Industrial Minerals and Rocks

5th Edition

Quartz crystals grown by Western Electric Co. (see chapter on materials for Electronic and Optical Uses beginning on p. 213). In the photograph are two loads of crystals from 6-in. ID vessels, one from 10-in. ID vessel

	6-In. Vessel	10-In. Vessel
Lb per crystal	1.75	1.75
Number of crystals	40	100
Run length, weeks	3	4
Natural crystal equivalent, lb	700	1750

Industrial Minerals and Rocks

(Nonmetallics other than Fuels)

Fifth Edition

Volume 1

Editorial Board

STANLEY J. LEFOND, Editor-in-Chief
Consultant, Evergreen, Colorado

ROBERT BATES
Ohio State University
Columbus, Ohio

JAMES C. BRADBURY
Illinois Geological Survey
Urbana, Illinois

B.F. BUIE
Florida State University
Tallahassee, Florida

RICHARD M. FOOSE
Amherst College
Amherst, Massachusetts

ROBERT B. HOY
Consultant
Applewood, Colorado

J.E. HUSTED
Georgia Institute of Technology
Atlanta, Georgia

H.N. MCCARL
University of Alabama in Birmingham
Birmingham, Alabama

L. ROE
Roeco, Inc.
Downers Grove, Illinois

L.F. ROONEY
US Geological Survey
APO, New York, New York

S. STOKOWSKI
Vulcan Materials
Birmingham, Alabama

Published by the
Society of Mining Engineers
of the
American Institute of Mining, Metallurgical, and Petroleum Engineers, Inc.
New York, New York • 1983

Printed in the United States of America
by Port City Press, Baltimore, Maryland

Library of Congress Catalog Card Number 82-071993
Volume 1 ISBN 0-89520-417-7
Set ISBN 0-89520-402-9

Preface

In 1980, after two printings of the fourth edition of *Industrial Minerals and Rocks,* the members attending a business meeting of the Industrial Minerals Division of the Society of Mining Engineers voted to publish a fifth edition. The original authors of the fourth edition were invited to update or revise their chapters. If the original author was unable to revise his chapter, another author did so. Some chapters were completely rewritten. We added a chapter on Alunite and divided Talc and Pyrophyllite into two chapters. *Industrial Minerals and Rocks* has also become international, with two chapters being written by authors in England and Australia. We hope future editions of *Industrial Minerals and Rocks* will include more chapters by authors from foreign countries.

As in the fourth edition of *Industrial Minerals and Rocks,* this edition is divided into four sections. However, the fifth edition breaks slightly with tradition in that it is printed in two volumes. As with the fourth edition, we made an attempt to be consistent in chapter content. However, we did not edit the papers to the point that they became stylized. Each author was permitted to write what he thought was important and to use his own words. We felt this made for more readable and interesting material. We attempted to incorporate the most recent prices and production data. Due to circumstances beyond our control, this was not possible in some instances.

Sadly, during the preparation of the fifth edition of *Industrial Minerals and Rocks,* we lost two editors of the fourth edition, Robert Grogan and Tom Murphy, and one editor of the fifth edition, Gerry V. Henderson. The fifth edition of *Industrial Minerals and Rocks* is dedicated to them.

The Industrial Minerals Division of SME is grateful to each of the authors and editors who gave so generously of their time and experience. We also appreciate the efforts of the many unknown secretaries who typed and retyped the various manuscripts.

Above all, however, we give our sincerest appreciation to Carmel A. Huestis for her diligence in preparing the various manuscripts for publication and to Marianne Snedeker for her patience in bringing the fifth edition through to publication.

Stanley J. Lefond
Editor-in-Chief

Contents

Volume 1

Frontispiece: Cultured Quartz Crystals ii

Preface v

Part 1. Introduction 1

Introduction 3

Part 2. Industrial Minerals Grouped by Uses 9

Abrasives 11
Ceramic Raw Materials 33
Chemical Industry 41
Construction Materials
Aggregates—Introduction 59
Crushed Stone 60
Lightweight Aggregates 81
Sand and Gravel 96
Slag 111
Cement and Cement Raw Materials 133
Dimension and Cut Stone 161
Gypsum and Anhydrite 183
Insulating Materials—Thermal and Sound 199
Roofing Granules 203
Electronic and Optical Uses 213
Fertilizer Minerals 233
Fillers, Filters, and Absorbents 243
Fluxes 259
Foundry Sand 271
Gem Materials 279
Glass Raw Materials 339
Mineral Pigments 349
Refractories 373
Well Drilling Fluids 387

Part 3. Sources of Information for Industrial Minerals 391

Sources of Information 393

Part 4. Commodities ... 415

Alunite ... 417
Asbestos ... 435
Barium Minerals ... 485
Bauxitic Raw Materials ... 503
Bituminous Materials ... 529
Boron and Borates ... 533
Bromine ... 561
Chromite ... 567
Clays ... 585
Diamonds ... 653
Diatomite ... 677
Feldspars ... 709
Index ... I-1

Volume 2

Frontispiece: Open Pit Boron Mine ... ii
Preface ... v

Part 4. Commodities, continued

Fluorspar and Cryolite ... 723
Glauconite ... 745
Graphite ... 757
Gypsum and Anhydrite ... 775
Iodine ... 793
Kyanite and Related Minerals ... 799
Lime ... 809
Limestone and Dolomite ... 833
Lithium Raw Materials ... 869
Magnesite and Related Minerals ... 881
Manganese ... 897
Meerschaum ... 909
Mica ... 915
Nepheline Syenite ... 931
Nitrogen Compounds ... 961
Olivine ... 989
Perlite ... 997
Phosphate Rock ... 1017
Potash ... 1049
Pumice, Pumicite, and Volcanic Cinders ... 1079
Pyrophyllite ... 1085
Rare Earths and Thorium ... 1109
Salt ... 1119
Sand and Gravel ... 1151
Silica and Silicon ... 1167
Sodium Carbonate Deposits ... 1187
Sodium Sulfate Deposits ... 1207
Staurolite ... 1225

Strontium 1229
Sulfur 1235
Talc 1275
Titanium Minerals 1303
Tripoli 1363
Vermiculite 1375
Wollastonite 1383
Zeolites
  Introduction 1391
  Synthetic Zeolites: Properties and Applications 1399
  Zeolites in Sedimentary Rocks 1413
  Commercial Utilization of Natural Zeolites 1418
Zirconium and Hafnium Minerals 1433

Index I-1

1

Introduction

Introduction

ROBERT L. BATES *

Definition

An industrial mineral, says the *Glossary of Geology*, is "any rock, mineral, or other naturally occurring substance of economic value, exclusive of metallic ores, mineral fuels, and gemstones; one of the nonmetallics." Probably most geologists and engineers would accept this simple and straightforward definition, at least as a general statement.

On analysis, however, certain anomalies crop up. For example, it may be considered arbitrary to include industrial diamond and corundum and to exclude their precious varieties, gem diamond and sapphire. So the present volume has a section on gems and gem materials, in spite of the definition. Another problem arises when we try to define precisely what is metallic and what is nonmetallic. Bauxite and ilmenite certainly qualify as metallic ores, but both are also raw materials of important nonmetallic substances; hence they are included in the book. A third anomaly is that, by common consent of geologists, engineers, mineral economists, and others concerned, the term industrial minerals may include not only rocks and minerals but also certain manufactured products like cement and lime, and even mixed groups of natural and artificial products, such as abrasives and refractories. Thus the central realm of the industrial minerals is clear enough, but its boundaries tend to be indefinable. Since a definition that allowed for every irregularity in the outer fringes would be excessively unwieldy, the one quoted above may be considered satisfactory.

The expression *industrial minerals* is commonly equated with *nonmetallics,* and with the somewhat more precise phrase *industrial minerals and rocks* as in the title of this book.

* Professor Emeritus, Dept. of Geology and Mineralogy, Ohio State University, Columbus, OH.

Characteristics of the Field

The industrial minerals as broadly interpreted are a highly diverse group of materials; indeed, we may say paradoxically that diversity is about all they have in common. At one end of the spectrum are ordinary earth materials such as gravel and crushed stone, with a value of perhaps $3 per st; at the opposite end are industrial diamonds, which sell (by the carat) at the rate of roughly $18,000,000 per st. Neither the metals nor the fuels, and few if any categories of manufactured products, include items that differ in value by more than six orders of magnitude. Most gravel and stone are of no value unless close to a market; diamonds and other exotic substances possess such unique and desirable properties that their occurrence in remote corners of the earth is no deterrent to their use. The broad range of industrial minerals falls between these extremes of unit value and place value.

The industrial minerals penetrate the entire fabric of modern industrialized society, from such fundamental operations as drilling for oil and smelting iron ore to the manufacture of virtually the whole array of consumer products. There is no industrial-minerals industry; there are many *industries*. The simple term *field* is convenient as an all-inclusive expression. An industry is considered to belong in the industrial-minerals field if its operations consist of extraction and processing, but not manufacturing. Thus the mining and beneficiation of feldspar is a mineral industry, but the manufacture of ceramic products is not. In the cement and lime industries, processing merges into manufacture, and the line is less distinct. It disappears altogether in a plant that extracts salt as artificial brines, and makes sodium and chlorine chemicals, all under the same roof.

An aspect of some industrial minerals that differentiates them from the metals is that their physical properties carry over into ultimate use of the product. Such minerals as mica, asbestos, diatomite, kaolin, and graphite—not to mention architectural stone and crushed stone—owe their value to physical properties that persist from quarry face to final application. This fact means, among other things, that geologists and engineers are likely to work closely together, since they are both concerned with the same material.

Some industries in the field are faced by problems of substitution and synthesis. Natural abrasives have been essentially replaced by artificial ones; dimension stone is hard-pressed by brick, tile, metal, and concrete. Replacement of a natural mineral by its artificial analog, made synthetically, faces producers of natural industrial diamond and quartz crystal. Some industrial minerals, on the other hand, are essentially immune from attacks of this kind. For example, because the humble diatom did things with silica—by the billions of billions, and down to the micron level—that man cannot do, incidentally producing a material of great industrial versatility, diatomite is not likely to suffer eclipse through replacement by other materials or by a synthetic equivalent.

The great diversity of the industrial minerals gives the field as a whole a rather acute problem of identity. The Perlite Institute speaks for perlite, and the National Sand and Gravel Association for the producers of sand and gravel; but no agency can speak for the entire field. This is probably not a very serious matter, except in the departments of geology and mineral engineering in our universities. There the mineral industries have traditionally meant metal mining, and economic geology has been equated with metallic ore deposits. Many schools are still dominated by what has been called the "metallic mentality," and the image of the immense field of the industrial minerals seems destined to continue somewhat vague and unfocused. It sometimes helps to remind the uninformed that the value of industrial minerals consumed annually in the United States is twice that of the metals.

Classification

The wide range and numerous peculiarities of the industrial minerals make their classification at once challenging and difficult. Five classifications proposed in the past two decades are summarized below.

Bates, 1960

To serve as the basis for a book on geology of the industrial minerals, a combined end-use and genetic classification was developed by Bates (1960). In this, the nonmetallics are first divided into the industrial rocks and the industrial minerals, on the basis of such economic criteria as bulk produced, unit value, place value, and international trade. Each group is then subdivided on geologic criteria: igneous, metamorphic, and sedimentary for the rocks, and pegmatite, vein and replacement, metamorphic, and sedimentary for the minerals. This classification has proved satisfactory for its purpose, which is systematic geologic study.

Wright and Burnett, 1962

In a paper on the search for industrial minerals in California, Wright and Burnett (1962) utilize a threefold classification, on the basis of unit price and bulk produced: (1) low-price, large-volume commodities, primarily construction materials; (2) high-price, large-volume commodities, such as raw materials for industrial chemicals; and (3) high-price, small-volume commodities, like feldspar, fluorspar, and talc. This simple plan is adequate for its purpose, which is to give the prospector and the potential small-mine operator an insight into the field of industrial minerals.

Fisher, 1969

A more elaborate classification has been set forth by Fisher (1969). Noting that the nonmetallics "represent a disparate and heterogeneous group," he utilizes unit value, place value, and representative volume of production as criteria for distinguishing six groups. These are (1) bulk constructional and building materials; (2) bulk ceramic raw materials, lime, and commodities of or from diversified industries; (3) specialty building materials and principal refractories; (4) major industrial, chemical, and fertilizer raw materials; (5) industrial rocks and minerals; and (6) specialty grade and precious rocks and minerals. Fisher discusses each group and gives much supplementary information, such as number of plants, processing costs, and marketing data. This is the most detailed classification that has appeared.

Kline, 1970

In a paper pointing out the great importance of the industrial minerals, Kline (1970) divides them into two main groups, each with similar end uses: chemical minerals and physical minerals. The former, used as sources of chemical elements, are altered in processing, mainly in the chemical, fertilizer, ceramic, and metallurgical industries. Physical minerals, not changed in composition by the user, are valued for particle-size distribution, brightness, surface area, or other physical properties. In his discussion, Kline points out among other things that many of the chemical minerals are produced by the end users—salt by chloralkali producers, limestone by steel companies, and so on—whereas the physical minerals have a wide range of purchasers and there is much less captive production.

Dunn, 1973

A "matrix classification" has been presented by Dunn (1973). In chart form, it is an attempt to relate economic and geologic factors. One coordinate is devoted to uses and processes, and the other to rock types; minerals that are referred to by numbers within appropriate matrix squares are listed in a separate column. This is an ingenious attempt at reconciling facts of geologic occurrence and end use, and should prove of interest and value.

Recent Developments

The field of industrial minerals has changed greatly since appearance of the 4th edition of *Industrial Minerals and Rocks* in 1975. Some of the more significant changes are reviewed in the following paragraphs.

Emergence of a World Community

As the world's population has increased, so has the demand for earth materials, and today these materials come from all corners of the globe. We find Canadian taconite ores pelletized with bentonite from Wyoming and Greece, Gulf Coast drilling muds weighted with barite from Ireland and Peru, and paint manufactured with titanium oxide derived from Australian beach sand. In 1981 the United States imported about 5% of its net consumption of cement, or some 3.6 Mt (4,000,000 st); large suppliers were Norway, Spain, the United Kingdom, and France. By no means is such traffic one-way. Kaolin, borax, and phosphate rock are among American products that enter widely into international trade. Concurrent with expanded trade is increased movement of venture capital across national borders and overseas—into the United States as well as out of it. Dramatic shifts of capacity and realignments of supply are more than ever a part of the world picture, as illustrated in the sulfur and potash industries. Today the whole fraternity of producers, processors, and consumers of industrial minerals operates on a worldwide scene.

Increased Government Influence

Another inevitable result of growing population is an expansion in the influence of government. This is felt by the industrial minerals from the national to the local level, abroad as well as at home. Nationalization, or strong moves in that direction, have affected the producers of Sri Lanka graphite, Indian mica, Turkish colemanite, and Guyana bauxite. The provincial government of Saskatchewan is in the potash business, and the government of Quebec is eyeing the asbestos industry. In this country, government's role in environmental protection has had a profound effect. At every stage of a mineral industry—from obtaining permission to mine a newly discovered deposit to meeting stringent new requirements at a long-established operation—companies must deal with regulatory agencies of federal, state, or local government. Another problem, that of land use and zoning, faces producers of sand, gravel, and crushed stone, who of necessity must operate in or near areas of high population density. In the years since 1960, governments have become progressively more involved, some would say more restrictive and high-handed, owing in the last analysis to the pressure of a constantly growing population on a planet with finite resources of air, water, land, and minerals.

Newcomers to the Field

During the 1970s, many firms from other fields became interested in industrial minerals. Oil companies established "solid minerals" divisions; metal-mining companies set up subsidiaries to find and produce nonmetallics. Concerns that had traditionally been in other lines of business entirely, such as chemicals or manufacturing, established industrial-minerals branches. "Conglomerates" acquired properties ranging from sand and gravel to high-value minerals. Today, a quarry operation may be owned by a corporation whose primary business is mining nickel or making pharmaceuticals.

Improved Communications

Communication and exchange of information in the field are appreciably better today than 12 or 15 years ago. An event of major significance was the appearance in 1967 of a new monthly magazine, *Industrial Minerals*. Published in London, it is dominantly but not exclusively a trade magazine. Its coverage is worldwide, and it has become essential reading for those seriously interested in the field. The annual review of industrial minerals in *Mining Engineering* has developed into another source of authoritative information. The Forum on Geology of Industrial Minerals has met annually since 1965; proceedings have been published and the meetings have generated considerable interest. The Forum has focused on practical problems of geological occurrence and how these problems affect extraction and use. The first Salt Symposium, sponsored by the Northern Ohio Geological Society, convened in Cleveland in 1962; later meetings were held in the same city in 1965 and 1969, in Houston in 1973, and in Hamburg in 1978. These meetings, and their published proceedings, have provided a focus for discussion of the geology, geochemistry, mining, and rock mechanics of salt and other evaporites.

About 400 geologists, engineers, chemists, and economists, mostly from the United Kingdom and western Europe, met in London on July 8–9, 1974, at the first International Industrial Minerals Congress. They heard papers on a wide range of topics, from geologic aspects of an undeveloped deposit to the testing of final products, and also had opportunity for informal exchange of views and information. Several field trips followed the meeting. Later congresses were held in Munich in 1976, Paris in 1978, Atlanta in 1980, and Madrid in 1982. More than 650 people attended the Atlanta meeting. Proceedings of the congresses are published by the sponsoring organization, the publisher of *Industrial Minerals* magazine. In September 1980, the same sponsor organized a conference on minerals and chemicals in drilling muds, which was held in Houston.

Developments in Exploration

Traditionally, the geologist has had little to do with the discovery of industrial mineral deposits, most of which were found by the prospector or by the driller for oil or water. Some progress was made during the 1960s in changing this picture. McKelvey and others of the US Geological Survey developed a geological model for the occurrence of phosphorite, involving such seemingly academic topics as paleogeography, oceanic upwelling, and continental drift. When applied, this model led to the discovery of large phosphorite deposits in Turkey, and immense deposits in Queensland, Australia. Bromine geochemistry was developed to the point where it can be utilized in locating potash deposits. Concurrent with these developments in geological science were improvements in various instruments for remote sensing, data handling and enhancement, atomic absorption, electron microscopy, and related techniques.

Changes in Technology

Perhaps the most outwardly visible changes that have taken place since the fourth edition of *Industrial Minerals and Rocks* are in mining, processing, handling, and transportation. The trend has been toward mechanization, larger size of units, simplified handling, and lower costs per ton. This trend is indicated by such underground equipment as the continuous mining machine and the high-capacity conveyor belt. In the open pit, trucks with as much as 200-ton capacity are loaded by shovels or draglines that seem to get larger every year. Material is moved across country by unit trains and in slurry pipelines, and across the oceans in large, self-unloading cargo vessels. Much of the mining and quarry equipment is of American manufacture, but Japanese, German, and Scandinavian machinery is also in widespread use.

Summary

The many companies in the industrial-minerals field, from the local gravel producer to the mining and processing complex owned by a major diversified firm, are all parts of a world community. Information, investment capital, and mineral commodities move among nations more readily today than at any time in the past. Governmental regulation has become and will remain a fact of corporate life. Geological, geophysical, and geochemical methods of prospecting are slowly being developed to the stage where they can be utilized to predict the existence of mineral deposits. Techniques of handling materials, from mine face to point of use, are evolving in the direction of greater size and capacity.

All developments are in response to the world's need for more and more earth materials at the lowest possible cost. One can only conclude a survey of the field by emphasizing the themes of energy, activity, and growth.

References

Absalom, S.T., "Cement," *Mineral Commodity Summaries 1982,* US Bureau of Mines, pp. 28–29.

Bates, R.L., 1960, "Classification," *Geology of the Industrial Rocks and Minerals,* 1st ed., Harper, New York, pp. 15–19.

Dunn, J.R., 1973, "A Matrix Classification for Industrial Minerals and Rocks," *Proceedings, 8th Forum on Geology of Industrial Minerals,* Public Information Circular 5, Iowa Geological Survey, pp. 185–189, chart.

Fisher, W.L., 1969, "The Nonmetallic Industrial Minerals: Examples of Diversity and Quantity," *Mining Congress Journal,* Vol. 55, No. 2, Feb., pp. 120–126.

Kline, C.H., 1970, "Industrial Minerals Are Big Business," *Mining Engineering,* Vol. 22, No. 12, Dec., pp. 46–48.

Wright, L.A., and Burnett, J.L., 1962, "The Search for Industrial Minerals," *Mineral Information Service,* California Div. of Mines and Geology, Vol. 15, No. 1, Jan., pp. 1–8.

2

Industrial Minerals Grouped by Uses

Abrasives

RICHARD P. HIGHT *†

Abrasives include the substances, natural or artificial, that are used to grind, polish, abrade, scour, clean, or otherwise remove solid material, usually by rubbing action but also by impact (pressure blasting for example). They do not include abrasive tools, for instance, lathe tools and files—or polishing agents such as waxes, which act by filling pores.

Detergents and cleaners whose action is chemical rather than physical are omitted although some chemical-action polishes and cleaners may also contain solid abrasives, for example, many automobile and metal polishes.

General Considerations

The most important physical properties of materials that qualify them for use as abrasives are hardness, toughness (or brittleness), grain shape and size, character of fracture or cleavage, purity or uniformity. For making bonded abrasive products such as grinding wheels, additional important factors are stability under high heat and bonding characteristics of grain surfaces. The economic factors of cost and availability are always important.

No one single property is paramount for any use. For some uses extreme hardness and toughness are needed, as in diamonds for drill bits; for others, the factors of greatest importance are hardness and ability to break down slowly under use, to develop fresh cutting edges when grains become worn—for example: in garnet for sandpaper neither highly cleavable or friable grains nor extremely tough grains are wanted. For still other uses, great hardness is objectionable; for example, abrasives for dentifrices and for glass-cleaning soaps. For the most efficient use in the more critical applications, the different types of abrasives are rarely completely interchangeable; thus, while crushed quartz and garnet are both used in sandpaper, the papers are not at all interchangeable in their use applications.

In the last analysis, the choice of a high grade abrasive depends upon the quality and quantity of work done by the abrasive per unit of cost. Initial cost of an artificial abrasive may be much greater than that of a natural abrasive but the artificial abrasive may do so much better work than the natural one, and do it so much faster that the ultimate cost will be less. It is for this reason that artificial abrasives have largely replaced natural abrasives.

Abrasive Value

Mineralogical hardness or "scratch" hardness as expressed in Mohs' scale is an important property in evaluating abrasive materials, but, as noted before, it is only one of several essential properties. The mineral hardness of pure crystal almandite garnet is about 7.5, but if the crystal is crossed by incipient fracture planes, or if it contains inclusions of other minerals, the apparent or useful hardness may be much lower. While the quartz grains in a sandstone have a hardness of 7, the bond holding the grains together may be so weak that the stone is valueless as a commercial abrasive. In artificially bonded wheels and stones, the hardness, strength, and character of the bond are fully as important as the hardness of the abrasive grains. Thus, in an overall consideration of abrasive hardness of loose abrasive grains, both "scratch" hardness and toughness must be considered. In naturally or artificially bonded abrasive stones, bond characteristics are a third factor, which is most important.

The problem of abrasive hardness is further complicated by the inadequacies of methods of testing hardness and of expressing relative values. The Mohs' scale is inadequate both because the methods of testing are very crude and

* Senior Mineralogist, Analytical Laboratory, Research & Development Dept., Norton Co., Worcester, MA.

† This chapter, authored by Raymond B. Ladoo in the third edition of this book, has been extensively rewritten, revised, and updated by R. P. Hight.

because the intervals between steps in the scale are not uniform; thus there is far less difference between hardnesses of 6 and 7 than between 9 and 10. Numerous attempts have been made to remedy these deficiencies. Ridgway, Ballard, and Bailey (1933) proposed an extension of Mohs' scale to include artificially prepared substances. Knoop, Peters, and Emerson (1939) devised a diamond indentation method of measuring hardness, which gives reproducible results on a wide variety of materials and has the merit of giving definite numerical values over a wide range up to the hardness of the diamond.

Table 1, after Thibault and Nyquist (1947), shows relative hardnesses on the Knoop scale.

Another consideration, perhaps even more important than true hardness in appraising abrasive value, is toughness or friability. Very hard materials are not necessarily tough or resistant to fracture. Diamond is an excellent example of this. It is the hardest material known to man yet it can fracture easily under impact.

Several techniques for testing friability have been developed and are in widespread use in the quality control of abrasive grain. The basic operating principle common to all strength-of-grain tests is to impact a given sample of abrasive grain under a set of standardized conditions followed by measurement of the amount of breakdown. Naturally, the higher the survival the tougher the grain, other factors being equal.

FIG. 1—*Wig-L-Bug mill.*

A simple and inexpensive device for performing this test is an ordinary ball mill. A detailed description of this test is given in the "American Standard Procedure for Ball Mill

TABLE 1—Knoop Hardness Numbers Determined at 100-G Load (K_{100}) for Various Hard Substances. Random Crystal Orientation

Substance	Knoop No. 100-G Load, (K_{100}) Average	Range
Primary boron carbide ("Norbide")	2800	2670-2940
Molded boron carbide ("Norbide")	2760	2580-2900
Gray silicon carbide (Regular "Crystolon")	2460	2250-2680
Gray silicon carbide (Regular "Crystolon" [check])	2550	2320-2760
Green silicon carbide (Green "Crystolon")	2480	2230-2740
Green silicon carbide (Green "Crystolon" [check])	2480	2130-2620
Titanium carbide	2470	2350-2620
Tungsten titanium carbide	2190	2050-2320
Alpha-aluminum oxide ("38 Alundum")	2050	1860-2200
Alpha-aluminum oxide (synthetic boule)	1950	1680-2100
Primary (unbonded) tungsten carbide (WC)	1880	1570-2140
Cemented carbide ("Kennametal K6")	1800	1700-1940
Cemented carbide ("Kennametal KM")	1500	1390-1600
Cemented carbide ("Kennametal K12")	1410	1280-1500
Garnet (Barton Mines, New York)	1360	1240-1440
Topaz (Thomas Range, Utah)	1340	1240-1500
Synthetic blue spinel (Linde)	1270	1190-1460
Quartz	820	760-880
Hardened tool steel, Rockwell C-60.5	740	730-760

Source: Thibault and Nyquist, 1947.

FIG. 2—*Strength-of-grain machine.*

Test for Friability of Abrasive Grain" (Anon., 1965).

The Wig-L-Bug Mill (Fig. 1) is also used for the testing of small samples. This feature makes it particularly attractive for testing the strength of diamond grain.

With this machine a small, closely sized weighed sample is placed in a capsule along with a milling sphere. The capsule and contents are vibrated vigorously for a given length of time. The sample is then removed (often by washing), rescreened, and the portion remaining on the screen is expressed as a percent survival. On a "Standard" sample, typical of the materials to be tested, it has been found best to adjust the running conditions so that approximately 50% breakdown is achieved.

Various other devices have been developed for impacting abrasive grain. A machine built and used by Norton Co. is one of these (Fig. 2). A distinct advantage that this machine has over the previously mentioned ones for ordinary abrasives is its capacity to test a relatively large sample 0.23 kg (½ lb). This machine ejects abrasive out of a rapidly whirling tube against the walls of the chamber. After striking the wall the grains accumulate in a collecting cylinder and are later screened to determine the percentage of surviving grain.

The grain shape is another fundamental physical property which must be considered in selecting the best abrasive for a particular application. Blocky and nearly equidimensional grains are called strong-shaped grains and are less friable than slivery or flakey ones which are considered weak-shaped grains. In grinding wheels strong-shaped grains are preferred in extra heavy duty applications involving high heat and pressure while weaker-shaped grains are selected for precision grinding to close tolerances and where high surface finishes are required. Strong- and weak-shaped grains, respectively, are illustrated in Figs. 3 and 4.

In modern abrasive technology close control of abrasive shape is necessary. The determination of the bulk density of an abrasive is a dependable indicator of shape inasmuch as bulk density is a function of shape. A sample of weak-shaped grain will not pack as efficiently and consequently will have a lower bulk density than the same abrasive but of a strong shape.

The following techniques are commonly used for the measurement of the bulk density (shape) of abrasive grain.

The "loose-pack-density test," as it is often called, is fully described in "American Standard Test for Bulk Density of Abrasive Grain" by the American Standards Assn. (Anon., 1964). The test apparatus is illustrated in Fig. 5. In performing this test, vibration is purposely avoided. In practice, the container is filled to overflowing and then carefully leveled with a

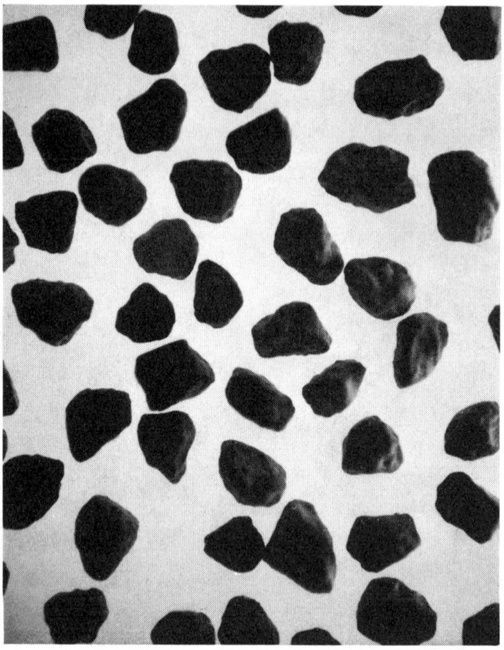

FIG. 3—*Strong-shaped abrasive grain.*

straight edge. Great care must be exercised so as to avoid any vibration. The weight of the grain filling the container is then obtained and the bulk density calculated using the traditional

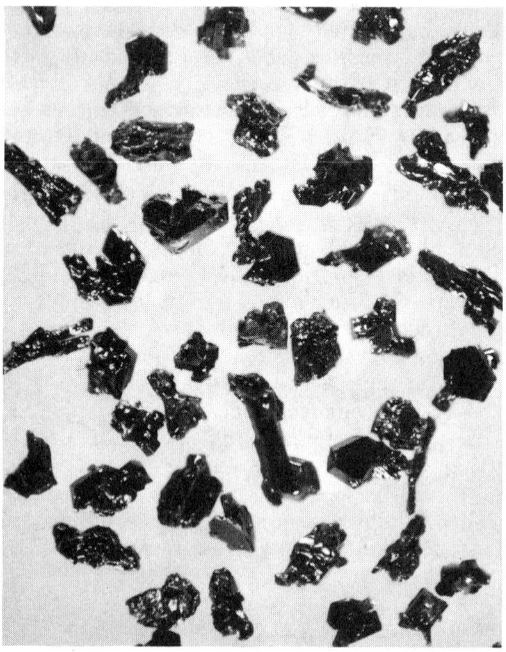

FIG. 4—*Weak-shaped abrasive grain.*

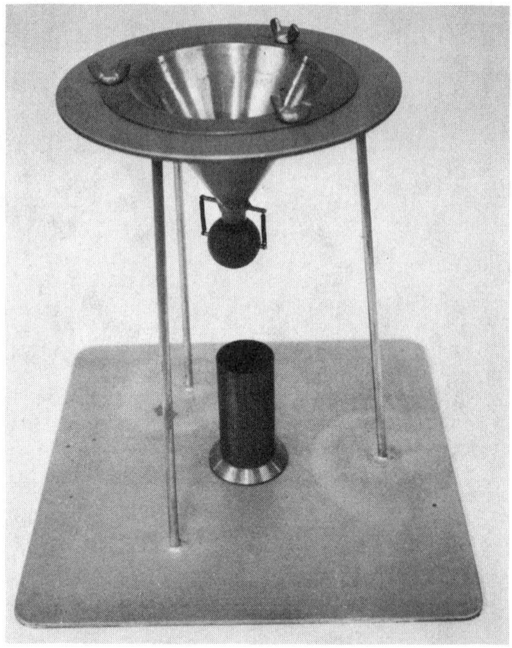

FIG. 5—*Loose-packed-density (LPD) apparatus.*

formula $D = W/V$ in which D is the bulk density of the sample in grams per cubic centimeter, W is the weight of the grain in grams, and V is the volume of the cylinder in cubic centimeters.

In the following test, however, controlled vibration is a necessary part of the determination. This second shape test is often referred to as "the weight per cubic foot" determination.

Fig. 6 shows equipment designed and built by Norton for the measurement of the weight per cubic foot of abrasive grain. In this test, as in the loose-pack-density test, it is absolutely essential that the sample to be tested be as closely sized and as free of dust and fines as possible, otherwise spurious results can be expected.

Before making a run with this equipment, the grain to be tested is placed in the hopper and a receiving container of suitable capacity is installed. A container with a capacity of 0.0035 m^3 (1/8 cu ft) is normally used but other containers—for example, 0.00088 to 0.00044 m^3 (1/32 or 1/64 cu ft)—may be used if the sample to be tested is small. To start a run, the motor is turned on and the storage hopper containing the sample is opened. When the motor starts, a gear-driven cam causes the collecting container to bump up and down, this being the controlled source of vibration, which causes

FIG. 6—*Weight-per-cubic-foot (WPCF) machine.*

settling or packing of the grains. Abrasive is allowed to flow continuously into the collecting container until it overflows and during the run more abrasive is added as needed to keep the container full-to-overflowing. The machine is so constructed as to collect any abrasive which overflows. Each bump cycle of the collecting cylinder is counted on an automatic counter. In practice the machine is run until a predetermined total number of bumps has been accumulated. At this point the motor is shut off and the excess abrasive removed by carefully drawing a straight edge across the top of the collecting cylinder. The grain is then weighed and expressed as "pounds-per-cubic foot."

Characteristics other than shape influence the strength of abrasive grains. The manufacture and processing of aluminous abrasives permit control of some of these important characteristics. For example, it has been clearly established by laboratory and field tests that, other factors being equal, the finer the crystal size of an abrasive the stronger that abrasive will be. The cooling history of a fused alumina melt is the dominant factor influencing the crystal size of the finished product. Perhaps at the cost of oversimplification it may be said that rapid cooling will result in fine crystal size and slow cooling will result in coarse crystals. As a frame of reference, a slow-cooled product might typically result in a product the crystal size of which would be in the order of 1500 m and a rapidly cooled product 400 m.

A rapid and inexpensive quality control check for crystal size is "the monocrystalline count." An immersion mount of the grit to be tested is prepared. Any convenient grit size can be used but 24-grit has been found by experience to be quite ideal. Each grain is then examined on the petrographic microscope to determine whether it is made up of a fragment of a single crystal or perhaps two or more crystals. The former grain is a monocrystalline grain while the latter is polycrystalline. A count is made of, say, 100 grains and that percentage which is found to be monocrystalline is by definition "the monocrystalline count." Generally speaking the higher the monocrystalline count the weaker the grain. It is theorized that a polycrystalline grain is less apt to disintegrate because a fracture upon starting through the grain might pass through one crystalline unit making up the grain but then be stopped or deflected upon encountering another crystalline unit not in crystallographic continuity with the first.

A secondary treatment which has been found to increase the strength of fused aluminum oxide abrasive grain is heat treatment or roasting. The theory is that the heated slag (impurities between crystals) softens and flows out onto the surface of the grain, cementing imperfections and healing incipient fractures.

Types of Abrasive Products

Loose Abrasive Grains: Abrasive grains are produced from a wide variety of materials, both for use as grains and for incorporation into other products. Thus abrasive grains are not only important products of themselves but are the starting point for making bonded shapes, coated abrasives, abrasive tools, polishes, cleaners, grinding pastes, and compounds.

For pressure blasting, in addition to silica sand, other natural mineral grains are used such as corundum, garnet, flint, and chert as well as manufactured products such as fused alumina, silicon carbide, and steel shot. All except the last have a hardness of 7 or more. Besides hardness the physical properties of toughness, grain shape, grain size, uniformity, and specific gravity are important. Grains should be tough, so that they will not disintegrate readily under impact. For some uses rounded grains are considered desirable, for others sharp cutting edges are favored. Uniformity of grain size and other physical properties is always desired. The higher the specific gravity, the greater the force of impact for grains of equal size. Each use has its own special requirements.

Common uses for pressure blasting are: cleaning stone and concrete, cleaning metal castings, preparing surfaces for painting, etching of glass, and plastics.

Loose grains of relatively coarse sizes are used for sawing stone, rough grinding plate glass, and surfacing stone. Materials used are quartz sand, garnet, corundum, emery, aluminum oxide, silicon carbide.

Grains of somewhat smaller sizes of the same materials are used for grinding lenses, rough polishing of building stone and plate glass, rough polishing of gems and ornamental stones, and dressing and polishing wood surfaces.

Fine-grained powders are used for polishing and lapping of a variety of materials including glass, building and decorative stone, metals, plastics, tile and artificial stone flooring, gems and semiprecious stones, and wood surfaces. The abrasive materials used are rouge, crocus, tin oxide, aluminum oxide, chromium oxide, cerium oxide, diamond dust, feldspar, garnet, rottenstone, pumice, diatomite, tripoli, ground silica, clay, whiting, zirconium oxide. Most automobile body cleaners and polishes contain diatomite. A familiar bathroom and kitchen scouring powder for glass and porcelain bodies uses feldspar or volcanic glass.

In one method for drilling small holes in glass, porcelain, gem stones, and similar materials, a brass tube is used, charged with diamond, corundum, emery, aluminum oxide, silicon carbide, or boron carbide dust in a vehicle of water, oil, or grease. Diamond-tipped drills are also used for this purpose.

Ground feldspar, fuller's earth, and other materials are used as nonskid dusting agents for oily floors. Emery, magnetite, and silicon carbide are cast into the treads of steps and ramps to increase "footing" and coarse or gritty slag and volcanic rocks are rolled into asphalt pavements to improve traction.

Bonded Abrasives: Abrasive grains, closely sized, are bonded and pressed or molded into a wide variety of bonded abrasives such as grinding wheels. Originally the abrasives in such products were corundum or emery but with the introduction of the electric-furnace abrasives, aluminum oxide, and silicon carbide, the use of the natural abrasives has very greatly declined. There are five main types of bonded abrasives, depending on the type of bond and method of manufacture: (1) Vitrified wheels with a clay-feldspar bond, which is vitrified in ceramic kilns. This bond type accounts for over 75% of the grinding wheels manufactured today. Because of their high rigidity and dimensional stability, vitrified bonded wheels are preferred for precision grinding operations. They are unaffected by water, acids, oils, and ordinary temperature variations. (2) Resinoid wheels with a hard synthetic resin bond. These are high speed wheels used in foundries, welding, and billet conditioning shops. They are also used in cutoff and thread grinding operations. (3) Rubber wheels, being bonded with natural or synthetic rubber, are somewhat elastic in nature. They are used for grinding ball races, most centerless feed wheels, and for portable snagging operations where finish is an important consideration. Rubber-bonded cutoff wheels can be made very thin. (4) Shellac wheels are used for producing high finishes on such items as camshafts and paper mill rolls. (5) Silicate bonds find application in operations where heat generated in grinding must be kept to a minimum. Silicate bonded wheels are mild acting and are used in grinding edge tools of all kinds.

Wheels are made that vary in at least five physical properties, aside from size and shape: (1) type of abrasive grain (aluminum oxide, silicon carbide, corundum, emery, diamond, etc.); (2) grit (size of abrasive grains, 8 to 1500-mesh); (3) grade (strength of bond, 18 to 20 grades); (4) structure (grain spacing, 10 to 12 spacings); (5) bond (vitrified, etc., five bond types).

Fig. 7 shows a wheel marking system which is widely used in the abrasive industry (Anon., 1970).

When all these variables are combined with an almost infinite number of sizes, shapes, and spindle diameters that are standard or can be had on special order, the complexity of the abrasive wheel business is apparent.

In addition to grinding wheels, bonded abrasives are made up into blocks, bricks, and sticks for sharpening and polishing stones such as oil stones, scythe stones, razor and cylinder

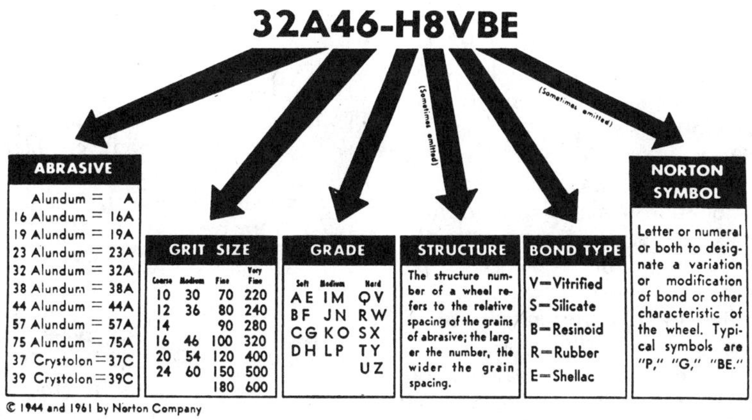

*Structure numbers are not always shown in the marking. However, such products may nevertheless be made to a definite structure.

FIG. 7—*Grinding wheel marking system.*

hones. Curved blocks and segments are made, which can be combined into large wheels such as pulpstones or used mounted or unmounted for grinding or polishing the interior or exterior of curved surfaces.

Rubber pencil and ink erasers contain abrasive grains and similar soft rubber wheels, sticks, and other forms, are made for finishing soft metals.

Coated Abrasives: Coated abrasives consist of sized abrasive grains cemented to paper or cloth backing. Originally "sandpaper" was perhaps coated with sand but today crushed quartz is used on silica-type paper because its grains have much sharper cutting edges. The use of crushed glass to surface "glass" paper is now obsolete. The principal abrasives used today for this purpose are crushed quartz, garnet, aluminum oxide, and silicon carbide. Most "emery" cloth today is coated with silicon carbide or aluminum oxide. "Flint" paper in the United States is coated with crushed quartz but true flint is used in Europe.

With the exception of sintered abrasives, all abrasive grains, natural or artificial, are made by graded crushing and close sizing. Crushing usually is done in rolls to avoid making extreme fines (for which there is little demand) with close screening between sets of two rolls each. Products are washed in classifiers to remove slimes, then are dried, passed through magnetic separators to remove iron-bearing material, and again very closely sized on screens. Such careful sizing is essential to prevent contamination of grades by coarser grains. The finest powders are sized by air or water flotation and sedimentation.

In earlier practice, grade numbers were more or less arbitrarily assigned to grain sizes and unfortunately the garnet, flint, and emery scales were all different. Artificial abrasives were given numbers based on bolting-cloth mesh designations, which roughly correspond with but are not exact equivalents of standard testing sieve meshes. Gradually garnet and "flint" paper grade numbers have come to include both the old and the new numbers.

TABLE 2—Size Grades for Abrasives

Modern Mesh Designation	Old Garnet Scale	Flint Scale	Emery Scale
400			
320			
280	8/0		
240	7/0*	5/0	
220	6/0	4/0	
		3/0	
180	5/0		3/0
150	4/0	2/0	2/0
120	3/0		1/0
		1/0	
100	2/0		
		1/2	1/2
80	1/0		1
		1	
60	1/2		1 1/2
50	1	1 1/2	2
		2	2 1/2
40	1 1/2		
		2 1/2	
36	2		3
30	2 1/2	3	
24	3	3 1/2	
	3 1/3†		

* Approximately 220-mesh Tyler.
† Approximately 20-mesh Tyler.

Table 2 shows the relationship between the different scale sizes.

The bond for coated abrasives may be either glue or a synthetic resin adhesive. The backing is either paper or cloth.

Garnet and "flint" papers and cloths are most widely used for wood, leather, hard rubber, plastics, felt, and rubbed paint and varnish finishes on metals. Abrasives coated with silicon carbide and aluminum oxide are used chiefly in the metal-working industries. To date, diamond has found but limited use in coated abrasives.

More detailed information on coated abrasives is contained in the section on garnet later in this chapter.

Grains and Powders For Soap, Cleaners, and Polishes: Many different materials, mostly natural but some manufactured, are used in making soaps, cleaners, and polishes.

Feldspar, pumice and pumicite, sand, ground quartz, tripoli, diatomaceous earth, clay, and wood flour are used in hand and scouring soaps. Often low price is the primary factor in the selection of the abrasive, in spite of extravagant claims by some manufacturers. The commercial success of a cleanser depends largely upon advertising and sales ability rather than on superiority over competitive products, provided a fairly good product is made. Occasionally the abrasive is a really important factor. A household cleanser should not contain quartz or any mineral of equal or greater hardness, as glasses and enamels are nearly all 6 or lower in hardness and are scratched by any grain of equal or greater hardness. The abrasive should not contain calcium carbonate, calcium sulfate, or other easily reactive lime compounds because lime reacts with most soaps to form insoluble substances, which are very difficult to remove from glass or enamel surfaces.

An ideal mineral abrasive for this use might be a mineral or rock between 3 and 5 in hardness occurring in abundance in easily mined deposits, close to rail transportation, and within a reasonable freight-haul distance of important markets. The material should be uniform in texture and physical properties and contain no impurities harder than 6 and no easily reactive impurities such as lime compounds.

The ultimate grain size of abrasives in this group depends upon the finished product. It ranges from extremely fine air-separated and water-floated particles in the micron sizes, much finer than 325-mesh, up to 100-mesh or even coarser for heavy-duty scouring soaps.

Classification

Abrasives * may be divided into two general classes, natural and manufactured. The former includes all rocks and minerals used for abrasive purposes without chemical or physical change other than crushing, shaping, or bonding into suitable forms. Manufactured or artificial abrasives are made either by heat or chemical action from metals or mineral raw materials.

Table 3 lists most of the important abrasives, classified as to inherent types and the forms in which they are used industrially.

For most types of use there are manufactured products that can be substituted for the natural products, usually at higher initial cost but with higher efficiency. This is not always true; for example, there is no satisfactory manufactured substitute for garnet for making coated abrasive paper and cloth. For some abrasives whose use is gradually lessening—for instance, chaser stones—the making of manufactured substitutes has not been economically attractive, but for even such a low-priced commodity as pressure blasting sand there are substitutes such as steel shot, fused aluminum oxide, and silicon carbide grains.

The decline in the use of most natural abrasives and their replacement by manufactured abrasives has not been a net loss to the mineral industry, however, for virtually all manufactured abrasives are made from mineral raw materials.

Natural Abrasives

Corundum and Emery: Corundum and emery have become of relatively little importance both from the standpoint of domestic production and consumption and on a worldwide basis. No corundum is mined today in the United States or Canada and there are only two relatively small producers of emery, both in the Peekskill, NY, area. At this locality, a spinel emery occurs in veins in an igneous complex of hornblende and olivine pyroxenite. It is associated with mica schist in rocks containing sillimanite, cordierite, garnet, and quartz. Most of the domestic emery is used in tumbling

* In this book separate chapters are devoted to diamonds, diatomite, tripoli, lightweight aggregates (including pumice), silica sand, and quartz. Hence in this chapter only brief reference will be made to these minerals.

TABLE 3—Classification of Abrasives

Natural Abrasives

- Superior Hardness (above 7 in Mohs' scale)
  - Diamond H-10
  - Corundum H-9
  - Emery H-7 to 9
  - Garnet H-6.5 to 7.5
  - Staurolite H-7.0 to 7.5
- Intermediate Hardness (H-5.5 to 7)
  - Silica Abrasives
    - Buhrstone
    - Chalcedony
    - Chert
    - Flint
    - Novaculite
    - Quartz
    - Quartzite
    - Sandstone
    - Silica sand
  - Other Rocks and Minerals
    - Argillaceous limestone
    - Basalt
    - Feldspar
    - Granite
    - Mica schist
    - Perlite
    - Pumice and pumicite
    - Quartz conglomerate
- Inferior Hardness (H-under 5.5)
  - Apatite
  - Calcite
  - Chalk
  - Clay
  - Diatomite
  - Dolomite
  - Iron oxides
  - Limestone
  - Rottenstone
  - Siliceous shale
  - Silt
  - Talc
  - Tripoli
  - Whiting

Manufactured Abrasives

- Boron carbide
- Boron nitride
- Calcium carbonate (pptd.)
- Calcium phosphate
- Cerium oxide
- Chromium oxide
- Clay (hard burned)
- Diamond
- Fused alumina
- Glass
- Iron oxides
- Lampblack
- Lime
- Magnesia (pptd.)
- Manganese dioxide
- Periclase (artif.)
- Silicon carbide
- Tantalum carbide
- Tin oxide
- Titanium carbide
- Tungsten carbide
- Zirconium oxide
- Zirconium silicate

Metallic abrasives, including steel wool, steel shot, angular steel grit, brass wool, and copper wool

Porcelain blocks for mill liners and grinding pebbles

Types of Abrasive Products

- Abrasive grains and powders, loose
- Abrasive grains bonded into wheels, blocks, and special shapes
- Coated abrasives; grains bonded to paper and cloth
- Abrasive grains and powders; paste form; oil or water vehicles
- Abrasive grains and powders; brick and stick form; grease, glue, and wax binders
- Natural rocks shaped into grindstones, pulpstones, chaser stones, millstones, etc.
- Natural rocks shaped into sharpening stones, such as oil stones, whetstones, scythe stones, razor hones, etc.
- Natural stones shaped into rubbing and polishing stones such as holystones and pumice scouring blocks
- Natural stones shaped into blocks for tube-mill and pebble-mill liners
- Pebbles, natural and manufactured, for grinding mills

barrels and for various types of nonslip floors and stair treads. Production in recent years has come from two producers and has averaged about 9 kt/a (10,000 stpy) valued at about $20 per ton.

Diamonds—Industrial: Diamonds are reviewed in detail in a separate chapter in the "Commodities" section; therefore only their uses as abrasives are outlined here.

In 1981 the total world production of natural industrial diamonds of all types was about 31,100,000 carats, with US imports of 21,600,000 valued at $5.34 per carat.

There are three major types of natural industrial diamonds: (1) bort, which includes off-color, flawed, or broken fragments of diamonds unsuitable for gems, (2) carbonado, or black diamond, which is a very hard and extremely tough aggregate of very small diamond crystals, and (3) ballas, a very hard, tough, globular mass of diamond crystals radiating from a common center. Carbonadoes come only from Bahia, Brazil; ballas chiefly from Brazil but a few from South Africa. Bort comes from all diamond-producing centers. In addition, there is a considerable production of diamond dust and powder, waste from cutting gem diamonds. Artificial diamonds are discussed in Part 4, Commodities, "Diamonds."

The industrial diamond has become one of the most important and essential materials in modern industry. Diamond drilling, once used only for locating metallic ores, is now widely used also for exploring nonmetallic mineral deposits; for exploring geologic structures; for testing foundations for dams, buildings, and heavy machinery; for exploring internal condition in heavy concrete structures such as dams; for stope mining; explosive demolitions under special conditions; and other purposes.

Some of the most important uses are: diamond-drill bits for drilling rock and concrete; diamond dies for wire drawings; diamond-tipped tools for truing abrasive wheels and for turning and boring hard rubber, fiber, vulcanite, hard plastics, etc.; diamond-toothed (segmental) saws and rim-impregnated (continuous rim) saws for sawing stone, glass, quartz, metals, slicing expansion joints in concrete highways, etc.; wheels, both for grinding and for cutoff work, in which the working face consists of diamond grit bonded with a resinoid, metal, or ceramic product; diamond-tipped tools for cutting glass and for engraving gems; diamond powder for cutting gems. High-speed tool steels, cemented carbides, and other exceedingly hard, tough alloys can be cut and shaped efficiently with diamond tools, and diamond-tipped tools are essential for the rapid and accurate shaping, truing, and dressing of abrasive wheels.

A diamond wheel marking system (Fig. 8), analogous to that previously shown for ordinary grinding wheels, is used by the abrasives industry (Anon., 1970).

With diamond, as with ordinary abrasives such as aluminum oxide and silicon carbide, the previously mentioned properties of size, shape, and friability must be closely controlled so that diamond with the proper physical properties may be selected for a particular application.

Resinoid or vitrified bonded diamond wheels require a fairly friable, weak-shaped diamond abrasive grain; metal bonded wheels require a more durable, strong-shaped grain.

Akin to the weight-per-cubic foot apparatus (see page 15) used for the ordinary abrasives a similar device (Fig. 9) was designed and built by Norton Co. (Hight, 1971) to determine the bulk density of diamond abrasive samples. It consists of a sample collecting tube which will hold a 20-g sample, funnel, vibrator, rheostat, depth gage, and timer. The cardinal rule in any determination of this type is close reproducibility of conditions from one test to another.

Another shape test is a microscopic technique, developed at the Diamond Research Laboratory, Johannesburg. It is fast and inexpensive but is quite subjective. This method was further refined by Custers and Raal (1959).

In this method four shape categories are recognized. Shape No. 1 is a near-perfect cube or sphere while shape No. 4 is a thin flake or a needlelike grain. Shapes No. 2 and No. 3 are intermediate with No. 2 nearer No. 1 and No. 3 nearer No. 4. The arithmetic mean of 100 grains is determined and this becomes the "shape count" for that particular sample.

A common and inexpensive friability test for measuring the strength of diamond is the Wig-L-Bug friability test mentioned on page 13.

A friability test which is similar in principle but much more sophisticated as to controlled running conditions was devised by Belling and Dyer (1965) at the Diamond Research Laboratory, Johannesburg.

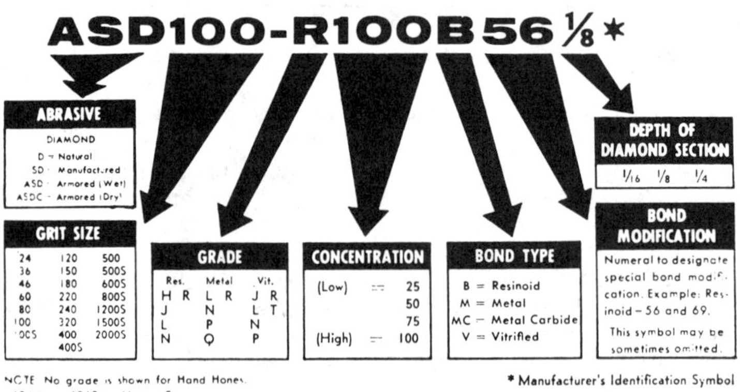

FIG. 8—*Diamond wheel marking system.*

FIG. 9—*Tap-density apparatus.*

Garnet: † The name "garnet" is given to a group of iron-aluminum silicate minerals having similar physical properties, crystal forms, and general chemical formula.

Composition—In garnet, general formula $3RO \cdot R_2O_3 \cdot 3SiO_2$, the bivalent element may be calcium, magnesium, ferrous iron, or manganese, the trivalent element, aluminum, ferric iron, or chromium, rarely titanium; further, silicon is also replaced by titanium in some examples. There are three prominent groups and various subdivisions under each, many of these blending into each other. These are as follows:

1) Aluminum Garnet
   A. Grossularite, calcium-aluminum garnet ($3CaO \cdot Al_2O_3 \cdot 3SiO_2$).
   B. Pyrope, magnesium-aluminum garnet ($3MgO \cdot Al_2O_3 \cdot 3SiO_2$).
   C. Almandite, iron-aluminum garnet ($3FeO \cdot Al_2O_3 \cdot 3SiO_2$).

† This portion of the chapter was originally prepared by H. H. Vogel, former Vice President of Barton Mines Corp. (now retired), and has been updated by Barton Mines Corp. staff.

   D. Spessartite, manganese-aluminum garnet ($3MnO \cdot Al_2O_3 \cdot 3SiO_2$).
2) Iron Garnet
   E. Andradite, calcium-iron garnet ($3CaO \cdot Fe_2O_3 \cdot 3SiO_2$).
3) Chromium Garnet
   F. Uvarovite, calcium chromium garnet ($3CaO \cdot Cr_2O_3 \cdot 3SiO_2$).

Properties

COLOR—Colors vary greatly but generally are as follows:

Grossularite—white, pale green, or yellow.
Pyrope—deep red to black.
Almandite—deep red, brownish red to black.
Spessartite—brown to red.
Andradite—black, green, and yellow green.
Uvarovite—emerald green.

Hardness ranges from 6 (grossularite) to 7.5 (almandite); some almandite has a hardness of between 8.0 and 9.0.

CRYSTAL SYSTEM—Cubic, commonly as rhombic dodecahedrons or tetragonal trisoctahedrons, or in combination of the two.

CLEAVAGE—Occasionally an indistinct dodecahedral cleavage is observed; some species of almandite possess a pronounced laminated structure, these are planes of weakness along which the mineral separates, this *parting* has no relation to the crystal form and is not a true cleavage.

FRACTURE—Garnets having a glassy structure usually have a marked conchoidal fracture but sometimes the mineral tends to break into thin flakes. In other varieties the fracture is sharp and uneven.

INDEX OF REFRACTION—The index of refraction of the garnet group ranges from 1.735 to 1.94.

TENACITY—Aggregates of crystal composed of many small individuals are brittle and shatter readily. Massive garnet and well formed crystals are remarkably tough and shatter with difficulty.

FUSIBILITY—Garnets having a high iron content, such as almandite, fuse at a temperature of about 1200°C. White garnets containing a considerable percentage of chromium are infusible.

OTHER PROPERTIES—Specific gravity of the garnet group ranges from 3.5 to 4.2; luster vitreous, resinous, or dull; transparent to opaque.

Production—Production of abrasive garnet in the United States, as published by US Bureau of Mines, is shown in Table 4.

Occurrence—Garnet commonly occurs as accessory minerals in a large variety of rocks,

TABLE 4—Production and Value of Garnet in the United States

Year	Short Tons*	Value, $
1973	22,772	2,380,000
1974	24,684	2,551,000
1975	17,204	2,092,000
1976	24,565	3,568,000
1977	20,022	3,136,000
1978	22,058	3,678,000
1979	23,303	4,335,000
1980	26,550	4,573,000
1981	27,850	4,300,000

Source: Adams, 1977; Baskin, 1979–1980; Smook, 1981.
*Metric equivalent: 1 st × 0.907 184 7 = t.

more particularly in gneisses and schists but also in contact metamorphic deposits, in crystalline limestones, pegmatites, and in serpentines. It occurs as gangue in ore veins formed at high temperatures. As most varieties of garnet are resistant to chemical and mechanical erosion, they tend to be concentrated in the sands of present-day or preexisting beaches, streams, or other alluvial deposits.

Deposits of garnet are found in many foreign countries and in nearly every state. An estimated 95% of the world's production of technical abrasive garnet comes from a small area in the Adirondack Mountains, New York State. Eighty percent of current usage of garnet is accounted for by United States industry.

The superior abrasive quality of Adirondack garnet currently being mined for technical applications is due to a combination of crystal properties including hardness (which is at the top of the garnet hardness range). This garnet, basically a combination of almandite and pyrope, exhibits incipient lamellar parting planes which break under pressure into sharp chiseledged plates. When crushed to a very fine size it retains its natural sharp irregular grain shape. These features, particularly important in coated abrasives, are also desirable for other abrasive applications. Since Adirondack garnet occurs in large crystals a complete range of sizes results from the crushing and grinding operations necessary for liberating and separating the garnet from matrix minerals.

United States—In recent years garnet production in the United States has been confined almost exclusively to two states, New York and Idaho.

NEW YORK. Although there were at one time several technical abrasive-grade garnet producers in New York, present production is confined to one company, the Barton Mines Corp., Gore Mountain, near North Creek. This company supplies garnet to coated abrasive, glass, and metal lapping industries throughout much of the world. It has been in continuous operation since 1878 and, through technical advances enabling it to utilize leaner ores, has constantly proved new ore reserves. This mine is believed to be the country's second oldest continuous operating mine under one management and, geologically, is one of the world's most interesting ore deposits.

The almandite-bearing rock, which is an igneous-metamorphic rock of uncertain origin, lies at the surface and is quarried. The ore body is over 2 km (1¼ miles) long and ranges in width from 15.2 to 91.4 m (50 to 300 ft). The principal gangue minerals are hornblende and plagioclase feldspar which constitute from 40 to 80% of the rock. Less abundant minerals include hypersthene, magnetite, biotite, apatite, and pyrite. The garnet content of the ore body ranges from 20 to less than 5% and averages slightly less than 10%. The garnet occurs as crystals, mostly imperfectly developed, known locally as "pockets." These range in size from a fraction of an inch to more than a foot in diameter. Occasionally crystals up to 914 mm (36 in.) in diameter have been seen; however, their average size is less than 102 mm (4 in).

Nearly every crystal of garnet is surrounded by a rim of coarsely crystalline hornblende. The quarry faces present a striking appearance, showing crimson-red garnet crystals set in a coal black background. Specific gravity of the garnet averages about 3.95, while the hornblende ordinarily is 3.07 to 3.24, with some specimens of very dense hornblende having been found with a specific gravity as high as 3.40. The general matrix is diorite.

The ore is mined in benches about 12.2 m (40 ft) in height, hauled in trucks, and dumped into a crushing system designed for initial liberation of garnet particles and reduction of particle size to dimensions required for separating the garnet from gangue minerals and other impurities. Separation is accomplished through a combination of concentrating methods including heavy media, magnetic, flotation, screening, tabling, and air and water separation. Processes are interconnected and are continuous or semicontinuous until a concentrate of 98% minimum garnet content has been achieved for the various grade sizes.

Elsewhere in New York, garnet is produced as a byproduct of wollastonite mining at Willsboro by Interpace Corp. and is sold in graded form for sandblasting operations in the Northeast. This garnet occurs as loosely cemented fine crystals that appear to be remnants of weathering and water deposition within preexisting stream channels or along shorelines. Separation is made electromagnetically after crushing and drying of the wollastonite ore. Separation is not complete and varying amounts of wollastonite accompany the garnet in its packaged form.

IDAHO. The only other important producer of abrasive garnet is the Sunshine Mining Co.'s subsidiary, Idaho Garnet Abrasive Co., whose deposit is on Emerald Creek and whose plant is at the creek's mouth near Fernwood, Benewah County, Idaho. This company works an alluvial deposit of almandite garnet, derived from the erosion of soft mica schists in which the garnets have a maximum grain size of about 4.8 mm (3⁄16 in.).

The garnet bearing gravel is mined by dragline and the garnet is concentrated on jigs and tables. It is then crushed and screened into various sizes.

This garnet is used mainly for sandblasting and spark plug cleaning.

FOREIGN COUNTRIES—Garnet is or has been commercially mined in Madagascar, Japan, Argentina, India, and Tanzania. Garnet mining in Russia has been reported but details regarding mining are not known.

Preparation For Market—GRADING AND SIZING—Practically 100% of the garnet sold today is a sized and graded product. Sizings are governed by mining techniques and end use. They generally conform to standards established in cooperation with the National Bureau of Standards for quality criteria, standard methods of test, rating, certification so as to provide a uniform basis for fair competition. Grading standards for coated abrasive use and micron sizing for grades used in glass grinding and metal lapping have exceedingly close tolerances. Wheel grades, polishing grades, and sandblast grades each have different standards and much greater tolerances. Government specifications can be found in Product Standard PS8–67, Grading of Abrasive Grain on Coated Abrasive Products, and in Commercial Standard CS–271–65, Grading of Abrasive Grains for Grinding Wheels. Also there are innumerable special grades for tumbling and miscellaneous uses.

HEAT TREATMENT—Practically all garnet grain used today is heat treated as a processing function and not for the purpose of improving inherent abrasiveness. Impurities picked up during processing adhere to garnet particle surfaces and destroy cleanliness and capillarity needed in later processing to give adhesion for bonding. Early in the 1930s it was thought that "roasting" temperatures improved the hardness, toughness, and fracture of the garnet. In the late 1940s it was thought higher temperatures would further increase the abrasive quality. Temperatures approaching 1000°C were used for some products. In the early 1960s scientific investigation determined that too high a temperature could destroy the abrasive quality of garnet and that heat treating solely accomplished the cleansing of particle surfaces or at most it produced superficial heating of microscopic fractures on the grain surface. In the mid-1960s the coated abrasive industry accepted uniform color standards assuring cleanliness of the grain surface and presenting a color standard most closely approaching the natural red color of abrasive garnet. This degree of heat treatment is known as RT treatment.

TEST FOR ABRASIVE QUALITY OF GARNET—There is no exact method for testing the abrasive quality of garnet or any loose abrasive except by practical application. However, several rough tests and examinations serve to indicate their abrasive possibilities. The fracture, sharpness and shape of the grain, the character of the grain structure, and the determination of presence of inclusions of other minerals which would weaken the grain structure can be studied under the microscope. Hardness and toughness of grains can be determined roughly by placing a definite grade, such as #60, between two glass microscope slides and rubbing them together. The relative scratch hardness of the grains and the amount of breakdown can be indicated in this manner.

The US Bureau of Standards has developed an abrasive tester designed primarily for testing abrasive quality of corundum. This can be adapted for testing the abrasive quality of any loose abrasive grain. Loose abrasive grain is also tested on various production or laboratory machines for evaluating glass removal on flat plates. These data can be correlated with the anticipated performance of the material to be abraded as an evaluation of actual abrasive performance. In all these methods actual stock removal is determined either by weighing or measuring. In the coated abrasive industry, test belts are made and stock removal is measured on test blocks. These tests measure stock removal only, and are a valuable indicator when other important variables such as bond, backing, coating, flex, grit size, etc. are closely controlled.

In judging the abrasive quality of garnet as well as all other abrasives, stock removal, while important, is not the only factor in judging the value of an abrasive. Surface finish in most cases is perhaps a more important factor. An abrasive used on some materials can be too hard and cause deep scratches in the substructure which cannot be removed in subsequent operations. Also, if the abrasive does not have the proper grain shape or becomes dull in use and does not break down to yield new sharp cutting edges while in use, it will tend to burn or gouge the material being ground or surfaced, thus rendering it unusable in subsequent operations.

Uses And Markets—The major industrial uses of garnet are: manufacture of coated abrasive products; grinding and lapping of glass, metals, ceramics, plastics; and for sandblasting. There are also a number of miscellaneous uses.

Abrasive Papers and Cloth—Garnet-coated papers and cloths are used primarily for wood sanding but also for finishing of leather, hard rubber, plastics, glass, and the softer metals. Coated abrasives have been improved in recent years mainly by using resin bonds in place of glue bonds for certain applications. Also, electrostatic coating is now in general use. In this process an electrostatic charge causes the grains to imbed upright in the wet bond on the backing. In effect the sharp cutting edges of the grain are bonded perpendicular to the backing. It also causes the individual grain particles to be spaced more evenly due to individual grain repulsion.

Garnet paper and cloth products comprise approximately 10 to 11% of the total volume of coated abrasives produced for the past several years. This percentage has remained fairly constant notwithstanding the general growth of the economy and substitutions of abrasive papers and cloths for other grinding methods.

Glass Surfacing—During World War II when imported corundum became scarce for grinding of optical lenses the War Production Board searched for substitute abrasives. Garnet was the logical mineral, possessing desirable properties and availability, although it had never before been processed for this application. After development it fulfilled the critical need and by the time the war ended the desirable qualities of garnet were widely known, particularly because it decreased lens scratching to a considerable extent. Its use improved the quality and reduced rejections, both important factors in the economy of lens grinding. Garnet also produced a finish which reduced polishing time. These same properties which made garnet a competitive abrasive for this application continue to produce new markets in the glass, plastics, ceramics, and softer metals industries.

Loose Grains—A major nontechnical use of garnet in loose form is for sandblasting. Garnet has the advantage of having no free silica and hence cannot cause silicosis. It is heavier than quartz sand and, thus, grains of equal size deliver harder blows. Practically all production of Idaho garnet is sold to West Coast aircraft industries as a sandblast agent because of its price and location.

Gem Garnet—Some domestic garnet is marketed for cutting into semiprecious gems. The dollar value of this is small and unimportant. Ordinarily gem garnet is produced in parts of the world having low labor costs; there are numerous such sources of suitable quality garnet.

Miscellaneous Uses—In grain sizes garnet is used for the manufacture of grinding wheels, as tumbling media, for nonskid applications, and as filter media. In powder sizes it is made into lapping and buffing compounds.

Markets And Prices—The problems of the established producers of garnet are meeting costs of technological developments, maintenance, and increasing sales in the face of competition from other natural and manufactured abrasives. In general, every abrasive has an application in which it surpasses other abrasives. In most applications these qualifications overlap and it is in these overlapping fields where competition is keenest and is based on relative cost, considering grinding and finish abilities.

Garnet has kept its relative position in the expanding abrasive market and present producers have solved problems of maintaining necessary ore reserves through exploration and application of new mining technologies which permit economical use of leaner ores.

Although garnet is common in most states and countries throughout the world, most of it is of a quality that would be uncompetitive. However, there are potential new sources and much technical quality byproduct garnet available. Relatively large capital investment is required to produce garnet and this factor combined with the relatively small and specialized market discourages the opening of new deposits. Furthermore, the present market is adequately supplied by present producers.

Market prices of garnet grains and powders vary widely depending on garnet type, purity, particle size, grade control, shipping quantities, and packaging.

Silica: Silica sand and quartz are reviewed in detail elsewhere in the volume, so that only their abrasive use and applications are given here.

Silica Sand—Silica sand is extensively used for pressure blasting, for the initial grinding or surfacing of plate glass, and as a cutting medium for gang saws on stone.

For pressure blasting the main centers of production in the United States are Ottawa, IL, and Cape May, NJ. The Ottawa material occurs as a friable sandstone, which is broken down to its natural grain, washed, and screened. The grains are spherical in shape. The Cape May sand is subangular and uneven in shape but larger grain sizes are obtainable, more so than with the Ottawa sand.

In Canada, pressure blast sand is obtained from the decomposed rock containing friable quartz and china clay at Lac Remi, north of Montreal, Que.; from a similar type of material near Smoky Falls, north of Lake Nipissing, 40 km (25 miles) west of North Bay, Ont.; also from a friable quartzite at East Templeton, Que., a few miles northeast of Ottawa, Ont. The grain of the Canadian sands is sharp. The ranges of grain size are approximately: No. 1 between 20 and 35-mesh for light work; No. 2 between 10 and 28-mesh; No. 3 between 6 and 10, and No. 4 between 4 and 8-mesh. The last two are used for the heavy cast-iron work and steel work.

Pure, clean beach and river sands and Illinois sand are used for the preliminary or coarse surfacing of plate glass. The crushed sand is water-graded into a number of grades at the glass plants and fed to the surfacing machines. Approximately three tons of sand are required to surface one ton of plate glass.

Cutting sand, composed of sharp, solid quartz grains, is used as an abrasive for sawing stone. It is usually ungraded and about equivalent to a No. 1 pressure blasting sand.

Burnishing sand is a fine, rounded-grain silica sand of uniform size between 65 and 100-mesh, used in rolling down and burnishing gold decorations on porcelain.

Evans (1980) gives the following quantities and values for industrial sand sold or used by producers in the US: for grinding and polishing 190.5 kt (210,000 st) at $1,710,000 and for pressure blasting 1.7 Mt (1,868,000 st) at $13,645,000.

Although production figures are not available at this time, it should be noted that blast furnace slags have been processed and used as substitutes for pressure blasting sand. Increased use of such materials may be expected in the future because of the possible health and environmental advantages over silica products.

Quartz—Crushed and graded quartz is used for the abrasive backing of "flint" sandpapers. Almost any deposit of massive white quartz is suitable. Being the cheapest of all the abrasive-coated paper, it still is sold in fair amount, mainly in hardware stores and by small jobbers. It is made only in the form of paper, not as cloth. True chalk flint from England and France is extensively employed for this purpose in Europe and has better cutting qualities and longer life than ordinary quartz.

Powdered quartz and silt are sometimes used for scouring compounds and for the harsher metal polishes.

Sand and sandstone are also crushed and ground and used for abrasive purposes, as in hand soaps, scouring compounds, metal polishes, etc. In 1979, Tepordei (1978-1979) reported a sale for abrasive use of 366.5 kt (404,000 st) of ground sand valued at about $5 million.

Soft Siliceous Powder Abrasives: Many natural highly siliceous materials either occur as a powder or are used only in the powder form for mild abrasives. For the majority of these, use as an abrasive is of minor importance as compared with their principal applications. Among these, diatomite, pumice, and tripoli are reviewed elsewhere in this volume. Their abrasive uses are therefore given only briefly here.

DIATOMITE—Diatomite production in the United States comes primarily from California and Nevada with small outputs from Oregon and Arizona (Table 5).

The amount used for abrasives is insignificant in comparison with its other applications —most important of which are filtration, fillers and insulation. Its abrasive uses include metal (silver) polishes, dental powders and pastes, and occasionally it is used as a friction agent in the manufacture of matches.

PUMICE—Under the name of pumice are included lump pumice and pumicite or volcanic

TABLE 5—Diatomite Sold or Used by Producers in the United States

Year	Domestic Production (Sales), Short Tons*	Average Value Per Ton, $
1973	609,000	59.26
1974	664,000	76.31
1975	573,000	80.01
1976	631,000	87.08
1977	648,000	98.56
1978	651,000	111.26
1979	717,000	125.97
1980	689,000	146.02
1981	707,000	152.48

Source: Meisinger, 1982.
*Metric equivalent: 1 st × 0.907 184 7 = t.

dust, the natural powder. Lump pumice is used by manufacturers of furniture and musical instruments; for dressing the wood and metal surfaces; by silver platers for preparing their metal surfaces: by lithographers for cleaning stone surfaces; for rubbing down and polishing fine tools and instruments; by restaurants for scouring grills and cooking utensils, and for domestic and toilet uses, such as hand cleaners. Pumicite or ground pumice is mainly used as a cleanser, the thin, sharp, and striated grains being particularly suitable.

The use of pumice as an abrasive has declined greatly in the last decade; the largest drop was in pumice, mostly pumicite, used in cleaning and scouring compounds and in hand soaps. Abrasive use of pumice now accounts for less than 1% of total production. Minor uses for abrasive pumice are in polishing compounds; polishing powders for bone, celluloid, and hard rubber; for dental use; and in some rubber erasers.

The 1980 consumption of pumice for abrasive purposes was about 24 kt (27,000 st), with a value of $568,000.

The consumption of pumice for "other uses" which includes abrasive use has declined from approximately 907 kt (1,000,000 st) in 1955 to 492.6 kt (543,000) in 1980.

TRIPOLI, MICROCRYSTALLINE SILICA, AND ROTTENSTONE—The fine-grained, porous materials, tripoli, microcrystalline silica, and rottenstone, are known to the trade as "sift silicas."

Tripoli, which in the United States comes from southwest Missouri and northeast Oklahoma, is mainly used in the form of made-up tripoli grease bricks or tripoli compositions for buffing and polishing. The compound is applied to a rapidly revolving belt or canvas wheel and used for the finishing or buffing of metals, plated products, and so forth. It is used also to a small extent in the manufacture of some scouring and cleaning powders and soaps; for the rubbing down of painted surfaces, such as automobile bodies. A similar but finer-grained material occurring in the northwest corner of Arkansas, about 80 km (50 miles) southeast of the Missouri deposits, is used mainly for oil-well drilling mud.

Microcrystalline (sometimes erroneously termed "amorphous") silica, which comes mainly from southwestern Illinois but to some extent from Wayne County, TN, also is used for buffing and polishing compounds. These compounds are termed "silica" by the trade and are much in demand for white "coloring" operations on high-class work. Chemically precipitated amorphous silicas also are used in polishing and buffing compositions.

Both tripoli and microcrystalline silica have been mined from deposits at Harrisburg, northwestern Georgia.

Rottenstone, a fine-grained gray-buff siliceous-argillaceous limestone, comes from Antes Forte, Lycoming County, PA, and is used as a polish base, for instance, for automobile polishes.

The domestic production of tripoli, rottenstone, and amorphous silica for abrasive use in the past few years has ranged from 72.5 to 116 kt (80,000 to 128,000 stpy) with a crude value of about $6 per st. Processed tripoli has values which range from $60 to $80 per st.

Nonsiliceous Soft Abrasives: *Ground Feldspar* is extensively used in scouring and cleaning compounds and for a window cleaner.

Chalk (calcium carbonate) is a soft, compact, fine-grained, white limestone composed of the calcareous remains of small marine shells. A small amount of this chalk—mainly from England and France—known as "whiting" is used as a very mild abrasive for hand polishing of nickel, gold, silver or plated ware, buttons, and similar materials.

China Clay (kaolin) and some pipe clays have been used successfully in polishing powders. Pipe clay at one time was the standard polish for naval and military tunic buttons.

BATH BRICK, used for scouring steel utensils, is made from a very fine-grained, quartzose clay found along the banks of the Parrot River in England.

STAUROLITE, a complex iron aluminum silicate, has been commercially recovered from various river placer deposits in the southeastern US. DuPont operates a deposit in Clay County, FL. It is used mostly for pressure blasting.

Sharpening Stones: Hand-used sharpening stones include scythestones, whetstones, oil stones, water stones, razor hones, holystones, and rubbing stones. While, in general, stones made from bonded artificial abrasive grains have largely replaced these natural abrasives there is still some small production of oilstones and whetstones in Arkansas, whetstones in Indiana, and scythestones in New Hampshire. Arkansas novaculite stones are still preferred for sharpening fine-edged tools for surgeons, carvers, and engravers. It is claimed that these stones give a smoother and longer-lasting edge.

Natural abrasive stones are made from a wide variety of material, including sandstone, novaculite, mica schist, siliceous argillite, shale, slate, and pumice. The superior cutting quality of some of these stones is due to the inclusion of well disseminated fine-grained inclusions of garnet or other minerals of superior hardness.

Grinding Pebbles and Tube-Mill Linings: Tube-mills, conical-mills, and cylindrical batch mills are used more often than any others for the fine grinding of hard ores, minerals, paints, chemicals, ceramic bodies and glazes and enamels, portland cement clinker, and similar materials.

Most efficient grinding can be done in mills with iron or steel liners and iron or steel balls, slugs, or rods, but where contamination of the color or chemical purity of the product by metals and metal oxides must be avoided, the grinding surfaces consist of blocks or bricks for the mill linings and natural pebbles or artificially prepared balls.

These materials should (1) have high abrasive hardness; (2) have great toughness and freedom from flaws, which will result in breaking and chipping; (3) produce a white powder when abraded; and (4) contain no dark-colored metallic impurities such as iron or manganese. While most of the natural and artificial mill linings and pebbles have a specific gravity close to that of quartz (2.6), a greater density in pebbles has obvious advantages. The heavier the pebbles per unit of size, the greater are the impact and abrasive grinding forces. Thus, with heavy pebbles less mill volume is needed for a given pebble load (by weight), resulting in greater mill capacity; or with equal pebble volume, faster grinding can be done. Small balls grind finer than large ones (because of more surface area per pound), but with small pebbles the unit weight is low unless high-density pebbles or balls are used.

The most favored natural mill-lining material for most purposes is Belgian silex, which is a very hard, tough, more or less cellular quartzite resembling French buhrstone. This is imported in rectangular blocks more or less shaped to fit the curve of a mill. During World Wars I and II, when imports were cut off, and to a much lesser extent at other times, domestic substitutes have been used. These have consisted chiefly of quartzite from near Jasper, MN; Iron City, TN; and Baraboo, WI; and granites from Salisbury, Lilesville, and Faith, NC. Some of these materials are reported to give service equal to that of Belgian silex. Special hard, dense, porcelain blocks are used in some mills for grinding paint, ceramic, and chemical materials.

For grinding media, Danish flint pebbles, when available, have long been standard because of their superior hardness, toughness, and uniformity. These pebbles are found on the shores of Greenland but are marketed through Denmark. Other foreign sources of similar pebbles exist along the seacoasts of Belgium, France (between Havre and St. Valery-sur-Somme), Norway, and England. Seven sizes of Danish pebbles are marketed, ranging from 25.4 to 203 mm (1 to nearly 8 in).

Domestic substitutes for grinding media are natural pebbles of flint, quartz, and quartzite, as well as artificially rounded (by tumbling small blocks in rotating cylinders) pebbles made from quartzite, granite, chalcedonized rhyolite, and other rocks. During World War II, pebbles were shipped from Encinitas Beach, CA (true flint); from Jasper, MN (quartzite); from the Austin chalk beds in south-central Texas (true flint); from Salisbury and Faith, NC (granite); from Baraboo, WI (quartzite), and several other points in the United States. Extensive deposits of quartzite pebbles in southwestern Saskatchewan have been reported. Beach pebbles were reported nonuniform in hardness and requiring careful sorting.

Ceramic pebbles are made from very dense, tough porcelain, alumina, and sillimanite (mullite). A dense, heavy pebble is made from zircon, having a specific gravity of 3.7. For certain purposes blocks of ore are used as grinding media; thus nepheline syenite blocks are used instead of pebbles for grinding this material.

Table 6 shows production statistics for special silica-stone products.

TABLE 6—Special Silica-Stone Products Including Grinding Pebbles, Grinastones, Oilstones, Tube-Mill Liners, and Whetstones

	1975	1976	1977	1978	1979	1980	1981
Short tons*	2,953	2,696	2,200	2,175	2,094	2,131	1,978
Value (thousand $)	1,061	1,404	3,236	2,630	2,064	2,233	2,660

Source: Baskin, 1978–1979; Smook, 1982.
*Metric equivalent: 1 st × 0.907 184 7 = t.

Manufactured (Artificial) Abrasives

The discovery of the electric-furnace method for making silicon carbide by Edward G. Acheson in 1891 began the revolution in the high-grade abrasives industry. Prior to that time, all abrasives were natural minerals and rocks, except for minor chemically prepared materials such as rouge and Vienna lime. Today, as noted before, artificial abrasives, with a few notable exceptions such as diamond and garnet, dominate if not monopolize the field of high-grade abrasives.

Manufactured abrasives are essential to modern industry because they are not only superior to natural abrasives but they are also much more uniform in quality and their properties can be varied to meet differing needs. Modern high-speed streamlined production methods require very accurate, specialized tools with a degree of uniformity and dependability that cannot be met with natural abrasives.

Manufactured abrasives may be divided into three main groups: (1) electric furnace products, (2) chemical precipitates, with (3) several miscellaneous additions.

Table 7 (Baskin, 1978-1979) gives production statistics for certain types of artificial abrasives.

Electric-Furnace Abrasives: Electric-furnace abrasives include silicon carbide, aluminum oxide, and boron carbide. Tungsten carbide, tantalum, and titanium carbides are used in making cemented-carbide high-speed cutting tools but are not considered as abrasives in customary usage.

Silicon Carbide—Silicon carbide, with the chemical formula SiC, is commonly known by the trade names Carborundum, Crystolon, and Carbolon. It is made by fusing a mixture of high-grade silica sand and carbon in an electric furnace. The preferable form of carbon used today is petroleum coke, but anthracite and coke made from coal low in ash have been used. High iron and alumina in the ash are objectionable because they tend to assist in the formation of elemental graphite and metallic silicon during the reaction. High-ash carbons make impure grades of silicon carbides.

"Silicon carbide is made by charging a resistance type electric furnace with pure glass sand . . . finely ground petroleum coke, saw-

TABLE 7—Crude Artificial Abrasives Produced in the United States and Canada (Thousand St[‡] and Thousand $)

Kind	1975	1976	1977	1978	1979	1980	1981
Silicon carbide*	134	159	192	182	196	170	159
Value	$31,842	$45,953	$53,814	$51,371	$62,702	$64,346	$70,941
Aluminum oxide (abrasive grade)*	141	191	185	142	225	193	199
Value	$28,368	$43,356	$48,819	$46,633	$67,511	$63,881	$67,253
Aluminum zirconium oxide	17	20	20	23	28	19	W
Value	$8,506	$11,383	$11,281	$14,668	$14,893	$8,438	W
Metallic abrasives†	236	250	243	204	264	235	247
Value	$72,864	$75,372	$72,740	$59,882	$84,918	$81,935	$90,471
Total	528	620	640	551	713	617	605
Value	$141,580	$176,064	$186,654	$172,554	$230,024	$218,600	$228,665

Source: Baskin, 1980; Smook, 1982.
*Figures include material used for refractories and other nonabrasive purposes.
†Shipments for US plants only.
W = Withheld.
‡Metric equivalent: 1 st × 0.907 184 7 = t.

dust and common salt. The silicon of the sand combines with the coke, which is carbon, to form silicon carbide. The sawdust makes the mixture porous so that the carbon monoxide gas formed in the course of the chemical reaction can escape freely" (Heywood, 1942). The salt (sodium chloride) combines with the various impurities in the sand and coke to form chlorides, which can then be eliminated from the mix, by volatilization. The temperature at the core of the furnace is about 2200°C. If the temperature is too high, silicon carbide decomposes, the silica being volatilized and the carbon becoming graphitized. The time of a run is about 60 hr (36 hr heating and 24 hr cooling). At the end of the run the core consists of loosely knitted silicon carbide crystals surrounded by unreacted or partially reacted raw materials. The quality of the product depends largely on the raw materials used and the accuracy of control and operation of the furnace.

Several types of products are made in which the toughness of the grains is varied. A special surface treatment is given to grains for coated abrasives to increase adhesion of the glue.

The lump material from the furnace is crushed, first in jaw crushers then in rolls, hammer mills, or ball mills, passed through a magnetic separator to remove magnetic impurities, and then washed to remove dust. The grains are then dried and screened mechanically into 20 or more standard sizes, from 8 to 240-mesh. The finer flour sizes are made by air or water flotation and sedimentation.

Fused Aluminum Oxide: Aluminum oxide abrasives are made in electric-arc furnaces. Bauxite, the crude raw material, should have the following typical physical and chemical properties:

Typical Chemical Analysis	
Al_2O_3	85.0% to 87.0%
TiO_2	3.0% to 4.5%
SiO_2	3.0% to 5.5%
Fe_2O_3	6.0% to 10.3%
L.O.I. plus combined water	1.0% to 2.0%

The ore as mined is crushed, then calcined in rotary kilns to remove both free and combined water. The calcined ore is mixed with iron borings (about 2%) and ground coke (about 3%) and charged into the furnace. The coke reduces the impurities which combine with the iron and sink to the bottom of the furnace.

At the end of the melting period, the charge is allowed to cool under controlled conditions to obtain desired texture. The finished, crystallized alumina consists of about 95% Al_2O_3, 3% TiO_2, 1.5 % SiO_2, and 0.5% Fe_2O_3. For certain purposes, Bayer-process alumina is used instead of calcined bauxite, making a product with about 99% Al_2O_3.

The cooled mass from the furnace is crushed, screened, and cleaned in about the way that silicon carbide is treated. Here also several types of grain are made, varying in toughness, type of fracture, glue adhesion, and other properties.

Two new heavy-duty abrasives have been developed recently. One is an electric furnace fusion product made up primarily of an intimately crystallized mixture of aluminum oxide and zirconia. Although other proportions have been successful, those containing approximately 25% and 40% zirconia, respectively, have found most widespread commercial application.

These fused alumina-zirconia abrasives are noted for their extreme durability and therefore are particularly well suited for foundry snagging and steel conditioning operations.

The other family of new abrasives is the sintered product. This differs drastically in method of manufacture from the more traditional fused product in that the raw materials are heated in kilns only to the sintering temperature (1200°C±) rather than being carried all the way to the fusion point (2000°C±). Although other basic starting materials have been used, bauxite with or without additives is formed while still in the green (unfired) state into predetermined sized and shaped granules by certain granulation techniques or by extrusion through an appropriately shaped orifice. The resultant presized and preshaped granules are then sintered to make the finished abrasive grain. Process steps, such as crushing, shaping,

Typical Screen Analysis	
+1/2 mesh	4%
+3 mesh	11%
+14 mesh	42%
+65 mesh	24%
+100 mesh	6%
+200 mesh	9%
+200 mesh	4%

sizing, washing, and drying common to manufacturing fused products, are thus eliminated in making a sintered product.

Boron Carbide: Boron carbide is an artificial abrasive introduced in 1934 by the Norton Co. under the name of Norbide. Its chemical formula is B_4C. Although considerably harder than silicon carbide it is far from the hardness of the diamond (see Table 1). It is made from boric oxide, B_2O_3, and carbon in the form of petroleum coke in a carbon resistance furnace at a temperature of about 2600°C. The finished product is crushed and ground to make a range of sizes of grains and powders. In grain form it is used for grinding and lapping operations for jobs previously possible only with diamond dust. Pulverized Norbide may be molded under very high heat and pressure to form extremely wear-resistant products such as pressure blast nozzle liners, thread guides, extrusion dies, and all types of extremely accurate plug, snap, and ring gages. It is not manufactured into bonded wheels or sharpening stones.

Boron Nitride: Boron nitride in cubic form is a new, manufactured abrasive discovered by General Electric Co. laboratories in 1957 and trademarked "Borazon."

It, unlike manufactured diamond, has no known counterpart in nature. It is made at temperatures and pressures comparable to those required for diamond manufacture. The thermal stability of cubic boron nitride exceeds that of diamond. It is stable at temperatures above 1371°C (2500°F) while diamond reverts to graphite at about 816°C (1500°F). The Knoop hardness (K_{100}) is 4700 which places it well above the ordinary abrasives but inferior to diamond (General Electric Co., 1970).

In resinoid bonded grinding wheels Borazon has been found to be effective in grinding hard and tough tool steels.

Tungsten Carbide: Tungsten carbide is classed as a cutting medium rather than as an abrasive. However, it could be argued that the action of the rock bit tipped with tungsten carbide on the end of the steel shank of a jackhammer is as much abrasive as it is cutting. Certainly the almost universal use of tungsten carbide tip on such bits has revolutionized drilling blastholes in mining and quarrying operations. Costs have been reduced because the penetration rate is increased, a smaller hole can be drilled permitting the use of sticks of dynamite with reduced diameter, giving more economical use of the explosive.

Mention has been made in an earlier paragraph of the use of steel impregnated with tungsten carbide in saw blades.

Manufactured Diamond: In February 1955, General Electric Co. announced that its laboratories had succeeded in producing diamond from carbonaceous material subjected to pressures of around 5.5 GPa (800,000 psi) at very high temperatures for prolonged periods. Crystals up to 1.6 mm (1/16 in.) long were made in 16 hr but smaller ones were produced in much shorter times. The press is capable of maintaining temperatures above 2760°C (5000°F) at pressures up to about 11 GPa (1.6 million psi). Costs were reported to be about twice that of comparable natural diamonds and sizes were small; nevertheless economical commercial production was anticipated. The diamonds so far made have not been of gem quality.

In October 1957, the company announced it had produced well over 100,000 carats and that by the end of the year a "substantial quantity" would be in industrial use. The company also stated that diamonds could now be made in a relatively few minutes and that their performance equals that of natural diamonds.

The use of manufactured diamond has steadily increased. In 1969, domestic production of manufactured industrial diamond amounted to approximately 13 million carats; by 1981 this quantity had soared to approximately 57.0 million carats.

Today there are several types of manufactured diamond on the market, of both domestic and foreign manufacture, each with unique characteristics which make them suitable for the specific end-use applications previously discussed under "Industrial Diamond (Natural)."

In resinoid bonded grinding wheels, the adherence of the bond to the abrasive seems to be improved if the diamond is metal-coated. Hence, metal-coated (nickel, copper) diamond is still another commercially available diamond type.

Metallic Abrasives: Metallic abrasives include crushed steel, steel shot, angular steel grit, steel wool, brass wool, and copper wool. The following description of the first three of these products is by Johnson and Schauble (1939):

> Crushed steel is made from high-carbon and crucible sheet steel specially treated to impart brittleness. It is then crushed to sizes ranging from 2 to 200-mesh. After screening, each batch is heat-treated and separated into 25 sizes ranging from 20 to 200-mesh. Sizes from 70-mesh upward are screened on silk bolting cloth, and the finer sizes in powder form are used in steel cement, various chemical compounds, and fireworks sparklers.
>
> Steel shot is merely chilled cast iron. Only raw materials of the highest grade, including selected scrap and charcoal iron, are used in its manufacture; these are melted in a cupola. During the casting period, the molten metal is separated into small spheri-

cal globules by directing high-pressure steam or heated compressed air against the stream of metal. The globules are blown into water and cooled. The shot, made brittle by the rapid cooling, is heat-treated to impart a temper of hardness and graded by mechanical means into 15 sizes that range from 4 to 90-mesh. Coarse material left on the 4-mesh screen is granulated by specially designed crushing equipment and in this form is known as angular steel grit. The crushed material is heat-treated to impart toughness and durability, then graded into 15 sizes ranging from 7 to 100-mesh. Steel shot and angular steel grit are used for many abrasive purposes.

Steel wool, made from both ordinary and stainless steel, in various finenesses, is used for finishing wood and soft metals such as aluminum and for scouring and cleaning. It is made in several types of machines by shaving or scraping continuously moving wire with a fixed serrated cutting tool. Brass and copper wool are used chiefly for household cleaning.

Chemical Precipitates: Chemical precipitates, mainly oxides, all have a very fine grain size and are used as final polishing agents.

Iron Oxides—ROUGE, a ferric oxide, has its largest use in the polishing of plate glass, but is also widely used in powder, paste, coated cloth, and stick form for polishing precious metals, stone, and other materials to give a final high luster.

CROCUS, similar to rouge in composition but a purplish red in color, is used in finishing cutlery and some brass work.

BLACK ROUGE, a precipitated black magnetic iron oxide, is used primarily in plate printing inks and in paints, but has small abrasive applications.

Other Precipitates—CHROMIUM OXIDE or "Green Rouge" is used chiefly for platinum and stainless steels.

TIN OXIDE is used largely for polishing glass and precious stones.

PUTTY POWDER is a mixture of tin oxide and oxalic acid, used for about the same purposes as tin oxide.

CERIUM OXIDE is a polishing agent for polishing glass and a substitute for tin oxide for some other uses.

MANGANESE DIOXIDE is reported as having limited abrasive use.

MAGNESIA in the soft precipitated form has been noted as a mild polishing powder. The hard-burned variety, artificial periclase, also has been used as an abrasive.

CALCIUM CARBONATE, precipitated, is a mild abrasive used in some dentifrices.

CALCIUM OXIDE OR LIME, under the names of Vienna lime or Sheffield lime, has important abrasive uses. This so-called Vienna lime (originally from Vienna, Austria), which is used on the American continent in grease-brick buffing composition, is obtained from certain beds of dolomite at Francis Creek and Manitowoc, WI. Vienna lime is made by calcining the dolomite and cleaning and grinding to a certain fineness. It is packed in carefully sealed containers. As soon as the lime becomes hydrated it ceases to function and also attacks the grease compositions. It is used for the buffing of brass, copper, bronze, steel, pearl, celluloid, and similar materials but its main use is for the coloring of nickel after plating, as it gives nickel a deep undersurface blue peculiar to the metal. Lime attacks aluminum, therefore it is not used on that metal.

The use of porcelain blocks and pebbles for grinding mills has been noted. This "porcelain" may consist largely of zirconium oxides or silicate, or of mullite converted from andalusite, kyanite or dumortierite.

Miscellaneous Manufactured Abrasives: *Glass,* crushed and screened, has been used for coated abrasives to make "glass" paper.

Lampblack, known as "satin rouge," finds some use for polishing celluloid and bone.

Clay, very hard burned and finely pulverized, has been reported as used as an abrasive for polishing metal.

Bibliography and References

Anon., 1964, "American Standard Test for Bulk Density of Abrasive Grain," American Standards Assn., B74.4.

Anon., 1965, "American Standard Test for Ball Mill Test for Friability of Abrasive Grain," American Standards Assn., B74.8.

Anon., 1970, *Abrasives and Grinding Wheels,* Norton Co., Worcester, MA, 36th printing, Mar., pp. 22–62.

Anon., 1970a, Technical Publication SMD-B-9-500, General Electric Co., Dec.

Adams, W.T., 1980, "Abrasive Materials," *Minerals Yearbook, 1977,* US Bureau of Mines, Vol. I, pp. 107–118.

Baskin, G.D., 1978–1979, "Abrasive Materials," Preprint, US Bureau of Mines, 1978–1979, 15 p.

Baskin, G.D., 1980, Diamond (Industrial), *Mineral Commodity Summaries, 1980,* US Bureau of Mines, pp. 46–47.

Belling, N.G., and Dyer, H.B., 1965, *Impact Strength Testing of Diamond Abrasive Grit,* Diamond Research Laboratories, Johannesburg, South Africa.

Clarke, R.G., 1972, "Abrasive Materials," *Minerals Yearbook 1972,* US Bureau of Mines, pp. 123–134.

Custers, J.F.H., and Raal, F.A., 1959, "A New Method of Determining the Average Shape of Diamond and Other Particles," *Industrial Diamond Review,* Vol. 19, p. 72.

Evans, J.R., 1980, "Sand and Gravel," *Minerals Yearbook, 1977,* US Bureau of Mines, Vol. I, pp. 797–819.

Harben, P., 1978, "Abrasives—Taking the Rough with the Smooth," *Industrial Minerals,* No. 134, Nov., pp. 49–73.

Heywood, J., 1942, *Grinding Wheels and Their Uses,* Penton Pub. Co., New York, 436 pp.

Hight, R.P., 1971, "Diamonds, Industrial," *Encyclopedia of Industrial Chemical Analysis,* Vol. 11, John Wiley, New York, pp. 462–484.

Johnson, B.L., and Schauble, 1939, "Abrasive Materials," *Minerals Yearbook 1939,* US Bureau of Mines, pp. 1225–1240.

Knoop, F., et al., 1939, National Bureau Standards Research Paper 1220, *Journal of Research,* 23 pp.

Meisinger, A.C., 1982, "Diatomite," *Mineral Commodity Summaries, 1981,* US Bureau of Mines, pp. 46–47.

Oskam, J., 1980, "Garnets, Abrasive and Gem," Preprint 80-382, SME-AIME Fall Meeting, Minneapolis, Oct., 5 pp.

Ridgeway, R.R., Ballard, A.H., and Bailey, B.L., 1933, "Hardness Values for Electrochemical Products," *Journal of the Electrochemical Society,* May.

Smook, J.F., 1982, "Abrasive Materials in 1981," *Mineral Industry Surveys,* US Bureau of Mines, 3 pp.

Tepordei, V.V., 1978–1979, "Sand and Gravel," Preprint, US Bureau of Mines, 1978–1979, 29 pp.

Thibault, N.W., and Nyquest, H.L., 1947, "The Measured Knoop Hardness of Hard Substances and Factors Affecting its Determination," *Transactions,* American Society for Metals, Vol. 38, pp. 271–330.

Ceramic Raw Materials

LANE MITCHELL *

A ceramic product or processed material is a solid composed of materials which have been subjected to heat above 468.3°C (875°F). The raw materials, which are blended together (or in some cases used singly), are selected and utilized for various reasons such as ease of fabrication, aid to processing, to yield desired properties after processing, lack or controllable degree of toxicity or hazard, availability, cost, and perhaps other considerations.

Almost all solid inorganic, some organic, and some liquid and gaseous materials have the potential to be used individually in some ceramic product. There are several fundamental functions to be fulfilled in every ceramic product and a number of secondary functions are desirable in specified products. These functions are generally met by the most widely occurring minerals and rocks of the earth and their debris. This fact characterizes ceramics as a field utilizing the raw materials most available to man. Of course, specified products utilize synthetics, organics, and rare inorganics to produce desired results.

Any solid nonmetallic product processed by subjection to temperature above a red heat may be called a ceramic product. Widely used construction materials and units, domestic utensils and units, industrial parts to facilitate manufacture, technical components, and aesthetic creations are very often of a ceramic nature because of their durability, inertness, hardness, ease of fabrication, cost, and availability of the ingredients. Often, organics or metals cannot meet desired specifications for some of these products.

The rocks and minerals and mineral debris of the earth are largely siliceous. Oxygen and silicon comprise roughly 75% of the earth's crust with aluminum, iron, calcium, and the alkalies following in abundance. Silica (SiO_2) and silicates can be melted to form glasses. Alumina (Al_2O_3), present in a mineral compound, gives greater durability, hardness, and strength, but lessens the glass-forming property. Iron, calcium, and the other alkalies flux silica and reduce melting points. Thus sand, which is generally almost pure SiO_2, and clay, which is generally hydrated alumina silicate, are the most widely used of all ceramic raw materials—principally because of the glass-forming propensities of silicates.

In ceramic formulations, clay is generally a refractory filler or skeletal material, having in the raw state a plasticity derived from its colloidal grain size and its water-loving nature. On the other hand, silica sand is a glass former or nonplastic filler. Clay and silica sand in pure forms are quite refractory, i.e., not easily melted. Clay, although siliceous, is less of a glass former than pure silica and is more refractory.

Natural contaminants, such as iron oxide or limestone, adulterate many clays and some sands. These contaminants act as fluxes and reduce the fusion points of the mineral or blends.

Glass is made from silica sand by fluxing the sand with strong bases, usually soda ash and limestone. Red building bricks are hard and durable because the sand and clay have been partially fused with help of natural iron oxide serving as a flux. The iron oxide performs a dual function, serving as a pigment as well as a flux. In a blend of clay and sand and iron oxide, the sand is more susceptible to fusion than is the clay.

Although silica sand and clay are the most widely used raw materials in most ceramic products, many other minerals, and some nonminerals are utilized in products where these materials can produce or aid in achieving desired results. Limestone and dolomite, feldspars, pigmentary minerals, refractory oxides or carbides, and special property producers can

* Director and Professor Emeritus, School of Ceramic Engineering, Georgia Institute of Technology, Atlanta, GA.

all contribute to the final composition of the product after firing.

On the other hand, there are many possible batch ingredients which are eliminated in the firing process: water or other liquids, sawdust, straw, organics, soaps, textile patterns, coal facings, woody additives, greases and oils, starches, and to some extent deflocculants, plaster of paris, and other materials.

In addition to the ingredients actually in the batch at the outset, there are many other possible materials which are necessary for, or contribute to, fabrication. The fuels of the firing phase; the gaseous atmospheres; the lubrication for dies and presses; materials for environmental control; and chemicals for suspension, flotation, or wetting may all play a prominent role in the manufacture of the product, while never being a part of the batch itself.

The inorganic constituents which leave residues in the final product, the materials which are eliminated from the batch or initial formula during firing, and the controllers or modifiers of the environment of production may all be classed as ceramic raw materials.

Functions of Ceramic Raw Materials

Although the complete behavior of any one material in a ceramic composition may not be limited to a single action, the principal functions of each material in a given formulation are rather clearly discernable. Broken down into discrete categories, the functions to be met are:

1) Skeleton formers or fillers
2) Glass formers and bonding agents
3) Fluxes
4) Special property producers or modifiers
5) Processing or fabrication aids
6) Environmental or fabrication conditioners
7) Secondary and overlapping functions

Skeleton Formers or Fillers

The principal skeleton formers are clay, silica, other refractory oxides, carbon, and rarer inert refractory compounds. A skeleton former might be thought of as an inert refractory material which leaves a durable residue after firing has removed gaseous radicals. While all silicates have glass-forming tendencies, the absence of a flux and the inclusion of alumina radicals in the material counter this tendency. Clay, therefore, which is hydrous aluminosilicate, is very refractory and difficult to turn into glass. Thus, it is the principal skeleton material. Other refractory oxides such as lime, magnesia, titania, zirconia, diaspore, bauxite, cordierite, sillimanite and andalusite, kyanite, mullite, wollastonite, talc, zircon, fosterite, fluorspar, and rarer materials can also serve as skeleton formers, particularly if there is not a eutectic formed with other ingredients. Of all these materials, clay and silica sand are by far the most abundant. Silica sand does serve as a filler in many products, but the glass-forming tendency probably prevails as the principal activity of this material in a ceramic product.

Clay: Clay is the name given to the naturally colloidal, hydrated, inorganic material of the earth. Disintegration of the aluminosilicate rocks and similar oxide materials of the earth's crust by the agencies of weathering leads to the formation of various types of clays. Some of these clays may be altered by geologic conditions, and some are more stable under surface conditions in one climate than in another.

The principal clay minerals are of four types, with many subtypes. The most abundant clay mineral at the surface in temperate humid climates is kaolinite. In deeper deposits, montmorillonite and illite are the more abundant clay minerals. Under special conditions, particularly in tropical climates, laterite, bauxite, and opaline tripoli are formed, and with exception of the tripoli, these are relatively nonsilicious. Kaolinitic clays predominate as ceramic fillers with relatively little utilization of montmorillonites and laterites. Illites are widely used for certain structural products since these clays contain their own flux and glass former.

Kaolinitic clays are found widely dispersed over the earth's surface, and the deposits of the clays reach a depth limited by deposition or weathering. The various clays are named according to their purity, their geologic formation, their history of transport or no transport, their adulteration, or their alteration by geologic forces. A pure white kaolinitic clay is called kaolin and is primary or saprolitic or residual if in the place of the mother rock or sedimentary if transported. If significantly contaminated with an admixture of iron oxide in various stages of oxidation, the clay is generally designated as "red burning clay" and is reddish in the raw state in warm climates and black or dark in colder climates. This color distinction in warmer climates is due to complete oxidation of humus with its consequent elimination, and complete oxidation of the iron oxide remaining. The greatest deposits of red kaolinitic clays are sedimentary deposits along river banks and on flood plains and in basins and swamps. The sedimentary clays so deposited are called alluvial clays. A red clay in the place of the mother rock is called red residual clay. Such latter clays are not very deep or extensive, and are more contaminated than are the alluvial clays which have been processed partially by nature during transport.

Kaolins which have become contaminated with organic matter (but not with iron) during deposition in a swampy environment are called ball clays in consequence of their great plasticity and tendency to "ball up."

Clays modified by organic reagents during coal formation or subsequent oxidation of coal in place are called coal measure fire clays. The organic reagents have played a role in purification of the clay, leaving a kaolinitic residue of high refractoriness. Also, the organic reagents have produced greater plasticity and moldability in such clay, and these properties enhance its value.

In addition to kaolins, alluvial clays, and coal measure fire clays, the only other clays of very great utilization in ceramic wares are shales, which mineralogically are usually illites rather than kaolinites. Such clays vary widely in properties, but in general are red burning, much more fusible, relatively hard, nonswelling, and easily processable. These filler clays contain their own flux due to high iron and potash content. Some schists are chemically like shales and have similar uses. Fuller's earth and bleaching clays, such as montmorillonitic clays, have a very great affinity for water and have a tendency to swell and are most difficult to dry without damage to a formed unit. These clays, therefore, have little use in the manufacture of shaped products when better clays are available. They do have uses as plasticizers, bonding clays, adsorbents, and in making grains of a vesicular texture.

Laterites of a composition of hydrated iron oxide are little used in America, although more widely used in India and other places. Bauxite, a hydrated alumina, has many ceramic uses because of its very great refractoriness, hardness, and durability. Tripoli and opaline silica are used occasionally in products where silica is a reagent, but rarely where the silica serves as a skeleton.

Other Refractory Filler Materials: Any material leaving a refractory inert residue may serve as a skeletal refractory material, if particular advantages outweigh a cost higher than clay or silica. Some refractory oxides are reactive with silica and cannot be used in combination with silica where inertness and rafractoriness must be preserved. Such oxides are those of calcium and magnesium. However, some natural combinations of these alkalies with silica such as wollastonite, fosterite, and talc have great filler uses.

Alumina has extreme refractoriness, inertness, stability, and hardness, and is a very valuable skeletal material, both singly and combined with other oxides. Pure alumina is processed from bauxite or diaspore and is now available in quantity for uses where the cost is justified. Aluminosilicates are also processed from bauxite and from other minerals such as sillimanite, kyanite, dumortierite, and andalusite and available to the trade in the form of mullite.

Zirconia (ZrO_2) and zircon (zirconium silicate) are each used as refractory inert skeletal materials mainly in refractory units. A synthetic, calcium zirconate, is also available, and is used similarly.

Likewise, magnesite, olivine, fosterite, cordierite, or synthetic oxides of magnesium are used in basic refractories where the function served is that of inert refractory skeletal material.

Talc and wollastonite are used for special purposes, but often those materials serve as clay substitutes and their functions are then mainly as inert refractory fillers.

Limestone, calcite, and aragonite are used mainly as cementitious materials where their functions are specific, but there are some uses of these materials as refractory fillers, especially when silica is absent. Dolomite is used as a refractory material, and in an expanded form as an insulating refractory. Small amounts of lime-bearing materials are fluxing in presence of silica.

Fluorspar is similar to limestone in its performance and is fluxing in presence of silica. Some refractory filler uses of fluorspar are recognized.

Carbon is a distinctively different ceramic raw material, but in carbon products, one function of the carbon is as a filler. The carbon may be in the form of graphite, of amorphous carbon from coke or petroleum coke, or synthesized pyrolytic carbon.

Chromite and refined chromium compounds, pyrophyllite, magnesite, magnesia, periclase, cordierite, barite, tin oxide, beryl, and a number of rare chemicals have refractory filler uses as well as other functions. Higher priced materials are utilized for specific functions rather than for filler use.

Glass Formers and Bonding Agents

Silica in its various forms is by far the principal glass former of ceramic materials. Most of the silica used is processed from relatively iron-free natural quartz sand, but some is ground from chert, beach pebbles, hard sandstone, gannister, novaculite, quartzite, and similar rocks. Silica may be combined with compounds having basic elements or fluorine in proportions that will form a eutectic fusion which will not readily recrystallize, thus forming glass on cooling.

Other forms of silica are sometimes used: diatomaceous earth, natural or synthetic colloidal silica, and fused silica. These forms are somewhat more costly and are used where the particular properties of these forms give an added advantage—often for functions other than glass forming.

Synthetic frits, which are largely glassy and silicious, are other contributors to the final structure. They are used as bonding agents.

There are relatively few glass formers other than silica, but several are used in special products where absence of silica is preferred. Arsenic oxide, boric oxide, phosphorus oxide, and selenium oxides are used as glass formers in certain specific products.

Glass is the main bonding agent in most final ceramic products, but a few products have crystalline bonds rather than glassy bonds. Thus, the materials which promote crystallization may be thought of as bonding agents. Examples are the mullite formers: bauxite, kyanite, sillimanite, and andalusite, dumortierite, and the special combinations of materials such as are used in pyroceram products.

Fluxes

Ceramic fluxes are mainly those materials which will form eutectics with silica and silicates. The basic elements lithium, sodium, potassium, calcium, magnesium, boron, iron, lead, and the acid element fluorine are the elements which are specifically utilized for such action. Such elements are found in a variety of materials, natural and synthetic. The fluxing power is usually dependent on the amount of the specific element used and the relative absence of counteracting elements such as aluminum and even silicon. The latter two elements tend to make a compound more refractory and less soluble. Hence, when maximum fluxing power is required and solubility is not a determining factor, as in commercial glass preparation, compounds lacking alumina and silica are used for fluxes such as soda ash, pearl ash, salt cake, borax, colemanite, fluorspar, white lead, red lead, litharge, and some phosphates. The principal fluxes of commercial glass are the carbonates of sodium and potassium and the less soluble carbonates of calcium and magnesium. A balance of the very soluble carbonates and the less soluble carbonates is desirable in order to achieve maximum fluidity of the melt consistent with stability of the final product. Too much sodium or potassium will tend to make a soluble product, whereas too much calcium or magnesium will tend to cause crystallization or devitrification of the product.

Salt cake is not usually favored because of trouble caused by the sulfate radical. Borax and boron compounds are used in combinations with other materials in glazes and enamels. In refractory glasses, boron forms a borosilicate which has higher fusion points and lower coefficient of thermal expansion. Hence, in this instance, the boron compound is not a flux.

Where a ceramic formulation uses water as a processing aid, soluble fluxes cannot be significantly tolerated. Hence, less fusible fluxes such as the potassium feldspars, orthoclase, microcline, and the sodium feldspar albite, or perthitic mixtures are used for such products as earthenware, porcelain, and stoneware. Nepheline syenite has somewhat higher fluxing power and is used as a feldspar substitute, as are synthetic frits. Lithium minerals such as spodumene and derivatives are also used as flux materials, sometimes blended with feldspars. In England, Cornwall Stone is used as a general filler-flux mixture.

The high calcium feldspar, anorthite, is not widely used as a flux because of certain disadvantages in processing.

Lead compounds are sometimes altered by combining with silica and the resulting synthetic lead silicate is sold to the trade as a less hazardous flux from standpoint of toxicity.

Iron oxide is rarely added to formulas to serve as a flux, but natural mixtures of clays with iron oxide achieve their vitrification from the content of iron oxide acting as a flux.

When metallic compounds are added to a formulation for other reasons, the metal element often shows some fluxing power.

Water initially used in processing often partially remains as steam in a kiln atmosphere and has a fluxing action on the ceramic unit.

Special Property Producers or Modifiers

The desirable properties in ceramic products other than inertness, durability, hardness, and refractiveness vary according to the use for which the product is intended. Such properties may be color, texture, insulating character, machinability, light sensitivity, conductivity, radiation modification and shielding, cementing capability, and many others.

Color Pigments: Colors in ceramic products are produced by use of certain inorganic compounds, processed in specific manners. Organic colors, such as are generally used in paints, dyes, and textiles, are destroyed in ceramic processing.

Color is generally produced by modification of the electron structure surrounding the nucleus of an atom having unsaturated inner orbits. The transition elements and the rare earth elements are the usual color producers. If the valence of such an element can be changed, modification or change of color results. The actual raw material for pigmentation depends on availability and cost, and is often an oxide, silicate, carbonate, or sulfate. Occasionally, a nonpigment oxide will modify the color of a pigment oxide, as will Al_2O_3 or SnO_2 in presence of chromium oxides.

A few colors are achieved by deposition of colloidal metals in specific grain sizes. Such pigments are gold and selenium.

The usual colors are achieved using compounds (or rarely the element itself) of the following elements:

Red:	Au, Se, (Cr+Al) (Cr+Sn) Fe
Orange:	U, Mo
Yellow:	Cd, Mo, Pr
Green:	Cr, Cu, Fe
Blue:	Co, Cu, Fe
Purple:	Mn
Brown:	Ni, Fe
Black:	Combinations of dark pigments, Mn, C
White:	Sn, Ti, Zr, P, Sb

Blends of these produce different shades and tones. Most pigments are processed and modified from original rocks and minerals and are therefore synthetics.

Texture and Insulating Character: Surface texture of a ceramic product may be modified by smoothing or roughing the surface. Silica sand and calcined clay and mixtures are used for roughing the surface, whereas glassy coatings, glazes, and enamels are used to smooth the surface. The raw materials for these vitreous coatings are the same as the glass formers plus pigments.

The microstructure of a product can be dense or vesicular. A porous microstructure enhances the insulating properties. To achieve a porous structure, raw materials such as diatomaceous earth, gas-forming minerals such as sulfur or fluorine-bearing shales, fluorspar, gypsum, sodium silicofluoride are often used. Burn-out materials such as sawdust, nut hulls, straw, bagasse, and combustible waste are used to leave voids.

Machinability: Ceramics are usually of a poorly machinable nature. Some very high alumina products have better machinability. Cermets, which are metal-ceramic combinations, have machinability approaching that of metals. The metals used in cermets are often nickel, chromium, and tin and the ceramics are often Al_2O_3, Cr_2O_3, and NiO_2. Other combinations are possible.

Radiation Moderating and Shielding Properties: For use in nuclear processes, oxides, carbides, and nitrides of uranium, processed from uranium ores, are used. Some thorium processed from monazite is also used.

Beryl and BeO, carbon as graphite or pyrolytic carbon, are moderators.

Heavy materials as massive concrete, barite, and lead glass are used for shielding.

Hafnium, cadmium, boron, and gadolinium compounds are used for high neutron absorption.

Cementing Properties: Portland cements are made from sand and clay by sintering the mixture. Tricalcium silicate, dicalcium silicate, dicalcium aluminate, and calcium iron aluminate are the cementing compounds produced. Gypsum is added to retard the setting rate. Magnesium in limestone cannot be tolerated in significant amount and the tolerances for different products are included in the Limestone chapter.

Lime is prepared by calcining limestone. The resulting quicklime is often hydrated to make slaked lime.

Plaster of paris is made by partial dehydration of gypsum. Some used plaster is recycled.

There are other cements prepared from special materials.

Fly ash, colloidal silica, or siliceous clay are sometimes added to a cement to create a pozzolana-type material.

Light-Sensitive Materials: The main light-sensitive materials are cadmium sulfide and silver chloride. Several other materials show light sensitivity to some degree.

Dielectric Materials: Natural steatite-talc, fosterite, and synthetic barium titanate, lead niobate and certain mixtures are used as high frequency dielectrics.

Processing or Fabrication Aids

Water, oils, organic binders, ball clay, bentonite and fuller's earth, starch, sawdust and burn-out material, textile and paper patterns and cores, plaster of paris, glue, sodium silicate, soda ash (sodium carbonate) phosphate electrolytes, waxes, greases, and dyes are each used as processing aids in certain products. Ball clay, which is a kaolinitic clay with natural organic content and little or no iron contamination, is widely used as a plasticizer in plastic formation and as a buffer in slip formation. The sulfate content of the ball clay can require certain controls. Bentonite and fuller's earth are plasticizers but cannot be used in large proportions in shaped products since their water absorption causes excessive shrinkage. Where adsorptivity is the desired end and the form is granular, these clays may be used successfully.

Environmental or Fabrication Conditioners

Control of the furnace atmosphere is the principal environmental material control which affects the finished product. Normally, a furnace atmosphere has a mixture of gases, derived from the burning of hydrocarbons from natural gas, fuel oil, or coal, mixed with air in a balance which is oxidizing.

For certain results such as the change in color of a normally red brick, a reducing atmosphere is required. This is obtained by reducing the admixed air content in gas or oil-fired furnaces or by burning oxygen consumers such as slack coal, wet grass or straw, waste rubber, zinc scraps, sawdust, or wood shavings in coal or other solid fuel furnaces. Usually the addition of oxygen consumers or the reduction in excess air is brought about near the end of the firing cycle. When firing carbon products, a continuous reducing atmosphere is generally required.

Nitrogen, argon, neon, helium, hydrogen, and carbon dioxide are all used to provide certain nonoxidizing atmospheres, dependent on process requirements.

Steam injection into a furnace atmosphere is occasionally used and here, the steam exerts a fluxing action on silicates.

Sodium chloride is vaporized in a kiln atmosphere when the product is to be salt glazed. Some borax or other materials are at times mixed with the salt.

Secondary and Overlapping Functions

Secondary functions of materials are often of major importance. Many materials used for a primary purpose exhibit important secondary functions which must be included in specifications. For instance, the choice of a kaolin or china clay for whitewares may depend on its plasticity, its whiteness, or its viscous behavior. These properties probably stem mainly from grain size, but they may control the desirability of the material for a given product or process.

Many ceramic clays are not pure kaolins but contain many natural adulterants which greatly affect the final product. The properties of moldability, of early susceptibility to vitrification, of good red or brown or buff color, and of convenient and cheap availability may appear to overshadow the prime function of clay as a refractory filler.

Natural kaolinitic clay mixtures often contain sand and silt and lime and iron-bearing pebbles or grains, but the clay substance still performs a filler function, whereas natural additives may alter the final state so as to completely mask the identity of the kaolinite. Iron oxide serves as a flux and as a pigment, sand as a nonplastic, part filler and part glass former, and lime as a flux and color modifier.

Clays not of a kaolinitic base, such as shales, phyllites, schists, and fullers earths, have basic chemical radicals in their compositions, causing quite different behavior in firing. Some clays undergo sudden fusion accompanied by bloating during the firing process, making them suitable for lightweight aggregate. Others form vesicular structures which enhance adsorption useful in products such as floor sweep and pet litter. Here, the filler function of the clay is almost entirely sacrificed to the secondary qualities given by the fluxes and contaminants.

Ball clays, mentioned previously as kaolinitic clays contaminated with organic matter but not with iron, are very useful in ceramic processing because of enhancement of moldability, bonding strength, and viscosity control. Coal measure fire clay has somewhat similar properties. These clays are used mainly because of their secondary functions.

Clays are silicates and under proper heat treatment will fuse to form glass, and this secondary function of clay as a glass former varies considerably. In most ceramic products, silica sand or potters flint is the principal glass former, but the clay may also participate in such a role.

Silica has nonglass forming functions as well. As a nonplastic, it opens up a body unit in a manner which facilitates drying, counters shrinkage and stickiness, and reduces water adsorption. Here the sand is a filler and special property producer. In glass, silica can have the additional properties of increasing refractoriness, reducing coefficient of thermal expansion, and increasing transparency to ultraviolet light. In insulating refractories, silica, as diatomaceous earth or tripoli, can enhance insulation properties because of its refractoriness and its porous structure.

Sources of Information—Ceramic Raw Materials

Other chapters in this book give descriptions of many mineral and rock raw materials mined in various localities and available to the ceramic manufacturer. Synthetics are available from numerous manufacturers and waste product additives from industries having such waste. There are numerous references to sources of both natural, synthetic, and waste materials. Excellent general listing of sources can be found in chemical, mining, and ceramic magazines (*Ceramic Age,* July issues; *Ceramic Industry,* January issues; *Ceramic Products Cyclopedia; Journal of the American Ceramic Society, American Ceramic Society Bulletin; Ceramic Abstracts; Industrial & Engineering Chemistry; Chemical Abstracts;* and *Mining Engineering*).

The book, *Ceramic Raw Materials* (Chas. Kline & Co.), is an invaluable reference to sources. Production figures, analysis of supply and outlook for the future are covered in the US Bureau of Mines *Mineral Facts and Figures* and *Minerals Yearbooks.* Current price ranges may be found in such publications as *The Chemical Marketing Reporter, American Metal Market,* and *Engineering & Mining Journal.*

Bibliography and References

Adams, P.J., 1961, *Geology and Ceramics,* H.M. Stationery Office, London, 27 pp.

Ceramics Data Book, Biennial 1922–1934, Annual from 1935 on, J.J. Suecy, ed., 1981, Cahners Publishing Co., Chicago.

Ceramics Monograph—A Handbook of Ceramics, published monthly in Interceram-International Ceramics Review, Germany.

Jones, J.T., and Bernard, M.F., 1972, "Ceramics," *Industrial Processing and Testing,* Iowa State University Press, Ames, 213 pp.

Kingery, W.D., et al., 1976, *Introduction to Ceramics,* 2nd ed., Science and Technology of Material Series, John Wiley, New York, 781 pp.

Norton, F.H., 1974, *Elements of Ceramics,* 2nd ed., Addison-Wesley, Reading, MA, 311 pp.

Pampuch, R., 1976, *Ceramic Materials,* Elsevier, Amsterdam, NY, 344 pp.

Parmelee, C.W., 1973, *Ceramic Glazes,* C.G. Harmon, ed., 3rd ed., Cahnew Books, Boston, 612 pp.

Ryan, W., 1968, *Properties of Ceramic Raw Materials,* 1st ed., Pergamon Press, New York, 107 pp.

Shaw, K., 1971, *Ceramic Glazes,* Elsevier, Amsterdam, NY, 166 pp.

Watson, I., 1981, "Minerals for Frits and Glazes—Value in Variety," *Industrial Minerals,* No. 165, June, pp. 23–35.

Chemical Industry

ROBERT B. FULTON, III *

The objective of this chapter is to discuss the interrelationship between industrial minerals and chemical manufacturing. It is intended to supplement rather than duplicate the commodity chapters. Particular emphasis is given to the pertinent chemical element and to market factors.

Condensing this broad subject into a few pages of this handbook permits treating only the most important elements derived from industrial minerals. Hydrocarbons, which quantitatively dominate as raw materials for the chemical industry, are omitted, as are the metallic elements and the minerals covered in other "use" chapters such as phosphorous, potassium and nitrogen for fertilizers, and titanium dioxide for pigments. The remaining six elements of major importance are: boron, bromine, chlorine, fluorine, sodium, and sulfur. These elements are treated individually under separate headings.

Table 1 (Jones, 1973) affords an overview of the main industrial minerals, the chemical products derived from them, and end uses of the products.

Salt brines have particular importance as raw material sources for the chemical industry. Table 2 (Anon., 1967) is a chart of the chemical compounds derived from four types of brines: (1) Owens Lake-type brines, which are sources of boron and sodium compounds; (2) Midland-type brines, from which bromine, iodine, and chlorides of calcium, magnesium, potassium and sodium are derived; (3) Searles Lake-type brines, yielding boron, bromine, lithium, magnesium, potassium, and sodium compounds; and (4) Silver Peak-type brines, produced mainly for lithium.

* Basic Materials Planning, Materials and Logistics Dept., E. I. du Pont de Nemours & Co., Wilmington, DE.

This chapter was reviewed by L.E. Mannion, Chief Geologist, Stauffer Chemical Co., Richmond, CA.

Market Attributes

Some of the important market traits common to industrial minerals used by the chemical industry are listed below.

1. They are international commodities, such as fluorspar and sulfur, which largely move to foreign consumers.

2. Grade, and freedom from deleterious elements are important factors affecting their usability in chemical processes. An example is salt (NaCl) used in electrolysis where ultrapure evaporated salt is required to meet rigid specifications.

3. Purified products take on the characteristics of specialty items and command a distinctly higher price than the basic commodity from which they are derived.

4. In practically all cases, chemical users require some sort of cleaning or beneficiation of the naturally-occurring mineral to bring it to specification, and individual specifications may vary from user to user for essentially the same use.

5. In some instances it is necessary to strike a balance between what the vendor can supply and what the buyer requires, with the result that specifications have to be eased to afford the needed materials in marginal cases.

6. Because they tend to be bulk commodities, low cost for handling and transportation are important and such costs may limit the area from which a chemical user can draw his supply. For example, fluorspar from China and Thailand normally goes to Japan, while South African fluorspar can go either to Japan or to the US.

7. Shipments are usually in bulk and frequently in multiple-car, full-trainload or full-shipload lots to reduce transport costs, which in turn may require large terminal investment facilities.

TABLE 1 — From Industrial Mineral to Inorganic Chemical

Commodity	Chemical Formula	Reaction Type	Reagents	Products: Formula	Products: Name	Uses for End Products
Spodumene	$LiAl(SO_4)_3$	Acid decomposition	Sulfuric acid	Li_2SO_4	Lithium sulfate	Ceramics, glass, fluxes, flame colorant
		Alkali decomposition	Limestone	LiOH	Lithium hydroxide	Absorption refrigeration, greases, high strength glass
Rock salt	NaCl	Electrolytic	Electricity, water	NaOH	Caustic soda	*Miscellaneous chemicals (30%), rayon (17%), pulp and paper (8%), metals (8%)
				Cl_2	Chlorine	Miscellaneous chemicals including organics (83%), pulp and paper (13%)
				H_2	Hydrogen	Ammonia synthesis, miscellaneous chemicals, metallurgy
		Ammoniation	Ammonia, lime, carbon dioxide	Na_2CO_3	Soda ash	Miscellaneous chemicals (37%), glass (30%) metallurgical (10%), soaps (16%)
				$CaCl_2$	Calcium chloride	Road treatment, dust suppression
Sylvite	KCl	Electrolytic	Electricity, water	KOH	Potassium hydroxide	General chemicals, fertilizers, detergents
				Cl_2	Chlorine	See above
				H_2	Hydrogen	See above
Beryl	$3BeO \cdot Al_2O_3 \cdot 6SiO_8$	High temperature fluorination	Sodium fluoroferrate	$Be(OH)_2$	Beryllium hydroxide	Fluorescent tubes, TV tubes, glass, ceramics
Limestone	$CaCO_3$	Calcination	—	CaO	Quicklime	Caustic soda manufacture, effluent treatment, water treatment
				CO_2	Carbon dioxide	General chemicals, soda ash
Anhydrite	$CaSO_4$	Carbothermic reduction	Natural gas, coal, air	H_2SO_4	Sulfuric acid	Fertilizers (36%), chemicals (26%), petroleum refining (12%), pigments (9%), metallurgical
				CaO	Quicklime	See above
Seawater	$MgCl_2$	Precipitation, calcination	Dolime	MgO		
Magnesite	$MgCO_3$	Calcination	—	MgO		
Brine	$MgCl_2$	Steam decomposition		MgO	Magnesium oxide	Refractories, general chemicals, fertilizers, cements
				HCl	Hydrochloric acid	Miscellaneous chemicals, metal pickling, organic chemicals
Celestite	$SrSO_4$	Carbothermic reduction	Coal	$SrCO_3$	Strontium carbonate	Glass, ceramics, pyrotechnics
Barytes	$BaSO_4$	Carbothermic reduction	Coal	$BaCO_3$ $BaCl_2$	Barium carbonate Barium chloride	Glass, ceramics, TV tubes, pyrotechnics, oxygen source, ferrite magnets, medicine, pigments, bricks, metal soaps

Bastnaesite Monazite Xenotime	Ce, La(PO_4)	Acid decomposition	Sulfuric acid	Ce, La O	Rare earth oxides	Glass colorants, polishing powders, TV tubes, lighter flints, arc carbons, catalysts, phosphors, ceramic stains, camera lenses
Zircon	$ZrSiO_4$	Chlorination	Chlorine, coke	ZrOCl	Zirconium oxychloride	Metal production, special chemicals
Rutile (natural & synthetic)	TiO_2	Reductive chlorination	Chlorine, coke	$TiCl_4$	Titanium tetrachloride	Pigments, titanium metal, waterproofing agents
Chromite	Fe_3CrO_4	Alkali decomposition	Soda ash	Na_2CrO_4	Sodium chromate	Chromium chemicals, electroplating baths, oxidants
Pyrolusite	MnO_2	Acid decomposition Carbothermic reduction	Hydrochloric acid Coal, nitric acid	$MnCl_2$ $Mn(NO_3)_2$	Manganous chloride Manganous nitrate	Dry batteries, soaps (dryers), fertilizers, glass, enamels, catalysts
Colemanite	$Ca_2B_4O_{11}\cdot 5H_2O$	Alkali decomposition	Soda ash	$NaBO_2$	Sodium borate	Detergents, fluxes
Bauxite	$Al_2O_3\cdot 2H_2O$	Alkali solution Acid decomposition	 Sulfuric acid	$Al(OH)_3$ $Al_2(SO_4)_3$	Aluminum hydroxide Aluminum sulfate	Water treatment chemicals, flocculants, dye mordants, refractories, catalysts
Quartz, silica	SiO_2	Alkali reaction	Soda ash	Na_2SiO_3	Sodium silicate	Adhesives, cements, deflocculants, soaps
		Reductive chlorination	Chlorine, coke	$SiCl_4$	Silicon tetrachloride	Organic chemicals (silicone fluids and rubbers)
Caliche	$NaNO_3$	Acid decomposition	Sulfuric acid	HNO_3	Nitric acid	Fertilizers, general chemicals
Phosphate rock	$Ca_5F(PO_4)_3$	Acid decomposition	Sulfuric acid	H_3PO_4	Phosphoric acid	Fertilizers, detergents, foodstuffs, organic chemicals, water treatment
		Electric furance reduction	Silica, coke	P	Phosphorus	See phosphoric acid
Brimstone	S_6	Burning	Air	SO_2	Sulfur dioxide	Fungicide, insecticide, reducing agent, solvent
		Catalytic oxidation	Air	H_2SO_4	Sulfuric acid	See above
Iron pyrite	FeS_2	Roasting, catalytic oxidation	Air	H_2SO_4	Sulfuric acid	See above
Fluorspar	CaF_2	Acid decomposition	Sulfuric acid	HF	Hydrofluoric acid	Organic chemicals, fluxes, etchants, metallurgy, isotope separation
Seawater, brines	NaBr	Chlorination	Chlorine	Br_2	Bromine	Organic synthesis, pharmaceuticals, petroleum additives
Seawater, brines, caliche	NaI	Chlorination Chemical reduction	Chlorine Sodium bisulfite	I_2	Iodine	Foodstuffs, sanitizers, organic chemicals

* Recent US consumption percentages in parentheses.

Source: Jones, 1973.

TABLE 2—Salt Brine Derivatives

Brines	Products	Derivatives	Uses
(Owens Lake type) →	Borax decahydrate Sodium sesquicarbonate Sodium chloride Sodium sulfate	Sodium carbonate	
(Midland type) →	Bromine Calcium chloride Iodine → Magnesium chloride Potassium chloride Sodium chloride	Hydriodic acid	
		Potassium iodide →	Food additive Photography Engraving and lithography
		Lead iodide →	Generator brushes
		Nickel iodide →	Catalyst
		Sodium iodide →	Food additive Photography Engraving and lithography
		Sodium iodate →	Improves bread quality
		Silver iodide →	Photography Dyes Cloud seeding for rain
		Titanium tetraiodide →	Sterospecific polymerization
		Resublimed iodine Organic iodine compounds Catalyst Medicine Disinfectant Iodophors (iodine-surfactant complexes)	
(Searles Lake type) →	Borax decahydate Lithium chloride Lithium carbonate Magnesium hydroxide Potassium chloride Sodium borate Sodium bromide → Sodium carbonate Sodium chloride Sodium fluoride Sodium phosphate Sodium sulfate	Bromine	
(Silver Peak type) →	Lithium carbonate → Potassium chloride	Lithium bromide Lithium chloride Lithium hydroxide Organolithium compounds	

Source: Anon., 1967 (courtesy Stanford Research Institute).

8. Purchases are generally by contract of one year or longer term, with spot buying playing only a minor role.

9. Contract prices are usually fixed in short term commitments but may vary according to assay, with premiums and penalties for content above or below the norm; however, general practice is for specifications to be fixed in the contract with minimums being set for the desired material and maximums for undesired elements. In longer term contracts, prices are often escalated on labor, fuel, and other vendor processing costs.

10. Suppliers of individual commodities to the chemical industry tend to be limited in number and are generally medium- to large-size producers that supply a few major consumers.

11. The bulk of the mineral volume is for basic chemical uses, sulfur supplies to sulfuric acid producers and fluorspar for hydrofluoric acid producers being typical examples. These basic chemical products then are used for the production of other products.

12. A significant part of the chemical industry's consumption of industrial minerals is for products that are rapidly growing in market volume, which places strain on supply sources to keep up with the demand.

13. Shortage of a supply of adequate quality leads consumers to seek substitutes. In the case of fluorspar, much work is being done on recovery of fluorine from phosphate rock. Success in the form of fluorosilicic acid production could, in time, affect the hydrofluoric acid chemical industry.

14. Though demand and price are generally following a strong upward trend, markets in the past have been characterized by cycles of shortage followed by oversupply, with attendant wide price fluctuations.

15. Barriers to trade can have an adverse effect on the necessary movement of industrial minerals used by the chemical industry in international trade. Antidumping laws, quotas, and tariffs can disrupt or dislocate normal markets.

16. Chemical industry consumers may back-integrate for security of supply or for favorable economics, sometimes by joint ownership and often with experienced mining partners.

Elements of Major Importance

Boron

Boron is used by the chemical industry mainly as borax, (disodium tetraborate decahydrate). It is derived principally from desert evaporite deposits. Major commercial boron minerals include: borax ($Na_2B_4O_7 \cdot 10H_2O$), colemanite ($Ca_2B_6O_{11} \cdot 5H_2O$), tincalconite ($Na_2B_4O_7 \cdot 5H_2O$), ulexite ($NaCaB_5O_9 \cdot 8H_2O$) and kernite ($Na_2B_4O_7 \cdot 4H_2O$). Commercially, sodium borates predominate and are supplied by refineries in a variety of grain sizes in either the purified borax form or as fused, crushed, anhydrous borax. Boric acid and boric oxide are also available. The industry defines boron content in terms of B_2O_3; e.g., granular borax decahydrate contains 36.5% B_2O_3, anhydrous borax, 69.1% and boric acid, 56.3%. These products are sold at 99.5% purity. Raw and semiprocessed ores are also sold.

Borates are used as fluxes in metal processing, in ceramics, as a constituent of heat-resistant glass, as an aid in fiber glass manufacture, as cleaning agents, and in textile manufacture. Borax and boric acid are used as extenders in dye making. In elemental form, boron acts as a neutron absorber in shielding of radioactivity. Organic boron compounds are used as gasoline additives and in insecticides and pharmaceuticals. The tri-fluoride etherate is used for manufacture of fluorinated hydrocarbons and in production of antiknock compounds for gasoline. Cubic boron nitride has extreme hardness and stability at high temperature. Fuels containing boron and hydrogen, outstanding for high energy yield, have not been widely used, mainly because of pollution effects and high cost. Boron starting-materials and end products are depicted in Table 3 (Anon., 1970).

The US produces 70% of the world's boron minerals and consumes 35%. According to the US Bureau of Mines (Lyday, 1982), production of boron minerals and compounds in the US in 1981 was 680 kt (750,000 st) valued at $414 million compared to a 1975 production of 1.06 Mt (1.17 million st) at $123 million, with especially high demand by glass and fire-retardant users. Annual growth of US consumption is expected to be 3.5 to 4.6%. Approximate fourth quarter 1982 f.o.b. prices quoted per short ton in bulk were:

Boric acid (99.9%, Tech.)	$550–597
Refined pentahydrate borax	$201

Opportunities for wider applications of boron compounds, in addition to the growth in established fields of use, lie in such uses as glass fibers for reinforced plastics used for weight reduction in automobiles, fuels, and as a possible replacement of fluorspar as a fluxing agent in steel making. Boron's capability of forming a large number of coordination compounds

TABLE 3—Boron Minerals and Chemicals

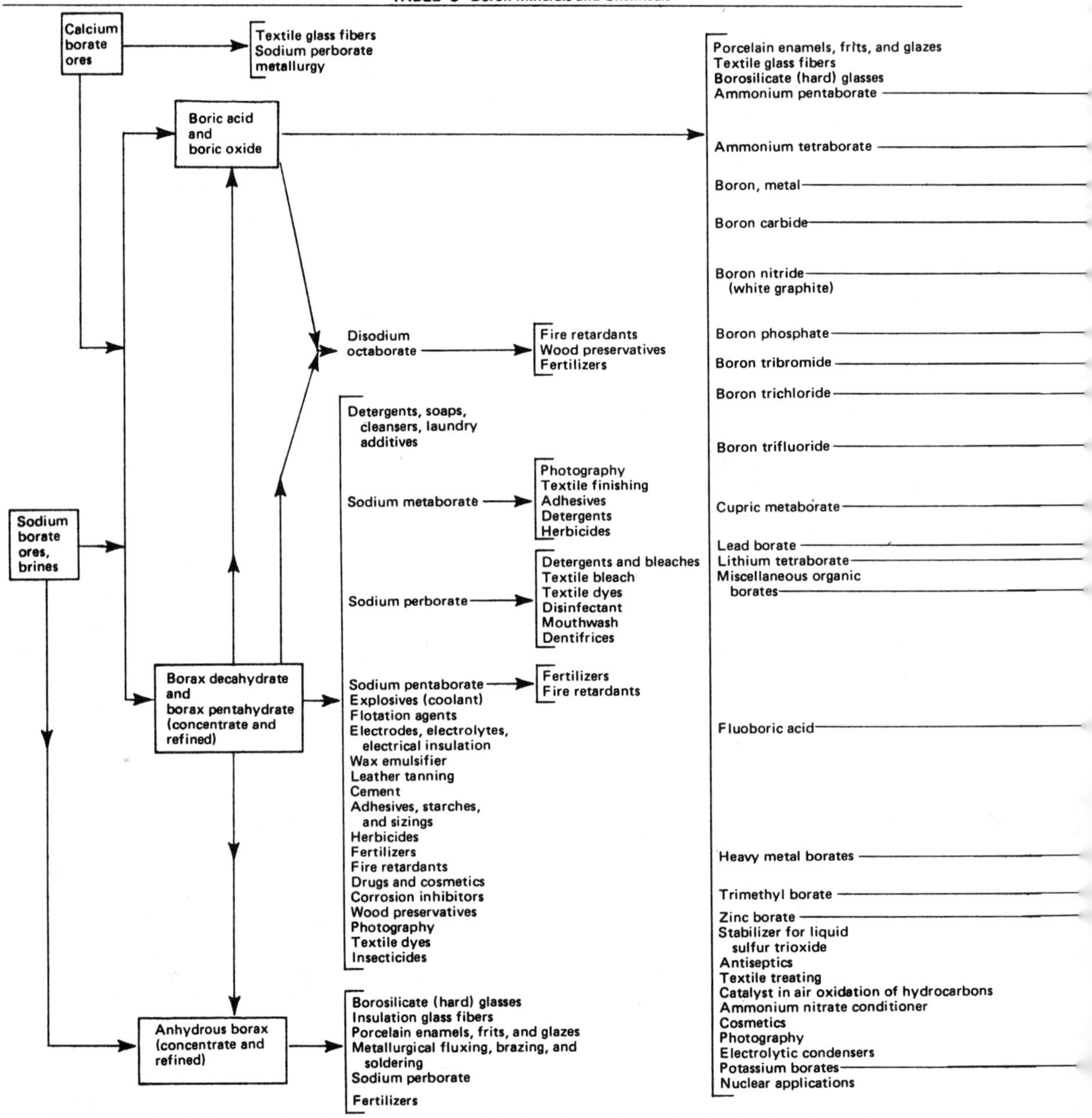

Source: Anon., 1973 (courtesy Stanford Research Institute).

TABLE 3—Continued

- → Electrical condensers; Fireproofing; Paper coating
- → Neutralizing agent; Fireproofing
- → Delay fuses; Flare igniters; Boron phosphide → High temperature transistors; Hexaboron silicide, Tetraboron silicide → Refractory material; Getter in manufacture of vacuum tubes; Abrasives; Grain refiner in aluminum manufacture; Nuclear shielding; Coating material for solar batteries; Semiconductors
- → Abrasives; Control rods for nuclear reactors
- → High-temperature lubricant; Refractory; Fibers; Thermionic devices
- → Ultra low loss dielectric
- → Diborane
- → Boron trifluoride; Diborane → [Pentaborane → Ethyl pentaborane; High energy fuels; Higher boron hydrides → Carborane polyesters, Rayon delustrant, Rubber vulcanization, Fluxing agent; Polymerization catalyst for ethylene; Copolymerization of styrene and butadiene]; Organo-boron compounds; Trichloroborazole → Gelling agent, Catalyst; Depolymerization of paraldehyde; Flux; Polymerizing agent for driers; Stabilizer for liquid sulfur trioxide
- → Neutron analysis; Catalyst
- → Fireproofing agent for wood; Fungicides → Wheat rust compounds; Oil pigments; Pigments for ceramics
- → Galvanoplastic work; Paint and varnish driers; Lead glass
- → Ceramics
- → Glycol borates, Alcohol borates → Lubricant and hydraulic fluic, Gasoline antiknock agent; Jet fuel deicer; Epoxy resin curing agents; Nuclear shield via neutron-detecting; Bactericide in fuels; Trimethoxyboroxine → Extinguishing metal fires, Epoxy foams
- → Ammonium fluoborate → Molding sand additive in magnesium casting; Cadmium fluoborate, Cupric fluoborate, Lead fluoborate, Nickel fluoborate → Electroplating baths; Potassium fluoborate → Binder and cooling agent in abrasive wheels, Electrolytic brightening of aluminum; Sodium fluoroborate → Electrochemistry, Heat treating magnesium alloys, Nonferrous flux; Stannous fluoborate, Zinc fluoborate → Electroplating baths
- → Fireproof and mold resistant paints
- → Sodium borohydride → Lithium borohydride; Potassium borohydride; Pharmaceuticals; Reduction of aldehydes and ketones; Vat dyes; Blowing rubber and epoxy resin foams; Paper bleach and pulp brightener; Hair waving formulations
- → Ceramics; Medicines; Fire-retardant; Fungicides
- → Inorganic boron compounds; Photography; Casein solvent; Metallurgical welding and brazing; Diazotype developer solutions

offers avenues for investigation which are being pursued to yield a variety of diverse products; some of which could have material future impact. Any immediate, large-volume impact from these new compounds seems unlikely, however.

Bromine

Bromine production in the US, normally in the range of 181.4 to 226.8 kt (400 to 500 million lb) per year, is derived from deep well brines in Michigan and Arkansas (Midland type) and from Searles Lake in California. At Searles Lake three of the seven producers account for 85% of the total. Worldwide, Israel is second as a bromine producer, at about 45.4 kt (100 million lb) per year. Other producers are France, Japan, United Kingdom and the USSR.

Bromine is used, mainly in the form of ethylene dibromide (EDB), by the chemical industry in making tetraethyl lead mixture, an antiknock additive for gasoline. The bromine compound inhibits deposition of lead when the gasoline burns. Methyl bromide has similar application. A variety of organobromides are used in agriculture as fumigants for controlling nematodes in soil and insects in storage of grain and seed. Elemental bromine is consumed in disinfectants and bleaching agents. Sodium, potassium and ammonium bromides have minor uses in textile fiber manufacturing, photography, and medicines. Bromine compounds are used in small amounts in making dyes, resins, leather, and rubber products and in production of textile intermediates and finishing agents. Newer uses are in the field of flame retardants and extinquishants and for gas- and oil-well completion fluids, the latter as calcium bromide.

US production in 1980 was about 170.1 kt (375 million lb), down substantially from about 226.8 kt (500 million lb) in 1979, the drop being mainly in the use of EDB for antiknock. In 1981, however, production increased to 195.9 kt (431.8 million lb). Domestically, 45% was used for EDB, 21% was used in the manufacture of flame retardants, 15% was for the production of drilling and completion fluids, 8% was sold as methyl bromine, and 11% was for the other bromine compounds described above. US exports reached 43.1 kt (95 million lb) in 1981, up from about 4.5 kt (10 million lb) in 1970. US capacity to produce is about 299 kt/a (660 million lb per year).

Bromine prices in 1980–1981 were 26 to 30¢ per lb, having increased from 22.5¢ in 1965–1968 and 18.8¢ in 1969–1970. The price outlook is for further increases due to increased cost of chlorine (used in the ratio of one lb of chlorine to recover two lb of bromine) and to increased cost of process energy. Future demand will depend on the usage of lead antiknock compounds, and on environmental considerations, but is expected to be 3 to 4% per year through 1986.

Chlorine

Chlorine production in the US in 1979 was about 11 Mt (12 million st), representing about 83% of installed capacity. Large chlorine producers, with plants based on the electrolysis of natural salt brines, are located in the New York-Ohio-Michigan region and on the Gulf Coast. Capacity increases of about 2.7 Mt (3 million st) in 1979–1980 were partly offset by the shutdown of small inefficient manufacturing units.

A number of principal consumers in the chemical industry are situated in the producing regions and in many cases are back-integrated with obvious advantages. Chlorine consumers who are not back-integrated either purchase chlorine or buy chlorine compounds such as hydrochloric acid, chloroform, ethyl chloride, ethylene dichloride, carbon tetrachloride and the like, depending on process requirements and economics. Choice of purchase mix is determined by relative costs; for example, the choice between either buying carbon tetrachloride, or buying chlorine and hydrocarbons for in-house production of the chlorocarbon. This choice has ramifications, such as the internal generation of hydrochloric acid, produced as a byproduct of hydrocarbon chlorination and of fluorination. The internally-produced hydrochloric acid can then be oxidized back to chlorine in a closed circuit, reducing the intake of purchased chlorine and thereby reducing costs.

Principal chemical industry uses for chlorine are for intermediates and for synthesis of organic and inorganic chlorides used in plastics, solvents, agricultural chemicals, automotive fluids, sanitation products, etc. Among the fastest growing is the use of ethylene dichloride intermediate for vinyl chloride, which increased from about 453.6 kt to 2721.6 kt (1,000 to 6,000 million lb) of ethylene dichloride in the 1960s and to nearly 4.5 Mt (10,000 million lb) in the late 1970s. Other uses of ethylene dichloride are for making solvents, such as trichloro- and perchloro-ethylene, and as a cheaper, though less effective, lead scavenger than ethylene dibromide.

An important inorganic intermediate is titanium tetrachloride; found in TiO_2 pigment manufacture by the so-called chloride process, now supplanting the older sulfate process.

Chlorine is also required for the manufacture of organic intermediates such as adiponitrile from butadiene. In this instance both chlorine and caustic are consumed, fortunately in the same ratio as they are produced by electrolysis of brine. This ratio, known industrially as one electrochemical unit (ECU), consists of one ton of chlorine to 1.1 ton of caustic soda. Neither the chlorine nor the caustic appears in the end product, but other routes to adiponitrile exist, affording the manufacturer an option to use or not use chlorine.

Growth of elastomeric products, made either by using chlorine, butadiene, and caustic or by using hydrochloric acid and acetylene, has contributed significantly to world demand for chlorine.

Product obsolescence can play an important role in the chlorine market, as in the case of byproduct chlorine from metallic sodium manufacture. Sodium production is expected to decrease if demand for use in producing lead antiknock compounds for gasoline declines, reducing the amount of chlorine available from this source.

Phosgene is another chlorine chemical important to chemical processing, e.g., in the production of agricultural chemicals, isocyanates, and other fiber intermediates and elastomers.

Demand for chlorine in the United States grew at a rate of about 8.5% per year during the 1960s, spurred by increased production of vinyl chloride and other chloro-organics. The rise was followed by a modest decline to about 5% from 1970 to 1971, mainly due to the curtailed use of chlorinated pesticides. In 1972–1973 demand recovered, exceeding the 1970 record and causing a tight-supply situation which was alleviated by adding to capacity in the late 1970s. Profitability has since suffered because rising costs have not been adequately compensated for by price increases, due in part to the weak economy in 1980.

Production growth of 3% per year from 1974 through 1979 was supported by a growth in exports, particularly in 1978 and 1979. Actual domestic growth was only about 1.5% per year and will continue at that low rate due to: (1) nonchlorine based processes supplementing some important chlorine requirements, (2) slackened growth in cyclic organic detergents which have been attacked by environmentalists, (3) leveling-off in growth of other major products containing chlorine such as vinyl chloride, and (4) the shortage of petroleum reflected both in process fuel and feedstocks.

The economics of the chlor-alkali industry was characterized in the 1960s by the ability to compensate for sharply rising costs by means of technical improvements and economies-of-scale in the manufacturing process, but the industry has now achieved whatever could be gained through these routes. Major consumers react now by seeking opportunities for back-integration or by making long-term contracts for purchases at the lowest negotiable price. The choice between these alternatives is determined by capital costs and contract terms and prices.

Chlorine-caustic market pricing in the 1960s and 1970s provides insight to the pricing mechanism at work. Chlorine pricing in the early 1960s was characterized by vendor willingness to make long-term contracts at a set price because, while raw material costs were slightly rising, technical improvements were offsetting costs. Such contracts, with terms of five to seven years, provided the seller with an adequate, though modest, return on investment. ECU prices, equally divided between chlorine and caustic, were about \$70. With the decline in chlorine demand along with the general economy in 1968–1971, chlorine prices could not be increased. The manufacturers took advantage of the tightness in the caustic market by raising caustic prices, thus compensating for the weakness in chlorine and giving a slight increase in overall profitability. Industry return on investment is not so great as to give strong incentive to accelerated expansion. As the economy recovered after 1971, the chlorine market improved and the industry was operating at capacity by the beginning of 1973, giving the opportunity for further price increases. These increases started in 1972 when an ECU delivered Gulf Coast was in the low \$80s and rose to the level of about \$100 early in 1974. Prices rose to the \$100 plus range in the late 1970s and into 1980, though spot prices weakened sharply late in 1980 due to the recession.

Since most of the chlorine produced in the United States is for captive use (about 60%), merchant sales have been used by the manufacturers to achieve high rates of operation, thereby minimizing unit costs of their captive requirements. These merchant sales are largely influenced by proximity of vendor to purchaser.

The overall market demand must be viewed in terms of chlorine chemicals; some of the ostensibly captive chlorine requirement of manufacturers is truly just one move downstream along the route to a final product. In the case of chlorinated hydrocarbons, the chlorine manufacturer takes the next step of treating hydrocarbons and selling the chlorinated compound.

About three-quarters of US salt production is used in the chemical industry where industrial capacity for salt brine electrolysis has barely kept ahead of rising demand. Individual new plants and additions to existing ones are now being sized at 700 tpd compared to about 200 tons in the mid-60s. Each major addition to capacity is thus 2 or 3% of the industry capacity. Assuming the forecast of 1.5% domestic growth rate is correct, one addition will be required every year or two for chlorine. If these are not forthcoming, shortages will persist.

Table 4 lists the US chlorine producers' approximate annual capacities.

An interesting change of technology in chlorine manufacture is the decline of mercury cells in favor of diaphragm cells due to environmental control measures, even though diaphragm cells have higher operating costs, albeit lower initial investment. Prior to this restructuring, the trend was toward larger and larger mercury cells, up to cells producing 20 tons of chlorine per day. Through the impact of pollution control, the emphasis has changed so that more than 70% of the chlorine produced by brine-electrolysis comes from diaphragm cells, in part also stimulated by development of the dimensionally stable electrode. This electrode eliminates the problem of gradual chlorination of the organic binders used in graphite anodes which resulted in inefficient cell operation as the anodes disintegrated. The new electrode made of titanium metal protected by a film of inert metal oxide permits construction of larger cells with higher current densities and results in a three-to-five year electrode life compared to three to six months for graphite electrodes. Diaphragm cells with a productive capacity of 3 tons of chlorine per day have a life of nine months to a year, being limited by diaphragm problems. Solution of these diaphragm difficulties is centered on new materials and techniques of construction.

TABLE 4—Annual Capacity of US Chlorine Producers

Company	Chlorine Capacity (Thousand st)*
Dow	4,600
PPG	1,570
Diamond	1,230
Hooker	1,160
Olin	930
Georgia-Pacific	510
Vulcan	490
Linden	490
BASF	490
Du Pont	480
Pennwalt	465
Stauffer	390
FMC	350
Kaiser	205
Alcoa	200
Ethyl	170
ICI	170
Weyerhaeuser	140
IMC	130
Goodrich	110
Shell	105
Others	1,170
Total	15,555

*Metric equivalent: 1 st × 0.907 184 7 = t.

A further cell modification is a bipolar cell, in which anodes are directly connected to adjacent cathodes to reduce the voltage drops now experienced in bussbar connections between cells. Development of bipolar membrane cells will permit higher concentrations of caustic in the cell, eliminating subsequent costly evaporation.

The long range market outlook is for continued modest, steady growth in use of chlorine by the chemical industry, despite adverse factors cited. A trend can be expected toward a greater percentage of foreign manufacturing based on chlorine and petrochemicals.

Fluorine

Fluorine's role in the chemical industry is mainly as a constituent of fluorocarbons and of inorganic fluorides. Hydrofluoric acid is the intermediate for practically all fluorine chemicals. It is produced by reacting fluorspar with sulfuric acid, yielding hydrogen fluoride gas and calcium sulfate. The gas is usually liquefied as anhydrous hydrogen fluoride, but may be absorbed in water to make aqueous hydrofluoric acid, usually 70% HF. In commercial practice about 2.2 tons of acid-grade fluorspar (97% CaF_2) are required to make one ton of hydrofluoric acid. Contaminants found in commercial, acid-grade fluorspar are calcium carbonate, silica, and sulfur; maxima are usually set.

Estimated production of hydrofluoric acid is shown in Table 5. Hydrofluoric acid producers in the United States are listed in Table 6.

Table 7 gives the estimated US consumption of hydrofluoric acid, by use, in 1980.

The fluorocarbons comprise a large family of useful compounds; the first to obtain commercial acceptance were used as refrigerants and still constitute a major part of the market. Because of environmental concern their use for aerosol propellants has been discontinued in the US. They are used for the manufacture of plastics characterized by chemical inertness, toughness, and electrical resistivity. Other fluorocarbons are used as fire-extinguishing agents, dielectrics, surfactants, anesthetics, and medicinals.

Inorganic fluorides are produced by reacting hydrofluoric acid with the appropriate base or basic salt. Principal uses of inorganic fluorides are outside of the chemical industry, such as: 1) the very large use of synthetic cryolite (Na_3AlF_6) used in the production of aluminum metal, 2) the relatively minor uses of sodium and potassium silicofluorides for fluxes and as constituents of enamels and glass, 3) aluminum fluoride as brazing flux, 4) sodium-aluminum-fluoride in resin-bonded grinding wheels, and 5) the use of sodium silicofluoride in water supply systems to control human tooth decay.

Hydrofluoric acid is used as a source of fluorine in reactions with uranium to produce gaseous uranium hexafluoride from which U^{235} is diffusion-separated from the more abundant

TABLE 5—Estimated Production of Hydrofluoric Acid and Consumption of Fluorspar, 1980 (Thousand Metric Tons)

	Hydrofluoric Acid Production*	Fluorspar Consumed†
United States	232	510
Western Europe	300	654
Japan	98	215

Source: Anon., 1982, "Product Review on Fluorine Minerals and Inorganic Compounds," *Chemical Economics Handbook,* SRI International, Menlo Park, CA.

* Includes hydrogen fluoride not withdrawn from the system, but used within it to produce other fluorine chemicals.

† Acid-grade fluorspar.

TABLE 6—Hydrofluoric Acid Producers in the United States

Company and Plant Location	Annual Capacity as of January 1982 (thousands of metric tons)	Remarks
Allied Chemical Corp.		
Baton Rouge, LA	17	
Baypoint, CA	6	
Geismar, LA	41	Increasing capacity to 86 kt/a by early 1983.
Nitro, WV	12	
Aluminum Co. of America		
Point Comfort, TX	41	Captive use.
E.I. du Pont de Nemours & Co., Inc.		
La Porte, TX	68	
Essex Chemical Corp.		
Paulsboro, NJ	10	
Gulf Oil Corp., Harshaw Chemical Co., subsidiary		
Cleveland, OH	14	
Pennwalt Corp.		
Calvert City, KY	23	
Total	232	

Source: Anon., 1982, "Product Review on Fluorine Minerals and Inorganic Compounds," *Chemical Economics Handbook,* SRI International, Menlo Park, CA.

Note: This table lists primary producers only. Reynolds Metals Co. operates a plant in Arkansas.

TABLE 7—Estimated US Consumption of Hydrofluoric Acid in 1980, by Use

Use	HF, 1000 Metric Tons	% of Total
Fluorocarbons	135	42
Aluminum production	105	33
Gasoline alkylation catalyst	16	5
Stainless steel pickling	10	3
Other	56	17
Total	322	100

Source: Anon., 1982, "Product Review on Fluorine Minerals and Inorganic Compounds," *Chemical Economics Handbook,* SRI International, Menlo Park, CA.

U^{238}. Hydrofluoric acid is also widely used as a catalyst in the alkylation of petroleum during high-octane gasoline production. A minor use of hydrofluoric acid is for etching glass, such as electric light bulbs.

The outlook for fluorine is reflected in projections published by the National Materials Advisory Board (Anon., 1970b), which indicate that US production of all fluorocarbons was about 1.9 billion lb in 1980, and that of the rest of the free world was 1.1 billion lb. Assuming that one lb of hydrofluoric acid is required for each three lb of fluorocarbon and that all hydrofluoric acid is produced from fluorspar, the amount of fluorspar to meet the fluorocarbon production requirement in 1980 is 725,000 net tons.

Although the demand for fluorspar requirements will probably continue to rise, the rate of increase will likely be moderate. The price-outlook for acid-grade fiuorspar is also a generally upward trend, with intermediate fluctuations largely dependent on: hydrofluoric acid demand, the success of aluminum manufacturing process now being tested that does not require fluorine, and the effect of substitution of alternate sources of fluorine. The main candidate for an alternate fluorine source is co-product fluorine from phosphate fertilizer manufacture. For hydrofluoric acid the trend is similar, but its demand and consequent price-pressure downside are dependent on the success of alternate aluminum manufacturing processes and upside on development of new fluorocarbon applications.

Sodium

Caustic soda, soda ash (sodium carbonate) and salt cake (sodium sulfate) are useful to the chemical industry mainly as sources of alkalinity and of the sodium ion.

Demand for caustic is in part related to chlorine, particularly with respect to adiponitrile production from butadiene, but is unrelated to chlorine in many of its applications. In cases where chlorine and caustic are related in direct proportion to the electrochemical unit (ECU) (described earlier in this chapter under *Chlorine*) a price advantage for the total ECU is possible because the vendor has no surplus of either commodity. The general case, however, is that pricing of the two are independent.

For the manufacture of certain sodium compounds, such as bicarbonate, cyanide, perborate, nitrate, and silicate, various sodium raw materials may be used. In making sodium bicarbonate and sodium silicate, soda ash is the usual starting material, but in the case of the silicate, caustic is sometimes used, while for nitrate, caustic is preferred.

Caustic is used in making such diverse products as cellophane, nylon intermediates, and pigments. It has important applications in waste-treatment processes, and pulp, paper and alumina manufacture. Increased demand for these uses has periodically led to tight supply and rising prices, partly because chlorine demand failed to keep pace with demand for caustic. Posted list price rose in early 1971 from $71 to $76 per ton with the concurrent elimination of discounts. These prices were followed by increases to $81 per ton in 1972, to $86 per ton in 1973, and to $100 in 1974. The price continued its rise to a plateau at $140 which held until 1978, then rose again to $160 by mid-1980, as the tight market persisted.

Recent shutdown of all except one Solvay-process plant in the US has been countered by an increase in production of mined soda ash. The overall growth of soda ash productive capacity was from 6.0 Mt (6.6 million st) in 1960 to 9.4 Mt (10.4 million st) in 1981. Production in 1981 was 7.5 Mt (8.3 million st) and domestic consumption about 6.5 Mt (7.2 million st).

The principal use of soda ash is in the glass industry (54%). Chemicals use 20%, pulp and paper 4%, soaps and detergents 7%, and water treatment 4%. Other domestic uses account for 11% of consumption.

Caustic will become more plentiful as the chlor-alkali industry responds to growing demand, but is currently tight while chlorine is plentiful. Caustic is expected to grow in international trade, where US exports have more than doubled since the mid-60s, mainly due to the growing demand for use in alumina production.

In looking to the future on prices, it has been noted that caustic soda pricing in the United States in the 1960s followed a downward trend that encouraged market-penetration. Because manufacturing costs remained high successful inroads were made into the soda-ash market. Technical improvements in manufacture can no longer be relied upon to offset rising costs. Since market penetration was achieved, prices have increased. Increases in the inflationary spiral are resulting in even higher prices. Further price increases will occur due to tight supply if the chlor-alkali industry fails to add the required new productive facilities in time to meet anticipated increased demand.

Sulfur

Mined sulfur is normally converted to sulfuric acid for use in fertilizer manufacture and chemical processing. About 50% of sulfur production is used for fertilizer manufacture and 20% for chemicals, followed by less than 5% each for pigments, textiles, steel and petroleum. Nonacid uses of sulfur are in pulp and paper manufacture (<5%), carbon disulfide (<5%) and insecticides (<5%).

Sulfur's usefulness is so widespread and pervasive in the chemical industry that its consumption closely follows the pattern of industrial activity. Tables 8 and 9 (Anon., 1967) provide a remarkably clear condensation, succinctly portraying the role of sulfur and sulfuric acid derivatives in organic and inorganic reactions and process applications. The history of sulfur in its relation to the chemical industry is interestingly told by Haynes (1959).

The following publications are recommended for current and comprehensive coverage of sulfur matters: (1) the magazine *Sulphur*, published by the British Sulphur Corp. Ltd., (2) the half-yearly *Statistical Supplement*, published by the British Sulphur Corp. Ltd. in conjunction with *Sulphur*, and (3) publications of The Sulphur Institute, Washington, DC.

Frasch-process sulfur mining has traditionally satisfied most sulfur requirements. Other important supplies of sulfur are coproduct sulfur from natural gas and petroleum, and sulfur from sulfide minerals, such as iron pyrite. A new supply of sulfur may be the byproducts of pollution control. The amount of this new supply is difficult to quantify but may be large enough to give incentive for new uses such as road paving, and to have an impact on sulfur and sulfuric acid prices.

Marketing of sulfur reflects the well-established preference of bulk consumers for deliveries in liquid form. High labor and energy costs for remelting, together with environmental aspects, continue to stimulate the preference for transport and use of liquid, yet distribution equipment is in short supply. Both producers and consumers are reluctant to make the financial commitments required for the new transport equipment and terminals required to increase shipping capacity.

Sulfur pricing is notoriously volatile. The present period of rising prices is expected to continue. The price in the Tampa area was $140 per lt in Jan. 1981, having risen from $31 in 1974. Canadian producers' price for recovered sulfur in 1973 went from 5 to $8 per lt to 12 to $15 per lt (rail) in Alberta and has continued sharply upward to $60 per lt, as of Dec., 1980. Delivered prices overseas ranged from 26 to $36 per lt for liquid and 24 to $33 for solid in northwest Europe in the early 1970s; and rose continuously to $75 in 1980. Prices were spurred in that year by delays in deliveries from Poland due to labor unrest and by cutoff of about one million lt per year supply due to the Iran/Iraq war. Sulfur is clearly an international commodity, with a market strongly influenced by supplies and demand outside the United States.

Future pricing, as in the past, will depend on relatively small excess or shortfall in supply being reflected by exaggerated price movements. Near-term emphasis is on upward spiraling prices until point-source recoveries and new sour-gas discoveries in the US Rocky Mountain Overthust Belt and elsewhere take effect.

Bibliography and References

Anon., 1967, *Chemical Origins and Markets,* Stanford Research Institute, Menlo Park, CA, 99 pp.

Anon., 1970, "The Adequacy of Acid-Grade Fluorspar Supplies Questioned," *Industrial Minerals,* No. 34, July, pp. 11–19.

Anon., 1970a, "Trends in the Usage of Fluorspar," PB 198 339, National Technical Information Service, National Materials Advisory Board, US Dept. of Commerce, 54 pp.

TABLE 8—Sulfur Derivatives

- Sulfur (Frasch process, recovered, mined) →
  - Ammonium hydrosulfide →
    - Ammonium polysulfides → Insect sprays
    - Photographic developer
    - Textile industry
    - Coloring brasses
  - Ammonium sulfite, Ammonium thiosulfate →
    - Reducing agent
    - Hair wave permanent
    - Photographic fixer
    - Liquid fertilizer additive
  - Ammonium thiocyanate →
    - Naphthylthiourea (antu)
      - Rodenticide
    - Thiourea
    - Guanidine sulfocyanate
    - Fertilizers
    - Photography
    - Freezing solution (liquid rocket propellant)
    - Weed killer/defoliant
    - Electroplating
  - Barium sulfate →
    - Barium sulfide → Depilatory; Vulcanizing
    - Paint pigment
    - Plastics
    - Rubber
    - Medicine
    - X-ray indicator
  - Calcium polysulfide →
    - Sodium hydrosulfide
    - Agricultural fungicide
  - Carbon disulfide →
    - Viscose rayon
    - Cellulose film
    - Ferbam → Fungicide
    - Fluorocarbons
    - Carbon tetrachloride
  - Hydrogen sulfide →
    - Ammonium sulfide → Textiles; Photo developer; Color brass/bronze
    - Mercaptoethanol → Dye solvent; Pharmaceuticals; Insecticide; Flotation agent
    - Sodium sulfide
    - Sodium hydrosulfide
    - Purify hydrochloric and sulfuric acid
    - Vulcanize synthetic rubber
  - Mercaptans →
    - Methyl mercaptan
      - Methionine → Protein food additive
      - Dimethyl sulfide
      - Dimethyl sulfoxide
      - Pharmaceuticals
    - Propyl and butyl mercaptans → Insecticides
    - Dodecyl mercaptan → Synthetic rubber; Fungicides; Surfactants
  - Molybdenum disulfide →
    - Molybdenum pentachloride
    - Solid lubricants
  - Mercaptobenzothiazole (Captax) → Rubber vulcanizing accelerator
  - Phosphorus pentasulfide → Lube oil additive; Rubber additive; Flotation agent; Insecticides
  - Phosphorus sesquisulfide → Matches
  - Potassium thiocyanate →
    - Thiourea
      - Aminothiazole → Sulfa drugs
      - Photography
      - Rubber accelerator
    - Freezing mixtures
    - Dyeing textiles
  - Sodium polysulfide →
    - Electroplating
    - Dye
    - Petroleum additive
    - Polysulfide polymers (Thiokol compounds)
  - Sodium thiosulfate
  - Sulfur dioxide →
    - Chlorine dioxide
    - Hydrochloric acid (Hargreaves process)
    - Hydroxylamine sulfate
    - Potassium bisulfite
    - Potassium hydrosulfide
    - Potassium sulfate (Hargreaves process)
    - Potassium thiosulfate
    - Sodium bisulfite
    - Sodium persulfate
    - Sodium sulfate (Hargreaves process)
    - Sodium sulfite
    - Sodium thiosulfate
    - Sulfur from hydrogen sulfide or from sour natural gas
    - Sulfur trioxide
    - Sulfuric acid (Chamber process)
    - Sulfurous acid
    - Sulfuryl chloride
    - Thionyl chloride
    - Zinc sulfide
    - Zinc hydrosulfite
    - Zinc sulfoxylate
  - Sulfur hexafluoride → Dielectric insulating gas for transformers, coaxial cables and other high-voltage electrical equipment
  - Sulfur monochloride →
    - Chlorinating agents
    - Complexes with polyvalent metal chlorides
    - Mustard gas → Chemical warfare agent
    - Organic acid anhydrides
    - Sulfur dichloride →
      - Sulfur tetrachloride
      - Thionyl chloride →
        - Alkyl chlorosulfonates
        - Isoamyl chloride
        - Sulfonyl chlorides
        - Thionyl bromide
        - Antihistamine compounds
        - Fatty acid chlorides
        - Organic acid anhydrides
        - Organic acid chlorides
        - Dyes
        - Nitriles
        - Pharmaceuticals
        - Catalyst (chlorobenzene from sulfuryl chloride)
        - Desiccant to remove water of hydration from
        - Synthetic pyrethrum
      - Chlorine carrier
      - Vulcanize rubber
      - Purify sugar juices
      - Organic chemicals
      - Insecticides
    - Thionyl chloride
    - Catalyst (chlorobenzene from benzene)
    - Rubber vulcanizing
    - Purifying sugar juices
    - Textile finish
    - Vulcanized oils
  - Sulfuric acid →
    - Hydrochloric acid, Nitric acid → Sodium bisulfate →
      - Sodium sulfate
      - Sodium sulfide
      - Mineral decomposing flux
      - Disinfectant
      - Dyeing
      - Carbonizing wool
      - Magnesia cements
      - Soap
  - Sulfur dyes
  - Sulfur nitrides
  - Agricultural fungicide
  - Polymerization inhibitors → Styrene; Vinyl acetate
  - Rubber vulcanizer

Source: Anon., 1967 (courtesy Stanford Research Institute).

TABLE 8—Continued

→ Caprolactam

→ Reduce organic compounds; Antiseptic; Food preservative; Fiber bleach; Tanning

→ Separation of heavy metals

Potassium persulfate → Bleach; Antiseptic; Pharmaceuticals; Synthetic rubber polymerizer

→ Potassium sulfide → Dimethyl sulfide → Dimethyl sulfoxide (Polyacrylonitrile solvent; Industrial cleaner; Paint stripping; Hydraulic fluid; Pesticides; Wool felt treatment); Odor donor natural gas; Catalyst impregnator

Potassium sulfide → Dipilatory; Medicine

Alum manufacture; Fertilizer; Medicine

→ Sodium hydrosulfite; Dye; Perfume; Wood pulp; Food preservative; Medicine; Pesticide; Fermentation industry

Sodium hydrosulfite → Sulfamic acid; Reducing agent; Dye stripping; Bleaching sugar, soap, minerals

→ Fat/oil bleach

→ Sodium thiosulfate; Dyes; Organic chemical; Textile bleach; Reducing agent; Latex treating; Food preservative; Photographic developer

Ammonium sulfamate → Flameproof textile/paper; Weed killer; Electroplating

Chlorosulfonic acid → Cyclamates; Sulfated surfactants; Saccharin; Poison gas; Sulfamic acid

Dioxane sulfotrioxide → Alkyl hydrogen sulfates

Dodecylbenzene sulfonate → Synthetic detergents using phosphate builders

Fluosulfonic acid → Organic catalyst; Electropolishing

Lauryl sulfate — Synthetic detergents

→ Oleum → Anthraquinone-disulfonic acids → Dyes; Dimethyl sulfate → Methylating amines and phenols; Naphthalenetrisulfonic acids; Sulfamic acid; Sulfanilic acid → Dyestuffs, Organic synthesis, Medicine; Sulfonic acids of saturated paraffin hydrocarbons → Sodium alkane sulfonates → Emulsifying agents → Cutting oils, Fungicides, Insecticides, Lubricating oil additives (→ Oil-soluble petroleum sulfonate, barium salt; Oil-soluble petroleum sulfonate, calcium salt; Oil-soluble petroleum sulfonate, sodium salt), Paper, Polishes, Textiles; Thionyl chloride

Pyridiniumsulfonic acid → Naphthylamine sulfonic acid; Naphthylamine disulfonic acid → Dyes

Pyrosulfuryl chloride → Synthetic processes

Sulfur doixide (high purity)

Sulfuric acid (contact process)

Thionyl chloride

Thioxane sulfotrioxide → Naphthalenesulfonic acid → Disinfectant soaps

→ Organic synthesis; Bleach straw, paper, textile; Wine, food, beer preservative

→ Alkyl sulfonyl chlorides; Aryl sulfonyl chlorides; Benzal chloride; Benzyl chloride; Chlorobenzene; Chlorophenols → Disinfectant; Rayon; Zinc sulfoxylate → Dye stripping; Military poison gases; Pharmaceuticals; Solvent for chlorination or bromination of dyes and dye intermediates; Dye stripping

→ Paint pigment; Glue and gelatine; Lithopone; X-ray/tv phosphor; Rubber; Plastic

→ Brighten paper pulp; Sugar juices; Bleach textiles, oils, straws

→ Dye stripping

→ Alkyl chlorides; Dialxyl sulfites

→ Surface active agents

→ Surfacants; Peroxides

→ Aluminum chloride; Magnesium chloride; Titanium trichloride

→ Leather dipilatory; Organic chemicals; Dyes; Rayon; Flotation process; Engraving/printing; Photographic reagent

(Right column, from top line:)

Corn starch

Natural glues

Pulp wood digestion (semichemical and sulfite processes)

Annealing of glass

Antichlor → Kraft pulp bleaching; Water purification

Bleach → Beet sugar solutions; Crude straw; Silk; Wool

Fertilizer

Fumigant → Grain; Silage

Heat treatment of magnesium alloys

Ore flotation agent (depressant of zinc blende during flotation of copper ores)

pH control

Preservative → Beer; Fresh and dried fruits; Meat; Wine

Refrigerant

Solvent extraction of lubricating oils

Solvent for aromatic and sulfur compounds in petroleum feed stocks (Edeleanu process)

Soybean protein

Sterilizing fermentation equipment (citric acid)

Tanning

Manufacture of → Activated carbon; Bromine; Iodine from oil well brines; Selenium from copper slimes

TABLE 9—Sulfuric Acid, Derivatives

Sulfuric acid* (from elemental sulfur; from pyrites; from smelter gases; from sludge acid; refortified acid; from H_2S) →

Inorganic reactions

- Aluminum sulfate → Sizing paper; Water purification; Dye mordant; Sewage treatment; Fire fighting; Fat/oil clarifier; Aluminum sodium sulfate → (Textile waterproofing; Tanning; Paper precipitant; Sugar refining)
- Iron free aluminum sulfate
- Hydrogen peroxide
- Ammonium bisulfate → Ammonium persulfate
- Ammonium sulfate (from coke-oven gas) → Fertilizer
- Ammonium sulfate (from synthetic ammonia) → Ammonium chloride → Dry cells; Cobaltous ammonium sulfate; Nickel ammonium sulfate → Electroplating; Fertilizer; Flame retardant in fabrics; Industrial ethyl alcohol from molasses (nutrient for yeast culture); Viscose rayon
- Antimony sulfate → Explosives
- Barium sulfate (synthetic blanc fixe) → Barium oxide; White pigment for paints
- Beryllium sulfate → Beryllium oxide
- Boric acid
- Cadmium sulfate → Cadmium electric cells; Medicine; Vacuum tube manufacture
- Chlorosulfonic acid → Sulfamic acid; Sulfonated oils → Textile finishing agents; Sulfuryl chloride; o-Toluene sulfonyl chloride → (Saccharin; Dyestuffs); Sulfated sufractants
- Chromic acid → Zinc chromate; Andizing aluminum and aluminum alloys; Ceramic glaze; Chrome plating baths; Copper stripping; Dyestuffs; Metal surface treatment; Pharmaceuticals; Rubber pigment
- Chromic sulfate → Green paint/varnish; Leather tanning; Textiles
- Cobaltous sulfate → Ceramics; Pigments; Glazes; Plating
- Copper sulfate → Fungicide
- Ferric sulfate → Aluminum etching; Pigment; Soil amendment; Coagulant in water and sewage treatment
- Ferrous sulfate
- Fluoboric acid → Electroplating
- Fluosulfonic acid → Aryl sulfonyl fluorides → Lubricating oil additives; Catalyst (alkylation of isoparaffins); Electropolishing; Wood preservative
- Hydrochloric acid (from sodium chloride)
- Hydrofluoric acid (from fluorspar)
- Hydrocyanic acid (from sodium cyanide) → Acrylonitrile; Acrylates; Adiponitrile; Dyes; Fumigants; Rubbers; Plastics; Chelating agents
- Hydrogen sulfide
- Lead sulfate, basic → Ceramics; Leaded zinc oxide; Pigment; Stabilizer in vinyl plastics
- Lithium sulfate → Pharmaceuticals
- Magnesium sulfate → Mineral water; Magnesium tungstate → Luminescent paint; Fireproofing formulations; Leather tanning; Medicine; Paper sizing
- Mercuric sulfate → Medicine; Gold/silver extractant; Galvanic batteries
- Mercurous sulfate → Laboratory batteries
- Nickel sulfate → Nickel-ammonium sulfate → Electroplating; Nickel catalyst; Nickel plating; Printing textiles
- Oleum
- Phosphoric acid (wet process) → Monoammonium phosphate → Diammonium phosphate → Fertilizer; Enriched superphosphate, Triple superphosphate → Fertilizers; Rust proofing metals; Soft drink; Sugar refining; Water treatment
- Potassium bisulfate → Food preservative; Tartrates
- Potassium dichromate
- Potassium sulfate (from potassium chloride) → Potassium aluminum sulfate (potash alum) → (Medicine; Dye mordant; Baking powder); Potassium bisulfate; Potassium persulfate; Potassium sulfide → (Medicine; Depliatory); Antiflash agent in smokeless powders; Fertilizers; Manufacture of gypsum wall board
- Silica gel
- Silicon tetrafluoride (by product of phosphate rock acidulation) → Fluosilicic acid
- Sodium bisulfate → Metallurgical flux; Pickling metal; Leather bleach; Cleaning agents
- Sodium dichromate
- Sodium sulfate →
- Sulfamic acid →
- Sulfuryl chloride
- Titanium dioxide → Pigment for → (Lacquers; Paint; Printing inks; Varnish); Delustering synthetic fibers; Single crystals for transducers; Filler for → (Coated fabrics and textiles; Floor coverings; Paper; Rubber); Welding rod
- Tungstic acid → Tungsten, metal; Textiles; Plastics
- Vanadyl sulfate → Mordant; Aniline black; Ceramic coloring
- Zinc sulfate

* Varying strengths of sulfuric acid are used as processes require (sulfuric acids = 100% H_2SO_4, or oleum 100% H_2SO_4)

Source: Anon., 1967 (courtesy Stanford Research Institute).

TABLE 9—Continued

Organic Reactions

- Acridine → Dyes
- Anthraquinone → Dye
- Chloroanthraquinone → Medicine
- Benzene (from coke-ovens)
- Benzene sulfonic acid → Phenol; Resorcinol; Organic synthesis; Detergent hydrotropes
- Chloral → Dichlorodiphenyltrichloroethane (DDT)
- Citric acid → Beverage; Pharmaceuticals; Effervescent salts
- Cresols and cresylic acid → Creosote; Disinfectant; Lube oil additive
- Cyanamide → Dicyandiamide
- Dichlorophene → Fungicide; Dermatological applications; Veterinary medicine
- Diethyl sulfate
- Dimethyl sulfate
- Dioxane → Solvent for cellulose acetate; Paint/lacquer; Dye dispersant; Cosmetics; Fumigant; Glues
- Dodecylbenzene sulfonic acid
- Ethyl hydrogen sulfate (from ethylene) → Diethyl sulfate; Ethyl alcohol; Ethyl ether → Organic solvent; Anesthetics; Medicine; Smokeless powder; Organic synthesis; Fumigant
- Fatty alcohols → Alkyl sulfates → Anionic detergents
- Formic acid → Formate esters; Dyeing; Laundry sour; Fumigant; Food preservative; Rubber coagulant; Ore flotation; Electroplating
- Hydrazine sulfate → Reducing agent; Condensation reactions; Acetate fiber catalyst; Blood tests; Metal and mineral analysis; Fungicide; In adhesive
- Lacquers
- Isopropyl alcohol → Acetone; Solvent; Deicing agent; Lacquers; Perfumes; Preservative
  - Acetone → Acetic anhydride; Diacetone alcohol; Methyl isobutyl ketone
- Ketoglutaric acid
- Lactic acid
- Mannitol hexanitrate (MHN) → High explosive
- Methyl methacrylate
- Naphthalene sulfonic acids → β-naphthol
- Naphthol sulfonic acids → Dyes
- Naphthylamine sulfonic acids → Naphthionic acid → Dye intermediate
- Oxalic acid (from calcium oxalate)
- Phenol (from coke-ovens, petroleum, and sodium phenolate)
- Phenanthroline → Dryer catalyst; Indicator
- Phenol sulfonic acid → Picric acid → Ammonium picrate → High explosive
- Salicylic acid
- Stearic acid
- Tall oil
- Toluene sulfonic acid, sodium salt; Xylene sulfonic acid, sodium salt → Detergent hydrotropes; Dyes; Organic synthesis
- Vanillin
- Mixed acid (nitric and sulfuric) →
  - Dipentaerythritolhexanitrate (DPEHN) → High explosive
  - Nitrocellulose → Guncotton → Single-base propellant; Double-base propellant
  - Nitrocellulose → Pyrocotton → Coating material
  - Nitrocellulose → Pyroxylin → Low energy propellant
  - Nitroglycerin; Pentaerythritoltetranitrate (PETN); Trinitrotoluene (TNT) → High explosives
- Synthetic petroleum sulfonates → Water soluble sulfonates for emulsifying agents; Oil soluble sulfonates for lube oil additives
- Manufacture of → Bromine dinitrobenzene → Trinitrobenzene; Kaolin; Naphthalene (from coal distillation); Nitrobenzene; Pyridine (from coal distillation); Xylene (from coal distillation)

Process Applications

- Additive to sodium peroxide baths for bleaching groundwood pulp
- Alumina silica catalysts
- Assistant in dyeing wool with acid dyes
- Catalyst in manufacture of → Acrylic esters by alcoholysis; Amyl acetate from amyl alcohol; Butyl acetate from butyl alcohol; Caprolactam; Cellulose acetate from cellulose by acetylation; DDT from chloral; Dialkyl phthalates from phthalic anhydride; Diphenylethane; Ethyl acetate from ethyl alcohol; Petroleum alkylate from petroleum feed stocks; Vinyl acetate from acetylene
- Chlorine drying
- Coagulant for SBR elastomers
- Corrosion inhibitor in phosphoric acid
- Cracked gasoline treating
- Creping agent for cotton
- Deodorizing tall oil
- Desulfonization
- Dessicant in manufacture of → Chlorine from sodium chloride; Nitric acid
- Desulfonization
- Electrolyte in lead-acid storage batteries
- Kraft pulping
- Mixed fertilizers
- Oat hulls → Furfural
- Ore flotation agent (pH modifier)
- Parchment paper
- Production of electrolytic copper and zinc
- Pickling of iron and steel
- Polymerization
- Regenerating bath for cellulose
- Sec and tert butyl alcohol
- Synergist for phenolic antioxidants
- Textile finisher
- Treating kerosine
- Water treatment/clarification
- Rubber coagulant

→

- Kraft paper production
- Lead sulfate → Storage batteries; Paint pigment
- Molybdate chrome orange
- Ceramic glaze
- Assistant in dyeing wool with acid dyes
- Synthetic detergents diluent
- Freezing mixture
- Glass
- Nickel smelting
- Spinning baths for viscose rayon
- Tanning
- Wood pulp (sulfate process)

→

- Ammonium sulfamate → Fire retardant
- Ceramic cleaning
- Electroplating
- Fire retardant
- Formaldehyde
- Metal pickling
- Plasticizer
- Weed killer

Anon., 1970b, et seq., *Chemical Economics Handbook,* Stanford Research Institute, Menlo Park, CA.

Anon., 1971, "Fluorspar, Cryolite and Major Fluorine Chemicals," Roskill Information Services, London, 140 pp.

Anon., 1972, "Du Pont Halon 1301 Fire Extinguishant," Organic Chemicals Dept., E.I. du Pont de Nemours & Co., Wilmington, DE, 14 pp.

Anon., 1973, "Alcoa Claims Aluminum-Smelting Process If Developed May Cut Electricity Use 30%," Aluminum Co. of America Press Conference report, *Wall Street Journal,* Jan. 12, last page.

Farmer, M.H., and Bertrand, R.R., 1971, "Long Range Sulfur Supply and Demand Model," PB 208 993, Esso Research and Engineering Co., National Technical Information Service, US Dept. of Commerce, 69 pp.

Fulton, R.B., 1971, "World Economics of Selected Industrial Minerals," *AIME Centennial Volume,* AIME, New York, pp. 74–87.

Gould, R.F., ed., 1961, *Borax to Boranes,* Advances in Chemistry Series No. 32, American Chemical Society, Washington, DC, 244 pp.

Haynes, W., 1959, *Brimstone, the Stone that Burns,* Van Nostrand, New York, 308 pp.

Johnstone, S.J., and Johnstone, M.G., 1961, *Minerals for the Chemical and Applied Industries,* 2nd ed., John Wiley, New York, 788 pp.

Jones, G.K., 1973, "Industrial Minerals and the Chemical Industry," *Industrial Minerals,* No. 73, Oct., pp. 9–29.

Kliff, E.J., et al., 1971, *World Supply and Demand for Fluorspar and Fluorine Products,* Charles H. Kline & Co., Fairfield, NJ, 466 pp.

Lyday, P.A., 1982, "Boron," *Mineral Commodity Summaries, 1982,* US Bureau of Mines, pp. 22–23.

Lyday, P.A., 1982, "Bromine," *Mineral Commodity Summaries, 1982,* US Bureau of Mines, pp. 24–25.

Raymont, M.E.D., 1972, "Sulphur Sources and Uses—Past, Present and Future," *Canadian Mining & Metallurgical Bulletin,* Nov., pp. 49–54.

Voskuil, W.H., 1955, *Minerals in World Industry,* McGraw-Hill, New York, 324 pp.

The US Bureau of Mines annual *Minerals Yearbook* and the *Mineral Industry Surveys* are also very helpful.

Construction Materials

The Construction Materials section contains the following Chapters:

Aggregates
- **Crushed Stone**
- **Lightweight Aggregates**
- **Sand and Gravel**
- **Slag**

Cement and Cement Raw Materials
Dimension and Cut Stone
Gypsum and Anhydrite
Insulating Materials—Thermal and Sound
Roofing Granules

Aggregates—Introduction

HENRY N. McCARL *

Mineral aggregates are those natural and manufactured industrial mineral and rock materials that provide bulk and strength in portland cement concrete, bituminous concrete mixes, and plaster or stucco surfaces. They may also provide special characteristics such as thermal and acoustical insulation, weight, surface textures, abrasion resistance, and impermeability to various concrete products and mixes. Most aggregates are relatively inexpensive materials, but despite their low unit value, the tonnages consumed each year in the United States and throughout the world make mineral aggregates by far the leading industrial mineral and rock materials both in terms of volume (tonnage) produced and value consumed.

The major use of mineral aggregates is in concrete. The low cost, high bulk aggregates tend to keep concrete construction costs at levels competitive with other building materials. Other major uses of aggregates include highway and railroad base or ballast materials, graded fill, and various industrial uses such as metallurgical and chemical fluxes and raw materials—these latter uses are often considered nonaggregate uses of mineral aggregate materials. The per-ton value of mineral aggregate materials at the production plant in 1980 and 1981 varied from $2 to $5 for most sand and gravel and crushed stone, $3 to $6 for air-cooled slag, $7 to $20 for expanded slag and expanded clay or shale aggregate, to as much as $165 to $320 for expanded perlite or vermiculite. The expanded lightweight aggregates have been dramatically affected by increasing fuel costs. It is ironic that the specialized aggregates used as insulating materials to conserve energy cost so much more to produce due to the high proportion of energy consumed in their production.

The first four parts of "Construction Materials" deal with the utilization of four major mineral aggregate types: Crushed Stone, Lightweight Aggregates, Sand and Gravel, and Slag. While the focus will be on the use of these industrial minerals and rocks, there will also be some coverage of sources, mining and production, especially in those cases where they are not discussed in the Commodity section of this book.

* Associate Professor of Economics and Geology, School of Business, The University of Alabama in Birmingham, Birmingham, AL.

Construction Materials

Aggregates—Crushed Stone

GEORGE H. K. SCHENCK *

THOMAS F. TORRIES †

Crushed and broken stone is used directly in construction as an aggregate and accounts for about half the value and quantity of natural aggregates consumed in the United States. Sand and gravel provide most of the balance of aggregates except for some lightweight materials such as pumice and expanded shale. Conditions necessary for production of crushed stone from a deposit are:

1) Quality—should pass rigid specifications for strength and durability.
2) Cost—low average total delivered cost.

Construction aggregates are hard, essentially inert materials suitable for being formed into a stable mass either (1) by the addition of cementing or binding materials that produce a concrete or (2) by compaction or by natural weight to produce a road base or foundation. Aggregates usually provide about 80 to 100% of the material volume of the construction masses in which they are used. The construction industry utilizes as aggregates about 80% of all crushed stone produced in the United States according to statistics provided by the US Bureau of Mines (USBM), Table 1, using definitions as shown below. Nonconstruction uses of crushed stone, Table 2, are included elsewhere in *Industrial Minerals and Rocks* because such uses require the special physical or chemical characteristics which are discussed in those chapters.

Principal aggregate mineral categories are (1) crushed stone and (2) sand and gravel. In 1980, there were 31 states in which over 9.0 Mt (10 million st) of sand, gravel, and crushed stone aggregate were produced or used and 9 states in which over 45.4 Mt (50 million st) of both categories of aggregates were produced or used (USBM).

* Associate Professor of Mineral Economics, The Pennsylvania State University, University Park, PA.

† Mineral Economist, Torries and Associates, San Antonio, TX.

Definition by USBM of Principal Crushed Stone Aggregate Functional Uses [Sizes as specified by American Society for Testing & Materials, ASTM]

Code	*Use and Product*
304	Concrete aggregate—nominal gradations from 88.9 mm (3.5 in.) maximum to 12.7 mm (0.5 in.) maximum. Follow American Society for Testing and Materials (ASTM) specification C33–78.
305	Bituminous aggregate—nominal gradations from 88.9 mm (3.5 in.) maximum to 12.7 mm (0.5 in.) maximum. Follow ASTM specification D602–79.
306	Macadam aggregates—nominal gradations from 88.9 mm (3.5 in.) maximum to 12.7 mm (0.5 in.) maximum. Follow ASTM specification D693–77.
307	Dense graded roadbase stone—nominal gradations from 38.1 mm (1.5 in.) maximum. Follow ASTM specification D2940–74.

308 Surface treatment aggregates—nominal gradations from 25.4 mm (1 in.) maximum. Follow ASTM specification D1139–79.

309 Unspecified construction aggregate & roadstone.

310 Riprap and jetty stone—heavy irregular rock chunks used chiefly for river and harbor work.

311 Railroad ballast—nominal gradations of crushed stone from 63.5 mm (2.5 in.) to No. 4 sieve used to form railway beds or substratum for new roads.

312 Filter stone—any porous stone through which water is filtered. Follow Federal (GSA) specification SS744a.

313 Manufactured fine aggregate (stone sand)—100% passing 9.5 mm (⅜ in.). Follow ASTM specification C33–78.

314 Terrazzo and exposed aggregate—small chips or pieces of crushed stone about 12.7 to 19.1 mm (0.5 to 0.75 in.) in diameter.

Purchasers of aggregate desire a reliable supply of specification stone at lowest delivered cost relative to alternative sources. Delivered cost of stone includes total production costs, distribution costs, and profit for the producer and hauler. It is common for f.o.b. prices and shipping cost to be quoted separately.

Suitable Locations

Many rock types meet the basic physical requirements established for aggregates,

TABLE 1—Crushed Stone Aggregate Use in the United States in 1980 (in Millions of Short Tons*)

Uses	Limestone[†] Dolomite	Granite, Syenite Gneiss	Traprock[‡]	Sandstone, Quartzite	Shell	Other and Specified	Total[§]	$ Value Total	$ Value Per Ton
Road base	151.9	32.2	20.2	7.1	2.6	5.1	219.1	649.2	2.96
Concrete	98.2	18.1	7.6	2.4		0.4	126.7	453.5	3.58
Unspecified aggregate	116.6	22.7	25.1	7.4	5.0	3.5	180.3	566.0	3.14
Bituminous	57.8	16.7	11.6	3.7		0.6	90.4	339.4	3.75
Surface treatment	36.4	3.4	3.9	1.2		0.3	45.2	156.3	3.46
Macadam	19.9	1.9	2.6	0.2		0.7	25.3	79.5	3.14
Riprap and jetty	15.3	2.8	3.7	1.1		0.6	23.5	75.8	3.23
Railroad ballast	13.0	12.3	3.4	1.4		0.1	30.2	91.3	3.02
Fill	2.1	0.3	W	0.2		0.5	3.1	8.1	2.61
Manufactured sand	15.2	3.0	1.0	1.0			20.2	80.1	3.97
Filter stone	3.5	1.5	0.4	0.2			5.6	19.4	3.46
Total	529.9	114.9	79.5	25.9	7.6	11.8	769.6	2518.6	3.27
(% of total)	(69)	(15)	(10)	(3)	(1)	(2)			
Average ton value	$3.20	$3.55	$3.69	$3.55	N/A	$2.92	$3.32		

Source: Singleton, 1980.
* Metric equivalent: st × 0.907 184 7 = t.
† Including marble and calcareous marl.
‡ Including gabbro, basalt, and diabase.
§ May not add due to rounding and inclusion of items withheld (W).
W-Withheld by US Bureau of Mines and no quantity included in total.

TABLE 2—Crushed Stone Use in the United States in 1980 (Units in Millions)

	Short Tons*	$ Value Total	$ Value Per St
	Physical Uses		
Aggregates, Subtotal	771.00 (79%)	2518.6	3.27
Other (granules, etc.)	8.3	38.0	4.58
Special (fillers, etc.)	5.7	80.9	14.19
Terrazzo and decorative	1.1	13.0	11.82
Nonaggregates, subtotal	15.1 (2%)	131.9	
Physical uses, subtotal	786.1	2650.5	
	Chemical Uses		
Cement, Subtotal	99.1 (10%)	234.6	2.37
Agriculture	36.6	155.4	4.25
Lime	30.3	95.1	3.14
Flux	16.1	60.1	3.73
Alkali works	1.9	5.7	3.00
Glass	2.1	15.8	7.52
Dead-burned dolomite	2.0	6.3	3.15
Ferrosilicon	0.1	1.0	10.00
Refractory	1.0	4.8	4.80
Noncement, subtotal	90.1 (9%)	344.2	
Chemical uses, subtotal	189.2	578.8	
	Unclassified Uses		
Not specified	3.4	14.1	4.15
Special uses	1.4	11.1	7.93
Unclassified, subtotal	4.8	25.2	
Total, all crushed stone	980.1	3254.5	3.32

Source: Singleton, 1980.
*Metric equivalent: st × 0.907 184 7 = t.

(Table 3). Limestone and dolomite account for 69% and granite for 15% of crushed stone aggregates as shown in Table 1.

Deposits in which it is possible to produce the large quantities required are common in most areas, and as a result, crushed stone, at an average 1981 price below $3.69/t ($3.35 per st), is one of the lowest unit cost items of commerce. Extraction of stone is usually conducted by open pit methods rather than by relatively more expensive underground methods because, on a national basis, surface exposures of acceptable rock are abundant.

The most economical rock to produce and use varies greatly from region to region. Georgia, North and South Carolina, and Virginia account for 75% of the production of crushed granite. Illinois, Indiana, Ohio, New York, and Pennsylvania account for about 30% of the limestone and dolomite produced while Missouri, Florida, Texas, and Tennessee account for another 30%. Shell is used in declining quantities in the Gulf states and over half is produced in Louisiana. Traprock production is scattered in pockets across the northern, western, southwestern, and eastern portions of the country. Production, in decreasing order, in Oregon, New Jersey, Washington, Connecticut, and California accounts for nearly 60% of the production (Singleton, 1980).

Rocks suitable for aggregate uses are well distributed across the United States and may be exposed or close to the surface for many miles. However, certain regions such as portions of the Gulf Coast, Florida, the Atlantic Coastal Plain, and the Great or High Plains are deficient in readily accessible deposits that meet typical national quality specifications. In

TABLE 3—Classification of Rocks Commonly Used for Crushed Stone

Group	General Classification	Rock	Approx. Specific Gravity	Los Angeles Abrasion Test* Avg.	Los Angeles Abrasion Test* Mid-Range†
Igneous	Intrusive	Gabbro	2.9	18	
		Granite	2.6	38	27-49
		Syenite	2.7	24	15-27
		Diorite	2.8		
		Peridotite	2.9		
	Extrusive	Felsite‡	2.6	18	
		Traprocks:			
		Basalt	2.8	14	10-17
		Diabase	2.9	18	13-21
Sedimentary	Calcareous	Dolomite	2.7	25	18-31
		Limestone	2.6	26	19-30
		Coquina			
		Coral	<2.6		
		Shell			
	Siliceous	Chert	2.5	26	
		Sandstone	2.6	38	24-48
		Graywacke	2.6	17	14-20§
Metamorphic	Foliated	Amphibolite	3.0	35	22-46
		Gneiss	2.7	45	33-57
		Schist	2.8	38	26-49
	Nonfoliated	Marble	2.7	47	26-64
		Quartzite	2.7	28	20-35

* Source: Woolf, 1953.
† Highest and lowest value after excluding top one-fifth and lowest one-fifth of tests reported by D. O. Woolf.
‡ Including andesite, dacite, rhyolite, and trachyte.
§ Private correspondence.

these areas rocks of lower quality have to be utilized. Where suitable alluvial deposits are available, their lower production costs will favor them over crushed stone.

Consumption

The uses of crushed stone are divided into two broad categories: (1) those uses in which the physical properties of the stone are more important, and (2) those uses in which the chemical properties of the stone are utilized. Almost 100% of the stone used for its physical properties and about 50% of the stone used for its chemical properties (primarily cement) are directly used in the construction sector of the national economy. The total quantity of crushed stone aggregates used in construction grew nationally at about 3.5% annually from 1960 to 1973. Since then production stagnated and did not exceed the record production of 960.7 Mt (1059 million st) in 1973 until 1979 when production reached 978.9 Mt (1079 million st). Production has dropped again; crushed stone production in 1981 was 815.6 Mt (899 million st). The average per ton price on a constant dollar basis has declined since 1960.

Construction uses of crushed stone can be divided into four end-use categories: highway, residential, nonresidential (commercial), and government. Highways account nationally for the largest share of total use though the average for an individual producer may be different. The quantities of crushed stone used regionally in each classification can be estimated by using historical usage factors or by using engineering estimates based upon a set of assumptions. All engineering estimates in this chapter are based upon a specific gravity of 1.66 for *crushed* limestone.

Highway Construction

Historical usage factors of stone used in the construction of highways are given in Table

FIG. 1—*Selected profile of crushed stone used for primary and interstate highways including later resurfacing.*

Item	Depth In.	Material	Crushed Aggregate*
1. Resurfacing	4	bit. conc.	Required
2. Surface	10	p.c. conc.	Required
	8	bit. conc.	Required
3. Subbase	14	coarse aggr.	Required
4. Select material	8	local g.d.m.	None
5. Subgrade	Varies	dirt	None
6. Shoulder	4	coarse aggr.	Required

* May be either stone or gravel origin.

Abbreviations: aggr., aggregate; bit., bituminous; conc., concrete; p.c., portland cement; g.d.m., good draining materials. Metric equivalent: in. × 25.4 = mm.

4. The percentage of stone costs relative to the overall cost of highway construction has been decreasing because the cost of stone is relatively stable compared to the rising trend of total highway construction costs.

The following usage factors were derived based on engineering estimates and the assumptions given for each type of highway:

Interstate Highway Construction: Assume that highways are constructed with 254 mm (10 in.) of concrete, 355.6 mm (14 in.) of subbase, have paved shoulders (Fig. 1), and that 8.1% of total aggregate used is natural sand, which is excluded from the use factors:

1) Six-lane interstate highway: 95.2 kt (105,000 st) of aggregate per mile.

TABLE 4—Use of Aggregate in Highway Construction (Three-Year Moving Averages)

A. Thousands of st* of aggregate used per million $ of highway construction cost†

	1962–1964	1968–1970	1974–1976	1976–1978	1978–1980
Federal aid—interstate					
Rural	98	83	50	41	37
Urban	46	45	26	23	23
Rural and urban	77	69	39	33	32
Federal aid—noninterstate					
Rural	127	76	54	44	39
Urban	63	55	35	29	23
Rural and urban	99	68	45	37	32
All federal aid	85	68	42	35	32

B. Cost of purchased aggregate as a % of federal-aid total highway construction cost†

	1968–1970	1974–1976	1976–1978	1978–1980
Interstate				
Rural	12	9	10	12
Urban	10	8	8	10
Total	12	9	9	11
Noninterstate primary				
Rural	13	13	12	12
Urban	14	12	10	9
Total	14	12	11	11

Source: US Federal Highway Administration, "Highway Construction Use Factors for Aggregates," Dept. of Transportation, Washington, DC, annual.

* Metric equivalent: st × 0.907 184 7 = t.

† All costs exclude right-of-way and engineering costs.

2) Four-lane interstate highway with bridges and interchanges as would be typical of an urban setting: 77 kt (85,000 st) of aggregate per mile.

3) Four-lane rural interstate highway with no bridges or interchanges: 59 kt (65,000 st) of aggregate per mile.

Primary Two-Lane Highway Construction: Assume that highways are paved with 101.6 mm (4 in.) of bituminous concrete and 203.2 mm (8 in.) of subbase, and that 94% of the total aggregate used must be coarse or graded aggregate including subbase:

1) Two-lane highway with bridges and interchanges: 21.7 kt (24,000 st) of aggregate per mile.

2) Two-lane highway without bridges and interchanges: 11.3 kt (12,500 st) of aggregate per mile.

Secondary Two-Lane Highway Construction: Assume that secondary roads are paved with a maximum of 101.6 mm (4 in.) of bituminous concrete and a maximum of 203.2 mm (8 in.) of subbase:

1) Municipal roads: 11.3 kt (12,500 st) of aggregate per mile.

2) County roads: 9 kt (10,000 st) of aggregate per mile.

Highway Maintenance: Based on operating statistics of the Pennsylvania Dept. of Highways in eastern Pennsylvania in 1969 (Torries, 1972), the per capita use of aggregate used in highway maintenance in this particular state is 0.079 t (0.087 st) per person. Approximately 65% of this total is classified as coarse aggregate. Bituminous concrete accounts for almost all repaving.

Residential and Nonresidential Construction: Regression analysis on data from eastern Pennsylvania (Torries, 1972) indicated that a per capita average of $80 of residential and nonresidential construction was spent in the region each year from 1964 to 1969. On a 1981 basis this would amount to approximately $201.50. Use factors of 0.014 and 0.013 t (0.015 and 0.014 st) of aggregate per dollar of residential and nonresidential construction, respectively, have been reported (McCarl, 1969). About 72% of the total quantity of aggregate used in residential and nonresidential construction is coarse aggregate.

Government Construction: Per capita use factors are not applicable for government construction projects because these projects are few in number and are usually very large heavy-construction projects, such as dams, that have little relation to the population of the area in which they occur. Aggregate use on such projects can best be estimated from the bidding plans.

Specifications

Specifications for crushed stone to be used for construction are set by agencies such as the American Society for Testing & Materials (ASTM), American Association of State Highway and Transportation Officials (AASHTO), US Dept. of Transportation (Federal Highway Administration), the US Corps of Engineers, and the various state highway departments. Any one specification for a particular use may differ somewhat among the various agencies. The specifications set by the individual highway departments are often adopted as the standard specification for all private construction within each state because the highway departments are the largest single user of aggregates in a state.

Different areas in the United States have different specifications for crushed stone to be used in similar applications. In addition, specifications are continually being changed by the various agencies in an attempt to guarantee a satisfactory material for a particular application without being overly restrictive or incurring unnecessary costs.

Standardized tests have been designed to evaluate the physical properties of an aggregate. A summary of the different types of tests commonly used, the ASTM and AASHTO code numbers for the tests, and sample AASHTO specifications are given in Table 5. For the details on how each test is executed, see the appropriate ASTM or AASHTO test identified by the code number.

Abrasion Tests—Tests that measure the resistance of an aggregate to abrasion are conducted on a Los Angeles machine, and the test is commonly called the "Los Angeles Rattler Test" (ASTM method C131 or C535).

Materials in which the individual grains are weak or are loosely cemented are not desirable as aggregate because these materials break down under pressure from use or during the handling and mixing procedures. The breakdown of the coarse aggregate causes a deviation from the desired gradation and may generate an excess of unwanted fine particles. Examples of rock types that commonly fail the Los Angeles test are weakly cemented sandstones, some argillaceous limestones, and weathered granites. It should be noted that different types of rock may have higher abrasion losses in the test

TABLE 5—Some Common Tests and Specifications for Crushed Stone

Physical Test	ASTM Test No.	AASHTO Test No.	AASHTO Spec. M79 Aggregate for Bituminous Concrete Surface Course	AASHTO Spec. M80 Coarse Aggregate for Portland Cement Concrete
Standard Specification	C33			
Abrasion				
Los Angeles Machine	C131 C535	T96	<40% loss	<40% loss
Soundness				
Sodium sulfate	C88	T104	<12% loss (5 cycles)	<12% loss (5 cycles)
Freeze and thaw	C682	T103		
Gradation				
Sieve Analysis	C136	T27	No. 7 and No. 57*	No. 357, 467 (blended from sizes 3 or 4 and 57 or 67), or size No. 57
Alkali reactivity	C227 C289 C586			
Deleterious materials				
Organic impurities	C40	T21		
Clay lumps	C142	T112		<0.25%
Soft fragments	C235	T189		<2.0%
Coal and lignite	C123	T113		<0.25%
Chert	See Alkali reactivity			
Excess fine material				
−200 mesh	C117	T11		<0.5%
Specific gravity and absorption				
Coarse aggregate	C127	T85		
Slag			70 lb per cu ft†	70 lb per cu ft
Fine aggregate	C128	T84		
Crushed count				
Retained on No. 4 Sieve, 1 frac. face			>50%	
Particle shape				
Thin or elongated				Commonly <15%
Petrographic analysis	C295			
Sampling method	D75	T2		

Source: Anon., 1980.
* See Table 6 of this chapter.
† Metric equivalent: lb per cu ft × 16.018 46 = kg/m³.

than other types but still perform equally well in service.

Soundness Tests—Two soundness tests commonly used are the sodium sulfate (or magnesium sulfate) soundness test and the freeze-thaw test. Repeated cycles of freezing and thawing or immersing the stone in a sodium or magnesium sulfate solution cause some rock to weaken as a result of the expansion of ice or salt crystals in the pores of the rock. These tests give a rough indication of porosity, permeability, and tensile strength of the rock. Rock types that commonly fail the soundness test include porous sandstones and shaly sedimentary rocks (refer to ASTM method C88).

Gradation Tests—Crushed stone is manufactured in a variety of sizes and gradations to meet the specifications set by the different users of crushed stone. Table 6 lists the gradations required by ASTM for several types of crushed stone used in highway construction. Fig. 2 illustrates the gradation included in a selected size, No. 357. In general, aggregate to be used in concrete should be graded to give as dense a mixture as possible and still retain the desired characteristics to make the concrete workable. The coarser sizes are used in more massive structures and the finer sizes where spaces between reinforcing steel and forms are small (Severinghaus, 1960). Conversely, aggregate used for drainage purposes should be uniformly sized to have as great a volume of voids among the pieces as possible and should be relatively free of fine particles. Several sizes of stone are blended with manufactured or natural

TABLE 6—Grading Requirements for Coarse Aggregates

Size No.	Nominal Size (Sieves with Square Openings)	Amounts Finer than Each Laboratory Sieve (Square Openings), Wt %												
		100 Mm (4 In.)	90 Mm (3½ In.)	75 Mm (3 In.)	63 Mm (2½ In.)	50 Mm (2 In.)	37.5 Mm (1½ In.)	25.0 Mm (1 In.)	19.0 Mm (¾ In.)	12.5 Mm (½ In.)	9.5 Mm (⅜ In.)	4.75-Mm (No. 4)	2.36-Mm (No. 8)	1.18-Mm (No. 16)
1	90 to 37.5 mm (3½ to 1½ in.)	100	90 to 100	...	25 to 60	...	0 to 15	...	0 to 5	...	...	...	...	...
2	63 to 37.5 mm (2½ to 1½ in.)	...	...	100	90 to 100	35 to 70	0 to 15	...	0 to 5	...	...	...	...	...
3	50 to 25.0 mm (2 to 1 in.)	...	...	...	100	90 to 100	35 to 70	0 to 15	...	0 to 5	...	...	...	...
357	50 to 4.75 mm (2 in. to No. 4)	...	...	...	100	95 to 100	...	35 to 70	...	10 to 30	...	0 to 5	...	...
4	37.5 to 19.0 mm (½ to ¾ in.)	...	...	...	...	100	90 to 100	20 to 55	0 to 15	...	0 to 5	...	...	...
467	37.5 to 4.75 mm (1½ in. to No. 4)	...	...	...	...	100	95 to 100	...	35 to 70	...	10 to 30	0 to 5	...	...
5	1 to ½-in. (25.0 to 12.5-mm)	...	...	...	...	...	100	90 to 100	20 to 55	0 to 10	0 to 5	...	...	...
56	1 to ⅜-in. (25.0 to 9.5-mm)	...	...	...	...	...	100	90 to 100	40 to 85	10 to 40	0 to 15	0 to 5	...	...
57	25.0 to 4.75 mm (1 in. to No. 4)	...	...	...	...	...	100	95 to 100	...	25 to 60	...	0 to 10	0 to 5	...
6	¾ to ⅜-in. (19.0 to 9.5-mm)	...	...	...	...	...	...	100	90 to 100	20 to 55	0 to 15	0 to 5	...	...
67	19.0 to 4.75 mm (¾ in. to No. 4)	...	...	...	...	...	...	100	90 to 100	...	20 to 55	0 to 10	0 to 5	...
7	12.5 to 4.75 mm (½ in. to No. 4)	...	...	...	...	...	...	...	100	90 to 100	40 to 70	0 to 15	0 to 5	...
8	9.5 to 2.36 mm (⅜ in. to No. 8)	...	...	...	...	...	...	...	...	100	85 to 100	10 to 30	0 to 10	0 to 5

Source: ASTM C33, for additional sizes ASTM D448, printed by permission of ASTM.

sand to maintain uniformity in gradation of the aggregate used in blacktop and concrete mixes.

Deleterious Materials—Deleterious materials include organic matter, chert, clay, coal and coke, shale, and soft particles. The amount of deleterious material allowed depends upon the type of deleterious material and the use to which the stone is to be put. It is usually desirable to limit the amount of chert in concrete aggregates because chert may be susceptible to alkali reactions in the concrete and may cause premature failure.

Color—Crushed stone used for decorative purposes must be of specific colors and hues. The colors must be permanent and not change with exposure to the weather or to atmospheric conditions.

Test for Excess Fine Material—A wash test is generally used to determine the amount of 74μ (−200 mesh) material in the aggregate. An excess of certain types of fine material prevents the aggregate from being freely drained, may form an unwanted coating in the coarse particles, and may require excessive amounts of binder to coat each particle.

Specific Gravity and Absorption Tests—Specific gravity of the crushed stone is used to determine the solid volume occupied by the aggregate so that the portland cement or bituminous concrete mixture can be designed correctly. Absorption is a measure of the porosity and permeability of a rock and is used to determine the proper amount of water to be added to portland cement concrete mixes. If the absorption of water by the stone is not taken into account, the resulting concrete mixes would be too dry. The ability of a rock to absorb moisture is also an important consideration in the manufacture of bituminous concrete mixes. All aggregate used in bituminous concrete must be dry, and water in the interior of more porous stone particles is difficult and expensive to remove.

Particle Shape—The shape of the crushed stone particles should be as cubical as possible with a minimum of flat and elongated pieces. Some AASHTO specifications are based on the percentage of particles that have a long dimension five times greater than the average thickness. Some structurally deformed rock deposits have pronounced platy cleavage and are difficult to crush without producing an unacceptable quantity of flat and elongated particles.

Crushed Count—This specification for the percentage of crushed particles in aggregate applies only to crushed gravel. Crushed count specification can be an important factor in choosing between a crushed stone or a crushed gravel source for a particular use.

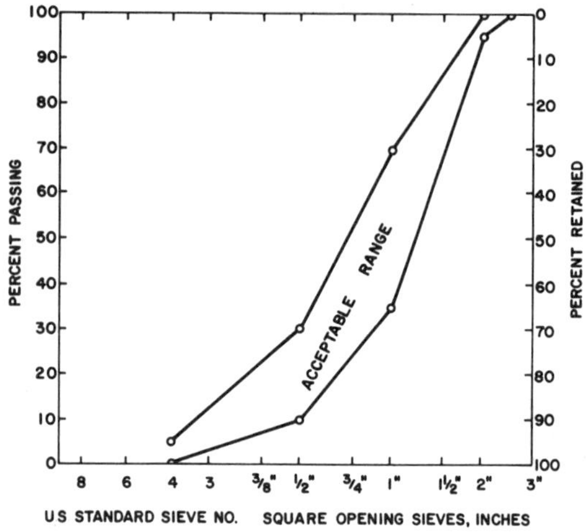

FIG. 2—*Range of acceptable gradation for ASTM size No. 357, from Table 6. Metric equivalent: in. × 25.4 = mm.*

Beam and Cylinder Tests for Concrete—In addition to meeting all other requirements for concrete aggregate, specimens of concrete made with the aggregate in question must be subjected to strength tests. In this manner, the concrete mix design, the cement, and the concrete aggregate are all tested simultaneously. If an aggregate meets all other specifications for concrete aggregate and the strengths of the resulting concrete are too low, the trouble probably is due to an improper mix design or improper curing of the concrete specimen rather than to the aggregate.

Skid-Resisting Qualities of Aggregate—Much research has been done concerning the type of aggregate to be used in the surface courses of highways to prevent accidents caused by skidding vehicles (Anon., 1972). In theory, a soft, homogenous rock that polishes easily and quickly should form a slick surface which would promote skidding, especially when the surface is wet. On the other hand, the skid-resisting properties of a highway surface should be improved if the aggregate used retains a rough surface in use. The subject is complex, and it is not yet possible to measure any selected parameters and predict accurately if a particular aggregate has skid-resisting properties or not. Skid performance tests on existing highway surfaces permit evaluation of material in the surface course of these highways.

Alternative Choices: It is often possible for the construction project designer or the crushed stone producer to meet a certain need with more than one type of material. The final choice is made on the basis of cost, but sometimes it is difficult to quantify the total real costs and benefits. Gravel is the most frequently used alternative to crushed stone.

The quantity of crushed stone used in building construction can be reduced by substituting other materials such as sand, gravel, steel, wood, lightweight aggregate, brick, oyster shell, shale, glass, aluminum, or dimension stone. Less aggregate usually is used in the construction of a highway paved with portland cement concrete than one paved with bituminous concrete (see Fig. 1). In addition, gravel can be used in lieu of crushed stone in portland cement concrete.

It is sometimes possible to blend materials in various proportions to make an acceptable product. Crushed gravel that does not meet the crushed count for a particular use can sometimes be blended with a small quantity of crushed stone to make a blended aggregate that meets all specified requirements. The sodium sulfate and the Los Angeles test results can be similarly upgraded by blending higher quality materials.

Whenever two or more types of materials are blended to make one aggregate, care must be taken to insure that the individual types are compatible. Great differences in specific gravities may cause segregation. Difference in thermal expansion coefficients may result in differential expansion of the final product which may, in turn, contribute to premature failure of the product.

Natural sand for use in concrete is scarce in some areas of the United States and the supply of concrete sand is augmented by manufactured sand, a product produced by crushing stone to meet specifications established for sand. Whether it is more economical to transport natural sand from greater distances or to manufacture sand from stone sources closer to the demand depends upon many factors in addition to the relative transportation and production costs. The rough particles of some manufactured sands may make the concrete hard to work and more labor might be required to pour and finish the concrete. Specifications must also be considered because a sand manufactured from limestone may not be acceptable by many agencies for use in portland cement concrete highway surfaces.

Production and Transportation

Production and distribution of crushed stone encompass all activities that contribute to the total cost of the delivered stone. This includes acquiring the deposit, stripping, drilling, blasting, crushing, screening, stockpiling, transporting the finished product, and meeting all pollution and reclamation regulations.

The following discussion focuses on the series of interrelated decisions concerning the location of the quarry and the methods by which the stone is to be quarried, processed, and transported. The manager's task is to see that the numerous preproduction decisions result in the optimal production system over the life of the quarry.

Plant Location

There must exist an identifiable market for crushed stone and workable deposits of suitable materials at a location where zoning, pollution, and other public-related constraints permit the quarrying and crushing of stone and desired related activities. The choice among several possible locations is based upon production and transportation costs, price, and risks over time. The possible inclusion of related activities, such as bituminous or portland cement concrete production, must be included in the analysis. The residual value of the land after the depletion of the deposit should also be considered when a production life of generally 15 years or less is anticipated.

Production Capacity of the Operation

The expected size of the market over time, recoverable reserves of stone, and economies of scale in production need to be considered in determining optimal plant capacity. All present and expected property setbacks and zoning and reclamation restrictions must be included in calculating recoverable reserves.

Economies of scale in production do exist in the production of crushed stone. Only 4% of the quarries operating in the United States produced over 907 kt (1.0 million st) in 1980; however, these accounted for 33% of total output. The average quarry produced about 214 kt (236,000 st) in 1980 (Singleton, 1980).

The size of a plant is determined by the interrelation between increased transportation costs required to serve a larger market area and operating and capital costs.

In general, it is not economical to transport stone by truck over 32.2 to 48.3 km (20 to 30 miles), because at this distance the transportation costs approximately equal the production cost. At transportation distances greater than 32.2 to 48.3 km (20 to 30 miles) it is often cheaper to open a new quarry unless shipment by rail or water is feasible. However, increased areas of urbanization and local zoning restrictions have caused aggregates to be transported ever-increasing distances because of the lack of available alternate sources closer to the markets.

General Plant Considerations

The site of initial production, future expansion of the quarry, temporary and permanent site screening, and pollution control should be planned at the time a new quarry is started. The nature and type of overburden (for quarry operations), land-use controls such as zoning, setback and depth restrictions, the thickness and attitude of the deposit, planned use of the land after the deposit is depleted, and the presence and quantity of ground water encountered influence decisions related to quarrying and reclamation procedures. It must be decided if the operation is to be worked by open pit, underground, or by a combination of methods. Quarrying below water level may result in increased reserves of clean stone with no additional overburden removal, but pumping cost, costs of disposing of the water, and the final end use of the depleted quarry (or cost of reclaiming the land) must be considered.

General Plant Arrangement: The relative locations of the primary crusher, crushing and screening plant, stockpiles, and ancillary equipment determine how efficiently materials can move through the plant. Items to be considered in designing a site layout include the quarry plan, topography, zoning and setback requirements, the types and locations of possible portland cement or bituminous concrete mixing plants, flooding problems, and intraplant material handling rates.

It is desirable to locate the primary crusher close to the quarry face to minimize the pit haul. This scheme may not be possible when more than one face is being worked or where there are serious water problems and the danger of flooding exists. It is also desirable to screen the plant from the sight of neighbors and motorists with landscaping such as planted embankments.

Operating Permits, Regulations and Mine Plans: It is necessary to acquire mine permits to open or continue quarry or mine operations. This permit usually requires that other permits, designed to assure against water and air pollution, be acquired. In particularly sensitive locations, an environmental impact statement may be utilized. Mine plans and land reclamation plans are a necessary part of the submissions included with a mine permit application. Permits and plans are site specific. Blasting and highway use permits along with licenses for engineers, blasters, and weigh scales are other necessary documents before operations can be started.

Bonding: Surface mining regulations require the posting of substantial financial assets as a bond that land reclamation will be done. The bond will be for a specific tract of land, can amount to as much as $5,000 per acre, and will be in the form of collateral such as US Treasury bills or a surety bond. The bond must exist before mining can start.

Stripping Plan: The size of the area to be stripped at any one time, whether to hire a contractor for the stripping, and what type of equipment to use must be decided by the quarry operator. Primary factors influencing these decisions include pollution, zoning, and reclamation considerations; the amount and type of material to be stripped; the distance the overburden is to be moved; equipment available; and time constraints. In some areas, the amount of land that can be stripped and left bare at any one time is regulated by law. If large quantities of materials are to be stripped, it may be more profitable for the quarry operator to contract the stripping to another company that specializes in moving large quantities of dirt.

The type of equipment used for stripping is influenced by the equipment already owned and any special conditions existing at the site. Trucks and shovels or front end loaders are commonly used where the overburden must be moved long distances [over 609.6 m (2000 ft)] or over public roads. Self-loading units such as scrapers are used for shorter hauls off public roads. Draglines and hoe-type excavators are commonly used where the overburden is sticky or is under water or where the rock surface is channeled and irregular. Under certain conditions draglines and dozers can be used to strip and place the dirt into previously quarried portions of the quarry so that stripping, quarrying, and reclaiming are accomplished simultaneously.

Drilling and Blasting Procedure: If a quarry operator does not contract the drilling and shooting to another firm, he must decide upon the type of drill to use, hole size, type of explosive, drill hole configuration, and the size and frequency of the blasts. Generally, large diameter holes (requiring large drills) are less expensive per ton of stone produced than small diameter holes. However, a quarry operation of less than 362.8 kt/a (400,000 stpy) does not generally require the full-time services of a large drill. A contract driller who services two or three quarries with a large drill can often do so at a lower unit cost than if each of the quarries did its own drilling.

Safety, zoning, and other public-related constraints; thickness of the deposit and bench heights; the geologic attitude of the deposit (inclination and thickness of the beds and the amount and type of fractures in the deposit); desired degree of rock fragmentation; and the number and configurations of the working faces must be considered when drilling and blasting procedures are selected.

Prevention of damage and public apprehension caused by blasting and its attendant noise is of major concern to the quarry operator. Ground vibrations are commonly measured by a recording seismograph to insure that the velocity and amplitude of the shock waves are within acceptable levels. The generally accepted level at which ground vibrations cause no structural damage is 50.8 mm/s (2 ips) (Nicholls, 1971). However, people near a quarry [less than 914.4 m (3000 ft) from the blast] may notice the ground vibrations and the airblast even though no property damage is done. Weather conditions have a profound effect on airblast which can be minimized by use of proper techniques (Schenck, 1971).

Types of Processing Required: Crushed stone plants can be generally classified as wet or dry plants. The preferred type of plant depends upon the classification of stone in the deposit, the types of products required, the amount and types of contaminants in the deposit, the availability of pond space and water, and zoning and pollution problems. Washing generally is used to remove undesirable amounts of clay from certain types of stone. Once the clays are removed, problems arise in the disposal of the tailings water and the settled solids in the tailings ponds. On the other hand, wet plants do not have the serious air pollution problems that are often associated with dry process plants. Air pollution controls require dust collectors which may present problems of waste disposal similar to those of wet plants. Larger plants will optimize particle size by controlling the circulating load between sizing screens and crushers.

Permanent Versus Portable Plants: A quarry operator may use a permanent (Fig. 3), a portable (Fig. 4), or a combination of the two types of plants to process quarried stone. Factors to be considered in making the choice include the desired capacity over time, the life of the deposit, the types of products required, labor rates, equipment available, impact of unionization and time constraints in obtaining equipment, setting up the plant, and obtaining production.

The equipment of a portable plant is often mounted on wheels and is of such a size and weight that each piece of machinery can be transported over public roads. Since portable plants are designed to be used under a variety of conditions, a portable plant may not be as efficient as a comparable permanent plant for any one location. Making a wide range of products at one time can sometimes be awkward in a portable plant.

Portable plants are often used to supply

FIG. 3—*Permanent plant for dolomitic-limestone in Virginia has 0.91 Mt/a (1-million stpy) capacity. The receiving hopper and primary jaw crusher are at left; primary screen and secondary cone crusher at left center; secondary screen enclosed at right with tertiary cone crusher in closed-loop to the left of it with manufactured sand and concrete block plants in background (courtesy* Rock Products).

FIG. 4—*Quarry and portable crushed stone plant processing 635t/h (700 stph) of Hopkinton limestone, Manchester, Iowa. Land-use and environmental regulations have tended to reduce the use of portable plants in favor of permanent plants (courtesy Iowa Manufacturing Co.).*

stone for a specific project such as the construction of a section of a highway or a dam. Since these plants are operated as an adjunct of the primary construction project, the labor rates paid in the quarry must follow the same schedule as the labor rates used on the main construction project. Permanent quarries supplying stone to a number of customers may have lower hourly wage rates than those paid on unionized heavy construction projects.

The delivery and set up time for a new permanent 272 to 363 t/h (300 to 400 stph) crushing and screening plant may be as much as six to nine months. If a stone producer has a portable plant available, the portable plant can often be set up and be in production within a few days.

Portable plants can be used in conjunction with a permanent plant to meet surge demands and to process materials in the deposit that cannot be processed efficiently in the permanent plant.

Types of Crushers: The selection of the proper size, type, and number of crushers depends upon the compressive strength and abrasiveness of the rock, the products desired, the quantity and type of deleterious materials in the deposit, the screening capacity, and the capital cost of the various types of crushers vs. the operating and maintenance costs associated with each type.

In general, crushers fragment rock by rapid impact, by slow compression, or by a combination of the two actions. Impact type crushers have greater capacity-to-cost ratios than do compression-type crushers, but abrasive stone (usually greater than 8% SiO_2) may cause accelerated wear on impact-type crushers. Crusher designers and manufacturers should be consulted to determine the proper types and numbers of crushers required to process a particular stone.

Screening, Classifying, and Washing Plant: Gradation specifications of crushed stone are met by screening, classifying, and washing the crushed stone. Many types and arrangements of screens and classifiers are available and the choice usually is made upon the recommendation of the equipment manufacturers. Factors to be considered in designing the screening, classifying, or washing plant include the number and types of products required, the nature of the material to be processed (amount of wet or sticky material, particle shape, amount of oversize and fine particles, and the densities of the various materials), and capital cost vs. the operating and maintenance costs of alternate equipment.

Stockpile Methods: The quarry operator must decide if the stockpile and its reclaim system should be automated or not. Automated systems require less manpower and fewer trucks and loaders but are generally not economical in small plants. Additional factors to be considered include the storage space available, the method used to transport the product, and the desired sustained and peak capacity of the plant. It is especially important to avoid segregation of stone sizes within each product during stockpiling and loading.

Pollution Controls: Air pollution control systems are classified as dry, wet, or combination systems. Dry systems include cyclones, air separators, and baghouses. The recovered dry fines may present a disposal problem or, as in a few cases, may yield a salable product.

Wet systems can utilize sprays for control of air pollution or can be an integral part of a conventional washing plant. Spray systems require relatively small amounts of water, and use of chemical wetting agents can further reduce water consumption. In a well-monitored spray system, the water that is used is combined with the dust and both are entrained with the crushed stone product. Caution must be exercised in considering if this entrained dust will exceed standards set for permissible amounts of 74μ (—200 mesh) particles in the finished product. Conventional washing systems use large quantities of water, resulting in an effluent carrying significant amounts of fines.

Water pollution, usually observed as turbidity, can be produced by these fines. Under standard water pollution regulations, wash water and runoff from the quarry requires impoundment in settling ponds. It is most desirable to recycle the water from the settling ponds within the operation.

An integrated approach to pollution control is required for the entire operation from quarry face to transport of the product to consumers. The type of control program selected depends upon the amount and type of dust generated, the availability of pond space and water, zoning, air and water pollution regulations, the type of processing plant utilized, amount of dust acceptable in the product, and capital cost vs. operating and maintenance costs. Special regulations are imposed on siliceous dusts because of the danger of silicosis.

Pollution regulations and limits vary, depending upon location and the enforcing agency. All in all, solutions to pollution problems have required substantial increases in the capital investment required to operate a quarry and crushing plant.

Transportation Methods

Most crushed stone used as aggregate is transported by truck for relatively short dis-

TABLE 7—Transportation Costs

Mode of Transportation	Cost (C) per Ton* for Haul Distance (M)		Additional Considerations
Truck	C = \$0.05	when: M ≤ 1	
	C = \$0.40 + \$0.10M	: M > 1, ≤ 3	
	C = −\$0.08 + \$0.20M$^{0.6}$	: M > 3	
1. Contract trucking provides flexible capacity to all delivery sites. 2. Backhaul arrangements may reduce haul cost to \$0.04 per ton per mile. 3. Stone can be delivered anywhere with short notice. 4. Traffic and road conditions greatly influence the cost. 5. Truck transport may get cheaper with improved highway network. 6. Maximum weight limits for travel on public highways must be observed.			
Rail	C = \$2.70 + \$0.016M		
1. A siding and loading tipple is required. 2. Stone may have to be reloaded into trucks at the destination for final delivery. 3. Delivery schedules are inflexible. 4. Best method available to transport railroad ballast.			
Water	C = \$2.90 + \$0.008 to C = \$5.75 + \$0.013M, depending on size of units		
1. Stone must be reloaded into trucks at the destination for final delivery. 2. Service may be interrupted during the winter months. 3. Delivery schedules are inflexible. 4. Dockside loading and unloading facilities are required.			

* Metric equivalent: st x 0.907 184 = t

tances [less than 48.3 km (30 miles)]. However, the possibility of rail or water transportation is an important consideration in many cases. Not only must the various direct transportation costs of each mode of transportation be compared, but each mode also has hidden advantages and disadvantages as is shown in Table 7. In addition, federal and state transportation regulations, which may severely constrain the feasible alternatives, must be observed with all modes of transportation.

Production Costs

It is not possible to make general statements about the production costs associated with crushed stone because the types of stone processed and the conditions under which the stone is crushed are so variable. However, it is possible to state a set of assumptions and then quantify production costs based on the assumptions, past practice figures, and engineering cost estimates. Production costs for the quarrying and processing of three different types of materials are shown in Table 8. The costs are based on the labor requirements and rates shown in Table 9 and on the following assumptions:

1) Only "normal" products are manufactured and these products meet all specifications for use in heavy construction.

2) There are "sufficient" reserves of materials available.

3) Open pit quarrying is possible with a level quarry floor and bench heights of 12.2 m (40 ft).

4) Ground water, overburden, pollution, and reclamation problems do not require any unusual expenditures.

In cases where underground mining must be considered, it can be assumed that total average mining costs in 1980 range from about \$4.96/t (\$4.50 per st) for room-and-pillar mining to about \$9.92/t (\$9 per st) for stope mining of limestone at large-scale operations.

The size and gradation of the stone produced influences production costs. In general, the finer the size, the greater the cost, as in the manufacture of stone sand. An exception might be the production of riprap in which the "gradation" is obtained by blasting, by use of the drop ball, and expensive "manual" selection of the proper mix of large stone sizes.

A given crushed stone plant and its associated stone deposit will have a limited com-

TABLE 8—Representative Open Pit Production Cost,*
272.2 T/h (300 stph) Plant Producing 453.6 Kt/a (500,000 stpy), All Day Operation

	Limestone-Dolomite, \$	Sandstone, Gabbro, Granite, \$	Traprock, Diabase, Basalt, \$
Stripping	0.03	0.03	0.03
Drilling	0.07	0.08	0.13
Shooting	0.10	0.10	0.11
Loading	0.48	0.48	0.48
Hauling	0.06	0.06	0.06
Processing	0.13	0.19	0.23
Stockpiling	0.13	0.13	0.13
Maintenance	0.54	0.67	0.87
Dust control	0.04	0.04	0.04
Royalty	0.10	0.10	0.10
Selling	0.06	0.06	0.06
Total average variable cost	1.30	1.50	1.80
Depreciation	0.38	0.58	0.58
Office overhead	0.10	0.10	0.10
Total average fixed cost	0.48	0.67	0.67
Total average cost	1.78	2.18	2.47
Total plant equipment Cost, × 1000	1925	2885	2885

*Based on 1972 costs escalated at a rate of price increases from 1972 through 1980. Totals may not add because of rounding.

TABLE 9—Labor Required for a 300 Tph* Crushed Stone Plant†

Item	No. Required Pit	Plant	Total
Shovel operator	1		1
50-ton pit vehicle operator	3		3
Driller	1		1‡
Driller helper	1		1‡
Blaster	1		1‡
Blaster helper	1		1‡
Mechanic	0.5	0.5	1
Oiler	0.5	0.5	1
Welder	0.5	0.5	1
Crusher operator		1	1
Conveyor and bin operator		1	1
Loader operator		1	1
Stock truck driver		3	3
Superintendent		1	1
Lab man		1	1
Scale man		1	1
Laborer		1	1
Drop ball operator	1		1
Total	10.5	11.5	22
Productivity per man @300 tons per man-hour	28.5	26.0	13.6

*Metric equivalent: st × 0.907 184 7 = t.
†Assumes stripping is done by contract work, limestone crushing is done by conventional open pit operations, and there are no unusual problems.
‡May be contracted.

bination of product sizes and alternative gradations that are feasible to produce. Every such combination will have an associated proportion of fine sizes that cannot be avoided. The production of such fines can often exceed demand. The operator must then pay special attention to their disposal by finding a market such as manufactured sand at some price above his next best alternative, by stockpiling for some future sale, or by wasting them in an environmentally sound manner and absorbing the cost of such disposal.

Economic and Commercial Considerations

Users of crushed stone are informed by project specifications as to the characteristics required of the mineral aggregates for a particular construction project. There is usually a choice of sources or rock types available in a market region and individual buyers will select that aggregate that will meet specifications at the lowest in-place cost for the materials.

Demand and Competition

The crushed stone industry is highly competitive and buyers of crushed stone are extremely price sensitive. Information as to prices and transportation costs from different sources is easily obtained. A stone producer has little freedom to differentiate his product from that of competitors on any basis other than service, which is the prompt dependable delivery of material that consistently meets all specifications. Occasionally stone from a particular source exceeds minimum specifications required by the purchaser and is preferred over stone that only meets the minimum requirements. In addition, many buyers purchase aggregate in quantities that might justify their own entry as aggregate producers should prices rise to a level that would justify such action.

Demand for aggregates is derived from the demand for other goods such as highways, buildings, and public works. The aggregate producer can change the timing or the magnitude of the demand for aggregates only to a limited extent. Transportation cost plus the general availability of aggregate sources limit the market area that can profitably be serviced by each stone quarry.

A major construction project near a quarry can create a temporary surge in orders. The producer must be prepared with stockpiles or excess capacity to supply such above-normal demand when it arises or it will be satisfied from some other source.

Industry Structure

There were about 4,150 crushed stone aggregate plants in the United States in 1980 (Singleton, 1981). Commonly, only a few producers are conveniently located to serve a particular market, resulting in an industry concentrated in a few firms on a local basis. However, pricing is often competitive because of the potential competition caused by the relative ease with which additional firms can enter the stone business, Table 10. As population centers increase in size, those quarries already in existence may enjoy higher prices due to locational advantages because new quarries can be opened only at larger distances from urban areas. During 1980, 171 quarries in the US produced 907 kt (one million st) of stone per year or more, for a total output of 289.2 Mt/a (318.8 million stpy) or 33% of the US production (Singleton, 1980). Conversely, half the quarries operating during 1980 produced less than 90.7 kt/a (100,000 stpy) and collectively accounted for only 7% of total production.

The structure of the crushed stone industry generally is competitive, localized, free of significant barriers to entry (with important local exceptions), and is not dominated by firms of national scope. Continuation of this industry structure is favored primarily by: (1) economies of scale reached at a size far below 1% of national demand, (2) a high ratio of transportation cost to production cost, (3) widespread availability of a close substitute (gravel), and (4) easily acquired managerial and financial requirements. The present trends to increased governmental involvement with the industry and increased limitations on availability of sites in urban areas in the future may inhibit entry and provide an incentive for multiplant growth of firms that possess the skills required to satisfy public demands.

Vertical integration that forecloses certain markets for independent stone producers exists to some degree. Stone users, such as construction firms, portland cement, and steel producers have entered the industry in order to increase their overall profitability and to stabilize income by backward integration. A few large nationwide stone companies have been organized by conglomerate industries. Forward integration exists for a few firms that market crushed stone aggregates as a joint product of other mineral production such as mining of chemical limestone or sale of waste rock from mining.

Member firms of the National Crushed Stone Association annually produce over two-thirds of all crushed stone. This national group is augmented by 22 state associations.

Prices and Profitability

Price per ton for sales in excess of 4.5 to 13.6 kt (5,000 to 15,000 st) is normally negotiated somewhat below the list prices a producer has established for spot sales. As a result, net sales prices may be considerably below that indicated by list prices. This does not neces-

TABLE 10—Factors That Affect Competition In Crushed Stone Markets

Factors Tending to *Promote* Competition	Factors Tending to *Inhibit* Competition
1. Widely understood technology and management.	1. Lack of available quarry sites. a. Reserves geologically absent b. Reserves depleted c. Buildings covering reserves d. Legal restrictions (zoning and pollution control)
2. Equity requirement in medium six figures, balance of financing in loans from equipment manufacturers.	2. Restrictive material specifications
3. Many alternative undeveloped quarry sites.	3. Quarry ownership by major portion of consumers.
4. Roving demand.	4. Dominant producer.
5. Costs and profits easily estimated.	5. Restrictive practices (favoritism) by labor, management, or government officials.
6. Consuming firms often with many times the financial strength of producers.	6. Inadequate transport capacity.
7. Availability of developed quarries due to completion of heavy construction projects.	7. Stable market.
8. Cheaper substitute (gravel) often available.	
9. Surplus transport capacity.	
10. Easily portable plants with operating costs competitive with those of stationary plants.	
11. Expanding market.	
12. Competitive bidding for large contracts.	
13. Substitutes available as byproducts from other sources.	

sarily mean a decrease in profitability since some costs decrease when filling large stone orders. Entry into a local market should only be made after a survey has established at what prices transactions actually take place, as price lists only establish an upper limit in such a study.

National f.o.b. average value of crushed stone produced in 1980 was $3.64/t ($3.30 per st) (Singleton, 1980). Although the reported value of crushed stone has been increasing, the deflated prices (adjusting for inflation) have been decreasing, as is shown in Fig. 5. The decrease in the real price of crushed stone is an indication of productivity gains in the industry. Because the crushed stone industry has been highly competitive over the time of the decline, much of the savings from the productivity gains have been passed on to the consumers.

Public Factors

Government agencies have been increasingly involved in the activities of crushed stone producers through regulation and taxation. Environmental controls, land use regulations, and worker health and safety standards have had a significant impact on capital and operating costs of existing plants and on the location and costs of new plants. Numerous permits are generally required. Examples of government programs include USBM safety regulations, state pollution control permits and reclamation regulations as well as the more common commercial requirements such as highway permits and business licenses.

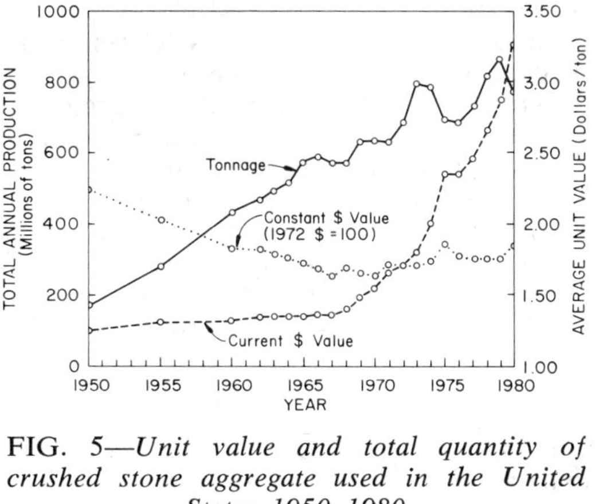

FIG. 5—*Unit value and total quantity of crushed stone aggregate used in the United States, 1950–1980.*

Noise, dust, water pollution, waste heaps, changes in the water table, traffic, and an unsightly landscape all may be the undesired by-products of quarrying where controls are lacking. An unregulated quarry can also create major public costs, for example, by silting a water course, by lowering nearby property values, or by startling neighbors with random blasts of explosives.

Adverse effects on the public can be minimized and many stone operators favor reasonable regulations with equitable enforcement to limit the external effects of stone production. The public also needs to be aware of the consequences of inhibiting the production of crushed stone because production of alternative materials could incur even greater environmental damage. Most quarry sites can be used after all reserves have been depleted, environmental damage as well as increased costs. Most quarry sites can be used after all reserves have been depleted, including uses that directly benefit the public such as parks or landfills. The public benefits from the present competitive organization of the crushed stone aggregates industry through lower cost public and private construction than would be possible if a philosophy inhibiting mineral production were to be imposed.

It is now becoming recognized by governments of metropolitan areas that preservation of aggregate reserves near consuming areas reduces costly transportation of aggregates that are necessary for construction and economic growth. Several states and cities have enacted statutes designed to preserve aggregate resources. Fairfax County, VA, perhaps the first to recognize the necessity of resource conservation, enacted appropriate zoning ordinances in 1961. Similar actions have been taken in California, Denver, CO, and Vermont.

Land Use

The land requirements for a quarry will include at least 40 470 to 80 940 m^2 (10 to 20 acres) for processing and stockpiling activities and at least another 80 940 m^2 (20 acres) of stone reserves. A quarry in which 272.2 to 453.6 kt/a (300,000 to 500,000 stpy) are produced commonly utilizes 809 400 m^2 (200 acres) of land. Such a parcel of suitably zoned land on the fringes of metropolitan areas will usually be valued at $5000 to well over $15,000 per acre. In addition, a quarry can only be situated where there is sufficient stone that meets specifications.

The result of such limitations in urban areas is that new operations are generally forced to the rural and semirural areas where good rock, permissive zoning, and good highway connections are present so as to minimize delivered cost to consumers. Suitable acreage in such a location would cost, in the early 1980s, from $2,000 to $3,500 per acre. Much semirural land can command a higher price and its use is therefore foreclosed to stone producers. Stone can also be purchased on a royalty basis with the royalty reflecting the value of the land. The normal range for royalty prices for aggregate stone is from $0.11 to $0.66/t ($0.10 to $0.60 per st) produced.

Zoning now exists for much of the semirural land near metropolitan areas so that the legal uses for the land are clearly specified. Almost all urban and suburban land is now zoned. Many presently active quarries were established prior to the imposition of zoning and either have been specifically zoned industrial or exist as nonconforming prior uses within areas where mining would otherwise be excluded. In either case, when the operator exhausts the acceptable stone within his parcel of suitably zoned land, he probably will have to move to a distant site.

Crushed stone operators generally desire to have their operations located close to large urban markets which are the areas most likely to have restrictive zoning regulations. There are in general five types of zoning regulations that pertain to quarrying:

1) Specific exclusion of quarrying or processing of stone.

2) Quarrying and processing of stone permitted in specified areas, not by right, but subject to approval by the zoning board.

3) Quarrying and processing of stone permitted by right in specified areas.

4) No mention of quarrying or processing of stone in the zoning ordinance.

5) Quarrying and processing of stone, either by right or subject to approval, but with stringent regulations on the quarry development and reclamation that effectively prohibit the operation of a stone quarry.

From the classification of zoning types, it can be seen that although zoning regulations may consider production of crushed stone, the decision as to whether a quarry may or may not be allowed for a specific area can still be open to question. In addition, zoning ordinances could be in various stages of development and the ordinances can change over time.

Taxation

In contrast to some other minerals, federal income tax accounting for aggregates is straightforward. The major decisions are whether any product can be construed to be specially classified for depletion such as chemical grade limestone, or whether cost depletion is more favorable than percentage depletion. Percentage depletion for construction aggregate and chemical limestone is 5% and 14% of gross sales, respectively.

State and local severance taxes and local property taxes on minerals should be checked before opening a quarry in an unfamiliar area. Property taxes on land can increase when a quarry is opened.

A few localities impose ad valorem or severance taxes of a set amount per ton on mineral production. For example, in 1972 Ohio instituted a severance tax on all minerals, which for crushed stone aggregates amounts to 1¢ per ton. Such taxes are easy for the taxing authority to administer, do not require the services of a highly skilled assessor, and are difficult to challenge once their legality is established under the state constitution. Sales taxes can also encompass some or all of a crushed stone producer's sales with the tax based on shipping point price.

Future Trends

The greatest uncertainty for producers of crushed stone will be the level of government regulation and restrictions associated with production and consumption. This uncertainty will be especially great for proposed new stone producing facilities that must compete with older, more favorably situated facilities. The response of the industry to inflationary pressures on costs will be to raise prices. The use of constant dollar conversions therefore will be necessary before valid analyses can be made of future trends in prices and value.

Demand

Demand for crushed stone should continue to parallel construction activity, which has been relatively depressed since 1973. Market areas for the individual producer will continue to be delineated by the interrelationships of the cost of truck transportation, production costs, and the location of producing plants. Depletion of sand and gravel deposits adjacent to certain metropolitan markets could cause a shift to increased use of crushed stone where quarry sites are available. Shipments of manufactured sand, which doubled in the decade ending in 1971, and tripled from 1971 to 1980, should increasingly penetrate the fine aggregate market.

Specifications for aggregates can be expected to change slowly as researchers continue to develop improved material performance tests such as that for frost susceptibility (ASTM C682) and to develop modified requirements for special aggregate applications such as skid-resistant surfaces for highways.

The structure of the industry may gradually change if external forces such as governmental regulations tend to inhibit competition. The short-run effect of federal, state, and local regulations in quarries and crushed stone operations is to decrease the number of present and new operations. This will increase the price of stone to the consumer because fewer quarries mean longer average truck hauls and decreased competition. The costs of government regulations will be passed on to the consumers of crushed stone because the profitability of the average crushed stone plant is already quite low.

If regulations help to control the undesirable features of quarries, quarry operation may become more acceptable to the public in the long run. Such a change in attitude could result in more permissive zoning that would tend to increase the number of quarries operating in the future. This could promote competition and produce a relative decline in the delivered cost of stone to the consumer.

Supply

There is no significant physical limit to reserves of stone suitable for supplying aggregate that meets quality specifications as they are presently established throughout the country. Locally, however, land use regulations may foreclose for the foreseeable future the use of major portions of reserves.

Despite anticipated productivity gains, total average cost of producing stone should continue to increase, probably at a rate higher than warranted by inflation. The primary causes will be the need to install expensive pollution control systems, the provision of idle land as buffer zones surrounding quarries, increases in taxation, and increased transportation costs. An increase in the number of states and localities imposing severance taxes can be anticipated as a result of the growing public interest in the operation of natural resource industries. Productivity gains due to technical improvements in capital equipment should continue to offset most of any normal increases in labor costs, on a constant dollar basis.

Firms that develop the specialized managerial talent required to respond to governmental activities affecting the crushed stone aggregates industry will tend to expand their share of the industry. The scope of such multiplant firms will be increased and their profit performance as a percentage of financial assets should exceed that of smaller, single-plant firms that are without the resources to respond to the changed environment of government-corporate relations.

Note: This chapter does not contain information directly concerning economic evaluation of a stone deposit or plant flowsheets. These subjects are treated in the Commodity section of this book.

Bibliography and References

Anon., 1966, *Standard Specifications for Highway Materials and Methods of Sampling and Testing,* Pt. I, "Specifications," 463 pp.; Pt. II, "Methods of Sampling and Testing," 9th ed., American Assn. of State Highway Officials, Washington, DC, 693 pp.

Anon., *Highway Construction Usage Factors for Aggregates,* US Federal Highway Administration, US Government Printing Office, Washington, DC, annual issues.

Anon., 1972, "Skid Resistance," *Synthesis of Highway Practice No. 14,* National Cooperative Highway Research Program, National Research Board, Washington, DC, 66 pp.

Anon., 1974, "Quality Concrete with Crushed Stone Aggregate," National Crushed Stone Assn., Washington, DC, 31 pp.

Anon., 1980, *1980 Book of ASTM Standards,* "Pt. 14, Concrete and Mineral Aggregates," American Society for Testing & Materials, Philadelphia, PA, 586 pp.

Anon., 1982, "The Seven Best Crushed Stone Plants," *Rock Products,* series of articles available as a reprint, 32 pp.

French, R.R., 1968, "Transportation of Mineral Aggregates in Indiana," *Indiana Academy of Science Proceedings for 1968,* Vol. 78, pp. 348–354.

McCarl, H.N., 1969, "The Mineral Aggregate Industry in the Vicinity of Baltimore, Maryland," unpublished Ph.D. Thesis, The Pennsylvania State University, University Park, Dec., 257 pp.

Nicholls, H.R., et al., 1971, "Blasting Vibrations and Their Effects on Structures," Bulletin 656, US Bureau of Mines, 105 pp.

Ozol, M.A., and Locke, S.R., 1974, *Study to Determine the Feasibility of an Experiment to Transfer Technology to the Crushed Stone Industry*, National Science Foundation, Contract NSF–C826, 94 pp.

Robertson, J.L., 1978, "Iowa Firm's Portable Plants Blanket Market Area," *Rock Products,* Vol. 81, No. 7, July, pp. 62–65.

Schenck, G.K., 1971, "Who Needs a Private Weather Service," *Mining Engineering,* Vol. 23, No. 4, Apr., pp. 54–57.

Schultz, G., 1981, "Aggregate Plant Design: The Planned Approach," *Rock Products,* series of articles available as a reprint, 32 pp.

Schwochow, S.D., 1980, "The Tortuous Trail to Permit Approval," *Rock Products,* Aug., pp. 50–53.

Severinghaus, N., 1960, "Crushed Stone," *Industrial Minerals and Rocks,* 3rd ed., J.L. Gillson, ed., AIME, New York, pp. 285–302.

Singleton, R.H., 1980, "Directory of Principal Stone Producers in the United States in 1980," *Mineral Industry Surveys,* US Bureau of Mines, 10 pp.

Singleton, R.H., 1980, "Stone," *Minerals Yearbook,* US Bureau of Mines, pp. 765–793.

Singleton, R.H., 1980, "Stone," *Mineral Facts and Problems,* US Bureau of Mines, pp. 853–868.

Singleton, R.H., 1982, "Stone," *Mineral Commodity Summaries, 1982,* US Bureau of Mines, pp. 146–147.

Soles, J.A., 1979, *Petrography in the Evaluation of Aggregates and Concrete,* CANMET, Report 80–3E, Energy Mines and Resources Dept., Ottawa, Dec., 14 pp.

Stearn, E.W., 1978, "A Look at Two Challenges to Quarrying," *Rock Products,* Vol. 81, No. 9, Sept., pp. 71–76.

Torries, T.F., 1972, "Economic Descriptions and Simulations of the Construction Aggregate Industry in the Allentown-Bethlehem-Easton Area, Pennsylvania," unpublished Ph.D. Thesis, The Pennsylvania State University, University Park, Sept., 188 pp.

Woolf, D.O., 1953, "Results of Physical Tests of Road Building Aggregates," Physical Research Branch, Bureau of Public Roads, US Dept. of Commerce, Washington, DC, 225 pp.

Also Useful:

Industrial Minerals magazine (London) is a good source of additional information.

Rock Products, Concrete Products, and *Pit and Quarry* magazines often contain informative articles on crushed stone and aggregates.

Construction Materials

Aggregates—Lightweight Aggregates

HENRY N. McCARL *

The production of lightweight aggregates in the United States has declined in recent years from about 18 Mt/a (20 million stpy) in the early 1970s to an average of between 10.9 and 11.8 Mt/a (12 and 13 million stpy) in the period 1979–1981. This has been caused by significantly higher costs and prices due to the large amounts of fuel and energy used in the production of most types of lightweight aggregate, and lower levels of construction activity. During the same period, however, the value of the smaller output has increased substantially so that the value of total lightweight aggregate production is roughly double what it was ten years ago.

Lightweight aggregates include a variety of mineral and rock materials used to provide bulk in concrete building units (block), lightweight structural concrete, and precast concrete units, as plaster aggregate, insulating fill, and other structural and/or insulation purposes. They are distinguished from other mineral aggregate materials by their lighter unit weight—1281 to 1602 kg/m³ (80 to 100 pcf) or less for loosely packed lightweight aggregates as compared to 2002 kg/m³ (125 pcf) or more for crushed stone, sand, gravel, air-cooled slag, and other more common mineral aggregates.

The various lightweight aggregate materials may be classified into four groups as follows:

1) *Natural Lightweight Aggregate Materials*—prepared by crushing and sizing natural rock materials such as pumice, scoria, tuff, breccia, and volcanic cinders.

2) *Manufactured Structural Lightweight Aggregates*—prepared by pyroprocessing clay, shale, or slate in rotary kilns or on traveling grate sintering machines.

3) *Byproduct Lightweight Aggregates*—prepared by crushing and sizing foamed and granulated slag, cinders, and coke breeze. With the exception of processed slag, this group is becoming less important as time passes.

4) *Manufactured Insulating Ultralightweight Aggregates*—prepared by pyroprocessing ground vermiculite or perlite.

This chapter will focus on the use of natural lightweight aggregates and manufactured insulating ultralightweight aggregates, and both the production and use of manufactured structural lightweight aggregates with some comments on byproduct materials. Detailed discussion of the production and use of both expanded and air-cooled slag is presented in a separate chapter in the utilization section of this volume, and there are separate commodity chapters on vermiculite, perlite, pumice, and volcanic cinders. Discussion of the commodities covered elsewhere in the book will be limited to observations on their utilization and relation to other lightweight aggregate materials.

All materials considered as lightweight aggregates are derived from some initial natural raw material. While historical definitions have included such organic materials as straw and peat as lightweight aggregates, the discussion in this chapter will restrict itself to lightweight aggregates derived from mineral or rock materials. Thus while the actual lightweight material in use may be a manufactured product, it is derived from an ore or raw material that is obviously a mineral resource. In this context, then, the reserves of raw materials for lightweight aggregate use may include iron ore and flux stone (the components of blast furnace slag) as well as coal (the raw material from which cinders are derived) as well as those clay, shale, and slate deposits that display the characteristics of bloating or controlled expansion when subjected to elevated temperatures [871 to 1316° C (1600 to 2400° F)].

* Associate Professor of Economics and Geology, School of Business, The University of Alabama in Birmingham, Birmingham, AL.

The principal properties desired in materials used as lightweight aggregates vary according to end use, but normally include:

1) *Light Weight.* Reduces dead load (weight in place) and facilitates the physical handling of the materials and resulting products.

2) *Thermal and Acoustical Insulation Properties.* Normally the result of air spaces, voids, or pores in the lightweight aggregate materials as finally used.

3) *High Fire Resistance.* Low probability of physical breakdown at temperatures below the melting point of the aggregate.

4) *Toughness.* Minimum tendency to crack or break if nailed or stressed in construction use.

Other properties that are normally considered desirable include:

5) Substantial compressive strength.
6) Low water absorption.
7) Good resistance to freezing and thawing.
8) Low shrinkage on drying and minimum thermal expansion.
9) Good bonding with cement.
10) Chemical inertness.
11) Good elastic properties.
12) Abrasion resistance.

The actual specifications and tolerances vary according to the end use or final product in which the lightweight aggregate is consumed. Specifications for various lightweight aggregate materials by use have been compiled by the American Society for Testing & Materials (ASTM) and their standards may be consulted as required. ASTM standards C330 (lightweight aggregate for structural concrete), C331 (lightweight aggregate for concrete masonry units), C332 (lightweight aggregate for insulating concrete) are the most relevant for most lightweight aggregate materials.

The variety and use of materials as lightweight aggregate is extensive. They are mostly bulky and of low unit value, and normally, they cannot be transported any great distance in final form and still remain competitive with alternative building materials. This means that few types of lightweight aggregate are used nationwide and most serve local and regional markets. Products such as pumice or expanded slag may be used in substantial quantities in some localities and may not even be available (at competitive prices) in others. Perlite and vermiculite serve national markets due to their relatively high value and because their "ores" are shipped prior to pyroprocessing and physical expansion. The more specialized uses enable higher market prices that lead to extensive geographic distribution and use.

Production

Naturally occuring lightweight aggregate raw materials such as pumice and volcanic cinders are normally mined by open pit or quarry methods, depending on the degree of consolidation of the raw materials. Mining costs are similar to those for sand and gravel except where substantial blasting and crushing are required—in which case the production costs parallel crushed stone. Selling prices are similar to sand, gravel, and crushed stone and have ranged from $2.50 to $4.25 per ton over the period 1976–1980. Output has increased in recent years despite price increases, and in 1981 was roughly 3.4 Mt (3.79 million st) (Meisinger, 1982). More details about the production of pumice and volcanic cinders may be found in a separate chapter in the Commodity section of this book.

Clay, shale, and slate are used as raw materials in the manufacture of building brick as well as lightweight aggregate. Most production plants for brick, however, do not seek the heat-expanding or pyroplastic properties desired by producers of lightweight aggregate. As lightweight aggregate raw materials, these substances should reach a pyroplastic state at elevated temperatures and then generate a gas that will expand or bloat the resulting product to a porous substance that will retain its physical strength despite its lighter unit weight when cooled.

Clay, shale, and slate are normally mined by open pit and quarry methods, then dried in large sheds or open stockpiles to control water content in the raw feed prior to high-temperature pyroprocessing in either rotary kilns or sintering machines. The resulting "clinker" may then be screened or crushed before screening to yield proper gradation mixes for final use as an aggregate in concrete masonry units, structural concrete, or other applications.

Figs. 1 and 2 are views of a modern rotary kiln lightweight aggregate plant in the eastern United States. Fig. 3 shows the control panel of the plant.

Table 1 summarizes the annual output of lightweight aggregate in the United States in the period 1980–1981 by quantity and esti-

FIG. 1—*Aerial view of Hercules Inc. 1.6 kt/d (1800 tpd) lightweight aggregate plant at Snowden, VA (photo courtesy of Kennedy Van Saun Corp.).*

FIG. 2—*Closeup view of rotary kilns at Hercules Inc. lightweight aggregate plant at Snowden, VA (photo courtesy of Kennedy Van Saun Corp.).*

FIG. 3—*Control panel of Hercules Inc. 1.6 kt/d (1800 tpd) lightweight aggregate plant at Snowden, VA (photo courtesy of Kennedy Van Saun Corp.).*

mated value. The significant price increases for expanded perlite and exfoliated vermiculite in recent years now give them the major portion of the market in terms of value despite contributing only about 6% of the volume (tonnage). Expanded clay, shale, and slate are still very important, but are not as dominant as they had been in the early 1970s when they accounted for about 60% of quantity and value (McCarl, 1975).

While raw material availability is important, most plants require a proximity to producers of concrete blocks and lightweight structural concrete, the major consumers of manufactured structural lightweight aggregates. Most producers of expanded clay, shale, and slate will find long-run economic survival rather precarious if they do not engage in substantial raw material testing, geologic and mining studies, and marketing research investigations

TABLE 1—1980–1981 Production of Lightweight Aggregates*

Material	Share of Market	
	By Quantity, %	By Value, %
Pumice and volcanic cinders	33	7
Expanded slag, cinders, fly ash aggregate, and other byproducts	18	5
Expanded clay, shale, slate	43	31
Expanded perlite	4	38
Exfoliated vermiculite	2	19
	100	100
Total industry†	13 million short tons††	$290 million

* Based on published statistics and industry estimates; data on value of expanded clay, shale, and slate are not normally published.
† Estimates (accuracy ± 10%)
†† Metric equivalent: 1 st ×0.907 184 7 = t.

prior to the location of a new plant (Conley et al., 1948; Reedy, 1971; Schwarzkopf, 1966; and Shildeler, 1961).

Production costs for expanded clay, shale, and slate have changed dramatically in the past ten years. Table 2 presents a cost comparison for a typical rotary kiln producer in 1970 and 1980. This is a composite based on several actual producers during this period as well as published reports (Harwell, 1965) and revisions of earlier work (McCarl, 1975). The most striking changes during this period were in rising fuel costs despite conservation measures leading to higher use efficiency, and other energy-related items such as refractories and electric power. The most severe fuel cost increases were at those plants utilizing oil and gas. Coal-fueled plants did not witness quite as dramatic (price) cost increases. Maintenance expenses were also higher, largely due to the older physical plant and necessity for maintenance of pollution control equipment. The magnitudes of the production cost increases are even more significant when one realizes that the Consumer Price Index, the major measure of inflation in the US, only increased by a factor of 2.0 (200%) in this same time period. This means that all direct production cost categories rose in real dollars as well as in relative amounts.

Expanded slag is a byproduct of iron and steel production. It is manufactured by spraying a stream of water through molten blast furnace slag as it is drawn from the furnace. The resulting foamed slag is then crushed and screened for use in concrete block or structural concrete. Further discussion of expanded slag may be found in the chapter on Slag in this book.

Cinders are a fast-disappearing byproduct of the burning of lump coal. Most coal-consuming processes now utilize pulverized coal in combustion and produce fly ash rather than cinders. Fly ash production has increased dramatically with the increased use of pulverized coal in electric power generation. Roughly 54.4 Mt (60 million st) of fly ash are produced annually, but utilization is only about 25% (Huhta and Stearn, 1980). While the production of lightweight aggregate from fly ash was once heralded as a major potential market, very little is currently being utilized as a lightweight aggregate raw material, and most fly ash is still a waste product rather than a byproduct from electric power generation.

Perlite and vermiculite are covered separately in the Commodity section of this book. The reader is directed to those separate chapters for a complete discussion of their occurrence and production.

Consumption

The quantity of pumice and volcanic cinders consumed in the United States has varied from 3.8 to 4.4 Mt/a (4.2 to 4.8 million stpy) during

TABLE 2—Comparative Operating Costs, Typical Rotary Kiln Production of Structural Lightweight Aggregate, 1970 and 1980*
(Cost in $ Per St of Output)

Item	1970	1980	Amount of Increase
Raw materials	$0.50	$ 1.25	+ 2.5
Labor	0.40	1.20	+ 3.0
Fuel	1.20	6.00	+ 5.0
Electric power	0.65	1.60	+ 2.5
Miscellaneous supplies and services	0.50		+ 7.0
(a) refractories		0.60	
(b) maintenance and repairs		1.50	
(c) other		1.40	
Direct production costs	$3.25	$13.55	+ 4.2
Depreciation (capital redemption)	0.50	0.85	+ 1.7
Sales, administration and general expenses	0.75	1.50	+ 2.0
Total production costs	$4.50	$15.90	+ 3.5
Overhead costs	1.25	1.20	—
Total costs	$5.75	$17.10	+ 3.0

* Based on production cost data both published and unpublished, principally: McCarl, 1975; Harwell, 1965; and miscellaneous producer data.

the period 1976–1980. This includes small amounts of imported pumice varying from 54.4 to 226.8 kt/a (60,000 to 250,000 stpy). Road construction accounted for 60% of the annual consumption and 25% was used as concrete admixtures and aggregates, the other 15% was consumed as railroad ballast, abrasives, and miscellaneous other uses. Average prices rose from $2.53 per ton in 1976 to an estimated $4.25 per ton in 1981. (Meisinger, 1982).

Consumption of expanded clay, shale, and slate has declined from about 10.9 Mt/a (12 million stpy) in the early 1970s (McCarl, 1975) to less than 5.4 Mt (6 million st) in 1980. The increasing production costs have forced producers to significantly increase prices, and this has led to a sharp decline in the amounts consumed annually. 1980 witnessed an approximate 18% decline in the sales of expanded shale attributed largely to lower levels of construction activity, high interest, and the general state of the economy (Berger, 1981; Robinson, 1981). Prices have roughly doubled in the past ten years from $6 to $10 per ton in the early 1970s (McCarl, 1975) to $11 to $20 per ton in 1980–1981. Many producers have also disappeared from the industry during this period of rapid increases in energy and other operating and capital costs. Raw materials for these structural lightweight aggregates have also increased in cost during this period. Typical values for lightweight aggregate raw materials (ores) and the expanded products at the production plant are shown in Table 3.

Expanded slag consumption has varied from 1.2 to 1.7 Mt/a (1.3 to 1.9 million stpy) in the last ten years with the output near the 1.7 Mt (1.9 million ton) level in both 1970 and 1978. Low production years were 1975 and 1980 due largely to low levels of construction activity in those recession periods. Over 95% of all expanded slag is used in lightweight concrete and concrete products. In recent years the consumption has been greater in lightweight concrete, reversing the pattern of the early 1970s when more was consumed in concrete products. In fact the proportions have reversed from 2:1 in concrete products (mostly blocks) to roughly 2:1 in lightweight structural concrete in 1978–1980 (Zlobik, 1981). The value of expanded slag has increased from about $3.35 per ton in 1970 to an average of about $7 per ton in 1980.

Expanded perlite consumption in the period 1978–1980 ranged from 454 to 499 kt/a (500,000 to 550,000 stpy) and was valued at from $106 to $120 million. The average unit value of the expanded product varied from roughly $150 per ton to over $300 per ton. The higher unit values were for expanded perlite treated with silicone. The average unit value for expanded perlite was approximately $200 per ton during the period 1978–1980, and has risen slightly more recently in response to rising fuel costs. A comparison of perlite end uses in 1972 and 1979 is presented in Table 4 to indicate relative shifts in market segments over the decade 1970–1980. 1972 end uses are representative of the early 1970s and 1979 end uses reflect the consumption

TABLE 3—Average Unit Value of Lightweight Aggregate Materials at Production Plants in 1980–1981*

Type of Material	$ Per St
Exfoliated vermiculite†	182.00–284.00
Crude vermiculite (ore)	76.32
Expanded perlite†	167.00–317.00
Crude perlite (ore)	30.15
Expanded clay, shale, and slate††	11.00–20.00
Clay and shale used for expansion (raw material)	1.50–3.00
Slate used for expansion (raw material)	4.50–6.00
Expanded slag	7.00
Pumice and volcanic cinders	4.25

*Specific prices are from US Bureau of Mines, *Mineral Commodity Summaries, 1982;* price ranges from trade journals and industry sources.

†Price ranges are for various grades; treated and specification-grades normally in higher portion of range.

††Prices for expanded clay, shale, and slate vary substantially, largely due to fuel and other energy costs.

TABLE 4—Perlite Uses in 1972 and 1979

Use	Share of Market, % 1972	1979
Filter aid	16	18
Plaster aggregate	12	4
Concrete aggregate	8	8
Horticultural products	3	7
Low temperature insulation	4	1
Formed products	8	14
Masonry and cavity fill insulation	4	4
Fillers	1	2
Other uses	44	42
	100	100

Sources: Perlite Institute; Milanese, 1981.

patterns in the later 1970s and into the early 1980s.

Exfoliated vermiculite in the period 1978–1980 was used in quantities of from 245 to 249 kt/a (270,000 to 275,000 stpy). This material was valued from \$50 to \$55 million. Unit values for exfoliated vermiculite during this period ranged from \$160 to just under \$300 per ton. The average value of exfoliated vermiculite in 1980 and 1981 was just over \$200 per ton, roughly the same as the unit value for expanded perlite, a major substitute material for many insulating uses. In 1981, 36% of all exfoliated vermiculite was used for construction aggregate and premixes, 27% in insulation products, 32% in horticulture, soil conditioning, and fertilizer carriers, and 5% in miscellaneous uses (Meisinger, 1982).

More detail on the use of slag, pumice, and volcanic cinders may be found in other chapters of this book. Separate commodity chapters may also be found on perlite and vermiculite.

Specifications

The consumption of lightweight aggregate materials is closely tied to construction codes and other specifications dealing with unit weight, size gradation, deleterious substances, and concrete-making properties. More detailed specifications may be obtained from the American Society for Testing & Materials (ASTM), Perlite Institute, Vermiculite Institute, National Slag Assn., Expanded Shale, Clay and Slate Institute, Lightweight Aggregate Producers Assn., American Concrete Institute, National Concrete Masonry Assn., National Ready-Mixed Concrete Assn., and various federal, state, county, and local government building codes.

Unit Weight—Dry loose weight may vary from 881 to 1121 kg/m^3 (55 to 70 pcf) for structural lightweight aggregates to 96 to 192 kg/m^3 (6 to 12 pcf) for exfoliated vermiculite and expanded perlite. Specific dry weights are given in Table 5.

Size Gradation—The size of particles relates to various other characteristics such as compressive strength, thermal and acoustical insulation values, heat resistance, and cement content in concrete mixes. It is generally expressed as ranges of percent by weight passing certain sieve or screen size openings. This gives an idea of the gradation of particle size in the raw material being used as aggregate. While similar to grading in other aggregates, there are some variations. Size gradation specifications vary according to use as shown in Table 6.

Deleterious Substances—The amount of contaminants in lightweight aggregates must be limited by weight so that the qualities of the final product (normally concrete or concrete block) will not be adversely affected. Organic impurities and iron oxides must be limited to avoid discoloration and staining. Clay lumps are limited to less than 2% by dry weight. Loss on ignition should be less than 5% except in the case of cinders where ignition loss should be less than 35%.

Concrete-Making Properties—A number of tests must normally be made with the lightweight aggregate in various concrete mixes. Specific quantities of fine and coarse aggregate are mixed with cement and water. This mixture is cast into cylinders and bars for a series of tests. The tests include compressive strength and unit weight (of concrete), drying shrinkage, visual inspection for popouts, and various durability tests such as freezing and thawing or sulfate reactivity or soundness tests.

Specifications vary depending on the final product to be made. ASTM or other standards should be consulted for specific details on testing and product requirements.

Other Specifications—In specialized uses such as insulating fill or as an aggregate in insulating concrete, other standards such as thermal conductivity, fire rating, and physical flow characteristics become important in judging the suitability of various lightweight aggregates. The actual requirements for any specialized application should be reviewed before any material is used for that purpose.

TABLE 5—Dry Weights for Various Lightweight Aggregates

Aggregate Type and Size Designation	Dry Loose Weight, Maximum, Pcf*
Structural lightweight aggregate	
Fine aggregate	70
Coarse aggregate	55
Combined fine and coarse aggregate	65
Vermiculite	10 (6 pcf minimum)
Perlite	12 (6 pcf minimum)

*Metric equivalent: 1 pcf × 16.018 46 = kg/m^3.

TABLE 6—Grading Requirements for Lightweight Aggregates, Percentages (by Weight) Passing Sieves Having Square Openings

Size Designation	1 In. (25.4 mm)	3/4 In. (19.0 mm)	1/2 In. (12.7 mm)	3/8 In. (9.51 mm)	No. 4 (4.76 mm)	No. 8 (2.38 mm)	No. 16 (1.19 mm)	No. 50 (0.29 mm)	No. 100 (0.149 mm)
Lightweight Aggregates for Structural Concrete (ASTM C330-69)									
Fine aggregate									
No. 4 to 0	—	—	—	100	85-100	—	40-80	10-35	5-25
Coarse aggregate									
1 in. to 1/2 in.	95-100	—	0-10	—	—	—	—	—	—
1 in. to No. 4	95-100	—	25-60	—	0-10	—	—	—	—
3/4 in. to No. 4	100	90-100	—	20-60	0-10	—	—	—	—
1/2 in. to No. 4	—	100	90-100	40-80	0-20	0-10	—	—	—
3/8 in. to No. 8	—	—	100	80-100	5-40	0-20	—	—	—
Combined fine and coarse aggregate									
1/2 in. to 0	—	100	95-100	—	50-80	—	—	5-20	2-15
3/8 in. to 0	—	—	100	—	65-90	35-65	—	10-25	5-15
Lightweight Aggregates for Concrete Masonry Units (ASTM C331-69)									
Fine aggregate									
No. 4 to 0	—	—	—	100	80-100	—	40-80	10-35	5-25
Coarse aggregate									
1/2 in. to No. 4	—	100	90-100	40-80	0-20	0-10	—	—	—
3/8 in. to No. 8	—	—	100	80-100	5-40	0-20	—	—	—
Combined fine and coarse aggregate									
1/2 in. to 0	—	100	95-100	—	50-80	—	—	5-20	2-15
3/8 in. to 0	—	—	100	90-100	65-90	35-65	—	10-25	5-15
Lightweight Aggregates for Insulating Concrete (ASTM C332-66)									
Group I (ultra-lightweights)									
Perlite	—	—	—	—	100	80-100	40-85	5-25	0-10
Vermiculite (Coarse)	—	—	—	100	98-100	60-100	40-85	2-20	0-10
Vermiculite (Fine)	—	—	—	—	—	100	85-100	2-40	0-10
Group II (expanded clay, shale, slate, and other materials)									
Fine aggregate									
No. 4 to 0	—	—	—	100	85-100	—	40-80	10-35	5-25
Coarse aggregate									
1/2 in. to No. 4	—	100	90-100	40-80	0-20	0-10	—	—	—
3/8 in. to No. 8	—	—	100	80-100	5-40	0-20	—	—	—
No. 4 to No. 8	—	—	—	100	90-100	0-20	—	—	—
Combined fine and coarse aggregate									
1/2 in. to 0	—	100	95-100	—	50-80	—	—	5-20	2-15
3/8 in. to 0	—	—	100	90-100	65-90	35-65	—	10-25	5-15

Source: Courtesy ASTM.

Substitutes

Virtually any naturally occurring lightweight material with the desired physical and chemical characteristics may be considered a lightweight aggregate. Small quantities of diatomite have been used for such purposes, and even meerschaum (sepiolite) has found use as a lightweight building material. Perhaps the only significant commercial substitutes are chemical foams and plastics such as styrofoam or polyurethane that compare in strength to exfoliated vermiculite. However the restriction on raw material supplies for the manufacture of these plastics brought about by the energy "crisis" in the early 1970s and substantially higher prices for petrochemicals are likely to limit the substitution of foamed plastics for most lightweight aggregate applications in the foreseeable future. Mineral wool and other insulating materials are also unlikely to make a significant impact on the growth in use of lightweight aggregates.

Substitution of manufactured lightweights for byproduct materials such as cinders and slag is likely to continue. Since lightweight aggregates often substitute for materials such as stone, sand, and gravel in concrete, about the only competitive threat from these natural materials could occur if fuel prices and availability of fuel supplies severely limit the ability to produce lightweight aggregates or render them uneconomic.

Some indications that higher manufacturing costs are curtailing structural lightweight aggregate markets are apparent. The drop in output of structural lightweights from roughly 11.4 Mt (12.6 million st) in 1972 to roughly 5.0 Mt (5.6 million st) in 1980–1981 is more than the drop in construction activity would indicate. Some of the decreased use is undoubtedly due to loss of market share to materials such as crushed stone and gravel, as the costs of producing structural lightweight aggregate have tripled from 1970 to 1980, largely due to higher energy costs and general inflation (see Table 2). Prices for crushed stone and sand and gravel have also risen, but the differences in unit cost between natural aggregates and manufactured lightweight aggregates are much greater today than they were in the early 1970s.

Economic Factors

Price Trends—Prices for manufactured lightweight aggregates have risen dramatically in the past ten years due to general inflation and higher energy costs. Fuel costs have risen about five times over their levels in the early 1970s, and other energy costs such as electric power are continuing to increase more rapidly than inflation. Cost of refractories has become a significant cost item as prices reflect higher energy costs in their manufacture. Energy and energy-related costs have increased as a proportion of total costs from roughly 25% in the early 1970s to roughly 50% in 1980 (see Table 2). Similar increases have been influencing the costs of expanded perlite and exfoliated vermiculite.

The prices of lightweight aggregates will continue to reflect the rising cost of energy, and prices are likely to parallel energy costs.

Transportation—All forms of surface transport are used in moving lightweight aggregates. Truck movement on highways and streets is especially important in the delivery of expanded aggregates either in bulk form or in containers. Economies of lightweight aggregate haulage are very similar to transportation of crushed stone and other aggregates. Due to the higher value of lightweights, however, some additional haulage can be tolerated with proportionately less delivered price impact. Transportation costs are likely to amount to from 15 to 35% of the delivered price for most structural lightweights. The impact of transportation costs on delivered prices of vermiculite and perlite is naturally less due to their higher unit value.

A good rule of thumb is that distance of movement of aggregate materials is likely to be directly proportional to unit price and inversely proportional to unit weight and bulk. Perlite and vermiculite ores are transported by rail from mines to delivery points throughout the United States and by ocean freighter throughout the world. Some vermiculite ores come to the United States by boat from South Africa, and rail haulage from the western US to the East Coast is common. However, the exfoliated or expanded products are not likely to be distributed outside local and regional markets due to their great bulk.

Rail transportation of other lightweight aggregates is largely regional in nature, and is normally limited by the proximity of competing producers. Trucking may be local or regional depending on the availability of rail transport and relative costs. Trucks are normally most economical for hauls of 80.5 km (50 miles) or less with railroads best suited for hauls in the 161 to 483-km (100 to 300-mile) range. The transportation method used is dictated by the access to other competitive modes as well as haulage costs. The object is to keep delivered price down to maximize returns to the producer.

Marketing—Most producers of natural and manufactured lightweight aggregates maintain their own sales staff and technical service representatives. Products are sold on the basis of specifications and cost (in-place) of the final product. In many cases the ability to reduce dead loads through the use of lighter aggregates will permit reductions in the cost of structural steel. Insulation properties of lightweights may enable the construction of thinner walls or cut down on the need for other insulation. These factors often permit cost saving in the construction of buildings despite higher unit prices for lightweight aggregate. Product yield often can be improved by the use of manufactured lightweights. Surface textures and uniformity are often superior to naturally occurring or byproduct aggregates.

Byproduct materials are normally handled by wholesale and retail dealers. The actual producers often handle slag sales, but cinders, fly ash, coke breeze, and other byproducts are usually sold through brokers.

Producers of vermiculite and perlite ores generally license or franchise brokers for the exfoliated or expanded product. Trademarks, promotional materials, specifications, and other sales aids are normally provided to the licensed or franchised dealer by the raw material supplier. The dealer pyroprocesses the ore and packages the product in bags or ships in bulk. The dealer acts as wholesaler in the distribution to large users and retail building material outlets.

Marketing research studies are usually important in the location of a new lightweight aggregate plant. The actual use and potential use of lightweight aggregate materials in the intended market must be measured or estimated from planned and prospective construction trends. The proximity of competitors and transportation costs in the market area must be carefully studied to determine the economic feasibility of a new production plant. Most lightweight aggregate producers do not maintain their own marketing research department, but this should never lead to the initiation of a new plant location without some market analysis. The choice of a consultant or specialized marketing research firm should be based on both the cost of the service and past experience in similar or related evaluations. The most prominent or prestigious consulting firms are not always the most qualified to do marketing studies for construction materials.

Government Considerations—Taxes, depreciation, and depletion allowances for lightweight aggregate producers are similar to those for the producers of other construction materials. Zoning and land-use regulation are restrictive in urban and other heavily populated areas, and are closely related to the situations encountered by producers of crushed stone or sand and gravel. However, the more limited occurrence of lightweight aggregate raw material deposits and the smaller number of producers generally leads to the location of plants at greater distance from the more heavily populated market centers and consequently less problems with zoning and land-use regulation. In cases where plants are simply pyroprocessing raw materials mined at distant locations or handling byproducts, they are in the same catagory as heavy industry and are located in industrial areas.

To date there have been no significant reductions in government regulations on air and water pollution. If some relaxation in environmental standards is achieved there will be some cost saving, but this is unlikely to create any significant reduction in production costs. The same observation can be made for safety and health regulations. Once implemented, the additional costs must be included in operating costs. Any relaxation in government regulations could, of course, hold down future cost increases. The public is slowly recognizing that business has little choice but to pass increased costs along to its customers as higher prices.

Outlook

Total consumption of lightweight aggregates declined roughly 35% during the 1970s. This was largely the result of significantly lower production and sales of expanded clays, shale, and slate. Some of the decline was also caused by sharply lower levels of construction activity in 1975, 1979, and 1980. The decline in construction has continued into 1981 and construction is likely to be depressed as long as interest rates remain relatively high.

Some increases have been noted in markets for pumice and volcanic cinders as these materials become relatively more attractive in terms of prices when compared to manufactured lightweights. Markets for insulating materials have increased as energy costs have increased, leading to higher production and use levels for expanded perlite and exfoliated vermiculite. Perlite use has increased roughly 3% per year while vermiculite consumption has grown at about 2% per year over the last five years (Meisinger, 1981, 1981b). Continued growth in markets for these materials is likely to continue into the foreseeable future.

The decline in markets for expanded clay, shale, and slate has stabilized in the period 1978–1980 with no significant growth likely in the near future. Some increase in construction activity would cause related growth in demand for these structural lightweight aggregates, but that turnaround in construction is unlikely until interest rates fall significantly.

In the long run, modest growth in demand for lightweight aggregates is unlikely to exceed 2% per year, and this will come mostly from natural lightweights and insulating materials due to favorable price considerations and continued high energy costs.

The ominous clouds of the energy crisis have priced expanded clay, shale, and slate out of some markets. However, those manufacturers who have remainded in the business have converted to coal as a fuel in larger proportions than was thought possible in the early 1970s. This has meant more reliable supplies of fuel and more modest increases in fuel costs. Most producers of expanded perlite and exfoliated vermiculite have not converted to alternative energy supplies and continue to use fuel oil and natural gas. The main hope for those who use petroleum and gas may come from the increased use of byproduct gas from coal mining and coke production. New discoveries of oil and gas are unlikely to solve the supply problems of the future.

As was the case in the past, it must be emphasized that the final products of the lightweight aggregate industry are energy-efficient buildings. This means that the necessity to conserve scarce energy resources will continue to create demand for lightweight aggregates as well as contribute to their cost of production. The insulation qualities of lightweight aggregates will continue to provide markets for these materials, even though the relative shares of the market will continue to shift in future years based on the costs and values of the final products compared to the costs and qualities of the raw materials.

Acknowledgments

The author wishes to express his appreciation to the Kennedy Van Saun Corp., Vulcan Materials Co., W. R. Grace & Co. Construction Products Division, Coronado Stone Products, *Pit and Quarry* Publications, *Rock Products* magazine, the US Bureau of Mines, the National Slag Association, and other organizations that provided data and advice on the material presented in this chapter.

In addition, the assistance of Ms. Jayne Handlin, my graduate research assistant, and Mrs. Rose McCoy, my secretary, is acknowledged with gratitude.

Bibliography and References

Anon., 1963, "New Boston and New Concrete (Federal Reserve Bank of Boston)," *New England Business Review,* Apr., pp. 5–7.

Anon., 1963a, "Some Things You Should Know About Lightweight Aggregates," reprint from *Concrete Products,* Vol. 66, No. 7, July, No. 8, Aug., No. 9, Sep., No. 10, Oct., 27 pp.

Anon., 1964, "Technical and Economic Feasibility of Establishing a Lightweight Aggregate Plant Using Marine Clays in York County, Maine," Area Redevelopment Administration, US Dept. of Commerce, Washington, Nov., 75 pp.

Anon., 1965, "Expanded Clay and Shale Lightweight Aggregate in the TVA Region," Div. of Water Control Planning, Geologic Br., Tennessee Valley Authority, Knoxville, TN, 22 pp.

Anon., 1973, "Lightweight Aggregates," *Rock Products,* Vol. 76, No. 12, Dec., pp. 48–49.

Anon., 1979, "Clay in 1979," *Mineral Industry Surveys Annual Advance Summary,* US Bureau of Mines, Washington, DC, 38 pp.

Aitcin, P.C., and Poulin, C., 1974, "Bubble Aggregate," *4th Proceedings, Mineral Waste Utilization Symposium,* Chicago, IL, May 7–8, pp. 265–268.

Barneyback, R.S., Jr., 1964, "The Effect of Lightweight Aggregate Physical and Chemical Characteristics on the Properties of Portland Cement Concrete Mixtures," *Trans. SME/AIME,* Vol. 229, p. 392.

Bauman, E.W., 1971, "Slag—Nonmetallic Aggregate from Steel-Making Process," *Rock Products,* Vol. 74, No. 10, Oct., pp. 116, 168.

Berger, R.L., 1974, "Properties of Concrete with Cement Clinker Aggregate," *Cement and Concrete Research,* Vol. 4, No. 1, Jan., pp. 99–112.

Berger, T.R., 1971, "Lightweight Aggregate—Newest Concrete Material," *Rock Products,* Vol. 74, No. 10, Oct., pp. 118, 170, 174.

Berger, T.R., 1972, "Trends in Lightweight Aggregate," paper at SME Fall Meeting, Birmingham, AL, Oct., 9 pp.

Berger, T.R., 1981, "Lightweight Aggregates Reflects Residential Construction," *Pit and Quarry,* Vol. 73, No. 7, Jan., p. 84.

Berry, E.E., 1979, "Concrete Made with Supplementary Cementary Materials," CANMET Report, No. 79–32, Aug., 22 pp.

Berry, E.E., Malhotra, V.M., 1980, "Fly Ash for Use in Concrete—A Critical Review," *Journal of the American Concrete Institute,* Vol. 77, No. 2, Mar.–Apr., pp. 59–73.

Blank, H.R., and Ledbetter, W.B., 1968, "Synthetic Aggregates," reprint from *Texas Highways,* Apr., 6 pp.

Boltz, C.L., 1977, "Everlasting Concrete from Pulverized Fuel Ash," *Indian Concrete Journal,* Vol. 51, No. 5, May, p. 159.

Burnett, J.L., 1960, "Expansible Shale," *Mineral Information Service,* California Div. of Mines, Vol. 13, No. 5, May, pp. 1–7.

Bush, A.L., 1973, "Lightweight Aggregates," *United States Mineral Resources,* Professional Paper No. 820, D.A. Brobst, ed., US Geological Survey, pp. 333–355.

Caldwell, D.C., 1959, "Glacial Lake and Glacial Marine Clays of the Farmington Area, Maine—Origin and Possible Use as Lightweight Aggregate," Special Geological Studies Series No. 3, Dept. of Economic Development, Maine Geological Survey, Augusta, June 1, 48 pp.

Chesterman, C.W., and Schmidt, F.S., 1956, "Pumice, Pumicite and Volcanic Cinders in California," Bulletin 174, Dept. of Natural Resources, California Div. of Mines, Dec., 119 pp.

Chitharanjan, N., Santhakumar, A.R., 1978, "Lightweight Aggregate Concrete Seismic Zones," *6th Symposium on Earthquake Engineering,* Vol. 1, Oct. 5–7, pp. 305–310.

Coda, F.M., 1973, "Perlite," *Mining Engineering,* Vol. 25, No. 1, Jan., pp. 41–42.

Cole, W.A., and Zetterstrom, J.D., 1954, "Investigation of Lightweight Aggregates of North and South Dakota," Report of Investigations 5065, US Bureau of Mines, July, 43 pp.

Cole, W.A., et al., 1961, "Lightweight Aggregates: Expansion Properties of Clays, Shales and Precambrian Rocks of Wisconsin," Report of Investigations 5906, US Bureau of Mines, 26 pp.

Conley, J.E., et al., 1948, "Production of Lightweight Concrete Aggregates from Clays, Shales, Slates, and Other Materials," Report of Investigations 4401, US Bureau of Mines, Nov., 121 pp.

Cook, D.J., 1973, "Expanded Polystyrene Beads as Lightweight Aggregate for Concrete," *Precast Concrete,* Vol. 4, No. 12, Dec., pp. 691–693.

Cooper, J.D., 1970, "Clays," *Mineral Facts and Problems,* Bulletin 650, US Bureau of Mines, pp. 923–938.

Coxon, P., Moss, D.W., 1975, "Sintering Machine Converts Fly Ash to Lightweight Aggregate," *Rock Products,* Vol. 78, No. 6, June, pp. 68, 70, 75.

Davis, J.F., 1963, "Lightweight Aggregate in New York State," *Empire State Geogram,* New York State Geological Survey, Vol. 2, No. 1, pp. 7–10, 12.

Diamond, W.G., et al., 1964, "Expanded Clay and Shale Lightweight Aggregate Industry in the South-Central United States," Information Circular 8233, US Bureau of Mines, 50 pp.

Doyle, R.G., 1962, "Results of New Research Investigation into the Use of Glacial Marine Clay as a Source of Lightweight Aggregate," Research Progress Report, Dept. of Economic Development, Maine Geological Survey, Augusta, Apr. 17, 9 pp.

Eckel, E.B., 1960, "Pumice and Pozzolan Deposits in the Lesser Antilles," Preliminary Report for the International Cooperation Administration, US Geological Survey, 58 pp.

Erskine, F.G., 1971, "Lightweight Aggregate Accelerates Systems Construction in Building," *Rock Products,* Vol. 74, No. 10, Oct., pp. 120, 178, 180, 182.

Everhart, J.O., et al., 1958, "A Study of Lightweight Aggregates," *Bulletin No. 169,* Engineering Experiment Station, The Ohio State University, Vol. 27, No. 3, May, 69 pp.

Faber, J.H., et al., 1967, "Fly Ash Utilization," Information Circular 8348, US Bureau of Mines, 345 pp.

Fairweather, V., 1975, "Flyash Pavements, Runways to Take Off?" *Civil Engineering,* Vol. 45, No. 8, Aug., pp. 57–58.

Fisher, W.L., and Garner, L.E., 1965, "Bloating Characteristics of East Texas Clays," Geological Circular 65–1, Bureau of Economic Geology, The University of Texas, Austin, Mar., 20 pp.

Fuess, D.A., 1961, "Survey of the Literature Pertaining to Expanded Clay and Shale Aggregates," *Monthly Progress Report No. 298,* College of Ceramics at Alfred University, State University of New York, Vol. 26, No. 4, Apr., 103 pp.

Grancher, R.A., 1969, "The Lightweight Aggregate Boom," *Rock Products,* Vol. 72, No. 12, Dec., pp. 59, 60, 62, 94.

Grancher, R.A., and Stearn, E.W., 1979, "Lightweight Aggregates, Gypsum and Slag," *Rock Products,* Vol. 82, No. 12, Dec., pp. 61–62.

Gromko, G.J., 1975, "Preliminary Investigation of the Feasibility of Spent Oil Shale As Road Construction Material," Transportation Research Board. Transportation Research Record, No. 549, pp. 47–54.

Grosh, W.A., and Hamlin, H.P., 1963, "Lightweight Aggregates: Expansion Properties of Clays, Shales and Argillites of Minnesota," Report of Investigations 6313, US Bureau of Mines, 30 pp.

Hamlin, H.P., and Templin, G., 1962, "Evaluating Raw Materials for Rotary-Kiln Production of Lightweight Aggregate," Information Circular 8122, US Bureau of Mines, 23 pp.

Harris, H.M., et al., 1962, "Resources for Making Expanded Aggregate in Western Washington and Oregon," Report of Investigations 6061, US Bureau of Mines, 41 pp.

Harrison, W.H., 1974, "Synthetic Aggregate Sources and Resources," *Concrete,* London, Vol. 8, No. 11, Nov., pp. 41–44.

Harwell, R.S., 1964, "The Economics of a Lightweight Aggregate Plant," paper at Annual Meeting of the Expanded Clay and Shale Assn., Miami Beach, FL, Jan. 9, 17 pp.

Harwell, R.S., 1965, "A Lightweight Aggregate Plant—Is It Feasible for Your Operation?" *Rock Products,* Vol. 68, No. 2, Feb., pp. 69–71.

Herod, B.C., 1962, "Colorado-New Mexico Teamed in Perlite," *Pit and Quarry,* Vol. 55, No. 2, Aug., pp. 96–101.

Herod, S., and Levine, S., 1980, "Aggregates, Other Major Nonmetals," *Pit and Quarry,* Vol. 72, No. 7, Jan., pp. 96–100.

Hollenbeck, R.P., and Tyrrell, M.E., 1969, "Raw Materials for Lightweight Aggregate in the Appalachian Region, Alabama and Georgia," Report of Investigations 7244, US Bureau of Mines, 21 pp.

Horler, D.B., 1977, "Some Advantages of Lightweight Aggregates," *Precast Concrete,* Vol. 8, No. 3, Mar., pp. 137–140.

Huhta, R.S., and Stearn, E., 1980, "The Aggregate Scene Around the Country" and "Lightweight Aggregates—New Markets Help Fill Construction Void," *Rock Products, Forecast,* Vol. 83, No. 12, pp. 46–52, 60–61.

Jain, K.K., Nauhyal, B.D., and Jain, O.P., 1975, "Compressive Strength and Modulus of Elasticity of Flyash Concrete," *Journal of the Institution of Engineers (India), Civil Engineering Division,* Vol. 56, Pt. C1 I., July, pp. 32–37.

Jardine, D.J., Wolhuter, C.W., 1977, "Some Properties of High-Slag Concrete," *Civil Engineer in South Africa,* Vol. 19, No. 11, Nov., pp. 249–251.

Jaster, M.C., 1956, "Perlite Resources in the United States," Bulletin 1027–I, US Geological Survey, pp. 375–403.

Kadey, F.L., 1963, "Petrographic Techniques in Perlite Evaluation," *Trans. SME/AIME,* Vol. 226, pp. 332–336.

Karni, J., and Carmel, D., 1974, "Lightweight Concrete Made with Artificial Lime," *Building International,* English ed., Vol. 7, No. 4, July–Aug., pp. 349–354.

Kasami, Hideo, Mishima, Tomio, et al., 1976, "Study on Artificial Super Lightweight Aggregate for Precast Concrete," *Takenaka Technical Research Report,* No. 16, pp. 107–122.

Klinefelter, T. A., 1960, "Lightweight Aggregates," *Industrial Minerals and Rocks,* 3rd ed., J.L. Gillson, ed., AIME, New York, pp. 487–495.

Knapton, J., 1976, "Pumice Aggregate for Concrete Blocks," *Precast Concrete,* Vol. 7, No. 5, May, pp. 247–248.

Krasnov, M.N., Norkin, I.A., and Fillipov, V.M., 1971, "Investigating the Features of Sampling Shales at High Capacity Power Stations," *Thermal Engineering,* Vol. 18, No. 8, Aug., pp. 67–72.

Lal, Kishan, Singh, Ranbir, and Vpadhyaya, V.B., 1976, "Bloated Clay: An Unconventional Artificial Lightweight Aggregate," *Indian Concrete Journal,* Vol. 50, No. 6, pp. 174–180, 186.

Lamb, D.R., and Hanna, S.J., eds., 1973, *Workshop on Expansive Clays and Shales in Highway Design and Construction, Proceedings, 1972,* 3 volumes.

Lapham, D.M., and Hoover, K.V., 1971, "Preliminary Evaluation of the Factors Affecting the Use Potential of Clays and Shales in Pennsylvania," SME Preprint 71H18, AIME Centennial Annual Meeting, New York, Mar., 18 pp.

Levine, S., 1961, "Which Is Better? Horizontal or Vertical Furnace for Processing Vermiculite," *Nonmetallic Minerals Processing,* Vol. 2, No. 3, Mar., p. 30.

Levine, S., 1962, "Sintering Control for Lightweight Aggregate Processing at Masslite," *Nonmetallic Minerals Processing,* Vol. 3, No. 10, Oct., pp. 20–24.

Levine, S., 1971, "The Rock Products Industry—An Overview: Lightweight Aggregate," *Rock Products,* Vol. 74, No. 10, Oct., pp. 84–85.

Levine, S., 1971a, "The Rock Products Industry—An Overview: Slag," *Rock Products,* Vol. 74, No. 10, Oct., p. 83.

Levine, S., and Stearn, E.W., 1972, "Lightweight Aggregates: Increased Production Forecast for All Types," *Rock Products,* Vol. 75, No. 12, Dec., pp. 63–64.

Lodding, W., 1964, "Evaluation of Raw Materials for Lightweight Aggregate," SME Preprint 64H71, AIME Annual Meeting, New York, Feb., 12 pp.

Lydon, F.D., 1973, "Some Criteria for the Choice and Use of Lightweight Concrete," *Building International,* English ed., Vol. 6, No. 3, May–June, pp. 321–338.

McCarl, H.N., 1963, "New England's Lightweight Aggregate Industry," *Minerals Processing,* Vol. 4, No. 7, July, pp. 31–34.

McCarl, H.N., 1963a, "The Lightweight Aggregate Industry in New England," Research Report No. 21, The Federal Reserve Bank of Boston, Boston, MA, 108 pp.

McCarl, H.N., 1964, "Lightweight Aggregate: A Growing Industry in New England," SME Preprint 64H31, AIME Annual Meeting, New York, Feb., 18 pp.

McCarl, H.N., 1964a, "Pennsylvania's Lightweight Aggregate Industry," *Minerals Processing,* Vol. 5, No. 12, Dec., pp. 18, 20, 25.

McCarl, H.N., 1975, "Lightweight Aggregates," *Industrial Minerals and Rocks,* 4th ed., S.J. Lefond, ed., AIME, New York, pp. 85–96.

McGrain, P., 1957, "Sources of Shale in Kentucky for Lightweight Aggregate Production," Report of Investigation No. 12, Kentucky Geological Survey, 23 pp.

McGrain, P., 1958, "Sources of Shale in Kentucky for Lightweight Aggregate Production No. 2," Report of Investigations No. 15, Kentucky Geological Survey, 30 pp.

Meisinger, A.C., 1980, "Vermiculite," *Mining Engineering,* Vol. 32, No. 5, May, pp. 588–589.

Meisinger, A.C., 1981, "Perlite," *Minerals Yearbook 1978–1979,* US Bureau of Mines, pp. 671–675.

Meisinger, A.C., 1981a, "Vermiculite," *Mining Engineering,* Vol. 33, No. 5, May, p. 592.

Meisinger, A.C., 1981b, "Vermiculite," *Minerals Yearbook 1978–1979,* US Bureau of Mines, pp. 977–980.

Meisinger, A.C., 1982, "Pumice and Volcanic Cinder," *Mineral Commodity Summaries 1982,* US Bureau of Mines, pp. 118–119.

Milanese, R.S., 1980, "Perlite," *Mining Engineering,* Vol. 32, No. 5, May, p. 579.

Milanese, R.S., 1981, "Perlite," *Mining Engineering,* Vol. 33, No. 5, May, p. 583.

Milas, J.E., 1970, "Sintering Machine Expands Clay at 50-Tph Rate," *Rock Products,* Vol. 73, No. 6, June, pp. 48–51, 96.

Millar, W.T., and Hamlin, H.P., 1964, "Examining and Testing Clay from Hartford County, Conn., for Lightweight Aggregate Use." Information Circular 8228, US Bureau of Mines, 21 pp.

Moss, D.W., 1976, "Lightweight Aggregate from Fly Ash in the United Kingdom," *Precast Concrete,* Vol. 7, No. 8, Aug., pp. 409–411.

Myers, J.B., 1960, "Vermiculite," *Industrial Minerals and Rocks,* 3rd ed., J.L. Gillson, ed., AIME, New York, pp. 889–899.

Myers, J.W., et al., 1964, "Production of Lightweight Aggregate from Washery Refuse," Report of Investigations 6449, US Bureau of Mines, 16 pp.

Nester, G.E., 1973, "Structural Mineral Aggregates," *Mining Engineering,* Vol. 25, No. 1, Jan., pp. 47–48.

North, O.S., 1960, "Vermiculite Sparkles in Modern Industry," *Rock Products,* Vol. 63, No. 10, Oct., pp. 94–97.

O'Neill, B.J., et al., 1965, "Properties and Uses of

Pennsylvania Shales and Clays," Mineral Resources Report M 51, Pennsylvania Bureau of Topographic and Geologic Surveys, 448 pp.

Owens, P.L., 1979, "Fly Ash and Its Usage in Concrete," *Concrete*, London, Vol. 13, No. 7, July, pp. 21–26.

Pallister, H.D., et al., 1964, "Clay and Shale for Lightweight Aggregate in Alabama," Circular No. 26, Geological Survey of Alabama, 45 pp.

Parks, W.S., et al., 1964, "Survey of Lightweight Aggregate Materials in Mississippi," Mississippi Geological, Economic and Topographical Survey, State Industrial and Technological Research Commission, Bulletin 103, 115 pp.

Peters, J.W., et al., 1968, "A Survey of the Location, Magnitude, Characteristics and Potential Uses of Pennsylvania Refuse," Special Research Report No. SR-67, Dept. of Mining, The Pennsylvania State University, University Park, Jan. 23, 180 pp.

Price, J.D., Troop, P., and Gershman, H.W., 1978, "Potential for Energy Conservation Through the Use of Slag and Fly Ash in Concrete, Final Report." Prepared for DOE by Gordian Assoc., Inc., Washington, DC.

Protze, H.G., and Levine, S., 1962, "Lightweight Aggregate for Structural Concrete and Products," *Nonmetallic Mineral Processing*, Vol. 3, No. 10, Oct., pp. 26–27.

Ravina, D., 1980, "Optimized Determination of PFA (Fly Ash) Fineness with Reference to Pozzolanic Activity," *Cement and Concrete Research*, Vol. 10, No. 4, July, pp. 573–580.

Reedy, R.W., 1971, "Lightweight Aggregate: Raw Materials, Plant Design, Selection of Equipment, Rotary Kiln Operation," reprint from *Pit and Quarry*, Vol. 63, Nos. 7, 10; Vol. 64, Nos. 2, 5, 16 pp.

Reichard, T.W., 1964, "Creep and Drying Shrinkage of Lightweight and Normal-Weight Concretes," Monograph 74, National Bureau of Standards, US Dept. of Commerce, Mar. 4, 30 pp.

Renfroe, O.S., ed., 1979, "Building Materials from Solid Wastes," Pollution Technology Review No. 61, Noyes Data Corp., Park Ridge, NJ, 275 pp.

Rhoads, D.C., Gordon, R.B., and Vaisnys, J.R., 1975, "Conversion of Marine Muds to Lightweight Construction Aggregate," *Environmental Science and Technology*, Vol. 9, No. 4, Apr., pp. 360–362.

Roberts, M.J., 1958, "Bloated Clay Aggregate, Its Production and Marketing Potential in Florida," Bulletin Series No. 98, Engineering & Industrial Experiment Station, University of Florida, 51 pp.

Robinson, H.C., 1981, "Expanded Shale Improves Efficiency," *Pit and Quarry*, Vol. 73, No. 7, Jan., pp. 84–85.

Ross, C.S., and Smith, R.L., 1961, "Ash-Flow Tuffs: Their Origin, Geologic Relations, and Identification," Professional Paper 366, US Geological Survey, 81 pp.

Schroeder, H.J., 1970, "Perlite," *Mineral Facts and Problems*, Bulletin 650, US Bureau of Mines, pp. 1129–1137.

Schroeder, H.J., 1970a, "Pumice," *Mineral Facts and Problems*, Bulletin 650, US Bureau of Mines, pp. 1171–1178.

Schroeder, H.J., 1970b, "Vermiculite," *Mineral Facts and Problems*, Bulletin 650, US Bureau of Mines, pp. 1283–1291.

Schwarzkopf, F., 1966, "Rotary-Kiln Processing of Lightweight Aggregate," *Minerals Processing*, Vol. 7, No. 4, Apr., pp. 34–41.

Sherwood, P.T., 1975, "Use of Waste and Low-Grade Materials in Road Construction-2 Colliery Shale," Transports and Road Research Laboratory, Great Britain, TRRL Report, LR 649, 21 pp.

Shildeler, J.J., 1961, "Manufacture and Use of Lightweight Aggregates for Structural Concrete," Bulletin D40, Research and Development Laboratories; Portland Cement Assn., Jan., 17 pp.

Stewart, G.W., 1959, "Lightweight Aggregate Raw Materials in New Hampshire," Mineral Resources Survey, Pt. 17, New Hampshire State Planning & Development Commission, Concord, 30 pp.

Strand, P.R., 1973, "Vermiculite," *Mining Engineering*, Vol. 25, No. 1, pp. 52–53.

Swamy, R.N., and Ibrahim, A.B., 1973, "Shrinkage and Creep Properties of High Early Strength Structural Lightweight Concrete," *Proceedings of the Institution of Civil Engineers, London*, Vol. 55, Part 2, Sept., pp. 635–646.

Sweeney, J.W., and Hamlin, H.P., 1965, "Lightweight Aggregates: Expansion Properties of Selected Illinois Shales and Clays," Report of Investigations 6614, US Bureau of Mines, 34 pp.

Sweeney, J.W., and Hamlin, H.P., 1965a, "Lightweight Aggregates: Expansion Properties of Selected Indiana Shales," Report of Investigations 6574, US Bureau of Mines, 28 pp.

Taeler, D.H., 1964, "Perlite Popping in the Nation's Capitol," *Minerals Processing*, Vol. 5, No. 7, July, pp. 31–32.

Teng, T.C.P., and Clisby, M.B., 1975, "Experimental Work for Active Clays in Mississippi," *Transportation Engineering Journal*, American Society of Civil Engineers, Vol. 101, No. 1, Feb., pp. 77–95.

Timms, A.G., 1967, "Lightweight Concrete and Aggregates, Parts I-III," *Modern Concrete*, Vol. 31, Pt. I, No. 2, June, pp. 28–30, 32; Pt. II, No. 3, July, pp. 29–30, 32, 35; Pt. III, No. 4, Aug., pp. 28, 30, 32.

Toles, G.E., 1975, "Lightweight Aggregates for Highways," *Quarry Management and Products*, Vol. 2, No. 6, June, pp. 157–158.

Trauffer, W., 1972, "New Lightweight Aggregate Plant of Hercules, Inc.," reprint from *Pit and Quarry*, Vol. 65, No. 4, Oct., 8 pp.

Utley, R.W., et al., 1964, "The Utilization of Coal Refuse for the Manufacture of Lightweight Aggregate," Special Research Report No. SR-46, Dept. of Mineral Preparation, The Pennsylvania State University, University Park, Sep. 1, 110 pp.

Utley, R.W., et al., 1965, "The Preparation of Coal Refuse for the Manufacture of Lightweight Aggregate," *Trans. SME/AIME*, Vol. 232, p. 346.

Venkatasubramanian, V., and Venkataraman, T.S., 1973, "Study of Mineral Fillers for Sheet-Asphalt Mixtures," *Highway Research Record*, No. 468, pp. 109–114.

Weinheimer, C.M., 1962, "Flyash Sintering for Aggregate," *Coal Utilization*, June, pp. 16–19.

White, W.A., 1960, "Lightweight Aggregate from Illinois Shales," Circular 290, Illinois State Geological Survey, Urbana, 29 pp.

Williams, E.G., et al., 1974, "Properties and Occurrence of Bloating Shales and Clays in the

Pennsylvanian of Western Pennsylvania," *Trans. SME-AIME,* Vol. 256, pp. 237–240.

Wills, M.H., Jr., 1974, "Lightweight Aggregate Particle Shape Effect on Structural Concrete," *Journal of the American Concrete Institute*, Vol. 71, No. 3, Mar., pp. 134–142.

Wilson, H.S., 1962, "Development of Canadian Lightweight Aggregate Industry," Information Circular 137, Dept. of Mines and Technical Surveys, Mines Branch, Ottawa, June, 22 pp.

Wilson, H.S., 1979, "Lightweight Aggregates: Properties, Applications and Outlook," Canada Centre for Mineral and Energy Technology (CANMET) Report, No. 79–33, Aug., 27 pp.

Zetterstrom, J.D., and Cole, W.A., 1956, "Expansion of Clays and Shales from North and South Dakota in a Rotary Kiln," Report of Investigations 5202, US Bureau of Mines, Feb., 13 pp.

Zlobik, A.B., 1981, "Slag—Iron and Steel," *Minerals Yearbook, 1978–1980,* US Bureau of Mines, pp. 821–833.

Additional Sources of Information

American Society for Testing & Materials (ASTM), "Lightweight Aggregate for Structural Concrete," Specification C330, revised periodically; "Lightweight Aggregate for Insulating Concrete," Specification C332, revised periodically; examples: ASTM C330–60T, ASTM C332–56T.

Cahners Publications, *Brick and Clay Record*, published monthly.

Concrete Facts, Journal of the Expanded Shale, Clay and Slate Institute (ESCI), published quarterly, National Press Bldg., Washington, DC, 20204.

"Lightweight Aggregate Outlook and Forecast," annual review in *Pit and Quarry* and *Rock Products,* December issues.

Pit and Quarry Publications, *Lightweight Aggregate Producing Plants in the US and Canada* (Map), Chicago, IL, 1963 ed., 1 p.; 1965 ed., 1 p.; 1969 ed., 1 p.

Construction Materials

Aggregates—Sand and Gravel

JAMES R. DUNN *

Major decisions in the sand and gravel industry are often influenced by a number of sociological, economical, and technological constraints. Demand and production trends, costs, and other historic data all show effects of such constraints as regulations, zoning, product specifications, and economic conditions.

"Sand" as most often used by industry is not sand in an engineering or geologic sense, but technically should be called "natural fine aggregate." The expression "fine aggregate" is preferable to "sand" because fine aggregate includes grain sizes which, according to engineering or geological classifications, are defined as pebble or grit size. Natural fine aggregate is distinguished from manufactured fine aggregate in that the manufactured material is crushed and screened to produce the required sizes, whereas natural fine aggregate usually is merely screened. Natural fine aggregate may also be mixed with the finer fraction of crushed gravel from a deposit. Coarse aggregate is generally defined as aggregate which will not pass a No. 4 or 4.76-mm (3/16-in.) sieve. Coarse aggregate may be natural (gravel) or manufactured (crushed stone, crushed gravel, slag, or coarse lightweight aggregate). Gravel is usually considered sufficiently crushed to qualify as manufactured coarse aggregate when the average size particle has two or more broken sides.

Consumption

Sand and gravel for construction are used as aggregate for portland cement and asphaltic concretes, as subbase under pavings, and as fill, particularly where good drainage is important. About 96% of the sand and gravel produced in the United States goes into construction, with 43% of the total 1981 demand used for portland cement concrete aggregates for general construction (Tepordei, 1982); of the remaining 57%, nearly 14% was used for asphaltic concrete mixtures; 24% for roadbases and coverings; 16% for construction fill; 3% for other uses. The production of sand and gravel in the United States in 1980 for construction purposes was 693 Mt (764 million st) with the breakdown by uses shown in Table 1.

The annual growth rate for total consumption since 1921 has averaged 6.4%, but, if depression and war years were eliminated, the growth rate would be about 9.4%. According to Cooper (1970), the growth rate from 1948 to 1968 was about 5.5% annually, but he considered this value as possibly higher than normal because of the interstate highway construction program. Judging by the low annual growth in recent years, Cooper's conclusion appears to be accurate.

Specifications

Specifications define the tests or critical observations that must be made on sand and gravel and the limiting values that must be achieved before an aggregate is considered suitable for a particular use. Tests for aggregates are an attempt to predict from simulated conditions or other observations what an aggregate might do in actual use, although tests may not uniformly achieve this purpose.

The specifications controlling the qualities of sand and gravel are variable depending upon the purpose for which aggregate is to be used, specifying agency, and the availability of materials. Specifications are normally concerned with: the reactions of the aggregates to alternate cycles of freezing and thawing or wetting and drying, both with and without salts present; chemical reactivity; resistance to abrasion and impact; particle shape; gradations; and deleterious constituents. Water absorption, specific gravity, color strength in fabricated concrete, and other characteristics of sand and gravel may be important in particular areas or for particular purposes.

When testing sand and gravel, standard test procedures must be followed. Duplicate tests on

* Chairman, Dunn Geoscience Corp., Latham, NY.

TABLE 1—Construction Sand and Gravel Sold or Used in the United States by Major Use

	1977			1978			1979			1980		
	Quantity, 1000 st	Value, $1000	Value per ton, $	Quantity 1000 st	Value, $1000	Value per ton, $	Quantity, 1000 st	Value, $1000	Value per ton, $	Quantity, 1000 st	Value, $1000	Value per ton, $
Concrete aggregate (including concrete sand)	347,447	811,866	2.34	369,500	898,600	2.43	357,100	923,300	2.59	287,600	847,100	2.95
Plaster and gunite sands	—	—	—	9,068	23,760	2.62	10,950	30,400	2.78	8,408	28,140	3.35
Concrete products (blocks, bricks, pipe, decorative, etc.)	53,042	129,260	2.43	41,950	104,000	2.48	32,780	86,900	2.65	30,970	87,130	2.81
Asphaltic concrete aggregates and other bituminous mixtures	132,237	284,204	2.15	141,700	323,800	2.29	142,000	343,000	2.42	111,800	321,800	2.88
Roadbases and coverings	188,843	343,666	1.82	212,500	412,900	1.94	222,400	458,800	2.06	181,100	436,100	2.41
Fill	158,691	210,805	1.33	164,100	232,800	1.42	155,500	240,500	1.55	122,000	217,100	1.78
Snow and ice control	—	—	—	6,974	13,990	2.01	8,207	16,670	2.03	6,621	14,500	2.19
Railroad ballast	1,203	2,751	2.29	1,478	3,460	2.34	1,190	2,849	2.39	946	2,499	2.64
Other uses	16,430	34,804	2.12	16,000	39,350	2.46	15,430	41,500	2.69	14,390	43,150	3.00
Total	897,900	1,817,000	2.02	962,300	2,053,000	2.13	945,500	2,144,000	2.27	764,000	1,998,000	2.62

Metric equivalent: 1 st × 0.907 184 7 = t.
Source: Modified from Tepordei (1980). Data may not add to totals shown because of independent rounding.

split samples, made by the specifying agency where possible, are strongly recommended.

The most generally used national guidelines for specifications and test procedures for mineral aggregates are outlined by the American Society for Testing and Materials (ASTM) and the American Association of State Highway and Transportation Officials (AASHTO). Most specifics of tests and procedures will not be discussed in this chapter nor will specific references to ASTM, AASHTO, or other agency procedures or specifications be made. Although specifying agencies usually follow guidelines set by ASTM and AASHTO, variations in such things as handling and calculating may cause significant variation in results.

Tests

Magnesium and Sodium Sulfate Soundness Tests: Sand and gravel are normally evaluated for their potential frost resistance on the basis of sulfate soundness tests, with some specifying agencies stressing the magnesium sulfate soundness test and others the sodium sulfate soundness test. The tests start with drying, screening, and weighing the aggregate, and then alternately soaking the aggregate in saturated sulfate brine and drying under carefully controlled conditions for a specified number of cycles (usually 4 to 10). The aggregate is then washed free of the salt, dried, back-screened, and weighed. The measure of the deterioration of the aggregate in the test (the test loss) is generally taken as the percent of aggregate which passes particular screens after the test related to that percent which passed before the test. The permissible loss is variable and depends largely on what test is used and on the anticipated use for the aggregate.

The sulfate soundness tests were devised in the early 1800s and have been frequently criticized as being old-fashioned and as not truly indicating the soundness of aggregate in actual use. For a comprehensive analysis see Garrity and Kriege (1935) and for a short summary see Bloem (1966). In many cases sulfate soundness tests are the only tests for soundness which are designated. Although there are some questions regarding their effectiveness, sulfate soundness tests continue to be the most common method specified for determining frost resistance of sand and gravel.

Durability Test: The durability test procedure consists of placing the aggregate in a wash pot, covering with water, and agitating for several minutes. The height of the layer of —200 mesh sediment in a sand equivalent graduate is the "durability factor." The test is a measure of abrasion resistance; in addition, the wetting of the aggregate deteriorates some shales which might not break down in sulfate soundness tests.

Freeze-Thaw Tests: Freezing and thawing tests may be made on fabricated concrete bars as a means of testing frost resistance of the total concrete system which includes aggregate. The amount of deterioration of the concrete may be taken as a function of change in the modulus of elasticity of the concrete, or expansive change in the length of the bars. This type of test has the theoretical advantage of being a more direct measure of frost resistance than the sulfate soundness tests.

Some porous aggregates may pass all soundness and freeze-thaw tests when unconfined and still cause destruction of concrete. The principles involved in this have been presented by many authors, but principally by workers in the Portland Cement Association (Powers, 1955; Verbeck and Landgren, 1960).

A particularly unfavorable situation is that of having a water-saturated, highly porous aggregate (over 3 to 4% water) in a dense, relatively impermeable portland cement paste. When water freezes directionally, hydrostatic pressures in the unfrozen water ahead of the ice at the interface of the aggregate and paste may often become excessive and create deleterious forces. The intentional entrainment of air bubbles in portland cement has largely minimized the effect of such forces by giving water many small voids into which to migrate, thus dissipating the hydraulic forces.

Two additional factors are critical in relating freeze-thaw test results to the actual performance of aggregate in use. The frequent use of NaCl or other salts on highways in winter radically increases deterioration (paradoxically, while decreasing the amount of freezable water); and much water—up to 100%—in many types of aggregate is not freezable (Dunn and Hudec, 1965, 1966, 1972; Larson and Cady, 1969). The water apparently is not freezable because the electrostatic forces bonding the water to mineral surfaces are stronger than the bond in ice. For these reasons, the relevance of freeze-thaw tests may be questioned.

New York State requires a freeze-thaw test on unconfined coarse aggregate in a 10% NaCl solution to indicate better what an aggregate will do under winter salting conditions. The test results on carbonate aggregate correlate to a remarkable degree with deterioration under natural weathering conditions in outcrops, but do not necessarily correlate with the natural weathering of the aggregate in concrete.

Wet-Dry Tests (Sorption Sensitivity): Shales, siltstones, many argillaceous limestones and dolomites, and certain other rocks basically deteriorate by wetting and drying or by humidity changes. In fact, a case can be made that most sedimentary rocks which have been observed to cause deterioration of concrete and deteriorate rapidly in nature are actually wet-dry (or sorption) sensitive, not freeze-thaw sensitive (Dunn and Hudec, 1972). Other authors have observed wet-dry deterioration of various rocks in various parts of the world (Garrity and Kriege, 1935; Rhoades and Mielenz, 1948). Many such rocks contain mostly water which is so tightly bound to internal mineral surfaces that it cannot freeze (Dunn and Hudec, 1965). Sodium chloride and other electrolytes in water in rock enhance wet-dry deterioration while reducing the amount of ice which forms. Salts may also greatly increase the quantity of water which may be absorbed by rock materials (Hudec and Rigber, 1976).

No standard tests which will detect sorption-sensitive rocks have been nationally promulgated, and sorption sensitivity is not uniformly detected by other soundness tests. However, petrographic analyses may be use to detect possibly deleterious argillaceous rocks. Dunn and Hudec (1972) suggest an absorption-adsorption-void space, three-component plot for detection of potential sorption sensitivity in carbonate rocks and suggest that the diagram may be valuable for some other argillaceous rock types, such as shales or siltstones.

Gravels occurring in unglaciated areas do not generally contain sorption-sensitive materials, but in glaciated northern areas and in areas where alpine glaciated materials are in outwash deposits, the problem may be common. Generally, sorption sensitivity is likely to be encountered in crushed stone.

Chemical Reactivity Tests: Aggregates are often chemically reactive under weathering conditions either alone or in combination with other components in the concrete system. The reactions are summarized as follows:

Alkali-Silica Reactions—Some mineral aggregates containing amorphous to finely divided silica have been found to react chemically with high-alkali portland cement, the so-called alkali-silica reaction first observed by Stanton (1940). Opal, chert, chalcedony, volcanic glasses, tridymite, cristobolite, some siltstones, and other siliceous materials have been found to be reactive.

Potentially reactive substances can be detected by a quick chemical test in which chemical resistivity is taken as a function of the reduction in alkalinity and the amount of silica dissolved from the aggregate in one-normal NaOH.

Bars fabricated of aggregate and portland cements of known alkali content may also be used to test alkali reactivity. The bars are stored in containers at specific temperatures over water to maintain 100% humidity. Alkali reactivity is taken as a function of the expansion of the mortar bars after specific time intervals, usually up to 12 months.

Alkali-silica reactivity can be controlled by using portland cement with less than 0.6% total alkali, by increasing or decreasing the amount of reactive aggregate away from the proportion of maximum reactivity (the pessimum proportion), or spreading the reaction by neutralizing the alkali quickly by adding such reactive substances as diatomite or fly ash.

Alkali-reactive substances are particularly prevalent in the sands and gravels of the western United States, but potentially reactive materials probably occur in most areas.

Alkali-Dolomite Reactions—Alkali-rich portland cements have been found to react with some argillaceous calcareous dolomites (Swanson and Gillott, 1960). Reaction products are largely various hydrous alkali carbonates. Expansion of the aggregate in the concrete results, and the concrete deteriorates. No general test is accepted to measure potentially alkali-reactive carbonate rocks. However, expansion of cylinders of suspected aggregate in NaOH solution seems to indicate reactivity, and the petrographic identification of argillaceous calcareous dolomites are indicators of potential reactivity. In New York many of the alkali-reactive carbonate rocks have been found to weather quickly on exposure, to be sorption-sensitive, and to fail the New York State Department of Transportation freeze-thaw test.

Avoiding sorption-sensitive carbonate rocks or argillaceous carbonate rocks in general is probably the best way of minimizing such chemical reactions. In general, gravels and sands rich in carbonate materials are not common, and those which do exist naturally contain a minimum quantity of argillaceous carbonate rocks because of the sorption-sensitivity of such rocks. Alkali-carbonate reactions between the components of sand and gravel and portland cement may be presumed to be rare.

Other Chemical Reactions—Oxidation of sulfides and the dissolving of certain soluble minerals may also be a problem with some aggregate, but generally sands and gravels contain little of such components.

Abrasion Tests: Gravel (and crushed stone) used in concretes usually must meet requirements for resistance to abrasion. The most general test is the Los Angeles abrasion test. Screened and weighed coarse aggregate is typically tumbled with steel balls in a barrel which is 28 ± 0.2 in. ID and 20 ± 0.2 in. inside length (ASTM C 131–76). The barrel contains flanges inside which, on rotation, raise the charge to the top of the barrel so that the charge drops to the bottom. After a specified number of rotations, usually 500, the aggregate is back-screened and the abrasion resistance is taken as a function of the percent of material which passes a given screen size.

The Deval abrasion test is similar to the Los Angeles abrasion test in screening and back-screening measurements, but the aggregate (with or without additional steel ball charge) is slid back and forth in small cylinders whose axes are set at a 30° angle to a rotating shaft. The Deval test is rarely used, primarily because the percent losses are so low that a few fortuitous weak or fractured particles can give unrealistic results.

Both the Los Angeles and Deval abrasion tests are really impact tests and more a measure of brittleness than abrasion resistance. Coarse grained rocks (many metamorphics and granites) tend to do poorly in the Los Angeles abrasion test, in particular because of their brittleness, yet they may be satisfactory in use.

Skid-Resistance Tests: In recent years the resistance of surface courses of concretes to the abrasion by vehicle tires has become critical because certain aggregates have been found to polish and cause dangerously slippery surfaces. Basically, the polishing of the surface of portland cement concrete is related to the hardness of the fine aggregate, while that of asphaltic concrete is related to the tendency of the coarse aggregate to polish.

Several methods of testing the amount of actual wear of surfaces in the laboratory and in the field have been devised (Gray and Renninger, 1966).

Wear of surfaces of concretes made with carbonate rocks has been found to vary directly with the quantity and size of the components which are insoluble in acid. Some specifying agencies require a minimum quantity of insoluble material of a grain size greater than 200 mesh.

No generally accepted tests are in use to measure skid resistance, although many specifying agencies employ locally standardized test methods. Generally, natural sands and gravels meet skid resistance requirements.

Miscellaneous Requirements—Nearly all agencies require sand and gravel used in concretes to be clean and to contain a minimum quantity of deleterious substances. Included in the category of deleterious substances are such things as excessive silt or clay, shale, clay balls, coal, lightweight material, and wood.

Aggregates must always have a specified range of particle sizes or gradations depending upon the use and, to some extent, the preference of the specifying agency. Some specifying agencies require that for some uses gravel must be broken so that for each particle there are a minimum number of fractured surfaces.

Substitutes

The average at-plant price of between $2 and $3 a ton sharply restricts possible alternate materials largely because of what might be called an energy barrier; any synthetic products require energy to grind or melt, and production costs rise rapidly with the addition of such energy. However, a review of the high delivered prices in some markets indicates that under certain local conditions high transportation energy may balance against high production energy. In addition, certain materials with special characteristics may replace some sand and gravel for special uses.

The major substitutes for gravel are crushed stone and slag. Generally, the prices for gravel and for stone and slag tend to be comparable in most markets, and for many uses gravel or crushed gravel may be used interchangeably with stone or slag, provided that the materials are of comparable quality in specified tests. However, for fill or for subbase items, poorer quality gravel is often used in preference to either concrete-grade gravel or crushed stone because the poorer quality gravel is less expensive.

Screenings from crushed stone or fine aggregate manufactured by crushing stone may replace natural sand or fine aggregate to varying degrees in market areas where natural material is in short supply. Crushed fine aggregate has the disadvantage of making stiff portland cement concrete which, although satisfactorily strong, is hard to finish.

Another substitute is oolitic aragonite, a natural calcium carbonate precipitate which has been dredged in the Bahamas and sold in some East Coast and Caribbean markets.

Some special manufactured products may replace sand and gravel to varying degrees in certain uses. Coarse lightweight aggregate is used in some concretes, and lightweight aggregate screenings may replace normal fine aggregate where lightweight concrete is needed. Waste slags from burned rubbish and waste glass have been suggested as possible replacements for sand and gravel, but their price is prohibitively high in most areas.

Economic Factors

Price Trends

The value for sand and gravel is shown in Table 1. Although the price of sand and gravel per ton has risen steadily since 1921 (Table 2), the average price is still less than $3. According to Evans (1978), considering inflation and taking 1976 dollars as the reference, the per-ton price of sand and gravel in 1976 was $2 compared to an adjusted price of $1.95 in 1957. Production efficiency in which each worker produces increasing tons per man-hour is the most important factor in keeping the price low. However, although production statistics on tons of sand and gravel produced per man-hour are not available, the National Crushed Stone Assn. (1981) showed that production of crushed stone per man-hour rose from 2.5 in 1949 and peaked at 8.8 in 1977, then dropped to 8.4 in 1978 and to 8.2 in 1979. The downward trend was the first since 1949. Whether sand and gravel production shows a similar trend is not known, but a downward trend is certainly likely because sand and gravel production trends tend to follow those of crushed stone.

Sand and gravel are high-bulk, low-value commodities and, as such, transportation/energy costs mean that prices actually paid at distribution points, ready-mix plants, or points of use are considerably higher than the at-plant prices. This is a particularly critical factor in metropolitan areas. Fig. 1 shows the average per-ton delivered price for sand, gravel, and crushed stone, and the prices of ready-mix concrete for 19 major cities in the United States. The figures are for the month of July for the years 1962 through 1980 and are compiled by *Engineering News-Record* (ENR) (McGraw-Hill Publishing Co.). The average prices of sand, gravel, and crushed stone as

TABLE 2—Annual Production and Value of Sand and Gravel for All Uses Since 1921

	Millions per year of	
Years	Tons	$
1921–1925	130*	83*
1926–1930	203	117
1931–1935	125	69
1936–1940	200	99
1941–1945	238	146
1946–1950	310	237
1951–1955	485	420
1956–1960	667	659
1961	752	751
1962	777	795
1963	822	847
1964	868	893
1965	908	957
1966	934	985
1967	906	981
1968	917	1,020
1969	937	1,070
1970	944	1,116
1971	920	1,149
1972	914	1,201
1973	984	1,359
1974	905	1,417
1975	789	1,340
1976	885	1,774
1977	929	2,028
1978	996	2,302
1979	979	2,427
1980	794	2,302
1981	760	2,356

* Average per year for the five-year intervals from 1921–1960, f.o.b. plant.

summarized by the US Bureau of Mines are also plotted for reference purposes (Fig. 2).

Local average delivered prices for sand and gravel vary widely depending upon the local supply and demand relationships. For instance, *ENR* market quotations show that gravel was $13 a ton in Kansas City and $11.60 to $11.30 a ton in Pittsburgh in April 1981; at the same time, a ton of gravel cost $3.67 in Los Angeles. The same index quotes the price of fine aggregate in April 1981 at $18 per cubic yard for New York City but only $2.30 per ton in Kansas City.

An excellent example of the local interplay of supply, demand, and price is seen in the fine aggregate market in the Greater New York metropolitan area. The price per cubic yard (delivered at dockside) held constant from January 1965 to January 1969 at $2.65, then jumped to $4.10 in May 1970. The jump coincided with the inability of major producers to replenish their reserves, causing 10 compa-

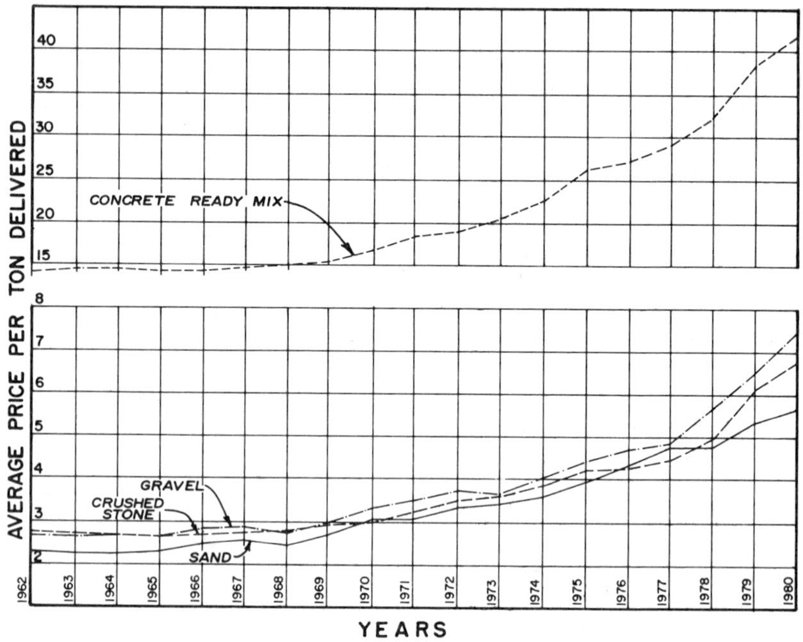

FIG. 1—*Average delivered price of sand, gravel, crushed stone, and concrete ready mix for 19 major US cities (July figures).*

nies to stop producing sand entirely (Hudec, 1970).

As noted, fine aggregate for New York City was $18 per cubic yard in April 1981, and has been even higher, which seems paradoxical since Long Island is one of the largest sand and gravel deposits in the world and much of eastern New Jersey is a sand and gravel deposit. Such shortages amid plenty are created largely by social pressures of zoning and construction.

The actual cost of getting sand and gravel to a job is still higher than the foregoing figures, because quotations are for delivery to ready-mix plants or other distribution points (f.o.b. city).

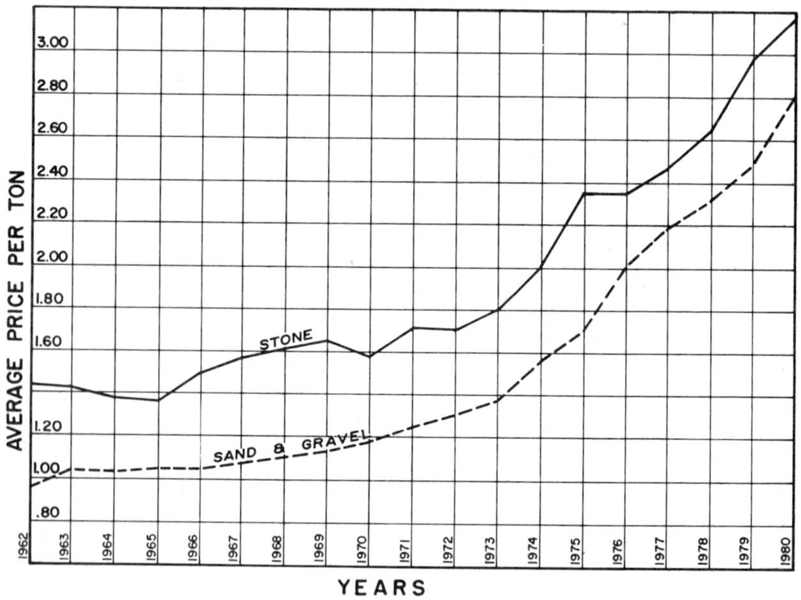

FIG. 2—*Variations in annual average price of sand, gravel, and crushed stone in the US.*

The significance of these figures is multiple:

(1) Production of sand and gravel has generally become more and more efficient because at-plant or f.o.b. prices have not risen comparably with reduction in dollar value.

(2) The price paid for sand and gravel by the consumer varies considerably depending upon local supply, demand, and transportation costs.

(3) Prices of delivered sand and gravel have risen sharply in some areas in recent years because of increased costs of transportation and amelioration of social conflicts.

(4) The distance of sand and gravel producers from competitors and markets is obviously a critical factor in the evaluation of a deposit.

Transportation

Methods and costs of moving sand and gravel to markets are particularly critical in the aggregate industry. In fact, it has been suggested that a way to compare deposits and give them value is to use transportation cost differentials and a discounted cash flow method for evaluation (Dunn, et al., 1970).

The methods used to haul mineral aggregate, in ascending order of cost, are: Great Lakes and oceangoing ships, barges, railroad cars, and trucks. Prices are usually figured in terms of cost per ton-mile, ranging from less than 2¢ per ton-mile for some of the very large bulk carriers to over 20¢ per ton-mile for trucks in metropolitan areas. In fact, the cost of hauling per ton-mile is highly variable for all modes of transportation, and there is considerable cost overlap between the various methods. Some costs are negotiated and vary considerably depending upon many variables of supply, demand, haul distances, quantities involved, size of carriers, and other competitive factors.

Most sand and gravel is moved by truck, at least for the final delivery. Truck haulage costs are commonly figured at so much per ton-mile for the first mile or other unit distance and a lesser amount for each additional mile. For example, the costs have been taken as 35¢ for the first ton-mile and 10¢ for each additional ton-mile, so that the cost of a 7.7 km (11-mile) haul would be $1.35 per ton. Using large haul units in nonurban driving conditions may generally result in lower than average costs, while trucking costs in metropolitan areas are likely to be much higher and often related to haul time, because travel may be very slow. In the latter case, a plant will usually have haul-cost zones radiating outward from the plant, the exact boundary for such zones being related to experience about the time it takes a truck to deliver its load and return. At the interface of competitive areas with other producers or when bidding major construction jobs, prices may be more variable.

Marketing

Production for major metropolitan areas is usually by fixed plant; but where market conditions shift rapidly, because of sudden changes in the amount of construction, production may be from portable plants.

In rural areas, semimobile plants to completely mobile plants equipped with rubber tires may move in, mine a deposit for a particular job such as a dam or highway, and move out when the work is completed. In such cases the producer of sand and gravel may be the contractor who was low bidder, or a subcontractor for the job. This situation is common in some parts of the Midwest and West where population centers are relatively far apart.

Metropolitan areas, on the other hand, are fixed market centers and consequently are served by relatively fixed sand and gravel or stone sources. In many market areas, asphalt plants and concrete ready-mix plants are located on the property of, or very close to, the producer of aggregate. It follows that producers of sand and gravel are often contractors and are normally in the concrete business. Thus much marketing is internal.

In major metropolitan areas, sand and gravel may be delivered by unitized train, barge, or other bulk carriers to distribution centers where portland cement ready-mix and asphaltic concrete plants are located. Sand and gravel normally leaves such distribution centers in a mixed form ready for laying or pouring within the market area.

In recent years, a strong trend has been toward consolidation of companies, and it has become common for sand and gravel companies to be part of a company with interests in crushed stone, cement, ready-mix plants, contracting, and other peripheral fields. Thus the tendency has been toward more internal marketing.

Governmental Considerations

Taxes and Subsidies

Sand and gravel producers selling to the construction industry have the minimum federal depletion allowance of 5% based on gross income; they have no subsidies. There is also no import duty on sand and gravel for construction purposes. (Actually, because of the extreme transportation cost sensitivity of sand and gravel, and because of its common occurrence, very little is either exported or imported.)

Other taxes are typical of those for any business. However, some political entities have imposed severance taxes on aggregate produced within their jurisdiction. Such taxes commonly range from a fraction of a cent to ten cents per ton produced. Whether this is a trend cannot be determined at present.

Zoning and Land Use

Sand and gravel operations often may be in conflict with their environment. The conflict often is manifested as zoning regulations requiring permits to operate or, in some cases, prohibiting production of sand and gravel. To a greater extent, permits are required before an operation may start. Often such permits are difficult to obtain, particularly along most of the east and west coasts. In most cases permits, when obtained, are contingent upon the producer meeting certain requirements for rehabilitation, work hours, trucking routes, and so on. Bonding of producers to insure that rehabilitation will be done satisfactorily is very common.

In some areas, such as California and parts of Virginia, mineral resource zoning, which sets aside land exclusively for sand and gravel production until the mineral resource has been depleted, is in effect.

Several court decisions against producers have served to reduce their reserves significantly or to completely shut down operations. In most cases producers started operating in rural areas only to find city populations gradually encompassing them. Their new neighbors then complained about the operations, with the complaints often leading to litigation. The result of such pressures on the aggregate industry, particularly the sand and gravel industry, has been to reduce dramatically the reserves available in many metropolitan areas. Shortages and subsequent sharp price changes have resulted. Some of the shortages, strikingly, occur in areas of abundant sand and gravel resources.

Environmental Regulations

Coinciding with the increased sociologic pressures of zoning and litigation, sand and gravel producers are required to meet more and more stringent environmental requirements. The requirements are moving toward tighter control of noise, vibrations, particulate matter (silica and dust in general), water pollution, and "visual pollution." Local rules and restrictions are highly variable, too numerous to review, but usually they have an exclusionary effect on the producer.

The following federal environmental laws regulate the sand and gravel industry in varying degrees (Dunn, et al., 1980):

- The National Environmental Policy Act
- Clean Air Act of 1970
- Energy Supply and Environmental Coordination Act of 1974
- Resource Conservation and Recovery Act of 1976
- Federal Water Pollution Control Act Amendments of 1972
- The Rivers and Harbors Act of 1899 (The Refuse Act)
- Water Resources Research Act of 1964
- Water Resources Planning Act
- Fish and Wildlife Coordination Act
- Ports and Waterways Safety Act of 1972
- Water Pollution Acts Amendments 1972
- Safe Drinking Water Act
- Deepwater Port Act of 1974
- Noise Control Act of 1974
- Coastal Zone Management Act of 1972
- Marine Mammal Protection Act
- Endangered Species Act of 1973
- Executive Order 11870 on Animal Damage Control
- Wild and Scenic River Act
- Soil and Water Resources Conservation Act of 1977

Air: At the federal level, regulation of particulate matter is under the control of the Environmental Protection Agency (EPA) under authorization of the Clean Air Act of 1970. Visible emissions, diesel engine exhausts, and fugitive dust are all under control of the EPA. However, at this time precise standards have not been set for ambient air quality. Implementation and enforcement are to be by the states with the EPA stepping in if enforcement is not satisfactory.

Water: Regulation of effluent water from sand and gravel operations is covered by the Federal Water Pollution Control Act Amendments of 1972. The regulatory system is called the National Pollutant Discharge Elimination System. The system is administered federally by EPA. Administration may be by state agencies when a state has a federally approved program.

Any company with effluent which goes directly or indirectly into navigable waters needs a discharge permit. As part of the permit, the operator needs to locate and show on a map all points of water discharge, the quantity discharged, and must analyze each discharge for turbulence, organic materials, pH, and various dissolved salts. The limits for the various materials checked have not been firmly established to date.

Noise: Local and state standards limiting hours of operation and noise level may apply to the industry. In addition, EPA has an Office of Noise Abatement which was authorized to establish standards under Title IV of the Clean Air Act Amendments of 1970. Noise affecting people not on a plant's work force currently is not federally regulated.

Rehabilitation: Federal laws controlling rehabilitation of extraction industry sites are not yet in effect, but several are before Congress. Boyd, et al. and Dunn, et al. (1980) in the National Academy of Sciences' study on the applicability to the rest of the mining industry of federal strip mining reclamation laws, concluded that the coal mining regulations were not applicable to the rest of the mineral industry. Furthermore, the committee concluded that new federal laws to control reclamation of mined land would be of minimal value. This was particularly true of sand and gravel and stone quarries because "about 40 states now have mined land reclamation laws. . . . thirty-one states have such laws which apply to the mineral aggregate industry" (Dunn, et al., 1980). The study also showed that over three-quarters of the construction aggregate industry was already regulated by state reclamation laws. In addition, because of the usual proximity of construction mineral operations to metropolitan areas and the pressure to use all land in such areas, even with no reclamation laws, some later use of mined land is virtually assured (Dunn, 1980). Largely based on this evidence, the President's Council on Environmental Quality has concluded that a new federal mined land reclamation law would not be appropriate.

Safety Regulations

Safety is regulated federally by the Mine Safety and Health Administration (MSHA), Department of Labor, mandated under the 1977 Mine Safety and Health Act. Inspections under MSHA are conducted by law twice annually and any safety violations observed must be cited. Fines are mandatory. As pointed out by NCSA (1981), an inspector cannot function in the capacity of safety advisor or consultant to operators. H.R. 1603 in the House of Representatives and S. 351 in the Senate, if passed, will provide some relief from MSHA by putting all mine safety under the Occupational Safety and Health Administration (OSHA) in the Department of Labor.

Future Prospects

Demand

The annual per capita consumption of sand and gravel in the United States increased from about 0.68 t (¾ st) in 1920 to about 4.5 t (5 st) in 1970, and is currently about 3.6 t (4 st). The total per capital annual consumption of mineral aggregates is grossly related to the affluence and rate of growth of the population. If increasing affluence and growth is the trend in the United States, the per capita consumption of mineral aggregates would be expected to increase and, therefore, the anticipated rising population curves could coincide with a more sharply rising consumption curve for mineral aggregates.

Cooper (1970), assuming a statistically probable long-range annual growth rate of between 3.9 and 4.7% for new construction, estimated the total consumption of sand and gravel by the year 2000 to be between 2.86 and 3.6 Gt/a (3.153 billion and 3.990 billion stpy). Using Cooper's figures, if US population reaches 300 million in the year 2000, the average annual per capita consumption will be between 9.1 and 11.8 t/a (10 and 13 stpy). However, where a population is relatively stable and has been affluent for a relatively long period, demand per capital tends to decrease. Thus per capita consumption in the northeastern US tends to be lower than the national average.

Although forecasting the average per capita demand for sand and gravel in the US is not totally accurate, current growth and population trends suggest average demand will not be as high as the previously estimated 9.1 to 11.8 t (10 to 13 st) per year per person. In fact, Pajalich (1975) estimated that demand for sand and gravel by the year 2000 will be 1.90 Gt (2.090 billion st), but just three years later Evans (1978) estimated the demand by the year 2000 to be 1.0 Gt (1.150 billion st), or about one third of Cooper's estimate.

Clearly, estimating future demand is subject to a number of influences, including current trends and the degree of optimism of the forecaster. One factor not previously considered is that as pressure on the American forests increases due to the use of wood for fuel, stone, concrete, and brick increasingly will be used in new construction (Dunn, et al., 1980). The increased use of stone products historically has been the trend in nations where the wood resource has been depleted, and the trend toward greater use of stone products for building construction is inevitable, given a steadily increasing population. However, the rate of increased new construction is not known because it is likely to be related to energy prices. The cost of energy is, in turn, complexly related to domestic energy production, conservation, and international pricing.

Supply Considerations

From a strictly geological view, the quantities of sand and gravel (as well as crushed stone) are essentially inexhaustable. Vast areas of the coastal plain from New Jersey through Texas are underlain by sand and gravel deposits. The deposits of glacially formed sand and gravel have just been tapped. And across the country fluvial terraces, dredgeable sand and gravel in waterways, and intermontane outwash and alluvial fan deposits are huge. Enormous sand and gravel deposits also occur on the continental shelves.

Cooper (1970) estimated the cumulative demand for sand and gravel to be very high—51.5 to 59.0 Gt (56.8 to 65.0 billion st) to the year 2000—with about 14.5 Gt (16 billion st) of sand and gravel currently under control of operators. Despite their geological abundance, filling the needs for sand and gravel to the year 2000 may be exceptionally difficult. Exploitable resources of natural fine and coarse aggregate should become increasingly less common because of sociological pressures.

Most of the demand for sand and gravel is in the urban areas and, as building accelerates and urban areas coalesce, natural deposits are directly covered over, or adverse zoning restrictions or other negative environmental factors cause vast reserves to become inaccessible. Much of New Jersey and all of Long Island are underlain by sand and gravel. Yet opening up a new deposit of sand and gravel is almost impossible in both areas. Ten operators who supplied sand and gravel to the New York metropolitan market have ceased production within the past decade, largely because of their inability to replenish reserves. The continental shelf deposits also may be difficult to operate for various reasons even if there is enabling federal legislation. The difficulty of obtaining suitable shore-based plant facilities to process the aggregate and the possibility that the silt produced may damage the fishing industry or beaches may be difficult problems to solve.

The great metropolitan area centering around New York City is experiencing a shortage of natural sand. Paradoxically, as noted previously, this shortage exists in an area of abundant natural fine aggregate, but the resource has been culturally nullified. Natural sand will be replaced to an ever greater degree by manufactured fine aggregate, and gravel will be replaced by crushed stone in the New York City market. Several other metropolitan areas such as Pittsburgh and Philadelphia have depleted or lost availability of nearby natural sand and prices are rising accordingly as haul distances increase or as natural fine aggregate is replaced by manufactured fine aggregate.

The change primarily to crushed fine aggregate in some areas will put stress on reserves of crushed stone. Yet in many areas new sites for the production of crushed stone will also be exceptionally difficult to locate for reasons discussed under the section, Governmental Considerations. Large areas of the East Coast are already in this category and new reserves of high quality rock near the sea are very difficult to find from Portland, ME to Brownsville, TX. Cultural nullification is one major factor, and south from approximately Norfolk, VA, stone is rare in coastal areas.

Other areas of the United States are experiencing similar reserve depletion. For instance, Sheridan (1967) points out that from 1935 to 1967 the sand and gravel reserves readily available to Denver, CO dropped from 816.5 to 90.7 Gt (900 million to 100 million st), with most of the loss caused by cultural nullification.

The trend toward mining out deposits and cultural nullification of sand and gravel near metropolitan areas will accelerate, and prices of sand and gravel at the point of use will be driven sharply upward as haul distances to more remote supplies increase. Resource planning on a regional level could alleviate the problem, but even if changes occur now, it could be too late to use much of our natural sand and gravel resources.

Plant Size Trends

For a time there was a trend in the aggregate industry toward fewer plants with higher production per plant. However, this may no longer hold true because the economic problems in the construction industry in the early 1980s are causing severe problems in the sand and gravel industry. The total number of sand and gravel operations declined from 7,447 in 1974 to 6,000 in 1981. Plants producing over 453.6 kt/a (500,000 stpy) supplied 41.1% of production in 1974 and 35.6% in 1980, while plants producing less than 90.7 kt/a (100,000 stpy) supplied 16.6% of production in 1973 and 18.1% in 1980. In that same period, the number of plants producing more than a million tons dropped from 100 to 66, reflecting the poor economy of the early 1980s.

Technology

As is the case for the whole mining industry, changes in technology are inevitable and occur in virtually every item of equipment. Excavating equipment has been changing toward larger units, and the great mobility of rubber-tired loaders has caused them to become more and more appealing. Pit trucks and trucks for hauling over highways have been increasing in size, and this trend is expected to continue. Plant automation will probably become more widespread, especially near metropolitan areas.

Beneficiation of sand and gravel materials undoubtedly will become more prevalent as deposits of higher quality materials become depleted. Methods of upgrading mineral materials may be based on differences in color, specific gravity, magnetic susceptibility, electrical conductivity, elasticity, particle shape, friability, and dielectric surface constant. Current beneficiation techniques for sand and gravel are not highly sophisticated and take such forms as heavy media separations, log washing, and gravity separation by jigging. Other methods, such as methods of separating flat particles from glacial deposits or the elastic fractionation of gravel, may become more common.

Chemical upgrading of aggregate may also become important as supplies are depleted locally.

Specification changes are expected to become more frequent and generally to move toward relaxation of requirements for the specific purpose of stretching depleting reserves. Specifications should also move toward being more positively related to potential service requirements.

Conservation of Sand and Gravel Resources

Mineral resource conservation is defined here as those things which must be done to assure a flow of mineral commodities to the market centers in reasonable quantities at reasonable prices, all done with reasonable environmental constraints.* Conservation, in general, would include those actions which lead to sustained yield of renewable resources, efficient use of nonrenewable resources (including recycling), and to the production of multiple values.

Increasing Resource Availability

The extent to which good conservation is practiced is most likely to be related to public understanding of the nature of the problem. To date, no long-range public policy about sand and gravel—or even for minerals in general—exists. Decisions about opening new aggregate deposits are made largely on an ad hoc basis and likely to be made in the emotional atmosphere of a public hearing. Persons who attend such hearings are largely objectors and live in the immediate vicinity of the proposed operations. The prospective sand and gravel operator in such a conflict is usually ill equipped to handle the emotions of the local populace. The net result of many such conflicts is that enormous sand and gravel resources have been rendered unavailable and haul distances and prices may increase spectacularly.

The general public is the largest loser in virtually all such conflicts in which the producer is the theoretical loser. If, for instance, the average producer of sand and gravel were five miles further from his market than he might otherwise be (because of being forced outward) the annual cost to the US would be about $450,000,000. Translated to fuel, if haul trucks average 5 mpg and carry 22.7 t (25 st) per load, the extra fuel would be 2,000,000 barrels annually. In addition, the number of trucks met on roads greatly increases for the general population and the road wear increases over a larger area. Therefore, the real conflict should not be between a local special-interest public and a producer, but between the local

* Reasonable price is the price an average user is willing to pay; reasonable quantity is the quantity needed to satisfy the major demands of the population; reasonable environmental constraints are those constraints needed so that the average person does not find the operation or its effects environmentally objectionable.

group and the general public. Yet the general public is rarely, if ever, represented in any mineral producer-public conflicts. The cost and the psychology are similar for conflicts involving any mineral or energy development.

One suggested answer to the dilemma has been to set aside resource zones to be reserved for production of sand and gravel. Colorado and Wisconsin have variations of mineral resource zones (Thomason, 1980) as have Fairfax County, Virginia, and the San Fernando Valley of California, among others. In such mineral resource zoning situations, it is not always clear how a new zone is to be set up if the originally zoned resources are depleted. And such zoning implies knowledge of the location of all or most economical deposits in an area and of future possible beneficiation methods, economics, and specifications.

An alternate has been suggested by Dunn and Moffett (1975). The concept is summarized in Dunn, et al., (1980) as follows:

> If it is in the interest of the general public to produce construction minerals efficiently, but is not felt by the local public to be in their interest, can the general public compensate the local public for any inconveniences or costs incurred?
>
> It would seem that a compensation system could be cost-justified, since (1) payments remain within the United States; (2) construction minerals may be produced closer to markets, resulting in a cost saving; and (3) minerals may be produced in the United States that might otherwise be bought from foreign sources.
>
> To the best knowledge of the panel, there is no precedence for such a system, and it may be legally or politically difficult to establish. Further, it is not known how large such compensation should be nor whether it might actually have the desired effect of reducing frictions and facilitating mineral production.
>
> The method of community compensation has the advantage of being relatively insensitive to technologic change and geologic ignorance, but is sensitive to community economic level, i.e., the less wealthy communities will find the compensation most alluring. The method has the further virtue that it will work for any area and for any mineral.

Multiple Land Use

In the best conservation practice, the act of using one earth resource should develop another, i.e., the multiple land-use concept in which concurrent and sequential multiple uses are planned for the land. The creation of planned lakes from excavation of sand and gravel, so prevalent in the Midwest, is a good example. Commonly, where conditions are suitable, excavations may be used for the systematic disposal of waste material and, once filled, the land used for still other purposes. In the Los Angeles basin, excavations have been filled with garbage, covered with soil, drilled to tap the gas (largely methane and carbon dioxide) for domestic or industrial use, with the surface used for a park or other purposes.

Mineral Resource Zoning

Setting land aside in major population areas is a way to assure sand and gravel availability. Cities usually must rely upon their surrounding rural areas, for such things as power, raw materials, water, food, etc.; in short, those things needed for survival. Currently, however, cities have little or no control over the political entities which surround them. Thus towns and counties, through zoning regulations, have the power to cut the supplies of inexpensive mineral aggregate materials to the cities. The resultant costs can be very high, as in the case of New York City where the cost of sand alone jumped about $5 million in a single year (Dunn, et al., 1971). For the greater New York metropolitan area, including coastal New Jersey, Westchester County, Long Island, and Connecticut, the annual cost of "nonplanning" has been estimated at $30 million (Bronitsky, 1973). Hewitt (1962) estimated that for the Toronto area the annual cost increase of forcing producers ten miles further from the market center would be $6.5 million.

Managing sand and gravel resources on a regional basis through planning departments has been tried. The mechanism is sand and gravel zoning. Such zoning for sand and gravel resources starts with a precise definition of the conditions. Realistically available reserves and the possibility of obtaining new reserves of comparable material must be determined. The details of management can be computerized (Wallace, et al., 1969; Wallace, et al., 1969a; and Dunn, et al., 1971) but the political implementation is exceptionally difficult. The Federal Enabling Act for Land Use Planning Studies is a step toward the sort of regional work that may be done. In recent years, however, the possibility of planners moving in the direction of enlightened mineral resource management appears to be remote.

Specifications

Specifications have a great influence on mineral reserves. Recent relaxation and alterations in specifications for sand and gravel in New York State have had the effect of multiplying reserves greatly. The major changes were: (1) raising the limits of the magnesium sulfate soundness test for fine and coarse aggregate, allowing many deposits to become suddenly acceptable; (2) changing the specified gradation of fine aggregate for portland cement concrete to the finer ASTM recommended gradation, allowing reserves short of the +16-mesh gradation (which had been pushed aside for years) to be acceptable; and (3) sampling of the finished gravel product rather than in the bank as formerly, thus encouraging beneficiation. These changes and many others were made only after research indicated that they would not cause inferior concrete products.

Much more research should be done by specifying or other agencies to determine whether aggregates are being overspecified. Overly stringent specifications create the tendency toward "high-grading" mineral aggregate resources. Sound conservation practice demands that the maximum possible quantity of material compatible with the requirements of good engineering practice be acceptable.

In addition, a thorough review of all nationally recommended tests for mineral aggregate may be in order, for, as previously indicated, many tests may not serve their expressed purposes sufficiently.

Bibliography and References

Anon., 1981, *Stone News*, National Crushed Stone Assn., Mar., p. 6.

Bloem, D.L., 1966, "Soundness and Deleterious Substances," NSGA Circular No. 97, National Sand and Gravel Association, pp. 497–512.

Boyd, J., et al., 1979, "A Study of Mineral Mining from the Perspective of the Surface Mining Control and Reclamation Act of 1977," Committee on Surface Mining and Reclamation, National Academy of Science, 339 pp.

Bronitsky, L., 1973, *The Economics of Construction Mineral Aggregates With an Analysis of the Industry in the Greater New York Metropolitan Area*, Ph.D. Thesis, Rensselaer Polytechnic Institute, Troy, NY.

Cooper, J.D., 1970, "Sand and Gravel," *Mineral Facts and Problems*, Bulletin 650, US Bureau of Mines, pp. 1185–1199.

Dunn, J.R., and Hudec, P.P., 1965, *The Influence of Clays on Water and Ice in Rock Pores*, Physical Research Report RR 65–8, New York State Dept. of Public Works, 150 pp.

Dunn, J.R., and Hudec, P.P., 1966, "Water, Clay and Rock Soundness," *Ohio Journal of Science*, Vol. 66, No. 2, 153 pp.

Dunn, J.R., and Hudec, P.P., 1972, "Frost and Sorption Effects in Argillaceous Rocks," *Proceedings*, Highway Research Board, No. 393, pp. 65–78.

Dunn, J.R., and Moffett, H.L., 1975, "Mineral Exploration in the U.S.: Problems and an Answer," *Mining Congress Journal*, Vol. 61, No. 11, pp. 22–26.

Dunn, J.R., et al., 1970, "How Valuable Are Your Mineral Reserves?" *Rock Products,* Vol. 73, No. 9, pp. 83–87, 132–133.

Dunn, J.R., et al., 1971, "Mineral Resource Valuation in the Public Interest," *Trans.-SME/AIME*, Vol. 250, pp. 280–284.

Dunn, J.R., et al., 1980, "Sand and Gravel, Crushed Stone Quarries & Blasting, & Other Construction Minerals," Appendix II, Committee on Surface Mining and Reclamation, National Academy of Science, 222 pp.

Evans, J.R., 1978, "Sand and Gravel," *Mineral Commodity Profiles*, MCP–23, US Bureau of Mines, 22 pp.

Garrity, L.B., and Kriege, H.F., 1935, "Studies of Accelerated Soundness Tests," *Proceedings*, Highway Research Board, Vol. 15, 237 pp.

Gray, J.E., and Renninger, F.A., 1966, "The Skid Resistance Properties of Carbonate Aggregates," *Highway Research Record 120*, Highway Research Board, pp. 18–34.

Hansen, W.C., 1963, "Crystal Growth as a Source of Expansion in Portland Cement Concrete," *Proceedings,* American Society for Testing and Materials, Vol. 63, p. 932.

Hewitt, D.F., 1962, *Urban Expansion and the Mineral Industries in the Toronto-Hamilton Area,* Industrial Report No. 8, Ontario Dept. of Mines, 11 pp.

Hudec, P.P., 1970, "Long Range Planning for Aggregate Materials in Metropolitan New York Area," *Proceedings*, 5th Forum on Geology of Industrial Minerals, Bull. M–64, Harrisburg, PA, pp. 165–177.

Hudec, P.P., and Rigbey, S.F., 1976, "The Effect of Sodium Chloride on Water Sorption Characteristics of Rock Aggregate," *Bull. of Assoc. of Eng. Geols.*, Vol. XIII, No. 3, pp. 199–211.

Lang, F.C., 1931, "Deleterious Substances in Concrete Aggregates," *Bulletin*, Cir. No. 10, National Sand and Gravel Assn., 17 pp.

Larson, T.D., and Cady, P.D., 1969, *Identification of Frost-Susceptible Particles in Concrete Aggregates*, Report 66, Highway Research Board.

Lenhart, W.B., 1960, "Sand and Gravel," *Industrial Minerals and Rocks*, 3rd ed., J.L. Gillson, ed., AIME, New York, pp. 733–758.

Pajalich, W., 1972, "Sand and Gravel," *Minerals Yearbook 1972*, US Bureau of Mines, pp. 1103–1121.

Pajalich, W., 1973, "Sand and Gravel," *Minerals Yearbook, 1973*, US Bureau of Mines.

Pajalich, W., 1975, "Sand and Gravel," *Minerals Yearbook, 1975*, Bulletin 667, US Bureau of Mines.

Powers, T.C., 1949, "The Air Requirement of Frost-Resistant Concrete," *Proceedings*, Vol. 29, Highway Research Board, pp. 184–211.

Powers, T.C., 1954, "Void Spacing as a Basis for Producing Air-Entrained Concrete," *Proceedings*, Vol. 50, American Concrete Institute, pp. 741–760.

Powers, T.C., 1955, "Basic Considerations Pertaining to Freezing-and-Thawing Tests," *Proceedings*, Vol. 55, American Society for Testing and Materials, pp. 1132–1155.

Powers, T.C., and Helmuth, R.A., 1953, "Theory of Volume Changes in Hardened Portland Cement Paste During Freezing," *Proceedings*, Vol. 32, Highway Research Board, pp. 285–297.

Rhoades, R., and Mielenz, R.C., 1948, "Petrographic and Mineralogic Characteristics of Aggregates," Symposium on Mineral Aggregates, American Society for Testing & Materials, 20 pp.

Sheridan, M.J., 1967, "Urbanization and Its Impact on the Mineral Aggregates Industry in the Denver, Colorado, Area," Information Circular 8329, US Bureau of Mines, 51 pp.

Stanton, T.E., 1940, "Expansion of Concrete Through Reaction Between Cement and Aggregate," *Proceedings*, Vol. 66, American Society of Civil Engineers, pp. 1781–1811.

Swanson, E.G., and Gillott, J.E., 1960, *Concrete Quality Control, Aggregate Characteristics, and the Cement-Aggregate Reaction*, Bulletin 275, Highway Research Board, 18 pp.

Tepordei, V., 1981, "Sand and Gravel," *Minerals Yearbook, 1980*, US Bureau of Mines, pp. 695–720.

Tepordei, V., 1982, "Sand and Gravel," *Mineral Commodity Summaries 1982*, US Bureau of Mines, pp. 132–133.

Thomson, R.D., 1980, "Mining Mineral Aggregates in Urban Areas," AAG Annual Meeting, Kentucky, 45 pp.

Verbeck, G., and Landgren, R., 1960, "Influence of Physical Characteristics of Aggregates on Frost Resistance of Concrete," *Proceedings*, Vol. 60, American Society for Testing & Materials, pp. 1063–1079.

Wallace, W.A., et al., 1969, "Planning for Construction Minerals in the Urban Environment," *Proceedings*, Joint National Meeting of the American Astronautical Society/Operations Research Society, Colorado, 253 pp.

Wallace, W.A., et al., 1969a, "Planning for Non-Renewable Common Mineral Resources in Urban-Suburban Environs," Open File Report 25–69, US Bureau of Mines, available from Clearinghouse for Federal Scientific and Technical Information, Springfield, VA 22151, as PB 188 458, 253 pp.

Wallace, W.A., et al., 1970, "Planning for Common Mineral Resources—A Computer Simulation Approach," *The Quarry Manager's Journal*, The Institute of Quarrying, London, pp. 107–111.

Construction Materials

Aggregates—Slag

HENRY N. McCARL *
H. K. EGGLESTON †
WILLIAM R. BARTON ‡

Slag is a byproduct of iron and steel production and its output is closely tied to the volume of hot metal output. However, slag was largely a waste product until a little more than 60 years ago, with an associated disposal cost. Today, slag actually contributes an economic supplement to the income of iron and steel producers.

The commercial utilization of iron and steel slag represents man's most successful attempt to utilize an industrial solid waste. Its consumption not only represents a solution to a potential solid waste problem, but also substitutes for large tonnages of natural aggregates.

Huge piles of waste slag were generated in the early years of iron and steelmaking. Occasionally, small amounts were used as road base. The earliest known use was in the days of the Roman Empire when slag from Catalan forges was used as a base material for Roman roads. The first sustained commercial use was, as is often the case in human history, in the manufacture of military armaments. Some profit-conscious munitions maker in the 16th century realized cannon balls could be made from slag rather than the more valuable iron. Other small volume uses developed over the years: cast concrete building blocks, slag-lime mortars and cements, slag wool, ballast, road base, agricultural slag (soil conditioner), and finally, as concrete aggregate. An early use of slag aggregate was in concrete-hulled ships built for the US Navy in the early part of the 20th century.

* Associate Professor of Economics and Geology, School of Business, The University of Alabama in Birmingham.

† President, National Slag Association, Washington, DC.

‡ Retired, formerly with the US Department of the Interior, Bureau of Mines, Durham, NH.

Not until after World War II was the true potential of blast-furnace slag as a construction aggregate recognized. Now, virtually all blast-furnace slag produced is used, and almost all of the earlier stockpiles of waste slag have been consumed.

Utilization of steelmaking slags developed much more slowly due to widely varying characteristics at different sources. In the past 10 years, however, the qualitative uniformity has increased significantly, techniques to eliminate the expansion potential of steel slags have been developed, and slags have been used successfully in many construction applications. These developments have resulted in a general acceptance of steelmaking slags for many purposes and a steady demand, so that all steel slags now are being used for some constructive purpose.

Although slag represents only about 2% of construction aggregate materials used in the US, its production involves certain features that make its contribution far more important than the national tonnage figures would indicate. Due to its byproduct relationship with iron and steel, its availability is restricted to those areas where iron and steel furnaces have been or are now located. In these areas, slag often represents a major proportion of the mineral aggregate market because of the special properties that make it highly desirable for certain uses. An example of this would be as an aggregate in road surfacing. Its resistance to polishing and wear make both blast-furnace and steel furnace slags far superior to most other natural aggregates as a highway surface material whenever traction is important.

The location and number of slag plants in the United States in 1980 is shown in Fig. 1. Prior to World War II, most blast furnaces were located around the Great Lakes—in

FIG. 1—*Location of slag processing plants in the United States, 1980. Multilocation symbols shown by numbers; locations may have several plants. (Source: Minerals Yearbooks 1978–1979, US Bureau of Mines.)*

Buffalo, Pittsburgh, Youngstown, Cleveland, Toledo, Detroit, Gary, and Chicago, with exceptions at Bethlehem, Philadelphia, Baltimore, and Birmingham. Since then, iron and steel plants have been established in other areas such as California, Texas, Colorado, Utah, Kentucky, and Illinois. In addition, many "minimills" producing steel in electric furnaces now produce steel slags used in many construction applications. In 1980, 107 slag plants in 20 states provided employment for about 5,200 people. These plants recovered nearly 3.6 Mt (4 million st) of slag-encrusted iron to be recycled into the nearby blast furnaces.

Much of the success in developing uses for slag can be attributed to the efforts of the National Slag Association, founded in 1918. Since that time the NSA has conducted numerous field and laboratory investigations to demonstrate the properties and uses for slag and slag products. These studies have resulted in widespread public acceptance of slag as an important construction raw material.

The American Society for Testing and Materials (ASTM) Designation C125 recognizes blast-furnace slag as "the nonmetallic product consisting essentially of silicates and aluminosilicates of lime, and other bases, which is developed simultaneously with iron in a blast furnace," but to most consumers it is simply thought of as the all-purpose construction aggregate.

Steel slag is the nonmetallic material consisting primarily of calcium silicates, calcium aluminoferrites, and fused oxides of calcium, iron, magnesium, and manganese produced in steelmaking furnaces simultaneously with the steel.

Consumption

Slag consumption in the United States is measured in terms of slag sold or used by iron and steel slag processing companies for construction purposes. Table 1 is a summary of the amount of iron and steel slag used for

TABLE 1—Slag Processed (Sold or Used) for Construction Purposes in the United States, By Type, 1970–1979 (Thousand St‡ and Thousand $)

	Iron Blast-Furnace Slag											
	Air-Cooled		Granulated		Expanded		Total Iron Slag†		Steel Slag		Total Slag†	
Year	Quantity	Value*	Quantity	Value	Quantity	Value	Quantity	Value	Quantity	Value	Quantity	Value
1970	22,252	42,135	1,936	2,134	1,959	6,570	26,147	50,839	7,539	8,832	33,686	59,671
1971	21,444	42,352	1,787	2,445	1,581	4,887	24,812	49,684	8,488	9,719	33,300	59,403
1972	21,878	44,787	1,657	3,059	1,518	5,529	25,053	53,375	10,162	11,023	35,125	64,398
1973	24,971	52,249	1,999	3,667	1,852	6,936	28,822	62,852	9,739	10,765	38,561	73,617
1974	26,226	57,227	2,081	4,442	1,573	6,461	29,880	68,130	8,862	11,195	38,742	79,325
1975	22,242	53,386	1,780	4,335	1,302	5,934	25,324	63,655	7,302	8,965	32,626	72,620
1976	22,899	59,813	1,618	3,529	1,492	6,610	26,009	69,952	6,588	9,728	32,597	79,680
1977	22,753	61,270	1,488	3,579	1,475	6,414	25,716	71,262	6,668	10,850	32,384	82,112
1978	25,119	73,148	1,372	3,608	1,914	9,641	28,404	86,398	8,457	14,510	36,861	100,908
1979	25,009	78,415	855	3,037	1,648	10,794	27,512	92,246	8,252	18,476	35,764	110,722
1980	17,113	65,313	772	2,938	1,156	8,028	19,041	76,279	6,158	16,270	25,199	92,549

*Value based on selling price at plant.
†Data may not add to totals shown because of independent rounding.
‡Metric equivalent: 1 st × 0.907 184 7 = t
Source: Zlobik, 1981; Singleton, 1981.

construction purposes in the United States from 1970 through 1979 (Drake, 1971, 1972, 1973, 1974; Evans, 1976, 1977; Palalich, 1975). Not included in these totals are about 3.6 Mt (4 million st) of blast-furnace slag on crusted iron processed by the slag plants and returned to the blast furnaces as a source of iron and flux. Also not included are about 6.4 Mt (7 million st) of steel slag used for agricultural purposes and as a feed for sintering plants and recycling as flux and other uses by the steel industry.

The amount of blast-furnace slag available for use each year is almost totally dependent on the amount of pig iron produced. The amount of new slag that becomes available each year has remained approximately constant during the most recent 10-year period due to changes in the raw material mix and quality of the iron ore. Fluctuations in the amount of blast-furnace slag processed and sold reflects the utilization of stockpiles which accumulated from earlier production at some locations and variations in the demand for slag as a construction aggregate.

Steelmaking slag being produced now is completely utilized either for construction purposes or as a portion of the burden in the blast furnace (consumption for this purpose has increased significantly in the past 10 years). This latter use, not included in the Bureau of Mines totals, accounts for the reduction in the use of steel slag for construction purposes since 1972. The recent reduction in the amount of granulated slag produced is due to reduction in the amount of this type of slag used by the cement industry; due to excess clinker production capacity and not to any quality deficiency in the slag. It is expected that the amount of slag used as a cementitious material, using either water-granulated or a new revolving drum air granulation method (see Technology), will greatly increase in the future. Atlantic Cement is constructing a facility at Sparrows Point, MD, which will have the capacity to produce 725.7 kt (800,000 st) of cement per year. This new processing method will also increase the amount of expanded slag produced in the future as it can produce either granulated or expanded slag.

Data on slags processed in the rest of the world are not available, but France, West Germany, and Great Britain together process slightly more than the US total, and when other nations are also considered the quantity used elsewhere must be at least twice that in the United States. Foreign uses are less diversified within each individual nation, and as a whole, foreign slags may be said to be less intensely utilized. In Great Britain consumption patterns are similar to those in the United States, while cement manufacture seems to be the overwhelming use in France, Germany, Italy, Japan, South Africa, and USSR. In the United States, slag is used in more than 40 different applications, although many of these end uses represent relatively small volume. Among the more important small volume end uses are: glass (lowers the cost, improves melting, results in a less brittle product), oyster beds (to provide a rough surface that young oysters can cling to), terrazzo-type floors, exposed aggregate panels, oil well pellets (propping agent), and landscaping (eye appeal in pathways and borders). The major uses are, of course, construction-oriented as can be seen in Table 2, which presents a breakdown of slag sold in the United States in 1980, for construction purposes.

The transition of iron blast-furnace slag from a waste product of iron production to an essentially fully utilized byproduct was a slow and difficult process. Because most of the potential end uses were of an engineering nature, it was necessary to prove conclusively that slag could provide economical, safe, and dependable service with durability at least equal to the material it was to replace. It required a long-term research, testing, and demonstration effort to develop designs, specifications, and construction practices that convinced engineers, architects, contractors, and buyers that slag was a useful material with benefits for them.

This same procedure was required for steelmaking slags but was not started until about 1965. Actual field performance has now demonstrated that steelmaking slags, properly processed, can be used for many construction purposes. The only limitations are that it is not suitable as an aggregate in cement concrete or cement stabilization, or as an untreated material in a constrained location (such as a floor fill), due to its potential to expand. It makes a very stable, high skid-resistant, bituminous concrete; and a high load-bearing base when not constrained.

The three forms in which blast-furnace slags are sold (air-cooled, granulated, and expanded or foamed) and the use of steel slags are discussed in the following sections of this chapter. In 1980, approximately 90% of blast-furnace slag sold or used was air-cooled, roughly 4% was granulated, and 6% was expanded or foamed.

Air-Cooled Blast-Furnace Slag

Air-cooled blast-furnace slag is specifically defined in ASTM specification C125 as "the material resulting from solidification of molten blast furnace slag under atmospheric conditions. Subsequent cooling may be accelerated by application of water to the solidified surface." Air-cooled blast-furnace slag is by far the most important slag product by virtue of tonnage and wide range of applications. Most of the research effort, laboratory investigations, and promotional impetus have been devoted to air-cooled slag during the 63-year history of the NSA. Over 75% of the more than 2,500 NSA technical reports, papers, and re-

TABLE 2—Slag Sold or Used in the United States in 1980, by Use and Type (Thousand St*)

Use	Air-Cooled Blast-Furnace Slag	Granulated Blast-Furnace Slag	Expanded Blast-Furnace Slag	Steel Slags	Total
Asphaltic concrete aggregate	2,928	—	—	662	3,590
Concrete aggregate	1,516	—	—	—	1,516
Lightweight concrete aggregate	—	—	369	—	369
Concrete products	390	—	527	—	917
Cement manufacture	1	—	—	—	1
Roadbases	5,881	644	—	3,231	9,756
Fill	2,362	—	—	1,251	3,613
Railroad ballast	2,151	—	—	644	2,795
Mineral wool	680	—	—	—	680
Roofing, built-up and shingles	234	—	—	—	234
Sewage treatment	59	—	—	—	59
Soil conditioning	—	—	—	—	—
Glass manufacture	W	—	—	—	—
Ice control	—	—	—	—	—
Fire protection	—	—	—	—	—
Other†	911	128	260	371	1,670
Total	17,113	772	1,156	6,159	25,200

W = Withheld to avoid disclosing company proprietary data.
*Metric equivalent: 1 st × 0.907 184 7 = t
† Includes airport runway use, drainage, miscellaneous, and uses indicated by W.
Source: Singleton, 1981.

TABLE 3—Air-Cooled Iron Blast-Furnace Slag Sold or Used in the United States, by Use* (Thousand St§ and Thousand $)

	Air-Cooled Blast-Furnace Slag							
	1977		1978		1979		1980	
Use	Quantity	Value	Quantity	Value	Quantity	Value	Quantity	Value
Asphaltic concrete aggregate	4,024	12,798	3,916	13,272	3,421	12,365	2,928	12,587
Concrete aggregate	2,553	7,535	2,192	6,529	2,367	8,199	1,516	6,743
Lightweight concrete aggregate	—	—	—	—	—	—	—	—
Concrete products	738	2,283	764	2,551	421	1,676	390	1,601
Cement manufacture	146	351	148	456	W	W	1	5
Roadbases	7,289	19,157	8,875	24,863	8,452	25,435	5,881	22,582
Fill	3,097	6,894	2,677	5,254	3,861	9,512	2,362	6,813
Railroad ballast	3,400	7,191	2,417	5,243	2,505	6,591	2,151	7,415
Mineral wool	525	1,457	987	3,390	826	3,374	680	3,354
Roofing, built-up and shingles	255	816	234	912	247	1,324	234	1,311
Sewage treatment	46	88	89	203	W	W	59	180
Soil conditioning	2	5	W	W	W	W	—	—
Glass manufacture	217	1,399	187	1,695	21	W	W	W
Ice control	27	76	W	W	W	W	—	—
Fire protection	—	—	—	—	—	—	—	—
Other†	433	1,119	2,632	8,779	2,889	9,939	911	2,724
TOTAL‡	22,753	61,270	25,119	73,148	25,009	78,415	17,113	65,313

W—Withheld to avoid disclosing company proprietary data.
*Value based on selling price at plant.
†Includes airport runway use, drainage, miscellaneous, and uses indicated by W.
‡Data may not add to totals shown because of independent rounding.
§Metric equivalent: 1 st × 0.907 184 7 = t.
Source: Zlobik, 1981; Singleton, 1981.

leases are concerned with some phase of air-cooled slag use.

Air-cooled slag normally is used interchangeably and competitively with natural mineral aggregates. It does, however, possess certain characteristic properties which set it apart from natural heavy construction aggregates and influence consumption patterns. Its compacted unit weight is 1121 to 1249 kg/m^3 (70 to 78 lb per cu ft), it has a rough vesicular surface, and tends to break into cubical pieces when crushed. Its hardness (near 7 on Mohs' scale) and toughness with its abrasive, vesicular surface are problems for conveyors, chutes, and liners. But these very properties make it an outstanding performer in many end uses. It exhibits excellent bonding in portland cement concrete and high stability in bituminous mixtures, and it gives high skid-resistance values to bituminous surfaces.

Air-cooled slag sold by processors for construction purposes in the United States, by use, in 1977, 1978, 1979, and 1980 is shown in Table 3. The bulk of this tonnage is used as aggregate in portland cement concrete, bituminous mixes of all types, base material, and for railroad ballast.

The use of slag as a raw material in the manufacture of glass is significant, but has stabilized in recent years. It is estimated that roughly 80% of amber glass—the familiar brown beer-bottle glass—contains slag as a major ingredient. Slag used in rock-wool insulation and fiberglass is of growing importance as more emphasis is placed on insulation to achieve energy conservation.

Granulated Slag

Granulated iron blast-furnace slag is defined in ASTM C125 as: "The glassy, granular material formed when molten blast-furnace slag is rapidly chilled, as by immersion in water." Molten slag, solidified rapidly in water, changes into amorphous glassy particles resembling natural sand. Because of its natural cementitious properties, it will slowly set over a period of time into something resembling concrete when compacted damp in layers of bases and fills. The damp, loose weight of granulated slag is 831 to 949 kg/m^3 (1400 to 1600 lb per cu yd). Compacted, it will weigh 1365 to 1661 kg/m³ (2300 to 2800 lb per cu yd).

This natural hydraulic property leads, as one might surmise, to a principle use in highway construction, mostly for bases, subbases, and fill. As a fill it results in little or no settlement to subsequent overlays; it is especially useful as fill beneath pavements, around bridge abutments, and also within building foundation walls where floor slabs will subsequently be constructed. Other important uses are in all types of cement, as a fine aggregate in concrete floors and in concrete masonry units, and in landscaping and agriculture. Slag containing a high percentage of lime is effective in reducing soil acidity and also contributes minor and trace elements which are beneficial to crop growth. Because of its grain size and structure, granulated slag is also useful for conditioning (lightening) heavy clay soils.

Granulated iron blast-furnace slag sold or used by processors in the United States in 1977–1980, by use, is shown in Table 4. It is anticipated that the tonnage of granulated slag produced each year will increase significantly in the near future due to new production facilities (specifically at Sparrows Point in Baltimore) and increased demand for granulated slag as a cementitious raw material.

Expanded Slag

Expanded iron blast-furnace slag is defined by ASTM C125 as: "The lightweight cellular material obtained by controlled processing of molten blast-furnace slag with water, or with water and other agents such as steam or compressed air or both." Crushed expanded slag is angular and roughly cubical in shape with a minimum of flat or elongated pieces. Its weight depends on the source, exact production method, and size gradation but lies in the 801 to 1041 kg/m³ (50 to 65 lb per cu ft) range for loose fine aggregate and 561 to 881 kg/m³ (35 to 55 lb per cu ft) range for loose coarse aggregate.

Most expanded slag is used as a lightweight aggregate in concrete block. The main advantages attributed to expanded slag concrete blocks are light weight, very high fire resistance, and very low shrinkage during block curing. Most expanded slag production goes for concrete block because of the constant heavy demand by block manufacturers for this high quality, economical aggregate. Expanded slag is also very suitable for use in lightweight structural concrete. When properly proportioned, air-entrained, and mixed, a high quality structural concrete can be obtained with unit weights in the range of 1762 to 1922 kg/m³ (110 to 120 lb per cu ft). An outstanding example of the use of expanded slag lightweight structural concrete is in the Prudential Building in Chicago, IL.

Expanded iron blast-furnace slag sold or used by processor in the United States in 1977–1980, by use, is shown with granulated slag in Table 4. The utilization of the new drum pelletizer (see Technology) is rapidly growing at slag plants in the US. This new machine will increase expanded slag production capability significantly and much larger quantities should be sold in the near future.

Steel Slag

Steel slags are used for base, fill, highway shoulders, railroad ballast, in bituminous mixes, and for agricultural purposes as a soil conditioner. Although steel slags do not have the extensive appeal of blast-furnace slags, their antiskid (rough texture) properties and ability to provide good color contrast with light-colored shoulder materials suit them particularly well for bituminous mixes, especially for road surfacing. Agricultural use of steel slags today is based largely on contained free lime and the presence of trace elements important to agriculture. In years past, steel slags also contained significant phosphorous, but the use of very low phosphorous iron ores has led to a reduction of this element in steel slags.

The weight of steel slag, its angularity, and its hardness make it an excellent railroad ballast material. It is not suitable for use as an aggregate in portland cement concrete nor as compacted fill beneath concrete slab floors because of its possible expansion characteristics.

The use of steel slags has stabilized in recent years, largely because of increased recycling of these materials as flux in the iron and steel furnaces. This has tended to limit availability of the steel slags for processing and sale. The NSA does not promote steel slag as much as blast-furnace slag because the quality and properties of steel slag vary significantly from one location to another and quality control practices differ from one steel plant to another. Past problems with expansion and deterioration of steel slags were caused by free lime or partly burned limestone in open-hearth slags. While the partial burning of limestone is disappearing with the replacement of open-hearth steel furnaces by basic oxygen process furnaces, free lime still causes problems in some applications.

Although uncontrolled expansion in service

TABLE 4—Granulated and Expanded Iron-Blast-Furnace Slag Sold or Used in the United States, by Use*
(Thousand St§ and Thousand $)

	Granulated Blast Furnace Slag								Expanded Furnace Slag							
	1977		1978		1979		1980		1977		1978		1979		1980	
Use	Quantity	Value	Quantity	Value	Quantity	Value	Quantity	Value	Quantity	Value	Quantity	Value	Quantity	Value	Quantity	Value
Asphaltic concrete aggregate	—	—	—	—	—	—	—	—	—	—	—	—	—	—	—	—
Concrete aggregate	—	—	—	—	—	—	—	—	—	—	—	—	—	—	—	—
Lightweight concrete aggregate	—	—	—	—	83	W	—	—	509	2,265	1,030	5,799	931	6,388	369	3,420
Concrete products	119	447	107	W	—	—	—	—	916	4,047	652	3,089	607	3,865	527	3,203
Cement manufacture	60	211	W	W	W	W	—	—	50	101	78	457	W	W	—	—
Roadbases	1,064	2,264	939	2,201	631	W	644	2,149	—	—	—	—	—	—	—	—
Fill	151	404	214	578	W	W	—	—	—	—	W	W	W	W	—	—
Railroad ballast	—	—	—	—	—	—	—	—	—	—	—	—	—	—	—	—
Mineral wool	—	—	—	—	—	—	—	—	—	—	—	—	—	—	—	—
Roofing, built-up and shingles	—	—	—	—	—	—	—	—	—	—	—	—	—	—	—	—
Sewage treatment	—	—	—	—	—	—	—	—	—	—	—	—	—	—	—	—
Soil conditioning	45	115	48	W	25	—	—	—	—	—	—	—	—	—	—	—
Glass manufacture	—	—	—	—	—	—	—	—	—	—	—	—	—	—	—	—
Ice control	8	31	1	W	—	—	—	—	—	—	—	—	—	—	—	—
Fire protection	—	—	—	—	—	—	—	—	—	—	—	—	—	—	—	—
Other†	40	107	67	828	109	3,037	128	789	—	—	153	295	110	541	260	1,405
TOTAL‡	1,488	3,579	1,372	3,608	855	3,037	772	2,938	1,475	6,414	1,914	9,641	1,648	10,794	1,156	8,028

W—Withheld to avoid disclosing company proprietary data.
*Value based on selling price at plant.
† Includes airport runway use, drainage, miscellaneous, and uses indicated by W.
‡ Data may not add to totals shown because of independent rounding.
§ Metric equivalent: 1 st × 0.907 184 7 = t.
Source: Zlobik, 1981; Singleton, 1981.

TABLE 5—Steel Slag Sold or Used in the United States, by Use*
(Thousand St§ and Thousand $)

Use	1977		1978		1979		1980	
	Quantity	Value	Quantity	Value	Quantity	Value	Quantity	Value
Asphaltic concrete aggregate	396	776	607	1,264	822	2,481	662	2,259
Concrete aggregate	—	—	—	—	—	—	—	—
Lightweight concrete aggregate	—	—	—	—	—	—	—	—
Concrete products	—	—	—	—	—	—	—	—
Cement manufacture	—	—	—	—	—	—	—	—
Roadbases	2,929	5,207	4,793	8,451	4,237	8,818	3,231	7,499
Fill	2,330	3,072	1,784	2,355	1,882	4,305	1,251	3,552
Railroad ballast	354	569	549	1,072	530	1,243	644	1,990
Mineral wool	—	—	—	—	—	—	—	—
Roofing, built-up and shingles	—	—	—	—	—	—	—	—
Sewage treatment	—	—	—	—	—	—	—	—
Soil conditioning	44	114	W	W	—	—	—	—
Glass manufacture	—	—	—	—	—	—	—	—
Ice control	—	—	—	—	—	—	—	—
Fire protection	35	76	—	—	8	W	—	—
Other†	580	1,036	723	1,368	773	1,629	371	970
TOTAL‡	6,668	10,850	8,457	14,510	8,252	18,476	6,158	16,270

W—Withheld to avoid disclosing company proprietary data.
*Value based on selling price at plant.
† Includes airport runway use, drainage, miscellaneous, and uses indicated by W.
‡ Data may not add to totals shown because of independent rounding.
§ Metric equivalent: 1 st × 0.907 184 7 = t.
Source: Zlobik, 1981; Singleton, 1981.

is still a threat, ways have been found to regulate it—principally wet aging for at least six months (to complete hydration prior to use), use of smaller sizes [−19.1 mm (−¾ in.)], or "open" gradings [larger sizes such as 38.1 or 50.8-mm (1½ or 2-in.) without sufficient fines in the mix to completely fill the voids]. At one plant, spent pickle liquor is applied to the slag to produce an acceptably stable product for use as a road-building material. The great bulk of steel slag is treated magnetically to recover contained metal. Coarser sizes of recovered metal are returned to the steel furnaces; finer sizes are recycled into the blast furnaces.

Steel slag sold or used by processors in the United States in 1977–1980, by use, is shown in Table 5.

Prices

Slag, because it must compete with actual and potential alternate materials for every end use, must be priced competitively in the marketplace. The various market values, by end use, reported by slag processors in 1978 and 1979 (Table 6) present a general picture of present prices and recent price trends. Because some firms do not keep records of material prices by end use, these figures really are approximations based on a processor knowing what his customers do with most of the slag they purchase.

Geographically, more detail is available for air-cooled slag construction aggregates. Prices for construction aggregates are regularly quoted in the periodical *Engineering News-Record* (ENR) for a group of US and Canadian cities. Slag generally commands a substantial premium in price over natural mineral aggregates for most construction uses.

Specifications and Properties

Blast-furnace slag generally is in all national specifications and specifications promulgated by states and localities in those areas where slag supplies are available. However, steel slags generally are not provided for in specifications, other than railroad ballast. They are accepted

TABLE 6—Average Value of Slag Sold or Used in the United States, by Use ($ per st)

Use	Air-Cooled			Granulated			Expanded			Steel Slag		
	1978	1979	1980	1978	1979	1980	1978	1979	1980	1978	1979	1980
Concrete aggregate	2.98	3.46	4.45	—	—	—	—	—	—	—	—	—
Lightweight concrete aggregate	—	—	—	—	W	—	5.63	6.86	9.27	—	—	—
Concrete products	3.34	3.98	4.11	W	—	—	4.74	6.37	6.08	—	—	—
Cement manufacture	3.08	W	4.36	W	W	7.49	5.86	W	6.11	—	—	—
Asphaltic concrete aggregate	3.39	3.61	4.30	—	—	—	—	—	—	2.08	3.02	3.41
Roadbases	2.80	3.01	3.84	2.35	W	3.34	—	—	—	1.76	2.08	2.32
Fill	1.96	2.46	2.88	2.70	W	6.13	W	W	5.04	1.32	2.29	2.84
Railroad ballast	2.17	2.63	3.45	—	—	—	—	—	—	1.95	2.35	3.09
Mineral wool	3.44	4.08	4.93	—	—	—	—	—	—	—	—	—
Roofing, built-up and shingles	3.90	5.36	5.60	—	—	—	—	—	—	—	—	—
Sewage treatment	2.28	W	3.05	—	—	—	—	—	—	—	—	—
Soil conditioning	W	W	—	W	W	—	—	—	—	W	—	—
Glass manufacture	9.06	W	W	—	—	—	—	—	W	—	—	—
Ice control	W	W	—	W	—	—	—	—	—	W	W	—
Other	3.34	3.42	2.99	3.71	3.55	—	1.93	7.95	—	1.89	2.09	2.61

W—Withheld to avoid disclosing company proprietary data; included with "other."
Source: Zlobik, 1981; Singleton, 1981.

on a case-by-case basis for individual projects. Where it is determined that steel slag can be used properly and effectively on a project, authorization apparently has been obtainable for its use. The Los Angeles abrasion test requirement for slag aggregates has been deleted from ASTM specifications both for portland cement concrete and for highway construction of all types. The change was adopted when it was shown that the Los Angeles abrasion test was not a valid method for judging the durability of slag or predicting its performance under conditions of actual service. This change in testing requirements did not affect stone or gravel specifications.

Because the basic chemistry for producing iron in blast furnaces has changed little over the years, the chemical composition of slag has remained quite uniform. Most slag produced will fall in the range of chemical compositions presented in Table 7.

Oxides are not present in their free form in slag, but are combined in various silicate and aluminosilicate minerals such as melitite, wollastonite, and merwinite with sulfur primarily present as calcium sulfide.

The chemical composition of blast-furnace slag is very similar, then, to the constituents that comprise portland cement. In fact, finely ground slag has definite hydraulic or cementitious properties which add to the strength of any portland cement concrete in which such slag is used. As a consequence, blast-furnace slag has been used extensively for some time in cement manufacturing in many European and other countries. Blast-furnace slag is chemically stable, homogeneous, basic in composition, and highly alkaline in the presence of moisture.

Slag is created in the blast furnace and is drawn off in molten form at a temperature range of 1371° C (2500° F). From this point on, the physical properties of the eventual slag product are determined by the methods used to

TABLE 7—Chemical Composition and Physical Properties of Air-Cooled Blast-Furnace Slags

Constituent or Property	Range
SiO_2 %	32-42
Al_2O_3 %	7-16
CaO %	32-45
MgO %	5-15
Fe_2O_3 %	0.1-1.5
MnO %	0.2-1.0
S %	1-2
Unit wt lb/ft³ *	70-85
Absorption %	1-6
Bulk dry specific gravity	2.0-2.5

* Metric equivalent: 1 lb/ft³ × 16.018 46 = kg/m³.
Source: National Slag Association.

TABLE 8—Slag Concrete Recommended Trial Mix Proportions

Coarse Slag Aggregate			Sand: Dry Wt per Cu Yd of Concrete for Cement Contents of No. Sacks per Cu Yd*			
Nominal Size and % Air Content	Dry Rodded Unit Wt, Lb per Cu Ft*	Dry Weight Per Cu Yd of Concrete, Lb*	5	6	7	Water, Gal per Cu Yd*
2 in.* to	70	1452	1355	1277	1199	
No. 4	80	1660	1281	1202	1124	30
(4%)	90	1867	1217	1139	1061	
1½ in. to	70	1395	1381	1303	1225	
No. 4	80	1594	1309	1231	1153	32
(4½%)	90	1794	1247	1169	1091	
1 in. to	70	1280	1517	1439	1361	
No. 4	80	1464	1448	1370	1290	34
(5%)	90	1646	1389	1311	1232	
¾ in. to	70	1185	1568	1490	1412	
No. 4	80	1354	1503	1425	1346	36
(6%)	90	1523	1447	1369	1291	
½ in. to	70	994	1714	1636	1558	
No. 4	80	1136	1658	1580	1502	39
(7%)	90	1278	1611	1533	1455	

*Metric equivalents: 1 in. × 25.4 = mm; 1 lb per cu ft × 16.018 46 = kg/m³; 1 lb per cu yd × 0.593 276 = kg/m³; 1 gal × 3.785 412 = L; 1 cu yd × 0.764 554 9 = m³.
Source: National Slag Association.

convert the slag to a solid state—air cooling, granulating, or expanding.

Air-Cooled Slag

Many concrete bridges, buildings, miles of pavements, and various concrete products use slag-bearing portland cement concrete. Slag is often the preferred and specified aggregate where a combination of high strength, durability, light weight, and economy are important factors. Under normal conditions slag concrete with a 28-day strength of 27.5 to 41 MPa (4000 to 6000 psi) in compression and flexural strengths of 4.1 to 5.1 MPa (600 to 750 psi) are readily obtained with a 5 to 7 sack per cu yd cement mix (Table 8). Concrete made with air-cooled slag aggregate will weigh about 113.4 kg (250 lb) less per cu m (cu yd) than concrete made with crushed stone or gravel. Because of the rough vesicular surface of the slag particles, excellent aggregate-cement bonding and concomitant excellent performance is provided. Compatible slag and cement coefficients of expansion assure continuation of the superior bond regardless of temperature variations during service.

TABLE 9—Bituminous Paving, Characteristic Mixes

Coarse Slag Aggregate		Mix	
Nominal Maximum Size	Compact Unit Weight, Lb per Cu Ft‡	Asphalt Content, Wt % of Total Mix*	Unit Wt. of Compacted Mix, Lb per Cu Ft‡
¾ in.‡	70-90	7.5-9.0	125-140
⅜ in.	70-90	8.0-9.5	127-142
No. 4	†	9.0-11.0	130-145

*Asphalt contents shown are for all-slag mixes. For mixes with slag coarse aggregate and natural sand fine aggregate the asphalt content should be reduced about 1¼%.
† No coarse aggregate, uses slag fine aggregate only.
‡ Metric equivalents: 1 in. × 25.4 = mm; 1 lb per cu ft × 16.018 46 = kg/m³.
Source: National Slag Association.

In bituminous paving products (Table 9) air-cooled slag provides for high stability, good adhesion, and durable nonskid surfaces. Because of these properties, it is extensively used in bituminous paving of many types. In many jurisdictions it is specified as a skid-resistant aggregate and in some localities is required in surface courses. It has been used in paving many automobile speedways including Indianapolis and Daytona.

The qualities that make air-cooled slag desirable for bituminous and portland cement concrete mixes also favor its use in base courses of any type. Such bases are highly stable. An important consideration favoring slag base or fill is the almost complete absence of settling after initial placement and compaction.

Slag can be charged with only minor additions of other ingredients as the furnace feed when producing mineral wool, but general practice is to add 15% or more SiO_2 or other similar materials to adjust the viscosity of the melt. The desirable properties of the end product make slag wool a preferred insulating product.

Where available, slag is the principal aggregate used in built-up tar or asphalt-bonded roofs (Table 10). It is resistant to weathering, stable, and its high opacity protects the bituminous bond from deterioration caused by exposure to sunlight. About 136 kg (300 lb) of slag is applied per 9.3 m² (100 sq ft) of roof as compared to 181 kg (400 lb) of gravel or stone, a 25% weight saving. Slag also is used extensively as the granular material coating composition shingles and roofing.

Slag is used extensively as railroad ballast. Its unit weight, particle shape, and rough surface permit it to interlock and to hold rails in alignment more securely than less stable ballast.

The high relative surface area and high voidage resulting from the vesicular nature of slag make it an efficient trickling filter medium for use in both industrial filtration plants and public sanitary sewage treatment installations. It is also often specified for subsurface absorption fields, seepage pits, and absorption beds in small residential and institutional septic systems. For these latter uses the slag should always be free of large amounts of fines that might cause clogging of the percolation bed. Crushed slag in sizes for typical coarse concrete aggregate is frequently used as coarse filter media.

TABLE 10—Gradation Requirements, Mineral Aggregate for Built-Up Roofs (ASTM D1863)

Sieve Size	Total % Passing, by Wt
¾ in.*	100
½ in.	90-100
⅜ in.	40-70
No. 4	0-15
No. 8	0-5

*Metric equivalent: 1 in. × 25.4 = mm.
Source: National Slag Association.

As can be surmised from the foregoing, specifications recommended as a guide for assuring reasonable performance for air-cooled slag products include those of ASTM, The American Assn. of State Highway and Transportation Officials (AASHTO), and the various state highway departments.

Granulated Slag

While not as widely available as air-cooled slag, the properties of granulated slag lend it to useful construction purposes. Its properties are affected by the production method used. The pit method, in which a large flow of slag is run directly into a water-filled pit, results in a product containing coarser sized material. A finer product is attained with the jet process in which streams of high-pressure water impinge on the slag at the end of the furnace slag runner. The water jet breaks the slag stream in droplets, yielding greater fine granulation. Slag made either way yields material that affords good consolidation in construction base or fill. The new drum method of processing slag, described in the "Technology" section, will be widely used in the near future and will greatly increase production potential for granulated slag.

Because of its excellent hydraulic properties, when granulated slag is compacted in the presence of moisture it will set like cement. A sample core taken from a granulated slag base one year after emplacement was tested by the Ohio Department of Highways and showed a compressive strength of 4.3 MPa (630 psi). In addition to its nonsettling and cementitious properties, granulated slag also is desirable for use as a base or fill because of its relatively light unit weight [1365 to 1661 kg/m³ (2300 to 2800 lb per cu yd) rolled in place] and its thermal insulation value which tends to minimize frost heaving. Of special note is its use as a pipeline backfill around gas and water lines. Granulated slag is the ideal material for backfill around pipes because its high alkalinity in the presence of moisture (pH 9–11) prevents a corrosive acid condition after the pipe is in the ground.

Three types of cement are made with granulated slag. It may be used as part of the kiln feed in making all types of portland cement, it may be interground with portland cement clinker to make portland blast-furnace slag cement, or it forms the principal ingredient of slag cement. The latter is an hydraulic cement in which slag may be used interground only with an air-entraining agent or mixed with lime and fly ash.

Other applications are in concrete, concrete masonry units, and as an agricultural liming material and for soil conditioning.

The ASTM specifications are suggested as guides to the use of granulated slag for all uses except base courses. The Ohio State Highway Department specifications should be used as a guide for base courses using granulated slag.

Expanded Slag

While the supply of expanded slag has gradually reduced in the recent past due to environmental constraints on production methods, it is anticipated that the installation of the new revolving drum device for processing slag described in the "Technology" section will permit the production of large quantities of pelletized expanded slag, as well as granulated, in the future and greatly increase the use of this type of slag.

Expanded slag is a preferred aggregate for use in lightweight structural concrete and concrete masonry units. Its properties also are suited for use as a loose fill insulating material. Its weight, depending upon the exact method used in expansion and particle size gradation, ranges from 641 to 961 kg/m³ (40 to 60 lb per cu ft). It is normally produced in two sizes: coarse, from 12.7 mm (½ in.) to No. 4 sieve, and fine, from No. 4 sieve down. Like other blast-furnace slag products, it is noncorrosive. Its thermal conductivity is exceptionally low.

Lightweight concrete mixes are best designed by following the trial mix procedures recommended by the American Concrete Institute (ACI) because the exact specific gravities of lightweight aggregates, and the net water-cement ratios of concrete mixes made with them, are difficult to determine accurately. Air entrainment is recommended for optimum workability but should be kept below 10% except in the case of low-strength insulating concretes. Natural sand is used as part or all of the fine aggregate in mixes to yield a higher strength-cement ratio.

Moduli of elasticity and bond and shear strength of expanded slag structural concretes are generally well within the limits of the ACI Building Code requirements. Its thermal insulating properties and fire resistance are outstanding (Table 11).

A typical 8 x 8 x 16-in. expanded slag concrete block will weigh 11 to 15 kg (25 to 33 lb). They are readily made to comply with all strength, shrinkage, and durability specifications. Thermal insulating, sound absorption and transmission properties, and fire resistance are outstanding. The latter property can be further enhanced by filling the cores of masonry units with loose expanded slag.

Among useful specifications and procedures that can be used as guides for the use of expanded slag products are those of ASTM and ACI.

Steel Slags

Steel slag is a byproduct of open hearth, electric, and most importantly, basic oxygen steel furnaces. It consists principally of a fused mixture of oxides and silicates. At times in specifying this material, it is referred to as "BOP" (Basic Oxygen Process) or "BOF" (Basic Oxygen Furnace) slag because the basic oxygen process is currently the most commonly used method of producing steel throughout most of the economically developed world.

Steel slag, because of its durability and stability, is excellent for ballast, base, certain types of fill, and highway shoulders. It is especially desirable for bituminous mixes due to its high skid-resistance. Most construction use of steel slag is on the basis of individual project approval. For most construction purposes, it may be used alone or combined with iron blast-furnace slag. In all such service it is specified to meet the minimum performance specifications set for any alternatively available aggregate. Specifications covering ballast use of steel slags have been published by the American Railway Engineering Association.

Substitutes

Talking about substitutes for slag is a case of putting the cart before the horse. Slag, in addition to being called the all-purpose construction aggregate, could just as readily be called the substitute extraordinaire. Every end use for slag existed before slag entered com-

TABLE 11—Fire Resistance Ratings, Reinforced Concrete Floors and Ceilings, Cast-In-Place

Aggregate	Slab Thickness, In.*	Minimum Protection to Reinforcement, In.*	Fire Rating, Hr
Expanded slag	4½	¾	4
Expanded shale	5¼	¾	4
Air-cooled slag	4¾	¾	2½
Air-cooled slag	6	1	4
Traprock or siliceous gravel	4¾	¾	2
Calcareous gravel or limestone	4¾	¾	2
Traprock, calcareous or siliceous gravel	6	1	3

* Metric equivalent: 1 in. × 25.4 = mm.
Source: National Slag Association (based on information from the American Insurance Association).

merce. And to enter commerce, slag had to displace some alternate material that was already serving a need more or less adequately. Reverse substitution, the original for the substitute, is always possible should necessity, economics, or technology dictate. Should slag supplies ever be curtailed, almost inexhaustible natural materials would be available as raw material supply supplements. Crushed stone, gravel, and sand can supply most of the heavy constructions aggregate needs. Impure limestone (wool rock), glass, and mixtures such as granite and limestone are all capable of being made into mineral wool. In Rhode Island, mineral wool was even made, experimentally, from limestone and meta-anthracite with fuel-grade gas as a coproduct (in this case they were making a slag as an intermediate step in a gas producer furnace). Alternative lightweight aggregates and agricultural liming products exist in abundance. The one thing of note is that no one material can substitute for slag in all its varied uses. In addition, none of them recover metallic iron as a useful and high market value byproduct. The slag industry recovers almost 3.6 Mt (4 million st) of slag-encrusted iron each year during its crushing and screening operations that previously was thrown out with the then-wasted slag. Now the industry saves the baby *and* the bath water instead of discarding them.

The good and substantive economic and technical reasons for which slag was able to displace more classic materials in the marketplace still pertain. Since that time, environmental objections to sand and gravel pits and stone quarries have added a new reason to discourage reverse substitution.

Price Trends

The slag industry, in common with other enterprises, is faced with a "cost-push" situation resulting in a rising price trend. In the period 1970–1980, slag prices have roughly doubled. Prices for slag used for fill, the lowest value use, went from $0.74 to almost $2.00 per ton during this period. The average value per ton for all slag sold in 1970 was $1.77, according to the US Bureau of Mines. In 1981, this had increased to $3.64 per ton. 1980 prices for air-cooled slag ranged from $3 to over $5 per ton.

In addition to "cost-push" inflation in historical areas such as labor, equipment, capital, and energy, the slag industry is faced with significant environmental protection costs, costs which are different than those faced by the competing crushed stone and sand and gravel producers. Because of slag pouring and cooling operations, the slag industry faces environmental problems in controlling gaseous emissions similar to those faced by the steel industry itself. Control of particulates, water pollution, and noise will add to production costs. All aggregate prices will also be affected, in addition to ecological consideration, by the requirements of worker health and safety laws. The slag industry is regulated under the Occupational Health and Safety Act that is not applied uniformly to all aggregate producers because stone and gravel come under regulations of the Mine Safety and Health Administration (MSHA). Slag already is at an economic disadvantage because of the depletion allowances permitted for stone and gravel, and new ecological and safety-health considerations also are projected as placing more severe burdens on slag. Even though many of these costs are unique to the slag industry, slag must remain competitive with natural materials in terms of price.

Slag is generally claimed to be better for most uses than competing natural products. However, the additional quality cannot be quantified sufficiently in the marketplace to sustain the use of slag at a higher unit price when a contractor can find a less expensive substitute that will satisfy minimum specifications. To permit slag to remain competitive, the iron and steel industry and the slag industry must cooperate in instituting all possible cost reduction practices or face the unhappy prospect of higher slag prices and dramatically reduced sales. In some cases, it may become necessary for the iron and steel industry itself to absorb some of the costs involved to permit the slag processor to remain competitive.

Transportation

Slag, like other aggregates, is traditionally a locally consumed product used near the steel facilities where it is produced. As is the case with most short-haul bulk items, the truck is now the normal method of delivery to the consumer—86% leaves the processing plant by truck. In 1979, another 11% was hauled by rail and the remaining 3% by barge (water). The decline in rail shipments has been dramatic over the past 15 years from 27% shipped by rail in 1966 to 12% in 1980.

The decline in rail shipments is not caused by shrinking market areas for slag, but by the increasing demand for truck delivery. In fact, slag availability has been the major constraint on slag sales. This also has contributed to the decline in what are normally longer distance hauls by rail. Heavy load demand and declining stockpiles now leave less slag available for shipment to more distant potential customers.

There are three additional reasons for the decline in rail shipments. The first and most important is the high cost of rail shipping (greatly increased in recent years); the second is the nonavailability of cars to ship at the right time to meet promised delivery schedules; and the final reason is the poor condition and uncleanliness of some cars when received for loading from the railroads. It is costly both in money and time to have to clean the cars properly; at times some cars leak so badly that much aggregate may be lost en route to the customer.

Tonnages of iron blast-furnace slag shipped in the United States in the period 1977–1979 by method of transportation are shown in Table 12.

Marketing

Slag is marketed regularly in the many areas where the iron and steel industry is located throughout the United States, Canada, Mexico, and other countries.

Slag processing firms generally are located within iron and steel producing industrial complexes and may or may not be subsidiaries of the iron and steel firms. In 1979, the US slag industry consisted of 107 processing plants located in 20 states. Table 13 lists some individual state production totals for blast-furnace slag for the period 1977–1979. Table 14 lists geographic regions of the United States and the combined amount of both iron and steel slags sold or used in those regions in 1979.

The industry supports the National Slag Association headquartered at 300 S. Washington St., Alexandria, VA, through corporate mem-

TABLE 12—Shipments of Iron and Steel Slag in the United States, by Method of Transportation

	1977		1978		1979		1980	
Method of Transportation	Quantity (Thousand St)	% of total	Quantity (Thousand St)	% of total	Quantity (Thousand St)	% of total	Quantity (Thousand St)	% of total
Rail	4,105	13	4,913	13	3,832	11	3,086	12
Truck	26,047	80	27,437	75	29,519	82	20,797	83
Waterway	1,329	4	1,879	5	1,126	3	682	3
Not transported (used at plant site)	903	3	2,632	7	1,287	4	634	2
Total	32,384	100	36,861	100	35,764	100	25,199	100

Metric equivalent: 1 st × 0.907 184 7 = t.

Source: Zlobik, 1981.

TABLE 13—Iron Blast-Furnace Slags Sold or Used in the United States, by State*
(Thousand St and Thousand $)

Year and State	Air-Cooled, Screened and Unscreened Quantity	Value	Total of all Types Quantity	Value
1977				
Alabama	1,232	3,814	1,232	3,814
California	547	991	547	991
Illinois	1,131	3,128	1,131	3,128
Ohio	5,039	15,210	6,030	17,753
Pennsylvania	4,481	14,784	5,314	17,325
Colorado, Texas, Utah	1,707	3,877	1,707	3,877
Other States†	8,616	19,466	9,755	24,374
Total	22,753	61,270	25,716	71,262
1978				
California	648	1,328	648	1,328
Illinois	1,000	2,564	1,000	2,564
Ohio	4,451	16,217	5,340	18,997
Pennsylvania	5,686	19,619	6,670	23,213
Colorado, Texas, Utah	2,185	5,590	2,185	5,590
Other states‡	11,149	27,831	12,561	34,706
Total §	25,119	73,148	28,404	86,398
1979				
California	663	1,482	663	1,482
Indiana	4,230	6,106	W	W
Kentucky	549	W	549	W
Ohio	4,448	17,580	5,157	W
Pennsylvania	5,518	20,424	6,042	24,390
Utah	1,028	2,273	1,028	2,273
West Virginia	843	W	843	W
Other #	7,729	30,551	13,230	64,101
Total #	25,009	78,415	27,512	92,246
1980				
Illinois, Indiana, Michigan	3,519	10,245	W	W
Ohio	3,210	14,740	W	W
Pennsylvania	4,299	17,885	W	W
Maryland, New York, West Virginia	1,603	4,996	W	W
Colorado, Texas, Utah	2,446	8,751	2,446	8,751
Alabama, Kentucky	1,509	7,298	1,509	7,298
California	528	1,398	528	1,398
Total §	17,113	65,313	19,041	76,279

W—Withheld to avoid disclosing company proprietary data; included with "Other."
*Value based on selling price at plant.
† Includes Indiana, Kentucky, Maryland, Michigan, New York, and West Virginia.
‡ Includes Alabama, Indiana, Kentucky, Maryland, Michigan, New York, and West Virginia.
§ Data may not add to totals shown because of independent rounding.
# Includes Alabama, Colorado, Illinois, Maryland, Michigan, New York, Texas, and categories indicated by symbol W.
Source: Zlobik, 1981, Singleton, 1981.

TABLE 14—Iron and Steel Slag Sold or Used in the United States, by Geographic Region, 1979 (Thousand St)

Region	Quantity
Middle Atlantic	11,002
New England	—
North Central	18,197
Pacific*	1,363
South	2,010
West/Midwest	3,192
Total	35,764

* Includes Alaska and Hawaii
Source: Zlobik, 1981.

berships. NSA disseminates information on the industry and its products and acts as a clearinghouse for inquiries about the commodity. NSA also conducts and stimulates research and helps in developing specifications and test procedures. Members of the NSA with their home office locations are:

United States

American Materials Corp., Hamilton, OH
Atlantic Cement Co., Inc., Stamford, CT
Bethlehem Steel Corp., Bethlehem, PA
Duquesne Slag Products Co., Pittsburgh, PA
Clarence A. Hackett, Inc., Tonawanda, NY
Gifford-Hill & Co., Inc., Shrevesport, LA
Heckett Engineering Co., Butler, PA
Houston Slag Materials Co., Houston, TX
Hurock, Inc., Houston, TX
Illinois Slag & Ballast Co., Chicago, IL
International Mill Service, Philadelphia, PA
Edw. C. Levy Co., Detroit, MI
The Levy Co., Chesterton, IN
Maryland Slag Co., Baltimore, MD
St. Louis Slag Products Co., St. Louis, MO
Sheridan Corp., Lebanon, PA
Standard Slag Co., Youngstown, OH
U.S. Steel Corp., Pittsburgh, PA
Vulcan Materials Co., Birmingham, AL
Vulcan Materials Co., Chicago, IL
Warner Co., Philadelphia, PA

Australia

Broken Hill Pty. Co., Ltd., Melbourne, Victoria
Blue Metal & Gravel Pty, Ltd., Wentworthville, N.S.W.
Blue Circle Southern Cement, Inc., Sydney, N.S.W.

Canada

National Slag Ltd., Hamilton, Ont.
Standard Slag Cement, Milton, Ont.

France

Gagneraud Pere Et Fils, Paris

Japan

Nippon Steel Co., Tokyo

Great Britain

Clugston Slag Co., Ltd., Scunthorpe, Lincolnshire
Appleby Slag Co., So. Humberside

South Africa

Pretoria Portland Cement Co., Johannesburg
Slagment, Ltd., Braamfontein, Transvaal

Associate Members

R. M. Chambers, Pittsburgh, PA
The Calumite Co., Trenton, NJ
McGraw Construction Co., Middletown, OH
Pennsylvania Slag Assoc., Pittsburgh, PA

Slag is processed and marketed for aggregate use in sizes and gradations to conform to the standard categories of the Simplified Practice Recommendation adopted by ASTM and AASHTO and also in other sizes and gradations as specified by the user. Additional detail on market products is included in the sections titled "Consumption" and "Specifications and Properties" and, in much greater detail, in the technical publications of NSA. Local market availability of slag types and size ranges can be obtained from the companies previously mentioned.

The slag processors are companies in the construction aggregates industry. They usually sell their product directly to the end user, that is, a ready-mix manufacturer, contractor, etc. Only two American slag processors are directly associated with the steel industry; the rest are independent aggregate producers. Some also produce gravel or crushed stone. A few sell only slag. The amount of slag used by slag processors themselves is very small, although a few own slag-consuming operations such as ready-mix concrete plants.

Government Considerations

The slag industry does not receive depletion allowances or any special subsidies. Because slag processing plants are generally located within iron and steel plant complexes, zoning and land use regulation are not problems.

The fact that slag is not a raw mineral product places it at a disadvantage compared with stone or sand and gravel, which are granted a 5% depletion allowance.

The fact that slag is a potential solid waste that is being usefully recovered should be considered in the future because such utilization is approved policy under the Federal Resources Recovery Act. Slag recovery also saves land from being despoiled by slag heaps and lessens the amount of land disturbed by aggregate mining operations.

Environmental Considerations

The slag industry has recognized its obligation to minimize particulate, gaseous, and aqueous emissions and has reduced these problems to acceptable levels from the regulatory viewpoint. These efforts required the expenditure of millions of dollars but are yielding substantive results in the control of undesirable side effects from slag processing.

Particulate emissions are controlled by the use of water sprays throughout the processing plants to preclude the dust becoming airborne. Dust control is a more difficult problem at slag plants than at stone plants because of the heat of the slag when it is delivered for screening and crushing. The slag is usually delivered warm, above 37.8° C (100° F), which tends to enhance the ability of dust to become airborne. At a large slag-processing plant in Detroit, a combination of infrared to sense the heat of the slag and water sprays are used to reduce the slag temperature to an acceptable level before it becomes plant feed. Throughout the plant, light density sensors scan the belts to measure the amount of dust in the air above belt transfer points. The sensors control the addition of water to the slag at the transfer points to provide sufficient moisture to allay dust but not clog the plant screens.

Basically, the slag industry is in compliance with the regulations of the Environmental Protection Administration (EPA) at the present time, and does not seem to face any significant new environmental problems.

Future Prospects

Supply-Demand

The demand for blast-furnace slag is such that the slag industry processes and markets all the slag now being produced and has eliminated practically all of the large slag banks generated during past years of blast-furnace operations.

The slag industry is totally predicated on iron and steel production—as pig iron production goes, so goes blast-furnace slag. There have been no recent significant changes in the ratio of slag produced per ton of hot metal. This currently varies from 227 to 272 kg/t (500 to 600 lb per ton) of hot metal. Its geographic availability is determined by the location of iron and steel production.

The drum pelletizer described in the "Technology" section, will greatly increase the capability of the slag industry to produce premium products. By adjusting drum speed and rate, direction, and pressure of water application, two totally different products can be made: one air-granulated material suitable for use as a cement, and another a pelletized expanded slag suitable as a lightweight aggregate. The cement product can be either ground to cement fineness and used to replace portland cement at the mixer, or blended with portland cement at the cement mill. Slag has been used as a cementitious material in Europe, Russia, Japan and South Africa for many years, but has not been used extensively in the US because the kiln capacity of the cement industry has been adequate for the demand. Because it takes much less energy to produce cement from slag than from portland clinker, there is a definite trend in the US now to use slag as a cementitious material. Energy and ecological constraints on expanded shale and clay production have reduced the availability of this type of lightweight aggregate—a shortage which can be alleviated by expanded slag.

Steelmaking slags also are fully used although their use as a construction aggregate is not as broad as blast-furnace slag due to the potential of some steel slags to expand (it is not used in cement concrete or in applications where it is "contained," as in a building floor fill). A large portion is recycled to the blast furnace by steel companies after sizing. Large quantities also are used as railroad ballast, and steel slag is the aggregate preferred by some jurisdictions for highway bituminous concrete because of its superior skid-resistance characteristics.

Because of their excellent performance in a broad range of applications for long periods of time, both blast-furnace and steelmaking slags are preferred for many uses. Consequently, in most locations more slag could be sold if it were available during normal construction level periods.

Technology

No major changes in steel industry technology are anticipated that will affect either the composition or quantity of blast-furnace or steel slags. The volume ratio, ton of slag per ton of iron, varies according to furnace burden which consists of ore, fuel, and flux. In past years, prior to much special enhancement of iron ores, the usual ratio was one-half to three-quarters of a ton of slag to each ton of hot metal produced. Now, with much of the ore refined and pelletized before charging as part of the furnace burden, the slag metal ratio is considerably reduced and is in the 227 to 272 kg/t (500 to 600 lb per ton) range. Theoretically, blast furnace burdens could be reformulated to reduce the amount of slag produced still further, but experimentation and tests have proven that it is more economical to run the furnaces as they are being operated now. The furnaces still using low iron ores and less than optimum pretreatment will gradually change to higher quality ores or be phased out of production. However, slag losses from this technological progress will be more than offset by increases in total iron and steel production.

The "direct reduction" method of producing steel has been tested and is feasible. However, it is not anticipated that this method of steel manufacture will be widely adopted in the next decade because of the huge capital expenditure required and the relatively higher energy consumption of the process. In fact, this higher energy consumption, accompanied by recent significant increases in energy prices, is making the economics of this process less attractive as time passes. The process also is hampered by a relatively low production capability. It is significant to note that large blast furnaces continue to be constructed at major steel installations. This indicates the steel industry does not anticipate the demise of blast furnaces in the immediate future.

A mechanical device known as a pelletizer, developed and made operational about 1970, is now recognized as a practical and environmentally acceptable device for expanding slag. The apparatus consists of a means of feeding a uniform flow of expanding molten slag to the surface of a horizontally-mounted, internally water-cooled rotating drum. A series of vanes welded to the surface of the drum throws the slag into the air where surface tension causes the individual particles to assume a semi-rounded shape.

Among the advantages of this process are:

1) Very low release of hydrogen sulfide compared to other slag-expanding systems.

2) The semisealed rounded surface of the aggregate requires minimum cement contents to produce lightweight concrete of the required strength.

3) Pelletizing is the only operating process which produces lightweight aggregate directly at the furnace.

4) The process also can produce a granulated (glassy) material of excellent quality for cement. Due to the limited amount of water used, the moisture content is generally low and due to its cellular structure, the product is generally easier to grind than conventional water granulated slag.

More than 30 machines, including 12 in the US, are currently operating in eight different countries.

The major technological thrust of the slag industry is toward new or higher quality products. Two of these are the use of blast-furnace slag as a cementitious material and as a lightweight construction aggregate, both made practical by the success of the drum pelletizer.

Continued improvement in processing equipment and methods may be expected to continue. During the past 10 years major improvements have been made in the capability of processing plants to remove magnetics (slag-encrusted iron and steel) from both blast-furnace and steelmaking slags through the installation of more sophisticated and efficient magnets. This has improved the quality of the slag as well as produced a valuable product for return to the steel industry.

Lime-fly ash-slag mixtures and mixtures of lime-granulated slag-natural aggregates are widely used in French highway construction. In the US, pelletized slag has the same potential for use in these stabilized-base mixtures, and in the stabilization of sludges produced by some environmental control processes. An example of this is the ground granulated blast-furnace slag used as the principal stabilizing agent in the *calcilux* process, which solidifies fluid wastes.

The ARMCO Steel Corp. has developed a solid waste recycling installation at Middletown, OH, that proportionately mixes steel slags with other plant wastes such as sludges, scale, and dusts with ore, limestone, and coke screenings. This mixture is then fused, cooled, crushed, and used as a blast-furnace feedstock.

Because the great age of iron and steel is rapidly being diluted into an age of iron plus aluminum plus plastics plus recycled materials, it is predicted that supply, and hence consumption, of slag will grow very slowly in the foreseeable future. The production of iron and steel has stabilized in recent years and in the current recession has actually declined. This means stable or declining near-term supplies of slag of all types. An average longer-term growth rate, in the range of 1 to 2% per year, is projected for both iron blast-furnace and steel slags.

One may be certain that above-average

growth rates can be anticipated for those slag products that command higher prices and slower-than-average rates, or even declines for low value products. Products such as expanded or granulated slag for roofing granules and cement, slag wool, and insulation materials all appear to have a bright future. The quantity of slag used in concrete and bituminous mixes will continue to be determined by the rate of building and highway construction. However, as the highway expenditures turn more to maintenance rather than new construction, the demand for slag for surface courses will continue to be strong. Demand for railroad ballast and lower quality base material and fill is likely to decline. In total, the future is still likely to be supply constrained by slag availability and higher value uses will continue to get top priority.

Bibliography and References

Anon., 1969, "Vulcan Materials Solves Slag Handling Problem at Fairfield, Alabama Plant," *Pit and Quarry*, Vol. 62, Sep., pp. 123–124.

Anon., 1978, "Skid Resistance of Bituminous Pavement Test Sections, Highway 401, Toronto By-Pass," IR–67, Ministry of Transportation and Communications, Ontario, Canada, Apr.

Anon., 1980, "Do It With Slag," *The Economist,* Vol. 275, No. 7139, June 28, p. 103.

Abrams, M.S., 1980, "Fire Tests of 3x3-Ft. Specimens Made With Pelletized Expanded Slag Aggregate," Construction Technology Laboratories, Skokie, IL, 17 pp.

Absalom, S.T., 1982, "Slag—Iron and Steel," *Mineral Commodity Summaries 1982,* US Bureau of Mines, pp. 140–141.

Atwell, J.S.F., 1974, "Some Properties of Ground Granulated Slag and Cement," *Proceedings*, Institution of Civil Engineers, Vol. 57, Part 2, London, June, pp. 233–250.

Barnett, S.C.C., 1979, "Methods and Economics of Slag Cleaning," *Mining Magazine*, Vol. 140, No. 5, May, pp. 408–409.

Barton, W.R., 1967, "Raw Materials for Cement Manufacture," *Proceedings*, 1st International Fly Ash Symposium, Information Circular 8348, US Bureau of Mines, pp. 46–51.

Barton, W.R., 1967a, "Raw Materials for Manufacture of Cement," *Pit and Quarry*, Vol. 60, No. 1, July, pp. 139–141.

Barton, W.R., 1975, "Construction Materials: Aggregates—Slag," *Industrial Minerals and Rocks*, 4th ed., S.J. Lefond, ed., AIME, New York, pp. 109–127.

Bauman, E.W., and Barton, W.R., 1967, "Utilization of Blast Furnace Slag as an All-Purpose Construction Aggregate," SME Preprint 67H48, AIME Annual Meeting, Los Angeles, 7 pp.

Blunk, G., 1979, "Utilization of Blast Furnace Slag and Steelmaking Slag in the Federal Republic of Germany," *Utility of Steelplant Slag*, Symposium paper, Feb., pp. 35–39.

Blunk, G., and Geissler, J., 1980, "Use of Steelmaking Slags, Represented by Selected Examples," *Stahl and Eisen*, Vol. 100, No. 3, Feb. 11, pp. 118–123.

Bosselman, K., et al., 1975, "New Slag Granulation and Dewatering System with Automatic Slag Unloading," *Proceedings*, Consistent Iron and Technology Meeting, London, Apr., pp. 91–95.

Bransford, T.L., 1973, "Analysis of Pavement Slipperiness in Alabama," ASTM Special Technical Publication 530, 22pp.

Chung, C.H., Mizuno, T., and Mackenzie, J.D., 1978, "Iron Recovery and Glass Fiber Production from Copper Slag," *Proceedings*, 6th Mineral Waste Utility Symposium, May 2–3, pp. 144–148.

Clarkson, J.F., et al., 1978, "Utilization of Smelter Slags at White Pine Copper Division," *Proceedings*, 6th Mineral Waste Utility Symposium, May, pp. 93, 101.

Clixby, G., 1977, "Experimental Study of Blast Furnace Bosh Slags and Its Application to Blast Furnace Practice," *Proceedings,* Ironmaking, Iron and Steel Division, The Metallurgical Society of AIME, Vol. 36, Apr., pp. 308–325.

Colussi, I., and Meriani, S., 1980, "Beneficiation of a Zinc-Oxide Rich Phase by Leaching Refired Imperial Smelting Process Slags," *International Journal of Mineral Processing*, Vol. 7, No. 1, May, pp. 43–55.

Cotsworth, R.P., 1978, "Use of Pelletized Slag in Concrete Masonry Units," *Journal of Testing and Evaluation*, Vol. 6, No. 2, Mar., pp. 148–152.

Drake, H.J., 1971, "Slag—Iron and Steel," *Minerals Yearbook*, US Bureau of Mines, pp. 1087–1093.

Drake, H.J., 1972, "Slag—Iron and Steel," *Minerals Yearbook*, US Bureau of Mines, pp. 1143–1146.

Drake, H.J., 1973, "Slag—Iron and Steel," *Minerals Yearbook*, US Bureau of Mines, pp. 1139–1143.

Drake, H.J., 1974, "Slag—Iron and Steel," *Minerals Yearbook*, US Bureau of Mines, pp. 1201–1205.

Dussart, J., 1979, "Blast Furnace and Steel Slags in France," *Utility of Steelplant Slags*, Symposium Paper, Feb., pp. 25–34.

Eggleston, H.K., 1970, "The Successful Utilization of Iron and Steel Slags," *Proceedings*, 2nd Mineral Waste Symposium, Chicago, US Bureau of Mines and ITT Research Institute, Mar. 18–19, pp. 15–22.

Eggleston, H.K., 1979, "Use of Blast Furnace and Steel Slags as a Rail Ballast Material in U.S.A.," *Utility of Steelplant Slags*, Symposium Paper, Feb., pp. 75–80.

Eggleston, H.K., 1981, "Slag Industry Hopeful About 1981," *Pit and Quarry*, Vol. 73, No. 7, Jan., p. 83.

Emery, J.J., 1976, "Slag," *Proceedings*, 5th Mineral Waste Utility Symposium, Chicago, IL, Apr., pp. 291–300.

Emery, J.J., 1979, "Canadian Developments in the Use of Wastes and By-Products," *CIM Bulletin,* Vol. 72, No. 812, Dec., pp. 88–94.

Emery, J.J., 1980, "Pelletized Lightweight Slag Aggregates," Concrete International 1980, London, pp. 37–47.

Evans, J.R., 1976, "Slag—Iron and Steel," *Minerals Yearbook*, US Bureau of Mines, pp. 1239–1249.

Evans, J.R., 1977, "Slag—Iron and Steel," *Minerals Yearbook,* US Bureau of Mines, pp. 845–855.

Forester, E.F., and Weston, P.L., Jr., 1967, "Heat Content of Some Blast-Furnace and Synthetic Slags," Report of Investigations 6886, US Bureau of Mines, 21 pp.

Fowler, E.W., and Lewis, D.W., 1963, "Flexural and Compression Tests of High Strength Air-Entrained Slag Concrete," *Journal of the American Concrete Institute*, Vol. 60, No. 1, Jan., pp. 113–128.

Frolov, A.E., et al., 1973, "Complex Treatment of Open-Hearth Slags," *Metallurgist (USSR),* Vol. 17, No. 7–8, July–Aug., pp. 572–574.

Fujimoto, Y., 1979, "Cooperative Studies in Japan on Granulated Slag Sands for Concrete," *Utility of Steelplant Slags*, Symposium Paper, Feb., pp. 65–74.

Gee, K.H., 1979, "Slag Cements Make Sense!" *Rock Products*, Vol. 82, No. 4, Apr., pp. 60–62.

Glazier, W.M., 1980, "The Use of Slag in Highway Construction," 49th Annual Meeting of the Highway Research Board, 22 pp.

Gorbachev, V.P., et al., 1974, "Viscosity of Iron-Bearing Magnesia-Aluminum Blast-Furnace Slags," *Steel in the USSR*, Vol. 4, No. 6, June, pp. 438–440.

Grancher, R.A., and Stearn, E.W., 1979, "Lightweight Aggregates, Gypsum and Slag," *Rock Products*, Vol. 82, No. 12, Dec., pp. 61–62.

Grieb, W.E., and Werner, G., 1961, "Final Report of Tests of Concrete Containing Portland Blast-Furnace Slag Cement," *Public Roads*, Vol. 31, No. 9, Aug., pp. 183–193.

Gutt, W., and Harrison, W.H., 1974, "Use of Lighter Weight Blast Furnace Slag as Dense Coarse Aggregate in Concrete," *Magazine of Concrete Research*, Vol. 26, No. 88, Sep., pp. 123–143.

Gutt, W., et al., 1967, "Blast Furnace Slag as Aggregate for Concrete," Research Series 70, Building Research Station, England, pp. 71–82.

Heaton, B.S., 1979, "Steel Furnace Slag as an Aggregate in Asphaltic Concrete," *Utility of Steelplant Slags*, Symposium Paper, Feb., pp. 81–85.

Hegmon, R.R., and Ryan, P.F., 1969, *The Polishing Resistance of Selected Slag Aggregates*, The Pennsylvania State University, University Park, PA, 14 pp.

Higginbotham, G., 1979, "Use of Blast Furnace Slag in the Manufacture of Glass," *Utility of Steelplant Slags*, Symposium Paper, Feb., pp. 105–114.

Hillnhuetter, F.W., and Kister, H., 1979, "Hoesch Ajo System for Granulation of Blast Furnace Slags," *Utility of Steelplant Slags*, Symposium Paper, Feb., 6 pp.

Hubbard, F., 1936, "Studies on Relation Between Characteristics of Blast-Furnace Slag and Other Coarse Aggregates and the Properties of Resultant Concretes," *Proceedings*, American Society for Testing and Materials, Vol. 36, Part 1, pp. 297–326.

Hubbard, F., and Williams, H.T., 1943, "Strength of Concrete as Related to Abrasion of the Blast-Furnace Slag Used as Coarse Aggregates," *Proceedings*, American Society for Testing and Materials, Vol. 43, pp. 1088–1094.

Huhta, R.S., and Stearn, E.W., 1980, "Lightweight Aggregates: New Markets Fill Construction Void," *Rock Products*, Vol. 83, No. 12, Dec., pp. 60–61.

Jefferson, R.H., Hattman, E.A., and Kusler, D.J., 1974, "Rapid Wet Chemical Methods for the Analysis of Blast Furnace Slags," Information Circular 8625, US Bureau of Mines, 15 pp.

Jones, J.B., Jr., 1968, "Granulated Slag for Liming Soils," *Ohio Report on Research and Development,* Ohio Research and Development Center, Wooster, OH, July–Aug., pp. 62–63.

Joseph, R.W., and Haddad, M., 1975, "Utilization of Steelmaking Slags in Blast Furnace Burdens," *Open Hearth Proceedings*, The Metallurgical Society of AIME, Apr., pp. 140–159.

Josephson, G.W., et al., 1949, "Iron Blast-Furnace Slag: Production, Processing, Properties and Uses," Bulletin 474, US Bureau of Mines, 148 pp.

Kashio, S., et al., 1980, "Raman Spectroscopic Study on the Structure of Silicate Slag," *Transactions*, Iron and Steel Institute of Japan, Vol. 20, No. 4, pp. 251–253.

Keil, F., 1949, "Hochofenschlacke (Blast-Furnace Slag)," *Verlag Stahleisen M.B.H.*, Dusseldorf, W. Germany, 346 pp.

Kleiger, P., and Isberver, A.W., 1967, "Laboratory Studies of Blended Cements—Portland Blast-Furnace Slag Cements," *Journal of the Portland Cement Association*, Vol. 9, No. 3, Sep., pp. 2–22.

Kokubi, M., et al., 1976, "Blast Furnace Slag Used as Cement Raw Material," *Journal*, Japan Society of Civil Engineers, Vol. 61, No. 4, Apr., pp. 20–29.

Kubodera, S., et al., 1979, "Approach to the Full Utilization of LD Slag," *Transactions*, Iron and Steel Institute of Japan, Vol. 19, No. 7, pp. 419–427.

Lanigan, P.G., 1979, "Processing of Steel Works Slags for the Building and Construction Industries," *Utility of Steelplant Slags*, Symposium Paper, Feb., pp. 95–104.

Larrabee, C.P., and Coburn, S.K., 1961, "Experience with Blast-Furnace Slag as an Aggregate in Reinforced Concrete," *Corrosion*, Vol. 17, Apr., pp. 155–156.

Levine, S., 1963, "Slag Pyroprocessing for Mineral Wool Producing," *Minerals Processing*, Vol. 4, No. 2, Feb., pp. 28–30.

Levine, S., 1968, "Open-Hearth Slag Produces New Aggregate Blends," *Rock Products*, Vol. 71, No. 5, May, pp. 122–128.

Levine, S., 1979, "Cement Industry Anticipates Steady Growth," *Pit and Quarry*, Vol. 72, No. 1, July, pp. 140–158.

Lewis, D.W., 1950, "Lightweight Concrete Made with Expanded Blast Furnace Slag," *Journal of the American Concrete Institute, Proceedings,* Vol. 47, No. 5, May, p. 84.

Lewis, D.W., and Hubbard, F., 1958, "Flexural and Compressive Strength Properties of Air-Entrained Concrete with Air-Cooled Blast-Furnace Slag Aggregate," *Proceedings*, American Society for Testing and Materials, Vol. 58, pp. 1143–1156.

Marek, C.R., et al., 1971, "Promising Replacements for Conventional Aggregates for Highway Use," Final Report, Report UILU-ENG-71-2010, National Cooperative Highway Research Program

Project 4–10, University of Illinois, Urbana, IL, Mar., 192 pp.
Mather, B., 1957, "Laboratory Tests of Portland Blast Furnace Slag Cements," *Journal of the American Concrete Institute, Proceedings*, Vol. 54, No. 3, Sep., pp. 205–232.
Matyas, A.G., 1978, "Utilization of Steelmaking Slag," *Iron and Steel Engineering*, Vol. 55, No. 8, Aug., pp. 29–30.
Mikhail, R., et al., 1977, "Air Entrainment in Portland Blast Furnace Slag Cement Pastes: Effects on Strength and Pore Structure," *Cement and Concrete Research*, Vol. 7, No. 5, Sep., pp. 515–522.
Mikhail, R., Mourad, W.E., and Gouda, V.K., 1974, "Hardened Portland Slag Cement Pastes . . . (1) Effects of Additives on Surface Area and on Pore Structure," *Cement and Concrete Research*, Vol. 4, No. 5, Sep., pp. 807–820.
Mills, R.H., 1979, "Significance of Latent Energy in Hydraulic Slags to the Cement and Concrete Industry," *Utility of Steelplant Slags*, Symposium Paper, Feb., pp. 51–57.
Minty, E.J., 1979, "Evaluation of Blast Furnace Slag as Base Material in Road Pavements," *Utility of Steelplant Slags*, Symposium Paper, Feb., pp. 17–23.
Munn, R.L., 1979, "Blast Furnace Slag as an Aggregate in Concrete," *Utility of Steelplant Slags*, Symposium Paper, Feb., pp. 41–50.
Murrie, S.J., and Jones, D.E., 1974, *Five-Year Corrosion Study of Steel Pipe in Various Backfill Materials*, Australian Iron and Steel Pty. Ltd., Port Kembla Corrosion and Processes Research Department, 6 pp.
Myasnik, A.A., et al., 1976, "Raising the Quality of Blast-Furnace Slag Aggregate and the Efficiency of its Utilization," *Steel in the USSR*, Vol. 6, No. 4, Apr., pp. 181–183.
Palalich, W., 1975, "Slag—Iron and Steel," *Minerals Yearbook*, US Bureau of Mines, pp. 1297–1301.
Pepper, L., and Mather, B., 1957, "Effectiveness of Mineral Admixtures in Preventing Excessive Expansion of Concrete Due to Alkali-Aggregate Reaction," *Proceedings*, American Society for Testing and Materials, Vol. 59, pp. 1178–1202.
Ray, H.S., 1979, "Measurement of Melting Temperatures of Blast Furnace Slags by Temperature Gradient Melting," *Metallurgical Transactions B*, Vol. 10B, No. 4, Dec., pp. 677–679.
Ray, H.S., Kanpur, B.S., and Lal, A., 1974, "Note on a Slag-Ceram," *Journal of the Institution of Engineers (India)*, Mining and Metallurgy Division, Vol 54, Part MM3, July, pp. 69–71.
Rellermeyer, H.M., and Mass, H.G., 1975, "Utilization of Bos Slags at the August Thyssen Huette AG," *Open Hearth Proceedings*, The Metallurgical Society of AIME, Vol. 58, Apr., pp. 131–137.
Renichi, K., et al., 1979, "Latent Hydraulic Property of Granulated Blast Furnace Slag by Various Activators," *Tetsu to Hagane*, Journal of the Iron and Steel Institute of Japan, Vol. 65, No. 13, Nov., pp. 1825–1829.
Shiomi, S., et al., 1979, "Separate Recovery of Iron and Phosphorous from Bof Slags by Using Tin," *Tetsu to Hagane*, Journal of the Iron and Steel Institute of Japan, Vol. 65, No. 14, Dec., pp. 2032–2037.
Singleton, R.H., 1981, "Slag—Iron and Steel," *Minerals Yearbook 1980*, US Bureau of Mines, pp. 747–756.
Smith, J.W., 1980, "Processing and Utilization of Steelmaking Slags," *Proceedings*, 3rd Mineral Waste Utilization Symposium, US Bureau of Mines, ITT Research Institute, Chicago, IL, Mar., pp. 52–54.
Spellman, L.U., 1979, "The Transition of Blast Furnace Slag as a Cementitious Material," Presented to 15th International Cement Seminar, Chicago, IL, 15 pp.
Spellman, L.U., 1980, "New Generations of Slag Cements Show Promise," *Rock Products*, Vol. 83, No. 2, Feb., pp. 45–48.
Taylor, I.F., 1974, "Cutting Costs with Slag Cement," *Civil Engineering*, Vol. 813, Apr., pp. 45, 47, 49.
Thomas, A., 1979, "Metallurgical and Slag Cements, the Indispensable Energy Savers," IEEE Cement Indiana Conference paper, May, 59 pp.
Thomas, G.H., and Stephenson, I.M., 1979, "Development of L.D. Slag as an Aggregate for Road Construction," *Utility of Steelplant Slags*, Symposium Paper, Feb., 10 pp.
Trauffer, W.E., 1962, "Levy's New Slag Plant Features," *Pit and Quarry*, Vol. 54, No. 7, Jan., pp. 135–143.
Trauffer, W.E., 1975, "Fountain's New Slag and Asphalt Plants in Pueblo, Colorado," *Pit and Quarry*, Vol. 67, No. 9, Mar., pp. 56–59.
Van der Vlist, A.A., 1967, "Robur Makes 330,000 Tons Per Year of Slag Cement," *Minerals Processing*, Vol. 8, No. 5, May, pp. 20–23.
Walter, P.D., Simpson, J.F., and Murrie, S.J., 1979, "Development of Blast Furnace Slag as an Aggregate on Road Surfacing," *Utility of Steelplant Slags*, Symposium Paper, Feb., pp. 87–93.
Whitehurst, E.A., and Goodwin, W.A., 1955, "Pavement Slipperiness in Tennessee," *Proceedings*, Highway Research Board, Vol. 34, pp. 194–209.
Zlobik, A.B., 1980, "Slag—Iron and Steel," *Minerals Yearbook*, US Bureau of Mines, pp. 821–833.
Zlobik, A.B., 1980a, "Slag—Iron and Steel," *Mineral Commodity Summaries 1980*, US Bureau of Mines, pp. 148–149.
Zlobik, A.B., 1981, "Slag—Iron and Steel," *Mineral Commodity Summaries 1981*, US Bureau of Mines, pp. 144–145.

Construction Materials

Cement and Cement Raw Materials

JOHN A. AMES *

WILLIAM E. CUTCLIFFE †

Webster's dictionary nearly equates portland cement with its current primary definition of cement. While such equation may be a triumph of common usage, the confusion between the terms *cement* and *concrete* unfortunately is very widespread. Full definition of cement properly would extend to many substances, but for purposes of this chapter cement is restricted to "varieties" of portland cement including "portland-pozzolan" cement.

Among the usually recited historical notes concerning cement is that it was developed by the Romans. Their use of cement in the great structures of Rome, and even in the far corners of their Empire, such as Hadrian's Wall in the north of England, testifies to the antiquity of a major cement industry. From English sources we know of John Smeaton's carefully proportioned hydraulic cement mixed in 1756 and of Joseph Aspdin's patented cement to which, in 1824, he gave the name portland cement. (He claimed that his cement mortar looked like the famous dimension limestone quarried from the Isle of Portland on England's south coast.) The year 1971 marked the centennial of portland cement manufacture in the United States. A fair historical account of the many real contributors to the cement industry since 1756 unfortunately would be too extensive for this chapter. Development of rotary kiln practice and burning at higher temperatures clearly was the major move into today's technology.

As part of the 5th edition of *Industrial Minerals and Rocks,* this chapter on utilization focuses on inferred interests of the anticipated readers. The main period for which statistics are cited generally is that since the 4th edition (1975) of this same volume.

Cement manufacture is the processing of selected and prepared mineral raw materials to produce the synthetic mineral mixture (clinker) that can be ground to a powder having the specified chemical composition and physical properties of cement. Cement making incorporates a number of distinct steps, but it is characterized by the key pyroprocessing (or "burning") step, which brings about the necessary changes in the raw materials. This pyroprocessing is a chemical process. Conversion of large volumes of mineral raw materials into specification-controlled cement at relatively low cost is the objective.

Statistics on the cement industry are of uneven quality. The US Bureau of Mines (USBM) is the principal collector of available data. These statistics can be only as good as the quality of information supplied by the reporting companies. The data perhaps most often questioned relate to productive capacity of the industry. The questions submitted by USBM are valid, nevertheless, and the information gained from them, as good as is available, is what is generally used. The Portland Cement Association also is a source of much quality information.

The year 1972 witnessed a test of the adaptability of cement industry statistics. This was the year of conversion from barrels and sacks of cement as the units of measure to tons and hundredweights. The barrels and sacks related to volumetric measures in that a sack was judged to be 1 cu ft (weighing a "standard" 94 lb). A barrel was 4 sacks. Volumes of concrete are stated in cubic yards, and the number of cubic feet of cement in a cubic yard of concrete was a sensible approach to the mensuration involved. However, cements do not weigh the same amount from plant to plant or from type to type, and the buyer was more interested in getting an assured weight for his money than in "buying air." Thus, a sack, in accepted practice, became 94 lb instead of 1 cu ft, and a "barrel" became 376 lb, regardless of whether it would fit into 4 cu ft of space.

* Consultant, Pittsburgh, PA.

† President, Dunn Geoscience Corp., Latham, NY.

With the *de facto* dealing in weight rather than in volumes, the awkward numbers made little sense except as common usage. The ton, now adopted as standard, is the short ton of 2000 lb, and the conversion factor from barrels to tons is 5.319. The sales unit of a hundredweight is subdivided readily out of the ton unit and is useful for pricing smaller or fractional parts of sales. Going metric remains to be accomplished.

Raw Materials

Concrete is so widely used as a construction material that cement plants geographically are spread to serve most areas. The major parameters that orient plant locations within the cement industry are markets, transportation, raw materials, available fuel, and power. Standards are determined by each company for quality and quantities of raw materials it deems requisite for its needs; 40 to 100 years' supply is the usual amount sought for a new plant. Sites are dictated primarily by market and/or transportation considerations. Long-distance transport of raw materials is avoided when possible. A number of exceptions to each such rule can be cited. Selection among raw materials that potentially could be moved to or obtained at each given site is determined by the economic and chemical advantages of each.

Objectives and Needs

The primary requirement for cement manufacturers is a source of lime (CaO) generally available as calcium carbonate ($CaCO_3$). Calcite, a crystalline form of $CaCO_3$, is the mineral raw material most often used and is found naturally as limestone in many varieties or in limestone's close relatives, marble and marl. Other sources include shell (oyster or coquina) in which the organically secreted calcium carbonate is present mainly as aragonite. Recently, use of oolitic aragonite sands (Bahamian) as a cement raw material added another source material of aragonite mineralogy. Lime-rich slag and other industrial waste products occasionally have been adapted as primary sources of lime for cement in the United States. Anhydrite (or/and gypsum) has long been used at a few plants in the world to produce both cement clinker and sulfuric acid from calcium sulfate. In Russia, feldspars (calcium alumino-silicates) have been the primary source of lime in cement raw material for at least one plant.

Secondary raw materials are required to supply the balance of silica and alumina when the lime content of the primary material is higher than needed. Except in the case of white cement manufacture, iron in the raw materials also is sought or is added. Although the sources of these chemical components are diverse, both from natural deposits and waste products, the ratio of silica/alumina (often also evaluated as silica/R_2O_3 with R_2O_3 including alumina plus the ferrous oxide value of the iron source) has to be closely controlled. Among the naturally occurring sources of silica are the sediments, sand, silt, clay, and loess, or their corresponding rocks, sandstone, siltstone, shale, or mudstone. Alumina sources generally are mud, clay, loess, and related rocks. Wastes that supply secondary components include slag, fly ash, red mud (from bauxite processing), and wash plant or mill tailings. Ash from the fuel (especially coal) used for processing the cement raw materials adds secondary components, sometimes in significant amounts. Table 1 lists types of raw materials used in the United States and provides a generalized statement of volumes used.

The considerable flexibility in selection of raw materials for a blended kiln feed lies mainly in the fact that cement making is a chemical process. Nevertheless, mineral and physical characteristics of the selected raw materials have a great deal to do with the costs and control of recovery, preparation, and processing of those materials. Powerful arguments can be made that impure limestone called cement rock, which requires virtually no addition of secondary raw materials, is preferable to the artificially blended materials, each of which in its own right may be relatively pure. However, the proverbial "cement rock" or limestone that contains the correct balance of lime, silica, alumina, and iron for even Type I cement is rare.

In addition to those raw materials that are blended to provide an acceptable kiln feed, a source of SO_3 is required. The SO_3 material is ground with clinker to provide predictable control of the setting times for concrete made with the cement. While the clinker itself normally contains SO_3, and can be produced near levels required by specification, natural gypsum or a gypsum-anhydrite source of SO_3 is preferred; and, in some quantity, gypsum is mandatory. The total amounts used in the United States for cement in 1977, 1978, 1979, and 1980 were 3.6, 3.9, 3.9, and 3.5 Mt (3,930,000, 4,260,000, 4,324,000, and 3,859,000 st), respectively.

Portland cement masonry mortar includes additional materials interground with clinker, generally dolomite or limestone, more rarely

TABLE 1—Raw Material Used In The Manufacture of Portland Cement Clinker in United States

Types of Raw Materials	
Calcium Carbonate Sources	**Alumina Sources**
Limestones, including:	Shale
Lithified limestone	Clay and mud
Chalks	Loess
Marbles	Slag
Marls	Fly ash
Oyster shell (reef)	Bauxite
Coquina shell	Alumina process waste (red mud)
Aragonite sands	Staurolite
Slag	Mill fines (granite)
Silica Sources	**Iron Sources**
Sand, sandstone, or quartzite	Iron ores
Clay and claystone	Blast furnace flue dusts
Shale	Pyrite cinder
Loess	Mill scale
Slag	Fly ash
Mill fines	
Fly ash	

	Volumes Reported 1000s St†				
	1976	1977	1978	1979	1980
Calcium carbonates (primary sources)	103,891	108,670	114,945	115,491	106,668
Primary alumina materials	10,851	11,009	11,382	11,667	10,726
Primary silica materials	2,682	2,710	3,016	2,936	2,662
Primary iron materials	825	870	1,037	1,063	1,175
Blast furnace slag	435	464	479	483	132
Miscellaneous*	272	392	505	515	772

Source: 1977, 1978, 1979 US Bureau of Mines *Minerals Yearbooks,* "Cement" chapter, Table 6, and Table 5, 1980. Calcareous, argillaceous, siliceous, ferrous, and blast furnace slag extracted from "other."

* Includes fluorspar, pumicite, calcium chloride, soda ash, borax, staurolite, fly ash, bauxite, diatomite, air-entraining compounds, and grinding aids.

† Metric equivalent: 1 st × 0.907 184 7 = t.

lime or other materials. Portland pozzolan cements and slag cement, of course, contain interground pozzolanic and slag materials.

Raw materials and fuels introduced into the cement kiln contain components other than those sought. These components are permitted, within specified limits, and certain of them, in some ways, are beneficial. If present above defined levels of concentration, these unsought components may constitute "intolerable impurities." Magnesium compounds perhaps are the most familiar and ubiquitous of these. At low levels of concentration magnesia's (MgO) role as a fluxing agent is beneficial. As long as it is present in low concentrations, magnesia is regarded to be innocuous. As its concentration in the cement increases, however, it becomes an intolerable impurity because, above certain levels of concentration, the usual magnesium compounds cause expansion and result in disruption of concrete. Others of the "uninvited" components are comparable in cement in that they are acceptable at lower but intolerable at higher concentrations.

Evaluation

In the areas of the United States where new cement plant sites are considered, the general geology of surface deposits is well enough known so that the process of exploration perhaps would be better characterized as evaluation. To be sure, the job of "location" still must be accomplished, but prudent selection among alternatives is a better description of the task and the objective than the term exploration implies. The complex set of factors to be evaluated importantly includes plant-design requirements necessary for the recovery, handling, and processing of the specific raw materials available to each site. Zoning and environmental considerations must be understood and their requirements satisfied at every operating location. Potential profitability still must stand as the final arbiter. Decisions concerning opening new plants or expanding old ones obviously must be based on hard, reliable data indicating adequate raw materials available at affordable costs. Cement plants are closed for many reasons, including the depletion of local raw materials, but that factor perhaps is the one most controllable by prudent selection of the plant site.

Recovery Steps

The winning or recovery of the mineral raw materials for cement manufacture adds up to a surprisingly large volume of mining activity. Roughly 114.3 Mt (125,994,000 st) of raw feed (expressed in dry terms) went into US kilns to produce cement in 1980. This feed weight does not include fuel, water, or air. In the processes of recovery and reclamation of cement raw materials, the tonnage of overburden handled each year may well equal the amounts of raw materials reported used. Open quarry methods typically are employed to produce the required cement raw materials. Usual steps in quarry operation include:

- Clearing and stripping
- Drilling
- Blasting and breaking
- Loading
- Transportation (hauling) to the crushing department
- Reclamation

Exceptions to this sequence can be cited for various operations, and the relative cost of each step may differ greatly from one operation to another. Some quarries have so little overburden that stripping is not necessary. In other quarries as much as 6 tons of overburden must be stripped for every ton of limestone recovered. Drilling, blasting, and breaking may be required to win hard rock in conventional limestone quarries or, at some plants, these methods may be eliminated because softer materials can be recovered by ripping or/and "scooping" operations. Loading is accomplished by a variety of equipment ranging from electric or diesel shovels or draglines to the numerous types of "end loaders" and "scooploaders." Truck hauling is the most common means of transporting stone out of or within the quarry, although belts, skip hoists, rail cars, and wheeled loaders ("pick and carry" use) are important among the wide variety of equipment used for transportation to the crusher.

The relatively few plants that depend on underground mining have dwindled in number. Compared to most open quarries, underground mining predictably is more costly. Plants whose lime requirements are met by materials other than limestone (limestone defined broadly to include rock types from marble through sedimentary limestone to marl) continue to decline. Shell beds (mostly oyster reefs up to many thousands of years old) are sources of cement raw material. Although some of the shell reefs are buried in mud and support no live oysters, some reefs do, and recent pressure by environmentalists to force dredgers to stop disturbing living oysters has made shell less available and more costly as a cement raw material. Operation methods range from dredging to modified quarrying to recover shell or other poorly or nonconsolidated carbonate materials.

Processing

Fig. 1 illustrates steps in the manufacture of cement. Plant flow descriptions illustrate that approximately every other step in the manufacturing of cement is transportation and is not a process step in the sense of this discussion. Process steps in manufacturing generally are visualized as being those which physically or chemically change the materials as they progress from their raw condition to the finished product. From the first step of recovery to the finished kiln feed, much of the processing involves the physical changes of size reduction and proper blending and mixing. The chemical changes brought about in the kiln most importantly include calcining of calcium carbonate to calcium oxide (lime) in an environment that promotes the lime's reaction with other components to form the synthetic minerals and glasses that comprise *clinker*. While size and shape of clinker particles are of great range, a typical range is "buckshot" to "golf ball" size; shapes of larger particles are rounded by rolling in the kiln and tend to be roughly spherical. The processes of clinker grinding primarily are physical although chemical changes are involved.

The whole scope of processing is difficult to describe in generalities. Each plant is uniquely designed in an effort to apply the best process art and equipment available at the time of that plant's construction to make, economically, products of competitive quality from its pe-

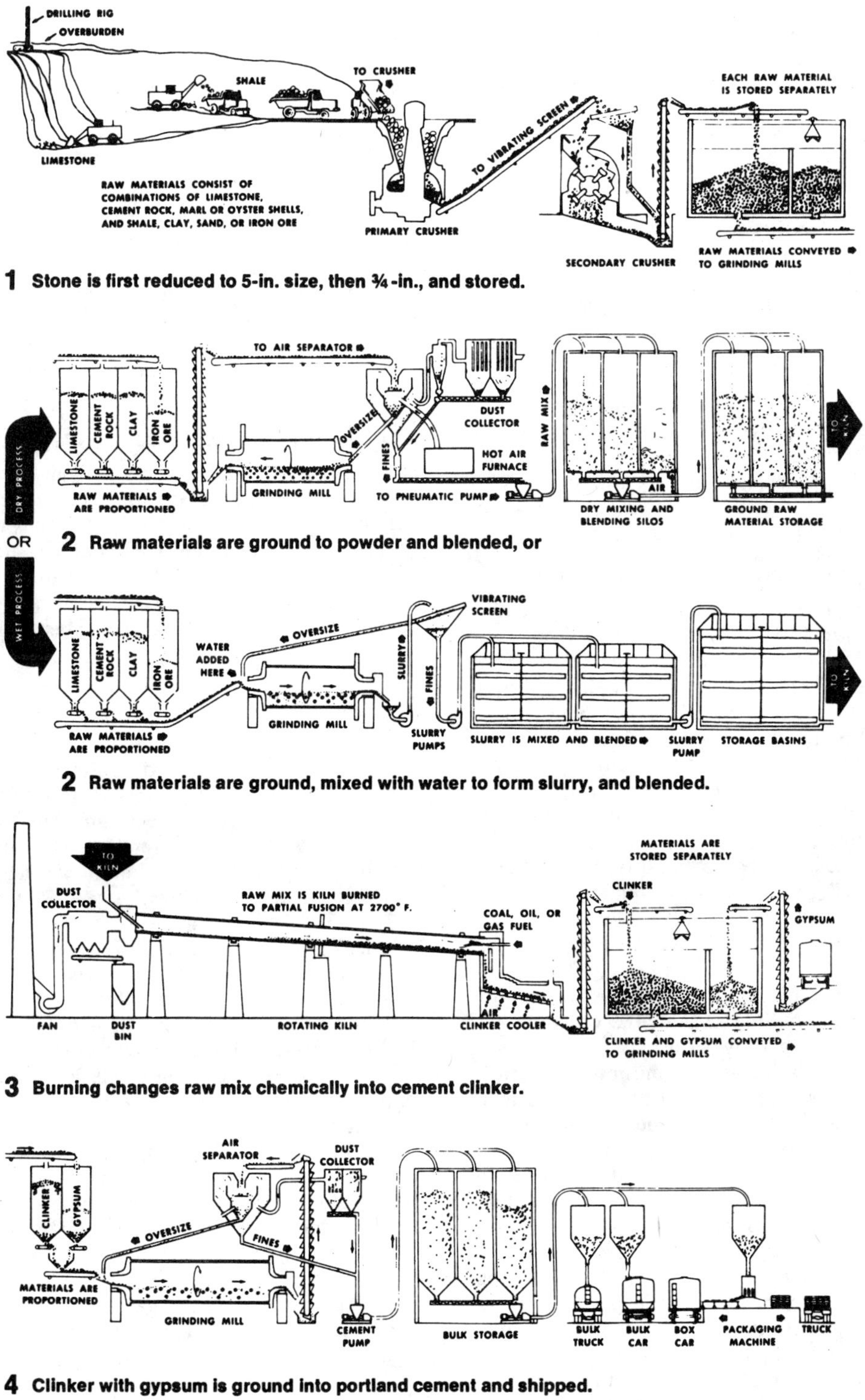

FIG. 1—*Steps in the manufacture of portland cement (Source: Portland Cement Assn.).*

culiar raw materials. Cost and opinions of what is best, as well as environmental considerations, obviously also influence which process flow and equipment are designed into a given plant.

Processing During Recovery

Recovery of the raw materials is a first step in processing. In contrast to preparing various size grades of crushed-stone aggregates, the objective in processing cement raw materials is to arrive at the fine kiln feed sizes as quickly and economically as possible. Drilling, blasting, and breaking are done so as to achieve size-reduction results consistent with the objectives of "fines" production and economy. Beyond these objectives, quarry practice is limited mainly by considerations of the community environment. "Let the explosives do the work" is an often cited rule; however, use of explosives is modified by safety, noise, and vibration standards. *Work* is the term descriptive of the energy applied to do the job of size reduction, and the processes through which work is applied again are a matter of plant design and practice. Quarrying, in the cement industry, most often is described to extend from breaking ground to the delivery of quarried materials into the crushing department. In many quarries blending for quality control begins with selection of working faces because the chemistry of the stone in each face is to some extent known. A variety of methods of quality monitoring is applied. Practice at some quarries includes close diamond-drilling surveillance as quarry faces progress; other deposits are more predictable.

Depending upon the raw material and the design of the milling system, the crushing system might be elaborate or it may not be present at all. In the usual case of limestone, the crushing process is expected to produce material of a given maximum size. Screens generally determine this crusher-run size, which, for various plants, typically is 9.5 mm (3/8 in.) up to 63.5 mm (2½ in.). Primary and secondary crushing stages, often followed by tertiary crushing, produce the crushed product for the mills. Virtually all varieties of crushing machines found in crushed stone plants are also found in the crushing departments of cement plants. Size grading is accomplished mainly to recycle coarse fractions for further crushing, to remove properly sized material from the stream and move it on to the next process, or to facilitate transportation. In a few instances discard of some or all of certain screened fractions is indicated. Crushed stone may move directly to bins from which mills are fed, but more often it moves to storage buildings and then to the bins. The stone storage and the bins provide additional opportunities for blending.

Beneficiation is employed at some plants. Beneficiation may be defined as a process step by which an undesirable component of the raw material is taken out through flotation or by an equally sophisticated technique. That type of beneficiation is employed at very few cement plants. However, the elimination of undesirable components from the recovered materials may be as simple as drying to eliminate water or as complex as roasting to drive off kerogens, and this latter case has to be regarded as a type of beneficiation. Screening, washing, or other methods of segregating unwanted materials for discard also is beneficiation. The term beneficiation may be expanded from the general upgrading of raw material to include the addition of relatively pure materials needed to bring the mix to the desired quality. A large number of cement plants employ this type of beneficiation.

Raw Milling

The primary objective of milling is to prepare sizes and mixtures of raw material for kiln feed that effectively and economically can be processed in the kiln to produce the desired cement. Standards of maximum size and percent through a given screen (say 90% through a 200 mesh screen with none coarser than 50 mesh) are set for the finished raw mill product. Determination of these standards usually is empirical. A sizable fraction of the mill-feed stream may occur in nature in qualifying finenesses, but no plants eliminate milling as a process. The mill acts as a superior mixing device in addition to accomplishing size reduction.

Historically, the early plants all were dry process. As the technology progressed a number of factors led to the development of wet-process plants, prime among them being wet raw materials, the efficiency of wet blending, and the then low cost of fuel. Wet marl, shell, and other raw materials can be handled and processed more conveniently without a drying stage. For many years the wet process dominated the design of new plants. Recent upward surges of fuel costs provide strong incentive for the designing of new plants as dry process. Fairly recent developments in dryer-grinding mills and heat-recuperating preheaters, added to available, efficient dry-blending systems, go far toward removing historical objections to dry-process installations. Conversion of existing wet plants to dry plants poses more difficult problems, and economic justification of this move is not automatic.

Raw-grinding practice varies rather widely, with the biggest difference between wet and dry milling. Nearly all wet plants employ mills ranging from rod or rod-peb to ball and tube mills. Water is added directly to the mills or may be introduced as a clay-wash mill product (clay slip). Wet milling produces a slurry of the ground kiln feed in which the water content is kept as close as possible to the minimum (usually 30 to 45%) that can be pumped and handled successfully. Nearly all wet milling systems employ positive sizing equipment from conventional vibrating screens to DSM screen installations, hydroclassifiers, hydroseparators, and thickeners. Coarse fractions are returned for further grinding and finished fractions are moved out of the milling system. Circulating loads are large. The slurry is pumped to storage tanks or basins, where blending is achieved. Mechanical and/or air agitation are main methods of keeping the slurry mixed. Closely controlled and variable kiln feeding is very important. Because the largest kilns are used in wet-process plants, the equipment employed to accomplish this feeding also is surprisingly large.

In traditional dry plants, the materials are dried between the stone-storage area and the raw mill-feed bins. Dryers may include their own heat sources or may use recuperated kiln heat. During grinding the material moisture content is usually 1% or less. Rotary dryers are the most widely used; other types include combined drying-grinding machines and traveling-grate installations. The variety of mills is greater in dry-process plants, where vertical, roller, and ball race mills are employed as well as ball mills and tube mills. Air separators classify the milled product and return the coarse fraction to the milling system. The circulating loads typically are large.

Dry blending is accomplished in a number of types of equipment, and the cost of compressing air for these operations often is a significant item. Positive feeding of the prepared dry kiln feed provides control out of feed-storage tanks or from preheater installations. Preheaters generally use heat recuperated from the kiln whether the kiln gases are directly in contact with the prepared raw feed or not. Auxiliary heat may be supplied. Among the more widely used designs are those in which incoming feed is exposed to kiln exit gases in large installations of "stacked cyclones" or related designs. European innovation and the wide acceptance of dry preheater were developments which matured in the 1970s.

Although preheating generally results in partial calcining of carbonates, the 1970s also witnessed the development of true precalcining systems. Enough heat is applied in the precalcining equipment to drive off nearly all of the CO_2 before the then partially burned material is introduced into the rotary kiln. A number of precalcining units are installed or under consideration in US plants. These precalcining units are appreciably more complex than traditional clinker-producing equipment, but they promise higher rates of production for the same invested capital (also less expense for refractories). However, preheating and precalcining systems may cause complicated process-imposed alkali problems because of the chemistry of certain raw materials. Understanding the alkali-processing relationships becomes more critical to the evaluation and selection of raw materials and fuels for successful precalcining operations. A primary objective in new designs is the more efficient use of fuel.

Another approach to preparation of kiln feed is to blend dry feed, then pelletize it (requiring addition of some water). The pellets are preheated, usually on a traveling grate prior to being introduced into the kiln. In many plants collected kiln dust is part of the "raw" material. Dust leaching equipment has been installed in some plants to remove undesirable soluble components from the dust. The collected dust may be reintroduced with the fuel stream (insufflation) or through other closely controllable feeding devices.

Burning

Pyroprocessing (burning) is the key process in the manufacture of cement. Burning at high temperatures causes properly constituted and prepared raw materials to react and combine to produce clinker containing a balance of synthetic compounds (minerals and glasses) that can be ground into the desired cement. Rotary kilns produce all domestic clinker.* The kiln-

* Worldwide practice includes vertical kiln installations of many varieties and a number of other types of plants. Raw material variations range from use of gypsum (or anhydrite) in plants that make coproduct cement and sulfuric acid to use of calcitic feldspar or lime, rather than calcium carbonate, as a primary raw material.

cooler department or system is the heart of the cement plant. Basically everything else is sized to serve it, and it should be run as nearly at capacity as is feasible. Plant-capacity expansion or reduction implies changes in kiln capacity.

Rotary kilns essentially are steel cylinders, ranging up to 7.6 m (25 ft) in diameter and 231.6 m (760 ft) in length and lined with fairly thick refractories (mostly "brick"). One function of the refractories is to develop a coating of adhering raw materials, especially in the hot zone, so that finished materials are processed over similar materials. Kilns are inclined slightly from the horizontal so that their rotation (usually 50 to 90 rph) moves the materials being processed from the feed end to the discharge end of the kiln at the desired rate of speed. Retention time of the processed material in the kiln is several hours although the burning zone occupies only about 15% of the kiln length and burning temperatures certainly are more important than the travel time of the material. Variable factors that influence kiln performance are legion, and their interaction perhaps is becoming better understood and more predictable. The effects of each variable and of changes in variables often cause kiln performance to be cyclic. A balanced, smooth operation is an objective in kiln operation.

Fuel is introduced under pressure through a burner pipe. The burner pipe is positioned (variable) in the product-discharge end of the kiln, and the fuel blown through the pipe is ignited into a flame, which extends well up into the kiln. The kiln zone in which the chemical combination of clinker compounds occurs normally begins about 12.2 to 15.2 m (40 to 50 ft) from the discharge end of the kiln. Powdered coal (usually milled in the kiln vicinity), oil, or gas are used depending on local costs and conveniences. As the materials being processed move from the feed end of the kiln toward and into the zone of the flame, they pass through a series of "stages." These stages include heating to lose water, calcination of the carbonates (i.e., the driving off of CO_2), and the fusion, melting, and reacting to form clinker compounds. Following calcination, the main reactions of hot lime with iron, alumina, and silica begin. Burning of the fuel is regulated so that the hottest kiln zone is from 1427 to 1649° C (2600° to 3000° F), much higher than required for calcination. Part of the kiln gas heats incoming raw material and the kiln beyond the burning zone. In kilns with chain sections (both wet and dry), the chain links are important exchangers of heat from the exiting gases to the solids as they move enroute to the burning zone. Gas-exit temperatures from dry-process kilns are much higher than exit temperatures from wet kilns. The thermal economy of dry kilns may thus be improved markedly by the addition of a preheater. Exit temperatures may range from 204 to 871° C (400° to 1600° F). Yet the range in exit-gas temperatures of a given kiln operating properly usually is quite narrow.

The soaring cost of fuel has encouraged widespread experimentation with waste materials, which range from wood, paper, and garbage to used oils, solvents, and chlorinated hydrocarbons. Although cement kilns have proven to be superior incinerators (especially wet kilns), notable successes using waste materials for fuel hardly can be described to have had an impact on industry practice. Use of petroleum coke blended into certain fuel streams continues to be important locally.

The processing of the minerals introduced into the kiln as raw feed changes them to synthetic products, which emerge from the kiln as clinker. Clinker compounds have been intensively investigated and described.* The brief statement descriptive of these compounds by Clausen (1960) is repeated here as Table 2. Successful kiln burning produces clinker that may form the classic rounded clinkers or that may be sand or dust sized. The clinker leaving the kiln is cooled in a wide variety of ways—from primitive pits to rotary units or fan-aerated reciprocating grate units of considerable sophistication. An objective of clinker-cooler design is the efficient reuse of the escaping heat. Cooling and "tempering" of clinker continue to gain stature as to their quality implications. The value of heat recuperation is self-evident.

Finish Grinding

Beyond the cooler, clinker generally is moved by cranes or conveyors into storage, where it may be segregated, tested, blended, and moved into bins for feeding to the clinker-grinding (finish) mills. The type of clinker storage available and the timing of clinker grinding both have quality implications. Clinker that was ground hot and old clinker that has be-

* The Onoda Cement Co. (Japan) was a leader in developing a unique method of analyzing clinker microscopically; although the Onoda method is approximate it is quite practical.

TABLE 2—Probable Cement Compounds

Compound	Abbreviation	Common Abbreviation	Oxide Composition	Stoichiometric Composition	Approx. Content in Type I Portland Cement, %†
Tricalcium silicate (Alite)	3 CS	C_3S	$(CaO)_3SiO_2$	Ca_3SiO_5	45
Dicalcium silicate (Belite)	2 CS	C_2S	$(CaO)_2SiO_2$	Ca_2SiO_4	27
Tricalcium aluminate	3 CA	C_3A	$(CaO)_3Al_2O_3$	$Ca_3Al_2O_6$	11
Tetracalcium-aluminoferrite*	4 CAF	C_4AF	$(CaO)_4(Al_2O_3)$-(Fe_2O_3)*	$Ca_4Al_2Fe_2O_{10}$	8

Source: Adapted from Clausen, 1960.

*This composition of the iron-containing phase is an approximation and may range from $(CaO)_2Fe_2O_3$ to $(CaO)_6(Al_2O_3)_2(Fe_2O_3)$.

†Commercial cements contain 4-6% gypsum or anhydrite (for regulation of the "setting time" of the concrete), approximately 0.5% each of alkali oxides (Na_2O and K_2O) and uncombined CaO, plus a few percent impurities, largely MgO.

come hydrated each produce cement with certain deficiencies.

The SO_3 content of finished cement is critical in providing the desired setting time for concrete made with that cement. Gypsum (3–6%) traditionally has been interground with clinker to provide this SO_3. Anhydrite and other sources of SO_3 have been found satisfactory substitutes for part, but not all, of the gypsum.

Clinker grinding may be accomplished in one type of mill or in a system including more than one type of mill. Rod, ball, roller, race, and tube mills are found in various finish-grinding installations. Nearly all finish-grinding systems are closed-circuit systems in which air separators provide classification. Fine finished product is sent to storage and coarser fractions are returned for further grinding. Careful proportioning of clinker and gypsum feed to the mills is a requisite, and the additions of air-entraining agents or other small amounts of ingredients must be accurately metered.

Good practice includes cooling of cement en route to storage. The product is sampled and its quality is verified before it is loaded for shipment. Concrete-storage silos generally are the major structures of the storage-and-shipping department, and that department is a significantly large capital cost factor in the whole plant.

Dust

Cement processing requires the creation and handling of fine mineral materials. Problems arise through the escape of these materials as dust. Dust collection, for many years, has required elaborate and costly equipment. Within the plant most dust-generating points can be equipped with adequate dust collectors; however, the big problem is the cleaning of kiln-exit gases. Mechanical collectors (including glass-bag installations), electrostatic precipitators, and wet scrubbers all have their own advantages and disadvantages. With legal requirements for collection becoming more demanding, technology gives little assurance that even new plants will operate fully in compliance with the standards. Particulate emissions standards expressed in terms of performance requirements need definition and standardization.

Manufacturing Control

In the broad sense, manufacturing control might be related to nearly every calibration, screen size, speed of drives, and so on, but the term now has come to imply the systems of electric devices that have been developed to control cement processing. Control capabilities have greatly improved in the last 20 years, and controls perhaps more refined than are justified may be imposed on some process steps. Safety of men and equipment has high priority in the establishing of safeguards of a control nature.

Historically, cement-manufacturing control developed through the use of instruments to help guide the operators. Methods of installing the instruments and watching, recording, and reacting to what they sensed were improved, along with the understanding of the information that they provided. From these bases came improvements in instruments, their housing, and finally, in centralized control places and practices. Experience within and outside the cement industry prompted the reaching for automation, perhaps prematurely, and the imposition of computer control.

Objectives

The manufacturer must justify expenditures for manufacturing control on the basis that it improves his costs and his products. Process control objectives relate to specification limits, which define chemical and physical product characteristics, and to the processes used to achieve them. Plant standards further narrow the limits for each specific operation. These parameters provide the reference framework for each process step and define the limits that the control sensors bring under surveillance and within which they call for action. Along with quality improvement, savings on labor, fuel, and power cost, as well as on equipment maintenance, are prime objectives. Effective controls are expected to accomplish smoother and more continuous operations to produce more cement as well as improve brick and chain life.

Modern laboratories can be equipped and staffed to provide quick information on sampled materials. Samples may be taken by continuous, on-line devices or be collected by more traditional hand methods. Computer evaluation of sample analyses (often determined by X-ray spectrophotometer), at many plants, provides information in unprecedented depth.

Devices and Systems

Control devices and sensors usually function to determine temperatures, pressures, speeds, rates of flow, weight, fullness (or emptiness), presence or absence of a component, and so on. A great variety of instruments is used. Closed circuit television in some plants provides better-than-direct views of processes. Analog and digital computers are part of some systems. Computerized evaluation of sensings and printouts of results can be achieved, although computer determinations more frequently signal desired action or reaction through other controls. Mill loads and efficiencies can be deduced by sonic sensing and feed added or withheld accordingly. Flow of electricity, radioactive-beam interruption, viscosity, and many other conditions can be sensed; action or reaction can be initiated as desired. Stopping or starting equipment often is automated; sequenced "off or on" reactions are routine; and variable feeding or withdrawal are part of the control provided for the operation of most plants.

Individual circuits are fitted together in systems, which often become elaborate and complex. Servicing of the systems generally can be boiled down to routine practice although troubleshooting may be demanding. Reliability and miniaturization of equipment continue to improve rapidly, and design of unitized components makes servicing less complicated at the plant control-room level.

Products

Cement has been selected by some to exemplify a one-product industry, but the product and the market are more complex than a superficial view reveals. Within the primary product line, pertinent specifications describe the characteristics of cement.

Five basic types of cement have been developed:

Type I—The "general-type" cement, qualified for a wide range of uses, mainly construction.

Type II—Cement qualified for general concrete construction that requires moderate resistance to sulfates and moderate heats of hydration.

Type III—Cement developed for high early strength uses.

Type IV—Cement developed for uses in which low heats of hydration particularly are desired.

Type V—Cement developed for uses requiring high sulfate resistance.

These five types are modified further or combined with other materials to qualify better for various uses. Cement into which an air-entraining agent has been introduced will produce concrete containing dispersed, tiny air bubbles, which act as buffers against certain types of deterioration (mainly those induced by freeze-thaw). Specification-controlled amounts of air in concrete, due to these bubbles, is sought mainly as modification of cement Types I, II, and III; these variations are designated as Types IA, IIA, and IIIA. The air-entraining agent is interground with the particular type of clinker to produce the "A" varieties of cements. The agents are numerous and many are proprietary chemicals offered under a "market name" rather than a compositional formula.

A number of cements sold under specifications are known by names descriptive of their uses or compositions. Some of these are masonry, oil well, portland-pozzolan, portland blast-furnace slag, waterproofed, and white cements. Natural cements, cement-containing and cementlike products also are prepared and sold under the manufacturer's guarantee whether they conform to accepted specifications or not. Premixed dry concrete or mortar is widely offered to the public; nearly all of it is packaged. Calcium aluminate cement is being produced in larger quantities than ever before.

Portland blast-furnace slag cement usually contains, by weight, 35 to 45% granulated blast-furnace slag although current specifications permit a range as great as from 25 to 65% of slag. Granulated blast-furnace slag of suitable quality is produced from a hot stream [approximately 1371°C (2500°F)] of slag by rapidly quenching it in water or steam and air. Quenching essentially produces a glass that chemically resembles a low-lime clinker. This granulated slag is somewhat more difficult to grind than clinker. When interground to fine sizes with Type I clinker, the slag imparts certain qualities to the finished Type I S cement. Granulated slag generally is cheaper than clinker.

Portland-pozzolan cement has proved itself in large western US dams. An important advantage is cost savings. Portland-pozzolan cement is between 15 and 40% by weight pozzolan. Pozzolans include some diatomaceous earths, opaline cherts and shales, tuffs, volcanic ashes or pumicites, clays, and fly ashes. The ASTM definition of pozzolan is, "A siliceous or siliceous and aluminous material, which in itself possesses little or no cementitious value but will, in finely divided form and in the presence of moisture, chemically react with calcium hydroxide at ordinary temperatures to form compounds possessing cementitious properties." Pozzolans may be natural, prepared, or artificial so long as they have the desired pozzolanic reaction in concrete as it sets. Portland cement hydration, in the concrete environment, liberates calcium hydroxide, which then reacts with the pozzolan, when it is available, to form compounds that are cementitious. Tested alone, pozzolans have little or no cementitious properties. Pozzolans in concrete counteract or prereact with aggregates, which may later cause alkali-aggregates problems if soda or potash are present above specified levels. Noncrystalline or cryptocrystalline substances qualify as pozzolans and include hydrous or opaline silicas, natural glasses such as pumicites, and synthetic glasses such as fly ashes. Clays, shales, diatomites, and many other materials seem to have pozzolanic properties enhanced by calcining.

Specifications

As technology replaces art in cement and concrete manufacture, use, and evaluation, specifications become more discriminating. Product quality requirements are formalized. Definitions of current practice or "state-of-the-art" standards generally are condensed into specifications. The importance of construction and its materials dictates procedures for arriving at meaningful specifications that serve to establish product quality. Quality consistency and reliability generally are predictable over large geographic areas because of specifications control.

The primary framework of specification guidance is that evolved by the American Society for Testing and Materials (ASTM). This society annually publishes formal specifications that result from considerable work and debate of the various committees of specialists, who are the key members of the ASTM. Membership of the committees is balanced with respect to producer and nonproducer interests, and the resulting specifications might be viewed as compromises. However, the standards are better characterized as representing what is possible and acceptable. Consumer interests generally are effectively protected. Other groups importantly involved in formulating specifications or standards (mainly for concrete) include the American Association of State Highway and Transportation Officials (AASHTO), the American Concrete Institute (ACI), the US Corps of Engineers, Bureau of Reclamation, and other government agencies and authorities. A notable trend is that private customers continue to become more knowledgeable and demanding buyers, who dictate specific requirements. Table 3 is a compact statement summarizing cement specifications.

As more complete standards and specifications are accepted for control of a product, regardless of its place of manufacture, a curious set of problems arises out of this uniformity. The cement industry wrestles with problems of a "single product" image, wherein quality becomes less effective as a selling device. Practices of cement marketing resulting from this situation invite charges of oligopoly against the industry because of the "single, uniform-product" image. Informed observers realize that the imposition of uniformity of product by outside agencies bears little relationship to

TABLE 3—Comparison of Requirements as Determined by the American Society for Testing and Materials (ASTM) and the American Association of State Highway and Transportation Officials (AASHTO)[a]

	Type of Cement					
	Type I		Type IA		Type II	
Item	ASTM	AASHTO	ASTM	AASHTO	ASTM	AASHTO
Standard Chemical Requirements						
Silicon dioxide (SiO_2), min. %	—	—	—	—	20.0	21.0
Aluminum oxide (Al_2O_3), max. %	—	—	—	—	6.0	6.0
Ferric oxide (Fe_2O_3), max. %	—	—	—	—	6.0	6.0
Magnesium oxide (MgO), max %	6.0	6.0	6.0	6.0	6.0	6.0
Sulfur trioxide (SO_3)[b], when:						
$3CaO \cdot Al_2O_3$[c] is 8% or less, max. %	3.0	3.0	3.0	3.0	3.0	3.0
$3CaO \cdot Al_2O_3$ is 8% or more, max. %	3.5	3.5	3.5	3.5	NA	NA
Loss on ignition, max. %	3.0	3.0	3.0	3.0	3.0	3.0
Insoluble residue, max. %	0.75	0.75	0.75	0.75	0.75	0.75
Tricalcium silicate ($3CaO \cdot SiO_2$), max. %	—	—	—	—	—	55
Dicalcium silicate ($2CaO \cdot SiO_2$), min. %	—	—	—	—	—	—
Tricalcium aluminate ($3CaO \cdot Al_2O_3$), max. %	—	15	—	15	8	8
Tetracalcium aluminoferrite + twice the tricalcium aluminate ($4CaO \cdot Al_2O_3 \cdot Fe_2O_3$) + $2(3CaO \cdot Al_2O_3)$ or solid solution ($4CaO \cdot Al_2O_3Fe_2O_3$ + $2CaO\ Fe_2O_3$) applicable, max. %	—	—	—	—	—	—
Optional Chemical Requirements (Note: These optional requirements apply only when specifically requested.)						
Tricalcium aluminate ($3CaO \cdot Al_2O_3$)						
for moderate sulfate resistance	—	—	—	—	—	—
for high sulfate resistance	—	—	—	—	—	—
Sum of tricalcium silicate and tricalcium aluminate[e], max. %	—	—	—	—	58	58
Alkalies[f] (Na_2O + 0.658 K_2O), max %	0.60	0.60	0.60	0.60	0.60	0.60
Standard Physical Requirements						
Air content of mortar[g], volume						
max. %	12	12	22	22	12	12
min. %	—	—	16	16	—	—
Fineness[h], specific surface, m^2/kg[i]						
Turbidimeter test:						
Average value, min.	160	160	160	160	160	160
Min. value, any one sample	150	150	150	150	150	150
Average value, max.	—	220	—	220	—	220
Max. value, any one sample	—	230	—	230	—	230
Air permeability test						
Average value, min.	280	280	280	280	280	280
Min. value, any one sample	260	260	260	260	260	260
Average value, max.	—	400	—	400	—	400
Max. value, any one sample	—	420	—	420	—	420
Soundness, autoclave expansion, Max. %	0.8	0.8	0.8	0.8	0.8	0.8
Time of setting[j], alternate methods						
Gillmore test						
Initial set, minutes, not less than	60	60	60	60	60	60
Final set, hr, not more than	10	10	10	—	10	—
Vicat test						
Initial set, minutes, not less than	45	45	45	45	45	45
Final set, hours, not more than	8	8	8	8	8	8
Strength, not less than the values shown for the ages indicated below[k]:						
Compressive strength, psi (MPa)						
1 day	—	—	—	—	—	—
3 days	1800 (12.4)	1800 (12.4)	1450 (10)	1450 (10)	1500 (10.3) 1000[l] (6.9)[l]	1500 (10.3) 1000[l] (6.9)[l]
7 days	2800 (19.3)	2800 (19.3)	2250 (15.5)	2250 (15.5)	2500 (17.2) 1700 (11.7)	2500 (17.2) 1700 (11.7)
28 days	—	—	—	—	—	—

TABLE 3—Continued

Type of Cement

Type IIA		Type III		Type IIIA		Type IV		Type V	
ASTM	AASHTO	ASTM	AASHTO	ASTM	AASHTO	ASTM	AASHTO	ASTM	AASHTO
20.0	21.0	—	—	—	—	—	—	—	—
6.0	6.0	—	—	—	—	—	—	—	—
6.0	6.0	—	—	—	—	6.5	6.5	—	—
6.0	6.0	6.0	6.0	6.0	6.0	6.0	6.0	6.0	6.0
3.0	3.0	3.0	3.0	3.5	3.5	3.5	3.5	3.5	3.5
NA	NA	4.5	4.5	4.5	4.5	NA	NA	NA	NA
3.0	3.0	3.5	3.5	3.5	3.5	2.5	2.5	3.0	3.0
0.75	0.75	0.75	0.75	0.75	0.75	0.75	0.75	0.75	0.75
—	55	—	—	—	—	35	35	—	—
—	—	—	—	—	—	40	40	—	—
8	8	15	15	15	15	7	7	5	5
—	—	—	—	—	—	—	—	20[d]	20[d]
—	—	8	8	8	8	—	—	—	—
—	—	5	5	5	5	—	—	—	—
58	58	—	—	—	—	—	—	—	—
0.60	0.60	0.60	0.60	0.60	0.60	0.60	0.60	0.60	0.60
22	22	12	12	22	22	12	12	12	12
16	16	—	—	16	16	—	—	—	—
160	160	—	—	—	—	160	160	160	160
150	150	—	—	—	—	150	150	150	150
—	220	—	—	—	—	—	220	—	220
—	230	—	—	—	—	—	230	—	230
280	280	—	—	—	—	280	280	280	280
260	260	—	—	—	—	260	260	260	260
—	400	—	—	—	—	—	400	—	400
—	420	—	—	—	—	—	420	—	420
0.8	0.8	0.8	0.8	0.8	0.8	0.8	0.8	0.8	0.8
60	60	60	60	60	60	60	60	60	60
10	10	10	10	10	10	10	10	10	10
45	45	45	45	45	45	45	45	45	45
8	8	8	8	8	8	8	8	8	8
—	—	1800	1800	1450	1450	—	—	—	—
		(12.4)	(12.4)	(10)	(10)				
1200	1200	3500	3500	2800	2800	—	—	1200	1200
(8.3)	(8.3)	(24.1)	(24.1)	(19.3)	(19.3)	—	—	(8.3)	(8.3)
800[f]	800[f]								
(5.5)[f]	(5.5)[f]								
2000	2000	—	—	—	—	1000	1000	2200	2200
(13.8)	(13.8)					(16.9)	(16.9)	(15.2)	(15.2)
1350[f]	1350[f]								
(9.3)	(9.3)								
—	—	—	—	—	—	2500	2500	3000	3000
						(17.2)	(17.2)	(20.7)	(20.7)

TABLE 3—Comparison of Requirements as Determined by the American Society for Testing and Materials (ASTM) and the American Association of State Highway and Transportation Officials (AASHTO)[a]

	Type of Cement					
	Type I		Type IA		Type II	
Item	ASTM	AASHTO	ASTM	AASHTO	ASTM	AASHTO
Optional Physical Requirements (Note: These optional requirements apply only when specifically requested.)						
Heat of hydration, cal/g (kJ/kg)						
7 days, maximum	—	—	—	—	70 (290)[m]	70 (290)[m]
28 days, maximum	—	—	—	—	80 (335)[m]	80 (335)[m]
False set, final penetration, min. %	50	50	50	50	50	50
Strength, not less than the values shown						
Compressive strength, psi (MPa)						
28 days	4000 (27.6)	3500 (24.3)	3200 (22.1)	2800 (19.3)	4000 (27.6) 3200[m] (22.1)[m]	3500 (24.1) 2800[m] (19.3)[m]
Sulfate expansion[n], 14 days, max. %	—	—	—	—	—	—

[a] ASTM C-150; AASHTO M85-79I.
[b] Sulfur trioxide. ASTM specifications state that there are cases where optimum SO_3 for a particular cement exceeds the limit in this specification. Where it has been demonstrated that by Method C563 that this condition exists, an additional amount of SO_3, in no case more than 0.5 by weight percent of cement, is permissible provided that, when the cement with the additional calcium sulfate is tested by Method C265, the calcium sulfate in the hydrated mortar at 24± ¼ hr expressed as SO_3 does not exceed 0.5 g/L. When the manufacturer supplies cement under this provision he will, upon request, supply supporting data to the purchaser.
[c] The expressing of chemical limitations by means of calculated assumed compounds does not necessarily mean that the oxides are actually or entirely present as such compounds.
[d] Does not apply when the sulfate expansion limit under "Optional Physical Requirements" is specified.
[e] This limit applies when moderate heat of hydration is required and tests for heat of hydration are not requested.
[f] Alkalies. This limit may be specified when the cement is to be used in concrete with aggregates that may be deleteriously reactive. Reference should be made to ASTM Specification C33 for suitable criteria of deleterious reactivity.
[g] Air content. Compliance with the requirement of this specification does not necessarily ensure that the desired air content will be obtained in concrete.

whether industry members act in concert in the market.

Consumption

End Uses—Quantities and Trends

The total production of cement is enormous (Table 4). Moreover, cement is peculiar in that nearly all of it actually finds its use as a minority component of concrete. Uses of concrete are diverse, and many do not lend themselves to being as readily defined as do the markets for cement. The status of construction activity largely determines how much concrete is used and how much cement is produced and sold. Identifying the general type of buyer provides one "classification" useful in thinking about consumption (Table 5).

Substitutes—Alternative Materials

In the context of our high technologic competence, the product cement seems vulnerable to substitutes. Capital investment for a single new plant is of a great enough magnitude to prompt the investor to evaluate the long-term threat of substitutes prior to commitment. The current situation appears to be that those materials that compete on the basis of performance cost too much, while those that compete economically do not perform so well in most applications.

Basic and product research by individual cement companies and by the Portland Cement Association for years have suffered substantial cutbacks due to low profits. This situation stimulated predictions that substitute materials were poised to take large segments of the market. Time will tell, but threatened long-range shortages of some of the basic substitute materials weigh in favor of cement. A number of countries place considerable emphasis on substituting concrete products for many of the materials that the United States industry is concerned will competitively replace concrete.

Economic Factors

The cement-industry statistic most meaningful to the customer is the price he pays for delivered cement. The components of that price tend to be lumped into the categories of cost, transportation, discounts and allowances, and profit. Costs may be accounted in detail or they may be so generalized as to be viewed

TABLE 3—Continued

Type of Cement									
Type IIA		Type III		Type IIIA		Type IV		Type V	
ASTM	AASHTO	ASTM	AASHTO	ASTM	AASHTO	ASTM	AASHTO	ASTM	AASHTO
70 (290)[m]	70 (290)[m]	—	—	—	—	60 (250)	60 (250)	—	—
—	80 (335)[m]	—	—	—	—	70 (290)	70 (290)	—	—
50	50	50	50	50	50	50	50	50	50
3200 (22.1) 2560[m] (17.7)[m]	2800 (19.3) 2240[m] (15.4)[m]	—	—	—	—	—	—	—	—
—	—	—	—	—	—	—	—	0.045	0.045

[h] Fineness. Either of two alternative fineness methods may be used at the option of the testing laboratory. However, when the sample fails to meet the requirements of the air permeability test, the turbidimeter test shall be used and the requirements for the turbidimetric method shall govern.
[i] These values were revised in ASTM specifications because of the change in unit from cm^2/g to m^2/kg.
[j] Time of setting. The purchaser should specify the type of setting time test required. In case he does not so specify the requirements of the Vicat test only shall govern.
[k] Strength. The strength at any specified age shall not be less than that attained at any previous testage.
[l] When the optional heat of hydration or the chemical limit on the sum of the tricalcium silicate and tricalcium aluminate is specified.
[m] When the heat of hydration requirements are specified, the sum of the tricalcium silicate and tricalcium aluminate shall not be specified. These strength requirements apply when either heat of hydration requirements or the sum of tricalcium silicate are specified.
[n] When the sulfate expansion is specified, it shall be instead of the limits of C_3A and $C_4AF + 2C_3A$ listed under "Standard Chemical Requirements."

TABLE 4—Annual Production, Shipments, and Capacity of the Cement Industry of the United States and Puerto Rico (1000s of st) §

Year	Production: Portland	Production: Masonry†	Production: Total	Shipment from Mills*: Portland	Shipment from Mills*: Masonry†	Shipment from Mills*: Total	No. of Active Portland Cement Plants	Capacity: Finish Grinding Capacity‡	Capacity: Clinker Capacity
1958	58,556	2,798	61,354	57,729	2,809	60,538	168		75,724
1960	59,974	2,843	62,817	58,711	2,781	61,492	176		81,393
1962	63,260	2,903	66,163	62,383	2,875	65,258	178		88,167
1964	69,303	3,150	72,453	68,865	3,189	72,054	181		90,168
1966	72,311	3,222	75,533	71,570	3,152	74,722	184		93,092
1968	74,243	3,264	77,507	74,720	3,260	77,980	183		95,703
1970	73,168	2,948	76,116	73,407	2,978	76,385	181		83,675
1971	77,007	3,309	80,317	79,005	3,392	82,397	176		85,791
1972	80,744	3,812	84,556	81,432	3,850	85,282	175		85,798
1973	83,476	4,022	87,498	86,399	4,130	90,529	173	100,413	86,882
1974	79,486	3,402	82,888	79,482	3,432	82,914	176	106,223	90,874
1975	66,796	2,925	69,721	67,776	2,908	70,684	174	106,111	92,264
1976	71,227	3,268	74,495	71,922	3,304	75,226	169	104,146	85,419
1977	76,340	3,720	80,060	77,850	3,764	81,614	163	103,880	88,237
1978	81,378	4,102	85,480	83,849	4,150	87,999	164	104,968	89,121
1979	82,071	3,833	85,904	83,390	3,762	87,152	164	106,446	89,694
1980	73,657	3,052	76,709	74,674	3,050	77,724	163	106,902	91,978

Source: US Bureau of Mines *Minerals Yearbooks.*
* From 1973 includes cement imported and distributed by domestic producers only.
† Masonry figures before 1973 include natural and pozzolan.
‡ Not available before 1973.
& Metric equivalent: 1 st × 0.907 184 7 = t.

TABLE 5—Cement Use in the United States (Percent by Type of Customer)

Designation	1963	1966	1968	1971	1972	1973	1974	1975	1976	1977	1978	1979	1980
Ready-mixed concrete	57	59	61	63	64	66	66	64	65	66	67	66	68
Concrete products	13	14	13	13	14	14	15	14	14	14	14	14	13
Highway contractors	12	11	10	9	8	7	7	8	8	6	5	4	4
Building materials dealers	11	8	8	9	8	8	7	8	7	8	8	9	7
Other uses or users	7	8	8	6	6	5	5	6	6	6	6	7	8

Source: US Bureau of Mines, *Mineral Yearbooks.*

as mill (plant) costs and overhead costs. Obviously the more costs are generalized for marketing purposes (arriving at a price), the less the interrelationships of those costs may be understood. The cement companies' usual practice of paying the transportation bill, and then including it in the price to the customer, becomes far more complex than one would suppose. Discounts and/or allowances amount to price-cutting devices although discounts are justified on the basis that they provide early payment, and allowances are the logical means to compete for the large-volume jobs or customers. The remaining component built into price should be profit. For the cement industry, profit has been an elusive goal in recent years in many market areas. Historically, the imposition of government price controls has created special problems with respect to profitability. Multiplant companies often must carry unprofitable operations by realizing sufficient profit in their other plants. Permanent closing of a cement plant has become a very costly move.

Costs

Elements of the cost of cement manufacture may be organized and accounted for in many ways. Changes in systems and practices of accounting are encouraged by new office equipment, tax laws, and evolving methods. Classifying expenditures, selecting categories into which costs fall, and relating them to each other in meaningful terms are formidable tasks. Operating personnel tend to prefer unit-cost accounting so that their surveillance of "jobs being done" is measured in terms that permit comparisons and reflect changes. Unit-cost examples would be: cost per ton, cost per foot drilled, cost per million Btu's, etc. General management may prefer having costs lumped into categories such as costs of men (labor), power, fuel, maintenance supplies, and so on. Discrimination between fixed and variable costs becomes most useful in understanding what happens to costs as a result of plant shutdowns or as a result of sales over a break-even level of operation. Many heroic efforts to lower manufacturing costs are overshadowed by whatever can be saved by operating the plant nearer to its rated capacity. Compared to the effect of raising prices or lowering transportation costs, the savings from lowering manufacturing costs are small.

Each plant has its own unique combination of initial capital, operating, and transportation costs; those costs partly are predetermined by the design, location, and manageability of the operation. In this capital-intensive industry the initial investment is so large that the cost of money is very significant. Particularly notable in the 1960s was the soaring cost of labor. Approximate doubling of the cost of a worker brought the pay of cement industry workers to a new, relatively high level compared to that of workers in other industries.

The 1970s witnessed further escalation of manning costs, but these years were particularly marked by the explosive increase in energy costs. General inflation of capital costs made questionable the wisdom of investment in sorely needed new plants in the US, while larger and more numerous plants characterized the cement industry's moves in other countries.

Transportation

As is often pointed out, essentially every other step in the manufacture of cement is transportation. Transportation of fuels and purchased raw materials is a sizable cost to most plants, as is transportation of raw materials to many plants. Clinker-transportation costs are important in some plants. But, unless otherwise qualified, the industry's transportation costs almost automatically are interpreted to be the cost of transporting finished cement to the consumer.

Cement is heavy in relation to its value so that the cost of moving it is proportionally high. Methods of transport are critical determinants of cost. For cement they importantly include various means of shipment by truck, rail, and barge or other waterborne vessel. Direct delivery to the customer by a single type of carrier predominates although cement that is moved through distribution terminals often moves the first leg of its journey via water or rail and then out of the terminal by truck. The 1960s witnessed schemes of cement transport more complicated than before, and the product often was moved greater distances. Handling of cement through distribution terminals reached its peak in the 1960s, but the trend is now back to direct shipment.

Cement moved from US producing plants is reported by the US Bureau of Mines to be transported by the means given in Table 6. The statistics in Table 6 show that when change is indicated, the cement industry is not slow to move into rather major changes in practice.

Marketing

The direct efforts of selling cement probably do virtually nothing toward increasing the total amount sold within a defined geographic area and within a limited period of time, such as a year. Achievement in the selling of cement long has been measured by a company's success in capturing or improving "its share" of the cement sold in any given market. (The location of US cement plants is given in Table 7.) This type of market change takes place much more dramatically upon the closing of a producing plant than upon the building of a new plant. Important as it is to push volume of sales to or beyond a given *percentage-of-market* goal, it is far more significant to sell the product for prices that will produce a reasonable return on the investment. During the 1960s failure to be reasonably profitable severely eroded the management personnel of many companies. The "soaring '70s" predicted by enthusiastic prognosticators turned out to be very spotty. Lack of the promise of reasonable returns on investment held back new plant construction in the US. During brief cement shortages, imported cement and clinker met the marginal needs.

The percent of capacity at which a cement plant must be operated to break even, say in a given year, is high and orients the marketing of cement toward high volumes of sales. During the 1960s the effects of industry overcapacity were disastrous. In some years prices were dropped to levels at which relatively few plants were profitable. While the economics of each plant and each market area is unique, still those areas overlap, and the problems generated by the condition of overcapacity spread throughout the whole country during the decade of the 1960s. During the 1970s the limited amount of new capacity which came onstream because investment was not justified now is inadequate to cope with brief spurts of high demand.

Selling tools have included the touting of color differences, offering technical assistance, custom-manufacturing of cement, providing facilities or special services, and providing aggregates to be sold along with the cement. Personal relationships of sales people, extension of credit terms, allowance of "customer pickup" (that is, the customer may send his own trucks into the cement plant or distribution terminal to pick up his cement), and price concessions also are among devices used to win (or buy) customers. In retrospect, many of the competition-generated schemes to establish better service for the customer were countered by competitors' lowering the price. Cement company profits sank from the best profit performance of all major industries in 1955 to the poorest in 1967. This experience was damaging.

Government proscriptions generally are shared by all heavy industry, but the cement industry also has received attention from the Federal Trade Commission. Presumably monopolistic practices justify this close surveillance, but this point becomes more debatable

TABLE 6—Percentage of Portland Cement Produced in the United States Transported in Bulk, in Paper Containers, by Truck, by Railroad, and By Water Carrier

Cement Transported	1965	1970	1971	1972	1973	1974	1975	1976	1977	1978	1979	1980
In bulk	88.6	90.8	91.4	91.4	92.0	92.2	92.0	92.0	91.8	92.2	92.2	93.0
In containers (paper)	11.4	9.2	8.6	8.6	8.0	7.8	8.0	8.0	8.2	7.8	7.8	7.0
By truck	68.6	79.8	82.7	82.7	83.5	84.3	86.3	86.6	87.5	89.2	89.8	90.8
By railroad	29.4	18.1	15.7	15.8	15.2	14.3	12.6	11.5	10.6	9.3	8.5	7.0
By water carriers	1.9	2.1	1.6	1.4	1.1	1.3	1.0	1.5	1.6	1.1	0.9	0.9

Source: US Bureau of Mines, *Minerals Yearbooks.*

* Includes Puerto Rico. Figures do not total 100% because a minor percent designated "not specified" was omitted.

TABLE 7—Distribution of US Cement Manufacturing Plants, 1981

West of Mississippi River

Producing District or Area	Specific State and Plant Location	Approx. Capacity, 1000 stpy*
Iowa	Iowa—Davenport, Des Moines, Mason City(2)	2,727
Greater St. Louis	Missouri—Hannibal, Clarksville, Selma, Cape Girardeau	4,000
Western South Dakota	South Dakota—Rapid City	1,140
Kansas City—Omaha	Missouri—Independence Kansas—Bonner Springs Nebraska—Louisville, Superior	2,000
Tri-State	Oklahoma—Tulsa, Pryor Kansas—Humboldt, Chanute, Fredonia, Independence	2,230
Ouachitas	Oklahoma—Ada Arkansas—Foreman, Ada	1,870
Greater Dallas-Ft. Worth	Texas—Dallas, Fort Worth, Midlothian(2), Maryneal, Waco	4,130
Greater Houston	Texas—Houston(3), Orange, Corpus Christi	1,121
Greater San Antonio	Texas—Buda, San Antonio(3), Balcones, Zorn	3,180
West Texas and New Mexico	Texas—Odessa, Amarillo, El Paso New Mexico—Tijeras	1,300
Front Range	Colorado—Boettcher, Lyons, Portland Wyoming—Laramie	1,850
Northern Mountain States	Montana—Trident, Montana City Utah—Salt Lake City, Devils Slide Idaho—Inkom Washington—Metaline Falls	1,850
Arizona	Arizona—Clarkdale, Rillito	1,720
Coastal Washington	Washington—Seattle(2), Bellingham	1.070
Oregon	Oregon—Lake Oswego, Dundee	860
Western Nevada	Nevada—Fernley	400
Northern California	California—Redding, San Andreas, Permanente, Santa Cruz	3,270
Southern California	California—Monolith, Mojave, Oro Grande, Lucerne Valley, Victorville, Lebec, Colton, Riverside(2)	7,380
Hawaii	Hawaii—Ewa Beach, Nanakuki	770
		42,868

East of Mississippi River

Producing District or Area	Specific State and Plant Locations	Approx. Capacity, 1000 stpy
New England	Maine—Thomaston	480
Hudson Valley	New York—Glen Falls, Ravena, Catskill, Cementon	3,200
Lehigh Valley	Pennsylvania—Stockertown, Nazareth(2), Bath, Northampton(2), Whitehall, Evansville	6,120
Extended Capital District	Pennsylvania (SE)—York Maryland—Union Bridge, Lime Kiln, Hagerstown West Virginia—Martinsburg	3,120
Blue Ridge	Virginia—Roanoke	1,200
Carolina's	North and South Carolina—Holly Hill, Harleysville(2), Castle Hayne, NC	3,150
Georgia	Georgia—Atlanta, Clinchfield	1,450
Florida	Florida—Tampa, Brooksville, Hialeah, Miami(2)	3,690
Upper Ohio River Valley	Pennsylvania—W. Winfield, Bessmer, Wampum, Pittsburgh Ohio—Middle Branch, Zanesville, Superior	3,200
Upper Michigan	Michigan—Alpena, Charlevoix, Petoskey	3,460
Detroit-Toledo	Michigan—Detroit, Dundee Ohio—Paulding	2,250
Greater Cincinnati-Louisville	Ohio—Fairborn Kentucky—Kosmosdale Indiana—Speed, Mitchell	3,390
Tennessee	Tennessee—Knoxville, Chatanooga, Richard City, Nashville, Kingsport	1,820
Alabama-Mississippi (Inland)	Alabama—Ragland, Leeds, Roberta, Demopolis, Birmingham Mississippi—Artesia	3,434
Mobile—New Orleans	Alabama—Mobile Louisiana—New Orleans	2,180
Greater Chicago	Wisconsin—Manitowoc Indiana—Buffington, Logansport, Greencastle Illinois—Dixon, Oglesby, LaSalle	3,420
Southern Illinois	Illinois—Joppa	1,300
		46,864
Puerto Rico	Puerto Rico—San Juan, Ponca, Dorado	2,310
		49,174

*Metric equivalent: 1 st × 0.907 184 7 = t.

as time goes by. Increased ownership of domestic cement producers by foreign companies certainly complicates the picture.

The question often is asked as to what government subsidies, direct or indirect, particularly favor the cement industry. Tax provisions of such wide effect as investment credit are held by some to favor particularly capital-intensive industries, and the cement industry certainly is in that group. However, special benefits for the cement industry as a whole due to this tax provision were not large because the industry hardly was in financial condition to make the investments from which these tax advantages might flow. Of course, some companies were engaged in capital programs that happened to be well-timed, by coincidence, with the credit provisions. Pollution control equipment, which required substantial investment, was being installed when the credits were offered. Local bond financing for law-imposed pollution control devices has helped make possible prompt compliance for installations that otherwise may have imposed a difficult financial burden on the companies involved.

Perhaps the subsidy granted the mineral industries that is most often pilloried is the familiar percentage depletion allowance granted by the US Congress to compensate for the physical using-up of nonrenewable mineral resources. The cement industry scarcely is a front-running beneficiary of this eminently sound tax relief measure. Because profitability is necessary to the realization of this source of tax relief, many plants have not benefited significantly from this source through many years. The biggest potential source of this benefit is the calcium carbonate raw materials valued at the kiln feed stage; 14% of this value is allowable. Import tariffs are regarded to be an item of government subsidy by some, although the present level of protection is not high.

Price Trends and Profits

The price of finished cement, a relatively sophisticated product, has had a rather remarkable recent history. Prices of cement decreased from the 1960 high to lows in the middle 1960s, and recovery was delayed until the 1970s. Average prices per short ton sold between 1960 and 1979 are indicated in Table 8.

Contract selling of cement generally offers little advantage to the cement companies. Customers rather freely break contracts, if it is convenient to them, and the producers usually

TABLE 8—Average Prices for Cement Sold in the United States, 1960–1979 (Dollars per short ton)

Year	Engineering News-Record*†	Actual Mill Value‡	Based on Constant‡ 1978 Dollars
1960	21.15	17.80	39.41
1962	21.68	17.48	37.67
1963	21.25	17.03	36.17
1964	21.04	16.98	35.51
1965	20.51	16.75	34.27
1966	21.09	16.59	32.86
1967	21.20	16.72	32.17
1968	21.41	16.80	30.94
1969	22.37	17.04	29.88
1970	22.68	17.69	29.44
1971	24.06	18.74	29.67
1972	25.44	20.31	30.88
1973	25.99	21.88	31.45
1974	30.73	26.52	34.76
1975	36.71	31.09	37.18
1976	40.53	33.86	38.50
1977	43.21	36.36	39.02
1978	46.89	40.70	40.70
1979	53.58	46.24	42.49

* 1960–1972 Converted from price per barrel listed in Table 8, 4th edition of *Industrial Minerals and Rocks* at 5.3 barrels per short ton.
† Yearly average of monthly cement prices in 20 US cities.
‡ From "Portland Cement," *Mineral Commodity Profile,* US Bureau of Mines.

regard themselves bound to honor such agreements. Cost-escalation provisions in such sales contracts are rare. Requirements of the environment-protective laws also force up prices because of the cost of compliance. How rapidly and how far cement prices will move up certainly is a complex question, but they should advance at a rate related to inflation of cost items.

Listening to similar arguments for years, and viewing precious little by way of results, cement company shareholders and boards of directors understandably have become arch skeptics. When will the industry become attractively profitable? Perhaps that day is still remote and unpredictable for the industry as a whole, but certain companies are perceived to be attractive both to acquisitors and to investors. The Red Queen's sage advice to Alice, "it takes all the running you can do to keep in the same place" unfortunately is all too appropriate to describe the experience of many segments of the cement industry. The entry of industrial conglomerate companies into the industry and the diversification of many companies which have their roots in the industry now has a significantly long history. Some criticism has been leveled at US cement producers for devoting some of their resources to diversification, rather than plowing those resources all back into new or modernized cement production facilities. Surely a careful study of actual investments in "bigger and better" cement plants in the 1960s and 1970s would enlighten the critics. Financial results generally have been disastrous. Conversely, some of the diversified companies have realized a rewarding flow of profits from their new venture activities.

Environmental Considerations

How Real and How Binding

The impact on the environment by cement making is assessable and visible. Even a small cement plant is a large industrial operation that processes a large amount of raw material into a large volume of product. As plants increase in size, they are expected to lower their unit cost, and the product partly relies upon its low cost to be marketable in our economy. But concentration of manufacturing volume into larger plants may concentrate pollution problems. Without belaboring the subject of environmentalists' unrealistic standards versus the ingredients of today's culture, informed people should strive to be aware of what can be done and where costs might be driven by standards that crowd the limitations of technology.

It is a fair statement that the industry recognizes that improvements can be made at every plant, and that many have been deferred because of tight money and squeezed profits. This situation is not the crux of the problem. Areas and intensity of environmental impact still need definition and study. Among the fundamental questions critical to decisions that lead to investment are many not within the control of the industry, such as:

1) Will the economy support a steady growth rate and continue construction activity in its approximate percent of the total gross national product?

2) Because CO_2 from cement plants often is cited as a possible pollutant, what precisely is its ecologic impact? Is this problem one that recognizes no national boundaries? If it is found to be critical, how will it be controlled?

3) Will federal, state, and local governmental pollution control agencies select pollution standards so demanding that they can be met in only the newest plants?

4) Will other types of controls be coordinated sensibly with environmental control requirements? For example, will local zoning that arbitrarily denies installations of heavy industry prevail if new, dust-free installations are proposed in those localities? Will price controls continue to be a threat and, when imposed, arbitrarily permit only a low percentage increase in price, when all other factors, including massive, pollution-control expenditures, combine to cause high percentages of increase in costs?

Other factors of environmental concern perhaps are more amenable to solution and should be debated so that progress is made toward realistic actions. For example:

1) Will foreign cement continue to be welcomed regardless of its quality and the impact of such imports on the US cement industry?

2) Practices historically employed in the strip mining of coal, combined with the acid water problems that almost characterize that activity, deservedly have been attacked because of damage caused in some areas. Partly as a result of this situation, all stripping and quarrying have become targets of control legislation. Laws requiring reclamation of strip-mined coalfields, and designed to deal with that type of damage, arbitrarily have been extended to cover other quarry activities. Some types of

quarries, including the deep limestone operations of the cement industry, are so different from coal strip mines that the extended laws are nonsense. Man's occupancy of the areas in which he lives causes and usually accelerates changes of land surfaces. Recovery of large volumes of stone will remove rock in noticeable amounts; hill forms may be modified, open pits dug, or underground mines excavated. In the normal course of man's active use of his environment, other excavations often are made for the purpose of modifying a hill so some flat surface is created on it for use; or an open pit is made for a reason; or an underground cavern is created for storage, or to accommodate some installation. Because these things are so, why not charge responsible planners with devising advantageous uses for holes in the ground that result from normal recovery of raw material? Such planning would constitute multiple use of resources that more nearly is true conservation than is nonuse. We have no choice but to recover minerals from their present locations.

3) The economics of transportation should be determined more precisely so that costs of transportation are known, not only in terms of freight rates, but in consumption of fuels and other resources and in the impact on the environment of the various types of transport used. With this information, the real cost effects would be assessable more accurately. FTC-guided competition might come to be recognized as being costly in terms of the environment as well as the economy.

4) Particularly in pursuit of the relative economics of energy, a vast amount of study has been devoted to process-energy balances of different industries. An attempt perhaps should be made to quantify the value of encouraging relatively efficient cement-producing plants capable of achieving low unit costs, compared to the real cost of creating and maintaining production of an equivalent amount of substitute materials.

5) What should be the goal for new construction, reconstruction, and provision of necessities and amenities for the people of the country in decades to come? What will these programs require by way of cement? What additional US capacity will be needed to meet these requirements?

6) For the cement industry, the cost of compliance with environmental protection standards should be spelled out for various stringencies of standards. As the costs for improving the environment increase, prices of the product must rise accordingly.

Pollution-control equipment in a new plant probably ranges from 10 to 20% of the total plant cost. Adding this type of equipment to existing systems, or replacing it, may be proportionately more expensive. These expenditures produce virtually no profit, and the equipment will increase operating costs and energy requirements. Generally the cement industry is adding the necessary pollution-control devices. Progress obviously will be guided by laws at a number of lawmaking levels, and the need for economic and technical understanding by lawmakers is emphasized. Adequate statements are needed as to present conditions, how the problems will be solved, and how rapidly solutions will be required. Most of the attention has been concentrated on air pollution by particulate matter although other air contamination and water and noise pollution are moving into positions of prominence.

The Environmental Protection Agency is the arm of federal administration now empowered to control the main problems as it defines them. EPA standards necessarily are being developed and changed as are those of many of the state level agencies. One pattern is clear, however; standards will become more and more restrictive. State regulations mostly recognize federal guidelines and can be no less demanding. State and local enforcement will be overridden by federal control mechanisms if the states are slow to move. In any case standards should be applied equitably to all industries. Some plants in some areas may find compliance virtually impossible. Standards for enforcement should as clearly take into account what can be done as what should be done. The environmentalists may ask for controls that are too stringent to be met by today's real competence in pollution control. Such demands would be consistent with their goal of forcing the highest possible degree of pollution control.

Effects on Supply and Demand

Environmental considerations increase the pressure on the cement industry to:

1) Abandon old plants rather than invest heavily in pollution-control devices in usually difficult circumstances.

2) Rebuild and add to those plants that can be modified to come into compliance with the new laws and still be profitable.

3) Critically scrutinize investment in new plants with pollution control equipment which promises to meet present and future environmental requirements. However, these uncertain

new burdens further cast doubt on the wisdom of building new plants. The last two decades of financial performance of the US cement industry probably will continue to discourage domestic companies from investing in new plants. Only the most optimistic are encouraged to venture into new plant construction, if they can afford it.

The net effect of the foregoing, and of all other pertinent factors, can be translated into increased costs. The precise weight of these factors in the decisions to close cement plants in the last decade may not be determinable. Costs imposed by environmental requirements may well have been the last straw in bringing about the decision to shut down some plants, although the move may have been considered for some time. In any case, the abandonment of some plants has an important impact on the financial performance of the surviving cement industry. Relatively frequent plant closings of the 1960s, 1970s, and 1980s without a surge of new capacity coming onstream should bring supply and demand into a better balance. The effect of 1979 price firmness in those market areas where shortages developed spread briefly to help prices in much of the country. Since 1980 the industry has suffered increasingly in the weak economy.

A Summary View

Within the 1960s and 1970s, a comprehensive discussion of changes important primarily to the cement industry are beyond brief review. The dismal financial performance of most of the US cement industry in the recessionary 1980s reemphasizes how dependent cement sellers are on a healthy economy. Preview of the later years of the 1980s promises to be more hazardous than prognosticating the last two decades. Energy sensitivity of cement manufacturing costs is a threatening factor. Cement use is projected to average somewhat above its historic growth rate beyond 1980 into the later years of this century, barring unforeseen changes. Some specific items of development which may be of summary interest are shown in Table 9. Although these are developments of the US industry, the origins of many were in other countries. Precision as to dates is unnecessary and the decades shown roughly allude to the period during which a development or advance generally matured into importance or is expected to do so.

The 1970s witnessed vast changes in the world and the US cement industries. Internationalization is a word which may be appropriate for what resulted, but scarcely any term can describe the problems which started with the Arab oil embargoes. Impacts on Third World countries, which are more dependent on cement than the industrialized nations, are just beginning to be appreciated.

The Canadian cement industry rapidly and preponderantly (more than 70% of its capacity) became internationalized (French, Swiss, Belgian, Saudi, and Lebanese) in the 1960s and 1970s. Part of the effort of the Canadian companies soon was directed toward selling in the US and by 1980 it was far ahead of Japan as the leading exporter to the US. Canadian companies have moved also to buy and build capacity in the US, as have German, Swiss, French, Italian, and other companies. Imports continued to rise. In some years this rise was spectacular, from 3.6 Mt (4,016,000 st) in 1977 to 8.5 Mt (9,413,000 st) in 1979, or 11.1% of plant shipments. In 1981, however, imports dropped to 3.6 Mt (4,000,000 st). Exports in 1981 were insignificant at 272.2 kt (300,000 st).

Equipment and Practice

The broad areas of knowledge from which new and modified cement technology are generated have been developing at remarkably accelerating rates. Cement-oriented technological advance has been broader and deeper than most casual observers realize. *Evolution* better describes cement industry progress than "breakthroughs." Although some breakthroughs have been claimed in the press, we have yet to witness successful, full-scale developments in the US that could be so described. Nevertheless, a fair statement may be that nearly every new operation incorporates features that represent modifications or advancements in equipment, controls, process flow, or/and practice. Some represent significant changes.

Worldwide developments in plant design and practice in the 1960s and 1970s have outdistanced the US industry in many ways. German and Japanese innovations are proving their worth in major cement plants. Particularly of note are the improved preheater kilns, improved roller-dryer mills, and precalciner plants. Those who despair of cement's ability to compete with other materials and disparage the industry's research posture need also to study the bullish picture of cement in the rest of the world. The US is a prime market for whatever is developed by way of new equipment and practice by aggressive, competent individuals, companies, and organizations throughout the

TABLE 9—Some Notable Developments in the United States Cement Industry during the 1960s and 1970s

1960s	1970s through 1980
Manufacturing	
Wet process installations probably reached their peak with giant kiln plants of the 1960s, which will be active for years.	Dry process plant installations have characterized new construction due to rapidly increasing fuel costs.
Quarry – Crushing	
Elevating pan scrapers were adapted for stripping where they can handle the materials and the volume is large. Smaller drill holes, closer patterns, and lower quarry faces became common practice and probably improved safety and decreased nuisance. Dropball breakage mostly replaced secondary blasting of large pieces. Most blasting converted to use of ANFO, but success with other slurries and water-gels indicate blasting agents and methods probably will continue to change. Wheeled loaders and other end loaders were displacing shovels for the loading of trucks both with shot rock and stripping.	Wheeled loaders increasingly have been used as "pick and carry" units to service mobile primary crushers located in the quarries with their crushed product being belted to secondary crushing plants—generally located out of active quarry areas. A higher percentage of impact crushing seems to have prevailed in cement rock quarries. In larger capacity quarries, some return to larger diameter blasthole practice has occurred.
Raw Milling – Drying	
Increase in size of wet grinding mills in the 1960s was spectacular, and overall efficiencies generally were achieved; nearly all were wet process, and some plants included rod or rod-peb mills. DSM screen positive classification systems installed in closed circuit became widely accepted and usually a source of savings or/and capacity relief. Pipeline transportation of slurry over relatively long distances was adapted to cement raw materials.	Roller-dryer milling systems more often have been employed where the raw material permits.
Preheating – Burning – Clinker Cooling	
The ever-increasing size of wet kilns probably reached its maximum. Both wet and dry kiln design effectively were committed to computers with generally good results. Chain use and effect still seems to generate almost as much theory as new results in the 1960s, with many variations being tried.	The dry counterflow preheaters are not new, but today's improved units are in majority of installations of this period. Conversion to the use of coal as the primary fuel has been wide spread.
Finish Milling	
The trend to giant mills with ever higher horsepower characterized finish mill installations in the 1960s. Sonic mill controls came into common use. Cement cooler installations became common.	Mill inefficiencies invited innovations in configuration, grinding media, and grinding environment.
Storage – Distribution	
Sales of cement in paper dwindled to still lower percentages. Long distance water transport increased and trucking made further drastic inroads on rail haulage. Nevertheless, railroads' pressure-flow cars did much to hasten the demise of many cement distribution terminals (terminals continue to be sold or abandoned). Managements more intimately involved themselves in trucking problems than they did in rail carrier problems in prior years.	Belt transport of cement (vs. air pumps) increased for short-distance steady moves. Local shortages of cement permitted some cement plants to choose closer, presumably more profitable markets in which to sell; paradoxically, the same situation created also the willingness of more distant customers to pay more for the transportation of their cement.

TABLE 9—(Continued)

1960s	1970s through 1980
Controls	
Centralized controls were an objective during the 1960s with development of far more discriminating sensors, better input intelligence, and long steps made toward automating responses to some of the input.	Controls rapidly continued to be refined and miniaturized, and whole groups of responses were automated after computerized consideration.
Pollution Control	
Sizing of dust collectors advanced rapidly to keep up with rapidly increasing plant sizes. Efficiency demands were higher than ever. Glass bag and wet scrubber installations came into prominence and electrostatic precipitators were improved to meet stringent requirements.	Costly efforts requiring an impressive array of equipment continued to be made to solve the very difficult technical problem of cleaning stack gases, and other sources of air pollution.
Product	
Shrinkage compensated and rapid setting concretes became available through the developments in cements, and the strong push for the use of fly ash in concrete spread the use of that type of pozzolan. Ferrocement received perhaps more press than use, but may be ready for further development.	Organic compound additions to cement along with curing of concrete through radiation continued to be developed as far as needs and costs justified, because the products had remarkable properties; plastics and cement are being joined to produce better materials rather than struggling to hold their positions separately.
Use Applications	
The 1960s was the second decade, and a far more successful one, in the remarkable growth of the prestressed field. The established technology of shotcrete was improved, and applications, particularly underground in rock, matured in the 1960s. Continuous slip forming of concrete in roads and many varieties of structures moved into new prominence. Lightweight concrete generally matured in its many tested and accepted uses.	Modular systems construction with concrete, so widely used in Euorpe, found a more effective place in the US in the 1970s. Metal fiber reinforcement of concrete found some specialized uses. Intense concentration by government agencies and others on the potential use of fly ash and a slag in finished cement has resulted in the major commitment to develop granulated slag at an East Coast location.
Organizational	
Cement companies entered the 1960s as a diversified group but often characterized by operations-engineering personnel dominating, or having a strong voice, in management. As sales situations became desperate a swing toward more sales-oriented managements took place; however, the move toward financially expert managers was the indicated direction by the end of the decade. With growing diversification, the professional manager, who might not have had an intimate cement background, more often was in charge.	With the vast array of available equipment to provide quick, accurate accounting, and computers aiding decision-making, it seems almost paradoxical that decentralization of management was the established trend of the 1970s. Widespread profit centers with considerable autonomy were either in effect or under formation, in many companies, as the decade ended. Although management's clear charge is to accomplish a proper return on investment, rather than develop fixation on this or that product, social and environmental awareness also will more become an asset to cement company management.

world. While this situation may be regarded by some to be a point of embarrassment, it actually is a positive factor working to the benefit of all. The acceptance of concrete as the primary construction material of the future, as well as of the present, seldom is seriously questioned beyond our borders. Table 10 gives some information on world cement production.

In the future we may see development of ground cements out of materials that have not been pyroprocessed or that have been much less intensively heat-treated than clinker. The adaptation in the US of the European practice of intergrinding high-quality granulated blast-furnace slag with portland cement clinker seems on the threshold of blossoming into a large-scale operation. Cement-metal and cement-plastic combinations will be developed for more specialized and demanding uses. Multiproduct or byproduct plants, which will include cement as one of the products, are expected to be developed. These plants likely would recycle materials, or, in larger proportions than today, be fed certain recycled materials. These materials range from byproduct gypsum waste to concrete itself. Underground locations may be sought and planned rather than avoided, not only for the recovery of the raw materials but for major parts of the plant.

Inefficiencies must increasingly become the target of technology. Better use of fuels and economy of heat may dictate a range of solutions from oxygen-fuel combinations to elaborate multipurpose plants in which cement making is but one step in heat use. Fluosolids reactors may displace rotary kilns, and new ways to apply energy to comminution of materials may make present milling practice obsolete.

General Comment

As technological inefficiencies must be attacked, so also must economic inefficiencies. Such inefficiencies range from plant and company organizational matters to prudent use of national resources. Selection of the best course of action may be most difficult. Apparent solutions so often create new problems of even greater magnitude than the original problems. The responsibility of finding better ways of doing things is clear, but oversimplified solutions invite trouble.

Assuming the experience of the 1960s provided valid lessons, we should know that a rush to construct new cement plants is the sure way to diminish the amplitude and shorten the duration of the positive upswing in this cyclic business. Acceptance in the 1960s that the basics of the business were on the brink of vast changes led to construction of giant wet plants designed to ship great distances and to force competition out of business. Implementations of these premises, at least partly erroneous, are now recognized as exceptionally costly mistakes. Permitting or planning long-distance transpor-

TABLE 10—Hydraulic Cement Production, World and by Country (Top 20 countries as of 1980—1000s of short tons)

	1960*	1970	1971	1972	1973	1974	1975	1976	1977	1978	1979	1980
USSR	50,314	104,993	110,596	114,970	120,703	126,925	134,482	136,957	140,055	139,944	135,597	138,000
Japan	24,933	63,040	65,515	73,120	86,007	80,588	72,220	75,742	80,620	93,550	96,787	97,144
US	63,043	74,619	80,316	84,556	87,498	82,888	69,721	74,495	80,058	85,480	85,904	76,709
China	17,701	11,000	25,300	31,240	33,880	27,560	33,100	44,100	59,250	71,898	81,461	89,728
Italy	17,214	36,460	35,052	36,882	39,961	40,024	37,738	40,044	41,580	41,621	44,250	47,206
West Germany	27,719	42,246	45,209	47,559	45,040	39,658	36,945	37,646	35,544	37,433	39,101	39,460
France	15,676	31,978	31,910	33,339	33,863	35,649	32,615	32,401	31,779	30,892	31,773	32,082
Spain‡	6,151	18,411	18,916	21,495	24,511	24,514	26,422	27,780	30,859	33,327	30,768	31,370
Brazil	4,916	9,923	10,806	12,545	14,709	16,440	19,221	21,105	21,123	24,361	26,800	29,200
Poland	7,293	13,426	14,420	15,417	17,143	18,480	20,393	21,826	23,479	23,920	21,138	20,330
India	8,574	14,929	16,418	17,306	16,535	15,745	17,895	20,547	21,010	21,561	20,133	19,300
South Korea	477	6,418	7,575	7,150	9,008	9,747	11,165	13,087	15,648	16,681	18,092	17,230
United Kingdom	14,932	18,802	19,508	19,894	22,037	19,603	18,625	17,394	17,037	17,540	17,791	16,323
Romania	3,379	8,958	9,395	10,154	10,856	12,340	12,699	14,427	15,295	16,191	17,196	17,200
Mexico	3,425	7,915	8,113	9,482	10,788	11,679	12,800	13,871	14,580	15,494	16,600	17,640
Turkey	2,255	7,024	8,320	9,286	9,868	9,846	11,966	13,576	15,248	16,676	15,199	15,650
East Germany	5,567	8,804	9,340	9.763	12,125	11,132	11,743	12,500	13,334	13,802	14,330	13,717
Greece	1,767	5,344	6,106	6,986	7,117	7,738	8,565	9,640	11,667	12,434	13,336	14,440
Taiwan	1,309	4,745	5,559	6,272	6,586	6,802	7,491	9,644	11,376	12,634	13,114	15,501
Canada	5,718	7,945	9,066	9.976	10,884	11,436	10,763	10,609	10,588	11,638	13,046	11,571
World	350,633	629,645	679,948	728,601	780,344	775,183	774,277	822,418	872,894	938,095	957,791	877,623

Source: US Bureau of Mines, *Minerals Yearbooks*

Converted from barrels at 5.3 barrels per short ton.

† Includes Puerto Rico

‡ Includes Canary Islands

tation of a commodity like cement eventually may be equated to planning to be inefficient. Ringing pronouncements in the late 1960s that plant obsolescence dictated a new round of plant construction amounted to the suggestion that the maximum effort to make cement slightly cheaper was in some way beneficial. Perhaps the industry's inability to pay for another round of new plants was the only reason another surge of then-unneeded capacity was avoided. What really was needed was enough price relief to provide resources to improve plants, to attract investors, and to meet sensibly the needs of the times rather than to dodge from one crisis to another.

Cement-company financial troubles of the 1960s relatively increased the strength of the involved unions. The weakened companies were unable to resist strike threats and were prone to unreasonable labor-contract concessions. When the closing of old plants seems prudent, if not mandatory for economic reasons, high costs of closings due to the union contracts tend paradoxically to force continued operation of old plants. Individual pay levels for cement workers moved in the decade from a modest position relative to other industries to a position near the top.

Many other cost factors inflated at more accelerated rates than those of the general economy. Competition intensified and the depressed prices that resulted caused low return on investment, mortality of management, and disorderly churning of organizations. As all of the factors came into play the cycle bottomed out, and the early 1970s promised a new era for cement prices and for the industry. The oil embargo and many other factors combined to make a shambles of hopes and predictions. Actually, real shortages developed, with 1973 and 1978 usually cited as peaks of the demand. The outcome of the shortages seemed more clearly to be an invitation to foreign cement producers than to provide any general benefit to the traditional US cement industry. Some observers point out the changing patterns of ownership and apparent strengthening of a few old-line US companies. The 1980s may prove to be a rigorous period of testing these companies, and the outcome is beyond prediction. Internationalization of the US cement industry will play a major role in the foreseeable future.

The pressures caused by cement shortages of the early 1970s may be repeated if the economy improves. Historically, cycles in industry performance have been the rule, and their peaks and valleys progressively seem shorter-term and more pronounced. One of the marketplace's answers to the shortages is to encourage importation of more and more foreign-made cement and clinker. The industry condition that promoted this situation, again, was overcapacity in the recent past and resultant depressed prices. Even today's inflation in virtually every cost area has not been adequately compensated by price increases. The temporary prosperity of exceptionally high-capacity utilization is vulnerable to even modest dips in this cyclic business, unless prices can be improved.

Bibliography and References

Literature and information sources dealing with the subject matter of this chapter are vast beyond assessment here, even if they were limited to writings in English and mainly generated in the United States. The magnitude of sources may be hinted by the fact that over 60 national "associations" exist that have something to do with cement and concrete. Government publications related to the subject fill a small library. This bibliography includes representative and major sources pertinent to the subject. Bibliographies of the literature cited here contain references that would comprise an adequate library on the subject. Publications of associations, societies, and institutes, along with other "trade journals," are important current sources of information, as are knowledgeable people in cement-related companies, design engineers, materials and equipment suppliers, and product users.

This bibliography is organized under headings intended to be descriptive.

Standard Reference "Classics"

Blanks, R.F., and Kennedy, H.L., 1955, *The Technology of Cement and Concrete,* John Wiley & Sons, New York, 422 pp.

Bogue, R.H., 1955, *The Chemistry of Portland Cement,* 2nd ed., Reinhold Publishing Corp., New York, 793 pp.

Clausen, C.F., 1960, "Cement Materials," *Industrial Minerals and Rocks,* 3rd ed., J.L. Gillson, ed., AIME, New York, pp. 203–231.

Eckel, E.C., 1928, *Cements, Limes and Plasters,* 3rd ed., John Wiley & Sons, New York, 401 pp.

Harben, P., 1979, "The Clamor for Cement," *Industrial Minerals,* No. 143, Aug., pp. 52–53.

Lea, F.M., 1971, *The Chemistry of Cement and Concrete,* 3rd ed. (1st American ed.), Chemical Publishing Co., New York, 760 pp.

LeLonde, W.S., and Janes, M., 1961, *Concrete Engineering Handbook,* McGraw-Hill, New York, 1216 pp.

Lesley, R.W., 1924, *History of the Portland Cement Industry in the United States,* Chicago International Trade Press, 330 pp.

Meade, R.K., 1926, *Portland Cement,* Chemical Publishing Co., New York, 707 pp.

Witt, J.C., 1947, *Portland Cement Technology,* Chemical Publishing Co., New York, 346 pp.

Some "Trade Journals" Include

Cement and Concrete Research, published bimonthly, Pergamon Press, New York.
Concrete Construction, published monthly, Box 555, Elmhurst, IL 60126
Concrete Products, published monthly, 300 W. Adams St., Chicago, IL 60606
Engineering News-Record, published weekly, McGraw-Hill Bldg., 1221 Avenue of the Americas, New York, NY 10020
Minerals Processing, published monthly, 865 Busse Highway, Park Ridge, IL 60068
Modern Concrete, published monthly, 105 W. Adams St., Chicago, IL 60603
Pit & Quarry, published monthly, 105 W. Adams St., Chicago, IL 60603
Rock Products, published monthly, 300 W. Adams St., Chicago, IL 60606

Some Organizations and Some of Their Publications

American Association of State Highway and Transportation Officials, 917 National Press Building, Washington, DC 20004
American Ceramic Society, 65 Ceramic Drive, Columbus, OH 43214; *Journal of the American Ceramic Society.*
American Concrete Institute, Box 4754 Redford Station, Detroit, MI 48219; *Journal of the American Concrete Institute,* published monthly, *ACI Manual of Concrete Practice* (3 parts), current edition, ACI Special Publications.
American Institute of Mining, Metallurgical, & Petroleum Engineers, Inc., 345 E. 47th St., New York, NY 10017; *Mining Engineering*, published monthly.
American Society for Testing and Materials, 1916 Race St., Philadelphia, PA 19103, *1980 Annual Book of ASTM Standards,* "Part 14, Concrete & Mineral Aggregates."
US Dept. of Defense, Corps of Engineers, various publications; e.g., *Handbook for Concrete and Cement,* 1949, with loose leaf revisions, 928 pp.
US Dept. of the Interior, Bureau of Mines; *Minerals Industry Survey* (Cement Monthly), *Minerals Yearbook* (Annual Cement Chapter).
US Dept. of the Interior, Bureau of Reclamation various publications; e.g., *Concrete Manual* (current edition), usually 600 to 700 pp.

Other Publications Since 1960

Hansen, W.C., and Offutt, J.S., 1969, *Gypsum and Anhydrite in Portland Cement,* 2nd ed., United States Gypsum Co., 81 pp.
MacDonald, L.P., et al., 1977, "Burning Waste Chlorinated Hydrocarbons in a Cement Kiln," Environmental Protection Service Report No. EPS–4–WP–77–2, 221 pp.
Peray, K.E., and Waddell, J.J., 1972, *The Rotary Cement Kiln,* Chemical Publishing Co. Inc., New York, 194 pp.
Queen, D.M., 1971, *The U.S. Cement Industry* (*A Memorandum*), Evans & Co., Inc. (investment brokers), New York, 50 pp.
Randall, R., Dec. 1961, "A Reference for Manufacturing Process Department Literature," MP-100, Portland Cement Assn., 172 pp.
Taylor, H.F.W., et al., 1964, *The Chemistry of Cements,* 2 vols., Academic Press, London and New York, 442 pp.
Troxell, G.E., Davis, H.E., and Kelly, J.W., 1968, *Composition and Properties of Concrete,* 2nd ed., University of California, Berkeley, Civil Engineering Series, 513 pp.

Construction Materials

Dimension and Cut Stone

W. ROBERT POWER *

Dimension stone is considered by many the premium material for beauty and durability in institutional and monumental construction. Nonetheless in the United States it commands an ever decreasing share of the total market. At the beginning of the century dimension stone accounted for more than half the stone produced in the United States, but by 1981 it accounted for little more than 0.1% and the number of producing quarries and finishing plants had greatly reduced.

This decrease in use is accounted for in part by competition from cheaper materials, but another factor seems to be the demand by many designers for uniformity and exact repetition of modular units in construction. Even where stone is used in large buildings modern architects want blandness of pattern and uniformity of color in the stone. Because most natural stone contains variations in color and pattern great strain is put on the quarriers to overcome this and produce a uniform product. This results in increased costs, increased waste at the quarries, and consequently increased price in the market. In spite of this, stone is still looked on as the most prestigious of building materials and its aesthetic appeal is born out by the many simulated stone patterns found in synthetic materials.

End Use and Specifications

The principal uses of dimension stone are in building construction and monuments. Other uses include flagging, curbing, paving, millstock slate, roofing slate, laboratory furniture, and refractory brick. Because many of these uses overlap, an orderly classification of dimension stone uses is most difficult. Various classifications have been used by Currier (1960), Barton (1968), the US Bureau of Mines yearbooks, and the Building Stone Institute (Anon., 1972).

Building Stone

Building stone includes all natural stone used in building construction, whether it be for structural support, curtain walls, veneer, floor tile, or strictly ornamental use. The US Bureau of Mines (USBM) includes all forms of building stone under the general classification of dimension stone. However, in the industry the only true dimension stone is that which is cut and finished to predetermined size and finish (cut stone of this chapter). For purposes of this chapter all building stone will be included as dimension stone.

Natural stone is probably man's oldest building material and for centuries has been the ultimate standard for quality construction. Other materials have made deep inroads into the use of natural stone, but the aesthetic appeal of stone is attested to by the attempts to simulate it in the manufacture of synthetic materials. Today, natural stone is still considered by many to be the ultimate in aesthetic appeal, durability, and ease of maintenance.

Rubble and Rough Construction Stone: Large blocks of rough-hewn granite or other stone have long been used in such structures as retaining walls, seawalls, and bridge work. Requirements for such stone are that it resist weathering and erosion and that it break naturally into rectangular shapes that are easily set. Granite fits these requirements admirably and is the most widely used stone for this purpose.

Smaller sizes of broken stone have recently become very popular as wall facing material in building construction, both for homes and small commercial buildings. The primary purpose is aesthetic—it replaces brick or other veneer—but ease of setting, weather resistance, light weight, and the ability to bond well to mortar are also important. The stone can be

* Professor, Georgia State University, Atlanta, GA.

produced as a byproduct from other stone operations or it can be produced from small pits with a minimum of capital investment by small operators. The only requirements are that it find a market and that it can be transported easily and cheaply to that market.

Because of the ease of production there are many small producers supplying local markets. In 1981, some 260 companies produced dimension stone in 425 quarries located in 39 states (Singleton, 1982). Some of the sellers quarry their own stone, but some also contract for the exclusive right to sell stone produced by a small operator in return for supplying advertising and distribution services.

The rock types advertised by the Building Stone Institute include all the standard dimension stone types as well as "obsidian" (which is actually pumice) and "specialty stones." Some of these have national distribution.

Rubble is set in random patterns. The size of pieces ranges from about 0.05 to 0.4 m^2 (½ to 4 sq ft) of exposed rock. A few years ago slabs of thin-bedded natural flagstone were commonly used with the bed set parallel to the wall, but recently the three-dimension effect has become more popular and natural flagstones are more commonly set in courses with the bed perpendicular to the wall (see next section). Nonetheless slab-shaped rocks have an advantage in that they cover more surface with less volume and weight. Low density rocks such as pumice and corkstone have several advantages. Shipping costs are lower, setting is easier for the stone mason, and fewer or no anchors are required to tie the stone veneer to the wall.

Ashlar: Ashlar consists of rectangular pieces of stone of nonuniform size that are set randomly in a wall. The exposed surface of each piece is generally less than 0.4 m^2 (4 sq ft). The surface may be sawed (smooth face) or rough-hewn (rock face). Ashlar blocks are prepared to specified thickness, usually about 76.2 mm (3 in.). Ashlar may be sawed from quarry blocks that are not suitable for finishing to cut dimension stone or it may be made from rocks that break naturally into rectangular pieces. Although smooth face ashlar was very popular in the past the rough-hewn look of rock face is more popular today.

Split-faced ashlar consists of brick-shaped blocks of stone that are laid in courses. Blocks of marble or limestone are sawed into modular thicknesses and then split with a power-driven wedge called a guillotine. The blocks are trimmed and laid in courses with the broken face showing. Natural flagstones (sandstone, slate, limestone) are similarly broken into brick-shaped pieces differing only in that the thickness is not precisely controlled. In the limestone and marble industry, ashlar consumes much stone that is considered "waste" in the production of cut dimension stone.

Cut Stone: Cut stone includes all building stone that is cut to precise dimensions on all sides. The surface may be textured, smoothed, or polished, but is always controlled and finished to specifications. In the industry this is known as dimension stone whereas rubble and ashlar are not. Cut dimension stone may be quarried and finished by a single producer or it may be finished by manufacturers who purchase quarry blocks or sawed slabs from a quarrier.

Accurate data on production of cut dimension stone are difficult to obtain. The USBM publishes annual reports on the production and value of various lithologies, but these are deceptive because they are reported in volume units whereas most cut stone used in construction is sold by the square foot. The production and finishing of cut stone is essentially a value-added business where actual cost of raw material is a small portion of total cost.

Specifications for cut dimension stone are stricter than for other categories of building stone. The most important physical tests as specified by the American Society for Testing & Materials are listed in Table 1 (Anon., 1966). In addition to the tests listed in Table 1, the coefficient of expansion and thermal conductivity are important in some applications.

Most cut dimension stone is used as curtain wall or veneer. The use of natural stone as structural members in buildings is almost nonexistent in modern architecture, but columns, lintels, and other structural members may be faced with natural stone.

Granite, limestone, and sandstone are generally used in slabs 76.2 to 127 mm (3 to 5 in.) thick. The stone is set as a nonload-bearing veneer with horizontal support at each floor and appropriate anchors to the structural wall behind. The load-bearing requirements of the stone are therefore generally limited to self support for a floor height despite the rather high requirements for compressive strength called for in the ASTM designations.

Marble more commonly is cut to a thickness of 22.2 to 31.8 mm (⅞ to 1¼ in.). The stone may be used as a facing on precast concrete or other structural backing material, or it may be

TABLE 1—ASTM Specifications for Building Stone

C97-47	Absorption and bulk specific gravity
C99-52	Modulus of rupture
C119-50	Definitions of terms relating to natural building stones
C120-52	Flexure testing of slate, modulus of rupture, and modulus of elasticity
C121-48	Water absorption of slate
C131-55	Abrasion resistance by Los Angeles test
C170-50	Compressive strength of natural building stone
C217-58	Weather (acid) resistance of natural slate
C218-48T	Combined effect of temperature cycles and wash salt solutions on natural building stones
C241-51	Abrasion resistance of stone subjected to foot traffic
C406-58	Roofing slate
C503-62 (R64)	Exterior marble
C543-64T	Slate blackboards
C568-67	Dimension limestone
C615-68	Structural granite
C616-68	Building sandstone
C629-68	Structural slate

hung using special systems developed by the Georgia Marble and Vermont Marble companies consisting of an arrangement of struts and angles which fit into grooves cut into the edge of the slabs. All of the methods rely on mechanical anchors to hold the stone and elastic caulking to seal the joints. They all have the advantage of easy placement on the building; some have the additional advantage of light weight.

Thin panels offer two advantages: savings in the cost of the stone, and savings in weight resulting in the use of less structural steel to support the stone. Because of the increased cost of structural steel, a recent development is to cut granite into 31.8-mm (1¼-in.) slabs, thereby saving weight.

Slate generally is used in natural thicknesses, commonly ranging from 6.4 to 50.8 mm (¼ to 2 in.) and with natural cleft or face showing. Honing the surface destroys the natural beauty of the stone and increases the chance of deleterious weathering. Slabs too thick for a particular application may be ground on the reverse side to specification. Slate is used in curtain wall construction, as facing material, and for flooring.

The systems approach in which natural stone is used in conjunction with other structural materials is a most significant recent development in the use of dimension stone. The systems approach developed naturally in conjunction with the use of very thin slabs in the marble industry; now limestone, granite, and slate are being individually suspended on the face of the building so as to allow movement at the interface. Elastic jointing compounds have made possible weathertight seals. Limestone is commonly preassembled into units such as column covers and window frames using epoxy glue and metal supports. The units are furnished with brackets for easy attachment to hangers built into structural members of the building.

A trend in the application of all kinds of dimension stone is toward use of larger slabs. It is not uncommon for limestone and granite panels 3.7 m (12 ft) long to be used in single floor to floor spans. Marble slabs are more limited in size, seldom exceeding 1.8 m (6 ft), but marble-faced panels backed with precast concrete or asbestos-cement material may be produced in single units for floor to floor installation. Because many of the large slabs are supported at top and bottom only, producers and producer associations have adopted wind load and other design specifications relating size of the panel to the thickness of the stone in order to give adequate strength. Furthermore, these requirements have put additional burdens on the quarrier because no physical flaws in the stone along which separation might occur can be tolerated. Whereas smaller sized pieces could be cut between natural flaws in a block, the quarrier is now called on to produce large, virtually flaw-free blocks of stone.

Color and pattern are inherent qualities of stone, but they can be influenced by the finish. Polished finishes tend to darken color and bring vivid contrasts. Rough textured surfaces tend to lighten color and subdue patterns. Only granite and serpentine will keep a polished surface in exterior work. Polished marble is used extensively on interior walls where the natural color and veining of many exotic marbles are used to advantage. In exterior work uniform color and subdued pattern are currently in favor; hence rough textured finishes are currently fashionable. These surfaces are achieved by flame treating granite. In the marble industry a textured surface is achieved with a steel rolling machine that rubs steel shot across the surface of the slab.

Granite and serpentine marble (verde antique) are favored stone for use in the base course and entrance features of large buildings. These are the only natural stones that will hold a polish in exterior use and polished surfaces are easily maintained as well as attractive.

Although seldom actually specified, the long-term availability of a particular stone may be an important factor in its selection for some uses. This is because of the possibility of additions or repair work which must match the original structure. A related factor is that many architects seem to prefer established building stones with which they are familiar. These considerations give long established quarries an advantage over new or untried stone despite any other specifications.

Monumental Stone

Monumental stone encompasses a variety of products ranging from simple grave markers, to statues, mausoleums, and elaborate structures such as the Lincoln and Jefferson memorials. In construction and fabrication the larger memorials and mausoleums differ little from building construction except that they may contain more elaborate carving.

Granite and marble, particularly white marble, are the preferred stones for monumental use. According to USBM statistics more than half the granite quarried is used for monumental purposes. Georgia, Vermont, and Alabama are the largest producers of monumental marble, but the producing companies produce more building than monumental stone.

Standard specifications for building stones do not apply to monumental stone unless the purchaser chooses. The principal requirements are for uniformity of texture and color, freedom from flaws, suitability for polishing and carving, and resistance to weathering. These qualities either are aesthetic, and hence not amenable to standard specifications, or are qualities best determined by experience. So far no one has developed a satisfactory test to predetermine the resistance of a stone to weathering.

Flagging

Flagging consists of thin slabs of stone used as paving. Slate and fine-grained sandstone are the most used rock types, but any naturally fissile stone can be used provided it has sufficient abrasion resistance.

Flagstone slabs are generally 25.4 to 76.2 mm (1 to 3 in.) thick but thinner slabs may be used where they are placed on an exceptionally firm base and subjected to only light traffic. Flagging may be sawed into rectangular shapes or may be produced in naturally irregular shapes. The largest production of cut flagging is produced from bluestone and slate. Much slate flagging is produced from quarry blocks that are not suitable for mill stock or other higher priced products. Random-shaped flagging is produced by small producers for local markets all over the country. A standard for abrasion resistance of stone subjected to foot traffic is given in ASTM designation C241–51.

Roofing Slate

Roofing slate was once used much more widely than at present. The annual dollar volume of roofing slate declined from $5,186,000 in 1908 to $2,302,000 in 1978. It is still used in some new construction and for repairs and restoration of older buildings. It is considered by many to be the most permanent type of roof covering. Standard specifications for roofing slate are given in ASTM designation C406–58.

Mill-Stock Slate

Smooth-finished slabs of slate are used for electric switchboard panels, billiard tables, blackboards, counter tops, and similar purposes. Electric slate must have high electrical resistance and be free of magnetite and other conducting minerals. Specifications for slate blackboards are given in ASTM designation C543–67. Either cleaved or smooth-finished slate may be used as hearthstones, mantels, steps, sills, or other structural purposes. Standard specifications for structural slate are given in ASTM designation C629–68.

Curbing and Paving Blocks

Granite curbing is used in large quantities because of its superior resistance to abrasion and weathering. In areas of abundant freeze-thaw cycling where deicing salts are used, granite has proved an especially resistant material. Granite curbing is broken out along the rift and grain of the stone, but it is also possible to break out curved pieces by close drilling.

Small rectangular blocks of granite were formerly used to pave roads, docks, freight yards, and other areas subjected to heavy traffic. This market has all but disappeared owing in part to the substitution of rubber-tired for steel-rim wheels. A small amount of granite paving is used for ornamental paving and in a similar application much granite is used to cover slopes near bridge abutments and other areas where slope stability and erosion is a problem.

Miscellaneous Uses

Other uses of dimension stone include laboratory furniture and sinks in which soapstone is considered the premium material. Sandstone, quartzite, and granite are used to line tube mills for grinding ore or other material. Precision surface plates for mounting optical instruments and fine machine work are made from granite. Sandstone and soapstone have been used as refractory "brick." In the past large numbers of grindstones and millstones were made from sandstone or granite. Arkansas novaculite, a fine quartzite, is fashioned into hones. Granite cylinders 6.1 to 9.1 m (20 to 30 ft) long and 1.5 m (5 ft) or more in diameter are used to grind pulp in paper mills.

Lithologic Classification

Almost every variety of rock can or has been used as dimension stone. The suitability of a particular stone for use as dimension stone is governed primarily by physical properties and aesthetic appeal. Mineralogy and chemistry are, for the most part, important only indirectly as they bear on the physical properties. The exception to this statement is the resistance of a rock to weathering which is profoundly affected by mineralogy and chemistry as well as by physical properties. Standard geologic nomenclature and classification of rocks is needlessly cumbersome for an industry which is dominated by nonprofessional, sales-oriented people and architects.

The American Society for Testing & Materials (ASTM) has adopted standard definitions for granite, limestone, marble, greenstone, sandstone, slate, and certain subdivision and varieties of each (ASTM designation C 119–50). The Building Stone Institute also gives a glossary of words including rock types in its annual *Stone Catalog*. These commercial definitions recognize and respect history and use within the industry. Although an attempt to keep the commercial definitions compatible with scientific terminology has been made, discrepancies still remain. The following sections discuss both scientific and commercial definitions.

Granite

Commercial Definition: Commercial granite includes all feldspathic rocks of visibly granular or gneissic texture. It thus includes granite family rocks as defined below plus syenite, gabbro, anorthosite, and other plutonic igneous rocks. In commercial usage fine grained igneous rocks such as diabase or basalt may be called "black granite."

Commercial granites are commonly named for the quarry or location where they are found, with names modified by adjectives giving color, texture, or some exotic description.

Geology: To the petrologist granite is a visibly crystalline rock with interlocking texture composed essentially of alkali feldspar and quartz. It is a member of the granite family of rocks which include all rocks with granitic texture and composed of essential quartz and feldspar. The granite family is subdivided into specific members such as *granite* (proper) and *granodiorite* on the basis of the kind and proportion of the three major feldspar varieties in the rock. All members of the family contain quartz and most contain minor accessory minerals such as biotite, muscovite, hornblende, and pyroxene.

Granitoid rocks containing essential feldspar, but no quartz, are called *syenite* or *monzonite* if alkali feldspar predominates, *diorite* if lime feldspar predominates. These rocks typically contain considerable amounts of ferromagnesian minerals such as biotite, hornblende, and pyroxene as accessories. Rocks with approximately equal proportions of lime feldspar and pyroxene are classed as *gabbro* or *norite*. These are included in the typical black granites of commercial usage.

Rocks with mineral composition of the granite family, but with a texture that shows distinct planar or linear properties owing to the parallel alignment of mineral grains are called *gneiss*.

All of the rocks discussed in this section are classed as plutonic by the geologist, meaning that they formed deep within the earth's crust. All but gneiss are typically igneous in origin meaning that they formed by consolidation from molten material. Most gneisses and some granite family members, however, are thought to have been formed by metamorphism; that is, by transformation from preexisting sedimentary or other rocks.

Granite family rocks may occur as dikes, sills, or large subjacent masses of irregular shape. The latter may be considered bottomless for practical economic purposes. The mode of emplacement may be by forceful injection where the molten magma shoulders aside the country rock, or more passively by stoping where the rising magma engulfs fragments of the country rock which either sink or are resorbed into the magma. Most granites are emplaced in the roots of geosynclinal mountains. Granite emplacement may be contemporaneous with orogenic folding (synorogenic) or may follow the principal orogenic movements (postorogenic). Postorogenic granites, passively emplaced by block stoping, are more apt to be free of internal strain and usually make better dimension stone.

Sandstone

Commercial Definition: Commercial sandstone is "a consolidated sand in which the grains are composed chiefly of quartz and felspar, of fragmental texture, and with various interstitial cementing materials, including silica, iron oxides, calcite, or clay" (ASTM designation C119–50). Quartzitic sandstone must contain at least 90% free silica; quartzite must contain 95% free silica (ASTM designation C616–68).

Common commercial varieties of sandstone quarried in the United States include *bluestone,* "a dense, hard, indurated, fine-grained feldspathic sandstone, most of which splits easily into thin, smooth slabs" and of various colors; and *brownstone,* "a sandstone of characteristic brown or reddish-brown color that is due to a prominent amount of iron oxide as interstitial material" (ASTM designation C119–50). Typical brownstones are arkosic sandstones quarried from Triassic basins in eastern United States. *Freestone* is sandstone or limestone that splits with equal ease in any direction.

Geology: Sandstone is a clastic sedimentary rock composed of indurated sand grains, most of which fall within the size range 1/16 to 2 mm. The interstices typically contain cementing material, but enough voids may remain to give the rock considerable porosity and permeability. Rocks composed of clastic fragments larger than sand size are called *conglomerate* if the pieces are rounded, *breccia* if angular. Rocks composed of clastic particles smaller than sand size are siltstone or shale.

Sandstones are deposited by waning currents, mostly in water, rarely in air by wind. They are the end product of a weathering and erosion cycle that starts with the disintegration and decomposition of source rock, followed by erosion and transportation of the resulting material to a final resting place where the sediment is deposited in horizontal layers. Quartz is the only abundant sand-size material in the source rocks that resists alteration or decomposition. Feldspar and other labile materials are gradually depleted during the cycle, leaving a concentrated residue of quartz. The amount of quartz concentration is a measure of the maturity of the sand. Quartz-rich sandstones, sometimes called *orthoquartzites,* tend to be well sorted and to be cemented with secondarily precipitated calcite or silica.

Immature sandstones containing abundant feldspar are called *feldspathic sandstones* (5 to 25% feldspar) or *arkoses* (more than 25% feldspar). Alternatively, immature sandstone containing abundant rock fragments is called *lithic sandstone, subgraywacke,* or *graywacke.* Immature sandstones tend to be poorly sorted with the interstices commonly filled with clay, iron oxide, and other detritus.

Perhaps the most conspicuous feature of sandstone is bedding or stratification which has formed by successive influxes of sediment. Bedding planes commonly are parting planes. The direction of bedding is called *rift* in the industry. Those bedding planes along which the rock splits with greatest ease are called *reeds.* Some sandstones have a second direction of easy splitting that is nearly perpendicular to bedding. The second direction, called *run,* can be of great aid to the quarrier.

Sandstones with thin, even, and regular bedding along which the rock easily splits are natural *flagstones.* Sandstones required for sawing into cut dimension stone must have massive beds as the saw cuts are generally made perpendicular to bedding. Some sandstones have internal *cross-bedding* that is not parallel to the principal stratification. If the stone parts along the cross-bedding, the stone is unsuitable for dimension stone.

Limestone

Commercial Definition: Limestone is "a rock of sedimentary origin . . . composed principally of calcium carbonate or the double carbonate of calcium and magnesium" (ASTM designation C 119–50). *Calcite limestone* contains less than 5% $MgCO_3$; *magnesian (dolomitic) limestone* contains 5 to 40% $MgCO_3$;

dolomite contains more than 40% $MgCO_3$. *Travertine* is a cellular limestone in which the cellular voids are generally concentrated in layers.

Although defined by the ASTM, the terms calcite limestone, dolomite, and magnesian limestone are seldom used commercially and the importance of the chemical distinction for dimension stone purposes is questionable. Commercial limestones are commonly designated by location and texture, for example, "Indiana Oolitic Limestone." The color may be included in a name such as "Cream Kasota Vein," or such other descriptive or exotic terminology may be used as salesmanship dictates.

The Building Stone Institute defines an *oolitic limestone* as a stone formed of shells and shell fragments. The fragments are of more or less uniform size and generally 2 mm or less in greatest dimension. It is considered noncrystalline in that calcite crystals do not form an interlocking texture, but rather the original fossil texture is well preserved. Many commercial oolitic limestones are composed in whole or part of oolites so would also fit the scientific definition of oolitic limestone.

Shell limestones generally contain larger unbroken fossil shells enclosed in a fine grained matrix. *Coquinas* are composed of loosely cemented shells and shell fragments.

Geology: *Limestone* is a sedimentary rock composed of one or both of the two polymorphs of calcium carbonate—calcite and aragonite. Aragonite is common in limestones of recent age, but as it is unstable and inverts to calcite, it is rare or absent in ancient limestones. Sedimentary rocks composed of the mineral dolomite (calcium-magnesium carbonate) are called dolomite or *dolostone*. Dolomitic limestones contain calcite and dolomite; magnesian limestones contain magnesium carbonate, but not necessarily as dolomite. Many limestones and dolostones contain abundant sand or clay in which case they may be called sandy or argillaceous limestone (or dolostone).

The vast majority of limestones are composed of shells or shell fragments. When deposited in turbulent water the fragments get broken, sorted, and reworked. Resulting features such as size sorting and bedding are the same as for sandstones and the rocks have been called *calcarenites*, where the word "arenite" designates sand-sized particles. In quieter water, the sturdier shells may remain intact but other shells readily disintegrate into ultrafine particles which may wash out leaving a residue of larger shells sorted as to size. In very quiet water the ultrafine material plus the shells of one-celled microorganisms may form a matrix in which larger fossils may or may not be imbedded. Rocks composed of finely commuted calcite are called *calcilutites*. Limestones composed predominantly or entirely of the shells of microorganisms are called chalk.

The only limestones known to form by direct chemical precipitation in open seas are in the form of oolites. Oolites are tiny (1 mm or less) spheres of calcite or aragonite, commonly formed with a concentric structure around a nucleus consisting of a tiny fossil or other fragment. Modern studies show that oolites form in shallow, turbulent, warm seas where loss of carbon dioxide to the atmosphere causes supersaturation of calcium carbonate. Oolitic limestones may be composed entirely of oolites or of a mixture of oolites and shell fragments.

Modern petrologic studies show that much of the finely commuted material in limestones is reworked by burrowing organisms and pelletized by filter feeders. Modern carbonate classifications are based on the texture of particles, be they fossil fragments, pellets, oolites, etc., and the degree of sorting of the particles. For further information the reader is referred to various texts on sedimentary petrology.

Marble

Commercial Definition: Commercial marble is any crystalline rock composed predominantly of calcite, dolomite, or serpentine that is capable of taking a polish. Commercial marble thus encompasses true marble in the geologic sense as well as many crystalline limestones, travertines, and serpentine. Onyx marble is a dense, crystalline form of limestone, deposited from cold water solutions and commonly found in caves. Verde antique is a massive serpentine, commonly crisscrossed by veinlets of calcite, and capable of taking a high polish. The Building Stone Institute glossary uses the term "crystalline limestone" synonymously with marble.

The ASTM recognizes calcite marble, dolomite marble, and magnesian marble as chemical varieties, but in commercial practice these distinctions seem to have little importance. Commercial trade names commonly give the quarry location, color, and a descriptive adjective describing the pattern or texture of the stone.

Where a pattern is referred to as "veined" the stone is generally cut across the bedding or layering. "Fleuri" patterns are achieved by cutting parallel to layering.

Geology: Marble in scientific usage is a metamorphosed limestone or dolostone which is so thoroughly recrystallized that much or all of the sedimentary and biologic textures are obliterated. An interlocking, mosaic texture results. Bedding may be partially preserved in the form of compositional layering or banding.

Impurities present in the original carbonate sediment commonly form a characteristic suite of magnesium and lime silicate accessory minerals. Most limestones are slightly magnesian to dolomitic and the chief mineral impurities are quartz and clay. Common accessory minerals are talc, chlorite, amphibole, and pyroxene. Accessory minerals are colorless to golden brown magnesian varieties if iron is absent, but green if iron is present. Organic material entrapped within the rock forms graphite that is commonly accompanied by finely disseminated pyrite. Pure calcite marble is white, but tiny amounts of impurities will color it significantly. Graphite and pyrite commonly color marble gray. Pyrite in the rock will produce a sulfurous odor when the rock is struck with a hammer. Finely disseminated hematite will color marble pink.

Most true marbles are deformed rocks in which original bedding planes have been folded, crumpled, or otherwise distorted. The crumpling is generally most intense along the axes of folds. The layering may remain straight and planar on the limbs. Such is the case in the West Rutland, VT, deposits where the character of the marble is very consistent parallel to the layering, but changes markedly across the layering. On the limbs of folds the layers may be stretched rather than crumpled. In some of the Alabama deposits brittle interbeds of dolomite marble have broken and pulled apart during deformation and the more plastic calcite marble squeezed into the spaces around the broken pieces. At Tate, GA, a combination of intense crumpling and stretching has practically obliterated all sense of original bedding and the entire deposit is a nearly homogeneous mass of swirly "veined" marble.

Crystalline limestones that take a polish are classed as marble commercially. Such limestones are not considered marble by petrologists, particularly if sedimentary and biologic textures are well preserved. The rocks may show incipient recrystallization of the matrix and be thoroughly cemented with secondary calcite. They are dense, nonporous rocks. Many of the more decorative marbles are of this type.

Serpentine, classed as marble commercially, is not so classed by petrologists. Most serpentine results from metamorphism of peridotite and related rock. Peridotites, originally composed of pyroxene and olivine, are recrystallized at relatively low metamorphic temperatures in the presence of water to form serpentine minerals (hydrous magnesian silicates) and fibrous amphibole. Iron released in the reaction commonly forms magnetite; and lime released from pyroxene, or introduced from outside, commonly forms calcite veinlets. Serpentine commonly forms in tabular or lens-shaped bodies associated with other metamorphic rocks.

Slate

Commercial slate is defined as a microgranular metamorphic rock derived from argillaceous sediments, and is characterized by an excellent parallel cleavage entirely independent of original bedding. The cleavage permits the rock to split easily into relatively thin slabs (ASTM designation C119–50). This definition applies for both commercial and scientific purposes. Slates are formed by dynamic metamorphism. Beds of differing composition that form color stripes on cleaved surfaces are called *ribbons*. Ribbons are considered deleterious in high grade slate. The best slate therefore comes from rocks having thick, massive, original bedding.

Quartz is usually the dominant mineral, but extremely fine-grained mica and other platy minerals are always present. The platy minerals tend to be aligned with the slaty cleavage and impart a sheen to the surface. Most slates contain pyrite or iron oxide and many contain graphite. The color of slate is determined principally by the oxidation state of the iron and the amount of graphite or other coloring material present. Graphite and pyrite impart a dark gray or black color; ferric iron oxides color the slate red. Green slates owe their color to the presence of chlorite, other iron silicates, or ferrous iron oxides.

Other Stone

Other rock types sold as dimension stone include *greenstone, obsidian,* and *traprock* or *basalt*.

Greenstone is a metamorphic rock containing "one or more of the following minerals: chlorite, epidote, or actinolite" (ASTM designation C119–50). It is derived by metamorphism of basalt, diabase, or other mafic igneous rocks.

Obsidian is volcanic glass. Nearly all the commercial obsidian sold as building stone in the United States is actually *pumice*, a highly vesicular or frothy volcanic glass that is the product of explosive volcanic eruptions. The rock has very low density and is abrasive. However it is easily shaped with hand tools. It is also sold as rubble.

Basalt is a microcrystalline volcanic rock composed of pyroxene and calcic plagioclase. *Traprock* is a common term for dike rock of the same composition. The correct petrologic term for traprock is *diabase* or *dolerite*. Diabase, unlike basalt, does not have columnar jointing nor flow structures and is coarser grained. When sold for dimension stone it is commonly called *black granite*.

Distribution of Deposits and Reserves

The ultimate reserves of stone must be considered inexhaustable. There can only be shortages of particular types or colors that can be discovered, quarried, finished, and transported to market at a competitive price. The distribution of the various lithologies is a result of geologic factors. The location of producing quarries or districts is governed by occurrence and proximity to market. The major market areas in the United States traditionally have been the northeast and upper midwestern states. However, as the country grows and the economy expands, major market areas are created in the new localities. The production of rubble stone which requires little equipment or capital investment follows the market, but the production of cut dimension stone has tended to remain concentrated in areas where the geology is favorable and traditional markets are close.

Space does not permit a detailed worldwide survey of producing areas and geologic provinces. The following sections will therefore treat mainly the various United States provinces. A brief section on other producing countries follows.

It is convenient to divide the United States into five geologic provinces for the purpose of discussing the distribution of lithologies (Fig. 1). Two of these are wholly or in part within the major traditional market area and provide or have provided the great bulk of dimension stone production in the United States.

Atlantic and Gulf Coastal Plain

The Atlantic and Gulf coastal plain (Fig. 1, No. 1) is underlain by poorly consolidated sedimentary rocks that lie in nearly horizontal strata. The rocks range in age from Cretaceous

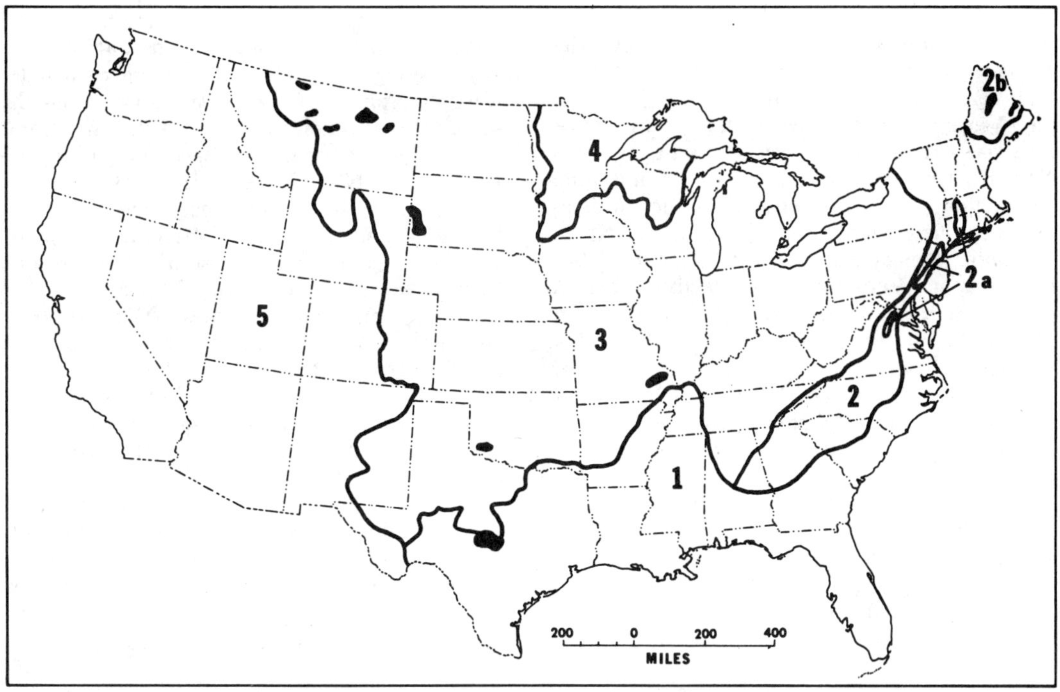

FIG. 1—*Dimension and cut stone—geologic provinces, United States.*

to Recent. Because the rocks are poorly consolidated, the province has yielded very little dimension stone. Soft coquina that hardens when exposed to air has been used as construction stone in Florida since the 17th century Spanish colonial times. Some production of limestone has recently been reported from Manatee County, FL.

Cretaceous limestone is quarried and finished to cut dimension stone near Liberty Hill, TX. Two varieties, one a uniform fine-grained cream-colored oolite and a second containing mega fossils, are quarried.

Appalachian Crystalline Province

The Appalachian crystalline province (Fig. 1, No. 2) extends from Alabama to Maine and beyond through the Maritime region of Canada. The province is sometimes called the "older Appalachians" because it is the exposed root zone of an ancient mountain chain that formed during mid-Paleozoic time. Granite, gneiss, marble, serpentine, and slate are exposed in these mountain roots. It leads the rest of the United States by far in the production of granite, slate, marble, and other crystalline rocks. The high production results from proximity to major markets.

Virginia, Pennsylvania, New York, and Vermont have long been the principal slate-producing states. Georgia, Vermont, and Alabama are leading marble producers. Serpentine and verde antique are produced in Vermont and Virginia. Almost every state in the province has produced dimension granite at some time. In 1969 Georgia was the leader in granite production, but Connecticut, Maine, Massachusetts, New Hampshire, New York, North Carolina, Pennsylvania, Rhode Island, South Carolina, and Vermont also had active quarries.

Triassic Basin Subprovince (Fig. 1, No. 2a): Down-faulted basins containing tilted, but otherwise undeformed sedimentary rocks of Triassic age, lie within the Appalachian crystalline province. These basins contain red arkosic sandstones that are sold as "brownstone." Brownstone was very popular near the turn of the century for facing town houses in large eastern cities. It has since lost favor and production is intermittent.

Northern Maine Subprovince (Fig. 1, No. 2b): Northern Maine is underlain by Paleozoic sedimentary rocks which have been intruded by granite in several places. The area is sparsely populated, has little industry, but no dimension stone production is reported.

Interior Sedimentary Province

The Interior sedimentary province (Fig. 1, No. 3) extends from the Appalachian Mountains to the Rockies. The sedimentary rocks are generally flat lying throughout the province. However they are strongly folded and faulted in the Appalachian Valley and Ridge physiographic province in the east and in the Ouachita Mountain region in Oklahoma and Arkansas. They are also tilted and faulted locally around uplifted domes where Precambrian rocks come to the surface. Crystalline rocks are absent except in the center of several domal uplifts where Precambrian basement rocks are exposed through the sedimentary cover in Montana, South Dakota, Oklahoma, Texas, and Missouri.

Rocks of Paleozoic age underlie the eastern two-thirds of the province, but Mesozoic and Cenozoic rocks cover older rocks in the western plains states. The Paleozoic rocks have yielded more than 80% of the domestic production of sandstone and limestone plus a considerable amount of crystalline limestone (i.e., commercial marble). The high production of this area results from its proximity to major markets.

During most of Paleozoic time the interior province was the site of a vast, shallow, inland sea that was bordered on the east by a large land mass—the older Appalachians. Clastic sediments poured into the area from this land mass forming near-shore and nonmarine sandstones along the eastern margin and carbonates were precipitated further west away from the source of sediment. As a result, the major sandstone producing states—Ohio, Pennsylvania, and New York—are located in the eastern part of the province whereas major limestone and crystalline limestone producing states—Indiana and Missouri—are located farther west. More than 60% of domestic limestone production comes from Indiana. However, most states in the province produce both limestone and sandstone. This overlapping distribution of sandstone and limestone reflects in part shifting shorelines and hence shifting sites of deposition of the different rock types during the Paleozoic Era. Tennessee in particular is a major producer of both sandstone and crystalline limestone (commercial marble).

Granite is produced from Precambrian rocks in Missouri, Oklahoma, South Dakota, and Texas.

Lake Superior Crystalline Province

The Lake Superior crystalline province (Fig. 1, No. 4) is part of the greater Canadian Shield where vast areas of Precambrian crystalline rocks are exposed. Parts of Michigan, Minnesota, and Wisconsin lie within the province. The province ranks second to the Appalachian crystalline province in the production of granite.

Western Province

The Western province (Fig. 1, No. 5) includes a wide variety of subprovinces that encompass nearly every conceivable geologic environment. Nearly every type of rock can be found. Probably the chief limiting factor in its development for dimension stone is the distance to major markets.

Granite is widespread and is produced in California, Colorado, and Washington. Sandstone is produced in a number of states, particularly those lying within the Colorado Plateau. Only California reports recent limestone production, but production was more widespread in the past. Onyx marble is produced in several southwestern states and travertine is produced in Idaho. This is the only province that produces obsidian (pumice) which comes principally from California and Nevada.

The small production from the province is probably because of lack of industrial development and distance from major markets rather than lack of suitable stone. Most production in the province is at locations near major population centers in California.

Foreign Sources

International trade in dimension stone is in the form of finished cut stone ready for installation, in sawed slabs, or in rough blocks. Granite, marble, and travertine are the principle rock types. Granite still is imported as blocks but the trend in marble and travertine is toward importing sawed and polished slabs. The great bulk of international trade is within the countries of Western Europe and North America.

Italy dominates international trade, being the leading importer of rough blocks and the leading exporter of finished stone. Italy also is the leading quarrier of marble and travertine. The Carrara district in northern Italy produces what is perhaps the world's most famous marble. Travertine has been produced from Tivoli since classic times. Marble in many patterns and colors had been quarried since Roman times and the classic names for these have come into generic use in much of the rest of the world.

Portugal is second in production of marble, mostly of colored varieties. Other Mediterranean countries also are important producers, as are Belgium and Norway.

Granite is produced extensively in northern European countries. Laurvikite, an iridescent syenite from Norway, is particularly popular in the United States. The black "petite granite" from Belgium actually is marble, but Belgium produces real granite as well.

Few countries lack potential for quarrying some sort of dimension stone. Sporadically, special varieties of marble, granite, and travertine come on the market from Latin American, Asian, or other countries. The bulk is not large and an important obstacle to the importation of more is the lack of consistency of the material and inability to produce enough stone to establish a market. Many underdeveloped countries encourage the production of dimension stone for domestic use, but their domestic markets usually have different requirements than US markets. For example, Mexico and other Latin American countries use large quantities of marble floor tile; in the United States carpeting has become the preferred flooring material. In Mexico this marble tile in variegated patterns is used on the exterior of buildings; US architects prefer uniformity of pattern and color and large slabs. In Mexico marble tile is used on interior walls in applications where in the United States ceramic tile is used. Part of these differences may be ascribed to the high setting costs in the United States, but some undoubtedly stem from architectural preferences.

Preparation for Markets

Quarry Operations

Quarry operations typically involve isolating a mass of stone by cutting it free on all sides but one from the parent mass. The isolated mass is then "lifted" or separated from the parent mass by breaking it free. The freed stone may be a block of size suitable for sale or further processing—a mill block—or it may be much larger—a quarry block. Quarry blocks are subdivided by breaking into smaller blocks that can be removed from the quarry and sent to the mill. Typical mill blocks range in size from 13.6 to 27.2 t (15 to 30 st), but may be as large as 63.5 t (70 st). Quarry operations thus involve cutting the stone, breaking the **stone, and removing the stone from the quarry.**

Quarry Plan: The quarry plan is determined **by the geometry of the deposit, the amount of overburden, and the necessity of handling rec-**tangular blocks weighing up to 63.5 t (70 st).

Most quarries are the open pit type (Fig. 2). They have relatively small surface area and they are worked downward in a pit below the general surface level. Open pit quarries have several advantages: they can be located on outcrops or where overburden is thin; they are within reach of fixed derricks with movable booms; electric service, air compressors, loading docks, and other service facilities can be more or less permanently fixed near the working area. Traditionally the quarry size is limited by the reach of the derrick boom. Larger open pit quarries generally require several derricks, both for lifting out blocks and for emplacing and moving quarry machinery. In recent years many operators have used large tractors, forklifts, and other machinery to move equipment and quarry blocks across the quarry floor thus extending the "reach" of the derrick boom.

Open pit quarries are particularly well suited to massive deposits such as granite and some marble formations. Steeply dipping layered deposits are also quarried by this method and the hanging wall is commonly undercut. Wall support is generally not required where a single opening by a single derrick is made. However, if successive quarries are opened along the strike, a web between quarries may be left for wall support.

The depth of open pit quarries is limited by the thickness or depth of good stone, by spalling or so-called "pressure breaks," or by excessive time loss in entering and leaving the quarry and in lifting out blocks. Open pits seldom reach depths of more than 60.1 m (200 ft).

Flat-lying, layered deposits may be worked as open shelf quarries in a hillside. In this case the quarry floor is level with the ground on one side. If the deposit is thick, the quarry may be changed into an open pit as deeper layers are removed. The quarry may also be extended under the hill as an underground quarry if the overburden is too thick for removal on the high side.

Underground quarries are room-and-pillar mines. Generally no more than 25% and commonly as little as 15% of the stone is left for roof support. Underground derricks are set for loading blocks. These may be of fixed boom type as blocks are generally skidded under them for loading onto trucks or other

FIG. 2—*General view of a dimension stone quarry.*

conveyances. In recent years forklifts strong enough to lift quarry blocks have become available, eliminating the need for underground derricks in flat-floored quarries.

Blocks are nearly always cut parallel to layering. Hence, where layering is tilted the quarry floors are not flat. This increases the difficulty of moving equipment and removing blocks. Movable equipment such as tractors and forklifts cannot be used and enough derricks must be set to reach the entire working area of the quarry.

After the initial opening is made quarry methods are generally the same as for open quarries except for block removal. The initial opening may be made by cutting out blocks in a manner similar to removing key blocks in an open quarry. The cost of removing these key or "tunnel" blocks is several times that for removing floor blocks. Alternatively, some underground marble quarries in Vermont are opened by removing an overlying layer of rock with conventional mining methods (i.e., blasting and mucking). In order to prevent blast damage to the underlying dimension stone the overlying material is first separated from the valuable stone by cutting a kerf at the top of the marble layer. This kerf cut also provides a good flat working floor for quarrying the underlying marble.

In Vermont, marble layers dipping 30° to 45° are quarried underground. The initial opening is made by cutting tunnel blocks and roof support is by the room-and-pillar method. Blocks are cut out along the bedding; hence working floors are inclined at the dip angle.

Cutting the Stone: One of the oldest methods for cutting marble, limestone, and sandstone is with channeling machines. These cut with the chopping action of several chisel-edged steel bars clamped together. The machine travels back and forth on tracks and a channel 50.8 to 63.5 mm (2 to 2½ in.) wide is cut. As the channel is deepened longer bars are substituted in the machine. The method is slow and the channel bars require frequent sharpening so that the quarrier is practically obliged to have a blacksmith shop. Channeling is gradually being replaced by other quarry methods.

Drilling and broaching is another long established method for cutting stone. Drill and broaching tools are mounted on a quarry bar built into a rigid frame. A series of closely spaced holes is drilled and the web between them is chipped out by the chiseling and chopping action of the broaching tool. The method is suitable for all kinds of stone, but is slow and, like the channeling steel, the broaching steel requires frequent sharpening. The method has largely been replaced by line drilling in modern quarries.

In line drilling, a drill is mounted on a quarry bar similar to that used in drilling and broaching (Fig. 3). A series of overlapping holes is drilled. The quarry bar serves to align the row of holes and hold the drill rigidly in position. The key to successful line drilling is a guide attached to the drill that is inserted into the last complete hole in the line, which serves to insure that successive holes overlap.

Wire saws have been used in some granite quarries in the United States and for many years they have been used to cut marble in Italy. They also have been used sporadically in the US to cut marble and limestone, but they have not replaced line drilling and their use in granite is all but eliminated by better methods.

The wire saw consists of a single, double, or triple strand of wire that runs over sheaves and is fed under tension into the stone. Sand, aluminum oxide, silicon carbide, or other abrasive is fed to the wire and wears a groove or channel in the stone. The sheaves are mounted on a tower or track so that they can be moved into the cut. In order to press the wire into the cut there must be two free and parallel sides to the block of stone being cut or else a large hole to admit the sheaves must be drilled. In some cases the hole need only be large enough to admit the axel, hub, and supporting arm of the sheave. The sheave is designed to cut its own groove as it is pressed into the stone. In some operations other quarry cutting methods are used to cut parallel channels between which the stone may be wire sawed.

The wire used may be many thousands of feet long and a single wire may be used to make many parallel or other cuts by guiding it with appropriate sheaves. At least one movable sheave must be set up with appropriate counterweights in order to keep tension on the wire.

One advantage of wire sawing is that it produces a smooth cut that minimizes subsequent finishing in the shop.

Jet channeling is the most common method of cutting granite. The method, also known as flame-cutting, consists of disintegrating the rock with intense heat from the combustion of fuel oil and oxygen. The thermal shock causes the rock to spall and the fragments are blown from the channel (Fig. 4). The torch, which is passed back and forth over the rock, is commonly attached to a frame in order to keep the cut true.

FIG. 3—*Quarry bar used in line drilling.*

FIG. 4—*Jet channeling in rock under high pressure.*

In some granite quarries large blocks are freed using this method and then are subdivided by using cheaper methods of breaking the rock. Jet channeling works well in granite and other hard brittle rocks that spall. It does not work in limestone or marble which calcine or fuse rather than spall.

Other methods have been used or tested. In some European quarries, huge circular diamond saws are used in the quarries. A chain sawlike device using carbide teeth is used in some Virginia soapstone quarries and similar machines have been built and reportedly are in use in some foreign limestone and marble quarries. Experiments have been made using diamond-set segments on wire saws. Other experiments are continuing and, judging from trends in the mills and shops, it seems probable that experiments with the use of diamond saws will continue.

One of the most promising new developments in cutting granite is the water jet. A pulsating jet of water at pressures ranging from 96.5 to 275.8 MPa (14,000 to 40,000 psi) is directed at the granite and causes it to disintegrate. It is less noisy, produces less dust, and promises to be more economical than jet piercing. The Elberton Granite Association in Georgia is currently testing two experimental machines.

Breaking and Lifting the Stone: In open pit quarries vertical cuts are made by dividing the floor of the quarry into appropriate sized blocks. An initial key block is removed by drilling a hole in the center to the depth of the vertical cuts. An explosive charge in this hole frees the block with or without damage to the block. Once the key block is removed a series of closely spaced holes is drilled underneath the remaining blocks and they are broken free with the use of feathers and wedges. Feathers are pieces of half-round steel that are inserted into the hole. Steel wedges are driven between the feathers to produce the break (Fig. 5). When a block of stone is broken free, it may be further subdivided by the same technique.

Granite typically has two directions of easy breakage called *rift* and *grain* or *rift* and *run*. In some quarries large blocks containing many thousand cubic feet of stone are cut free by jet channeling. These blocks are then broken into mill-size blocks along the rift and run directions. Some granite ledge quarries rely entirely on breakage to produce dimension blocks. Because of easy splitting along the rift and run, holes drilled for feathers and wedges in granite commonly need penetrate only a few

FIG. 5—*Driving feathers and wedges in order to "lift" block by breaking it free.*

inches of stone. When the break is produced, it runs true for the entire length of the block (Fig. 6).

Marble and limestone are typically cut into pieces of appropriate height and width for mill blocks, but of much greater length. These are then "lifted" (broken free from the quarry floor), using feathers and wedges. They are also broken into appropriate lengths by the same technique. Drill holes for the feathers and wedges generally penetrate the entire width of the block.

In order to speed the process of breaking or lifting the stone some companies have experimented with explosives, using various presplit breakage techniques. Time and money can be saved, but there is still some concern over possible damage to the stone. Dimension stone quarriers have traditionally shied away from use of explosives, but recent experience indicates that if used properly explosives can be used successfully and economically.

A serious problem in some quarries is the failure of stone to break along predetermined directions. This may happen in stone that does not have a natural grain or in which the quarry direction does not parallel the natural grain. In some deep marble quarries in Georgia, stone that is cut free on the sides has a tendency to break loose from the floor along arch-shaped fractures rather than parallel to the quarry floor. This not only ruins a mill block, but presents a costly cleanup problem in order to keep the quarry floor flat. These breaks, called "rainbow" breaks by the quarrymen, are most

FIG. 6—*Quarry worker wedging out granite blocks. Jackhammers, sledgehammers, and wedges are used to size and free blocks.*

common in large, deep quarries and where a long time has elapsed between cutting the block free on the sides and lifting it from the quarry floor. Some deep quarries have been abandoned because of rainbow breaks even though the stone was good otherwise.

Georgia Marble recently introduced a new method of breaking quarry blocks free with air pressure. A quarry block is cut free on all sides but the bottom. The bottom is drilled as though for using feathers and wedges. A large reinforced bladder is inserted in the vertical kerf that separates the block from the quarry. Air pressure applied to the bladder breaks the block free and sets it out from the quarry face. Usually 22.7 kg (50 lb) of pressure or less is sufficient.

The Georgia Marble Co. has also experimented with wire sawing vertical slabs with very large surface area. These slabs are broken free, turned over on their side, and broken into appropriate sized mill blocks. This procedure not only eliminates much expensive cutting, but allows the quarrier more flexibility in choosing the size and position of the quarry blocks to be cut from the slab. The slab can often be subdivided in such a manner that a minimum number of blocks contain natural flaws, whereas with a set floor pattern the quarrier has less control.

Removing Blocks from the Quarry: Quarry blocks are most typically removed from the quarry by fixed derricks. The procedure is slow and all other operations under the derrick boom must be halted while blocks are removed. The length of the boom also limits the size of some quarries. Blocks may be skidded from greater distance with tractors and crawler-type derricks which can also add flexibility. The most efficient new innovation is the use of forklifts which are now available in sizes large enough to handle mill blocks.

Sawing

Primary saw cuts may be made with a gang saw, wire saw, or circular saw.

Gang saws consist of a group of flat steel blades set in a frame that moves with a reciprocating motion. The frame is automatically fed into the block at a rate that depends on the type of stone and the abrasive or cutting edge used. A sand-water mixture may be fed into the saw for cutting soft stone. For harder stone such as granite or quartzite, heavier notched blades are used and the cutting media is steel shot. For faster cutting the blade may be fed with aluminum oxide, carborundum, or other abrasive. The latest development is the use of set teeth. These may be carbide for softer limestones, but harder rocks such as marble and granite require diamonds. Diamonds are set in segments which are brazed to the gang-saw blade. Set teeth generally cut faster and smoother than loose abrasives. As many as 40 blades may be set in a single gang saw when thin stock is being cut.

In most gang saws the blade is horizontal

and the frame is pressed downward through the block. The size of the block that can be sawed is thus limited by the size of the frame and blades. A newer development is vertically set gangs through which the block moves horizontally. Longer slabs can be produced.

One of the problems in gang saws is keeping equal tension on the blades. If a blade loosens it may wander or buckle and jam the saw or ruin the cut. The traditional method of holding a blade in position is with wedges driven through a hole in the blade outside the frame. Considerable skill is required to maintain in a single frame a large number of blades under equal tension. A more modern method involves the use of hydraulic tensioners.

Wire saws are commonly used to make primary cuts in granite blocks. These may be grouped into multiple or "ganged" wire saws. Silicon carbide is the most commonly used abrasive.

Wire saws are also used to make irregular or curved cuts. The saw is set vertically and the stone is fed into it as wood is fed into a band saw. Wire saws have also been used to turn large columns. Here the saw is usually set horizontal and the stone is slowly turned into it.

The most recent development in the granite industry is the use of large [2.7-m (9-ft) diam] circular saws to make primary cuts. The saw makes repeated passes across the block cutting 6.4 to 9.5 mm (¼ to ⅜ in.) at each pass. The resulting cut is remarkably smooth so that increased sawing costs are more than compensated for.

Circular saws are used for secondary and later cuts. Some European manufacturers use very large circular saws for primary cuts but that practice is very unusual in the United States. Most circular saws use impregnated diamond rims for their cutting edge. These are so efficient and economical that their use is nearly universal.

Shaping, Grinding, and Polishing

A wide variety of methods is used for finishing the stone. Shaping may be done by hand-used pneumatic tools, by wire saws that cut as a band saw cuts wood, or by carborundum wheels that grind the stone to a reciprocal shape just as a wood-router bit shapes a board. For softer stones large milling machines using carbide teeth may be used to gouge shapes into the stone. Cylindrical shapes are made on a lathe that slowly turns the stone into a carbide bit or into a wire saw. Boring machines may be set with diamonds or with carbide.

Flat surfaces may be ground on a large horizontal cast-iron wheel called a rubbing bed. A water-sand mixture is used as abrasive. More modern plants have replaced rubbing beds with automatic surfacing machines using impregnated diamonds. The stone is fed under a series of successively finer horizontal wheels on a belt. Some machines fine grind the edge of the stone as well as the surface. Much polishing is still done by a conventional grinding-polishing machine that moves a spindle over the stone surface. Wheels using successively finer abrasives are set on the spindle.

Inevitably much of the finishing is handwork by skilled craftsmen. Despite many recent efforts at labor saving the production of cut dimension stone remains a labor-intensive industry.

Evaluation of Deposits

In order to properly evaluate a potential dimension stone deposit, one must determine the marketability of the stone, its soundness, its size and shape, the overburden, the distance to market, the kinds of transportation available, the availability of labor, the availability of power and other utilities, and so on. Most of these must be determined in the evaluation of any mineral deposit, but several are unique to dimension stone deposits.

Marketability

The marketability of a stone is governed largely by aesthetic factors not subject to quantitative measurement. Aesthetic factors include color, pattern, and texture. Unless the deposit is an extension of a proved quarry or an exact duplicate of a popular stone the judgment should be left to an experienced architect or marketing man.

It is important that a reliable sample be obtained. The best sample is a block from which slabs 0.09 m^2 (1 sq ft) or larger can be sawed. These slabs are used to evaluate color and pattern and also can be used for making required physical tests. The more varied the color and pattern the larger the sample should be. Drill cores are commonly deceptive, but may be suitable for fine-grained uniform textured stone or for extensions of known deposits.

It is the responsibility of the examining geologist or engineer to determine the size, extent, and uniformity of the deposit and to obtain samples that are truly representative. In the case of layered deposits, samples from each separate stratum should be obtained and compared for uniformity.

Soundness

Soundness or freedom from flaws is one of the most critical factors in evaluating a dimension stone deposit and one of the most difficult to determine. Every joint and fracture should be mapped. This should be supplemented by core drilling parallel to the direction in which the stone is to be sawed. If at all possible the deposit should be drilled in two intersecting directions in order to pick up all possible weakness directions that will limit the size of blocks and slabs. If mapping shows a pronounced joint direction, the drilling should be perpendicular to it. The drilling must be done with extreme care and nothing smaller than BX size should be used. If the core breaks on anything but a natural weakness, it will have failed in its purpose of proving "sound" stone.

Any directions of natural splitting that would aid in quarrying should be determined. Granite and many limestones and sandstones have such directions that are known as rift, run, grain, etc. These can be beneficial, whereas close-spaced open fractures are deleterious. The investigator should remember that the ultimate object is to free large blocks that are without internal fractures that will appear only after the block is sawed.

Some rocks that have been subjected to great stress are still strained. In many cases such stone will develop new fractures and split after blocks are quarried. Some hidden fractures show up only after the block has been sawed. This is costly as the flaw appears only after a large investment has been made in the stone. This type of fracture is probably less common in horizontally layered sedimentary rocks and in postorogenic granites that have intruded passively than in rocks that have been subjected to tectonic stress.

Other Factors

The investigator should remember that quarrying blocks for cut and sized dimension stone is still a high labor input industry. He should also keep in mind that each salable unit—a quarry block—will weigh 9.1 t (10 st) or more and will require special handling methods to extract and transport to market. The total output of a quarry is small as compared to other industrial rocks and minerals. It takes a large operation to extract more than a few blocks a day and few if any deposits would justify building new roads, railroads or expensive utility lines just to service a quarry.

The physical tests required by ASTM specifications (see "End Use and Specifications") require a sample from which blocks up to 102 × 203 × 57 mm (4 × 8 × 2¼ in.) in size can be cut.

Depletion

The Internal Revenue Code allows a depletion allowance of 14% on most dimension stone. The allowance is on net sales of mill blocks without sawing or any other finishing. However, most dimension stone is quarried, sawed, and otherwise finished by single vertically integrated companies and a problem arises in assigning costs and profits to quarry operations alone. Because the value of a particular stone depends on aesthetic appeal, ease of working, and many other factors, it is virtually impossible to set a standard value on quarry blocks. Each company case is individual and must be negotiated with the Internal Revenue Service in order to establish a price and profit on quarry blocks.

Special Supplement on Stone Preservation and Stone Preservatives by Erhard M. Winkler *

Stone decay is a natural process, not new, but accelerated in the urban environment. Acid-producing sulfate and carbon dioxide produced by the burning of wood, coal, and peat caused deterioration of buildings in the Middle Ages and attempts were made to stem this decay by the application of linseed oil and waxes. The crumbly and porous limestones used in southern England were treated with lime water and whitewash, which produced a temporary retardation of decay and saved the

* Professor, Department of Earth Sciences, Notre Dame University, Indiana.

gothic statues at the Wells Cathedral from crumbling. Stone decay is accelerated today owing to increasing population and industrialization. Recently the introduction of plastics and acrylics has opened up new possibilities for preventing stone decay. However, long-lasting improvements by surface applications have, in many instances, failed to materialize.

Moisture, even though barred from surface access, is the chief culprit in stone decay. Close analysis reveals four main sources of moisture in masonry: rain, rising ground moisture (often loaded with salts), leaky plumbing, and condensation from the interior. The problem of stone preservation is to prevent the introduction of this water.

Stone preservatives may be classed in two categories: surface sealers which repel water at the surface, and consolidants which fill pore space within the decayed stone. A combination of both may be used and the chemicals used may be inorganic or organic (Table 2).

Surface Sealants

Many sealers made for concrete and brick masonry are also used for stone. These form an impervious skin which, while preventing surface moisture from entering, may allow the entrapment of moisture within. Pressure from the trapped moisture may cause scaling of the surface. This may be accelerated if salts build up within the skin. Protection by an impervious or semipervious skin will be successful only if the stone is dense and impervious to other moisture. Oils, hot waxes, and silicones are relatively easy to apply, but generally last only 5 to 10 years.

Stone Solidifiers

The purpose of solidifiers is to fill pore space and recement crumbling stone. Whitewashing of limestone has been practiced since the Middle Ages for this purpose and is still carried out in England. Today a large number of chemicals, protected by numerous patents, is available. Few of these are compatible with the original minerals, and they cannot guarantee permanence. The following properties of solidifers should be closely observed:

1) Depth of penetration and readiness to penetrate.

2) Similarity of coefficient of expansion during heating and cooling cycles.

3) Degree of bonding between mineral grains and consolidants.

4) Degree of sealing of pores.

5) Development of byproducts harmful to the stone, e.g., efflorescents.

Clifton (1980) classifies the consolidants into four major groups according to their chemistry: inorganic materials, alkoxysilanes, synthetic organic polymers, and waxes.

Inorganic consolidants: Lime milk $Ca(OH)_2$ is the oldest remedy attempted for limestone. Limited protection is achieved if it reacts with

TABLE 2—Stone Preservation and Stone Preservatives

Type of Preservative	Materials	Results
Surface sealants	Organic oils and waxes	Impervious skin or semipermeable to permit vapors to pass
Consolidants		
Inorganic	Milk of lime Ca $(OH)_2$	Formation of calcite with CO_2 of atmosphere at or near surface
	Barium hydroxide with urea	Formation of white $BaSO_4$ or $BaCO_3$
	Fluoridate	To form insoluble CaF_2
Organic	Alkoxysilanes with moisture	Precipitates quartz
	Polymers, acrylic polymers	Plexiglass-like fillers and binders
Stone cleaning	Washing with water	Removal of dust
	Soaking with water and pressure jets	Removal of grime and soot
	Abrasives, hard and soft, dry or with water jet	Removal of hard crust
	Microsandblasting	Removal of hidden patches on sculptures
	Chemicals and acids	Effective removal, but discoloring is common, efflorescence

carbon dioxide in the atmosphere to form calcite. Clark and Ashurst (1972) discuss the effect at Wells Cathedral. Attempts to replace calcite with barium carbonate or fluorite have often produced changes in color and molecular volume.

Organic consolidants: The silification of crumbling sandstone with alkoxysilanes, which react with moisture to form quartz, has achieved some success. However, some of the preservative may be unable to contact water which does not penetrate as deeply as the preservative. Polymers and acrylic polymers tend to penetrate well and form a strong bond. Winkler (1979) has questioned the permanence of the plexiglass substance and its contact with inorganic material.

Many new preservatives have been used with inadequate performance records. Clarke and Ashurst (1972) documented the performance of stone preservatives used in London and southwest England. After 4 to 10 years of observation they concluded that most preservatives had very little beneficial effect or else accelerated stone decay. New preservatives and techniques brought onto the market should not be used until adequate field tests by qualified and disinterested people have been made. Accelerated laboratory tests do not substitute for field experience.

Protection Against Low Plants and Bacteria

Algae and other low plants frequently coat the weather side of masonry in humid climates or along the permanent wet line at the top of rising groundwater. Though most algae do not extract nutrients from the stone, they cover the open ends of capillaries and prevent the stone from drying properly. Dead plant material on the surface contributes humic acid, a powerful chelating agent. Bacteria populate stone after incipient decay. Sulfate bacteria such as *Thiobacilus* metabolize sulfur from the atmosphere, converting calcite to undesirable gypsum. Although some people believe the effect of bacteria to be overrated, opinions are divided. Bactericides may rid the surface of damaging algae and bacteria temporarily. Van de Molen, et al. (1979) describe a UV radiation-emitting device for destroying bacteria.

Cleaning Outside Masonry

Careful consideration as to the nature of the accumulated dirt and its method of attachment should be made before cleaning. The cleaning itself may damage the stone. In many cases the preservation of a certain patina or slight discoloration adds a feel of antiquity.

Torraca (1976) gives main requirements for cleaning of stone: (1) It should be possible to control the cleaning process and graduate the cleaning process at will; (2) The cleaning process must not produce materials which may be the cause of future deterioration, e.g., soluble salts; and (3) The cleaned surface must be, as far as possible, smooth and free from cracks and other defects that would accelerate weathering.

Different techniques are available to rid masonry surfaces of undesirable deposits. Pure water appears to be simplest and the most popular method in use today. Generally, slow soaking precedes final washing with high-pressure jet streams. Most of the oily and sooty grime can be removed this way, and patina-like coatings of ferric hydroxide can be controlled by the adjustment of the water nozzle pressure.

Sandblasting using quartz or milder, often synthetic material is common. The type of abrasive and nozzle pressure determine the severity of cleaning action. Some communities prohibit the use of quartz. Selective microblasting with adjustable nozzles has been used, sometimes in combination with nebulized water, to clean undesirable deposits which accumulate in areas protected from cleansing rain.

Cleansing acids and chemicals, most with secret formulas, have been looked on with suspicion because of the potential for corrosive action. Hydrofluoric acid has caused more damage than any other chemical. Chemicals may cause efflorescence, either by the actual chemicals left on (or in) the stone, or by their reaction with the stone to create other deleterious products.

All methods for cleaning, especially high-pressure water jets, may result in undesirable penetration of water beneath the surface of the stone.

Moisture and Stone Preservation

Moisture is the chief cause of stone decay. Inscriptions on ancient Egyptian monuments engraved in sandstone and granite are still visible in their original sharpness and clarity beyond the reach of ground moisture. Many preservatives tend to form a strong surface skin, often stronger than the stone itself. Moisture tends to work from behind (within) the treated surface and cause scaling or honeycomb weathering. In many cases the rate of decay is accelerated by the use of surface treatments. A full understanding of the origin and movement of moisture and the location of its source should precede any decision regarding selection of a stone preservative. Sealing a surface from acid rain may permit entrapment behind the surface of moisture that comes from the ground, from condensation within the building, or from leaky plumbing. Each of these sources may have its own chemical peculiarities. A total and permanent removal of all moisture from a building or monument is the only sure protection against weathering.

Bibliography and References

Anon., 1961, "American Standard Specifications for Thin Exterior Marble in Curtain or Panel Walls." Marble Institute of America, American Institute of Architects, File 8-B-1, 26 pp.

Anon., 1966, American Society for Testing & Materials, *Book of ASTM Standards,* Pt. 12, Philadelphia, PA, 468 pp.

Anon., 1972, "Stone Catalog," Building Stone Institute, 1972–1973.

Anon., 1979, "Water Jet Machines Tested in Elberton," *Elberton Graniteer*, Vol. 23, pp. 4–5.

Barton, W.R., 1968, "Dimension Stone," Information Circular 8391, US Bureau of Mines, 147 pp.

Bowles, O., 1939, *The Stone Industries,* 2d ed., McGraw-Hill Co., Inc., New York, 519 pp.

Bowles, O., 1955, "Memorial Stone," Information Circular 7720, US Bureau of Mines, 6 pp.

Bowles, O., 1955a, "Slate," Information Circular 7719, US Bureau of Mines, 12 pp.

Bowles, O., 1956, "Granite as Dimension Stone," Information Circular 7753, US Bureau of Mines, 18 pp.

Bowles, O., 1956a, "Limestone and Dolomite," Information Circular 7738, US Bureau of Mines, 29 pp.

Bowles, O., 1958, "Marble," Information Circular 7829, US Bureau of Mines, 31 pp.

Bowles, O., 1960, "Dimension Stone," *Industrial Minerals and Rocks,* 3rd ed., J.L. Gillson, ed., AIME, New York, pp. 321–337.

Bowles, O., and Barton, W.R., 1963, "Sandstone as Dimension Stone," Information Circular 8182, US Bureau of Mines, 30 pp.

Clarke, B.L., and Ashurst, J., 1972, "Stone Preservation Experiments," Dept. of the Environment, Garston, Watford, England, 78 pp.

Clifton, J.R., 1980, "Stone Consolidating Material—A Status Report," Technical Note 1118, National Bureau of Standards, 46 pp.

Cooper, J.D., 1970, "Stone," *Mineral Facts and Problems*, Bulletin 650, US Bureau of Mines, pp. 1219–1235.

Currier, L.W., 1960, "Geologic Appraisal of Dimension-Stone Deposits," Bulletin 1109, US Geological Survey, pp. 7–14.

Drake, H.J., 1970, "Stone," *Minerals Yearbook 1970*, US Bureau of Mines, pp. 1039–1057.

Drake, H.J., 1971, "Stone," *Minerals Yearbook 1971*, US Bureau of Mines, pp. 1097–1117.

Drake, H.J., 1972, "Stone," *Minerals Yearbook 1972*, US Bureau of Mines, pp. 1153–1173.

Hockman, A., 1953, "Physical Properties of Currently Produced Marbles," Circular LC1010, National Bureau of Standards, 13 pp.

Ladabouche, J.J., 1973, "Dimension Stone," *Mining Engineering*, Vol. 25, No. 1, Jan., p. 32.

Lewin, S.Z., and Baer, N.S., 1974, "Rationale of the Barium Hydroxide-Urea Treatment of Decayed Stone," *Studies in Conservation*, Vol. 19, pp. 24–35.

Meggers, E.R., 1956, "Selected Bibliography on Building Construction and Maintenance," Building Materials and Structures Report 140, National Bureau of Standards, 40 pp.

Power, W.R., 1973, "An Evaluation of Building Dimension Stone Deposits," *Mining Engineering,* Vol. 24, No. 6, June, pp. 42–44.

Shadmon, A., 1970, "The Modern Stone Age," *Philippines Arts and Architecture,* Vol. 7, No. 4.

Shadmon, A., 1972, "Stone in Israel," Ministry of Development, Natural Resources Research Organization, Stone Technology Center, Technion, Haifa, 64 pp.

Shelton, J.E., and Drake, H.J., 1975, "Stone," *Mineral Facts and Problems,* Bulletin 667, US Bureau of Mines, pp. 1031–1048.

Singleton, R., 1982, "Stone," *Mineral Commodity Summaries 1982,* US Bureau of Mines, pp. 146–147.

Torraca, G., 1976, "Treatment of Stone in Monuments: A Review of Principles and Processes," *The Conservation of Stone,* R. Rossi-Manares, ed., Centro per la Conservazione delle Sculture all'Aperto, Bologna, pp. 297–316.

Van der Molen, J.H., et al., 1980, "Growth Control of Algae on Historical Monuments by a Mobile UV (MUVU)," *Studies in Conservation,* Vol. 24, pp. 71–77.

Winkler, E.M., 1973, *Stone: Properties, Durability in Man's Environment,* Springer-Verlag, New York, 230 pp.

Winkler, E.M., 1978, "Stone Preservation: The Earth Scientist's View," *Bulletin,* Association for Preservation Technology, Vol. X(2), pp. 118–122.

Wood, S.O., Jr., and Carrales, M., Jr., 1969, "Stone," *Minerals Yearbook*, Vol. I–II, US Bureau of Mines, pp. 1019–1040.

Construction Materials

Gypsum and Anhydrite

FRANK C. APPLEYARD*

Gypsum, the dihydrate form of calcium sulfate, has a history of usefulness to man dating back several thousand years, and a worldwide industry has been built on the mining and processing of this versatile industrial mineral. However, anhydrite, the other principal calcium sulfate mineral, has very limited use, so much so that separate statistics for it are not always kept, but are usually included with those for gypsum.

Although often found in close association, the two minerals have important chemical differences in that anhydrite is the anhydrous form of calcium sulfate ($CaSO_4$) whereas gypsum is the dihydrate ($CaSO_4 \cdot 2H_2O$). These two molecules of combined water result in significant differences in hardness, density, and solubility of the two minerals. However, of most importance is the difference in thermal properties wherein gypsum, upon the application of a modest amount of heat, can be converted to the hemihydrate of calcium sulfate ($CaSO_4 \cdot \frac{1}{2} H_2O$) or plaster of paris, an intermediate product upon which is based over 90% of the value of all calcium sulfate end products. Anhydrite, on the other hand, is inert at these temperatures and its uses have been limited to only those few cases where calcium sulfate in the anhydrous form may have an advantage.

The earliest known use of gypsum plaster dates back some 5000 years to the time of the Pharaohs of Egypt who had discovered that gypsum rock, when exposed to fire, broke down into a powder which when mixed with water would form a putty that could be "plastered" on a rough mud brick or stone wall to make a smooth finish. It was used down through the centuries in a limited way for ornamental purposes, but did not achieve wide acceptance because of its quick (25 to 30 min) setting or hardening time which made it difficult to use. The first real understanding of gypsum chemistry was developed in France about 1755, and the gradual growth of its present utilization dates from that time.

In the United States, the first utilization of gypsum was to grind it to a fine material for use as a soil conditioner at about the time of the Revolutionary War, and the term "land plaster" for ground gypsum has remained in use. Another cementitious material—lime or calcium oxide—had become the usual plastering material, but by about 1870 a method for retarding the set of gypsum plaster was discovered and today it has almost completely replaced lime as a plastering material.

In recent years, gypsum products sold in the United States have reached a total valuation of approximately $1.4 billion per year as illustrated in Table 1.

Structure of the Industry

In the technologically more advanced countries such as the United States, Canada, Western Europe, and Japan where the industry is well developed, it tends to be organized into large, well financed, vertically integrated companies. On the other extreme, in those countries where the usage of gypsum is not extensive, its production is usually accomplished by small operators with limited objectives and capabilities. Several North American and western European companies operate on an international basis through subsidiaries or other arrangements (equity or otherwise) but in general the percentage of total business done outside their home country is small (Anon., 1972).

There are a total of 45 companies (1981) who mine and/or process gypsum in the United States, ranging in size from large, multiplant, multiproduct organizations to small, one-plant, one-product companies who serve a local market for portland cement retarder or for agricultural gypsum. Of these, 15 companies have calcining and wallboard manufacturing facilities, and the remaining 30 ship only uncalcined prod-

* Retired; formerly Vice President, Raw Material Resources, U.S. Gypsum Co., Chicago, IL.

TABLE 1—Gypsum Products (Made from Domestic, Imported, and Byproduct Gypsum) Sold or Used in the United States, By Use

Use	1978 Quantity, 1000 t	1978 Value, $1000	1979 Quantity, 1000 t	1979 Value, $1000	1980 Quantity 1000 t	1980 Value, $1000
Uncalcined:						
Portland cement	3 817	36,897	3 648	38,223	3 524	41,440
Agriculture¶	1 367	11,315	1 541	14,064	1 504	19,121
Other	148	4,298	112	3,846	122	4,353
Total*	5 332	52,510	5 303	56,133	5 150	64,914
Calcined:						
Industrial plaster	347	24,177	331	23,663	357	28,296
Building plaster:						
Regular base coat	200	10,580	121	6,733	130	7,657
Mill-mixed base coat	100	6,854	89	6,725	81	6,985
Veneer plaster	97	8,719	89	8,591	72	7,942
Other†	80	4,693	78	5,385	79	6,427
Total*	477	30,846	377	27,434	362	29,011
Prefabricated products‡	13 417	1,140,480	13 783	1,284,763	11 816	1,119,728
Total calcined	14 241	1,195,503	14 491	1,335,860	12 535	1,177,035
Grand total	19 573	1,248,013	19 794	1,391,993	17 685	1,241,949

Source: Pressler, 1979, 1981.
* Data may not add to totals shown because of independent rounding.
† Includes gaging, molding, and Keene's cement, roof deck concrete, and other uses.
‡ Includes weight of paper, metal, or other materials.
¶ Includes 606 530 t of byproduct gypsum in 1978, 750 680 t in 1979, and 601,452 t in 1980.

ucts. Of the 15 companies with calcining and manufacturing facilities, seven are multiplant operations with anywhere from two to 23 different locations, with the remainder being single-plant operations. In 1981 there were 72 active calcining and manufacturing plants.

The larger companies are completely integrated from mining to the marketing and distribution of finished products. As the importance of wallboard has grown, some of the companies have also built paper mills to supply part or all of their needs for the face and back paper used on the board. This trend towards vertical integration has resulted in strong competition, and in a continuing need to control all phases of their business, one result being that there is practically no market for crude gypsum rock to the calcining-manufacturing companies.

World Trade

World trade in gypsum consists primarily in the movement of crude gypsum rock, although to a very limited extent, certain manufactured products do move across international boundaries. The United States is at the same time the world's largest producer and the largest importer of gypsum rock. Canada is the largest exporter of gypsum, with almost all of it being in the form of crude rock to the United States which is mined by subsidiaries of US companies.

In Europe, France is both the largest producer and the largest exporter of gypsum, with its exports being in the form of crude rock. Spain, Poland, and West Germany also export substantial quantities of gypsum rock (Anon., 1978).

The primary reason for the limited amount of world trade in gypsum or its products is found in the widespread occurrence of deposits, and in the relatively low ratio of manufacturing cost to product weight. This results in freight being a major segment of delivered cost, a fact which discourages long, international movements. For further discussion of occurrence and the concept of "place value," and for statistics regarding United States imports, the reader is referred to the chapter on Gypsum and Anhydrite in the "Commodities" section of this book.

Basic Processing Methods

The processing of gypsum can be divided into three basic steps: (1) rock preparation; (2) calcining; and (3) formulating and manufacturing. This is illustrated graphically in Fig. 1. The specific means used to accomplish each of these steps will vary with the quality of the gypsum being processed and with the type of product to be made, and in the following discussion generally used methods and equipment are reviewed.

Rock Processing

Primary crushing is accomplished by gyratory, jaw, or impact crushers, with the type selected depending upon the size of the mine-run rock, the desired throughput, and the type of subsequent processing. Secondary crushing is done with a variety of standard units, but hammer mills and cone-type crushers are most commonly utilized. Fine grinding of both uncalcined and calcined gypsum is generally accomplished by air-swept roller mills fitted with integral air separators for better control of particle size, although in the past few years high-energy impact mills plus air classifiers also have been used.

Both primary and secondary crushing steps are usually conducted with vibrating screens in the circuit, in part to maximize crushing efficiency and to reduce the production of ultrafines, but also to recover portland cement rock, this being the first marketable product of gypsum processing. The particle size of this product will vary with the requirements of the cement plant using it, but in the United States it most often is controlled within a range of 38.1 to 50.8-mm (1½ to 2-in.) top size by 6.4 to 12.7-mm (¼ to ½-in.) bottom size.

Drying (the removal of free moisture) may or may not be used either before or after the secondary crushing stages, with its need being a function of the amount of free moisture in the mine run rock. Crushed gypsum is difficult to handle in that −4 mesh material is not free-flowing, particularly if wet, so a drying step is often added to rock preparation to assure the free flow of material in subsequent steps. This most often is accomplished in rotary dryers, and must be carefully controlled so that the temperature of the rock does not exceed 49°C (120°F), this being the point at which dissociation of the combined water begins to take place.

In many cases, the purity or quality of gypsum as mined is sufficiently high that it can be used without upgrading. However, the use of beneficiation techniques is becoming more common as the quality of final products is improved to meet the higher requirements of the marketplace. The most common form of beneficiation is simple classification of particle size by dry screening, air separation, or other means wherein a size fraction containing a significant portion of impurity is discarded. In general, this method is used to reduce clay or sand ("dirt") impurities, although in some cases, impurities which are harder than gypsum, and hence tend to concentrate in the coarser size fractions after crushing, can also be reduced by screening.

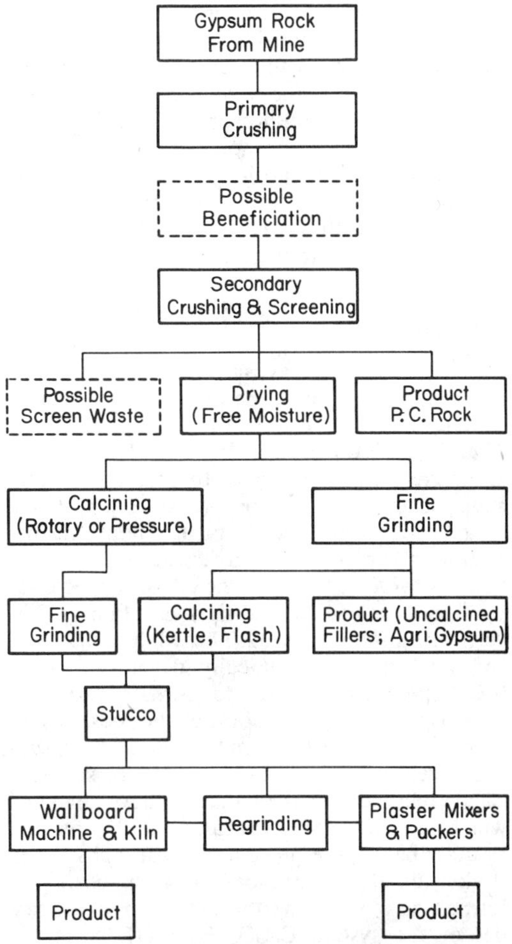

FIG. 1—*Generalized flow diagram of gypsum processing.*

Washing and/or wet screening is used in a few cases, particularly where the need for a white color exists. And the use of heavy media sink-float separation for the removal of impurities is currently employed at only one operation in the United States and one in Canada. In many cases, gypsum is amenable to other gravity methods of beneficiation, or to froth flotation or to color sorting, and some laboratory work has been reported on these approaches (French, 1967; MacPherson, 1950). However, the relatively high cost of these methods and the availability of deposits which do not require this type of beneficiation prevent commercial applications, with the exception of an electronic color sorter used at one location in England.

Calcining

As applied to the gypsum industry, calcining is the step of reducing the dihydrate of calcium sulfate ($CaSO_4 \cdot 2H_2O$) to the hemihydrate ($CaSO_4 \cdot \frac{1}{2}H_2O$) or anhydrous ($CaSO_4$) forms. In the very early history of the making of gypsum plaster, pieces of rock were simply heated in an open wood fire to bring about the dehydration process. Then crude ovens were built in which "fist" or "head" size pieces of gypsum were placed, and in a variation of this, vertical kilns were used. These latter may still be found in less technically developed countries, operating with crude burners and with little or no prior preparation of the rock. However, where the utilization of calcined gypsum has developed on a broad scale, more sophisticated methods of calcination have evolved which require prior preparation of the gypsum as described in the preceding paragraphs.

Considerable theoretical work has been performed as regards the thermodynamic properties of the system $CaSO_4 \cdot H_2O$ (Kelly et al., 1941; Riddell, 1950) and the literature is quite complete in this regard. In its practical application, four calcined products are made on a commercial basis: the two forms (alpha and beta) of hemihydrate; a soluble anhydrite; and an insoluble or "dead-burned" anhydrite. All hemihydrates, no matter how made, are usually called **"stucco"** within the industry, but this term is not to be confused with the mixture of portland cement and sand stucco sometimes used for exterior application on buildings. The term **"plaster of paris"** is also often used for hemihydrate, and had its origin from the large deposits of gypsum under and near the city of Paris, France.

It is the consensus that hemihydrate is the only lower hydrate of calcium sulfate whose identity has been established with any degree of certainty (Riddell, 1950). Two forms of hemihydrate—alpha and beta—have been identified, both with the same crystalline form; however, the beta form has a definitely higher energy content and a higher solubility rate, and the two forms can only be distinguished one from the other by highly sophisticated analytical methods. The alpha form of hemihydrate is more stable, or less reactive, than the beta form, and has a slower rate of strength development. This is a disadvantage in many uses for stucco, but nevertheless rehydrated alpha hemihydrate makes a denser, stronger plaster which is advantageous in other uses. The principal characteristics of these two forms of hemihydrate are given in Table 2.

Alpha hemihydrate can be prepared by dissociation of gypsum in a water saturated environment above a temperature of 97°C, and in practice this is usually accomplished at elevated pressure in an autoclave in the presence of steam. *Beta hemihydrate* can be prepared by the dissociation of gypsum in a vacuum at 100°C, but in practice is accomplished in a less than saturated environment at atmospheric pressure.

The greatest utilization of calcined gypsum is in the manufacture of wallboard and the formulation of plasters for building markets, uses for which early strength development is necessary, and the more reactive beta hemihydrate is most suitable for this purpose. On the other hand, less reactive alpha hemihydrate, with slower strength development, makes a denser, higher strength cast, and may be preferred for industrial uses where these characteristics are important. Other factors to be considered in the utilization of these two forms of hemihydrate are cost (beta is considerably less expensive to make) and water demand (alpha requires less excess water—over and

TABLE 2—Characteristics of Alpha and Beta Hemihydrates

	Normal Consistency*	Setting Range In Min, Vicat	Typical Set Expansion, In. per In.†	Avg. Compressive Strength, Psi, Dry†	Reaction
Alpha	40-43	20-30	0.003	5500	Neutral
Beta	64-66	10-20	0.0018	2000	Neutral

* Parts water to 100 parts stucco by weight to make a pourable slurry.
† Metric equivalents: 1 in. × 25.4 = mm; 1 psi × 6.894 757 = kPa.

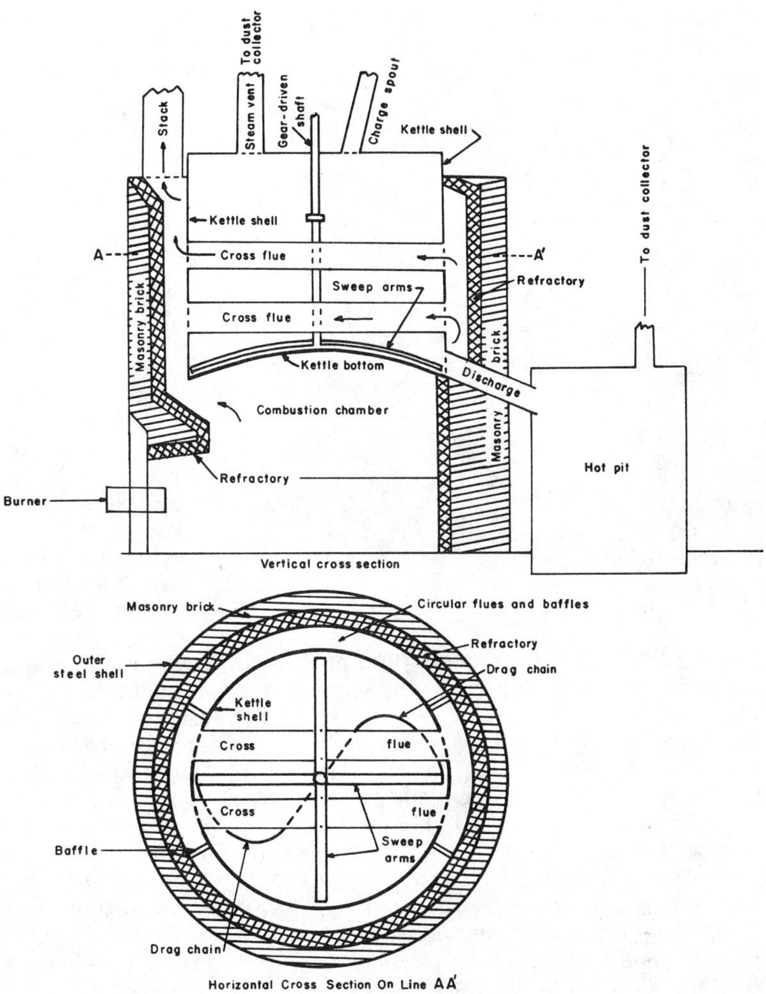

FIG. 2—*Generalized sections of a calcining kettle.*

above that necessary for rehydration—to make a slurry of equivalent consistency or viscosity).

By far the most commonly used vessel to calcine gypsum is the kettle, the basic design of which originated about 1870. Its general features are illustrated in Fig. 2. In its simplest form, the kettle is a cylindrical steel vessel with a height greater than its diameter which is enclosed in a refractory shell (Fig. 3). Heat is introduced in a firebox below the vessel, and flows upward around it. To improve heat transfer to its pulverized gypsum contents, four or more horizontal cross-flues are installed, and agitation is added by means of a vertical, rotating shaft with sweeps or rabble arms which stir the contents. The kettle is filled from the top, and an opening at the bottom with a quick operating gate provides for fast "dumping" at the desired moment.

This is a batch process, with the full cycle from filling to dumping requiring 2 to 3 hr. Kettles have been built with 1.8 to 27 t (2 to 30 st) capacity of finished product, but the usual size is 13.6 to 18 t (15 to 20 st). The normal cycle requires from 20 to 30 min to fill the kettle, and 90 to 120 min to bring the mass up to the desired temperature after filling is complete. Dumping is usually accomplished in 5 min or less, with the finished stucco falling by gravity into a vented "hot pit" or holding bin where it is kept for an hour or more to allow as much as possible of the entrapped steam to escape before it is transferred to a storage bin at the point of use.

For kettle calcining, gypsum **(land plaster)** is usually ground to a fineness of 90% −100 mesh, and stored in an appropriate bin. Approximately 791.3 MJ (750,000 Btu) are re-

FIG. 3—*Exterior view of gypsum kettle.*

quired per ton of finished stucco in a modern, well-designed kettle, and may be provided by coal, oil, or gas, using a burner so designed that quick changes in heat can be made.

In the calcining process, gypsum begins to lose its water of crystallization when it reaches a temperature of 43° to 49°C (110° to 120°F). During the filling part of the cycle, the firing rate is controlled to keep the temperature of the kettle contents at about 104°C (220°F). When filling has been completed, the firing rate is increased, with different rates producing different properties in the finished stucco. As the temperature of the mass rises, water of crystallization is released as steam, and at 116°–121°C (240°–250°F) the mass boils vigorously as the vapor pressure of the released steam equals atmospheric pressure. After boiling ceases and little or no more steam is given off, the contents of the kettle settle to 12–15% less than the original volume. This quiet period indicates that most of the gypsum particles are dehydrated to the hemihydrate. Heating is continued to dry the mass until its temperature reaches 149° to 166°C (300° to 330°F) at which point it is dumped into the hot pit and the cycle is complete.

The product of the foregoing operation is known as "first settle" stucco, and contains from 5 to 6% combined water as against 20.9% contained in 100% pure gypsum, i.e., approximately 75% of the water of crystallization has been driven off. Because this reaction takes place at atmospheric pressure in a less than water saturated environment, most of the resulting hemihydrate is in the beta form, although varying amounts of alpha hemihydrate may also be present depending upon the design of the kettle as regards the removal of steam vapor and the temperatures used.

If desired, a "second settle" stucco can be made after reaching the "first boil" and "first settle" stages. This is done by increasing the temperature to about 177°C (350°F), at which point the gypsum again starts to boil and give off steam. This second boil is of shorter duration and less vigorous than the first because there is a smaller percent of water of crystallization to be removed. At about 204°C (400°F) the material is dumped, at which point almost all of the water of crystallization has been removed, and the material is usually referred to as **soluble anhydrite**. It differs from first settle stucco in that it has less plasticity, but after rehydration will make a product with greater density and strength. Also, it has a high affinity for moisture, and is primarily useful for its desiccant properties.

A recent innovation to the batch process has been the "continuous kettle" method, wherein after an initial charge of land plaster is brought approximately to the finished hemihydrate stage, instead of being dumped, fresh land plaster is fed into the kettle as a continuous stream of stucco is removed. This technique requires carefully controlled conditions, the details of which are considered proprietary, but is gaining in use by certain segments of the industry.

Other methods also may be used to produce beta hemihydrate with variations of the flash drying concept being perhaps the second most common calcining method. To a lesser degree hollow flight screw conveyors are used as a calcining medium, with heat being applied by circulating hot liquid or gas through the hollow flight and/or by heated jackets around the conveyor box. The rotary kiln is still used for calcining in some parts of the world, but has been almost totally replaced in the United States with either kettles or flash calciners.

Grinding and calcining can be done more or less simultaneously in impact mills swept by heated air. However, the flash calciners, hollow flight conveyor, and impact mill methods are neither as versatile nor as readily controlled as the kettle, and not nearly as widely used.

Aridized stucco refers to first-settle kettle stucco to which a small percentage of a soluble salt has been added during the filling part of the cycle. In general practice, approximately 0.68 kg (1½ lb) of calcium chloride is used per ton of stucco, with the exact amount depending upon the purity of the gypsum rock and the kind and amount of impurities present. This treatment increases the vapor pressure within the kettle so that calcining takes place at a somewhat lower temperature, and the resulting stucco, which contains a relatively high percent of alpha hemihydrate, is used in the formulation of many industrial plasters.

Low water demand stucco can also be produced by adding water to the kettle after the batch calcination cycle is complete. The stucco is quenched to about 93°C (200°F) and then reheated to drive off excess water before dumping to the hot pit. Alternatively, low water demand stucco can be made by adding a measured amount of water to the stucco after it has left the kettle.

Pressure calcining in steam autoclaves, with or without the addition of a crystal modifier, is used to make superstrength plasters for certain industrial uses. The gypsum may be introduced into the vessel either as a crushed, sized rock or as a slurry, with the stucco being dried as soon as calcination is completed. This method produces all alpha hemihydrate, is relatively high cost, and its use is limited to just those products requiring this type of specialized plaster. Alpha hemihydrate may also be produced by dehydration of gypsum in hot, salty water.

Dead-burned gypsum or **insoluble anhydrite** is made in beehive kilns, rotary calciners, or flash calciners using much higher temperatures [up to 482°C (900°F)] than are required for other calcining methods. The resulting product is an anhydrous calcium sulfate which does not take up water (rehydrate) at any appreciable rate, and is the least soluble of any of the forms of calcium sulfate (Riddell, 1950). The primary uses of this material are to make Keenes cement by the addition of chemical accelerators to restore its setting properties, and to produce a mineral filler by fine grinding.

Hydraulic gypsum refers to lump (head size) gypsum which has been heated to about 899°C (1650°F) at which temperature the material is at a red heat and calcium sulfate begins to break down to release sulfur as sulfur dioxide gas. This has the effect of increasing the calcium oxide content, and results in a material which when mixed with water is slow setting as compared to other calcined gypsum. However, it provides a very hard, dense product that is quite durable, and is sometimes referred to as Estrich gypsum (a German term). Although it has been used in Europe as a flooring material, it has not found application in the United States.

Formulation And Manufacturing

The direct product of gypsum calcining is almost never used without some further processing such as grinding, mixing with additives, or rehydrating and casting it into block or wallboard. Kettle stucco, the most common calcined material, is not just a single form of calcium sulfate, but instead is a mixture of particles in varying states of dehydration from the dihydrate to anhydrous forms. That is, each particle is not dehydrated to the same degree, a factor which varies with the method of calcination. The amount of variation can be minimized by careful control of the calcining process, but it may also be a deciding factor in considering the type of subsequent treatment to be given the stucco.

Probably the earliest attempts at formulating calcined gypsum to improve its utility were the addition of fibers and/or aggregates. However, the single most important discovery was that its setting time could be either retarded or accelerated to very exact time limits by mixing certain materials with stucco (plaster of paris). Properly calcined gypsum when mixed with water will have a natural setting (hardening) time of 15 to 25 min, and this is usually too fast for satisfactory application on walls or ceilings. However, in the middle 1870s a method was discovered to retard the set to 2 or 3 hr (or longer). This led to the manufacture of "slow set plaster," which was the key to the rapid growth in the use of gypsum in the building industry.

Retarders are made from organic compounds and several different materials can be used. For the most part, these are formulated and manufactured by the major gypsum companies for their own use. **Accelerators** can be natural or synthetic salts, such as potassium sulfate, or gypsum itself. Finely ground raw gypsum, or rehydrated stucco cast into a block and then finely ground, are both commonly used to accelerate the set of stucco for the manufacture of wallboard. By the proper use of these materials, gypsum plaster can be accelerated to set within 3 to 4 min, or be retarded so that its set does not occur for several hours. However, adjustment of the set can cause variations in other properties, especially strength, so that care is advised in the use of these materials. The type of stucco and the characteristics of the plaster being formulated can be controlled by other additives mixed in precise proportions prior to bagging, and much research has been conducted to determine the formulas used. Among other products, plasters premixed with sand, expanded perlite, or exfoliated vermiculite aggregates are offered for the building trades, and a wide range of formulated plasters for industrial use are also available.

Gypsum is often reground after calcining to impart special qualities to the stucco. The classic buhrstone mill once used for this purpose has been replaced by roller mills operating in closed circuit with air separators. In recent years, high energy impact mills have successfully been employed. When stucco is to be used with aggregate (as for sanded or perlited wall plaster), a tube mill is used for regrinding.

One of the early uses of gypsum as a construction material was to cast it into blocks or tile. Continuous casting machines were developed for this operation, along with kilns to dry the product by hastening the removal of excess moisture. Blocks are cast either as solid masses or as hollow tiles, and can be made in any desired dimension. Metal reinforcing is used for some products, and some are cast in a metal frame to produce a "metal edged" tile or plank. The addition of fiber for reinforcing is a common practice, and use of an accelerator hastens the initial set of the stucco in the mold.

However, the largest end use for gypsum today is in the manufacture of **wallboard,** and approximately 65 to 70% of the gypsum consumed in the United States is used for this purpose. In its simplest form, wallboard is a thin, flat slab of gypsum made by casting a water slurry of stucco in a paper "envelope," i.e., it consists of a gypsum core with specially prepared paper bonded to both sides and the edges. This is a sophisticated product, made on high speed, highly automated machines to a wide range of precise specifications.

The bond between the paper and the gypsum core is produced by the growth of gypsum crystals locking themselves into the fibers of the paper as the stucco is rehydrated, rather than by the use of adhesives. Extensive research has gone into the formulation of the gypsum slurry used in wallboard, but essentially it consists of beta hemihydrate, an accelerator, fibers, starch, and a foaming agent, with stucco making up at least 95% of the materials used prior to mixing with water. Fig. 4 shows a general process flow diagram of a wallboard plant.

In the manufacturing process, the dry components are brought together with water in a high energy mixer designed for a continuous flow-through of material, and the resulting slurry is dropped on a moving strip of paper, the edges of which have been turned up to form a wide, shallow trough. This mass then moves under a second strip of paper, and the roughly formed board moves between edge guides and under a 609.6-mm (24-in.) diam roller called a "master roll" (Fig. 5) or, in some cases, a stationary steel "forming plate," which imparts the exact desired width and thickness of the board. The edges of the bottom paper are turned over to form square or rounded (as desired) corners and an adhesive is added so that the top paper is bonded along its edges to the lower. The thickness of the board can be varied by raising or lowering the master roll or forming plate, and the now fully formed board moves along on a flat conveyor belt several hundred feet in length.

The speed of this "board line" varies with the design of the machine and the product being made, but usually ranges between 0.3 and 1.0 m/s (60 and 200 fpm). The length of the board line is designed to provide sufficient time (5 to 6 min) for the stucco to set, at which point the moving strip is cut at desired intervals by a rotating knife to produce individual boards ready for drying to remove excess free moisture. This is accomplished by moving them through a continuous, multideck kiln heated directly by hot air or indirectly by steam coils, in which close control is required to remove the free moisture without starting to re-calcine the gypsum core.

The most popular size of wallboard is 1.2-m × 2.4-m × 12.7-mm (4-ft × 8-ft × ½-in.) thick, but a wide variety of sizes is made and offered to the market. After emerging from the kiln, the board moves automatically through devices that smooth the ends and cut to the precise length desired. Usually, one face is made with light colored, smooth texture paper which forms the finished wall surface, and two boards are then placed with these faces together to form a "bundle" for shipping. Or if desired, the board can move on to a finishing machine where a special facing is added such as vinyl or imitation wood veneer paper to make a pre-decorated panel.

Gypsum sheathing is made in the same general manner except that asphalt-impregnated paper is used, and asphalt may also be added to the core to improve its water resistant qualities. Plasterboard is a 406.4-mm (16-in.)-wide board used as a base for plastered walls instead of wood or metal lath, and is made on the same machine as wallboard. Shipment of finished material is by truck or rail, and most wallboard

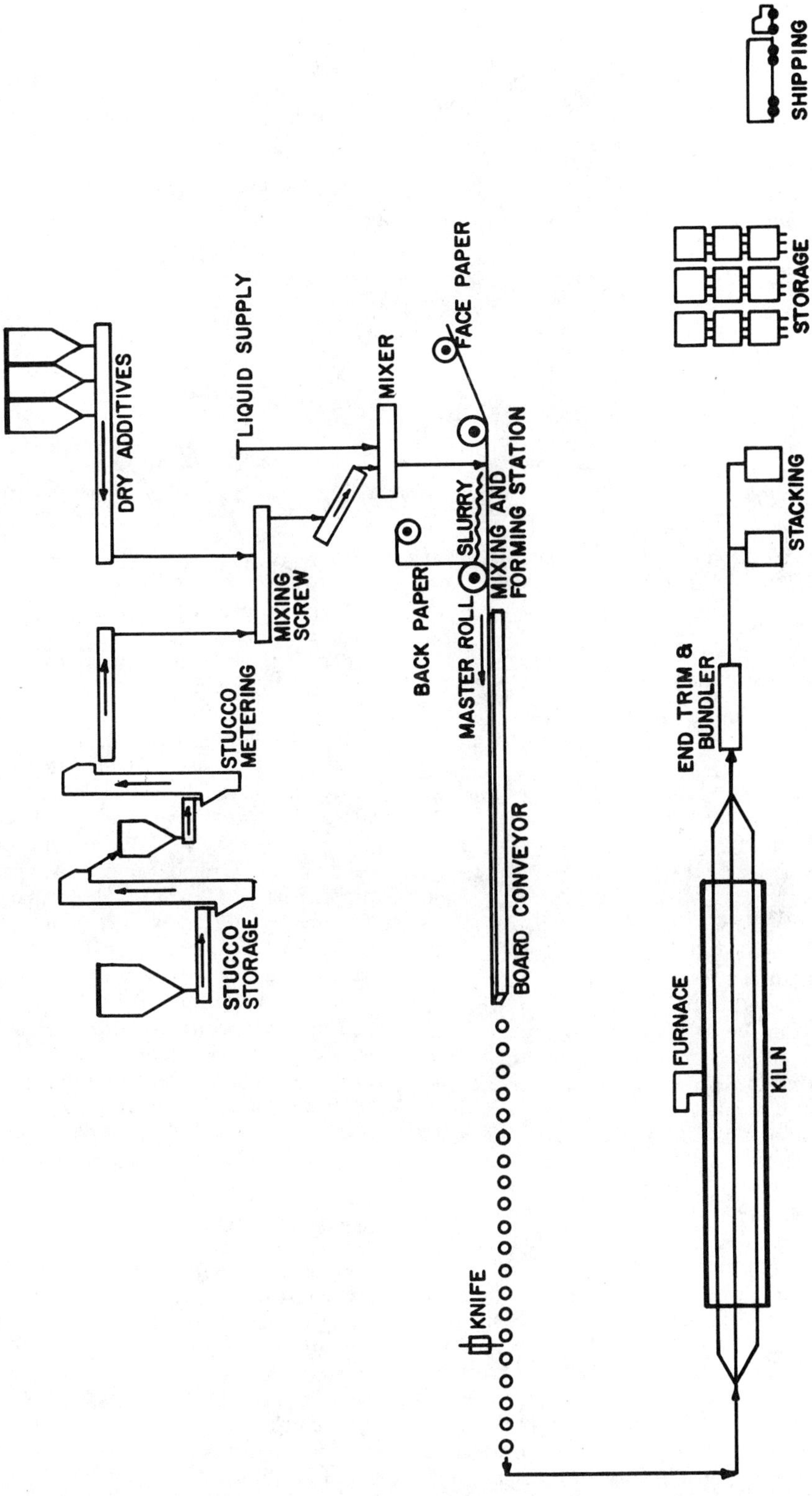

FIG. 4—*Simplified flow diagram for complete small gypsum plant.*

FIG. 5—*Mixing and forming section of wallboard machine showing stucco slurry feeding onto bottom paper at master roll.*

plants have large warehousing space adjacent to the manufacturing area in which to carry adequate stocks of various types and sizes of products.

Energy Considerations

Gypsum is an energy-intensive industry, particularly as regards fuel which is used in the rock drying and calcining operations, as well as in the manufacture of paper and the drying of excess moisture from wallboard. In total, an average 3.7 to 3.8 GJ (3.5 to 3.6 million Btu) are required per 92.9 square meters (thousand square feet) of 12.7-mm (½-in.) thick wallboard for these four operations. The drying of both rock and wallboard is usually accomplished in direct fired kilns (although steam coils can be used for wallboard) in which a clean-burning fuel is required to prevent contamination of the product, and natural gas is the preferred fuel for this purpose. The calcination step, particularly when carried out in a kettle, requires the ability to change the intensity of the heat quickly, and either natural gas or fuel oil is preferred for this property, although coal can also be used. Intermediate pressure steam is used for drying paper, with the choice of boiler fuel being based upon economic and environmental considerations.

The rapid rise of energy cost in recent years has encouraged extensive development and use of conservation methods, as well as investigation of process technology to reduce the energy requirement. Also of interest are fuel substitution possibilities which may be required as a result of changing availability and pricing policies.

Major Categories of End Uses

Gypsum product lines are divided into several categories based upon end uses and/or processing methods. Three major groups are generally recognized: (1) construction, i.e., all those products used directly as a construction or building material; (2) industrial—all those products used in the manufacture or processing of other materials; and (3) agricultural. As indicated in Table 3, the construction group utilizes about 70% of the gypsum rock used in the United States, and produces approximately 94% of total product valuation. The industrial group uses about 23% of the rock, but yields only 5 to 6% of the product value, and the balance is accounted for by agricultural gypsum.

The proportions produced for each use will vary considerably from country to country, and in many instances the only use for gypsum will be in the industrial category as portland cement retarder. In the following discussion, a few of the more important end uses will be reviewed.

Construction Group

Gypsum products are used in the building and construction industry as covering and finishing materials. They are not structural components in the sense that they provide the frame or structure of a building; instead, this requirement is met with wood, steel, concrete, concrete block, and fired or adobe brick. It is in the covering of a structural frame (Fig. 6), or in the finishing of a masonry wall, that gypsum products find their greatest utilization, and they

TABLE 3—Estimated Tons Gypsum Used and Value of Products By Major End Use Categories in the United States.*

Category	1977 Quantity, 1000 t	1977 Value, $1000	1978 Quantity, 1000 t	1978 Value, $1000	1979 Quantity, 1000 t	1979 Value, $1000
Construction	13 169	842,767	14 053	1,171,326	14 286	1,312,197
% of total	70	93	71	94	71	94
Industrial	4 174	54,583	4 428	65,372	4 201	65,732
% of total	22	6	22	5	21	5
Agricultural	1 519	13,176	1 367	11,315	1 541	14,064
% of total	8	1	7	1	8	1
Grand totals	18 862	910,526	19 848	1,248,013	20 028	1,391,993

*Source: Pressler, 1979 (modified).

compete in these applications with other covering or finishing materials, such as wood products, plastics, other cementitious products (cement, lime, or in primitive regions, even mud).

By far the largest market for gypsum is in wood frame buildings, and because wood has historically been the basic structural material in the United States, most of the development for this usage has taken place in this country. Gypsum construction products are also compatible with steel or aluminum framing, and may be used in conjunction with masonry construction, although their economics in this latter application are more difficult to justify.

FIG. 6—*Installing gypsum wallboard partition using steel studs.*

The basic characteristics of gypsum which give it value in the manufacture of construction materials are its fireproofing qualities, the versatility which comes from being able to use it in many forms, and its economy of application. It is used entirely in its calcined form, with stucco being formulated and/or manufactured to make plasters, block or tile, and wallboards. In recent years, approximately 90% of the gypsum used in the United States for construction materials is in the manufacture of various sizes and types of wallboard with a total surface area of some 1486 km^2 (16 billion sq ft) per year.

Gypsum is noncombustible and is one of the most effective fireproofing materials known, a property which has its basis in the combined water content of the product. When the hemihydrate form of gypsum (stucco) is rehydrated to make a covering material it reverts back to the dihydrate, so that when heat from fire hits a plaster wall (or gypsum wallboard) it again begins to lose its combined water in the form of steam which dampens the transmission of heat. Extensive research into the utilization of this property has resulted in a number of products with a high fire rating, and this is an important reason for the wide acceptance of gypsum products by the construction industry, particularly when wood framing is used.

The versatility of gypsum is also a major advantage. Plaster can be used in many forms, as can wallboard and cast block or tile, and by the use of different calcining methods and/or additives, physical properties can be varied to meet a wide range of requirements. These characteristics give the architect and builder several choices of design, and enable maximum flexibility in producing a structure to meet the desired specifications.

The economics of application of many building materials are usually measured in terms of cost per square yard of finished wall, or cost per square foot of floor space, or of cubic feet of content, which in turn is a function of the cost of material delivered to the job plus the cost of installation, all to provide a desired finished structure. In the final analysis, the wide use of gypsum construction products is based upon the industry's ability to keep its manufacturing costs low, and to minimize the use of labor in their installation. The relatively low processing cost of gypsum is due to its non-abrasiveness and low calcining temperature. Also, the wide geographical distribution of deposits and the resulting low cost of transportation contributes to economical delivery.

Plastered walls and ceilings, which in the early 1900s were widely used in the United States, have to a major extent given way to "dry wall" construction employing gypsum wallboard. However, one-coat "veneer-type" plasters have been developed for application over 1.2-m (4-ft)-wide plaster base, and many plasters are now formulated for machine mixing on the job and pumping to the point of use. Also, plasters continue to be used where special effects are desired, and their flexibility enables wide latitude in finished design.

Gypsum wallboard was originally developed in the United States, and the first product, made about 1890, consisted of at least four sheets of felt or paper with thin layers of gypsum stucco between them. By 1914, the concept of just a top and bottom sheet of paper with a single layer of gypsum had been perfected, and the use of this product received a big boost in the construction of buildings during World War I. Following World War II, as the increase in hourly wage rates raised the in-place cost of plastered walls, the use of gypsum wallboard expanded rapidly because it also provided fireproof construction, but with a lower labor requirement than the traditional two-coat plaster wall.

The present dominant position of wallboard stems not only from its economy of application, but also from the variety of installation systems which the industry has developed. For example, one such installation is the moveable partition; another is a method for enclosing elevator shafts in high rise buildings; and still other types are designed to produce optimum acoustical properties in multiple unit housing.

The use of gypsum partition and roof tile has declined in recent years, although they are still preferred for some applications. Poured gypsum roof decks are often specified for industrial buildings, in which a specially formulated plaster slurry is pumped in place over a wallboard base to provide a completely fireproof roof. Other construction uses include water-resistant sheathing, plasterbase board, and fireproofing building steel.

Industrial Group

Gypsum finds a wide variety of markets in industrial applications which bear little or no relationship to the construction products discussed previously. Industrial gypsum can be divided into three broad categories: calcined, anhydrous, and uncalcined, each of which utilizes distinct properties of the mineral.

The industrial segment of the gypsum industry is well developed in the United States; however, markets are widely scattered geographically, many product lines tend to be specialized, and with the exception of portland cement retarder rock, individual customers are likely to want tens or hundreds of tons of a product, rather than thousands of tons.

Calcined Industrial Gypsum: Perhaps the largest use of calcined gypsum by industry is for mold making in the manufacture of sanitary ware, pottery, metal casting, and decorative objects.

Molding plaster is made from high purity gypsum (95%) at select locations and is marketed on the basis of its water demand, strength, white color, setting time, and closely controlled expansion/shrinkage qualities. It may be made from either beta or alpha hemihydrate, or a combination of the two.

The qualities which make gypsum an excellent material for mold making also contribute to its use as a **casting plaster,** and gypsum casts are widely used by many industries. Included in this category is dental plaster, a highly specialized material made for dental and orthopedic work which is accelerated to produce a setting time of only 3 to 4 min.

Another important industrial use is as a cementing agent in well drilling, particularly in the oil and gas industries. This usually is formulated from low consistency alpha hemihydrate (pressure calcined) and is very effective for sealing off porous or cavernous rock stratas often encountered in the drilling of oil, gas, or water wells.

A small but growing use of high density, high strength plasters is in the maintenance of floors, and for quick patching high use surfaces (such as highways or airport runways) that cannot be closed down for more conventional repairs. In both this and the well drilling application, gypsum plaster may be mixed with other cementitious materials to obtain optimum results.

Anhydrous Gypsum: In addition to the hemihydrate industrial plasters described previously, there are several products made from calcium sulfate with no water of crystallization, i.e., from "second settle" stucco made when the calcining temperature has been carried up to about 204°C (400°F). This material, often referred to as "soluble anhydrite," has a high affinity for water, making it an efficient drying agent, and is sold in various particle size gradations for use as a desiccant in laboratory and commercial applications. Also, a finely ground soluble anhydrite is used as a carrier for certain types of insecticides which must remain absolutely dry to maintain their toxicity.

When the calcination process is carried to a temperature of 482°C (900°F), a dead-burned or insoluble anhydrite product is formed which is used for industrial filler applications or for the manufacture of Keenes cement. Dead-burned gypsum is preferred for some filler uses over uncalcined gypsum, particularly where the temperature of the product in which the filler is used may go over 49°C (120°F), the point at which natural gypsum begins to release water of crystallization. Also, dead-burning usually produces a whiter product, and is preferred where color (i.e., lack of color) is important. Other uses for dead-burned fillers are a source of calcium in food products, and for yeast and beer processing. They also serve as a diluent or extender in such compositions as pills, rubber, artificial wood, plastics, paper, and pigments.

Keenes cement is a generic name for dead-burned gypsum which by the use of additives can be made to set and harden after being mixed with water. The usual set specifications fall in the range of 4 to 12 hr, and its major use is as a wall plaster where extra density, strength, and hardness are desired. It is made in only a few locations, and then only in small quantities.

Uncalcined Industrial Gypsum: This category, sometimes referred to as "raw" gypsum, includes the largest volume use for the mineral outside that for wallboard, i.e., as a retarder for portland cement. In 1980 (see Table 1) 3.5 Mt (3,900,000 st) or 20% of all the gypsum used in the United States was consumed by this market.

It has long been known that calcium sulfate compounds will control the setting time of portland cement. They also will control the rate at which the cement paste develops strength, and the shrinkage of cement products during drying. The use of calcium sulfate compounds in cement has been the subject of extensive research, and is closely regulated by ASTM standards. The amount of gypsum (or gypsum/anhydrite blend) used depends upon: (1) the SO_3 content of the gypsum, (2) the type of cement, and (3) the mix of raw materials used to make the cement clinker. In practice, the amount of gypsum used will vary from 4 to 6%, averaging approximately 5% by weight of finished cement. It is mixed with the clinker as it is fed into the grinding mill, and the two materials are then ground together.

The setting and strength development of portland cement is based upon several complex chemical reactions. As these are better understood, raw material specifications, including those for gypsum, are being tightened, and particular emphasis is being placed upon the uniformity of all constituents. There also is a trend towards the use of more SO_3 per unit of cement, with the result that the gypsum industry is being asked to provide increasing quantities of uniform gypsum and anhydrite products for portland cement manufacture.

The foregoing statistics on portland cement retarder usage include anhydrite, and in fact, this is the largest market in the United States for this mineral. However, it is a highly fragmented market because not all cement plants want anhydrite in their retarder rock. Where used, it is generally blended with gypsum, the usual blend containing 40% of the mineral anhydrite. This means that any one cement plant might use only 4.5 to 9.0 kt/a (5,000 to 10,000 tpy) of anhydrite, and because many plants prefer straight gypsum to a gypsum/anhydrite (GA) blend, it is estimated that probably no more than 5% of the 3.5 Mt (3,900,000 st) of gypsum reported used for portland cement retarder in 1980 was the mineral anhydrite.

Terra alba is a term applied to high purity, finely ground raw gypsum. Usual specifications call for a 95% −325 mesh, very white color material whose exact specifications as to color and particle size vary with the end use. It has the same general utilization as dead-burned (anhydrous) gypsum, that is, as an inert filler or diluent, and/or as a source of soluble and biologically available calcium. However, as noted under that discussion, the two products do have different characteristics, and are not completely interchangeable. When used in bread and bakery products, or in pharmaceuticals, terra alba must meet Food Chemicals Codex specifications, and generally is made from +97% pure gypsum. In one of its more interesting applications, the presence of calcium sulfate fillers in water for the brewing of beer has been found to be very desirable because it assists in the development of proper yeast fermentation, helps the settling out of the yeast, and aids in the clarification of the beer.

Glass batch gypsum is the uncalcined mineral ground and sized to be comparable to the fineness of sand, and is used in the manufacture of container glass. The gypsum, which must be fairly dust free, is thoroughly mixed with the other ingredients of a batch prior to charging the furnace, and performs three functions: as an oxidizing agent, as a fining agent, and it aids in removal of scum.

Agricultural Group

The oldest use of gypsum in the United States, dating back to pre-Revolutionary War times, is as a soil conditioner. Generally called "land plaster," this material can be produced from either gypsum or anhydrite, and is usually ground in an air-swept mill to a fineness ranging between 75 and 90% −100 mesh. A screened and air separated granular-type product is available in certain locations in limited quantities which will average only 25% through 100 mesh. Also, in a few western states, particularly California, gypsite (an impure, unconsolidated form of gypsum) is sold in large quantities, and generally requires no processing except a hammer mill to break down the lumps.

The principal purposes for applying gypsum on land are to:

1) Improve the structure or physical conditions of soil by breaking up compacted clays; and by increasing porosity, aid in drainage.

2) Supply neutral, soluble calcium for peanuts and other crops.

3) Supply quickly available sulfate sulfur.

4) Neutralize sodium compounds in alkali soils.

5) Correct high sodium irrigation waters.

6) Improve availability and utilization of nitrogen (S-N relationship).

7) Clear up muddy farm ponds.

8) Stimulate soil microorganisms.

In other agricultural related uses, ground calcium sulfate, either in the dihydrate or anhydrous form, is often added as an ingredient in formulating feeds and feed premixes for beef cattle, dairy cows, and sheep. In this use, it supplies total sulfur requirements in safe, easy to mix form, increases the efficiency of nonprotein nitrogen in urea feeds, is an ideal supplement for ensilage enhancers, and is an effective regulator for self-feeding on the range. Also, terra alba gypsum is used as a carrier for insecticides and for micronutrients, with granulated and pelletized gypsum being produced in limited quantities for the same purpose.

Minor Uses for Calcium Sulfate

A minor use at the present time for gypsum and anhydrite is for its sulfur values. It has often been pointed out that these two calcium sulfate minerals constitute the world's largest reserve of sulfur, and considerable research has gone into the technology of this use. However, the economics of utilizing either gypsum or anhydrite for this purpose are rarely competitive with other sources of sulfur. In the late 1960s when demand for sulfur exceeded its supply, pushing the price up into the $40 per ton range, a number of new investigations were undertaken, but none resulted in a viable operation. Since that time sulfur prices have gone full cycle, including the establishment of new highs of over $100/t, but no new plans to extract sulfur or sulfuric acid from calcium sulfate have been announced.

The most common approach has been to bring about the thermal dissociation of calcium sulfate to produce calcium oxide and sulfur dioxide, converting the sulfur dioxide into sulfuric acid and the calcium oxide into cement clinker (Bedwell, 1952; Hull, 1957). This process was originally developed by Müller and Kühne in Germany in the 1920s. It was improved upon by Imperial Chemical Industries in England in the 1930s, and is based upon burning (calcining) a mixture of anhydrite, shale, silica, and coke in rotary kilns. Approximately one t of cement clinker is produced per ton of sulfuric acid from 1.4 t (1.6 st) of anhydrite [or 1.8 t (2.0 st) of gypsum].

In 1969, seven plants utilizing this process were operating in England, East Germany, Poland, and Austria, most of them using anhydrite as the principal raw material. However, ten years later in 1979 only four such plants were operating, these being in Poland, Austria, East Germany, and South Africa. The Austrian and South African plants utilize the OSW-Krupp process (an offshoot of the Mueller-Kuhne process) and use phosphogypsum plus some natural anhydrite or gypsum as a raw material (Wheelock, 1980).

Although the process was offered to and considered in the United States during the sulfur shortage of the late 1960s, no plants were built. The problem was that a plant of sufficient size to be economic in cement production would produce too much sulfuric acid. Capital requirements would be correspondingly high, and the risk of balancing two markets—cement and sulfur—which historically have wide price swings was considered to be too great.

A variation from portland cement-sulfuric acid route is production of high quality lime and sulfuric acid (Wheelock and Boylan, 1968). In this process, gypsum or anhydrite is heated in a reducing atmosphere (without the addition of alumina, silica, and iron required for portland cement) to produce a relatively pure calcium oxide. Although considerable experimentation has been performed, no commercial operation has yet been developed. In both the cement-acid and lime-acid processes, anhydrite is the preferred mineral because of the higher yield of sulfur per ton, and elimination of the Btus needed to drive off the water of crystallization in gypsum.

Another use of gypsum for its sulfur value is in the production of ammonium sulfate. Several plants have been built in sulfur-poor countries, and the technology is well developed (Habashi, 1969). In 1968–1969 a plant for this purpose was built in California; however, it was closed down about a year later.

A serious attempt was made in Texas in the late 1960s to produce elemental sulfur from gypsum, but technical and economic problems prevented successful operation. The US Bureau of Mines has developed a process to produce elemental sulfur from gypsum with soda ash as a co-product, but no known commercial application has been made (Robinson and McFarlane, 1969).

Prices

In Table 4, average prices of five representative gypsum products are shown for 1979, and are typical of the spread in value between different products. These figures represent mill net prices, and are averages across the United States. Prices at any given location will vary widely from these averages, and to them must be added freight and other distribution costs to the ultimate consumer. Delivered costs, based upon list prices, are quoted for several cities in *Engineering News-Record*.

TABLE 4—Average Gypsum Prices (1980)*

Portland cement retarder	$10.67 per st
Agricultural gypsum	11.53 per st
Basecoat plaster	53.55 per st
Industrial, molding & casting plasters	72.00 per st
Wallboard (½-in. thickness)	83.08 per st

* Source: Based upon total quantities and value of these products as shown in US Bureau of Mines Annual Advance Summary (Pressler, 1980).

The recent trend of gypsum prices is shown in Table 5, and reflects the drop in building and construction activity in 1968–1970 when the industry was operating below capacity, and the subsequent improvement beginning in 1971.

TABLE 5—Price Index for Gypsum Products (1967 = 100)*

1967	100.0
1968	103.6
1969	106.7
1970	100.0
1971	107.0
1972	114.7
1973	120.9
1974	137.6
1975	144.0
1976	154.4
1977	183.5
1978	236.0
1979	254.9
1980	251.8

* Source: Based upon Standard Industrial Classification 3275 (gypsum products: block, board, plaster, rock, tile, etc.) as reported by the Bureau of Labor Statistics, US Dept. of Labor.

Bibliography and References

Anon., 1970, "World Trade in Gypsum," *Industrial Minerals*, Oct., pp. 9–25.

Anon., 1978, "The Economics of Gypsum, Second Edition," *Roskill Information Services Ltd.*

Ashizawa, R.Y., 1969, "Gypsum," *Minerals Yearbook*, US Bureau of Mines, pp. 547–553.

Bedwell, W.L., 1952, "The Production of Sulphuric Acid From Calcium Sulphate," The Royal Institute of Chemistry, Lectures Monographs, No. 3, 21 pp.

Dawes, R.E., 1970, "Gypsum," *Minerals Yearbook*, US Bureau of Mines, pp. 559–566.

Duda, W.H., 1966, "Simultaneous Production of Cement Clinker and Sulfuric Acid," *Minerals Processing*, Aug., pp. 10–13, 26.

Fisher, H.C., 1953, "The Setting of Gypsum Plaster," *ASTM Bulletin*, Sep., pp. 43–47.

French, R.R., 1967, "Beneficiation of Low Grade Gypsum by Electronic Color Sorting," Preprint 67-H-318, SME Fall Meeting, 15 pp.

Habashi, F., 1969, "Processes For Sulfur Recovery From Ores," *Mining Congress Journal,* Vol. 55, No. 6, June, pp. 38–46.

Hull, W.Q., et al., 1957, "Sulfuric Acid From Anhydrite," *Industrial & Engineering Chemistry,* Vol. 49, No. 8, Aug., pp. 1204–14.

Kelley, K.K., et al., 1941, "Thermodynamic Properties of Gypsum and Its Dehydration Products," *Technical Paper 625*, US Bureau of Mines, 73 pp.

Lane, M.K., 1968, "Disintegration of Plaster Particles In Water," *Rock Products,* Mar., pp. 60–63, 108; Apr., pp. 73–75, 116–17.

MacPherson, A.R., 1950, "Recent Investigations Into The Beneficiation of Canadian Gypsum," Canada Department of Mines & Technical Surveys, Memorandum Series No. 111, Nov., pp. 1–28.

Moyer, F.T., 1939, "Gypsum and Anhydrite," *Information Circular 7049*, US Bureau of Mines, Feb., 45 pp.

Pressler, J.W., 1979, "Gypsum in 1979," Annual Advance Summary, US Bureau of Mines, 17 pp.

Pressler, J.W., 1981, "Gypsum," *Minerals Yearbook, 1980,* US Bureau of Mines, pp. 385–396.

Riddell, W.C., 1950, "Physical Properties of Gypsum," *Rock Products,* May, pp. 68–71, 102.

Riddell, W.C., 1954, "Effect of Some Inorganic and Organic Compounds on the Solubility, Setting Time and Tensile Strength of Calcined Gypsum," *Rock Products*, Oct., pp. 109, 113, 117.

Robinson, V., and McFarlane, J.D., 1969, "Recovery of Sulphur Values From Gypsum and Anhydrite," *CIM Bulletin,* Vol. 62, No. 689, Sep., pp. 967–971.

Smith, F.H., 1963, "Aging of Calcium Sulfate Hemihydrate," *Nature*, Vol. 198, pp. 1055–1056.

Wheelock, T.D., and Boylan, D.R., 1968, "Production of Sulfuric Acid From Calcium Sulfate By Reductive Reaction," American Institute of Chemical Engineers, Preprint 6B, Annual Meeting, Dec., 23 pp.

Wheelock, T.D., 1980, "Desulfurization of Phosphogypsum," International Symposium on Phosphogypsum, Nov. 5–7.

Wilder, F.A., 1923, "Gypsum: Its Occurrence, Origin, Technology and Uses," *Iowa Geological Survey,* Vol. 28, 1917–18, pp. 47–537.

Wyman, R.A., 1969, "The Floatability of Twenty One Non-Metallic Ores," Canada Department of Energy, Mines & Resources, TB 108, Jan., 51 pp.

Construction Materials

Insulating Materials—Thermal and Sound

RAYMOND J. KUJAWA*

For general purposes, insulating materials may be any of those mineral substances that provide a barrier between a desired human environmental feature and an unwanted condition. In this sense, radiation shields, electrical insulations, cryogenic insulation, and acoustical barriers would qualify. For this chapter only the thermal and acoustic insulations used for the median range of normal comfort control will be covered. Because of their magnitude and complexity the other fields are left to other chapters. In another sense, this section deals only with the ultralightweight aggregates and insulations, rather than the multitude of lightweight structural materials.

Definitions and Uses

The insulating materials covered in this chapter are the three groups of mineral products that provide thermal or acoustic barriers through the median temperatures encountered in home and industry. The three mineral categories are perlite, vermiculite, and the silicate wools whether they be categorized as glass, mineral, or rock.

Perlites are those amorphous aluminum silicates with minor impurities that take the form of a felsitic, glassy lava possessing a pearl-like luster and concentric cracking. In a more commercial sense the term is used to describe either the natural or processed acidic, volcanic glass, containing from 2 to 5% combined water, that may be converted to a frothy mass by heating rapidly to the softening point.

Vermiculites are micaceous, hydrated magnesium iron aluminum sheet silicates of varying composition. Commercial vermiculites are any of these sheet silicates that are capable of sufficient expansion on rapid heating to make an acceptable, highly porous product.

Mineral wools, rock wools, silicate cottons, glass fiber are generic names for manufactured fibers in which the fiber forming substance may be slag, certain rocks, or glass. The fibers may be made by a blowing or spinning process. The resultant fibers may be felted or matted or gathered in the manner best suited to the use.

The three groups may be used competitively to insulate ceilings and walls. The fibers have strong advantages in this area as they may be manufactured in batts that are stable in residential wall construction. The perlites and vermiculites are more suited to insulate cores and cavities of masonry walls and as plaster and concrete aggregates for thermal insulations. Vermiculite is widely used for insulating attic floors. Because of manufacturing techniques and controls, the glass fibers have made great strides as composites in the plastics industry but only a few of these uses classify in the insulation category.

Raw Materials

Deposits

In the commercial sense, the perlites are any volcanic glass that may be bloated to an acceptable degree by rapid heating. Thus, many of the pumicite occurrences may be interlisted as perlite. As perlite and pumice are particularly susceptible to alteration by geologic processes, commercial deposits of these materials are restricted to areas of relatively recent volcanism (Tertiary and Quaternary), such as the western United States. The chapters on perlite and pumice in this book cover the details of geology, reserves, and beneficiation of the mineral material.

As mentioned, the vermiculite category covers a number of specific sheet silicates, many of which are closely related to the micas. The vermiculites are of sufficiently similar composition and character for them to be concentrated by flotation. All of the commercially important

* Chief Geologist, Construction Products Div., W.R. Grace & Co., Libby, MT.

Note: This chapter was revised by Research Dept., Construction Products Division, W.R. Grace & Co., Cambridge, MA.

vermiculite deposits are associated with basic or ultrabasic rock types. The chief domestic deposits are in Montana and South Carolina. The largest known foreign deposit is in Transvaal Province, Republic of South Africa (Schroeder, 1970a). The details of geologic occurrence are covered in the vermiculite chapter.

The mineral wools may be made from melts of specific argillaceous limestones, particular smelter slags, or glasses of a particular composition. The many manufacturing plants available as sources of molten material and the recent development of the multitude of special compositions has reduced the importance of natural sources. However, the natural raw materials are still of great importance and attention should be directed to the specific sections—limestones, etc.

Economic and Environmental Factors in Raw Material Supply

Because all of these related mineral insulators are low value, bulk commodities, it has been desirable to develop those deposits that were closest to markets or close to cheap transportation. As our individual demands increase, the utilization of more remote occurrences will probably follow.

Any area with adequate facilities may well be a valuable location provided the questions of transportation and political climate can be answered. With insulating materials the historic problem has been transportation but the awakening worldwide interest in environmental matters is adding the new factors of air pollution control, water pollution control, control of strip mining, etc. These environmental factors may well add to the attraction of a more remote location and thus some of the transportation costs may be offset.

The perlite producers are approaching an era that will require some greater utilization of beneficiation. While perlite occurrences are widely distributed the best locations with respect to marketing are being mined. These properties will begin to use new techniques to enhance the quality of the crude material being shipped to expanders.

Traditionally, as with many of the older metallic properties, deposits of insulating materials were opened by mining the highest grade ores possible with the least amount of concentrating equipment. Naturally this "high-grading" allowed maximum market entry and development with minimum capital and presented a product at a most attractive price, today's "introductory offer." Since that time the producers of all the different insulating materials have introduced quality control programs, specific standards, and to some degree they have all developed beneficiation methods. These techniques allow utilization of lower grade materials but the capital requirements are increasing rapidly.

By way of example, the major vermiculite producers have engaged in extensive expansion programs. While the basic reason is the natural increase in production at lower costs, a large portion of the capital outlay has gone into pollution control equipment and the recovery of smaller particles. These expansion programs are trending toward wet methods that allow higher recoveries at higher cost but with less ecologic impact.

In a more important sense thermal insulations, by definition, restricted many deposits and their potential. More recently the same minerals began to find use for sound control. Most recently these minerals have developed areas of use in agriculture. Consequently, mineral deposits that had marginal values for one use must now be considered in a broader aspect. Virtually no commodity in the basic mineral industry survives on one use.

In spite of the projected excess stocks in all parts of the mineral insulation field, there is wide interest in these materials. This general interest is causing much private examination of mineral deposits. However, because of the nature of the industry, an individual is frequently frustrated by the apparent lack of market interest in a mineral discovery of this sort. There are few obvious clues in corporate names and, frequently, slow response to inquiry. The best source of information and assistance lies in the smaller governmental agencies. This may be such as a state or provincial department of mines or a community business bureau. This route will probably satisfy a prospector quicker, and direct a prompt reply.

Processing and Manufacturing

Vermiculite and perlite undergo milling procedures near the mine sites, to prepare them for exfoliation. Milling consists of crushing, drying, screening, and separation, with ores of specific gradation stored until transported to processing plants. Gradation is based on accepted standards such as ASTM or federal specifications. The light, bulky nature of the finished products dictates that the output can only be shipped limited distances without freight rendering the product noncompetitive. Hence, the trend among major producers is to establish multiple processing plants, each of which serves a limited geographic area.

When received at an expanding plant, ore is exfoliated in furnaces and discharged—typically into multiwall paper bags or plastic containers. Furnaces may be fitted with dust collectors, de-stoners, or equipment to introduce other ingredients into the package, i.e., emulsions to improve water-repellency.

Exfoliation results in an increase in bulk in the range of 8 to 12 times, with an inverse reduction in density.

Mineral wools (rock wool, glass wool) are derived from fiber-forming, or fabricating procedures that result in production of a fibrous material with properties that have led to their extensive use as thermal insulation for homes, commercial buildings, and various industrial uses.

Loose, or pouring wool, is generally a byproduct of the manufacturing process, accumulated and packaged for use as attic insulation. Much of the mineral wool produced is modified by adhering it with facings of kraft paper or foil to form batts or blankets to render it more suitable for application.

Density and thickness of mineral wool blankets can be predetermined during manufacture, as dictated by performance and market requirements.

Industry Trends

During the decade of the 1970s, the markets for insulating materials expanded greatly due to the skyrocketing cost of energy. The pressure of energy costs created an awareness in the consuming public that a measure of cost control was to add insulation to presently constructed buildings and to increase the volume in new construction.

The demand for residential insulation is expected to level off while the demand for nonresidential insulation is expected to increase at a rate dependent on nonresidential construction. Insulation will be a larger fraction of the cost of nonresidential construction in the 1980s than it has been in the past.

The single most important impetus in improving the thermal integrity of new construction and equipment has been the inclusion of energy conservation in building design. Historically, the domestic construction industry, because of its high sensitivity to first costs, has ignored the full-scale improvement of certain qualities that in the long run would be beneficial to fuel savings. Since the Arab oil embargo, however, there has been a slow but steady movement in this country to establish mandatory regulations for energy conservation. The US Department of Housing and Urban Development (HUD) has had a Minimum Property Standard for residential construction for a number of years that was applicable to construction involving federal funding, i.e., FHA and VA mortgages. The American Society of Heating, Refrigerating and Air Conditioning Engineers (ASHRAE) has also developed standards to regulate the design of new construction. In addition to these two standards, many states have either adopted these standards or set up their own requirements for achieving energy savings in new construction within their jurisdiction.

Generally, the proposed HUD standards are more or less in agreement with the thermal insulation requirements set up by ASHRAE 90–75. There is a strong movement among the various states and model code authorities to adopt ASHRAE 90–75 for new construction. Even though the HUD requirements would relate only to those buildings constructed using federal funds, the standards themselves would, by custom, become adopted by most conventional mortgage builders, particularly if state building codes follow suit.

Mineral insulations now face strong competition from synthetic organic foam insulation products in some of their traditional markets. The competition will be based on the relation of cost to the efficiency of the insulation materials and will be greatest in markets where fire resistance of high temperature applications is not critical. The key for continued growth in the mineral insulation industry will be to develop markets and applications which make use of the advantages mineral insulations offer, namely, fire resistance, performance at high and very low temperatures and permanence as installed.

Bibliography and References

Anon., 1972, *What Mining Means to the United States*, American Mining Congress, Washington, DC, 24 pp.

Bush, A.L., 1973, "Lightweight Aggregates," *U.S. Mineral Resources*, D.A. Brobst and W.P. Pratt, eds., Professional Paper 820, US Geological Survey, pp. 333–355.

Dana, H.J., and Lyle, R.E., 1935, "Comparative Heat Loss on Insulated and Uninsulated Buildings," Bulletin 45, Engineering Experiment Station, State College of Washington, Vol. 17, No. 10, Mar., 26 pp.

Schroeder, H.J., 1970, "Perlite," Bulletin 650, US Bureau of Mines, pp. 1129–1137.

Schroeder, H.J., 1970a, "Vermiculite," Bulletin 650, US Bureau of Mines, pp. 1283–1291.

Construction Materials

Roofing Granules

CLIFFORD L. JEWETT *

ROBERT C. COLLINS †

LEONARD W. WEAVER ‡

THOMAS McSHEA §

The development and production of artificially colored roofing granules have been key factors in the success of the asphalt roofing industry. They provide protection for the asphalt from solar radiation and satisfy the homeowner's desire for a wide range of attractive colors. As a result, the asphalt shingle has become the most commonly used roofing material today. In addition to providing a stimulus to the asphalt roofing industry, artificial methods of coloring granules completely changed the nature of granule production from one of quarrying, crushing, and screening to a complex technical operation. Today's roofing granules are the result of many years of research, and they are produced in modern plants to exact specifications.

History

Asphalt roofing materials available today are the product of a long evolution of technological development. Forerunners of the present bituminous products, appearing as early as 1780, were primitive constructions of burlap saturated with crude tar or pitch. An early improvement to this basic effort was the dusting of talc, mica, or other finely ground material on the coated surfaces to keep the roofing from sticking to itself in storage. At times some pigmentation was added with the dusting material for initial color, but this was far from satisfactory. In fact, this early product possessed none of the qualifications of a good roof surfacing material. It was not economical; it was difficult to handle and apply; it was not attractive; it provided no resistance to fire; and it would not withstand the elements as well as other products then in use.

Methods and machines were soon devised to saturate endless webs of felt with asphalt. Even though saturated felt became available, this product of rags and wood fiber had almost the same limitations as burlap and tar. Even with heavier coatings of asphalt applied to both sides of the sheet for rigidity and longer life, the product still lacked color and fire resistance. Further, it was found that constant exposure to the sun soon caused the asphalt to become brittle, and it eventually cracked and leaked. This led to the first important use for crushed mineral products on asphalt-coated shingles. When a layer of crushed mineral was embedded in the surface exposed to the weather, the asphalt coating was protected from ultraviolet light and retained its elasticity over a much longer period of time. In addition to this protection, the mineral surfacing provided some resistance to fire. Fire resistance in the past had been found only in the expensive slate or tile roofs and the unattractive and expensive metal roof coverings which were then available.

The most common mineral granule applied to the early asphalt roofing products after 1900 was screened waste from slate quarries. Slate granules, being available in dull shades of red,

* Retired; formerly Technical Director, Industrial Mineral Products Div., 3M Company, St. Paul, MN.

† Former Manager, CUSTM-Roofing Granules, Industrial Mineral Products Div., 3M Company, St. Paul, MN.

‡ Retired; formerly Manager, Technical Center, Bird and Son, East Walpole, MA.

§ Geologist-Chemist, Technical Center, Bird and Son, East Walpole, MA.

The chapter on Roofing Granules in the 4th edition (1975) of *Industrial Minerals and Rocks* was prepared by Clifford Jewett and Robert C. Collins. This chapter in the 5th edition has been extensively modified and updated by Leonard Weaver and Thomas McShea.

green, black, purple and a few other miscellaneous colors, did provide a degree of color. Since slate has a flat platelike cleavage, it provides a good protective cover for the asphalt-coated sheets. Records show that coarse slate screenings were used for asphalt roofing as early as 1906. As the asphalt roofing industry developed, many other natural colored rocks and minerals were used to coat the surfaces of the roofing. The list included silica sand, quartz, novaculite, feldspar, mica, greenstone, talc, glass, slag, basalt, granite, rhyolite, and oyster shells. In addition, a number of manufactured granular mineral materials, such as crushed brick and tile, fired clay, and white porcelain were used. The latter was introduced in 1912, but never became popular because of its poor adhesion and its tendency to stain with oils from the asphalt.

Development of Artificially Colored Granules

Records indicate that S. W. Osgood produced the first artificially colored roofing granules in 1914 by running molten slag into a solution of sodium or potassium silicate which was absorbed by the porous slag. Various colors were obtained by adding colored pigments such as chrome oxide or iron oxide to the silicate coating and by subsequently heat treating the granules. In 1922 H. Fisher developed another technique in which green slate granules were saturated with a copper sulfate solution and heated to 760°C (1400°F) to form copper oxide. The hot granules were then quenched in a ferrous sulfate solution and refired. This resulted in a copper ferrite glaze on the surface of the granules which was variable from yellow to red, depending upon the proportion of the oxides used.

Many investigators developed a variety of granule coloring processes. More than 150 granule patents have been issued. These included "painted" granules by Schrader and Sproat, metal salt processes by Pelton and Fisher, and colored slags by Finney and Van de Mark. Probably the most significant of the processes invented in the early period were the sodium silicate coatings introduced about 1925 by H. C. Fisher of the Richardson Co., the phosphate coatings of H. R. Gundlach of the Central Commercial Co., the fused glaze coatings of W. H. Alton, L. Brown, E. H. Nichols, and 3M Co.'s G. W. Swenson, and the silicate-clay coatings of C. L. Jewett of the 3M Co. and M. R. Beasley of Bird & Son. The silicate-clay process is the basis for much of today's roofing granule technology. In this process the granules are coated with a mixture of sodium silicate, clay, and inorganic colored pigments, followed by firing in rotary kilns.

A serious problem was soon discovered. It was found that the granule base must be almost completely opaque to protect the asphalt adequately from the sun's rays. If the short, actinic rays of the sun (ultraviolet light) pass through the granules, these rays will deteriorate the underlying asphalt, with the result that adhesion is lost and the granules come off. In the mid-1930s, the roofing industry generally recognized the inadequacy of translucent granules and turned to opaque base materials. This led to the end of the fused glaze coatings which were discolored by the iron and manganese in the opaque bases.

The important developments since about 1940 consist mainly of modifications of the silicate process. Some of these modifications allow for lower firing temperatures to permit the use of heat-sensitive pigments and to improve other properties. Others provide improvements in color and weatherability. Many of the recent patents involve the development of improved light-colored and white granules. In the late 1960s a new algae-fungus resistant granule was developed and offered to the trade. This new white granule incorporates an inhibitor which dissolves slowly and retards the growth of algae and fungus which discolor the granule surface, particularly in the southern US.

Specifications for A Granule Base Rock

Twenty-five or more years ago the specifications for roofing granules were much looser than they are today. No single type or family of rock or minerals can be considered as a class to be a good and acceptable granule base, but rather a mineral deposit (Table 1) must be selected which meets rigid physical and chemical specifications. In recent years, these specifications have been so tightened by advanced technology and competition that only a few rocks can be processed acceptably and economically. Even after a geologic investigation has located a suitable rock formation, quarry operations must follow definite development programs. Quality control procedures must start at the quarry site. Therefore, the costs of geologic exploration, quarry development, and operation are usually higher than those found in normal commercial rock and stone operations.

A rock intended for use in manufacturing

TABLE 1—Rock Types Used As Granule Base

West Coast	
	Basaltic river gravels
	Dacite porphyry
Midwest	
	Andesite
	Argillite (graystone)
	Nepheline syenite
East Coast	
	Rhyolite (partially metamorphosed)
	Diabase
	Greenstone
	Arkosic quartzite
Canada	
	Basalt

colored roofing granules must have the following properties:

A) It must resist weathering.
B) It must be adaptable to the coloring process.
C) The deposit must be uniform.
D) The deposit must contain sufficient tonnage.
E) It must have low porosity.
F) It must be almost completely opaque to ultraviolet light.
G) It must be tough.
H) It must provide equidimensional fracture upon crushing.

A discussion of these properties follows:

A) *It must resist weathering.* The weathering properties of a roofing granule base are of primary importance. If a granule base does not adequately resist exposure to the elements, the granules do not provide the necessary protection to the asphalt product nor do they give color permanence. Granule disintegration due to weathering is of two types, mechanical and chemical.

Mechanical weathering is caused chiefly by differential thermal expansion of the various rock-forming minerals and by moisture which seeps into the voids around mineral crystals, causing rock distintegration upon freezing.

Chemical weathering is caused chiefly by the solution, hydration, or oxidation of certain mineral substances. This action is usually the result of the action of rain water or dew conveying dissolved gases, often acidic. Note: carbonates and sulfates not present in suitable granule rock.

B) *It must be adaptable to the coloring process.*

1) Surface characteristics—after crushing, many of the rocks do not have the surface characteristics which are desirable for a granule base. For example, quartz, when crushed, has a smooth conchoidal fracture and it is difficult to adhere a coating to such a surface. Also the mineral constituents must be sufficiently inert to prevent their reacting chemically with the processing chemicals in manufacture.

2) Color change upon processing—many of the naturally occurring rocks contain high percentages of iron compounds. When this type of base rock is heated, it becomes brownish-red in color and changes the appearance of the coated granule. The basalt deposits generally have this property, some to an excessive degree.

3) Spalling—many rocks will spall or flake off in the heating process due to the presence of mineral constituents of different expansion characteristics and the presence of molecular water. This condition alters the screen grade and leaves uncolored faces on the granule. Granite is particularly susceptible to this mode of failure.

C) *The deposit must be uniform.* It is essential that uniform standards be maintained in the roofing industry. Therefore, after granule manufacturing facilities have been established, production becomes basically a duplication process. In order to conform to the specifications, it is of the utmost importance that the rock base be of uniform quality and homogeneous in color as well as in physical properties. It must not have small variable seams nor contaminations which make it impossible to maintain uniform color and processing conditions.

D) *The deposit must contain sufficient tonnage.* Successful roofing granule operations cannot be sporadic and temporary. The capital investment required to construct a granule plant today is so great that the operator must plan to produce steadily over a long period. To justify such an investment, a producer should have ample rock of uniform quality to protect his investment and return a profit over a long period of time (e.g., 25 to 50 years).

E) *It must have low porosity.* Granule rock should be massive, dense, and nonporous. Partially weathered surface rock, particularly in nonglaciated areas, is often porous. High porosity is detrimental, because it allows the penetration of moisture and will, in some cases, permit the granule to act as a wick, allowing the seepage of water into the interior of the asphalt shingle, thereby causing shingle blistering, cracking, and general deterioration. As the coloring process involves applying a pigmented coating on the surface of the granules, a porous base will increase costs substantially by the absorption of coating chemicals.

F) *It must be opaque to ultraviolet light.* The asphalt roofing industry has found through experience and research that the actinic rays of the sun (ultraviolet light) rapidly deteriorate roofing asphalt. A mineral surface is applied to protect the asphalt from this radiation. In the early days, before this factor was fully understood, very little attention was paid to the transparency of the mineral surfacing grain. Roofing manufacturers have placed strict specifications on ultraviolet light transmission of granules. Therefore, it is of the utmost importance that a base rock be opaque.

G) *It must be tough.* Base granules must be naturally hard, but more important is their toughness. True, it is more expensive to crush a tough, hard rock to granule grade before the coloring process, but this very toughness assures that there will be a minimum of attrition breakdown in handling. It must be realized that there is some unavoidable attrition of the granule by conveyors, chutes, screens, and elevators, both before and after coloring. Then, too, granules must withstand additional handling during subsequent processing at the roofing plant.

H) *It must provide equidimensional fracture upon crushing.* Since roofing granules are applied to a hot moving asphalt-coated sheet and pressed into the asphalt with embedding rolls, flat or splintery granules will be embedded directionally, and light will be reflected from the shingle in a manner which gives the impression of distinct color differences when viewed from different angles. Use of granules made from rock which has a cubical fracture helps to eliminate this problem.

Granule Processing

Plant Site Location

Detailed geological investigations combined with a technical laboratory evaluation and various business aspects, including environ-

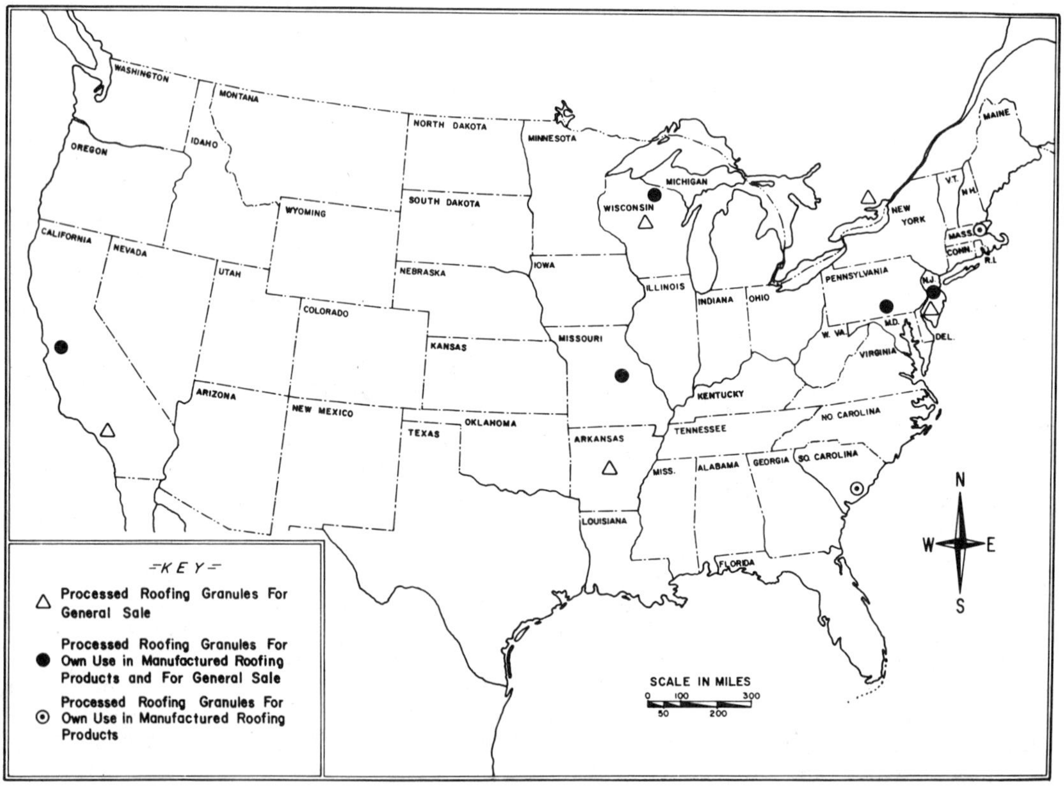

FIG. 1—*Location of ceramic-coated roofing granule plants (1981) in the United States.*

mental considerations, will generally determine a granule plant site location. Proximity to markets is a very important factor. Accessibility to transportation, water, electricity, gas, and labor supply are factors in the granule industry. Aspects of zoning and regulations of local municipalities should be thoroughly investigated. Fig. 1 shows the locations of ceramic-coated colored roofing granule plants in the United States (1981).

Quarrying

The operation of a granule quarry is in many respects similar to the operation of a quarry opened for crushed rock aggregate. The same quarry, in certain instances, may be used for both purposes. The main difference in these two types of operation is to be found in the quality problems encountered in a granule quarry. It is frequently necessary to conduct considerable diamond drilling in order to develop efficient quarry operations. Roofing granules are produced principally in open-cut operations. In general these steps are followed: first, removal of overburden; second, drilling and blasting followed by secondary breakage (many operations now prefer the drop-ball method of secondary breakage rather than secondary blasting); third, loading the rock by power shovel into quarry trucks and hauling it to the primary crusher.

Crushing and Screening

Primary crushing is done by either jaw or gyratory crushers of sufficient size for the desired capacity. A washing operation usually is not necessary, but occasionally quarry conditions are such that contaminating material must be removed by passing the rock over grizzly bars or Ty-Rock screens. Secondary crushing ordinarily is performed by cones, rolls, or hammer mills in circuit with vibrating screens.

In the modern granule plant, (Fig. 2) elaborate dust control systems are necessary, not

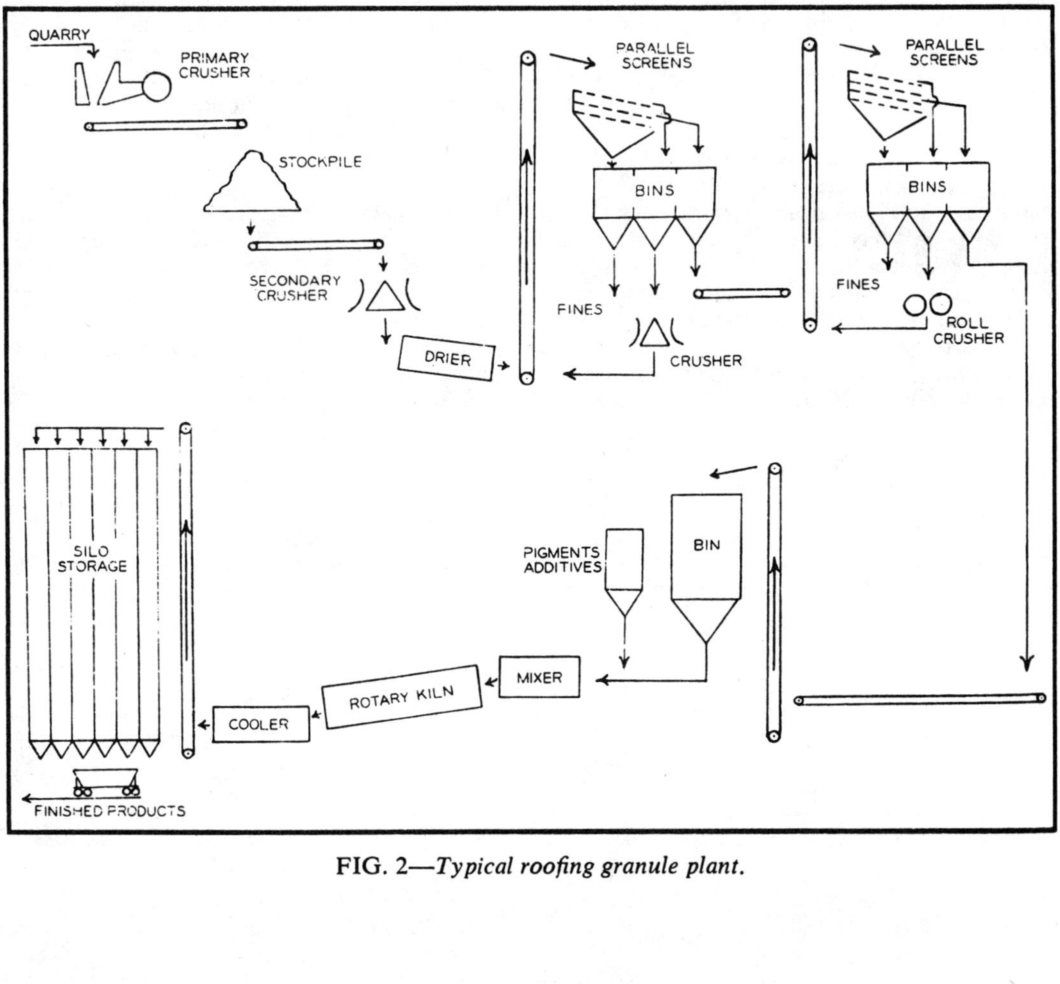

FIG. 2—*Typical roofing granule plant.*

only to protect the health of the employees but also to comply with local and state laws and zoning ordinances. In dust control systems, cyclones, electric precipitators, bag filters, or combinations of these are usually used.

The fines are stockpiled or reprocessed to meet specifications for mineral fillers, stabilizers, dusting compounds, or other products.

It is desirable to have several hundred tons of storage tank capacity for base granules to eliminate delays in the manufacturing process.

Coloring

The raw granules are trucked or conveyed by belts and elevators to the coloring section of the plant. Here they are mixed with pigments and other coating materials in rotary type mixers. The coated granules are fed into a rotary kiln where they are dried and undergo heat treatment to insolubilize the binder. Granules are discharged from the kiln into a rotary type cooler or similar device. Rescreening on vibrating screens is usually necessary after the firing and cooling process to maintain granule grade. This is done either before storage or in the shipping and loading process.

Surface treatments of oil and various other organic and inorganic materials are applied either in the rotary cooler or along the flow line before shipment. A system of bins or tanks is used for storage and for blending the finished granules before shipment. The smaller granule plants may have only a few bins, but the larger companies can store up to 200 carloads of finished material. In most cases, roofing granules are shipped in bulk, either in box or hopper-bottom cars. Some roofing companies which do not have bulk storage capacity for small tonnage items receive shipments in bags. Railroad cars or trucks are loaded either by gravity flow systems or mechanical carloaders.

Quality Control

Roofing companies have established rigid specifications over the years. It is now necessary for granule producers to maintain effective quality control laboratories as well as research facilities. Quality control procedures have been developed to test the important properties of granules desired by the roofing companies. These tests were developed to evaluate the following properties:

Base Rock: The durability of the granule base; its resistance to chemical and mechanical breakdown on handling and exposure.

Grading: The granule grading affects the appearance, color, and surface coverage of mineral surfaced roofing.

Shape and Size: Granule shape is determined primarily by crushing characteristics of the base rock. A wedge or blocky granule is preferred by most manufacturers. Complete coverage is obtained by controlling the distribution of the particle size within the grade limits. In this way, the spaces between the larger particles are filled by smaller granules. Excessive fines hinder the embedding of the larger size granules. Table 2 lists the typical grading specifications for roofing granules.

Color: The initial color as produced as well as color permanence on exposure.

Adhesion: The initial granule adhesion to asphalt as well as the resistance to shedding of granules in handling, packaging, and on outdoor exposure.

Opacity: Ultraviolet light decomposes asphalt. Opaque granules protect the roofing from this photochemical reaction.

TABLE 2—Typical Grading Specifications for Roofing Granules

Sieve Number		Percentage Retained					
		No. 8 Grade*		No. 9 Grade*		No. 11 Grade*	
U S Standard	Tyler	Min.	Max.	Min.	Max.	Min.	Max.
6	6	0	2	0	0	0	0
8	8	15	30	0	8	0	0
12	10	30	50	35	50	0	2
16	14	18	37	25	45	30	40
20	20	2	12	10	20	30	45
30	28	0	5	0	7	15	30
40	35	0	0	0	0	2	9
50	48	0	0	0	0	0	2

*Standard Specification Sieve Analysis of Granular Mineral Surfacing for Asphalt Roofing and Shingles, ASTM Designation: D 451.

Efflorescence: A white crust of soluble salts deposited on granules under certain atmospheric conditions.

Staining: The absorption of asphaltic oils from the asphalt which changes the appearance of the granules.

Sooting: The retention of dust and dirt on the granule surface. This factor is primarily found in industrial areas and is affected by numerous conditions, such as electrostatic charges.

Blistering: The expansion of water vapor within the shingle resulting in bubbles within the coating asphalt. Accelerated weathering tests of various types such as freeze-thaw tests, sodium sulfate tests, the weatherometer, etc., have all been evaluated and, in some cases, have demonstrated merit for evaluating certain properties. It has been necessary for some of the larger granule producers to maintain weather decks in various parts of the United States for the purpose of collecting data on the performance of granules and coated roofing.

The two major granule producers maintain research and technical service groups, which study the problems of roofing fabrication as a service to the roofing companies.

Auxiliary Mineral Products

So far this discussion has dealt with ceramic-coated colored roofing granules, but there are other mineral commodities directly related to the asphalt roofing industry.

Natural or Headlap Granules

During the latter part of the 1950s and early 1960s the asphalt roofing industry modified some of its shingle designs. Prior to this time most of the asphalt shingles were completely surfaced with a premium or processed granule. For economic reasons the roofing companies began to use a cheaper uncoated granule for the unexposed portion of the shingle referred to in the trade as the headlap. These natural or headlap granules must possess many of the qualities of a surfacing granule used on the exposed portion of the shingle, since the headlap granule is exposed to the weather through the cutout portion of the overlying shingle. A few of these quality requirements are:

1) Opaqueness to ultraviolet light.
2) Uniformity of gradation.
3) Natural color uniformity.
4) Weatherability.

Many types of material are currently used for this headlap granule such as slags, slates, shales, and uncoated roofing granule bases.

Asphalt Stabilizer and Backing Material

Finely ground materials were first used to stabilize the asphalt coating, that is, prevent the asphalt from sliding at normal roof temperatures. It was soon learned that some of these stabilizers also made the asphalt coating more resistant to weathering and less susceptible to shattering in cold weather. Much research has been conducted on the desirability of various amounts, grades, and types of stabilizers. At the present time, asphalt compositions containing up to 60% stabilizers are being used.

Sieve analyses of stabilizers vary greatly although they usually range from 60 mesh to finer than 325 mesh. The most common mineral types presently used are rhyolite, granite, trap rock (basalts and diabase), slate, limestone, and silica.

The asphalt shingle industry also uses a different mineral material for backing and, on some products, surfacing in place of granules. This is applied on the back of the asphalt-coated sheet as it runs through the roofing machine. The primary purpose of this application is to prevent the roofing from sticking together in a package. In making roll roofing on which granules are not used, commonly referred to as smooth roll roofing, both sides of the asphalt-coated roll are surfaced with finely divided flaky talc or mica because the flakes cover relatively large areas and also provide a slippery surface. Other materials occasionally used are schist, pyrophyllite, and finely ground rock dust, or fine sand.

Built-Up Roofing Aggregate

Mineral material used for roofing includes that group of coarse aggregates used on built-up roofs. These aggregates include miscellaneous rocks and gravels. Their selection, properties, and qualities are not to be confused with roofing granules which require different properties.

The principal roof surfacing on commercial structures is the bitumen and gravel type. This type of application is limited to roofs which have a slope of 88.9 mm per 0.3 m (3½ in. per ft) or less. It is constructed of repeated layers of felt and hot bitumen topped with mineral aggregate. The principal purpose of the mineral topping is to protect the bitumen from the sun's actinic rays. The most common mineral surfacing used on this type of roof is locally obtained pea gravel and, in a few cases, crushed stone or slag. The aggregate is generally washed and delivered to the job in bags or bulk quantities with little attention paid to color or other properties.

Distribution and handling of the numerous natural colored aggregates presented a problem to the jobbers and distributors. Currently, the practice is for roofing material distributors to handle the more popular types, and the material is delivered to the job in 36.3- to 45.4-kg (80- to 100-lb) bags. Many natural occurring colored rocks are shipped for considerable distances, but only for specialty uses, for instance, in the case of marble chips where extreme whiteness is needed.

Asphalt

Of course, one of the principal mineral products used in asphalt roofing is the asphalt itself. It will not be discussed in detail here, but some of the problems of the roofing industry related to the use of asphalt should be mentioned. Not only must roofing manufacturers locate near a refinery area because of shipping costs, but they must consider the various types of asphalts available. Crudes from different parts of the country produce different types of asphalts, and tests conducted by the Asphalt Roofing Manufacturers Association have shown that the properties and durability of the asphalt coating is largely a function of the asphalt source. This makes the selection of the asphalts very important. Fig. 3 shows the shipment of mineral surfaced asphalt roofing and siding.

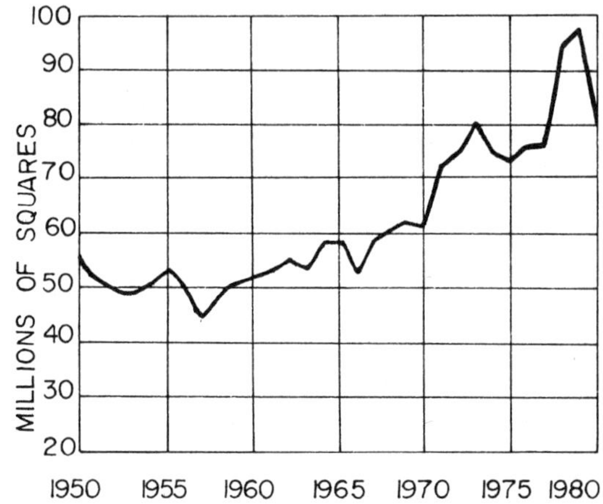

FIG. 3—*Shipments of mineral-surfaced asphalt roofing and siding. Value for 1977 does not include glass mat; 1980, estimated by Bird and Son. Source: US Dept. of Commerce, Bureau of Census "Facts for Industry" and Current Industrial Report Series M29A.*

Marketing

Production Volume

In 1978 there were approximately 5.4 Mt (6 million st) of roofing granules produced; in 1979 the figure was approximately 4.8 Mt (5.3 million st). Although there was no report on the breakdown between ceramic coated and natural granules, 75% were ceramic coated in many previous years. The other 25%, referred to in the trade as natural granules, are divided between crushed slate, slag, and uncoated granule base rocks. These natural granules now find utilization chiefly in the headlap portion of asphalt shingles. For economic reasons the natural granules are usually secured from sources close to the roofing manufacturing operation, while artificially colored granules are often shipped for considerable distances. Statistics of granules production (Table 3) have been compiled by the US Bureau of Mines.

Transportation and Price

Roofing granules are sold by the ton, usually in carload lots weighing from 27.2 to 90.7 t (30 to 100 st) each, and are shipped by rail in either covered hoppers or boxcars. A few bag shipments are made by trucks. Virtually all of this granule production is consumed by the asphalt roofing industry with minor quantities being sold as decorative material for surfacing tennis courts, private driveways, etc. Currently there are approximately 25 companies manufacturing asphalt roofing products in approximately 100 manufacturing plants spread throughout the US.

The selling prices in 1979 for natural and for processed roofing granules ranged respectively between \$26.85 to \$31.00 per st (natural) and

TABLE 3—Roofing Granules Shipped in the United States*
(Natural and Processed Granules Combined)

	Total, 1000 st†	Value, \$1000‡
1974	4,353	11,267
1975	4,381	11,236
1976	4,394	11,798
1977	5,099	14,702
1978	6,070	18,762
1979	5,264	18,221
1980	4,460	17,295

* As reported to the US Bureau of Mines.
† Metric equivalent: 1 st × 0.907 184 7 = t.
‡ Values are only for natural granules, not for processed.
Source: *Minerals Yearbook,* US Bureau of Mines.

$42.65 to $85.60 per st (processed) f.o.b. at the manufacturing plant. Various colors sell at different prices because of the varying costs of pigments and processing. Roofing granules are sold with close color control from processing plants in different sections of the country in order to allow for nationwide color standards.

Granule Surfaced Roofing Industry Trends

In the early days of asphalt roofing production, black, red, and green mineral surfaced roofing were the most popular. This was due to the historic use of black, red, and green slates on premium roofing materials at that time. In the residential building boom which followed World War II, the architectural design of the ranch style home made the roofing material much more important to the overall appearance of the home. This required more attention from the homeowner. During this period ceramic-coated artificially colored roofing granules became the dominant mineral surfacing on asphalt roofing products. In the early and mid-1950s white and pastel roofing became the vogue. During the mid-1960s roofing colors became more subdued with less color contrast in the shingle blends. The granule manufacturers met these needs from time to time with additional colors so that now most granule manufacturers offer a complete line of colors ranging from the light pastels to the dark rustic and earthtone colors. White still remains a popular color in the southern states due to its heat reflectance. It has been determined that the temperature under a shingle surfaced with white granules may be as much as 19.4°C or 35°F lower than the temperature under a shingle surfaced with black granules when exposed to full sunlight. With the introduction of the algae-fungus resistant white granule in the late 1960s which will stay white longer, white roofing products will probably remain favored in these areas.

Many homeowners like the appearance of a thicker, more rugged looking roof. This desire is being met in two ways. First, the granule manufacturers are producing a coarser grade roofing granule than the standard material used on conventional shingles, and this is being used by the shingle manufacturer in color shadow lines and other blend patterns to give greater visual depth to the product. Secondly, many roofing companies are fabricating so-called double shingles; that is, they are laminating tab on tab to give a greater thickness and rugged appearance. Such asphalt shingles closely resemble wood shingles in appearance yet give the fire resistance required by many building codes today. These new types of shingles are finding acceptance where a "premium" appearance is preferred.

Bibliography and References

Abraham, H., 1937, *Asphalts and Allied Substances,* 4th ed., D. Van Nostrand Co., New York, pp. 632–34, 1335–1338.

Behre, C.H., Jr., 1949, "Slate," *Industrial Minerals and Rocks,* 2nd ed., S.H. Dolbear, ed., AIME, New York, pp. 927–944.

Berger, E.H., 1935, "Ceramic Roofing Granules," *Ceramic Age,* Feb., pp. 49–51.

Bowles, O., 1934, *The Stone Industries,* McGraw-Hill Book Co., New York, pp. 229–289.

Bowles, O., 1955, "Slate," US Information Circular 7719, June, 12 pp.

Brown, J.A., 1960, "Granule," *Industrial Minerals and Rocks,* 3rd ed., J.L. Gillson, ed., AIME, New York, pp. 443–454.

Dale, T.N., 1914, "Slate in the United States," Bulletin 586, US Geological Survey, pp. 1–220.

Drake, H.J., 1972, "Stone," *Minerals Yearbook 1972,* US Bureau of Mines, pp. 1153–1173.

Eardley-Wilmot, V.L., 1946, "The Roofing Granule Industry," *Transactions, Canadian Institute of Mining and Metallurgy,* pp. 49, 379–394.

Foose, R.M., 1945, "Three Roofing-Granule Plants in Pennsylvania," *Mining Technology,* AIME, Jan., TP 1787.

Greenfield, S.H., 1956, "Effects of Mineral Additives on the Durability of Coating-Grade Roofing Asphalts," Building Materials and Structures Report 147, Sep., US Dept. of Commerce.

Greenfield, S.H., 1969, "Natural Weathering of Mineral Stabilized Asphalt Coatings on Organic Felts," Building Science Series 24, Oct., National Bureau of Standards.

Hirschwald, J., 1912, *Handbuch der bautechnischen Gesteinsprufung,* Berlin, Gebr. Borntraeger, pp. 1–359, 592–645.

Hughes, H.H., 1937, "Granule," *Industrial Minerals and Rocks,* 1st ed., S.H. Dolbear, ed., AIME, New York, pp. 347–352.

Ingram, S.H., 1949, "Roofing Granules in the Southwest," *California Journal of Mines and Geol.,* July, pp. 455–458.

Josephson, G. W., 1949, "Roofing Granules," *Mining Technology,* AIME, Mar., TP 1725.

Josephson, G.W., 1949a, "Granules," *Industrial Minerals and Rocks,* 2nd ed., S.H. Dolbear, ed., AIME, New York, pp. 404–416.

Ladoo, R.B., and Myers, W.M., 1951, *Non-metallic Minerals,* 2nd ed., McGraw-Hill, New York, pp. 476–483.

Singleton, R.H., 1981, "Stone," *Minerals Yearbook, 1980,* US Bureau of Mines, pp. 765–793.

Strahan, J.L., 1970, *Manufacture, Selection and Application of Asphalt Roofing and Siding Products,* 10th ed., Asphalt Roofing Manufacturers Association, New York, 97 pp.

US Patents

Osgood, S.W., 1–25–16, US Patent 1,169,506, "Process for Treating Mineral Substance."

Fischer, H.C., 2–9–26, US Patent 1,572,425, "Process for Coloring Mineral Matter."

Gundlach, H.R., 10–20–36, US Patents 2,057,677, "Colored Roofing Granule," 2,057,678, "Roofing Granules and Method of Coloring Same;" and 2,057,679, "Colored Roofing Granule and Method of Manufacturing the Same."

Alton, W.H., 7–17–28, US Patent 1,677,701, "Granule and Method of Producing Same."

Brown, L., 11–10–31, US Patent 1,831,784, "Glazed Rock Granule Surface and Roofing Surfaced Therewith."

Nichols, E.H., 4–10–34, US Patent 1,954,778, "Roofing Granule and Method of Producing Same."

Beasley, M.R., 5–14–35, US Patent 2,001,448, "Manufacture of Artificially Colored Granules."

Swenson, G.W., 10–12–43, US Patent 2,331,356, "Colored Surface Coated Glazed Granules, Useful as Roofing Granules, and Process of Making the Same."

Jewett, C.L., 6–26–45, US Patent 2,379,358, "Colored Granulated Material."

McMahon, W.A., 3 M Co., 4–21–70, US Patent 3,507,676, "Zinc Containing Algicidal Surfacing."

Skadulis, A., 3M Co., 9–15–70, US Patent 3,528,842, "Copper Compound-Containing Algicidal Surfacing and Process."

Electronic and Optical Uses

Original by DANFORTH R. HALE *
Revised by R. E. BLAIR †

Minerals for electronic and optical uses divide easily into two sections: (1) quartz and (2) minerals other than quartz.

Quartz

Quartz, having a great usefulness discovered by the radio communication enthusiasts about 1918, has become a subject of amazingly expanded interest in the US over the past decade as a man-made mineral and a multimillion dollar crystal-growing business.

An economic summary of the commercial growing of quartz crystals has a place in a handbook directed to the mineral engineering industry when it is remembered that quartz crystals have long been an important commercial mineral, and that the raw material for "cultured quartz"—that is to say, quartz crystals grown through the ingenuity of man—is still natural quartz.

Nearly all the natural crystals come from Brazil. The larger pieces which meet rigorous standards of quality are used for electronic and, to a lesser extent, optical components. Smaller pieces and fragments are used for vitreous silica. The need for high quality material in quantity led to US government-sponsored research programs in the 1940s that have resulted in the factory growing of beautiful crystals of prescribed shape, size, and quality.

The development of the cultured quartz crystal is typical of man's success in adapting the product of the mine to increasingly sophisticated uses. A remarkable achievement perhaps, but foreshadowed by experiments in 1908 by Giorgio Spezia (1908), an Italian geologist studying the relative effects of temperature and alkaline environment on the solubility of quartz.

Modern radio equipment is most often controlled as to frequency by the presence in the circuit of a separately added "crystal"—the 1918 discovery responsible for the existence and growth of the quartz industry. The "crystal" is quartz, but this component is a carefully oriented and prepared slice from a crystal, but not a crystal as recognized by a rock hound or seen in a museum.

How quartz operates to control frequencies is not a proper subject for a handbook on industrial minerals, and references should be consulted (Cady, 1964; Mason, 1964). Quartz belongs to that class of materials called dielectrics: those that do not conduct an electric current but permit electric fields to exist and act across them. Quartz shows the "piezoelectric effect," which means that when a quartz plate is mechanically deformed against its natural stiffness, one of its surfaces becomes negatively charged, the other, positively charged. When the plate is released from the stress quickly, the charges disappear as the plate regains its original shape, but because of mechanical momentum the plate deforms in the opposite direction (to a lesser amount) and the surfaces correspondingly become charged in the opposite direction. By coating the two surfaces thinly with metal and attaching flexible wires, these charges may be brought into an electronic circuit. If the surfaces are suddenly electrically charged by movement of current through the wires, the "converse piezoelectric effect" occurs —the plate deforms. Carry the thought further and it is realized that an alternating current flowing through the wires responds to the mechanical oscillation of the plate and is keyed to this oscillation. By controlling the thickness of the plate its mechanical vibration frequency can be varied through a wide range.

One type of quartz plate, called the AT-cut, having a precisely defined orientation with respect to the crystallographic axes of the crystal, vibrates on a microscopic scale much as a book would deform when placed flat on a table and the top cover moved parallel back and forth with the hand. There are at least 17 other orientations which have been studied, some of

* Senior Scientist, Clevite Corp., Retired.
† Consultant, Arvada, CO.

which have preferred uses in various applications (Cady, 1964).

The quartz crystal industry is composed of three main segments (excluding fused quartz and quartz used for optical purposes):

1. *Natural electronic-grade quartz crystals.* Mined quartz suitable for fabrication into piezoelectric units. Zlobik (1981a) estimates the waste to ore ratio at 1:1000 to 1,000,000, depending upon deposit.

2. *Lasca.* Mined quartz usable as feedstock in the production of cultured quartz. Approximately 0.63 kg (1.4 lb) of lasca are required to produce 0.45 kg (1.0 lb) of cultured quartz.

3. *Cultured quartz.* Cultured quartz is produced from lasca feedstock in a process of crystal growth in an autoclave under conditions of heat, pressure, and time. It is estimated that 0.45 kg (1.0 lb) of cultured quartz is equivalent to 1.4 to 4.5 kg (3 to 10 lb) of natural quartz crystal regarding yield of commercial quartz suitable for slicing into piezoelectric units. The chronology of the development of the quartz crystal industry both natural and cultured follows:

Date	Comment
1918	Discovery of the piezoelectric effects of quartz crystal
1921	Application of the piezoelectric effects of quartz crystal in the circuitry of radios
1948	Establishment of the quartz crystal commodity stockpile by the US government
1952	US consumption of quartz crystal at an all time high of 228 t (503,000 lb) natural crystal only
1958	First commercial production of cultured quartz crystal
1970	Cultured quartz crystal production exceeds imports of natural quartz crystal
1971	Cultured quartz crystal consumption surpasses natural quartz crystal consumption
1976	Surge in the use of quartz crystal, both natural and cultured, due to citizen band radio sales and then decline due to change in technology in citizen band radio circuitry
1977–1982	Continued replacement of natural quartz crystal by cultured quartz crystal, large overcapacity in the cultured quartz crystal industry and new developments in the use of quartz crystal in watches and in microprocessors for automobiles and toys

Production and Consumption

In 1980, three companies in the United States produced some 181 t (400,000 lb) of quartz crystal which included lasca, specimen, and jewelry material and quartz suitable for direct fabrication into piezometric units. Some 370 t (816,000 lb) of quartz were imported, primarily from Brazil. This quartz includes specimen and jewelry material as well as lasca. Some 465 t (1,026,000 lb) of lasca were used by seven companies in five states to produce an estimated 343 t (757,000 lb) of cultured quartz in 1980. The US government stockpile of natural quartz crystal was 1 kt (2,420,000 lb) at the end of 1980, of which 826 t (1,820,000 lb) was classified as excess. Stockpile sales of excess poundage were suspended pending reevaluation of stockpile goals. US government stockpile sales had been, for the most part, exported due to the preference by crystal cutters for using cultured quartz.

The flow of natural electronic-grade quartz from mine to finished units is shown in Fig. 1 (Zlobik, 1981a). The US lasca supply demand for 1978 is shown in Fig. 2 (Zlobik, 1981a).

Exports and Imports

As noted, Brazil is the world's major producer and exporter of natural electronic-grade quartz crystal and lasca. The United States is the major importer and consumer of lasca as well as the major producer, consumer, and exporter of cultured quartz crystals. Japan, the Federal Republic of Germany, Hong Kong, and Switzerland received most of the exports of cultured quartz from the United States.

Prices and Trends

Prices of quartz material vary considerably from \$0.60/lb (0.45 kg) for lasca to over \$35/lb (0.45 kg) for cultured quartz.

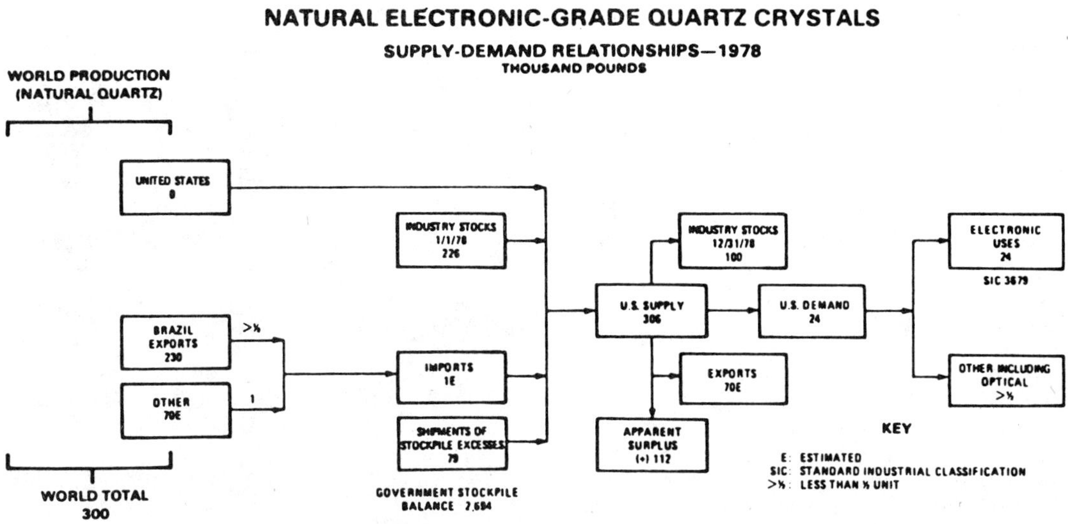

FIG. 1—*US natural electronic-grade quartz crystal supply-demand* (*Zlobik, 1981a*).

The electronic-grade quartz crystal industry encompasses the fully integrated producer/consumer from the base mineral, cutting of blanks for piezoelectric unit production, piezoelectric unit production and fabrication into a finished electronic unit, such as a radio, to companies that are only involved in segments of the industry. Pricing in the industry reflects this segmented or captive structure depending on where the transfer of the mineral takes place.

Future trends in the industry are expected to be:

1. Continuing displacement of natural electronic-grade quartz crystal with cultured quartz crystal.
2. Local development of quartz production for lasca use as an alternative to Brazilian lasca.
3. Growth in the exports of US cultured quartz to locations that, due to favorable labor rates, produce piezoelectric-containing electronic units; development of cultured quartz crystal production or expansion of existing quartz crystal production in these locales.

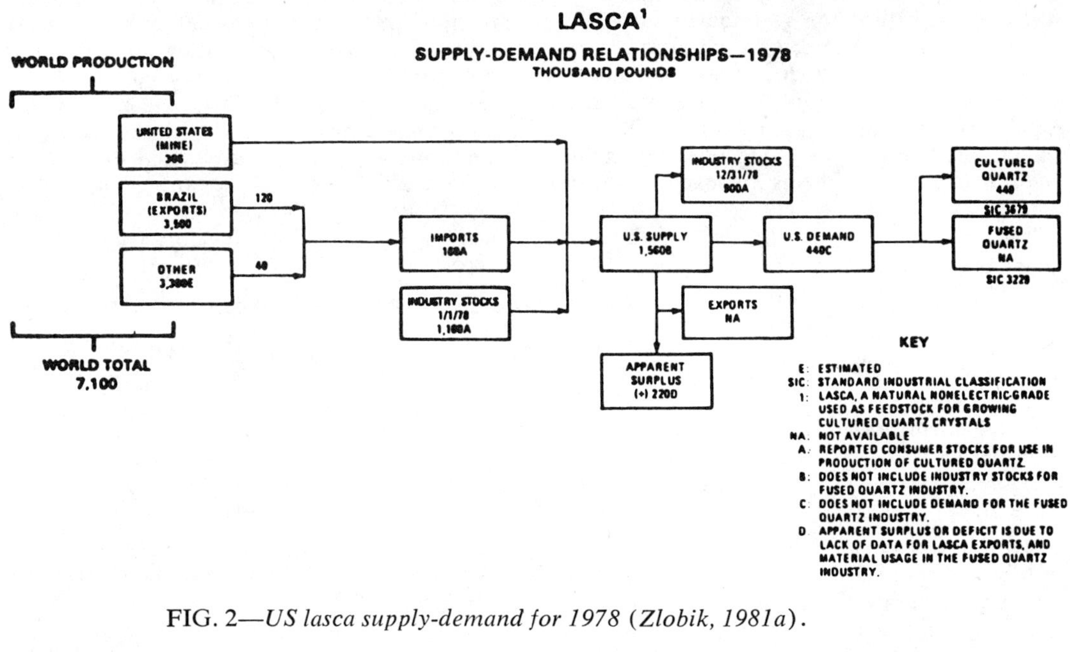

FIG. 2—*US lasca supply-demand for 1978* (*Zlobik, 1981a*).

4. Expansion of quartz crystal-containing electronic units used as timing devices in microprocessors, for example, in automobile production.

Production of Cultured Quartz

The early history of producing large, high-quality quartz crystals in the laboratory and in the factory, and the general manufacturing process, have been described by Waesche (1960).

Today there are seven operations in the US, and operations are reported in at least three other countries: England, Japan, and Russia.

The manufacturing process makes use of "hydrothermal growth," so named because it involves an aqueous solution at elevated temperature, and is similar to the "hydrothermal action" that develops ore bodies and crystal formations in the crust of the earth. The synthetic process may be briefly described as follows: A vertical steel vessel is provided with suitable internal fittings and pressure gage, safety blow-off valve, thermocouples, and a source of electrical heating. The vessel is charged with (1) nutrient material (lascas) to about ¼ of its volume, (2) dilute alkaline solution to about ¾ full, and (3) an array of seed plates. The vessel is closed and brought to a temperature of about 350°C, producing a high pressure. The heat source is adjusted to provide a temperature difference between the area of dissolution and the area of deposition and growth. Conditions of constant temperature and pressure are maintained for a period of weeks or longer. The status of the growing crystals can be monitored using radiography with cobalt-60.

Vessels up to 254 mm (10 in.) ID, and capable of operating at up to 275.8 MPa (40,000 psi), are in use. Fig. 3 shows man-made quartz crystals being removed from an autoclave or reaction vessel.

The conditions of growth used by the two major US producers are here compared:

FIG. 3—*Extracting man-made crystals from an autoclave or reaction vessel at Western Electric plant, North Andover, MA (courtesy Western Electric Co.).*

The Sawyer Research Products Co., providing quartz crystals for the optical industry, reports maximum size crystals grown as follows: 7.7 kg (17 lb) in 1965, 9.8 kg (22 lb) in 1971, and 190.9 kg (42 lb) in 1972 (private communication).

Although there are no deposits of natural electronic-grade quartz in the US, (Waesche, 1960), high purity US quartzite has been used successfully in producing cultured quartz.

Seed: The earliest seed plate that had much attention was a thin plate cut parallel to the minor rhombohedral face, often identified as *r* or as $\{01\bar{1}1\}$ by Bravais symbols, a face which makes an angle of 38°13′ with the *c*-axis. This seems a reasonable choice since the plate should grow parallel to itself, and the AT-cut plate (Cady, 1964), whose surface makes an angle of 35°15′ with the *c*-axis, is perhaps the most used for frequency-control oscillators. Indeed, this choice of seed orientation was

	Sawyer Research Products *	Western Electric Co.†
Solution	0.83 molar Na_2CO_3	1.00 molar NaOH
Percent of fill ‡	76	82
Growing temperature, °C	350	350
Dissolving temperature, higher by	5 to 20°C	about 40°C
Pressure, psi §	12,000	40,000
Time, days	42	24

* Private communication.
† Rudd et al., 1966.
§ Metric equivalent: psi × 6.894 757 = kPa.
‡ Percent of fill: percent of the internal volume of the vessel not occupied by solid substances when it is being charged.

patented (Freedman and Tuttle, 1951), particularly, however, as promoting rapid crystalline growth. The patent goes on to say that this orientation grows "much more rapidly than seed plates cut at other angles."

However, it turns out that in the alkaline solutions used for growth, the minor rhombohedral face soon grows itself out of existence! The major rhombohedral face grows much slower, the prism faces grow at a negligibly slow rate, and the crystal ultimately becomes a more or less square bar (depending on how the seed was supported) elongated in the *c*-direction, lengthening slowly.

In 1954 a newer seed type called the "*Y*-bar seed" (Jaffe and Turobinski, 1960), was invented and studied: "an elongated quartz crystal seed bar having its length extending substantially perpendicular to a crystallographic *X*-axis and substantially perpendicular to the *Z*-axis of the crystal, said seed bar being small in cross-sectional dimension perpendicular to its length." In piezoelectric engineering the *X*-, *Y*-, and *Z*-axes are mutually perpendicular, the *X*-axis being one of the *a*-axes, and the *Z*-axis being the *c*-axis, in the hexagonal system to which quartz belongs. This seed type became so successful that it is now the standard seed in the cultured quartz industry.

On this seed the growth takes place rapidly in the *c*-direction, easily 1 mm per day under good conditions, and less rapidly in the *a*-direction at right angles. The growth in the *c*-direction yields, as the mineralogist would expect, no flat face but, on the contrary, a surface composed of many rounded small mounds occasionally showing growth spirals. The *c*-face or basal face, (0001), is seen sometimes on the tiny crystals studied as micromounts, but it is not a plane.

Growth on a Y-bar seed consists of growth in both the *c*-direction and the *a*-direction. Studies have shown that the highest quality is the *c*-direction growth; hence a modified Y-bar is often chosen as a seed, being long in the direction perpendicular to an *a*-axis and of moderate dimensions parallel to that *a*-axis (Fig. 4).

Growth in the *c*-direction has been found to be highly satisfactory for producing crystals destined for the relatively small AT-cut oscillator plates and various other units, but the need for longer lengths for frequency filters has led to the successful use of relatively large seeds oriented parallel to the minor rhombohedral face, $\{01\bar{1}1\}$, a cyclic sweep to the early seed (Lias and Rudd, 1969). Fig. 5 shows representative man-made quartz crystals.

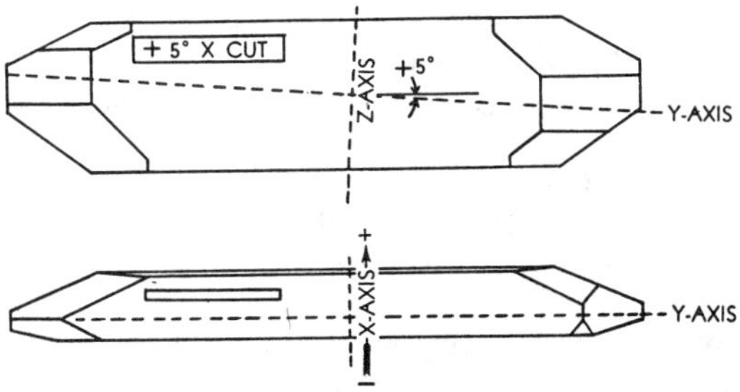

FIG. 4—*Drawing of cultured quartz crystal slightly modified from standard Y-bar, the seed being cut at 85° to the c-axis, as shown, instead of 90° to the c-axis. The drawing shows elevation and plan views (Sawyer, 1971).*

The Q of Quartz: How does one judge the quality of a find of quartz crystal? A material with a subtle use may require a subtle test of its useful value. The "Q" of quartz, a dimensionless quotient, is the mark of electronic quality. It is often called "mechanical Q." It is highest when there is lowest energy loss in a quartz oscillator due to internal friction. Given a resonator vibrating at resonance frequency while connected to a source of ac of that frequency, the instant the current is stopped, the vibration amplitude decays. The Q expresses the slowness of the decay. The dying sound of a struck bell is a close analogy. A bell having a long tintinnabulation might have a Q of 250,000; a good quartz resonator has a Q of about one million.

Practically, the Q of a resonator can be determined as a ratio,

$$Q = \frac{2\pi fL}{R}$$

where the quantities are measured for the

FIG. 5—*Representative quartz crystals. Left to right, the seeds on which the crystals were grown are: minor rhombohedral, CT-cut, DT-cut, and two cut at 5° off the basal plane. The growing run was three weeks long, made in 1970. The three large crystals weigh about 0.79 kg (1¾ lb) each (courtesy Western Electric Co.).*

equivalent electrical circuit (a resonating electronic circuit of the same frequency), f being the resonance frequency, L the inductance, and R the resistance (Heising, 1946).

A direct method for observing Q displays the decay of the amplitude of a specially made 5 MHz oscillator on the screen of a cathode-ray tube (King, 1959). Still another method involves the examination of the graph of resistance as a function of frequency (Adams and Kusters, 1968).

The Q of the best natural quartz (measured at room temperature) lies in the range 1.0×10^6 to 3.0×10^6, and the usual good cultured quartz may be in the range of $\frac{1}{10}$ of these. In the early days of quartz growth study, Q was much lower, and although the crystals withdrawn from the pressure vessels were clear and sparkling in their beauty, it was discouraging that the men in white coats could not improve a little on nature.

At Western Electric Co., quartz finally attained a coveted high value of Q by the use, rather surprisingly, of $LiNO_2$, lithium nitrite, as an additive to the growth solution (Ballman et al., 1966). Lithium as, say, LiF, had been found to improve the Q, perhaps by decreasing the infrared absorption at 2.86μ (or the wave number, 3500 cm^{-1}) (King et al., 1962). The addition of the nitrite raised the Q, at slow growth rate, into the region of natural quartz values. The anion may be adsorbed onto the growing crystal and may decrease the presence of hydroxyl ion in the quartz.

Sawyer Research Products has developed means of growing what it labels as "Premium Q Cultured Quartz" through slow growth in sodium carbonate solution which in general yields a slower rate than sodium hydroxide solutions (Rudd et al., 1967; Sawyer, 1972). "Quality control on all pertinent aspects of the growth process has been tightened with the result that a series of runs has been produced whose 5 MHz 5th overtone Q's were consistently higher than those yielded by natural quartz" (Sawyer, 1971).

Compared to glass, quartz has a considerably widened spectral window. Hence the use of vitreous silica for optical components and the use of quartz crystal in spectrometer prisms. However, quartz shows a number of absorption peaks in the infrared, certain of which have been identified with impurities—elements and radicals. In particular, hydrogen, which is present in great quantity in the alkaline growth solutions in the hydroxyl ion, as well as in aqueous solutions in the earth's crust, is known to be responsible for a characteristic high absorption peak in the wave-number region, 3000 to 3700 cm^{-1} (Katz, 1962).

A relationship was determined experimentally between the infrared absorption at 3500 cm^{-1} and the mechanical Q (Cady, 1964). Use of this information has led Western Electric (Rudd and Lias, 1967) and Sawyer Research Products (Sawyer, 1972), to inspect their production of crystals for Q by measuring this absorption with an infrared spectrophotometer. This simple and rapid method has made it easier to monitor the research programs leading to improved Q as well as to provide quality control.

Defects in Quartz

For much of the product of mine and smelter, microscopic and submicroscopic details need no attention. On the contrary, quartz destined for piezoelectric oscillators must be reasonably free from various types of defects down to those affecting the crystalline structure. Defects include twinning, structural imperfections such as "slip" and "screw" dislocations, impurity ions, and microscopic inclusions. For a review of defects as affecting the quality of quartz for electronic and optical use, see Waesche (1960) and National Stockpile Purchase Specification, P-43-R1, 1968 (Department of Commerce). For a more general mineralogical view see Frondel (1961).

Quartz may show two types of twinning: electrical or Dauphiné twinning and/or optical or Brazil twinning. These are beautifully displayed under reflected light on slices etched in hydrofluoric acid, a standard method for checking the quality of natural quartz during the course of oscillator manufacture (resonator plates must not show twin boundaries). In cultured quartz, the seed being carefully chosen, twinning is very rarely seen.

At the start of the crystal-growing process, when pressure vessel, solution, and seeds are heating up, the alkaline solution is likely to dissolve holes and tubes along dislocation lines such that when growth sets in, the opening is covered and a microscopic bubble results. If, instead of continuing the process, the vessel is cooled and opened, the seed plates will show etch pits of characteristic shapes, and suggest the density of the defects.

Studies using a special double-crystal X-ray spectrometer technique have disclosed much information on structure defects. The technique produces a "topography" of internal structure, and assists in relating defects in structure to others such as substitutional impurities (Lang, 1967).

Considering growth on the basal plane, yielding the cobble texture, the basic mechanism of growth seems to be along screw dislocations, often resulting in spiral pattern visible to the eye or under the microscope. Examination by light interference has produced pictures of great interest and beauty (Joshi and Vagh, 1964).

Quartz often darkens in certain areas when irradiated with X-rays, high-voltage electrons, or gamma rays from cobalt-60. This was traced to the presence of Al^{3+} substituting for some of the Si^{4+}. The charge compensation for the trivalent aluminum is in general provided by Na^+ or H^+ or Li^+ if the latter has been added intentionally as an impurity (Ballman et al., 1966). A broad study of hydrogen in quartz has illuminated some of this behavior (Katz, 1962).

A powerful method of studying a variety of defects, substitutional, interstitial, and structural, is to place a quartz block between metal electrodes, heat the assembly, and apply a dc voltage for a period of time (Pfenninger, 1961). This reference is a detailed study using electrodes of silver, gold, copper, and platinum; voltages up to 670; and temperatures to 800°C. Observation means included the microscope, the electron microscope, and infrared spectrometry. A beautiful series of photographs shows veils of defects entering quartz, leaving it, "decorating" lines of structural dislocations.

Regretfully this investigation did not encompass cultured quartz. However, for preparing quartz especially stable in a radiation environment, cultured blocks are now "swept" of various defects by this general procedure (private communication).

A defect occasionally seen—mostly in earlier, less well controlled growing operations—are microscopic white particles identified as clumps of acmite crystals, $NaFeSi_2O_8$.

Developments in Quartz Growth and Technology

Low Optical Transmission Loss: Cultured quartz has been examined as a material for transmitting infrared laser light of 1.06μ wavelength (from the YAG:Nd laser). It was found that growth in a silver tube to diminish iron contamination yields a quartz equivalent in low optical loss to the best commercial vitreous silica (Lias and Rudd, 1967). An interesting correlation was found between H^+ concentration in the quartz and the absorption at 1.06μ, the hydrogen ions seemingly providing charge-compensation for the Fe^{2+} or other divalent impurity ions associated with absorption in that region.

Growth Rate 2 mm/d on Basal (0001) Plane: Stemming from an eagerness to grow the crystals faster—an economic advantage if sufficiently high Q and other desirable characteristics can be maintained—growth studies at 374°C and at a pressure of about 275.8 MPa (40,000 psi), led to a rate of about 2.6 mm/d and a Q of the order of 1.4×10^6 (Lias et al., 1972). The high rate due to higher expected solubility was anticipated; the excellent Q was a bonus, and may signify that at the higher temperature H^+ goes less readily into the growing crystal.

Measurement of Viscoelastic Materials: A device successfully used to study the shear modulus of viscous materials employs a "delay line" of suitable mechanically rigid material, e.g., vitreous silica or aluminum, having mounted on the end a piezoelectric transducer to generate shear waves. This is typically quartz, providing frequencies in the range 1 to 7 MHz, and may be a piezoelectric ceramic. The delay line is inserted in the sample and the system brought to constant temperature to within 0.005°C. Pulses of waves are sent down the delay line, and the echoes are measured as to their attenuation and phase shift, from which an accurate calculation can be made of the shear modulus (Hunston et al., 1972).

An "ultrasonic impedometer" has been made to measure the dynamic mechanical properties of viscoelastic materials (adhesives, paints). It employs Y-cut, shear-wave resonators of 5 MHz frequency attached to the ends of a 10-cm bar of vitreous silica. The bar ends are cut at 79°, producing a shape suggesting a bathtub. The bulk shear wave is reflected off the top surface of the bar on which is spread the material to be studied. The reflected wave returns to the transmitting transducer and an amplified signal brought out on an oscilloscope. The impedance of a film of material can be monitored as it is drying (Myers and Schultz, 1962). For further general information on ultrasonics see Bradfield (1970).

Pressure Transducer: Quartz being piezoelectric, i.e., being a material which produces an electric field as a response to stress, the question arises, how useful a pressure gage can be made from it? A gage having a range of 0 to 69 MPa (0 to 10^4 psi) and an accuracy of about 0.07 kPa (0.01 psi) is made using a periphery-stressed resonator as a pressure transducer. The transducer unit is a circular 5 MHz resonator whose edge is integral with a surrounding cylinder of quartz, the axis of the cylinder being identical with the normal through the resonator center. The cylinder and resonator are fashioned by grinding techniques from a single crystal of quartz! The resonator may be at AT-cut or other cut to provide that the temperature coefficient of frequency at zero pressure is zero at a given temperature in the range 0 to 100°C (Karrer and Leach, 1969).

Pressure Gage for High Impact Conditions: A serious study, largely at Sandia Laboratories (Albuquerque, NM) has produced a gage technique for studying the stresses due to shock-wave loading to pressures as high as 6×689 MPa (6×10^5 psi). Based on the piezoelectricity of quartz, the device has been used to study the behavior of quartz and also of other solids under great impact (Ingram and Graham, 1972). A valuable use of an outgrowth of this technique is the measurement of forces produced in underground nuclear explosions.

While stresses of 4 GPa (6×10^5 psi) are routinely measured, a tungsten facing on the quartz disk extends the upper stress limit of the experiments to 12 GPa. Ordinarily the stress in an oscillating quartz resonator is probably of the order of 600 kPa.

A projectile in a cylinder is faced with a disk of quartz, and another disk is provided at the closed end of the cylinder. In this "Sandia Quartz Gage" for impact experiments, the cylinder is a 27-m compressed gas gun which can accelerate the projectile to 510 m/s. This impact velocity can be measured to 0.01%.

Disks of *X*-cut (the face of the disk being perpendicular to an *a*-axis) cultured quartz are used, which may be 126 mm in diameter and 6.3 mm thick—these dimensions yielding about a quarter of an ampere of piezoelectric current at a stress of 1 GPa.

During the study of strains up to 4.3×10^{-2} in these experiments, the analysis of data provided unusually accurate values for the linear and nonlinear piezoelectric stress constants, and the longitudinal elastic constants up to the fourth order (Graham, 1972).

Quartz Thermometer: A quartz resonator having its resonance frequency unvarying with temperature has not been made, but the behavior with temperature has been thoroughly studied, so that, for a given temperature range, a resonator may be made having a reasonably low, and accurately known, variation. Advantage has been taken of this property to measure temperature difference (Smith and Spencer, 1963). A 5° *Y*-cut (Cady, 1964), operating at 5 MHz, has a temperature-frequency coefficient of about 80 ppm per °C and responds to a temperature deviation of 3.8×10^{-6} °C, indicating a very sensitive device.

A commercial development of the quartz thermometer led successfully through an analysis of plate orientations to find one yielding a linear frequency-temperature relationship. The available instrument has a range of −80°C to 250°C, with a maximum resolution of 0.0001°C. A digital readout is used, and with optional amplifiers the sensory probe may be as far as 1372 m (4500 ft) from the measuring equipment (Hammond et al., 1965).

Quartz Microbalance: A quartz resonator can be brought to a slightly lower frequency by loading it with a little extra mass, e.g., of evaporated metal. Hence the change in frequency on loading makes the resonator a mass-measuring device. This idea has led to the construction of microbalances reading to about 10^{-12} g for certain unusual applications, for example the measurement of gases that can be adsorbed in a layer of material coating the resonator: a hydrophilic substance for weighing water vapor. The frequency change is directly proportional to the change in mass. A 5-mm-diam, 15 MHz resonator has a mass sensitivity of 2600 Hz change per microgram.

The resonator is usually an AT-cut. The frequency is of course also a function of temperature. In one method of assisting in this problem, there is applied to one side of the resonator a zigzag strip of evaporated-on nickel which serves as heater, thermocouple, and radio-frequency electrode (King, 1969; Saubrey, 1964; Stockbridge, 1966).

Surface Waves on Quartz: Elastic solids may display bulk waves or surface waves. Bulk waves are found in quartz oscillators themselves, and in "buffer rods" or other attachments for conveying a vibration from an oscillator along a length of rigid elastic solid. A use of bulk waves in such a delay line is mentioned under "Measurement of Viscoelastic Materials." Bulk waves in quartz have been used also to modulate light.

Surface waves—those Lord Rayleigh waves which now find standard application in earthquake study—can be produced on various crystalline solids (Viktorov, 1968; White, 1970). They can be launched in the *X*-direction (i.e. along an *a*-axis) on a quartz bar by means of electrical impulses applied to the surface through two evaporated electrodes having the appearance of two interleaved combs, or the illustrative diagram of a multiplate capacitor. They can be received and converted into electrical signals by similar electrodes at the far end. The bar or plate of quartz may be cut parallel to a plane containing an *a*- and the *c*-axis—the *Y*-cut of the piezoelectric engineer. In one application the bar is 254 mm (10 in.) long.

The interesting advantage of using elastic waves in certain electronic circuits is the slowness with which they move: about 1.5×10^3 to 15×10^3 m/s instead of the 3×10^8 m/s which is the velocity of electric signals and light. The surface waves move in the lower third of that range. A further advantage of surface waves is that they can be sampled or modified. If they are accompanied by electric or magnetic fields (when they are produced in piezoelectric or magnetic materials), the passage of the wave and its characteristics can be picked up by a sensing device not in contact with the surface.

Another synthetic crystal is also of much interest in surface wave applications: lithium niobate (see synthetic crystal section).

Miscellaneous Uses: Further uses of quartz resonators have appeared in recent years in color TV circuits and in watches. The resonator used in a watch design is not the face-shear mode most used for high frequency, 0.5 to 50 MHz, but a flexural, or wave-shaped, mode operating in the range 2 to 20 kHz (Cady, 1964).

Minerals Other Than Quartz

Crystals of a size to hold in the palm of your hand are of enormous interest to the technology of the 1980s because of the astonishing things they can be made to do. Not many mined mineral crystals have been among these, or are likely to be. However, a scientist in a modern physical laboratory, dreaming up a circuit (electronic, optical, magnetic, or any combination) does not ask whether an existent *mineral* might help, but asks whether the professional crystal grower can provide a material which the solid-state physicist says might have some of the properties to make the dream realizable.

Few minerals are found having sufficiently high quality to be employed directly. There is one special exception, calcite; and optical calcite, highly useful, is becoming so hard to find in the sizes demanded that investigations are under way to synthesize it.

Natural quartz is still important in the frequency-control oscillator industry; mica and selenite are used, in very small amounts, in making light retardation plates as microscope accessories.

If a mineral is observed to have a useful property, the crystal grower can probably reproduce it in an improved crystal: larger, freer of optical flaws, purer or conversely doped with impurities—in either case to provide it with an effective property.

The synthetic crystals are, in any case, derived from mined ores, which undergo various refining and purifying processes—some of them rather sophisticated, leading then to the growing and cutting and polishing into usable shapes.

Many of the currently exploited properties of light and crystals have been known since early in the 18th century! Recently developed techniques have been needed to study and apply them, e.g., intense light beams, high optical quality crystals, sensitive measuring equipment. The technology in optical communication is now on the edge of such an expansion as radio and television experienced in the 1940s. The frequency-control resonator, the diode, and the transistor were among those early achievements. Radio frequencies have given way, along the growing front of research, to ultra high frequencies, and now to frequencies close to and within the visible range of the spectrum.

Minerals of Technological Interest

Table 1 is a listing of several minerals studied earlier or in current commercial use or under present investigation. There are others not listed, some of only casual past interest, some which have been unintentionally overlooked.

The list is alphabetical, and includes quartz, the subject of the previous section. The mineral name is preceded in most instances by the Dana identifying number. The chemical formula and the crystal system are given. It should be noted that this is the system for the crystal at ordinary temperature: at lower temperature a crystal may move into a less symmetric system, and vice versa. Where appropriate, related or replacing crystals are mentioned; the latter may have enhanced, or developed, new and exciting properties.

The "useful property," the "application," and the "device" are not easily differentiated in every case, depending on the state-of-the-art. Sometimes they are, e.g.:

TABLE 1—Minerals for Electronic and Optical Uses

Dana No.	Mineral	Chemical Formula	Crystal System	Related or Replacing Crystal	Useful Property and Reference	Application	Device, Equipment	Commercial Growers (Keyed to Table 2)	Growth Studies, References
2613	Altaite	PbTe	Iso	PbS	Photoconductivity (Bube, 1960)	Pyrometry	Photoconductors, IR spectroscope (Bube, 1960), semiconductor laser (Harman, 1971)		
	Almandite	$Fe_3Al_2Si_3O_{12}$	Iso	$Y_3Fe_5O_{12}$ "YIG" $Y_3Al_5O_{12}$ "YAG"	Magnetooptic Fluorescence Hardness, n = 1.8 (high)	Microwaves Laser, YAG:Nd (Geusic et al., 1964) Gem, YAG	Laser, YAG:Nd (Geusic et al., 1964)	1,3,14	(Makram, 1968) (Rudness and Kebler, 1959) (Tolksdorf, 1968)
	Apatite	$Ca_5F(PO_4)_3$	Hex		Fluorescence with Mn (Ryan, 1970) IR spectra (Knubovets et al., 1970)		Laser, high gain		(Mazelsky, 1968)
38.4.3	Berlinite	$AlPO_4$	Hex	Quartz	Piezoel. (Stanley, 1954)	Probably none; piezoel. low			
26.1.7	Boracite	$Mg_3B_7O_{13}Cl$	Iso Ort (Cady, 1962)		Piezoel. (Cady, 1964) Pyroel. (Cady, 1964) Ferroel. (Jona and Shirane, 1962)				(Nassau and Shiever, 1973)
9.1.1.5	Bromyrite	AgBr	Iso		IR transmission 0.5-35μ		IR spectrophotometer	7	
39.2.1	Brushite	$CaHPO_4$	Mon	KH_2PO_4?	Piezoel. (Cady, 1964)				
14.1.1.1	Calcite	$CaCO_3$	Hex	$NaNO_3$	Strongly birefringent	Polarization of light. (Waesche, 1960)	Nicol prism, Glan-Thompson prism, etc.	9,15	(Hatta et al., Henisch, 1970 Liaw and Faust, 1972)
9.1.1.4	Cerargyrite	AgCl	Iso		IR transmission 0.4-30μ		IR polarizer by reflection	7.	
269	Cinnabar	HgS	Hex		Optical rotation, IR transmission 1-13μ		Photoconductors		(Toudic and Aumont, 1971)
2616	Clausthalite	PbSe	Iso	PbS	Photocond. in IR to 5.6μ	Detect IR radiation.	IR spectrophotometer (Bube, 1960)		
25.1.9	Colemanite	$Ca_2B_6O_{11}\cdot 5H_2O$	Mon		Ferroel. (Cady, 1962)				

TABLE 1 (Continued)

Dana No.	Mineral	Chemical Formula	Crystal System	Related or Replacing Crystal	Useful Property and Reference	Application	Device, Equipment	Commercial Growers (Keyed to Table 2)	Growth Studies, References
8321	Columbite	$(Fe,Mn)(Nb,Ta)_2O_2$	Rho		Piezel. Ferroel. Luminescence (Hordvik and Schlossberg, 1972)	Electroop. modulation. Hologram recording (Laudise, 1968) (Stepke, 1972)		7,8,14	(Laudise, 1968)
4411	Corundum	Al_2O_3	Hex	(Sapphire, ruby)	High ref. index,	Substrate for Si (Cullen, 1971) Basis for "star sapphires"		2,7,10,13,14	(Pollock et al., 1972)
9.2.1	Fluorite	CaF_2	Iso		Fluorescence Ref. index low Dispersion low Transmission 0.12-9.0μ	Windows, prisms, lenses.	Scintillation counter, CaF_2:Eu. Laser (Garbuny, 1965) (Nassau, 1965)	7,10,11	(Nassau, 1964)
2611	Galena	PbS	Iso	Note altaite and claust-halite	Photocond. to 2.8μ	IR detection (Bube, 1960)	IR spectroscopes. Semiconductor laser (Harman, 1971)		
2642	Greenockite	CdS	Hex		Piezoel. Photoc., Photovoltaic, Pyroel., Electrooptic, Linear compressibility (44)	Mixing light beams.	Photoconductors, solar cells. Acousto-electric oscillator (Fochs, 1970)	5,6	(Fochs, 1968) (Nassau and Shiever, 1972) (Parker and Pinnell, 1968)
9.1.1.1	Halite	NaCl	Iso	NaI	Transmission, 0.2-15μ	Windows, prisms, lenses.	IR spectrometers, X-ray spectrometers.	7,10,11	
28.4.3.1	Langbeinite	$(NH_4)_2Cd_2(SO_4)_3$			Ferroel. (Jona and Shirane, 1962)				
	Leushite	$NaNbO_3$	Orth	Other niobates, see columbite	Antiferroel. (Jona and Shirane, 1962)	(See Columbite)			
7216	Magnetite	Fe_3O_4	Iso	$ZnNiFe_2O_4$ $BaFe_2O_4$	Low coercive force High remanence	Ferromagnetic applications; replaced by better crystals.	Memory devices, and permanent magnets		
	Muscovite	$KAl_3Si_3O_{10}(OH)_2$	Mon		Birefringent, good dielectric coef. high diel. strength.	Insulation.	Capacitors, 1/4-wave plates.	15	

TABLE 1 (Continued)

Dana No.	Mineral	Chemical Formula	Crystal System	Related or Replacing Crystal	Useful Property and Reference	Application	Device, Equipment	Commercial Growers (Keyed to Table 2)	Growth Studies, References
9.1.3.1	Nantokite	CuCl	Iso		Electrooptic, transmission 0.4-20μ	Electrooptic.	Light modulator (Murray, 1964)		
4211	Periclase	MgO	Iso			Insulator		8	(Butler, 1969)
7421	Perovskite	$CaTiO_3$	Orth	$BaTiO_3$ (Tet) $PbZrTiO_3$	High diel. coef. Ferroel. (Jona and Shirane, 1962)	Modulation of light phase (Kaminow, 1965)	Phono pickups, many piezoel. devices (Wheeler, 1972)		(Nassau and Broyer, 1962) (Von Hippel, 1963)
				$SrTiO_3$ (Iso)	High ref. index	Gem			
3212	Proustite	Ag_3AsS_3	Hex		Pyroel., photocond., Electrooptic.	Electroop. modulation			(Bardley and Jones, 1968) (Gentile and Staufsudd, 1968) (Hulme, 1972)
	Quartz	SiO_2	Hex	No equivalent crystal	Piezoel. (Cady, 1964) (Mason, 1964) Transmission, 0.15-3.5μ Birefringent. Color grown in (Crowningshield, 1972) (Pough, 1970)	Frequency control (Cady, 1964) surface waves (Klerk, 1972) (White, 1970) cultured gems (Crowningshield, 1972) (Klerk, 1972)	Oscillator or "crystal", delay lines, transducers, filters.	12	See Sec. 38 on "QUARTZ" IM & R, 3rd ed.
4411	Ruby	Al_2O_3:Cr	Hex	Corundum	Fluorescence, high hardness.	Holography, communications.	Laser, (Garbuny, 1965) (Maiman, 1968)	14	(Nassau and Broyer, 1962a)
4511	Rutile	TiO_2	Tet		High diel. coef. 173 perp. c-axis			10	(Farrell and Ling, 1972) (Nassau and Broyer, 1962) (Rosenberger and DeLong, 1972).
	Sapphire–	see Corundum							
48.1.3.1	Scheelite	$CaWO_4$	Tet		Fluorescence, electroluminescence.		Laser: $CaWO_4$:Nd (Garbuny, 1965) (Johnson and Nassau, 1961) $CaWO_4$:Sm	8	(Nassau and Broyer, 1962a)

TABLE 1 (Continued)

Dana No.	Mineral	Chemical Formula	Crystal System	Related or Replacing Crystal	Useful Property and Reference	Application	Device, Equipment	Commercial Growers (Keyed to Table 2)	Growth Studies, References
29.6.3	Selenite	$CaSO_4 \cdot 2H_2O$	Mon		Birefringent.	Petrography	1/2-wave plates		
2621	Sphalerite	ZnS	Iso		Piezoel. (Cady, 1964) Birefringent (Yu and Cardona, 1973) Optical rotation	Light modulation (Buhrer et al., 1963)		5	(Parker and Pinnel, 1968)
7211	Spinel	$MgAl_2O_4$	Iso		Good transmission, diel. coef.: 8.4	Substrate for Si (Cullen, 1971)		10,14	(Wood and White, 1968)
9.1.1.2	Sylvite	KCl	Iso	KBr KI	Transmission 0.38-21μ	Study of color centers		7,10,11	
8322	Tantalate	$(Fe,Mn)(Ta,Nb)_2O_6$		Various tantalates $LiTaO_3$ (Jona and Shirane, 1962)	Ferroel. Low biref., transmission 0.35-4.0μ	Light modulation (Laudise, 1968)		4	
454	Tellurite	TeO_2	Tet		Piezoel., low velocity shear wave.	Acoustooptic (Bonner et al., 1972)	Light deflector, modulator.	8	(Bonner et al., 1972)
	Tourmaline	$NaFe_3B_3Al_3$-$(Al_3Si_6O_{27})(OH)_4$ Schorl variety	Hex	Quartz	Piezoel., pyroel. (Cady, 1964) Slightly electrooptic.				
	Vaterite	$CaCO_3$	Hex		Strong pos. birefringence (Jona and Shirane, 1962)	No use noted.			
48.1.4.1	Wulfenite	$PbMoO_4$	Tet	$CaMoO_4$	High polarizability (Pinnow, 1969) Transmission, 0.45-3.9μ	Acoustooptic	Light deflector, modulator.	8,14	(Pinnow et al., 1968)
4221	Zincite	ZnO	Hex		Piezoel.	Acousto-electric (Kolb, 1967)		10	(Nielsen, 1968) (Kolb, 1967)

Mineral	*Useful Property*	*Application*	*Device*
Quartz	Piezoelectricity	Frequency control	Oscillator or "crystal"
Ruby	Fluorescence	Light amplification	Laser

Some of the listed properties are obtained in enhanced value from "water-soluble" crystals—hence not minerals. While halite and sylvite, for example, are minerals soluble in water, the crystal most phenomenal as a ferroelectric and piezoelectric crystal is Rochelle salt, a rather complex and very soluble organic compound. Other important synthetic crystals also not found as minerals include germanium and silicon, the single-crystal chemical elements from which come diodes, transistors, integrated microcircuits, etc.

"Almandite" garnet is listed as the precursor of yttrium iron garnet and yttrium aluminum garnet. "Perovskite" is so listed for the much valued barium titanate and lead zirconium titanate. "Columbite" is listed for the family of niobates. The growers and users of optical crystals have been using "niobium" instead of "columbium" for the element of atomic number 41.

Often there are recognizable grower-suppliers for the synthetic crystals, as indicated; two listed suppliers however are middlemen. Other crystals may still be in the laboratory, and references to "growth studies" would indicate where.

Table 2, keyed to Table 1, is a partial list of commercial producers of mineral crystals.

Glossary

The following glossary contains some of the terms applying to the use of minerals and various near-mineral synthetic crystals in electronic and optical technology. Optics texts are mentioned in the general bibliography for further coverage of that limited field.

Absorbance. The value of the expression, $\ln(I_0/I)$, where I_0/I is known as the *absorptance* and is a numerical statement of the absorption of radiant energy by a transparent material: the ratio of the intensity (W/m^2) of the incident light to the intensity leaving. The constitutive equation is

$$I = I_0 e^{-\alpha d}$$

where α is the absorption coefficient and d is the length of light path in the material.

Acoustooptic. In general, a light-modulating device in which an acoustic wave, ultrasonic in frequency, passing through a suitable crystal plate, sets up strained layers with its wave fronts such that an entrant light beam is deflected. Ideally the deflection can be controlled through an important angle which can be varied at frequencies up into the MHz range.

Beam Splitter. A thin sheet of glass or clear plastic having its surface treated to reflect about half the light falling on it at approximately 45° off the normal to the surface. In a more advanced type, two right-angle prisms are cemented along the hypotenuse faces with a material having selected reflection properties which may control the ratio of transmitted to reflected light, and/or the proportion of color.

Delay Line. A device used to delay an elec-

TABLE 2—Commercial Producers of Mineral Crystals (Keyed to Table 1)

1. Allied Chemical Corp., Synthetic Crystal Products, Box 1021R, Morristown, NJ 07960
2. Carmen Sapphire Corp., Reseda, CA 91335
3. Crystal Optics Research, Inc., 3680 S. State St., Ann Arbor, MI 48104
4. Crystal Technology, Inc., 2510 Old Middlefield Way, Mt. View, CA 94040
5. Eagle Picher Industries, Box 47, Joplin, MO 64801
6. Gould, Inc., Gould Laboratories, Crystal Section, 540 East 105th St., Cleveland, OH 44108
7. The Harshaw Chemical Company, 6801 Cochran Rd., Solon, OH 44139
8. Isomet Corp., 103 Bauer Drive, Oakland, NJ 07436
9. Karl Lambrecht Corp., 4318 N. Lincoln Ave., Chicago, IL 60618
10. Materials Research Corp., Orangeburg, NY 10962
11. Optovac, Inc., E. Brookfield Rd., North Brookfield, MA 01535
12. Sawyer Research Products, Inc., 35400 Lakeland Blvd., Eastlake, OH 44094
13. Tyco Laboratories, 16 Hickory Drive, Waltham, MA 02154
14. Union Carbide Corp., Crystal Products Department, 8888 Balboa Ave., San Diego, CA 92123
15. Ward's Natural Science Establishment, Inc., Box 1712, Rochester, NY 14603

Note: 9, Lambrecht, and 15, Ward's, are suppliers only.

trical signal for a short time, usually of the order of milliseconds. In one method, a piezoelectric transducer is mounted on an elongated piece of rigid material, e.g., vitreous silica, aluminum. A signal is introduced by the transducer, and picked up on reflection, or by a transducer at the far end.

Dielectric. A substance which is an electrical insulator as contrasted with an electrical conductor. The term, of course, includes the case of no substance—vacuum—and many gases and liquids. Many solid dielectrics, mineral and nonmineral, are opaque or absorb all nonreflected light in a very thin layer. The dielectrics useful in optical technology are transparent in the visible or near-visible range of radiation.

Ferroelectric Crystal. A member of a group characterized by being naturally electrically polarized in a restricted temperature range, and (hence) showing hysteresis in the relationship between polarization and applied electric field. If the crystal is heated to the transition temperature (Curie point) the crystal becomes nonpolar. *Piezoelectric* crystals, to be sure, become electrically polarized when subject to a deforming stress.

Fluorescence. The emission of radiation from a substance which is absorbing radiation of higher energy, shorter wavelength. (But there are interesting cases of the [apparent] reverse energy relationship.) Often large crystal size is not important, and powdered fluorescent and luminescent crystals, called phosphors, are used in cathode ray tubes, television kinescopes, fluorescent lamps. On the contrary, a large fluorescent crystal bar is used in the solid-state laser: ruby, scheelite, fluorite. Crystals do not have the property of fluorescence if they are strictly pure, and the variation in amount and type of impurity yields useful variations in behavior.

Harmonic Generation. The combining of two sets of wave motions having different, but not far removed, frequencies, so as to produce an ongoing wave whose frequency is a multiple or submultiple of the input frequency. Long known in radio technology, this effect in optics uses crystals as the medium in which the action occurs. For further information—without, however, but slight mention of minerals, see Schlossberg and Kelley (1972).

Holography. A type of "whole-writing," often caught on a photographic film, involving a laser beam divided by means of a beam-splitter so that the continuing beam strikes the film and the reflected beam strikes and is reflected back by the object onto the film. The two beams, direct and twice reflected, meet, interfere, and record their interference pattern in the film. A three-dimensional view of the object can be seen in the processed film illuminated by monochromatic light. Remarkable features are: no lens is used, the view is not simply two-dimensional, a piece of the film will display the same view as the whole, a large amount of information is stored in the film. A crystal of lithium niobate has been used as the recording medium.

Laser. The device which shoots out an intense, narrow beam of coherent, monochromatic light as a result of light amplification by the stimulated emission of radiation (Maiman, 1960). Originally a ruby rod energized by an intense source of light nearby, the laser, after a few years of investigations, can now be of glass, liquid, dye solution, or gas. The light from the ruby, at 9630Å, is at the infrared edge of vision, and now lasers can produce radiation of both longer and shorter wavelengths. The power of lasers covers the range of about 1 MW to 1 kW, with the pulsed type capable of emitting at much higher levels. Through modulation at the source and demodulation at a receiving station, the laser can be used as a means of transmitting, theoretically, up to 10^{10} channels each 10 kHz wide (Laudise, 1968, and Nassau, 1965).

Magnetooptic. In certain crystals through which a beam of polarized light is passing, the application of a magnetic field with its lines of force parallel to the beam, causes a rotation of the plane of polarization: Faraday effect, related to the Zeeman effect. This magneto-optic effect can be used to modulate a light beam.

Modulation. The process and the effect of adding information to a radio carrier wave or to a light beam. In radio, AM and FM are well known. Modulation of a light beam may be accomplished by altering its amplitude, frequency, phase, or plane of polarization. The modulation can take a digital form, so as to transmit simply 0 and 1, or an analog form, which includes voice and music. Modulation of the amplitude (or intensity) of a light beam can be had, for example, by applying a varying electric field to an electrooptic crystal carrying the beam. The crystal is sandwiched between crossed polarizers to provide the rest of the optical requirement.

Nonlinear. Behavior that does not follow a straight-line or proportionality relationship when one variable is altered and a dependent variable is observed. In electronics, for example, the behavior of current through a diode with respect to applied voltage is nonlinear. The optical properties of a medium are linear when the intensity of the transmitted light is low, as is normally the case. Nonlinear refers to changes in the optical properties as the intensity of transmitted light is greatly increased, or when two or more strong light beams are combined in the medium. The study of nonlinear phenomena is the basis of much current progress in electronic and optical technology: solar power, transistor, laser, holography (Schlossberg and Kelley, 1972).

Piezoelectric Effect. The reaction of a plate or bar of substance to a stress (compression or bending or twisting) by the appearance of an electric field across two opposite faces. The converse effect obtains also—the element tends to deform when an electric field is applied. Actual motion or strain is, of course, usually involved. The substance, often a macrocrystal, may be an assemblage of microcrystals as in a piezoelectric ceramic. A necessary crystallographic condition is that the crystal has no center of symmetry.

Applications include frequency-control resonators of quartz, and phonograph pickups of Rochelle salt (not a mineral) and of lead zirconium titanate (related to perovskite).

Property. Any of the principal characteristics of a substance, especially as determined by the senses, but also as determined by any variety of sophisticated sensing equipment, e.g., infrared transparency, laser capability, semiconduction.

Pyroelectric. The property of a crystal such that electric charges appear on certain faces of it when it is heated or cooled. Strontium niobate, a synthetic crystal related to columbite, shows such a strong effect that it can serve as a practical infrared detector.

Q-Switch, or Q-Spoiler. A fast-acting light valve used with lasers to hold back, arbitrarily, the light emission and let it burst forth at the flick of the switch in a high energy flash. Electrooptic crystals, and the Kerr cell using nitrobenzene and a high voltage pulse, have been used for such shutters, and require the presence of polarizers on each side, in a crossed position. Acoustooptic crystals and solutions of certain dyes are also used.

Resonator. A device or system in which a characteristic vibration may occur involving a continuous exchange between potential and kinetic energy.

A piezoelectric resonator is a rectangular plate (sometimes a disk or bar) of piezoelectric material, often quartz, which will vibrate characteristically when placed in an alternating voltage field having the same frequency as the natural mechanical frequency of the plate. The motion may take various types or modes: compression, shear, or flexural. Piezoelectric resonators of quartz for frequency control cover the range of about 1 kHz to 55 MHz. A piezoelectric plate can be driven at frequencies different from the resonance frequency; an application of such an oscillator is found in the cutting of phonograph records.

An example of an acoustic energy cavity resonator is the organ pipe. In microwave technology a resonator is a metal box or cavity in which electromagnetic energy may be "stored," the resonance frequency being determined by the geometry of the enclosure.

An example of an optical resonator is the Fabry-Perot resonator or interferometer: two parallel, partially reflecting mirrors cause the light entering in one direction to oscillate before exiting. In a common application the light is passed through a converging lens to form on a screen an interference figure of highly contrasting fringes. The laser may be called a Fabry-Perot resonator.

Quartz
Bibliography and References

Anon., 1956 on, *Proceedings,* Annual Frequency Control Symposium, US Army Electronics Command, Ft. Monmouth, NJ (printed from 1956, 10th Symposium, and annually since).

Anon., 1971, "Premium Q Cultured Quartz," Sawyer Research Corp., Advertising literature.

Adams, C., et al., 1968, "Measurement Techniques for Quartz Crystals," *Frequency,* Vol. 6, No. 8, pp. 22–25.

Ballman, A.A., and Laudise, R.A., 1963, "Hydrothermal Growth," *Art and Science of Growing Crystals,* J.J. Gilman, ed., John Wiley, New York, pp. 231–251.

Ballman, A.A., et al., 1966, "Synthetic Quartz with a Mechanical Q Equivalent to Natural Quartz," *Applied Physics Letters,* Vol. 8, No. 2, pp. 53–54.

Bradfield, G., 1970, "Ultrasonic Transducers, 1. Introduction to Ultrasonic Transducers, Part A," *Ultrasonics,* Vol. 8, No. 2, April, pp. 112–123; "Part B," July, pp. 177–189.

Cady, W.G., 1964, *Piezoelectricity,* Rev. ed., Dover Publications, New York, especially Chaps. 16, 17.

Dodd, D.W., and Fraser, D.B., 1965, "The 3000–3900 cm^{-1} Absorption Bands and Anelasticity in Crystalline Alpha-Quartz," *Journal of the Physics and Chemistry of Solids,* Vol. 26, pp. 673–686.

Friedman, I.I., and Tuttle, O.F., 1951, "Method of Selection of Oriented Seed Plate for Quartz Synthesis," US Patent 2,558,745, July 3.

Frondel, C., 1961, *The System of Mineralogy of J. Dana and E. Dana,* Vol. 3 "Silica Minerals," 7th Ed., 6th printing, John Wiley, New York, 334 pp.

Gerber, E.A., and Sykes, R.A., 1967, "Quartz Frequency Standards," *Proceedings,* Institute of Electrical and Electronics Engineers, Vol. 55, No. 6, June, pp. 783–791.

Graham, R.A., 1972, "Strain Dependence of Longitudinal Piezoelectric, Elastic, and Dielectric Constants of X-Cut Quartz," *Physical Review B,* Vol. 6, No. 12, Dec. 15, pp. 4779–4792.

Hammond, D.L., et al., 1965, "A Linear, Quartz-Crystal, Temperature-Sensing Element," *Transactions,* Instrument Society of America, Vol. 4, Oct., pp. 349–354.

Heising, R.A., 1946, *Quartz Crystals for Electrical Circuits,* D. Van Nostrand, New York, p. 458.

Hunston, D.L., et al., 1972, "Viscoelastic Measurements of Poly(dimethylsiloxanes) and a Polystyrene Solution Using the Layered Waveguide Technique," *Transactions,* Society of Rheology, Vol. 16, No. 1, pp. 45–57.

Ingram, G.E., and Graham, R.A., 1972, "Quartz Gauge Technique for Impact Experiments," *Proceedings of the 5th Symposium on Detonation,* or Sandia Laboratories (Albuquerque, NM) Report No. SC-R-72 2722.

Jaffe, H., and Turobinski, T.J., 1960, "Method of Growing Quartz Single Crystals," US Patent 2,923,605, Feb. 2.

Joshi, M.S., and Vagh, A.S., 1964, "Growth Spirals on Prism Faces of Cultured Quartz," *American Mineralogist,* Vol. 49, Nov., pp. 1771–1773.

Karrer, H.E., and Leach, J., 1969, "A Quartz Resonator Pressure Transducer," *Transactions on Industrial Electronics and Control Instrumentation,* IEEE, Vol. IECI-16, No. 1, July, pp. 44–50.

Katz, A., 1962, "Hydrogen in Alpha-Quartz," *Philips Research Reports,* Vol. 17, pp. 133–279 (N.V. Philips Co. Eindhoven, Netherlands).

King, J.C., 1959, *Bell System Technical Journal,* Vol. 38, p. 573.

King, J.C., et al., 1962, *Journal of Physics and Chemistry of Solids,* Vol. 23, p. 1019.

King, W.H. Jr., 1969, "Using Quartz Crystals as Sorption Detectors," *Research/Development,* Pt. 1, April, pp. 28–34; Part 2, May, pp. 28–33.

Kolb, E.D., et al., 1972, "Low Optical Loss Synthetic Quartz," *Materials Research Bulletin,* Vol. 7, pp. 397–406.

Lang, A.R., 1967, "Fault Surfaces in Alpha Quartz," *Crystal Growth,* supplement of *Physics and Chemistry of Solids,* Pergamon Press, pp. 833–838.

Lias, N.C., and Rudd, D.W., 1969, "Growth of Synthetic Quartz for Use in High Frequency Monolithic Crystal Filters," *Western Electric Engineer,* Vol. 13, No. 2, April.

Lias, N.C., et al., 1973, "The Growth of High Acoustic Q Quartz at High Growth Rates," American Association for Crystal Growth Conference, Princeton, NJ, July 30, *Journal of Crystal Growth,* Vol. 18, No. 1, pp. 1–6, Amsterdam, Netherlands.

Mason, W.P., ed., 1964, *Physical Acoustics,* Vol. 1, Pt. A, Academic Press, New York.

Meisinger, A.C., 1981, "Quartz Crystal" section, Minor Nonmetals Chapter, *Minerals Yearbook, 1980,* Vol. I, US Bureau of Mines, pp. 941–942.

Myers, R.R., and Schultz, R.K., 1962, "A Kinetic Study of Film Drying via Changes in Mechanical Properties," *Official Digest,* Federation of Societies for Paint Technology, Vol. 24, 801 pp.

Petkof, B., 1970, "Quartz Crystal," *Mineral Facts and Problems,* Bulletin 650, US Bureau of Mines, pp. 1179–1184.

Pfenninger, H., 1961, "Diffusion von Kationen und Abscheidung von Metallen in Quarz unter elektrischer Feldeinwirkung," Ph.D. Thesis, University of Zurick, 130 pp.

Rudd, D.W., et al., 1966, "Mechanical Q of Alpha-quartz Rapidly Evaluated," *Western Electric Engineer,* Vol. 10, p. 22.

Rudd, D.W., and Lias, N.C., 1967, "Recent Advances in Synthetic Quartz Production," *Chemical Engineering Progress Symposium Series,* Vol. 63, No. 76, p. 33.

Saubrey, G., 1964, "Messung von Plattenschwingen sehr kleiner Amplitude durch Lichtstrommodulation," *Zeitschrift für Physik,* Vol. 178, p. 457; see also his paper in *Proceedings,* 15th Annual Frequency Control Symposium, Ft. Monmouth, NJ.

Sawyer, B., 1972, "Q Capability Indications from Infrared Absorption Measurements for Na_2CO_3 Process Cultured Quartz," *Transactions on Sonics and Ultrasonics,* IEEE, Vol. SU-19, No. 1, Jan., pp. 41–44.

Smith, W.L., and Spencer, W.J., 1963, "Quartz Crystal Thermometer for Measuring Temperature Deviations in 10^{-8} to 10^{-00} C Range," *Review of Scientific Instruments,* Vol. 34, p. 268.

Sosman, R.B., 1965, *The Phases of Silica,* Rutgers University Press, New Brunswick, NJ, p. 148.

Spezia, G.R., 1908, "Sull'accrescimento del quarzo," *Atti della R. Accad. delle Scienze di Torino,* Vol. 44, pp. 95–105.

Stockbridge, C.D., 1966, "Resonance Frequency Versus Mass Added to Quartz Crystals," *Vacuum Microbalance Techniques,* Vol. 5, Plenum Press, New York, pp. 193–205.

Viktorov, I.A., 1968, *Rayleigh and Lamb Waves,* Plenum Press, New York.

Waesche, H.H., 1960, "Quartz Crystal and Optical Calcite," *Industrial Minerals and Rocks,* 3rd ed., J.L. Gillson, ed., AIME, New York, pp. 687–695.

Watson, I., 1979, "Quartz Crystal," *Industrial Minerals,* No. 140, May, pp. 21–29.

White, R.M., 1970, "Surface Elastic Waves," *Proceedings,* Institute of Electrical and Electronic Engineers, Vol. 58, No. 8, Aug., pp. 1238–1276.

Zlobik, A.B., 1981a, "Quartz Crystal," *Mineral Facts and Problems,* Bulletin 671, 1980 ed., US Bureau of Mines, pp. 729–736.

Zlobik, A.B., 1981b, "Quartz Crystal, Industrial," *Mineral Commodity Summaries, 1981,* US Bureau of Mines, pp. 120–121.

Minerals Other Than Quartz Bibliography and References

Bardsley, W., and Jones, O., 1968, "On the Crystal Growth of Optical Quality Proustite and Pyrargyrite," *Journal of Crystal Growth,* Vol. 3, No. 4, pp. 268–271.

Bonner, W.A., et al., 1972, "High Quality Tellurium Dioxide for Acousto-optic and Non-Linear

Applications," *Journal of Electronic Materials,* Vol. 1, No. 1.

Bube, R.H., 1960, *Photoconductivity of Solids,* John Wiley, New York, p. 428.

Buhrer, C.F., et al., 1963, "Electro-optic Light Modulation with Cubic Crystals," *Applied Optics,* Vol. 2, Aug., p. 845.

Butler, C.T., et al., 1969, "Growth of Pure Single Crystals of MgO by Arc Furnace," American Association for Crystal Growth, Gaithersburg, Md., Aug. 11–13.

Cady, W.G., 1964, *Piezoelectricity,* Dover Publications, New York, pp. 134, 201, 229, 230, 435, 484, 704.

Crowningshield, R., 1972, *Gems and Gemology,* Spring issue.

Cullen, G.W., 1971, "The Preparation and Properties of Chemically Vapor Deposited Silicon on Sapphire and Spinel," *Journal of Crystal Growth,* Vol. 9, pp. 107–125.

Farrell, E.F., and Ling, A., 1972, "Chemical Vapor Grown TiO_2 Single Crystals," American Association for Crystal Growth, Conference II, Princeton, NJ, July.

Fochs, P.D., et al., 1968, "Growth of Cadmium Sulfide Single Crystals of Controlled Composition from the Vapor Phase," *Journal of Crystal Growth,* Vol. 3, No. 4, pp. 122–125.

Fochs, P.D., 1970, "The Cadmium Sulfide Acoustoelectric Oscillator," *Journal of Science and Technology,* Vol. 37, No. 4, pp. 171–181.

Garbuny, M., 1965, *Optical Physics,* Academic Press, New York, 371 pp.

Gentile, A.L., and Stafsudd, O.M., 1968, "Czochralski-Grown Proustite and Related Compounds," *Journal of Crystal Growth,* Vol. 3, No. 4, pp. 272–274.

Geusic, J.E., et al., 1964, *Applied Physics Letters,* Vol. 4, 182 pp.

Harmon, T.C., 1971, *Journal of the Physics and Chemistry of Solids,* Supplement, Vol. 32, p. 363.

Hatta, J., et al., "Product Engineering for Growth of Synthetic Calcite Polarizing Material," Contract DAAB-05-71-C-2636, US Army Electronics Command, 225 S. 18th St., Philadelphia, PA.

Henisch, H.K., 1970, *Crystal Growth in Gels,* Pennsylvania State University Press, 63 pp.

Hiroshi, Y., 1972, "Synthetic Crystals of Japan," *Japan Electronic Engineering,* June, pp. 41–47.

Hordvik, A., and Schlossberg, H., 1972, "Luminescence of Lithium Niobate," *Applied Physics Letters,* Vol. 20, No. 5, pp. 197–199.

Hulme, K.F., 1967, *Applied Physics Letters,* Vol. 10, p. 133.

Johnson, L.F., and Nassau, K., 1961, *Proceedings,* Institute of Radio Engineers, Vol. 49, p. 1704.

Jona, F., and Shirane, G., 1962, *Ferroelectric Crystals,* Macmillan Co., New York, pp. 108, 216, 226, 338, 361, 363, 371.

Kaminow, I.P., 1965, "Barium Titanate Light Phase Modulator," *Applied Physics Letters,* Vol. 7, p. 123.

Klerk, J.de, 1972, "Elastic Surface Waves," *Physics Today,* Vol. 25, No. 11, pp. 32–39.

Knubovets, R.G., et al., 1970, "Long Wave Infrared Spectra of Some Apatites," *Optics and Spectroscopy,* Vol. 29, No. 1, July, pp. 32–39.

Kolb, E.D., 1967, "Hydrothermal Growth of Low Carrier Concentration ZnO at High Water and Hydrogen Pressures," *Materials Research Bulletin,* Vol. 2, pp. 1099–1106.

Laudise, R.A., 1968, "The Search for Nonlinear Optical Materials for Laser Communication," *Bell Laboratories Record,* Vol. 46, Jan., pp. 2–7.

Liaw, H.M., and Faust, J.W.Jr., 1972, "The Growth of Calcite in Aqueous Solution," American Association for Crystal Growth, 2nd National Conference, Princeton, NJ, July 30.

Maiman, T.H., 1960, "Stimulated Optical Radiation in Ruby," *Nature,* Vol. 187, p. 493.

Makram, H., 1968, "Improvement in Yield of Yttrium Iron Garnet," *Journal of Crystal Growth,* Vol. 3, No. 4, pp. 447–451.

Mason, W.P., ed., 1964, *Physical Acoustics,* Vol. 1, Pt. A, Academic Press, NY.

Mazelsky, R., et al., 1968, "Czochralski Growth of Calcium Fluophosphate," *Journal of Crystal Growth,* Vol. 3, No. 4, pp. 360–364.

Montalvo, R.A., and Langer, D.W., 1970, "Linear Compressibilities of II–VI Compound Single Crystals," *Journal of Applied Physics,* Vol. 41, No. 10, Sept., pp. 4101–4104.

Murray, L.A., 1964, "Modulation Crystals for Lasers," *Electronic Industries,* Feb., pp. 83–85.

Nassau, K., and Broyer, A.M., 1962, "Application of Czochralski Crystal-Pulling Technique to High-Melting Oxides," *Journal of the American Ceramic Society,* Vol. 45, Oct., pp. 474–478.

Nassau, K., and Broyer, A.M., 1962, "Calcium Tungstate: Czochralski Growth, Perfection, and Substitution," *Journal of Applied Physics,* Vol. 33, Oct., pp. 3064–3073.

Nassau, K., 1965, "Lasers and Laser Materials," *Materials Research and Standards,* Vol. 5, Jan., pp. 3–11.

Nassau, K., 1964, "Growing Synthetic Crystals," *Lapidary Journal,* Vol. 18, No. 3, June, pp. 386–389.

Nassau, K., 1964, "Growing Synthetic Crystals," *Lapidary Journal,* Vol. 18, No. 4, July, pp. 474–477, 483.

Nassau, K., and Shiever, J.W., 1972, "Modified Technique for the Growth of Boracite Crystals," *Journal of Crystal Growth,* Vol. 16, No. 1, pp. 59–61.

Nassau, K., and Shiever, J.W., 1972, "Vapor Growth of II-VI Compounds and the Identification of Donors and Acceptors," *Journal of Crystal Growth,* Vol. 13/14, pp. 375–379.

Nielsen, K.F., 1968, "Growth of ZnO Single Crystals by the Vapor Phase Reaction Method," *Journal of Crystal Growth,* Vol. 3, No. 4, pp. 141–145.

Parker, S.G., and Pinnell, J.E., 1968, "Molten Flux Growth of Cubic ZnS Crystals," *Journal of Crystal Growth,* Vol. 3, No. 4, pp. 490–495.

Pinnow, D.A., et al., 1969, "Lead Molybdate," *Applied Physics Letters,* Vol. 17, pp. 83–86.

Pollock, J.T.A., et al., 1972, "Edge-Defined, Film-Fed Growth of Sapphire," American Association for Crystal Growth, 2nd National Conference, Princeton, NJ, July 30.

Pough, F., 1970, "Colored Synthetic Quartz from Russia," *Lapidary Journal,* June 24, pp. 444–446.

Rosenberger, F., and DeLong, M.C., 1972, "Growth of Rutile by Chemical Vapor Transport," American Association for Crystal Growth,

2nd National Conference, Princeton, NJ, July 30.

Rudness, R.G., and Kebler, R.W., 1959, "Growth of YIG by the Flame Fusion Process," *Linde Industrial Crystal Bulletin,* Linde Co., 30 East 43rd St., NY, 10017, Sept. 21.

Ryan, F.M., et al., 1970, "Optical Properties of Divalent Manganese in Apatite," *Physical Review B,* Vol. 2, No. 7, Oct., pp. 2341–2352.

Schlossberg, H.R., and Kelley, P.L., 1972, "Using Tunable Lasers," *Physics Today,* Vol. 25, No. 7, July, pp. 36–44.

Stanley, J.M., 1954, *Industrial and Engineering Chemistry,* Vol. 46, p. 1684.

Stepke, E.T., 1972, "Optical Mass Memories," *Electrooptical Systems Design,* Vol. 4, No. 11, Oct., pp. 12–17.

Tolksdorf, W., 1968, "Growth of Yttrium Iron Garnet Single Crystals," *Journal of Crystal Growth,* Vol. 3, No. 4, pp. 463–466.

Toudic, Y. and Aumont, R., 1971, "Croissance par voie Hydrothermale de Monocristaux de Cinabre de Dimensions Importantes," *Journal of Crystal Growth,* Vol. 10, pp. 170–174.

Von Hippel, H., 1963, *Technical Reports,* Laboratory of Insulation Research, Cambridge, MA, Vol. 44, March.

Wheeler, D., 1972, "Titanates," *Ceramic Industry Magazine,* July, pp. 28–29.

White, R.M., 1970, "Surface Elastic Waves," *Proceedings,* Institute of Electrical and Electronics Engineers, Vol. 58, No. 8, Aug.

Wood, J.D.C., and White, E.A.D., 1968, "Growth of Stoichiometric Manganese Aluminate Spinel Crystal by Flux Evaporation," *Journal of Crystal Growth,* Vol. 3, No. 4, pp. 480–484.

Yu, P.Y., and Cardona, M., 1973, "Intrinsic Piezobirefringence of Several Semiconducting Chalcogenides," *Journal of the Physics and Chemistry of Solids,* Vol. 34, pp. 29–56.

Fertilizer Minerals

Original by JOHN P. BRYANT *
Revised by SAMUEL KINCHELOE †

Plant nutrients are obtained by plants from both the air and the soil. Carbon dioxide, a gaseous form of carbon and oxygen, supplies the carbon which usually makes up 50% or more of plant structure. Plants are capable of absorbing carbon dioxide directly from the air. Most of this nutrient comes from the air, although some carbon dioxide is absorbed by soil water and taken into the plant as liquid.

Growing plants use more water than any other substance. Plant nutrients in the soil dissolve in soil water; soil water entering the plant carries nutrients into the plant structure. Water is then given off into the air as the plant breathes or transpires.

Inorganic or mineral nutrients essential to plants are supplied through the soil. Inorganic or mineral composition of plants usually is less than 5% of their weight. In most crop-growing areas, much of this nutrition is inadequately supplied from the soil and must be provided by fertilizers. Proper fertilization of a soil to improve its mineral level and productivity is the role of the fertilizer industry.

The 13 essential plant nutrients which often must be added to soils or plants and are generally considered fertilizer minerals are categorized as primary minerals (nitrogen, phosphorus, and potassium); secondary minerals (calcium, magnesium, and sulfur); and micronutrients, or trace minerals (boron, iron, manganese, copper, zinc, molybdenum, and chlorine). These classifications are based on quantity of nutrients utilized by plants and have no implication as to function or degree of necessity.

One point of major significance to the fertilizer industry is that not all nutrient carriers supply that nutrient in a chemical form which will be available to the plant. Thus, a product which may be considered a fertilizer mineral may require substantial processing or treatment before it becomes useful as a nutrient carrier.

* Agronomic Services, International Minerals & Chemical Corp., Skokie, IL.

† Manager, Agronomic Services, International Minerals & Chemical Corp., Atlanta, GA.

Fertilizer Industry Scope

Until about 1900 demands for higher crop yields were met simply by bringing new agricultural lands into cultivation. Since that time great progress has been made in increasing yields through fertilization and improved varieties and cultural practices. The importance of the fertilizer industry to the economy and welfare of the US is widely recognized. Fig. 1 illustrates the relationship of lime and fertilizers to crop yields.

In terms of tons of fertilizer products, the growth of the industry has been phenomenal, particularly between 1950 and 1980. Table 1 shows the increase in fertilizer usage.

Primary Plant Nutrients in Fertilizers

Primary minerals (nitrogen, phosphorus, and potassium) are designated as such because

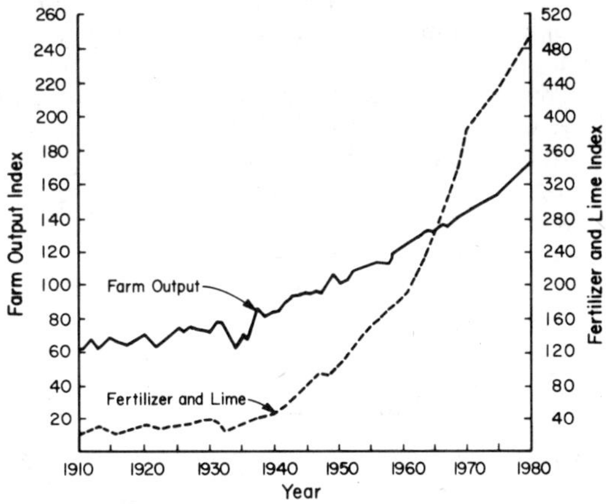

FIG. 1—*Indices of farm output and input of fertilizer and lime, US agriculture 1910–1980 (1947–1949 = 100). (US Dept. of Agriculture).*

TABLE 1—Total Consumption of Plant Nutrients in the United States, St*

Year	N, P_2O_5, K_2O
1920	1,145,000
1930	1,526,000
1940	1,766,000
1950	4,058,000
1960	7,464,000
1970	16,068,000
1980	23,083,000

Source: US Dept. of Agriculture.
* Metric equivalent: 1 st × 0.907 184 7 = t.

of their significance to plant growth. Practically all soils are deficient in nitrogen. Most soils are deficient to varying degree in phosphorus and potassium. Consequently, these nutrients comprise the vast majority of fertilizers applied to soils.

Nitrogen

Although nitrogen is not a fertilizer mineral in that it is not a mined product, it is perhaps the most important fertilizer nutrient or constituent. It accounts for about half of the total nutrients applied as fertilizer in the US. Table 2 shows consumption of nitrogen in North America and its growth here and in other world markets in recent years.

The manufacturing process for nitrogen involves the fixation of atmospheric nitrogen.

Nitrogen is absorbed from the soil by plants most commonly as nitrate (NO_3^-) but may be taken up also as ammonium (NH_4^+). After it enters the plant, nitrogen is converted into complex compounds and ultimately into protein.

Commercially available sources of nitrogen include both organic and inorganic forms. Regardless of the original source, however, all nitrogen must be converted to the nitrate or ammonium form before it can be absorbed by the plant.

In years past, virtually all nitrogen was applied to soil in the form of animal manures. Certain soils in Europe have a "man-made" surface layer more than 0.9 m (3 ft) thick due to centuries of manuring. One of the earliest commercial sources of nitrogen (also organic) was guano, excreta and remains of sea fowl found in deposits on certain arid islands. The material often contains as much as 13% nitrogen and about 9% phosphorus. Neither manure nor guano is an important commercial source of fertilizer today; animal manure is applied to soils primarily because of the need for disposal. The major organic sources of nitrogen in use today are urea (NH_2CONH_2) and urea formaldehyde ($NH_2CONH \cdot CH_2OH$); both are synthetic products.

Commercially important inorganic sources of nitrogen include ammonium nitrate (NH_4NO_3) available in both solid (33–34% N) and liquid (19–21% N) forms; ammonium sulfate [$(NH_4)_2SO_4$—21% N] calcium nitrate [$Ca(NO_3)_2$—15% N]; potassium nitrate (KNO_3—13–15% N); urea (NH_2 $CONH_2$—45–46% N); anhydrous ammonia (NH_3—82% N); and nitrogen solutions (28–32% N).

Nitrogen is applied to soils either prior to planting, at planting, or after planting as a sidedress. Since all inorganic sources of nitrogen convert to the usable nitrate form rather rapidly in warm, moist soils, and since nitrate leaches rapidly, either method is agronomically and economically sound. The crop, soil, and weather conditions may dictate which materials and application methods are best.

In 1970 about 48 kg (106 lb) of nitrogen was applied to each acre of corn, accounting for approximately half the total US consumption. The average rate increased to about 54 kg (120 lb) per acre in 1978, and accounted for approximately 38% of US consumption of nitrogen.

TABLE 2—World Nitrogen Fertilizer Consumption (Mt, Nitrogen)

Year	North America	Western Europe	Eastern Europe	Oceania and Japan	Asia	Africa	Latin America	Communist Asia	Total
1955	1.83	2.10	0.92	0.56	0.43	0.18	0.25	0.24	6.51
1960	2.55	3.02	1.62	0.62	0.73	0.22	0.29	0.58	9.63
1965	4.38	4.28	3.37	0.80	1.36	0.51	0.74	0.96	16.40
1970	7.04	6.00	6.85	1.08	2.66	0.74	1.18	3.18	28.73
1975	8.34	7.24	10.78	0.90	4.38	1.07	1.82	2.27	36.80
1980	11.18	9.58	12.44	1.06	7.79	1.51	2.67	11.05	57.28

Source: TVA, UN-FAO.

Phosphorus

Of the three major or primary nutrients in fertilizers, phosphorus is the least used. In 1980 total phosphorus (P) used in the US was 2.38 Mt. And, although its growth has not paralleled that of nitrogen, its world consumption has increased from 3.32 Mt in 1955 to 8.27 Mt in 1970 and to 13.21 Mt in 1980 (Table 3).

Phosphorus is absorbed from soils by plants principally as the $H_2PO_4^-$ ion. Once inside the plant it enters into the composition of phospholipids and of nucleic acids. A major metabolic role of phosphorus in the cells of both plants and animals is the providing of energy which results from the high energy phosphate bonds.

In growing plants, phosphorus is most abundant in the growing tissue where it is utilized in the synthesis of nucleoproteins. As the plant matures, phosphorus accumulates in the seed and fruit.

Fertilizer phosphorus is highly reactive with clay and organic constituents in soils and forms relatively insoluble compounds. As a result it does not move downward with percolating ground waters but tends to accumulate in the surface layer of soil. Thus after several years of high phosphorus fertilization, farmers are able to reduce their phosphorus fertilization without loss in yield. However, high crop yields usually require a maintenance application of fertilizer.

The development of the modern fertilizer phosphate industry dates back to 1840 when Justus von Liebig, a German scientist, demonstrated that the fertilizer value of bones was significantly increased by treatment with sulfuric acid. Two years later, the process by which phosphate rock (apatite) was treated with sulfuric acid to produce ordinary superphosphate (20% P_2O_5) was patented in England by John Bennett Lawes. Commercial production began in the US in 1852.

In today's multimillion-dollar fertilizer phosphate industry, nearly all products originate from phosphate rock (see the chapter on Phosphate Rock). It is still acidulated with sulfuric acid using essentially the same principle employed in England more than 100 years ago. Due to its low cost, ordinary superphosphate dominated the phosphate market until recent years when new materials began to replace it, primarily because of their higher analysis and lower transportation costs. The three products currently dominating the world phosphate market include normal superphosphate [$Ca(H_2PO_4)_2\ H_2O + CaSO_4\ 2H_2O$]; concentrated superphosphate [$Ca(H_2PO_4)_2\ H_2O$]; and complex fertilizers including monoammonium phosphate—$NH_4H_2PO_4$; diammonium phosphate—$(NH_4)_2HPO_4$; and nitric phosphate of varying analyses. Fig. 2 illustrates how the world phosphorus product mix is changing.

Nearly all commercially available sources of phosphorus are highly water-soluble. However, when the calcium phosphates, such as normal superphosphate or concentrated superphosphate, are ammoniated to make other complex fertilizer products, the high pH of the manufacturing system often causes further reaction of calcium with the phosphorus, resulting in precipitation of insoluble calcium phosphate compounds. These products are more effective on acid soils than on alkaline soils.

A major consideration in phosphorus application to soil and its utilization by plants is soil pH. The influence of soil pH on phosphorus solubility or availability to plants is illustrated in Fig. 3.

In acid soils the phosphorus can be complexed by iron and aluminum compounds. At alkaline pH levels, phosphorus reacts with calcium, forming insoluble calcium phosphates. These problems are partially overcome by liming acid soils and applying only high water-soluble phosphate sources where soils are alkaline.

Because of its immobility in soils, phos-

TABLE 3—World Phosphorus Fertilizer Consumption (Mt, Phosphorus)

Year	North America	Western Europe	Eastern Europe	Oceania and Japan	Asia	Africa	Latin America	Communist Asia	Total
1955	1.00	1.25	0.47	0.40	0.06	0.08	0.05	0.01	3.32
1960	1.13	1.54	0.70	0.52	0.10	0.10	0.07	0.03	4.19
1965	1.51	1.91	1.16	0.76	0.22	0.11	0.20	0.17	6.04
1970	1.94	2.43	1.78	0.84	0.28	0.22	0.34	0.44	8.27
1975	2.02	2.22	2.72	0.71	0.71	0.33	0.66	0.73	10.10
1980	2.44	2.59	3.54	0.95	1.32	0.41	1.01	0.95	13.21

Source: TVA, UN-FAO.

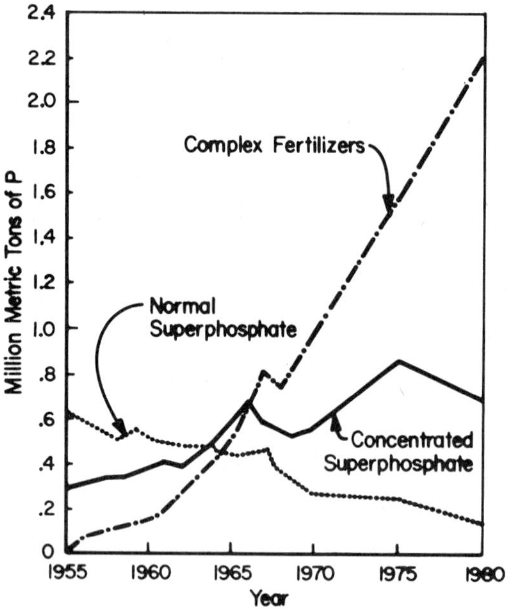

FIG. 2—*Production of major phosphate fertilizer materials in the US. (I.M.C.)*

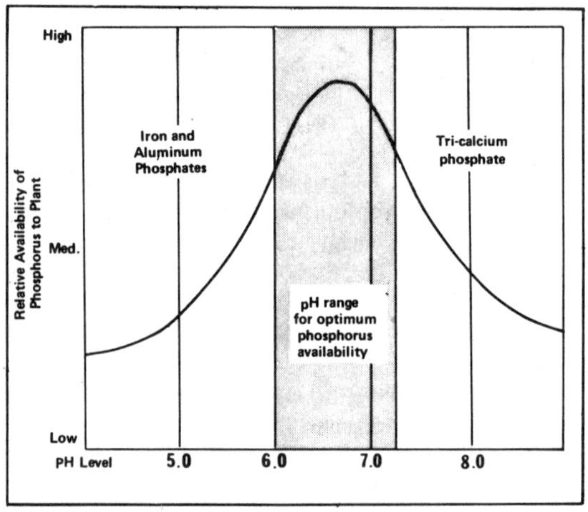

FIG. 3—*Diagram showing relationship of soil pH to phosphorus availability.*

phorus must be incorporated into the soil for optimum effectiveness. Broadcasting phosphorus fertilizers with subsequent incorporation is an effective method of application, although recovery of applied phosphorus is usually higher where it is placed in a band beside the planted seed—particularly when soil phosphorus is quite low. Also, small amounts of certain phosphorus products have been effectively applied in direct contact with the seed. Broadcasting is by far the dominant method of phosphorus fertilizer application in the US.

Potassium

In the decade from 1960 to 1970, world potassium consumption nearly doubled. Table 4 shows that it increased from 6.73 Mt to 12.59 Mt in that period. By 1980 consumption increased to 19.89 Mt.

Growing plants absorb potassium as K^+ ion. Potassium is unlike all other mineral constituents in that its exact metabolic role in the plant is unknown. It occurs in the ionic state and is not a component of compounds within the plant. However, it is indispensable as a nutrient and is required in large quantities by most plants. Although they are not fully understood, the fundamental roles of potassium in the plant are undoubtedly regulatory or catalytic.

Muriate of potash (KCl) is the dominant source of fertilizer potassium throughout the world. Other important, but much less significant, sources are sulfate of potash (K_2SO_4) and sulfate of potash-magnesia ($K_2SO_4 \cdot 2MgSO_4$) (see the chapter on Potash).

Potassium application methods are very similar to those of phosphorus; most is broadcast with a lesser amount applied banded alongside the seed at planting time. Muriate of potash is

TABLE 4—World Potassium Fertilizer Consumption (Mt, Potassium)

Year	North America	Western Europe	Eastern Europe	Oceania and Japan	Asia	Africa	Latin America	Communist Asia	Total
1955	1.47	2.06	1.28	0.34	0.05	0.04	0.09	0.00	5.33
1960	1.69	2.71	1.56	0.49	0.10	0.06	0.11	0.01	6.73
1965	2.23	3.24	2.45	0.61	0.19	0.13	0.23	0.08	9.16
1970	3.18	3.63	3.97	0.72	0.39	0.18	0.46	0.06	12.59
1975	3.52	3.82	6.08	0.80	0.76	0.29	0.76	0.19	16.22
1980	5.00	4.54	6.44	0.81	1.15	0.29	1.32	0.34	19.89

Source: TVA, UN-FAO.

usually mixed with phosphorus carriers and the two nutrients are applied simultaneously. Almost two-thirds of the potassium used in US is applied to soils of the Midwestern corn belt.

Secondary Plant Nutrients in Fertilizers

Calcium, magnesium, and sulfur—the secondary nutrients—are required by plants in smaller amounts. In many areas, sulfur and magnesium are becoming increasingly important as fertilizer nutrients. Increasing crop yields, sandy soils, and leaching due to high rainfall or to irrigation are factors which influence the use of magnesium and sulfur in fertilizers.

The use of fertilizer calcium strictly as a nutrient is very limited. Its use is dictated by special crop needs and unusual soil conditions.

Common fertilizer materials for secondary nutrients are potassium sulfate, sulfate of potash-magnesia, gypsum, and normal superphosphate.

Calcium

Soil calcium is important not only as a plant nutrient but also because of its very favorable influence on maintaining the desired physical and biological conditions in the soil. Of the exchangeable bases found in soil (potassium, magnesium, calcium, and sodium), calcium is dominant, normally accounting for 50 to 80% of the total base exchange.

Calcium is absorbed by plants as the Ca^{++} ion. Inside the plant calcium tends to accumulate in the older tissue. Much of the calcium in plants is permanently fixed in the cell walls as a calcium salt of the pectic compounds in the middle lamella, giving cell walls their strength. Calcium also is important in the reduction of nitrates as they make their way toward protein formation.

Most soils, even acid ones, contain sufficient calcium for plant growth; calcium-deficient plants are seldom observed. However, many soils are too acid for optimum plant growth, and the soil pH can be raised to a desirable level most economically with finely ground agricultural lime ($CaCO_3$). Table 5 shows lime consumption in the US.

Soils of the humid regions east of the Mississippi River generally require higher rates and more frequent applications of lime since the soil calcium is leached from those soils by relatively high rainfall, resulting in the lowering of the soil pH.

TABLE 5—Use of Agricultural Limestone in the United States

Year	Limestone, st*
1930	3,587,700
1940	14,406,000
1950	29,842,100
1960	22,613,600
1970	25,900,800
1980	34,402,400

Source: National Limestone Institute.
*Metric equivalent: 1 st × 0.907 184 7 = t.

Besides creating a desirable soil pH level for optimum plant growth and resulting in a more desirable soil physical condition, a secondary effect of lime is that soil pH influences availability of certain other nutrients. The favorable influence of soil pH on phosphorus availability is illustrated in Fig. 3. Conversely, the micronutrients boron, manganese, iron, copper, and zinc unfortunately are made less available (Fig. 4) as the soil pH is increased toward neutrality. Thus these micronutrients must often be added back to the soil as fertilizer.

The need for lime is tremendous; some southeastern states cite need by more than half the agricultural soils. Lime supplies are virtually unlimited and commercial deposits have been found in most agricultural areas. Common lime materials are calcite ($CaCO_3$) and, to a lesser extent, basic slag which is a calcium silicate material which also contains some phosphorus and magnesium. These materials are sometimes used as fillers in mixed fertilizers although more than 95% of agricultural lime is applied directly to soils.

Certain crops such as peanuts and tomatoes

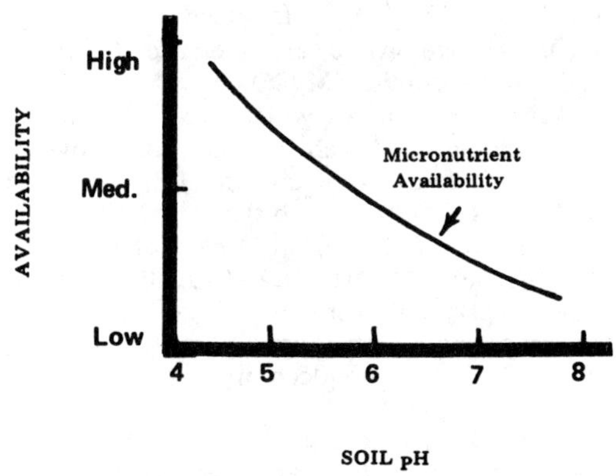

FIG. 4—*Diagram showing relationship of soil pH to micronutrient (manganese, iron, copper, zinc, boron) availability.*

require a large amount of soluble calcium. Gypsum ($CaSO_4 \cdot 2H_2O$) is often used where these crops are grown. It is usually applied directly rather than as a component of mixed fertilizer. Other carriers of much less significance for their calcium content are normal and concentrated superphosphate and calcium nitrate.

Magnesium

Magnesium, like calcium, is required as both a nutrient and a necessary soil component. The desired exchangeable or available level of magnesium in soils is usually about one-tenth that of calcium. The use of high-calcium lime materials and the natural tendency of many soils to be low in magnesium has stimulated the practice of including soluble magnesium carriers in fertilizers. They are an important component of premium fertilizer products. These soluble magnesium materials have increased the crop production efficiency of the major nutrients.

Magnesium is absorbed by plants as the Mg^{++} ion. Its role inside the plant is of major significance in that it is a constituent of the chlorophyll molecule. It is also important in phosphorus metabolism.

Magnesium carriers either added to fertilizers or applied directly can be classified by their chemical nature—or the degree of water-solubility. The magnesium source must either be water-soluble initially or convertible to a water-soluble form before the magnesium can be absorbed by the plant.

Magnesium sources which are soluble or readily available are hydrated magnesium sulfate and sulfate of potash-magnesia (langbeinite: $K_2SO_4 \cdot 2MgSO_4$). Included in the magnesium sulfate group are kieserite ($MgSO_4 \cdot H_2O$) and epsomite ($MgSO_4 \cdot 7H_2O$).

Magnesium sources which are only slightly water-soluble but which become soluble over a period of 2 to 4 years are (1) dolomite ($CaCO_3 \cdot MgCO_3$); (2) brucite ($MgO \cdot H_2O$); (3) magnesite ($MgCO_3$); (4) basic slag (which contains various nutrients and usually from 3–12% MgO); (5) periclase (MgO); and (6) the magnesium-ammonium phosphates ($MgNH_4PO_4 \cdot H_2O$) produced in fertilizer manufacture.

There are also some slowly soluble magnesium materials that are sold as fertilizer materials in some areas. These include serpentine $(Mg,Fe)_3Si_2O_5(OH)_4$ and olivine $(Mg,Fe)_2SiO_4$.

TABLE 6—North American Consumption of Magnesium Fertilizers

Source	Tons/Mg
Magnesium sulfate	3,300
Sulfate of potash-magnesia	100,000
Liming materials	2,385,400
Other	100,000

Source: International Minerals & Chemical Corp.

Although there is no accurate record of magnesium consumption, Table 6 lists the approximate tonnage of magnesium on an elemental Mg basis.

Although much of the soluble magnesium materials are mixed into other fertilizers, nearly all other magnesium carriers are directly applied to soils. Dolomite, by far the most significant carrier, is applied to soils at rates of 1 to 3 tons per acre.

Sulfur

Sulfur has been recognized as an essential plant nutrient for more than 130 years. Further, it has been basic to the fertilizer industry since its inception due to the treatment of phosphate rock with sulfuric acid in manufacture of superphosphate. In that process the sulfuric acid reacts with calcium in the rock, resulting in the formation of gypsum ($CaSO_4 \cdot 2H_2O$). The gypsum is not removed from the normal superphosphate so that the product used by farmers contains 12% sulfur.

Until recently normal superphosphate was the chief carrier of sulfur in fertilizers and nearly all fertilizers contained some normal superphosphate. In recent years, however, increasing importance has been placed on higher analysis phosphates. In these products more sulfuric acid is actually used initially to treat the rock, but the product is further processed so that most of the sulfur content is removed.

Some plants require almost as much sulfur as they do phosphorus. Sulfur enters the plant as the $SO_4^{=}$ ion. Inside the plant sulfur becomes a necessary component of protein. It is also a constituent of the mustard oil glycosides which impart characteristic odors and flavors to such species as mustard, onions, and garlic.

The major carriers of sulfur in today's fertilizers are normal superphosphate, potassium sulfate, langbeinite, and gypsum. In chemically manufactured mixed fertilizer, sulfuric acid is still in wide use for neutralization of the free ammonia used in formulation.

In certain areas where response to sulfur is

obtained, langbeinite ($K_2SO_4 \cdot 2MgSO_4$) is increasing in importance as a sulfur carrier. Elemental sulfur is also commonly used, but its oxidation to the available sulfate form is known to proceed slowly and it tends to reduce soil pH.

Ammonium thiosulfate [$(NH_4)_2S_2O_3$] is perhaps the most widely used sulfur carrier in the fluid fertilizer industry. It is made by reacting SO_2 and aqueous ammonia with further reaction with elemental sulfur. It has about 26% sulfur in the sulfate ($SO_4^{=}$) form and is completely soluble in liquid N-P-K mixtures. It is important to the fluid fertilizer industry because fluids are normally sulfur-free and are not compatible with most other sulfur carriers.

Sulfur is increasingly important not only to farmers because of more widespread soil deficiencies but also to the phosphate industry. Table 7 shows the increase in sulfur consumption in the US fertilizer industry from 1950 to 1980.

TABLE 7—US Sulfur Consumption for All Fertilizer Industry Purposes

Year	Consumption, t
1950	1,300,000
1960	2,000,000
1970	4,600,000
1980	9,600,000

Courtesy The Sulphur Institute.

Micronutrients in Fertilizers

Micronutrients are essential to plants, but are used only in minute quantities. They are measurable in plants only in parts per million (ppm). They almost always occur in soils in very small quantities. Micronutrients which must be added to certain soils as fertilizers include boron, copper, iron, manganese, molybdenum, and zinc. Chlorine is also an essential nutrient but there are no known areas of soil deficiency.

Increased crop production, high yield goals, and diminished soil reserves of micronutrients have combined to increase the consumption of micronutrient minerals as fertilizers, particularly in recent years. Partially estimated micronutrient consumption data show that the tonnage of these minerals consumed in the US increased more than five-fold in the 30-year period 1950-1980 (Table 8).

Most micronutrient carriers are incorporated with multinutrient mixed fertilizers. They may be mixed in chemically manufactured fertilizers, in blends, or in fluids. Lesser amounts of micronutrients are applied directly to soils or used as foliar sprays.

TABLE 8—US Consumption of Micronutrients (Manganese, Iron, Zinc, Boron, Molybdenum)

Year	Consumption, t
1950	12,300
1970	37,800
1980	67,500

Source: US Dept. of Agriculture, I.M.C. estimates.

Micronutrients, just as primary and secondary nutrients, must be water-soluble before they can be absorbed by plants. The relationship of soil pH to solubility or availability of micronutrients to plants is illustrated in Fig. 4. Micronutrients are shown in Table 9.

Boron

The nutrient that is probably added to more fertilizers than any other of the micronutrients is boron (see the chapter on Boron and Borates). It is necessary for the growth of many economic crops such as alfalfa, cotton, clover, and corn. Many plants will not produce seed without boron fertilization. A large number of plants will not tolerate more than traces of soluble boron. It is added to a fertilizer as water-soluble borax, modified borax (agricultural borate), colemanite, and frits. The common boron mineral carriers for fertilizer use are borax ($Na_2B_4O_7 \cdot 10H_2O$) and colemanite ($Ca_2B_6O_{11} \cdot 5H_2O$) and impure synthetic borosilicate glasses.

Iron

Iron is an important micronutrient closely associated with chlorophyll formation and a catalyst in plants. Lack of sufficient iron in plants often results in chlorosis or yellowing of the plant. This element is often supplied as an incidental impurity in the phosphate and potash carriers. Some sulfuric acids used in ammoniation also carry definite quantities of iron. Although most soils contain sufficient iron, it often is not available to the plants, particularly in high pH soils. Therefore, supplemental additions of specific iron carriers are made to fertilizers.

Manganese

Another essential micronutrient closely associated with iron and having somewhat similar

TABLE 9—Categories of Micronutrients

Chemistry	Compound	Solubility/Availability
Sulfates	Inorganic	Water-soluble; readily available
Oxides	Inorganic	Insoluble; becomes available only in acid soils
Oxy-sulfates	Inorganic	Variable solubility
Frits*	Inorganic	Insoluble; compound becomes slowly available
Chelates †	Organic	Soluble; availability is controlled
Complexes	Organic/Inorganic	Soluble

* Frits are made by fusing a mixture of the micronutrients with a phosphate, silicate, or other matrix in a smelting furnace.
† Chelated micronutrients are manufactured by reacting the micronutrient metal with a chelate such as ethylenediaminetetraacetic acid (EDTA), nitrilotriacetic acid (NTA), or other suitable chelating agents.

functions in the plant is manganese (see the chapter on Manganese). It will not take the place of iron. Either its absence or excess reduces the effectiveness of iron. Manganese along with iron is probably the main oxidation-reduction regulator in plants. It is often deficient in soils, particularly those that have been recently limed.

This element is used as a trace element supplement in fertilizers. It is supplied in the form of soluble sulfates, various insoluble oxides, and even as chelated manganese. Usually the less soluble oxides are less effective, although this depends upon other factors. The chief minerals which supply these manganese materials for fertilizer are pyrolusite (MnO_2), psilomelane (mixed Mn-oxides), and byproduct reduced manganous oxide (MnO).

Copper

Although it stimulates crop production, the total functions of this element in plant growth are not fully understood. It is known to stimulate seed production. It probably also functions as catalyst and regulator. Since plants absorb copper readily, its application with fertilizer is one of the important ways of getting this essential animal nutrient into the feed of animals.

Many soils—especially the leached, sandy, and excessively cropped ones—need copper in the fertilizer for good crop production. High organic soils are also often copper-deficient. It is usually added to the fertilizer mixture as a sulfate, oxide, and even as a chelated copper. Chalcocite (Cu_2S), chalcopyrite ($CuFeS_2$), and cuprite (Cu_2O) are probably the most important mineral sources from which other products originate.

Zinc

A micronutrient closely associated with copper is zinc. It cannot take the place of copper. Its role is usually described as that of a catalyst. In its absence, most plants are stunted or make improper growth. Zinc deficiency occurs in wide areas of the US.

Zinc usually is applied with the fertilizer mixture as a sulfate, sulfide, oxide, and even in the chelated form. Most of these carriers have been prepared from the zinc mineral sphalerite (ZnS) and byproduct oxides from zinc processing operations.

Molybdenum

Of all the elements definitely considered to be essential to plants, molybdenum is required in smallest amounts. Little is known regarding the exact function of molybdenum although it is known to be associated with the reduction of nitrates. There are many soils on which little growth can be sustained without molybdenum additions. It very often stimulates the growth of legumes and may have something to do with nitrogen changes in plant tissue. In animals it is closely associated with copper. Excesses of molybdenum can be just as injurious as inadequate amounts. It is used in many fertilizers in small quantities. This may be as part of a general micronutrient mixture or alone. The carriers are usually sodium molybdate and ammonium molybdate. The chief source of these molybdenum compounds is the mineral molybdenite (MoS_2).

Bibliography and References

Anon., 1979, "Agricultural Statistics," US Department of Agriculture.

Anon., 1980, *World Survey of Phosphate Deposits,* 4th ed., British Sulphur Corp., London.

Anon., 1981, "Commercial Fertilizers Consumption for Year Ended June 30, 1981," Nov., Economics & Statistical Service, Crop Reporting Board, US Dept. of Agriculture.

Anon., 1981a, "Commercial Fertilizer Consumption by Class for Year Ended June 30, 1981,"

Dec., Economics & Statistical Service, Crop Reporting Board, US Dept. of Agriculture.

Anon., 1981b, "FAO Monthly Bulletin of Statistics," *FAO Fertilizer Yearbook,* UN-FAO, Vol. 3, March, various years.

Bryant, J. P., 1975, "Fertilizer Minerals," *Industrial Minerals and Rocks,* 4th ed., S.J. Lefond, ed., AIME, New York, pp. 225–233.

Cummings, R.W., and Gleason, R.N., 1971, "The Role of Fertilizer in Agricultural Development," *Fertilizer Technology and Use,* Soil Science Society of America, Madison, WI, pp. 1–25.

Drake, H.J., 1972, "Stone," *Minerals Yearbook 1972,* US Bureau of Mines, pp. 1153–1173.

Harre, E.A., et al., 1971, "The World Fertilizer Market," *Fertilizer Technology and Use,* Soil Science Society of America, Madison, WI, pp. 27–55.

Omasta, S., 1981, *Tonnage Report,* National Limestone Institute, Inc.

Parker, F.W., 1962, "Fertilizer and Economic Development," *Fertilizer Technology and Use,* Soil Science Society of America, Madison, WI, pp. 1–22.

Fillers, Filters, and Absorbents*

NELSON SEVERINGHAUS, JR. †

Mineral fillers have been defined as inert materials that are included in a composition for some useful purpose. Because this definition and the scope of this review are not specific—indeed, exceptions abound—accurate total statistics are unavailable. One source estimates the total production in the United States in 1979 at 11.3 Mt (12.5 million st) valued at $534 million, averaging $49.19 per ton. Included are a broad group of minerals—asbestos, barite, bentonite, ground limestone and dolomite, clays, diatomite, feldspar, micas, silica, talc. Additional references will be made to synthetic mineral-derived fillers such as aluminum hydrate, precipitated calcium carbonate, and pyrogenic and precipitated silicas and silicates.

A section on related uses at the end of this chapter covers absorbents and filters.

Mineral fillers can be included in compounds to accomplish a variety of final results. Hardness, brittleness, impact strength, compressive strength, softening point, fire resistance, surface texture, electrical conductivity, and a host of other physical characteristics may be modified by the inclusion of fillers.

These effects are the result of properties of the fillers, including chemical activity, hardness, particle size, shape and distribution, surface structure, color, density, and refractive index. To facilitate an overview of the mineral filler industry, this chapter will review the general usage of fillers, the inherent characteristics that intensify their utilization, and similarly their manufactured characteristics. Major using industries will then be examined in detail. Specific information on deposits and processing for industrial minerals will be found in the appropriate commodity chapter. All of these same minerals have additional significant uses based on their chemical composition or other characteristics. References will be found in other use chapters and summarized in commodity chapters.

The chart in Table 1, updated from the 4th edition of *Industrial Minerals and Rocks,* illustrates the major mineral fillers (both natural and manufactured) and their more important end uses.

Use of Mineral Fillers

Archaeologists have found that more than 5000 years ago, bitumen used for mortar and waterproofing contained mineral contaminants that acted as fillers. Since that time man has added mineral products to compounds to improve certain characteristics. Perhaps the most common objective of filler usage is reduction of cost of the end product. Certainly the inclusion of a filler that costs 2¢ per 0.45 kg (lb) will reduce the cost of a plastic product whose resin costs 50¢ per 0.45 kg (lb). But the decision to use a specific filler is not that simple. It may "soak up" binder, making a compound more viscous and perhaps preventing its flow into a mold as an injection molding compound, but similarly will give good anti-sag characteristics to putty, caulking compound, or undercoating.

Fillers may be incorporated to modify:

1) Cost
2) Physical characteristics
3) Flow characteristics
4) Fire resistance
5) Density
6) Heat conductivity
7) Color or brightness and opacity
8) Hardness, brittleness, impact strength
9) Deformity, viscosity, softening point
10) Processing problems
11) Electrical conductivity
12) Surface texture
13) Thermal expansion characteristics
14) Abrasion resistance

* Space prohibits a complete listing of people who have assisted in compiling this chapter. Those who made major contributions include: F.C. Appleyard, R.E. Blair, Garth Coombs, Wharton Jackson, Mitchell Koppelman, P.E. Kummer, W.B. Midgette, R.W. Sparks, Thomas Thompson, K.F. Weitz.

† President, Franklin Limestone Co., Nashville, TN.

TABLE 1—The More Important Mineral Fillers

	Major Filler Uses	Distinguishing Properties	Remarks
Asbestos* (Chrysotile)	Floor tile, plastics, "black line" products, asb-cement composition, building and insulating compounds.	The most fibrous mineral filler. Uses based on reinforcing and viscosity control effects.	Consumption declining, possibly related to carcinogenic concerns. Filler uses take mostly "fines" and "shorts" (7 group) but some 8, 6, and 5 group uses are akin to fillers. 1979 est. 93,000 tons, $310 per ton.
Barite	Rubber, paint, and urethanes.	The highest density mineral filler; also inert.	Ground barite used for filler purpose. Short supply. 1979 est. 101,000 tons, $174 per ton.
Bentonite	Pesticides, pellets, asphalt, detergents.	Not important as filler, except as insecticide dust diluent.	Wyoming grades growing as binder for iron ore pellets. 1979 est. 100,000 tons, $24 per ton.
Diatomite	Paints, paper, plastics, pesticides, cementitious mixtures and most filler uses.	An inert or reactive opaline silica depending on use, unique microscopic structure.	Natural or calcined, milled or processed grades are used as filler. Much difference in grades. 1979 est. 150,000 tons, $118 per ton.
Feldspar and nepheline syenite	Paints, plastics.	Translucency in plastics. Tint retention in paint. Chemical stability.	Major use is glass and ceramics. 1979 est. 30,000 tons, $45 per ton.
Fuller's earth	Pesticides principally, but some in greases and paper.	Important as pesticide dust diluent; absorptive capacity relatively good; price low.	Primary uses involve its absorbent character. 1979 est. 410,000 tons, $52 per ton.
Gypsum	Paints, paper, cotton goods, pesticides.	Low cost. A minor filler material used raw or calcined.	60% as calcined, 40% as uncalcined. 1979 est. 20,000 tons, $31 per ton.
Kaolin	Paper, paint, adhesives, oil cloth, plastics, rubber, ink, pesticides.	Good white color, low cost low absorption, abundant.	Paper industry consumes 70% of total production. 1979 est. 3,900,000 tons, $66 per ton.
Limestone†	Asphalt, fertilizers, insecticides, paints, rubber, plastics, and many others.	Abundant and cheap, particles soft, soluble in acids; has largest tonnage use of all fillers.	Large tonnages of low grade limestone and dolomite used in bitumens, fertilizers, and pesticides; quality grades employed in paints, rubber, and other filler uses. 1979 est. 5 million tons, $25 per ton.
Mica (muscovite)‡	Dry ground-roofing, joint cement and paint. Wet ground-paint, rubber and wallpaper.	Micaceous structure unique, valuable for electrical properties, useful in specialties.	Substantial volume is by-product of lithium and feldspar production. 1979 est. 122,000 tons, $120 per ton.
Perlite	Fines used in paint, loose fill, drilling muds, plastics, and misc.	Lightweight vesicular structure. Filler uses as yet not important.	1979 est. 12,000 tons, $150 per ton.
Portland cement	Asphaltic concrete and pavements and other compositions.	Cheap and abundant; may have some bonding action.	Filler uses of portland cement greatly secondary to its uses in construction. 1979 est. 100,000 tons, $45 per ton.
Pumicite	Stuccos, plasters and paint.	Natural vesicular structure, harsh. Filler uses minor.	Main uses as abrasive, concrete admixture, loose fill, and lightweight aggregate. 1979 est. 25,000 tons, $25 per ton.
Pyrophyllite	Insecticides, asphalt roofing, paint, rubber, battery boxes.	Like talc, but more refractory.	Volume and usage usually reported with talc. 1979 est. 10,000 tons, $23 per ton.
Rock dusts	Asphaltic compounds and cheap filler uses.	Low cost; chemical stability, strength.	Usually low cost ground carbonates, but dust collector fines also used. 1979 est. 500,000 tons, $8 per ton.
Silicas, tripoli	Paints, bitumens, foundry facings, etc.	Hardness, inertness, strength, low cost; low soluble salts.	Silica used in various related filler fields. Differentiation between amorphous and crystalline is not clear. 1979 est. 150,000 tons, $24 per ton.

TABLE 1—Continued

	Major Filler Uses	Distinguishing Properties	Remarks
Slate	Roofings, sealing compounds, asphalt linoleum, paints, hard rubber.	Cheap, abundant, inert and compatible with bitumens.	Statistics included in Rock Dusts.
Talc	Paints, insecticides, roofing, rubber, asphalt, paper, cosmetics, textiles.	Soft platey structure, good adhesion.	Talc has important ceramic, foundry and miscellaneous uses. 1979 est. 500,000 tons, $65 per ton.
Vermiculite	Fertilizers, pesticides. Potting soils	Filler grades are fines from expanded aggregate; filler uses not large.	1979 est. 84,000 tons, $187 per ton.

*Minor tonnage of anthophyllite and amphibole asbestos are used as fillers.
†Includes dolomite and dolomitic limestone; chemically precipitated calcium carbonates are also of considerable importance as fillers.
‡Phlogopite, biotite, and indeterminate micas are also used as fillers.
Note: Accurate statistical information for filler use is not readily available. Estimates were used in all categories. Prices of individual commodities have a broad range, depending on many variables; amounts shown are averages. Metric equivalent: 1 st × 0.907 1847 = t.

Seldom is the desired modification achieved without some sacrifice of other qualities. Some plastics may become more brittle when cost-reducing fillers are added. Some paints may tend to settle out in the can when anti-chalking fillers are added. Each using industry has found an acceptable solution to problems similar to these usually through optimizing the degree of modification that is acceptable. Frequently, combinations of fillers are utilized to achieve a number of desirable modifications. Hence, a quality paint may contain kaolin, calcium carbonate, talc, and perhaps barite, each included to assist in achieving a specific favorable effect.

The major natural fillers are asbestos, clays, limestone, talcs, diatomite, and siliceous minerals. The chemically modified or synthetic products constitute a second important group. Others less commonly used include gypsum, vermiculite, perlite, wollastonite, feldspar, tripoli, nepheline syenite, coal, and pulverized waste products such as slag and fly ash. While most choices will be based on price and performance coupled with availability, a surprising number of applications continue as a result of traditions or habit. The natural mineral filler industry experiences competition from processed minerals including portland cement, lime, and other man-made fillers, as well as organic materials including wood flour, sisal fiber, cork, nut hull meal and bagasse.

While some of the filler products are mined for their filler applications only, most have other uses also; and, for some, filling is a minor utilization. Hence, more talc and feldspar are produced for the ceramic industry than for filling. Substantially more limestone is used for construction aggregate and calcination than as fillers.

One filler may be substituted for another in some uses, but as technology advances this becomes more difficult. Replacement frequently requires total reformulation, a task the consumer is not likely to begin unless cost or performance improvements are apparent.

Inherent Physical Characteristics

One classification of fillers is performance in use, which, in turn, is a result of both the inherent characteristics of the mineral itself and the characteristics imparted by processing. Natural characteristics include such physical properties as hardness, natural particle size and shape, color, refractive index, chemical composition, chemical inertness, or reactivity. Physical characteristics of the deposit—including purity, ease of separation, and geographic location—are also a factor of economics and are covered in that portion of this chapter. It should be noted that some trace elements may totally disqualify an otherwise acceptable filler mineral deposit. Hence, a calcium carbonate deposit containing a substantial amount of magnesium carbonate may not produce an acceptable rubber filler due to the chelating effect of the magnesium ions. Other trace minerals can affect viscosity and aging.

Hardness affects the utilization of a filler as well as its production cost. Many silica minerals or silicates may score rolls, spreader blades, or knives, forcing the choice of a talc or calcium carbonate alternative. This adverse effect of hardness may be mitigated by using finer particle sizes. Conversely, these same hard minerals add abrasion resistance to rubber, scrubability to paint, and other desirable results.

Particle size, as it decreases, tends to fill smaller voids, give smoother surfaces, and increase the viscosity of systems. Particle size effects may result from the average size, which affects viscosity and binder demand; the size distribution, which affects packing and resultant density and film integrity; the top size, which may affect gloss and become the origin of fracture failures. Small particle sizes occur naturally in some minerals, primarily the clays and diatomite. Most other minerals must be ground to provide low particle-size effects while clays and diatomite are usually sized to separate their fine particles into desirable groupings, though milling may be required to free some particles.

Particle shapes strongly affect utilization. The acicular asbestos particles are regularly called on for reinforcing while plates of mica or kaolin provide a tough leaflike, surface film.

Where color is important in any end product, the color of the filler is similarly so. Paints and paper are obvious examples. Less so are plastics; white or colored rubber products; and even dry wall joint cement, which, if too dark, will show through a one-coat paint. Most fillers become more acceptable as they become whiter. In addition, the presence of tint may preclude utilization since correcting a distinctive blue, green, or yellow tint may result in an unacceptably gray end product. Color is usually measured as reflectance from a prepared flat cake of the ground filler. While natural color is basic, it may be modified by various processing techniques as we will see later.

Index of refraction is important in some applications. If the filler's index of refraction is very nearly that of the substance it is filling, the filler will show very little, giving an appearance of translucency. Vinyl floor tile that one apparently "looks into" achieves the effect through the use of talc as a filler. Nepheline syenite exhibits similar characteristics in many resins.

Chemical composition affects the utilization of many fillers. Calcium carbonate will tend to react chemically with acid paper manufacturing processes while kaolin will not. Battery boxes for lead-acid batteries must be compounded with nonreactive fillers. Low water solubility is a necessity for almost all filler applications.

Scores of other physical characteristics have a bearing on intended uses and, indeed, will make possible markets for fillers that would not otherwise exist. Table 2 (Cummins, 1960) lists some of the important major mineral fillers.

Processed Physical Characteristics

The most important filler physical characteristics result from particle size distribution and particle shapes. Whether the sizing be simply separation of naturally occurring particles or the fracturing and classification of discrete mineral particles, size itself will govern utilization and price more than any other single factor.

Size reduction may involve grinding by such mills as roller, pebble, fluid energy, vibrating, burr, impact, pin, or other grinding devices. It may involve exfoliation, as with vermiculite, or delamination, as with kaolin and mica.

In some applications, a maximum of fines is desirable, but for others, size reduction may be performed in a manner that minimizes the production of fine particles. As a general rule, the more fine particles the higher the binder demand. In many cases, this may be desirable, even necessary, to incorporate the required amount of binder per unit weight of finished product. In other instances, the objective may be to minimize binder requirements in which case fine particles may be removed from the filler.

Classification, even to low micron sizes, may be done in a liquid or air medium. Screen sizes seldom are used for production below 100 mesh for typical filler applications. As previously mentioned, sharpness of top size cut may be important in controlling wear or streaks in finish or gloss in a finish.

Color may be process-modified through grinding—the finer a filler is ground the whiter it will appear; or classification—some objectionable coloring materials tend to concentrate in various size ranges; or physical beneficiation—flotation, electrostatic, electromagnetic, or optical separation may remove objectionable color material.

Many filler industries continue to utilize hand sorting and most use some form of pit or mine selection of color control as well as other properties. Still other products use chemical processes for color improvement. The two most popular are calcination, as is used for kaolin, talc, and diatomite, and chemical leaching, as is used in some processing of barite, talc, and kaolin.

In many cases, impurities, other than color, must be removed from a filler before it is ac-

TABLE 2—Some Properties of the Major Mineral Fillers

	Theoretical Chemical Composition	Specific Gravity	Bulk Density Lb per Cu Ft*	Hardness, Mohs' Scale	Refractive Index	Reaction, pH	Oil Absorption, Cc per 100 G	Particle Characteristics
Asbestos (chrysotile)	$3MgO \cdot 2SiO_2 \cdot 2H_2O$	2.5-2.6	10-40	2.5-4.0	1.51-1.55	8.5-10.3	40-90	Fibers fine, easily separable. Fibrils hexagonally close packed 100-300Å.
Barite	$BaSO_4$	4.3-4.6	80-150	2.5-3.5	1.64	7	6-10	Generally equi-dimensional.
Bentonite	$(Mg,Ca)O \cdot Al_2O_3$ $5SiO_2 \cdot xH_2O$	2.3-2.8	50-60	1.5+	1.55-1.56	6.2-9.0	20-30	Porous microaggregates, irregular shapes; ultimate plate structure.
Diatomite	$SiO_2 \cdot xH_2O$	2.0-2.35	6-20	4.5-6.0	1.42-1.49	6-8.5	100-300	Unique diatom structure; micro and ultramicro porosity.
Fuller's earth	$(Mg,Ca)O \cdot Al_2O_3$ $5SiO_2 \cdot xH_2O$	2.2-2.6	27-38	4	1.50	7.5-8.2	30	Apparently equi-dimensional. Electronmicroscopically fibrous, lath-like.
Gypsum	$CaSO_4 \cdot 2H_2O$	2.3	25-40	1.5-2.0	1.52	6.5-7	17-25	Irregular, roughly equi-dimensional.
Kaolin	$Al_2O_3 \cdot 2SiO_2 \cdot 2H_2O$	2.6	20-40	2.0-2.5	1.56-1.58	4.5-7	25-50	Thin, flat hexagonal plates, 0.05-2μ size and stacks of same.
Limestone	$CaCO_3$	2.7	40-100	3	1.63-1.66	7.8-8.5	6-30	Variable size particles; ultimate rhombs.
Mica (muscovite)	$H_2KAl_3(SiO_4)_3$	2.7-3.0	12-20	2.0-3.0	1.59±	7.4-9.4	25-50	Platelike particles.
Nepheline syenite	$K_2O \cdot 7d\text{-}_2O \cdot 4.5Al_2O_3 \cdot$ $20SiO_2$	2.61	50-80	5.5-6.0	1.53	9.9	21-29	Nodular and irregular
Perlite	Like rhyolite	2.5-2.6	4-20	5.0	1.48-1.49	9	50-275	Expanded "glass" bubbles and fragments.
Portland cement	Essentially Ca silicates and aluminates	2.9-3.15	90-100	5.6	17.2±	11.0-12.6	20	Variable, smooth, rounded, angular and flake particles.
Pumicite	A silicate, like rhyolite	2.2-2.63	40-50	5-6	1.49-1.50	7-9	30-40	Vesicular.
Pyrophyllite	$Al_2O_3 \cdot 4SiO_2 \cdot H_2O$	2.8-2.9	25-30	1-2	1.57-1.59	6-8	40-70	Minute foliated plates or scales and extralong particles.
Rock dusts	Variable	2.6-3.3	50-100	4-6.5	Variable	Usually above 7	20-40	Variable.
Silicas, crystalline and microcrystalline	SiO_2	2.60-2.65	50-80	6.5-7.0	1.53-1.54	6-7	20-50	Variable sized, angular and equidimensional particles, or minute particles to porous masses.
Slate	Mixture of mineral silicates	2.7-2.8	40-80	4-6	—	6.8	20-25	Flat or wedge-shaped, or spherical grains.
Talc	$H_2Mg_3(SiO_3)_4$	2.6-3.0	26-60	1-1.5	1.57-1.59	8.1-9.0	20-50	Lamellar, foliated, or microfibrous.
Vermiculite	$(Mg, Fe)_3(Si, Al)_4$ $O_{10}(OH)_2 \cdot 4H_2O$	2.2-2.7	6-10; fines-20	1.5	1.56	Pract. neutral	—	Platelets or lamellar structure.

Modified from Cummins, 1960.
*Metric equivalent: 1 lb per cu ft × 16.018 46 = kg/m^3.

ceptable in a given application. Hard silica may be separated from the desired soft talc.

Where desirable, qualities can be added by coatings or by chemical treatment. Some coatings reduce the natural tendency for finely ground particles to adhere together or agglomerate. Others improve processing such as extrusion or ease of wettability in water, oil, solvent, and other systems. Still others result in added strength characteristics (reinforcing fillers) or improved electrical characteristics in the end product.

Test Methods

Test methods to evaluate the aforementioned characteristics are as different as the minerals themselves and tend to be specialized for each using industry.

Screen mesh sizes, commonly to 325 mesh and occasionally to 5μ, are used. Particle size curves are determined through air permeability, gaseous absorption, micrograph counts, settling in water, and specialized devices, such as the Coulter Counter, the Sedigraph, and the Microtrac. Color measurements are accomplished by measuring the light reflected from a carefully prepared powder cake surface using one or more colors as the light source and, in some cases, reducing the data to a continuous curve indicating the light reflected through most of the visible spectrum.

Moisture present in fillers may have a great effect on the end product; hence, this is a commonly measured and reported item.

Other properties sometimes measured and reported include bulk volume resistivity for electrical applications; pH of the filler in an aqueous or other solution; bulk density in varying stages of compaction; effective volume displacement in a liquid. Abrasiveness is measured for the paper industry. Oil absorption is a crude test for binder demand and is generally an indicator of particle size, particle size distribution, and surface area.

The requirements for accurately controlled physical characteristics and, of necessity, accurate measurements continue to increase in the filler industry. To be assured of the acceptability of the end product, the processor must know that his specified filler will perform in a predetermined manner. Economics becomes a substantial factor since a filler valued at $50 can render a batch of plastic valued at $5000 essentially useless. The requirement for improved performance, however, tends to bring about more exact processing and testing with resultant higher prices.

Since each major consuming industry has problems unique to that industry, many of the previously described physical properties and tests will be elaborated on as they apply to that industry. Additionally, where alternate fillers are commonly used, these possible substitutions will be pointed out.

Applications

Bituminous Compositions

It is estimated that 816 kt (900,000 st) of fillers were used in asphalt shingles and roll roofing alone in 1979. The f.o.b. plant cost ranged from $7 to $15 per ton. The volume and price figures dictate that the sources be relatively close to the consumer and that operating costs be low. Other major consuming industries include filling road paving asphalt, flooring material, joint fillers and expansion joints, waterproof coatings, and battery boxes. Loadings up to 65% are not uncommon in some of these applications.

Fillers for this industry have basic requirements including inertness in bitumastic compositions, physical resistance to process degradation, water insolubility, absence of hygroscopic tendencies, and opacity to active sunlight. Color is not a usual factor.

Many sources then can supply adequate fillers. These include, in addition to the common ground limestone, almost any stone that can be economically pulverized. Almost all have been used at one time or another. Dust collector discharges have become increasingly popular sources for these finely divided products.

Fillers are added to bituminous compositions to elevate the softening point. Asbestos is exceptionally effective (and generally more expensive) than common pulverized minerals. Weathering durability is improved through crack resistance imparted by asbestos and sunlight resistance derived from opaque ground stone fillers. Fire resistance is achieved by diluting the asphalts with noncombustible fillers, using fillers that puff up when heated, using fillers that reduce flow (melting) when hot and tend to reinforce the ash if charred, and fillers that assist in heat dissipation. Asbestos is a major contributor of all these properties.

Fillers modify the physical properties of bituminous compositions by increasing cold flow resistance, slide and indentation resistance, and viscosity. By using carefully controlled combinations of limestone and short asbestos fiber, a bitumastic compound can be made to flow easily for brush application, yet not sag or drip after application.

Water absorption is detrimental to bituminous products; therefore, fillers should be chosen which minimize this effect. Impact resistance and physical strengths are improved by the addition of fillers. Pliability (bending strength) may be increased to a point with fillers; but other strength characteristics may decrease with excessive filler bodies.

Roofings and sidings, including cement compounds and coatings, use blown asphalt with heavy (20 to 60%) loadings of filler. Most specifications include only screen gradation (with sufficient points to control void volume) and moisture content. A typical screen analysis is:

% Retained on US Standard 20 mesh— 0
% Retained on US Standard 60 mesh— 3 max
% Retained on US Standard 100 mesh—25 max
% Passing a US Standard 200 mesh—40–90

In paving compounds only the −200 mesh fraction is considered as filler, and some of this may come from the fine aggregate. The filler may affect the void volume and may modify the binder or both. The proper filler will increase deflection resistance and stability. Common fillers are limestone, portland cement, and silica, with the choice usually made on price. The primary specification is based on size.

Pipeline protection is obtained through coating pipelines with bitumastic compounds. Fillers are used to eliminate cold flow, provide impact resistance, resist water absorption, and permit fast penetration of the compound into the wrapping mat where one is used. Most of these fillers must pass a 200 mesh sieve.

Special coatings and cements include large uses such as damp-proofings, insulating mastics, sound deadeners, and automotive undercoatings. Asbestos is commonly required for low-temperature strength and anti-slump qualities, but the quantities of high-bulking fillers used are relatively small.

Plastics

Fillers, as applied to the plastic industry, should be referred to as "functional inorganics" since these materials contribute far more to the finished product than merely filling up space to gain cost savings in the finished product.

Plastics are usually classified as thermoplastics—materials which in their final molded shape can be repeatedly softened by the application of heat and remolded into the original or another shape—and thermosetting plastics—materials which when molded are permanently shaped and cannot be softened by heat for remolding.

Plastic resins are generally formulated with filler, stabilizer, colorants, and additives to achieve required physical properties. Resins may be dry or liquid. The dry system components may be dry mixed and fed to an extruder producing an end product. Much more commonly, both dry and wet systems are mixed, milled at a doughlike consistency with the addition of some heat (fusion), cooled, and ground to a granule or powder. This material is then extruded, compressed, or injected to yield the desired final shape. Some extremely liquid systems may be cast, but even these usually require heat for rapid curing.

Performance tests for filler materials are unique to the plastics industry. They include:

1) Dispersability of a filler in a fused composition.
2) Abrasiveness of filler particles as related to wear on metal extrusion dies.
3) Flow properties of a filler as related to the ease of moving it from storage hoppers to points of use.
4) Flow characteristics of filler-loaded plastics.

There are many filler materials which have wide acceptance and are used in large quantities in the plastic industry. These are outlined with some of their most important applications in the following paragraphs.

CALCIUM CARBONATE—Ground calcium carbonates are the most common fillers and have the greatest usage in this application. They may be wet or dry-ground products produced in a 0.1 to a 20μ size category. These are particularly attractive to the plastic industry in that they are low in cost, low in oil absorption, easily and quickly dispersed in conventional mixing equipment, high in brightness range, soft in hardness value, and are low in abrasion rating. The ultrafine variety may be used to impart mar resistance, produce a high gloss on finished surfaces, and help to reduce the common "whitening" on flexed materials. In addition, when properly formulated, they may give improved dielectric properties, impact resistance, cold flow properties, low and high temperature part behavior, weatherability, control part shrinkage, molded part crazing, and common plasticizer migration reduction.

There is an increased use of coated calcium

carbonate products. The coatings are chemically bonded to the filler and enhance various processing or end product performance criteria.

Some of the major plastic applications for calcium carbonate are:

Thermoset molded application—20 to 200 parts $CaCO_3$ per 100 parts resin.

Auto front sections, dashboards, auto parts, bodies.
Pleasure and commercial boats.
Building panels, plumbing equipment, tubs, shower enclosures.
Furniture parts, electrical appliances, pipe.

Thermoplastic applications—10 to 50 parts $CaCO_3$ per 100 parts resin.

House sidings, gutters, moldings, wall covering, upholstery, floorings, carpets, structural forms, electrical wire coatings, plastic sheetings, auto parts, lumber replacements, shoe soles and heels, and housewares.

Other applications

Adhesives, sealants, caulking compounds, latexes, phonograph records, etc.

ASBESTOS—Asbestos is an ideal reinforcement-type filler additive to achieve a final product which has high mechanical strength, excellent heat resistance, reduced mold shrinkage, improved molding characteristics, good surface appearance, and excellent electrical properties. It also aids in lowering unit cost. It is used in both thermoplastic and thermosetting molding applications, generally in the shorter fiber lengths.

Some of the most common thermoplastic applications are in vinyl/asbestos floor tile, electrical application for moldings which require high temperature in use and good dielectric properties, application of parts which need friction properties, sealant compounds, mastic compounds, pipe, and high temperature panels and sheeting.

Some of the most common thermosetting applications for asbestos are in the auto industry: fan shrouds, heater duct work, heater air-conditioner unit housings, and auto firewall components. Other uses include fabrication of marine engine housings, tractor housings, lawn mower covers and shrouds, heat duct work for the building industry, high temperature pipe, tubing, and special parts requiring high strength at high temperatures.

TALC—Talc is processed similarly to calcium carbonate. It does not have as wide usage in plastics as calcium carbonate. It does, however, have some unique physical properties which are quite valuable to the plastic industry. It is very white in color, has a high degree of softness and smoothness, providing excellent coverage, good luster, and high sheen to finished parts. Talc particles may be described as a series of very small plate-shaped particles which, when aligned, tend to make possible a finished part with improved physical properties. The nature of the particle size and shape will tend to produce "slip" or a form of lubricity. It is somewhat difficult to disperse in conventional plastic mixing equipment due to overall tendency for more bulking. The material is less sensitive to pH changes; therefore, it can be used in applications where minor acidity is a problem.

Talc, as a filler for plastic, has some special applications. Generally, it is not used as the sole filler in molding formulas but in combination with other filler materials. It facilitates removal from molds, helps achieve a better sheen or gloss on parts, and lowers the extrusion die wear through its lubricating effect. Talc also promotes better fire retardancy and aids in processing and final application of various plastic sealants and mastics. It is also useful as an aid in certain color applications, as a filler for parts in contact with acids, and as a viscosity modifier to accomplish some improvement of overall finished product physical properties.

BARITE—Barite, as a filler for plastic, has somewhat limited usage. It is supplied as a finely ground powder, white to shades of gray-black, with varying degrees of hardness, and is insoluble in water and normal acids. In this application, it can be considered as a chemically inert material. It is basically crystalline in structure. It is very heavy with a specific gravity of 4.5 when pure. It does impart some desirable properties for special applications where additional weight is required. It is used in the furniture industry, in combination with other mineral fillers, to achieve weight and density duplicating that of hardwoods. It has wide usage in molded parts that are subjected to acid conditions, and is used in brake lining materials.

CLAYS—Clays, especially the calcined products, are used quite extensively in the thermoplastic wire coating applications. They improve the insulation properties of the coating and exhibit good dielectric properties. The nature of the particle promotes fast extrusion with little die wear. Clays have very high oil and plasticizer absorption; therefore, except in special applications, they are generally not used as a single filler in formulations. They may be combined with calcium carbonate or other mineral filler products. They are used in thermoplastic applications where density is important and good opaque qualities are required. They are economical fillers for general use and will generally not seriously deteriorate physical properties. Their usage in sizable quantities in a formulation generally will require increased plasticizer to compensate for their high absorption rate. Clays, when combined with PVC, will have a greater stiffening effect than calcium carbonate; however, greater hardness will result.

Clay, combined with calcium carbonate, may be used in thermosetting polyester compounds to lower the exotherm, reduce the cost of resin, and modify the mechanical and surface properties. Clay will aid the thickening mechanism of sheet molding polyester compositions. Clays are used in other thermosetting applications to lower cost where final parts are to have good electrical properties and surface appearance.

Rubber

The art and science of rubber compounding are complex but are highly developed with much technical literature in which the function of fillers is exhaustively covered. Most researchers relate the action of reinforcing fillers to the surface active characteristics of the filler particles. Whether for natural rubber or many of the more recent synthetics, including thermoplastic urethanes, fillers are added to achieve varied results.

Most fillers must be fine particle size both for total dispersion in the compounding state and for avoiding large contact surfaces which may originate tensile failures. In white goods, such as rubber footwear and mechanical goods, color may be important; but, in black or dark colored products, filler color may be insignificant.

Major rubber products using mineral fillers include footwear and shoe products, wire and cable, and mechanical goods such as industrial, construction and houseware products. It should be noted that tires contain little mineral filler. Man-made carbon black and silicas are the principle fillers.

Kaolin, usually air-floated, is the most commonly used mineral filler. The fine particle-size products have a pronounced stiffening effect on the modulus of rubber goods. Additionally, kaolin adds hardness, abrasion resistance, and some tensile strength. Calcium carbonate, also used in major quantities, allows large loadings without loss of softness, elongation, or resilience.

Talc is commonly used as an anti-stick dusting compound throughout the rubber industry. It also lubricates rubber compounds for improved extrusion performance and provides reinforcing. Mica is also used occasionally for anti-stick or mold-release purposes.

It should be noted that rubber products are called on to perform in a broad range of physical characteristics. The chemistry involved is highly complex and affects, to an extent, the filler. In rubber a number of specialized additives are used, most of them man-made. Some of these might be considered fillers because of the quantities involved or the similarity to the natural mineral fillers. Those products are beyond the scope of this review.

Paper

More than 3.4 Mt (3.75 million st) of mineral fillers, at an average cost of approximately \$65 per ton, were used by the paper industry in 1977. Most are used as fillers or coatings though some talc is utilized for specialized purposes such as pitch control. A mineral filler is used in combination with bonding (sizing) agents, such as starch, rosin, or high molecular weight polymers to fill the voids between the fine intermeshed and felted cellulose fibers which comprise the paper sheet. Filled paper when calendared has the opacity, smoothness, ink receptivity, body, and feel that is essential for printing. The filler content of uncoated papers ranges from 4 to 30% by wt (usually 8 to 17%). Kaolin is by far the most common filler supplying more than 80% of the market. Calcium carbonate usage is increasing with an industry shift toward chemically neutral or alkaline size from the almost universal acidic rosin size. Other filler pigments include man-made products such as alumina hydrate and precipitated silica. The fillers are usually added at the beaters as an essential part of the mixture and are selected so as to give good working qualities on the paper machine (high speeds, low abrasion, and rapid drainage), as well as to provide the desired properties in the paper sheet.

Coated paper has the surface of the prefinished and filled sheet improved through the application of a mineral filled system to one or both sides of the sheet. This may be done on the paper-making machine at the time of manufacture or subsequently in an off-machine step. The coating particles are suspended in an adhesive mixture such as starch, casein, or synthetic latex plus other ingredients. Specially processed kaolin of good white color is the principal coating filler, but calcium carbonate is increasingly employed. TiO_2 is added for brightness and opacity, with talc, diatomite, alumina hydrate, synthetic silicates, delaminated kaolin, and ultrafine calcium carbonate being added to extend the TiO_2 and add other properties. The coating mixture is affected by the particle size, specific gravity, rheological effects, and cost of the pigment-filler combination used.

The art of paper making is much concerned with its filler uses. Among the properties required in varying degrees for paper filling and coating purposes are high brightness, a high refractive index, softness, freedom from gritty particles, chemical inertness, satisfactory ink receptivity, high retention in the paper web, and rheological properties that permit easy flow and application. With this wide range of desired properties, it is obvious no narrow set of specifications will apply. Common tests include color measurements, particle size curves, viscosity determinations, and sieve retention on very fine mesh screens. To meet the wide variety of physical requirements, producers have created as many as 40 distinctive specifications for the various kaolin products.

Protective and Decorative Coatings

Technological developments in these fields have required increasingly specialized and precisely processed extender pigments. A broad range of coatings are carefully formulated to meet many different applications. In addition to suitable vehicles made up from oils, resins, and solvents with prime pigments, extender pigments are a basic requirement in almost all types of coatings.

Fillers and extender pigments are used to balance flow characteristics; control penetration; reduce requirements for high cost prime pigments; provide flatness; improve sheen and color uniformity; aid in controlling chalking, pigment dispersion, and mildew; reduce film tack; control consistency; enhance the toughness, hardness, scrubbability, chemical resistance, or weatherability of the applied paint film; and to reduce costs. The role of the paint chemist is to derive the optimum balance of these various properties, and in so doing he finds the judicious use of mineral fillers most important.

The coatings industries have come to recognize these mineral products as white extender pigments. The minerals coming within this category vary in particle characteristics and physical properties, but all are white or near white, and have refractive indices substantially below 1.75. Calcium carbonates are the most widely used extender pigments. Others are kaolin, both natural and calcined; talc; pyrophyllite; diatomite; feldspar; mica; barite; silicas; and extensive quantities of the man-made fillers.

Extender pigments are employed in most types of coatings, including interior and exterior house paints of oil or water base; industrial paints, both primers and top coats; and traffic paint plus a host of minor applications.

The practice of utilizing two or more extenders is common, since most coatings require the benefits offered. The desirable effect of one is sometimes reinforced by the presence of another of the extender pigments. The loading is usually stated in terms of pigment volume concentration (PVC) which is the volume percentage of the coating occupied by pigment. This is not to be confused with the plastic PVC (polyvinyl chloride).

The durability of outside paints is related to the structure of the film and its porosity, since vapor transmission is necessary to provide blister resistance. Extender pigments also have a bearing on the chalking characteristics of the paint film. Semigloss and flat paints are formulated with extender pigments which not only aid in pigment suspension but also increase opacity and give the desired degree of flatness. Calcium carbonate is used in these paints; but diatomite and some types of talcs are the most effective flattening pigments. Emulsion paints contain a disperse phase of polymer binder and a continuous aqueous phase using mineral extenders and pigments that are as essential as in the oil base coatings.

The ASTM rating system evaluates the performance of exterior paints through analysis of a number of factors. These include general appearance, blistering, chalking, cracking, checking, dirt retention, fading, flaking, mildew resistance. Mineral extender pigments are a major factor in favorably meeting these criteria.

Other Filler Uses

The extensive use of fillers of all types in fertilizers is diminishing as the trend toward high performance granular and liquid fertilizers increases. Some fertilizers utilize an inert filler to adjust the desired ratio of fertilizer elements. Soil conservationists are encouraging the use of dolomite or other products which will also benefit the soil.

Kaolin and diatomite are used as anti-caking agents for ammonium nitrate by coating the freshly formed prills. An anionic surfactant coating on the grades of kaolin commonly used assists in the distribution of relatively minute amounts of surfactant over all surfaces.

Adhesives are sometimes compounded with mineral fillers where they reinforce, control the viscosity and penetration, and lower the cost of the end product. Since fillers tend to reduce the strength of the bond, they are found more frequently in those compounds which require a low bond strength. One example of heavy mineral loading is the mastic used to secure asphalt or vinyl floor coverings.

Textile sizings are made up of a binder, starch, glue, or synthetic polymer together with mineral powders referred to as weight-adding agents. Kaolin, talc, and barite are the most commonly used agents. Finishing agents are employed to improve web impregnation, appearance, dye receptivity; to provide temporary sizing; and to add weight.

The carpet industry uses filler in the latex or other binder used to secure the loop of tufted carpets and the secondary backing, when one is used. Urethane or other polymers used as foam backings or as a separate underlay are also filled with up to 600 parts of filler per 100 parts of polymer. Ground limestone is the predominant filler. The use of alumina trihydrate for reducing flammability is declining. Approximately 653 kt (720,000 st) of limestone were used in 1980 at an average price of $14 per ton.

Fillers are extensively utilized in many facets of the construction industry. Asbestos cement products and specialty concrete blocks utilize ground silica to add strength and accelerate the curing process. Dry wall joint cements are heavily filled, usually with calcium carbonate. Mica is extensively used as a reinforcing medium in these compounds, and attapulgite provides sag control. Caulking compounds, putties, and mastics likewise may contain up to 90% or more mineral filler. Though most fillers in this industry are relatively coarse ground, whiteness is important since these products are frequently left exposed without painting or must be completely covered in one coat.

Since our original definition of mineral fillers was at best indefinite, any cutoff point in this summary would be similarly so. A number of related uses should be mentioned for members of the mineral filler group.

Coal mine rock dust, usually limestone, is used extensively to dilute the naturally occurring coal dust underground to mitigate the explosive hazard. It is applied wet or dry from shaft stations right up to the working face. Standard requirements are that the total free and combined silica content be less than 5% and that the dust be ground finer than 83% passing a US Standard 200-mesh sieve.

Athletic field line marker, usually white calcium carbonate ground somewhat finer, is utilized because it presents no hazard to the athlete.

Welding rod coatings, crayons, chewing gum, lubricants, match heads, cleansers, polishes, wood and concrete filling compounds, soap, paper maché, oilcloth, window shades, oil well drill muds, cosmetics, pharmaceuticals, and animal feed nutrition also use one or more of the common mineral fillers in notable quantity. Compared to the major uses covered in detail these are relatively minor but may present additional marketing outlets. It must be noted that many of the mineral fillers discussed have additional applications, as insulators, soil conditioners, chemical raw material, ceramic constituents, abrasives, bonding agents, and numerous others that here are arbitrarily considered outside the field of fillers.

Filter Media

The removal of solid particles from a liquid medium has taxed man's ingenuity for centuries. Mineral filter mediums are extensively utilized, especially in those applications requiring the filtration of high volumes of liquids.

Sand is perhaps the most common. Most sand filters are horizontal beds of carefully graded grains beginning with coarse particles to form the bed and becoming progressively finer toward the top layer. Cost, durability, and availability are factors in the choice of sands. Specifications include freedom from organic matter, silt, minerals which fracture or deteriorate, and acid-soluble components. The grains must be rounded or angular and very durable. Sizing is a factor of the viscosity of the filtrate, the type contaminant, and the flow rate desired. Uniformity coefficient of the sand and absence of −50 mesh sizes are critical. The American Water Works Assn. has prepared standard specifications to cover a wide variety of water filtration applications. Size specifications vary for filtration sand for swimming pool, industrial, and sewage disposal sludge bed filters.

A broad range of more difficult filtration problems exist. To a major degree they have been overcome through the use of diatomite. Particles of diatomite occur in a variety of forms, affording a shape factor combined with the rigidity of the particles, that make an excellent raw material for the production of filter aids. The natural diatomite filter aids are practically pure amorphous silica and are essentially inert. However, as most diatomite filter aids are generally processed by flux calcination to improve flow rate characteristics, the amorphous silica is partially or largely converted to cristobalite which is essentially inert.

A good filter aid for industrial filtration must possess several qualifications: form a very porous cake, have a low surface area, have correct particle size distribution, have low retention, resist collapse under pressure, and be available in a number of grades. Performance of a good filter aid is measured in filtrate clarity, the filtration flow rate, and the filter cycle length.

A. B. Cummins (1960) categorized diatomite filtrations as shown in Table 3.

Greatest commercial pressure to diatomite as a filter aid comes from expanded perlite, which has invaded markets in sugar, alginates, and pharmaceutical filtration.

Testing methods are unique in the diatomite filter industry, and the basic device used is the bomb filter leaf, which is essentially a small pressure vessel, fitted with an internally adjustable filtering area and connected to an air source. The bomb filter will rapidly determine the best balance of variables to obtain required clarity, maximum throughput, and at minimum filter aid dosage. Other test methods that miniaturize full scale filtration include the leaf filter and the Walton filter. Additional evaluation methods include the wet density test, microscopic evaluation, screen tests, pore size distribution, turbidity, particle size, surface area, and absorption.

Some specialized applications utilize asbestos as a filter medium. In the petroleum industry, fuller's earth (attapulgite) is primarily used in the refining of petroleum products

TABLE 3—Classification of Diatomite Filtrations

FOOD PRODUCTS	Inorganic-general
Fruit products (apple, grape, citrus)	Organic-general
Sugars (cane, beet, plantation, etc.)	
Corn products (sugar, glucose, oil)	INDUSTRIAL
Oils and fats (vegetable, animal)	Varnishes and lacquers
Beverages (beer, wine, whisky)	Metalworking industries (coolants, cutting oils, etc.)
Packing house products	Electroplating
Miscellaneous (casein, dairy products)	Alginates
	Soaps and glycerine
METALLURGICAL	Nonedible oils and greases
Cyanidation	Cellulose liquors
Uranium processing	Nylon and rayon liquids
Recovery of metal values	Synthetic resins
Metal extraction (Be, W, others)	Catalytic reactions (hydrogenation)
PETROLEUM	WATERS
Dewaxing	Municipal
Still residues	Industrial
Petrochemical	Military (potable, engine condensate)
Contact filtration (clay removal)	Swimming pools
Used oils (crank case, transformers)	Super-clarified (electronics, beverages)
Oil well pressurizing	
DRUGS AND PHARMACEUTICALS	MISCELLANEOUS
Pharmaceuticals	Dry cleaning
Antibiotics	Laboratory filtrations
Enzymes, serums, vitamins, etc.	Jet fuels
Cosmetics	Yeast and waxes
CHEMICALS	WASTE DISPOSALS
Dyestuffs and intermediates	Paper mill effluents
Acids (sulfuric, phosphoric, organic)	Sewage
Titanium pigments	Radioactive (underground injections)
Sulfur	Laundry waste waters
Solvents	Industrial wastes, generally

(percolation and contact filtration), though activated bauxite and magnesium silicate have been used as substitutes in recent years. The action is considered to involve selective adsorption as well as filtration.

Absorbents

Several minerals are highly sorptive in their natural or modified form. These include attapulgite, bentonite, montmorillonite, diatomite, and other minor minerals. A large void volume appears to be the leading factor in their high absorption. Sorption is generally considered to include two separation mechanisms. Absorption is a process readily reversible by simple low temperature heating. Absorbents may be restored to their original absorptive character by such processing. Adsorption is a selective process and one that is not readily reversible by low temperature heating.

Attapulgite and montmorillonite, also categorized as fuller's earths, are the most widely used adsorbers. They are capable of absorbing twice their own weight in water. Both are processed by drying and in some cases by calcining, with the degree of preparation having a major effect on their sorptive properties. They are used to purify, by selective adsorption, oils, fats, waxes, resins, brewery products, and many other industrial items. Attapulgite has supplanted most other carriers for insecticides, herbicides, and soil fumigants, since it has the best overall qualities of high absorption, moderate density, and good flowability. Pet litter and industrial absorbents have become other major markets in recent years for these products, with montmorillonite predominating; an estimated 1.4 Mt (1,500,000 st) was shipped in 1980, with 75% used as absorbents.

Economic Summary

Each of the mineral filler products reviewed has established its position in the economy because of the characteristics it imparts to some end product. While a specific filler (or filter or absorber) will do one job best, economic forces generally determine utilization, with the user balancing a cost-to-performance equation. Factors involved include:

1) Geographic location, with resultant transportation costs.
2) Purity of physical state, with resultant processing costs.
3) Physical characteristics, with resultant handling and processing costs.

All of these must be balanced against value of performance. Throughout this chapter, the particular characteristics that favor the utilization of a given filler in a given application have been noted. It would be impossible to equate the various fillers economically except within the framework of a specific application.

Similarly, attempting to summarize specifications would accomplish little since there are perhaps 40 grades of kaolin, 25 grades of calcium carbonate, 8 grades of barite, 10 grades of mica, 20 grades of talc, and so on, each with characteristics usually optimized for one or two industries. Coated, bleached, or treated products multiply the possibilities. One must carefully evaluate a given market to determine those specifications most important for that market. Price ranges are as broad as specifications. The *Minerals Yearbook* summarizes price spans for most products annually. A comprehensive market analysis would be required to pinpoint specific industry prices. Extreme caution must be used in the economic evaluation of deposits based on published commodity statistics, as the values shown will include substantial processing costs and may include expensive additives and packaging, and in some cases, reflect transportation factors.

The trend in the filler industry has been to do more processing with the ultimate aim of making the product performance more desirable, usually in a specific application. This increased processing can only lead to increased prices. It should be noted that, as consumption rates increase, the economics of bulk handling becomes more evident. The trend toward bulk rail and truck shipments is anticipated to increase rapidly outside the kaolin industry where it is already entrenched. Major quantities of water-processed kaolin are shipped as water slurries, and the shipping of ground limestone slurries is increasing. Containerized shipments have become commonplace for the overseas transport of mineral fillers whose values will support an export market.

Bibliography and References

Those references listed in prior editions of *Industrial Minerals and Rocks* are most important in developing the history and forecasting trends in the mineral filler industry. Some of these references are classics and have not been superseded in their field. In addition to the references, various trade associations can be most helpful in providing related information. These are also listed.

General

Anon., 1980, *Mineral Facts and Problems,* Bulletin 671, US Bureau of Mines, 1060 pp.

Clark, L., Faith, W.L., and Keyes, B., 1965, *Industrial Chemicals,* John Wiley, New York, 852 pp.

Cummins, A.B., 1960, "Mineral Fillers," *Industrial Minerals and Rocks,* 3rd ed., J.L. Gillson, ed., AIME, New York, pp. 567–584.

Fisher, W.L., 1969, "The Nonmetallic Industrial Minerals," *Mining Congress Journal,* Feb., pp. 120–126.

Grim, R.E., 1962, *Applied Clay Mineralogy,* McGraw-Hill, New York, 422 pp.

Harrell, G.O., and Harrell, D.E., 1971, "Survey of Mineral Fillers in Selected Industries," School of Engineering, North Carolina State University at Raleigh, 106 pp.

Kollonitsch, V., and Kliff, E.J., 1969, "Functional Mineral Pigments," SME Preprint 69H81, AIME Annual Meeting, Washington, DC, Feb., 24 pp.

Patton, T.C., ed., 1973, "Extender (Filler) Pigments," *Pigment Handbook,* Vol. I, John Wiley/Interscience, New York, pp. 109–321.

Robertson, R.H.S., 1960, *Mineral Use Guide* (Robertson's Spider Webs), Cleaver-Hume Press, London, 44 leaves.

Robie, E.H., ed., 1964, *Economics of the Mineral Industries,* 2nd ed., AIME New York, 787 pp.

Skow, M.L., 1962, "Mica—A Materials Survey," Information Circular 8125, US Bureau of Mines, 241 pp.

Weitz, K.F., 1970, "Applications of Mineral Fillers in the Paint and Plastics Industries," *The Canadian Mining & Metallurgical Bulletin,* Vol. 63, No. 697, May, pp. 600–606.

Uses in Bituminous Compositions

Abraham, H., 1960, *Asphalts and Allied Substances,* Vol. 3, 6th ed., D. Van Nostrand, Princeton, NJ, 969 pp.

Cook, J.P., 1970, *Construction Sealants and Adhesives,* John Wiley, New York, 269 pp.

Greenfeld, S.H., 1954, "Laboratory Studies of Several Factors Affecting the Behavior of Some Minerals as Stabilizers for Asphalt Roofing," Report 3616, National Bureau of Standards, 11 pp.

Greenfeld, S.H., 1956, "Effects of Mineral Additives on the Durability of Coating-Grade Roofing Asphalts," Building Materials and Structures Report 147, National Bureau of Standards, Sep., 14 pp.

Damusis, A., ed., 1967, *Sealants,* Reinhold, New York, 382 pp.

Hoiberg, A.J., 1965, *Bituminous Materials: Asphalts, Tars, and Pitches,* 3 vols., John Wiley/Interscience, New York; "General Aspects," Vol. 1, 1964, 432 pp.; "Asphalts," Vol. 2, 1965, 638 pp.; "Coal, Tars, and Pitches," Vol. 3, 1966, 585 pp.

The following ASTM publications are also of interest:

"Symposium on Accelerated Durability Testing of Bituminous Minerals," 1949, ASTM Special Technical Publication No. 94.

"Symposium on Bituminous Waterproofing and Roofing Materials," 1959, ASTM Special Technical Publication 280.

Uses in Rubber

Anon., 1961, *Compounding Ingredients for Rubber,* 3rd ed., compiled by eds. of *Rubber World,* New York, 722 pp.

Grim, R.E., 1968, *Clay Mineralogy,* 2nd ed., McGraw-Hill, New York, 384 pp.

Lamar, R.S., Mulryan, H.T., and Warner, M.F., "Reinforcing EPR and EPT with Ultra-Fine Talc," *Rubber World,* Vol. 147, No. 5, Feb., pp. 60–64.

Pollack, L., 1980, *The Rubber Redbook,* 32nd ed., pub. by Palmerton Publishing Co., New York, 778 pp.

Uses in Plastics

Hosford, R.C., 1965, "A Guide to Fillers and Reinforcements for Thermosets," *Plastics Technology,* Feb., pp. 34–44.

Hunt, R.E., 1969, "How to Select Fillers for Thermoplastics," *Plastics Technology,* Nov., pp. 38–43.

Katz, H.S., and Milewski, J.V., eds., 1978, *Handbook of Fillers and Reinforcements for Plastics,* D. Van Nostrand-Reinhold, New York, 652 pp.

Miles, D.C., and Bristoh, J.H., 1965, *Polymer Technology,* Chemical Publishing Co., New York, 444 pp.

Mohr, G., and Oleesky, S.S., 1973, *Handbook of Technology and Engineering of Reinforced Plastics,* 2nd ed., Van Nostrand-Reinhold, New York, 608 pp.

Rauch, J., ed., 1978, *Kline Guide to the Plastics Industry,* Chas. H. Kline & Co., Inc., Fairfield, NJ, 181 pp.

Simonds, H.R., and Church, J.E., 1967, *The Encyclopedia of Basic Materials for Plastics,* Van Nostrand-Reinhold, New York, 500 pp.

Thomson, R.D., 1960, "Nonmetallic Mineral Fillers in Plastics," SME Preprint 60H105, AIME Annual Meeting, New York, Feb., 18 pp.

Wells, G.W., 1980, "Barytes Market-Filler-Extender and Chemical Uses," SME Preprint 80–29, AIME Annual Meeting, Las Vegas, Feb., 4 pp.

Plastics Compounding has published (1978–1982) a series of articles on various mineral filler applications in plastics (1129 E. 17th Ave., Denver, CO, 80218).

Uses in Paints

Anon., no date, *Paint Formulary Using Wet Ground Mica,* Wet Ground Mica Assn., Inc.

Bradbury, J.C., and Ehrlinger, H.P. III, 1980, "Tripoli," SME Preprint 80–389, SME-AIME Fall Meeting, Minneapolis, Oct., 4 pp.

Gaynes, N.I., 1967, *Formulation of Organic Coatings,* Van Nostrand, Princeton, NJ, 386 pp.

Hall, R.F., 1959, "Effect of Particle Size Distribution of Pigment and Pigment Volume Concentration on Flat Wall Paint Properties (Solvent

Type)," *Official Digest,* Federated Societies of Paint Technology, June, pp. 4–15.
Martens, C.R., ed., 1968, *Technology of Paints, Varnishes and Lacquers,* Van Nostrand-Reinhold, New York, 744 pp.
Parker, D.H., 1965, *Principles of Surface Coating Technology,* John Wiley/Interscience, New York, 817 pp.
Patton, T.C., 1964, *Paint Flow and Pigment Dispersion,* John Wiley/Interscience, New York, 479 pp.
Payne, H.F., 1961, *Organic Coating Technology,* John Wiley, New York; Vol. 1, 1954, 674 pp.; Vol. 2, 1961, 725 pp.
Rauch, J., ed., 1979, *Kline Guide to the Paint Industry,* Chas. H. Kline & Co., Inc., Fairfield, NJ, 89 pp.

Uses in Paper

Casey, J.P., 1960, *Pulp and Paper,* 3 vols., 2nd ed., John Wiley/Interscience, New York.
Huber, J., ed., 1980, *Kline Guide to the Paper Industry,* Chas. H. Kline & Co., Inc., Fairfield, NJ, 343 pp.

The following TAPPI (Technical Association of the Pulp and Paper Industry, Atlanta) publications are also of interest:

"Paper Coating Additives," 1963, Monograph Series No. 25.
"Paper Coating Pigments," 1966, Monograph Series No. 30.
"TAPPI Standards and Suggested Methods," 1960.

Other Uses

Cummins, A.B., 1973, "The Development of Diatomite Filter Aid Filtration," *Filtration & Separation,* Pt. 1, Jan./Feb., Pt. 2, Mar./Apr., Pt. 3, May/June.
Fahn, R., 1979, "Acid Activated Clays and Their Adsorption Properties," SME Preprint 79–328, SME-AIME Fall Meeting, Tucson, Oct., 13 pp.
Haas, C.Y., 1971, "Attapulgus Clays for Future Industrial Minerals Market," SME Preprint 71H54, AIME Centennial Annual Meeting, New York, Mar., 13 pp.
Haden, W.L., Jr., and Schwint, I.A., 1967, "Attapulgite, Its Properties and Applications," *Industrial & Engineering Chemistry,* Vol. 59, No. 9, Sep., pp. 59–69.
Key, W.W., 1965, "Mineral Fillers for the California Pesticide Industry," Information Circular 8260, US Bureau of Mines, 39 pp.
Lang, W.J., 1971, "Bentonite: The Demand and Market of the Future," SME Preprint 71H29, AIME Centennial Annual Meeting, New York, Mar., 9 pp.
Parker, D.C., 1979, "Porters Creek Clay, Geology, Processing, and Marketing of a Unique Mineral Assemblage," SME Preprint 79–339, SME-AIME Fall Meeting, Tucson, Oct., 6 pp.
Tooker, D.J., 1980, "How a 'Tee-Hee' Mineral Got the Last Laugh," SME Preprint 80–36, AIME Annual Meeting, Las Vegas, Feb., 10 pp. (Pet Litter).
Torok, A., and Thompson, T.D., 1972, "Activated Bleaching Clay for the Future," *SME/AIME Trans.,* Vol. 252, pp. 15–17.

Industrial Minerals magazine has published many extensive articles on most of the minerals used as fillers, and is helpful for reference on these minerals and their applications.

The following professional associations are also sources of information on mineral fillers:

The Adhesive and Sealant Council, Inc.
1600 North Wilson Blvd.
Arlington, VA 22209

American Assn. of Textile Chemists & Colorists
P.O. Box 12215
Research Triangle Park, NC 27709

American Assn. for Textile Technology, Inc.
1040 Avenue of the Americas
New York, NY 10018

American Paper Institute, Inc.
260 Madison Ave.
New York, NY 10016

American Petroleum Institute
2101 L Street, NW
Washington, DC 20037

The Asphalt Institute
Asphalt Institute Building
College Park, MD 20740

Asphalt Roofing Manufacturers Assn.
1800 Massachusetts Ave.
Washington, DC 20036

Cultured Marble Institute
435 North Michigan Ave.
Chicago, IL 60611

Expansion Joint Manufacturers Assn.
707 Westchester Ave.
White Plains, NY 10604

Federation of Societies for Coatings Technology
1315 Walnut St.
Philadelphia, PA 19107

The Fertilizer Institute
1015 18th St., NW
Washington, DC 20036

National Assn. of Glue Manufacturers, Inc.
663 Fifth Ave.
New York, NY 10022

National Assn. of Printing Ink Manufacturers
550 Mamaroneck Ave.
Harrison, NY 10528

National Paint and Coatings Assn.
1500 Rhode Island Ave., NW
Washington, DC 20005

Plastic Pipe and Fittings Assn.
355 Lexington Ave.
New York, NY 10017

Rubber Manufacturers Assn.
1901 Pennsylvania Ave.
Washington, DC 20006

The Society of the Plastics Industry, Inc.
355 Lexington Ave.
New York, NY 10017

Technical Assn. of the Pulp and Paper Industries
1 Dunwoody Park
Atlanta, GA 30338

Wallcovering Manufacturers Assn.
66 Morris Ave.
Springfield, NJ 07081

Fluxes

FREDERICK V. LAWRENCE, JR.*

Broadly speaking, *fluxes* are substances which promote wetting and spreading or enhance the fluidity and manipulative properties of materials in joining, fusion, and smelting operations. The term most strictly applies to substances used to facilitate soldering and brazing; but in the discussion which follows, the definition will be broadened to include materials used to promote fusion in arc welding and which aid in the smelting of metals. In each of the mentioned applications, the role of flux and its composition differs considerably; consequently in the discussion which follows, fluxes for soldering, brazing, arc welding, and smelting applications will be considered separately.

Soldering and Brazing Fluxes

Soldering and brazing are metallurgical joining processes in which a joint is formed using a filler metal of composition dissimilar to that of the joined pieces (base metal). The filler metal, solder or braze, is always of lower melting point than the base metal. Consequently with these processes there is no melting of the base metal. The successful creation of the joint depends upon the wetting and spreading of the solder or braze around or throughout the joint. The obstacles to successful soldering and brazing are contaminants and tenacious oxide films on the metal surfaces. In soldering and brazing applications, fluxes act to remove and exclude oxides and other impurities from the joint.

There is no fundamental difference between soldering and brazing. The distinction between them is made on the basis of temperature. Joining operations carried out at temperatures above 500° C are termed brazing. The higher temperatures used in brazing enable higher melting point filler metals to be used which generally provide stronger joints. The higher temperatures of brazing processes also require fluxes which are substantially different from those used in soldering and which will remain effective at high temperatures. Consequently the distinction between soldering and brazing can also be made on the basis of the fluxes used.

Soldering Fluxes

Even thoroughly cleaned and degreased, metal surfaces still possess tenacious oxide films which inhibit the soldering of the joint. The reactions which form these films increase with increasing temperature. The rate of oxide formation and the tenacity of the film varies with each base metal. Aluminum, magnesium, stainless and high alloy steels, aluminum and silicon bronzes when exposed to air form hard, adherent oxide films. These metals require highly active and corrosive fluxes to remove and prevent the reformation of oxide films during soldering. Metals such as copper and silver, on the contrary, form less tenacious films at a lower rate, so that mild fluxes are sufficient.

Soldering fluxes must dissolve or remove oxides and promote rapid wetting and spreading of the solder. The flux removes the oxide film by loosening and floating it off into the main body of the flux. The oxide removal is often aided by the alloying of the solder with the base metal, undermining small areas of the oxide and detaching them (Anon., 1959). The flux coagulates and suspends the oxide particles which have been loosened. The flux forms a protective cover preventing the reformation of the oxides, permitting the spreading of the solder and the formation of a metallic bond between the solder and the base metal.

Constituents of Soldering Fluxes: Soldering fluxes are commonly grouped into three categories: highly corrosive, intermediate, and noncorrosive. One normally would choose to use the least corrosive flux which will give satisfactory performance.

* Associate Professor, Depts. of Metallurgy and Mining Engineering and Civil Engineering, University of Illinois, Urbana-Champaign, IL.

Corrosive Fluxes—The most active and therefore the most corrosive fluxes consist of inorganic acids and salts. They may be applied as pastes or as dry salts. The corrosive fluxes are stable at the higher soldering temperatures but also remain active after soldering and consequently must be removed. Table 1 gives the common constituents of corrosive soldering fluxes.

TABLE 1—Soldering Flux Constituents

(a) Zinc chloride
(b) Ammonium chloride
(c) Stannous chloride
(d) Sodium or potassium chloride
(e) Lithium chloride
(f) Aluminum chloride
(g) Sodium or potassium fluoride
(h) Boron trifluoride
(i) Hydrochloric acid
(j) Hydrofluoric acid
(k) Orthophosphoric acid
(l) Fluoboric acid

Corrosive Flux Ingredients

Zinc Chloride. The main ingredient in the majority of corrosive fluxes is zinc chloride. It can be prepared by adding an excess of zinc metal to concentrated hydrochloric acid or can be purchased as fused zinc chloride which is readily available and more convenient to use. Zinc chloride has a melting temperature well above the solidus temperature of most commercial tin-lead solders. For this reason, if zinc chloride is used alone, unmelted salt particles may be entrapped in the joint. These inclusions will corrode the joint and make it weak. It is good practice, therefore, to mix other organic chlorides with zinc chloride to lower the melting temperature of the flux.
Ammonium Chloride. A water solution of ammonium chloride may be used as a flux. When the solvent evaporates, the ammonium chloride sublimes as a white fume without melting. It is less effective as a flux than zinc chloride because the protective action of a molten salt is not present and the base metal may reoxidize before it reaches soldering temperature. A combination of one part of ammonium chloride to three parts of zinc chloride forms a eutectic flux mixture which melts at 177°C (350°F). This mixture takes advantage of the excellent oxide reducing properties of the ammonium chloride and the excellent protective action of the zinc chloride to form a flux which is considerably more effective than either constituent when used alone. A 1:9 ratio of the salts (1 part ammonium chloride to 9 parts zinc chloride) is commonly used without fear of flux inclusions in the soldered joint.
Stannous Chloride. Stannous chloride is formed by dissolving metallic tin in hydrochloric acid. It is commercially available in the anhydrous and hydrated forms. Stannous chloride is a highly effective flux when used alone in powder, paste or molten form. It is also effective when mixed with zinc and ammonium chlorides.
Sodium or Potassium Chloride. Sodium chloride is not effective as a flux when used alone. It is effective in diluting zinc chloride and lowers its melting point. A low melting flux can be obtained by mixing 9 parts of

TABLE 1—Continued

zinc chloride with 2 parts of sodium chloride. A ternary eutectic mixture, melting at 203°C (397°F), is obtained by mixing 75 parts zinc chloride, 11 parts sodium chloride, and 14 parts potassium chloride.
Other Chlorides and Fluorides. Lithium and aluminum chlorides and fluorides are seldom used alone but they are used effectively as fluxes when mixed with other compounds.
Hydrochloric Acid. Hydrochloric acid has limited use when used alone as a flux. When hydrochloric acid is applied to galvanized iron, the zinc coating is dissolved to form zinc chloride which acts as the flux. Hydrochloric acid is used to activate the zinc chloride type fluxes. Mixtures of inorganic salts and hydrochloric acid are the basis of stainless steel fluxes.
Hydrofluoric Acid. Hydrofluoric acid is extremely corrosive. It is added to zinc chloride base fluxes for the purpose of dissolving silicon inclusions on the surface of cast iron.
Orthophosphoric Acid. Orthophosphoric acid is an effective flux for steel, copper, and brass. It leaves a glassy residue which serves as a protective coating. A diluted solution is especially effective on high tensile manganese bronze.

Source: Courtesy of American Welding Society (Anon., 1959).

Intermediate Fluxes—The intermediate fluxes are less active than the corrosive fluxes and generally consist of mild organic acids and bases. These materials are most active at the soldering temperatures but tend to decompose under prolonged exposure to high temperatures. Their tendency to volatilize limits their usefulness with flame heating. The residue of these fluxes is relatively inert and can be easily removed. A list of intermediate flux compounds is given in Table 2.

Noncorrosive Fluxes—The least active fluxes are composed of a solution of rosin, which consists largely of abietic acid. Rosin solders become active at soldering temperatures up to 316°C (600°F) after which they begin to decompose. The residue is noncorrosive. Rosin by itself is not a highly active flux. Many proprietary rosin fluxes contain additives to enhance the fluxing action while retaining the other advantages. *Stabilized rosin fluxes* contain additives which allow the anhydride structure of the rosin to become fully effective. *Activated rosin fluxes* depend upon the additives for increased activity. Additives such as hydrazine hydrohalide, glutamic and biguanide hydrochlorides, benzoic acid, succinic acid, cetyl pyridinium bromide, and aniline compounds may be used in amounts of 0.2 to 5%.

Typical Soldering Flux Composition: Commercially available soldering fluxes are proprietary formulations. Typical fluxes similar to those obtainable are given in Tables 3, 4, and 5.

Brazing Fluxes

As was the case with soldering fluxes, *brazing fluxes* are materials which promote the brazing

TABLE 2—Typical Intermediate and Noncorrosive Flux Constituents

Organic Acids
- Lactic acid
- Citric acid
- Oleic acid
- Stearic acid
- Glutamic acid
- Benzoic acid
- Oxalic acid
- Phthalic acid
- Abietic acid

Organic Hydrohalides
- Glutamic acid hydrochloride
- Aniline hydrochloride or phosphate
- Hydrazine hydrobromide or hydrochloride
- Cetyl trimethyl ammonium bromide
- Ethyl dimethyl detyl ammonium bromide
- Cetyl pyridinium bromide

Amines and Others
- Urea
- Diethylene di- or triamine
- Glycerol

Source: Courtesy of American Welding Society (Anon., 1959).

TABLE 3—Corrosive General Purpose Fluxes

These fluxes are effective on low-carbon steel, copper, brass, and bronze. Applications are in the production of auto radiators, air conditioning and refrigerating equipment, body soldering, and sheet metal assembly.

1. Zinc chloride	40 oz*
Ammonium chloride	4 oz
Water to make	1 gal
2. Zinc chloride	36 oz
Sodium chloride	10 oz
Ammonium chloride	1/2 oz
Hydrochloric acid	1 oz
Water to make	1 gal
3. Zinc chloride	21 oz
Sodium chloride	6 oz
(Dry flux for molten solder cover in dip soldering)	
4. Zinc chloride	25 oz
Ammonium chloride	3-1/2 oz
Petroleum jelly	65 oz
Water	6-1/2 oz
(Paste flux)	

These fluxes can be used for soldering stainless steel, alloy steel, nickel alloys, silicon and aluminum bronzes, zinc coated sheet, cast iron, and aluminum.

1. Zinc chloride	85 oz
Ammonium chloride	6-1/2 oz
Stannous chloride	9 oz
Hydrochloric acid	2 oz
Water to make	1 gal
Wetting agent (optional)	0.1% by wt
(For stainless steel and galvanized iron)	
2. Zinc chloride	48 oz
Ammonium chloride	5 oz
Hydrochloric acid	3 oz
Water to make	1 gal
Wetting agent (optional)	0.1% by wt
(For stainless steel)	
3. Zinc chloride	16 oz
Ammonium chloride	16 oz
Glycerin	16 oz
Water (For Monel)	1 pt
4. Orthophosphoric acid (85%)	34 oz
Water	16 oz
(For high tensile manganese bronze, copper, brass)	
5. Zinc chloride	32 oz
Ammonium chloride	4 oz
Sodium chloride	8 oz
Hydrochloric acid	8 oz
Water to make (for cast iron)	1 gal
6. Zinc chloride	40 oz
Ammonium chloride	4 oz
Hydrofluoric acid	
Water to make (for cast iron)	1 gal
7. Stannous chloride	83 oz
Zinc dihydrazinium chloride	7 oz
Hydrazine hydrobromide	10 oz
Water	10 oz
(Paste flux for soldering aluminum)	
8. Cadmium fluoboride	5 oz
Zinc fluoboride	5 oz
Fluoboric acid	6 oz
Diethanol amine	20 oz
Diethanol diamine	4 oz
Diethylene triamine	10 oz
(For soldering aluminum)	
9. Potassium chloride	45 oz
Sodium chloride	30 oz
Lithium chloride	15 oz
Potassium fluoride	7 oz
Sodium pyrophosphate	3 oz
(For soldering aluminum)	
10. Triethanolamine	25 oz
Fluoboric acid	3 oz
Cadmium fluoborate	2 oz
(Chloride-free organic flux for soldering aluminum. Fluxing range 350° to 525° F. The viscous liquid can be dissolved with water or alcohol to any desired concentration.)	
11. Stannous chloride	44 oz
Ammonium chloride	5 oz
Sodium fluoride	1 oz
(Reaction-type flux for soldering aluminum. Fluxing range 540° to 720° F or higher. It may be used as a dry powder mixture or it may be suspended in alcohol.)	
12. Zinc chloride	44 oz
Ammonium chloride	5 oz
Sodium fluoride	1 oz
(Reaction-type flux for soldering aluminum. Fluxing range 620° to 720° F or higher. It may be used as a dry powder or mixed with water or alcohol.)	

Source: Courtesy of American Welding Society (Anon., 1959).

*Metric equivalents: oz × 28.349 520 = g; gal × 3.785 412 = L; pt × 0.47 = L; (°F − 32)/1.8 = °C.

TABLE 4—Intermediate Fluxes

These fluxes contain organic compounds which decompose at soldering temperatures. When properly used, the mildly corrosive elements in the flux volatilize, leaving a residue relatively inert and easily removed with water. They are effective on all materials which are solderable with mild fluxes. Typical compositions are as follows:

1. Glutamic acid hydrochloride	19 oz*
Urea	11 oz
Water	1 gal
Wetting agent	0.2% by wt
(For copper and Inconel)	
2. Hydrazine monohydrobromide	10 oz
Water	90 oz
Nonionic wetting agent	1/20 oz
3. Lactic acid (85%)	9 oz
Water	42 oz
Wetting agent	1/10 oz
(For beryllium copper)	

Source: Courtesy of American Welding Society (Anon., 1959).

* Metric equivalents: US fl oz × 29.573 530 = cm³; gal × 3.785 412 = L.

of a joint by the removal of oxide films on the base metal. Unlike soldering fluxes, brazing fluxes must remain active at the higher temperatures normally encountered in brazing operations. Most brazing fluxes are metallic salts which must be melted before they become active. It is desirable that the flux melt at a temperature about 50° C below the melting point of the filler metal and remain active up to the maximum brazing temperature. The viscosity of the molten flux is important and influences the ease with which the flux is displaced by the filler metal. Joints having entrapped flux deposits are usually inferior. Finally, the flux must be easily removed in order to avoid corrosion of the joint during service.

TABLE 5—Noncorrosive Fluxes

The rosin-base fluxes—plain, stabilized, and activated—belong in this class. For all electronic and critical soldering applications, water white rosin dissolved in an organic solvent (item 1 below) is the safest known flux. Activators added to the rosin increase the activity, but the flux residue from these fluxes should pass tests for noncorrosivity and nonconductance when used on electronic applications. These fluxes are effective on clean copper, brass, bronze, tinplate, terneplate, electro-deposited tin and tin alloy coatings, cadmium, nickel, and silver.

1. Water white rosin	10-25% by wt
Alcohol, turpentine, or petroleum spirits	Balance
2. Water white rosin	40% by wt
Glutamic acid hydrochloride	2% by wt
Alcohol	Balance
3. Water white rosin	40% by wt
Cetyl pyridinium bromide	4% by wt
Alcohol	Balance
4. Water white rosin	40% by wt
Stearine	4% by wt
Alcohol	Balance
5. Water white rosin	40% by wt
Hydrazine hydrobromide	2% by wt
Alcohol	Balance

Source: Courtesy of American Welding Society (Anon., 1959).

Constituents of Brazing Fluxes: Most commercially available brazing fluxes are the result of considerable development and often contain a large number of components each of which plays a different role. The chemistry of the flux at brazing temperature is complex and highly transient. The most common constituents of brazing fluxes are: *borates, fluoborates* (potassium, sodium, etc.), *fluorides* (sodium, potassium, lithium, etc.), *chlorides* (sodium, potassium, lithium, zinc), *acids* (boric, calcined boric acid), *alkalies* (potassium hydroxide, sodium hydroxide), wetting agents, and water. A discussion of the role of each of these constituents is given in Table 6.

Brazing fluxes can be separated into three main categories: (1) borax fluxes, (2) fluoride fluxes, and (3) alkali-halide fluxes.

Borax Fluxes—Borax fluxes have been used for centuries and are cheap and effective for metals which melt above 800° C. They are an effective brazing flux for metals such as steel, copper, and brass but are unsatisfactory for metals with refractory oxides such as aluminum, titanium, etc.

Boric acid is seldom used alone as a flux because of its lesser effectiveness as an oxide solvent. However, because molten boric acid does cover the work well, it is used in combination with borax. Borax-boric acid mixtures maintain a continuous coating over metal surfaces at brazing temperatures. Table 7 gives the classification system currently in use by the American Welding Society (AWS). Borax-based fluxes are AWS Type 5.

Fluoride Fluxes—Borax fluxes are not useful for filler metals which melt below about 800° C. With the introduction of silver-base brazes (silver solder) lower melting point fluxes were required. Alkali fluorides and bifluoride additions to borax fluxes lower their melting point, increase their oxide removal rate, and decrease their viscosity. Because sodium salts produce a glare when heated, there has been a tendency to substitute the more costly potassium salts for sodium, resulting in an even greater reduction in melting point and a further increase in oxide reduction powers. Fluoride-based fluxes may initially consist of potassium

TABLE 6—Brazing Flux Constituents

Borates. These compounds are useful in formulating the higher melting fluxes. They have good oxide dissolving power and provide protection against oxidation for long periods of time. Most of the borates melt and are effective at temperatures around 760°C (1400°F) or higher. Even in the molten condition they have a relatively high viscosity and therefore must be used with other salts.
Fused Borax. Fused borax is another high melting material that is active at high temperatures. However, it is little used in the lower melting brazing fluxes.
Fluoborates. Fluoborates react similarly to the other borates in many respects. However, they do not provide protection from oxidation to the same extent as the other borates. They flow better in the molten state and have greater oxide-dissolving properties than the other borates. Fluoborates are used with other borates or with alkaline compounds, such as carbonates, to make a flux. Another related class of compounds is fluosilicoborates. These are somewhat higher in melting point than the fluoborates and provide good coverage and surface adherence. The high melting point limits their use.
Fluorides. The fluorides react readily with most metallic oxides at elevated temperatures and therefore are used in fluxes as cleaning agents. They are particularly important in applications where refractory oxides, like chromium oxides, aluminum oxides, etc., are encountered. In addition to cleaning, the fluorides increase the fluidity of molten borates thereby improving the capillary flow of the molten brazing filler metal.
Chlorides. The chlorides function in a manner similar to that of the fluorides. They are less costly but less effective. Chlorides must be used judiciously since, at elevated temperatures, they tend to oxidize the work.
Boric Acid. Boric acid is a principal constituent used in brazing fluxes. It is employed in its conventional form or sometimes in the calcined form. The calcined form reacts similarly to boric acid but has a somewhat higher melting point. Boric acid is employed principally as a cleaning agent. However, it is used in a flux because it improves the viscosity of the flux residue. Further, it has the property of facilitating the removal of the glasslike flux residue after brazing. The melting point of boric acid is best described as being below that of the borates but higher than that of the fluorides.
Alkalies. Potassium and sodium hydroxide are used sparingly, if at all, because of their deliquescent properties. Even small amounts involved with other flux agents can create difficulty in humid weather. When used, the function is to elevate the useful working temperature of the flux.
Water. Water will be present in brazing fluxes either as water of crystallization in the chemicals used in formulating the flux or as a separate addition to the flux for the purpose of making it a paste or liquid. If the water of crystallization is undesirable in a chemical, about the only way to remove it is to calcine that particular ingredient. Water which may be used to form a paste must be checked as to its suitability, and excessively hard waters should be avoided. In mixing fluxes, if the available water is unsuited, then it is a wise precaution to employ alcohol.

Source: Courtesy of American Welding Society (Anon., 1955).

borate, potassium fluoride, and boric acid. Upon mixing with water a number of reversible chemical reactions may occur resulting in the formation of new compounds. Typical fluoride fluxes are AWS Type 3A and 3B listed in Table 7.

Alkali Halide Fluxes—Although the fluoride-type fluxes are satisfactory for the removal of most refractory oxides, a special type of flux must be used for aluminum and its alloys and titanium and zirconium. These reactive metals require very aggressive fluxes. Fluxes which are suitable for these metals are basically mixtures of the alkali metal chlorides and fluorides. A typical flux composition follows, which is suitable for well-cleaned aluminum surfaces (Brooker and Beatson, 1953):

Constituent	Wt %
Sodium chloride	31
Potassium chloride	44
Lithium chloride	15
Potassium fluoride	7
Potassium bisulfate	3

When it is not possible to thoroughly clean an aluminum surface prior to brazing, or where the joint will be heated to temperature slowly, it is desirable to use a *reaction flux* which contains zinc salts, that alloy with the base metal and promote spreading and oxide removal. Additions of approximately 7% by wt of zinc metal are the safe upper bound. Further additions weaken the joint by forming excessively large intermetallic layers. The alkali halides are relatively expensive materials (with the exception of sodium chloride) and the relative cost of the materials tends to determine the compositions used. The ranges of typical compositions are (Brooker and Beatson, 1953):

Constituent	Wt %
Sodium chloride	25–30
Potassium chloride	30–55
Lithium chloride	15–30
Lithium fluoride	8–10
Zinc chloride	7–12

Economy of Soldering and Brazing Fluxes

Because relatively small amounts of flux are used in brazing and soldering operations the cost of the flux itself is usually not the cost determining factor in the operation. The cost of the filler metal (for instance, silver brazes) usually is far greater than the cost of the flux. However, one economic consideration associated with the use of a flux is the cost of its removal. Most active fluxes remain corrosive

TABLE 7—Applications for Commercially Available Brazing Fluxes*

AWS Brazing Flux Type No.	Metal Combinations for Which Various Fluxes Are Suitable		Effective Temperature Range of Flux, °C	Major Constituents of Flux	Physical Form	Methods of Application†
	Base Metals	Filler Metals (B = Braze)				
1	Aluminum and aluminum alloys	BAlSi	370-600	Fluorides; chlorides	Powder	1,2,3,4
2	Magnesium alloys	BMg	480-650	Fluorides; chlorides	Powder	3,4
3A	Copper and copper-base alloys (except those with aluminum); iron base alloys; cast iron; carbon and alloy steel; nickel and nickel base alloys; stainless steels; precious metals (gold, silver, palladium, etc.)	BCuP BAg	620-870	Boric acid Borates Fluorides Fluoborate Wetting agent	Powder Paste Liquid	1,2,3
3B	Copper and copper-base alloys (except those with aluminum); iron base alloys; cast iron; carbon and alloy steel; nickel and nickel base alloys; stainless steels; precious metals (gold, silver, palladium, etc.)	BCu BCuP BAg BAu RBCuZn BNi	730-1150	Boric acid Borates Fluorides Fluoborate Wetting agent	Powder Paste Liquid	1,2,3
4	Aluminum-bronze; aluminum-brass§	BAg BCuZn BCuP	570-870	Borates Fluorides Chlorides	Powder Paste	1,2,3
5	Copper and copper-base alloys (except those with aluminum); nickel and nickel-base alloys; stainless steels; carbon and alloy steels; cast iron and miscellaneous iron base alloys; precious metals (except gold and silver)	BCu BCuP BAg (8-19) BAu BCuZn BNi	760-1200	Borax Boric acid Borates (No fluorides)	Powder Paste Liquid	1,2,3
6	Titanium and zirconium based alloys	BAg	370-880	Fluorides Chlorides Wetting agents	Paste Powder	1,2,3

* This table provides a guide for classification of most of the proprietary fluxes available commercially. For additional data consult AWS specification for brazing filler metal A5.8 ASTM B260; consult also AWS Brazing Manual, 1963 ed.

† 1—Sprinkle dry powder on joint; 2—dip heated filler metal rod in powder of paste; 3—mix to paste consistency with water, alcohol, monochlorobenzene, etc.; 4—molten flux bath. Some Type 3A fluxes are specifically recommended for base metals listed under Type 4.

§ In some cases Type 1 flux may be used on base metals listed under Type 4.

Source: Courtesy of American Welding Society (Anon., 1955).

after use and consequently must be *completely* removed. Measures necessary to insure complete flux removal therefore are important and contribute to the overall cost of the joining operation.

Fluxing Materials Used In Arc Welding

A variety of minerals are used to promote the joining of metal by arc welding. Strictly speaking these materials are not fluxes but rather deoxidants, slag formers, and arc stabilizers. The molten covering material does aid in the manipulation of the weld pool and consequently is often considered a flux.

The materials used in the coatings of welding electrodes serve a variety of purposes:

1) To provide a suitable gaseous atmosphere for the arc and the weld pool,

2) To produce a slag which will shield the weld pool, deoxidize the metal in the weld pool, and aid in weld pool manipulation,

3) To act as a vehicle for alloying agents, iron powder, and deoxidants,

4) To improve arc stability and metal transfer.

Three representative applications of fluxing materials in arc welding are the shielded metal arc, flux-cored arc, and submerged arc processes.

Fluxing Materials Used in Shielded Metal Arc Welding

In the shielded metal arc process the welding electrode which is manipulated by hand in a holder, is coated with a complex mixture of minerals each of which plays a different role during the manufacture, storage, and use of the electrode. The materials used may be classed as:

Shielding gas formers
Slag formers
Arc stabilizers
Extrusion aids
Deposition rate augmentors
Deoxidants
Alloying agents
Binders
Fluxing agents

There are many different electrode covering compositions in use as can be seen in Table 8. These different formulations find application in particular welding situations where either deep penetration, high deposition rate, fast freezing, or hydrogen-free conditions are desired.

Fluxing Materials Used in Flux-Cored Arc Welding

In the flux-cored arc welding process, a tubular filler metal wire is fed through the welding torch and is melted by an arc struck between the end of the wire and the base metal. Flux materials which are trapped in the core of the wire act in much the same way as the external coverings on shielded metal arc electrodes. The flux provides the shielding gas in the self-shielding method or may be supplemented by an auxiliary shielding gas, usually carbon dioxide.

The fluxes used in flux-cored electrodes for use with auxiliary carbon dioxide shielding gas are of three basic types: (1) high-titania, (2) lime-titania, and (3) lime.

The high-titania type produces little spatter, a smooth weld bead, and has good penetration. The lime-titania type of flux possesses good arc stability and fairly low spatter. The lime type flux produces a basic slag.

The constituents of the fluxes used and slags formed with auxiliary carbon dioxide shielding gas are given in Table 9. The compounds in this table can be incorporated into the flux in many forms. Silicon dioxide may be added as any one of the many natural silicates, as sand, as ground glass, or as ferrosilicon. The main difference between the three types of flux is the titanium dioxide and silicon dioxide contents in the titania type and the high calcium oxide content in the lime-titania and lime types.

In the self-shielding flux-cored process the fluxes must provide the shielding gas. In this respect the self-shielding flux-cored process resembles the shielded metal arc process electrodes turned inside out. However, because the gas forming constituents of the flux are on the inside of the electrode, they cannot be used as effectively without producing some spatter.

There are four types of self-shielding flux-cored wires:

1) Fluorspar—alumina.
2) Fluorspar—titania.
3) Fluorspar—lime-titania.
4) Fluorspar—lime.

Types 1 and 2 are in the most common usage. Type 4 may be used as a self-shielding electrode or with auxiliary carbon dioxide shielding gas. The constituents of self-shielding fluxes and the slags formed are given in Table 10.

TABLE 8—Typical Functions and Composition Ranges of Constituents of Coverings on Mild Steel Arc Welding Electrodes

Constituent of Covering	Function of Constituent		Composition Range, %, in Covering on Electrode of Class:								
	Primary	Secondary	E6010, E6011	E6012, E6013	E6020	E6027	E7014	E7016	E7018	E7024	E7028
Cellulose	Shielding gas		25 to 40	2 to 12	1 to 5	0 to 5	2 to 6	—	—	1 to 5	—
Calcium carbonate	Shielding gas	Fluxing agent	—	0 to 5	0 to 5	0 to 5	0 to 5	15 to 30	15 to 30	0 to 5	0 to 5
Fluorspar	Slag former	Fluxing agent	—	—	—	—	—	15 to 30	15 to 30	—	5 to 10
Dolomite	Shielding gas	Fluxing agent	—	—	—	—	—	—	—	—	5 to 10
Titanium dioxide (rutile)	Slag former	Arc stabilizer	10 to 20	30 to 55	0 to 5	0 to 5	20 to 35	15 to 30	0 to 5	20 to 35	10 to 20
Potassium titanate	Arc stabilizer	Slag former	*	*	—	—	—	—	0 to 5	—	0 to 5
Feldspar	Slag former	Stabilizer	—	0 to 20	5 to 20	0 to 5	0 to 5	0 to 5	0 to 5	—	0 to 5
Mica	Extrusion	Stabilizer	—	0 to 15	0 to 10	—	0 to 5	—	—	0 to 5	—
Clay	Extrusion	Slag former	—	0 to 10	0 to 5	0 to 5	0 to 5	—	—	—	—
Silica	Slag former	—	—	—	5 to 20	—	—	—	—	—	—
Asbestos	Slag former	Extrusion	10 to 20	—	—	—	—	—	—	—	—
Manganese oxide	Slag former	Alloying	—	—	0 to 20	0 to 15	—	—	—	—	—
Iron oxide	Slag former	—	—	—	15 to 45	5 to 20	—	—	—	—	—
Iron powder	Deposition rate	Contact Welding	—	—	—	40 to 55	25 to 40	—	25 to 40	40 to 55	40 to 55
Ferrosilicon	Deoxidizer	—	—	—	0 to 5	0 to 10	0 to 5	5 to 10	5 to 10	0 to 5	2 to 6
Ferromanganese	Alloying	Deoxidizer	5 to 10	5 to 10	5 to 20	5 to 15	5 to 10	2 to 6	2 to 6	5 to 10	2 to 6
Sodium silicate	Binder	Fluxing agent	20 to 30	5 to 10	5 to 15	5 to 10	0 to 10	0 to 5	0 to 5	0 to 10	0 to 5
Potassium silicate	Arc stabilizer	Binder	*	5 to 15 *	0 to 5	0 to 5	5 to 10	5 to 10	5 to 10	0 to 10	0 to 5

* Used (in place of constituent on line above) in E6011 and E6013 electrodes to permit welding with alternating current.

Source: Courtesy of the American Society for Metals (Anon., 1971).

TABLE 9—Typical Flux and Slag Compositions for the Three Types of Carbon-Dioxide-Shielded Flux-Cored Electrodes

Compound or Element	Type 1 High-Titania (Nonbasic)		Type 2 Lime-Titania (Basic or Neutral)		Type 3 Lime (Basic)	
	Flux, %	Slag, %	Flux, %	Slag, %	Flux, %	Slag, %
Silica	21.0	16.8	17.8	16.1	7.5	14.8
Aluminum oxide	2.1	4.2	4.3	4.8	0.5	—
Titanium dioxide	40.5	50.0	9.8	10.8	—	—
Zirconium dioxide	—	—	6.2	6.7	—	—
Calcium oxide	0.7	—	9.7	10.0	3.2	11.3
Sodium oxide	1.6	2.8	1.9	—	—	—
Potassium oxide	1.4	—	1.5	2.7	0.5	—
Carbon dioxide	0.5	—	—	—	2.5	—
Carbon	0.6	—	0.3	—	1.1	—
Iron	20.1	—	24.7	—	55.0	—
Manganese	15.8	—	13.0	—	7.2	—
Calcium fluoride	—	—	18.0	24.0	20.5	43.5
Manganese oxide	—	21.3	—	22.8	—	20.4
Ferrous oxide	—	5.7	—	2.5	—	10.3
Flux	13.0	—	13.0	—	27.0	—
AWS classification	E70T-1 or E70T-2		E70T-1		E70T-1 or E70T-5	

Courtesy of American Society for Metals (Anon., 1971).

Fluxes Used in the Submerged Arc Process

In the submerged arc welding process the arc is struck between bare filler metal wire and base metal. The arc is protected from the atmosphere by a blanket of granular, fusible flux which is laid over the joint prior to welding and removed afterwards.

Fluxes used in the submerged arc process usually consist of manganese oxide, silicon oxide, titanium oxide, aluminum oxide, calcium oxide, zirconium oxide, magnesium oxide, and other compounds. The flux is melted by the welding arc and, in the molten state, blankets the weld metal and protects it from the atmosphere. Submerged arc fluxes are generally neutral and must not evolve large amounts of gas during the welding operation. Submerged arc welding fluxes are supplied in three forms: prefused, bonded, and agglomerated.

TABLE 10—Typical Flux and Slag Compositions for the Four Types of Self-Shielding Flux-Cored Electrodes

Compound or Element	Type 1 Fluorspar-Aluminum		Type 2 Fluorspar-Titania		Type 3 Fluorspar-Lime-Titania		Type 4 Fluorspar-Lime	
	Flux, %	Slag, %	Flux, %	Slag, %	Flux, %	Slag, %	Flux, %	Slag, %
Silica	0.5	—	3.6	0.2	4.2	1.8	6.9	0.2
Aluminum	15.4	—	1.9	—	1.4	—	—	—
Aluminum oxide	—	11.8	—	6.5	—	6.0	0.6	12.8
Titanium dioxide	—	—	20.6	27.0	14.7	33.5	1.2	2.3
Calcium oxide	—	—	—	—	4.0	—	3.2	4.1
Magnesium oxide	12.6	9.2	4.5	4.5	2.2	6.0	—	0.9
Potassium oxide	0.4	—	0.6	1.8	—	—	—	—
Sodium oxide	0.2	—	0.1	1.0	—	—	0.6	0.9
Carbon	1.2	—	0.6	—	0.6	—	0.3	—
Carbon dioxide	0.4	—	0.6	—	2.1	—	1.3	—
Iron	4.0	—	50.0	—	50.5	—	58.0	—
Manganese	3.0	—	4.5	—	2.0	—	7.9	—
Nickel	—	—	—	—	2.4	—	—	—
Calcium fluoride	63.5	76.1	22.0	53.0	15.3	47.5	22.0	73.7
Manganese oxide	—	0.4	—	1.1	—	2.8	—	0.7
Ferrous oxide	—	2.5	—	1.9	—	3.6	—	3.0
Flux	18.0	—	18.0	—	26.0	—	26.0	—
AWS classification	E70T-4, E60T-7 or E60T-8		E70T-3		E70T-6		E70T-5	

Source: Courtesy of American Society of Metals (Anon., 1971).

Prefused fluxes are produced by melting the ingredients in an electric furnace. After melting the flux is quenched and crushed. The resulting product is a glass which has the following advantageous properties: good chemical homogeneity, the fine particles can be reused without changing chemical composition, they are nonhygroscopic, unfused portions of the flux can be reused without changing the chemical composition, and they are suitable for high welding speeds. The largest disadvantage of prefused fluxes is the segregation and/or loss of the ferroalloys and deoxidizers which occur during the high temperature prefusing operation. *Bonded fluxes* are produced by grinding the constituents together with a potassium silicate or sodium silicate binder, which results in the pelletization of the flux. The advantages of this method are: the retention of the deoxidizers and ferroalloys; a lower density flux results, which allows thicker flux blankets to be used; and the resulting slag is more easily removed after welding. The disadvantages of bonded flux are that the fines cannot be removed without altering its composition and that the flux is more hygroscopic. *Agglomerated fluxes* are similar to bonded fluxes except that a ceramic binding agent is used and requires higher processing temperatures. The constituents used in submerged arc fluxes and the composition of the resulting slag are given in Table 11.

Fluxes Used In Smelting

Smelting is a metallurgical operation in which a metal is separated or concentrated from its ore by heat. The waste materials resulting from this separation, termed *slags,* must have certain attributes for efficient smelting. The slag produced must not be costly, must have a low formation and melting temperature, must be fluid at smelting temperatures, must have an appreciably lower specific gravity than the metal, and must not dissolve appreciable amounts of the metal being smelted (Bray, 1947; Anon., 1964).

The slags of certain ores are termed "self-fluxing," that is, they naturally have all the desirable attributes mentioned previously. In most cases, however, it is desirable to add materials, *fluxes,* which modify the properties of the slag either by lowering the melting point and specific gravity or by increasing the slag fluidity. A secondary role played by fluxes is to provide a substance which removes (or slags out) chemical species which are combined or may recombine with the metal being purified. In this connection, fluxes are termed *acidic, neutral,* or *basic.* This classification arises from the acidity or basic nature of solutions of these materials in water and more directly from the fact that they will react with basic or acid metal compounds freeing the metal and forming a fusible slag. Basic fluxes react with acidic impurities, etc.

Some of the more common fluxing materials

TABLE 11—Typical Compositions of Manganese Silicate Fluxes for Submerged-Arc Welding

Substance	Prefused Flux, %	Bonded Flux, %
Manganese oxide	42.0	36.5
Manganese dioxide	—	5.2*
Silica	45.0	38.0
Calcium fluoride	6.9	3.9
Calcium oxide	1.2	0.8
Magnesium oxide	0.3	2.7
Barium oxide	0.1	0.3
Aluminum oxide	2.0	1.1
Ferric oxide	1.5	—
Ferrous oxide	—	2.7
Titanium oxide	0.1	0.1
Potassium oxide	0.4	—
Sodium oxide	0.4	1.5
Lead oxide	0.1	0.1
Ferrosilicon (50%)	—	7.1†
Ratio manganese oxide/silica	$\frac{42.0}{45.0} = 0.93$	$\frac{40.7}{45.6} = 0.89$

* At welding temperatures, manganese dioxide reacts with silicon in ferrosilicon to yield additional manganese oxide equal to 4.2% of the total flux (0.815 × 5.2% = 4.2%).

† In reaction with manganese dioxide at welding temperatures, silicon in ferrosilicon forms additional silica equal to 7.6% of the total flux (2.14 × 7.1% ÷ 2 = 7.6%).

Source: Courtesy of American Society for Metals (Anon., 1971).

used in smelting are limestone, silica, and fluorspar.

Limestone is one of the most common basic fluxes in both ferrous and nonferrous metallurgy. Limestone is composed principally of calcium carbonate ($CaCO_3$) (see Table 12). Limestone decomposes at high temperatures into calcium oxide and carbon dioxide. Most copper and lead ores are acidic and must be fluxed with a basic material such as limestone. Calcium oxide decreases the specific gravity and fusion temperature of these slags. Limestone is usually used as a flux in conjunction with iron oxide, promoting fluidity and further decreasing the melting point of the slag. In steelmaking limestone or dolomite (calcium-magnesium carbonate) may be used. When large amounts of sulfur are to be removed, limestone is preferred. The dolomitic fluxes are objectionable to a degree, since magnesium oxide in large enough quantities gives rise to viscous slag. Calcium oxide is added as calcined or "burnt" lime in the basic open-hearth steelmaking process. The effectiveness of a natural basic flux is impaired by the "acidic" impurities it contains. The effectiveness of a basic flux can be expressed in terms of "available base" which is the amount of basic material remaining after neutralizing its own impurities.

Silica is one of the most common fluxes and one of the cheapest. Silica for fluxing is available as sandstone, quartzite, and quartz. Silicates containing feldspar, hornblende, and mica are not favored since they do not aid in fluxing and only burden the furnace in melting them. Silica is the only substance used strictly as an acid flux. Silica is seldom charged into steelmaking blast furnaces, however, since the lining of the furnace usually supplies all that is needed. Silica may be used in basic steelmaking when too much lime has been charged.

TABLE 13—Typical Composition of Fluorspar (Dry Basis)

Ingredient	%
CaF_2	81.0
SiO_2	4.75
Al_2O_3, Fe_2O_3	1.00
S	1.00
$CaCO_3$	Remainder

Source: Courtesy United States Steel Corp. (Anon., 1964).

Fluorspar is considered a neutral flux and is used to make slags more fusible and fluid. The principle compound in fluorspar is calcium fluoride, as can be seen in Table 13.

Sources of Smelting Fluxes

The sources of fluxes are inexpensive materials such as mentioned earlier. Silica is commonly available and may be used in the form of sand, gravel, and quartz. Limestones of fluxing quality underlie the areas drained by the Mississippi and Ohio Rivers. Large portions of the limestone used for steelmaking in Pittsburgh originates in western Pennsylvania; while in the Great Lakes area it is obtained from lower Michigan, northern Ohio, and southern Illinois. Dolomite is obtained in eastern West Virginia, western Pennsylvania, and most of the Great Lakes states. Domestic fluorspar comes principally from western Kentucky and southern Illinois.

TABLE 12—Representative Compositions of Limestone, Burnt Lime, Dolomite, and Burnt Dolomite (Natural State Determinations)

Ingredient	Limestone, %	Burnt Lime, %	Washed Dolomite, %	Burnt Dolomite, %
$CaCO_3$	95.06	–	54.74	–
$MgCO_3$	0.54	0.76	39.61	–
Fe_2O_3 / Al_2O_3	0.70	0.93	0.43	1.57
SiO_2	1.73	2.55	0.74	1.53
Sulfur	0.049	0.07	0.026	0.037
Phosphorus	0.020	0.03	0.006	0.009
H_2O	1.70	–	4.00	–
CaO	–	81.36	–	–
MgO	–	–	–	56.35
Loss on ignition	–	14.00	–	1.60

Source: Courtesy United States Steel Corp. (Anon., 1964).

Bibliography and References

Anon., 1955, *Brazing Manual,* American Welding Society, Reinhold, London, pp. 33–47.

Anon., 1959, *Soldering Manual,* American Welding Society, New York, pp. 25–42.

Anon., 1964, *The Making, Shaping, and Treating of Steel,* United States Steel Corp., Pittsburgh, pp. 210–215.

Anon., 1971, "Welding and Brazing," *Metals Handbook,* 8th ed., Vol. 6, American Society for Metals, Metals Park, Aug., pp. 6–77.

Bray, J.L., 1947, *Non-Ferrous Production Metallurgy,* John Wiley and Sons, New York, pp. 17–25.

Brooker, H.R., and Beatson, E.V., 1953, *Industrial Brazing,* Iliffe and Sons, London, pp. 117–126.

Lancaster, J.F., 1965, *The Metallurgy of Welding, Brazing and Soldering,* American Elsevier, New York, pp. 130–134.

Foundry Sand

H. E. WILBORG *

Revised by G. V. HENDERSON †

This chapter deals with those sands employed by foundries for the manufacture of cores and molds used in the casting of such common metals as steel, gray iron, ductile iron, aluminum-based alloys, and copper-based alloys.

Most metal is cast in "green sand" molds in which the sand mixture, consisting primarily of silica, clay, and water, forms a plastic material. This plastic material is rammed around a pattern in a split box called a flask, then the box is separated, the pattern is withdrawn, and the metal is poured. These green sand molds are broken apart when the metal has solidified. The sand is then remixed and a new mold made for the next casting. To a lesser extent silica sand is mixed with resins or oils and, after forming, the mixture is hardened by either heat or chemical reaction. In any case, the foundry mold or core is used just once before being broken up and remixed for manufacturing another core or mold.

By far the greatest consumption in the foundry is of those types of sand high in silica, low in clay (under 0.5%), and occurring in numerous locations thoughout the United States.

Consumption

End Uses

Foundry sands fall into four distinct categories: silica sand, lake sand, bank sand, and natural molding sand. *Silica sand* is a general term used to describe washed, graded, dried, and cooled clay-free sands which are found in many parts of the United States. They generally contain 90 to 100% silica but vary greatly in chemical analysis and grain structure. *Lake sand* is the sand dredged from Lake Michigan or dug from the dunes bordering that lake. *Bank sand* is composed primarily of silica grain but containing up to 5% clay. It is shipped in both the damp and dry state but the normal processing of bank sand consists only of drying and scalping. Occasionally two different grades of bank sand are blended prior to or during loading. *Natural molding sand* is a term used to describe a sand containing sufficient clay as mined to be suitable for molding purposes. Clay content may vary from 10 to 24% with the balance of the material consisting primarily of silica sand. These sands range in fineness from one peaked on the 20 mesh to one which contains more than 50% of the particles passing the 270 mesh.

Lake sands are relatively coarse and are used in great quantities for the production of automotive castings. Bank sands and natural molding sands are gradually phasing out as the production of castings becomes more a science and less an art. The vast majority of foundry sands used today are those known as the silica sands and lake sands.

Foundry sands are used in two basic ways. The majority are used in the green or uncured state with the balance used in the cured or set state. Green sand is grain bonded with clay and contains varying amounts of water to plasticize the clay in order to bond the grains together. Green sand is either naturally bonded or "synthetic" sand. The naturally bonded molding sands are generally found in the Hudson River Valley of New York State, Cumberland County in New Jersey, and in Tennessee and Mississippi. Synthetic sand is molding sand compounded in the individual foundry. It is made up primarily of graded silica sand, western or southern bentonite, water, and carbonaceous and/or cellulosic materials.

Green sand is used primarily in making molds. The use of naturally bonded green sand has been declining for many years, primarily because naturally bonded sands are more difficult for the foundryman to find and control. Synthetic green sand, on the other hand, can be

* Formerly Sales Manager (Retired), Whitehead Brothers Co., Florham Park, NJ.

† Deceased.

quickly altered in the foundry to change the physical properties of the sand as need arises. Also, good naturally bonded molding sands are mined at just a few locations in the United States so the logistics problem has also influenced the trend toward synthetic sands based on washed, graded, and dried silica sands and lake sands.

Cured sands are combinations of silica sand and resins or oils. These sands are cured by either baking or by chemical reaction. Cured sands are used primarily for the manufacture of cores (that portion of the completed mold which forms the voids in a casting) but are being used more and more to make the complete mold.

About 7.9 Mt (8.7 million st) of sand at an average price of about $8.50 per ton were used by the foundry trade in 1980 (Tepordei, 1980).

Prices

The following net ton prices are those effective in late 1980 and pertain to dried bulk sand in carload lots f.o.b. shipping point.

Northeast $7.00–$9.00
Southeast $6.00–$8.50
Lake sand $4.50
Illinois sandstone $6.00–$9.00
Nevada $11.00–$16.00
West Coast $10.00–$13.00

Most sands in the northeast and southeast are graded by wet classification. The difference in price reflects: (1) higher labor costs in the northeast, and (2) harsh northern winter weather which increases the unit labor cost for classifying and handling. Lake sand is relatively low priced because little processing is required other than drying. The Illinois sandstone reflects a lower price than those of the northeast, primarily because most sand producers in this area sell large tonnages to the glass melting industry and because of this large overall volume their unit cost is lower. Specifications for foundry sand are different than for glass sand. Therefore the market allows higher cost for foundry sand.

The price range within a given district reflects the price from coarse to fine. Fine sands in northern regions are at a premium, while in southern regions coarse sands are at a premium. Two basic factors account for the higher price of the finer sands: (1) the drying cost is higher for finer sands because of considerably more surface area and therefore more water must be driven off, and (2) the finer sands are either scarcer or require additional processing and classifying to get them to the desired grade.

Specifications

Medium grade washed eastern silica sands (Table 1) are used predominantly in the casting of steel and heavy gray and ductile iron. With silica content over 98% and a fusion point over 1682°C (3060°F), these sands have excellent refractoriness and are of a gradation which presents a suitable finish to the larger castings. The eastern silica sands are produced in a wide range of fineness. Very fine grades of washed eastern silica sand (Table 1) are generally used for making precision castings and as a base sand for aluminum castings and light copper-based alloy castings. The grade shown in Table 1 is probably as fine a sand as is produced commercially anywhere.

Silica sands are found in many parts of the United States but a popular sand is that produced in the midwest from the St. Peter sandstone (Table 1). Because of its round grain shape, particularly in the coarser screen fractions, this sand has excellent flowability and relatively smooth surface that requires less bonding material than angular or subangular sands.

Nevada silica sand (Table 1) is used predominantly in West Coast foundries.

Fine bank sand (Table 1) is used as a base for molding sand in the casting of medium-to-small gray iron castings. Typical of such castings are those used in hardware and hand tools. These fine sands are also used as a base for molding sands for casting aluminum and copper-based alloys, producing excellent finish and fine detail.

Fine Albany naturally bonded molding sands (Table 1) are used as shipped in the casting of aluminum, light copper-based alloys, and light gray iron. Like all natural molding sands, these contain enough clay as mined to provide the necessary bond for green sand molding. This sand is still popular in smaller foundries that have little mechanization or mixing equipment and performs very well in the hands of skilled molders.

Heavy, naturally bonded molding sands and gravels (Table 1) are used in the molding of large iron castings. These sands are generally coarse and high in fire clay content. The mold is frequently dried prior to casting of such items as paper rolls and ingot molds for the steel mills.

Lake sands (Table 1) are used in large quan-

TABLE 1—Typical Analyses of Various Foundry Sands

Medium Grade Washed Eastern Silica Sand		Fine Grade Washed Eastern Silica Sand		Midwestern Round Grain Silica Sand, St. Peter's Sandstone		Silica Sand from Nevada	
US Series	% Retained	US Series	% Retained	US Series	% Retained	US Series	% Retained
20	0.4	50	0.4	40	2.0	30	0.5
30	1.6	70	5.2	50	24.0	40	7.5
40	6.4	100	14.6	70	40.0	50	19.5
50	22.6	140	23.4	100	22.0	70	42.0
70	40.4	200	25.8	140	8.0	100	18.5
100	23.6	270	16.2	200	3.0	140	10.0
140	4.6	Pan	14.4	270	1.0	200	1.5
200	0.6	AFS Fineness #	148	Pan	0.0	270	0.5
270	0.0			AFS Fineness #	60	Pan	0.0
Pan	0.0					AFS Fineness #	57
AFS Fineness #	53						

Fine Bank Sand—New York State		Fine Grade Naturally Bonded Albany Sand		Naturally Bonded Molding Gravel		Lake Sand	
US Series	% Retained	US Series	% Retained	US Series	% Retained	US Series	% Retained
30	0.2	12	Trace	6	3.0	30	Trace
40	0.2	20	Trace	12	8.8	40	2.4
50	1.0	30	0.2	20	16.0	50	22.6
70	4.4	40	0.2	30	13.8	70	44.4
100	12.4	50	1.0	40	17.6	100	24.0
140	20.6	70	3.4	50	10.8	140	6.0
200	24.8	100	10.2	70	5.6	200	0.6
270	18.2	140	12.0	100	2.0	270	0.0
Pan	18.0	200	13.6	140	0.8	Pan	0.0
AFS Fineness #	157	270	13.0	200	0.4	AFS Fineness #	55
		Pan	33.0	270	0.2		
		AFS clay	13.4	Pan	1.0		
		AFS Fineness #	191	AFS clay	20.0		
				AFS Fineness #	29		

tities for the production of automotive and farm machinery castings. These sands, while lower in silica content than "silica sands," are quite round in shape. They contain only small percentages of fine particles, so are economical to use because of relatively low binder demand.

Alternative Materials

Minerals other than silica are also used in the foundry industry for the manufacture of molds and cores. These other minerals are used for their peculiar characteristics and generally for special applications. Since silica is by far the most prevalent and commonly found mineral in the earth's crust, silica is also by far the cheapest refractory material available for foundry use.

Zircon is widely distributed in nature, being found in small quantities almost everywhere. Workable commercial deposits are found in Florida and Australia. Concentrated zircon in the Australian deposit is about 15% whereas that in Florida runs only 1.5% but the market price is about the same because of the higher shipping costs from Australia. Typical price is about \$165 per ton bulk 65% ZrO_2 f.o.b. Florida or port of entry.

The properties of zircon which make it highly desirable as a molding medium and a mold coating are high refractory value, high conductivity, high density, low expansion, and resistance to being wetted by molten metal. The melting point of commercial zircon is close to 2538°C (4600°F). The high conductivity and high density of zircon tend to chill or freeze the molten metal more rapidly. Zircon can absorb about twice the amount of heat twice as fast as silica sand. These properties give the foundryman a very useful tool for controlling directional solidification.

Zircon is used both as a molding medium

and as a core sand. In addition, however, zircon finds wide use as a refractory in the manufacture of core and mold washes. For this purpose the zircon is ground in a ball mill, usually to pass a 325-mesh screen. This fine material is made into a wash by combining it with a suspending agent and a binder. Such a wash is brushed or sprayed onto the mold or core surface and is effective in stopping penetration of the liquid metal into the interstices of the core or mold. This is particularly important in certain high leaded bronzes and steel castings.

Olivine sand is produced in the state of Washington and, to a lesser extent, in North Carolina. Olivine is a magnesium-iron-orthosilicate and contains no free silica. Olivine sand as used in the foundry has a fusion point of approximately 1816°C (3300°F). It has a lower thermal expansion rate than silica and therefore exhibits less tendency toward such defects as rattails, scabs, and buckles. Olivine sand has approximately 25% greater specific gravity than quartz. Because it contains no free silica, manganese steel castings made in olivine come out much cleaner than those made in normal silica sand. Olivine sand is also used successfully in the manufacture of copper-based alloy castings. Its price, however, has limited its application. Olivine sand is priced at approximately $56 per ton in bags f.o.b. Washington State.

Chromite sands used in the United States are imported from South Africa. The foundry application is similar to that of zircon sand in that it is more dense than silica and serves to chill the metal and provide directional solidification. It has a low rate of expansion and is not easily wetted by molten metal. The use of chromite is particularly prominent in steel foundries and medium-to-heavy gray iron foundries. Chromite sands are priced at $80 to $90 per ton in bags f.o.b. port of entry.

Staurolite is a naturally occurring mineral with the general formula $FeAl_5Si_2O_{12}(OH)$. A foundry grade of staurolite has been developed and is mined in north-central Florida. The price in late 1980 was $60 per ton in bags in carload lots f.o.b. shipping point. Its bulk density is 2050 kg/m^3 (128 lb per cu ft) and it begins to melt at 1538°C (2800°F) which is too low for use in some steel applications. However, it would not be affected by molten iron or nonferrous metals. The product also has a low rate of thermal expansion as compared to silica, as do the other substitute materials mentioned previously.

Economic Factors

Price Trends

Recent prices of foundry sands indicate an increase of approximately 7 to 12% per year. Federal regulations on wage and price control have had a large effect on this but competition among sand producers probably would not have allowed any greater increases anyway. Lessening demand for all products by foundries in what might well be called a "foundry recession" during the year 1971 and the first half of 1972 resulted in a condition of overcapacity in foundry sands in most areas. This, together with the ancient law of supply and demand, restricted price increases at a time when labor, power, and other costs were going up. The price trend, however, is upward and will probably continue so because of increased capital expenditure as well as increased costs for labor and energy.

Transportation

Until 50 years ago the greatest volume of foundry sands moved by water. Most shippers were located near water where they loaded the sand on barges for transportation via a complex system of canals, lakes, and rivers. No washed silica sand was used at that time as molding sands were all naturally bonded and the core sands consisted of unwashed bank sands. These sands were often dug by hand because of their occurrence in thin veins and carried by horse and wagon to narrow gage railways which delivered the sand to the loading docks. Today there is still some water movement. Large quantities of lake sand are delivered to various ports on the Great Lakes but most shipments are by railroad or truck.

The trend is strongly swinging toward truck delivery for shipments up to a 241.4-km (150-mile) radius of the sand plants. Since most of the sand consumed is dried and free-flowing, pressure trucks similar to those used for hauling cement are now used in the foundry sand trade. In pressure truck delivery the driver hooks up a hose to a standpipe and in 2 hr or so discharges 18 t (20 st) or more of sand, blowing it into a silo. This type of delivery eliminates the need for any unloading labor at the consumer's plant. An alternative is to use covered hopper rail cars but this method requires mechanical unloading equipment which is justified only for the larger user. Another reason for the strong trend toward trucking is the often erratic service given by the railroads as evidenced by unavailability and poor condition of cars, and the wide variation in the time needed to get the car of sand from the producer's plant to the consumer. Rail rates are causing concern within the industry with a 5 to 7% price increase twice a year.

Marketing—Distribution Channels

Foundry sands are sold by direct salesmen and by jobbers and manufacturers' representatives. Most sand producers have direct salesmen located in the vicinity of their largest customers. These salesmen can be considered sales engineers since they must of necessity be qualified to enter into technical discussions with their customers. Unfortunately, many of the new people are not students of the industry. The foundry sand salesman needs to know not only his own product but the total application of that product, whether used for molding or core sand. He is expected to help diagnose casting defects, instruct in the application of new products, and in general provide strong technical assistance to his customers.

Some sand producers operate sophisticated laboratories, not only for quality control of their own product but to assist customers in determining the cause of problems in a foundry. This help can go beyond the realm of mold and core practice to encompass the area of metallurgy also. All this assists the foundry sand salesman in selling his specialized product as well as his own service. Many manufacturers' representatives who sell on a commission basis also have the same capability as the sand producer's direct salesmen and can be particularly helpful when they cover areas beyond the boundaries of the sand producer's territory.

Government Considerations

Taxes and Subsidies

There are no peculiar taxes or subsidies applying to foundry sands. Those taxes, such as state franchise, unemployment, property tax, etc., applying to all businesses apply to the operation of sand plants as well. Although there is no outright subsidy, the foundry sand producers of industrial sands do fall into the rule of 14% depletion allowance—a factor which can be quite significant.

Zoning and Land Use

The use of building zoning codes has spread beyond the city and the suburb, extending to even small isolated townships in which sand plants are often located. Zoning codes per se are not harmful to sand producers but plant managers are urged to attend municipal government meetings discussing such codes. Neglect of this phase might well result in imposition of restrictions concerning the size of pit that can be mined at one time and many other limiting factors which could be enacted into law and result in an economic blow to the producer.

Many municipalities are requiring permits to continue mining and in most cases request detailed plans and maps for reclamation of mined-over land before issuing such permits. Bonds are usually posted to insure reclamation of the land. A "good neighbor" policy toward a municipality is essential to be sure that unnecessarily restrictive laws are not enacted. One move in this direction is to improve the appearance of a sand plant. This can be particularly important when the plant is viewed from a main road. Many sand producers are working in this direction by constructing sand berms and/or planting trees and shrubbery to hide the structure of the plant. Offering use of the sand producer's land for recreational purposes of the township inhabitants can go a long way toward insuring good relations between the company and the municipality.

Environmental Considerations

The time has passed when a foundry sand producer, or any other manufacturer, can ignore the effect that his plant operation has on the water, air, noise level, and esthetics of the community. Federal, state, and municipal governments have passed or are formulating laws governing all forms of nuisance or danger to the community. Today a producer is hard pressed to determine what regulations to follow but he is wise to attempt to comply with all legislation.

The effect of these environmental considerations upon supply and demand is hard to determine. A marginal operation may well not be able to afford the necessary capital investment for air pollution control equipment, etc., and therefore be forced to close. However, those producers who are in a good position vis-a-vis the foundry market will conform to all the laws and pass the cost on to their consumers through price increases.

Water Pollution

Most silica sand producers use a great deal of water for washing and in some cases for classifying their sands. The source of this water is usually natural streams but occasionally driven wells are used. In Nevada, wells must be dug to provide a source of water. All this water, with the exception of that lost by evaporation, must eventually be discharged into some waterway which then becomes a navigable stream and therefore is under the jurisdiction of various authorities.

Since foundry sands are not usually acid-treated there is not likely to be a chemical change to the water during the processing of the sand and the big problem is to cleanse the water of clay washed from the raw sand. This clay varies from one geographical area to another in its ability to settle out of water prior to discharge into a stream. Usually large settling beds are provided for this purpose but in some cases this method alone is not adequate. Under more difficult conditions chemicals, often alum, are added to the water to build a floc and thus settle out fine clay particles. In effect, then, the silica sand producer who washes sand is, of necessity, in the water treatment business. In at least one instance, the producer is maintaining daily, weekly, monthly, and semiannual records of test results for seven or eight physical characteristics of the water entering and discharging from the plant. Initial tests show discharge water to be better quality than that entering the plant.

Air Pollution

Air pollution plagues the silica sand producer in three areas: stockpiles, plant roads, and sand dryers. Most plants have large stockpiles of raw or graded sands in the open and the finer grades are subject to blowing under windy conditions. The probable requirement will be to eliminate this blowing by sprinkling water on stockpiles to keep the surface damp. Plant roads, which are usually unpaved, will require sprinkler trucks to allay the dust.

A more complicated problem centers around the sand dryer. Some sands are dried by passing damp sand over steam pipes but in modern practice the majority of foundry sand plants use rotary kiln or fluid-bed dryers. The latter two methods require huge quantities of air to support combustion and this air carries a considerable amount of particulate matter out through the stack. These fines must be trapped and either returned to the commercial product or disposed of in some manner. Both wet and dry methods of dust control are used in this industry. Either method requires considerable maintenance to keep in good operation. Testing for the amount of particulate matter exhausted is difficult and standards have not yet been formally established.

Noise Pollution

Noise pollution results in a nuisance factor to neighbors and a health factor to employees. Trends toward correcting excessive noise include better muffling of engines and mechanized equipment, dampening and baffling techniques for other equipment, isolation booths for plant operators, and ear protection devices for individuals who cannot be isolated from the noise.

Future Prospects

Demand

It is pretty safe to assume that silica sand will be the main molding medium during the next ten years. There will, however, be a continued and perhaps accelerated growth in specialty sands for mold facings and cores.

Sand reclamation equipment installed in foundries is on the increase. Reclaiming units are particularly efficient in processing cores and molds made with no-bake resin-bonded sands. The reclaimer need only crush the sand down to grain size and exhaust the fines. The finished sand can be reused for no-bake cores and molds. The reclaimer has become more popular of late because of increasing difficulty and expense incurred in disposing of used foundry sand. The recent development in the industry would indicate a lower demand for sand but that is not necessarily the case. Foundries using the no-bake resin sand usually add 20 to 30% new silica sand to each batch, which is considerably more than the amount of new sand required for each batch of green molding sand. The anticipated increase in casting production will help maintain good sand sales. Resin cost for chemical-bonded sand is $0.02 per pound, plus sampling cost.

Currently in this country there is a strong trend for the smaller foundry to close. There will always be a place for the jobbing foundry that can make short runs, but the small or medium size foundry can no longer compete with the larger foundries for long-run production work. The capital cost of mechanization in order to reduce the unit cost of labor has had considerable influence on the closing of foundries for perhaps 20 years. The economic slowdown in 1971 and part of 1972 accelerated this trend and the need to install expensive air pollution control equipment with no sign of an increase in profits sounded the death knell for additional foundries. While many small foundries have gone out of business, those remaining appear larger and stronger. There has actually been a percentage increase of 43% in the amount of castings produced from 1962 to 1972. At the same time the number of foundries in operation was reduced by approximately 16%. As the foundries became larger they applied more technical knowledge to the operations and also became more scientific in their management approaches. The decade of 1962–1972 appeared to be a decade of great advances in high production molding machines and methods.

This trend toward fewer and larger foundries exerts great influence on the competitive position of sand producers. Each customer becomes more and more important to the competitive sand producers in that area. This greater competition tends to restrict price increases and require additional capital investment to meet tighter specifications.

Supply-Demand Considerations

Just as the smaller foundries have closed because of lack of capital or know-how needed to upgrade their plants and make them competitive in the new markets, so too will many sand plants be closed in the future. The ability to purchase the equipment which would reduce unit labor cost is only one item of expense facing the sand producer. A great deal of additional capital will be required to put in necessary equipment to control air and water purity to the new standards being established.

As the remaining sand producers become larger and fewer they will be able to meet their customers on a more equal footing with respect to the producers' demand and the customers' denial of equitable price increases to cover increased costs of manufacture.

Technology

The modern foundryman is accepting new ideas as a way of life. Typically, we have seen the advent of a new series of core systems, including air set, no-bake, hot box, and cold box, which are now moving into molding. We have seen an increase in mechanized and automated systems to mix, prepare, and reclaim the various types of sand mixes. Automatic molding is developing from a specialized method applicable only to high production foundries to one that can be adapted to many kinds of casting manufacture. Flaskless molding has become commonplace and more variations are coming on the market every year. Casting shakeout, cleaning, and material handling have changed almost beyond recognition. Blast cleaning is a mainstay in most foundries. Materials handling was revolutionized by the forklift truck, the hoist, the crane, and a multiplicity of conveyors. At the same time, of course, melting and metal handling are undergoing widespread change; plants are being cleaned up to look more like modern manufacturing facilities than the old, dark shops of the past. We will see changes that add up to an almost entirely new industry. Metal will still be melted and poured in some sort of mold but the similarity ends there.

The new technology not only will permit castings to compete with other components in the marketplace but will also enable foundries to cope with ecological, occupational, and health regulations, and latter-day problems including labor availability, quality and costs, material shortages, liability requirements, and others. The "seat of the pants" foundryman is going to fade away gradually and will be replaced with manager/specialists who presumably are better equipped to cope with the increasing sophistication of the technology and the overall proliferation of management problems. The foundry of the future will not only be mechanized but automated with built-in quality control checks at numerous positions along the line of production. Quality control checks will also extend to more inspection of incoming raw materials, such as sand. The supplier will be responsible for providing sand to meet the new requirements of the new processes and equipment. As an example, even today silica sand must be cooled to below 37.8°C (100°F) before shipment to a foundry if the sand is to be used in some of the newer, more exotic sand binder systems.

Projections and Forecasts

The demand for foundry sands in the United States will probably continue to grow at a rate of 2 to 6% per year for the next several years. It is estimated that by 1985 the United States foundry industry will be shipping 32 Mt (35 million st) of castings. The trend for castings will be more in the direction of using aluminum rather than steel as the automobile industry, the largest user of castings, moves to lighter weight metals. Aluminum castings do not require refractory sands and therefore more high quality silica sands will be used.

Reclaiming used foundry sands will become more prevalent in the industry as the cost of raw materials increases. Higher costs are attributed primarily to higher transportation and energy costs. Thermal reclamation is presently the most efficient method of sand reclaiming. The present cost per ton of reclaimed sand by this method varies from $12 to $16, depending on the efficiency of the process. Mechanical or attrition type of sand reclaiming will continue as well, but this method does not appear to be as popular as the thermal process. As sand reclaiming becomes more prevalent, the lower grade, less expensive sands will probably be in less demand.

Olivine sand is expected to increase in consumption because it has the unique ability of producing a better, more consistent finish on the metal than does regular silica sand. It also eliminates the problems of silicosis attributed to quartz sands.

Another reason for the expected increase in olivine sand consumption is the fact that finer mesh material, required for the better, cleaner finishes on metal castings, can be used and still maintain good permeability. Silica sands often require coarser, less desirable sizes to maintain permeability requirements.

Bibliography and References

Anon., 1962, *Molding Materials and Methods,* 1st ed., American Foundrymen's Society.

Dietert, H.W., 1966, "Core Sands," *Foundry Core Practice,* 3rd ed., American Foundrymen's Society.

Drake, H.J., 1970, "Sand and Gravel," *Minerals Yearbook, 1970,* US Bureau of Mines, pp. 991–1011.

Laux, J.P., Jr., and Lipton, J.M., 1972, "Introduction to Staurolite Sand," *Foundry,* May 1972, pp. 158–160.

Miske, J.C., Gibbs, D.D., and Maddox, R.S., 1972, *Charting Foundry Trends to 1982,* Penton Publishing Co.

Pajalich, W., 1971, "Sand and Gravel," *Minerals Yearbook, 1971,* US Bureau of Mines, pp. 1043–1066.

Pajalich, W., 1972, "Sand and Gravel," *Minerals Yearbook, 1972,* US Bureau of Mines, pp. 1103–1121.

Tepordei, V.V., 1978–1979, "Sand and Gravel," US Bureau of Mines, Preprint, 29 pp.

Tepordei, V.V., 1980, "Sand and Gravel," US Bureau of Mines, Preprint, 26 pages.

Wood, S.O., Jr., 1969, "Sand and Gravel," *Minerals Yearbook,* US Bureau of Mines, pp. 969–989.

Gem Materials

RICHARD H. JAHNS *

Gem materials, comprising those minerals and closely allied natural substances used for personal adornment, for the fashioning of ornamental objects, or for other decorative purposes, have been valued in human commerce for thousands of years. Amber, jet, natural glasses, and at least 15 mineral species were sought and used by primitive man, and most of the materials in common use today were known well before the time of Christ. Their variety is remarkably great, and the traceable relationships among their respective sources, modes of occurrence, properties, and histories of production, preparation, and valuation are extremely complex. Their uses have been affected by numerous shifts in fashion, and from time to time by superstition, by decrees of church and state, by personal interpretations of religious significance, and by various presumptions concerning medicinal value and influence on human behavior (Ball, 1935, 1950; Kunz, 1913; Pogue, 1915; Whitlock, 1946).

Thanks to their durability, portability, and relative scarcity, some gem stones have been employed as media of exchange, and the investment value of the more precious stones has long been recognized and exploited, especially in Asiatic and European countries. Demands for gem materials have stimulated trade among various peoples for centuries, and they also were responsible for much early geographic exploration. During more recent times they have prompted numerous geological studies and a considerable amount of research on the growth and properties of crystals.

Apart from decorative uses, some naturally occurring gem materials also are employed directly in numerous fields of industry. Substances that are manufactured for gem or industrial uses are termed *synthetic gem materials;* most of these correspond in composition and crystal structure to certain minerals found in nature. *Imitation materials* are those whose use is based upon their resemblance to substances of different composition and greater intrinsic value.

Gem stone is a designation commonly restricted to those materials suitable for personal adornment, and the complementary term *ornamental stone* is applied to the remaining materials not used industrially. This distinction is arbitrary and somewhat inconsistent, owing to numerous geographic differences and historical shifts in end uses. Further, there is a complete gradation between the category of ornamental stones and that of dimension stone. Substances that lose their identity in use are excluded from the gem category; thus a transparent crystal of beryl would not be classified as a gem stone if it were processed as a source of beryllium metal.

The adjectives *precious* and *semiprecious* have been employed as a means for distinguishing different gem materials on the basis of recognized value.

Diamond, emerald, ruby, and sapphire consistently have been regarded as precious stones, and in cut form they commonly are referred to as the *noble gems.* Opal, pearl, turquoise, and chrysoberyl also have been classed as precious at various times. In recent years, however, it has been increasingly recognized that no satisfactory distinction can be based upon assessed value or any group of specific properties, and now all gem stones normally are referred to as precious. Indeed, the tendency is to extend this term to include many of the so-called ornamental stones, and the designation semiprecious has been essentially discontinued.

Unlike most mineral commodities of economic value, gem materials are little altered in processing for end uses. Some stones are used in essentially their natural form, and most are reshaped and polished without significant changes in their fundamental properties. Thus the preparation of gem and ornamental stones is intended chiefly to emphasize or enhance

* Professor of Geology and Applied Earth Sciences, School of Earth Sciences, Stanford University, Stanford, CA.

desirable characteristics that are present initially. The worth of most synthetic gem stones is founded upon the identity between their major properties and the natural characteristics of the corresponding minerals, just as the pleasing qualities of imitation stones depend upon how well certain of their properties simulate those of more valuable substances.

Specifications

The most important specifications for gem materials represent four basic factors: *beauty, durability, rarity,* and *portability.* Beauty, whether expressed as splendor, "purity," attractiveness, or in some other way, is mainly a function of personal response to optical properties such as color, transparency, indices of refraction, and dispersion. Freedom from visible imperfections is important for most transparent stones. For some substances beauty depends also upon asterism, opalescence, textural and structural features, or other special characteristics. *Permanence* of all these properties is essential to the highest valuation of any stone.

Durability, or resistance to abrasion, pitting, chipping, and splitting, determines the extent to which a finished stone retains its details of form and beauty during normal use. Superior durability is most desirable for stones mounted in personal jewelry. It is dependent chiefly upon hardness, and to various degrees upon tenacity and the occurrence and nature of cleavage, parting, or internal strains in the material. Insolubility in water and other common fluids, along with resistance to chemical alteration under normal atmospheric conditions, also are necessary elements of durability in its broadest sense.

Rarity is an essential qualification, and for some stones it outweighs even their physical characteristics. Thus, emerald commands much higher prices than other, more common varieties of beryl that are equally durable and, in the eyes of many viewers, equally beautiful. Similarly, purple spinel is more favorably regarded than the relatively abundant garnet of the same color, whereas green garnet has a much higher value than the more abundant green spinel. However, rarity can be undesirable if it is extreme, as it can lead to uncertainty of supply and even to ignorance or lack of appreciation of the gem stone on the part of potential users. Transparent andalusite, euclase, phenakite, and staurolite, for example, are so rare that they have attracted little attention in the market, despite their durability and the great beauty of some specimens.

Portability, though desirable for obvious reasons, is of lesser significance in governing the value of gem materials. The stones that best meet the other three qualifications are generally small in size, and hence are readily transported from place to place. Those that occur in part as large crystals or masses (e.g., quartz, feldspars) are of lesser stature in terms of their beauty, durability, or rarity.

A fifth factor must be added as a basic measure of worth among cut gems. Known in the trade as *"make,"* it includes perfection of surface finish, the symmetry and accurate positioning of facets or faces, and the degree to which the general proportioning of the stone permits the most effective display of its optical properties. It therefore is a measure of what the cutter, lapidary, or other craftsman has done to enhance the natural attributes of the raw material, or "rough." Similar qualifications apply to the artistry and workmanship reflected in other ornaments and decorative products made from gem materials.

The value of industrial gem materials is essentially independent of their beauty, and instead is based mainly upon shape, hardness and toughness, various optical properties, special pigmenting effects, or combinations of these features. The broad distinction between materials used for decoration and those used industrially is largely a matter of grading according to certain elements of quality, but specifications tend to shift considerably with time in response to changes in supply, market conditions in the jewelry trade, the demands of industry, and other factors.

General Classification and Compositional Variety

Inasmuch as physical properties are expressions of chemical composition, gem materials are most logically classified on the basis of their respective identities as elements and compounds, or various aggregates of these. At least 150 different substances have been used for gem and ornamental purposes, and the most important of these are listed in Table 1. A few are organic in composition and origin, and nearly all the rest are minerals in the strict sense of the term. So great is the variety of these minerals that a consideration of gem materials amounts to a survey of mineralogy itself; in general, however, attention can be focused upon those substances whose chemical bonding is strong and in which the bonding and

TABLE 1—Properties of Gem Minerals and Other Gem Materials

Mineral or other Substance*	General Composition	Chief Color or Colors	Transparent†	Nontransparent†	Mohs' Hardness	Specific Gravity	Index or Indices of Refract.‡	Other Features
Actinolite	Ca, Mg, Fe, OH silicate	Green, yellowish green		x	5–6	3.0–3.2	Mod.	Compact bundles and mats of prisms or fibers
[Allanite (Orthite)]	Ca, Na, Al, Fe, Mn, Mg, rare-earths, OH aluminosilicate	Brown, brownish green, black		x	6	3.5–4.2	High	Bright, pitchy luster
AMBER	Hydrocarbon	Yellow; grayish, reddish, brownish, greenish	x	x	2–2½	1.0–1.1	Low	Commonly contains small inclusions of other organic matter
[Amblygonite]	Li, Na, Al, F, OH phosphate	White, pinkish, reddish, greenish, bluish, yellowish, brownish	x	x	5½–6	3.0–3.2	Mod.	Color generally very pale
[Anatase]	Ti oxide	Yellow, blue, brown, black	x		5½–6	3.8–4.0	Extreme	Adamantine luster; color zoning in some crystals
Andalusite	Al silicate	Gray, pink, green, yellowish, brown, red, violet	x		6–7	3.1–3.2	Mod.	Regularly disposed carbonaceous inclusions in chiastolite
Chiastolite				x				
Apatite	Ca, F, Cl, OH phosphate	Colorless, yellow, green, blue, pink, violet	x		4½–5	3.0–3.2	Mod.	Uneven color and color zoning common
[Axinite]	Ca, Mn, Fe, Mg, Al, OH borosilicate	Gray, blue, yellow, yellowish-reddish brown, green, violet	x		6½–7	3.3	Mod.	Distinct pleochroism common
Azurite	Cu, OH carbonate	Deep blue		x	3½–4	3.8	High	Commonly with malachite
Benitoite	Ba, Ti silicate	Colorless, blue, purple	x	x	6–6½	3.6	High	Uneven color in many crystals
BERYL								
Aquamarine	Be, Al silicate	Blue, greenish blue	x	x	7½–8	2.6–2.9	Mod.	Mild dispersion; pleochroism recognizable in some varieties; some crystals yield cat's-eye stones
Emerald		Green	x					
Golden beryl, Heliodor		Yellow, orangy	x					
Goshenite		Colorless	x					
Morganite		Pink, peach, rose-red	x	x				
Common beryl		Green, yellowish green	x	x				
[Beryllonite]	Na, Be phosphate	Colorless to pale yellow	x		5½–6	2.85	Mod.	Some crystals yield cat's-eye stones

TABLE 1—Properties of Gem Minerals and Other Gem Materials (Continued)

Mineral or other Substance*	General Composition	Chief Color or Colors	Transparent†	Nontransparent†	Mohs' Hardness	Specific Gravity	Index or Indices of Refract.‡	Other Features
[Brazilianite]	Na, Al, OH phosphate	Yellow-green	x		5½	3.0	Mod.	
CALCITE								
Satin spar	Ca carbonate	White, pink, reddish, bluish, greenish, brownish, etc.		x	3	2.7	Low to Mod.	Satin spar chatoyant; marble mottled; Mexican onyx variegated in layers
Marble				x				
Mexican onyx				x				
[Cassiterite]	Sn oxide	Colorless, yellow, red, brown, gray, black	x		6–7	6.5–7.1	V. high	Adamantine luster; color zoning in some crystals
[Celestite]	Sr sulfate	Blue, gray, yellowish	x		3½	4.0	Mod.	Pleochroism recognizable in deeply colored specimens
[Chromite]	Fe, Cr oxide	Brownish black to black		x	5½–6	4.6–5.2	V. high	Pitchy, submetallic to metallic luster
CHRYSOBERYL								
Alexandrite	Be aluminate	Green (daylight), red (artificial light)	x		8½	3.5–3.8	High	Strong pleochroism; mild dispersion
Cat's-eye (Cymophane)		Yellowish green to green		x				
Chrysolite		Yellowish, bluish, and brownish green	x					
Chrysocolla	Cu silicate, hydrous	Green, blue		x	2–2½	2.0–2.4	Low to Mod.	Commonly impregnated with silica
[Cinnabar]	Hg sulfide	Red, brownish to grayish red	x	x	2–2½	8.1	Extreme	Adamantine luster
[Cobaltite]	Co, As sulfide	Pinkish silvery white		x	5½	6.0–6.4	Extreme	
CORAL								
Black coral	Ca carbonate	Black		x	3½–4	2.6–2.7	Mod.	Branching forms common
Common coral		White, gray, green, etc.		x				
Precious coral		Orange-red to red		x				
Cordierite (Iolite, dichroite)	Al, Mg, Fe aluminosilicate	Gray, blue, yellowish, reddish, greenish, brown	x	x	7–7½	2.5–2.8	Low	Strong pleochroism common

TABLE 1—Properties of Gem Minerals and Other Gem Materials (Continued)

Mineral or other Substance*	General Composition	Chief Color or Colors	Transparent†	Nontransparent†	Mohs' Hardness	Specific Gravity	Index or Indices of Refract.‡	Other Features
CORUNDUM Ruby	Al oxide	Pink, red, orangy to purplish red, purple	x	x	9	3.9–4.1	High	Color zoning common; distinct pleochroism; mild dispersion
Sapphire		Colorless, blue, yellow, pink, green, violet, etc.	x	x				
Star ruby and star sapphire				x				
[Covellite]	Cu sulfide	Blue to blue-gray		x	1½–2	4.6–4.8	Extreme	Submetallic luster
[Danburite]	Ca borosilicate	Colorless, yellow, yellowish brown	x		7	3	Mod.	Very strong dispersion
Datolite	Ca, OH borosilicate	White, gray, yellowish, reddish, greenish, brownish		x	5–5½	2.9–3.0	Mod.	Streaking or uneven color distribution common
DIAMOND	C	Colorless, blue-white, yellow, red, brown green, blue, gray	x		10	3.5	Extreme	Adamantine luster; very strong dispersion
Diopside	Ca, Mg, Fe silicate	Colorless, gray, yellow, green, blue	x		5–6	3.2–3.4	Mod.	Color zoning in many crystals
Dioptase	Cu silicate, hydrous	Green	x		5	3.3	Mod.	Resembles emerald
[Dumortierite]	Al, B, OH aluminosilicate	Blue, greenish blue, reddish violet		x	7	3.3	Mod.	Very strong pleochroism; commonly in dense aggregates with quartz
Enstatite Bronzite	Mg silicate	White, gray, yellowish, brownish, greenish		x	5–6	3.2	Mod.	Bronzite variety fibrous, with chatoyance and bronzy luster
Epidote (Pistacite)	Ca, Al, Fe, OH aluminosilicate	Gray, yellow, yellow-green, green, reddish, brown	x	x	6–7	3.3–3.5	High	Strong pleochroism; mild dispersion
[Euclase]	Be, OH aluminosilicate	Colorless, green, blue	x		7½	3.0–3.1	Mod.	Colors generally pale; mild pleochroism
[Euxenite]	Ca, rare-earths, U, Th, Cb, Ta, Ti oxide	Brown to black		x	5–6	4.8–5.9	V. high to Extreme	Bright, pitchy luster

TABLE 1—Properties of Gem Minerals and Other Gem Materials (Continued)

Mineral or other Substance*	General Composition	Chief Color or Colors	Transparent†	Nontransparent†	Mohs' Hardness	Specific Gravity	Index or Indices of Refract.‡	Other Features
FELDSPARS								
Adularia, Orthoclase	K, Na aluminosilicate	Colorless to yellow	x		6–6½	2.5–2.8	Low to Mod.	Moonstones opalescent, chatoyant; sunstone red spangled; aventurine green or red spangled; nontransparent gem labradorite and oligoclase gray with marked play of colors
Amazonstone		Green, bluish to yellowish green		x				
Precious moonstone		White to bluish gray		x				
Aventurine	Na, K, Ca aluminosilicate	White or gray		x				
Moonstone				x				
Sunstone				x				
Labradorite, Oligoclase	Ca, Na aluminosilicate	White, gray	x	x				
[Fergusonite]	Rare-earths, Fe, Cb, Ta, Ti oxide	Gray, yellow, brown, black		x	6	4.0–7.0	V. high	Bright, pitchy luster in darkest varieties
Fluorite	Ca fluoride	Colorless, white, yellow, green, blue, red, purple, brown	x	x	4	3.1–3.3	Low	Uneven color and color zoning common
[Gadolinite]	Fe, Be, rare-earths silicate	Brown, greenish brown, black		x	6½–7	4.0–4.6	High	Bright to moderately dull luster
GARNETS								
Almandine (Carbuncle)	Fe, Al silicate	Red, violet, brownish, black	x	x	7½			
Pyrope	Mg, Al silicate	Red to black	x		7–7½			
Rhodolite	Mg, Fe, Al silicate	Rose-red to purple	x		7–7½	3.4–4.3		
Grossularite	Ca, Al, Fe silicate	White, yellow, red, green, brown	x	x	7		High	Color zoning common in some varieties; dispersion moderate to high; some varieties show asterism
Spessartine	Mn, Fe, Al silicate	Red, orangy, yellowish, brown	x		7–7½			
Essonite (Hessonite)	Ca, Mn, Fe, Al silicate	Yellowish orange to orange-brown	x		7–7½			
Demantoid		Yellow to green	x					
Melanite	Ca, Fe, Al silicate	Black		x	6½–7			
Topazolite		Yellow	x					
Uvarovite	Ca, Cr silicate	Green	x		6½–7			

TABLE 1—Properties of Gem Minerals and Other Gem Materials (Continued)

Mineral or other Substance*	General Composition	Chief Color or Colors	Transparent†	Nontransparent†	Mohs' Hardness	Specific Gravity	Index or Indices of Refract.‡	Other Features
GLASSES (manufactured)	Wide range of silicates, borosilicates, etc.	Colorless, any color	x	x	2½–5	1.3–5.5	Low to Extreme	Wide range in luster, dispersion, etc.
Gold	Au	Golden yellow		x	2½–3	16–19		Commonly in quartz
GYPSUM								
Alabaster	Ca sulfate, hydrous	White		x	2	2.2–2.4	Low	Satin spar fibrous, silky, chatoyant
Satin spar			x	x				
[Hambergite]	Be, OH borate	Colorless	x		7½	2.4	Mod.	
[Haüynite]	Na, Ca sulfate-aluminosilicate	Blue, green, red, yellow		x	5½–6	2.4–2.5	Low	Uneven color common
[Helvite]	Mn, Fe, Zn, Be, S silicate	Gray, yellow, red, green, brown	x		6	3.2–3.7	High	
HEMATITE	Fe oxide	Steel gray to black		x	5–6½	4.9–5.3	Extreme	Bright metallic luster
[Hemimorphite (Calamine)]	Zn, OH silicate, hydrous	Colorless, white, blue, green, gray, brown	x	x	5	3.4–3.5	Mod.	Strong dispersion
[Howlite]	Ca, OH borosilicate, hydrous	White		x	3½	2.6	Mod.	Commonly in matrix of other minerals
Hypersthene Bronzite	Mg, Fe silicate	Yellow, brown, green		x	5–6	3.3–3.5	Mod. to High	Bronzite variety fibrous to lamellar, with chatoyance and bronzy luster
Idocrase (Vesuvianite)	Ca, Mg, Fe, Al, OH silicate	Green, brown, yellow, blue		x	6½	3.3–3.5	High	Color uneven in some specimens
JADE								
Jadeite	Na, Al silicate	White, green, yellowish, brownish, violet		x	6½–7	3.3	Mod.	Compact and tough; color commonly uneven, with mottling in white
Nephrite	Ca, Mg, Fe, OH silicate	Yellowish green		x	6½–7	3.0–3.2	Mod.	
JET	C, with H, O, etc.	Black		x	2–2½	1.1–1.4		Tough and compact; resinous luster

TABLE 1—Properties of Gem Minerals and Other Gem Materials (Continued)

Mineral or other Substance*	General Composition	Chief Color or Colors	Transparent†	Nontransparent†	Mohs' Hardness	Specific Gravity	Index or Indices of Refract.‡	Other Features
Kornerupine	Mg, Fe, Al borosilicate	Colorless, yellow, green, brown	x		6½	3.3–3.5	Mod. to High	
Kyanite	Al silicate	Colorless, gray, blue, green, brown	x	x	4–7	3.5–3.7	High	Uneven color common; strong pleochroism
Lazulite	Mg, Fe, Al, OH phosphate	Blue		x	5–6	3.1–3.4	Mod.	Distinct pleochroism
LAZURITE (Lapis lazuli)	Na, S aluminosilicate	Deep blue to greenish blue		x	5–5½	2.4–2.9	Low	Commonly with matrix of pyrite and other minerals
[Lepidolite]	K, Li, Al, F, OH aluminosilicate	Colorless, gray, pink, red, violet, purple, yellow		x	2½–4	2.8–3.3	Low to Mod.	Mainly in dense crystalline aggregates
Malachite	Cu, OH carbonate	Green		x	3½–5½	3.9–4.0	Mod. to High	Commonly impregnated or interlayered with other minerals
[Marcasite]	Fe sulfide	Bronze yellow		x	6–6½	4.9		
[Microlite]	Ca, Na, F, OH, Ta, Cb oxide	Yellow, red, green, brown, black	x		5–5½	5.5–6.4	V. High	Color zoning common
Moldavite (Tektite)	Aluminosilicate glass	Green, yellowish green, brown, light to very dark gray	x	x	5½	2.3–2.5	Low	Possibly of meteoritic origin
[Nepheline]	Na, K aluminosilicate	Colorless, gray, yellowish, pinkish, reddish		x	5½–6	2.6	Low	Commonly with haüynite, sodalite
Obsidian	Aluminosilicate glass	Colorless, red, green, brown, black	x	x	5–5½	2.3–2.6	Low to Mod.	Some types variegated
OLIVINE								
Chrysolite	Mg, Fe silicate	Yellow, greenish yellow, brownish	x	x	6½–7	3.2–3.5	Mod. to High	Mild dispersion
Peridot		Slightly yellowish green	x					
OPAL	Si silicate, hydrous	Colorless, many colors		x	5½–6½	1.9–2.3	Low	Play of colors; many varieties impure

TABLE 1—Properties of Gem Minerals and Other Gem Materials (Continued)

Mineral or other Substance*	General Composition	Chief Color or Colors	Transparent†	Nontransparent†	Mohs' Hardness	Specific Gravity	Index or Indices of Refract.‡	Other Features
PEARL	Ca carbonate	White, gray, black, many pale colors		x	2½–4	2.5–2.7	Mod.	Various natural shapes used
[Pectolite]	Na, Ca, Mn, OH silicate	Colorless, white, gray, pinkish		x	4½–5	2.7–2.9	Mod.	Commonly fibrous
Phenakite	Be silicate	Colorless, yellow, rose-red, brown	x		7½–8	3.0	Mod.	
[Pollucite]	Cs aluminosilicate	Colorless, white, gray, pinkish	x	x	6½	2.9	Low	
[Polycrase]	Ca, rare earths, U, Th, Ti, Cb, Ta oxide	Brown to black		x	5–6	4.8–5.4	Extreme	Bright, pitchy luster
Prehnite Chlorastrolite	Ca, Al, OH aluminosilicate	White, gray, yellowish green, green		x	6–6½	2.8–3.2	Mod.	Chlorastrolite green with white mottling; chatoyant
PYRITE	Fe sulfide	Brass yellow		x	6–6½	5.0		Some crystals used in natural form
[Pyrochlore]	Na, Ca, rare-earths, Th, U, Mg, Fe, Mn, Cb, Ta, F oxide	Reddish brown to black		x	5–5½	4.2–5.7	V. high to Extreme	Bright, pitchy luster
QUARTZ Megascopically crystalline varieties:								
Amethyst	Si silicate	Purple, violet	x	x	7	2.7	Low	Aventurine spangled, glistening; cat's-eye and tiger's-eye fibrous, chatoyant; gold quartz with native gold; rutilated quartz with many needles of rutile
Aventurine		Yellow, red, green, brown		x				
Cat's-eye		Gray, green, brown		x				
Citrine		Yellow	x					
Gold quartz		Milky white		x				
Rock crystal		Colorless	x					
Rose quartz		Rose-pink	x	x				Color commonly uneven; some varieties show asterism
Rutilated quartz (Sagenite)		Colorless to smoky yellow	x	x				

TABLE 1—Properties of Gem Minerals and Other Gem Materials (Continued)

Mineral or other Substance*	General Composition	Chief Color or Colors	Transparent†	Nontransparent†	Mohs' Hardness	Specific Gravity	Index or Indices of Refract.‡	Other Features
Sapphire quartz (Siderite)		Blue	x					
Smoky quartz (Cairngorm)		Smoky yellow to black	x	x				
Tiger's-eye		Red, blue, brown		x				
Cryptocrystalline varieties:								
Agate	Si silicate, commonly somewhat impure	Varicolored		x	7	2.6	Low	Agate with curved or angular bands; onyx and sardonyx with straight bands; bloodstone with red spots and veins; jasper with bands or irregular markings; plasma with white or yellowish spots; many combinations are known
Bloodstone		Dark green		x				
Carnelian		Red, orange-red		x				
Chalcedony		White, gray, bluish, black		x				
Chrysoprase		Light green		x				
Jasper		Yellow, red, brown, green, bluish gray		x				
Onyx		Varicolored		x				
Plasma		Green		x				
Prase		Green		x				
Sardonyx		Red, brown, white		x				
[Realgar]	As sulfide	Red to orange-yellow		x	1½–2	3.6	Extreme	Commonly with matrix of other minerals
[Rhodizite]	Na, K, Li, Al, Be borate	Colorless, yellow	x		8	3.4	Mod.	
Rhodonite	Mn, Fe, Ca silicate	Pink, rose-red, gray, yellowish		x	5½–6½	3.4–3.7	High	Commonly with black spots and veins
Rutile	Ti oxide	Yellow, red, violet, brown, greenish, bluish, black	x	x	6–6½	4.2–4.3	Extreme	Adamantine luster
[Samarskite]	Ca, Fe, rare-earths, Pb, Th, Cb, Ta, Ti, Sn oxide	Brown to black		x	5–6	5.6–6.2	Extreme	Bright, pitchy luster
Scapolite (Wernerite)	Na, Ca, Cl, CO_3, SO_4 aluminosilicate	Colorless, white, gray, pink, reddish, bluish, greenish	x	x	5–6	2.5–2.8	Low to Mod.	Chatoyance in some varieties

TABLE 1—Properties of Gem Minerals and Other Gem Materials (Continued)

Mineral or other Substance*	General Composition	Chief Color or Colors	Transparent†	Nontransparent†	Mohs' Hardness	Specific Gravity	Index or Indices of Refract.‡	Other Features
[Scheelite]	Ca tungstate	Colorless, white, gray, yellow	x		4½–5	6.1	V. high	
Sepiolite (Meerschaum)	Mg, OH aluminosilicate, hydrous	White, creamy yellow, pink, greenish		x	2–2½	1–2	Low	
Serpentine	Mg, OH silicate	Yellow, light to very dark green		x	2–4	2.5–2.8	Mod.	Fibrous to massive; color mottling and streaking in some varieties
Sillimanite	Al silicate	White, gray, greenish, brownish		x	6–7½	3.2	Mod.	Commonly in tough, fibrous aggregates
Smithsonite	Zn carbonate	White, yellow, green, blue		x	5	4.1–4.5	Mod. to High	
Sodalite	Na, Cl aluminosilicate	Colorless, gray, yellow, pink, blue, green		x	5–6	2.2–2.4	Low	Commonly with nepheline, haüynite
Sphalerite	Zn sulfide	Colorless, yellow, red, green, brown, black	x		3½–4	3.9–4.1	Extreme	Resinous adamantine luster; extreme dispersion
Sphene (Titanite)	Ca titanosilicate	Gray, yellow, green, brown, red, black	x		5–5½	3.4–3.6	High to V. high	Distinct pleochroism; strong dispersion; color zoning common
SPINELS	Mg, Fe, Zn, Al oxide	Yellow, pink, red, orange-red, violet, purple, blue, green	x	x	8	3.5–3.7	High	Moderate dispersion; asterism in some varieties
SPODUMENE								
Hiddenite	Li, Al silicate	Yellow-green to green	x		6–7	3.0–3.2	Mod.	Distinct pleochroism; mild dispersion
Kunzite		Pink to lilac	x					
Triphane		Yellow to colorless	x					
Staurolite	Al, Fe, OH silicate	Reddish to yellowish brown, brownish black	x	x	7–7½	3.6–3.8	High	Cruciform twins common; moderate dispersion
Steatite (Soapstone) Agalmatolite	Mg, OH silicate	White, gray, green, reddish, brownish		x	1½–2½	2.7–2.8	Low to Mod.	Compact aggregate of talc and other minerals
Thomsonite	Na, Ca aluminosilicate, hydrous	Colorless, white, yellow, pink, red, green		x	5–5½	2.3–2.4	Low	Color mottling and banding common

TABLE 1—Properties of Gem Minerals and Other Gem Materials (Continued)

Mineral or other Substance*	General Composition	Chief Color or Colors	Transparent†	Nontransparent†	Mohs' Hardness	Specific Gravity	Index or Indices of Refract.‡	Other Features
TOPAZ	Al, F, OH silicate	Colorless, pink, yellow, blue, purplish, reddish, orangy, brownish	x		8	3.4–3.6	Mod.	Mild dispersion
TOURMALINE								
Achroite	Na, Li, Fe, Al, OH borosilicate	Colorless	x		7–7½	2.9–3.2	Mod.	Color zoning very common; some varieties yield cat's-eye stones; strong pleochroism; mild dispersion
Emeralite		Green	x	x				
Indicolite		Blue	x					
Rubellite		Rose-red to purplish	x	x				
Schorlite		Black	x	x				
Siberite		Violet	x					
Other varieties		Green, yellow, brown, etc.	x	x				
TURQUOISE	Cu, Al, OH phosphate, hydrous	Green, blue, yellowish, greenish gray		x	5–6	2.6–2.8	Mod.	Commonly in matrix of other minerals
Variscite	Al, Fe phosphate, hydrous	White, yellow, blue, green		x	3½–5	2.5–2.8	Mod.	Commonly in matrix of other minerals
[Willemite]	Zn silicate	White, gray, yellow, red, green, blue, brown	x		5–6	3.9–4.2	Mod. to High	Deeply colored varieties pleochroic
ZIRCON								
Blue zircon	Zr silicate	Blue, greenish blue	x		6–7½	4.0–4.8	V. high	Strong dispersion; commonly metamict, wholly or in part
Hyacinth, Jacinth		Yellow, orange, red, brown	x					
Jargon		Gray, green, smoky brown, etc.	x					
White zircon		White	x					
Zincite	Zn oxide	Red, orange, yellow	x		4	5.4–5.7	V. high	
Zoisite	Ca, Al, OH aluminosilicate	Colorless, gray, green, blue, brown	x	x	6–6½	3.2–3.4	Mod. to High	Colored varieties strongly pleochroic
Thulite		Pink, rose-red		x				

* Those substances in most common use are shown in capital letters; those rarely used for gems or for ornamental purposes are shown within brackets.

† As ordinarily used for gem or ornamental purposes.

‡ Designated as follows:

Low—<1.56 Mod.—1.56-1.70 High—1.70-1.90 V. high—1.90-2.10 Extreme—>2.10

nature of the constituent ions or atoms produce desirable optical effects.

Silicate minerals constitute about one-third of all gem materials, aluminosilicates nearly one-fifth, and oxides nearly one-seventh. The remaining compositional groups, listed in order of decreasing numerical representation, are the sulfides, phosphates, borosilicates, carbonates, native elements, sulfates, titanosilicates, borates, halides, and aluminates.

Twenty-two minerals and mineral groups, together with five other substances, have been most widely used as gem materials. Among these, the elemental mineral diamond is by far the most important in terms of value. The other minerals can be classified as follows:

Silicates—beryl, garnets, jadeite, nephrite, olivine, opal, quartz, spodumene, topaz, zircon.
Aluminosilicates—feldspars, lazurite.
Oxides—corundum, hematite, spinel.
Aluminate—chrysoberyl.
Borosilicate—tourmaline.
Carbonate—calcite.
Phosphate—turquoise.
Sulfate—gypsum.
Sulfide—pyrite.

The principal substances of organic origin are amber, ivory, jet, and calcium carbonate in the form of coral and pearl. The natural glasses complete the short list of common gem materials not ordinarily regarded as minerals.

Further classification of gem materials commonly is made on the basis of specific properties that are responsible for beauty and durability. Data on these features are summarized in Table 1.

Influence of Properties on Preparation and Use

Decorative Uses

The dominant use of gem materials today lies in the manufacture of jewelry. Typical finished articles are rings, bracelets, brooches, charms, clips, cuff links, earrings, necklaces, hair ornaments, pendants, tiepins, jewel-decorated watches, and jewel flowers and other specially designed objects. Of lesser but substantial total value are the materials normally fashioned into book ends, chimes, desk ornaments, lamp bases, various mosaics, pipe bowls and other smokers' articles, polished spheres, statuettes, trays, vases, special bowls, cups, and dishes, and a host of other articles not intended for personal wear. Nearly all these are primarily ornamental in purpose, but many are utilitarian as well. The variety among jewelry and other objects made from gem materials has been limited solely by man's own imagination. Numerous summary descriptions of these products have been published (Baerwald and Mahoney, 1949; Ball, 1950; Rogers and Beard, 1940), and information on both past and current developments in the trade is provided by several of the periodicals listed near the end of this chapter.

Gems, ornamental objects, and rough stones also are displayed as such in numerous personal and institutional collections, which include nearly all known types of material. Indeed, several collections have served directly or indirectly as bases for books on gems or gemology (Merrill, 1922; Church, 1924; Merrill and Foshag, 1943; McLintock and Sabine, 1951; Desautels, 1979). The aggregate value of existing collections, many of them accumulated over long periods of time, is difficult to estimate but assuredly is very great.

Although fashion and other human factors have a highly significant influence upon the preparation and uses of gem stones, the respective physical properties of these materials are fundamentally more important because they impose practical limits on the finished forms and specific uses for which each kind of stone is well qualified. Only a few of the major properties can be considered here (Tables 1, 7); additional data are included in an extensive published record, and specially in definitive treatments by Bauer (1904, 1968, 1969), Bauer and Schlossmacher (1932), Kraus and Slawson (1947), Smith and Phillips (1958), and Gubelin (1979).

Hardness: The hardest minerals, such as diamond, corundum, chrysoberyl, spinel, topaz, and most varieties of beryl, garnet, tourmaline, and zircon, are superior to all others for use in personal jewelry, as they can be cut into faceted gems that take an excellent and lasting polish. These minerals lie above quartz on the Mohs' scale (Tables 1, 2) and hence are little affected by daily contact with fabrics, dust, and grit in the air and additional materials that contain tiny particles of quartz, feldspar, or other abrasive substances.

Moderately hard minerals that can be scratched by quartz also are much used in jewelry, but they generally become dull with prolonged wear unless they are handled carefully. Feldspars, olivine, opal, rutile, spodumene, and some garnets, tourmaline, and zircon belong in this category. Relative softness is one

TABLE 2—Relative Hardness of Selected Materials

Material	Relative Hardness Value*	Hardness Rank on Mohs' Scale
Gypsum	20-32	2
Calcite	75-130	3
Fluorite	100-150	4
Flint glass†	480	—
Apatite	500-530	5
High-alumina glass†	550	—
Orthoclase	560-600	6
Quartz	700-900	7
Topaz	1050-1250	8
Spinel†	1100-1250	8
Corundum†	1700-2200	9
Diamond	8000-8500	10

* Determined mainly through measurements, by several investigators, of indentations made by a diamond blade. Variations in hardness value for a single mineral species reflect differences in determinative techniques, as well as actual differences in hardness for different crystallographic directions.
† Synthetic material.

of the chief disadvantages of glasses that are used as gems. Loss of polish and sharpness through wear is particularly objectionable in facet-cut gems, and hence the softer stones prepared in this form are not satisfactory for rings that are worn continuously. They commonly are suitable, however, for pendants, brooches, and other types of jewelry that are less exposed to hard use. Amber, calcite, jet, lazurite, turquoise, and other materials of inferior hardness are used almost exclusively in rounded rather than faceted form.

The Mohs' scale indicates nothing more than hardness *rank,* and it is far from linear in terms of correlated *values* of hardness (Table 2). Thus corundum is at least twice as hard as quartz, and diamond is more than four times as hard as corundum. Further, many gem minerals show distinct *variations* in hardness with crystallographic direction. Such vectorial differences are of vital importance in the cutting and polishing of gem diamonds and other very hard stones, as well as in certain aspects of their industrial uses (Kraus and Slawson, 1939; Kraus et al., 1942, 1943, 1948; Long, 1951, 1954; Long and Slawson, 1952; Denning, 1953; Grodzinski, 1953).

Cleavage, Parting, and Tenacity: Many crystalline substances used as gems are cleavable in one or more directions, some can be separated along surfaces of parting ("false cleavage"), and a few exhibit both these properties. Cleavage and parting cracks commonly are present as flaws in rough material, and the portions of stones thus affected must be eliminated in the preparation of top-grade gems. Such imperfections are difficult to detect in certain nontransparent stones (e.g., cat's-eye, star ruby and sapphire), and sometimes are overlooked in cutting and polishing. Though they have little effect on the appearance of many gems, they do constitute surfaces of weakness along which splitting can occur during use.

Minerals with fairly ready cleavage, such as spodumene, topaz, and some olivine and spinel, require special precautions in cutting, polishing, and setting in jewelry, lest cracks and chips develop. Spodumene is particularly difficult to handle without inducing separation along its two directions of cleavage. On the other hand, the smooth but very difficult cleavages of diamond, quartz, some beryl, and a few other very hard minerals, as well as parting surfaces in corundum, diamond, and some garnets and pyroxenes, provide useful controls for the preliminary splitting of rough stones by the gem cutter.

Many gems fashioned from beryl, quartz, zircon, and several other minerals without ready cleavage or parting are easily chipped or shattered when knocked sharply against another object. This commonly is due to intrinsic *brittleness* of the mineral, to strains set up during cutting and polishing of the gem, or to strains resulting from heat treatment given the stone in order to improve its color. Contrasting with brittleness on the tenacity scale is *toughness,* which characterizes jadeite, nephrite, sillimanite, and the microcrystalline aggregates of diamond and other minerals known as carbonado (Table 3). Most materials with such high fracture strength are compact polycrystalline aggregates that are very resistant to wear, and hence they are as suitable for many forms of jewelry as other minerals of superior hardness.

Transparency, Brilliance, and Fire: Transparent stones ordinarily are cut in faceted forms, whereas nontransparent stones are finished in flat or rounded forms with polished or carved surfaces (Figs. 1, 2). Exceptions to this practice include the transparent ruby, sapphire, beryl, quartz, and tourmaline that are carved or shaped into spheres and other rounded objects, and the translucent varieties of opal and quartz that are facet-cut. Almandine, pyrope, and melanite garnets, as well as black tourmaline and a few other transparent but extremely dark-colored minerals, are prepared in both forms. The same is true of some opaque minerals, particularly hematite, marcasite, and pyrite.

The pleasing brilliance and sparkle, or gen-

TABLE 3—Orders of Hardness and Toughness of Selected Gem Materials

Order of Increasing Hardness (Numbers Indicate Rank on Mohs' Scale)	Order of Increasing Toughness*
Orthoclase (6)	Spodumene
Olivine	Orthoclase
Nephrite	Topaz
Spodumene	Zircon
Jadeite	Garnets
Sillimanite	Beryl
Quartz (7)	Spinels
Garnets	Olivine
Zircon	Diamond
Beryl	Quartz
Topaz (8)	Corundum
Spinels	Chrysoberyl
Chrysoberyl	Sillimanite
Corundum (9)	Jadeite
Carbonado†	Nephrite
Diamond (10)	Carbonado†

* Determined from tests of fracture strength and widespread lapidary experience (Parsons, 1969).
† Crystallite aggregate of diamond and other minerals.

eral "liveliness," shown by many faceted stones are related fundamentally to their indices of refraction. The effects of light refracted into such gems are more important than the effects of light reflected directly back from their surfaces. Maximum brilliance is obtained when the highest percentage of light entering the stone through its upper facets is totally reflected from its lower facets and ultimately emerges in upward directions, as shown in Fig. 3. For all practical styles of cuts, the downward or lateral "leakage" of light from a given gem can be held to a minimum if the index (or indices) of refraction of the mineral is high and the critical angle at the mineral-air interface is therefore low, because total reflection then takes place within the stone for light rays incident upon the lower facets over a wide range of angles. Thus diamond, rutile, zircon, and other minerals with very high indices of refraction (see Table 1) are distinguished by their sparkle and brilliance when properly cut.

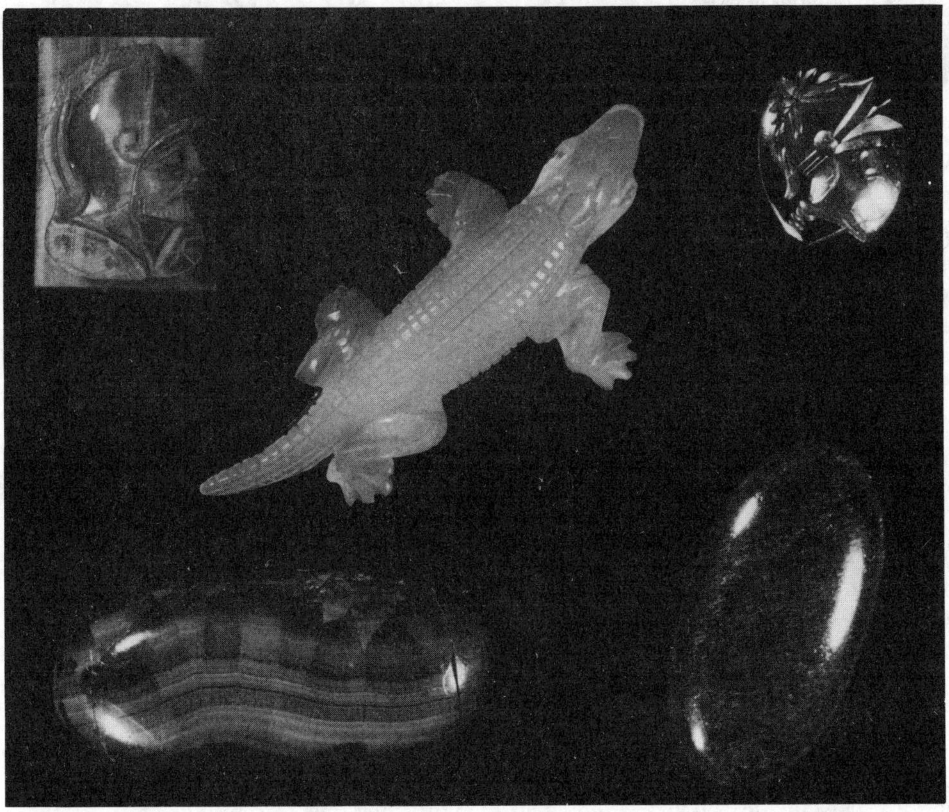

FIG. 1—*Typical nontransparent gems, shown slightly greater than natural size. Carved tiger's-eye, upper left; carved hematite, upper right; carved fire opal, center; cabochon-cut malachite, lower left; cabochon-cut obsidian showing opalescence, lower right. Pennsylvania State University Collection.*

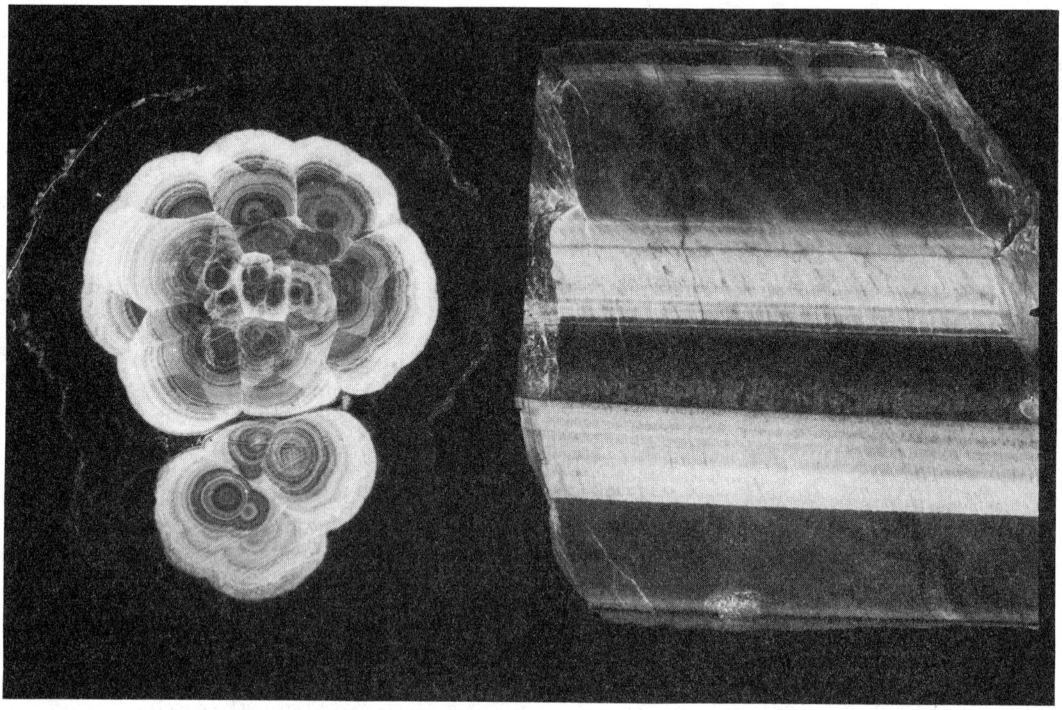

FIG. 2—*Polished slabs of nontransparent gem materials, shown slightly greater than natural size. Concentrically banded malachite (light) rimmed by azurite (dark) at left; tiger's-eye at right. Pennsylvania State University Collection.*

Many gem minerals are optically anisotropic, and hence show double refraction. This complicates the paths of light transmitted through them, but ordinarily contributes nothing to their brilliance. Notable effects, however, are shown by isotropic and anisotropic minerals that have strong *dispersion,* i.e., that have markedly different indices of refraction for frequencies of light corresponding to different parts of the visible spectrum (Fig. 3). These minerals, which include diamond, rutile, sphalerite, sphene, zircon, and some varieties of garnet (Table 1), yield facet-cut gems that show the shifting flashes of color known as *fire.*

Color and Luster: Color is an important property of all gem materials, and is by far the most desirable feature of those that are opaque (e.g., jade, lazurite, turquoise) and those that are transparent but have low or moderate indices of refraction (e.g., apatite, beryl, spodumene). Almost any color can be found among the gem minerals (Bauer, 1968; Bank, 1970; Nassau, 1974–1975), and some species such as beryl, corundum, quartz, tourmaline, and the garnets and spinels exhibit a wide range of colors (Tables 1, 7). Preferences for various colors tend to change with time, but red, blue, green, and violet to purple gems consistently have been more popular than those of yellow, yellowish green, or greenish blue color. Intense colors in medium to moderately dark tones tend to compensate for lack of brilliance and fire in many stones.

Evenness of color is sought in most material for facet-cut gems, but distinct color zoning also enhances the value of some ruby, sapphire, tourmaline, and garnets. Another desirable but fundamentally different type of variation is shown by anisotropic colored minerals in which the selective absorption of certain components of white light differs with the direction of light transmission. Such *pleochroism* leads to differences in hue, tone, or both these features when a gem stone is viewed in different directions by transmitted light. It is most strongly expressed by andalusite, axinite, benitoite, chrysoberyl, cordierite, dioptase, epidote, kyanite, spodumene, staurolite, tourmaline, zoisite, and some deeply colored varieties of beryl, pyroxenes, and quartz. In general, pleochroic stones are carefully oriented with respect to their crystallographic directions when they are

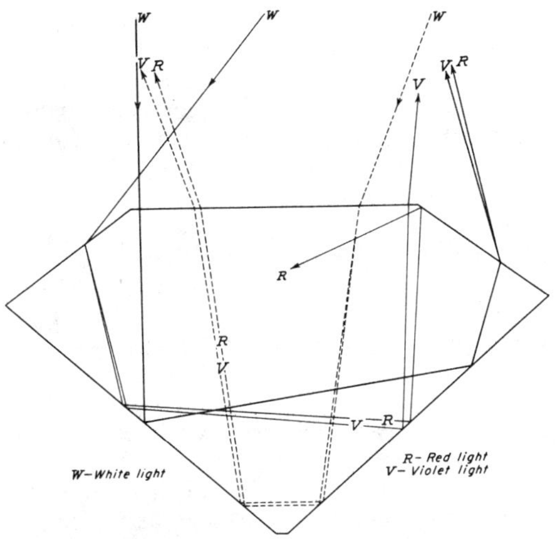

FIG. 3—*Section of a brilliant-cut stone with highly desirable optical properties. Typical reflection, refraction, and dispersion of incident white light are represented diagrammatically.*

cut, so that they can display the most pleasing colors and color tones in the normal directions of view.

The highly prized alexandrite variety of chrysoberyl and its synthetic corundum analogue strongly absorb yellow light and are pleochroic as well. These features, together with the spectral contrasts between natural and artificial light, produce the "alexandrite effect" in such gems, i.e., they appear green in daylight and red to purplish red in most other kinds of illumination. The green color can be attributed mainly to the presence of some chromium ions in the crystal structures, and the marked color changes to differential light absorption in spectral ranges to which the human eye is highly sensitive (Poole, 1964; Farrell and Newnham, 1965; White, Roy, and Crichton, 1967; Hassan and El-Rakhawy, 1974).

Color mottling, streaking, or banding is primarily responsible for the desirability of many nontransparent gem materials, which commonly are prepared in forms that yield the most attractive markings or patterns (Figs. 1, 2). Some of these color variations are due to differences in amounts of pigmenting impurities, as in agate, sardonyx, and other types of cryptocrystalline quartz; some to variations in texture within an aggregate of a single mineral, as in tiger's-eye (Fig. 2) and chalcedonic quartz; others to enclosures of one material in another, as carbonaceous inclusions in andalusite and inclusions and veinlets of manganese oxides in agate and rhodonite; and still others to aggregates of two or more minerals, as azurite and malachite (Fig. 2), lazurite and pyrite, and turquoise with various "matrix" minerals.

The luster of a gem material, or its surface appearance in reflected light, is independent of its color. Among the nontransparent minerals, the bright metallic luster of hematite and pyrite, the bright pitchy luster of allanite and several chemically complex multiple oxides (e.g., euxenite, fergusonite, pyrochlore), the bronzy luster of certain pyroxenes, and the delicately silky luster of some jades and feldspars are highly desirable. Many of these substances show vectorial differences in luster, and gems cut from them commonly are oriented to provide the most attractive possible appearance. The greasy, oily, resinous, waxy, or dull luster of other nontransparent materials (e.g., amber, lazurite, prehnite, turquoise) generally contributes little to their beauty, which is derived instead from other properties.

A splendent, or adamantine, luster characterizes most of the transparent gem stones with high indices of refraction, and enhances their brilliant appearance. The vitreous luster of many minerals with lower indices of refraction neither enhances nor strongly detracts from their appearance as facet-cut gems, but most transparent minerals with markedly greasy, oily, resinous, or dull surface appearance (e.g., some apatite, epidote, feldspars) are cut in rounded form because faceted gems with such luster tend to be unattractive and "dead" in appearance.

Among all the known gem stones, only the diamond combines great durability with adamantine luster, pleasing colors, and outstanding brilliance and fire. However, the most desirable deep blue, green, and red diamonds of good quality, known in the trade as "fancies," are so rare that even this mineral generally lacks one of the major attributes of the ideal gem stone. Excellent colors are provided by emerald, ruby, and sapphire, but these stones have limited brilliance and fire. Rutile, on the other hand, combines good color and fire with great brilliance when properly cut, but it lacks superior hardness. And sphalerite, with its very high index of refraction and by far the highest dispersion among all transparent gem minerals, probably would have very great value were it not for softness and ready cleavage that restrict its use to display in collections.

Other Properties: Asterism, as shown by star ruby and star sapphire, is due to tiny, regularly arranged inclusions; such stones therefore must be carefully oriented when cut, in order to center the starlike blades of light on the rounded upper surfaces of the finished gems. Asterism also is exhibited by gems cut from some types of chrysoberyl and quartz, and rarely, beryl, diopside, garnet, opal, and spinel. Reflections from abundant inclusions of microscopic size yield a faint whitish sheen, known as "silk," in much transparent and nearly transparent ruby and sapphire.

Proper orientation of cuts with rounded form is necessary to show a well-developed "eye" in gems made from chatoyant varieties of beryl, beryllonite, bronzite, calcite (satin spar), chrysoberyl (true cat's-eye), gypsum (satin spar), prehnite (chlorastrolite), quartz (cat's-eye, tiger's-eye), and tourmaline. The chatoyance is due to fibrous structure or to the presence of elongate cavities or solid inclusions of minute size. The sharpness and regularity of the "eye" effect is mainly a function of how uniformly the fibers or inclusions are distributed and aligned.

Finishing in various rounded forms enhances the opalescence and the attractive play of colors in precious varieties of opal. These colors, quite different in origin from the fire in transparent gems, are caused by interference of light along minute, regularly spaced voids and other internal inhomogeneities. Milky or pearly opalescence also characterizes moonstone varieties of the feldspars and some glasses (Fig. 1), and a distinct play of colors appears in some of the plagioclase feldspars. A glistening or spangled appearance is typical of sunstone feldspar and the aventurine varieties of feldspar and quartz.

Industrial Uses

Gem materials have served man for purely utilitarian purposes since the most ancient times. Cryptocrystalline quartz and natural glasses were fashioned into weapons long before metals were recognized, and subsequently jade, sillimanite, and zoisite were used for pestles and ax heads, serpentine and steatite for utensils, garnet and quartz for abrasive purposes, and hematite, malachite, and lazurite (lapis lazuli) as pigments. Centuries before Christ, the transparency of calcite, fluorite, colorless quartz (rock crystal), and some glasses prompted the shaping of the first crude lenses, the hardness and compactness of several materials underlay their widespread use in the form of seals, and the superior hardness of diamond and corundum was exploited in connection with the engraving of gems and other objects. These and numerous additional early uses (Ball, 1931) were gradually improved, expanded, and shifted in response to changes in human cultures and discoveries of new materials, a general trend that foreshadowed the tremendous growth and diversification of industrial uses during modern times.

Diamonds: The diamond is by far the most important of the industrial gem materials, with uses founded in various ways upon hardness, cleavage and parting, optical properties, presence of sharp points and edges, and capacity for taking and maintaining a high polish. During recent years industrial applications have absorbed more than 90% of the world's diamond production as high-quality individual stones (*fine industrials*), stones or crystal aggregates of lower grade (*bort*), extraordinarily tough microcrystalline to cryptocrystalline aggregates with minor amounts of other minerals (*carbonado*), and very hard, tough aggregates of small, radially disposed crystals (*ballas*). Some industrial stones are classed as *cuttables,* as they can be fashioned into sound gems. Most of them are off-color, and their gem use at any time depends mainly upon currently prevailing market conditions in the jewelry and industrial trades, and upon the respective supplies of rough stones representing better and poorer grades.

Specifications, selection, preparation, and uses of industrial diamonds have been discussed extensively in the published record (Tyler, 1932; Kraus and Slawson, 1941; Austin, Mercer, and Shipley, 1948; Long, 1951, 1954; Long and Slawson, 1952; Denning, 1953; Grodzinski, 1953; Tolansky, 1955; Jahns, 1960; Chandler, 1964), and an up-to-date summary by Reckling et al. appears in another chapter (Diamonds, Industrial and Gem) of the present volume.

Other Stones: Large quantities of hard and relatively tough gem materials have been shaped for use as bearings in watches, chronometers, and various timing devices, in gages and metering equipment, in compasses, and in analytical balances and a host of other scientific instruments. The many finished forms include knife edges, half-cylinders, plates, cups, and various pierced elements. Most are very small, and require precision methods in their manufacture.

Sapphire and ruby once were by far the most important raw materials for jewel bearings, and diamond, garnet, quartz, spinel, and a little chrysoberyl, topaz, zircon, and natural glass constituted nearly all the remainder. During the past five decades, however, they have been largely displaced by synthetic corundum, spinel, and hard glasses, which can be provided more cheaply and with excellent quality control. Agate, chalcedony, sardonyx, and garnet still are used for many of the large but relatively inexpensive types of bearings, and both diamond and natural ruby and sapphire of superior quality are used for other types when exacting specifications justify their higher cost.

Corundum, feldspars, garnets, hematite, quartz, topaz, natural glasses, and several other moderately to very hard gem materials can be broken into abrasive fragments with sharp corners and edges. The cleavage of topaz and the ready parting of some natural corundum, however, reduce the effectiveness of these minerals in many applications. The rough stones are crushed, and the fragments are then carefully size-classified and used in the form of a slurry or paste with water, oil, wax, or some other carrier. Some also are cemented as facings on paper or cloth, are bonded with metals and ceramics for grinding wheels, or are used in loose form for blast-abrading purposes that range from the cleaning of heavy castings to the most delicate of dental work. During the past two decades, several of these materials have been supplanted by synthetic corundum, various artificial oxides and carbides, and to some extent by less expensive natural materials not classed as gem stones.

Both synthetic and some natural ruby and sapphire are used in gages and surface-testing equipment, as dies, as needlepoints in phonographs and many recording devices, and as windows in combustion chambers. Cryptocrystalline quartz is the chief raw material for laboratory mortars and pestles. The chemical resistance, hardness, and limited response of this mineral to changes in temperature make it superior to glass and many other materials for optical flats, gage blocks, and precision balance weights. Large single crystals generally are used for such purposes.

Clear, unflawed pieces of rock crystal are fashioned into lenses and lens components, as well as wedges, plates, and prisms for numerous optical instruments. The rotatory effect of this mineral on plane polarized light is responsible for its use in saccharimeters and several special accessories for polarizing microscopes. Quartz crystals that meet certain tolerances with respect to the amount, kind, and distribution of twinning, solid inclusions, gas bubbles, and other flaws are cut into plates of specified thicknesses for frequency control in radios, quartz-clocks, wave filters, and many other electronic devices. The piezoelectric properties of this mineral also underlie its use in microphones and pressure indicators. Precise orientation is an absolute necessity for most uses, as both the optical and piezoelectric properties vary with crystallographic direction. Some other substances grown artificially from solutions in the form of large, clear crystals have partly displaced quartz for piezoelectric elements in which extreme thinness and shock resistance are not essential.

Crystals of tourmaline have been cut into piezoelectric components for pressure gages and other devices, but uncertainties with respect to the supply of suitable material have severely limited these applications. Owing to its strong pleochroism, tourmaline also is employed as a means for obtaining polarized light; it generally is mounted in tongs that constitute a simple but effective polariscope.

The very high birefringence of calcite underlies its many important optical uses. Clear, unflawed material (Iceland spar) is made up into plates and polarizing prisms for microscopes, photometers, spectrometers, devices for measuring birefrigence, dichroscopes for the determination of pleochoism in gem minerals, and many other instruments. It also has been used in special gun sights and bomb sights. Its vectorial differences in hardness and its ready cleavage must be considered in shaping and finishing pieces of this mineral, and precise orientation with respect to crystallographic directions is essential for obtaining the desired optical effects. Polaroid, an artificial material whose polarizing characteristics and light transmission were progressively improved after its commercial introduction some five decades ago, has been substituted for both calcite and tourmaline in many optical uses.

Fluorite, with its low index of refraction and very faint dispersion, is extensively used as components of spherically and chromatically corrected lenses. It also is highly transparent to light in both the ultraviolet and infrared ranges, and for this reason is employed as prisms and lenses in spectrographs and other special-purpose optical instruments. Fluorite must be colorless, transparent, and essentially flawless to meet most optical specifications, and fortunately much natural material can be decolored satisfactorily. Artificially grown crystals of both calcium fluoride and lithium fluoride have been substituted for some uses.

Finely divided fluorite is applied to the

surfaces of lenses and other optical elements in order to improve their light-transmitting qualities, and such "coated optics" are widely used in cameras, binoculars, telescopes, and many other devices. Some fluorite-coated gems also have appeared on the market.

Clear crystals of gypsum (selenite) have limited application in optical equipment, chiefly as plates for the sensitive detection of birefringence.

Lazurite once was greatly valued as a permanent blue pigment, especially by artists. The less expensive azurite was similarly used, but unfortunately this mineral alters very slowly to malachite when exposed in finely divided form to the atmosphere; thus the blues in many old paintings have become green with age. Malachite, turquoise, and hematite of gem quality also have been ground for pigmenting purposes, but these and other gem materials long since have been displaced by less costly and more satisfactory pigments.

Origin and Geologic Occurrence

Gem materials occur in nearly all kinds of rocks and mineral deposits, ordinarily as very minor constituents. Taken together in terms of known or inferred origin, they probably represent most of the chemical environments and ranges of temperature and pressure to be found in the earth's crust and upper mantle. They have been variously formed by crystallization from silicate melts, by crystallization or recrystallization under metamorphic conditions, by deposition from aqueous solutions, by organic action, and by combinations of these processes. Many also have been mechanically concentrated at later times to form parts of placer deposits. A substantially larger fraction of the world's total gem-stone production, in terms of value, has come from placer deposits than from primary occurrences, thanks mainly to numerous detrital accumulations of diamonds, sapphires, and rubies. Among the primary sources, the ultramafic intrusive rocks have been by far the most important because of the diamonds obtained from them. The relative importance of major primary sources of all gem materials other than diamonds and pearls is indicated in Fig. 4.

Most of the world's great gem-producing districts lie within or along the margins of the Precambrian shield areas. Here the chief occurrences are in ancient igneous and metamorphic terranes, and in thousands of satellite placer deposits that range in age from Precambrian to Holocene. The younger placer accumulations were largely swept away by Pleistocene glaciation from the Canadian and Fennoscandian shield areas, and their content of gem stones was scattered through the glacial drift. This accounts, at least in part, for the relatively low production of gem materials from these regions.

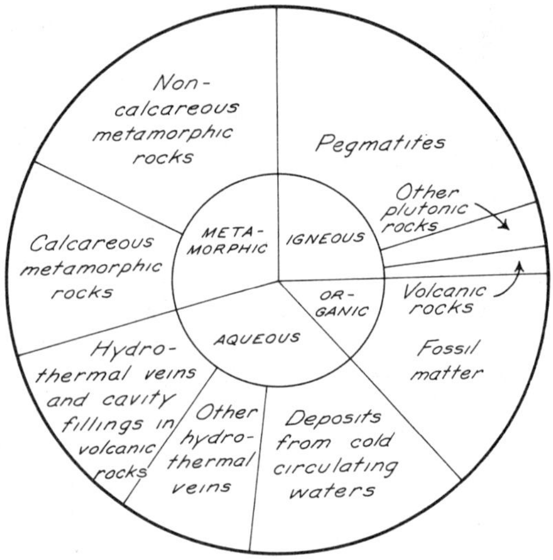

FIG. 4—*Estimated relative importance of major primary sources of gem materials (exclusive of diamonds and pearls), in terms of values of their respective contributions to the world's output since 1899.*

Most important among the world's younger primary deposits are those in Mesozoic and Cenozoic volcanic rocks, in Paleozoic and Mesozoic pegmatites, and in Cretaceous ultramafic rocks. Reckoned in terms of value, however, little more than one-quarter of all gem materials (other than pearls) produced to date was originally formed in rocks younger than Precambrian.

The extensive geologic literature on gemstone deposits and districts is highly varied in emphasis, scope, and detail. Individual contributions range from travelogues to highly technical reports, from brief locality descriptions to monographs, and from assemblages of crystallographic data to economic appraisals. Some are focused upon the gem materials themselves, others upon their occurrence and origin, and still others upon their exploitation. Treatment of gem materials in many reports is incidental to broader commodity studies, geologic investigations, or even to studies in such seemingly unrelated areas as paleontology (e.g., fossils in amber) and meteoritics (gem minerals in stony and iron meteorites). These and other contrasts are well illustrated, for example, by publications on the gem fields of India, Burma, and Sri Lanka (Ceylon), (Barrington Brown and Judd, 1896; Adams, 1926; Iyer, 1942, 1953), pegmatite deposits in California, Maine, North Carolina, and Malagasy (Hamlin, 1895; Kunz, 1905, 1907; Bastin, 1911; Lacroix, 1922; Jahns and Wright, 1951; Sinkankas, 1976), corundum deposits in Montana (Pirsson, 1900; Clabaugh, 1953), and occurrences of many other types (Pogue, 1915; Williams, 1932; Oppenheim, 1948; Staatz and Bower, 1952).

Igneous Rocks

Ultramafic Rocks: Dikes and plug- and pipe-like bodies of kimberlite (peridotite) and related ultramafic rocks yield diamonds and other gem minerals in many parts of the world. Most appear to be essentially vertical volanic necks, now deeply truncated by erosion, and their constituent rocks commonly are crowded with inclusions of material derived from both higher and lower parts of the adjacent sections. Evidences of explosive volcanic activity and composite injection of magma are widespread. Origin of the sparsely scattered diamonds and their numerous tiny inclusions of other minerals has long been a matter of controversy (Williams, 1932; O'Hara and Mercy, 1963; Meyer and Boyd, 1969; Frantesson, 1970; Dawson, 1971), but the sum of present evidence suggests high-temperature crystallization from ultrabasic magma at depths of 95 km (60 miles) or more, followed by upward movement in volcanic conduits (Kennedy and Nordlie, 1968; Harris and Middlemost, 1969; MacGregor, 1970; Meyer and Boyd, 1972).

Among the other gem materials obtained from mafic and ultramafic igneous rocks are sapphire, ruby, diopside, enstatite, hypersthene, olivine, and pyrope, rhodolite, and uvarovite garnets. The pyrope and sapphire are most important commercially. Some occurrences of nephrite and jadeite have been interpreted as intrusive masses of basic rocks that have been highly metamorphosed or otherwise altered.

Pegmatites: Of all rock types, the pegmatites contain the greatest variety of gem minerals, and they constitute the major igneous source of gem minerals other than diamond (Fig. 4). They have yielded anatase, apatite, axinite, emerald and other varieties of beryl, chrysoberyl, cordierite, danburite, diopside, feldspars, fluorite, garnets, kornerupine, kyanite, lazulite, phenakite, quartz, rutile, sapphire and ruby, scapolite, sphene, spinels, spodumene, topaz, tourmaline, zircon, and numerous species that are uncommon or absent among all other types of geologic occurrence.

Gem materials are present in only a small percentage of known pegmatite bodies, and occur chiefly in those of granitic and syenitic to quartz dioritic composition. Such bodies range considerably in size and shape, but no correlation has been established between dimensional features and content of gem minerals. The crystals of best quality ordinarily are concentrated in well-defined "pockets" or "paystreaks" that are ellipsoidal to highly irregular in form (Fig. 5) and rarely constitute large parts of the containing pegmatite bodies.

Some pockets are encountered as true cavities with crystal-studded walls, but most are partly or completely filled with compact to spongy aggregates of quartz, alkali feldspars, and various micas and clay minerals. Commonly scattered through these aggregates are superbly formed crystals of other minerals, many of them transparent and some of them essentially flawless, as well as broken fragments or residual parts of such crystals (Fig. 6). Some gem minerals also are disseminated or sporadically scattered through essentially solid pegmatite, but the yield of high-quality material from such "frozen" crystals generally is low. Milky quartz and rose quartz are obtained from aggregates of very large anhedral crystals that occur chiefly in the central parts of pegmatite bodies without pockets. Compact, scaly to waxy aggregates of attractively colored

FIG. 5—*A typical mined-out gem pocket in pegmatite, Pala district, California. This occurrence yielded spodumene, beryl, and quartz of gem quality, along with waste materials consisting mainly of albite, potash feldspar, quartz, muscovite, lepidolite, cookeite, clay minerals.*

FIG. 6—*Two large crystal fragments of clear, flawless purple kunzite in a matrix of clay minerals and quartz, Pala district, California. The piece at upper left is about 51 mm (2 in.) long. Similar material was obtained from the pocket shown in Fig. 5.*

muscovite and lepidolite are obtained from both pockets and pocket-free pegmatite.

Other Igneous Rocks: Some granitic intrusive rocks furnish small quantities of gem materials, notably garnets, sphene, and zircon that occur as accessory constituents. Quartz, feldspar, and topaz are obtained from miarolitic pods and cavities, which can be likened to miniature bodies of pegmatite within the enclosing granitic rock. Perhaps somewhat analogous are the beryl- and topaz-bearing cavities in a few volcanic rocks of rhyolitic composition.

Felsic to mafic intrusive rocks of alkaline affinities yield sapphire and ruby, feldspars, feldspathoids, melanite and grossularite garnets, and zircon. Volcanic rocks of intermediate to basic composition are sources of sapphire, ruby, plagioclase feldspars, olivine, and pyrope garnet. Diamond occurs in the matrices of some intermediate to basic igneous breccias that appear to have formed as volcanic throat rocks. Gems have been fashioned directly from obsidian and other volcanic glasses, especially those that contain attractively distributed inclusions or clusters of tiny crystals.

Metamorphic Rocks

Calcareous Rocks: Crystalline limestones and other calcareous metamorphic rocks constitute the chief source of lazurite and gem spinels. They also have yielded substantial quantities of ruby, sapphire, and emerald, along with some apatite, axinite, cordierite, diopside, epidote, grossularite and andradite garnets, fluorite, idocrase, rhodonite, scapolite, sphalerite, sphene, steatite, tourmaline, willemite, and zincite. Some occurrences are in regionally metamorphosed terranes, but most are contact-metamorphic, associated with intrusive rocks of basic to silicic composition, and range lithologically from nearly pure marble to skarn, tactite, and other silicated rocks.

The gem materials occur mainly as large, irregular crystals and crystalline masses (e.g., scapolite, lazurite), as small individuals disseminated through the host rock (e.g., garnets, ruby, spinels, tourmaline), and as groups of well-formed crystals that line primary cavities (e.g., emerald, epidote). Considerable variations in distribution, concentration, and quality are characteristic, and the significance of most deposits lies in their role as sources of richer placer accumulations.

Noncalcareous Rocks: Schists and gneisses are sources of emerald, ruby, sapphire, and most of the other gem minerals that occur in metamorphosed carbonate rocks. In addition, actinolite, anatase, andalusite, aquamarine and other varieties of beryl, chrysoberyl, garnets (demantoid, almandine, and spessartine), jadeite and nephrite, quartz (e.g., jasper, tiger's-eye), sillimanite, staurolite, topaz, and zoisite have been obtained commercially. The gem minerals typically form metacrysts and crystal clusters in finer-grained host rocks such as slate, phyllite, schist, and hornfels. Those in some metasedimentary rocks appear to have been present prior to the metamorphism, and

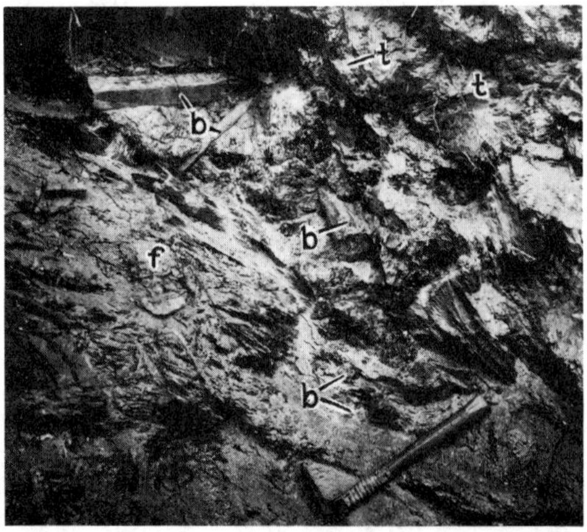

FIG. 7—*Crystals of beryl (b), fluorite (f), and topaz (t) in mica-rich schist, Petaca district, New Mexico. Some of the beryl is aquamarine of gem quality.*

hence may well represent ancient placer occurrences.

Beryl, chrysoberyl, phenakite, and other typical pegmatite minerals appear as metacrysts and cavity linings in some schists and and gneisses (Fig. 7). Nearly all these occurrences are closely associated with masses of pegmatite or other igneous rocks, and probably are genetically related to them. The principal deposits of jadeite, nephrite, serpentine, and steatite are found in metamorphosed igneous rocks of basic to ultrabasic composition, and in vein- or dikelike masses associated with such rocks.

Deposits Formed by Aqueous Solutions

Disseminations, cavity fillings, veins, and stockworks of opal, variscite, and many forms of quartz occur in igneous rocks, especially volcanic rocks of intermediate to acid composition, and generally have been ascribed to the action of hydrothermal solutions. Other aqueous fluids have been responsible for the deposition, at low to high temperatures, of calcite, datolite, epidote, pectolite, prehnite, thomsonite, and both coarse-grained and cryptocrystalline varieties of quartz. These minerals ordinarily occupy fissures and cavities in basic volcanic rocks.

Typical vein deposits, found in a wide variety of host rocks, have yielded amethyst, rock crystal, citrine, agate, jasper, and many other varieties of quartz, as well as barite, benitoite, calcite, celestite, emerald, epidote, feldspars, fluorite, gypsum, spessartine and uvarovite garnets, hematite, kyanite, opal, topaz, and tourmaline. Some of these deposits were formed by cold waters, and others evidently are of hydrothermal origin. Vein occurrences of several minerals, notably emerald, topaz, and tourmaline, generally show clear evidence of magmatic affinities.

Hydrothermal ore deposits contain pyrite, sphalerite, and other primary minerals that have been used as gems, and the oxidation of such mineralized ground has led to formation of important concentrations of secondary gem minerals, especially azurite, malachite, chrysocolla, dioptase, hemimorphite, smithsonite, and turquoise.

Organic Accumulations

Jet, which can be taken to include several compact varieties of lignite and coal, is one of the few gem materials indigenous to the sedimentary rocks in which they are found. Also of vegetable origin is amber, a fossil resin derived from extinct species of coniferous trees. It is obtained from sedimentary deposits that in a sense are fossil placer concentrations.

Pearls are composed of very fine-grained aragonite and a little amorphous organic matter deposited in concentric layers by certain species of mollusks at points within their shells. Mother-of-pearl, obtained from the innermost parts of the shells themselves, also consists of very fine-grained aragonite in thin layers. Commercial production of these materials is from sources of modern age. Both modern and fossil corals have been used as gem material, the former very extensively. Some varieties of oolitic hematite, deposited through the action of ancient organisms, also have been obtained for gem purposes from sedimentary sections of various ages.

Pseudomorphs of turquoise after bone, pyrite and opal after the shells of various invertebrates, and quartz after both plant and animal matter are the chief inorganic gem materials that represent earlier organic processes.

Placer Deposits

Gem-bearing placer deposits are widely distributed, and many of them are richer, more accessible, and more easily worked than typical lode deposits. The gem stones in detrital accumulations, having survived the rough treatment of mechanical concentration, are generally of much better average quality than those in primary source occurrences.

Gem materials that are resistant to chemical decay under normal conditions at the earth's surface are preserved in soft, weathered parts of many lode deposits, and both saprolitic and eluvial concentrations are formed as other constituents are dissolved, washed, or blown away. Thus the diamond content of the weathered "yellow ground" at the outcrops of the South African peridotite pipes has proved to be five to more than ten times greater than that of the fresh "blue ground" at depth, and many surface concentrations of beryl, spodumene, topaz, and tourmaline in Brazil are substantially richer than the pegmatite bodies whence these minerals were derived. Extreme local concentration is illustrated by several Burmese deposits, where crystals of ruby and spinel have been weathered from metamorphosed limestone and have accumulated in solution cavities. Some South American occurrences of emerald are similar. Residual concentrations of gem stones, especially those of pegmatitic origin, also are known from the deeply weathered terranes of the Guianas, Madagascar, the southeastern United States, and many other parts of the world.

Specific gravity (Table 1) and physical resistance are primarily responsible for the preservation and distribution of the detrital gem materials. Thus large fragments of amethyst in the modern stream gravels of Brazil are found at considerable distances from their sources, whereas the accompanying cleavage fragments of topaz are much smaller, even though derived from original crystals of comparable size; spodumene, with two directions of fairly ready cleavage, ordinarily appears in such localities as small splinters. Gem stones of relatively high specific gravity typically are concentrated at or near the bottoms of gravel sections, mainly in pockets and elongate "streaks" of sinuous form. Some of the richest concentrations are found in crevices, potholes, and other depressions in the immediately underlying bedrock. Many placer sections include layers and lenses of relatively fine-grained impervious sediments that constitute "false bottoms," above which perched accumulations of gem materials are sometimes encountered.

Particularly well known among gem-bearing placer deposits are the stream gravels of Burma, Sri Lanka, Indochina, Thailand, Brazil, Australia, and southern, central, and western Africa. Modern beach gravels also have yielded much gem material, including diamond in South-West Africa; ruby in Burma; jasper, prehnite, and thomsonite along the shores of Lake Superior; and nephrite and jadeite along the California coast. Gravels, dune sands, and playa sediments in desert regions, where the ordinary processes of concentration are augmented by the winnowing action of the wind, have been sources of diamond, olivine, pyrope garnet, turquoise, jade minerals, opal, cryptocrystalline quartz, and other gem stones, chiefly in Australia, Afghanistan, Egypt, central Africa, South-West Africa, central Asia, and the western United States.

Somewhat older placer deposits are identified with fluviatile and lacustrine terraces, uplifted or drowned bars and beaches, and numerous surficial features of glacial origin. Still older deposits are known from sedimentary and metamorphic sections ranging in age from Cenozoic to Precambrian. Many of these are rocks as hard and well consolidated as those containing typical primary deposits. Examples of fossil placer occurrences include Cretaceous conglomerates that yield much of the garnet obtained from central Europe, diamond-bearing glaciofluviatile deposits of probable Precambrian age in Brazil, and Precambrian conglomerate that contains diamonds in India.

Some of the world's best-known placer deposits represent more than one cycle of erosion and deposition. Thus diamond-bearing sedimentary rocks of Precambrian, late Paleozoic, and Mesozoic age occur in parts of Africa, where diamonds also are present in Quaternary terrace deposits and Holocene marine and nonmarine gravels. Each type of occurrence may well have contributed diamonds to any of the younger deposits. Recycling has been most significant for diamond, corundum, and other very durable minerals, but it has been by no means restricted to them. Amber, for example, occurs along the present beaches of the Baltic coast in Europe, where it has been derived from fossil placer deposits of Tertiary age.

Geographic Distribution

World Deposits

Gem deposits of major economic importance are present in every continent. The minerals best representing Asia are ruby, sapphire, spinels, and zircon from Burma, Sri Lanka, and Indochina; lazurite from Afghanistan; jade from Burma and China; and diamond and many forms of quartz from India. Siberia may well contain important deposits of diamonds, chrysoberyl, and other gem stones. The chief occurrences in Europe, including the Ural Mountains region, have yielded amber in the countries fringing the Baltic Sea; chrysoberyl, diamond, emerald and other varieties of beryl, garnets, and quartz in Russia; pyrite and marcasite in France and Italy; garnets in Czechoslovakia; and quartz in the Alpine region.

South America is best known for diamonds, emerald and other kinds of beryl, quartz, topaz, and tourmaline from Brazil, lazurite from Chile, emerald from Colombia, and diamonds from British Guiana and Venezuela. Opal and sapphire feature Australia's deposits. Most important in North America are occurrences of sapphire, garnets, jade minerals, spodumene, tourmaline, turquoise, and cryptocrystalline varieties of quartz. In the African continent, diamonds have been obtained mainly from the Republic of South Africa, Ghana, Angola, Botswana, South-West Africa, Sierra Leone, Zaire, Tanzania, and central Africa; chrysoberyl, emerald, garnets, sapphire and ruby, tiger's-eye, tourmaline, and zircon from Rhodesia, South Africa, Kenya, and Tanzania;

olivine and turquoise from Egypt; and tourmaline from Malagasy and South-West Africa.

Burma, Sri Lanka, India, Malagasy, South-West Africa, the United States, the Alpine region of Europe, and the Ural Mountains of Eurasia are famous for the remarkable variety of gem materials they have yielded. Additional information on geographic distribution of deposits is presented in Table 4.

Deposits in the United States

Occurrences of gem stones in the United States are distinguished more for their great variety and domestic popularity than for their impact on world economy. Those of principal interest include the lode and placer deposits of sapphire in Montana and jade in Alaska, California, Washington, and Wyoming; the lode deposits of turquoise, variscite, opal, and chrysocolla in the southwestern states; the vein and other lode occurrences of amethyst in Arizona and quartz crystal in Arkansas, California, New York, and North Carolina; various garnet deposits in Alaska, Arizona, California, Idaho, New Mexico, New York, and North Carolina; the gem-bearing pegmatites of California, North Carolina, and the New England States; widely distributed placer accumulations of agate, chalcedony, jasper, petrified wood, and other forms of cryptocrystalline quartz in the western states; and coral in Hawaiian waters.

About seventy-five different minerals have been produced and sold as gem stones in the United States. Occurrences of deep blue benitoite (California), the rose-red to purple rhodolite garnet (North Carolina), and the greenish chlorastrolite (Lake Superior region) are found nowhere else in the world. The superior color tones and other qualities of amazonite from Colorado, Pennsylvania, and Virginia; amethyst from Arizona; beryl, garnet, and hiddenite from North Carolina; beryl from Colorado and Maine; tourmaline from Maine; star garnet from Idaho; variscite and beryl from Utah; black opal from Nevada; jade from Wyoming; azurite, malachite, moss agate, and chrysoprase from several western states; and beryl, garnet, kunzite, tourmaline, and gold quartz from California have established significant demands for these domestic gem materials in the world market.

Diamonds have been found in almost all the states, chiefly in stream gravels and glacial drift. Most have been recovered from glacial deposits in the Great Lakes region, and presumably were derived from bedrock sources somewhere in Canada. Many of these stones are of excellent quality, but their number and total weight have been trivial on the world scene. A small pipe of altered peridotite near Murfreesboro, AR, has yielded some diamonds since it was discovered in 1906.

The principal domestic occurrences of gem materials are summarized by states in Table 4. More detailed locality lists have been assembled by Henry (1948), Dake (1950), MacFall (1951), Pearl (1951), Ransom (1955), Schlegel (1957), Sinkankas (1959, 1976), and many others, in large part for the use of amateur collectors. Additional information has appeared in several of the journals and periodical reports listed near the end of this chapter.

Prospecting, Mining, and Recovery

During the past half century the search for deposits of gem materials and the working of known deposits have required the simultaneous efforts of 90,000 to 160,000 persons, at least four-fifths of them non-Caucasians. The number of individuals thus occupied at any one time has depended mainly upon the economic health of the trade, the availability of unmined reserves, various political controls, and the condition of certain international relationships.

Although some deposits are very rich, the concentration of gem stones within any large block of ground ordinarily is low, regardless of the type of occurrence. Thus extraction has been based either upon highly selective mining of relatively rich pockets and shoots, or upon more systematic operations in which large amounts of low-grade materials are handled by bulk methods.

Exploration and Discovery

The original discoveries of most gem materials have been happy accidents of observation, many of them attending searches or mining operations for other substances. Excellent examples of serendipitous findings are those of kunzite from tourmaline prospecting in California, turquoise from copper prospecting and mining in Arizona, and rhodolite garnet from sapphire prospecting in North Carolina. Direct search for gem materials, largely a matter of examining natural exposures for uncommon but distinctive stones, has been focused mainly upon unconsolidated stream gravels and other placer accumulations, and to a lesser extent upon certain bedrock masses. Many lode deposits have been found through the tracing of placer occurrences and fragments of surface float to their sources.

TABLE 4—Principal Geographic Sources of Gem Minerals and Other Gem Materials

Mineral or Other Substance	World Occurrences*																	
	Europe								Asia									
	Austria–Hungary	British Islands	Czechoslovakia	France	Germany	Italy–Sicily	Norway–Sweden	Switzerland	USSR	Afghanistan	Burma	Sri Lanka	China	India	Indochina	Thailand	Tibet	Turkey
Amber		x		x	⊕	X	x	x	⊕		x		x	x				
Andalusite	x				x				x			⊕						
Apatite	x	x	x		⊕	x	x	X	x			⊕						
Axinite		x		⊕	x			x										
Azurite and malachite				x		x			⊕									
Beryl																		
Emerald	x								⊕									
Other varieties	x	x		x	x	x			X			x		x				
Chrysoberyl			x						⊕			⊕	x	x				
Chrysocolla		x							⊕									
Cordierite					x		x					⊕						
Corundum																		
Ruby									x	X	⊕	x	x	x	X	x		
Sapphire								x	X	x	X	⊕	x	X	⊕	X	x	
Diamond									⊕					X				
Feldspars																		
Moonstone						x		X			X	⊕						
Other varieties	x	x	x	x	x	x	x	x	⊕									
Fluorite	x	⊕	x		X			x	x									
Garnets																		
Pyrope			⊕		x				x		x	X	x					
Other varieties	X		⊕	x	X	x	x	x	⊕			⊕		X			x	
Gypsum	x	⊕	x	x	x	X		x	X	x								x
Hematite		⊕		x	x	x	X		X				x	⊕				
Jade																		
Jadeite								x	x		⊕		X	x	x	x	X	x
Nephrite					x			x	⊕	x	x		⊕	x	x	x	x	
Jet	x	⊕	x	X	X	X			x						X	x		
Lazulite	⊕		x				x	x										

TABLE 4—Principal Geographic Sources of Gem Minerals and Other Gem Materials (Continued)

Africa								South America				North America					
Angola–Zaire-Congo	Egypt	Madagascar	Rhodesia	South-West Africa	Tanzania	R. of So. Africa	Australia	Brazil	Chile	Colombia	Uruguay–Paraguay	Canada	Cent. Amer. Countries	Mexico	United States	Other Countries†	Chief Occurrences in the United States‡
														x		*Denmark, Poland, Romania,* Dominican Republic	MD, MA, NJ, NY, NC, WY
		x					x	⊕							x	Spain	CA, ME, MA, PA
		x						X						X	x	Japan, Spain	CA, CT, *ME*, NH, NJ, NY, NC
														x	x	Tasmania	*CA*, NJ, PA
x	⊕		X	X			X		X					x	X		*AZ*, NV, *NM*, UT, etc.
	X					x	x	x		⊕					x		*NC*, SC
	x	X		X	x	x		⊕							X	Mozambique	*CA, CO*, CT, *ME,* MA, NH, UT
		X	x					X							x	Tasmania	*CO*, CT, *ME*, NY
x	x		X	x			x		X					x	⊕		*AZ*, CA, *NV*, *NM*, PA, UT, WY, etc.
		x						x							x	Greenland	CT
		x					x								x		*MT, NC,* SC
		x				x	⊕					x			x		CA, GA, *MT*, NY, NC, SC
⊕			x	X	X	⊕	x	X								Borneo, *Chad,* Guyana, *Ghana, Sierra Leone*, etc.	*AR*, CA, MI, etc.
	x			x				X							x		*CA, CO*, NC, NM, PA, etc.
	x	X		x				X				x		X	X	*Japan*	*CA, CO*, ME, MA, NM, NY, NC, *PA, TX, VA*
				X	x		x					x		x	X		CA, CO, *IL, KY, NM, NY*, etc.
				x		⊕									x		*AZ*, CA, CO, MT, *NM*, NC
	X	X	x	x	x	x	x	x			x	x		x	X	Finland, Greenland, Kenya, Romania	*AK*, AZ, *CA*, CO, NV, NH, *NY, NC*, VA, etc.
	⊕											x	x	⊕	x	Greece, Spain	CO, *MI*, NY, NM, TN, etc.
								X	x			x		x	X	Algeria, Morocco, Venezuela	*MI, MN*, MO, NY, PA, *WI*, WY, etc.
								x			x		x	x	X	New Zealand,	*AK, CA*, MT,
							x	x		x		x	x	x	x	*Pakistan*	NV, NC, OR, *WY*
												x			X	*Spain*	CO, NM, *PA*, UT
		X						X							X		GA, NC

TABLE 4—Principal Geographic Sources of Gem Minerals and Other Gem Materials (Continued)

	World Occurrences*																	
	Europe								Asia									
Mineral or Other Substance	Austria–Hungary	British Islands	Czechoslovakia	France	Germany	Italy–Sicily	Norway–Sweden	Switzerland	USSR	Afghanistan	Burma	Sri Lanka	China	India	Indochina	Thailand	Tibet	Turkey
Lazurite									X	⊕								
Olivine											X	⊕		x				
Opal	x		X						x	x			x	X			x	x
Phenakite				x			x	x	⊕									
Prehnite	x	x		⊕	x	x		x					x					
Pyrite and marcasite	x		x	⊕	X	⊕	x	x										
Quartz																		
Amethyst	x	x	x					x	X			X		X				
Rock crystal	x	X	x	X	x	X	x	⊕	X	x	x	X	x	X	x		x	
Other crystalline varieties	X	x		X	X	x	x	X	X	x	x	X	x	X			x	x
Agate, chalcedony, and other cryptocrystalline varieties	x	x	X	x	X	x			X	x			X	⊕			x	x
Scapolite							x		X									
Sepiolite			x															⊕
Serpentine		X		x	x		x		x	⊕		x	⊕					
Smithsonite	x	x		x	x	x			⊕									
Sphene	x			x		x	x	⊕										
Spinels					x	x	x			X	⊕	⊕	x	X	⊕	X	x	
Spodumene																		
Topaz		X	x		x		x		X			X		X				

TABLE 4—Principal Geographic Sources of Gem Minerals and Other Gem Materials (Continued)

Africa								South America				North America					
Angola–Zaire-Congo	Egypt	Madagascar	Rhodesia	South-West Africa	Tanzania	R. of So. Africa	Australia	Brazil	Chile	Colombia	Uruguay–Paraguay	Canada	Cent. Amer. Countries	Mexico	United States	Other Countries†	Chief Occurrences in the United States‡
									X			x		x	x	*Iran*	CA, CO, MO
	⊕		x	x	x	x	X	X	x	x	x		x	x	X		*AZ*, CA, *NM*, NC
x	x			x			⊕	X				x	x	X	x	Japan, New Zealand	AZ, CA, CO, *ID, NV,* NM, OR, UT, WA, etc.
		x			x			⊕						x	x		CA, *CO, ME*, NH, NC, VA
						x									X		CT, MA, *MI, MN, NJ,* PA
x	x		x					x	x					x	x	Japan, Romania, Spain	AZ, CA, CO, IL, NY, PA, etc.
	X	X						⊕			X	x		X	x	*Iran*, Romania	*AZ*, CA, NH, ME, MI, NC, PA, VA, WY
	X	X		x				⊕			x	x	x	X	X	*Japan* ⊕	*AR*, CA, CO, *NY*, NC, etc.
	x	X	x	x	x	⊕	x	⊕			x	x	x	x	X	*Japan*	AK, *CA, CO,* ME, MT, NH, NC, *SD,* VT
x	X	x	x	x		x	X	X			X	x	x	X	⊕	*Arabia, Argentina,* Iran, Poland	AZ, *CA,* CO, ID, *MI, MT, NV,* NM, *OR,* UT, *WA,* WI, WY, etc.
		⊕													x	Finland	MA, NJ, NY
															x	*Greece*, Morocco, Spain	CA, NM, PA, UT
						x						x			x		CA, MA, NJ, PA, RI, VT, VA, WA
				X		x	x							x	x	*Greece* ⊕, Romania	*AR, MO, NM,* PA, UT, WI
		X										x		x	x		AR, CA, ME, PA
		X					X	X				x		x	x		CA, GA, MT, NJ, NY, NC
		X						⊕							⊕		*CA,* CT, *NC*
	x	x		x			x	⊕						x	X	Japan, Nigeria, Tasmania	*CA, CO,* CT, ME, NH, *TX, UT*

TABLE 4—Principal Geographic Sources of Gem Minerals and Other Gem Materials (Continued)

Mineral or Other Substance	World Occurrences*																	
	Europe									Asia								
	Austria–Hungary	British Islands	Czechoslovakia	France	Germany	Italy–Sicily	Norway–Sweden	Switzerland	USSR	Afghanistan	Burma	Sri Lanka	China	India	Indochina	Thailand	Tibet	Turkey
Tourmaline	x	x	x	x	x		x	x	X		X	X		X				
Turquoise				x	x				X				x					
Variscite					x													
Zircon				x	x		x		x		X	⊕			⊕	X		
Zoisite					X	X	X		X		X	X			X	X		

* Countries included in the tabulation are designated, with respect to estimated total production of a given mineral or material, as follows:
x—appreciable commercial output ⊕—major fraction of world output, in terms of value
X—substantial fraction of world output, in terms of value

The earliest subsurface prospecting was confined to soft ground in which test pits and crude borings could be developed most readily, and even today much of this work is done in relatively primitive and unsystematic ways. The native of Burma, Sri Lanka, or Indochina, for example, makes simple excavations in placer ground or weathered rock, the basis for his locations ranging from shrewd reasoning to pure chance. Or he may thrust an iron bar into the ground, then withdraw it and examine its surface for scratches that might indicate the presence of very hard gem stones. The native prospector in Arizona, with an even simpler approach, searches for garnet and peridot by examining the granular debris of ant hills. But methods of search also have been improved with time, especially by large companies and government organizations, and in many regions they have evolved into well-planned campaigns that include test pitting, drilling, and geophysical work aimed not only at discovery but also at determining the extent and grade of deposits that might be present.

Modern prospecting by technically trained personnel has profited from the imaginative application of air photography and more remote imagery, geologic mapping, geophysical and geochemical techniques, careful sampling, and reasoning based upon analogy and inferred genesis of various deposits. Thus, geomorphic studies have led to major discoveries of diamond-bearing terrace sections in Africa and to delineation of deeply weathered and hence easily worked pegmatite-bearing terranes in the United States. The combining of subsurface data with modern theories of stream behavior has led to prediction, discovery, and tracing of "threads of deposition" and associated "pay streaks" in numerous placer accumulations. Detailed sampling and petrographic studies have established correlations between the frequency of gem minerals and that of much more common species in some alluvial terranes, thereby providing a useful method for estimating grade. Bodies of kimberlite and other igneous rocks have been located by careful geologic mapping and complementary geophysical studies, but extensive sampling still is required to determine the grade of these rocks.

Nearly all present-day prospecting for gem materials in the United States is done by individuals, most of them amateur collectors appropriately known as "rock hounds." More than 250,000 such persons visit the field each year, and as many as 50,000 of them are active members of nearly a thousand more or less organized mineral societies throughout the country. That they have constituted a potent

TABLE 4—Principal Geographic Sources of Gem Minerals and Other Gem Materials (Continued)

Africa								South America				North America					
Angola–Zaire-Congo	Egypt	Madagascar	Rhodesia	South-West Africa	Tanzania	R. of So. Africa	Australia	Brazil	Chile	Colombia	Uruguay–Paraguay	Canada	Cent. Amer. Countries	Mexico	United States	Other Countries†	Chief Occurrences in the United States‡
	x	X		⊕		x		⊕							⊕		*CA,* CO, CT, *ME,* MA, NJ, NY, PA
	⊕						X		x				x	x	⊕	*Iran* ⊕	*AZ, CA, CO, NV, NM,* UT, WI, WY
															⊕	Spain	*UT*
		X					x	x					x		x	New Zealand, Tasmania	CO, ME, MA, NY, NC, TX
		x		x	⊕	x	x	x							x		CA, CO, NV, NM

† Italicized name indicates estimated total production amounting to a substantial fraction of the world output, in terms of value, for the corresponding mineral or material.
‡ Italicized name indicates estimated total production amounting to a substantial or dominant fraction of the United States output, in terms of value, for the corresponding mineral or material.

factor in the discovery and working of significant occurrences of garnet, jade, quartz, topaz, tourmaline, and many other minerals is not surprising, for they combine great energy, enthusiasm, and mobility with a willingness to seek out and explore any rumor or suggestion of a new locality. Many have made discoveries by reviving the ancient technique of "just going out and looking around."

Historic trends in important discoveries of gem-stone deposits are pointed up in Table 5. Discovery rates have increased in all continents with the passage of time, reaching maxima in Asia and Europe prior to 1900 and in the other continents since then. The very high rates in Africa, Australia, and North America during the twentieth century are attributable to vigorous exploration and to the improvement of prospecting techniques, and it is interesting to note that, of about 600 major discoveries thus far known in the world, nearly one-third can be assigned to the present century. The overall rate of future discoveries, however, is likely to decline significantly for most of the continents before the year 2000, and final figures for the present century almost certainly will be substantially less than those enclosed by parentheses in Table 5.

TABLE 5—Distribution of Important Discoveries of Gem Deposits Since 3500 B.C.*

Continent	Average Number of Discoveries per Century: 3500 B.C.- 0 A.D.	1- 1500	1501- 1800	1801- 1900	1901- 2000†	Total Number of Discoveries since 3500 B.C.
Africa	0.8	0.4	1.0	8.0	(120)	129
Asia	1.5	2.3	7.0	24.0	(17)	144
Australia	?	?	1.0	11.0	(43)	44+
Europe	0.5	2.0	10.7	13.0	(4)	95
North America	?	1.7	4.0	30.0	(66)	113+
South America	?	0.4	9.7	14.0	(20)	72+
						597+

* Based in part upon data compiled by Ball (1931).
† Discoveries per century computed on basis of number of discoveries during the period 1901-1980.

Most important among the world's early discoveries (Ball, 1922, 1931) were those of amber in Europe (prior to 25,000 B.C.); turquoise in Egypt and lapis lazuli in Afghanistan (prior to 3500 B.C.); emerald in Egypt (about 1900 B.C.) and South America (date unknown); and diamond, ruby, and sapphire in India and Sri Lanka (about 800 B.C.). Diamonds were first discovered in Borneo about 600 A.D. and in Brazil in 1720, and the lode occurrences in South Africa were first recognized in 1867. Important diamond finds subsequently were made in British Guiana (1891), Rhodesia (1903), South-West Africa (1908), Zaire (1906-1913), Tanzania (1908, 1924), Congo (1915, 1928), Ghana (1919), South Africa (1926, 1927), Sierra Leone (1930), and Asiatic Russia (1938).

Mining Methods

Lode Deposits: Nearly all lode mining for gems has begun with development of open cuts in promising parts of the rock body, most simply by the digging of soft material with hand tools and the removal of harder rock by prying and barring along joints and other discontinuities. Larger masses have been broken mainly by drilling and wedging or blasting, although the ancient method of heating and water-quenching still is employed in some parts of the world. Equipment and techniques used in current mining depend in various ways upon the location and size of the deposit, the nature of the gem stone concentrations, and the skill and resources of the operator. Present gem mining ranges from the highly systematic and mechanized removal of diamondiferous kimberlite from large surface and underground workings in South Africa to almost casual surface scraping or potholing of garnetiferous schist in Alaska or topaz-bearing rhyolite in Utah. Apart from such extremes, most of the lode mines consist of irregular open cuts, with or without appended shafts, inclines, tunnels, and stopes of rather unsystematic distribution. Individual workings rarely are large, as they are confined to local concentrations of desirable materials or are aimed at such concentrations through barren rock. Soft, altered rock is most easily handled, and yields gem stones with a minimum of difficulty and damage. The extent of weathering thus can be a very significant factor. Many gem mines have been profitable only near the surface, and relatively few include extensive workings in hard rock.

Drilling, blasting, and timbering generally are held to an absolute minimum. Equipment typically is simple, and the most effective tools for removing gem-bearing material have proved to be the screwdriver, chisel, small pick, bar, and hammer. In the most primitive operations, the broken rock is moved in small baskets or trays, often from person to person. Elsewhere buckets, wheelbarrows, small cars, and skips are drawn, pushed, and lifted by hand or by means of windlasses or whims. Mechanized hauling and hoisting is confined to a few relatively large mines. Rough sorting is done within the workings of some mines, but all final separation of gem stones is accomplished outside.

Placer Deposits: Most placer operations for gem stones are small in scale and simple in procedure. Digging ordinarily is done by hand, and some overburden is sluiced away by crude methods. Where the overburden is thick, pits or shafts are sunk to the pay horizons, which then are worked laterally until caving becomes imminent. Further operations are carried out from new shafts. Timbering is confined to the simplest of shoring. Drifts, commonly referred to as "swallows' nest workings," are driven from naturally exposed bases of many terrace sections.

Hoisting and haulage methods are similar to those employed in most small lode mines. Standard equipment in many Asiatic and African operations is the empty tin or large can, which is variously used for carrying gravel, bailing water, and storing concentrates. Desirable materials are recovered by divers from some river and lake beds, and this long-established technique most recently has been applied along the coasts of California and Hawaii, where skin divers have gathered fragments of jade and coral, respectively, from the ocean bottom.

Bulk handling of material characterizes many of the larger placer operations. Sluicing is common practice, and bulldozers, dragline scrapers, dredges, power shovels, and other heavy items of equipment also are used. Such large-scale operations depend for their success on the blocking out of workable ground in advance of mining.

Concentration and Recovery

Lode Deposits: Most material removed from typical vein and pocket occurrences is broken or crushed where necessary, and is concentrated by washing, winnowing, screening, handpicking, and by various combinations of these methods. Several stages of treatment commonly are employed, and the most valuable materials often are repicked as many as six times. Old mine dumps and spoil heaps have been reworked in some districts, and amateur collectors in the United States have repeatedly screened and picked such accumulations at hundreds of localities.

Mechanical methods of separation and recovery are emphasized in large-scale operations such as those in the South African diamond pipes, but they are rare elsewhere. Various attempts have been made to recover garnets, turquoise, and other gem minerals in large mills operated for bulk products such as feldspars and copper concentrates, but the recovery generally has been effected by handpicking at an early stage in the flowsheet. It often has proved more satisfactory to separate the gem-bearing rock during mining rather than milling at such localities.

Placer Deposits: Gem stones are extracted from loose placer materials and the thoroughly weathered rock of many lode deposits by washing, dry winnowing, screening, panning, jigging, tabling, and handpicking, generally in some combination of stages. Equipment ranges from baskets, hand pans, and simple log washers to belts, tables, and electrostatic separators. Large companies depend mainly on efficient standard bulk concentrating techniques; individual miners and collectors, at the other extreme, derive their return more through careful observation and patient hand work than from the use of complex tools.

Special Losses: Owing to the high values of most gem materials, particular efforts generally are made to insure maximum recovery from the deposits that are worked. The same high values, coupled with the portability of most stones, also invite attempts at theft, and even the beauty of the stones themselves can induce unfortunate reactions in otherwise honest persons. Idle properties in all parts of the world have been visited and worked by individuals or groups without knowledge or consent of the owners. The large output from these informal activities has appeared in collections, in various commercial transactions, and even in the regular trade.

Operating properties are equally liable to theft by visitors and employees. Access therefore is carefully regulated, and if the mine output is of high unit value, special precautions are taken against the many ingenious attempts at "high grading," which range from the swallowing of gem stones by entrepreneural individuals to various elaborately coordinated exchanges of stones for other, less valuable materials. As pointed out by Ball (1949), thefts do occur in the face of the most careful monitoring, generally "in direct ratio to the primitiveness of the mining methods."

Synthetic and Imitation Gem Materials

Corundum and Spinel

The synthesis of gem stones began more than a century ago, and rubies large enough for gem use were manufactured as early as 1877. Fragments of natural ruby also were sintered to form *reconstructed* ruby, but many of these stones were so unsound that they cracked or shattered during subsequent cutting and use. Development of the flame-fusion (Verneuil) process in 1900 provided the basis for commercial synthesis of corundum. In this relatively simple method, finely ground aluminum oxide of great purity is dropped through the flame of an inverted oxyhydrogen blowpipe that operates within a combustion chamber. The fused material collects continuously on a rod or spindle placed beneath the flame, and there forms a carrot- or bottle-shaped *boule* that is a single crystal of corundum, in general with polysynthetic twinning. Typical boules are 51 to 127 mm (2 to 5 in.) long and weigh 75 to 500 carats; considerably larger ones have been made, including a 28,000-carat white sapphire 203 mm (8 in.) long and 89 mm (3½ in.) in diam.

A boule is grown in 2 to 5 hr, after which it is cooled slowly for a short time and then removed from the furnace and allowed to continue its cooling more quickly. The entire cooling process can require no more than an hour, but this is so rapid that considerable strain is set up within the boule. Most of the strain must be relieved before the material can be used, which is accomplished by lengthwise splitting. Improvements in manufacturing techniques during recent decades have made it possible to control the crystallographic orientation of boules within narrow limits, and hence to obtain better splitting and relief of strain. Another outgrowth of modern technology is the production of slender rods, some as long as 0.9 m (3 ft), and single crystal filaments as long as 30.5 m (100 ft). These have primary shape advantages in connection with certain industrial uses, and short segments of the rods also are rounded and flame polished for use in jewelry and jewelry design.

Corundum rods commonly are grown by modified Verneuil methods or by a process known as crystal drawing or "pulling," in which a small seed crystal is placed in contact with a pool of alumina melt and then is slowly raised, with or without rotation about a vertical axis. A large single crystal is formed upon the seed and becomes increasingly elongate as the upward withdrawal continues. Alternative melt-growth methods include crystallization on seeds placed in slowly cooling crucibles of fused alumina, single-crystal pattern casting from seeds placed in molds of tungsten or other materials with very high melting points, vacuum growth in crucibles, crystallization from controlled domains of melt formed by intense electron heating, and growth from melts comprising alumina and one or more fluxes. These different methods yield a remarkable variety of product shapes: in widest current demand are single-crystal rods and filaments, discs, boules of many forms, and bodies cast to specified forms.

Some corundum has been synthesized commercially by hydrothermal methods, which are relatively expensive but provide certain advantages of controllable crystal growth and a yield of material with relatively little internal strain. The feed stock, either corundum, alumina, or aluminum hydroxide, typically is placed at one end of an elongate pressure-resistant autoclave, or "bomb," and a small seed crystal of corundum is placed near the other end. After an aqueous solution of sodium carbonate is added, the autoclave is sealed and raised to temperatures commonly in the range 400 to 460° C. An elevated pressure, generally within the range 69 to 415 MPa (10,000 to 60,000 psi), and an end-to-end temperature differential of 20° to 40°C are maintained within the vessel for a given crystallization run, during which the nutrient material at the hotter end is progressively dissolved into the aqueous vapor and transferred through this fluid to the other end for deposition on the seed crystal.

Spinel also is synthesized in substantial quantities by means of the Verneuil process. Colored varieties of both spinel and corundum are obtained by adding metallic oxides to the raw alumina. In the manufacture of corundum, for example, 2 to 7% of chromium oxide in the batch yields various tones of red and lesser amounts yield pink and very pale red, 2% of iron oxide with lesser amounts of titanium oxide yield blue, a mixture of cobalt and vanadium oxides green, a mixture of iron and chromium oxides orange, and so on. The colors and color changes characteristic of alexandrite can be duplicated by adding about 3% of vanadium oxide. This variety of synthetic corundum has been widely marketed under the misleading names *alexandrine*, *synthetic alexandrite*, and even *alexandrite*. Alexandrite and numerous other natural stones of superior value also are simulated by colored varieties of synthetic spinel.

Both synthetic corundum and spinel are popular in their own right for gem uses, as their price is relatively low and their composition and properties do not differ significantly from those of the corresponding natural stones. Further, they can be produced in masses that are relatively free from flaws, and that have attractive and evenly distributed colors. Some of the material is sufficiently strained and brittle to develop internal cracks during and after cutting, but this objectionable feature is not characteristic of stones made during recent years. Here it is interesting to note that colorless synthetic corundum now can be effectively applied as a hard protective coating on some softer natural and synthetic gem stones with superior optical properties, Surface hardnesses near 9 on the Mohs' scale are achieved, generally via high-temperature deposition of a corundum film from gaseous mixtures containing an aluminum alcoholate.

Synthetic star ruby and star sapphire also are made by flame-fusion and crystal–drawing methods. The asterism is obtained through controlled exsolution of impurities, generally rutile in the form of tiny, needlelike crystals that assume preferred orientations within the host crystals of corundum. These asteriated gem stones are in considerable demand, but have not affected the higher values of the corresponding natural stones, A variety of synthetic spinel recently has been produced to resemble lapis lazuli. The German manufacturers can add specks of gold to match the appearance of the pyrite that commonly occurs in natural material.

Synthetic corundum is employed mainly in the manufacture of jewel bearings and for other industrial uses already mentioned. It is sounder and internally more homogeneous than typical natural ruby and sapphire, and also has a lower coefficient of friction. Raw boule will yield four to seven watch jewels or one industrial jewel per carat (Thompson, 1955). Synthetic spinel has broadly similar uses; although it is softer than corundum, it can be surface hardened by special heat treatment. Thanks to their shape and uniform diameter, rods of corundum and spinel are preferred over boules for some types of bearings and pointed instruments. They can be heated and bent in certain directions, and commonly are fashioned into hooks and loops for use as thread guides. The rough exterior, sawed, or ground surfaces of such material can be effectively smoothed or "glossed" by carefully controlled "skin melting" and annealing. This commonly improves the strength of both rods and thin plates. Surface smoothing also is accomplished by thermal bonding of a thinly applied coating of magnesia, silica, or other oxide that forms a solid solution, eutectic, or peritectic with the host material.

Emerald

Emerald crystals of modest size were first synthesized in Germany in 1930, and small crystals were grown a few years later by the Chatham Research Laboratories in San Francisco. Two general procedures, each quite different from the flame-fusion process, have since evolved for commercial production. One is flux-fusion (melt-flux), and the other crystallization in a closed container under hydrothermal conditions. Not all details have been made public (Nassau, 1976; Yaverbaum, 1980). The flux-fusion method has been much improved during recent years, and crystals weighing 1000 carats or more have been produced. The material of medium to best quality matches the properties of natural emeralds almost perfectly. Though not inexpensive, it has found a ready market.

Diamond and Boron Nitride

The long-sought synthesis of diamond required development of a vessel capable of maintaining extremely high pressure and temperature conditions for a sufficiently long period of time to permit satisfactory crystallization. This difficult problem was solved in 1954, when the General Electric Co. succeeded in producing diamonds from graphite at temperatures near 2760°C (5000°F) and under pressures of more than 1×6.9 GPa (1×10^6 psi). Three years later, the same basic techniques, known as the Hall process, were successfully applied to an analogous synthesis of cubic boron nitride from the hexagonal form of that substance. The cubic form, now known in the trade as *borazon,* has properties resembling those of diamond, as long ago predicted by crystal chemists, and it is much more stable at very high temperatures.

As detailed elsewhere in this volume, synthetic diamond is now widely accepted for certain industrial applications, and additional methods of synthesis are under investigation. The Hall process itself has served as a base for extensive subsequent research in areas such as vessel design, catalysts, physical conditions of crystallization, use of seed crystals, and control of size, habit, purity, and surface characteristics of the grown crystals (Yaverbaum, 1980).

Crystals of gem quality also have been made in the laboratory, but at costs yet too high to permit competition with natural diamonds. Borazon has been entering the industrial market as a substitute for synthetic diamond, but its potentialities as a gem material with an attractive range of colors remain to be fully determined.

Other Synthetic Materials

Synthetic rutile, prepared by the flame-fusion process and marketed under names such as *titania, titangem,* and *miridas,* is cut into transparent gems with great brilliance and fire, but light-colored material free from an objectionable yellowish tinge is difficult to obtain. Some splendid blue, red, and deep yellow stones have been produced. Synthetic star rutile of nearly white color has been prepared, but it is not yet marketed in significant quantities. The hardness of titania is slightly less than that of quartz, so that finished stones must be handled carefully when worn in jewelry.

Both barium titanate and strontium titanate, substances not found in nature, have been synthesized by the Verneuil process in the form of clear crystals. Their optical properties are similar to those of rutile, but their hardness is inferior and hence their use as gems has been limited. Some attractive stones with transparent protective coatings of corundum have appeared in the gem trade from time to time. Clear and essentially colorless single crystals of aluminum titanate, calcium titanate, yttrium titanate, and zinc titanate have been grown during recent years. These substances show promise for gem use, as they have high indices of refraction, high dispersion, moderate to superior hardness, and potential for synthesis in attractive colors. Aluminum titanate, nearly as hard as corundum and highly resistant to thermal shock, also could serve for many industrial uses.

Chrysoberyl has been made synthetically, some of it with the color changes characteristic of natural alexandrite. It has not appeared in quantity on the market owing to high manufacturing costs and some superior features of synthetic corundum and spinel that can be used for the same purposes. Other naturally occurring gem materials successfully synthesized by various methods during recent years include amethyst, citrine, cat's-eye, and sapphire quartz, as well as jade, lapis lazuli, opal, and turquoise. Excellent crystals of quartz in several colors have been grown hydrothermally, whereas the polycrystalline cat's-eye material represents the drawing of fused silica through dies or tubes. Artificial jade in several colors has been derived from finely ground "ceramic-type" mixtures of lead oxide, nepheline syenite, zinc oxide or carbonate, potash feldspar, quartz sand, kaolin, and chromophoric oxides. Such mixtures are calcined and fused in crucibles; the cooled products have been judged suitable for gem, decorative, and industrial uses.

To form lapis lazuli, crystals of sodalite or nepheline, along with sulfur, polysulfides, or sulfur-rich mixtures, typically have been vacuum calcined and then heat treated in autoclaves under moderately high pressures of helium or other inert gases. Partial or complete conversion of the starting crystals to blue lazurite occurs in the temperature range 500 to 1200° C, with full conversion generally at temperatures of 800° C and higher.

Opal of high quality has been produced by a complex process that involves growth of uniform tiny spheroids of silica, sedimentation of these particles on a silica substrate in geometrically regular light-diffracting arrays, and careful fixing of the arrays through various combinations of drying, sintering or welding of the particles, and additions of silica in sol form (Darragh, Gaskin, and Sanders, 1976; Yaverbaum, 1980). The best of the synthetic material, which is marketed as both white and dark-colored stones, is comparable to natural precious opal in most respects but generally exhibits more vivid colors, shows a delicate mosaic or "lizard-skin" structure, and is somewhat softer and more porous.

In response to greatly increased demands for turquoise, synthetic stones were introduced to the market in 1972. This material, which is attractive in texture and color, appears to consist of powdered natural turquoise that is tightly bonded. Large quantities of relatively soft and porous natural turquoise also have been impregnated with plastics or other enhancing agents; such material is referred to by reputable dealers as *stabilized turquoise*.

Hydrothermally grown crystals of clear, colorless quartz have been produced in increasingly large numbers for industrial applications during the past four decades, and the present annual output is very substantial. Crystals of fluorite, halite, tourmaline, and several other minerals also have been synthesized hydrothermally for technological uses.

Recent additions to the family of synthetic gem stones are YAG (yttrium-aluminum garnet), YGG (yttrium-gallium garnet), and GGG (gadolinium-gallium garnet). Unlike natural garnets, these substances are silicon-free. The crystals can be grown by flame-fusion and crystal-drawing methods, but most are results of flux-fusion processing. YAG is used in laser, microwave, ultrasonic, and other technical devices. Thanks mainly to its superior hardness, moderately high dispersion and index of refraction, relatively low cost, and availability as clear, colorless, or variously colored material, it has been successfully marketed as a substitute for diamond under names such as *diamonair, diamite, triamond,* and *simulated diamond.* YGG is slightly less hard but otherwise similar in characteristics. GGG, though even less hard, has a high index of refraction, high dispersion, and commercial promise as a gem diamond simulant.

Synthetic lime-silicate garnets have been produced for a wide variety of ornamental and abrasive uses, but in the gem market they are not competitive with harder and more brilliant crystalline materials. Superior products with a different crystal structure recently have been obtained through fusion and cooling of carefully controlled mixtures of inexpensive raw materials. Self-fluxing mixtures of alumina, calcium carbonate, and silica hydrate, for example, readily yield clear crystals with hardness approaching that of corundum. They can be cut into gems of exceptional brilliance.

The most recent and perhaps the most attractive synthetic diamond substitute to appear on the market is cubic zirconia, ZrO_2, which is sold chiefly under the names *djevalite, phianite,* and *zirconia.* It has become very popular, and world production in 1978 amounted to about 15 million carats of rough material. The hardness of cubic zirconia generally is greater than 8 on the Mohs' scale, like that of YAG, and its index of refraction and dispersion are very high. Single clear crystals, colorless or in various hues, are grown by flux-melting or by zone-melting ("skull melting") techniques (Nassau, 1977), with starting mixtures of zirconium oxide for nourishment, fluxes and coloring oxides as needed, and scandium oxide or a rare-earth oxide to stabilize the cubic structure.

For relatively recent and more detailed

summaries of synthetic gem materials, reference can be made to Webster (1970), Nassau and Nassau (1980), and Yaverbaum (1980).

Imitation Stones

The term *imitation* applies generally to substances that are quite different in composition from the gem materials they are intended to resemble (e.g., green garnet simulating emerald; titania, YAG, and white sapphire simulating diamond; black cryptocrystalline quartz simulating jet). In the trade, however, this designation ordinarily is restricted to manufactured noncrystalline substances that are compositionally different from any found in nature. Imitations of glass, known as *paste* or *strass,* are by far the most widely used, owing to their low cost and their great range in color and other properties. Glasses best simulate beryl, jade, quartz, topaz, and tourmaline, and are somewhat less effective as imitations of diamond, ruby, sapphire, and zircon.

All glass stones are objectionably soft, and are suitable only for costume jewelry or special displays. Many of the glasses are lead-bearing types that have high indices of refraction and can be fashioned into sparkling stones, but these are particularly soft and hence liable to rapid wear. The lower facets of some glass stones are silvered or backed with foil in order to increase their brilliance. These are sold under the general name *rhinestone*. The cheapest type of imitation gem is the *molded* stone, which consists of glass or a plastic that has been pressed into a carefully designed mold, with or without subsequent polishing of the facets on the solidified material.

Bakelite, lucite, cellulose acetate, and other plastics are widely used to imitate amber, jet, hematite, turquoise, and other nontransparent materials. Increasing quantities of transparent plastics are being employed in the manufacture of costume jewelry.

Pearls are remarkably well imitated by solid spheres of translucent glass and by hollow glass spheres that are filled with wax. The characteristic luster is obtained by repeated surface applications of finely ground fish scales in ammonia. Cheaper imitations are made of translucent glasses whose own luster is pearly.

Production and Consumption

Production of gem materials rarely is equivalent to consumption during the same year, but an approximate balance generally is struck over periods ranging from 3 to 12 years. At times the trade has been seriously affected by overproduction of certain stones, especially during periods following important discoveries or the recognition of new markets. Examples include emerald from South America (1565–1790), diamond from Brazil (1725–1740, 1850–1860) and South Africa (1872–1889), ruby and sapphire from Asia and Australia (1800–1910), and tourmaline from the United States (1903–1912). In such instances production ultimately has dropped off in response to normal economic factors, various governmental regulations, or exhaustion of the richest and most easily worked portions of major deposits.

In the past, demands for some industrial stones have exceeded available supplies, but these problems have been resolved by new discoveries, improvements in use technologies and recovery of material, and especially by the substitution of other grades or kinds of materials. Thus synthetic corundum and spinel have largely supplanted natural stones, and polaroid is now extensively used in place of optical calcite. By far the largest segment of the diamond-producing industry is controlled by one organization, which maintains considerable flexibility of supply by accumulating stocks of rough material. The merit of this policy was clearly demonstrated during World War II, when demands greatly exceeded the available output from all mines.

Materials for gem uses seldom have been in serious short supply relative to their normal long-term availability. Mining of most stones is traditionally sporadic, and is rather sensitive to shifts in demand and general economic conditions. Known reserves of unmined ground are small for most deposits other than those yielding diamonds, but this is a reflection more of operating policies than of true availability. Finished gems and ornaments are essentially permanent articles that constitute a steadily growing reserve of salable material. Some have been destroyed or lost through accident, to be sure, but this attrition has been insignificant. Thousands of old stones appear on the market each year, either in original or recut form, and, as indicated by Ball (1949), "many an Egyptian or Roman gem is worn today by an American woman or a Wall Street tycoon." Perhaps the most extreme example of recirculation is furnished by peridot; large stones of this gem olivine have not been mined in appreciable quantity for many centuries.

World Production

Leadership in gem stone production has shifted during historic time from Europe to

Africa, thence to Asia, thence to South America, and finally back to Africa (Ball, 1922, 1931, 1935). The Baltic amber mines dominated the industry from 25,000 to 3400 B.C., after which they were surpassed by operations for turquoise and then for emeralds in Egypt. From 800 B.C. to 1725 A.D. the production of diamond, ruby, and sapphire maintained India and Sri Lanka in the top position. Brazil then assumed the lead through its output of diamonds, but in turn was displaced by South Africa in 1872. Diamonds produced in South Africa and the Belgian Congo (Zaire) have made these countries dominant during recent decades. Brazil and several European countries have continued as significant sources of many gem stones, but Egypt, India, and Sri Lanka together account for less than 1% of the world's current output.

The total production of gem materials in the world to date can be valued at approximately $45 billion. Nearly three-fourths of this total is represented by cut diamonds now in the possession of mankind, and about four-fifths by all existing cut gems. Thanks mainly to its contribution of diamonds, Africa is the leading continent in total output of gem materials. Asia has been the leader with respect to the aggregate value, in rough form, of all gem stones other than diamonds produced to date, and has been followed by Europe, South America, Australia, North America, and Africa in that order. Respective rankings for two different periods are shown in Fig. 8.

Production and consumption of gem stones were fairly stable during the eighteenth century, when the value of an average year's output was slightly less than $5 million. The annual value gradually rose during the following century, and increased spectacularly after 1870. The $20-million level was first reached in 1880, the $80-million level in 1913, and a $97-million peak was reached in 1928. As shown in Fig. 13 (p. 329), production during the first half of this century was violently affected by world conflicts and economic depressions. The trend since 1943 has been vigorously upward, reaching annual values exceeding $200 million by the late fifties, $1 billion by the late sixties, and $4 billion by the late seventies. The impressive increases in production of natural gem materials during the past century are ascribable principally to growth in the demand for diamonds, approximately 45 million carats of which were produced from the world's deposits in the peak year 1976.

The Republic of South Africa, long a prime source of natural gem diamonds, accounted for 35% to 45% of the world's annual output during the period 1960 to 1980. While South Africa's large production has been remarkably steady, that from Russian deposits and from occurrences in several other African countries has been growing in response to vigorous programs of exploration and development. For the five-year period 1976–1980, diamond-producing countries were led by the Republic of South Africa (35% of world gem output, 14% of world industrial-diamond output), the USSR (21%, 29%), Zaire (6%, 37%), and South-West Africa (17%, <1%). Following them were Botswana (4%, 8%), Angola (4%, <1%), Sierra Leone (3%, 2%), Ghana (2%, 6%), Venezuela (2%, 2%), Tanzania (2%, 1%), Central African Empire (2%, <1%), and Liberia (1%, 1%). Additional minor sources included Brazil, from which all carbonado and most of the world's ballas have been produced in the past, along with Guinea, India, Lesotho, Ivory Coast, Guyana, and Indonesia.

The rankings are little changed by expression of outputs in terms of value, although the position of South-West Africa is thereby

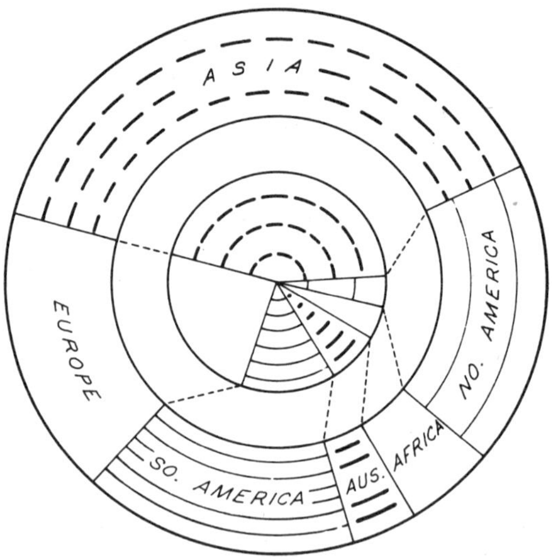

FIG. 8—*Estimated relative importance of the continents as sources of gem materials (exclusive of diamonds and pearls). Outer part of diagram indicates respective contributions to the value of the world's total output during the 50-year period since 1920, inner part respective contributions to world's total previous output. Inner part based mainly on data from Ball (1922) and Meisner (1929).*

markedly improved and the spreads between the other leading countries are increased. Of the diamond production from South-West Africa, 95% is gem material, as compared with an average of 26% for recent world production. Corresponding recent gem fractions have been 48% for the Republic of South Africa, about 20% for the USSR, and 5% for Zaire. In general, relatively small fractions characterize those countries in which lode production exceeds that from placer deposits.

Since World War II, African nations have accounted for nearly three-fourths of the world's output of all natural gem materials in terms of value. Along with the major diamond-producing countries of other continents, Brazil, Colombia, and Australia have supplied much of the remainder, each accounting for more than 1% of the total during recent years. Countries producing significant fractions of 1% include Burma, Czechoslovakia, Germany, India, Indochina, Japan, People's Republic of China, Sri Lanka, Thailand, the United States, and Zambia. Operations in many countries are financed in part by United States capital.

Fig. 9 illustrates the estimated relative importance, in terms of value at the mine, of gem materials other than diamonds and pearls that

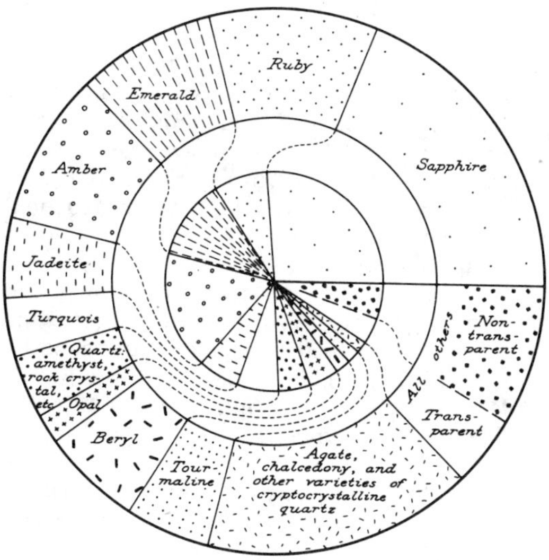

FIG. 9—*Estimated relative contributions of major gem materials to the value of the world's total output (exclusive of diamonds and pearls). Outer part of diagram represents the output during the 50-year period since 1920, inner part the previous known output. Inner part based mainly on data from Ball (1922) and Meisner (1929).*

have been produced prior to 1920 and in the 50-year period dating from 1920. During the first half of the present century the value of the world's total output in normal years changed little from an average of about $5 million, but progressive and dramatic increases have occurred since the time of World War II. These increases reflect both rising demands and widespread inflationary trends in costs and prices. Significant shifts in production have occurred among individual minerals. The exhaustion of some deposits and the effects of depressions, international strife, political unrest, and changes in fashion were reflected in decreased outputs of amber, emerald, and sapphire during several decades prior to 1960, after which production of emeralds and sapphire began to recover strongly. Increased outputs of beryl, tourmaline, cryptocrystalline varieties of quartz, turquoise, and other nontransparent materials since World War II have been due largely to increased demands in the United States, especially by amateur lapidaries.

Placer deposits account for more of the total production than lode occurrences, although the dominance is no longer as great as it once was. As in the case of diamonds, the percentage of first-quality gem stock tends to be higher in placers than in lode deposits, but the yields vary greatly from one occurrence to another, even for a single material. About 20% of the amber mined in the Baltic region is suitable for ornamental uses. Higher figures apply to most nontransparent minerals, and much lower figures to minerals that typically are facet-cut. About 4% of the mine-run Colombian emerald, for example, is of superior color and slightly enough flawed to be cut into gems of good to excellent quality. Most yields of sapphire, ruby, spinel, beryl, tourmaline, and many other transparent minerals have been at similar or even lower levels.

Production of Colombian emerald in recent decades has averaged more than 12,000 carats per year for periods when the government-owned mines in the Muzo and Chivor districts were worked, but formal operations often were brought to a standstill by local disturbances and illicit dealings. The mines were shut down from 1973 until 1977, when activities were resumed on a large scale under government leases and military surveillance. Resolution of most difficulties is attested to by a record of annual production valued at only $7.2 million in 1972 but nearly $50 million in 1978 and more than $100 million in 1979.

Other countries with significant outputs of emerald during the seventies include the USSR, India, Brazil, Australia, Republic of South Africa, and Zambia.

Since 1935 the output of ruby, sapphire, and spinel from Burma has averaged slightly more than 100,000 carats per year, that from Sri Lanka and India about 150,000 carats per year, and that from Thailand and Indochina about 75,000 carats per year. Production trends have tended downward for most of these countries during the past two decades, but uniformly reliable data are not available. Australia, long known as the world's predominant source of gem opal, has also become the world's leading producer of sapphire since the sixties. The USSR recently has supplied modestly increasing amounts of sapphire, while significant amounts of ruby and spinel have been produced in Afghanistan, ruby in Kenya and Pakistan, and spinel in Madagascar, Australia, and Brazil. The output of beryl, topaz, and tourmaline from Brazil has averaged about 6 million carats per year since World War II.

Owing in part to increasing outputs and in large part to inflationary trends, the values of gem production from many countries have risen almost startlingly during the seventies. Thus the average value of gem stones produced in Sri Lanka was about $500,000 per year for the period 1945–1973 and $700,000 for the period 1974–1980; in Burma it was about $350,000 and $600,000; in Australia about $900,000 and more than $15 million; and in the United States $900,000 and $8 million for the same respective periods.

The world's annual output of gem coral was at the million-dollar level for many years, but in general dwindled after the twenties. Sharply upward shifts occurred during the seventies, when the value of production exceeded $5 million in most years. The average annual output of pearls, chiefly cultured material from Japan, has been increasing since World War II, with yearly values greater than $10 million in 1973, $20 million in 1977, and $45 million in 1979. The relatively small production of natural pearls has come from widely distributed fisheries in the Pacific Islands, Japan, Australia, and in countries bordering the Indian Ocean, Red Sea, Persian Gulf, and Carribean Sea.

Synthetic diamonds are produced at a current rate of about 60 million carats per year, mainly in the United States, Western Europe, Russia, and South Africa. Synthetic corundum and spinel are manufactured chiefly in Europe and the United States, and to a much lesser extent in Japan, Israel, and India. The principal centers are in Germany, Switzerland, France, and England. Switzerland is by far the most important source of jewel bearings; others, some of them intermittently active, include Canada, England, France, Germany, Italy, Israel, Japan, Russia, and the United States. Synthetic emerald, entirely for gem uses, is produced in Germany, the USSR, Switzerland, and the United States. Imitation stones are manufactured on a very large scale, chiefly in Austria, Czechoslovakia, Germany, Hong Kong, and Japan. Imitation pearls are produced mainly in Hong Kong, Japan, Spain, and Taiwan.

United States Production

The aggregate value of the many kinds of gem materials obtained from deposits in the United States since 1880 is approximately $106 million, which corresponds to less than 1% of the world output during the same period. Domestic production reached its first major peak in 1909, with a value of $534,400 for that year, and its second peak in 1921, with a value of $518,300. The depression year 1934 marked the all-time low, when the value was only $3,200. Since then a strong upward trend, interrupted only by World War II, has led to annual values that reached $1 million in 1958, $2 million in 1968, and an average of $8.7 million in the period 1975–1980 (Fig. 10). For most of the years since 1880, US production constituted from 0.2 to 1.3% of the world's total output in terms of value.

The most significant domestic mining for gem stones prior to 1910 stemmed in part from new discoveries, such as that of tourmaline in Maine (1820), gold quartz in California (1849), sapphire in Montana (1865), tourmaline in California (1872), hiddenite in North Carolina (1879), and kunzite in California (1902), and in part from reoccupation of old Indian diggings for allophane, chlorastrolite, garnets, peridot, turquoise, variscite, and agate, petrified wood, and other varieties of quartz. Production of beryl, feldspar, and coarsely crystalline quartz, mainly from California, the Carolinas, Colorado, South Dakota, and the New England states, also dates from the nineteenth century. The principal occurrences that most recently have yielded these and many other minerals are summarized in Table 4. The leading states, in terms of value

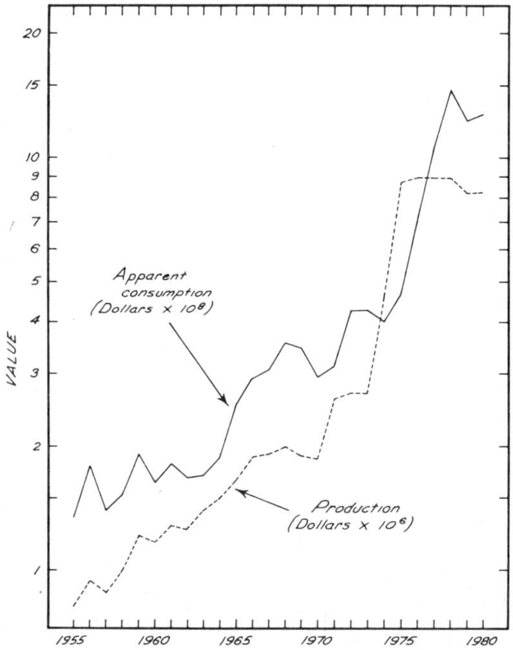

FIG. 10—*United States production and apparent consumption of gem materials, according to values, during the period 1955–1980. Based on data compiled by US Bureau of Mines. Apparent consumption is reckoned as domestic production plus imports, minus exports and reexports. Note logarithmic scale for values.*

of their all-time production, are Arizona, Montana, Oregon, Nevada, and California, in that order, and leaders during the past ten years of burgeoning production have been Arizona, Nevada, Oregon, Maine, and California. Most important among the other producing states are Colorado, Idaho, Michigan, New Mexico, North Carolina, Texas, Utah, Washington, and Wyoming. In most of the years since World War II, fewer than a dozen states have accounted for more than four-fifths of the total domestic output.

Trends in United States production of gem materials from 1900 to 1980 are indicated by five-year totals in Fig. 11. The prominent trough of the mid-thirties separates two periods of basically different characteristics. The earlier of these was typified by the professional operations of both mining companies and individuals, the dominance of sapphire and ruby, tourmaline, and turquoise in their output, and the marketing of this output chiefly through regular trade channels. The earliest formal mining organization was the Emerald and Hiddenite Mining Co., which obtained $22,500 in gem stones from the pegmatite deposits at Stony Point, NC, during the period 1880–1889. Many other companies subsequently were active, the largest of them in the sapphire fields of Montana.

The more recent period of relatively large production has been characterized by the activities of a few small companies, some professional collectors and individual operators, and many thousands of amateurs engaged in part-time collecting and fee mining. Jade, turquoise, and the quartz-family minerals have dominated the output, whereas the production of sapphire, ruby, and tourmaline remained small until the seventies (Fig. 11). The record of the seventies indicates greatly increased contributions of tourmaline from Maine and California, along with a spectacular growth in turquoise production that accounts for recent leadership of Arizona and Nevada among the gem-producing states.

Since 1940 substantial amounts of jade have been obtained from deposits in Wyoming and California, and from occurrences in Alaska that were worked by the Eskimos centuries ago, and since 1960 coral from Hawaii, peridot from Arizona, and obsidian from Arizona, California, and Oregon have become significant components of total production. Most of the gem-stone output has remained in private hands, and much of the rest has been sold through numerous local dealers and roadside "rock shops." Reliable statistics on these materials are not available, and the aggregate production of domestic gem stones during recent years probably has been substantially greater than that indicated in Figs. 10 and 11.

About 87% of the sapphire and ruby produced commercially in the United States was exported to Switzerland for industrial use prior to World War II. Large quantities of pink tourmaline and jade have been exported to China, chiefly for the preparation of polished or carved gems and ornaments. Shipments of many other minerals have gone abroad from time to time, but the bulk of the production has been absorbed within the United States.

Domestic output of synthetic corundum and spinel, amounting to millions of carats per year and with early peaks of more than 100 million carats per year during World War II, comes mainly from plants of the Linde Air Products Co. in the Chicago area, and also from the Union Carbide Corp. at Niagara Falls, NY and the Bell Telephone Laboratories, Inc. at

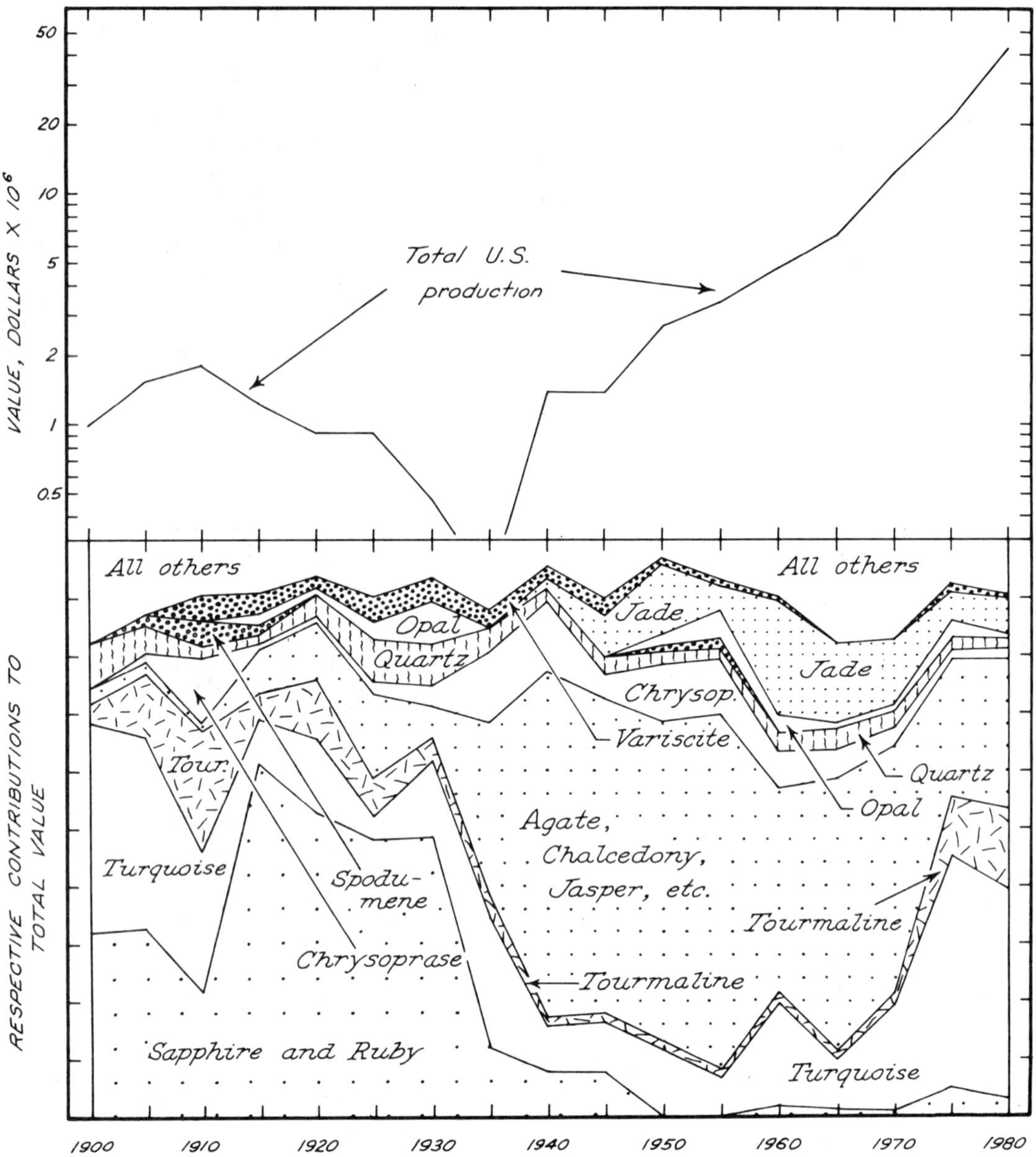

FIG. 11—*Trend, expressed by five-year totals, in value of gem materials produced in the United States since 1895, and temporal shifts in relative contributions of selected materials to the totals.*

Murray Hill, NJ. This material is widely used for both gem and technical purposes. Large quantities of diamond, along with boron nitride and other synthetics, are produced by the General Electric Co. in Schenectady, NY, and an impressive amount of quartz crystal is manufactured each year by Western Electric Co. in North Andover, MA and by several other large companies. Synthetic emerald is produced mainly by the Chatham Research Laboratories in San Francisco, during recent years at an annual rate of more than 60,000 carats. Approximately 10% of the output, obtained by a flux-fusion process, is transparent rough of excellent quality. Emerald also has been produced by other methods in laboratories elsewhere, chiefly by the Linde Co. and Union Carbide Corp.

Substantial quantities of synthetic rutile, titanates, and other gem materials have been manufactured by the Linde Co., National Lead Co. (NL Industries) of New York, Union Carbide Corp., and the American Cyanamid Co. of Wayne, NJ. At present the Ceres Corp. of Waltham, MA is the only major domestic producer of cubic zirconia. Additional organizations that have been active in the field of gem synthesis include Creative Crystals, Inc., Crystal Systems, Inc., North American Rockwell Corp., and Westinghouse Electric Corp.

United States Consumption and Imports

Demands in the United States for both gem materials and industrial gem stones are greater by far than those of any other nation. Since 1950, for example, imports of industrial diamonds have amounted to about 85% by weight of concomitant world production, and imports of gem diamonds, in both rough and cut form, have constituted nearly one-half of the world output of rough stones. The United States consumes about three-fourths of the diamonds that are cut, and provides comparable demands for most other relatively valuable gems. Considerably more than half of the cultured pearls grown in Japanese waters since World War II has been imported by the United States.

Diamonds have accounted for 85 to 89% of the total annual value of gem materials imported for consumption in the United States during the past three decades. Trends in value of the other gem materials, as shown in Fig. 12 for imports during the period 1955–1972, reflect a growing total demand that has been highly sensitive to the general prosperity of the country. They also reflect the increasing popularity of emerald and other beryls, jade, opal, ruby and sapphire, topaz, tourmaline, and several varieties of synthetic gems. A partial breakdown of gem-stone imports for seven contrasting years is provided in Table 6.

Retail jewelry sales in the United States amounted to less than $200 million in 1932, a typical year of the Great Depression, but rose to points near the $1-billion level in 1943 and 1953, years of wartime and postwar prosperity. They broke through the $2-billion level during the seventies. Reflected in the figures of Table 6 are expanding domestic markets for both cut and uncut diamonds, increases in funds available for luxury purchases and the growing tendencies of purchasers to regard gems as a hedge against inflation, the cutting-off of Japanese sources of cultured pearls and

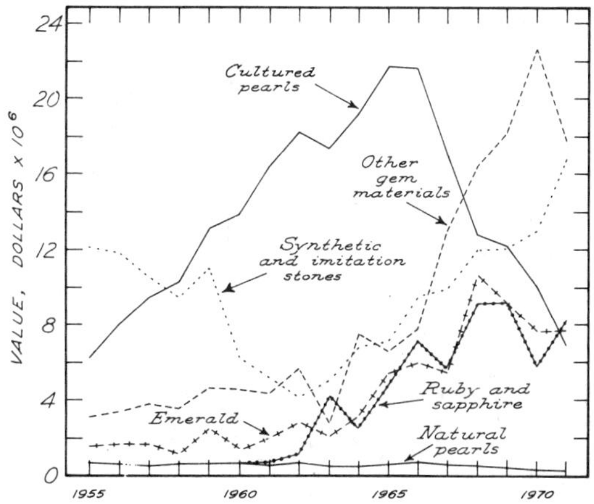

FIG. 12—*Trends in values of gem materials (exclusive of diamonds) imported for consumption in the United States during the period 1955–1971.*

central European sources of imitation stones during World War II, the increasing demands for emerald, ruby, and many other precious stones despite some uncertainties in the supply of newly mined material, and the diminishing popularity of "marcasites," which include pyrite, hematite, and some true marcasite. Imports of imitation stones at times have been very large, and in 1953 they accounted for more than 95% of the imitation and synthetic stones noted in Table 6. Imports of precious stones other than diamonds since World War II have come chiefly from Australia, Brazil, Colombia, India, Indochina, South Africa, and Thailand, and also from nearly 30 other countries that include Belgium, Burma, Sri Lanka, France, Israel, Italy, Japan, Switzerland, the United Kingdom, and West Germany.

Political and Commercial Control

The original gathering of gem stones by individuals and local specialist groups continued with little change among most peoples of the western hemisphere until essentially modern times (Kunz, 1907; Ball, 1941), whereas nearly all gem mining in Europe, Africa, and Asia came under the control of rulers and ruling groups centuries before Christ. Some later operations also were sponsored by the Church, especially in Europe and parts of Asia. The gem mining company has been a relatively recent development, starting in 1715 with the organization of a group to work the garnet deposits of Bohemia (Ball, 1922). Hundreds

TABLE 6—Gem Materials Imported for Consumption in the United States, by Values in Thousands of Dollars*

(Excluding Materials Not Suitable for Jewelry)

Commodity	1943	1953	1958	1963	1968	1973	1978
Diamonds (gem quality):							
Rough or uncut	37,443	57,001	72,563	129,870	252,653	460,198	848,651
Cut but not set	31,454	50,572	68,068	93,977	222,478	360,987	1,112,907
Emerald†	33	349	1,100	2,081	10,644	32,865	82,237
Rubies and sapphires†	< 1	< 1	< 1	4,115	9,175	19,336	46,858
Pearls and parts, not strung or set:							
Natural	167	265	597	479	525	368	840
Cultured or cultivated	< 1	3,770	10,347	17,427	12,865	9,232	27,152
Other precious stones:							
Rough or uncut	48	204	717	1,708	5,062	5,859	15,888
Cut but not set	2,590	2,219	2,904	2,710	11,412	26,575	51,760
Imitation and synthetic stones, including pearls	279	15,718	9,621	5,067	12,378	20,467	30,892
Marcasites	96	97	26	< 1	1	28	40
Total	72,110	130,195	165,943	257,454	537,193	935,915	2,217,225

* Extracts from data compiled mainly by US Bureau of Mines.
† Including very small fractions of rough or uncut material.

of companies, most of them small, have been formed since. Although a large fraction of the world's output has stemmed from their activities, many of them have enjoyed little or no financial success. Even Burma Ruby Mines, Ltd., a British corporation that recovered large quantities of gem stones from the Mogok tract in Burma during the period 1887–1931, realized but slight profit from these extensive operations. Other companies have flourished briefly before ceasing operations, and only a few have continued to grow and prosper. Outstanding in this last category is DeBeers Consolidated Mines, Ltd., which has played a dominant role in supplying the world's diamonds since 1888. The history of this organization has been both complex and fascinating (Chilvers, 1939).

State control of gem-stone sources has continued to the present time in most parts of the world. It ranges from outright ownership of the deposits (Colombia, Namaqualand in South-West Africa, Russia), which are mined by the government or are leased to operating companies, to the imposing of various regulations, restrictions, and even embargoes on exports. Except during periods of national emergency, the United States Government has exercised little direct control over the mining of deposits not on federal land.

Owing to the profound geographic differences between regions of production and centers of demand, international flow of gem materials is essential to the existence of the industry. This flow is channeled through the tariff structures of most countries, commonly for both incoming and outgoing material. Export duties typically are at the 10 or 20% level, but are considerably higher in some parts of the world, especially for rough stones in countries attempting to foster or protect domestic cutting industries. Import duties are more widespread, and generally range from 2 to 100% ad valorem. Many export-import relationships are complicated by various exchange controls. High tariffs and the great value per unit weight of most precious stones have prompted countless attempts at smuggling.

Most duty rates on imports to the United States date from the Tariff Acts of 1922 and 1930. Statutory rates typically have been 10% ad valorem for gem diamonds and other natural precious stones in cut form, 20% for marcasites, 10% for natural and cultured pearls, 40 to 90% for imitation pearls (depending on make and size), 20% for cut or faceted nonopaque imitation stones, and 60% for most other kinds of imitation materials. Some of the rates have been lowered in recent years;

cut but unset diamonds, for example, are now dutiable at only 1% if less than one-half carat and 2% if greater than one-half carat, and emeralds are duty-free. Rough or uncut natural stones for both gem and industrial uses also are duty-free. The import duty on other materials such as synthetic corundum boule, jewel bearings, and diamond wire-drawing dies generally ranges from 10 to 30% ad valorem, in some instances depending on the country of origin. A 10% excise tax was levied on jewelry in 1941, was increased to 20% in the wartime year 1944, was lowered to 10% in 1954, and was removed in 1965.

Marketing

Organization and Methods

Most of the world's diamond output is marketed through channels that are effectively controlled by the DeBeers syndicate, but brokers also can obtain rough gem material from various independent sources. Subsequent distribution often is highly complex, and by the time a stone has been cut, shipped to areas of demand, and sold at the retail level, it ordinarily has passed through many hands in the trade. Industrial diamonds can be purchased directly from a central DeBeers-controlled source, from independent producers, from various brokers, or from cutters who have accumulated stocks of reject material from the preparation of gems.

The marketing structure for most other kinds of gem stones is both varied and flexible. At one extreme is the direct mail-order distribution of rough material from the mine to the ultimate consumer, generally an amateur lapidary. At the other is a long series of transactions involving the producer, several brokers, and one or more cutters before a stone or parcel of stones reaches the jobber or wholesale jeweler for further distribution. During this time the material may well have passed one or more times through a well-established marketing center such as Antwerp, Bombay, London, New York, or Paris. It will have been graded and regraded, both before and after cutting, but it may or may not have been inspected by the office-based brokers who have dealt with it.

The bases for individual transactions range from formal contracts to auctions at both wholesale and retail levels. Old gems and jewelry commonly are sold at auction, often with no more than sight identification of the stones by the bidders. Large quantities of both old and new gems also are sold each year in completely informal ways, generally to tourists and other persons who are attracted by what seem to be unusually good bargains. Such purchasers rarely have opportunity to determine the real value of the material in question, and hence are particularly liable to the cozenage and outright fraud that are widespread outside of well-established sales channels.

Units, Nomenclature, and Identification

The standard unit of weight now used for the more valuable gem stones and cut gems is the *carat,* or *metric carat.* One carat is equivalent to 0.2 g, 200 mg, or 3.0865 grains; 155.52 carats equals 1 troy oz, 141.75 carats 1 avoirdupois oz. The weight of small cut diamonds also is indicated in terms of *points* (one point equals 0.01 carat). During recent years crushing bort has been sold by the *kilogram* (1 kg equals 5000 carats).

Most of the less costly gem stones, such as amazonite and moonstone, amethyst and citrine, azurite and malachite, lapis lazuli, fire opal, and turquois, are sold in rough form by the *gram, kilogram, avoirdupois ounce, troy ounce,* or *pound.* Both weight and dimensional terms, generally in the metric system, are used to define finished gems made from the more bulky materials of intermediate and lesser value (cf. feldspars, jade, opal, and turquoise in Table 7). For pearls the standard unit is the *pearl grain,* which is equivalent to 0.05 g or 0.25 carat. This differs from the standard grain, which is 0.0648 g.

Some gem names are identical with the names of the minerals themselves, and other well-established names either are different from the corresponding mineral names or are applied to different varieties of the same mineral (Tables 1, 7). Still others are applied to more than one mineral; the marcasites ("sulfur diamonds") of the trade, for example, comprise pyrite and hematite, along with some marcasite and other opaque minerals.

Much confusion stems from the widespread use of more than a hundred trade names that range from meaningless to misleading. These generally resemble or even are identical with the names of valuable gem stones, and are applied to materials of lesser intrinsic worth. Thus idocrase has been marketed as "California jade;" red varieties of garnet as "Adelaide ruby," "American ruby," "Arizona ruby," and "Australian ruby;" and green tourmaline as "African emerald," "Brazilian emerald," "Brazilian peridot," and "Peridot of Ceylon."

TABLE 7—Comparative Retail Values* of Cut But Unset Gem Stones, Exclusive of Diamonds, Correlated with Color, Size, and Other Features (For Stones Finished in Well-Oriented Cuts of Standard Styles and Fair to Good Quality; Transparent Stones Clear and Free or Nearly Free from Flaws, Except Where Otherwise Indicated)

Mineral	Gem Stone	Trans-parent	Semitrans-parent to Opaque	Finished Form: Facet-Cut	Finished Form: Other Forms	Color, Color Value, and Other Features		Weight, Carats	Most Common Price or Price Range, $ per Carat	Additional Features Responsible for Highest Prices
BERYL	Aquamarine	×		×		Blue to greenish blue†	Light to medium	3 5 10	15–90 25–175 45–375	Even color, with minimum of green component
							Medium dark	3 5 10	60–650 75–800 120–1,400	
	Emerald	×		×		Light to medium green	Moderately to heavily flawed	1 8	20–450 50–1,000	Even color, green or very slightly bluish green, with minimum of yellow component
							Slightly flawed	1 8	110–750 800–3,000	
							Flawless or nearly so	1 4 8	750–5,200 1,300–7,500 1,700–13,000	
						Moderately dark green	Moderately to heavily flawed	1 8	100–750 275–2,200	
							Slightly flawed	1 8	850–2,800 2,800–9,000	
							Flawless or nearly so	1 4 8	3,000–20,000 4,800–40,000 6,500–60,000	
	Morganite	×		×		Pink, peach, and rose-red	Light to medium dark	3 10	8–75 15–300	Even color, with deepest and warmest tones
	Goshenite	×		×		Colorless		5-10	3–12	Even color, with deepest tones
	Common beryl	×		×		Yellow green†	Light to medium	3 10	10–20 15–45	
	Golden beryl, Heliodor	×		×		Yellow to pinkish or orangy yellow	Light to medium dark	3 10	8–45 15–70	
	Red beryl	×		×		Rose-red	Medium to medium dark	1 3	600–2,600 2,800–4,500	Even, deep color
CHRYSOBERYL	Alexandrite	×		×		Green (daylight), red to purplish red (artificial light)	Light	3 10	180–1,800 450–3,000	Sharp and intense distinction in colors, as viewed in natural versus artificial light
							Medium to medium dark	3 5 10	700–1,250 2,500–25,000 3,600–40,000	
	Cat's-eye		×		×	Green, opalescent	Light	3 10	150–650 400–4,800	Sharply defined eye with regular form
							Medium to medium dark	3 5 10	650–2,700 2,200–7,500 4,500–18,000	
	Chrysolite	×		×		Yellowish green	Medium to medium dark	3 10	50–125 110–550	Even, deep color

TABLE 7—Comparative Retail Values* of Cut But Unset Gem Stones, Exclusive of Diamonds, Correlated with Color, Size, and Other Features (Continued)

Mineral	Gem Stone	Trans-parent	Semitrans-parent to Opaque	Finished Form: Facet-Cut	Finished Form: Other Forms	Color, Color Value, and Other Features		Weight, Carats	Most Common Price or Price Range, $ per Carat	Additional Features Responsible for Highest Prices
CORUNDUM	Ruby	x		x		Purple to purplish red†	Medium or dark	1 10	700–2,400 3,200–11,000	Even color, very slightly purplish red
							Medium dark	1 5 10	1,200–13,000 4,500–20,000 12,000–47,000	
						Very slightly purplish or orangy red†	Medium or dark	1 10	600–12,000 3,600–45,000	
							Medium dark	1 5 10	1,800–17,000 4,500–40,000 11,000–200,000	
	Star ruby		x		x	Slightly purplish red	Light to medium	1 5 10	250–2,400 750–11,000 1,250–32,000	Even color or distinctive color zoning; regular, well-defined star
							Medium dark	1 5 10	450–14,000 2,500–40,000 3,000–80,000	
	Sapphire	x		x		Blue with slight violet cast	Light or very dark	1 10	100–600 250–2,800	Even color, moderately deep
							Medium dark	1 5 10	500–6,500 1,200–10,000 2,000–18,000	
						Blue	Light or very dark	1 10	60–160 100–250	
							Medium dark	1 5 10	500–600 750–2,200 800–3,000	
						Green, greenish blue, yellow, pink, colorless,† etc.	Light to medium dark	1 10	20–350 80–1,100	
	Star sapphire		x		x	Blue with slight violet cast	Light, medium, or dark	1 10	25–180 100–2,500	Even color or distinctive color zoning; regular, well-defined star
							Medium dark	1 5 10	200–1,800 750–9,000 1,400–12,000	
						Slightly grayish blue†	Light to dark	1 10	30–400 250–2,400	
						Pink	Light	1 10	40–300 65–700	
						Bluish gray, purple, black	Light to dark	1 10	30–350 40–1,800	

TABLE 7—Comparative Retail Values* of Cut But Unset Gem Stones, Exclusive of Diamonds, Correlated with Color, Size, and Other Features (Continued)

Mineral	Gem Stone	Transparent	Semitransparent to Opaque	Finished Form: Facet-Cut	Finished Form: Other Forms	Color, Color Value, and Other Features		Weight, Carats	Most Common Price or Price Range, $ per Carat	Additional Features Responsible for Highest Prices
FELDSPARS	Moonstone		x		x	White to gray	Opalescent	<10 >10	3–80 3–50	Even color tone, with bright blue reflections
	Amazonstone		x		x	Green, bluish green	Medium to medium dark	<10 >10	1–12 1–65	Deepest tones of bluish green
	Sunstone, aventurine		x		x	White or gray, with red or green spangles		>5	1–25	Abundantly spangled
	Transparents	x		x		Colorless to yellow		>2	3–25	Warm color tone
GARNETS	Almandine, pyrope	x		x		Red to purple	Light to dark	1 10	10–150 25–175	Even color in warm, medium deep tones
			x	x	x	Brownish red to purple	Very dark	<10 >10	3–35 10–70	
	Rhodolite	x		x		Rose-red to purple	Light to medium dark	1 10	50–300 75–800	
	Essonite	x		x		Yellowish orange to orange-brown	Light to medium dark	1 10	50–200 120–1,400	
	Demantoid	x		x		Yellow to green	Light to medium dark	1 10	600–1,800 1,500–6,000	
	Melanite		x		x	Black	Essentially opaque	>5	2–35	
JADE MINERALS	Jadeite		x		x	Green	Medium to dark	For stones averaging 15 X 20 mm in plan	800–20,000 per stone	Intense, even color, with minimum of white mottling; minimum opacity
						Green	Light to medium		120–800 per stone	
						Yellowish green	Medium		100–700 per stone	
						Violet	Light to medium dark		10–700 per stone	
						Brownish green	Medium		80–150 per stone	
						White or reddish brown	Light to medium		5–80 per stone	
	Nephrite		x		x	Yellowish green	Medium		10–90 per stone	

TABLE 7—Comparative Retail Values* of Cut But Unset Gem Stones, Exclusive of Diamonds, Correlated with Color, Size, and Other Features (Continued)

Mineral	Gem Stone	Trans-parent	Semitrans-parent to Opaque	Finished Form: Facet-Cut	Finished Form: Other Forms	Color, Color Value, and Other Features		Weight, Carats	Most Common Price or Price Range, $ per Carat	Additional Features Responsible for Highest Prices
OLIVINE	Peridot	x		x		Slightly yellowish green	Light to medium dark	1 10	30–250 90–600	Even color, with maximum deepness of tone
						Greenish yellow		1 10	25–100 65–420	
						Yellowish green		1 10	25–130 60–750	
OPAL	Black opal		x		x	Very dark gray or blue to black		1 10	160–5,500 600–8,000	Intense play of several colors among large, even-size patches, each of uniform color
	White opal		x		x	Light to very light body color		1 10	15–800 140–1,800	
	Fire opal	x	x		x	Yellow, orange, or red		1 10	12–50 40–150	
QUARTZ	Amethyst	x		x	x	Reddish to purplish violet†	Medium to dark	5-20	15–120	Even color, distinct dichroism
						Violet,† purple†	Light to dark	5-20	8–50	
	Citrine	x		x		Yellow†	Light to medium dark	>3	5–20	Even color or attractive variations in color; distinct dichroism in some of colored varieties
						Orange-yellow†		>3	8–40	
	Green quartz	x		x		Green‡	Medium to dark	>3	3–18	
	Rock crystal	x		x	x	Colorless		>3	0.1–4	
	Rose quartz	x	x	x	x	Rose-pink	Light to medium	>3	0.5–10	
	Smoky quartz	x	x	x	x	Smoky yellow to black	Light to very dark	>3	0.5–20	
SPINELS	Blue spinel (Sapphirine)	x		x		Blue	Medium to dark	5-20	15–80	Intense, even color, medium to medium deep
	Green spinel (Chlorospinel)	x		x		Yellowish green	Light to dark	5-20	10–60	
	Orange spinel (Rubicelle)	x		x		Yellow to orange-red	Medium	5 20	35–500 35–2,000	
	Purple spinel (Almandine)	x		x		Violet, purple	Light to dark	5-20	10–100	
	Red spinel (Ruby spinel, Balas ruby)	x		x		Pink to purplish red	Light to dark	5 20	15–80 40–280	
SPODUMENE	Hiddenite	x		x		Green	Light	5-20	3–25	Even color of maximum depth
							Medium	5-20	6–120	
	Kunzite	x		x		Pink to lilac	Light	5-20	3–35	
							Medium	5-20	5–150	
	Triphane	x		x		Yellow to colorless		5-20	3–35	

TABLE 7—Comparative Retail Values* of Cut But Unset Gem Stones, Exclusive of Diamonds, Correlated with Color, Size, and Other Features (Continued)

Mineral	Gem Stone	Trans-parent	Semitrans-parent to Opaque	Finished Form: Facet-Cut	Finished Form: Other Forms	Color, Color Value, and Other Features		Weight, Carats	Most Common Price or Price Range, $ per Carat	Additional Features Responsible for Highest Prices
TOPAZ	Precious topaz	×		×		Purple red†	Light	5-15	10–40	Even color in deepest warm tones; high luster
							Medium	5 15	80–250 120–1,400	
						Pink†	Light	5-15	8–40	
							Medium	5 15	70–1,600 150–3,500	
						Orangy to brownish yellow,† reddish brown†	Light	5-15	60–220	
							Medium	5-15	100–850	
						Blue, yellow	Light	5-15	5–50	
							Medium	5-15	40–800	
TOURMALINE	Tourmaline (Achroite, Emeralite, Indicolite, Rubellite, Siberite, etc.)	×	×	×	×	Pink, red, purplish to violet-red	Light to medium	5-15	25–120	Even, medium to medium deep color; combinations of color in some stones
							Medium dark	5 15	80–250 125–450	
						Blue, green,† bluish to yellowish green†	Light	5-15	20–60	
							Medium to dark	5-15	45–140	
						Colorless, brown, yellow, etc.		5-15	10–50	
TURQUOISE	Turquoise		×		×	Blue		For stones averaging 15 × 20 mm in plan	10–140	Even color; minimum of matrix material
						Green to greenish blue			5–50	
						Yellowish green to greenish gray			2–25	
ZIRCON	Blue zircon	×		×		Blue,‡ very slightly greenish blue‡	Medium	1-15	4–40	Even color of purest blue; highest indices of refraction
							Medium dark	1-5 10-15	6–75 20–175	
						Very slightly yellowish or brownish blue‡		1-15	3–60	
	Green zircon	×		×		Green†	Medium	1-15	5–25	
							Medium dark	1-15	5–50	
	Hyacinth, Jacinth, Jargon	×		×		Yellow, orange, red, violet, brown, etc.	Medium	1-15	10–35	
							Medium dark	1-15	15–80	
	White zircon	×		×		Colorless†		1-15	3–40	
ZOISITE	Tanzanite	×		×		Blue†		5 10	400–1,200 750–2,800	Even deep blue color.

* Based upon updated informal reports from 22 jewelers and gem brokers. Indicated values represent sales at the retail level in the United States, each sale involving not more than a few individual stones. No established scale of prices is indicated or implied.

† Color of some stones obtained or improved by heat treatment.

‡ Color of all stones obtained by heat treatment.

Colored varieties of synthetic corundum have been referred to as amethyst, aquamarine, emerald, danburite, diamond, garnet, kunzite, topaz, and zircon, with or without designation as synthetic. The term "scientific," in combination with the names of valuable gem minerals, has been used to dignify many glass imitations. The sale of yellow quartz (citrine) for topaz has long been promoted by the use of such deceiving names as "Bohemian topaz," "Brazilian topaz," "Jewelers' topaz," "Spanish topaz," "quartz topaz," or, worst of all, "topaz." Probably more than four-fifths of the "topaz" now in the hands of ultimate users of jewelry actually is quartz.

Data on both proper and improper naming of gem stones have been published (Bauer and Schlossmacher, 1932; Briggs, 1944; Pough, Bowman, and Hoke, 1945; Kraus and Slawson, 1947; Shipley, 1951; Hurlbut and Switzer, 1979), and the American Gem Society, the Precious Stone Dealers Assn. of New York, and other responsible organizations have made vigorous efforts to standardize the nomenclature for the trade. An approved scheme was put forth three decades ago (Kraus, 1951), but much still remains to be accomplished in its application.

Independent identification of gem stones, though desirable in any legitimate transaction, has long been neglected at the retail and consumer levels. During the past four decades, however, both jewelers and the general public have become increasingly well informed concerning the recognition and testing of gem materials, mainly from publications (Anderson, 1942, 1947; Kraus and Slawson, 1947; Webster, 1952, 1971; Gubelin, 1953; Liddicoat, 1975) and through training provided by the Gemological Institute of America, members of University staffs, and a few independent gemologists. Large numbers of retail jewelers, for example, are now familiar with the use of instruments and the application of criteria for distinguishing natural, synthetic, and imitation stones, as well as specific varieties within each of these groups. Particularly difficult identifications and nearly all detailed appraisals of value, however, still are left to a relatively few experts in this specialized area.

Value and Prices

The basic worth of the noble, or most precious, stones is mainly a function of their intrinsic properties and the "make" of the gem fashioned from them. For gems of progressively lesser stature, value also is influenced to an increasing extent by fashion, superstition, royal and church sponsorship, national pride, fear of imitations that cannot be distinguished satisfactorily, publicity campaigns, new discoveries, and variations in supply, national controls, and world economic conditions. These factors have been discussed in some detail by Ball (1935). The demand for pearls and cut diamonds, chrysoberyls, emeralds, rubies, and sapphires responds mainly to changes in national prosperity and attendant luxury-purchasing power of consumers. The established price of diamonds, however, remains essentially independent of this factor, owing to the controlled release of cuttable stones to the market.

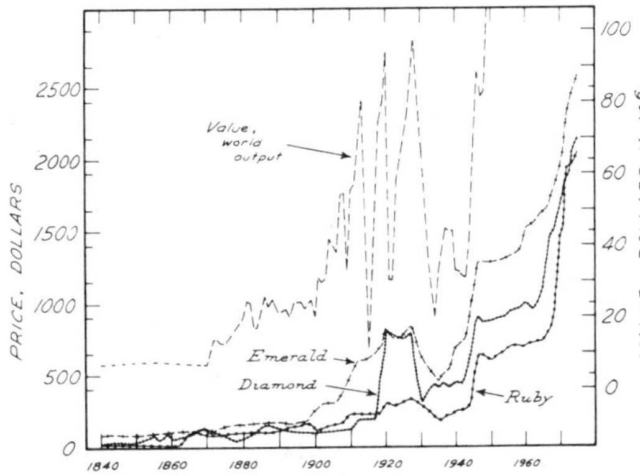

FIG. 13—*Trends in average retail prices for emerald, diamond, and ruby in the form of one-carat, well-cut gems of superior quality during the period 1840–1973. Total value of the world's production of gem materials (exclusive of pearls) is shown for 1840–1948.*

For finished gems in the ½ to 3 carat range, diamond was the most valuable stone from about 26 A.D. to 1500, after which ruby assumed the lead. Diamond again led during part of the nineteenth century, but was displaced at least twice by emerald. Emerald generally has been the most expensive stone since 1871. Trends in the prices of these three gem stones are shown in Fig. 13 for the period 1840–1973. Based largely upon data assembled by Ball (1935, 1949) and reports during the past 25 years by numerous distributors, each of the curves applies to "a one-carat, well-cut, rather fine" gem, and hence to a gem of better than average quality.

Average ranges in retail prices of cut but unset gem stones, which are two to four times the respective wholesale ranges, are listed in

Table 7 for the major minerals other than diamond. The indicated price ranges, expressed in dollars *per carat,* are based upon sales of individual stones and very small groups of stones during the period 1978–1980. They point up the relative prices of different gem minerals, along with the dependence of price upon size, color, color value, and other features. Though reasonably accurate, they represent no standard scale of prices established in the trade.

Relations between retail price and the size of finished gems are shown graphically in Fig. 14. The curves, which represent market conditions in the period 1970–1973, would be displaced considerably upward if they were to reflect 1980 markets (cf. Table 7). They apply only to stones of exceptionally fine color and quality, for which there is no real fixed price, but the indicated values can be regarded as maximal in the early seventies. The curve for ruby crosses those representing emerald and diamond, reflecting the scarcity of large rubies of top quality. In the lower price ranges, similar relationships are characteristic of star ruby, cat's-eye, spinel, and some tourmaline. For purposes of comparison, average 1973 retail prices for cut diamonds of good intrinsic quality and "make" were about $600 for a ½-carat stone, $1800 for a 1-carat stone, and $12,000 for a 3-carat stone. Corresponding prices for stones of exceptional quality and ideal "make" were about $2500, $8000, and $75,000; by 1979 these respective retail prices had risen to $4500, $65,000, and $500,000 or more.

Transparent andalusite, diopside, epidote, kyanite, and sphene, in the form of cut gems of good to excellent quality, now command retail prices of $6 to about $90 per carat, depending upon size and color of the individual stone. Some sphene of fine green color is marketed at the $150-per-carat level. Benitoite and zoisite (tanzanite) are currently in great demand, and best-color stones larger than one carat are being sold for $200 to $3000 per carat. Most of these minerals are rare in transparent crystals, and for some of them finished gems in sizes above three to five carats can be extremely difficult to obtain in any color or quality. Prices for 10 to 15-carat stones of most nontransparent minerals, including datolite, feldspathoids, hematite, idocrase, marcasite, prehnite, pyrite, rhodonite, serpentine, smithsonite, variscite, and zeolites, range from $5 to $65 per stone according to quality. Most amber is in the same price category, lapis lazuli is somewhat more expensive, and natural glasses are less expensive.

The retail price of cut synthetic emerald of good color varies according to the number and distribution of flaws, and ranges from $20 to $300 for most 1-carat stones. The corresponding range for most 2-carat stones is $25 to $350 per carat. Cut synthetic rutile commands prices of $8 to $40 per carat in the 1 to 10-carat size range; the most expensive stones are those that are least yellowish in color. Boules of transparent ruby, sapphire, and spinel are sold for less than a dollar per carat, so that the cost of finished gems is mainly that of cutting and handling. Prices are about $2 to $25 per stone, and among those of like color and quality the per-carat price decreases appreciably with increasing size. Finished gems of synthetic star ruby and star sapphire are being sold for $12 to $35 per carat in the 1 to 10-carat size range. The typical per-carat price for strontium titanate gems is $15 to $30, and for those of synthetic yttrium-aluminum garnet (YAG) is $15 to $35.

Pearls are sold in parcels of undrilled individuals and in both uniform and graduated strands. Prices vary greatly according to the size, color, luster, sphericity, soundness, and surface quality of the pearls, and according to the excellence of their matching in a given lot. Thus a single undrilled cultured pearl might be sold at retail for less than a dollar or as much as $600, and a graduated 432 mm (17-in.) strand for as little as $5 or as much as $5000 or more. Corresponding price ranges are considerably higher for natural pearls.

Preparation of Finished Stones

Cutting

The lapidary art was practiced thousands of years ago, when rough gem materials were crudely shaped by means of percussion chipping. The appearance of finished stones subsequently was improved through the application of techniques for grinding, polishing, carving, and engraving; bruting, or rubbing with heavy pressure against a very hard stone, also was found to be effective in the preliminary shaping of many gems. Facet-cutting originated in India perhaps 2000 years ago (Ball, 1949), and beryl gems were prepared in this form by Roman lapidaries shortly thereafter. Diamonds were not faceted until the fifteenth century, when appropriate techniques were developed in India and soon were transmitted eastward to Borneo and westward to Europe.

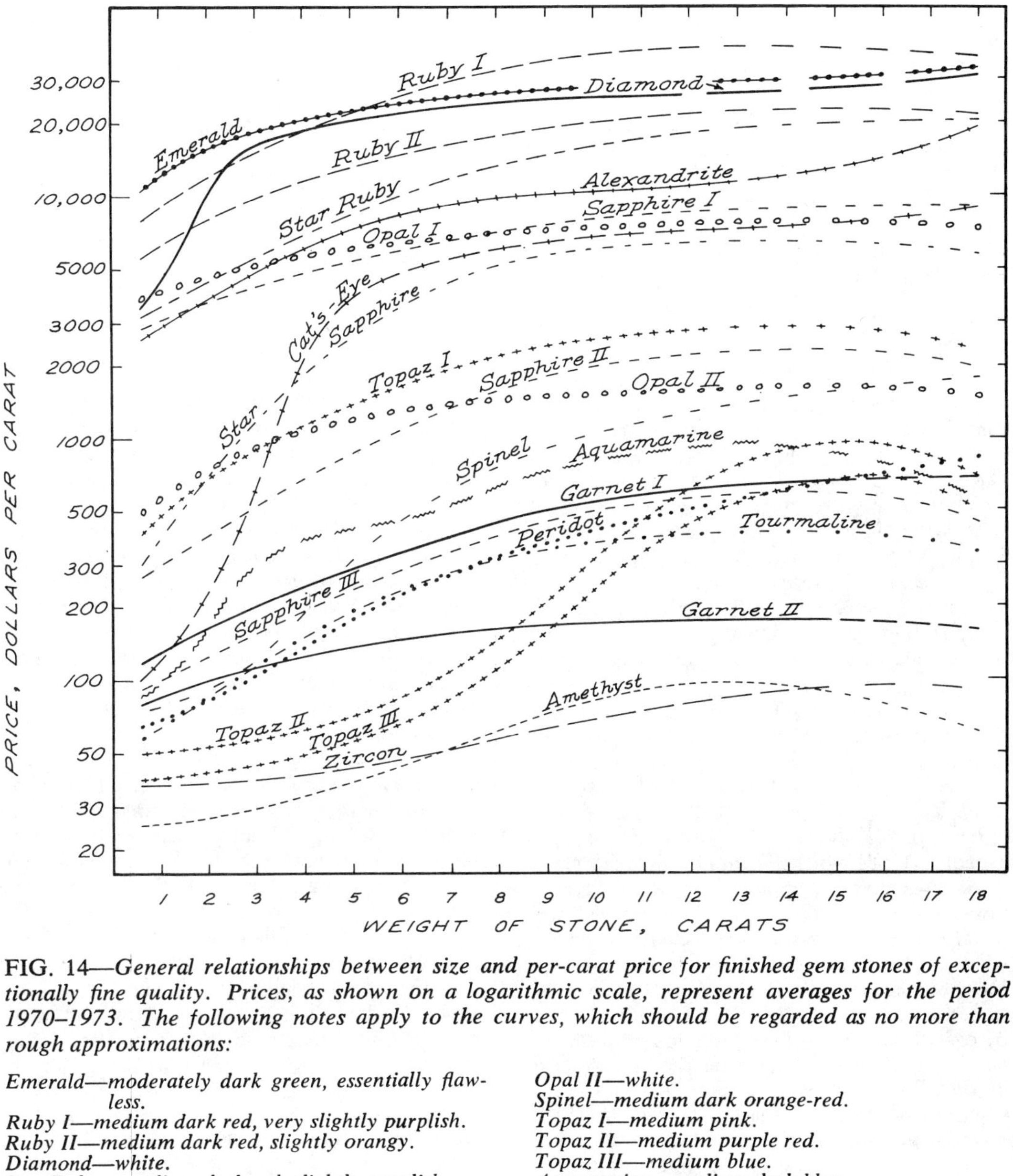

FIG. 14—*General relationships between size and per-carat price for finished gem stones of exceptionally fine quality. Prices, as shown on a logarithmic scale, represent averages for the period 1970–1973. The following notes apply to the curves, which should be regarded as no more than rough approximations:*

Emerald—moderately dark green, essentially flawless.
Ruby I—medium dark red, very slightly purplish.
Ruby II—medium dark red, slightly orangy.
Diamond—white.
Star Ruby—medium dark red, slightly purplish.
Alexandrite—medium dark, intense colors, with sharp color change.
Sapphire I—medium dark blue, slightly violetish.
Sapphire II—medium dark blue.
Sapphire III—medium dark pink or greenish blue.
Star Sapphire—medium dark blue, slightly violetish.
Cat's-eye—chrysoberyl, medium dark green.
Opal I—dark blue to very dark gray.
Opal II—white.
Spinel—medium dark orange-red.
Topaz I—medium pink.
Topaz II—medium purple red.
Topaz III—medium blue.
Aquamarine—medium dark blue.
Tourmaline—medium dark pink, red, purplish red, or violet-red.
Peridot—medium dark green.
Zircon—medium dark blue.
Garnet I—medium dark green.
Garnet II—medium dark purplish red.
Amethyst—medium dark violet, reddish to purplish.

Belgium, Germany, the Netherlands, Israel, and the United States are now the principal cutting countries; others include Borneo, Brazil, Cuba, England, India, and South Africa.

Emerald, ruby, sapphire, and other stones of high value are cut mainly in Bangkok, Bombay, Colombo, London, New York, and Paris, and at various localities in Germany, India, and Russia. Cutting of gems and ornaments from the less expensive materials has flourished in the Idar-Oberstein and Freiburg districts of Germany, the Jura Mountains region of France, the Turnov district of Czechoslovakia, and at many places in Russia and the United States. Jade and several other nontransparent materials are finished mainly in China and Japan. No significant amount of facet cutting has been done in this part of the Orient, which long has been famous for work in the round.

The cutting of a gem normally begins with careful inspection of the rough stone, after which it is divided by cleaving or sawing in order to obtain one or more pieces of desirable shape and good quality. After a given piece is further shaped by hand or lathe bruting, it is subjected to coarse and fine grinding against abrasive-impregnated laps and wheels, polishing with abrasives on metal or wooden disks and wheels, and drilling, carving, and engraving with diamond tools. The chief abrasives employed are diamond and silicon carbide. Special gages and mechanical devices generally are used in the proportioning of facet-cut stones.

In general, the softer gem materials are roughly shaped with a hammer, further separated by means of diamond or silicon carbide saws, ground against sandstone wheels and metal laps and wheels, and polished on wooden wheels. Many kinds of abrasives are employed. Most of the work is done freehand, and the stones generally are proportioned and finished by eye. *Tumbling,* a bulk technique developed nearly a century ago, lately has been much used in the United States for the preparation of stones with irregular shape. Pieces of rough material are placed in a revolving mill, commonly of cylindrical form, and are batch ground and polished in several stages with successively finer abrasives.

Thousands of small gem cutting and lapidary shops are being operated throughout the United States at the present time, some by professionals who sell most of the product to collectors, but the majority by individual hobbyists who work on rough material they have collected or purchased. Numerous publications (Baxter, 1938; Dake, Fleener, and Wilson, 1938; McKinley, 1939; Von Bernewitz and Hess, 1940; Dake and Pearl, 1945; Howard, 1946; Willems, 1948; Grodzinski, 1953; Sinkankas, 1955; Parsons, 1969), ranging from popular handbooks to highly technical summaries, have appeared in response to a burgeoning demand for information on the finishing of gem materials, and have contributed to the remarkable skill developed by many self-trained amateurs.

Finished Forms

The sparkle and fire in many transparent gem stones are most effectively displayed by means of the *brilliant* cut (Fig. 3), which has been popular since its introduction in 1660. The standard form of this cut comprises 58 facets and has a circumferential outline, or girdle, of circular shape (Fig. 15); various modifications are distinguished by greater or lesser numbers of facets or by elliptical, triangular, drop-like, or otherwise noncircular girdle shapes (e.g., marquise cut, Fig. 16). The proportioning of the *brilliant* cut repeatedly has been modified to improve the liveliness of gems, especially diamonds, but only in recent decades have the relationships between optical effects and the disposition and orientation of facets been widely appreciated within the jewelry trade.

Many stones of potentially great brilliance still are cut too deep or too flat in order to reduce wastage from the original pieces of rough material. Although this practice may seem economically sound, it does not yield gems of top quality. If either the crown or the pavilion (Fig. 15) of the stone is overly thick with respect to its diameter at the girdle, light is lost by refraction through the pavilion facets and the stone appears somewhat glassy and dark. Undesirable leakage of light also takes place if either the crown or the pavilion is too thin or shallow with respect to the girdle diameter, and such a gem is so lifeless that it is said to have a "fisheye" or "tapioca" appearance. In addition, brilliance is sacrificed in stones whose facets are properly oriented but whose crowns are made too thin in order to obtain a large top facet, or table. Although relatively "dead," such gems tend to appear larger than they really are, and the term "swindled" is widely and aptly applied to them. They find a ready market among persons to whom size is the most important criterion of a good bargain.

The brilliant cut often is used to enhance the appearance of unevenly colored stones. If the

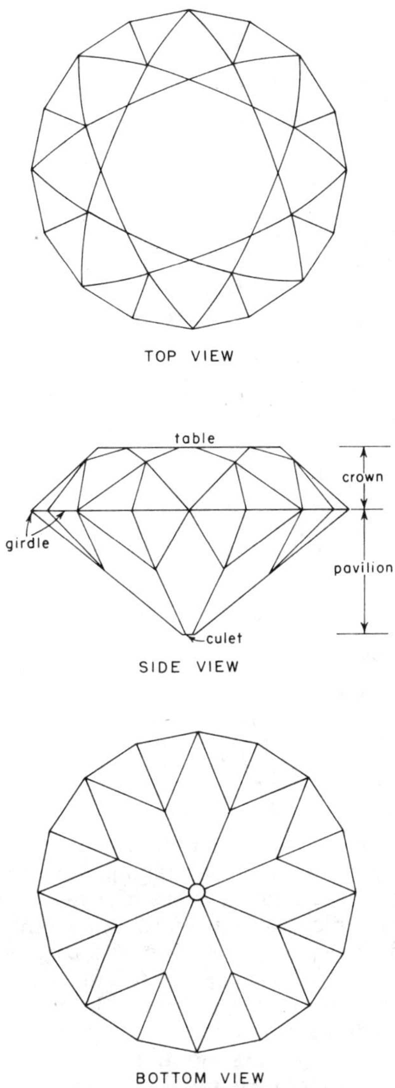

FIG. 15—*The standard brilliant cut.*

portion with the best, or deepest, color lies immediately above the culet (Fig. 15) in such a gem, the entire stone appears to be more deeply and uniformly colored than it actually is. Other types of cuts, generally square or rectangular in plan (Fig. 16), are widely used for stones with distinct, uniformly distributed color and relatively low indices of refraction. *Emerald, step, baguette, square,* and a large number of special cuts vary considerably in their proportions, which may emphasize brilliance, color tone, economy of the original rough material, or other factors. Some of these cuts, especially the baguette and the marquise, are also applied to diamonds and other stones with superior brilliance. The beauty of many stones with deep or unevenly distributed colors is best shown by means of various *mixed* cuts, comprising brilliant-faceted crowns and step-faceted pavilions.

Of all the rounded cuts applied to gem stones, the *cabochon* is the oldest and has remained the most popular to the present time. It generally has a convex upper surface (Fig. 1) and a flat, convex, or concave lower surface, and it appears in many different forms. Other rounded cuts range from *polished spheres* to *"baroques"* (polished stones of irregular form), and from *cameos, intaglios,* and *curvettes* (Fig. 1) to the most intricately carved, engraved, and pierced gems and ornaments.

Special Treatment

Careful heating under various conditions improves the color of many gem stones. Thus sapphire can be changed from yellow to colorless and from violet to pink, beryl from green to blue, quartz from pale violet to orange brown and from smoky to yellow, chalcedony from gray to red, topaz from yellow to pink or colorless, tourmaline from dark blue and dark green to medium green, and zircon from reddish and brownish to blue or colorless. Some of the color changes extend throughout the stones, others are only "skin deep." Most are essentially permanent under all conditions to which gems and jewelry are normally exposed, and many heat-treated stones therefore have a respected position in the trade (Table 7). Such treatment, however, tends to set up strains within most gems, and hence can induce brittleness or the formation of tiny cracks.

Color changes also can be developed in many minerals by exposure to X-rays, by bombardment in a cyclotron or nuclear reactor, or by irradiation in other ways (Kraus and Slawson, 1947; Pough and Rogers, 1947; Hurlbut and Switzer, 1979). Thus colorless or pale-colored diamonds can be changed to green, blue, or brown, colorless quartz to brown or smoky gray, purplish spodumene to green, colorless topaz to purple or brown, pink topaz to orange or brown, blue beryl to green, colorless and pink sapphire to orange, and yellowish scapolite to purplish. Some of the new colors appear to be essentially permanent, but many fade spontaneously or are readily changed to the original colors by heating, long-term exposure to sunlight, or other kinds of treatment. Irradiated stones have been marketed in in-

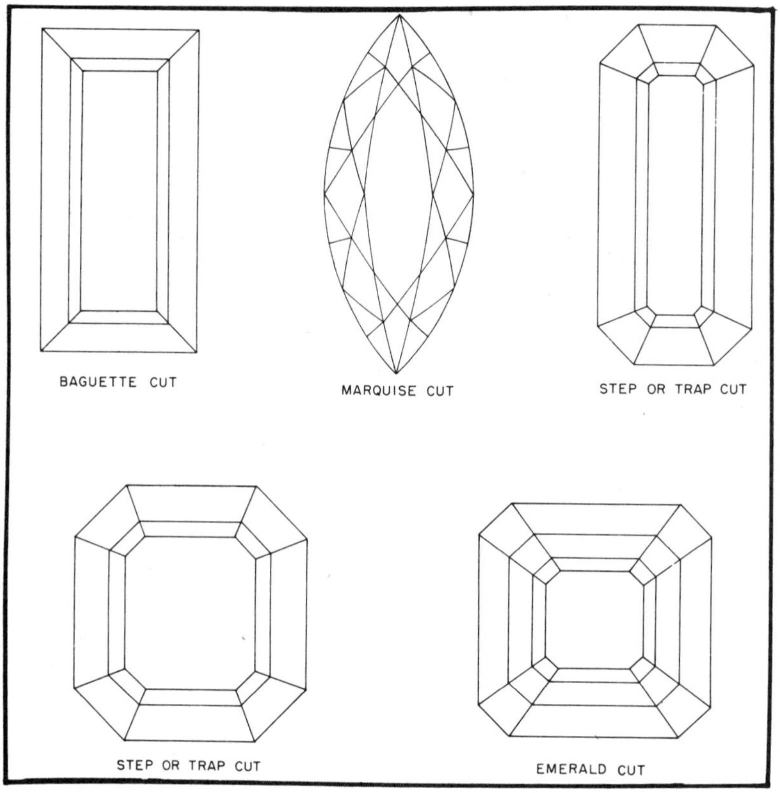

FIG. 16—*Typical cuts used for transparent stones that have attractive, evenly distributed color but relatively little brilliance.*

creasingly significant numbers during recent years.

The natural colors or color tones of some gem stones are not permanent. The lilac to purple kunzite, for example, fades appreciably during years of exposure to sunlight. The attractive plays of color in many opal gems become subdued whenever some of the contained water is lost from the material. To forestall this partial desiccation, which often is accompanied by cracking, some dealers immerse their opals in water or protect them with films of oil until they are ready for sale.

Numerous types of calcite, gypsum, and cryptocrystalline quartz are artificially colored or stained by means of various solutions, with or without subsequent heat treatment. The colors of jade, opal, and turquoise commonly are intensified by dyeing, and some of the techniques have been extended to transparent minerals. Dyed quartz and calcite have been marketed as imitations of jade, lapis lazuli, and turquoise.

Dyes, colored foils, enamels, finely divided fluorite, and other coatings are variously applied to the surfaces of many gems, mainly to improve color and other optical properties. Star sapphires and star rubies have been simulated by the cementing of appropriately colored foils or mirrors to the backs of gems cut from milky asteriated quartz. Further, a milky stone that does not show asterism can be made to yield a star if its underside (or the surface of an applied mirror) is scribed with fine grooves or scratches arranged in a regular pattern.

Composite or assembled stones, consisting of two or more parts that have been cemented or fused together in order to increase their size or improve their color, have been widely marketed. Those that comprise two or three sections of genuine material are known respectively as *true doublets* or *true triplets. False doublets,* in contrast, generally comprise a crown or partial crown of genuine material, such as diamond, ruby, sapphire, or garnet, and a pavilion of colored glass or an inferior natural material (Fig. 17). Most *false triplets* contain a slice of colored glass, pocket of liquid, or some type of colored film between two sections of hard material. Fortunately, most composite, dyed, coated, and specially backed stones can be identified as such through careful inspection or the application of simple tests.

Sources of Additional Information

Gem materials are extensively described in numerous books, some of them handsomely illustrated, that have been published during the past century. The treatment ranges from essentially nontechnical to that found in standard texts on mineralogy. Features of occurrence, as well as trends in use, classification, value and prices, have been summarized by Ball (1922, 1931, 1935, 1941, 1949, 1950), Meisner (1929), and more recently in somewhat abbreviated form by the US Bureau of Mines (Thompson, 1955). The Bureau also has provided more detailed summaries relating to individual gem materials.

Useful comments, economic data, and references to the current literature appear from time to time in *Mineral Trade Notes* and annually in the *Minerals Yearbook*, both published by the US Bureau of Mines. Until 1921 similar data, in many instances accompanied by information on geologic occurrence, were provided by the US Geological Survey in the yearbook *Mineral Resources of the United States*. Current developments within the industry are reported in the *Jewelers' Circular-Keystone* and other trade journals, some of which also contain annual summaries. Statistical data were drawn from all these sources in the preparation of Figs. 8–12 for this chapter.

Other sources of information have been cited specifically in the foregoing pages. To these can be added the following partial list of journals that together serve amateur and professional mineralogists, crystallographers, collectors, cutters, and lapidaries, as well as those persons variously engaged in the production and marketing of gem materials.

The American Mineralogist, Mineralogical Society of America, Ann Arbor, MI.
The Desert Magazine, El Centro, CA.
Die Deutsche Goldschmiede-Zeitung, Wilhelm Diebener, Leipzig.
Earth Science, Midwest Federation of Mineralogical Societies, Chicago, IL.
Gems and Gemology, Gemological Institute of America, Los Angeles, CA.
Gems and Minerals, California Federation of Mineralogical Societies, Mentone, CA.
The Gemmologist, National Assn. of Goldsmiths Press, London.
Industrial Diamond Review, National Assn. of Goldsmiths Press, London.
The Jewelers' Circular-Keystone, Chilton Press, New York.
Journal of Gemmology, Gemmological Assn. of Great Britain.
The Lapidary Journal, Lapidary Journal, Inc., Hollywood, CA.
Mineralogical Magazine, The Mineralogical Society, London.
The Mineralogical Record, Mineralogical Record, Inc., Bowie, MD.
The Mineralogist, The Mineralogist Publishing Co., Portland, OR.
National Jeweler, Chicago, IL.
Rocks and Minerals, Rocks and Minerals Assn., Peekskill, NY.

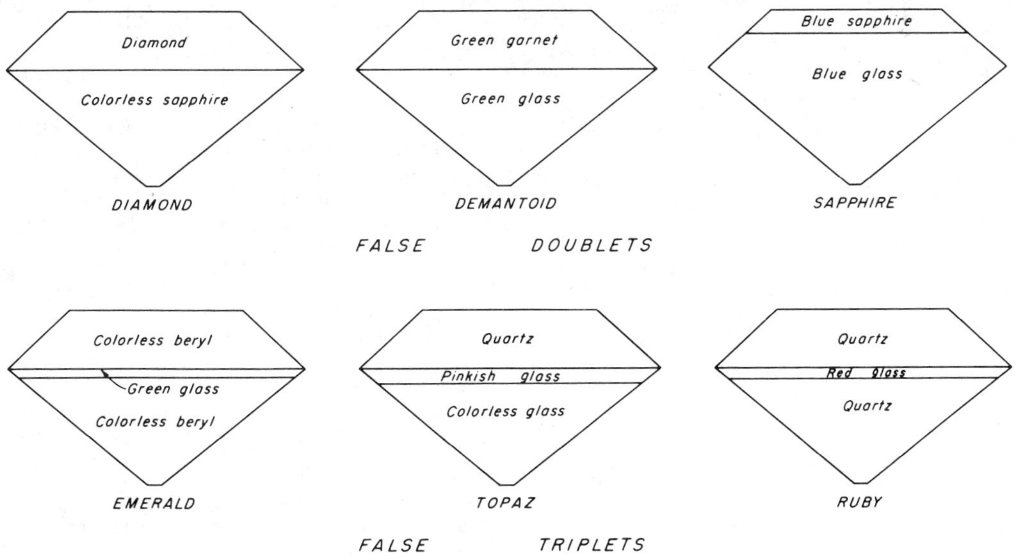

FIG. 17—*Typical composite or assembled gems consisting of two or more parts cemented or fused together to increase their size or improve their color.*

Bibliography and References

Adams, F.D., 1926, "A Visit to the Gem Districts of Burma and Ceylon," *Bulletin,* Canadian Institute of Mining & Met., Vol. 19, pp. 213–246.

Aitkens, I., 1931, "Emeralds," Information Circular 6459, US Bureau of Mines, 18 pp.

Aitkens, I., 1931a, "Rubies and Sapphires," Information Circular 6471, US Bureau of Mines, 11 pp.

Aitkens, I., 1931b, "Turquoise," Information Circular 6491, US Bureau of Mines, 17 pp.

Aitkens, I., 1931c, "Opals," Information Circular 6493, US Bureau of Mines, 9 pp.

Aitkens, I., 1931d, "Topaz," Information Circular 6502, US Bureau of Mines, 11 pp.

Aitkens, I., 1931e, "Garnets (Gem Stones)," Information Circular 6518, US Bureau of Mines, 11 pp.

Aitkens, I., 1931f, "Feldspar Gems (Amazon Stone, Moonstone, Sunstone)," Information Circular 6533, US Bureau of Mines, 10 pp.

Aitkens, I., 1931g, "Tourmaline," Information Circular 6539, US Bureau of Mines, 8 pp.

Aitkens, I., 1932, "Quartz Gem Stones," Information Circular 6561, US Bureau of Mines, 15 pp.

Anderson, B.W., 1942, *Gem Testing for Jewellers,* Heywood and Co., Ltd., London, 220 pp.

Anderson, B.W., 1947, *Gem Testing,* 4th ed., Heywood and Co., Ltd., London, 252 pp.; 1948, Emerson Books, Inc., New York, 256 pp.

Austin, A.C., et al., 1948, *The Story of Diamonds,* 3rd ed., Gemological Inst. of America, Los Angeles, CA, 101 pp.

Baerwald, M., and Mahoney, T., 1949, *Gems and Jewelry Today,* Marcel Rodd Co., Inc., New York, 303 pp.

Ball, S.H., 1922, "The Geologic and Geographic Occurrence of Precious Stones," *Economic Geology,* Vol. 17, No. 7, Nov., pp. 575–601.

Ball, S.H., 1931, "Historical Notes on Gem Mining," *Economic Geology,* Vol. 26, pp. 681–738.

Ball, S.H., 1935, "A Historical Study of Precious Stone Valuation and Prices," *Economic Geology,* Vol. 30, No. 6, Sep.-Oct., pp. 630–642.

Ball, S.H., 1941, "The Mining of Gems and Ornamental Stones by American Indians," Bulletin 128, Bureau of American Ethnology, pp. 1–77.

Ball, S.H., 1949, "Precious Stones," *Industrial Minerals and Rocks,* 2nd ed., AIME, New York, pp. 714–747.

Ball, S.H., 1950, *A Roman Book on Precious Stones,* Gemological Institute of America, Los Angeles, CA, 338 pp.

Bank, H., 1970, *Precious Stones and Minerals,* Frederick Warne and Co. Ltd., London, 125 pp.

Barrington Brown, C., and Judd, J.W., 1896, "The Rubies of Burma and Associated Minerals," *Philosophical Transactions,* Royal Society of London, Vol. 187A, pp. 151–228.

Bastin, E.S., 1911, "Geology of the Pegmatites and Associated Rocks of Maine, Including Feldspar, Quartz, Mica, and Gem Deposits," Bulletin 445, US Geological Survey, 152 pp.

Bauer, M., (1904) 1968, *Precious Stones* (English trans., with additions, by L.J. Spencer), Dover Publications, Inc., New York, 627 pp.; 1969, Charles E. Tuttle Co., Rutland, VT, and Tokyo, Japan, 647 pp.

Bauer, M.H., and Schlossmacher, K., 1932, *Edelsteinkunde,* 3rd ed., Bernhard Tauchnitz, Leipzig, 871 pp.

Baxter, W.T., 1938, *Jewelry, Gem Cutting, and Metalcraft,* McGraw-Hill, New York, 334 pp.

Berman, R., ed., 1965, *Physical Properties of Diamond,* Oxford University Press, Oxford, 443 pp.

Briggs, H.E., 1944, *An Encyclopedia of Gems,* Gemological Institute of America, Los Angeles, CA, 83 pp.

Chandler, H.P., 1964, "Industrial Diamond—A Materials Survey," Information Circular 8200, US Bureau of Mines, 149 pp.

Chilvers, H.A., 1939, *The Story of De Beers,* Cassell and Co., Ltd., London, 350 pp.

Chudoba, K.F., and Gubelin, E.J., 1953, *Schmuck- und Edelsteinkundliches Taschenbuch,* Verlag Bonner Univ.—Buchdruckerei, Gebr. Scheur, Bonn, 158 pp.

Church, A.H., 1924, *Precious Stones,* Victoria and Albert Museum, London, 164 pp.

Clabaugh, S.E., 1953, Corundum Deposits of Montana: Bull. 983, US Geological Survey, 100 pp.

Dake, H.C., 1950, *Northwest Gem Trails; A Field Guide for the Gem Hunter, the Mineral Collector, and the Tourist,* The Mineralogist Publishing Co., Portland, OR, 80 pp.

Dake, H.C., et al., 1938, *Quartz Family Minerals,* Whittlesey House, New York, 304 pp.

Dake, H.C., and Pearl, R.M., 1945, *The Art of Gem Cutting,* 3rd ed., The Mineralogist Publishing Co., Portland, OR, 128 pp.

Darragh, P.J., Gaskin, A.J., and Sanders, J.V., 1976, "Opal," *Scientific American,* Vol. 234, No. 4, pp. 84–85, 88–95.

Davies, G., 1977, "The Optical Properties of Diamond," *Chemistry and Physics of Carbon,* Vol. 13, pp. 1–143.

Dawson, J.B., 1968, "Recent Researches on Kimberlite and Diamond Geology," *Economic Geology,* Vol. 63, No. 5, Aug., pp. 504–511.

Dawson, J.B., 1971, *The Genesis of Kimberlite, Diamond Research for 1971,* Industrial Diamond Information Bureau, London, pp. 2–7.

Denning, R.M., 1953, "Directional Grinding Hardness in Diamond," *American Mineralogist,* Vol. 38, No. 1–2, Jan.-Feb., pp. 108–117.

Desautels, P.E., 1979, *Treasures in the Smithsonian; the Gem Collection,* Smithsonian Institution Press, Washington, DC, 77 pp.

Eppler, W.F., 1934, *Edelsteine und Schmucksteine,* 2nd ed., Wilhelm Diebener, Leipzig, 559 pp.

Eyles, W.C., 1964, *The Book of Opals,* Charles E. Tuttle Co., Rutland, VT, Tokyo, Japan, 224 pp.

Farrell, E.F., and Newnham, R.E., 1965, "Crystal-field Spectra of Chrysoberyl, Alexandrite, Peridot, and Sinhalite," *American Mineralogist,* Vol. 50, No. 11–12, Nov.-Dec., pp. 1972–1981.

Farrington, O.C., 1903, *Gems and Gem Minerals,* A.W. Mumford, Chicago, IL, 229 pp.

Fisher, P.J., 1966, *The Science of Gems,* Charles Scribner's Sons, New York, 189 pp.

Frantsesson, E.V., 1970, *The Petrology of Kimberlites* (English trans. by D.A. Brown), Publication 150, Australian National University, Dept. of Geology, 195 pp.

Grodzinski, P., 1953, *Diamond Technology,* 2nd ed., National Assn. Goldsmiths Press, Ltd., London, 784 pp.

Gubelin, E.J., 1953, *Inclusions as a Means of Gemstone Identification,* Gemological Inst. of America, Los Angeles, CA, 220 pp.

Gubelin, E. J., 1979, *Internal World of Gemstones,* 2nd ed., Butterworth and Company Ltd., London, 234 pp.

Hamlin, A.C., 1895, *The History of Mount Mica, Maine,* Augustus Choate Hamlin, Bangor, 72 pp.

Harris, P.G., and Middlemost, E.A.K., 1969, "The Evolution of Kimberlites," *Lithos,* Vol. 3, pp. 77–88.

Hassan, F., and El-Rakhawy, A., 1974, "Chromium III Centers in Synthetic Alexandrite," *American Mineralogist,* Vol. 59, No. 1, pp. 159–165.

Henry, D.J., 1948, *California Gem Trails; A Field Guide for the Gem Hunter, the Mineral Collector, and the Tourist,* The Mineralogist Publishing Co., Portland, OR, 63 pp.

Howard, J.H., 1946, *Revised Lapidary Handbook,* J. Harry Howard, Greenville, SC, 220 pp.

Hurlbut, C. S., Jr., and Switzer, G. S., 1979, *Gemology,* John Wiley and Sons, New York, 253 pp.

Iyer, L.A.N., 1942, "Indian Precious Stones," Bulletin 6, Geological Survey of India, 54 pp.

Iyer, L.A.N., 1953, "The Geology and Gem-stones of the Mogok Stone Tract, Burma," Memoir 82, Geological Survey of India, 100 pp.

Jahns, R.H., 1960, "Gem Stones and Allied Materials," *Industrial Minerals and Rocks,* 3rd ed., J.L. Gillson, ed., AIME, New York, pp. 383–441.

Jahns, R.H., 1975, "Gem Materials," *Industrial Minerals and Rocks,* 4th ed., S.J. Lefond, ed., AIME, New York, pp. 271–326.

Jahns, R.H., and Wright, L.A., 1951, "Gem- and Lithium-bearing Pegmatites of the Pala District, San Diego County, California," Special Report 7-A, California Division of Mines, 72 pp.

Kennedy, G.C., and Nordlie, B.E., 1968, "The Genesis of Diamond Deposits," *Economic Geology,* Vol. 63, No. 5, Aug., pp. 495–503.

Kraus, E.H., 1951, "Nomenclature of Gems," *Gems and Gemology,* Vol. 7, pp. 58–67.

Kraus, E.H., et al., 1942, "Symposium on Diamonds," *American Mineralogist,* Vol. 27, No. 3, Mar., pp. 162–191; 1943, Vol. 28, No. 3, Mar., pp. 141–150.

Kraus, E.H., et al., 1948, "Third Symposium on Diamonds," *American Mineralogist,* Vol. 31, No. 3–4, Mar.-Apr., pp. 135–167.

Kraus, E.H., and Slawson, C.B., 1939, "Variation of Hardness in the Diamond," *American Mineralogist,* Vol. 24, No. 11, Nov., pp. 661–676.

Kraus, E.H., and Slawson, C.B., 1941, "Cutting of Diamonds for Industrial Purposes," *American Mineralogist,* Vol. 26, No. 3, Mar., pp. 153–160.

Kraus, E.H., and Slawson, C.B., 1947, *Gems and Gem Materials,* 5th ed., McGraw-Hill Book Co., New York, 332 pp.

Kunz, G.F., 1892, *Gems and Precious Stones of North America,* 2nd ed., The Scientific Publishing Co., New York, 367 pp.; reprinted 1968, Dover Publications, Inc., New York.

Kunz, G.F., 1905, "Gems, Jewelers' Materials, and Ornamental Stones of California (2nd ed.), Bulletin 37, Calif. State Mining Bureau, 171 pp.

Kunz, G.F., 1907, "History of the Gems Found in North Carolina," Bulletin 12, North Carolina Geological Survey, 60 pp.

Kunz, G.F., 1913, *The Curious Lore of Precious Stones,* J.B. Lippincott Co., Philadelphia, PA, 406 pp.

Lacroix, A., 1922, "Minéralogie Appliquée, Lithologie," *Minéralogie de Madagascar,* Vol. 2, Augustin Challamel, ed., Paris, 694 pp.

Liddicoat, R.T., Jr., 1975, *Handbook of Gem Identification,* 10th ed., Gemological Institute of America, Los Angeles, CA, 440 pp.

Long, A.E., 1951, "Diamond Orientation in Diamond Bits. Procedures and Preliminary Results," Report of Investigation 4800, US Bureau of Mines, 16 pp.

Long, A.E., 1954, "Diamond Orientation in Diamond Bits. Effects of Orienting High-grade Drill Bort in Diamond-coring Bit Crowns," Report of Investigation 5015, US Bureau of Mines, 13 pp.

Long, A.E., and Slawson, C.B., 1952, "Diamond Orientation in Diamond Bits. A Method of Identifying Hard Vectors for Setting Purposes," Report of Investigation 4853, US Bureau of Mines, 6 pp.

MacFall, R.P., 1951, *Gem Hunters' Guide,* Science and Mechanics Publishing Co., Chicago, 187 pp.

MacGregor, I.D., 1970, "An Hypothesis for the Origin of Kimberlite," Special Paper 3, Mineralogical Society of America, pp. 51–62.

McKinley, W.C., 1939, *American Gem Cabochons, an Illustrated Handbook of Domestic Semiprecious Stones Cut Unfaceted,* Lightner Publishing Co., New York.

McIver, J.R., 1967, *Gems, Minerals, and Rocks in Southern Africa,* American Elsevier Publishing Co., New York, 268 pp.

McLintock, W.F.P., and Sabine, P.A., 1951, *A Guide to the Collection of Gemstones in the Geological Museum,* 3rd ed., H.M. Stationery Office, London, 75 pp.

Meisner, M., 1929, *Weltmontanstatistik,* Vol. 1, Die Versorgung der Weltwirtschaft und Bergwerkserzeugnissen, Stuttgart, pp. 348–366.

Merrill, G.P., 1922, "Handbook and Descriptive Catalogue of the Collections of Gems and Precious Stones in the United States National Museum," Bulletin 118, US National Museum, 225 pp.

Merrill, G.P., and Foshag, W.F., 1943, "Minerals from Earth and Sky," Pt. 2, *Gems and Gem Materials,* Vol. 3, Smithsonian Science Series, pp. 159–312.

Meyer, H.O.A., and Boyd, F.R., 1969, "Mineral Inclusions in Diamond," *Year Book No. 67,* Carnegie Institution, Washington, DC, pp. 130–135.

Meyer, H.O.A., and Boyd, F.R., 1972, "Composition and Origin of Crystalline Inclusions in Natural Diamonds," *Geochimica et Cosmochimica Acta,* Vol. 36, pp. 1255–1273.

Nassau, K., 1974, "Natural, Treated, Synthetic, and Imitation Gems," *Gems and Gemology,* Vol. 14, No. 11, pp. 322–326.

Nassau, K., 1974–1975, "The Origins of Color in Gems and Minerals," *Gems and Gemology,* Vol. 14, No. 12, pp. 354–361; Vol. 15, No. 1, pp. 2–11; Vol. 15, No. 2, pp. 34–43.

Nassau, K., 1976, "Synthetic Emerald: the Confusing History and the Current Technologies," *Journal of Crystal Growth,* Vol. 35, pp. 211–222.

Nassau, K., 1977, "Cubic Zirconia, the Latest Diamond Imitation and Skull Melting," *Lapidary Journal,* Vol. 31, pp. 899–904, 922–926.

Nassau, K., and Nassau, J., 1980, "The Growth of Synthetic and Imitation Gems," *Crystals: Growth, Properties, and Applications: Vol. 2. Growth and Properties,* H.C. Freyhardt, ed., Springer-Verlag, Berlin, pp. 1–50.

O'Hara, M.J., and Mercy, E.L.P., 1963, "Petrology and Petrogenesis of some Garnetiferous Peridotites," *Transactions,* Royal Society of Edinburgh, Vol. 65, pp. 251–314.

Oppenheim, V., 1948, "The Muzo Emerald Zone, Colombia, S.A.," *Economic Geology,* Vol. 43, No. 1, Jan.-Feb., pp. 31–38.

Parsons, C.J., 1969, *Practical Gem Knowledge for the Amateur,* Lapidary Journal, Inc., San Diego, CA.

Pearl, R.M., 1948, *Popular Gemology,* John Wiley, New York, 316 pp.

Pearl, R.M., 1951, *Colorado Gem Trails,* Sage Books, Golden, CO, 125 pp.

Pearl, R.M., 1964, *Gems, Minerals, Crystals, and Ores,* The Odyssey Press, New York, 320 pp.

Petar, A.V., 1934, "Amber," Information Circular 6789, US Bureau of Mines, 13 pp.

Petar, A.V., 1936, "Jade," Information Circular 6844, US Bureau of Mines, 16 pp.

Pirsson, L.V., 1900, "Petrography of the Igneous Rocks of the Little Belt Mountains, Montana," 20th Annual Report, US Geological Survey, Pt. 3, pp. 463–581.

Pogue, J.E., 1915, "The Turquoise; A Study of Its History, Mineralogy, Geology," Memoir 12, National Academy of Science, 206 pp.

Poole, C.P., Jr., 1964, "The Optical Spectra and Color of Chromium Containing Solids," *Journal of Physical & Chemical Solids,* Vol. 25, pp. 1169–1182.

Pough, F.H., Bowman, J.J., and Hoke, C.M., 1945, *The Jeweler's Dictionary,* Jeweler's Circular-Keystone, New York.

Pough, F.H., and Rogers, T.H., 1947, "Experiments in X-ray Irradiation of Gem Stones," *American Mineralogist,* Vol. 32, No. 1–2, Jan.-Feb., pp. 31–43.

Ransom, J.E., 1955, *Arizona Gem Trails and the Colorado Desert of California; A Field Guide for the Gem Hunter, the Mineral Collector, the Uranium Hunter,* The Mineralogist Publishing Co., Portland, OR, 96 pp.

Rogers, F., and Beard, A., 1940, *5000 Years of Gems and Jewelry,* J.B. Lippincott Co., New York, 309 pp.

Rossman, G.R., 1981, "Color in Gems: the New Technologies," *Gems and Gemology,* Vol. 17, No. 2, pp. 60–71.

Schlegel, D.M., 1957, "Gem Stones of the United States," Bulletin 1042-G, US Geological Survey, pp. 203–253.

Shipley, R.M., 1951, *Dictionary of Gems and Gemology,* 5th ed., Gemological Institute of America, Los Angeles, CA, 261 pp.

Sinkankas, J., 1955, *Gem Cutting; A Lapidary's Manual,* D. Van Nostrand, Princeton, NJ, 413 pp.

Sinkankas, J., 1959, *Gemstones of North America,* D. Van Nostrand, Princeton, NJ, 675 pp.

Sinkankas, J., 1961, *Gemstones and Minerals,* D. Van Nostrand, Princeton, NJ, 387 pp.

Sinkankas, J., 1968, *Van Nostrand's Standard Catalog of Gems,* D. Van Nostrand, Princeton, NJ, 286 pp.

Sinkankas, J., 1976, *Gemstones of North America,* Volume II, Van Nostrand-Reinhold Co., New York, 494 pp.

Smith, G.F.H., and Phillips, F.C., 1972, *Gemstones,* 14th ed., Pitman Publishing Corp., New York, 580 pp.

Staatz, M.H., and Bauer, H.L., Jr., 1952, "Virgin Valley Opal District, Humboldt County, Nevada," Circular 142, US Geological Survey, 7 pp.

Thompson, R.D., 1955, "Gems," *Mineral Facts and Problems,* Bulletin 556, US Bureau of Mines, pp. 299–307.

Tolansky, S., 1955, *The Microstructures of Diamond Surfaces,* National Assn. Goldsmiths Press, Ltd., London, 67 pp.

Tyler, P.M., 1932, "Abrasive and Industrial Diamonds," Information Circular 6562, US Bureau of Mines, 25 pp.

Von Bernewitz, M.W., and Hess, F.L., 1940, "Cutting and Polishing Stones," Information Circular 7107, US Bureau of Mines, 23 pp.

Webster, R., 1952, *Practical Gemology; A Study of the Identification of Gem-stones, Pearls, and Ornamental Minerals,* 2nd ed., National Assn. Goldsmiths Press, Ltd., London, 180 pp.

Webster, R., 1970, "Modern Synthetic Gemstones," *Journal of Gemology,* Vol. 12, pp. 101–148.

Webster, R.A., 1971, "Comprehensive Compendium of Modern Synthetic Gem Stones, Pts. 1 and 2," *Lapidary Journal,* Vol. 25, pp. 275–280, 304–317.

Webster, R., and Hinton, V.V., 1945, *Introductory Gemology,* Gemological Institute of America, Los Angeles, CA, 181 pp.

Weinstein, M., 1946, *Precious and Semi-precious Stones,* 4th ed., Sir Isaac Pitman and Sons., Ltd., London, 185 pp.

White, W.B., Roy, R., and Crichton, J.M., 1967, "The Alexandrite Effect, an Optical Study," *American Mineralogist,* Vol. 52, No. 5–6, May-June, pp. 867–871.

Whitlock, H.P., 1946, *The Story of the Gems,* Emerson Books, Inc., New York, 206 pp.

Willems, J.D., 1948, *Gem Cutting,* The Manual Arts Press, Peoria, IL, 224 pp.

Williams, A.F., 1932, *The Genesis of the Diamond,* 2 vols., Ernest Benn, Ltd., London, 636 pp.

Wodiska, J., 1909, *A Book of Precious Stones,* G.P. Putnam's Sons, New York, 365 pp.

Yaverbaum, L. H., ed., 1980, *Synthetic Gems, Production Techniques,* Noyes Data Corp., Park Ridge, NJ, 353 pp.

Youngman, E.P., 1931, "Zircon (The Gem)," Information Circular 6465, US Bureau of Mines, 20 pp.

Glass Raw Materials

H. N. MILLS*†

The glass industry is a major user of many industrial minerals in the manufacture of its product. It is the intent of this chapter to: (1) acquaint the reader with the glass industry by including a few production statistics, and (2) cover the various aspects of the major raw materials used to make glass.

The matter of glass compositions, properties, and batches is considered beyond the scope of this chapter which deals only with raw materials. For more details on the particular minerals and rocks used in glassmaking, the reader is referred to chapters in the commodity section of this book.

Glass Industry Statistics

Glass manufacture in the United States can be divided into three major product classifications as follows:

1) Glass containers
2) Pressed and blown glass
3) Flat glass

Glass containers include bottles and jars for the packaging of food and beverages, as well as drugs, chemicals, and cosmetics. Pressed and blown glass includes table, kitchen, art and novelty ware, lighting and electronic glassware, and fiber glass. Flat glass covers sheet or window glass, plate, laminated, and safety glass.

The US Department of Commerce statistics, as recorded in the *1977 Census of Manufacturers,* compares the dollar value of glass shipments for each product category. These statistics are shown in Table 1 (Anon., 1980) and are self-explanatory.

A comparison of the tonnage output in each product classification is difficult, as production data is reported in "gross" for glass containers, "pounds" for fiber glass, "pieces" for glassware, and "square feet" for flat glass. In the case of glass containers, an average "pounds per gross" factor can be applied to the total production figures reported in thousands of gross in Bureau of Census industrial reports. Table 2 presents the calculated annual production tonnage for glass containers for the years 1974–1979. Except for 1979, there was a yearly increase in production. The tonnage figures of Table 2 are further influenced by decreasing weight per gross which has occurred in the glass container industry since 1975. This reflects the gradual shift to lighter weight containers for beverages.

TABLE 1—1977 Dollar Value of Glass Shipments in the United States

Product	$	% of Total
Glass Containers	3,618,000,000	48.0
Pressed and Blown	2,100,700,000	27.9
Flat Glass	1,817,600,000	24.1
Total	7,536,300,000	100.0

Source: 1977 Census of Manufacturers, US Dept. of Commerce.

Other segments of the glass industry have experienced growth, but the figures in tons are not published information. Fiber glass growth has been influenced by the heavy demand for insulation as part of energy conservation efforts on a national basis. The production of flat glass by the float process has continued to grow with several new plants entering production. All portions of the glass industry experienced the effects of the general economic decline of 1979–1980, but the outlook for the future is bright.

Batch Materials For Glass Use

Materials and Function

Soda-lime-silica glass represents the largest proportion of glass manufactured for commercial use. It is produced by the fluxing and

* Section Head—Glass and Furnace Technology, O-I Glass Container Div., Owens-Illinois, Inc., Toledo, OH.

† Acknowledgment is made to Owens-Illinois, Inc. for use of unpublished data from the writer's personal files.

TABLE 2—Glass Container Production Statistics 1974 to 1979

Year	Annual Production, Short Tons*	% Increase/Decrease Over Previous Year
1974	12,337,468	—
1975	12,720,645	+3.1
1976	13,370,500	+5.1
1977	13,291,198	−0.6
1978	13,857,861	+4.3
1979	13,411,258	−3.2

Source: Bureau of the Census Industrial Reports, 1975-1979, US Dept. of Commerce and Glass Packaging Institute.
*Metric equivalent: 1 st × 0.907 184 7 = t.

melting of pure silica sand in the glass melting furnace. Silica (SiO_2) is the glass-forming oxide, soda (Na_2O) is the fluxing agent, and lime (CaO or CaO + MgO) is the stabilizing material to provide chemical durability. Various other oxides may be added to the glass formula to provide other desired properties. For example, Al_2O_3 and B_2O_3 improve chemical durability; color may be imparted by adding oxides of chromium, cobalt, iron, or nickel; and the melting and refining process in glass manufacture can be improved by additions of carbon, sulfur or sulfur oxides.

Glass compositions must take into account the end use of the product and the type of process used in forming the product—poured, pressed, drawn, rolled, molded, etc. The glass technologist must consider these, as well as cost, in developing the optimum glass composition. As might be expected, a variety of modifications can be made to the basic soda-lime-silica composition to achieve the desired objectives. The reader is referred to the literature on glass for more information in this area.

The raw materials for the manufacture of glass come primarily from mined minerals and rocks that are selected for their purity and stable composition, or those that may be beneficiated to a suitable product. The relatively low cost of glass as a material can be attributed to the general availability, plentiful supply, and moderate cost of raw materials.

Table 3 lists various glassmaking raw materials and the major oxide or element for which they are used in the glass batch formulation. This list is by no means complete, but does contain the major materials. Other oxides or metals may be added to the batch formula to provide special properties for particular applications.

Typical Compositions

Table 4 lists typical oxide compositions (or elements) of some major glass raw materials. These so-called "raw material factors" are used to calculate the batch mixture needed to produce a certain glass composition. Normally, the best average analysis is used to obtain these factors for each raw materials source.

Two sands are shown in the table—white sand for the manufacture of flint or white glass and yellow sand for the manufacture of colored glass. The major difference in the two sands involves the iron and chrome oxide content, with the yellow sand having six times the Fe_2O_3 and ten times the Cr_2O_3 of the white sand. Iron and chrome oxides impart shades of green color to glass, the depth of color being dependent on the concentration. Flint or white glass cannot be made if the concentration of coloring oxides becomes too great.

The three columns of data to the far right in Table 4 are also important to the glass technologist or batch and furnace engineer. The column headed "% Glass Made" is the sum of the oxide composition for each material and represents the amount of glass produced. The column headed "% Fusion Loss" represents the difference between 100% and "% Glass Made." This is the portion of each raw material lost in the melting process. These volatiles

TABLE 3—Raw Materials for Glassmaking Oxides or Elements

Raw Material	Oxide or Element
Silica sand	SiO_2
Soda ash	Na_2O
Limestone	CaO
Dolomite (raw or burned)	CaO + MgO
Feldspar, aplite, nepheline syenite	$Al_2O_3 + Na_2O + K_2O$
Blast furnace slag	$Al_2O_3 + S^=$
Borax, razorite	B_2O_3
Salt cake	$SO_3 + Na_2O$
Gypsum	SO_3 + CaO
Barite	SO_3 + BaO
Fluorspar	F_2
Arsenic	As_2O_3
Iron chromite	Cr_2O_3
Iron pyrite	Fe_2O_3 + S
Nitre	$Na_2O + O_2$
Selenium	Se
Carbon	C

Source: Compiled by H. N. Mills, Owens-Illinois, Inc.

TABLE 4—Typical Compositions of Some Glassmaking Raw Materials

Material	Oxide Composition, %										% Glass Made	% Fusion Loss	Bulk Density* lb per cu ft†
	SiO_2	Al_2O_3	Fe_2O_3	Cr_2O_3	TiO_2	CaO	MgO	Na_2O	K_2O	Other			
White sand	99.7	0.08	0.025	0.0002	0.02	0.01	0.01	0.004	0.005	—	99.9	0.1	95
Yellow sand	99.1	0.25	0.150	0.0020	0.05	0.01	0.01	0.015	0.050	—	99.6	0.4	87
Soda ash	—	—	0.002	—	—	—	—	58.0	—	—	58.0	42.0	63
Calcite limestone	0.7	0.1	0.075	0.0010	—	54.7	0.60	—	—	—	56.2	43.8	91
Dolomite limestone	0.5	0.3	0.075	0.0005	—	31.0	21.0	—	—	—	52.9	47.1	90
Feldspar	68.0	19.4	0.075	0.0003	0.01	1.4	0.02	6.9	4.0	—	99.8	0.2	89
Aplite	60.5	23.0	0.41	0.0003	0.20	5.6	0.05	6.1	2.9	—	98.8	1.2	90
Nepheline syenite	59.5	23.9	0.085	0.0003	—	0.2	—	10.5	5.0	—	99.2	0.8	95
Blast furnace slag	38.0	7.4	0.30	0.0030	0.40	38.8	12.5	0.4	0.4	MnO 0.7; $S^=$ 0.7	99.0	1.0	99
Salt cake	—	—	0.003	—	—	—	—	43.5	—	SO_3 56.2	43.5	56.5	85
Fluorspar	2.5	—	0.19	—	—	65.8	—	—	—	F_2 43.5	68.5	31.5	91
Borax (anhyd.)	—	—	—	—	—	—	—	30.8	—	B_2O_3 69.2	100.0	—	62
Sodium nitrate	—	—	—	—	—	—	—	36.5	—	—	36.5	63.5	—
Arsenic	—	—	—	—	—	—	—	—	—	As_2O_3 100.0	—	—	—
Selenium	—	—	—	—	—	—	—	—	—	Se 100.0	—	—	—
Iron pyrite	3.1	—	63.2	0.0050	—	—	—	—	—	$S^=$ 50.3	66.3	33.7	125
Iron chromite	1.4	15.0	28.0	45.0	—	—	10.4	—	—	—	99.8	0.2	88
Gypsum	3.5	0.9	0.33	—	—	30.9	1.3	—	—	SO_3 42.7	36.9	63.1	75
Barite	—	—	—	—	—	—	—	—	—	BaO 65.7; SO_3 34.3	65.7	34.3	120

Source: Compiled by H. N. Mills, Owens-Illinois, Inc.
* Loose Packing
† Metric equivalent: lb per cu ft × 16.018 46 = kg/m³.

may be CO_2, SO_3, H_2O, etc., depending on the chemical nature of the raw material.

The last column in Table 4 lists the bulk densities for each material. This data is useful in calculating bin capacities for storage of each raw material. The data given is for loose packing of the material. The bulk density for each specific material used at the glass plant can be determined by ASTM procedure and is needed for accurate inventory data.

Pricing

The majority of the raw materials used in making glass are in abundance in the earth's crust. The price of the materials varies with the cost of mining and subsequent processing, as well as total tonnage produced. In recent years, substantial added costs have been incurred due to the high price of energy and the need to meet environmental requirements. Table 5 contains a current (1981) listing of the range of f.o.b. prices of some of the most common glass raw materials.

Of the large-tonnage raw materials, soda ash, a chemically pure material, is the most expensive, accounting for over 55% of the raw material cost per ton of glass in most soda-lime compositions. The major source of soda ash in the United States is the trona ore beds of Wyoming. Here soda ash is produced by a solution and recrystallization process. Only one ammonia-soda (Solvay) process plant for manufacture of soda ash remains in operation in the United States.

TABLE 5—Pricing of Glassmaking Raw Materials, 1981

Material	F.O.B. Price Range per Ton, $
Glass sand	5–13
Soda ash	86–123
Limestone (calcite, dolomite)	5–18
Feldspar, aplite, syenite	20–31
Blast furnace slag	22–23
Borax (anhydrous)	152–270
Salt cake	54–83
Gypsum	18–22
Barite	66–72
Fluorspar	140–208
Arsenic	480–920
Iron chromite	148–191
Iron pyrite	86–115
Nitre	146–160
Selenium	6.50–15 (per lb)
Carbon	92–120

Source: Corporate Purchasing, Owens-Illinois, Inc.

The cost of shipping raw materials from their source to the glass plant is another major item. Freight costs often exceed the f.o.b. price of raw materials and have been increasing at a steady rate because of high energy and other costs. It is most desirable that the raw material source be close to the glass plant to minimize freight cost. This is not always possible, however, as the glass plant should also be close to its market to minimize the even higher freight on finished glass products.

The glass industry is fortunate that at least two of the basic raw materials—sand and limestone—may be found in most areas of the country. By selective mining and modern mineral beneficiation techniques, material from these deposits can be upgraded to meet glass raw material specifications.

The shipment of bulk glass raw materials is normally by closed hopper bottom rail cars or trucks. Normal capacity is 63.5 to 90.7 t (70 to 100 st) for rail cars and 18 to 22.7 t (20 to 25 st) for trucks. Long-distance hauls are usually by rail; for the shorter haul the choice may be between rail and truck, on the basis of competition for the business.

US Consumption of Some Glass Raw Materials

Table 6 contains data from the US Bureau of Mines *Minerals Yearbooks* relative to US consumption of materials used in glassmaking for the years 1972–1977. Figures for total usage are shown, followed by the percent used only in glass. For example: in 1972, there were 26.8 Mt (29,530,000 st) of sand used for molding, grinding, blasting, engine, filtration, glass, and other industrial uses, *not* including construction. Of this tonnage, 37%—about 9.8 Mt (10.9 million st)—was used in glassmaking.

Use in glass of the major raw materials—sand, soda ash, and limestone—has remained at a relatively fixed percentage of the total consumption. Of the total crushed limestone and dolomite consumed in the US, the percentage used in glass is extremely small compared to the quantities used in road building and construction. Approximately half the soda ash and about 40% of the industrial sand is used in glass.

The soda ash figures point out the changes in that industry resulting from the shift from Solvay process to natural process soda ash. By 1977, natural process production amounted to approximately 80% of the total consumption as opposed to a little over 40% in 1972.

Fluorspar usage in glass continued to decline as glassmakers tried to minimize fluoride compound emissions from glass furnaces to meet air quality standards. Selenium, used in the decolorization of glass and classified as a hazardous material, has also shown decline in consumption for glass.

Raw Material Specifications

Chemical

The chemical specifications for glassmaking raw materials must recognize the type of glass in which the material is to be used, and the capability of the processed material to meet these requirements. A *consistent* raw material composition is the most important requirement.

Chemical specifications on glassmaking raw materials can be set in great detail, with plus or minus limits on each oxide. This is not always realistic, as the raw material supplier has no way of controlling each oxide in a naturally occurring mineral. Ideally, the specification should guarantee a certain minimum purity of the raw material and provide limits on critical oxides over which the supplier has control.

Iron oxide content is perhaps the most critical chemical component for all glassmaking raw materials. Practically all glass raw material processors must include some type of equipment in their process to minimize and control iron, or it must be done through selec-

TABLE 6—US Consumption Statistics of Some Glass Raw Materials

	1972		1973		1974		1975		1976		1977		1978		1979	
Material	Thousand Tons	% in Glass	Thousand Tons	% in Glass	Thousand Tons	% in Glass	Thousand Tons	% in Glass	Thousand Tons	% in Glass	Thousand Tons	% in Glass	Thousand Tons	% in Glass	Thousand Tons	% in Glass
Industrial sand	29,530	37	28,974	38	29,070	38	27,283	39	29,914	41	31,360	38	32,850	42	33,510	41
Soda ash																
Mfg.	4,305	47	3,813	51	3,507	40	2,802	47	2,344	47	1,812	47	1,500	55	W	55
Natural	3,218		3,722		4,059		4,328		5,216		6,228		6,790		8,253	
Crushed limestone and dolomite	671,496	0.27	774,397	0.22	751,520	0.23	666,320	0.30	662,880	0.32	706,520	0.25	785,734	0.26	812,054	0.26
Industrial lime	17,534	2.1	18,131	2.0	11,970	2.6	10,666	2.4	11,717	1.8	12,007	1.8	12,952	1.7	13,446	1.4
Feldspar	581	50	589	53	548	36	542	51	731	67	735	56	735	54	740	55
Nepheline syenite	456	Major	474	Major	505	Major	425	Major	499	Major	502	Major	638	Major	NA	Major
Salt cake—refined	801	NA	907	NA	783	NA	796	NA	766	NA	741	NA	630	NA	514	NA
Boron oxide	607	50	664	50	619	50	603	50	630	50	735	46	778	50	799	50
Fluorspar	1,352	1.5	1,352	0.8	1,525	1.1	1,245	0.7	1,274	0.9	1,162	0.8	1,204	0.9	1,136	1.0
Arsenic	13.6	NA	13.5	NA	13.7	NA	12	NA	4.3	NA	6.0	NA	10.3	5	12.3	5
Selenium	(lb) 985,000	33	(lb) 1,368,889	34	(lb) 1,565,466	34	(lb) 1,062,372	28	(lb) 989,831	30	(lb) 871,161	30	(lb) 799,853	30	(lb) 683,903	30

Source: US Bureau of Mines *Minerals Yearbooks,* Vol. I, 1972-1979.
Thousand tons = Thousand short tons of material sold, imported, or manufactured for US consumption. Metric equivalents: 1 st × 0.907 184 7 = t; lb × 0.453 592 4 = kg.
NA = Data not available.

tive mining. Removal and control of iron are accomplished through such processes as washing, scrubbing, classification, flotation, magnetic or electrostatic separation, screening, acid leaching, etc. Iron control is important because of its effect on glass color, particularly white or flint glasses. In colored glass, iron fluctuation affects heat transmission during melting and forming operations.

Column A of Table 7 lists some typical chemical specifications for ten major glassmaking raw materials. Maximum levels of impurities allowed and specification limits on the major oxides are shown.

Physical

The grain size distribution of glassmaking raw materials is very critical, and the raw material supplier must exercise diligence in

TABLE 7—Typical Chemical and Physical Working Specifications on Glass Raw Materials

Material	Chemical Specifications	Physical Specifications*
White sand	SiO_2 over 99% Fe_2O_3 - 0.030% max. Cr_2O_3 - 0.0003% max.	+20 mesh - 0 +30 mesh - 1% max. −100 mesh - 15% max.
Yellow sand	SiO_2 over 98.5% Fe_2O_3 - 0.25% max. Cr_2O_3 - 0.005% max.	+20 mesh - 0 +30 mesh - 1% max. −100 mesh - 15% max.
Soda ash	Na_2CO_3 over 99% NaCl - 0.5% max. Fe_2O_3 - 0.002% max.	+16 mesh - 0 +30 mesh - 3% max. −200 mesh - 3% max.
Limestone	CaO + MgO over 54% Fe_2O_3 - 0.10% max.	+16 mesh - 1% max. +20 mesh - 15% max. −100 mesh - 20% max.
Feldspar	Al_2O_3 over 19% Alkali - over 11% Fe_2O_3 - 0.10% max.	+16 mesh - 0 +20 mesh - 1% max. −100 mesh - 25% max.
Nepheline syenite	Al_2O_3 over 22% Alkali - over 13% SiO_2 - 62% max. Fe_2O_3 - 0.10% max.	+30 mesh - 0 +40 mesh - 3.5% max. −100 mesh - 35% max.
Aplite	Al_2O_3 over 22% (±0.5%) Fe_2O_3 - 0.10% max.	+16 mesh - 0 +20 mesh - 2% max. +30 mesh - 20% max. −100 mesh - 10% max.
Blast furnace slag	Al_2O_3 - AV. ±0.5% SiO_2, CaO, MgO - AV. ±1.0% Fe_2O_3 - 0.33% max. $S^=$ - AV. ±0.14%	+8 mesh - 0 +16 mesh - 5% max. −100 mesh - 30% max.
Salt cake	Na_2SO_4 over 99% NaCl - 0.20% max. Fe_2O_3 - 0.002% max.	+16 mesh - 0 +20 mesh - 1% max. +30 mesh - 2% max. −100 mesh - 54% max.
Gypsum	Fe_2O_3 - 0.25% max. % CaO - AV. ±1.0% % SO_3 - AV. ±2.0%	+16 mesh - 0 +20 mesh - 0.5% max. +30 mesh - 12% max. −100 mesh - 25% max.

Source: Compiled by H. N. Mills.
* Physical specification on grain size distribution based on % retained on US Series screens.

meeting size specifications. With regard to the effects of sizing in the glass process, the following are of concern:

Meltability: Materials such as sand, nepheline syenite, aplite, and feldspar are the most difficult to melt in the glass furnace. It is important that coarser fractions be eliminated from these materials to aid melting. Physical sizing normally is −30 mesh; the trend is −40 mesh for improved meltability.

Mixing and Segregation: Because glass batches are mixes of the various raw materials in proper proportions, it is important that physical sizing of the materials be similar, in order to improve blending of the materials and diminish chances of segregation in handling the mixed batch. A uniform batch mix helps melting and improves glass homogeneity. Over 50% of the grain size distribution on major glassmaking raw materials falls between −30 and +100 mesh using US Series screens.

Dusting: Too many fines in the raw materials cause dust in handling and within the glass furnace itself. The former is detrimental to housekeeping, equipment, and personnel, while the latter increases erosion of furnace refractories. The size distribution below 100 mesh is critical from the dusting standpoint and specifications are set to control this.

Column B of Table 7 lists some typical sizing specifications applied to ten major glassmaking raw materials. Specifications on screens other than those listed may be desirable for specific materials as required by the end use and nature of the material.

Other Raw Material Specifications

The glass technologist is concerned about refractory impurities which might be contained in certain raw materials in addition to normal chemical and physical characteristics. Refractory particles are those which may survive the glass melting process and be present in the finished glass as a solid inclusion or "stone" defect. The larger the particle, the greater the chance it will come through the glass furnace.

These refractory particles may be naturally occurring minerals contained in trace quantities in the raw materials or a result of either contamination through the raw material processing or chemical interaction in such processes as calcination. Table 8 lists some of the refractory impurities which have been found in various raw materials.

Specifications on refractory heavy minerals (RHM) are usually based on size and quantity. The particles +60 mesh or larger are the ones of major concern in glass melting. The quantity may be set on the basis of weight percent or actual particle count from a given size sample. Typical working specifications for a glass sand might be set as follows:

Sample Size = 400 g

Weight Basis

Maximum +60 mesh RHM = 0.0003%

Particle Count Basis

+40 mesh RHM = 2 particles maximum

−40 +60 mesh RHM = 20 particles maximum

In the laboratory, the quantity of refractory heavy minerals in a particular sample is determined by a heavy liquid density separation, followed by soda ash fusion, and petrographic count and identification of the mineral species.

Byproduct Materials In Glass

Byproduct materials containing desired glassmaking oxides are excellent candidates for glass raw materials. Each byproduct must be

TABLE 8—Refractory Impurities Found in Some Glassmaking Raw Materials

Material	Sand	Limestone	Burned Dolomite	Feldspar	Nepheline Syenite	Aplite
Refractory Impurities	Sillimanite Kyanite Andalusite Zircon Spinel Corundum Chromite Kaolin	Corundum Spinel Chromite Copper metal	Clay Spinel	Sillimanite Corundum Zircon Chromite Spinel Copper metal Cassiterite Chrysoberyl	Corundum Zircon	Sillimanite Kyanite Zircon Spinel Corundum

Source: Jasinski, 1970.

evaluated for its chemical and physical properties to determine its suitability for use in the glass batch. Use of byproducts can in many cases be a cost-effective means of obtaining glass oxides while aiding in the disposition of these secondary materials. Some of the byproduct materials used in glass are:

Feldspathic sand from mineral beneficiation processes can be utilized as a source of silica, alumina, and alkali to the glass composition. Such material can come from sand beneficiation or recovery of lithium minerals, where the feldspathic material is separated. The alumina may be 10 to 12%, the silica 80 to 82%, and the alkali 7 to 8%. Iron is normally low.

Clay or *kaolin* can be a source of alumina in colored glass compositions, so limited because of its high iron content. The clay must be sized fine to melt properly. Dusting is controlled by adding wetting agents to the glass batch. Clay is a byproduct of silica sand mining in some areas, notably the Southwest.

Blast furnace slag is a byproduct of the ironmaking industry, and contains many of the glassmaking oxides in a relatively low melting composition. It is used as a source of alumina and sulfide sulfur in amber glass, and as a refining agent in lesser quantities in flint and green glasses. Iron and chrome are the major contaminants in slag, but usually can be controlled by proper selection of material and magnetic separation processing. Table 4 lists typical chemical and physical working specifications.

Sodium sulfate is produced as a byproduct from some chemical processes as well as the manufacture of paper. In addition, it may be collected in electrostatic precipitators on the exhaust system of the furnace. Properly handled and with attention to chemical quality control, such material can be a source of sodium sulfate to the glass batch as a refining agent and melting aid.

Caustic soda, or sodium hydroxide, might be considered a byproduct from the manufacture of chlorine. It is also a basic chemical. Caustic can be used in the glass batch as a source of soda to replace part or all of the soda ash, or as a wetting agent to control furnace dust. Caustic has not been widely used in glass, however, because of cost and handling considerations. Caustic normally is supplied as a 50 or 73% solution, or in solid form as flakes or pellets.

Cullet, or recycled glass, is a byproduct of the glass industry itself. Clean and crushed cullet can be readily recycled through the glass furnace and it is normal to use 15 to 20% of this material in most glass batches. Higher percentages can be used if the cullet can be handled as a raw material and its composition used in calculating the glass batch. Cullet is beneficial in lowering glass melting temperatures, improving melting capability, and decreasing glass furnace emissions. The glass container industry has very active programs to recycle as much cullet as possible.

Future Prospects

Environment and Health

The glass industry is vitally involved in efforts to minimize environmental pollution and improve occupational health and safety. From the raw material standpoint, actions have been taken to eliminate or make substitution for materials considered hazardous, and those which contribute to emissions from glass furnaces. Dust is controlled by use of wetting agents or mechanical dust collectors. Respirators are required for workers in a batch dust exposure area at the glass plant.

Environmental and health considerations will continue to exert a major influence on materials used in making glass and on the specifications, both chemical and physical, requested.

Technology and Growth

Growth in all phases of glass production is expected to continue, with ups and downs related to changes in the product mix and the national economy. Glass technology continues to progress in increasing productivity, decreasing costs, and making the most efficient use of energy.

The higher glass melting rates of the 1980s will increase the demand for quality in raw materials. Those raw materials most difficult to melt—sand, nepheline syenite, etc.—show a trend to less coarse material while maintaining a low level of fines to minimize dusting.

Cullet will be increasingly used as the glass melting process requires more output at lower temperatures and fewer furnace emissions.

Glass batch agglomeration continues to offer promise as a means to increase the melting rate of raw materials, save energy, and possibly serve as an emission absorbant. As the cost of energy has increased, batch agglomeration (pelletizing, briquetting, etc.) becomes more attractive from the economic standpoint.

The glass and mineral industries are closely allied in that the products of the latter are vital to the former. As the technology of each industry develops, requirements and capabilities change. These must be taken into account in long-term planning in the glass raw materials area.

Bibliography and References

Anon., 1972, *Glass Containers,* Glass Container Manufacturers Institute, New York, NY.

Anon., 1973, "An Introduction to the Glass Industry," *Industrial Minerals,* No. 74, Nov., pp. 9–23.

Anon., 1980a, *1977 Census of Manufacturers,* Glass Containers SIC3221, Flat Glass SIC3211, Pressed & Blown Glass, NEC SIC3229, US Dept. of Commerce, Bureau of the Census.

Anon., 1980b, "Glass Containers," *Current Industrial Reports, 1974–77,* Series M32G, US Dept. of Commerce, Bureau of the Census.

Anon., 1980c, *1979 Annual Report,* Glass Packaging Institute, Washington, DC.

Jasinski, J., 1970, "Refractory Impurities in Raw Materials," 1970 Conference on Glass Problems, Ohio State University, Columbus, OH.

Scholes, S.R., 1946, *Modern Glass Practice,* Industrial Publications, Inc., Chicago, IL, 289 pp.

Tepordei, V.V., 1978–1979, "Sand and Gravel," Preprint, US Bureau of Mines, 29 pp.

Tooley, F.V., 1953, *Handbook of Glass Manufacture,* Ogden Publishing, New York, NY, 506 pp.

Also Useful:

Minerals Yearbooks, Vol. 1, "Metals and Minerals," 1965–1980, US Dept. of the Interior, Bureau of Mines.

Mineral Pigments

KENNETH R. HANCOCK *

Iron oxides are unique in that they are the only significant *colored* mineral found in a natural state suitable for use as a pigment after being pulverized to pigmentary size. The current world production of iron oxide pigments is estimated by the author to be between 499 and 590 kt (550,000 and 650,000 st) based on 1979 data. The fact that highly colored natural deposits of iron ore are found throughout the world accounts for its use in prehistoric man's artistic cave paintings. These early artists did not realize that they had established man's first paint test fence which now has some 20,000 years of exposure.

In addition to abundance, constituting about 7% of the earth's crust, iron oxides have the advantages of low cost, permanency, and are not toxic. Through the centuries, succeeding civilizations have utilized iron oxides as a major source for decoration and protection when coloring was desired. In the last century the chemical industry has improved on nature by developing a complete range of synthetic iron oxide pigments which surpass the pigments produced from natural iron ores in uniformity, color quality, and chemical purity. In the United States alone, the combined production of natural and synthetic iron oxides produced sales of over $28,000,000 in 1970 (Stipp, 1970) and increasing to $97 million by 1980 (Spinrad, 1980).

Nomenclature

With the increasing importance of the synthetic iron oxides it is necessary to make a distinction between the natural or mineral pigments and the synthetic pigments. Natural pigments are those products which are derived from selected ores and should not be confused with iron ores mined for steel production. Iron ores that are mined for steel must be capable of being mined and reduced to iron on a commercial basis. These ores are selected on the basis of iron content and processing economics. It is therefore unusual when iron ores for steel production are suitable for use as mineral pigments. Natural pigment ore sources are selected for their special physical-chemical properties and can command a premium price.

Synthetic pigments, and in this instance iron oxides, are those pigments produced from basic chemicals. Through chemical synthesis pigmentary sized particles are produced as opposed to comminution, the major procedure common to all natural iron oxide pigments:

Classification

An important characteristic of a pigment is its color and, therefore, a logical separation can be made on this basis.

A. Yellow Iron Oxide Pigments
   1. Natural mineral origin
      a) Goethite
      b) Lepidocrocite
      c) Ochres
      d) Siennas
      e) Limonite
   2. Synthetic pigments
      a) Goethite

B. Red Iron Oxide Pigments
   1. Natural mineral origin
      a) Hematite
      b) Siderite (calcined)
      c) Pyrites (calcined)
   2. Synthetic pigments
      a) Hematite

C. Brown Iron Oxide Pigments
   1. Natural mineral origin
      a) Umbers
      b) Limonite (calcined)
      c) Siderite (calcined)
      d) Goethite (bog ore or sulfur mud)
   2. Synthetic pigments
      a) Blends of hematite, goethite, and magnetite

* Director of Quality Assurance, Mineral Products, Pfizer, Inc., New York, NY.

b) Coprecipitated hematite-magnetite
c) Maghemite

D. Black Iron Oxide Pigments
1. Natural mineral pigments
a) Magnetite
b) Slate (mixed minerals)
2. Synthetic pigments
a) Magnetite

Pigmentary minerals, natural or synthetic, have physical-chemical properties that are important and distinguish them from metallurgical, cementitious, construction, agricultural, and other types of industrial minerals.

The chemical composition of natural minerals, with respect to iron content, is somewhat secondary to the type and form of the companion minerals. For example, an iron ore possessing good pigmentary qualities but associated with appreciable quantities of quartz would be undesirable due to the difficulty in grinding quartz. High percentages of organic matter would eliminate simple grinding and would require prior calcination to decompose the organics. The converse is true concerning synthetic oxides where a high iron content is indicative of chemical synthesis and minimal impurities.

Chemical Properties

Most of the pigmentary iron ores are nonreactive (Casey, 1966; Mellor, 1953), chemically inert, and contain only traces of heavy metals and toxic elements, usually chemically bound in the form of complex silicates. Synthetic pigments produced from basic raw materials have the advantage that purification steps can be employed to reduce the heavy metal content: lead, antimony, arsenic, cadmium, mercury, and selenium of the starting raw materials. If subsequent precautions are followed avoiding processing contamination, the heavy metal impurities in synthetic oxides will be about one-tenth that of the best naturals.

Iron oxide pigments have a high degree of lightfastness. This is a term which refers to the ability of a pigmented film to resist exposure to radiant energy with minimal physical and/or chemical change.

Physical Property Definitions

The physical properties of iron oxide pigments are usually more important than chemical properties in view of their inertness. Several physical properties require definition to provide a clear understanding of a pigment's merit and to distinguish between naturals and synthetics or pigments of other families.

Oil Absorption: Oil absorption is defined as that weight of vehicle required to wet a definite weight of pigment to form a paste. There are numerous methods (Anon., 1972; Stewart, 1942; Von Fisher, 1948) and devices for determining oil absorption. Values given in this article will be by the ASTM spatula rub-up procedure. Oil absorption, or binder demand, is an important property since it relates to particle size, shape, and surface characteristics and places certain limitations on paint formulation or other end uses where solid-liquid interactions are important.

Surface Area: Two pigments of essentially the same mean particle size may have widely divergent surface areas. This may be due to shape and/or surface characteristics such as porosity, pore size, smoothness, and convolutions. Since spectral and surface energy properties depend on the surface, its measurement becomes an important factor. The surface area of a particle may be defined as the area of the total external surfaces including pores but excluding *internal* (isolated) pores. When external pores are smaller than the molecule of the media being absorbed, they will not be measured. The most widely accepted and fundamental method for determining surface area is by the adsorption of a gas, usually nitrogen (Brunauer, et al., 1938; Herdan, 1953) and is designated as the BET method. The value obtained expresses surface area in square meters per gram.

Particle Size: No single method or instrument currently in existence will adequately measure particles ranging in size from 100 to 0.01 μ. Complete books have been devoted to this subject, which is probably the most important physical property of pigments. It is not sufficient to state an average particle size since the size distribution of particles is equally important. Unless otherwise specified, particle diameter will be the diameter of a sphere having the same terminal velocity as the particle. This is commonly called a Stokes' diameter (Cadle, 1955, p. 198) and is derived from Stokes' equation for particles settling under the action of gravity through an infinite fluid.

Terms which are quite often confused in particle technology are:

1) *Discrete Particle:* A single unit of matter which can change in size only by the breakdown or fracture of chemical bonds within its structure.

2) *Crystallite:* A minute single crystalline

unit i.e., that part of a particle that has its atoms and molecules arranged in a perfect crystal lattice. A *discrete* particle is composed of an assemblage of a large number of crystallites.

3) *Aggregate:* A group of discrete particles bound together by strong chemical bonds.

4) *Agglomerate:* A group of discrete particles and/or aggregates bound together by weak physical bonds.

5) *Flocculate:* Groups of aggregates or particles usually formed in suspensions and readily dispersed with minimal forces.

Shape: In some applications particle shape can have a more profound effect than size. Micaceous iron oxide (lamellar platelets) and acicular particles contribute completely different characteristics to a pigmented system than their equivalent spherical counterpart. Various techniques have been devised to apply numerical values for shape factors (Cadle, 1955, p. 23; Herdan, 1953, p. 37–39). Pigments may be classified according to shape into five categories: (1) spheroidal, (2) cubical, (3) nodular, (4) acicular, and (5) lamellar.

Optical Properties: The optical properties of pigments are dependent on the molecular and crystal structure and are influenced by the size and shape of the discrete particles. Since pigments are rarely used in their powder state, the ultimate effect depends on the system. Light entering a pigmented system may be absorbed, reflected, or refracted through particle-to-particle interactions. The manner and intensity of the light reflected results in spectral reflectance which can be measured by a spectrophotometer.

Magnetic Properties: Iron oxides are usually antiferromagnetic (Schieber, 1967) but they can also be ferromagnetic as produced synthetically and in certain natural mineral species. The magnetic properties of ferromagnetic materials may be determined by various techniques. A few of the more common are: magnetic moment balance, vibrating sample magnetometer, 60-Hz B-H meter, dc hysteresisgraph and permeameters.

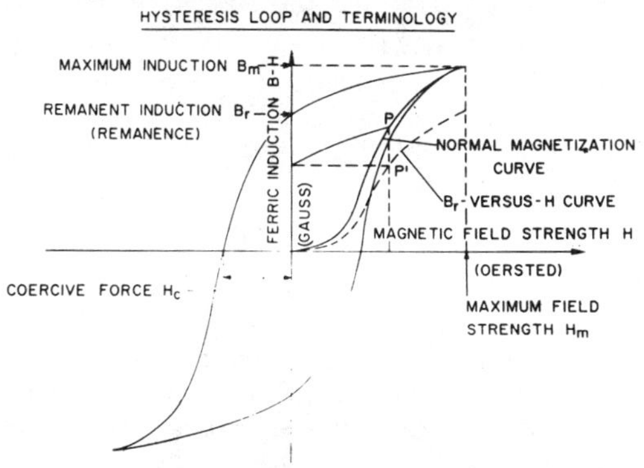

FIG. 1—*Hysteresis loop used to characterize magnetic iron oxides.*

Magnetic materials are best described by measuring the response to a magnetic field of varying intensity. This produces a magnetization curve or *BH* curve as shown in Fig. 1. Ferromagnetic materials may then be characterized with values of coercive force (H_c), remanence (B_r), and saturation (B_m) at a specified field strength in oersteds (H).

Natural Iron Oxides

Natural iron oxide minerals as mined for pigment applications normally require more than simple grinding to attain commercial interest. The major minerals are listed along with their chemical and physical properties. It should be noted that the chemical compositions, especially those of goethite and hematite, do not represent the pure mineral but rather the range as received by the pigment manufacturer.

Hematite

This important iron-bearing mineral, besides providing a source for steel manufacture, is a major source for mineral pigments.

Occurrence: Major iron ore sources are sedimentary as exemplified by the Lake Superior deposits. Hematite also occurs as igneous, volcanic, hydrothermal, metamorphic, and weathered deposits. Several deposits are unique for their high color value, such as those on the Island of Ormuz in the Persian Gulf or Spanish deposits near Malaga. Micaceous hematite, unique for its lamellar structure, is mined in Austria and England.

Physical-Chemical Properties:

1) *Chemical Formula*—Fe_2O_3

2) *Chemical Composition of Ore*—50 to 95% Fe_2O_3, associated with other iron or silicate minerals

3) *Crystal Class*—hexagonal; scalenohedral

4) *Habit*—thick to thin tabular, as rosettes, rhombohedral, sometimes platy, also earthy

5) *Physical*—no cleavage, fracture subconchoidal to uneven. Hardness—5 to 6

6) *Other Properties:*

Melting point—1475° to 1565° C
Boiling point—decomposes
Solubility—insoluble in water and common organic solvents
Index of refraction—3.042
Specific gravity—5.26

Magnetic—usually antiferromagnetic, exists as ferromagnetic synthetically and naturally

7) *Reaction Mechanism*—will form by solid state reactions a wide variety of ferrites of the composition $M \cdot Fe_2O_4$ where M represents one or more metals such as manganese, zinc, magnesium, nickel, barium, yttrium, etc. These ferrites have important electrical and magnetic properties and to a lesser extent pigmentary properties.

Goethite

The second most common iron-bearing mineral is goethite. This includes other hydrated mineral iron oxides such as the limonites, ochres, and siennas.

Occurrence: Goethite is typically formed under oxidizing conditions as a weathering product of ferriferrous minerals, siderite, pyrite, and magnetite, and is found widespread, as deposits in bogs. It is commonly associated with hematite, pyrolusite, manganite, calcite, quartz, lepidocrocite, clay minerals, and limonite. Hydrous iron oxide is also called limonite. By X-ray examination most limonites are actually cryptocrystalline goethite with absorbed or capillary water (Berman, 1946, p. 685). Several excellent pigment ore sources are found in India. Ochre is mined on Cape Province, South Africa, and traditionally in France. Domestic ochre is mined in Georgia, Virginia, and Colorado.

Physical-Chemical Properties:

1) *Chemical Formula*—$Fe_2O_3 \cdot nH_2O$

2) *Chemical Composition of Ore*—50 to 88% Fe_2O_3, 9 to 10% H_2O, balance silicate or clay minerals

3) *Crystal Class*—orthorhombic, dipyramidal

4) *Habit*—prismatic and striated, also in tablets or scales, grading into acicular and long prismatic forms; also earthy

5) *Physical*—cleavage [010] perfect, [100] less so, fracture uneven, hardness 5 to 5½

6) *Other Properties:*

Melting point—decomposes: naturals (finely ground) start to lose chemically combined H_2O at 200° C and essentially dehydrated at 315 to 320° C, synthetics start at 200° C and are dehydrated at 300° C (private communication).

Boiling point—decomposes

Solubility—considered insoluble in water and common organic solvents, 0.0015 gpl cold water

Index of refraction—2.26 to 2.27

Specific gravity—3.3 to 4.0

Magnetic—antiferromagnetic

7) *Reaction Mechanisms*—will form by solid state reactions ferrites as noted under hematite; the removal of the combined water precedes the reaction.

Can also be converted from $Fe_2O_3 \cdot nH_2O$ to αFe_2O_3 by heat and to ferromagnetic γFe_2O_3 by reduction and oxidation.

Lepidocrocite

This natural mineral has essentially the same chemical composition as goethite but has different crystallographic constants.

Occurrence: Formed under similar circumstances as goethite and often associated with it.

Physical-Chemical Properties:

1) *Chemical Formula*—$Fe_2O_3 \cdot nH_2O$ or FeO (OH)

2) *Chemical Composition*—85 to 89% Fe_2O_3, 10 to 11% H_2O; minor amounts of silicates

3) *Crystal Class*—orthorhombic; dipyramidal

4) *Habit*—as flattened elongated scales sometimes striated; usually as isolated crystals or plumose groups on a matrix

5) *Physical*—cleavage [010] perfect, [100] less perfect, [001] good; hardness—5.0

6) *Other Properties:*

Melting point—decomposes, similar to goethite

Boiling point—decomposes

Solubility—considered insoluble in water and common organic solvents

Index of refraction—1.94

Specific gravity—4.05 to 4.10

Magnetic—antiferromagnetic

7) *Reaction Mechanisms* — lepidocrocite converts to ferromagnetic γFe_2O_3 on heating at low temperature and to αFe_2O_3 at high temperatures; will enter into other reactions similar to goethite.

Magnetite

The mineral magnetite occurs abundantly and is found worldwide in many locations. The magnetic nature of this mineral gives it considerable historic interest going back into antiquity.

Occurrence: Magnetite deposits are found as a magmatic segregation deposit, as an accessory mineral in igneous rocks, in metamorphic deposits in limestones and in chlorite schists with pyrite, in sulfide vein deposits, and in meteorites. Commercial deposits are found in US, Norway, Sweden, USSR, South Africa, and India.

Physical-Chemical Properties:

1) *Chemical Formula*—Fe_3O_4

2) *Chemical Composition*—95.0 to 99.0% Fe_3O_4 associated with minor amounts of silicate minerals; purity can be high for a natural mineral

3) *Crystal Class*—isometric (cubic); hexoctahedral

4) *Habit*—usually octahedral, well-formed nearly perfect crystals are common; occurs in coarse to fine grained masses.

5) *Physical*—octahedral parting on [111] especially good; conchoidal to uneven fracture. [001], [011], [138] also reported as parting planes in magnetite; hardness 5.5 to 6.5

6) *Other Properties:*

Melting point—decomposes
Boiling point—decomposes
Solubility—insoluble in water and common organic solvents
Magnetic—ferromagnetic
Index of refraction—2.42
Specific gravity—5.16 to 5.17

7) *Reaction Mechanisms*—synthetic magnetites start to decompose to αFe_2O_3 at 105° and are completely oxidized at 385° C, respectively; natural magnetite, if finely ground, will start to decompose at 124° C and be completely oxidized at 550° C (private communication); will form spinel compounds of the general formula $M \cdot Fe_2O_4$ where M represents metals such as manganese, zinc, barium.

Pyrites and Siderites

These minerals do not fall into the class of iron oxides since they require transformation by heat to convert to the oxide. As such, they are potential raw materials rather than end products.

Natural Mineral Statistics

Based on United States production figures, the utilization of *natural* mineral sources for the production of pigments has increased over the last ten years. A major change occurred in the mid-1960s when the volume of naturals increased dramatically. Production was lower in 1970 due to adverse domestic economic conditions, but production has continued to increase through 1979 (Tables 1 and 2).

Manufacturing

The manufacturing procedures for the naturals must be designed around the starting ore base. For example, siderite ($FeCO_3$) requires calcination to convert to the oxide whereas pyrites cinder (calcined pyrite), although essentially Fe_2O_3, must be freed of its acidic residues. Also, certain ores, although suitable for use by simple grinding, develop more desirable colors if calcined. Other natural ores, siennas, and umbers are produced in raw and calcined form. Figs. 2 and 3 show typical flowsheets for both ore types.

Some of the hematites and the magnetites

TABLE 1—Salient Statistics of Iron Oxide Pigment Materials in the US 1961–65 (Average) and 1973 to 1980

	1961–65 Average	1973	1974	1975	1976	1977	1978	1979	1980
Mine Production*	55	W	W	43	67	59	85	88	49
Crude pigments sold or used*	56	W	W	40	60	56	76	75	63
Value $†	464	931	1,429	1,093	1,626	2,143	2,799	2,578	4,043
Iron oxides from steel plant wastes*	NA	NA	W	19	21	21	21	25	21
Value $†	NA	NA	W	1,102	1,258	1,644	1,396	1,703	1,394
Finished pigments sold*	117	149	147	105	136	141	153	156	136
Value $†	21,163	43,514	60,612	46,206	64,506	73,851	81,830	94,175	97,270
Exports*	4	10	10	9	6	6	7	5	5
Value $†	1,287	3,101	3,466	2,523	3,353	4,065	6,649	7,359	9,132
Imports for consumption*	14	51	54	28	50	59	71	55	39
Value $†	1,561	12,005	16,367	9,184	16,554	20,596	24,706	24,341	20,035

Source: Stipp, 1961–65; Collins, 1973–79; Spinrad, 1980.
*Thousands of short tons. Metric equivalent: 1 st × 0.907 184 7 = t.
†Thousands of dollars.
W—Withheld to avoid disclosing company confidential data.

TABLE 2—Finished Iron Oxide Pigments Sold by Processors in the US By Type (Value Rounded Out to Nearest Thousand $)

Pigment	1970		1978		1979		1980	
	St	Value	St	Value	St	Value	St	Value
Natural								
Black: magnetite			10,707	1,219	8,075	906	5,402	635
Brown:								
Iron oxide*	11,823[1]	2,058[1]	8,151	2,260	10,075	3,481	8,123	2,026
Umbers:								
Burnt	4,258	1,014	5,546	2,868	4,495	2,665	3,954	2,583
Raw	1,087	262	2,040	936	1,782	970	1,383	873
Red:								
Iron oxide†	26,847	1,813	40,476	3,596	40,618	3,953	33,136	3,379
Sienna, burnt	963	352	665	416	647	464	544	401
Yellow								
Ochre§	4,718	312	7,588	913	6,865	945	5,214	850
Sienna, raw	619	201	745	380	683	399	630	395
Total natural	50,315	6,012	75,918	12,588	73,240	13,782	58,386	11,143
Synthetic								
Black: Magnetic	6,262	1,379	[2]	[2]	[2]	[2]	—	—
Brown: Iron oxide	4,656	1,731	11,351	10,364	11,404	11,319	10,328	10,820
Red:								
Pure red iron oxides	—	—	26,433	25,267	33,344	32,540	31,998	34,791
Calcined copperas	14,199	4,602	—	—	—	—	—	—
Other processes¶	14,461	3,786	8,969	7,951	10,291	12,053	11,044	17,367
Venetian red	592	124	—	—	—	—	—	—
Yellow: Iron oxide	25,075	8,004	25,916	22,725	24,550	22,651	21,703	21,424
Total manufactured	65,245	19,626	72,669	66,307	79,589	78,563	75,073	84,402
Mixtures of naturals and synthetics	8,428	2,559	3,923	2,935	3,205	1,830	2,877	1,726
Total	123,988	28,197	152,510	81,830	156,036	94,175	136,336	97,270

Source: Stipp, 1970; Collins, 1978, 1979; Spinrad, 1980. Metric equivalent: 1 st × 0.907 184 7 = t.
* Includes Vandyke brown.
† Includes pyrite cinder.
§ Includes yellow iron oxide.
¶ Includes other manufactured red iron oxide.
[1] Includes magnetite.
[2] Included with brown 1978–79.

are magnetically beneficiated to increase the chemical purity and remove undesirable siliceous minerals. This is particularly true if the end use has chemical purity requirements such as ferrites (Anon., 1965). Magnetic beneficiation is most efficient if the mesh size is controlled within fairly narrow limits. Therefore, wet magnetic beneficiation usually follows the crushing and screening operation and prior to fine grinding or calcination. The critical variables are: magnetic field strength, pole and plate designs, solids concentration and feed rate, demagnetizing field strength, and wash water feed rates. The process produces a superconcentrate with purities in excess of 99.5% Fe_2O_3 or Fe_3O_4 from lower grade concentrates.

The use of fluid energy mills, micronizers, and improved methods of classification has enabled the manufacturer to produce natural iron oxide pigments that compete with synthetics in many applications where chemical purity and the ultimate discrete particle size is not a factor.

Products

The broad selection of manufacturing procedures coupled with the wide variety of natural ore results in over 200 different products being marketed in the United States alone. In Table 3 some of the major ore sources are tabulated along with a *single* typical product prepared from the ore. Selling price and end use considerations result in a range of products being

TABLE 3–Iron Oxide Pigments From Natural Ores–Typical Physical and Chemical Properties

Mineral Class:	Hematite	Hematite	Hematite	Hematite	Hematite (Specular-) (Micaceous)	Hematite	Goethite (Ochre)	Goethite (Ochre)	Goethite (Sienna)	Goethite (Sienna)	Umber	Umber	Magnetite
Ore Source:	US	Venezuela (Calcined)	Persian Gulf	Spanish Oxide	Austria	Pyrites Cinder	Georgia	France	Italy	Italy (Calcined)	Cyprus	Cyprus (Calcined)	US
Chemical Properties													
% Fe_2O_3	63.0	96.5	73.0	85.0	90.0	82.0	50.0	20.0	62.0	68	45.0	55.0	98.5*
% SiO_2	12.6	1.50	21.0	6.5	3.5	Balance—Mainly Al Silicates	14.0	50.0	Balance—Mainly Al Silicates	Balance—Mainly Al Silicates	Balance—Mainly Al Silicates	Balance—Mainly Al Silicates	Balance—Mainly Al Silicates
% Al_2O_3	4.8	1.50	3.0	1.0	2.4		6.00	18.0					
% MgO	3.2	0.02	0.33	1.0	1.0		0.20	0.25					
% CaO	7.5	0.03	0.80	3.0	0.5		0.20	0.25					
% MnO_2	–	0.02	0.06	–	–		0.62	–			11.0	10.0	
% Loss on ignition													
@ 1000° C	8.2	0.50	2.50	3.0	0.15	–	9.00	9.00	15-20	–	10-15	–	–
% H_2O soluble salts	0.8	0.20	<3.0	1-2	0.08	3-3.5	0.50	–	0.45	–	0.8	0.4	0.10
Other Properties													
Color	Brownish red	Reddish brown	Bright red	Bright red	Brownish red	Red	Brownish yellow	Dull yellow	Brownish yellow	Reddish brown	Greenish brown	Brown	Black
Oil absorption†	15	14	13	17	11	21	24	32	36	42	44	54	15
Particle size‡													
Range (10-90%)	0.3-7	0.3-8	0.3-7	0.25-15	8-65¶	0.2-6	0.2-15	0.2-40	0.1-8	0..05-6	0.6-7	0.13-3	1-10§
median	0.8	3.0	0.6	1.5	25¶	2.0	1.0	4.0	0.6	0.7	3.0	0.6	–
% retention +325	<0.20	<0.01	<0.20	<0.50	30	<0.01	<2.0	<3.0	<1.0	<0.20	<0.20	<0.20	<0.10
Surface area#	8.0	6.1	4.7	5.8	0.08	8.73	15.4	–	33.4	36.0	101.7	86.9	4.3
Specific gravity	3.78	4.90	4.35	4.67	4.8	4.48	3.47	2.88	3.58	4.03	3.24	3.74	5.03

Source: Technical bulletins and reports of domestic iron oxide pigment producers.

* Contains 30% FeO.

† Oil absorption, pounds of oil per 100 lb of pigment.

‡ In microns as determined by Andreasen pipette or Micromeritics sedigraph, except as noted. Range is the percentage by weight—between 10% to 90%. 50% by weight point is median size.

¶ Values by microscopy due to lamellar structure.

§ Magnetic—estimated from electron micrographs.

Square meters per gram—by BET.

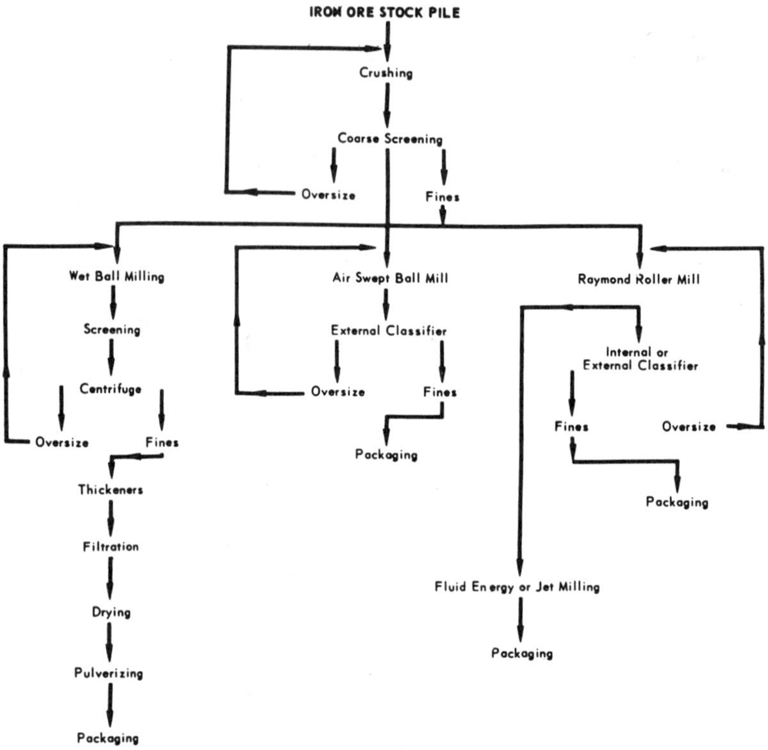

FIG. 2—*Simplified flowsheet showing alternate processing routes for natural iron oxide pigment ores requiring grinding.*

produced from a single ore body. For example, a natural iron oxide for a paint primer must be finely ground but has wide latitude in its color specifications. Concrete coloring pigments, however, have stringent color requirements but moderate fineness requirements.

All natural products, without exception, are subjected to some type of final grinding. Comminution, as opposed to particle synthesis, produces characteristic fragmented particles covering a wide size distribution. In Fig. 4 a synthetic iron oxide is compared with a natural ground iron ore showing a major difference in size and fragmented nature of the natural ore.

Fig. 5 shows the difference in particle size distribution between a synthetic and a natural ground product.

Spectral reflectance curves are shown in Figs. 6 and 7 which compare a synthetic yellow and a synthetic red with similar natural ground products. The spectral reflectance curves show the naturals to have weaker tinting strengths and lower color purity.

Table 4 lists the price range of natural iron oxide pigments marketed in the United States and the approximate varieties available within the various categories.

Applications

Many of the applications for natural iron oxides derive from their low cost and permanency (Pfizer, Technical Bulletin). In considering applications and end uses there is considerable overlap between the naturals and synthetics; both, for example, are used extensively in paints. Wherever possible, a distinction will be made to highlight the advantages and limitations of natural pigments.

Paints and Coatings: The variety and types of industrial coatings are too numerous to mention individually. There are several general types of paints wherein natural iron oxides provide excellent properties and represent major uses. *Primers, surfacers,* and *primer-surfacers* vary in the degree of pigment loading and function as a base for various top coats. The primer pigments are usually hematites and can be highly loaded due to high density, fairly large particle size, and low oil absorption.

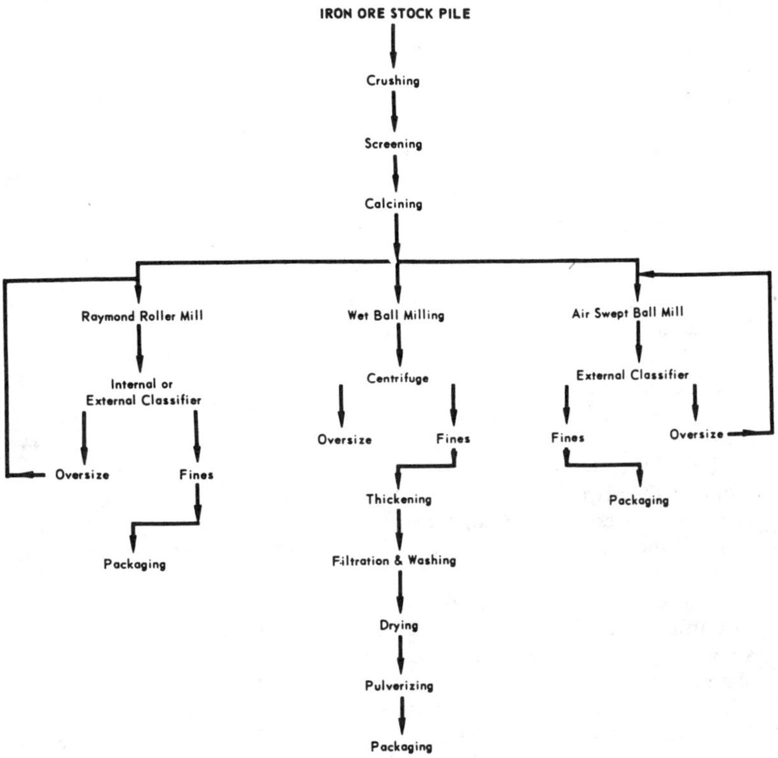

FIG. 3—*Simplified flowsheet showing alternate processing routes for natural iron oxide pigment ores requiring calcining.*

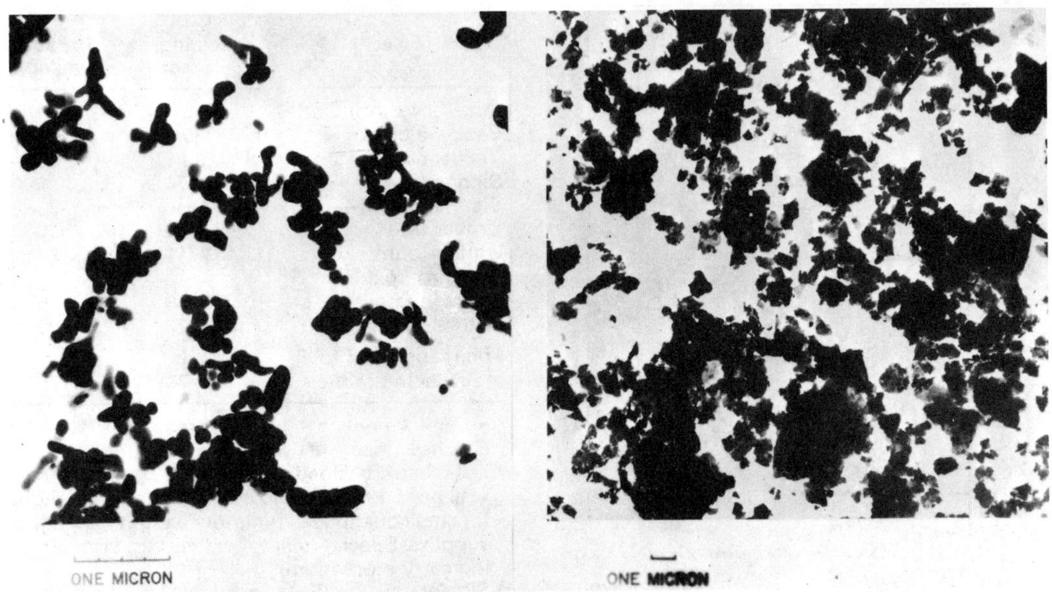

FIG. 4—*Comparison of particle size between A (right) synthetic iron oxide and B (left) ground natural iron ore.*

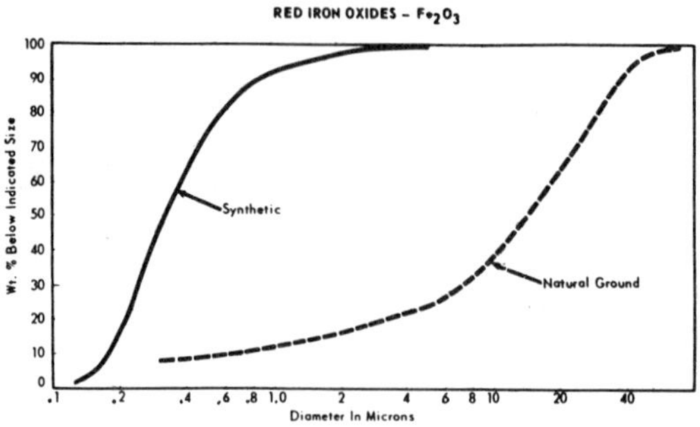

FIG. 5—*Particle size distribution comparing a fine particle synthetic with a natural iron oxide.*

Industrial maintenance finishes for railway equipment and metal structures largely employ hematites. Micaceous iron oxide (Anon., 1972c), due to its platelike structure, has found extensive use in Europe in railway maintenance coatings, bridge paints, industrial maintenance finishes, and other exterior applications.

Micaceous iron oxide is not a primer or a rust inhibiting pigment but is used as part of a corrosion control system. To capitalize on its platelike structure, the paint formulator must grind the paint properly to secure a lamellar film structure (Anon., 1971).

Natural umbers and siennas have unique color characteristics which are not easily matched with synthetics and hence are the

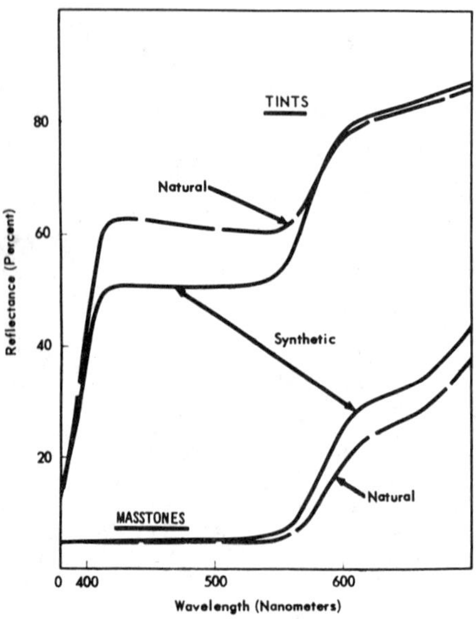

FIG. 6—*Spectral reflectance curves of masstone and 1:50 TiO_2 tint of a synthetic copperas red vs. a natural red iron oxide.*

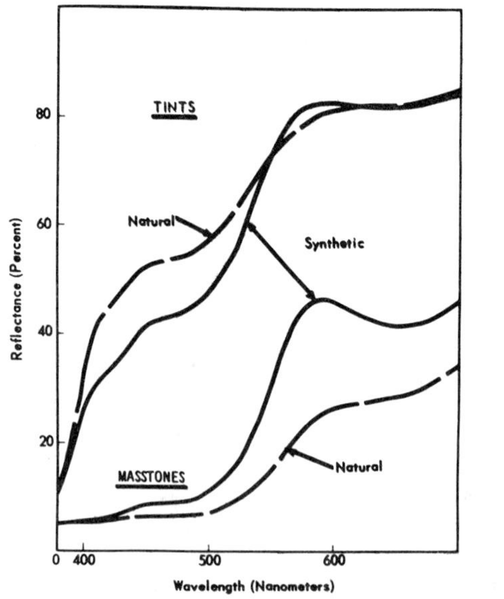

FIG. 7—*Spectral reflectance curves of masstone and 1:10 TiO_2 tint of a synthetic yellow oxide vs. a natural domestic ochre.*

major pigments in stains. With the increased use of prestained wall panels, this is an important use.

The naturals are also used in general interior

TABLE 4—1980 Price Range for Natural Iron Oxide Pigments Marketed in the United States (Value Rounded to Nearest Cent Per Lb)

	F.O.B.* Selling Price	Products Available†
Micaceous iron oxide	55	1
Vandyke brown	36	4
Yellow ochre	23	6
Sienna (raw)	33–70	8
Sienna (burnt)	37–56	20
Umber (raw)	31–39	16
Umber (burnt)	31–41	34
Black iron oxide (magnetite)	25–30	6
Natural hematites	6–32	30
Price range and total available products	6–70¢/lb	125

Source: *American Paint Journal* and suppliers published price.

* Less than truckload price in 22.6-kg (50-lb) multiwall bags. Price about 2¢ to 10¢ per pound higher for small quantities, minimum weight depends on supplier. Special milling, palletizing, and packaging carry a premium.

† Similar or duplicate products are carried by suppliers, increasing the number of apparent varieties.

and exterior house paints, but to a lesser extent than synthetics.

Electronic: Natural iron oxides are used to some extent in the preparation of barium ferrite ($BaFe_{12}O_{19}$) and to a lesser extent in the preparation of manganese zinc ferrite ($Mn_xZn_y \cdot Fe_2O_4$). Manganese zinc ferrites are very sensitive to impurities and hence the use of naturals has been limited. Highly beneficiated magnetite or hematite can be used to prepare barium ferrites. Beneficiated magnetite when finely ground is used in magnetic inks (Love, 1961). The magnetic energy (remanence) of an ink is proportional to the volume of iron oxide in the ink. Due to the low binder demand of the dense, natural magnetites, high loadings of 60 to 70% by weight of magnetite pigment can be obtained.

Rubber, Plastics, and Related Compounds: Natural pigments are used in rubber and many of the more common plastics such as vinyls, phenolics, polyurethanes, and epoxies (Woernle, 1967). Umbers contain a high percentage of manganese which can cause undesirable effects such as accelerating the aging of rubber and retarding the curing of polyesters. The general trend has been toward the synthetics based on color value per unit of price.

Concrete Products and Building Materials: Natural oxides are suitable in all respects for use in coloring concrete, mortar, asbestos shingles, and similar building materials (Love, 1968). In the coloring of concrete and shingles, the tint is the important characteristic and synthetic pigments have replaced many of the naturals because of their higher tinting strength.

Rouge: Prior to the introduction of the float glass process, large tonnages of natural iron oxides were used in polishing plate (flat) glass. This use has declined dramatically in recent years due to the float glass process. Opthalmic rouge and metal polishing rouge represent a much smaller but more constant market. Specially processed grit-free hematites having high glass removal rates are required. A significant market outlet is in coloring glass and is referred to as "batch rouge."

Miscellaneous: Natural iron oxides are used in numerous other industrial applications: foundry core washes, ceramics, paper, fertilizer, and as the iron source in glassmaking.

Synthetic Iron Oxides

There are four major types of synthetic iron oxides: reds, yellows, browns, and blacks. All can be prepared from basic chemicals, usually iron salts and alkalis. The advantage of chemical synthesis is that it provides precise control over two important pigment parameters: particle size (shape) and chemical purity.

Raw Materials

Iron Salts: There are only two iron salts of commercial significance, ferrous sulfate and ferrous chloride, that are used in the manufacture of synthetic iron oxides. Ferrous sulfate ($FeSO_4 \cdot 7H_2O$), or copperas, as it is commonly called, was the historical base for red oxides produced by calcination. Subsequently, ferrous sulfate was used as the iron salt for the production of yellows, reds, browns, and black by precipitation. In recent years, the increased use of hydrochloric acid for the pickling of steel resulted in an excess of ferrous chloride. The chloride pickle liquor produced, however, far exceeds the utilization capabilities of synthetic iron oxide producers.

Ferrous sulfate for pigment production is available from three sources. The first consists of neutralizing scrap iron with sulfuric acid. A second is byproduct sulfate pickle liquor from the steel mills. This is a mixture of about 15% $FeSO_4$ and 10% H_2SO_4, the balance water. The excess acid is converted to $FeSO_4$ by neutralization with iron. The third major source is the titanium pigment industry which produces byproduct $FeSO_4 \cdot 7H_2O$. If utilized in crystal form, it must be redissolved and purified. Ferrous sulfate liquor containing free sulfuric acid is also available from the titanium industry. In this case the processing is the same as steel mill pickle liquor.

Ferrous chloride has rarely been utilized for the production of color pigments by domestic pigment manufacturers. Its use limitations are more fully discussed under *Byproducts*.

Iron: The iron source is generally stampings and punchings of light gage soft black iron. Pigment producers avoid scrap that is plated (galvanized) or contains alloying elements due to the adverse affect of these impurities.

Alkalis: The major alkali used in the United States is sodium hydroxide (NaOH) followed by ammonium hydroxide (NH_4OH). The least important is lime (CaO). Both sodium hydroxide and liquid NH_3 can be purchased with a high degree of chemical purity, which is not true of commercial limes.

Byproducts: Most of the byproducts require further processing either as raw materials or end products. The spray roasting of ferrous chloride pickle liquor produces Fe_2O_3 and HCl. Both can be returned to the steelmaking processes, but some steelmakers divert a portion of the Fe_2O_3 to pigmentary applications where color is not important. The spray roasted oxides are poor in color quality. Even poorer in quality are the dust collector recovered oxides produced from the basic oxygen steel lancing process. The thousands of tons of these two waste products have depressed pigment prices in a few areas where quality is not a major consideration.

Mill scale, although abundant in supply, has found only very minor usage in the iron oxide industry. Depending on the steel mill, the type steel, the stage of the steel when it is rolled, and the recovery system, all can cause a wide variation in chemical composition. One of the major contaminants of this product is refractory material due to spalling of furnaces and linings. Also, mill scale is not a completely oxidized product and cannot be used for pigmentary purposes without further processing.

Organic reduction processes which utilize pulverized iron as the reductant produce byproduct black iron oxide sludges. These sludges must be separated from the organic reaction products and converted to Fe_2O_3 to have any commercial value.

Other waste products that have attracted attention over the years but have not resulted in commercial pigments are: red mud from bauxite processing, sponge iron from the cement copper mining operation, and sulfur mud or "yellow boy," a precipitate formed from acid coal mine waters.

Manufacturing Processes

There are only three major methods for the manufacture of synthetic iron oxides: thermal decomposition of iron salts or iron compounds, precipitation of iron salts usually accompanied by oxidation, and organic reduction processes utilizing iron.

Thermal Decomposition of Iron Salts: There are numerous combinations of processing techniques depending on the starting basis. Only one flowsheet (Fig. 8) will be outlined, starting with scrap iron and sulfuric acid (Ayers, 1946). If the starting base is steel mill pickle liquor or titanium copperas, the process is essentially the same with the exception that steps may be introduced to remove undesirable impurities.

Fig. 8 illustrates the production of iron oxides by the calcination of ferrous sulfate. The major chemical reactions that are involved in this process are (Hoffman, 1972):

Chemical Reactions Involved In The Thermal Decompositions of Ferrous Sulfate:

Neutralization $$H_2SO_4 + Fe + H_2O \longrightarrow FeSO_4 + H_2\uparrow + H_2O$$

Evaporation Crystallization $$FeSO_4 + 7H_2O \longrightarrow FeSO_4 \cdot 7H_2O$$

Dehydration $$FeSO_4 \cdot 7H_2O \xrightarrow{\Delta} FeSO_4 \cdot H_2O + 6H_2O$$

Decompositions Above 1200°F

Without air
$$6FeSO_4 \cdot H_2O \xrightarrow{\Delta} 2Fe_2O_3 + Fe_2(SO_4)_3 + 6H_2O + 3SO_2$$
$$Fe_2(SO_4)_3 \xrightarrow{\Delta} Fe_2O_3 + 3SO_3$$
$$3SO_3 \xrightarrow{\Delta} 3SO_2 + 1\tfrac{1}{2}\, O_2$$

With air
$$6FeSO_4 \cdot H_2O + 1\tfrac{1}{2}O_2 \xrightarrow{\Delta} Fe_2O_3 + 2Fe_2(SO_4)_3 + 6H_2O$$
$$2Fe_2(SO_4)_3 \overset{\Delta}{\rightleftharpoons} 2Fe_2O_3 + 6SO_3$$
$$6SO_3 \overset{\Delta}{\rightleftharpoons} 6SO_2 + 3O_2$$

The last two chemical reactions are reversible at temperatures below 650°C (1200°F) down to about 316°C (600°F).

One major advantage of this process is that the products of combustion and waste products are reclaimed or reused.

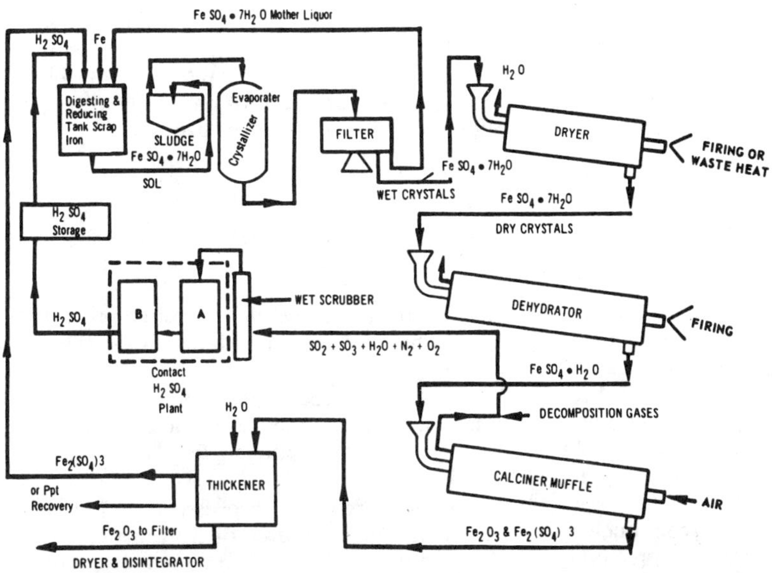

FIG. 8—*Simplified flow diagram for synthetic iron oxide (copperas red type).*

One of the earliest synthetic iron oxides and a variation of the calcination of ferrous sulfate are the "Venetian reds." These oxides are produced by calcining an intimate mixture of ferrous sulfate and lime or preferably hydrated lime ($Ca(OH)_2$). The chemical reactions may be expressed as follows (Love and Ayers, 1942):

$$4FeSO_4 \cdot 7H_2O + 4CaO + O_2 \xrightarrow{\Delta} 2Fe_2O_3 + 4CaSO_4 + 28H_2O \quad (1)$$

or

$$2FeSO_4 \cdot 7H_2O + CaO \xrightarrow{\Delta} Fe_2O_3 + CaSO_4 + SO_2 + 14H_2O \quad (2)$$

The difference between Reactions 1 and 2 is the amount of oxygen present in the calcining operation. Generally speaking, both reactions are involved and an end product of about 40% Fe_2O_3 is obtained. The calcination product must be ground to be suitable for pigmentary applications. The process has one advantage in that the kiln product does not require washing. "Venetian reds" have been supplanted by other synthetic reds.

The amount of hydrochloric acid that is used in the pickling of steel was 136 kt (150,000 st) in 1965 and is estimated to be doubling about every four years in the United States alone (Anon., 1966). Between 45.4 and 54.4 Mt (50 and 60 million st) of steel are pickled yearly (Buckley, 1967). As the steel mills convert to hydrochloric acid in order to recover the oxide and reuse the hydrochloric acid, the amount of available byproduct iron oxide will exceed the total present production of iron oxide pigments by five or six times. The byproduct oxide therefore is usually pelletized and recycled into the steel process. Depending on the equipment and the conditions, the oxide has little color value and contains undesirable chlorides.

Precipitation Reactions: Precipitation processes were developed in the early 1900s (Penniman and Zoph, 1921) but did not achieve commercial importance until the 1930 to 1940s. Since that time the major growth in United States' synthetic iron oxide manufacture has been by precipitation technology. The precipitation process will produce yellows, reds, browns, and blacks and additional products can be produced by subsequent procedures, such as calcination, reduction, etc.

In the manufacture of yellow oxides by the Penniman-Zoph process, the initial reaction is one of nucleation, wherein a seed nucleus is formed by the reaction of an iron salt and an alkali. The nucleation reaction is normally conducted in an excess of $FeSO_4$, and can be described by the following equations:

$$4NaOH + 2FeSO_4 + H_2O \longrightarrow 2Fe(OH)_2 + 2Na_2SO_4 + H_2O$$

$$2Fe(OH)_2 + \tfrac{1}{2} O_2 \longrightarrow \underset{\text{nucleus}}{\underline{Fe_2O_3 \cdot H_2O}} + H_2O$$

The nucleus is transferred to a precipitator containing $FeSO_4$, H_2O, and iron, usually in the form of thin-gage stampings. The seed particles are continuously circulated over the bed of iron in an oxidizing environment at elevated temperatures [60° to 87.8°C (140° to 190°F)] and, as the iron oxidizes or dissolves, the reaction products precipitate on the nuclei causing the particles to grow in size. The chemical reactions involved are:

$$2Fe^{++} + \tfrac{1}{2}\,O_2 + 3H_2O \rightleftharpoons Fe_2O_3 \cdot H_2O + 4H^+$$

also

$$2FeSO_4 + 3H_2O + \tfrac{1}{2}\,O_2 \longrightarrow Fe_2O_3 \cdot H_2O + 2H_2SO_4$$

$$H_2SO_4 + Fe \longrightarrow FeSO_4 + H_2$$

From the simplified chemical reactions, the precipitation process would not appear to be complex but the actual gas-liquid-solid chemical interactions are not completely known. A host of variables affect the reaction and the products. Some of the more important variables are: (1) rate of circulation, (2) quantity of oxygen, (3) temperature, (4) quantity of initial seed nucleus, (5) size and shape of seed nucleus (Hancock, 1970). In Fig. 9, a typical precipitation flowsheet is shown which will produce a wide color range of yellow iron oxides.

A class of pigments that has attracted considerable attention in recent years is the transparent or low opacity pigments. These pigments are produced by precipitation techniques (Anon., 1971a) and are characterized by very fine particle size, usually less than 0.10μ.

Precipitated reds can be produced by processes similar to and utilizing equipment that is essentially the same as that employed for yellow oxide manufacturing. The difference lies in the preparation of the nucleus which contains ultrafine hematite (Fe_2O_3) nuclei (Ayers, 1960; Bennetch, 1957, 1961; Marsh, 1955; Toxby, 1952).

Subsequent processing conditions are also much more critical since normal precipitation reactions tend to produce $Fe_2O_3 \cdot H_2O$ as opposed to Fe_2O_3.

The two preceding processes employ scrap iron as the major source of iron in the growth of the nuclei. Both yellow and red can be produced by nuclei growth by simultaneously precipitating alkali and iron salts in the presence of the nuclei (Martin, 1960). If conducted properly, the nuclei will grow to pigmentary size in the same manner as the Penniman-Zoph process.

Another class of pigments produced by precipitation are the browns and blacks. The blacks are Fe_3O_4 and browns are $x\,FeO \cdot y\,Fe_2O_3$. Commercially, they are prepared by reacting ferrous sulfate with caustic at temperatures close to boiling and at an alkaline pH (Ayers, 1938).

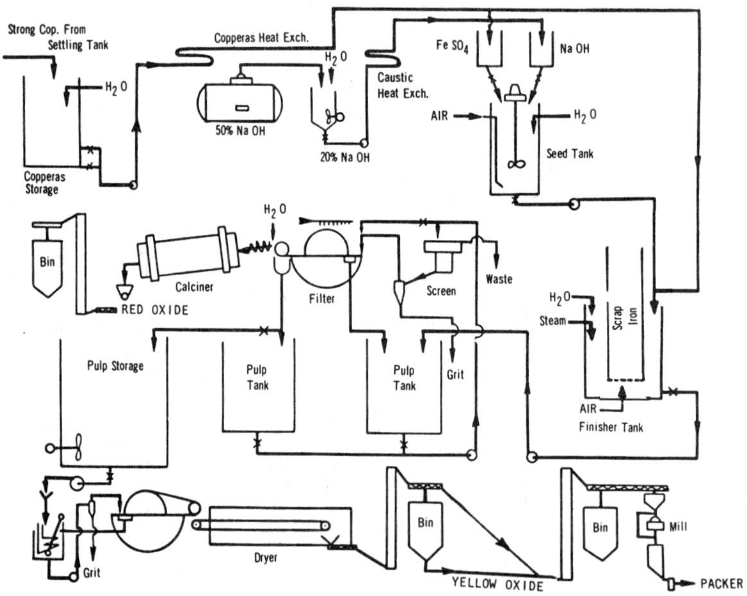

FIG. 9—*Simplified flow diagram of a synthetic yellow iron oxide plant.*

The precipitation processes have a further advantage in that their end products are suitable for additional processing to produce different pigments. The yellows, blacks, and browns all may be calcined to produce red and brown pigments. If the temperature is below 649° C (1200° F), the shape of the precursor is usually retained. Synthetic browns are also made by producing intimate blends of synthetic reds, yellows, and blacks.

The most sophisticated product to be manufactured by a combination of precipitation and controlled calcination techniques is magnetic γFe_2O_3 from nonmagnetic $\alpha Fe_2O_3 \cdot H_2O$ (Camras, 1954). In this process particle size control is exercised in the selection of the yellow oxide, which is acicular. The yellow is processed in a rotary kiln at temperatures between 315° C (600° F) and 538° C (1000° F). The reactions are:

$$\underset{\text{(yellow)}}{\alpha Fe_2O_3 \cdot H_2O} \xrightarrow{\Delta} \underset{\text{(red-nonmagnetic)}}{\alpha Fe_2O_3 + H_2O}$$

$$\underset{\text{(red)}}{3Fe_2O_3 + H_2} \xrightarrow{\Delta} \underset{\text{(black-magnetic)}}{2Fe_3O_4 + H_2O}$$

$$\underset{\text{(black)}}{2Fe_3O_4 + \tfrac{1}{2}\, O_2} \xrightarrow{\Delta} \underset{\text{(brown-magnetic)}}{3Fe_2O_3}$$

Reduction of Organic Compounds: The reduction of organic compounds, using iron as the reducing agent, resulting in iron oxides has been practiced in Germany for over 50 years. Most organic chemical manufacturers, however, use different reducing agents or, if they use iron, are more concerned with the organic compound than with the iron oxides produced. Consequently, most iron oxide sludges from the reduction of organic compounds are an impure waste product.

If nitrobenzene is reduced to aniline in the presence of various chemicals, it will produce specific colors of iron oxides (Love, 1946). $AlCl_3$ may be used to produce yellow oxides but is omitted when blacks are desired. Reds can be produced by calcining either yellow or blacks. In each the choice of iron source is carefully controlled and usually consists of iron filings or iron powder.

Miscellaneous Manufacturing Processes: One major domestic pigment producer manufactures tan pigments that are actually ferrites (Downs and Martin, 1950, 1959). These are formed by interaction of iron oxides with other metallic oxides to form ferrites. The reactions are as follows:

$$ZnO + Fe_2O_3 \xrightarrow{\Delta} Zn \cdot Fe_2O_4$$

$$MgO + Fe_2O_3 \xrightarrow{\Delta} Mg \cdot Fe_2O_4$$

A final example of a product which is between a synthetic and a natural, with respect to its variable nature and very high impurity level, is the byproduct oxide from the basic oxygen furnace (BOF) steel production. The byproduct oxide is collected from the gasses evolved during the lancing process. The 1970 world production of BOF steel was estimated at 272 Mt (300 million st) (McGannon, 1968; Taylor, 1966). Each ton of steel requires between 0.1 and 0.05 tons of oxygen. The oxygen in addition to oxidizing the carbon and entering into the slag formation reactions produces about 4.5–5.4 Mt/a (5–6 million stpy) of oxide fumes or dust. This is roughly 10 to 15 times the world production of pigment grade oxides.

Products and Properties

The *synthetic* manufacturing approach opens up an almost infinite number of distinct product possibilities. An iron oxide producer, if he wishes to cover all major *synthetic* marketing applications, should have over 100 oxides available for sale.

In Table 5 a cross tabulation is made listing the synthetic reds, yellows, browns, blacks, intermediate colors, and some specialty items such as the transparent iron oxides, magnetic recording oxides, and a few that are not color oxides but still are classed as synthetics (byproduct steel mill oxide).

In the major classifications, such as the synthetic yellows and reds, the finest and largest discrete particles are tabulated. Within any one classification, for example, synthetic yellow oxides, there can be as many as 10 to 12 intermediate colors.

Fig. 10 shows the particle size range of the copperas red family. The typical shape of a calcined copperas red (Fig. 11) may be compared to a precipitated red oxide (Fig. 12). Fig. 13 shows the wide size distribution of a Venetian red. The very large particles are the $CaSO_4$.

Fig. 14, a yellow iron oxide is shown before and after being subjected to high temperatures of about 1038° C (1900° F).

Fig. 15 shows a synthetic, precipitated black iron oxide. In Fig. 16, a magnetic recording oxide retains the acicular shape of its yellow precursor.

In Table 6, the price range and number of products are listed.

TABLE 5—Synthetic Iron Oxide Pigments—Typical Physical and Chemical Properties

Chemical Formula:	Fe_2O_3	Fe_2O_3	Fe_2O_3	Fe_2O_3	Fe_2O_3	Fe_2O_3	Fe_2O_3	Fe_2O_3	Fe_2O_3 $CaSO_4$	Fe_2O_3
Mineral Class:	Hematite	Hematite	Hematite	Hematite	Hematite	Hematite	Hematite	Hematite	Mixed	Hematite
Method of Mfg.:	Decomp. $FeSO_4$	Decomp. $FeSO_4$	Ppt.	Ppt.	Calc. of $Fe_2O_3 \cdot H_2O$	Calc. of Fe_3O_4	Pickle Liq. Decomp. (HCl)	B.O.F. Flue Dust	Decomp. $FeSO_4$ & $Ca(OH)_2$	Organic Reduction (Calcined)
Chemical Properties										
% Fe_2O_3	99.5	99.5	97.0	98.0	97.5	99.3	98.4	82.2	40.0	96.5
% FeO	–	–	–	–	–	–	–	–	–	–
% SiO_2	0.03	0.03	0.05	0.05	0.10	0.01	0.07	9.10	–	{2.5}
% Al_2O_3	0.02	0.03	0.01	0.01	0.01	0.05	0.005	3.10	–	
% MgO	0.01	0.005	0.01	0.01	0.01	0.015	0.03	1.00	–	–
% CaO	0.02	0.03	0.10	0.10	0.05	0.006	0.05	0.40	25.0	–
% Mn	0.02	0.25	0.02	0.02	0.02	0.10	0.30	0.36	–	0.10
% SO_3	0.05	0.02	0.80	0.80	0.60	0.10	–	–	15.0	–
% Cl_2	–	–	–	–	–	–	–	–	–	–
% LOI @ 1000° C	0.20	0.05	2.50	1.50	0.70	0.60	–	2.50	–	0.50
% Soluble salts	0.10	0.10	0.10	0.10	1.00	0.12	0.10	0.22	–	0.30
Other Properties										
Color	Light red	Dark red	Light red	Dark red	Medium yellow red	Red	Dark red (poor color)	No value	Light red	Light red
Oil absorption*	24	13	24	22	48	–	24	–	20	26
Particle size†										
Range (10-90%)	0.1-1	2-6	0.1-.6	0.5-1.5	0.5-1	0.3-1.2	0.3-7	Varies widely	0.3-5	–
median, μ	0.25	3.0	0.35	0.90	0.75	0.50	1.5	Varies	1.2	0.3‡
Particle shape	Nodular	Nodular	Spheroidal	Spheroidal	Acicular	Cubical	Spheroidal	Spheroidal	Mixed system	Spheroidal
% Retention +325	<0.10	<0.10	<0.05	<0.10	<0.10	–	<0.10	20.0	<0.50	0.05
Surface area¶	10.0	2.2	9.4	3.7	9.1	5.4	6.3	–	8.8	–
Specific gravity	5.15	5.15	4.90	4.9	4.9	5.15	5.18	–	3.55	5.2

Source: Technical bulletins and reports of pigment producers.

Federal Regulations covering paints and coatings: ANSI (American Natural Standards Institute) and FDA 1973 and 1974. Source: "Coatings," National Paint and Coatings Assn., Washington, D.C., Vol. 24, No. 17, Sep 1972, p. 77.

* Oil absorption—lbs of oil per 100 lbs of pigment.

† As determined by Andreasen pipette or Micromeritics sedigraph, except as noted. Range is the weight percentage between 10% and 90%. 50% weight point is median size.

‡ By electron microscopy, acicular, width × length.

¶ Square meters per gram—by B.E.T.

§ Magnetic, by electron microscopy.

\# Total iron expressed as Fe_2O_3.

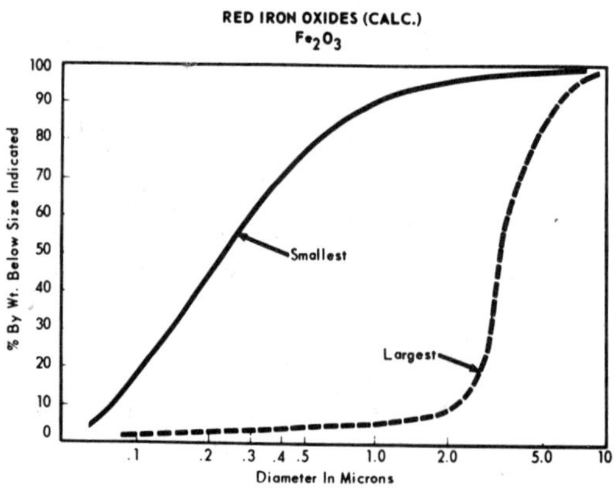

FIG. 10—*The range in particle size distribution for synthetic red iron oxides of the copperas red type.*

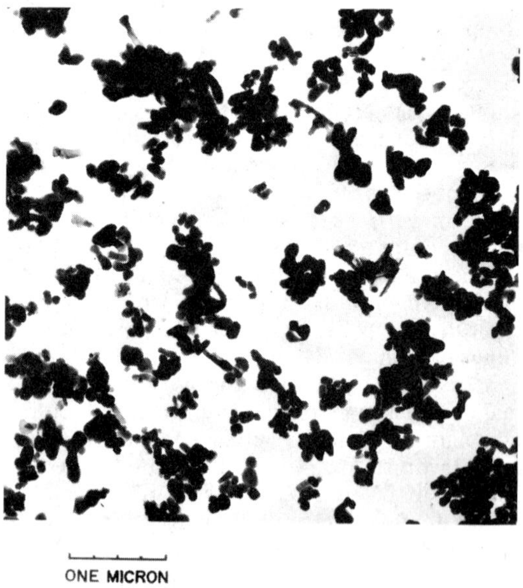

FIG. 11—*Light synthetic red iron oxide, copperas red type.*

TABLE 5—Cont'd.

$Fe_2O_3 \cdot xH_2O$	$Fe_2O_3 \cdot H_2O$	$Fe_2O_3 \cdot H_2O$	$Fe_2O_3 \cdot H_2O$	$Fe_2O_3 \cdot xH_2O$	$ZnFe_2O_4$	$xFeO \cdot yFe_2O_3$	–	γFe_2O_3	Fe_3O_4
–	Goethite	Goethite	Goethite	–	Zinc Ferrite	Hematite Magnetite	Goethite Hematite Magnetite	Gamma Hematite	Magnetite
Ppt.	Ppt.	Ppt.	Organic Reduction	Ppt.	Calcination	Ppt.	Blend	Reduction & Oxidation	Ppt.
94.0	87.0	87.5	86.5	83.0	(99.6 As) ($ZnO \cdot Fe_2O_3$)	93.3#	94.0	98.5	98.8#
–	–	–	–	–		9.0	Balance depends on blended components	–	23.2
0.10	0.05	0.05	{2.0}	–	0.15	0.01		0.06	0.12
–	0.04	0.04		–	0.002	0.01		0.01	0.07
–	0.02	0.02	–	–	–	1.12		0.03	0.03
–	–	–	–	–	–	0.04		–	–
0.14	0.05	0.05	–	0.18	0.01	0.19		0.05	0.20
–	1.00	1.00	–	–	0.03	0.20		1.00	0.60
–	–	–	–	–	–	–			–
5.50	11.00	11.00	12.5	16.2	0.10	–		1.00	–
–	0.15	0.10	0.30	–	0.07	–		0.40	0.15
Transparent red	Light yellow	Dark yellow	Yellow	Transparent yellow	Tan	Brown	Brown	Brown	Black
84	55	52	42	55	34	28	26	60	27
–	0.1-0.4	1-2.5	–	–	0.4-1.2	0.2-0.8	–	–	0.2-0.6
<0.10	0.20	1.5	0.2 x .7‡	<0.10	–	0.4	0.5§	0.2§	0.4§
Acicular	Acicular	Acicular	Acicular	Acicular	Acicular	Cubical	Mixed systems	Acicular	Cubical
0.20	<0.10	<0.10	0.10	0.16	<0.10	0.20	<0.20	<1	<0.10
–	15.2	7.4	–	–	–	11.8	–	15.4	7.2
4.56	4.03	4.03	3.97	3.88	5.24	4.77	4.70	4.60	4.96

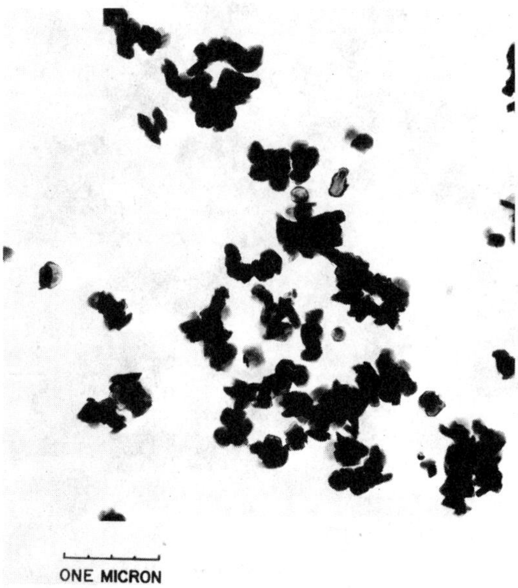

FIG. 12—*Light synthetic red iron oxide, precipitated type.*

FIG. 13—*Light synthetic red iron oxide, Venetian red type.*

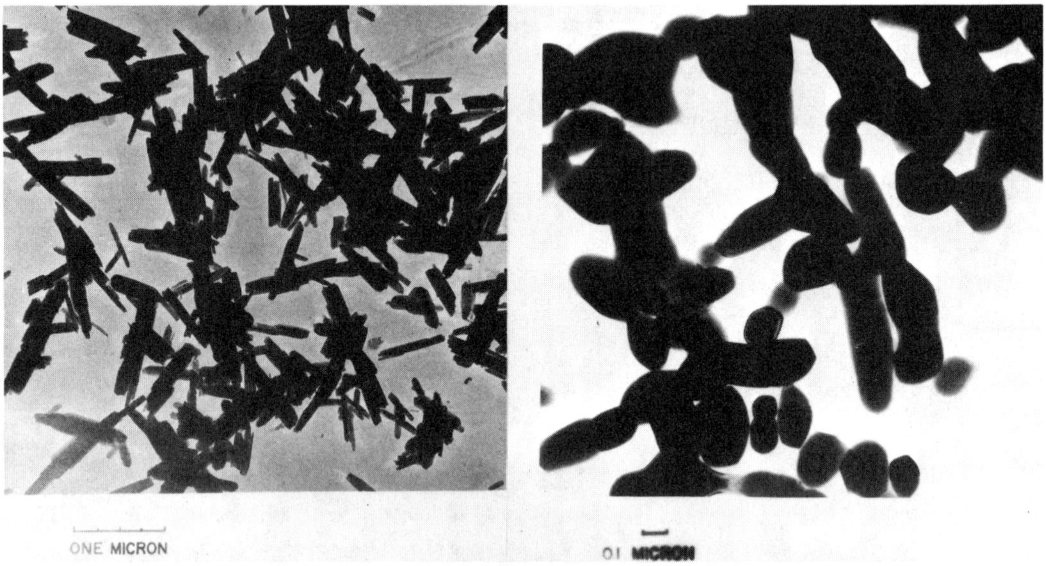

FIG. 14—*Change in morphology in synthetic yellow iron oxide ($Fe_2O_3 \cdot H_2O$) by heat treatment to red oxide (Fe_2O_3).*

Specifications

Specifications vary widely based on the end use application. A paint manufacturer will be most concerned with color and top size whereas a manufacturer of ferrites requires chemical purity.

Typical Color Pigment Specifications:

Mass color	3 ΔE FMC-2 units (max)
Tint tone	3 ΔE FMC-2 units (max)
Tint strength	±5% of standard (max)
% Fe_2O_3	Min. 99.0% (copperas red type)
	Min. 96.0% (precipitated red type)
325 mesh sieve retention	0.10% max
Moisture at 110°C	0.25% max
H_2O soluble salts	0.15% max
Dispersion for specially milled reds	6.0 min. Hegman fineness
Oil absorption	±10% of standard (max)

Natural pigments will have much wider tolerances and in some cases are not controlled for color and tinting strength.

Applications

As previously noted there is an overlap between natural and synthetic iron oxides. There

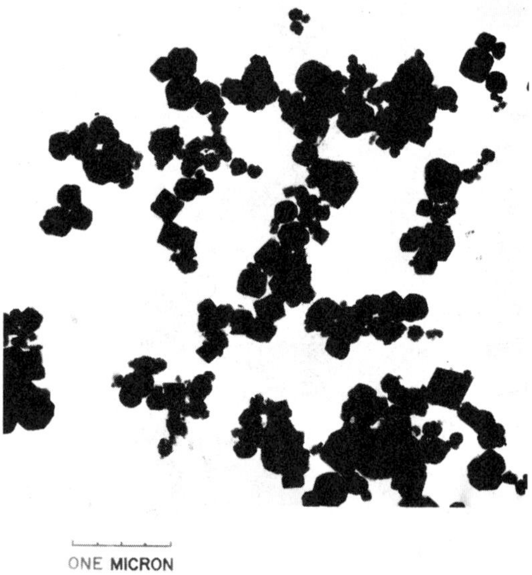

FIG. 15—*Synthetic black iron oxide (precipitated magnetite).*

FIG. 16—*Magnetic iron oxide (γFe_2O_3) retains shape of yellow oxide precursor.*

are some applications, either by custom or necessity, where only synthetics are used (Pfizer, Technical Bulletin).

Paint: Synthetic oxides have higher "chroma" or color purity and greater tinting strength than natural oxides. The color variation from lot to lot can be closely controlled, which is not always true of a natural. They have excellent suspending properties. Hence they are ideally suited for universal tinting systems, powder coatings, automotive finishes, appliance enamels, interior and exterior paints (Anon., 1968; Parfitt, 1969; Parker, 1965; Patterson, 1967). They are nontoxic and meet existing federal regulations.

A trend in automotive finishes to the copper and bronzes has created a market for ultrafine iron oxides referred to as "transparent" or "low opacity" pigments. These pigments are combined with aluminum pigments to produce the metallized automotive finishes (Anon., 1972a). The present market is estimated to be about 1 to 2 kt/a (3 to 5 million lb per year), worth about $6 million.

Electronic Applications (Ferrites): The traditional synthetic reds had the major share of the ferrite raw material market until about 1970. In recent years highly beneficiated naturals and the steel mill ferrous chloride oxides have cut into this market primarily in the less critical barium ferrite market. The soft ferrite market remains as a large consumer of synthetic reds. Copperas reds and calcined yellows are preferred due to chemical purity, high density, and minimum volatiles (Stephens, 1959). The 1972 estimated shipments of permanent magnets, TV yokes, memory cores, and ferrite parts was $192 million (Anon., 1972b). This market currently is about $260 million (Anon., 1981). The iron oxide portion of the total ferrite dollar sales is less than 5% (Kline, 1981). The future for ferrites looks bright and ferrite shipments of all types are expected to exceed 45 kt (50,000 st).

Magnetic Recording: Magnetic tape recording and magnetic ink oxides represented a US market in excess of 6 kt/a (15,000,000 lb per

TABLE 6—1980 Price Range for Synthetic Iron Oxides Marketed in the United States (Values Rounded to Closest Cent Per Lb)

Synthetic Oxide Type	F.O.B.* Selling Price	Products† Available
Red Iron Oxides		
Copperas reds	52–59	27
Precipitated reds	53–55	6
Calcined yellows	45–71	18
Organic reduction (calcined)	51–58	20–56
Yellow Iron Oxides		
Regular	47–52	35
Low oil	48–55	30
Tans		
Browns & Tans		
Synthetic blends	51–57	138
Precipitated	56–58	2
Blacks		
Precipitated	56–58	7
Organic reduction	50–55	21
Magnetic Oxides		
Gamma ferric oxide	102–540	16
Magnetite	170–450	4
Transparent		
Yellows	150–290	6
Reds	175–320	5
Electronic (Ferrite) Grades		
Red iron oxides	10–58	10
Pet Food And Cosmetic Grades		
Reds	52–57	16
Blacks	57–66	4
Price range and total available products	10–540¢/lb	365–401

Source: *American Paint Journal* and suppliers published prices.

* Less truckload price in 22.6-kg (50-lb) multiwall bags. Price about 3 to 10¢ per pound higher for small quantities. Minimum weight depends on supplier. Special milling, palletizing, and packaging carry a premium.

† Similar or duplicate products are carried by suppliers, increasing the number of apparent varieties.

year) in 1962 (Anon., 1968) and has continued to grow at a highly accelerated growth rate in recent years due to the market penetration of video cassette recorders. CrO_2 has made some inroads, but the major part of the market is gamma ferric oxide.

The requirements for audio, video, instrumentation, and disk recording vary and special oxides are produced for these applications. Slow recording speeds and high fidelity require ultrafine gamma ferric oxides. The gamma ferric oxides cover a fairly wide range of (coercive force) H_c values from 250 to 400 oersteds. This can be extended to 500 to 800 oersteds by modification with cobalt. Typical σ_s (saturation magnetization) values are 73 to 80 emu/gr. Magnetic inks for check data processing, credit cards, and railway tickets use synthetic magnetite which provides high pigment concentrations in the binder system.

Concrete Products and Building Materials: Synthetics have the advantage of greater tinting strength and uniformity and are replacing naturals in many applications. This is a growing market consuming between 13 and 18 kt/a (30 and 40 million lb per year) domestically and estimated to be in the 11 kt/a (25 million lb per year) range in the United Kingdom.

Plastics, Floor Tile, Linoleum, Rubber: Higher chemical purity, higher chroma masstones, and stronger tints result in a lower cost per color unit, accounting for greater usage of synthetics. Many plastics require stabilization which is more easily accomplished with the consistent chemical purity of the synthetics.

Miscellaneous: Synthetic iron oxides are used singly or in combination with other elements as catalysts in various conversions for ammonia, styrene, butadiene, hydrogen, and formaldehyde (Mills, 1964). The US production of styrene butadiene catalysts was 1.8 kt (4 million lb) in 1963. The production of styrene type plastic materials increased from slightly over 454 kt (500,000 st) in 1960 to 1 Mt (1.2 million st) in 1966 (Anon., 1967) and 3.4 Mt (3.8 million st) in 1978 (Anon., 1978).

Iron oxide is added to animal feeds as a source of iron nutrient and also as a coloring agent for dog foods.

Iron oxides are considered to have a generally recognized as safe (GRAS) rating by the Food and Drug Administration (FDA). Certain grades that meet the maximums of 3 ppm arsenic, 10 ppm lead, and 3 ppm mercury are approved for drug and cosmetic use. Iron oxides for pet foods must not exceed 5 ppm arsenic, 20 ppm lead, and 3 ppm mercury. High-purity iron oxides meeting the heavy metal requirements carry a premium of about 4 to 6¢ per lb to cover purification and analytical costs.

Synthetic oxides are used for opthalmic polishing and as the iron source, batch rouge, for coloring glass and providing ultraviolet absorptive properties.

An interesting application—but as yet of no commercial significance—is in the field of medicine. Magnetic oxides have the ability to absorb myxoviruses and can then be magnetically separated from the system (Warren, 1969).

Marketing

The demand for iron oxides is closely related to domestic and worldwide requirements of the construction, coatings, automotive, plastics, electronics, and chemicals industries. With the exception of electronics and automotive, where miniaturization and imports have cut into domestic demands, the remaining industries were projected to double in dollar sales from 1970 to 1980 (Anon., 1971b).

Iron oxide prices were very stable from 1960 through 1970. The inflationary costs of labor and supplies plus the impact of high energy costs necessitated significant price increases during the period 1970 to 1980. Table 7 shows the ten-year price trend of a selected range of natural and synthetic iron oxides.

Most oxide suppliers offer discounts of 3 to 10¢ for truckload or carload orders. Special packaging such as shrink or stretch wrap may carry a premium. A broad product line is necessary to offer pool car or full truckload discounts. The lighter synthetic shades are usually 1 to 3¢ more expensive per pound, and special processing such as fluid energy milling carries a premium. The majority of the iron oxides are packed in multiwall paper bags. Extra-ply bags, barrels, drums, or other special packaging requirements carry a special price.

The marketing of iron oxides varies with the size of the producer. Major oxide producers maintain their own sales force and warehouses. The smaller producers rely heavily on sales agents or representatives.

The transportation of iron oxides within the United States is accomplished primarily by truck and rail. In 1968 as much as 50% of

TABLE 7—Current Iron Oxide Prices and Ten-Year Trend

	1970 ¢ per lb*	1976 ¢ per lb	1977 ¢ per lb	1978 ¢ per lb	1979 ¢ per lb	1980 ¢ per lb
Blacks:						
Synthetic	16–19	31–38	31–38	31–38	43–49	43–49
Micaceous	—	38–40	38–40	55	55	55
Browns:						
Ground iron ore	7–10	7–9	7–9	10–13	10–13	10–13
Metallic	10	11–13	11–13	—	—	18–20
Pure synthetic	19–22	36–45	40–47	47–50	50–52	47–55
Burnt sienna (Italian)	13–28	30–40	34–47	34–48	34–48	70
Burnt umber (Turkey)	11–15	24–28	24–26	26–29	26–29	31–36
Red:						
Domestic primers	7–10	17–23	16–22	20–23	20–23	22–29
Pure synthetic	18	38–41	43–45	44–46	44–47	45–52
Spanish	12	—	26	26	26	28–32
Yellow:						
Synthetic	16–18	37–40	39–42	43–44	47–48	52
Ochre (French)	10–12	18–23	—	—	—	23
Ochre (Domestic)	6	8–9	13	13	14	13

Source: *American Paint Journal,* July issues for years shown.
*Prices are f.o.b. per pound in less than carload lots unless indicated.

iron oxides moved via rail. This figure was drastically reduced over the next several years due to damage in transit, shorter lead time on orders, lower average inventories, and the increase in freight rates. The amount of oxide moved by rail bottomed out in the late 1970s at between 5 and 15% of the total. Since that time railroads have taken action to reduce the trend and regain the lost business.

Deregulation of the transportation industry has presented new opportunities for the reduction of freight costs by the use of co-op carriers, contract carriers, aggregate tenders to common carriers, and private fleet operations. By combining different modes of transport, iron oxide producers have been and should in the future be able to keep transportation costs, as a percent of delivered price, relatively low.

Environmental Considerations

Perhaps the greatest single factor affecting iron oxide manufacturing today has been federal and state regulations regarding ecology and all its implications. Whereas the major steel and titanium processors were content to dispose of their iron salts by dumping, they are now forced to recover and process these salts prior to disposal or close operations. This has resulted in huge tonnages of cheap byproduct iron oxides. The iron oxide manufacturers have also been required to recover and process any soluble iron salts which were previously discharged, resulting in increased manufacturing costs. On the one hand they are beset by cheap byproducts and on the other hand by rising costs.

A very favorable outgrowth of environmental regulations concerns the *nontoxic* characteristic of iron oxides. The FDA has issued concentration limits of seven elements, and other elements are under fire in various state regulations. Table 8 lists the federal specification covering *coatings* for heavy metal and lead concentrations (Stephenson, 1972). Since iron oxides constitute only part of pigmented systems, it is obvious that their use would meet federal requirements. The single exception shown is a natural iron oxide that is high in lead. Other naturals are below the 600 ppm maximum.

Future Outlook for Iron Oxides

With the projected population growth and its attendant demands for construction, wood products, appliances, paints, plastics, chemicals, and related industrial products, the future for the permanent nontoxic iron oxides is very good.

TABLE 8—Concentrations of Heavy Metals in Light or Medium Shade Iron Oxides in Parts Per Million

Element	Federal Regulation*	Copperas Reds	Precipitated Reds	Precipitated Yellow	Synthetic Black	Natural Red	Raw Sienna	Burnt Umber
Arsenic		1-5	30-60	10-40	5-15	50-100	70-120	100-200
Antimony		5-15	5-20	5-20	10	40-80	†	†
Cadmium		<20	<20	<20	<20	<50	<50	<50
Selenium		<20	<20	<20	<20	<100	<100	<100
Mercury		<0.1	<0.1	<0.1	<0.1	<0.1	<0.1	<0.1
Total maximum	500	60	120	100	65	330	~370	~450
Lead	5000 1973 600 1974	1-15	10-40	10-50	50-100	6000+	100-500	100-300
Barium (total)		10-100	30-120	100-1000	100-1000	100-200	100-200	400-700
Barium (soluble)		N.P.	N.P.	N.P.	N.P.	–	–	–

Source: Stephenson, W., 1972.

* Federal regulation covering paints and coatings, 1973 and 1974, American Natural Standards Inst. (ANSI) and FDA (also NPCA, 1972).

Code < indicates if the element is present at all, it is below the detection limit of analytical method used.

Hg determined by flameless atomic absorption. As by U.S.P. colorimetry method. Remaining elements by X-ray fluorescence.

N.P.—not present.

† Potassium spectral interference

Two major consumers of iron oxides are paint and plastics. Sales of paints and allied products increased from \$2.4 billion to \$3.6 billion from 1963 to 1971. A \$7 billion market was projected for 1980. The plastic resins market doubled during this same period and was projected to reach \$12 billion by 1980 (Anon., 1971b). The major business upturn in 1972 has resulted in worldwide shortages of iron oxides (Banov, 1972). Shortages of synthetic yellow iron oxide were projected to remain critical into mid-1973.

The last 20 years produced a few iron oxide innovations—transparent iron oxides, high-purity ferrite oxides, grades meeting special FDA requirements for heavy metals, magnetic ink pigments, new magnetic iron oxides, and new types of doped magnetic oxides. Due to the high cost of energy and a shift to water-based coatings systems, iron oxides have recently become available in a high concentration slurry form. This simplifies handling and, in some applications such as concrete coloring, results in an increase in tinting strength. The number of pigments available as easy dispersing types increased during the last ten years. The use of spectrophotometers coupled with computers to monitor and control iron oxide pigments gained wide acceptance in recent years.

Despite the serious effects of inflation, recession, and high energy costs, which have depressed two major iron oxide markets—automotive and housing production—the advantages of iron oxides (nontoxic, compatible in many systems, and a good cost-color value) should assure continued growth of production in the years ahead.

Bibliography and References

Anon., 1960, "Iron Oxide Resurgence, *Oil, Paint, and Drug Reporter,* June 13.

Anon., 1965, "Wet Magnetic Separation Comes of Age," *Canadian Chemical Processing,* Feb.

Anon., 1966, "Trends in Steel Pickling and Waste Treatment," *The Magazine of Metals Producing,* Mar., pp. 65–76.

Anon., 1967, *Business Statistics,* US Dept. of Commerce, p. 129.

Anon., 1968, "Inorganic Color Pigments," *Federation of Societies for Paint Technology,* Unit 8, Sec. 2, Mar.

Anon., 1971, "Iron Oxide Pigments," *Industrial Minerals,* No. 50, Nov., pp. 9–19.

Anon., 1971a, "Trans-Oxide Pigments," Technical Bulletin, SC 22—3/71, Hilton Davis Chemical Co.

Anon., 1971b, *U.S. Industrial Outlook 1971,* US Dept. of Commerce, pp. 188, 199.

Anon., 1972, "Oil Absorption by Spatula Rub Out," *ASTM Annual Book of Standards,* Vol. 20, p. 170.

Anon., 1972a, "More Sparkle for Auto Finishes," *Chemical Week,* Aug. 2, p. 29.

Anon., 1972b, "Sales Estimates 1972," *Ceramic Age,* Jan., p. 25.

Anon., 1972c, "Technical Report "316" MIOX," Reichard-Coulston Inc., Aug. 18.

Anon., 1978, "Manville Chemical Products Synopsis," Reporting Service, Manville, New York, Oct.

Anon., 1981, Market Forecast for Electronics, Vol. 54, No. 1, Jan. 13, pp. 130–133.

Ayers, J.W., 1938, "Manufacture of Black Iron Oxide," US Patent 2,133,267, (assigned to Pfizer Inc.), Oct. 18.

Ayers, J.W., 1946, "Process for the Production of Iron Oxide," US Patent 2,394,579, (assigned to Pfizer Inc.), Feb. 12.

Ayers, J.W., 1960, "Production of Red Oxide of Iron Pigments," US Patent 2,937,927 (assigned to Pfizer Inc.), May 24.

Banov, A., 1972, "Paint Companies Find Some Yellows and Whites Tight," *American Paint Journal Convention Daily,* Oct. 26, pp. 12, 31.

Bennetch, L., 1957, "Preparation of Red Iron Oxide," US Patent 2,785,991, (assigned to Pfizer Inc.), Mar. 19.

Bennetch, L., 1961, "Red Oxide of Iron," US Patent 3,009,821, (assigned to Pfizer Inc.), Nov. 21.

Berman, H., et al., 1946, *Dana's System of Minerology,* Vol. 1, 7 ed., John Wiley, New York, pp. 685–687.

Brunauer, S., et al., 1938, "Adsorption of Gases in Multi-Molecular Layers," *Journal,* American Chemical Society, Vol. 60, p. 309.

Buckley, J.A., 1967, "Liquor Regeneration Slashes Cost of Steel Pickling," *Chemical Engineering,* Jan. 2, pp. 56–58.

Cadle, R.D., 1955, *Particle Size Determination,* Interscience Publishers Inc., New York, pp. 23–25, 198.

Camras, M., 1954, "Magnetic Impulse Record Member," US Patent 2,694,656, (assigned to Armour Research Foundation), Nov. 16.

Casey, R., 1966, "Iron Compounds, Iron (III) Oxide," *Encyclopedia of Chemical Technology,* Kirk-Othmer, ed., Vol. 12, 2nd ed., Interscience, New York, p. 39.

Collins, C.T., 1975, "Iron Oxide Pigments," *Minerals Yearbook,* US Bureau of Mines, pp. 751–758.

Collins, C.T., 1976, "Iron Oxide Pigments," *Minerals Yearbook,* US Bureau of Mines, pp. 673–680.

Collins, C.T., 1977, "Iron Oxide Pigments," *Minerals Yearbook,* US Bureau of Mines, pp. 499–505.

Collins, C.T., 1979, "Iron Oxide Pigments," *Minerals Yearbook,* US Bureau of Mines, pp. 457–464.

Downs, C.D., and Martin, J., 1950, "Preparation of Magnesium Ferrite Pigment," US Patent 2,502,130, (assigned to Cities Services Inc.), Mar. 28.

Downs, C.D., and Martin, J., 1959, "Method for Producing Zinc Ferrite Pigment," US Patent 2,904,395, (assigned to Cities Services Inc.), Sep. 15.

Hancock, K.R., 1970, "Preparation of Ultrafine Particles of Metal Oxides," *Ultrafine Grain Ceramics,* Syracuse University Press, pp. 39–59.

Herdan, G., 1953, *Small Particle Statistics,* Elsevier Publishing Co., New York, pp. 37–39, 230–240, 267–269.

Hoffman, M., 1972, "Iron Oxide Technology," Videotape Presentation, Pfizer Inc.

Jolly, J.L.W., and Collins, C.T., 1980, "Natural Iron Pigments," Information Circular 8813, US Bureau of Mines, 79 pp.

Kline, C., 1981, "Ceramic Raw Materials," Charles Kline & Co. Report, Fairfield, NJ, In preparation.

Love, C.H., 1946, "German Production of Some of the More Important Inorganic Pigments," FIAT Final Report 814, June 24, Office of Military Government for Germany (US), pp. 7–19.

Love, C.H., 1961, "Magnetism, Magnetic Recording and Magnetic Inks," *American Ink Maker,* Vol. 39, No. 5, pp. 50–52, 55–57.

Love, C.H., 1968, "Guidelines for Coloring Concrete Products," *Concrete Products,* Vol. 71, No. 7, July, pp. 34–37.

Love, C.H., and Ayers, J.W., 1942, "The Iron Oxide Pigments," *Protective and Decorative Coatings,* J.J. Mattiello, ed., Vol 2, John Wiley and Sons, New York, pp. 317–318.

Marsh, B., 1955, "Manufacture of Red Hydrous Ferric Oxide," US Patent 2,716,595, (assigned to Pfizer Inc.), Aug. 30.

Martin, J., 1960, "Manufacture of Iron Oxides," US Patent 2,939,767, (assigned to Cities Services Inc.), June 7.

Mattiello, J.J., 1942, "Protective and Decorative Coatings," in *The Iron Oxide Pigments,* Vol. II, John Wiley & Sons, New York, pp. 317–318.

McGannon, H.E., 1968, "Steel, Oxygen Steel Making Process," *Encyclopedia of Chemical Technology,* Kirk-Othmer, ed., Vol. 18, 2nd ed., Interscience, New York, pp. 741–747.

Mellor, J.W., 1953, *A Comprehensive Treatise on Inorganic and Theoretical Chemistry,* Vol. 13, Lowe and Brydone Ltd., London, pp. 805–817.

Mills, G.A., 1964, "Catalysis," *Encyclopedia of Chemical Technology,* Kirk-Othmer, ed., Vol. 4, 2nd ed., Interscience, New York, pp. 534–586.

Parfitt, G.D., 1969, *Dispersion of Powders in Liquids,* Elsevier Publishing Co. Ltd., pp. 81–121, 285–324.

Parker, D., 1965, *Principles of Surface Coating Technology,* Interscience, New York, 817 pp.

Patterson, D., 1967, *Pigments, an Introduction to their Physical Chemistry,* Elsevier Publishing Co. Ltd., pp. 50–65.

Penniman, R., and Zoph, N., 1921, "Process of Manufacturing Iron Compounds and Product," US Patent 1,368,748, (assigned to National Ferrite Co.), Feb. 15.

Schieber, M.M., 1967, *Experimental Magnetochemistry,* E.P. Wohlfarth, ed., Chap. 4, Vol. 8, North Holland Pub. Co., Amsterdam, pp. 157–243.

Spinrad, W.I., Jr., 1980, "Iron Oxide Pigments," *Minerals Yearbook, 1980,* US Bureau of Mines, pp. 429–436.

Stephens, R.A., 1959, "Iron Oxides as Raw Materials for the Manufacture of Magnetic Ferrites," *American Ceramic Society Bulletin,* Vol. 38, No. 3, Mar. 15, pp. 106–109.

Stephenson, W., 1972, "Project Cope," presentation to Federated Societies of Paint Technology, May 2.

Stewart, J.R., 1942, *The National Paint Dictionary,* 2nd ed., Stewart Research Laboratory, Washington, DC, p. 95..

Stipp, H.E., 1970, "Iron Oxide Pigments," *Minerals Yearbook 1970,* US Bureau of Mines, pp. 627–630.

Stipp, H.E., 1971, "Iron Oxide Pigments," *Minerals Yearbook 1971,* US Bureau of Mines, pp. 653–656.

Stipp, H.E., 1972, "Iron Oxide Pigments," *Minerals Yearbook 1972,* US Bureau of Mines, pp. 683–687.

Stipp, H.E., 1973, "Iron Oxide Pigments," *Minerals Yearbook, 1973,* US Bureau of Mines, pp. 673–678.

Taylor, A.H., 1966, "Oxygen Uses," *Encyclopedia of Chemical Technology,* Kirk-Othmer, ed., Vol. 14, 2nd ed., Interscience, New York, pp. 402–403.

Toxby, T., 1952, "Method of Making Iron Oxide Pigment," US Patent 2,620,261, (assigned to Pfizer Inc.), Dec. 2.

Von Fisher, W., 1948, *Paint and Varnish Technology,* Reinhold Publishing Corp., New York, pp. 56–57.

Warren, J., 1969, "The Adjuvant Effect of Powdered Ferric Oxide: Enhancement of Response to Mycoplasma Pneumonia in Respiratory Synctal Virus Vaccines," *The Journal of Immunology,* Vol. 102, No. 5, pp. 1200, 1308.

Woernle, A.K., 1967, "Iron Oxide Pigments," *The Encyclopedia of Basic Materials for Plastics,* Reinhold Publishing Co., New York, pp. 261–264.

The following company technical bulletins are also of interest:

Cities Service Co., Technical Bulletin, "Low Opacity Colors."

Pfizer, Inc., "Technical Bulletins on Natural and Synthetic Iron Oxides:" TR–7–4(1971), TR–8–4(1971), TR–108–5(1971), TR–154–8 (1971), TR–5–4(1972), TR–77–5(1972), TR–166–4(1972).

Refractories

JAMES A. CROOKSTON *

WILLIAM D. FITZPATRICK †

Committee C-8 of the American Society for Testing and Materials defines *refractories* as "material, usually nonmetallic, used to withstand high temperature," and it defines the term *refractoriness* thus, "in refractories, the capability of maintaining the desired degree of chemical and physical identity at high temperatures and in the environment and conditions of use."

The ability to withstand high temperatures is critical to classifying a material as a refractory. However, as the foregoing definition of "refractoriness" implies, the refractory material may be subjected to and is expected to resist not only high temperatures but a variety of other destructive forces incidental to its use. These include abrasion, impact, thermal shock, chemical attack, and high level loads at high temperatures.

The various industrial applications of refractories represent an infinite variety of combinations and degrees of severity of these destructive forces, including high temperatures. Naturally, such a variety of application conditions leads to a need for a wide variety of refractory products to satisfy those conditions. The number of refractory products available is great enough that some form of classification is necessary to a discussion of refractories.

Refractories may be classified in several ways, but the most suitable one is based on their general compositions and to lesser degrees on their physical condition and distinctive properties. The primary types of brick include aluminosilicate brick (fire clay and alumina), basic brick (magnesia and chrome ore, alone and in various combinations), silica brick, insulating brick, and special refractories (carbon, SiC, ZrO_2, zircon, borides, and nitrides). Brick is, of course, preformed and usually is heat treated at high temperatures before use, but in some instances is unburned and/or chemically bonded, including the relatively new class of tar-bonded magnesia and/or dolomite brick.

In addition to preformed refractories, another broad category includes the so-called specialty refractories, which are usually made in the same types as itemized for brick—bonding mortars and high temperature cements, hydraulic setting castables or refractory concrete, plastic firebrick, and ramming mixes. Refractory grains, such as fire clay and high alumina calcines, dead-burned magnesite and periclase, and dead-burned dolomite are also considered refractory products.

Bricks are also classified on the basis of physical dimensions, as standard sizes and special shapes. The standard sizes are those which are of definite shape and design and are designated as 228.6 x 114.3 x 63.5 mm (9 x 4½ x 2½ in.) and 228.6 x 114.3 x 76.2 mm (9 x 4½ x 3 in.) straights and series. Bricks of nonstandard sizes are referred to as special shapes. These come about as a result of a particular need for a given furnace or for a particular application.

Types and Classes of Refractories

Fire Clay Refractories

Classification: Fire clay refractories, as the name implies, are composed essentially of fire clay and/or calcined fire clay in varying proportions. The ASTM Standard Classification of Fireclay and High Alumina Refractory Brick, C27–70 (reapproved 1976), gives five classes of fire clay brick designated as super-

* Vice-President and Director of Research, A.P. Green Refractories Co., subsidiary of U.S. Gypsum Co., Mexico, MO.

† Senior Research Consultant, A.P. Green Refractories Co., subsidiary of the U.S. Gypsum Co., Mexico, MO.

duty, high duty, semisilica, medium duty, and low duty. The semisilica class is based on its load-bearing ability at high temperatures and its silica content, while the other four classes use as the primary criterion for classification their degree of refractoriness or pyrometric cone equivalent (PCE) (Table 1).

Pouring pit refractories are those used in the handling of molten steel from the furnace to the ingot molds or to the casting machines and include ladle brick, stopper rod sleeves, hot tops, teeming nozzles, runner brick, and certain other shapes, such as ingot mold plugs. These pouring pit refractories are made principally from fire clay and account for a significant proportion of the fire clay brick and tile sold. Acid-resistant brick and tile are generally classified as fire clay refractories as most of them are made from special types of fire clays. The bulk of the fire clay refractories sold are in the super and high duty classes.

While not a separate classification by ASTM standards, a recent development in the superduty class of fire clay refractories is a brick characterized by significantly enhanced load-bearing properties. This development is based on the availability of raw materials of extremely low alkali content which allows the achievement of this desirable property. Another recent development in fire clay refractories has been the technique of pitch-impregnation. The resulting carbon in the pores renders the brick more resistant to wetting and corrosion by metals and slags.

Raw Materials: Fire clays include all clays having a fusion point above PCE 15, 1424°C (2595°F). They are mainly composed of hydrated aluminum silicates with minor amounts of accessory minerals. The clays of greatest importance as refractory raw materials are flint and semiflint fire clays, plastic and semiplastic fire clays, and kaolin.

Flint and semiflint clays are found in sedimentary facies of Mississippian and Pennsylvanian age. They may be derived from the weathering of pegmatites or as a result of a clay material weathering in place, and occur in rather limited areas of Pennsylvania, Ohio, Maryland, Kentucky, Colorado, and Missouri. Physically, these flint clays are hard dense clays which exhibit a conchoidal fracture and will not slake in water. They are resistant to

TABLE 1—Fire Clay and High-Alumina Refractory Brick Classified According to Classes and Subdivided Into Types.

Class	Type	Pyrometric Cone Equivalent, Min.	Panel Spalling Loss, Max, %	Hot Load Subsidence, Max, %	Reheat Shrinkage, Max, %	Modulus of Rupture, Min. Psi (MPa)	Other Test Requirements
Fire Clay Brick							
Superduty	Regular	33	8 at 3000°F (1650°C)	. . .	1.0 at 2910°F (1600°C)	600 (4.14)	. . .
	Spall resistant	33	4 at 3000°F (1650°C)	. . .	1.0 at 2910°F (1600°C)	600 (4.14)	. . .
	Slag resistant	33	. . .	. . .	. . .	1000 (6.89)	Bulk density, min, 140 lb per cu ft (2243 kg/m³)
High duty	Regular	31½		. . .	. . .	. . .	. . .
	Spall resistant	31½	10 at 2910°F (1600°C)	. . .	. . .	500 (3.45)	. . .
	Slag resistant	31½	. . .	. . .	. . .	1200 (8.27)	Bulk density, min, 137 lb per cu ft (2194 kg/m³) or max porosity 15%
Semisilica		. . .	. . .	1.5 at 2460°F (1350°C)	. . .	300 (2.07)	Silica content, min, 72%
Medium duty		29	. . .	. . .	. . .	500 (3.45)	. . .
Low duty		15	. . .	. . .	. . .	600 (4.14)	. . .
High-Alumina Brick							
50% alumina		34	. . .	. . .	. . .	. . .	Alumina content, 50 ± 2.5%
60% alumina		35	. . .	. . .	. . .	. . .	Alumina content, 60 ± 2.5%
70% alumina		36	. . .	. . .	. . .	. . .	Alumina content, 70 ± 2.5%
80% alumina		37	. . .	. . .	. . .	. . .	Alumina content, 80 ± 2.5%
85% alumina		. . .	. . .	. . .	. . .	. . .	Alumina content, 85 ± 2.0%
90% alumina		. . .	. . .	. . .	. . .	. . .	Alumina content, 90 ± 2.0%
99% alumina		. . .	. . .	. . .	. . .	. . .	Alumina content, min., 97%

Source: ASTM Standard Classification C27-70 (reapproved 1976).

weathering. They are essentially pure, extremely fine-grained kaolinite, $Al_2O_3 \cdot 2SiO_2 \cdot 2H_2O$, and the best refractory grade flint clays have PCEs of cone 33 to cone 35 and are low in impurities such as iron and alkalies. The drying and firing shrinkages of flint clays are quite variable, but would be considered medium to high.

The plastic and semiplastic clays are found in sedimentary facies of Mississippian to Tertiary age. Deposits of these clays occur in many states with the principal ones located in Pennsylvania, Ohio, Kentucky, Maryland, West Virginia, Alabama, Missouri, Texas, and Colorado. These clays vary considerably in plasticity, refractoriness, and bond strength, and they will readily work up with water into a plastic mass showing high drying and firing shrinkage. The top quality plastic clays have PCEs of cone 26 to cone 33 and are generally relatively low in iron and alkali impurities.

Kaolins are found in two types of deposits, residual and sedimentary. The residual deposits are those in which the kaolin is in the location where it was altered from the parent rock. Kaolins for use in refractories, however, are generally sedimentary in origin. The majority of the kaolin deposits in the United States are located below the fall line in North Carolina, South Carolina, Georgia, and Alabama. The refractory kaolins have a PCE ranging from cone 28 to cone 35. While they are usually moderately plastic, kaolins have low green strengths and high firing shrinkages. The best refractory grade kaolin must have a low iron and alkali content and should exhibit a relatively high fired density. The extremely low alkali contents of some kaolins, such as from the Alabama area, make them valuable for the production of superduty brick having enhanced load bearing properties at high temperatures.

In addition to the raw materials described above, some fire clay brick, especially those of the low-duty class of pouring pit refractories, are made of clays which exhibit particular properties, such as the ability to bloat or expand when exposed to high temperatures.

Manufacture: There are several forming methods used for the production of fire clay brick and tile. The mixes from which these brick and tile are made vary from about 10% clay and about 90% fire clay calcines or crushed fire brick to 100% clay. The mixes for each manufacturing process are grain sized to impart definite properties to the finished brick and tile.

The clays and calcines or grogs are crushed by various means. The crushed materials are incorporated into the batch for forming by either of the following two procedures: (1) the crushed raw materials are batched, ground to the desired grain size, and then tempered to the desired consistency, or (2) the crushed raw materials are ground and sized separately, batched, and then tempered to the desired consistency. Equipment for crushing and batching and tempering are of the commonly accepted types.

Most of the fire clay brick and tile are made by the dry press process. The tempered batch is fed to the mold cavity of the press, and the brick or tile are formed by the use of high pressure, derived mechanically or hydraulically. Some brick is formed on vibrating presses in which the pressure is less but the vibratory action aids compaction.

In order to impart specific properties or to make brick and tile whose configuration makes it difficult to form them by the dry press process, some brick and tile are made by the stiff mud process. In this process, the mix is tempered to a plastic consistency and extruded through a die as a plastic column. The column is cut into individual units, which may be "repressed" to give them sharp corners and edges or formed into intricate special shapes by hydraulic or runner presses.

Large and intricate shapes are sometimes made by hand molding, but more often, these shapes are made by ramming the mix into the mold with an airhammer. The use of an airhammer makes it possible to use a stiffer mix consistency and to more closely duplicate machine made refractories.

Slip casting is used to make some special purpose tile. In this process, the ground materials are formed into a slurry or slip with water and electrolytes, which give the slip free-flowing characteristics with a minimum amount of water. The molds used in the slip casting process are made of plaster, which absorbs much of the water and permits early removal from the mold.

The formed brick and tile made by any of these processes are dried in either periodic or tunnel dryers, with the most difficult to dry tile being dried under controlled conditions of humidity and temperature. The dried brick or tile are then fired in periodic or tunnel kilns using appropriate firing schedules designed to impart the desired properties to the finished product. The firing schedules used for the various products will vary widely depending upon the properties desired in the final product. For instance, the pouring pit refractories generally are fired at a low temperature and on a relatively long schedule to impart the desired property of reheat expansion. During the firing process, mechanical and chemically combined water is driven off and iron, sulfur, and organic impurities are oxidized. The mineral transformations and volume changes caused by the application of high temperatures are accomplished, and as a result of firing, brick or tile of the desired mineralogical and physical properties is produced.

Uses: Fire clay brick and tile are widely used because of their generally good overall properties and relatively low costs. Among the largest consumers of fire clay brick are blast furnaces and blast furnace stoves in which the blast furnace combustion air is preheated to improve blast furnace efficiency. Other large consumers of fire clay brick and tile are the ladles and accessory equipment used to transfer molten metal between furnaces and between the finishing furnaces and ingot molds or casting machines.

Fire clay brick and tile are also widely used to build furnaces for baking the carbon electrodes used in the aluminum industry, as general purpose steel mill brick, fireplace brick, ceramic kiln construction, ceramic kiln furniture, glass tank regenerator wall and checker support brick, and for the linings of heat-treating furnaces.

High Alumina Brick

Classification: High alumina brick and tile are those alumina-silica refractories containing 50% or more of alumina. This brick is classified in accordance with its alumina content as shown in Table 1, which is the ASTM Standard Classification C27–70. In addition to the high alumina brick and tile covered by this specification are those classified as mullite refractories. This class is described in ASTM Standard Classification C467–72 (reapproved 1977) as having an alumina content of 56 to 79%, a maximum amount of metal oxides other than alumina and silica of 5%, a maximum deformation under 172 kPa (25 psi) load and 1593°C (2900°F) for 90 min of 5%, and consisting predominately of mullite ($3Al_2O_3 \cdot 2SiO_2$) crystals.

Raw Materials: The more common high alumina raw materials used in the high alumina refractories include bauxite [$Al(OH)_3$], diaspore ($Al_2O_3 \cdot H_2O$), kyanite ($Al_2O_3 \cdot SiO_2$), sillimanite ($Al_2O_3 \cdot SiO_2$), andalusite ($Al_2O_3 \cdot SiO_2$), and fused and sintered alumina (Al_2O_3).

Bauxite for refractory purposes is considered to be a mixture of gibbsite [$Al(OH)_3$] and kaolinite ($Al_2O_3 \cdot 2SiO_2 \cdot 2H_2O$), and most of the bauxite used for this purpose originates in either South America or mainland China. There are also commercial deposits in Arkansas, Alabama, and Georgia. The Alabama material is particularly suited to the manufacture of high quality high alumina brick by virtue of its high purity and high fired density. Because of the high shrinkage of bauxite on firing, the raw material is nearly always calcined at high temperatures before use.

Diaspore has lost its former widespread use because of the depletion of better quality reserves.

Kyanite, sillimanite, and andalusite all form mullite and silica when heated above 1554°C (2830°F). India is the largest source of kyanite and sillimanite, although large reserves of kyanite occur in Virginia, Georgia, and South Carolina. The domestic kyanite does not burn to as high density as the Indian material.

Synthetic mullite may also be considered a raw material for high alumina brick. It is made by calcining either natural or synthetic mixtures of alumina and silica to high temperature to form grain composed predominately of the mineral mullite.

Manufacture: High alumina brick and tile are made primarily by the dry press process. The principal ingredients used in these mixes have normally been calcined to take care of shrinkage and convert them to either mullite, mullite and glass, or mullite and corundum. The calcined portion of the mixes is crushed, ground, and grain sized prior to batching with the bonding clays and tempering in a suitable mixer. However, some large and intricate shapes are made by the hand mold or airhammer processes, and for these processes, the mix is prepared in the same manner as for dry pressing. In addition, some special tile are made by either the stiff mud or slip casting process.

The formed brick and tile are dried in either periodic or tunnel dryers and then fired in tunnel or periodic kilns to prescribed firing temperatures designed to impart the desired properties to the finished product. Firing schedules for the various high alumina brick are designed for their particular needs. Some high alumina brick is impregnated with pitch after firing, using procedures described later for pitch-impregnated periclase brick.

Uses: High alumina brick and tile are more resistant to the effects of high temperatures, and therefore can be used at higher temperatures than the fire clay brick and tile. In recent years, the service conditions for many refractory applications have become more severe, and there has been a steady trend toward the use of high alumina brick and tile to replace fire clay.

Among the large consumers of high alumina brick and tile are rotary kilns for the production of cement and lime products, roof brick for electric steel-melting furnaces, blast furnace lining and blast furnace stove brick, ladle brick, and aluminum contact refractories in aluminum-melting furnaces.

Silica Brick

Classification: ASTM designation C416–70 (reapproved 1976), the Standard Classification of Silica Refractory Brick, covers all silica brick having an alumina content of less than 1.5%, titania content less than 0.2%, Fe_2O_3 content less than 2.5%, calcium oxide less than 4.0%, and an average modulus of rupture not less than 3 MPa (500 psi). Silica brick is further classified on the basis of its "flux factor" which equals the percent of alumina plus twice the percent of the total alkalies. Type A silica brick has a flux factor of 0.50 or less, and Type B includes all other silica brick meeting the general requirements.

Raw Materials: Gannister, a true quartzite, is a commonly used silica raw material for the manufacture of silica brick. The principal deposits in the United States are the Medina Sandstone of Pennsylvania, the Baraboo Quartzite of Devil's Lake Region, Wisconsin, and the quartzite of Alabama and Colorado. Pebble rock such as the Sharon Conglomerate of Ohio is also widely used for brick. Not all silica raw materials of these types are suited for silica brick manufacture. Criteria for judging raw material suitability include chemical purity and burning characteristics conducive to giving brick of adequate strength.

Manufacture: Because of the harmful effect of alumina and alkalies, some of the raw materials used for the manufacture of silica brick must be crushed, if necessary, and then washed to remove clay and other impurities. The washed silica is then ground in either a muller-type mill or in a rod mill and grain sized into the coarser fractions, and a portion of the silica is ball milled.

The several grain size fractions of the raw material are batched to give the specified grain sizing, and tempered in a muller-type mixer with about 3% lime hydrate, water, and other bonding agents added. The brick and tile are formed either at high pressures on mechanical presses or at slightly lower pressures on impact presses where vibration aids compaction, or by hand molding in which the prepared mix is pounded into the mold by the use of air-hammers or heavy mauls.

The formed brick or tile are dried either in batch or continuous dryers, and are then fired at a temperature of approximately 1482°C (2700°F) in continuous tunnel or periodic kilns. In firing silica brick, a rigid firing schedule must be followed to accommodate the volume changes accompanying the several mineral conversions and inversions which silica undergoes when heated. The cooling rate must also be closely controlled in the lower temperature ranges to accommodate the cristobalite inversion, which is accompanied by a rapid volume change. Well-fired silica brick show most of the quartz to have been converted to the lower density forms of silica, cristobalite and tridymite, and consequently, the fired brick show a permanent volume expansion of about 13–15% and a corresponding decrease in density.

Uses: Silica brick and tile are relatively inexpensive, but, because of their rapid change of size through the cristobalite inversion range, 121–288°C (250–550°F), silica brick tends to crack when heated rapidly at low temperatures and this limits its area of application. In the past, silica brick was widely used as roof brick for open hearth and electric furnaces melting steel. Changes in steelmaking technology have reduced the number of operating open hearths and raised the open hearth and electric furnace operating temperatures above the service temperature of silica brick, and as a consequence, the amount of silica brick sold has shown an eightfold decrease since the middle 1950s.

Because of its excellent hot load resistance at temperatures close to its melting temperature, silica brick or tile continues to be used in coke ovens, ceramic kilns, glass tank crowns, and as a checker in the regenerators of some glass tanks.

In European and Japanese steel mills, silica brick is used as a blast furnace checker in some instances. This practice has been adopted for some American blast furnaces and, should it become widespread, the market for silica brick would show a considerable increase.

Insulating Brick

Classification: Insulating firebrick is characterized by low thermal conductivity, low heat capacity, low density, and high porosity. It is classified in the ASTM Standard Classification

TABLE 2—Grouping of Insulating Firebrick

Group Identification	Reheat Change Not More Than 2% When Tested at °F (°C)	Bulk Density Not Greater Than, Lb per Cu Ft (Mg/m³)
Group 16	1550 (845)	34 (0.54)
Group 20	1950 (1065)	40 (0.64)
Group 23	2250 (1230)	48 (0.77)
Group 26	2550 (1400)	54 (0.86)
Group 28	2750 (1510)	60 (0.96)
Group 30	2950 (1620)	68 (1.09)
Group 32	3150 (1730)	95 (1.52)
Group 33	3250 (1790)	95 (1.52)

From: ASTM Standard Classification C155-70 (reapproved 1976).

of Insulating Firebrick, C155–70 (reapproved 1976), as shown in Table 2, which lists groups 16, 20, 23, 26, 28, 30, 32, and 33. By multiplying the group numbers by 100, the ultimate use temperature in degrees Fahrenheit is derived. The table also indicates that the permissible linear reheat change is 2% maximum for each group when exposed for 24 hr to a temperature 50°F less than the ultimate use temperature. In addition to these groups, other lightweight firebricks, such as insulating silica brick and insulating zircon and zirconia brick, are available commercially.

Raw Materials: Insulating firebrick is generally made of normal high alumina and fire clay raw materials such as alumina, bauxite, kyanite, kaolin, and fire clay, but in some instances special materials such as zirconia or silica are used. Raw materials added to provide the lightweight characteristic of insulating firebrick include: (1) combustible materials, such as sawdust, coal, coke, paper, napthalene, anthracene, and plastic beads which leave voids in the brick after burning; (2) lightweight inorganic aggregates or fibers such as diatomaceous earth, expanded perlite, expanded vermiculite, expanded clay, foamed aggregate, alumina bubbles, mineral wool, and synthetic alumina-silica fibers; and (3) foaming agents to form voids in the formed brick through release of gases.

Manufacture: Insulating brick is generally made by either extrusion, pressing, or casting. The extrusion process normally uses combustible materials to achieve light weight, although lightweight aggregates are also used in this process. The pressing process uses some of the combustible materials, but only those which will not compress and later expand when the pressure is released. Lightweight aggregates are commonly used in the pressing process. The casting process generally incorporates the use of foaming agents and may also include the use of other raw materials to achieve light weight.

After forming by any of these processes, drying and firing is accomplished under closely controlled conditions. Most of the insulating brick requires grinding after firing to provide uniformly sized brick, although some brick made by pressing and using some of the lightweight aggregates does not require final grinding to size.

Uses: Insulating brick is used to reduce heat flow through refractory walls and to reduce the amount of heat required to heat the walls, thus serving to reduce the fuel consumption and time to bring a furnace or other enclosure to temperature.

Special Refractories

This group of refractories includes zircon, zirconia, ultra high alumina, silicon carbide, fusion cast products, chromic oxide, and carbon and graphite.

Zircon: Zircon is a highly refractory zirconium silicate, quite resistant to thermal shock and susceptible to attack by basic slags and iron oxide. There are two types of zircon brick classified by ASTM C545–70 (reapproved 1976)—Type A (Regular) having a bulk density of less than 3.8 Mg/m³ (240 pcf) and Type B (Dense) having a bulk density greater than 3.8 Mg/m³ (240 pcf). Both types contain not less than 60% ZrO_2 and not less than 30% SiO_2.

Zircon brick is normally made by dry pressing followed by drying and firing at high temperatures. In addition to dry pressing, relatively large as well as smaller special shapes are made by isostatic forming. Normally, in this process, the extremely fine zircon is prepared in slip form, spray dried to form free-flowing granules, formed under high pressures while contained in a flexible mold in an isostatic pressure chamber, and fired at high temperatures. This process provides a product that may be characterized by extremely high uniformity and density. Slip casting is also used to make large and/or intricate shapes.

Zircon brick and shapes are used in the glass industry and are generally limited to low soda borosilicate glasses. "E" fiber glass tanks and forehearths use this refractory extensively.

Zirconia: Zirconia, ZrO_2, is highly refractory and, when properly made, strong, hard, and resistant to chemical attack, particularly by molten silicates. Because of its unusual thermal expansion characteristics, it is normally modified by additions of calcia, magnesia, or yttria for practical industrial use. The manufacture of this product in brick and shapes is normally by dry pressing, isostatic forming, or slip casting.

Ultra High Alumina: This product is characterized by extreme purity, above 99.5% Al_2O_3, and is of fine grain size. It is used for extreme resistance to abrasion and chemical attack and for processing some glasses which are compatible with alumina. Various processes are used in manufacture, including slip casting, dry pressing, and isostatic forming. Densities achieved may approach theoretical.

Silicon Carbide: Silicon carbide grain used in the manufacture of brick is prepared in electrical resistance furnaces by the reaction of carbon (coke), silica, and a small amount of soda ash. It does not soften or melt below its dissociation temperature of 2510°C (4550°F). SiC is not attacked by most acids, but is deteriorated by basic slags, molten iron, and oxidizes when heated in air. SiC refractories are formed by dry pressing, airhammer, and vibration (jolting); some are also prepared by hot pressing to form pieces used for exotic applications such as turbine blades. SiC refractories find use as ceramic setter tile; muffles; in boilers; in heat treating, annealing, and forging furnaces; in electrical resistance elements; and in gas producers.

Nitride bonded silicon carbide bricks have been used successfully in the linings of Japanese blast furnaces and recently this practice has been adopted by some European and American steel mills. The use of ramming mixes containing a significant amount of silicon carbide grain for blast furnace tapholes, troughs, and runners has recently begun to find American applications, especially for larger furnaces.

Fusion Cast Products: A number of refractories are formed by fusing the refractory mixture and casting into required shapes; these products include alpha and beta alumina, basic compositions, and a variety of alumina-zirconia-silica compositions. Uses of these refractories vary but include applications involving glass contact, high load stresses, abrasive conditions, and limited areas of high vulnerability in the steel industry.

Chromic Oxide: These products are made predominately of chromic oxide (Cr_2O_3) either by slip casting or isostatic forming. They are fine grained, extremely dense bodies, and are used primarily in fiber glass tanks to reduce glass corrosion. Chromic oxide is a colorant in glass, but the rate of solution of these refractories may be low enough to permit its use in selected areas of glass tanks.

Carbon and Graphite: The largest application of carbon refractories is in the hearth, sidewalls, and bosh sections of blast furnaces. The primary raw material for most carbon brick is anthracite-based carbon. Graphite is used in the manufacture of crucibles and ladle stopper heads, and some graphite is used in plastic firebrick. Oxygen, alkalies, water vapor, and carbon dioxide will attack carbon at elevated temperatures, but under reducing conditions, carbon can withstand temperatures above 3316°C (6000°F) without melting or softening.

Refractory Specialties

Refractory specialties are those refractory products which are sold as nonformed products including mortars, castables, plastics, and ramming mixes. Mortars are normally used for bonding together brick of the same type as the mortar. The mortar joints used in refractory construction are quite thin so the refractory mortar must be capable of forming a thin joint. Some mortars are used in a thicker than usual consistency as a coating on refractory brick.

Many castables are used in the same manner as construction concrete to form a monolithic refractory. A trend toward gun application of castables is apparent, since this type of application has the advantage of requiring much less forming than the casting procedure. The installation costs and good properties of castables have led to their wide use in the petrochemical industry, as well as for repairing refractory linings including blast furnaces.

Plastics and ramming mixes are used to repair refractory installations where the repairs can be installed against the existing refractory structure. Because of their capability of being molded in place, they also are used for refractory installations that would otherwise require intricate—and therefore expensive—preformed shapes.

Mortars: Refractory mortars are of two general types. Air setting mortars have a bonding agent which imparts a set to the mortar on drying. With further heating the strength given by the bonding agent decreases in some cases, while the bond derived from heat-induced ceramic reactions increases. Heat setting mortars have only moderate strength on drying and develop a strong ceramic bond on exposure to high temperatures. Refractory mortars vary greatly in composition depending on the refractory raw materials used in them and are generally of similar composition to the brick with which they are used. They are of fine grain size and use plasticizing agents to give them the proper and desired workability. Their manufacture is relatively simple, comprising crushing and milling the refractory raw materials to the desired sizes, batching with the required bonding and plasticizing agents, mixing either wet or dry, and packaging accordingly.

Castables: Refractory castables are produced from refractory aggregates and bonding agents such that when mixed with water and poured into a mold they will set in a reasonable time due to chemical reaction. Most refractory castables are boned with hydraulic-setting calcium aluminate cements. They generally are divided into heavy and insulating castables, and each of these two types is further classified according to its service temperature and, for the insulating castables, density. This is shown in Table 3 from ASTM C401–77, which is the Standard Classification of Castable Refractories. A majority of refractory castables are of the fire clay and high alumina type, but other aggregates may be used such as chrome ore, tabular alumina, and chrome-magnesite mixtures. When intended for gunning application, special additives are used.

The manufacture of refractory castables consists of crushing and sizing the refractory aggregate, batching with the required bonding agent and additives, mixing dry in a blade or concrete-type mixer, and packaging in bags or drums.

Plastics: Plastics are products sold in a stiff plastic condition which are then formed into shape or placed at the installation site. Various types of plastic firebrick are made, depending primarily on the aggregate used, such as fire clay grog, crushed fire clay brick, bauxite, tabular alumina, and chrome ore. These products are divided into air-setting and heat-setting types. The quality or refractoriness of a given plastic is dependent on the aggregate used as well as the type and amount of bonding and plasticizing agents used. The manufacture of plastics consists of crushing and sizing the aggregate, batching to give the desired mix grain sizing, mixing and tempering, and generally extrusion to form a clot which is then sliced for ease of installation. Packaging is usually in cartons with protective wrapping to avoid premature drying, although some are sold without extrusion and are packaged in drums.

Ramming Mixes: Ramming mixes consist primarily of ground, sized refractory aggregates with minor amounts of other materials

TABLE 3—Regular Castable Refractories Applying to Both Normal-Strength and High-Strength Types

	Classes of Alumina-Silica-Base Castable Refractories						
Test Requirements	Class A	Class B	Class C	Class D	Class E	Class F	Class G
Permanent linear shrinkage, not more than 1.5% when fired for 5 h at	2000°F (1095°C)	2300°F (1260°C)	2500°F (1370°C)	2700°F (1480°C)	2900°F (1595°C)	3100°F (1705°C)	3200°F (1760°C)

Insulating Castable Refractories

	Classes of Insulating Castable Refractories				
Test Requirements	Class N	Class O	Class P	Class Q	Class R
Permanent linear shrinkage, not more than 1.5% when fired for 5 h at	1700°F (925°C)	1900°F (1040°C)	2100°F (1150°C)	2300°F (1260°C)	2500°F (1370°C)
Maximum bulk density, lb/ft³ (Mg/m³) after drying at 220 to 230°F (105 to 110°C)	55 (0.88)	65 (1.04)	75 (1.20)	90 (1.44)	95 (1.52)
	Class S	Class T	Class U	Class V	
Permanent linear shrinkage, not more than 1.5% when fired for 5 h at	2700°F (1480°C)	2900°F (1595°C)	3000°F (1650°C)	3200°F (1760°C)	
Maximum bulk density, lb/ft³ (Mg/m³) after drying at 220 to 230°F (105 to 110°C)	95 (1.52)	100 (1.60)	105 (1.68)	105 (1.68)	

Source: ASTM Standard Classification C401-77.

added to impart workability and to act as a bond. The refractory aggregate used in these products includes silica, tabular alumina, calcined fire clay, calcined bauxite, silicon carbide grain, zircon, periclase, and chrome ore. The amount of clay, bentonite, or other plasticizers and bonding agents is normally less than that contained in plastics. Some ramming mixes are supplied dry for preparation with water at the installation site while others are supplied at the required ramming consistency. In the latter case, they are packaged in drums, while in the dry form they are normally packaged in bags. Ramming mixes may be of either the air-setting or heat-setting type, and when rammed into place, they harden to form well-bonded monolithic refractories.

TABLE 4—Classification of Chrome-Magnesite Brick, Magnesite-Chrome Brick, and Magnesite Brick According to MgO Content

Class	MgO Content, %*	
	Nominal	Minimum
Chrome-Magnesite and Magnesite-Chrome Brick		
30	30	25
40	40	35
50	50	45
60	60	55
70	70	65
80	80	75
Magnesite Brick		
90	90	86
95	95	91

*Ignition-free basis.
Source: ASTM Standard Classification C455-76.

Basic Brick

Classification: ASTM defines basic refractories thus, "Refractories whose major constituent is lime, magnesia, or both and which may react chemically with acid refractories, acid slags, or acid fluxes at high temperatures." A note to this definition states, "Commercial use of this term also includes refractories made of chrome ore or combinations of chrome and dead-burned magnesite." In this chapter, the broad definition is used.

The ASTM Standard Classification C455–76 classifies basic brick, composed of chrome ore and dead-burned magnesite as end members and combinations of them, on the basis of the MgO content (Table 4). In addition to this classification, basic brick is further divided into chemically bonded brick and burned brick. Burned brick is further divided into that burned at normal temperatures and that burned at high temperatures, generally exceeding 1649°C (3000°F), to effect the so-called direct bonding microstructure. The latter brick is primarily characterized by its enhanced strength at high temperatures. Both chemically bonded and burned brick are further classified as metal encased brick when partially enclosed with soft steel plates which, when exposed to high temperatures, act to bond the brick.

The advent of basic oxygen steelmaking and its rapidly expanding use over the past several years has led to an entirely new class of basic refractories. These are the pitch-bonded and pitch-impregnated brick, and this group includes: (1) pitch-bonded periclase, (2) pitch-bonded dolomite, (3) pitch-bonded dolomite periclase, and (4) pitch-impregnated periclase brick. Pitch-bonded brick is sometimes subjected to a low temperature heat treatment, and this is referred to as pitch-bonded tempered brick. It is characterized by resistance to peeling or slumping during the burning-in of a basic oxygen furnace (BOF) lining. The pitch-bonded and pitch-bonded tempered periclase bricks are generally of two types, one of high purity (about 96% MgO plus) and one of lower purity (nominally 90% MgO). The dolomite bricks are of two types, the regular dead-burned dolomite and a low flux dolomite characterized by lower amounts of oxides other than CaO and MgO. The pitch-impregnated periclase bricks are generally of two types, a regular burn and a direct bonded (high fired) type.

Raw Materials: *Chrome Ore*—Chromite has the theoretical composition $FeCr_2O_4$, but commercial refractory chrome ores contain substantial amounts of alumina, magnesia, and silica. They are characterized by a chromic oxide (Cr_2O_3) plus alumina (Al_2O_3) content not less than 60% and a silica content less than 6%. They will generally range from 30–50% Cr_2O_3, 12–16% FeO, 13–30% Al_2O_3, 14–20% MgO, 1–6% SiO_2, and up to 1% CaO. High quality refractory ores, such as that from the Philippines, which is the primary source, have about 32% Cr_2O_3 and 30% Al_2O_3, while other excellent ores have as much as 42% Cr_2O_3 and correspondingly lower alumina contents. Other sources of chrome ore include Turkey, Zimbabwe, and South Africa. Albania and the USSR also mine chromite.

Magnesite and Periclase—The term *magnesite grain* is defined by ASTM as "dead burned magnesia in granular form of size suitable for refractory purposes." The term *magnesite, seawater* is defined as, "Dead-burned magnesia made by a chemical process using seawater or other saline solutions." These are, of course, refractory terms and differ from the mineralogical term *magnesite,* which refers to a mineral composed of $MgCO_3$. The term *periclase* is often used in refractories to describe dead-burned magnesia derived from seawater or brines, while mineralogically *periclase* refers to the high temperature form of magnesia.

Most of the dead-burned magnesite grain derived from natural deposits uses magnesium carbonate as its source, although some brucite ($Mg(OH)_2$) deposits have been used. The largest deposits of magnesite in North America are in Washington, Nevada, western Canada, and Quebec. Other dead-burned magnesite grains used in the United States are derived from natural deposits in Greece, Austria, and Yugoslavia.

Some of the natural deposits provide grain of excellent quality and high purity; however, the need for periclase grain of high purity and closely controlled properties has led to a significant growth in the production of such grains from seawater and brines. This chemical process involves precipitation of magnesium hydroxide from the solutions using slaked lime, slaked dolomite, or caustic, followed by conversion of the magnesium hydroxide to periclase by calcining in rotary or shaft kilns at extremely high temperatures. The grain derived from either natural deposits or from magnesium-containing solutions is rendered relatively inert to reaction with moisture or carbon dioxide of the atmosphere, or in other words, is considered "dead-burned."

Periclase grains derived from magnesium-containing solutions can be modified in their chemical composition during manufacture, and grains having MgO contents in excess of 98% are commercially available.

Dolomite—Dolomite grain is obtained by treating the mineral dolomite $Ca,Mg(CO_3)_2$ at high temperatures, thus driving off the CO_2 and rendering the resultant material more dense. Extensive deposits of high purity dolomite are readily available. Two types of grains are produced for refractory use, the regular grain and a low flux grain of higher purity.

Manufacture: In the manufacture of basic brick, the chrome ore and periclase or magnesite ingredients are crushed, ground, and carefully grain sized into several fractions. Generally, two separate systems are used for these two ingredients in order to minimize contamination during processing. Because the mineral placement, i.e., the size fractions in which the periclase and chrome ore occur, has such a strong influence on the properties of these brick, it is important that the grain size of each fraction be controlled within its specified limits. The grain size fractions of each mix ingredient are carefully batched and then mixed in a muller-type mixer with the tempering water and additives. Most basic brick is formed on mechanical or hydraulic presses at high pressure, but intricate shapes are formed by the airhammer process. The chemically bonded category of this brick is dried at a sufficiently high temperature so that the chemical bond will develop. The burned brick is dried and then fired at sufficiently high temperatures to develop a ceramic bond, and the direct bonded brick is fired at about 1760°C (3200°F) in order to develop the direct or crystal-to-crystal bond. Either the chemically bonded or the burned brick of both types may be further processed to metal-encased brick by attaching soft steel plates by various means.

The periclase used in the manufacture of brick for the basic oxygen furnace (BOF) is crushed, ground, and sized into fractions. The coarse sized fractions of periclase or magnesite are batched and heated to about 204°C (400°F). The heated aggregate is then mixed with the pulverized periclase fraction, heated pitch, and carbon particulates in a blade or pan-type mixer. The hot mix is promptly conveyed to a mechanical press and the brick is formed by the use of high pressure. The pitch-bonded brick is then cooled prior to shipment. The tempered equivalents of this brick are made by heating the pitch-bonded brick to about 316°C (600°F) in order to remove some of the volatiles and develop some carbon bonding. Tempered brick shows better strength during the early heat-up of the BOF than the regular pitch-bonded type.

The pitch-bonded and pitch-bonded tempered dolomite and dolomite-periclase brick are made in the same manner but with the coarse size fractions being dolomite grain.

Burned-impregnated periclase brick is made by the same process as conventional basic brick, except after firing the brick is heated to about 204°C (400°F), subjected to vacuum, and then impregnated in a pressure chamber with pitch under a pressure of several atmospheres.

Uses: Basic brick finds its greatest use in the steel industry in open hearths and electric furnaces, but also finds considerable use in cement kilns and the furnaces of the copper and nickel processing industries, as well as in glass tank regenerators. At present, most of these applications use the burned type of brick and in the case of the steel, cement and nonferrous industries, direct bonded brick is beginning to find greater favor. The brick used in the glass tank regenerators is of the ceramic and direct-bonded type, but in the basic checkers, chemically bonded brick of various qualities also finds use. Pitch-bonded, pitch-bonded tempered, and pitch-impregnated brick find their principal uses in the basic oxygen furnace in various areas. The trend is toward zoned linings in these furnaces; for example, pitch-impregnated high fired periclase brick is used in the charge pad area where it is subjected to the impingement of scrap and hot metal. It is interesting to note, however, that brick of this type is being tried in other applications, such as in electric furnaces for steel melting.

Economics and Future Trends

Table 5 presents selected statistical data for refractory shipments during the past ten years as compiled by the US Bureau of Census. These data are shown in 1967 dollars in order to eliminate the distorting effect of inflation. From these data it is obvious that certain categories of refractories have increased in dollar volume. High alumina brick, extra high alumina brick, fire clay specialties, and nonclay specialties have shown a marked increase from the beginning of the decade, while special refractory brick and basic brick have shown upward trends. Silica brick and shapes showed an increase from their low point of the early 1970s because of their increased use in blast furnace stoves and an increase in coke oven construction. Other categories remained at relatively constant levels. The largest single dollar category throughout the decade has been basic brick and shapes.

These trends are better understood when one considers certain trends in the usage of refractories during this period and at present. Increasingly, users of refractories have been more willing to accept higher first costs provided that the refractory item gives increased life or otherwise contributes to lower refractory cost per unit of product being produced. Thus, for example, high alumina refractories are gradually replacing fire clay refractories in some applications. Another important consideration for the refractories user is production loss due to equipment downtime, and to minimize this the use of refractory specialties as materials for patching or partial rebuild has found increasing favor.

TABLE 5—Shipments of Refractories in the United States (Millions of 1967 Dollars)

	1970	1971	1972	1973	1974	1975	1976	1977	1978	1979
Clay Refractories										
Brick and shapes										
Fire clay	79.7	69.4	72.5	71.0	73.6	72.7	70.8	70.0	75.2	75.6
High alumina	36.9	34.9	43.5	47.5	51.0	46.9	43.1	46.8	58.7	62.5
Pouring pit	42.0	37.9	44.4	44.1	43.7	34.5	37.3	41.5	45.2	44.4
Insulating brick	15.7	11.7	12.4	13.6	14.3	15.5	14.9	13.5	16.4	15.2
Fire clay specialties*	58.0	55.6	61.8	66.9	73.8	80.2	87.3	92.8	105.0	121.9
Total clay refractories	232.3	209.5	234.6	243.0	256.3	249.9	253.3	264.6	299.7	319.6
Nonclay refractories										
Brick and shapes										
Silica	13.8	10.6	10.8	11.3	12.5	18.1	19.8	16.0	18.8	19.3
Basic	114.1	102.3	112.6	130.9	136.4	108.9	118.3	125.5	157.0	148.3
Extra high alumina	18.2	16.6	14.7	16.0	19.3	19.0	20.4	21.9	25.0	29.3
Other (special refractories)	80.4	73.2	78.4	84.1	93.1	87.0	85.6	93.9	110.0	118.0
Nonclay specialties*	75.2	82.2	88.6	93.2	93.5	91.4	95.0	95.6	117.9	131.7
Total nonclay refractories	301.7	284.9	305.1	355.6	354.8	324.4	339.2	352.8	428.8	446.7
Total refractories	534.0	494.4	539.7	578.5	611.1	574.3	592.5	617.4	728.5	766.2

Source: US Bureau of Census *Report on Refractories,* series MQ-32C.
* Mortars, castables, plastics, ramming mixes, and sized refractory grains.

Although the dollar value of refractories shipments has tended to increase over the years, the number of units of product shipped has had a tendency to decrease. This trend is also related in part to changes in practice by refractories users. For example, steelmaking vessels since the advent of the BOF process have tended to grow smaller in volume, and yet these smaller vessels produce the same or greater quantities of steel annually than the larger furnaces they have replaced. Refractories themselves in every category have undergone upgrading and improvement during the last decade, and these improved products tend to outlast the old. In turn, the new products sell at higher prices. Many improvements in metallurgical and minerals processing during the decade were made possible largely or entirely due to these new and improved refractories.

The trend, therefore, seems clearly established toward refractories of improved properties and often higher chemical purity and cost but in decreasing unit volume. With this in mind, a few comments will be made concerning the present principal sources of refractory raw materials.

Fire clays and fire clay calcines for use in refractories come almost entirely from domestic sources and the mining and processing of these materials are largely carried out by the refractories producers themselves. Sources of high quality fire clays having low levels of contamination have diminished with years of mining and before the end of this century it may be necessary to mine and beneficiate lower grade clays. The cost of beneficiation could significantly increase product costs. One factor in the wide usage of fire clay brick and specialties in the past has been their lower first cost vs. many other categories of refractories, and a significant increase in cost would undoubtedly cause some refractories users to reconsider using other types of refractories.

No refractory grade chrome ore deposits exist in the United States, and the refractories industry, like the metallurgical industry, is dependent on importation of chrome ores. Of equal, if not greater, importance is the importation of bauxite, for, although sources of domestic bauxite exist and are being worked, the volume of bauxite required for high alumina refractories demands importation. Both chrome ore and bauxite are found in areas of the world which are politically sensitive, and an uninterrupted source of supply must be a point of future concern to refractories manufacturers and users. It should be noted with regard to chrome ore that the change in steelmaking toward the BOF process has decreased the requirement for refractory grade chrome.

Magnesite, dolomite, and periclase present a much more favorable outlook. Not only are natural magnesite and dolomite available domestically, but the reserves of periclase derived from seawater are theoretically limitless. In connection with seawater periclase it is of interest that the processing facilities already in operation in the United States have more than enough capacity to supply present needs.

Other factors that may have a strong influence on future trends in refractory usage are: (1) changes in metallurgical practices, (2)changing sources of available energy, and (3) ecological considerations.

One change in metallurgical processing presently being practiced on a limited basis is the direct reduction of iron ore. Where presently practiced, the economic conditions or available raw materials make these processes more attractive than reduction of iron ore by the more normally used blast furnace. Should these direct reduction processes be improved significantly to allow their more widespread use, the markets for fire clay and other refractories used in blast furnaces and stoves would be unfavorably affected.

The diminishing sources of available energy may have a strong influence on furnace and process design. Changes designed to conserve energy may increase the consumption of insulating and other refractories as they will call for less dissipation of heat by water and air cooling.

Ecological considerations at present are exerting an influence on the choice of processing furnace, with consideration being given to the use of electric furnaces in place of fuel-fired furnaces, because of their lesser emission of fumes. In the future, a stronger trend to electric furnace processing may evolve, which could influence the refractory consumption pattern.

Bibliography and References

Anon., 1961, *Modern Refractory Practices,* 4th ed., Harbison-Walker Refractory Co., Pittsburgh, PA, 607 pp.

Anon., 1978, "Refractories—Changing Patterns of Consumption," *Industrial Minerals,* Pt. 1, No. 125, Feb., pp. 15–35.

Anon., 1978a, "Refractories, Producers Progress," *Industrial Minerals,* Pt. 2, No. 128, May, pp. 21–49.

Budnikov, P.P., 1964, *The Technology of Ceramic Refractories,* M.I.T. Press, Cambridge, MA, 647 pp.

Chesters, J.H., 1963, ***Steel Plant Refractories, Testing, Research and Development,*** 2nd ed., United Steel Co., Sheffield, Eng., 779 pp.

Granitzki, K.E., and Hail, W., 1970, "Jahresuebersicht Feuerfeste Baustoffe, 6" (Sixth Annual Summary of Recent Literature on Refractory Materials), *Giesserei,* Vol. 57, No. 7, Mar. 26, pp. 175–81 (German).

Grim, R.E., 1968, *Clay Mineralogy,* 2nd ed., McGraw-Hill, New York, 596 pp.

Herold, P.G., 1942, "Relation of Permeability and Specific Gravity of Insulating Refractories," University of Missouri, Rolla, School of Mines and Metallurgy, Technical Series, Vol. 14, No. 4, 51 pp.

Havranek, P., 1970, "Naga Utvecklingslinjer pa de Eldfasta Material ens Omrade under 1960-Talet (Some trends in the developments of refractories in the 1960s), *Jernkonterets Annales,* Vol. 154, No. 1, pp. 1–8 (Swedish).

Jaffee, R.I., 1959, "A Brief Review of Refractory Metals," Battelle Memorial Institute, Columbus, Ohio, Defense Metals Information Center, Memo. 40, 34 pp.

Levin, E.M., et al., 1959, *Phase Diagrams for Ceramists,* The American Ceramic Society, Columbus, Ohio, P. 2, 286 pp.

Levin, E.M., et al., 1969, ***Phase Diagrams for Ceramists,*** The American Ceramic Society, Columbus, Ohio, Supplement.

Miska, W.S., 1968, "Refractory Use Patterns in the Iron and Steel Industry of the United States." Information Circular 8382, US Bureau of Mines, 67 pp.

Norton, F.H., 1968, ***Refractories,*** 4th ed. McGraw-Hill, New York, 450 pp.

Schwarzkopf, P., 1953, ***Refractory Hard Metals, Borides, Carbides, Nitrides, and Silicides,*** Macmillan, New York, 447 pp.

Schwarzkopf, P., et al., 1960, *Cemented Carbides,* Macmillan, New York, 349 pp.

Searle, A.B., 1948, ***Refractories for Furnaces, Kilns, Retorts, Etc.,*** 2nd ed. rev., C. Lockwood, London, 121 pp.

Also useful:

1980 Annual Book of ASTM Standards, Pt. 13, American Society for Testing & Materials, Philadelphia, PA.

"Refractories," *Current Industrial Reports,* Bureau of the Census, US Department of Commerce, Washington, DC.

Refractories Bibliography, 1928–1947 inclusive, 1947–1956 inclusive, American Iron & Steel Institute, The American Ceramic Society.

Well Drilling Fluids

NEAL DAVIS *
STANLEY J. LEFOND †

Drilling an oil well or most other types of drilling or coring is no longer a simple and uncomplicated operation. Drilling today at depths which exceed 9144 m (30,000 ft) is hazardous and requires personnel with long and varied experience in drilling. It also requires extensive experience and knowledge in the use of materials which are added to the drilling fluid or "mud," more often than not materials utilizing an industrial mineral. The drilling fluid can vary from a simple mixture of water and "mud" to a very sophisticated, scientifically compounded drilling fluid.

Industrial Minerals Used in Drilling Fluids

Industrial minerals which are commonly used in oilwell drilling and completion are listed in Table 1. The most important minerals, determined by sales, are bentonite and barite. The other minerals listed are either used to achieve a specific purpose or are used regularly but in limited quantities. The main suppliers of drilling fluids internationally are Magcobar Div., Dresser Industries, Inc.; Baroid Div., National Lead Co.; IMC Drilling Mud Co., (a subsidiary of International Minerals and Chemicals Corp. and Halliburton Co.); and Milchem, a division of Baker Industries. Most of the industrial minerals are found in North America, with barite mines and processing plants located in Nevada, Arkansas, Missouri, and Arizona, and bentonite mines near Graybull, WY. Other mines, such as those in Nova Scotia, are capable of shipping to South America and to the European and Middle East countries.

Most of the materials processed by any of the suppliers must conform to specifications set by the American Petroleum Institute (API) or the Oil Companies Material Organization (OCMA). These specifications provide for uniform physical and chemical properties of the materials to be sold on the open market.

* Technical Service Engineer, Magcobar Operations, Oilfield Products Div., Dresser Industries Inc., Houston, Texas.

† Consultant, Industrial Minerals, Evergreen, CO.

Function of Drilling Fluids

In the early days of rotary drilling, the primary function of drilling fluids was to bring the cuttings from the bottom of the hole to the surface. Today it is recognized the drilling fluid has at least ten important functions:

1) To remove the cuttings from the bottom of the hole and carry them to the surface.
2) To cool and lubricate the drill string and drill bit.
3) To build an impermeable cake on the wall of the hole.
4) To control subsurface pressures.
5) To hold cuttings and weight material in suspension when circulation is interrupted.
6) To release sand and cuttings at the surface.
7) To support part of the weight of drill pipe and casing.
8) To reduce to a minimum any adverse effects upon the formation adjacent to the hole.
9) To insure maximum information about the formation penetrated.
10) To transmit hydraulic horsepower to the bit.

In order to accomplish some or all of these functions while penetrating formations of various characteristics, several types of drilling fluids systems have been designed.

Types of Drilling Muds

There are two basic drilling fluid systems—water-base muds and oil-base muds. There is a wide variety of groups within these two main

TABLE 1—Industrial Minerals Used in Well Drilling

Additives	Form	Specifications	Uses
Asbestos	Fiber	Shorts, 6-7 grade	Improves cutting recovery
Attapulgite	Ground to −200 mesh	Apparent viscosity: 25 g clay in 250 cc of saturated saltwater, 30 mi. dial reading Test at 600 rpm in rotational viscometer Wet screen analysis residue, 8.0 max Moisture, 20% max	Imparts suspension properties in saltwater
Barite	Finely ground −325 mesh	Specific gravity 4.2 min. Wet screen analysis, 200 mesh (US) residue, 3.% max; 325 mesh residue, 5.%; soluble alkaline earths, 200 ppm max	To control specific gravity in water- and oil-base muds
Bentonite	Ground to −200 mesh	Plastic viscosity, 10 g bentonite in 350 cc water, 8.0 g per min Filter loss, 10 g bentonite in 350 cc water, 14 cc max Wet screen analysis, 200 mesh (US) residue, 2.5% max; moisture content, 12% max	Used to increase suspending power of water-based muds To form a mud cake on the wall of the hole
Bicarbonate of soda	Ground		Same use as soda ash
Diatomaceous earth	Ground		Cement additive Lost circulation
Galena	Finely ground		Weighting agent
Graphite	Flake		Lubricant
Gypsum	Ground		To coagulate clay and thin the mud in "Gyp" mud; also to treat for carbonate contamination
Lime	Ground powder		To control alkalinity in freshwater and/or mud; to treat contamination
Limestone or dolomite	Finely ground		Low density weighting agent
Mica	Flakes		Lost circulation prevention
Perlite	Expanded		Cement additive
Quartz	Sized grains		Propping agent in hydrofracing
Salt	Ground crystal		To control hole size when drilling through salt horizons; to minimize the dispersion of sticky clay cuttings in the mud
Soda ash	Ground		Useful in controlling flow properties of freshwater muds and primarily to overcome thickening due to calcium contamination
Volcanic ash	Ground		Cement additive

divisions; each has its advantages and disadvantages within a certain area. The type and composition of the drilling mud depends upon the requirements of the particular drilling operation. Holes must be drilled through different type formations requiring different types of fluids and ingredients. Economics, contamination, available makeup water, pressures, downhole temperatures, and many other factors all are significant in the choice of the drilling fluid.

Water-Base Muds

The majority of drilling fluids may be classified as water-base muds. The term water-base mud means any drilling fluid having a continuous liquid phase of water, in which clay or claylike materials are in suspension. There are numerous mud additives used to obtain special properties, but basically all water-base muds have three basic components or phases.

1) Water phase—water is the continuous phase of the mud.
2) Reactive solids phase—commercial clays and hydratable drilled solids compose the reactive solids in the mud. These solids are subject to chemical treatment to control the properties of the drilling fluid.
3) Inert solids phase—barite is the principal commercial inert or nonreactive solid.

Makeup Water: The water used for makeup and maintenance of water-base muds is classified as either fresh or salt base. Freshwater muds contain little or no salt concentration in the form of sodium, calcium, or magnesium. If the makeup water does contain high concentrations of calcium or magnesium, the water may be treated with soda ash or bicarbonate of soda to reduce the concentration.

Salt-Base Mud

Saltwater drilling fluid contains a relatively high concentration of salt ranging from brackish or seawater up to and including saturated saltwater. The salt-base mud system is usually designed for a specific purpose—i.e., drilling salt-contaminated formation or to prevent shale hydration—or for economy—i.e., availability of seawater in offshore drilling.

Oil-Base Muds

An oil mud is defined as a drilling fluid having a continuous liquid phase of oil. There are two types of oil mud available. Oil-base muds are mixtures of oxidized asphalt, organic acids, alkali, stabilizing agents, and high flash diesel oil. Such mud contains 3 to 5% of water emulsified in the oil. The second type is an inverted emulsion oil mud, containing up to 50% water emulsified in the oil. Various additives are used to emulsify the water and stabilize both systems.

Oil muds are used when there are special economic advantages over water-base systems. Oil muds can be used justifiably:

1) To drill troublesome shales.
2) To drill deep, hot holes.
3) To minimize formation damage in possible production zones.
4) To drill contaminated zones.
5) To drill hydrogen sulfide (H_2S) gas zones and carbon dioxide gas (CO_2), minimizing drill pipe corrosion.
6) As a packer or workover fluids.

The foregoing are the main advantages of the use of an oil mud over a water-base mud. Economics and the availability of diesel play a great importance in the limited use of an oil mud.

Industrial Mineral Usage

There are many products used in the makeup and treatment of the various drilling fluids. The basic products utilized in the drilling industry are barite and bentonite, a sodium montmorillonite. Most of the other products can be classified as to the service they perform.

Weighting Materials

Barite, barium sulfate, is a natural ore mined domestically and internationally and is used to increase the density of all drilling fluids up to a maximum of 2.6 kg/L (22 lb per gal). It is used to hold back high formation pressures and troublesome shales. Other materials used to increase the density of the mud are ground limestone, used in oil muds to minimize formation damage, and galena, used in preparation of muds requiring densities of 2.6–3.6 kg/L (22–30 lb per gal) for completion or kill operations in high pressure formations.

Viscosifiers

Viscosifiers are used to produce drilling fluid viscosity for the rapid removal of cuttings from the bit and carry the cuttings to the surface and to provide adequate support of cuttings and weight material when circulation is interrupted. The main viscosifiers are:

1) Wyoming bentonite (sodium montmorillonite) is used in freshwater mud systems.

2) Attapulgite clay is used in water containing high salt concentration.

3) Asbestos is also used to achieve viscosity either in a freshwater or saltwater mud either as a replacement for one of the water types or as a supplement.

4) Low-yield clays (usually composed of calcium montmorillonite) used in some drilling operations because of availability or to achieve a special purpose. This clay does not have the "yield" characteristics of the Wyoming bentonite.

Thinning and Dispersing

To control the rheology or flow properties of a water-base mud system, organic and inorganic chemicals are added to a mud system which change the electrochemical attraction between the colloidal particles in the system and also to treat and eliminate contaminants. Among the commonly used thinners and dispersants are lignins, lignites, lignosulfonates, and polyphosphates.

Fluid-Loss Control Agents

To prevent extensive filtrate invasion of the formation from the fluid of the mud system, some means of controlling fluid loss is needed. Although bentonite and water can build an acceptable filter cake around the borehole in the surface drilling, more control is needed as drilling progresses into formations containing contaminants and as temperature increases. Thinners and dispersants give additional help, but the complete control can be achieved with fluid-loss control agents. Most of these agents are organic colloids such as pregelatinized starch, carboxymethyl cellulose, polyanionic cellulose, resins, and other high molecular polymers.

Commercial Chemicals

There is a large line of commercial chemicals used in all types of drilling fluid to control alkalinity, treat contamination, alter the water phase of the mud system, and help alter the viscosity and thixotropic properties of the mud system. To control alkalinity, caustic soda and lime (calcium hydroxide) normally are used. Contaminates, especially calcium, are treated with sodium bicarbonate, soda ash, and gypsum. Salt (sodium chloride) and calcium chloride are used to alter the salt concentration in the water phase in fresh and saltwater, and oil mud systems; sodium chromate and chrome alumchromic chloride are also used in limited quantities to help control viscosity under unusual circumstances.

Lost Circulation Materials

The loss of the drilling fluid to highly fractured and porous formations is a common occurrence. Many materials are used to help seal off the thief zones. There is actually no one type of material which has a high success ratio. Among the many used to some degree of success are mica (processed muscovite), diatomaceous earth, processed cellophane flakes, ground paper, and organic products such as redwood fibers, ground almond shells, sawdust, cane fiber, and cottonseed hulls.

Oil Mud Products

Most oil muds use organic colloids to control viscosity and stability. However such materials as barite, gilsonite, blown asphalts, and ground limestone give additional properties desired by the mud engineer.

Specialty Products

In many instances special mud properties are desired when unusual circumstances are encountered while drilling. There are many specialty products available, most of them originating from organic compounds. Among those industrial minerals used are blown asphalt, graphite, silicates, aluminum stearates, and polymers used for flocculation and viscosity.

As the demand for more oil production grows, so does the demand for more drilling. In the United States, the quest for increased oil exploration is limited at the present time (1981) because of shortages: shortages of drilling personnel, drilling rigs, tubular goods such as drill pipe and casing, and a shortage of some of the industrial minerals listed previously.

Bibliography and References

Anon., 1972, "Drilling Fluids Engineering Manual," Magcobar Operations, Oilfield Products Div., Dresser Industries Inc., Houston, Texas, Jan., 204 pp.

Jones, G.K., 1972, "Industrial Minerals in Oilwell Drilling," *Industrial Minerals,* No. 9, Sep., pp. 9–31.

3

Sources of Information for Industrial Minerals

Sources of Information*

ROBERT B. HOY †

If the reader finds that the basic information in a commodity chapter is insufficient, he can consult the appropriate sources in this chapter to find more detailed or more up-to-date information.

In these days, technology develops so rapidly that this 5th edition will be somewhat out of date even by the time it is distributed. A comparison of this edition with the 4th edition (1975) of *Industrial Minerals and Rocks* will clearly demonstate how rapidly the scene changes.

One of the best sources of new information, of course, is attendance at technical meetings. For example, the papers presented at the national and sectional meetings of AIME usually discuss new developments in the respective divisions of the society. Frequently this information is available in *Preprints*. In addition some are printed in *Mining Engineering* and/or the *Transactions of the Society of Mining Engineers of AIME*. Furthermore, other professional and trade journals contain articles pertinent to the field. A journal introduced in 1967, *Industrial Minerals,* published by the Metal Bulletin, Ltd., London, is an excellent source of information. The *Mining Journal,* Ltd., also in London, publishes a weekly journal, a *Mining Magazine* (monthly), and a *Mining Annual Review*. The latter covers the field from three aspects: (1) by commodities, (2) through technical developments, and (3) geographically. *The Industrial Minerals Directory 1977,* 1st ed., Metal Bulletin Books, Ltd., London, provides addresses of organizations in the field.

Other useful magazines and periodicals include *Engineering and Mining Journal, Pit and Quarry, Rock Products,* and *World Mining,* all published in the United States.

In addition many new books related to industrial minerals have been published since the 4th edition of *Industrial Minerals and Rocks*. Most of these are included in the bibliography of each chapter in this volume.

Another excellent information source is the proceedings of the *Forum on Geology of Industrial Minerals*. This group, which held its 18th annual meeting in 1982, is sponsored by a state geological survey or another mineral agency on a rotational basis. The proceedings have been published by the various sponsors and are available as indicated in Table 1.

Two major sources of basic industrial mineral and rock data in the United States, and also a good source for many foreign countries, are the US Geological Survey and the US Bureau of Mines. The two agencies publish many reports and maps on a variety of geological and mining subjects. In addition, each agency has an indexed list of all their publications with a free service announcing new publications: *New Publications of the Geological Survey,* US Geological Survey, 329 National Center, Reston, VA 22092 and *New Publications—Bureau of Mines,* Publications Distribution Branch, US Bureau of Mines, 4800 Forbes Ave., Pittsburgh, PA 15213.

The US Bureau of Mines (USBM) publishes annually *Minerals Yearbooks* which provide statistical data on mineral products including production, consumption, imports and exports, as well as information on new developments. Another very useful USBM series is called *Mineral Facts and Problems*. Editions were published in 1960, 1965, 1970, and 1975. Four major series of publications by the Bureau are *Information Circulars, Reports of Investigations, Mineral Trade Notes,* and *Minerals and Materials*. The last two are particularly good for news of foreign developments. The Bureau of Mines Employee Directory, available at all USBM offices, lists addresses and telephone numbers for all offices as well as same for all commodity and area specialists.

* Sam Patterson, US Geological Survey, Reston, VA, made the initial suggestion for inclusion of Section 4, "Sources of Information" in the 4th edition of *Industrial Minerals and Rocks*.

† Consultant, Golden, CO.

TABLE 1—Publications of the Forum on Geology of Industrial Minerals

Date & Place of Meeting	Title of Publication	Where It Can Be Ordered and Price
1st: Feb. 26-27, 1965, Columbus, Ohio	"A Symposium on Geology of Industrial Limestone and Dolomite," *The Ohio Journal of Science,* 1966, Vol. 66, pp. 97-191	Ohio Academy of Science 505 King Ave., Columbus, Ohio 43201, $2.00.
2nd: Mar. 31-Apr. 1, 1966, Bloomington, IN	*Proceedings of the Second Forum on Industrial Minerals,* Indiana Geological Surv., 1966, 197 pp. (out of print)	Publications Section Indiana Geol. Survey, 611 N. Walnut Grove Bloomington, IN 47401, $3.00.
3rd: Apr. 6-7, 1967 Lawrence, KS	Angino, E. E., and Hardy, R. G., eds., *Proceedings of the Third Forum on Geology of Industrial Minerals,* State Geological Survey of Kansas, Spec. Distrib. 34, 174 pp.	State Geological Survey of Kansas, The University of Kansas, Lawrence, KS 66044, $2.00.
4th: Mar. 14-15, 1968, Austin, Texas	Brown, L. F., Jr., ed., *Proceedings of the Fourth Forum on Geology of Industrial Minerals*: Bureau of Economic Geology, The University of Texas, 174 pp.	Bureau of Economic Geology, The University of Texas, University Station, Box X, Austin, Texas 78712, $2.00.
5th: May 1-2, 1969, Harrisburg, PA	Hoover, K. V., ed., *Proceedings of the Fifth Forum on Geology of Industrial Minerals*: Pennsylvania Geological Survey, 4th ser., Mineral Resources Dept., M64, 1970, 278 pp.	Bureau of Publications, Tenth & Market Sts., Harrisburg, PA 17125, $2.35.
6th: Apr. 2-3, 1970, Ann Arbor, MI	Kneller, W. A., ed., *Proceedings of the Sixth Forum on Geology of Industrial Minerals*: Michigan Dept. of Natural Resources, Geological Survey Div. Miscellany No. 1, 1970, 160 pp.	Publications Room, Michigan Dept. of Natural Resources, Geol. Survey Div. Stevens T. Mason Bldg., Lansing, MI 48909, $3.00.
7th: Apr. 28-30, 1971, Tampa, FL	Puri, H. S., ed., *Proceedings of the Seventh Forum on Geology of Industrial Minerals*, Florida Bureau of Geol. Special Publication 17, 1971	Bureau of Geology, Dept. of Natural Resources, P. O. Box 631, Tallahassee, FL 32302, $3.00.
8th: Apr. 12-14, 1972, Iowa City, IA	Dorheim, F. H., ed., *Proceedings of the Eighth Forum on Geology of Industrial Minerals*, Iowa Geological Survey, 1972	Iowa Geological Survey, Iowa City, IA 52240, $2.00.
9th: Apr. 26-28, 1973, Paducah, KY	Hutcheson, D. W., ed., *Proceedings of the Ninth Forum on Geology of Industrial Minerals*, Kentucky Geological Survey, 1973	Kentucky Geological Survey, 120 Graham Ave., Lexington, KY 40506, $2.00.
10th: Apr. 17-19, 1974, Columbus, Ohio	Bates, R. L., and Collins, H. R., eds., *Proceedings of the Tenth Forum on Geology of Industrial Minerals*, Div. of Ohio Geological Survey, 1974	Div. of Ohio Geological Survey, 1207 Grandview Ave., Columbus, Ohio 43212, $2.00.
11th: June 18-20, 1975, Kalispell, MT	R. B. Berg, ed., *Eleventh Industrial Minerals Forum*, Montana Bureau of Mines and Geology, Spec. Pub. 74, 186 pp.	Montana College of Mineral Science & Technology Room 203-B Main Hall, Butte, MT 59701, $4.50.
12th: April 22-24, 1976, Atlanta, GA	*Twelfth Forum on the Geology of Industrial Minerals,* Compiled by P. A. Platt, 1978, 78 pp.	Georgia Geol. Survey Dept. of Natl. Resources, Rm. 400 Agriculture Bldg. 19 Martin Luther King Jr. Dr. SW, Atlanta, GA 30334, $2.10 + $1.00 postage and handling.
13th: May 12-14, 1977, Norman, OK	Kenneth S. Johnson and Judy A. Russell, eds., *Proceedings of the Thirteenth Annual Forum on the Industrial Minerals*, Circ. 79, 107 pp.	Oklahoma Geol. Survey 830 Van Vleet Oval, Room 163, Norman OK 73069

TABLE 1—Publications of the Forum on Geology of Industrial Minerals (Continued)

Date & Place of Meeting	Title of Publication	Where It Can Be Ordered and Price
14th: May 4-6, 1978, Albany, NY	Robert H. Fakunding, ed., *Geology of the Industrial Minerals in the 1970's*, N.Y.S. Museum Bull. 435, 1980	New York State Geol. Survey, Rm. 3136, Cultural Ed. Center Albany, NY 12230
15th: June 13-15, 1979, Golden, CO	Stephen D. Schwochow, ed., *Industrial Minerals in Colorado and the Rocky Mtn. Region.* (Resource Series 8)	Colorado Geological Survey, 1313 Sherman Rm. 715, Denver, CO 80203, $5.00.
16th: April 9-11, 1980, St. Louis, MO	James A. Martin, ed., *Geology of Barite, Fireclay, and Stone Deposits; Planning for Industrial Mineral Development,* Information Circular 28.	Missouri Dept. of Natural Resources, Div. of Geol. & Land Survey, Box 250 Rolla, MO 65401
17th: May 13-15, 1981, Albuquerque, NM	G. S. Austin, ed., *Industrial Minerals and Rocks of the Southwest,* Circular 182.	New Mexico Bureau of Mines, Campus Station Socorro, NM 87801 $13.00
18th: April 14-16, 1982, Bloomington, IN	Donald Carr, ed., *Construction Materials,* Occasional Paper 37.	Ohio Ministry of Natural Resources Columbus, OH $4.50

Professional Paper No. 820 of the US Geological Survey, 1973, summarizes data on the mineral resources of the entire country.

The USGS, in cooperation with USBM, other federal and various state agencies, has also prepared summary reports on mineral and water resources of given areas which include pertinent information on industrial minerals.

Another series of valuable reports prepared by USGS is the state *Mineral and Water Resources* publications. These were prepared mostly at the request of members of the US Senate Interior and Insular Affairs Committee. The states for which these reports were published are listed in Table 2.

The US Bureau of Mines has major offices in Washington, DC (Administrative) and (Division of Publication), and Pittsburgh, PA (Branch of Production and Distribution). These and many other offices in the United States are valuable sources of information on such subjects as health and safety, mineral resources, and environmental development.

The US Geological Survey maintains the public inquiries offices, listed in Table 3, to provide convenient contacts for obtaining information concerning the work of the USGS.

Other excellent sources of information are the state and country geological surveys. Almost any state or country with a significant industrial rock and mineral industry has a state or federal geological survey. Most of these have geologists and engineers who are well informed regarding the mining activities of their particular area. In addition, they usually have good ideas regarding the potential for new mineral developments. The names of the state and country agencies and addresses are listed in Table 4.

Trade associations, listed in Table 5, are also good sources of data and information on their particular industrial mineral interest.

TABLE 2—List of State Mineral and Water Resources Publications

State	Agency	No.	Year Published
Alaska			1964
Arizona	State Bureau of Mines	Bull. 180	1969
California			1966
Colorado	Mining Industrial Development Board		1964, 1968
Idaho			1964
Missouri	Division of Geological Survey and Water Resources	2nd Series	1967
Montana			1963, 1965, 1968
Nevada	State Bureau of Mines	Bull. 65	1964
New Mexico	State Bureau of Mines	Bull. 87	1965
Oregon			1969
S. Dakota			1964
Utah	State Geological and Mineralogical Survey	Bull. 73	1964, 1969
Washington			1966
Wyoming			1960

TABLE 3—Material Available from Public Inquiries Offices of the US Geological Survey*

Location	Topographic Maps	Thematic Maps	Book Reports
Alaska Anchorage 99501 108 Skyline Bldg. 508 2nd Ave. Phone: (907) 277-0577	Alaska	Alaska	Alaska
California Los Angeles 90012 7638 Federal Bldg. 300 North Los Angeles St. Phone: (213) 688-2850	Alaska, Arizona, California, Hawaii, Nevada, Oregon, Washington	Alaska, Arizona, California, Hawaii, Nevada, Oregon, Washington	Alaska, Arizona, California, Hawaii, Nevada, Oregon, Washington
Menlo Park 94025 Rm. 122, Bldg. 3 (MS 33) 345 Middlefield Road Phone: (415) 323-8111, ext. 2817	Alaska, Arizona, California, Hawaii, Idaho, Nevada, Oregon	Alaska, Arizona, California, Hawaii, Idaho, Nevada, Oregon	Alaska, Arizona, California, Hawaii, Idaho, Nevada, Oregon
San Francisco 94111 504 Custom House 555 Battery St. Phone: (415) 556-5627	Alaska (selected), Arizona, California, Hawaii, Idaho, Nevada, Oregon, Washington	Alaska (selected), Arizona, California, Hawaii, Idaho, Nevada, Oregon, Washington	Alaska, Arizona, California, Hawaii, Idaho, Nevada, Oregon, Washington
Colorado Denver 80294 169 Federal Bldg. 1961 Stout St. Phone: (303) 837-4169	Alaska (selected), Arizona, Colorado, Kansas (selected), Montana (selected), Nebraska (selected), New Mexico, North Dakota (selected), South Dakota (selected), Utah, Wyoming	Alaska (selected), Arizona, Colorado, Kansas, Montana, Nebraska, New Mexico, North Dakota, South Dakota, Utah, Wyoming	Alaska (selected), Arizona, Colorado, Kansas (selected), Montana, Nebraska (selected), New Mexico, North Dakota (selected), South Dakota (selected), Utah, Wyoming
Texas Dallas 75202 Room 1 C45 1100 Commerce St. Phone: (214) 749-3230	Arkansas, Louisiana, New Mexico, Oklahoma, Texas	Arkansas, Louisiana, New Mexico, Oklahoma, Texas	Arkansas, Louisiana, New Mexico, Oklahoma, Texas
Utah Salt Lake City 84111 8102 Federal Bldg. 125 South State St. Phone: (801) 524-5652	Arizona, Colorado, Idaho, Montana (selected), Nevada, New Mexico (selected), Utah, Wyoming	Arizona, Colorado, Idaho, Montana, Nevada, New Mexico, Utah, Wyoming	Arizona, Colorado, Idaho, Montana, Nevada, New Mexico, Utah, Wyoming
Washington Spokane 99201 678 US Court House West 920 Riverside Ave. Phone: (509) 456-2524	Alaska (selected), Idaho, Western Montana, Oregon, Washington	Alaska, Idaho, Montana, Oregon, Washington	Alaska, Idaho, Montana, Oregon, Washington

Note: For information on states east of the Mississippi, contact Eastern Distribution Branch, USGS. For books, the address is 604 S. Pickett St., Alexandria, VA 22304. For maps, 1200 S. Eads St., Arlington, VA 22202.
* Also see "A Guide to Obtaining Information," from the USGS, 1979, compiled by Paul F. Clarke, Helen E. Hodgson, and Gary W. North, Geological Survey Circular 777.

TABLE 4—Sources of Information: Geological Surveys, Universities, and Publications

The selected list of sources of information on industrial minerals and rocks in this section is provided to facilitate locating the major governmental and private entities in position to furnish such information, including societies and institutes; colleges, universities, and technical schools. In the case of publications, it should be noted that, in addition to those listed, most of the governmental and private organizations publish a wide variety of reports, bulletins, monographs, surveys and the like, including, in the case of governmental agencies, production statistics.

A major portion of the data for the preceding edition (4th) was adapted from Section 35, *SME Mining Engineering Handbook*, A. B. Cummins and I. A. Given, eds., AIME, New York, 1973. The same format has been followed in the fifth edition; however, many modifications have been required because of reorganizations of governmental agencies, moves to new locations, and other changes.

The names of the mining schools and colleges in the following, with some exceptions, were taken from the *International Handbook of Universities and Other Institutions of Higher Learning*, and *Commonwealth Universities Yearbook*, with the kind permission of the International Association of Universities, Paris, and the Association of Commonwealth Universities, London, owners of the copyrights.

The Foreign Mineral Specialists of the US Bureau of Mines, Robert Evans (Assistant Director of the USGS), Samuel Shephard (US Bureau of Mines), and other individuals too numerous to mention assisted in updating respective parts of the tables.

The *E/MJ International Directory of Mining* was useful as a source of changes of addresses.

TABLE 4—Continued

Abbreviations in the listings include the following: A. P., B. P. and C. P., Apartado Postal, Boite Postal and Caixa Postal (post office boxes); Br., Branch; and F., Faculty.

NORTH AMERICA

CANADA

Dept. of Energy, Mines & Resources

Deputy Minister, 588 Booth St., Ottawa, Ont.; branch office, 2136 Foy Rd., Quebec City, Que.

Geological Survey of Canada, 601 Booth St., Ottawa; 326 Howe St., Vancouver, B.C. Mines Br., 555 Booth St., Ottawa.

Metals Reduction & Energy Center, c/o Research Council of Alberta, P. O. Sub 11, Calgary, Alta.

Surveys and Mapping Br., 615 Booth St., Ottawa.

Mining Research Centre, 3303 33d St. N.W., Calgary, Alta.

Mineral Resources Br., No. 8 Temp. Bldg., Ottawa.

Provincial Agencies

Alberta—Dept. of Mines and Minerals, 9915 108th St., Edmonton.

British Columbia—Dept. of Mines and Petroleum Resources, 406 Douglas Bldg., Victoria.

Manitoba—Dept. of Mines, Resources and Environmental Management, 310 Legislation Bldg., Winnipeg.

New Brunswick—Dept. of Natural Resources, Centennial Bldg., Fredericton.

Newfoundland—Dept. of Mines & Energy, Confederation Bldg., 95 Bonaventure Ave., St. Johns.

Nova Scotia—Dept. of Mines, P. O. Box 1087, Halifax.

Ontario—Ministry of Natural Resources, Whitney Block, Queen's Park, Toronto.

Prince Edward Island—Dept. of Industry and Commerce, P. O. Box 2000, Charlottetown.

Quebec—Dept. of Natural Resources, 1620 Blvd. de l'Entente, Quebec City.

Saskatchewan—Dept. of Mineral Resources, Legislative Bldg., Regina.

Yukon and Northwest Territories—Dept. of Indian and Northern Affairs, Ottawa.

Mining and Geological Societies and Institutes

The Canadian Institute of Mining & Metallurgy, 400–1130 Sherbrooke St. West, Montreal, Que H3A 2M8.
In affiliation—The Mining Society of Nova Scotia, Halifax.

Geological Association of Canada, Dept. of Earth Sciences, U. of Waterloo, Waterloo N2L 3Gl.

Colleges and Schools of Mines

University of Alberta, F. of Engrg., Edmonton, Alta.

University of British Columbia, F. of Applied Science, Vancouver, B.C.

Dept. of Mining Engrg., Nova Scotia Technical College, P. O. Box 1000, Halifax, N.S.

Queen's University, F. of Applied Science, Kingston, Ont.

Ecole Polytechnique, 2500, Ave. Marie-Guyard, Montreal 26, Que.

Laval University, F. of Science, Quebec City, Que.

Dept. of Mining Engrg. & Applied Science, McGill University, Montreal 2, Que.

Magazines & Periodicals

CIM Bulletin, The Canadian Institute of Mining & Metallurgy. 400–1130 Sherbrooke St., Montreal, Que. West, H3A 2M8.

Canadian Mining Journal, 310 Victoria Ave., Suite 201, Westmount, Que. H32, 2M9.

Canadian Pit and Quarry, Don Mills, Ont.

Mining in Canada, Don Mills, Ont.

Northern Miner, Flabatt Ave., Toronto, Ont., M5A 3P2.

Western Miner, 1200 W. Pender St., Vancouver, B. C. V6E ZT4.

GREENLAND

Gronlands Geologiske Undersogelse. (Geological Survey), Ostervolgade 7, Kobenhaven K, Denmark.

MEXICO

Direccion General de Minas Arcos de Belen, No. 30 Z. p. 7, Mexico, D.F.

Comision de Formento Minero, Puente de Tecamachalco 26, A.P. 10762 y 41505, Mexico 10, D.F.

Regional Offices: Aldama No. 510, Chihuahua, Chih; Bruno Martinez No. 403–314, Durango, Dgo.; Palacio Federal, 5° piso Ala Sur, Guadalajara, Jal.; Bvd. Miguel Hidalgo j Costilla No. 15–2° piso, Hermosillo, Son.; Ave. Juarez No. 2318–401, Puebla, Pue.; Vergara No. 8, Queretaro, Qro.; Ignacio Allende Nte. 254 2° piso, Saltillo, Coah.; Juarez No. 19 2° piso, Zacatecas, Zac.

Agencias de Mineria: Morelos No. 202–2° piso, Aguascalientes, Ags; Palacio Municipal, Alamos, Son., Hidalgo No. 20, Altar, Son.; Villada No. 413, Arteaga, Mich.; Santos Degollado No. 14, Batopilas Chih.; Av. Juarez No. 202A, Cananea, Son.; 1° de Mayo No. 440, Ciudad Guzman, Jal.; Edificio Clouthier Desp. 111, Culiacan, Sin.; Palacio Municipal, Cumpas, Son.; Palacio Federal, Chihuahua, Chih.; Ave. Juarez No. 13–D, Chilpancingo, Gro.; Bruno Martinez No. 403–219, Durango, Dgo.; Macheros No. 325, Ensenada, B.C.N.; Palacio Federal, 5° piso Ala Sur, Guadalajara, Jal.; Donato Guerra No. 1, Gpe. y Calvo, Chih.; Palacio Federal, Guanajuato, Gto.; Bvd. M. Hgo. y Costilla No. 15 2° piso, Hermosillo, Son.; Ojinaga No. 7–101, 102, Hidalgo de P., Chih.; Constitucion No. 3, Mascota, Jal.; Arcos de Belen No. 30 P.B., Mexico, D. F.; Palacio Federal, Monterrey, N.L.; Palacio Federal, Morelia, Mich.; Palacio Federal Oaxaca. Oax.; Avenue La Cumbre S/N, Ocampo, Chih.; Vergara No. 8, Queretaro, Qro.; Independencia No. 121, Sabinas, Coah.; Aldama Pte. 397 Mezzanine, Saltillo, Coah.; Palacio Federal, San Luis Potosi, SLP; Casa Blanca Mesa Francia, S/N, Sta. Rosalia, B.C.S.; Ave. Hidalgo No. 196, Sombrerete, Zac.; Domicilio Concocido, Temoris, Chih.; Pedraza No. 40 Nte, Tepic, Nay.; Palacio Federal, Torreon, Coah.; Ave Juarez No. 19 2° piso, Zacatecas, Zac.

Fidelcomiso de Gobierno Federal de Mexico Para al Fomento de la Exploracion, Explotacion y Beneficio de Minerales no Metalicos, Paseo de la Reforma 77, Mexico, D.F.

Asociacion de Ingenieros de Minas, Metalurgistes y Geologos de Mexico, A.P. 1260, Mexico, D.F.

TABLE 4—Continued

Instituto de Geologia, Universidad Nacional Autonomo de Mexico, Ciudad Universitaria, Mexico 20, D.F.
Universidad de Coahuila, S. of Engrg. & Metallurgy, Edo. Ateneo Fuente, 2 piso Saltillo, Coah.
Universidad de Guanajuato, S. of Mining, Apt. 24, Guanajuato, Gto.
Universidad Nacional Autonomo de Mexico, F. of Engrg.; Ciudad Universitaria, Mexico 20, D.F.
Mineria y Metalurgia, Asociacion de Ingenieros de Minas, Metalurgistes y Geologia de Mexico, address as above.

UNITED STATES

The Energy and Minerals Division of the Department of the Interior encompasses the US Bureau of Mines, US Geological Survey, Ocean Mining Administration, Office of Minerals Policy and Research Analysis, and Office of Surface Mining Reclamation and Enforcement.

Dept. of the Interior, Bureau of Mines

Director Assistant Director Deputy Directors Deputy Assistant Director Management Services Officer Division of Publication	Columbia Plaza 2401 E Street NW Washington, DC 20241

Branch of Production and Distribution,
4800 Forbes Ave.,
Pittsburgh, PA 15213

Minerals Information and Analysis

Deputy Director Office of Mineral Information Systems Division of Field Operations	Columbia Plaza 2401 E Street NW Washington, DC 20241
Alaska Field Operations Center	P. O. Box 550, Juneau, AK 99802
Eastern Field Operations Center	4800 Forbes Ave., Pittsburgh, PA 15213
Intermountain Field Operations Center	Building 20, Denver Federal Center, Denver, CO 80225
Western Field Operations Center	E. Montgomery Ave., Spokane, WA 99207
Division of Minerals Availability Division of Mineral Land Assessment Division of Production Consumption, Data Collection and Interpretation Branch of Domestic Data Branch of Foreign Data Division of Analytical Studies	Columbia Plaza 2401 E Street NW Washington, DC 20241

Minerals Research

Deputy Director Division of Minerals Environmental Technology Division of Minerals Health and Safety Technology Division of Research Center	Columbia Plaza 2401 E Street NW Washington, DC 20241

Operations

Albany Research Center	1450 Queen Ave., SW, P. O. Box 70, Albany, OR 97321
Avondale Research Center	4900 La Salle Rd., Avondale, MD 20782
Denver Research Center	Bldg. 20, Denver Federal Center, Denver, CO 80225
Pittsburgh Research Center	Cochrans Mill Rd., P. O. Box 18070, Pittsburgh, PA 15236
Reno Research Center	1605 Evans Ave., Reno, NV 89520
Rolla Research Center	P. O. Box 280, Rolla, MO 65401
Salt Lake City Research Center	1600 E. First South, Salt Lake City, UT 84112
Spokane Research Center,	E. 315 Montgomery Ave., Spokane, WA 99207
Tuscaloosa Research Center,	P. O. Box L, University, AL 35486
Twin Cities Research Center,	P. O. Box 1660, Twin Cities, MN 55111

Dept. of Interior, Geological Survey

US Geological Survey, National Center, Sunrise Valley Drive, Reston, VA 22092.
Director, Associate Director, Assistant Director for Eastern, Central, & Western Regions, Assistant Directors for Research; Engineering Geology; Resource Programs; Programs & Administration.
Office of Earth Sciences Applications—Chief
Chief, Office of Earth Resources Observation Systems; Chief, Office of Earth Sciences Assistance; Chief, Office of Environmental Affairs; Chief, Office of Resource Planning Analysis; Chief, EROS Data Center.
Conservation Division—Chief
Deputy Division Chief for Onshore Minerals; Deputy Division Chief for Offshore Minerals.
Geologic Division—Chief Geologist
Chief, Office of Mineral Resources; Chief, Office of Resource Analysis; Chief, Office of International Geology; Chief, Office of Energy Resources; Chief, Office of Marine Geology; Chief, Office of Environmental Geology; Chief, Office of Earthquake Studies; Chief, Office of Geochemistry and Geophysics; Chief, Office of Scientific Publications.
National Mapping Division—Chief
Asst. Division Chief, Research; Asst. Division Chief, Plans and Operations; Asst. Division Chief, Information and Data Services.

TABLE 4—Continued

Water Resources Division—Chief Hydrologist
Asst. Chief Hydrologist, Operations; Asst. Chief Hydrologist, Research and Technical Coordination; Chief, Office of Water Data, Coordination; Chief, Office of International Activities.

Offices of District Chiefs: Rm 202 Oil & Gas Board Bldg., Univ. of Alabama, Tuscaloosa, AL 35486; 218 E. Street, Anchorage, AK 99501; 301 W. Congress Street, Tucson, AZ 85701; 700 West Capitol Avenue, Little Rock, AR 72201; 855 Oak Grove Avenue, Menlo Park, CA 94025; Bldg. 53, Denver Federal Center, Denver, CO 80225; 135 High Street, Hartford, CT 06103; 325 John Knox Road, Tallahassee, FL 32303; 6481 Peachtree Industrial Bldg., Doraville, GA 30360; 300 Ala Moana Blvd., Honolulu, HI 96850; 550 West Fort Street, Boise, ID 83724; 605 N. Neil Street, Champaign, IL 61820; 1819 North Meridian Street, Indianapolis, IN 46202; 400 South Clinton Street, Iowa City, IA 52244; Univ. of Kansas, 1950 Avenue A, Lawrence, KS 66045; 500 Federal Place, Louisville, KY 40202; 6554 Florida Blvd., Baton Rouge, LA 70896; 150 Causeway Street, Boston, MA 02114; 8600 La Salle Road, Towson, MD 21204; 6520 Mercantile Way, Lansing, MI 48910; 702 Post Office Bldg., St. Paul, MN 55101; 100 West Capitol Street, Jackson, MS 39201; 1400 Independence Road, Rolla, MO 65401; Federal Bldg., Helena, MT 59601; 100 Centennial Mall North, Lincoln, NE 68508; 705 North Plaza Street, Carson City, NV 89701; 402 East State Street, Trenton, NJ 08607; 505 Marquette NW, Albuquerque, NM 87125; 236 US Post Office & Court House, Albany, NY 12201; Rm 436 Century Sta. P. O. Bldg., Raleigh, NC 27602; 821 E. Interstate Avenue, Bismarck, ND 58501; 975 West Third Avenue, Columbus, OH 43212; 215 N. W. 3rd, Oklahoma City, OK 73102; 820 NE Holladay Street, Portland, OR 97232; 228 Walnut Street, Harrisburg, PA 17108; Fort Buchanan, San Juan, PR 00936; 1835 Assembly Street, Columbia, SC 29201; 200 4th St., S.W., Huron, SD 57350; Federal Bldg., US Courthouse, Nashville, TN 37203; 300 East 8th St., Austin, TX 78701; 1745 West 1700 South, Salt Lake City, UT 84104; 200 West Grace St., Richmond, VA 23220; 1201 Pacific Avenue, Tacoma, WA 98402; 1815 University Avenue, Madison, WI 53706; 2120 Capitol Avenue, Cheyenne, WY 82001.

Principal US Geological Survey Libraries:
2255 North Gemini Drive, Flagstaff, AZ 86001; 345 Middlefield Rd., MS 55, Menlo Park, CA 94025; Denver West Office Park, Bldg. 3, 1526 Cole Blvd., Golden, CO 80401; EROS Data Center, Sioux Falls, SD 57198; 950 National Center, Room 4–A–100, 12201 Sunrise Valley Drive, Reston, VA 22092.

State Agencies—Mine Inspection, Geology

Alabama—Div. of Safety & Inspection, Dept. of Industrial Relations, 1816 8th Ave., North, Birmingham 35203; Geological Survey of Alabama, P. O. Box O, University 35486; Surface Mining Reclamation Commission, P. O. Box 1027, Jasper 35501.

Alaska—Dept. of Natural Resources, Pouch M, Juneau 99811; 3001 Porcupine Dr., Anchorage, 99501 and P. O. Box 80007, College 99708; Arctic Environmental Information and Data Center, 707 A Street, Anchorage 99501.

Arizona—Bureau of Geology and Mineral Technology, U. of Arizona, Tucson 85721; Department of Mineral Resources, Mineral Building, State Fairgrounds, McDowell and 19th Aves., Phoenix, 85007; State Mine Inspector, Capitol West Wing, Suite 705, 1700 West Washington St., Phoenix 85007.

Arkansas—Arkansas Geological Commission, Vardelle Parham Geology Center, 3815 W. Rossevelt Rd., Little Rock 72204; Dept. of Labor, 247 Central Mall, Fort Smith 72901; Occupational Safety Division (Dept. of Labor), Capitol Hill Bldg., 6th Floor, Little Rock 72201.

California—Division of Mines and Geology, Dept. of Conservation, Resources Bldg., 1416 9th St., Room 1341, Sacramento 95814; Div. Office, Ferry Bldg. San Francisco 94111; Division of Occupational Safety and Health, Mine and Tunnel Unit, Dept. of Industrial Relations, Agriculture, and Services Agency, 455 Golden Gate Ave., Room 7237, San Francisco 94101.

Colorado—Geological Survey, 1313 Sherman St., Room 703, Denver 80203; Division of Mines, 1313 Sherman St., Denver 80203; Division of Mined Land Reclamation, 1313 Sherman St., Denver 80203.

Connecticut—Geological and Natural History Survey, D.E.P., State Office Bldg., Room 561, 165 Capitol Ave., Hartford 06115; Department of Labor, 200 Folly Brook Blvd., Wethersfield 06109.

Delaware—Geological Survey, University of Delaware, 10 Penny Hall, Newark 19711; Dept. of Labor, 506 W. 10th St., Wilmington 19801.

Florida—Bureau of Geology, Dept. of Natural Resources, 903 W. Tennessee St., Tallahassee 32304; Industrial Safety Section, Bureau of Workmen's Compensation, Dept. of Commerce, 216 Ashley Bldg., 3121 Executive Cir. Dr. East, Tallahassee 32304.

Georgia—Geologic Survey, Environmental Protection Division, Dept. of Natural Resources, 19 Martin Luther King, Jr. Dr., S. W., Atlanta 30334.

Hawaii—Div. of Water and Land Development, Dept. of Land and Natural Resources, P. O. Box 373, Honolulu, 96809; Dept. of Labor and Industrial Relations, 825 Mililnai St., Room 203, Honolulu 96813.

Idaho—Bureau of Mines and Geology, Moscow 8384; Inspector of Mines, State Capitol, Boise 83701.

Illinois—Dept. of Mines and Minerals, William G. Stratton Bldg., Room 704, Springfield 62706; Geological Survey, 121 Natural Resource Bldg., Urbana 61801.

Indiana—Bureau of Mines and Mining, 119 Wabash Ave., Terre Haute 47807; Division of Labor, 1013 State Office Bldg., 100 North Senate Ave., Indianapolis 46204; Dept. of Natural Resources, Indiana Geological Survey, 611 North Walnut Grove, Bloomington 47401; Dept. of Natural Resources, Div. of Reclamation, 613

TABLE 4—Continued

State Office Bldg., 100 North Senate Ave., Indianapolis 46204.

Iowa—Geological Survey, Geological Survey Bldg., 123 North Capitol, Iowa City 52242; Dept. of Soil Conservation, Div. of Mine Land Reclamation, Wallace State Office Bldg., Des Moines 50319.

Kansas—Geological Survey, University of Kansas, 1930 Avenue A, Campus West, Lawrence 66044; Industrial Safety Division, Dept. of Human Resources, 510 West 10th St., 2nd Floor, Topeka 66612.

Kentucky—Dept. of Mines and Minerals (coal), P. O. Box 680, Lexington 40501; Dept. of Labor, (metal and nonmetal), US Highway 127, South, Frankfort 40601; Geological Survey, University of Kentucky, Room 311, Breckenridge Hall, Lexington 40506.

Louisiana—Div. of Safety, Dept. of Labor, P. O. Box 44094, Capitol Station, Baton Rouge 70804; Geological Survey, P. O. Box G, University Station, Baton Rouge 70893.

Maine—Geological Survey, Dept. of Conservation, State House Station #22, Augusta 04333; Div. of Industrial Safety, Bureau of Labor, Dept. of Manpower Affairs, State Office Bldg., Augusta 04333.

Maryland—Geological Survey, Merryman Hall, Johns Hopkins University, Baltimore 21218; Bureau of Mines, (coal), City Bldg., Westernport 21562; Dept. of Labor and Industry (metal and nonmetal), 301 W. Preston St., Baltimore 21201.

Massachusetts—Dept. of Labor and Industry, 100 Cambridge St., Boston 02202; Dept. of Environmental Quality Engineering, Room 532, 100 Nashua St., Boston 02114.

Michigan—Dept. of Natural Resources, Geological Survey Division, 4th Floor, Mason Bldg. (Box 30028), Lansing 48909; Dept. of Natural Resources, Air Quality Division, 3rd Floor, General Office Bldg. (Box 30028), Lansing 48909; Dept. of Labor, Bureau of Safety and Regulation, 7159 Harris Dr. (Box 30015), Lansing 40909; Dept. of Public Health, Bureau of Environmental and Occupational Health, 3500 N. Logan (Box 30035), Lansing 48914.

Minnesota—Geological Survey, University of Minnesota, 1633 Eustis St., St. Paul 55108; Dept. of Natural Resources, Minerals Division, Box 45, Centennial Office Bldg., St. Paul 55155; Dept. of Labor and Industry, Occupational Safety and Health, Space Center Bldg., 444 Lafayette Rd., St. Paul 55101.

Mississippi—Geological, Economic and Topographic Survey, P. O. Box 4915, Jackson 39216; State Board of Health, Division of Occupational Safety and Health, P. O. Box 1700, Jackson 39205.

Missouri—Dept. of Natural Resources, Division of Geology and Land Survey, P. O. Box 250, Rolla 65401; Dept. of Labor and Industrial Relations, P. O. Box 449, Jefferson City 65101.

Montana—Bureau of Mines and Geology, Montana College of Mineral Science and Technology, Butte 59701; Bureau of Safety and Health, Workmen's Compensation Division, 815 Front Street, Helena 59601.

Nebraska—Geological Survey, Conservation and Survey Division, University of Nebraska, 113 Nebraska Hall, Lincoln 68588; Dept. of Labor, Division of Safety, 550 South 16th St., Lincoln 68509.

Nevada—Bureau of Mines and Geology, University of Nevada, Reno 89557; Industrial Commission, 515 E. Musser St., Carson City 89714.

New Hampshire—Office of State Geologist, James Hall, Durham 03824; Dept. of Labor, 19 Pillsbury St., Concord 03301.

New Jersey—Bureau of Geology and Topography, P. O. Box 1390, Trenton 08625; Bureau of Engineering and Safety, John Fitch Plaza (P. O. Box 709), Trenton 08625.

New Mexico—Bureau of Mines and Mineral Resources, Campus Station, Socorro 87801; Mining and Minerals Division, P. O. Box 2860, Santa Fe 87501; Energy and Minerals Dept., P. O. Box 2770, Santa Fe 87501.

New York—Museum and Science Service, Geological Survey, State Education Bldg., Room 973, Albany 12224; Dept. of Environmental Conservation, Bureau of Mineral Resources, 50 Wolf Road, Albany 12233.

North Carolina—Geological Survey Section, Division of Land Resources, Dept. of Natural Resources and Community Development, P. O. Box 27687, Raleigh 27611; Mine and Quarry Division, Dept. of Labor, P. O. Box 27407, Raleigh 27611.

North Dakota—Geological Survey, University Station, Grand Forks 58201; Reclamation and Plant Siting Division, Public Service Commission, Capitol Bldg., Bismarck 58501, Workmen's Compensation Bureau, Capitol Bldg., Bismarck 58501.

Ohio—Geological Survey, Dept. of Natural Resources, Fountain Square, Bldg. 6, Columbus 42334; Division of Reclamation, Dept. of Natural Resources, 1930 Belcher Dr., Columbus 43224; Division of Mines, 2323 W. Fifth Ave., Columbus, 43204.

Oklahoma—Geological Survey, University of Oklahoma, 830 Van Vleet Oval, Room 163, Norman 73069; Dept. of Mines, 4040 N. Lincoln, Suite 107, Oklahoma City 73105.

Oregon—Dept. of Geology and Mineral Industries, 1069 State Office Bldg., 1400 S. W. Fifth Avenue, Portland 97201; Workmen's Compensation Board, Accident Prevention Division, Labor and Industries Bldg., Salem 97301.

Pennsylvania—Topographic and Geologic Survey, Dept. of Environmental Resources, P. O. Box 2357, Harrisburg 17102; Dept. of Environmental Resources, Office of Deep Mine Safety, P. O. Box 2063, Harrisburg 17120.

Rhode Island—Associate Geologist for Marine Affairs, Graduate School of Oceanography, University of Rhode Island, Kingston 02881; Dept. of Labor, 220 Elmwood Avenue, Providence 02907; Statewide Planning Program, 265 Melrose Street, Providence 02903; Asst. Director for Regulation, Dept. of Environmental Management, 75 Davis St., Providence 02908.

South Carolina—Division of Geology, Geological Survey, Harbison Forest Road, Columbia 29210; Land Resources Conservation Commission, 2221

TABLE 4—Continued

Devine St., Suite 222, Columbia 29205; Dept. of Labor, Division of Occupational Health and Safety, Landmark Center, P. O. Box 11329, Columbia 29204.

South Dakota—Geological Survey, Western Field Office, 301 West Blvd., Rapid City 57701; Geological Survey, Science Center, University of South Dakota, Vermillion 57069; Dept. of Natural Resource Development, Office Bldg. No. 2, Pierre 57501; Dept. of Agriculture, Division of Conservation, (Surface Mining), Office Bldg. No. 1, Pierre 57501.

Tennessee—Dept. of Conservation, Division of Geology, G–5 State Office Bldg., Nashville 37219; Dept. of Labor, Division of Mine Inspection, 618 Church Ave., S. W., Knoxville 37902; Dept. of Conservation, Division of Surface Mining, 2611 West End Ave., Nashville 37203; Dept. of Conservation, Division of Soil Conservation (Mine Land Reclamation), 1720 West End Ave., Nashville 37203.

Texas—Bureau of Economic Geology, University of Texas at Austin, University Station, Box X, Austin 78712; Division of Surface Mining and Reclamation, Texas Railroad Commission, P. O. Drawer 12967, Capitol Station, Austin 78711.

Utah—Geological and Mineral Survey, 606 Black Hawk Way, Salt Lake City 84801; The Industrial Commission of Utah, 350 E. Fifth South St., Salt Lake City 84111.

Vermont—Office of the State Geologist, Agency of Environmental Conservation, Heritage 11 Office Bldg., 79 River St., Montpelier 05602; Dept. of Labor and Industry, State Office Bldg., Montpelier 05603.

Virginia—Dept. of Conservation and Economic Development, Division of Mineral Resources, P. O. Box 3667, Charlottesville 22903; Dept. of Labor and Industry, Division of Mines and Quarries, Big Stone Gap 34319; Dept. of Labor and Industry, P. O. Box 1814, Richmond 23214.

Washington—Dept. of Natural Resources, Division of Geology and Earth Resources, 1404 Jefferson St., Olympia 98504; Dept. of Labor and Industry, Division of Industrial Safety and Health, 308 E. Fourth Avenue, Olympia 98504.

West Virginia—Geological and Economic Survey, P. O. Box 879, Morgantown 26505; Dept. of Mines, Room E–151, State Capitol Bldg., Charleston 25305.

Wisconsin—Geological and Natural History Survey, University of Wisconsin, 1815 University Ave., Madison 53706; Dept. of Industry, Labor and Human Relations, Safety and Buildings Division, 201 E. Washington Ave., Madison 53703.

Wyoming—Geological Survey, P. O. Box 3008, University Station, University of Wyoming, Laramie 82071; State Inspector of Mines, P. O. Box 1064, Rock Springs 82901.

Mining & Geological Associations, Societies, and Institutes

American Geological Institute, 5205 Leesburg Pike, Falls Church, VA 22041.

American Institute of Professional Geologists, 622 Gardenia, Golden, CO 80401.

American Mining Congress, Suite 1100, 1200 18th St., NW, Washington, DC 20036.

Association of American State Geologists, No. Carolina Dept. of Resources and Commercial Development, P. O. Box 27687, Raleigh, NC 27611.

The Geological Society of America, Inc., 3300 Penrose Place, Boulder, CO 80301.

Society of Economic Geologists, Inc., 185 Estes St., Denver, CO 80226.

Society of Mining Engineers of AIME, Caller No. D, Littleton, CO 80123.

Colleges and Schools of Mines

Dept. of Civil and Mineral Engineering, University of Alabama, University 35486.

Dept. of Mining & Metallurgy, College of Earth Sciences & Mineral Industry, University of Alaska, College 99735.

Division of Geosciences, University of Alaska, Fairbanks, AK 99701.

College of Mines, University of Arizona, Tucson 85721.

Dept. of Mineral Technology, University of California, Berkeley 94720.

School of Earth Sciences, Stanford University, Stanford, CA 94305.

Mining Engrg. Dept., Colorado School of Mines, Golden 80401.

Dept. of Earth Sciences, Dartmouth College, Hanover, NH 03755.

Mineral Engrg. Br., Georgia Institute of Technology, Atlanta 30332.

Dept. of Mining Engrg. and Metallurgy, University of Idaho, Moscow 83843.

Dept. of Metallurgy, University of Illinois, Urbana 61801.

Mining Engrg., Civil Engrg. Dept., University of Kentucky, Lexington 40506.

Dept. of Mining Engrg., Michigan Technological University, Houghton 49931.

Dept. of Civil & Mineral Engrg., University of Minnesota, Minneapolis 55455.

School of Mines & Metallurgy, University of Missouri, Rolla 65401.

Montana College of Mineral Science & Technology, Butte 59701.

Dept. of Metallurgical and Materials Engineering, University of Nevada-Reno, Reno 89507.

Mackay School of Mines, University of Nevada, Reno 89507.

Dept. of Petroleum & Mining Engineering, New Mexico Institute of Mining & Technology, Socorro 87801.

Henry Krumb School of Mines, Columbia University, Seeley W. Mudd Bldg., New York 10027.

Dept. of Mineral Engineering, College of Earth and Mineral Sciences, Pennsylvania State University, University Park 16802.

Dept. of Mining Engrg., South Dakota School of Mines & Technology, Rapid City 57701.

Dept. of Mining & Metallurgy, Texas Western College, El Paso 79902.

Dept. of Geological Sciences, University of Texas at El Paso, Texas 79968.

College of Mines and Mineral Industries, University of Utah, Salt Lake City 84112.

TABLE 4—Continued

Div. of Minerals Engineering, Virginia Polytechnic Institute & State University, Blacksburg 24061.
Dept. of Mining, Metallurgical and Ceramic Engineering, University of Washington, Seattle 98195.
Dept. of Mining, Washington State University, Pullman 99164.
College of Mineral and Energy Resources, West Virginia University, Morgantown 26506.
Dept. of Metallurgical & Mineral Engineering, University of Wisconsin, Madison 53706.
Dept. of Mining Engrg., University of Wisconsin–Platteville, Platteville 53818.

Magazines & Periodicals

Engineering & Mining Journal (*E/MJ*) 1221 Ave. of the Americas, New York, NY 10020.
Mines Magazine, Guggenheim Hall, Golden, CO 80401.
Mining Congress Journal, American Mining Congress, 1100 Ring Bldg., Washington, DC 20036.
Mining Engineering, Society of Mining Engineers of AIME, Caller No. D, Littleton, CO 80123.
Pit and Quarry, 105 W. Adams St., Chicago, IL 60603.
Rock Products, 300 W. Adams St., Chicago, IL 60606.
World Mining, 500 Howard St., San Francisco, CA 94105.

CARIBBEAN ISLANDS

CUBA

Institute Nacional de Investigaciones Cientificas, Cerro 827, Havana.

DOMINICAN REPUBLIC

Direccion General de Mineria, Secretaria de Estado de Industria y Comercio, Santo Domingo.

GUADELOUPE

See French Guiana, South America.

HAITI

National Institute of Mineral Resources, Port-au-Prince.
Dept. of Mines and Energy Resources, Port-au-Prince.

JAMAICA

Mines and Geology Division, Ministry of Mining and Natural Resources, P. O. Box 189, Hope Gardens, Kingston 6.

MARTINIQUE

See French Guiana, South America.

PUERTO RICO

Mineralogy and Geological Sec., Industrial Research, Economic Development Administration, Box 38, Roosevelt Sta., Hato Rey.

TRINIDAD & TOBAGO

Ministry of Petroleum and Mines, P. O. Box 96, Port-of-Spain, Trinidad.

CENTRAL AMERICA

COSTA RICA

Direccion de Geologia, Minas y Petroleo, Ministerio de Industria y Comercio, San Jose.
Asociacion de Geologos de Costa Rica, Apt. 6153, San Jose.
Escuela Centroamericana de Geologia, Ciudad Universitaria.

EL SALVADOR

Centro de Estudios de Investigaciones Geotecnics, Ministry of Public Works, A.P. 109, San Salvador.

GUATEMALA

Direccion General de Mineraria e Hidrocarburos, 10a. Calle 11–46, Zona 1 Guatemala City.

HONDURAS

Departamento de Minas, Direccion General de Recoursos Naturales, Comyaquela, D.C.

NICARAGUA

Instituto Nicaraguense de Recursos Naturales y del Ambiente, Managua. Codemina (Corp. for Mining Development), Managua.

PANAMA

Departamento de Recursos Minerales, A.P. 1631, Panama City.

SOUTH AMERICA

ARGENTINA

Secretaria de Estado de Mineria, Diag. Julio A. Roca 651, 9° Piso, Capital Federal, Buenos Aires.
Direccion Nacional de Geologia y Mineria, Diag. Julio A. Roca 651, 10° Piso, Capital Federal, Buenos Aires.
Colegio Argentino de Ingenieros de Minas, Avda. del Libertador 1109, Oeste, San Juan.
Universidad Nacional de Cuyo, F. of Engrg. & Exact, Physical & Natural Sciences, Rivadavia 65, Mendoza.
Revista, Direccion Nacional de Geologia y Mineria, above address.
Mineria, Avda. de Mayo 1370, 9° Piso, Capital Federal, Buenos Aires.

BOLIVIA

Ministerio de Mineria y Metalurgia, Avda. 16 de Julio 1769, La Paz.

TABLE 4—Continued

Servicio Geologico de Bolivia, Casilla 2729, La Paz.
Sociedad de Ingenieros de Bolivia, Avda. 20 de Octubre, La Paz.
Sociedad Geologica Boliviana, Casila 2308, La Paz.
Universidad Tecnica de Oruro, F. of Engrg.–Civil & Mining, Avda. 6 de Octubre 1207, Oruro.
Instituto Tecnologico Boliviana, School of Mining, Avda. Mariscal Santa Cruz, Esquina Ayacucho, Casilla 2977, La Paz.

BRAZIL

Companhia de Minerais Pesguisa de Recursos Minerais (CPRM), Rio de Janeiro.
Departamento Nacional da Producao Mineral, Avda. Pasteur 404-ZC-82, GB. Brasil. Divisao de Geologia e Mineralogia.
Sociedade Brasileira de Geologia, C.P. 20897, Sao Paulo.
Sociedade Mineira de Engenharia, Belo Horizonte, Minas Gerais.
Universidad do Recife, School of Engrg., Rua do Hospicio 619, Recife, Pernambuco.
Escola de Minas de Ouro Preto, Praca Tiradentes, Ouro Preto, Minas Gerais.
Mineracao e Metalurgia, Avda. N.S. Copacabana 637, Con. 501, GB. Brasil.
Geologia e Metalurgia, Praco Cel. Fernando Prestes 74, Sao Paulo.

CHILE

Servicio de Minas del Estado, Ministerio de Mineria, Casilla 1347, Santiago.
Instituto de Investigaciones Geologicas, Augustinas 785, 5° Pisa, Casilla 10465, Santiago.
Instituto de Ingeneiros de Minas de Chile, Clasificador 671, Santiago.
Instituto de Geofisica y Sismologia, Universidad de Chile, Casilla 2777, Santiago.
Universidad de Chile, Casilla 10D, Alamedo Bernardo O'Higgins 1058, Santiago.
Universidad Tecnica del Estadi, F. of Mining, Engrg., Metallurgy & Chemical Engrg., Avda. Ecuador 3467, Santiago.
Minerales, Instituto Ingeneiros de Minas de Chile, above address.

COLOMBIA

Instituto Nacional de Investigaciones Geologico–Mineria, Bogota.
Ministerio de Minas y Petroleos, Bogota.
Division de Minas.
Servicio Geologico Nacional.
Universidad Nacional de Colombia, F. of Engrg., Bogota.
Universidad Pedagogica y Tecnologia de Colombia, Tunja.

ECUADOR

Direccion General de Geologia y Minas, Ministerio de Recursos Naturales y Energeticos, Guayaquil.
Direccion General de Hidrocarburos, Ministerio de Recursos Naturales y Energeticos, Quito.
Universidad Central de Ecuador, F. of Engrg., Ciudad Universitaria, Quito.
Escuela Superior Politeonica del Litoral, Dept. of Geological, Mining & Petroleum Engrg., Guayaquil.

FRENCH GUIANA

Arrondissement Mineralogique de la Guyane Competence: Guyane-Guadeloupe-Martinique, B.P. 333, Cayenne.
Bureau de Recherches Geologiques et Minieres, 74, Rue de la Federation, Paris, 15e, France.

GUYANA

Mines Div., Ministry of Energy and Natural Resources, Upper Brickdam, Georgetown.
Geological Survey, Ministry of Energy & Natural Resources, Georgetown.

PARAGUAY

Direccion de Recursos Minerales, Asuncion.

PERU

Direccion General de Mineria, Ministerio de Energia y Minas, Mza. J.9 Av. de las Artes, Urbanizacion San Borja, Lima.
Servicio Nacional de Geologia y Mineria, Av. Republica de Chile 250, Lima.
Instituto de Ingenieros de Minas del Peru, Las Camelias 555, San Isidro, Lima.
Sociedad Geologica del Peru, Camilio Carrilo 300, Lima.
Capitulo Ingenieros de Minas y Capitulo de Ingenieros Geologos, Colegio de Ingenieros del Peru, Av. Arequipa 4947, Lima.
Universidad Nacional de Ingenieria, F. of Geological, Mining & Metallurgical Engrg., Casilla 1301, Lima.
Boletin Bimensual, Instituto de Ingenieros de Minas, above address.
Boletin Bimensual, Sociedad Nacional de Mineria y Petroleo, Av. Nicolas de Pierola, Lima.

SURINAM

Geological and Mining Service, Ministry of Development, Paramaribo.

URUGUAY

Administracion Nacional de Combustiables, Alchohol y Portland, Montevideo.
Instituto Geologica del Uruguay, Montevideo.
Ministerio de Industria y Energia, Montevideo.

VENEZUELA

Direccion de Geologia (and Direccion de Minas), Ministerio de Energia y Minas, Torre Norte, Piso 9, Centro Simon Bolivar, Caracas.
Asociacion Venezolana de Geologicia, Apt. 4400, Caracas.

TABLE 4—Continued

Sociedad Venezolana de Geologos, Apt. del Este 5199, Caracas.
Universidad Central de Venezuela, Caracas.
Universidad del Oriente, Cerro Colorado, Comana, Sucre.

EUROPE

ALBANIA

Ministry for Heavy Industry and Mining, Tirana.
Universiteti Shteteror i Tiranas, Tirana.

AUSTRIA

Bergebehord, Bundesministerium fur Handel, Gewerbe und Industrie, Stubenring 1, 1011 Vienna.
Geologische Bundesanstalt, Rasumofskygasse 23, 1013 Vienna.
Bergmannischer Verband Osterreichs (Mining Institut), 8700 Leoben.
Leoben Mining University, Franz Josef Strasse 18, 8700 Leoben.
Berg und Huttenmannische Monatshefte (Mining & Smelting), Springer-Verlag, Molkerbastei 5, Vienna.

BELGIUM

Administration des Mines, Ministere des Affaires Economiques et de l'Energie, Rue J. A. Demot, 24–26, 1040-Brussels.
Institut National des Industries Extractives, Bois du Val-Benoit, Rue du Chera, Liege.
Service Geologique de Belgique, Rue Jenner, 12, 1040-Brussels.
Union Professionnelle des Ingenieurs des Mines, Rue J. A. Demot, 24–26, 1040-Brussels.
Annales de Mines de Belgique, Rue du Chera, 200, B-4,000-Liege.

BULGARIA

Ministry of Chemistry & Metallurgy, Bansky Sq. 1, Sofia-Centre.
Geologic Committee (geology, prospecting, geologic research), G. Dimitrov St. No. 22, Sofia-Centre.
State Economic Trust for Nonmetallic Minerals, Haidushka Poliana St. No. 5, Sofia.
Scientific Technical Union of Mining, Geology & Metallurgy, Rakovsky St. 108, Sofia-Centre.
National Committee for Geology, Georgi Dimitrov Boul. No. 22, Sofia.
Viss Minno-Geologki Institut, F. of Mining Engrg., Sofia-Darvevetzi.
Vuglishta (Coal), Slaveikov Sq. No. 4, Sofia-Centre.
Rudodobiv i Metalurgia (Ore Mining & Metallurgy), Bansky Sq. No. 1, Sofia-Centre.
Annual, Research & Designers' Institute for Ore Mining & Dressing (NIPRORUDA), Al. Stamboliisky Boul. No. 205, Sofia.
Annual, Mining Research Institute, MINPROEKT, address as above.
Annual, Higher Mining-Geological Institute, Durvenitza Quarter, Sofia.

CZECHOSLOVAKIA

Cesky Bansky Urad (Czech Bureau of Mines), Kozi 1, Prague 1.
Slovensky Bansky Urad, Leningradska 24, Bratislava.
Ustredni Geologicky Ustav, Malostranske Nam. 19, Prague 1.
Mining Sec., Cesky Syaz Vedecko-Technikya Spolecnosti (Czech Union of Scientific & Technological Societies), Siroka 5, Prague.
Slovenska Banska Spolecnost (Slovak Mining Society), Stefanovka 39, Bratislava.
Vysoka Skola Banska (College of Mining), Dvorakova 7, Kosice.
Rudy (Ores), Krakovska S., Prague 1.
Geologicky Pruzkum (Geological Survey), Malostranske Nam. 19, Prague 1.
Ustar Prievidza (Research Institute for Mining), Prievidza, CSSR.

DENMARK

Danmarks Geologische Undersogolese, Oster Volgade 11, DK-1350, Copenhagen.
Dansk Geologisk Forening (Association), Oster Volgade 10, DK 1350, Copenhagen.
Mineralogical-Geological Museum, University of Copenhagen, Oster Volgade 7, Copenhagen.

FINLAND

Bureau of Mines, Ministry of Commerce & Industry, Hallituskatu 4, Helsinki.
Geologinen Tutkimuslaitos, Otaniemi.
Geological Society of Finland, Otaniemi.
Teknillinen Korkeakoulu (Inst. of Technology), Bersmannavagen 2A, SF 02150, Esboo 14.
Vuorimiesyhdistys R. Y. (Soc. of Engineers and Geologists of the Mining Industry), c/o Tekn. Dr. Mattiketola, Outokumpo Oy, PB 27, SF-02201, Esbo 20.
Geologi, Otaniemi.
Skandinavian Journal of Metallurgy, Box 244, SF-00131, Helsinki 13.
Vuoriteollisuus-Bergshanteringen Otberget, 2 B 19, SF-02150, Esbo 15.

FRANCE

Conseil Generale des Mines, 3, Rue Barbet-de-Jouy, 75700 Paris.
Direction Generale de Energie et des Matieres Premieres, 99, Rue de Grenelle, 75700 Paris.
Bureau de Documentation Miniere, 4 Rue Las Cases, 75007 Paris.
Bureau de Recherches Geologiques et Minieres (BRGM), Siege Social: 6–8, Rue Chasseloup-Laubat, 75737 Paris Cedex 15.
Bureau de Recherches Geologiques et Mineres, Centre Scientifique et Technique: Orleans-la-Source (Loiret).
Bureau de Recherches Geologiques et Mineres, Publications, *Chronique de la Recherche Miniere Geologic Appliquee*
Societe de L'Industrie Minerale, Paris office: 35, Rue des Petites-Champs, 75001 Paris; Administration: 19, Rue du Grand-Moulin, 42029 Saint-Etienne Cedex.
Societe de L'Industrie Minerale, Publications *Annuaire* (Yearbook) with Organigrammes de *L'Industrie Extractive Francaise Industrie Minerale.*

TABLE 4—Continued

Union National des Industries de Carrieres et Materiaux de Construction (UNICEM), 3 Rue Alfred-Roll, 75849 Paris Cedex 17.
Ecole National Superieure des Mines de Paris, 60, Boulevard Saint-Michel, 75272 Paris Cedex 06. *Addresse a Corbeil*: Centre des Materiaux-Enceinte SNECMA, BP 87, 91005 Evry; *Addresse a Fontainebleau:* Ecole des Mines, 35, Rue Saint-Honore, 77305 Fontainebleau; *Addresse a Sophia-Antipolist:* Ecole des Mines, Sophia-Antopolis, 06560 Valbonne.
Ecole Nationale Superieure des Mines de Saint-Etienne, 158 Cours Fauriel, 42023 Saint-Etienne.
Ecole Nationale Superieure de la Metallurgie et de L'Industrie des Mines de Nancy, Pare de Saurupt, 54042 Nancy.
Ecole Nationale des Techniques Industrielles et des Mines de Douai, 941, Rue Charles-Bourseul, BP 838, 59508 Douai.
Ecole Nationale des Techniques Industrielles et des Mines D'Ales, 6, Avenue de Clavieres, 30107 Ales.
Ecole Nationale Superieure de Geologie Appliquee et de Prospection Miniere, 94, Avenue de Lattre-de-Tassigny, 54001 Nancy.
Societe Geologiques de France, Maison de la Geologie, 77, Rue Claude Bernard, 75005 Paris.
Annales des Mines, 4, Rue Las-Cases, 75007 Paris.
Mines et Metallurgie, 16, Place Sainte-Claire, BP 385, RP 38015 Grenoble.

GERMANY, FEDERAL REPUBLIC OF

Bundesanstalt fur Geowissenschaften und Robstoffe (Geol. and Mining Survey), 3000 Hannover 51, Alfred-Bentz Haus, P. O. Box 5101 53, Hannover-Buchholz.
GDBM, Gesellschaft Deutscher Metallhutten und Bergleute EV (Society of Mining and Metallurgical Engineering) 3392 Clausthal Zellerfeld, Paul-Ernst-Str. 10, P. O. Box 210, Postfach 51.
Deutsche Geologische Vereinigung E.V., 3 Hannover-Bucholz, Stillweg 2, Postfach 54.
Rheinisch-Westfalische Technische Hochschule Aachen, Fachabteilung fur Bergbau (Mining School) 5100 Aachen Templergraben 55.
Technische Universitat Berlin, Fachbereich 16: Bergbau und Geowissenschaft, 1000 Berlin 12, Strasse des 17. Juni 135.
Technische Universitat Clausthal, Abteilung Bergbau und Minealische Rohstoffe (Faculty of Mining and Minerals), 3392 Clausthall-Zellerfeld, Adolph-Romer-Strasse 2A.
Bergbauwissenschaften (Mining Sciences), Erzstrasse, 3392 Clausthal-Zellerfeld.
Bergfreiheit, Alfredstrasse 77–79, 4300 Essen.
Gluekauf (Mining Magazine), Verlag Gluchauf GmbH, Franz-Fischer-Weg 61, Postfach 10 39 45, D-4300, Essen 1.
Jahrbuch fur Bergbau, Energie, Mineralol, und Chemie (Yearbook), Verlag Gluckauf GmbH, 4300 Essen 1, Postfach 10, 39, 45, Franz-Fischer-Neg. 61.

GERMAN DEMOCRATIC REPUBLIC

Bergakademie Freiberg, Sektion Geotechnik und Bergbau, 92 Freiberg Sachs.
Bergakademie, VEB Deutsche Verlag fur Grundstoffenindustrie, Karl Heine Str. 27, Leipzig W. 31.

GREECE

Ministry of Industry, Direction of Mines, 80 Michalakopoulou St., Athens.
Institute of Geological and Mineral Research (IGME), 70 Messogion Street, Athens.
Institute for Geology and Subsurface Research, 5 Amerikis Str., Athens.

GREENLAND

Gronlands Geologiske Undersogelse (Geol. Survey), Oster Volgade 10, DK-1350, Copenhagen, Denmark.

HUNGARY

Mining Sec., Ministry of Heavy Industries, V Marko-ut 16, Budapest.
Magyar Allami, Foldtani Intezet, Nepstadion-ut 14, Budapest XIV.
Orsznagos Magyar Banyasyati es Kohaszati Egyesulet (Hungarian Mining & Metallurgical Society), Szbasbag Ter 17, Budapest.
Geological Society of Hungary, Magyar Tudomenyos Ekadamie, Budapest.
Nehezepari Muszaki Egyetem (University of Heavy Industry), Egyetemi Varos, Miskolc.
Felosofodu Banyagazdasagi Technikum (College of Mining), Tatabanya.
Felosofoku Vegyipari Gepezeti Technikum (Chemical & Mining Engrg.), Esztergom-Kenyermezo.
Banyaszati Lapok (Mining Journal), Hungarian Mining & Metallurgical Society, address as above.

ICELAND

Dept. of Geology & Geography, Museum of Natural History, P.O. 532, Reykjavik.

IRELAND (EIRE)

Geological Survey of Ireland, 14 Hume St., Dublin 2, Erie.
Geological Survey of Northern Ireland, Belfast.

ITALY

Direzione Generale delle Miniere, Via Molise, 2-00100-Rome.
Servizio Geologica, Salita San Nicola da Telentino, 1/B-00187-Rome.
Associazione Mineraria Italiana, Via Sardegna, 14-Rome.
Associazione Nationale Geologi Italiani, Via Cesare Battisti, 11, 20100-Milan.
Societa Geologica Italiana Presso, Instituto di Geologia, Citta Universitaria, 00187-Rome.
Instituto di Geologica e Paleontologia, Universita di Roma, Citta Universitaria, 00187-Rome.
Universita degli Studi di Bologna, F. of Engrg., Via Zamboni 33, Bologna.

TABLE 4—Continued

Universita degli Studi di Cagliari, F. of Engrg., Via Universita 40, Cagliari.

Politecnica di Torino, F. of Engrg., Coras Duca degli Abruzzi 24, Torino.

L'Industria Mineraria, Via Sardegna, 14-Rome.

Relazione sul Servizio Minerario e Statische delle Industria Estrattive in Italia (annual), Via Molise, 2-Rome.

LUXEMBOURG

Inspector du Travail et des Mines, (Mines Inspectorate), Luxembourg.

Service Geologique, Direction des Ponts et Chassees, 13 Rue J. P. Koenig.

NETHERLANDS

Inspector General of Mines, Apollolaan 9, Heerlen.

Rijks Geologische Stichting, Spaarne 17, P.O. Box 157, Haarlem.

Koninklijk Nederlands Geologisch Mijnbouwkundig Genootschap (Royal Netherlands Geological & Mining Society), P.O. Box 285, Delft.

Technische Hogeschool Delft, Dept. of Mining, 134 Julianalaan, Delft.

Geologie en Mijnbouw, P.O. Box 285, Delft.

Jaarboek van de Mijnbouwkundige Vereniging, Mijnbouwstraat 20, Delft.

NORWAY

Bergverks-NYTT, Post boks 1438, N-7001, Trondheim.

Bergverkskontoret (Dept. of Mines), Industridepartmentet, Oslo-Dep, Oslo 1.

Den Norsk Ingeniorforening, Bergingeniorenes Avdeling, c/o Bergavdelingen, 7034 Trondheim NTH.

Norsk Geologisk Foroning, c/o Geologisk Museum, Sars Gate 1, Oslo 5.

Norges Geologisks Undersokelse (Geological Survey), Post boks 3006, 7001 Trondheim.

Selskapet for Industiell og Teknisk Forskning (SINTEF), Ved Norges Tekniske Hogskole, N-7039, Trondheim-NTH.

Universitet i Trondheim, Norges Tekniske Hogskole, N-7034, Trondheim-NTH.

Teknisk Ukoblad, Kronprinsensgate 17, Oslo 1.

Tidsskrift for Kejmi, Bergvesen og Metallurgi, Drammansveien 10, Oslo.

POLAND

Ministerstwo Gornicta i Energetyki (Ministry of Mining & Energy), Ul. Krucza 36, Warsaw.

Glowny Instytut Gornicta (General Mining Institute), Plac Gwarkow 1, Katowice.

Stowarzyszenie Inzynierowi Technikow Gornictwa (SIT Gornictwa), U1, Powstancow 25, Katowice.

Central Mining Institute, Katowice, Pl. Gwarkow.

Centraly Urzad Geologii (Polish Geological Society), Ul. Jasna 6, Warsaw.

Akademia Gorniczo-Hutnicza (Mining & Metallurgy), Mickiewicza 30, Krakow.

Politechnica Slaska im Wincentego Pstrowskiego, F. of Mining, Ul. S. Konarskiego 23, Gliwice.

Przeglad Gorniczy, Ul. Armii Czerwonej 51, Katowice.

Widomosci Gornicze, same address.

PORTUGAL

Institutio Nacional de Estatistica, Avenida Antonio Jose de Almeida, Lisboa.

Servicio de Fomento Minieiro, Rua da Amiera, S. Mamede de Infesta.

Servicos Geologicos de Portugal, Rua Academia das Ciencias 19-2°, Lisboa-2.

Ordem des Engenheiros, Avda. Antonio Augusto de Aguiar, 3-D, Lisboa-1.

Sociedade Geologica de Portugal, Faculdade de Ciencias, Rua Escola Politenica, Lisboa-2.

Universidade de Lisboa, F. of Sciences, Cidade Universitaria, Lisboa.

Universidade do Porto, F. of Engrg., Praca Gomez Texeira, Porto.

Boletin de Minas (quarterly), Rua Antonio Enes, 7, Lisboa-1.

ROMANIA

Ministerful Industriei Miniere si Geologiei, Str. Mendeleev 34–36, Bucuresti 1, O.P. 22.

Institutul Geologic, Soseaua Kiseleff 55, Bucuresti 1.

Institut de Mine Petrosani, Strada Institutui 20, Petrosani.

Revisita Minelor, Splaiul Unirii 127, Bucuresti 4.

SPAIN

Instituto Nacional de Industria, Ministerio de Industria, Plaza de Salamanica, 8, Madrid.

Direccion de Minas, Serrano 35, Madrid 1.

Instituto Geologico y Minero de Espana, Rios Rosas 23, Madrid 3.

Associacion Nacional de Ingenieros de Minas, General Goded 38, Madrid.

Instituto de Investigaciones Geologicas, Paseo de la Castellana 84, Madrid 6.

Escuela Tecnica Superior de Ingenieros de Minas, Calle Independence, Oviedo.

Instituti Politecnica Superior de Madrid, School of Mining Engrg., Ciudad Universitaria, Madrid.

Mineria y Metallurgia, Principe No. 9, Madrid 12.

Industria Minera, General Goded 38, Madrid.

Mine & Quarry Directory, Consejo Superior del Ministerio de Industria, Cristobal Bordiu 32, Madrid.

SWEDEN

Jernkontoret, Box 1721, 11187 Stockholm.

Sveriges Geologiska Undersokning (SGU), Uppsala 670, 75128 Uppsala.

Svenska Gruvforeningen (Swedish Mining Association), Box 5501, 11485 Stockholm.

Kungliga Tekniska Hogskolan, School of Mining & Metallurgy, FALK 10044, Stockholm 70.

SWITZERLAND

Department of Interior, Koniz Strasse 74, Bern.

Schweizerische Geologische Kommission, Bernoullianum, CH 40,000, Basle.

TABLE 4—Continued

UNION OF SOVIET SOCIALIST REPUBLICS

Ministry of Geology & Conservation of Natural Resources, Ul. Bolshahya Gruzinskaya 4/6, Mockba (Moscow) D-242.

Scientific-Technological Mining Institute, Karetni Sq. No. 10/18, Mockba K-6.

Doneckij Ordena Trudovogo Krasnogo Znameni Politehniceskij Institut, F. of Mining, Ul. Artema 58, Doneck, Ukrainskaja SSR.

Har'kovsky Institut Gornago Masenostroenja, Avtomatiki i Vujeislitl'noj Tekniki (Institute of Mining Machinery, Automation & Computer Technology), Pr. V. I. Lenina 14, Har'kov, Ukrainskaja SSR.

Karagandskij Politechniceskij Institut, F. of Mining, Bel'var Mira 22, Karaganda, Kazanskaja SSR.

Irkutskij Politehnkeskij Institut, F. of Mining, Ul. Lermontova 83, Irkutsk, Rossiskaja SFSR.

Kazanskij Politehniceskij Institut, Universitetskaja Ul. 22, Alma-Ata, Kazanskaja SSR.

Noril'skij Vecernij Industrial'nij Institut, F. of Mining, Juznaja Linia 7, Noril'sk, Rossiskaja SFSR.

Novocerkasskij Ordena Trudovogo Krasnogo Znameni Politehniceskij Institut im Sergo Ordzonikidze, F. of Mining & Geology, Ul. Prosvescenija 132, Novocerkassk, Rossiskaja SFSR.

Permskij Politehniceskij Institut, F. of Mining, Komsomolskij Pr 29a, Perm, Rossiskaja SFSR.

Tallinna Polutehniline Institut, F. of Mining & Chemical Engrg., Ul. Kalenina 101 Tallin, Estonskaja SSR.

Taskentskij Politehniceskij Institut, F. of Mining, Ul. Jakuba Kolasa 16, Tashkent, Uzbekskaja SSR.

Dnepropetrovskij Ordena Trudovogo Krasnogo Znameni Gornyj Institut im Artema, Inst. of Mining, Pr. Karla Marska 21, Dnepropetrovsk, Ukrainskaja SSR.

Kemerovskij Gornyj Institut, F. of Mining, Vesennjaja 28, Kemerovo Rossiskaja SFSR.

Krivorozskij Gornyj Institut, F. of Mining, Puskinskaja 37, Krivoi Rog, Ukrainskaja SSR.

Leningradskij Orden Leninia i Ordena Trudovogo Kransnogo Znameni Gornyj Institut im G. V. Plehanova, Vasil'evskij Ostrov, 21-ia Linija 2, Leningrad, Rossiskaja SFSR.

Moskovskij Institut Radioelektroniki i Goroj Elektromehaniki (Inst. of Radio-Electronics & Mining Electrical Engrg.), F. of Mining, Leninskij Pr. 6, Mockva, Rossiskaja SFSR.

Sverdlovsk Gornyj Institut im V. V. Vahruseva, F. of Mining, Ul. Kujbyseva 30, Rossiskaja SFSR.

Kommunarskij Gornometallurgiceskij Institut (Institute of Mining & Metallurgy), Pr. Mira 48 Kommunarsk, Luganskoj Obl., Ukrainskaja SSR.

Krasnojorski Institut Cvetnyh Metallou im M. I. Kalinina (Institute of Nonferrous Metals), F. of Mining & Metallurgical Engrg., Vuzovskij Pr. 3, Krasnojorsk 25, Rossiskaja SFSR.

Magnitogorskij Gornometallurgiceskij Institut im G. I. Nosova, F. of Mining & Metallurgical Engrg., Pr. V. I. Lenina 38, Magnitogorsk, Rossiskaja SFSR.

Sibirskij Metallurgiceskij Institut im Sergo Ordzonikidze (Institute of Metallurgy), F. of Mining, Rudokoprovaja Ul. 48, Novo-Kuzneck, Kemerovoskoj Obl., Rossiskaja SFSR.

Gorni Zhournal (Mining Journal), Pr. Vlademcrova 6, Mockba K-12.

UNITED KINGDOM

Dept. of Trade and Industry, Thames House So., Millbank, London S.W. 1, England.
Divs.: Coal, Iron and Steel, Petroleum, Safety and Health, Mines and Quarries Inspectorate.

Geological Survey of Northern Ireland, 20 College Gardens, Belfast, Northern Ireland.

Institute of Geological Sciences, Exhibition Rd., So. Kensington, London S.W. 7, England.

The Institution of Mining Engineers, 3 Grosvenor Crescent, London, S.W. 1, England.

The Institution of Mining and Metallurgy, 44 Portland Place, London, W. 1, England.

The Mining Institute of Scotland, Royal College of Science and Technology, George St., Glasgow, C. 1, Scotland.

Geological Society of London, Burlington House, Piccadilly, London W1V OJU.

Geologists Association, 278 Fir Tree Rd., Epsom, Surrey, England.

Yorkshire Geological Society, Dept. of Earth Sciences, The University, Leeds, LS2 9JT, England.

Royal School of Mines, Imperial College of Science & Technology, London, S.W. 7, England.

Bulletin, The Institution of Mining & Metallurgy, 44 Portland Place, London, W. 1, England.

Industrial Minerals, Metal Bulletin Ltd., 45–46 Lower Marsh, London, SE 1, 7RG, England. U.S. address: *Industrial Minerals,* 708 3rd Ave., New York City, NY 10017.

The Mining Engineer, The Institution of Mining Engineers, address as above.

Mining Magazine, 15 Wilson St., London, EC 2M 27R, England.

Mining Journal, (weekly), address as above.

Mining & Minerals Engineering, 10 Laystall St., London, EC 1, England.

YUGOSLAVIA

Univerzitet U Beogradu, Faculty of Mining and Geology, Dusina 7., 11000 Beograd.

Geoloski Institut Srpske Akademije Nauka.-Biograd. Knez Mihajlova, Beograd.

Sveuciliste U Zagrebu (F. of Mining, Geology and Petroleum Engineering) Trg. Marsala Tita 14, Zagreb.

Universitet U Sarajevu. F. of Mining, Obala 7, Sarajevo.

Mining and Metallurgical Quarterly, Univerza V Ljubljani, F. of Natural Sciences and Technology, Ljubljana.

Rudarski Glasnik, Quarterly, Rudarski Institut-Beograd, Batajnicki Put 2, Zemun.

Rudarstvo i Metalurgija, Drustvo Rudarskih i Metallurskih Inzenjera i Tehnicara, Beograd.

AFRICA

ALGERIA

Societe Nationale Sonatrach, 80 Avenue Ghermoul, Algiers.

TABLE 4—Continued

Direction des Mines et de la Geologie, Ministere de l'Industrialization et de l'Energie, Immeuble Mauretania-Agna, Algiers.

ANGOLA

Ministry of Industry and Power, National Directorate for Geology and Mining, Luanda.
National Fuels Company (SONANGOL), Luanda.

BOTSWANA

Ministry of Mineral Resources and Water Affairs, Gaborone.
Geological Survey, P. O. Box 94, Lobatse.

BURUNDI, REPUBLIC OF

Dept. de Geologie et Mines, Ministere des Affaires Economiques et Financieres, B. P. 745, Bujumbura.

CAMEROON

Director of Mineral Resources, Ministry of Commerce and Industrial Development, B. P. 1004, Yaounde.

CENTRAL AFRICAN REPUBLIC

Ministere de Travaux Publics, des Transports, des Communications, et des Mines, B.P. 737, Bangui.

CHAD REPUBLIC

La Service des Mines et de la Republique des Tchad, Service des Travaux Publics, N'Djamena.

CONGO, PEOPLE'S REPUBLIC OF

Service des Mines et de la Geologie, B.P. 12, Brazzaville.

DAHOMEY

Service of Mines & Geology, Ministry of Public Works, Transport & Telecommunications, Box 249, Contonou.

DJIBOUTI

Service des Travaux Publics de Territoires Francais, Djibouti.

EGYPT

See United Arab Republic.

ETHIOPIA

Ministry of Mines, Imperial Ethiopian Govt., P. O. Box 486, Addis Ababa.
The Secretariat, Economic Commission for Africa, Africa Hall, Addis Ababa.
Ministry of Mines, Energy and Water Resources, P. O. Box 486, Addis Ababa.

GABON

Direction des Mines, B.P. 576, Libreville.
Ministry of Mines, Energy and Hydraulic Resources, Libreville.

GHANA

Mines Dept., P. O. Box 254, Takoridi.
Geological Survey Dept., P. O. Box M80, Accra.
State Diamond Mining Corp., P. O. Box 2751, Accra.

GUINEA

Service des Mines et de la Geologie, Conakry.
Institut Polytechnique de Conakry, B.P. 1147, Conakry.
Minister de Mines, Conakry.

IVORY COAST

Direction des Mines Service Geologique, B.P. 1368, Abidjan.

KENYA

Mines & Geological Depts., Ministry of Natural Resources, P.O. Box 30009, Madini House Nairobi.

LIBERIA

Liberian Hydrological Services, Ministry of Lands & Mines, Monrovia.
Liberian Bureau of Mines & Liberian Geological Survey, Ministry of Lands and Mines, P. O. Box 9024, Monrovia.

LIBYA

Geological Division, Ministry of Industry, Tripoli.

MALAGASY, REPUBLIC OF

Direction des Mines et de l'Energie (including Service Geologique), Ministere de la Industries et de la Mines, B.P. 280, Tananarive.

MALAWI

Geological Survey, The Geological Survey Department, Ministry of Natural Resources, Zamba.
Director of Mines and Geology, Ministry of Industrial Development & Tourism, Bamako.
Director General de la Societe Nationale de Recherches et de l'Exploitation des Resources Minieres du Mali (SONAREM), B.P. 2, Kati.

MAURITANIA

Direction des Mines et de la Geologie, B.P. 199, Nouakchott.

MOROCCO

Office Cherifen des Phosphates, Rabat.

TABLE 4—Continued

Moroccan Bureau de Recherches et de Participations Minieres, Rabat.
Direction des Mines et de la Geologie, Ministere de l'Industrie et Mines, Rabat (Publication: *Mines et Geologie,* Quarterly).
Societe Nationale des Produits Petroliers, Rabat.

MOZAMBIQUE

Direccao dos Servicos de Geologia e Minas, Rui dos Santos Garcia, Maputu.
Estadios Gerais Universitarios de Mozambique, C.P. 257, Maputu.

NAMIBIA

See Southwest Africa.

NIGER

Direction des Mines et de la Geologie, Minister des Mines et de l'Hydraulique, B.P. 257, Niamey.
Niger Bureau of Mines, B.P. 257, Niamey.
Societe Miniere du Niger (SMN), Naimey.
Office National des Resources Minieres (ONAREM), B.P. 210, Naimey.

NIGERIA

Ministry of Mines and Power, P.M.B. 12574, Lagos.
Nigerian Chamber of Mines, P. O. Box 454, Jos.
Nigerian Mining Corporation, c/o Ministry of Mines and Power.
Nigerian Coal Corporation, c/o Ministry of Mines and Power, Yakubu Gowon Street, Lagos.
Nigerian Geological Survey Dept., Ministry of Mines and Power, Kaduna South, North State.
Nigerian National Petroleum Co., P.M.B. 12650, Lagos.
Nigerian Mining, Geological & Metallurgical Society, Dept. of Geology, University of Ibadan, Ibadan.

REUNION

Service de Travaux Publics, St. Denis.

RHODESIA

See Zimbabwe.

RAWANDA

Service Geologique, B.P. 15, Ruhengeri.

SENEGAL

Direction des Mines et de la Geologie, B.P. 1238, Dakar.

SIERRA LEONE

Ministry of Lands and Mines, Freetown.

SOMALIA

Geological Survey Dept., Ministry of Mining and Water Development, Magadiscio (Mogadishu).
Geological Survey Dept., Box 41, Hargeisa.

SOUTH AFRICA, REPUBLIC OF

The Director, Geological Survey, Private Bag X112, Pretoria, Transvaal.
The Government Mining Engineer, P. O. Box 1132, Johannesburg, 2000, Transvaal.
The Director, Minerals Bureau, Private Bag X4, Braamfontein, 2017 Transvaal.
National Institute for Metallurgy, Private Bag X3015, Randsburg, 2125 South Africa, (Publication: *Minerals Science and Engineering*).
The Pithead Press (Pty) Ltd., 91 Mool Street, Johannesburg (Publication: *Coal, Gold & Base Metals of Southern Africa*).
The South African Institute of Mining and Metallurgy, P. O. Box 61019, Marshalltown 2107 (Publication: *Journal of the South African Institute of Mining and Metallurgy*).
Geological Society of South Africa, 77 Marshall Street, Johannesburg.
South African Mining and Engineering Journal, 66 Commissioner Street, P. O. Box 8308, Johannesburg 2000.

SOUTHWEST AFRICA (NAMIBIA)

Mines Division, Southwest Africa Administration, Windhoek.
Geological Survey, P. O. Box 2168, Windhoek.

SPANISH SAHARA

See Morocco.

SUDAN

Geological Survey Dept., Ministry of Mining & Industry, P. O. Box 410, Khartoum.

SWAZILAND

Geological Survey and Mines Department, P. O. Box 9, Mbabane.

TANZANIA

Mineral Resources Division, Ministry of Commerce & Industries, P. O. Box 903, Dodoma.
Water, Energy & Minerals, Dar es Salaam.

TOGO, REPUBLIC OF

Direcion des Mines et de la Geologic, Ministere des Travaux Publics, Mines, Transportes, des Postes et Telecommunications, B.P. 356, Lome.

TUNISIA

Geological Survey of Tunisia, 95 Ave. Mohamed V. Tunis.

TABLE 4—Continued

UGANDA

Geological Survey & Mines Dept., P. O. Box 9, Entebbe.

UNITED ARAB REPUBLIC

Mines & Fuel Dept., 19 Badrawi Ashour St., Dokki, Cairo.
Geological Survey, Abbasiya P. O., Cairo.
Geological Society of Egypt, 1 Elhamy St., Kasr el Doubara, Cairo.
University of Cairo, F. of Engrg, Cairo.
Higher Industrial Institute, Dept. of Mining Engrg., Aswan.
Chamber of Petroleum Mines and Quarries Industries, Federation of Egyptian Industries, 26a, Sherif Street, Cairo.

UPPER VOLTA

Direction de la Geologie et des Mines, Ministry of Industry & Mines, Ougadougou (Ouazadougou).

ZAIRE (CONGO, REPUBLIC OF)

Service Geologiquc National de la Republique du Congo, B.P. 898, Kinshasa.
Institut National des Mines, B.P. 3131, Bukavu.
Universite Officielle du Congo, B.P. 1825, Lumumbashi.
Ministry of Mines and Energy, Brazzaville.

ZAMBIA

Geologica Survey Dept., Ministry of Mines, P. O. Box RW 135, Lusaka.
Geological Society of Zambia, P. O. Box RW 135, Lusaka.
Mines Development Dept., Ministry of Mines, P. O. Box 1969, Lusaka.
National Council for Scientific Research, Ministry of Mines, P. O. Box Ch 158, Lusaka.
Copper Industry Service Bureau, P. O. Box 2100, Kitwe (Publication: *Zambia Mining Yearbook*).
Institution of Mining and Metallurgy, Zambian Section, P. O. Box 450, Kitwe.
Mindeco Limited, Zimco House, Cairo Rd., P. O. Box 90, Lusaka.

ZIMBABWE

Ministry of Mines & Lands (including Geological Survey Office), Causeway, P. O. Box 8039 (Private Bag 7709 Causeway), Salisbury.
Institute of Mining Research, P. O. Box MP 167, Mount Pleasant, Salisbury.

MIDDLE EAST

BAHRAIN

Ministry for Development and Industry, Al Manamah (Manama).

CYPRUS

Ministry of Commerce and Industry, Mines Dept. & Geological Survey Dept., 3 Electra Street, P. O. Box 809, Nicosia.

IRAN

General Directorship of Mines, Ministry of Industry and Mines, Tehran.
Geological Survey of Iran, P. O. Box 1964, Tehran.
Mining and Metallurgical Corp. of Iran, 467 Takhte Jamshed, Tehran.
Society of Iranian Engineers, Behesht Ave., Tehran (Publication: *Journal of the Society of Iranian Engineers,* Quarterly).

IRAQ

Ministry of Oils and Minerals, Baghdad (Includes Geol. Survey).

ISRAEL

Ministry of Energy and Infrastructure, 30 Malchei Israel St., 95–501 Jerusalem.
Geological Survey of Israel, Ministry of Energy and Infrastructure, 30 Malchei Israel St., 95–501 Jerusalem. (Publication: *The Geological Survey Bulletin*).
Israel Society for Rock and Mineral Engineering, Mineral Engineering Dept. Technion City, 32 000 Haifa.
Israel Geological Society, P. O. Box 1239, 91 012 Jerusalem.
Israel Journal of Earth Sciences, P. O. Box 801, 91 007 Jerusalem.

JORDAN

Mines and Geological Survey Dept., Natural Resources Authority, Box 2220, Amman.

KUWAIT

Kuwait Oil Affairs Dept., Ahmadi 94.
Kuwait National Petroleum Co., P. O. Box 70, Kuwait (Al Kuwayt).

LEBANON

Direction General des Travaux Publics, Beirut.
Ministry of Industry and Petroleum, Beirut.

OMAN

Oman Mining Co., Sohar.
Petroleum & Mineral Resources Ministry, Muscat.

QATAR

Qatar Petroleum Production Authority (QPPA), Oil and Finance Ministry, Doho.

TABLE 4—Continued

SAUDI ARABIA

Directorate General of Mineral Resources, Riyadh.
Ministry of Petroleum & Mineral Resources, Riyadh.
General Petroleum and Mineral Organization, Box 757, Riyadh.
College of Petroleum & Mining, Dhahran.

SYRIA

Dept. of Geological Research & Mineral Prospecting, General Establishment for Geology and Mineral Resources, Jsir al Abiad, Damascus.

TURKEY

Maden Tetkik ve Arama Enstitusu, Posta Kutusu 116, Ankara.
General Director, Mineral Research and Exploration Institute, P.K. 116, Ankara.
Mining Engineers Society, Salenek St. 19/3, Yenisehir, Ankara (Publication: *Madencilik,* bimonthly).
Istanbul Teknik Universitesi, Dept. of Mining Engineering, Gummussuyu Caddesi 87, Beyoglu, Istanbul.

UNITED ARAB EMIRATES

Ministry of Petroleum and Mineral Resources, Abu Dhabi.

YEMEN ARAB REPUBLIC

Office of Mineral Resources, Ministry of Public Works, Sana.
Oil and Minerals Organization, Board of Petroleum and Minerals, Geology & Mineral Exploration Dept., Sana.

ASIA

AFGHANISTAN

Mines & Geological Survey Dept., Ministry of Mines & Industry, Kabul.

BORNEO

Geological Survey Dept., Borneo Region, Kuchin, Sarawak, Malaysia.

BRUNEI

The Government Geologist, Brunei Town.

BURMA

Ministry of Mines, Rangoon.
Geological Dept., 226 Mahabandoola St., Box 843, Rangoon.
Government Technical Institute, Insein.
University of Rangoon, Rangoon.

CAMBODIA

Service des Mines, Phnomh-Penh.
Institut Technique Superier de l'Amitie Khmero-Sovietique, Phnomh-Penh.

CHINA, REPUBLIC OF

Mining Research & Service Organization, Ministry of Economic Affairs (Mining & Geology), 261 Roosevelt Rd., Sec. 3, Taipei, Taiwan.
Geological Survey of Taiwan, 1 Kaifeng St., Sec. 1, Taipei, Taiwan.
Taiwan Mining Institute, 38 Hwa Yin St., Sec. 2, Taipei, Taiwan.
The Chinese Institute of Mining & Metallurgical Engineers, 38-1 Chi Nan Rd., Sec. 2, Taipei, Taiwan. (Publication: *Mining & Metallurgy,* quarterly).
Geological Society of China, 245 Chilung Rd., Sec. 3, Taipei, Taiwan.
Taiwan Provincial Taipei Institute of Technology, Dept. of Mining & Metallurgical Engrg., 3 Shin-Sheng Rd., Sec. 1, Taipei, Taiwan.
Mining Technical Digest, 261 Roosevelt Rd., Sec. 3, Taipei, Taiwan.

CHINA, PEOPLE'S REPUBLIC OF

Chinese Ministry of Geology, Peking.
China University, Canton, Kwangtun.
Chungking Technical College, Chungking, Szechuan.
Kunming Technical College, Kunming, Yunnan.
Kweiyang Technical College, Kweiyang, Kweichow.
Northwest China Polytechnical University, Sienyang, Shensi.
South-Central China College of Geology, Changsha, Hunan.
South-Central China College of Mining & Metallurgy, Changsha, Hunan.
Peking College of Iron & Steel Technology, Peking, Hopeh.
Hofei College of Mining Engineering, Sian, Shensi.
Northeast China Technical College, F. of Mining Engrg., Shenyang, Liaoning.
Northwest China College of Mining Engineering, Taiyuan, Shensi.
Taiyuan College of Mining Engineering, Taiyuan, Shensi.
Tsinan College of Coal Mining Engineering, Tsinan, Shantung.

INDIA

Bureau of Mines, Ministry of Petroleum & Chemicals & Mines & Metals, Nagpur-440001.
Geological Survey of India, 27 Chowringhee Rd., Calcutta 16.
Geological, Mining & Metallurgical Institute of India, Presidency C, Calcutta 12.
Indian Institute of Technology, Mining Engineering D., Kharagpur-721 302.
Oil and Natural Gas Commission, Dehra Dun, Uttar Pradesh.
Hindustan Copper Ltd., P.O. Khetri Nagar, Rajasthan.

TABLE 4—Continued

Presidency College, Calcutta 12.

Cement Corp. of India, Ltd., Nehru Place, New Delhi-110 019.

Rajasthan State Minerals, Ltd. (Phosphates), Udaipur, Rajasthan.

Indian Journal of Mining, Geology, Metallurgy & Fuel, E. Ry. N.B. Rd., Raniganj P.O., Burdwan, W. Bengal.

Journal of Mines, Metals & Fuels, 6/2 Maden St., Calcutta-700 072.

Mines & Minerals (quarterly) and *Indian Minerals Year Book,* Indian Bureau of Mines, Nagpur.

INDONESIA

Bureau of Mineral Resources Development, Ministry of Mines, Merdeka Selatan 18, Djakarta.

Dept. Pertambangan, P.N. Aneka Tambang, Bungur Besar 24/26, Djakarta.

Directorate of Mines, Dept. of Basic Industry & Mining, 8 Djalan Gadja Mada, Djakarta.

Geological Survey of Indonesia, Djalan Diponegoro 57, Bandung.

Universitatis Kalimatan Timur, F. of Mining Engrg., Kalimatan, Timur.

Institut Teknologi Bandung, F. of Mineral Technology, Djalan Tasmansari 64, Bandung.

JAPAN

Mine Bureau, Ministry of International Trade & Industry, 3-1 Kasumigaski, Chiyoda-ku Tokyo.

Geological Survey of Japan, 135 Hisamoto-cho, Kawasaki, Kanagawa.

Mitsubishi Metal Mining Co., Ltd., No. 6 1-Chrome, Ohte-Machi, Chiyoda-ku, Tokyo.

Mitsu Mining & Smelting Co., Ltd., 2-Chrome Nihonbashi-Miromacht, Chuo-ku, Tokyo.

Nippon Mining Co., Ltd., 3 Akasaka Alo-cho, Minato-ku, Tokyo.

Japanese Mining Industry Association, c/o Shin Hibiya Bldg., No. 3-6, 1-Chrome, Uchisaiwai-cho, Chiyoda-ku, Tokyo.

The Mining & Metallurgical Institute of Japan, 24, 1-Chrome, Suda-cho Kando, Chiyoda-ku, Tokyo.

Akita Daragaku (Akita University) Mining College, 1 Fukada, Tegata, Akita.

Kozan (Mine), Japan Mining Industry Assn., 3-Chome Kanda Nishiki-cho, Chiyoda-ku, Tokyo.

KOREA, REPUBLIC OF

Bureau of Mines, Ministry of Commerce & Industry, 149 Sejong-Ro, Jongro-ku, Seoul.

Korean Institute of Geoscience and Mineral Resources (KIGAM) 219–5, Garibong-dong, Youngdung-Po-Gu, Seoul, 150–06.

The Korean Institute of Mining, c/o Dept. of Mining, College of Engineering, Seoul National University, 445 Kongneung-Dong, Sungpuk-ku, Seoul.

The Korean Institute of Mining & Geology, c/o Yunse University, 134 Sinchon-Dong, Sudaimun-ku, Seoul.

Geological Society of Korea, c/o Dept. of Geology, College of Literature & Science, Seoul National University, 21 Dongsoong-Dong, Jongro-ku, Seoul.

Seoul Dae Hak Kyo (Seoul National University), College of Engrg. (including Mining), see address, Geological Society of Korea.

Journal of the Korean Institute of Mining, address as above.

Journal of the Korean Institute of Mining & Geology, address as above.

Journal of the Geological Society of Korea } address as above.
Geology and Ore Deposits }

Bulletin of Mining, Mining Association of Korea, 34-24 Tongsi-Dong, Jongro-ku, Seoul.

Mining Industry, The Institution of Mining Productivity of Korea, 231-57 Heuksuk-Dong, Youngdeungpo-ku, Seoul.

KOREA, PEOPLE'S DEMOCRATIC REPUBLIC OF

Tchunjin Institute of Mining & Metallurgy, Tchunjin.

LAOS

Dept. des Mines, Ministere du Plan, B.P. 46, Vientiane.

MALAYSIA (WEST)

Dept. of Mines, Govt. Offices, Jalan Gurney, Kuala Lumpur.

Geological Survey of Malaysia, Scrivenor Rd., Tiger Lane, Ipoh, Perak.

The Malaysian Association of Mining Engineers, Jalan Gurney, Kuala Lumpur.

The Institution of Mining & Metallurgy (Malaysian Sec.), P.O. Box 300, Kuala Lumpur.

The Geological Society of Malaysia, c/o Dept. of Geology, University of Malaysia, Kuala Lumpur.

NEPAL

Geological Survey, Ministry of Industry & Commerce, Lainchaur, Kathmandu.

Nepal Bureau of Mines, Ministry of Industry & Commerce, address as above.

PAKISTAN

Directorate of Mineral Development, 11 Lytton Rd., Lahore.

Geological Survey of Pakistan, P.O. Box 15, Quetta.

Pakistan Mineral Development Corp., PIDC House, Zrauddin Ahmid Rd., Karachi-4.

Oil and Gas Development Corp., Central Hotel Bldg., Club Road, Karachi.

PHILIPPINES

Bureau of Mines, Dept. of Agriculture & Natural Resources, P.O. Box 1595, Manila.

Philippine Society of Mining, Metallurgical & Geological Engineers, P.O. Box 1595, Manila.

Adamson University, Dept. of Mining Engrg., 959 San Marceleno, Ermita, Manila.

University of the Philippines, College of Engrg., Quezon.

University of South Philippines, College of Engrg., Mabine St., Cebu City.

Asia Mining, 1176 Pasong Tamo, Makati, Rizal.

TABLE 4—Continued

SRI LANKA (CEYLON)

Geological Survey Dept., 48 Sri Jinaratana Rd., Colombo 2.

Consejo de Mineraia y Metalurgia, Cristobal Bordice 34, Madrid, Spain.

Institute Geologico y Minero de Campana, Rio Rosas 23, Madrid Spain.

THAILAND

Royal Dept. of Mineral Resources, Rama VI Rd., Bangkok, 4.

VIETNAM, SOCIALIST REPUBLIC OF

Truong Dai-Hoc Bach Khoa (Polytechnic Inst.), F. of Mining & Geology, Hanoi.

Truong Dai-Hoc Mo Diachat (Inst. of Mining & Geology), Hanoi.

Geological Services, 31 Rue Han-Thuyen, Ho Chi Minh (Saigon).

Direction General des Mines de l'Industrie et de l'Artisanat, Service Geologique, 59, Rue Gia Long, Ho Chi Minh.

AUSTRALIA AND THE PACIFIC ISLANDS

AUSTRALIA

Bureau of Mineral Resources, Geology & Geophysics, London Circuit, P.O. 378, Canberra City, A.C.T.

State Government Organizations

New South Wales—Dept. of Mines, 11 Loftus St., Sydney.

Northern Territory—Mines Br., Northern Territory Administration, Darwin.

Queensland—Dept. of Mines, Geological Survey of Queensland, Mineral House, 2 Edward St., Brisbane.

South Australia—Dept. of Mines, 169 Rundle St., Adelaide.

Tasmania—Dept. of Mines, G.P.O. Box 124B, Hobart.

Victoria—Dept. of Mines, Treasury Pl., Melbourne C. 2.

Western Australia—Dept. of Mines, Treasury Bldgs., St. George's Terrace, Perth; Geological Survey, 26 Francis St., Perth.

New Guinea, Papua—Dept. of Lands, Surveys and Mines, Konedobu, Papua.

Societies-Schools-Publications

The Australasian Institute of Mining & Metallurgy, Clumes Ross House, 191 Royal Parade-Parkville 3052, Melbourne, Victoria.

Mining Dept., University of Melbourne, Parkville, N. 2, Victoria.

School of Mines of Western Australia, P.O. Box 62, Kalgoorlie.

Australian Mining, 47 Chippen St., Box 65, Chippendale 2008, N.S.W.

Australasian Mining, Box 5416 CC, C.P.O., Melbourne, Victoria.

Bulletin, Australasian Institute of Mining & Metallurgy, address as above.

Mine & Quarry Mechanisation (annual), P.O. Box 55, Avalon Beach, N.S.W.

Quarry Mine & Pit (quarterly), G.P.O. Box 4138, Sydney, N.S.W.

BORNEO

Geological Survey, Borneo Region (E. Malaysia), Kuching, Sarawak.

BRITISH SOLOMON ISLANDS

Geological Survey Dept., P.O. Box G-24, Honiara, Guadalcanal.

FIJI

Director of Geological Survey, P. M. B., G.P.O., Suva.

Bureau of Mines, Dept. of Lands, Mines and Surveys, Suva.

NEW CALEDONIA

Service des Mines, Noumea.

NEW HEBRIDES

Geological Survey, New Hebrides Condominium, British Residence, Vila.

NEW ZEALAND

Dept. of Mines, Wakefield St., Wellington.

Geological Survey, Dept. of Scientific & Industrial Research, Andrews Ave., P.O. Box 30368, Lower Hutt, Wellington.

Australian Institute of Mining & Metallurgy, N.Z. Br., P.O. Box 6342, Wellington C. 2.

New Zealand Geological Society, c/o Canterbury University, P.B., Christchurch.

TABLE 5—List of American and Foreign Trade Associations Which Specialize in Industrial Minerals

AMERICAN

American Mining Congress, 1200 18th St., NW, Washington, DC 20036
American Ceramic Society, 65 Ceramic Drive, Columbus, OH 43214
American Road Builders Association, 525 School St., SW, Washington, DC 20024
American Society for Testing & Materials, 1916 Race St., Philadelphia, PA 19103
Asbestos-Cement Products Association, 521 Fifth Ave., Room 2000, New York, NY 10017
Asphalt Institute, Asphalt Institute Bldg., College Park, MD 20740
Ceramic Lightweight Aggregate Association, Camino De Los Mares, San Clemente, CA 92672
Gypsum Association, 1603 Orrington Ave., Evanston, IL 60201
Lightweight Aggregate Producers Association, 546 Hamilton St., Allentown, PA 18101
Midwest Crushed Limestone Institute, P.O. Box 184, Decatur, IL 62525
Mineral Aggregates Association, P.O. Box 203, Parkersburg, WV 26101
Missouri Limestone Producers Assn., 124 E. Capitol Ave., Jefferson City, MO 65101.
National Ash Association, 1819 H St., NW, Washington, DC 20006
National Asphalt Pavement Association, 6811 Kenilworth Ave., Riverdale, MD 20840
National Crushed Stone Association, 1415 Elliott Place, NW, Washington, DC 20007
National Lime Association, 5010 Wisconsin Ave., NW, Washington, DC 20016
National Limestone Institute, 1315 16th St., NW, Washington, DC 20036
National Mineral Wool Insulation Association, 211 E. 51st St., New York, NY 10022
National Quartz Producers Council, P.O. Box 771, Golden, CO 80401
National Sand & Gravel Association, 900 Spring St., Silver Spring, MD 20910
National Slag Association, 300 S. Washington St., Alexandria, VA 22314
Perlite Institute, 45 W. 45th St., New York, NY 10036
Portland Cement Association, Old Orchard Rd., Skokie, IL 60076
Pulverized Limestone Association, 412 Delaware Ave., Towson, MD 21204
Salt Institute, 206 N. Washington St., Suite 300, Alexandria, VA 22314
Sorptive Minerals Institute, 111 E. Wacker Drive, Chicago, IL 60601
Southern California Rock Products Assn., P.O. Box 40, S. Pasadena, CA 91030
Sulphur Institute, 1725 K St., NW, Washington, DC 20006
Vermiculite Association, 52 Executive Park South, Atlanta, GA 30329

FOREIGN

Aggregate Producers Association of Ontario, P.O. Box 483, London, Ont., Canada
Agrupacion de Fabricantes de Cemento de Espana, Alfanso XII, 26, Madrid 14, Spain
Associacoa Brasileria de Cimento Portland, Caixa Postal 8071, Sao Paulo, Brazil, S.A.
Associacoa Tecnia da Industria do Cimento, Av. 5 de Outubro, 54, 2°-D, Lisboa 1, Portugal
Associated Portland Cement Manufacturers Ltd., Portland House, Tothill St., Westminister S.W., London, England
Association of the Greek Cement Industry, 8, Dragatsaniou St., Athens 122, Greece
Auckland Sand Suppliers' Association, P.O. Box 1882, Auckland, New Zealand
British Ceramic Research Assn., Queens Rd., Penkhull, Stoke-on-Trent, ST4 7LQ
British Granite & Whinstone Federation, 18 Berkeley St., London W. 1, England
British Quarrying & Slag Federation, 16 Berkeley St., London WIX 6 AP, England
Bundesverband der Deutschen Zementindustrie, e.V., 5000 Koln 16, Riehler Strasse 8, Postfach 4, Germany
Camera Nacional del Cemento, Paseo de la Reforma 398 (202), Mexico 6, D.F., Mexico
Canadian Institute of Mining and Metallurgy, 906, 1117 Catherines St. W., Montreal, Que. H3B 1J3, Canada.
Canadian Potash Producers Assn., 14th Floor, 8 King St. E, Toronto, Ont. M5C 13C, Canada.
Cembureau, 2 Rue St. Charles, Paris (15), France
Cement & Concrete Association, 52 Grosvenor Gardens, London, S.W. 1, England
Cement & Concrete Association of Australia, 147-151 Walker St., North Sidney, N.S.W., Australia
Cement Manufacturers Association, Express Bldg., Churchgate, Bombay 1, India
Cement Research Institute of India, M-10, South Extension 2, Ring Road, New Delhi 11, India
Centre National de Recherches Scientifiques et Techniques pour l'Industrie Cimentiere, Rue Cesar Franck 46, B 1050, Brussels, Belgium
Danish National Institute of Building Research, 20 Borgergade, Copenhagen K, Denmark
Engineering Institute of Canada, 2050 Mansfield St., Montreal 2, P.Q., Canada
Forschungsinsitut der Zementindustries, 400 Dusseldorf-Nord, Tannenstrasse 2/4, Germany
Gypsum Plasterboard Development Association, G.P.O. Box 321, London, N.W. 1, England
Instituto del Cemento Portland Argentina, San Martin 1137, Buenos Aires, Argentina, S.A.
Instituto Mexicano del Cemento y del Concreto, A. C. Insurgentes Sur 1846, Mexico 20, D.F., Mexico
Japan Cement Engineering Association, No. 1, Akasaka-Daimachi Minatoku, Tokyo, Japan
Limestone Federation, Manfield House, 376-9, Strand, London W.C. 2, England
Portland Cement Institute, 18 Kew Road, Richmond, Johannesburg, Rep. of South Africa
Quebec Asbestos Mining Assn., 320, 580 Grandee Allee E, Quebec, Que., G1R 2K2, Canada
Sand & Gravel Association of Great Britain, 48 Park St., London, W. 1, England
Sociedad Boliviana de Cemento, La Paz, Bolivia, S.A.
Societe des Ciments, Libanais, Lebanon
Syndicat National des Fabricants de Ciments et de Chaux, 41, Avenue de Friedland, Paris (8), France
Verein Deutscher Zementwerke, e.V., 400 Dusseldorf-Nord, Tannenstrasses 2/4, Germany
Verein Schweiz, Zement, Kalk und Gipsfabrikanten, Talstrasse 83, 8001 Zurich, Switzerland
Verkoopassociatie Nederland Cement, Enci-Eemij-Robur, N.V., Herengracht 507, Amsterdam C, The Netherlands
Wellington Sand, Gravel & Metal Producers Association, P.O. Box 17778, Wellington, New Zealand

4

Commodities

Alunite

ROBERT B. HALL *
CHARLES W. BAUER †

Alunite, $KAl_3(SO_4)_2(OH)_6$, was used from the 15th century until early in the present century as a source of potash alum and aluminum sulfate. Since early in this century its use for this purpose has declined to virtually nil with acid treatment of bauxite or clay replacing alunite as a source of alum. During the First World War alunite served as an emergency source of potassium sulfate fertilizer in the United States and Australia. More recently, alunite has been investigated as a nonbauxite ore of aluminum, with potassium sulfate and sulfuric acid as recoverable byproducts. The Soviet Union established a commercial-scale plant in the mid-1960s producing cell-grade alumina from alunitic ore and recovering by-product potassium sulfate and sulfuric acid. Interest in nonbauxite ores, including alunite, is expected to continue in countries that presently are dependent on imports of bauxite or alumina to support their aluminum industries. The US has the largest aluminum producing and fabricating industry in the world but imports more than 90% of its aluminum-bearing raw materials. This chapter reviews the status of alunite as a potential source of aluminum with byproduct fertilizer, and the resources available to support this industry, whenever this becomes economically feasible.

Historical Background

Alunite has received little attention until recently, yet for centuries it held a higher place than many of the industrial minerals considered essential today. The deposits of alunite at Tolfa, northwest of Rome, were mined almost continuously from 1462 until about the middle of the present century (DeLaunay, 1907; Lombardi, 1977). The potash alum made from the alunite, called Roman alum, was a valued commodity in international trade, being used in the making of paper, in tanning leather, and as a mordant in textile dyeing. The industry at Tolfa is said to have been controlled by the Vatican and was a source of revenue for the Papal State until fairly recent times (Lombardi et al., 1977).

Since early in the present century, the preferred method for making alum has been by acid treatment of bauxite, clay, or other aluminous material, so that alunite has fallen into near-total disuse as a source of alum chemicals. Appreciable quantities of mixed alunite-kaolin rock still are quarried in the Tolfa district for use by the Italian cement industry (Lombardi and Mattias, 1979), but the role of this material in the making of Italian cement is not clear. A large deposit of alunite-bearing rock in Chekiang province, mainland China, is said to have yielded a "great quantity" of alum over a period of several centuries (Yih, 1931). During the 19th century, alunite deposits in France, Spain, Australia, and other countries were exploited as raw material for making potash alum and aluminum sulfate.

Interest in alunite was revived during the First World War, not for alum, but as an emergency source of potassium sulfate fertilizer. Large veins of nearly pure alunite had just been discovered on Alunite Ridge, 11 km southwest of Marysvale in Piute County, Utah (Butler and Gale, 1912), and these were quickly brought into production for making potash fertilizer to replace supplies cut off from the famous deposits at Stassfurt, Germany. An estimated 225 kt of Marysvale alunite was mined and processed at that time (Callaghan, 1973). Similarly, Australian alunite deposits were mined as a source of potash fertilizer during both world wars (Hall, 1978).

* Geologist, Branch of Central Mineral Resources, US Geological Survey, Denver, CO.
† Senior Project Manager-Geologist, Earth Sciences, Inc., Golden, CO.

Shortly before entry of the US into World War II, a comprehensive review of the US alunite resources was made by the US Bureau of Mines (USBM) (Thoenen, 1941). Thoenen recognized the deposits at Marysvale as the most promising then available to American industry and, after the entry of the US into the war, efforts were accelerated to explore the deposits and to develop a method for economically processing the ore to produce aluminum (Fleischer, 1944; Hild, 1946; Callaghan, 1973). Japan also experimented with alunite as a wartime aluminum ore (Allen, 1947). With the end of the wartime emergency, alunite was relegated to the status of an unpromising, noneconomic resource (Anon., 1970) and was all but forgotten. The Soviet Union, as a matter of national policy aimed at raw material self-sufficiency, established a 200 kt/a alumina-from-alunite plant in the state of Azerbaijan in the mid 1960s. The Soviet Union, like the United States, is deficient in bauxite resources (Shabad, 1976). US interest in alunite was revived in 1970 when large resources of alunitic rock were discovered 140 km west of Marysvale by personnel of Earth Sciences, Inc., Golden, CO. The deposits, located in the southern Wah Wah Mountains of Beaver County, Utah, 87 km northwest of Cedar City, were explored by trenching and drilling. A proven-plus-probable reserve of 136 Mt of rock containing about 38% alunite, with an additional inferred reserve of 500 Mt of similar grade, was announced (Anon., 1976). Pilot-plant testing was conducted at Golden, CO from November 1973 until the end of 1976 by the Alumet consortium comprising Earth Sciences, Inc., National Steel Co. of Pittsburgh, PA, and the Southwire Co. of Carrollton, GA. The preferred process is generally similar to that used by the Soviet Union (Thompson, 1976).

The Alumet partnership announced intent to establish an alumina-from-alunite industry in southwest Utah with a capacity to produce 454 kt/a of cell-grade alumina with 227 kt/a of fertilizer-grade potassium sulfate and 408 kt/a of industrial-grade sulfuric acid as byproducts. The plan called for calcined phosphate rock to be shipped by rail to the Utah plant site from deposits in southeastern Idaho and to be treated with the acid to produce phosphate fertilizer. Sale of two fertilizer byproducts would considerably enhance the economics of alumina production from low-grade alunite-bearing rock (Walker and Stevens, 1974; Parkinson, 1974; Thompson, 1976).

Construction of the Alumet project still had not been scheduled as of 1982, but may proceed if technological breakthroughs are achieved or if costs of imported bauxite and alumina rise to levels at which use of alunitic rock becomes more attractive economically. Obsolescence and aging of conventional Bayer plants designed to operate on imported bauxite also may become a factor contributing to development of nonbauxite ores. Costs of new greenfield plants fed by domestic alunite may compare favorably with new Bayer plants fed by imported bauxite, when all factors are considered. Alunite one day may be regarded by the aluminum industry in much the same way that taconite now is regarded by the steel industry.

Geology

Alunite occurs in a variety of deposits ranging in size from mere nodules and lenticles a few centimeters across to huge masses comprising several hundreds of millions of tons of altered rock containing 30 to 40% alunite. Nearly pure alunite occurs in hypogene veins.

Classification of Deposits

Three main classes of alunite deposits are commonly recognized (Hall, 1978, 1980):

Veins: The veins at Alunite Ridge, Marysvale, UT are steeply dipping fissure fillings with sinuous trend enclosed in strongly altered volcanic lava flows, mud flows, and ash-flow tuffs (Butler and Gale, 1912; Callaghan, 1973). DeLaunay (1907) described veins at the classic deposits at Tolfa, Italy that are similar to those at Marysvale. Veins several meters wide and several hundred meters long provided the bulk of the alunite production at Tolfa (DeLaunay, 1907). Lombardi (1977) reports an estimated total production at Tolfa of 18 Mt during a span of five centuries. Callaghan (1973) reports an estimated production of 225 kt from veins at Marysvale during the First World War. Veins smaller than those at Marysvale occur in Lincoln, Mineral, and Humboldt counties, Nevada, and Yuma County, Arizona, with tenor ranging from 75 to nearly 100% alunite. Alunite veins may be white to yellow cryptocrystalline, or may contain coarse pink crystals 10 to 20 mm long. Although the high-grade alunite in veins almost certainly would be an acceptable substitute for bauxite, the total resource available in veins is too small to constitute a raw material base for industry.

Nodules and seams in argillaceous sedimentary rocks: Nodules and thin discontinuous layers and seams of alunite or natroalunite are surprisingly common and widespread geographically (Hall, 1978), occurring in shales, micaceous schists, or clayey beds. They appear to have been formed supergenically or diagenetically by the action of sulfate-rich acid ground water on illitic or mica-rich argillic sediments. Oxidation of pyrite disseminated in adjacent or superjacent rock provides the necessary acid; potassium is derived from illite or muscovite mica in the alunitized host sediment. The purity of alunite nodules may approach that of veins, but these sedimentary occurrences are, for the most part, confined to thin discontinuous layers commonly intermixed with kaolin, and do not form bodies large enough to serve as an aluminum resource.

An unusual sedimentary deposit at Lake Campion, near Chandler, Western Australia, was mined during World War II as a source of potassium sulfate fertilizer. The lake sediment contains 60% alunite. The balance is mostly detrital quartz, mica, and iron oxide. Reserve was estimated at 12 Mt (Fitzgerald, 1945). The origin of this deposit is not clear, but it may have been formed by a diagenetic mechanism similar to that postulated by King (1953) to explain alunite-rich lake sediment at Pidinga, South Australia. The Australian lake deposits, although much larger than most sedimentary alunite deposits, do not constitute a long-term aluminum-resource base.

Replacement bodies in solfatarically and hydrothermally altered volcanic, subvolcanic, and hypabyssal intrusive rocks: This class of deposit constitutes the bulk of alunite resources, both in the United States and elsewhere in the world, and would serve as the resource base for any new aluminum industries designed to use alunite as a raw material. Although these deposits are lower in grade than veins or sedimentary deposits, the alunitized rock occurs in huge bodies amenable to open-pit mining. The tonnages are sufficient to feed an economic-scale alumina plant for 20 years or more, which is long enough at least to amortize the plant. The characteristics and origin of replacement deposits are discussed more fully in the following section.

Mineral and Chemical Composition and Other Characteristics

Pure alunite ($KAl_3(SO_4)_2(OH)_6$) theoretically contains 11.37% K_2O, 36.92% Al_2O_3, 38.66% SO_3, and 13.05% H_2O. Analyses of some crystals may approach this composition, but most natural alunites contain some sodium substituting for potassium. If the atomic ratio Na:K is equal to or greater than 1, the mineral is called natroalunite. If the Na:K ratio is greater than 1:3, the material may be called sodian alunite, although this name is sometimes misinterpreted as a synonym for natroalunite, and no consistent rule seems to have been followed in the literature. The K-rich variety is more common in nature, and is preferred because relatively soda-free K_2SO_4 fertilizer is a more valuable byproduct than salt cake, Na_2SO_4.

The replacement-type alunite ore bears little resemblance to the museum specimens of alunite described in mineralogy texts and journals. Typically, the alunite occurs in fine-grained altered volcanic or subvolcanic rock, or coarser hypabyssal intrusive porphyries. The altered rock is composed principally of microcrystalline quartz and alunite with minor amounts of hematite and rutile or anatase. Clays and other caustic-soluble silica minerals are common alteration constituents and may be troublesome in the Alumet process—or any caustic-leach process if present in more than small amounts. Identification of alunite-bearing rocks in the field is not easy because altered volcanics look very much alike whether rich in alunite or in kaolinite, sericite, or other alteration minerals. Alunite has a specific gravity of 2.82, slightly higher than quartz and clays, and so a greater *heft* commonly is a good clue to presence of high alunite content in the rock. Vein alunite commonly is flesh-pink, but color is a poor criterion in identifying alunitized rocks, which may be off-white, gray, yellowish, brownish, or pinkish, usually mottled or variegated, and iron-stained. A yellowish-orange tint generally is a clue to presence of jarosite. Microcrystalline quartz is a major constituent in typical ore-grade rock, which is tenacious, hard, and abrasive, with dull luster and an uneven to subconchoidal fracture; deceptively crumbly or powdery material may be encountered occasionally, but is comparatively rare. Cristobalite-bearing alunitic rock commonly has a subvitreous waxy luster and good conchoidal fracture. Most alunitized rocks are so fine-grained that their composition is best determined by powder X-ray diffractometry supplemented by thin-section examination and whole-rock chemical analysis. Table 1 shows the composition of theoretically pure $KAl_3(SO_4)_2(OH)_6$, one sample of high-grade vein from Marysvale, UT, and representative silicified and alunitized vol-

TABLE 1—Analyses of Alunite and Alunite-bearing Rocks

Sample Type*	1	2*	3*	4*	5*	6*	7*	8	9
SiO_2		3.30	38.00	54.00	50.00	48.00	28.00	55.77	37.55
Al_2O_3	36.92	34.00	23.00	15.00	14.00	16.00	25.00	17.03	22.52
Fe_2O_3		0.54	0.40	1.20	0.08	3.10	6.20	3.05	4.16
FeO								0.29	
MgO		0.01†	0.02†		0.02†	0.01†	0.02†	0.22	0.22
CaO		0.33	<0.10	0.10	0.11	0.11	0.50	0.33	0.58
Na_2O		0.69†	0.27†	0.50†	0.22†	1.21†	0.82†	0.34	2.21
K_2O	11.37	5.30	5.50	4.10	3.60	2.20	3.90	4.17	3.56
TiO_2		0.08	0.20	0.76	0.83	0.47	0.20	0.22	0.53
P_2O_5		<1.00	<1.00	<1.00	<1.00	<1.00	<1.00		
MnO		0.05		<0.05	<0.05	<0.05		0.01	
SO_3	38.66	37.50	23.00	16.00	12.50	17.00	21.75	13.25	20.51
H_2O^+	13.05							5.52	8.44
H_2O^-								0.35	
TOTAL	100.00	Inc.‡	Inc.‡	Inc.‡	Inc.‡	Inc.‡	Inc.‡	100.55	100.28

Sample Type and Source

1. Theoretical $KAl_3(SO_4)_2(OH)_6$.
2. Vein alunite, Alunite Ridge, Marysvale, Piute County, Utah.
3. Alunite ore, Alumet lease exploration pit, southern Wah Wah Mtns., Beaver County, Utah. Approximately 60% alunite.
4. Alunite ore, Red Mountain, Santa Cruz County, Arizona. Approximately 39% alunite.
5. Alunite ore, Red Mountain, Hinsdale County, Colorado. Approximately 34% alunite.
6. Alunite ore, Saddleback Mountain, Grant County, New Mexico. Approximately 40% alunite.
7. Alunitic rock, The Elbow, Lyon County, Nevada; ridge crest 3 km north of Mineral County boundary. Approximately 50% alunite, 5% jarosite, 4% hematite. Silica dominantly cristobalite.
8. Alunitic rock, ore-grade, from Allumiere near Tolfa, Italy (Lombardi, 1977).
9. Average of 63 analyses of alunitized volcanic rock at Dashkesan (Zaglik), Azerbaijan, USSR (Kashkai, 1961).

* SiO_2, Al_2O_3, Fe_2O_3, CaO, K_2O, TiO_2, MnO, SO_3 analyzed by X-ray fluorescence, US Geological Survey laboratories, Denver. Determinations approximate; only first two digits are significant.

† Na_2O, MgO determined by atomic absorption, US Geological Survey laboratories, Denver.

‡ Inc., incomplete analysis, total not given.

canic rocks from various localities, rich enough in alunite to qualify as alunite ore.

Chemical analyses of alunite-bearing rocks can be misleading if not interpreted carefully. Chemical data should be supplemented with X-ray identification of mineral constituents to insure a reasonably accurate assessment of mineral composition. For example, the presence of kaolinite ($Al_2Si_2O_5(OH)_4$) will contribute a certain amount of Al_2O_3 over that attributable to alunite, and of SiO_2 over that attributable to quartz. In this case, the calculation of alunite content based on Al_2O_3, and of quartz content based on SiO_2, will be erroneously high. The presence of jarosite ($KFe_3(SO_4)_2(OH)_6$) will increase SO_3 and K_2O in the analysis so that calculation of alunite content based on SO_3 or K_2O will yield an erroneously high figure. Fortunately, a good powder X-ray diffraction pattern, properly interpreted, reveals minerals present in amounts great enough to significantly affect chemical analyses. Our experience has shown that X-ray patterns, when compared to sets of patterns of materials of known composition, can be used to estimate alunite content to within 5% of true value. Properly standardized computer-controlled automatic-loading X-ray diffractometers could be used to control ore grade and quality in a large-scale alunite ore mining and processing plant.

Zoning of Replacement Deposits

Mineralogical zoning is characteristic of large replacement-type alunite deposits in the western United States. Identical or very similar zoning also has been reported in the geological literature of other countries.

Four principal zones generally are recognized: siliceous core or cap; quartz-alunite; argillic; and propylitic.

The siliceous core or cap (Zone 1) is central and uppermost, succeeded outward and downward by the quartz-alunite zone (Zone 2), the argillic zone (Zone 3), and finally the propylitic zone (Zone 4). These four zones may be recognized laterally on the surface and also vertically, although deep drilling generally is necessary to reveal an in-depth zoning relationship.

The following description represents an ideal case, seldom realized in nature. Most natural deposits are irregular and inhomogeneous with enclaves of one zoning assemblage enclosed in another. The alteration may be telescoped so that one or several zones are very narrow or not apparent to casual observation. Moreover, zonal boundaries are subtle, difficult to map visually, and are drawn arbitrarily based on X-ray powder diffraction determinations.

Siliceous Core or Cap Zone: Rock here is highly siliceous and may resemble chert or opalite, and if formed at or very near the surface it may be porous like siliceous sinter. Native sulfur commonly is present in pores and cavities. Quartz typically is the dominant silica phase, but cristobalite is not uncommon, and amorphous silica and tridymite have been identified in some areas. This zone is thought to mark the main vent for, and final stage of alteration caused by, strongly acid volcanogenic hydrothermal fluids and gases that leached alkalis, alkaline earths, alumina, and other elements from the volcanic rocks leaving a silica residue. Some of the silica is introduced from the underlying zone of alunitic alteration and the rock may contain 90% or more of silica.

Quartz-Alunite Zone: The siliceous zone grades outward to a zone of highly altered rock composed mainly of microcrystalline quartz and alunite. This constitutes the *ore* zone of a commercial alunite deposit. Tenor of alunite may be 30% or higher. Other minerals normally present in small amounts include hematite, rutile, and anatase. Kaolinite (or its polymorph, dickite) and other phyllosilicate minerals, as well as opaline cristobalite, are common, but if these siliceous minerals are present in more than very minor amounts they have a deleterious effect on ore quality as explained under discussion of the Alumet process (p. 428). Pyrite may be found in quartz-alunite rock, particularly at depth. Jarosite, the iron-bearing sulfate analog of alunite, may be present as either a hypogene or a supergene constituent. Alum minerals occur as incidental weathering efflorescences, but rarely are found at depth.

containing appreciable amounts of kaolin clay minerals (hence *argillic*), usually kaolinite, but occasionally dickite. Halloysite, both hydrated and nonhydrated, is reported to be common in some Italian alunite-bearing districts (Lombardi and Mattias, 1979); however, halloysite does not seem to be as common as kaolinite in the western US deposits. Microcrystalline quartz still is dominant, and alunite may be subequal in amount to kaolin. Material in this zone is not as desirable for processing by the Alumet process as the quartz-alunite material because clays and other phyllosilicate minerals are deleterious. (This is not necessarily true for acid-leach processes.) Minerals other than the kaolin group also occurring in the argillic zone include sericite, diaspore, pyrophyllite, and more rarely, zunyite. Hematite and rutile or anatase are ubiquitous in this as in all of the other zones. Jarosite commonly is associated with alunite but may be supergene rather than hypogene. Jarosite and iron oxide minerals impart yellowish and reddish hues to the altered rocks.

The argillic alteration assemblage of Zone 3 changes gradually, progressing outward from the alunite-rich zone (Zone 2) toward the propylitic zone (Zone 4). Kaolin-group minerals become less abundant, with sericite commonly subequal to quartz. Where sericite becomes the dominant alteration mineral in the outer or central part of Zone 3, this part may merit the informal designation, *phyllic zone,* but this is not so common as to justify formal designation as part of the zoning pattern described here. Smectite clays and chlorite become prominent in the more distal part of the argillic zone, and alunite is sparse or absent.

Propylitic Zone: The argillic zone grades imperceptibly into the propylitic zone, characterized by dull grayish-green color and usually well-preserved primary rock texture. Quartz is less prominent than in the preceding zones and is likely to be primary rather than a secondary alteration mineral. Characteristic members of the alteration suite in this zone include epidote, chlorite, zeolites, pyrite, and calcite—minerals which are unstable and could not exist in the more strongly acid environment of the quartz-alunite zone. Primary plagioclase may be saussuritized and the mafic minerals uralitized, indicating effects of a milder less acidic alteration than that prevailing toward the center.

Genesis of Replacement Alunite Deposits

The replacement deposits are believed to have been formed by sulfuric acid-charged hydrothermal fluids and solfataric gases emitted during the late stage of a volcanic episode. Soluble elements are leached from the permeable rocks, typically pyroclastics of rhyolitic to andesitic or dacitic composition, although flows and hypabyssal porphyries are susceptible also. Silica released by destruction of feldspar, mica, and other rock-forming minerals recrystallizes into very fine-grained quartz or cristobalite. Microscopic textures suggest that fine-grained alunite crystallizes more or less simultaneously with the silica phase (Fig. 1). Metasomatic volume-for-volume replacement may take place in a way (admittedly not well understood) that preserves original rock texture, sometimes remarkably well (Fig. 2). These silica-rich rocks have been called *secondary quartzites* by Soviet geologists (Kirova, 1959; Naboko, 1963; Kashkai, 1970).

Alteration is controlled by texture and composition of the rock, porosity and permeability of the rock, chemistry of the fluids, temperature and pressure, rate and continuity of flow of the fluids, duration of the episode, and structures such as beds, faults, and joints. The fluids may rise along a localized vent, in which case the alteration zoning will be rudely concentric or target-like in plan, grading from highly silicified rock at the *bullseye* through alunitic and argillic rings to propylitic at the perimeter. In many cases, the vent seems to have been a large fracture or fault, so that the alteration pattern is elongated with strongly silicified rock along a main axis flanked successively by alunitic, argillic, and finally propylitic zones on each side of the siliceous axial zone. The chemical system is dynamic; the fluids cool and become less acid as they migrate outward from the central channel and react with the rock. Accordingly, the alteration assemblages in the respective zones reflect the stability fields of the constituent minerals, as elucidated by Hemley et al. (1969).

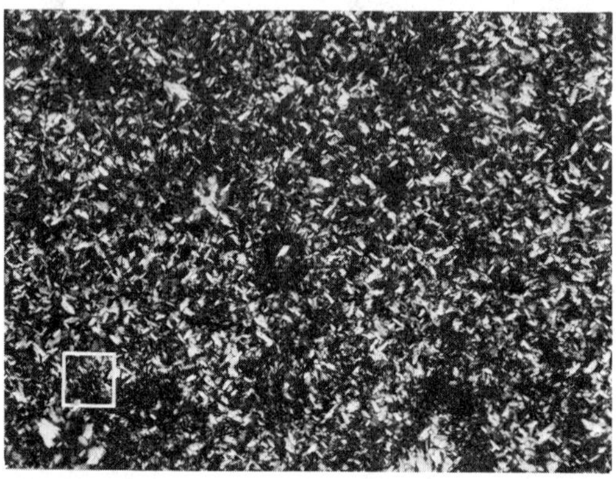

FIG. 1—*Photomicrograph of high-grade alunite rock from the east slope of Red Mountain, Hinsdale County, Colorado, 5 km SSW of Lake City. Square 0.1 mm across. Crossed polarizers. Elongate shreddy flakes are alunite; dark matrix (low birefringence) is microcrystalline quartz. Curious runic arrangement of elongated alunite crystallites is characteristic of metasomatically altered alunite-rich rock.*

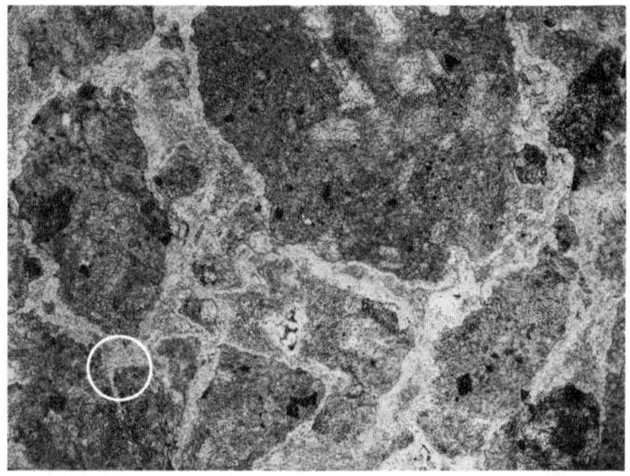

FIG. 2—*Photomicrograph of lithic ash tuff metasomatically altered to high-grade alunite rock, Marble Mountain, Rio Grande County, Colorado, 10 km east of Summitville. Plane light. Circle 0.1 mm across. Lithic ash fragments (gray) composed of extremely fine-grained microquartz and alunite; matrix (light gray), very fine-grained alunite with subordinate microquartz. Original texture is well preserved in spite of profound change in chemical and mineral composition by hydrothermal-solfataric alteration.*

The alunitization process is a near-surface phenomenon, but essentially is hypogene rather than supergene. Supergene alunite can form by downward percolation of acidic surface water as postulated by Schoen, White, and Hemley (1974) at Steamboat Springs, NV, but multimillion-ton bodies of alunitic rock are formed by an essentially hypogenic process—that is, by uprising, boiling hydrothermal solutions. The sulfuric acid of the system is derived from volcanogenic H_2S rising from the lower part of the volcanic system, oxidizing near the surface where boiling occurs and exposure to atmospheric oxygen takes place above the water table. A small amount of the water in the hydrothermal system may be primarily magmatic, or in some cases perhaps connate water also, but numerous isotopic studies made since the late 1960s suggest that by far the greater part of the water in these systems is deeply circulating meteoric water in the upsurging part of a convection cell, driven by heat from underlying magma or a still-hot intrusive body (Taylor, 1974; White, Muffler, and Truesdell, 1971). Multimillion-ton deposits of quartz-alunite rock probably were formed by descent of the water table from higher to lower levels with passage of time, thus forming successively deeper alunitized rock beneath earlier-formed alunitized rock. Alunitic rock is found at considerable depth in pervasively altered districts; for example, a drillhole at a deposit in Beaver County, Utah, penetrated 300 m of alunite-bearing rock (Parkinson, 1974). Hypogene alunite was formed to a depth of several hundred meters at El Salvador porphyry copper deposit in northern Chile and at the Mi Vida porphyry copper deposit in northwestern Argentina (Gustafson and Hunt, 1975; Khoukharsky and Mirré, 1976).

The role of bacteria in forming sulfuric acid by oxidation of elemental sulfur in volcanic hot-spring environments has been pointed out by some investigators (Ivanov et al., 1968; Brock and Mosser, 1975; Zinder and Brock, 1977). It is doubtful, however, that microbial processes were dominant in forming huge masses of alunitic rock.

From the foregoing, it is apparent that volcanic terrains are the natural habitat of large replacement-type alunite bodies, which constitute the principal resource of alunite worldwide. Calderas are especially favorable because they provide the elements essential for formation of alunite-rich rock: volcanic rock cover susceptible to alunitization; an underlying magma chamber, heat from which can drive a hydrothermal convection cell; ring fractures or faults that provide vents for uprising hydrothermal fluids; and a source of sulfur.

Resources

The alumina-from-alunite plant in Azerbaijan, USSR has shown that alunitic rock can be used as an ore of aluminum with fertilizer byproducts. Future establishment of similar plants in the United States and elsewhere in the world depends on the demonstration of economic feasibility. In general, for alunitized rock to be a suitable ore, it should contain at least 30% of alunite with the remainder mostly microcrystalline quartz. Soluble silica in the form of clay, mica, or opaline cristobalite are undesirable contaminants in the Alumet caustic leach process, and are intolerable if present in more than a minor amount. Iron and titanium oxides are tolerable in the metallurgical process, if not excessive. A deposit, even one meeting the 30% alunite criterion, cannot be classed as an ore body unless it contains sufficient ore-grade rock to feed a plant of economic size long enough to fully amortize it, generally about 20 years. A deposit containing 90 Mt of ore-grade (30% alunite) rock is considered to be potentially exploitable.

Principal domestic and foreign resources are summarized briefly here. More comprehensive discussions are given in Hall (1978) and Bauer (1980).

US Alunite Resources

Large alunite deposits have been identified in some western states during the last decade. The best explored deposits are in southwestern Utah, but deposits in Arizona and Colorado also have been tested. Other prospects having large-tonnage potential are known in Nevada and New Mexico. Fig. 3 shows location of major alunite districts in the United States.

Utah: A group of four closely spaced deposits in the southern Wah Wah Mountains of Beaver County, Utah, has been mentioned already. This discovery was made by Earth Sciences, Inc. in 1970 (Walker and Stevens, 1974; Parkinson, 1974). Drilling at one deposit identified a reserve of 135 Mt of material containing at least 30% alunite. Additional potential resources of 500 Mt of the same grade have been estimated in three adjoining deposits. Several other deposits lie within a few kilometers of the group of four (Anon., 1976). The deposits have been formed in Tertiary age volcanic rocks of andesitic to rhyolitic composition. The individual deposits are elongated in plan, and at least one is known to extend to a depth of 300 m, although bottoms of alunitized ground have not been completely delimited. The elongated form suggests local control by faults or major fractures.

The Marysvale district, Piute and Sevier counties, was fairly well explored during World War I and World War II and is the only district in the United States to have produced more than token amounts of alunite (Callaghan, 1973). However, the resource potential at Marysvale, estimated to be on the order of 43 Mt, clearly appears to be well below that of

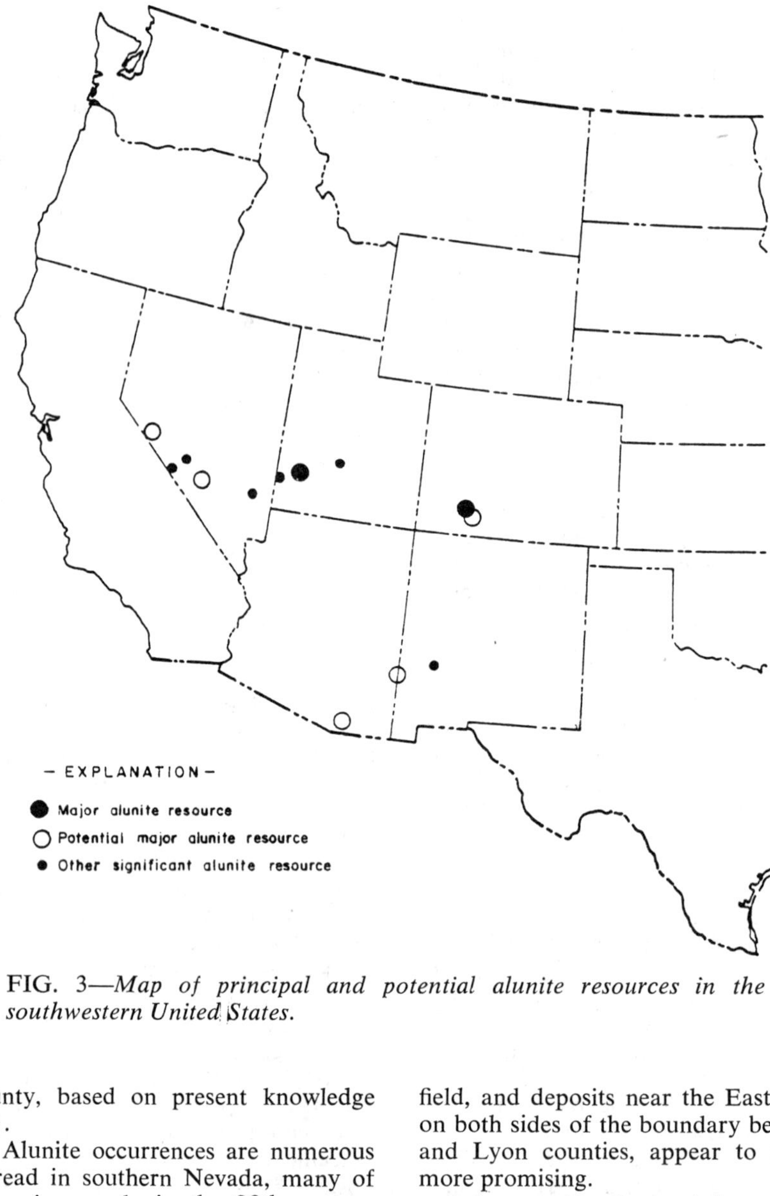

FIG. 3—*Map of principal and potential alunite resources in the southwestern United States.*

Beaver County, based on present knowledge (Hall, 1978).

Nevada: Alunite occurrences are numerous and widespread in southern Nevada, many of them known since early in the 20th century (Hall, 1978). Sporadic attempts were made to develop vein-type deposits in Lincoln (Hewett et al., 1936) and Humboldt counties (Clark, 1918; Vanderburg, 1938). Reconnaissance sampling by Earth Sciences, Inc. during the mid-1970s revealed several large deposits; however, based on preliminary data, the Nevada deposits are generally lower grade and less homogeneous than the deposits in Utah. Exploration has not progressed sufficiently to allow presentation of reserve-resource estimates for Nevada. Deposits in Esmeralda County, south and west of Goldfield, and deposits near the East Walker River on both sides of the boundary between Mineral and Lyon counties, appear to be among the more promising.

Arizona: Alunite in Arizona was reported first in the Patagonia district, Santa Cruz County, south of Tucson (Schrader, 1913, 1915). An attempt to mine alunite veins at Sugarloaf Peak, Yuma County, was described by Heineman (1935). One deposit with commercial potential has been recognized at Red Mountain, 5 km southeast of Patagonia. The deposit is a complex assemblage of rhyolitic latitic to andesitic flows, tuffs, breccias, and porphyries that have been pervasively alunitized. Exploratory drilling over the summit was carried out in 1975. A resource estimate of

200 Mt is conservative, but the Red Mountain deposit presently is considered to be less promising than deposits in Utah because of questionable ore quality (Hall, 1978). Low-grade copper mineralization was discovered by deep drilling (Corn, 1975) prior to drilling for alunite. This interesting association of alunite with another kind of ore has been observed in many places (see *Relation of Alunite to Hypogene Ore Deposits*).

New Mexico: The earliest reported occurrence of alunite in New Mexico is at Alum Mountain, Grant County, on the south side of the Gila River (Hayes, 1907). Additional occurrences are in Sandoval, Sierra, and Luna counties. An interesting deposit at Saddleback Mountain, western Grant County, has a potential resource tentatively estimated at 60 Mt of alunitized rhyodacitic tuffs and breccias of Tertiary age, containing about 30% alunite (Hall, 1978). Further exploration is needed to estimate tonnage and grade more precisely at this and other deposits.

Colorado: Alunite was reported in the Rosita Hills, Custer County, and at Calico Peak, Dolores County, in the late 19th century (Cross, 1891, 1896; Cross and Spencer, 1900). Additional occurrences were noted later in Hinsdale County (Larsen, 1913), and in Rio Grande County (Gardner, 1943; Steven and Ratté, 1960). Some of these deposits attracted interest during the First and Second World Wars but no significant development ensued. The most promising deposit of alunite recognized in Colorado so far is that at Red Mountain, 5 km south of Lake City in Hinsdale County (not to be confused with the previously mentioned Red Mountain in Santa Cruz County, Arizona, nor with other Red Mountains in Colorado). The Red Mountain deposit was explored by two drillholes in 1976, one to a depth of 210 m, the other to 99 m. A third hole to a depth of 93 m was drilled in 1978. Estimates by Earth Sciences Inc. show demonstrated reserve of 55 Mt with grade at or near 40% alunite. Potential resources may be as great as 1.5 Gt, making the Red Mountain deposit one of the largest in the world. Metallurgical testing of ore compares favorably with tests made on ore from the Beaver County, Utah, deposits which have been proposed for commercial development. However, there are considerable environmental and operational obstacles to developing Red Mountain; accordingly, development of this deposit does not appear likely until feasibility of an alumina-from-alunite industry is proven elsewhere.

Other States: Additional alunite deposits occur in California, Washington and Texas, but none appear to have a potential for commercial development comparable to those mentioned above. A comprehensive review of domestic alunite deposits in other states, including many of only mineralogical or scientific interest, has been published (Hall, 1978).

Summary of US Alunite Resource Position

A demonstrated reserve base of 650 Mt of alunite-bearing rock in 23 deposits has been suggested for the United States, with an additional inferred reserve base of 2.5 Gt and an additional potential resource * of nearly 5 Gt (Bauer, 1980). In contrast, an earlier estimate published by the US Geological Survey (USGS) suggests a reserve base of 252 Mt of alunite-bearing rock and an additional 1.4 Gt of potential resources (Hall, 1978). The latter more conservative estimates were made without access to extensive surface sampling and drilling data available to Bauer in 1980.

The most thoroughly explored deposits are those in the southern Wah Wah Mountains of Utah, where tonnages and grades are fairly well established. In spite of uncertainty and difficulty in assigning an absolute number to the reserve base and to potential resources at the present time, it is clear that the volume of domestic alunite resources is adequate to support an alumina-from-alunite industry (with fertilizer and acid byproducts), granted the proviso that rock containing 30% or more of alunite can be mined and processed profitably. This invalidates the conclusion reached by the National Materials Advisory Board (Anon., 1970) that "alunite has little potential for being a raw material of aluminum in this country because all known deposits are either small and

* Demonstrated reserve base and inferred reserve base as used here conform generally to definitions of these terms as given in US Geological Survey Circular 831 (Anon, 1980). The term *potential* resource as used here includes resource as defined in USGS Circular 831, p. 1, not otherwise included in demonstrated or inferred reserve base. It includes hypothetical but not speculative resources as defined in USGS Circular 831. The Soviet Union uses alunite-bearing rock as an ore of aluminum with potassium sulfate and sulfuric acid byproducts, but we designate the material as alunite-bearing rock rather than as ore because it has not yet been demonstrated on a commercial scale that this material can be mined and processed at a profit in an open-market economy.

scattered or have the mineral disseminated through volcanic rock."

Foreign Alunite Resources

Alunite occurs in many countries and can truly be said to have global distribution, although deposits large and rich enough to be potentially exploitable as sources of aluminum and potassium fertilizer are comparatively rare. If a successful alumina-from-alunite industry eventually is established in the US or in another industrialized nation operating under a free-market economy, it is virtually certain that further investigation will prove the existence of additional alunite resources in other parts of the world. So far, the alunite industry operating since 1966 in the USSR has not stimulated the establishment of similar operations outside of the Soviet Union. Reports concerning the Soviet alunite operation have been sparse and vague, but apparently it has had operational and economic difficulties that have clouded prospects for its expansion (Shabad, 1976). The worldwide subeconomic resources of alunitic rock in a number of countries are known to be large. A summary of worldwide alunite resources more comprehensive than that given here may be found in Hall (1978).

USSR: The excellent treatise on alunite by Kashkai (1970), which unfortunately is not available in English translation, contains the most comprehensive summary known to us of alunite resources in the Soviet Union. Although the author does not give specific data on reserves and resources, he presents details of mineralogy, geologic setting, and genesis for a large number of deposits. From Kashkai's work, one may conclude that the Soviet Union probably is better endowed with alunite resources than any other nation in the world. This is true both because of its vast size and because it encompasses large regions of volcanism where hydrothermal-solfataric processes have profoundly altered feldspathic rocks of various kinds to produce large bodies of silicified and alunitized rock. The alumina-from-alunite plant at Kirovabad in Azerbaijan is fed by alunitized tuffs of Late Jurassic age in deposits located near Zaglik, a few kilometers northwest of Dashkesan. Alunite content of the rock apparently is about 40%.

There are more than 80 other deposits and occurrences in the Soviet Union, including those in Kazakhstan, Armenia, Georgia, the Ural Mountains, Altai, Uzbek, Tadzhik, Kirgiz, the Sikhote Alin cordillera, and elsewhere. It is unlikely that all of these deposits are of a size and grade that would encourage development.

Asia: The People's Republic of China has a very large deposit of alunite-rich rock in the Pinyang-Fanshan district in Chekiang province (Yih, 1931) and other deposits aggregating hundreds of millions of tons in Fukien and adjoining provinces (Ikonnikov, 1975).

Japan experimented with alunitic rock as a source of aluminum and potash during the Second World War (Allen, 1947), using material from deposits in Japan and South Korea, which was under control of Japan at that time. Apparently the Japanese deposits in Shizuoka and Hyogo prefectures (Iwao, 1949, 1953) and the Korean deposits in South Cholla province (Cho and Moon, 1978) are comparatively small, not especially promising for commercial development.

Europe: The long-exploited deposits at Tolfa, north of Rome, have been mentioned previously. A short review of alunite as a potential nonbauxite ore of aluminum with fertilizer byproducts was published recently by the Commission of the European Communities with headquarters in Brussels (Landi, 1978). The geology of Italian deposits has been described in detail by Lombardi and Mattias (1979). Although historically important, the Italian resources do not appear to be large enough to support an alumina-from-alunite industry of economic scale; moreover, except for certain localized areas, the alunitic rock generally is contaminated with inordinately high amounts of clay, mica, and other forms of soluble silica, including amorphous silica and cristobalite (Lombardi and Mattias, 1979).

Alunite deposits in the Puy-de-Dôme region of central France were mined during the 19th century for making alum (Charrin, 1940, 1948; Gautier, 1940; Lacroix, 1962). It is not clear whether resources there could support an alumina industry. France is the only country in western Europe to have appreciable bauxite resources, so that alunite has received little attention there as a potential ore of aluminum.

Spain is modestly endowed with alunite resources (Caballero et al., 1974), and a deposit near Segovia was proposed for development during the mid-1970s (Kühnel et al., 1975; Galan and Lopez-Aguayo, 1977). A deposit of volcanogenic origin on the island of Milos, Greece, has been considered for development, however, details are unavailable. Quartz-alunite zones in hydrothermally altered andesitic volcanic rocks of Upper Cretaceous age in Bulgaria are thought to be potentially exploitable (Velinov, 1973).

Mexico and South America: Alunite was recognized as a common constituent in hydrothermal kaolin deposits in Mexico during the 1940s, according to Knížek and Fetter (1950). During the 1950s, a nationwide investigation of alunite deposits in Mexico was undertaken, and a comprehensive report has been published (Yris Rovirosa, 1964). Experiments were conducted at the University of Guanajuato, and several processes were developed on a laboratory scale for extraction of potassium fertilizers and alumina from Mexican alunitic rock (Lopez, 1977). In the 1970s an alunite-processing plant was projected for construction at Salamanca, Guanajuato, to be fed by material from deposits in the Juventino Rosas district northwest of Celaya, Guanajuato, but the project was suspended. Probably the most promising alunite resources recognized in Mexico so far are in Guanajuato, but occurrences are numerous in other states (Keller and Hanson, 1968; Hanson, 1975; Yris Rovirosa, 1964), although no single deposit has yet been proven to be of a size and grade that would support a plant of economic scale.

On the South American continent, deposits of an unusual character have been proposed for commercial development in Chubut Province, southern Argentina. Alunite occurs there in *bochones* (large nodules or round boulderlike masses) 100 mm to 1.5 m long in bentonitic tuffs of Paleocene age (Angelelli, Schalamuk, and Arrospide, 1976). Originally investigated as a potential source of alum chemicals for domestic water-supply treatment (Catalano and Fernandez Segura, 1953; Bertello, 1957), the Chubut deposits were considered during the 1970s as a possible nonbauxite ore of aluminum. However, the available reserve was determined to be insufficient for this purpose (Macchiaverna, 1977). Moreover, the *bochones*, although consisting mainly of alunite and microcrystalline quartz, also contain appreciable opaline cristobalite, an undesirable contaminant. Alunite is associated with dickite in a zoned deposit of hydrothermal origin at Cerro Bayo in the Argentine Andes (Angelelli, Schalamuk, and Arrospide, 1976), and as a hypogene alteration mineral in an altered rhyolite breccia pipe above the Mi Vida copper orebody in Catamarca Province, northwestern Argentina (Koukharsky and Mirré, 1976). Alunite is part of the alteration assemblage in porphyry wallrock of the silver-tin veins at the Potosí and Oruro mining districts, southwest Bolivia (Turneaure, 1960), and also at the sulfide veins of Cerro de Pasco, Peru (Graton and Bowditch, 1936). Alunite-bearing altered volcanics cap the El Salvador porphyry copper deposit in the Andes of northern Chile (Gustafson and Hunt, 1975). Hydrothermally altered terrains of Mexico and the Andean *cordillera* are good exploration targets for potentially exploitable deposits of alunite.

Australia: Alunite deposits in Australia were mined as a source of alum in the 1890s and the pre-World War II period, and during both world wars as a source of potassium sulfate fertilizer (Pittman, 1901; Anon., 1917; Fitzgerald, 1945). The deposit at Alum Mountain near Bullahdellah, New South Wales consists of hydrothermally altered alunitized rhyolite interlayered with sedimentary rocks of Carboniferous and Permian age. Deposits in South Australia consist of nodules of alunite in decomposed slate (Anon., 1917). Bedded lacustrine deposits of an unusual type, the origin of which is enigmatic, occur at Pidinga on the Nullarbor Plain, South Australia (Armstrong, 1950; King, 1953) and at Lake Campion near Chandler, Western Australia (Fitzgerald, 1945). Mining of alunite for potassium sulfate fertilizer was suspended at Lake Campion at the end of World War II; however, experiments were continued in an attempt to develop a feasible method of extracting alumina from the alunite-rich sediment (Bayliss, Ewers, and Miles, 1951). Interest in alunite as a nonbauxite ore of aluminum waned with development, in the 1950s, of Australia's enormous bauxite resources.

Other Countries: A project to develop a billion-ton deposit of alunite near Takestan, 150 km west of Tehran, Iran, was proposed (Anon., 1977) but seems unlikely to materialize in the foreseeable future. Several deposits in Turkey have been suggested as a possible source of potash fertilizer raw material (Tolun, 1974). Alunite is reported in northwest Pakistan (Ahmad, 1953; Schmidt, Clark, and Bernstein, 1975), but the potential for commercial exploitation is not known. Other occurrences, most of which apparently are of scientific but not commercial interest, are reported in Israel (Bentor, 1966; Goldbery, 1978, 1980), Egypt (Gad and Barrett, 1949), Morocco (Destombes and Lucas, 1956; Martin Vivaldi, 1963; Dietrich, 1965), Tanzania (Bassett, 1954), Niger (Faure, Greigert, and Martinet, 1959), New Zealand (Hutton, 1947; Steiner, 1953), Sumatra (Leinz, 1933), and the Philippines (Comsti, 1969).

Technology

The technique for making potash alum (*Roman alum*) and aluminum sulfate from alunite was developed at Tolfa, Italy, during the 15th century and was used with little change into the 20th century. Presently, the acid treatment of bauxite or clay is preferred for making alum, and alunite no longer is used for this purpose. Summary descriptions of the Tolfa method for making alum from alunite and of generally similar methods used in France during the 19th century and in England and Australia during the late 19th and early 20th centuries are given by Butler and Gale (1912).

The importance of alunite today lies not in its historical use as a raw material for making alum, but in its potential for becoming a nonbauxite ore of aluminum with fertilizer byproducts.

Alumina-From-Alunite Processes

Kalunite Process: During the Second World War a pilot plant for testing methods of extraction of alumina and byproduct potash fertilizer from Marysvale alunite was erected at Salt Lake City, a project sponsored by the Defense Plant Corp. (Fleischer, 1944). The Kalunite process involves reaction of dehydrated alunite (metalunite) with a solution of sulfuric acid and potassium sulfate to produce potassium alum ($K_2SO_4 \cdot Al_2(SO_4)_3 \cdot 24H_2O$). The potassium alum is heated under pressure to make water-insoluble basic alum ($K_2SO_4 \cdot 3Al_2O_3 \cdot 4SO_3 \cdot 9H_2O$), with liberation of K_2SO_4 and H_2SO_4 in solution. Calcination of the basic alum in a multiple muffled-hearth furnace produces a mixture of K_2SO_4 and Al_2O_3. The sulfate is precipitated from solution as potash fertilizer; the insoluble aluminous residue is calcined to Bayer-grade alumina. Details of the process and difficulties associated with it were described by Fleischer (1944). The Kalunite process represented a significant technological development in the utilization of alunite as an ore of aluminum, but was plagued with problems of materials handling, high costs, high energy use, and impurities in the alumina. It has not been used by alunite-processing experimenters since closure of the Salt Lake Lake City pilot plant at the end of the Second World War.

UG Processes: Three methods for extracting alumina from alunitic rock with simultaneous production of fertilizer byproducts have been developed on a laboratory and small pilot-plant scale at the University of Guanajuato, state of Guanajuato, Mexico (Lopez, 1977). The three methods differ only in detail. The basic UG process commences with dehydroxylation of crushed alunite ore at temperatures below 700°C, followed by reaction with ammonia gas in a boiling aqueous suspension. The filtrate from this treatment is a solution of potassium and ammonium sulfates; the filter cake is said to be boehmite, an aluminum monohydrate (Lopez, 1977). The synthetic boehmite ($Al_2O_3 \cdot H_2O$) is reacted with SO_2 in an aqueous suspension at temperatures below 75°C, forming soluble aluminum sulfite ($A1_2O_3(SO_2)_3(H_2SO_3)_n$, $n \geqq 1$. Treatment of unreacted sludge with added H_2SO_4 makes a solution of aluminum sulfate ($Al_2(SO_4)_3$), which is reacted with the previously obtained aluminum sulfite solution, precipitating a mixture of nearly pure aluminum hydrate and aluminum sulfate; SO_2 offgas is recycled in the process. The mixed aluminum sulfate-hydrate precipitate then is calcined, driving off SO_3, SO_2, and O_2, leaving a Bayer-grade alumina residue. The offgases are recycled to make the H_2SO_4 used in the previous steps (Lopez, 1977). Valuable byproducts are potassium and ammonium sulfate fertilizers. The second UG process is reported to be a simplification of the first (Lopez, 1977) intended for use on higher-grade ores, where several reaction and filtration steps can be bypassed. The third UG process entails treatment of the impure synthetic boehmite ($Al_2O_3 \cdot H_2O$), produced as filter cake in the first or basic UG process, by a Bayer-type alkaline (NaOH) leach instead of with sulfur dioxide gas. The resulting sodium aluminate solution is seeded to precipitate synthetic gibbsite ($Al_2O_3 \cdot 3H_2O$) as in the Bayer process. The alumina trihydrate is washed and calcined to produce Bayer-grade alumina sand.

Alumet Process: The process developed by the Alumet partnership (which comprises Earth Sciences Inc. of Golden, CO, National Steel Corp. of Pittsburgh, PA, and the Southwire Co. of Carrollton, GA) was tested in a pilot plant at Golden, Colorado, from 1973 until the end of 1976 (Thompson, 1976) and is generally similar to the process employed at the Soviet alumina-from-alunite plant in Azerbaijan. Principal steps in the process are: crush and grind quartz-alunite ore to minus 20 mesh (0.841 mm); dehydrate in a fluidized-bed reactor at 500 to 600°C; roast dehydrated ore in reducing atmosphere in fluidized bed reactor, driving off SO_2, which is passed to a sulfuric acid plant for conversion to H_2SO_4; leach roasted product in an aqueous solution to remove K_2SO_4; filter, pass K_2SO_4 liquor to crystallizer-compactor-dryer unit for production of fertilizer-grade potassium sulfate; pass filter cake, which consists mainly of fine quartz sand and amorphous alumina to Bayer-type caustic (NaOH) leach (alumina goes into solution as sodium aluminate); filter and pass silica sand residue to tailing; desilicate and filter sodium aluminate solution; seed and precipitate aluminum trihydrate; and calcine precipitate to Bayer-grade alumina.

Fig. 4 is a generalized flowsheet of the Alumet process. The Alumet project plan would react the sulfuric acid, produced by roasting the dehydrated ore, with calcined phosphate rock hauled by rail from southeast Idaho to the proposed Utah plantsite, to produce phosphoric acid for phosphate fertilizer. Credits for phosphoric acid and potassium sulfate fertilizer by-products would offset in great part the disadvantage of low-alumina alunitic rock. The modified Bayer process to treat the alumina-silica residue following removal of K_2SO_4 has two significant advantages over conventional Bayer processing: alumina is dissolved in a caustic leach at about 95°C at atmospheric pressure; and alunitic rock contains much less iron oxide than most bauxites and so produces very few "red mud" tailings. In contrast, most conventional Bayer processes require temperatures of 130° to 200°C and pressures on the order of 1.4 MPa (200 psi); disposal of caustic red-mud tailings constitutes a significant cost at conventional Bayer plants.

Possible disadvantages of the Alumet process include: fluidized bed roast must be carried out under narrow and complex restraints on both temperature and velocity of feed-flow (Thompson, 1976); 30% alunite ore contains only 11% alumina in the alunite as contrasted with most bauxites containing 45 to 50% Al_2O_3; fully two-thirds of the raw feed is useless gangue, adding greatly to materials handling costs; and ore quality must be carefully monitored. If the principal gangue is microcrystalline quartz, there is little difficulty because quartz resists attack under the relatively mild conditions of the Alumet modified-Bayer caustic leach. However, clays, micas, opaline cristobalite, and other silica-bearing minerals present in more than very small amounts can be deleterious because they are attacked by caustic; an inordinately high amount of silica may be dissolved in the sodium aluminate solution. Loss of alumina and soda during the desilication step can make such material uneconomical to use, even if the tenor of alunite is 30% or more in the original feed. Clays, micas, and cristobalite

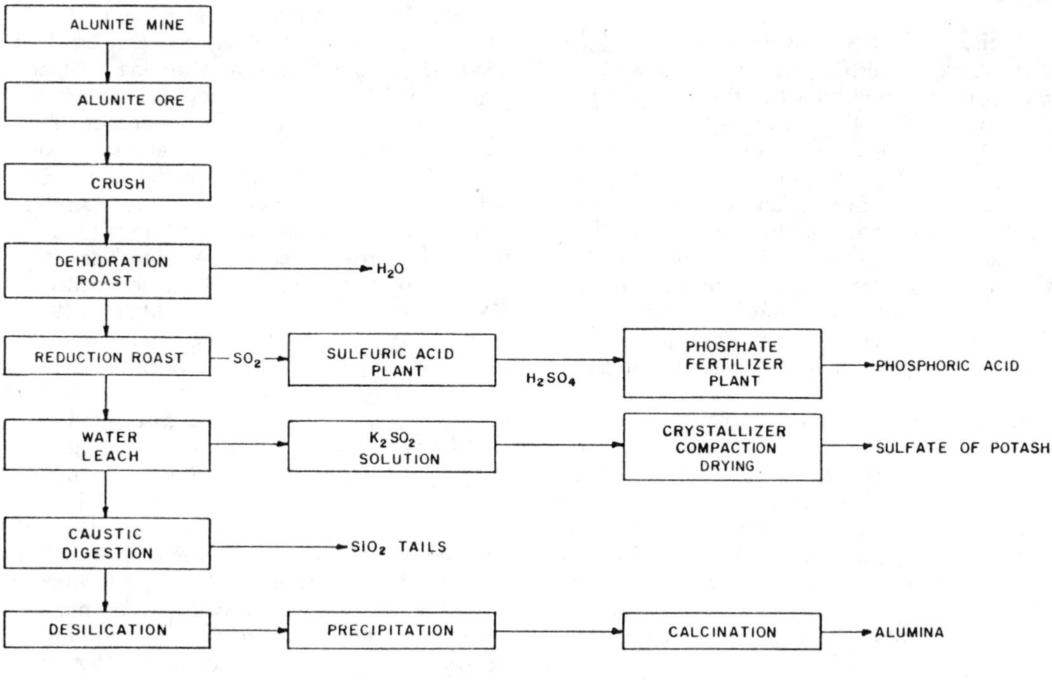

FIG. 4—*Alumet process.*

are common in hydrothermally and solfatarically altered terrains where alunite is formed. This limitation must be evaluated carefully when applying the Alumet process (or any other caustic-leach process) to alunite ore (Hall, 1978, 1980).

Other Processes: Details are lacking, but apparently the process used at the Soviet plant—the only operating alumina-from-alunite plant in the world—is similar to the Alumet process.

A number of processes besides those already mentioned have been proposed and dozens of patent applications have been submitted over the years. Most of these involve a variation of either an acid or an alkali leach to extract alumina from the residue, after K_2SO_4 has been removed from the alunite molecule. Various techniques of heat treatment and materials handling also have been proposed. Several processes designed to produce only potassium sulfate fertilizer, without regard to recovery of alumina or acid, were employed at two small commercial plants at Marysvale, UT during the First World War. However, no process for extraction of alumina, other than that presently used in the Soviet Union, has been taken beyond laboratory or pilot-plant scale.

Mining

Alunite veins at Marysvale, UT were mined by underground drifts and adits and also by small open pits and trenches during the First World War, when the high-grade vein matter was used to provide potassium sulfate fertilizer (Callaghan, 1973). Alunite at Tolfa, Italy, was mined for nearly five centuries by large open trenches and quarries (Butler and Gale, 1912; Lombardi and Mattias, 1979). During World War II, several disseminated replacement deposits at Marysvale, including the White Horse (Willard and Proctor, 1946), were quarried to obtain ore for the Kalunite plant at Salt Lake City.

Although alunite is not presently mined in the United States, open pits on a scale comparable to the porphyry copper mines in America's Southwest would appear to be the only economically feasible mining method at the +100 Mt low-grade alunite deposits. There is little barren overburden at these deposits, but savings here could be offset by the need to selectively remove subore-grade material to waste.

Relation to Alunite to Hypogene Ore Deposits

Apart from consideration of alunitic rock as a possible ore of aluminum is the association of alunite with hypogene deposits of metals other than aluminum. Following F. L. Ransome's observation (1907, 1909) of alunite associated with gold at Goldfield, NV, many other examples have been reported of alunite associated with metalliferous deposits. We shall not discuss theoretical concepts or offer models to explain these associations, but merely list some examples to emphasize the point.

Alunite occurs with auriferous veins at Rodalquilar in southeastern Spain (Lodder, 1966) and at the old Rattlesnake Jack prospect near Lead, SD (Grout and Schwartz, 1927). The huge Red Mountain alunite deposit near Lake City in Hinsdale County, Colorado, is flanked on the north and east by numerous base- and precious-metal deposits, especially along Henson Creek (Slack and Lipman, 1979). Extensive alunitization in the volcanic complex of the Marysvale district, Piute and Sevier counties, Utah, appears to be related genetically to base- and precious-metal deposits (Steven et al., 1978). Also in Utah, alunite is prominent in the alteration assemblages associated with polymetallic deposits of the East Tintic district in Juab County (Lovering et al., 1949), and at the old Horn Silver mine in the Frisco mining district, Beaver County (Stringham, 1967). Alunite occurs with lead-silver ore at the Flathead Mine, northwestern Montana (Shenon, 1935), with opalite mercury (cinnabar) deposits in Nevada (Bailey and Phoenix, 1944; Knopf, 1916), and also with mercury deposits in the Cordero-McDermitt district at the Oregon-Nevada state line (Rytuba and Glanzman, 1979). Pervasive alunitization in the upper parts of porphyry copper systems has been reported by Corn (1975), Gustafson and Hunt (1975), and Koukharsky and Mirré (1976). Sillitoe (1973) mentions the common occurrence of alunite in altered volcanic caprock above porphyry copper systems in the Andean Cordillera, and also alunite associated with lead-silver deposits in northwest Argentina (Sillitoe, 1975). Kashkai (1970) cites numerous examples of the association of alunite with metal sulfides. Many additional examples could be cited. The presence of alunite, especially in a large pervasively altered area of volcanic rock, may be interpreted as a *red flag* marking a favorable target for further exploration for possible metalliferous deposits, nearby or at depth. Wallace (1979) has cited such altered areas in western Nevada as possible *signatures* of hidden porphyry copper deposits.

Prospecting for Alunite Deposits

Any pervasively altered and bleached iron-stained area in a volcanic setting may contain alunite. Whitish bleached zones normally grade outward into characteristically dull greenish-gray propylitized rock. It is difficult to identify alunite-bearing rock by visual examination alone, but slightly greater heaviness or *heft* compared to similar appearing rock can be a clue. A simple test using pH paper to measure acidity of condensed water inside a test-tube in which a powdered sample has been strongly heated detects alunitic rock, but should be used with caution (Cunningham and Hall, 1976). Powder X-ray diffractometry is the most reliable identification method.

Geologists of Earth Sciences, Inc. have been very successful in identifying large potentially alunitized bodies from aircraft flying over rugged volcanic terrains. An entire region can be reconnoitered in a few hours compared to days or weeks needed for ground reconnaissance. Favorable targets can be quickly identified for follow-up ground examination and sampling. Bleached and iron-stained zones (color anomalies) are recognized more easily from the air than from ground level. However, it should be emphasized that areas showing high albedo (high reflectivity) do not always contain alunite. Recent research intended to improve and refine remote-sensing techniques using the visible and near-infrared part of the electromagnetic spectrum offers hope that alunite-bearing ground eventually can be detected directly by satellites or high-flying aircraft equipped with infrared remote-sensing instruments (Hunt and Ashley, 1979; Hunt and Hall, 1981).

Acknowledgments

The authors acknowledge with thanks the helpful suggestions of John G. Wells and Charles G. Cunningham of the US Geological Survey and Douglas N. Stevens of Earth Sciences, Inc.

Bibliography and References

Ahmad, M.I., 1953, "Volcanoes and Sulphur of Western Baluchistan," *Records*, Geological Survey of Pakistan, Vol. 4, Part 3, pp. 1–16.

Allen, G.L., 1947, "Aluminum Metallurgy in the Japanese Empire," US Bureau of Mines, Mineral Trade Notes, Special Supplement No. 19 to Vol. 25, No. 4, Oct. 1947, 76 pp.

Angelelli, V., Schalamuk, I.B., and Arrospide, A., 1976, "Los Yacimientos No Metalíferos y Rocas de Aplicación de la Region Patagonia-Comahue," Ministerio de Economía, Secretaria de Estado de Minería, Republica Argentina, pp. 17–19.

Anon., 1917, "The Alunite Deposits of Australia and Their Utilization," Bulletin 3, Australia Advisory Council of Science and Industry, 38 pp.

Anon., 1970, *Processes for Extracting Alumina from Nonbauxite Ores,* National Research Council, National Materials Advisory Board NMAB–278, 88 pp.

Anon., 1976, "Annual Report for 1975," Earth Sciences, Inc., Golden, CO, 24 pp.

Anon., 1977, "Soviet Union and Iran Agree to Build Large Alumina-from-Alunite Plant," *World Mining*, Vol. 30, No. 1, Jan., p. 57.

Anon., 1980, "Principles of a Resource/Reserve Classification for Minerals," Circular 831, US Bureau of Mines and US Geological Survey, 5 pp.

Armstrong, A.T., 1950, "Pidinga Alunite Deposit," South Australia Department of Mines Mining Review No. 89 (1948), pp. 126–136.

Bailey, E.H., and Phoenix, D.A., 1944, "Quicksilver Deposits in Nevada," *Nevada University Bulletin*, Vol. 38, No. 5, 206 pp.

Bassett, H., 1954, "New and Old Minerals in Tanganyika," Tanganyika Geological Survey Records, Vol. 1 (1951), pp. 60–67.

Bauer, C.W., 1980, "Alumina from Alunite—Domestic Resources," Preprint No. 80–119, SME-AIME Annual Meeting, Las Vegas, Feb. 24–28, 3 pp.

Bayliss, N.S., Ewers, W.E., and Miles, G.L., 1951, "The Production of Alumina from Alunite," *Australian Journal of Applied Science*, Vol. 2, No. 2, pp. 267–275.

Bentor, Y.K., 1966, "The Clays of Israel," *Guidebook to the Excursions*, International Clay Conference, Jerusalem, Israel Program of Scientific Translations, 121 pp.

Bertello, L.F., 1957, "Utilización de Alunita Nacional Para la Fabricación de Sulfato de Aluminio," *Industria y Quimica*, Vol. 18, pp. 25–30, 35–36.

Brock, T.D., and Mosser, J.L., 1975, "Rate of Sulfuric-Acid Production in Yellowstone National Park," *Geological Society of America Bulletin*, Vol. 86, pp. 194–198.

Butler, B.S., and Gale, H.S., 1912, "Alunite, a Newly Discovered Deposit near Marysvale, Utah," Bulletin 511, US Geological Survey, 64 pp.

Caballero, M.A., Menendez del Valle, F., and Martin Vivaldi, J.L., 1974, "Yacimientos Españoles de Bauxitas y Alunitas," *Boletín de Geología y Mineralogía*, Vol. 85, pp. 32–42.

Callaghan, E., 1973, "Mineral Resource Potential of Piute County, Utah, and Adjoining Area," Bulletin 102, Utah Geological and Mineralogical Survey, 135 pp.

Catalano, E., and Fernandez Segura, A., 1953, "Alunita Silicificada de Camarones," *Revista de Administración General de Obras Sanitarias*, Buenos Aires, Vol. 17, No. 152, pp. 137–151.

Charrin, V., 1940, "L'Alunite du Puy-de-Dôme," *L'Argile*, Vol. 198, pp. 1–5.

Charrin, V., 1948, "Le Gisement d'Alunite du Mont Dore," *L'Industrie Chimique*, Vol. 35, p. 26.

Cho, H.I., and Moon, H.S., 1978, "Alunite Deposits in Korea," special publication of Korea Research Institute of Geoscience and Mineral Resources, 101 pp. [In Korean, English abstract.]

Clark, I.C., 1918, "Recently Recognized Alunite Deposits at Sulphur, Humboldt County, Nevada," *Engineering and Mining Journal*, Vol. 106, No. 4, pp. 159–162.

Comsti, F.A., 1969, "Kaolin Deposits in the Philippines," *Proceedings*, 23rd International Geological Congress, Symposium 1, Kaolin Deposits of the World. B-Overseas Countries, Prague, 1968, pp. 39–42.

Corn, R.M., 1975, "Alteration-Mineralization Zoning, Red Mountain, Arizona," *Economic Geology*, Vol. 70, pp. 1437–1447.

Cross, W., 1891, "On Alunite and Diaspore from the Rosita Hills, Colorado," *American Journal of Science*, Series 3, Vol. 41, No. 246, pp. 466–475.

Cross, W., 1896, "Geology of Silver Cliff and the Rosita Hills, Colorado," 17th Annual Report, Part 2, US Geological Survey, pp. 263–403.

Cross, W., and Spencer, A.C., 1900, "Geology of the Rico Mountains, Colorado," 21st Annual Report, Part 2, US Geological Survey, pp. 7–165.

Cunningham, C.G., and Hall, R.B., 1976, "Field and Laboratory Tests for the Detection of Alunite and Determination of Atomic Percent Potassium," *Economic Geology*, Vol. 71, No. 8, pp. 1596–1598.

DeLaunay, L., 1907, "La Metallogénie de l'Italie," 10th International Geological Congress, Comptes Rendus, Part 1, pp. 679–686. [English translation of the part discussing alunite is given in Butler and Gale, 1912, pp. 51–55.]

Destombes, J., and Lucas, J., 1956, "Une Occurrence de Natroalunite dans le Gothlandien de Tazzarine (Jebel Sahro-Sud)," Morocco Service Géologique Notes et Mémoirs No. 133, pp. 117–123.

Dietrich, J.E., 1965, "Zunyite d'Azrou-Melloul," Morocco Service Géologique Notes et Mémoirs No. 183, pp. 105–107.

Faure, H., Greigert, J., and Martinet, B., 1959, "Occurrences d'Alunite au Soudan et au Niger," Societé Géologique France, Comptes Rendus, 1959, Ser. 7, pp. 175–176.

Fitzgerald, M. L., 1945, "Production of Potash from Alunite, Lake Campion, Western Australia," *Chemical Engineering and Mining Review*, Vol. 37, No. 440, pp. 241–248.

Fleischer, A., 1944, "The Kalunite Process," *Trans.*, AIME, Vol. 159, pp. 267–279.

Gad, G.M., and Barrett, L.R., 1949, "The Constitution of Some Egyptian Clays," *Mineralogical Magazine*, Vol. 28, pp. 587–597.

Galan, E., and Lopez-Aguayo, F., 1977, "Occurrences of Alunite in Spain," *Proceedings*, 8th International Kaolin Symposium and Meeting on Alunite International Geol. Correlation Program, Working Group, Genesis of Kaolins Madrid-Rome, Sept. 7–16, Paper No. A–1, 4 pp.

Gardner, L.S., 1943, unpublished data, US Geological Survey.

Gautier, P., 1940, "Note sur le Gisement d'Alunite de Madriat (Puy-de-Dôme)," *Revue des Sciences Naturelles d'Auvergne*, Vol. 7, pp. 96–98.

Goldbery, R., 1978, "Early Diagenetic Nonhydrothermal Na-Alunite in Jurassic Flint Clays, Makhtesh Ramon, Israel," *Geological Society of America Bulletin*, Vol. 89, pp. 687–698.

Goldbery, R., 1980, "Early Diagenetic Na-Alunite in Miocene Algal Mat Intertidal Facies, Ras Sudar, Sinai," *Sedimentology*, Vol. 27, No. 2, pp. 189–198.

Graton, L.C., and Bowditch, S.I., 1936, "Alkaline and Acid Solutions in Hypogene Zoning at Cerro de Pasco," *Economic Geology*, Vol. 31, No. 7, pp. 651–698.

Grout, F.F., and Schwartz, G.M., 1927, "An Alunite Gold Ore in the Black Hills," *Economic Geology*, Vol. 22, No. 4, pp. 369–373.

Gustafson, L.B., and Hunt, J.P., 1975, "The Porphyry Copper Deposit at El Salvador, Chile," *Economic Geology*, Vol. 70, No. 5, pp. 857–912.

Hall, R.B., 1978, "World Nonbauxite Aluminum Resources—Alunite," Professional Paper 1076–A, US Geological Survey, 35 pp.

Hall, R.B., 1980, "Alunite—A Potentially Economic Source of Aluminum and Potash?" *Proceedings*, 15th Forum on Geology of Industrial Minerals, Colorado Geological Survey Resource Series 8, pp. 77–88.

Hanson, R.F., 1975, "The Geology of Some Clay Deposits in the Trans-Mexico Volcanic Belt and on the Mexican Central Plateau," Guidebook to Field Trip FT–1, Hydrothermal Kaolins International Clay Conference, Mexico City, July 24–30, 121 pp.

Hayes, C.W., 1907, "The Gila River Alum Deposits," Bulletin 315, US Geological Survey, pp. 215–224.

Heineman, R.E., 1935, "Sugarloaf Butte Alunite," *Engineering and Mining Journal*, Vol. 136, No. 3, pp. 138–139.

Hemley, J.J., et al., 1969, "Some Stability Relations of Alunite," *Economic Geology*, Vol. 64, No. 6, pp. 599–610.

Hewett, D.F., et al., 1936, "Mineral Resources of the Region Around Boulder Dam," Bulletin 871, US Geological Survey, 197 pp.

Hild, J.H., 1946, "Exploration of Alunite Deposits, Marysvale, Piute County, Utah," Report of Investigations 3972, US Bureau of Mines, 74 pp.

Hunt, G.R., and Ashley, R.P., 1979, "Spectra of Altered Rocks in the Visible and Near Infrared," *Economic Geology*, Vol. 74, No. 7, pp. 1613–1629.

Hunt, G.R., and Hall, R.B., 1981, "Identification of Kaolins and Associated Minerals in Altered Volcanic Rocks by Infrared Spectroscopy," *Clays and Clay Minerals*, Vol. 29, No. 1, pp. 76–78.

Hutton, C.O., 1947, "Contributions to the Mineralogy of New Zealand, Part 2," *Transactions and Proceedings*, Royal Society of New Zealand, Vol. 76, Part 3, pp. 358–361.

Ikonnikov, A.B., 1975, "Mineral Resources of China," Microform Publication 2 [6 microfiche sheets], Geological Society of America, 552 pp.

Ivanov, M.V., Grinenko, V.A., and Lein, A.Y., 1968, "Exogenetic Formation of Sulfuric Acid in Fumarole Fields of Volcanoes of the Kuril Islands," *Geochemistry International*, Vol. 5, pp. 1133–1140.

Iwao, S., 1949, "The Alunite Deposits of Japan," Report No. 130, Japan Geological Survey, 100 pp. [In Japanese with English Resumé.]

Iwao, S., 1953, "Alunite Deposits in Japan," *Proceedings*, 7th Pacific Science Congress, Pacific Science Assoc., Vol. 2, pp. 229–234.

Kashkai, M.A., 1961, "Alunite Deposits, Their Classification and Associated Processes," *Izvestiya, Academy of Sciences of the USSR*, Geologic Series, No. 7, July 1961. [English translation published by American Geological Institute, July 1962], pp. 57–63.

Kashkai, M.A., 1970, *Alunites, Their Genesis and Utilization*, Moscow, Nedra Publishing House, Vol. 1, 400 pp., Vol. 2, 317 pp. [In Russian, with English summary in Vol. 2, pp. 231–239.]

Keller, W.D., and Hanson, R.F., 1968, "Hydrothermal Alteration of a Rhyolitic Flow Breccia Near San Luis Potosí, Mexico, to Refractory Kaolin," *Clays and Clay Minerals*, Vol. 16, No. 3, pp. 223–229.

King, D., 1953, "Origin of Alunite Deposits at Pidinga, South Australia," *Economic Geology*, Vol. 48, No. 8, pp. 689–703.

Kirova, T.V., 1959, "Some Secondary Quartzites of the Ore-Bearing Part of the Altai and the Chemical Nature of Their Formation," Leningrad Gornogo Institute Zapiski, Vol. 35, No. 2, pp. 214–226. [In Russian; English abstract in *Chemical Abstracts*, 1960, Vol. 54, Column 8499, item h.]

Knížek, J.O., and Fetter, H., 1950, "The Refractory Properties of Alunite," *Transactions*, British Ceramic Society, Vol. 49, pp. 202–285.

Knopf, A., 1916, "Some Cinnabar Deposits in Western Nevada," Bulletin 620, US Geological Survey, pp. 59–68.

Koukharsky, M., and Mirré, J.C., 1976, "Mi Vida Prospect: A Porphyry Copper-Type Deposit in Northwestern Argentina," *Economic Geology* Vol. 71, No. 5, pp. 849–863.

Kühnel, R.A., Koster, F., and Roorda, H.J., 1975, "Alunite and Associated Clay Minerals from Riazaa, Province Segovia, Central Spain," *Abstracts*, International Clay Conference, Mexico City, July 24–30, pp. 158–159.

Lacroix, A., 1962, *Mineralogie de la France et de ses Anciens Territoires d'Outre Mer*, Vol. 4, Paris, A. Blanchard, 923 pp. ["Groupe de l'Alunite," pp. 138–143].

Landi, M.F., 1978, "Leucite-Bearing Rocks and Alunites and Their Utilization as Aluminium Raw Materials," *Raw Materials, Research and Development, III. Aluminium*, Dossier published by Commission of the European Communities, Brussels. [Specialized Report II on Alternative Aluminium Ores], 73 pp.

Larsen, E.S., 1913, "Alunite in the San Cristobal Quadrangle," Bulletin 530, US Geological Survey, pp. 179–183.

Leinz, V., 1933, "Alunite-Diaspor Neubildungen in Liparitischen Tuffen von Sumatra," *Geologische Rundschau*, Vol. 23a, pp. 18–20.

Lodder, W., 1966, "Gold-alunite Deposits and Zonal Wall-rock Alteration near Rodalquilar, SE Spain," *Memoirs*, Amsterdam University Geological Institute, 93 pp.

Lombardi, G., 1977, "Alunite in Italy, Occurrences and Genesis," *Proceedings*, 8th International Kaolin Symposium and Meeting on Alunite [International Geological Correlation Program, Working Group, Genesis of Kaolins] Madrid-Rome, Sept. 7–16, Paper No. A–3, 19 pp.

Lombardi, G., and Mattias, P., 1979, "Petrology and Mineralogy of the Kaolin and Alunite Mineralizations of Latium (Italy)," *Geologica Romana*, Vol. 18, pp. 157–214.

Lombardi, G., Mattias, P., and Uras, I., 1977, *Guidebook for the Excursions in Italy*, 8th International Kaolin Symposium and Meeting on Alunite [International Geological Correlation Program, Working Group, Genesis of Kaolins], Madrid-Rome, Sept. 7–16, 45 pp.

Lopez, W.X., 1977, "Three Methods to Produce Alumina from Alunite," *Light Metals 1977*, Vol. 2, TMS-AIME Annual Meeting, Atlanta, pp. 49–58.

Lovering, T.S. (in collaboration with 11 other authors), 1949, "Rock Alteration as a Guide to Ore—East Tintic District, Utah," *Monograph 1, Economic Geology*, 64 pp.

Macchiaverna, E.G., 1977, Aluminio Argentino, personal communication.

Martin Vivaldi, J.L., 1963, "The Bentonites of Cabo de Gata (Southeast Spain) and of Guelaya Volcanic Province (North Morocco)," *Clay and Clay Minerals*, Proceedings, 11th Conference on Clays and Clay Minerals, Ottawa, Aug. 1962, Macmillan, pp. 327–357.

Naboko, S.I., 1963, "Conditions of the Present-Day Hydrothermal Metamorphism of Volcanic Rocks," *International Geology Review*, Vol. 5, No. 7, pp. 850–858.

Parkinson, G., 1971, "Low-grade Alunite Yields Alumina and Fertilizers Too," *Chemical Engineering*, Vol. 78, pp. 83–85.

Parkinson, G., 1974, "Golden Pilot Plant Points the Way to 500,000-tpy Alumina-from-Alunite Mine and Plant in Utah," *Engineering and Mining Journal*, Vol. 175, No. 8, pp. 75–78.

Pittman, E.P., 1901, "Alunite or Alumstone in New South Wales," *The Mineral Resources of New South Wales*, New South Wales Geological Survey Report, Sydney, pp. 415–420.

Ransome, F.L., 1907, "The Association of Alunite with Gold in the Goldfield District, Nevada," *Economic Geology*, Vol. 2, No. 7, pp. 667–692.

Ransome, F.L., 1909, "The Geology and Ore Deposits of Goldfield, Nevada," Professional Paper 66, US Geological Survey, 258 pp.

Rytuba, J., and Glanzman, R.K., 1979, "Relation of Mercury, Uranium, and Lithium Deposits to the McDermitt Caldera Complex, Nevada-Oregon," in *Papers on Mineral Deposits of Western North America*, Report 33, Nevada Bureau of Mines and Geology, pp. 109–117.

Schmidt, R.G., Clark, B.B., and Bernstein, R., 1975, "A Search for Sulfide-Bearing Areas Using Landsat-1 Data and Digital Image-Processing Techniques," *Proceedings*, NASA Earth Resources Survey Symposium, Houston, TX, June, US National Aeronautics and Space Administration Technical Memorandum X–58168, Vol. 1–B, pp. 1013–1027.

Schoen, R., White, D.E., and Hemley, J.J., 1974, "Argillization by Descending Acid at Steamboat Springs, Nevada," *Clays and Clay Minerals*, Vol. 22, No. 1, pp. 1–22.

Schrader, F.C., 1913, "Alunite in Patagonia, Arizona, and Bovard, Nevada," *Economic Geology*, Vol. 8, pp. 752–767.

Schrader, F.C., 1915, "Mineral Deposits of the Santa Rita and Patagonia Mountains, Arizona, with Contributions by James H. Hill," Bulletin

582, US Geological Survey, 373 pp. [Section on alunite, pp. 361–364.]

Shabad, T., 1976, "Soviet Experiment in Aluminum Fails," *New York Times*, May 8, Section 3, p. 5.

Shenon, P.J., 1935, "Genesis of the Ore at the Flathead Mine, Northwestern Montana," *Economic Geology*, Vol. 30, No. 6, pp. 585–603.

Sillitoe, R.H., 1973, "The Tops and Bottoms of Porphyry Copper Deposits," *Economic Geology*, Vol. 68, No. 6, pp. 799–815.

Sillitoe, R.H., 1975, "Lead-Silver, Manganese, and Native Sulfur Mineralization Within a Stratovolcano, El Queva, Northwest Argentina," *Economic Geology*, Vol. 70, No. 7, pp. 1190–1201.

Slack, J.F., and Lipman, P.W., 1979, "Chronology of Alteration, Mineralization, and Caldera Evolution in the Lake City Area, Western San Juan Mountains, Colorado," in *Papers on Mineral Deposits of Western North America*, Report 33, Nevada Bureau of Mines and Geology, pp. 151–157.

Steiner, A., 1953, "Hydrothermal Rock Alteration at Wairakei, New Zealand," *Economic Geology*, Vol. 48, pp. 1–13.

Steven, T.A., Cunningham, C.G., and Rowley, P.D., 1978, "Geology and Mineralization in the Marysvale Mining Area, West-Central Utah," *Guidebook to Mineral Deposits of Southwestern Utah* (International Association on the Genesis of Ore Deposits, Aug., Excursion C–2), Publication 7, Utah Geological Association, pp. 67–70.

Steven, T.A., and Ratté, J.C., 1960, "Geology and Ore Deposits of the Summitville District, San Juan Mountains, Colorado," Professional Paper 343, US Geological Survey, 70 pp.

Stringham, B., 1967, "Hydrothermal Alteration Near the Horn Silver Mine, Beaver County, Utah," Special Studies 16, Utah Geological and Mineralogical Survey, pp. 35.

Taylor, H.P., Jr., 1974, "The Application of Oxygen and Hydrogen Isotope Studies to Problems of Hydrothermal Alteration and Ore Deposition," *Economic Geology*, Vol. 69, No. 6, pp. 843–883.

Thoenen, J.R., 1941, "Alunite Resources of the United States," Report of Investigations 3561, US Bureau of Mines, 48 pp.

Thompson, D.L., 1976, "Alumina and Byproducts from Utah Alunite—The Alumet Project," *World Mining and Metals Technology*, Proceedings of the joint Mining and Metallurgical Institute of Japan (MMIJ)-AIME meeting, Denver, CO, Sept. 1–3, AIME, Vol. 2, pp. 681–684.

Tolun, R., 1974, "Possibilities for the Evaluation of Turkish Alunites," *CENTO Symposium on the Mining and Beneficiation of Fertilizer Minerals*, Symposium held in Istanbul, Turkey, Nov. 19–24, 1973. Central Treaty Organization, Ankara, Turkey, pp. 286–291.

Turneaure, F.S., 1960, "A Comparative Study of Major Ore Deposits of Central Bolivia," *Economic Geology*, Vol. 55, No. 2, pp. 574–606.

Vanderburg, W.O., 1938, "Reconnaissance of Mining Districts in Humboldt County, Nevada," Information Circular 6995, US Bureau of Mines, 54 pp.

Velinov, I.A., 1973, "Mineralogical-Petrological Analysis of the Hydrothermally Altered Upper Cretaceous Volcanics from the Western Part of the Srednogorian Zone," [Abstract of thesis to be presented for the degree of Doctor of the Geological and Mineralogical Sciences], Bulgarian Academy of Sciences, Ministry of Chemical Industries and Power Production, 26 pp. [English version.]

Walker, W.W., and Stevens, D.N., 1974, "The Earth-Sciences-National-Southwire Alunite-to-Alumina Project," *Light Metals 1974*, Vol. 3, TMS-AIME Proceedings of the AIME annual meeting, Houston, TX, pp. 683–688.

Wallace, A.B., 1979, "Possible Signatures of Buried Porphyry-Copper Deposits in Middle to Late Tertiary Volcanic Rocks of Western Nevada," *Papers on Mineral Deposits of Western North America*, Report 33, Nevada Bureau of Mines and Geology, pp. 69–76.

White, D.E., Muffler, L.J.P., and Truesdell, A.H., 1971, "Vapor-Dominated Hydrothermal Systems Compared with Hot-Water Systems," *Economic Geology*, Vol. 66, No. 1, pp. 75–97.

Willard, M.E., and Proctor, P.D., 1946, "White Horse Alunite Deposit, Marysvale, Utah," *Economic Geology*, Vol. 41, No. 6, pp. 619–643.

Yih, L.F., 1931, "The Alunitization and Pyrophyllitization of the Rhyolite and Tuff in Some Maritime Districts of Southeastern China," *Academica Sinica*, Geological Memoir 11, National Research Institute of Geology, 58 pp.

Yris Rovirosa, N., 1964, *La Alunita*, Mexico City, Nacional Financiera, S.A., Fideicomiso de Minerales No Metalicos, 190 pp.

Zinder, S., and Brock, T.D., 1977, "Sulfur Dioxide in Geothermal Waters and Gases," *Geochimica et Cosmochimica Acta*, Vol. 41, pp. 78–79.

Asbestos *†

E. L. MANN ‡

Asbestos is the generic name given to a group of fibrous mineral silicates found in nature. They are all incombustible and can be separated by mechanical means into fibers of various lengths and cross sections, but each differs in chemical composition from the others. The commercial grades of asbestos consist of a spectrum of lengths and sizes and are not definitive as are most organic fibers.

It is generally recognized that there are two main groups of asbestos. The first only contains the fibrous serpentine called chrysotile and comprises about 94% of the world production of asbestos. The second group contains five minerals in the amphibole series—crocidolite, amosite, anthophyllite, tremolite, and actinolite. The latter two have no commercial importance.

* Previous editions—1937, 1949, 1960, and 1975—have developed a format describing the general physical properties of asbestos; the distribution of deposits throughout the world; description of mining and milling methods; tests and specifications; grades produced; prices, uses, and markets.

In the 4th and 5th editions certain phases are discussed in more detail. The sections relating to world occurrences, geology of deposits, and methods of evaluation are covered more fully. Statistics have been augmented and updated whenever possible and the bibliography enlarged.

† The 4th edition was coordinated by R. W. Winson formerly Mine Project Manager, Canadian Johns-Manville Co., Ltd., Asbestos, Quebec, Canada (now deceased) with assistance from J. W. Axelson (Properties), E. L. Mann (Geology; World Deposits; Production; Exploration and Evaluation), J. N. Gauthier (Mining), E. B. McKenna (Milling), D. P. R. Smyth (Tests and Specifications; Uses) and R. T. Hughes (Markets).

Grateful acknowledgement for assistance in reviewing the latest edition is given to the following: W. C. Streib, Research and Development Center, Manville Products Corp., Denver (Properties), F. J. Evelegh (Deposits and Production) and E. J. Day (Exploitation), both of Johns-Manville Canada Inc., and Margaret Cole (Production and Uses) and D. M. Kelleher (Markets), both of Manville Products Corp., Denver, CO.

‡ Chief Geologist and Exploration Manager, Exploration Department, Manville Products Corp., Denver, CO.

Properties

Physical

The structures of asbestos fibers have been studied intensively over a number of years. X-ray diffraction patterns have been used as a means of identification and classification. Low angle X-ray scattering techniques have shown that chrysotile fibers are "hexagonally close packed" and parallel to each other, having cross-sectional diameters varying from 180Å to 300Å units, while the amphibole fibers are many times larger in cross section, have considerably smaller surface areas, and the fibers are less symmetrical than those of chrysotile. Electron micrographs of the more common types of asbestos are shown in Figs. 1–4. Fig. 1 illustrates the softness and flexibility of chrysotile which make it suitable for textile-type products.

For a number of years there was considerable controversy over the exact nature of chrysotile fiber, but the photographic evidence in Figs. 5–7 clearly show that it is tubular in structure. The curvature is the result of the mismatch in the dimensions of the silica [Si_2O_5] and brucite [$Mg(OH)_2$] layers which are on the inside and outside, respectively, of a double sheet. This curvature had been predicted by Pauling (1930) years ago based on the molecular dimensions of the two component structures.

In contrast to this, the structure of the amphiboles consists of two ribbon chains based on Si_4O_{11} units separated by a band of seven cations with two hydroxyl groups attached to

FIG. 1—*Electron micrograph of Jeffrey chrysotile. X8000. Reduced approximately 33% in reproduction.*

the central cation of each unit cell. These stack together to form a sandwich of ribbons in the final structure with a marked cleavage developed between the ribbons.

Asbestos fibers are extensively used to provide reinforcement in composite materials. Consequently, the tensile strength of the fiber is an highly desirable physical property. However, the extremely fine diameters and variations in fiber lengths make this very difficult to measure. Different varieties of asbestos exhibit a wide range of tensile strengths, both one from another as well as a single variety tested from different occurrences. Crocidolite has the greatest tensile strength followed by chrysotile, amosite, anthophyllite, tremolite, and actinolite in descending order.

Chemical Composition

The theoretical compositions of the six grades of asbestos are given in Table 1. It can readily be seen that the basic structure of all five minerals in the amphibole series is the same. The only difference is in the amount and type of metallic cations present. Consequently, all the amphiboles are quite similar but do differ substantially from chrysotile.

The idealized formula for the different types of asbestos are given in Table 1, but the actual compositions are different because of various impurities and differences in the chemical composition for different deposits. Table 2 shows the range of compositions for the most important components in commercial asbestos. In addition, there will be many trace elements present but these are rarely present in amounts greater than 0.20%.

Chrysotile is the result of hydrothermal reactions under certain conditions within serpentine rock, and it is only natural that there will be imperfections within the crystal structure and the inclusion of impurities between fibers and at the ends of the fibers near the wall rock. It is very difficult to remove these impurities by mechanical means, so they are present in the final product. Generally, they do not have any significant effect on the use of asbestos; but in a few cases, such as electrical applications or some filtration systems, the amount of impurities has to be limited.

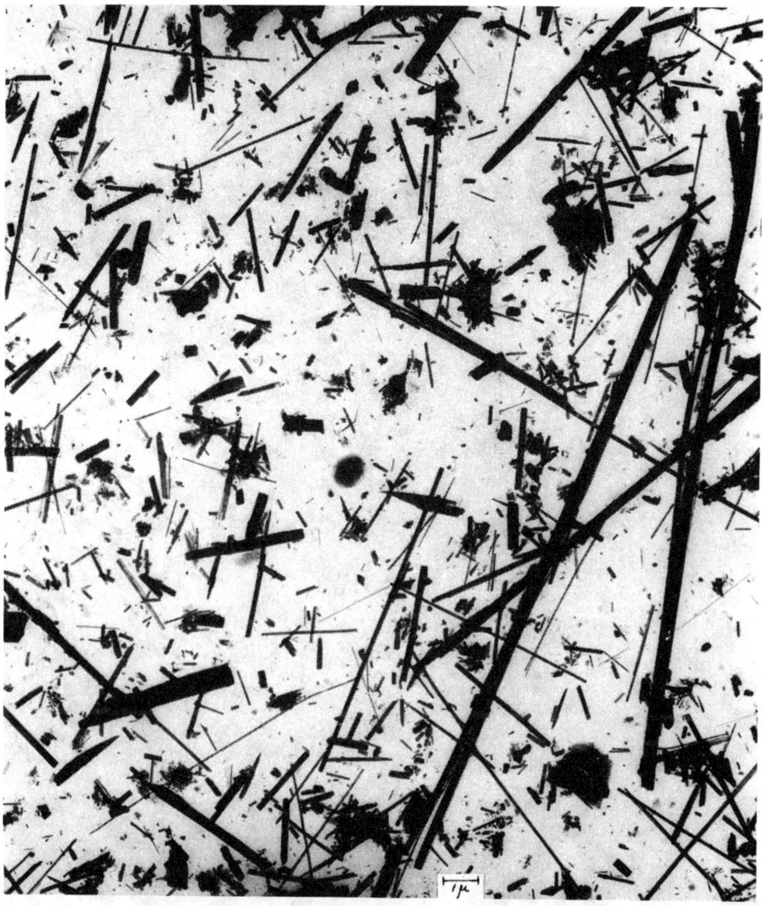

FIG. 2—*Electron micrograph of crocidolite. X8000. Reduced approximately 33% in reproduction.*

Crocidolite and amosite occur in metamorphosed siliceous-ferruginous sediments. Amphibole fibers are generally harsher, more springy and more brittle than those of chrysotile, making them faster draining and more bulky in various manufacturing uses. The fibers are often comparatively long, ranging up to several inches. These long lengths have no advantage for most uses and may detract from the value, since the greatest use is in asbestos cement products where fiber lengths of ⅛ to ⅜ in. in length are more suitable.

Anthophyllite (Rabbit, 1948) has a low ignition loss comparable to that of tremolite. It is never known to occur with fibers sufficiently flexible to be spun nor with the tensile strength required to make them of value to the asbestos cement industry. Resistance to chemical reaction opens a specialized field for this fibrous mineral as a filter medium and as a filler. A comparison of the general properties (Badollet, 1951) of the six varieties of asbestos is given in Table 3.

Thermal analysis (Pundsack, 1955) of chrysotile shows a broad endothermic peak between 600° and 720°C, which corresponds to the loss of structural water and an exothermic peak at 800°C associated with a phase change to forsterite.

The hydrogen ion concentration of chrysotile has been reported to vary from 9.0 to 10.3, depending on occurrence and method of measurement. A value of 9.7 will be most typical. The hydrolysis of chrysotile in distilled water free from carbon dioxide resembles the action of magnesium hydroxide. Chrysotile is also a selectively adsorptive material, taking in up to 2 to 3 wt% moisture from saturated air. This is due to its ultra small fiber diameters, its relatively reactive surface, and its high specific surface area. Chrysotile also adsorbs iodine from solutions thereby providing an excellent

FIG. 3—*Amosite, Penge, South Africa. X5000. Reduced approximately 33% in reproduction.*

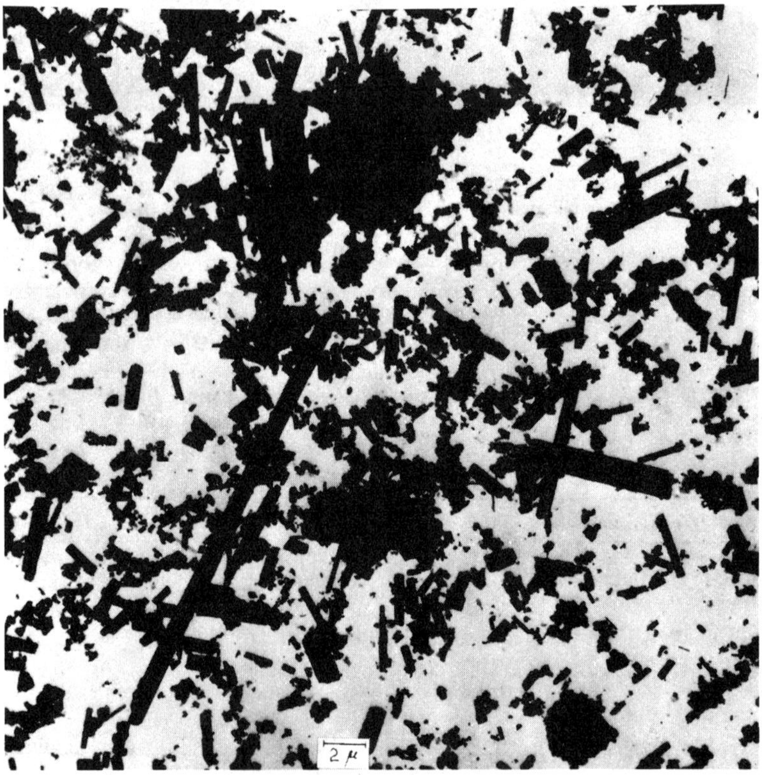

FIG. 4—*Anthophyllite, Georgia. X5000. Reduced approximately 33% in reproduction.*

FIG. 5—*Chrysotile, X150,000. Reduced approximately 33% in reproduction.*

staining technique for the detection of chrysotile fibers.

Chrysotile is unique as compared to the other types of asbestos and most materials of a similar nature such as clays and talcs, since it carries a positive charge or has a plus zeta potential in a water slurry. The magnitude will vary with chrysotile from different deposits. Similarly, the isoelectric point or pH for zero potential will be different for chrysotiles from various sources. The isoelectric point for chrysotile will vary from 10.5 to 11.3 depending

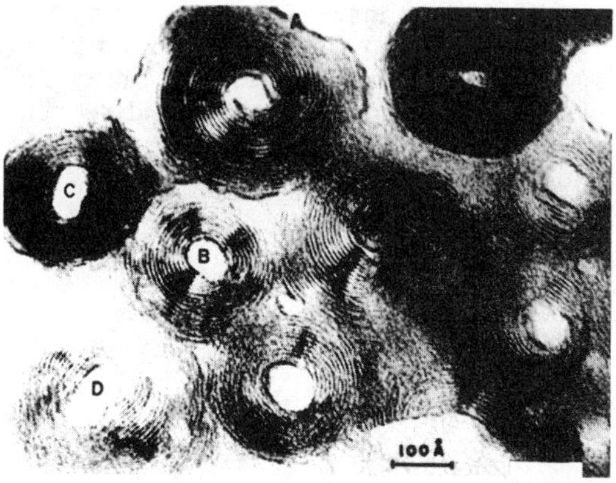

FIG. 6—*High resolution electron micrograph of chrysotile. Reprinted with permission from "Acta Chrystalogica," (Yada, 1967). Reduced approximately 38% in reproduction.*

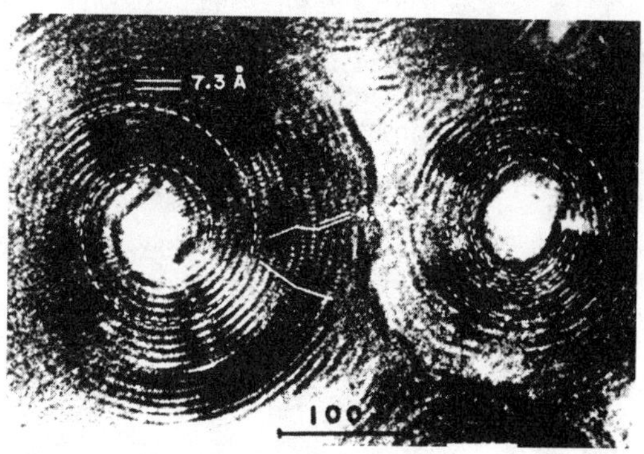

FIG. 7—High resolution electron micrograph of chrysotile. Reprinted with permission from "Acta Chrystalogica," (Yada, 1967). Reduced approximately 38% in reproduction.

TABLE 1—Types of Asbestos

Type	Theoretical Formulae
Chrysotile Hydrated magnesium silicate	$Mg_6[(OH)_4Si_2O_5]_2$
Crocidolite Complex sodium iron silicate (riebeckite) commonly called Blue Asbestos	$Na_2Fe_5[(OH)Si_4O_{11}]_2$
Amosite Iron magnesium silicate (grunerite)—same as anthophyllite except for high iron content	$MgFe_6[(OH)Si_4O_{11}]_2$
Anthophyllite Magnesium silicate with varying amounts of iron in the lattice	$(Mg,Fe)_7[(OH)Si_4O_{11}]_2$
Tremolite Calcium magnesium silicate	$Ca_2(Mg,Fe)_5[(OH)Si_4O_{11}]_2$
Actinolite Calcium magnesium silicate with varying amounts of iron	$Ca_2(Mg,Fe)_5[(OH)Si_4O_{11}]_2$

on the source of the fiber and the method of measurement.

Careful density measurements (Pundsack, 1956) of chrysotile have shown values ranging from 2.53 to 2.58 with 2.56 most typical.

The surface area determination of chrysotile depends upon the type of measuring apparatus employed. The industry generally uses an air permeability method (Badollet, 1951) for measuring the degree of openness of the fiber. If the true surface area is desired, it is necessary to employ gas adsorption techniques. Specific areas by air permeability will range from about 0.4 to 5 m^2/g whereas comparable gas adsorption areas will vary by a factor of 10 from 4 to 50 m^2/g depending on the source of the fiber and the degree of openness.

Effect of Heat on Asbestos

The weight loss for different types of asbestos when heated for 2 hr each at successive temperatures from 400° to 1800°F is shown in Table 4 (Badollet, 1951; Badollet and Streib, 1955). As expected, the weight loss for the amphiboles is considerably less than for chrysotile because of much less available ^-OH to be driven off as water. In spite of this, the amphiboles after heating at 800°F will still be more brittle than a chrysotile heated at 100° to 1100°F.

Table 5 shows the effect of heating chrysotile on the tensile strength of the fiber. The fiber bundles tested for tensile strength were very small so heat transfer would be effected in the 3-min heating time. Even after heating at 1200°F chrysotile retains 32% of its strength.

Structural changes in chrysotile can occur under conditions of intense grinding. This causes the structure to become amorphous with a loss of recognizable X-ray diffraction or electron microscope images. This apparently occurs because of localized temperature surges within the fibrils which produces dehydroxylation as the tremendous impact energies are absorbed.

Amphibole asbestos fibers also decompose at higher temperatures. This is particularly true when ferrous iron is present as it causes complex decomposition or thermal analysis determinations which are extremely dependent on atmospheric compositional variations.

TABLE 2—Approximate Chemical Analysis of Asbestos

	Chrysotile	Crocidolite	Amosite	Anthophyllite	Tremolite
SiO_2, %	37-44	49-53	49-53	56-58	53-62
MgO, %	39-44	0-3	1-7	28-34	0-30
FeO, %	0-6	13-20	34-44	3-12	1.5-5
Fe_2O_3, %	0.1-5	17-20	—	—	—
Al_2O_3, %	0.2-1.5	—	2-9	0.5-1.5	1-4
H_2O, %	12-15	2.5-4.5	2-5	1-6	0-5
CaO, %	tr-5	—	—	—	0-18
Na_2O, %	—	4.0-6.5	—	—	0-9
CaO + Na_2O, %	—	—	0.5-2.5	—	—

TABLE 3—Properties of Asbestos Fibers (Badollet, 1951)

Property	Actinolite	Amosite	Anthophyllite	Chrysotile	Crocidolite	Tremolite
Structure	Reticulated long prismatic crystals and fibers	Lamellar. Coarse to fine fibrous and asbestiform	Lamellar. Fibrous asbestiform	Usually highly fibrous fibers fine and easily separable	Fibrous in ironstones	Long. Prismatic and fibrous aggregates
Mineral association	In limestone and in crystalline schists	In crystalline schists, etc.	In crystalline schists and gneisses	In altered peridotite adjacent to serpentine and limestone near contact with basic igneous rocks	Iron rich silicious agrillite in quartzose schists	In Mg limestones as alteration products of highly magnesian rocks. Metamorphic and igneous rocks
Origin	Results of metamorphism	Metamorphic	Metamorphic. Usually from olivine	Alteration and metamorphism of basic igneous rocks rich in magnesian silicates	Regional metamorphism	Metamorphic
Veining	Slip or mass fiber	Cross fiber	Slip. Mass fiber unoriented and interlacing	Cross and slip fibers	Cross fiber	Slip or mass fiber
Essential composition	Ca•Mg•Fe•Silicate water up to 5%	Silicate of Fe and Mg higher iron than anthophyllite	Mg silicate with iron	Hydrosilicate of magnesium	Silicate of Na and Fe with some water	Ca and Mg silicate with some water
Crystal structure	Long and thin columnar to fibrous	Prismatic. Lamellar to fibrous	Prismatic. Lamellar to fibrous	Fibrous and asbestiform	Fibrous	Long and thin columnar to fibrous
Crystal system	Monoclinic	Monoclinic	Orthorhombic	Monoclinic (pseudo-orthorhombic?)	Monoclinic	Monoclinic
Color	Greenish	Ash gray or brown	Grayish white. Brown-gray or green	White. Gray. Green. Yellowish	Lavender-blue. Metallic-blue	Gray-white. Greenish Yellowish. Bluish
Luster	Silky	Vitreous. Somewhat pearly	Vitreous to pearly	Silky	Silky to dull	Silky
Hardness	6±	5.5-6.0	5.5-6.0	2.5-4.0	4	5.5
Specific gravity	3.0-3.2	3.1-3.25	2.85-3.1	2.4-2.6	3.2-3.3	2.9-3.2
Cleavage	110 perfect	110 perfect	110 perfect	010 perfect	110 perfect	110 perfect
Optical properties	Biaxial negative extinction inclined	Biaxial and positive extinction parallel	Biaxial positive extinction parallel	Biaxial positive extinction parallel	Biaxial ± extinction inclined	Biaxial negative extinction inclined
Refractive index	1.63± weakly pleochroic	1.64±	1.61±	1.51-1.55	1.7 pleochroic	1.61±

TABLE 3—Continued

Length	Short to long	2 to 11 in., varies	Short	Short to long	Short to long	Short to long
Texture	Harsh	Coarse but somewhat pliable	Harsh	Soft to harsh. Also silky	Soft to harsh	Generally harsh. Sometimes soft
Specific heat, Btu per lb per °F	0.217	0.193	0.210	0.266	0.201	0.212
Tensile strength, psi	1000 and less	16,000 90,000	4000 and less	80,000 100,000	100,000 300,000	1000 8000
Temperature at maximum ignition loss, °F	—	1600 to 1800	1800	1800	1200	1800
Filtration properties	Medium	Fast	Medium	Slow	Fast	Medium
Electric charge	Negative	Negative	Negative	Positive	Negative	Negative
Fusion point, °F	2540	2550	2675	2770	2180	2400
Spinnability	Poor	Fair	Poor	Very good	Fair	Poor
Resistance to acids and alkalies	Fair	Good	Very good	Poor	Good	Good
Magnetite content	—	0	0	0-5.2	3.0-5.9	0
Mineral impurities present	Lime, iron	Iron	Iron	Iron, chrome, nickel, lime	Iron	Lime
Flexibility	Poor	Good	Poor	High	Good	Poor
Resistance to heat	—	Good. Brittle at high	Very good	Good. Brittle at high	Poor, fuses	Fair to good
Ionizable salts, µmhos (relative electrical conductance)	—	1.34	0.58	1.82	0.84	—

TABLE 4—Effect of Temperature on Loss in Weight of Asbestos Fibers

°F	Time, Hr	% Loss in Wt Amosite	Anthophyllite	Chrysotile	Crocidolite	Tremolite
400	2	0.23	0.05	0.30	0.08	0.04
600	2	0.57	0.24	0.85	0.25	0.08
700	2	0.80	0.30	1.78	0.49	0.13
800	2	0.98	0.38	2.17	0.73	0.22
900	2	1.07	0.41	2.83	0.83	0.26
1000	2	1.16	0.44	3.99	0.86	0.29
1100	2	1.36	0.52	10.38	1.00	0.37
1200	2	1.39	0.54	12.75	1.04	0.37
1400	2	1.43	0.54	13.43	1.03	0.47
1500	2	—	0.64	—	—	0.56
1600	2	1.52	1.12	13.62	0.93*	0.67
1700	2	—	1.73	—	—	1.40
1800	2	1.53	2.39	13.77	0.77*	2.18

* Iron changing in weight due to oxidation.

Chemical Resistance

Asbestos is often used in chemical service where resistance to chemicals is of paramount interest. Table 6 gives data on the effect of various acids and caustic soda on asbestos after long exposure at room temperature and for short exposure at boiling temperatures. Because of the basic nature of chrysotile, it is seriously attacked by strong acids and should not be exposed to them. It is resistant to 25% caustic but recent data show that it is attacked by strong caustic (70%) at its boiling temperature.

TABLE 5—Effect of Heat on Tensile Strength of Canadian Chrysotile Crude

	Tensile Strength, Psi	% of Original Tensile Strength
Original crude, no heat	131,000	—
Heated 3 min at 600°F	120,000	91.6
Heated 3 min at 800°F	96,000	73.3
Heated 3 min at 1000°F	78,000	59.5
Heated 3 min at 1200°F	42,000	32.0

TABLE 6—Solubility of Asbestos

% Loss in Wt, Refluxing 2 Hr, 25% Acid or Caustic

	HCl	CH_3COOH	H_3PO_4	H_2SO_4	NaOH
Actinolite	20.31	12.28	20.19	20.38	9.25
Amosite	12.84	2.63	11.67	11.35	6.97
Anthophyllite	2.66	0.60	3.16	2.73	1.22
Chrysotile	55.69	23.42	55.18	55.75	0.99
Crocidolite	4.38	0.91	4.37	3.69	1.35
Tremolite	4.77	1.99	4.99	4.58	1.80

% Loss in Wt, Room Temperature 26°C for 528 Hr, 25% Acid or Caustic

	HCl	CH_3COOH	H_3PO_4	H_2SO_4	NaOH
Actinolite	22.55	12.14	20.10	20.60	9.43
Amosite	12.00	3.08	11.83	11.71	6.82
Anthophyllite	2.13	1.04	3.29	2.90	1.77
Chrysotile	56.00	24.04	56.45	56.00	1.03
Crocidolite	3.14	1.02	3.81	3.48	1.20
Tremolite	4.22	1.41	4.89	4.74	1.65

Most amphibole fiber varieties are more acid resistant than those of chrysotile. However, they are not entirely acid resistant and can exhibit weight losses of 2 to 20% when subjected to boiling concentrated acids.

Geology

General

Chrysotile constitutes about 94% of the current world production of asbestos and, of this amount, all but a fraction of a percent is derived from deposits whose host rocks are ultrabasic in composition (Berger, 1963; Bates, 1969; Keith and Bain, 1932). This other fraction of the chrysotile production is derived from serpentinized dolomitic limestone. Among the other varieties of asbestos, amosite and crocidolite are found in certain metamorphosed ferruginous sedimentary formations and, together, account for some 5.3% of world production. Tremolite and anthophyllite make up the balance of the production and are generally found in association with highly metamorphosed ultrabasic rocks.

The bulk of world chrysotile production comes from the USSR and Canada. Southern Africa, including Zimbabwe, Swaziland, and South Africa, is the third largest producer with the balance coming from other parts of the world.

Of the Canadian production, by far the greatest amount comes from the Eastern Townships of Quebec (Fig. 8) with northern British Columbia, Ungava, Newfoundland, and Ontario contributing the balance. Russian production comes mainly from the Bazhenovo district in the central Urals, the Dzhetygara area of northwest Kazakhstan, the Kiembay area in the southern Urals and from Ak-Dovurak near the Yenesei River west of Lake Baikal.

The age of asbestos deposits varies greatly from earliest Precambrian in Zimbabwe and Swaziland to Upper Jurassic in California. The Ontario, Ungava, and Brazil deposits are all Precambrian, while those of the Eastern Townships, Vermont, and Newfoundland are all ascribed a Mid-Paleozoic age, associated with early folding in the Appalachian Mountain belt. The deposits of western Canada are connected with mountain building in the Late Paleozoic while the Russian deposits vary from Early Paleozoic to Late Paleozoic or Triassic in age.

Structures

General deformation in the form of faulting, folding, or shearing evidently plays a major role in the localizing of asbestos deposits (Cooke, 1937). The introduction of swarms of small intrusive dikes of acidic to intermediate composition also contributes to the general fracturing and opening up of the rock which, in turn, leads to serpentinization and the formation of chrysotile fiber.

The majority of chrysotile ore deposits tend to occur in serpentinized peridotite rather than dunitic host material. This is probably due to the fact that serpentinized dunite bodies tend to flow when subjected to stress, thereby aiding in the development of fracturing in the surrounding more brittle peridotite.

Asbestos Veining

Chrysotile asbestos is referred to as cross fiber where the fibers lie transverse to the vein, and as slip fiber where the fiber lies in the plane of the vein (Gold, 1967; Cooke, 1937). In the case of cross fiber, the orientation may be anywhere from normal to sharply inclined to the vein walls, in which case it can be referred to as oblique fiber. The fiber in the veins may be straight, gently curved, or contorted. Fiber containing sharp flexures is apt to break easily.

The quality of the fiber ranges from soft and silky to harsh; from highly flexible to brittle; and from strong to weak.

Partings are often present in the veins, either as microscopic discontinuities or as irregular shaped inclusions composed of picrolite, serpentine, brucite, and magnetite, or combinations of these. The veins may be relatively persistent or short and lenticular, occurring in the form of stockworks or in parallel systems known as ribbon fiber. They may be fissure fillings, replacements, or stress-relief features, and are often the result of a combination of these processes.

Some very short fiber is derived from a variety referred to as "mass" fiber. Mass fiber consists of either the complete replacement of most of the original mineral constituents to fiber, each grain of which has a different fiber orientation, or a dense network of fine reticulating replacement veins. Both varieties result in a very high content of short fiber and the rock, which appears to be largely fibrous, is sometimes referred to as "fur" rock.

Other varieties of mass fiber include "platy"

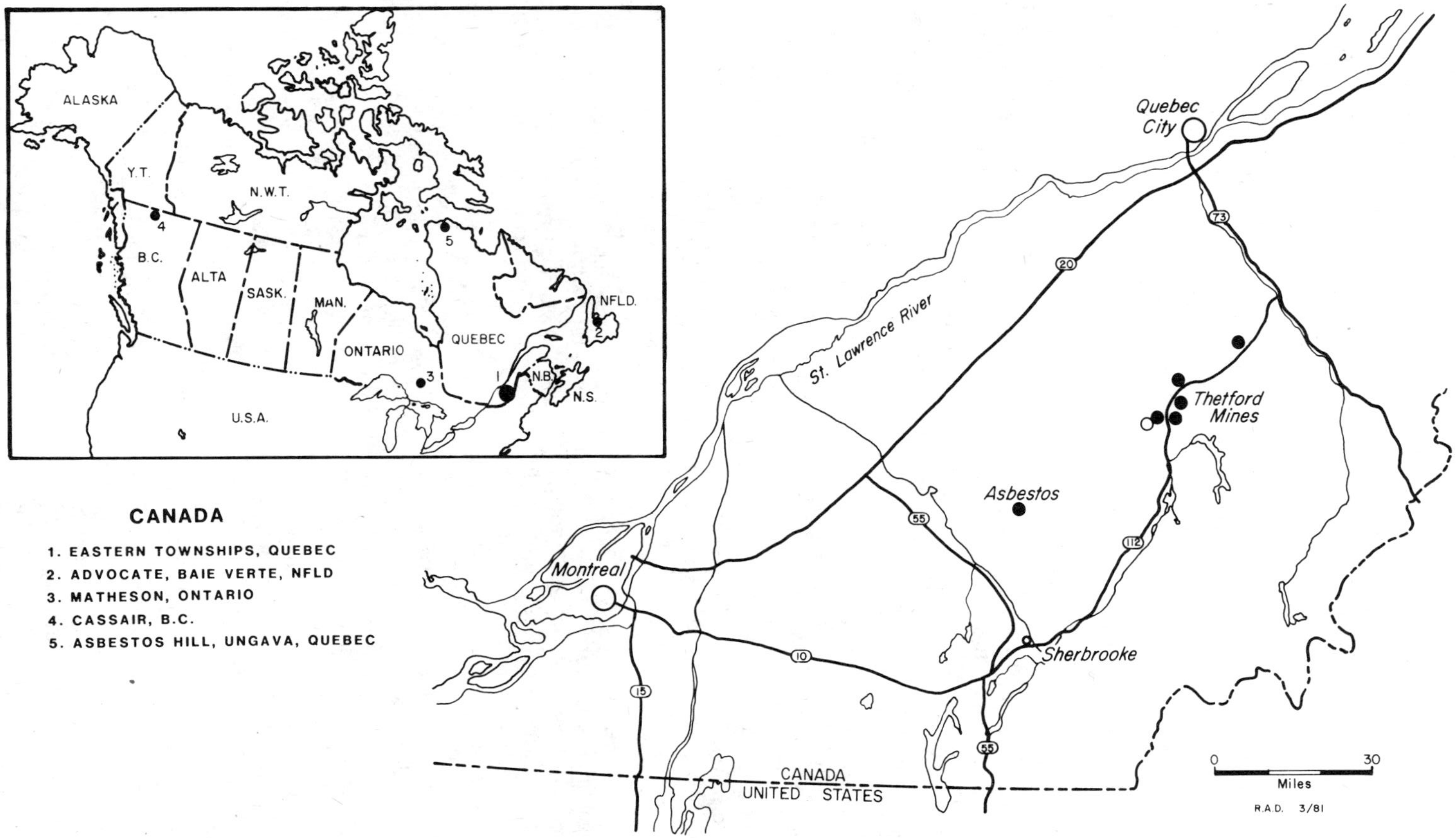

FIG. 8—*Asbestos mining areas in Canada. At right, the Eastern Townships of Quebec.*

fiber, of which the Coalinga deposit is unique. A similar, but less important, variety is known as "mountain leather."

World Asbestos Deposits (Fig. 9)

North America

Canada: *Eastern Townships, Quebec*—Chrysotile asbestos was discovered in two localities in the Eastern Townships in 1878 and has been actively mined at Thetford Mines and Asbestos for over 100 years. The chrysotile deposits of Quebec's Eastern Townships occur intermittently along a major serpentine belt which arcs northeastward into the Gaspé Peninsula in one direction and southward into the Appalachian Mountain belt of Vermont in the other (see Fig. 8).

The area yields over 80% of Canada's fiber production and some 30% of the world total, thereby vying closely with the Bazhenovo district of the Soviet Union for first place in world production. Most of the major occurrences are located along a sector 55 miles long which lies parallel to, and 50 miles southeast of, the St. Lawrence River extending from a point near East Broughton, 50 miles south of Quebec City, southwestward past Thetford Mines and Black Lake to Asbestos and Danville which lie approximately 85 miles east of Montreal.

This serpentine belt is a typical ophiolitic complex of gabbroic and dioritic rocks, and of pyroxenite, peridotite and dunite, serpentinized to various degrees with associated granite, rodingitic, and talc-carbonate rocks. This belt has been described in the past as a partially serpentinized ultrabasic intrusive emplaced in the crust along faulted zones of weakness. More recent study would indicate that this giant complex has been extruded onto or just under unconsolidated aluminous and silicious sediments in a eugeosynclinal ocean basin. Magmatic differentiation within the ultramafic magma is thought to be responsible for the formation of the various pyroxenite, peridotite, and dunite phases found in the complex.

The small granitic masses found within the ultramafic complex have been considered intrusive but may be the result of the ingestion and digestion of fragments of underlying sea floor sediments. Subsequent faulting and shearing has resulted in numerous dislocations within and adjacent to the ultramafic complex.

Consideration of this complex as a submarine extrusive simplifies the problems of water for serpentinization, allowance for large volume change, the occurrence of fiber as open fracture fillings in stockworks, and banded fiber in ribbon zones. It also explains the presence of "pillows," usually gradationally serpentinized from the surface inward.

The fiber occurs mainly in highly serpentinized peridotite (harzburgite containing 10–15% enstatite) and is found in numerous occurrences along the strike of the complex.

There are presently six deposits being mined in the Eastern Townships and numerous others are either dormant or completely mined out. For purposes of description the deposits may be grouped on the basis of their distribution within the belt, i.e., the Pennington dike deposits, those of Thetford Mines area, the Black Lake deposits, and those of Asbestos.

Pennington Dike, Thetford Mines-East Broughton Area (Anon., 1972; Merrill, 1957; Riordan, 1957; Rowbotham, 1970)—Separated from the main Thetford ultrabasic intrusive and extending northeastward for some 17 miles or more is the Pennington "dike" which has supported a number of asbestos mining operations, both past and present, over much of its length. This serpentinite body is actually tabular and more sill-like in form, being closely conformable with the country rocks which dip up to 60°SE. Shearing is conspicuous throughout its length. This is interpreted as being a major thrust fault, the plane of which occurs at the base of the sill. Where the sill is narrow, shearing extends from wall to wall and much of the rock is steatized, but the wider portions which reach a thickness of up to 800 ft are less affected. It is these wider bulges which contain most of the major ore zones. Slip fiber predominates in all the deposits, though some cross fiber is found in the wider, less sheared sections, particularly toward the southwest.

The narrowness of the ore bodies and the attitude of the sill are obstacles to mining any appreciable depth. Only two mines, National Asbestos and Carey-Canadian, are presently operating on the Pennington dike.

Thetford Mines Area (Cooke, 1937; Riordon, 1957, 1957a, b; Riordon and Laliberté, 1957)—A large mass of periodotite is host to the Thetford Mines and Black Lake deposits. The Thetford group lies at the northeast end of this mass and contains four closely connected deposits, all of which lie in a well-defined zone along the hanging-wall side of a prominent fault structure marked by the presence of intensive talc-carbonate alteration. The zone has a length of 6000 ft and a width of 1500 ft with the largest and deepest deposits, the King-Bell-Johnson complex, extending to a depth of 1500 ft or more. These deep zones are mined by underground methods. The other deposits are the Bennett-Martin, the Beaver-Consolidated, and the Beaver "C." The first two are now exhausted.

A number of subsidiary shears, most of which have dips in the opposite direction to the main fault and strikes that roughly parallel it, tend to break the ore zone into segments. Most of these subsidiary shears are accompanied by numerous bodies of syenite, and evidently developed along the original zones of weakness into which these acid rocks were introduced. Some large bodies of serpentinized dunite situated within the ore zone contributed to the fracturing in the surrounding peridotite by behaving as relatively incompetent masses which tended to flow under stress.

Black Lake Area (Anon., 1972; Riordon, 1957a, c, d; Riordon and Laliberté, 1957)—There are three major ore bodies in the general vicinity of Black Lake. The British Canadian-Megantic mine is over 1 mile long and ½ mile wide, and occupies a position adjacent to a pronounced flexure in the peridotite contact immediately east of the town of Black Lake. This ore body is cut up into ore shoots by a series of roughly parallel shear zones and syenite dikes striking tangentially to the folded contact. Large bodies of serpentinized dunite are also found within the ore body and produce low grade to barren sections.

Southwest of the town of Black Lake are the ore bodies which comprise the Lake Asbestos mine, the lake having been drained to permit mining of the deposit beneath. The easternmost of these ore zones is situated along the hanging-wall side of a northwest-southeast transverse fault which dips to the northeast. Like the major fault in Thetford this zone has also undergone intensive steatization. Another ore body to the west was located beneath a large, flat-lying acid dike which evidently localized the fracturing in its vicinity and no doubt was responsible for the formation of fiber in the vicinity.

The Normandie mine is located about a mile to the west of Lake Asbestos and two miles southwest of the town of Black Lake. This deposit consists of at least three separate ore bodies lying along a northeasterly plunging fold axis which manifests itself in a pronounced *Z* bend in the peridotite-Caldwell schist contact in the vicinity. The old Vimy Ridge deposit which is now completely worked out is the most southerly of these and occurred close to surface, with much of the upper portions of the ore body truncated and removed by erosion.

The central zone lay completely buried beneath a capping of drift and overburden, and was only discovered by regular grid pattern diamond drilling in the peripheral areas surrounding the old Vimy Ridge mine. The Normandie open pit mine, a very productive operation in its day, has now been mined out.

Further deep drilling on the downdip extension of the Normandie ore body led to the discovery of a deep ore zone lying well beneath the overfolded capping of Caldwell quartzites. This deposit, the Penhale ore body, has been explored by underground methods and is now being considered for production.

The origin of these three ore bodies is closely related to the folding but may also be dependent on a fault zone trending southwest which parallels the fold axis to the east of the ore bodies.

Whereas the other deposits in the Eastern Townships are made up of a stockwork of asbestos veins, the Normandie, Vimy Ridge, and Penhale ore bodies are composed predominantly of a prominent system of parallel veins of ribbon fiber. These veins form a conjugate pattern but dip predominantly to the northeast with the rake of the ore zone.

The Vimy Ridge and Normandie ore bodies are separated by numerous large syenite dikes and some shearing, whereas the Normandie and Penhale deposits are separated by a zone of low grade to barren material.

Asbestos-Shipton Area (Allen et al., 1957; Bourassa, 1957; Rowbotham, 1970; Riordon and Laliberté, 1957)—The Jeffrey and the now-exhausted Nicolet ore bodies occur along the north and footwall side of a peridotite-dunite intrusive. The Jeffrey deposit, discovered over 100 years ago, has been producing since the early 1880s and is now the largest asbestos mine outside the USSR. The ore body itself has an ellipsoidal to cylindrical or roughly pipe-like form, having an average diameter of about 2000 ft and its major axis plunging about 55° to the southwest. The ore zone is bounded on both the footwall and hanging-wall contacts by major faults or zones of shearing which dip south-southeast at about 65° to 70°. The ore body is also cut by several irregular shears which trend roughly parallel to the footwall contact and divide the deposit into five major ore zones. Numerous acid dikes of dioritic to

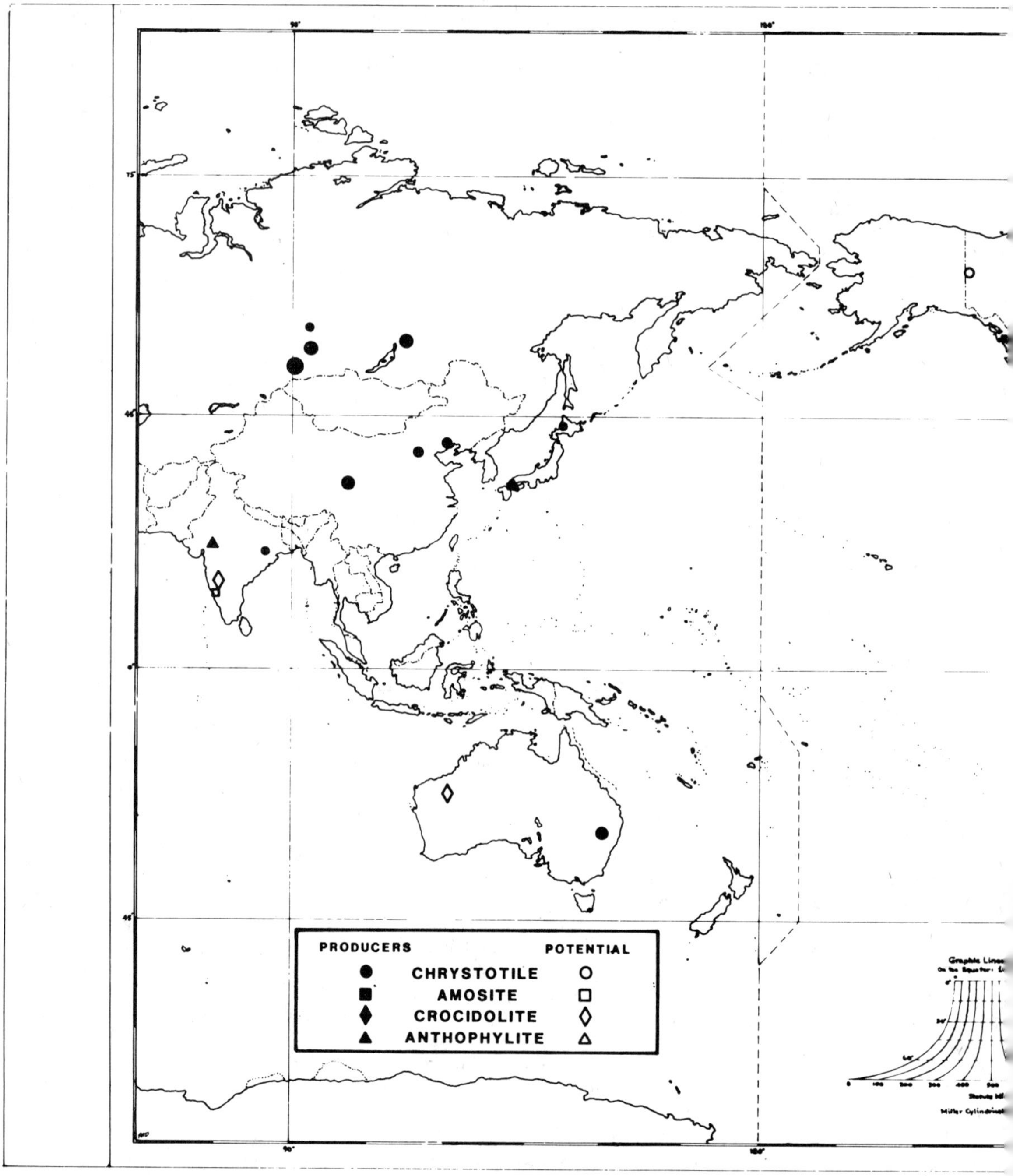

FIG. 9—*World asbestos deposits.*

ASBESTOS DEPOSITS
of the
WORLD
R.A.D. 3/81

syenitic composition tend to follow the shear zones but a few exhibit later cross-cutting tendencies. The fiber is generally of good- to medium-grade cross fiber, though sections of slip fiber and mass fiber also occur.

The Nicolet ore body occurred as a narrow zone along the footwall contact, being bounded on its south side by a prominent zone of shearing. This deposit has now been exhausted.

Baie Verte, Newfoundland—Asbestos is currently mined near Baie Verte (Straw, 1961) on the north shore of Newfoundland. The Advocate ore body forms a continuous fiber zone within the highly serpentinized periphery of a massive elliptical pyroxenite body, 1 mile long by ¼ mile wide. The inward dipping fiber zone surrounds the massive pyroxenite core and is in turn surrounded by highly sheared serpentinite. The serpentinite is part of the 55-mile belt of ultrabasic rocks that separate the schists of the Ordovician Fleur de Lys group from the volcanics to the east.

The massive pyroxenite core shows strong parallel, nearly vertical bands of pyroxene-rich and pyroxene-poor phases, suggesting that it is a fragment of differentiated oceanic crust caught up in a subsequent ultramafic flow. The fiber probably formed in fractures on the highly stressed, highly serpentinized periphyry of this peridotite core.

The fiber occurs as a stockwork of good quality fiber veins averaging ¼ to 5⁄16 in. in length. After mining and milling the fiber is loaded directly at the nearby terminal for shipment to overseas markets.

Reeves Township, Timmins, Ontario—Chrysotile asbestos occurs at Reeves (Douglas, 1970; Rowbotham, 1970; Hendry and Conn, 1957) 40 miles southwest of Timmins. This deposit lies near the northern tip of a large serpentine mass in contact with Precambrian metavolcanics. The south or footwall side of the ore body is bordered by a major east-west fault that dips 45°N in the serpentine. Extensive talc-carbonate alteration is common along this fault. Two north-south diabase dikes cut the ore zone. These have caused pronounced alteration of the fiber along their contacts.

The deposit was mined for a number of years but finally closed in 1974 due to marginal grades and pervasive alteration of the fiber.

The deposit in Munro Township in eastern Ontario has long been exhausted by mining. This deposit occurred within a narrow ultrabasic dike bounded by faults containing pronounced talc-carbonate alteration. The ore zone was also offset by cross faulting.

Cassiar, Northern British Columbia—The Cassiar asbestos deposit (Douglas, 1970; Smitheringale, 1957) occurs at an elevation of over 6000 ft near the peak of Mount McDame in the Cassiar Range. Fiber in the Cassiar deposit is long and of exceptional grade and quality and is associated with an ultrabasic sill intruded into a folded succession of volcanic flows and sediments of Devonian age. The asbestos occurs as cross fiber veins in dark green serpentine commonly ½ to 1 in. wide and up to 15 ft long. This fiber is shipped in special containers by road to Whitehorse, then by narrow-gauge railway to Skagway and from there by ship to overseas markets.

Clinton Creek, Yukon Territory—Mining of asbestos at Clinton Creek (Douglas, 1970) near the Canada-Alaska boundary was short-lived and the relatively small deposit is now closed.

The deposit lies just south of the Yukon River within 150 miles of the Arctic Circle. Here cross fiber averaging close to ¼-in. in length occurred in a small ultrabasic body enclosed by phyllites, argillites, and quartzites of Paleozoic age. A portion of the deposit has also undergone strong talc-carbonate alteration. This remote location necessitated costly road, rail, and shipping transportation similar to that of Cassiar asbestos which decreased the margin of profit considerably.

Asbestos Hill, Putunig, Ungava, Quebec—Situated some 40 miles from tidewater on Quebec's Hudson Strait (Mann, 1962; Stewart, 1978) on the western margin of a large ultrabasic to basic intrusive, is one of Canada's newest and richest asbestos producers, Asbestos Hill. This deposit, about 2500 ft long and up to 200 ft wide, occurs at a point where the dip of the serpentine changes from gently inward-dipping through vertical to steeply outward dipping from the intrusive complex. There is also evidence of a small drag fold where the dip changes most abruptly that appears to form the locus of the ore zone. Talc alteration is present in some zones. The ore is exceptionally high grade with the bulk of the fiber ranging from 3⁄16 to 7⁄16 in. in length and a minimum amount of short fiber. This produces a good Group 4 asbestos-cement fiber. The ore near the surface was, until recently, mined by open pit but the deeper parts are now being considered for development by underground methods. The fiber is partially concentrated at the mine site before being transported to Deception Bay and thence by sea during a 3½ to 4-month shipping season to Nordenham, Germany, for final processing.

Other Potential Deposits—Several other deposits have been considered at various times for possible production. These include the Abitibi deposit situated 50 miles north of Amos; the McAdam deposit 17 miles east of Chibougamau, both in Quebec; the Lloyd Lake Midlothian Township deposit 40 miles south of Timmins, Ont.; and the Garrison deposit 70 miles east of Timmins. An attempt was made to develop the Midlothian deposit at Lloyd Lake but the project encountered serious financial problems and has been halted.

United States of America: Chrysotile asbestos is mined near Eden, VT, and in two localities in California. Small quantities of chrysotile are also recovered from dolomitic limestones in Arizona.

Pacific Asbestos, Copperopolis, California—The Pacific Asbestos deposit (Leney and Loeb, 1972) is located between Copperopolis and Sonora in the foothills of the Sierras about 125 miles east of San Francisco. This deposit occurs within a belt of ultrabasic rocks which trend northwest for 20 miles and reach a maximum width of 4 miles. Strike faulting is common and the peridotites and dunites have been variously altered to serpentine. The ore body occupies a zone about 2000 ft wide between two bands of antigorite schist which are believed to represent zones of shearing within the intrusive. Fiber occurs as cross fiber up to 1 in. in length but most of the fiber is ¼ in. or less. The ore body is cut by narrow dioritic dikes. Talc-carbonate alteration is also prevalent in some portions of the pit.

New Idria-Coalinga, California—The recognition of "platy chrysotile" in the New Idria ultrabasic intrusive led to the successful mining of this fiber in two or three operations northwest of Coalinga (Munro and Reim, 1962) in Fresno County. The ultrabasic body trends northwestward and covers an area some 15 miles in length and 3 to 5 miles wide, having been intruded into Jurassic sediments in Late Jurassic or Early Cretaceous times. The intrusive is composed of highly sheared serpentinite characterized by its extremely platy, slickensided nature. "Boulders" of massive serpentinized material occur scattered through the loose platy serpentine. There is also evidence of abundant landslide material, a factor which may have contributed to the extreme deformation of the serpentine.

The ore contains abundant short chrysotile fiber similar to the Canadian variety but low prices and severe safety and health regulations caused the closing of two of the three producers in the area.

Eden, Vermont—The southern extension of the Eastern Townships serpentine belt of Quebec continues intermittently southward down the Appalachian chain into northern Vermont (Rowbotham, 1970). Fiber occurrences have been known for many years over a considerable area in Lamoille and Orleans Counties. The deposit which has been producing for a number of years lies near the village of Eden. Here a substantial ore body, some 3000 ft wide, is being mined at the foot of Mount Belvidere between elevations of 1000 and 1500 ft.

Slip fiber is the more predominant variety mined but some good quality cross fiber is also recovered. The host rock is a highly serpentinized peridotite with the fiber occurring along a contact between gneiss and amphibolite schist.

Mexico: *Cuicatlan, Oaxaca*—A large occurrence of slip fiber has been discovered at Concepcion Papalo east of Cuicatlan, and about 50 miles north-northwest of Oaxaca in Mexico. The Pegaso occurrence is exposed on two sides of a ridge in highly dissected terrain and appears to be part of a tabular serpentinite body which forms the core of a highly folded and sheared sequence of rocks.

The occurrence is presently under study for possible production but the lower tensile strength and the fact that the fiber is slip variety render the deposit very marginal.

Ciudad Victoria, Tamaulipas—Occurrences of chrysotile fiber are known to occur in the Sierra Madre Oriental Mountains west of Ciudad Victoria (Rowbotham, 1970) in the State of Tamaulipas. These are generally small, discontinuous zones of ribbon fiber and are of little economic significance.

South America

Cana Brava, Goiás, Brazil: The Cana Brava asbestos deposit is located some 120 miles north of Brasilia and 60 miles northeast of Uruaçá in the State of Goiás. Little is known about the deposit except that it occurs in a large ultrabasic body which intrudes gneisses of the Brazilian Precambrian Shield. The ore is apparently a good quality cross fiber of medium length which occurs in two separate zones 1800 ft long and roughly 200 ft wide. The bulk of this fiber is utilized by local Brazilian manufacturers.

Las Brisas, Antioquia, Colombia: A deposit

of chrysotile asbestos is at present being readied for production at Las Brisas, Antioquia, in central Colombia (Harris, 1973). This property lies about 80 miles north of Medellin or roughly midway between Bogota and the Caribbean port of Cartagena. The deposit is reported to comprise at least two separate zones containing fiber of distinctly different grade.

Europe

Balangero, Northern Italy: The bulk of the Italian chrysotile production comes from Balangero (Rowbotham, 1970), a large, low grade slip fiber deposit located near San Vittore about 20 miles north-northwest of Turin. The ore zone occurs in a highly sheared serpentine body about 1500 ft wide which trends north-south and is intrusive into paragneisses in the foothills of the Alps in Western Italy.

Zidani, Kozani, Northern Greece: Situated about 75 miles southwest of Thessaloniki and 20 miles due south of Kozani is a large, low grade slip fiber deposit which currently is being developed for production. The deposit occurs in a highly sheared and fractured serpentinite, probably of dunitic composition which is flanked by crystalline marble to the west. The property was in production in 1981.

Troodos, Cyprus: Cyprus (Bear, 1963; Vokes, 1964) is an ancient source of asbestos, as it was known and exploited by the ancient Greeks and Romans. These deposits are some of the largest occurrences of chrysotile in the European area.

Asbestos has been mined for the past 70 years near Amiandos in the Troodos Mountains in the south-southwestern portion of the island. The chrysotile occurs in a body of bastite-serpentinite covering an area some 3½ miles north-south and 2½ miles wide, being roughly oval in plan. The best fiber is confined to the southern part of the area near the faulted eastern margin of the peridotite where deformation was most intense. The fiber occurs in small discontinuous lenticles or in larger composite veins of cross fiber and picrolite up to ½ in. wide and several feet long. The fiber seldom exceeds ¼ in. in length. These deposits are probably Upper Cretaceous in age.

The Troodos mine is at an altitude of 5000 ft and the crude ore is transported to Limassol by a 19-mile-long aerial ropeway for further processing.

Yugoslavia: Short fiber chrysotile has been produced from a number of mines in Yugoslavia (Rowbotham, 1970) for many years. Many of these mines have undergone expansion recently and this is reflected in the latest production figures. Deposits are mined at Korlace in Serbia and near Tuzla, Bosnia. Details of the geology of these deposits are not available.

USSR: Russia (Nalivkin, 1960; Rowbotham, 1970; Smirnov, 1971; Stenho and Fronek, 1977; and Petrov and Znamensky, 1978) is now considered to be the largest producer of asbestos in the world with an estimated production of 3,197,000 short tons of fiber in 1978 and 2,723,000 tons in 1979.

The USSR's known reserves are vast and are considered to surpass those of any other country.

There are four major areas of asbestos production in the USSR: the Bazhenovo deposits of the Central and Northern Urals, the Dzhetygara district in the Southern Urals of western Kazakhstan, the newly opened Kiembay district also in the southern Urals and the Ak-Dovurak deposits in Tuva ASSR some 600 miles west of the southwest end of Lake Baikal.

Bazhenovo (Bajenovsk) District—According to its production and reserves, the Uralian deposits of Bazhenovo occupy first place in the USSR and rival the Eastern Townships of Quebec for the leading asbestos district of the world. These deposits are affiliated with the Bazhenovo-Alapajevsk or eastern serpentine belt which extends for a distance of about 110 miles in the Central Urals. Most of the deposits are centered around the new town of Asbest which is 50 miles northeast of Sverdlovsk and 900 miles east of Moscow.

The deposits are associated with fault zones in serpentinites which formed predominantly as the result of alteration of peridotites and, to a lesser extent, pyroxenites. Formation of the veins of chrysotile is thought to be by filling of fissures in a massif of ultrabasic rocks.

Mines in the Asbest area and others at Alapajevsk, Novaja and Lesnoje, together with the deposits at Rez (East Aninsk) Krasnouralsk, Lukosk and Krivisk have an annual capacity of at least 1.5 million tons of fiber.

Besides the Bazhenovo group, a number of other deposits of asbestos are known.

Dzhetygara District, Northwest Kazakhstan—This constitutes the second largest area of production in the USSR and has been in production since 1965. The district has a planned annual production of about 800,000 tons.

The Dzhetygara district lies some 140 miles southwest of Kustenay. Little is known of the geology of the deposits except that they occur in serpentinized peridotites of Early Hercynian (Middle Devonian) age.

Kiembay (Kijembai), Southern Urals—This deposit, the third largest in the USSR, is located some 90 miles east of Orsk in the Orenburg Administrative Region. The deposit occurs in a completely serpentinized ultrabasic intrusive. Severe weathering 150 to 200 ft deep has caused the surface fiber to become weaker and relatively brittle. The deposit has been developed recently by a collective combine with Czechoslovakia, Poland, Hungary, East Germany, Romania and Bulgaria participating with the USSR to bring it to its planned productive capacity of 550,000 tpy.

Ak-Dovurak (Aktovrak), Tuva ASSR—Important deposits of asbestos have recently been developed near Ak-Dovurak, Tuva, near the Yenesei River to the west of Lake Baikal. These deposits are associated with a belt of serpentinized ultrabasic rocks of the Salair complex which forms part of the Lower Hercynian of the eastern Altai-Sayan Province. The ultrabasic intrusions occur intermittently in belts which extend along major faults.

At least six significant deposits have been discovered and the district is reported to contain vast reserves near the village of Kyzyl-Mazalyk and the Kedrovsk deposit 30 miles to the south.

Other Deposits—Chrysotile asbestos deposits are known at Molodeznoje (Molodjezhnoje) some 130 miles east of the northern end of Lake Baikal. The deposit is of excellent grade but will have to await development until the Baikal-Amur railway is constructed.

The Ilcirsk and Sajan deposits in the same general area also have good potential for future production.

A peculiar fiber deposit is located near Pechenge close to the Norwegian border. Although the deposit is large, the high iron content is very detrimental to the value and it is doubtful whether the deposit will ever prove to be economic.

Africa

Shabani, Zimbabwe: The Shabani mine (Laubscher, 1964; 1968; Oldham, 1968; Wilson, 1968), situated some 100 miles east of Bulawayo, has the largest reserves of fiber in southern Africa. The deposit occurs near the base of a lenticular ultrabasic sill 9 miles long and 1½ to 3 miles wide which intrudes Early Precambrian gneisses. The sill strikes northwest and dips about 40° SW and is composed of dunite at the base, overlain by peridotite, pyroxenite, and gabbro. A swarm of younger diabase dikes cut the ore zone but these are themselves truncated by strong zones of shearing, both parallel and oblique to the basal contact. The intrusion and shearing has caused strong serpentinization of the dunite and heavy talc-carbonate alteration, particularly along the footwall shear zone. The fiber occurs as a stockwork of cross fiber veins described as "stress-controlled dilation seams." The quality is good and the length of the fiber up to 1½ in. in places.

Fiber is also mined at other localities in Zimbabwe, notably in the Mashaba, Belingwe, and Filabusi districts.

Msauli, South Africa: The Msauli mine (Hall, 1930; Pelletier, 1964; Van Biljon, 1964) is located 16 miles south-southeast of Barberton in the eastern Transvaal and only six miles southwest of the Havelock mine across the Swaziland border. The serpentine in which this ore body occurs is very similar to that of the Havelock deposit in what appears to be the faulted continuation of the same sill-like ultrabasic intrusive. The serpentine body averages 450 ft wide and dips about 65° to the east. The best fiber is reported to occur to the north in association with major faulting. The fiber occurs as a stockwork of veins from 1/32 to ¾ in. and is of the same general high quality as that of Havelock. However, some sections contain a higher percentage of magnetite.

Havelock, Swaziland: The Havelock mine (Pelletier, 1964; Van Biljon, 1964) is located within ½ mile of Swaziland's western border with the Republic of South Africa, and lies some 12 miles south-southeast of Barberton in the eastern Transvaal. Here, a sill-like body of serpentine of the ancient Jamestown Complex is intercalated in the Early Precambrian Fig Tree sedimentary sequence, all of which dip south at 55°. The ore zone has a strike length of 5000 ft and widths from 60 to 350 ft averaging about 110 ft. The ore body is traversed by a longitudinal fault zone with associated transverse faulting. Talc-carbonate alteration is prevalent in the western portion of the mine. The fiber occurs as a stockwork of high quality cross fiber seams, which average ½ in. or more in length and comprise 3% to 4% of the rock. Mining of the upper portion of the ore body was by open cast workings, the deeper portions now being mined by underground methods. The ore is transported by aerial cableway to Barberton for processing.

Other Deposits: Many small deposits of

chrysotile asbestos occur at many locations in northern and eastern Transvaal but their combined production has never been appreciable. Most of these are elongated ribbon fiber zones which are in ultrabasic rocks or associated with dolomitic limestones.

Occurrences of chrysotile fiber are also reported from the Ingessana Hills and Qala en Nahl districts of eastern Sudan. Both areas have been explored extensively and several small deposits have been indicated, but the total potential remains relatively small.

Australia

Woodsreef (Transpacific Asbestos), New South Wales: The Woodsreef (Straw, 1964; Butt, 1978) asbestos deposit is located just east of Barraba approximately 340 miles north of Sydney in northeastern New South Wales. The deposit occurs in a lens of serpentinite approximately 5 miles long and 1½ miles wide which forms part of the Great Serpentine Belt of New South Wales. The ore body occurs in a layered peridotite-dunite complex which has intruded sediments of Mid-Paleozoic age, and is approximately 2300 ft long and 250 ft wide. It is bounded on the western margin by prominent faulting and shearing. Strong shear zones are also prevalent within the fiber zone. The fiber occurs mainly as a stockwork of cross fiber seams and is generally short to medium in length. Although production commenced in 1972, the mine has encountered serious fiber recovery problems and the long-term viability of the deposit is still in question. Most of the fiber is exported to Japan but some is consumed locally.

New Zealand

Pyke Asbestos, South Island: A small occurrence of chrysotile fiber is reported near the headwaters of Pyke River (Babcock, 1978) high in the Southern Alps some 50 miles northeast of Milford Sound on the South Island of New Zealand. Due to its isolated location and relatively low grade, it is not expected to be exploited in the near future.

Asia

China: Little is known about the geology of Chinese asbestos deposits apart from a few short reports written by various mining and trade delegations (Mamen, 1973). However, various production estimates indicate that China ranks fourth in world production with 231,000 tons of fiber produced in 1978.

The Lai-yüan district of Hopeh, situated approximately 100 miles southwest of Peking, was for many years the only source of chrysotile in China.

Emphasis has recently shifted to the Shihmien area which is located about 250 miles west of Chungking in southwest Szechwan (the word *Shihmien* is translated as *rock cotton*). At least a dozen projects now contribute to the production from this area which totals more than 50,000 tpy of long and short chrysotile fiber.

A small amount of fiber is also recovered from the Chin-chou mine in Liaoning which lies some 265 miles east-northeast of Peking.

India: Indian asbestos production (private communication) is small and includes chrysotile of both ultrabasic and limestone origin, together with some amphibole varieties.

The small Roro deposit, located in southern Bihar State, about 175 miles west of Calcutta, is an example of chrysotile production. Here the ribbon fiber occurs in five separate discontinuous zones along highly serpentinized contact zones between massive serpentinized peridotite bands and massive to sheared "clot" peridotite. One of these, the Upper Zone, is economic and contains from 7% to 15% fiber over minimum mining widths. The ore is hand-cobbed and sorted for milling.

Japan: Chrysotile asbestos is mined at two localities in the Yamabe district of Hokkaido, Japan (Rowbotham, 1970). The fiber is generally short and all of the production is utilized locally.

Chrysotile in Dolomite

In addition to the large deposits of chrysotile asbestos found in serpentinized peridotite there are comparatively small tonnages mined from serpentinized dolomitic limestones (duToit, 1946; Hall, 1930; Pelletier, 1964; Rowbotham, 1970; Van Biljon, 1964). Such fiber is often of high quality and is free of the magnetite found associated to a varying degree with deposits where the original host rock is of ultrabasic igneous origin.

Chrysotile of this type has been mined from the Carolina district in the Transvaal and in the Salt River and Sierra Ancha regions in Arizona where it is found in narrow bands associated with diabase sills. The only chrysotile of this variety known in the USSR occurs near Aspagash just south of Krasnovask on the Yenesei River.

Crocidolite and Amosite

Both crocidolite and amosite occur in South Africa (Cilliers and Genis, 1964; Cilliers et al., 1964; duToit, 1946; Frankel, 1953; Hall, 1930; Keep, 1961; Pelletier, 1964) in bedded metamorphosed sedimentary formations known as banded ironstones, that may be classed as ferruginous quartzite or probably more exactly as an iron-rich silicified argillite. Cross fiber veins are found as closely spaced ribbons roughly conformable with the bedding which in some localities is distorted and steeply dipping. Crocidolite veins are generally less than 2 or 3 in. thick with the bulk of the fiber between ¼ to ¾ in. in length. Amosite veins on the other hand often range up to 10 or 12 in. thick.

Crocidolite is found over a large area of Cape Province occurring in a belt of the Lower Griquatown series of the Transvaal system. One of the oldest producing areas is in the vicinity of Prieska on the Orange River with notable production coming from Westerberg and Koegas, located a few miles down the river. Another producing area of blue asbestos in the Cape Province is farther north in the Kuruman district, and at Pomfret, just south of the Botswana border.

Approximately 200 miles northeast of Johannesburg both crocidolite and amosite are found in similar formations in the vicinity of Pietersburg in northern Transvaal. In some places here the two varieties have been known to occur side by side in the same vein. Such a vein, having fibers composed of crocidolite at one end and amosite at the other, is referred to as a "doublet." It is interesting to note that there is generally no weakening of the fibrils at the point of contact of the two varieties of fiber.

In the Pietersburg area a quite silky-fibered phase of amosite called *montasite* is found which is often a dusty blue in color and has sometimes been confused with crocidolite.

Widespread occurrences of amosite and crocidolite are known to occur in the Bababudan Hills north of Chikmagalur in Mysore, India. Both varieties of fiber occur in an assemblage of banded ironstones and shales similar to those of northeastern Transvaal but the amosite in particular is noticeably shorter. However, one area of amosite has been intensively investigated for possible production.

Deposits of crocidolite are found in the Hammersley range of western Australia, as well as in Bolivia and elsewhere.

Amosite-bearing banded ironstone crops out for a distance of over 20 miles near Penge in the Lydenburg district of the Transvaal. The fiber-bearing "reef" is being worked at three points along the strike, at Penge, and at the Weltevreden and the Kromellenboog mines. A good cross section of the country rock has been exposed by diamond drilling to 1000 ft in depth. The fiber-bearing banded ironstones and associated sedimentary members dip southwestward at about 20°. This sequence has been intruded by thin persistent sills of dolerite which are conformable with the bedding.

The following column is from a 1200-ft diamond drill hole intersecting the fiber-bearing horizon at a depth of 1000 ft below the surface at the "Amosa" mine which is part of the Penge property:

Shales
Dolerite sills
Shales
Bevetts conglomerate
Banded ironstone (main fiber zone)
Dolerite sill
Banded ironstone (sometimes fiber-bearing)
Dolomite

The Bevetts conglomerate is considered an important "marker" for the fiber-bearing formation.

In addition to the rock series in the foregoing, diabase cuts vertically across the whole formation at intervals. Evidence as to possible relationship of these dikes to the genesis of the amosite mineralization is inconclusive.

With the steady depletion of known amosite ore reserves, extensive exploration for additional ore was initiated a few years ago. Considerable success was achieved by utilizing a novel structural approach developed by the Cape Asbestos Group (private communication).

Careful mapping of the main marker horizons with respect to detailed topographic contours disclosed a system of broad, gentle north-south folds having amplitudes of about 50 ft and wavelengths of 4000 to 5000 ft between crests. A second, superimposed pattern of east-west cross folds was also detected having similar amplitude and a wavelength approaching 7000 ft. Careful structural mapping further indicated that each of the known or producing deposits was located on the loci of intersections of either anticlinal or synclinal structures.

Consequently, by drilling similar loci, at least five new fiber occurrences were discovered and the sixth failed only because the precipitous terrain in the area prevented drilling directly on target. Nevertheless, a drill hole angled toward the target from the nearest accessible location did disclose traces of fiber on the periphery of the target.

This record of successful drilling gives considerable weight to the hypothesis of a structural control for the genesis of amosite fiber. Similar patterns of cross-folding have been noted in many crocidolite deposits in the Northwest Cape and elsewhere.

It should be remembered, however, that the main prerequisite for any fiber development is to have the necessary chemical components for a specific fiber variety. Without the required chemical constituents, no amount of folding, faulting, or alteration can produce any fiber.

It is interesting to consider this structural origin for the genesis of chrysotile fiber as well, because the effects of faulting, folding, and dilation by shearing or intrusion almost invariably occur in close proximity to fiber occurrences.

Anthophyllite, Tremolite, and Actinolite

Production of amphibole asbestos (Rowbotham, 1970), other than crocidolite and amosite, includes significant amounts of anthophyllite, a small quantity of tremolite, and token amounts of fibrous actinolite.

Of the world's known anthophyllite deposits, Finland has been by far the most important producer. However, the famous deposits at Paakkila in the parish of Tuusniemi near Outokumpu in eastern Finland, as well as those of Maljasalmi in Kuusjarvi Parish some 10 miles to the east have now been mined out and no further production is anticipated.

The anthophyllite occurred as a series of lenses of amphibolitized and serpentinized ultrabasic material originally thought to be dunitic in composition. These lenses appear to have become detached from larger bodies during an early period of severe deformation. The serpentinized amphibolite lenses occur in a biotite gneiss which is intruded by Late Karelian granites and pegmatite.

The size of these bodies was generally small, varying from 125 to 200 ft in length and 30 to 80 ft in width with a potential of 40,000 to 100,000 tons of anthophyllite asbestos in each pod.

Exploration for these isolated bodies is difficult and often costly as the area is drift-covered and most of the ore lenses are not exposed at surface.

Present mining of these ore bodies is by open pit with the ore being processed at a centrally located mill.

Other amphibole asbestos is reported from many countries including Yugoslavia, Bulgaria, Brazil, Japan, India, and Taiwan. Little is known of the mode of occurrence of these deposits or their present production status.

A small quantity of anthophyllite is mined near Greenmountain, Yancey County in North Carolina. Production has been small and erratic but is reported to be increasing slightly. Similar material is known to occur in the nearby state of Georgia.

Italy continues to produce a limited quantity of long-fibered tremolite from small deposits located at Val Malenco, in the Sondrio district some 60 miles north of Milan. Reports of tremolite fiber are also known from the Aosta district north of Turin in the Italian Alps.

The Mavita anthophyllite mines of Mozambique, closed since 1975, are expected to reopen shortly pending the testing of a pilot plant (*Asbestos*, 1980).

Although some occurrences of fibrous actinolite are reported, production is extremely limited and of negligible value.

Principal Deposits

In the foregoing sections, the major asbestos deposits of the world have been briefly described. Global distribution of these deposits appears to be rather haphazard, but closer examination reveals that their occurrence generally coincides with the earth's major chains of mountain building of all ages. This is particularly true of the Urals, the Appalachians, and Rocky Mountains, but the relationship is more difficult to recognize in Precambrian fold belts.

Production

Of the world's two leading producers, both the USSR and Canada (Hendry, 1972) have vast reserves of chrysotile fiber; though, from the limited information available, it would appear that Russia has more reserves than Canada. Elsewhere in the world, reserves seem to be fairly evenly distributed among those countries presently producing, or those in which potential deposits are known (Table 7).

The production of fiber, including both

chrysotile and the various types of amphibole asbestos, increased steadily to a peak in 1978, but has declined since (Fig. 10). Of particular note is the lead taken by the Soviet Union in the mid-sixties. However, it is interesting to note that some authorities consider this figure to be biased by the inclusion of a greater proportion of very short fiber grades not presently recovered in Canada.

Russian production of medium and short fiber chrysotile is concentrated in the Central Urals where the large "Asbest" combine, together with smaller mines and mills at Alapajevsk, Nizhnitagil, Kiyembay, and south of Sverdlovsk, are capable of producing about 1.5 million tpy of fiber. Production has been steadily increasing over the last 25 years, both at existing mines and at new mills at Dzhetygara in northwest Kazakhstan and Ak-Dovurak in the Tuva ASSR. Other deposits such as Molodezhnoe and Ilchirsk in the Buryat SSR and those discovered at Bugetsaysk in Kazakhstan await development.

There are no less than six large mills in the "Asbest" combine, the sixth being commissioned recently. Combined capacity of these mills is now well over 1 million tpy. The first stage of the Dzhetygara combine in Kazakhstan was completed in 1965 with a productive capacity of 500,000 tpy. Since then a second stage of development and expansion has been completed.

The first stage of the "Tuvaasbest" combine at Ak-Dovurak came into operation in 1966. This was followed by further expansion to give a potential capacity of over 500,000 tpy of fiber.

The graph showing Canadian production has been climbing steadily over the last 20 years. This has been brought about by the enlarged capacities of many plants and the opening of several new mines such as Advocate, Reeves, Clinton Creek, and Asbestos Hill. During this same period several deposits have been depleted resulting in the closing of several properties including Munro, Nicolet, Flintcote, Reeves, Normandie, and Clinton Creek.

According to the latest statistics, the Eastern Townships of Quebec (which have been producing asbestos for the past 100 years) and Ungava now account for 1.482 million tpy of fiber. This represents approximately 89% of Canadian production, with the balance coming from British Columbia (7.8%) and Newfoundland (3.2%). The Jeffrey mine at Asbestos is the largest single producer, accounting for some 600,000 tpy of fiber. The balance of the production, amounting to approximately 750,000 tons, comes from a number of mines located in and around Thetford Mines.

African production ranks third in world production statistics. It should be remembered, however, that this production comprises 430,000 tons of chrysotile produced in Zimbabwe, the Republic of South Africa, and Swaziland, as well as 207,000 tons of crocidolite and 64,000 tons of amosite produced in South Africa. Zimbabwean production has increased appreciably in the last few years since the completion of a large expansion program at Shabani.

Recent mining of these ore bodies was by open pit with the ore being processed at a centrally located mill.
and 231,000 tons in 1978.

European production has shown only minor fluctuations over the past 15 years. Most of the mines are relatively small and production tends to fluctuate. Italy leads European production with 146,000 tons of fiber, most of which is mined at Balangero. French production has all but ceased with the closing of the Corsican asbestos mines.

The United States ranks eighth in world production with 103,000 tons in 1978 and a similar amount in 1979. Most of this production comes from California and Vermont with a small amount from chrysotile in dolomites mined in Arizona. Overall production took a noticeable increase in 1963 and 1964 when the Coalinga-New Idria deposits were first developed but this has dropped again with the closing of two of the mines in this area.

Most other countries show little change in overall production. Brazil, however, has risen to seventh in world production due largely to the opening of Cana Brava mine north of Brasilia at Uruaçú.

Australia ranks ninth in world production with all of its tonnage coming from the Woodsreef deposit near Barraba in New South Wales.

The production of crocidolite and amosite comes solely from the Republic of South Africa. Most of the crocidolite production comes from a number of small mines in the Prieska, Griquatown, and Kuruman districts of the Northern Cape but some production is also mined in the Pietersburg district of the northeastern Transvaal.

Varieties of amphibole asbestos, other than crocidolite and amosite, have a comparatively limited production. Of this, anthophyllite dominates the field, leaving a very limited quantity of tremolite and negligible amounts of fibrous actinolite.

Of the world's known anthophyllite deposits,

TABLE 7—Asbestos—World Production (1000 St)

	1954	1955	1956	1957	1958	1959	1960	1961	1962	1963	1964	1965
North America:												
Canada	924	1064	1014	1046	925	1050	1118	1174	1216	1277	1421	1388
United States	48	45	41	44	44	45	45	53	53	66	101	118
Mexico	—	—	—	—	—	—	—	—	—	—	—	—
Total	972	1109	1055	1090	969	1095	1163	1227	1269	1343	1522	1506
South America:												
Argentina	Tr	1	Tr	Tr	Tr	Tr	—	—	—	Tr	1	Tr
Bolivia	Tr	—	Tr	Tr	—	Tr	Tr	Tr	Tr	Tr	Tr	—
Brazil	3	3	4	3	4	4	4	3	5	1	—	1
Venezuela	1	2	5	8	9	5	4	1	—	—	—	—
Total	4	6	9	11	13	9	8	4	5	1	1	1
Europe:												
Austria	—	—	—	—	—	—	Tr	1	1	1	—	—
Bulgaria	1	1	1	1	1	1	1	1	1	1	1	1
Finland	8	19	8	10	8	10	11	10	11	10	12	13
France	14	14	9	16	21	23	29	31	28	27	24	11
Greece	Tr	Tr	Tr	Tr	—	—	—	—	—	—	—	—
Italy	26	35	39	38	40	53	61	63	61	63	76	79
Portugal	Tr	Tr	Tr	Tr	Tr	Tr	Tr	Tr	—	Tr	—	Tr
Spain	Tr	—	—	—	—	Tr	Tr	Tr	—	—	—	—
USSR[e]	375	450	500	500	550	600	660	880	1100	1200	1300	1500[e]
Yugoslavia	4	4	4	5	6	5	6	7	7	9	9	11
Total	428	523	561	570	626	692	768	993	1209	1311	1422	1615
Africa:												
Botswana	1	1	1	2	2	1	1	2	2	2	2	1
Eritrea	—	—	—	Tr	Tr	—	—	—	—	—	—	—
Kenya	Tr	Tr	Tr	Tr	Tr	Tr	Tr	Tr	Tr	Tr	Tr	Tr
Morocco	1	1	Tr	Tr	—	—	—	—	—	—	—	Tr
Mozambique	Tr	Tr	Tr	Tr	Tr	Tr	Tr	Tr	Tr	Tr	—	—
Zimbabwe	90	105	119	132	127	120	134	162	142	142	153	185[e]
South Africa	109	120	137	157	176	182	176	195	221	206	216	241
Swaziland	30	33	30	31	25	25	32	31	33	33	40	41
United Arab Republic	Tr	—	—	Tr	Tr	1	Tr	Tr	1	Tr	2	3
Total	231	260	287	322	330	329	343	390	399	383	413	471
Asia:												
Afghanistan	—	—	—	—	—	—	—	—	—	—	—	—
China[e]	15	23	26	33	66	90	90	100	100	110	130	140
Cyprus	15	15	15	15	16	14	23	16	22	20	14	18
India	Tr	2	1	2	1	1	2	2	2	3	3	5
Iran	—	Tr	2	Tr	Tr	—	—	—	—	—	—	—
Japan	7	7	10	13	11	14	17	19	15	18	18	16
Korea	Tr	Tr	Tr	Tr	Tr	Tr	1	Tr	1	2	1	2
Philippines	—	—	—	—	—	Tr	Tr	Tr	1	Tr	1	—
Taiwan	Tr	Tr	Tr	Tr	Tr	Tr	Tr	Tr	1	1	1	1
Turkey	Tr	Tr	1	Tr	1	Tr	Tr	Tr	1	Tr	1	1
Total	37	47	55	63	95	119	133	137	143	154	169	183
Australia	5	6	10	15	16	18	16	17	18	13	14	12
New Zealand	—	Tr	Tr	Tr	Tr	1	Tr	Tr	Tr	Tr	—	—
Total	5	6	10	15	16	19	16	17	18	13	14	12
Total World[e]	1677	1951	1977	2071	2049	2263	2431	2768	3043	3205	3541	3788

Source: 1954–1969, 1979 compiled from *US Minerals Yearbook,* US Bureau of Mines; 1970–1971 compiled from British Institute of Geological Sciences, and various trade journals; 1972–1978 compiled from Johns-Manville records and various trade journals.

1966	1967	1968	1969	1970	1971	1972	1973	1974	1975	1976	1977	1978	1979[e]
1489	1452	1510	1577	1662	1635	1629	1810	1824	1260	1664	1598	1568	1655
126	123	121	126	126	131	132	150	112	99	115	109	103	103
—	—	—	—	—	—	—	—	—	—	—	—	—	—
1615	1575	1631	1703	1788	1766	1761	1960	1936	1359	1779	1707	1671	1758
Tr	Tr	Tr	Tr	Tr	Tr	Tr	—	Tr	Tr	Tr	Tr	Tr	Tr
—	—	—	—	—	—	—	—	—	—	—	—	—	—
2	2	5	10	14	22	36	48	71	77	77	103	135	132
—	—	—	—	—	—	—	—	—	—	—	—	—	—
2	2	5	10	14	22	36	48	71	77	77	103	135	132
—	—	—	—	—	—	—	—	—	—	—	—	—	—
1	2	2	3	4	3	3	14	27	24	Tr	Tr	Tr	Tr
13	12	14	15	15	15	7	7	7	3	—	—	—	—
Tr	Tr	—	1	1	—	Tr	—	—	—	—	—	—	—
—	—	—	—	—	—	—	—	—	—	—	—	—	—
91	111	114	124	131	132	146	165	163	162	184	165	154	143
Tr	Tr	Tr	Tr	Tr	—	Tr	—	—	—	—	—	—	—
—	—	—	—	—	—	—	—	—	—	—	—	—	—
1720	1980	1980	2000	2100	2200	2425	1698	2205	2955	3080	3197	3197	2723
8	10	11	13	13	17	12	9	12	15	13	9	11	12
1833	2115	2121	2156	2264	2367	2593	1893	2414	3159	3277	3371	3362	2878
—	—	—	—	—	—	—	—	—	—	—	—	—	—
—	—	—	—	—	—	—	—	—	—	—	—	—	—
Tr	Tr	—	—	—	—	—	—	—	—	—	—	—	—
1	1	Tr	1	—	—	Tr	—	—	—	—	—	—	—
—	—	—	—	—	—	—	—	—	—	—	—	—	—
175	107[e]	95[e]	88[e]	88[e]	88	88	259	264	270	182	221	221	259
277	268	261	285	317	352	356	369	364	391	402	408	386	275
36	40	43	43	36	39	39	39	41	41	43	47	41	42
2	NA	3	—	—	—	—	—	—	—	—	—	—	—
491	416	402	417	441	479	483	667	669	702	627	676	648	576
—	—	—	—	—	—	—	—	—	—	15	14	14	4
150	165	165	180	190	176	220	138	198	199	232	231	231	276
24	19	21	24	28	25	31	35	31	35	37	39	39	40
8	8	10	11	11	12	14	8	8	5	5	5	8	22
—	—	—	—	—	—	—	—	—	—	—	—	—	—
21	27	24	23	24	24	16	9	5	5	9	8	7	9
1	2	4	7	2	—	2	7	7	7	5	7	15	15
1	Tr	Tr	Tr	1	—	—	—	—	—	—	—	—	—
1	1	1	3	3	—	3	5	4	2	2	3	2	3
1	2	4	6	2	2	4	5	18	18	11	9	11	11
207	224	229	254	261	239	290	207	271	271	316	316	327	380
13	1	1	1	1	1	1	49	37	61	66	56	69	77
—	—	—	—	—	—	—	—	—	—	—	—	—	—
13	1	1	1	1	1	1	49	37	61	66	56	69	77
4161	4333	4389	4541	4769	4874	5164	4824	5398	5629	6142	6229	6212	5801

NA Not Available
e estimate
Tr Trace

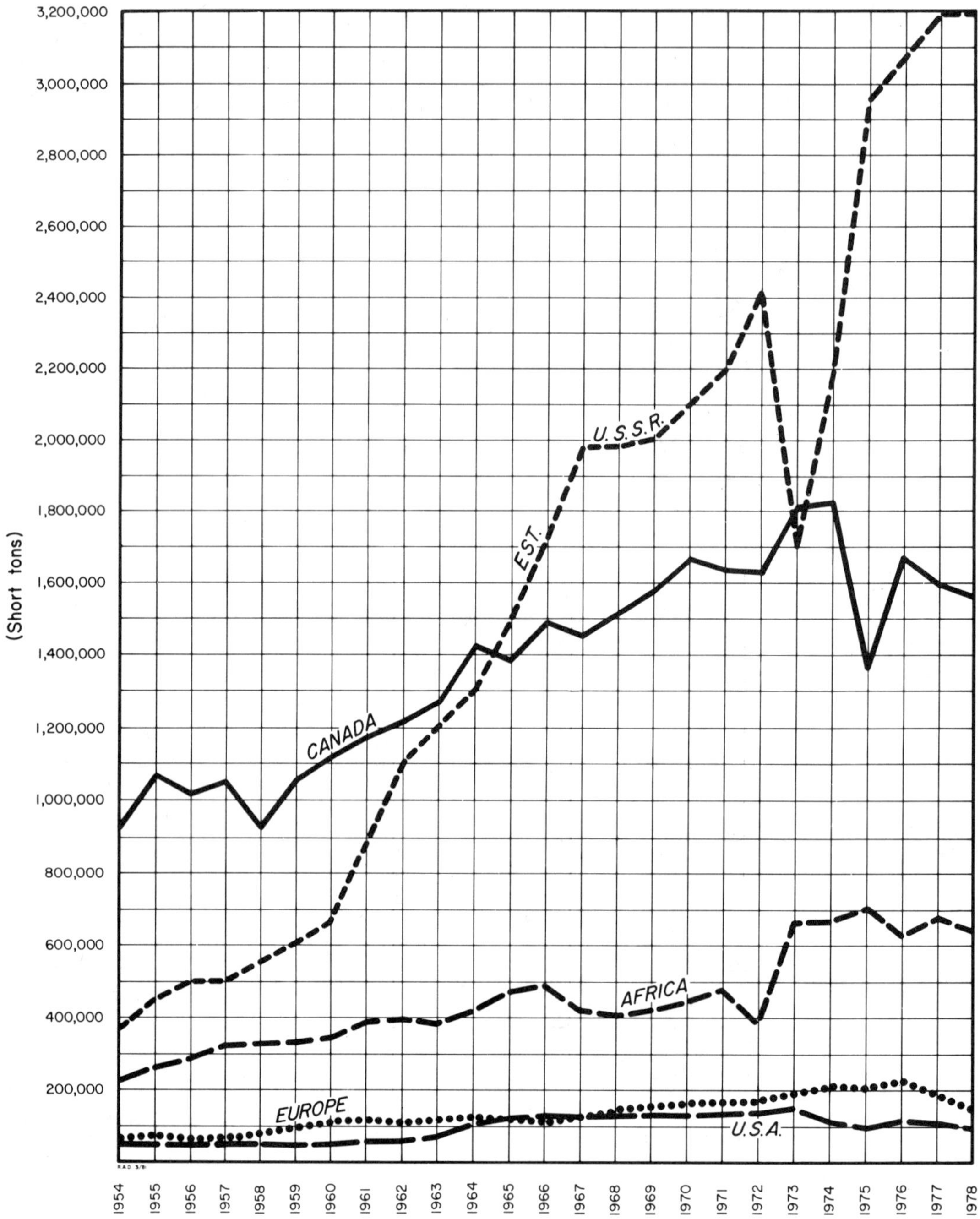

FIG. 10—*Asbestos production, major areas.*

Finland dominated the field with an annual production of 15,000 tons in 1971, but over the next four years production dropped progressively until the mines closed in the mid-seventies. Yugoslavia, Bulgaria, Brazil, Japan, and India are all reported to have some amphibole asbestos production, but details of these deposits and pertinent statistics are unavailable.

A small quantity of long-fibered tremolite is produced from small deposits in the Sondrio and Aosta districts in northern Italy.

Table 8 gives the total world production of all types of asbestos by countries for the years 1977–1980. Similar statistics for crocidolite and amosite are also shown but comparative figures for anthophyllite production are unavailable.

Exploration and Evaluation

The following observations concerning exploration and evaluation apply mainly to chrysotile deposits and to those, in particular, which occur in ultrabasic rocks, and are essentially a review of current methods employed in Canada.

Geophysical

Owing to the lack of outcrop within areas of ultrabasic rocks, or within a belt in which these rocks are expected to occur, both aeromagnetic and ground magnetic surveys (Conn, 1967; Low, 1951) are often employed in the early stage of exploration for asbestos. Ground magnetic surveys may be used to check and define

TABLE 8—Approximate Production of All Types of Raw Asbestos by Countries (St)

	1977	1978	1979	1980
USSR	2,899,000	2,689,600	2,524,300	2,645,500
Canada	1,663,400	1,571,900	1,654,600	1,410,900
Republic of South Africa	419,100	284,400	274,500	295,400
People's Republic of China	275,600	275,600	275,600	275,600
Zimbabwe	300,900	274,500	286,600	289,900
Italy	165,300	154,300	148,800	134,500
Brazil	102,500	135,600	152,100	159,800
United States	101,400	102,500	98,100	97,000
Australia	56,200	69,400	87,100	91,500
Swaziland	49,600	40,800	37,500	37,500
Cyprus	40,800	37,500	39,700	44,100
Korea.	6,600	15,400	16,500	15,400
Afghanistan	15,400	15,400	4,400	*
Yugoslavia	9,900	11,000	11,000	11,000
Turkey	12,100	14,300	11,000	11,000
Japan	7,700	6,600	4,400	4,400
India	5,500	6,600	6,600	6,600
Taiwan	3,300	2,200	3,300	2,400
Total	6,134,300	5,707,600	5,636,100	5,532,500
Amphibole Asbestos Production Included in Above Tabulation				
Amosite				
Republic of South Africa: Transvaal	73,800	44,700	43,100	56,900
Crocidolite				
Republic of South Africa: Cape Province, Transvaal	221,600	151,300	130,400	120,800
Tremolite, Actinolite, and Anthophyllite				
Republic of South Africa	600	—	—	—
Finland, Yugoslavia, US, Japan, Brazil, Bulgaria, India, Italy	16,500[e]	16,500[e]	13,200[e]	13,200[e]
Total, All Amphiboles	312,500	212,500	186,700	190,900

* No production figures available.
[e] Estimate.

in more detail anomalies obtained by an airborne survey of a large area, or the ground survey alone may be used for the purpose of exploring a small area.

In the case of an airborne survey, flight lines are normally spaced at ¼-mile intervals, and flown as close to 500-ft elevation as possible. Where the terrain is rugged, helicopters are used which are capable of maintaining a constant altitude of 300 ft above ground level. The spacing between profiles on a ground survey is usually 200 or 300 ft and readings are taken at intervals of 50 or 100 ft.

Magnetic surveys are used to locate and define the areas of ultrabasic rocks and, within these, the areas which have been subjected to extensive serpentinization. This is possible because this type of alteration produces a higher content of secondary magnetite. Asbestos deposits in ultrabasic rocks are a result of intensive serpentinization, and for that reason asbestos veining is usually accompanied by a higher concentration of magnetite than is normally found in the barren serpentine. It follows, therefore, that magnetic anomalies obtained over an area of ultrabasic rocks are favorable places to explore for asbestos. Figs. 11 and 12 show the results of ground magnetometer surveys carried out over two known ore bodies.

Modern instrumentation has made great strides in recent years, and it is now possible to conduct precise surveys with small lightweight magnetometers, in contrast to the cumbersome equipment used in the past.

Diamond Drilling

Diamond drilling is normally employed to probe beneath the overburden to assess and define the limits of an asbestos deposit. As asbestos ore bodies are usually large in volume, it is customary to drill vertical holes on a grid pattern. In the initial stages of exploration an interval of 400 ft and sometimes more may be used, filling in to an interval of 100 ft or even less, where an asbestos-bearing zone is encountered. In cases where a deposit is elongated in one direction, holes are generally spaced at closer intervals across the strike. Narrow, tabular targets are best explored by angle holes planned to give the attitude and true thickness of the body.

Care should be exercised to test the area with one or two preliminary drill holes to determine

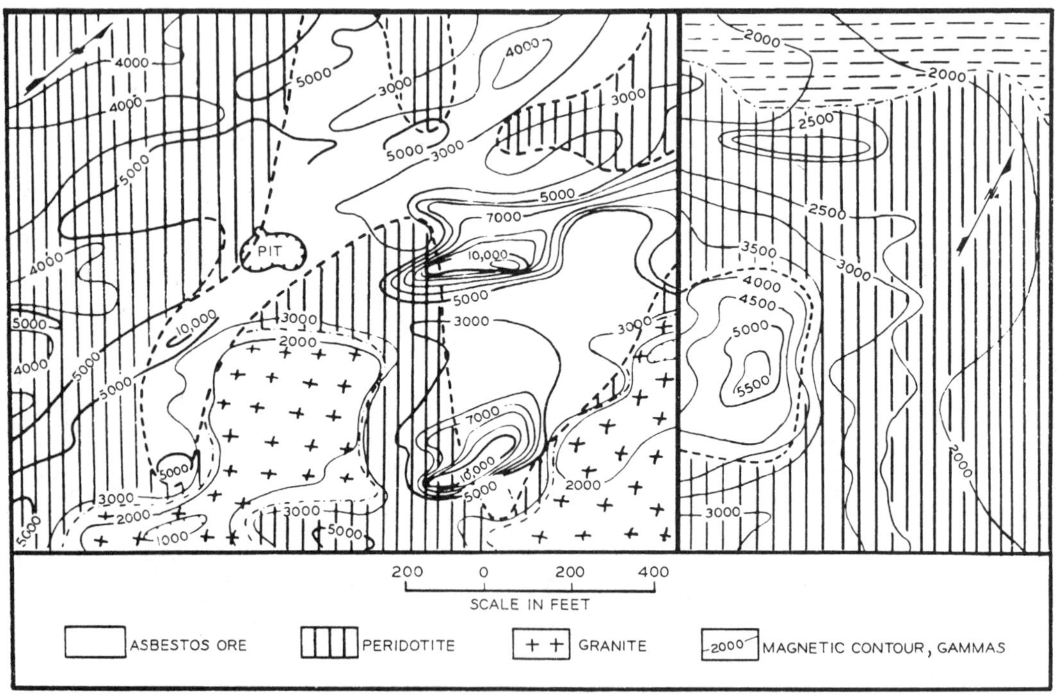

FIGS. 11 and 12—*Results of ground magnetometer surveys of two known ore bodies.*

whether the fiber has a preferential vein angle or not. If the angles of the vein intersections appear random, no changes need be made to the program. If, on the other hand, the deposit exhibits prominent vein angles which diverge from the average 45°, then the attitude of all later drill holes should be changed to correct for this variation.

In the past, drill evaluation programs frequently used small diameter, AX (1⁹⁄₃₂ in.) or EX (¹⁵⁄₁₆ in.), core sizes for evaluating any deposits where shearing was minimal and where holes did not exceed 300 ft. Experience has shown, however, that large core sizes such as NX (2⅛ in.) or BX (1⅝ in.), are preferable for better depth penetration, more geologic data, and to aid in fiber logging and dry milling of the core.

Wire-line drill equipment and the use of non-rotating core barrels are also recommended to minimize fiber loss by grinding of the core during drilling.

Steps are taken to recover the sludge only where core recovery is poor, which is generally the case with slip fiber occurrences. Because of the tendency of the fiber to fluff up and remain in suspension, much greater settling tank capacity is required than is the case when recovering sludge from other minerals. Care must also be taken to avoid contamination by grease and vegetable matter as these cannot be burned off without damaging the fiber.

In regions where permafrost conditions are expected, every precaution should be taken to avoid freezing the string of drill rods down the hole. The use of suitable low-freezing drill lubricant media such as a concentrated brine solution is adequate to allow drilling in most areas of permafrost. However, due to the high percentage of fiber cuttings in the water return, the brine solution has to be renewed at frequent intervals—a factor which considerably increases the cost of drilling asbestos prospects in permafrost areas. A number of successful drill ventures under permafrost conditions have been completed, including Asbestos Hill which showed the frozen ground to persist to at least 1000 ft below surface.

In areas where drilling is impractical, exposure by trenching or exploration beneath the surface by adit or shaft and lateral workings may offer the only means of assessing a deposit.

Evaluation

The evaluation (Conn and Mann, 1971; Dean and Mann, 1968; Oughtred, 1952) of any asbestos deposit entails the determination of its size, as well as the grade and quality of its fiber. Dimensions of a mineral zone are established by conventional methods, such as mapping, magnetic surveys, trenching, and diamond drilling. The value of the contained asbestos fiber is dependent on numerous physical properties such as fiber length, strength, flexibility, harshness, and color, besides the actual amount of fiber present.

The determination of grade cannot be based on a simple chemical analysis as both the fiber and the wall rock have essentially the same chemical composition. To avoid the complete crushing of the rock and physical separation of the fiber into different lengths, a method of visual evaluation has been developed which requires careful enumeration of the total number of fiber veins, together with the average length of fiber in each vein. These fiber lengths, if expressed in sixteenths of an inch, give an approximation of the grade, each ¹⁄₁₆ in. vein of fiber in a 5-ft section being approximately 0.1% fiber.

As the price of fiber varies considerably depending on its length, the use of grade based solely on its percentage fiber content is of little significance. Instead, the product of these two variables, expressed in dollars per ton of rock, offers a far more meaningful value which can be used for direct comparison in the final evaluation.

Drill Core: The method of visual evaluation is ideally suited to the evaluation of drill core.

Each vein of cross fiber is logged and the length carefully measured and recorded in multiples of ¹⁄₁₆ in. (Adoption of the metric system based on multiples of 1 mm is also acceptable.) As veins are often of irregular width, the average width of each vein should be estimated. Some veins are of a composite nature carrying partings, or the fiber may have kinks which cause it to break into shorter lengths. The fiber in some veins may be at right angles to the vein walls, in others sharply inclined. Allowance must be made for these conditions in arriving at the true lengths of fiber.

The more precise determination of fiber content calls for the measurement of each vein angle in the core to permit individual volume corrections. However, in practice it is usually assumed that the random angle is 45° and a factor of 1.414 (inverse of the sine of 45°) is used. However, this can be quite misleading if the average angle is rather small, so logging results should be checked against laboratory recovery and bulk sampling wherever possible.

As a rule, the visual reading gives an indicated lower yield of a higher value fiber than a corresponding mill test. This is to be expected as the visual readings disregard fiber lengths of less than 1/16 in. which are, of course, recovered in milling and are important in grading. Furthermore, a certain amount of pulverized host rock adheres to the fiber, further reducing the grade and increasing the yield by 20% or more. On the other hand, there may be some loss of veins in the core and the veins themselves contain foreign material such as nonfibrous serpentine, picrolite, and magnetite. There is also a tendency, on the one hand, for some breaking of fibers to occur in the mining and milling process while, on the other hand, the fluffed up milled product tends to remain on a screen which an unopened fiber would pass through. For most practical purposes, the discrepancies are compensating and, when a suitable volume correction factor is used, the ore values found by the two methods are generally quite close.

Slip fiber is often associated as a minor constituent in deposits made up mainly of cross fiber, and in others the slip fiber may be the predominant type. It is not easy to determine the slip fiber length by the normal visual methods of logging core; however, its presence may be recorded separately by vein widths and the percentage slip fiber determined in the same manner as used for cross fiber. Laboratory assistance is almost essential in evaluating slip fiber deposits.

Regardless of the method used in evaluating asbestos ore, the final answer should give the value of the fiber, the yield and, from this, the value of the ore in dollars per ton. A bulk sample properly milled and graded readily supplies the answer, whereas a laboratory test normally gives only the yield of an ungraded fiber and, therefore, of unknown value. The visual method, after due allowance for vein angles, gives the yield only and further calculations are required in order to approximate the fiber and ore value.

In order to place a value on the ungraded fiber resulting from laboratory extraction a graph is used (Fig. 13). The curve on this graph is derived from the Quebec Standard Tests (described elsewhere in this chapter) for average test production of all standard grades by weighting the fractions found in each box to give each grade a "point rating." The prices for Quebec asbestos are based on a minimum shipping test but for practical purposes the average test of production is about 20% higher. Weighting factors for the fraction are 100 for +½ in., 40 for −½ + 4 mesh, 10 for −4 + 10 mesh, and 1 for −10 mesh. These have no real significance and have been chosen only to control the shape of the curve.

As an example, a fiber testing 5.7 oz on the ½ in., 7.5 on the 4 mesh, and 2.3 on the 10 mesh with 0.5 in the pan would have a point rating of:

$$\begin{aligned} 5.7 \times 100 &= 570 \\ 7.5 \times 40 &= 300 \\ 2.3 \times 10 &= 23 \\ 0.5 \times 1 &= - \\ \hline & 893 \text{ points} \end{aligned}$$

and from the curve the fiber value would be about $455 per ton (1973 prices).

In applying this same procedure to the visual method, it is necessary to correlate the observed lengths with the corresponding Quebec Standard Test. In order to do so, it is assumed that fibers of ½ in. length and more will stay on the ½ in. screen, ¼ to 7/16 on the 4-mesh screen, and ⅛ and 3/16 in. on the 10-mesh, whereas the 1/16 is assumed to end up in the pan. A synthetic Standard Test can then be arrived at by simple arithmetic.

To illustrate this procedure, assume that a core length of 85 ft contains four veins of ½ in. length, two of 9/16 in., one of ⅝ in., two of ⅞ in., and one of 1 in.; 16 of ¼ in., three of

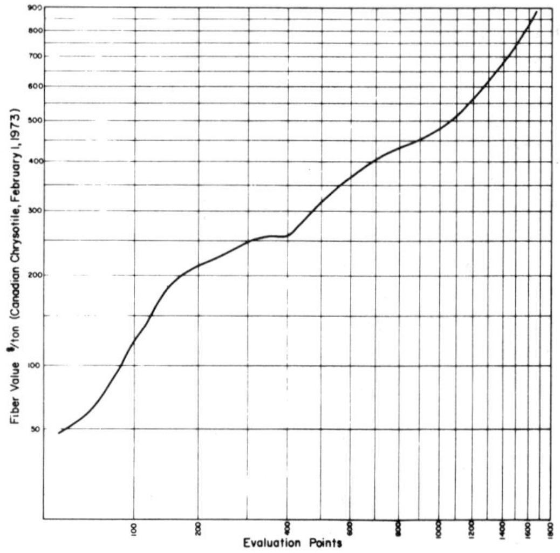

FIG. 13—*Graph for evaluating ungraded fiber based on 1973 prices. (Reproduced by permission of Asbestos Corp. Ltd.)*

$\frac{5}{16}$ in., two of $\frac{3}{8}$ in., and two of $\frac{7}{16}$ in.; 23 of $\frac{1}{8}$ in. and 10 of $\frac{3}{16}$ in.; 74 of $\frac{1}{16}$ in.

Reducing everything to a common denominator of $\frac{1}{16}$th, there would be 104 such sixteenths of $\frac{1}{2}$ in. and over, 105 of $\frac{1}{4}$ to $\frac{7}{16}$ in., 76 of $\frac{1}{8}$ and $\frac{3}{16}$ in., and 74 of $\frac{1}{16}$ in., totaling 359 in all.

In 85 ft there are $85 \times 12 \times 16$ sixteenths; therefore, the uncorrected yield is

$$\frac{359}{85 \times 12 \times 16} \times 100 = 2.2\%.$$

A correction must be made next for the average vein angle which here will be assumed at 45°:

$$\text{Corrected yield} = \frac{2.2}{\sin 45°} = 3.1\%.$$

Next to arrive at the synthetic Standard Test:

½ In.	4 Mesh	10 Mesh	Pan	Total
104	105	76	74	359 sixteenths
104/359 × 100	105/359 × 100	76/359 × 100	74/359 × 100	100%
4.6	4.7	3.4	3.3	16.0 oz

Then to find the fiber value from the graph:

$$\begin{aligned} 4.6 \times 100 &= 460 \\ 4.7 \times 40 &= 188 \\ 3.4 \times 10 &= 34 \\ 3.3 \times 1 &= 3 \\ \hline & 685 \text{ points} \end{aligned}$$

From the graph the fiber value is found to be \$400 a ton, and from this the indicated ore value is calculated to be 3.1% of \$400 or \$12.40 per ton (1973 prices).

Face Readings

On the surface or in underground workings, channel sampling may be employed. On the other hand, it is also possible to log these surfaces in a similar manner to that used for drill core. One method is to take a linear reading along either wall of a drift or crosscut and another is to take cross sections at intervals of 5 ft on the back and both walls to ensure that veins running parallel to the drive are not excluded.

An alternative method is to record all the veins in the face, walls, and back after each round. In order to arrive at a percentage, a factor based on the area involved is applied to the reading for each surface, and an average percentage for the round is determined in this manner.

The fact, however, that the rock tends to break along one or more of the numerous fault planes present rather than across the fault bounded blocks to expose the veins within them, makes it difficult to obtain representative results.

Bulk Sampling

The various methods of logging and sampling outlined seldom give entirely dependable results. The visual methods of logging will usually produce dependable results only where there is a relatively low content of shorts. Slip fiber, as already noted, presents a problem in this connection. Even laboratory results, which are dependent on complete extraction, usually give an appreciably higher fiber value than that obtained in a conventional mill. It is not easy to simulate in a laboratory the conditions to be found at an operating mine where the fiber from the time of blasting to the final product is subjected to a good deal of handling, some of which is rather severe. The fibers, as a consequence, suffer some breakage in the process.

Bulk sampling is often resorted to as a means to check and to arrive at a suitable factor to be applied to drill core data. This may be done by diamond drilling a block of ground at close intervals prior to mining and milling. The core is then read visually and treated in the laboratory. Provided precautions are taken to avoid contamination and the sample is sufficiently large, results should be reasonably reliable despite the erratic distribution of fiber in the rock.

A 12,000-ton bulk sample from one deposit indicated that a yield factor of two would be necessary to bring the "visual" ore value in line with that found by milling. In this deposit there were approximately equal amounts of slip and cross fiber and it is observed that the visual determined yield is low and the fiber value exaggerated.

	Fiber Value per Ton, $	%	Ore Value per Ton, $	Yield Correction Factor
Mill recovery	111	6.80	7.53	1.00
Laboratory recovery	164	4.59	7.70	0.98
Visual analysis	202	1.87 *	3.78	1.99

* No correction for vein angles.

A 50,000-ton sample assessed in the same manner also indicated a factor of two but showed that the fiber value determined visually was fairly close to that found in milling. This is attributed to the absence of slip fiber which, when present, makes the estimation of fiber value more difficult.

	Fiber Value per Ton, $	%	Ore Value per Ton, $	Yield Correction Factor
Mill recovery	153	2.75	4.20	1.00
Laboratory recovery	160	2.52	4.03	1.04
Visual analysis	148	1.37 *	2.03	2.07

* No correction for vein angles.

Tensile Strength

In considering the preceding paragraphs, it is important to point out that the measurements made and the evaluations so obtained are based entirely on the length of the fiber which is today a secondary factor in fiber value. Most asbestos-cement fiber grades sold today are valued for the strength they lend to cement or other mixes and a standard scale of Strength Units has been established.

It is imperative then that any evaluation of a chrysotile deposit include Strength Unit evaluations which are laboratory tests involving the testing of an asbestos-cement tile made with the subject fiber. Once the inherent strength of the fiber from a particular ore body has been determined it is usually possible to equate strength to length distribution and dust content measurements.

Tonnage and Grade

To estimate the tonnage and grade of a deposit from diamond drill core data, individual drill holes may be weighted according to their interval using the polygonal method. An alternative procedure is to use cross sections, or groups of cross sections and, by weighting the individual holes in each section, determine the average grade for each section.

A third method employs contoured cross sections, wherein the contouring is based on a reasonable interpolation of the intervening area between drill holes. This method permits the estimator to make use of all available geological information in his interpolation. In open pit operations contoured horizontal sections may be prepared in this manner to correspond with expected mining level intervals, and these serve as a useful guide to mining. Separate horizontal sections contoured for rock value and fiber value per ton permit the mine operator to produce a more balanced mill feed with respect to both fiber content and grades.

Exploitation

Mining

The development of asbestos mining in Canada has been accompanied by a succession of mining practices and equipment. The simplest hand methods were, quite naturally, first employed in shallow open pit workings. During one period, when pits were reaching considerable depth, quite elaborate overhead cableway derricks were popular. Modern power shovels loading into heavy-duty trucks have now supplanted all other loading and transporting equipment where open cast methods are employed. Underground methods which have been used in the past include glory holes, shrinkage and sublevel stoping, and block caving.

In some cases, scraper hoists drag the broken ore from the underground drawpoints and in others it is passed directly through grizzlies. Large primary crushers may be located underground. The trend that prevailed in Canada, in the late 1940s and early 50s when most of asbestos was mined by underground methods, has been reversed and now open pit mining prevails.

It is generally considered that surface mining is more advantageous in recovery, grade control, economy, and safety. The reversing of that trend is mostly due to the technological innovations in mining equipment: shovel size varies from 2½ to 15-cu-yd capacity, truck size goes from 25 to 200-ton payload, rock drilling equipment has improved considerably, and new blasting agents and techniques have been introduced. These innovations permit increased waste to ore ratio up to 3:1 or more, thus lengthening the economic life of open pits.

In Africa a large part of all the chrysotile is mined by underground methods. The ore bodies are generally tabular in shape with a pronounced dip, with the result that the economic limit for quarry mining was reached at a comparatively early stage. Ore widths in the larger mines commonly range from 80 to 200 and up to 400 ft. Some ore bodies, notably in the Shabani district of Zimbabwe, are quite extensive in length; in one case, development has extended for three miles along the strike, and is being developed or diamond drilled to over a thousand feet in depth. Several underground methods have been used including cut and fill. Presently, sublevel stoping and caving accounts for most of the tonnage extracted. Caving may be initiated by blasting holes drilled upward from sublevel crosscuts, starting first on the hanging–wall side and retreating over a considerable width toward the footwall, or the development and retreat may be along the strike of the ore. In some cases high pressures have developed from an arching effect but these pressures have been largely released by mining a vertical slot which may extend through to surface and toward which the ore is allowed to cave. In the sublevel stoping method a slot may also be opened across the center of the ore body. The long holes which are fanned out from the sublevel drifts are then blasted toward the slot, and mining proceeds as a systematic retreat in two directions away from the opening thus made.

Drilling in the softer portions of these South African chrysotile mines is now frequently accomplished by rotary drills employing bits in which tungsten carbide cutting lugs are set. In drilling long holes, which may be up to 70 ft or more for sublevel mining, short lengths of jointed rods are used. With percussion type drills, which are generally used for harder rock than the rotary drills, flexible ribbon steel with a chisel bit of tungsten carbide has proven useful and efficient.

Much of the amosite and crocidolite in Africa has been mined from small, narrow open cuts or adits following the fiber-bearing band. In the Penge area the operations have been developed for larger scale and deeper mining. The main ironstone "reef" runs quite consistently at around 54 in. thick with a dip of about 20°. A sloping face is advanced along the strike by drilling rounds of about 6 ft with jackhammers. The broken material is shoveled back from the face into temporary pillars for roof support. The remaining material is hand-loaded into small cars, hoisted to a haulage level, and transported to a sorting plant for further removal of waste rock before being sent to the cobbing plant and mill. Formerly the fiber-bearing portion was hand-sorted underground which resulted in excessive values being left behind in the pillars. The ironstone fractures into rectangular blocks which lend themselves well to being built up in this way. The main fiber-bearing reef commonly contains up to 30% asbestos, so that the proportion of waste to barren works out well for the system used. The method is locally referred to as shrinkage stopage but should not be confused because of this nomenclature with the latter system as commonly practiced in other areas.

Milling

Mill flow lines are designed to suit the characteristics of a given ore body. The basic steps are similar for most flow lines. The raw fiber is recovered using the dry process which has been used almost exclusively since the inception of asbestos milling in Canada some 90 years ago. However, in certain cases, wet processing of at least some of the fiber has occurred. This method of processing may be examined more closely in the future.

There have been few major changes in the basic process since the last volume of this series was published in 1975. The major improvements have evolved around material handling, packaging, and shipping methods. The improved methods developed by the asbestos mining industry have made it possible for both the producer and customer to store, ship, receive, and use fiber in a relatively dust-free environment.

The major portion of the production is being packaged in woven polylaminate 50 kg pressure-packed bags for the domestic and export markets.

Fiber in water-soluble unsized kraft paper bags is also provided to some customers who consume the bag, along with the fiber in wet paper-making production processes.

Recent developments in packaging equipment have permitted the packaging of fiber in high density blocks, approximately 1 cu ft in volume, containing 45 kg. Some producers are supplying small tonnages of limited grades in these blocks which, because of their size, offer some economics in transportation and storage.

Unitized bags, with coverings of stretch or shrink wrap plastic film for additional stability and protection in rail and container shipments, are supplied by some producers. The present trend is toward 100% mechanical handling of the asbestos at both the mines and point of use.

The flowsheet shown in Fig. 14 is a generalization of the essential steps and equipment used in a Canadian chrysotile asbestos mill producing the full range of raw asbestos fiber grades Group 3 to Group 7.

Fig. 15 shows a section of a Canadian asbestos mill primary rock and fiber cleaning circuit.

Fig. 16 illustrates a schematic flow line of a two-product Canadian asbestos mill.

Fig. 17 illustrates a recent Canadian installation of a primary and secondary crushing, drying, and ore concentration system.

The description that follows of the essential steps of a milling operation is based on Figs. 14 and 17.

Mill feed is derived from the underground or pit operation. Primary crushing may be done in underground stations in the case of an underground mine or surface plant when treating open pit ore.

Large jaw crushers (the 48 x 60-in. size is the most popular) are used in the majority of the Canadian plants either underground or on surface. Heavy-duty pan feeders have proven quite practical and comparatively trouble-free, particularly for handling open pit ore containing large unbroken pieces which are likely to cause blockages. Lately the trend has been toward large capacity gyratory crushers which eliminates the need for feeders and bins.

In countries where relatively cheap labor is still available a certain amount of hand sorting continues to be done before or after (or both) primary crushing for the purpose of removing barren rock and also to recover pieces of the larger asbestos veins to be used in the production of No. 1 or No. 2 crudes. Reduction in primary crushers of the size mentioned is normally to about 7 in., and is followed quite commonly by 7-ft or smaller cone crushers with settings ranging from ¾ to 2½ in.

Ore concentration is an important step in the milling of asbestos ore and is particularly important to the lower grade ore bodies. It is not uncommon to discard up to 40% of the mine ore through selective impacting and screening in the primary and secondary crushing circuits. Nearly all producers are able to accomplish beneficiation of this type to some degree. The use of magnetic pulleys is a relatively new approach being used by some producers for upgrading the mine ore. Not all asbestos ore bodies are amenable to this type of separation.

The next step is to dry the ore. It is considered good practice to provide a storage large enough for one shift's production of mine ore to assist in regulating the flow rate to the dryers.

Most of the moisture is contained in the −1½ in. fraction. At some plants this fraction is screened and dried separately from the oversize (1½ to 7 in.).

In all stages of beneficiation it is important to minimize abrasion and cutting of the fiber particularly in the earlier stages before the longer and more easily released fiber fractions are separated from the gangue.

The selection of a dryer is largely based on local preference. The two more commonly used are the rotary and vertical dryers. Recently an installation has been made using fluid bed dryers.

The chief advantage of the vertical tower and fluid bed over the rotary kiln-type dryer is normally a lesser degree of mechanical damage to the fiber. On the other hand, the rotary dryer is preferable and is also more effective for open pit ores containing snow and pieces of ice.

Where conditions permit, oversize (say, +1½ to 3 in.) material may be separated before drying, and bypassed to lessen the pebble-milling action of rotary dryers. Where vertical dryers are used the feed should not be much coarser than a 1½ in. grizzly opening size in order to avoid blockages in the gridwork of arresting bars with which the dryers are fitted.

There is a limit to the air velocity which may be used in the counterflow vertical-type dryer where the products of combustion from an oil or coal-fired furnace rise through the falling ore. Near the bottom of the tower, where the hot gases enter, fluffed particles of fiber may be temporarily held in suspension in a damaging temperature environment.

One solution to this objectionable feature is to have at least part of the hot gases introduced at the top of the tower and to be drawn downward with the falling material.

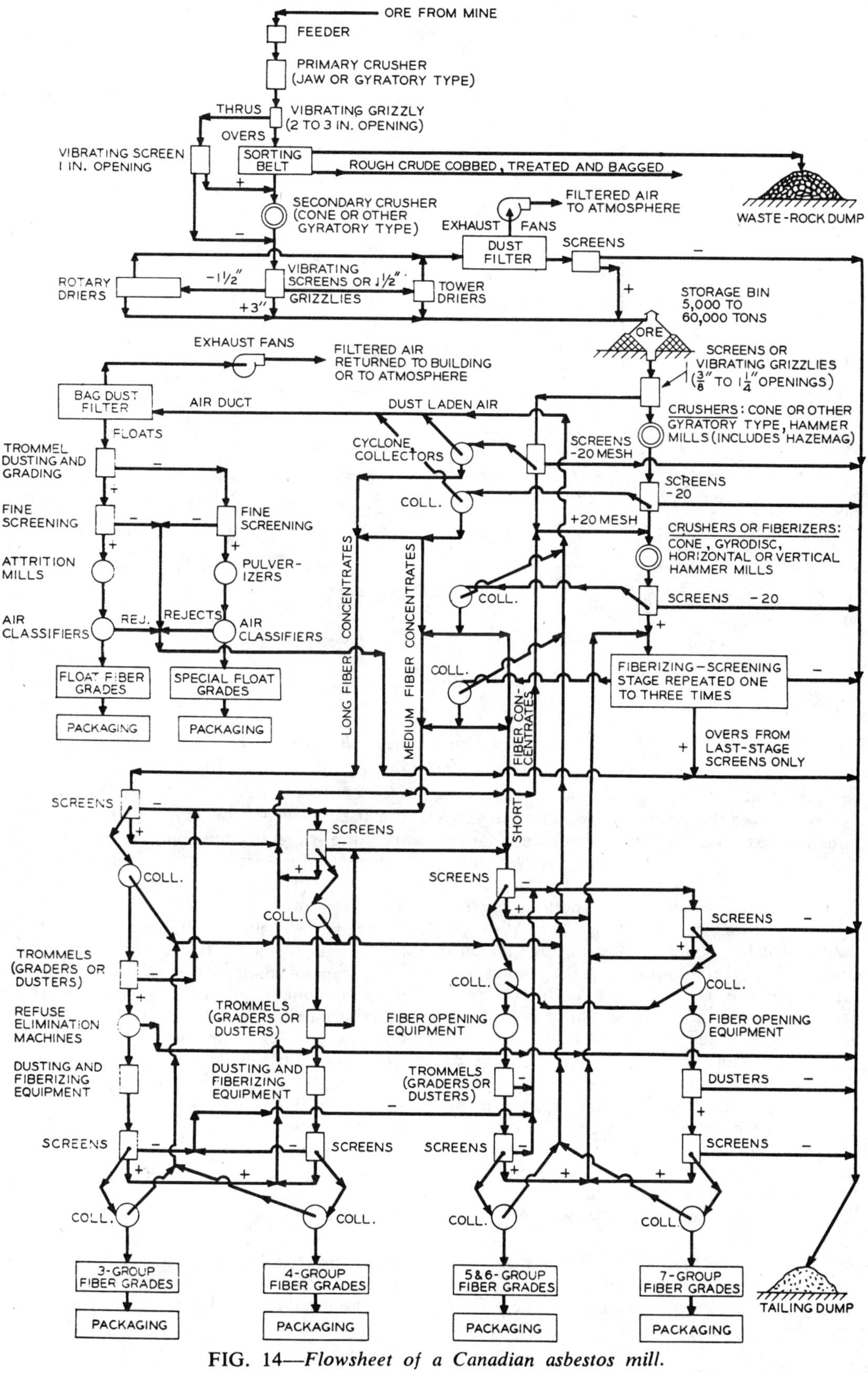

FIG. 14—*Flowsheet of a Canadian asbestos mill.*

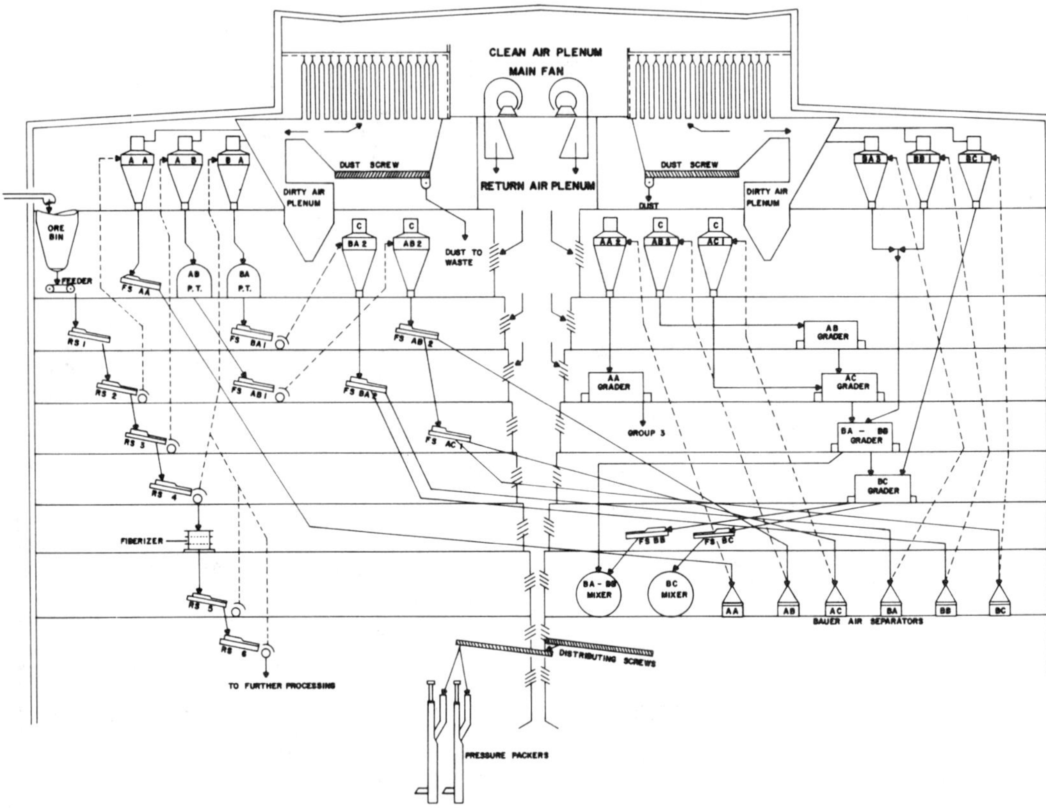

FIG. 15—*Section of a Canadian asbestos mill showing the primary rock line and fiber cleaning circuit.*

Dryers using the practical application of this principle have the discharge gas takeoff at a point midway up the height of the dryer, thus keeping air velocities and temperature within permissible limits toward the bottom of the tower. An after-cooling device may also be used in which cool air is blown through the discharged hot ore to remove further moisture before the ore is discharged onto conveyor belts. The air from this operation may be used as preheated makeup air in the dryers.

The air and products-of-combustion as discharged from the dryers carry a considerable volume of dust. It is desirable, therefore, to provide filtering equipment in order to keep the plant area clean. Bag filters using various fabrics have operated successfully and are certainly the most positive means of removing the dust. Care must be exercised, however, to guard against condensation.

An alternative filter which has been used is known as the Impingo. This employs an ever-changing layer of cold moist aggregate, or the ore itself screened to, say, +1-in. size. Some moisture is precipitated from the gases and dust adheres to the surfaces of the crushed stone or pebbles. This system has proven ineffective in most applications.

It has been found desirable to have substantial storage capacity between drying and the main stage of milling. Besides assuring an adequate and more uniform supply of ore to the mill, further drying (up to 1% moisture loss) takes place during the storage. The size of storage facilities will vary with mill capacity.

As will be apparent from the flowsheet (Fig. 14) the release and separation of fiber from gangue is accomplished by successive stages of crushing or comminution by impact. Impactors are designed to release the fiber from the host rock and at the same time produce a minimum of fines.

Fiber in the form of cross-fiber veins or slip fiber thus released is at the same time sufficiently "teased out" so that the portion so freed may be lifted by air suction, leaving most of the

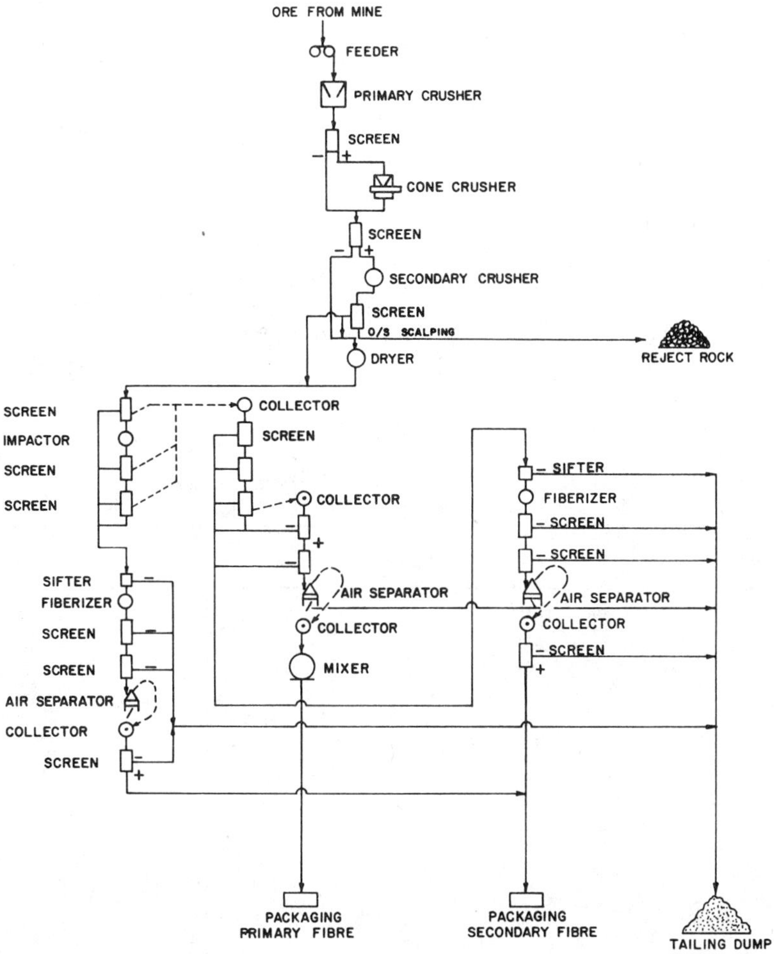

FIG. 16—*Schematic flow of a two-fiber-product Canadian asbestos mill.*

rock as a reject to go to the next stage of impacting and eventually to tailings.

Finer fractions are generally screened out prior to air separation. Otherwise a large portion of rock fines would be aspirated with the fiber. Rotary aspirators designed to present a uniform layer of material to the aspirating hood are commonly used to improve the aspirating efficiency. The products of these first stages of separation may be considered as concentrates. They contain a large percentage of rock particles and must be further treated in the fiber grading division of the plant.

Here the concentrates are further graded and undergo a series of cleaning operations for the purpose of removing sand and dust. Screens, trommels, specific gravity air separators, and modified forms of these such as graders and dusters, further clean the fiber and separate it into standard grade lengths.

In the grading mill the fiber is further separated into the approximate quality brackets required for specific grades and is subjected to several stages of screenings using shaking screens, rotating screens, conventional trommels, trommel-like graders, and rotary dusters.

When well opened or fluffed out grades are called for, the fiber is subject to special treatment in one or more of a variety of machines ranging from graders or Willows (fixed shell trommels having a rotating center shaft to which beater arms are attached) to one of several types of high-speed hammer mills, disk grinders, or pulverizers. The type of machine or machines used depends upon the length and type of fiber to be processed and the degree of opening or fluffing up required. This additional treatment is generally given to the shorter fibers.

A large volume of air is used in an asbestos

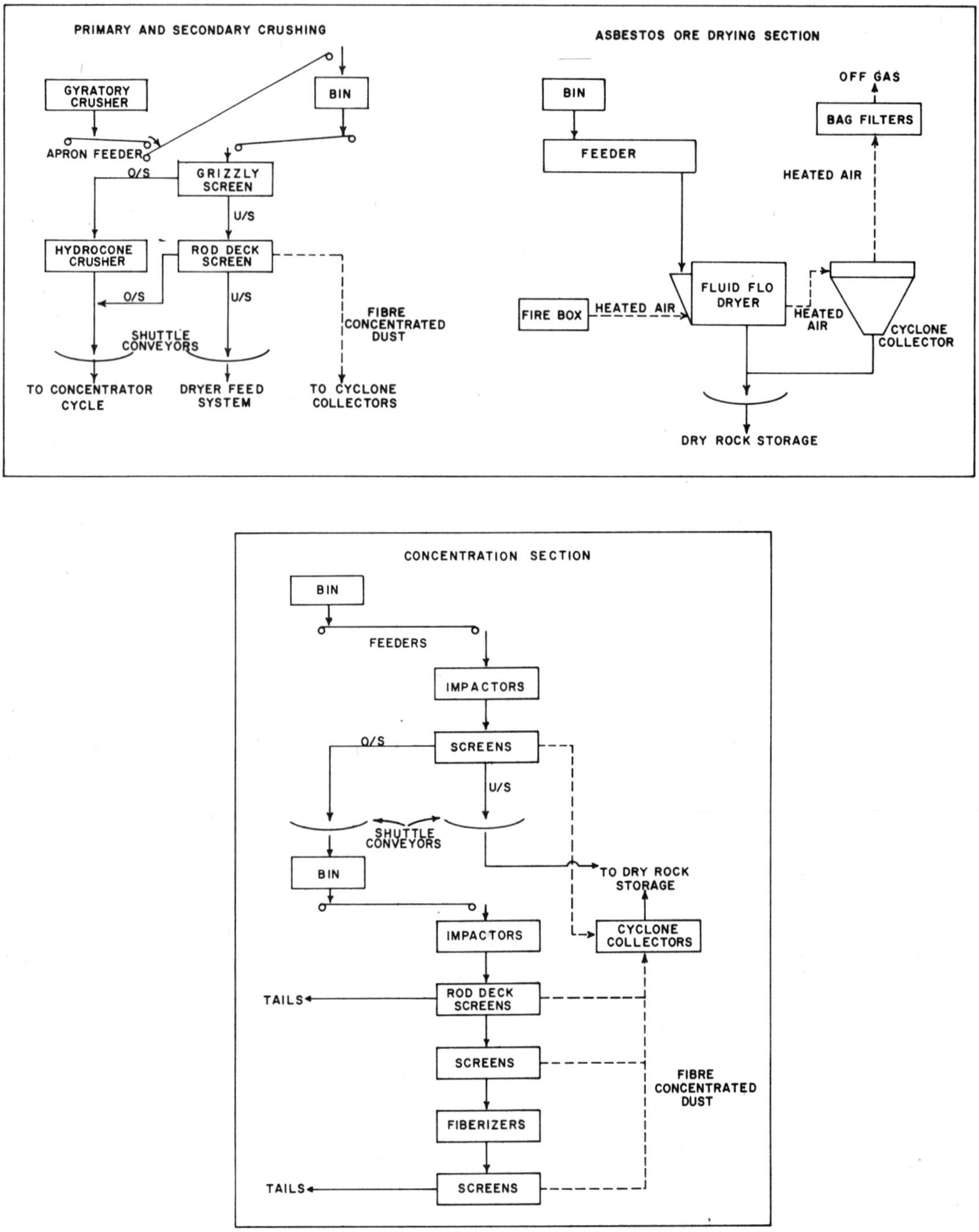

FIG. 17—*Recent Canadian installation of a primary and secondary crushing, drying, and ore concentration system.*

mill, most of which is for the primary purpose of separating fiber from rock, and dust by air suction and classification.

An additional quantity is also required for dust elimination at many points within the building to keep the dust content of air in the building to a minimum. It is estimated that mills having a capacity of, say, 5000 tpd of ore, use from 500,000 to over 800,000 cfm of free air.

The average consumption in Canadian mills is something in the order of 7 to 14 tons of air per ton of ore treated (Kula and Wiser, 1970; Rozovsky, 1957).

Modern practice requires all this air to be filtered before being discharged from the mill into the open. This has led to extensive bag filter units being installed in nearly all mills in recent years.

In Zimbabwe, South Africa, and Swaziland, where there are several modern mills, one finds that the general approach to the beneficiation of chrysotile does not differ greatly from Canadian practice.

One feature of African milling is careful control of moisture content as this may actually fall too low, particularly during the dry season. Fiber may be more easily broken or damaged under extremely dry conditions.

In Zimbabwe the small pan crusher or Chilean mill was one in extensive use. This is no longer widely used for general milling but has been adopted to some extent in Canada for the preparation of crudes.

African practice, for the preparation of crocidolite and amosite, is generally very simple, consisting to a large degree in hand sorting, crushing, and screening as with the "crude" grades in Canada. However, in the Penge area of the Transvaal, milling practice is somewhat more elaborate and several milled grades of amosite are produced.

In addition to the more or less conventional sorting and some very difficult cobbing due to the extreme toughness of the rock, the fiber is treated in different mill circuits variously equipped with jaw crushers, cone crushers, rolls, hammer mills, and trommels. As an example the longest grade, D_3, is prepared essentially as follows:

3 x 14 in. grizzly
↓
Reduction to 1½ in. in small jaw crusher
↓
Crushing rolls (24 x 48 in.) to ¾ in.
↓
Trommel—1½ in.—undersize to next shorter grade circuit
↓
Crushing rolls (12 x 36 in.) to ½ in.
↓
Trommel—¾ in. (as above)
↓
Degritting trommels (2 in parallel)—¼ in.
↓
Bagging

In the circuits for shorter grades, cone crushers, hammer mills, and other fiberizing and cleaning equipment is now used. Trommels have proven effective as a cleaning device for medium as well as the long grades.

For the very short fibers high-speed vibrating screens are found more efficient. Air separation is practiced in various forms such as air traps, suction hoods, and screens, depending on the fiber grade being treated.

Tests and Specifications

The only test that is generally recognized by the government in Canada is the Quebec Asbestos Mining Association classification known as the Quebec Standard (Q.S.) Test. (This test is performed in the Quebec Standard Testing Machine which consists of a nest of three sieve boxes with ½ in. screen, 4-mesh, and 10-mesh screens, respectively, and a bottom box serving as a pan. A test is made with 16 oz of asbestos and the whole mechanically shaken for 600 revolutions at 328 rpm. The equipment is made to exact standard specifications, and can be purchased through the secretary of the Quebec Asbestos Mining Assn., Quebec City, Prov. of Quebec, Canada, G1R 2K2. The Quebec Asbestos Mining Assn. has set specifications for the various fiber grades showing the minimum limits for the quantities of fiber that should remain on each screen. It is used as a production control and serves as a specification of the grade of fiber for sale.

This type of test has also been partially adopted by other countries, and in some cases the results are expressed in percentages instead of ounces.

Other variations of the Quebec Standard Test at some localities involve the Ro-Tapping of the Q.S. fractions for further evaluating purposes, because it was believed that the Ro-Tap would give a better distribution of the fiber lengths.

The Ro-Tap test is used by many asbestos mill operators to obtain a fiber distribution impossible to get by the Quebec testing machine. The use of the Ro-Tap differs with operators at different plants. It is not a standardized test accepted by the industry, and therefore the type and size of screens vary, as well as the lengths of time of Ro-Tapping and the quantity of fiber used in making a test.

In many cases, a cubic-foot box or a fraction of a cubic-foot box is used for obtaining the density of fibers. When the box is filled with the fiber, it is weighed and the values expressed as pounds per cubic foot. Many customers make density measurements as a guide for the introduction of the fiber in a specific product.

Wet classification and screening have been employed to an increasing extent in recent years. Equipment such as the McNett or the Clark wet testing machine may be used for production purposes permitting the operator a much closer check on fiber lengths, nonfibrous fractions, and dust content.

Many tests are made at the special request of customers who in turn may evolve their own best methods to evaluate the fiber in terms of their finished products. This has resulted in tests such as: viscosity, absorption, grit, troweling, bulk, penetration, soluble salts, magnetite content, compressibility, flexing, tensile strength, color, strength units, surface area, and others being made.

Fiber producers in other countries such as Zimbabwe, USSR, Australia, Cape Province, and Transvaal, South Africa, have evolved their own testing methods for grading and classifying fibers, but do use the Quebec Testing Machine to a certain degree.

The classifications for various grades from the more important producing areas are given in Tables 9–14.

Uses of Asbestos

The uses of asbestos are numerous and at least 3000 different products are being manufactured in the world today which take advantage of its unique properties. Some of the major

TABLE 9—Quebec Grading as Set Forth by the Quebec Asbestos Mining Assn.

SPECIFICATION

Group No. 1

No. 1 Crude-cross-fiber veins having 3/4-in. staple and longer.

Group No. 2

No. 2 Crude-cross-fiber veins having 3/8-in. staple up to 3/4-in. Run-of-Mine Crude consists of unsorted crudes. Sundry Crudes—consist of crudes other than above specified.

	Guaranteed Minimum Shipping Test			
Group No. 3 (Commonly referred to as textile or shipping fibers)	1/2 In., Oz	4 Mesh, Oz	10 Mesh, Oz	Pan, Oz
3F	10.5	3.9	1.3	0.3
3K	7	7	1.5	0.5
3R	4	7	4	1
3T	2	8	4	2
3Z	1	9	4	2
Group No. 4 (Commonly referred to as asbestos cement fibers)				
4A	0	8	6	2
4D	0	7	6	3
4H	0	5	8	3
4H	0	5	7	4
4K	0	4	9	3
4M	0	4	8	4
4R	0	3	9	4
4T	0	2	10	4
4Z	0	1.5	9.5	5
Group No. 5 (Often referred to as paper stock grades)				
5D	0	0.5	10.5	5
5K	0	0	12	4
5M	0	0	11	5
5R	0	0	10	6
5Z	0	0	8.6	7.4
Group No. 6 (Paper and shingle fibers)				
6D	0	0	7	9
6F	0	0	6	10
Group No. 7 (Shorts and floats)				
7D	0	0	5	11
7F	0	0	4	12
7H	0	0	3	13
7K	0	0	2	14
7M	0	0	1	15
7R	0	0	0	16
7T	0	0	0	16
7RF and 7TF Floats	0	0	0	16
7W	0	0	0	16
Group No. 8 & 9 (Sands and gravels)				
8S	0	0	0	16
	Minimum 50 lb per cu ft			
8T	0	0	0	16
	Minimum 75 lb per cu ft			
9T	0	0	0	16
	More than 75 lb per cu ft			

TABLE 10—Cassiar Asbestos Grades, Northern British Columbia

C-1	Crude 3/4-in. staple and longer.	
AAA	Extra long spinning fiber	– Canadian Group 3
AA	Long spinning fiber	– Canadian Group 3
A	Spinning fiber	– Canadian Group 3
AC	Spinning fiber	– Canadian Group 3
CC	Spinning fiber	– Canadian Group 3
AK	Asbestos-cement fiber	– Canadian Group 4
CP	Asbestos-cement fiber	– Canadian Group 4
AS	Asbestos-cement fiber	– Canadian Group 4
CT	Asbestos-cement fiber	– Canadian Group 4
AX	Asbestos-cement fiber	– Canadian Group 5
AY	Asbestos-cement fiber	– Canadian Group 5
CY	Asbestos-cement fiber	– Canadian Group 5
AZ	Asbestos-cement fiber	– Canadian Group 6
CZ	Asbestos-cement fiber	– Canadian Group 6

TABLE 11—Classification of African Chrysotile*

Grade	Source
C&G1 Long, crudy textile fiber	From mines in the Shabani district of Zimbabwe
C&G2 Textile fiber	
C&G3 Long shingle fiber	
C&G4 Shingle fiber	
C&G5 Short shingle fiber or paper stock	
VRA-2 Textile fiber	From the Mashaba district of Zimbabwe
VRA-3 Long shingle fiber	
VRA-4 Shingle fiber	
HVL2 Textile fiber	From the Havelock mine in Swaziland
HVL3 Long shingle fiber	
HVL4 Shingle fiber	
HVL5 Short shingle or paper stock	
Msauli – Grade 4, Msauli – Grade 5 Shingle fibers	From the Msauli mine near Swaziland border in the Transvaal
Amianthus 1 and 2 – Textile fibers	From the Barberton district of the Transvaal
F – Long shingle fiber	
AA – Shingle fiber	
Munnik-Myburgh M1 – Textile fiber	From the Barberton district near Nelspruit of the Transvaal
M3 – Shingle fiber	
M4 – Short shingle fiber	

*** In the absence of a current listing of Zimbabwean, South African, and Swaziland grades, this list from the 1975 edition is included to indicate type and source.**

ones are listed, along with a brief discussion of the products involved (Badollet, 1948).

Crudes No. 1 and No. 2—Chrysotile crudes are usually processed by the customer to produce a long spinning fiber for use in textiles. A desirable fiber for textiles is one that has good flexibility, is soft, low in soluble salts and magnetite, and easily carded without an excess drop in shorts. It should also be free from wood and blasting wire or fuse wire! Fiber of this quality would be desirable for all textile uses, including those for the electrical industry. In some cases, this grade of fiber is used as felts in laminates along with resins to form a strong molded sheet for use in airplanes, boats, etc.

Crocidolite crudes must be carefully processed to produce long fibers which can be used in textiles, gaskets, ropes, or in laminates with resins.

Amosite crudes, after reprocessing, will give a long bulky fiber that is used in blanket insulation or in products requiring a low density and good insulation value. Spinning of amosite can be accomplished, but is difficult.

Group 3 Milled Fibers—Chrysotile fibers that meet this classification are generally used in textiles. Some are used in long fiber asbestos papers, packings, gaskets, brake linings, clutch facings, electrolytic diaphragms, pipe coverings, and insulating blocks. Some of the fibers of this group are now being used in laminates with resins.

Crocidolite may be used for most of the purposes listed previously, providing color is not an objection. It should generally be carefully prepared in a well-opened condition.

Group 4 Milled Fibers—Many grades of fibers in this classification are used in asbestos-cement processes to produce pipe, jackets, boards, sheets, and a variety of hand-molded articles. Other uses are for papers, pipe coverings, packings, gaskets, millboards, and plastics.

Crocidolite, similar to Group 4 chrysotile, is used in asbestos-cement pipes, in some packings and in some gaskets.

Amosite is used in magnesia blocks, pipe coverings, and other insulation compositions when a light density is desired. It is also used in acetylene cylinders to give strength to the calcium silicate mix during curing.

Group 5 Milled Fibers—Fibers of this group sometimes are used as replacements for Group 4 fibers, and therefore the products made with these fibers would include asbestos-cement sheets, corrugated or flat boards, pipe, electrical panels, papers, millboards, pipe coverings, gaskets, packings, brake linings, and plastics.

Crocidolites of this classification are used in asbestos-cement pipes.

Group 6 Milled Fibers—Chrysotile fibers of Group 6 are used in asbestos-cement shingles, flat sheets, corrugated sheets, boards, brake lining, papers, millboards, and plastics.

Crocidolite of this grading is used in the

TABLE 12—Classification of Russian Chrysotile (Condensed)

Grade	Type	Texture	Mark according to USSR Standards, 1972
0	Spinning fiber	Harsh, crudy	DV-0 80, DV-0 55
1	Spinning fiber	Harsh, crudy Semi-crudy	J-1-50, J-1-38 PRJ-1-75, PRJ-1-50
2	Spinning fiber	Harsh, crudy Semi-crudy Semi-open	J-2-20 PRJ-2-30, PRJ-2-15 P-2-30, P-2-15
3	Asbestos-cement fiber	Harsh, crudy Semi-open	J-3-40 P-3-70, P-3-60, P-3-50 M-3-70, M-3-60
4	Asbestos-cement fiber	Semi-open Open, soft	P-4-40, P-4-30, P-4-20, P-4-5 M-4-40, M-4-30, M-4-20, M-4-5
5	Paper fiber	Semi-open Open, soft	P-5-67, P-5-65, P-5-52, P-5-50 M-5-65, M-5-50
6	Paper and shingle Fiber and shorts	Semi-open Open, soft	P-6-45, P-6-30 M-6-40, M-6-30 K-6-45, K-6-30, K-6-20, K-6-5
7	Unguaranteed		7-300 7-370 7-450 7-520

production of asbestos-cement ML6 products. The balance would be classed as a filler and used wherever its physical properties could be employed to advantage.

Amosite of this classification is considered quite short for most purposes, and would probably be used as a cheap filler to free up asbestos-cement slurries prior to pressing.

Group 7 Milled Fibers—These fibers find usage in certain papers, cements, asphalt roof coatings, paints, welding rods, floor tile, and plastics.

TABLE 13—Classification of African Amosite

Grade	Approximate Range Average Fiber Length, In.
S 11	1–1 1/2
W 3	1/2–1 1/2
K 3	1/2–1 1/2
SK	3/16–3/4
S 33	1/8–1/2
S 33/65	1/8–1/2
GW	1/8–1/4
GK	1/8–1/4
S 44	1/8–1/4
RK	1/8–1/4
6605	1/16–1/8

The grades listed above are produced in the Penge area from the Penge mine and the Weltevreden and Kromellenboog mines. The fiber lengths shown are not necessarily exact, but have been included in order to give some indication of the relative lengths of fibers of the different grades.

Crocidolite and amosite fibers of a length equivalent to Group 7 chrysotile are not known to be used in commercial products unless as a cheap filler where color is not objectionable.

Floats—Chrysotile floats (Badollet, 1952, 1956) are used extensively in plastics and coatings.

The selection of asbestos fiber for a particular application will depend upon the processing method, as well as the desired properties of the end product. The following examples outline some of the major uses for asbestos fiber and illustrate pertinent factors which influence the choice of fiber grade and type.

Asbestos-Cement Products—Pressure pipe which must conform to hydrostatic test speci-

TABLE 14—Cape Blue (Crocidolite)

Grade	Approximate Range Average Fiber Length, In.	Typical Values Surface Area (Rigden), Cm^2/g
C	1 1/4–1 3/4	1,500
S	1/4–3/4	5,500
S 80	1/4–3/4	9,500
P 25	1/4–3/4	8,300
H	1/8–1/2	7,500
H 80	1/8–1/2	10,000
713	1/8–1/2	13,000
WDS	1/8–3/8	9,000

Cape Blue Mines (Pty) Ld. produce most of the Blue asbestos. Among other producers grading practice is to refer to "long" and "short" for the bulk of their production.

fications is produced from high quality Group 4 fiber, usually a blend of chrysotile and crocidolite, to ensure a good modulus of rupture. On the other hand, the flexural and impact strength requirements for asbestos cement sheets can usually be met by using a Group 6 fiber. Formulations for corrugated sheets generally include some Group 5 material to improve adhesion of the wet sheets during the forming process.

In all asbestos cement products made by the wet machine process, drainage is an important fiber characteristic since it has direct bearing on the production rate. For this reason, preference may be given to fast filtering fibers and amosite may be included in the formulation as a filter aid.

The requirements for shingles made by a dry process are not so exacting from either the strength or drainage point of view and can be satisfied with a lower quality 6 or 6–7 blend.

Asbestos Paper—Traditionally asbestos fiber in conjunction with an organic binder has been used for the manufacture of paper and millboard which, in turn, were converted to roofing felt, pipe coverings, electrical insulations, and many other products. Various blends of Groups 4, 5, and 6 fibers are used for this application, depending upon the desired strength and porosity of the paper.

More recently, a latex-asbestos process has been developed in which a long clean Group 7 fiber having a high surface area is coated with latex rubber by a chemical precipitation method and the resulting finish formed into a continuous sheet on a paper machine. A large proportion of this latex asbestos paper is used as an underlayment for vinyl-rolled floor covering. A lesser amount is used in the manufacture of gaskets.

Friction Materials—This product line cannot be related to any particular fiber group, since it spans the complete spectrum from Group 3 spinning grades to the shorter Group 7.

The explanation lies in the wide variety of products, which fall within the general classification of "friction materials," and the equally diversified manufacturing processes involved.

Clutch plates are made from an asbestos open-weave cloth impregnated with resin and bonded to a steel disk. A similar product can be manufactured by molding a dry resin-fiber blend under conditions of high temperature and pressure onto a packing plate. For the first method, a Group 3 fiber is required, whereas the molding process utilizes a Group 5.

Automobile brake linings bonded to a steel shoe are usually made from Group 7 fiber in a semi-wet extrusion process while heavy blocks for railcars and large vehicles use Group 5 or 6 fiber dry-molded and machined to finished dimensions.

Group 5 fiber is also used extensively in disk brake pad formulations.

Sheet Packing—Latex asbestos paper made from Group 7 fiber can be densified and used for gasketing, but most sheet packing material is formed on a sheeter machine by a calendering process. This latter method requires a longer fiber in the Group 4 to 5 range, which has been cleaned and opened. The fiber is blended with natural or synthetic rubber, plasticizers, and other ingredients in a high shear mixer to form a dough which is later calendered into sheets of various thicknesses.

Floor Tile—A large volume of Group 7 fiber is supplied to manufacturers of vinyl floor tile. This product requires a short, clean, well opened Group 7 fiber having a high degree of uniformity in such properties as length distribution, absorption, and color. This latter feature is particularly important to the industry because of the need for matching of shades from different production runs.

Asphalt Products—Group 7 asbestos fiber in combination with asphalt and various solvents form the basis of a wide variety of products often classified under the catchall heading of "Blackline."

These include spray or brush-on roof coatings, sound deadeners for automobile body panels, and caulking components.

In recent years, automobile underbody protective coatings applied by airless spray equipment have provided an outlet for Group 7 fiber. Since the finished compound must pass through an orifice 0.021 to 0.028 in. in diameter under high pressure during application to the automobile, the fiber used must meet strongest specifications on fineness and viscosity building properties.

Short Group 7 chrysotile asbestos added to hot asphalt paving mix helps to improve characteristics of toughness, flexibility, and water impermeability. Asbestos modified pavements have proved successful in high traffic density areas, such as busy street intersections, bus stops, and bridge decks.

Caulking Compounds—Combinations of long asbestos with cement and other ingredients, along with waterproofing resins, are used to produce special types of caulking compounds.

In many cases, short asbestos and floats are also combined with various types of resins and

TABLE 15—Asbestos Prices—Canada, US

Standard Grade Designation		
Quebec, Canada	Per Ton of 2000 Lb, f.o.b. Mine Canadian Dollars	
Grades 3Z to 3F spinning fiber	$ 1202 to	$ 1995
Grades 4T to 4A asbestos cement fiber	850	1256
Grades 5Z to 5D paper fiber	497	687
Grade 6D paper and shingle fiber	408	426
Grades 7TS to 7D shorts	125	253
Paperbestos No. 1 to 5 H.D. Blocks	237	782
Cassiar, B.C. Canada	Per Ton of 2000 Lb, f.o.b. North Vancouver, B.C. Canadian Dollars	
AAA nonferrous spinning fiber: Canadian Group 3		$ 2000
AA nonferrous spinning fiber: Canadian Group 3		1600
A nonferrous spinning fiber: Canadian Group 3		1265
AC nonferrous spinning fiber: Canadian Group 3		1155
AK asbestos cement fiber; Canadian Group 4		1075
AS asbestos cement fiber; Canadian Group 4		1000
AX asbestos cement fiber; Canadian Group 4		930
AY asbestos cement fiber; Canadian Group 5		660
AZ asbestos cement fiber; Canadian Group 6		430
Specialty Grades AK100, AS100, AX100 and AY120 are increased an additional $25 per ton above list price of base grade.		
Vermont, US	Per Ton of 2000 Lb f.o.b. Morrisville, VT US Dollars	
Grade 3Z fiber		$ 1135
Grades 4T to 4D fiber	$ 754 to	833
Grades 5R to 5K fiber	495 to	584
Grade 6D waste		363
Grades 7R to 7D shorts	113 to	215
Hooke No. 1		1520
Hooke No. 2		880
California, US (Calidria Asbestos)	Per Ton of 2000 Lb, f.o.b. King City, CA US Dollars	
SG-103 pellets in bags; vinyl-asbestos floor tile		$ 136
SG-104 pellets in bulk; vinyl-asbestos floor tile		116
SG-130 open fiber in bags; general		194
SG-200 pellets in bags; rubber		176
SG-210 open fiber in bags; general		250
HPP pellets in bags; rubber reinforcement		224
HPO open fiber in bags; asphalt		250
RG-110 open fiber in bags; roofing		270
RG-144 open fiber in bags; resin viscosity control		700
RG-244 open fiber in bags; resin viscosity control		3300
Arizona, US	Per Ton of 2000 Lb, f.o.b. Globe, AZ US Dollars	
No. 1 crude (soft)		$ 3000
AAA		2000
Grade 3Z nonferrous filtering-plastic	$ 1350 to	1400
Grade 4T nonferrous filtering-plastic	700	1350
Group 7 white shorts	50	200

Source: Asbestos, January 1981.

TABLE 16—Annual Exports of Asbestos from Canada by Areas*
(St)

	1977	1978
United States	545,500	544,200
Europe including United Kindgom	1,006,400	898,700
South and Central America	39,200	37,700
Africa and Asia	472,100	392,600
Others	417,900	220,100

*Dollar value not available.

other materials to produce a soft plastic caulking compound that remains soft or it may be controlled so as to set up as a hard mass.

Plastics—Structural materials using plastics reinforced by asbestos or as a combination of asbestos and glass are now of considerable importance commercially. The asbestos may be in the form of a mat, or as paper or cloth to form laminates with resins such as polyesters, phenolics, thermosetting silicones, melamines, and furanes.

The use of long fiber chrysotile, crocidolite, and in some cases amosite, in the form of felts or papers and impregnated with resins produces a tough product of high strength and good heat resistance. These products have been used in aeroplane wings in England and in small sailing boats, radar scanner aerials, aircraft tanks, automobile bodies, and other products including rocket tubes, missile nose cones, and other parts.

In some cases, asbestos in the form of cloth or a millboard type is impregnated by resins to obtain a strong sheet for structural use.

Short Group 7 fiber and floats are also used extensively as a fibrous filler for the production of molded phenolic resin and polyester parts, such as automobile heater and air conditioner housings, electric kettle bases, and other appliance parts. In such applications, freedom from abrasive particles is especially important to hold die wear at minimum level.

Joint Filler—Another interesting use for short Group 7 fiber and floats is in the manufacture of joint filler cements and texture paints. Here again, fineness, and whiteness are critical properties of the fiber. Also, since the viscosity of the mix after the addition of a prescribed amount of fluid must be consistent from batch to batch, the absorptive capacity of the asbestos must be controlled within very narrow limits.

Prices for fibers from the major Canadian and American producing areas are listed in Table 15. Russian fiber prices vary with fiber

TABLE 17—Russia: Exports of Asbestos to Countries Importing over 1,000 t in Any One Year, 1974, 1975

	1974		1975	
	t	1,000 Roubles	t	1,000 Roubles
Austria	7.9	0.79	10.0	1.54
Belgium-Luxembourg	14.9	1.62	14.2	1.98
Bulgaria	32.6	4.23	28.8	4.55
Cuba	14.0	2.08	8.8	1.40
Czechoslovakia	28.0	4.15	31.2	5.53
France	54.4	5.86	63.2	9.23
East Germany	44.7	5.14	59.4	9.35
West Germany	18.5	2.19	17.9	2.60
Hungary	26.0	3.32	27.9	4.44
India	35.4	5.09	29.2	4.62
Italy	1.2	0.15	13.4	2.38
Japan	66.5	8.01	68.3	10.68
Poland	74.9	9.61	76.2	12.50
Romania	21.9	2.94	34.3	5.80
US	—	—	10.7	2.56
Yugoslavia	32.0	4.49	33.7	6.31
Others	55.3	6.89	89.3	11.05
	528.2	66.56	616.5	97.63

TABLE 18—Supply-Demand Relationships for Asbestos, 1979

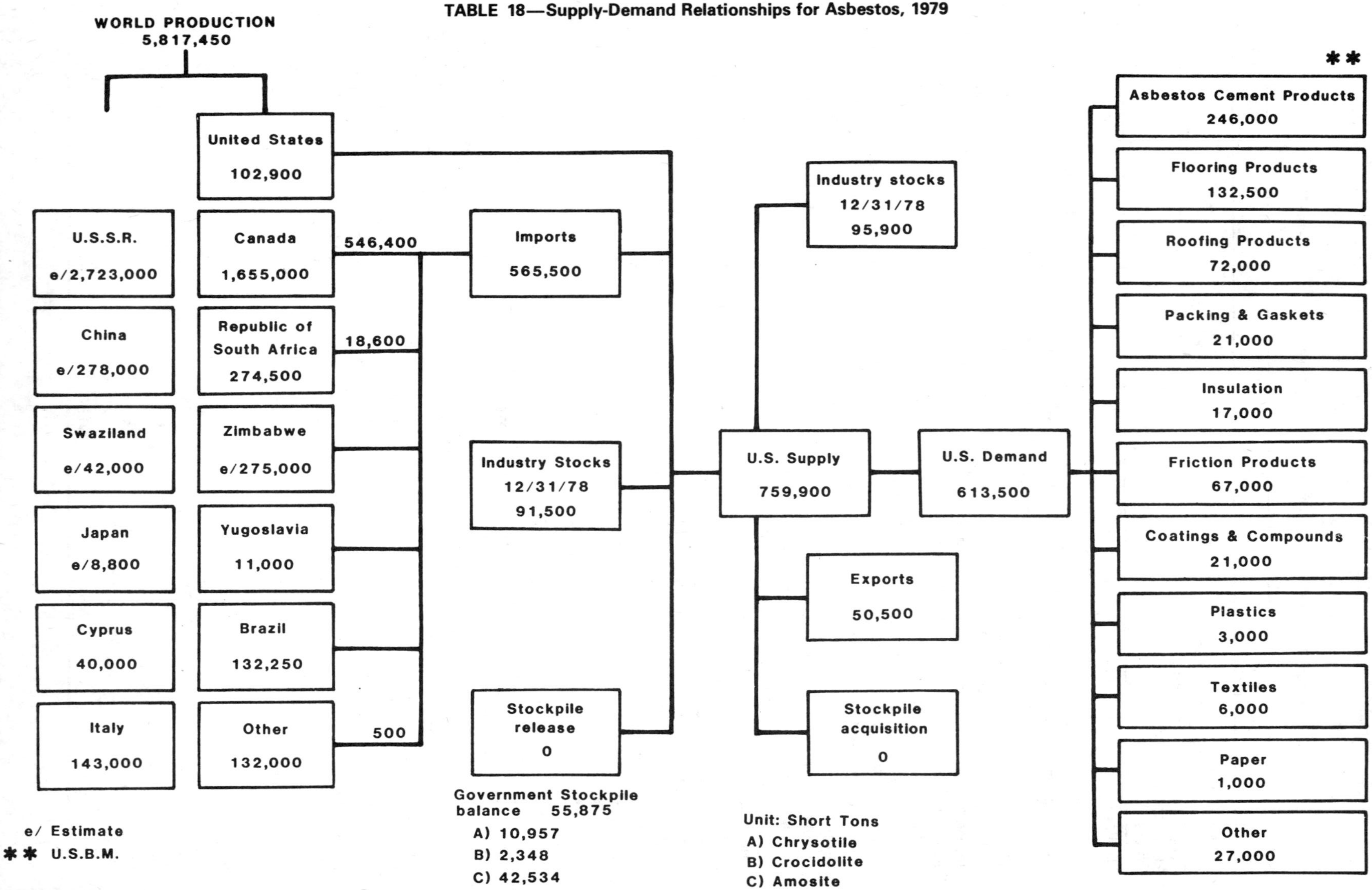

availability, and only certain grades are offered for sale.

Markets

In the late 1960s and early 1970s the consumption of asbestos fiber increased at the rate of 3 to 4% yearly. The US, Western Europe, and Japan have remained the major outlets for asbestos fiber exports. In recent years some of the developing countries have become a major factor in the rate of asbestos fiber consumption.

On a worldwide basis, estimated consumption is approximately 6 million tons, of which Canada supplies about 28% of world requirements. As indicated in Table 16, Western Europe accounts for 40% of total Canadian exports.

The largest producer is Russia which produces 50% of the world requirements. Russia exports from 330,000 to 660,000 short tons per year (see Table 17).

The major end uses for asbestos fiber are asbestos-cement sheet, asbestos-cement pipe, cushion vinyl floor backing, floor tile, friction materials, rolled roofing, asbestos-filled asphalts, plastics, coatings and other lesser uses. A breakdown of US 1978 end use consumption shows asbestos-cement products consuming 41% of total fiber, and floor tiles 20% (Table 18).

However, due to the slower increase in consumption for asbestos-cement pipe and sheet and continued increase in the use of substitutes, the overall world growth rate will decline in the coming years.

Health Hazards

In the area of health hazards, the asbestos industry has taken measures to answer statements on the usage and safe application of asbestos by setting up the Asbestos Information Assn./North America, whose basic objectives are to: (1) provide an authoritative channel of communication, (2) rebut irresponsible statements, (3) disseminate information on the safe uses of asbestos in our modern technological society.

Bibliography and References

Anon., 1966, *Testing Procedures for Chrysotile Asbestos Fiber,* 2nd ed., Quebec Asbestos Mining Assn.

Anon., 1972, *Geological Guide to the Asbestos-Mining Region of Southeast Quebec,* Quebec Asbestos Mining Assn., p. 32.

Allen, C.C., Gill, J.C., and Koski, J.S., 1957, "The Jeffrey Mine of Canadian Johns-Manville Company, Limited," *The Geology of Canadian Industrial Mineral Deposits,* 6th Commonwealth Mining and Metallurgical Congress, pp. 27–36.

Anderson, H.V., and Clark, G.L., 1929, "Application of X-Rays in the Classification of Fibrous Silicate Minerals Commonly Termed Asbestos," *Industrial & Engineering Chemistry,* No. 10, pp. 924–933.

Aruja, E., 1944, "Displacement of X-Ray Reflections," *Nature,* Vol. 154, p. 53.

Aruja, E., 1944a, "An X-Ray Study of the Crystal Structure of Antigorite," *Mineralogical Magazine,* Vol. 27, pp. 65–74.

Asbestos, monthly periodical, P.O. Box 471, Willow Grove, PA 19090.

Avery, R.B., Conant, M.L., and Weissenborn, H.F., 1958, "Selected Annotated Bibliography of Asbestos Resources in the United States and Canada," Bulletin 1019-L, US Geological Survey, pp. 817–865.

Babcock, R.C., 1978, "The Pyke Asbestos Deposits, New Zealand," SME Preprint No. 78–H–84, AIME Annual Meeting, Denver.

Badollet, M.S., 1937, US Patent 2,068,219, Jan. 19.

Badollet, M.S., 1948, "Asbestos," *Encyclopedia of Chemical Technology,* Vol. 2, Interscience, New York, pp. 134–142.

Badollet, M.S., 1951, "Asbestos, A Mineral of Unparalleled Properties," *Transactions,* Canadian Institute of Mining & Metallurgy, Vol. 54, pp. 151–160.

Badollet, M.S., 1952, "Asbestos Floats," *Transactions,* Canadian Institute of Mining & Metallurgy, Vol. 55, May, pp. 185–189.

Badollet, M.S., 1956, "The Role of Asbestos in Plastics," *Transactions,* Canadian Institute of Mining & Metallurgy, Vol. 59, pp. 283–288.

Badollet, M.S., 1963, "Asbestos," *Encyclopedia of Chemical Technology,* Vol. 2, Kirk-Othmer, ed., Interscience, New York, pp. 734–747.

Badollet, M.S., and Streib, W.C., 1947, US Patent 2,616,801, Sept. 20.

Badollet, M.S., and Streib, W.C., 1955, "The Heat Treatment of Chrysotile Asbestos Fibers," *Transactions,* Canadian Institute of Mining & Metallurgy, Vol. 58, pp. 33–37.

Bates, R.L., 1969, "Metamorphic Minerals—Asbestos," *The Geology of Industrial Rocks and Minerals,* Dover Publications, New York, pp. 317–328.

Bates, T.F., et al., 1950, "Tubular Crystals of Chrysotile Asbestos," *Science,* Vol. 3, pp. 512–513.

Bear, L.M., 1963, "The Mineral Resources and Mining Industry of Cyprus," Bulletin No. 1, Geological Survey, Cyprus, 208 pp.

Berger, H., 1963, *Asbestos Fundamentals,* Chemical Publishing Co., New York, 171 pp.

Bourassa, P.J., 1957, "The Asbestos Mine of Nicolet Asbestos Mines Limited," *The Geology of Canadian Industrial Mineral Deposits,* 6th Commonwealth Mining and Metallurgical Congress, pp. 26–27.

Bowles, O., 1955, "The Asbestos Industry," Bulletin 552, US Bureau of Mines, 122 pp.

Bragg, W.L., 1937, "The Pyroxene and Amphibole Groups," Atomic Structure of Minerals, Univ. Cornell Press, p. 184.

Brindley, G.W., and Zussman, J., 1957, "A Structural Study of the Thermal Transformation of

Serpentine Minerals to Forsterite," *American Mineralogist,* Vol. 42, No. 7–8, pp. 461–474.

Butt, B.C., 1978, "Exploration Forecasts and Exploitation Realities of the Woodsreef Mine, New South Wales, Australia," SME Preprint No. 78–H–71, AIME Annual Meeting, Denver.

Carroll-Porczynski, C.Z., 1956, *Asbestos,* The Textile Institute, Manchester, England.

Chidester, A.H., and Shride, A.F., 1962, "Asbestos in the U.S., Exclusive of Alaska and Hawaii," Mineral Investigation Research Map MP 17, US Geological Survey.

Cilliers, J.J.le R., 1964, "Amosite at the Penge Asbestos Mine," Vol. 2, *The Geology of Some Ore Deposits of Southern Africa,* Geological Society of South Africa, pp. 579–591.

Cilliers, J.J.le R., and Genis, J.H., 1964, "Crocidolite Asbestos in the Cape Province," *The Geology of Some Ore Deposits of Southern Africa,* Vol. 2, The Geological Society of South Africa, pp. 543–570.

Cilliers, J.J.le R., et al., 1961, "Crocidolite from the Koegas-Westerberg Area, South Africa," *Economic Geology,* Vol. 56, pp. 1421–1437.

Conn, H.M.K., 1967, "Geophysics and Asbestos Exploration," *Mining and Groundwater Geophysics,* Economic Geology Report No. 26, Geologic Survey of Canada, pp. 485–491.

Conn, H.K., and Mann, E.L., 1971, "Evaluation of Asbestos Deposits," SME Preprint No. 71–H–27, AIME Annual Meeting, New York, 9 pp.

Cooke, H.C., 1937, "Thetford, Disraeli and Eastern Half of Warwick Map Areas, Quebec," Memoir 211, Geological Survey of Canada, pp. 86–140.

Dean, A.W., and Mann, E.L., 1968, "The Evaluation of Chrysotile Asbestos Deposits," *Ore Reserve Estimation and Grade Control,* Special Vol. 9, Canadian Institute of Mining & Metallurgy, pp. 281–286.

Douglas, R.J.W., 1970, *Geology and Economic Minerals of Canada,* Economic Geology Report No. 1, Geological Survey of Canada, Dept. of Energy, Mines and Resources, 838 pp.

duToit, A.L., 1946, "The Origin of the Amphibole Asbestos Deposits of South Africa," *Transactions,* Geological Society of South Africa, Vol. 48, pp. 161–206.

Fankuchen, I., and Schneider, M., 1944, "Low Angle X-Ray Scattering from Chrysotiles," *Journal of American Chemical Society,* Vol. 66, No. 3, Mar., pp. 500–501.

Frankel, J.J., 1953, "South African Asbestos Fibres," *Mining Magazine,* London, Nos. 2 and 3, pp. 89, 73–83; 142–149.

Genis, J.H., 1964, "The Formation of Crocidolite Asbestos," *The Geology of Some Ore Deposits in Southern Africa,* Vol. 2, Geological Society of South Africa, pp. 571–578.

Gold, D.P., 1967, "Local Deformation Structures in a Serpentinite," *Ultramafic and Related Rocks,* P.J. Wyllie, ed., John Wiley, New York, pp. 200–202.

Graham, R.P.D., 1944, "Serpentine Belt, Eastern Townships," *Geology of Quebec,* Geology Report 20, Dept. of Mines, Quebec, Vol. 2, pp. 439–443.

Hall, A.L., 1930, *Asbestos in the Union of South Africa,* Memoir 12, 2nd ed., Geological Survey of South Africa, p. 324.

Harris, H.I., 1973, "How Nicolet Proved and Evaluated Columbian Asbestos Deposit," *World Mining,* Vol. 26, No. 13, pp. 43–46.

Hendry, N.W., 1972, "The Outlook for Asbestos in Canada," *Bulletin,* Canadian Institute of Mining & Metallurgy, Vol. 65, No. 724, Aug., pp. 40–44.

Hendry, N.W., and Conn, H.K., 1957, "The Ontario Asbestos Properties of Canadian Johns-Manville Company, Limited," *The Geology of Canadian Industrial Mineral Deposits,* 6th Commonwealth Mining and Metallurgical Congress, pp. 36–44.

Hillier, J., and Turkevich, J., 1949, "Electron Microscopy of Colloidal Systems," *Analytical Chemistry,* Vol. 21, No. 4, Apr., pp. 475–485.

Hodgson, A.A., 1965, "Fibrous Silicates," Lecture Series No. 4, Royal Institute of Chemistry.

Keep, F.E., 1961, "Amphibole Asbestos in the Union of South Africa," *Transactions,* 7th Commonwealth Mining and Metallurgical Congress, Vol. 1, pp. 90–120.

Keith, S.B., and Bain, G.W., 1932, "Chrysotile Asbestos: 1. Chrysotile Veins," *Economic Geology,* Vol. 27, pp. 169–188.

Kula, J., and Wiser, J.P., 1970, "Msauli Asbestos Mill," *World Mining,* Sep., pp. 26–29.

Lamarche, R.Y., 1972, "Ophiolites of Southern Quebec," *Canadian Contributions 1–11 to the Geodynamics Project—A Symposium,* Earth Physical Branch, Dept. of Energy, Mines and Resources, Ottawa.

Laubscher, D.H., 1964, "The Occurrence and Origin of Chrysotile Asbestos and Associated Rocks, Shabani, Southern Rhodesia," *The Geology of Some Ore Deposits of Southern Africa,* Vol. 2, The Geological Society of South Africa, pp. 593–624.

Laubscher, D.H., 1968, "The Origin and Occurrence of Chrysotile Asbestos in the Shabani and Mashaba Areas, Rhodesia," Symposium on Rhodesian Basement Complex, *Transactions,* Geological Society of South Africa Annexure, Vol. 71, pp. 195–204.

Leney, G.W., and Loeb, E.E., 1972, "The Geology and Mining Operations at Pacific Asbestos Corporation," *Asbestos,* Vol. 54, No. 4, pp. 4–14.

Low, J.H., 1951, "Magnetic Prospecting Methods in Asbestos Exploration," *Transactions,* Canadian Institute of Mining & Metallurgy, Vol. 54, pp. 388–395.

Mamen, C., ed., 1973, "China's Mineral Industry," *Canadian Mining Journal,* Vol. 94, No. 1, pp. 21–31.

May, T.C., and Lewis, R.W., 1970, "Asbestos," *Mineral Facts and Problems,* Bulletin 650, US Bureau of Mines, pp. 851–863.

Merrill, R.J., 1957, "The Carey-Canadian Asbestos Deposit," *The Geology of Canadian Industrial Mineral Deposits,* 6th Commonwealth Mining and Metallurgical Congress, pp. 45–49.

Miles, K.R., 1942, "The Blue Asbestos-Bearing Banded Iron Formations of the Hammersley Range, Western Australia," Bulletin No. 100, Geological Survey of Western Australia, Pt. 1, pp. 5–37.

Mining Magazine, 1980, Vol. 143, No. 6, p. 537.

Munro, R.C., and Reim, K.M., 1962, "Coalinga Asbestos Fiber—A Newcomer to the Asbestos Industry," *Canadian Mining Journal,* Vol. 83,

No. 8, Aug.; *Mining Engineering,* Vol. 14, No. 9, pp. 60–62.

Nalivkin, D.V., 1960, *The Geology of the U.S.S.R. —A Short Outline,* trans. by S.I. Tomkeiff, J.E. Richey, trans. ed., Pergamon Press, 170 pp.

Oldham, J.W., 1968, "A Short Note on the Recent Geological Mapping of the Shabani Area," Symposium on Rhodesian Basement Complex, *Transactions,* Geological Society of South Africa Annexure, Vol. 71, pp. 189–194.

Parry, J., 1980, "Zimbabwe Opens Its Doors," *Industrial Minerals,* No. 158, p. 55.

Pauling, L., 1930, "The Structure of the Chlorites," *Proceedings,* National Academy of Science, Vol. 16, p. 578.

Pelletier, R.A., 1964, *Mineral Resources of South-Central Africa,* Oxford University Press, 277 pp.

Petrov, V.P., and Znamensky, V.S., 1978, "Asbestos Deposits of the USSR," SME Preprint No. 78–H–106, AIME Annual Meeting, Denver.

Pundsack, F.L., 1955, "The Properties of Asbestos. I. The Colloidal and Surface Chemistry of Chrysotile," *Journal of Physical Chemistry,* Vol. 59, No. 9, Sep., pp. 892–895.

Pundsack, F.L., 1956, "The Properties of Asbestos. II. The Density and Structure of Chrysotile," *Journal of Physical Chemistry,* Vol. 60, No. 3, Mar., pp. 361–364.

Pundsack, F.L., and Reimschussel, G., 1956, "The Properties of Asbestos. III. Basicity of Chrysotile Suspensions," *Journal of Physical Chemistry,* Vol. 60, Sep., pp. 1218–1222.

Rabbit, J.C., 1948, "A New Study of the Anthophyllite Series," *American Mineralogist,* Vol. 33, May-June, pp. 263–323.

Rice, S.J., 1963, "California Asbestos Industry," Mineral Information Service, California Div. of Mines, Vol. 16, No. 9, pp. 4–6.

Riordon, P.H., 1952, "Geology of the Thetford-Black Lake District of Quebec with Particular Reference to the Asbestos Deposits," Ph.D. Thesis, McGill University, unpublished.

Riordon, P.H., 1955, "The Genesis of Asbestos in Ultrabasic Rocks," *Economic Geology,* Vol. 50, No. 1, pp. 67–81.

Riordon, P.H., 1957, "The Structural Environment of the Thetford-Black Lake Asbestos Deposits," *Proceedings,* Geological Assn. of Canada, Vol. 9, pp. 83–93.

Riordon, P.H., 1957a, "The Asbestos Belt of Southeastern Quebec," *The Geology of Canadian Industrial Mineral Deposits,* 6th Commonwealth Mining and Metallurgical Congress, pp. 3–8.

Riordon, P.H., 1957b, "The Asbestos Deposits of Thetford Mines, Quebec," *The Geology of Canadian Industrial Mineral Deposits,* 6th Commonwealth Mining and Metallurgical Congress, pp. 9–17.

Riordon, P.H., 1957c, "The British Canadian Mine," *The Geology of Canadian Industrial Mineral Deposits,* 6th Commonwealth Mining and Metallurgical Congress, pp. 17–21.

Riordon, P.H., 1957d, "Normandie and Vimy Ridge Mines," *The Geology of Canadian Industrial Mineral Deposits,* 6th Commonwealth Mining and Metallurgical Congress, pp. 21–26.

Riordon, P.H., and Laliberté, R., 1957, "Asbestos Deposits of Southern Quebec," *Excursion B-08 Guidebook,* 24th International Geological Congress, Canada, pp. 1–21.

Robinson, K., and Shaw, E.R.S., 1952, "Summarized Proceeding of a Conference on Structures of Silicate Minerals (November 1951)," *British Journal of Applied Physics,* Vol. 3, Sept., pp. 277–282.

Rosato, D.V., 1959, *Asbestos, Its Industrial Applications,* Reinhold, New York, pp. 198–199.

Rowbotham, P.I., ed., 1970, "World Asbestos Industry," *Industrial Minerals,* No. 28, Jan., pp. 17–29.

Rozovsky, H., 1957, "Air in Asbestos Milling," Ventilating Conference, Keller Center for Continuing Education, Michigan State University.

Smirnov, V.I., 1971, *Essays on Metallogeny,* trans. by E.A. Alexandrov, Queens College Press, 96 pp.

Smitheringale, W.V., 1957, "The Mine of Cassiar Asbestos Corporation Limited, Cassiar, B.C.," *The Geology of Canadian Industrial Mineral Deposits,* 6th Commonwealth Mining and Metallurgical Congress, pp. 49–53.

Speil, S.S., and Leineweber, J.P., 1969, "Asbestos Minerals in Modern Technology," *Environmental Research,* Vol. 2, No. 3, Apr., pp. 166–208.

Stehno, J., and Fronek, R., 1977, "The Soviet Asbestos Deposits. A Guarantee for the Continuing Development of the Czechoslovakian Asbestos Cement Industry," *Stavivo* (Building Materials), No. 11, pp. 430–432. (Translation from Czechoslovakian technical journal.)

Stewart, R.V., 1978, "Geology and Evaluation of the Asbestos Hill Orebody," SME Preprint No. 78–H–69, AIME Annual Meeting, Denver.

St-Julien, P., 1967, "Tectonics of Part of the Appalachian Region of Southeastern Quebec," Special Publication 10, Royal Society of Canada, pp. 41–47.

Straw, D.J., 1955, "A World Survey of the Main Chrysotile Asbestos Deposits," *Canadian Mining & Metallurgical Bulletin,* Vol. 48, pp. 610–630.

Streib, W.C., 1978, "Asbestos," *Kirk-Othmer: Encyclopedia of Chemical Technology,* Vol. 3, 3rd ed., John Wiley & Sons, Inc., pp. 267–283.

US Bureau of Mines, *Minerals Yearbook,* annually.

Van Biljon, W.J., 1964, "The Chrysotile Deposits of the Eastern Transvaal and Swaziland," *Geology of Some Ore Deposits in Southern Africa,* Vol. 2, Geological Society of South Africa, pp. 625–669.

Vokes, F.M., 1964, "Asbestos Bearing Claims on Troodos," US Special Fund Project, Cyprus, United Nations unpublished report.

Warren, B.E., 1932, "Structure of Asbestos—An X-Ray Study," *Industrial & Engineering Chemistry,* Vol. 24, No. 4, pp. 419–422.

Warren, B.E., 1942, "X-Ray Study of Chrysotile Asbestos," *American Mineralogist,* No. 27, p. 235.

Warren, B.E., and Bragg, W.L., 1928, "The Structure of Diopside," *Zeitschrift fuer Krist,* Vol. 69, pp. 168–193.

Warren, B.E., and Hering, K.W., 1941, "The Random Structure of Chrysotile Asbestos," *Physical Reviews,* No. 59, p. 925.

Whittaker, E.J.W., 1952, "The Unit Cell of Chrysotile," *Acta Chrystalogica,* Vol. 5, pp. 143–144.

Whittaker, E.J.W., and Zussman, J., 1956, "The Characterization of Serpentine Minerals by

X-Ray Diffraction," *Mineralogical Magazine,* No. 31, pp. 107–126.

Wilson, J.F., 1968, "The Mashaba Igneous Complex and Its Subsequent Deformation," *Symposium on Rhodesian Basement Complex,* Geological Society of South Africa Annexure, Vol. 71, pp. 175–188.

Yada, K., 1967, "Study of Chrysotile Asbestos by a High Resolution Microscope," *Acta Chrystalogica,* Vol. 23, pp. 704–710.

Yada, K., 1971, "Study of Microstructure of Chrysotile Asbestos by High Resolution Electron Microscopy," *Acta Chrystalogica,* Vol. A27, pp. 659–664.

Zussman, J., Brindley, G.W., and Comer, J.J., 1957, "Electron Diffraction Studies of Serpentine Minerals," *American Mineralogist,* No. 42, pp. 133–153.

Barium Minerals

DONALD A. BROBST *

The minerals barite ($BaSO_4$—barium sulfate) and witherite ($BaCO_3$—barium carbonate) are the chief commercial sources of the element barium and its compounds whose many uses are nearly hidden among the technical complexities of modern industrial processes and products.

Barite, the major ore mineral, is extremely vital to the petroleum industry which in 1980 consumed at least 90% of the world's production of 8.8 million tons as a major ingredient of the heavy fluid, called mud, that is circulated in rotary drilling of oil and gas wells. The remaining 10% of the barite production was consumed chiefly in the manufacture of barium chemicals and glass and as a pigment, filler, and extender. Barite occurs throughout much of the world and is available from three major geologic types of deposits—vein and cavity filling, residual, and bedded—in sufficient quantity at competitive prices to meet current demands. The world's demand for barite is expected to increase, and geologic circumstances are favorable for the discovery of new deposits of commercial value.

Witherite is much less common and abundant than barite, although it is more desirable in many ways as a raw material for the production of barium chemicals. The United States has not produced witherite since about 1950; England is currently the chief producer of witherite.

End Uses

Most of the world's barite production since 1926 has been used as a weighting agent for the muds circulated in rotary drilling of oil and gas wells. The muds fundamentally are mixtures of water, clay, and barite in different proportions that vary according to local reservoir conditions. Muds with a specific gravity as great as 2.5 are used. The mud is pumped down the hollow drill stem, passes through the bit at the bottom of the hole, and rises to the surface in the space between the drill stem and the wall of the hole. In the course of this circulation the drill bit is cooled, cuttings are removed, the drill stem is lubricated, the walls of the hole are sealed, and the hydrostatic head of the column of weighted fluid helps to confine high oil and gas pressures. The latter feature aids in prevention of gushers, thus reducing both environmental pollution and waste of oil and gas resources as well as conserving the natural reservoir pressure for greater production and rate of recovery of the products contained in the rocks. Barite is particularly well suited for drilling mud because it is clean, easy to handle, soft (nonabrasive), heavy, virtually inert chemically, and relatively inexpensive compared to many other available heavy materials.

Barite is used in the manufacture of glass in continuous tanks. The addition of barite homogenizes the melt and gives greater brilliance and clarity to the finished glass.

Ground barite, both unbleached and bleached by sulfuric acid, is a common industrial filler, extender, and weighting agent. The rubber industry is a major consumer of barite as a filler. Barite also is added to bristolboard, heavy printing paper, playing cards, rope finishes, brake linings, clutch facings, plastics, and linoleum, to name but a few uses. Bleached barite has long been an extender in white lead paint because of its weight. The low index of refraction of barite, however, makes its ability to cover marks poorer than some other substances, but its low capacity for absorption of oil is a good feature. Off-color or unbleached barite can be used as a filler in colored paints.

In the construction industry, some lump barite is used in concrete aggregate to weigh down pipelines buried in marshy areas and to shield nuclear reactors. Because barite absorbs gamma radiation well, its use reduces the amount of expensive lead shielding otherwise

* Consulting Geologist, Reston, VA.

necessary. A mixture of rubber and asphalt that contains about 10% barite has been used successfully as a durable paving material for parking lots, roads, and airport runways. Recently the tires of heavy road construction equipment have been partly filled with a ground barite mixture to add weight as an aid in packing areas of fill.

Barium ceramics have some important industrial uses. Unusual electromechanical properties make barium titanate ceramics useful in the electronics industry. Barium ferrate ceramics are used in permanent magnets.

As a pigment, barite is used as the natural or bleached white sulfate, as blanc fixe, and as lithopone. Blanc fixe is a white pigment of reprecipitated chemically pure barium sulfate that is used where higher purity is required than found in natural barite. Blanc fixe is probably best known as an indicator in medical X-ray photography. Lithopone is another chemically prepared white pigment that consists of 70% barium sulfate and 30% zinc sulfide.

A survey of the uses of barium chemicals in 1961 listed more than 2000 specific industrial applications in 17 major classifications (Anon., 1961). Various manufactured barium compounds, including the carbonate, chloride, nitrate, oxide, peroxide, hydroxide, sulfide, sulfate, and some organic varieties are widely used as reagents and catalysts, as in sugar refining; sizing, waterproofing, and fireproofing for textiles; drugs, for the manufacture of vitamins, hormones, and blood coagulants; various pyrotechnics, especially green flares; stabilizers for glue and starch, coagulation of synthetic rubber; plastics; insecticides, herbicides, and germicides; case hardening of steel; phosphors for fluorescent lamps; bonding phosphors to glass for black and white TV tubes; reagents for beneficiation of ores and paper manufacture; welding fluxes; additives for oil and grease; the melting and refining of magnesium; the recovery of indium; and the recovery of zinc from dross, and as a scum preventative in the ceramic and brick industries.

Barium metal is used as a "getter" (deoxidizer) in degasifying TV and other vacuum tubes. Barium is alloyed with other metals such as aluminum, magnesium, lead, and calcium. An alloy of barium, lead, and calcium is used for bearings.

Product Specifications

The barite product for use in drilling mud should have a specific gravity of at least 4.2 (about 92 to 94% $BaSO_4$) and contain less than 0.1% soluble salts. A few percent iron oxide is permitted. At least 90 to 95% of the ground barite should pass 325 mesh.

Glassmakers have their own rigid specifications for the barite used in their products. These requirements include: a minimum of 96 to 98% $BaSO_4$; no more than 0.1 to 0.2% Fe_2O_3; and only a trace of TiO_2. Some specifications allow no more than 1.5% SiO_2 and 0.15% Al_2O_3. Sizing is one of the stringent requirements set by individual users. Most manufacturers want a product that will pass a 16-mesh screen but with as little as 5 to as much as 40% passing a 100-mesh screen. Material of finer size is undesirable because it tends to ball up in the melt.

Specifications for barite to be used as filler, extender, and weighting agent are set by various users, but most manufacturers want a fine-grained product that is virtually all —325 mesh. Color is important to many users.

The American Society for Testing and Materials presents specifications for barite used in pigments (D602–42) that require a minimum of 94% $BaSO_4$ and no more than 0.05% Fe_2O_3, 0.2% water-soluble materials, 0.5% moisture, and 2% foreign matter. Specifications for blanc fixe are similar, except that a minimum of 97% $BaSO_4$ is required and a maximum of only 0.02% iron oxide is permitted. The manufacture of lithopone requires 95 to 98% $BaSO_4$ and no more than 1% iron oxide.

Manufacturers of barium chemicals want a raw material containing at least 95% $BaSO_4$, less than 1% iron oxide and 1% strontium sulfate, and only traces of fluorine. The barite for use in chemical plants generally is sized between 4 and 20-mesh, although some producers buy crude lump ore and grind it to their own specifications. Barite that is too finely ground is lost as flue dust and barite that is too coarse mixes poorly with carbon and yields poor results on roasting.

Barite for heavy aggregate in concrete generally is gravel-sized material.

Geology

Barite occurs in many geological environments in sedimentary, igneous, and metamorphic rocks. Some recent references to the geology of barite deposits are cited in this report and many more are available in an annotated bibliography (Dean and Brobst, 1955) and in other reports on barite reserves and resources of the United States and the world (Brobst, 1958, 1965, 1970, 1973; Anon., 1970).

Mineralogy

Barite ($BaSO_4$—barium sulfate), also called barytes, tiff, cawk, or heavy spar, is the most common and abundant ore mineral of barium. Pure barite contains 58.8% Ba (or 65.7% BaO and 34.3% SO_3) and has a calculated specific gravity of 4.5, but inclusions in natural barite may reduce this value considerably. Barite crystallizes in the dipyramidal class of the orthorhombic system; well-formed crystals are generally tabular and have three cleavages, the most perfect of which is parallel to (001). The barite of many deposits, however, has an uneven fracture and an apparent cleavage caused by separation along planes between successively deposited layers. The barite in most commercial deposits occurs as irregular masses, concretions, nodules, rosettelike aggregates, and in laminated to massive beds of fine crystallinity. The mineral occurs in many colors, although shades of white to dark gray and black are most common in commercial deposits. The hardness is 2.5 to 3.5 on Mohs' scale (in which the hardness of a fingernail is about 2). Commercially, the terms "hard" and "soft" refer to the ease with which barite can be ground. Barite is relatively insoluble in water and acid, and thus can be used as a chemically inert material.

Barite is most commonly associated with quartz, chert, jasperoid; calcite, dolomite, siderite, rhodochrosite, celestite; fluorite; various sulfide minerals, such as pyrite, chalcopyrite, galena, sphalerite, and their oxidation products. Ferruginous clays make up a large part of the residual deposits of barite. Barite is a common gangue mineral in many types of ore deposits, especially veins that are mined principally for other types of commodities, including lead, zinc, gold, silver, fluorite, and rare-earth minerals. Barite is a common constituent of some siderite-rich iron deposits in Europe, but deposits of this type have not been found in the United States.

Minor amounts of other metals commonly occur in barite (Brobst, 1958). Barite in vein and residual deposits may contain as much as several percent of strontium which substitutes for barium isomorphously in the crystal structure because of the close similarity in the ionic radius of the two metals in their bivalent state.

Witherite ($BaCO_3$—barium carbonate) contains 70% Ba (or 77.7% BaO and 22.3% CO_2), and it has a calculated specific gravity of 4.2 and a hardness of 3 to 3.5 on Mohs' scale. Like barite, this mineral also crystallizes in the dipyramidal class of the orthorhombic system. The cleavage parallel to {010} is distinct, {110} is imperfect. The fracture is uneven and the luster is vitreous to resinous. Crystals or complex masses are colorless to milky white or gray or tints of yellow, brown, or green. Witherite is a minor accessory mineral in some barite deposits. In the United States, only the mine at El Portal, Mariposa County, CA, has produced witherite in large amounts; operations ended there about 1950. Witherite is highly desirable as a source for the production of barium chemicals because of its solubility in acid. It seems, however, that witherite does not commonly occur in commercial concentrations and that the mineral probably constitutes only a small part of the world's resources of barium.

Sanbornite ($BaSi_2O_5$—barium silicate) contains 50% barium and is of potential interest as a source of barium for chemicals. Unlike many silicate minerals, sanbornite is soluble in acid which permits the recovery of barium by further chemical processing. The mineral, formerly considered rare, has been found in abundance with other barium silicate minerals in contact zones with metamorphic rocks, mostly quartzite, in roof pendants of the Sierra Nevada batholith in Fresno County, CA (Matthews and Alfors, 1962). The resource potential of sanbornite is virtually unstudied.

Barium occurs in other minerals, but in concentrations so low that none of these minerals is likely to be a potential source of the element soon.

Types of Deposits

The commercial deposits of barite may be geologically classified by their mode of occurrence into three major types—the vein and cavity-filling deposits, the residual deposits, and the bedded deposits.

Vein and Cavity-Filling Deposits: The vein and cavity-filling deposits are those in which barite and associated minerals occur along faults, gashes, joints, bedding planes, breccia zones, and solution channels and in various sink structures. The deposits in solution channels and sink structures are most common in limestone. The host rocks are of Precambrian to Tertiary age in the United States. Many deposits in the Western states are associated with igneous rocks of Tertiary age although notable exceptions are vein deposits of barite and rare-earth minerals associated with igneous rocks of Precambrian age at Mountain Pass, San Bernardino County, CA, and in the Wet Mountains, Custer County, CO. In most of the Eastern and Midwestern states an association with igneous activity is not obvious.

Most of the barite in the vein and cavity fillings is dense and gray to white, and is associated with other minerals including those of the metallic ores as already mentioned. The grade of the barite ore differs by deposit and within deposits. The deposits commonly have sharp contacts with wall rocks; large-scale replacement of the host rocks beyond ore-controlling structures is rare. Within individual districts, the deposits commonly are scattered and irregular, and range in thickness from a few inches to a few feet and in length from tens to hundreds of feet. Barite cementation of breccia zones forms thin deposits of small extent in the eastern United States.

Barite deposits in collapse and sink structures are common in central Missouri, where they are known as circle deposits, and in the Appalachian states. The deposits have yielded rich ore, but the individual bodies tend to be small.

On weathering, vein and cavity-filling deposits may form valuable bodies of residual ore.

The barite and other minerals of the vein and cavity-filling deposits are typical of an epithermal suite precipitated from low-temperature hydrothermal solutions. Many geologists agree that most vein deposits containing barite in the Western states had such an origin. A hydrothermal origin also is postulated for many deposits in the Midcontinent region and in the Eastern states where the term "telethermal" has been applied to indicate that the solutions traveled farther from their source and, therefore, were somewhat cooler than those generally termed "epithermal." Evidence has been accumulating recently to suggest that some barite and minerals of base metals may form vein deposits from circulating connate and ground waters (Sawkins, 1966).

Residual Deposits: Residual barite deposits occur in unconsolidated material and are formed by weathering from preexisting deposits. Many residual deposits of commercial value in the United States lie within the clayey residuum derived from Cambrian and Ordovician limestone and dolomite. Classic examples of these deposits occur in southeastern Missouri (Brobst and Wagner, 1967) and in the Appalachian region from Pennsylvania to Alabama (Brobst and Hobbs, 1968), but especially in the Sweetwater district, TN (Maher, 1970), and the Cartersville district, GA (Kesler, 1950).

Most of the residual barite is white, and translucent to opaque; it occurs in mammillary, fibrous, or dense fine-grained masses, but subhedral to euhedral crystals also are found. The barite in these deposits ranges in size from microscopic particles to irregular boulders weighing hundreds of pounds. Most of the masses, however, are 1 to 6 in. in diam. Chert, jasperoid, and drusy quartz are common in many deposits. Small amounts of pyrite, galena, and sphalerite occur in or on some of the barite, and locally the lead and zinc minerals are recovered as a byproduct. Incompletely weathered rock fragments and yellow, red, or brown illitic clays typically make up the rest of most deposits.

The grade of these residual barite ores varies greatly. Much residuum in commercial deposits contains less than 10% to about 20% barite.

The size and shape of the residual barite deposits varies greatly. Some of the larger deposits in Georgia, Missouri, and Tennessee extend over several hundred acres. The shape of many residual deposits reflects that of the original deposit. Deposits derived from solution channels and vein systems tend to be elongate and those derived from sink structures tend to be circular.

The depth of residual deposits differs according to local conditions. Some residual deposits in Georgia terminate at the contact with bedrock as deep as 150 ft beneath the surface. The residual deposits in Missouri terminate at bedrock that in many places is only 10 to 15 ft beneath the surface. The barite may be distributed randomly through the residuum or concentrated near the contact with bedrock. The contact of the residuum and the bedrock generally is irregular. Pillars or ribs of bedrock that protrude into, or even through, the residuum are common in many deposits.

The bedrock in many areas of residual barite deposits contains some veins filled with barite, fluorite, calcite, and quartz and locally some sulfide minerals including pyrite, chalcopyrite, galena, and sphalerite. In other deposits, only thin veinlets or disseminated lumps and pods of barite occur in the carbonate bedrock. The source of the barite in residual deposits generally has been assumed to be epigenetic (introduced) hydrothermal vein material left after the host rock was dissolved.

Today's knowledge of mineralogy and crystallography suggests that any barium that was trapped in precipitating carbonate at the time the host rocks were deposited will fit into the structure of aragonite which seems to be a common primary precipitate of calcium carbonate. Because of the differences in the ionic radii of the cations and the crystal structure, barium will not fit into the structure of calcite or dolomite. Thus, when aragonite is converted to calcite or dolomite, any structurally included barium is expelled. Any sulfate ions in the pore fluids might then fix the barium ions as barite in veinlets or nodules in the carbonate rocks. If the pore fluid is chloride-rich and sulfate-poor, the expelled barium ions might be converted to the very soluble chloride, in which case the barium could be held in solution to be precipitated later, in place or elsewhere, as barite when sulfate ions became available. Connate water may play an important role in the redistribution of barium in the carbonate environment. The foregoing is perhaps an oversimplification of a complex chemical process, but it does offer an explanation of why calcite and dolomite-rich rocks generally contain only small amounts of barium. Just as important perhaps is the largely unevaluated roll of organisms, especially bacteria, in the mobilization and redistribution of elements such as barium in the sedimentary environment under conditions of low temperature and pressure.

Bedded Deposits: Bedded deposits include those in which barite occurs as a principal mineral or cementing agent in stratiform bodies in layered sequences of rock. The major deposits of commercial value in this type are bedded concentrations of fetid, generally dark-gray to black, fine-grained barite. Some of these deposits contain millions of tons of barite easily beneficiated for use in drilling mud and chemical plants. In 1972 deposits of this type were mined commercially in Arkansas, California, and Nevada. Geologic descriptions are available for some of the deposits in Arkansas, including the major deposit in the Chamberlain Creek syncline near Magnet Cove (Scull, 1958; Zimmermann and Amstutz, 1964); in Nevada, including the deposits in the Toquima Range (Shawe et al., 1969) and the Shoshone Range (Ketner, 1965; Zimmermann, 1969); in California, near Castella, Shasta County (Weber and Matthews, 1967); and in eastern Washington (Mills et al., 1971). The bedded deposits of barite are worthy of considerable discussion because of their current economic value and the good potential for finding more of them in many parts of the world.

The dark beds of barite are a few inches to more than 50 ft thick and extend over many acres. The beds of barite commonly are intercalated with dark chert and siliceous shale and siltstone. Some barite-rich zones of these rocks are more than 100 ft thick. Some individual beds of barite are massive, but commonly the beds are laminated. The grains of barite generally are less than 0.1 mm in diameter. Barite rosettes and nodules make up a substantial part of some beds, and a smaller part of some silty or shaly beds. Some nodules have concentric rings of barite, and the rosettes have a radially bladed structure. Both structures suggest that room for free growth existed during deposition. Conglomeratic beds consisting of nodules and fragments of barite, chert, phosphate (apatite), and fragments of rock in a fine-grained matrix of barite have been found in the Toquima Range, Nevada (Shawe et al., 1969). Many small-scale sedimentary textures and structures, as well as rhythmic alternations of beds, also have been described (Zimmermann, 1969; Zimmermann and Amstutz; 1964).

Many beds of ore in these deposits consist of 50 to 95% barite. The chief impurity is fine-grained quartz whose abundance is inverse to that of barite. Small amounts of clay minerals and pyrite are common. Carbonate minerals are rare. Spectrographic analyses of many samples indicate that the composition of these barite deposits is extremely low in calcium and magnesium—the total of these two elements generally is less than 1%. The most abundant minor element is strontium; it occurs in amounts up to about 7000 ppm (0.7%). Both manganese and vanadium content generally are less than 50 ppm. The content of cobalt, copper, chromium, nickel, yttrium, and zirconium is less than 20 ppm each. The trace element suite is limited, and the abundance of each element is usually small as compared to the suite and abundance of the trace elements in vein deposits.

The beds of black barite contain several percent organic matter and characteristically give off the odor of hydrogen sulfide (H_2S) when struck with a hammer. The hydrogen sulfide is believed to be the product of the decay of organic matter and reduction of sulfate by anaerobic bacteria. Fatty acids, hydrocarbons, and other organic compounds have been extracted from samples collected in Arkansas and Nevada (Miller et al., 1977). The fatty acids identified are considered to be biological markers for the cell walls of bacteria that probably once lived on or within crystallizing barite masses.

Rocks containing the black bedded deposits commonly are of mid-Paleozoic age even though they are widely scattered geographically. The relation is not fully understood, but it is perhaps more than coincidental. Most of the deposits in Arkansas are of Devonian and Mississippian age; those in California and Nevada are mostly of Devonian age, although some in Nevada may be of Ordovician age. Somewhat similar deposits at Meggen, Westphalia (Ehrenberg et al., 1954), and in the Saarland (Krebs, 1970) of western Germany also occur in Devonian rocks.

Earlier workers have proposed that the black, fetid beds of barite in Arkansas (Scull, 1958) and Nevada (Ketner, 1965) originated by replacement of carbonate rocks by barium-bearing hydrothermal solutions from intrusive igneous activity. The appeal of this hypothesis is greatly reduced by: the absence of igneous rocks in the vicinity of many deposits, especially in western Arkansas, except for the deposit near Magnet Cove; the absence of anomalously large concentrations of such elements as copper, lead, zinc, silver, molybdenum, and mercury expected in and around hydrothermal ore deposits; and the absence of abundant carbonate rock available for replacement in the formations that contain rich beds of barite.

The black bedded barite deposits probably are of sedimentary origin and formed virtually at the same time as the enclosing rocks by organic and inorganic chemical processes of concentration and deposition. A sedimentary origin for deposits in the Shoshone Range, NV (Zimmermann, 1969) and in the Chamberlain Creek syncline, Arkansas, was postulated from structural and textural evidence (Zimmermann and Amstutz, 1964). A sedimentary origin for the deposits in the Toquima Range, NV, was proposed from textural, structural, and chemical evidence, including the effects of contact metamorphism on barite beds (coarsening of the grain size by recrystallization and leaching of the dark color) from emplacement of a stock (Shawe et al., 1969).

The source of the barium in these deposits has not been established, although some proposals have been made. Hydrothermal solutions are considered by Dunham and Hanor (1967) to be the ultimate source of all barite deposits. Sources in submarine volcanic emanations and springs were suggested for the barite and sulfide beds at Meggen, Germany (Ehrenberg, 1954). A barium source in seawater with extraction related to biological activity has been proposed (Shawe et al., 1969). The recycling of barium from preexisting rocks should not be overlooked as a source for the barium in the bedded deposits. Connate brines seem to be good concentrators and transporters of barium (Sawkins, 1966). Barium-rich chloride waters reaching sulfate-rich marine or brackish water could mix locally to yield deposits of barite.

White to clear barite occurs as a cement in some sandstones and conglomerates, such as the Pike gravel, the basal formation of the Trinity Group of Cretaceous age in Howard and Sevier Counties, AR (Scull, 1958). The Pike gravel marks the Cretaceous strandline and unconformably overlies folded rocks of Paleozoic age. About 20 miles to the north, the Paleozoic rocks contain deposits of black bedded barite, which perhaps are the source of the barium for the barite cement in the Pike gravel.

Distribution of Deposits

United States: *Missouri*—The major deposits of Missouri are clustered in the Washington County area, about 50 airline miles southwest of St. Louis (Brobst and Wagner, 1967); Wharton, 1972). This area, commonly considered part of the Southeast Missouri Lead district, has yielded about 11.5 million tons of barite through 1972—more than any other district in the United States. The barite is concentrated in residuum overlying the Potosi and Eminence dolomites of Cambrian age. Single areas of many acres may be ore-bearing, but the richest parts of the deposits form "runs" or "leads" 10 to 20 ft wide and several hundred feet long that generally lie within 15 ft of the surface. Some deposits contain as much as 25% barite, but much ground contains less than 10%.

The barite deposits of the Central district lie mostly in Cole, Miller, Moniteau, and Morgan counties. Several hundred deposits are known in the Gasconade and Jefferson City dolomites of Ordovician age and the Burlington limestone of Mississippian age, and a few occur in the Roubidoux formation of Ordovician age and in the Cherokee rocks of Pennsylvanian age. The most important deposits, called circle deposits, are bell, flask, or cone-shaped in the vertical dimension and are approximately circular in plan. Most of the deposits are small, the largest being 200 to 250 ft in diam, and some were worked to a depth of 125 ft. The grade of the ore is extremely irregular.

Arkansas—The largest single deposit of barite mined in the United States is in the Chamberlain Creek syncline, near Malvern, Hot Spring County (Scull, 1958). Since 1942, this black bedded barite deposit has yielded virtually all of the 8.5 million tons of barite produced in Arkansas through 1972. The barite occurs in a zone about 60 ft thick in the lower part of the Stanley shale of Mississippian age in a southwest-plunging syncline. The ore body is about three-quarters of a mile long and nearly one-half mile wide. The grade of the ore varies, but the richest parts of the ore body contain more than 70% $BaSO_4$. The associated minerals are quartz, chert, pyrite, iron oxides, clay minerals, and some small amounts of calcite, alstonite ($CaBa(CO_3)_2$), and benstonite ($MgCa_6Ba_6(CO_3)_{13}$). Other bedded deposits and a few vein deposits are known in the Stanley shale and the underlying Arkansas novaculite (Devonian and Mississippian) in Montgomery, Pike, and Polk counties. Barite locally cements parts of the Pike gravel, the basal formation of the Trinity Group of Cretaceous age in Howard and Sevier counties.

Georgia—For many years barite has been mined from the residual deposits of the Cartersville district, Bartow County (Kesler, 1950). The deposits are associated with the residuum of the Weisner, Shady, and Rome formations of Cambrian age. The general features of the deposits are similar to those other residual deposits already described, except that some barite has been recovered from pits as much as 150 ft deep. Most of the more than 4 million tons of barite mined through 1970 in Georgia has come from the Cartersville district; however deposits in Polk, Floyd, Cherokee, Gordon, Murray, and Whitfield counties also have been worked from time to time (Butts and Gildersleeve, 1948).

Nevada—Many barite deposits of the bedded and vein types are widely distributed in 11 counties of Nevada (Horton, 1963, 1964). Much of the three million tons of barite mined in Nevada from 1907 through 1972 has come from the black bedded deposits in the Shoshone Range, Lander, and Eureka counties. Most of the bedded deposits occur in the Slaven chert, a formation of Devonian age that consists chiefly of black chert and black carbonaceous shale and siltstone and minor amounts of intercalated dark limestone and limy, brown sandstone. Some of the deposits may occur in similar rocks of Ordovician age. The deposits generally are lenticular and some beds contain as much as 97% $BaSO_4$, and others are of sufficiently high grade for use in drilling mud with no beneficiation other than grinding. The major mines in the Shoshone Range include those in the Argenta, Bateman Canyon, Greystone, Mountain Springs, Pleasant View, and Sierra Chemical Co. deposits. Other bedded deposits occur at the Rossi mine and elsewhere in Elko County; in the Toquima Range, Nye County (Shawe et al., 1969); and in the Osgood Mountains, Humboldt County (Hotz and Willden, 1964).

Tennessee—An estimated 1.7 million tons of barite has been mined through 1972 from the deposits of Tennessee, mostly from the residual deposits of the Sweetwater district that lies in parts of Loudon, McMinn, and Monroe counties. The barite is associated with the residuum overlying the upper part of the Knox Group of Ordovician age. The bed rock contains thin veins and shatter zones filled with barite, fluorite, pyrite, quartz, and calcite that are considered the source of the material concentrated in the residuum. The deposits occur in three separate belts, each bounded by a thrust fault of southeasterly dip. The large deposits cover several acres and are as much as 60 to 80 ft thick. Other residual deposits are known in the Fall Branch-Greenville district in northeast Tennessee. The Pall Mall mine, Fentress County, is a unique deposit in clay that overlies limestone in the St. Louis limestone of Mississippian age. The source beds for the barite are not obvious.

In the Del Rio district, Cocke County, barite occurs with pyrite, fluorite, chert, quartz, hematite, ankerite, and dolomite in veins and replacements of fault breccia associated with thrust faults involving sedimentary rocks of Precambrian and Cambrian age.

Vein deposits containing barite, fluorite, and calcite with smaller amounts of sphalerite, galena, pyrite, and local traces of chalcopyrite and greenockite are associated chiefly with host rocks of Ordovician age in the Central Tennessee district.

The geology of the barite deposits of Tennessee has been described by Maher (1970).

California—Barite is known from more than 150 localities in California, but more than 90% of the nearly one million tons of barite produced in the state through 1970 has been mined from five deposits in the Sierra Nevada region: the Nine Mile Canyon mine, Tulare County; the El Portal mine, Mariposa County; the Democrat and Spanish mines, Nevada County; and the Almanor mine, Plumas County (Weber, 1966). In recent years most of the state's production has come from the dark bedded deposit in rocks of Devonian age on Girard Ridge in the Klamath Mountains, Shasta County (Weber and Matthews, 1967). Barite also occurs in the silver-bearing veins of the Calico district, San Bernardino County. In the Mountain Pass district, San Bernardino County, rare-earth minerals are mined from large reserves in carbonate bodies that also contain 20 to 25% barite which could be recovered as a byproduct (Olson et al., 1954).

Alaska—Barite deposits are known in the limestone belt of southeastern Alaska including those on Castle Island in Duncan Canal near Petersburg, at Lime Point on Prince of Wales Island, and Kuiu Island and the adjacent Keku Islands in the vicinity of Cornwallis Peninsula (Plafker and Berg, 1964). The deposits at and below tidewater on Castle Island have yielded most of the state's production of barite through 1972. Lode deposits of barite with silver and base metals near Glacier Creek (Skagway B-4 quadrangle) have been described by Mackevett (1971).

Deposits containing barite, fluorite, and base and precious metals occur in marble and schist along the road to Teller about 20 miles north of Nome on the Seward Peninsula (Brobst et al., 1971).

Descriptions of barium-bearing siliceous rocks of mid-Paleozoic to possibly Cretaceous age in the Brooks Range suggest that bedded barite deposits may occur in the region (Tailleur, 1970).

Other States—In addition to the deposits already mentioned in Georgia and Tennessee, many scattered residual and some vein deposits of barite occur in other parts of the Appalachian region (Brobst and Hobbs, 1968), including Pennsylvania, Virginia, Kentucky, North and South Carolina, and Alabama. In the central region, barite occurs in the Illinois-Kentucky fluorspar district, in the lead-zinc veins of the southwestern Wisconsin district, and at scattered localities in Oklahoma and Texas. Barite deposits mostly of the vein and cavity-filling type occur throughout the Rocky Mountain region in Arizona (Stewart and Pfister, 1960), New Mexico (Williams et al., 1964), Utah (Brobst, 1964b), Colorado (Brobst, 1964), Idaho (Brobst, 1964a), and Montana. In Washington, barite has been produced from vein and bedded deposits, chiefly in Stevens and Pend Oreille counties (Mills et al., 1971; Moen, 1964).

Foreign Deposits: Canada became a large producer of barite in 1942 when large deposits associated with faulted and folded sedimentary rocks of Mississippian and Triassic age were opened in the vicinity of Walton, Nova Scotia (Boyle, 1963). In 1978 flooding closed the Walton mine near Pembrooke. Mining had been conducted since 1941 and the ore body had essentially been exhausted (Morse, 1980). Large deposits also occur in veins and in limestone, especially in British Columbia (Morrow, et al., 1978). Large bedded deposits of mostly barite alone or associated with fine-grained sulfides of base metals occur in shales of Paleozoic age in the Selwyn basin in the Yukon Territory (Carne, 1976; Blusson, 1976). Other resources of barite are accumulating in Canadian tailings ponds at the estimated annual rate of 225,000 tons (Ross, 1960). Mountain Minerals, Ltd. is reworking several old tailings ponds from lead-zinc operations to recover the barite.

Mexico is a large producer of vein barite largely from the Muzquiz district, Coahuila and the Galeana district, Nuevo Leon. Large deposits of black bedded barite similar to those of Nevada have been reported in northern Mexico, but detailed descriptions have not as yet been published.

The vein deposits of the Rimac Valley, Peru and the bedded deposits of Camamu Bay region in northern Brazil have been the chief sources of barite in South America. Argentina, Chile, Columbia, and Cuba also have recorded barite production.

Germany supplied nearly half of the world's output of barite prior to World War II, but production has declined in recent years. At Meggen in Westphalia, West Germany, large deposits of bedded barite associated with sphalerite, pyrite, galena, and tetrahedrite in marine rocks of Devonian age recently have been exhausted (Krebs, 1981). Barite has been mined at Rammelsberg in the western Harz mountains. Bedded deposits also occur in rocks of Devonian age in the Saarland (Krebs, 1970). Other barite deposits mostly of the vein type are in Hesse, Baden, and Bavaria. In East Germany, the barite deposits are concentrated in the Erzgebirge, Harz Mountains, Vogtland, and Thuringia (Klaus and Schröder, 1967).

Barite is produced in many other nations of Europe, especially in the Union of Soviet Socialist Republics, from vein deposits in the Transcaucusus region of Georgia, Armenia, and Azerbaijan; in Italy, from the veins with base metals in Sardinia; in France, chiefly from vein deposits in the Central Massif and southern regions; in Yugoslavia, from vein and bedded deposits; in Greece, from vein deposits on the Aegean Islands of Milos, Mykonos, Polyvos, and Krimlos; in Great Britain, from vein deposits in northern England and Scotland; in Ireland, from large bedded deposits in County Tipperary; and in Romania, Bulgaria, Poland, Czechoslovakia, Austria, Spain, and Portugal, from various deposits, but mostly of the vein type.

Most of the barite mined in Africa has come from large vein deposits in Morocco and Algeria. Barite veins in gneiss of Precambrian age in the Gibi area, Liberia, are of commercial interest.

Vein deposits of barite have been mined in India, Iran, the Democratic Peoples Republic of Korea (North Korea), the Republic of Korea (South Korea), Japan, and the Philippines. Mainland China is a major producer of barite in Asia, but little is known about the geology of the deposits. In Thailand, barite deposits termed bedded replacements of sedimentary rocks of Devonian and Carboniferous age recently have come into production.

Commercially significant deposits of bedded barite have been found in Proterozoic sedimentary rocks in the Mangampeta area, Cuddapah district, Andhra Pradesh, India (Kurien, et al., 1977; Bose and Vaidyanathan, 1979).

In Oceania, barite deposits are known in Australia, New Zealand, and the Fiji Islands. The major production in the region has been from vein deposits in rocks of Precambrian and Cambrian age in the Flinders and Mount Lofty Ranges, South Australia (Gourlay, 1965).

The nations of the world that have recorded production of barite from 1945 through 1979 are listed in Table 1.

Reserves

The present status of knowledge of world barite reserves is summarized in Table 1. A broad use of the term *reserves* is necessary here because of the many shades of meaning the word expresses when used by authors throughout the world. The term *reserves* is used here to include bodies of barite-rich rock of sufficient size and grade to be mined now or to warrant serious consideration for mining under slightly more favorable economic conditions. Resources of barite, a category that includes reserves plus other barite-rich materials that will have to wait for greater changes in economics and technology before being utilized, are large and include such unconventional materials as the barite nodules of the sea floor (Brobst, 1973).

Exploration

Highly sophisticated technical methods for prospecting barite deposits have not yet been developed, chiefly because new deposits have been fairly easy to find in known districts.

The success of future prospecting will be increased if prospecting is guided by a knowledge of the geologic association and the role of barium in various geologic environments, especially in the sedimentary environment (Brobst, 1980). Deposits should be sought in residuum overlying carbonate-rich rocks, especially those of Cambrian and Ordovician age. Other areas of well-weathered rocks that have been hosts for veins or pods of barite also should be examined for residual deposits. The fetid black beds of barite occur abundantly in sequences of dark siliceous rocks of mid-Paleozoic age, especially those of Devonian and Mississippian age. These black beds may be overlooked because the rocks are easily confused with dark impure limestones, except for the noticeably greater weight of the barite-rich beds. The widespread, but probably mistaken, idea that bedded deposits must be of hydrothermal origin has diverted the attention of prospectors from search in sedimentary basins away from known igneous rocks. If, indeed, the apparent positive correlation between bedded barite deposits and organic materials discussed in the geologic sections of this chapter is correct, then the possibility of finding barite deposits at sites of ancient and modern lakes and swamps containing sufficient sulfate, especially in coastal areas, must be considered seriously.

Geophysical techniques to explore for barite deposits have not been used widely. Gravity surveys for residual barite deposits have been undertaken in Missouri (Uhley and Scharon, 1954) where they indicated that large concentrations of barite could be outlined and the tonnage computed with an expected error of 35% or less. The gravity study must be followed, however, with test pitting to determine the cause of gravity anomalies which reflected not only the presence of ore but also of pinnacles of bedrock, flint bars, and unknown concentrations of mass and/or probable errors. In East Germany, the use of geophysical techniques to search for hidden deposits of barite and fluorspar was described as less than ideal (Klaus and Schröder, 1967).

Useful geochemical techniques are available for exploration. A relatively simple and inexpensive turbidimetric test for barium was applied in a region of bedded barite deposits

TABLE 1—Barite: World Production, 1945 through 1979, and World Reserves, 1980

	Production, Thousands of Short Tons	Reserves, Millions of Tons
North America		
United States	35,292	78
Mexico	7,244	15
Canada	5,407	50
West Indies and Central America		
Cuba	39	*
Jamaica	—	*
Belize	—	*
South America		
Peru	4,875	5
Brazil	1,451	5
Argentina	787	*
Chile	559	5
Colombia	219	1
Bolivia	—	0.5
Europe		
German Federal Republic (West Germany)	12,468	4
USSR	7,533	10
Italy	4,851	6
Ireland	3,861	3
France	3,341	5
Yugoslavia	2,600	12
Greece	2,557	4
Romania	2,369	2
United Kingdom	2,117	2
Poland	1,251	2
Spain	893	5
German Democratic Republic (East Germany)	614	3
Bulgaria	150	33
Czechoslovakia	99	2
Austria	94	0.3
Portugal	31	0.6
Sweden	7	—
Belgium	—	0.6
Africa		
Morocco	2,522	5
Algeria	1,570	5
Tunisia	130	2
Republic of South Africa	84	3
Egypt	62	*
Zimbabwe	10	1
Swaziland	10	1
Kenya	5	*
Liberia	—	2
Malagasy Republic	—	0.5
Zambia	—	*
Tanzania	—	*
Uganda	—	*
Somalia	—	*
Near East		
India	2,680	65
Iran	1,641	3
Turkey	1,294	4
Pakistan	177	4
Burma	275	0.2
Afghanistan	47	3
Saudi Arabia	—	0.1
Far East		
People's Republic of China	5,005	30
Democratic People's Republic of Korea (North Korea)	2,117	3
Thailand	1,811	12
Japan	1,307	3
Republic of Korea (South Korea)	28	3
Philippines	59	*
Oceania		
Australia	569	5
Fiji	—	*
Total	122,112	406

* No reserve figures available, but some reserves probable; —, no production data available. Production data from US Bureau of Mines *Minerals Yearbooks*. Reserve data revised 1981 from Brobst (1970, Table 4).

in western Arkansas (Brobst and Ward, 1965). It was concluded from the study that barium anomalies could be found and that targets for further study could be outlined. Geochemical studies can be carried out with the use of emission spectrographic and X-ray fluorescence techniques, but initial cost of the equipment is high. The use of a portable radioisotope fluorescence analyzer in geochemical prospecting for barite and celestite in the west of England has been described by Ball, et al. (1979).

Evaluation of Deposits

Evaluating the commercial potential of a specific barite prospect is a complex process that requires careful study of the relation of pertinent geologic, technologic, and economic factors. Geologic factors to consider are the type of deposit and its structure, size, grade, and mineralogy. Knowledge of zoning, texture, grain size, and chemical composition of the ore minerals and the associated gangue is also critical to the geologic evaluation. Geologic information puts limits on technical factors such as the choice of mining and beneficiation methods to be used. Economic factors of major importance include: location of the deposit with respect to markets, current and future, and transportation to them; costs of land acquisition; costs of labor and equipment to mine, beneficiate, and process the ore; and problems of financing and taxes.

The summaries of the characteristics of barite deposits and references to more detailed descriptions of deposits already given should be of value in the geologic appraisal of the commercial potential of deposits.

Barite prospects generally are evaluated geologically by detailed geologic mapping and surveying as well as by digging test pits and trenches for further examination and sampling of the rock body. Drilling is commonly used for evaluation of vein and bedded deposits, and also has been used successfully in residual deposits (Kesler, 1949).

Beneficiation tests must be made to assure that the anticipated mineral separation and recovery are attainable and that barite products can be made profitably to meet the specifications of the purchaser. In some cases, barite is only one of the salable products recoverable from a deposit. Some barite users do not like to buy byproduct barite. They fear that the availability of the barite is unsteady because its production is tied to the demand for the mine's primary product. Byproduct barite probably will find greater acceptance in future markets; its potential value should not be ignored when estimating the ultimate worth of an ore deposit.

Preparation for Market

Mining and Beneficiation

Residual barite deposits generally are mined by open pit methods. After the topsoil is stripped away, the barite-bearing residuum is dug with 1.5 to 1.75-yd power shovels, although draglines and front-end loaders are used in some mines. The ore is loaded into 15 to 30-ton off-the-road dump trucks for hauling to the washer plant, rarely more than three miles from the operating pit.

The ore-bearing residuum is dumped into a hopper from which it is fed to rotary breakers using water at high pressure. The softer barite breaks up more readily than the gangue materials; the barite moves on through the circuit, but the coarse gangue is diverted to the dump. The barite and remaining gangue move on to the log washers which separate the clay from the gravel and the barite. The clay particles flow away, by gravity, to the tailings pond. A secondary rotary breaker or trommel screen receives the product of the log washer. Oversized material from the trommel is byproduct used for road building. The baritic material that passes the trommel screen is moved to jigs which separate, by gravity, barite from gravel. The gravel-sized waste from this step also goes to use as road building material. The barite product is then ready for grinding or other processing. A typical flowsheet for this process is shown in Fig. 1.

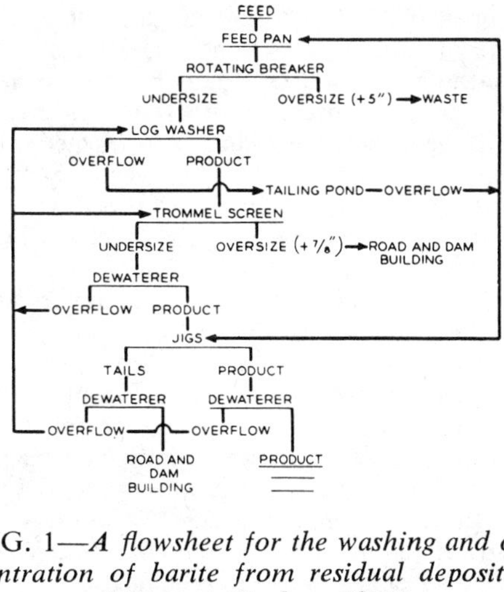

FIG. 1—*A flowsheet for the washing and concentration of barite from residual deposits in Missouri (Sackett, 1962).*

Washer plants vary in size, but the capacity generally ranges from 70 to 120 cu yd per hr of residuum. As much as 5000 gpm of water is used in large washers. It is said that 100 to 200 lb of barite is recovered per cubic yard of residuum measured in the ground, but the efficiency of the plants varies greatly. The recovery of the barite in the jig product may range from 65 to 85% (Wharton, 1972).

According to Wharton (1972), conventional crushing equipment is not currently used in the beneficiation of residual barite in Missouri because the beneficiation process itself accomplished this mission. The major component of these ores is still clay, but the solid fragments other than the barite and dolomite are hard, abrasive forms of silica-chert, chalcedony, and drusy quartz. In the normal agitation of the beneficiation process, the soft and brittle barite is dislodged from the other materials. Considerable amounts of barite are lost with the coarse gangue, but about two-thirds of the total losses are believed to travel with the slimes to the tailings ponds. Perhaps 20 to 30% of barite in the original ore is lost and comes to rest in the tailings ponds. Wharton (1972) estimated that 1.9 million tons of barite might be available in 67 tailings ponds in the Washington County district, Missouri, an amount equal to about 17% of the recorded production of the district through 1971.

Bedded and vein deposits may be mined by either open pit or underground methods, depending on the characteristics of each deposit. Open pit methods are easier and less costly than methods generally employed underground. At Chamberlain Creek, near Magnet Cove, AR, the large bedded deposit has been mined by both open pit and underground methods. The bedded deposits of the Shoshone Range, NV, are mined in open pits where blasting is not always required because the rocks in many deposits have been tectonically deformed and broken.

Barite from many vein deposits requires concentration to meet the buyer's specifications. Depending upon the suite of minerals in the vein deposit, beneficiation by the simple, inexpensive gravity separations is ineffective and the more complicated and expensive flotation techniques are needed. Intergrowths of barite and fluorite are common, but the two minerals were impossible to separate commercially until recently when flotation methods were developed (Eddy and Browning, 1964). Many vein deposits containing sulfide minerals could yield barite to meet commercial specifications as a coproduct. In many places, however, this operation is still an unprofitable venture, but more for economic than technical reasons.

The ore from some bedded deposits requires no beneficiation to meet certain user's specifications, especially the requirements for drilling mud, but ore from many deposits needs some concentration. The bedded ore from the deposit near Magnet Cove, AR, is beneficiated by double crushing, jigging, wet grinding in ball mills to pass 325 mesh followed by flotation of barite with anionic reagents, according to the general scheme shown in Fig. 2 (Sackett, 1962).

Processing

Barite is available from producers as crude lumps and as a jigged or flotation concentrate. For many uses the lump or jigged concentrate must be ground to meet the user's specifications. Grinding is done either wet or dry, but much barite is ground wet because it is still wet from prior steps of beneficiation or will be wet again in further stages of processing, as in bleaching with sulfuric acid to remove objectionable iron stains. Barite is ground in ball mills, if iron contamination is unimportant. Raymond mills have been used for grinding in many places. Most ground barite is packed in paper bags holding 100 lb, except bleached barite which may be packed in 50-lb bags.

Preparation of Chemicals: The first step in the chemical preparation of barium compounds is the roasting of "hard" lump barite with carbon in a rotary kiln at a temperature of about 2400°F. This process reduces the barium sul-

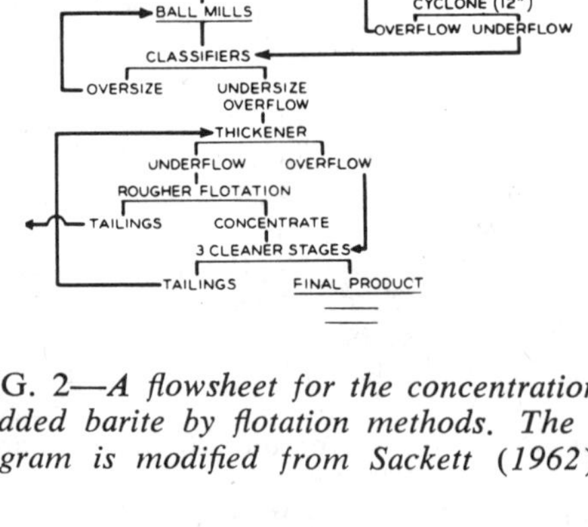

FIG. 2—*A flowsheet for the concentration of bedded barite by flotation methods. The diagram is modified from Sackett (1962).*

fate that is insoluble in water to barium sulfide (black ash) that is soluble in water. The barium sulfide product is leached to yield a solution that, because of hydrolysis, contains $Ba(OH)_2$ and $Ba(SH)_2$. The solution is filtered to remove excess carbon and impurities. High purity barite is needed for the roasting process because iron oxide, silica, and alumina can react with barium to form insoluble compounds. One percent iron oxide can convert as much as 4% barium sulfide to insoluble compounds.

The solution leached from the black ash is the basic raw material for making lithopone, blanc fixe, and other barium chemicals. If a solution of zinc sulfate is added, the resulting precipitate is a mixture containing about 30% zinc sulfide and 70% barium sulfate. This precipitate is filtered, washed, dried, and calcined under carefully controlled conditions, quenched in water, wet ground, and then dried to yield the white pigment known as lithopone. Approximately one ton of barite is required to manufacture each ton of lithopone, even though the product is only 70% $BaSO_4$.

If sodium sulfate is added to the black ash solution, the chemically pure precipitate of barium sulfate, called blanc fixe, is formed.

If sodium carbonate is added to the black ash solution, the precipitate is barium carbonate. The precipitate may be filtered, washed, dried, and ground for use as the carbonate or for conversion to other barium compounds. Heating the carbonate in the presence of carbon yields barium oxide which then can be oxidized to barium peroxide or hydrated to barium hydroxide. The nitrate and chloride of barium generally are prepared by treating barium carbonate with the corresponding acids.

Barium metal may be produced by electrolyzing barium chloride in the fused state or by the Guntz process (Johnstone and Johnstone, 1964) in which barium oxide is reduced by metallic aluminum or silicon at about 1200°C under reduced pressure in an electric furnace.

Barium chemicals are currently produced by the J. T. Baker Co., Phillipsburg, NJ; Chemetron Corp., Huntington, WV; Chemical Products Corp., Cartersville, GA; The Great Western Sugar Co., Johnstown, CO; Inorganic Chemicals Div., F M C Corp., Modesto, CA; Mallinckrodt Chemical Works, St. Louis, MO; Ozark Smelting and Mining Co., Coffeyville, KS; and Sherwin-Williams Co., Ashtabula, OH.

Transportation

Barite is a commodity of small unit value and, therefore, the cost of transportation to market is an important consideration. Only temporary demands can be supplied by higher than normal costs of transport. Marine transport is sufficiently low in cost to allow long hauls of foreign barite to US ports on the Atlantic and Gulf Coasts at prices competitive with that from domestic sources, despite the additional costs of tariffs.

Crude barite may be shipped in bulk in open railroad cars. Ground barite generally is shipped in 100-lb bags or in bulk in covered hopper cars. Long rail and truck hauls generally are prohibitive in cost, although an alteration in rail-freight rates recently has opened Gulf Coast markets to barite mined in Nevada.

Marketing and Consumption

The principal marketing areas for barite in the United States are the Eastern states, the Midwestern states, the Gulf Coast, and California. All of these areas use ground and crude or crushed barite. Most of the barite for markets in the Eastern states is purchased from deposits in Georgia, Tennessee, and South Carolina, and additional supplies from abroad are received at East Coast ports. Most of the barite used in this region is used for chemicals and as a raw material in various manufacturing processes.

Barite for drilling mud in the Gulf Coast area is supplied from Arkansas, Missouri, and Nevada. Domestic supplies in this major consuming region were augmented in the last decade by imports of crude barite, chiefly from Canada, Ireland, Mexico, and Peru. Much of the imported barite is ground at the plants of the major companies at New Orleans and Lake Charles, LA, and at Houston and Brownsville, TX.

Barite is mined in Missouri for the chemical and manufacturing industries of the Midwestern states, located principally in Missouri, Illinois, and Kansas.

Rising demands for barite in the oil and chemical industry on the West Coast are met by the mines of Nevada and California.

Mud-grade barite for the Rocky Mountains and the northern plains has come from mines in Idaho, Montana, Nevada, and Washington. Some barite is imported from British Columbia, which is the chief supplier for the oil fields of Alberta.

Some mud-grade barite has been produced in Alaska for use by the petroleum industry north of the Brooks Range.

The total domestic consumption of barite fluctuates annually with the rise and fall of drilling in the petroleum industry. Statistics available in the *Minerals Yearbook* compiled annually by the US Bureau of Mines recount the history in detail. From 1960 to 1973, the US consumed annually about 1.5 million tons of barite, of which it produced nearly 1 million tons and imported the remainder. About 80% was used in drilling mud; 10% in chemicals; 5% in filler, mostly paint and rubber; and 5% in glass and other uses. By 1980, domestic production had risen to an estimated 2.7 million tons and imports to 1.8 million tons from an estimated 8.8 million tons world production. Production in 1981 was estimated at 2.4 million tons with an estimated 1.9 million tons imported out of a total world production of 8.3 million tons.

Current prices for barite are available regularly in the monthly issues of the *Engineering & Mining Journal.* In October 1982, lump barite suitable for the chemical and glass industries was quoted at $90 per st for hand-picked material. A flotation product for these uses was listed at $105 per st. Imported mud-grade barite (specific gravity 4.2 to 4.3) c.i.f. Gulf ports sold for $32 to $59 per st. Wet-ground barite (—325 mesh, in 50-lb bags) sold for $80 to $155 per ton in carload lots f.o.b. shipping point. Dry-ground, mud-grade barite (specific gravity 4.2 to 4.3) sold for $87 to $120 per ton f.o.b. shipping point. Imported ground mud-grade barite sold for $65 to $75 f.o.b. shipping point. Bulk shipments of crude barite can compete easily with domestically produced barite despite the tariff of $1.27 per lt at the 1982 rates.

The quoted prices in many cases are flexibly structured, depending on the desired specifications and the quantity needed by the purchaser. Most of the finished mud-grade product is sold with engineering services by the major companies, the largest of which are the Baroid Div., NL Industries, Inc.; Dresser Minerals Div., Dresser Industries, Inc.; Minerals Div., Milchem, Inc.; and IMCO Services Div., Halliburton Co. Small producers commonly sell their barite to major users according to specially negotiated terms that generally vary somewhat from published prices.

Future Considerations and Trends

The current upward trend of world barite production will probably continue, but the rate of increase will depend mostly on the rates of drilling for oil and gas which seem likely to continue at high levels for two to three more decades. Annual world production of barite doubled from about two million tons to four million tons between 1950 and 1970. Production doubled again to 8 million tons annually by 1980, only one decade later, when the cumulative production for that period amounted to more than 60 million tons. The 400 million tons of barite in the reserve do not appear especially large when considering the current production rates and status of detailed knowledge of some of the deposits which are remote from current markets. Those observations come sharply into focus when compared with the estimated total world production of about 95 million tons from 1850 to 1970, of which 70 million tons were mined from 1945 to 1970. New deposits must yet be sought, found, and developed, a process which takes place slowly.

Geologic factors suggest that world barite production can be increased in the years ahead and that the possibilities for discovering new barite deposits throughout the world are favorable. Among the most worthy targets are deposits of bedded barite similar to those in Arkansas and Nevada. Such deposits probably have not yet been sought seriously or recognized in many areas. These deposits tend to be large and are amenable to large-scale, low-cost operations.

Vein deposits mined chiefly for barite are yielding major amounts of ore in many countries of the world, and substantial amounts will continue to be obtained from vein deposits of all sizes. Smaller vein deposits, however, will lose their economic significance in the more advanced stages of industrialization as has already happened in the United States. The search will have to be pressed for larger deposits with great volumes of ore.

An increasingly important source of barite may well be as a coproduct from the mining of iron, base-metal, fluorspar, and rare-earth ores in deposits already known in many parts of the world. Larger volumes of barite also may be recovered from mine dumps and tailings ponds, where the barite already is partly ground and concentrated.

Too few data are available to evaluate the possibilities of increasing the world's reserves of barite in residual deposits. Prospecting and assessing residual barite deposits are difficult. Despite the problems, much barite has been mined from residual deposits in the United States, where the deposits have been large, although relatively low grade, and easy to mine by open pit methods. Ore from residual deposits generally can be beneficiated easily and inexpensively.

Substitutes

Substitutes for barite have been tried in drilling mud, but all have generally been found unsatisfactory. Celestite ($SrSO_4$) has been tried and rejected because of its higher cost and lower specific gravity (3.95 to 3.97). Iron ores have been found more abrasive and unpleasant to handle because of the color. Galena (PbS) has been tried, but found too expensive.

The use of lithopone as a pigment has nearly ended because of greater use of titanium dioxide pigments.

The industrial uses of most barium compounds will continue because of their cost advantages compared to satisfactory substitutes.

If a less expensive method can be devised to prepare barium metal, it may come to wide use in a compact energy cell that also utilizes air and chlorine (Lewis, 1970).

Barium chloride will find wider application in water purification and pollution abatement, if the cost of manufacture can be reduced.

Taxes and Tariffs

Barite producers have a depletion allowance of 14% on both foreign and domestic production. Some states have severance taxes.

Tariffs set by the "Kennedy Round" on imported barite and barium products for 1972 include: \$1.27 per lt on crude barite; \$3.25 per lt on barite ground or otherwise manufactured; and a range of \$0.60 to \$3.25 per lb on various barium chemicals. In 1972, crude witherite was importable free of duty, but ground witherite had a 6% ad valorem tax.

Barite and barium compounds have not been included in the list of strategic materials to be stockpiled because the United States is potentially self-sufficient. Barite deposits and chemical plants are scattered around the nation. The nations of North America possess a substantial part of the world's known reserves of barite.

Recycling and Ecology

Almost none of the barite now mined finds its way to recycled use. The drilling mud residual is generally abandoned in waste ponds after the well is completed. Most of the barite used annually thus is diluted, scattered, and lost. The waste ponds are easily rehabilitated. Most of the other uses are such that the contained barite or barium compound is virtually impossible to reclaim.

The barite industry has no abnormally difficult ecology problems to solve other than those attendant to the mining and beneficiation of ores. Barite flotation plants and washers need tailings ponds to store their generally inert wastes. Some toxic gases are generated in chemical plants, especially hydrogen sulfide in the manufacture of barium chloride. The toxic gases can be withheld from discharge, and the hydrogen sulfide is converted to other sulfide products.

Bibliography and References

Anon., 1961, *Barium Bibliography,* Mineral Products Div., Food Machinery & Chemical Corp., New York, 433 pp.

Anon., 1970, "The World's Barite Industry," *Industrial Minerals,* London, No. 32, May, pp. 15–32.

Anon., 1978, "Barytes—Its Uses Outside Drilling Muds," *Industrial Minerals,* No. 133, Oct., pp. 33–43.

Anon., 1978a, "Barytes—Producers Review," *Industrial Minerals,* No. 130, July, pp. 33–47.

Ball, T.K., et al., 1979, "Geochemical Prospecting for Barite and Celestite Using a Portable Radioisotope Fluorescence Analyser," *Journal of Geochemical Exploration,* Vol. 11, pp. 277–284.

Bates, R.L., 1960, *Geology of the Industrial Minerals and Rocks,* Harper & Brothers, New York, pp. 287–296.

Blusson, S.L., 1976, "Selwyn Basin, Yukon and District of Mackenzie," Paper 76–A1, Geological Survey Canada, pp. 131–132.

Bose, R.N., and Vaidyanathan, N.C., 1979, "Gravity Surveys around Mangampeta, Cuddapah District, Andhra Pradesh," *Journal* Geological Survey of India, Vol. 20, No. 11, pp. 540–547.

Boyle, R.W., 1963, "Geology of the Barite, Manganese, and Lead-Zinc-Copper-Silver Deposits of the Walton-Cheverie Area, Nova Scotia," Paper 62–65, Canada Geological Survey, 36 pp.

Brobst, D.A., 1958, "Barite Resources of the United States," Bulletin 1072-B, US Geological Survey, pp. 67–130.

Brobst, D.A., 1964, "Barite," *Mineral and Water Resources of Colorado,* Report for Committee on Interior and Insular Affairs, US Senate, pp. 157–160.

Brobst, D.A., 1964a, "Barite," *Mineral and Water Resources of Idaho,* Special Report 1, Idaho Bureau of Mines & Geology, pp. 46–48.

Brobst, D.A., 1964b, "Barite," *Mineral and Water Resources of Utah,* Bulletin 73, Utah Geological & Mineralogical Survey, pp. 154–157.

Brobst, D.A., 1965, "Barite in the United States," Mineral Investigations Resources Map MR-43, US Geological Survey.

Brobst, D.A., 1970, "Barite: World Production, Reserves, and Future Prospects," Bulletin 1321, US Geological Survey, 46 pp.

Brobst, D.A., 1973, "Barite," in Brobst, D.A., and Pratt, W.P., "United States Mineral Resources," Professional Paper 820, US Geological Survey, pp. 75–84.

Brobst, D.A., 1980, "Barite Resources in Their Geologic Framework," *Proceedings of Minerals and Chemicals in Drilling Muds—The 80's and Beyond*, P.W. Harben, ed., *Industrial Minerals*, New York, pp. 30–39.

Brobst, D.A., and Hobbs, R.G., 1968, "Barite" in "Mineral Resources of the Appalachian Region," Professional Paper 580, US Geological Survey, pp. 270–277.

Brobst, D.A., et al., 1971, "Geology and Geochemistry of the Sinuk River Barite Deposit, Seward Peninsula, Alaska," *Geological Survey Research*, Professional Paper 750-D, US Geological Survey, pp. D1-D8.

Brobst, D.A., and Wagner, R.J., 1967, "Barite," *Mineral and Water Resources of Missouri,* Vol. 43, Missouri Div. of Geological Survey and Water Resources, pp. 99–106.

Brobst, D.A., and Ward, F.N., 1965, "A Turbidimetric Test for Barium and Its Geological Application in Arkansas," *Economic Geology,* Vol. 60, No. 5, pp. 1020–1040.

Butts, C., and Gildersleeve, B., 1948, "Geology and Mineral Resources of the Paleozoic Area in Northwest Georgia," Bulletin 54, Georgia Geological Survey, pp. 83–85.

Carne, R.C., 1976, "Stratabound Barite and Lead-Zinc-Barite Deposits in Eastern Selwyn Basin, Yukon Territory," Canada Indian and Northern Affairs Open-File Report EGS–1976–16, 41 pp.

Dean, B.G., and Brobst, D.A., 1955, "Annotated Bibliography and Index Map of Barite Deposits in the United States," Bulletin 1019-C, US Geological Survey, pp. 145–186.

Dunham, A.C., and Hanor, J.S., 1967, "Controls on Barite Mineralization in the Western United States," *Economic Geology,* Vol. 62, No. 1, pp. 82–94.

Eddy, W.H., and Browning, J.S., 1964, "Selective Flotation of a Barite-Fluorspar Ore from Tennessee," Report of Investigations RI-6491, US Bureau of Mines, 8 pp.

Ehrenberg, H., Pilger, A., and Schröder, F., 1954, "Das Schwefelkies-Zinkblende-Schwerspatlager von Meggen (Westfalen)," Beihefte No. 12, Geologisches Jahrbuch, 352 pp.

Gourlay, A.J.C., 1965, "Barite," *Australian Mineral Industry,* McLeod, ed., Bulletin 72, Australia Bureau of Mineral Resources, Geology & Geophysics, pp. 61–72.

Harben, P.W., 1978, "Raw Materials for the Oil Well Drilling Industry: An *Industrial Minerals* Consumer Survey," *Industrial Minerals,* London, 144 pp.

Horton, R.C., 1963, "An Inventory of Barite Occurrences in Nevada," Report 4, Nevada Bureau of Mines, 18 pp.

Horton, R.C., 1964, "Barite," *Mineral and Water Resources of Nevada,* Report, US Geological Survey and Nevada Bureau of Mines, 18 pp.

Hotz, P.E., and Willden, R., 1964, "Geology and Mineral Deposits of the Osgood Mountains Quadrangle, Humboldt County, Nevada," Professional Paper 431, US Geological Survey, 128 pp.

Johnstone, S.J., 1954, "Barium," *Minerals for the Chemical and Allied Industries,* 1st ed., John Wiley, New York, pp. 64–74.

Johnstone, S.J., and Johnstone, M.C., 1961, "Barium," *Minerals for the Chemical and Allied Industries,* 2nd ed., John Wiley, New York, pp. 59–69.

Joseph, J., 1979, "Mud Suppliers Face Baryte Squeeze," *Industrial Minerals,* No. 146, Nov., pp. 33–38.

Kesler, T.L., 1949, "Occurrence and Exploration of Barite Deposits at Cartersville, Georgia," *Trans. AIME,* Vol. 184, pp. 371–375.

Kesler, T.L., 1950, "Geology and Mineral Deposits of the Cartersville District, Georgia," Professional Paper 224, US Geological Survey, 97 pp.

Ketner, K.B., 1965, "Economic Geology" *in* Gilully, J., and Gates, A., "Tectonic and Igneous Geology of the Northern Shoshone Range, Nevada," Professional Paper 465, US Geological Survey, pp. 129–144.

Klaus, D., and Schröder, N., 1967, "Erfahrungen beim Einsatz geophysikalischer Messverfahren in der Flusspat—und Schwerspaterkundung," *Zeitschrift für Angewandte Geologie,* Vol. 13, No. 11–12, pp. 610–618.

Krebs, W., 1970, "Nachweis von Oberdevon in der Schwerspat-Gube Eisen (Saargebiet) und die Folgerungen für Paläogeographie und Lagerstättenkunde des linksrheinischen Schiefergebirges," Neues Jahrbuch für Geologie und Palaontologie Monatsheft, Jahrgang, Heft 8, pp. 465–480.

Krebs, W., 1981, "The Geology of the Meggen Ore Deposit," *Handbook of Strata-Bound and Stratiform Deposits,* K.H. Wolf, ed., Elsevier Scientific Publishing Co., Amsterdam, pp. 509–549.

Kurien, T.K., et al., 1977, "Baryte Mineralization and Its Origin, Mangampeta Cuddapah District, Andhra Pradesh," *The Indian Mineralogist,* Vol. 18, pp. 1–6.

Lewis, R.W., 1970, "Barium," *Mineral Facts and Problems,* Bulletin 650, US Bureau of Mines, pp. 865–877.

Mackevett, E.M., Jr., 1971, "Analyses of Samples and Preliminary Geologic Summary of Barite-Silver-Base Metal Deposits near Glacier Creek, Skagway B-4 Quadrangle, Southeastern Alaska," Open File Report, US Geological Survey, 8 pp.

Maher, S.W., 1970, "Barite Resources of Tennessee," Report of Investigations 28, Tennessee Div. of Geology, 40 pp.

Matthews, R.A., and Alfors, J.T., 1962, "Sanbornite from Rush Creek, Fresno County," *Mineral Information Service,* California Div. of Mines and Geology, Vol. 15, No. 6, pp. 1–3.

Miller, R.E., et al., 1977, "The Organic Geochemistry of Black Sedimentary Barite—Significance and Implications of Trapped Fatty Acids," *Organic Geochemistry,* Vol. 1, pp. 11–26.

Mills, J.W., et al., 1971, "Bedded Barite Deposits of Stevens County, Washington," *Economic Geology,* Vol. 66, No. 8, pp. 1157–1163.

Moen, W.S., 1964, "Barite in Washington," Bulletin 51, Washington Division of Mines and Geology, 112 pp.

Morrow, D.W., et al., 1978, "A Hypothesis Concerning the Origin of Barite in Devonian Carbonate Rocks in Northeastern British Columbia," *Canadian Journal Earth Science,* Vol. 15, No. 9, pp. 1391–1406.

Morse, D.E., 1980, "Barite," Preprint, *Minerals Yearbook, 1978–1979*, US Bureau of Mines, 10 pp.

Olson, J.C., et al., 1954, "Geology of the Rare-Earth Deposits of the Mountain Pass District, San Bernardino County, California," Professional Paper 261, US Geological Survey, 75 pp.

Plafker, G., and Berg, H.C., 1964, "Nonmetallic Mineral Resources" *Mineral and Water Resources of Alaska,* Report for the Committee of Interior and Insular Affairs, United States Senate, pp. 125–148.

Ross, J.S., 1960, "The Barium Minerals Industry in Canada," Mines Branch Information Circular IC-126, Canada Dept. of Mines & Technical Surveys, 60 pp.

Sackett, E.L.H., 1962, "Barite-Little-Known Industry That Means 'Mud' to Oil Men," *Mining Engineering,* Vol. 14, No. 5, May, pp. 46–49.

Sawkins, F.J., 1966, "Ore Genesis in the Northern Pennine Orefield in the Light of Fluid Inclusion Studies," *Economic Geology,* Vol. 61, No. 2, pp. 385–401.

Scull, B.J., 1958, "Origin and Occurrence of Barite in Arkansas," Information Circular 18, Arkansas Geology & Conservation Commission, 101 pp.

Shawe, D.R., et al., 1969, "Newly Discovered Bedded Barite Deposits in Northumberland Canyon, Nye County, Nevada," *Economic Geology,* Vol. 64, No. 3, pp. 245–254.

Stewart, L.A. and Pfister, A.J., 1960, "Barite Deposits of Arizona," Report of Investigations RI-5651, US Bureau of Mines, 89 pp.

Tailleur, I.L., 1970, "Lead-, Zinc-, and Barite-Bearing Samples from the Western Brooks Range, Alaska," Open File Report, US Geological Survey, 16 pp.

Uhley, R.P., and Scharon, LeR., 1954, "Gravity Surveys for Residual Barite Deposits in Missouri," *Mining Engineering,* Vol. 6, No. 1, pp. 52–56.

Weber, F.H., Jr., 1966, "Barite" *Mineral and Water Resources of California,* Pt. 1, Bulletin 191, California Div. of Mines and Geology, pp. 94–98.

Weber, F.H., Jr., and Matthews, R.A., 1967, "Prospecting For Barite in Northern Shasta County," *Mineral Information Service,* Vol. 20, No. 9, California Div. of Mines and Geology, pp. 107–114.

Wharton, H.M., 1972, "Barite Ore Potential of Four Tailings Ponds, Washington County Barite District, Missouri," Report of Investigations 53, Missouri Geological Survey and Water Resources, 91 pp.

Williams, F.E., et al., 1964, "Barite Deposits of New Mexico," Circular 76, New Mexico Bureau of Mines & Mineral Resources, 46 pp.

Zimmermann, R.A., 1969, "Stratabound Barite Deposits in Nevada," *Mineralium Deposita,* Berlin, Vol. 4, pp. 401–409.

Zimmermann, R.A., and Amstutz, G.C., 1964, "Small Scale Sedimentary Features in the Arkansas Barite District," *Developments in Sedimentology,* Vol. 2, G.C. Amstutz, ed., Elsevier, New York, pp. 157–163.

Bauxitic Raw Materials

JAMES W. SHAFFER *

Aluminum is the most abundant metallic element and forms 8% of the earth's crust. Because of its chemical activity it does not occur in nature in a metallic form but principally in the silicates, feldspar, mica, shale, and clay. Bauxitic raw materials—bauxitic clay, aluminous laterite, and bauxite—develop from these aluminum silicate sources. Bauxite and aluminous laterite are naturally occurring mixtures of aluminum hydroxides and mineral impurities, and they are the major sources of aluminum.

End Uses

Bauxite is the basic ore of the aluminum industry. It is refined into aluminum hydroxide and calcined for feed stock to electrolytic smelters. Over 90% of the tonnage mined is refined and smelted into aluminum. The total tonnage consumed by other industries is small in comparison, but bauxite and alumina are important for the manufacture of refractories, ceramics, chemicals, abrasives, pharmaceuticals, catalysts, cements, dessicants, and flame retardants. Some aluminum-containing industrial products are made from clays and other materials but the majority originate from bauxite directly or from aluminum hydroxide and alumina refined from bauxite. Lateritic type bauxite is also used as building and decorative stone.

Geology

Mineralogy and Properties

Bauxite is a mineral raw material composed principally of one or more aluminum hydroxide minerals and impurities of silica, iron oxide, titania, and other elements in minor or trace amounts. The important aluminum hydroxide minerals are gibbsite, boehmite, and diaspore. Bayerite, nordstrandite, and tucanite are naturally occurring aluminum hydroxides but are rare and not commercially important. Mixtures of gibbsite and boehmite are common, boehmite and diaspore less common, and gibbsite and diaspore rare. Gibbsite is aluminum trihydroxide. Boehmite and diaspore are aluminum oxide hydroxides and have essentially the same composition, but diaspore has a greater hardness and a more tightly bonded and denser atomic structure.

Although not technically correct, the commercial terms *trihydrate* and *monohydrate* are frequently used to distinguish types of bauxite. If bauxite is all or nearly all gibbsite, it is referred to as a trihydrate ore; if boehmite or diaspore or a mixture of both are the dominant minerals, the bauxite is called a monohydrate ore. If substantial amounts of both gibbsite and boehmite are present, the bauxite is frequently referred to as a *mixed ore*. Properties of major bauxite minerals are shown in Table 1.

Silica, iron oxides, and titania are major impurities always present in deposits of bauxitic raw materials.. Silica impurities usually occur as clay minerals, micaceous minerals, or quartz, but cristobalite and feldspar are found in some deposits. The most common clay minerals are kaolins; however, illites and chlorites are often found in small quantities. Varying proportions of aluminum, iron, and clay minerals make bauxite, aluminous laterite, ferruginous laterite, and bauxitic clay deposits gradational with each other.

Hematite and goethite are the common iron mineral impurities in many bauxites. Magnetite and ilmenite occur in deposits derived from basic igneous rocks. Other iron minerals associated with bauxite are maghemite, pyrite, siderite, and marcasite.

The titanium minerals found in bauxites are anatase, leucoxene, rutile, ilmenite, titaniferous

* Consultant, Orinda, CA.

TABLE 1—Properties of Major Bauxite Minerals

	Gibbsite	Boehmite	Diaspore
Chemical Formula	$Al(OH)_3$ or $Al_2O_3 \cdot 3H_2O$	$AlOOH$ or $Al_2O_3 \cdot H_2O$	$AlOOH$ or $Al_2O_3 \cdot H_2O$
Al_2O_3 Content, %	65.4	85	85
H_2O Content, %	34.6	15	15
Crystal System	Monoclinic	Orthorhombic	Orthorhombic
Hardness, Mohs' Scale	2.3–3.5	3.5–5	6.5–7
Density, g/cm^3	2.3–2.4	3.01–3.06	3.3–3.5

magnetite, and sphene. In addition to iron and titanium, beryllium, calcium, carbon, cerium, chromium, columbium, fluorine, gallium, magnesium, manganese, phosphorous, potassium, thorium, uranium, vanadium, zinc, and zirconium have been found as minor impurities in bauxites.

Bauxite is one of the most variable mineral raw materials in physical appearance. Color variations include white, gray, pink, brown, yellow, tan, green, and nearly black. Bauxite may be earthy, clayey, massive, oolitic, pisolitic, nodular, fine-grained, vermicular, brecciated, saccharoidal, vesicular, botryoidal, or platy. The in-situ dry bulk density of most bauxites is between 1.3 and 2.4 but some are known to be as high as 3.7. This wide range is due to variations in porosity and the amount of iron minerals present.

Classification of Deposits

Bauxite deposits may be classified according to geologic origin, chemical composition, mineral composition, texture, topographic position, underlying rock type, and mode of occurrence. Examples of various types of deposits are shown in Fig. 1A–D.

Blanket Deposits: Bauxite and aluminous laterite deposits occurring as flat or nearly flat-lying layers near the surface are classified as blanket deposits. In plan they may extend for kilometers. Thicknesses may vary from a meter or less to 20 m, but 4 to 6 m is most common. Some deposits are at low elevation, covering many square kilometers of a gently undulating surface. Other deposits occur at the tops or upper slopes of hills, ridges, and plateaus. At some localities, the surface on which the bauxite formed has been eroded and dissected and the plateaus capped with bauxite are now outliers and remnants of an older, more extensive surface. Where the bauxite formation is massive and rocky, escarpments frequently form along the edges of plateaus, and bauxite boulders are abundant along the slopes.

Overburden on blanket deposits may be absent or range up to 10 m in thickness. Where overburden is absent, the surface is often rough and irregular with an abundance of loose bauxite.

A layer of massive bauxite containing abundant iron is common in the uppermost parts of blanket deposits. Where the deposit is made up of bauxite concretions distributed in a clay matrix, iron enrichment in the upper parts is not as common. Horizontal grade variations are usually gradual, but there are examples of high-grade bauxite surrounded by or associated with siliceous and ferruginous laterite. Large blanket deposits occur in South America, Australia, West Africa, and India and some of them are extremely high grade.

Pocket Deposits: On the Caribbean Islands of Jamaica and Hispaniola and in southern Europe, bauxite occurs in karst depressions. These occurrences are classified as pocket deposits, and their size and shape are extremely variable. Thicknesses of pocket deposits range from less than one to more than 30 m. In some areas the bauxite-filled depressions are isolated while in other areas the depressions coalesce and enclose large and extensive deposits. A common feature of pocket deposits in carbonate rocks is the abrupt and irregular contact between the bauxite and the underlying rock along with the upward extension of limestone pinnacles into the bauxite. Pocket deposits also occur as lenses within clay or as masses on igneous and metamorphic rocks.

Interlayered Deposits: Interlayered deposits are those which are overlain by younger sedimentary rocks or volcanics. They formed at the surface as one of the other types of deposits and subsequently were buried. Interlayered

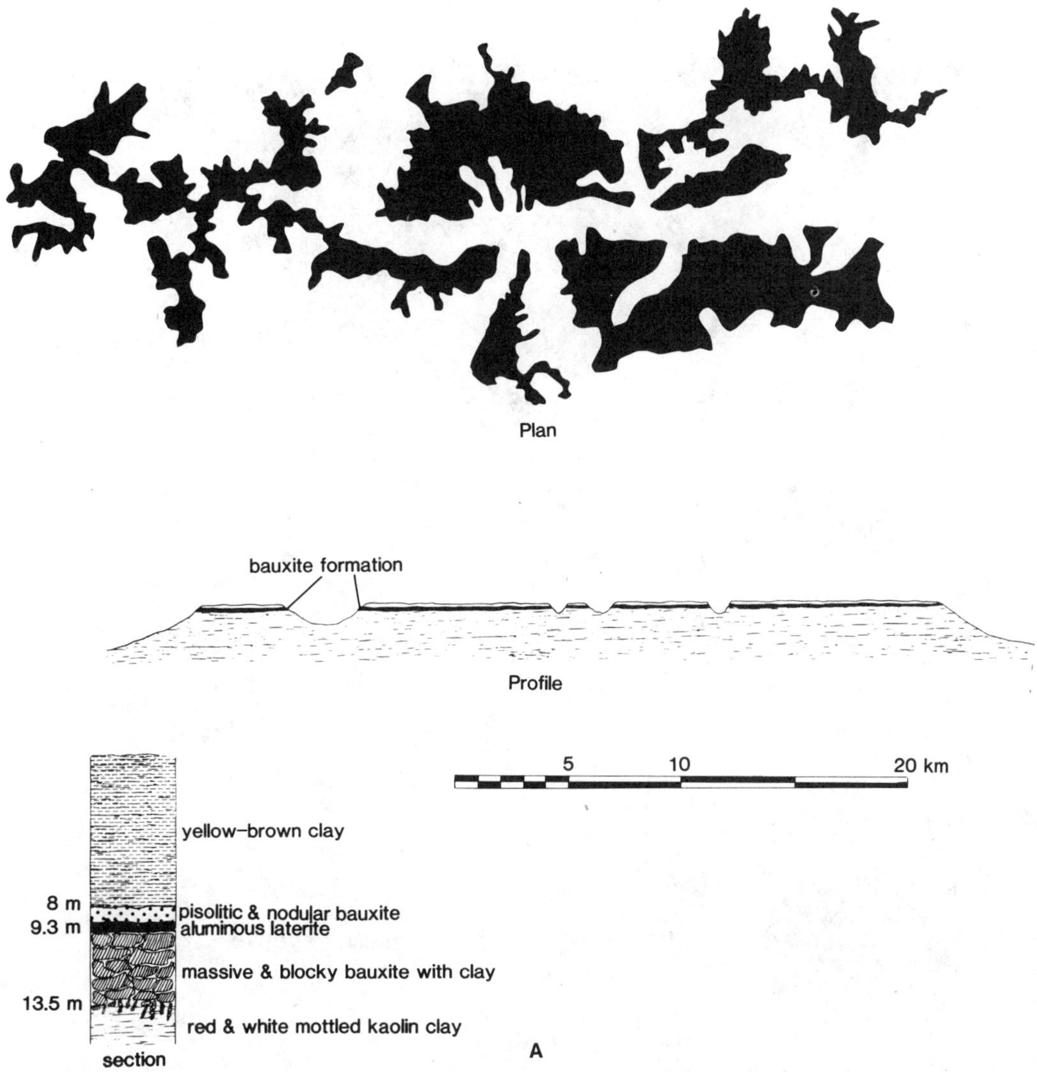

FIG. 1—*Examples of bauxite deposits. A. Plateau blanket deposits, Brazil.*

deposits are known in the United States, Guyana, Suriname, the USSR, China, Ireland, and southern Europe. They may be either large or small, depending in part on the amount of erosion prior to their burial. The interlayered deposits of Suriname and Guyana are overlain by unconsolidated sands and clays.

Interlayered deposits are usually more compact than surface deposits due to the weight of overlying rocks. In some localities in southern Europe and Turkey the rocks enclosing the bauxite have been folded and faulted. Where deformation of the enclosing rocks has been extreme, bauxite deposits are found on the limbs of steeply dipping and overturned folds. This deformation has resulted in partial alteration of gibbsite and boehmite to diaspore and in some cases to corundum.

Detrital Deposits: Detrital deposits result from the accumulation of transported bauxite eroded from other deposits. Deposits of this type are found in Arkansas.

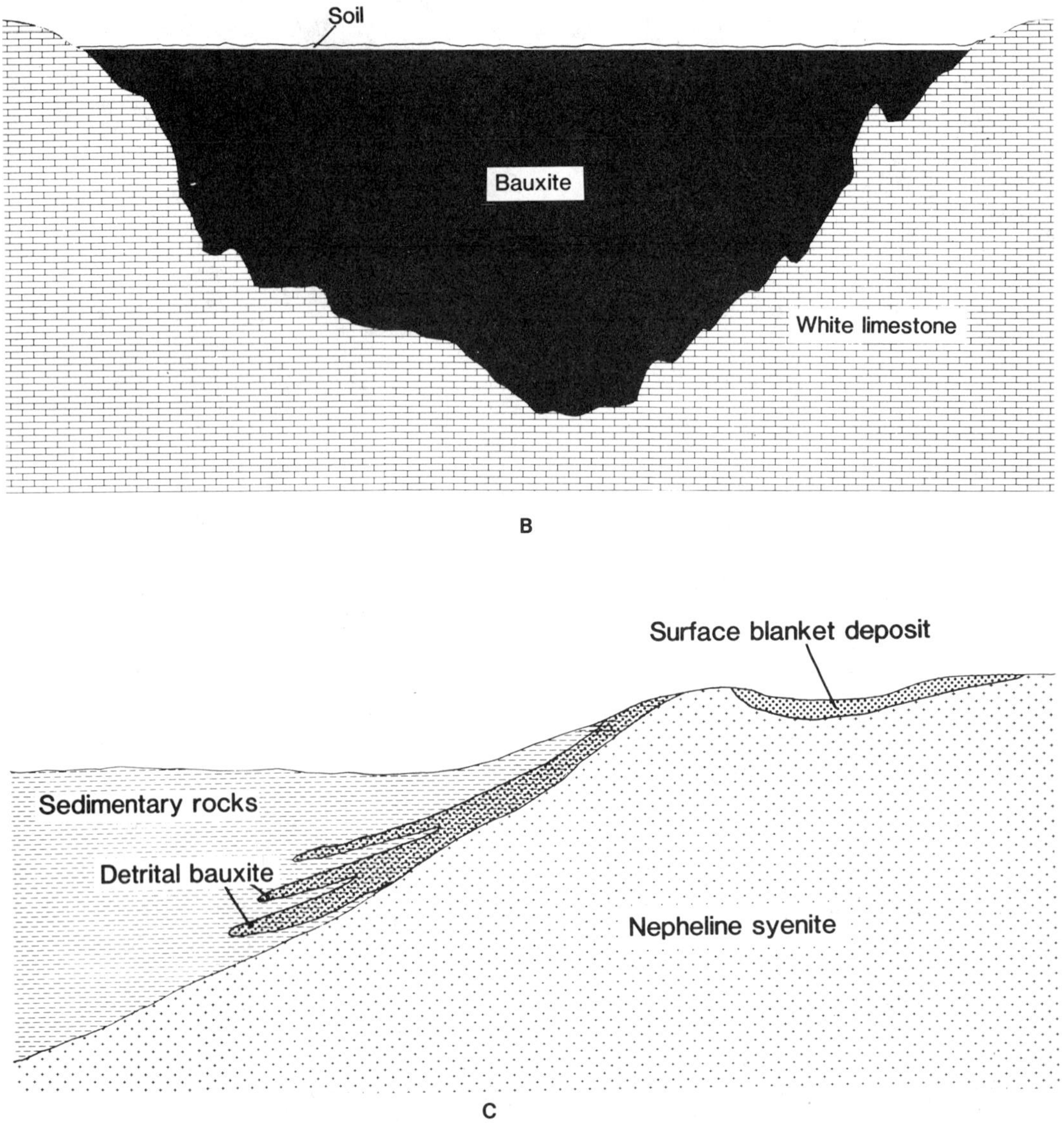

FIG. 1—*B. Surface pocket deposit, Jamaica. C. Blanket and detrital deposits, Arkansas, United States (Source: Gordon, 1958).*

Origin

There are different opinions regarding the origin of bauxite but general agreement that bauxite is a product of chemical weathering of rocks containing aluminum. The different theories concern the original aluminous source, the mechanics, and the chemistry involved in aluminum enrichment and desilication during weathering. Conditions and interrelated factors that occurred were probably interrupted and became unbalanced at times. Consequently, it is difficult to simulate in laboratory experiments all natural weathering conditions.

One theory proposes that aluminum and iron compounds were weathered and dissolved from rocks, transported as molecular or colloidal solutions to swampy basins and lagoons, then precipitated as aluminous and ferric gels. This

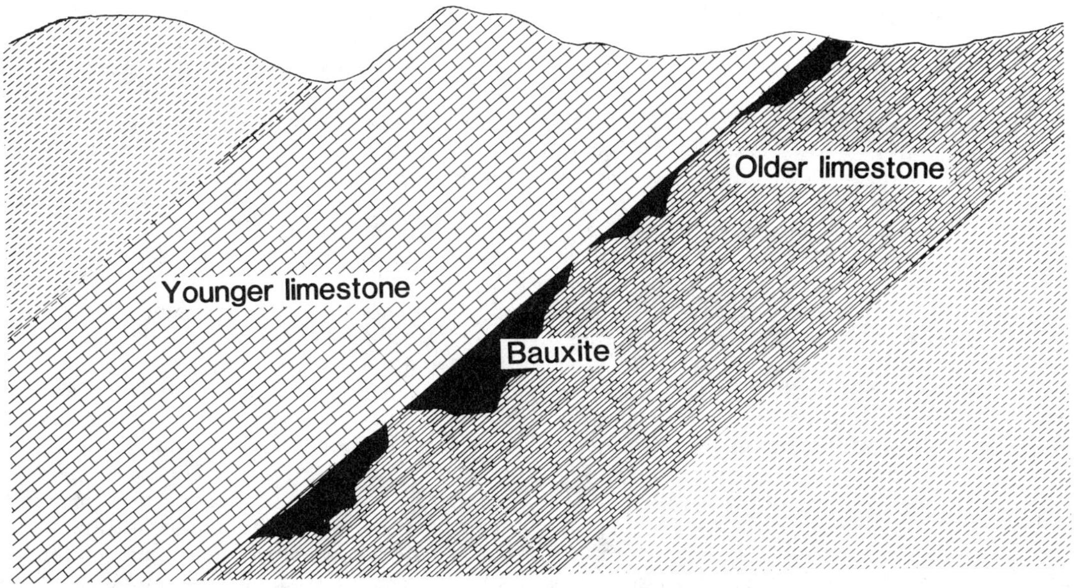

FIG. 1—*D. Interlayered pocket deposits, Southern Europe.*

theory rejects the idea of residual bauxite formation by tropical weathering.

Another theory suggests that bauxite resulted from differences in chemical mobility of the rock-forming elements while exposed to the downward percolation and movement of meteoric water during weathering. The pH of ground waters usually ranges between 3.5 and 8.0. Within this range and with variations in oxidizing and reducing conditions minerals are altered. The alkalies, alkaline earths and silicon are chemically mobile within pH ranges wherein aluminum and iron are stable or only slightly mobile. The lesser mobility of aluminum and iron results in residual bauxite formation, with aluminum and iron being taken into solution and reprecipitated as hydroxides and oxides. Aluminum enrichment below the residual bauxite may occur if the pH of the ground water changes to allow aluminum and iron or aluminum alone to be taken into solution and then reprecipitated with a further pH change.

Considerable disagreement exists regarding the origin of bauxite occurring on carbonate rocks. One group of mineralogists and geologists proposes that insoluble residue remaining from the leaching of limestone and dolomite was the aluminous source. Opposed to this are those who believe that the bauxite was derived from material other than limestone and suggest that volcanic ash or alluvium from sedimentary, igneous, or metamorphic rocks was deposited in karst depressions and subsequently altered to bauxite.

Favorable conditions often listed for bauxite formation are (1) source rocks rich in aluminum; (2) permeability of rocks or clays to permit desilication by groundwater movement downward and laterally; (3) tropical or semitropical climate, with abundant rainfall and alternating wet and dry periods; (4) topographical relief to allow drainage and fluctuations in groundwater levels; (5) a rate of chemical weathering exceeding the rate of erosion to permit the accumulation of weathered products; (6) sufficient time; and (7) presence of vegetation. While vegetation may be a factor by contributing to humic acids, bauxite formed during the Precambrian era, a geologic time at which land plants had not yet developed.

Bauxite has formed from clays and nearly every type of rock containing aluminum. These rocks include granite, nepheline syenite, anorthosite, phonolite, basalt, andesite, dolerite, gabbro, hornfels, schist, slate, kaolinitic sandstone, siltstone, and shale. Apparently, a high aluminum content in the parent rock is not as important in bauxite formation as the intensity and duration of the bauxitization and laterization process. The extensive blanket deposits along the west coast of the Cape York Peninsula (Australia) were derived from kaolinitic sands having an alumina content as low as 4% (Loughnan and Bayless, 1961).

Analyses of parent rocks and bauxites which formed on them (Table 2) illustrate the chemical changes during bauxitization. Because of the variable chemical composition of the parent rocks and the extent of chemical weathering, a wide range in the composition of bauxites has resulted (Table 3).

The geological age of some deposits cannot be precisely determined because they formed during long, quiet intervals and the enclosing rocks may be of widely differing ages. Bauxite has formed as early as the Precambrian and the conditions for its formation exist at present. During the Tertiary period conditions favorable for laterization were widespread, and this resulted in the formation of many high-grade bauxite deposits. Many deposits formed during the Tertiary period were derived from much older Precambrian crystalline and metamorphic rocks. The deposits of Tertiary age tend to be predominantly gibbsitic in composition with lesser amounts of boehmite.

During the Mesozoic era, bauxite developed on carbonate rocks in many areas in southern Europe. Boehmite is the principal bauxite mineral, although locally there are substantial amounts of diaspore. Gibbsite is also present in small amounts.

Deposits of Paleozoic age have been found in Asia, Europe, and North America. Diaspore and boehmite are the principal minerals. The diaspore deposits of Missouri are dated as Pennsylvanian in age.

Some of the oldest known deposits are of Precambrian age in Siberia. Possibly other older bauxites formed in the geologic past but were subsequently removed by erosion or remain undiscovered under overlying formations. It has been suggested that corundum-bearing rocks of early Proterozoic and Archean metamorphics in the USSR represent metamorphosed bauxite (Goretzky, 1959). General age relationships of various bauxitic raw material deposits are listed in Table 4 illustrating the many periods in geologic time when conditions were favorable for bauxite formation.

Description of Deposits

Bauxite is known to occur on all continents except Antarctica, and production was reported

TABLE 2—Analytical Comparisons of Parent Rocks and Bauxites

	Al_2O_3, %	SiO_2, %	Fe_2O_3 Plus FeO, %	TiO_2, %	Loss on Ignition, %
Arkansas					
Nepheline syenite	21.1	56.5	3.0	0.5	—
Bauxite	59.8	5.3	1.9	2.0	30.6
Australia					
Kaolinitic sandstone	5.2	91.1	0.5	0.3	1.8
Bauxite	50.0	9.6	9.2	2.2	26.6
Brazil					
Kaolinitic clay	37.9	30.4	14.5	1.5	15.8
Bauxite	51.3	3.0	17.5	1.1	27.1
Phonolite	20.9	53.1	5.3	0.4	1.7
Bauxite	55.1	5.5	9.6	1.8	27.7
Ghana					
Shale	18.2	61.0	4.7	1.0	7.5
Bauxite	60.5	1.4	9.8	2.2	25.6
Guinea					
Diabase	12.4	51.3	9.5	0.7	0.4
Bauxite	60.2	2.7	3.9	1.6	32.0
Guyana					
Epidiorite	18.9	49.1	6.4	0.9	0.4
Bauxite	56.4	3.1	11.6	0.8	27.8
Dolerite	17.3	52.0	11.2	0.5	—
Bauxite	59.0	9.8	1.1	0.7	28.1
India					
Basalt	16.8	47.4	14.9	1.7	2.2
Bauxite	56.4	1.7	7.1	8.9	25.4

TABLE 3—Range of Major Chemical Constituents of Bauxites

	Al_2O_3, %	SiO_2, %	Fe_2O_3, %	TiO_2, %	Loss on Ignition, %
Australia					
Cape York	53–60	2–10	5–13	2.1–3.1	21–29
Gove	52 available	2–6 reactive	13–28	3.4	26.3
Darling Ranges	25 available	0.5–2.5 reactive	15–20		
Kimberly Region	47–50	2.5–3.5			
Brazil					
Minas Gerais	55–59	1.6–5.6	6.9–9.6		
Amazon	50–61	3.7–9.0	1.7–14	1.1–2.0	25–30
China	50–70	9–15	1–13	2	
Dominican Republic	46–49	1.6–5.2	19–21		
France	55–70	3–16	4–25	2–3.5	
Ghana					
Yenahin	41–63	0.2–3.1	1.2–30.9	1.5–5.3	20–29
Awaso	48–61	0.4–2.4	4–22	0.8–2.1	26–33
Kibi	32–60	0.3–2.9	6–45	2.0–6.2	13–30
Greece	35–65	0.4–3.0	7.5–30	1.3–3.2	
Guinea	40–65	0.5–5	2–30	3–5	22–32
Guyana	51–61	4–6	1–8	2–3	25–32
Haiti	46.8	3.4	21.9	2.8	24.1
Hungary	50–60	1–8	15–20	2–3	13–20
India	45–60	1–5	3–20	5–10	22–27
Indonesia	37–56	4–16	4–26		
Jamaica	49–51	0.7–1.6	19–21	2.5–2.7	25–27
Malaysia	38–60	1–13	3–21	1–2	
Sierra Leone	51–55	1.5–2	10–18	1.5	27–31
Suriname	50–60	2–6	2–15	2–3	29–31
United States					
Arkansas	45–57	5–24	2–12	1.6–2.4	22–28
Oregon & Washington	31–35	5–11	33–35	5–6	16–20
Southeastern states	51–56	12–15	1–5	1.5–3.5	22–30
USSR	26–52	2–32	1–45	1.4–3.2	
Yugoslavia	48–60	1–8	17–26	2.5–3.5	13–27
Romania	55	5	22	1–2	
Turkey	55–60	5–7	15–20	2–3	12–14

for 28 countries in 1980. The largest reserves or potential resources are in tropical or subtropical areas of Australia, Africa, Asia, South America, and the Caribbean. Bauxite reserves are sufficiently widespread and no single country has been able to control world supply.

Reserves

The listing of world reserves and potential resources of bauxitic raw materials (Table 5) is more of an order-of-magnitude guess than an accurate estimate. Some figures are based on actual exploration, but others are based on generalized observations and represent, in part, personal opinions. *Reserve* usually refers to material that can be profitably mined, processed, and marketed under prevailing conditions and economics at the time the estimate was made. *Potential resource* is a category used for material that could possibly be used economically at some future time. In the last ten years there have been discoveries of bauxite that can be measured in hundreds of millions or billions of tons. An increase in bauxite demand or political changes disrupting bauxite supply could rapidly change many deposits from a potential resource to a reserve category, and the reliability of predictions taking these events into account is limited.

In preparing a reserve estimate there is also the problem of compensating for ore reserve additions and depletion in mining areas. Allowances for new discoveries and depletion have been made in the estimate in a general way from available data.

The political stability or investment climate of countries influences the decision of investors in new mining ventures. Large tonnages of bauxite are known in developing countries and

TABLE 4—General Age Relationships of Various Bauxitic Raw Material Deposits

Millions of Years	Era	Period	Epoch	Areas
1	Cenozoic	Quaternary	Recent and Pleistocene	Central America: Costa Rica, Panama South America: Colombia Pacific Islands: Hawaii, Fiji, Solomons
		Tertiary	Undifferentiated	Africa: Ghana, Guinea, Ivory Coast Australia: Western South America: Brazil, Venezuela
12			Pliocene	
23			Miocene	Caribbean: Jamaica, Haiti, Dominican Republic North America: U.S.—Oregon, Washington Australia: Victoria
35			Oligocene	Europe: Northern Ireland
55			Eocene	Asia: India, Deccan, Kashmir-Jammu Australia: Queensland, Northern Territory, Tasmania Europe: Italy, Yugoslavia North America: U.S.—Arkansas, Alabama, Georgia, Mississippi South America: French Guiana, Guyana, Suriname
70			Paleocene	Europe: Yugoslavia
90	Mesozoic	Cretaceous	Upper	Asia: Turkey Europe: Austria, Greece, Hungary, Italy, Yugoslavia
120			Middle	
135			Lower	Asia: Turkey, Saudi Arabia Europe: France, Greece, Hungary, Italy, Romania, Yugoslavia
180		Jurassic		Europe: Yugoslavia, USSR
		Triassic	Upper	Europe: Yugoslavia
			Middle	Europe: Yugoslavia
220			Lower	Asia: China
270	Paleozoic	Permian		Asia: Turkey
320		Pennsylvanian		North America: U.S.—Missouri, Pennsylvania
350		Mississippian		Asia: China—Yunnan, Kweichow Europe: USSR—Tikhvin
		Devonian	Upper	
			Middle	Asia: USSR—Urals Europe: Spain
400			Lower	
		Silurian		
490		Ordovician		
600		Cambrian		Asia: USSR
		Precambrian	Upper	Asia: USSR—Bokson, East Sayan Mtns.

Source: Modified from Patterson, 1967.

TABLE 5—Estimated Reserves Plus Potential Resources of Bauxite, Mt

Region / Country	Mt
North America	
United States	
Arkansas	35
Southeastern states	10
Oregon & Washington	50
	95
Central America	
Costa Rica	150
Caribbean	
Dominican Republic & Haiti	40
Jamaica	900
	940
South America	
Brazil	4,550
Colombia	20
French Guiana	170
Guyana	1,160
Suriname	400
Venezuela	500
	6,800
Europe	
France	40
Greece	900
Hungary	400
Italy	20
Romania	15
USSR & Soviet Asia	300
	1,675
Africa	
Cameroon	1,000
Chad	30
Ghana	410
Guinea	8,200
Guinea Bissau	10
Ivory Coast	20
Malagasy	150
Malawi	60
Mali	820
Mozambique	60
Sierra Leone	140
South Africa	70
Zaire	50
Zimbabwe	5
	11,025
Asia	
China	1,000
India	2,400
Indonesia	1,200
Malaysia	25
Pakistan	15
Taiwan	200
Turkey	120
	4,960
Australia	5,760
New Zealand	20
Oceania (Admiralty Islands, Babelthuap, Fiji & Solomon Islands)	70
World Total, rounded	31,500

the distinction between economic reserves and potential resources based on politics has not been made.

Another unknown factor that relates to reserve classification is the increasing environmental concern regarding mining in general, but more specifically surface mining. Future legislation may prohibit or be so restrictive as to prevent some bauxite deposits from ever being mined. Most tonnage estimates of world bauxite reserves include tonnages of only a few million tons for some areas and countries. When the high cost of mine development—especially in remote locations—is taken into consideration it is improbable that some of these smaller deposits will ever be mined. This is due to the limited future mine life for recovering capital investment. Consequently, some of the tonnages shown in Table 5 should be discounted as being viable reserves unless they are related to operating mines.

Principal Producing Areas

World bauxite production in 1980 was approximately 90 Mt. As shown in Table 6, three countries, Australia, Guinea and Jamaica, supplied 60% of this total.

Australia

Before 1963, Australia's production was less than 100 kt per year or less than 1% of world total. By 1980 production had expanded to 27 Mt or 30% of world total. The active districts are the Cape York Peninsula of Queensland, Gove Peninsula of Northern Territory and the Darling Ranges of Western Australia.

Queensland, Cape York Peninsula: The ore in the vicinity of Weipa occurs as a flat-lying to a gently undulating formation from 1 to 10 m thick. Overburden consists of a meter of soil. The upper part of the bauxite deposits

TABLE 6—World Bauxite Production by Country, 1980*

Country	Production Rank	Production, kt	% of World Total
Australia	1	27,580	30.7
Guinea	2	13,780	15.3
Jamaica	3	12,260	13.6
Suriname	4	4,700	5.2
USSR	5	4,600	5.1
Brazil	6	3,970	4.4
Yugoslavia	7	3,140	3.5
Hungary	8	3,020	3.4
Greece	9	2,950	3.3
Guyana	10	2,350	2.6
India	11	1,740	1.9
France	12	1,660	1.8
United States	13	1,560	1.7
China	14	1,500	1.7
Indonesia	15	1,220	1.4
Malaysia	16	920	1.0
Romania	17	710	0.8
Dominican Republic	18	600	0.7
Sierra Leone	19	590	0.6
Haiti	20	450	0.5
Turkey	21	350	0.4
Ghana	22	230	0.3
Others combined†	23	50	−0.1
		89,930	100.0

* USBM figures rounded to nearest 10,000 t.
† Includes Italy, Pakistan, Spain, West Germany, and Zimbabwe.

contains spherical pisolites from 1 to 20 mm in diameter, which occur in a loose, reddish-brown, silty sand matrix. Below the pisolitic bauxite there is a distinct change to a zone of nodules and concretions usually high in iron and silica. Under this nodular zone are kaolinitic and sandy clays (Evans, 1960).

After separation of the matrix and pisolites by wet screening, the final product averages about 54 to 57% Al_2O_3 and 5.5% SiO_2. The deposits contain mixtures of gibbsite and boehmite, but gibbsite is the predominant alumina mineral. Weipa bauxite is exported and is also refined into alumina at Gladstone in southeastern Queensland. A special grade of bauxite is calcined for abrasive use.

Northern Territory, Gove Peninsula: Bauxite occurs on a plateau of laterized sediments. A pisolitic layer extends from the surface to a depth of 1 to 5.5 m. Pisolites are approximately 5 mm in diameter, and a small quartz grain in the center is common. For the first meter they are loosely cemented, but below 1.2 m they are strongly cemented and the bauxite breaks across the pisolites. Below the pisolitic layer is a layer of irregularly shaped nodules up to 25 mm in length formed by the cementation of several pisolites. A reported grade is 48.7% Al_2O_3, 3.6% reactive silica, and 17% Fe_2O_3, but there are substantial tonnages of higher grade bauxite in the district. Most of the alumina is present as gibbsite, but about 3% is boehmite (Bland, 1972; Dunn, 1965). Bauxite is refined into alumina in a nearby plant and is also exported.

Western Australia, Darling Ranges: The deposits are aluminous laterites developed on remnant hills and ridges of a dissected surface approaching peneplanation. Bauxite deposits occur in an area approximately 320 km long and 40 km wide. Most ore bodies are smaller than those in other Australian bauxite districts, and they occur in areas of ferruginous laterite. The deposits mined range in thickness from a meter to more than 6 m, with an average thickness of 3.2 m. The overburden thickness averages 0.75 m. In the upper part of the laterite section there is a hard capping with a friable zone below.

Grade varies from 25 to 48% available Al_2O_3. Although quartz is abundant in the deposits it is nonreactive in the caustic liquors used in the alumina refining process, and remains in the red mud waste. The amount of reactive SiO_2, which occurs mainly in clay minerals, is 1 to 2%. All of the production is refined into alumina at refineries relatively close to the mines.

There are extensive undeveloped deposits in the Kimberley area of Western Australia.

Africa

Guinea: Reserves are the largest known for any country. Mining began on a steady basis at Los Islands in 1952. These deposits are nearly depleted. At the present time mining is active at Fria, Kindia, and Sangaredi.

Fria bauxite is composed of red gibbsitic earth with hard concretions, lumps, and crusts. Gibbsite is the bauxite mineral. The ore is about 8 m thick and the overburden is 100 to 200 mm. Ore analysis is 48% total Al_2O_3 and 2.5% SiO_2 of which 30% is fine quartz. After drilling and blasting, ore is loaded into trucks and hauled to a nearby alumina refinery. Alumina is transported by rail 145 km to Conakry and exported.

The Debele deposit near Kindia is 6 to 8 m thick and overlain by 0.3 to 0.5 m of soil. Ore grade is about 48% Al_2O_3, 1 to 3% SiO_2. Following drilling and blasting, ore is loaded into trucks and hauled to a crushing and screening plant where it is reduced to —60 mm. The crushed ore is hauled 100 km by rail to Conakry and exported to the Soviet Union.

The largest mine is at Sangaredi in western Guinea. This is the richest single bauxite deposit known, with Al_2O_3 59 to 60% and SiO_2 less than 1.5%. Most of the Al_2O_3 is present as gibbsite but up to 3 to 6% is present in boehmite. The ore consists of a rock formation 30 m thick at the top of a plateau. There is no overburden. Blasted ore is loaded directly into railroad cars for the 142-km haul to Kamsar. Track mining is used throughout. At Kamsar the bauxite is crushed to —100 mm, dried to about 5% free moisture, and stockpiled for eventual export to Europe and North America. In addition to metal grade, calcined abrasive grade bauxite is produced.

There are a number of other large deposits in the general vicinity of Sangaredi. Although they are lower in grade than Sangaredi they are suitable for metal grade use. In the Tougue and Dabola areas there are extensive undeveloped deposits containing 47 to 52% Al_2O_3 and 2 to 4% SiO_2.

Ghana: The principal reserves are in the Kibi, Yenahin, and Awaso areas, but Awaso is the only area in production. All production is exported to Europe.

Sierra Leone: Bauxite is mined in the Makanji Hills area for export to Europe and North America. A new mine is planned near Port Loko.

Caribbean

Jamaica: The commercial possibilities of Jamaican bauxite were first realized in 1942. Production began in 1952 and within a few years Jamaica became a major world producer. The bauxite is of the terra rossa type occurring as concentrations of soft earthy material in limestone sinkholes and channels in karst depressions. Particle size varies from less than 1 to 40 μm. The chemical composition varies from 40 to 50% Al_2O_3, 1 to 8% SiO_2, and 17 to 20% Fe_2O_3. Field moisture is about 19 to 24%. Overburden is only 0.3 to 0.6 m of soil. Deposits vary in size from hundreds to millions of tons. Gibbsite is the predominant alumina mineral but some areas contain as much as 20% boehmite. Bauxite is exported and is also refined locally into alumina.

Dominican Republic and Haiti: Both countries are small producers compared to Jamaica and export bauxite to the United States.

South America

Suriname: Discoveries of bauxite were made in the coastal plain in 1915 and shipments began in 1922. These deposits have long been one of the major sources of high-grade gibbsitic bauxite. The surface deposits with little overburden occur in the coastal plain on the tops of low hills and ridges dipping gently northward toward the coast. Thicknesses range from 3 to 6 m. Buried deposits are found below clay and sand overburden having thicknesses up to more than 30 m. High-iron bauxite forms a capping on many of the deposits. Grade ranges from 50 to 60% Al_2O_3 and 2 to 4% SiO_2. Suriname has an aluminum industry integrated from bauxite to metal. Special grades of bauxite are also produced.

Bauxite and lateritic bauxite also occur on hills and plateaus in the interior parts of the country but these deposits have not yet been developed.

Guyana: Since 1917 Guyana has been a producer of high-grade gibbsitic bauxite. Part of the production is refined into alumina in Guyana and part is exported. The unusually low iron content of Guyana bauxite makes it desirable for the manufacture of aluminous chemicals, refractories, and abrasives, and Guyana is the leading country in the production of these special grades.

The main deposits currently being mined lie within a belt roughly parallel with the coastline and ranging from 24 to 110 km inland. Deposits occur as cappings on an old peneplained surface (Hose, 1959), most of which is now overlain by unconsolidated sediments of sand and clay. Ore thicknesses range from 4.5 to 10 m and overburden thickness varies to more than 60 m. Undeveloped deposits also occur on ridges and plateaus inland from the bauxite belt.

Brazil: Brazil has a domestic aluminum industry integrated from bauxite to finished products. Bauxite for the alumina refineries is supplied by mines in the state of Minas Gerais.

Discoveries of large reserves have been made in the state of Para along the Amazon and its tributaries and south of Belem since the late 1950s. Bauxite was also discovered in the Carajas area while work was being done on the iron deposits. The largest areas are along the south side of the Trombetas River above Oriximina and in the vicinity of Paragominas south of Belem. Initial production from the Trombetas area began during 1979. The other Para deposits had not been developed as of 1980. Planning is being done for expansion of the Brazilian aluminum industry based on these large reserves and the development of hydroelectric power.

Amazon bauxites occur on plateaus which are the erosional remnants of a dissected peneplain. A section through the deposits consists of clay overburden, nodular bauxite from several millimeters to 2.5 m thick, zero to 2.5 m of ferruginous bauxite, 1 to 7 m of massive bauxite with clay seams and underlying kaolin clay. Gibbsite is the major bauxite mineral; less than 1% is boehmite. After beneficiation to remove clay, the massive bauxite averages about 50% available Al_2O_3 and 4.6% reactive SiO_2.

Venezuela: Commercial grade bauxite was discovered in the western part of Bolivar in 1976. When these deposits are developed, the domestic aluminum industry will be integrated from ore to metal.

Europe

Greece: Production began in the 1930s and remained on a small scale until the early 1950s. Since that time production has steadily increased, and in 1980 Greece produced 3.0 Mt. One type of ore is a mixture of boehmite and diaspore. Most of the deposits occur as pockets or lenses between layers of carbonate rocks.

Hungary: Bauxite deposits in Hungary have been worked since 1925. The ore is used by the domestic aluminum industry and is also exported. Major producing areas are Nyirad, Halimba, Hislod, Iszkaszentgyory, and Gant. Most of the deposits occur as pockets in buried karst depressions, and about 83% of the production is from underground mines (Staudinger, 1971). Many deposits are mixtures of gibbsite and boehmite with 1 or 2% of diaspore, but other ore bodies are mainly gibbsitic. There are also deposits containing mixtures of boehmite and diaspore.

Yugoslavia: Since about 1915 Yugoslavia has been a bauxite producer, and ore is refined into alumina or is exported. Deposits occur in a belt along the Adriatic coast for a distance of about 560 km. The belt is approximately 60 km wide in the northwest and widens to 160 km in the southeast at the Albanian border. Deposits are irregular fillings of karst depressions overlain by younger carbonate rocks. The formations enclosing the bauxite are deeply folded and bauxite extends to depths of more than 600 m. The age of the Yugoslav bauxite varies from Triassic to Eocene. Younger deposits contain gibbsite and boehmite, and the older deposits contain mostly boehmite. In some localities diaspore is the major alumina mineral (Patterson, 1967).

France: In 1821 the French chemist P. Bertier discovered that material which had been considered a red clay contained practically no silica and was not clay at all. To this substance the name *bauxite* was given, after the community of Les Baux near Arles in southern France where the first samples were obtained (Fox, 1932). With the development of the aluminum industry, France became a major bauxite producer. However, the French aluminum industry has gradually become more dependent on imports of bauxite and alumina. Most French bauxite is refined into alumina but it is also used in the production of chemicals, cements, abrasives, and refractories.

The major deposits are in a belt a few kilometers north of the Mediterranean coast extending from Spain nearly to Italy. They are irregularly shaped beds, lenses, and pockets between limestone and dolomite formations that have been folded and faulted. Some deposits are nearly horizontal while others are nearly vertical on the limbs of synclines. Boehmite is the major mineral but various amounts of gibbsite and diaspore are also present.

USSR: The main bauxite-producing area is in the northern Urals. Other important areas are Kazakhstan and Turgay, and additional mine developments are scheduled for Krasnyoktyabrskoye and Belinskoye in north Kazakhstan. Production has increased steadily in the Soviet Union since World War II, and in 1980 total output was on the order of 4.6 Mt. The Soviet aluminum industry uses nepheline and alunite in addition to bauxite as aluminum ores. Some Russian bauxites are low-grade in comparison to the quality of ore used by other countries. Alumina content varies from 26 to 52% (Strishkov, 1969). Alumina minerals in the bauxites are gibbsite, boehmite, diaspore, and corundum (Gladkovsky and Ushatinsky, 1963).

Asia

China: Production is related to the requirements of the domestic aluminum industry and the export of low-iron bauxite suitable for refractories. Most of the bauxite produced is diaspore. In recent years China has increased its share of the world refractory grade market.

India: Bauxite production is directly related to the domestic aluminum industry since very little bauxite is exported. Power shortages in recent years have reduced metal production to less than installed capacity resulting in a corresponding decrease in bauxite output. There have been recent discoveries of large deposits in the Eastern Ghats.

Indonesia: Bauxite has been mined on Bintan and adjacent islands since 1935. All of the production is exported, principally to Japan. The better quality reserves are nearly depleted and feasibility studies have been made to utilize higher silica material within Indonesia.

In western Kalimantan there have been discoveries of large tonnages averaging about 40% available Al_2O_3 and 3% reactive silica. The Kalimantan and also the Bintan bauxites are gibbsitic and contain little or no boehmite. Studies are being made to develop the Kalimantan deposits for export and refining into alumina in Indonesia.

Malaysia: Mines are located in the state of Johore and all of the production is exported, mostly to Japan.

Turkey: Bauxite is produced for the Turkish aluminum industry and for export to USSR. The alumina refinery near Seydisehir operates on boehmitic and gibbsitic bauxite mined nearby. Large reserves of diasporitic bauxite are in Mugla Province and the central Taurus region.

North America

United States: Production of bauxite in the United States began in 1888 in north Georgia, and about 1895 in Arkansas. Production expanded steadily with the growth of the aluminum industry. The United States for many years was one of the leading world producers until the period immediately following World War II. At that time the US became increasingly dependent on foreign bauxite sources.

Bauxite has been mined in Alabama, Arkansas, Georgia, Mississippi, Tennessee, and Virginia. About 85% of US production is now mined in Arkansas and the remainder in Alabama and Georgia.

In Arkansas bauxite occurs as residual deposits on kaolinized nepheline syenite and kaolin or as detrital deposits. All production is from surface mines.

In the coastal plain of Alabama and Georgia, deposits of bauxite and bauxitic clay occur as pockets and lenses within deposits of kaolin. These pockets are small and contain up to several tens of thousands of tons. The ore contains abundant silica but little iron, although ferruginous deposits have been found. Most of the material is used in the manufacture of aluminous refractories with lesser amounts being used for chemicals and as filter media. In recent years deeper exploration drilling has resulted in new discoveries.

Diaspore occurs within deposits of burley and flint clay in limestone sinkholes in the southern clay district of central Missouri. It was used primarily in the manufacture of refractory products but at one time was also used in the production of abrasives and cements. New discoveries did not keep pace with production and the reserves of diaspore are mostly depleted. Diaspore also occurs in west central Pennsylvania.

Ferruginous bauxite or aluminous laterite resulting from the laterization of Columbia River basalt occurs in northwestern Oregon and southwestern Washington. These deposits have not yet been developed.

On the islands of Hawaii, Maui, and Kauai bauxite has resulted from the weathering of basaltic lavas. Many of the deposits occur in cultivated and residential areas having higher value as real estate than as a source of bauxite. Exploitation in the foreseeable future is unlikely.

Exploration

In the search for bauxite, areas where geological evidence indicates that conditions at one time were favorable for laterite or bauxite development are investigated. Because bauxite is believed to have formed under tropical weathering conditions, efforts are directed to areas where the climate was tropical or subtropical in the not-too-distant geological past. In tropical areas, land forms which have remained stable and have been preserved from total erosional destruction are considered prospecting targets. Such land forms are mesas, plateaus, peneplains, base leveled areas, gently rolling areas, or gradual slopes—all of which have or did have adequate drainage.

Field Techniques

Topographic maps and aerial photographs, when available, are useful in the study of land forms in areas considered for exploration. Aerial reconnaissance flights are frequently made to study topography and ground access routes. Often a cloud cover interferes with aerial photography in the tropics. Airborne side-looking radar is capable of penetrating cloud cover and making possible the production of mosaics which show land forms distinctly in detail.

Areas considered favorable are examined on the ground for outcrops, float, and stream gravel. Float tends to become case-hardened and durable. When a discovery has been made, holes are drilled or pits are dug to obtain samples. During reconnaissance work in remote areas where supplying field parties is difficult, hand tools for drilling and test pitting are generally used. The shallow depth of many deposits also makes this possible. Initially, holes and pits are widely spaced to get a general impression of the deposit.

Once a deposit has been found, drilling or test pitting is usually done on a rectangular grid spacing. Hole spacing depends in part on the size and attitude of the deposits and the amount of reliable information desired. Spacings of 150 x 150 m and 300 x 300 m have been used on large blanket deposits, while spacings of 15 x 30 m and 30 x 30 m have been suitable for smaller pocket deposits.

Prospecting for buried deposits involves the mapping of geologic structures or regional dips for the selection of specific areas for drilling. In order to minimize drilling and increase its effectiveness, geophysical surveys are sometimes made.

Geophysical

The physical characteristics of bauxite are such that geophysical surveying cannot clearly distinguish between bauxite and surrounding rocks, but geophysical methods are used to search for areas where bauxite might occur.

Types of geophysical surveys that have been made in the search for buried bauxite include seismic refraction, gravimetric, magnetic, and resistivity. These have met with varying degrees of success. Some investigators of the bauxite of Guyana and Suriname believe that buried deposits in the coastal plain areas are in some way related to higher parts of the basement relief (Van der Hammen and Wijmstra, 1964). Geophysical surveys have been made to search for these basement highs and to determine depth to basement.

In the Bicske Basin of Transdanubia, Hungary, bauxite occurs in karst depressions in basement carbonate rocks. The basement rocks are buried under sediments up to 425 m thick and faults occur in the area. Geophysical methods have been used to determine basement depth and dip of the basement surface, to trace main faults, and to outline structural units. More efficient location of drill holes has resulted (Ottlik and Szabadvary, 1971). Portable seismographs such as those used by excavation contractors have been used to measure thickness of overburden with reasonable success.

For the study of areas before beginning field work, LANDSAT data should be helpful in studying land forms, alteration patterns, and geomorphological characteristics.

Drilling

The variable physical nature of bauxitic materials frequently poses unique drilling problems. Equipment suitable for drilling the soft terra rossa deposits of the Caribbean area is unsatisfactory for drilling hard massive types of bauxite. Nearly every type of drill has been used in bauxite exploration.

Both percussion and rotary types are used. The simplest percussion hand drill consists of steel rods connected to a chopping bit. It is used with or without casing, depending on the type and thickness of overburden. Truck-mounted churn drills are also used where it is feasible to build the necessary access roads. Churn drills are used to make and recover cuttings or for drive tube sampling.

If the bauxite formation consists of a mixture of bauxite nodules and clay or of bauxite with seams and streaks of clay, it is essential that a sample for metallurgical testing be obtained. Some bauxites on an as-mined basis are high in reactive silica and require scrubbing to reduce the reactive silica to a level acceptable for Bayer plant feed. Drilling methods that produce fine particle sizes in cuttings or cause a washing-out of clay from the sample are not suitable. In some formations test pitting is necessary for reliable sampling.

Rotary core drills using diamond or tungsten carbide bits are suitable in bauxitic formations where variations of hard and soft material are not great. In material which is difficult to core, modifications in core barrel design, variations in water usage or the use of compressed air instead of water frequently improves core recovery.

Noncoring rotary drills include hand augers and mechanically driven types. Hand augers, though primitive, have definite advantages in areas where access is difficult and labor is unskilled and inexpensive. Noncoring drills using compressed air to raise drill cuttings to the surface are used in Jamaica but they are unsuitable in wet formations.

The reverse cycle or vacuum rotary drill has been used for bauxite sampling. A partial vacuum is maintained at the bit and cuttings pass through the drill rods into a cyclone and are discharged into a collecting tube. The drill works best when the formation is dry or nearly so. It is unsuitable when metallurgical testing is required on samples.

Rotary drills using continuous flights of augers for transporting cuttings to the surface are satisfactory in drilling deposits of clay and bauxitic materials. The sample becomes somewhat mixed, but when there are not great vertical variations in grade, this is not a serious problem. This type of drill is high-speed and requires no water.

A bucket-type drill used mainly in construction work for drilling large diameter holes has been used in bauxite drilling. It is capable of drilling holes from 0.3 to 1 m or more in diameter and removing all cuttings with a drilling bucket attached to the kelly. It is a high-speed drill capable of producing samples sufficiently large for metallurgical testing.

Preliminary Evaluation

Exploration samples are generally analyzed for silica, iron as Fe_2O_3, titania, loss on ignition, and alumina. Samples are often analyzed for available alumina and reactive silica. Available alumina is that which can be recovered in an alumina plant at specified digestion conditions. Because these conditions vary among operators, the meaning of available alumina is not uniform. Reactive silica is the silica that combines with caustic soda and alumina in the alumina refining process. It is usually in the form of clay minerals or finely divided quartz.

In new areas or in reconnaissance work, samples are frequently first evaluated on the basis of reactive silica content. If in excess of a desired limit on reactive silica no further analyses are made.

During the accumulation of drill hole and analytical data, tonnage and grade estimates are made. Ore volume is converted to tonnage either by estimating bulk density or making actual bulk density measurements. Because density is a direct multiplier in converting volume to tonnage, large errors in tonnage estimates have resulted when densities differing widely from the true value have been used.

Evaluation of Deposits

Specifications

Metal Grade: In practically all uses of bauxite, a high alumina content is desired. When used for metal grade, bauxite is refined into alumina and the alumina is electrolytically smelted into aluminum. For bauxite used in the Bayer process—the most commonly used process for refining bauxite into alumina—a reactive silica content not exceeding 5% is desirable, although bauxites containing more than this amount are used. A sintering process is used on some bauxites containing 8 to 15% reactive silica. The use of bauxite containing such a high percentage of reactive silica is justified by economic conditions unique to certain operations. There are no generally accepted maximum limits on iron and titania impurities.

The lower limit of alumina content for metal grade bauxite is governed by economic factors such as mining, transportation, and processing costs; bauxite taxation; geographical location of alumina plants in relation to bauxite sources and other raw materials used in alumina refining; and the availability and cost of alternate or higher grade sources of bauxite. The interrelationship of all or many of these factors may permit the use of bauxite containing as little as 25% available alumina as is presently being done in Western Australia.

Refractory Grade: The general requirements for refractory grade are that it must be low in

Fe_2O_3 and TiO_2 and after calcining have maximum density and minimum porosity. There is a wide range in the Al_2O_3 and SiO_2 content of bauxites used for refractories due to the varying requirements for and applications of alumina refractory products. For the highest grade of refractories made from calcined bauxite, Al_2O_3 usually is 88% or more, Fe_2O_3 2% or less, and TiO_2 3.8% or less. A very low K_2O plus Na_2O content is required and a low CaO plus MgO is desirable.

Guyana for many years has been the major source of high Al_2O_3 calcined refractory bauxite, and, as a result, the grade produced frequently is used as a reference for the comparison of bauxite from alternate sources. In recent years China has increased exports of calcined refractory grade.

Abrasive Grade: Specifications for calcined abrasive grade are not as stringent as those for the high-alumina refractories. More bauxites are able to meet abrasive specifications mainly because more iron is tolerated. A titania content of about 3 to 5% is desirable as it imparts greater toughness to abrasive products.

Calcined bauxite is sold with guaranteed and typical analyses. In Table 7 are shown typical analyses of various grades of calcined refractory bauxite. Table 8 shows the range of calcined bauxite use for abrasive manufacture.

Chemical Grade: Several grades of bauxite are produced for the manufacture of aluminous chemicals. The iron content should be low and the alumina should be readily soluble in sulfuric acid. The range of chemical grade bauxite is shown in Table 9.

Cement Grade: Bauxite suitable for the manufacture of high-alumina cements should not have a silica content greater than 6%. If the silica exceeds this amount, the rapid hardening properties of the cement will be decreased. The iron content of the bauxite can be relatively high and is actually desirable because the iron compounds in the cement contribute to its hardening properties and act as fluxes in some cement manufacturing processes. If the iron content of the bauxite is too high, however, the refractoriness and strength development of the cement will be less.

Testing

Physical Tests: Tests include scrubbing, screening, crushing, grinding, and pumping determinations. Some bauxites require beneficiation before being acceptable for plant feed to alumina refineries. Silica in the form of clay minerals is a common impurity. Clay may surround bauxite concretions and blocks in the deposits or be intimately mixed within bauxite, and before a scrubbing plant can be designed, testing of crude ore is required. Batch testing is frequently followed by work on larger bulk samples for equipment selection and power requirements. Heavy liquid and heavy media testing are made on ores that can be improved by iron removal.

Transport of bauxite from mine or beneficiation plant to an alumina plant or shipping point is an important cost element. When pipeline slurry transport is being considered, bauxite samples are sent to testing laboratories for pumping and pipeline loop testing.

When being evaluated for refractory use, samples are calcined at different temperatures to determine at what points optimum bulk density and porosity are reached. If desired density and porosity are achieved, samples are fired and tested for high temperature stability, hot strength under load, spalling, and porosity. The temperature range of these firing tests approximates the temperature range of refractories in industrial service.

TABLE 7—Typical Analyses of Calcined Refractory-Grade Bauxites

	Source				
	China	US	China	China	Guyana
Grade	60	70	80	+80	+80
Al_2O_3, %	63.5	70.5	84.5	89	88.3
SiO_2, %	32.4	25.3	9.9	6	6.5
TiO_2, %	2.2	2.7	3.8	3.3	3.2
Fe_2O_3, %	1.5	1.4	1.6	1.2	1.8
Loss on ignition, %	0.15	—	0.20	0.05	0.25
Bulk sp gr	2.70	2.85	2.80	3.15	3.10

TABLE 8—Analysis of Calcined Abrasive-Grade Bauxite

Al_2O_3	85–87%
Fe_2O_3	6.0–10.3%
SiO_2	3.0–5.5%
TiO_2	3.0–4.5%
Loss on ignition	1.0–2.0%

Preparation For Markets

Mining Methods

Bauxite mining is much like the mining of other minerals and has no unusual or unique features. Approximately 85 to 90% of world production is from surface mines and the balance is from underground operations.

The wide variation in physical characteristics of bauxite deposits results in different types of mining methods. At some mines there is little or no overburden, whereas at others overburden may be 70 m or more, which can cause slope stability problems with water-saturated sands and clays. Indurated bauxite requires blasting. Bauxite formations composed of loose pisolites, oolites, or concretions in a clay or sandy matrix are also usually easy to mine, but sometimes ripping is done to facilitate digging. Loading is usually into trucks or directly into rail cars for transport to crushing or washing plants or to stockpiles. The relatively thin nature of some deposits permits mining in a single lift. Other bauxite formations are thick or deep and several lifts or benches are necessary during mining.

Types of equipment used for stripping are scrapers, draglines, shovels, and bucket wheel excavators. Hydraulicking and dredging has also been used. Shovels, draglines, and end loaders are the common loading machines.

Underground bauxite mines are in southern Europe and Russia. The European deposits are pockets, lenses, or beds between layers of folded carbonate rocks. Mine access is by adit, shaft, or decline. Mining methods used are top slicing, sublevel stoping, block caving combined with shrinkage stoping, longwall, and room-and-pillar. Excessive water is a problem at most operations, especially in workings below the karst water table. Where there are possibilities of encountering strong water inrushes in limestone and dolomite, the active water level is lowered by drilling dewatering shafts before the start of mining operations (Mecsnober and Rosta, 1971).

TABLE 9—Analysis of Dried Chemical Grade Bauxite

Al_2O_3	56.5–60.5%
SiO_2	4.3–9.0%
TiO_2	2.3–3.5%
Fe_2O_3	1.5–3.0%
Loss of ignition	28–31%

Milling Techniques

Bauxite requires simple processing as compared with the intricate milling and mineral dressing techniques in the base metal industry. Bauxite processing is simple because most ore as mined is of an acceptable grade—or can be beneficiated by removing clay—and does not require elaborate and costly treatment. Also, the impurities of some bauxites are associated with aluminous minerals in such a way that they do not yield to mechanical removal or cannot be removed economically.

Many bauxites can be substantially upgraded by the removal of a large part of the silica-rich clay in crude ore by scrubbing and wet screening. Heavy media separation has been used for the removal of iron minerals in lateritic bauxite. At some mines in India iron is removed from lateritic bauxite by hand sorting and picking.

Siderite is an impurity in Arkansas bauxite. This ferrous carbonate mineral reacts with Bayer plant liquor reducing the alumina extraction capability of the liquor. Removal of siderite is accomplished by spiral concentrators and high intensity magnetic separators (Hunton and Dale, 1973).

Transportation

Transportation charges are a major cost element of bauxite production; however, some is transported thousands of miles to consuming areas. Barge and ocean vessel shipping at bulk rates is commonly used. Rail transportation is usually limited to short distances because of the higher costs.

Processing Techniques

Alumina: The refining of bauxite into alumina by the Bayer process is accomplished by digesting bauxite in caustic soda to dissolve the Al_2O_3 in the bauxite and to remove impurities. The dissolved Al_2O_3 is precipitated as aluminum hydroxide and calcined to alumina. Aluminum hydroxide is also referred to as alumina trihydrate, alumina hydrate, or hydrate. This nomenclature is considered technically incorrect by some writers but the terms are in common usage.

In the Bayer process bauxite is taken from storage, ground (not required for finely-divided Caribbean bauxite), and mixed with recycled caustic soda plant liquor to form a slurry. The slurry is pumped to digestion vessels and combined with additional caustic soda liquor at elevated temperatures and pressures and is agitated. Temperature is maintained by high pressure steam. Al_2O_3 in the bauxite is dissolved by the caustic solution to form sodium aluminate. Silica in the form of clay minerals in the bauxite reacts with the caustic soda and Al_2O_3 dissolved from the bauxite and precipitates as a desilication product which is a complex sodium aluminum silicate. The other impurities—which are mostly iron oxides, titania, and quartz, if present in the bauxite—are insoluble and remain in suspension.

After digestion, temperature and pressure are reduced to atmospheric conditions and the caustic slurry containing Al_2O_3 in solution and impurities in suspension is pumped to a clarification section of the plant. Impurities are separated in settling tanks, washed with water, and pumped as a red mud to disposal. The semiclarified caustic liquor containing the dissolved Al_2O_3 is pumped to filters to further remove impurities.

The clarified caustic liquor is then pumped to vertical tanks to which is added previously precipitated alumina hydrate. As the agitated liquor is cooled, alumina hydrate acts as seed or nuclei for the formation and crystalline growth of precipitating alumina hydrate. The separated and washed hydrate is calcined to alumina or used in other industrial applications.

The aluminum hydroxide minerals as well as the mineral impurities present in bauxite affect the processing techniques necessary to produce a commercially acceptable alumina product. Boehmite is a less soluble mineral in caustic soda than gibbsite and requires higher digestion temperatures and/or concentrations of caustic soda. Diaspore is less soluble than boehmite and requires additional treatment.

Silica minerals can be as important to processing as the aluminum hydroxide minerals. The desilication product formed from the reaction of caustic soda with alumina and with silica in minerals of the kaolin group results in an approximate loss of 1.2 t of caustic soda expressed as sodium carbonate and 1 t of alumina for each ton of silica. Some micaceous minerals are partly solubilized. Coarse quartz is not solubilized, but quartz of fine particle size becomes increasingly soluble under higher digestion temperatures. A reactive silica content not exceeding 5% is desirable to prevent excessive caustic soda consumption and alumina losses.

Hematite is usually not a processing and settling problem. Goethite and goethite with some of the Fe_2O_3 replaced by Al_2O_3 in solid solution causes red mud settling and clarification problems. This lattice-bound Al_2O_3 in goethite can be partially extracted, but is usually retained in the red mud. Siderite is not a common mineral in bauxite but, if not removed, it reacts with Bayer plant liquor, reducing the Al_2O_3 extraction capability of the liquor.

Titania present in bauxites is usually only a few percent or less. Some bauxites do have TiO_2 as high as 12% but this is uncommon. Titania is not considered a serious processing problem, as it tends to remain in the red mud wastes.

Any element, even in minor amounts, can be a problem if it results in alumina contamination. Phosphorous is controlled by adding lime to precipitate calcium phosphate, which remains in the red mud. Highly soluble phosphate minerals, such as variscite and crandallite, contribute to poor red mud settling. Potential contamination by zinc is controlled by adding sodium sulfide to form zinc sulfide, which is removed from the process liquor during clarification.

Refractories: Alumina refractories are produced from clays, bauxitic clays, bauxite, alumina, and fused alumina. Alumina content ranges from about 50% on a calcined basis in bauxitic clays to nearly 100% in fused alumina. Bauxitic clays are calcined at temperatures between 1600 and 1760°C and are converted to mullite for refractories in the 50 to 70% Al_2O_3 range. Refractories containing 70 to 85% Al_2O_3 are made from higher grade bauxite, principally from Guyana and China, and are calcined in the range of 1400 to 1800°C. After calcining, bauxite is mixed with binders and pressed, extruded, cast, and shaped into a variety of forms which are then dried and transferred to kilns for firing. Calcined bauxite is also used in unfired monolithic refractories.

The highest Al_2O_3 refractory is made from Bayer process alumina. Tabular alumina is high purity alumina that has been converted to the corundum state by heating to a temperature just below the fusion point of 2300°C. This produces very hard and dense tabletlike crystals which have good thermal conductivity and resistance to thermal shock. It is used for the manufacture of refractory bricks, castables, ramming mixes, electrical insulators, high grade ceramics, and catalyst bed supports, and in chemical processes requiring a material resistant to chemical attack. Fused alumina is also used in the manufacture of refractories containing 90 to nearly 100% alumina.

Abrasives: Two main varieties of fused aluminum oxide abrasives are made: white and brown. The white product is made from calcined alumina produced by the Bayer process, and the brown product is made from calcined bauxite. About 80% of total world production of aluminum oxide abrasives are of the brown variety.

In making the brown variety, bauxite, coke, and iron borings are charged to an electric-arc furnace. The iron and silica impurities in the bauxite are reduced by the coke, and the iron borings help in slagging. A slag of ferrosilicon forms and settles to the bottom of the melt. The melted bauxite is cooled, crushed, and ground and is the crude abrasive from which a variety of products are made (Anon., 1971).

Chemicals: Aluminum chemicals made from bauxite are aluminum sulfate, aluminum chloride, alumina trihydrate, and sodium aluminate. Aluminum sulfate is produced by digesting bauxite in sulfuric acid. It is used by the paper industry to fix the size which is added to the pulp. It is also used to coagulate suspended matter in water, as a waterproofing agent for concrete, and in petroleum refining.

Aluminum chloride is produced by treating heated bauxite with chlorine gas. It is used in refining mineral oils and in the manufacture of certain organic compounds.

Alumina trihydrate and sodium aluminate are both produced in the Bayer alumina process. Trihydrate is used by the glass and ceramics industry. Sodium aluminate is used in water purification, in the manufacture of milk glass, in the hardening of building stones, as a setting-up agent for acid-resistant enamel slips, and as a substitute for sodium silicate and sodium carbonate in whiteware slips.

Alumina trihydrate is the starting material for the manufacture of aluminum chemicals purer than those originating from bauxite. Iron-free aluminum sulfate is made from the trihydrate and is used by the paper industry and to produce other aluminum chemicals such as aluminum stearate, resinate, acetate, and pharmaceuticals. Aluminum fluoride and synthetic cryolite are produced from the trihydrate and are essential in smelting aluminum. Trihydrate is used as a flame-retardant filler in the backing of carpets. Chemically bound water is released as vapor around 200°C in sufficient quantities to extinguish or retard the spread of flames.

Activated bauxite is used as a drying agent and absorbant in the clarification and decolorization of oils and raw sugar solutions. It is prepared by partially calcining special chemical grades of bauxite at temperatures of 120 to 180°C to reduce the chemical bound water to 6 to 8%.

Activated alumina is prepared from alumina trihydrate in a similar way by partial calcination. It is used as a powder, granules, or porous balls to dry gases and organic liquids.

Cement: When ordinary aggregates and high-alumina cement are used, concrete can be made that is slow-setting, rapid-hardening, and resistant to chemical attack, especially from sulfates. Concrete made from refractory aggregates and high-alumina cement is capable of withstanding high furnace temperatures.

High-alumina cement is produced by fusing bauxite and limestone in rotary kilns and reverberatory furnaces. It is also produced in blast furnaces. The fused product is cooled and ground to cement fineness (Robson, 1962). A higher grade of cement is made using alumina instead of bauxite.

Markets—Consumption

As with most raw materials, the greatest markets for bauxite and alumina are the industrialized areas of North America, Europe, and Japan. Consumption of bauxite is related to the economy and growth of the aluminum industry, which is committed to the use of bauxite as the basic aluminum ore. Fig. 2 shows the relationship of the US bauxite and alumina supply.

Prices

Bauxite for metal grade use does not have a quoted or recognized price like some mineral commodities. Most bauxite is sold in one of the following ways: (1) transfers or sales by subsidiaries or affiliates to parent companies for their own use; (2) under negotiated agreements between buyer and seller for specified tonnages, grades, and delivery periods; and (3) occasional spot sales.

Aluminum companies mining bauxite through

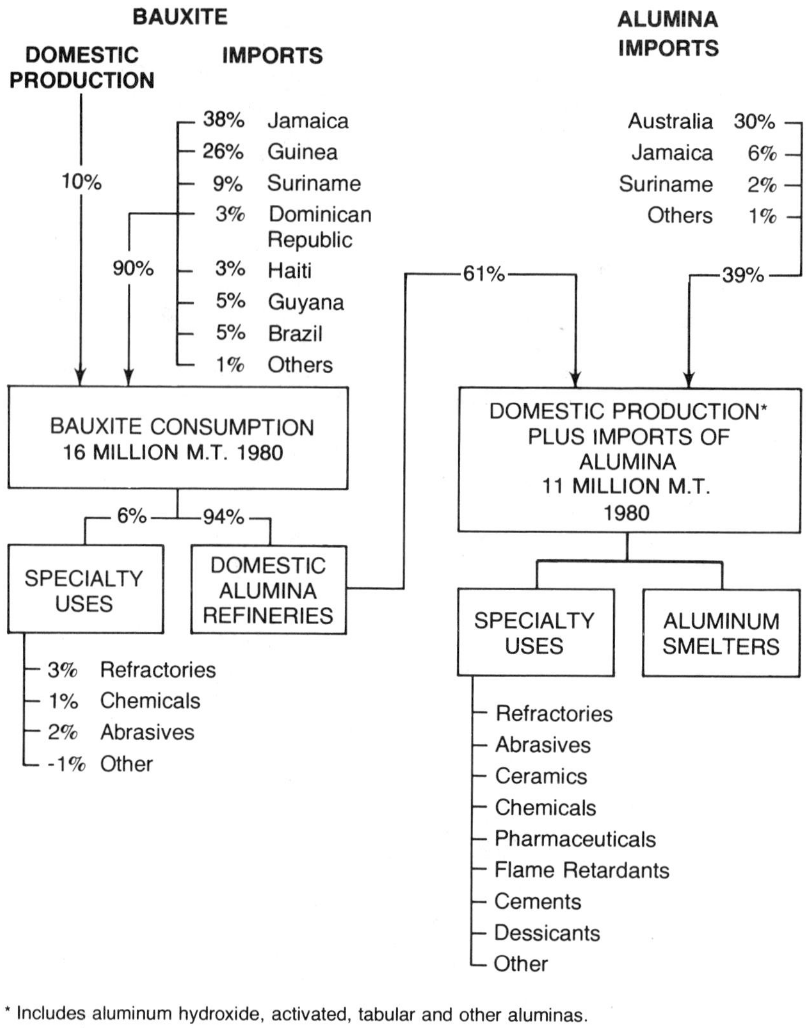

FIG. 2—*US bauxite and the alumina supply (1980). Includes aluminum hydroxide, activated, tabular, and other aluminas. Source: US Bureau of Mines.*

subsidiaries or affiliates for internal use establish their own bauxite value or price for tax or cost accounting purposes. Bauxite sold on contract in arms-length agreements is usually based upon a formula using an f.o.b. base price, base grade, and escalation. Bauxite differing from the base grade stated in percent Al_2O_3 and SiO_2 content involves price adjustments. If the Al_2O_3 content exceeds the base grade, the price is adjusted upward, and if it is lower, the price is adjusted downward. In a similar way, price adjustments are made for variations in the SiO_2 content from base grade. Downward adjustments from base price are made for increasing amounts of SiO_2, and upward price adjustments are made for decreasing amounts of SiO_2. Contracts frequently carry escalator clauses based on wage and price indices and petroleum prices. The f.o.b. price for bauxite sold on contract is extremely variable in different producing areas. In 1980 the f.o.b. price at shipping point generally ranged between \$15 and \$35/t. The amount of bauxite moved in spot sales is small compared to intercompany transfers and contracted sales. Prices for such sales usually are not revealed.

Prior to 1974 revenue to governments from bauxite production was from royalties, severance taxes, and income taxes on profits. In 1974 a group of bauxite producing countries organized and formed the International Bauxite Association, IBA, having as their stated objective ". . . the equitable distribution of benefits which accrue from the exploitation of bauxite resources while taking into account the interest of consumers." Some—but not all—of the member countries of the IBA have levied bauxite taxation indexed to the price of aluminum metal, thereby greatly increasing the total cost of bauxite. These bauxite taxes are not uniform among countries imposing them, but in 1980 were in the range of \$12 to \$17/t, or more than the f.o.b. selling price in some countries not imposing levies. In 1980 member countries of the IBA were Australia, Dominican Republic, Ghana, Guinea, Guyana, Haiti, Indonesia, Jamaica, Sierra Leone, Suriname, and Yugoslavia.

In 1980 calcined refractory grade bauxite from Guyana was priced at \$240 to \$250/t c.i.f. US ports. The price of Chinese refractory grade was about \$165 to \$185/t depending on Al_2O_3 content. Calcined abrasive grade was quoted at approximately \$140 to \$160/t c.i.f. and dried chemical grade bauxite was approximately \$60/t c.i.f.

Future Considerations and Trends

Substitutes

Among the raw materials other than bauxite that have been considered as sources of aluminum are clay, alunite, saprolite, aluminous shale, dawsonite, and igneous rock with a high aluminum content. Resources of nonbauxitic raw material containing aluminum far exceed the reserves of bauxite, are virtually unlimited, and are far more widely distributed. Technology has been developed and is being advanced for the use of bauxite substitutes. During World War II nonbauxite substitutes were used by several countries when supplies of bauxite were curtailed or reduced. The Soviet Union has for some years processed nepheline rock and alunite to fill part of its alumina requirements.

The heavy dependence of the US aluminum industry on imported bauxite, the rapid increase of prices through production levies and taxes in 1974 by some member countries of the IBA, and increasing ocean freight costs brought to attention the vulnerability of the industry regarding bauxite supplies. A program was sponsored by the US Bureau of Mines (USBM), with participation by aluminum producers, to investigate the feasibility of producing alumina from raw materials other than bauxite. Two processes were studied in detail: a nitric acid-clay process and a hydrochloric acid-clay process. The hydrochloric acid-clay process was considered more promising and process designs were developed for a demonstration plant. Research has continued on the program but by the end of 1980 no announcements to build a demonstration plant had been made. Some companies independently have developed processes and are conducting research to produce alumina from nonbauxitic raw material. Present technology does not permit the conversion of a Bayer alumina plant to the use of nonbauxitic ores, so entirely new plants would be required.

That future time when substitutes for bauxite become more economically attractive could be brought closer by a significant technological breakthrough; general disruption of bauxite supplies; increased taxation in bauxite-supplying countries; demands by governments of bauxite-supplying countries that the industry would find difficult or impossible—or be unwilling—to meet; and development within countries of economic and political climates that would discourage venture capital from the development of new bauxite sources. It is improbable that all or most of these possibilities would occur simultaneously or within a short period of time or throughout the bauxite-supplying countries. The major aluminum producers have protected themselves against such occasional occurrences by obtaining ore from more than one source.

Government Control

There is an increasing trend in overseas countries toward government ownership, participation, or control of extractive industries. State ownership of mines in countries with centrally planned economies has long existed, but a more active state role is now being played in various countries with market economies. The export of bauxite by foreign-owned companies began early in the history of the aluminum industry. As the industry expands and new bauxite mines are proposed in developing countries, the host governments frequently insist on government participation as a condition for mine development. At some existing operations there has been a shift from total private ownership to a partnership with the government.

Refinery Location

Another trend has been the locating of alumina refineries at or near bauxite mines. This meets, in part, the objectives of the government of the bauxite-producing country to produce a higher value commodity for export. A contributing factor for on-site processing is the changing economies of transporting bauxite long distances to refineries versus refining bauxite at the mine and shipping alumina.

Consortia and Partnerships

Most of the known readily accessible bauxite deposits have been developed. Development of new mines in more remote locations has rapidly raised the capital cost per annual ton of mine capacity in several ways. Transporting workmen, construction equipment, and supplies to a remote location is more costly. An entire town to house construction and operating personnel may be required along with a transportation system and deep-water port. All of these major items can contribute to make the total capital cost amount to hundreds of millions of dollars. In order to achieve a reasonable economy of scale an initial production rate of several million tons per year is often necessary. This much initial production may exceed the ability of a single company to finance the project or to use or market all of the mine production. To overcome these problems there is a continuing trend to form a consortium or partnership of bauxite-producing and consuming companies for new mine development. There is a wide range in capital cost of new mines depending on whether infrastructure and service facilities already exist or must be constructed. Estimated costs of new mines in 1980 ranged from \$50 to \$100 per annual ton of mine capacity.

Environmental Considerations

Environmental regulations for bauxite mining range from none in some countries to rigid in others. There is a growing concern regarding land disturbance and the clearing of timber and brush for mine development, resulting in delays and modifications of mining plans. Replacement of topsoil and reforestation may be required. Jamaica for many years has required the restoration of mined areas to a level of productivity or utilization that existed prior to mining. This includes the reshaping and grading of pits, replacement of topsoil that had been previously stripped, and planting trees and grasses to prevent erosion. In Western Australia there was considerable resistance to the expansion of bauxite mining and refining, causing concern about future growth of the industry. Land reclamation in the Weipa area of Queensland, Australia, includes partial filling of mined areas and overburden replacement. Trial plantings of various tree species have been made in attempts to replace the indigenous trash trees with more useful varieties. France maintains strict regulations of mine openings and buildings because of the proximity of tourist attractions.

Inasmuch as bauxite is a product of chemical weathering, practically all the leachable products have already been removed and the drainage from bauxite mines and dumps contain no pollutants that are toxic. Many surface bauxite mines are in tropical or subtropical areas where rainfall is abundant and the growth of vegetation is rapid, which naturally promote the reestablishment of plant life.

Environmental regulations in the United States are strict concerning the discharge of industrial wastes into streams and open bodies of water. Red mud resulting from refining bauxite into alumina must be impounded. Other countries have or are developing similar regulations.

Acknowledgments

The manuscript for this chapter was reviewed by W. L. Barham, K. B. Bengtson, C. P. Bingham, R. G. Breuer, H. M. Mikami, and E. O. Strahl of Kaiser Aluminum and Sam H. Patterson of the US Geological Survey. Their comments and suggestions are gratefully acknowledged.

Bibliography and References

Anon., 1970, "Weipa Bauxite," Comalco, Melbourne, 32 pp.

Anon., 1971, "Abrasives," *Industrial Minerals*, No. 46, July, pp. 9–28.

Anon., 1972, "An Introduction to Refractories," *Industrial Minerals*, No. 58, July, pp. 9–23.

Anon., 1972a, "CE Minerals: Georgian Bauxite and Kaolin Calcined for Refractory Grog," *Industrial Minerals*, No. 56, May, pp. 17–22.

Adams, J.A.S., and Richardson, K.A., 1960, "Thorium, Uranium and Zirconium Concentrations in Bauxite," *Economic Geology*, Vol. 55, No. 8, pp. 1653–1675.

Allen, S.A., 1949, "Bauxite Investigations, Eufaula District, Barbour and Henry Counties, Ala.," Report of Investigations 4521, US Bureau of Mines, 85 pp.

Allen, V.T., and Sherman, G.D., 1965, "Genesis of Hawaiian Bauxite," *Economic Geology*, Vol. 60, No. 1, Jan.-Feb., pp. 89–99.

Altschuler, Z.S., et al., 1955, "The Aluminum Phosphate Zone of the Bone Valley Formation, Florida, and Its Uranium Deposits," Professional Paper 300, US Geological Survey, pp. 495–504.

Anderson, E., 1966, "Geophysical Survey Operations of the Demerara Bauxite Co. in Connection with the Search for Bauxite in Guyana," *Proceedings*, 7th Inter-Guiana Geological Conference, Paramaribo, 10 pp.

Auvert, R., and Barisain, J., 1969, "New French Bauxite Mine," *Mining Magazine*, Vol. 121, No. 4, Oct., p. 351.

Baker, R., 1972, "Del Park Bauxite Project," *Mining Magazine,* Vol. 127, No. 4, Oct., pp. 351–355.

Bardossy, G., 1971, "Bauxite Deposits of Hungary," *Proceedings*, 2nd International Symposium of ICSOBA, G. Varhegyi, ed., Vol. 2, Research Institute for Non-Ferrous Metallurgy, Budapest, pp. 9–20.

Bland, F., 1972, "Electronic Process Control in Aluminium Production at Gove," *Australian Chemical Processing & Engineering*, Mar., pp. 20–23.

Bleakley, D., 1964, "Bauxites and Laterites of British Guiana," *Bulletin 34*, Geological Survey of British Guiana, Georgetown, 153 pp.

Brocewell, S., 1962, "Bauxite, Alumina and Aluminium," Overseas Geological Surveys, Mineral Resources Div., London, 235 pp.

Brucy, J., and Auvert, R., 1971, "The Mazaugues—Aval Mining Works—Sinking of an Incline," *Proceedings*, 2nd International Symposium of ICSOBA, G. Varhegyi, ed., Vol. 2, Research Institute for Non-Ferrous Metallurgy, Budapest, pp. 189–195.

Burst, J.F., 1972, "Genetic Relationship of the Andersonville, Georgia and Eufaula, Alabama Bauxitic Kaolin Areas," SME Preprint 72H350, SME Fall Meeting, Birmingham, AL, Oct., 20 pp.

Bushinsky, G.I., 1964, "Types of Karst Bauxite Deposits and Their Genesis," *Symposium sur les Bauxites, Oxydes et Hydroxydes D'Aluminium,* M. Karsulin, ed., Vol. 1, Academie Yougoslave Sci. et Arts, Zagreb, pp. 93–105.

Calhoun, W.A., and Hill, T.E., 1962, "Metallurgical Testing of Hawaiian Ferruginous Bauxites," Report of Investigations 6003, US Bureau of Mines, 43 pp.

Clarke, E.W., 1924, "The Data of Geochemistry," Bulletin 770, US Geological Survey, 841 pp.

Clarke, O.M., Jr., 1966, "The Formation of Bauxite on Karst Topography in Eufaula District, Alabama and Jamaica, West Indies," *Economic Geology*, Vol. 61, No. 5, Aug., pp. 903–916.

Cochet, Y.R., 1971, "Origin of the Bauxite Deposits in the Padurea Craiului Mtns., Rumania," *Proceedings*, 2nd International Symposium of ICSOBA, G. Varhegyi, ed., Vol. 2, Research Institute for Non-Ferrous Metallurgy, Budapest, pp. 63–68.

Cooper, W.G.G., 1958, "The Bauxite Deposits of the Gold Coast," Bulletin 7, Gold Coast Geological Survey, Accra, 33 pp.

Dana, E.S., 1932, *A Textbook of Mineralogy*, 4th ed., John Wiley, New York, pp. 503, 508.

Deschamps, J., 1977, "Bauxite Mining at Les Baux de Provence, France," Preprint 77H392, SME Fall Meeting, St. Louis.

Denisevitch, A.A., 1959, "Bauxites of the Carpathians and Their Genesis," (in Russian, reviewed by V. A. Zans), *Economic Geology*, Vol. 54, No. 5, Aug., p. 963.

Dennen, W.H., and Norton, H.A., 1977, "Geology and Geochemistry of Bauxite Deposits in the Lower Amazon Basin," *Economic Geology*, Vol. 72, pp. 82–89.

De Weisse, G., and Louvain, R., 1956, "How French Bauxite is Mined," *Mining World*, Vol. 18, No. 13, Dec., pp. 51–53, 76.

Dunn, P.G., 1965, "Bauxite Deposits of Gove Peninsula, Arnhem Land," *Geology of Australian Ore Deposits*, 8th Commonwealth Mining and Metallurgical Congress Australia and New Zealand, J. McAndrew, ed., 2nd ed., Vol. 1, Australasian Institute of Mining & Metallurgy, Melbourne, pp. 216–218.

Edwards, J.D., 1949, "The Combination Process for Alumina," *Trans. AIME*, Vol. 182, pp. 9–13.

Evans, H.J., 1960, "High Alumina Laterites of North Queensland," Preprint, Laterites and Associated Minerals Symposium, Brisbane, 11 pp.

Eyles, V.A., 1952, "The Composition and Origin of the Antrim Laterites and Bauxites," *Memoirs*, Geological Survey, Government of Northern Ireland, Belfast, 90 pp.

Fox, C.S., 1932, *Bauxite and Aluminous Laterite*, 2nd ed., Crosby, Lockwood and Son, London, p. 1.

Franotovic, D., 1955, "Dalmatia Leads Yugoslavia's Growing Bauxite Industry," *Engineering and Mining Journal*, Vol. 156, No. 12, Dec., pp. 78–84.

Gladkovsky, A.K., and Ushatinsky, I.N., 1963, "Genesis and Alteration of Aluminous Minerals in Bauxites," *Symposium sur les Bauxites, Oxydes et Hydroxydes d'Aluminium*, M. Karsulin, ed., Vol. 1, Academie Yougoslave des Sci. et Arts, Zagreb, pp. 153–170.

Goldich, S.S., and Bergquist, H.R., 1947, "Aluminous Lateritic Soil of the Sierra de Bahoruco Area Dominican Republic, W.I.," Bulletin 953-C, US Geological Survey, 84 pp.

Gordon, M., and Tracey, J.I., 1952, "Origin of the Arkansas Bauxite Deposits," *Problems of Clay and Laterite Genesis*, AIME, New York, pp. 12–34.

Gordon, M., et al., 1958, "Geology of the Arkansas Bauxite Region," Professional Paper 299, US Geological Survey, 268 pp.

Goretzky, J.K., 1959, "Principles of Distribution of Bauxite Deposits and the Conditions of Their Formation," (in Russian reviewed by V.A. Zans), *Economic Geology*, Vol. 54, No. 5, Aug., p. 960.

Greig, E.W., 1977, "Trombetas and Other Amazon Bauxites, Brazil," Preprint 77H92, SME-AIME Annual Meeting, Atlanta, 34 pp.

Grubb, P.L.C., 1963, "Critical Factors in the Genesis, Extent, and Grade of Some Residual Bauxite Deposits," *Economic Geology*, Vol. 58, pp. 1267–1277.

Grubb, P.L.C., 1971, "Mineralogical Anomalies in the Darling Ranges Bauxites at Jarrahdale, Western Australia," *Economic Geology*, Vol. 66, pp. 1005–1016.

Grubb, P.L.C., 1979, "Genesis of Bauxite Deposits in the Lower Amazon Basin and Guiana's Coastal Plain," *Economic Geology*, Vol. 74, pp. 735–750.

Grubic, A., 1964, "The Stratigraphic Position of Bauxites in the Yugoslav Dinarids," *Symposium sur les Bauxites, Oxydes et Hydroxydes D'Aluminium*, M. Karsulin, ed., Vol. 1, Academie Yougoslave des Sci. et Arts, Zagreb, pp. 51–79.

Guicherit, R., 1966, "Recent Shallow Refraction Surveys in the Coastal Area of Suriname," Informatie 25/1966, Geologisch Mijnbouwkundige Dienst, Paramaribo, 6 pp.

Harder, E.C., 1952, "Examples of Bauxite Deposits Illustrating Variations in Origin," *Problems of Clay and Laterite Genesis*, AIME, New York, pp. 35–64.

Hoppe, R.W., 1977, "Sangaredi—an African Plateau of Bauxite," *Engineering and Mining Journal*, Vol. 178, No. 8, August, pp. 83–90.

Hose, H.R., 1959, "The Origin of Bauxites in British Guiana and Jamaica," *Proceedings*, 5th Inter-Guiana Geological Conference, Georgetown, pp. 185–198.

Hunton, B.W., and Dale, N.W., 1973, "Siderite Removal by Spiral Concentrators and Magnetic Separators from Arkansas Bauxite," *Light Metals 1973*, A.V. Clack, ed., Vol. 2, TMS/AIME, New York, pp. 843–860.

Jackson, R.L., 1971, "Description of the Ferruginous Bauxite Ore Profile in Columbia County, Oregon," Vol. 33, No. 12, Dec., *The Ore Bin*, Oregon Dept. of Geology & Mineral Industries, pp. 223–229.

Kear, D., et al., 1961, "Bauxite Deposits in Northland," Information Series 32, New Zealand Dept. of Scientific & Industrial Research, Lower Hutt, 58 pp.

Kerbyson, J.D., and Schandorf, J.R.H., 1966, "Chemical Analyses of Ghana Rocks, Ores and Minerals," Bulletin 42, Ghana Geological Survey, Accra, pp. 138–140.

Lea, F.M., 1956, "High Alumina Cement," *The Chemistry of Cement and Concrete*, 2nd ed., Edward Arnold, London, pp. 426–460.

Lee, J.W., 1971, "Exploration and Development Drilling for Bauxite in Jamaica," *The Journal of the Geological Society of Jamaica*, Bauxite/Alumina Symposium, Jan., pp. 37–40.

Libby, F.W., et al., 1945, "Ferruginous Bauxite Deposits in Northwestern Oregon," Bulletin 29, Oregon Dept. of Geology & Mineral Industries, 97 pp.

Lindsley, A.W., 1965, "Aluminium," *The Australian Mining, Metallurgical and Mineral Industry*, 8th Commonwealth Mining and Metallurgical Congress Australia and New Zealand, J.T. Woodcock, ed., Vol. 3, Australasian Institute of Mining & Metallurgy, Melbourne, pp. 143–156.

Loughnan, F.C., and Bayliss, P., 1961, "The Mineralogy of the Bauxite Deposits near Weipa, Queensland," *The American Mineralogist*, Vol. 46, Jan.-Feb., pp. 209–217.

Lyons, L.A., 1972, "Gove Bauxite," *World Mining*, Vol. 8, No. 9, Aug., pp. 78–81.

McQueen, H.S., 1943, "Geology of the Fire Clay Districts of East Central Missouri," Vol. 28, Missouri Geological Survey & Water Resources, pp. 201–214.

Mecsnober, M., and Rosta, F., 1971, "Experiences of Shaft Sinking by Drilling in Hungarian Bauxite Mining," *Proceedings*, 2nd International Symposium of ICSOBA, G. Varhegyi, ed., Vol. 2, Research Institute for Non-Ferrous Metallurgy, Budapest, pp. 211–218.

Moses, J.H., and Michell, W.D., 1963, "Bauxite Deposits of British Guiana and Surinam in Relation to Underlying Unconsolidated Sediments Suggesting Two-Step Origin," *Economic Geology*, Vol. 58, No. 2, Mar.-Apr., pp. 250–262.

Nia, R., 1971, "Genesis of Boehmite and Diaspore in Greek Upper Cretaceous Bauxites of the Parnassos—Kona Zone," *Proceedings*, 2nd International Symposium of ICSOBA, G. Varhegyi, ed., Vol. 2, Research Institute for Non-Ferrous Metallurgy, Budapest, pp. 69–98.

Norton, S.A., 1973, "Laterite and Bauxite Formation," *Economic Geology*, Vol. 68, pp. 353–361.

Ottlik, P., and Szabadvary, L., 1971, "Geophysics in Bauxite Prospecting," *Proceedings*, 2nd International Symposium of ICSOBA, G. Varhegyi, ed., Vol. 2, Research Institute for Non-Ferrous Metallurgy, Budapest, pp. 157–168.

Overstreet, E.F., 1964, "Geology of the Southeastern Bauxite Deposits," Bulletin 1199–A, US Geological Survey, 19 pp.

Patterson, S.H., 1967, "Bauxite Reserves and Potential Aluminum Resources of the World," Bulletin 1228, US Geological Survey, 176 pp.

Pearson, T.G., 1955, "The Chemical Background of the Aluminium Industry," No. 3, The Royal Institute of Chemistry, London, 103 pp.

Petersen, U., 1971, "Laterite and Bauxite Formation," *Economic Geology*, Vol. 66, pp. 1070–1071.

Ratmiroff, Gregor de, 1979, "Discovery and Development of de Las Pijiguaos Bauxite Deposit, Venezuela," Preprint 79–87, SME-AIME Annual Meeting, New Orleans, 10 pp.

Roberts, A.L., 1971, "Mining of Bauxite in Jamaica," *The Journal of the Geological Society of Jamaica*, Bauxite/Alumina Symposium 1971, Special Issue, pp. 41–51.

Roberts, R.H., 1977, "Bauxite Mining—Weipa 1967–1977," Preprint 77H328, SME Fall Meeting, St. Louis, 40 pp.

Robson, T.D., 1962, *High-Alumina Cements and Concretes*, John Wiley, New York, 1962, 263 pp.

Roy Chowdhury, M.K., 1965, "Bauxite in India," Ser. A, No. 25, Geological Survey of India, Calcutta, 127 pp.

Ryan, C.W., 1960, "A Guide to the Known Minerals of Turkey," Apr., Mineral Research & Exploration Institute of Turkey, Ankara, pp. 178–180.

Sechiari, P.G., and Sangster, K.J., 1979, "Bauxite Project at Paragominas, Brazil," Preprint 79–11, SME-AIME Annual Meeting, New Orleans, 15 pp.

Sherman, G.D., 1952, "The Genesis and Morphology of the Alumina-Rich Laterite Clays," *Problems of Clay and Laterite Genesis*, AIME, New York, pp. 154–161.

Sinclair, I.G.L., 1967, "Bauxite Genesis in Jamaica: New Evidence From Trace Element Distribution," *Economic Geology*, Vol. 62, No. 4, June-July, pp. 482–486.

Sinkovec, B., 1971, "Geological Features and Origin of the Vrace Bauxite Deposit, Croatia,"

Proceedings, 2nd International Symposium of ICSOBA, G. Varhegyi, ed., Vol. 2, Research Institute for Non-Ferrous Metallurgy, Budapest.

Skillings, D.N., Jr., 1979, "Trombetas Bauxite Operation Starting up in Brazil," *Skillings Mining Review*, Vol. 68, No. 14, April 7, pp. 8–14.

Staudinger, J., 1971, "Some Problems of Technology in Underground Bauxite Mining," *Proceedings*, 2nd International Symposium of ICSOBA, G. Varhegyi, ed., Vol. 2, Research Institute for Non-Ferrous Metallurgy, Budapest, pp. 253–261.

Strahl, E.O., 1971, "The Mineralogy of Jamaican Bauxite and its Effect on Bayer Process Technology," *The Journal of the Geological Society of Jamaica*, Bauxite/Alumina Symposium, Special Issue, Jan., pp. 62–69.

Taylor, G.R., and Hughes, G.W., 1975, "Biogenesis of the Rennell Bauxite," *Economic Geology*, Vol. 70, pp. 542–546.

Valeton, I., 1972, *Bauxites*, Elsevier, Amsterdam, 226 pp.

Van der Hammen, T., and Wijmstra, T.A., 1964, "A Palynological Study on the Tertiary and Upper Cretaceous of British Guiana," *Leidse Geologische Mededelingen*, Vol. 30, pp. 216–232.

Wefers, K., and Bell, G.M., 1972, "Oxides and Hydroxides of Aluminum," Technical Paper No. 19, Alcoa Research Laboratories, East Louis, IL, 51 pp.

White, A.H., 1976, "Genesis of Low Iron Bauxite, Northeastern Cape York, Queensland, Australia," *Economic Geology*, Vol. 71, pp. 1526–1532.

Wyllie, R.J.M., 1976, "OBK, Guinea's National Bauxite Company," *World Mining*, Vol. 29, No. 4, April, pp. 72–75.

Van Kersen, J.F., 1956, "Bauxite Deposits in Surinam and Demerara," *Leidse Geologische Mededelingen*, Vol. 21, pp. 249–375.

Zans, V.A., 1952, "Bauxite Resources of Jamaica and Their Development," *Colonial Geology and Mineral Resources*, London, Vol. 3, No. 4, pp. 307–333.

Bituminous Materials

Original by JOHN M. BOSTWICK *
Revised by JAMES C. BRADBURY †

Bitumen is a general term for a group of materials composed of mixtures of hydrocarbons that are fusible and soluble in carbon disulfide. Included in the group are petroleum, asphalts, asphaltites, and mineral waxes.

Asphalt is a dark-colored, cementitious solid or semisolid, which generally softens on heating and resumes a more viscous or solid state when cooled. It is found in native form in fissures and pore spaces in rocks and as lakes. It is also obtained from the residues produced in petroleum refining.

Asphaltites are black, naturally occurring solid bitumens. They are generally shiny and break with a conchoidal fracture. Common asphaltites are gilsonite (uintahite), grahamite, and glance pitch.

Mineral waxes are solid or semisolid, relatively nonvolatile bitumens of various colors, having an unctuous feel and waxy luster. Of the native waxes, ozokerite is the best known. Montan wax is also classed as a natural wax, although it never occurs in the pure state and must be extracted from lignite or cannel coal (Ladoo, 1920). Mineral waxes may also be obtained by pyroprocessing of petroleum, lignite, and peat.

Pyrobitumens are dark-colored, solid hydrocarbons that are infusible and relatively insoluble in carbon disulfide. They are so-called because they will yield liquid and gaseous hydrocarbons only when heated. *Asphaltic pyrobitumens* are defined as "substantially free of oxygenated bodies" and include wurtzilite (elaterite), albertite, and impsonite (Thrush, 1968). *Nonasphaltic pyrobitumens* contain oxygenated bodies; examples are coal and peat. The nonasphaltic type will not be discussed further.

* Buyer, Eastman Kodak Co., Greeley, CO.
† Principal Geologist, Illinois State Geological Survey, Champaign, IL.

Petroleum Asphalts

Asphalts refined from crude petroleum have almost completely replaced the native types. Refineries can produce asphalts in every desirable grade more economically and in greater quantity than the native types and with a variety of properties that meet the requirements of the various manufacturers using them. For all practical purposes, it is estimated that more than 95% of the asphalts consumed in the United States are refined from crude petroleum (private communication).

Native Asphalts

Native asphalts are dark brown or black, semisolid or solid substances that, in theory, formed by slow, natural fractionation of crude petroleum at or near the surface of the earth (Goulston, 1938). They are colloidal systems of high molecular weight components dispersed in a medium consisting of low molecular weight components (Pracy, 1954). Native asphalts may occur as lakes, as impregnations of rock (rock asphalt) or as deposits consisting of mixtures of natural asphalt and mineral matter in various proportions (Stern, 1960).

Practical use of native asphalt dates to about 3000 B.C. It was mined from surface pits or lakes and used, then as now, for masonry mortar, waterproofing aqueducts and tunnels, binders for road construction, and as an article of commerce (Miller, 1938). Generally the native asphalts find little commercial use today except where they are geographically abundant, easily mined and transported, or have specific or special application. Petroleum-refined asphalt is produced by unusually economical techniques and is much less costly to use.

The largest and best known of the lake deposits is that on the Island of Trinidad, where the lake covers an area of 114 acres and is 285 ft deep at its center (Stern, 1960). The lake asphalt is refined by heating to evaporate included water and is used in street paving, asphaltic roofing, mastic flooring, waterproofing compounds, and many other uses.

Deposits of asphalt mixed with mineral matter and exploited on a limited commercial scale are found in Albania, on the Island of Boeton in the Dutch East Indies, and in Iraq along the Euphrates River near Baghdad (Stern, 1960). The Albanian material, marketed under the name of "Selenitza," contains 10 to 20% finely divided mineral matter. The Boeton asphalt is admixed with microscopic coral shells and clay. The Iraq deposits are about 26% mineral matter. Their chief use is in paving mixtures, but the Albanian product has also been used in bituminous coatings, mastic, and paints.

Rock asphalt is found as deposits of bituminous sandstone and limestone. The impregnating asphalt was probably emplaced in the host rock as petroleum and converted to its present form by natural polymerization. The chief deposits in the United States are in Kentucky, Texas, Oklahoma, Louisiana, Utah, Arkansas, California, and Alabama; in Europe it is found in France, Germany, Italy, and Switzerland (Kretchman, 1957). In mining, the overburden of earth, sand, or other material is stripped off, and the rock asphalt is blasted, crushed, graded, and blended. The bitumen content ranges from 3 to 15% and is blended to about 6.5 to 7.5%.

Rock asphalts are graded as to their origin (Miller, 1938).

Limestone rock asphalt consists of natural limestone rock asphalt containing 9–12% bitumen and the balance relatively contaminant free.

Rock asphalt consists of natural sandstone rock asphalt containing about 7% bitumen with the remainder being free of clay and foreign material.

Processed rock asphalt consists of natural sandstone rock asphalt which is blendable with various proportions of asphalt cement. The blend depends on the application intended.

Rock asphalt uses consist primarily of various paving applications with some specialty uses in flooring, roofing, and waterproofing (Anon., 1973). In Europe the material tends to be used locally and in more varied applications. When mixed with purer native or petroleum asphalts it is applied hot (375°F) to damp-proof walls and roofs, and acid-proof tanks and work surfaces. Concrete and wood leaching tanks and acid vessels are commonly lined by hand application to make them waterproof and/or acid-proof in both Europe and the United States (Miller, 1938). Most American rock asphalts contain more bitumen than do the European types and can be used for paving without heating (Miller, 1938).

Asphaltites

Gilsonite

Gilsonite, also called uintahite, is a dark-colored, brittle solid with low specific gravity and brown streak. It was discovered in 1862 on the Uintah Indian Reservation at Fort Duchesne, Utah. The deposit extends from 5 miles east of the Colorado border to 60 miles west in Utah (Richardson, 1916). It occurs in long, nearly vertical, narrow fissures as much as 40 miles long and up to 20 ft wide.

Mining of asphaltites, such as gilsonite, is done hydraulically as they tend to produce a fine explosive dust when mechanically mined (Jones, 1939; Kretchman, 1957). Hydraulic mining uses either 2000 psi water jets or water-flooded rotary cutters which produce a slurry. The slurry is washed to the floor of the drifts, flows to a sump, and then is pumped to a surface preparation plant. The material is processed to certain particle sizes, graded, packaged, and shipped.

Gilsonite is graded into *ordinary, selects,* and *screened selects.* The grades are defined by melting point, fracture, and degree of fines.

Ordinary gilsonite has a melting point of 325°F, a coarse fracture, and is easily reduced to fines.

Select gilsonite has a melting point of 288°F, a lustrous fracture, and is not easily reduced to fines.

Screened select gilsonite has generally the same properties as select but nearly no fines.

Gilsonite uses include varnishes, battery boxes, flooring materials, mineral rubber, brake linings, paints, and inks (Jones, 1939).

Grahamite

Grahamite was first discovered in Ritchie County, WV, in the early 1860s. Grahamite differs from gilsonite in that it is harder, heavier, more brittle, less soluble, and has a much higher melting point. It occurs in Cuba, Trinidad, West Virginia, Colorado, and Oklahoma. Grahamite may be the result of metamorphic changes in gilsonite (Richardson, 1916). It is blacker than gilsonite and exhibits a conchoidal, hackley fracture and has been mined to depths of 360 ft in veins which are more or less horizontal. Grahamite varies widely in composition from fairly pure to as high as 50% mineral matter.

Grahamite is no longer mined and has not been used in many years (Stern, 1960). It was used for waterproofing, coatings, and mortar.

Glance Pitch

Glance pitch (Miller, 1938) has been called *manjak*. It is intermediate in properties between gilsonite and grahamite, being heavier and higher melting than gilsonite, but less so than grahamite. It was first discovered on Barbados but is also found in Utah, Cuba, Colombia, Russia, Argentina, and Mexico.

Glance pitch uses include black varnishes, japans, thermoplastics, paints, and compounding.

Mineral Waxes

Ozokerite

Ozokerite (Pracy, 1954; Miller, 1938) is found in limited amounts in Utah, Russia, Romania, and in somewhat larger amounts along the Carpathian Mountains in Poland, Czechoslovakia, and Russia. It is a waxy, solid bitumen of paraffinic origin. It occurs both pure and mixed with mineral matter. The material is hand-mined and skimmed from a hot flotation cell. Acid and alkali washing followed by filtering yields "ceresine," which is light brown to white in color. Ozokerite uses include insulation materials, paints, wood fillers, floor polishes, rubber mixtures, and high grade candles (Stern, 1960).

Montan Wax

Montan wax, although classed as a natural wax, is not found in the pure state but must be extracted by means of solvents from lignite or cannel coal. Pyropissite, a type of cannel coal once mined near Halle in Saxony, contained 50–70% montan wax, although most lignites used contain 1–20% (Ladoo, 1920). The processed wax varies from white to nearly dark brown in color and has a melting point of 170–200° F. Its uses are similar to those of ozokerite. A company in California is believed to be producing montan wax.

Pyrobitumens

Wurtzilite, Elaterite, Albertite, Impsonite

Pyrobitumens include wurtzilite (elaterite), albertite, and impsonite. Wurtzilite is mined in Uintah County, Utah, from veins that are 1 to 25 in. thick and as much as 3 miles in length (Fene, 1928). It is black in color but streaks to light brown, is hard and brittle.

Crude wurtzilite must be cracked in a still, recondensed, and graded. Grading is according to still residence time, which affects the degree of hardness or softness. Generally, longer heating yields a lower softening point (Kretchman, 1957).

Albertite and impsonite are in the same class of pyrobitumens as wurtzilite (elaterite). They are also cracked in a still, recondensed, and graded. Albertite contains about 10% mineral matter (twice that of wurtzilite). They are distinguished from one another by specific gravity and carbon content, elaterite having the lowest gravity and fixed carbon and impsonite the highest. All four are used to some degree in rubber compounds, paints, varnishes, and insulating and waterproofing compositions (Kretchman, 1957).

Production of Native Asphalts in US

Native asphalts, including asphaltites, were produced in four states in 1979. Volume totalled 1.6 million st with a value of $25,600,000. Bituminous limestone was marketed by Whites Uvalde Mines and Uvalde Rock Asphalt in Uvalde County, Texas; Southern Stone Co., Colbert, AL; and Barton County Rock Asphalt Co., Barton County, MO. The product was used mainly for street and road repair. The asphaltite gilsonite was produced in Utah by American Gilsonite Co., Uinta County, and Zeigler Chemical and Mineral Co., Weber County. Gilsonite was used in a variety of industrial products other than road materials (Singleton, 1978-1979).

References

Fene, W.J., 1928, "The Mining of Gilsonite in Utah," Information Circular 6069, US Bureau of Mines, 6 pp.

Goulston, W.W., 1938, "The Manufacture of Asphaltic Bitumen," *The Science of Petroleum,* Vol. 4, Oxford University Press, London, pp. 2690–2699.

Jones, F.R., 1939, "Gilsonite," *Chemistry and Industry,* Vol. 58, p. 800.

Kretchman, H.F., 1957, "The Story of Gilsonite," American Gilsonite Co., Salt Lake City, Utah, pp. 51–54, 73–74.

Ladoo, R.B., 1920, "The Natural Hydrocarbons; Gilsonite, Elaterite, Wurtzilite, Grahamite,

Ozokerite and Others," Report of Investigation, US Bureau of Mines, 12 pp.

Miller, J.S., Jr., 1938, "Native Asphalts and Bitumens," *The Science of Petroleum,* Vol. 4, Oxford University Press, London, pp. 2710–2727.

Pracy, H.E.F., 1954, "Manufacture of Asphaltic Bitumen," *Modern Petroleum Technology,* 2nd ed., Pt. 3, The Institute of Petroleum, London, pp. 325–331.

Richardson, C., 1916, "Gilsonite and Grahamite, The Result of Metamorphism of Petroleum Under a Particular Environment," *Journal of Industrial and Engineering Chemistry,* Vol. 8, p. 493.

Singleton, R.H., 1978–1979, "Minor Nonmetals," Preprint, *Minerals Yearbook, 1978–1979,* US Bureau of Mines, 1 p.

Stern, K., 1960, "Native Bitumens, Pyrobitumens, and Asphaltic Type Petroleum Bitumens," *Industrial Minerals and Rocks,* 3rd ed., J.L. Gillson, ed., AIME, New York, pp. 631–637.

Thrush, P.W., ed., 1968, *A Dictionary of Mining, Mineral, and Related Terms,* US Bureau of Mines, Washington, DC, 1269 pp.

Boron and Borates

ROBERT B. KISTLER *

WARD C. SMITH †

The borate industry is one of the few sectors of the mining and mineral-processing industry which the United States still dominates. Since about 1927, the United States has supplied over half of the world's total boron and borate raw materials, and many of the chemical derivatives as well. Chiefly responsible for this record are companies producing sodium borates from two large deposits in southern California, at Boron (Kramer) and Searles Lake. The sodium borates, or products derived from borax and borax-containing brines, are the preferred borates for industry because they are soluble in water and combine readily with most other chemicals. Calcium borates are required for certain end uses, and these are produced from smaller colemanite-ulexite-probertite bodies. Sodium-free borate compounds also are produced from boric acid, which can be manufactured from either sodium or nonsodium borate minerals.

During the past hundred years, the use of boron compounds throughout the world has risen from a few tons per year to the equivalent of over a million tons of B_2O_3. Leadership in supplying borates shifted from country to country and from region to region during this period, generally as the result of the development of newly discovered deposits that were better in quality, more accessible, or more economically exploitable than previous sources (Table 1).

* United States Borax and Chemical Corp., Los Angeles, CA.

† School of Earth Sciences, Stanford University, Stanford, CA.

Primary Products and End Uses

The principal borate products offered by primary producers in the United States are listed in Table 2. In general, these products are utilized worldwide by almost all consumers, regardless of the original source of the primary borates. In the United States, the primary borates all originate in southern California. Colemanite is produced as crude or processed crude ore from the Death Valley area and is utilized by the fiber glass and ceramic industries without further treatment. Production from Searles Lake, CA is from brines that are processed in chemical plants where borax and other compounds are recovered and the principal borate products are produced. At Boron, CA, the world's leading producer is mining a deposit of mineral borax and operating a chemical refinery. The mineral borax is mined together with included clay, and refined borax or "10-mol" is one of the refinery products. The mineral and the chemical compound are identical, $Na_2B_4O_7 \cdot 10H_2O$. In addition to shipping all major refined borate products from

TABLE 1—World Production of Boron Materials and Compounds (1000 st)

	1976	1977	1978	1979	1980
United States	1246	1469	1554	1590	1545
Turkey	1005	1211	1455	1036	990
USSR*	200	200	220	220	220
Argentina/Chile/Peru	103	117	191	206*	212*
Mainland China*	25	30	30	30	30

Source: Lyday, 1981.
* Estimated.

TABLE 2—Principal Borate Products, United States Primary Producers

Product	Composition	B_2O_3 Content, Wt %	Price per st,* Carload Lots, $
Borax decahydrate (technical grade)	$Na_2B_4O_7 \cdot 10H_2O$	36.5	147.00
Borax pentahydrate†	$Na_2B_4O_7 \cdot 5H_2O$	48	169.00
Borax anhydrous‡	$Na_2B_4O_7$	69	446.00
Boric acid	H_3BO_3	56	448.00
Colemanite concentrate #	$Ca_2B_6O_{11} \cdot 5H_2O$	37	203.00
Colemanite, calcined #	$Ca_2B_6O_{11} \cdot 5H_2O$	43	268.75
Ulexite/probertite‖	$NaCaB_5O_9 \cdot 8H_2O / NaCaB_5O_9 \cdot 5H_2O$	26	70.20

Source: Price schedules, US Borax, May 1, 1980, and ABC, Jan. 1, 1980.
* Colemanite, f.o.b. Las Vegas, NV; others, f.o.b. producing plants, California.
† Also sold as an ore concentrate, "Rasorite 46" (registered name), averaging 47% B_2O_3.
‡ Also sold as an ore concentrate, "Rasorite 65" (registered name), averaging 66% B_2O_3.
# Colemanite concentrate produced by ABC is crushed and floated, with both a float concentrate and a calcined product sold.
‖ Ulexite/probertite produced by ABC is crushed and screened to −200 mesh.

Boron, the company ships semiprocessed ore concentrates to treatment plants elsewhere.

Borates over the years have evolved from a high-priced commodity used for medicines, food preservatives, ceramic glazes, and metallurgic fluxes to a relatively low-priced chemical raw material used in large tonnages, in this country mostly in the glass and ceramic industries, in Europe mostly in the manufacture of detergents. The principal end uses of borates produced in the United States are listed in Table 3. Of the 49% exported in 1980, a large part is shipped as ore concentrates. The glass, fiber, and insulation industries account for 68% of domestic consumption. Miscellaneous (9%) includes the uses in pharmaceuticals and cosmetics, antifreeze compounds, adhesives, solvents, tanning agents, fire retardants and wood preservatives, welding flux, and specialty metals and alloys. It also includes such minor uses as gasoline and other fuel additives, extra hard abrasives and gems, embalming fluid, and nuclear reactor control rods and shielding. Research is continuing in many areas, including synthetic organic borane compounds which are finding uses in both medicine and metallurgy; titanium diboride coatings for heat- and wear-resistant valves and cutting tools; various borosilicate glasses and chemical compounds for solar heating and cooling applications; new

TABLE 3—Utilization of Borates Produced in the United States, 1979

	Sodium Borates					
	Decahydrate %	Pentahydrate %	Anhydrous %	Boric Acid %	Colemanite and Other %	Total %
Domestic end uses						
Glass	—	—	5	1	—	6
Glass fibers	—	—	10	—	2	12
Textile fiber	—	—	—	4	4	8
Enamel	—	—	2	—	—	2
Cleaners	4	1	—	—	—	5
Fertilizers	—	—	2	—	—	2
Cellulosic insulation	—	—	—	5	2	7
Misc/unreported	—	1	1	7	—	9
Total	4	2	20	17	8	51
Exports	2	21*	20	6	—	49
Total	6	23	40	23	8	100

Source: Absalom, 1980.
* Includes ore concentrates produced for export.

uses for boron filaments in thermoplastic resin matrix and in fiber optics; and as possible compounds (amino acid analogs) which show promise as therapeutic agents (Absalom, 1979). While research continues on the substitution of borates for fluorspar in the basic oxygen furnace steel manufacturing process, industry acceptance over the past five years has been relatively negative.

Geology

Mineralogy

Although there are at least 150 minerals known to contain the element boron (B), only three are currently utilized in any quantity for their borate content. These are:

Borax (tincal)	$Na_2B_4O_7 \cdot 10H_2O$	Primary world source
Colemanite	$Ca_2B_6O_{11} \cdot 5H_2O$	Produced in U.S. and Turkey
Szaibelyite	$MgBO_2(OH)$	Principal Russian source

For descriptions of minerals in detail, readers are referred to mineralogy texts. Here, mineral names and compositions are taken from *Dana's System of Mineralogy* (Palache, Berman, and Frondel, 1951). Some names of long standing are persistently used in the borate industry: "tincal" is an old name for borax, often used to distinguish the naturally occurring mineral from the identical manufactured compound.

Borax is by far the most important mineral for the borate industry. It crushes freely and it dissolves readily in water, its solubility and rate of solution increasing with water temperature. In contrast, kernite, which is present in large tonnages in the borax deposits at Boron and at Tincalayu, Argentina, is less suitable, although it has a higher content of B_2O_3. Having excellent cleavage, kernite crushes into fibers that tend to mat and clog handling equipment and being rather slowly soluble in water, it requires autoclaves for efficient treatment. However, the new boric acid plant just completed at Boron by US Borax will utilize kernite and mixed borax-kernite ore exclusively.

Colemanite is the preferred calcium-bearing borate. It is very slowly soluble in water, although it dissolves readily in acid. Some colemanite is used in southern Europe in chemical plants to produce boric acid and perborate, because the supply from Turkey provides B_2O_3 at lowest cost. In the United States it is used in the fiber glass industry. In Turkey, much colemanite is beneficiated simply by hand-sorting and washing, with minimum crushing to avoid losses in fines. In the United States, colemanite's property of decrepitating on heating has been utilized to separate it from matrix, usually claystone, producing a semirefined product. Current practice is to separate the colemanite by the sink-float process. The coexisting ulexite/probertite is crushed, screened, and sold without further treatment.

Szaibelyite is the principal Russian source of borate material. It is a magnesium borate and, like colemanite, is very slowly soluble in water. Although it is less satisfactory than either borax or colemanite, the Russians utilize substantial tonnages rather than import foreign borates. Recent reports also suggest that the principal source of Chinese production is now szaibelyite from the Liaoning area of Manchuria.

The list of boron minerals in Table 4 includes some that were utilized at one time even though they are difficult to treat, because they were available where deposits of borate minerals easier to refine had not been found. Ulexite is the primary marsh or playa borate and was used extensively worldwide until larger deposits of borax and colemanite were discovered. Priceite (pandermite) was mined in Turkey and hydroboracite in the USSR. Others such as inyoite, howlite, meyerhofferite, and kurnakovite occur intimately associated with the major ores. Boracite was used in Germany, where it and minor magnesium borates were recovered as a byproduct of potash mining. Still lower in boron content are ludwigite-paigeite and the borosilicates datolite, axinite, and tourmaline; their potential as sources of boron has been studied extensively only in the USSR. Sassolite has only mineralogic interest at most occurrences, for the quantity found is generally very small. In the Lardarello region of Italy, however, natural steam carries boric acid recoverable as sassolite, and for a long period prior to 1965 several thousand tons a year were produced.

Among studies of boron minerals in recent years, two kinds have special value. One, studies of crystal structures have illuminated many of the interrelationships of borates in groups such as borax-kernite-tincalconite and inyoite-meyerhofferite-colemanite (Christ, Truesdell, and Erd, 1967). These also lead to a satisfac-

TABLE 4—Composition of Principal Boron Minerals

Mineral	Chemical Composition	B_2O_3 Content,* Wt %
Sassolite	$B(OH)_3$	56.4
Borax	$Na_2B_4O_7 \cdot 10H_2O$	36.5
Tincalconite	$Na_2B_4O_7 \cdot 5H_2O$	47.8
Kernite	$Na_2B_4O_7 \cdot 4H_2O$	51.0
Ulexite	$NaCaB_5O_9 \cdot 8H_2O$	43.0
Probertite	$NaCaB_5O_9 \cdot 5H_2O$	49.6
Priceite	$Ca_4B_{10}O_{19} \cdot 7H_2O$	49.8
Inyoite	$Ca_2B_6O_{11} \cdot 13H_2O$	37.6
Meyerhofferite	$Ca_2B_6O_{11} \cdot 7H_2O$	46.7
Colemanite	$Ca_2B_6O_{11} \cdot 5H_2O$	50.8
Hydroboracite	$CaMgB_6O_{11} \cdot 6H_2O$	50.5
Kurnakovite	$Mg_2B_6O_{11} \cdot 15H_2O$	37.3
Szaibelyite (Ascharite)	$MgBO_2(OH)$	41.4
Boracite	$Mg_3B_7O_{13} \cdot Cl$	62.2
Datolite	$CaBSiO_4(OH)$	21.8
Ludwigite-Paigeite	$(Mg,Fe)_2FeBO_5$	13.5-17.8
Howlite	$H_5Ca_2SiB_5O_{14}$	44.5
Axinite	$HCa(Mn,Fe)Al_2B(SiO_4)_4$	5.0-6.5(approx.)
Tourmaline	$(Na,Ca)_3B_3Al_6Si_6O_{27}(OH,F)_4$	9.0-11.0(approx.)

* Boron content is about one-third the B_2O_3 content, or 1:3.22.

tory classification of borate minerals based on crystal chemistry (Tennyson, 1963; Christ, 1969, 1972). Second are the field studies yielding notable reports on Death Valley colemanite deposits (McAllister, 1970); Searles Lake evaporites (Smith and Haines, 1964; Smith, 1979); Boron borax deposit (Barnard and Kistler, 1966; Morgan and Erd, 1969; Siefke, 1980); Tincalayu borax deposit (Aristarain and Erd, 1971; Hurlbut, Aristarain, and Erd, 1973); and Turkish colemanite deposits (Özpeker, 1969; Helvaci and Firman, 1976; Wendel, 1978). The field studies have added many new minerals to the list of identified borates and, more importantly, have defined more sharply how various boron minerals occur in their natural geologic environments.

Classification and Origin

Boron mineral deposits often are classified simply according to the main mineral of value —borax, ulexite, colemanite, etc.—because this identifies the feature most important in utilization. Much more is indicated, particularly about such features as size and the mineral and rock associations so important in exploration and development, if the deposit is also classified according to the geologic environment in which it originated (Meixner, 1953; Aristarain and Hurlbut, 1972).

Borate deposits that are mined or potentially minable contain a small part of the total boron in the world. Most of the boron is dispersed in trace amounts in geologic situations from which it cannot be recovered economically. Seawater contains 4.6 ppm, the earth's crust averages about 3 ppm, and the continental crust, predominantly granitic rocks, averages about 10 ppm (Parker, 1967). Among ordinary rocks, marine shales are unusually rich in boron, with 100 to 230 ppm. In the deposits sought for commercial use, boron has been concentrated enormously above these background amounts. Concentrating boron into the minable deposits involved geologic processes and conditions that are noted as their classification is reviewed.

Paragenesis of Hydrated Borates: Boron does not occur in nature as the element. It is found combined with oxygen in about 150 minerals and with fluorine in two. The more numerous minerals are hydrated borates that form under conditions that prevail at or near the earth's surface. The sequence in which borates form has been established by field observation and laboratory experiment. Among the sodium borates, for example, borax is the primary mineral formed in a lake or elsewhere at ordinary temperatures. Borax dehydrates readily in dry air to tincalconite, and in moist air the conversion reverses readily. Kernite, stable at somewhat higher temperatures than borax, formed from the metamorphism of borax at Boron, CA, and Tincalayu, Argentina. A depth of about 2500 ft and a temperature of about 58°C was calculated for the kernite at Boron (Christ and Garrels, 1959).

Ulexite is the common primary sodium-calcium borate mineral formed in borate marshes or playas under normal circumstances. The lower hydrate, probertite, has been found only in deposits which were probably subjected to elevated temperatures and pressures, as at Boron and in some of the Death Valley deposits. For the calcium borates, inyoite-meyerhofferite-colemanite, the lowest hydrate is the common mineral, although it may not form under normal surface conditions. The highest hydrate, inyoite, has been found as a primary mineral in a few small deposits in South America (Aristarain and Erd, 1971); however, sodium is usually available and consequently ulexite is likely to form (Foshag, 1921). The origin of colemanite is somewhat complex, as it is generally found as a "secondary" mineral with or replacing ulexite, suggesting that the sodium is removed and/or replaced by calcium-rich waters (fluids) under conditions where inyoite is not the stable mineral. Other transformations of the hydrated borates, by change in water content or chemical composition, are demonstrated by minor occurrences. In general, they show that borates reflect sensitive response to their geochemical environment (McAllister, 1970).

Boron Deposits in Nonmarine Basins: The largest known boron deposits originated as evaporites or chemical precipitates interbedded with clays, mudstones, tuffs, limestones, and similar lacustrine sediments. There is evidence that most of the deposits were close to thermal (volcanic) springs at the time of their formation and these are generally regarded as the source of the boron (Waring, 1965). Numerous hot springs and associated basic volcanics throughout the world today contain boron in concentrations of 50 ppm (B) or greater, and several South American springs are actively depositing borates. The first borax discovered in the United States was found in the muds of a warm spring at Borax Lake, CA (Hanks, 1883). The Italian (Tuscany) steam vents from which sassolite was recovered may also represent an active volcanic source. In addition to the concentrated source of the borates and a "basin" in which they can collect, an arid to semiarid climate also seems to be an essential requirement during deposition and concentration of economic amounts of the soluble borates. These concentrations of soluble borates can, in the long run, only be preserved by burial; however, the lack of deposits of soluble borates older than mid-Tertiary may indicate that even burial is not adequate for their protection over long periods of geologic time.

Hydrated borates may accumulate in several ways within a nonmarine basin. Borates may be deposited in layers in a *spring apron* around a borate spring, with ulexite, borax, or inyoite as the primary borate mineral (Muessig, 1966). In a *pool* dominantly fed by a borate spring, borax crystals may grow in bottom mud or at the intermittently dried margins, as at Borax Lake, CA (Hanks, 1883; White, 1957), and near Rudok, Tibet (Hutchinson, 1957). If the central salt flat or *mud flat* of an arid basin receives boron by seepage, intermittent springs or surface flow, evaporation develops a surface efflorescence or precipitate, or an accumulation of crystals just below the surface; these are the "marsh" or playa deposits of early mining in California and Nevada and some of the salar deposits of South America (Papke, 1976). Finally, there are what we may properly call *lake* deposits, because their deposition required much more than seasonal flooding; here we refer to the borax deposits such as Boron and Kirka, formed by chemical precipitation. There is some disagreement as to whether the thicker ulexite deposits, such as those in Death Valley for example, are spring apron or lake deposits.

There is probably a complete gradation between spring apron, pool, and small lake deposits, and a study of the recent South American deposits suggests that many "mud flat" accumulations probably originated as spring aprons or pool deposits. Mud flat deposits, however, are generally thinner, lower grade, and spread over a wider area than the others; they are also generally mixed or interlayered with either evaporite minerals or clastic sediments ranging to conglomerate.

Searles Lake provides the example, possibly unique, of an important boron concentration formed in a continental basin by the evaporation of deep lake waters (Gale, 1914; Smith, 1966; 1979). The borates there consist of borax crystals interbedded with other crystalline saline minerals, mostly sodium chloride, sodium-calcium carbonates, and sulfates. Thermal springs in the distant headwaters have been identified as adequate sources of the boron. The other dissolved constituents gathered in the large drainage area reached Searles Lake separated from most of the detrital materials which settled in a series of five higher basins. Stratification of the denser borate containing salts occurred during several periods in the lake's 40,000-year history, while the fresher waters flowed on to Death Valley, the final lake in the chain. The borax from these salt layers is now recovered by pumping the entrapped brines (see Borate Producing Regions).

The deposits of borates not accompanied by an array of evaporite minerals originated in lakes that were not dessicated during borate deposition; these are chemical precipitates. The large borax deposits at Boron and Kirka clearly are in this category. Some deposits of ulexite-colemanite (with ulexite the primary borate) may also form in borate lakes rather than on playas. Several conditions favoring the development of a borate lake can be cited:

1) An unusually abundant supply of boron from a nearby spring. A warm spring would be most favorable, because it sets up the possibility of precipitation by cooling.

2) A small lake with a small drainage area. This would permit the source of boron to dominate over precipitation or stream inflow and might also minimize dilution with detritus.

3) Contrasting inflows, such as sodium borate in thermal waters and calcium in stream waters. This would cause precipitation by mixing.

4) The lake basin should be deep enough to take the deposit, yet have overflow to prevent buildup of saline mineral constituents.

5) An even local climate. A lake is sensitive to temperature changes and its circulation is affected by wind changes; this may have a role in precipitating borates or other minerals.

Boron Minerals in Marine Evaporite Basins: The magnesium borate, boracite, and a number of less common borates are known to occur in evaporite beds of Permian age formed by extended evaporation of seawater. No deposits of these minerals have been mined for boron alone, but they have been used in Germany, where they were recovered as a byproduct during potash mining. There are also some borates in the associated gypsum-anhydrite beds. In the USSR, borates have been found in Permian potash beds that occur in a series of domes. Borates mined in the Inder region are described as secondary veins and replacements containing boron leached from salt dome borates of marine origin (Godlevsky, 1937). These deposits may represent an unusual concentration of boron in seawater (the "normal" concentration of boron in seawater is 4.6 ppm), but more probably an additional surge of boron from a more concentrated source such as submarine volcanic vents is needed to explain these apparently unique deposits.

Boron Deposits at Granitic Contacts: Some pegmatites and contact metamorphic rocks contain assemblages of various boron-containing minerals such as tourmaline, ludwigite-paigeite, or datolite. These represent "concentrations" of boron that relate more or less directly to the crystallization of intrusive granitic magma. Analyses show that granites average about 10 ppm boron with a few exceptions ranging up to 300 ppm. However, boron does not enter readily into the crystal structure of the common rock-forming minerals; hence when magma crystallizes the boron tends to leave with released water. The conditions of high temperature and fluids under high pressure at a contact also open the possibility that boron may be extracted from the country rocks. In either case, if the boron is not fixed in the contact zone, there is the possibility of its being transported to the surface for release at thermal springs, steam vents, and volcanic fumaroles.

Borate Producing Regions and Their Reserves

Borate deposits are known to occur in North and South America, Europe, Asia, and Asia Minor (Fig. 1). Within this worldwide distribution, however, commercially exploitable deposits are very limited in areal extent. Seven countries (Table 5) supply almost all the world's borates with the United States producing about 50% of this total, all of which comes from southern California. In the western hemisphere, the major deposits of hydrated borates are known in two areas, the southwestern United States and the altiplano of the Andes of South America. Considering with them the minor deposits and active boron-bearing springs and volcanic vents, an abundance of boron is found along the circum-Pacific belt, where minable deposits of soluble borates are preserved in areas which have had a history of relatively arid climate since the mid-Tertiary period. In the eastern hemisphere, boron follows a similar tectonic belt that extends from the Mediterranean through the major deposits of Turkey to Tibet. In the borate area of Turkey the present climate is too humid for boron accumulation, and all the known deposits are of Tertiary age.

The United States Bureau of Mines and others have estimated world reserves of boron at 1 billion st (Table 6, Absalom, 1979).

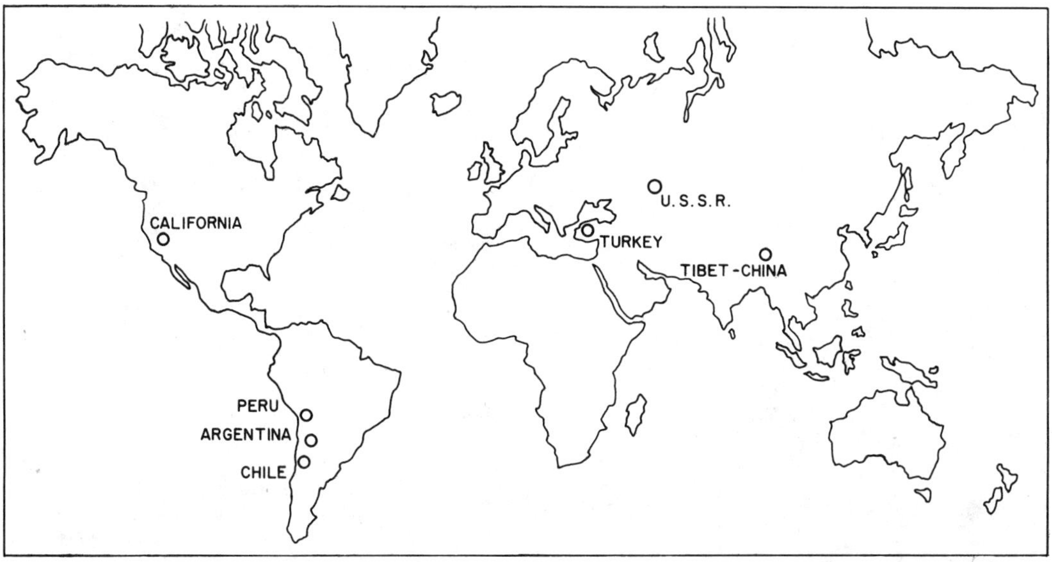

FIG. 1—*Current world sources of borates, 1980.*

United States

In 1856 Dr. J. A. Veach discovered borax in the muds of a mineral spring at what is now Borax Lake, Lake County, CA. Production began there in 1864 and soon spread to the desert areas of western Nevada and southeastern California. The first 63 years of domestic production saw several changes in both the type and location of the mineral deposits mined; but in general, production increased and the price of borax decreased as larger and better deposits were found and refinery techniques improved. The year 1927 was a benchmark in US borate history as both the great borax deposit at Boron and the Searles Lake brines with their recoverable borax content were brought into production and largely replaced all other domestic sources. Borates are presently produced at three locations in Southern California: Boron, Searles Lake, and the Death Valley area (Fig. 2).

Boron, CA (Kramer): The Boron (Kramer) deposit is located between the towns of Mojave and Barstow in the Upper Mojave Desert, about 145 km (90 air miles) northeast of Los

TABLE 5—Capacities of Boron Mineral Producers

Country	Location	Producer	Product	Capacity, 1000 st B_2O_3
United States	Boron	US Borax	Sodium borates	795
	Searles Lake	Kerr-McGee	Sodium borates	122
	Death Valley	ABC	Colemanite	75
Turkey	Emet	Etibank	Colemanite	450
	Kirka	Etibank	Sodium borates	400
	Banderma	Etibank	Colemanite	400
	Bigadic	Etibank	Colemanite, ulexite	50
Argentina	Tincalayu	Boroquimica	Sodium borates	64
Chile/Peru	Misc.	Misc.	Ulexite	54
USSR	Alga (Processing plants)		Magnesium borates	227
China	Chaerhan (Processing plants)		Magnesium borates	35

Source: Fajans, 1980.

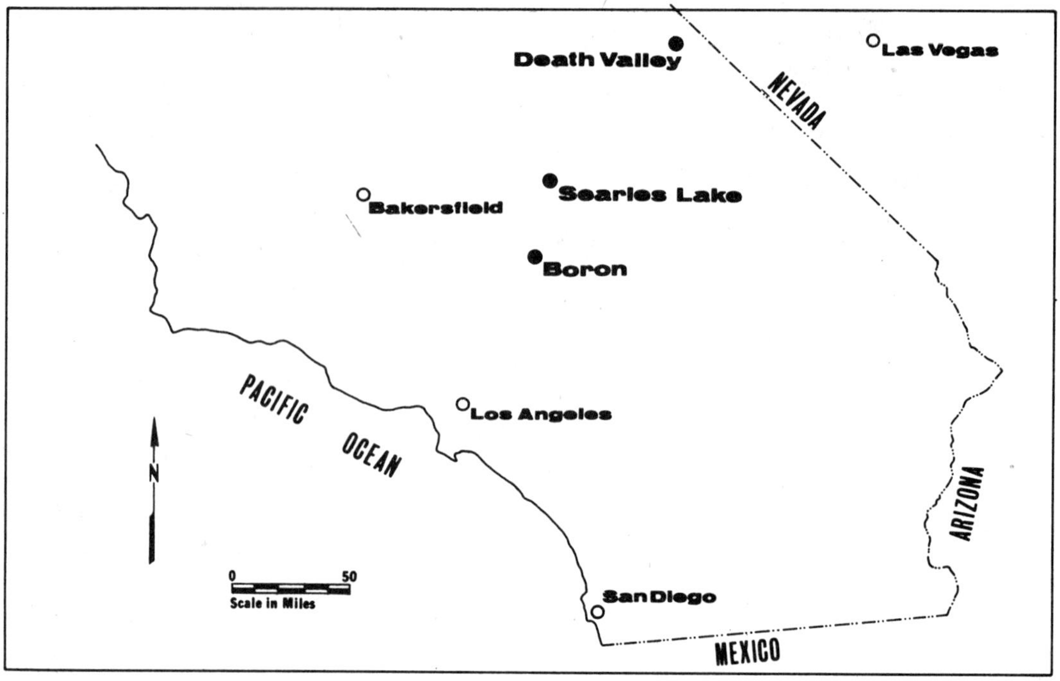

FIG. 2—*Index map of Southern California borate producers, 1980.*

Angeles. Geologically, the deposit lies near the edge of a large Tertiary basin. The "basement" rocks consist of Jurassic and Cretaceous granitics and metavolcanics; the lower Tertiary sequence consists of continental sediments, primarily arkosic sands and silts apparently derived from the earlier granites, with occasional freshwater limestones and volcanics. Fig. 3 shows a stratigraphic section at Boron, CA. The borates are underlain by Tertiary basalt flows, and overlain by a series of continental sands similar to those below the basalt. Diagnostic fossils are rare in the area; however, the Boron deposit has been dated as Upper Middle Miocene on the basis of mammalian fossils found above the borates (Whistler, 1965).

TABLE 6—Estimated World Reserves of Boron

	Million St B
North America	20
Turkey	40
USSR	20
South America	10
China	10

Source: Absalom, 1979.

The Boron deposit consists of a lenticular mass of borax and interbedded clays. It is approximately 1.6 km (1 mile) long, 0.8 km (½ mile) wide, and 100 m (300 ft) thick. Outward from this central borax core, the deposit shows facies changes to ulexite and clay. This ulexite facies also exists above and below the borax. The deposit is thought to have formed in a shallow "permanent" lake fed by sodium- and boron-rich thermal (volcanic) springs which accompanied late stage volcanic activity. Several tuff and tuffaceous clay beds, some of which exhibit ripple marks and other evidence of shallow water deposition, occur interbedded with the borax. The original lake basin was controlled by faulting contemporaneous with the deposit; continued movement tilted and dropped the lake sediments and buried them under at least 762 m (2500 ft) of late Miocene and Pliocene arkosic sediments (Christ and Garrels, 1959). The more deeply buried portions of the borax deposit were metamorphosed to kernite at this time. Renewed movement in late Pliocene-Pleistocene time uplifted the deposit, removed a portion of it by erosion, and caused the outer portions of the kernite to rehydrate to borax. The deposit was then covered by 9 to 21 m (30 to 70 ft) of Pleistocene and Recent sands and gravels (Barnard and Kistler, 1966; Smith, 1968; Siefke, 1980).

The Boron deposit did not crop out; it was discovered in 1913 by Dr. Suckow, a homesteader, who struck colemanite while drilling a water well. However, the sodium borates were not found in the area until 1925. Production of kernite and borax began in 1927 and mining, soon entirely in borax, continued by underground methods until 1957, when production began from the present open pit. US Borax presently mines the deposit by truck-shovel methods with the borax ore being removed from the pit by conveyor and processed in the adjacent refinery (see Frontispiece, Vol. 2). Reserves of the Boron deposit have been estimated by the US Geological Survey to range from 6.3 to 13.6 Mt (7 to 15 million st) B or about 87 Mt (96 million st) of ore grading 25% B_2O_3 (Smith, 1975; Wendel, 1978).

Searles Lake, CA: Searles Lake occupies the central portion of a small desert valley midway between Death Valley and Boron, CA. The "lake" consists of a central salt flat overlying a mass of admixed salts, borax, clays, and interstitial brine, approximately 103 km^2 (40 square miles) in extent. These salts and brines are thought to be the end results of the desiccation and concentration in a series of late Quaternary lakes extending along the east front of the Sierra Nevada. The boron contained within the salt bodies originated in the hot springs along the Sierra Nevada mountain front faults in Long and Owens Valleys (Smith, 1979; Mumford, 1954).

There are two producing horizons in the

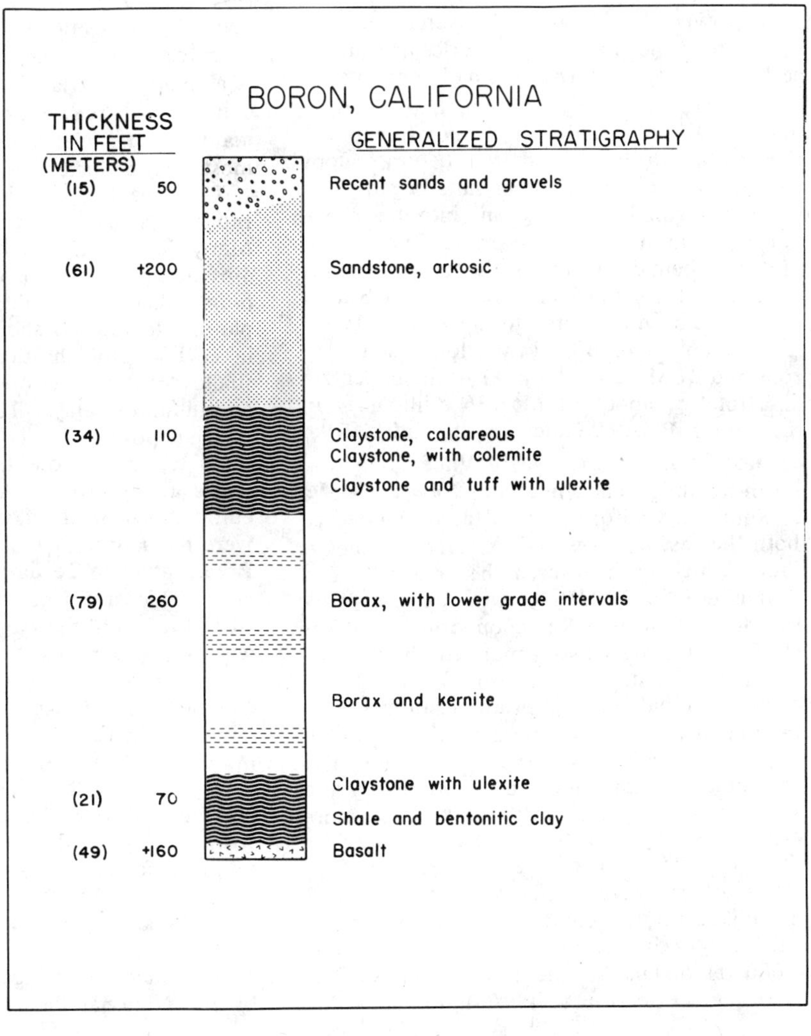

FIG. 3—*Stratigraphic section, Boron, CA.*

lake: the "upper salt," which ranges from 9 to 27 m (30 to 90 ft) thick, averaging 15 m (50 ft), with a brine content averaging 1% B_2O_3, and the "lower salt," which is 8 to 14 m (25 to 45 ft) thick, averaging 12 m (39 ft), with a brine content averaging 1.2% B_2O_3. The upper salt is overlain by a Recent surficial mud and salt crust about 7 to 9 m (22 to 30 ft) thick and separated from the lower salt by 3 to 6 m (12 to 20 ft), averaging 4 m (13 ft), of relatively impervious clays called the "parting muds." The interstitial multicomponent brine which makes up almost 45% by volume of these salt horizons is pumped from a series of wells in the middle of the lake. A relatively constant brine content is maintained by natural recharge and the effluent from the adjacent refineries. The two producing salt horizons are pumped separately as they differ in chemical makeup, and therefore in plant treatment.

Playa-type borax was discovered on the shores of Searles Lake by D. Searles and others perhaps as early as 1863. Small amounts of borax were produced from the playa surface until 1919 when, after extensive potash exploration drilling during World War I, borax along with trona and various other salts began to be recovered from brine wells sunk into the central portion of the lake. At the present time, Kerr-McGee Chemical Corp. is the only company producing from the lake. Reserves in Searles Lake have been estimated to range from 16 Mt (18 million st) of B_2O_3 in the lower salt horizons and 27 Mt (30 million st) in the upper salt to a total of about 33 Mt (36 million st) of recoverable B_2O_3 (Wendel, 1978; Smith, 1979).

Death Valley, CA: Death Valley is a long north-trending valley near the Nevada border of Southern California. It is fault-bounded on both the east and west sides by recently active faults which have lowered the area near Bad Water to 86 m (282 ft) below sea level while raising the Panamint Range opposite to 3367 m (11,045 ft). The Valley contained several lakes during the Pleistocene and over 300 m (1000 ft) of the salts, halite and gypsum, accumulated in the Bad Water area at that time. Minor amounts of the playa borates, borax and ulexite, also accumulated in the muds and seeps nearby and were gathered in the early 1880s. The more productive playa borates near Harmony, 32 km (20 miles) to the north near the present Park Headquarters, were at least partly derived from the adjacent borate-containing Tertiary beds. In late Tertiary time (Mio-Pliocene), over 2130 m (7000 ft) of lake sediments were deposited in what is now Furnace Creek Wash, to the east of Death Valley. It is these lake sediments of the Furnace Creek Formation which contain bedded borates. Fig. 4 shows a stratigraphic section of the Furnace Creek Formation near Death Valley, CA.

"The main deposits (Tertiary colemanite-ulexite deposits) are low in the Furnace Creek Formation, starting in the top of the lowest conglomerate and continuing to about 500 feet higher. The deposits form elongate lenses as much as 200 feet thick in which the borates are interlayered with mudstone, limestone, conglomerate, and gradational variations of these lithologies. The three minable borate minerals, where found in the same deposit, are distributed in three zones in the following order of increasing abundance: probertite in the core or deepest zone, ulexite next, and colemanite in the surrounding outermost zone [Fig. 5]. The boundaries of the zones are generally transitional, but colemanite veins sharply transect some of the ulexite. Probertite is a significantly large part of only one of the deposits, whereas ulexite is a large part of many, and colemanite is predominant in most." (McAllister, 1970.)

The area has undergone several periods of faulting since the borates were deposited. Many deposits are fault bounded and bedding dips of up to 45° are common. Recent mapping indicates that the deposits are elongate parallel to the old shore lines of the Tertiary lake. Laterally, the deposits are equivalent to shallow water limestones or limey mudstones, and tuffaceous clays. These are probably spring apron deposits.

As noted previously, the Death Valley deposits are of two types, the Recent playa accumulations on the floor of the Valley which were the source of the original 20 Mule Team Borax, and the Tertiary colemanite-ulexite deposits which furnished the bulk of US supplies from 1907 until the mines closed in 1927 along with similar deposits at Calico, Lang, and others in western Nevada and southern California. The commercially worked deposits range in size from 181 kt (200,000 st) to over 13.6 Mt (15 million st) averaging 18 to 24% B_2O_3. Present production by the American Borate Co. (ABC) is from both open pit and underground mines, with underground production by continuous mining machines. The ore is trucked to a refinery near Death Valley Junction (Evans, 1976; Blakely, 1980).

Further mining of colemanite in this area has been limited by stringent regulations imposed by the National Park Service upon those de-

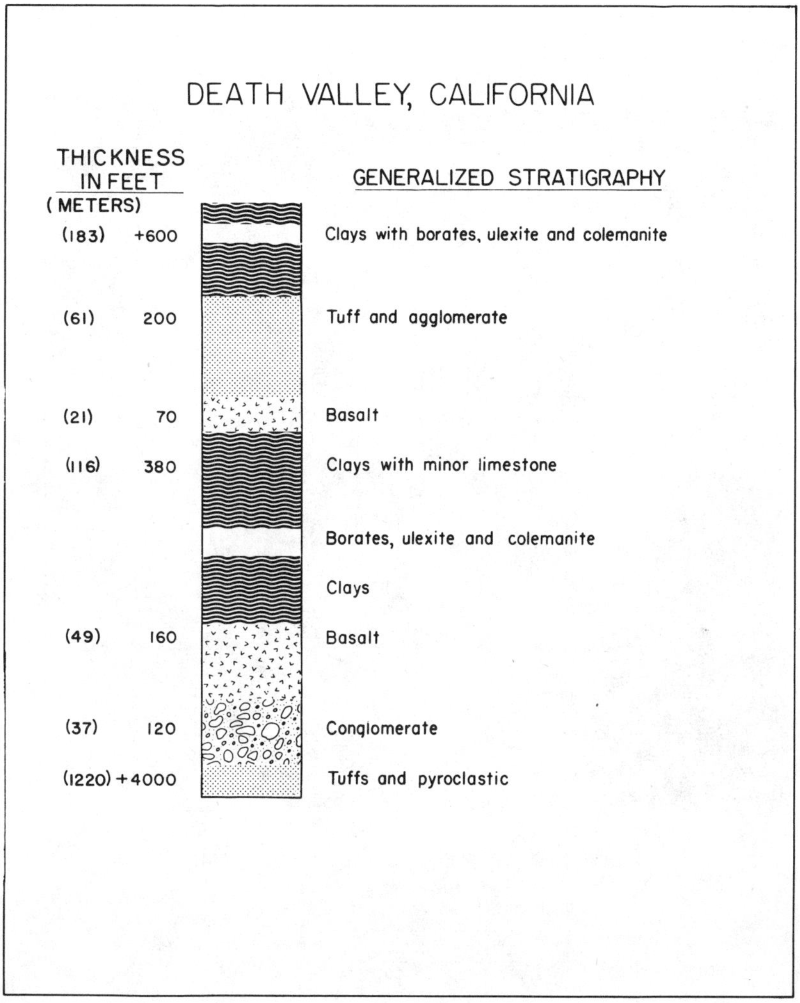

FIG. 4—*Stratigraphic section, Death Valley, CA.*

posits within the Death Valley National Monument, which includes the bulk of the known but undeveloped colemanite-ulexite reserves in the United States. Reserves in Death Valley are estimated at 41 Mt (45 million st) of ore (Absalom, 1980). Wendel (1978) lists the colemanite/ulexite reserves in Death Valley at 28.8 Mt (31.8 million st) of ore. During 1981, two borate companies began developmental drilling of colemanite claims in the National Monument, the first active effort to evaluate reserves in the Monument since 1976 (Lyday, 1981a).

Europe—Asia

Turkey

Turkish borates have been known since at least the 5th century, when they were used by Byzantine Empire artisans for their glazes and for fluxing precious metals. The borates probably came from the Sultan Cayir area, later the site of borate mining from 1865 to 1954. In 1950, a specimen of colemanite was identified from the Bigadic area, some 45 km south of Sultan Cayir. Subsequent prospecting resulted in the discovery of many deposits of colemanite in the 1950s and to the finding of an enormous borax deposit in the 1960s. Turkey is now recognized as having the largest known reserves of borate minerals in the world (Absalom, 1979). Almost all Turkish production is exported. Major mining districts are in the Bigadic and Emet basins of western Anatolia (colemanite) and in the Kirka area south of Eskisehir (borax), Fig. 6. The borates are shipped by truck and rail from the mines to Bandirma, where the government has built a

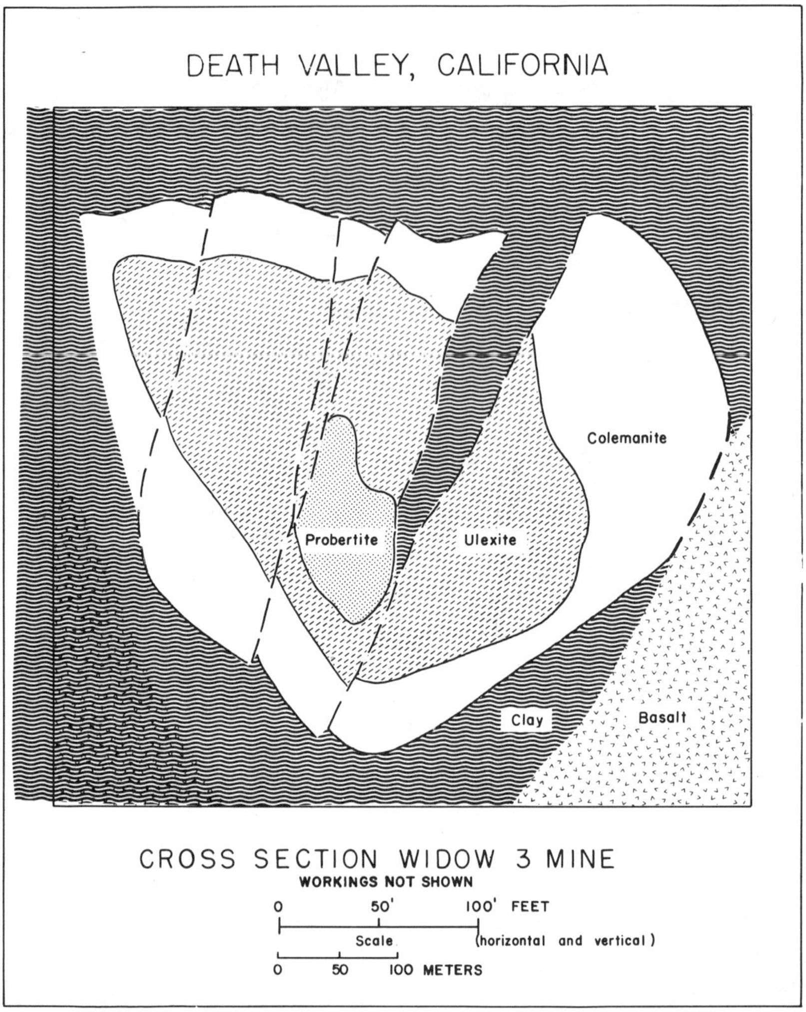

FIG. 5—*Cross section, Widow 3 mine, Death Valley, CA.*

boric acid plant and port facilities. Most of the Turkish borate mines are now run by Etibank following nationalization. Etibank recently has opened up two colemanite deposits, one at Balikesir and one at Kestelek, both located to the north of the town of Bigadic.

Geologically, the borate deposits of Turkey are like those in the Tertiary rocks of California (Meixner, 1953). They are lenticular beds in moderately deformed sections of sedimentary rocks that were deposited in continental basins during a late Tertiary period of tectonic-volcanic activity. The priceite at Sultan Cayir is in highly gypsiferous beds, but the colemanite elsewhere is in shale and marl, and the borax at Kirka is intimately interlayered with dolomitic clay and, in places, ulexite. Limestone, thick-bedded in part, makes up much of the sections, and tuff is abundant. The lack of halite or other saline minerals (except gypsum) and the presence of lignite (Kestelek, Emet, and Kirka) suggest that in the Tertiary the climate there was not extremely arid. The borates that accumulated, presumably during episodes of thermal spring activity, should be considered chemical precipitates rather than evaporites (Özpeker, 1969; Brown and Jones, 1971; Watt, 1973).

Bigadic: The Bigadic region of Turkey includes at least 12 mines in the vicinity of the towns of Bigadic and Susurluk which are south and east of the city of Balikesir in Balikesir Province. The deposits consist of priceite (pandermite) and/or colemanite-ulexite interbedded with late Tertiary marls, clays, and gypsum. These relatively small deposits, 54.4 to 453.5 kt

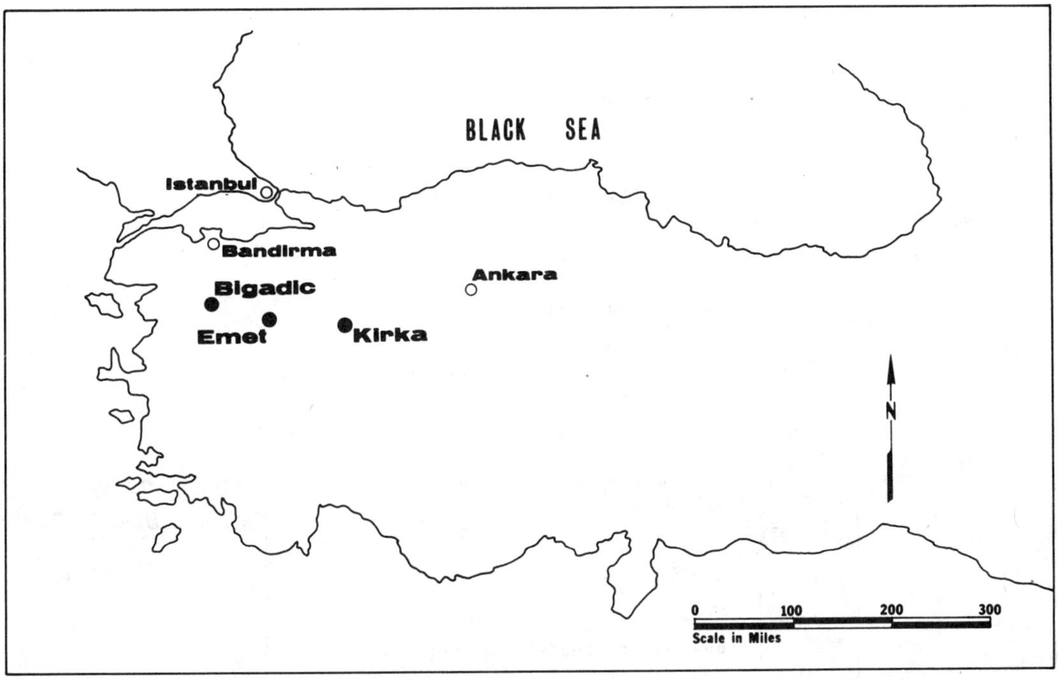

FIG. 6—*Index map of Turkish borate producers.*

(60,000 to 500,000 st), are lens-shaped, and contain numerous clay partings (Fig. 7). The rocks are moderately folded with dips ranging from 10° to 40°. Faulting is also common with a number of the mines fault-bounded. Much of the mining is by relatively simple methods utilizing hand labor. After mining, the colemanite ore is washed or allowed to weather and then is hand-cobbed to produce a direct shipping product. The deposits in the Bigadic region appear to be similar to those in Death Valley. Several deposits have undergone rapid erosion and redeposition so that the borates now occur as nodules and boulders in conglomerates from which they are recovered (Meixner, 1956). Reserves in the Bigadic district are estimated to be about 30 Mt of ore, containing about 10 Mt B_2O_3 (Wendel, 1978).

Emet: The Emet borate district is located between Gediz and Tavsanli in the western part of the province of Kutahya, Turkey. There are at least eight mines in the area both to the north and south of the town of Emet. The deposits consist of colemanite interstratified with late Tertiary marls, clays, limestones, and tuffs (Fig. 8). The largest colemanite deposit is the Hisarcik mine operated by Etibank near Emet. The ore consists of nodular colemanite closely packed in shale. The upper part of the shale averages about 14 m thick and is three-fourths colemanite, the lower part is about 8 m thick and about one-fourth colemanite. These are mined in an open pit now about 1.5 km long, where there is an overburden of limestone about 40 m thick. The shale zones are taken in succession on two benches. After light blasting, the colemanite nodules are hand-picked and stacked for drying or are washed and screened prior to shipment. The washing plant has a capacity of 1360 t/d (1500 stpd) (Wendel, 1978). The ore is characterized by its large rounded masses, some as much as 2 m in diam, and by the relatively large amounts of realgar (AsS) associated with the ore. Probable ore reserves are estimated to be 110 Mt averaging 45% B_2O_3 (Helvaci and Firman, 1976).

Kirka: The Kirka deposit, located about 60 km south of the city of Eskisehir in the Province of Eskisehir, was discovered in 1960. Initial work concentrated on the near surface colemanite, but later drilling disclosed sizable thicknesses of borax and this is now being brought into open pit production by Etibank utilizing modern truck-shovel methods. The deposit lies along the east side of a Tertiary basin, roughly 3 x 7 km in extent and adjacent

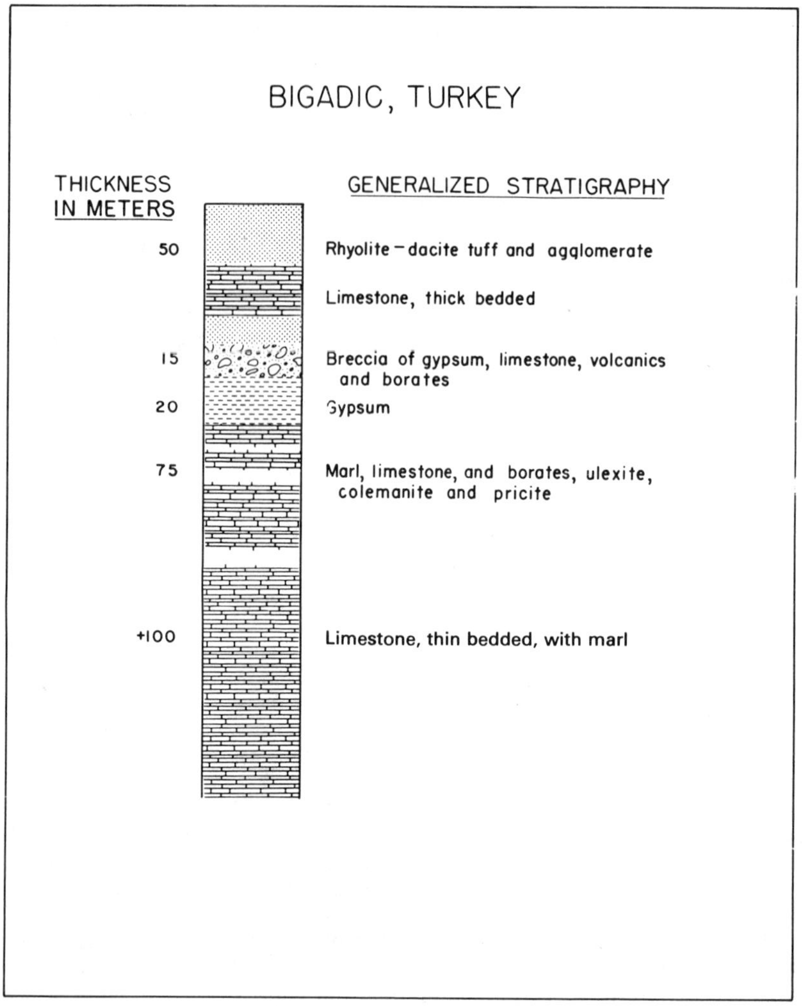

FIG. 7—*Stratigraphic section, Bigadic, Turkey.*

to a large travertine cone. Tertiary marls underlie and overlie the borates. As exposed in the open pit, the ore is mostly monomineralic borax, thinly interlayered and intergrown with dolomitic clay and in places, ulexite (Fig. 9). The ore body has an indicated thickness of up to 145 m and the reserves are estimated at 100 Mt of run-of-mine ore containing an average of 20 to 25% B_2O_3. Wendel (1978) also states these reserves should be adequate for about 35 years at the present rate of production.

USSR

Recent data (Absalom, 1979) indicate that the USSR is now producing about 90 kt/a of B_2O_3. There are little or no recent data published in English on the Russian sources for this production, though it is assumed to come mainly from deposits in the Inder District of Southern Russia. The mineral deposits there are reported to be magnesium borates, primarily szaibelyite and associated hydroboracite, with gypsum, carbonates, and reddish clays; borax is also recovered from the brines of Lake Inder, a large salt lake east of the Ural River between the Caspian Sea and the City of Uraljsk. These sources were discovered in the mid-1930s as a result of a large-scale prospecting effort by Soviet geologists. The "bedded" borates crop out in numerous locations in the district as lenses and "pockets" several meters thick above a large salt dome with a surface area of about 250 km². The salts in the core of the dome

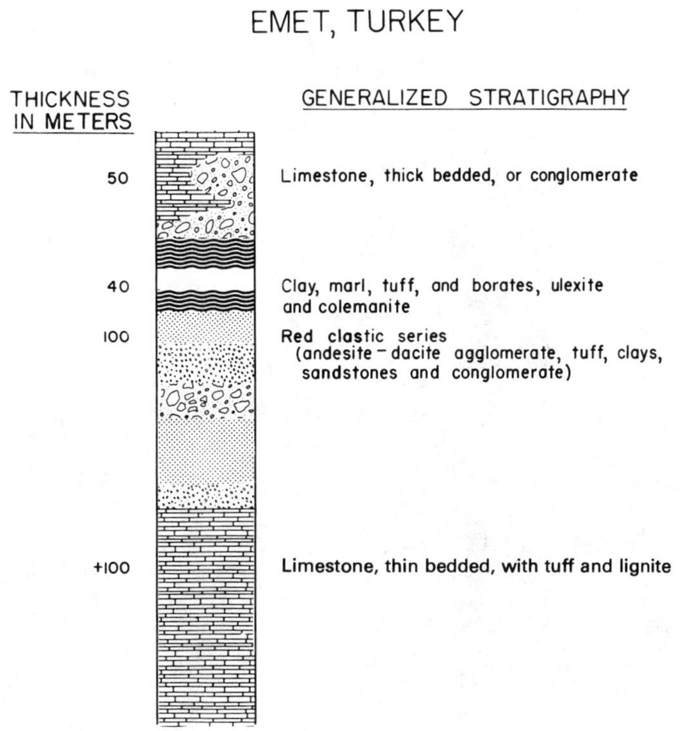

FIG. 8—*Stratigraphic section, Emet, Turkey.*

consist of Permian marine evaporites which have intruded Mesozoic and Tertiary sediments. The dome is capped by a series of gypsum and clay beds and the borates occur along fractures near the top of the salt; they are said to replace the salt in some areas and the clay and gypsum in other areas. Reserves are estimated at 350 kt to 7 Mt of ore having a grade of 20% B_2O_3.

Lake Inder, which resembles Searles Lake to some degree, consists of a surface salt crust, below which is a brine-saturated salt section 36 m thick containing potassium chloride, bromides, and borax; the brine averaging 0.23% B_2O_3. As the lake receives drainage from the salt dome area, the borates there are assumed to be the source for the boron in the lake brines.

The Russians have also done considerable research on obtaining borate minerals and/or producing boric acid from skarn deposits of borosilicates or anhydrous borates. These deposits may account for a minor amount of the Russian production (Valyashko, 1970; Utekin, 1965; Strishkov, 1971).

Italy: Although Italy is no longer a producer of borate raw materials, the Tuscany District was the leading world source of borates from 1828 until about 1880. Production continued thereafter at a rate of several thousand tons a year, but ended in 1965. Boric acid (or sassolite, H_3BO_3), with some ammonium borate and ammonium sulfate, was recovered by controlled evaporation of condensed steam from natural vents (soffioni) and drilled wells. The steam ranges in temperature from 140 to 230°C, and contains about 0.1% B_2O_3. According to White (1957) its properties indicate that the steam

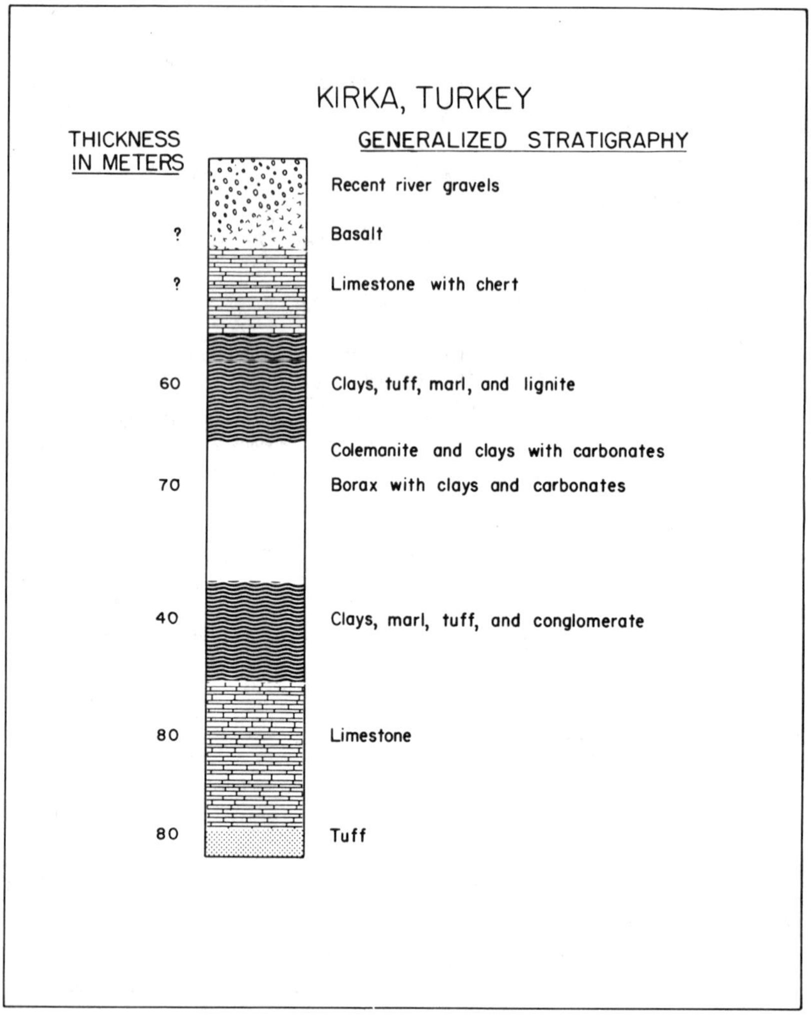

FIG. 9—*Stratigraphic section, Kirka, Turkey.*

brought boron from a volcanic source. Other students of the district consider it possible that the steam picked up boron from the rocks it passed through in rising to the surface. The rocks are chiefly sediments of Cretaceous to Pliocene age. No borates have been found in them, but they may contain traces of boron. Having developed a sizable industry based on using natural boric acid from Tuscany, Italy remains a leader in the European production of boric acid and perborates by processing imported raw materials.

South America

There are over 40 borate deposits in an 885-km (550-mile) stretch of the high Andes along the common borders of Argentina, Bolivia, Chile, and Peru (Fig. 10). This is an arid segment of the tectonic-volcanic belt, characterized by many closed basins with playas or salt flats, called *salares*. Reserves are said to total millions of tons, but the figures lack supporting data. For many years (1852–1967) Chile was a major producer of borates, but industry gradually shifted to the larger and richer deposits found in adjacent areas of Argentina. At present, most production from South America is from the Salta Province of Argentina.

Borate spring deposits are better developed in South America than anywhere else in the free world. These deposits consist of cones and aprons of ulexite and in one case borax, built up around vents from which hot to cool

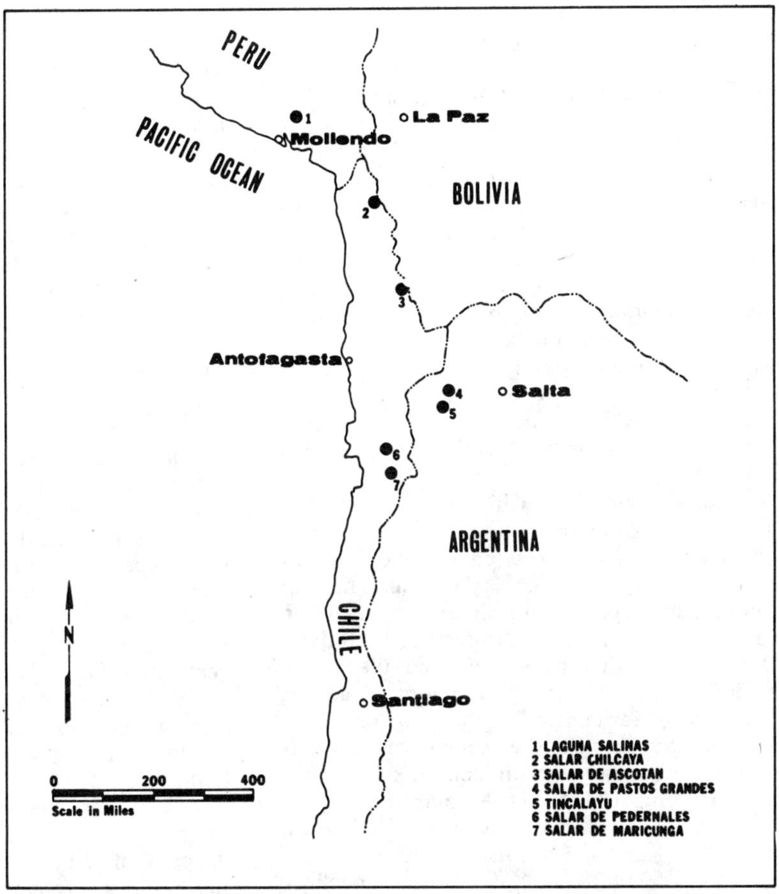

FIG. 10—*Principal South American deposits.*

waters and gas are still issuing in some cases. Most deposits are associated with calcareous tufa which occurs as a late-stage capping over the borates, and sometimes with halite and gypsum as well. Recent volcanic activity is indicated by basaltic to rhyolitic flows in adjacent areas and a volcanic source for the borates is presumed. Old rock in the vicinity of the deposits varies greatly from Precambrian to Tertiary and appears to have no genetic relationship to the deposits. The deposits currently forming are quite small, ranging from a few hundred to approximately 9 kt (10,000 st) of borates averaging 20% B_2O_3 (Muessig, 1966).

The salar deposits of South America consist of beds and nodules of ulexite with occasional borax or inyoite, associated with Recent playa sediments, primarily muds, silts, halite, and gypsum. The borate-bearing playas occupy individual basins which range in size up to several hundred square kilometers. Springs and seeps appear to be the source of the borates. Weathering of adjacent areas and leaching of buried sediments seem to be improbable sources according to knowledgeable observers. These salar deposits contain up to several million tons of crude borates and support a small local industry. Mining conditions are difficult due to the extreme climatic conditions and high altitude and to the wetness of the playas which hinders heavy equipment. Most deposits are mined by hand and the borates transported to the railheads by small trucks (Muessig, 1966).

Argentina: In 1979 Argentina produced about 132.4 kt (146,000 st) of boron minerals and compounds, mostly from the Tincalayu borax deposit (Lyday, 1981). The borax is in Pliocene rocks, and occurs as a pluglike lens of very fine crystalline borax about 100 m in diameter and more than 30 m thick which was apparently pushed upward along with associated playa sediments (Hurlbut, et al., 1973). The borax rests on massive halite, and is partially overlain by about 50 m of reddish salty detrital sediments that include thin layers containing, separately, ulexite, inyoite, kurnakovite, and some kernite. The deposit and the surrounding sediments are folded into an anticline, truncated by a fault at one end. The deposit is mined by open pit methods with relatively modern equipment by Boroquimica SAMICAF. Other borate deposits of Tertiary age are known in Argentina. One large deposit at the Salar de Pastos Grandes consists mostly of bedded colemanite and hydroboracite. The more common salar deposits of borax and ulexite known in the area are worked intermittently by small operators (Anon., 1979). However, the general conditions of high altitude and remote location confront operators with formidable problems (Catalano, 1927).

Chile: Ulexite from Chile was important in the world market prior to the 1950s. About three-fourths of this production came from the Salar de Ascotan, which is large and has the special advantage of being situated on the international railroad from Antofagasta to Bolivia. Other salares that have produced ulexite include: Surire (Chilcaya), Pintados, Atacama, Pedernales, Maricunga. Except Pintados, all are at high altitudes near the Andean crest, where transportation and working conditions are difficult (Leiding, 1942). At Ascotan, mining began in 1883, reached a peak in 1913 when exports from all the Chilean deposits amounted to about 50,000 tons of mineral, declined abruptly in the 1920s, and became inactive in 1967. The ulexite is mostly just below the surface where it occurs as irregular masses and in beds that attain a meter thickness, averaging perhaps 300 mm. Mining was by hand-dug trenches. The ulexite was piled for air drying prior to rail shipment, then roasted at the port for additional drying before shipping overseas. Borates from Chile have again entered the world market in significant amounts, with production in 1979 estimated to be almost 27.2 kt (30,000 st). Most of this production originates from the Atacama and Ascotan salares.

Peru: Peru exported ulexite from about 1894 until the 1920s and production is again being reported. Total exports prior to 1930 were about 100,000 tons. The source was the Laguna de Salinas, a large mountain basin east of Arequipa, reached by 56 km (35 miles) of mountain road that crosses a mountain pass at nearly 5200 m (17,000 ft). The basin, altitude 4328 m (14,200 ft), has a seasonally flooded floor, and the ulexite was recovered along the lake margins in the dry season, from shallow pits dug by hand. Ulexite occurs as irregular masses and lenses 100 to 1000 mm thick in fine sandy detritus, below the surface layer of sulfate and halite that crystallizes from the lake waters each year. The selected ulexite was air-dried or roasted on iron plates over open fireplaces fueled with yareta (a common plant in the Andes). In the old days, the ulexite was taken to Arequipa on llamas, burros, or mules. The deposit is reported to contain 9 Mt (10 million st) of ulexite. Bolivia has several similar salares in the southwestern part of the country, and production from that area is also being considered.

Other Areas

Europe: Boracite ($Mg_3B_7O_{13}\cdot Cl$) was discovered in northern Germany in the potash mines near Stassfurt (East Germany), where it occurs as crystals and nodules. It also has been reported from the edges of a large salt dome near Hamburg where it occurs associated with gypsum and anhydrite. These borates, which are Upper Permian in age, are almost certainly of marine origin. They were recovered in small quantities as a byproduct of the potash mining operations and are not considered of commercial importance in today's markets (Kuhn, 1962).

Asia: There are almost no data available on the borates of Mainland China and Tibet. The deposits of Tibet, which reportedly furnished the borates brought back from "China" by Marco Polo in the 13th century, occur as cones and crusts around thermal springs and in the muds of brine pools. The best known of these deposits are in the Puga Valley of Kashmir in Western China. Current production is from the Chaerhan and Yuncheng salt lakes of Central China which are in some ways similar to Searles Lake, and from a newly reported deposit in Liaoning province in northeast China where szaibelyite is the principal mineral. Mining of the Liaoning Deposit is reported to be by open-cut methods with an expected output of 90 kt/a (100,000 stpy) (Anon., 1980b; Wang, 1979). China exports some boric acid and borax decahydrate to neighboring countries and would appear to be self-sufficient in borates.

Central America: An occurrence of the mineral howlite, a borosilicate, was reported 8 km southeast of Magdalena in northern Sonora, Mexico, and has stirred some industry interest (Lefond and Barker, 1979; Barker and Lefond, 1979; Lyday, 1981).

The Mexican Geological Survey has discovered colemanite and howlite in a similar setting about 30 km northwest of Tubutama, Sonora. These could be the first significant borate finds in Mexico and Central America (Gomez, 1980).

Exploration

Successful exploration for borates in recent years consisted of core drilling favorable areas selected after detailed geologic surveys in the vicinity of known borate occurrences in nonmarine sedimentary rocks of Cenozoic age. Some exploration guides are recognized. Because the bedded borates, borax, and ulexite-colemanite accumulate in lakes, sections of lake beds are studied closely, particularly if they include saline minerals. Tuffaceous and other volcanic materials associated with lake beds are also considered encouraging, although not a direct guide to borates. Some borate deposits seem to be closely associated with basic volcanics, others with tufa cones and similar spring deposits, so indications of centers of volcanism may be favorable. In arid regions, near-surface accumulations of borates at springs or on playas are possible indicators of older borates in the adjacent ranges or subsurface. The combination of a salt flat having marginal borates and surrounding hills having ancient high shorelines suggests the possibility that extensive evaporation may have formed saline beds containing valuable brines. In humid areas, typical associated minerals such as gypsum, or secondary calcite, may provide guides to areas worth testing for borates that do not show at the weathered surface.

Field Techniques

There are two field tests applicable to the common borates. In the flame test, the mineral is dissolved in sulfuric acid, mixed with alcohol, and ignited. "She burns green, Rosie . . .!" Or, dissolved in hydrochloric acid, borate will turn turmeric paper reddish or pinkish brown. Borax or other borates which will dissolve rapidly in water can sometimes be detected by tasting, but this test is not considered reliable. Tests of physical properties are given in mineralogic texts, but it should be noted that borates in older bedded deposits are much harder and denser than playa varieties usually cited. Some Turkish colemanite and associated secondary calcite are said to have a distinctive yellow-white color under ultraviolet light, but this is not reported for other colemanite areas.

Geochemical Methods

Geochemical surveys for boron have been reported in Russian literature, presumably as a prospecting method, but none have been described for other areas. A number of elements that often accompany boron might be considered as potentially useful indicators or pathfinder elements in a regional geochemical program, but boron itself would seem to be the most reliable indicator. Boron is highly mobile, so water sampling would appear to be useful in a search for the soluble borates. As little as 0.5 ppm of boron in water can be detected. Agricultural studies show that some plants are highly sensitive to boron, but surveys of vegetation have not been pursued as a prospecting technique except in Russia (Brooks, 1979).

Geophysical Surveys

Geophysical surveys using gravity, magnetic, and resistivity methods have been used in borate exploration mostly to determine the shape of buried structures or the profile of basin bedrock beneath sedimentary fill. No method that indicates borates or borate bodies directly has been reported. Shallow seismic methods have been tried with some success as a means of tracing limestone-colemanite zones, or lenses, that have a much higher velocity than enclosing shales. Similar tests on borax showed no indication of the ore. Various well-logging techniques will indicate borate zones in a sequence of sands and shales and show, indirectly, the borate content (Fertl, 1971). However, it is generally easier and more reliable to examine the core or cuttings for borate minerals and analyze for B_2O_3.

Drilling

Drills of 460 m (1500 ft) or less rated capacity are generally used for borate exploration as borates are not considered economic below these depths. Drilling can be with either air or fluid depending upon ground conditions and whether cores are being cut or not. In areas of recent basalt cover, a combination of pneumatic down-the-hole equipment followed by casing and conventional coring has been tried with some success. Cores of BX size or larger are generally cut if the core is to be split and assayed. Double-tube core barrels are preferred to help prevent excessive wearing and dissolving of the core. In areas where the borates are interbedded with sticky muds, chrome plating of the inner barrel has helped improve daily footage. Coring bits are chosen which direct most of the fluid away from the core to prevent washing and dissolving the softer beds. Where very soluble borates are involved, short core runs are recommended. Short core runs are also recommended for some colemanite areas as the hard ore nodules tend to roll within the core barrel if the colemanite is interbedded with soft shales. Generally, however, borates core rather well and recoveries of 90% or better are the rule.

Evaluation of Deposits

Under current economic conditions, bedded borates at depths greater than 460 m (1500 ft) probably cannot be exploited at a profit. Borax-bearing brines might be extracted from greater depth. As is true of most nonmetallic commodities, special advantages such as closeness to markets, or new uses for a particular mineral, may change the minimum requirements as shown in Table 7.

Testing

Preliminary chemical tests are normally confined to determining the grade of the ore and the impurities in it. For the grade, both water-soluble and acid-soluble B_2O_3 are determined. The impurities are determined by chemical analysis and mineral identification; they include chlorides, sulfides and sulfates, iron, and other metallic elements. Depending upon the borate mineral involved and its prospective use, the specifications and tolerances can vary.

Physical tests bearing on the separation of the borates from the gangue, which generally is a clay or limestone, include the standard milling tests to determine the size at which the mineral is most efficiently liberated, and the crushing strength of the ore. For the soluble borates, the temperature and rate of dissolving are important, as are the settling rates of the nonsolubles and the rate at which borax recrystallizes from the liquor. Standard mining tests are run on material from the prospect to determine the best mining method at its location.

Preparation for Markets

Mining Methods

Mining methods are chosen which will meet the demand for the mineral by the company and which will fit the particular ore body involved. Two current examples from the United States will illustrate this point.

In Death Valley, the American Borate Co. (ABC) is mining a colemanite/ulexite/probertite ore body by underground methods. The body extends down to 366 m (1200 ft) below the Furnace Creek wash just within the eastern boundary of Death Valley National Monument. Primarily due to Monument restraints, ABC sank its concrete-lined shaft [ID 4.9 m (16 ft)] outside the Monument some 610 m (2,000 ft) from the ore body. The shaft was completed in late 1979 at a depth of 378 m (1240 ft). Drifts were driven to the ore body at the 232-m (760-ft) and 341-m (1120-ft) levels. The ore body is developed by room-and-pillar methods with the rooms roughly 9 m (30 ft) wide and up to 152 m (500 ft) long. Development and mining currently is being done with four mechanical miners of the roadheader type; two Dosco TB 600s and two AEC 330s. The ore is moved to ore passes by 3 m^3 (4 cu yd) Scooptrams, then loaded onto 18-t (20-st) diesel trucks by 3.8 m^3 (5 cu yd) load-haul-dump machines for transport to the shaft for hoisting. Stopes are backfilled with shale waste. Produc-

TABLE 7—Minimum Requirements for a Minable Borate Deposit, United States

Mineral	Size, Millions of Tons	Minimum Grade, Wt % B_2O_3	Maximum Depth, Ft	Remarks
Borax	1 to 5	10	1500	Present supply excellent
Colemanite	0.5 to 2	15	1500	Market Eastern US and Europe
Complex brine	Large	1	1500	May depend upon recovery of other salts
Ulexite	0.5 to 1	20	1500	Present market very small

tion currently is about 1 088 t/d (1200 stpd) (Blakely, 1980).

At Boron, US Borax mines a large, thick, relatively flat-lying borax ore body by open pit methods. The overburden, consisting of semi-consolidated sands, consolidated arkosic sands, and lacustrine shales, is drilled and blasted in 9-m (30-ft) lifts. The ore, consisting of borax and admixed borax and kernite, is also drilled and blasted. The overburden is loaded by two P&H 2100 BL electric shovels with 14.5 m^3 (19 cu yd) buckets onto 154-t (170-st) capacity Lectra-Haul M-36 trucks. The ore is loaded onto a fleet of Lectra-Haul M-100 trucks by a P&H 2100 shovel and three Marathon-LeTourneau front-end loaders, all with 11.5 m^3 (15 cu yd) buckets. The loaders and the 90.7-t (100-st) trucks also assist in overburden loading and removal as required. The ore is trucked to an in-pit primary crusher and then conveyed to the adjacent refinery, as described in "Milling." A 4.5:1 stripping ratio currently is used at the Boron pit, with some of the fringe ore slated for later underground mining. The pit was 168 m (550 ft) deep as of September, 1980; future pit extensions will be somewhat deeper as the pit advances to the south and east.

Milling

Milling of the borax ores at Boron consists of crushing to —203.2 mm (—8 in.) in a 1.4 X 1.8-m (54 X 70-in.) Jeffery hammer mill located in the open pit. The ore is transported from this primary crusher to a surface ore stockpile by a 1.06-m (42-in.) by 427-m (1400-ft) inclined belt conveyor which feeds a traveling twin-boom stacker which feeds two coarse-ore stockpiles, one for the borax refinery and one for the new kernite-feed boric acid plant. Ore from these coarse-ore stockpiles is moved by a wheel reclaimer onto a 762-mm (30-in.) belt conveyor which feeds a Jeffery hammer mill. Ore from this secondary crusher is screened to —25.4 mm (—1 in.) and transported by belt to a series of ore bins which provide surge and holding capacity for the refinery.

In Death Valley, the colemanite ore is rough screened at the mine and then trucked some 48 km (30 miles) to the plant site, located outside of Death Valley. There the ore is reduced to —9.8 mm (—3/8 in.), fed to a washing plant to remove the clay, then fed to a ball mill to produce a —35 mesh, +10 μm feed to a bank of six float cells. Float concentrates from the cleaner stage are thickened, filtered, and dried, then run through cyclones to remove the —325 mesh material. A portion of this float product is sold directly while the remainder is calcined. The calciner product is also run through a series of cyclones, cooled, and classified, and the cleaned product is bagged or bulk loaded for shipment (Smith and Walters, 1980).

Processing Techniques

The conditions required for crystallizing the various borates have been studied in detail (Teeple, 1929; Kemp, 1956; Garrett, 1960). Likewise, the literature describing advances in technology is extensive. The following are brief descriptions of current industry practices in the United States.

Sodium Borate Ores: US Borax presently mines at Boron an ore composed of borax and clay with few other impurities (Fig. 11). The ore is crushed and screened to —25.4 mm (—1 in.) as noted previously, then dissolved in hot, weak, recycled borax liquor. The dissolved ore solution is passed over vibrating screens which separate and remove the larger insoluble pieces. The liquor and fines are then pumped into a series of six large countercurrent thickeners. Clarified borax liquor from the thickeners is filtered and pumped to vacuum crystallizers. Crystals of borax from the crystallizers are separated from the spent liquor in automatic centrifuges and then dried as refined borax decahydrate, or borax pentahydrate. A portion of the product is used as feed to make fused products, while another portion is used to make boric acid. The fused products, anhydrous boric acid (ABA), and anhydrous borax are formed in gas-fired furnaces which melt the borax and form a glass which is cooled, rolled, and broken, then screened to produce the final product. The formation of both ABA and boric acid require the addition of concentrated sulfuric acid and produce sodium sulfate as a byproduct (Watt, 1973). In the company's new boric acid plant, crushed kernite and borax-kernite ore are mixed with sulfuric acid, the insolubles are removed, and boric acid is crystallized from the clarified liquor.

Brines Containing Sodium Borates: Sodium borates (borax) are recovered from the brines of Searles Lake, CA by one of two processes—evaporation or carbonation—by the Kerr-McGee Corp. at its lakeside refineries at West End and Trona.

In the carbonation process, carbon dioxide from lime kiln gases or boiler flue gases is bubbled through the brine to crystallize out

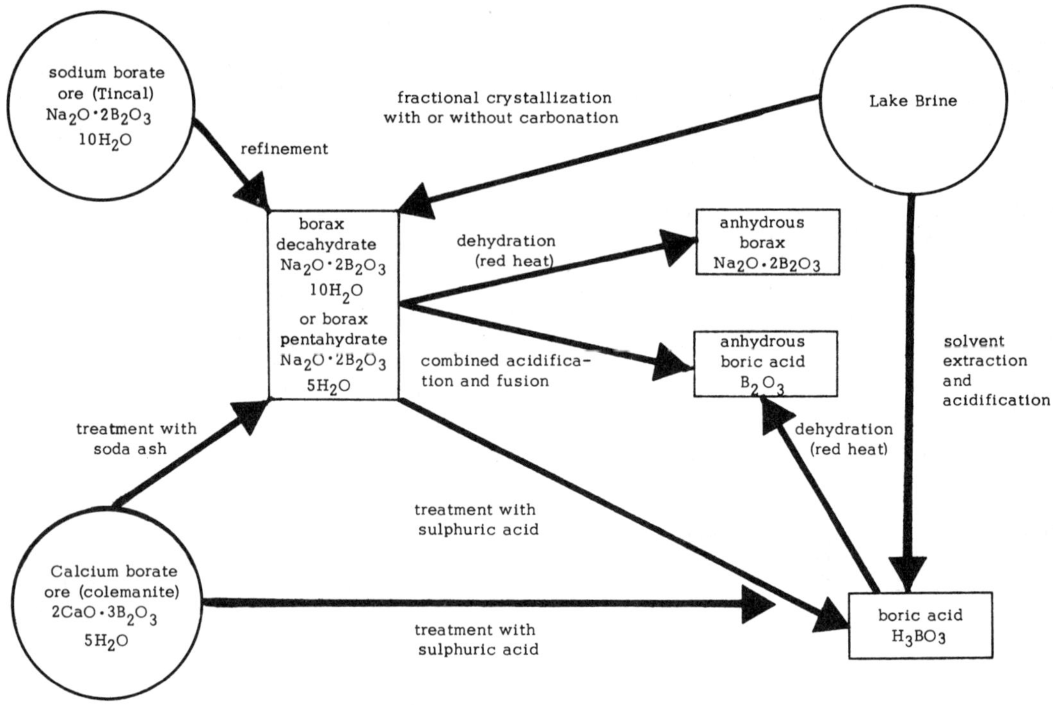

FIG. 11—*Refining and production of boron compounds (Watt, 1973).*

sodium bicarbonate. Borax is then crystallized by neutralization and cooling in vacuum crystallizers. Further cooling crystallizes the sodium sulfate. Borax from the crystallizers is dewatered in automatic centrifuges and dried to produce borax decahydrate and borax pentahydrate.

The evaporation process used by Kerr-McGee consists of rapid, controlled cooling which selectively "drops out" halite (NaCl), sodium carbonate (Na_2CO_3), and potash (KCl). The remaining borax-containing liquor is fed into crystallizer tanks containing borax seed crystals. The resultant borax crystals are separated from the liquor in cyclones, filtered, washed, redissolved, and recrystallized in vacuum crystallizers. Product from these crystallizers is also separated in cyclones and then dried to produce the final products (Barrett, 1960; Weisman and Tandy, 1975). Kerr-McGee also makes boric acid from the brines, utilizing a solvent extraction process. The process uses a kerosene solution of a chelating agent which is then acidified with sulfuric acid. The aqueous phase is treated with activated carbon to yield an organic-free aqueous phase which contains boric acid. Evaparation and crystallization produce a final boric acid product (Anon., 1980).

Future Considerations and Trends

Exploration and Development

The trends in the United States include exploring for bedded sodium and calcium borates and developing more efficient methods for utilizing lower grade deposits. Research continues on the possible commercial uses of ulexite. Connate brines that contain very small amounts of boron also have been studied, although the recovery and refining of borate from them is not economically feasible at present. As noted previously, the Russians have done much research directed toward the utilization of the common boron-containing silicate minerals. At present, the supplies of borax and colemanite in the free world are adequate to the point where little research has been directed to this possibility outside Russia.

The recovery of boron from secondary

sources is regarded as impractical. In most cases, it is obvious that the expense of reclaiming the relatively minor amounts of borates involved far outweighs the value of an equivalent primary borate product.

There are few substitutes for borate products. In most applications they provide unique chemical properties at a reasonable price. Soda ash can be used in some applications as a substitute, but in general its use results in products of inferior quality.

The known and probable reserves of boron minerals in the areas now productive are so large that all of the world's anticipated demand for established uses can be met for at least one hundred years. These requirements probably will rise about 4% a year (Absalom, 1979). An extraordinary jump in the rate of consumption, resulting from some unexpected new use, would be required to shorten significantly the inferred long period of abundant supply.

Transportation

The trend in the borate industry has been away from bagged, small lot shipments to large volume bulk transport. It is estimated that two-thirds of the domestic sales and three-fourths of the foreign shipments from this country are moved in bulk. Most domestic shipments are made in specially constructed 90-t (100-st) capacity railroad hopper cars; most of the exported borates are moved in large bulk shipments. Transport of export products has been greatly helped by the construction of bulk terminal silos by US Borax at its Wilmington, CA, port facilities and in the Netherlands at Rotterdam, the main port of importation for Europe. Kerr-McGee utilizes similar facilities in the United Kingdom and in the Netherlands. In Europe, much of the borate product moves by barge to river ports in Germany and neighboring countries; the remainder moves by lighters, rail, and truck in both bag and bulk.

Most of the borates imported to the US are shipped in bags, some directly from the Turkish mines. The main port of entry for colemanite is Charleston, SC. There is no duty on imported crude borates. Several companies in South Carolina take the crude Turkish ore and crush, screen, store, and rebag it for use by certain fiber glass manufacturers. This material moves by both rail and truck transport to the eastern markets.

Governmental Controls

In the United States, deposits of borax and kernite on federal lands, as well as brines that contain sodium borate, can be exploited only under provisions of the Mineral Leasing Act of 1917 and 1920 applicable to sodium minerals. The federal government's position is that ulexite is also a "sodium mineral." Colemanite can be claimed under the Mining Laws of 1892. Boron minerals carry a 14% depletion allowance. Most of the known colemanite deposits and some favorable prospecting areas lie within Death Valley National Monument and the newly created California Desert Conservation Area.

Environmental Considerations

The borate industry has expressed its interest in conservation increasingly over the past decade, chiefly by mining lower grade ores and weaker brines while increasing efficiency of the refineries. A major problem at the plants has been control of the fine light dust they generate. Recent engineering advances have cut dust emissions by more than 90%, and further reductions are sought and will be required under ever-tighter governmental regulations.

Plants vary in their tolerance for boron, and boron is used in several compounds and different concentrations for both plant growth and for soil sterilization. The concentration of B_2O_3 must exceed 4.0 mg/L to be harmful to most plants. The detection of boron pollution has been hampered somewhat by the lack of historical data on boron distribution. In the few areas where the data are known, boron from natural sources far exceeds that added by man. Borates are not considered toxic to humans and boric acid is not an industrial poison; however, absorption of large amounts may produce pathological disturbances (Sprague, 1972; Anon., 1980).

Markets: Consumption and Outlook

Consumption of domestic borates is described under "End Uses" (See Table 3), and the trend of US borate production is shown in Fig. 12. The United States is the world's largest user of borates and borate products, consuming about 32% of the world production. The manufacture of various kinds of glass, insulating material, and glass fiber accounts for 68% of the domestic consumption (Absalom, 1979). Other major end uses are listed in Table 3 and Fig. 13. The growth rate for domestic use has averaged

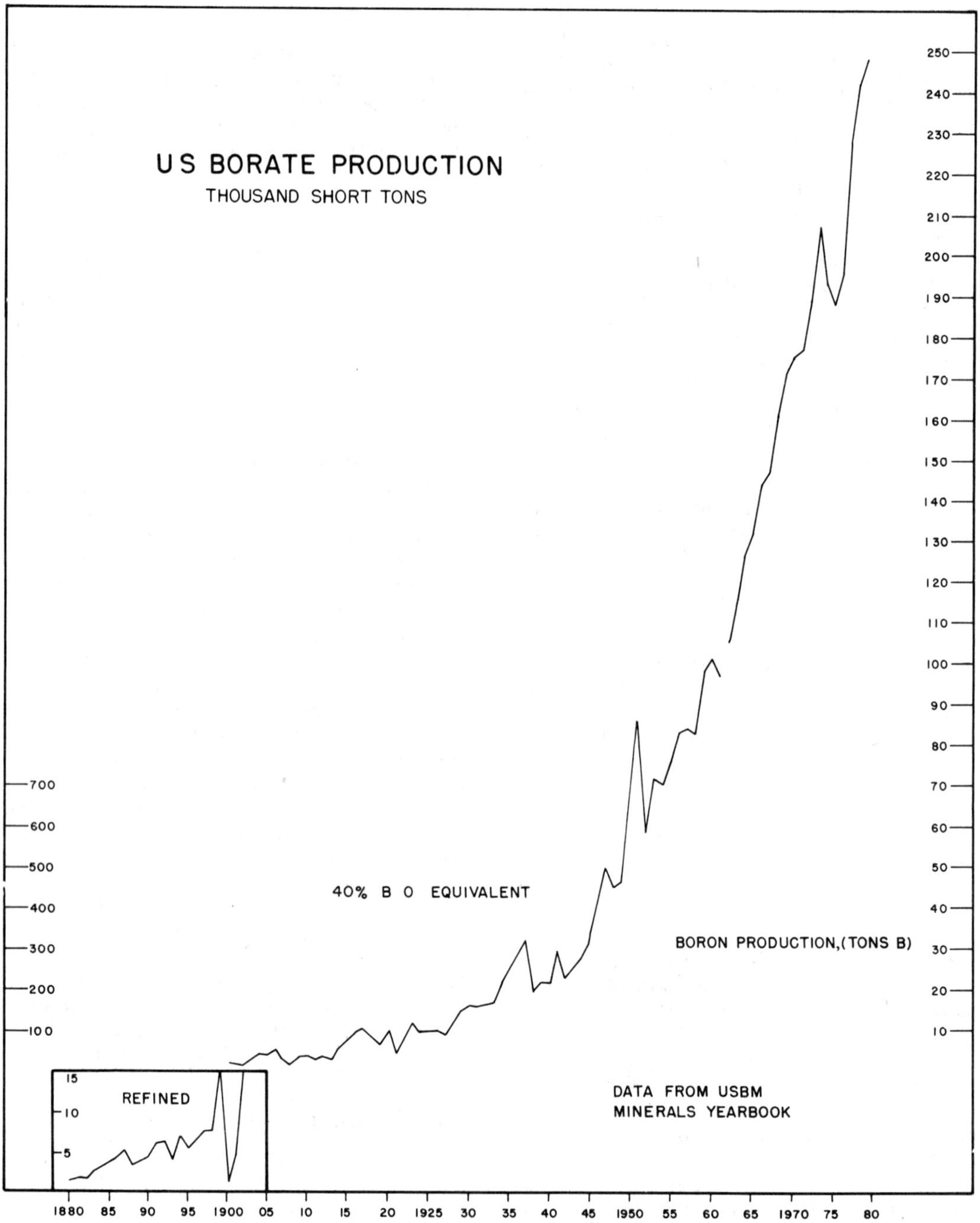

FIG. 12—*US borate production (data from US Bureau of Mines Mineral Surveys, Boron).*

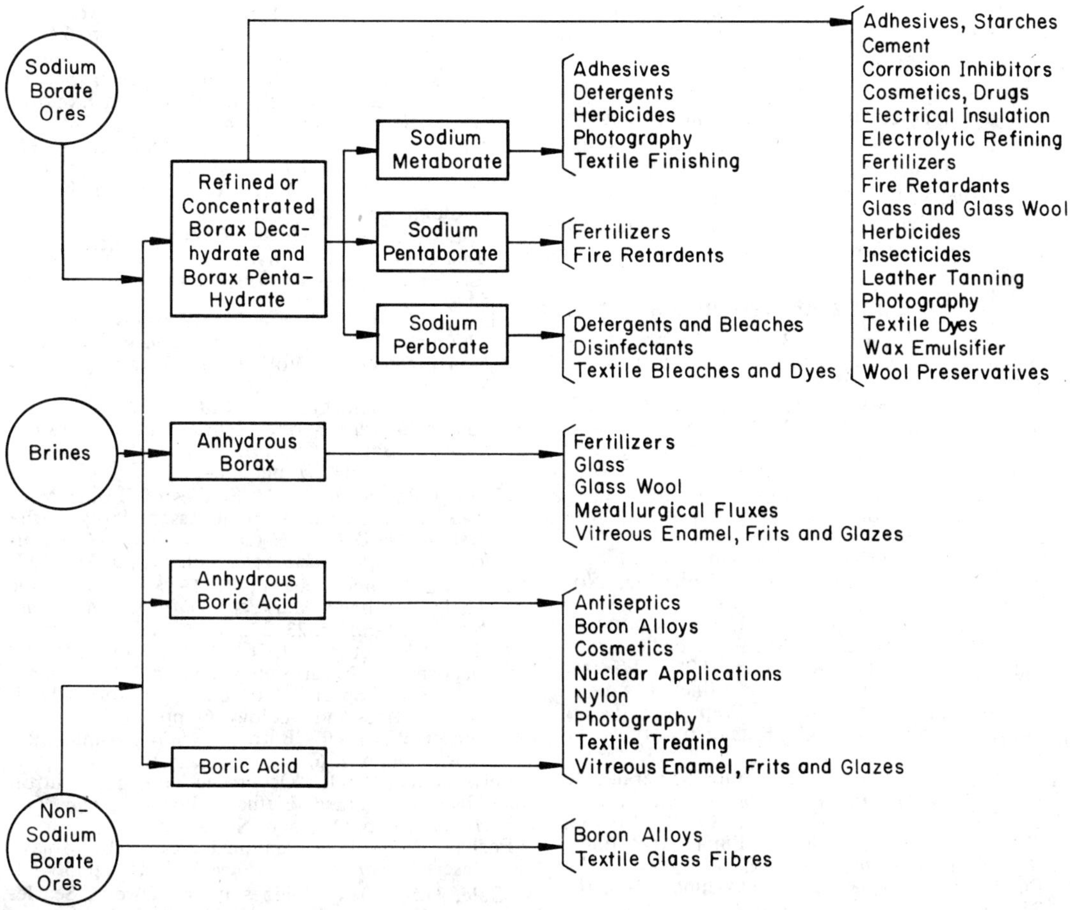

FIG. 13—*Main uses of boron compounds (Watt, 1973).*

4.7% over the 19-year period 1960-1978, with a most likely annual growth rate for the world of about 3.5% through the year 2000. The projection for the United States is 3.8% per year for the same period (Hall, 1980). Marketing experts foresee no dramatic change in the domestic market, either in growth rate or kinds of end uses, in the next few years (Anon., 1977).

In Europe, the major use for borates is in the soap and detergent field, accounting for over a third of their consumption. Glass, enamels, and ceramics also account for about a third of the European market, in marked contrast to the United States where these comprise only 16% of the market. The glass and textile fiber market is growing, but at a slower pace than in this country (Absalom, 1979).

In the Far East, there is a smaller but expanding market with increased use of better grade glass products by the Japanese and some increase in trade with other countries. Exports from the United States probably will be reduced somewhat when Turkish sodium borates become available in quantity at lower cost. South American production is expected to decline with the eventual exhaustion of the sodium borate deposit at Tincalayu although there is increased interest in producing boric acid from ulexite for export by both Chile and Peru (Anon., 1980a). Future trade with the USSR and China is unpredictable. Russia seems to be producing enough for its domestic uses, but apparently lacks the large high-grade ore bodies required for a significant impact on the international market. China has been a recent importer of borates as well as exporting minor amounts of borate products. However, there is increased interest in foreign trade in China, and its new deposit at Liaoning may be utilized with an export market in mind. Some Turkish borate goes to the USSR at present, and this trade is expected to increase. India recently erected a boric acid plant which utilizes imported feed, and this supplies part of the Near East market. Markets for refined borates in the nations of Africa and the Far and Near East are not expected to change greatly in the near future.

Bibliography and References

Anon., 1977, "Comments—Borates: A Forecaster's Dilemma," *Industrial Minerals,* No. 120, Sept., p. 7.

Anon., 1979, "High Hopes for Borate Production," *Industrial Minerals,* No. 126, Mar., p. 9.

Anon., 1980, "Chemical Briefs 1: Boric Acid," *Chemical Purchasing,* June, pp. 27–32.

Anon., 1980a, "Chemical Supplies for Insulation," Boric Acid Imports, Working Paper (preliminary), US Dept. of Commerce, July.

Anon., 1980b, "Production Begins at Chinese Boron Mine," *Mining Journal,* Vol. 295, No. 7563, Aug. 1, p. 95.

Absalom, S.T., 1979, "Boron," *Mineral Commodity Profiles,* US Bureau of Mines, 17 pp.

Absalom, S.T., 1980, "Boron in 1979," *Mineral Industry Surveys,* US Bureau of Mines, 13 pp.

Absalom, S.T., 1980a, "Boron," Preprint, *Minerals Yearbook 1978–1979,* US Bureau of Mines, 11 pp.

Ahlfeld, F., 1954, "Los Yacimientos Minerales de Bolivia," *Banco Minero de Bolivia,* p. 227.

Aristarain, L., and Erd, R.C., 1971, "Inyoita, $2CaO \cdot 3B_2O_3 \cdot 13H_2O$, de la Puna Argentina," *Anales de la Sociedad Cientifica Argentina (La Plata),* Vol. 191, Pt. 5–6, May-June, pp. 191–211.

Aristarain, L.F., and Hurlbut, C.S., Jr., 1972, "Boron Minerals and Deposits," *The Mineralogical Record,* Vol. 3, No. 4, July-Aug., pp. 165–172; Vol. 3, No. 4, Sep.-Oct., pp. 213–220.

Barker, J.M., and Lefond, S.J., 1979, "Some Additional Borate and Zeolites from the Mesa Del Alamo Borate District, North Central Sonora, Mexico," Preprint 79-367, SME Fall Meeting, St. Louis, 12 pp.

Barnard, R.M., and Kistler, R.B., 1956, "Stratigraphic and Structural Evolution of the Kramer Sodium Borate Body, Boron, California," *2nd Symposium on Salt,* J.L. Rau, ed., Vol. 1, Northern Ohio Geological Society, Cleveland, Ohio, pp. 133–150.

Bekisoglu, K.A., 1962, "Boron Deposits in Turkey," *Turkish Economic Review,* Vol. 3, No. 4, June-July, pp. 12–34; Union of Chambers of Commerce, Industry and Commodity Exchanges of Turkey, Ankara.

Blakely, J.W., 1980, "American Borate's Billie Mine in Death Valley," *World Mining,* Apr., pp. 57–60.

Blue, T.A., and Treskon, D.J., 1970, "Boron Minerals and Chemicals," *Chemical Economics Handbook,* Stanford Research Institute, pp. 717.5011–717.5020.

Brooks, R.R., 1979, "Indicator Plants for Mineral Prospecting—A Critique," *Journal of Geochemical Exploration,* Vol. 12, No. 1, Nov., pp. 67–78.

Brown, W.W., and Jones, K.D., 1971, "Borate Deposits of Turkey," *Geology and History of Turkey,* A.S. Campbell, ed., Petroleum Exploration Society of Libya, Tripoli, pp. 483–492.

Catalano, L.R., 1927, "Geologia Quimica de los Boratos," *Anales del Ministerio de Agricultura,* Publication No. 28, Direccion General de Minas, Geologia e Hidrologia, p. 101.

Christ, C.L., 1969, "Crystal Chemistry of Boron," *Handbook of Geochemistry,* K.H. Wedepohl, ed., Vol. 2, Springer-Verlag, Berlin, pp. 5A1–5A6.

Christ, C.L., 1972, "Some Contributions of Crystal Chemistry to Geochemistry," *Journal of Geological Education,* Vol. 20, No. 5, Oct., pp. 235–245.

Christ, C.L., and Garrels, R.M., 1959, "Relations among Sodium Borate Hydrates at the Kramer Deposit, Boron, California," *American Journal of Science,* Vol. 257, pp. 516–528.

Christ, C.L., et al., 1967, "Studies of Borate Minerals XIII. Borate Mineral Assemblages in the System, $Na_2O \cdot CaO \cdot MgO \cdot B_2O_3 \cdot H_2O$," *Geochimica et Cosmochimica Acta,* Vol. 31, pp. 313–337.

Dickson, T., and Harben, P., 1983, "Borates and Their Becalmed Markets," *Industrial Minerals,* No. 184, Jan., pp. 23–27.

Evans, J.R., et al., 1976, "Mines and Mineral Deposits in Death Valley National Monument, California," Special Report 125, California Division of Mines and Geology, 60 pp.

Fajans, E.W., 1980, "Boron," *Mining,* Annual Review, June, p. 126.

Fertl, W.H., 1971, "Occurrence of the Neutron Absorbing Trace Element Boron," *The Log Analyst,* Sep.-Oct. and Nov.-Dec.

Foshag, W.H., 1921, "Saline Lakes of the Mojave Desert," *Economic Geology,* Vol. 21, pp. 56–64.

Gale, H.S., 1914, "Salines in the Owens, Searles, and Panamint Basins, Southeastern California," Bulletin 580–L, US Geological Survey, pp. 251–323.

Gale, W.A., 1964, "History and Technology of the Borax Industry," *Boron, Metallo-boron Compounds and Boranes,* R.M. Adams, ed., Chap. 1, Interscience Publishers (John Wiley), New York, pp. 1–27.

Garrett, D.E., 1960, "Borax Processing at Searles Lake," *Industrial Minerals and Rocks,* 3rd ed., J.L. Gillson, ed., AIME, New York, pp. 119–122.

Godlevsky, M.N., 1937, "Mineralogical Investigation of the Inder Borate Deposits," *Memoires de la Societe Russe de Mineralogie,* Vol. 66, No. 2, pp. 345–368.

Gomez, A., et al., 1980, "Estudio Geologico del Deposito de Boratos del Area de Tubutama, Sonora," VIII Seminario Interna Consejo de Recursos Minerales, pp. 367–403.

Hall, W.B., 1980, "Bureau of Mines Forecasts of Supply and Demand for Major Industrial Minerals," Preprint, 4th Industrial Minerals International Conference, Atlanta, GA, 7 pp.

Hanks, H.G., 1883, "Report on the Borax Deposits of California and Nevada," *California State Mining Bulletin,* 3rd Annual Report, Part 2.

Helvaci, C., and Firman, R.J., 1976, "Geological Setting and Mineralogy of Emet Borate Deposits, Turkey," *Applied Earth Science,* Vol. 85, Section B, Institution of Mining and Metallurgy, pp. B141–B152.

Hurlbut, C.S., Jr., et al., 1973, "Kernite from Tincalayu, Salta, Argentina," *American Mineralogist,* Vol. 50, Nos. 3, 4, Mar.-Apr., pp. 308–313.

Hutchinson, G.E., 1957, *A Treatise on Limnology,* Vol. 1, John Wiley, New York, pp. 570–571.

Inan, J., et al., 1973, "Turkey: Mineralogy, Chemistry and Origin of Kirka Borate Deposit, Eskisehir Province," *Transactions/Section B,* Institution of Mining & Metallurgy, Aug., pp. B114–B123.

Kemp, P.H., 1956, "The Chemistry of Borates: Part I," Borax Consolidated Ltd., London, pp. 1–90.

Kuhn, R., 1962, "Geochemistry of the German Potash Deposits," Special Paper 88, Geological Society of America, pp. 460–466.

Lefond, S.J., and Barker, J.M., 1979, "A Borate and Zeolite Occurrence near Magdalena, Sonora, Mexico," *Economic Geology,* Vol. 74, No. 8, Dec., pp. 1887–1889.

Lefond, S.J., and Barker, J.M., 1980, "Borates—Past, Present and Future," Preprint 80–101, SME-AIME Annual Meeting, Las Vegas, NV, 6 pp.

Leiding, B., 1942, "Inventario de Reservas de Minerales y Metalicos en Chile," *Anales de Primer Congreso Panamericano de Ingenieria de Minas y Geologia,* Santiago de Chile, pp. 1665–1667.

Lyday, P.A., 1981, "Boron in Mineral Commodity Summary—1980," US Geological Survey, pp. 22–23.

Lyday, P.A., 1981a, "Boron," *Mineral Commodity Summaries, 1982,* US Bureau of Mines, p. 22.

MacMillan, R.T., 1970, "Boron," *Mineral Facts and Problems,* US Bureau of Mines Bulletin 650, pp. 879–891.

Maden T.veA.E., 1965, "Borate Deposits of Turkey," Publication No. 125, Mineral Research and Exploration Institute of Turkey, Ankara, pp. 1–11.

Matkovich, V.I., 1977, *Boron and Refractive Borides,* Springer-Verlag, Berlin, 656 pp.

McAllister, J.F., 1970, "Geology of the Furnace Creek Borate Area, Death Valley, Inyo County, California," Map Sheet 14, 1970, California Div. of Mines and Geology, 1 map, pp. 1–9.

McAnulty, W.N., and Hoffer, J.M., 1972, "A New Howlite Occurrence in Sonora, Mexico," Bol. Soc. Geologia Mexicana, Vol. 33, pp. 21–24.

Meixner, H., 1953, "Neue Turkische Boratlagerstatten," *Berg- und Hüttenmännische Monatshefte,* Vol. 98, No. 5, pp. 86–92.

Meixner, H., 1956, "Die Neue Turkische Boratprovinz um lskelekoy bei Bigadic im Viayet Balikesir," *Kali und Steinsalz,* Vol. 2, pp. 43–47.

Miller, W.C., 1965, "Boron," Bulletin 630, *Mineral Facts and Problems,* US Bureau of Mines, pp. 149–157.

Morgan, V., and Erd, R.C., 1969, "Minerals of the Kramer Borate District California," Mineral Information Service, California Div. of Mines, Vol. 22, Nos. 9, 10, pp. 143–154; 165–172.

Muessig, S., 1958, "First Known Occurrence of Inyoite in a Playa, at Laguna Salinas, Peru," *American Mineralogist,* Vol. 43, Nov.-Dec., pp. 1144–1147.

Muessig, S., 1966, "Recent South American Borate Deposits," *2nd Symposium on Salt,* J.L. Rau, ed., Vol. 1, Northern Ohio Geological Society, Cleveland, pp. 151–159.

Mumford, R.W., 1954, "Deposits of Saline Minerals," Chapter VIII, *Geology of Southern California,* Bulletin 170, Division of Mines, Dept. of Natural Resources, State of California, pp. 15–22.

Murdock, T.G., 1958, "The Boron Industry in Turkey," *Mineral Trade Notes,* Vol. 46, No. 5, Special Supplement No. 53, US Bureau of Mines, May, pp. 1–47.

Özpeker, I., 1969, "Bati Anadolu Borat Yataklarinin Mukayeseli Jenetik Etudu," *Istanbul Teknik Universitesi Maden Fakultesi,* pp. 1–116.

Palache, C., et al., 1951, *"Dana's System of Mineralogy,"* 7th ed., Vol. 2, John Wiley, New York, 1951, pp. 320–389.

Papke, K.G., 1976, "Evaporites and Brines in Nevada Playas," Bulletin 87, Nevada Bureau of Mines and Geology, 33 pp.

Parker, R.L., 1967, "Composition of the Earth's Crust," Professional Paper 440-D, US Geological Survey, pp. D1–D19.

Rosenfelder, W.J., 1978, "The Industrial Uses of Boron Chemicals," *Chemistry and Industry,* June 17, pp. 413–416.

Schaller, W.T., 1930, "Borate Minerals from the Kramer District, Mojave Desert, California," Professional Paper 1581, US Geological Survey, pp. 137–170.

Siefke, J.W., 1980, "Geology of the Kramer Borate Deposit, Boron, California," Fall Field Trip, South Coast Geological Society, Preprint AIME, 8 pp.

Smith, G.I., 1966, "Geology of Searles Lake—A Guide to Prospecting for Buried Continental Salines," *2nd Symposium on Salt,* J.L. Rau, ed., Northern Ohio Geological Society, Cleveland, pp. 167–180.

Smith, G.I., 1979, "Surface Stratigraphy and Geochemistry of Late Quaternary Evaporites, Searles Lake, California," Professional Paper 1047, US Geological Survey, 130 pp.

Smith, G.I., and Haines, D.V., 1964, "Character and Distribution of Nonclastic Minerals in the Searles Lake Evaporite Deposit, California," Bulletin 1181–P, US Geological Survey, pp. P1–P58.

Smith, G.I., et al., 1973, "Evaporites and Brines," *US Mineral Resources,* Professional Paper 820, US Geological Survey, pp. 207–208.

Smith, P.R., and Walters, R.A., 1980, "Production of Colemanite at American Borate Company's Plant Near Lathrop Wells, Nevada," *Mining Engineering,* Feb., pp. 199–204.

Smith, W.C., 1962, "Borates in the United States," Mineral Investigations Resource Map MR14, US Geological Survey, 1 map, 4 pp.

Smith, W.C., 1968, "Borax Solution at Kramer, Calif.," *Economic Geology,* Vol. 63, No. 8, Dec., pp. 877–883.

Sprague, R.W,. 1972, *The Ecological Significance of Boron,* Ward Ritchie Press, 58 pp.

Strishkov, V.V., 1971, "The Mineral Industry of the USSR," *Minerals Yearbook 1969,* Vol. 4, US Bureau of Mines, p. 726.

Teeple, J.E., 1929, "The Industrial Development of Searles Lake Brines," Monograph 49, American Chemical Society, 182 pp.

Tennyson, C., 1963, "Eine Systematik der Borate auf Kristall Chemischer Grundlage," *Fortschritte der Mineralogie,* Vol. 43, Sep., pp. 64–91.

Utekhin, G. M., 1965, "The Problem of Distribution Features and Conditions of Boron Concentration in Skarns," (E.A. Alexandrov, Review), *Economic Geology,* Vol. 60, No. 8, Dec., p. 1750.

Valyashko, M.G., 1970, "Genesis and Exploration of Borate Deposits Related to Marine Salt Deposits," *International Geology Review,* Vol. 12, No. 6, June, pp. 711–719.

Wang, K.P., 1979, "China," *International Minerals/Metals Review,* McGraw-Hill Publishing Co., New York, pp. 18–19.

Waring, G.A. (rev. by Blankenship and Bentall), 1965, "Thermal Springs of the United States and Other Countries of the World—A Summary," Professional Paper 492, US Geological Survey, 383 pp.

Watt, K.A.L.G., 1973, "The Borate Industry," *World Minerals and Metals,* British Sulphur Corp., Ltd., London, Mar.-Apr., No. 12, pp. 5–12.

Weisman, W.J., and Tandy, C.W., 1975, "Sodium Sulfate," *Industrial Minerals and Rocks,* 4th ed., S.J. Lefond, ed., AIME, New York, pp. 1081–1093.

Wendel, C., 1978, "Special Report of Borate Resources," Mining and Minerals Division, NPS/US Department of Interior, Apr., 138 pp.

White, D.E., 1957, "Magmatic, Connate, and Metamorphic Waters," *Bulletin,* Geological Society of America, Vol. 68, pp. 1659–1682.

White, D.E., 1957, "Thermal Waters of Volcanic Origin," *Bulletin,* Geological Society of America, Vol. 68, Dec., pp. 1637–1657.

Whistler, D.P., 1965, "A New Hemingfordian (Middle Miocene) Mammalian Fauna from Boron, California—Its Stratigraphic Implication within the Western Mojave Desert," Ph.D. Thesis, University of California, Riverside.

Bromine

Original by J. H. JENSEN *
Revised by W. E. BRECKOFF
and
A. P. ANDERSON †

Bromine is the intermediate member of the halogen family of elements between iodine, a solid, and chlorine, a gas. The name is derived from the Greek "bromos," meaning stench. Bromine is the only nonmetallic element that is liquid at room temperatures with a melting point of −7.2° C and a boiling point of 58.78° C. Reddish brown in color, it dissolves easily in water to yield a red solution. Bromine volatilizes at ordinary temperatures to a red vapor that is very irritating to the eyes, nose, and throat. Density of the liquid is 3.12 and gas weighs 7.59 gpl at standard conditions. As the element is highly active, it exists in nature only as bromide compounds. Bromine was discovered by Antoine Jerome Balard in seawater bitterns in 1826 and first prepared in quantity in 1860.

Distribution

Bromine is widely distributed in nature and compounds occur in solution in natural brines, in seawater, saline lakes, oil and gas well brines, and in evaporite chloride minerals. The concentration of bromine salts ranges from 60–65 ppm in seawater to several thousand ppm in oil well brines and saline lakes. The distribution of bromine in marine evaporite minerals and in the solutions from which they crystallized was determined many years ago by H. E. Boeke (1908). Theoretical aspects regarding the geochemistry of bromine have been published in numerous papers (Raup and Mite, 1969). Bromine occurs in solid solution in carnallite and sylvite associated in potash deposits; in halite in salt beds; salt in oil shales; and in trona deposits. Its presence is useful as a guide in geochemical exploration.

The analysis of small amounts of bromides in the presence of large amounts of chlorides is discussed by W. Schwerdtner (1963). Manganese, iron, and iodine may interfere with the determination. Other analytical methods are discussed at length in the literature (Stenger and Atchison, 1964).

Sources and Reserves

The sources and reserves of bromine, given in Table 1, appear ample to meet present and foreseeable requirements in all producing countries.

With the occurrence of bromine salts in seawater and other brines, the potential for bromine production is worldwide.

It has been estimated that 99% of all commercially recoverable bromine is in the ocean. However, actual subterranean brines having a higher bromine content than seawater will continue to be the preferred sources of bromine for the foreseeable future.

TABLE 1—Sources and Reserves of Bromine

Country	Source	Bromine Concentration, Ppm
US	Arkansas—oil well brines	5000
	California—Searles Lake	85
	Michigan—brines	1300
	Texas—brines,	—
	seawater	60-65
France	Alsatian potash fields	—
Israel	Dead Sea	10,000
Japan	Seawater	60-65
U.K.	Seawater	60-65
W. Germany	Stassfurt Potash	1500-2500

* Western Knapp Engineering Div., Arthur G. McKee Co., San Francisco, CA.

† Ethyl Chemicals Group, Ethyl Corp., Baton Rouge, LA.

Markets and Prices

About 85 to 90% of all bromine generated is consumed at the producing site by the bromine producer. The balance is shipped to a large variety of chemical processors who convert elemental bromine into such products as flame retardants, dyes, and pharmaceuticals and into chemicals used in the photographic and agricultural industry. The single largest compound produced is ethylene dibromide. Ethylene dibromide consumes about 50% of the bromine produced in the US, a dramatic drop from 1972, when it represented about 67%. Restrictions on the use of lead antiknock compounds in gasoline have led to a much smaller consumption of ethylene dibromide, which is used in leaded fuels to convert lead oxides into volatile compounds. Ethylene dibromide has a second significant, although smaller, use as a soil fumigant (nematocide). Ethylene dibromide serves also as a raw material for the production of vinyl bromide, which is used as a comonomer in certain acrylic fibers to impart flame retardancy.

The next largest use area is represented by flame retardants, which consume about 20% of the bromine consumed in the US. The bulk of these are ring-bromined aromatic compounds that impart flame retardancy to a variety of plastics.

Methyl bromide, a soil and space fumigant, consumes another estimated 10% of the US bromine production.

A new and formidable use for bromine has developed over the last five years in oil and gas well completions, workovers, and as packer fluids in high pressure geological formations. Estimated consumption is about 10% of the US bromine used.

The balance of the bromine consumed is used in miscellaneous applications: fire extinguishing agents, anesthetics, pharmaceuticals, dyes, and refrigeration systems, and water treatment chemicals (Anon., 1980).

The consumption figure of 387 million lb for 1979 represents the use of virgin bromine only. The actual generation of bromine is considerably higher, for two reasons: many brominations generate byproduct streams that contain bromides and are recycled, and many products use bromine compounds as intermediates, but do not contain bromine in the final product. These bromine values are recycled also.

Table 2 gives the uses of bromine in the United States in 1979.

TABLE 2—Usage of Bromine in the US, 1979

Manufacture of	Thousands of lb	Percent
Gasoline antiknock additive	198,000	51
Flame retardants	73,000	19
Oil field fluids	43,000	11
Agricultural chemicals	50,000	10
Miscellaneous uses	23,000	9
Total	387,000	100

Source: Anon., 1980.

Product Specifications

Although there are ACS specifications for various grades of bromine, they have no real significance. All merchant bromine in the US contains less than 100 ppm of impurities and is therefore 99.99% pure.

The price of bromine varies with the type of container. Bromine shipped in bulk (truck trailer or tank car) carries a list price of 26.5¢ per lb f.o.b. plant in 1981. Bromine in containers sells for 75¢ per lb delivered, east of the Rockies (Anon., 1981).

Production of Bromine

Production in the United States began about 1867 as a byproduct of salt manufacture in the Pomery district along the Ohio River. Output increased steadily to more than a million pounds at the beginning of the 20th century.

The most dramatic change in the world's bromine supply is the emergence of Israel as a major bromine producer, recovering bromine from Dead Sea brines. Table 3 (Lyday, 1981) shows the production of bromine in 1979-1980.

US exports of elemental bromine and bromine compounds totaled 94 million lb in 1980.

Preparation for Market

Six companies (Table 4) operated a total of nine bromine plants in the US in 1980 for an estimated production of 375 million lb of virgin bromine.

Bromine occasionally is recovered with one or more other products. Iodine and calcium, magnesium, sodium, and potassium chlorides are recovered from brines in Michigan. In West Germany potash is produced as a coproduct while in Israel bromine is obtained from the waste liquors of a potash plant at the

TABLE 3—World Production of Bromine, 1980 (Millions of lb)

Country	Production
France	35.0
West Germany	8.6
India	NA
Israel	100.0*
Italy	1.3
Japan	26.0
Spain	0.9
United Kingdom	55.0
United States	375.0
USSR	33.0

*Since 1979 bromine capacity has increased to about 150 million lb.
Note: It is not clear whether recycle streams are included in these figures. Most likely they are included at least for the US.

Dead Sea. Another bromine operation on the Jordanian side of the Dead Sea is being planned for a production of 66 million lb per year (Absalom, 1979).

Processing Techniques

Bromine is extracted from brines by a steaming-out process and from sea water by a blowing-out process. However, in both processes, the essential steps are (Lyday, 1980):

1. Treatment with chlorine to oxidize the bromide to bromine.
2. Removal of bromine vapor from the solution.
3. Condensation of bromine vapor or reacted to form a salt or acid.
4. Purification of bromine.

TABLE 4—Bromine Producing Operations in the US, 1980

Company	Location	Brine
Arkansas		
Arkansas Chemicals*	El Dorado	Well
Dow	Magnolia	Well
Ethyl	Magnolia	Well
Great Lakes Chemical	El Dorado	Well
Velsicol†	Marysville	Well
	El Dorado	Well
Michigan		
Dow	Ludington	Well
Dow	Midland	Well
Morton	Manistee	Well

*Jointly owned by Great Lakes Chemical and PPG.
†The bromine related assets of Velsicol (formerly Michigan Chemical), a subsidiary of North-West Industries, are being purchased by Great Lakes Chemical.

A process for recovering bromine from the relatively low concentration in seawater was developed in 1924 by the S.S. Ethyl, a floating chemical research plant. The process precipitates bromide with aniline to form tribromoaniline. Other methods use air instead of steam and in other cases the bromine is absorbed in sodium carbonate solutions. Should the market and price justify it, seawater can be an exhaustless source of bromine.

Elemental bromine is contained and transported, in order of decreasing quantities, as follows:

Glass or lead-lined tank cars and trucks.
10-gal lead-lined steel drums.
Case of nine 6½-lb glass bottles with lead caps.
1-lb jars, 40–50% bromine absorbed in 10–20 mesh charcoal.

The shipment of large quantities is covered under the requirements of ICC section 252 (i).

Subsidies, Depletion, and Tariffs

There are no US government incentive and procurement subsidies.

Tariffs on bromine and its compounds as of 1981 are given in Table 5.

The depletion allowance on brine wells, both domestic and foreign, was 5% in 1981.

Future Consideration and Trends

The decline of ethylene dibromide usage in gasoline because of the restrictions on lead antiknocks has prevented any growth of bromine over the last five years. The void is slowly being filled by increased consumption in flame retardants, fumigants, and specifically and to the largest extent by the use of heavy brines prepared from calcium bromide and zinc bromide in the oilfield. The fluids are being used as completion, workover, and packer fluids in specific geological formations to control pressure and prevent formation damage.

Another possible use for bromine that could emerge is in the form of bromine chloride as a disinfectant for the treatment of power plant cooling waters and of municipal waste water.

TABLE 5—Tariffs on Bromine and Its Compounds (1981)

Item	TSUS No.	Rate of Duty ¢/lb	1982	1983	1984	1985	1986	1987
Bromine	415.05	4.3	3.9	3.5	3.1	2.8	2.4	2.0
Potassium bromide	420.02	1.9*	1.8	1.7	1.7	1.6	1.5	1.5
Sodium bromide	420.82	6.7*	6.1	5.5	4.9	4.2	3.6	3.0
Ethylene dibromide	429.28	1.0 + 5.4% ad valorem*						0.7 + 3.5% ad valorem

*Valid for 1981. Under the Tokyo round agreement tariffs will be reduced annually until 1987.

The US Bureau of Mines has investigated the use of heavy liquid bromine compounds for mineral separation on a continuous basis (Llewellyn and Browning, 1971; Tippin et al., 1970). Acetylene tetrabromide, specific gravity 2.95, has been used for some time in laboratories in batch testing to provide information for design of heavy media separation plants.

Health and Safety Factors

Bromine presents serious hazards of injury by contact with the liquid and inhalation of the vapor. The liquid rapidly attacks the skin and comparatively low concentrations of the vapor are highly irritating to the eyes and the entire respiratory tract (Stenger and Atchison, 1964). Vapor concentrations of 500–1000 ppm (by volume) are rapidly fatal on short exposure; 40–60 ppm are dangerous to life on exposure for 0.5 to 1 hr. The maximum concentration considered safe for 8-hr exposure is less than 1 ppm. The ACGIH (accepted) limit is 0.1 ppm in air. At this level bromine can be detected by its odor.

Plants designed for the use of bromine should be designed for the rapid disposal of spilled liquid and for adequate venting of vapors.

Bibliography and References

Absalom, S.T., 1981, "Bromine," Preprint, 1978–1979, US Bureau of Mines, 7 pp.

Anon., 1967, "Bromine Chemical Profile," *Oil, Paint and Drug Reporter,* Vol. 191, No. 4, Jan. 23, p. 9.

Anon., 1968, "Bromine, Its Properties and Uses," Michigan Chemical Corp., Chicago, IL, 62 pp.

Anon., 1968a, "How to Handle Bromine," Michigan Chemical Corp., Chicago, IL, 36 pp.

Anon., 1968b, "The Dow Chemical Co., 80 Years of Leadership . . . and Still Pioneering, Bromine and Brominated Compounds," Midland, MI, 89 pp.

Anon., 1969, "Dow Products 1968–69," Midland, MI, pp. 9, 16.

Anon., 1972, "Extraction of Chemicals from Seawater, Inland Brines and Rock Salt Deposits," United Nations, New York, 59 pp.

Anon., 1980, "A Jump in Bromine Sales Orders Signals a Record Year," *Chemical Week,* Vol. 127, No. 6, Aug. 6, p. 53.

Anon., 1981, *Chemical Marketing Reporter,* Vol. 219, No. 20, May 18, p. 43.

Boeke, H.E., 1908, "Uber das Kristallisationsschema der Chloride, Bromide, Jodide von Natrium, Kalium und Magnesium, sowie uber das Vorkommen das Broms und das Fehlen von Jod in der Kalisalzlagerstatten," *Zeitschrift fuer Kristallographie,* Vol. 45, pp. 346–391.

Collins, A.G., 1966, "Here's How Producers Can Turn Brine Disposal Into a Profit," *Oil and Gas Journal,* Vol. 64, No. 27, July 4, pp. 112–113.

Jolles, Z.E., 1966, *Bromine and Its Compounds,* Academic Press Inc., New York, 940 pp.

Keiser, H.D., 1960, "Bromine" in "Minor Industrial Minerals," *Industrial Minerals and Rocks,* J.L. Gillson, ed., AIME, New York, pp. 607–609.

Llewellyn, T.O., and Browning, J.S., 1971, "Continuous Heavy Liquid Concentration of Kyanite," Report of Investigations No. 7481, US Bureau of Mines, Mar., 11 pp.

Lyday, P.A., 1981, "Bromine," *Mineral Commodity Summaries 1982,* US Bureau of Mines, pp. 24–25.

MacMillan, R.T., 1970, "Bromine," *Mineral Facts and Problems,* Bulletin 650, US Bureau of Mines, pp. 893–901.

Raup, O.B., and Mite, R.J., 1969, "Bromine Distribution of Paleosalinities from Wall Cuttings, Paradox Basin, Utah and Colorado," *3rd Symposium on Salt,* Northern Ohio Geological Society, pp. 40–47.

Schwerdtner, W., 1963, "Analysis of Small Amounts of Bromides in Presence of Large Amounts of Chlorides," *1st Symposium on Salt,* Cleveland, 1962, pp. 247–248.

Stenger, V.A., and Atchison, G.J., 1964, "Bromine," *Encyclopedia of Chemical Technology,* Vol. 3, 2nd ed., R.F. Kirk and D.F. Othmer, eds., Interscience Publishers, New York, pp. 750–783.

Tallmage, J.A., et al., 1964, "Minerals From Sea Salt," *Industrial and Engineering Chemistry,* Vol. 56, No. 7, July, pp. 44–65.

Tippin, R.B., et al., 1970, "Continuous Heavy Liquid Concentration of Spodumene," Report of Investigations 7379, US Bureau of Mines, May, 24 pp.

Chromite

HARRY M. MIKAMI *

Chromite is the only ore mineral of metallic chromium and chromium compounds and chemicals. Because of this fact, *chromite* and *chrome ore* are used synonymously in trade literature. In commercial market quotations, chrome ore is the term commonly used. Chromite per se, because of properties imparted by its chromium content, is used in refractories and in special purpose molding sands for metal casting. There are many other minerals containing some, generally minor, amounts of chromium, but none are commercial sources of the element (Thayer, 1956).

Chromium and chromite have many diverse uses that, directly and indirectly, critically affect vast segments of our modern industrial system. Most important are probably the metallurgical applications wherein chromium is a component of heat-, abrasion-, corrosion- and oxidation-resistant and high-strength alloys of many types. Stainless steels are the largest volume category of chrome-bearing alloys. Chromium chemicals are used in leather tanning, in pigments, dyes and mordants, printing, chemical process industries, photography, pharmaceuticals, metal plating, pure chromium metal production, and for many other purposes. Chromite is a necessary constituent in basic refractories indispensable for the production of steel, copper, cement, and glass.

Under present conditions, the US and North America, with the possible exception of Cuba, have no commercial ore reserves. There are chromite deposits, as will be described, but they would be used only under highly abnormal conditions if all outside sources were cut off for an extended period. All chromite used in the US is imported from the eastern hemisphere in which are contained the principal world reserves. The main suppliers in 1980 were the Republic of South Africa (43%), the USSR (21%), the Philippines (12%), Albania (11%), Turkey (8%), and others (5%). The strategic nature of chromite is obvious.

* Formerly Program Manager, Kaiser Aluminum and Chemical Corp. Now Consultant, Pleasanton, CA.

End Uses—Grades and Specifications

As noted before, the three principal uses for chromite are metallurgical, refractories, and chemical with a fourth minor use as foundry molding sand. For *metallurgical* applications chrome ore is first reduced in electric carbon arc furnaces to various types of ferrochromium alloys. There are high, medium, and low carbon ferrochromiums with carbon ranging from 0.01 up to 10%. Chromium itself ranges from 50 to 73%. Ferrosilicon chromium and ferrochromium-silicon grades have chromium from 33 to 67% and silicon from 3 to about 48%. Since iron is reduced with chromium, the Cr:Fe ratio of the ore will determine the ultimate grade of the ferrochromium whereas carbon and silicon can be controlled during processing. MgO, Al_2O_3, CaO, and excess SiO_2 are removed in slag. A small amount of pure chromium metal is prepared by aluminothermic and electrolytic processes from essentially iron-free chrome components that have been made via the chemical route. The higher grades of ferrochromium with chromium in the 66 to 73% range can only be made from ores with not only high chromium content, but low iron, i.e., high Cr:Fe ratio. These ores, conventionally termed metallurgical grade, have Cr_2O_3 contents of 45 to 56% with Cr:Fe ratios of 2.5 to 4.3. A standard composition is generally 48% Cr_2O_3 and 3:1 Cr:Fe ratio. Very high grade ores are often blended with lower grades even with the high iron chemical grade. For some ferrochromiums, additions of high alumina refractory grade are used as slag conditioners to facilitate smelting while still contributing chromium values. A lower grade ferrochromium, the high carbon charge chrome, has become the predominant grade in use today for making chromium-bearing steel alloys. This material can have as low as 50% chromium and can be made from chemical grade ore with 44% Cr_2O_3 and a Cr:Fe ratio of 1.5:1.

The widespread use of charge chrome was made possible by the argon oxygen decarburization (AOD) and the vacuum oxygen decarburization (VOD) methods for making stainless steel (Saccamano, et al., 1969; Anon., 1972; Anon., 1971; Robiette, 1972; Krivsky, 1972). These new processes, especially AOD, now make most of the stainless steels produced in the world today. They have converted the enormous South African chrome ores, long known as *chemical* grade, into *metallurgical* grade reserves. The old grade designations will undoubtedly linger, but will be understood in the present context.

Chromite is used in basic refractories that range from essentially all chromite to various mixtures of chromite and magnesia. The average content is about 45% chrome ore to 55% magnesia. An increase in the magnesia-chrome ratio during the past 15 to 20 years is one reason for the decline in refractory chromite usage. Another reason is that the basic oxygen steelmaking process uses mostly magnesia and dolomite refractories without chromite. That process, and its modifications, known as basic oxygen furnace (BOF, BOP, LD, Q-BOP, OBM, etc.), is now the predominant steelmaking method in the US, Japan, and Western Europe. Magnesia-chrome refractories are still the main refractories for open hearths and electric arc furnaces, although in electric arc furnaces increasing application of magnesia-carbon brick and watercooled wall panels is cutting down use of magnesia-chrome brick. About 80% of basic refractories are consumed in steelmaking. The remainder goes into the portland cement, nonferrous, glass, power generation, and paper industries. So-called *refractory grade* chrome ores traditionally have low Cr_2O_3 (30–40%), relatively high Al_2O_3 (25–32%), and, although a low Cr:Fe ratio (2.0–2.5:1), they have relatively low iron (total Fe as FeO <15%). Principal source of this type for the US is the Masinloc district in the Philippines. Just as all grades of chrome ores can be used, to some extent, for metallurgical purposes, so can all grades be used for refractories although not for all classes. High iron chromites have generally been successful in magnesia-chrome refractories with 50% or more magnesia. They are virtually the only chromites employed in fused cast or rebonded fused grain basic refractories in the US. A major development in chromite-containing refractories occurred in 1962 with the introduction of direct bonded magnesia-chrome brick. For this brick fine granular, low silica (1–3%) concentrates became preferable, or rather mandatory, instead of lump ores that contained ~5–6% SiO_2.

Chromite is used in sintered prereacted grain or coclinker in which the chromite is reacted with magnesia at high temperatures in briquettes, pellets, or other agglomerated forms. This grain serves as an intermediate raw material for high performance refractory brick.

A possible problem for chromite-bearing refractories in the portland cement and glass industries is that some of the chromium is oxidized to hexavalent Cr and made water-soluble by reaction with alkalies. This creates a hazardous waste disposal problem for the used brick. At this writing nonchrome brick is being investigated for this application. Other possible solutions are reduction treatment of the wastes and safe disposal methods.

Chemical grade chrome ore is used to produce sodium dichromate (also called bichromate) which, in turn, becomes the feedstock for a long list of chromium chemicals used for electroplating chromium metal; as pigments for paints, dyes, inks; textile mordants; photography; metal corrosion control; leather tanning; wood preservation; lithography; and various other chemical processes. Chemical grade ore in the US generally means $Cr_2O_3 \geqq 44\%$, Cr:Fe about 1.5, and silica under 3.5%, although generally under 2.5%. All of this ore comes from the Transvaal of South Africa. Most of this ore is friable and breaks down to a granular sandlike size. Silica in the mined ore is fairly low, <5%, but can be further reduced by simple washing, wet classification, or gravity concentration.

Foundry molding sand chromite used in both ferrous and nonferrous metal casting where special surface effects are sought is specially concentrated and sized Transvaal chemical grade ore.

Industry uses of chrome ores in terms of percentages and tonnages in the US for the last 40 years are shown in Figs. 1 and 2. Metallurgical and refractory chromites are strongly steel industry-oriented and are subject to that industry's cyclical nature though not necessarily to its long-term trends. Decline in refractory chrome ore usage is mainly accounted for by the increasing predominance of BOF steelmaking. In 1955, open hearths made 90% of all US steel, whereas in 1980 its percentage was down to 11.8%. In the same interval, BOF output rose from 0.3 to 60.6%. BOFs use vir-

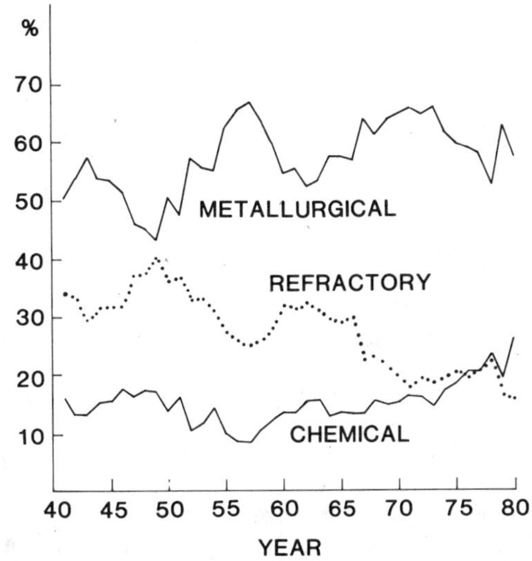

FIG. 1—*Percent usage of chrome ores by type in the United States, 1941–1980.*

tually no chrome ore. Electric furnace steel, which does use chromite refractories, rose from 6.8 to 27.6% of the total during the same period. During the period when steel production changed from mainly open hearth to predominantly basic oxygen process, chrome ore used in steelmaking refractories fell from about 365,000 st annually to an estimated 100,000 to 120,000 st range (depending on actual steel

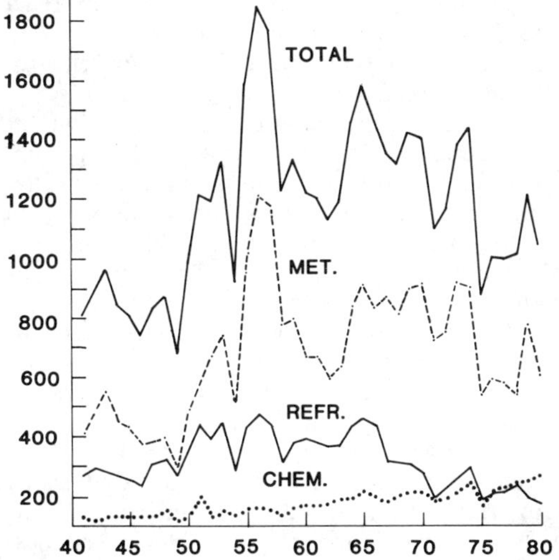

FIG. 2—*Consumption of chrome ores, total and by grade, in the United States, 1941–1980.*

production, which fluctuates considerably). Chromite usage will continue to decrease in this industry, although at a somewhat slower rate.

Geology and Mineralogy

Mineralogy

Chromite is a variety, or more properly stated, a composition range, of the spinel group of minerals (Stevens, 1944; Ulmer, 1970). This range can be expressed as $(Mg, Fe^{2+})O \cdot (Cr, Al, Fe^{3+})_2O_3$.

The composition also can be generalized as $R^{2+}O \cdot R^{3+}{}_2O_3$. Molecular ratio of divalent to trivalent parts theoretically should be, and generally is, close to one. A lower than one ratio usually indicates oxidation of some original FeO to Fe_2O_3 (Stevens, 1944).

Chromite compositions can be represented graphically by a prism of six theoretical end members as shown in Fig. 3a. The curved pie-shaped volume within the prism is where the natural chromites occur. Extending this volume down to the $FeFe_2O_4$ corner may allow for chromian magnetites but chrome ore chromites are all near the $(Mg, Fe^{2+})(Cr, Al)_2O_4$ side as shown by the shaded area of the mid-plane projection, Fig. 3b. A further idea of composition range is gained from the projection of Fig. 3c. (All diagrams show molecular proportions.) The constituents shown by these diagrams account, in most chromites, for $99^+\%$ of their compositions. But titanium, vanadium, manganese, and nickel are present in traces up to slightly more than 1% in most chromites.

TiO_2 is the most prominent of those accessory oxides and occurs from a few hundredths to about 1%. Reports of TiO_2 of 1.4 to 2.0% may indicate a separate titanium mineral, such as rutile or ilmenite. Unusual Ti-Cr spinels (not ores) have been reported (Cameron and Glover, 1973). Vanadium is seldom analyzed but accounts for around ½% (as V_2O_5) in Bushveld chromites (Cameron and Emerson, 1959). In most chromites vanadium amounts to less than 0.1%; MnO and NiO are very persistent constituents and generally each amounts to a few tenths of a percent. The latter three elements can be accommodated into the spinel structure.

Chromites of economic interest have a Cr_2O_3 content of 25 to about 65%. Predominantly, the Cr_2O_3 of commercial ores ranges from 30 to about 60%. This refers to the chromite mineral free of gangue—which, of course, it never is. The always present interstitial gangue minerals are generally orthopyroxene, olivine, calcic plagioclase and their hydrous alteration products, serpentine, chlorites, and talc. In fact, the hydrous alteration magnesian silicates are probably more commonly present than the original minerals (although they may reveal, by microstructure, the original silicates from which they were derived). Other less common to rare gangue minerals that have been found in various deposits include dolomite, magnesite, brucite, sepiolite, chrome tourmaline, uvarovite (chrome garnet), kämmererite, quartz, kaolin, pyrite, mica, and goethite. These minerals have been formed by metasomatism. A spectacular, though exceptional, example of this process is the large-scale alteration of original ultramafics to talc-quartz and talc-carbonate host rocks of the Selukwe, Zimbabwe, chromite deposits (Cotterill, 1969).

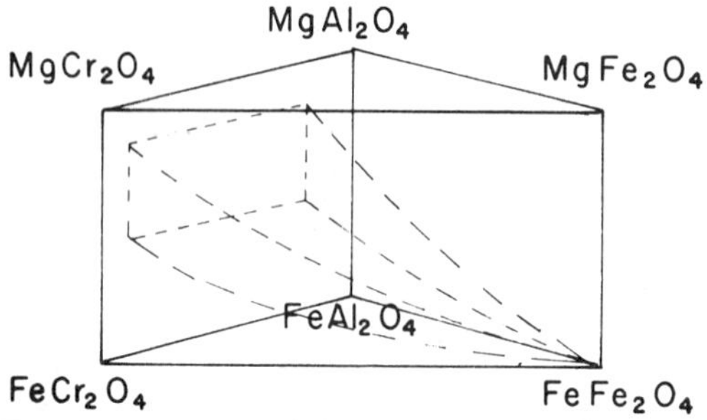

FIG. 3a—*Prism representing molecular compositions of chromites in terms of six end members. Dashed lines outline volume containing most natural chromites.*

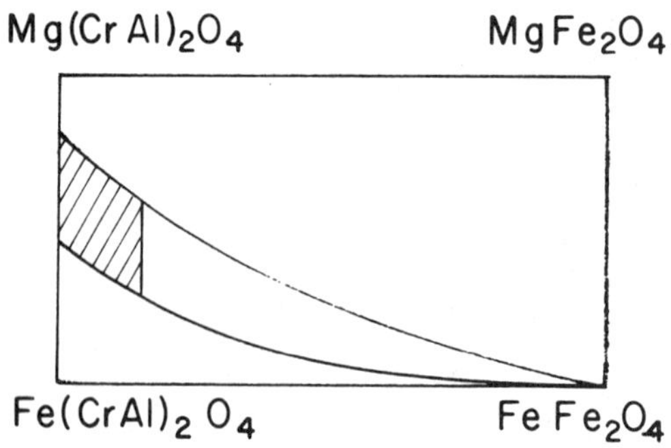

FIG. 3b—*Cross-section through Fig. 3a. Shaded area contains the commercial chromites indicating relatively low amount of ferric iron.*

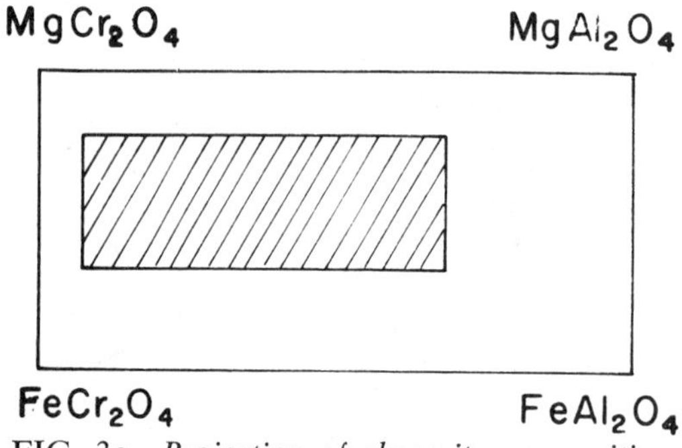

FIG. 3c—*Projection of chromite compositions on (Mg, Fe^{2+}) (Cr, Al)$_2$O$_4$ face.*

The geologist, engineer, or industry consumer will practically always be dealing with chrome ores or concentrates that are mixtures of pure chromite and essentially chrome-free silicate gangues. Enrichment of ore in chromium physically can only be accomplished by separating chromite from gangue, but the Cr_2O_3 content will never be higher than that of the chromite. Physical beneficiation methods as discussed later are the only economical methods of upgrading chrome ore. Chemical methods for removing iron from chromite are known but not practical at present.

Table 1 lists some typical pure chromite analyses accompanied by analyses of the corresponding ores showing the diluent effects of the gangue minerals. SiO_2 and CaO are exclusively contained by the gangue and Cr_2O_3 by the chromite whereas MgO, Al_2O_3, and FeO are distributed in both minerals.

Physically, chromite is generally jet to dark grayish black with luster from shiny vitreous to resinous to dull. Hardness is 5.5 to 6.5 and specific gravity, 4.1 to 4.7. It is brittle, without cleavage, and fractures unevenly with conchoidal tendencies. Some chromites may react weakly to a hand magnet, but most will not. But all can be moved in a strong electromagnetic field.

Chrome ore generally occurs as massive aggregates of chromite grains 0.05 to 30 mm, more commonly 1 to 5 mm, in size. Interstitial gangue in shippable massive ore amounts from as little as 2 or 3% to as much as 25% for very high chromium chromite. The amount of gangue permissible before concentration becomes necessary depends, of course, on the chromite composition. As gangue increases, chrome ore grades from massive to disseminated ore. Gangue in crude ore in some mining operations can go well beyond 50% if beneficiation economics allow. Chromite grains range from near euhedral octahedrons to anhedral rounded to irregular blocky shapes. Cracks and fractures are common in some deposits. Microbrecciation to mylonitization is

TABLE 1—Analyses of Typical Chromites and Corresponding Chrome Ores

	No. 1		No. 2		No. 3		No. 4		No. 5		No. 6	
	C*	O*	C*	O*	C*	O*	C*	O*	C*	O*	C*	O*
Cr_2O_3	47.41	44.52	36.24	32.10	61.44	54.91	57.70	47.00	58.32	48.50	55.90	49.19
Al_2O_3	14.82	15.50	31.86	30.20	11.41	9.92	13.44	12.65	11.06	10.01	11.92	11.37
Fe_2O_3	9.21		2.97		nil		3.42		4.10		2.18	
FeO	16.86		11.32		12.53		11.66		11.10		18.80	
Tot. Fe as FeO†	(25.26)	24.72	(13.99)	12.72	—	12.41	(14.70)	11.93	(14.75)	13.28		18.27
MgO	11.40	10.10	17.10	18.06	13.66	14.92	13.29	15.46	14.23	18.83	10.36	12.35
CaO		0.30		0.44		0.70		1.77		0.40		0.48
SiO_2		2.24		5.00		5.02		5.71		6.94		6.58
TiO_2	0.41	0.43	0.38	0.30	0.17	0.20	0.39	0.32	0.06	0.05	0.76	0.67
MnO	0.09	0.07	0.11	0.08	0.16	0.14	0.24	0.06	0.15	0.12	0.30	0.26
NiO			0.12	0.10					0.20	0.18	0.09	0.08
V_2O_5	0.33	0.30							0.07	0.05	0.18	0.16
L.O.I.				0.35		0.92		3.95		1.20		
Total	100.53	98.17	100.10	99.35	99.37	98.64	100.24	98.85	99.29	99.56	100.49	100.44
Cr:Fe	1.67	1.58	2.30	2.22	4.31	3.90	3.45	3.47	3.45	3.25	2.37	2.37

*C = chromite, O = ore or concentrate.

†Separate FeO and Fe_2O_3 are generally not determined. Generally only Cr_2O_3, Total Fe, Al_2O_3, MgO, SiO_2, and CaO are available in out-turn analyses. These are given on a dried basis. CaO and SiO_2 ranged from nil to 0.17 in separated chromites and were omitted.

No. 1—Transvaal Steelport. 2—Philippine Masinloc. 3—Kempirsai (USSR). 4—Selukwe, Zimbabwe (talc-dolomite gangue). 5—Turkish. 6—Great Dyke.

prominent in the Selukwe, Zimbabwe ores (Cotterill, 1969). Lineation, foliation, and series of aligned partings, the so-called "pull-apart" textures, are found in some deposits (Thayer, 1964). Another common texture is nodular ore, also called grape, shot, pebbles, etc. This consists of spherical to ovoid aggregates mostly 5 to 40 mm in diameter. The nodules consist predominantly of chromite grains, generally in the 1 to 5 mm size range, in a silicate (commonly olivine-serpentine) matrix.

In the Transvaal stratiform deposits nodules of a texture and mineralogy somewhat different from those commonly seen in podiform deposits are found (Thayer, 1969). The Transvaal (Bushveld) nodules are spherical to ovoid bodies 5 to 50 mm in diameter. Cameron and Desborough (1969) state:

> Each nodule consists of one or more crystals of bronzite or plagioclase poikilitically enclosing abundant fine-grained chromite grains. Nodules are generally set in a matrix of coarser-grained chromite. They are high in silica compared to the matrix material, and portions of certain seams in which nodules are abnormally abundant are unworkable. At certain mines, light crushing of the crude ore disintegrates the matrix material, the nodules are then removed by simple screening, and the ore is correspondingly upgraded.

In podiform deposits, nodules generally are not excessively high in silica and may comprise good ore if sufficiently concentrated in the host rock matrix. For a further description of nodules and orbicules in podiform deposits, see Thayer (1969).

In thin section chromite ranges from dark yellow through orange reds to opaque. Polished sections in vertically reflected light are light to medium gray. Chromite is isometric; hexoctahedral, in the space group Fd3m. Unit cell size *a* ranges continuously from about 8.200Å for high aluminum, low iron and chromium to ~8.330 for high chromium and iron chromites.

Geology

All primary chromites, and this means nearly all presently economic chrome ore deposits, occur in ultramafic rocks or their altered derivatives. The primary ultramafic rocks are peridotite, dunite, and pyroxenite and the principal derived rock is serpentine (properly called serpentinite). Serpentine is, in fact, the most common gangue mineral associated with chromite. Chlorite and talc schists are also alteration products of the primary silicate minerals. The ultramafic rocks are parts of peridotite or pyroxenite-gabbro complexes. Gabbros (including norite) and anorthosite are present near many chromite ore bodies in lesser amounts than the ultramafic rocks. In some chrome ore districts, e.g., Masinloc (Philippines) and Camagüey (Cuba), the ore bodies are found in peridotite near the peridotite-gabbro contact (Stoll, 1958; Davis et al., 1957). Gabbroic rocks occur in some ore bodies as cross-cutting dikes and inclusions. In some districts, e.g., the Bushveld, chrome ores may occur within norite or anorthosite, but mainly they occur in pyroxenite or peridotite.

Types of Deposits

Chrome ore deposits are divided by structural types into two major categories: stratiform and podiform. The *stratiform* deposits, exemplified by the chromite of the Bushveld complex of the Transvaal, are characterized by occurrence as parallel layers of great lateral continuity. In this stratiform class, the ore layers tend to be coextensive with the containing layered rock sequence much as are coal seams within sedimentary formations. The containing rock layers are themselves recognizably persistent, continuous, and extensive so that the occurrence and placement of the ore layers are amenable to conventional geologic mapping methods. Other examples of stratiform deposits are the Great Dyke of Zimbabwe (Worst, 1960); Kemi in Finland (Kahma, et al., 1962); Campo Formoso, Brazil (Thayer, personal communication); the Stillwater Complex, Montana (Jackson, 1964); Fiskenaesset, Greenland (Ghisler and Windley, 1967); and Bird River Complex, Manitoba (Jackson, 1964). The last three named are not economic ore deposits at the present time.

The other major type of chrome ore deposit, which includes all known deposits besides those already named, has been variously designated as sackform, piecemeal, discontinuous, or podiform. Following Wells, et al. (1946) and Thayer (1964), the term *podiform* will be used here. In contrast to the stratiform, the podiform deposits occur as pods, lenses, sackforms, slabs, and all sorts of irregular shapes. The ore bodies may also be tabular and occur in layers, but the important feature that distinguishes this type from the stratiform is that the ore bodies are not continuous, and in many districts follow no recognizable systematic distribution within the containing country rock. In fact, the distribution may be erratic, random, and unpredictable.

As mentioned previously, in some districts ore bodies seem to be concentrated in peridotite near its contact with gabbro, a fact that serves as an exploration guide, but since the designated areas are still very large in proportion to the size of the ore bodies, this criterion cannot locate individual ore bodies. Ore bodies may occur in strings or clusters, but these assemblages may give out without warning. The size of these podiform chrome ore bodies range from a few tons (depending on economics of mining) to one that contained more than 14 million tons (Coto mine, Philippines). The majority of bodies range from a few tens of thousands to a few hundred thousand tons. Single ore bodies of even one million tons are quite rare, e.g., Gölolan mine, Turkey.

The contacts between massive chromite and country rock may range from very sharp to gradational in single ore bodies. By gradational is meant grading into rock by increasing amount of silicate and corresponding decrease of chromite. If the gradational zone is wide, the practical mining limit defines the boundary of the ore body. In some deposits, ore with high silicate content may be mined as disseminated ore and be subject to beneficiation. Float, or alluvial, chromite may contribute to production in some districts but generally is a minor factor. Such deposits derived from adjacent primary deposits are mined in Zimbabwe and Brazil. During 1980 a small chromite beach sand mine on Palawan Island in the Philippines was started up.

Mining and Beneficiation

Chromite is mined by both conventional open pit and underground methods. Where relatively large bodies are near the surface so that stripping of overburden is economically feasible, of course, open pit mining will be used. A good example of this is the Coto deposit in Luzon, probably the largest single podiform body ever discovered, which was an irregular plate forming the top of a hill. The majority of bodies are of such size or shape that underground mining is required. For example, the seams of the Transvaal that are mined are at most about 5 ft thick and dip 10° to 40°, making stripping impossible. The parallel seams of the Great Dyke range from 2 to 18 in., but average only 7 in. and the distance between seams is some hundreds of feet, so they can only be mined separately rather than two or more in one opening. The seams dip from 15° to 50°, depending on their location in the synclinal structure of the Great Dyke. But, detrital and eluvial chromite of the Great Dyke has been mined and constitute a substantial reserve although minor in comparison to the seam deposits. Stratiform seam ore bodies have been mined downdip several thousand feet without thinning of the seams. They have been followed for tens of miles along strike. The only constraints are economic.

Probably the majority of podiform chrome ore bodies are mined underground. Many of these tend to be elongate and to plunge steeply. Good examples are the Eskisehir and Guleman-Soridağ districts of Turkey and the Camagüey district of Cuba.

Although large segments of most ore bodies of the major chrome ore districts of the world are direct shipping ore, beneficiation is becoming increasingly important. Selective mining, hand cobbing and sorting, picking belts, wash screening, and wet classification have all been used for many years, more or less, depending on ore grade and labor costs. But more intensive beneficiation methods now used include heavy media (sink-float), shaking tables, spirals, hydroclones, electromagnetic and electrostatic separations, and flotation (Kurochin, 1971). At the Palawan beach sand operation, concentration is by wet screening, spirals, and shaking tables in optimum combinations. All of these are methods that separate chromite from silicate gangue which leave chromium, iron, aluminum, and magnesium fixed by their content in the chromite. There are several chemical methods to enrich the chromite in Cr_2O_3 by reducing the iron, thereby increasing the Cr:Fe ratio which has been an important consideration for metallurgical use. One method involves selectively reducing iron oxides to metallic iron and leaching this out with acid (Harris, 1964). Another removes iron selectively by chlorination (Severns, 1969). Neither of these methods has been used on a significant commercial basis and probably will not be in the foreseeable future because of developments that will be discussed later.

Exploration, Prospecting, and Evaluation of Reserves

The most fundamental basis for searching for chromite is its virtually exclusive occurrence in ultramafic rocks that are, in turn, parts of ultramafic-gabbro complexes. But there are a number of large areas of ultramafic rocks in which no significant chromite ore bodies have ever been found, although there is no known basis for saying that they contain none. The second important point is identifying a chromite deposit as stratiform or podiform. The differences are so dramatic that there should be no confusion between the clearly recognizable stratiform deposits such as the Bushveld, Great Dyke, Stillwater, and Bird River on the one hand, and those presently designated podiform which encompasses most of the rest of the known commercial deposits. These include the Balkans-Asia Minor region, USSR, Zimbabwe exclusive of the Great Dyke, Caribbean, Philippines, India, New Caledonia, Pakistan, Iran, Sudan, and possibly others.

New deposits classified as stratiform in the past two decades are Kemi, Finland (Kahma, 1962); Campo Formoso in Bahia, Brazil (Thayer, personal communication); and Fiskenaesset, Greenland (Ghisler and Windley, 1967). The Brazilian deposit had been thought to be podiform but more detailed mapping has identified it as a sheared-out stratiform traceable for ten miles in heavily weathered terrane.

Prospecting and estimation of reserves in the stratiform complexes are subject to conventional geological field methods because there is a rational genetic basis for explaining and predicting the distribution of the chromite layers. The theory is that the chromite and containing ultramafites originated from a common liquid magma by differentiation and crystal settling or what has been termed *magmatic sedimentation*. Furthermore, these formations have been comparatively little disturbed from their original locations or attitudes. There has been some faulting and tilting, but no reemplacement. The structural displacements are still generally amenable to solutions by stratigraphic methods. The theory has difficulty in explaining every mineralogical and chemical detail, variation, and nuance found in the deposits, but overall it appears to be the best available and—more importantly for the mining geologist—it works.

The podiform deposits present a drastically different picture. Although the petrogenetic relations between chromite and host rocks appear to be similar, i.e., explainable by differential crystallization from a common magma, their size, shape, and, especially, their spacial arrangement and distribution are completely different from the stratiform. The ore bodies are separate, discrete, and generally appear to have been thoroughly scrambled. A currently favored theory of podiform origin proposed by Thayer (1969) is that these ore bodies were once layers in stratiform complexes but have since been broken up by tectonic forces within the earth's mantle and reemplaced tectonically, or by solid or quasi-solid flow into crustal rocks. The ultramafic complexes of which podiform chrome districts are a part, are called alpine-type complexes because they are always characterized by intense regional deformation (Jackson and Thayer, 1972). It is postulated that the settled chromite layers (chromitite) were relatively well consolidated and cohesive as compared to the surrounding semisolid mixture of crystals and liquid, or mush, which was able to physically entrain and support the higher specific gravity chromite masses by virtue of motion and high effective viscosity. Broken segments of the chromite layers were jumbled and carried along in the moving stream and were frozen in place in varying attitudes and locations as the crystal mush finally solidified. In periods of relative quiescence, the chromite agglomerates would sink even in the mush in a manner simulating Stokes law according to which the velocity of descent is proportional to the square of the diameter of the body. This could explain the extreme paucity of very large podiform bodies (1 to 15 million tons) and the comparative abundance of small bodies.

The Selukwe district of Zimbabwe described by Cotterill (1969) possesses characteristics of both podiform and stratiform deposits. The district appears to consist of sliced-up remnants of an originally stratiform complex. Within some slices, ore zones (though not bodies) can be followed for the length of the slice. The actual ore bodies are most generally described as lenses or "pipelike" and are discontinuous or irregularly distributed. Zones of occurrence of ore bodies tend to extend the full strike distance of a slice but in the dip direction ore distribution is mostly restricted to much less than half the extent of the ore zone. Much of the chromitite is strongly sheared, i.e., microbrecciated or mylonitized. Cotterill (1969) believes that the chromitite lenses represent original cumulus chromite that settled in local furrows or depressions caused by buckling of silicate layers during tectonic movements. He does not believe the lenses were reemplaced through a semisolid silicate matrix but owe their present distribution to the segmentation of an original ultrabasic complex into slices. Jackson and Thayer (1972) classify the Selukwe district as a stratiform complex which it may be, viewed broadly petrogenetically. But from the practical viewpoint of finding ore bodies and calculating reserves, the Selukwe district is more closely akin to podiform districts than it is to classic stratiform deposits such as the Bushveld or Great Dyke. As more knowledge is gained, the transitional events between original chromite accumulation and final emplacement will become better known. In the Selukwe district these events seem somewhat more clear, than say, in Turkey or Masinloc but still far from obvious.

The implications of the petrogenetic hypotheses on exploration and prospecting for podiform ore bodies could be profound. If the podiform bodies are broken up remnants of originally stratiform deposits, only a tiny fraction of the original ore bodies have ever been found. How can the number and size of the original layers be estimated from the remnants? How much remains at economically exploitable levels, and do the fragments have a decipherable pattern of distribution? These are unanswered questions and the estimation of ultimate reserves of any podiform district is largely speculation. The fact that some of these districts have, in the past, managed to maintain stated reserves at, say, 10 times annual production for 10 or 20 years indicates the estimators had no real idea of the reserves, but gave developed ore plus a reasonable inference based on experience, intuition, and an optimistic outlook.

Although there is no present scientific methodology for extrapolating podiform reserves, there are guides to prospecting that increase the probability of finding new ore bodies.

A great deal of interest has been shown recently in methods of exploration and prospecting for chrome ores. As mentioned previously, stratigraphic methods are best applied to stratiform deposits. But geophysical methods have had limited success as aids to exploring these deposits as at Kemi, Finland, where the extension of ore was traced under glacial deposits. The main interest lies in ore finding methods for podiform deposits. The relationships of structural and textural features of ore bodies to those of the host rocks in some districts give clues to the probable locations of other ore bodies. These are empirical relationships and do not assure the occurrence of new ore bodies. Of geophysical methods, magnetic methods have been of little use. Gravimetric methods have had limited success in a few cases (Yüngül, 1956; Davis, et al., 1957). Hosking (1964) describes the possibilities of using geochemical methods for chrome ore prospecting mainly to narrow the areas for drilling. Drilling, drifting, or trenching near known ore bodies is the usual current method of discovering new bodies.

World Chromite Resources

The estimation of reserves of stratiform deposits is on an entirely different basis from those of the podiform districts. In the stratiform districts, the chromite layers can be extrapolated laterally and vertically within layered host formations which are relatively undisturbed tectonically. Estimation of podiform reserves on the basis of only mined or drilled ore bodies would lead to unrealistically low estimates for many of these districts if past experience is considered. Generally this experience, coupled with current activity, is used to project or extrapolate the discovery of new reserves for at least a few years. These estimates are only order of magnitude approximations by which to compare the various deposits. It is unlikely that the proportion of estimated chrome ore in the two great stratiform complexes, the Bushveld and Great Dyke, to that of all the rest will change significantly since they contain over 99% of the world total (Thayer and Lipin, 1978). But, among the podiform producing districts, the relations could change by the discovery of comparatively small bodies.

North America

United States: In the early 19th century the southeastern Pennsylvania-Maryland area was the major chromite producer in the then small world chromite trade, most of which was in chrome chemicals with no metallurgical or refractories usage. Significant production did not last beyond the 1800s. Chrome ore was produced in Montana, Oregon, and California under government subsidies during and after World War II, but none has been produced since 1961 when the last government program was phased out. At the present time, the US has no commercial chrome ore deposits nor is there much probability that any will be discovered in the foreseeable future. But there are many chromite occurrences in the US substantial enough to be designated potential resources and these could conceivably be transformed into actual ore reserves by highly abnormal circumstances that would prevent or threaten delivery of overseas material. These potential resources are estimated to amount to about ten million st of chrome ore or concentrates averaging about 40% Cr_2O_3. Of this amount, 85% is in the Stillwater stratiform complex of Montana, about 10% in Oregon, and 5% in California, Alaska, Wyoming, and Washington (Thayer, 1962). About half of the Oregon resource would be attributed to very low grade chromite-bearing beach sands.

Canada: Canada, like the US, presently has no commercial reserves, but it has a low grade resource in the stratiform Bird River complex of Manitoba, with an estimated nine million tons of high-iron potential chrome ore.

Cuba: Cuba is the only country in North America currently with significant reserves. There is little available data on current production but it is believed to be minor. Most of this ore is the high alumina refractory grade in scattered podiform deposits. During the 1940s, Cuban ore was the principal refractories chromite used in the US. Average annual production in the 1940s was 210,000 st, during the 1950s it was 78,000 st, and in the 1960s, the average fell to 35,000 st (estimated). In the period 1965–1968 an average of 19,000 st was exported annually to Poland. Because of the political situation, there have been no US imports from Cuba since 1961. There may still be potential resources of a million or so tons but there is insufficient information to make any statement about reserves.

Greenland: A large high-iron, low-grade stratiform deposit has been discovered in west Greenland that, although incompletely mapped, may add several hundred million tons to the world's potential resources (Ghisler and Windley, 1967). Because of its location, it is of interest to the US and Canada as a possible strategic source. Actual exploitation is not probable in the foreseeable future.

Central America

Guatemala has shown up in chromite production statistics, the last time in 1961, at a few hundred tons per year. But the deposits are small, scattered, and do not constitute significant reserves at this time.

South America

Brazil is the only other western hemisphere country with significant chrome ore production. Chrome deposits are in Bahiá near Campo Formoso. Originally considered podiform, these deposits are now recognized as parts of a broken up, sheared-out stratiform complex, segments of which have been traced for 10 miles (Thayer, personal communication). Massive ore occurs in layers up to 2 m thick and weathered disseminated ore can be mined to depths of 20 m or more. Both high chrome metallurgical and high iron chemical grade ore are found. Preliminary estimates of shippable ore or concentrates are on the order of 15 million st.

The foregoing review of known commercially available western hemisphere chromite reserves shows that, at present, they are small and inadequate for US needs. Minor amounts of chromite have been produced in Colombia and Mexico. This explains the total dependence on eastern hemisphere countries.

Europe

Albania: Albania has had a steadily growing production, making it the world's third largest chromite producer. The deposits are in the Balkan-Asia Minor ultramafic province in the Dinaride extension of the Alpine mountain chain. Principal deposits are near the border with Yugoslavia. As in many podiform areas with active mining and accompanying exploration, reserve estimates are continually exceeded by production as the years go on. Albanian reserves were estimated at 500 kt in 1960, but through 1979 that country produced over 10 Mt. Annual production in the 1960s averaged 280 kt/a. Since then it has increased every year, reaching 1.2 Mt in 1979. The deposits are podiform bodies of moderate size. Estimating reserves is inaccurate, but a figure of 5 to 10 Mt seems reasonable. Cr_2O_3 content ranges from 40 to 49%, averaging about 40%. Cr:Fe ratio is 2:6, which classifies it as metallurgical grade chromite. It is used mainly in ferroalloys.

Finland: The high-iron stratiform type deposit at Kemi, Finland is the most recently discovered producing mine. Estimated reserve is 37 Mt, which translates into about 12 Mt of usable concentrates. The significance of this smallish, rather low-grade deposit is that it is serving as the raw material base of a domestic ferrochromium industry. Ore exports reached a peak in 1977 but have since fallen off as more chromium is exported in ferrochromium. The Kemi deposit was entirely hidden by glacial cover. Accidentally discovered by an excavation, its evaluation was aided by magnetometer and gravimeter (Kahma, et al., 1962).

USSR: Outside Russia, very little is known about the reserves or potential resources of this country except to judge by the record of past and current production and exports, which are comparatively large for podiform deposits. There are two main areas: Sarany in the southern Urals has medium to low grade ore; and the Kempirsai district of northern Kazakhstan has some of the world's highest grade metallurgical ore, much of it 55% Cr_2O_3 and 4:1 Cr:Fe ratio. Total reserves have been put at about 100 Mt. Both the production rate and export histories of Soviet Union chrome ore would appear to support this number (Mikami, 1982).

Russia was the world's largest chrome ore producing country for about 25 years through 1975, and during that period, was one of the largest chrome ore exporting nations. Because ore production has increased every year since 1953 and exports have been at high levels, one can believe that reserves must be substantial.

Yugoslavia: Yugoslavia has the same type of random podiform deposits as the other Balkan-Asia Minor countries (Albania, Greece, Turkey, Cyprus) and from 1952 to 1963 produced over 100 kt/a, but output has declined in the past ten years to under 40 kt. This would appear to indicate that production may be overtaking reserves. If true, it is a situation that could occur in other podiform areas, even if active mining and exploration are taking place.

Africa

Madagascar: Madagascar had negligible to very small, intermittent production until 1969 when intensive development of a moderate size deposit was undertaken. Production for 1979 was about 150 kt. Reserves are placed at 3 to 4 Mt of metallurgical grade concentrates.

Republic of South Africa: This country is now the world's major chromite producer, having passed the USSR in 1976. It unquestionably possesses the largest reserves, estimated at over 2 Gt recoverable under present conditions. Inferred potential resources are placed at 20 Gt (von Gruenwaldt, 1977). Chromite occurs in the stratiform layers of the Bushveld complex in the Transvaal. Its amount can be estimated with vastly more assurance than that of the podiform districts. This Transvaal ore is of the high-iron, so-called chemical grade with Cr_2O_3 mostly 44 to 46% but some up to 48%. The Cr:Fe ratio is mainly ~1:5 but the relatively small deposits of 48% Cr_2O_3 have ratios up to 2:1. South Africa is the source of most of the ore used in Western Europe, the US, and Japan to make sodium dichromate, the feedstock for all other chromium chemicals. It has also become one of the largest producers and exporters of ferrochromium, most of which is charge grade. South African chromite is the principal ore used in foundry sand in the US. It is used in refractories, but in a more limited way than the Masinloc ore from the Philippines.

Zimbabwe: Zimbabwe is the only country with major reserves of both stratiform and podiform types. In the Selukwe district outside the Great Dyke, which is regarded as stratiform by some authorities, are important podiform reserves of high-grade metallurgical ore, estimated at 5 Mt. Two other small deposits may add another 2 Mt. But these deposits are dwarfed by the seams of the stratiform Great Dyke, reserves of which are estimated at about 1 Gt (in which are included 60 Mt of eluvial ore resulting from weathering in place). Ores and concentrates range in grade from chemical (44% Cr_2O_3, 2 to 2.3 Cr:Fe ratio) to metallurgical (48% Cr_2O_3, 3.0 Cr:Fe ratio). The reserve is estimated to be one-half metallurgical grade. The Zimbabwe reserve alone would, theoretically, supply world needs at current rates for nearly 100 years. One calculation, on geological grounds, of the occurrence of chromite in the Great Dyke, disregarding mining feasibility, has come up with a figure of 71 Gt (Stanley, 1961). This interesting figure has no status for reserve or even potential resource purposes at the present time. Zimbabwe has had a turbulent history in recent years that has interrupted mining and exports from time to time, as will be discussed.

Asia

India: There are extensive ultramafic areas in India (Bala Sundaram, 1972) with a wide range of chrome ore grades from refractory to metallurgical and some chemical. Numerous occurrences have been reported but significant production has come from Orissa, Mysore, Bihar, and Madras. The first two are the most important. The government has been active in exploration and is encouraging a domestic ferroalloy industry. The Geological Survey of India in 1969 estimated the proved and probable reserves of Orissa and Mysore at 7.2 Mt. If mining and exploration continue, ultimate reserves may prove to be substantially more than that. Thayer and Lipin (1978) classify India's deposits as "severely deformed stratiform" and estimate reserves at 5 Mt and resources at 20 to 25 Mt. Intensive exploration recently is said to have doubled or tripled Indian reserves (Lipin, 1981).

Iran: Iran has two commercially important areas, Froumad in the north, said to have possibly 1 Mt, and a southern area near the Persian Gulf port of Bander Abbas. The southern area is the most important and has the most potential. In this latter area are the Esfandegheh concession with about 1 Mt inferred and the Faryab concession which is said to have possible reserves of 50 Mt. Ore and concentrates are metallurgical grade. If transportation is improved the economics of mining would be sufficiently improved to permit more beneficiation which, in turn, would increase actual reserves considerably. Ore bodies are relatively small lenticular, tabular, and irregular shapes, but the outcropping and near surface bodies are numerous (Haeri, 1961).

Philippines: The main chromite region in the Philippines is in Zambales province on Luzon Island about 100 miles northwest of Manila. The Coto ore body and adjacent deposits near Masinloc have been the principal source of refractory grade ore for the US since 1947. These ores generally are designated Masinloc and are characterized by high alumina (Table 1). The Coto ore body originally contained about 14 Mt (Stoll, 1958) of massive ore, making it the largest single podiform body ever found as far as published data are concerned.

About 10 miles north of Masinloc and within the same ultramafic-gabbro complex, high Cr, high Cr:Fe ratio metallurgical ore occurs and is being mined in the Acoje, Trident, and some smaller mines. From 1946 to 1980, about 14 Mt of the Masinloc refractory ore and 3 Mt of metallurgical ore were produced in the Philippines. Most of the original massive ore of the Coto body has been mined, but much concentratable ore remains, and some other smaller ore bodies have been found. There is probably enough ore for ten years at current production rates. This is based on reserves that have been stated each year and production rates. Again reserve estimates in podiform deposits are difficult. Mining could well be going on for 20 or more years in the Masinloc area (Dela Cruz, 1982).

There are chromite occurrences elsewhere in the Philippines. A small chromite recovery operation on Palawan Island, an alluvial delta deposit with sands and gravel containing 1 to 10% chromite, was started late in 1980. About a million tons of chromite concentrates with ~42% Cr_2O_3, Cr:Fe = 1.7, SiO_2 ~1 may be recoverable. Concentration is effected by classification, spirals, and shaking tables. Initial production is 20 kt/a.

Turkey: Turkey has been one of the world's major sources of chromite for over 50 years. The ore occurs in numerous deposits in two serpentine belts that are continuations of the Balkans ultramafic province. There have been several large ore bodies in the million ton class, but at present most of the ore is in smaller podiform bodies. Hundreds of occurrences are scattered throughout Turkey, but there are about half a dozen important areas of which the Elazig (Guleman-Soridağ-Kefdağ region), the Eskisehir, and the Mugla areas are notable examples. Although ultimate reserves are impossible to estimate, the Turkish Mineral Exploration Institute in 1972 placed total reserves at 11.5 Mt. Disseminated ore makes up 47% and massive ore 53% of this total. Present reserves are probably about 5 Mt, and potential resources about 15 Mt.

China: A small chromite mine has been reported (Yoshino, et al., 1981) operating in the Sinkiang (Xinjiang) province about 400 km from the capital of Urumchi, which is near the border with the USSR. The chemical composition is similar to that of the Masinloc deposit in the Philippines. Production is stated to be "several tens of thousands of tons" annually, all of which has been used internally, except for recent small exports to Japan.

Other Countries

Other countries with minor current production and small reserves are Cyprus, Greece, Japan, Pakistan, and Sudan. Countries not currently producing but which have chromite occurrences and have reported minor production at some time in the past 25 years are Afghanistan, Colombia, Egypt, Mexico, Portugal, Australia, Guatemala, and Sierra Leone. Reserves for these countries can be assumed to be quite minor. North Vietnam, North Korea, Bulgaria, and Rumania are known to have small producing deposits but data on them are not available. New Caledonia had a modest production of high grade ore until 1963, when the Tiebaghi mine ceased production. Very minor production resumed in 1975. It has been reported that chromite deposits are being investigated in Indonesia and Papua-New Guinea (Anon., 1980).

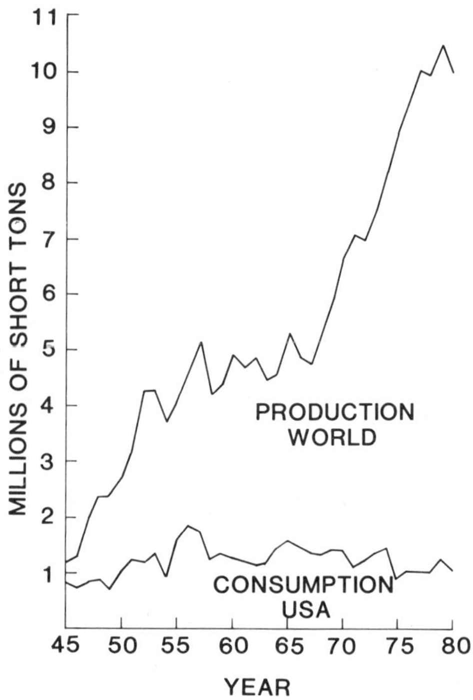

FIG. 4—*World production and United States consumption, mostly imported, of chrome ore, 1945–1980.*

Production, Consumption and Trade

World chrome ore production has been in a strong uptrend since 1945 (Fig. 4). The outputs of various countries in the period 1967 to 1979 are listed in Table 2. It can be seen that the six largest producing countries, each with more than half a million tons, accounted for 90% of world production in 1979. Production of several important countries (Albania, Zimbabwe, USSR) is estimated by the US Bureau of Mines (USBM). According to a published report (Anon., 1971b), the USSR produced 3 million st in 1970 and 2.78 million st in 1965 instead of the 1.93 and 1.56 million st figures, respectively, estimated by USBM.

Table 2 shows a steady uptrend in world production with only two down years in the 1967 to 1980 period. The average annual increase was 8½%. On the other hand, US imports for consumption in the same period (Table 3, Fig. 4) present a sawtooth curve with a flat to slightly downward trend. Part of the difference in the two trends is accounted for by increased importation of chromium as ferrochromium rather than as chromite. More importantly, the world outside the United States has been increasing its use of chromite at a substantially greater rate than the US. South Africa contributed the greatest amount to world production, followed by the USSR and Albania. Brazil, Finland, and Turkey had significant increases during the period. The six leading countries out of 21 produced 90% of the total. Of the principal consumers of chromium—the US, Western Europe, Japan, and USSR—only the last is a significant producer of chromite.

Prices

Fig. 5 shows prices of three principal types of chrome ore imported into the US over an

TABLE 2—World Production of Chrome Ore, All Grades
(Thousands of Short Tons)

Country	1967	1968	1969	1970	1971	1972	1973	1974	1975	1976	1977	1978	1979	1980[e]
Albania[e]	360	360	473	516	553	630	720	788	859	915	970	1025	1120	1190
Brazil	8	15	17	30	31	33	33	97	191	205	209	209	375	385
Colombia[e]	—	—	—	—	—	—	—	13	13	6	6	6	6	6
Cuba[e]	*na*	*na*	*na*	*na*	22	22	22	22	35	35	35	35	31	35
Cyprus	24	28	26	37	46	26	33	37	30	10	16	17	17	18
Finland	7	40	79	133	123	107	130	170	160	193	186	196	195	193
Greece	13	14	27	29	16	24	24	11	39	38	46	44	60	68
India	121	227	250	299	288	325	295	435	551	443	389	293	341	352
Iran[e]	165	176	165	220	194	198	200	193	190	176	257	218	150	90
Japan	50	31	33	36	35	27	26	29	26	24	20	10	13	15
Madagascar	—	—	88	144	165	123	173	172	214	245	182	152	141	165
New Caledonia	2	—	—	—	—	—	—	—	2	11	9	9	14	14
Pakistan	29	29	25	32	27	36	24	11	11	12	9	12	3	3
Philippines	463	446	517	624	474	385	640	584	573	475	593	592	618	631
South Africa	1277	1271	1320	1573	1812	1635	1818	2069	2288	2656	3372	3466	3634	3764
Sudan	20	33	28	52	23	25	35	22	17	24	19	20	31	28
Turkey[e]	409	665	500	572	665	710	617	734	790	782	760	750	500	440
USSR[e]	1731	1820	1874	1930	1980	2040	2100	2150	2290	2300	2400	2500	2650	2700
Vietnam[e]	*na*	*na*	*na*	*na*	*na*	*na*	*na*	*na*	*na*	10	11	13	15	17
Yugoslavia	52	52	43	45	38	31	4	1	2	2	2	2	2	—
Zimbabwe[e]	*na*	*na*	400	400	600	600	600	650	650	952	747	527	597	611
Total	4720	5206	5865	6672	7093	6977	7507	8187	8931	9372	10172	9920	10513	10725

e = estimated. *na* = not available.
Other countries producing small amounts, <1000 stpy, are Egypt, Argentina, Thailand (latter reported 6000 st in 1979). Possible commercial deposits have been reported in Indonesia and Papua-New Guinea.
Source: US Bureau of Mines.

extended period. South African Transvaal ore, type A, had a rather flat price curve from 1945 to 1973 reflecting the very large supply situation. The sudden jump in 1974 was triggered by stepped-up inflation from the oil price rise and other factors. After 1974 (shown by the dotted line in Fig. 5) prices are quoted at the loading point rather than at East Coast US ports for types A and B. Therefore the price escalation was even higher than shown because ocean freights have to be added. Prices for type C (USSR) were quoted in metric tons f.o.b. loading point from 1971 and after. This ore was not available in 1978 to 1979 but limited imports resumed in 1980. No 1978 through 1980 prices were available. Lower grades, ~40%, continued to be imported from the USSR during the suspension period of type C. Prices for the Philippine refractory grade have not been published but they were also subject to sharp escalation after 1974 in addition to a decline in quality (higher silica).

For metallurgical ore the base price applied to a standard, generally 48% Cr_2O_3 and 3:1 Cr:Fe ratio with differentials as analyses depart from these figures either plus or minus. For chemical grade, minimum Cr_2O_3 and maximum SiO_2 are generally specified.

Grain size, i.e., lump vs. fines, and degree of fineness also affect price. With the increasing need for beneficiation, the proportion of lump is decreasing.

Because of its large supply, Transvaal ore will probably remain the most stable in price. Some escalation from shipping costs can be expected. The high Cr metallurgical grade ore price probably will be tempered by decreasing demand for the highest Cr qualities of ferrochromium for the reasons previously cited. Because of US dependence on imports from the eastern hemisphere, political stability is a major factor in the supply-demand equation.

Economic and Political Factors

Since the end of World War II the supply situation has been generally favorable. Prices were relatively stable until 1974 when they surged upward. Interruptions from some sources were readily made up from others. During and after the Korean War in the 1950s, the USSR cut off shipments to the US but these were resumed in the 1960s. Ore shipments from Cuba ceased in 1962 and have not been resumed. Because of declining production, Cuba would not have been an important source. Zimbabwe (Rhodesia) ore was cut off by UN sanctions from 1967 until 1972. This was more serious than Cuba, but the deficit was made up by increased imports from Turkey and the USSR at sharply higher prices. Today a lack of Zimbabwe ore would be less serious because of the extensive use of low-grade charge chrome permitted by the AOD. Under the Byrd

TABLE 3—Imports of Chromite into the US for Consumption, Thousands of Short Tons

	1970		1971		1972		1973		1974		1975		1976		1977		1978		1979		1980	
	Ore	Cr_2O_3*	Ore	Cr_2O_3	Ore	Cr_2O_3	Ore	Cr_2O_3	Ore	Cr_2O_3	Ore	Cr_2O_3	Ore	Cr_2O_3	Ore	Cr_2O_3	Ore	Cr_2O_3	Ore	Cr_2O_3	Ore	Cr_2O_3
Source:																						
Albania	—		—		—		—		7	3	51	23	18	8	25	11	43	19	106	46	86	38
Finland	—		—		—		—		—		19	4	169	71	269	71	69	23	29	9	47	16
India	—		—		—		—		—		12	7	18	7	—		—		—		—	
Iran	31	15	12	5	14	6	—		—		11	5	—		—		—		—		—	
Madagascar	—		—		13	4	—		—		—		—		—		—		—		43	20
Pakistan	31	15	35	17	27	13	22	11	—		—		—		—		—		—		—	
Philippines	210	71	159	53	131	43	190	65	264	85	210	70	166	55	172	56	185	60	194	62	138	45
So. Afr., Rep. of	407	182	422	189	249	113	303	136	341	153	289	134	438	194	659	306	490	219	353	158	405	178
Turkey	257	116	338	150	96	51	131	59	120	56	173	75	212	89	124	53	70	26	71	32	93	37
USSR	469	252	307	164	434	226	242	120	298	150	349	175	189	89	80	39	157	61	251	98	170	76
Zimbabwe†	—		26	12	92	43	43	21	72	34	138	66	35	15	—		—		—		—	
Other	—		—		—		—		—		—		30	5	7	2	—		—		—	
Total, st	1405	651	1299	590	1056	499	931	412	1102	481	1252	559	1275	533	1336	538	1013	408	1004	405	982	410
Total, $ Value‡	32		32		28		21		28		61		70		69		51		54		57	

Source: US Bureau of Mines, *Minerals Yearbooks* to 1977 and *Mineral Industry Surveys*, 1978–1980.
* Cr_2O_3 content
† Rhodesia before 1979.
‡ Millions of dollars.

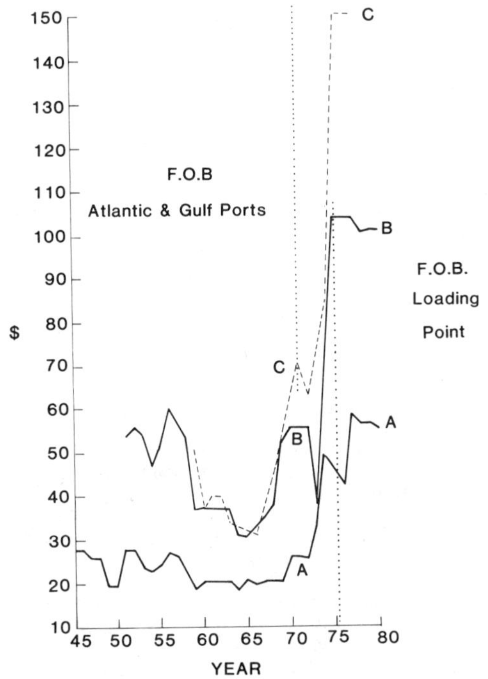

FIG. 5—*Chrome ore nominal prices. A* = *44%* Cr_2O_3*, no ratio. South African Transvaal. B* = *48%* Cr_2O_3*, 3:1 Cr:Fe ratio, Turkish. Comparable Zimbabwe when quoted had similar prices. C* = *54 to 56%* Cr_2O_3*, 4:1 Cr:Fe ratio, USSR. Prices subject to penalties or premiums on variance from standards. Ore A and B are quoted per long dry ton f.o.b. cars US Atlantic or Gulf ports through 1973. From 1974 on they are quoted f.o.b. loading point as indicated by the dotted line. Ore C is per long dry ton, f.o.b. same ports through 1970. Starting in 1971 C was quoted in metric tons f.o.b. loading point.*

amendment, Zimbabwean imports resumed in 1972 and continued until 1976 when sanctions were reimposed. After Zimbabwean independence, sanctions were again lifted in December 1979. No ore has been imported in 1979 to 1980 but some ferrochromium has been.

After Mozambique received independence in 1975 there was a considerable slowdown in export shipments of South African chrome ore for about a year because all the ore had gone by rail through Mozambique to the port of Maputo. Conditions have since improved substantially. An alternate route entirely within South Africa to Richards Bay is available but the South African government prefers to ship the bulk of the ore through Maputo at the present time. Political factors have not interrupted the steady flow of Philippine chrome ore but a severe typhoon wrecked the plant and interrupted shipments in 1976.

Logically, there have been no import duties on chrome ore, but in 1980 the government placed a charge of $1.56 per ton of contained Cr on all imported chrome ore for the so-called superfund that will go toward secure hazardous solid waste disposal sites.

Technological Developments and Future Outlook

In the 4th edition of this book the AOD and VOD, especially the former, were cited as tremendously significant in their effects on chrome ore usage. Now virtually all stainless steel, many high alloys, and even many low alloys are made by these methods. While this has decreased our dependence on high-Cr, high Cr:Fe ratio ores, it has increased our dependence on South African chromites or ferrochromium produced from it. The high-grade metallurgical ores from the USSR, Albania, Zimbabwe, and Turkey are still valuable and will continue to command a price premium over Transvaal but they are no longer indispensable.

A development that is beginning to increase is agglomerating fine ores or concentrates by briquetting or other means in order to increase the yield of these materials (Madronic, 1970). Briquetting of fine chrome ore is being practiced by some, but not all, plants in South Africa, Zimbabwe, Germany, and India. Prereduction by sintering briquettes with reducing agents before the electric furnace is also beginning to be used by some plants (Rath, et al., 1981; Nafziger, et al., 1981).

Production and export of ferrochromium by chrome ore-producing countries has continued strong growth. In 1980 more than half the chromium imported into the US came in the form of ferrochromium rather than chrome ore (Watson, 1980). This trend will continue because of the greater profit margins chromite-producing countries can realize on ferrochrome compared to ore. South Africa and Zimbabwe are major ferrochromium producers and, because of their enormous reserves, could become predominant factors in this market.

The United States is completely dependent on foreign sources at the end of long supply lines, so chrome ore and ferrochromium are important elements in the US strategic materials stockpiling program. The most important noncommunist world sources are South Africa, Turkey, the Philippines, and potentially Zimbabwe. These countries have current or potential political instability problems. Second and third largest US suppliers are the USSR and Albania, both of which have in the past banned shipments to the US. This potentially dangerous problem has prompted a number of studies of alternatives to chromium (Anon., 1978), but no satisfactory solutions are known. As a possible future domestic source in addition to the Stillwater complex, chromite-bearing (2 to 10%) ultramafic rocks, of which there are billions of tons in the US, are being investigated. Large-scale mining and concentration, similar to porphyry copper, would be required, and chromite concentrate costs would be many times higher than current prices (Lipin, 1981).

Refractory grade chrome ore consumption in the US and probably the rest of the world will continue its downward trend relative to other uses (Figs. 1 and 2) as open hearths continue to be replaced by BOFs. Electric arc furnaces have and will continue a moderate growth and these use chromite-bearing refractories. But increasing application of watercooled walls and of magnesia-carbon brick will restrict the growth of chromite usage in these furnaces if they do not cause an actual decrease. Refractories for the glass, copper, nickel, and cement industries still use chromite refractories. But in glass and cement the oxidation of trivalent to soluble hexavalent chromium poses a hazardous waste problem that could threaten chromite usage unless a solution can be found.

There appear to be no radical technical developments or marked trends in chrome ore for chemical usage as there have been for metallurgical and refractories usages. Production of chromium chemicals has been relatively stable in the US for the past decade. Because of wide diversity of use, they are not subject to as severe fluctuations as chromium usage in steel or refractories. Chrome chemicals may continue a slow growth that is less than the growth of the general economy of the US.

Acknowledgments

The author wishes to express thanks to Thomas P. Thayer, John M. Warde, John L. Morning, Bruce R. Lipin, and Edward C. Peterson for valuable assistance in preparing this chapter.

Bibliography and References

Anon., 1961, Symposium on Chrome Ore, Central Treaty Organization, Ankara, Turkey, Sept. 1960, *Min. Res. Expl. Inst. Turkey*, 272 pp.

Anon., 1962, *Chromium—Materials Survey*, US Dept. of Commerce, Business and Defense Service Administration, Iron and Steel Div., 96 pp.

Anon., 1969, "New Chromite Mine and Washing-Plant Readied for Production," *World Mining*, Vol. 5, No. 3, March, p. 17.

Anon., 1971, "Chromium," *Philippine Mining Yearbook*, Vol. 13, No. 7, pp. 26, 34.

Anon., 1971a, *International Chromium Report*, American Metal Market, Sec. 2, June 25, pp. 1A-8A.

Anon., 1971b, "Soviet Reveals Chromite Output," New York Times, April 17.

Anon., 1971c, "Trends in the Use of Ferroalloys by the Steel Industry of the United States," National Materials Advisory Board, Publ. NMAB 276, National Academy of Sciences—National Academy of Engineering, Washington, DC, July, 116 pp.

Anon., 1972, "AOD—Significant Advance in Stainless Steel Making," *33 Magazine*, June, pp. 40–45.

Anon., 1972a, *Chromium*, National Conference on Materials Policy, Interim Report, April, pp. 14–16.

Anon., 1972b, "Chromium Minerals, Ferrochrome, Chromium, and Chromium Chemicals: World Survey of Production and Consumption," Roskill Information Services, Ltd., London, 236 pp.

Anon., 1972c, *Chromium Section*, American Metal Market, Sec. 2, Oct. 18, pp. 1A–12A.

Anon., 1977, "Chromium," *Mineral Commodity Profiles*, US Bureau of Mines, 14 pp.

Anon., 1978, "Contingency Plans for Chromium Utilization," National Materials Advisory Board Publ. NMAB 335, National Academy of Sciences, Washington, DC, 347 pp.

Anon., 1980, "Chromium," *Mineral Commodity Summaries*, US Bureau of Mines, pp. 34–35.

Anon., 1980a, "Papua-New Guinea Chromite Prospects," *Industrial Minerals*, No. 152, May, p. 19.

Anon., 1980b, "Indonesia: Developments in Chrome Production," TIZ-Fachberichte Rohstoff-Engineering, Heft 3, 104 Jahrgang, p. 145.

Bala Sundaram, M.S., 1972, "A Review of Reserves and Demand Pattern of Iron, Chromium, Manganese Ores of India," *Eastern Metals Review*, Annual Number, Vol. 25, pp. 39–49.

Bosum, W., 1964, "Theoretical Limits of the Application of Geophysical Methods in Prospecting for Chromite," Chap. 8, *Methods of Prospection for Chromite*, R. Woodtli, ed., OECD, Paris, pp. 209–224.

Brantley, F.E., 1970, "Chromium," *Mineral Facts and Problems*, 1970 ed., US Bureau of Mines, Bulletin 650, pp. 247–262.

Broad, W.J., 1980, "Resource Wars: The Lure of South Africa," *Science*, Vol. 210, No. 4474, pp. 1099–1100.

Buros, J., 1976, "Sowjetunion-Chromit," Rohstoffwirtschaftlichte Landerberichte VIII, Bundesanstalt fur Geowissenschaften und Rohstoffe, Hannover.

Cameron, E.N., 1963, "Structure and Rock Sequences of the Critical Zone of the Eastern

Bushveld Complex," Special Paper No. 1, Mineralogical Society of America, pp. 93–107.

Cameron, E.N., 1975, "Postcumulus and Subsolidus Equilibration of Chromite and Coexisting Silicates in the Eastern Bushveld Complex," *Geochimica et Cosmochimica Acta,* Vol. 39, pp. 1021–1033.

Cameron, E.N., 1980, "Evolution of the Lower Critical Zone, Central Sector, Eastern Bushveld Complex and Its Chromite Deposits," *Economic Geology,* Vol. 75, No. 6, pp. 845–871.

Cameron, E.N., and Desborough, G.A., 1969, "Occurrence and Characteristics of Chromite Deposits—Eastern Bushveld Complex," in "Magmatic Ore Deposits," H.D.B. Wilson, ed., Monograph 4, *Economic Geology,* pp. 23–40.

Cameron, E.N., and Emerson, M.E., 1959, "Origin of Certain Chromite Deposits of the Eastern Part of the Bushveld Complex," *Economic Geology,* Vol. 54, No. 7, pp. 1151–1213.

Cameron, E.N., and Glover, E.D., 1973, "Unusual Titanian-Chromium Spinels from the Eastern Bushveld Complex," *American Mineralogist,* Vol. 58, No. 3–4, Mar.-Apr., pp. 172–188.

Cotterill, P., 1969, "The Chromite Deposits of Selukwe, Rhodesia," in "Magmatic Ore Deposits," H.B.D. Wilson, ed., Monograph 4, *Economic Geology,* pp. 154–186.

Davis, W.E., et al., 1957, "Gravity Prospecting for Chromite Deposits in Camaguey Province, Cuba," *Geophysics,* Vol. 22, No. 4, pp. 848–869.

Dela Cruz, S.D., 1982, "Masinloc Chromite Operation of Benguet Corp., Republic of the Philippines," *Proceedings,* Raw Materials for Refractories Conference, sponsored by US Bureau of Mines at the University of Alabama, Tuscaloosa, Feb. 8–9, pp. 264–285.

Dickson, T., 1980, "Chromite: Southern Africa Holds Sway," *Industrial Minerals,* No. 150, Mar., pp. 53–73.

Engin, T., and Aucott, J.W., 1971, "A Microprobe Study of Chromites from the Andizlik-Zimparalik Area, SW Turkey," *Mineralogical Magazine,* Vol. 38, Mar., pp. 76–82.

Engin, T., and Hirst, D.M., 1970, "The Alpine Chrome Ores of the Andizlik-Zimparalik Area, Fethiye, SW Turkey," *Trans./Sect. B,* Institution of Mining and Metallurgy, Vol. 79, pp. B15–B30.

Fernandez, N.S., 1960, "Notes on the Geology and Chromite Deposits of the Zambales Range," *The Philippine Geologist,* Vol. 14, No. 1, pp. 1–8.

Fourie, G.P., 1959, "The Chromite Deposits in the Rustenburg Area," Bulletin 27, Geological Survey of Union of South Africa, 45 pp.

Ghisler, M., and Windley, B.F., 1967, "The Chromite Deposits of the Fiskenaesset Region, West Greenland," Report No. 12, Geological Survey of Greenland, Copenhagen, Den., 39 pp.

Golding, H.G., and Johnson, K.R., 1971, "Variation in Gross Chemical Composition and Related Physical Properties of Podiform Chromite in the Coolac District, NSW Australia," *Economic Geolgoy,* Vol. 66, No. 7, pp. 1017–1027.

Greathead, H.G., 1963, "Chromium," Minerals Bureau of the Government Mining Engineer's Div., Republic of South Africa, 18 pp.

Haeri, Y., 1961, "Geology of Iran's Chromite Deposits," *Symposium on Chrome Ores,* Central Treaty Organization, Ankara, Sep. 1960, Min. Res. Expl. Inst. of Turkey, pp. 21–26.

Harris, D.L., 1964, "Chemical Upgrading of Stillwater Chromite," *Trans. SME/AIME,* Vol. 229, pp. 267–281.

Heiligman, H.A., and Mikami, H.M., 1960, "Chromite," *Industrial Minerals and Rocks,* 3rd ed., J.L. Gillson, ed., AIME, New York, pp. 243-258.

Helke, A., 1962, "The Metallogeny of the Chromite Deposits of the Guleman District, Turkey," *Economic Geology,* Vol. 57, No. 6, pp. 954–962.

Homer, D., and Wilshire, B., 1980, "Tonnage Demand for Steel Plant Refractories, 1975–2000," *Refractories Journal,* Vol. 2, Mar.-Apr., pp. 17–22.

Hosking, K.F.G., 1964, "The Application of Applied Geochemical Methods to the Search from Chrome Ore Deposits," *Methods of Prospection for Chromite,* R. Woodtli, ed., OECD, Paris, pp. 149–174.

Jackson, E.D., 1964, "Primary Features of Stratiform Chromite Deposits," Chap. 8, *Methods of Prospection for Chromite,* R. Woodtli, ed., OECD, Paris, pp. 111–134.

Jackson, E.D., and Thayer, T.P., 1972, "Some Criteria for Distinguishing between Stratiform, Concentric and Alpine Peridotite-Gabbro Complexes," 24th International Geological Congress, Sec. 2, pp. 289–296.

Jankovic, A.S., 1964, "Prospecting for Chromite Deposits in Yugoslavia," *Methods of Prospection for Chromite,* Chap. 12, R. Woodtli, ed., OECD, Paris, pp. 203–208.

Kahma, A., et al., 1962, "On the Prospecting and Geology of the Kemi Chromite Deposit, Finland," *Bull. Comm. Geol. Finlande,* No. 194, 91 pp.

Krivsky, W.A., 1973, "The Linde Argon-Oxygen Process for Stainless Steel; A Case Study of a Major Innovation in a Basic Industry," *Trans. TMS/AIME,* Vol. 4, June, pp. 1439–1447.

Kurochin, M.G., 1971, "Rational Scheme for Beneficiating Kimpersai Chromite Ores," *Ogneupory* (*Refractories,* Engl. trans., Consultant Bureau. NY), Vol. 14, Sept.-Oct., pp. 504–506.

Laming, J., 1970, "Sintered and Chemically Bonded MgO-Chrome Ore Refractories," *High Temperature Oxides,* Pt. I, A.M. Alper, ed., Academic Press, New York, pp. 143–189.

Lipin, B., 1981, Private communication.

Madronic, I., 1970, "Use of Chromium Ore Briquettes in the Manufacture of Ferrochrome Silicon," *AIME Electric Furnace Conference Proceedings,* Vol. 28, pp. 174–178.

Mikami, H.M., 1975, "Chromite," *Industrial Minerals and Rocks,* 4th ed., S.J. Lefond, ed., AIME, New York, pp. 501–517.

Mikami, H.M., 1982, "Refractory Chromites from Southern Africa and Other Non-Masinloc Sources," *Proceedings,* Raw Materials for Refractories Conference, sponsored by US Bureau of Mines at the University of Alabama, Tuscaloosa, Feb. 8–9, pp. 286–308.

Morning, J.L., 1972, "Chromium," *Minerals Yearbook 1972,* US Bureau of Mines, pp. 289–299.

Morning, J.L., 1977, "Chromium," Preprint, *Minerals Yearbook 1977,* US Bureau of Mines, 11 pp.

Nafziger, R.H., et al., 1981, "Prereduction and Melting of Domestic Chromites," *AIME Electric Furnace Conference Proceedings,* Vol. 38, Dec.

Parasnis, D.S., 1964, "Some Aspects of Geophysical Prospecting for Chromite," *Methods of Prospection for Chromite,* Chap. 14, R. Woodtli, ed., OECD, Paris, pp. 225–233.

Pearre, N.C., and Heyl, A.V., 1960, "Chromite and Other Mineral Deposits in Serpentine Rocks of the Piedmont Upland in Maryland, Pennsylvania, and Delaware," 1082-K, US Geological Survey, pp. 707–833.

Peatfield, M., and Spencer, D.R.F., 1979, "Chromite Usage in Refractory Brick Production," presented at the Extended General Meeting of the Institution of Mining and Metallurgy on Chromium, Dec., 21 pp.

Peterson, E.C., 1981, "Chromium," *Minerals Yearbook 1980,* US Bureau of Mines, pp. 189–200.

Rath, and Rottman, 1981, "Use of Briquetted Feed for the Production of Ferrochrome," *AIME Electric Furnace Conference Proceedings,* Vol. 38, Dec.

Robiette, A.G.E., 1972, "Steelmaking Trends and the Role of Ferrochrome in Stainless Steel,—A World Survey," *Metal Bulletin,* London, pp. 63–71.

Romagnoli, E., 1972, "Chrome Ore," *Engineering & Mining Journal,* Vol. 173, No. 3, Mar., pp. 116, 135–137.

Roskill, O.W., 1972, "Chromium Minerals, Ferrochrome, Chromium, and Chromium Chemicals," *World Survey of Production and Consumption,* Roskill Information Services, Ltd., London, 236 pp.

Rossman, D.L., et al., 1959, "Chromite Deposits on Insular Chromite Reservation No. One, Zambales, Philippines," *Chromite,* Publication No. 19, Philippine Bureau of Mines, 12 pp.

Saccomano, J.M., et al., 1969, "Making Stainless Steel in the Argon-Oxygen Reactor at Joslyn," *Journal of Metals,* Vol. 21, No. 2, pp. 59–64.

Severns, W.H., Jr., 1969, "Process for Beneficiating Chrome Ores," US Patent 3,473,916, Oct. 21, pp. 1–10.

Stanley, R., 1961, "Chromium in Southern Rhodesia," Mines Dept. of Southern Rhodesia, 21 pp.

Stevens, R.E., 1944, "Compositions of Some Chromites of the Western Hemisphere," *American Mineralogist,* Vol. 29, No. 1–2, pp. 1–34.

Stoll, W.C., 1958, "Geology and Petrology of the Masinloc Chromite Deposit, Zambales, Luzon, Philippine Islands," *Bulletin of the Geological Society of America,* Vol. 69, No. 4, pp. 419–448.

Sully, A.H., and Brandes, E.A., 1967, "Chromium," *Metallurgy of the Rarer Metals,* 2nd ed., No. 1, 373 pp.

Thayer, T.P., 1956, "Mineralogy and Geology of Chromium," *Chromium,* Chap. 2, M.J. Udy, ed., American Chemical Society Monograph 132, Reinhold, New York, Vol. 1, pp. 14–52.

Thayer, T.P., 1962, *World Resources, Chromium Minerals,* Chaps. 2, 3, Business and Defense Services Administration, US Dept. of Commerce, Washington, DC, pp. 3–47.

Thayer, T.P., 1964, "Principal Features and Origin of Podiform Chromite Deposits and Some Observations on the Guleman-Soridag District, Turkey," *Economic Geology,* Vol. 59, No. 8, pp. 1497–1524.

Thayer, T.P., 1969, "Gravity Differentiation and Magmatic Re-Emplacement of Podiform Chromite Deposits, Magmatic Ore Deposits," H.D.B. Wilson, ed., Monograph 4, *Economic Geology,* pp. 132–146.

Thayer, T.P., and Lipin, B.R., 1978, "A Geological Analysis of World Chromite Production to the Year 2000 A.D.," *Proceedings,* Council of Economics of AIME.

Udy, M.J., 1956, ed., "Chromium," *Chemistry,* Vol. 1, 433 pp., and *Metallurgy,* Vol. 2, 402 pp., American Chemical Society Monograph 132, Reinhold, New York.

Ulmer, G.C., 1970, "Chromite Spinels," *High Temperature Oxides,* Chap. 6, A.M. Alper, ed., Academic Press, New York, pp. 252–314.

Von Gruenwaldt, G., 1977, "The Mineral Resources of the Bushveld Complex," *Minerals Science and Engineering,* Vol. 9, No. 2, pp. 83–95.

Warde, J.M., 1981, Private communication.

Watson, G.A., 1980, "Chromium, Today and Tomorrow," *Iron and Steel Maker,* May, pp. 19–22.

Wechsler, P., et al., 1980, "Chromium," *American Metal Market,* Dec. 19, pp. 11–25.

Wells, F.G., et al., 1946, "Chromite Deposits of Del Norte County, California," California Div. of Mines, Pt. 1, Ch. 1, pp. 1–76.

Wilson, H.D.B., ed., 1969, "Magmatic Ore Deposits," Monograph 4, *Economic Geology,* 366 pp.

Woodtli, R., ed., 1963, *Methods of Prospection for Chromite,* OECD, Paris, 242 pp.

Worst, B.G., 1960, "The Great Dyke of Southern Rhodesia," Bulletin 47, Southern Rhodesian Geological Survey, 234 pp.

Yoshino, S., et al., 1981, "Materials from China," *Taikabutsu,* Vol. 33, No. 278, pp. 25–28.

Yüngül, S., 1956, "Prospecting for Chromite with Gravimeter and Magnetometer over Rugged Topography in East Turkey," *Geophysics,* Vol. 21, No. 2, pp. 433–454.

Zednicek, W., 1969, "Mineralogische Untersuchungen von Chromerzen Verschiedener Lagerstätten der Erde," *Radex Rundschau,* Heft 4, pp. 650–680.

See Also

American Iron and Steel Institute, *Annual Statistical Report,* issued annually.

Engineering and Mining Journal, Markets, Chrome Ore Prices, New York, issued monthly.

US Bureau of Mines, *Mineral Trade Notes,* issued monthly.

US Bureau of Mines, *Minerals Yearbook,* issued annually.

American Metal Market, New York, issued daily.

Clays*

Original by SAM H. PATTERSON †
and
HAYDN H. MURRAY ‡
Revised by HAYDN H. MURRAY

The term *clay* is somewhat ambiguous unless specifically defined, because it is used in three ways: (1) as a diverse group of fine-grained minerals, (2) as a rock term, and (3) as a particle-size term. Actually, most persons using the term clay realize that it has several meanings, and in most instances they define it. As a rock term, clay is difficult to define because of the wide variety of materials that compose it; therefore, the definition must be general. Clay is a natural earthy, fine-grained material composed largely of a group of crystalline minerals known as the clay minerals. These minerals are hydrous silicates composed mainly of silica, alumina, and water. Several of these minerals also contain appreciable quantities of iron, alkalies, and alkaline earths. Many definitions state that a clay is plastic when wet. Most clay materials do have this property, but some clays are not plastic; for example, halloysite and flint clay.

As a particle-size term, clay is used for the category that includes the smallest particles. The maximum-size particles in the clay-size grade are defined differently on various grade scales. Soil investigators and mineralogists generally use 2 micrometers as the maximum size, whereas the widely used scale by Wentworth (1922) defines clay as material finer than approximately 4 micrometers.

Some authorities find it convenient to use the term clay for any fine-grained, natural, earthy, argillaceous material (Grim, 1968). When used this way, the term includes clay, shale, or argillite, and some argillaceous soils.

* Publication authorized by the Director, US Geological Survey. For other Acknowledgments, see pp. 642–643.

† US Geological Survey, Reston, VA.

‡ Georgia Kaolin Co., Elizabeth, NJ; presently, Dept. of Geology, Indiana University, Bloomington, IN.

Even though no standard definition of the term clay is accepted by geologists, agronomists, engineers, and others, the term is generally understood by those who use it. Clay is an abundant natural raw material, and it has an amazing variety of uses and properties which will be discussed in this chapter.

Clays rank among the leading industrial minerals in both tonnages produced and total value. Their importance is indicated in Table 1, which lists the 1978, 1979, and 1980 production and values according to the US Bureau of Mines (Ampian, 1980).

Statistical data summarizing world production of kaolin, bentonite, and fuller's earth are compiled by the US Bureau of Mines (Ampian and Polk, 1981). The total world bentonite production in 1980 was 6,625,204 st of which the United States produced 4,184,619 st. The total fuller's earth produced in the world in 1980 was 1,952,600 st of which 1,534,000 st was produced in the United States. The total world kaolin production in 1980 was estimated at 22,971,000 st. Approximately 13 million tons of this total was produced in two countries, the United States and the United Kingdom. World production data on ball clay, fire clay, and miscellaneous clays are not available.

As industrial minerals, clays are a complex group that consists of several mineral commodities, each having somewhat different mineralogy, geologic occurrence, technology, and uses. In this chapter these commodities are organized as follows: (1) bentonite and fuller's earth; (2) kaolin, ball clay, halloysite, and refractory clays; and (3) miscellaneous clay and shale. Bentonite and fuller's earth are grouped together because they are closely interrelated. Though bentonite is a term based on mineral composition and fuller's earth is a term based on use, the two are virtually inseparable, because much, but by no means all, clay sold as

TABLE 1—Clay Sold or Used in the United States, 1978, 1979, and 1980

	1978		1979		1980	
Clay*	Tonnage	Value $	Tonnage	Value $	Tonnage	Value $
Bentonite	4,467,605	88,425,737	4,422,075	106,528,550	4,184,619	115,234,996
Fuller's earth	1,529,619	73,430,515	1,568,247	81,761,964	1,533,803	79,724,017
Kaolin	6,973,314	367,782,408	7,760,600	462,320,088	7,878,993	527,098,609
Ball clay	936,251	21,549,559	987,012	26,119,965	893,624	26,844,126
Fire clay†	3,125,953	42,561,432	2,932,343	47,178,993	2,095,361	36,022,134
Miscellaneous clay	40,074,738	124,068,443	37,278,803	122,735,117	32,494,837	114,700,246
Total‡	57,107,480	717,818,034	54,949,080	846,644,677	49,081,237	899,624,128

Source: Ampian, 1980; Ampían and Polk, 1981.
* Definitions of classes of clay listed are given in the appropriate places in the text.
† Fire clay is used with essentially the same meaning as refractory clay in this chapter. Fire clay is also the only type of clay having a sharply declining production in recent years. This is due mainly to the method of reporting, in which some types of fire clay used for other than refractory products are now classified as miscellaneous clay, and to the increasing use of higher quality and more durable refractory materials.
‡ Includes Puerto Rico.

fuller's earth is actually bentonite. The overlapping of the two terms is particularly evident where both bentonite and nonbentonite fuller's earth are used for the same purposes or products, such as in drilling muds, bleaching or clarifying fats and oils, and carriers for insecticides and fertilizers. Kaolin, ball clay, halloysite, and refractory clays are grouped together because they consist mainly of minerals of the kaolin group. Miscellaneous clay and shale is a grouping of several fine-grained materials of the type referred to as common clay in some reports.

The scientific and technological publications pertaining to clays are so numerous that to cover all the material in the format used for other chapters would require a book. Therefore, the authors have attempted to summarize the subject matter and wherever possible have indicated by bibliographic reference where more detailed information is available. For example, the reference to Bicker (1970) is to a bulletin containing only brief summary information on the economic geology of bentonite in Mississippi, but that bulletin contains references to virtually all published reports on bentonite in that state.

The remaining parts of this introduction will be mainly an outline of some of the major advancements in the broad field of clays since the 4th edition of *Industrial Minerals and Rocks* (Patterson and Murray, 1975) and acknowledgment to those who have helped in preparing this chapter. Our intent is to note where some of the contributions, mainly in clay mineralogy, have been made, so that those who are interested can find detailed information not included in this report.

A vast amount of new information on the mineralogy, geology, and technology of clays has become available in recent years. One of the principal sources is The Clay Minerals Society. This professional and scientific society was organized in 1963 from an informal group sponsored and partially supported for the preceding years by the National Academy of Sciences-National Research Council. For five years after that date, the papers presented at the society's annual meetings were published in volumes by Pergamon Press. Since 1968, The Clay Minerals Society has published its own journal under the title *Clays and Clay Minerals.* A second major source of information on clays is the publications of the Association Internationale pour l'Étude des Argiles (AIPEA); these publications are principally volumes containing the papers presented at the triennial meetings of the association. A third excellent source of information on clays is *Clay Minerals,* the journal of The Clay Minerals Group of The Mineralogical Society (London). In addition, clay minerals groups or societies in France, Spain, Italy, Japan, Argentina, West Germany, and other countries have periodic publications. Considerable information on kaolin resources in several countries is available in three volumes resulting from a symposium held at the International Geological Congress in Czechoslovakia in 1968 (Malkovský and Vachtl, 1969a, 1969b, 1969c) and in the report of another symposium on kaolin held at the 1972 International Clay Conference (AIPEA) in Spain in 1972 (Serratosa, 1972).

Several major textbooks and reference volumes on clays have been published in the last few years. They include one concerned mainly with mineralogy and another on applied mineralogy of clays by Grim (1962, 1968). The Clay Minerals Group of the Mineralogical Society (London) has a classic series on clay minerals, including a volume concerned with differential thermal analysis, edited by Mackenzie (1957), one on the X-ray identification and crystal structures of clay minerals edited by Brown (1961), and a third on the electron-optical investigation of clays edited by Gard (1971). A volume on the electron microscopy of clay minerals has also been prepared by Beutelspacher and Van Der Marel (1968). A Special Paper of The Geological Society of America by Carroll (1970) outlines guides to the X-ray identification of clays. An excellent volume on colloid chemistry of clays has been authored by Van Olphen (1963). In 1967, Zvyagin (1967) published a book concerned with the electron diffraction analysis of clays. Gillott's (1968) volume outlines the role of clays in engineering geology. Recent volumes concerning clays used in ceramic materials include books by Grimshaw (1972) and Clews (1969). An excellent volume on the geology and origin of clays is one by Millot (1964), which has been translated from French into English. A recent Special Paper of The Geological Society of America by Weaver and others (1972) contains considerable information on the geochemistry of clays related to diagenesis and metamorphism. Another recent volume by Weaver and Pollard (1973) contains information on the chemistry of clays, including oxide data and structural formulae.

A great deal of information on the geology, technology, and uses of clay has also been published by various provincial, state, and federal geological surveys, bureaus of mines, and other agencies. Most of these organizations and some of the major reports are listed by Hoy in Part 3 of Volume 1 and will not be repeated here; however, reference to some of the more important articles will be made in discussions in the following sections of this chapter.

Many changes have taken place in the various clay industries in recent years. Several of these have come about because of technological advancements, changing economic conditions, new uses, shifts in demand, and increases in both domestic and export markets. The changing conditions, particularly increasing freight rates, the energy crisis, other production and marketing costs, and the pressures brought about by governmental controls and mining laws resulting from requirements and restrictions related to environmental considerations, etc., have motivated the formation of organizations to look after the interests of some of the clay-producing industries. One of these organizations formed recently is the Absorptive Minerals Institute, having headquarters at No. 1 Illinois Center, 111 E. Wacker Drive, Chicago, IL 60601. This institute was formed mainly to represent the producers of absorbent granules and other fuller's earth products. A similar organization, the Bentonite Producers Association, was organized in 1971 in Casper, WY, and represents the interests of the producers of bentonite. The kaolin industry organized the US Clay Producers Traffic Association which was incorporated in 1953 in New Jersey. In addition to these organizations dealing specifically with clays, the clay industry and other mining industries formed mining associations in many states, including Georgia and Texas in 1972, in order to protect their interests against restrictive state and federal legislation.

Bentonite and Fuller's Earth

Definitions and Classifications

The term *bentonite* was first proposed in 1898 by Knight (1898), a year after he had named this clay taylorite; taylorite was found to be preoccupied. The name *taylorite* was after the Taylor ranch, the site of the first mine, near Rock River, WY, and the name *bentonite* is from the Benton Shale in which the clay was thought at that time to occur. The Benton Shale was, in turn, named after Fort Benton, MT, located more than 400 miles north of Rock River.

Early in the 20th century, several geologists recognized that bentonite, mainly in beds in Cretaceous and Tertiary rocks, originated from transported volcanic materials. This recognition led to definitions based on origin, the one most widely quoted and generally accepted by geologists is the following definition by Ross and Shannon (1926):

> Bentonite is a *rock* composed essentially of a crystalline clay-like mineral formed by devitrification and the accompanying chemical alteration of a glassy igneous material,

> usually a tuff or volcanic ash; and it often contains variable proportions of accessory crystal grains that were originally phenocrysts in the volcanic glass. These are feldspar (commonly orthoclase and oligoclase), biotite, quartz, pyroxenes, zircon and various other minerals typical of volcanic rocks. The characteristic clay-like mineral has a micaceous habit and facile cleavage, high birefringence and a texture inherited from volcanic tuff or ash, and it is usually the mineral montmorillonite, but less often beidelite.

The difficulty in applying the foregoing definition to bentonite as an industrial mineral commodity is that it is based on origin and is restrictive to an ash, tuff, or volcanic glass parent material. Therefore, bedded deposits consisting of the clay minerals required by this definition but having uncertain origin or parent materials cannot properly be called bentonite. Furthermore, many deposits in the western United States and in other countries that have formed from rocks other than the types required by the definition are being mined and sold as bentonite.

Perhaps the best definition of bentonite as an industrial mineral is one given by R. E. Grim in a plenary lecture at the International Clay Conference (AIPEA) at Madrid, Spain, June 27, 1972. According to this redefinition, which will be used in this chapter, bentonite is a clay consisting essentially of smectite minerals (montmorillonite group of some usages), regardless of origin or occurrence. This definition solves the problem of the difference between the geologic and industrial usages of the term and overcomes the difficulty in assigning a name to smectite clay that formed from igneous rock other than ash, tuff, or glass, or those of sedimentary or uncertain origin. However, bentonite, when used with this meaning is still a rock term (consisting of more than one mineral), and it will not be possible to distinguish it from fuller's earth in many instances.

One way of classifying bentonite is based on its swelling capacities when wet or added to water. Bentonite having sodium (Na^+) as either the dominant or as an abundant exchangeable ion typically has very high swelling capacities and forms gel-like masses when added to water. Bentonite in which exchangeable calcium (Ca^{++}) is more abundant than other ions has much lower swelling capacities than sodium varieties. Some calcium types swell little more than common clay, and most crumble into granular masses in water. Intermediate calcium-sodium bentonites, the so-called mixed types, tend to swell moderately and to form gel of lesser volumes than equal masses of the sodium type. Bentonite, because of the general relationship of swelling and exchangeable ion characteristics, is commonly divided into the high-swelling or sodium, low-swelling or calcium, and moderate-swelling or intermediate types. The term *subbentonite* (Davis, Vacher, and Conley, 1940) is used inconsistently in industry for the low or moderate-swelling varieties. The authors believe the use of this term should be discouraged because of its implication of a low quality or low value and the lack of a mineralogical or usc basis for it.

In the United States bentonite is also classified by geographic location and the uses for which it is sold. Inasmuch as most of the low-swelling calcium type occurs in states bordering the Gulf of Mexico, this variety is commonly called *Southern bentonite*. The largest high-swelling sodium bentonite deposits and the major producing districts are in Wyoming and adjacent states. Therefore, this bentonite is commonly called *Wyoming* or *Western* type. Such terms as drilling mud bentonite, foundry bond bentonite, and taconite bond bentonite relating directly to use are applied in marketing. Other terms including high and low-yield bentonite, high and low-gel bentonite, and high and low-strength bentonite are also used to distinguish different grades.

The varied classifications notwithstanding, bentonite occurs in so many different varieties that some cannot logically be classified according to any of the foregoing groupings. One of these types is hectorite, which is a high-swelling lithium-bearing variety of smectite occurring mainly in California and adjacent states. It is, therefore, not a Wyoming type, and it occurs even farther west than the so-called western bentonites. Others are bentonites having magnesium (Mg^{++}) or hydrogen (H^-) as the most abundant or dominant exchangeable ion. These types are neither sodium nor calcium varieties, and apparently some are low swelling, whereas others have rather high-swelling capacities.

Still another type of bentonite, the so-called potassium type, K-bentonite, or metabentonite, occurs in Ordovician and other Paleozoic rocks at many places in Appalachian and Mississippi Valley regions and elsewhere. This bentonite, which is generally thought to have formed from volcanic ash, consists mainly of illite and mixed-layer minerals. Smectite minerals are ordinarily present in only minor quantities. Because of its mineral composition, this bentonite contains appreciably more potassium than most other types.

The meta prefix originates from the idea that this bentonite was altered by low-grade metamorphism or diagenesis. Apparently, the only attempt to use K-bentonite in the United States was when five or six carloads were mined near Dayton, Rhea County, TN, about 1928; this bentonite was used for purifying lard (Gildersleeve, 1946), and no further mention of this type will be made.

The term *fuller's earth* is more or less a catchall for clay or other fine-grained earthy material suitable for bleaching and absorbent and certain other uses. It has no compositional or mineralogical meaning. The origin of the term dates back into antiquity; it was first applied to material used in cleansing and fulling wool, thereby removing the lanolin and dirt from it. When in the latter half of the last century it was found that some earths used for fulling would also serve in decolorizing and purifying mineral, vegetable, and animal oils, the term *fuller's earth* was modified to include this usage.

Extensive use of fuller's earth in processing mineral oils in the first half of this century and the virtual end of its use in fulling fiber led to the general application of the term, fuller's earth, as meaning primarily earth used in petroleum processing. Further modification of the meaning came about as other uses developed for the earth and replaced oil processing as the major use. The term fuller's earth is retained, however, for fine-grained materials used for many purposes. Most of these uses require absorbent properties in one form or another, but the properties required for some uses, such as certain drilling muds and fillers, are other than absorbency and, therefore, result in further modification of the term.

The classification and understanding of fuller's earth are also complicated by several terms having more or less duplicate or overlapping meanings. When applied in the oil-processing sense, fuller's earth has the same meaning as *naturally active clay*. Fuller's earth and other clays which are treated with acid or otherwise altered to improve their desirable properties are called *activated clay* (Torok and Thompson, 1972). The terms *bleaching clay* and *bleaching earth* are applied mainly to both naturally active and activated clay, but they also include activated bauxite (Rich, 1960). Both naturally active and activated clays are included under the term *absorbent clays* (Nutting, 1943); therefore, this term has nearly the same meaning as bleaching clay. The term *absorbent clay* is applied to fuller's earth used for a wide variety of absorbing purposes which are different from that of processing oils.

History and Use

Bentonite: Bentonite mining began on the Taylor ranch near Rock River, WY, in 1888. This bentonite reportedly was shipped crude to Philadelphia where it sold for $25 a ton and was used in making cosmetics. A mine was opened in the Upton, WY, district on the western side of the Black Hills in 1903. The early method of drying was to rake the bentonite by hand over steel plates heated by wood or coal fires.

The date of the first mining of southern bentonite is somewhat conjectural because of the confusion between the terms bentonite and fuller's earth. One fuller's earth deposit near Gonzales, Texas, that is almost certainly bentonite was mined as early as 1906, and another deposit in that same state mined during the Civil War (Lang et al., 1940) may also have been bentonite. Large-scale mining of calcium-type bentonite began in Mississippi in the early 1930s (Bicker, 1970), and bentonite deposits were used as fuller's earth in several states about this time.

The value of bentonite as foundry-sand bond was recognized in the 1920s, and the iron and steel foundries have since been major consumers of bentonite. Bentonite was first used as drilling mud in the late 1920s or early 1930s, and bentonite is still one of the most efficient materials for drilling muds where the rocks penetrated contain only fresh water. The high-swelling Wyoming bentonite is the most efficient type for drilling mud, but the high-viscosity properties of hectorite make it useful in muds (Larsen, 1955). Some Texas bentonite is also used for drilling muds, particularly in shallow wells, and commonly it is treated with soda ash, polymers, or other chemicals to make it suitable for this use. Low-quality bentonite when sold for use in drilling mud is commonly classified merely as clay rather than bentonite (Anon., 1972d).

Bentonite production remained small during the first few decades of mining, and the first yearly total sufficiently large to be separated from miscellaneous clays in the US Bureau of Mines statistics was in 1930 when production was only a few thousand tons. Demands for bentonite increased during World War II, but a yearly total of 1 million tons was not reached until 1950. More than 2 million tons was produced in 1966, and the peak production year to date was 1978, when 4,467,605 tons were sold or used by producers. Bleaching clay (oil

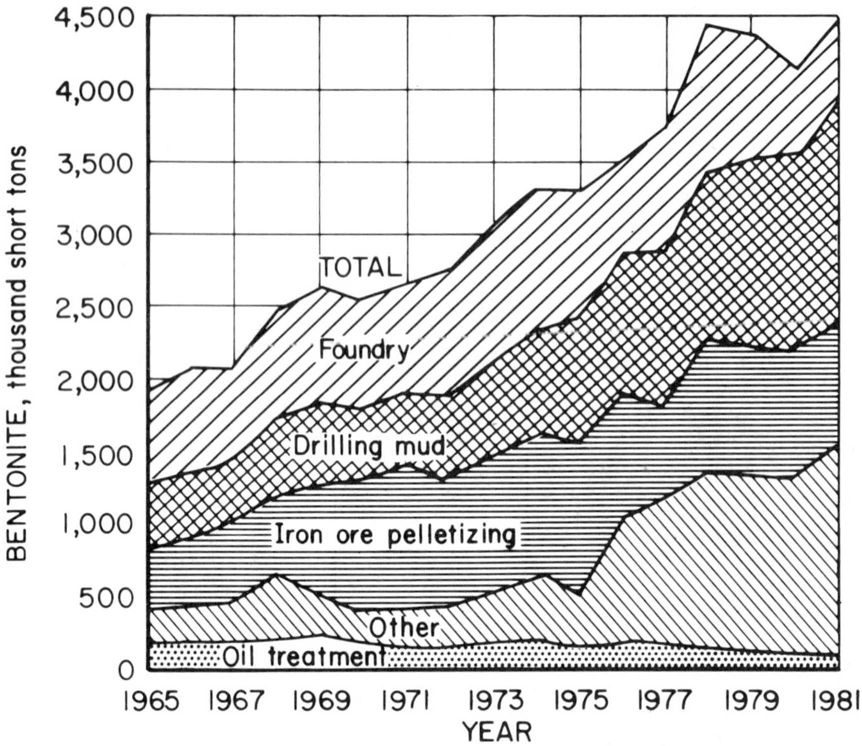

FIG. 1—*Bentonite sold or used by domestic producers for specified uses (Ampian, 1980, 1983).*

refining, filtering, clarifying, and decolorizing), drilling mud, iron-ore pelletizing, and foundry-sand bonding have been the major uses of bentonite since World War II (Fig. 1). Beginning about 1950, steel companies used Wyoming-type bentonite as a bond for pelletized taconite ore. This use grew rapidly, and in recent years approximately one-third of the national production has been used for this purpose.

In addition to the major uses, bentonite is used in many miscellaneous products, and hundreds of patents for speciality uses have been issued or applied for. The speciality uses include filtering agents (one of which is a high-value product for clarifying wine, and another is a less costly one for treating waste water), water impedance (preventing seepage loss from reservoirs, irrigation ditches, and waste-disposal ponds, and seepage through basement walls, tunnel walls, and other structures), ingredients in cosmetics, animal feed, pharmaceuticals, colloidal fillers for certain types of paints, an additive to ceramic raw materials to increase plasticity, fire-retarding materials, and for many other purposes (Papin, 1964). The white bentonites occurring in Texas and Nevada and imported from Italy are particularly suitable for many of the speciality products, and a great deal of research evaluating them for many uses has been done. One of the unique uses of high-swelling bentonite that is likely to increase is in the "slurry trench" or "diaphragm wall" method of excavation in construction in areas of unconsolidated rock or soil (Blackman, 1969; Lang, 1971). In this method, the trench or hole being excavated is filled with bentonite slurry and the earth being excavated is removed through it. A thin filter cake on the walls of the excavation prevents loss of fluid, and the hydrostatic head of the slurry prevents caving and running of loose soils, which makes costly shoring unnecessary. Considerable quantities of bentonite were once used in making catalysts for petroleum refining, but the market for these types of catalysts in the United States declined after World War II. However, sizable quantities of catalysts have been produced from bentonite and similar clays in Japan, the United Kingdom, and elsewhere in recent years. Acid-activated bentonite is used for bleaching oils and in making multiple-copy paper requiring no carbon paper.

Fuller's Earth: Published reports give rather confusing accounts of the discovery of fuller's earth in the United States. Several state that fuller's earth was discovered in this country in 1893 near Quincy, FL, by the Owl Commercial Co. (predecessor of the present makers of White Owl cigars) during an attempt to burn brick from clays on tobacco property. The clay was not suitable for making brick, but an Alsatian immigrant employed as a farm worker recognized that it was similar to fuller's earth mined in Germany. His observation led to development of the first mine near Quincy two years later and to the use of this clay in processing mineral oils. Fuller's earth had been mined on a small scale, however, in Arkansas in 1891 (Miser, 1913), and tested for use in the refining of cottonseed oil.

Both of these reported discoveries are related to the use of fuller's earth in processing oils and fail to note its earlier use for other purposes. Clays and other earthy materials were undoubtedly used by the early settlers from Europe in cleansing wool and other materials. Little effort was made to search the historical records and to document this, but it was found that soldiers stationed near Perth Amboy, NJ, used Woodbridge fire clay for cleansing buckskins during the Revolutionary War (Cook and Smock, 1878). Fuller's earth associated with an iron-ore bed near Kent, CT, had been mined in the early 1800s (Silliman, 1820), and fuller's earth near Falls City, Texas, had been used to bleach sugar during the Civil War (Lang et al., 1940). Also, American Indians were reported to have used bentonite in cleaning blankets, and they probably dug it for other cleansing purposes before the Columbian period.

The use of fuller's earth in the refining of oils continued to be the major one for many years. The demand for fuller's earth for processing mineral oil increased rapidly in the first part of this century and reached a peak of approximately 317,000 tons in 1930; this was 91.1% of the total United States production that year. The production of fuller's earth for this purpose began to decrease thereafter, especially when activated bauxite was introduced in 1937 and magnesium silicate in 1940 as more efficient substitutes. The market for this use was also depressed by improvements in refining methods, which produce oils requiring little purification or bleaching. In recent years, the production of fuller's earth for use in refining mineral oils has maintained a rather uniform rate of 35,000 to 40,000 tpy. The use of fuller's earth in processing animal and vegetable oils has never been a major one and has decreased to the point where it is included as a part of the miscellaneous uses in the US Bureau of Mines *Minerals Yearbooks*.

Palygorskite (attapulgite) fuller's earth was first sold for drilling mud in 1941. The market for this use expanded slowly and has maintained a level of 7 to 10% of the total United States production during the last few years. Most of the fuller's earth sold for drilling mud comes from the southern part of the Meigs-Attapulgus-Quincy district of Georgia and Florida. Palygorskite clays produced in this area are superior to most other fuller's earth for muds used in drilling salt formations, but because of high water loss, they are inferior to bentonite where the rocks drilled contain no saltwater.

Fuller's earth was used in significant quantities as a carrier for insecticides and fungicides by 1950, and the market for this use has grown at a rather uniform rate since that year. In 1979, nearly 14% of the fuller's earth produced was used for this purpose.

The use of fuller's earth granules for absorbent purposes began during the 1930s, but this use did not expand significantly until the World War II period when fuller's earth was used as an absorbent for greases, oil, water, chemicals, and other undesirable substances on the floors of factories, filling stations, canning plants, aircraft hangers, decks and engine rooms of ships, and other installations. The absorbent granules are porous and ordinarily weigh less than 30 lb per cu ft. Because of their light weight, size, porosity, and absorbent properties, the granules are suitable for many uses, and since World War II many different markets for them have developed (Fig. 2). Among other uses, they are now sold for litter and bedding for poultry, pets, and other animals, and as a soil conditioner in greenhouses and for golf courses. In 1982, 895,700 tons of fuller's earth were sold for absorbent and filtering uses; this tonnage was nearly 53% of the total fuller's earth produced in the United States that year.

The foregoing discussions apply only to those uses consuming sufficient quantities of fuller's

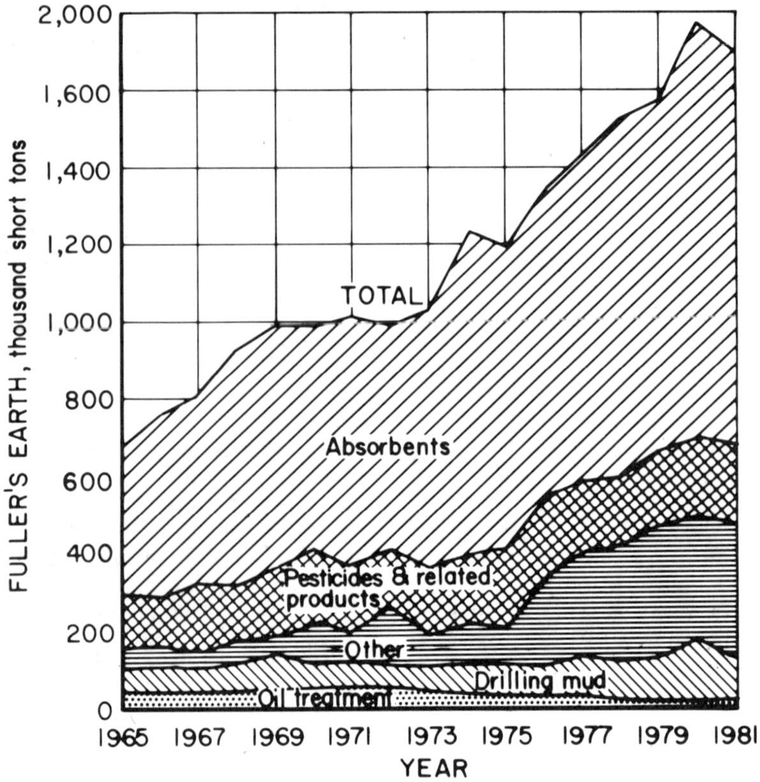

FIG. 2—*Fuller's earth sold or used by domestic producers for specified uses (Ampian, 1980, 1983).*

earth to be classified separately by the US Bureau of Mines in the *Minerals Yearbooks*. In addition, smaller quantities of fuller's earth are or have been produced for many miscellaneous uses. According to Oulton (1965), more than 90 different grades of fuller's earth are produced. Some of these grades are used for pharmaceuticals designed to absorb toxins, bacteria, and alkaloids; for treatment of dysentery; for purifying water and dry-cleaning fluids, dry-cleaning powders and granules; for the manufacture of NCR (no carbon required) multiple-copy paper; for the manufacture of wallpaper; and as extenders or fillers for plastic, paint, and putty. Fuller's earth mined near Ellenton, FL, was used for making lightweight aggregate for the construction of concrete barges during World War II (Calver, 1957; Greaves-Walker, et al., 1951). Still other uses of fuller's earth and its suitability for uses in new products are outlined by Haden and Schwint (1967), Haden (1972), and Haas (1970). One special use of fuller's earth is as a carrier of platinum catalysts that are made in the United Kingdom from sepiolite clays mined in Spain. Other uses of sepiolite fuller's earth (Chambers, 1959) are similar to those of the palygorskite (attapulgite) type mined in the United States.

Some reports also note that fuller's earth is used in making cement. Probably the basis for these reports is the mining of the Twiggs Clay Member of the Barnwell Formation near Clinchfield, GA, and the use of the clay in making portland cement. It is used for this purpose mainly to obtain the desired chemical composition in the clinker from which the cement is made. The classification of this clay as fuller's earth results from the fact that the Twiggs Clay Member is mined for fuller's earth, and this unit is widely known in Georgia as a source of fuller's earth.

Product Specifications

Product specifications have been tentatively standardized for bentonite and (or) fuller's earth sold for drilling mud, foundry-sand bond, absorbent granules, and oil bleaching and will be outlined briefly. Specifications for most of the many other products made from these clays are prepared to meet the requirements of the individual customer, and they commonly vary with each.

Drilling Muds: The most critical specifications for bentonite and palygorskite (attapulgite) fuller's earth for drilling muds are the suspension properties, wet-screen analysis, and moisture as shipped.

Suspension Properties—The test for suspension properties involves the preparation of a suspension consisting of 22.5 g of bentonite in 350 cc of distilled water. The suspension is aged and the viscosity determined and yield point calculated from dial readings at 300 and 600 rpm with a direct-indicating viscometer, according to procedure and equipment requirements outlined in American Petroleum Institute specification API STD 13A, 5th edition, 1969. Another important suspension requirement is the filtrate test, which is a measure of the volume of water lost from the prepared suspension when tested in a pressurized filter press according to procedure outlined in American Petroleum Institute specification API RP 13. According to these specifications, to be drilling-mud quality, a bentonite must have a dial reading at 600 rpm of 30 minimum, a yield point pounds per 100 sq ft of 3 times plastic viscosity maximum, and a filtrate volume of 13.5 maximum. Palygorskite (attapulgite) prepared for drilling mud must fulfill the same viscosity specification as bentonite, but the suspension is made with water containing 40 g of salt (NaCl) per 100 cc of water. Yield point and filtrate requirements are not ordinarily specified for palygorskite-type drilling muds. Many companies still use a yield specification for bentonite. Yield is a term used in earlier American Petroleum Institute specifications for the number of barrels of 15 centepoise viscosity mud that can be made from a ton of bentonite. The yield requirement is ordinarily 90 bbl per ton minimum.

Wet-Screen Analysis—Wet-screen analysis (grit test) is a measure of the material in bentonite or palygorskite (attapulgite) mud that is coarser than 200-mesh US Series sieve. The specified test is made by mixing 10 g bentonite in 350 cc water containing 0.2 g dispersing agent, stirring, aging, stirring, and washing through a sieve with a specified spray system (API STD 13A). The residue on the sieve is then dried, weighed, and the percentage of the original bentonite is determined. Bentonite to fulfill the specifications for drilling mud must have no more than 2.5% residue (grit) and palygorskite no more than 8% residue.

Moisture—The maximum moisture content of bentonite when shipped from the plant where it is processed is 10% and that for palygorskite is 16%.

Foundry Sand Bond: Tentative specifications for Western bentonite for bonding foundry sand are outlined in Steel Founders Society of America SFSA Designation 13T-65 issued in 1965. This specification requires the following characteristics and properties for Western bentonite, as tested by methods described in the specification: (1) water content shall not exceed 12% or be less than 6%; (2) pH value shall be equal to or greater than 8.2, (3) calcium oxide content shall not exceed 0.70%, and (4) the liquid limit shall not be less than 600 or greater than 850. Each foundry has its own green, dry, and hot-strength specifications for bentonite, which vary with the type of metal, size of castings, and foundry production procedures. Methods of testing strengths of bentonite-bonded sands are outlined in detail in publications of the American Foundrymen's Society (Anon., 1962a, 1963). The tests require the preparation of a specified mixture of bentonite, standard testing sand, and water; the preparation of test cylinders; and testing strengths with approved apparatus, according to controlled procedures.

Iron-Ore Pelletizing: Specifications for bentonite used for pelletizing taconite-type iron-ore have not been standardized, and several tests are used (Sastry and Fuerstenau, 1971; Wakeman, 1972). As green pellets must be capable of withstanding handling, compaction, and drying and the dry pellets must be even stronger, mixtures of bentonite, iron ore, and water are commonly tested for wet-drop strength, wet-compression strength, plastic deformation, and dry-compression strength.

Absorbent Granules: Most absorbent granules marketed are prepared to fulfill the requirements outlined in Federal Specification P-A-1056A, Absorbent Material, Oil and Water (for floors and decks), outlined for purchases by the US General Services Administration. A similar set of specifications is outlined in the American Society for Testing & Materials (ASTM) Standard: C 431–65 Standard Methods for Sampling and Evaluation of Sorptive Mineral Products Used as Floor Absorbents. The size and absorbent characteristics required of granules as outlined in the federal specification are listed in Table 2.

The specifications require that absorbent granules consist of a uniform mixture of minerals of the silicate type. They also must be clean, uniform, and free of lumps or foreign

TABLE 2—Characteristics of Absorbent Granules

Characteristic	Maximum	Minimum
Matter retained on a No. 6 US Standard sieve	1.0%	0.0%
Matter retained on a No. 30 US Standard sieve	99.0%	52.0%
Matter retained on a No. 40 US Standard sieve	99.8%	73.0%
Matter retained on a No. 60 US Standard sieve	100.0%	90.0%
Ability to absorb lubricating oil (per gram of sample)	—	0.8 ml
Ability to absorb distilled water (per gram of sample)	—	0.9 ml
Solubility in distilled water	1.5%	—

matter, and no more than 10% of them can pass through an 80-mesh sieve in the attrition resistance test. The attrition-resistance test is made by shaking with steel balls on a screen, according to a specified procedure.

Fuller's Earth for Bleaching Oils: Test methods for evaluating bentonite and other types of fuller's earths for bleaching soybean and cottonseed oils are outlined in the American Oil Chemists Society AOCS Official Method Cc 8b-52, revised April 1952, and AOCS Official Method Cc 8a-52, corrected 1958. These specifications contain instructions on bench-type tests, including stirring time, heating rates and temperatures; approved equipment; quantities of raw oil and clay required; methods of color determinations; and so forth. They also require the comparison of the material tested with an official natural bleaching earth approved by the American Oil Chemists Society. The color determination of the bleached oils and their comparisons with the oil bleached by the official earth in the same test are the basis for purchase specifications.

Geology

Mineralogy: The principal clay mineral in bentonite is smectite, according to the definition used in this chapter, and many fuller's earth deposits also consist chiefly of this mineral. The term *smectite* is applied as a group name, and *montmorillonite* is a mineral species name. This usage conforms to the growing acceptance of the term smectite and does away with the confusing use of montmorillonite as both mineral species and group names. Montmorillonite, including both sodium and calcium varieties, is the most common member of the smectite group occurring in bentonite. However, saponite, a magnesian smectite, and hectorite, a lithium-bearing magnesian variety, are the major minerals in some bentonites.

Smectite minerals occur in extremely small particles (Fig. 3); therefore, detailed information on them is difficult to obtain and is, in part, based on theoretical considerations. As the mineralogy of this group of clays is complex, the reader is referred to authoritative books and other references given in the introduction for detailed information on them, and only the following brief summary will be given here.

According to the most generally accepted structure, smectite consists of two silica tetrahedral sheets with a central octahedral sheet. The theoretical structural formula is $(OH)_4Si_8Al_4O_{20}\cdot nH_2O$, and the theoretical composition without interlayer material is SiO_2, 66.7%; Al_2O_3, 28.3%; and H_2O, 5%. However, smectite always differs from the foregoing theoretical formula and composition (Table 3) because of substitutions of various ions for silicon in the tetrahedral coordination and for aluminum in the octahedral sheet. Furthermore, the substituted ions are thought to be commonly of different valence than the theoretical ion replaced, which results in an unbalancing of the charge in a unit of smectite. This charge deficiency is balanced by exchangeable ions, which vary considerably and result in further differences in the compositions of smectite. Cations which are exchangeable in bentonite include sodium, calcium, potassium, magnesium, lithium, and hydrogen. The exchange capacity of most bentonite is within the range of 60 to 150 milliequivalents per 100 g.

Unit layers of smectite are thought to be stacked with the oxygen layer of one silica tetrahedral sheet adjacent to the similar layer in the neighboring unit. Only a very weak bond exists between neighboring units, and water or other polar molecules can enter between unit layers causing the lattice to expand in the c direction. This expansion by polar molecules and the collapse of expanded units with heat, which can be recognized by X-ray diffraction methods, is applied in studying smectite and distinguishing members of this group from other clay minerals.

The ion-exchange characteristics have an important role in controlling or influencing the physical properties of bentonite and the bentonitic fuller's earths. In general, those bentonites

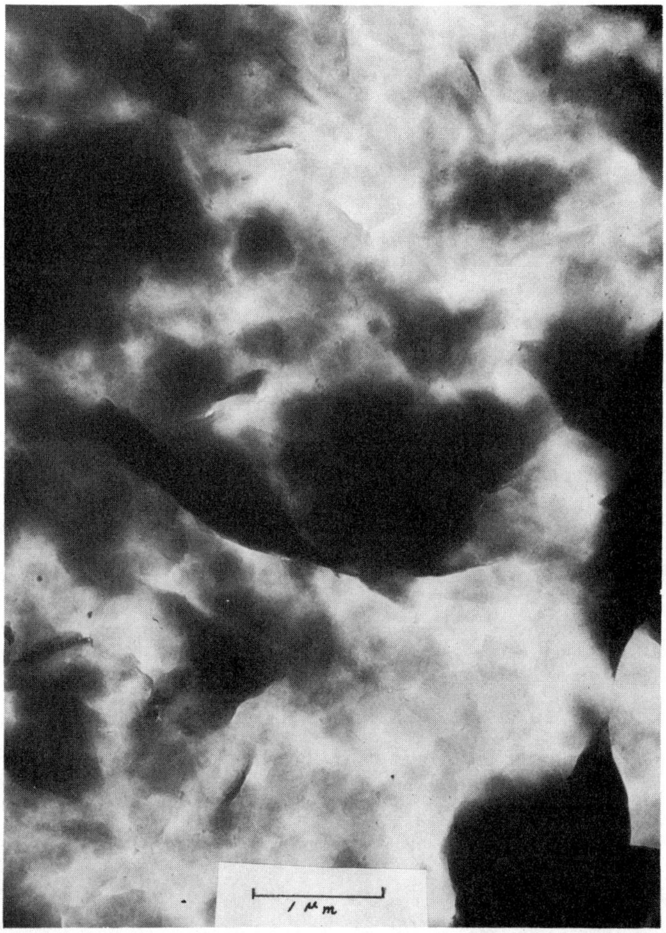

FIG. 3—*Electron micrograph of bentonite from Clay Spur, WY (API sample 26). Micrograph by E.J. Dwornik, US Geological Survey.*

that have Na^+ as the dominant exchangeable ion have very high swelling capacities and colloidal properties, and those in which Ca^{++} is the dominant ion tend to swell little more than other clays. Because of their high colloidal properties, high-swelling sodium bentonites are valuable as drilling muds and other uses requiring thixotropic suspensions. The calcium varieties usually have little value for such uses without beneficiation by the addition of soda ash or other chemicals, and even when so treated, are not ordinarily as efficient as the sodium types. The type of exchangeable ion also has a major role in controlling the bonding characteristics. As pointed out by Grim (1962), sodium bentonites have very strong foundry sand bonding properties after drying (dry strength), but only moderate strengths when moist or wet (green strength). Calcium bentonites, on the other hand, have high green strengths and low to moderate dry strengths. However, part of these differences in bond strengths may be due to particle size or other characteristics, and there is some evidence that the dry strength of some calcium bentonites can be increased by lengthening the mulling (mixing) time in the preparation of foundry sand.

All bentonites contain mineral impurities, which vary considerably in type and quantity present. A few are contaminated with minor quantities of clay minerals other than smectite, such as kaolinite and illite. The common nonclay minerals in bentonite include those listed in the older definition by Ross and Shannon (1926), and minor quantities of most of the accessory minerals in volcanic rock also may be present. Selenite is abundant in many deposits, and some beds contain carbonate veins and concretions. Zeolite minerals and opal, cristobalite, or other forms of poorly ordered

TABLE 3—Chemical Analyses of Clays

	1	2	3	4	5	6	7	8	9	10
SiO_2	54.34	54.62	50.20	49.72	49.91	53.95	53.96	43.98	45.20	46.77
Al_2O_3	19.92	20.08	16.19	16.39	17.20	0.14	8.56	38.46	37.02	37.79
Fe_2O_3	1.11	2.36	4.13	3.10	2.17	0.03	3.10	—	0.27	0.45
FeO	2.17	1.03	—	1.00	0.26	—	0.19	0.03	0.06	0.11
TiO_2	0.11	0.11	0.20	0.58	0.24	Tr	0.24	0.01	1.26	0.02
CaO	0.33	0.40	2.18	4.19	2.31	0.16	2.01	0.32	0.52	0.13
MgO	2.21	2.15	4.12	2.56	3.45	25.89	10.07	Tr	0.47	0.24
Na_2O	2.65	2.26	0.17	0.66	0.14	3.04	0.03	0.14	0.36	0.05
K_2O	0.45	0.35	0.16	1.55	0.28	0.23	0.39	0.48	0.49	1.49
MnO	0.01	0.01	—	0.01	0.04	—	—	—	—	Tr
SO_3	0.18	0.16	—	—	—	—	—	—	—	—
H_2O-	11.73	11.78	15.58	12.21	15.77	9.29	9.79	2.58	1.55	0.61
H_2O+	4.50	4.58	7.67	5.71	7.70	5.61	11.51	14.59	13.27	12.18
CO_2	—	—	—	1.71	—	—	—	—	—	—
P_2O_5	—	—	—	0.06	—	—	—	—	—	0.10
S	—	—	—	0.73	—	—	—	—	—	—
BaO	—	—	—	—	—	—	—	—	—	—
ZnO	—	—	—	0.03	—	—	—	—	—	—
C	—	—	—	—	Tr	—	—	—	—	—
LiO_2	—	—	—	—	—	1.22	—	—	—	0.03
F	—	—	—	—	—	—	—	—	—	0.24
Total	99.71	99.89	100.50	100.21	99.47	99.56	99.85	100.59	100.47	100.21
Less O for S				0.37						
				99.84						

1. Bentonite, blue-gray, clay spur bed, Belle Fourche, SD (Foster, 1953)
2. Same as 1 but oxidized to olive green (Foster, 1953).
3. Bentonite, Polkville, MS (Ross and Hendricks, 1934).
4. Fuller's earth, Combe Hay, Somerset, United Kingdom (Kerr et al., 1950).
5. Bentonite, Chambers, AZ [Cheto] (Kerr et al., 1950).
6. Hectorite, Hector, CA (Kerr et al., 1950).
7. Palygorskite (attapulgite), Attapulgus, GA (Kerr et al., 1950).
8. Halloysite, Eureka, Utah (Kerr et al., 1950).
9. Kaolinite, Macon, GA (Kerr et al., 1950).
10. Kaolinite, St. Austell, United Kingdom (Kerr et al., 1950).

silica are present in some bentonite deposits in western United States and at several other places in the world. Some deposits are contaminated with incompletely altered parts of the rock from which they formed. The vitric latitic tuff remaining in the bentonite at Cheto, AZ (Sloane and Guilbert, 1967), is an example of this type of material.

The principal fuller's earth deposits other than the bentonite type consist of palygorskite (attapulgite) and sepiolite. According to Bradley (1940), palygorskite has an ideal formula of $(OH_2)_4(OH)_2Mg_5Si_8O_{20}\cdot 4H_2O$, but substitution of Al^{3+} for either Mg^{++} or Si^{4+} takes place. It consists of two silica chains linked in amphibolelike structures and has both monoclinic and orthorhombic symmetry (Christ et al., 1969). Sepiolite, $(Si_{12})(Mg_9)O_{30}(OH)_6(OH_2)_4\cdot 6H_2O$, is thought to have three pyroxene-type chains (Grim, 1968), and, therefore, the unit cell is somewhat larger than for palygorskite. Both palygorskite and sepiolite typically occur in fibrous and elongate lathlike particles. Sepiolite is also the mineral forming meerschaum (see "Meerschaum).

Other nonbentonite fuller's earths include an uncommon type of rock called opal claystone (Heron, et al., 1965). This rock is more than three-fourths opal or another form of poorly ordered silica, and the remainder is montmorillonite and other mineral impurities. It was mined in South Carolina in the early part of this century and used for bleaching oil, and it is the raw material to be mined near Rimini, SC, to supply a fuller's earth plant under construction in 1972. Nonbentonite clays that have been used as fuller's earths in the past include the Anna kaolin in Illinois, halloysite in Utah (Schroter and Campbell, 1940), and glacial clays in Massachusetts. In addition to these nonbentonite clays, there are extensive deposits of fuller's earth consisting of very impure montmorillonite that are not ordinarily thought of as being bentonite. Two such deposits are the Porters Creek Clay (Paleocene) in the Mississippi Embayment region and the

Twiggs Clay Member of the Barnwell Formation (Eocene) in Georgia.

Occurrence: Though bentonite deposits of Jurassic age are extensive in western United States and elsewhere in the world, virtually all deposits mined to date are of Cretaceous age or younger. All but a very minor tonnage of the high-swelling Wyoming-type bentonite produced has been mined from beds of Cretaceous age, and more than three-fourths of the total has come from a single formation—the Mowry Shale. The Southern or calcium bentonite is mainly in formations of Cretaceous, Eocene, and Miocene age. The abundant miscellaneous bentonite deposits, other than the high-swelling type, in the western states are mainly of Tertiary age. Deposits scattered through Nevada, California, Oregon, and Idaho are mainly in upper Tertiary rocks, and many are thought to be Late Miocene or Pliocene in age. The widely known Cheto bentonite near Chambers, AZ, is in the Bidahochi Formation of Pliocene age, and deposits mined in western Oklahoma are also in Pliocene rocks. Deposits in the Ash Meadows district, NV, occur in rocks mapped by Denny and Drewes (1965) as Pleistocene in age, and if this age assignment is correct, they are among the youngest bentonites in the world.

Most bentonite occurs in beds or lenticular bodies aligned along a definite stratigraphic zone. Some beds extend for more than 200 miles, and other deposits, particularly the lenticular ones, are only a few hundred yards in diameter. Deposits mined range from about 1 to more than 30 ft in thickness. Bedded deposits ordinarily have a characteristic sharp contact with underlying rocks and a gradational one with overlying strata. Deposits formed by hydrothermal processes tend to be irregularly shaped and to grade into the host rock in all directions.

Most bentonite, including both the high and low-swelling types, has a characteristic waxy or soaplike texture. Parts of deposits near the surface tend to be light-yellowish green or gray when the natural moisture is present and to become lighter in color when they dry. Deposits under considerable overburden tend to be bluish gray. The change to lighter shades near the surface is related to the oxidation of iron. Typical outcrops of high-swelling bentonite have a popcornlike or frothy texture caused by alternate swelling and drying of the bentonite. Outcrops of low-swelling types commonly have a cracked appearance thought by some to resemble alligator hide.

Most bentonite in North America and elsewhere occurs in sedimentary rocks and was formed in place from volcanic ash or tuff. Some deposits have been formed by hydrothermal alteration of volcanic or other igneous rock, and other deposits of nearly pure smectite may be accumulations that were transported and deposited in marine or alkaline lake water with only minor postdepositional alteration.

The extensive palygorskite (attapulgite) fuller's earth deposits in the southern part of the Meigs-Attapulgus-Quincy district, Georgia and Florida (Fig. 4), occur as tabular lenses and discontinuous beds in the Hawthorn Formation of Miocene age. This clay is green and bluish gray, and most of it has a waxy or soaplike texture.

In addition to the deposits in the United States, palygorskite fuller's earth is also mined in the following regions: (1) Pout district near Mbour, Senegal (Wirth, 1968); (2) Ukrainian district, USSR (Ovcharenko et al., 1967); and (3) Mudh district, India (Siddiqui, 1968). The Senegal deposits are in marine beds of Early Eocene age. The Ukrainian palygorskite is interbedded with bentonite. Deposits in India occur between fossiliferous limestones and apparently are marine in origin.

Insofar as the authors are aware, the only sepiolite fuller's earth mined at the time this report was written is in the Vallecas and another district in Spain (Anon., 1972b). However, economic sepiolite deposits have been discovered recently in the Ash Meadows district, Nevada (private communication). The Spanish deposits occur in an evaporite sequence of Tertiary age. The Nevada deposits are closely associated with calcium, sodium, and magnesium bentonites of Pleistocene age.

Origin: The evidence for most bedded bentonite having formed from volcanic materials transported considerable distances in the atmosphere was presented by several geologists early in the century, including Hewett (1917), Wherry (1917), and recently by Slaughter and Earley (1965). This conclusion is based primarily on the purity of smectite, on the fact that the nonclay minerals in bentonite are angular and of the type occurring in volcanic rock, rather than mixtures of rounded grains characteristic of detrital sediments, and on the presence of relict shard textures in many bentonites (Ross and Shannon, 1926). Ample evidence supports the conclusion that the parent ash of most bentonite was deposited under marine conditions, but a few deposits apparently accumulated in alkaline lakes (Papke, 1969). As would be expected, different bentonites have

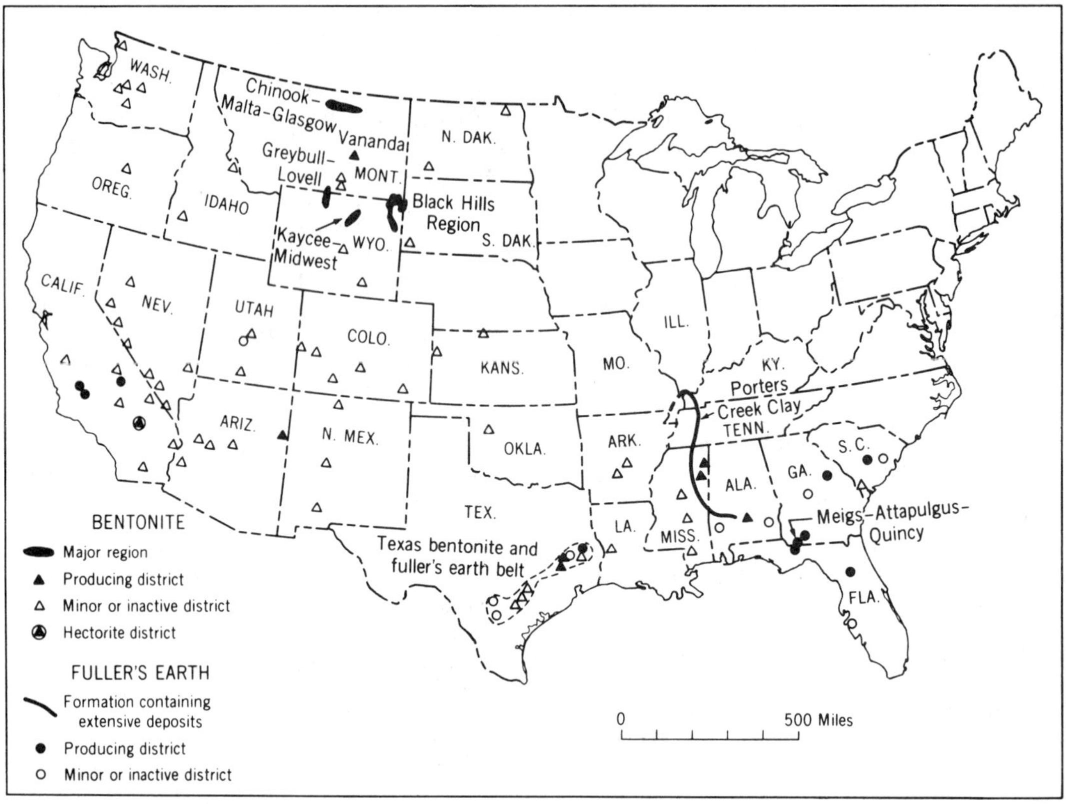

FIG. 4—*Map showing bentonite and fuller's earth districts in the United States.*

formed from volcanic rock of somewhat different types, and the most common parent materials range from andesite to rhyolite in composition. Opinions differ concerning the process and time of alteration of the ash. Several authors have noted that change in ash would have begun with its contact with water, but when the alteration was completed is questionable. Papke (1969) believes that one type of bentonite deposit formed in alkaline lakes soon after deposition and another type formed more slowly as the result of alteration by ground water. Alteration by ground water requires burial by younger rocks and probably regional uplift and a considerable interval of geologic time. In the authors' opinion, changes have taken place in bentonite deposits throughout virtually their entire geologic history. Some deposits may have been altered extensively before burial beneath younger rocks. However, the prominent cherty zone below several beds points to the downward transport of silica by ground water from altering deposits; this must have taken place after burial.

Bentonite formed by hydrothermal activity occurs in Nevada (Papke, 1969), California (Kelley, 1966), Spain (Anon., 1972b), and elsewhere. Deposits of this type commonly are associated with other minerals known to have formed by hydrothermal processes. They also tend to occur in irregularly shaped bodies rather than in beds, and they ordinarily grade into the host rock through partially altered zones. The distribution and shape of bentonite formed hydrothermally are commonly controlled by fault zones, joints, and other geologic features providing access for heated water.

Alteration of volcanic ash or tuff in restricted alkaline lakes heated by hot-spring activity is also thought to have formed bentonite (Ames, et al., 1958). The hectorite deposits in California probably formed this way, and this clay passed through an intermediate stage in which most of the ash was altered to zeolite. A magnesium bentonite called amargosite formerly mined near Shoshone, CA, also formed by the alteration of vitric volcanic material by hot springs (Sheppard and Gude, 1968).

Though the authors have doubts about the theory, some bentonite deposits are thought to have been deposited mainly in the form of smectite minerals. According to this explanation of origin, smectite minerals, having formed by weathering of volcanic or other igneous rocks, were transported, separated from impurities, and deposited. This theory has been applied to explain the origin of deposits in India (Siddique and Bahl, 1965), the USSR (Tazhibayeva and Galiyev, 1972), and Nevada (Papke, 1969). Bentonitic fuller's earth deposits in the United Kingdom were thought by Robertson (1961) to have been transported from smectite-rich soils on the European landmass. Such soil would have formed mainly on nonigneous rocks.

As most fuller's earth deposits are bentonites, they have the same origins; however, some fuller's earths consist of minerals other than smectite and may be of quite different origin. The palygorskite (attapulgite) fuller's earth deposits in Georgia and Florida are one type that probably formed in a different way. These deposits have been investigated by several geologists and mineralogists, and no one has found convincing evidence of any volcanic materials having ever been present. The first author has investigated these deposits in detail and believes that the palygorskite in them formed in place from seawater evaporating in a tidal-flat environment. This theory is essentially the same as the "neoformation" idea of Millot (1964).

Distribution of Deposits

North America: *United States*—SOUTHERN BENTONITE AND FULLER'S EARTH—Southern or calcium bentonite is now mined in Texas, Mississippi, Alabama, Oklahoma, and Louisiana (Teague, 1972). Undeveloped bentonite deposits occur in South Carolina (Robinson, Buie, and Johnson, 1961). At present, at least 17 different bentonite mining operations are active, and 11 plants are processing bentonite. One of these plants is in Alabama, four are in Mississippi, one is in Louisiana, and five are in Texas.

The bentonite deposits in Texas (Fig. 4) are mainly in an extensive belt in the Coastal Plain region. Valuable deposits occur in formations ranging in age from Late Cretaceous to Miocene (Hagner, 1939; Maxwell, 1962; Fisher et al., 1965). The bentonite is mined intermittently near Mineral Springs, Lincoln Parish, LA (Gann, 1971). Deposits also were formerly mined in the Vernon Parish (Welch, 1942), which is in the west-central part of Louisiana. The Vernon Parish deposits are in the Fleming Formation (Miocene) which also contains bentonite in Texas. Most of the bentonite produced in Mississippi is mined near Aberdeen, Monroe County, where the American Colloid and International Minerals & Chemical Corps. operate plants. Bentonite is also mined intermittently in Smith County, and clay that is presumably bentonite is produced in Pearl River County. The deposits mined in Monroe County are in the Eutaw Formation of Late Cretaceous age. Deposits mined in Smith County are in Oligocene rocks, and the deposits in Pearl River County are of Miocene age (Bicker, 1970). The bentonite processed near Sandy Ridge, AL, occurs in the Ripley Formation [Upper Cretaceous] (Monroe, 1941).

Fuller's earth is produced mainly in the southern states; and the Meigs-Attapulgus-Quincy district, Georgia and Florida (Fig. 4), is the leading fuller's earth-producing district in the United States. The deposits mined in this district occur in the Hawthorn Formation of Miocene age. Those in the northern part of the district consist chiefly of diatomaceous montmorillonite containing minor quantities of other clay minerals and clastic nonclay mineral impurities. Some of the deposits in the southern part of the district are the purest palygorskite (attapulgite) mined in the world. The fuller's earth mined in central Peninsular Florida also occurs in the Hawthorn Formation, and this clay consists chiefly of montmorillonite.

Large deposits of fuller's earth occur in the Twiggs Clay member of the Barnwell Formation of Eocene age. These deposits crop out in a belt extending across central Georgia seaward from the fall line. The Twiggs Clay member is now mined and processed by the Georgia Tennessee Mining Co. at Wrens, GA (Patterson, 1972), and it was formerly mined at several other places in this state. The Twiggs Clay member consists mainly of montmorillonite and cristobalite (Brindley, 1957) or opal (Carver, 1972).

In 1973, five plants were actively producing fuller's earth from the Porters Creek Clay of Paleocene age. This formation extends from west-central Alabama across the northeastern part of Mississippi, the western parts of Tennessee and Kentucky, southernmost Illinois, and into southeastern Missouri. Most of the Porters Creek Clay consists of a mixture of montmorillonite and other clay minerals, and much of it contains appreciable quantities of cristobalite and nonclay mineral impurities (Sims, 1972).

Fuller's earth used mainly in bleaching animal and vegetable oils is now produced in Texas, and historically this state has ranked with the leading producers. The fuller's earth produced in Texas is mined from bentonite beds.

WYOMING OR HIGH-SWELLING BENTONITE DEPOSITS—The Northern Rocky Mountains-High Plains region has led the world in the production of Wyoming or high-swelling bentonite since the first mining of this clay. In 1973, 16 plants were processing bentonite in this region. The major producing districts (Fig. 4) are as follows: (1) the northern and western Black Hills districts, Wyoming, Montana, and South Dakota; (2) Kaycee-Midwest, Wyoming; (3) Greybull-Lovell, Wyoming and Montana; (4) Vananda, Montana; and (5) Chinook-Malta-Glasgow, Montana.

Virtually all the bentonite produced in the Northern Rocky-High Plains region occurs in sedimentary formations of Cretaceous age. Most of the bentonite produced in the Black Hills district has been mined from the Clay Spur Bentonite Bed, which is in the uppermost part of the Mowry Shale (Davis, 1965; Knechtel and Patterson, 1962). High-swelling bentonite deposits in the Newcastle Sandstone have also been mined, and a small tonnage has been produced from beds in the Belle Fourche Shale. A bentonite in the upper part of the Pierre Shale was mined on the southern flank of the Black Hills before and during World War II and processed for use as water softener. A very low swelling bentonite having the properties of the southern-type bentonite is now mined on a small scale from a bed in the Pierre Shale in the northern Black Hills.

Most of the bentonite mined in the Kaycee-Midwest and Greybull-Lovell districts occurs in the Mowry Shale, but some beds in younger formations are also mined in these districts (Anon., 1969h). The bentonite produced mainly for taconite-bonding purposes in the Vananda and Chinook-Malta-Glasgow districts, Montana, occurs as extensive beds in the Bearpaw Shale (Berg, 1969, 1970), which is considerably younger than the Mowry Shale.

HECTORITE—Virtually all the hectorite mined to date has been near Hector, CA, the type locality. The hectorite occurs as a waxy soft nodular layer ranging from 6 to 8 ft in thickness. It is associated with sandstone and clay beds of Tertiary age that accumulated in an alkaline lake environment. The beds below the hectorite contain travertine formed as the result of hot-spring activity (Ames, et al., 1958). The valuable hectorite deposits are overlain by Quaternary basalt.

A large deposit of hectorite has been discovered recently in the Amargosa Valley, California (private communication). Plans for mining and processing this clay were in an advanced stage of preparation when this report was written.

In addition to the Hector and the Amargosa districts, hectorite also occurs 16 miles north of Amboy, CA, and in Yavapai County, AZ (Norton, 1965). Deposits not known to the authors probably occur at other localities, and there are many areas where the geologic conditions are favorable for more discoveries.

OTHER DEPOSITS IN WESTERN STATES—Bentonite (other than the Wyoming type and hectorite) and fuller's earth deposits are scattered throughout the western states (Fig. 4). In general, bentonite production from these deposits has been small, as compared with that from the Wyoming and Southern types, because many of these clays are of lower quality and all of them are more distant from the major consumers. One district in which bentonite has been mined continuously for a long time is in the vicinity of Chambers [Cheto], AZ (Kiersch and Keller, 1955). This bentonite is a low-swelling clay used as bleaching earth and in making desiccants. Bentonite used for preparing livestock feed is mined in southwestern Idaho, southern California, and probably elsewhere. Deposits in central Oregon are used in the preparation of retardants used in fighting forest fires. Some of the white bentonite mined in Nevada is used in pharmaceuticals and beauty aids. A bentonite in the Ash Meadows district composed of saponite (Papke, 1969) was mined extensively several decades ago, and other deposits in the state have been mined for use in water impedance and for other purposes. Low-swelling or calcium-type bentonite also occurs and is produced in the region containing the Wyoming or high-swelling bentonite. Low-swelling bentonites are mined from deposits in the northern Black Hills and the flanks of the Big Horn Mountains. This bentonite is marketed primarily for foundry-sand bonding.

The most active fuller's earth-producing districts in the western states are in Kern County, CA; plants operated by Excel Minerals are at McKittrick and Taft, and the Panamint Marketing plant is at Maricopa. These plants process a clay mined from the Monterey Shale of Miocene age. Some beds in the Monterey Shale consist chiefly of diatoms and montmorillonite, and probably the raw material mined is of this composition. Absorbent granules, insecticide carriers, and other fuller's earth products are prepared for market. In addition to the deposits in Kern County, bentonite near Olancha and probably deposits at other localities in California have been processed for filtering and decolorizing agents in recent years.

RESOURCES—Resources of bentonite and fuller's earth in the United States are very large. The reserves of bentonite are at least one billion tons, and the total resources, including all bentonite that will be eventually suitable for one use or another, are considerably larger. This reserve estimate of one billion tons includes very large deposits that have a quality suitable for binding iron-ore pellets but that would be classed as submarginal for drilling mud and several speciality uses. The reserves of very high quality drilling-mud bentonite are limited and are much in demand. Resources of fuller's earth are tentatively estimated at two billion tons. However, almost any figure could be used here, if all the clays equal in quality to the lowest grade material used as bleaching clay in the past were included. A large part of the fuller's earth resources is in the extensive Porters Creek Clay and the Twiggs Clay Member of the Barnwell Formation. These deposits are suitable for use in absorbent granules and certain other fuller's earth products, but they are not used in drilling mud and several of the speciality products.

Canada—Bentonite occurs in beds of Cretaceous and Tertiary ages at many places in western Canada (Ross, 1964; Spence, 1924). The major producing districts (Fig. 5) are Onoway (Baroid Canada, Ltd.) and Rosalind (Dresser Industries, Inc.) in Alberta and in the Pembina district (Pembina Mountain Clays, Ltd.) in Manitoba. Bentonite has also been mined near Princeton, B.C., and deposits near Avonlea, Sask. (Anon., 1971), and along the Mackenzie River in the vicinity of Inuvik, N.W.T., have been investigated by industry.

The deposits mined in Alberta are in the Edmonton formation, the uppermost Cretaceous unit in this region. Those mined at Rosa-

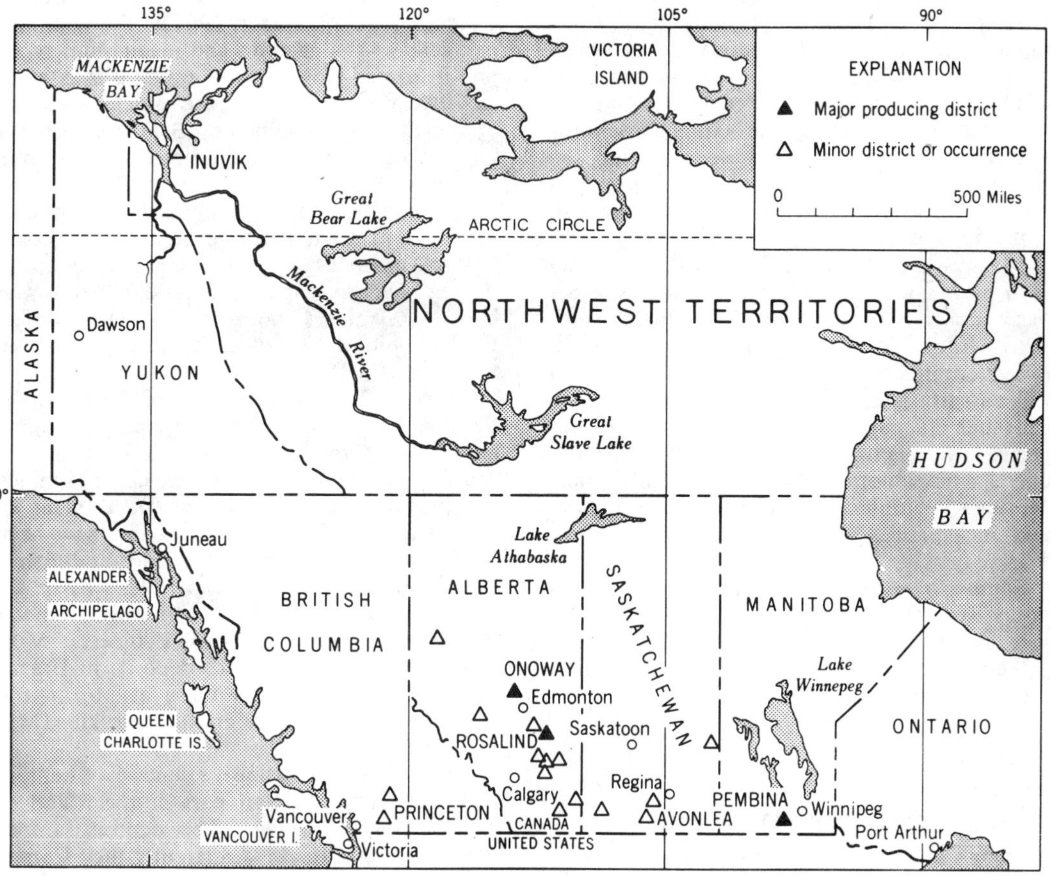

FIG. 5—*Map showing bentonite districts in Canada.*

lind are nearly flat lying and are 8 to 10 ft thick. The deposits mined at Onoway are 2 to 2½ ft thick, and the overburden removed from them has an average thickness of about 7 ft. Bentonite deposits mined in the Pembina district are in the Vermillion River formation (Upper Cretaceous), which consists mainly of black shale. The bentonite near Princeton, B.C., is of Tertiary age, and it is interbedded with lignite (Spence, 1924).

Several grades of bentonite are produced in Canada. The swelling bentonite mined in Alberta is sold for use in drilling mud, foundry-sand bond, fire retardant, and stock feed pelletizing (Ross, 1964). The bentonite produced in the Pembina Valley is a low or nonswelling variety. It is processed in a plant at Morden, Man., and some is activated with sulfuric acid in a plant at Winnipeg. This bentonite is used for foundry-sand bond, insecticide carriers, and pelletizing animal feed. The activated bentonite is used for bleaching mineral lubricating oils and waxes, re-refining used lubricating oils, and decolorizing animal and vegetable oils.

According to Ross (1964), only a very few million tons of bentonite reserves have been proved in Canada. However, the potential resources of bentonite in the country must be very large, considering the large areas now known to contain scattered deposits and the general similarities of strata to Cretaceous and Tertiary rocks in the United States known to contain bentonite.

Mexico—The German firm Süd-Chemie AG with its Mexican affiliate Tonsil Mexicana SA has operated a bentonite plant in Puebla since 1967 (Anon., 1969k), and another plant is active intermittently at Monterrey. Bentonite and fuller's earth also occur in Querétaro, Michoacán, and Guanajuato, and a few of these deposits have been mined on a small scale (Esquivel and Zamora, 1958). Apparently, most Mexican bentonite deposits occur in sedimentary rocks, and some of these are of the nonmarine type.

South America: *Argentina*—Bentonite deposits are scattered throughout several provinces in Argentina (Bordas, 1947). The principal deposits mined and the 1969 production (Anon., 1971b) are as follows:

Principal Deposits	Province	Tons Produced
Rio Chico	Chubut	5,286
El Alamo y Las Hegueras	Mendoza	19,799
El Catalán	Neuquén	4,622
Del Lago Camino y Rob	Río Negro	14,427
La Emilia y Cristina	San Juan	18,005
	Total	62,139

A few thousand tons of fuller's earth are also produced each year. The fuller's earth deposits mined are in La Rioja and Río Negro provinces.

Brazil—A plant processing montmorillonite clays for use as foundry-sand bond was built at Sacramento, Minas Gerais, in 1962 (Anon., 1962). Bentonite has also been found at several other localities in the country.

Peru—Bentonite and fuller's earth occur in the provinces of Pisco, Canete, Paita, and Contralmirante Villar, and elsewhere (Caberra, 1963). Bentonite has been mined and used in iron-ore pelletizing in Peru for several years (Anon., 1967a), and present production is estimated to be 20,000 tpy (Anon., 1973a).

Europe: *Cyprus*—A low-swelling calcium bentonite is mined in the Troulli district near Kambia (Anon., 1969l). A few thousand tons are produced annually. Part is processed in a small plant at Vassiliko, and part is shipped to Israel for beneficiation and use in several products (Anon., 1970e).

Czechoslovakia—Bentonite occurs in the northwestern, central, and eastern parts of Czechoslovakia (Gregor, 1967). The bentonite is chiefly of the calcium and magnesium types, and it formed from rhyolite tuffs. It is as much as 70% montmorillonite; the nonclay minerals present include cristobalite. Bentonite reserves in Czechoslovakia were estimated to be 10 million tons.

Apparently, the most productive bentonite district is in the northwestern part of the country, inasmuch as a bentonite processing plant having a yearly capacity of 100,000 tons was built at Zelenice (Anon., 1969l). One-third of the bentonite produced in this plant is reported to be available for export.

France—The company Société Française des Bentonites et Dérives SarL. markets several grades of bentonite, mined principally in central France (Anon., 1969j). The deposits mined are presumably those in the Department of Vienne and the Limousin region that have been worked for many years (Déribére and Esme, 1943). Other bentonite and fuller's earth deposits are in Vaucluse (Mormoiron, Sainte-Radegonde, and Apt) and Pyrénées. Most bentonite and fuller's earth deposits in France are chiefly smectite, but those at Mormoiron are mixtures of smectite and palygorskite (Millot, 1964).

Greece—Rather extensive deposits of bentonite occur in the eastern part of the island of Milos (Anon., 1969l). The deposits are as much as 100 ft thick. Apparently they formed by the alteration of ash or tuffs of Pliocene age in a marine environment (Wetzenstein, 1972). One plant near deep water operated by the

Greek company, Silver and Barytes Ores Mining Co., has a yearly capacity of 200,000 tons. Three other companies are also mining bentonite on this island (Wayland, 1971). The bentonite producers on Milos have been particularly successful in selling bentonite for iron-ore pelletizing, mainly because of the nearness of deposits to deep-water shipping. Bentonite produced on this island is also sold for drilling mud and foundry-sand bond, and for other purposes.

In addition to the deposits on Milos, bentonite also occurs on Mikonos where three companies have been active in recent years. With production on both islands, Greece ranks among the leading producers of bentonite.

Hungary—Bentonite occurs at scattered localities in Hungary. Recently, bentonite-processing facilities were concentrated at Mod, under the supervision of personnel of the National Ore and Mineral Mining Enterprise (Sondermayer, 1967). The new plant is reported to have a capacity for processing 100,000 tpy of raw bentonite. Both natural calcium and sodium-exchanged bentonites are produced for several different uses (Anon., 1970d).

Italy—Italy is one of the leading European bentonite producers. The deposits mined are on the island of Ponza, off the western coast, and on Sardinia. The Ponza deposits are near Cala dell Acqua. They are in volcanic rock and were formed by hydrothermal alteration of rhyolite tuffs (Lupino, 1954). The average thickness of these deposits is about 10 m, and 2.9 million tons is estimated to be present (Anon., 1969i). Several grades of bentonite are produced on Ponza, and one of them is a colloidal white bentonite that brings a high price for speciality uses. This white bentonite is one of the few bentonites imported by the United States.

The deposits on Sardinia are mined by Industria Chimca Laviosa SpA, an Italian company (Anon., 1970b), and NL Industries, Inc. (Wayland, 1971). The Italian company is mining the Pedra de Fogu deposits near Alghero, and grades suitable for drilling mud, iron-ore pelletizing, civil engineering applications, water treatment, and wine purification are produced. The deposits mined occur in altered rocks (Pietracaprina and Novelli, 1972). The bentonite mined is the swelling type. It is approximately 70% montmorillonite, and the impurities are chiefly free silica, illite, and calcite. The deposits mined by NL Industries are near Nurallao (Anon., 1969g). These deposits are in Miocene marine sediments, and they are underlain by Jurassic limestone and trachyte flows. This bentonite is sold primarily for drilling muds in countries bordering the Mediterranean Sea and in the Near East.

Poland—Bentonite deposits of Eocene and Miocene age occur in southern Poland; these are processed for use as bleaching clay (Nowacki and Ciechomska, 1964) and probably for other uses.

Romania—Romania has been one of the major East European bentonite-producing countries for several years. The bentonite is mined underground in the Alba-Iulia-Ocana Mures district (Neacşu, 1969), and apparently it is produced elsewhere. The yearly production of bentonite in Romania is estimated to be more than 100,000 tons (Huvos and Sondermayer, 1968).

Spain—Bentonite and fuller's earth occur at several places in Spain. The most productive bentonite deposits are in the Cabo de Gata region in Almeria. These deposits have formed by the hydrothermal alteration of rhyolite and andesite igneous rocks (Anon., 1972b; Martin Vivaldi and Linares, 1968, 1969). The more intensely altered parts of deposits contain kaolinite, alunite, and jarosite. Other parts contain minor amounts of palygorskite and sepiolite, which presumably indicates an excess of magnesium at the time of alteration. Fuller's earth deposits consisting mainly of sepiolite occur at Vallecas and elsewhere in the Tagus basin (Anon., 1972b; Huertas, et al., 1971). The deposits are in an evaporite sequence of Tertiary age. The annual output of fuller's earth in the Vallecas district is approximately 14 kt. Absorbent granules, insecticide carriers, and materials for other uses are produced.

Switzerland—Bentonite is mined in Switzerland and used in bonding foundry sands (Grim, 1962).

United Kingdom—Calcium montmorillonite clays occur at many places in the United Kingdom, and fuller's earth has been mined since the Roman period nearly 2000 years ago (Anon., 1969f). Fuller's earth districts now active are at Redhill in Surrey, Combe Hay south of Bath in Somerset, and Woburn in Bedfordshire. In addition to these districts, large deposits of calcium montmorillonite have been discovered in the Swindon-Abington district in Berkshire by the Institute of Geological Sciences (Poole and Kelk, 1971; Poole et al., 1971), and other deposits occur near Clophill in Bedfordshire, and Maidstone in Kent.

The deposits in Surrey and Bedfordshire are in the Lower Greensand, which is the upper formation of the Lower Cretaceous Series (Holmes, 1950). These deposits consist of hard bluish-gray clay having a soapy feel. They are in beds of irregular thicknesses, and at places as many as eight beds of clay are present in the Lower Greensand. Thicknesses of clay mined range from 5 to 8 ft. Most mining is by stripping methods, but deposits in the Combe Hay district are mined underground.

The fuller's earth at Combe Hay district occurs in the Great Oolite series of Middle Jurassic age. These fuller's earth deposits are commonly 6 to 7 ft thick. As in several bentonite deposits in the United States, the fresh clay in the Combe Hay district is bluish green, and the weathered parts of the deposits are yellowish.

Fuller's earth production in the United Kingdom has quadrupled since 1945, and 176,000 tons was produced in 1970 (Anon., 1972c). Probably the biggest market is for foundry-sand bond, and some of the clay is treated with soda ash for this use. Sodium-exchanged fuller's earth is also used in pelletizing animal feeds, treating domestic water supplies, and for several uses requiring swelling or thixotropic properties, including civil engineering uses, drilling mud, and for water-impeding purposes. Approximately a fifth of the fuller's earth produced is acid activated; this clay is sold mainly for alkylation catalysts.

USSR—The principal bentonite deposits in the USSR are in the Volga Region, western and central parts of the Ukraine, Transcaucasia Region, Kazakhstan, and Central Asia (Lebedinskii and Kirichenko, 1972; Anon., 1970a). Smaller bentonite resources also occur in eastern and western Siberia, Crimea, and in the northern Caucasus.

Several large bentonite deposits of Lower Miocene age occur in the Cherkassy district (Ovcharenko et al., 1967), which is located along the border between the Cherkassy and Kiev regions in the Ukraine. The deposits consist of five layers having maximum thicknesses ranging from 1.5 to 8 m. Though most of this bentonite consists chiefly of montmorillonite containing minor quantities of detrital mineral impurities, one layer as much as 2 m thick is mainly palygorskite (attapulgite). This palygorskite was discovered during geological studies in 1954, and it is the first fuller's earth deposit of this composition found in sufficient quantities for mining in Russia.

Most bentonite produced in Azerbaijan in the Transcaucasia Region is mined from beds of Cretaceous age (Seidov and Alizade, 1966), but younger deposits are also present in this region. The bentonite in Azerbaijan is interbedded with tuff and calcareous sedimentary rocks and is thought to have formed mainly by the decomposition and alteration of volcanic ash in a marine environment. However, some deposits may have been altered from other rocks by hydrothermal solutions.

The scattered bentonite deposits in Kazakhstan occur mainly in Lower Tertiary rocks (Akhmed, 1963; Tazhibayeva and Galiyev, 1972). They are thought to have altered from hydromica and related minerals in an alkaline and deoxydizing environment in inland seas in regions of arid climate. The mica minerals apparently were transported from weathered rocks in land areas.

The principal uses of bentonite in USSR are in the manufacture of heavy clay products, drilling mud, and iron-ore pelletizing, and minor quantities are used in purifying oils, water, and wine. In 1967 about 1.5 million tons was produced (Lebedinskii and Kirichenko, 1972), and the total consumption was about 3.4 million tons. Imports to fulfill requirements came from Italy, Czechoslovakia, and Hungary.

West Germany—Lower Bavaria has been a major producer of smectitic fuller's earth and bentonite for many years. The two major producers are Süd Chemie AG, which has a plant at Mooseberg (Anon., 1969k), and Erbslöh and Co., which operates a plant at Landshut (Anon., 1969e). The deposits are mined mainly by open-pit methods, but some underground mining is still done. Both companies produce sodium-exchanged bentonite for use in foundry-sand bonding and for use in several products requiring colloidal properties. Smectite clays are also acid activated and exported to several countries for processing oils and waxes.

Yugoslavia—Calcium-type bentonite occurs in many places in Yugoslavia, and the total resources in the country are very large (Anon., 1969m). The bentonite is processed, including some sodium exchanging, at several places. The three largest producers have a total annual plant capacity of 100,000 tons, of which 30,000 can be acid-activated. Yugoslavia exports bentonite mainly to East European countries, but minor tonnages are also shipped to West European markets. The quantity exported in 1968 was 23,745 tons.

The best quality bentonite in Yugoslavia is in the Ginovci deposits, 81 km northeast of Skopje (Konta, et al., 1971). These deposits are in lacustrine beds of late Pliocene age, and they formed by alteration of dacite tuffs. The bentonite is reported to be more than 85% smectite and contains 1.2% quartz and 0.5% cristobalite. Reserves of the three best grades of bentonite in these deposits are estimated to be 1.4 million tons. The Ginovci bentonite is converted into a sodium type in the modern Bentomak plant near Kriva Palanka.

Africa: *Algeria*—Calcium-type bentonite is mined in the Marnia and Mostaganem districts of Algeria (Anon., 1969l). Algerian bentonite production in recent years has ranged from 25,000 to 38,000 tpy. Part of this clay is processed in a plant at Marnia, and part is shipped across the Moroccan border to a plant at Oudja.

The bentonite in the Marnia district occurs in both altered rhyolite masses and as beds in sedimentary rock of Lower Miocene age (Sadran, et al., 1955). Much of this bentonite is essentially pure montmorillonite, but some of the deposits altered from volcanic rock contain traces of kaolinite and some of the transported deposits are 10 to 20% kaolinite.

Kenya—Bentonite in the Athi River valley has been mined intermittently for use as foundry-sand bonding (Thompson, 1952).

Malagasy—Bentonite is produced on a small scale in this country, inasmuch as 99 tons were reported to have been exported from that country in 1969 (Anon., 1969c).

Morocco—Extensive deposits of bentonite occur in the Camp Berteaux district and other localities near Taourirt in eastern Morocco. This bentonite is the product of the transformation of the glassy materials in volcanic ash. It occurs in beds interstratified with Miocene lagoonal or lacustrine deposits (Millot, 1964). Production in recent years has ranged from 20,000 to 30,000 tpy (Anon., 1969l).

Another type of clay consisting mainly of magnesium montmorillonite called *ghassoul* (rhassoul in French) is produced from deposits in the Tamdafelt district in the middle Moulaya valley (Jeannette, 1952; Eyssautier, 1952) and sold for several uses requiring absorbent clay. Ghassoul is a brown to black waxy montmorillonite containing 26 to 28% MgO. It occurs in five separate beds ranging in thickness from 15 to 80 cm.

Mozambique—A bentonite processing plant near Impamputo having a capacity of 24 to 26 tpd began production in 1963. High-quality bentonite also has been produced at Lourenço Marques in recent years, and some of it is marketed in Europe (Anon., 1973a).

Senegal—Large fuller's earth deposits consisting principally of palygorskite (attapulgite) occur in the Pout region and elsewhere in Senegal (Wirth, 1968). The deposits are in lower Eocene sedimentary beds of marine origin. According to the published test data, the deposits in Senegal are not as high grade as some of the best fuller's earth mined in the United States, but they are used for several of the same products. Most of the fuller's earth produced in the Pout region is shipped to European markets.

South Africa—Bentonite is mined in Transvaal and Orange Free State, and it is reported to have been discovered near Plettenberg Bay, Cape Province (Anon., 1972j; Woodmansee and Murchison, 1969). Production has increased in recent years and now virtually all of South Africa's bentonite requirements are fulfilled by the domestic output. Apparently, the deposits mined consist chiefly of smectite minerals. The authors are unaware of any development of the palygorskite (attapulgite) deposits at Springbok Flats that are widely referred to in the geologic literature.

Tanganyika—The production of a few tons of bentonite in Tanganyika in 1958 was reported by Ross (1964).

Tanzania—Slatick (1969) noted the production of 203 tons of bentonite in Tanzania and small quantities mined in other years.

Asia: *India*—Major fuller's earth deposits occur in at least five states in India, and bentonite occurs at several localities. Probably the best known deposits are in the Barmer district, Rajasthan (Siddique and Bahl, 1965). They are believed to be transported clays deposited in a marine embayment during late Tertiary time, and the clay minerals formed by the weathering of igneous rock. Other deposits occur in Tertiary beds in Gujarat and in Miocene rocks in Jammu and Kashmir; those in Madras and Bihar are of Jurassic age. Fuller's earth mined in the Mudh district is of Eocene age, and it is chiefly palygorskite (Siddiqui, 1968). Indian bentonite is used for drilling mud, foundry-sand bonding, water sealant, and other uses. The deposits in Gujarat are near the coast, and some of this bentonite is sold in Europe and elsewhere. Apparently most of the fuller's earth produced is used domestically for purifying oils and processing fibers.

Japan—Two rather distinct types of smectite mineral deposits occur in Japan (Takeshi and Kato, 1969); one is called bentonite and the other, acid clay. The bentonite is composed chiefly of sodium montmorillonite, and the acid clay is montmorillonite in which H^- has substituted for exchangeable Mg^{++} and Ca^{++}. Deposits of both types occur at several places in Japan. Bentonite altered from bedded rocks called *Green Tuff* of Miocene age occur in Hokkaido and several places in Honshu (Takeshi and Kato, 1969; Urasima, Syoya, and Suzaki, 1966). Large deposits formed by hydrothermal alteration of rhyolite and rhyolitic tuffs occur in Higashikambara-gun district, Niigata Prefecture (Takeshi, 1963). The Teikoku deposit, one of several in this district, is principally sodium montmorillonite. Bentonite formed by diagenic alteration of tuff beds occurs at Usui-gun, Gunma Prefecture, and several places in Yamagata Prefecture (Takeshi and Kato, 1969; Omori, et al., 1961). Several of these deposits contain appreciable quantities of zeolite, cristobalite, and detrital minerals, which contaminate the montmorillonite.

Large deposits of acid clay formed by diagenic processes occur at Tsuruoka-shi district, Yamagata Prefecture, at Kitakanbara-gun, Niigata Prefecture, and elsewhere. Acid clay formed hydrothermally also occurs at several places in Niigata.

The yearly bentonite production in Japan is estimated to be approximately 300,000 tons, and demands are increasing (Takeshi and Kato, 1969). Several companies are now active and are processing bentonite in different ways for domestic markets and export. Japanese bentonite is used for foundry-sand bond, drilling mud, several pelletizing and briquetting bond uses, agricultural engineering, soil-improving agent, additives for ceramic raw materials, carriers for agricultural chemicals, paper making, filler for water-based paint, clarifying and coagulation liquids, making organic-clad bentonite, and for other uses. Acid clay is used mainly to produce activated clay by a sulfuric acid-treatment process. Aluminum sulfate and synthetic gypsum, made with the addition of calcium, are recovered as byproducts of this process. The activated clay is used in decolorizing lubricating oils and other oils and fats, for hygroscopic and absorptive applications, and as petroleum catalyst. Acid clay is also used in the manufacture of globular alumina gel, silica gel, fine powdery silica, filler for noncarbon duplicating paper, and for other uses. The exported Japanese bentonite and acid clay are shipped primarily to countries in southeastern Asia, and petroleum catalysts are apparently the principal products shipped.

Pakistan—Bentonite occurs at several places in Pakistan. Deposits in the Bhimber district of Azad Kashmir of probable middle Pliocene age have been mined since 1958 (Ali and Shahn, 1962). The bentonite is used for several purposes, including the preparation of surgical dressings. A plant for processing activated bentonite at the rate of 20 tpd was under construction at Gujrat in 1962 (Nawaz, 1962).

South Korea—The Dong Bo Clay Industrial Co. operates a bentonite plant in Seoul (Anon., 1973a). The mines are located in North Kyongsang Province.

Turkey—Bentonite and fuller's earth occur at several places in Turkey. Bentonite has been produced on a small scale for domestic use as foundry-sand bond, wine purification, insecticide carriers, and other uses for several years. In 1973, plans were underway for expansion of mining and processing of sodium bentonite (Anon., 1973). Fuller's earth presumably has been used locally for centuries.

The older productive deposits apparently are at Killik, Eskişehir Province, and Kursunlu and Küçük, Çankiri Province (Anon., 1965). The sodium bentonite is near Resadiye, Tokat (Anon., 1973).

Oceania: *Australia*—A few hundred tons of bentonite have been produced in Australia in recent years (Kalix, 1971), and there has been intermittent mining of fuller's earth for a long time. The bentonite produced in 1970, the last year for which figures are available, was 136 tons from Nanango Mineral Field and 100 tons from Ipswich in Queensland, and 116 tons from lake deposits at Marahagee and Woodanilling in the southwestern part of Western Australia. In that year, 58,244 tons of bentonite was imported from the United States, and the average f.o.b. value was $23.08 per ton. Considerable interest in bentonite in Australia has developed because of its possible use in pelletizing iron ore and other uses, and it seems likely that valuable deposits will be found. Fuller's earth was mined on a small scale in the Dubbo district, New South Wales, as recently as 1969.

New Zealand—Bentonite mining began in New Zealand at Porangahau, Hawke's Bay, in 1940, and in recent years it has also been produced in the Harper Hills region of Canterbury. The Harper Hills deposits mined are in lower Eocene beds, and apparently they formed by the alteration of volcanic materials deposited in a marine environment (Ritchie, et al., 1969). Bentonite also occurs in beds of Miocene and probable Late Cretaceous ages in this region. The bentonite mined in the Harper Hills region is a low-swelling iron-rich variety, containing as much as 10% kaolinitic clay. New Zealand bentonite is used mainly for foundry-sand bond and as an additive to ceramic material, but it may be suitable for other uses, including pelletizing taconite iron ore.

Evaluation of Deposits

Exploration: The exploration for bentonite and fuller's earth deposits, as for virtually all minerals, requires first an understanding of the geologic occurrence of deposits, then sampling, testing, and appraising of results. An understanding of the geology of the deposits is essential in the early stages and is gained by first reviewing all published and other information on deposits under consideration or on similar rocks in the region. The second step is geologic field investigations, which may require mapping, but reconnaissance studies have been adequate for many deposits. The Wyoming bentonite beds are generally easy to find in the field, because of their tendency to swell when wet and to develop a "popcorn" or frothy texture caused by shrinkage when drying. The calcium or low-swelling bentonite and some fuller's earth deposits are not as easy to find as the swelling-type bentonite, and extensive drilling is commonly required.

Once located, deposits are explored by augering or drilling. The auger method has been widely used for bentonite, mainly because the overlying rocks can be penetrated by such equipment and the bentonite sampled satisfactorily, particularly if the hole is cleaned when the top of the bentonite is reached. Power augers are also used in some fuller's earth exploration, but they have generally been found to be unsatisfactory in searching for palygorskite (attapulgite) type deposits. These deposits are commonly explored with drilling equipment, and a fishtail bit is used to penetrate the soft sandy overburden, the cuttings being flushed with water. A core barrel with a toothed cutting head is used when the top of the clay is approached, and the clay is sampled with such equipment. A double-tubed barrel has been found to be effective in obtaining more complete recovery in some deposits. Core drilling is also used in exploring for hectorite, because of hard rock in the overburden.

Testing: The testing of bentonite and fuller's earth is handicapped by the general absence of universally accepted procedures and specifications. As a result, these clays are tested and their valuable properties are expressed in many different ways. Some of the reasons for this are the widely varying properties of the clays and the many different specifications outlined by purchasers for many uses. The purchaser of a single grade of bentonite may require tests of a slurry of different concentrations, preparation procedures, aging, temperature specifications, and even measurements of liquid and plastic limits. Clearly, the absence of standardized test procedures indicates the need for an effort on the part of the American Society for Testing & Materials or another organization, with the cooperation of industry, to establish useful procedures having wide application. Rather than go into details on the many empirical and tentative test procedures that are now in use in evaluating bentonite and fuller's earth and that are likely to be superseded by more scientific and technically accurate methods, only a few generalized comments on tests will be made in the following paragraphs.

Bentonite and fuller's earth are tested for use in drilling mud by adding water and mixing slurries of specific concentrations and measuring viscosity, yield, yield point, and filtrate (a measure of wall-building properties), and making a wet-screen analysis, according to standard procedures recommended by the American Petroleum Institute (Anon., 1969, 1969a). The Oil Companies Materials Association (Anon., 1969b) in the United Kingdom also has specifications for testing drilling muds. Tests for foundry sand-bonding properties require preparation of standard mixtures of bentonite and sand and measurements of compression strengths of carefully prepared test pieces in a manner similar to the methods outlined by Fisk (1946), the American Foundrymen's Society (Anon., 1962a; 1963), and the Steel Founder's Society of America (Anon., 1965a). Tests for taconite pelletizing bonds vary considerably (Davison, 1969), but some require certain strengths in pellets measured by the number of times they can be dropped in specified intervals onto a hard surface without breaking (Stone, et al., 1971; Sastry and Fuerstenau, 1971; Wakeman, 1972). Oil-bleaching tests of fuller's earth and bentonite are commonly made according to procedures described by the American Oil Chemists' Society (Anon., 1958). Ross (1964) and Nutting (1943) also have given procedures for evaluating fuller's earth for use in bleaching oils. Tests for a variety of different insecticide, dust diluents, and carrier uses have been described by Weidhaas and Brann (1955). Most tests for absorbent granules are made according to procedures in Federal Specifications P-A-1056A, Absorbent Material, Oil and Water (for floors and decks), available from the US General Services Administration. No standard procedures have been developed for testing bentonite for use in water impedance, but rather detailed procedures for testing bentonites for canal and pond sealants have been described by the Colorado State University Experiment Station (Dirmeyer and Skinner, 1968).

As swelling bentonites are generally more

suitable for uses requiring colloidal properties and low-swelling ones are used in other ways, a preliminary indication of the potential value of bentonite can be determined by a swelling test. Perhaps the most common test is the simple procedure of adding 2 g of dried and ground bentonite to a 100-ml graduate filled with distilled water and determining the volume of the settled bentonite in milliliters (Ross, 1964).

Mineralogical Investigations: In addition to the testing of the type outlined in the foregoing paragraphs, information on the mineralogical and chemical composition of bentonite and fuller's earth is also required. Mineralogical composition, including the purity and undesirable contaminants, is determined in some company laboratories. Other companies obtain information of this type from consultants who are experienced in the use of X-ray, electron microscope, differential thermal analysis, petrographic microscope, and other methods of investigating clays.

Appraisal of Field and Laboratory Results: The final step in the evaluation of bentonite and fuller's earth deposits involves the synthesis of data collected during the field investigations, sampling and testing, and other laboratory programs. Maps are prepared locating drill holes and showing the thicknesses and quality of clay and thicknesses of overburden. Reserves of various grades of clay are calculated, and the mining cost is determined from such maps and from geologic and engineering notes taken during the fieldwork.

Preparation for Markets

Mining Methods: In the United States, virtually all bentonite and fuller's earth are mined by stripping methods. Bulldozers and scrapers or pans are most commonly used in removing overburden, but one company stripping near Attapulgus, GA, operates a large walking dragline powered by electricity for stripping. In the typical pit, the overburden is removed in panels, the clay mined by loading trucks with dragline or endloaders, and the overburden from an adjacent panel shifted to the mined-out area. One company hauling on privately owned roads operates trucks having 100-ton capacity, but most hauling requires use of public highways and lighter trucks. Thicknesses of overburden removed vary considerably. Most Wyoming-type bentonite mined is under less than 30 ft of overburden, but in a few places overburden is as much as 40 ft thick. The thickness of overburden removed from some southern bentonite deposits is as much as 100 ft (Teague, 1972), and beds of the same thickness have been stripped from low-swelling bentonite in one district in Arizona. The overburden stripped from fuller's earth deposits ranges from a few feet to more than 75 ft.

The rocks overlying most bentonite and fuller's earth deposits are soft and can be loaded directly by self-propelled or pushed scrapers, but in places they must be loosened by bulldozers and rippers before loading. Some of the overburden above fuller's earth near Attapulgus, GA, and Quincy, FL, contains limestone and cemented sandstone that must be blasted. Volcanic rocks above hectorite deposits in California also must be blasted where stripping is done. Bentonite is mined underground by the square-set method in the New Discovery mine near Beatty, NV (Papke, 1969), and insofar as the authors are aware, this is the only bentonite now mined underground in North America. Hectorite in California and bentonite in the Ash Meadows district, Nevada (Papke, 1969), and the Chambers [Cheto], Arizona, district, were formerly mined underground. Early in this century, overburden was stripped from fuller's earth near Quincy, FL by hydraulic methods, and the clay was transported by trams.

Most bentonite and fuller's earth produced in other countries are also strip mined, but underground methods are used in a few places, including the Combe Hay district in the United Kingdom (Anon., 1969f). Deposits in this district are selectively mined by hand, using compressed air picks. The clay is extracted in lumps 3 to 18 inches across, and it is hauled to underground storage facilities in mine cars drawn by electric locomotives. It is loaded as needed from the storage dumps onto trucks and hauled to the plant.

Because of variable physical properties, most bentonite and some fuller's earth deposits are selectively mined. Bentonite from a single pit or bed may be separated into as many as three stockpiles at the plant, each with different physical properties. Various grades of bentonite suitable for different uses are then prepared from the separate piles and by blending clay from more than one pile. Blending bentonite as it is dumped on stockpiles, using earth-moving or cultivating equipment to obtain a uniform clay, is also common practice. The properties of fuller's earth, and particularly the palygorskite (attapulgite) type, vary considerably, but deposits tend to be more uniform than bentonite. Commonly, an entire deposit will be suitable for one product, and another deposit in the same district may have the properties required for a different use.

Beneficiation and Processing Techniques: Both bentonite and fuller's earth are processed by simple milling techniques that involve removal of water and, in some instances, other volatile matter, and grinding to suitable sizes. The Wyoming or high-swelling bentonite, when delivered to the plant, contains approximately 30% moisture; the free water in the southern or calcium-type bentonite is about 25%. Some fuller's earth contains as much as 50% volatile matter and 10% undesirable impurities. The processed bentonite ordinarily contains only 7 or 8% water, but because it is hygroscopic, it may contain considerably more when used. Moisture contents of most freshly processed fuller's earth are approximately the same as those of bentonite. Most absorbent granules, however, have been heated beyond the temperature required to drive off hydroxyl water in the minerals present, and any moisture in the granules when consumed has been absorbed.

In most plants, the raw clay is passed through some sort of a clay "slicer" to break up the large chunks before drying. One plant processing fuller's earth uses an extruding process before drying to improve properties desired in certain products. In this process the crushed clay is fed into a pugmill, where water is added, and the clay is then extruded as rods approximately ½ in. in diam. Drying in most plants is done by gas or oil-fired rotary dryers as large as 10 ft in diam and 65 ft long. A fluid-bed dryer is in operation in one bentonite plant at Colony, WY, and a similar dryer is used in processing fuller's earth at Meigs, GA. The temperature in dryers varies with the intended use of the clay. The desirable properties of both bentonite and colloidal grades of fuller's earth are reduced greatly if the clay is heated too much. The temperature in dryers processing bentonite is likely to be approximately 800°C at the inlet, 100° to 200°C at the outlet, and 400° to 500°C in the main drying zone. The bentonite itself is kept at temperatures of less than 150°C (Ross, 1964). Temperatures used in processing fuller's earth range from 150° to 650°C. The lower part of the range is used for colloidal grades and the higher part for absorbent granules.

The dried clay is ground and sized in several ways. In some plants, rods in rotary dryers do much of the grinding, but most of the powdered product is ground with roll and hammer mills or other pulverizers, and screened. Most bentonite and much of the fuller's earth used is ground to approximately 90% finer than 200 mesh (US Standard sieve size). Some fuller's earth for special markets is reported to be 95% finer than 10-micrometer particle size. Granular grades of bentonite are also sold, and the wide use of absorbent granule-type fuller's earth requires the preparation of a coarse product. The preparation of coarse absorbent granules is, in effect, a beneficiation process, because much of the fine and medium-grained sand impurities in the fuller's earth are in the size fraction discarded.

Hectorite and some southern bentonite, in addition to being processed like other bentonite, are beneficiated by a hydroclassification process. In this process, the clay is dispersed in water and pumped into a multistage concentrating circuit where nonclay and other undesirable materials are separated and removed. The slurry is then passed through drum or spray-type dryers, and the concentrate is ground. A high-value material is prepared in one plant by centrifuging a dilute slurry.

Transportation: Several types of packaging are used for both bentonite and fuller's earth. For other than bulk shipments, multiwalled paper bags, some of which have plastic liners, are most common. Fiber drums are used for specialty products sold in rather small quantities. Metal drums of 100-lb capacity are required for some uses, particularly absorbent granules used aboard ships and where lengthy periods of storage or repeated handling are necessary. One French company uses the modern containerized method for export shipments (Anon., 1969j). Most of the granules sold for animal litter are packaged in 10 to 30-lb paper bags, and packaging is a major item in the cost of the product.

The method of transporting bentonite and fuller's earth depends on use and location of markets. Most of the processed bentonite is shipped by rail in bulk hopper cars, but boxcars and bulk trucks are also used. Bagged bentonite is commonly palletized for shipment by either rail or truck. Trains carrying only bentonite are used for large shipments to distant points and for overseas consignments. Ore carrier-type ships having bentonite as the only cargo are commonly used for Great Lakes and overseas shipments. The bentonite transported this way is only partly dried at the point of origin and is shipped semicrude. This method makes it possible to load and unload it by bulk-handling equipment, avoiding the costs of bagging, but it requires further drying and grinding at or near the place used. Much of the bentonite used in pelletizing taconite ore, as well as most of that sold in the European markets, is handled in this way.

Fuller's earth is transported in a similar manner, except that orders are not sufficiently large to permit trainload or some of the other bulk-handling methods. Though some fuller's earth is transported unbagged in special cars, most is bagged for rail transportation. Trucking is also used more extensively in the transporting of fuller's earth than for bentonite, because orders tend to be smaller and are delivered directly to the warehouses of distributors. Granules for this use are commonly delivered by truck to warehouses of the distributing chain store.

Synthetic and Organic-clad "Bentonite"

A plant (Fig. 6) operated by Baroid Div., NL Industries, Inc., at Houston, Texas, making synthetic mica-montmorillonite began production in 1970. An early announcement of this plant indicated that one of the products was hectorite (Anon., 1970). Published mineralogical reports related to the research on the development of this synthetic material refer to a randomly interstratified alumina montmorillonoid (Granquist and Pallack, 1967; Granquist, et al., 1972) and SMM, which stands for synthetic mica-montmorillonite (Wright, et al., 1972). The principal product made in 1972 was advertised in a company pamphlet as BARASYM SMM. It consists of small stacks of thin, parallel, irregular platelets, having an equivalent diameter on the order of 1000Å and a thickness of about 10Å. This material is recommended for usc in catalytic cracking, hydrogenation/dehydrogenation, double-bond isomerization, and disproportionation, and it may be a base for reforming catalysts and a component in hydrotreating catalysts. As a petroleum-cracking catalyst, BARASYM SMM is reported to be intermediate in activity between silica-alumina and zeolitic catalysts, to have a high selectivity for gasoline, and product distribution similar to silica-alumina catalysts.

FIG. 6—*Synthetic "bentonite" plant, Houston, Texas (courtesy Baroid Div., NL Industries Inc.).*

The company's advertising pamphlet indicates it has typical properties as follows:

Appearance: White, free-flowing powder
Particle size: 20–80 micron microspheres
Density:

Apparent: 0.6–0.7 g per ml
True: 2.5 g per ml

Typical analysis: 48% SiO_2
(as received)
37% Al_2O_3
2.6% NH_3
1.6% F
12.0% H_2O

Trace analysis: 10 ppm Ni
190 ppm Fe
800 ppm Na
100 ppm Ca
250 ppm Mg
0 ppm SO_4

Surface area: 110–160 sq m per g, dependent on calcination temperature
Saturation volume (calcined 1300°F)
N_2 sorption: 0.27 ml per g
Hg porosimeter, 50,000 psi: 0.62 ml per g (excluding macropore volume)
Cation exchange capacity: 1.5–1.6 meq per g
Accessible * surface acidity: 1.8×10^{19} sites per g

---

* To pyridine vapor at 200°C

Synthetic hectorite is also made in the United Kingdom by Laporte Industries, Ltd., and sold under the brand name *Laponite CP* (Green, et al., 1970). This material is purer than naturally occurring varieties, as it contains no fluoride. It presumably is suitable for some of the same uses that synthetic bentonite was developed for in the United States.

For several years an organic-treated bentonite has been on the market under the registered trade name *Bentone* (Grim, 1962). This product is processed in such a way that the original inorganic exchangeable ion on the bentonite is replaced by an alkyl amine organic cation. This reaction produces a hydrophobic clay because the inorganic ions that can be hydrated are removed, and a large part of the clay-mineral surface formerly capable of adsorbing water is coated by hydrocarbon chains. Organic-clad bentonites are used in making paint; to gel organic liquids; to produce greases having superior properties for adherence to metal, ability to repel water, and resistance to high temperatures; and in other products.

Organic-clad bentonite is also made in the United Kingdom and France (Anon., 1970) and in Japan (Takeshi and Kato, 1969). Some of the Japanese product is sold under the brand name *Nikkagel.* It is prepared by treating betonite with alcohol under pressure and heat followed by ion exchange with an organic ammonium salt. A product called *Organite* is prepared by further exchanging Nikkagel with an amine having a benzene nucleus.

Future Considerations and Trends

The overall demands for both bentonite and fuller's earth seem destined to increase at a rate approximately equal to the growth in Gross National Product. According to Cooper (1970), in the year 2000, production of bentonite in the United States is likely to be three to five times the 1968 amount. The growth rate in fuller's earth requirements will probably exceed the rate forecast for bentonite.

Both bentonite and fuller's earth are used in so many different products and in so many ways that there are many possible substitutes for specific uses, and both have lost markets to other materials in the past. One of the most obvious losses of fuller's earth to other materials occurred when the use of synthetic materials in petroleum refining increased before World War II, and more efficient methods of refining were introduced. Bentonite sales have also suffered mildly from the use of oil-based and speciality muds; however, bentonite is likely to remain the cheapest and most efficient material for mud in certain types of drilling for years to come. Bentonite is in some danger of a market loss which may be caused by a breakthrough in the use of another material for bonding taconite pellets. Though the proportion of bentonite added in taconite bonding is small, it is still a contaminant, and some type of bonding material that will burn off completely or even add heat in the steel furnace will continue to be sought.

The manufacture of synthetic hectorite and other montmorillonitelike mineral products in the United States and United Kingdom marks a technological breakthrough. Worldwide markets for organic-clad bentonite and washed or otherwise purified white bentonite for speciality products are also likely to increase with general economic growth and the development of new uses and products.

Rising costs of transportation and fuel are two of the major pressures on the bentonite and fuller's earth industries that are likely to have a significant influence on future trends. The cost spread between rail and ocean transportation is the principal reason for the loss to Greek producers of some of the Wyoming-type bentonite market for bonding taconite ore in eastern Canada. The comparatively low cost of ocean transportation is also the primary motivation behind an overseas search for bentonite reserves near deep water. Progressive increases in fuel costs, particularly for the low-sulfur type required for ecological reasons, seem to be almost a certain forecast. Fuel shortages have caused temporary shutdowns of some plants, and problems related to fuel supply seemed ominous at the time this article was written.

Kaolin, Ball Clay, Halloysite, and Refractory Clay

Definitions and Classifications

Kaolin: The name kaolin is derived from the Chinese term *Kauling* meaning high ridge, the name for a hill near Jauchau Fu, China, where the clay was mined centuries ago. Apparently the first to use the name kaolinite for the mineral of kaolin were Johnson and Blake (1867). The term kaolin is now variously used as a clay-mineral group, a rock term (consisting of more than one mineral), an industrial mineral commodity, and interchangeably with the term china clay. The following mineralogical definition by Ross and Kerr (1931) is probably the most widely accepted one.

> By kaolin is understood the rock mass which is composed essentially of a clay material that is low in iron and usually white or nearly white in color. The kaolin-forming clays are hydrous aluminum silicates of approximately the composition $2H_2O \cdot Al_2O_3 \cdot 2SiO_2$, and it is believed that other bases if present represent impurities or adsorbed materials. Kaolinite is the mineral that characterizes most kaolins, but it and the other kaolin minerals may also occur to a greater or lesser extent in clays and other rocks that are too heterogeneous to be called kaolin.

When applied as a term for an economic clay commodity, the foregoing definition must be modified to include some indication of use and to account for the fact that most kaolin now marketed is beneficiated to improve purity and whiteness. The following definition will apply to further discussions of kaolin in this chapter. Kaolin is a clay consisting of substantially pure kaolinite, or related clay minerals, that is naturally or can be beneficiated to be white or nearly white, will fire white or nearly white, and is amenable to beneficiation by known methods to make it suitable for use in whiteware, paper, rubber, paint, and similar uses. The term is applied without direct relation to purity of deposits. Many very large kaolin deposits are essentially pure, and they require little concentration during preparation for market. Most are slightly off-color and require bleaching, and others contain as little as 10% clay that must be washed and concentrated to recover marketable kaolin.

China clay is an ancient term originating from the use of clay, later found to be kaolinitic, in porcelain tableware and art objects in China. This term will be used rarely in this chapter.

Ball Clay: *Ball clay* is a term apparently originating in the United Kingdom many decades ago from the mining practice of prying out a lump of clay, rolling it into a crude ball, and loading it into a horse-drawn cart or wagon. Ball clay is defined by the American Society for Testing & Materials (Anon., 1971a) as a secondary clay commonly characterized by the presence of organic matter, high plasticity, high dry strength, long vitrification range, and a light color when fired. Kaolinite is the principal mineral constituent of ball clay, and it typically makes up more than 70% of this type of clay. The minor differences between the light fired color required by this definition and the white required for kaolin provide little practical basis for distinguishing these two types of clay. Therefore, ball clay is actually a variety of kaolin, but it is treated as a separate type of clay because of its wide application in marketing and as a use term in the ceramic industries.

The term ball clay is used in the United States, United Kingdom, India, Republic of South Africa, and a few other countries. Similar clay or clay used in the same products as ball clay is included with kaolin or china clay in most countries.

Halloysite: Halloysite occurs in two forms; one has the composition $(OH)_8Si_4Al_4O_{10} \cdot 4H_2O$ and the other $(OH)_8Si_4Al_4O_{10}$. The hydrated form is generally a dense porcelainlike hard clay that dehydrates at surface temperatures or slightly above to a white or light-colored porous, friable, or almost cottony-textured material. The dehydrated form is similar to kaolinite in composition and mineral structure, and the hydrated form has a *c*-axis spacing greater than that of kaolinite. The two forms of halloysite have caused considerable nomenclature problems. The system most commonly used now is to restrict the name *halloysite* to the hydrated form and to call the dehydrated form *metahalloysite*. However, some reports use the name *endellite* for the hydrated form and the name *halloysite* for the dehydrated form. As virtually all this clay below the surface occurs in the hydrated form, only the name *halloysite* will be used further in this chapter, except in the mineralogy discussion.

Halloysite, like ball clay, is a variety of kaolin, and it is included with kaolin in most production statistics. However, the term halloysite is accepted worldwide, and it is used in the clay industries, as well as by mineralogists, geologists, and soil scientists.

Refractory Clay: Refractory clay includes several varieties of kaolinitic clay that are used in the manufacture of products requiring resistance to high temperatures. The qualities and properties of refractory clays are expressed in terms of pyrometric cone equivalents (PCE) which is a method of designating fusion points. Pyrometric cone equivalent is defined as the number of that standard cone whose tip would touch the supporting plaque simultaneously with that of a cone of the material being investigated, when tested in accordance with the Standard Method of Test for Pyrometric Cone Equivalent (PCE) of refractory materials (ASTM Designation C24) of the American Society for Testing & Materials.

The fusion points of products made from refractory clays range from just above PCE 19, the lower cutoff according to accepted definitions (Norton, 1968), to as high as cone 37 (Table 4). The refractory properties of kaolinitic clays are ordinarily classified, according to suitability for heat service, as low (PCE 19 to 26), moderate (PCE 26 to 31½), high (PCE 31½ to 33), and superduty (PCE 33 to 34). A few essentially pure kaolinite clays have fusion points as high as cone 35, and some refractory kaolins and fire clays containing diaspore, boehmite, or gibbsite have PCE's as high as cone 37.

TABLE 4—Temperature End Points of American Pyrometric Cones

No. of Cone	End Point* °F	End Point* °C
18	2714	1490
19	2768	1520
20	2786	1530
23	2876	1580
26	2903	1595
27	2921	1605
28	2939	1615
29	2984	1640
30	3002	1650
31	3056	1680
32	3092	1700
32½	3137	1725
33	3173	1745
34	3200	1760
35	3245	1785
36	3290	1810
37	3308	1820
38	3335	1835

* Temperatures given are approximate and are applicable only when heated at specified rates (Norton, 1968).

The principal types of clay included in the refractory clay are fire clay, kaolin, and ball clay. Definitions of refractory kaolin and refractory ball clay are the same as given for these two types of foregoing definitions, except that the adjective "refractory" is added to indicate use. The definition of fire clay used in this chapter is essentially the one by Norton (1968), which assigns this term to all clays that are not white burning and have a PCE above 19. The term *fire clay*, therefore, excludes most kaolin and ball clay because they burn white, but it does include kaolin and ball clay that are colored when fired. This classification of refractory clay as a group name and the threefold subdivision are by no means universally applied. Other reports, including earlier ones by the authors themselves, have used the terms *refractory* and *fire clay* interchangeably.

Kaolin

History and Uses: The sedimentary kaolins of the Coastal Plain of Georgia and South Carolina and the residual kaolins of North Carolina have been known since Colonial times. Smith (1929) quotes from Sholes in his chronological history of Savannah, "1741—Porcelain Clay was discovered in or near Savannah by Mr. Duchet, and china cups made." In 1767, Josiah Wedgwood sent Thomas Griffiths to Macon County, North Carolina, to send a shipment of china clay to England (Goff, 1959). Minton (1922) recorded that, "As early as 1766 American clays from Georgia, Florida, and the Carolinas were being sent to England in considerable quantities." These clays were regularly imported and used by Wedgwood until the clays of England were available. In 1768 (Barton, 1966), the English kaolins in Cornwall were discovered, which virtually ended the mining of the Georgia kaolins for more than a century. In 1876 (Smith, 1929), the mining of the sedimentary kaolins of Georgia was started again by Riverside Mills of Augusta in Richmond County. Mining of the Georgia kaolins has been continuous since that time. In addition to the original pottery in Savannah, a second pottery was established near Bath, Aiken County,

SC, at an early date. This plant was destroyed during Sherman's march to the sea (Lang et al., 1940). However, kaolin mining was active again in Aiken County shortly after the Civil War and has continued with only minor interruptions until the present. South Carolina has ranked as the second leading kaolin-producing state for several decades. Kaolin mining in North Carolina started in 1888 in Jackson County (Parker, 1946), and in 1904, the earliest mines opened in the Spruce Pine district. The first kaolin operation in Vermont (Ogden, 1969) was in the early 1800s, and most of the product was sold in Troy and Albany, NY.

Production figures for kaolin from most areas in the late 1800s and early 1900s are difficult to trace. It is estimated that about 3 million tons were produced in Georgia before 1932 (Kesler, 1956), after which much better statistics have been available. In 1980, total production in Georgia alone was over 6 million tons, and South Carolina ranking second produced more than 657 thousand tons.

Kaolin has many industrial applications (Murray, 1963a), and new uses are still being discovered. It is a unique industrial mineral because it is chemically inert over a relatively wide pH range, is white, has good covering or hiding power when used as a pigment or extender in coated films and filling applications, is soft and nonabrasive, has low conductivity of heat and electricity, and costs less than most materials with which it competes. Some uses of kaolin require very rigid specifications including particle-size distribution, color and brightness, and viscosity, whereas other uses require practically no specifications; for example, in cement, where the chemical composition is most important. The better grades of kaolin make up most of the tonnage sold and, of course, have the highest value. Many grades of kaolin are specially designed for specific uses, in particular for paper, paint, rubber, plastics, and ceramics.

The paper industry is by far the leading consumer of kaolin (Fig. 7), and approximately 50% of the total production is used in paper products. Much of the paper used in today's color picture magazines contains as much as 30% by weight of kaolin. Kaolin used in paper fills the interstices of the sheet and coats the surface, imparting smoothness, gloss, brightness, opacity, and printability (Bundy, et al., 1965). Kaolin is less expensive than paper pulp and thereby effectively lowers paper-production costs, and it is cheaper or more efficient than other materials used in filling and coating. It is inert to the other ingredients in paper, is retained well between the paper fibers, is available in large quantity, and the platy structure of kaolinite lends itself to the production of high-gloss papers. The low viscosity of kaolin at both high and low shear rates is a very important property, because at today's very high speed production rates the coating must be applied at high solids content and still give the correct coating thickness and opacity to the paper. High-quality paper coating kaolin flows readily when applied with very high speed coating equipment, giving the paper a smooth and even film.

The rubber industry uses large tonnages of kaolin as a filler or extender in both natural and synthetic rubber (Fig. 7). Kaolin is incorporated in the latex mix to improve properties such as strength, abrasion resistance, and rigidity; furthermore, it lowers the cost. Through the years, two terms, "hard-kaolins" and "soft-kaolins" (Klinefelter et al., 1943), have become commonplace in describing types of kaolin used by the rubber industry. The hard kaolin, though distinctly softer than flint kaolin or flint clay, is a very fine particle-size kaolin that tends to improve the tensile strength of rubber and its resistance to tear and abrasion. The principal products in which large quantities of the hard varieties are used are footwear and cable covers. Soft kaolin lacks the reinforcing properties of the hard kaolins and is used to lower elasticity and improve abrasion resistance, particularly in such products as floor tile and soft rubber goods.

Kaolin is used as an extender in paints because it is chemically inert, has a high covering power, gives desirable flow properties, is low in cost, is white, and reduces the amount of expensive pigment required. In addition, it has very excellent suspension properties and is available in a wide range of particle sizes which can be used in many types of paints. For example, coarse-particle kaolins are used in paints where a dull or flat finish is required, and fine-particle kaolins are used in high-gloss paints. Large quantities of calcined kaolin and water-beneficiated kaolin are used in interior flat wall paints and in metal primers. Water-beneficiated grades of kaolin disperse easily in water and are, therefore, particularly suited to latex paints. Some kaolin is chemically treated to make it organophilic or hydrophobic and thereby is suitable for use in exterior oil-base paints. Calcined kaolin, because of its resistance to abrasion and dry covering properties, is being used in increasingly large quantities by paint manufacturers.

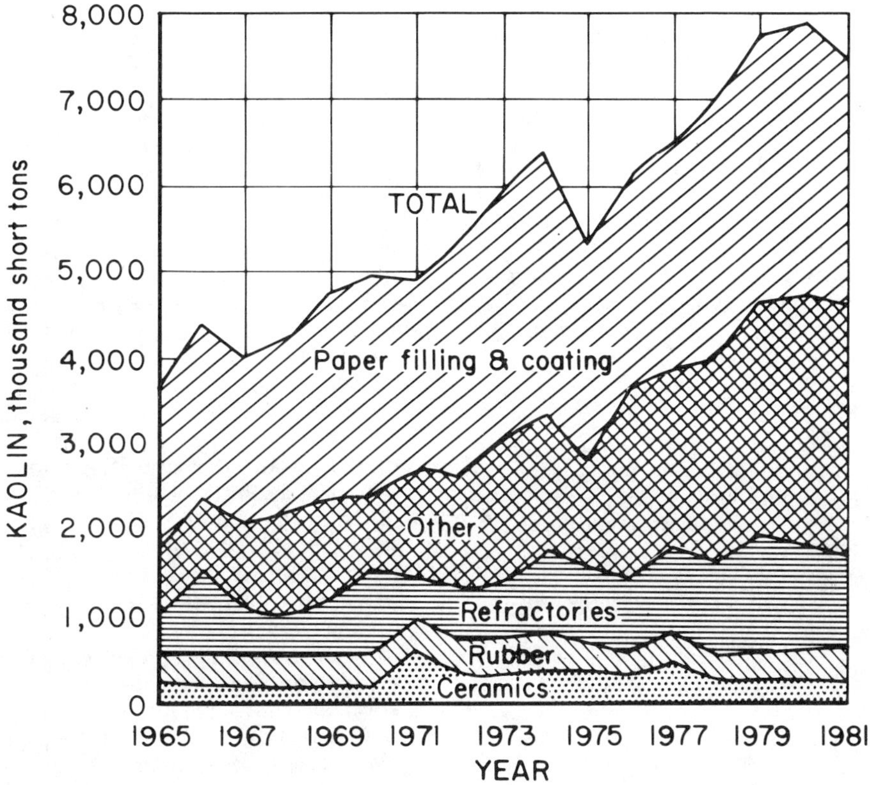

FIG. 7—*Kaolin sold or used by domestic producers for specified uses (Ampian, 1980, 1983).*

Calcined kaolin is an excellent extender for TiO_2, the leading paint pigment, particularly in latex types, and its use reduces costs and simplifies the formulation of the paint.

Kaolin is used extensively as a filler in plastics. The desirable characteristics in plastics resulting from the use of kaolin include smoother surfaces, more attractive finishes, dimensional stability, and resistance to chemical attack. Manufacturers of polyvinyl chloride (PVC) use kaolin as a reinforcing agent, and it makes the plastic more durable. Calcined and partially calcined kaolins are used as a filler in PVC wire insulation to improve electrical resistivity. In manufacturing glass-reinforced polyesters, kaolin has helped eliminate flow problems which hindered production of large products, such as automobile body parts and boat hulls. The kaolin actually results in a stronger, more uniform part or body. Plastics are a major area of future growth for kaolin products. They are used extensively in vinyl floor coverings, and surface-modified kaolin gives improved dispersibility in certain plastic systems, which will greatly increase their usage.

Historically, kaolin was first used in ceramics, and this is perhaps still the best known application, even though the total tonnage sold for this use is small compared with that marketed for paper, paint, and rubber manufacturing (Fig. 7). It is used in the manufacture of whitewares, wall tile, insulators, refractories, and in some face brick when a white color is desired. Specially designed grades with high green and dry strengths are available for use by whiteware manufacturers. The most important properties for this use are plasticity, strength, and fired color. Kaolin is used in insulators because of its properties of low conductivity, high dielectric constant, low power loss, plasticity, and fired strength. Kaolin is used in casting operations because of its excellent flowability, suspension properties, color, and water-release characteristics. Kaolin is also used in porcelain, again because of its excellent suspension characteristics and good color.

Kaolin has many other important applications, but in terms of tonnages required compared with the aforementioned uses, they are rather small. Although the following list is not complete, it gives an idea of the extremely wide range of products containing kaolin as a suspending agent, filler, extending agent, or a main constituent because of its chemical or mineral composition, whiteness, particle size, or other properties:

ink	cement
adhesives	fertilizers
insecticides	plaster
medicines	filter aids
food additives	cosmetics
catalyst preparations	chemicals
bleaching	crayons
adsorbents	pencils
	detergents

porcelain enamels
paste
roofing granules
sizing
foundries
linoleum
floor tiles
textiles

Production Specifications: *Paper Filler and Coater*—The most important specifications for kaolin used in paper applications are brightness, particle size and particle-size distribution, viscosity, and screen residue. Some purchasers also require tests of abrasiveness. The Technical Association of the Pulp and Paper Industry [TAPPI] (Lyons, 1966) has given detailed information regarding the properties and test procedures for kaolin used in paper, and only brief descriptions of the more important tests will be outlined.

BRIGHTNESS—This test consists of drying, pulverizing, forming a plaque, comparing with a standard of known brightness, and calculating the brightness measured at 457 millimicrons with a reflectance meter constructed, calibrated, and operated according to the requirements of The Technical Association of the Pulp and Paper Industry Standard T452M-48. This test must be very carefully done by an experienced and competent technician in order to obtain accurate and reproducible results. In general, a minimum brightness of 80% is required for filler clay and 85% for coating clays.

PARTICLE SIZE.—Many methods of determining particle size and particle-size distribution are used, but the most accurate is based on the Stokes principle. Size determinations can be made by the application of this principle, because settling rates in fluids are controlled by particle diameters when the attraction of gravity, densities of particles and fluids, shape of particles, and the viscosity of the fluids are identical. Coarse particles, therefore, settle more rapidly than finer ones, and determinations of a given particle size and distribution can be made by the time-rate principle. The time-rate principle is equally valid for settling by gravity or when settling is augmented by centrifugation. A rapid test for plant control consists of the preparation of a standard, dilute deflocculated slip of the clay in question, settling by controlled centrifugal force for predetermined short-time intervals under such conditions as to effect reproducible packing of the sediment, and measuring the apparent volume of sediment at each time interval. Typical particle-size analyses of filler and coating grades are shown on Fig. 8.

VISCOSITY—Flow properties of kaolin are very important to the paper coater because they affect the functioning of the coating operation as well as the final quality of the paper product. Two viscosity tests are used in the kaolin industry, one to measure high shear viscosity and the other to measure low shear viscosity. The measurements are generally made on slurry suspensions at 71% solids. A Brookfield four-speed viscometer is used to measure low shear viscosity. A Hercules viscometer is used to measure high shear viscosity. Standard procedures and conditions are rigorously followed in making these tests.

SCREEN RESIDUE—Material held on a 325-mesh screen is regarded as grit or residue. A 100-g sample is thoroughly mixed and dispersed with a dispersing chemical. After carefully determining the weight and percent solids of the dispersed slip, the amount of residue retained on the screens can be accurately weighed and percentages calculated.

Filler for Paint and Plastics—In addition to several of the properties required for use in paper, the kaolin used in paint and plastics must meet certain electrical resistivity specifications. The resistivity test is designed to give an indication of the amount of residual salts that might be contained in the clay. The higher resistivity values reflect a low soluble-salt content. A certain weight of kaolin is added to a specified weight of distilled water. The suspension is boiled for 5 min and reweighed. The loss of

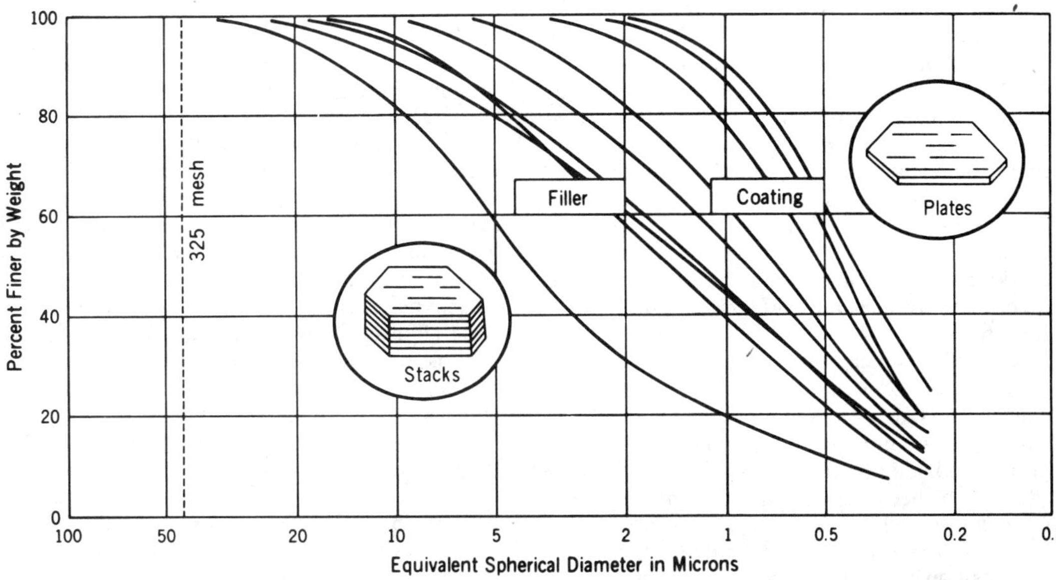

FIG. 8—*Graph showing typical particle sizes of paper filler and coating grades of kaolin.*

distilled water from boiling is replaced, and the conductivity is determined using an industrial conductivity bridge.

For certain paint applications, the Hegman fineness-of-grind test specification is important. This test is intended to measure the degree of dispersion (Anon., 1970c) and consists of a fineness gage and a scraper. The gage is calibrated from 0 to 8, and the fineness is reported by numbers. The higher the number, the finer the particles and the better the dispersion. Some paint customers specify that a pigment or extender must be finer than a particular Hegman number.

Ceramics—The most important tests for kaolin sold for ceramic products are modulus of rupture, casting rate, pyrometric cone equivalent (PCE), fired color, and shrinkage. The measurement of strength of clay bodies is a basic requirement by the ceramic industry. The modulus of rupture (MOR) is the fracture strength of a material under a bending load. The MOR measurement is made on a long bar of either rectangular or circular cross section, supported near its ends; a load is applied to the central part of the supported span. ASTM has standard methods of test for MOR of ceramics.

Casting rate is important, and fine-grained bodies cast more slowly than coarse ones. Viscosity of a slip must be carefully controlled, for if it is too viscous the slip will not properly fill the mold or drain cleanly. Therefore, viscosity is measured and controlled on shipments of kaolin used in the casting process.

For certain applications, particularly for refractory purposes, the pyrometric cone equivalent (PCE) is measured. The pyrometric cone measures the combined effects of temperature and time (Anon., 1952; Klinefelter and Hamlin, 1957; Norton, 1968). The cones consist of a series of standardized unfired ceramic compositions molded into the shape of triangular pyramids. The sample of kaolin is molded into the standard cone shape and is heated along with several standard cones so that its end point can be determined in terms of an equivalent cone number. ASTM test method C24–56 gives the procedure of making this PCE test (see Table 4).

The fired color of kaolin is also important because some burns white and some burns buff or slightly pink. This is a visual test, and the test specimen, fired to a predetermined temperature, is compared with a standard.

Shrinkage is an important property. Ceramic articles undergo shrinkage at several different points in the manufacturing sequence. During drying, the article will shrink in varying amounts depending upon its composition and the amount of water present. The article shrinks further during firing. Therefore, it is important to know both the drying shrinkage and the firing shrinkage. Linear shrinkage and volume shrinkage can both be measured, although linear shrinkage is more commonly reported (Jones and Berard, 1972).

Miscellaneous Uses and Products—Because of the many different uses of kaolin, the industry is plagued by many different purchase specifications and product requirements. Virtually none of these, except those outlined in foregoing discussions, have been standardized, and purchase specifications are frequently changed. However, the grades of kaolin now marketed by the major producers, those grades described in sales and advertising brochures, can, with or without blending, fulfill a wide variety of specifications.

Geology: *Mineralogy*—By definitions, the clays discussed in this section consist chiefly of kaolin-group minerals. The members of this group are kaolinite, nacrite, dickite, and halloysite. With the exception of the hydrated form of halloysite, which has somewhat more water, all these minerals have essentially the same composition. Their overall crystal structures are also similar, but minor differences in arrangements of ions in octahedral positions, stacking of unit layers, crystal habit, etc., are sufficient to cause the use of separate names. Kaolinite is, of course, the most common member of the group, and halloysite obviously forms halloysite deposits. The other two members of the group rarely occur in economic deposits.

Structurally, kaolinite consists of an alumina octahedral sheet and a silica tetrahedral sheet. These sheets form triclinic crystals. The theoretical structure formula of kaolinite is $(OH)_8$-$Si_4Al_4O_{10}$, and the theoretical composition is SiO_2 46.54%, Al_2O_3 39.5%, and H_2O 13.96%. There is relatively little ionic substitution in the mineral lattice, although there is evidence suggesting minor substitution of iron for aluminum in some kaolinite. The perfection or degree of ordering of kaolinite crystals varies considerably. By interpretation of X-ray diffraction data, Murray and Lyons (1956) have shown that the crystallinity of Georgia kaolin ranges from poorly ordered to very well ordered forms. Other investigators (Brindley and Robinson, 1946; Range, et al., 1969) have found similar variations in ordering in kaolinite from several localities. The more perfectly ordered forms of kaolinite commonly occur in hexagonal or subhexagonal crystals. Growths of crystals along the *c*-axis into accordionlike or vermicular forms are also common. Fig. 9 is an electron micrograph which shows both the hexagonal thin plates and vermicular stacks of kaolinite.

The clays described in this section contain, in their natural occurrence, a wide variety of mineral impurities. Georgia kaolin generally is 85 to 95% kaolinite, the remainder is mainly quartz, minor muscovite, biotite, smectite, ilmenite, anatase, rutile, leucoxene, goethite, and traces of zircon, tourmaline, kyanite, and graphite. Kaolin deposits in Latah County, Idaho, contain considerable ilmenite (Hosterman et al., 1960). Kaolinite and halloysite make up from 10 to 40% of residual deposits in the Spruce Pine district, North Carolina. The bulk of these deposits consists of angular quartz, muscovite, microcline, and partly altered plagioclase (Parker, 1946). The well-known English kaolins, which are 10 to 40% kaolinite, consist mainly of quartz, mica, and potash feldspar; tourmaline is also present (Bristow, 1969).

Occurrence—Most kaolin is soft and plastic when natural moisture is present or where water has been added to dried material. One way of describing the natural texture is by the colloquial phrase—a good whittling clay—which refers to the ease with which the clay can be carved with a knife. Dry kaolin is ordinarily friable, and a common but not a diagnostic test for it is its stickiness when touched to the tongue. Some types of kaolin, which have apparently been under considerable overburden or which contain introduced silica, are hard and are sometimes referred to as flint kaolin. Though much more lithified than plastic varieties, this so-called flint kaolin is softer than most flint clay referred to in the section on fire clay, which also consists of kaolinite.

Kaolin and related clays occur in several different types of deposits. Many kaolin deposits throughout the world are in the form of tabular lenses and discontinuous beds in sedimentary rock. Most deposits of this type are of Cretaceous age or younger. The shapes and lateral extent of such deposits vary appreciably. Deposits as much as 60 ft thick are common in some districts, and bodies of kaolin are known to extend laterally for more than a mile. Extensive sedimentary deposits of this type occur in the Georgia-South Carolina kaolin belt (Anon., 1970i), Arkansas bauxite region (Gordon, et al., 1958), and Ione district, California (Johnson and Ricker, 1948) [Fig. 10].

Residual kaolin deposits occur in weathered rock, and they are at or very near the surface, except where they formed during ancient weathering cycles and were later buried by younger rocks. Most residual kaolin deposits have formed from igneous rock. The age of kaolin formation is commonly controlled by the age of the present land surface rather than the age of the parent rock. Residual kaolin deposits are

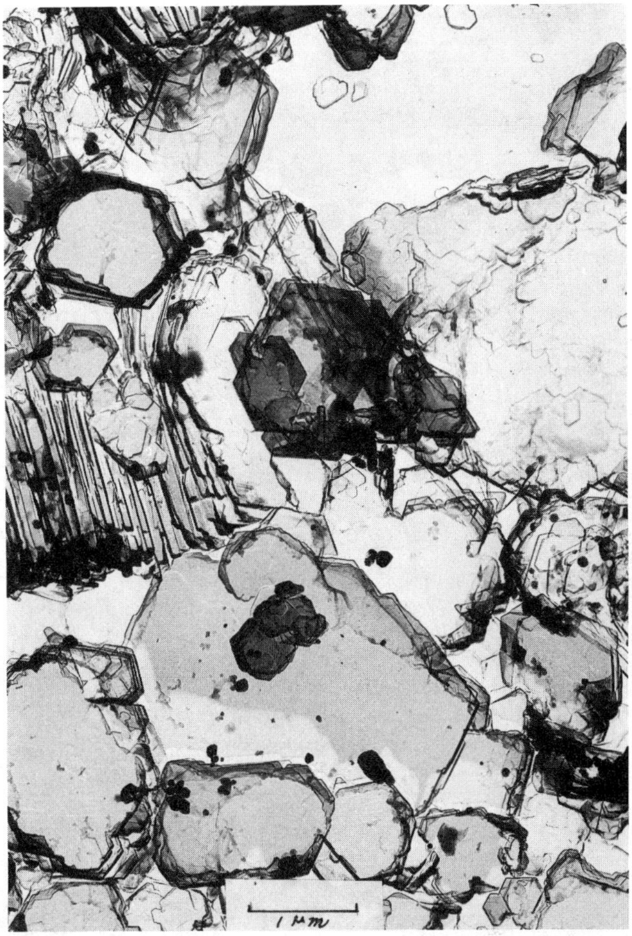

FIG. 9—*Electron micrograph of Georgia kaolin showing hexagonal thin plates and vermicular forms of kaolinite.*

generally irregularly shaped and grade downward into the parent rock through partly altered weathered saprolite zones. They also commonly contain rounded boulders and irregular masses of fresh or only partly altered rock that were originally in the centers of joint blocks or otherwise protected from percolating water. Residual kaolin deposits occur in the Spruce Pine district, North Carolina (Parker, 1946); the Arkansas bauxite district (Gordon, et al., 1958); Latah County, Idaho (Ponder and Keller, 1960); Alberhill district, California (Cleveland, 1957); and elsewhere in the United States (Fig. 10). They also occur in many other countries and are particularly common in Czechoslovakia (Kuzvart, 1969) and Brazil (de Souza Santos and de Souza Santos, 1972).

Some kaolin deposits throughout the world occur in rocks that have been hydrothermally altered. Such deposits are ordinarily in irregularly elongate pods or pipelike bodies aligned along joints, faults, or other permeable zones that allowed movement of warm or hot aqueous solutions. Some deposits of this type also occur in altered tuffs and are bedded. As with the residual type, the age of hydrothermal kaolin formation is not necessarily related to the age of the host rock. Most hydrothermal clay deposits are probably no older than Tertiary, though some occur in rocks that are much older. Insofar as the authors are aware, deposits in the Little Antelope Valley, California (Cleveland, 1957), contain the only hydrothermal kaolin now mined in the United States. It is used for refractory clay. However, kaolin formed by this process occurs at many places in the western states, and deposits were formerly mined on a small scale in several mining districts.

Origin—The two fundamental geologic proc-

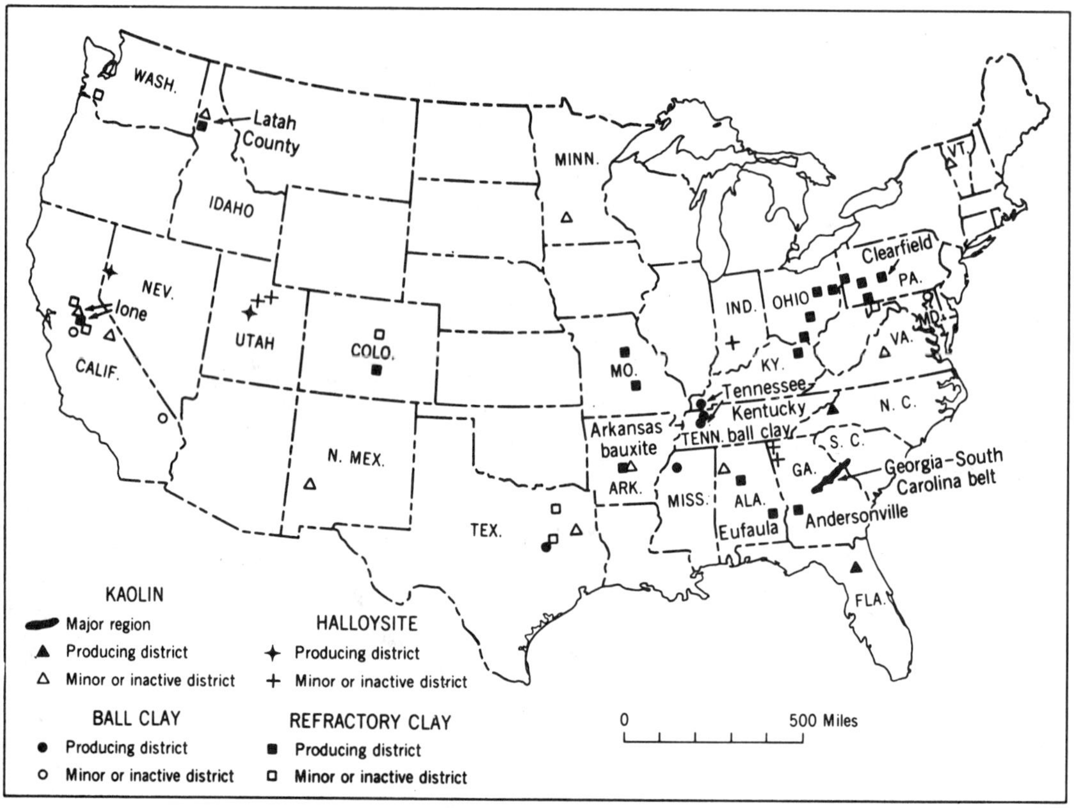

FIG. 10—*Map showing kaolin, ball clay, halloysite, and refractory clay districts in the US.*

esses that form most of the clay minerals of the kaolin group are weathering (Keller, 1964) and hydrothermal alteration (Sales and Meyer, 1949). These processes under certain conditions remove elements other than silicon, aluminum, oxygen, and hydrogen, the constituents of these clays. The clays form as the other elements are removed from preexisting minerals and possibly by the growth of new atomic arrangements from materials in solution. Though most kaolin minerals form in these ways, the origin of valuable deposits is quite another matter. Certainly several commercial clay deposits throughout the world have formed by hydrothermal processes, and others have formed in place by the weathering of other rocks. However, many of the large high-grade deposits of kaolin occur in sedimentary rocks and have either been (1) transported and deposited as kaolin; (2) altered from previously transported rocks by processes related to surficial weathering, submarine or subaqueous alteration, or postburial (diagenetic) changes; or (3) combinations of these possibilities.

Knowledge regarding the origin of kaolin deposits in the United States varies considerably. The origins of some of the deposits are reasonably well established, but geologists have widely differing opinions on how other deposits formed. Clearly, many of the scattered kaolins in western states formed by hydrothermal processes, but few such deposits have been significant sources of kaolin. In the Arkansas bauxite region (Gordon, et al., 1958) and in Latah County, Idaho (Hosterman, et al., 1960), are districts containing both transported clay and residual kaolin formed by weathering of igneous rock. In both districts, the kaolin grades downward through saprolitic zones into parent rock, which is nepheline syenite in Arkansas and granodiorite in Idaho. Also, in both districts, kaolin transported from the residual material occurs in sedimentary rocks—in marine shale in Arkansas and in lacustrine beds in Idaho. However, this explanation of origin is inadequate to account for all the kaolinitic clays in the regions containing these districts. The transported kaolin in Arkansas is of Wilcox age (early Eocene), and kaolinitic clay occurs in rocks of this age throughout a much more extensive region than can be related to the nepheline syenite in Arkansas. Similarly, the Miocene clays in Idaho cannot all be related to granodiorite. Another type of kaolin deposit that is reasonably well understood occurs as sinkhole fillings in the Valley and Ridge province in the eastern part of the United States. This type of kaolin apparently formed by the leaching of other aluminous minerals by downward-moving water in sinkholes. Probably the parent aluminous materials were mainly residuum from the solution of carbonate rock. This explanation is in part based on the presence of cores of gibbsitic bauxite in the centers of such deposits and the enclosing envelopes of cherty clay separating the kaolin from the carbonate rock in the walls of the sinkholes.

Strange as it may seem, the deposits whose origins are least understood include those in the extensive belt in Georgia and South Carolina which have been studied by many geologists and mineralogists and which lead the world as a source of commercial kaolin. Virtually all who have investigated these deposits agree that either the kaolin or the material from which it formed has been transported. Areas of disagreement center about explanations of how such very large deposits of fine-grained white clay could have formed in environments in which some sediments being deposited consisted of sand containing much more iron than does the clay. Furthermore an adequate explanation of origin of the deposits must account for the following facts: (1) some deposits of kaolin are finer grained than others; (2) some contain abundant accordionlike booklets and some do not; (3) gibbsite is present in some deposits and not in others; (4) lignitic layers or layers rich in organic matter are present in some deposits; and (5) smectite is present in some deposits in variable quantities.

One of the first geologic studies of the Georgia kaolin was by Ladd (1898); he described the deposits but did not discuss their origin. Veatch (1909) concluded that the lowermost member of the Cretaceous system, the Tuscaloosa, is the most important clay-bearing formation in Georgia. He indicated that the age of the formation in Georgia was determined only by its stratigraphic position. Veatch believed that the material making up the Tuscaloosa Formation was derived from disintegrated and decomposed rocks of the Piedmont Plateau. The greater part of the Piedmont region was a land surface for a very long period of geologic time. Just before the beginning of the Cretaceous, the Piedmont region was uplifted and tilted southeast. Streams rapidly carried the weathered residue to the Cretaceous sea. Veatch postulated that an enormous quantity of sediment was dumped rapidly at the stream mouths, forming extensive sand flats. Fine clay particles were deposited in nonpersistent and lenticular beds of white clay in the deeper and quieter waters near the stream mouths. He considered the Tuscaloosa in Georgia a nonmarine deposit.

Smith (1929) concluded that the basal Cretaceous sediments in Georgia were Upper Cretaceous and correlated them with the Middendorf Formation of eastern South Carolina. Smith believed that Veatch's explanation of origin was essentially correct, except that he agreed with Neumann (1927) that the kaolin was deposited in saltwater. Cooke (1943) assigned the basal Cretaceous sediments to the Upper Cretaceous under the name Tuscaloosa, which is still used. Kesler (1963) believes that in east-central Georgia there is no basis for subdividing the Upper Cretaceous, and he found that the kaolin lenses have no preferred stratigraphic position within the formation. Kesler theorized that feldspathic sands were deposited as delta sediments above sea level and were subsequently altered to kaolin. The kaolin was separated from the coarser sediment and deposited at points of lowest elevation. As the kaolin accumulated in the ponds, sand was washed from the margin into and over the edge of each deposit, forming gradational and interfingering contacts.

Bates (1964) reviewed the origin of the Georgia kaolins and summarized the mineralogic and geologic data, some of which support a marine origin and some a freshwater origin. Hinckley (1961) suggested that face-to-face flocculation of kaolin particles in a marine environment would result in the properties that distinguish a so-called "hard" kaolin, whereas edge-to-face flocculation in freshwater would produce a less dense, "soft" clay. Buie (1964) proposed the possibility that volcanic ash was first altered to montmorillonite and then to kaolinite. This is a radical departure from previously proposed theories of origin, and much more detailed field and laboratory work must be done before this hypothesis can be adequately tested. Jonas (1964) concluded from a detailed petrologic study that the originally deposited sediment was muscovite, feldspar, and quartz, and that postdepositional alteration and recrystallization produced the kaolin deposits now mined.

Recent studies of spores, pollen, and invertebrate fossils in and associated with kaolin indicate that many of the deposits thought to be Cretaceous are of Eocene age (Buie and Fountain, 1967; Cousminer and Terris, 1972; Scrudato, 1969), a possibility suggested by Eargle (1955). The first announcement of the Eocene age for any of the deposits was by Buie and Fountain, who on the basis of paleontological evidence established the presence of middle Eocene strata beneath some of the commercial kaolin deposits and above others. In one of the studies, Cousminer and Terris (1972) concluded through palynological work that some, if not all, of the sedimentary kaolin is middle Eocene in age. Results of palynological studies by R. H. Tschudy of the US Geological Survey on samples collected by Patterson have confirmed that some of the kaolin deposits are of Late Cretaceous age or older, and that some are no older than Claiborne (middle Eocene). Tschudy's work also revealed that sand beds of Paleocene age are present in the kaolin belt. This raises the possibility that some of the kaolin deposits may be of Paleocene age and that kaolin deposition took place during several different intervals of geologic time.

The recent findings on the range in age of Georgia kaolin deposits reinforce our opinion that the deposits have had a very complex history and origin, and the theories advanced to date do not provide adequate explanations. Probably the purity of the kaolin is the result of some sedimentary winnowing process, an idea favored by Kesler (1963) and others, as evidence now available suggests that the ultimate source was aluminum silicate minerals in the crystalline rocks to the northwest. The environment in which the deposition took place is yet to be established, but it was probably similar to the present coast of Georgia and South Carolina, where there are abundant offshore sand bars and swampy, muddy lagoons and winding estuaries. How iron was removed from the kaolin remains a problem; removal of iron may have been due mainly to chelating by organic acids from decaying plant materials, and oxidizing pyrite may have contributed to the process. Some of the differences in physical properties of the kaolin may be due to the fact that parts of the deposits, particularly the younger ones, have been reworked. Other differences in physical properties are probably due to varying degrees of diagenetic alteration, as most of the deposits now known to be Cretaceous in age contain more vermicular crystal growths and coarser particles than do the deposits known to be of Eocene age.

Major Kaolin-Producing Countries: *United States*—GEORGIA-SOUTH CAROLINA KAOLIN BELT—The principal reserves and the major kaolin-producing centers in the United States are in a belt extending from Twiggs County in the central part of Georgia to Lexington County in west-central South Carolina. This same belt extends southwest from Macon to Andersonville, GA (Zapp, 1965), and west to Eufaula, AL (Warren and Clark, 1965). The Andersonville and Eufaula districts contain appreciable kaolin resources, but both are centers of production of refractory kaolin and refractory bauxitic materials, and no kaolin, as defined in this report, is now produced in either district.

The deposits in the Georgia-South Carolina belt occur in sedimentary rocks a short distance seaward from the crystalline rocks of the Piedmont region. The major mining and beneficiation activities are in Twiggs, Wilkinson, and Washington Counties, Georgia, and in the vicinity of Aiken, SC. Large reserves of fine-particle-size kaolin have been discovered in the last 10 to 15 years in the vicinity of Wrens, GA, which is in the east-central part of the belt. Other reserves are scattered throughout the belt. Very large potential resources of off-color high-iron clay and deposits under overburden that is too thick to be stripped profitably, at present, are scattered throughout the seaward parts of the belt.

ALABAMA—Kaolin is mined from the Tuscaloosa Group in Marion County, Alabama (Clarke, 1964). It has been mined in this area for 40 years, and approximately 1 million tons has been shipped. This kaolin is relatively pure and is easily beneficiated by removing quartz and mica. Most of the production from this district has gone into various ceramic products.

CALIFORNIA—Filler-type kaolin is mined and processed near Ione and Bishop, CA. The Ione clay is sedimentary in origin (Bates, 1945), whereas the kaolin near Bishop is hydrothermal (Cleveland, 1957).

FLORIDA—The Florida kaolin is found in sand and gravel deposits of the Citronelle Formation (Calver, 1949). The formation ranges from 20 to 55 ft in thickness, and the recoverable kaolin averages about 15%. The only operating mine is in Putnam County. Most of the production is sold to the ceramic industry. Recently, these kaolinite sediments were named the Fort Preston Formation and dated middle Miocene (Calver, 1949).

IDAHO—Filler-type kaolin is mined and beneficiated in Latah County, Idaho. These deposits occur as residual weathered products of granodiorite and as lacustrine sediments (Hosterman, et al., 1960; Ponder and Keller, 1960). Mineralogical analyses show that some of the deposits contain halloysite in addition to kaolinite and that the deposits are variable, some being almost all kaolin and others containing kaolin, quartz, orthoclase, mica, and minor quantities of other minerals.

NORTH CAROLINA—Near Spruce Pine, a primary kaolin (the term *primary kaolin* is used here as commonly applied in the kaolin industry for hydrothermal-type deposits and for residual deposits of kaolin formed by in-situ weathering) altered from pegmatite and aplite (Parker, 1946) is mined and beneficiated. This kaolin is a mixture of kaolinite and halloysite and is used primarily by the ceramic industry. These deposits generally contain approximately 40% kaolin and halloysite, so that the mica, quartz, hornblende, and other minerals must be removed by wet beneficiation methods to produce a usable product.

TEXAS—Near Kosse, Texas, a kaolin deposit of early Eocene age in the Wilcox Group is mined and beneficiated as a filler and ceramic clay (Fisher et al., 1965). Pence (1954) described these deposits as nonmarine in origin. They contain 40 to 70% quartz, and the kaolin is a byproduct of a glass-sand operation.

OTHER STATES—Kaolin has been mined in the past or has been reported present in commercial quantities in Arkansas (Tracey, 1944), Illinois (Grim, 1934), Minnesota (Parham, 1970), Mississippi (Conant, 1965), New Mexico (Glassmire, 1957), Vermont (Ogden, 1969), and Virginia (Ries and Sommers, 1920).

RESOURCES—Resources of kaolin in the United States of a quality equal to or better than the lower grade clay now mined are at least 1 billion tons. The one grade of kaolin about which there is concern for long-range supply is the very high grade fine-particle-size white clay suitable as a paper coater and for certain other speciality uses (Buie, 1972). Reserves of this type of clay probably total 100 to 200 million tons. Potential resources of kaolin, including discolored and very sandy clay and that below more overburden than can be profitably stripped at present, are several times the estimated reserves.

United Kingdom—The United Kingdom is second only to the United States as a producer of kaolin or china clay and is the largest exporter of kaolin in the world (Anon., 1972g). China clay is the United Kingdom's chief export, and almost the entire output comes from the St. Austell area of Cornwall. Historically, the china clays in Cornwall were discovered by William Cookworthy (Barton, 1966) in the mid-1700s to be suitable for making porcelain. In Cornwall, an extensive region of kaolinized granite has been formed by hydrothermal alteration of feldspar (Bristow, 1969). The granite masses were emplaced in Permian time and are now exposed over large areas of Devon and Cornwall. The china clay deposits typically occur in funnel or troughlike bodies narrowing downward. Hot acidic solutions are thought to have migrated upward, guided by structural weaknesses such as faults and joints, and then to have been trapped under a roof of relatively impermeable rock. The hot solutions attacked the granite, altering the feldspar and mica to kaolinite. Quartz, tourmaline, and mica are the principal impurities. Drilling has shown kaolinization to exist at depths of more than 800 ft (Bristow, 1969).

In most places, the clay can be mined using high-pressure water jets, but sometimes blasting or ripping is required to loosen the rocks. Selective mining is very important because of variability in the various deposits. Much blending is done to produce a uniform product. As the rock face is washed, the slurry containing the kaolin in suspension flows to the lowest point in the pit. The slurry is then pumped into spiral classifiers which remove the coarse material. After this initial classification, the slurry is pumped to the surface of the pit where hydrocyclones remove finer sand and mica particles. The slurry is then thickened in large-diameter tanks. A series of refining tanks is used to remove the heavier coarser particles. After primary classification, the slurry is fed to centrifuges to separate the coating kaolin from the filler grade. The kaolin is dewatered by filter presses and is dried. Two new processes have been designed—the tube filter press (Clark, 1968), which utilizes extremely high pressures and produces a relatively dry filter cake that needs little if any additional drying, and a process catalogued as micromineral separation (British Patent No. 1,222,508), which produces a rather clean pure kaolin.

More than 2.7 million tons of kaolin were produced in England in 1970, of which 2.1 million tons were exported (Anon., 1972f). The exports are handled through the ports of Par and Fowey. The kaolin is transported to port by rail or truck. Most ships that transport kaolin are in the 400 to 500-ton range, but 1000-ton vessels can be accommodated at Par and 10,000-ton vessels at Fowey.

USSR—Kaolin is mined and processed in several districts in the USSR (Petrov, 1969). Residual, hydrothermal, and sedimentary deposits are worked. One kaolin deposit in the Transcarpathian region consists of virtually pure dickite (Rusko, 1973), and other deposits in this same district consist of mixtures of kaolinite, dickite, and halloysite. The Gluhovetski district, southwest of Kiev, approximately midway between the cities of Zhitomir and Vinitsa, is one of the major centers of production. The best grades of kaolin are produced from deposits formed in weathered granite. Production from this district is about 700,000 tpy, and both wet and dry beneficiation methods are used. Another major producing district is in the vicinity of Prosyanovski in the Dnepropetrovski region. The deposits in this area also occur in weathered granite and gneiss. Nearly 1 million tons of kaolin were produced from this region in 1966. Reserves of kaolin in the Prosyanovski district are estimated at 160 million tons. Other large kaolin deposits occur in Kazakhstan, where a large kaolin beneficiation plant was reported to be in the advanced stages of planning in 1968. The largest sedimentary kaolin is the Pologa deposit on the Konok River near the Pologa Station. Many other kaolin deposits have been described by Petrov (1969). The Russian kaolin is used in the paper, ceramic, rubber, and chemical industries.

Minor Kaolin-Producing Countries: *North America*—MEXICO—Most of the kaolin in Mexico is in primary deposits originating by hydrothermal alteration of Tertiary rhyolitic rocks (Pasquera, et al., 1969). The deposits are found in almost every state, but the largest commercial deposits are in the central part of the country. All the kaolin in Mexico is dry processed for use in the paper, ceramic, refractories, rubber, chemical, and pharmaceutical industries. Reserves are estimated to be more than 500 million tons. At present, more than 100,000 tpy is mined and processed. Kaolin is imported from the United States for paper-coating applications.

South America—ARGENTINA—The kaolin production in Argentina in 1969 was 80,905 tons (Anon., 1971b). Most of this kaolin was mined in the Blaya Dougnac, Don Carlos, and Maruja districts, Chubut. Districts in Río Negro and San Juan also had active mines in recent years.

BRAZIL—Kaolin deposits in weathered granites, pegmatites, and other crystalline rocks occur at many localities in Brazil, and recently very large transported deposits have been discovered. The deposits in weathered rocks have been mined for domestic use for many years. The large transported deposits were not developed at the time this report was prepared, but exploration by American interests and Brazilian government agencies suggests that production is likely. One region containing significant deposits of transported kaolin is along the Jari River (Colligan, 1973), a tributary of the Amazon. The deposits along this river are of Pliocene age (Klammer, 1971). They are as much as 35 m thick and extend for several kilometers. Reserves are considerably more than 50 million tons. Another region in which the kaolin deposits are being evaluated is along the Capim River approximately 200 km south of Bélem (White, 1973).

GUYANA—Very large transported kaolin deposits are associated with bauxite in Guyana and Surinam (Moses and Mitchell, 1963). These deposits are in sedimentary rocks underlying the Coastal Plain seaward from a shield area of ancient crystalline rocks. The regional geology is, therefore, similar to that of the Georgia-South Carolina kaolin belt. Japanese interests are reported to be mining 300 tons of kaolin a month and shipping it to Japan (Ampian, 1973).

OTHER COUNTRIES—Both residual and transported kaolin deposits occur at several places in Chile, Colombia, and Venezuela (Malkovský and Vachtl, 1969c). Hydrothermal deposits are also present in Chile, as indicated by the association of kaolin with alunite. Small tonnages of kaolin are also produced in Peru, Paraguay, and Ecuador (Ampian, 1973).

Europe—The kaolin operations in continental Europe are mostly small scale because of the scarcity of high-grade deposits. Most of the kaolin produced is still used by the ceramics industry and as filler for paper, and only minor quantities are sold for other applications. West Germany produces nearly 500,000 tpy kaolin, the main production being from Bavaria. The kaolin deposits in Brittany in France are being worked to a much greater extent, and production from this area will increase substantially. Czechoslovakia, where the production is controlled by the state, is becoming an important producer, and exports kaolin to other European countries.

AUSTRIA—Economically important kaolin deposits of primary origin occur on the *Bohemian Massif*, a deeply eroded complex of granitic and metamorphic rocks (Holzer and Wieden, 1969). Most of this kaolin is used in the ceramic industry and as filler in the paper and rubber industry. In 1966, more than 450,000 tons were mined in Austria, but less than 100,000 tons of this was processed using wet beneficiation techniques.

BULGARIA—Sedimentary kaolin deposits of probable Pliocene age occur in northwestern Bulgaria (Karanov, et al., 1969). The individual deposits are small because the kaolin is associated with sand which fills sinkholes in a karst area underlain by Lower Cretaceous limestones. The average kaolin content in these sands is about 18 to 25%. The best kaolin has an iron content of 0.6 to 0.8%. The kaolin is used locally in the ceramic and paper industries.

CZECHOSLOVAKIA—Czechoslovakia is the principal producer of kaolin in Eastern Europe. The major deposits are in the vicinity of Karlovy-Vary, Plzeň (Pilsen), and Podborany in west Bohemia (Kuzvart, 1969). The kaolin is primary, is of high quality, and is used for the production of fine ceramics, paper, rubber, and refractories. More than half a million tons has been mined, and by 1975, the annual production will probably be as much as 1 million tons. Reserves are estimated to be more than 1 billion tons (Kuzvart, 1969). Production is state controlled, and the product is sold at a low price. The largest export trade is with other Eastern European countries, but in the last two years a significant tonnage has been shipped into Western Europe, particularly West Germany (Anon., 1972f).

DENMARK—A small tonnage of primary kaolin is mined on the Island of Bornholm primarily for the Danish ceramic industry in Copenhagen (Almeborg, et al., 1969). Most of the output goes into the manufacture of low-alumina refractories and glazed heavy-clay products. The quality of the kaolin is not high enough to be used in whitewares or as filler for paper.

EAST GERMANY—Both primary and secondary kaolin deposits are worked in East Germany (Störr, et al., 1969). The kaolin is mined from an area north of Bautzen, Dresden, and Colditz. The iron content of the various commercial deposits ranges from as low as 0.2% to as high as 2%. Most of the kaolin is used for ceramic purposes, but shipments for paper and rubber filler are increasing. Reserves are estimated to be considerably more than 100 million tons.

FRANCE—Many small companies produce kaolin from districts in Brittany and around the Massif Central in central France. The deposits in Brittany are of approximately the same age as those in Devon and Cornwall in England (Anon., 1972f). In Brittany, the deposits are all primary in origin; and the deposits around the Massif Central are both primary and secondary (Damiani and Trautmann, 1969). About 80% of the French production is from Brittany and is near the coast. Production is more than 250,000 tpy. Most kaolin is used by the ceramic industry, but its use as a filler for paper, paint, and rubber is increasing because of improved processing techniques now being implemented. France imports about 250,000 tpy, mainly from England and the United States to meet its requirements for kaolin.

GREECE—Low-grade deposits of kaolin suitable for use in cement, ceramics, refractories, and paper filler occur on the Aegean islands of Milos, Polyaegos, and Mytilene (Anon., 1972f). Reserves are small and production is limited.

HUNGARY—Most of the commercial kaolin deposits in Hungary are confined to an area in the northeast, in the Tokaj Mountains. These deposits are primary and have been derived from rhyolite (Varju, 1969). Secondary kaolin-bearing sandstones are mined and washed at Sarisap in the Visegrad Mountains west of Budapest. The output is small, probably less than 60,000 tpy and is used in ceramics, paper, rubber, and plastics. Reserves are relatively small, so that the kaolin industry in Hungary will probably continue to produce primarily for domestic usage.

ITALY—Both primary and secondary kaolin deposits are mined in Italy, primarily for the ceramic industry. Production is small and has never reached 90,000 tpy (Moretti and Pieruccini, 1969). The most important areas are in northwest Italy near Brella, on southern Sardinia, and near Rome.

POLAND—In the Lower Silesia area of Poland southwest of Wrocaw, primary kaolin derived from granite and gneiss is mined for local consumption mainly for the ceramic industry (Anon., 1970h; Gawronski and Kozydra, 1969). Reserves of individual deposits range from 2 million to more than 10 million tons, and total reserves are indicated to be several tens of millions of tons.

SPAIN—Spain has the fourth largest production of kaolin in Western Europe; in 1970, the total was 285,000 tons (Anon., 1972f). By 1976, according to the National Mining Plan, the production will be more than 500,000 tons. Both primary and secondary deposits are mined in Spain (Martin Vivaldi, 1969). The deposits are widely scattered, but the commercially important production is concentrated in the northwest near Galicia-Asturias and in the east near Cuenca-Teruel-Valencia. The primary deposits of Galicia were formed by the hydrothermal alteration of mylonitized granite; reserves exceed 300 million tons. The secondary deposits in eastern Spain are Cretaceous in age and are principally kaolinitic sands ranging from 10 to 20 m in thickness. Reserves are estimated to be more than 1 billion tons. Most of the production is used by the ceramic industry, primarily in refractories, but as beneficiation techniques improve, more will be consumed in the paper, paint, rubber, and other industries that need filler clays.

West Germany—The largest kaolin mining district is in the townships of Hirschau and Schnaittenbach in Bavaria (Anon., 1972f). The deposits are sedimentary in origin, derived from decomposed granite in the Naab basin mountains, and are Triassic in age. These deposits consist of a mixture of quartz, feldspar, and kaolin and are about 40 m thick. The material is wet processed to recover about 10 to 20% kaolin, and the quartz and feldspar are also salable products. Because of the high content of quartz and feldspar, sophisticated hydrocyclone-processing stages are used to make very precise particle separations (Anon., 1972f). Other areas in West Germany where kaolin is worked are Oberneisen near Frankfurt, Hesse, and Westphalia (Lippert et al., 1969).

Other Countries—Small tonnages of kaolin are produced in Romania, Sweden, Portugal, Yugoslavia, and Belgium (Ampian, 1973; Malkovský and Vachtl, 1969b).

Africa—Republic of South Africa—The kaolin deposits of South Africa have been derived from the weathering of granite, shale, and tillite (Coetzee, 1969). Kaolin is mined near Capetown, Grahamstown, and Durban. With the exception of the large deposit of altered tillite near Grahamstown (Murray and Smith, 1973), deposits are small but of good color and quality. The Grahamstown deposit is very large, with reserves in excess of 50 million tons. but the quality ranges from excellent to very poor and it changes in very short intervals both horizontally and vertically. Approximately 50,000 tons are used locally in South Africa, the ceramic industry taking more than 50% of the production. Other uses include insecticides, rubber, paint, plastics, and pharmaceuticals. The large deposit at Grahamstown could be potentially valuable in the future as advanced beneficiation techniques become available and water supply to the area is improved. A severe water-supply problem for wet beneficiation of kaolin exists throughout South Africa.

Swaziland—Kaolin deposits of economic importance are found in the Malilangatsha Mountains of the Manzini district (Hunter and Urie, 1969). Kaolinized dikes ranging in width from 3 to 10 m, 75 m in depth, and as much as 620 m in length have been studied. The kaolin ranges from white to pale red, depending on the iron content. The entire production of some few thousand tons is exported to South Africa.

Other Countries—Other African countries having minor kaolin production include Angola, Arab Republic of Egypt, Ethiopia, Kenya, Nigeria, Tanzania, Malagasy, Mozambique, and Morocco (Ampian, 1973; Malkovský and Vachtl, 1969c). The deposits in the Pugu Hills district near Dar es Salaam, Tanzania, are among those having economic potential. The kaolin is in upper Tertiary sandstone which is 30% white clay (Harris, 1969). Attractive features of these deposits include a location only 50 km from deep water and a possible glass sand coproduct. However, an attempt in the early 1950s to develop and export this kaolin failed.

Asia—Sri Lanka—The best known kaolin deposit on Sri Lanka is the Boralesgamuwa field, south of Colombo. The kaolin is probably Pliocene or Pleistocene in age, is sedimentary, and occurs in layers and pockets 4 to 19 ft thick (Pattiaratchi, 1969). The percentage of kaolin ranges from 20 to 85, and the major impurity is quartz. The kaolin is used for ceramics, paper, rubber, paint, and as an insecticide carrier. The production is small, about 3000 tpy, and the estimated reserves are about 8 million tons.

India—Kaolin and related clay materials are mined in many locations in India (Arogyaswamy, 1968). The major producing area is in the vicinity of Chaibasa. The deposit mined in this district occurs in weathered granite. High-quality kaolin is also produced in the vicinity of Kundra, Chattanur, and Quilon, Kerala. The kaolin in this region occurs in altered feldspathic granite, and the altered zones are as much as 10 m thick.

Indonesia—Large residual kaolin deposits altered from granite are known to occur on the islands of Belitung and Banka (private communication). Although the deposits are large, little is known about the quality of the kaolin. There are small wet-processing operations which produce china clay for the ceramic industry in Indonesia.

IRAN—The largest kaolin deposit in Iran is in the central part of the country near Simiron (Khadem, 1969). This deposit is the result of deep weathering at a diastem in a Cretaceous limestone sequence. The material is used for fine ceramic and refractory purposes. The reserves are about 15 million tons.

JAPAN—Large primary and secondary kaolin deposits occur in Japan (Fuji, Okamo, and Shimazaki, 1969). The deposits are scattered throughout the islands. The primary deposits were formed by hydrothermal alteration of volcanic and granitic complexes and from in-situ weathering of granite, pegmatite, and volcanic rock. The secondary deposits are lacustrine, of lower Pliocene age, and range from 20 cm to 20 m in thickness. About 200,000 tpy is produced in Japan (Anon., 1972f). The reserves in many deposits range from 5 to 20 million tons. The major uses of the kaolin are in fine whiteware, pottery, refractories, and as paper fillers. Mining in Japan becomes more difficult each year because of increasing land values and the difficulty in purchasing enough property to sustain reserves required for the operation of a wet-beneficiation plant.

KOREA—Primary and secondary kaolins occur in Korea (Heikes and Kim, 1965). Both dry and wet beneficiation methods are used to process the kaolin for the ceramic, refractory, paper, and rubber industries. More than 100,000 tons are mined for domestic and export markets; over half of this is shipped crude to Japan for use as paper filler. The largest deposits, consisting of mixtures of kaolinite and halloysite, are in the Hadong-Sanchong area of southernmost Korea.

MALAYSIA—Residual kaolin altered from granite is being mined and wet processed to make paper-filler clay (Anon., 1970f) east of Kuala Lumpur. The reserves are large, but present production is only about 5000 tpy.

PEOPLES REPUBLIC OF CHINA—Kaolin not only was used first in China, but advertisements in recent trade journals indicate that it is still being mined on a large scale and is available for export. Historically, deposits in the Ching-Tae-Chen district in Fuliang, Kiangsi Province, were the principal source of kaolin used in making fine ceramic ware (Pai, 1945). These deposits are associated with Ordovician and Silurian phyllite. The clay is mostly light gray to white, and it is 70% pure kaolinite. Presumably, this district is still in production, but this was not verified. Other kaolin-producing regions in China include Singtze, Loping, Yukan, and other districts in Kaingsi Province, and Tzuhsien, Kaiping, Fengyn, and Chinghsing, Hopeh Province. Kaolin in deeply weathered granite occurs at several localities in Kwangtung Province which surrounds Hong Kong. This kaolin formed from granite may be the clay advertised for export because it is much closer to the ocean than most other deposits.

In addition to the kaolin mentioned in the foregoing paragraph, kaolinitic clays associated with diasporic and bauxitic clays in Carboniferous, Permian, and Mesozoic coal measures occur in Szechuan, Honan, Kupeh, and Fukian Provinces and elsewhere. Several of these clays are called kaolins and are used in ceramic products. Residual kaolin formed by the weathering of arkosic sandstone occurs in Szechuan Province, and deposits formed from weathered feldspar in granite and pegmatite occur at several places in China.

PHILIPPINES—Philippine kaolin deposits are primary and are generally small and scattered. Most are hydrothermal in origin, although one deposit called Bukidnon is believed to have formed from in-situ weathering of a dacite (Comsti, 1969). The major deposits are on Luzon and Mindanao. The bulk of the kaolin is used in refractories, tiles, and sanitary wares. Very little is used in the paper and rubber industry. The reserves are limited and transportation is costly; these factors and the variable quality of the kaolin make the mining operations very marginal.

Turkey—Both primary and secondary kaolin deposits are mined in Turkey, primarily for the Turkish ceramic industry. Approximately 100,000 tons of kaolin are used annually in Turkey by the ceramic industry (Jeyhan, 1969). Reserves, estimated at 2.5 million tons, are mainly in the western half of the country.

OTHER COUNTRIES—The production of kaolin has been reported in recent years (Ampian, 1973) in Pakistan, Taiwan, Thailand, and South Vietnam.

Oceania—AUSTRALIA—Most of the kaolin deposits in Australia are primary and have been formed as a result of deep weathering of both silicic and mafic rocks (Gaskin, 1969). Secondary deposits in Western Australia are associated with the coal measures of Cretaceous age. Some of the Australian kaolin is very white and can be used readily as filler and for ceramic purposes (Anon., 1970g), but the viscosity of almost all the kaolin is very high, which prohibits its use for paper coating. The bulk of the kaolin produced is used for refractories; the rest is consumed by the ceramic, paper, paint, plastics, and rubber industries. Although no precise estimates of reserves have been made, the total amount of kaolin in Australia is known to be very large. At present, more than 100,000 tons is processed in Australia primarily for local consumption, although some is being exported to Japan as paper filler. Most of the deposits mined are within 200 km of the major cities of Melbourne, Sydney, Adelaide, and Perth. Extensive exploration for kaolin is being carried out in Australia by several major companies, primarily because of the lucrative Japanese export market.

NEW ZEALAND—The only kaolin being produced in New Zealand at present is in the Northland some 120 km north of Auckland (Bowen, 1969). The deposits are hydrothermal and are mixtures of kaolinite and halloysite. Reserves are estimated to be approximately 20 million tons. The kaolin is wet beneficiated for use as a ceramic whiteware component and as filler for paper and paint. The kaolin is very bright and white but has a high viscosity and is somewhat abrasive, which prohibits its use in paper coating. At present, the entire output of some 5000 tpy is utilized in New Zealand.

Exploration, Evaluation

Exploration in the United States: The principal productive deposits of kaolin were found in the Carolinas and Georgia in the late 19th and early 20th centuries. Most discoveries were in outcrops in stream beds and road cuts. Today, exploration is carried out in geologically favorable areas, taking into account topography and elevation. Geologic and topographic maps, if available, and aerial photographs are used to lay out exploration drill-hole patterns. Most exploration holes are drilled to depths of as much as 200 ft and on 400- to 800-ft centers. Many deposits were missed in the past because only random spot drilling instead of pattern drilling was used. Also, because drilling was limited to 50 or 60 ft, many good deposits at greater depths were not discovered. Auger or fishtail drills are used to the top of the kaolin bed, and then a special core barrel is used to core the kaolin. Cores are difficult to recover, and special care by expert drillers is necessary to get good core recovery. In many areas, two or three kaolin beds separated by sand are penetrated where drilling is as deep as 200 ft.

Many geophysical methods of exploration for kaolin have been attempted, but to date none has been successful (Gross, 1960). Geochemical techniques also have been unsuccessful. The only positive method to locate and delineate kaolin deposits to date is by pattern drilling.

Evaluation of Deposits: After coring the kaolin and recording the thickness of the various strata above the kaolin and the thickness of the kaolin bed, a large-scale map is prepared. On this map, the drill holes are located and the thickness of the over-burden and the kaolin are recorded. The cores are sent to the crude evaluation laboratory where the kaolin is prepared and tested.

The cores are cleaned and divided into sections 2 ft or less in length for testing. The reason for splitting the cores into short sections is to be able to spot any lens or section that may be off-color, have high grit or high viscosity, or not meet specifications in some way. The 2-ft sections are dried and pulverized. The pulverized material is then evaluated using the following tests for paper-type kaolin:

1) Percent grit or screen residue
2) Particle-size distribution
3) Low shear viscosity
4) High shear viscosity
5) Brightness
6) Leachability

Grit percentage is determined as explained in the section on end uses and specifications and is an important crude evaluation test. A good rule of thumb is that as much as 5% grit can be handled easily using air-flotation techniques. Much higher percentages of grit can be handled using wet beneficiation methods, but the higher the grit percentage the lower the recovery of kaolin and the higher the cost. Normally in the United States, a maximum of 10 to 15% grit is set for the upper limit of grit percentage, but if the kaolin is of unusually high quality, a higher grit percentage can be tolerated.

Particle-size distribution is important because from this test the percent recovery of coating and filler clays is determined. For many years an important delineation point has been the percent of particles less than 2 micrometers in size (Lyons, 1966). This percentage can vary from as low as 50% to as high as 90% in the Georgia deposits. This is an important test because brightness, viscosity, and other physical properties are closely related to fineness. Also, a predetermined fineness range must be supplied to the beneficiation plant to meet the filler and coating requirements and to provide for uniform process procedures. It is, therefore, necessary to blend kaolins from various deposits to regulate the particle size of the material being fed to the plant. At this point in the testing, one or more coating clay fractions are prepared. Normally this fraction will contain 80% of the particles less than 2 micrometers in size. If special evaluation is necessary, a coating clay fraction containing 92% of the particles less than 2 micrometers in size is prepared.

Both low shear and high shear viscosity tests are run on the degritted crude clay and on the coating fraction or fractions that are prepared. The upper limit of low shear viscosity that can be tolerated is 800 cps at 10 rpm and a high shear of 300 rpm at 1800 reciprocal seconds. The low shear viscosity is a measure of thixotropy, and high shear viscosity gives an indication of dilatancy (Murray, 1970).

Brightness is measured on the degritted crude clay and on the coating clay fractions. For No. 2 coating clay (80% finer than 2 micrometers) a minimum brightness of 85.5% is required, and for No. 1 coating clay (92% finer than 2 micrometers), 87.0% brightness is required. Brightness determinations are made following procedures of TAPPI Standard T646.

Leachability is measured on the coating clay fractions to determine the ultimate brightness that can be attained using standard reducing and oxidizing leach procedures. These determinations are very important because it establishes the level of treatment and whether the kaolin can be utilized alone or must be blended to meet brightness specifications.

For air-float clays only grit and brightness are determined. For ceramic clays, grit, brightness, green and dry strength, and fired color are determined. Chemical analyses for iron and alumina are commonly required for kaolin used in a refractory product. Other factors that can be important are the particle shape, pH, percentage of particles less than 0.25 micrometers in size, percentage of nonkaolin minerals, particularly smectite, in some cases total surface area, and the quantity of dispersing agent necessary to attain the lowest viscosity at 71% solids.

After all the tests are performed and evaluated and the overburden thicknesses and the thicknesses of the deposit have been determined, a decision is made whether the deposit is of a quality suitable for any of the present uses for kaolin or a potential aluminum resource, or whether the property should be dropped. The economic limits of the ratio of overburden to kaolin that can be mined economically range between 6 and 8 to 1, depending on the type of overburden, and the reclamation requirements imposed by federal and/or state law.

If the decision is made to keep the deposit for future mining, then additional drilling is done on 200- or 100-ft centers. From this drilling information, additional maps and cross sections are prepared, from which a stripping and mining plan is devised.

The classification of a deposit as potential aluminum clay is of relatively recent practice as a result of the interest in using kaolin as a raw material to produce aluminum. The United States is dependent on foreign sources of bauxite to supply approximately 90% of its aluminum requirements. The purchase of bauxite represents a significant outlay of dollars which contributes to our balance of payment deficits. Kaolin clays in Georgia and South Carolina contain 35 to 39% Al_2O_3 and represent a very sizable resource from which aluminum could be recovered if the economics become favorable. A report by the National Materials Advisory Board (Anon., 1970i) reviewed the various processes for extracting alumina from nonbauxite ores. An article (Anon., 1973b) also described a process to make aluminum using high-alumina clay which is chlorinated and reacted with manganese to form aluminum. Coal is used as the fuel.

Preparation for Markets

Mining and Beneficiation: The mining of kaolin is a highly sophisticated mechanized operation (Murray, 1963b). After extensive testing to insure that the deposit can be beneficiated to meet the necessary quality standards of the various process industries, mining plans are prepared which include clearing of timber, removing the overburden by either motorized scrapers or large draglines, establishing a land-reclamation schedule and program that must be approved by officials in some states, such as Georgia, and establishing approved environmental controls on water drainage and dust emanation. After the overburden is removed and the top of the kaolin deposit is exposed, additional drilling is done in order to provide detailed information of quality variations within the deposit so that a predetermined mining program can be followed that will meet the requirements needed by the beneficiation plant. Detailed maps showing the elevations of the top of the kaolin, thickness of the deposit, all quality determinations, and lateral extent are prepared which are used to control the mining operation properly.

The principal beneficiation steps are shown in Fig. 11. After the kaolin is exposed and tested, the kaolin is mined by shovels, draglines, motorized scrapers, or front-end loaders. The kaolin is either loaded in trucks or dropped into a blunger which disperses it in water with the aid of a dispersant chemical to form a clay-water slip or slurry. This slurry is pumped from the blunger to a degritting station, where, by the use of settling boxes, screens, and/or hydrocyclones, the coarse foreign particles called grit are removed. Grit is defined as that material coarser than about 44 micrometers. After degritting, the kaolin slurry is collected in large storage tanks and then pumped through pipelines to the beneficiation plant.

Two basically different processes are used to produce a salable kaolin product—a dry process and a wet process. The dry process is simple and yields a lower cost and lower quality product than the wet process. In the dry process, the properties of the finished kaolin reflect generally the properties originally found in the crude kaolin. Therefore, deposits containing the desired qualities of brightness, low grit content, and particle-size distribution must be located by drilling and testing. In the dry process, the kaolin is crushed to approximately egg size, dried in rotary dryers, pulverized, and air-floated. The air-floating process removes most of the grit.

The wet process (Fig. 11) has been highly developed in recent years to produce kaolin products of uniform and predetermined physical and chemical properties. Crude kaolin of variable quality can be wet processed to produce a uniform high-quality product.

The kaolin slurry is pumped from the mines to the plant, where it is stored in large tanks. Kaolins of various qualities are blended at the mine tanks or in the terminal tanks at the plant to provide a feed for the beneficiation plant that will produce products to meet specifications. The first step in the wet process is to fractionate the kaolin slurry into coarse and fine fractions by using continuous centrifuges, hydrocyclones, or hydroseparators (Fig. 11). After fractionation, the kaolin is leached to remove ferruginous coloring compounds. In this step the kaolin is acidified with sulfuric acid to a pH of about 3.0 to dissolve the iron. A strong reducing agent such as a hydrosulfite is then added, and the iron reacts and forms a soluble sulfate, which is removed during the dewatering step. The dewatering of the flocculated kaolin slip is accomplished by using high-speed centrifuges, rotary vacuum filters, or filter presses. Sometimes the flocculated slip is thickened by using thickeners or centrifuges before filtration. After filtration, the kaolin filter cake can be extruded to noodle form and dried in apron dryers or rotary dryers or it can be redispersed to the fluid state and spray dried, drum dried, or shipped in slurry form as 70% solids in tank cars.

The foregoing paragraph is a simplified description of the wet process. Other processes are used to produce brighter and whiter kaolins and to reduce viscosity. Flotation (Greene and Duke, 1962), selective flocculation (Bundy and Berberich, 1969), and magnetic separation (Iannicelli, et al., 1969) are used to remove iron and titanium impurities to produce kaolin in the range of 90% brightness. This kaolin has a blue-white shade.

The most recent development for improving brightness and whiteness is a high-intensity magnetic separator which removes substantial quantities of iron and titanium mineral impurities. The intensity of these newly developed units ranges from 18 to 22 kilogauss. The main feature of the separator system (Fig. 12) is a canister filled with a special type of fine steel wool which provides a uniform magnetic field through which the kaolin slurry is pumped and the magnetic particles retained in the matrix. Periodically, the magnet is deenergized, and a high volume of water under pressure is back flushed through the matrix to flush out the impurities (Oder and Price, 1973).

Delaminated kaolin is produced by splitting apart the larger books or stacks of kaolin (Fig. 13); this yields a product in which the individual plates tend to be wider, thinner, and whiter. The processes used for delamination include wet attrition grinding (Gunn and Morris, 1965) and extrusion (Lyons, 1959). Delaminated kaolin has special application in lightweight paper coatings, in paint films, and in special rubber applications.

Calcination is another widely used process to produce special products from kaolin. When kaolin is heated to approximately 1050°C, it is converted to mullite, silica-alumina spinel, and cristobalite. The product is whiter, brighter, has better hiding properties when used in thin film applications, and is more abrasive.

Other special processes are used to modify

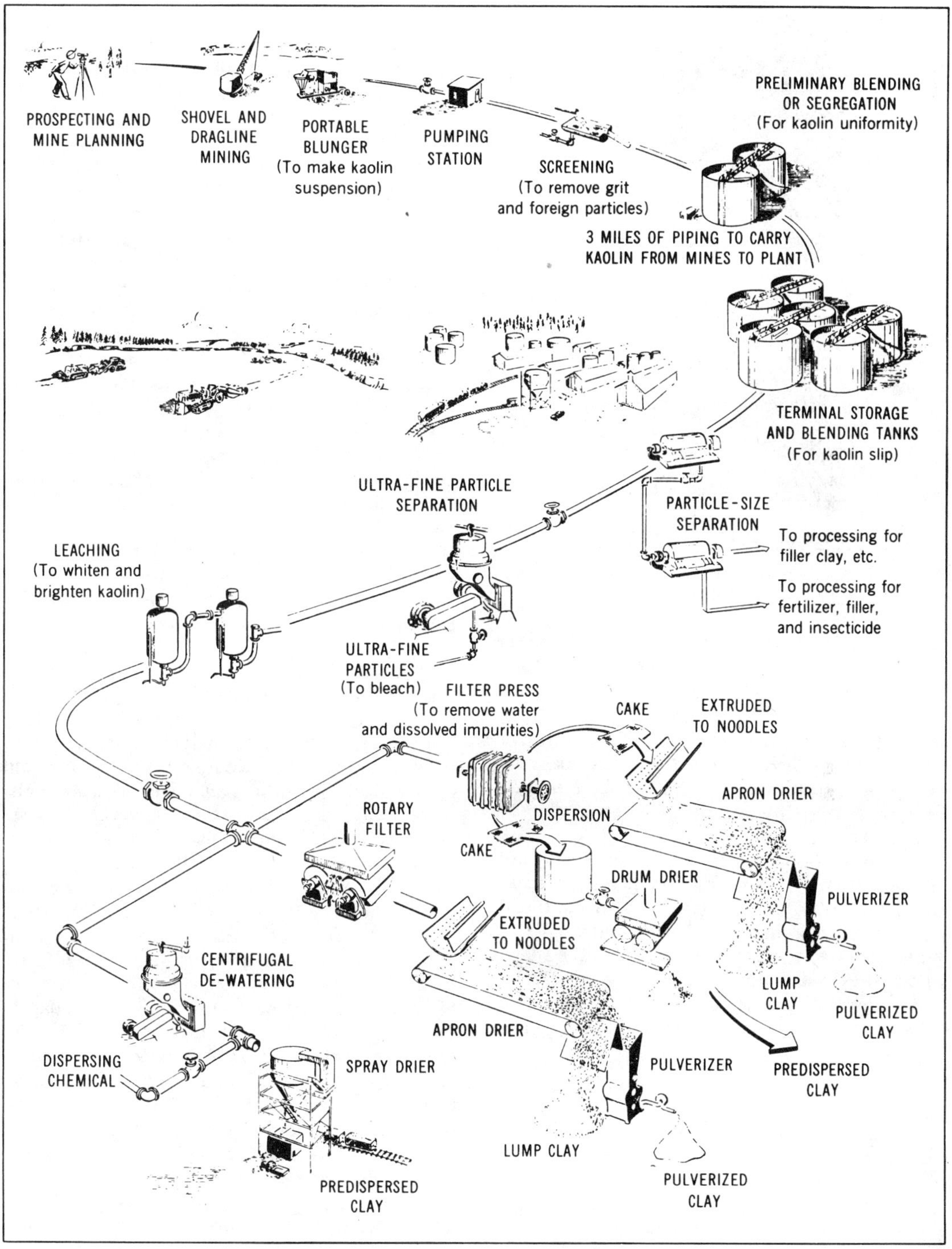

FIG. 11—*Illustrated flow diagram showing kaolin beneficiation methods* (*Murray, 1963b*).

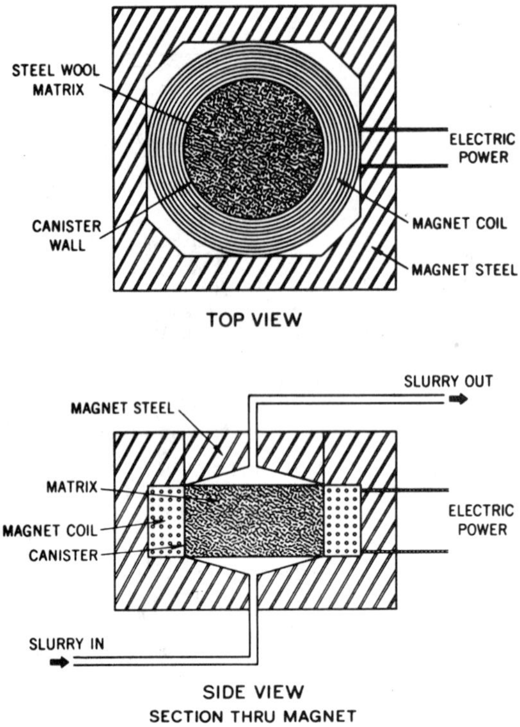

FIG. 12—*Diagrammatic illustration of magnetic separator used in removing iron and very fine-grained titanium minerals from kaolin.*

the surface of the kaolin chemically to improve its function in many systems, such as ink, rubber, paint, and plastics. These surface-modified kaolins can be hydrophilic, hydrophobic, or organophilic.

Transportation: The kaolin products are shipped from the beneficiation plants in bulk, in slurry form at 70% solids, and in multiwall paper bags. Kaolin is supplied commercially in pulverized form, in lump, generally in the shape of small extruded noodles, in spray-dried bead form, in small pellets, or in flake form. The weight per cubic foot of the kaolin will vary

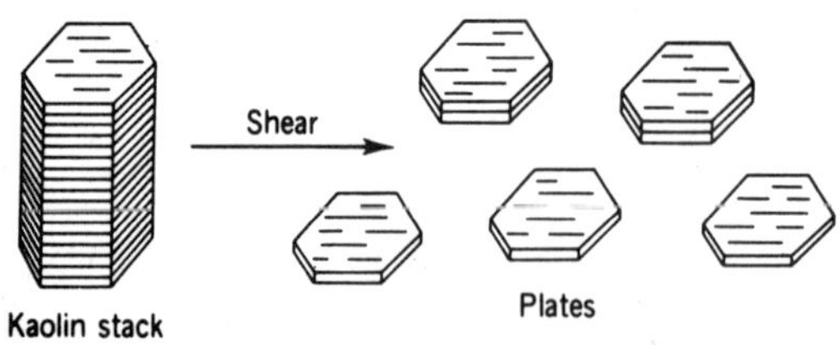

FIG. 13—*Diagrammatic illustration of the splitting of kaolinite booklets or stacks into thin plates by the delamination process.*

widely depending on whether it is in pulverized, lump, bead, or in some other form. Lump clay will generally weigh 60 to 65 lb per cu ft. Spray-dried clay in bead form will weigh 40 to 50 lb per cu ft. Depending on the fineness of pulverization and moisture content, pulverized kaolin will range from about 35 lb per cu ft for air-floated kaolin to as low as 20 lb per cu ft for some very fine particle size finely pulverized kaolin.

Kaolin is shipped in bulk form in boxcars, in covered hopper cars, and in slurry form in tank cars or in tank trucks. Kaolin is also shipped in 50-lb multiwall bags in boxcars or in trucks when the customer cannot handle the material in bulk. Export shipments are made in bulk ore-type carriers in 2,000- to 15,000-ton lots, in either lump or spray-dried form. Some export tonnage is also shipped in special paper bags of 25- to 50-kilo capacity.

Markets and Consumption: The various markets to which kaolin is supplied were discussed in the section on end uses. In 1980, 7,878,993 tons of kaolin were produced and sold in the United States (Table 1). Fig. 7 shows the tons of kaolin sold for use as paper filling and coating, rubber, ceramics, refractories, and other uses for the years 1929–1979. Since 1930, the market for kaolin has increased tenfold. In the 1930s, practically all the kaolin sold to the paper industry was used as filler, whereas today the proportion of filler clay to coating clay sold to the paper industry is approximately 1 to 2. Coating clay has a much higher value per ton than filler clay, and this increase in the use of coating clay has had more effect on the kaolin industry than any other single factor over the past 40 years. In 1980, the total value of kaolin sold was reported at \$527,098,609 (Ampian and Polk, 1981).

Exports of kaolin to Europe, Japan, and South America have increased significantly from 1965 to the present largely because of the increasing use of coating clay by the paper industry. In 1980, 1,392,000 tons of kaolin having a value of \$133.7 million, was exported (Ampian and Polk, 1981). The largest export tonnages of kaolin are shipped from Savannah, GA, and Port Royal, SC.

Future Considerations and Trends: Cooper (1970) has estimated that the demand for kaolin from the United States in the year 2000 will be five to six times greater than the amount now being sold. He has estimated a high of 23,680,000 tons and a low of 16,710,000 tons. He has assumed a technological and economic breakthrough that will allow kaolin to be used for the production of alumina and aluminum. If this projection is even partly correct, the future of the kaolin industry is indeed exciting and dynamic.

Competitive products for kaolin are ground calcium carbonate, talc, precipitated silica and silicates, alumina hydrate, and plastic pigments. Ground calcium carbonate is cheaper, has better color, and has lower oil absorption. Kaolin, however, has better flow characteristics and gives better hiding or opacity. In paper uses, ground calcium carbonate has several disadvantages with respect to kaolin: (1) it is reactive in acid finishes, (2) the ground calcium carbonate particles are too large to produce high gloss, (3) it gives poorer flow characteristics at high solids, and (4) it is more abrasive. Precipitated calcium carbonate is used in coating paper to give higher brightness and better ink receptivity.

Talc competes with kaolin mostly in paint, caulking compounds, and drywall compounds. In paints, talc gives better brightness and flatting in interior paints.

Precipitated silica and silicates give better reinforcement to rubber, produce higher tensile strength, modulus of elasticity, and resistance to abrasion. These materials are also added in paper to improve brightness and to extend the TiO_2 used in paper. However, kaolin is only one-third to one-half as expensive; consequently the precipitated silicas and silicates will be used only in conjunction with kaolin to achieve specific requirements.

Alumina hydrate is used in paper manufacturing to give higher brightness and to extend TiO_2. It is seldom used alone but is blended with kaolin. Alumina hydrate is two or three times more expensive than kaolin.

The drive to produce very lightweight coated paper has led to the development of plastic pigments which are uniform in size, lightweight, and spherical (Anon., undated; Heiser and Shand, 1973). The pigments give some superior properties such as gloss, opacity, and improved printability to coated paper but are much more expensive than kaolin. These pigments will be blended with kaolin to give certain improved properties for particular applications but will not replace kaolin as a coating pigment.

Ball Clay

History and Uses: Probably the first deposits of the material now known as ball clay mined in the United States were those worked near Paris, TN in 1860 for use in a local pottery (Hosterman, 1973). The first clay shipped to out-of-state consumers was that produced in west Tennessee in 1894 by a Mr. I. Mandle. Later, the industry expanded into Kentucky, and the ball clay districts in the western parts of these two states competed with districts in the United Kingdom (Anon., 1969d) for the world leadership in production. In 1981, more than 916 thousand tons of ball clay was produced in the United States (Ampian, 1982). More than 80% of this tonnage was produced by companies active in Tennessee and Kentucky. The other states producing ball clay are Mississippi, Texas, and California (Fig. 10). Ball clay is also produced intermittently in Maryland.

Ball clay is used principally in the manufacture of vitreous china sanitary ware, electrical porcelain, floor and wall tile, dinnerware, and artware (Phelps, 1972). It is also used in refractory products, ceramic glazes, and porcelain enamel slips.

Geology: *Mineralogy*—Kaolinite is the principal mineral in the ball clay deposits in Tennessee and Kentucky, but other clay minerals are present and nonclay minerals vary appreciably in abundance. The kaolinite in these deposits is fine-grained, and much of it is poorly ordered. The main clay minerals other than kaolin are illite, smectite, chlorite, and mixed-layer clay. Quartz is by far the most abundant nonclay mineral; in some deposits it makes up as little as 5% of the clay, but in others, as much as 30% (Olive and Finch, 1969). The other nonclay minerals ordinarily present occur in very minor or trace amounts and include plagioclase, potassium feldspar, and calcite. Organic matter in the form of black leaf imprints and disseminated lignitic and peaty materials are common in most ball clays.

Occurrence and Origin—Ball clay is extremely plastic. Deposits range from very light gray to nearly black. Most deposits are lenticular bodies varying considerably in size and shape. Some are more than 30 ft thick and extend over areas 1000 ft wide and 2500 ft long, but most deposits are smaller. The physical properties of a typical deposit vary considerably, and layers and pockets mined from a given deposit are commonly separated in different stockpiles suitable for a particular use.

Most deposits in Tennessee and Kentucky are in beds of Claiborne age (middle Eocene), but a few, including one mined, are in Wilcox beds (lower Eocene) (Olive and Finch, 1969). The deposits mined in Texas near Henry's Chapel, northeastern Cherokee County (Fisher et al., 1965) are also in Wilcox beds. The ball clay mined in Panola County, Mississippi (Bicker, 1970), is in a formation of Claiborne age, and the deposits are approximately the same age as younger ones in Tennessee and Kentucky. The deposits mined in Stanislaus County, California, are in beds of Paleocene or Eocene age (Cleveland, 1957). Those mined on a small scale in the Hart district in eastern San Bernardino County, California, are in hydrothermally altered Tertiary rhyolites (Kelley, 1966).

Most ball clay occurs in transported sedimentary deposits, but one minor producing district in California contains hydrothermally altered clay. The deposits in the Tennessee, Kentucky, and Mississippi districts are in sedimentary formations cropping out along the eastern edge of the Mississippi embayment. According to one theory of origin (Olive and Finch, 1969), they accumulated in an elongate broad valley. The abundance of plant remains and organic matter associated with this clay indicates that the depositional basins probably were swampy.

Distribution of Deposits: *United States*—Any attempt to estimate the resources of ball clay is based on piecemeal information. The reserves, primarily in Tennessee and Kentucky, total probably as much as 100 million tons, and are owned by eight or ten different companies; they are thought adequate to sustain the industry for 100 years (Phelps, 1972). Additional reserves are in the producing districts in Mississippi, Texas, and California. Potential resources of low-grade ball clay, which probably would require beneficiation for market, occur in the vicinities of all of these districts. The prospects are also favorable for the discovery of more deposits of kaolinitic clay having the physical properties of ball clay in several southern states. Furthermore, if the term ball clay is applied to such deposits as the hydrothermally formed clay at Hart, CA, many areas in the western states are likely to contain large deposits.

Other Countries—The principal ball clay deposits in the United Kingdom occur in the Bovey Basin southeast of Dartmoor in Devonshire, the vicinity of Petrockstow in north Devonshire, and in the Wareham and Poole regions of Dorsetshire. These ball clays occur in sedimentary rocks of Tertiary age, consisting mainly of lenticular units of sand, clay, and lignite of probable lacustrine origin (Scott, 1929). The clay is very fine grained plastic kaolin ranging in color from off-white to dark brown and gray. The darker colors, as in the ball clay in Kentucky and Tennessee, apparently are caused by carbonaceous impurities, and the clay burns white. The deposits are believed to have been transported from altered granite and other igneous rocks. Watts, Blake, Bearne, and Co. is the major producer of British ball clay.

Ball clay occurs at several localities in India (Arogyaswamy, 1968), but production is limited, and in recent years India has been an importer of ball clay. The best quality ball clay produced is in the Than district of Gujarat. Other deposits are mined in two districts in Kerala.

Evaluation of Deposits: The prospecting techniques for ball clay are very similar to those outlined for kaolin. The most commonly used tests are very similar to those used for evaluating the ceramic properties of kaolin and heat-resistant properties of refractory clays. Chemical analysis for soluble sulfate in ball clays is commonly required because this material can be tolerated in only very minor amounts in most ceramic products.

Preparation for Markets: Ball clay is mined by stripping methods and trucked to plants. Most plant processing involves primarily drying and grinding to a suitable size. A few plants air-float part of their output. Most ball clay is packaged in paper bags, but some is shipped in bulk carload lots in both crude and processed form. In recent years, some ball clay has been shipped in tank cars. This clay is prepared for shipment by deflocculation in water, and it is shipped as 60 to 62% solids (Phelps, 1972). Such slurries are used by plants manufacturing ceramics by the slip-casting method.

Future Considerations and Trends: The production of ball clay in the United States has grown rather consistently for the last several decades, and continuation of this trend is forecast. Nationwide demands are likely to be 2 to 2.3 million tons in the year 2000 (Cooper, 1970). In general, the ball clay industry faces little competition from foreign sources, and exports are appreciably greater than imports. However, the major districts in Tennessee and Kentucky are in the central part of the country and as rail transportation costs increase, a greater proportion of the demands in the populous coastal region may be satisfied by imports.

Halloysite

History and Uses: Apparently the first halloysite used in the United States was that mined on a small scale near Rising Fawn, GA in the late 1800s. This clay is reported to have been sold as a filler or extender in food (Veatch, 1909). Deposits near Gore, GA were mined and used for making alum about 1912 (Broadhurst and Teague, 1954). A few hundred tons of halloysite was mined from deposits in Indiana and used for pottery and alum during and after World War I (Callaghan, 1948). The major source of halloysite in this country has been the approximately 1 million tons mined from the Dragon mine, Tintic district, Utah (Fig. 10), by The Anaconda Co. It is processed into petroleum-cracking catalyst at the Filtrol Corp. plant in Salt Lake City, Utah. Mining to supply this plant began in 1949. Minor tonnages of halloysite were produced from this same mine before 1949 for other uses. Halloysite occurring at several other places in Utah has also been used in making light-colored brick, firebrick, tile, and as a paper filler (Van Sant, 1964). Halloysite in Nevada has been used in portland cement in recent years (Papke, 1971). Deposits in Idaho consisting of mixed kaolinite and halloysite are used in refractories (Hosterman et al., 1960), and deposits of similar clay-mineral composition in North Carolina (Parker, 1946) are mined for use in ceramics and other products.

Geology: *Mineralogy*—Halloysite is so similar to kaolinite that the two cannot be distinguished without careful mineralogical work (Brindley, et al., 1963). Very fine tubular structures commonly observable in electron micrographs of dehydrated halloysite (metahalloysite) are ordinarily diagnostic of this mineral. The shortcoming of this criterion is that not all metahalloysite occurs in the tubular form. A reliable criterion is that the basal (001) spacing of halloysite is at approximately 10Å; as halloysite converts irreversibly to metahalloysite, the spacing collapses to about 7.2Å, which is essentially the same spacing as for kaolinite. Halloysite, therefore, can be identified by the greater spacing in X-ray diffractograms of samples containing natural moisture.

Nonclay minerals and other impurities in halloysite vary considerably in different deposits. Some deposits, such as those in the Dragon mine, Utah, are exceptionally pure and contain only minor quantities of contaminants, consisting chiefly of pyrite, silica, dolomite, iron-bearing minerals and noncrystalline materials, and manganese oxides (Morris, 1968). In most deposits, silica is the principal impurity; it occurs in the form of quartz, cristobalite, chert, and poorly crystalline forms. Many halloysite deposits are intermixed with kaolinite, and several in western United States are closely associated with alunite. Gibbsite and manganese minerals also occur in some halloysite deposits.

Occurrence and Origin—Small deposits of halloysite occur at many places in the United States, but known deposits large enough to mine profitably are rare. The deposits in the Dragon mine, Utah, the major producer, are two large pipelike bodies replacing lower Paleozoic limestone near the contact with monzonite porphyry (Kildale and Thomas, 1957). The two bodies are separated by a zone of iron-oxide deposits extending along the Dragon fissure (Morris, 1968). The occurrence of these deposits and the minerals associated with them clearly indicate that they formed by hydrothermal activity. Other deposits thought by Papke (1971) to have formed hydrothermally occur in the Terraced Hills, Washoe County, Nevada. These deposits are bedded and apparently altered mainly from tuff. The halloysite in the Gore district, Georgia, is bedded and occurs in the Armuchee Chert of Devonian age (Broadhurst and Teague, 1954). An adequate explanation of these deposits has never been advanced, but the extent of the beds suggests that they are of sedimentary origin. Halloysite mixed with kaolinite in the Spruce Pine district, North Carolina (Hunter and Hash, 1953; Parker, 1946), and the Bovill district, Idaho (Hosterman et al., 1960), is mainly residual and formed by the weathering of igneous rocks. Several deposits in foreign countries are thought to have formed hydrothermally, and some are the result of surficial weathering.

Distribution of Deposits: *United States*—Virtually all major productive halloysite deposits in the United States have been referred to in the foregoing discussions. Most of the many other known scattered deposits are either too small or too impure to be considered as significant sources of halloysite. Possible exceptions include the following: (1) the halloysite mixed with kaolin in weathered nepheline syenite masses in the Arkansas bauxite region (Gordon, et al., 1958); (2) the body of halloysite reported to have been exposed for 350 ft in the adit of a mine in the Bullion (Railroad) district, Nevada (Olson, 1964); and (3) rather extensive deposits reported near Bartow, GA (Anon., 1959). Possibilities for the discovery of new large high-grade deposits are not easy to evaluate. The likelihood of major discoveries does not seem promising, but rocks in many large areas in western United States have been altered hydrothermally, and several of these could contain large deposits of halloysite. Also, resources of halloysite in the old Gore mining district in Georgia are estimated at more than a million tons (Broadhurst and Teague, 1954), and some use may again be found for these deposits.

Other Countries—Halloysite occurs in many countries, but, as in the United States, most deposits are small; the authors are aware of small-scale mining only in Morocco, Japan, Korea, Czechoslovakia, and New Zealand. The deposits mined in Morocco are in the Maaza district about 6 km west of Melilla (Hilali et al., 1969; Martin Vivaldi and Vilchez, 1959). These deposits occur above travertine and are overlain by pyroclastic materials. They are in a bed of Pliocene age, ranging in thickness from 1.5 to 2.5 m. Halloysite mined in Japan includes deposits that formed by hydrothermal alteration of granite and quartz porphyry and others that were transported and deposited in sedimentary basins (Takeshi and Kato, 1969). Hydrothermal deposits occur at the Taishu mine, and white sedimentary halloysite is in the Tajimi and Arikabe deposits. The halloysite mined in Korea is presumably from the San Chong and Tan Song deposits described by Kim and Kim (1964). The halloysite deposits mined in Czechoslovakia are in the Michalovce district in the eastern part of the country (Kuzvart, 1969). These deposits are thought to have formed by the hydrothermal alteration of dacite. They extend over an area 500 m square and are as much as 36 m thick. They occur below overburden 30 to 40 m thick. The halloysite mined is reported to be suitable for use in ceramics, refractories, paper and rubber fillers, and in refining sugar. The deposits in New Zealand occur in the Maungaperua district on the North Island. This clay is mined by New Zealand China Clays, Ltd., and sold for use in paper filler, ceramics, paint extenders, and filler for adhesives and cosmetics. The halloysite is in massive deposits formed by hydrothermal alteration of andesite of middle Cenozoic age (Bowen, 1969).

Several of the kaolin deposits in other countries consist of mixtures of halloysite and kaolinite. One such mixed deposit occurs in altered felsitic volcanic rocks in the Burela district, Spain (Anon., 1972b), the principal kaolin-producing center in that country. The kaolin produced contains appreciably more halloysite than kaolinite, but the clay must be considered to be a mixed type. Another mixed halloysite-kaolinite deposit is in the Bukidon district in the Philippines (Comsti, 1969).

Future Considerations and Trends: Halloysite will probably be in demand in the United States for use in petroleum catalysts and other products as long as suitable high-quality deposits can be mined at a reasonable cost. As mentioned previously, known high-grade deposits are limited, and though additional deposits may be found, the point may be reached when it will be cheaper to use substitute materials than to prospect for new deposits. Catalysts made from halloysite are in competition with those manufactured from kaolin and synthetic materials containing alumina and silica.

The products other than catalysts made from halloysite in recent years are all low-value or high-weight items lacking the broad market base of catalysts. All these other products can be made from other clays or other materials, but halloysite will continue to be used for them until reserves are exhausted, cheaper substitutes found, or local markets disappear.

Refractory Clays

History and Uses: The first plant in the United States for making refractory products from clays began operation at Salamander, NJ, in 1825. The refractory industry grew rapidly and expanded from New Jersey to centers in Pennsylvania, the Ohio River valley, Missouri, and to western states.

Refractory clays are used mainly in making firebrick and block of many shapes, insulating brick, saggers, refractory mortars and mixes, monolithic and castable materials, ramming and air gun mixes, and other products. A product called *chamottes* in Europe (Anon., 1972, 1972i) and elsewhere is made by calcining high-grade fire clays and other kaolinitic clays. A similar material produced in the United States is called merely calcined clay or calcined kaolin. A related material called mullite refractory is made by calcining bauxitic clay or clayey bauxite. Further fabrication into finished refractory products is required for chamottes and calcined clay. Considerable tonnages of fire clay also have been used in the past in the United States in the manufacture of light-colored face brick, tile, stoneware, and other products.

Product Specifications: The specifications for refractory clays are as many as the different uses. As resistance to heat is the most essential property, many specifications include reference to pyrometric cone (Table 4), indicating the heat duty required. Resistance to shrinkage, warping, cracking, and abrasion is also very important in many products, and expressions of the requirements for these properties appear in some specifications. Many of the end products must fulfill rigid specifications on dimensions. Restrictions on raw clay commonly include cutoffs on plasticity, impurities, and content of alkalies, alkaline earths, and other constituents that reduce fusion points.

Geology: The mineralogy, occurrence, and origin of refractory kaolin and refractory ball clay have been discussed in the foregoing sections. Therefore, the following discussions will be concerned with these subjects only as they apply to fire clay.

Mineralogy—The degree of ordering of the kaolinite, the principal mineral in fire clay, varies considerably. Most of the kaolinite in the harder and purer varieties of fire clay (flint clay) is very well ordered, and it is even more perfectly crystalline than the kaolinite in many kaolins. The kaolinite in the hard fire clay also tends to occur in coarser particles, and in the harder varieties the grains are interlocking. The kaolinite in the plastic varieties of fire clay, however, is ordinarily characterized by imperfectly crystalline structures. The more disordered form of this kaolinite has been widely referred to as fire clay mineral in an effort to distinguish it from the more perfectly crystalized forms.

Some of the better grades of refractory clays contain minerals that are richer in aluminum than kaolinite, and, therefore, higher alumina products can be made from them. Some refractory kaolin is a mixture of kaolinite and gibbsite, $Al_2O_3 \cdot 3H_2O$. The better grades of fire clay, that are exceptionally rich in aluminum, consist of kaolinite and diaspore, $Al_2O_3 \cdot H_2O$, and minor quantities of boehmite, $Al_2O_3 \cdot H_2O$.

Occurrence and Origin—The physical characteristics of fire clay vary considerably; the clays range from soft and plastic to flintlike. Fire clay, therefore, is divisible into plastic, semiplastic, semiflint, and flint types. Flint clay has unique properties for a clay; it lacks plasticity, breaks with a conchoidal parting into shardlike particles, and most of it is as hard as limestone. Typical diaspore-bearing flint clay is hard and has an oolitic, pisolitic, or nodular texture. Such clay is referred to as burley, birdseye, or nodular clay.

Most fire clay occurs in sedimentary rocks, and deposits range in age from Pennsylvanian to Tertiary. Fire clay is particularly common in rocks of Pennsylvanian age, and deposits of this age occur as underclays (seat earths) immediately below or closely associated with coalbeds. Fire clay of Cretaceous and Tertiary age occurs mainly in lenticular bodies. Some deposits of these ages are associated with lignite and are probably underclays, and others have apparently been transported and deposited in local basins in a nearshore, swamp, or floodplain environment.

Opinions differ considerably on the origin of the extensive fire clays occurring below coalbeds. Some geologists (Keller, 1970; Patterson and Hosterman, 1963), including the authors, believe that most underclay deposits formed mainly by the alteration of aluminous sediments in a swampy environment. Some mineralogists, however, believe that these clays were transported and some sort of a sedimentary winnowing process caused the rather pure accumulations of kaolinite in fire clay. Some clays of Cretaceous and Tertiary age have been transported and formed by a process similar to those thought to have formed the transported kaolin deposits.

Distribution of Deposits: *United States*—Fire clay of Pennsylvanian age is widely scattered throughout the Appalachian region and parts of the Mississippi Valley (Anon., 1967). The very high quality diaspore-bearing clay, as well as other grades of fire clay, occur in the Clearfield district, Pennsylvania (Anon., 1964; Bolger and Weitz, 1952) (Fig. 10), and the Ozark region in Missouri (Keller, 1952; Keller et al., 1954). High grade kaolinitic flint clay and semiflint clays are mined in the Olive Hill district, Kentucky (Patterson and Hosterman, 1963); Oak Hill district, Ohio (Stout, et al., 1923); and Somerset district, Pennsylvania (Hosterman, et al., 1968). Major districts producing fire clay suitable for low- and moderate-heat-duty refractory products include the Allegheny Valley and Beaver Valley districts, Pennsylvania; Cordova district, Alabama; East Liverpool district, Ohio and West Virginia; and the Tuscarawas Valley and Hocking Valley districts, Ohio (Hosterman, et al., 1968); and a region including parts of Monroe, Audrain, Callaway, and Montgomery Counties, Missouri (McQueen, 1943).

The largest and best quality fire clay deposits in the Rocky Mountain region are in Fremont, Pueblo, Custer, Huerfano, Jefferson, and Las Animas Counties, Colorado. These deposits are in the Purgatoire Formation and Dakota Sandstone of Cretaceous age (Waagé, 1953). Fire clay in these deposits occurs as isolated tabular lenses ranging in thickness from 3 to 20 ft.

In addition to the fire clay mentioned in the foregoing discussions, other deposits are scattered throughout the western states (Anon., 1967; Mark, 1963; Van Sant, 1959, 1964).

These deposits are of several types, and they include transported, residual, and hydrothermally formed clay. Districts in which these scattered deposits occur include several areas in King County and the Castle Rock area in Cowlitz County, Washington; the Molalla and Hobart Butte areas, Oregon; and the Alberhill area, California. Most of these deposits are of Cenozoic age.

Refractory kaolin is produced primarily in the following districts: (1) Georgia-South Carolina kaolin belt; (2) Andersonville, Georgia (Anon., 1972a); (3) Eufaula, Alabama; (4) Arkansas bauxite region; (5) scattered districts in Texas; (6) Latah County, Idaho; and (7) Ione, California (Anon., 1972e). Refractory ball clay is produced primarily in the western parts of Kentucky and Tennessee, but some of the ball clay produced in Texas may be used in refractory products.

The total refractory clay resources in the United States suitable for low-heat-duty refractory products may be as much as 7 billion tons (Hosterman, 1973). However, the reserves—those clays that can be mined and processed now at competitive costs and are suitable for low-heat-duty products or better—are probably no more than 1 billion tons. The reserves are primarily in the fire clay and refractory kaolin and ball clay districts outlined previously. The remaining resources of high-grade diaspore clay are mainly under considerable overburden in Pennsylvania, and only small resources are present in Missouri. The bauxitic kaolin resources are primarily in the Arkansas bauxite region, and the Eufaula, Alabama, and Andersonville, Georgia, districts. Some deposits also occur in other parts of Alabama and Georgia.

Other Countries—Refractory clays are produced in many countries; however, the information available to the authors on worldwide production is sketchy and incomplete. This is partly because the distinction between fire clay and miscellaneous clay is not made in some countries, and clay used for refractory products is lumped with kaolin in others.

Those countries that produce more than 1 million tons of refractory (fire) clay annually include the United Kingdom, Federal Republic of Germany (West Germany), and Japan. The production in France exceeded a million tons a few years ago but has apparently dropped below this figure in recent years. Other countries producing 100,000 to 1 million tons annually include Argentina, Australia, India, Italy, Mexico, New Zealand, Sweden, the United Arab Republic, Uruguay, and Yugoslavia. Refractory clay is exported from the Peoples Republic of China, which is apparently a major producer. Fire clay is also mined in the USSR, Czechoslovakia, Hungary, Poland, and other East European countries.

As in the United States, fire clay in other countries is in several types of deposits. Kaolinitic fire clay associated with coalbeds of Carboniferous age is mined in the United Kingdom, France, Federal Republic of Germany, and elsewhere. Very high alumina diaspore-bearing flint clays are sources of refractory materials in Scotland, South Africa, Israel, and the Peoples Republic of China. Refractory kaolin is mined in Japan, India, Hungary, Italy, Czechoslovakia, Poland, Sweden, Yugoslavia, Mexico, Argentina, Brazil, Chile, Peru, Republic of South Africa, United Arab Republic, Australia, New Zealand, Iran, and other countries, and most such deposits are no older than Cretaceous.

Exploration and Evaluation: The methods used in exploring and evaluating fire clay deposits are in general similar to those used for bentonite and kaolin, except that different drilling and testing procedures are required. Drilling fire clay deposits, particularly those of Pennsylvanian age, ordinarily requires diamond bits and core barrels of the type used for minerals in hard rock. This is because the clay itself and the rocks overlying the deposits are hard. The most common test in the evaluation of fire clay is the determination of PCE (pyrometric cone equivalents, Table 4, p. 613).

In addition to the PCE requirement, fired test pieces also may be tested for such properties as high-temperature stability, hot strength, porosity, and spalling (Anon., 1972h; Norton, 1968). Chemical analyses are also required for some evaluations, and the contents of alkalies, alkaline earths, iron oxides, and a few less common elements are critical, because they act as fluxes when clay refractories are heated.

Preparation for Markets: Fire clay and other refractory clays are mined by methods similar to those used for other types of clay. However, a much higher percentage of fire clay mining in the past has been underground. The underground mining has been required because much of the higher quality clay occurs at considerable depths, and the overlying rocks are too hard to strip profitably.

Firebrick and related products are prepared for markets by grinding the clay, shaping or forming the firebrick or other products, and firing. Commonly several varieties of fire clay and grog (crushed previously fired clay) are blended to make a mix desired for a certain product, and blends are varied considerably to prepare different grades of firebrick. The harder clays are ordinarily the most refractory, and as they lack plasticity, small proportions of kaolin or plastic fire clay or both are added to aid in forming and bonding the brick until firing. Most firebrick is formed by hydraulic or other mechanical presses, but several types of products are intricately shaped and must be formed by hand.

Most firebrick is fired in tunnel kilns heated by gas or oil. A few plants, particularly those making specially shaped products or having a limited market, use downdraft kilns. Finished products are palletized and shipped by rail or truck. Heavy crating with padded packaging is required for overseas shipments. Calcined clay is ordinarily fired in rotary kilns and is shipped in both paper bags and in bulk carloads.

Future Considerations and Trends: The peak output of refractory (fire) clay in the United States was 11.8 million tons in 1957. Production has decreased, with only minor yearly reversals of the trend, since that date. In 1980, the total refractory clay produced was 2 million tons valued at $36 million (Table 1). This trend in decreasing production of fire clay has been due mainly to the following changes: (1) the development of more durable aluminous refractory materials; (2) the shift to the basic oxygen process in steelmaking, which requires basic rather than alumina-type firebrick in furnaces; (3) changes in statistical reporting, whereby fire clay used in light-colored facebrick, stoneware, and ceramic products is now classified as miscellaneous clay rather than fire clay.

The decrease in demands for refractory clays brought about by the factors outlined in the previous paragraph has probably bottomed out, and markets for refractory clay and products made from it can be expected to increase consistently. Demands forecast for refractory clay in the year 2000 are expected to be 12 to 17 million st (Cooper, 1970).

The availability of high-grade and very high grade refractory clays at competitive costs is one of the long-range problems faced by the refractories industry. Deposits of very high grade diaspore clay in Pennsylvania and Missouri that can be mined by stripping methods are nearing exhaustion, and underground mining increases costs considerably. The cost of underground mining is one of the reasons why the production of refractory bauxitic and kaolinitic clay has increased in the Andersonville, GA and Eufaula, AL, districts in recent years. The trend for the increasing use of these bauxitic clays is expected to continue. Another trend may be an increase in the beneficiation of kaolin both to improve purity and to recover a coproduct, such as glass sand, to make the cost competitive. Kaolin and glass sand are now produced as coproducts from impure deposits in Texas, Idaho, and California.

Miscellaneous Clay and Shale

Miscellaneous clay and shale includes a wide variety of clay and other fine-grained rocks that are used in many ways. Most products made from them are fired and include such things as structural and face brick, drain tile, vitrified pipe, quarry tile, flue tile, conduit, pottery, stoneware, and roofing tile. Very large tonnages of these materials are bloated by firing to form lightweight aggregate, and large quantities are used in making portland cement (see the specific chapters in this volume). Uses not requiring firing of a finished product include filler for paint and other products and shale and clay used for packing dynamite in blastholes and for plugging oil and gas wells that are no longer in use. Adobe building blocks are also used in unfired form in several southwestern states.

Properties

Clay and shale are used in so many different structural clay products that they necessarily have a wide range in physical properties. The properties are plasticity, green strength, dry strength, drying and firing shrinkage, vitrification range, and fired color. The properties desired vary with the structural clay product made. For example, a clay used in making conduit tile must be very plastic and have high green and dry strengths and uniform shrinkage, but for drain tile or common brick these properties do not have to be controlled so closely.

Most clays are plastic when naturally wet or mixed with water. Plasticity can be defined as the property of a material to undergo permanent deformation in any direction without rupture under stress beyond that of elastic yielding. Some clay materials are very plastic and are called *fat* clay; others having little plasticity are called *lean* clay. The type of clay mineral, particle size, particle shape, organic matter, soluble salts, adsorbed ions, and the amount and type of nonclay minerals influence the plastic properties of clay.

Green-strength and dry-strength properties of clay and shale are important because most structural clay products are handled at least once and must be strong enough to maintain shape. Green strength is the strength of the clay material in the wet or plastic state. Dry strength is the strength of the clay after it is dried. Plasticity and green strength are closely related and are influenced mainly by the same variables. Dry strength is dependent on the proportion of fine particles present, the shape of the individual particles, the degree of hydration of the clay fraction, the method of forming the ware, and the rate and thoroughness of the drying. The presence of a small amount of smectite, which occurs in very fine hydrated particles, generally increases dry strength.

Both drying and firing shrinkages are critical properties of clay used for structural clay products. Shrinkage is the loss in volume of a clay as it is dried or fired. Drying shrinkage is dependent on the water content, the character of the clay minerals, and their particle size. Most plastic clays shrink appreciably with drying, which tends to produce cracking and warping. Sandy clay or clay having low plasticity has low shrinkage properties but tends to produce a weak porous body. Smectite minerals in rather large amounts (10–25%) commonly cause excessive shrinkage, cracking, and slow drying. Firing shrinkage depends on the density of the clay, volatile materials present, the types of crystalline phase changes that take place during firing, and the dehydration characteristics of the clay minerals.

The temperature range of vitrification or glass formation during firing is a very important property of clay and shale used in structural clay products. Vitrification is a process of gradual fusion in which some of the more easily melted constituents produce increasing amounts of liquid as the temperature is increased. This liquid produces glass as the material is cooled, and the glass is the bonding material in the final fired product. Some clay has a short vitrification range, and when such clay is fired the temperature of the kiln must be very closely regulated. Clay consisting chiefly of illite, smectite, or chlorite has lower vitrification temperature than kaolinite-rich clay. Some mineral impurities in clay, such as calcite and feldspar, lower vitrification temperatures by acting as fluxes. The degree of vitrification attained also depends on the duration of firing and the temperature. Commonly the degree of vitrification is regulated by the amount of shrinkage and porosity needed in the final product.

Uniform color is an essential property of many structural clay products. The color of a product is influenced by the state of oxidation and particle size of the iron minerals; the firing temperature and degree of vitrification; the proportion of alumina, lime, and magnesia in the clay material; and the composition of the gases in the kiln during the burning operation. High-grade white-burning clay contains less than 1% Fe_2O_3, buff-burning clay ordinarily contains 1–5% Fe_2O_3, and red-burning clay contains 5% or more Fe_2O_3. Other constituents also affect the color, but finely divided iron-bearing minerals generally are the principal materials causing color in fired products.

Occurrence

Miscellaneous clay and shale occur in many types of rocks ranging in age from Precambrian to Holocene. They include glacial clay, soils, alluvium, loess, shale, weathered and fresh schist, slate, and argillite. Some fire clay and kaolin are also included in the miscellaneous clay group, particularly when used in the manufacture of structural-clay products. This group consists of so many different types of rocks that only a few broad generalizations about their mineral makeup are possible. The most common mineral in many of them is one of the members of the mica group, and the dominant one in a given deposit may be illite, sericite, or one of the micas normally occurring in coarser grain sizes, such as muscovite and biotite. In addition to illite, the clay minerals present commonly are kaolinite, smectite, mixed-layer varieties, and chlorite. Some materials in this group actually contain more quartz and other detrital minerals than clay minerals.

Miscellaneous clay and shale are now mined in every state in the United States, except Alaska and Rhode Island, according to US Bureau of Mines *Minerals Yearbooks*. In 1980, 32.4 million tons of these materials having a value of $114.7 million was produced (Table 1). The states producing more than 1 million tons were Alabama, California, Georgia, Illinois, Indiana, Iowa, Louisiana, Maryland, Michigan, Missouri, New York, North Carolina, Ohio, Pennsylvania, South Carolina, Texas, and Virginia. Texas was the leading state in production in 1980, with 3.4 million tons, and North Carolina was second, with 2.8 million tons.

Testing, Processing, and Markets

The testing of miscellaneous clay and shale for most products requires the preparation of test pieces of specific dimension, drying, and firing (Clews, 1969; Grimshaw, 1972; Klinefelter and Hamlin, 1957). Fired test pieces are tested for such properties as shrinkage, warping, cracking, permeability, modulus of rupture, crushing or compression strengths. Color and soluble salt content are also important for some products.

Miscellaneous clay and shale is dug from open pits. Pits must be near plants and the digging must be cheap, because both the raw material and finished products are heavy, and the margin of profit is ordinarily low. Most products made from these materials are processed and marketed much in the manner described for refractory clays.

Future Considerations and Trends

The long-range trends in demands for some products made from miscellaneous clay and shale can be expected to keep pace with the growth in the gross national product, but demands for others are likely to fall short of this growth rate. Demands for clay and shale required for portland cement and lightweight aggregate are increasing, and this trend is likely to continue. Though demands for heavy structural-clay products will continue, these materials are facing strong competition from cement, wood, glass, plastics, aluminum and other metals. Growth in demands for structural-clay products is also hampered by the heavy weight of these products, which limits their market range, and by increasing production costs. The rise in fuel costs, which has been forecast, seems likely to be a major problem faced by the industry.

Another strong trend in the industry using miscellaneous clay and shale has been the shift to rural areas. For a long time, plants were near population centers because of transportation costs. Suburban growth in many population centers has forced plants to close or to relocate because of ecological pressures and raw-material supply problems. This trend seems certain to continue until virtually all active plants are in rural areas, which, in turn, tends to favor a substitute material because of increased transportation costs and other factors.

Government Controls and Influences

Bentonite and Kaolin on Public Domain Lands

Most of the bentonite and part of the kaolin resources in the United States are on public domain lands. At the time this chapter was written, bentonite and kaolin deposits on federal lands for which mineral rights were not withdrawn or otherwise not obtainable were subject to the general mining laws. Mining claims could be filed on valid bentonite deposits, and under certain conditions such claims could be patented. The staking and patenting of mining claims on public domain lands are primarily under the jurisdiction of the Bureau of Land Management, an agency of the US Department of Interior. Further information on the staking and patenting of mineral claims on federal lands can be found in Bureau of Land Management pamphlets 0–377–927 and 0–377–930.

Depletion Rates

The following depletion rates were in effect in 1973 (Anon., 1972k): kaolin, ball clay, bentonite, fuller's earth, and fire clay, 14% domestic and 14% foreign; clay and shale used for making brick, tile, and lightweight aggregate, 7½%, both domestic and foreign. The depletion rate for clay used for the extraction of alumina or aluminum compounds was set at 22%, but because no clays have yet been used for this puropse in the United States the rate was not applied.

Tariffs

The statutory rates of duty per long ton for the various classes of clay in effect in 1980 (Ampian, 1981) are listed in Table 5.

Ecology

All segments of the clay industry, like all other mineral industries, face major problems related to the environmental controls, and compliance with regulations now in effect has added significant production costs. Virtually all the major plants producing clay involving a dry-grinding process have installed equipment to control dust emissions in the atmosphere. Not all such equipment has proved effective, and some companies are faced with further expenditures for improving dust control. The control of turbidity and other contamination in water pumped from pits and discharged from washing and other types of plants is also required. This regulation has forced mining to be shut down at least temporarily in flooded fuller's earth pits and adds the costs of maintaining settling ponds to remove the suspended materials in the discharge from kaolin-washing plants. As most clays are strip mined, land

TABLE 5—Tariffs per Long Ton for Various Classes of Clay, 1980

Item	TSUS Number	Rate of Duty (per Long Ton, $) Most Favored Nation (MFN) 1/1/81	Most Favored Nation (MFN) 1/1/87	Non-MFN 1/1/81
China clay and kaolin	521.41	33.0¢	33.0¢	$2.50
Fuller's earth, not beneficiated	521.51	25.0¢	25.0¢	1.50
Wholly or partly beneficiated	521.54	50.0¢	50.0¢	3.25
Bentonite	521.61	40.0¢	40.0¢	3.25
Ball clays, not beneficiated	521.71	41.0¢	38.0¢	1.00
Wholly or partly beneficiated	521.74	83.0¢	77.0¢	2.00
Other clays, not beneficiated	521.81	Free	Free	1.00
Wholly or partly beneficiated	521.84	50.0¢	50.0¢	2.00

reclamation costs are now a major item. Commonly, as much as $500 per acre is spent in restoring mined bentonite lands in Wyoming, where the value for the surface alone is only about one-tenth of that amount. The restoration of some of the strip-mined land in the Georgia kaolin belt costs as much as $800 per acre, but the average cost is about $300 per acre. The high costs of reclamation in Georgia are due to state laws governing mining that are even more stringent than the federal regulations.

Acknowledgments

Much of the authors' knowledge of the various clay industries has been gained by conversations and other communications with many professional and management personnel. Several of these persons contributed to the information required by one or both of the authors for the preparation of this chapter, and others have aided them by contributing to their general background knowledge in the past. Among those to whom we wish to express thanks are John McKenzie and Kefton H. Teague of International Minerals and Chemical Corp.; William T. Granquist and John W. Jordon of NL Industries; Arthur G. Clem and E. P. Weaver of American Colloid Inc.; W. L. Otwell, D. B. Crosten, and T. E. Shufflebarger, Jr., of Pennsylvania Glass Sand Corp.; Finus Turner of Dresser Magcobar; Jack W. Williamson, W. L. Haden, Jr., and T. D. Oulton of Engelhard Minerals and Chemicals Corp.; James L. Rodgers of Georgia-Tennessee Mining Co.; R. C. Valenta and Richard M. Jaffee of Oil Dri Corp. of America; Robert P. Smith, Waverly Mineral Products Co.; J. F. Burst and Glen P. Jones of General Refractories Co.; J. F. Wescott of A. P. Green Refractories; Edward Ruh and O. M. Wickem of Harbison-Walker; P. E. Turbett and H. C. Spinks of Clays Corp.; D. W. L. Spry, Clive W. Gronow, and William Mallory of Anglo-American Clay Co.; James D. Cooper and Sarkis G. Ampian of the US Bureau of Mines; Jack Rogers, E. F. Oxford, Tijs Volker, and Ralph B. Hall of J. M. Huber Corp.; R. V. Colligan, Fred Gunn, H .H. Morris, Paul Sennet, James Olivier, and Robert Weaver of Freeport Kaolin Co.; Paul Thiele and Owen Etheridge of Thiele Kaolin Co.; Wayne M. Bundy, John M. Smith, and Doral Mills of Georgia Kaolin Co.; George Eusner of Mulcoa-CE Minerals; R. E. Grim, University of Illinois; W. D. Keller and W. D. Johns, University of Missouri; George W. Brindley and Thomas F. Bates, Pennsylvania State University; William F. Bradley and Edward Jonas, University of Texas; C. E. Weaver and W. E. Moody, Georgia Institute of Technology; John Koenig and Malcolm McLaren, Rutgers University; N. O. Clark and C. M. Bristow, English China Clays; B. F. Buie, Florida State University; and Walter Parham, Minnesota Geological Survey.

Consulting geologists and engineers who have assisted the authors in many ways include Thomas L. Kesler, W. B. Beaty, Padriac Partridge, E. W. Adams, Richard H. Olson, G. W. Phelps, William J. Lang, Richards A. Rowland, and Wayne Fowler.

The authors were also aided in the preparation of this chapter by having access to information gathered by John W. Hosterman (1973) during the preparation of the clay chapter in an overall review of the mineral resources of the United States published by the US Geological Survey. E. C. Freshney of the U.K. Institute of Geological Sciences supplied references to material on some of the clay deposits in the United Kingdom discussed in the previous pages.

M. Kuzvart of Charles University, Prague, organized the Kaolin Symposium at the 23rd International Geological Congress in 1968 from which much of the information on foreign kaolin deposits was taken. Martin Vivaldi, Ciudad University, Madrid, Spain, organized the AIPEA Meeting of 1972 and personally provided information concerning the clays in Spain. R. E. Grim and B. F. Buie read the manuscript and both made several helpful suggestions. K. Y. Lee translated the sections of a report on which the discussion of kaolin in the Peoples Republic of China is based.

Bibliography and References

Anon., undated, "Plastic Pigments for Paper," Technical Bulletin No. 6314, Monsanto Co., 15 pp.

Anon., 1952, *Manual of ASTM Standards on Refractory Materials,* Committee C-8, American Society for Testing & Materials, 294 pp.

Anon., 1958, "Bleaching Tests," A.O.C.S. Official Method, Cc 8a 52, American Oil Chemists Society, 2 pp.

Anon., 1959, "Halloysite in the Cartersville District," *Georgia Mineral Newsletter,* Georgia Geological Survey, Vol. 12, No. 2, pp. 42–43.

Anon., 1962, "Brazil," *Mining Journal,* Mar. 9, p. 244.

Anon., 1962a, *Moulding Methods and Materials,* American Foundrymen's Society, Chicago, IL, 619 pp.

Anon., 1963, *Foundry Sand Handbook,* 7th ed., American Foundrymen's Society, Chicago, IL, 200 pp.

Anon., 1964, "Map of the Mercer Clay and Adjacent Units in Clearfield, Centre, and Clinton Counties, Pennsylvania," 4th Series, No. 12, Pennsylvania Geological Survey.

Anon., 1965, "Bentonite in Turkey," *Mineral Trade Notes,* US Bureau of Mines, Vol. 60, No. 4, pp. 12–14.

Anon., 1965a, "Tentative Specifications for Western Bentonite," SFSA Designation 13T-65, Steel Founder's Society of America, 5 pp.

Anon., 1967, "Potential Sources of Aluminum," Information Circular 8335, US Bureau of Mines, 148 pp.

Anon., 1967a, "Area Reports: International," *Minerals Yearbook, 1965,* Vol. 4, US Bureau of Mines, p. 343.

Anon., 1969, "API Recommended Practice Standard Procedure for Testing Drilling Fluids," API RP 13B, 2nd ed., American Petroleum Institute, 19 pp.

Anon., 1969a, "API Specifications for Oil Well Drilling Fluid Materials," API STD 13a, 5th ed., American Petroleum Institute, 11 pp.

Anon., 1969b, "Bentonite," Specifications No. DFCP.4, Oil Companies Materials Assns., London.

Anon., 1969c, "Bentonite's Indispensible Role in Industry," *Industrial Minerals,* No. 25, Oct., pp. 9–14.

Anon., 1969d, "Buoyant Market for Ball Clays," *Industrial Minerals,* No. 17, Feb., pp. 9–15.

Anon., 1969e, "Erbsloh & Co.: Pioneer Producer of Sodium-Exchanged Bentonite," *Industrial Minerals,* No. 27, Dec., pp. 33–34.

Anon., 1969f, "Laporte's Lead in Fuller's Earth, Continued Expansion at Redhill," *Industrial Minerals,* No. 24, Sep., pp. 15–16.

Anon., 1969g, "National Lead Company, Huge Reserves Held by Baroid Division," *Industrial Minerals,* No. 24, Sep., pp. 25–27.

Anon., 1969h, *Review and Forecast,* A Report Prepared for the Wyoming Natural Resources Board and the State Water Planning Program, Book I, "Summary," pp. 129–137; Book III, "Non-Fuel Minerals," pp. 1–68, A38–A60, Cameron Engineers, Denver, CO.

Anon., 1969i, "SAMIP: Mineria Isole Pontine Spa," *Industrial Minerals,* No. 24, Sep., pp. 21–23.

Anon., 1969j, "Ste. Francaise des Bentonites, A Range of Quality Clays," *Industrial Minerals,* No. 24, Sep., p. 45.

Anon., 1969k, "Sud-Chemie AG, Bleaching Earths from Bavaria," *Industrial Minerals,* No. 24, Sep., p. 44.

Anon., 1969l, "The Bentonite Industry Expands Capacity," *Industrial Minerals,* No. 24, Sep., pp. 10–14.

Anon., 1969m, "Yugoslavia's Bentonite Industry," *Industrial Minerals,* No. 27, Dec., pp. 35–37.

Anon., 1970, "An Emphasis on Minerals Marketing, Berk Ltd.'s Mineral Products Division," *Industrial Minerals,* No. 36, Sep., pp. 40–41.

Anon., 1970a, "Bentonite Clays of the Volga Regions," *Transactions,* Kazan Geological Institute, Ministry of Geology of the USSR, 1175 pp.

Anon., 1970b, "Bentonite from Sardinia, Laviosa a Leading Supplier of Processed Clays," *Industrial Minerals,* No. 31, Apr., pp. 47–48.

Anon., 1970c, "Fineness of Dispersion of Pigment-Vehicle Systems," *Annual Book of ASTM Standards, 1970,* American Society for Testing & Materials, D1210–64, pp. 210–215.

Anon., 1970d, "Hungary's Mineral Industry, Istvan Soha Details the Non-Metallics Produced," *Industrial Minerals,* No. 29, Feb., pp. 35–37.

Anon., 1970e, "Israel, New Process Up-Grades Cyprus Bentonite," *Industrial Minerals,* No. 39, Dec., 37–38.

Anon., 1970f, "Malaysia, Sanyo Backs Clay Project," *Industrial Minerals,* No. 35, Aug., p. 31.

Anon., 1970g, "Pacific Plans Kaolin Production," *Industrial Minerals,* No. 29, Feb., p. 39.

Anon., 1970h, "Poland, Kaolin Processing Commences," *Industrial Minerals,* No. 33, June, p. 46.

Anon., 1970i, "Processes for Extracting Alumina from Non-Bauxite Ores," Report 278, National Materials Advisory Board, 88 pp.

Anon., 1971, "Canada, Industmin Branches Out into Bentonite," *Industrial Minerals,* No. 45, June, p. 30.

Anon., 1971a, "Designation C-242–60," *Annual Book of ASTM Standards, 1971,* Pt. 13, American Society for Testing & Materials, 198 pp.

Anon., 1971b, *Estadictica Minera Año, 1969,* Argentina Direccion Nacional de Promocion Minera, 255 pp.

Anon., 1972, "An Introduction to Refractories," *Industrial Minerals,* No. 58, July, pp. 9–23.

Anon., 1972a, "CE Minerals: Georgian Bauxite and Kaolin Calcined for Refractory Grog," *Industrial Minerals,* No. 56, May, pp. 17–22.

Anon., 1972b, "Field Trip Guides," International Clay Conference, Madrid, Assoc. International pour l'Etude des Argiles (AIPEA), pp. II-1 to II-10, III-59.

Anon., 1972c, "Fullers Earth," *Industrial Minerals,* No. 62, Nov., p. 42.

Anon., 1972d, "Industrial Minerals in Oil Well Drilling," *Industrial Minerals,* No. 60, Sep., pp. 9–31.

Anon., 1972e, "Ione Calcined Kaolin, Interpace's Refractory Aggregate," *Industrial Minerals,* No. 59, Aug., pp. 37–41.

Anon., 1972f, "Kaolin in Europe and Japan," *Industrial Minerals,* No. 53, Feb., pp. 9–16.

Anon., 1972g, "Kaolin in the UK, English China Clay Defends Its Lead in World Paper," *Industrial Minerals,* No. 52, Jan., pp. 9–29.

Anon., 1972h, "Kaolin Operations of Gebrüder Dorfner at Hirschau," *Industrial Minerals,* No. 53, Feb., pp. 17–19.

Anon., 1972i, "Refractory Raw Materials, the Producers Reviewed," *Industrial Minerals,* No. 59, Aug., pp. 9–30.

Anon., 1972j, "South Africa: An Explosive Mining Potential," *Engineering & Mining Journal,* Vol. 173, No. 11, Nov., pp. 101–213.

Anon., 1972k, "Tariff Schedule of the United States Annotated," Publication No. 452, US Tariff Commission.

Anon., 1973, "Bentonite in Europe and the Mediterranean," *Industrial Minerals,* No. 64, Jan., pp. 9–20.

Anon., 1973a, "Making Aluminum Without Electricity," *Business Week,* No. 2268, Feb. 24, pp. 58E-58H.

Akhmed, K.S., ed., 1963, "Bentonity Uzbekistan," *Akad. Nauk. Uzbekskoi Institute Khim. and Geology,* Taskent, 195 pp.

Ali, T.S., and Shahn, I., 1962, "The Bentonite Resources of Pakistan," *Symposium on Industrial Rocks and Minerals,* Central Treaty Organization, Lahore, Pakistan, Dec. 1962, The Mineral Research and Exploration Institute of Turkey, 1963, pp. 153–160.

Almeborg, J., et al., 1969, "Kaolin Deposits of Denmark," "Proceedings, Symposium I. Kaolin Deposits of the World. A. Europe," *23rd International Geological Congress, Prague, 1968, Report,* M. Malkovský, and J. Vachtl, eds., Vol. 15, pp. 75–84.

Ames, L.L., Jr., et al., 1958, "A Contribution on the Hector, California, Bentonite Deposit," *Economic Geology,* Vol. 53, No. 1, pp. 22–37.

Ampian, S.G., 1972, "Clays," *Minerals Yearbook 1971,* US Bureau of Mines, Vol. 1, pp. 301–327.

Ampian, S.G., 1980, "Clays," Preprint, 1978–1979, US Bureau of Mines, 41 pp.

Ampian, S.G., 1982, "Clays," *Mineral Commodity Summaries, 1982,* US Bureau of Mines, pp. 34–35.

Ampian, S.G., 1983, "Clays," *Mineral Commodity Summaries, 1983,* US Bureau of Mines, pp. 34–35.

Ampian, S.G., and Polk, D.W., 1981, "Clays," *Minerals Yearbook, 1980,* US Bureau of Mines, pp. 201–236.

Arogyaswamy, R.N.P., 1968, "Clays," Economic Geology Bulletin 29, India Geological Survey, Series A, 220 pp.

Barton, R.M., 1966, *A History of the Cornish China Clay Industry,* D. Bradford Barton, Ltd., Truro, 221 pp.

Bates, T.F., 1945, "Origin of the Edwin Clay, Ione, California," *Geological Society of America Bulletin,* Vol. 56, No. 1, pp. 1–38.

Bates, T.F., 1964, "Geology and Mineralogy of the Sedimentary Kaolins of the Southeastern United States—A Review," *Clays and Clay Minerals—Proceedings of the Twelfth National Conference . . . 1963,* W.F. Bradley, ed., Macmillan, New York, pp. 177–194.

Berg, R.B., 1969, "Bentonite in Montana," Bulletin 74, Montana Bureau of Mines and Geology, 34 pp.

Berg, R.B., 1970, "Bentonite Deposits of the Ingomar-Vananda Area, Treasure and Rosebud Counties, Montana," Special Publication 51, Montana Bureau of Mines and Geology, 5 pp.

Beutelspacher, H., and Van Der Marel, H.W., 1968, *Atlas of Electron Microscopy of Clay Minerals and Their Admixtures,* Elsevier, Amsterdam, 333 pp.

Bicker, A.R., Jr., 1970, "Economic Minerals of Mississippi," Bulletin 112, Mississippi Geological, Economic, and Topographical Survey, 80 pp.

Blackman, A.G., 1969, "Bentonites Major Market: Two, Civil Engineering," *Industrial Minerals,* No. 25, pp. 23-25.

Bolger, R.C., and Weitz, J.H., 1952, "Mineralogy and Origin of the Mercer Fireclay of Northcentral Pennsylvania," *Problems of Clay and Laterite Genesis,* AIME, New York, pp. 81–93.

Bordas, A.F., 1947, "Contribucion al Concimento de las Bentonitas Argentinos," *Industria Minera, Cãmera Argentina de Mineria,* Vol. 6, No. 68, pp. 43–48; No. 69, pp. 56–60; No. 70, pp. 96–100; No. 71, pp. 24–27; No. 72, pp. 51–53.

Bowen, F.E., 1969, "Kaolin Deposits of New Zealand," "Proceedings of Symposium I. Kaolin Deposits of the World. B. Overseas Countries," *23rd International Geological Congress, Prague, 1968, Report,* M. Malkovský and J. Vachtl, eds., Vol. 16, pp. 151–157.

Bradley, W.F., 1940, "The Structural Scheme of Attapulgite," *American Mineralogist,* Vol. 25, No. 6, pp. 405–410.

Brindley, G.W., 1957, "Fuller's Earth from near Dry Branch, Georgia, a Montmorillonite-Cristobalite Clay," *Clay Minerals Bulletin,* Vol. 3, pp. 167–169.

Brindley, G.W., et al., 1963, "Mineralogical Studies of Kaolinite-Halloysite Clays: Part I. Identification Problems," *American Mineralogist,* Vol. 48, Nos. 7, 8, pp. 897–910.

Brindley, G.W., and Robinson, K., 1946, "Randomness in Kaolin Clays," *Faraday Society Transactions,* Vol. 428, No. 4213, pp. 198–205.

Bristow, C.M., 1969, "Kaolin Deposits of the United Kingdom of Great Britain and Northern Ireland," "Proceedings of Symposium I. Kaolin Deposits of the World. A. Europe," *23rd International Geological Congress, Prague, 1968, Report,* M. Malkovský and J. Vachtl, eds., Vol. 15, pp. 275–288.

Broadhurst, S.D., and Teague, K.H., 1954, "Halloysite in Chattooga County, Georgia," *Georgia Mineral Newsletter,* Vol. 7, No. 2, pp. 56–61.

Brown, G., ed., 1961, *The X-Ray Identification and Crystal Structures of Clay Minerals,* Mineralogical Society (Clay Minerals Group), London, 544 pp.

Buie, B.F., 1964, "Possibility of Volcanic Origin of the Cretaceous Sedimentary Kaolin of South Carolina and Georgia," *Clays and Clay Minerals —Proceedings of the Twelfth National Conference . . . 1963,* W.F. Bradley, ed., MacMillan, New York, p. 195.

Buie, B.F., 1972, "Future of the Kaolin Industry in Southeastern United States," H.S. Puri, ed., *Proceedings, 7th Forum on Geology of Industrial Minerals, Tampa, Fla., April 28–30, 1971,"* Special Publication 17, Florida Bureau of Geology, pp. 103–107.

Buie, B.F., and Fountain, R.C., 1967, "Tertiary and Cretaceous Age of Kaolin Deposits in Georgia and South Carolina," Geological Society of America, Southeastern Section, 1967 Annual Meeting Program, p. 19 (G.S.A. Special Paper No. 115, p. 465).

Bundy, W.M., and Berberich, J.P., 1969, "Kaolin Treatment," US Patent 3,477,809, Nov. 11.

Bundy, W.M., et al., 1965, "Physico-Chemical Properties of Kaolinite and Relationship to Paper Coating Quality," *Technical Association Pulp and Paper Industry,* Vol. 48, pp. 688–696.

Caberra, A.LaR., 1963, "I. Bentonitas, II. Diatomitas," *Peru, Ministerio de Fomento y Obras Publicas,* Minerales No—metalicos Serie, Memoir No. 7, 107 pp.

Callaghan, E., 1948, "Endellite Deposits in Gardner Mine Ridge, Lawrence County, Indiana," Bulletin 1, Indiana Division of Geology, 47 pp.

Calver, J.L., 1949, "Florida Kaolins and Clays," Information Circular 2, Florida Geological Survey, 59 pp.

Calver, J.L., 1957, "Mining and Mineral Resources," Bulletin 39, Florida Geological Survey, 132 pp.

Carroll, D., 1970, "Clay Minerals: A Guide to Their X-Ray Identification," Special Paper 126, Geological Society of America, 80 pp.

Carver, R.E., 1972, "Absorption Characteristics of Opaline Clays from the Eocene of Georgia," *Proceedings Seventh Forum on Geology of Industrial Minerals,* H.S. Puri, ed., Florida Bureau of Geology, Special Publication 17, pp. 91–101.

Chambers, G.P.C., 1959, "Some Industrial Applications of the Clay Mineral Sepiolite," *Silicates Industrials,* Vol. 24, pp. 181–189.

Christ, C.L., et al., 1969, "Palygorskite: New X-Ray Data," *American Mineralogist,* Vol. 54, Nos. 1, 2, pp. 198–205.

Clark, N.O., 1968, "Tube Press," US Patent Application No. 739442.

Clarke, O.M., Jr., 1964, "Clay Deposits of the Tuscaloosa Group in Alabama," *Clays and Clay Minerals—Proceedings of the Twelfth National Conference in Clays and Clay Minerals . . . 1963,* W.F. Bradley, ed., Macmillan, New York, pp. 495–507.

Cleveland, G.B., 1957, "Clay," *Mineral Commodities in California,* L.A. Wright, ed., California Division of Mines, Bulletin 176, pp. 131–152.

Clews, F.H., 1969, *Heavy Clay Technology,* 2d ed., Academic Press, New York, 481 pp.

Coetzee, C.B., 1969, "Kaolin Deposits of the Republic of South Africa," "Proceedings of Symposium I. Kaolin Deposits of the World. B. Overseas Countries," *23rd International Geological Congress, Prague, 1968, Report,* M. Malkovský and J. Vachtl, eds., Vol. 16, pp. 61–65.

Colligan, R.V., 1973, "Kaolin—U.S. Capacity Exceeds Use, but Sales to Grow in '73," *Engineering & Mining Journal,* Vol. 174, No. 3, pp. 158–159.

Comsti, F.A., 1969, "Kaolin Deposits of the Philippines," "Proceedings of Symposium I. Kaolin Deposits of the World. B. Overseas Countries," *23rd International Geological Congress, Prague, 1968, Report,* M. Malkovský and J. Vachtl, eds., Vol. 16, pp. 39–42.

Conant, L.C., 1965, "Bauxite and Kaolin Deposits of Mississippi, Exclusive of the Tippah-Benton District," Bulletin 1199-B, US Geological Survey, 70 pp.

Cook, G.H., and Smock, J.E., 1878, *Report on the Clay Deposits of the Woodbridge, South Amboy, and Other Places,* New Jersey Geological Survey, Trenton, 381 pp.

Cooke, C.W., 1943, "Geology of the Coastal Plain of Georgia," Bulletin 941, US Geological Survey, 121 pp.

Coope, B.M., 1979, "Kaolin, a Review of Production and Processing," *Industrial Minerals,* No. 136, Jan., pp. 31–49.

Cooper, J.D., 1970, "Clays," *Mineral Facts and Problems 1970,* US Bureau of Mines, Bulletin 650, pp. 923–938.

Cousminer, H.L., and Terris, L., 1972, "Palynology of Paleocene Clays from Georgia," Abstracts of Papers, 5th Annual Meeting, American Association of Stratigraphic Palynologists, 1 p.

Damiani, L., and Trautmann, F., 1969, "Les Dépots de Kaolins Français," "Proceedings of Symposium I. Kaolin Deposits of the World. A. Europe," *23rd International Geological Congress, Prague, 1968, Report,* M. Malkovský and J. Vachtl, eds., Vol. 15, pp. 141–178.

Davis, C.W., et al., 1940, "Bentonite, Its Properties, Mining, Preparation, and Utilization," Technical Paper 609, US Bureau of Mines, 83 pp.

Davis, J.C., 1965, "Bentonite Deposits of the Clay Spur District, Crook and Weston Counties, Wyoming," Preliminary Report No. 4, Wyoming Geological Survey, 17 pp.

Davison, J., 1969, "Bentonites Major Markets: Three, Iron Ore Pelletising," *Industrial Minerals,* No. 25, Oct., pp. 27–31.

Denny, C.S., and Drewes, H., 1965, "Geology of the Ash Meadows Quadrangle, Nevada-California," Bulletin 1181-L, US Geological Survey, 56 pp.

Déribéré, M., and Esme, A., 1943, *La Bentonite, les Argiles Colloidales et Leurs Emplois,* Dunod, Paris, 175 pp.

de Souza Santos, P., and de Souza Santos, H., 1972, "Kaolin Clays of Brazil: Mineralogy and Properties," Program—1972 International Clay Conference (AIPEA), Madrid, June 25-30 (Paper presented without abstract or preprint).

Dirmeyer, R.D., and Skinner, M.M., 1968, "Evaluation of Colorado Clays for Sealing Purposes," Technical Bulletin 83, Colorado State University Experiment Station, 36 pp.

Eargle, D.H., 1955, "Stratigraphy of the Outcropping Cretaceous Rocks of Georgia," Bulletin 1014, US Geological Survey, 101 pp.

Esquivel, J.M., and Zamora, S., 1958, "Informe

Sorbre Minerales no Metalicos," Bulletin 44, Mexico Consejo de Recursos Naturales no Renovables, 152 pp.

Eyssautier, L., 1952 "L'Industrie Miniere du Moroc," Notes et Memoires, Morocco Service des Mines et de la Carte Géologique, 184 pp.

Fisher, W.L., et al., 1965, "Rock and Mineral Resources of East Texas," Report of Investigations 54, Texas Bureau of Economic Geology, 439 pp.

Fisk, H.G., 1946, "Bentonite, with Test Methods and Results of Tests of Wyoming Bentonites," Bulletin 2, Wyoming, University Natural Research Institute, 39 pp.

Foster, M.D., 1953, "Geochemical Studies of Clay Minerals: II. Relation Between Ionic Substitution and Swelling in Montmorillonites," *American Mineralogist,* Vol. 38, Nos. 11, 12, pp. 994–1006.

Fuji, N., et al., 1969, "Kaolin Deposits of Japan," "Proceedings of Symposium I. Kaolin Deposits of the World. B. Overseas Countries," *23rd International Geological Congress, Prague, 1968, Report,* M. Malkovský and J. Vachtl, eds., Vol. 16, pp. 29–37.

Gann, D.E., Jr., 1971, "Sedimentary Analysis of the Filtro Corporation Bentonite Quarry at Mineral Springs, Lincoln Parish, Louisiana," M.Sc. Thesis Northeast Louisiana State College, Monroe.

Gard, J.A., ed., 1971, *The Electron-optical Investigations of Clays,* Mineralogical Society (Clay Minerals Group), London, 383 pp.

Gaskin, A.J., 1969, "Kaolin Deposits of Australia," "Proceedings of Symposium I. Kaolin Deposits of the World. B. Overseas Countries," *23rd International Geological Congress, Prague, 1968, Report,* M. Malkovský and J. Vachtl, eds., Vol. 16, pp. 115–150.

Gawronski, O., and Kozydra, Z., 1969, "Kaolin Deposits of Poland," "Proceedings of Symposium I. Kaolin Deposits of the World. A. Europe," *23rd International Geological Congress, Prague, 1968, Report,* M. Malkovský and J. Vachtl, eds., Vol. 15, pp. 217–223.

Gildersleeve, B., 1946, "Minerals and Structural Materials of East Tennessee," Report B, Tennessee Valley Authority Regional Products Research Division, 26 pp.

Gillott, J.E., 1968, *Clay in Engineering Geology,* Elsevier, New York, 296 pp.

Glassmire, S.H., 1957, *Clay Mineral Potential of Northeastern New Mexico,* New Mexico Economic Development Commission, 12 pp.

Goff, J.A., 1959, "Thomas Griffiths, A Journal of the Voyage to South Carolina, 1767, to Obtain Cherokee Clay for Josiah Wedgewood, with annotations," *Georgia Mineral Newsletter,* Vol. 12, No. 3, pp. 113–122.

Gordon, M., Jr., et al., 1958, "Geology of the Arkansas Bauxite Region," Professional Paper 299. US Geological Survey, 268 pp.

Granquist, W.T., et al., 1972, "Clay Mineral Synthesis—III. Rapid Hydrothermal Crystallization of an Aluminian Smectite," *Clays and Clay Minerals,* Vol. 20, No. 5, pp. 323–329.

Granquist, W.T., and Pallack, S.S., 1967, "Clay Mineral Synthesis—II. A Randomly Interstratified Aluminian Montmorillonoid," *American Mineralogist,* Vol. 52, pp. 212–226.

Greaves-Walker, A.F., et al., 1951, "The Development of Lightweight Aggregate from Florida Clays," Bulletin 46, Florida Engineering and Industrial Experiment Station (*Florida University Engineering Progress,* Vol. 5, No. 9), 23 pp.

Green, J.M., et al., 1970, "Thermal Reactions of Synthetic Hectorite," *Clays and Clay Minerals,* Vol. 18, pp. 339–346.

Greene, E.W., and Duke, J.B., 1962, "Selective Froth Flotation of Ultrafine Minerals or Slimes," *Trans. SME/AIME,* Vol. 223, pp. 389–395.

Gregor, M., 1967, "Industrielle Forschung und Verwertung von Bentoniten in der Tschechoslovakei," *International Clay Conference 1966, Jerusalem, Israel, June 20–24,* L. Heller, ed., Vol. 2, pp. 271–279.

Grim, R.E., 1934, "Petrology of the Kaolin Deposits near Anna, Illinois," *Economic Geology,* Vol. 24, pp. 659–670.

Grim, R.E., 1962, *Applied Clay Mineralogy,* McGraw-Hill, New York, 422 pp.

Grim, R.E., 1968, *Clay Mineralogy,* 2d ed., McGraw-Hill, New York, 596 pp.

Grimshaw, R.W., 1972, *The Chemistry and Physics of Clays and Other Ceramic Raw Materials,* 4th ed., rev., Wiley-Interscience, New York, 1024 pp.

Gross, G.W., 1960, "Location of Clay Deposits by Combined Self-Potential and Resistivity Surveys," *Trans. SME/AIME,* Vol. 217, pp. 124–130.

Gunn, F.A., and Morris, H.H., 1965, "Delaminated Domestic Sedimentary Clay Products and Method of Preparation Thereof," US Patent 3,171,718, Mar. 2.

Haas, C.Y., 1970, "Attapulgite Clays for Future Industrial Mineral Markets" (Abstract), *Mining Engineering,* Vol. 22, No. 12, p. 63.

Haden, L.W., Jr., 1972, "Attapulgite," SME Preprint No. 72H327, SME Fall Meeting, Birmingham, Oct., 8 pp.

Haden, L.W., Jr., and Schwint, I.A., 1967, "Attapulgite, Its Properties and Applications," *Industrial & Engineering Chemistry,* Vol. 59, No. 9, pp. 58–69.

Hagner, A.F., 1939, "Adsorptive Clays of the Texas Gulf Coast," *American Mineralogist,* Vol. 24, pp. 67–108.

Harben, P., 1979, "Paper Expansion Spurs Kaolin in Georgia," *Industrial Minerals,* No. 147, Dec., pp. 23–33.

Harris, J.F., 1969, "Kaolin Deposits in Tanzania," "Proceedings of Symposium I. Kaolin Deposits of the World. B. Overseas Countries," *23rd International Geological Congress, Prague, 1968, Report,* M. Malkovský and J. Vachtl, eds., Vol. 16, pp. 75–78.

Heikes, G.C., and Kim, H.K., 1965, *The Kaolins of Korea,* issued by Mining Branch Industry—Engineering Division, US Operations Mission to Korea, Agency for International Development, 104 pp.

Heiser, E.J., and Shand, A., 1973, "Lightweight Polymeric Pigment," *Technical Association Pulp and Paper Industry* (TAPPI), Vol. 56, pp. 70–73.

Heron, S.D., Jr., et al., 1965, "Clays and Opal-Bearing Claystones of the South Carolina Coastal Plain," Bulletin 31, South Carolina Division Geology, 65 pp.

Hewett, D.F., 1917, "The Origin of Bentonite," *Journal,* Washington Academy of Science, Vol. 9, pp. 77–96.

Hilali, E.A., et al., 1969, "Les gîtes de Kaolin du Maroc," "Proceedings of Symposium I. Kaolin Deposits of the World. B. Overseas Countries," *23rd International Geological Congress, Prague, 1968, Report,* M. Malkovský and J. Vachtl, eds., Vol. 16, pp. 55–59.

Hinckley, D.N., 1961, "Mineralogical and Chemical Variations in the Kaolin Deposits of the Coastal Plain of Georgia and South Carolina," Ph.D. Thesis, Pennsylvania State University, University Park., 194 pp.

Holmes, R.J., 1950, "Reference Clay Localities—Europe," Preliminary Report No. 4, American Petroleum Institute, Project 49, Clay Mineral Standards, Columbia University, 101 pp.

Holzer, H.F., and Wieden, P., 1969, "Kaolin Deposits of Austria," "Proceedings of Symposium I. Kaolin Deposits of the World. A. Europe," *23rd International Geological Congress, Prague, 1968, Report,* M. Malkovský and J. Vachtl, eds., Vol. 15, pp. 25–32.

Hosterman, J.W., 1973, "Clays," Professional Paper 820, US Geological Survey, pp. 123–131.

Hosterman, J.W., et al., 1960, "Investigations of Some Clay Deposits in Washington and Idaho," Bulletin 1091, US Geological Survey, 147 pp.

Hosterman, J.W., et al., 1968, "Clay," Professional Paper 580, US Geological Survey, pp. 167–188.

Huertas, F., et al., 1971, "Minerales Fibrosos de la Arcilla en Cuencas Sedimentarias Españolas. I. Cuenca del Tajo," *Spain Inst. Geológico y Minero Boletin,* Vol. 82, No. 6, pp. 28–36.

Hunter, C.E., and Hash, L.J., 1953, "Halloysite Deposits of Western North Carolina," Bulletin 58, North Carolina Division Mineral Resources, 32 pp.

Hunter, D.R., and Urie, J.G., 1969, "Kaolin Deposits of Swaziland," "Proceedings of Symposium I. Kaolin Deposits of the World. B. Overseas Countries," *23rd International Geological Congress, Prague, Report,* M. Malkovský and J. Vachtl, eds., Vol. 16, pp. 67–74.

Huvos, J.B., and Sondermayer, R.V., 1968, "Mineral Industry of Romania," *Minerals Yearbook, 1968,* US Bureau of Mines, Vol. 4, p. 620.

Iannicelli, J., et al., 1969, "Process for Improving the Brightness of Clays," US Patent 3,471,011, Oct. 7.

Jeannette, A., 1952, "Argiles Smectiques et Rhassoul," *International Geological Congress 19th, Algiers, 1952,* Chap. 20, Monograph Regionales Ser. 3, Maroc, No. 1, pp. 371–383.

Jeyhan, I., 1969, "Kaolin Deposits of Turkey," "Proceedings of Symposium I. Kaolin Deposits of the World. B. Overseas Countries," *23rd International Geological Congress, Prague, 1968, Report,* M. Malkovský and J. Vachtl, eds., Vol. 16, pp. 43–51.

Johnson, F.T., and Ricker, S., 1948, "Ione-Carbondale Clays, Amador County, Calif.," Report of Investigations 4213, US Bureau of Mines, 6 pp.

Johnson, S.W., and Blake, J.M., 1867, "On Kaolinite and Pholerite," *American Journal of Science,* 2d ser., Vol. 43, pp. 531–561.

Jonas, E.C., 1964, "Petrology of the Dry Branch, Georgia Kaolin Deposits," *Clays and Clay Minerals—Proceedings of the Twelfth National Conference . . . 1963,* W.F. Bradley, ed., Macmillan, New York, pp. 199–205.

Jones, J.T., and Berard, M.F., 1972, *Ceramics: Industrial Processing and Testing,* Iowa State University Press, Ames, Iowa, 213 pp.

Kalix, Z., 1971, "Clays Including Bentonite, Fuller's Earth and Damourite," *Australian Mineral Industry, 1970,* Review, pp. 82–87.

Karanov, K., et al., 1969, "Kaolin Deposits of Bulgaria," "Proceedings of Symposium I. Kaolin Deposits of the World. A. Europe," *23rd International Geological Congress, Prague, 1968, Report,* M. Malkovský and J. Vachtl, eds., Vol. 15, pp. 33–46.

Keller, W.D., 1952, "Observations on the Origin of Missouri High-Alumina Clays," *Problems of Clay and Laterite Genesis,* AIME, New York, pp. 115–134.

Keller, W.D., 1964, "The Origin of High-Alumina Clay Minerals—A Review," *Clays and Clay Minerals—Proceedings of the Twelfth National Conference on Clays and Clay Minerals . . . 1963,* W.F. Bradley, ed., Macmillan, New York. pp. 129–151.

Keller, W.D., 1970, "Environmental Aspects of Clay Minerals," *Journal of Sedimentary Petrology,* Vol. 40, pp. 788–813.

Keller, W.D., et al., 1954, "The Origin of Missouri Fire Clays," *Clays and Clay Minerals—Proceedings of the Second National Conference on Clay Minerals . . .* 1953, A. Swineford, ed., US National Research Council Publication 327, pp. 7–46.

Kelley, F.R., 1966, "Clay," *Mineral and Water Resources of California, P. 1, Mineral Resources,* US 89th Congress, 2d Session, Senate Committee on Interior and Insular Affairs, Committee Print, pp. 126–134 (reprinted in Bulletin 191, California Division of Mines and Geology, pp. 126–134).

Kerr, P.F., et al., 1950, "Analytical Data on Reference Clay Materials," Report No. 7, American Petroleum Institute Project 49, Columbia University Preliminary, 160 pp.

Kesler, T.L., 1956, "Environment and Origin of the Cretaceous Kaolin Deposits of Georgia and South Carolina," *Economic Geology,* Vol. 51, No. 6, pp. 541–554.

Kesler, T.L., 1963, "Environment and Origin of the Cretaceous Kaolin Deposits of Georgia and South Carolina," *Georgia Mineral Newsletter,* Vol. 16, Nos. 1–2, pp. 2–11.

Khadem, N., 1969, "Kaolin Deposits of Iran," "Proceedings of Symposium I. Kaolin Deposits of the World. B. Overseas Countries," *23rd International Geological Congress, Prague, 1968, Report,* M. Malkovský and J. Vachtl, eds., Vol. 16, pp. 25–28.

Kiersch, G.A., and Keller, W.D., 1955, "Bleaching Clay Deposits, Sanders-Defiance Plateau District, Navajo County, Arizona," *Economic Geology,* Vol. 50, No. 5, pp. 469–494.

Kildale, M.B., and Thomas, R.C., 1957, "Geology of the Halloysite Deposits at the Dragon Mine," *Utah Geological Society Guidebook to the Geology of Utah,* No. 12, pp. 94–96.

Kim, J.H., and Kim, D.H., 1964, "Investigation Report of the San Chong and Tan Song Halloysite Deposits, San Chong Gun, Kyongsang Namdo, Korea," Bulletin 7, Korea Geological Survey, pp. 67–88.

Klammer, G., 1971, "Uber plio-pleistozäne und ihre Sedimente im unteren Amazonasgebiet." *Zeitschrift für Geomorphologie,* N.F., Vol. 15, No. 1, pp. 62–106.

Klinefelter, T.A., and Hamlin, H.P., 1957, "Syllabus of Clay Testing," Bulletin 565, US Bureau of Mines, 67 pp.

Klinefelter, T.A., et al., 1943, "Hard and Soft Kaolins of Georgia," Report of Investigation 3682, US Bureau of Mines, pp. 1–20.

Klinger, F.L., 1965, "Mineral Industry of Finland," *Minerals Yearbook,* US Bureau of Mines, Vol. 4, pp. 229–238.

Knechtel, M.M., and Patterson, S. H., 1962, "Bentonite Deposits of the Northern Black Hills District, Wyoming, Montana, and South Dakota," Bulletin 1082-M, US Geological Survey, pp. 893–1030.

Knight, W.C., 1898, "Bentonite," *Engineering & Mining Journal,* Vol. 66, No. 17, p. 491.

Konta, J., et al., 1971, "Mineral and Chemical Composition of Four Varieties of Bentonite from Ginovci," *Acta Universitatis Carolinae-Geologica,* No. 3, pp. 175–187.

Kuzvart, M., 1969, "Kaolin Deposits of Czechoslovakia," "Proceedings of Symposium I. Kaolin Deposits of the World. A. Europe," *23rd International Geological Congress, Prague, 1968, Report,* M. Malkovský and J. Vachtl, eds., Vol. 15, pp. 47–73.

Ladd, G.E., 1898, "A Preliminary Report on a Part of the Clays of Georgia," Bulletin No. 6, Georgia Geological Survey, pp. 81–91.

Lang, W.B., et al., 1940, "Clay Investigations in the Southern States, 1934–35," Bulletin 901, US Geological Survey, 346 pp.

Lang, W.J., 1971, "Bentonite: The Demand and Markets of the Future," SME Preprint No. 71H29, AIME Centennial Annual Meeting, New York, Mar., 9 pp.

Larsen, D.H., 1955, "Use of Clay in Drilling Fluids," *Clays and Clay Technology,* J.A. Pask and M.D. Turner, eds., California Division of Mines Bulletin 169, pp. 269–281.

Lebedinskii, V.I., and Kirichenko, L.P., 1972, "Concerning the Collection 'Bentonitic Clays of the Volta Regions' (A Review and Discussion trans. From *Lithologiya i Poleznye Iskpaemye,* No. 3, 1972)," *Lithology and Mineral Resources,* Consultants Bureau, New York, Vol. 7, No. 3, pp. 393–394.

Lippert, H.J., et al., 1969, "Die Kaolinlagerstätten der Bundesrepublik Deutschland," "Proceedings of Symposium I. Kaolin Deposits of the World. A. Europe," *23rd International Geological Congress, Prague, 1968, Report,* M. Malkovský and J. Vacht, eds., Vol. 15, pp. 85–105.

Lupino, R., 1954, "Sulla Bentonite dell'Isola di Ponza," *Ricera Science,* Rome, Vol. 24, No. 11, pp. 2326–2339.

Lyons, S.C., 1959, "Method of Treating Kaolinitic Clay," US Patent 2,904,267, Sept. 15.

Lyons, S.C., 1966, "Clays," "Paper Coating Pigments," *Technical Association Pulp and Paper Industry Monograph 30,* H.H. Murray, ed., pp. 57–124.

Mackenzie, R.C., ed., 1957, *The Differential Thermal Investigation of Clays,* Mineralogical Society (Clay Minerals Group), London, 1957, 456 pp.

Malkovský, M., and Vachtl, J., eds., 1969a, "Proceedings of Symposium I. Genesis of Kaolin Deposits," *23rd International Geological Congress, Prague, 1968, Report,* Vol. 14, 135 pp.

Malkovský, M., and Vachtl, J., eds., 1969b, "Proceedings of Symposium I. Kaolin Deposits of the World. A. Europe," *23rd International Geological Congress, Prague, 1968, Report,* Vol. 15, 322 pp.

Malkovský, M., and Vachtl, J., eds., 1969c, "Proceedings of Symposium I. Kaolin Deposits of the World. B. Overseas Countries," *23rd International Geological Congress, Prague, 1968, Report,* Vol. 16, 157 pp.

Mark, H., 1963, "High-Alumina Kaolinitic Clay in the United States," Mineral Resource Investigations Map MR-37, US Geological Survey.

Martin Vivaldi, J.L., 1969, "Kaolin Deposits of Spain," "Proceedings of Symposium I. Kaolin Deposits of the World. A. Europe," *23rd International Geological Congress, Prague, 1968, Report,* M. Malkovský and J. Vachtl, eds., Vol. 15, pp. 225–261.

Martin Vivaldi, J.L., and Girela Vilchez, F., 1959, "A Study of the Halloysite from Maazza (North Morocco)," *Silicates Industriels,* Vol. 24, No. 7–8, pp. 380–385.

Martin Vivaldi, J.L., and Linares, G.J., 1968, "Las Bentonitas de Cabo de Gata: II. Yacimiento de Palma del Muerto," *Spain Inst. Geológico y Minero Boletin,* Vol. 79, No. 6, pp. 65–71.

Martin Vivaldi, J.L., and Linares, G.J., 1969, "Las Bentonitas de Cabo de Gata, III, Consideracions sobre la Mineralógia y Genesis de los Yacimientos Estudiadon," *Spain Inst. Geológico y Minero Boletin,* Vol. 80, No. 1, pp. 74–80.

Maxwell, R.A., 1962, "Mineral Resources of South Texas," Report of Investigation 43, Texas Bureau of Economic Geology, 140 pp.

McQueen, H.S., 1943, "Geology of the Fire Clay Districts of East-central Missouri," *Missouri Geological Survey and Water Resources,* Vol. 28, 2d ser., 250 pp.

Millot, G., 1964, *Géologie des Argiles,* Masson, Paris, 320 pp. (trans. by W.R. Farrand and H. Paquet, published as *Geology of Clays,* Springer-Verlag, New York, 1970, 425 pp.).

Minton, L.H., 1922, "New Jersey's Part in the Ceramic History of America," *Ceramist,* Vol. 2, 271 pp.

Miser, H.D., 1913, "Developed Deposits of Fuller's Earth in Arkansas," Bulletin 530, US Geological Survey, pp. 207–220.

Monroe, W.H., 1941, "Notes on Deposits of Selma and Ripley Age in Alabama," Bulletin 48, Alabama Geological Survey, 150 pp.

Moretti, A., and Pieruccini, V., 1969, "Italian Kaolin Deposits," "Proceedings of Symposium I. Kaolin Deposits of the World. A. Europe," *23rd International Geological Congress, Prague, 1968, Report,* M. Malkovský and J. Vachtl, eds., Vol. 15, pp. 201–209.

Morris, H.T., 1968, "The Main Tintic Mining District, Utah," *Ore Deposits of the United States, 1933–1967,* Vol. II, J.D. Ridge, ed., AIME, New York, pp. 1043–1073.

Moses, J., and Mitchell, W.D., 1963, "Bauxite Deposits of British Guiana and Surinam in Relation to Underlying Unconsolidated Sediments Suggesting Two-Step Origin," *Economic Geology,* Vol. 58, No. 6, pp. 250–262.

Murray, H.H., 1963, "Industrial Applications of Kaolin," *Clays and Clay Minerals, Proceedings of the Tenth National Conference . . . 1961,* Vol. 10, A. Swineford, ed., MacMillan, New York, pp. 291–298.

Murray, H.H., 1963a, "Mining and Processing Industrial Kaolins," *Georgia Mineral Newsletter,* Vol. 16, pp. 3–11.

Murray, H.H., 1970, "China Clay," *Handbook of Pulp and Paper Technology,* K.W. Britt, ed., Van Nostrand Reinhold Co., New York, pp. 617–625.

Murray, H.H., and Lyons, S.C., 1956, "Correlation of Paper Coating Quality with Degree of Crystal Perfection of Kaolinite," *Clays and Clay Minerals—Proceedings of the Fourth National Conference . . . 1955,* A. Swineford, ed., National Research Council, Pub. 456, pp. 31–40.

Murray, H.H., and Smith, J.M., 1973, "The Geology and Mineralogy of the Grahamstown, South Africa Kaolin Deposit," Program and Abstracts, 22nd Annual Clay Minerals Conference, Clay Minerals Society, p. 48.

Nawaz, A.S., 1962, "New Pakistan Plant for the Manufacture of Activated Earth from Bentonite," *Symposium on Industrial Rocks and Minerals,* Central Treaty Organization, Lahore, Pakistan, Dec. 1962, The Mineral Research and Exploration Institute of Turkey, 1963, pp. 166–168.

Neacşu, 1969, *Bentonitele din Regiunea Alba-Iulia-Ocna-Mureşului,* Editura Academiei Republicii Socialista, România, Bucareşti, 203 pp.

Neumann, F.R., 1927, "Origin of the Cretaceous White Clays of South Carolina," *Economic Geology,* Vol. 22, pp. 380–386.

Norton, F.H., 1968, *Refractories,* McGraw-Hill, New York, 228 pp.

Norton, J.J., 1965, "Lithium-Bearing Bentonite Deposits, Yavapai County, Arizona," Professional Paper 525-D, US Geological Survey, pp. D163–D166.

Nowacki, J. and Ciechomska, B., 1964 "Krajowe Surowce do Produkcji ziem Bielacgch (Domestic Raw Materials for Production of Bleaching Earths)," *Przegląd Geologiczny,* Vol. 12, No. 11, pp. 442–443.

Nutting, P.D., 1943, "Adsorbent Clays, Their Distribution, Properties, Production, and Uses," Bulletin 928-C, US Geological Survey, pp. 127–221.

Oder, R.R., and Price, C.R., 1973, "A New Clay Technology—Its Implications for the Paper Industry," *TAPPI Coating Conference 1973,* Technical Association of the Paper Industry, Atlanta, pp. 23–27.

Ogden, D.G. ,1969, "Geology and Origin of the Kaolin at East Monkton, Vermont," *Economic Geology,* Vermont Geological Survey, No. 3, 40 pp.

Olive, W.W., and Finch, W.I., 1969, "Stratigraphic and Mineralogic Relations and Ceramic Properties of Clay Deposits of Eocene Age in the Jackson Purchase Region, Kentucky, and in Adjacent Parts of Tennessee," Bulletin 1282, US Geological Survey, 35 pp.

Olson, R.H., 1964, "Clays," *Mineral and Water Resources of Nevada,* US Geological Survey & Nevada Bureau of Mines, US Congress, 88th, 2d Session, Senate Document 87, pp. 185–189.

Omori, K., et al., 1961, "X-Ray Studies on Bentonite from Kaminoyama, Yamagata Prefecture," *Japan Association Mineralogists, Petrologists, and Economic Geologists Journal,* Vol. 45, No. 3, pp. 81–88.

Oultin, T.D., 1965, "Mining, Production and Uses of Attapulgite Clay Products," SME Preprint 65H39, AIME Annual Meeting, 10 pp.

Ovcharenko, F.D., et al., 1967, *Investigations of the Physicochemical Mechanics of Clay-Mineral Dispersions* [trans. from Russian by Z. Lerman], Israel Program Science Translations, Jerusalem, 146 pp.

Pai, C.C., 1945, "General Statement on the Mining Industry," Special Report No. 7, China Geological Survey, 772 pp.

Papin, R., 1964, "Bentonite," *Encyclopedia of Chemical Technology,* Kirk-Othmer, ed., Vol. 3, pp. 339–360.

Papke, K.G., 1969, "Montmorillonite Deposits in Nevada," *Clays and Clay Minerals,* Vol. 17, No. 4, pp. 211–222.

Papke, K.G., 1971, "Halloysite Deposits in the Terraced Hills, Washoe County, Nevada," *Clays and Clay Minerals,* Vol. 19, No. 2, pp. 71–74.

Parham, W.E., 1970, "Clay Mineralogy and Geology of Minnesota's Kaolin Clays," SP-10, Minnesota Geological Survey, 142 pp.

Parker, J.M., III, 1946, "Residual Kaolin Deposits of the Spruce Pine District, North Carolina," Bulletin 48, North Carolina Division Mineral Resources, 45 pp.

Pasquera, R., et al., 1969, "Kaolin Deposits of Mexico," "Proceedings of Symposium I. Kaolin Deposits of the World. B. Overseas Countries," *23rd International Geological Congress, Prague, 1968, Report,* M. Malkovský and J. Vachtl, eds., Vol. 16, pp. 105–110.

Patterson, S.H., 1972, "Fuller's Earth and Bentonite in the Southeastern States," *Proceedings Seventh Forum on Geology of Industrial Minerals,* H.S. Puri, ed., Special Publication 17, Florida Bureau of Geology, pp. 37–49.

Patterson, S.H., and Hosterman, J.W., 1963, "Geology and Refractory Clay Deposits of the Haldeman and Wrigley Quadrangles, Kentucky," Bulletin 1122-F, US Geological Survey, 113 pp.

Patterson, S.H., and Murray, H.H., 1975, "Clays," *Industrial Minerals and Rocks,* 4th ed., S.J. Lefond, ed., AIME, New York, pp. 519-585.

Pattiaratchi, D.B., 1969, "Kaolin Deposits of Ceylon," "Proceedings of Symposium I. Kaolin Deposits of the World. B. Overseas Countries," *23rd International Geological Congress, Prague, 1968, Report,* M. Malkovský and J. Vachtl, eds., Vol. 16, pp. 17–24.

Pence, F.K., 1954, "Wilcox Sand-Kaolins of Northeastern Central Texas," Publication No. 5416, Texas University, 99 pp.

Petrov, V.P., 1969, "Kaolin Deposits of the USSR," "Proceedings of Symposium I. Kaolin Deposits of the World. A. Europe," *23rd International Geological Congress, Prague, 1968, Report,* M. Malkovský and J. Vachtl, eds., Vol. 15, pp. 289–319.

Phelps, G.W., 1972, "The Ball Clays of Tennessee and Kentucky," SME Preprint No. 72H305, SME Fall Meeting, Birmingham, Oct., 11 pp.

Pietracaprina, A., and Novelli, G., 1972, "Bentonite Deposits at Uri, Sardinia, Italy," *Clay Mineralogy,* Vol. 9, No. 3, pp. 351–355.

Ponder, H., and Keller, W.D., 1960, "Geology, Mineralogy, and Genesis of Selected Fireclays from Latah County, Idaho," *Clays and Clay Minerals—Proceedings of the Eighth National Conference . . . 1959,* A. Swineford and P.C. Franks, eds., Pergamon Press, New York, pp. 44–62.

Poole, E.G., and Kelk, B., 1971, "Calcium Montmorillonite (Fuller's Earth) in the Lower Greensand of the Baulking Area, Berkshire," Report No. 71/4, Great Britain Institute of Geological Sciences, 56 pp.

Poole, E.G., et al., 1971, "Calcium Montmorillonite (Fuller's Earth) in the Lower Greensand of the Fernham Area, Berkshire," Report No. 71/12, Great Britain Institute of Geological Sciences, 65 pp.

Range, K.J., et al., 1969, "Fire-Clay Type Kaolinite or Fire-Clay Mineral? Experimental Classification of Kaolinite-Halloysite Minerals," *Proceedings of International Clay Conference, Tokyo, Japan, 1969,* Vol. 2, pp. 5–8.

Rich, A.D., 1960, "Bleaching Clay," *Industrial Minerals and Rocks,* 3d ed., J.L. Gillson, ed., AIME, New York, pp. 93–101.

Ries, H., and Sommers, R.E., 1920, "The Clays and Shales of Virginia West of the Blue Ridge," Bulletin 20, Virginia Geological Survey, pp. 20–21.

Ritchie, J.A., et al., 1969, "Bentonites of Canterbury," *New Zealand Journal of Geology and Geophysics,* Vol. 12, No. 4, pp. 583–608.

Robertson, R.H.S., 1961, "The Origin of English Fuller's Earth," *Clay Minerals Bulletin,* Vol. 4, No. 26, pp. 282–287.

Robinson, G.C., et al., 1961, "Common Clays of the Coastal Plain of South Carolina and Their Use in Structural Clay Products," Bulletin 21, South Carolina Division Geology, 71 pp.

Ross, C.S., and Hendricks, S.B., 1934, "Minerals of the Montmorillonite Group, Their Origin and Relation to Soils and Clays," Professional Paper 205-B, US Geological Survey, pp. 23–79.

Ross, C.S., and Kerr, P.F., 1931, "The Kaolin Minerals," Professional Paper 165-E, US Geological Survey, pp. 151–180.

Ross, C.S., and Shannon, E.V., 1926, "The Minerals of Bentonite and Related Clays and Their Physical Properties," *American Ceramic Society Journal,* Vol. 9, No. 2, pp. 77–96.

Ross, J.S., 1964, "Bentonite in Canada," Mines Branch Monograph 873, Canada Dept. of Mines and Technical Surveys, 61 pp.

Rusko, Yu. A., 1973, "Dickite Kaolin of the Beregovian Deposit," (Inst. Geokim. Fiz. Miner, Kiev USSR), *Akademiya Nauk. Ukrainskoy RSR Dopovidi,* Seriya B, Vol. 35, No. 1, pp. 43–46 (in Russian).

Sadran, G., et al., 1955, "Sur L'Origine des Gisements de Bentonites de Lalla Maghnia (Oran)," *Algéria Service de la Carte Géologique Bulletin* 5, New Series, pp. 212–234.

Sales, R.H., and Meyer, C., 1949, "Results from Preliminary Studies of Vein Formation at Butte, Montana," *Economic Geology,* Vol. 44, pp. 465–484.

Sastry, K.V.S., and Fuerstenau, D.W., 1971, "A Laboratory Method for Determining the Balling Behavior of Taconite Concentrates," *Trans. SME/AIME,* Vol. 250, No. 1, pp. 64–67.

Schroter, G.A., and Campbell, I., 1940, "Geological Features of Some Deposits of Bleaching Clay," Technical Publication 1139, AIME, Vol. 4, No. 1, pp. 1–31.

Scott, A., 1929, "Ball Clays," *Special Report of the Mineral Resources of Great Britain,* Memoir, Great Britain Geological Survey, Vol. 31, 73 pp.

Scrudato, R.J., 1969, "Origin of East-Central Georgia Kaolin Deposits [Abstract]," Abstracts with Programs, Geological Society of America, Vol. 1, Pt. 7, p. 203.

Seidov, A.G., and Alizade, Kh.A., 1966, "The Formation and Mineralogy of Bentonites in Azerbaijan," *Clay Minerals,* Vol. 6, No. 3, pp. 157–166.

Serratosa, J.M., ed., 1972, *Kaolin Symposium,* International Clay Conference, Madrid, 195 pp.

Sheppard, R.A., and Gude, A.J.,3d, 1968, "Distribution and Genesis of Authigenic Silicate Minerals in Tuffs of Pleistocene Lake Tecopa, Inyo County, California," Professional Paper 597, US Geological Survey, 38 pp.

Siddiqui, M.K.H., 1968, *Bleaching Earths,* Pergamon Press, New York, 86 pp.

Siddique, H.N., and Bahl, D.P., 1965, "Geology of the Bentonite Deposits of Barmer District, Rájasthan," Memoir, India Geological Survey, Vol. 96, 96 pp.

Silliman, B., 1820, "Sketches of a Tour in the Counties of New Haven and Litchfield in Connecticut, with Notices of the Geology, Mineralogy, and Scenery, etc.," *American Journal of Science,* Vol. 2, No. 2, pp. 201–235.

Sims, J.D., 1972, "Petrographic Evidence for Volcanic Origin of Part of the Porters Creek Clay, Jackson Purchase Region, Western Kentucky," Professional Paper 800-C, US Geological Survey, pp. C39–C51.

Slatick, E.R., 1969, "Mineral Industries of Kenya, Tanzania, and Uganda," *Minerals Yearbook, 1969,* US Bureau of Mines, Vol. 4, pp. 441–450.

Slaughter, M., and Earley, J.W., 1965, "Mineralogy and Geological Significance of the Mowry Bentonites, Wyoming," Special Paper 83, Geological Society of America, 91 pp.

Sloane, R.L., and Guilbert, J.M., 1967, "Electron-Optical Study of Alteration to Smectite in the Cheto Clay Deposit," *Clays and Clay Minerals —Proceedings of the Fifteenth Conference . . . 1966,* Pergamon Press, New York, pp. 35–44.

Smith, R.W., 1929, "Sedimentary Kaolins of the Coastal Plain of Georgia," Bulletin 44, Georgia Geological Survey, 482 pp.

Sondermayer, R.V., 1967, "Mineral Industry of Hungary," *Minerals Yearbook 1967,* US Bureau of Mines, Vol. 4, pp. 339–345.

Spence, H.S., 1924, "Bentonite," Report 626, Canada Mines Branch, 35 pp.

Stone, R.L., et al., 1971, "Muller Mixing of Bentonite and Iron Ore Concentrate at Cornwall," *Trans. SME/AIME,* Vol. 250, pp. 236–239.

Störr, M., et al., 1969, "Kaolinlagerstätten der Deutschen Demokratischen Republik," "Proceedings of Symposium I. Kaolin Deposits of the World. A. Europe," *23rd International Geological Congress, Prague, 1968, Report,* M. Malkovský and J. Vachtl, eds., Vol. 15, pp. 107–140.

Stout, W.E., et al., 1923, "Coal-formation of Clays of Ohio," Bulletin 26, Ohio Geological Survey, 4th Series, 588 pp.

Takeshi, H., 1963, "On the Bentonite Deposits in Higashi Kambara-gun, Niigata Prefecture," *Japan Geological Survey Bulletin,* Vol. 14, No. 1, pp. 29–38.

Takeshi, H., and Kato, C., 1969, "Minerals," *The Clays of Japan,* ed. by Editorial Subcommittee for 1969 International Clay Conference, Geological Survey of Japan, pp. 103–120.

Tazhibayeva, P.T., and Galiyev, M.S., 1972, "Genetic Features of South Kazakhstan Bentonites (Abstract)," *International Clay Conference, Madrid, June 25–30, 1972,* pp. 113–114.

Teague, K.H., 1972, "Southern Bentonite," SME Preprint No. 72H328, SME Fall Meeting, Birmingham, Oct., 7 pp.

Thompson, A.G., 1952, "Colonial Minerals Development—XII. Kenya," *Mining Journal,* London, Vol. 239, No. 6102, p. 120.

Torok, A., and Thompson, T.D., 1972, "Activated Bleaching Clay for the Future," *Trans. SME/AIME,* Vol. 252, pp. 15–17.

Tracey, J.I., Jr., 1944, "The High-Alumina Clays of Pulaski and Saline Counties, Arkansas," *American Ceramic Society Journal,* Vol. 27, No. 8, pp. 246–249.

Urasima, Y., et al., 1966, "Bentonite Deposits at Hattari, Hokkaido (I)," *Japan Association Mineralogists, Petrologists, and Economic Geologists Journal,* Vol. 55, No. 1, pp. 17–24.

Van Olphen, H., 1963, *An Introduction to Clay Colloid Chemistry for Clay Technologists, Geologists, and Soil Scientists,* Interscience, New York, 301 pp.

Van Sant, J.N., 1959, "Refractory-Clay Deposits of Colorado," Report of Investigations 5553, US Bureau of Mines, 156 pp.

Van Sant, J.N., 1964, "Refractory-Clay Deposits of Utah," Information Circular 8213, US Bureau of Mines, 176 pp.

Varju, G., 1969, "Kaolin Deposits of Hungary," "Proceedings of Symposium I. Kaolin Deposits of the World. A. Europe," *23rd International Geological Congress, Prague, 1968, Report,* M. Malkovský and J. Vachtl, eds., Vol. 15, pp. 179–200.

Veatch, O., 1909, "Second Report on the Clay Deposits of Georgia," Bulletin 18, Georgia Geological Survey, 53 pp.

Waagé, K.M., 1953, "Refractory Clay Deposits of South-central Colorado," Bulletin 993, US Geological Survey, 104 pp.

Wakeman, J.S., 1972, "Method for Evaluation of Data from the Batch Testing of Green Pellets," *Trans. SME/AIME,* Vol. 252, pp. 83–86.

Warren, W.C., and Clark, L.D., 1965, "Bauxite Deposits of the Eufaula District, Alabama," Bulletin 1199-E, US Geological Survey, 31 pp.

Wayland, T.E., 1971, "Geologic Occurrence and Evaluation of Bentonite Deposits," *Trans. SME/AIME,* Vol. 250, pp. 120–132.

Weaver, C.E., et al., 1972, "Clay Water Diagenesis during Burial: How Mud Becomes Gneiss," Special Paper 132, Geological Society of America, 99 pp.

Weaver, C.E., and Pollard, L.O., 1973, *The Chemistry of Clay Minerals,* Elsevier, New York, 224 pp.

Weidhaas, D.E., and Brann, J.L., Jr., 1955, *Handbook of Insecticide Dust Diluents and Carriers,* 2d ed., Dorland Books, Caldwell, NJ, 233 pp.

Welch, R.L., 1942, "Geology of Vernon Parish," Bulletin 22, Louisiana Geological Survey, 90 pp.

Wentworth, C.K., 1922, "A Scale of Grade and Class Terms for Clastic Sediments," *Journal of Geology,* Vol. 30, pp. 377–392.

Wetzenstein, W., 1972, "Die Bentonitlagerstätten im Ostteil der Insel Milos und Ihre Mineralogische Zusammensetung," *Geological Society of Greece Bulletin,* Vol. 9, No. 1, pp. 144–171.

Wherry, E.T., 1917, "Clay Derived from Volcanic Dust in the Pierre of South Dakota," *Journal,* Washington Academy of Science, Vol. 7, pp. 576–583.

White, M.G., 1973, "Probing the Unknown Amazon Basin, a Roundup of 21 Mineral Exploration Programs in Brazil," *Engineering & Mining Journal,* Vol. 174, No. 5, pp. 72–76.

Wirth, L., 1968, "Attapulgites du Sénégal Occidental," Rapport No. 26, Laboratorie de Geologie, Dakar, Université, Faculté des Sciences, 55 pp.

Woodmansee, W.C., and Muchison, R.C., 1969, "Mineral Industry of South Africa," *Minerals Yearbook, 1969,* US Bureau of Mines, Vol. 4, pp. 623–644.

Wright, A.C., et al., 1972, "Catalysis by Layer Lattice Silicates—I. The Structure and Thermal Modification of a Synthetic Ammonium Dioctahedral Clay," *Journal of Catalysis,* Vol. 25, No. 1, pp. 65–80.

Zapp, A.D., 1965, "Bauxite Deposits of the Andersonville District, Georgia," Bulletin 1199-G, US Geological Survey, 37 pp.

Zvyagin, B.B., 1967, *Electron-Diffraction Analysis of Clay Mineral Structures,* Plenum Press, New York, 364 pp.

Diamonds*

K. RECKLING †

R. B. HOY ‡

STANLEY J. LEFOND §

Reported world production of natural diamonds approximates 40,000,000 carats a year (1980). The Republic of Zaire is the leading producer, with an output which is primarily industrial rather than gem grade. The USSR ranks second, the Republic of South Africa third, and Botswana, fourth. Diamonds are also mined in South-West Africa (Namibia), Angola, Ghana, Sierra Leone, and other African countries. In the western hemisphere, Brazil and Venezuela are the principal producers (Fig. 1). A very small output of diamonds is mined today in India, which was the first source of commercial production. In the US, efforts at commercial diamond mining have been confined to a small area near Murfreesboro, AR. The first diamond was found in a kimberlite pipe there in 1906. Small-scale trial mining has not, however, proved economical.

Since diamonds were first discovered over 2,000 years ago, only about 260 tons have been mined. In order to obtain 1 g (5 carats) of diamonds, it is necessary to remove and process approximately 25 tons of rock. Recovering this small percentage involves a combination of highly developed techniques in mining, quarrying, and earth-moving, and extremely sophisticated recovery processes.

End Uses

Diamonds are used for two unrelated end uses; gem diamonds are jewels of great beauty while industrial diamonds are essential materials of modern industry. Although imitation stones are substituted for the gem diamond, none of these matches its properties sufficiently well to offer real competition. Synthetic industrial diamonds are now of a quality and size that permit them to be substituted for natural diamond in numerous industrial applications. For example, synthetic diamonds are available today in sizes up to 100 stones per carat (14/16 US mesh size). In addition, polycrystalline synthetic diamond inserts, such as De Beers Syndite®, General Electric's Compax® and Stratapax® and Megadiamond's Megapax® have replaced natural diamond in turning tools, mining and oil drilling bits and dressing tool applications.

Gem Diamonds

A diamond is a natural prism. It has the ability to bounce light rays (reflection), to bend them (refraction), and to break them into all colors of the rainbow (dispersion). This ability to modify light gives the diamond three characteristics for providing beauty. First is *brilliance,* the reflection of light back to the eye. Second is *fire,* dispersion of light into the colors of the spectrum. Third is *scintillation,* the "twinkling" that occurs when the diamond is moved (Bauer, 1969).

The skill of the diamond cutter is required to unlock the brilliant, fiery beauty of a gem diamond (Fisher, 1966). Diamond-cutting skills have gradually developed over the centuries. The first cutters, in India, learned how to grind one rough diamond against another to remove their "skins" and to give them various shapes. This revealed some fire and beauty, but little compared to that of the diamonds that are cut today.

* The authors wish to thank Industrial Diamonds Information Bureau, London, and Peter Silveri Associates, New York, for their help in supplying information and illustrations for this chapter.

† Diamond Abrasives Corp., New York, NY.

‡ Consultant, Golden, CO.

§ Consultant, Industrial Minerals, Evergreen, CO.

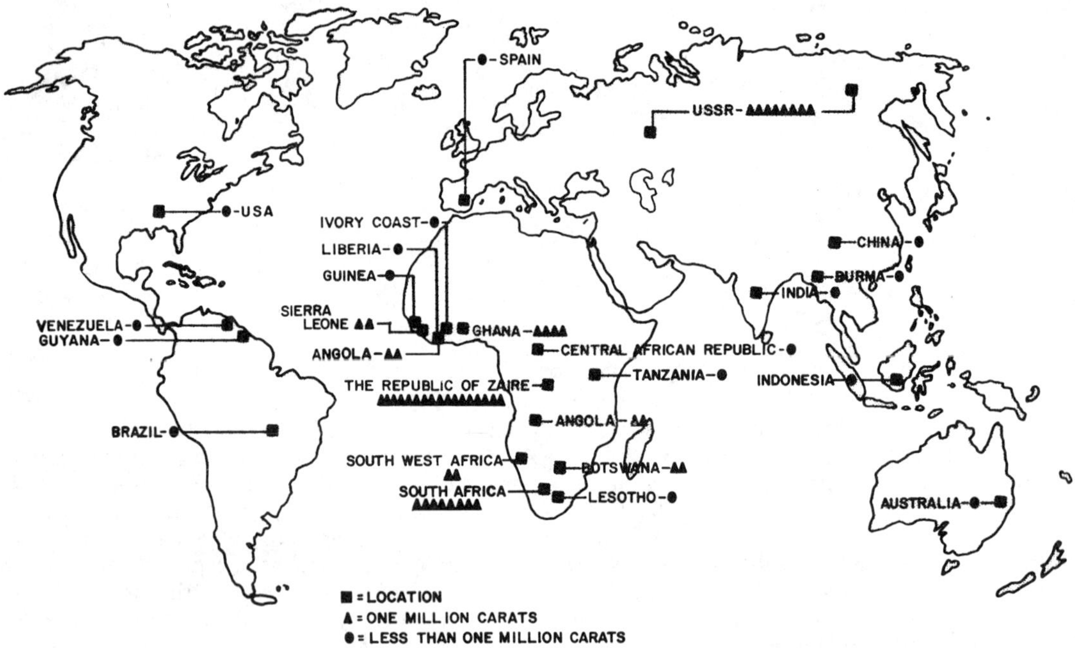

FIG. 1—*World diamond production from natural sources.*

About 1475, a Flemish cutter named Van Gerquem discovered that diamond dust on an iron wheel would polish facets on a diamond to provide more brilliance and fire. Cutters continued to experiment with facets, and shortly before 1700 a Venetian cutter named Peruzzi developed the 58-facet brilliant cut. With the development of the science of optics, cutters have learned more about the behavior of light within the diamond. Early in the 20th century, they learned how to angle Peruzzi's 58 facets precisely so that light enters the top of the diamond, reflects around inside, and comes back out of the top with maximum brilliance, fire, and scintillation. The full beauty of the modern diamond is unlocked only when the cutter follows this precise formula; any deviation robs the diamond of its full potential (Fig. 2). Even today, many diamonds are cut either too deep or too flat in order to reduce wastage from the original pieces of rough material. Although this practice may appear economically sound, it does not produce gems of top quality. Sometimes crowns of brilliant cuts are made too thin in order to maintain a large top facet. This cut provides "dead" stones, which tend to appear larger than they are. Such a gem appeals to persons to whom size is the most important criterion of a good bargain.

Industrial Diamonds

The diamond is by far the most important industrial abrasive. As recently as 40 years ago, consumption of industrial diamonds was less than that of gem diamonds, but since that time, industrial use has grown to a position of great dominance. During the six-year period 1929–1934, the material produced for industrial use amounted to about 74% by weight of the world's total output of diamonds. This ratio

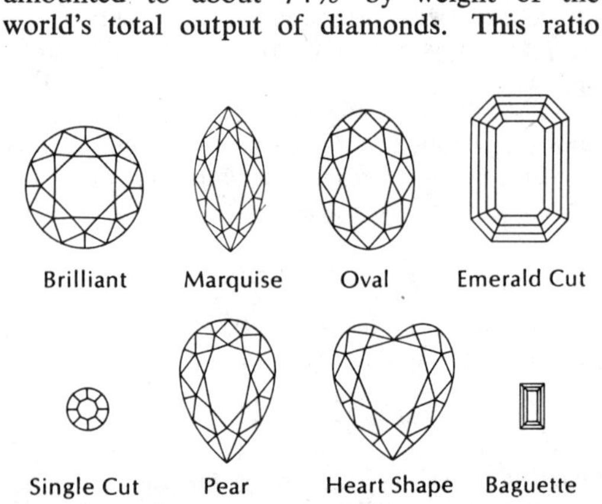

FIG. 2—*Diagram of the principal shapes into which diamonds are cut today. The single cut and baguette are often used as side stones for engagement rings and embellishment for jewelry.*

increased to 83% for the 1950–1955 period, to 87% in 1968, and to 89% in 1979. The many uses responsible for these significant increases are dependent on the properties of diamond, including hardness, cleavage, and parting; optical characteristics; presence of sharp points and edges; and capacity for taking and maintaining a high polish.

The utilitarian role of the diamond was confined primarily to lapidary products until the industrial revolution, which created the first demand for diamond as an industrial tool. In 1777, a British inventor and instrument maker, Jesse Ramsden, used a diamond to cut a precision screw for an engine that he had invented.

The first authentic description of industrial diamonds being set in a saw was recorded in 1854 by a Frenchman, Dumain. Eight years later a Swiss watchmaker, Jean Leschot, developed the first diamond drill bit for use in a hand-operated machine, which was employed to drill blastholes in rock.

In 1864, diamond bits were put to their severest test up to that time in the construction of the Mont Cenis tunnel in the Alps. A few years later a steam-powered diamond drill with a speed of 30 rpm was able to penetrate rock at the modest rate of 1 fph.

As the industrial revolution gained momentum on both sides of the Atlantic, metal replaced wood and machines replaced people. Thus the foundation was laid for precision engineering and the recognition of diamond as an indispensable tool of industry.

The next major demand for industrial diamond came after World War I with the development of cemented carbide cutting tools. Diamond was found to be the most effective medium for finishing and grinding the new ultrahard metal. This discovery rapidly increased the demand for industrial diamonds.

The availability of inexpensive diamond material inspired tremendous research into applications. By 1935, the first successful phenol-resin grinding wheel containing diamond had been marketed. Soon afterward, the metal-bonded and vitrified diamond wheels were produced, and, as the matrices and bonds that held the diamond grit in place began to improve, the popularity of diamond grinding wheels grew. The outbreak of World War II and the accompanying increase in use of hard metal tools in the munitions industry increased the demand for industrial diamonds.

Since about 1950, the development of ultrahard ceramics, semiconductor materials, plastics, and exotic metal alloys has further consolidated the diamond position as an indispensable tool of industry. Only diamond is hard enough to cut these superhard materials with the precision, speed, and economy that industry demands today.

Special machines equipped with industrial diamonds are used to remove bumps from concrete runways and highways. They are also used to modify highway surfaces in order to prevent skid accidents. Many skids are caused by hydroplaning, a phenomenon that occurs when the roadway is wet. Tires mount a film of water and virtually lose contact with the road surface. Diamond machines cut neat, narrow grooves in the pavement, thereby creating an escape route for the water.

Although dentists have used diamond drills for nearly a century, they have found a new use for diamonds as dental crowns in the form of very fine dust. The minute diamond crystals give wear and impact resistance to a dental resin employed to fabricate bridges, jackets, and other forms of dental restoration. Diamonds are also used in surgery. An ophthalmologist has developed a new technique in cataract surgery in which a perfect incision is made by a diamond knife with a microscopically perfect cutting edge. The blade's edge appears straight even when examined under 1000-power magnification. A New York plastic surgeon specializing in the repair of peripheral nerves uses precision forceps whose tips are coated with diamond dust. The ability of the diamond-tipped forceps to grip fine tissues firmly, but without damage, is an important technical advance in the field of microsurgery.

Resembling the forceps are diamond-tipped tweezers, which have gained favor with hobbyists and collectors who work with small parts or delicate items. Because they have from three to four times the gripping power of conventional instruments, they are excellent for plucking eyebrows. Nail files coated with diamond dust are another popular item because they are almost indestructible.

In Europe, many manufacturers of pasta have switched from steel to diamond saws to slice uncooked spaghetti into desired sizes. Steel blades dull in about 2 hr, while diamond blades hold their cutting edge for several months. For similar reasons, industrial diamonds are used to cut frozen fish to package size.

Although virtually unnoticed, industrial diamonds frequently surround us. In modern oil furnaces, some nozzles through which fuel is sprayed are made of diamond, since fuel oil is abrasive and tends to enlarge metal nozzles. Fine wires used in telephones, transistor radios, and television sets owe their precision diameters to diamond dies. Eye glasses and contact lenses are also shaped with industrial diamonds. In fact, the mirrors in your home and the glass in your automobile were most likely ground with diamonds. Because of their abrasion-resistance, diamond-tipped drills are also used for geological exploration and oil-well drilling. They also played a major role in the Mohole project in the Pacific Ocean.

Industrial diamonds have made contributions to America's space program, most notably in the production of the Gemini Apollo spacecraft. The thousands of yards of complex wiring, delicate instrumentation, ceramic fiber insulation, stabilizing gyroscope, nose cap, and reentry heat shield—all these components were produced with the aid of industrial diamond tools. Diamonds even played their part in man's spectacular step onto the moon—not only in the manufacture of the vehicles that got him there, but in the delicate instruments that helped him to probe the moon's surface to uncover the secrets that lie beneath.

Doulton, a name (Royal Doulton) synonymous with fine china for over a century, now is a mammoth international company manufacturing not only tableware and figurines, but also a whole host of other items ranging from sonar detection equipment to bullet-resistant glass windows. The company, having a reputation for use of up-to-date methods and equipment, uses metal-band diamond wheels for grinding and Hole-O-Matic diamond drills for making holes in such items as safety doors (Herbert, 1976).

New developments in the application of industrial diamonds are described in the publication *Industrial Diamond Review*. Some of these uses are new, others are improvements in established applications. They include precision drilling of small holes (Herbert, 1976), cutting of hard, brittle materials (Forner, 1978), drilling of concrete (Seversky, 1976), hard rock drilling (Hill, 1975), trueing of tungsten carbide gages (Herbert, 1977), use of diamonds in the electronics industry, and as an optical material (Seal, 1978).

Alternate materials are available for many of the applications served by diamond, but these materials are generally so much less efficient and adaptable that they are poor substitutes. No effective substitute has been found for use in drilling bits for hard rock or for coring concrete.

Classification: Natural industrial diamonds come in many shapes and sizes. They are generally classified according to use, in the following groups: tool and die stones, drilling material, grits and powders. Fig. 3 shows the pattern of industrial diamond consumption. It is expected that due to substitution of synthetic diamond in drilling bits and wire drawing dies that natural diamond usage will decline from 26% in 1973 to about 8% by 1983.

Industrial diamonds are further classified according to type of material, as follows:

1) *Industrial stones* (fine industrials): Stones of large size not suited for gem use because of shape, mechanical imperfections, or undesirable color.

2) *Bort* (*boart, boort, bortz, bowr*): Stones whose small size, irregular shape, content of flaws or inclusions, or occurrences in finely crystalline aggregates make them unsuitable for gem use. *Drilling bort* are those stones whose soundness permits their use in diamond drill bits. The more abundant crystals and crystal aggregates of lower grade are classed as *crushing bort,* which is suitable for being crushed into grit, powder, and dust. This latter category includes worn or broken stones from used diamond tools, and fragments formed during preparation of gems and shaped industrial stones.

3) *Carbonado* (*carbon, black diamond*): Compact, opaque, dark gray to black crystalline

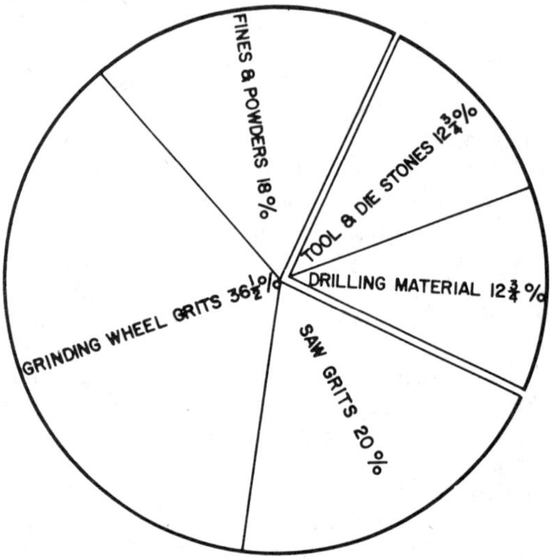

FIG. 3—*Pattern of world industrial diamond consumption. Grit products and powders constitute about 75% of the overall total.*

material composed of diamond, graphite, and possibly some amorphous carbon. It has no cleavage and is extremely tough. It occurs as rounded masses of average size greater than that of gem stones.

4) *Ballas* (*bort-ballas, short-bort*): Dense, globular aggregates of small radially oriented and extremely difficult cleavage. It is both very hard and very tough.

The average person tends to think that all rough diamonds are of the octahedron form, but many rough stones are not of this shape (Fig. 4).

Tool and Die Stones—The larger diamonds of higher quality are called tool and die stones. These are used as dressers (to true and dress conventional abrasive wheels), turning, boring, and milling tools, and wire-drawing dies.

To obtain maximum efficiency from a tool diamond, it is important to orient it so that the hard direction or the most abrasion-resistant face bears the work load. Orientation is also important in the manufacture of diamond wire-drawing dies.

Drilling Materials—Drilling accounts for 10% of all diamond use. Blast-hole drill bits are usually made with small crystals of regular shape. Mineral exploration bits require diamonds ranging in size from 15 to 100 stones per carat. Masonry and concrete bits make use of smaller diamonds ranging from 50 to 200 stones per carat. Oil well bits require larger sizes, usually 12 stones per carat and sometimes stones as large as ¾ carat.

To improve the supply of natural drilling material, stones with poor shapes are subjected to high-velocity air milling with diamond dust to improve their shape and strength. In addition, surfaces of some diamonds are polished because polished surfaces reduce friction during drilling operations.

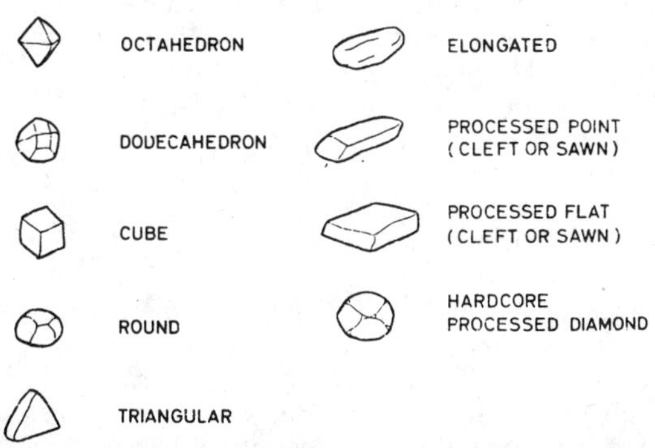

FIG. 4—*Industrial diamond shapes.*

Drilling material is also used in the manufacture of multiple-layer diamond dressers and rotary-form dressers. These are used for dressing and trueing surfaces of precision grinding wheels, cutting glass, engraving glass, and harder materials, and for sawing and shaping carbides, corundum, dimension stone, plastics, and various metals.

Grits and Powders—Approximately 75% of all industrial diamonds—both natural and synthetic—are used in the form of grit and powder (Fig. 4). Diamond grit is used primarily in grinding wheels and saw blades. There is a specific diamond grit type, shape, and size for each specific application. Diamond abrasive grains are sized by sieving on woven wire mesh or electroformed screens. Screens are designated according to the number of meshes to the inch, and screened grits are expressed in terms of the screens between which they fall. A −270 +325 grit is one which has passed through a 270-mesh screen and will not pass through a 325-mesh screen.

Diamond grit ranges in size from very coarse material (8/10 (+8 −10) mesh) to fine material (325/400 (−325 +400) mesh). Very fine diamond powder is referred to as graded micron powder and is used in the manufacture of lapping compounds and as fine finishing tools. Diamond micron powders are graded by a combination of sedimentation and centrifuging. Table 1 lists the standard grade numbers for these powders.

Geology

General

Diamonds are composed of a single element, carbon, crystallized in cubic form. Diamonds range from colorless through faint tinges of blue-white, yellow, red, brown, green, blue, and gray in gem form; and from yellow-brown to dark brown and black in industrial form. It is the hardest known material, listed at 10 on the Mohs' hardness scale, actually almost five times as hard as corundum listed at 9. The specific gravity is high (3.5) and index of refraction is very high (2.42). Color dispersion is exceptionally strong, producing the characteristic play of color in the gem. Table 2 lists many of the various properties of diamonds (Anon., undated).

Occurrences

Kimberlite Occurrences: The primary source of diamonds is a rock called kimberlite or

TABLE 1—Standard Grade Numbers for Diamond Powders*

Grade Nos.	μ Range	Approximate Mesh Equivalent
1/2	0-1	50,000
1	0-2	14,000
3	2-4	8,000
6	4-8	3,000
9	8-12	1,800
15	12-22	1,200
30	22-36	800
45	36-54	500/600
60	54-80	400/500

* Source: US Bureau of Standards, C/S 261/63.

"blue ground," which occurs in volcanic pipes and dikes. The pipes are generally vertical and range in shape from nearly circular to elliptical. They range in size from small pipes only a few feet in diameter to the well-known pipes in the Kimberley district of South Africa which have surface areas from 10 to more than 30 acres. Although the intrusions are called pipes, they are commonly funnel or carrot-shaped in three dimensions, and many become dikelike at depth. The kimberlite forming the pipes and dikes is a porphyritic ultramafic rock, usually altered to serpentine, containing a large quantity of accessory minerals. Not all kimberlites contain diamonds.

The most important accessory minerals are pyrope garnet, chrome diopside, ilmenite, and perovskite (Motti, 1969). These minerals, being more abundant than diamonds, have frequently served as indicators to geologists in their search for diamond-bearing kimberlite occurrences. Although there has been much controversy over the origin of diamonds, available evidence favors primary crystallization from ultramafic magma at considerable depth, followed by upward movement in volcanic conduits (Davidson, 1964). Kennedy and Nordlie (1968) ascribe an important role to CO_2 in both the genesis and emplacement. They suggest emplacement by an explosive CO_2 gas drive with the diamond-bearing rocks coming from a depth of no less than 200 km (120 miles).

Alluvial Occurrences: Diamond-bearing kimberlite in volcanic pipes served as the main source from which widely distributed alluvial diamond deposits were formed. Weathering processes disintegrated kimberlite and other rocks at the earth's surface. These processes are particularly effective in warm humid climates where near-surface rock formations, including kimberlite, were broken down and the products carried away by river systems and deposited over widely scattered areas, often far from their sources. Some geologists believe that a thickness of at least 3000 ft of rock has been eroded away in the Kimberley region of South Africa and distributed by streams.

Diamonds and other hard, weather-resistant minerals, released from the disintegrating kimberlite, were concentrated by flowing water and deposited in topographically favorable localities to form the alluvial deposits found today. Due to the high specific gravity of diamonds they tend to be concentrated at the bottom of alluvial deposits.

Reserves

World reserves of diamonds can be estimated only very roughly. Some figures have not been updated for over a decade. The estimate in Table 3 (Lampietti and Sutherland, 1978) indicates reserves do not appear to be sufficient to supply gem stones and industrial stones for more than about 30 years at current rates of consumption. The estimate for the Republic of South Africa has been adjusted to allow for production to 1976 and to include the significant increase in reserves resulting from reopening of the De Beers mine and discovery of the Finsch deposit. The USSR estimate is of low reliability.

Diamonds have been found in a pipelike structure of diamond-bearing kimberlite in the Lake Argyle area in Western Australia near the Northern Territory border. The prospect has yielded more than 5,000 carats (Garlick, 1979).

Producing Countries

World production of diamonds in 1979 was about 110 million carats, of which about 35% were natural stones. One-tenth of the total production was used for gems (Baskins, 1980). Table 4 lists output for major producing countries.

In recent years, production of natural diamonds has been concentrated in Africa and South America. Principal African production is in Zaire, the Republic of South Africa, Botswana, South-West Africa (Namibia), Ghana, Tanzania, and Sierra Leone. South American production is in Brazil, Guyana, and Venezuela (Baskins, 1980).

South Africa and South-West Africa (Namibia) are the main producers of gem stones. The USSR and African countries also are the world's main producers of natural industrial

TABLE 2—Properties of Diamonds

GENERAL PROPERTIES

I Chemical composition

Diamond is composed of the single element carbon.

Major impurities: Nitrogen, up to 0.2% in natural Type I diamond (see Section II).
Nickel, iron *etc.* up to 10% as inclusions in synthetic diamond (ppm or less in natural diamond).
Aluminum, up to 100 ppm in natural Type IIb diamond and 150 ppm in special doped Type IIb synthetic diamond. Boron, between 3 and 270 atomic ppm in specially doped synthetic diamond—now thought to be responsible for semiconducting properties of Type IIb diamond.
Others generally < 100 ppm
Inclusions: 22 mineral species have been positively identified.

II Classification

Type Ia diamond: Contains nitrogen as an impurity in fairly substantial amounts (of the order of 0.1%), and which appears to have segregated into relatively large sheets or platelets within the crystal. Most natural diamonds are of this type.

Type Ib diamond: Also contains nitrogen as an impurity but in dispersed form. Almost all synthetic diamonds are of this type.

Type IIa diamond: Effectively free of nitrogen impurity. Very rare in nature, these diamonds have enhanced optical and thermal properties.

Type IIb diamond: A very pure type of diamond which has semiconducting properties: generally blue in color. Extremely rare in nature. Semiconducting properties can be imparted to synthetic crystals by the incorporation of suitable impurities.

Note: Classification is based principally on optical properties. Further subdivisions can be made almost *ad infinitum* (indeed, almost every diamond is different in some way) but the above classification is universally accepted. Some diamonds have been shown to consist of more than one type, *eg* a complex interweaving of Type I and Type II material.

III Crystal structure

Diamond type.

Unit cell: Cubic, lattice constant a_0 = between $3.56683 \pm 1 \times 10^{-5}$ Å and $3.56725 \pm 3 \times 10^{-5}$ Å (25°C).

Space group: O_h^7.

IV Density

Value: Average of 35 diamonds: 3.51524 ± 0.00005 g per cc (25°C).

MECHANICAL PROPERTIES

I Hardness

(a) Scratch hardness (Mohs' Scale)

The Mohs' hardness is a scratch hardness test and is related to the indentation hardness of the solid (see below). If the Mohs' number is M and the indentation hardness in kg per sq mm is H, the relation between these quantities is approximately

$$\log H = 0.2\,M + 1.5$$

There is reasonable equality of intervals between the first 9 integers on the Mohs' scale, but the interval between 9 (corundum) and 10 (diamond) represents a much larger difference in indentation hardness than a single unit on the Mohs' scale would suggest.

(b) Indentation hardness (Knoop Scale)

Value ((001) diamond surface): 5,700–10,400 kg per sq mm depending on crystallographic direction and with normal loads of 500 g, 1 kg, and 2 kg.

Comment: There are many other scales, some appropriate to all materials, some only to metals. A Knoop indentor produces a wedge-shaped indentation in the form of a parallelogram with one diagonal at least seven times longer than the other diagonal, and this method is generally considered to be the most accurate for crystalline solids. On any scale diamond is the hardest known material.

It has recently been shown that the measured hardness value is significantly influenced by the normal load, the indentor shape, and its crystallographic relationship with that of the indented surface. It has also been shown that the indentation hardness of diamond decreases with increasing temperature.

II Elastic moduli and compressibility

Note: Young's modulus does not vary greatly with orientation as the condition for isotropy is $\frac{2C_{44}}{C_{11} - C_{12}} = 1$ (C. S. Smith, Vol. 6, *Solid State Physics*, p. 248) and using values given below $\frac{2C_{44}}{C_{11} - C_{12}}$ is 1.54, 1.14, 1.21, and 1.30, respectively. This is not far off unity except for the first case (the ratios for copper and lead are 3.21 and 3.89 respectively). Gilman (*Fracture*, p. 204, John Wiley & Son, New York, NY, 1959) quotes values $E_{100} = 10.5$, $E_{110} = 11.6$, $E_{111} = 12.0$ (all $\times 10^{12}$ dynes cm^{-2}), confirming little variation with orientation. Since diamond specimens are small and difficult to deform, numerical values of Young's modulus from direct deflectional

Table 2—Continued

methods are hardly obtainable. However, Young's modulus can be calculated from

$$E = \frac{(C_{11} - C_{12})(C_{11} + 2C_{12})}{C_{11} + C_{12}} = \frac{1}{S_{11}}$$

Using the values given below, $E = 7.24$, 9.48, 10.5, and 96.4, respectively (all $\times 10^{12}$ dynes cm^{-2}).

For *rough* calculations, Young's modulus can be taken as 10.5×10^{12} dynes cm^{-2} in any orientation.

(a) Elastic Moduli

Bhagavantam and Bhimasenachar:
$C_{11} = 9.5$, $C_{12} = 3.9$, $C_{44} = 4.3$
Prince and Wooster:
$C_{11} = 11.0$, $C_{12} = 3.3$, $C_{44} = 4.4$
McSkimin and Bond:
$C_{11} = 10.76$, $C_{12} = 1.25$, $C_{44} = 5.76$
Markham: $C_{11} = 10.76$, $C_{12} = 2.75$, $C_{44} = 5.19$
(all $\times 10^{12}$ dynes cm^{-2})

(b) Bulk Modulus

Calculated from elastic moduli by formula

$$K = \frac{1}{3}(C_{11} + 2C_{12})$$

Bhagavantam and Bhimasenachar: $K = 5.8 \times 10^{12}$ dynes cm^{-2}
Price and Wooster: $K = 5.9 \times 10^{12}$ dynes cm^{-2}
McSkimin and Bond: $K = 4.42 \times 10^{12}$ dynes cm^{-2}
Markham: $K = 5.42 \times 10^{12}$ dynes cm^{-2}

(c) Compressibility

Calculated from mean of Bhagavantam and Bhimasenachar and Prince and Wooster results.
Value: 1.7×10^{-7} sq cm kg^{-1}.

III Strength

(a) Tensile Strength

Diamond normally cleaves on the (111) plane, but cleavage has been observed on the (110) plane, and, to a lesser extent, some other planes (J. R. Sutton, *Diamond*, Murby & Co., London, 1928). Curved cracks can be produced by gradual loading with a spherical indentor. Ring cracks are found frequently on diamond surfaces and can be made by both slow and impact loading with indentors, not necessarily diamond.

Experimental values for strength show wide variations. This is partly a reflection of the difficulty encountered in making tests with diamonds, but also indicates that the strength of indivdual diamonds is significantly affected by the defects, inclusions and impurities which they contain. Small diamonds (or small loaded areas) usually exhibit greater strength than large diamonds (or large loaded areas), *i.e.* there is a size effect.

Value of tensile strength

4×10^{10} dynes cm^{-2} (Seal)
1.3 to 2.5×10^{10} dynes cm^{-2}
(Howes, indentation method, value depended on indentor material and size).

(b) Shear Strength

Theoretical value: 12×10^{11} dynes cm^{-2} (Tyson)
Experimental values: 8.7×10^{10} dynes cm^{-2} (Bowden & Hanwell, from a friction experiment)
3×10^{9} dynes cm^{-2} (Hull & Malloy, from a torsion test).

Note: Doubt must exist as to whether these strength values are genuine shear strengths; it is very likely that the diamonds still failed by cleavage. The variation is because of the very different method of test. In the friction experiment only a very small area of material would be stressed, while in the torsion test failure would probably start at a comparatively large defect at the specimen edge.

(ii) For synthetic diamonds with translucent inclusions or small concentrations of very fine inclusions:
Area of faces 50–75 sq mm $\times 10^{-4}$
Average 17.44×10^{10} dynes cm^{-2}
Maximum 24.50×10^{10} dynes cm^{-2}
Area of faces 660–730 $mm^2 \times 10^{-4}$
Average 6.17×10^{10} dynes cm^{-2}

(c) Compressive Strength

(i) Average value for natural octahedra containing no visible flaws or inclusions =
8.68×10^{10} dynes cm^{-2}
Maximum value =
16.53×10^{10} dynes cm^{-2}
(Area of faces over which load applied approximately 10^{-2} sq cm).

OPTICAL AND ELECTRICAL PROPERTIES

I Refractive index

Values (at H_g green 5461 Å) : $\mu = 2.4237$
(at Hα red 6563 Å) : $\mu = 2.4099$
Near cutoff in ultraviolet at 2265 Å : $\mu = 2.7151$

(d) Plastic Flow

Recent work by Gane and Brookes suggests that plastic deformation would appear to be possible in diamonds at room temperature, although the stress required to do this must be very high and close to the crystal's theoretical strength. However, it is very difficult to achieve these stress levels in practice without fracture first occurring. Earlier it had been established that there was no detectable plastic flow in three-point bending experiments up to critical stress for cleavage at temperatures less than 1600°C.

Note: Fits empirical formula:

$$\mu^2 - 1 = \frac{\epsilon_2\lambda^2}{\lambda^2 - \lambda_1^2} + \frac{\epsilon_2\lambda^2}{\lambda^2 - \lambda_2^2}$$

where $\epsilon_1 = 0.3306$
$\epsilon_2 = 4.3356$
$\lambda_1 = 1750$ Å
$\lambda_2 = 1060$ Å

II Dielectric constant

Value (at 27°C and 0–3 kHz) : 5.58 ± 0.03.

Table 2—Continued

III Optical transparency

Type IIa diamond : 2250 Å to 2.5μ. > 6μ
Type I diamond : 3400 Å to 2.5μ. > 10μ

Note: More restricted range in Type I diamonds due to nitrogen content.

Note: Visible region 4000–7000 Å.
Any coloration is due to absorption bands in this region, or in the case of Premier Overblue (Type IIb) to the tail of the infrared absorption spectrum associated with the acceptor center.

IV Resistivity

Value (at 20°C): Type I and most Type IIa : > 10^{16} ohm cm.
Type IIb : 10–10^{3} ohm cm.

Note: Figures for diamond only apply in the dark, for there is considerable photoconductivity in ultraviolet light.

Note: Diamonds exist which would normally be classified as Type IIa, but which have resistivities intermediate between Type IIb and Types I and IIa. Hence there is really a range from 10–10^{16} ohm cm at 20°C. However, most diamonds are insulators with a resistivity of > 10^{16} ohm cm.

Comment: Insulating diamond can have a very high resistivity, comparable with or better than many insulators in commercial use. The best insulator is probably fused silica (> 5×10^{18} ohm cm).

THERMAL PROPERTIES

I Thermal conductivity

Typical values (at 20°C):
Type I 9 watts deg^{-1} cm^{-1}
Type IIa 26 watts deg^{-1} cm^{-1}
Maximum at − 190°C:
Type I 24 watts deg^{-1} cm^{-1}
Type IIa 120 watts deg^{-1} cm^{-1}

Comment: Highest known value: value dominated by impurity content.

II Thermal expansion (linear coefficient)

Value (at 20°C) : $0.8 \pm 0.1 \times 10^{-6}$
(at − 100°C) : $0.4 \pm 0.1 \times 10^{-6}$
(at 100°C–900°C) : $1.5 - 4.8 \times 10^{-6}$

Note: Grüneisen's Law obeyed between 150°C and 950°C.

III Specific heat (constant volume C_v)

Value: (at 20°C : 1.478 cal g-$atom^{-1}$ deg^{-1}
Note:
Effective Debye temperature 0–800°C
$\theta_D = 1800 \pm 10$°K
Effective Debye temperature 0°K
$\theta_D = 2220 \pm 20$°K

Source: Anon., undated.

TABLE 3—Estimated 1977 World Resources of Natural Diamonds (Millions of Carats)

Country	Reserves*	Identified	Undiscovered	Total
Angolo	20†	—	25	45
Botswana	60	25	25	110
Central African Empire	0	2	50	52
Ghana	15	15	0	30†
Guinea	0	1	50	51
Ivory Coast	0	0	0	0
Lesotho	5	—	25	30
Liberia	0	—	25	25
Sierra Leone	6	—	25	31
Republic of South Africa‡	72	50	25	147
S.W.A. (Namibia)‡	15	20	0	35
Tanzania	2	0	0	2
Zaire	120†	330†	50	550
Brazil	1	5	50	56
Venezuela, Guyana, Colombia	0	2	100	102
India, Indonesia	0	0	25	25
USSR	25§	50	50	125
North America	0	0	50	50
Australia, Antarctica	0	0	25	25

Source: Lampietti and Sutherland, 1978.
*Use of the word *reserves* does not rigorously follow USGS/USBM classification of 1976 (Anon., 1976).
†After Bardet, 1973.
‡Including offshore.
§Clarke, 1975.

TABLE 4—World Production of Natural Diamonds, 1979

	Country	Total Production (Thousand carats)	Industrial %
1	Zaire	11,160	95
2	USSR	10,700	80
3	South Africa	7,640	52
4	Botswana	3,340	85
5	South West Africa (Namibia)	1,950	1
6	Ghana	1,500	90
7	Angola	750	25
8	Venezuela	750	62
9	Sierra Leone	710	60
10	Liberia	310	58
11	Central African Empire	300	40
12	Tanzania	290	50
13	Brazil	90	50
14	Guinea	85	68
15	Lesotho	70	80
16	Guyana	17	50
17	India	16	59
18	Indonesia	15	60
19	Ivory Coast	5	100

Source: Baskins, 1980.

diamonds. The US is the largest producer of synthetic stones.

About 70 million carats of synthetic diamonds are produced annually, mainly in the US, USSR, South Africa, and in the Shannon Free Zone of the Republic of Ireland.

Exploration

The first step in diamond exploration is selection of the locality to be explored. Naturally, favored areas are those where diamonds have been previously found in large quantities, such as the countries in Africa where diamonds are abundant in both pipe and alluvial deposits, and in South America in alluvial deposits. More speculative exploration has taken place in Ontario, Quebec, Labrador, and the Northwest Territories, Canada, where kimberlite pipes are located (Gunn, 1967; Brummer, 1978), and in mid-continental United States where a kimberlite pipe in Arkansas contains small quantities of diamonds. Diamonds were discovered in kimberlite diatremes in the State Line district in southeastern Wyoming along the Colorado border in 1975 (Hausel, 1979). More than 90 kimberlite occurrences are known in an area extending from west of Boulder, CO to the Iron Mountain district of Wyoming. Cominco American expects to start operation of a $2.5 million plant to test diamond-bearing rocks found in southern Wyoming and northern Colorado (Anon., 1981). Geologic structures favorable for diamonds are horsts and grabens. Host rocks are kimberlites, with rare occurrences in syenites, ijolites, melilites, and carbonatites (Dawson, 1962, 1964, 1968; Ilupin, 1977; Hausel, et al., 1979).

Many companies spend considerable sums exploring for diamonds. However, the largest consistent expenditure is by De Beers, according to their annual report, amounting to about $20 million per year (Lampietti, 1978).

Kimberlite is frequently intruded along zones of structural weakness in the earth's crust. For this reason, examination of LANDSAT or SLAR imagery, that defines major lineaments, may help select target areas for more detailed study (Lampietti, 1978).

Geophysical exploration, particularly airborne and ground magnetic surveying, is adaptable to diamond exploration. The kimberlite pipes, being ultramafic rocks, usually have more magnetite than the rocks surrounding them, and have a characteristic magnetic pattern reflecting their shapes, usually a circular or elliptical magnetic high (Gerryts, 1970). Electrical resistivity results reveal that weathered kimberlite is relatively conductive (80 to 250 ohm-ft) compared to intruded Precambrian granites (300 to 7,400 ohm-ft) in the Wyoming-Colorado area (Hausel, 1979). In areas known to contain kimberlite pipes, their locations can sometimes be pinpointed by panning alluvial material in streams. If characteristic heavy minerals of kimberlites are found (such as diamond, perovskite, ilmenite, chromite, chrome diopside, and pyrope garnet), these can be traced upstream to the source (Brown, et al., 1967). The careful determination of exact composition of pyrope garnets and ilmenites may help differentiate diamond-bearing from barren kimberlites (Lampietti, 1978).

Geochemical exploration makes use of nickel and columbium content of stream sediments and soils. Anomalous halos have been detected in surface soils several hundred feet downslope from kimberlite outcrops and stream sediment anomalies exist several thousand feet downstream from outcrops (Gregory and Tooms, 1969). However, geochemical methods are not widely used because heavy mineral sampling is more efficient and simpler (Lampietti, 1978; Hausel, et al., 1979c).

In the average commercial deposit, approximately 25 tons of rock must be mined to obtain one gram of diamond—a high ratio of concentration, on the order of 1:5,000,000 or higher. Because of this extremely sparse distribution, bulk sampling methods are best for evaluating both pipe and placer deposits. Pitting and trenching is better than drilling. If holes are drilled, their diameter should be as large as practical to provide representative samples. Holes are frequently drilled in alluvial deposits merely to determine depths to, and extent of, diamond-bearing gravels.

Evaluation of Deposits

Most diamond deposits contain both gem and industrial diamonds. Exploration should provide sufficient information to determine the quantity and quality of diamonds per unit volume, and the total volume of the deposit. Estimates are made of the cost of mining and recovery of the diamonds. A feasibility study can then be made to determine whether the deposit can be mined profitably.

Preparation for Markets

Mining Methods

During the second half of the 19th century before the diamond-bearing kimberlite pipes were discovered in South Africa, diamonds were produced entirely from alluvial deposits in open pits by use of very primitive tools and techniques. Picks and shovels were practically the only tools used for mining, and the hand-held washing pan was used for concentration, with hand sorting to recover the diamond from the washed concentrate.

The discovery of kimberlite pipes necessitated the development of completely new methods of mining and recovery. These pipes eventually have to be mined underground, with methods adaptable to treating large quantities of material.

Pipe Mining: A typical diamond pipe is roughly cylindrical in shape and from a few hundred feet to 3,000 ft in diameter. However, the pipe at Orapa, Botswana, oval in outline, is almost a mile long and 3,000 ft across (Hoppe, 1980). Pipes may be many shapes. For example, Jwaneng in Botswana is actually two cylinders connected at depth. Operations usually begin with surface excavation of weathered periodotite or "yellow ground." As mining proceeds over the entire outcrop of pipe and the pit becomes deeper, the walls begin to cave, filling the pit with rubble. To continue mining, it is necessary to mine underground. This is done by sinking shafts in the adjacent country rock and driving haulageways at 600 ft intervals to connect the shafts with the diamond-bearing ground.

From the shafts, raises are driven and intermediate levels are established at about 40-ft intervals. These are driven horizontally into the solid kimberlite below the open pit. This system is expanded to form a series of chambers through excavation of the diamond-bearing rock. This method combines shrinkage stoping and sublevel caving (Peele, 1952). The material is loaded into cars, hauled to the shaft, and hoisted to the surface. In some pipes, mining has reached great depth. The Kimberley mine, in South Africa, has been mined to a depth of nearly 4000 ft.

The mining method described is called chambering. Another variation is block caving, a system in which some material is extracted leaving inverted cone-shaped openings which eventually collapse and cave the block of "blue ground" from above. The material is drawn away through drifts by mechanical scrapers, then hoisted to the surface. Block caving is safer, faster, and more economical than chambering.

The first attempt at block caving in diamond mining was conducted by the De Beers at the Bultfontein mine in 1955. Jagersfontein mine was completely converted to block caving in 1958, and all the Kimberley pipes are now mined by this method (Bruton, 1978).

Open-Pit Mining: Open-pit mining methods

are relatively simple, differing from one mine to another mainly in the type of machinery used. All kimberlite mines are first operated as open-pits. Only when open-pit mining becomes so deep as to be inefficient and unsafe does underground mining begin. A brief description of the method used at the Finsch mine in South Africa will give a general idea of the technique used (Fig. 5).

The Finsch pipe is roughly circular, 1,750 × 1,500 ft, with almost vertical sidewalls of country rock. The overburden, consisting of topsoil and banded ironstone, from 10 to 40 ft thick, has to be removed to expose the weathered kimberlite, or "yellow ground," which extends to a depth of about 100 ft. The "blue ground" below continues to an indefinite depth, at least 900 ft. Blastholes 45 ft deep are drilled 25° from the vertical on levels. The holes are charged with explosives and blasted. The broken kimberlite is removed by mechanical shovels and trucked to the treatment plant. As mining proceeds, the waste rock from the sidewalls of the pipe is systematically cut back and removed to prevent it from falling into the pit. This waste is dumped outside the perimeter of the mine.

It is expected that open-pit mining at Finsch will continue for approximately 20 years reaching a depth of 900 ft (275 m) before a changeover to underground methods is required.

In other large-scale surface-mining operations in the Republic of Zaire, Botswana, Angola, South-West Africa (Namibia), and Sierra Leone other modern equipment is used, such as bucket-wheel excavators, draglines, front-end loaders, shovels, scraper-loaders, and giant trucks, to move overburden and to mine and transport the ore.

Benching: A combination of open-pit and underground mining is the method that has been used at the Premier mine in the Transvaal since 1945. The 2,800 by 1,500 ft (850 x 450 m) oval-pipe is divided into two parts by a

FIG. 5—*Aerial view of the Finsch open-pit mine and processing plant in South Africa.* (*Courtesy De Beers Consolidated Mines, Ltd.*)

200 ft (60 m) wide quartzite mass reaching to a depth of more than 1,000 ft (305 m). The blue ground on both sides of the quartzite divider is blasted by benching and dropped below into inverted cone-shaped openings cut into the blue ground at a current depth of 1,170 ft (355 m). The benches are mined to form an overall slope of 65°, steep enough to permit the ore to gravitate to the cone level below.

Placer Mining: Placer diamonds, which have been eroded from pipes, are found in both riverbed and beach deposits. At some places, diamonds are found in marine terraces elevated above the sea; elsewhere, in existing beaches or in beaches now submerged at some depth beneath the sea (Murray, 1970; Oostdam, 1970; Bruton, 1978).

The most extensive placer diamond mining is at Consolidated Diamond Mines (CDM) in South-West Africa (Namibia). CDM operates over a 50-mile stretch at Orangemund, reported to be the richest gem-producing area in the world, and further south at Kleinzee.

The first step in developing an ordinary placer deposit is to sink prospecting boreholes to determine the thickness of the sand overburden and find the geological character of the diamondiferous gravel. In a typical deposit, bulldozers remove the overburden and expose the gravel. Prospecting trenches about a yard wide and a mile long are then cut. These are sampled by simple hand sorting of diamonds from the gravel to determine the richness of the deposit.

If the deposit proves to be large and sufficiently rich, a formidable earth-moving operation begins. Giant excavators strip away the overburden at the rate of as much as 200 tph to expose the diamond-bearing gravel, which typically averages 3 to 4 ft thick. The gravel is either bulldozed into convenient stockpiles for mechanical loading or is loaded directly into dump trucks for delivery to the nearest screening plant. Because a high percentage of diamonds in gravel occurs on or near bedrock, it is essential to clean the rock very thoroughly. All crevices and holes in the bedrock are searched carefully either by hand labor or by mechanical devices that have been devised specifically for this purpose.

At the field screening plant, the diamond-bearing gravel is placed on vibrating screens, which reject 80% as waste and retain 20% as diamond-rich residue to be taken by truck or train to a central treatment plant.

Placer deposits in submerged ocean beaches off the coast of South-West Africa (Namibia) were mined for several years. The diamond-bearing gravel was hydraulically dredged aboard a ship and the diamonds were recovered. This ocean diamond mining proved uneconomic and was discontinued in 1970 (Bruton, 1978).

Processing Techniques

Lode Deposits: Material from lode deposits is crushed if necessary and concentrated by washing, winnowing, screening, and hand sorting, or by combinations of these methods. In some localities diamond-bearing material from dumps and waste heaps is frequently reworked several times.

Methods of concentrating diamonds in large-scale operations have become more mechanized with time. In the early years of shallow mining, the "yellow ground" and the partially weathered "blue ground" required little crushing. The material, when exposed to the atmosphere, gradually disintegrated if allowed to remain exposed in open cuts, or if spread out in nearby areas for as many as 18 months. Thus no crushing was required before processing. However, as mining has progressed deeper into unweathered "blue ground," the material is crushed in either jaw or gyratory crushers or by corrugated rolls, generally in two or more stages separated by screening. Figs. 6 and 7 (Anon., undated) show three general types of flowsheets.

Devices used for this purpose include the rotary washing pan, jigs, heavy-media separators, and hydrocyclones. Each of these machines makes use of differences in specific gravity to separate diamonds and other heavy minerals from the lighter minerals. For example, in the rotary washing pan—a low, circular pan of large diameter—the crushed material is mixed with mud to form a "puddle" of material with a density of about 1.25. Although far below the density of even the light mineral fraction of the kimberlite, the density of this "puddle" appears to play an important part in the process. The waste is moved off the top into a central rotating weir by rotary toothed blades. The diamonds and other heavy minerals settle to the bottom and are drawn off periodically from the outer rim of the pan. The concentrate is then classified, and either jigged or separated by heavy media to produce a rough concentrate usually amounting to about 1% of the feed from the mine. Larger, more modern plants have replaced the jigs with heavy-media separators (Fig. 6). In these, a

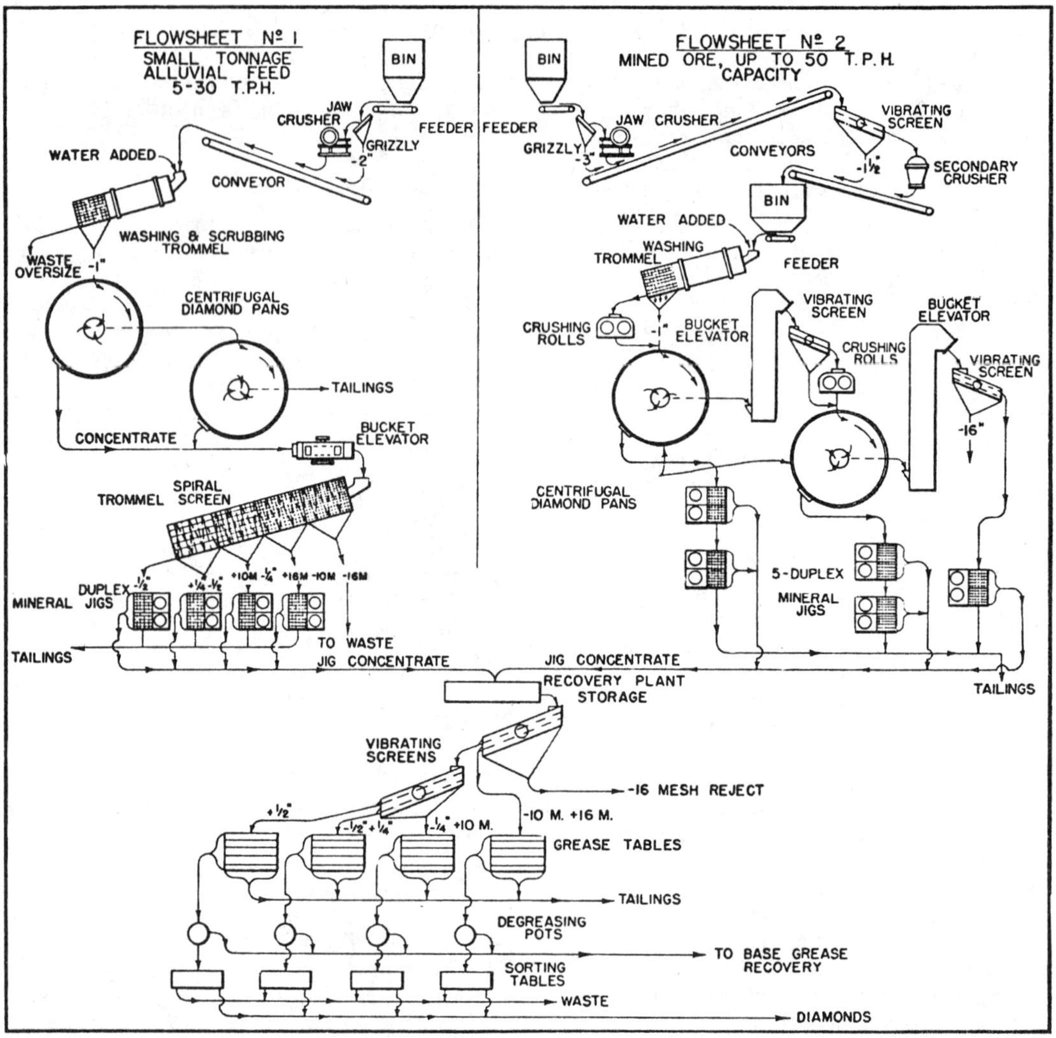

FIG. 6—*General flowsheets for small to medium operations.* (*Courtesy Denver Equipment Co.*)

slurry of ferrosilicon powder in water with a density of 2.7 to 3.1 is used to float the light particles, separating them from the heavier minerals, including diamonds, which sink and are collected as a final gravity concentrate for further treatment.

The final concentrate was formerly processed by separating the diamonds from the other heavy minerals by hand sorting. Today most operators use one of several greased-surface methods. The principle of greased-surface concentration was discovered at Kimberley in 1896 when it was observed that diamonds from freshly mined kimberlite adhered tenaciously to axle grease with which they accidentally came into contact, whereas the bulk of the associated heavy minerals did not. This important discovery greatly reduced the amount of hand sorting required to separate the diamonds from other heavy minerals.

One type of concentrator is the vibrating grease table, which consists of a stepped surface made of heavy-gage aluminum (Fig. 7). The surfaces of the steps are covered with a thick layer of a heavy grease. The table is vibrated while the diamond-bearing concentrate, mixed in water, is passed over the steps. The diamonds adhere to the grease; the waste minerals are washed away. The diamonds are collected by stopping the process and scraping them with some of the grease off the surface of the table. The table is regreased and the process continues. The scrapings are boiled in water to remove the grease from the diamonds.

At Orapa, Botswana, the diamonds do not adhere to the grease because of a surface coating. Instead of greased tables or belts, an X-ray unit spots the diamonds, activating a microsecond air jet that sweeps the diamond from the belt into a collector (Hoppe, 1980).

Placer Deposits: Diamonds are recovered from loose placer materials and from the weathered rock of lode deposits by washing, dry winnowing, screening, panning, jigging, tabling, and hand picking, generally in some combination. In partially consolidated gravels, some crushing is required. Usually simple devices, such as log washers, are sufficient to break up the gravels and separate the clay fractions. Other equipment used includes baskets, hand pans, grease belts, and tables.

The grease belt is the greased-surface concentration method best adapted to alluvial materials. In fact, it was developed to process diamond-bearing material from certain alluvial deposits. A modification of the normal process was necessary because alluvial diamonds, particularly those from marine terraces, frequently have a film of mineral salts on their surfaces, which renders them wettable in water. A wet diamond will not adhere to grease.

The De Beers Diamond Research Laboratory

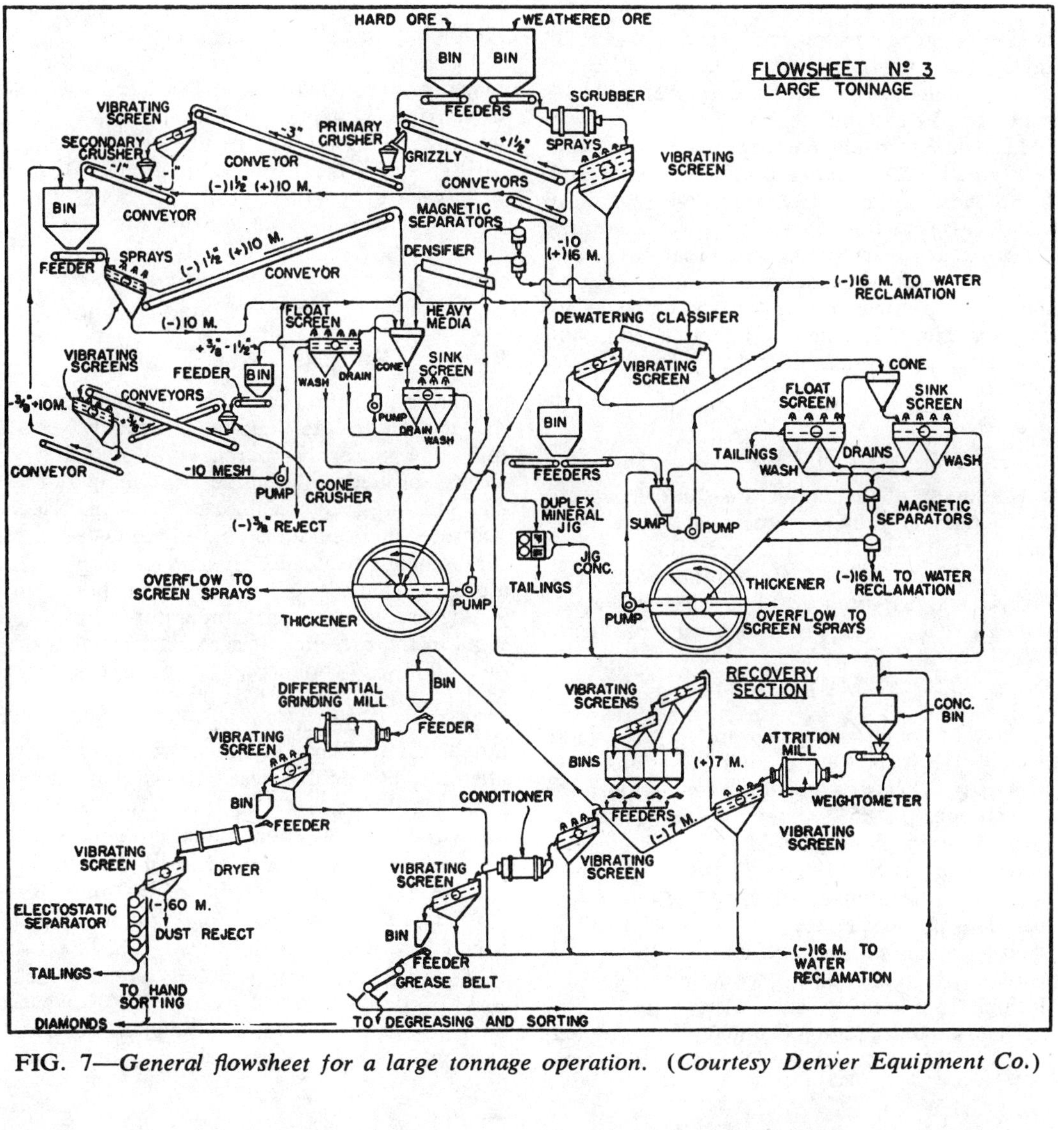

FIG. 7—*General flowsheet for a large tonnage operation.* (*Courtesy Denver Equipment Co.*)

developed a method using a soap solution made of corn-acid oil and caustic soda. Treatment with this solution produces a water-repellent surface on the diamond but not on the other minerals. The method also incorporates use of a greased belt, which permits a continuous rather than a batch process (Fig. 6). Both the soap solution and the grease are added to the belt surface as the belt is in operation. The method is in use at the Consolidated Diamond Mines of South-West Africa (Namibia); at the Finsch and Williamson mines in South Africa; at Luxilo plant in Angola; and at the State Alluvial Diggings in Cape Province.

Further treatment is required when the grease-belt process is used on finer sizes of alluvial material. In order to clean surfaces and disintegrate softer materials to slimes, the degreased concentrates are processed in ball mills. The material is screened, and the clean coarser diamond particles are further concentrated by electrostatic separation.

Material from alluvial deposits is also concentrated by X-ray separation. Diamonds tend to luminesce in an X-ray beam, whereas most of the associated minerals do not. Luminescence excites a photomultiplier that triggers a gate that diverts the diamonds from the path of the gravel passing through the machine. Calcite sometimes luminesces like diamond, but this effect is neutralized by use of a suitable filter.

Transportation

Essentially all diamonds are shipped via air which permits delivery almost anywhere in a matter of hours. The costs are usually small compared to the value of the diamond. This favorable transportation factor also applies to grinding wheels, saws, bits, and other products.

Marketing

Sales of rough diamonds are largely made through De Beers' Central Selling Organization (CSO). CSO has sold about $1 billion worth of diamonds, or 85% of world sales (Bruton, 1978). Although only about $200 million (the actual figure is not disclosed) is for industrial diamonds, the number of carats of industrial diamonds produced has ranged from 80 to 90% of total diamond production. Most of the world's rough-diamond production not sold through CSO finds its way to Antwerp.

The Diamond Trading Company was founded in 1934. Together with the Diamond Producers Association, it formed the nucleus of CSO, a group of marketing companies through which all the principal producers sell diamonds in cooperative arrangement.

Prior to formation of the CSO, the diamond business was subject to slumps about every eight years. In the mid-1920s, the discovery of high-quality alluvial diamonds at Lichtenburg, Namaqualand and South-West Africa (Namibia) flooded the market with diamonds that could be produced for significantly lower costs than those from deep mines at Kimberley (Bruton, 1978), resulting in extremely low earnings from the Kimberley production.

During the world depression in the early 1930s, the situation worsened. De Beers was unable to sell enough diamonds to cover expenses. The mines were all shut down and Kimberley was almost a ghost city.

Because of these conditions, Sir Ernest Oppenheimer, Chairman of De Beers, was able to convince a majority of producers to unite to control production and regulate sales (Bruton, 1978). Since the formation of CSO, all producers have benefitted from the stability that was brought to the marketplace.

Future Consideration and Trends

Diamond Synthesis

During the 19th century many attempts were made to snythesize diamonds. Broadly speaking, three approaches were made—precipitation from saturated solutions at high temperatures and pressures, chemical reaction at high temperatures and pressures, and the application of simultaneous heat and pressure to graphite in a press (Bundy, 1962). The last of these, subjecting graphite to high temperatures and pressures in the presence of a catalyst metal is the method now in commercial use (Ringwood and Major, 1966).

Moissan and Hannay are probably the most notable of the early scientists who made these attempts, but neither was successful in manufacturing diamonds. The reason for their failure was that the essential material for successful synthesis—tungsten carbide used in pressure chambers—was not available at the time. The first successful synthesis of diamond is attributed to a Swedish team of scientists (Liander, 1955) from Allmänna Svenska Elektriska Aktiebolaget (ASEA) in 1953. However, their process at first was not thought to be commercially viable, and no public announcement was made. Early in the 1950s, General Electric in the United States initiated a research program in superpressure reaction, which ultimately led to successful diamond synthesis in late 1954. Although this discovery was known in 1955, an order for secrecy was imposed on the General Electric process, and the apparatus patent applications restricted the publication of any results of their work until 1961.

Also in the mid-1950s, a team of scientists was assembled in the De Beers Diamond Research Laboratory in Johannesburg, South Africa for the purpose of devising a process for diamond synthesis, and they announced success in 1958 (Legrand, 1980).

In the early days of diamond synthesis scientists were only interested in producing diamonds, hence, little attention was given to the quality of the product. It was soon realized that improvements in quality were necessary. Research effort for this objective has continued relentlessly. The hydraulic presses in the De Beers synthetic diamond plant in Ireland (Fig. 8) are products of such research.

The relative merits of synthetic vs. natural diamond grits have been evaluated and discussed on many occasions. There is no doubt that there are specific applications in which the synthetic product has gained broad acceptance—most notably in resinoid grinding wheels.

A major difference between natural and synthetic diamonds is the metal content. Synthetic diamonds contain significant amounts of the metal solvent used in synthesis. The included metal is believed to exist in three forms, namely (1) macroscopic metal inclusions; (2) small platelike inclusions dominantly on the

FIG. 8—*These 10,000-ton hydraulic presses in the De Beers plant in Shannon, Ireland convert graphite into diamond by the simultaneous application of ultrahigh pressure and temperature.*

cleavage planes; and (3) substitutional metal atoms in diamonds synthesized from nickel or nickel-based alloys.

Due to differences in the coefficients of thermal expansion and the compressibilities of the metal inclusions in diamonds, the metal causes the diamonds to be stressed. The resulting strain causes birefringence which can be seen when synthetic diamonds are examined by use of a polarizing microscope. The strength of synthetic diamonds decreases with increasing the metal content.

Synthetic micron-size diamonds are also produced by an explosive-shock technique first achieved by Stanford Research Institute and developed commercially by E. I. du Pont de Nemours & Co., Inc. The product is a polycrystalline type of diamond, which can, at this time, be manufactured only in micron sizes (De Carli and Jamieson, 1961).

Synthetic diamonds are being produced today in several countries throughout the world. While De Beers Industrial Diamond Division and General Electric's Specialty Materials Division are still the world's leading producers, many new companies have entered the sales fray. In the US, there are four manufacturers of diamonds. There are also producers in the Soviet Union, Japan, Rumania, China, India, Czechoslovakia, Germany, Greece, Ireland, Finland, Sweden, and South Africa. While many of these producers are not able to turn out the high quality products that De Beers and General Electric make, they have caused an oversupply condition in the market place, resulting in substantial price reductions of most synthetic diamonds.

Recent Developments

Continued research and development have improved the quality and lowered the cost of synthetic stones. Temperatures and pressures required have been lowered. Crystals can be made with better shapes. Recently, methods have been developed to recrystallize small synthetic diamonds to form gem-quality diamonds of a carat and larger. However, this has not yet been done profitably.

Sintered masses of diamond material have been produced in predetermined shapes, but only in small quantities. This process is reported to have been developed in both the USSR and the US.

A most recent process reported in the US is the growing of diamonds from hydrocarbon gas at atmospheric pressure and moderate temperature.

A major technological development expected and overdue in diamond synthesis is commercial production of large industrial stones to replace the use of natural stones in such applications as drilling bits (Adams, 1979).

Today production of synthetic stones far exceeds the production of natural stones. Approximately 110 million carats are manufactured by De Beers in South Africa, Ireland, and Sweden; by General Electric in the US; by the USSR in Kiev; and by Japan. Small quantities are also produced in other countries (Legrand, 1980).

Following the successful manufacture of small diamond crystals, research continued on crystallization of larger stones suitable for jewelry. Finally in 1970 General Electric laboratories succeeded. A few stones of very high quality were produced, some as large as one carat. These can be seen on display at the Smithsonian Institute in Washington, DC. These results proved that it is possible to manufacture synthetic stones of fine quality large enough for use in jewelry. However, the process is so costly that the synthetic diamonds are far more expensive than natural stones of the same quality (Legrand, 1980; Bruton, 1978).

Crystal now can be made in larger sizes (100 stones per carat—i.e., 14/16 US mesh size) with better shapes and higher thermal stability.

Polycrystalline or composite diamond inserts are produced by sintering together small particles of diamond at temperatures above 1400°C and at pressures on the order of 60,000 atmospheres.

The polycrystalline tool itself consists of a dense sintered diamond layer which is bonded onto a tungsten carbide substrate. In natural diamond tools chipping occurs by the diamonds dividing along naturally occurring cleavage planes which are parallel to the octahedral faces of the diamond. Wear of this nature is absent from polycrystalline diamond tools because the diamond grains, which make up the sintered diamond layers, are randomly orientated.

These polycrystalline tools are ideally suited for use in cutting tools for machining of nonferrous materials, and in abrasive wheel dressers, wear parts, wire dies, and rotary drill bits.

The following chart gives the crystallographic data of the starting material, and of Syndite.

Material	*Plane hkl*	B_0 *[deg.]*	*L* [μm]
Synthetic diamond	111	0·13	—
	220	0·15	—
	311	0·18	0·5
Natural diamond	111	0·15	—
	220	0·18	—
	311	0·18	0·5
Syndite	111	0·18	0·088
	220	0·35	0·050
	311	0·54	0·037
	222	0·54	0·046

$$L = \frac{0 \cdot 9\,\lambda}{B_0 \cos\theta_B}$$

L = crystallite size
θ_B = Bragg angle
λ = wavelength of the radiation
B_0 = half-peak width of x-ray beam

Substitutes

The same technique used in diamond synthesis has been applied to the synthesis of cubic boron nitride. Cubic boron nitride is produced by De Beers Industrial Diamond Division under the trade name Amber Boron Nitride and by General Electric under the trade name Borazon. Today cubic boron nitride is available in several grades with varying physical properties which, when correctly applied, make it an excellent abrasive for grinding hardened and difficult-to-grind steels.

The chart below compares some properties

Property	*Amber Boron Nitride*	*Diamond*	Al_2O_3	*SiC*
Indentation Hardness (Knoop) (kg/mm^2)	4500	8000	2100	2500
Reactive (at high temperature) to	Water	Oxygen	Unreactive	Iron
Density (g/cm^3) Unclad	3.45	3.52	3.7/4.0	3.2
Density (g/cm^3) Clad	5.2±0.1	—	—	—

of Amber Boron Nitride with those of other abrasive materials, using average values.

With its excellent thermal stability, Amber Boron Nitride may be heated in air up to 1,400°C (2,552°F) before its cubic structure reverts to hexagonal boron nitride. However, cubic boron nitride can react with water at temperatures in excess of 800°C (1,472°F), which can be reached easily at the point of cutting. Care should be taken, therefore, in wet grinding operations where water is used as a coolant. Reducing the water content in the grinding fluid can substantially increase the life of the grinding wheel.

Polycrystalline CBN cutting tools are manufactured by De Beers Industrial Diamond Division under the trade name Amborite, by General Electric under the trade name BZN Compax, and by Sumitomo of Japan. The chart on the following page lists the properties of Amborite. Results were extracted from non-standard tests described in "AMBORITE—A Remarkable New De Beers Cutting Material," *Industrial Diamond Review,* June 1980.

The CBN polycrystalline cutting tools are finding increasing application for machining hardened steels (+45 Rc), high nickel content alloys (*aerospace alloys*), and difficult-to-machine ferrous materials.

Commercial synthetic rutile is marketed under the name *titania.* In appearance, titania passes as an imitation diamond, being transparent and possessing great brilliance and fire. However, it frequently has an objectionable yellow tinge, and its hardness is less than that of quartz, so that faceted stones are subject to wear and breakage when worn in jewelry.

A relatively new imitation diamond is yttrium aluminum garnet (YAG). YAG was advertized as being only 1.5 less hard than diamond (8.5 compared to 10 on the Mohs' hardness scale). However, tests by De Beers show that YAG is only one-tenth as hard as diamond by impact tests and abrades 3,000 times more readily (Caveney, 1976). YAG is produced by mixing ingredients in a platinum crucible. The "seed" thus formed grows as it is rotated at a temperature of approximately 3500°F. The index of refraction is considerably less than in diamond. The dispersion is reasonably high but lower than that of diamond. The color, hardness, and freedom from impurities make the stone an inexpensive substitute for gem diamond.

Tool Material	*Knoop Hardness (GPa)*	*Transverse Rupture Strength (Gpa)*	*Ultimate Tensile Strength (GPa)*	*Young's Modulus (GPa)*	*Modulus of Rigidity (GPa)*	*Bulk Modulus (GPa)*	*Poisson's Ratio —*
Amborite	40·5	0·57	0·45	680	279	405	0·22
Tungsten Carbide K10	17·9	1·70	1·01	630	258	375	0·22
Aluminum Oxide	17·1	0·26	0·24	365	147	234	0·24

Stocks, Prices, and Production

Government Stockpile

Crushing bort, dust and powder, and industrial stones have been stockpiled by the US government since 1946. As of December 31, 1978, the government stockpile contained 26.0 million carats of crushing bort, dust and powder; 20.0 million carats of industrial stones; and 25,473 carats of dies and small pieces (Adams, 1979).

Industry Stocks

At the end of 1977 inventories of bort and grit amounted to 17.3 million carats and industrial stones added about 5.0 million carats (Adams, 1979).

Prices and Costs

Prices for industrial diamonds have decreased over the years (Adams, 1979). From 1954 to 1977 average prices for bort and powder decreased from \$3.14 to \$2.08 per carat. Average prices for stones for the same period decreased from \$23.13 to \$6.20 per carat. This decrease is the result of greater efficiency developed through large-scale production and use of improved techniques. The time-price relationships, showing actual prices and constant-dollar values, are shown in Table 5 (Adams, 1979).

A recent trend in industrial stones, however, has been price increase. The average value reached a minimum in 1970 of \$5.62 per carat. Since then the price has increased to \$7.78 in 1978 as a result of anticipated shortage in supply. This is caused in part because about 50% of stones previously classed as industrial can be converted to gem stones by laser treatment (Adams, 1979).

US and World Production

US production, consisting of synthetic bort, powder and dust, was 42.0 million carats in 1980. World production of diamond stones was 16.8 million carats. Natural diamond, bort, powder and dust added 14.3 million carats and synthetic bort and grit added 40.1 million carats excluding the US production. The world total for all synthetic and natural diamond production was about 113.0 million carats.

Government Control

Government Programs and Strategic Considerations

Industrial diamonds are included on the list of materials eligible for exploration assistance by the Office of Minerals Exploration (OME) of the US Geological Survey. The government will pay 50% of qualifying exploration costs.

Taxes and Tariffs

No special taxes are placed on the diamond-producing or diamond-products industries. US companies receive a 14% depletion allowance for both domestic and foreign production of natural diamonds. There is no import duty on either natural or synthetic industrial diamonds (synthetic dust and powder). A tariff rate of 7.5% ad valorem was in effect in 1977 for other synthetic diamonds. Rough natural diamonds can be imported duty free. However, as of January 1, 1971, cut diamonds even though unset, were charged duty of 4.5% if less than ½ carat, and 6% if over ½ carat. Statutory duty for both was 10%.

Secondary Sources and Recovery

The main sources of recovery for diamonds are the diamond grit and powders salvaged from industrial wastes by users of diamond grinding wheels, saws, and lapidary compounds. This amounts to about 3 million carats annually. The other major source is the substantial quantity of stones recovered from broken or worn diamond tools and drilling

TABLE 5—Time-price Relationships in Industrial Diamond, 1954-1978 (Average Annual Price, Dollars per Carat)

	Bort and Powder		Stones	
Year	Actual prices	Based on constant 1977 dollars	Actual prices	Based on constant 1977 dollars
1954	3.14	7.43	9.77	23.13
1955	3.18	7.37	8.11	18.80
1956	3.08	6.92	9.24	20.76
1957	3.30	7.17	7.50	16.30
1958	3.28	7.02	6.64	14.20
1959	3.42	7.16	7.51	15.72
1960	3.11	6.40	6.85	14.10
1961	3.05	6.22	7.39	15.07
1962	3.06	6.13	7.26	14.54
1963	2.97	5.86	7.05	13.92
1964	2.80	5.44	6.38	12.40
1965	2.79	5.31	7.00	13.31
1966	2.57	4.73	5.78	10.64
1967	2.38	4.26	6.13	10.96
1968	2.18	3.73	8.65	14.80
1969	2.18	3.55	6.20	10.10
1970	2.14	3.31	5.62	8.69
1971	2.23	3.28	5.76	8.48
1972	2.14	3.02	5.80	8.20
1973	2.10	2.81	5.90	7.88
1974	2.07	2.52	5.95	7.25
1975	2.10	2.33	6.54	7.27
1976	2.11	2.23	6.59	6.96
1977	2.08	2.08	6.20	6.20
1978	2.20	—	7.78	—

bits. These are returned to the manufacturer for reuse.

Ecology

The diamond industry, being based in countries outside the United States, has been faced with few problems regarding ecologic restraints. However, environmentalists are becoming more active in Africa in recent years.

Future of Diamond

In 1967, the machining of cast iron and steel was a process virtually untested. De Beers DXDA-MC, a metal-clad synthetic, was specially formulated to grind certain types of steel. Wheelmakers have reported that their customers have had excellent results in grinding D series tool steels, Ferro-Tic, cast iron, titanium, stainless steels, and combination steel and carbide dies.

Scientists at the De Beers Diamond Research Laboratory feel that research indicates quite convincingly that the characteristics of natural grit can be altered in ways that will permit production of synthetic materials of virtually any type—hard or friable, octahedrons or slivers, particles of varying sizes and qualities.

The indicated future of industrial diamond is indeed bright. Already by 1976, the annual world consumption exceeded the 100 million carats predicted as being conceivable by the end of the decade. It is believed that research will pave the way for an almost incredible growth.

Bibliography and References

Anon., undated, "Properties of Diamonds," De Beers Industrial Diamond Div., London, 4 pp.

Anon., undated, "Concentration of Diamonds," Denver Equipment Co., Flow Sheet Study, Bulletin No. M7-F74, 8 pp.

Anon., 1976, "Principles of the Mineral Resource

Classification of the US Bureau of Mines and the US Geological Survey," Bulletin 1450A, US Geological Survey, pp. A1–A5.

Anon., 1981, "Cominco American Will Begin Diamond Operation," *The Mining Record,* July 1, p. 4.

Adams, W.T., 1979, "The History and Use of Diamond," Methuen & Co., Ltd., London, p. 166.

Anders, E., and Lipschutz, M.E., 1966, Reply (to N.L. Carter and G.C. Kennedy's reply, 1966, to critique of "Origin of Diamonds in the Canyon Diablo and Novo Urei Meteorites"), *Journal of Geophysical Research,* Vol. 71, No. 2, pp. 673–674.

Bardet, M.G., 1973, *Geologie du Diamant,* Memoires du B.R.G.M., No. 83, 3 volumes.

Baskins, G.D., 1980, "Gem Stones," Preprint, 1978–1979, US Bureau of Mines, 13 pp.

Bauer, M., 1969, *Precious Stones,* Charles E. Tuttle Co., Rutland, VT and Tokyo, Japan, 636 pp.

Berman, R., 1962, "Graphite Diamond Equilibrium Boundary," *1st International Congress on Diamonds in Industry,* Ditchling Press, Sussex, England, pp. 291–295.

Berman, R., and Simon, F., 1955, "On the Graphite-Diamond Equilibrium," *Zeitschrift fuer Electrochemie,* Vol. 59, pp. 333–338.

Bovenkerk, H.P., et al., 1959, "Preparation of Diamonds," *Nature,* Vol. 184, pp. 1094–1098.

Brown, D.D., et al., 1967, "The Source of Alluvial Kimberlite Indicator Minerals in the James Bay Lowland," Ontario Dept. of Mines Misc. Paper, MP-7, 35 pp.

Brummer, J.J., 1978, *Diamonds in Canada,* CIM Bulletin, Oct., pp. 64–79.

Bruton, E., 1978, *Diamonds,* 2nd ed., Chilton Book Co., Radnor, PA, p. 532.

Bundy, F.P., 1962, "Direct Conversion of Graphite to Diamond in Static Pressure Apparatus," *Science,* Vol. 137, pp. 1057–1058.

Bundy, F.P., et al., 1955, "Man-Made Diamonds," *Nature,* Vol. 176, pp. 51–55.

Carter, N.L., and Kennedy, G.C., 1966, "Origin of Diamonds in the Canyon Diablo and Novo Urei Meteorites—A reply" (to discussion of 1964 paper by E. Andres and M. E. Lipschutz), *Journal of Geophysical Research,* Vol. 71, No. 2, pp. 663–672.

Caveney, R.J., 1976, "The Hardness of Diamond and Gem Diamond Substitutes," *Industrial Diamond Review,* Feb., pp. 45–47.

Chandler, H.P., 1964, "Industrial Diamond—A Materials Survey," Information Circular 8200, US Bureau of Mines, 150 pp.

Clarke, R.G., 1975, "Diamond-Industrial," *Mineral Facts and Problems,* Bulletin 667, US Bureau of Mines, pp. 325–338.

Collins, R.S., 1972, "Diamonds, Mining, Annual Review," *Mining Journal,* London, pp. 121–126.

Cooper, J.D., 1970, "Diamond-Industrial," *Mineral Facts and Problems,* Bulletin 650, US Bureau of Mines, pp. 953–965.

Davidson, C.F., 1964, "On Diamantiferous Diatremes," *Economic Geology,* Vol. 59, No. 7, pp. 1368–1380.

Dawson, J.B., 1960, "Comparative Study of the Geology and Petrology of the Kimberlites of the Basuto Province," PhD. Thesis, Dept. of Geology, University of Leeds, England.

Dawson, J.B., 1962, "Basutoland Kimberlites," *Geological Society of America Bulletin,* Vol. 73, pp. 545–560.

Dawson, J.B., 1964, "An Aid to Prospecting for Kimberlites," *Economic Geology,* Vol. 59, No. 7, pp. 1385–1386.

Dawson, J.B., 1968, "Recent Researches on Kimberlite and Diamond Geology," *Economic Geology,* Vol. 63, No. 5, pp. 504–511.

De Angelis, W.M., 1974, "Diamonds in the Glass Industry," *Proceedings: Diamond: A Partner in Productivity,* Symposium, Nov. 11–12, Industrial Diamond Assn. of America, Washington, DC.

De Carli, P.S., and Jamieson, J.S., 1961, "Formation of Diamond By Explosive Shock," *Science,* Vol. 133, pp. 1821–1822.

Dickinson, S.K., Jr., 1966, "The Chemical Erosion of Diamond By Boron—An Analogy of Graphitization (abstract)," *American Mineralogist,* Vol. 51, Nos. 1–2, p. 262.

Fernandez, L.L., 1968, "Recent Developments in the Diamond Mining Industry in British Guiana," *Transactions,* 4th Caribbean Geologic Conference, 1968, pp. 415–417.

Fisher, P.J., 1965, *The Science of Gems,* Charles Scribner & Sons, New York, 1966, p. 189; originally published in England as *Jewels,* 1965.

Forner, H., 1978, "The Precision Cutting of Hard, Brittle Materials Using Diamond Sawblades," *Industrial Diamond Review,* Apr., pp. 136–141.

Frank, F.C., 1969, "Diamonds and Deep Fluids in the Upper Mantle," *The Application of Modern Physics to the Earth and Planetary Interiors,* S.K. Runcorn, ed., Wiley-Interscience, London and New York, pp. 247–250.

Garlick, H.J., 1979, "Australian Diamond Prospects—The Story So Far," *Industrial Minerals,* No. 137, Feb., pp. 17–26.

Gerryts, E., 1970, "Diamond Prospecting by Geophysical Methods; A Review of Current Practice," *Mining and Groundwater Geophysics/1967,* Economic Geology Report No. 26, Canadian Geological Survey, pp. 439–446.

Gleason, J.J., 1974, "The Evaluation of Diamond Uses in Industry: Past, Present, and Future," *Proceedings: Diamond-Partner in Productivity,* pp. 153–157.

Gold, D.P., 1968, "Natural and Synthetic Diamonds and the North American Outlook," *Earth and Mineral Science,* Vol. 37, pp. 37–43.

Gregory, P., and Tooms, J.S., 1969, "Geochemical Prospecting for Kimberlites," International Geochemical Exploration Symposium, *Colorado School of Mines Quarterly,* Vol. 64, No. 1, pp. 265–305.

Gunn, C.B., 1967, "The Origin of Diamonds in Drift of the North Central United States—A Discussion (of paper by H.P. Schwarez, 1965)," *Journal of Geology,* Vol. 75, No. 2, pp. 232, 233.

Gunn, C.B., 1968, "Provenance of Diamonds in the Glacial Drift of the Great Lakes Region, North America (abstract)," *Bulletin,* Canadian Petroleum Geology, Vol. 16, No. 3, p. 418.

Hausel, W.D., et al., 1979, "Exploration for Diamond-Bearing Kimberlite in Colorado and Wyoming: An Evaluation of Exploration Techniques," Report of Investigation No. 19, Geological Survey of Wyoming, 29 pp.

Hausel, W.D., et al., 1979a, "Update on Exploration for Diamonds in Colorado-Wyoming Kimberlite Province," *AAPG Bulletin,* Vol. 63, No. 5, p. 830.

Hausel, W.D., et al., 1979b, "Preliminary Report on Exploration for Diamondiferous Kimberlites, Colorado-Wyoming," *Mining Yearbook,* Colorado Mining Association, pp. 109–120.

Hausel, W.D., et al., 1979c, "Prospecting for Kimberlites in Wyoming Using Heavy Mineral Alluvial Sampling Methods," Open File Report 79–6, Geological Survey of Wyoming, 13 pp.

Hausel, W.D., et al., 1980, "Exploration for Diamond-Bearing Kimberlite in Colorado and Wyoming—An Update," Preprint No. 80–310, SME-AIME Fall Meeting, Minneapolis, 4 pp.

Herbert, S., 1976, "A New Production Plant Gets into Its Stride," *Industrial Diamond Review,* Sept., pp. 315–318.

Herbert, S., 1976a, "Superstrength Diamond Solves Small-Hole Drilling Problem," *Industrial Diamond Review,* May, pp. 170–172.

Herbert, S., 1977, "Best of Both Worlds," (Diamond Abrasive as Carbide Grinding Medium), *Industrial Diamond Review,* Dec., pp. 422–424.

Hill, B.S., 1975, "The Performance of Processed Drilling in Hard Rock," *Industrial Diamond Review,* Aug., pp. 282–288.

Hoppe, R., 1980, "Diamonds from the Kalahari," *Engineering and Mining Journal,* Vol. 181, No. 5, May, pp. 64–82.

Hunt, G., 1975, "Diamond Prospecting and Mantle Xenoliths of the Kimberlite Pipes of Elliott Co., Kentucky," GSA Abstracts with Programs, Vol. 7, No. 4, p. 503.

Hurlbut, P.K., 1967, "Diamonds (abstract)," *Guidebook of Defiance-Zuni-Mt. Taylor Region, Arizona and New Mexico,* New Mexico Bureau of Mines and Mineral Resources, pp. 226–227.

Ilupin, I.P., 1977, "The Mineral Composition of Kimberlites and the Genesis from Depth in Kimberlites," *Geologiya i Geofizika,* Vol. 1, No. 12, pp. 100–105.

Jones, R., 1979, "Diamonds," *Rock & Gem,* Vol. 9, No. 1, pp. 47–71 and No. 2, pp. 48–93.

Kennedy, G.C., and Nordlie, B.E., 1968, "The Genesis of Diamond Deposits," *Economic Geology,* Vol. 63, No. 5, pp. 495–503.

Kopecky, L., 1960, "Diamond Prospects in Czech Massif," *Izvest.* Akad. Nauk SSSR, Ser. Geol., Vol. 12, pp. 46–55.

Lampietti, F.M.J., and Sutherland, D., 1978, "Prospecting for Diamonds, Some Current Aspects," *Mining Magazine,* Aug., pp. 117–123.

Lang, A.R., 1970, "Diamond Accumulation in the Upper Mantle," *Nature,* Vol. 226, No. 5243, p. 345.

Lawrence, A.L., 1967, "Field Prospecting for Diamonds in Isolated Areas," *Bulletin,* Geological Society of Liberia, Vol. 2, pp. 1–3.

Legrand, J., et al., 1980, *Diamonds: Myth, Magic and Reality,* Crown Publishers, Inc., New York, p. 287. Comprehensive and authoritative, more than 420 excellent full-color illustrations.

Leighton, V.L., and McCallum, M.E., 1979, "Rapid Evaluation of Heavy Minerals in Stream Sediments of the Prairie Divide Area of Northern Colorado: A Tool for Kimberlite Exploration," Open File Report 79–761, US Geological Survey.

Leuria, B., 1966, "Diamond Prospecting in Lofa County," *Bulletin,* Geological Society of Liberia, Vol. 1, No. 1, pp. 27–35.

Leuria, B., 1967, "Notes on the Geology of the Lofa River South of Wea Sua, with a Comment on Diamond Dispersal and Recovery," *Bulletin,* Geological Society of Liberia, Vol. 2, pp. 4–8.

Leuria, B., and Stracke, K.J., 1966, "Diamonds and Their Occurrences in Liberia," *Bulletin,* Geological Society of Liberia, Vol. 1, No. 1, pp. 5–12.

Liander, H., 1955, "Artificial Diamonds," *ASEA Journal,* Vol. 28, p. 97.

Linari-Linholm, A.A., 1951, "Recovery of Diamonds by Electrostatic Separation," *Journal,* Chemical, Metallurgical, & Mining Society of South Africa, Vol. 52, No. 2, Aug., pp. 33–35.

Loewenstein, W., 1965, "Glaciation and the Origin and Distribution of Diamond," *Lapidary Journal,* Vol. 18, pp. 1204–1213.

Macnae, J.C., 1979, "Kimberlite and Exploration Geophysics," *Geophysics,* Vol. 44, No. 8, pp. 1395–1416.

Mathewson, W.F., et al., 1966, "An Analysis of the Coated Diamond Bond System," *Grinding and Finishing,* Vol. 12, No. 12, Dec., pp. 18–25.

McCallum, M.E., 1974, "Infrared Detection of Kimberlitic Diatremes in Northern Colorado and Southern Wyoming," *Contr. to Geology,* Vol. 13, No. 1, University of Wyoming, pp. 16–17.

McCallum, M.E., 1979, "Geochemical Prospecting for Kimberlite in the Colorado-Wyoming State Line District (Abstract)," *Abstracts for 1979,* Vol. 11, No. 6, Geological Society of America, p. 279.

McCallum, M.E., and Smith, C.B., 1978, "Minor and Trace Element Contents of Kimberlites of the Front Range, Colorado and Wyoming," Open File Report 78–1011, US Geological Survey, 20 pp.

Meyer, H.O.A., 1968, "Mineral Inclusions in Diamonds," Yearbook 66, Carnegie Inst., Washington, pp. 446–450.

Meyer, H.O.A., and Harris, J.W., 1969, "Crystalline Inclusions in Diamonds from Ghana and Venezuela," *Symposium on Volcanoes and Their Roots,* Oxford University, p. 16.

Meyer, H.O.A., and Boyd, F.R., 1970, "Crystalline Inclusions in Natural Diamonds (abstract)," *International Symposium on Mechanical Processes in the Mantle,* US National Academy of Sciences, p. 37.

Moore, M., 1976, "A Review of Professor Tolansky's Research on Synthetic Diamonds," *Industrial Diamond Review,* Nov., pp. 395–398. (Lists Tolansky's papers.)

Motti, E., 1969, "Prospecting for Diamonds and Heavy Minerals in the Alluvium of the Arabian Shield," *Mineral Resources Research,* Saudi Arabia, Min. Petrol. Miner. Resour., Jiddah, pp. 68–69.

Murray, L.G., 1970, "Exploration and Sampling Methods Employed in Offshore Diamond Industry," *Mining and Petroleum Geology,* 9th Commonw. Mining Met. Congr., Vol. 2, pp. 71–94.

O'Donovan, K.H., 1975, "Synthetic Diamond," *Annals of CIRP,* Vol. 24, No. 1, pp. 265–270.

O'Donovan, K.H., 1976, "Synthetic Diamond," *The South African Mechanical Engineer,* Vol. 26, Aug., pp. 311–319.

Oostdam, B.L., 1970, "Exploration for Marine Placer Deposits of Diamonds (with French abstract)," *Mining and Groundwater Geophysics/1967,* Economic Geology Report No. 26, Canada Geological Survey, pp. 447–461.

Paterson, N.R., et al., 1977, "Geophysical Exploration for Kimberlites, with Special Reference to Lesotho," *Geophysics,* Vol. 42, No. 7, p. 1531.

Peele, R., ed., 1952, "Practice at De Beers Diamond Mines" (L.J. Parkinson, H.T. Dickinson), *Mining Engineers Handbook,* 6th ed., Vol. 1, John Wiley, New York, pp. 10–392—10–398.

Pustovalov, L., and Yu, I., 1966, "Diamonds of the Russian Platform," *Mining and Minerals Engineering* (London), Vol. 2, No. 3, March, pp. 113–115.

Ringwood, A.E., and Major, A., 1966, "Synthesis of Diamonds," *Australian Journal of Chemistry,* Vol. 19 (IO), p. 19.

Schwarez, H.P., 1965, "The Origin of Diamonds in Drift of the North Central United States," *Journal of Geology,* Vol. 73, No. 4, pp. 657–663.

Seal, M., 1978, "The Increasing Application of Diamond as an Optical Material and in the Electronics Industry," *Industrial Diamonds Review,* Apr., pp. 130–134.

Sharma, S.P., 1975, "Production of Diamonds from Graphite Using Explosive-Driven Implosions," UTIAS Technical Note No. 196, Institute for Aerospace Studies, University of Toronto.

Swersky, G., and Van Biljon, J.J., 1976, "The Influence of Machine and Toll Parameters on the Drilling of Concrete with Diamond Bits," *Industrial Diamond Review,* March, pp. 82–89.

Tolansky, S., 1962, *The History and Use of Diamond,* Methuen & Co., Ltd., London, p. 166.

Tolansky, S., and Komatsu, H., 1967, "Abundance of Type II Diamonds," *Science,* Vol. 157, No. 3793, pp. 1173–1175.

Weatherby, E.W., 1952, "The Principles of Diamond Sorting and Valuation," *Optima* (Johannesburg), Vol. 2, No. 2, June, pp. 5–9.

Weaver, H.B., 1964, "Development and Application of Manufactured Diamond," SME Preprint 64H61, AIME Annual Meeting, New York, 8 pp.

Wedlake, R.J., 1977, "Diamond Synthesis," *Industrial Diamond Review,* June, pp. 196–199.

Wentdorf, R.N., and Bovenkirk, H.P., 1961, "The Origin of Natural Diamonds," *Astrophysics Journal,* Vol. 134, pp. 995–1005.

Wilks, E.M., and Wilks, J., 1978, "The Mechanical Strength and Reliability of Natural Diamonds," *Diamond Research,* pp. 2–10.

Williams, A.F., 1932, *The Genesis of the Diamond,* Vol. 2, E. Benn Ltd., London.

Diatomite

FREDERIC L. KADEY, JR.*

Diatomite is a siliceous, sedimentary rock consisting principally of the fossilized skeletal remains of the diatom, a unicellular aquatic plant related to the algae. Thus, it has been formed by the induration of diatomaceous ooze, and consists mainly of diatomaceous silica, a form or variety of opal which is first formed in the cell walls of the living diatom. Diatomaceous silica is not generally regarded as a synonym or the equivalent for diatomite, although it has been so used at various times. Accurately, diatomaceous silica is the preferred name for the principal mineral component of which the rock, diatomite, is composed. The terms diatomaceous earth and kieselguhr are used as synonymous with diatomite. The designations tripoli, tripolite, infusorial earth, etc., were used at one time but are now obsolete. With the changing nomenclature, these terms that were at one time correct when proposed and used for generations would be considered incorrect if used today in the light of current knowledge. The designation diatomite is reserved for those accumulations of diatomaceous silica that are of sufficient quality, size, and minability to be considered of potential commercial value.

Processed diatomite possesses an unusual particulate structure and chemical stability that lends itself to applications not filled by any other form of silica. Foremost among these applications is its use as a filter aid, which accounts for over half of its current consumption. Its unique diatom structure, low bulk density, high absorptive capacity, high surface area, and relatively low abrasion are attributes responsible for its utility as a functional filler and as an extender in paint, paper, rubber, and in plastics; and as an anti-caking agent; thermal insulating material; catalyst carrier; and chromatographic support; polish, abrasive, and pesticide extender to name a few representative applications.

The United States is the principal producing country, although diatomite is found in numerous other locations.

Geology

Composition and Morphology

Diatomaceous silica qualifies as a mineral of organic origin in much the same way that aragonite and collophane do. The silica of the fossilized diatom skeleton closely resembles opal or hydrous silica in composition ($SiO_2 \cdot nH_2O$).[4, 12, 203] The silica is of acute biological significance, not only for the cell wall component, but also for the basic life process.[45, 51] Without silica, cell development ceases.[31] In addition to bound water, varying between 3.5 and 8%, the siliceous skeleton may also contain, in solid solution, or as part of the SiO_2 complex, small amounts of associated inorganic components—alumina, principally—and lesser amounts of iron,[46, 49, 55] alkaline earths, alkali metals, and other minor constituents.[59, 63, 67] Boron is reported to be an essential element for diatom growth.[52, 58] Since diatomaceous silica is not pure hydrous silica but contains other intimately associated elements, there is good reason to consider it a distinct type or variety.[4] Associated with the diatomaceous silica, and integrated as part of the diatomite, may be variable amounts of organic matter, soluble salts, and particles of rock-forming minerals that were syngenetically deposited or precipitated with the diatom frustules. Sand, clay, carbonate, and volcanic ash are typical common contaminants. Other contaminating minerals may be present, such as feldspar, mica, amphiboles, pyroxenes, rutile, zircon—the result of weathering, then transporting, and subsequent redeposition of surrounding land masses. Commercial diatomite may also contain fragments and particles of

* Exploration Manager, Manville International Corp., Denver, CO.

other such organisms as silico-flagellates, radiolaria and siliceous sponges.

In a commercial diatomite, silica makes up the bulk of the chemical composition; usually over 86% and as high as 94%. Alumina and iron generally are at least 1.5 and 0.2%, respectively. This includes not only that believed to be incorporated as part of the skeleton but iron and alumina associated with many of the contaminants. Lesser amounts of other elements, a small part of which may be secreted in the diatom skeleton, comprise the balance of the total chemical composition. The manner in which many of these elements are associated is not presently known. Table 1 illustrates the chemical composition of diatomites from various areas.[4] Although diatoms appear amorphous under the light microscope, X-ray studies show untreated diatomite to have a broad halo in the region of the principal cristobalite peak, thus it has been referred to as "micro-amorphous."[4] The main X-ray line is an approximation and not identical with α-cristobalite.[66] Some researchers have reported β-cristobalite to be prevalent.[71] The crystalline impurities produce their own X-ray lines; hence they furnish an identification of their nature, to a greater or lesser degree, depending on the amounts present. The ultimate hardness of the diatom skeleton is between 4½ and 5 on the Mohs' scale. After calcination or flux calcination, the Mohs' hardness is increased to 5½ to 6. The friability, or the propensity of the skeleton to break down, rather than to abrade, renders a measurement of hardness meaningless without also a consideration of the particle size.[172] The specific gravity ranges from 1.95 to 2.3. In calculating settling velocities, bulking values, etc., an apparent specific gravity of 2.0 for natural milled powders and of 2.3 for flux calcined powders is generally used.[137] Refractive index is variable between about 1.40 and 1.46 for natural earth, and increases to 1.49 for flux calcined diatomite.

Taxonomically, diatoms are divided into two broad categories: Centricae (discoid) and Pennatae (elongate to filiform). The study of the various intricate shapes and structural patterns of individual siliceous skeletons is as old as the use of the light microscope itself. Each form consists of two valves that are bound together by a connecting band or girdle. In the living diatom, these encase the cell contents.

Each siliceous valve is punctated by a system or pattern of openings that are arranged in a consistent and orderly design (Figs. 1–3). Furthermore, each valve appears to consist of an inner and an outer platelike surface, separated by ribs that result in a chambered interior. The structure of each surface is different in that the nature of the openings from each surface into the chamber is not necessarily the same. It is on the basis of the valve structure that diatoms are classified. The openings in the skeleton, classified by diatomists and divided into primary, secondary, and in some species, tertiary structures, are believed to simply support the membranc of thc living diatom through which the nutrients pass by the process of osmosis. The valves vary between approximately 5 and 1000μ in diameter, or maximum dimension, depending on the genus (Fig. 4). Most species fall within the range of 50 to 150μ. It is not within the scope of this chapter to dwell further on the botanical aspects of the diatom, although a few of the numerous references on the subject are included in the bibliography for the interested reader.

Suffice it to say that outside the realm of mining and commerce, the diatom has its own nomenclature and scientific entity, the study of which distinguishes it as a source of enjoyment for amateur and professional microscopists alike; and as a scientific tool in the fields of limnology, stratigraphic correlation, and other similar noncommercial applications.[3, 75, 82, 95, 194] Certain properties of diatomite—physical and chemical—may be visualized as primary or fundamental in nature. The nature and configuration of the skeletal structure, specific gravity, refractive index, hardness and friability, and composition are a few. These are the properties that determine secondary or derived properties which also endow diatomite with the attributes that set it apart from other sources of silica. It is an accurate generalization to state that the skeletal structure or configuration of the diatom is the principal primary property that controls most of the derived or secondary properties. Low bulk density, low wet or cake density, and high surface area may be visualized as examples of derived or secondary properties. The loose weight and wet density, for example, are a function of, and depend on, the skeletal structure and specific gravity. The particle size and shape on which bulk and wet cake density depend can be altered to a degree by milling; but ultimately, they are determined by such primary properties as structure, density, and friability. Certain properties

TABLE 1—Chemical Composition of Natural Diatomites from Various Localities (Oven-Dried Basis)

Constituent	Lompoc, CA	Maryland Calvert Formation	Nevada	Idaho	Kenya Soysambu	Japan Nilgata Earth	Russia Kamy-shlov Urals	Spain Alba-cete	Mexico Jalisco	Algeria (Primo Grade)
Silica (SiO_2), %	89.70	79.55	86.00	89.82	84.50	86.0	79.92	88.60	91.20	58.40
Alumina (Al_2O_3), %	3.72	8.18	5.27	1.82	3.06	5.8	6.58	0.62	3.20	1.66
Iron Oxide (Fe_2O_3), %	1.09	2.62	2.12	0.44	1.86	1.6	3.56	0.20	0.70	1.55
Titanium (TiO_2), %	0.10	0.70	0.21	0.07	0.17	0.22	0.48	0.05	0.16	0.10
Phosphate (P_2O_5), %	0.10	—	0.06	0.13	0.04	0.03	—	—	0.05	0.20
Lime (CaO), %	0.30	0.25	0.34	1.26	1.80	0.70	1.43	3.00	0.19	13.80
Magnesia (MgO), %	0.55	1.30	0.39	0.54	0.39	0.29	0.98	0.81	0.42	4.57
Sodium (Na_2O), %	0.31	1.31	0.24	1.03	1.19	0.48	0.65	0.50	0.13	0.96
Potassium (K_2O), %	0.41		0.29	0.22	0.91	0.53	0.72	0.39	0.24	0.50
Ignition loss, %	3.70	5.80	4.90	4.02	6.08	4.4	4.91	5.20	3.60	17.48*
Total	99.98	99.71	99.82	99.35	100.0	100.05	99.23	99.37	99.89	99.22

* Includes 13.9% CO_2.

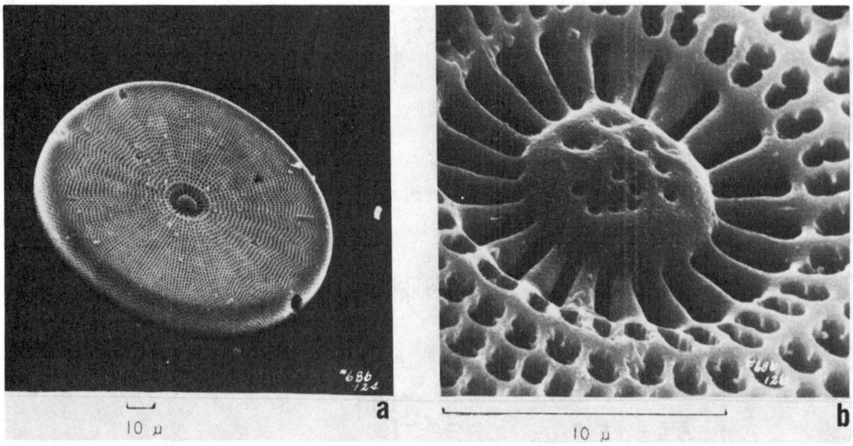

FIG. 1—*Scanning electron micrograph of Arachnoidiscus ornatus, Lompoc, CA.*

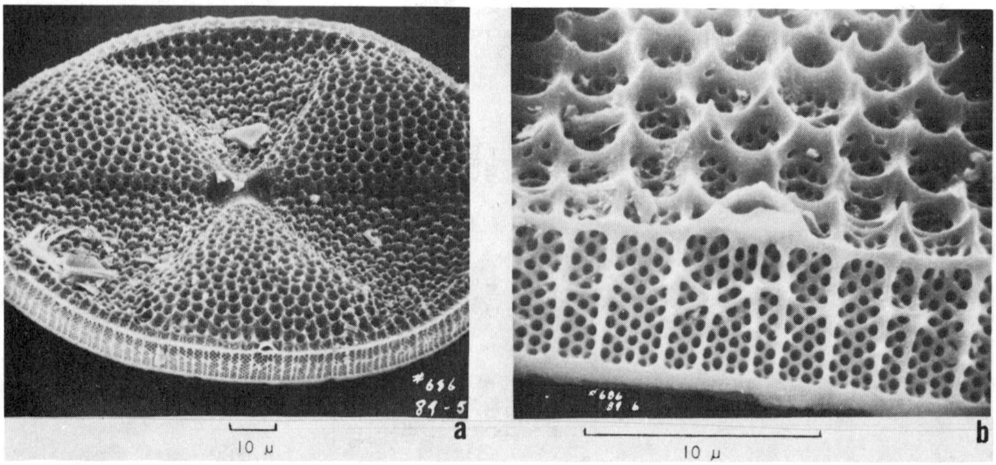

FIG. 2—*Scanning electron micrograph of Actinoptychus undulatus, Lompoc, CA.*

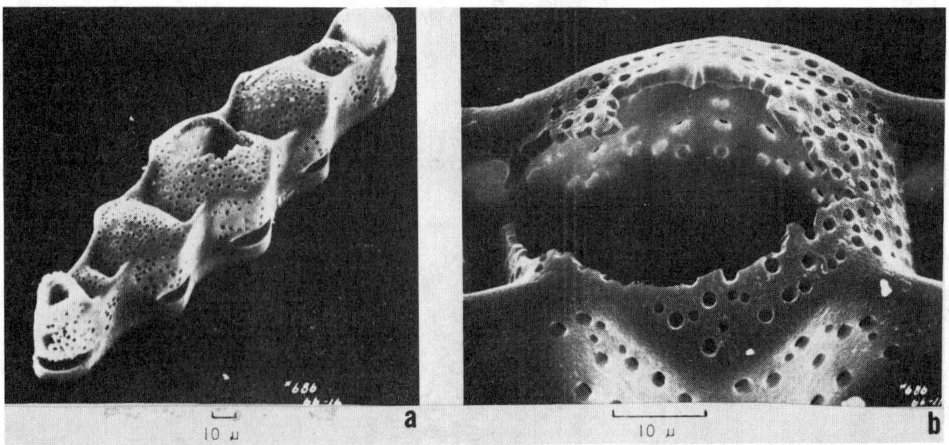

FIG. 3—*Scanning electron micrograph of Biddulphia Tuomeyi, Lompoc, CA.*

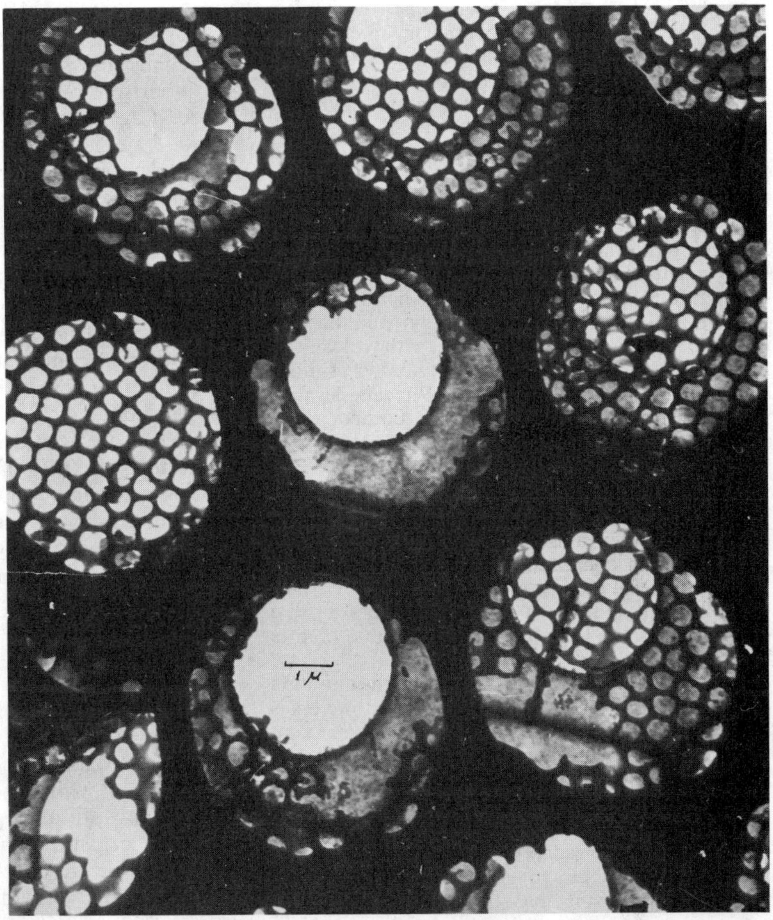

FIG. 4—*Electron micrograph of portion of Coscinodiscus radiatus valve. The primary structure includes the larger openings of approximately 3μ diam, occurring in one plate of the valve. The secondary structure is the meshlike openings of 0.5μ diam that cover the other plate. Vestiges of undissolved tertiary structure can be seen in a very few of the secondary openings.*

as pH, water solubility, and abrasiveness can be modified by extraneous material that was deposited syngenetically with the diatoms.

Mode of Occurrence and Origin

The frustule or siliceous skeleton of the diatom, a unicellular or noncellular microscopic algae, of the class Bacillariophyceae, and the order Bacillariaes, serves as the ultimate building block of which diatomite is composed. The group comprises over 300 genera and 12,000 to 16,000 species. In living form, accumulations of the diatom may be seen as the iridescent scum on ponds, the slippery gelatinous film on seaweed, on the bellies of certain species of whales, and other such varied habitats as oceanic ice floes, hot springs, moist soil, and particularly as masses of planktonic colonies on the open sea. Their natural function appears to be that of a food for other organisms of the sea. Furthermore, their role in controlling the geochemical balance of silica in marine and in lacustrine waters is scientifically recognized.[36, 37, 48, 68, 90]

Environmental conditions for growth include at least the five following major requirements:

1) Large shallow basins (preferrably 35 m or less in depth) for deposition, so that photosynthesis can occur. With regard to lacustrine deposits, a shallow lake provides sufficient actinic light for photosynthesis for not only pelagic diatoms, but also for benthonic forms attached to stones and to plants on the lake bottom. In the case of thick deposits of marine diatomites, there is evidence for a down warping of the basin of deposition, thus maintaining fairly shallow water for benthonic species. The open sea is reportedly the best environment for pelagic diatoms.

2) An abundant supply of soluble silica.[31, 36, 37, 45, 51] There is a worldwide correlation between the existence of thick diatomite deposits and proximity to volcanic ash occurrences.[203] While volcanic ash does not necessarily have to

accompany diatom deposition, some mechanism for increasing the silica content in marine and lacustrine bodies beyond the present day norm is necessary for the formation of commercially thick deposits. There are numerous examples wherein marine and nonmarine deposits meet this condition. Typical of these are at Lake Myvatn, Iceland; in the state of Jalisco, Mexico, where the deposits border the ancient shores of Lake Atotanilco; and those occurrences bordering Lake Rotorura, New Zealand; in Nevada the late Miocene Virgin Valley beds in Humbolt County, and the early Pliocene Esmeralda formation in Nye and Esmeralda Counties; the Payette formation in Idaho and Eastern Oregon. These all are associated with volcanism. Deposits of marine diatomite exhibit a similar correlation; the late Miocene-early Pliocene Sisquoc formation which comprises the Lompoc, CA diatomite; the middle and late Miocene Monterey formation of the Coast Ranges; and the middle Miocene Temblor formation east of Coalinga, CA, are examples of diatomites associated with contemporary volcanism. Particles of volcanic ash are a common contaminant of some diatomites.

3) An abundant supply of nutrients. In most lakes that are nontoxic to diatom proliferation the supply of nutrients is often more available than is the supply of silica.

4) The absence of toxic or growth-inhibiting constituents in the water. Although few lakes contain toxic water in the usual sense, many in which the rate of evaporation exceeds inflow during long periods of the year build up concentrations of soluble salts to the point of inhibiting diatom growth.[47, 60, 191]

5) A minimum supply of clastic sedimentary materials. While this, per se, is not a requirement for diatom growth, low nondiatomaceous contamination is paramount for the development of a commercially suitable deposit.

The effect of temperature, light, pressure, and other factors on diatom growth has been discussed.[34, 50, 64, 195, 204, 206] In addition to its unique siliceous skeleton, the living form has a nucleus, it produces certain protoplasmic substances by the process of photosynthesis, and incidental to its metabolism, it manufactures oil and vitamins. The rate of reproduction of diatoms varies with the species from between two or three times a day to once a week; and one diatom may have 100 million descendents in 30 days.[199, 200, 209] Deposition of the skeletal remains occurs after it has served its natural function. Thus, given the right conditions of environment and geologic location, tremendously thick deposits of diatomaceous ooze may build up on the floor of the containing body of water.

Potassium-argon dating of volcanic minerals and glass in North Pacific sediments has established Tertiary sedimentation rates of from less than 1 mm per 1000 years for deep sea red clay, to 1 cm per 1000 years for calcareous-siliceous ooze nearer the continent.[51, 80] In comparison, the rhythmic banding seen in Lompoc, CA diatomites suggests a rate considerably faster than that—probably of the order of 1 mm or more per year. Gross has calculated a sedimentation rate of 4 mm per year for a 25% diatom-75% silt sediment, deposited in Sannich Inlet, B.C.[105]

After deposition, such subsequent geologic forces as consolidation, burial under what will later be overburden, regional uplift, and partial erosion come into play to expose, yet protect, the deposit for later discovery and exploitation.

Because of the delicate nature of the diatom skeleton, deposits of diatomite to be useful to industry cannot undergo any great degree of regional metamorphism or chemical alteration. For this reason, geologic conditions that have not resulted in an appreciable degree of consolidation or of cementation are preferable. When orogenic forces are excessive, the resulting metamorphosis produces opaline cherts, porcelanites, and similar more indurated materials of noncommercial interest.

In place, diatomite is soft and "punky," and has a chalklike appearance. Color may vary from snow white in a pure, well bleached and dry deposit, to olive green or darker where substantial organic remains are still present and where moisture content is high. It may exhibit stratification, caused by either, or both, sedimentation of particularly flat beds or a preponderance of discoid diatoms, or by seasonally rhythmic deposition of clay and other impurities. On the other hand, it may be massive and show no stratification. It may be so loosely consolidated that when handled, a field sample will readily break down to a powder, or it may be hard enough to crack "brittley" when struck with a hammer. In addition to induration through consolidation, precipitation of carbonate for example, or "baking" by volcanic flows can destroy an otherwise good deposit. The better quality diatomite is lightweight, usually possessing a block density between 20 and 34 lb per cu ft.

Classification of Deposits

The various species of diatom thrive in either a marine or a lacustrine environment. Some forms live in brackish waters. Identification of the diatoms from an unknown deposit label it as having been laid down in either one environment or the other. It is, therefore, to be expected that a major criterion of deposit classification deals with whether it is of marine or of freshwater origin. Some investigators have added to the environments just mentioned, whether deposits are of modern lake, marsh, or bog origin.[5] These criteria are important because the diatom assemblages associated with marine environments are quite different from those that live only in freshwater habitats. The association of forms or the diatom assemblage, as seen by means of the microscope, not only serves to differentiate marine origin from freshwater, but also in many cases, to identify the deposit location from which an unknown sample may have come (Figs. 5–9). Diatom assemblages, like fingerprints, are specific to individual locations. Because of the structural differences related to origin, as has been pointed out, diatomites have a range of properties and produce a range of effects in the numerous uses to which they have been applied. Numerically, most of the known deposits in the world are of lacustrine origin. Generally however, those of marine origin, although less numerous, tend to be larger.

Distribution of Deposits

The occurrence of diatomaceous silica is widespread throughout the world. Although algae appeared quite early in geologic history, commercial deposits are generally restricted to sedimentary formations of Tertiary and of later age, and further limited ecologically by those conditions for formation that have been previously described. However, when one considers the numerous other limiting factors that must be taken into account before an occurrence qualifies as a commercial deposit—quality, minability, location, and size—then the numbers are few indeed. While a good portion of the California coast could be considered "diatomaceous in character," an area of hardly more than four square miles contains diatomite of high quality, commercial value.

North America

United States: Occurrences of diatomite have been reported from just about all of the east and the west coastal states, and indeed, from

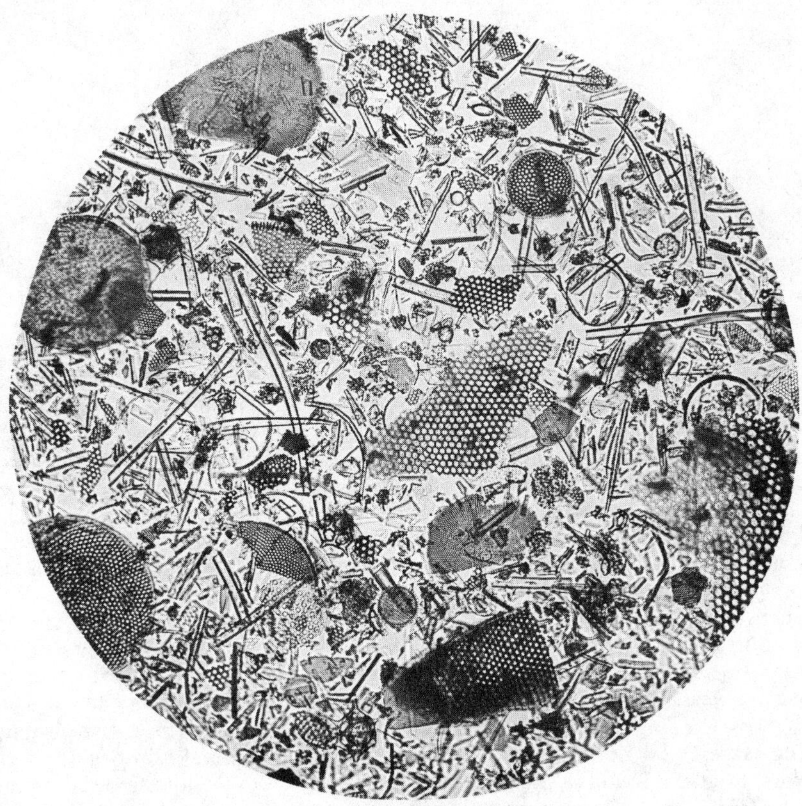

FIG. 5—*Micrograph of marine diatom assemblage from Lompoc, CA.*

100 μ

FIG. 6—*Micrograph of lacustrine diatom assemblage from Carlin, NV.*

many bordering these states in the interior of the country. Commercial production, however, has been limited to a few of these. First American production was from Maryland, where from 1884 until 1930, marine diatomite was extracted from the Fairhaven member of the middle Miocene Calvert formation.[129, 215] This diatomite outcrops along the banks of the Patuxent, Rappahannock, and Potomac rivers; and in cliffs along Chesapeake Bay. It is contaminated with varying amounts of loosely held silica sand, most of which can be removed in processing. More intimately held montmorillonite, illite, and kaolinite are also present. Removal of the clay on an experimental basis has been attempted with little practical success, for other than low quality applications.[129] Today, the remains of a diatomite enterprise of bygone years can be seen at such places as Kaylors Landing, MD, lying idly and in ruins. The largest and most uniform deposits in the world are to be found in the vicinity of Lompoc, CA (Fig. 10). The diatomite sequence of commercial significance is of the order of 1000 ft thick and is part of a thicker diatomite series of marine origin belonging to the Sisquoc formation of late Miocene or possibly early Pliocene age. Although the Lompoc diatomite may have been known since the time of the Spanish Conquistadors in the 1760s, it was not recognized as such until over a century later, and first mining was not started until about 1890.[2] The strata at Lompoc, which present a good example of rhythmic bedding, are mined by the Filtration and Minerals Div. of Manville International Corp. principally from a broad pitching syncline with related smaller anticlines and synclines. Quarries located at strategic places on the flanks and nose of the major syncline, and related structures, permit extraction of crude from various parts of the stratigraphic column, thus allowing for blending of crude types to fit the product demand. The Dicalite Div. of Grefco, Inc. also produces from an area in the general Lompoc district.

Diatomite was mined by the Dicalite Co. from a deposit in the Palos Verdes Hills near Los Angeles starting in the 1930s, but that de-

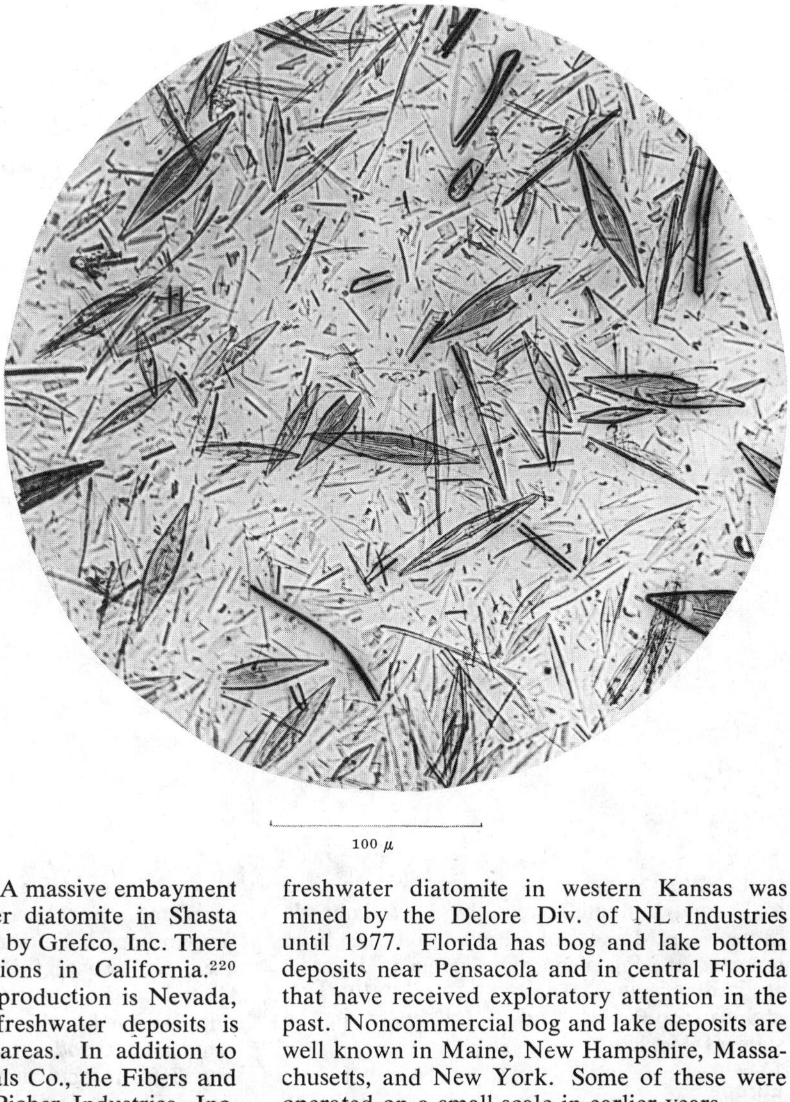

FIG. 7—*Micrograph of bog diatom assemblage from Pernambuco, Brazil.*

posit is now depleted.[230] A massive embayment of commercial freshwater diatomite in Shasta County has been outlined by Grefco, Inc. There are other minor operations in California.[220] Second to California in production is Nevada, where diatomite from freshwater deposits is mined in two principal areas. In addition to Cyprus Industrial Minerals Co., the Fibers and Minerals Div. of Eagle-Picher Industries, Inc. mine diatomite from deposits near Lovelock and at Clark, near Reno. These are of late Tertiary or Pleistocene age. The beds at Clark are close to the surface and are relatively low in moisture. Diatomite is also mined by the Dicalite Div. of Grefco from a freshwater deposit near Basalt, about 60 miles from Tonopah in southwestern Nevada. In the state of Washington, diatomite is produced from a lacustrine deposit near Quincy by the Kenite Div. of Witco Chemicals Corp.[7] At one time, there was production from a deposit at Kittitas, WA. Minor production has been reported from Oregon [115] and from Arizona.[116] Other states that have produced diatomite in the past include Idaho and Utah. A mixture of carbonate and freshwater diatomite in western Kansas was mined by the Delore Div. of NL Industries until 1977. Florida has bog and lake bottom deposits near Pensacola and in central Florida that have received exploratory attention in the past. Noncommercial bog and lake deposits are well known in Maine, New Hampshire, Massachusetts, and New York. Some of these were operated on a small scale in earlier years.

Canada: Production in Canada at present is limited to freshwater Miocene deposits near Quesnel, B.C., where Crownite Diatoms Ltd. and Pacific Diatomite Ltd. produce a fertilizer grade product. Eastern Canada has well-known occurrences in Nova Scotia and New Brunswick that are not profitable to exploit at the present time.

Mexico: Diatomite occurs in several states in Mexico, and among these are Tlaxcala, Colima, Jalisco, Michoacan, and Mexico. Some production has been known since 1927. Good quality earth is mined near Catarina in Jalisco (Fig. 11) by Diatomita San Nicolas, S.A. de C.V. Filter aids and fillers from this operation are shipped to Latin America, Europe, and Aus-

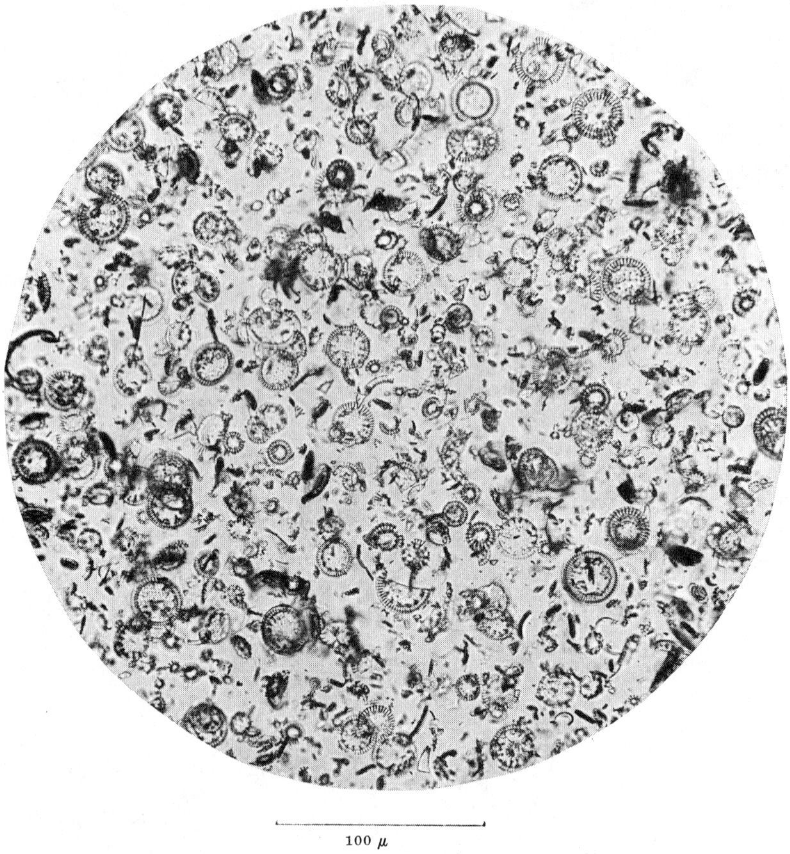

FIG. 8—*Micrograph of lacustrine diatom assemblage from Elche de la Sierra, Spain.*

tralia. Other commercial deposits are exploited at Zacapu and at Tuxpan by Kieselguhr de Mexico in Michoacan. Production has been reported at Magdalena and Cocula in Jalisco,[2] at La Blanca in Tlaxcala, at San Martin Texmelucan in Puebla, and at Ixtlahuaca in the state of Mexico.

Europe

The most significant European sources of diatomite are freshwater Tertiary and Quaternary deposits located in the Massif Central area of southern France. Deposits at Collandres, and at Saint-Bauzile (Ardeche) in the Privas area, are worked by Soc. CECA (Carbonisation et Charbons Actifs). A deposit near Murat, mined by Manville de France, is processed into filter aids. The Luneburger-Heide deposits of Western Germany were the first commercially mined deposits in the world, but have declined in importance in recent years. Some production is, however, reported from Tagebau and from Unterluss by Kieselguhr Industrie GmbH. In Italy, commercial deposits are located at Arcidosso and Santa Fiora and mined by Winkelmann Mineraria S.p.A. at Castel del Piano, all near Monte Amiata. Diatomiti Italiane A.P.E.S.-S.p.A. operate further south in the Viterbo area, and Diatom S.p.A. mine a deposit at Castiglione in Teverina. Medium grade diatomite is mined in the area of Tombolia in northern Italy. Good quality Spanish diatomite is mined from lacustrine deposits between Hellin and Elche de la Sierra by Manville Española S.A. in the southeastern part of the country. Some of these deposits were mined from underground galleries for over 50 years, but in recent years have been converted to open pit operations. Some of the earth is a remarkably white color, but chert lenses and carbonate beds necessitate highly selective mining methods. There are other, low quality occurrences in Spain, including a marine occurrence near Almeria. In Iceland, a diatomaceous ooze of Holocene Age is dredged from Lake Myvatn and pumped in slurry to the processing plant by Kisilidjan, H.F., where the volcanic ash is removed by hydroclones. The resulting slurry, after pumping to settling ponds, is dewatered and dried with geothermal steam prior to flux

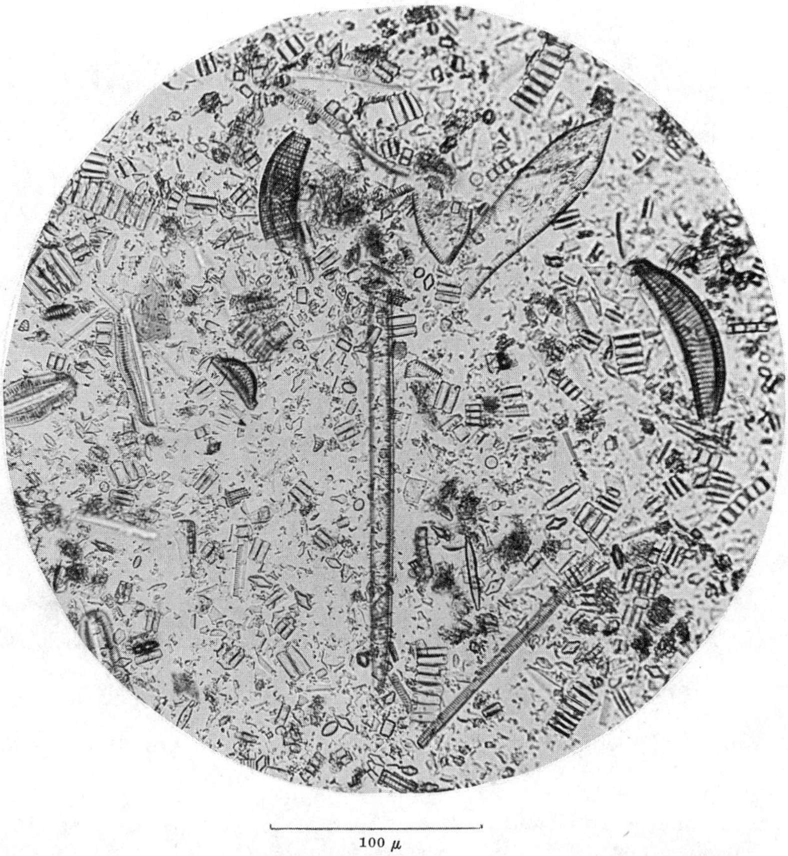

FIG. 9—*Micrograph of modern lake diatom assemblage from Myvatn, Iceland.*

calcining. The diatomite of Skye, Scotland, has been known for some time.[92] Some diatomite is mined by Cape Insulation Ltd., from a wet lake basin deposit near Kendal in Westmoreland, U.K. The earth is dark in color and is not suitable for filter aid manufacture. It is used mostly for thermal insulation.

An unusual, but commercially viable deposit, is the diatomite-clay mixture of Denmark called Moler. This earth is not suitable for filter aids because of the iron oxide and clay content, but over 200,000 tpy is produced and exported elsewhere in Europe for fertilizer coating and for insulating. Moler is of marine origin and of Tertiary age and occurs on the islands of Fur and Mors in northern Jutland. Some diatomite of good quality has been produced in recent years in southern Denmark.

There are numerous deposits in the USSR. Although little information is available, production is estimated to be of the order of 300,000 tpy. Freshwater diatomite deposits exist near Prilep, Jugoslavia, and in central Anatolia, Turkey. Romania reports 45,000 tpy production for fertilizer coating from a lacustrine deposit near Adamclisi. Small unproductive deposits occur in many other European countries.

Africa

Impure diatomite, suitable only for insulating materials, is mined by Marico Mineral Co. from a deposit in the Ermelco District, Transvaal, and by Witdraai Kieselguhr Co. from a deposit in the Prieska District, near Postmasburg, Cape Province.

East Africa produces both filter aid and filler grade products from deposits near Gilgil, Kenya. These deposits, mined by East African Diatomite Industries Ltd., were deposited in Pleistocene lakes in the Rift Valley. Several other smaller deposits are not of current economic interest.

Marine deposits of Miocene age occur near Sig (formerly St. Denis-du-Sig) in Algeria. These are mined by Sonarem, a government agency, from underground galleries and a large

FIG. 10—*Aerial view showing extent of Manville diatomite quarries at Lompoc, CA.*

portion of the production is shipped to France and Italy. There is a marked similarity in appearance between the Algerian and Lompoc diatomites when viewed under the microscope. The diatomite from Algeria however, is characterized by a noticeable amount of carbonate contamination, and is deficient in many of the diatom types that, in the Lompoc earth, provide a better balance for filtration. Numerous, intensely folded, thin beds occur near Mosteganum, Algeria.

South America

Production of the order of several thousand tons per year is mined from small, scattered deposits in Rio Negro Province in Argentina. Much of the earth occurs under a basalt cap, which necessitates mining from galleries and in one instance as an open pit operation after blasting away the basalt. There are also small impure occurrences in San Juan province near Calingasta, and in remote parts of Salta Prov-

FIG. 11—*View of one of the Diatomita San Nicolas quarries at Catarina, Jalisco, Mexico.*

ince. There are numerous bog deposits in Brazil. The states of Ceará and Bahiá contain the most deposits, although eight other states are reported to have smaller impure occurrences. Peru has occurrences at Pisco, Piurá, Chiclayo, and Arequipa; and Chile has deposits of high brightness earth at Arica and Chiloe. In Colombia, diatomite occurs at Tunja and Antioquia. Diatomite occurs to a lesser degree in other Latin American countries.

Asia

There is considerable potential for development of deposits in the Far East. Japan has a growing industry based on its own marine and lacustrine deposits operated by Showa Chemical Co. Deposits, possibly suitable for local markets, exist in Indonesia and in Korea. China has commercial deposits about which little is known. There are several small deposits in Australia and in New Zealand, but for filter aid quality, these countries depend on imports, principally from the US. The pressure from increasing transportation costs, however, is lending motivation to the exploration and possible development of shallow mediocre quality bog deposits in Western Australia near Perth and Geraldton.[228]

Elsewhere in the world are numerous other occurrences of diatomaceous silica which, for one reason or another, have not been exploited. Furthermore, accumulations of diatomaceous ooze are forming today that will constitute the deposits of "tomorrow."

Reserves

Present reserves are estimated to be adequate beyond the year 2000. Presently known deposits are stretching beyond forecasted depletion through the efforts of research and development in finding methods to process lower grade crude. Other deposits that are currently considered marginal by virtue of quality, minability, or accessibility will undoubtedly take on new importance should the present reserves begin to dwindle.

Exploration

Keeping in mind the previously described limiting criteria for formation, prospecting for diatomite entails reconnaisance of potentially suitable terrain. Because it is usually soft and easily eroded, white "showings" in stream banks or in road cuts should be investigated.

One would surmise that gravimetric surveying would be fruitful in the search for diatomite horizons in conjunction with higher density beds. So far, however, no successful operations of this kind have been reported. Geophysical refraction seismic surveys have successfully outlined the depth of shallow fresh water diatomite basins. This works particularly well where the soft (low velocity) diatomite is underlain by higher velocity basalt (C.M. Smith, Grefco, Inc., personal communication).

Geochemical methods so far have not been adapted to the prospecting for, nor exploration of, diatomite. However, narrow pass-band infrared imagery (3–4 and 4.5–5.5 micrometers) has been used to recognize diatomite from aircraft by its thermal characteristics.[79] Changes in vegetation have been noted over diatomite-bearing areas of bog deposits, and this might indicate the possible application of geobotanical techniques in prospecting.

When carried to the point of development, exploration of diatomite deposits is pursued in stages. After an occurrence has been recognized through prospecting, usually, the first step in exploration consists of a preliminary sampling of all visible outcrops. The nature of the material is noted and recorded by measuring the attitude of the beds, and by observing all other visible structural and stratigraphic features. Sampling intervals are divided into visible increments if bedding, textural, and color changes or stratification are evident. If thick enough and no visible divisions are apparent, the stratigraphic interval spanned by each sample should be no more than 150 cm for each channel cut. In this way, nonvisible characteristics—diatom assemblage, chemical changes, etc.—may be noticed and characterized during testing of the samples. In subsequent exploration stages, the 150-cm interval may be reduced, if required. Evidence of staining, degree of consolidation, judgment of color, bedding, and stratification, etc., are all field characteristics that are important, and which should be noted. Pending favorable results from the laboratory evaluation of the samples, the next stage of exploration is planned to delineate the reserves within the area and to further assess the quality. With horizontal beds under relatively thin overburden and in areas of gently rolling topography, the digging or augering of vertical exploratory shafts to expose the entire stratigraphic column has proven highly satisfactory. In countries with low labor rates and in locations where mechanical equipment is difficult to maintain, hand-dug shafts up to 50 m in depth have many advantages. When these are of the order of 1 to 1½ m in diam, ingenious "bird cage" types of sampling platforms, lowered by windlass into the hole from a tripod arrangement, have been used to support a geologist, who logs the hole and collects samples from the wall of the shaft. Whether hand-dug or machine-bored with a large auger, these openings in the deposit have the advantage that the structure and nature of the beds may be noted and correlated from hole to hole. Where overburden is minimal or nonexistent, holes dug by backhoe have been used. While these have the advantage of the speed of excavation that is provided by mechanical equipment, depth is limited to about 6 or 7 m at the most.

Core drilling in diatomite is specialized and requires special equipment and experienced drilling crews. Little, if any, good quality diatomite will produce satisfactory core with a diameter of less than 10 to 15 cm. Any formations that will, are usually too highly consolidated to be of much commercial value. A 10-cm core if possible to obtain, however, provides the amount of material that is required for testing. Where any amount of topographic relief is present, trenching by bulldozer on hillsides will expose the bedding and will permit subsequent channel sampling. The opening of trenches by bulldozer across the bedding of dipping strata will remove overburden and expose the strata for sampling.

The positioning of drill holes, shafts, or trenches to adequately cover a deposit, depends upon a number of factors. Sample positions are most commonly arranged systematically in a grid to cover the area to be explored. The distance between sample locations is dependent, among other things, on the lateral variation in important properties or characteristics of the diatomite. Where important properties are thought or known to change rapidly with lateral extent, holes must be placed more closely together than when a high degree of uniformity is experienced in preliminary study. (Sample holes, on as close as 30-m centers, have been used.)

The advantage of two and three-stage exploratory programs is that exploration can begin with a relatively economically wide spacing of sampling locations. Upon testing of the samples from such a stage, a judgment can be made whether, because of unfavorable findings, the program should be aborted, or if a subsequent program is needed in which the hole spacing is reduced. Subsequent stages of exploration consist of additional specialized sampling and of the selection of bulk samples for plant scale trials. Whereas the foregoing methods are suitable for dry deposits, specialized techniques must be developed for bog deposits and for those occurring under lakes or ponds.

In the case of shallow lakes, sample locations are marked with survey poles driven into the ooze. Peat bog samplers that extract a 30 to 60-cm incremental sample have been designed with long shafts so that a sample can be extracted from as much as 8 m below the surface. These may be used from a boat or raft that is floated into position and anchored. Some preliminary field testing can be incorporated into the exploratory phase of deposit evaluation. Field examination of samples by portable microscope can reveal considerable information. Other obvious field aids include an HCl test for carbonate, a grit test by grinding the crude between teeth, and the noting of appreciable water solubles by taste.

Evaluation of Deposits

A preliminary idea of quality can be gained from the foregoing field observations. The most obvious recognizable property is color. The higher the brightness of a diatomite, the more attractive its potential as a filler is likely to be. Another property that is evident in the field is block density. A low block density is indicative, among other things, of freedom from contaminating solids, such as sand and clay. Diatom type and degree of consolidation are also reflected in block density. A low degree of consolidation is desirable. Highly consolidated diatomites are difficult to mill and result in degradation of the skeletal structures. As previously mentioned, the most useful tool in the field is a portable microscope. When performed by an experienced operator, microscopic examination in the field can be used to ascertain diatom constitution and contaminants, to direct the course of exploration, to provide a stratigraphic correlation, and to eliminate the shipping of useless samples to the testing center.

While much useful information can be collected in the field, and can lead at that point to a firm recommendation by the geologist to *not* consider the deposit further, the ultimate judgment of quality, however, is formed from the results of usually extensive testing in the laboratory. The chemical analysis of a diatomite, while useful to some extent, is not an effective criterion in predicting the performance for most applications. It is to be expected of course that the lower the percent of nonsilica components, the better—to be sure their absence is essential in many cases. However, simply the indication of high SiO_2 content alone is not a sufficient sign of a suitable crude. Chemical purity, then, is an attribute that becomes more important only after other indications of suitability have been ascertained. The properties that are measured in the laboratory are designed to reflect the performance of the finished product in a specific, or similar grouping, of applications. The milling of the crude in the laboratory must, of course, simulate the treatment that it would receive in plant equipment. Such properties as microscopic constitution, loose weight, wet density, screen size, brightness, abrasion, water absorption, filtration flow rate and clarity, pH, and resistivity, and in specialized cases, porosity and acid soluble iron, calcium, and trace elemental content are just a representative few of the many characteristics that are of importance in various applications and which must be evaluated in the laboratory. Special applications generate the need and the design of additional special and often relatively expensive testing methods.

Preparation for Markets

Mining

By and large, the most economical and indeed the only method of extraction used in the United States is quarrying or open pit mining. In Europe, Africa, South America, and parts of Asia, commercial deposits are mined underground as well. In Iceland, the deposit at Lake Myvatn is dredged from beneath about 1 m of water and pumped in slurry form to the processing plant 2 km away.

In the larger open pit quarries of the western United States, nearly all use power equipment. Since blasting is not ordinarily required, only bulldozers, front-end loaders, and trucks are required to extract and transport the crude to the mill. In the Manville Celite® operations at Lompoc, bulldozer-rooter combinations loosen the consolidated strata which are loaded into diesel-powered bottom-dump trucks by Euclid belt loaders, by diesel and electric shovels, and by front-end loaders. The result is an endless procession of conveyance from the quarries to strategically located stockpiles over vertical storage shafts. These are connected to the processing plant through an underground transportation system, which permits considerabe flexibility in the blending of various crudes at the mill.

In smaller American operations, diesel-powered scrapers convey the earth to stockpiles where front-end loaders are used to fill the trucks for transportation to the mill. Where hard contaminants or thin strata are encountered, power shovel application may be required.

Although some operations outside the United States are carried on by large companies and use modern mining methods, diatomite mining in many parts of the world is still, for the most part, a small or family owned operation, often from underground tunnels and galleries, and of pick and shovel and wheelbarrow magnitude. Where the climate will allow, the ore is spread on the ground or on drying platforms, and the moisture reduced to less than 20%. Rotary dryers are often used rather than simultaneous milling-drying processes.

At the Manville Lompoc deposit, extensive sampling and testing precede the mining operation for quality control. Selective mining then consists of separating "critical horizons" which are sent to waste. In many smaller deposits, such as are prevalent elsewhere in the world, selective mining of narrow beds by hand to separate chert lenses and carbonate inclusions is widely used.

The attitude of the beds, their thickness, the distribution of intercalated impure horizons, and thickness of overburden all have a bearing on the method of mining that is best to use. Underground mining of horizontal beds consists simply of following each stratum as far into the deposit as is economical. Then recovery of part of the remaining approximate 44% that is required to support the roof is removed on the way out, if the mine is ever depleted to that extent.

Underground mining of inclined beds becomes increasingly difficult with increased dip. Depending on topography and on the arrangement of galleries, exploitation may not be economically feasible for any appreciable distance. Open pit quarrying is obviously preferable to underground mining because of lower mining costs and greater rates of recovery. When the beds are dipping gently, open pit mining offers considerable flexibility in that quarries can be located where beds of a particular quality outcrop, thus allowing for blending of several crude types. Horizontal beds in rolling topography offer similar advantages.

Milling and Processing

Because of the high moisture content of the crude and other processing losses, it is highly desirable to have the mill as near as possible to the mine. Under unusual conditions, however, crude diatomite has been trucked considerable

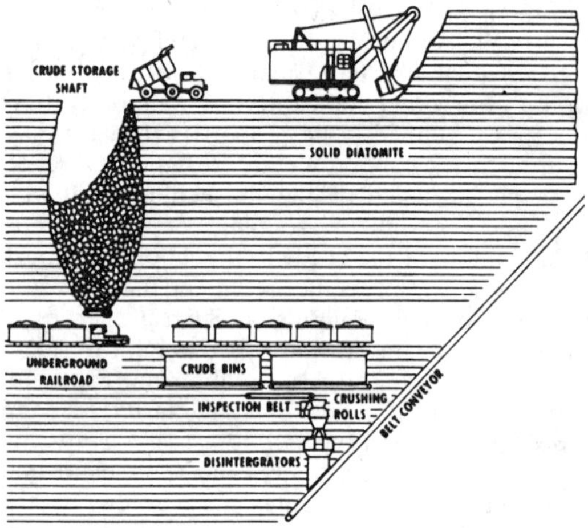

FIG. 12—*Schematic representation of the mining, transporting, processing, and packaging of Celite diatomite powders. Reprinted by special permission from* Chemical Engineering.

distances to a plant. The Manville Lompoc plant, for example, is adjacent to the quarries, whereas the Manville Española S.A. plant at Alicante, Spain, is situated over 100 miles from the source of crude.

Since the particulate shape and structure of the diatom skeleton is the physical property that most distinctly sets diatomite apart from other forms of silica, and for which its uniqueness is most responsible, great care is taken during milling and processing to preserve this structure. Such size reduction methods commonly used in the processing of other industrial minerals, as ball milling or grinding, would destroy the delicate structure and would render it useless for such applications as filtration or as a flatting agent in paint (Fig. 12).

Since crude diatomite commonly contains as much as 40% moisture, and in many cases over 60%, primary crushing to aggregate size is followed by a simultaneous milling-drying as the suspended particles of diatomite are carried in a stream of hot gases. Passage of the suspended particles in the hot gases through a series of fans, cyclones, separators, and a baghouse results in the separation of the powder into various sizes, in the removal of waste impurities, and in the expulsion of the sorbed water. Ratios of over three tons of crude in place, to one ton of filter aid product, are not unusual. Diatomite products, so processed without further treatment, are bagged or handled in bulk as "natural" milled products.

When the adjustment of particle size distribution is required for such applications as fast flow rate filter aids, the heating of the powder to incipient fusion in large rotary kilns, followed by further milling and classifying, results in straight calcined grades. Further adjustment of particle size is effected by the addition of a flux—usually soda ash—before the calcining step. The use of sodium chloride as a flux is still common outside the US, although in this country, its corrosive action is avoided. Such products are referred to as "flux calcined."

It should be pointed out that the term "calcination" is a misnomer when referring to the heat treatment of diatomite. The process is not, in the correct technical sense, one of calcination at all. Rather, it is the agglomeration of fines through incipient fusion or sintering—often with a flux. The incorrectly introduced term has persisted and has become entrenched in the particular nomenclature of the trade; and for historical reasons, its use is being continued here.

Simple calcining without a flux (straight calcining) results in a product with a pink cast. The color is caused by the oxidation of iron in the crude and becomes more intense with an increasing iron oxide content. Flux calcining produces a white product in good quality diatomite, in part believed to be caused by the conversion of the iron to complex sodium-aluminum-iron silicates, rather than to the oxide. A pinkish cast is often observed in flux calcined diatomite, if the amount of flux required for optimum filtration properties is insufficient to complex all of the available iron.

Calcining and flux calcining produce other changes in the diatom particle. Among these are the loss of the combined water that is part of the opaline structure, degradation of the tertiary and secondary structure of the diatom valve through incipient fusion, and conversion of portions of the otherwise amorphous silica to cristobalite (Figs. 13a-d).

Filter aid powders for special uses are pro-

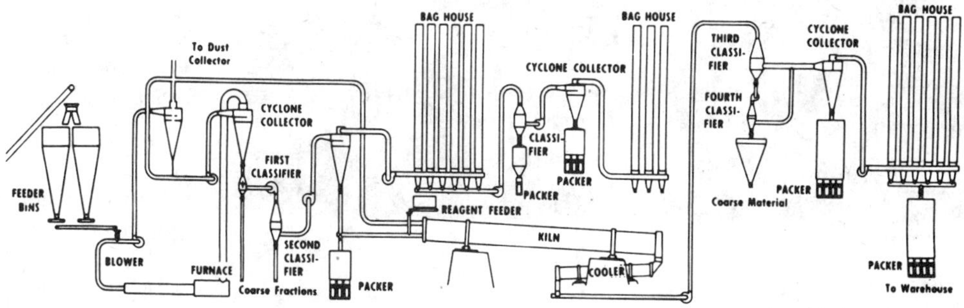

Chemical Engineering
Copyright McGraw-Hill Publishing Co

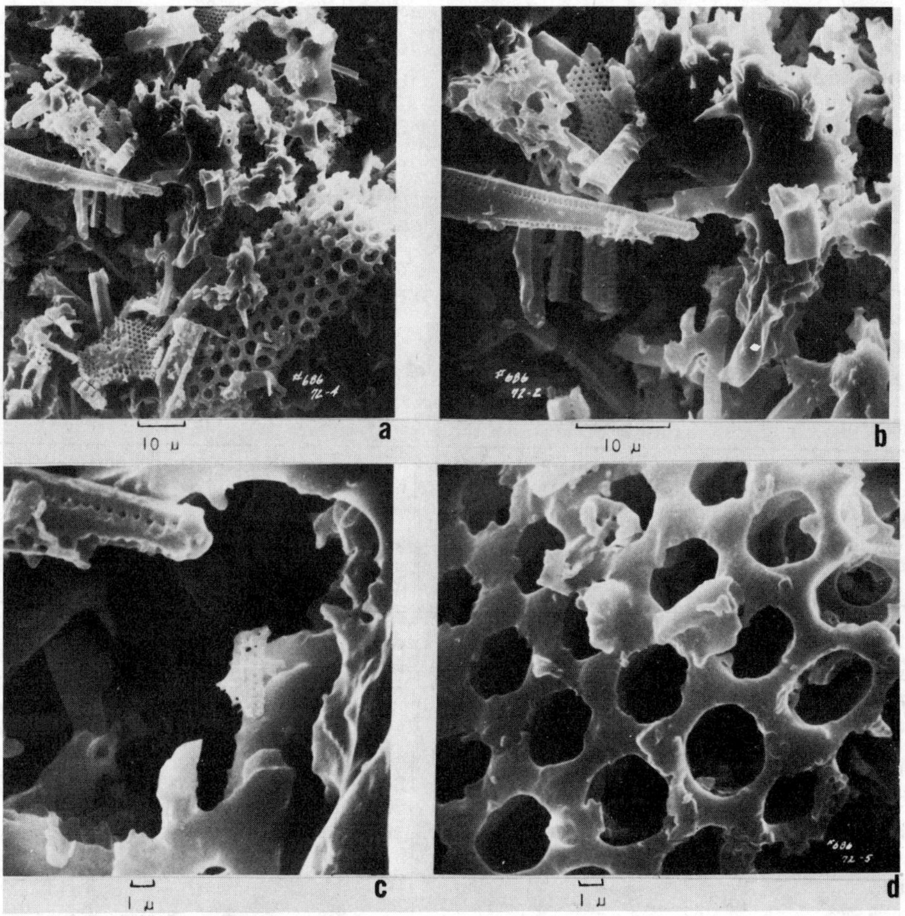

FIG. 13—*Scanning electron micrograph of flux-calcined Lompoc, CA, diatomite illustrating agglomeration of fine particles and incipient fusion of siliceous structure.*

FIG. 14—*An aluminum, pneumatic, dry bulk tanker on semi and full trailers for transporting diatomite powders. These are 1250 cu ft each (8–9 tons) in capacity. Each has its own low pressure air system for unloading to the customer's storage bin.*

duced by acid treatment of dried and milled material, in combination with conventional calcination and flux calcination. Specially prepared diatomite aggregates have also been produced for use as supports in gas liquid chromatography by special sizing to close tolerances, followed by acid treatment, and by special surface treatments with silanes or silicones to deactivate the support surface. The application of diatomite in brick and in extruded and aggregate forms has declined in recent years, hence the manufacture of product types in other than fine powders is less important. Special milling and classifying techniques have been employed in the control of particle size distribution for functional fillers. For the efficient control of gloss and sheen in paint, for example, grades of diatomite powders are available in which the nonfunctional fines, as well as the oversized particles, have been removed during processing.[164] This point is treated in more detail later.

While the greater portion of diatomite powders are packaged and shipped in 50-lb bags, or equivalent metric quantities, in recent years progress has been made in the pneumatic bulk handling of diatomite in conjunction with commercial shipments in bulk box cars, in bulk trucks, and in pressure differential special bulk compartmented cars [137, 164, 216] (Fig. 14).

Testing and Specifications

As is common with most industrial minerals, the testing procedures by which processed diatomite powders and aggregates are evaluated and standardized are designed to quantify an

TABLE 2—Trace Elemental Composition of a Typical Celite® Diatomite Product*

Element	Ppm	Element	Ppm	Element	Ppm	Element	Ppm
Antimony (Sb)	2	Gadolinium (Gd)	<1	Neodymium (Nd)	20	Tantalum (Ta)	20
Arsenic (As)	5	Gallium (Ga)	5	Nickel (Ni)	120	Tellurium (Te)	<2
Barium (Ba)	30	Germanium (Ge)	<10	Niobium (Nb)	5	Terbium (Tb)	<0.2
Beryllium (Be)	1	Gold (Au)	<0.5	Osmium (Os)	<0.5	Thallium (Tl)	<0.5
Bismuth (Bi)	<0.5	Hafnium (Hf)	<0.5	Palladium (Pd)	<1	Thorium (Th)	5
Boron (B)	100	Holmium (Ho)	<0.2	Platinum (Pt)	<2	Thulium (Tm)	0.2
Bromine (Br)	20	Indium (In)	<0.5	Praseodymium (Pr)	2	Tin (Sn)	<1
Cadmium (Cd)	2	Iodine (I)	1	Rhenium (Re)	<0.5	Tungsten (W)	<0.5
Cerium (Ce)	10	Iridium (Ir)	<0.5	Rhodium (Rh)	<0.5	Uranium (U)	5
Cesium (Cs)	5	Lanthanum (La)	10	Rubidium (Rb)	10	Vanadium (V)	200
Chlorine (Cl)	400	Lead (Pb)	2	Ruthenium (Ru)	<1	Ytterbium (Yb)	<0.5
Chromium (Cr)	100	Lithium (Li)	1	Samarium (Sm)	2	Yttrium (Y)	100
Cobalt (Co)	5	Lutetium (Lu)	<0.2	Scandium (Sc)	20	Zinc (Zn)	<10
Copper (Cu)	40	Manganese (Mn)	60	Selenium (Se)	10	Zirconium (Zr)	20
Dysprosium (Dy)	<1	Mercury (Hg)	0.3	Silver (Ag)	<0.5		
Erbium (Er)	<0.5	Molybdenum (Mo)	5	Strontium (Sr)	20		
Europium (Eu)	1						
Fluorine (F)	50						

® Registered Manville trademark.

* The symbol <, less than, indicates below detection limit of method used. Values are reported in parts per million (ppm).

attribute required in the performance of a function. Filter aids must produce clarity of varying degree, depending on the fluid to be filtered, and at a reasonable flow rate. A filter aid grade, then, may be subjected to a filtration test that measures flow rate through a filter cake under controlled conditions, as well as the clarity of the resulting filtrate. Since diatomite is purchased on a weight basis, and used on a volume basis, the cake or wet density is also important. To arrive at a suitable product, control of particle size distribution, as reflected in screen analysis is important. In certain filler applications, particle size distribution must be characterized quite precisely with sedimentation and related techniques. Lately, the Coulter Counter has been widely used. Particularly in food processing and conditioning, pH and resistivity of a water slurry are important. Recently, soluble trace elemental analysis to conform with Food Codex methods and limits has become important. Table 2 shows the trace elemental composition of a typical Celite® diatomite. In functional fillers for paper or paint, brightness of the powder, as reflected in a TAPPI or a General Electric brightness; fineness translated into Hegman reading; vehicle demand expressed in oil and water absorption; and abrasion for paper by the Valley Iron method are only a few examples of the properties that must be measured and controlled. These and other properties are the subject of specifications and of quality control procedures by which manufacturer and buyer agree on a product. Most of the tests are well known, whereas others are proprietary and confidential. Often, it is more practical to judge a particular characteristic by comparison against a standard sample. This is common and accepted practice, and is used particularly in evaluating properties that do not lend themselves readily to quantification.

TABLE 3—Domestic Production of Diatomite

Year	Sales, 1000 St	Average Value $ per Ton
1973	609	59.26
1974	664	76.31
1975	573	80.01
1976	631	87.08
1977	648	98.56
1978	651	111.26
1979	717	125.97
1980	689	146.02
1981	707	152.48

Source: Meisinger, 1982.

The following are specifications and test procedures currently in use by the ASTM:

D 604–42 (reapproved 1975)	Diatomaceous Silica Pigment
D 719–63 (reapproved 1976)	Analysis of Diatomaceous Silica Pigment
C 517–71	Diatomaceous Earth Block and Pipe Insulation

The following military specifications are currently in effect:

MIL–S–15191B	(Oct. 20, 1967) Silica, Diatomaceous (Flattening Extender Pigment)
MIL–D–20550B	(Aug. 8, 1968) Diatomaceous Earth
MIL–F–52637	(Feb. 20, 1969) Filter Aids, Water Purification
52–MA–522a	Diatomaceous Silica Pigment

The following TAPPI specification is current:

T658 OS–77 (1977)	Properties of Diatomaceous Silica

Production and Consumption

Although affected by changes in the economy, the production of diatomite has shown steady increase through 1974. A sharp drop in 1975 was followed by steady recovery that surpassed 1974, until 1980. According to US Bureau of Mines statistics, the average value per ton has increased steadily. This trend is reflected in Table 3. The breakdown of domestic diatomite consumption by applications is illustrated in Table 4. The exportation of diatomite has increased steadily except for a drop in 1975 that gradually recovered in subsequent years. The trend is illustrated in Table 5.

The United States leads world production; California has remained the predominant producing state, followed by Nevada, Washington, and Oregon. Four companies, Manville International Corp., Grefco, Inc., Eagle Picher Industries, and Witco Chemical Corp., account for approximately 90% of domestic production. Russian production, which is second to that of the United States, tends to run between 45 to 50% of US production. Other major producing countries are France, Germany, and if Moler is considered, Denmark. Minor world production also comes from over 20 other countries. Total world production had not, as of 1980, quite reached 2 million tons, having been of the order of about 1.75 million tons in 1979.

TABLE 4—Domestic Consumption of Diatomite by Principal Uses (% of Total Consumption)

Use	1973	1974	1975	1976	1977	1978	1979	1980
Filtration	61	60	60	60	59	63	63	66
Fillers	18*	19*	20*	21*	22*	23	21	21
Insulation	4	5	4	5	5	3	3	3
Miscellaneous	17	16	16	14	14	11	11	10

*The US Bureau of Mines included fillers with miscellaneous until 1978. The split between fillers and miscellaneous before 1978 is the author's estimate.

Marketing

The successful marketing of diatomite products by the major producers has depended, to a large degree, on their ability to furnish high caliber technical sales service to the customer. Technically trained sales engineers, supported by research and development organizations, have resulted in the solution of customer production problems with the consequent introduction of a particular diatomite grade to do the job. In technical service to the paint industry, for example, the development of cost saving formulations, through the introduction into the formulation of specially developed extender pigments and diatomite grades for control of gloss and sheen, has been of great help to paint manufacturers. There can hardly be a better foundation on which to base sales effort or to assure product performance. A similar approach in the selection of filter aids to meet a specific requirement, or to solve a unique problem, has entrenched trade or brand names in the minds of users who have come to demand a high degree of quality and of uniformity. While this approach has been of considerable assistance to industry, and has ultimately benefited the consumer, it has made marketing more difficult for minor producers with an inferior deposit or an untrained sales organization to compete in the more sophisticated or demanding applications of diatomite. The small producer, however, supplying to a local, or to a lower grade market has been able to compete successfully. This is particularly applicable to the small foreign producer against American imports.

Diatomite powders are traditionally sold as carload, l.c.l., and warehouse. American exports are to a large degree sold through distributors, as are diatomite chromatographic supports, both domestically and abroad.

Because of its low bulk, freight and containers constitute a substantial part of the cost to the consumer. The precise effect of the cost of transportation on the diatomite industry is being treated separately in this chapter.

Transportation

The low bulk density of processed diatomite presents unique transportation problems. Since all American commercial diatomite originates in the western states, and the predominant markets are located east of the Mississippi River, a substantial part of the eastern delivered cost consists of freight charges. The price per ton of a California processed filter aid delivered in New York consists, for example, of approximately 33% transportation costs. The successful exportation of American diatomite to Europe, Africa, Latin America, and the Far East has traditionally been dependent on a reputation and a need for superb quality.

With the cost of ocean freight increasing steadily, together with the rising costs of production at home, exports have been under continued pressure from the local sources. For this reason, the motivation for foreign exploration by American companies in the major market areas is apparent. The trend, therefore, toward decreasing exports in future years is to be anticipated. The following are typical current (May 1981) transportation costs (inland plus ocean freight) from Lompoc, CA, to representative foreign ports:

Port	Cost per Mt, $
Capetown, South Africa	253
Yokohama, Japan	121
Buenos Aires, Argentina	213
Sydney, Australia	282
Hamburg, West Germany	179
Hull, England	181

Uses

Filtration: By far, the widest use for processed diatomite is as a filter aid for the separa-

TABLE 5—Exports of Diatomite

	Quantity, 1000 St	Value, 1000 $
1974	186	17,541
1975	147	15,314
1976	149	16,832
1977	152	18,876
1978	153	21,463
1979	170	26,496
1980	173	32,238
1981	160	24,397

Source: US Bureau of Mines.

tion of suspended solids from fluids.[234] About 50% of all processed diatomite is channeled into this application. Indeed, the probability of economic success for a potential deposit is seriously reduced if it cannot be processed into a single, if not a range, of filter aid products. The requirements for filter aid suitability are subtly complex and many. Above all, diatom skeletal constitution and structure, density, and soluble impurities are principal considerations. The part that the skeletal structure plays in a filter aid is of paramount importance. In addition to the space between diatom particles in a filter cake, the interstices and chambers provided within the structure play their part in the trapping of impurities. The relationship and interplay between variously shaped particles in a multi-species assemblage is also important. A single-species assemblage of forms often has less to offer than an association consisting of numerous different species. The balance, for example, between pennate and discoid species found in the marine earth at Lompoc, together with its response to calcination and flux calcination for the increase in flow rate, is unmatched with regard to its flow rate-clarity relationship. Certain lacustrine earths, consisting of a completely different assemblage or of wholly one species, while lacking the advantages mentioned, possess other merits for certain applications. For example, the lacustrine earth mined for some years at Terrebonne, OR, (and now depleted) was a favorable fast flow rate filter aid for rotary precoat filtration in that the filter cake was less susceptible than that made with other earths to cracking on the rotary drum. The need for low density and particularly in filter aids sold for food processing for low solubility content is obvious.

Diatomite is processed into filter aids for all types of food and nonfood processing applications.[149] The selection of the proper grade of filter aid depends on the size of the suspended particles that are to be removed. It is axiomatic in the use of filter aids that as the particle size, and thus the flow rate increases, the ability of the filter aid to remove small particles of suspended matter decreases.[137] Conversely, as filter aid particle size and, therefore, flow rate decreases, the ability of the filter aid to remove small particles of suspended matter increases. The factors that govern which end of this flow rate-clarity relationship is to be emphasized will depend very much on the type and the particle size distribution of the undissolved solids being removed. The best filter aid is that grade that will result in the fastest flow rate (or the greatest throughput per dollars worth of filter aid) and yet will provide adequate clarity. The correct clarity must be determined and specified by the filter aid user. The calcination and flux calcination of diatomite powders are performed for the purpose of adjusting particle size distribution through incipient fusion and by agglomeration of the fines and structure to produce a range of flow rates, hence clarities.

The more commonly known applications are the use of processed diatomite powders in the filtration of dry cleaning solvents;[132] pharmaceuticals;[157] beer, whiskey, and wine; raw sugar liquors; antibiotics;[151] industrial,[146] municipal,[141, 152] and swimming pool waters;[133] fruit and vegetable juices; lube, rolling mill, and cutting oils; jet fuels; organic and inorganic chemicals; and varnishes and lacquers. Diatomite has also been mixed with asbestos, and with other minerals, and surface-treated and acid-washed to improve its effectiveness in special filtration applications. Table 6 lists the physical properties of some commercial filter aids.

Fillers: The second largest use of diatomite is that employed as a filler. Although natural milled powders were used to some extent in paint, paper, abrasives, and other uses before there were baghouse grades, later filler applications were the result of product development to use the baghouse fines that would otherwise be discarded as a byproduct of the filter aid manufacturing process. In recent years, however, while filler products are usually produced in conjunction with, and as an integral part of the manufacturing process for filter aids, their specifications and properties are tailored to meet the requirements of a demanding filler market. Indeed, the term "filler" was often used as the catchall term for those uses that were not directed toward filtration.

Lately, however, the expression, *functional filler*, has been set apart and reserved for those

TABLE 6—Physical Properties of Some Commercial Filter Aids

	Grade*	Color	Density, Lb per Cu Ft Dry	Density, Lb per Cu Ft Wet	Screen Analysis, % Retained, 150 Mesh	pH	Specific Gravity	Water Absorption	Relative Flow Rate	Typical Applications, Filtration
Natural	Filter-Cel	Gray	7.0	15.9	1.0	7.0	2.10	235	100	Vegetable oil catalyst
Calcined	Celite 505	Pink	8.0	21.0	–	7.0	2.15	170	135	Apple juice
	Standard Super-Cel	Pink	8.0	17.2	3.0	7.0	2.15	255	200	Beer and wine
	Celite 512	Pink	8.0	17.9	4.0	7.0	2.15	250	300	Sugar
Flux Calcined	Hyflo Super-Cel	White	9.0	17.2	5.0	10.0	2.30	245	500	Dry cleaning solvents, chemicals, etc. (most widely used filter aid)
	Celite 501	White	9.5	16.9	8.0	10.0	2.30	250	750	Grape juice
	Celite 503	White	9.5	17.2	9.0	10.0	2.30	240	900	Industrial and potable water
	Celite 535	White	12.0	17.6	9.0	10.0	2.30	245	1350	Industrial wastes
	Celite 545	White	12.0	18.0	12.0	10.0	2.30	240	2160	Swimming pools
	Celite 550	White	18.1	21.0	20.0	8.0	2.30	220	2380	Pharmaceuticals
	Celite 560	White	19.5	20.0	50.0	10.0	2.30	220	7500	Phosphoric acid

* Grades listed are Manville registered trademarks.

filler applications in which diatomite produces a desired effect or function that is more than simply the bulking or partial replacement of an otherwise more expensive component in the formulation. The larger part of the total filler market is that in which diatomite serves as a functional filler as defined, rather than as a mineral filler. The following discussion, rather than application-oriented, is designed to highlight those properties of diatomite that are important in its application as a functional filler, as well as an absorbent, mild abrasive, extender, insulation, conditioning agent, and catalytic or chromatographic support. In general, more than one property is important in any application. A key property, supported, however, by others, usually provides the specific function.

Particulate Structure—Much has already been stated regarding the unique nature of the diatom valve and pointing out its difference from other forms of silica. The use of diatomite as a functional filler in paint for the control of gloss and sheen, accounts for a major share of the dollar volume attributed to fillers. The small particles serve to roughen the otherwise smooth paint film and to produce a flatting effect.[162, 163] Precise particle size control is required during production of flatting agents, since the extreme fines contribute little to the flatting effect, but because of their absorption do increase the vehicle demand. Particles that are too coarse, on the other hand, will cause surface blemishes and unsightly spots. This is not to say that only the structure is important in this application. Structure is of primary importance, but brightness, absorption, pH, refractive index, and chemical stability are also important.

The unique structure of the diatom valve makes it useful as an antiblocking agent in polyethylene film production. The film is blown as an envelope, and while hot, has a tendency for the surfaces to stick together or "block." If a small amount of diatomite (0.05 to 0.5% by weight) of the right particle size distribution is incorporated into the film, the fine particles sticking through the plastic surface provide a mechanical separation or "antiblocking" effect that keeps the surfaces from sticking.[164]

As in the flatting of paint, such other supporting properties as absorption, brightness, refractive index, and particle size distribution are important and must be considered in the manufacture of antiblocking agents.

Liquid Absorption—Diatomite powders, because of their high surface area and low bulking value, have a high absorptive capacity. They can absorb up to 2½ times their weight of water. Specially processed diatomite powders are still "dry" and free flowing even after the addition of 50% by weight of water. This property contributes to their application as the liquid carrier in rug cleaners, pesticide carriers, and pitch control in paper manufacture. In the chemical industry, such hazardous materials as sulfuric and phosphoric acids are converted through the use of diatomite to dry powders for safer and easier handling and storage.

Inertness—A high quality diatomite has an exceptionally high silica content—as much as 94%. Because it is essentially silica, it is inert to most chemical reactions and is resistant to extremely high temperatures, with a softening point of about 2600°F. Coupled with its other attributes, inertness and temperature resistance are responsible for its utility as a catalyst carrier and as an insulation. Diatomite catalyst supports are ideal for the nickel catalyst used in hydrogenation processes, the vanadium catalyst used in the manufacture of sulfuric acid, and the phosphoric acid catalyst used in the petroleum industry.

Mild Abrasive—Because it is similar in composition to opaline silica, diatomite has a hardness sufficient to produce abrasion on metal surfaces. However, to this attribute is coupled the fact that the delicate skeletal structure has a friability and particle size that causes it to polish, rather than to scratch.[172] The highly refined (or low contaminant) natural milled product is ideal for incorporation into silver polish formulations, and the slightly increased strength or particle integrity produced by flux calcination makes this type of treatment suitable for producing the more abrasive effect required in automobile polishes.

Reinforcing Effect—The particulate structure produces a semireinforcing effect in such diverse materials as silicone rubber specialties, and in mechanical rubber goods.

Silica Source—Because of its high content of silica and its high surface area, diatomite is a particularly suitable and reactive form of silica in combination with lime for the hydrothermal manufacture of lime-silicate insulations and of highly absorptive calcium-silicate powders.

Chromatographic Support—The use of diatomite as a chromatographic support is a unique development, principally of the 1960s, and makes use of and illustrates essentially all the properties that have been previously described.[180] Structure and absorption account for its unique capacity to carry sufficient amounts of the liquid phase.[187] An inert surface is required to keep the support from reacting and interfering with the partitioning ability of the liquid phase. When properly processed and treated, diatomite chromatographic supports satisfy all these requirements.[184, 187]

Miscellaneous—Other applications that make use of the unusual properties listed above include anticaking agent on ammonium-nitrate prills, match head composition to control after glow, welding rod composition, use in battery box separators, pozzolan and concrete additive to improve workability and reduce bleeding, acetylene containers, and a stabilizer in explosives, drilling mud additive, and as a conditioner of animal foods. Table 7 lists the physical properties of some commercial fillers. The uses of diatomite are also covered in the chapter on "Fillers, Filters, and Absorbents," pages 243-257, Volume 1.

Further Considerations and Trends

Since the United States is self-sufficient, imports of diatomite have been negligible. However, depending on the course of foreign exploration, on processing costs, and on transportation charges, it is conceivable that eastern US markets could be penetrated to some degree by European or African sources.

The presence of eastern US occurrences of diatomite, while not a threat to western supremacy at present, could become competitive. For certain low quality applications, the marine diatomite outcroppings in Maryland and Virginia could have utility. Marked advances in processing are required before this earth could approach the quality of western deposits. The bog deposits of Florida, New England, and New Brunswick, because of their eastern location, are intriguing. Advances in dredging technology and in imaginative processing will be required to upgrade these to commercial potential.

Substitutes

While diatomite possesses properties and characteristics that are unmatched by other forms of silica—indeed by any other type of

TABLE 7—Physical Properties of Some Commercial Fillers

Application:	Paper		Paint		Abrasives			Conditioning Agent		Anti-Blocking Agent	Reinforcing Agent
	Cylinder Machine Production	Fourdrinier Production	High Flatting	Low Luster Stir-In	Silver Polish	Automobile Polish	Buffing Compound	Toxicant Extender	Fertilizer Coating	Polyethylene	Rubber
Brand*	Celite 321A	Celite 305	Celite 281	Celite 499	Snow Floss	Super Floss	Celite HSC	Celite 209	Celite 379	Celite White Mist	Celite 270
Loose wt, lb per cu ft	7.0	7.2	8.0	7.2	7.1	8.7	10.7	7.2	7.9	8.8	8.6
Wet density, lb per cu ft	20	22.5	20.0	19.0	22.0	26.0	17.0	22.5	19.2	20.8	24.0
Moisture content, %	4.0	3.0	0.5	0.1	4.1	0.1	0.5	4.0	3.3	0.1	0.3
Retained on 150 mesh, % wt	—	0.1	Trace	0	Trace	0	5.6	0.1	0.1	0	0.1
Retained on 325 mesh, % wt	8.0	0.4	1.2	Trace	0.3	0.1	19.6	0.4	3.0	Trace	0.7
Oil absorption, % wt	210	180	145	160	190	120	185	180	190	160	150
Color	Gray	Gray	White	White	Gray	White	White	Gray	Gray	White	Pink
Brightness, TAPPI	65	63	88	89	64	91	86	63	59	90	60
Hegman	—	—	1	4	4½	3¾	—	—	—	5	—
Sheen	—	—	12	15	50	50	—	—	—	30	—
Specific gravity (apparent)	2.0	2.0	2.3	2.3	2.0	2.3	2.3	2.0	2.0	2.3	2.2
pH	6.6	6.2	8.7	9.5	6.5	9.2	8.9	6.2	7.1	9.8	6.5
Resistivity, ohms	3000	3000	13,000	7400	3000	5000	9000	3000	4000	6500	30,000
Refractive index	1.42	1.42	1.48	1.48	1.42	1.48	1.48	1.42	1.42	1.48	1.44
Surface area, sq. m per g	20-30	20-30	0.7-3.5	0.7-3.5	20-30	0.7-3.5	0.7-3.5	20-30	20-30	0.7-3.5	4-6
Porosity, %	65-85	65-85	65-85	65-85	65-85	65-85	65-85	65-85	65-85	65-85	65-85
Particle size range, μ	8-1	4-1									
Valley Iron Abrasion, mg	440	170	850	495	140	375	1900	170	410	390	300

* Brands listed are Manville Registered Trademarks.

mineral powder—in certain applications, it has lost some ground to new and different materials. Where for decades it has occupied a place of preeminence in the field of filtration, a relative newcomer has appeared to challenge its position in specific applications—perlite. Perlite, when expanded and milled into a filter aid, has cut into the rotary precoat filtration field. It has taken over some markets for sugar, for alginates, and for pharmaceutical filtration. The percentage of total perlite going into filtration applications has increased steadily to where it was approximately 18% in 1979. Another area where inroads have been made into diatomite's role is in paper fillers and in paint where calcined clay and talc in recent years have partially replaced it. However, some of the major diatomite producers are also in the perlite and talc business.

Tariffs and Taxes

There is no tariff on the importation of diatomite into the United States. In those countries that have a customs tax, the Brussels Tariff designations are generally used. The classifications into which the various grades and types of diatomite are relegated are not uniform among countries, but depend on the interpretation of the responsible customs officials in each country. Of course, the rate of taxation is also variable among countries. In Japan, for example, natural and straight calcined grades are admitted duty free under the Brussels Tariff No. 25–12, which is defined to include "Siliceous fossil meals and similar siliceous earths (for example, kieselguhr, tripolite, or diatomite), whether or not calcined, of an apparent specific gravity of one or less." Flux calcined grades, on the other hand, are taxed 4% in Japan under designation 38–03–2, which applies to "activated carbon (decolorizing, depolarizing, or absorbent) activated diatomite, activated clay, activated bauxite, and other activated natural mineral products." Diatomite catalyst carriers imported into Japan are classed as "Laboratory, chemical, or industrial wares," and taxed 6% under designation 69–09–1.

The European Common Market countries use the designation 25–12–00 for natural and straight calcined products and 38–03–00 for flux calcined products. Some countries insert other figures for the last two digits for further classification. There is still no duty charged against these tariff numbers.

One legitimate point of contention among diatomite producers appears to be the classification of flux calcined diatomite by government authorities under designation 38–03–2, thus including it with "activated" mineral materials. It cannot be logically argued with any degree of scientific accuracy that the flux calcination of diatomite should be visualized as an *activating* process similar to that of clays or carbon.

Under the Tax Reform Act of 1969, the percentage depletion allowance for diatomite was reduced from 15 to 14%.

Health and Ecology

Considerable expenditures of money and effort by the major producers have been made to control dust both in mines and in plants. Since the calcined and flux calcined grades of diatomite contain cristobalite,[6] protection of workers against pneumoconeosis through the use of dust control equipment and through education has become common practice. The health of workers with major companies is periodically checked by physical examinations that include chest X-rays. The filtering and optional heating of air of mining equipment cabs is not uncommon among the larger producers. While untreated diatomaceous silica is not known to be hazardous, many in the industry regard all dusts to be a nuisance that should be held to a minimum and the comfort and protection of workers is provided for accordingly. Respirators are usually required of workers in parts of the plant, and environmental control departments periodically monitor the dust count at strategic stations within the processing plants. A modern, large diatomite plant is a model in dust control and cleanliness.

Acknowledgments

The author gratefully acknowledges the assistance of Arthur B. Cummins, Irene Crespin, Celia Kamau, and Ruth Patrick in the compilation of this bibliography, and of these and numerous other colleagues within Manville for their valuable suggestions.

Bibliography and References

Almost all of the references listed herein, which by no means includes all those available, are dated 1960 or later. The reader is referred to the previous editions of *Industrial Minerals and Rocks* for bibliographic information prior to 1960.

General

1. Anon., 1964, "Celite—The Story of Diatomite," Johns-Manville Corp. Bulletin FA-41A, 26 pp.

2. Anon., 1969, "Diatomite," *Industrial Minerals,* No. 18, March, pp. 9–27.
3. Black, J.P., 1971, "The Almost Perfect Kidnapping," *Readers Digest,* Vol. 98, No. 590, June, pp. 140–144.
4. Cummins, A.B., 1960, "Diatomite," *Industrial Minerals and Rocks,* 3rd ed., J.L. Gillson, ed., AIME, New York, pp. 303–314.
5. Durham, D.L., 1973, "Diatomite," in "United States Mineral Resources," Professional Paper 820, US Geological Survey, pp. 191–195.
6. Dutra, F.R., 1965, "Diatomaceous Earth Pneumoconeosis," *Archives Environmental Health,* Vol. 11, Nov., pp. 613–619.
7. Frankenhoff, C.A., 1970, "Kenite Diatomite," *Industrial Minerals,* No. 33, June, pp. 51–52.
8. Frondel, C., 1962, "Silica Minerals," *Dana's System of Mineralogy,* Vol. 3, 7th ed., p. 287.
9. Hartwell, J.W., and Schreck, V., 1962, "Diatomite," *Minerals Yearbook,* US Bureau of Mines, pp. 531–536.
10. Petkof, B., 1970, "Diatomite," *Minerals Yearbook,* US Bureau of Mines, pp. 501–503.
11. Schroeder, H.J., 1970, "Diatomite," *Minerals Facts and Problems,* Bulletin No. 650, US Bureau of Mines, pp. 967–975.

Structure

12. Ahn, Y.-P., and Lee, K.-G., 1960, "Electron Microscopical Observations on Fine Structure of Diatom Frustules Which are Collected in the East Coast of Kyung-Pook Province," *Korean Journal of Botany,* Vol. 3, No. 2, pp. 26–28.
13. Bens, E.M., and Drew, C.M., 1967, "Diatomaceous Earth: Scanning Electron Microscope of 'Chromosorb P'," *Nature,* Vol. 216, No. 5119, pp. 1046–1048.
14. Drum, R.W., and Pankratz, H.S., 1963, "Porous Plates in the Punctae of a Small Nitzschia," *Microscope and Crystal Front,* Vol. 14, No. 1, pp. 1–4.
15. Drum, R.W., et al., 1966, *Electron Microscopy of Diatom Cells,* Helmcke and Krieger, Diatomeen-Schalen im Electronmikroskopischen Bild, Vol. 6, 24 pp.
16. Drum, R.W., 1969, "Electron Microscope Observations of Diatoms," *Osterr. Bot. Zeitschrift.,* Vol. 116, pp. 321–330.
17. Gerloff, J., 1963, "Submicroscopic Structure of the Diatom Frustule and Modern Technical Considerations," *Sitzber. Ges. Naturf. Freunde Berlin,* Vol. 3, pp. 144–146.
18. Hasle, G.R., 1968, "The Valve Processes of the Centric Diatom Genus *Thalassiosira,*" *Nytt Mag. Botanik,* Vol. 15, No. 3, pp. 193–201.
19. Helmcke, J.G., 1961, "Electron Microscopy and Algae Structures-Versuch Eiaer Gestaltsanalyse un Diatomeenschalen," *Recent Advances in Botany,* Sec. 3, University of Toronto Press, pp. 216–231.
20. Helmcke, J.G., and Krieger, W., 1963, "Diatomeenschalen Im Elektronenmikroskopischen Bild. III. Teil," *Phycologia,* Vol. 3, No. 2, p. 84.
21. Jerković, L., 1972, "L'ultrastructure des frustules de quelques espèces endemiques des Diatomées de la Yougoslavie," *Arch F. Hydrobiol.,* Sup. 41 (1); Algolog. Stud. No. 6, pp. 1–10.
22. Mehta, S.C., et al., 1961, "The Fine Structure and the Cell Wall Nature of *Diatoma Hiemale* var. *mesoden* (Ehr.) Grun," *Revue Algol.,* Vol. 6, No. 1, pp. 49–52.
23. Mehta, S.C., and Venkataraman, G.S., 1961, "Fine Structure of Diatom Valves III," *Botanical Magazine of Tokyo,* Vol. 74, No. 875, pp. 248–253.
24. Okuno, H., 1960, "Fine Structure of Diatomaceous Earth Particle," *The World through the Electron Microscope,* by Suito, Society of Chemistry, Tokyo.
25. Okuno, H., 1960a, "Electron Microscopical Study on Fine Structure of Diatom Frustules, XVIII," *Botanical Magazine of Tokyo,* Vol. 73, No. 865–866, pp. 310–316.
26. Okuno, H., 1962, "Electron Microscopical Study on Fine Structures of Diatom Frustules XIX," *Botanical Magazine of Tokyo,* Vol. 75, No. 886, pp. 119–226.
27. Stoermer, E.F., and Pankratz, H.S., 1964, "Fine Structure of the Diatom *Amphipleura pellucida.* I. Wall Structure," *American Journal of Botany,* Vol. 1, No. 9, pp. 986–990.
28. Venkataraman, G.S., and Mehta, S.C., 1960, "Fine Structure of Diatom Valves—I," *Phytomorphology,* Vol. 10, No. 2, pp. 116–118.
29. Venkataraman, G.S., et al., 1961, "Fine Structure of Diatom Valves—II," *Journal of Indian Botanical Society,* Vol. 40, No. 2, pp. 271–273.
30. Venkataraman, G.S., et al., 1962, "Fine Structure of Diatom Valves IV," *Nova Hedwigia,* Vol. 4, No. 1–2, pp. 127–129.

Chemistry

31. Arehart, J.L., 1972, "Diatoms and Silicon," *Sea Frontiers,* Vol. 18, No. 2, Mar./Apr., pp. 90–94.
32. Barashkov, G.K., 1960, "The Chemistry of Diatomaceous Algae (Diatomese)," (in Russian), *Bot. Zhur.,* Vol. 45, No. 9, pp. 1350–1356.
33. Barashkov, G.K., 1962, "Chemistry of Some Marine Plankton Diatoms," *Trudy Murmansk Mersk. Biol. Inst.,* Vol. 4, No. 8, pp. 27–46.
34. Berger, L.R., 1971, "Effects of Hydrostatic Pressure on Photosynthesis and Growth of Unicellular Marine Algae and Diatoms," *National Technical Information Service,* AD–720, Vol. 401, 11 pp.
35. Bernhard, M., and Zattera, A., 1967, "A Comparison Between the Uptake of Radioactive and Stable Zinc by a Marine Unicellular Alga," *Symposium on Radioecology,* D.J. Nelson, and F.C. Evans, eds., Vol. 2, pp. 389–398.
36. Bogoyavlenskiy, A.N., 1966, "Distribution and Migration of Dissolved Silica in Oceans," *International Geological Review,* Vol. 9, No. 2, pp. 133–153.
37. Calvert, S.E., 1968, "Silica Balance in the Ocean and Diagenesis," *Nature,* Vol. 219, No. 5157, pp. 919–920.
38. Coombs, J., and Volcani, B.E., 1968, "Studies on the Biochemistry and Fine Structure of Silica Shell Formation in Diatoms. Silicon Induced Metabolic Transients in Navicula pelliculosa (Breb) Hilse," *Planta,* Vol. 80, pp. 264–279.

39. Coombs, J., and Volcani, B.E., 1968a, "Studies on the Biochemistry and Fine Structure of Silica Shell Formation in Diatoms. Chemical Changes in the Wall of Navicula pelliculosa during Its Formation," *Contributions,* University of California, Vol. 38, No. 2463, p. xiv.
40. Coombs, J., et al., 1968, "Studies on the Biochemistry and Fine Structure of Silica Shell Formation in Diatoms. VI. Fine Structure of Colchicine-induced Polyploids of Navicula pelliculosa (Breb) Hilse," *Contributions,* University of California, Vol. 38, No. 2461, p. xiv.
41. Coombs, J., et al., 1968a, "Studies on the Biochemistry and Fine Structure of Silica Shell Formation in Diatoms. VII. Fine Structure of Colchicine-Induced Polyploids of Navicula pelliculosa (Breb) Hilse," *Zeitschrift fuer Pflanzanphysiol.,* Vol. 59, No. 3, pp. 274–284.
42. Darley, W.M., 1970, "Silicon and the Division Cycle of the Diatoms Navicula pelliculosa and Cylindrotheca fusiformis," *Proceedings,* North American Paleontology Convention, Vol. 2, pp. 994–1009.
43. Desikachavy, T.T., and Dweltz, N.E., 1961, "The Chemical Composition of the Diatom Frustula," *Proceedings,* Indian Academy of Science, Vol. 53, No. 4, pp. 157–165.
44. Distanov, U.G., 1968, "Conditions of Silica Accumulation in the Paleocene Basin of the Volga and Caspian Sea Regions," *Doklady Akademii Nauk SSR,* Vol. 180, No. 1, pp. 204–207.
45. Dunn, J.H., 1968, "Utilization of Silica by Diatoms," *Dissertation Abstracts B.,* Vol. 29, No. 4, p. 1447.
46. Harvey, H.W., 1937, "The Supply of Iron to Diatoms," *Journal of Marine Biology Assn., U.K.,* Vol. 22, pp. 206–219.
47. Jorgensen, E.G., 1960, "The Effects of Salinity, Temperature, and Light Intensity on Growth and Chlorophyll Formation of Nitzschia ovalis." *Year Book,* Carnegie Inst., Washington, Vol. 59, pp. 348–349.
48. Kilham, P., 1971, "A Hypothesis Concerning Silica and the Fresh Water Planktonic Diatoms," *Limnol. Oceander* (Livca), Vol. 16, No. 1, pp. 10–18.
49. Kramer, D.D., and Ryther, J.H., 1960, "The Iron Requirement of Some Marine Plankton Algae," *Biological Bulletin,* Vol. 119, No. 2, p. 324.
50. Lanza, G.R., and Cairns, J. Jr., 1971, "Certain Effects of Sudden Temperature Increases Upon Diatom Populations," *ASB Bulletin,* Vol. 18, No. 2, p. 42.
51. Lewin, J.C., 1959, "Silicon as an Essential Element for Diatom Cultures," *Recent Advances in Botany,* 9th International Botanical Congress, Vol. 1, pp. 253–254.
52. Lewin, J.C., 1966, "Boron as a Growth Requirement for Diatoms," *Journal of Phycologia,* Vol. 2, No. 4, pp. 160–163.
53. Lewin, J.C., et al., 1966a, "Silica Shell Formation in Synchronously Dividing Diatoms," *Cell Synchrony—Studies in Biosynthetic Regulation,* 1966, pp. 169–188.
54. Lewin, J.C., 1966b, "Silicon Metabolism in Diatoms. V. Germanium Dioxide, a Specific Inhibitor of Diatom Growth," *Phycologia,* Vol. 6, pp. 1–11.
55. Lewin, J.C., and Chen, C.-H., 1971, "Available Iron: A Limiting Factor for Marine Phytoplankton," *Limnol Oceander.,* Vol. 16. No. 4, pp. 670–675.
56. Millington, W.F., and Gawlik, S.R., 1967, "Silica in the Wall of *Pediastrum,*" *Nature,* Vol. 216, No. 5110, p. 68.
57. Millot, G., and Fauck, R., 1971, "Origin of Silica in Climatic Silicification and Quaternary Diatomites of Sahara," C.R. Academy of Sciences Ser. D, Vol. 272, No. 1, pp. 4–7.
58. Neals, T.F., 1967, "The Boron Nutrition of the Diatom, *Cylindrotheca fusiformis,* Grown on Agar, and the Biological Activity of Some Substituted Phenylboronic Acids," *Australian Journal of Biological Science,* Vol. 20, No. 1, pp. 67–76.
59. Patrick, R., et al., 1969, "Temperature and Manganese as Determining Factors in the Presence of Diatom or Blue-Green Algal Floras in Streams," *Proceedings,* National Academy of Science, Vol. 64, No. 2, pp. 472–478.
60. Qasim, S.Z., et al., 1972, "The Influence of Salinity on the Rate of Photosynthesis and Abundance of some Tropical Phytoplankton," *Marine Biology,* Vol. 12, No. 3, pp. 200–206.
61. Reimann, B., et al., 1965, "Studies on the Biochemistry and the Fine Structure of Silica Shell-Formation in Diatoms. I. The Structure of the Cell Wall of *Cylindrotheca fusiformis,*" *Journal of Cell Biology,* Vol. 24, pp. 39–55.
62. Reimann, B., et al., 1966, "Studies on the Biochemistry and Fine Structure of Silica Shell Formation in Diatoms. II. The Structure of the Cell Wall of *Navicula pelliculosa* (Breb.) Hilse," *Journal of Phycologia,* Vol. 2, No. 2, pp. 74–84.
63. Riley, J.P., 1971, "The Distribution of Trace Elements in Some Species of Phytoplankton Grown in Culture," *Journal of Marine Biology Assn., U.K.,* Vol. 51, No. 1, pp. 63–72.
64. Ryther, J.H., and Guillard, R.R.L., 1962, "Studies of Marine Planktonic Diatoms. III. Some Effects of Temperature on Respiration of Five Species," *Canadian Journal of Microbiology,* Vol. 8, No. 4, pp. 447–453.
65. Shimazu, M., 1961, "A Study of the Crystallization of Diatomite. Crystallization of Quartz and Cristobalite after 900°C," *Journal of the Mineralogical Society of Japan,* Vol. 5, pp. 77–88.
66. Slisarenko, F.A., et al., 1965, "Physicochemical and Sorption Properties of Volga Diatomite," *Zhurnal Prikladnar Khimii,* Vol. 38, No. 11, pp. 2430–2439.
67. Steemann, N.E., and Wium-Andersen, S., 1971, "The Influence of Cu on Photosynthesis and Growth in Diatoms," *Physiological Plantology,* Vol. 24, pp. 480–484.
68. Strakhov., N.M., 1967, "Some Aspects of the Geochemistry of Silica," Introduction to *Geochemistry of Silica,* N.M. Strakhov, ed.; English translation in *International Geological Review,* Vol. 9, No. 1, pp. 25–27.
69. Wawrik, F., 1960, "Silicic Acid and Diatom Growth in the Lower Lake Lunz and Neighbouring Streams, including the Ybbs." (in German), *Wass. und Abwass.,* pp. 108–132.
70. Werner, D., and Pirson, A., 1967, "On the Reversible Storage of Silicic Acid in *Cyclo-*

tella cryptica," *Arch. Mikrobiol.,* Vol. 57, No. 1, pp. 43–50.

71. Zabelin, V.A., 1960, "The Crystal Structure of SiO_2 in Opal-Containing Rocks of the Volga Region," *Doklady Akademii. Nauk SSSR,* Vol. 125, pp. 1941–1944.
72. Zabelin, V.A., 1962, "Examination of Crystal Structure of Silica in Opals and Opal Containing Rocks," *Zap. Vses Mineralog. Obschchestva,* Vol. 91, pp. 343–350.

Geology

73. Abbott, W. H., 1971, "Diatom Investigations of Southern Ocean Deep-Sea Cores," *Antarctic Journal of United States,* Vol. 6, No. 5, pp. 171–172.
74. Arakyan, T.A., 1969, "Genesis of Diatomaceous Earth in the Sisian Deposit," *Izvestiya Nauki O Zemle,* Akademii Nauk SSR, Vol. 22, No. 5, pp. 91–96.
75. Bradbury, J.P., 1972, "Diatoms and Paleolimnology," Abstr. in: *Transactions,* American Microscopic Society, Vol. 91, No. 1, p. 84.
76. Bramlette, M.N., 1961, "Pelagic Sediments," *Oceanography,* A.A.S. Pub. No. 67, M. Searles, ed., pp. 345–366.
77. Calvert, S.E., 1964, "Factors Affecting the Distribution of Laminated Diatomaceous Sediments in the Gulf of California," *Marine Geology of the Gulf of California,* American Assn. of Petroleum Geologists, Memoir 3, pp. 311–330.
78. Calvert, S.E., 1966, "Origin of Diatom-Rich Varved Sediments from Gulf of California," *Journal of Geology,* Vol. 74, No. 5, pp. 546–565.
79. Carter, W.D., 1971, "ERTSA—A New Apogee for Mineral Finding," *Mining Engineering,* May, pp. 51–53.
80. Dymond, J.R., 1966, "Potassium-Argon Geochronology of Deep-Sea Sediments," *Science,* Vol. 152, May 24, pp. 1239–1241.
81. Emiliani, C., and Milliman, J.D., 1966, "Deep-Sea Sediments and Their Geological Record," *Earth-Science Reviews,* Vol. 1, No. 2–3, pp. 105–132.
82. Ichikawa, W., 1967, "Fossil Diatoms and Geology," *Journal of Geological Society of Japan,* Vol. 73, No. 2, pp. 53–62.
83. Karaeva, N.I., 1971, "The Diatoms of the Genus *Navicula Bory* in the Palaeogene and Neogene of the USSR," *Botan. Zhurnal,* Vol. 56, No. 7, pp. 953–962.
84. Lanskaya, L.V., 1961, "Rate and Conditions of Division of Marine Planktonic Algae under Culture Conditions," *Primary Production of Seas and Inland Waters,* pp. 328–332.
85. Lohman, K.E., 1961, "Geologic Ranges of Cenozoic Nonmarine Diatoms," Professional Paper 424-D, US Geological Survey, pp. 234–236.
86. Margolef, R., 1961, "Velocidad de Sedimentacion de Organisms Pasivos de Fitoplanction," *Barcelona Inv. Pesqueros 18,* pp. 3–8.
87. Schrader, H.-J., 1971, "Fecal Pellets: Role in Sedimentation of Pelagic Diatoms," *Science,* Vol. 174, Oct. 1, pp. 55–57.
88. Tasch, P., 1967, "The Problem of Primary Production in the Seas Through Geologic Time," *Review of Paleobotanical Palynology,* Vol. 1, No. 1/4, pp. 283–290.
89. Williamson, R., 1966, "Exploration for Diatomites," *Colorado School of Mines Industrial Minerals Bulletin,* Vol. 9, No. 3, May, pp. 1–14.
90. Zhuze, A.P., 1966, "Siliceous Sediments of Modern and Ancient Lakes," *Geochemistry of Silica,* Strakhov, ed., pp. 301–320.

Occurrences and Deposits

91. Anon., 1961, "Diatomaceous Earth in Kentmere Valley, North Lancashire Co., England," *Chemical Trade Journal & Chemical Engineering* (London), Vol. 149, No. 3882, Oct. 27, p. 912.
92. Anon., 1961a, "The Skye Diatomite Deposits," *Chemical Trade Journal & Chemical Engineering* (London), Vol. 148, No. 3850, Mar. 17, p. 594.
93. Alimen, H., et al., 1970, "Sediments Quaternaires a Diatomees du Sahara Nord-Occidental," *Bulletin of Society of Geology of France,* 7 Ser., Vol. 12, No. 1, pp. 103–107.
94. Anderson, F.W., and Dunham, K.C., 1966, "The Geology of Northern Skye," *Memoir,* Geological Survey of Scotland, Edinburgh, H.M. Stationary Office, p. 216.
95. Andrews, G.W., 1966, "Late Pleistocene Diatoms from Trempealeau Valley, Wisc.," Professional Paper 523-A, US Geological Survey, pp. A1–A27.
96. Andrews, G.W., 1970, "Late Miocene Non-Marine Diatoms from the Lake Kilgore Area, Cherry Co., Nebraska," Professional Paper 683-A, US Geological Survey, pp. A1–A24.
97. Andrews, G.W., 1970a, "Early Miocene Non-Marine Diatoms from the Pine Ridge Area, Sioux Co., Nebraska," Professional Paper 683-E, US Geological Survey, pp. E1–E17.
98. Archibold, N.L., 1966, "Industrial Mineral Deposits of Mineral Co., Nevada," Report 14, Nevada Bureau of Mines, p. 32.
99. Barber, H.G., and Carter, J.R., 1971, "An Account of Fossil Freshwater Diatomaceous Earth from Gordon Road Site, Auckland, New Zealand, Part 2," *Microscopy,* Journal of Quekett Micr. Club, Vol. 32, No. 1, pp. 24–28.
100. Barber, H.G., and Carter, J.R., 1972, "An Account of Fossil Freshwater Diatomaceous Earth from Gordon Road Site, Auckland, New Zealand," *Microscopy,* Journal of Quekett Micr. Club, Vol. 32, No. 5, pp. 141–147.
101. Beck, F.M., 1966, "Diatomite in Maine," *Contributions to the Geology of Maine,* Bulletin IE, Maine Geological Survey, p. 10–17.
102. Charrin, V., 1962, "Les Gisements Francais de Kieselguhr," *Genie Civil,* No. 139, pp. 84–88.
103. Endrihinskii, A.S., and Cheremisinova, E.A., 1970, "On Miocene Sedimentary Rocks from the Vitim Plateau," *Doklady Akademii Nauk SSR,* Seriya Geol., Vol. 191, Nos. 4–6, pp. 885–888.
104. Ermina, A.S., 1964, "Characteristics of Dia-

tomites in Area of the Gulf of Finland and the Lodoga Lake," *Gruntoved. i Inzhenernyi Geologii,* Leningrad University, pp. 89–111.

105. Gross, M., et al., 1963, "Varved Marine Sediments in a Stagnant Fjord," *Science,* Vol. 141, No. 3584, Sep. 6, pp. 918–991.

106. Gulyan, E.Kh. and Arutyunyan, A.A., 1967, "Wzhardosk Diatomaceous Earth Deposit in Armenia SSR", *Razvedka Okhrana Nedr,* Vol. 33, No. 4, pp. 8–11.

107. Herbert, C., 1968, "The Mineral Industry of New South Wales—Diatomite," *Geological Survey Bulletin of New South Wales,* No. 4, pp. 5–29.

108. Ichikawa, W., 1971, "Diatom Remains from the Surface Layer in the Bottom of the Japan Sea. 1. Core Samples Collected by Seifu-Maru," *Bulletin,* Japan Sea Research Institute, Kanazawa University, No. 3, pp. 1–34.

109. Ichikawa, W., 1971a, "Diatom Remains from the Surface Layer in the Bottom of the Japan Sea. 2. Core Samples Collected by Seifu-Maru," *Bulletin,* Japan Sea Research Institute, Kanazawa University, No. 4, pp. 1–42.

110. Khalfina, N.A., 1971, "Diatoms from the Type Locality of Karginsky Beds, Lower Reaches of the Yenisei River," *Morskoi Pleistotsen Sibirskikh Rvanin,* Trudy, Institut Geologii i Geofiziki Akademii Nauk SSR, Sibirskoe Otdelenie, Vyp., Vol. 104, pp. 100–112.

111. Knezevic, V., 1968, "Diatomaceous Earth Deposit near Barosevoc and its Economic Significance," *Tehnika* (Belgrade), Vol. 23, No. 9, pp. 1509–1521.

112. Kozyrenko, T.F., 1971, "New and Rare Diatoms of Neogene Sediments of Kuellsk, Australia," *Novosti Sistematiki Nizshikh Rnstenii,* Vol. 8, pp. 34–41.

113. Levin, J., 1968, "Preliminary Examination of Kieselguhr from Witdraai South of Postmasburg," Research Report No. 287, National Institute of Metallurgy, South Africa.

114. Lohman, K.E., and Andrews, G.W., 1968, "Late Eocene Non-Marine Diatoms from the Beaver Divide Area, Fremont Co., Wyo.," Professional Paper 593E, US Geological Survey, pp. E1–E26.

115. Mason, R.S., 1965, "Oregon's Mineral Industry in 1964," *The Ore Bin,* Vol. 27, No. 1, Jan., pp. 1–24.

116. McCrory, F.J., and O'Haire, R.T., 1961 "Map of Known Non-Metallic Mineral Occurrences of Arizona," Arizona Bureau of Mines, University of Arizona, Nov.

117. McLaughlin, R.B., 1972, "A Preliminary Account of Diatomaceous Deposits on Kenai Peninsula, Alaska," *Microscopy,* Journal of Quekett Micr. Club, Vol. 32, No. 5, pp. 151–154.

118. Okuno, H., 1967, "Diatomaceous Earth in Oki Island," *Journal of Japanese Botany,* Vol. 43, No. 1, Jan., pp. 17–25.

119. Olson, R. H., 1960, "A Survey of Industrial Mineral Possibilities in Nevada," 13B, Nevada Bureau of Mines, pp. 25–38.

120. Peterson, N., 1971, "Diatomite Occurrences in Klamath Co.," *The Ore Bin,* Vol. 33, No. 1, Jan., p. 14.

121. Pharr, R.F., 1961, "Diatomaceous Sediments in Virginia," *Virginia Minerals,* Vol. 2, No. 3, Aug., pp. 25–31.

122. Pironkov, S., and Kovacher, K., 1965, "Some Characteristics of the Diatomite from the Gurmen Deposit in the Blagsevgrad District," *Tr. Geol. Mino Dels,* Vissh. Mino-Geol. Inst., pp. 237–242.

123. Ruth, J.W., 1971, "Late Tertiary Diatoms and Silicoflagellates of the California Coastal Area," Abstract, Geological Society of America, Vol. 3, No. 2, p. 188.

124. Stall, W.C., 1961, "A Short Review of Peru's Mineral Resources," *Economic Geology,* Vol. 56, pp. 985–990.

125. Stall, W.C., 1962, "Notes on the Mineral Resources of Ecuador," *Economic Geology,* Vol. 57, No. 5, Aug., pp. 799–808.

Mining and Processing

126. Anon., 1967, "Diatomite Slurry Pipeline in B.C." *Minerals Processing,* Vol. 8, No. 6, June, p. 28.

127. Bogdau, I.G., 1963, "Treatment of Kieselguhr with Chlorine," *Sakhorn. Prom.,* Vol. 37, No. 2, pp. 22–23.

128. DeCastro, R.R., 1962, "Diatomaceous Earth," *Grasas Aceitas,* Vol. 13, pp. 134–138.

129. Knechtel, M.M., and Hosterman, J.W., 1965, "Outlook for Resumption of Diatomite Mining in Southern Maryland and Eastern Virginia," Prof. Pap. 525D, US Geological Survey, pp. 151–155.

130. Pironkov, S., and Kovacher, K., 1965, "Production of Kieselguhr from Bulgarian Raw Material," *Rudodobiv Met.* (Sofia), Vol. 20, No. 4, pp. 19–22.

131. Velihaza, E.J., and Maltsev, P.M., 1961, "The Effect of Various Methods of Treatment of Kieselguhr from the Kirovgrad Region on Its Chemical Composition," *Tr. Kievsk. Inst. Pischevoi Prom.,* No. 24, pp. 156–158.

Filtration

132. Anon., 1960, "Norge Launches Coin-Op Drycleaners," *Chemical & Engineering News,* Vol. 38, No. 42, Oct. 17, p. 38.

133. Anon., 1960a, "Swimming Luxury in Sweden," *Power Specialist,* Vol. 35, No. 2, pp. 10–11.

134. Anon., 1962, "Diatomite Filter Removes Iron," *Engineering News,* Vol. 169, Lowell, MA, 8 pp.

135. Anon., 1968, "Johns Manville Offers Latest ABS Removal System," *Chemical and Engineering News,* Vol. 40, No. 44, Oct. 29, p. 48.

136. Anon., 1970, "The Secret of the Fossil Filters," *Filtration and Separation,* Eagle-Picher Co., Vol. 7, No. 6, Nov.-Dec., p. 742.

137. Anon., 1972, *Filter Aid Filtration,* Johns-Manville Corp., Bulletin FA-84A, 27 pp.

138. Anderson, D.R., 1967, *Coated Diatomite for Treatment of Radioactive Wastes,* University Microfilms, Ann Arbor, Mich., Order 68–279E.

139. Bartolo, A.M., 1961, "Continuous Pipe-Line Defacation with Kieselguhr," Proceedings, Technical Sessions on Bone Char, pp. 145–153.

140. Baumann, E.R., et al., 1962, "A Theory of Diatomite Filtration," *Journal of the Ameri-*

can *Water Works Association,* Vol. 54, No. 9, Sep., pp. 1109–1119.

141. Baumann, E.R., 1965, "Diatomite Filters for Municipal Use," *Journal of the American Water Works Association,* Vol. 57, Feb., pp. 157–180.

142. Bell, G.R., 1961, "Coagulant Coatings Open New Applications to Filter Aids," *Proceedings,* 22nd Annual Water Works Conference of Western Pennsylvania, pp. 129–140.

143. Bell, G.R., 1961a, "Use of Coagulant Coated Filter Aids," Paper given at American Water Works Assn., California Section, Oct. 27, 11 pp.

144. Bell, G.R., 1965, "Removal of Soluble Iron by Filtration," *Journal of the American Water Works Association,* Vol. 57, No. 4, Apr., pp. 458–471.

145. Bell, G.R., 1965a, "Removal of Manganese by Controlled Precipitation and Filtration," *Journal of the American Water Works Association,* Vol. 57, No. 5, May, pp. 655–662.

146. Buck, F.W., 1966, "Diatomite Filtration of Industrial Wastes," *Water and Sewage Works,* Vol. 113, June, pp. 200–204.

147. Burns, D.E., et al, 1970, "Particulate Removal on Coated Filter Media," *Journal of the American Water Works Association,* Vol. 62, Feb., pp. 121–126.

148. Coogan, G.T., 1962, "Diatomite Filtration for Removal of Iron and Manganese," *Journal of the American Water Works Association,* Vol. 54, pp. 1507–1517.

149. Cummins, A.B., 1973, "Development of Diatomite Filter Aid Filtration," *Filtration and Separation* (London), Pt. I, Vol. 10, No. 1, Jan.-Feb., pp. 53–60; Pt. II, Vol. 10, No. 2, Mar.-Apr., 1973, pp. 215–219. Pt. III, Vol. 10, No. 3, May-June, 1973, pp. 317–320, 332.

150. Emmet, R.C., and Dahlstrom, D.A., 1961, "A Guide to Continuous Filtration of Precipitates," *Mining Engineering,* Jan., pp. 45–50.

151. Filipposyants, T.T., et al., 1961, "Use of Diatomaceous Earth in Filtration of Antibiotic Culture Media," *Med. Prom. SSSR,* Vol. 15, No. 11, pp. 46–50.

152. Harding, H., 1960, "Diatomite Filtration Proves Economical Standby for Water Plant," *Public Works Magazine,* Vol. 91, No. 11, Nov., p. 103.

153. Hernandez, L., et al., 1961, "A Comparative Study of Some of the Effects of Filter Media on Atropine Sulfate and Vitamin B_{12} Solutions," *Parental Drug Association Bulletin,* Vol. 15, No. 4, pp. 16–23.

154. Hunter, J.V., et al., 1966, "Coliform Organism Removals by Diatomite Filtration," *Journal of the American Water Works Association,* Vol. 58, No. 9, Sep., pp. 1160–1169.

155. Jackson, T.M., 1961, "Filter Aids Speed Up Difficult Filtrations," *Chemical Engineering,* Vol. 68, No. 6, Mar., pp. 141–146.

156. Jahreis, C.A., 1961, "Filter Performance under Field Conditions" *Chemical Engineering Progress,* Vol. 57, No. 7, July, pp. 60–62.

157. Midivnishivili, O.M., et al., 1961, "Comparative Study of USSR Diatomites as Filtration Material in the Production of Insulin," *Tr. Kovhazsk. Inst. Mineralin Syrya,* Vol. 3, pp. 231–237.

158. Peravalov, V.G., and Alehseeva, V.A., 1961, "The Use of Diatomaceous Earth for Purifying Water for Crude Oil Production," *Neftyanoe Khozyaistvo,* Vol. 39, No. 4, pp. 52–55.

159. Rankin, R.S., 1960, "Diatomite Water Filters for Municipal Suppliers," *Public Works,* Vol. 91, Feb., pp. 60, 68, 70.

160. Timen, J.C., 1970, "Diatomite Filtration Reduces Mercury Pollution," *Chemical Engineering,* Vol. 77, No. 27, Dec., p. 77.

Filler and Related Uses

161. Anon., 1960, "Jewels of the Sea Help Blue Coral Preserve Lustrous Finishes," *New York Herald Tribune,* Sec. 11, Mar. 20, p. 4.

162. Anon., 1961, "New Agent Gives Paint Makers Both Fast Dispersion and High Flatting," *Power Specialist,* Vol. 36, No. 4, p. 14.

163. Anon., 1961, "The Paint Industry," *Chemical & Engineering News,* Vol. 47, Dec., pp. 31–57.

164. Anon., 1969, "Celite and Micro-Cel Functional Fillers," FA-64A, Johns-Manville Corp., Apr., 27 pp.

165. Ammons, V., 1963, "The Dispersed Pigment Problem," *Industrial & Engineering Chemistry,* Vol. 55, No. 4, pp. 40–47.

166. Agronomov, A.E., and Mardashev, Vu.S., 1961, "Structure and Activity of Supported Nickel Catalysts I. Change in the Structure of the Carrieron Deposition of Nickel," *Zhurnal Fig. Khimii,* Vol. 35, pp. 1660–1670.

167. Bareja, H., and Ruskiewicz, M., 1972, "New Possibilities of Diatomaceous Earth in the Polish Industry," *Index of Geology,* Vol. 36, No. 7, Sec. 2, No. E–72–21630; or *Przett Geol.,* Vol. 20, No. 2, pp. 74–78.

168. Brody, D.E., 1966, "Pigment Technology," Pts. I, II, III, *Paint and Varnish Production,* May, pp. 59, 61, 63, 65, 67, 69, 92, June, pp. 57–69, July, pp. 51–62.

169. Brody, D.E., 1968, "Performance of Flatting Pigments in High Speed Dispersion," *Journal of Paint Technology,* Vol. 40, No. 525, Oct., pp. 439–445.

170. Coombs, G., 1970, "Pesticide Formulation as Seen by a Producer of Carriers," *Agricultural Chemicals,* Vol. 15, June, pp. 47, 49, 50, 103, 105.

171. Gorshkov, V.S., et al., 1957, "Determination of the Degree of Interaction of a Filler with the Resin in Poly (Vinylchloride) Compositions on the Basis of Their Solution Kinectics," *Sibirskii Tr. Vses Nauch-Issledovat,* Institut Novykh Stroit Mater, No. 15, pp. 30–32.

172. Kadey, F.L., et al., 1960, "Relationship of Hardness, Friability, and Particle Size to the Abrasive Performance of Diatomaceous Silica," *Trans. AIME,* Vol. 217, pp. 446–450.

173. Kadey, F.L., 1967, "Unique Properties of Diatomite and New Filler Uses," SME Preprint No. 67HI84, AIME Annual Meeting, Los Angeles, Feb., 8 pp.

174. Kranich, H., and Zilli, R.J., 1966, "Diatomite in Chemical Specialties," *Soap & Chemical Specialties,* Vol. 42, No. 12, pp. 73–74.

175. Kranich, H., 1973, "Diatomaceous Silica," *Pigment Handbook,* Vol. 1, T.C. Patton, ed., John Wiley, New York, pp. 141–155.
176. Kollonitsch, V., et al., 1970, "Functional Mineral Pigments," *Industrial Minerals,* Pt. 1, No. 38, Nov., pp. 35–40; Pt. 2, No. 39, Dec., pp. 46–51.
177. Ovcharenko, F.D., et al., 1960, "Activated Diatomite, a New Rubber Filler," *Dopovidi Akademii Nauk Ukrainskoi RSR,* No. 1, pp. 54–59.
178. Ovcharenko, F.D., et al., 1961, "Activated Diatomite as a Rubber Reinforcing Agent," *Dopovidi Akademii Novik Ukrainskoi RSR,* pp. 504–507.
179. Strong, R.G., and Sbur, D.E., 1963, "Protection of Wheat Seed with Diatomaceous Earth," *Journal of Economic Entomology,* Vol. 56, No. 3, pp. 372–374.

Chromatography

180. Anon., 1971, "Chromosorb—Diatomite Supports for Gas Liquid Chromatography," Johns-Manville Corp., FF-1010, Nov., 11 pp.
181. Ane, W.A., et al., 1971, "The Distribution of Support Bonded Silicones on Chromosorbs," *Journal of Chromatography,* Vol. 56, No. 2, Apr., pp. 295–310.
182. Arutyunyan, A.Kh., and Shakhbayyan, A.M., 1960, "Purification of Sewage Waters by Absorption Chromatography," *Transactions,* Erevansk. Med. Institute, No. 11, pp. 169–172.
183. Ayers, D.O., et al., 1961, "Principals of High Speed Gas Chromatography with Packed Columns," *Analytical Chemistry,* Vol. 33, No. 8, July, pp. 986–991.
184. Bohemen, J., et al., 1960, "A Study of the Absorptive Properties of Firebrick in Relation to its Use as a Solid Support in Gas-Liquid Chromatography," *Journal of Chemical Society,* pp. 2444–2457.
185. Bryzgalova, N.I., et al., 1969, "Preparation and Investigation of Macroporous Diatomite Support for Gas Chromatography," *Neftekhimiya,* Vol. 9, No. 3, May-June, pp. 463–469.
186. Hayashi, N., et al., 1970, "Studies on Surface Structures of Diatomaceous Earth Supports and Packing Materials," *Bunseki Kagoku,* Vol. 19, No. 7, July, pp. 907–918.
187. Ottenstein, D.M., 1963, "Column Support Materials for Use in Gas Chromatography," *Journal of Gas Chromatography,* Vol. 1, No. 4, Apr., pp. 11–22.
188. Strain, H.H., 1960, "Chromatography," *Analytical Chemistry,* Vol. 32, No. 5, Apr. pp. 3R–18R.
189. Vigdergauz, M.S., 1968, "Inert Solid Support for Gas Chromatography," *Khim. Tekhnol. Topl. Masel,* Vol. 13, No. 3, Mar., pp. 60–63.

Diatoms

190. Cheremissinova, H.A., 1971, "New and Interesting Diatoms from the Neogene of Pribaikal," *Nov. Sist. Niz. Rast.,* Vol. 8, pp. 52–66.
191. Chin, T.G., et al., 1965, "Influence of Temperature and Salinity on the Growth of Three Species of Plankton Diatoms," *Ocenol. Limnol. Sinica.,* Vol. 7, No. 4, pp. 384–394.
192. Conger, P.S., 1966, "How Diatoms Move," *Journal of Phycology,* Vol. 2 sup., p. 7.
193. Desikachary, T.V., 1962, "The Diatoms," *Current Science,* Vol. 31, Feb., pp. 43–45.
194. Donahue, J.G., 1970, "Pleistocene Diatoms as Climatic Indicators in North Pacific Sediments," Memoir No. 126, Geological Society of America, pp. 121–138.
195. Felfoldy, L.J.M., 1961, "Effect of Temperature on the Photosynthesis of a Natural Diatom Population," *Am. Institute of Biology,* Vol. 28, pp. 95–98.
196. Hellebust, J.A., 1970, "Plants," *Marine Ecology, Vol. 1, Environmental Factors,* O. Kunne, ed., pp. 125–158.
197. Hopkins, J.T., 1967, "The Diatom Trail," *Microscopy,* Vol. 30, No. 9, pp. 209–217.
198. Hopkins, J.T., 1969, "Diatom Motility: An Explanation and a Problem," *British Phycology Bulletin,* Vol. 3, No. 1, pp. 63–67.
199. Hughes, J.C., and Lund, J.W.G., 1962, "The Rate of Growth of Asterionella Formosa Hass. in Relation to its Ecology," *Arch. Mikrobiol.,* Vol. 42, pp. 117–129.
200. Jitts, H.R., et al., 1964, "The Cell Division Rates of Some Marine Phytoplankton as a Function of Light and Temperature," *Journal of Fisheries Research Board of Canada,* Vol. 21, No. 1, pp. 139–157.
201. Jouse, A.P., 1968, "Diatoms from Pleistocene and Late Pliocene Cores of the Northern Pacific Ocean," *Osnovnye Problemy Mikropaleontologii i Organogennogo Osadkonakopleniya v Okeanakh i Moryakh,* pp. 5–27.
202. Ko-Bayashi, T., 1963, "Variations on some Pennate Diatoms from Antarctica," Japanese Antarctic Research Expedition, 1960–1962, Seci. Repts. Ser. E. Biol., No. 18, p. 20.
203. Lohmann, K.E., 1960, "The Ubiquitous Diatom—A Brief Survey of the Present State of Knowledge," *American Journal of Science,* No. 258-A, pp. 180–191.
204. Patrick, R., 1971, "The Effects of Increasing Light and Temperature on the Structure of Diatom Communities," *Limnol. Oceander,* Vol. 16, No. 2, pp. 405–421.
205. Schluter, M., 1961, "The Importance of Littoral Diatoms in Our Waters," *Zeitschrift fuer Fisch.,* Vol. 10, pp. 351–359.
206. Ukeles, R., 1961, "The Effect of Temperature on the Growth and Survival of Several Marine Algal Species," *Biological Bulletin,* Vol. 120, pp. 255–264.
207. Wright, H.E., Jr., 1970, "Diatoms in Lakes and Lake Sediments as an Index to Environment," *University of Minnesota Technical Progress Report,* May 1, 1970, January 31, 1971.
208. Zabelina, E.K., 1968, "Diatoms in the Sediments of the Northern Part of the White Sea," *Osnovnye Problemy Mikropaleontologii i Organogennogo Osadkonakopleniya v Okeanakh i Moryakh,* pp. 85–95.
209. Zahl, P.A., 1961, "How the Sun Gives Life to the Sea," *National Geographic,* Vol. 119, No. 2, Feb., pp. 199–225.

Selected References Since 1974

210. Anon., 1974, "A Wide Range of Uses for Old Skeletons," *Industrial Minerals*, No. 86, Nov., pp. 9–25.
211. Anon., 1975, "The Industrial Minerals of Denmark," *Industrial Minerals*, No. 89, Feb., pp. 9–29.
212. Anon., 1979, "Pilot Project to Produce Oil from Diatomite," *World Mining*, March, p. 83.
213. Abbott, P.L., and Gastil, R.G., 1979, "Baja California Geology," Geological Society of America Annual Meeting, San Diego, Nov., publ. prepared for field trips 10, 12, 13, and 26, pp. 107–111.
214. Alciatore, A.F., et al., 1974, "Advances in Perlite and Diatomite Filter Aid Research," *Filtration and Separation*, March/April, pp. 121–126.
215. Andrews, G.W., 1976, "Miocene Marine Diatoms from the Choptank Formation, Calvert County, Maryand," Professional Paper P-910, US Geological Survey, 26 pp.
216. Basso, A.J., 1974, "Bulk Handling and Shipment of Diatomaceous Earth," *Filtration Engineering*, Vol. 5, No. 4, pp. 10–11, 14.
217. Baudrimont, R., and Degiovanni, C., 1974, "Sedimentoligie—Interpretation Paleo-ecologique des Diatomites du Miocene Superior de L'Algerie Occidentale," Academie des Sciences CR, Vol. 279, No. 16, Serie D, pp. 1337–1340.
218. Blake, E.E., and Paulsen, M.B., 1978, "Diatomite Filtration Saves Space and Time," *Water and Water Engineering*, May, pp. 27–30.
219. Bryant, E.A., and Bailey, D., 1979, "Ozone-Diatomaceous Earth Filtration for Treatment of New York City Croton Water Supply," *Proceedings*, Part I, AWWA 1979 Annual Conference, San Francisco, June 24–29, pp. 283–299.
220. Clark, W.B., 1978, "Diatomite Industry in California," *California Geology*, Jan., 9 pp.
221. Coombs, G., 1980, "Diatomite," *Mining Engineering*, May, p. 568.
222. Culver, R.H., 1975, "Diatomaceous Earth Filtration," *Chemical Engineering Progress*, Vol. 71, No. 12, pp. 51–54.
223. Erguvanli, K., et al., 1975, "Siliceous Rock Deposits and Its Economic Importance in Turkey," Congress of Earth Sciences, on the Occasion of the 50th Anniversary of the Turkish Republic," Dec. 17–19, 1973, published by Mineral Research and Exploration Institute.
224. Meisinger, A.C., 1978–1979, "Diatomite," *Minerals Yearbook*, US Bureau of Mines, pp. 313–315.
225. Meisinger, A.C., 1980, "Diatomite," Preprint, *Mineral Facts and Problems*, Bulletin No. 671, US Bureau of Mines, 7 pp.
226. Meisinger, A.C., 1982, "Diatomite," *Mineral Commodity Summaries, 1981*, US Bureau of Mines, pp. 46–47.
227. Patel, P.P., et al., 1979, "On the Occurrence of Diatomite from Bhavnagar Area, Gujarat," *Journal of the Geological Society of India*, Vol. 20, No. 3, pp. 114–117.
228. Patterson, K., 1977, "Australian Mineral Industry—Diatomite," *Minor Metals and Minerals—1975 Review*, Australian Bureau of Mineral Resources, Geology and Geophysics, pp. 364–366.
229. Pinto, T., 1976, "Developments in Beer Filtration," *The Brewer*, April, pp. 115–120.
230. Rowell, H.C., 1980, "Diatom Biostratigraphy of the Monterey Formation, Palos Verdes Hills, California," M.S. Thesis, University of Southern California, 123 pp.
231. Smith, G.R.S., 1975, "Filter Aid Regeneration and Recovery," *Chemical Engineering Progress*, Vol. 71, No. 12, pp. 37–39.
232. Stosur, J.J., and David, A., 1976, "Petrophysical Evaluation of the Diatomite Formation of the Lost Hills Field, California," *Journal of Petroleum Technology*, Oct., pp. 1138–1144.
233. Takayama, Y., 1979, "Preparation of Silyated Celite by a Vapour-Phase Method," *Journal of Chromatography*, Vol. 178, pp. 63–70.
234. Tiller, F.M., and Lloyd, P.T., 1974, "Theory and Practice of Solid-Liquid Separation," Chemical Engineering Dept., University of Houston, 12 chapters.
235. Timmermans, D.E.B., 1980, "Porous Old Skeletons Playing a Vital Role in Industry," *South African Mining and Engineering Journal*, Jan., pp. 103–107.

Feldspars

Original by CARROLL P. ROGERS, JR.*
and
J. PHILIP NEAL †
Revised by KEFTON H. TEAGUE ‡

Feldspars, the most abundant minerals of the igneous rocks, occur in numerous forms and mixtures. The feldspars of commercial significance are found in widely distributed pegmatites as large crystals usually free of iron-bearing impurities and thus suitable for hand-cobbing, and also in larger bodies where the ore contains various types of feldspar intermingled with quartz and relatively free of iron-bearing impurities, or at least readily unlocked from impurities present at relatively coarse mesh. The latter type of ore includes some pegmatites, and the "alaskite" bodies so common in the Spruce Pine district of North Carolina. On the west coast of the US and in Oklahoma there is—or has been—commercial recovery of natural feldspathic sand as a source of alumina in glass and ceramics. "Aplite" in commercial terminology is a light-colored igneous rock with a granitic composition and a fine sugary texture, often banded with readily removable iron impurities. Only one area in Virginia produces aplite for the glass industry; it is located near Montpelier in Hanover County.

According to Deer et al. (1966) the word feldspar derives from the Swedish *feldt* or *fält* (field) and "spath:" chunks of rock appearing in tilled ground overlying granite. Castle and Gillson (1960) indicate a Germanic origin, citing "spat," which is said to refer to any transparent or translucent material which is readily cleavable. The term "spar" has in the past been applied to minerals other than feldspar, such as barite, calcite, and fluorite. It is correct to call the latter "fluorspar." There is a true barium feldspar, but it is rare and of no present economic importance. What is recognized as feldspar or "spar" by present producers and consumers consists of three silicate minerals which, if pure, would have the formulas $KAlSi_3O_8$ (microcline or orthoclase), $NaAlSi_3O_8$ (albite), and $CaAl_2Si_2O_8$ (anorthite). These are almost never found in pure form in nature, but occur together in great abundance in a three-component system. A considerable number of different combined ratios of these three exists, with the exception that isomorphism between potassium spar and calcium spar is very limited. User specifications for feldspar products are largely built around the prevalent naturally occurring ratios. Passages following will refer to "K-spar," "Na-spar," or "Ca-spar."

Nomenclature and Descriptive Terms

Relating to both the mineralogy and the economics of feldspar are certain terms whose coverage it is useful to specify.

Commercially, *aplite* is a feldspathic rock mined and beneficiated in one location in Virginia, in which both titanium and feldspar minerals are present—the latter now being the major, if not the only, economic product. Technically, the term aplite when applied to this rock is of questionable accuracy. But, in light of accepted commercial usage, the term will be here used to designate it. Regarding the Virginia rock, Ross (1941) is cited.

Alaskite is another term requiring understanding if not precise use. A distinctive rock type customarily called by this name is found near Spruce Pine, Mitchell County, North Carolina. An important major feldspar source, it contains somewhat higher levels of plagioclase feldspar and quartz than alaskite as defined in the *Glossary of Geology and Related Sciences*, which states that "orthoclase, microcline, and subordinate quartz are the principal

* President, The Feldspar Corp., Spruce Pine, NC.
† Ore Dressing Specialist, Minerals Research Laboratory, North Carolina State University, Asheville, NC.
‡ International Minerals & Chemical Corp., Skokie, IL.

minerals" and "plagioclase may or may not be present." Brown (1962) notes a granitoid rock near Peakville in Bedford County, Virginia (a past commercial feldspar source) as being alaskite, but he also classifies the Spruce Pine rock as such. There is some professional disagreement as to whether the Spruce Pine rock is a granite or a pegmatite. In any event, the term "alaskite" will, in this chapter, pertain to the relatively coarse-grained granitelike Spruce Pine ore which is well known to the feldspar industry.

Graphic granite (also called corduroy spar) can be briefly described as a pegmatite rock predominating in K-spar, with the secondary mineral, quartz, forming a distinctive pattern in it. It is a source of feldspar where high K_2O assay is required.

Pegmatite is prevalent, widely distributed, generally coarse-grained igneous rock from which are obtained predominantly potash feldspar and a certain number of other economic minerals. They occur frequently in association with granites, and usually have major minerals in common with these, but frequently a higher percentage of some (such as K-spar) in large crystals. And, in some instances, there are also additional, often more exotic, minerals present.

Perthite is a microscopic intergrowth of plagioclase in K-spar, having a high K-spar to Na-spar ratio. In graphic granite and in pegmatites, perthite is of common occurrence.

Commercially, *soda spar* is a mixture assaying 7% Na_2O or higher while *potash spar* contains 10% K_2O or higher. The term soda spar has recently changed. Formerly it denoted a higher Na_2O assay than at present. *Pottery spar* is a feldspar product which goes generally into a shaped, fired body where the spar exerts a fluxing action. Such feldspar is usually ground to −200 mesh or finer. The presence of a given level of K-spar is important here, although potash spar is not required in all ceramic applications. *Glass spar* is—with very few exceptions—a soda spar ground to −20 mesh or sometimes as fine as −40 mesh. It is expected to assay a given level of Al_2O_3—the most important chemical value which the spar furnishes to the melt.

Feldspathic sand is a naturally occurring or processed mixture of feldspar and quartz.

Mineralogy and Geology

Feldspar, or feldspar-plus-quartz, is now being obtained from a limited number of rock types. Most of these have been referred to previously.

The following, as present feldspar sources, appear to merit attention from a mineralogical standpoint:

1) pegmatite, including graphic granite;
2) "alaskite" as recognized commercially;
3) "aplite" in the same sense;
4) granite;
5) feldspathic sand;
6) feldspathic quartzite.

Pegmatite is discussed only briefly here, but several good references are cited in the bibliography.

Generally speaking, feldspathic pegmatite bodies have the following notable characteristics:

1) derivation from residual magma solutions;

2) a wide variety of accessory minerals with frequent evidence of magmatic alteration and replacement;

3) considerable variability in shape, size, and extent of body;

4) a tendency toward a concentric sequence of mineral zones, with varying distributions and ratios of minerals; and

5) major minerals in common with associated granites.

Point 4 is commonly seen as the result of successive inward differential crystallization. This phenomenon follows a certain sequence of mineral assemblages (Cameron, et al., 1949).

With these characteristics, involving so many thermal and chemical variables, pegmatites contain feldspars and numerous other minerals in many combinations and a variety of crystal sizes. A list is cited of minor elements which can be found in various pegmatites (Castle and Gillson, 1960):

Sb	Sr	Cb	Zn	Sc	Sn
As	Be	Ta	Li	S	Ti
Bi	Cs	Cu	Mo	Th	Zr
Ba	Ru	Pb	W	U	Rare Earths

In addition, low concentrations of Cr, Co, Hg, Se, Te, V, and precious metals are occasionally found.

The following accessory minerals may be found in pegmatites in addition to the more usual ones (K-spar, Na-spar, quartz, and muscovite) (Williams, et al., 1954):

Lepidolite	Zircon
Spodumene	Tourmaline
Almandine	Topaz
Spessartite	Epidote

Allanite	Amblygonite	Columbite
Beryl	Fluorite	Tantalite
Apatite	Wolframite	Lithiophillite
Eudialite	Cassiterite	

To these could be added pollucite, petalite, and others.

Distribution of pegmatites in the United States, and their respective sizes and mineralogies, can be further studied by consulting Cameron, et al. (1949). The principal pegmatite regions of the US can be summarized as follows:

1) New England: southwestern Maine, New Hampshire, Massachusetts, and Connecticut.

2) Southeastern states from Virginia through the Carolinas to Georgia and Alabama.

3) The Black Hills of South Dakota.

4) The Rocky Mountain states, especially Colorado and New Mexico.

5) Small pegmatite districts in California, Arizona, Idaho, Texas, and Wyoming.

There are scattered pegmatites elsewhere, such as in southeastern Pennsylvania, southeastern New York, northern New Jersey, central Wisconsin, and central Washington.

Where pegmatite consists of large mineral crystals of K-spar, quartz, and not much else, the rock is—or has been—mined and hand-cobbed for potash feldspar (*block spar*). In other pegmatites, mineral substitutions and alterations have taken place, there are various minor and accessory minerals, and the rock is usually finer grained. Thus it requires beneficiation by relatively fine grinding (−20 mesh), and then by magnetic separation and/or froth flotation, to secure desired concentrates. In addition to feldspar, quartz, muscovite mica, spodumene, pollucite, and other economic minerals are obtained from the finer grained pegmatites. For commercial mining, the best zone or zones of any given pegmatite must be selected to yield the desired minerals in optimum availability and quality. Pegmatite containing spodumene occurs in contiguous bodies near Kings Mountain, North Carolina, and is an important source of feldspar and feldspathic sand—these being actually secondary to the spodumene content in economic importance. This pegmatite is the characteristic rock of the so-called tin-spodumene belt of the Carolinas. The belt can be briefly described as "a narrow, sinuous zone which strikes northeastward, paralleling in general the layering and foliation of the principal rock units of the area" (Broadhurst, 1956).

In considering theories regarding pegmatite formation, a concept involving mineral deposition from an aqueous or pneumatolytic environment may be of interest. This theory is intermittently referred to by Park, et al. (1964), wherein additional references are given, and by Turner, et al. (1960), dealing with related physical laws.

Graphic granite is cited by Cameron, et al. (1949), as one pegmatite manifestation. Williams, et al. (1954) describe graphic granite by the phrase "a cuneiform intergrowth of quartz and potash feldspar." This type of intergrowth is supposed to have come about through simultaneous or rapid crystallization from a viscous melt, along with a certain amount of replacement of massive K-spar by quartz. Like coarse-grained pegmatite, graphic granite will yield high-grade potash feldspar if beneficiated by grinding and flotation. If dry ground as is, the resulting feldspar product, being less pure, would be lower grade than fine-ground block spar.

The alaskite commercially mined in the area of Spruce Pine, North Carolina, is defined by Parker (1952) as a Paleozoic igneous intrusive and also a stocklike pegmatite mass of Carboniferous age. Parker divides the intrusives of this area (those covered by his designation "pegmatite") into (1) some unevenly distributed aplites in the true mineralogical sense, (2) pegmatites of coarse-grained crystalline texture in locations mostly peripheral to the principal finer grained rock, and (3) the principal type, which is characterized by various writers as alaskite or granite. As elsewhere, the coarse-grained pegmatites have yielded commercial hand-mined block spar.

The alaskite is composed, roughly, of the following principal mineral ingredients.

Plagioclase:	45%
Quartz:	25%
Microcline:	20%
Muscovite:	10%

These proportions may vary; there may be less microcline or muscovite and sometimes none. Plagioclase and quartz are nearly always present. The commercially mined rock of the principal mass contains plagioclase-quartz-perthite-muscovite, usually in that order of abundance. In addition there are present minor amounts of garnet, biotite, and apatite plus very small quantities of beryl, tourmaline, epidote, and others (Parker, 1952).

Except for varying textures, the mostly peripheral coarse-grained bodies easily classified as pegmatites differ little in their collective mineralogy from that of the overall central body, leading to the conclusion of derivation from a common magma. Moreover, the prevailing mineralogical ratios of the central body are quite atypical of granite. Grain size of the alaskite is, on the average, between 0.25 and 0.5 in.: considerably coarser than the average granite or granodiorite (Parker, 1952).

The Spruce Pine alaskite has the characteristics of exceptional mineral purity, uniformity, coarse-grained texture, and large size of deposit, which are sought in commercial ore of feldspar.

The feldspathic aplite of Virginia, previously mentioned, is described by Castle and Gillson (1960) as essentially an intrusive mass having variable texture. It has presumably undergone magmatic alteration from coarse feldspar and is a host rock for titanium minerals not found in true aplite. According to Williams, et al. (1954), aplite in the mineralogical sense developed from residual magmas and occurs, associated with pegmatite, as narrow, relatively homogeneous intrusives with such accessory minerals as garnet, zircon, and tourmaline. These factors, plus differing content of K-, Na-, and Ca-feldspars, indicate that the Virginia rock should not be classified as true aplite. (For an additional reference on this rock, see Ross, 1941).

Granites are extremely abundant, a certain number of them being good sources of high-quality feldspar. Granite is coarse-grained acid plutonic igneous rock. Quartz generally makes up 25 to 30% of the rock; feldspar is the other major component. There is a wide range of minor constituents, among them muscovite, biotite, apatite, zircon, allanite, hornblende, magnetite, epidote, zoisite, and garnet. The feldspar component of granite is principally of the alkali members, which is to say crystalline feldspar containing some ratio of the potassium and sodium feldspars. The remaining feldspar is of the plagioclase series, i.e., a varying ratio of sodium and calcium feldspar in solid solution.

As long as such ores as the pegmatites and the Spruce Pine alaskite are readily available, granite nearby is unlikely to be mined for feldspar. Being generally finer grained than pegmatites, granite must, of course, be also beneficiated by froth flotation. Where granite has been beneficiated for feldspar in the United States, the operation has been in conjunction with quarrying for production of road or dimension stone: fines from such an operation can offer an economic advantage in terms of reduced mining and grinding costs. The granite must be sufficiently coarse-grained so that there is substantial mineral liberation at 40 mesh or coarser. Inclusions of minor, iron-bearing minerals must be at a minimum. Near Pacolet, Spartanburg County, South Carolina, one such granite quarrying operation has in the past furnished tailings which were beneficiated into feldspathic sand. This granite happens to be mineralogically close to the optimum for feldspar beneficiation. The operations are discussed by Eddy, et al. (1972).

Various beach and river sands containing economic amounts of feldspar have apparently acquired the feldspar principally from granitic bodies, and possibly also from feldspathic metamorphic rocks. Basis for this statement is in the assay of feldspar concentrates from these sands. Normal decomposition of feldspar involves first the breakdown and kaolinization of Na-spar, leaving behind a component higher in K-spar. Thus, if an assay is run on spar concentrate from feldspathic sand and the K_2O is high (perhaps 10% or more), that would indicate either extreme decomposition or that the feldspar originated in a pegmatite. Most concentrates from feldspathic sand, however, usually do not assay over 7% K_2O: considerably below usual pegmatite feldspars, and much more typical of feldspar from partially weathered granite, with part or all of its plagioclase component weathered away. The feldspar in these sands, then, probably is mainly from the original alkaline feldspar of granite.

Since each of the rock types discussed has its own distinctive mineralogy, reflected by the types and ratios of feldspars present, it follows that varying commercial chemical standards for assorted feldspar uses will be best met by different mineralogical combinations found in aplite, alaskite, granite, feldspathic sand, or pegmatite. (These are cited here in more or less rising order of their K-spar to Na-spar ratio.)

The presence or absence of Ca-spar is usually of slight importance to the feldspar user. Since pure Ca-spar (anorthite) contains a theoretical 36.7% Al_2O_3, it is welcome in most glass applications. In certain ceramic uses it may need to be limited.

The geology and mineralogy of feldspar are thoroughly discussed by Barth (1969) and the physical and chemical characteristics by Berry, et al. (1959) and Deer, et al. (1966).

Distribution and Reserves

Feldspar is one of the most plentiful minerals in the earth's crust, occurring in various rock types within a short distance of almost anywhere. Thus, any absolute figure on reserve tonnage is academic. Any estimate of tonnage reserves would have to be relative to the economics of the mineral at the time. The feldspar market in the United States has been highly competitive for many years, and by now the surviving producers have located themselves on ore bodies of the best available quality, i.e., yielding a high percentage of feldspar easily concentrated as a quality product. Mineral purity, size of ore body, and coarse-grain liberation have been paramount considerations. Existing feldspar producers generally have available reserves which are calculated in quarter centuries, if not centuries. There may arise, of course, complications due to zoning or environmental limitations—or perhaps to shipping difficulties. Should such limitations preclude exploitation of many prime quality feldspar reserves, then the vast and widespread granite formations and the unworked alluvial sands may come into the picture. Cost and quality could well be variables of concern, but sheer physical reserves of marketable feldspar are on hand in copious quantity.

One rock type not previously discussed has recently emerged as an interesting potential reserve of potash spar. Located in San Bernardino County, California, this potential ore body is described by Sheppard and Gude (1965) as an altered tuff whose principal constituent is K-spar. It is part of the Barstow Formation and covers a wide area. Its chief disadvantage appears to be a high iron content in much of the feldspar.

Feldspar Production

Table 1 (Wells and Potter) shows world feldspar production by countries, 1959 through 1980. The columns denoting calculated four-year averages (1959–1974) do not show totals because changing compilation techniques and categories over the years introduced minor statistical variations which, if fully explained, would have necessitated a complex set of qualifying footnotes. Individual trends shown in the averages of these earlier years are here perhaps the most valuable type of data. More precise figures on this period may be found in the annual *Minerals Yearbook*. Table 2 shows tonnages of domestic feldspar shipped from merchant mills to destination states (1977 through 1980). Table 3 shows the derivation and use of ground feldspar sold by merchant mills in the United States, 1978 through 1980.

Table 1 shows that the US led all other nations in 1980 in feldspar production (710,000 st), followed by West Germany (408,000) and the USSR (est. 340,000). Other countries producing over 100,000 st of feldspar in 1980 included Italy, France, Spain, Mexico, and Brazil.

While nearly 200,000 tpy of aplite are produced in the US, Japan is by far the largest producer with production of between 400,000 and 500,000 tpy, according to the 1980 US Bureau of Mines *Minerals Yearbook*.

Producing Areas, US

In the United States, the state of North Carolina now holds a substantial lead in feldspar production. In 1979, it is likely that about 70% of all domestic feldspar product originated in that state. The state of Connecticut now appears to be second in feldspar production, followed by Georgia and California. Minor quantities of feldspar are shipped from Wyoming.

The Spruce Pine alaskite, measurable in terms of perhaps several square miles in area, irregularly shaped, with associated coarse pegmatites, is the principal feldspar ore of North Carolina. Around Bryson City in Swain County are pegmatites which have in the past yielded block feldspar. In recent years, North Carolina pegmatites in the area of Bessemer City (Gaston County), Lincolnton (Lincoln County), and Kings Mountain (Cleveland County) have been processed by flotation to yield feldspar concentrate. At these places, feldspar is a byproduct of spodumene production. One North Carolina company produces potash feldspar by flotation from a decomposed graphic granite.

South Dakota and Wyoming produce block spar from pegmatites. This type was formerly produced in Connecticut, but in recent years a shift has been made to froth flotation, working finer grained pegmatites. A deeply weathered pegmatite in Georgia is processed by flotation in order to concentrate potash spar. The Georgia producer is developing a feldspar source in a weathered potash granite in Greene County to supplement present mill feed.

Current Feldspar Producers, US

Table 4 gives company names and plant locations of the 11 US feldspar producers operating in 1982. Of the plant units mentioned, two utilize dry grinding, with hand-cobbed feldspar as a feed source for one and a feldspathic quartzite for the other; ten are froth flotation plants with pegmatite, alaskite, weathered granite, or river and beach sands as feed. The flotation plants have the option to produce both feldspar concentrates and feldspathic sand, and most of them probably do.

TABLE 1—World Production of Feldspar, by Countries, 1959-1980

Country*	1959-1962	1963-1966	1967-1970	1971-1974	1975	1976	1977	1978[p]	1979	1980
	(Average; lt)	(Average; lt)			St					
North America										
Canada	11,678	8,836	9,834	10,026†						
Guatemala				29,464‡	[e]33,000	[e]22,000	14,408	16,950	15,000	13,000
Mexico	ND	ND	76,320	118,133	80,732	80,732	[e]83,000	140,604	140,000	140,000
United States	510,014	604,050	651,334	676,311	669,898	739,684	733,963	735,000	[e]740,000	710,000
South America										
Argentina	6,885	15,393	22,692	46,230	63,934	75,204	85,274	51,034	52,000	47,000
Brazil				103,547§	73,167	92,742	[e]94,000	118,085	129,000	405,000
Chile	1,497	731	1,661	1,789	421	106	[e]110	995	1,000	—
Colombia	14,894	13,284	20,899	27,039	36,376	[e]36,500	36,500	29,162	30,000	33,000
Peru	505	613	1,585	1,958	[e]3,300	4,305	[e]4,500	10,759	9,900	10,000
Uruguay	659	1,029	996	1,019	1,939	1,262	1,700	2,199	3,000	3,000
Europe										
Austria	4,225	1,646	1,886	2,860	‖			3,181	3,000	8,000
Finland	11,624	16,232	55,287	60,483	75,593	75,192	[e]77,000	78,628	77,000	75,000
France¶	123,304	195,418	166,670	197,414	219,360	207,234	[e]215,000	[e]208,000	209,000	220,000
Germany, W.	243,694	293,183	325,662	352,782	436,331	462,944	[e]475,000	425,040	430,000	408,000
Italy	84,165	109,271	173,620	333,146	206,522	201,287	244,000	276,771	276,000	325,000
Norway**	71,330	73,966	120,156	225,781	49,557	41,546	44,000	78,264	78,300	77,000
Poland[e]	NA	26,720	28,750	29,464	33,000	33,000	33,000	44,000	44,000	44,000
Portugal	2,276	5,290	27,281	22,317	14,506	14,686	11,908	18,987	29,000	29,000
Romania[e]				NA	64,000	64,000	64,000	66,000	66,000	66,000
Spain¶	10,395	18,036	47,630	67,425	95,102	[e]100,000	[e]100,000	[e]110,000	99,000	138,000
Sweden	52,426	44,674	31,494	29,850	49,320	49,324	[e]50,000	[e]57,000	57,000	55,000
USSR[e]	192,500	220,000	240,000	260,714	310,000	310,000	320,000	330,000	340,000	340,000
United Kingdom (China Stone)[e]	ND	ND	ND	52,785	55,000	55,000	55,000	55,000	55,000	55,000
Yugoslavia	21,221	39,660	43,120	51,389	60,129	27,983	[e]27,600	[e]63,000	60,000	60,000
Africa										
Egypt	423	4,032	NA	2,870	937	2,346	[e]2,400	3,678	3,690	4,000
Ethiopia	1,460	3,939	ND	NA	NA	NA	NA	NA	NA	—
Kenya	ND	ND		2,194	1,781	[e]1,800	2,060	1,046	NA	1,000
Malagasy Republic (Madagascar)	ND	ND	NA	NA	753			[e]1	1	—
Mozambique[e]	ND	ND	7,000	409	950	950	950	1,000	1,000	—
Nigeria[e]				NA	5,500	5,500	5,500	NA	NA	6,000
South Africa (Republic of)	19,595	39,542	21,090	26,769	33,460	50,858	56,172	57,921	58,000	55,000
Zambia				NA	1,294	570	[e]900	368	550	—
Asia										
Burma	ND	ND	ND	607	840	981	1,356	2,205	2,000	2,000
Hong Kong	1,593	1,425	1,555	1,180	2,270	2,534	[e]2,800	3,480	3,300	20,000
India	12,212	24,442	30,141	46,360	46,817	58,878	[e]58,741	52,672	50,000	55,000
Japan‡‡	62,407	55,719	58,190	54,450	43,494	45,434	[e]47,000	[e]46,000	[e]45,000	40,000
Korea (Republic of)	6,085	14,012	21,989	24,218	22,198	28,889	54,425	76,280	75,000	88,000
Pakistan	NA	634	ND	1,250	2,981	3,299	[e]3,300	15,769	13,000	16,000
Philippines	8,858	8,766	NA	37,791	4,307	16,799	15,988	18,966	20,000	19,000
Sri Lanka	81	283	674	551	859	3,526	3,600	3,483	3,500	4,000
Thailand§				NA	14,358	13,511	[e]14,000	35,917	35,000	28,000
Turkey	NA	NA	NA	NA	NA	63,714	82,894	83,004	80,000	79,000
Oceania										
Australia	7,972	8,550	4,388	3,036	3,366	4,958	[e]3,300	3,053	2,000	5,000
Total	††	††	††	††	2,895,141	2,935,564	3,044,505	3,401,953	3,410,000	3,782,000

Source: Compiled from *Minerals Yearbooks,* US Bureau of Mines.

* In addition to the countries listed, People's Republic of China, Czechoslovakia, and South-West Africa (Namibia) produce feldspar, but available information is inadequate to make reliable estimates of output levels.

† No production after 1972.

‡ Production first reported 1974.

[e] Estimated

ND No data.

NA Not available.

§ Production first reported 1973.

‖ No production after 1973.

¶ Includes pegmatite.

** Described in source as lump feldspar; does not include nepheline syenite.

†† Total omitted for reasons stated in text.

[p] Preliminary

‡‡ In addition, the following quantities of aplite and other feldspathic rock were produced:

1963-65 (average) 250,694 lt, 1966-68: 325,575 lt, 1969-71: 444,251 lt, 1972-74: 472,912 lt, 1975: 357,056 st, 1976: 394,533 st.

TABLE 2—Destination of Shipments of Feldspar Sold or Used by Producers in the United States, by State*

(Short Tons)

State	1977	1978	1979	1980
Alabama	†	35,500	13,900	21,100
Arkansas	5,500	5,200	W	W
California	†	†	ᵉ15,000	—
Connecticut	†	23,800	21,600	18,400
Florida	†	20,000	23,600	32,800
Georgia	†	35,800	69,000	64,700
Illinois	37,000	47,600	43,700	36,600
Indiana	30,800	32,600	25,300	26,700
Kentucky	10,100	10,200	13,100	12,800
Louisiana	16,200	19,200	16,900	14,600
Maryland	5,000	6,500	7,600	5,100
Massachusetts	18,400	W	W	11,100
Michigan	800	2,500	4,000	2,700
Mississippi	20,800	22,000	17,600	15,600
Missouri	7,600	4,200	7,600	4,900
New Jersey	45,100	50,400	59,600	64,600
New York	20,600	21,400	22,000	23,100
Ohio	63,300	59,200	64,400	56,400
Oklahoma	34,300	33,600	31,700	31,000
Pennsylvania	53,700	55,400	52,900	46,200
South Carolina	NA	W	17,700	15,600
Tennessee	21,700	19,700	19,400	18,300
Texas	39,400	38,800	40,400	35,000
West Virginia	37,000	38,200	59,800	55,400
Other‡	267,200	153,200	97,200	97,300
Total	735,000¶	735,000	744,000	710,000

Source: Potter, 1979, 1980.

ᵉ Estimated. (Data are incomplete: Bureau of Mines estimate is 15,000 tons or more.)
W Withheld to avoid disclosing company proprietary data: included with "Other."
* Includes potash feldspar (8% K_2O or higher).
† Data are incomplete; included with "Other."
‡ Includes Kansas, Rhode Island, other states, states indicated by symbol W or footnote †, exports to foreign destinations, and unknown.
¶ Data do not add to total shown because of independent rounding.

Two plants in California use froth flotation to remove feldspar from feldspathic beach sands. The plant utilizing river sand as a feldspar source washes, dries, magnetically separates, acid leaches, and rewashes in the production of a feldspathic sand.

Table 4 emphasizes the shift away from hand-cobbed feldspar toward flotation concentrates. One of the contributing factors toward this shift has been the near exhaustion of easily available block feldspar in many of the producing districts and rapidly escalating freight costs which burden customers using feldspar from remote producing locations.

Feldspar Production Outside the US

A partial list of foreign producers would include the flotation plant of Bjorum Sibelco Norfloat A/S and Co. This plant has a capacity of approximately 65,000 tpy, and serves much of Europe and the British Isles. In West Germany, Franz Mandt and Sons operates three feldspar plants, using both imported hand-cobbed feldspar and local feldspathic pegmatites as sources of feed for fine grinding. West Germany also obtains considerable byproduct feldspar from local kaolin operations. In France, Denain-Anzin operates several dry grinding plants, using hand-cobbed spar from the Perpignan area. Maffei and Co. of Italy is one of the largest producers of feldspar in Europe, shipping over 200,000 tpy.

In Mexico, Materias Primas, with headquarters in Monterrey, operates a feldspar plant near Queretaro, using a feldspathic sand as feed; this raw material has been transported to the site by wind.

TABLE 3—Feldspar Sold or Used by Producers in the United States, by Use*
(Thousand short tons and thousand dollars)

	1978		1979		1980	
	Quantity	Value	Quantity	Value	Quantity	Value
Hand-cobbed:						
Pottery	W	W	W	W	W	W
Other	26	1,260	20	1,260	15	995
Total	26	1,260	20	1,260	15	995
Flotation concentrate:						
Glass	297	6,700	304	7,250	298	7,870
Pottery	W	W	W	W	W	W
Other	271	9,390	281	10,660	266	10,990
Total	568	16,090	585	17,910	564	18,860
Feldspar-silica mixture:†						
Glass	100	3,500	102	3,590	106	4,790
Pottery	W	W	W	W	W	W
Other	41	1,900	38	1,840	25	1,620
Total	141	5,400	140	5,430	131	6,410
Total:						
Glass‡	397	10,200	406	10,840	404	12,660
Pottery	284	9,925	312	12,220	276	11,390
Other¶	54	2,625	27	1,540	30	2,220
Total	735	22,750	§744	24,600	710	26,300

Source: Potter, 1979, 1980.

W Withheld to avoid disclosing company proprietary data: included with "Other."
* Includes potash feldspar (8% K_2O or higher).
† Feldspar content.
‡ Includes container glass, flat glass, and fiber glass.
¶ Includes soaps, abrasives, sanitary ware, fillers, electrical insulators, etc., and unknown; totals for "Quantity" and "Value" do not correspond to the sums of the subtotals of the three "Other" categories above.
§ Data do not add to total shown because of independent rounding.

Many countries in South America produce limited amounts of feldspathic products, the same being true of Africa and Asia.

With the exception of "iron curtain" countries in Europe, it is believed all production not specifically described is by hand selection, rather than by more sophisticated beneficiation techniques.

Aplite Production

The one remaining aplite plant in the United States is located at Montpelier, VA, and is operated by The Feldspar Corp. It produces a low-iron aplite suitable for flint glass and some other glass applications and employs wet methods.

The other former producer at Piney River, VA produced aplite by dry methods suitable for amber and green glass and window glass. The Piney River operation was closed July 1, 1980.

Domestic production of aplite in 1979 is estimated at 205,000 st. This figure has remained fairly constant over recent years and represents about 28% of the recorded 1979 feldspar production.

Exploration, Mining, and Processing

Exploration for feldspar and aplite deposits is fairly simple. A geologist with basic geologic experience in pegmatites, granites, and clays explores potential areas selected by a knowledgeable producer, seeking outcrops and small rock chips appearing as "float." Bulldozers and ditchers may be used to strip a relatively shallow overburden. Diamond drilling has been used to determine depth of a deposit, but is

TABLE 4—Domestic Feldspar Producers, 1982

Company	Plant Location(s)	Basic Plant Type	Type of Ore
IMCORE Division International Minerals & Chemical Corp.	Penland, NC	Flotation	Alaskite
Lawson-United Feldspar & Mineral Co.	Spruce Pine, NC	Flotation	Alaskite
Feldspar Corp.	Middletown, CT	Flotation	Pegmatite
	(2 plants) Spruce Pine, NC	Flotation	Alaskite
	Monticello, GA	Flotation	Pegmatite
	Siloam, GA (Mine)		Weathered granite
Kings Mountain Silica Co.	Kings Mountain, NC	Flotation	Graphic granite
Foote Mineral Co.	Kings Mountain, NC	Flotation	Pegmatite
Pacer Corp.	Custer, SD	Dry grind	Pegmatite
Lithium Corp. of America	Bessemer City, NC	Flotation	Pegmatite
Owens-Illinois Glass Co.	Mission Viejo, CA	Flotation	Beach sand
Calspar Corp.	Santa Fe Springs, CA Ord Mountain, CA (Mine)	Dry grind	Feldspathic quartzite
Arkhola Sand and Gravel Co.	Muskogee, OK	Wash, size Acid leach	River sand
Ottawa Silica Co.	Oceanside, CA	Flotation	Beach sand

often costly with relation to the information obtained.

In those deposits appearing sufficiently promising, representative samples are collected and refined in a pilot plant—this often being preceded by bench work. The product is then evaluated by chemical assay and possibly by physical testing: button firing, sample batch runs, etc.

The feldspar industry has learned that, within rather broad limits, feldspathic minerals are suitable for glass and ceramics, provided iron-bearing impurities and the surrounding country rock can be separated from the finished product. Generally the deciding factor is whether, at approximately 20 mesh, or at most 30 mesh, the pure feldspathic mineral can be unlocked from its associated impurities. If it can, the product derived from flotation or magnetic treatment is evaluated as to suitability for available markets.

The refined product is usually screen-tested for the particular mesh size range for which it is ground. Fusion tests show both meltability and color of the melted product. Chemical tests employed are conventional quantitative analyses: silica, alumina, iron oxide, calcia, magnesia, alkalies, and ignition loss are determined. Following these tests, the computation of tonnage from areas stripped and/or drilled completes the basis for decision on whether to mine the deposit.

Most pegmatite, alaskite, and granite ores of feldspar are mined by conventional open-pit methods: removal of overburden, drilling and blasting on 20- to 40-ft benches, loading, and removal by trucks.

Feldspar ores are mostly refined today by froth flotation. In a typical plant, after the raw material is crushed and ground it is screened at a suitable liberation mesh size and deslimed. Micaceous minerals are then removed with amine as collector in an acid environment, assisted by pine oil and fuel oil. Next, after dewatering of the feed in a classifier or cyclone for reagent removal, the pH is again lowered by sulfuric acid, and petroleum sulfonate (mahogany soap) is used as collector to remove iron-bearing minerals. In the last step, the machine discharge from the preceding flotation is again dewatered, and the feldspar floated away from the quartz in a hydrofluoric-acid environment, again using (cationic) amine as collector. The feldspar float concentrate is then dewatered either in drain bins or over a vacuum filter or centrifuge, and is finally dried in rotary dryers.

An additional process at a typical feldspar flotation plant today is the use of a high-intensity induced-field magnetic separator. Earlier use of this feature involved the dry rolls type of machine, but beginning in the early 1960s there emerged a new generation of wet separators, usually operating with slow-moving, horizontally rotating chambers in a turntable, cycling through pinning and flushing; concentrating the magnetics on shaped steel elements; and discharging them as each chamber rotates out of the magnetic field to be replaced by a clean one. A dry type of magnetic separator can, of course, be used only on the dried end product, whereas a wet separator can be employed in a number of spots in the course of the wet processing, depending on expediency, the characteristics of the plant, etc. Magnetic fields involved in either case are generally 15,000 gauss or higher; weaker fields are usually ineffective. A high-intensity magnetic separator may be used either as a primary remover of iron minerals, partially or completely replacing flotation, or as a backup device in case the froth flotation circuits did not work properly for some reason.

The process of electrostatic separation for feldspar beneficiation does not appear to have survived commercially. The last company employing this process went out of business in the early 1960s.

Although the major tonnage of feldspar is moved by rail, truck transportation is becoming increasingly popular where short distances are involved, because of convenience and economics.

Aplite is in part mined by conventional stripping, drilling, and blasting, and in part by using tractors with hydraulic rippers followed by elevator scrapers, which both load and haul the ore to the plant. At Montpelier, VA, where wet methods are used, the wetted ore is put through a large hammer mill, then a rod mill and screens, reducing it to 30 mesh; it then passes through scrubber and classifier circuits before spiraling to remove biotite, ilmenite, and rutile. The remaining material is then dewatered in drain bins and dried in a rotary dryer. After this it is passed over high-intensity magnetic separators and stored in silos for loading.

History of Domestic Feldspar Production

Significant commercial production of feldspar in the United States started early in the 20th century as a replacement for Cornwall Stone, which was imported from England and used by the English potters who settled around Trenton, NJ, and East Liverpool, Ohio. Pressure to reduce costs resulted in the testing of feldspar deposits in eastern Pennsylvania and in Connecticut. At first, milling was done by the individual potteries from hand-selected crude feldspar. Eventually commercial grinders opened feldspar deposits and turned out a product which, in the early days, was ground in stone chaser mills and screened. The screening technique was very inefficient, since screening at finer than 40 mesh has always been difficult. In time, continuous ball mills were developed, and sizing was accomplished with air separators. By 1919, there were mills in the Trenton, NJ area; in New England; and in East Liverpool, Ohio, using these methods. They secured their hand-cobbed, crude feldspar first from Connecticut and later from New Hampshire and Maine.

Because of the winter weather in New England, ore bodies were prospected and developed in the Spruce Pine area in North Carolina at about this time, and also in the area near Brookneal, VA. Soon there was a thriving milling industry in the South. Clinchfield Products Co. at Erwin, TN, owned by a Mr. Rockefeller, used Spruce Pine hand-cobbed spar. The Erwin Feldspar Corp. was also located in Erwin. A small plant at Bristol, TN, later moved to Spruce Pine and became the United Feldspar and Mineral Co.

By 1930, half the feldspar used in the country was produced in North Carolina and half in New England. Exploration was begun in areas around Denver, CO, and in South Dakota. Mills were subsequently built at Keystone and Custer, SD, and at Denver and Salida, CO. During all this period, the principal search was for potash feldspar, which was preferred by the potteries because of its excellent firing characteristics in vitrified ceramic pieces.

By the end of the 1930s, there were approximately 23 feldspar mills operating in the New England area; in North Carolina, Tennessee, and Virginia; and in the Colorado-South Dakota area. Ten of these plants had been combined by a Mr. Margerum into what was known as Consolidated Feldspar Corp. The other companies, operating one or two mills each, were as follows:

United Feldspar and Mineral Co., Spruce Pine, NC and W. Paris, ME
Eureka Flint and Spar Co., Portland, CT and Trenton, NJ
Golding-Keene Co., Keene, NH and Tocane, NC
Genessee Feldspar Co., Rochester, NY
Clinchfield Sand and Feldspar Co., Bedford, VA
Feldspar Mining Co., Burnsville, NC
North Carolina Feldspar Corp., Erwin, TN
F. E. Schundler Co., Custer, SD
Western Feldspar Milling Co., Salida, CO
Western Feldspar Co., Warroad, MN

At about the beginning of World War II, the US Bureau of Mines had worked out a technique for refining feldspar from pegmatites and coarse-grained granites that were low in iron, by the use of froth flotation. Use of F^- ion in an acid environment, plus cationic (amine) collectors synthesized from the meat-packing industry, allowed concentration of the feldspar as froth product—the quartz being depressed as machine discharge. This work was piloted jointly by Consolidated Feldspar Corp. at Erwin, TN and the US Bureau of Mines, and resulted in the construction of two flotation plants—one in Keene, NH, by the Golding-Keene Co. and one at Kona, NC, by Consolidated Feldspar Corp. The first of these plants operated successfully for only about five years, but the Kona plant became a great success. Continual efforts were made at process improvement, the feldspar being used both in the glass industry and in ceramics.

By 1949, Feldspar Milling Co. and North Carolina Feldspar Corp., in a joint venture at Spruce Pine, constructed a flotation plant with alaskite as ore, using improvements on the original Bureau of Mines process which had been developed at the North Carolina State College Minerals Research Laboratory in Asheville. From the start, this operation pointed the way toward both lower cost plant construction and lower cost flotation processing, and these improvements were used by most flotation plants built subsequently.

Although commercial feldspar production was first directed to ceramics, the glass industry had found that feldspar, if low enough in iron, is highly desirable as partial replacement for soda ash, and especially to introduce alumina into the glass batch. This addition of alumina to the batch, which traditionally had been made solely with sand, lime, and soda ash, improves the strength of the glass and its resistance to thermal shock; lengthens the period of workability during forming; and, perhaps most important, gives the glass resistance to "blooming" or frosting that develops on the outside of glass which is exposed to weather.

Economic and Competitive Factors

The feldspar industry and its chief competitors, the nepheline syenite and aplite industries, have had a history of alternating between overproduction and shortage during the last 45 years. Each period of shortage has resulted in a slight rise in price, followed by construction of new producing facilities, resulting in oversupply and depressed prices. Overcapacity and very low prices in the early 1930s were followed by some shortages and slightly rising prices during World War II, and these by a great expansion in the late 1940s and 50s. This brought on a serious slump about 1960. New and expanded facilities came into production in 1972, with some cloudiness in the short-term economic picture. Certainly there will be no shortage of feldspathic materials for the principal markets: glass, ceramics, and chemical fillers.

Feldspar, nepheline syenite, and aplite, being relatively inexpensive nonmetallic minerals, are strongly affected by freight rates, with the result that most grades of these materials move within fairly short distances, mostly under 1000 miles. Customers can always use substitutes when the cost becomes too high. The single possible exception to this is the customer who can use only high-potassium feldspar, who often pays substantial delivered prices to obtain the product he needs.

The glass industry, because of its ability to use a variety of alumina sources, can switch from feldspar to nepheline syenite, aplite, or feldspathic sand by relatively simple reformulation; and, since it is possible to calculate cost per pound of alumina in a given mineral—with certain additional credit for alkalies present—the glass industry usually uses the alumina source of lowest delivered price at each glass plant.

The ceramic industry has not been so flexible, because of the long time lapse between the mixing of raw materials and the final delivery of the product to the consumer; flaws might not show up for a considerable period. Also since, in most ceramics, raw materials are a small part of the cost, ceramic manufacturers have seldom switched materials solely because of cost factors. Only in recent years, when froth flotation of feldspar has made available considerable product uniformity, with resulting interchangeability, has it been possible for a ceramic customer to switch materials with any feeling of confidence that he would not run into trouble.

Similarly, the filler market for feldspathic materials has been somewhat less flexible than the glass market, because the function of a filler in products such as latex foam, urethane, acrylic, and paint is not well understood, and usage is on a "cut and try" basis. Consumers use a product which they know will work, and only switch to cheaper products after exhaustive testing.

Markets for feldspar, nepheline syenite, and aplite have not grown rapidly in comparison with other industries, and mortality among producers has been great. This has resulted from swings between overproduction and shortage, from exhaustion of some deposits, and from inefficient operations which failed to provide for modernization and improved techniques.

It is probable that, in the future, because of the high cost of a flotation plant, more thought will accompany the planning of additional operations, and only as demand can absorb products—with assured profit—will anyone risk the large investment. This may also have a stabilizing influence on prices, eliminating cyclical swings of overproduction and shortage.

Needless to say, a potential producer who fails to study costs and potential markets, or who does not analyze freight rates and their resulting distribution patterns, is heading for trouble. Availability of working capital, usually roughly equal to cost of facilities, is imperative —both to assure financial responsibility and, more important, to carry the operation through the period of startup expense plus the time required to reach a profitable volume of sales.

There are a few substitutes for the feldspathic family of minerals. One is pyrophyllite, which has replaced a substantial amount of feldspar in the manufacture of tile and similar items. Another is steel mill slag (Table 5) that has been magnetically treated to remove most of the iron; this is a source of some alumina and of considerable flux in glassmaking. It also has some advantages as a "fining" agent in the glass batch at time of melting.

As more sophisticated beneficiation techniques are developed for the recovery of feldspar, valuable co- and byproducts can and should be recovered. Producers using alaskite as ore produce byproduct quartz and muscovite mica. One producer, using as feed source weathered graphic granites, recovers clay for brick, muscovite mica, feldspar, and quartz. Two lithium producers recover byproduct feldspar, quartz, and muscovite. Not only does the recovery of co- and byproduct commodities enhance the value of the ore, cost of tailings disposal is correspondingly reduced.

Feldspar producing companies have no subsidies. The chief competitive material is nepheline syenite from Canada. Nepheline syenite is a low-iron rock composed of nepheline (a silica-deficient feldspar), albite, and microcline. No tariff is levied on the entry of nepheline syenite into the US, or the entry of feldspar into Canada. Nepheline syenite is fully dealt with elsewhere in this publication.

Most feldspar producers are granted percentage depletion. The rate is usually 14%, although on byproduct mica, often produced along with feldspar, the rate is 22%.

Producers of feldspathic minerals are, of course, among those industries under government control, both state and federal, not only in the matter of meeting stringent air and water standards, but also because of new mine-reclamation laws.

The Occupational Safety and Health Act (1972) has substantially affected mining. In many older plants, stairs, handrails, guards, and other safety devices had to be changed or newly installed.

Producers have installed dust collectors, cyclones, and scrubbers, and constructed filtration plants, settling ponds, or other systems of cleaning process water before returning it to local streams.

Mine reclamation includes replacement of overburden and also return of tailings to old mine sites before planting grass and trees. A rule of thumb is that it costs about as much "to fill a hole as to make it," and producers are beginning to look at costs with this figure in mind.

State and federal regulations covering ecologically oriented control of water and air pollution and mine reclamation eventually will create a cost exceeding that of labor in producing feldspathic minerals. Full effects of this are just becoming apparent, as most state laws are patterned on federal regulations—which the state then has the option of administering.

There is no present recycling of feldspathic products, although some products incorporating

TABLE 5—Typical Analyses, Feldspathic and Aluminous Materials*

Soda Flotation Feldspar, Spruce Pine, NC	
SiO_2	67.54%
Al_2O_3	19.25
Fe_2O_3	0.06
CaO	1.94
MgO	Trace
K_2O	4.05
Na_2O	6.96
Loss	0.13

Potash Flotation Feldspar, Kings Mountain, NC	
SiO_2	67.04%
Al_2O_3	18.02
Fe_3O_3	0.04
CaO	0.38
MgO	Trace
K_2O	12.10
Na_2O	2.12
Loss	0.30

Dry Ground Feldspar, Custer, SD	
SiO_2	71.84%
Al_2O_3	16.06
Fe_2O_3	0.09
CaO	0.48
MgO	Trace
K_2O	7.60
Na_2O	3.72
Loss	0.20

Feldspathic Sand, Bessemer City, NC	
SiO_2	79.20%
Al_2O_3	12.10
Fe_2O_3	0.06
CaO	0.52
MgO	Trace
K_2O	2.62
Na_2O	4.80
Li_2O	0.35
Loss	0.35

Low Iron Aplite, Montpelier, VA	
SiO_2	63.71%
Al_2O_3	21.89
Fe_2O_3	0.09
CaO	5.70
MgO	Trace
TiO_2	0.43
K_2O	2.37
Na_2O	5.60
Loss	0.21

Canadian Nepheline Syenite Nephton, Ont., Canada	
SiO_2	61.40%
Al_2O_3	22.74
Fe_2O_3	0.06
CaO	0.70
MgO	Trace
K_2O	4.95
Na_2O	9.54
Loss	0.60

Steel Mill Slag ("Calumite")	
SiO_2	38.8%
Al_2O_3	10.5
Fe_2O_3	0.3
MnO	8.3
CaO	38.5
MgO	1.4
K_2O	0.5
Na_2O	0.4
Sulfur cpds.	1.1

* From public merchandising data of the respective producers.

them are recycled: for example, container glass, which is collected, sorted, and reintroduced into the glass batch as "cullet."

Uses, Marketing, and Prices

Feldspar, feldspathic sand, and aplite are used in the manufacture of glass as a source of alumina and as a partial replacement for soda ash. Feldspar is used in the production of porcelain enamel, as a flux, and as a source of alumina. It is also used in a full spectrum of ceramic products which include pottery, plumbing fixtures, electrical porcelain, ceramic tile, dinnerware (both vitreous and semivitreous) and some minor uses such as englobes and frits for brick and structural ceramics, as well as art pottery and planters. When finely ground, feldspar has proved quite suitable as a filler in latex, in paint, and, in smaller quantities, in urethane and acrylics. A little is used in abrasive cleaners and polishes, but this market has been taken over almost entirely by volcanic ash or silica flour.

The foregoing uses apply to some degree to feldspathic sand as well, unless a formulation requires minimum silica or a given high K_2O assay.

About 65% of all feldspathic material is used by the glass industry; 30% in ceramics; and 5% in fillers and other applications. Aside from the glass industry, a very small tonnage of aplite "fines" is used as a flux in brickmaking: this is really a byproduct, high in iron, which has been removed prior to the magnetic separation step so as not to be detrimental to the process.

Certain feldspars are of value for ornamental reasons, either as gems or as decorative building stone. Both of these uses are of extremely minor importance in the total economic picture.

Table 5 sets forth the chemical specifications for some principal feldspathic or aluminous materials.

Engineering and Mining Journal of November 1982 quoted f.o.b. feldspar prices in North Carolina, Georgia, and Connecticut ranging from \$27.50 per ton for —20-mesh float concentrate suitable for glass, up to \$70.25 per ton for —200-mesh float concentrate assaying high in K_2O, produced for ceramics. A few special items may be priced slightly higher. Prices can logically be expected to increase at a measured pace, due to economic factors cited in preceding paragraphs. Feldspar commands a higher price when fine-ground, and producers find this process a welcome avenue to additional profits. There is also a rising price scale roughly related to rising K_2O assay, since feldspar high in K_2O value is a premium product in today's market.

Acknowledgment

In the organization and writing of the portions of this chapter relating to geology and mineralogy, the assistance of Leonard S. Wiener, geologist with the North Carolina Div. of Mineral Resources, is acknowledged with sincere thanks.

Bibliography and References

Barth, T.F.W., 1969, *Feldspars,* John Wiley and Sons, New York, 261 pp.

Berry, L.G., and Mason, B., 1959, *Mineralogy—Concepts, Descriptions, Determinations,* Wm. H. Freeman, San Francisco, 612 pp.

Broadhurst, S.D., 1956, "Lithium Resources of North Carolina," *Information Circular 15,* North Carolina Div. of Mineral Resources, 37 pp.

Brown, W. R., 1962, "Mica and Feldspar Deposits of Virginia," *Mineral Resources Report 3,* Virginia Div. of Mineral Resources, 195 pp.

Cameron, E.N., et al., 1949, "Internal Structure of Granitic Pegmatites," *Economic Geology,* Monograph 2, 115 pp.

Castle, J.E., and Gillson, J.L., 1960, "Feldspar, Nepheline Syenite, and Aplite," *Industrial Minerals and Rocks,* 3rd ed., J. Gillson, ed., AIME, New York, pp. 339–362.

Clark, W.B., 1977, "Feldspar Deposit in the Ord Mountains, San Bernardino County, CA," *California Geology,* Vol. 29, pp. 81–85.

Deer, W.A., et al., 1966, *An Introduction to the Rock-Forming Minerals,* John Wiley and Sons, New York, 528 pp.

Eddy, W.H., et al., 1972, "Recovery of Glass Sand from South Carolina Waste Granite Fines," Report of Investigation 7651, US Bureau of Mines, 11 pp.

Kesler, T.L., 1961, "Exploration of the Kings Mountain Pegmatites," *Mining Engineering,* Vol. 13, pp. 1062–1068.

Park, C.E. Jr., and MacDiarmid, R.A., 1964, *Ore Deposits,* Wm. H. Freeman, San Francisco, 475 pp.

Parker, J.M., III, 1952, "Geology and Structure of Part of the Spruce Pine District, North Carolina," Bulletin 65, North Carolina Div. of Mineral Resources, 26 pp.

Potter, M.J., 1975–1980, "Feldspar," *Minerals Yearbook,* US Bureau of Mines.

Ross, C.S., 1941, "Titanium Deposits of Amherst and Nelson Counties, Virginia," Professional Paper 198, US Geological Survey, 56 pp.

Sheppard, R.A., and Gude, A.J., III, 1965, "Potash Feldspar of Possible Economic Value in the Barstow Formation, San Bernardino, California," Circular 500, US Geological Survey, 7 pp.

Turner, F.J., and Verhoogen, J., 1960, *Igneous and Metamorphic Petrology,* McGraw-Hill, New York, 602 pp.

Wells, J.R., 1971–74, "Feldspar," *Minerals Yearbook,* US Bureau of Mines.

Williams, H., Turner, F.J., and Gilbert, C.M., 1954, *Petrography—An Introduction to the Study of Rocks in Thin Sections,* Wm. H. Freeman, San Francisco, 406 pp.

Index

A

Abandoned mines, use of, 851, 1132
Abietic acid, 260
Abrasives, 11–31, 521
  abrasive value, 11
  action of, 11
  alumina, 16, 29, 521
  analyses, 29
  artificial abrasives, 27–31
  bauxite, 35, 503, 515, 518, 521
  beach pebbles, 27
  bonded abrasives, 16, 17
  borazon, 30
  boron carbide, 12, 29, 30
  boron nitride, 29, 30, 43, 313
  brass wool, 30
  ceramic pebbles, 27
  cerium oxide, 16, 31
  chemical precipitates, 31
  chert, 15
  choice of, 11
  chromium oxide, 31
  classification of, 18–31
  clay, 31
  coated, 17, 18
  copper wool, 30
  corundum, 15, 16
    synthetic, 313–315
  crocus, 31
  crystal size, importance, 15
  definition, 11
  density, bulk testing, 13, 14
  diamonds, 16, 19, 20, 654–657
    see also Diamonds.
    manufactured, 19, 20, 30
    wheel marking system, 20
  diatomite, 16, 25, 26, 38, 699
  electric furnace, used in, 28
  emery, 16, 18, 19
  feldspar, 16, 18, 26, 721
  flint, 17, 18, 27
  fluorspar, 724
  friability test, 12, 13
  fuller's earth, 16
  fused alumina oxide, 29, 521
  garnets, 21–25
  glass, 31
  grade sizes, 17
  grinding pebbles, 27
    lime, 31
  iron oxide, 30, 31
Abrasives, hardness, 12
    Knoop, 12
    Mohs' scale, 292
  kaolin, 26
  lampblack, 31
  lime, 31
  loose-grains, 15–18
  loose-pack density test, 13, 14
  magnesia, 31
  manufactured abrasives, 28–31
  metal finishing, 16, 31
  metallic abrasives, 28, 30
  monocrystalline count, 15
  mullite, 27, 34, 376
  natural abrasives, 18–27
    corundum, 15, 16, 18, 19, 313
    diamonds, 19, 20, 654–656
    diatomite, 16, 25, 26, 38, 699
    emery, 16, 18, 19
    garnets, 21–25
    pumice, 26
    quartz, 25
    silica, 25, 26
    staurolite, 26
    tripoli, 16, 18, 26, 1364, 1371
  nonsiliceous, 26
  polishing, 30, 31
  production
    alumina oxide, 28
    metallic abrasive, 28
    putty powder, 31
    silicon carbide, 28
    special silica stone product, 28
  properties, physical, 11
  pumice, 16, 18, 26
  putty powder, 31
  rottenstone, 26
  rouge, 30
  sand, crushed, 25
  shape, grain, 13, 14
  sharpening stones, 26, 27
  silica, microcrystalline, 26
  silica sand, 16, 17, 25
  siliceous powder, 25, 26
  silicon carbide, 28, 29
Abrasives, soft, 25, 26
  steel shot, 30
  steel wool, 30
  strength of grains, 13
  tests, 12, 13, 14, 15
  tripoli, 16, 18, 26, 34, 39, 1364, 1371
  tube-mill linings, 27
  tungsten carbide, 30
  types of abrasives, 15–18, 19
  uses, 15, 16, 17, 18, 24, 1388
  value, abrasive, 11
  Wig-L-Bug, use of, 12, 13
  weight per cubic foot, test, 14, 15
  whiting, 16, 26
  zircon-alumina, 29
Absorbance, 226
Absorbents, minerals, 255
Accelerators, 189
Acrylic fibers, 962
Actinolite, 435, 440–443, 456, 473
Activated clay, 589
Addy quartzite, 1173
Adobe, 639
Afghanistan
  chromite, 578
  gems, 302
  lithium, 870
Africa
  See separate countries.
  gems, 302
  geological agencies, 407–410
Aggregates, 59–131, 1151–1166
  alternate choices, 69
  asphaltic, 1151
  barite, 485
  byproduct, 1042
  cinders, 81, 85, 86
  consumption, 64
  crushed stone, 60–80, 847, 852
    alternate design choices, 69
    capacity of plant, 69, 70
    chert, 63

Aggregates, crushed stone
- competition, 74–76
- consumption, 61–64
- costs, 64, 73–76
- crushing, 72
- definition, 60
- demand, 74, 75
- depletion allowance, 78
- factors affecting competition, 75, 76, 77
- future trend, 78, 79
- general plant considerations, 70–73
- government controls, 70, 77
- gravel, 1151–1166
- gradation, acceptable, 68
- highway construction, 63–65
- industry, 75, 76, 83
- labor requirements, 75
- land use, 77, 78
- limestone, 63, 68, 844
- location requirements, 61–63, 69, 70
- nonconstruction use, 62
- physical properties, rocks, 63
- plant considerations, 69, 70, 71–73, 75, 76
- prices, 77, 78
- production, 61, 69
- profitability, 77
- qualities, 60
- reserves, 79
- rock types used, 62, 63, 844
  - physical properties, 63, 844
- shape, particle, 68
- sand, 1151–1166
- shipments, 852
- slag, as an aggregate, 111
- specifications, 60, 61, 65–69
- structure, industry, 75, 76
- substitutes, 69
- supply, 79
- taxation, 78
- tests, 65–68, 100, 119, 1158
- transportation, 73, 74, 852
- use, 59, 61, 62
- value, 77, 78
- zoning regulations, 78

definition, 60, 61, 1151

demand, 107, 108

dolomite, 63

lightweight aggregates, 81–95
- bituminous paving mixes, 120

Aggregates, lightweight
- byproducts, 81, 85, 111–131
- calcilux process, 128
- cinders, 81, 85
- classification, 81
- clay, 81
- coke breeze, 81
- commodities, used for, 82
- consumption, 85–87
- costs, operating, 85
- definition, 81
- deleterious substances, 87, 1152
- economic factors, 89, 90
- expanded slag, 85
- government considerations, 90
- grading requirements, 88
- marketing, 90
- operating costs, 85
- outlook, 90, 91
- perlite, 59, 81, 86, 88
- prices, 86, 89
- production, 81–85, 86
- properties of, 82
- pumice, 81, 85
- scoria, 81
- shale, 82, 1152
- silicone treated, 86
- slag, 81, 85, 86, 111–131
- slate, 85
- specifications, 82, 87
- substitutes, 89
- transportation, 89, 90
- tuff, 81
- types, 81
- uses, 82
- value, average, 86
- vermiculite, 81, 86, 87, 1375–1381
- volcanic cinders, 81, 85, 1079–1184
- weight of, dry, 81, 87
- zeolite, use in, 1422

limestone, aggregate, 63

pumice, 81, 85

rock types, suitability of, 1152

sand and gravel, 96–110, 1151–1166
- abrasion tests, 65, 66, 100
- conservation, 105–107
- consumption, 96, 97, 1163
- control, governmental, 103, 104, 1065
- definition, 96
- demand, 106, 107
- deleterious material, 1152
- depletion allowance, 104, 1160
- deposits, 1153–1155
- dredging, 1160, 1161

Aggregates, sand and gravel
- economic factors, 101–103
- environmental considerations, 104, 107, 1165, 1166
- evaluation, 1157–1160
- exploration, 1156, 1157
- exports, 1156
- geologic occurrences, 1153–1155
- government considerations, 103, 104, 1065, 1160
- grading, 1158, 1160
- growth rate, 96, 1165
- imports, 1156
- land use, multiple, 106
- marketing, 103, 124, 126, 1162
- microscope examination, 1160
- mining, 1160, 1161
- prices, 101, 1163
- processing, 1161, 1162
- production, 101, 1151
- properties, 1151, 1152
- reactivity, 1160
- requirements, general, 1151
- reserves, 1155, 1156
- resource availability, 105, 106
- safety regulations, 107
- specifications, 96, 97, 109
- substitutes, 100, 101, 1165
- supply-demand considerations, 107, 108
- taxes, 103, 104
- technology, changes in, 105, 127–129
- tests, 98–100, 1152, 1157–1160
- transportation, 73, 103, 124, 1163
- transportation costs, 103, 1163, 1164
- trend, 101, 1163, 1165
- utilization, 1151
- uses, 96, 1151–1153
- value, 101
- zoning, 104, 106, 1165

shale, 81, 82

slag, 111–131
- air-cooled blast-furnace slag, 114, 115, 118–122
- aggregates, lightweight, 82–95
- ballast, railway, 121
- bituminous paving mixes, 120, 121

Aggregates, slag, blast furnace slag, production, 120
  chemical composition, 110
  coarse slag, 120
  concrete making properties, 87
  consumption, 112, 113
  composition, chemical, 119
  definition, 111
  deleterious substances, 87
  employment in, 112
  environmental considerations, 127
  expanded slag, 116, 117, 119, 122, 123
  fire resistance ratings, 123
  future prospects, 127
  government considerations, 126, 127
  gradation requirements, 121
  granulated slag, 115–117, 119, 121, 122
  granules, roofing, 203–212
  history, early, 111
  marketing, 124–126
  mixes, 120–122
  pavements, use in, 121
  physical properties, blast furnace slag, 119
  plants, 112
  pozzolan, 143
  price trends, 123, 124
  prices, 113, 115, 117, 123
  producers, 125
  production, 112, 113
  production centers, 82, 83, 109, 110
  properties, 119
  sales, 113, 124–126
  slag, use in cement, 143
  specifications, slag, 118–123
  steel slag, 118, 119, 122
  substitutes, 122, 123
  transportation, methods of, 124
  types of slag, 114–122
  uses, 114, 118, 119, 121
  value, slag, 114, 118, 119
  wool, slag, 115, 121
Air-cooled blast furnace slag, 114, 115, 117, 119, 120, 121
  analyses, 119
  production, 113, 115, 125
  prices, 113, 123
  sales, 114
  specifications, 118–120
  strength, 120, 122
  uses, 115, 120
  weight, 120
Aglime, 237
Air separation, zeolite used in, 1423, 1424
Akermanite, 1389
Alabama
  asphalt, 530, 531
  barite, 492
  bauxite, 515
  bentonite, 599
  clay, 374, 375
  fuller's earth, 599
  gannister, 377
  glauconite, 752, 765
  granite, 170
  graphite, 759, 760
  kaolin, 622, 638
  lime, 827, 828
  limestone, 853
  manganese, 898
  marble, 164, 170
  marl, 854
  mica, 918
  monument stone, 164, 854
  pegmatites, 711
  plastic clay, 375
  silica, 1171
  slag, 112, 125
  talc, 1288, 1289, 1296
  tripoli, 1368
  zircon, 1368
Alabaster, 776
Alaska
  barite, 492, 498
  chromite, 575
  fluorspar, 728
  jade, 303
  perlite, 1002
Alaskite, 711
Albania
  asphalt, 530
  chromite, 381, 576, 578
Alberta
  bentonite, 601
  silica, 1176
  sodium sulfate, 1210, 1218
  sulfur, 1251, 1253
  tar sand, 1253
Albertite, 530, 531
Alexandrite, 295, 312
  synthetic, 312, 316
Algeria
  barite, 493
  bentonite, 605
  celestite, 1233
  diatomite, 687
  phosphate, 1035
Alkali-dolomite reactions, 99
Alkali-flux, 262
Alkali-halide flux, 263
Allard Lake, 1337, 1345
Alluvial deposits, 1154
  processing, 1347–1351
  flowsheet, 1350
Almandite, 22, 226
Alpha-hemihydrate, 186, 189, 191
Alpha quartz, 1167
Alumina, 29
  abrasives, 35
  analysis, 29
  bricks, 376, 377
  cement, use in, 521
  ceramics, 33, 35
  chemical analysis, 29
  fused, 29
  hydrate, 633
  lime, used in, 820
  nepheline syenite, 932, 942–953, 954
  processing, 29, 519–523
  production, 28
  screen analysis, 29
  sintered, 376
Alumina hydrate, 633
Aluminum chemicals, 521
Aluminum
  kaolin, source of, 629
  nepheline syenite, source of, 932, 953, 954, 1042
Aluminum oxide, fused, 29, 417
Alunite, 417–434
  alunitization, 422
  Alumet process, 419–421, 428–430
  alumina source, 417, 427
  analyses, 420
  bacteria, role of, 423
  bochones, 427
  boehmite, synthetic, 428
  bullseye, 422
  composition, 419, 420
  deposits
    Argentina, 423
    Australia, 417, 419
    Chile, 423
    China, 417
    France, 417
    Italy, 417, 418
    Japan, 417
    Spain, 417
    types of, 418–422
    United States
      Arizona, 418, 423
      California, 425
      Colorado, 425
      Nevada, 418, 422
      New Mexico, 423
      Texas, 425
      Utah, 417–419, 423, 430
      Washington, 425
    USSR, 418
  fertilizer, 418
  genesis of, 421–423
  geology, 418–423
  historical background, 417, 418
  hypogene ore deposits, 430, 431
  jarosite, 419–421

Alunite, Kalunite process, 428
  Marysvale, 417–419, 425
  mining, 430
  natroalunite, 419
  photomicrographs, 422
  potash, source of, 417
  processes
    Alumet, 418, 419, 421, 428–430
    Kalunite, 428
    Soviet process, 430
    UG process, 428
  prospecting for, 431
  resources, 423–426
    Argentina, 427, 430
    Asia, 426
    Bolivia, 427
    Europe, 426, 427
    Iran, 427
    Pakistan, 427
    South America, 427
    Turkey, 427
    United States, 423–426
      Arizona, 424, 425
      Colorado, 423, 425
      Nevada, 424
      New Mexico, 425
      Utah, 423, 424
    USSR, 426
  Roman alum, 417, 418
  technology, 428–430
  UG process, 428
  uses, 417
  zoning, of deposit, 420, 421
Amargosite, 598
Amber, 279, 281, 291, 301, 315, 316
Amblygonite, 870, 877
  see also Lithium raw materials.
Amino resins, 963
Ammonia
  see also Nitrogen compounds.
  coke oven byproduct, 970
  synthetic, 963–970
Ammonia, anhydrous, 961
  fertilizer, 234
Ammonium nitrate, 234
Ammonium sulfate, 197, 234, 981
Ammonium thiosulfate, 239
Amorphous graphite, 758
Amorphous silica, 1363
  see also Tripoli.
Amosite, 437–443, 455–457, 461, 475, 476
Amphibole asbestos, 435, 444, 456
Analcime, 1393, 1394
Anatase, 1311, 1323
Anauxite, 1172
Andalusite, 34, 35, 376, 799, 800, 802, 803
Andreasen size determination, 946
Angola
  diamonds, 302
  fluorspar, 734
  gems, 302
  kaolin, 626
Angular steel grit, 30
Anhydrite, 183–199, 775–777
  portland cement material, 141
  sulfur source, 1256, 1257
  use, 775
Animal feed, urea, 963
Ankerite, 833, 834
Anorthite, 36
Antarctica, sodium sulfate, 1216
Anthrophyllite, 435, 439, 440, 441, 443, 444, 456, 483
  see also Asbestos.
Antigorite, 1282
Apache tears, 1007
Apatite, 1017, 1021
  see also Phosphate rock.
Aplite, 710, 716, 719–721, 955
  see also Feldspar.
  production, 644, 716
Aragonite, 35, 134, 301, 833, 834
Arcon, 912
Arc welding, 265, 266
Argentina
  barite, 492
  bentonite, 602
  borates, 549, 550
  celestite, 1233
  diatomite, 688
  fire clay, 638
  fluorspar, 732
  garnet, 23
  glance pitch, 531
  kaolin, 624, 638
  lithium, 869, 870, 873, 876
  perlite, 999, 1003
  rare earths, 1113
  refractory clay, 638
  salt, 1126
  soda ash, 1200
  sodium sulfate, 1207, 1215, 1221
  vermiculite, 1377
  zeolite, 1418
Aridized anhydrite, 188, 189
Arizona
  asbestos, 454
  barite, 492
  bentonite, 598
  diatomite, 25, 685
  feldspar, 711
  fluorspar, 728
  gem, 303, 319
  gypsum, 778, 783
  hectorite, 600
  kyanite, 801
  lime, 827, 828
Arizona, manganese, 898
  meerschaum, 911
  perlite, 997, 999, 1003
  potash, 1066
  salt, 1124
  silica, 1173
  sodium sulfate, 1209, 1213
  titanium, 1316
  vermiculite, 1377
  wollastonite, 1385
  zeolites, 1414, 1424, 1425
Arkansas
  asphalt, 530
  barite, 489, 491, 496, 497
  bauxite, 508, 515, 519
  bromine, 48
  diamonds, 303
  fuller's earth, 591
  halloysite, 635
  kaolin, 618–620, 623, 638
  kimberlite, 662
  manganese, 897, 898
  nepheline syenite, 508, 931, 942, 943
  novaculite, 27, 1369
  pyrophyllite, 1097
  rock asphalt, 530
  silica, 1170
  talc, 1298
  titanium, 1316, 1320
  tripoli, 26, 1367, 1369
  vermiculite, 1377
  whetstones, 27
Arkose, 166, 170
Artificial abrasives, 27–31
Asbestos, 435–484
  actinolite, 435, 440–444, 456, 473
  amosite, 437–443, 455–457, 461, 475, 476
  amphibole asbestos, 435, 444, 456
  anthophyllite, 435, 439–444, 456, 483
  bulk sampling, 465
  Cape Blue, 476
  chemical composition, 436–440
  chemical resistance, 443
  chrysotile, 435–437, 439–444, 446, 454, 455, 474–481
  classification, 474–476
  composition, 436–440
  consumption, 481
  crocidolite, 437, 440–443, 455–457, 475
  cross fiber, 444, 446, 462, 463
  definition, 435
  deposits, 435, 444, 446, 454, 455
  distribution, 446–454

Asbestos, dolomite, deposits in, 454, 455
drilling, 462, 463
electron micrographs, 436–439
exploitation, 466–473
exploration, 461–466
exports, 479
evaluation, 463–466
face reading, 465
fiber evaluation, 437, 444, 446, 462–464
filler, 250, 479
geology, 444–446
grading of, 474, 475, 476
health hazards, 481
heat, effect on, 440
markets, 481
mass fibers, 441, 442, 444
mill, flowsheets, 469, 470, 471, 472
milling, 467–473
mining, 466, 467
montasite, 455
occurrence, 446–454
ores, 437, 444, 446, 461–466
prices, 478
production, 456–461
properties, 435–444
prospecting, 461–466
Quebec grading, 474
Quebec standard, 473, 474
reserves, calculation of, 463–466
slip fibers, 444, 463, 464, 466
solubility of, 443
specifications, 473, 474
spinning, 441, 442
stabilizer, use as, 209
specifications, 473, 474
strength units, 466
structures, 444
supply-demand relationship, 480
tensile strength, 466, 473, 474
tests, 473, 474
tremolite, 435, 440–444, 456, 461, 1282
types of, 440
uses, 250, 479, 480
varieties, 435
veining, 444–446
zeta potential, 430
Ashlar, 162
Asia
see also individual countries.
geological agencies, 411–413
Asphalt, native, 529, 530
aspect ratio, 925
stabilizer, 190, 209
Asphalt roofing, development of, 204
Asphaltites, 530, 531
gilsonite, 530
glance pitch, 531
grahamite, 530, 531
petroleum asphalts, 529
Asterism, 295, 296, 312
AT-cut quartz, 213, 216, 217
Attapulgite
see also Fuller's earth.
absorbent, 255
analysis, 596
deposits, 591
palygorskite, 591
uses, 591–594
Australia
asbestos, 454
barite, 493
bauxite, 508, 511–513, 523
bentonite, 606
chromite, 578
coral, 318
crocidolite, 454
diamonds, 658
diatomite, 689
fire clay, 638
fluorspar, 736
fuller's earth, 606
gems, 302, 317, 318
geological agencies, 413
glauconite, 753
gypsum, 778
graphite, 767
kaolin, 627, 638
lithium, 870, 874
magnesite, 883, 884, 885
monazite, 1113
pearls, 318
perlite, 1006
phosphate, 1028
potash, 1071
pyrophyllite, 1093–1095
rare earths, 1111
refractory clay, 638
refractory kaolin, 638
sillimanite, 802
titanium, 1305, 1315, 1322, 1333, 1339, 1340, 1352
vermiculite, 1377
zeolites, 1416
zircon, 273, 1433, 1435, 1443
Austria
barite, 492
graphite, 761, 762, 764
iron oxide pigment, 355
kaolin, 624
kyanite, 801
magnesite, 883, 884, 885
meerschaum, 911
olivine, 889, 991
sulfur, 1256
Autoclave, quartz synthesis, 216, 217
Aventurine, feldspar, 296
Aztec sandstone, 1172

B

Bahco classifier, 946
Bacteria, 1249
Baddeleyite, 1433, 1435, 1436
Baquette, 33
Bahamas
aragonite, 101
salt, 1125
Balangero area, 452
Balds, 992
Ball clay, 35, 38, 612, 633, 634
ceramics, 35, 38
composition, 633
definition, 35, 633
depletion allowance, 641
deposits, 620, 633
distribution, 633, 634
evaluation, 634
future consideration, 634
geology, 633, 634
history, 633
impurities, 633
milling, 634
mining of, 634
mineralogy, 633
occurrence, 633, 634
origin, 633, 634
preparation for market, 634
production, 586, 633
reserves, 634
tariff, 642
trends, 634
uses, 633
Ballas, 19, 292, 296, 316
Ballast, railroad, 61
Bank sand, 271
Banka drill, 1178
Barbados, glance pitch, 531
Barilla, 1187
Barite. *See* Barium minerals.
Barium chemicals, 486, 496, 497
Barium minerals, 485–501
aggregate, 486
barium chemicals, 486, 496, 497
beneficiation, 495, 496
ceramics, 486
chemicals, 486, 496
consumption, 497, 498
depletion allowance, 499
deposits, distribution, 490–493
deposits, types of, 487–490, 495, 496, 498
ecological impact, 499
evaluation of deposits, 495
exploration, 493–495
filler, 244, 245, 247, 250
flowsheet, 495, 496
future considerations, 498, 499

Barium minerals, geology, 486–495
- imports, 498
- marketing, 497, 498
- milling, 495–497
- mineralogy, 487
- mining, 495–497
- plants, chemical, 497
- preparation of chemicals, 496, 497
- prices, 498
- processing, 496, 497
- production, 494
- recycling, 499
- reserves, 493, 494, 498
- shielding, 37, 485, 486
- specifications, 486
- substitutes, 499
- tariff, 499
- taxes, 499
- transportation, 497
- trends, 499
- uses, 35, 388, 389, 485, 486
- witherite, 485, 499

Barium titanate, 313

Barytes. *See* Barium minerals.

Basalt, 165, 168, 169, 205, 209

Base exchange, 745

Base granules, 206

Base, molding sands for, 272

Baseline sandstone, 1172

Basic brick, 381–383

Basic oxygen furnace, 122, 382, 819

Basic oxygen process, 122, 741

Bassanite, 776

Bastnaesite, 1110–1114, 1116, 1117

Bath brick, 26

Bath rouge, 359

Batteries
- graphite in, 768
- lithium, 878
- manganese, 902

Bauxitic raw materials, 503–527
- abrasive material, 35, 521, 522
- abundance, 503
- activated, 521
- age, of deposits, 508, 510
- alumina, 519, 520
- analyses, 508, 509, 518, 519
- bauxite, 504, 508, 509
- Bayer process, 418, 520
- boehmite, 504, 515
- cement, 521
- chemicals, 521
- classification of deposits, 504, 505
- composition, 508, 509, 518, 519

Bauxitic raw materials, consumption, 521
- deposits, 504–511
- diaspore, 504, 515, 637, 638
- environmental considerations, 524
- evaluation, 517–521
- exploration, 516, 517
- ferruginous bauxite, 515
- fused aluminum oxide, 29
- future considerations, 523, 524
- geology, 503–508
- gibbsite, 504, 515
- government control, 523, 524
- grades, 517–521
- high alumina bricks, 520, 521
- imports, 522
- impurities, 503
- milling, 519
- mineralogy, 503, 504
- mining, 519
- mixed ore, 503
- nepheline syenite, 506
- origin, 506–508
- preparation for market, 519–521
- prices, 521–523
- processing, 519–521
- producing countries, 511–516
- production, 512, 522
- properties, 503, 504
- refractories, 520, 521
- reserves, 509–511
- specifications, 35, 517, 518
- substitutes, 523
- testing, 517, 518
- transportation, 519
- uses, 35, 518, 519, 520, 521, 522

Bayer process
- aluminum process, 418, 519, 520
- lime used in, 820
- sulfuric acid, 1263

Beach pebbles, 27

Beach sand deposits, 1347, 1348

Beam and cylinder test, 68

Beam splitter, 226

Bearings, graphite for, 768

Belgium
- flint pebbles, 27
- glauconite, 752
- kaolin, 626
- lime, 828
- quartz, 27
- silex, 27

Beneficiation. *See* preparation under names of rocks and minerals.

Benitoite, 303

Bentone, 611

Bentonite, 38, 587–612
- absorbent granules, 255, 589, 594
- activated clay, 589
- analyses, 596
- beneficiation, 609
- bleaching clay, 589
- bond, foundry sand, 593
- claiming land, 641
- classification, 587–589
- definition, 587, 588
- demand, 611
- depletion allowance, 641
- deposits, distribution, 598–606
- drilling, 607
- drying, 609, 640
- electron micrograph, 595
- evaluation, 606
- exploration, 606, 607
- fillers, 244
- fuller's earth, 587–612
- future considerations, 611, 612
- geology, 594–599
- history, 589, 590
- impurities, 595
- industry growth, 589
- ion exchange, 594
- iron ore pelletizing, 593
- K-bentonite, 588
- metabentonite, 588
- milling, 609
- mineralogy, 594–597
- mineralogical investigation, 608
- mining methods, 608, 609
- mixed bentonites, 588
- montmorillonite, 594
- occurrence, 597, 599–606
- organic-clad bentonite, 610, 611
- origin, 588, 591–593, 597, 598
- plants, 598, 601, 602
- plasticizer, 38
- potash-bentonite, 588, 589
- preparation for market, 608, 609
- processing techniques, 609
- production, 585–587, 589, 590
- results, appraisal, 608
- size, ore body, 597
- smectite, 588, 594
- southern bentonite, 588
- specifications, 593, 594
- sub-bentonite, 588
- swelling bentonite, 588
- synthetic, 610, 611
- tariff, 643
- testing, 607, 608
- transportation, 609, 610
- trend, 611, 612
- use, 589–591, 593

Bentonite, western bentonite, 588
  Wyoming bentonite, 588
Beryl, 35, 37, 42, 292, 294, 296
BET, surface area, 350
Beta-hemihydrate, 186, 188
Bicarbonate of soda, 388, 390
Bidahochi formation, 1173
Biotite, 915, 916
Bitumens, fillers for, 248, 249
Bituminous material, 529–532
  albertite, 529, 531
  asphalt, 529
  asphaltic pyrobitumens, 529
  asphaltites, 530–532
  bituminous limestone, 530
  ceresine, 531
  definition, 529
  elaterite, 531
  first used, 529
  gilsonite, 530
  glance pitch, 531
  grahamite, 530, 531
  impsonite, 531
  limestone, rock asphalt, 530
  manjak, 531
  mineral waxes, 529
  mining, 530
  montan wax, 531
  native asphalt, 529, 530
  nonasphaltic pyrobitumens, 529
  ozokerite, 531
  pavement bituminous mixes, 120
  petroleum asphalts, 529
  processed rock asphalt, 530
  production, 531
  pyrobitumens, 529, 531
  rock asphalt, 530
  sands, bituminous, 529, 1176
  uses, 529
  uintahite, 529
  varieties, 530, 531
  wurtzilite, 529, 531
Black granite, 165, 169
Black marl, 745
Black powder, 963
Black rouge, abrasive, 31
Black sand, 1316
Blanc fixe, 486
Blast furnace slag, air-cooled, 114, 115, 135, 143, 344, 346
Blast sand, 1225
Bleaching clay, 523
  see also Fuller's earth.
Block mica, 915
Blockspar, 711
Blue ground, 301
Blue Mountain syenite, 937, 948
Bluestone, 166
Boehmite, 504
Bochones, 427
Boeton, 530
BOF. *See* Basic oxygen furnace.
BOP. *See* Basic oxygen process.
Bolivia
  bastnaesite, 1111, 1113
  borates, 548
  crocidolite, 455
  lithium, 869, 870, 871
  monazite, 1111, 1113
  soda ash, 1200
Bonaire
  salt, 1125
Bonded abrasives, 16, 17
Bonding clay, 589
Bone magnesite, 882
Bone phosphate of lime, 1017
Boracite, 550
Borax
  see also Boron and Borates.
  Boron, CA, 539, 541
  brazing flux, 36
  brines, 553, 554
  composition, 536
  deposit size, 552
  flux, 36
  growth potential, 48
  origin, 41, 536–553
  prices, 48
  processing, 553, 554
  production, 48
  Turkey, 543–546
  uses, 48
Borazon, 29, 30, 313, 671
Bordeaux mixture, 822
Boric acid, 547, 555
Boron and borates, 45–48, 533–560
  borax, 36, 41, 48, 539, 541, 553, 554
  boron chemicals, 41, 43, 44, 46, 47
  classification, 536
  colemanite, 5, 43, 536, 552
  composition, minerals, 536
  consumption, 45, 534, 555
  depletion allowance, 555
  deposits, 536–543, 552
  environmental considerations, 555
  evaluation, 552
  exploration, 550–552, 554, 555
  exports, 534
  fertilizer, 239
  flux, 45, 262, 263
  future considerations, trends, 48, 544, 555
  geology, 535–538
  government controls, 555
  imports, 555
  marine deposition, 537
  markets, 555, 556
  marsh deposits, 537
  micronutrient, 223, 239
Boron and Borates, milling, 553
  mineralogy, 45, 535, 536
  minerals, 536
  mining, 552, 553
  neutron absorption, 37
  nonmarine deposition, 537, 538
  ores, 43
  origin, 41, 536–538
  preparation for market, 552–554
  prices, 48
  processing, 553, 554
  producers, 539
  production, 48, 533, 556
  products, 533–535
  refining, 554
  regions
    Argentina, 549, 550
    Chile, 550
    China, 550
    Germany, 550
    Italy, 547
    Mexico, 550
    Peru, 550
    Tibet, 550
    Turkey, 543–546
    US, 539–543
      California, 539
      Nevada, 539
    USSR, 546, 547
  reserves, 539–546, 552
  resources, map, 539
  sodium sulfate from, 553
  test for borates, 551
  testing, 552
  transportation, 555
  trends, 554–558
  uses, 36, 38, 43, 44, 47, 533–535, 557
Boron, CA, 539–541
Boron carbide, 12, 29, 30
Boron nitride, 29, 30, 43, 313, 320, 671
Bort, 19, 292, 296
Botswana
  diamonds, 302
  soda ash, 1216
Boule, 311, 312
Brake linings, graphite for, 768, 769
Brass wool, abrasive, 30
Brazil
  asbestos, 451
  baddeleyite, 1436
  barite, 492
  bauxite, 508, 514
  bentonite, 602
  bort, 19
  carbonados, 19
  chromite, 575
  diamonds, 653, 658
  diatomite, 689

Brazil, fire clays, 638
fluorspar, 732
gems, 302
graphite, 761, 763, 767
kaolin, 624, 638
kyanite, 806
lithium, 870, 871, 873
magnesite, 882, 884, 885
manganese, 898
monazite, 1113
oil shale, 1253
nepheline syenite, 935, 936, 942, 1042
olivine, 991
phosphate, 1021, 1032, 1033, 1042
potash, 1051, 1066, 1067
pyrophyllite, 1096, 1097
quartz, 213
rare earths, 1112
salt, 1125, 1126
soda ash, 1200
sulfur, 1260
talc, 1282
titanium, 1315, 1323, 1334
tourmaline, 302
vermiculite, 1377
yttrium, 1113
zirconium, 1443
Brazil twinning, 218
Brazing, 259, 261–263
Breeder reactors, 1117
Breunnerite, 881, 885
Brick, bath, 26
Bricks, refractory, 373–378, 381–383
Brilliance, 292–294
Brimstone, 1235
see also Sulfur.
Brine, artificial, 42
Brines, natural, 41, 44, 871, 873, 875
borax source, 541, 542
bromine, 41, 45, 561
chart, uses, 44
chlorine, 41, 45
iodine, 41, 793, 796
lithium, 41, 869, 871, 873, 876
magnesium, 881, 889
potassium, 41
sodium, 41
types of, 44
Briquetting, 723, 725, 739, 740, 887
British Columbia
barite, 492
bentonite, 601
chrysotile, 450
diatomite, 685
fluorspar, 728
gypsum, 784
magnesite, 884
British Columbia, nepheline syenite, 936
perlite, 682, 685
silica, 1176
soda ash, 1188
wollastonite, 1386
zeolite, 1416
Bromine, 48, 561–565
brines, 41, 45
depletion allowance, 563
exploration guide for potash, 1060
prices, 48
production, 48
salt, content, 562
sources, 41
uses, 48
Brookfield viscosity, 616
Brownstone, 166
Bruce Williams process, 903
Brucite, 382, 881, 882, 888
Bryn Mawr gravel, 1711
Buckles, 274
Buhrstone, 27
Building stone, 161, 164
Built up mica, 915
Bulgaria
amphibole asbestos, 461
barite, 493
chromite, 578
fluorspar, 734
kaolin, 625
perlite, 999, 1004
zeolite, 1417, 1419, 1421, 1422
Bulk sampling, asbestos, 465, 466
Bullseye alunite, 422
Burkeite, 1189, 1199
Burma, gems, 19, 302
Burnishing sand, 25
Burundi, bastnaesite, 1111
Butyllithium, 877, 878

C

Calcareous marl, 854
Calcined clay, 614, 636
Calcined kaolin, 614, 628
Calcite
fertilizer, 233, 237, 238
optical, 221, 297, 834
Calcium, 237, 238
Calcium carbide, 724
lime, for use in, 821
limestone, for use in, 846
Calcium carbonate
abrasive, 19
precipitated, 31, 633
rubber filler, 251
substitute for, 1298, 1372
Calcium cyanamide, 962
Calcium nitrate, 962
Calcium zirconate, 35
Caliche, nitrates from, 971
California
asbestos, 451
asphalt, 530
ball clay, 633, 634
barite, 488, 489, 491, 492
bastnaesite, 1113, 1114, 1115
bentonite, 597
borax, 537, 539–543
bromine, 48
chromite, 575
clay, 633, 634, 638
diatomite, 25, 26, 682–685, 693
dimension dolomite, 854
dimension limestone, 854
feldspathic sand, 713
fire clay, 638
flint pebbles, 27
fluorspar, 729
gems, 303, 318, 319
glass sand, 639
granite, 171
gypsum, 778, 784
hectorite, 598, 600
iodine, 794
kaolin, 618, 619, 622, 638
kyanite, 801
lime, 828
limestone, 853
lithium, 870, 875
magnesite, 890, 891
pegmatites, 997, 999, 1003
perlite, 1003
phosphate, 1043
potash, 1073
pumice, 171, 1079, 1080
pyrophyllite, 1093
rare earths, 1111
salt, 1121, 1124
sandbornite, 487
silica, 1171, 1172, 1173
slag, 125
soda ash, 1187, 1188
sodium sulfate, 1212, 1220, 1290, 1291, 1293, 1294, 1298
sulfur, 1251, 1252
talc, 1275, 1283, 1284, 1285
tar sands, 1253
thorium, 1113–1115
titanium, 1316
tripoli, 1363, 1369
vermiculite, 1377
witherite, 487
wollastonite, 1383–1385
zeolites, 1414, 1416, 1421
Cameos, 333
CAN, 962
Cana Brava area, 451
Canada
asbestos, 445, 446–451, 456, 460
barite, 492
bentonite, 601, 602

Canada, blasting sand, 25
- *see also* individual provinces.
- brucite, 382
- celestite, 1032, 1229
- chromite, 575
- diatomite, 685
- fluorspar, 728
- graphite, 765
- gypsum, 184, 777, 779, 784
- kyanite, 803
- lime, 828
- lithium, 870, 871
- magnesite, 883, 884
- mica, 918
- monazite, 1113
- nepheline syenite, 931, 933, 937, 938, 947–949
- olivine, 991
- perlite, 1002
- potash, 1062
- pyrophyllite, 1096
- quartzite pebbles, 27
- silica sand, 25, 1174–1176
- sodium sulfate, 1207
- tar sand, 1442
- titanium slag, 1305
- vermiculite, 1377
- zeolites, 1422

Cap rock, 1243, 1244

Cape Blue. *See* Crocidolite.

Carat, unit of weight, 323

Carbolon, 28

Carbon
- *see also* Graphite.
- ceramic raw material, 35
- forms of, 757
- refractory, 379

Carbonada, 19, 292, 296, 316

Carbonate rocks. *See* Dolomite, Limestone, Marble.

Carbonatites
- apatite in, 1020
- fluorspar in, 727
- phosphates in, 1020
- rare earths in, 1111, 1112

Carborundum, 28

Carlsbad area, 1059–1066, 1071
- bromine, 1059
- potash, 1059, 1060
- stratigraphy, 1059

Caribbean Islands, geological agencies, 402

Carnelian, 1168

Carnallite, 1050
- *see also* Potash.

Carthage marble, 834

Castable refractories, 380

Casting plaster, 194

Catalyst, 636, 1425, 1426

Cat's eye, 296

Caustic, calcined magnesite, 882
- *see also* Magnesite.

Caustic soda
- *see also* Sodium hydroxide.
- production, 49
- use, 346

Cawk, 487

CCE, 845

Celestite. *See* Glass raw materials, raw materials composition; Strontium.

Cement, 37, 133–159
- aggregates for, 37, 59–131, 1151–1153
- alternate raw materials, 147
- capacity of industry, 146, 150
- chemical requirements, 144–146
- clinker, 139–141, 155
- composition, 144
- comment, general, 157, 158
- competition, foreign, 152
- compounds, 141
- consumption, 146–148, 150
- control, manufacturing, 141, 142, 156
- costs, 147, 148
- definition, 133
- deleterious material, 135
- depletion allowance, 151
- developments, notable, 155, 156
- dry blending, 138
- dry process plants, 138, 139
- drying, 155
- dust control, 141
- economic factors, 135, 148–151
- environmental considerations, 152–154, 156
- evaluation, 135, 136, 247
- gypsum, cement from, 1256
- high-alumina, 521
- history, 133, 134
- hydraulic cement, 157
- kilns, 135, 139, 140
- location, plants, 134, 135
- manufacture, steps in, 137–139, 141, 142, 155
- marketing, 149, 150, 155
- physical requirements, 144–146
- plants, 146, 150
- pollution control, 153
- processing, 136–141, 155
- price trend, 151, 152
- pozzolan cement, 143
- profits, 151, 152
- production, 146
  - US, 134
  - world, 157
- products, types of, 142, 143
- pyroprocessing, 139
- quality control, 138
- quarry operations, 136, 155

Cement, raw materials, 134–136, 141, 143, 147, 635, 784, 785, 931, 953, 1256, 1279, 1421
- requirements, 144–146
- rotary kiln, 140
- specifications, 144–146
- subsidies, 151
- substitutes, 147
- summary view, 154
- tests, 68
- transportation, 148, 149
- types, 142, 148
- underground mines, 136
- units, 133, 134
  - costs, 148
- use of, 147
- weight, 133, 134
- wet process plants, 138, 139
- white, 1087, 1088, 1103

Central African Republic, diamonds, 661

Ceramic, 35

Ceramic pebbles, 27

Ceramic raw materials, 33–39
- anadalusite, 34, 35, 376, 799, 802, 803
- barium ceramics, 35, 486
- dielectric materials, 38, 213
- feldspar, 33, 36, 719, 721
- fluorspar, 35, 724, 725
- function of, 34–39
- kaolin, specifications, 617, 618
- lime, 34, 37, 822
- nepheline syenite, 36, 932, 945, 946
- olivine, 35, 989
- silica, 33, 34, 36, 39, 1170
- specifications, 617, 618
- talc, 35, 1277
- titanium, 1312

Cerium, 1110, 1116

Cerium oxide abrasive, 16, 31

Cermets, 37

Ceylon. *See* Sri Lanka.

Chabazite, 1395, 1396

Chalk, 167
- abrasive, 26
- cement raw materials, 135
- flint, 25

Chamottes, 636, 1085

Chaser stones, 18

Chatham process, 313

Chatoyance, 296

Chemical industry, 41–58
- brines, salt, 41, 42, 44
- chlor-alkali industry balance, 48

Chemical industry, consumption, 43
  industrial minerals, 41–43
  major elements, 43
    boron, 36, 38, 43–48, 533–560
    bromine, 46–49, 561–565
    caustic soda, 52
    chlorine, 48–50
    consumption, 51, 52
    fluorine, 50–52, 723, 724
    hydrofluoric acid, 50–52, 723
    producers, 51
    sodium carbonate, 52, 1077, 1187–1206
    sodium chemicals, 52
    sodium hydroxide, 52, 53
    sodium sulfate, 52, 1207–1223
    sulfur, 41, 43, 53–55, 1235–1273
    sulfuric acid, uses, 41–43, 54, 55
Chemical precipitates, abrasives, 30, 31
Chemical reactivity tests, sand and gravel, 99, 100
Chert, 15, 68, 835
Chile
  barite, 492
  borates, 550
  diatomite, 689
  fluorspar, 732
  gems, 302
  iodine, 795
  kaolin, 624, 638
  lithium, 869, 871
  lazurite, 302
  lithium, 869, 871
  nitrates, 970, 971
  perlite, 1003
  potash, 1066, 1067
  salt, 1126
  soda ash, 1200
  sodium sulfate, 1207, 1216
  sulfur, 1251, 1260
  zeolite, 1418
China
  asbestos, 454
  barite, 493
  bastnaesite, 1113
  bauxite, 458, 515
  borates, 550
  calcined bauxite, 518
  chromite, 578
  diaspore, 515
  diatomite, 689
  flint clay, 638
  fluorspar, 735
  gems, 302
  graphite, 761
  kaolin, 627
China, magnesite, 884
  mica, 918
  oil shale, 1253
  potash, 1070
  refractory clay, 638
  salt, 1126
  talc, 1283
  titanium, 1336
  vermiculite, 1377
  zirconium, 1433
China clay, 26, 612
Chlorastrolite, 303
Chlor-alkali industry, 48–51
Chloride process, TiO_2 manufacture, 49
Chlorine, 48–50
  fertilizer, 233
Christmas Island
  guano, 1021
  phosphate, 1021
Chromite, 567–584
  analyses, 569, 571
  beneficiation, 572, 573
  ceramic, 35
  chemical grade, 568, 569
  composition, 569–571
  consumption, 569, 578
  deposits, types, 572
  definition, 567
  economic factors, 567, 579–581
  evaluation, 573, 574
  exploration, 573, 574
  foundry sand, 274, 568, 1440
  future trends, 581, 582
  gangue, 571, 572
  geology, 569–573
  grades, 35, 274, 567–569
  imports, 578, 579–581
  metallurgical grade, 567, 568
  milling, 572, 573
  mineralogy, 569–571
  mining, 572, 573
  molding sand, 274, 568
  olivine, competition, 994
  podiform deposits, 572, 576
  prospecting, 573, 574
  prices, 274, 578–581
  production, 578, 579
  properties, 274
  ratios, Cr:Fe, 567, 568
  refractories, 379–381, 568
  refractory filler, 35
  reserves, estimating, 574
  resources, world, 574–578
    Africa, 576, 577
      Madagascar, 576
      South Africa, 576
      Zimbabwe, 576
    Asia, 577, 578
      China, 578
      India, 577
      Philippines, 577
Chromite, resources, Asia, Turkey, 577, 578
    Central America, 575
    Europe, 576
      Albania, 576
      Finland, 576
      USSR, 576
      Yugoslavia, 576
    North America, 575
      Cuba, 575
      Greenland, 575
      United States, 575
    South America, 575, 576
  specifications, 567–569
  stockpiling, 567
  stratiform deposits, 572, 576
  uses, 35, 274, 567–569
Chromium oxide, abrasives, 31
Chrysoberyl, 279, 282, 291
  synthetic, 313, 314
Chrysophase, 1168
Chrysotile
  see also Asbestos.
  composition, 440
  electron micrograph, 436, 439
  geology, 444–446
  heat, effect on, 440
  occurrences, 454, 455
  origin, 436
  pH, 437
  production, 444
  properties, table, 441, 442
  solubility, 443
  structure, 435
  thermal analysis, 437
  uses, 474–481
  zeta potential, 439
Cinders
  as aggregate, 81
  coal, 85
  volcanic, 81, 85, 86, 1079
Circum-Pacific belt, 1001, 1002
Claus process, sulfur from H_2S, 1252, 1253
Clays, 585–651
  abrasive, 31
  absorbent granules, 593, 594
  analyses, table of, 596
  attapulgite. *See* Fuller's earth.
  ball clay, 35, 38, 586, 612, 620, 633, 634, 641, 642
  bentonite, 38, 587–612
  calcined clay, 614, 628, 636
  ceramic, 33, 34
  definition, 585
  depletion allowance, 641
  drying, 609
  ecological considerations, 641, 642
  electron micrographs, 595, 619
  fillers, 244, 250–252

Clays, fire clay, 34, 373–376, 576, 586, 641
  firing, shrinkage, 640
  fuller's earth, 16, 38, 587–612, 641
  glass, use in, 346
  granule, 209, 593, 594
  green strength, 640
  halloysite, 596, 612, 613, 620, 634–636
  hectorite, 594, 596, 598, 600, 607–609, 611
  illite, 34, 633
  kaolin, 26, 34, 35, 38, 375, 376, 586, 596, 599, 612–633, 635, 641, 642
  lightweight aggregate, 82
  opal claystone, 596
  palygorskite, 596
  porcelain, clays for, 614
  pottery, clays for, 639
  production, 586, 590, 614
    Argentina, 624
    Australia, 627, 628
    Austria, 624
    China, 627
    Czechoslovakia, 625
    Denmark, 625
    France, 625
    Guyana, 624
    Hungary, 625
    India, 626
    Italy, 625
    Japan, 672
    Korea, South, 627
    Malaysia, 627
    Mexico, 624
    New Zealand, 628
    Philippines, 627
    Spain, 625
    UK, 623, 624
    US, 622, 623
    USSR, 624
    West Germany, 626
  refractory clay, 586, 620, 636–639
  reserves
    Brazil, 624
    Bulgaria, 625
    Czechoslovakia, 625
    Iran, 627
    New Zealand, 628
    Poland, 625
    Sri Lanka, 626
    Turkey, 627
    US, 623
    USSR, 624
  saponite, 594, 600
  sepiolite, 596
    see also Meerschaum.
Clays, smectite, 594, 597, 599, 604, 605, 633, 640, 870
  societies, 586
  uses, 588–592, 599, 602, 606, 633–636
  vitrification range, 640
  whiteware, 1089
Clayton Valley, 871, 873
Clinker, 139–141
Clinoenstatile, 1279
Clinoptilolite, 1397, 1398
Coal, sulfur from, 1254
Coated abrasives, 17, 18
Ccated optics, 298
Coatings
  calcium carbonate, 633
  decorative, protective, 252
  kaolin, 616
  talc, 1278
Coesite, 1167
Coke, 35
Coke oven gases, 970
Colemanite, 5, 43, 536
  see also Boron and borates.
Coalinga area, 451
Colloidal silica, 1168
Colombia
  asbestos, 451, 452
  barite, 492
  chromite, 578
  diatomite, 689
  emeralds, 302
  glance pitch, 531
  kaolin, 624
  salt, 1126
  tar sands, 1253
Colorado
  barite, 492
  clay, 374, 375, 637
  cryolite, 741
  feldspar, 711
  fluorspar, 727, 729
  gannister, 377
  gems, 303, 319
  grahamite, 530
  granite, 171
  gypsum, 778, 782
  lithium, 873
  oil shale, 1253
  pegmatites, 711
  perlite, 997, 1003
  salt, 1123, 1124
  sand-gravel demand, 108
  slag, 125
  soda ash, 1188
  thorium, 1111
  titanium, 1321
  vermiculite, 1377
  zirconium, 1435
Composite gems, 334
Computers, use of, 142
Concrete
  aggregates, 60–131
  air-entrained, 142
Congo (Brazzaville), potash, 1070
Connecticut
  feldspar, 711
  fuller's earth, 591
  granite, 170, 711
  mica, 918
  pegmatites, 711
  talc, 1288
  trap rock, 62
Construction materials
  aggregates
    crushed stone, 60–80
    lightweight aggregates, 81–95
    sand and gravel, 96–110
    slag, 111–131
  cement, cement raw materials, 133–159
  dimension, cut stone, 161–181
  gypsum, anhydrite, 183–198
  insulation materials, 199–201
  roofing granules, 203–212
Contact process, sulfuric acid, 1266
Copper, 233, 240
Copper wool abrasive, 30
Copperas red, 361, 364
Copperopolis area, 451
Coquina, 167, 170
Coral, gem, 291, 301, 303, 318
Cordierite, ceramic, 34, 35, 1278
Corduroy spar, 710
Corkstone, 162
Cornwall stone, 36
Corundum, 15, 16, 18, 292
  synthetic, 312, 313, 319
Cosmetics, 52, 1277
Costs
  open pit, 74
  transportation, 73, 853
Cristobalite, 1167, 1168
Crocidolite, 437, 440–443, 457
  see also Asbestos.
  uses, 475
Crocus, abrasive, 31
Cross fiber, 444, 463
Crucibles, graphite for, 769
Crushed steel, 30
Crushed stone, 60–80, 847
  aggregates, 60–80
  capacity, plant, 69, 70
  chert, 63
  competition, factors affecting, 74–76
  consumption, 61–64
  costs, 64, 73, 75, 76
    transportation, 73, 74, 852, 853
  crushing, 72
  definition, 60

Crushed stone, deleterious material, 68
- demand, 74
- depletion allowance, 78
- distribution, 69
- future trend, 78, 79
- government controls, 70, 77
- gradation, acceptable, 68
- grading requirements, 67, 68
- highway construction, 63–65
- industry, 75, 76, 83
- labor requirements, 75
- land use, 77, 78
- limestone, 62, 63, 844
- location requirements, 61–63, 69, 70
- nonconstruction use, 62
- physical properties, rocks, 63
- plant considerations, 69, 70–73, 75, 76
- prices, royalty, 78
- production, 61–69
- profitability, 77
- qualities, 60
- reserves, 79
- rock types used, 62, 63, 844
- shape, particle, 68
- shipments, 852
- slag aggregate, 111
- specifications, 60, 61, 65–69
- substitutes, 69
- supply, 79
- taxation, 78
- tests, 65–69, 98, 100, 1157, 1158
- transportation, 73, 74, 852, 853
- uses, 59, 61, 62, 847
- value, 77, 78
- zoning regulations, 78

Crushing
- asbestos, 468
- borates, 553
- stone, 72

Cryolite, 741, 742, 1042
- wollastonite, substitute for, 1388

Crystalline graphite, 757

Crystallite, 350, 351

Crystals
- *see also* Electronic and optical uses.
- producers, 214, 226
- synthetic, 214, 226

Crystolon, 28

Cuba
- barite, 492
- chromite, 575
- glance pitch, 531
- grahamite, 530

Cuicatlan area, 451

Cullet, 346

Cultured quartz, 38, 210, 213, 214, 216–221
- *see also* Quartz.

Curbing block, 164, 165

Cured sands, 272

Curvettes, 333

Cut stone, 162, 163

Cuttables, diamonds, 19, 296

Cutting sand, 25

Cyprus
- asbestos, 452
- bentonite, 602
- chromite, 578
- iron oxide pigments, 355
- sulfur, 1254

Czechoslovakia
- barite, 493
- bentonite, 602
- clay, 625, 636, 638, 683
- magnesite, 884
- ozokerite, 531
- perlite, 1004

D

Danakil Depression, 1051

Darling Range, 512, 513

Daugherty process, 903

Dauphine twinning, 218

Dawsonite, 523

Dead-burned gypsum, 189

Dead-burned magnesite, 381, 882, 887, 893
- *see also* Magnesite.

Dead gems, 295, 332, 654

Dead perlite, 1010

Death Valley, 542, 543

Decorative coatings, 252

Delamination, 630, 631, 632

Delaware, glauconite, 752

Delay line, 226, 227

Denmark
- diatomite, 687
- flint pebbles, 27
- kaolin, 625
- quartzite, 27

Density test, loose grains, 20

Desulfurization, 1256

Deval abrasion test, 100, 844

Diabase, 314

Diamite, 314

Diamonair, 314

Diamond drilling, asbestos, 462, 463

Diamonds, 19, 20, 653–676
- *see also* Gem materials.
- abrasive, 15, 16, 19, 20
- accessory minerals, 658
- alluvial occurrences, 667, 668
- ballas, 19, 657
- black diamond, 656
- boart, 656
- boort, 656
- bort, 19, 656

Diamonds, bortz, 656
- bowr, 656
- brilliance, 653
- carbonado, 19, 656, 657
- classification, 656, 657
- composition, 659
- consumption, 656
- cuttables, 298
- cutting, 653
- dead stone, 295, 332, 654
- diamite, 314
- diamonair, 314
- discoveries, 653
- dispersion, 294, 653
- ecology, 673
- evaluation, 663
- exploration, 662, 663
- fancies, 295
- fire, 295, 653
- flowsheets, 666, 667
- future of, 673
- gems, 279, 283, 295, 332, 653, 654
- geology, 657, 658, 662
- government control, 672
- grade, 663
- grits, 657
- hardness, 291–293
- inclusions, 299
- index of refraction, 653–660
- industrial, 19, 296, 654–657
- kimberlite, 657, 658, 662
- manufactured, 16, 19, 20, 30, 662, 668–670
- marketing, 668
- metal coating, 30
- milling, 310, 311, 665–668
- mining, 663–665
- powders, grade of, 658
- price/time, 673
- processing, 665–668
- production, 19, 654, 662, 672
- producing controls, 668
- properties, 659–661
- reserves, 658, 661
- scintillation, 653
- secondary recovery, 672, 673
- shapes, 654
- short-bort, 657
- sources, 654
- stockpile, 672
- substitutes, 671, 672
- synthetic, 313, 318, 668–670
- tariffs, 672
- taxes, 672
- transportation, 668
- trends, 668
- trimond, 314
- value, 321
- uses, 653

Diaphragm cells, chlorine production, 50

Diaspore, 34, 376, 377, 504, 515, 637, 638, 1090
Diatomite, 16, 25, 26, 38, 677–708
  abrasive, 16, 25, 26, 38, 699
  absorbent, 699
  acid treatment, 692
  aggregate, lightweight, 89
  analyses, 678, 679, 694
  calcining, 692
  ceramic, 36, 39
  chromatographic support, 699
  classification of deposits, 682, 683
  composition, 677–681, 694
  consumption, 695
  definition, 677
  density, 678, 690
  depletion allowance, 701
  deposits, 682–691
  ecology, 701
  electron micrographs, 680, 681, 683, 684–687, 693
  evaluation, 690, 691
  exploration, 689, 690
  exports, 696
  filler, 254, 697–699
  filter, 254, 696, 697
  flow diagram, 692, 693
  flux calcining, 692
  geology, 677–683
  glass former, 36
  impurities, 677
  lacustrine deposits, 682
  marine deposits, 682
  marketing, 696
  microstructure, 678
  milling, 691–694
  mining, 691
  morphology, 677–681
  occurrences, 681, 682
  origin, 681, 682
  packaging, 694
  preparation, for market, 691–699
  processing, 692, 693
  producers, 658, 662
  production, 695
  properties, 659–661, 699, 700
  reserves, 689
  sampling, 663
  specifications, 694, 695
  structure, diatom, 677
  substitutes, 699–701
  taxes, 701
  tariffs, 701
  testing, 694, 695
  transportation, 696
  types of, 678
  uses, 244, 254, 388, 677
    cement, 143
    ceramic, 36, 39
    filler, 697–699
Diatomite, uses, filter, 696, 697
    properties of, 698
    well drilling fluids, 388
Didymium oxide, 1110
Dielectric materials, 38, 213, 227
Dimension stone, 161–181
  ashlar, 157, 158, 162
  basalt, 168, 169
  building stone, 161–181
  color, 161
  consumption, 162, 847
  curbing block, 164, 165
  cut stone, 162, 163
  cutting, 173–175
  definition, commercial, 165–167
  depletion allowance, 180, 181
  deposits, distribution, 169–171
  dolomite, 167
  end use, 161–165
  evaluation of deposits, 179, 180
  finishing, 164, 177
  flagstone, 166
  fleuri pattern, 167
  geologic provinces, 169–171
  granite, 161–166
  greenstone, 168
  limestone, 162, 163, 166, 167, 847
  lithologic classification, 165–169
  marble, 162, 167–173
  marketability, 180
  mill-stock slate, 164
  monument stone, 164
  obsidian, 167, 168
  onyx, 167
  pattern, quality of, 163
  paving blocks, 164
  polish, 163, 164
  preparation for markets, 171–177
  preservation of stone, 177, 178
  producers, 162
  production, 162, 847, 854
  pumice, 168
  quarrying, 171–177
  reserves, 169–171
  rock types, commercial definition, 165–167
  roofing slate, 164
  rough construction stone, 161, 162
  rubble, 161, 162
  sandstone, 166
    abrasive use, 25
    bluestone, 166
    brownstone, 70
    freestone, 166
    reeds, 166
    refractory brick, 165
Dimension stone, sandstone, rift, 166, 176
  sawing, 176, 177
  slate, 162–164, 167, 168
    filler, use as, 245
    granules, 209
  soundness, 180
  underground quarries, 172, 173
  use, 847
  zeolites, as, 1420
Discrete particle, definition, 350
Dispersion, 294
Dolomite, 167, 833–868
  aggregate, 61
  burnt dolomite, 269
  ceramic, 33, 35
  composition, 269
  fertilizer, 238
  flux, 33, 269
  refractories, 381, 382, 881
  smelting use, 269
  specifications, 770–772
  uses, 33, 238, 269, 342, 345, 381, 382, 881
Domes, salt, 1123–1125
Dominican Republic
  bauxite, 513, 523
  gypsum, 784
Dredges, 1155, 1160, 1328, 1348
Drilling
  asbestos, 462, 463
  bauxite, 456, 457
Drilling muds. *See* Well drilling fluids.
Dry press process, brick, tile, 375
Dumortierite, 35, 799
Dunite, 989
Durability test, sand, gravel, 98
Dysprosium, price, 1110

E

Earth diatomaceous, 653–676
  see also Diatomite.
Ecuador, kaolin, 624
Egypt
  chromite, 578
  clay, 626, 638
  gems, 303, 316
  nitrates, 970
  phosphate, 1031
  rare earths, 1112
  salt, 1126
  soda ash, 1197
  titanium, 1315, 1334
  vermiculite, 1377
  zirconium, 1433
Elaeolite, 933
Elastomic products, 49
Elaterite, 531
Electric furnace production, 28, 1018
Electrical twinning, 218

Electrochemical unit, 50
Electronic uses, 213–231.
 see also Quartz.
Emeralds, 313, 317, 318
Emery, 16, 18, 19
Endellite. *See* Halloysite.
Eneabba, 1332, 1333
Epsom salts, 893
Erbium, price, 1110
Erionite, 1395, 1396
Esker, 1154, 1155
Ethiopia
 kaolin, 626
 potash, 1051, 1070
Ethylene dichloride, 49
Eucryptite, 870
Eureka sandstone, 1172
Europe, geological agencies, 404–407
Europium, price, 1110
Expanded aggregates, 86
Expanded slag, 85, 116, 117, 119, 122
Extenders, 253
 see also Mineral fillers.

F

Fancies, 295
Fat clay, 639
Feathers, in quarrying, 175
Feldspar, 709–722
 abrasive, 16, 18, 26
 alaskite, 709–711
 aplite, 709, 712, 716, 721
 block spar, 711
 bonding agent, 16
 cement, use in, 134
 ceramics, use in, 33, 36, 716, 721
 depletion allowance, 720
 feldspathic sand, 710
 filler, 244
 glass, use in, 716, 719, 721
 government control, 721, 722
 granules, 204
 history, 718, 719
 markets, 719, 720
 milling, 717, 718
 mining, 717
 nepheline syenite, 719, 721
 pegmatites, 710, 711
 perthite, 710
 prices, 722
 processing, 716–718
 producers, 717, 719
 producing areas, 711, 713, 715, 716
 production, 713, 714, 716
 reclamation, 720
 reserves, 712, 713
 sand, 710
Feldspar, substitutes, 719, 720
 uses
  ceramic, 33, 36, 721
  fillers, 721
  glass, 344, 721
Feldspathic sand, 710
Ferrierite, 1417
Ferrites, 367, 1229
Ferroalloys, 904, 1116, 1173
Ferroelectric crystal, 227
Ferromanganese, grades, 904
Ferrotitanium, 1313
Fertilizer minerals, 233–241
 ammonia, 234
 boron, micronutrient, 239
 calcium, 233, 237, 238
  see also Limestone, Dolomite.
 consumption, 234
 copper, micronutrient, 233, 240
 importance, 233
 industry scope, 233
 iron, micronutrient, 239
 magnesium, 238, 239
  see also Limestone, Dolomite.
 micronutrients, 239, 240
 nitrogen, 233, 234, 977
  see also Nitrogen compounds.
 nutrient consumption, 234
 olivine, 989
 phosphorus, 233, 235, 236
  see also Phosphate rock.
 potassium, 233, 236, 237
  see also Potash.
 primary plant nutrients, 233–237
 secondary plant nutrients, 233, 236, 237
 sulfur, 233, 238, 239
  see also Sulfur.
 tagged nitrogen, 972
 trace elements, 233, 239, 240
 zinc, micronutrient, 240
Fiber glass, production, 339
Fibrolite, 802
Fiji, barite, 493
Fillers
 see also Mineral fillers.
 ceramic, 34
 clinoenstatite, 1279
 diatomite, 245, 247
 feldspar, 245, 721
 kaolin, 245, 246, 247
 nepheline syenite, 931, 934, 944
 perlite, 1013
 tripoli, 245
 wollastonite, 245
 zeolite, 1422
Filter aids
 see also Mineral filters.
 diatomite, 254
Filter stone, 61
Fining agent, 720
Finland
 anthophyllite, 461
 chromite, 576
 sulfur, 1254, 1255
 talc, 1282, 1283
 titanium, 1338
 vermiculite, 1377
 wollastonite, 1383, 1386, 1388
Fire clay, definition, 373, 374, 613
 see also Refractory clay.
Fire, of gem, 286, 292
Flagging, 164, 166
Flagstone, 162, 164
Flake graphite, 758
Flake mica, 915, 919–921
Flame cutting, 173–175
Flame treatment, 167, 339
Fleuri pattern, 167
Flint, 17, 18
Flint, abrasive, 17, 18
Flint clays, 374
Flint kaolin, 618
Flint paper, 17
Flint pebbles, 27
Florida
 attapulgite, 591
 clays, 591, 599, 622
 diatomite, 685
 kyanite, 800, 803
 limestone production, 62
 magnesite, synthetic, 890
 marl, 854
 phosphate, 1030, 1032–1034, 1039, 1043, 1045
 shells, 62
 silica, 1170, 1171
 staurolite, 26, 27, 1225
 titanium, 1303, 1304, 1316, 1324–1326
 zircon, 274, 1433, 1436–1438
Flour, silica, 1168, 1172
Flour, sulfur, 1238
Flue gas, 1256
Fluid-energy grinding mill, 1295
Fluorine, 50, 51, 723–725
 phosphates, from, 740, 1041
 tremolite, 1283
Fluorinated hydrocarbons, 723, 724
Fluorite. *See* Fluorspar.
Fluorocarbons, 51, 52
Fluorescence, 227
Fluorspar, cryolite, 269, 723–744
 beneficiation, 737, 738
 ceramics, use in, 34, 35, 734
 consumption, 725, 738, 739
  fluorocarbons, 52, 724

Fluorspar, consumption
fluxes, types of, 33, 267, 268, 739
glass, 724
hydrofluoric acid, 51, 724
steel, 724, 725
fluorides from phosphates, 740
flux, 33, 267, 268, 739
gem, 297
occurrences, mode of, 726–728
reserves
Alaska, 728
Italy, 733
Ontario, 728
world, 736
uses, 723, 724
abrasives, 724
ceramics, 34, 35, 724, 734
chemicals, 723
fluorinated hydrocarbons, 52, 724
flux, 33, 267, 268, 739
glass, 724
hydrofluoric acid, 56, 724
optics, 298
steel, 724, 725
welding flux, 269
Flux factor, 377
Fluxes, 33, 259–270
agents, 26, 36, 269
borates, 45, 46, 262, 263
ceramic fluxes, 36
economy of, 36
flux modifiers, 37
lithium, 36
nepheline syenite, 36
salt cake, 36
Fly ash, 85, 135, 209
Forsterite, 35
Forum on Geology of Industrial Minerals, 394, 395
Foundry facings, graphite, 768
Foundry sand, 271–278
olivine, 274–278, 989–996
staurolite, 274, 1225–1227
zircon, 273, 274, 1433–1446
France
andalusite, 802
asphalt, 530
barite, 492
bauxite, 514, 515
bromine, 48
buhrstone, 27
clay, 602, 625, 638
chalk, 25, 26
chrysotile, 457
diatomite, 686
feldspar, 715
flint pebbles, 27
fluorspar, 732, 733
gems, 302
gypsum, 184, 778
iodine, 795
iron oxide pigments, 355
France, kyanite, 803
meerschaum, 911, 912
mica, 918
perlite, 1005
potash, 1067
salt, 1126
slag, 113
sulfur, 1247, 1251
talc, 1283
zeolite, 1417, 1422
Fraccing sand, 1168
Francolite, 1022
Frasch process, 53, 1235, 1241, 1242, 1245–1247
Freestone, 166
Freeze-thaw test, 98
French chalk, 25, 26
Friability tests, 12, 13, 15
Frits, 36
Fuchite, 916
Fuller's earth, 12, 16, 587–612
abrasive, 16
absorbent, 255
ceramic, 35
distribution, 598
filler, 241
processing aid, 38
tariff, 642
uses, 38, 591, 592, 594
Fur, asbestos, 444
Fused alumina oxide, 29
Fusion cast products, 379

G

Gabon, manganese, 898
Gadolinium, price, 1110
Gadolinium-gallium garnet, 314
Galena, 388
Gannister, 363, 374, 379, 1168
Garnets, 17, 21–25
yttrium-aluminum, 314
Gas, natural, 976
Gehlenite, 1389
Geikelite, 1389
Gem materials, 279–338
see also names of stones.
ballas, 19, 296
bort, 10, 292, 296
carbonado, 19, 292, 296
control, political, 321–323
corundum, 15, 16, 18
synthetic, 312, 313, 319
diamonds, 19, 20, 314, 653–676
garnet, 24, 315
marketing, preparation for, 23, 323–330
quartz, 1167, 1168
uses
decorative, 291
gems, 653, 654
history, 279
industrial, 654, 655
Geological agencies
see also Sources of information.
Africa, 407–410
Asia, 411–413
Australia, 413
Caribbean Islands, 402
Central America, 402
Europe, 404–407
Middle East, 410, 411
North America, 393–402
Pacific Islands, 413
South America, 402–404
Georgia
attapulgite, 592
barite, 491, 492
bauxite, 515
clay, 592, 599, 613, 618, 621, 622, 638
feldspar, 711
granite, crushed, 62, 170
halloysite, 634, 635
iron oxide pigments, 355
kyanite, 376, 800, 801
manganese, 898
marble, 164, 170
mica, 918
monumental stone, 164
olivine, 888, 889, 991
pegmatite, 711
silica, 26, 1170, 1171
sillimanite, 802
titanium, 1316, 1324, 1326
talc, 1284, 1285, 1289, 1298
tripoli, 26, 1368
vermiculite, 1377
zircon, 1436
Germany, East
barite, 492
borates, 550
fluorspar, 733
kaolin, 625
potash, 1068
sulfur, 1256
Germany, West
asphalt, 530
barite, 492
bromine, 562
clay, 492, 604, 626, 638
diatomite, 686
feldspar, 715
fluorspar, 733
graphite, 760, 765, 768
gypsum, 778
iodine, 795
lime, 828
lithium, 876
perlite, 1005
potash, 417, 1068, 1069
quartz, 214
salt, 1126
slag, 113
sulfur, 1242, 1252

GGG, 314
Ghana
  bauxite, 508, 513, 523
  diamonds, 302
  gems, 302
  manganese, 897
Gibbsite, 376, 377, 504, 635
Gilsonite, 530
Glacial deposits, source of aggregates, 1154, 1155
Glance pitch, 531
Glass, 31, 33, 36
  raw materials for, 344, 902
Glass batch gypsum, 196
Glass paper, 31
Glass raw materials, 339–347
  abrasive, 17, 31
  batch requirements, 339, 340
  ceramic, 33, 36
  feldspar, 721
  fiber glass, 339
  formers
    diatomite, 36
    silica, 33, 36
  gypsum, used in, 196, 340
  imitation gems, 315
  industrial minerals, consumption, 33, 340, 341
  natural, 279
  processing, 33
  production, 340, 955
  raw materials composition, 340, 341, 344
    borax, 340, 341
    celestite, 1229, 1231
    clay, 33, 346
    dolomite, 33, 340
    feldspar, 340, 341, 344
    lime, 822
    limestone, 33, 341, 344, 846
    nepheline syenite, 341, 344, 955
    silica sand, 33, 36, 340, 341, 344
    soda ash, 340, 341, 344, 1217
Glass wool, 199–201
  see also Insulating materials.
Glauberite, 1209
Glauber's salt, 1207
Glauconite, 745–756
Glossary, electronic, optical, 226–228
Gneiss, 165
Goethite, pigment, 352
Gosiute, Lake, 1199
Gove Peninsula, 512
Grading requirements
  lightweight aggregates, 88
  slate, 164, 168
Grahamite, 530, 531
Grainers, 1136
Granite, 161, 164, 165, 167, 168
  aggregate, 61
  composition, 712
  crushed, 62
  cutting, 177
  definition, 167, 168, 712
  flame treatment, 173, 174
  graphic, 710, 711
Granulated slag, 115–117, 121, 122
  base rock, 204, 205, 208
Granules, roofing, 203–212
  absorbent, characteristics of, 594
Graphic granite, 710, 711
Graphite, 5, 757–773
  refractories, 379, 769
  reserves
    Alabama, 760
    Korea, 763, 764
    Norway, 765
  types of, 757, 758, 759
  uses, 388, 757, 768, 769
Gravel, definition, 1151
Gravel spar, 730
Graywacke, 166
Great Britain. *See* United Kingdom.
Great Dyke, 572
Great Salt Lake
  lithium in, 875
  potash, 1072
  sodium sulfate, 1212, 1220
Greece
  asbestos, 452
  barite, 492
  bauxite, 514
  chromite, 578
  clay, 602, 625
  fluorspar, 733
  magnesite, 382, 884, 885, 888
  meerschaum, 911
  olivine, 991
  perlite, 999, 1003
  pumice, 1084
  salt, 1126
  talc, 1283
  wollastonite, 1387
Green earth. *See* Glauconite.
Green River Basin, 1199, 1200
Green rouge, 31
Greenland
  chromite, 575
  cryolite, 741
  flint pebbles, 27
  nepheline syenite, 935, 936
Greensand. *See* Glauconite, Foundry sands.
Greenstone, 168, 205
Grinding pebbles, 27, 31
Grog, 638
Guano, 234, 1021, 1031
Guatemala
  chromite, 575
Guatemala, fluorspar, 732
Guillotine, 162
Guinea
  bauxite, 508, 513, 523
  diamonds, 661, 662
  gems, 302
Guyana
  bauxite, 5, 513, 514
  calcined bauxite, 458
  diamonds, 658
  kaolin, 624
Gypsite, 777
Gypsum, anhydrite, 183–199, 755–792, 1256, 1257
  ammonium sulfate, 968
  anhydrite, 775, 776, 1256
  anhydrous, 195
  cement raw material, 134, 141, 775, 784, 785, 1256
  fertilizer, 238
  glass batch gypsum, 196, 344
  major uses, 775
    agriculture, 196, 238, 968
    construction, 192–194
    industrial, 194–196
  phosphate, byproduct, 238
  portland cement material, 134, 141, 784, 785, 1256
  processing, 185–196
  secondary plant nutrients, 238
  sulfur, source of, 1256, 1257
  superphosphate, processing of, 238, 1041
  uses, 196, 197, 775
    agriculture, 196, 238, 1256, 1257
    cement, use in, 134, 141, 775
    construction group, 190, 194, 195
    fertilizer, 238
    filler, 244, 247
    industrial, 194–196
    stucco, 188, 189

H

Hafnium, 1441
  see also Zirconium.
Haiti, bauxite, 513, 523
Halloysite, 612, 613, 620, 634–636
  analyses, 596, 612
Hand molding, 375
Hanksite, 1199
Hard kaolin, 614
Hardness, scales
  Knoop, 12, 659
  Mohs', 292, 293
Hardness, selected materials, 19, 284, 285
Harmonic generation, 227

Hawaii
- bauxite, 515
- coral, 291, 301, 303

Headlap granules, 209
Hectorite, 594, 596, 598, 600, 607, 611, 875
Hegman test, 617
Helfferichs test, 1417
Hematite, pigment, 351, 352
Hercules viscosity, 629
Hiddenite, 870
- *see also* Lithium raw materials.

High alumina brick, 374, 520, 521
High alumina cement, 521
High grading, 311
Highway construction, aggregates, 63–65
Hispaniola, bauxite, 453, 504
Histogram, 1099
Holland
- glauconite, 753
- salt, 1126

Holography, 227
Honduras, fluorspar, 732
Hong Kong, graphite, 764
Horse, salt, 1122
Howlite, 550
Humphrey's spiral, 922
Hungary
- bauxite, 514
- clays, 603, 625, 638
- perlite, 999, 1004
- zeolites, 1417, 1419, 1421, 1422, 1426

Hydraulic cement, 1957
Hydraulic lime, 824
Hydrocarbons, 529–532, 1132
- *see also* Bituminous material.

Hydrochloric acid, flux, 260
Hydrofluoric acid, 50–52, 260
Hydrogel process, 1407–1409
Hydrogen sulfide, 489
Hysteresis loop, 351

I

Iceland
- diatomite, 686
- geothermal sources, 875
- lithium, 875
- perlite, 1001, 1005

Iceland spar, 297
Idaho
- andalusite, 803
- barite, 492
- clay, 597, 618–620, 623, 635, 638
- diatomite, 685
- feldspar, 711
- fluorspar, 729
- garnet, 23, 24
- gems, 303

Idaho, glass sand, 639
- kyanite, 801
- monazite, 1111
- pegmatite, 711
- perlite, 997, 1003
- phosphate, 1022, 1031, 1034, 1042
- salt, 1124
- silica, 1171, 1172, 1173
- thorium, 1111
- travertine, 171
- vermiculite, 1377
- wollastonite, 1385
- yttrium, 1112

Illinois
- barite, 492
- clay, 596, 599, 623
- dimension stone, 62
- fluorspar, 726, 729, 730
- lime, 827, 828
- limestone, 836
- limestone production, 62, 853
- silica, 1170
- slag, 112, 125
- tripoli, 26, 1363, 1364, 1366, 1367

Ilmenite, 1313, 1314
- *see also* Titanium.

Illite, ceramic, 34
Imitation gems, 315
Impsonite, 531
Inagua, salt, 1124
Inder Lake, 507
India
- amosite, 455
- asbestos, 454, 455
- barite, 493
- bauxite, 508, 515
- chromite, 577, 581
- clay, 597, 605, 626, 634, 638
- crocidolite, 455
- diamonds, 302
- fire clay, 638
- garnet, 28
- gems, 302
- graphite, 761, 764
- kyanite, 376, 799, 801, 803
- magnesite, 883, 884, 885
- manganese, 897
- meerschaum, 911, 912
- mica, 5, 917
- monazite, 1113
- olivine, 991
- pyrophyllite, 1097
- rare earths, 1113
- sillimanite, 802
- sulfur, 1256, 1260
- talc, 1283
- thorium, 1113
- titanium, 1333
- vermiculite, 1377
- wollastonite, 1383, 1386
- yttrium, 1113
- zircon, 1443

Indiana
- dimension stone, 62, 854
- dolomite, 62, 854
- gypsum, 125, 777, 782
- halloysite, 635
- lime, 828, 836
- limestone, 62, 854
- marl, 854
- slag, 125
- whetstones, 27

Indonesia
- bauxite, 515, 523
- chromite, 578
- diatomite, 689
- gypsum, 795
- iodine, 793
- kaolin, 626, 627
- monazite, 1113

Industrial minerals, 1–6, 42, 43, 393–414
Information, sources of, 292–414
Insoluble anhydrite, 189
Insulating materials, 199–201
- bricks, 377, 378

Intaglios, 333
Iodine, 41, 793–797
Ion exchange, clay, 594, 595
Ione formation, 1172
Iowa
- dimension stone, 854
- dolomite, 854
- gypsum, 778, 780, 781
- limestone, 854

Iran
- barite, 493
- chromite, 577
- kaolin, 627, 638
- meerschaum, 911, 912
- sodium sulfate, 1216
- sulfur, 1242, 1252

Iraq
- asphalt, 530
- sulfur, 1242, 1249, 1250, 1252

Ireland, Northern, perlite, 1001
Ireland, Republic of
- barite, 493
- bauxite, 505
- diamonds, synthetic, 662
- iodine, 795
- magnesite, synthetic, 890

Iron, 233, 239
Iron oxide
- abrasives, 30, 31
- ceramics, 30, 36, 38
- natural, 351–359
- synthetic, 359–368

Isotopes
- carbon, 1240
- iodine, 794
- sulfur, 1240

Israel
- bromine, 562
- diaspore, 638
- clay, 638

Israel, magnesium chloride, 891
phosphate, 1031, 1035
potash, 1070
Italy
asbestos, 452, 457, 461
asphalt, 530
baddeleyite, 1436
barite, 492
borates, 547, 548
celestite, 1233
clay, 603, 638
diatomite, 686
feldspar, 713
fluorspar, 728, 733
gems, 302
graphite, 761, 765, 767
gypsum, 778
iron oxide pigments, 355
kaolin, 625
magnesite, 890
marble, 171
olivine, 991
perlite, 1004
potash, 1069
pumice, 1084
pyrite, 1236, 1254
salt, 1126
slag, 113
sulfur, 1236, 1241, 1247, 1251, 1254
tremolite, 461
wollastonite, 1387
zeolite, 1417, 1419, 1420, 1422
Ivory Coast
diamonds, 661, 662
lithium, 874

J

Jade, 299, 302, 303, 314, 319
Jamaica
bauxite, 513, 523
gypsum, 778
Japan
asbestos, 454
barite, 493
bromine, 48
chromite, 578
clay, 605, 627, 636, 638
cultured quartz, 214
diatomite, 689
fluorspar, 735
graphite, 763
garnet, 23
iodine, 793, 795, 796
magnesite, synthetic, 890
olivine, 991
perlite, 999
pearls, 318
pyrite, 1236, 1254
pyrophyllite, 1085, 1091
refractories, consumption, 1086
Japan, silica brick, 377
slag, 113
sulfur, 1236, 1242, 1251, 1254, 1260
vermiculite, 1377
zeolites, 1414, 1419, 1420, 1421, 1422, 1423, 1424
Java, iodine, 795
Jet, 292, 301, 315
Jet channelling, granite, 175
Jetty stone, 61
Jordan
bromine, 563
potash, 1070
phosphates, 1035
Jugoslavia, diatomite, 687

K

K-bentonite, 588
synthetic, 610, 611
Kainite, 1070
Kames, source of aggregate, 1154
Kansas
diatomite, 685
gypsum, 783
vermiculite, 1377
Kaolin, 26, 34, 35, 38, 375, 376, 586, 594, 596, 599, 612–633, 635, 641
definition, 612
processing, 629, 630, 631, 632
production, 586, 590, 614
Argentina, 624
Australia, 627
Austria, 624
Czechoslovakia, 625
Denmark, 625
France, 625
Guyana, 624
Hungary, 625
India, 626
Italy, 625
Japan, 627
Korea, South, 627
Malaysia, 627
Mexico, 624
New Zealand, 628
Philippines, 627
Spain, 625
UK, 623, 624
US, 622, 623
USSR, 624
West Germany, 626
reserves
Brazil, 624
Bulgaria, 625
Czechoslovakia, 625
Iran, 627
New Zealand, 628
Poland, 625
Sri Lanka, 626
Turkey, 627
Kaolin, reserves
US, 623
USSR, 624
saponite, 594, 600
sepiolite, 596
see also Meerschaum.
smectite, 594, 597, 599, 604, 605, 633, 640
see also Bentonite, Fuller's earth.
specifications, 616–618
uses, 26, 34, 35, 38, 614–617
Kaolinite. *See* Kaolin.
Kara-Bogaz-Gol, 1209
Kasota vein, creamy, 122, 167
Keatite, 1167
Keenes cement, 189, 823
Kellogg, M.W., ammonia process, 964
Kentucky
asphalt, 530
barite, 492
clay, 374, 375, 599, 623, 624, 633, 634, **637**, **638**
fluorspar, 729, 730
lime, 827, 828
limestone, 853
rock asphalt, 530
silica, 1178
slag, 125
tar sand, 1253
Kenya
bentonite, 605
diatomite, 687
fluorspar, 734
graphite, 766
kaolin, 626
kyanite, 801
meerschaum, 911, 912
monazite, 1113
soda ash, 1187, 1197
vermiculite, 1377
wollastonite, 1383, 1386
zeolite, 1416
Kernite, 45, 536
Kettle, calcining, 63–65
Keuper, 1125
Kieselguhr. *See* Diatomite.
Kilns, 81–83, 85, 139, 140, 186–188, 812, 813
Kimberlite, 308
Knoop hardness number, 12
Kokoman, 734
Kop, 1183
Korea, North
barite, 493
chromite, 578
fluorspar, 735
magnesite, 884
talc, 1283
Korea, South
barite, 493
chromite, 578

Korea, South, fluorspar, 735, 736
  graphite, 761–763, 767
  halloysite, 636
  kaolin, 627
  pyrophyllite, 1092, 1107
  magnesite, 883
  talc, 1283
Kramer. *See* Boron, CA.
Kroll process, titanium, 1311, 1312
Kunzite, 303, 318, 870
Kyanite, 376, 799–807
Kyasill, 800

L

Lake sand, 271
Lampblack, abrasive, 31
Land plaster, 183, 186, 187
Langbeinite, 236, 238, 1071
Lapis lazuli, synthetic, 314
Laponite, 611
Larvikite, 171
Las Brisas, area, 451
Lasca, 214, 215
Laser, 227
Laterite, 35
Lazurite, 292, 298, 300
Lead, flux, 36
Lean clay, 639
Lepidocrocite, 352
Lepidolite, 915
Lechatelierite, 1167
Leucoxene, 1314
Liberia, barite, 493
Libya, 1070
Light-sensitive materials, ceramic, 37
Lightweight aggregates, 81–95
  bituminous paving mixes, 120
  properties, desirable, 82
  types, 81
  zeolite, use in, 1442
Lime, 809–831
  abrasive, 31
  analyses, 269, 811
  burned lime, 269, 801
  burnt dolomite, 269
  composition, 269, 824
  flux, 269
  uses, 31, 34, 37, 134, 267, 388, 821–824
Limestone, 62, 833–868
  see also Crushed stone, Dimension stone.
  aggregate, 60, 61
  agricultural use, 237
  bituminous, 530
  cement, use in, 135
  ceramic, 33, 35
  chalk, 167
  commercial definition, 163
  composition, 61, 269, 839, 855
Limestone, consumption, agricultural, 237
  crushed stone, 60–80, 847, 852
  cut stone, 162, 163
  definition, commercial, 163
  density, 63, 842, 843
  dimension stone, 166, 167, 847
  evaluation, 841–848
  filler, 244, 847
  flux, 35, 269
  glass, use in, 344
  oolitic, 167, 835
  production, 847, 849, 852, 853
  production, dimension stone, 162, 166, 854
  shells, consumption, 62, 855
  uses, 62, 846–848
    aggregates, 62, 847
    agriculture, 62, 237, 847
    calcium carbide, 846
    cement, 135, 847
    ceramic, 33, 35
    flux, 269
    glass, 344, 847
    well drilling fluids, 388
Limestone rock asphalt, 530
Lithium, 869–880
  batteries, 902
  brines, 41, 871, 873, 875
  flux, 36
  minerals, 870, 875, 877
  potential districts
    Chile, 875
    Utah, 875
    Zaire, 874
  producing districts, 871–874
    Argentina, 873, 874
    Australia, 874
    Brazil, 873
    Canada, 874, 875
    Chile, 875
    Mozambique, 874
    Namibia, 874
    Ruanda, 874
    South Africa, 874
    Uganda, 874
    US, 871, 872
  reserves, 871–874
    Argentina, 873, 874
    Australia, 874
    Brazil, 873
    Canada, 874, 875
    Chile, 875
    Namibia, 874
    US, 871, 872
    Zaire, 874
    Zimbabwe, 873
Lithopone, 486
Lively perlite, 1010
Loading materials, 243–257
  see also Mineral fillers.
Long Island, salt, 1124
Loose abrasive grains, 15–17
Loose-pack density test, 13, 14
Lorrain quartzite, 1174
Lord Rayleight waves, 220
Los Angeles rattler test, 65, 100, 119, 844, 1158
Louisiana
  asphalt, 530
  bentonite, 599
  iodine, 794
  salt, 1119, 1123, 1140
  shell, 62, 854
  silica, 1173
  sulfur, 1245
Lubricants, graphite for, 769
Lump graphite, 758
Luster, 294, 295
Lutetium, price, 1110

M

MacIntyre mine, 1317–1319, 1343, 1344
Magadi, Lake, 1187, 1197, 1198
Magnesia, 31, 34
Magnesite, 881–896
  ceramic, 35
  dead-burned, 882, 893
  deposits, 884–886
  flux, 35
  production, 892, 893
  seawater as source, 381
  synthetic, 889–892
  talc with, 1145
  uses, 35, 883, 893, 894
Magnesium
  brines, 881, 889–891
  secondary fertilizer, 238
Magnesium metal, 890
Magnesium sulfate, soundness test, 66, 98
Magnet Cove, 1320
Magnetic surveying, asbestos, 461, 462
Magnetite pigment, 352, 353
Magnetoptic, 227
Maine
  diatomite, 685
  feldspar, 711
  gems, 303, 319
  granite, 170
  graphite, 762
  manganese, 899
  marl, 854
  pegmatites, 639
Make, of a gem, 280, 329
Makgadikgadi, 1197
Malagasy Republic (Madagascar)
  bauxite, 515
  chromite, 576
  clays, 605, 626
  garnet, 23

Malagasy Republic (Madagascar)
  gems, 302
  graphite, 758, 759, 760, 761, 766, 768
  mica, 918
  monazite, 1112, 1113
  rare earths, 113
  thorium, 1112
Malawi
  fluorspar, 734
  monazite, 1113
  zirconium, 1433
Malaysia
  kaolin, 627
  titanium, 1333
  zirconium, 1433
Manchuria
  graphite, 763
  magnesite, 883
  soda ash, 1198
  talc, 1283
Manganese, 897–908
  deposits, 897, 898
  fertilizer, 239, 240, 902
  micronutrient, 233, 239
  uses, 233, 239, 901, 902
Manganese dioxide
  abrasive, 31
  battery use, 905
  synthetic, 904
Manganese silicate, flux, 268
Manitoba
  bentonite, 601
  chromite, 575
  gypsum, 784
  lithium, 874, 875
  silica, 1175
Manjak, 531
Man-made quartz
  see also Cultured quartz.
Manto, 726
Manufactured abrasives, 27–31
  boron carbide, 12, 29
  diamonds, manufactured, 16, 19, 20, 30
  silicon carbide, 16, 17, 29
Manufactured diamonds, 30, 668–670
  see also Diamonds.
Marble, 167, 168, 173, 835
  types, 166, 167
Marcasite, 321, 323
Marekanite, 1007, 1003
Marinas, salt, 1120
Marl, calcareous, 854
  cement raw materials, 135
Marsh borate deposits, 537
Maryland
  chromite, 575
  clay, 374, 375
  diatomite, 684
  glauconite, 752
  limestone, 853
  silica, 1171
Maryland, slag, 112
  talc, 1284, 1285, 1288
Mass fiber, 441, 442, 444
Massachusetts
  clay, 596
  diatomite, 685
  feldspar, 711
  granite, 170
  pegmatites, 711
  talc, 1287, 1288
Matrix, 1018, 1030
Medina sandstone, 1174
Meerschaum, 89, 909–914
Mercury cells, chlorine production, 50
Merseber process, ammonium sulfate from, 1256
Met-spar, 725
Meta-bentonite, 588
Metal finishing abrasive, 16, 31
Metahalloysite, 612
Metallic abrasives, 28, 30, 31
Metallurgical pebbles, 1171
Methyl bromide, use, 48
Mexico
  asbestos, 451
  barite, 492
  borates, 550, 551
  celestite, 1229
  chromite, 578
  clay, 602, 624, 638
  diatomite, 685, 686
  feldspar, 715
  fluorspar, 727, 728
  glance pitch, 531
  graphite, 761, 762, 766, 767
  gypsum, 777
  limestone, 840
  magnesite, synthetic, 890
  nepheline syenite, 942
  nitrates, 970
  perlite, 1003
  salt, 1121, 1124, 1125
  silica, 1176, 1177
  sodium sulfate, 1213, 1220
  sulfur, 1236, 1241, 1251, 1252, 1260
  tequesquite, 1121
  titanium, 1324
  vermiculite, 1329, 1377
  wollastonite, 1383, 1385
  zeolite, 1418–1420, 1422
Mica, 203, 209, 915–929
  deposits, 919, 920
  filler, 244, 252, 927
  flake, 915, 919, 920
  processing, 921–923
  scrap, 915, 918, 919, 920, 921
  substitute for talc, 1298
  synthetic, 915
  uses, 209, 388, 917, 927, 928
Micanite, 915
Michigan
  bromine, 48, 562
  dolomite, 85
  gypsum, 777, 781
  iodine, 795, 796
  lime, 827, 828
  limestone, 836
  magnesite, synthetic, 891
  marl, 854
  potash, 1066
  salt, 1140
  slag, 112, 125
Micrite, 837
Micro-Cel, 1389
Microcrystalline silica, 26
Microporite, 822
Midland type brine, 44
Mill scale, 360
Mill-stock slate, 164
Mine, salt, temperature, 1311
Mineral absorbents, 255
Mineral fillers, 243–257
  ceramics, 24, 25
  nepheline syenite, used as, 245, 247, 932, 934
  paints, 252, 931
  plastics, use in, 249, 250, 931
  properties, 244, 245, 246, 247
  rubber, fillers used in, 251, 931
  uses, 243–245
  zeolite, 1422
Mineral filters, 253–255
  properties, table of, 698
Mineral oils, bleaching clays, processing, 594
Mineral pigments, 349–372
  definition, 349
  processing, 353–356, 360–363
  synthetic, 359, 360, 366–368
  titanium, 1303, 1304
Mineral wax, 529, 531
Mineral wool, 199–201
  see also Insulating materials.
Minnesota
  dimension stone, 854
  dolomite, 854
  kaolin, 623
  limestone, 854
  manganese, 898
  quartzite, 27
  silica, 1173
  titanium, 1321
Mirabilite, 1207
Miridas, 313
Misch metal, 1110
Mississippi
  asphalt, 531
  bauxite, 515
  clays, 599, 633
  magnesite, synthetic, 890
  marl, 854

Missouri
barite, 490
bauxite, 515
clay, 374, 375, 599, 637
diaspore, 515
granite, 170
lime, 827, 828
limestone, 62
silica, 1170
tripoli, 26, 1361, 1364–1366, 1369
Modulation, 227
Modules of rupture, 617
Mohs' scale of hardness, 292, 293
Molded stone, 325
Molding plaster, 194
Molding sand
see also Foundry sand.
chromite, 568
natural, 271
olivine, 898, 990
silica, 271, 1168
Molecular sieves, 1400, 1407, 1408, 1411–1413
see also Zeolites.
definition, 1400
uses, 1411–1413
Moler, 687
Molybdenum, 233, 240
Monazite, 1109–1117
see also Rare earths.
deposits, 1112
uses, 1109, 1110
Mongolia
fluorspar, 734
rare earths, 1111
soda ash, 1198
Monocrystalline count, 15
Montana
barite, 492
chromite, 575
clays, 587, 600
fluorspar, 730, 731
gems, 303, 318
graphite, 762, 765, 767
gypsum, 778, 783
jade, 303
manganese, 898, 902
phosphate, 1022, 1031, 1042
silica, 1171
sodium sulfate, 1213
talc, 1283, 1284, 1285, 1291–1293, 1298
titanium, 1321
vermiculite, 200, 1376, 1377, 1378
Montasite, 455
Montmorillonite, 255
see also Clays.
Montan wax, 529, 531
Monumental stone, 164
MOR, modules of rupture, 617
Mordenite, 1395, 1396, 1419
Morocco
barite, 493
celestite, 1233
clays, 605, 626, 636
fluorspar, 734
meerschaum, 911
phosphate, 1031, 1032, 1034
potash, 1070
salt, 1126
vermiculite, 1377
wollastonite, 1386
Mortar, 823
Mother-of-pearl, 301
Mountain leather, 446
Mountain Pass, 1113, 1114, 1115
Mozambique
anthophyllite, 456
clays, 605, 626
fluorspar, 734
lithium, 871, 874
monazite, 1112, 1113
titanium, 1336
Muds, drilling, 387–390
see also Well drilling fluids, 387–390.
Muller-Kuhne process, 1256
Mullite, 24–27, 34, 376, 377, 520, 799, 800, 806, 1085
Muriate of potash, 236
Muscovite, 915, 916

N

Nahcolite, 1203
Namibia
amblygonite, 874
diamonds, 302, 658
fluorspar, 727, 735
gems, 302
graphite, 766
lepidolite, 870, 874
lithium, 874
petalite, 874
sillimanite, 799, 802
soda ash, 1063
sodium sulfate, 1197
wollastonite, 1383, 1386
Native asphalts, 529–532
Natroalunite, 419
Natron, 1188
Natural abrasives, 15, 16, 18–27
Natural glass, 279
Natural granules, 209
Natural iron oxides, 351
goethite, 352
hematite, 351
lepidocrocite, 352
magnetite, 352, 353
Natural molding sand, 271
Natural pigments, 351–359
Nauru, phosphate, 1021
Nebraska, soda ash, 1189
Neodymium, price, 1110
Nepal, magnesite, 883, 885, 886
Nepheline syenite, 506, 931–960
alumina, from, 932, 953, 954, 1042
deposits, 935, 936
flowsheets
IMC, Ont., 949
Indusmin, Ont., 947
Stjernöy, Norway, 951
granule, 205
reserves
Canada, 937
Norway, 937
Siberia, 937
USSR, 937
talc, substitutes for, 1298
uses, 931, 932
Nevada
barite, 489, 490, 491, 497
bentonite, 590, 597, 598
borates, 537
brines, 869, 871
brucite, 382
diatomite, 685
fluorspar, 727, 731, 739
gems, 319
gypsum, 783
halloysite, 635
lithium, 869, 870, 871, 876
magnesite, 382, 886, 887
manganese, 899, 902
perlite, 997, 999, 1003
phosphate, 1022
silica, 272, 1171, 1172
soda ash, 1189
sodium sulfate, 1209, 1213
tripoli, 1369
vermiculite, 1377
wollastonite, 1385
zeolite, 1414, 1417, 1426
New Brunswick
diatomite, 685
gypsum, 779
potash, 1059, 1072
New Caledonia
chromite, 578
olivine, 991
New Guinea, chromite, 578
New Hampshire
diatomite, 685
feldspar, 711
granite, 170, 711
mica, 917
pegmatites, 711
scythestones, 27
New Jersey
clays, 591, 636
feldspar, 711
foundry sand, 271
glauconite, 751
magnesite, synthetic, 890
manganese, 898
silica sand, 25

New Jersey, titanium, 1316, 1324, 1326, 1327
traprock, 62
New Mexico
barite, 492
bastnaesite, 1111
feldspar, 711
fluorspar, 727, 731
gems, 303
graphite, 761
gypsum, 778, 783
kaolin, 623
meerschaum, 910, 912
mica, 918
pegmatites, 711
perlite, 997, 999, 1002
potash, 1059, 1060, 1066, 1071, 1073, 1075
sodium sulfate, 1213
sulfur, 1247
talc, 1284, 1285
vermiculite, 1377
wollastonite, 1385
zirconium, 1435
New York
corundum, 18
diatomite, 685
emery, 18
feldspar, 711
foundry sand, 273
garnet, 18, 22, 23
gem, 303
granite, 763
gypsum, 781, 782
limestone, 62
salt, 1129, 1132, 1133, 1135, 1140
sand and gravel, 101, 102
sandstone, 170
silica, 1171
slag, 112, 125
slate, 170
talc, 1275, 1276, 1283–1285, 1286, 1287, 1293, 1296, 1298
titanium, 1303, 1316, 1317, 1336, 1342
wollastonite, 1384, 1387
New Zealand
asbestos, 454
barite, 493
clays, 606, 628, 636, 638
diatomite, 689
geothermal sources, 875
graphite, 767
lithium, 875
olivine, 991
perlite, 1006
sulfur, 1251
titanium, 1314, 1335
wollastonite, 1383
zeolites, 1415
Newfoundland
asbestos, 444, 450, 457
Newfoundland, fluorspar, 728
gypsum, 779
olivine, 992
pyrophyllite, 1094, 1105
salt, 1124
Nigeria
cryolite, 741
fluorspar, 734
kaolin, 626
thorium, 1112
Nikkagel, 611
Nitrates, Chilean, 961, 970, 971
Nitric acid, 963, 965, 966
Nitrile rubber, 962
Nitrogen compounds, 234, 961–988
ammonia, synthetic, 234, 963–965, 978, 982–987
application, 234
composition, 234
fertilizer, 234, 961, 968, 974, 977, 979
uses, 962, 963
Noble gems, 279
Norbide, 29
North Carolina
alaskite, 713
andalusite, 802
anthophyllite, 456
barite, 492
clays, 613, 618, 619, 623
feldspar, 711
gems, 303
granite, 27, 67
hiddenite, 870
kyanite, 800
lithium, 870, 871, 872, 876
marl, 854
mica, 918
olivine, 274, 888, 889, 991, 992
phosphate, 1028, 1032, 1034, 1043
pyrophyllite, 1092, 1105–1107
quartzite, 27
silica, 170
talc, 1284, 1285, 1288, 1289, 1298
titanium, 1315, 1320
vermiculite, 1377
North Dakota, sodium sulfate, 1213
Northwest Territories
bastnaesite, 1111
bentonite, 601
thorium, 1111
Norway
feldspar, 715
flint pebbles, 27
fluorspar, 733
graphite, 760, 765
iodine, 795
kyanite, 801
Norway, larvikite, 171
magnesite, synthetic, 890
mica, 918
monazite, 1113
nepheline syenite, 933, 935, 936, 938–942, 949–951
olivine, 888, 889, 991
pyrite, 1254, 1255
sulfur, 1254, 1255
talc, 1282, 1283
titanium, 1322, 1337, 1345
Nossen process, 903
Nova Scotia
andalusite, 802
barite, 492
celestite, 1229, 1232
diatomite, 685
fluorspar, 728
gypsum, 779
potash, 1066, 1072
salt, 1124
silica, 1175
Novaculite, 204, 1367
Nylon, 962, 963, 968

O

Obsidian, 168, 1001
Ocean floor deposits, manganese, 899
Ocean mining, 1161
Oceania, phosphate, 1039
Ohio
bromine, 562
clays, 374, 375, 637
gannister, 277
gypsum, 778, 782
lime, 828
limestone, 62, 836, 853
salt, 1140
sandstone, 170
silica, 1171
slag, 112, 125
Oil absorption, definition, 350
Oil shale, sulfur from, 1253, 1254
Oklahoma
asphalt, 530
barite, 492
clays, 599
feldspar, 709
grahamite, 530
granite, 170
gypsum, 778, 783
iodine, 795
limestone, 170, 853
silica, 1170, 1173
tripoli, 26, 1363, 1364–1366, 1369
Olivine, 33, 238, 274, 888, 889, 893, 894, 989–996
ceramics, 35

Olivine, deposits, 991
  fertilizer, 238, 989
  foundry sand, 274, 990
  gem, 318
  magnesium, source of, 238, 989
  processing, 993
  properties, 274
  sources, 274, 991
  uses, 989, 990
    ceramic, 35
    fertilizer, 238, 989
    foundry sand, 274, 990
    heat storage, 894, 994
    refractory, 893, 894, 921
Ontario
  asbestos, 450
  bastnaesite, 1111, 1113, 1115
  blasting sand, 25
  fluorspar, 728
  graphite, 763
  gypsum, 781, 782
  kimberlite, 662
  nepheline syenite, 936, 937, 938, 947–949
  salt, 1124
  silica, 25, 1174, 1175
Onyx, 167, 171
Oolites, 167, 835
Opal, 279, 296, 301, 314, 320, 1168
Opal claystone, 596
Opalescence, 296
Opaline quartz, 1167
Open pans, 1136
Optical calcite, 221
Optical materials, 221–231
Oregon
  chromite, 575
  clays, 597, 638
  diatomite, 25
  ferruginous bauxite, 515
  fluorspar, 732
  gems, 319
  perlite, 1003
  pumice, 1080
  silica, 1171
  sodium sulfate, 1211
  talc, 1298
  traprock, 62
  zeolites, 1421
Organoclad bentonite, 610, 611
Oriskany sandstone, 1170, 1174
Ornamental stones, 279
Otanmaki deposit, 1345
Otjevalunda pan, 1197, 1198
Owens Lake brine, 44
Oxychloride cement, 894
Ozokerite, 531

P

Packing, graphite for, 769
Paints, 614–616
  lime, use in, 822
  manganese, use in, 902
  pyrophyllite, use in, 1089
  talc, use in, 1277
  wollastonite, use in, 1383, 1387, 1388
Pakistan
  celestite, 1233
  chromite, 578
  clays, 606, 627
  fluorspar, 736
  potash, 1070
  vermiculite, 1377
Palygorskite, 591
Pandermite, 535
Pans, salt, 1120
Paper
  coatings, 614, 616
  fillers used in, 614, 616, 1422
  lime, used in, 822
  sodium sulfate, 1217, 1218
  talc, use in, 1277
  titanium, used in, 1305, 1306
  zeolite, 1422
Paradox Basin, 1123, 1124
Paraguay, kaolin, 624
Particle size, 350, 351
  paste, 315
Paving blocks, 164, 165
Pay streak, 308
Pearl grain, unit of weight, 323
Pearls, 301, 315, 322, 323
Pebbles, ceramic, 27
Pee-Dee gravel, 1171
Pegmatites, 299, 300, 710, 728, 869, 870, 919
  see also Feldspars, Aplite, Mica.
  gem stone source, 299, 300
  thorium in, 1111
Pencils, graphite for, 769
Pennsylvania
  barite, 492
  bauxite, 515
  chromite, 575
  clays, 374, 375, 637, 638
  diaspore, 515
  gannister, 374
  granite, 170
  graphite, 765
  lime, 827, 828
  limestone, 62
  pegmatites, 71
  refractory, 637
  rottenstone, 26
  sandstone, 170
  silica, 1170
  slag, 122, 125
  slate, 170
  talc, 1288
  vermiculite, 1377
Periclase, 381, 382, 384
  abrasive, 31
Peridot. *See* Olivine.
Perlite, 81, 86, 87, 199–201, 244, 378, 997–1015
  see also Lightweight aggregates.
  analyses, 999
  definition, 997
  deposits, 1001–1006
  evaluation, 1008
  processing, 1009, 1010
  uses, 997, 1013
Permanent vs. portable plants, 71, 72
Perovskite, 1313
Perthite, 710
Peru
  barite, 492
  borates, 548, 550
  clays, 602, 624, 638
  diatomite, 689
  phosphate, 1024
  potash, 1066, 1067
  soda ash, 1200
Petalite, 870, 877
Petite granite, 171
Petrochemicals, fluorspar, uses in, 724
Petroleum
  asphalt, 529
  source of sulfur, 1251–1253
Phianite, 314
Philippines
  barite, 493
  chromite, 381, 577
  clays, 627, 636
  manganese, 897
  perlite, 1006
  sulfur, 1251
Phillipsite, 1395, 1396
Pholgopite, 915, 917, 919, 922
Phosam process, 970
Phosgene, 49
Phosphate rock, 1017–1047
  deposits, 1018, 1020, 1021
  diammonium phosphate, 969
  evaluation, 1031
  fertilizers, 235, 236, 1035, 1036, 1038–1040
  fluorine in, 1042
  glauconite, source of, 745
  processing, 1018–1020
  rare earths in, 1112
  terminology, 1017, 1018
  yttrium, in, 1112
Phosphogypsum, 1256
Phosphoric acid, 1040, 1041
Phosphorus, 1018, 1031, 1036–1038
  see also Phosphate rock.
  fertilizer, 235, 236
  uses, 1035–1040
Phosphoria formation, 1022–1024
Piezoelectricity of quartz, 228

Pigments. *See* Mineral pigments.
Pigment volume concentration (PVC), 252
Pipe clay abrasive, 26
Pitch bonded refractories, 381
Pitch impregnated refractories, 381
Place value, 4, 788, 942, 946, 947, 1180, 1181
Plant nutrients. *See* Fertilizer minerals.
Plaster
  gypsum, 193, 194
  lime in, 823
  vermiculite, use in, 1380
Plaster of paris, 34, 183
Plastic clays, 380
Plastics, 249–251
  asbestos, 250
  barite, 250
  clay, 250, 251
  fillers used in, 249, 251
  kaolin, 616, 617
  talc used in, 1277
  titanium, in, 1305
  use as imitation gems, 315
  wollastonite used in, 1388
Playa lake, 1121, 1122
Pleochroism, 294
Plumbago, 757
Plumping hides, 822, 823
Poland
  barite, 493
  clays, 603, 625, 638
  glauconite, 753
  gypsum, 184
  ozokerite, 531
  potash, 1069
  salt, 1126
  sulfur, 1242, 1247, 1248
  wollastonite, 1385
Polaroid, 297
Polishing, abrasives for, 30, 31
Porcelain blocks and pebbles, 27
Porcelain, clays, 615
Portable vs. permanent plants, 71, 72
Portland cement. *See* Cement, Cement raw materials.
Portugal
  barite, 493
  chromite, 578
  kaolin, 625
  marble, 171
  salt, 1126
Potash, 236, 237, 1049–1078
  brines, 41, 1957
  bromine, use in exploration for, 1060
  Carlsbad district, 1059, 1060, 1066, 1071
  deposits, 1052–1057
Potash, fertilizer, 236, 237
  glauconite, source of, 745
  nepheline syenite, source of, 951–953
  processing, 1071, 1072
  reserves, 1065, 1066
    Africa, 1070
    Brazil, 1067
    Canada, 1062, 1066
    Morocco, 1070
    New Mexico, 1066
    Russia, 1069
    South America, 1066, 1067
    world, table, 1063–1065
  resources, 1061, 1063–1065
  uses, 236, 237, 1049
  Zechstein, 1068, 1069
Potash bentonite, 588
Potassium
  see also Potash.
  brines, 41
  carbonate, from nepheline syenite, 951–953
  fertilizer, 236, 237
  oxide. *See* Potash.
  sulfate, 419, 1066, 1072
Potsdam sandstone, 1174
Pottery, clays for, 639
Pottsville formation, 1171
Powders, abrasives, 16
Pozzolan, 134, 135, 143
  zeolite used in, 1421
Prairie Evaporite formation, 1060
Praseodymium, price, 1110
Precious stones. *See* Gem materials.
Pressure blasting, 16, 27
Priceite, 435
Prills, 967
Protective coatings, 252
Puerto Rico, zeolites, 1416
Pulaskite, 931
Pulp, paper, lime used in, 822
Pumice and pumicite, 199, 999, 1079–1084
  abrasive, 16, 18, 25, 26, 1081
  aggregate, 84, 1083
  consumption, 85
  lightweight aggregate, 82, 1083, 1084
  mining, 86, 1079
  uses, 1079
    abrasive, 16, 18, 25, 26, 1081
    aggregate, 1081
    building stone, 1081
    filler, 244
Pumicite. *See* Pumice.
Putty powder, abrasive, 31
Putunig area, 458
PVC (pigment volume concentration), 252
Pyke area, 454
Pyrite, 353
  source of iron, 135
  source of sulfur, 1254
Pyrobitumens, 531
Pyrolusite, 897, 902
Pyrometric cone equivalents (PCE), table, 613
Pyrophanite, 1313
Pyrophyllite, 1085–1108
  analyses, 1093, 1097, 1098, 1102, 1105
  deposits
    Argentina, 1096
    Australia, 1093, 1094
    Brazil, 1096
    Canada, 1094, 1095
    India, 1097
    Japan, 1091
    South Africa, 1097
    South Korea, 1092
    Turkey, 1097
    United States, 1092
  evaluation, 1104
  processing, 1105, 1106
  substitutes, 1085, 1108
    for feldspar, 720
  uses, 1085, 1086, 1087, 1280
Pyropissite, 531
Pyroprocessing, 139, 140
Pyrrhotite, source of sulfur, 1254–1256

Q

Quarrying, 136
  dimension stone, 171–175
  granules, 207
  underground, 172, 173
Quartz, 213–221
  see also Silica and silicon.
  abrasive, 25
  cultured, 38, 214
  dielectric material, 38, 213, 227
  gems, 1167, 1168
  glossary, 226–228
  synthesis, 216–218, 320
  uses, 219–221
Q-spoiler, 228
Q-switch, 228
Q of quartz, 217, 218
Quartzite, 27
  pebble, metallurgical, 1171
Quicklime. *See* Lime.
Quebec
  asbestos, 445, 446–451
  bastnaesite, 1111
  blasting sand, 25
  brucite, 382, 888
  chrysotile, 445, 446–451
  graphite, 763, 765
  kyanite, 801
  lithium, 874
  magnesite, 883, 884, 887, 888

Quebec, mica, 918
nepheline syenite, 936
olivine, 992
pyrophyllite, 1096
silica, 1050
talc, 1287, 1295
thorium, 1111
titanium, 1321, 1337, 1345
Quebec standard test, 464

R

Radiation shielding, ceramics, 37
Ramming, 375, 379
mixes, 380, 381, 799, 989
Rare earths, 1109–1118
bastnaesite, 1110
deposits, 1111, 1112
exploration, 1113, 1114
monazite, 1110
processing, 1114, 1115
uses, 1109, 1110
Rattails, 274
Reaction flux, 263
Reclamation costs, 642
Reconstituted mica, 915
Reconstructed gems, 311
Recovered sulfur from
coal, 1254
crude oil, 1251, 1252
oil sand, 1253
oil shale, 1253, 1254
sour gas, 1051, 1052
sulfides, 1254, 1255
tar sands, 1253, 1254
Red mud, 740
Reeds, 166
Reefs, 835, 836
Refractories, 373–385
alumina, 376, 377
chromite, 381
clay, 373–376
lime, 822
magnesite, 381, 382
mullite, 24–27, 34, 376, 377, 520, 799, 800, 1085
olivine, 989
periclase, 381, 382, 882
raw materials
dolomite, 167, 382
graphite, 379
mullite, 27, 376, 377, 799
olivine, 989
periclase, 381, 382
zircon, 378
talc, 1277, 1279
types, 373–378, 380, 381–383
Refractory clay, 636–639
ultrahigh alumina, 379
uses, 588–592, 599, 602, 606, 633–636
Resonator, 228
Retarders, in cement, 189, 190
Rhinestone, 315
Rhode Island
granite, 170
graphite, 761
talc, 1288
Rhodesia. *See* Zimbabwe.
Rift, 166
and grain, 175
Riprap, 61
Rochelle salt, 226
Rock asphalt, 530
Rock hounds, 308
Rock salt. *See* Salt.
Rocks used for aggregates, 1444
see also Crushed stone.
physical properties, 846–848
Rock wool, 199–201
Roman alum, 417, 418
Romania
barite, 493
chromite, 578
clays, 603, 626
diatomite, 687
fluorspar, 734
ozokerite, 531
salt, 1126
talc, 1283
Roofing materials, 203–210
see also Granules.
Rosin solders, flux, 260
Rottenstone, 1363, 1364, 1369
abrasives, 16, 26
composition, 26, 1364
deposits, 26, 1364
sales, 26
uses, 1369
Rouge, 31
Ruanda, amblygonite, 871, 874
Rubber
fillers in, 251, 614
wollastonite use in, 1383
Rubble, 161, 162
Ruby, 279
asterism, 295
occurrence, 300, 304
production, 320
synthetic, 311–313
use, 296
Runner sand, 1168
Russia. *See* USSR.
Rutile, 1314
see also Titanium minerals.
gem, 313
mineralogy, 1314
synthetic, 313, 321

S

Sabhka, 777
Salado formation, 1059, 1060
Salinas, 1120
Salt, 98, 1119–1149
brines, derivatives, chart, 44
cap rock, 1243, 1244
ceramic use, 38
Salt, deposits, 1120, 1121, 1123, 1124
domes, 1123, 1124, 1243–1245
cap rock, 777
evaluation, 1129
evaporated, 1135
mining, 1130, 1131, 1133–1135
origin, 1122, 1123
solar, 1133–1135
solution mining, 1131–1133
uses, 1142, 1143, 1144, 1145
Salt cake, 419, 1217, 1218
see also Sodium sulfate.
ceramic use, 36, 38
detergents, 1217
flux, 36
glass, use in, 344
paper, use in, 1217
Salt earth, 1126
Salt lakes, 1120, 1121
Salt pan, 1120
Samarium, price, 1110
Sampling, silica, 1178–1180
Sanbornite, 487
Sand and gravel, 96–110, 1151–1166
deposits, types of, 1154, 1155
evaluation, 1157–1160
sand equivalent, 1160
tests, 98, 99, 1157, 1158, 1160
uses, 96
Sand blasting, 1125
Sand, foundry. *See* Foundry sand.
Sandhills Belt, 1178
Sandpaper, 17
Sandstone, 61, 166
Sanford Lake, 1317–1319
Saponite, 594, 600
Sapphire, 279
asterism, 295
production, 312, 320
synthetic, 311, 312
uses, 296, 297
Sardinia
fluorspar, 733
talc, 1283
Saskatchewan
bentonite, 601
potash, 1059, 1061, 1071–1074
quartzite pebbles, 27
silica, 1175, 1176
sodium sulfate, 1210, 1218
Sassolite, 535
Satin rouge, 31
Satin spar, 296
Saudi Arabia, magnesite, 886
Saws, stone cutting, 173, 174
Scabs, 274
Scapolite, 883
Scattergram, 1101

Scintillation, 654
Scoria, 1079
 aggregate, 81
Scotland
 diatomite, 687
 flint clay, 638
 iodine, 795
 olivine, 991, 992
 perlite, 1001
Scrap mica, 915, 919, 920
Scratch test, 11
Screw dislocation, 218
Scythestones, 27
Sea foam, 909
 see also Meerschaum.
Searles Lake, 44, 48, 533, 537, 541, 542, 553, 873, 1072, 1121, 1187, 1199, 1201, 1212, 1220
Seawater, source of magnesia, 382, 891
Sechura Desert, 1024, 1067
Secondary plant nutrient, 237–239
Seed quartz, 216, 217
Selenite, 221, 298, 776
Selenitza, 530
Self-fluxing ore, 268
Semi-precious stones, 279
Senegal
 bentonite, 597, 605
 fuller's earth, 597
 wavellite, 1031
 zirconium, 1433
Sepiolite. *See* Meerschaum.
Sergipe, 1066, 1067, 1125
Serpentine, 446
 dimension stone, 168
Shale, 81, 82, 639–641
 aggregate, 81, 82, 86
 cement material, 134, 639
 properties, 639
Shape, particle, definition, 351
Sharpening stones, 26
Shells, 61–63, 204, 854
 use in cement, 134, 135
Shielded arc welding, 265
Shot steel, 30
Siberia
 bauxite, 508
 graphite, 764
 nepheline syenite, 936, 937
 soda ash, 1198
Sicily
 potash, 1071
 sulfur, 1241, 1247
Siderite, 353, 834
Sierra Leone
 bauxite, 513, 523
 chromite, 578
 diamonds, 302, 658
 titanium, 1333
Silex, 27
Silica, silicon, 1167–1185
 see also Tripoli.
 abrasive, 16, 17, 25, 27
 bricks, 377
 ceramics, 33, 34, 35, 39
 filler, 244
 flux, 269
 foundry sand, 271–278
 gannister, 363, 374, 377, 1168
 glass former, 33, 36
 glass, use in, 344, 1179
 granules, 204
 microcrystalline, 26
 molding sand, 271
 skeleton former, 34
 specifications, 344, 1168, 1169
 tripoli, 1363–1374
 uses, 1168, 1169, 1173
 zonal restrictions, 106, 107
Silicate cotton, 199
Siliceous powder, 25, 26
Silicon carbide, 12, 16, 17, 27–29, 378, 379, 1174
Silicon metal, 1167, 1173
Silicone, 86
Sillimanite, 376, 799, 800, 802
 ceramics, 34
 porcelain, use in, 27
 refractories, 376
Silver Peak Marsh, 44, 871–873
 see also Lithium.
Skeleton formers, 34
Slag, 81, 82, 89, 111–131
 cement, raw materal for, 134, 143
 deleterious substances, 87
 expanded slag, 85, 116, 117, 122, 123
 glass, raw material for, 344
 granules, roofing, 203–212
 steel, 118, 122
 types, 115–122
 uses, 120–122
Slate, 163, 164, 168
 filler, use as, 245
 granule, 209
 lightweight aggregate, 82
 mining, 82
 soundness, 180
 specifications, 162–164
 travertine, 167
 underground quarries, 172, 173
 uses, 164
Slip casting, 375
Slip dislocation, 218
Slip fiber, 464
Smackover formation, 869
Smectite, 594, 640, 870
 see also Bentonite, Fuller's earth.
Soap, scouring, 18
Soapstone, 165, 175
 see also Talc.
Soda ash, 52, 1187–1206
 see also Sodium carbonate.
 ceramic, 36
 flux, 36, 38
Sodium, 49
Sodium carbonate, 52, 1187–1206
 ceramic, 36
 flux, 36
 geology, 1188, 1189
 Green River, 1199, 1200, 1201, 1202
 nepheline syenite, 931
 Searles Lake, 48, 533, 537, 541, 542, 1072, 1121, 1187, 1199, 1201, 1212, 1220
 Solvay process, 52, 1201
 uses, 52, 344, 1188
Sodium chloride, 38, 98
 see also Salt.
Sodium hydroxide, 42, 48, 52, 99
Sodium sulfate, 52, 1207–1223
 glass use in, 346
 producers
  Canada, 1210, 1211, 1218
  Mexico, 1220
  US, 1211–1213, 1218–1220
 production, 1217
  Argentina, 1215
  Canada, 1210, 1211
  Chile, 1216
  Mexico, 1213, 1214
  Russia, 1214, 1215
  Spain, 1215
  Turkey, 1215
  US, 1207, 1211–1213
 soundness test, 98
 uses, 36, 38, 1217, 1218
Soffioni. *See* Sassolite.
Soft kaolin, 614
Soft silica, 1363
Soft siliceous powder, 25
Soil stabilization, lime in, 823
Solar salt, 1133–1135
Soldering, 259, 261, 262
Soluble anhydrite, 188
Solution mining, 902
 potash, 1072
 production, 1132
 salt, 1127, 1131
 wells, 1132
Solvay process, 52, 1201
Somali Republic, meerschaum, 911, 912
Sorel slag, 1345, 1346
Sorption sensitivity, sand, gravel, 99
Sour gas, 1251–1253

Sources of information, 393–414
 Africa, geological agencies, 407–410
 Asia, geological agencies, 411–413
 Australia, geological agencies, 413
 Caribbean Islands, geological agencies, 402
 Europe, geological agencies, 404–407
 Forum on Industrial Minerals publications, 394, 395
 North America, geological agencies, 397–402
 Pacific Islands, geological agencies, 413
 periodicals, 402
 Public Inquiries Office, USGS, 396
 publications
  Canada, 397
  US, 402
 schools of mines
  Canada, 397
  US, 401, 402
 societies, mining, geological, 401, 402
 South America, geological agencies, 402–404
 state agencies, US, 399–401
 state mineral, water resource publications, 395
 trade associations, 414
 US Bureau of Mines offices, 398
 US Geological Survey offices, 398, 399
South Africa
 amblygonite, 871
 amosite, 453, 455
 andalusite, 803
 apatite, 1021
 bastnaesite, 1111, 1113
 bort, 19
 carbonado, 19
 chromite, 274, 381, 576
 chrysotile, 453
 clays, 605, 612, 636, 638
 corundum, 806
 crocidolite, 455
 diamonds, 302
 diaspore, 572, 658
 fluorspar, 726, 727, 734
 gems, 302
 graphite, 766
 kimberlite, 658
 kyanite, 799, 803, 806
 lithium, 874
 magnesite, 884, 885
 manganese, 898
 meerschaum, 911
South Africa, mica, 918
 monazite, 1113
 nitrates, 970
 olivine, 991
 pyrite, 1236
 pyrophyllite, 1097
 rare earths, 1111, 1113
 silica, 1183
 sillimanite, 802
 slag, 113, 1305, 1315, 1334, 1335
 soda ash, 1193
 sulfur, 1260
 titanium, 113, 1305, 1315, 1334, 1335
 vermiculite, 200, 1376
 wollastonite, 1383, 1386
 zirconium, 1433, 1436, 1437
South America
 see individual countries.
 bauxite, 504
 diatomite, 688, 689
 feldspar, 715
 gems, 302
South Carolina
 barite, 492
 clay, 596, 599, 613, 618, 622, 638
 feldspar, 711, 713
 granite, 62, 170
 kyanite, 376, 800, 801
 marl, 854
South Carolina, mica, 918
 silica, 1170
 sillimanite, 802
 vermiculite, 200, 1376, 1379
South Dakota
 amblygonite, 873
 feldspar, 711
 granite, 170
 manganese, 898
 silica, 1173
 spodumene, 873
South West Africa. *See* Namibia.
Southern bentonite, 588
Spain
 barite, 493
 celestite, 1229
 clays, 597, 598, 603, 625, 636
 cryolite, 741
 diatomite, 686
 fluorspar, 726, 727, 733
 graphite, 765
 magnesite, 884, 885
 meerschaum, 911
 olivine, 991
 potash, 1069, 1070
 pyrite, 1236, 1254
 salt, 1126
 sepiolite, 592
 sodium sulfate, 1207, 1215, 1221
 sulfur, 1236, 1247, 1254
Spain, vermiculite, 1377
 wollastonite, 1386
Spanish Sahara, phosphates, 1029
Spar
 block, 711
 feldspar, 709–722
 heavy. *See* Barite.
 Iceland, 297
 satin, 296
Spath, 709
Specifications. *See* individual commodities.
Sphalerite, dispersion, 294
Spiegeleisen, 904
Spinel, synthetic, 311, 312
Splittings, 915
Spodumene, 870, 877
 see also Lithium raw materials.
 ceramics, 36
Spring apron, 537
Sri Lanka (Ceylon)
 baddeleyite, 1436
 gem, 302
 graphite, 5, 758, 762, 763, 764, 767, 768
 kaolin, 626
 monazite, 1113
 titanium, 1334
 yttrium, 1112
 zirconium, 1433, 1443
St. Lucie zone, 1170
St. Peter sandstone, 272, 1173
Stabilization, land, 823
Stabilizer, asphalt, 209
Staurolite, 27, 274, 1225–1227, 1325
 abrasive, 26
 foundry sand, 274
 use, 274, 1225
Steatite, 1275, 1281
Steel shot, abrasive, 16, 18, 30
Steel slag, 122, 126
Steel wool, 30
Stishovite, 1167
Stjernöy, Norway, 938–942, 949–951
Stockpiling
 fluorspar, 740
 iodine, 796
Stokes diameter, 350
Stone
 see also names of rocks.
 crushed stone, 60–80
 dimension stone, 161–181
 sand, gravel, 96–110
 solidifier, 178, 179
Stoneware, clays for, 612, 617
Storgangen, 1322
Strass, 315
Stratiform deposits, sulfur, 1247
Strontianite, 1229

Strontium, 1229–1233
 specifications, 1229
 titanate, gem, 313
 uses, 1229, 1231
Stucco, 188, 189
Submerged arc fluxes, 267, 268
Sudan
 chromite, 578
 fluorspar, 735
 lithium, 874
 wollastonite, 1383, 1386
Sugar industry, lime use in, 822
Sulfates, 1256, 1257, 1308–1310
Sulfide ores, sources of sulfur, 1238
Sulfur, 238, 239, 1235–1274, 1297
 alunite, with, 421
 ceramic, 35
 consumption, fertilizer, 239
 deposits, 1245–1251
 economic factors, 68, 1268
 fertilizers, 238, 239
 plant nutrient, 233, 238, 239
 secondary plant nutrient, 233, 238, 239
 sources of, 1235, 1236, 1238, 1243, 1251
 uses, 1261–1264
Sulfur dioxide emission control, 1255, 1256
Sulfuric acid, 1261–1266
 superphosphates, use of, 238
Sunstone, 296
Supai formation, 1172
Superphosphates, 238
Surface area, definition, 350
Surinam
 bauxite, 513, 523
 kyanite, 801
 lithium, 871
Surveying, airborne for asbestos, 462
Swallow nest workings, 310
Swaziland
 chrysotile, 453
 kaolin, 626
Sweden
 clay, 626, 638
 feldspar, 715
 fluorspar, 733
 lithium, 870
 olivine, 991
 petalite, 870
 pyrite, 1254
 sulfur, 1254
Switzerland
 asphalt, 530
 bentonite, 603
 fluorspar, 733
Sylvania sandstone, 1174
Sylvanite, 1072
 see also Potash.
Synopal, 1389
Synthetic materials
 corundum, 311, 312
 diamonds, 313, 318, 668–670
 gems, 311–315
 iron oxides, 359–368
 magnesite, 889–892
 mica, 915
 quartz, 213–221, 314
 rutile, 313
 zeolites, 1399–1414
Szaibelyite, 535, 536

T

Taiwan, kaolin, 627
Talc, 203, 209, 1275–1301
 ceramic use, 34, 35, 38, 1277
 deposits, 1283–1293
 filler, use as, 33, 35, 38, 250, 251, 1280
 paint, use in, 1277, 1278
 processing, 1294–1297
 properties, 1277, 1281
 specifications, 1276, 1277
Tanganyika
 bentonite, 605
 meerschaum, 912
Tanzania
 clays, 605, 626
 diamonds, 302, 658
 garnet, 23
 gems, 302
 meerschaum, 910, 911
 monazite, 1113
 soda ash, 1198
 vermiculite, 1377
 zeolites, 1416
 zirconium, 1433
Tap density, 20, 21
Tar sands, 1235
Taylorite, 587
Tecopa, Lake, 1416
Tejon formation, 1172
Tellnes, deposit, 1337, 1345
Tennessee
 barite, 491, 492
 bauxite, 515
 clays, 589, 633, 634, 638
 fluorspar, 732
 foundry sand, 271
 glauconite, 752
 lime, 827
 microcrystalline silica, 26
 phosphate, 1029, 1030, 1043
 quartzite, 27
 silica, 26
 titanium, 1316, 1324, 1327
 tripoli, 26, 1364, 1367, 1368
Tequesquite, 1121
Terbium, price, 1110
Terra alba, 195, 196
Tetraethyl lead, 48
Texas
 asphalt, 530, 531
 barite, 492
 clays, 375, 589, 590, 599, 623, 633, 634
 dimension stone, 170
 feldspar, 711
 flint, 27
 fluorspar, 732
 gems, 319
 glass sand, 639
 glauconite, 752
 granite, 170
 graphite, 750, 765
 gypsum, 778, 783
 lime, 827, 828
 limestone, 62, 854
 magnesite, synthetic, 890
 marl, 854
 pegmatite, 711
 rock asphalt, 530, 531
 salt, 1132, 1140
 sand, 108
 silica, 1173
 slag, 125
 sodium sulfate, 1212, 1219
 sulfur, 1236, 1248, 1250, 1252, 1256
 talc, 1275, 1283, 1286, 1287, 1290, 1298
 tripoli, 1368, 1369
 vermiculite, 1377
 zeolites, 1414
Thailand
 barite, 493
 fluorspar, 736
 gems, 302
 kaolin, 627
 potash, 1070
 pyrophyllite, 1097
 salt, 1126
 zirconium, 1433
Thenardite, 1207
Thermonatrite, 1188
Thetford area, 446, 447
Thin sections, use of, 944
Thorite, 1111
Thorium, 1109–1112, 1114, 1116, 1117
Thulium, price, 1110
Tibet, borates, 550
Tiff, 487
Tiger's-eye, 1168
Tile, 375
Tincal, 45, 535
 see also Borax.
Tincalconite, 45, 536
 see also Borax.
Tin oxide
 abrasive, 16, 31
 ceramic, 35, 37
Titangem, 313
Titania, 313, 520, 671

Titanium carbide, 1313
Titanium dioxide pigments, 1304
Titanium minerals, 1303–1362
  alluvial deposits, 1348–1351
  analyses, 1335
    Brazil, 1335
    Sanford Hill, 1318
    South Africa, 1336
  beach deposits, 1347–1351
  capacity, industry
    US, 1308
    world, 1309
  distribution of deposits, 1316–1336
    Australia, map, 1330
    Brazil, map, 1335
  evaluation, 1339–1342
  gem, 313
  processes
    chlorite, 149, 1310, 1311
    sulfate, 1305–1308
  reserves
    Australia, 1331
    world, 1317
  sponge, producers, 1312
  substitutes, 1278, 1351
  uses, 1303, 1306
Tofa, 417, 418
Topaz, 291, 292, 799
Tourmaline, 297
  synthetic, 314
  use, 297
Trace elements fertilizers, 239
Trade associations, 414
Trail Ridge, 1437, 1438
Transparency, 292
Transportation
  costs, 73, 103, 104
  lightweight aggregates, 89
  molding sands, 274, 275
Transvaal, South Africa
  amosite, 455, 457
  chromite, 576
  chrysotile, 455
  crocidolite, 455, 457
  diatomite, 687
  fluorspar, 727, 734
  graphite, 766
  thorium, 1111
Traprock, 61, 164
Travertine, 167, 171
Tremolite, 435, 440–444, 456, 461, 1281
Triamond, 314
Tridymite, 1167
Trinidad
  asphalt, 529
  grahamite, 530
Triphane, 870
Tripoli, 26, 1363–1374
  abrasives, 16, 18, 26, 1364, 1371
  ceramics, 34, 39
  occurrence, 26, 1364–1369
Tripoli, uses
    abrasive, 16, 18, 26, 1364, 1371
    ceramics, 34, 39
    fillers, 1371, 1372
Tripolite, 1363
Tritium, 878
Trona, 1188, 1197, 1198, 1199, 1200
  see also Sodium carbonate.
Troodos area, 452
Tube-mill linings, 27
Tuff, aggregates, 81
Tumbling, 332
Tungsten carbide, 30
Tunisia
  fluorspar, 735
  phosphate, 1031, 1034
  salt, 1123
Turilari playa, 876
Turkey
  bauxite, 515
  borates, 5, 535, 543–546
  celestite, 1229
  chromite, 381, 577, 578
  clays, 606, 627
  diatomite, 687
  fluorspar, 735
  magnesite, 884, 885, 888
  meerschaum, 910, 911
  perlite, 1004
  phosphate, 1028
  pyrophyllite, 1097
  sodium sulfate, 1215
  sulfur, 1236, 1251
Turquoise, stabilized, 314
Twinning defects, 218

U

Uganda
  apatite, 1021
  lithium, 874
  vermiculite, 1377
Uintahite, 530
Ulexite, 534, 536, 537, 542, 544, 548
Ultrahigh-alumina bricks, 376
Umber, 355
Underground mining, 849
Underground storage, 1132
Ungava, asbestos, 444
United Kingdom
  barite, 493
  bath brick, 26
  bromine, 48
  celestite, 1231, 1232
  chalk, 26
    flint, 25
  clays, 603, 608, 623, 624, 634, 638
  cultured quartz, 216
  diaspore, 638
  diatomite, 687
  flint clay, 638
  flint pebbles, 27
  fluorspar, 723, 726, 733, 734
  glauconite, 753
  gypsum, 778
  iodine, 795
  magnesite, synthetic, 890
  mica, 918
  potash, 1067
  quartz, 216
  salt, 1135
  slag, 113
  witherite, 485
  zeolites, 1422
United States
  see also individual states.
  andalusite, 802, 803
  bauxite, 515, 516
  bentonite, 599–601
  bromine, 561, 562, 563
  chromite, 384
  chrysotile, 451, 457
  clays, 585, 599–601, 612, 622–624
  diamonds, synthetic, 662, 670
  diatomite, 685, 686
  feldspar, 709, 713–715, 718
  fluorspar, 728–732
  gems, 303, 318–321
  gypsum, 184
  kyanite, 800–802
  lime, 827
  magnesite, 883
  mica, 981
  monazite, 1114, 1115
  nepheline syenite, 931, 942
  nitrates, 970
  olivine, 991, 992, 993
  perlite, 998
  phosphate, 1030, 1033, 1034
  potash, 1059, 1060, 1066, 1071, 1073, 1095
  pyrophyllite, 1092, 1093
  rare earths, 1114, 1115
  salt, 1125
  sillimanite, 799, 802
  slag, 111, 112
  soda ash, 1187, 1188
  sulfur, 1236
  vermiculite, 1377
  wollastonite, 1384, 1385
  zeolites, 1393–1399, 1414
Uranium in phosphate, 1041
Urea, 234, 962, 963, 966
Urite, 942, 952
Uruguay
  fire clay, 638
  monazite, 1113
  rare earths, 1113
  titanium, 1336
USSR (Russia)
  apatite, 1021
  asbestos, 452, 453, 479

USSR (Russia), barite, 492
bauxite, 515
borates, 546, 547
bromine, 48
chromite, 381, 576
chrysotile, 452, 453, 479
clays, 597, 604, 638
cryolite, 741
cultured quartz, 216
diamonds, 653, 662
diatomite, 687
fluorspar, 734
garnet, 23
gems, 303
glance pitch, 531
glauconite, 753
graphite, 767
iodine, 795
kaolin, 624
lithium, 870
magnesite, 883, 884, 890
manganese, 897, 898
mica, 918
nepheline syenite, 931, 935, 951–954
oil shale, 1253
olivine, 991
perlite, 1004
phosphate, 1020, 1027, 1031, 1032, 1037, 1039
potash, 1069
salt, 1126
slag, 113
sodium sulfate, 1207, 1209, 1214
sulfur, 1236, 1249, 1252, 1256
synthetic diamonds, production, 662
talc, 1283
thorium, 1112
titanium, 1336
vermiculite, 1377
wollastonite, 1383, 1385
yttrium, 1112
zeolites, 1416
zirconium, 1416
Utah
asphalt, 530
brines, 869
diatomite, 685
fluorspar, 732
gems, 303
gilsonite, 530
glance pitch, 531
gypsum, 778
halloysite, 635
hectorite, 875
lime, 827
lithium, 870, 875
oil shale, 1253
ozokerite, 531
perlite, 1003
phosphate, 1022, 1031, 1034, 1042
Utah, potash, 1066, 1073
salt, 1123
silica, 1173
slag, 125
sodium sulfate, 1209, 1212
tar sand, 1253
titanium, 1316
wollastonite, 1385
wurtzilite, 531
zeolites, 1415
zirconium, 1435

V

Valley abrasion test, 1278
Van Dyke zone, 1170
Venetian red iron oxide, 365
Venezuela
bauxite, 514
diamonds, 302, 658
gems, 302
iron oxide pigments, 355
kaolin, 624
sulfur, 1252
tar sand, 1253
Verde antique, 164, 167, 170
Vermiculite, 81, 85, 87, 90, 199, 200, 916, 1375–1381
consumption, 87
deposits, 1377
evaluation, 1377
expansion, 86, 1375
filler, 245
grading requirements, 88
lightweight aggregates, 81, 86, 87, 1380
marketing, 90
prices, 86, 1380
production, 84, 1379, 1380
specifications, 1379
uses, 81, 87, 1380
value, unit, 86
weight, dry, 87
Vermont
asbestos, 451
granite, 170
kaolin, 623
marble, 73, 164, 170
monument stone, 164
serpentine, 170
slate, 170
talc, 1275, 1276, 1282, 1286–1288, 1293, 1295, 1298
Verde antique, 170
Verneuil process, 311–313
Vienna lime, 31
Viet Nam, North, chromite, 578
Viet Nam, South, kaolin, 627
Virginia
aplite, 709, 716, 718, 719
barite, 492
bauxite, 515
Virginia, feldspar, 712, 716, 718, 719
gem, 303
glauconite, 752
granite, 170
crushed, 62
gypsum, 778, 782
kaolin, 623
kyanite, 376, 801
lime, 824, 827, 828
limestone, 853
marble, 164
marl, 854
pegmatites, 711
salt, 1119, 1124
serpentine, 170
silica, 1170
slate, 170
talc, 1286, 1287, 1288, 1298
titanium, 1319
vermiculite, 1377
Viscosity, 616, 629
Volcanic
cinder, 81–86, 1081
see also Pumice.
well drilling fluids, 388
dust, 26
glass, 997

W

Wallboard, 194, 195, 1089
Washington
barite, 492
brucite, 382
chromite, 575
diatomite, 685
feldspar, 711
ferruginous bauxite, 515
fire clay, 638
fluorspar, 732
granite, 171
magnesite, 382, 883, 887
olivine, 272, 274, 991, 993
perlite, 1003
silica, 1171, 1173
sodium sulfate, 1213
traprock, 62
zeolite, 1416
Waste treatment
glauconite, 746, 747
lime use, 820, 821
Wavellite, 1031
Waxes, mineral, 529
Weight-per-cubic-foot test, 14, 15
Weights, units, gems, 323
Welch process, 903
Welding rod coatings, 265, 268, 269
Well drilling fluids, 387–390
Western bentonite, 390

West Virginia
- clays, 374, 375, 637
- grahamite, 530
- limestone, 853
- salt, 1119
- silica, 1170
- white cement, 1087, 1088, 1103

Wheel abrasive, 16, 17
Whiting abrasive, 16, 26
Wicking, 1281
Wig-L-Bug mill, 12, 13
Winnepeg formation, 1175
Wire saws, 173, 174
Wisconsin
- barite, 492
- dolomite, 31, 854
- feldspar, 711
- gannister, 377
- granite, 171
- limestone, 854
- manganese, 898
- quartzite, 27
- silica, 1170

Witherite, 485, 487, 499
Wollanita, 1389
Wollastonite, 800, 1383–1390
- ceramics, 34, 35, 1388
- production, 22, 1387, 1388
- refractory filler, 35
- uses, 1383, 1388
  - ceramic, 34, 35

Wonderstone, 1089, 1097
Woodsreef area, 454
Wool
- blanket, 201
- brass, 30
- mineral, 200, 201, 1388
- slag, 115, 121
- steel, 30

Wurtzilite, 531
Wyoming
- bentonite, 589, 595, 600
- chromite, 575
- feldspar, 711
- fluorspar, 732
- gypsum, 778, 783
- jade, 303
- lithium, 873
- oil shale, 1253
- pegmatites, 711

Wyoming, phosphate, 1022, 1042
- salt, 1124
- silica, 1172
- soda ash, 1188, 1199, 1200, 1201
- sodium sulfate, 1213
- tar sand, 1253
- titanium, 1320
- trona, 1199, 1200
- vermiculite, 1377
- zeolites, 1423
- zirconium, 1435

X

Xenotime, 1111
X-ray diffraction
- clay identification, 594
- gems, color changes, 333
- zeolite identification, 1415

Xerogels, 1168

Y

Yadkin valley, 1320
YAG, 314, 671
Y-bar seed, 210, 217
Yellow ground, 301
Yellow iron oxide, flow diagram, 362
YGG, 314
Yttrium, 1112
- *see also* Rare earths, Thorium.
- growth rate, 1116
- price, 1110

Yttrium-aluminum-garnet, 314
Yttrocerite, 1111
Yttrofluorite, 1111
Yugoslavia
- asbestos, 452
- barite, 492
- bauxite, 513, 523
- chromite, 576
- clays, 604, 626, 638
- diatomite, 687
- fluorspar, 734
- magnesite, 884, 885
- perlite, 1004
- zeolites, 1417, 1419

Yukon, chrysotile, 450

Z

Zaire, diamonds, 302, 653, 658
Zechstein, 1049, 1067–1069, 1126
Zeolites, 1391–1431
- deposits, 1419–1421
- minerals, 1415
- processing, 1407–1413
- reserves, 1414, 1417, 1418
- structure of, 1400–1403, 1413
- synthetic, 1399–1413, 1414
- uses, 1411, 1412, 1420, 1425, 1426

Zeta potential, chrysotile, 439, 440
Zidani Kozani area, 452
Zimbabwe
- amblygonite, 871
- apatite, 1021, 1033
- asbestos, 453
- chromite, 576, 577, 579, 581
- eucryptite, 870
- ferroalloy, 581
- gems, 302
- lepidolite, 870
- lithium, 870, 871, 873, 874
- magnesia, 885
- olivine, 991
- rhodocrosite, 881
- vermiculite, 1377

Zinc, 38
- fertilizer, 233
- micronutrient, 240

Zinnawaldite, 916
Zircon, 27, 273, 314, 378
- alumina abrasive, 29
- ceramic use, 35
- distribution, 273, 274
- foundry sand, 273, 274, 1433
- prices, 1441, 1442
- refractories, 378, 1438
- uses, 273, 274

Zirconia, 34, 35, 314
Zirconium minerals, 273, 274, 1433–1446
- *see also* Titanium.
- abrasive, 29
- ceramics, use in, 35
- foundry sand, 273
- gems, 314
- refractories, 378
- uses, 1086, 1438, 1440

Zoning, 35, 271, 272
- crushed stone, 78
- sand and gravel, 106, 107
- silica, 1177

Industrial Minerals and Rocks

5th Edition

Aerial view of open pit boron mine at Boron, CA (courtesy of US Borax and Chemical Corp.).

Industrial Minerals and Rocks

(Nonmetallics other than Fuels)

Fifth Edition

Volume 2

Published by the
Society of Mining Engineers
of the
American Institute of Mining, Metallurgical, and Petroleum Engineers, Inc.
New York, New York • 1983

Printed in the United States of America
by Port City Press, Baltimore, Maryland

Library of Congress Catalog Card Number 82-071993
Volume 2 ISBN 0-89520-418-5
Set ISBN 0-89520-402-9

Preface

In 1980, after two printings of the fourth edition of *Industrial Minerals and Rocks,* the members attending a business meeting of the Industrial Minerals Division of the Society of Mining Engineers voted to publish a fifth edition. The original authors of the fourth edition were invited to update or revise their chapters. If the original author was unable to revise his chapter, another author did so. Some chapters were completely rewritten. We added a chapter on Alunite and divided Talc and Pyrophyllite into two chapters. *Industrial Minerals and Rocks* has also become international, with two chapters being written by authors in England and Australia. We hope future editions of *Industrial Minerals and Rocks* will include more chapters by authors from foreign countries.

As in the fourth edition of *Industrial Minerals and Rocks,* this edition is divided into four sections. However, the fifth edition breaks slightly with tradition in that it is printed in two volumes. As with the fourth edition, we made an attempt to be consistent in chapter content. However, we did not edit the papers to the point that they became stylized. Each author was permitted to write what he thought was important and to use his own words. We felt this made for more readable and interesting material. We attempted to incorporate the most recent prices and production data. Due to circumstances beyond our control, this was not possible in some instances.

Sadly, during the preparation of the fifth edition of *Industrial Minerals and Rocks,* we lost two editors of the fourth edition, Robert Grogan and Tom Murphy, and one editor of the fifth edition, Gerry V. Henderson. The fifth edition of *Industrial Minerals and Rocks* is dedicated to them.

The Industrial Minerals Division of SME is grateful to each of the authors and editors who gave so generously of their time and experience. We also appreciate the efforts of the many unknown secretaries who typed and retyped the various manuscripts.

Above all, however, we give our sincerest appreciation to Carmel A. Huestis for her diligence in preparing the various manuscripts for publication and to Marianne Snedeker for her patience in bringing the fifth edition through to publication.

Stanley J. Lefond
Editor-in-Chief

Contents

Volume 2

Frontispiece: Open Pit Boron Mine ii
Preface v

Part 4. Commodities, continued

Fluorspar and Cryolite 723
Glauconite 745
Graphite 757
Gypsum and Anhydrite 775
Iodine 793
Kyanite and Related Minerals 799
Lime 809
Limestone and Dolomite 833
Lithium Raw Materials 869
Magnesite and Related Minerals 881
Manganese 897
Meerschaum 909
Mica 915
Nepheline Syenite 931
Nitrogen Compounds 961
Olivine 989
Perlite 997
Phosphate Rock 1017
Potash 1049
Pumice, Pumicite, and Volcanic Cinders 1079
Pyrophyllite 1085
Rare Earths and Thorium 1109
Salt 1119
Sand and Gravel 1151
Silica and Silicon 1167
Sodium Carbonate Deposits 1187
Sodium Sulfate Deposits 1207
Staurolite 1225
Strontium 1229
Sulfur 1235
Talc 1275
Titanium Minerals 1303
Tripoli 1363
Vermiculite 1375
Wollastonite 1383

Zeolites
Introduction ... 1391
Synthetic Zeolites: Properties and Applications ... 1399
Zeolites in Sedimentary Rocks ... 1413
Commercial Utilization of Natural Zeolites ... 1418
Zirconium and Hafnium Minerals ... 1433
Index ... I-1

Volume 1

Frontispiece: Cultured Quartz Crystals ... ii
Preface ... v

Part 1. Introduction ... 1

Introduction ... 3

Part 2. Industrial Minerals Grouped by Uses ... 9

Abrasives ... 11
Ceramic Raw Materials ... 33
Chemical Industry ... 41
Construction Materials
Aggregates—Introduction ... 59
Crushed Stone ... 60
Lightweight Aggregates ... 81
Sand and Gravel ... 96
Slag ... 111
Cement and Cement Raw Materials ... 133
Dimension and Cut Stone ... 161
Gypsum and Anhydrite ... 183
Insulating Materials—Thermal and Sound ... 199
Roofing Granules ... 203
Electronic and Optical Uses ... 213
Fertilizer Minerals ... 233
Fillers, Filters, and Absorbents ... 243
Fluxes ... 259
Foundry Sand ... 271
Gem Materials ... 279
Glass Raw Materials ... 339
Mineral Pigments ... 349
Refractories ... 373
Well Drilling Fluids ... 387

Part 3. Sources of Information for Industrial Minerals ... 391

Sources of Information ... 393

Part 4. Commodities 415

Alunite 417
Asbestos 435
Barium Minerals 485
Bauxitic Raw Materials 503
Bituminous Materials 529
Boron and Borates 533
Bromine 561
Chromite 567
Clays 585
Diamonds 653
Diatomite 677
Feldspars 709

Index I-1

Fluorspar and Cryolite*

ROBERT B. FULTON †

GILL MONTGOMERY ‡

Fluorspar is the commercial name for fluorite, a mineral that is calcium fluoride, CaF_2. The name, derived from the Latin word *fluere* (to flow), refers to its low melting point and its early use in metallurgy as a flux. It is the principal industrial source of the element fluorine.

Cryolite, sodium aluminum fluoride, Na_3AlF_6, is a rare mineral which has been found in commercial quantities only in Greenland. The natural material has been supplanted by synthetic cryolite for its principal industrial use in the manufacture of aluminum.

History

Fluorspar was used by the early Greeks and Romans for ornamental purposes as vases, drinking cups, and table tops. Various peoples, including the Chinese and the American Indians, carved ornaments and figurines from large crystals. Its usefulness as a flux was known to Agricola in sixteenth century Europe.

Fluorspar mining began in England about 1775 and at various places in the United States between 1820 and 1840. Production grew substantially following the development of basic open-hearth steelmaking, wherein it is used as a flux. Use was stimulated by growth of the steel, aluminum, chemical, and ceramic industries, particularly during World Wars I and II. Fluorocarbons entered the picture in 1931. The use of anhydrous hydrogen fluoride (HF) as a catalyst in the manufacture of alkylate for high octane fuel began in 1942.

Differential flotation for separating fluorspar from galena, sphalerite, and common gangue minerals in the 1930s, and the application of heavy-media concentrating methods to the treatment of low-grade ores in the 1940s were outstanding technological advances that facilitated increased production.

Recently, pelletizing and briquetting of flotation concentrates for use in steel furnaces and the development of flotation schemes for beneficiating ores containing abundant dolomite and barite have been major improvements in the industry.

Uses of Fluorspar

Fluorspar is used to make hydrogen fluoride, also called hydrofluoric acid, an intermediate for fluorocarbons, aluminum fluoride, and synthetic cryolite. It is used as a flux in the steel and ceramic industries, in iron foundry and ferroalloy practice, and has many minor specialized uses.

Hydrogen fluoride (HF) is produced by reacting acid grade (97% CaF_2) fluorspar with sulfuric acid in a heated kiln or retort to produce HF gas and calcium sulfate. After purification by scrubbing, condensing, and distillation, the HF is marketed as anhydrous HF, a colorless fuming liquid, or it may be absorbed in water to form the aqueous acid, usually 70% HF.

Synthetic cryolite, organic and inorganic fluoride chemicals, and elemental fluorine are made from hydrofluoric acid. The acid itself is important in catalysis in the manufacture of alkylate, which is an ingredient in high-octane fuel for aircraft and automobiles, in steel pickling, enamel stripping, glass etching and polishing, and in various electroplating operations. The manufacture of one ton of virgin aluminum requires about 50 to 60 lb of fluorine content in synthetic cryolite and alumnium fluoride. This quantity, through improved technology and recovery practices, is being lowered significantly.

* Revised from Robert M. Grogan and Gill Montgomery, "Fluorspar and Cryolite," *Industrial Minerals and Rocks*, 4th ed., 1975.

† Basic Materials Planning, Materials and Logistics Dept., E. I. du Pont de Nemours and Co., Inc., Wilmington, DE.

‡ Minerals Consultant, Box 410, Eldorado, IL.

Elemental fluorine is prepared from anhydrous hydrofluoric acid by electrolysis. Gaseous at room temperature and pressure, fluorine is compressed to a liquid for shipment in cylinders or in tank trucks. Elemental fluorine is used to make uranium hexafluoride, sulfur hexafluoride, and halogen fluorides. Gaseous uranium hexafluoride is used in separating U^{235} from U^{238} by the diffusion process. Sulfur hexafluoride is a stable high dielectric gas used in coaxial cables, transformers, and radar wave guides. Halogen fluorides have important applications, mostly as substitutes for elemental fluorine which is more difficult to handle.

Emulsified perfluorochemicals, organic compounds in which all hydrogen atoms have been replaced by fluorine, are undergoing investigation as promising blood substitutes. They transport oxygen and, in conjunction with a simulated blood serum, perform many functions of whole blood. With further development, ultimately they may be useful in saving lives of animals and humans in emergencies during periods of acute shortages of natural blood.

Inorganic fluorides are used as insecticides, preservatives, antiseptics, ceramic additives, in electroplating solutions, as fluxes, antioxidants, and many other ways. Boron trifluoride is an important catalyst.

Organic fluorides are volume leaders in the fluorine chemical industry. Fluorinated chlorocarbons and fluorocarbons are prepared by the interaction of anhydrous HF with chloroform, perchlorethylene and carbon tetrachloride, and are characterized by low toxicity and notable chemical stability. They perform outstandingly as refrigerants, aerosol propellants, solvents, and cleaning agents and as intermediates for polymers such as fluorocarbon resins and elastomers.

Fluorocarbon resins are inert compounds with an unusually low coefficient of friction which have found a number of applications for parts that cannot be oiled, such as the bearings inside of automobile doors, for window raising equipment, in small electronic equipment, and for the manufacture of chemical resistant gaskets and valve parts, pipe and tank linings, flexible tubing and containers, and cookware.

In the steel industry, fluorspar is used as a flux in basic open-hearth, basic oxygen, and electric furnaces where it is added to the heats in amounts ranging from 2 to 20 lb per ton of steel produced. The average in the US is currently about 6 lb per ton. It promotes fluidity of the slag and thus facilitates removal of sulfur and phosphorus from the steel into the slag. It serves the same purpose in iron foundries, where it is added to the cupola charge in the proportion of around 15 to 20 lb per ton of metal melted.

In the ceramic industries, fluorspar is used to make flint glass, white or colored opal glasses, and enamels. Flint glass mixtures commonly contain 3% fluorspar. Opal glasses, containing 10 to 20% fluorspar, are used in containers for foods, drugs and toiletries, and in ornamental glassware and lavatory and restaurant fixtures. Opaque enamels are used to cover steel stoves, refrigerators, cabinets, bathtubs, and cooking ware and for facings on brick and tile, and other structural materials. Fluorspar makes up to 3 to 10% of the weight of the enamel. Many types of welding rod coatings incorporate fluorspar or fluorspar mixtures. Ceramic grade fluorspar is used in the manufacture of magnesium and calcium metals, and in the preparation of some manganese chemicals.

Lower grades of ceramic spar are used in making fiberglass insulation, in zinc smelting, as a clinkering aid in the manufacture of portland cement, as an inhibitor of vanadium green scumming in the manufacture of buff-faced brick, and as an abrasive on certain types of sandpapers. Various grades of fluorspar are used in electric furnace manufacture of calcium carbide, in making electrodes for arc lamps, and as a bonding material for abrasive wheels, among numerous minor uses.

Product Specifications

The three principal market grades of fluorspar are acid, ceramic, and metallurgical. Specifications for these are fairly well defined, although the requirements set by individual consumers may vary in detail.

Acid grade fluorspar is defined as not less than 97% CaF_2 but some manufacturers of hydrofluoric acid both in the US and in Europe can use 96%, or slightly lower if the remaining impurities are acceptable. Since the US tariff structure favors fluorspar of 97% or over, imports of acid grade or imports of material used for ceramic fluorspar purposes must be over 97% CaF_2. Users specify limits for silica, calcium carbonate, arsenic, lead, sulphide sulfur, phosphorous and other deleterious constituents. Moisture content of dried concentrates for acid and ceramic grades is usually specified to be not more than 0.10% H_2O, but in some ceramic uses up to 1 or 2% may be required to relieve dust problems in handling bulk materials.

There are two standard ceramic grades: No. 1 ceramic containing 95 to 96% CaF_2, and No. 2 ceramic which customarily contains 85 to over 90% CaF_2. A medium grade of about 93 to 94% CaF_2 is specified by some users. In some instances buyers of ceramic grades may specify not over 2.5 or 3.0% silica, limited amounts of calcite and ferric oxide, and only traces of lead and zinc sulfides. Practically every ceramic grade user has his own specifications, and producers tailor their products accordingly.

In the United States, metallurgical grade, called "metspar," contains a minimum of 60 "effective" percent of fluorspar and generally not over 0.30% sulfide sulfur and 0.25 to 0.50% lead. The "effective" percentage is calculated by multiplying the silica percentage in the analysis by 2.5 and subtracting this number from the calcium fluoride percentage in the analysis; the result is termed "effective units." Thus, a concentrate with 85% calcium fluoride and 6% silica would be 70% effective grade. To qualify as "lump" or "gravel" spar, it must pass a 1 to 1½ in. screen and contain less than 15% of material passing a 1/16 in. screen.

In markets outside the United States, metspar may contain a minimum of 80% CaF_2 and a maximum of 15% silica.

In fluorspar's use as a flux in the steel industry, the trend is toward briquets and away from metspar due to growing scarcity of satisfactory gravel fluorspar. Briquets are made from the fines accumulated during metspar preparation and from flotation concentrates. Most of the imported fluorspar concentrates used to make briquettes are acid grade, for tariff reasons. Domestically produced ceramic grade concentrates are also used. Most briquets are made on roll presses, from "peach seed" sizes up to two-inch square briquets. The most popular binders used are molasses and lime, the use of which does not require baking ovens. Pellets have been made on balling machines using sodium silicate binder, and extruded shapes have been made. There is a strong trend toward making diluted briquets containing steel mill wastes, such as mill scale, flue dust, shredded scrap, iron ore fines, and manganese ore fines. Some diluents have fluxing values. Briquets having as little as 25% CaF_2 contained are in common use, but briquets made from acid grade fluorspar grading above 90% are in demand. Users of the diluted briquet claim fluorspar economy plus a way to make use of steel mill wastes.

Geology

Composition and Properties

Theoretically pure fluorite contains 51.1% calcium and 48.9% fluorine. Substitution of small percentages of cerium and yttrium for calcium has been noted. Inclusions of gases and fluids, such as petroleum and water, and of solid minerals such as pyrite, marcasite, and other sulfides are common. Free fluorine is present in some crystals. The fluorspar of commerce contains attached and admixed mineral impurities, such as calcite, quartz, barite, celestite, various sulfides, or phosphates.

Fluorite tends to occur in well-formed isometric crystals, forming cubes and octahedrons. It also occurs in massive and in earthy forms, and as crusts or globular aggregates with radial fibrous texture. Crystalline fluorspar exhibits a great range of colors, from colorless and water-clear to yellow, blue, purple, green, rose, red, bluish and purplish black, and brown. The colors may occur in alternating bands parallel to cube faces. They may be altered by exposure to X-rays, heat, ultraviolet light, and pressure. Colors are caused by a variety of factors including the presence of trace impurities and displaced ions in the lattice. Long exposure to sunlight, such as occurs on mine dumps, frequently results in the fading of the original coloration.

Some varieties fluoresce blue or violet under ultraviolet light or cathode rays. Some specimens phosphoresce when heated or after exposure to sunlight or ultraviolet light, and some exhibit triboluminescence.

The mineral has a hardness of 4, and is the type-mineral of that hardness on the Mohs' scale. Its specific gravity is 3.18 when crystalline, but ranges from 3.01 to 3.6 in various forms. The luster is vitreous. The mineral has perfect octahedral cleavage, and "diamonds" made by knocking the corners off of cubic crystals are seen in many collections.

When pulverized and treated with sulfuric acid, fluorite decomposes into gaseous hydrogen fluoride and calcium sulfate, which is the fundamental reaction in the production of hydrofluoric acid. When added to metallurgical slags it imparts greater fluidity at lower temperatures making it valuable in steelmaking and ferrous and nonferrous foundry practice. In ceramic mixes it promotes crystallization around individual centers, and thus is useful in making opal glasses. Crystalline fluorite has a very low index of refraction ($n = 1.4339$), low dispersion, is isotropic, and has an unusual ability to transmit ultraviolet light. These are the properties which make it useful in optical systems as prisms and as components of high quality, special-purpose lenses. Synthetic fluorite has replaced the natural mineral for optical uses.

Modes of Occurrence

Fluorite occurs in a wide variety of geological environments, evidencing deposition under an extended range of physical and chemical conditions. At one extreme it is present as an accessory mineral in granites and related igneous rocks; while at the other extreme, it is found as crystals in geodes and as botryoidal linings in limestone caves.

From an economic standpoint, the most important modes of occurrence of the mineral are:

1) Fissure veins in igneous, metamorphic, and sedimentary rocks.

2) Stratiform replacement deposits in carbonate rocks.

3) Replacements in carbonate rocks along contacts with acid igneous intrusives.

4) Stockworks and fillings in shattered zones.

5) Deposits at the margins of carbonatite and alkalic rock complexes.

6) Residual concentrations resulting from the weathering of primary deposits.

7) Occurrences as recoverable gangue in base metal deposits.

Less common modes, in some places commercially important, include:

8) Fillings in breccia pipes.

9) Fillings in open spaces.

10) Pegmatites.

Least common but of potential importance are:

11) Deposits in lake sediments.

Fissure Veins: Fissure veins, usually along faults or shear zones, are the most readily recognized form in which fluorspar deposits occur the world over. Silica, calcite or other carbonates, iron, lead, and zinc sulfides, and barite are the typical associated minerals. In some vein deposits as in the Rosiclare district of southern Illinois, fluorite appears to have replaced a prior vein filling of calcite. Along some veins in carbonate rocks, fluorspar has replaced the wall rock at intersections with favorable beds, providing large minable tonnage mined from some deposits. Although vein structures are remarkably persistent, the fluorspar itself commonly occurs as lenses or ore shoots separated by barren or poorly mineralized portions of vein. Ore shoot widths of 2 to 30 ft and lengths of 200 to 1000 ft are common, but there is great variation from deposit to deposit. Vein systems may be up to several miles in length, and ore may be present to depths of 1000 ft or more below the surface. The CaF_2 content of minable portions of veins normally ranges from 25 to 80%, although grades above 90% occur in limited areas.

Some of the world's great vein deposits include the Osor deposit in northeastern Spain, the Torgola deposit in northern Italy, the Muscadroxiu-Genna Tres Montis vein system in Sardinia, the Longstone Edge-Sallet Hole deposit in England, and the Rosiclare-Goodhope vein system in southern Illinois.

Stratiform Deposits: Stratiform, manto, or bedded deposits occur in carbonate rocks. Certain beds are replaced along or adjacent to structural breaks such as joints or faults. This relationship to structural features is very clear in some deposits, but obscure in others. Frequently there is a sandstone, shale, or clay capping. Typically there is evidence of loss of volume in the replaced zones with attendant development of gentle synclinical structures in overlying strata or of collapse structures, sometimes pipelike in shape. In some districts, as in southern Illinois, no connection is recognized between the mineralization and any igneous activity, whereas in others, such as the Encantada district in northern Coahuila, Mexico, the presence of rhyolite plugs and sills in the general vicinity of the spar deposits, and the association of spar with rhyolite injections along bedding planes, suggests an association. Stratiform deposits are known in many parts of the world and are particularly well developed in the Cave in Rock district of southern Illinois, in the northern part of the State of Coahuila in Mexico, and in the Ottoshoop (Zeerust) District of Transvaal in South Africa. In Illinois the deposits range from a few inches to 20 ft thick, 50 to 500 ft wide, and are up to 4½ miles long. In Coahuila the individual ore bodies are smaller but relatively more numerous and widespread. In the Ottoshoop District bedded deposits occur in an area 10 miles long and 6 miles wide in a dolomite facies underlying a prominent chert bed.

In stratiform deposits, textural features of the parent rock such as sedimentary banding are commonly preserved. Frequently associated with the banded ore is a massive crystalline type which appears to have filled open spaces left from the dissolving of limestone by the ore-bearing solutions or their precursors. Minerals accompanying fluorspar in stratiform deposits are calcite, dolomite, quartz, galena, sphalerite, pyrite, marcasite, barite, and celestite. CaF_2 content in minable deposits ranges from 15% upward. A few ore bodies in Illinois yield direct shipping metallurgical grade fluorspar containing 85% or more CaF_2.

Replacement Deposits: Replacement deposits in carbonate rocks along contacts with intrusive rhyolite bodies are well developed in the Rio Verde, San Luis Potosi, and Aguachile districts in Mexico. They include some of the largest and highest grade fluorspar deposits. The fluorspar is not thought to be contact metamorphic in origin, but introduced later, following the contact zone as a conduit replacing the limestone outward from the contact, either massively or selectively along certain beds. At Aguachile, cross sections show ore shoots resembling one side of a Christmas tree.

Stockworks: Fluorspar often occurs as stockworks and fillings in shear and breccia zones. Many occurrences in the American West are of the stockwork type and, though wide, usually have low overall CaF_2 content. Deposits in the Zuni Mountains of New Mexico and near Jamestown, Colorado are examples. The Zwartkloof deposit in the Transvaal Province of South Africa consists of three vertical breccia zones in an east-west line in felsite containing stockworks of fluorite-carbonate veins. The largest zone is 200 by 600 ft in plan and persists to 3000 ft below the surface. The fluorspar grade is about 14%. The Buffalo deposit, near Naboomspruit in the Transvaal, consists of a network of fluorspar veinlets in sill-like bodies of fine-grained pink granite which are inclusions in coarse red granite of the Bushveld complex.

Carbonatite and Alkalic Rock Complexes: Fluorspar is a common mineral in carbonatite and alkalic rock complexes, rarely in sufficient abundance to be economic. The Okorusu deposit in South-West Africa (Namibia) of this type consists of a number of bodies of fluorspar in limestones, quartzites, and related rocks which have been intruded and metamorphosed by an alkaline igneous rock complex including a nepheline syenite stock. The fluorspar replaces bedded and brecciated limestone, marble, and quartzite, forming large lenticular masses. Apatite and quartz are abundant accessory minerals. At Amba Dongar, India, veins and replacement bodies of fluorspar occur in carbonate rocks bordering ankeritic carbonatite intrusives. Fluorite is also present in the carbonatite itself.

Residual Deposits: Concentrations of fluorspar in clayey and sandy residuum resulting from surficial weathering of fluorspar veins and replacement deposits in some places are sources principally of metallurgical spar. This category includes detrital deposits blanketing the apex of veins, as well as deeply weathered upper portions of the veins themselves extending to depths of 100 ft or more. Such deposits were of major importance in Illinois and Kentucky. Similar deeply weathered ore has been mined in England, Thailand, and the Asturias district of northwestern Spain. Weathered residuum, called *kokoman*, is mined in the Marico district in South Africa, where it is the result of the weathering of gently dipping replacement bodies.

Gangue Mineral: Fluorspar occurs as a major gangue mineral in lead-zinc veins in many parts of the world, and in some, averaging 10 to 20%, is economically recoverable. Acid grade fluorspar is produced on a large scale from lead-zinc mill tailings near Parral in Mexico.

Breccia Pipes: Fluorite occurs in breccia pipes in the Thomas Range, Utah and near Beatty, Nevada. Pipes in the Thomas Range are circular to oval in plan, up to 150 ft in diameter and over 200 ft in depth, and are formed in dolomite by replacement along shattered zones associated with faults and intrusive breccias. Fluorspar occurs as soft friable masses and in boxworks of fine-grained more resistant veinlets, and is nearly unrecognizable. At the Daisy mine near Beatty, fluorspar replaces brecciated dolomite in pipelike bodies bounded by gouge zones along two sets of intersecting faults.

Fillings in Open Spaces: Fluorspar occasionally partially fills open spaces, both in veins and stratiform deposits. Spectacular examples of this type occur in the San Vicente district of northern Coahuila, where fluorspar occurs in veins and mantos as pure massive incrustations of mamillary, stalactitic, and stalagmitic forms. The Fluorspar-Gero-Penber vein system of the Northgate mine in Colorado is a similar occurrence where fluorspar occurs in botryoidal layers on the walls of open fissures and as concretionary coatings surrounding loose fragments of country rock. Lower parts of open areas of these fissures are partly filled with concretionary pebblelike masses, in places cemented into porous, rubbly aggregates.

Pegmatites: Many pegmatites contain minor amounts of fluorspar. Grade was high enough to support a mining operation at the Crystal Mountain occurrence in Montana, where three large tabular bodies of massive fluorspar containing minor amounts of biotite, quartz, feldspar, and other igneous-type accessory minerals, occur in coarse-grained biotite granite.

Lake Sediments: Fluorspar occurs in unconsolidated clayey and sandy pyroclastic sediments in the beds of former lakes near Castel Giuliano, about 25 miles north of Rome, Italy. Fluorine of volcanic origin permeated the lake sediments resulting in deposition of minute disseminated crystals of fluorspar, which make up as much as 50 to 60% of the clayey parts and 15% of the sandy parts of the deposits. It is accompanied by barite, apatite, calcite, dolomite, and opal.

Distribution of Deposits

Fluorspar deposits occur on every continent. Major producers include the United States, Mexico, Spain, England, France, Italy, South Africa, China, Thailand, Mongolia, and the USSR. Lesser producers include Argentina, Brazil, Bulgaria, Czechoslovakia, Kenya, Tunisia, Morocco, West Germany, East Germany, Romania, India, and Korea.

North America: *Canada*—The only large production in Canada was from the St. Lawrence area of the Burin Peninsula in southern Newfoundland, which yielded over 3 million tons of ore from 1940 until operation ceased in 1978. Other deposits occur in Ontario at Madoc and near Wilberforce and Cobden; at Rock Candy and Birch Island near Kamloops and near Quesnel Lake in British Columbia; along the Lower Laird River in northern British Columbia; and at Lake Ainslie on Cape Breton Island in Nova Scotia.

On the Burin Peninsula, fluorspar occurs in veins in granite and rhyolite porphyry along steeply dipping faults. Average thickness of higher grade veins is 3 to 5 ft and of lower grade veins is 15 to 20 ft. Some veins have been traced for more than a mile.

Veins at Madoc are small and have been worked principally in wartime. The veins at Lake Ainslie are low grade fluorspar with associated abundant barite. A deposit at Birch Island remains undeveloped with reserves said to be 1.5 million tons of 29% CaF_2 grade. Bedded deposits on the Lower Laird River near Mile 497 on the Alaska Highway in northern British Columbia have not proved to be of current commercial interest. Two low grade fluorite deposits near Likely, BC, were reported to be under exploration early in 1981.

Mexico—Mexico is the world's largest producer of fluorspar. In 1981, production was 1 116 332 t, having held at that approximate level steadily since 1975, but sharply declined during 1982, totaling only 605 643 t in the first three quarters compared to 831 687 t in the first three quarters of 1981. Important deposits occur in San Luis Potosi, Coahuila, Chihuahua, and Guanajuato.

Many types of deposits are mined. In the Musquiz district of northern Coahuila, mantos in Cretaceous limestones are most important. Northwestern Coahuila has vein deposits and replacement deposits in limestone associated with rhyolite sills and plugs. Deposits in the Paila district of central Coahuila are veins. In San Luis Potosi and adjoining Guanajuato are several large replacement deposits in limestone associated with rhyolite intrusives. Numerous vein deposits occur in Chihuahua, but its principal production is acid grade concentrate recovered from sulfide mill tailings near Parral.

Eight flotation plants in Mexico have an annual production capacity of approximately 600 kt of acid grade concentrate. A plant just across the border at Eagle Pass, Texas, using Mexican ore, has a nominal capacity of 90 kt/a. Total acid grade production in 1981 was 510 kt; metspar and sub-met production was approximately 501 kt; and ceramic was 108 kt. Total sales in 1981 were 995 kt. While most is exported, Mexican consumption of acid grade has increased substantially with four hydrofluoric acid plants, including two relatively new ones (Quimica Fluor starting in 1975, the Fluorex in 1979), resulting in domestic acid grade sales of 177 kt in 1981.

United States—ALASKA—The Lost River deposit, located in the western portion of the Seward Peninsula 85 miles northwest of Nome, was worked intermittently for tin from 1904 to 1955 and has since been explored for fluorspar, beryllium, and tin. Fluorspar-tin-tungsten-beryllium mineralization occurs in acidic dikes, in skarn deposits along granite-limestone contacts, and brecciated limestone adjacent to thrust faults. Mineralization in one zone was reported to be 27 million tons grading 16.3% CaF_2, 0.15% tin, and 0.03% tungsten. Fluorspar also occurs in six other zones in the district (Anon., 1972), but there has been no production.

ARIZONA.—Although 56 occurrences of fluorspar were listed in one summary (Van Alstine, 1969) and 96 in another (Elevatorski, 1971), to date cumulative production has probably been less than 50,000 tons. Deposits are widely scattered in the southern half of the state, principally as small epithermal veins in fissures and brecciated zones along faults in intrusive, metamorphic, and sedimentary rocks. No large replacement deposits have been found. Fluorite occurs mostly with calcite and quartz in nonmetaliferous veins, although in the Castle Dome district it is a gangue mineral along with barite, quartz, and calcite in veins which were mined on a small scale for their lead-silver content.

The Castle Dome, Duncan, Sierrita Mountain, Tonto Basin, and Whetstone Mountain districts have been producers. The Lone Star mine in the Whetstone Mountain district yielded 10,000 tons of metallurgical spar in 1967, but then closed down.

Vein ore from the Packard and other nearby mines was processed into acid grade fluorspar in a flotation mill at Tonto Basin until these veins were exhausted. Ore from the Turkey mine on McFadden Peak north of Globe was processed at that mill before it was decommissioned in 1979. This mine and other localities in Arizona remain inactive.

CALIFORNIA—Although there are many occurrences of fluorspar in California, only small amounts have been produced for ceramic and metallurgical purposes. Metspar was produced on a small scale from the Clark Mountain area of San Bernardino County in the 1950s and 1960s. California occurrences have been described by Chesterman (1966) and Elevatorski (1968).

COLORADO—Colorado in some years ranked second in annual production, after Illinois. Aggregate production of crude ore to the present is estimated at about 2.5 million tons.

A total of 63 occurrences have been identified in the state (Van Alstine, 1964), including five major districts, all now idle. Northgate and Jamestown districts were the largest and most recently closed. The others are Browns Canyon, Poncha Springs, and Wagon Wheel Gap.

In the Jamestown district, fluorspar occurs in pipelike bodies and breccia zones in granite and granodiorite, and in associated veins. Output of the Burlington mine was treated at the former Valmont Mill near Boulder. Other mines include the Argo, Emmett, and Blue Jay, ore from which was concentrated to acid grade in a mill, located at Jamestown. Neither mill exists now.

At Northgate, fluorspar occurs in veins in shear and breccia zones in granite and schist, and has been mined open cut and underground. During World War II the district produced metspar. More recently the ore was concentrated to acid grade in a mill built in 1952, now idle. Exploratory drilling continues from time to time and may result in reopening as prices continue to rise.

The Browns Canyon district included a number of mines, developed on veins in rhyolite and in granite. Two flotation mills were operated at various times. In the Poncha Springs district a flotation mill operated for several years in the mid-1950s on low grade ore consisting of fluorspar veinlets in fractured and brecciated zones in granite and granite gneiss. The Wagon Wheel Gap mine was on a vein in a sheeted zone in rhyolite yielding metspar.

IDAHO—The principal deposits of fluorspar in Idaho are in the Bayhorse-Keystone Mt. area of Custer County, southwest of Challis, and in the Meyers Cove, Big Squaw Creek, and Stanley areas (Anderson, 1964).

The Bayhorse dolomite replacement deposits, as outlined by coredrilling and partially developed for mining, may exceed 3.2 million st of 36% CaF_2, with a 25% grade cutoff. Other deposits, both vein and replacement, occur in the Bayhorse district and on adjoining Keystone Mt. The district has potential if production economics can offset freight disadvantages.

A flotation mill operated briefly at Meyers Cove in the fifties, producing about 13,000 tons of acid grade concentrate from fluorspar mined in brecciated volcanics and fissure fillings. Later, a heavy media plant was built to make metallurgical gravel spar, but was unsuccessful due to excessive silica locked in the ore. This mill was moved to the Garden Creek area west of Challis to be used on free-milling vein ores, but is not in operation.

At Big Squaw Creek, a vein is exposed in the northeast canyon wall of the Salmon river west of Shoup. At Stanley, fluorspar occurs in shear zones in granite and volcanics.

ILLINOIS-KENTUCKY—Current production is focused at two flotation mills situated in an area of 700 square miles in southern Illinois and adjacent Kentucky which for a number of years was the world's largest producer (Baxter, et al., 1973) and continues as the major US producing district. Shipments of fluorspar concentrates of all grades during the period 1880 through 1980 are estimated to have totaled 11.5 million st, which is probably equivalent to mine production of approximately 33 million tons of crude ore.

Vein deposits are the most numerous, but since 1950 the bedded deposits near Cave in

Rock, IL, have been the source of most of the district's output, and outrank the vein deposits in terms of total all-time district production. Recently production from vein deposits has resumed importance as some of the older bedded deposits have been depleted.

In addition to the major district, a small production has been recorded from veins southwest of Lexington, in central Kentucky.

Vein deposits of the Illinois-Kentucky district occur along an extensive and intricate system of faults in sedimentary rocks of Mississippian age. Most of these faults are the steeply-dipping normal type, trending northeast, with displacement ranging from a few feet to over 1000 ft. The most important deposits are on faults of 50 to 500-ft displacement. Veins range in width up to 30 ft and to as much as 60 ft in exceptional cases. They have been mined for 900 ft vertically and for two miles in length. Veins pinch and swell horizontally and vertically with the proportion of fluorspar to total vein material ranging from 0 to 100% within short distances. Near the surface and to depths as great as 250 ft, the veins contain much clay owing to the dissolving of vein calcite and limestone wall rock by circulating ground water leaving the argillaceous material in the limestone behind. Strong walls are found at depth in most places. A large inflow of water at high pressure was characteristic of the deeper mines particularly at Rosiclare.

Major replacement deposits occur in Illinois northwest and north of the village of Cave in Rock, others occur in Kentucky south of Carrsville and near Hampton in Livingston County. The Cave in Rock group of mines occupies a zone of minor structural disturbance one mile wide and over six miles long, paralleling a major northeast trending fault which has maximum displacement of about 1000 ft. The deposits lie along one or both sides of minor fractures and small faults and are characteristically long and narrow. Major deposits occur along fracture-fault systems with less than one ft of displacement. Widths of 50 to 150 ft and thicknesses of 3 to 15 ft are typical. Maximum widths of 325 to 350 ft have been found. Lengths are mostly 200 to 1500 ft. One ore body is known to extend continuously for at least 4½ miles. The deposits occur through a vertical range of about 180 ft in four preferred sets of limestone beds and follow these beds down dip from the outcrops (which were discovered originally in a prominent bluff) to depths well over 1000 ft. Mining in the former Minerva mine (now Inverness) currently is below 1300 ft and drilling discloses ore at least 120 ft deeper in lower horizons. Three to four bedded horizons often exist superimposed along the same fracture system, and within a vertical interval of 150 ft. Fluorite is the principal mineral, accompanied by sphalerite and galena in many deposits. Quartz, barite, and calcite are common, along with pyrite, marcasite, witherite, and strontianite. Banding, which may be a relic from the replaced limestone or from rhythmic metasomatic deposition, is characteristic of some deposits, known locally as "coontail ore." Crude ores have 15 to 90% calcium fluoride, and some contain up to 3½% zinc and 5% lead.

Residual deposits have been less important producers, but have yielded high quality fluorspar. These "gravel spar" deposits are as much as 60 ft wide and extend to depths of a hundred feet or more. The fluorspar occurs as weathered fragments from the size of boulders down to sand grains dispersed in a clayey matrix.

Montana—Most of the fluorspar production has come from the Crystal Mountain deposits near Darby in Ravalli County. The deposits of massive fluorite in granite and metasediments were discovered in 1951 and developed into one of the nation's major metallurgical grade fluorspar mines.

The mine was a side-hill benched open pit, at the 7000 ft elevation on Sapphire Mt. Ore was mined, hand-sorted, and sized at the mine continuously from 1952 to 1973. This roughly concentrated ore was hauled 25 miles by truck to Darby, where finished concentration was accomplished in a 60-ton per hour heavy-media cone plant. The metspar was shipped by rail to steel plant consumers and government stockpiles.

Shipments from the accumulated fines at the heavy-media plant continued from 1973 to 1980 to briquetting plants serving the steel industry. Production from 1952 to 1980 totalled 566,000 tons of which 520,000 was metspar, 2000 tons was metspar-sized jig product (¼ x ½ in.), and 44,000 tons was sub-metgrade fines. Mining and shipment of fines resumed in 1981.

Numerous other fluorspar occurrences have been recorded in Montana but have no production history. Fluorspar occurs as gangue in metalliferous veins in marginal portions of the Boulder batholith in west-central Montana, and is present as veins and replacement seams in intrusive syenite and in limestone in several areas characterized by alkalic intrusives in the north-central part of the state.

The presence of fluorspar has been used as a

guide to gold deposits. In the Old Glory deposit in the Pryor Mountains, fluorspar occurs with the uranium mineral, tyuyamunite, replacing limestone in brecciated and cavernous portions of the Madison limestone.

NEVADA—Fluorspar became commercially important in Nevada in 1919 when the Daisy mine near Beatty began its long productive life. Nevada deposits are described in a recent publication by Papke (1979). Most of the 62 deposits he describes lie in a broad, arcuate band across the southern part of the state, and northward through the central part. Total production has been about 575,000 st, most from the Daisy (Crowell) mine in southern Nye County, the Baxter (Kaiser) mine in northeastern Mineral County, the Goldspar mine near the Daisy, and the Carp (Wells Cargo) mine in southeastern Lincoln County. About 19 other deposits have had some production, but only the Daisy mine is active. More than a fourth of the deposits are in the Quinn Canyon and Adaven district in the southeastern part of the state, but the small size of the deposits and remoteness of this area has hindered their development. Small metspar shipments have been made from Caliente.

Deposits are hydrothermal or pyrometasomatic. They occur as replacement bodies, veins, replacement and/or filling in jasperoid bodies, stockworks, and breccia pipes. The Daisy deposit is mined underground, with total past production of about 225,000 st. It consists of irregular, generally steeply dipping, structurally complex replacement bodies in dolomite. The ore is soft, friable, and relatively low in silica. Most has been used as metspar. Some has been shipped to a southern California cement producer. The Baxter deposit, a vein in andesitic volcanic rocks, developed to a depth of 600 ft, had a total mine production of 182,000 st, of which two-thirds was made into acid-grade concentrate in a mill operated from 1952 to 1957 by Kaiser Aluminum and Chemical Corporation at Fallon, Nevada. The Goldspar and nearby Mary deposits are breccia pipe and replacement bodies in dolomite. All the fluorspar from these open-pit mines was used in making cement. At the Carp mine an estimated 45,000 st of metspar was produced in four open pits. The ore came from nearly flat, manto-like replacement bodies, generally concordant with bedding of the dolomite host rock.

NEW MEXICO—There are numerous occurrences of fluorspar in New Mexico but production has been small. Van Alstine (1965) listed 65 areas in the central, north central, and southwestern parts of the state. The fluorspar map MR-60 of the USGS and accompanying tabulation in 1974, by Worl, Van Alstine, and Heyl, classified some 51 locations as to type, tonnage, and associated minerals.

Eight areas each produced at least 20,000 tons of crude ore and cumulative total statewide production has probably been about 700,000 tons. Production peaked in World War II, when New Mexico was the fourth largest producing state after Illinois, Kentucky, and Colorado.

Most production came from the Zuni Mts., Sierra Caballos Mts., Gila district, Burro Mts., Cooks Peak, and from Fluorite Ridge and Tortugas near Las Cruces. Flotation plants have operated at various times at Chise, Silver City, Gila, Lordsburg, Deming, Los Lunas, San Antonio, and near Hachita. A drum type heavy-media separator making metspar is operating at the Great Eagle Mine north of Lordsburg.

In the Zuni Mts., the "27" vein deposits were the most productive, with the Mirabal deposit contributing. Until closed in 1953, these had produced totally about 224,000 tons of ore which was concentrated to acid grade at the Zuni Milling Co. flotation mill at Los Lunas (Goddard 1966).

In the Cooks Peak district near Deming, White Eagle, Greenleaf and others in the Fluorite Ridge district, and Burro Chief and Shrine deposits in the Burro Mts. were important producers. All are vein deposits in brecciated or fractured zones along faults in igneous rocks. The Lyda K Mine, south of Truth or Consequences, is a siliceous vein in Precambrian granite. The Ruby mine north of Las Cruces contains veins in limestone near an igneous contact. At Chise small manto replacement bodies occur in limestone beds. The Bishops Cap deposit is a vein in limestone, with some adjacent replacement ore. In the Hansonburg district, east of San Antonio, fluorite occurs with lead and barite as vug filling in brecciated limestone.

The Salado deposit, southwest of Truth or Consequences, consists of tabular masses of fractured and brecciated jasperoid with the fluorspar forming a filling and cement. Between 2½ and 3 million tons of low-grade mineralization have been estimated from drilling results. A thickness of over 40 ft in places would permit open-pit mining, with little overburden. Difficult flotation metallurgy, low grade, and remoteness from market have inhibited development.

OREGON—Near Rome in the southeastern part of the state, fluorspar occurs as nearly spherical grains in volcanic tuff, tuffaceous mudstone, and mudstone of Tertiary lake deposits (Sheppard and Gude, 1969). The fluorspar content reaches 16% in some zones, but is generally much lower. The tonnage of contained fluorspar is estimated at 11 Mt. Because of very fine particle size, concentration to acceptable grade has not proved practical.

TENNESSEE—A small amount of metspar was produced from a small fissure vein in limestone in the central part of the state, where it occurs with barite and sulfides (Jewell, 1947).

Fluorite has been noted as a gangue mineral in zinc deposits near Carthage, locally comprising 12 to 15% of the ore but averaging much less.

A drilling program in the Sweetwater area indicated fluorspar-zinc-barite mineralization several hundred feet below the surface in buried karst breccia zones similar to the Mascot district, TN zinc deposits. Exploratory drifting was done from a 600 ft shaft sunk in 1979, now plugged.

TEXAS—Metallurgical grade fluorspar production began in 1971 in the Christmas Mts. south of Alpine from bedded limestone replacement mantos adjacent to intrusive rhyolite bodies. In late 1980, a 900 ft decline was sunk beneath a laccolithic structure where exploration and development work continue.

In the Eagle Mts. southwest of Van Horn, fluorspar occurs as replacement of limestone and as fissure deposits in rhyolite. In the period 1942 to 1950, about 15,000 tons of metallurgical gravel and acid grade were produced.

UTAH—Most of the deposits are located in the west-central part of the state in the Spor Mountain, Indian Peak, Pine Grove, and Star districts. The Spor Mountain district in the Thomas Range northwest of Delta has been the most consistent producer. The deposits are classified as pipes, veins, and disseminated bodies in dolomite and volcanic rocks. Most of the production has come from the pipes. The most productive mines have been the Lost Sheep, Fluorine Queen, and Bell Hill.

In the Indian Peak district veins occur along faults in volcanic rocks. The Cougar Spar and Blue Bell mines were World War II producers of metspar. Small production has also been recorded from the Pine Grove and Star districts, and from the Rain Bow mine in Millard County and the Silver Queen mine in Tooele County.

Cumulative Utah production of crude ore to date is estimated to have been 250,000 tons.

WASHINGTON—A small production of metspar has come from the Mitchem mine near Keller, in Ferry County.

WYOMING—There are reports of fluorspar in Crook and Albany counties.

Central America: Occurrences have been reported in Honduras and Guatemala, but details are not available.

South America: *Argentina*—Vein deposits have been mined intermittently in the Sierra Comechigones, 30 miles northwest of Cordoba and others developed for production in the southern part of Rio Negro province and the adjoining northern part of Chubut. In recent years metspar has been produced from a group of veins in the Sierra Grande district of Rio Negro, including the Delta vein, reportedly with 4.0 million tons of drill-indicated ore of 51% CaF_2 grade. A small flotation mill at Valcheta has produced concentrates for domestic use. In Chubut a number of veins have been discovered in areas west of Puerto Madryn.

Brazil—Principal deposits are in the vicinity of Criciuma in the Santa Catarina province of southern Brazil. Small-scale production has been reported from deposits in the provinces of Rio Grande do Norte, Paraiba, and Bahia.

In the Criciuma district, three flotation mills have an aggregate capacity of about 35 kt/a of acid grade concentrate. Known deposits are not large and continued output will depend on the discovery of new reserves and the resolution of water problems.

Chile—Small tonnage of fluorspar has been produced from veins in granite in the Province of Coquimbo. One deposit, the Mercedes mine, is near the village of Paihwano.

Europe: *France*—Important deposits occur in the Departments of Haute-Loire, Pyrenees-Orientales, Var, Tarn, Saone-et-Loire, and Puy-de-Dome. Details have been described in a series of articles by Chermette (1972–1973 and 1979).

Epithermal vein deposits occur in the Morvan, Auvergne, Limousin, and Albi districts of the Massif Central, the Maures and Esterel districts of the Mediterranean coastal area northeast of Toulon, in the eastern Pyrenees west of Peripignan, and in the Vosges. The vein deposits contain about 50% CaF_2, 20% silica, 5 to 10% barite, and 3 to 5% sulfides. Stratiform deposits with about 35% CaF_2 and 15% barite occur in the Morvan district. Depostis having stratiform characteristics are important producers in the Escaro district in the Pyrenees southwest of Perpignan. Deposits at Le Beix, Chavaniac, and Le Maine in the Massif Central are exhausted. La Charbonniere and

Langeac have also closed. Fonte Sante (near Cannes in Var), Le Burc (near Albi in Tarn), Montroc (near Albi), Escaro and Chaillac continue in operation. Fonte Sante is noteworthy for the economic occurrence of sellaite (MgF_2) which is blended with the fluorspar product and is used at the Lyons hydrofluoric acid plant.

Production of all grades of concentrates in 1979 was about 300 kt, divided 177 kt acid grade and about 116 kt metspar. Exports were 9 kt acid grade and 84 kt metspar.

Germany, East—East Germany is estimated to produce about 100 kt/a of fluorspar. Deposits are in Saxony, Thuringia, and Anhalt, but details are lacking.

Germany, West—Production in 1979 amounted to about 63 kt, derived from vein deposits in the Todtnau, Hesselbach, Oberwolfach, and Pforzheim districts in the Black Forest region in the southwest, and in the Naaburg and Sulzbach districts in Bavaria. Acid-grade concentrate is produced in four flotation plants. As a large consumer of both acid grade and metspar concentrates, West Germany imports a large part of its requirements. Imports in 1979 were from South Africa (53 456 t), Spain (33 910 t), Morocco (16 840 t), UK (13 351 t), and Kenya (10 986 t).

Greece—Fluorspar occurs as a gangue mineral in the silver ores of Laurium, but Greece imports its requirements, having produced only 400 t in 1979.

Italy—Italy produced about 183 kt in 1979. Mines in Sardinia are the principal producers. Smaller outputs have come from Brescia, Trento, Bolzano, and Bergamo in northern Italy. The Genna Tres Montis-Muscadroxiu-S'Acqua Frida vein system in southern Sardinia near the village of Silius is credited with 8 million tons of ore reserve averaging 40 to 45% CaF_2, making it one of the world's largest deposits. Vein widths are 5 to 8 m. Barite and galena are recoverable accessory minerals. Underground workings have demonstrated a strike length of about 3 km. Numerous other vein deposits and a skarn deposit occur in southern Sardinia.

Of two flotation plants at Assemini, near Cagliari, one is active with a nominal capacity of 200 kt/a of acid grade concentrate including metspar briquettes, plus barite and galena concentrates. Actual capacity is said to be about 150 kt/a due to restriction of labor regulations. The other, an older mill with acid grade capacity of 40 kt/a, is idle. In 1979, Sardinia produced 107 kt of acid grade from 320 kt of crude ore.

In northern Italy important deposits include Prestavel and Vallarsa near Bolzano, Torgola near Brescia, and Camerata Cornello near Bergamo. In 1979 Societa Fluormine produced 16 kt of acid grade from 71 kt of ore from Prestavel and Vallersa which also yielded 4.7 kt ceramic grade and 19.7 kt metallurgical grade.

The Pianciano deposit in the Castel Giuliano area 40 km northwest of Rome consists of fluorspar impregnating lake beds of volcanic ash (Spada, 1969). The fluorspar content, ranging from 20 to 55%, was derived from volcanic emanations along with barium, strontium, and phosphorous. The fluorspar is too fine-grained to be recovered by flotation, but tests indicate a possibility of producing metallurgical grade by hydrocyclone separation.

Norway—Veins are reported in the Buskerud, Telemark, Vest Agder, and Drammen regions west and southwest of Oslo. The Lassedal mine near Kongsberg was formerly a producer.

Spain—Spain is estimated to have produced about 200 kt of metgrade and acid grade fluorspar in 1980 mainly for export to Europe with smaller amounts to the US and Canada. In 1979 total production was 191 kt, divided 41 kt metspar and 150 kt acid grade. Principal deposits are in the Asturias region of northern Spain near the ports of Ribadesella and Aviles; in the southeast near the port of Almeria; and in south-central Spain near Seville and Cordoba. Deposits are both veins and replacement bodies. Acid-grade material is derived in the retreatment of old lead mine wastes in the Almeria-Berja area. In 1980, seven companies produced acid grade, four in Asturias, two in the Seville-Cordoba region, and one near Berja. Grades of ore range from 40 to 45% for some of the vein-mines, down to 15 to 20% in the case of old lead mine waste. The large Osor vein in Gerona is exhausted. Weakening market conditions in 1982 led to closure, at least temporarily, of all but three mills: two belonging to Minersa, one of which is in Asturias, the other in Almeria; and one belonging to Minas de Orgiva near Berja.

Sweden—The sole producer in Sweden, the government-owned Yxsjoberg mine near Ludvika, 200 km northwest of Stockholm, where fluorspar occurs with tungsten and copper, closed in 1977. Other deposits have been reported between Branteriks and Onslunda in Skane in southern Sweden. The Osterlen district was formerly a small producer.

Switzerland—Occurrences are known, but none of commercial importance.

United Kingdom—Among numerous fluorite occurrences in the United Kingdom, present operations are located in the Southern Pennine Orefield of north Derbyshire and in the Northern Pennine Orefield of west Durham. Production comes from fissure veins and bedded replacement deposits, together with associated old lead mine waste and tailings dumps. The deposits occur in rocks of the Carboniferous Limestone Series (Mississippian). Ore is mined both underground and open pit. The deposits contain variable amounts of galena, sphalerite, calcite, barite, quartz, and iron sulfides. Some were originally mined for lead, even before written record.

Installed capacity for the production of all grades of fluorspar is now in excess of 300 kt/a of which 245 kt is acid-grade capacity. Two acid-grade mill operations are in the Southern Pennine Orefields, and two in the Northern Pennine Orefield. Problems with ore supply have prevented capacity production.

Imports were 28 102 t in 1979 up from 3 783 t in 1978. South Africa provided 16 580 t followed by Morocco (5 100 t) and Mexico (3 203 t).

USSR and Soviet Bloc—USSR is estimated by the US Bureau of Mines to have produced about 560 kt of fluorspar in 1979, and the same in 1980. Details regarding the sources of this output are not available, but deposits are reported in the Transbaikal, Tashkent, and Yaroslav regions. Czechoslovakia produces about 90 kt/a from deposits in Bohemia. Mongolia produced about 410 kt in 1980, having steadily increased from 80 kt/a in 1972. Deposits are southwest of Ulan Bator in the Dorono Gobi district. Romania produces about 20 kt/a; Bulgaria about 25 kt/a, and Yugoslavia has a small production. The USSR imports about 100 kt/a of acid grade and about 400 kt/a of met grade to supply its requirements of more than 1 Mt/a. Sources include China, Kenya, and Thailand.

Africa: *Angola*—Deposits are reported in Angola, but details are lacking.

Kenya—A 100% government-owned acid grade mill, formerly jointly owned with IMC and Bamburi Portland Cement, produces concentrates relatively high in P_2O_5 from epithermal vein and replacement deposits in the Kerio Valley, about 130 km northwest of Nairobi. Ore ranging from 30 to 45% CaF_2 is mined open pit. The acid grade mill's output capacity is rated at 9,000 tons per month. Concentrates are trucked to rail at Kaptagat, near Eldoret, and loaded on vessel at Mombasa. Other similar deposits are reported along the Rift structure.

Malawi—Exploration for fluorspar deposits related to carbonatite structures has been reported; five such occurrences are noted by Van Alstine and Schruben (1980).

Morocco—The El Hammam deposit near Meknes is a nearly vertical vein on an east-west fault in sericite schist, grading about 45% CaF_2. A flotation plant with a capacity of 65 kt/a of acid grade concentrate is operated by Samine, owned 55% by Omnium Nord Africain, 10% by Pechiney, and 35% by the Moroccan government (BRPM). Three other deposits are recognized in this region. Other veins occur in the Djebel Tirremi area, 10 km northeast of Taourirt, in eastern Morocco (Van Alstine and Schruben, 1980).

Mozambique—Large vein deposits occur approximately 150 km southeast of Tete. Remote from transportation and from sources of power, this area has been a minor producer of metspar. Further development may result as the area is opened up as a consequence of the construction of the Cabora Bassa dam on the Zambesi River. Six localities are listed by Van Alstine and Schruben (1980).

Nigeria—Deposits containing fluorspar with galena are reported in limestones in the Arufu district of Gongola State near the border with Benue State, in the Benue Trough about 80 km east of Makurdi, and fluorspar in greisens near Jos (Van Alstine and Schruben, 1980).

South Africa—In 1978 the Republic of South Africa achieved its present (1981) rank as third largest fluorspar producing country, surpassed by Mexico and the USSR. Its large reserves indicate it will continue to be prominent in the world picture. Deposits occur in several areas, but production has come principally from mines in the western and northern parts of Transvaal.

Acid grade output of four flotation plants in 1980 was 469 682 t, up from 387 310 t in 1979. Other production consists of metallurgical grade fluorspar, some of which is hand-cobbed *kokoman* residual ore derived from small mines, and small amounts of ceramic grade. Acid grade production is exported to Japan, Europe, Australia, and the United States through the port at Durban. In 1980 total production of all grades was almost 523 kt.

Domestically, AECI consumes acid grade to make hydrofluoric acid and the steel industry consumes essentially all of the metgrade as well as about 20 kt/a of 80% CaF_2 briquettes.

The Buffalo open pit mine of Transvaal Mining & Finance, a subsidiary of General Mining and Finance, near Naboomspruit exploits 14 to 16% grade deposits consisting of a network of fluorspar veinlets in sills in a pink granite which is itself an inclusion in coarse Bushveld granite. Similar deposits occur nearby.

Near Warmbad, the Zwartkloof deposit consists of three breccia zones containing a stockwork of fluorspar veinlets in felsite. Operating difficulties and the low grade of the ore for underground operation, approximately 14% CaF_2, caused operations to close in 1973.

At the open pit Vergenoeg mine of Chrome Chemicals, a subsidiary of Bayer, near Pienaarsrivier, fluorspar ore grading 40% occurs with abundant iron oxides (up to 50%) in a massive deposit in felsite.

In the Marico-Zeerust district, fluorspar occurs as replacements in dolomite and as associated residual deposits. Much of the ore in dolomite is disseminated along intersecting fractures. For many years metspar was produced by screening coarse fragments from residual surface deposits. Exploration has disclosed the existence of large reserves of primary ore probably on the order of 60 million tons containing 15 to 30% CaF_2. Two open pit mines, the Chemspar mine of Phelps Dodge and the Marico mine of Barlow Rand, with capacity of about 120 kt/a, are located here. Armco Bronne, a subsidiary of Armco, Inc., has an active exploration program and expects to develop an underground operation by 1985.

Namibia (South-West Africa)—The Okurusu deposit near Ojiwarongo consists of replacements or segregations in carbonatite and veins in quartzite. After yielding a few thousand tons of metspar, it has been idle. Reserves are indicated to be between 6 and 7 million tons grading 40% CaF_2. Vein ore near Omaruru was mined open pit and concentrated to acid grade in 25 kt/a flotation mill from reserves in excess of 600,000 tons. Vein and replacement deposits are known in other parts of the country. The lack of water and transportation facilities has hindered development.

Sudan—Fluorspar deposits occur at Jebel Semeih and Jebel Dumbeir approximately 200 miles southwest of Khartoum and 830 miles by rail from Port Sudan. At Jebel Semeih a highly siliceous fluorspar vein occurs in a granitic host rock. At Jebel Dumbeir fluorspar occurs in metasediments as replacements of bands of marble, as thin quartz-fluorite veins, and as bodies of fluorite associated with red feldspathic rock. No production is reported.

Tunisia—A flotation mill at the Hamman Zriba deposit near Zaghouan, 54 km south of Tunis, produces about 30 kt/a of acid grade concentrate from a bedded deposit in Jurassic limestone. Proven reserves are said to be 4 million tons of ore containing 35% CaF_2 and considerable barite. In the same general area, replacement vein deposits occur at Djebel Staa, Hammam Djedida, Djebel Oust, and Djebel el Kohol. Other deposits are at Bourchiba, on Cap Bon, and Bou Jaber.

Turkey—Fluorspar deposits in Turkey have been described (Anon., 1965); there is a very small annual production.

Asia: *China*—China is emerging as a major producer with an annual production of about 410,000 t, mainly from deposits in the central and northern parts of Zhejiang Province, northeastern Hebei Province, northern Guangdong Province, northeastern Anhui Province, and eastern Shandong Province. Other production is at Taoling in Hunan Province, Ming Gang in Henan, at Kaiping in Liaoning Province, and at Yulin in Guangxi Province (a new discovery near rail). In all, fluorspar occurrences are reported in 17 provinces. China exported 105 020 t to Japan in 1978 and 205 489 t in 1979, mainly metallurgical grade. Acid grade production is increasing with the conversion of sulfide flotation mills to acid grade, including one in northeastern Hebei and one in eastern Shandong, and the expansion of existing acid grade mills in Henan and Zhejiang.

India—At Amba Dongar, 350 miles north of Bombay in Gujarat State, veins and replacement bodies occur in the carbonate wall rock around ankeritic carbonatite intrusives, similar to the occurrences at Okurusu in Namibia (South-West Africa) (Deans, et al., 1972). A heavy media and flotation concentration plant has been constructed nearby at Kadipani, with a nominal capacity of 37.5 kt/a of product, of which 30% is acid grade and 70% metgrade. Reserves are reported to be 11.6 Mt of 30% fluorite content, of which proved reserves are said to be 4.7 Mt.

Other mineralized areas are reported at Chandri-Dungri in the Drug district of Madhya-Pradesh and at Mando-ki-Pal in the Dungarpur district of Rajasthan.

Japan—Japan formerly produced less than 10 kt/a of fluorspar from the Island of Hokkaido but ceased production in 1972 and imports its requirements.

Korea, North—This country has an annual production of about 40 kt.

Korea, South—Fluorspar deposits occur in the northern, central, and southern parts of the country. Approximately 70 mines have been producers, formerly with annual production of about 50 kt, largely for export (Kim et al.,

1972), but production was down to 8 kt in 1979. Both vein and replacement deposits occur. Minable ore reserves have been estimated at 660 kt (Jee and Kye, 1971). Producers have included the Choonchon Shinpo, Dojon, Kumi, Busang, and Kumbo mines.

Pakistan—Minor production has been reported from the Kalat district 45 to 60 miles south of Quetta in Baluchistan. There are also deposits south of Quetta.

Thailand—After the initiation of commercial fluorspar production in 1960, Thailand advanced to the status of one of the largest producers in the world. In 1979 production was 260 kt, comprising 178 kt metspar and acid grade. Exports totaled 205 kt, largely to Japan (116 kt) and the USSR (50 kt). Acid grade exports were 49.6 kt of that total, of which 39 kt went to Japan; 5 kt to India and 4.5 kt to Australia. In 1980, production reached 306 kt of which 173 kt was metspar and acid grade and 133 kt was low grade. Exports totaled 214 kt, 117 kt to Japan and 47.5 kt to the USSR. Acid grade exports were 60 kt of which Japan received 40 kt; the USSR, 10 kt; India, 5 kt; and Australia 3.5 kt.

Fluorspar deposits occur in a region from the northern border with Burma southward for over 800 miles to a point about 400 miles southwest of Bangkok. Active mines are concentrated in three areas: Chiangmai, Lumphun, and Mai Hongsun Provinces in the north; Kanchanaburi, Petchaburi, and Ratchaburi Provinces southwest of Bangkok; and Krabi Province in the south (Gardner and Smith, 1965; Anon., 1972; and private communication, B. L. Hodge, 1981).

The northern deposits are mainly replacements in Paleozoic limestones which, so far, have been found suitable only for the production of metspar, derived by hand sorting with large resulting waste dumps containing low grade ore. Residual deposits are widespread. A heavy media concentrating plant, with a 60 kt/a production capacity originally installed at Ban Hong in Lamphun Province in the northern area, 320 miles north of Bangkok, has been moved a short distance to Mae Tha. The product is trucked to river barges at Bank Pa In. In the southern area, the deposits are veins in metamorphosed, fine-grained, fractured Paleozoic clastics and in granite. Two flotation plants are operating, one at Ban Lard in Petchburi Province, about 100 miles south of Bangkok, with an annual production of 50 kt of acid grade concentrate and another in Khlong Thom district, Krabi Province, about 60 miles northwest of Trang, with an annual production of about 15 kt of acid grade.

Australia: Deposits of fluorspar have been found in all the Australian states except the Northern Territory (Liddy, 1971), but are undeveloped because they are small and remote. There have been investigations in the Chillagoe-Mungana-Almaden and Forsayth areas of northern Queensland, and in Western Australia in the Pilbara district 135 miles southeast of Port Hedland, the Yinnietharra area 200 miles from Carnarvon, and the Speewah Valley area southwest in Wyndham.

World Activity in Fluorspar

Overall, reported reserves of fluorspar have been increasing for several decades along with increasing annual production as producer interest in fluorspar has grown, either to fill a need for captive raw material or as a marketing opportunity. In the late 1970s, rising prices spurred new mine development and the reopening of some of those previously closed. Growth was particularly noteworthy in South Africa and China and revitalization in Spain, but recession in the early 1980s led to retrenchment with mines being closed, at least temporarily, especially in Spain and Mexico.

Exploration

Because fluorspar resists chemical weathering, it can be traced in the soils overlying weathered veins. Cleavage fragments washed clean by rainwater or exposed in anthills or spoil piles from animal burrows are useful clues. Silicified veins resisting erosion may stand up like a "reef." Bedded deposits are less discernable, but in areas of sufficient topographic relief the outcropping edges of deposits or slumped fragments of ore may be found. Owing to its softness and cleavability, fluorspar does not survive in the beds of streams and ordinarily cannot be traced by panning.

The search for bedded deposits usually involves locating a mineralized horizon and following it with vertical drill holes drilled to intersect the favorable beds.

Geophysical prospecting methods are not applicable. Photogeology can be of assistance in locating and tracing structures and outcrops, as well as the areal distribution of stratigraphic units known to be favorable for veins or replacement deposits.

Geochemical methods have been employed with varying degrees of success. Fluorine anomalies have been found in ground waters and surface streams, in soil samples and in stream sediments, but with little practiced success in discovery of ore deposits. Various nonfluorine elements detectable geochemically can be guides to fluorspar deposits.

Once mineralization has been detected, prospect shafts, drifts, and core or churn drill holes are used to probe the structures.

Preparation for Markets

Mining Methods

In some areas, such as in the Illinois-Kentucky district and the Southern Pennine district of Derbyshire in England, veins may be weathered to depths of as much as 250 ft. Such weathered ore, a mixture of clay and fragments of fluorspar and detached wall rock, may be mined open pit with draglines, scrapers, or power shovels to depths of as much as 150 ft. Below that, underground mining methods are used, involving modified top slicing or overhead shrinkage stoping.

Until the late 1960s, most vein mining in the Illinois-Kentucky district was done by shrinkage stoping and open stoping where strong walls occurred. Closely spaced shrinkage stope bins have given way to widely spaced bins, with electric and air slushers being used in the tops of stopes to transport the overbreak to the ore pass. Air-operated rubber-tired muckhaul units have been adapted to in-stope work. Where shrinkage stoping is used, broken ore is commonly moved to the shaft by track haulage using battery locomotives and one or two-ton side dump cars.

With the development of diesel haul units of less than 5-ft width, one Illinois company changed from shrinkage stope mining to ramp subleveling in veins, serving the loaders which traversed the sublevels and ramps with small diesel trucks carrying three to four tons. Output per manshift improved, with the elimination of labor-intensive timbering, chute building, and track work. Ventilation for the diesel equipment is usually handled by lines of woven plastic tubing. In shafts, bucket hoisting has been supplanted by skips with guides in all instances. In larger mines, crushers are installed over skip loading pockets at the shaft bottom, improving skip loading.

In bedded deposits, room-and-pillar patterns are used, with the widths of rooms governed by roof conditions. Roof bolting has supplanted wooden supports. Most equipment is rubber-tired and diesel-operated, including the muckhaul units which have buckets ranging in size from 1 to 5 yd and rubber-tired diesel trucks ranging in size from 3- to 18-ton capacity. Drilling is done by diesel-propelled jumbos in the bedded ore mines, but the jack-leg drills are still used in narrower working places and drifts. In multileveled ore bodies, haulage ramps on 12 to 15% grades connect the levels. Vertical raises are used to facilitate ventilation requirements. Most drilling uses tungsten carbide bits, or the throwaway type hardened steel bits. In blasting the trend is increasingly toward the use of ammonium nitrate-fuel-oil mixtures. Mine crews, supervisors, and mechanics are usually provided with diesel personnel vehicles to facilitate mobility.

Wherever widths of ore bodies, depths of overburden and the strength of the sidewalls permit, it is common practice to mine by open pit methods, as is the case for all mines now producing in South Africa. This optimizes the advantages to be gained by using lower-cost explosives, rubber-tire earthmoving equipment, and economical quarry bench mining methods.

Beneficiation

Most fluorspar must be upgraded for marketing. It is sold in a range of calcium fluoride contents, and the beneficiation techniques used are tailored to meet the character of the ores and the specifications for particular uses.

Metallurgical spar with a calcium fluoride content in the range of 75 to 85% is often produced by the hand-sorting of high grade lump crude ore, followed by crushing and screening to remove most of the —10-mesh fraction. In the case of ores of lower grade, and/or ores with relatively coarse interlocking of minerals, gravity processes of concentration are used for producing metspar based on the specific gravity of 3.2 for spar and less than 2.80 for most gangue minerals.

Heavy media cone and drum separators are particularly effective in the size range of 1½ x 3⁄16 in., either for producing metallurgical gravel or for preconcentrating the crude ore for flotation feed. For the finer sizes the heavy media cyclone process is frequently used. The high tonnage capacity and low operating costs of heavy media methods give very satisfactory results. The "sink" can be sold as metspar, the "sands" for flotation feed, and the "float" frequently is marketed for road gravel and concrete aggregate. Ores as low as 14% of calcium fluoride are being preconcentrated to yield a flotation feed of 40% CaF_2 or more. Lead and zinc sulfides and barite, by reason of their high gravity, concentrate with the fluorspar to enrich the flotation feed in these valuable minerals. Many ores which would otherwise be too low grade to warrant mining and processing can be preconcentrated to acceptable flotation feed grade with a tailings loss of less than 12% of the CaF_2 in the crude ore. Washing plants are also used ahead of flotation to remove clay or wad (manganese oxides) in some areas.

Barite, a fairly common accessory mineral in fluorspar deposits, in most instances can be effectively separated from the fluorspar by flotation and made into a salable product.

As a consequence of the decreasing supply of natural metspar gravel, the agglomeration of fluorspar products by pelletizing and briquetting has expanded. Feed for agglomeration covers a variety of products such as gravel fines or screenings and various grades of flotation concentrates.

Ceramic and acid grades of fluorspar are produced by the froth flotation process, which has reached a high level of metallurgical efficiency. In many cases ceramic grades with a calcium fluoride content in the range of 90 to 96% can be made with good recovery from ores which otherwise would be difficult to concentrate to 97% CaF_2 acid-grade spar. Ore from the mine is crushed and ground to proper size, care being taken not to overgrind and cause fluorspar to be lost in the slimes. If sulfides are present they are floated off with xanthate collector in separate circuits ahead of the fluorspar, first the lead sulfide and then the zinc sulfide. Then all the easy-floating fluorspar is removed in a quick pass through a rougher flotation circuit and sent on to the cleaner circuit; the rougher tailing is then discarded. The middling product is reground to separate the more finely interlocked grains of fluorspar and gangue, and these are then separated in one or more cleaner circuits. The final products generally comprise an acid-grade concentrate and in some cases one or more concentrates of lower grade which are sold as ceramic grade or pelletized and sold as metallurgical grade. Fatty acids are used as collectors for the fluorspar. Quebracho or tannin is used to depress calcite and dolomite; sodium silicate to depress iron oxides and silica; and chromates, starch, and dextrin to reject barite. Cyanide is used to depress sulfides. Lime, caustic, or soda ash can be used for pH control. Flotation temperatures range from ambient to heating to 25 to 80°C.

A standard method for determination of calcium fluoride in fluorspar is being promulgated by ASTM under Designation E 815–81 wherein the sample is digested with nitric and perchloric acid and the fluorine is expelled by fuming. The residue is dissolved in dilute HCl, the solution made alkaline, and the calcium titrated with standard EDTA solution. Calcium present as carbonate is determined in a separate sample after extracting with dilute acetic acid and a correction for calcium fluoride, solubilized by dilute acetic acid digestion, is applied by determining the fluoride in the acetic acid extract by fluoride ion-selective electrode. The CaF_2 content is then calculated.

Neutron activation is used as an analytical method in determining calcium fluoride in ores and mill products having a content of less than 95%. The method is rapid, and is a particularly useful tool in analyzing ore samples and for mill control. The X-ray fluorescence analyzer is used to determine the abundance of other minerals occurring with the fluorspar such as barite, sphalerite, and galena.

Transportation

Most acid grade fluorspar in world markets is shipped in bulk by boat, rail, or truck in the form of damp filter cake containing about 8 to 10% moisture, because damp spar is easier to load and unload, and dust losses are minimized. This moisture is removed by heating in rotary drying kilns or other types of dryers before treating with sulfuric acid to make hydrofluoric acid because the fluorspar has to be practically bone dry when it goes into the acid reactors. Some acid spar is dried at the point of origin and shipped in covered hopper cars or tank trucks to consumers. A relatively small amount of acid grade spar is dried and bagged for shipment to purchasers of small quantities for specialty uses.

Metallurgical grade fluorspar is shipped as lump or gravel, in barges, ships, or railroad cars. Ceramic grades of fluorspar are usually marketed in bags; larger ceramic users, however, buy dried concentrates shipped in bulk by rail.

Markets, Production, and Consumption

Among market-economy countries, the eight principal producers in 1979 were, in descending order, Mexico, South Africa, France, Thailand, Spain, Italy, United Kingdom, and the US, each of which produced more than 100 kt. Total market-economy production was about 2.8 Mt. World production of all grades approached 4.6 Mt. The principal consuming countries in 1979 were the United States, the USSR, Japan, West Germany, the United Kingdom, Canada, Italy, and France. Because of the disparities in production vs. consumption, the principal importing countries in 1979 were the United States, Japan, the USSR, West Germany, and Canada. Major exporting countries were Mexico, South Africa, Spain, Thailand, Mongolia, France, and China.

The United States relies heavily on imports, especially of acid grade and metspar and does not produce enough ceramic grade to meet full requirements. In 1979, 369 279 t were mined and 322 650 t were treated to produce 96 253 t, compared to reported consumption of 1 030 086 t. Imports in 1979 amounted to 926 kt and declined about 15% to 784 kt in 1980. US consumption in 1980 was 861 kt, about 16% lower than the 1.03 Mt consumed in 1979. Figures of mined, treated, and finished production for 1980 were withheld by the US Bureau of Mines. World production in 1980 reached about 4.7 Mt, a modest increase over 1979.

In 1980, nearly 80% of the domestic US fluorspar production was mined and processed in Hardin and Pope counties, Illinois, in the mills of Ozark-Mahoning Co. and Inverness Company. Ozark's flotation mill, at Rosiclare, IL, is served by the Illinois Central Gulf railroad and by a barge loading facility on the Ohio River. Preconcentration by heavy media is done by Ozark-Mahoning adjacent to several bedded replacement mines northwest of Cave in Rock, IL, and at the new Knight shaft, west of Rosiclare. Metspar and flotation feed are produced at the heavy media separation facilities. In flotation, lead and zinc sulfide concentrates are produced in addition to acid grade fluorspar.

Inverness operates its principal flotation mill 5 miles north of Cave in Rock, adjacent to the No. 1 mine, the largest fluorspar mine in the US. Crude ore is preconcentrated by heavy media separation. The mill produces ceramic spar, lead and zinc sulfide concentrates, and a high grade barite concentrate. The smaller Crystal mill, located northwest of Cave in Rock, is closed. Concentrates are shipped by truck to rail-loading facilities at Rosiclare and Junction, IL, or to Cave in Rock, IL, for barge loading. Ore from Inverness' Spivey vein-mine in northern Hardin County is hauled to the No. 1 mill for treatment.

In Kentucky, Cerro Corp. constructed a heavy media and flotation plant in 1974, with a stated capacity of 60,000 stpy of acid grade flotation concentrate, at the Babb-Barnes property north of Salem, KY. It is now owned by Frontier Spar, a subsidiary of Marathon Oil, and is idle. The flotation mill of Pennwalt Corp., near Mexico, KY, with a capacity of about 35,000 stpy of flotation products, was idled in 1973 following mine exhaustion and was subsequently briefly leased to Armco but is now idle. There are several small mines in both Kentucky and Illinois, whose operations are intermittent, that sell ore to the established mills. Drill-indicated ore on several properties in Kentucky is not now being produced.

In Colorado, Ozark-Mahoning Co. operated a flotation mill until late 1973 adjacent to its Northgate mine, one of the nation's largest, near Cowdrey. This mill, with a capacity of 50,000 stpy of acid grade spar, is now idle. At various times, Ozark has operated other vein mines in the area to supplement Northgate production.

Roberts Mining Co. processed ore from its Crystal Mountain mine near Darby, Montana, in a heavy-media plant, producing only metspar. Reynolds Metal Co. operates a flotation plant at Eagle Pass, Texas. For mill feed it uses ores and metallurgical screenings of Mexican origin. Small flotation mills, with short operating histories, have been operated in Arizona, New Mexico, Colorado, Utah, and Nevada. J. Irving Crowell at Beatty, NV produces metspar.

Briquetting presses are operated by Ozark-Mahoning Co. at Rosiclare, IL; Cametco Corp. at Duquesne, PA; Mercier Corp. at Dearborn, MI; and Oglebay-Norton Co., Pennwalt Corp., and Delhi Foundry Sand Co. at Brownsville, TX.

Flux bricks containing varying amounts of fluorspar and other additives, usually patented formulations, are produced by Cleveland Flux Co. in Cleveland, OH, and Allen Park, MI; by Mercier Corp. in Dearborn, MI; and by Glen-Gerry Shale Brick Corp. at Reading, PA.

Most of the metspar screenings used in briquettes and flux brick are imported from Mexico. The imported flotation concentrates used for briquetting are usually filter cake of medium moisture content and are always acid grade because of the lower tariff on imported material assaying over 97% CaF_2. Domestic flotation concentrates used in metspar briquettes are normally below acid grade and contain only enough CaF_2 to achieve acceptable grades.

Binders used in making roll-compacted briquets include molasses-lime mixtures (with self-curing properties), tall oil derivatives and lignin paper mill byproducts, the latter two requiring oven curing. Other binders have practical applications, both in the US and foreign areas. Sodium silicate was used in making disk pellets. Most of the flux bricks use portland cement as a binder.

Future Considerations and Trends

Fluorspar Substitutes in Metallurgical Use

A number of substitutes for the traditional metspar gravel of natural origin have been proposed and investigated from time to time (Buxton and Sandaluk, 1972). From the worldwide viewpoint, high grade sources of natural metspar are declining and most of the crude ores currently available tend to be of lower CaF_2 content and require flotation to separate the fluorspar from the other minerals in the ore. Rising fluorspar prices and shipping costs have stimulated the search for substitutes, as has the need to reduce fluoride air and water pollution at points of consumption. Investigations have taken three general courses: (1) making artificial metallurgical gravel by forming flotation concentrates into pellets and briquettes, with or without the admixture of natural gravel spar screenings; (2) developing substitute nonfluoride fluxes; and (3) making pellets from mixtures of fluorspar and substitute fluxes.

Steel company research groups and governmental agencies have examined various aluminum-bearing materials including bauxite and the "red mud" residue from the manufacture of alumina, dolomite, olivine, serpentine, manganese ores, ferromanganese slags and dusts, titanium ores such as ilmenite and natural ilmenite-hematite mixtures, borates (including borax, colemanite, and rasorite), soda ash, and various iron residues including mill scale and dust.

Mixtures of iron, aluminum, and calcium oxides are used as extenders for fluorspar, to reduce the poundage of fluorspar required per ton of steel produced. In the event that a serious shortage of fluorspar did develop, the work done to date indicates substitutes would be available.

Phosphate-Derived Fluorides

Natural phosphate rock contains approximately 3.5% fluorine as a component of the mineral fluorapatite. In phosphate processing, the fluorine volatilizes as silicon tetrafluoride, which hydrolizes to fluosilicic acid when the gas is passed through water scrubbers. Currently most of this acid is neutralized with lime. However, some fluosilicic acid is converted to aluminum fluoride and cyrolite (Steininger, 1972) and several plants in the US are recovering fluorine values in this fashion. Alcoa's plant in Fort Meade, FL, completed in 1970, converts fluosilicic acid to aluminum fluoride, and at Mulberry, FL, Kaiser Aluminum and Chemical Co. converts fluosilicic acid to sodium fluosilicate, which is converted to aluminum fluoride at Kaiser's plant at Chalmette, LA.

The quantity of fluorine present in phosphate rock deposits is enormous (Worl, et al., 1973). Current US identified reserves of phosphate rock are estimated to be 10.5 billion tons, which at 3.5% would contain approximately 370 million tons of fluorine, equivalent to about 740 million tons of fluorspar. In existing operations the quantity of fluorine which is recovered varies widely, but in general may be in the range of 25% to 35% of the amount present in the processed rock. According to Johnson, et al. (1973), if all the recoverable byproduct fluorine from phosphate rock processed in the US in 1968 had been available for domestic consumption, it would have supplied 18% of the total demand. Present high prices for fluorspar which increased 60% in one year's time, January 1979 to January 1980, have accelerated interest in fluosilicic acid.

Depletion and Duties

The depletion allowance is 22% for domestically mined fluorspar and 14% for foreign operations of US companies.

The duty on imported fluorspar, is $2.10 per long ton on fluorspar which assays 97.0% CaF_2 or more (acid grade), and is 13.5% ad valorem for all lower grades and ores, most of which are metallurgical grade.

The Federal Emergency Management Agency, through the General Services Administration (GSA), maintains a strategic stockpile of fluorspar. National defense stockpile objectives as of September 30, 1980 were 1,400,000 st acid grade and 1,700,000 st of metallurgical grade compared to an inventory of 895,983 st acid grade and 411,738 st metallurgical, a shortfall of 504,017 st and 1,288,262 st respectively.

Changes in Patterns of Fluorspar Consumption

In recent years there has been a trend in the aluminum industry toward lower consumption of fluorine salts per ton of metal produced. Aluminum companies are recovering and recycling fluorine values from pot linings and fumes, larger pots are being used which absorb proportionately less fluoride electrolyte, more aluminum fluoride and cryolite are being manufactured from fluorsilicic acid, and greater overall operating efficiency is being achieved.

Looking ahead, even greater reductions may result from two developments announced by the Aluminum Co. of America. One is a patented process in which up to 99% of the fluoride values in fume and particulate material in the off-gas from the pot lines is recovered by adsorption on a fluidized bed of ordinary alumina which is subsequently used to make aluminum metal. The other is a process for making aluminum which requires no fluorides at all (Anon., 1973). In this system, alumina is reacted with chlorine, and the resulting aluminum chloride is electrolytically processed to yield aluminum metal and chlorine. This process is said to reduce electrical energy requirements by 30%, to be free of undesirable emissions, and if piloting is successful, may be used in new smelters.

Prior to the introduction and widespread adoption of the basic oxygen steel furnace (BOF), there had been a gradual decrease in consumption of fluorspar per ton of steel produced. With the arrival of the BOF, however, fluorspar consumption increased approximately two to three times above the consumption in basic open hearth furnaces. Use is growing of fluorspar concentrates of higher CaF_2 content, which permits closer control of the reactions in the furnace during the short heating cycle characteristic of the BOF process, and the introduction of smaller amounts of possible undesirable contaminants. Current average use is about six pounds of CaF_2 per ton of steel in the US.

Another new development in the steel industry is the bottom oxygen-blowing process which has been applied to both open hearth furnaces and oxygen converters. In both applications, provision is made for injecting hydrocarbon fuel, oxygen, and pulverized fluxes such as fluorspar and lime into the base of the steelmaking vessel below the surface of the molten metal. It has been claimed that fluorspar consumption per ton of steel may be reduced by as much as 50% in addition to various other benefits. If this method becomes widely adopted, the requirement for lump metallurgical spar will drop and specifications will call for a finely ground spar, either ceramic or acid grade.

Over the years ahead, increasing industrialization around the world will lead to increased fluorspar production and consumption, but changes can be expected in unit consumption in accordance with new developments, the desire to cut costs, and in response to environmental controls.

Ecology

Environmental restrictions and controls are affecting the fluorspar industry. Meeting official standards will require additional investment and cause significant increases in cost of production.

The disposal of mill tailings creates a problem because even after impounding and clarification the liquid effluents may contain traces of chemicals that are in violation of environmental standards. The trace amount of fluoride ion concentration resulting from the natural solubility of calcium fluoride is itself of concern.

Dust emissions from drying and dry handling equipment can be controlled, and it is desirable from an economic standpoint to reduce the loss of valuable product in the form of dust, as well as for environmental control.

Cryolite

Natural cryolite (Na_3AlF_6) is a colorless mineral containing 54.3% fluorine, 12.8% aluminum, and 32.9% sodium. It has a hardness of 2.5, a specific gravity of 2.96, and crystallizes in the monoclinic system. Its mean index of refraction is about 1.339, which is close to that of water.

Cryolite has been found in commercial quantities only at one place in the world, Ivigtut, West Greenland, although small economically unimportant occurrences have been found in the St. Peter's Dome District near Pike's Peak in Colorado, and in the USSR, Spain, Canada, and Nigeria.

The Greenland deposit is a pegmatite in porphyritic granite in which the cryolite is accompanied by fluorspar, siderite, pyrite, arsenopyrite, galena, topaz, molybdenite, and a number of rare aluminum fluoride minerals. Approximately 90 different minerals have been found in the deposit, several of which have not been found elsewhere. The deposit lies on the shore of the Arsuk Fiord on the southwest coast of Greenland. It was worked by open pit and was mined out in 1962 (Maersk, 1973) after approximately a century of operation. Cryolite was stockpiled at Ivigtut for shipment to a concentrating plant in Copenhagen operated by Kryolitselskabet Oresund, which is partly owned by the Danish government. In this plant the crude ore was concentrated by flotation to a product containing 98 to 99% cryolite. For 100 years crude ore also was shipped to the United States for processing by Pennsalt Chemicals Corp., but this arrangement terminated in 1965.

Originally cryolite was used mainly in the production of soda, alum, and aluminum sulfate, but this eventually became unprofitable, and in 1889 the mineral assumed a new and important role as the molten bath in which alumina was dissolved for electrolytic reduction to aluminum metal by the Hall process. Demand for cryolite in the aluminum industry outstripped the natural supply and the gap was filled by synthetic cryolite manufactured from hydrofluoric acid, sodium carbonate, and aluminum hydrate. In the ceramic industry, cryolite is used as a whitener for enamels and an opacifier in glasses. It is also used in bonding grinding wheels and abrasives, and as an ingredient in welding rod coatings. Very finely ground cryolite is an active agent in some insecticide mixtures.

Acknowledgment

The authors appreciate valuable comments on the manuscript by Dr. B. L. Hodge of B. L. Hodge and Partners.

Bibliography and References

Anon., 1965, "Barytes and Fluorite Deposits of Turkey," Maden Tetkik Arama Enstit. Yayinl, No. 126, Ankara, 11 pp.

Anon., 1972, Annual Report, Lost River Mining Corp.

Anon., 1972a, "Progress in Modernizing Thailand's Fluorspar Industry," *Industrial Minerals,* No. 56, May, pp. 31–33.

Anon., 1973, "Alcoa Announces Development of Pollution-Free Aluminum Smelting Process," *Engineering & Mining Journal,* Vol. 174, No. 2, Feb., p. 30.

Anon., 1973a, "Trade and Trends in Fluorspar," *Industrial Minerals,* No. 64, June, pp. 23–31.

Anon., 1973b, "Thailand Flotation Mill Now," *World Mining,* Vol. 20, June, pp. 24–25.

Anon., 1978, "Samuk Enters the Major Fluorspar League," *Industrial Minerals,* No. 135, Dec., pp. 39–43.

Agricola, G., 1546, *Bermannus Sive de Re Metallica; De Re Metallica,* A. Hoover and H. Hoover, trans., Dover Publications, New York 466 pp.

Anderson, A.L., and Van Alstine, R.E., 1964, "Fluorspar," *Mineral and Water Resources of Idaho,* 88th Congress, 2nd Session, US Government Printing Office, Washington, DC, pp. 79–84.

Argall, G.O., Jr., 1949, "Fluorspar," in "Industrial Minerals of Colorado," *Quarterly of the Colorado School of Mines,* Vol. 44, No. 2, pp. 179–208.

Bauxton, F.M., and Sandaluk, P.A., 1972, "Fluorspar Substitutes in Steelmaking," reprint, American Iron and Steel Institute, Regional Technical Meeting, Pittsburgh, Nov. 9.

Baxter, J.W., et al., 1973, "A Geologic Excursion to Fluorspar Mines in Hardin and Pope Counties, Illinois," Guidebook Series II, Illinois State Geological Survey, 28 pp.

Chermette, A., 1950, "L'Exploitation du Spath-Fluor en France de 1938 á 1946," *L'Echo des Mines et de la Metallurges,* Vol. 427, No. 3, pp. 547–549.

Chermette, A., 1964–65, "Les Ressources du Mexique en Spath-Fluor," *Mines et Metallurgie,* Oct., pp. 467–469; Nov., pp. 519–521; Dec., pp. 571–574; Jan., pp. 25–27; Feb. pp. 77–79.

Chermette, A., 1972–1973, "Un Demi-Siècle de Spath-Fluor Français (1922–1972)," *Mines et Metallurgie,* Oct., pp. 179–182; Nov./Dec., pp. 201–204, 213; Jan./Feb., pp. 9–12, 14; Mar., pp. 8, 11; Apr., pp. 10–13; May, pp. 9–16.

Chermette, A., 1979, "Le Spath Fluor est a un Tournant," *Mines et Metallurgie,* No. 145, pp. 14–19 and No. 146, pp. 20–24.

Chesterman, C.W., 1966, "Fluorspar," *Mineral Resources of California,* Bulletin 191, California Div. of Mines and Geology, pp. 165–168.

Coope, B., 1978, "Fluorspar—Down But Not Out," *Industrial Minerals,* No. 125, Feb., pp. 39–66.

Cornwall, H.R., 1972, "Geology and Mineral Resources of Southern Nye County, Nevada," Bulletin 77, Nevada Bureau of Mines and Geology, 49 pp.

Dasch, M.D., 1964, "Fluorine," *Mineral and Water Resources of Utah,* Bulletin 74, Utah Geological and Mineralogical Survey, pp. 162–168.

Deans, T., et al., 1972, "Metasomatic Feldspar Rocks (Potash Fenites) Associated with the Fluoride Deposits and Carbonatites of Amba Dongar, Gujarat, India," *Transactions, Sec. B,* Institution of Mining & Metallurgy, Vol. 81, Bulletin 783, p. B6.

Dunham, K.C., 1952, "Fluorspar," Memoir, Special Report on Mineral Research, Geological Survey of Great Britain, Vol. 4, 143 pp.

Elevatorski, E.A., 1968, "California Fluorspar," *Mineral Information Service,* State of California, Vol. 21, No. 9, pp. 127–130.

Elevatorski, E.A., 1971, "Arizona Fluorspar," Arizona Dept. of Mineral Resources, 51 pp.

Ford, T.D., and Ineson, P.R., 1971, "The Fluorspar Potential of the Derbyshire Ore Field," *Transactions, Sec. B,* Institution of Mining and Metallurgy, Vol. 80, pp. B185–205.

Funnell, J.E., and Wolff, E.J., 1964, "Fluorspar (Fluorite)," *Compendium on Nonmetallic Minerals of Arizona,* prepared for Arizona Public Service Co. by Southwest Research Institute, pp. 105–114.

Gardner, L.S., and Smith, R.M., 1965, "Fluorspar Deposits of Thailand," Report of Investigation No. 10, Dept. of Minerals Resources, Bangkok, 42 pp.

Gillerman, E., 1953, "Fluorspar Deposits of the Eagle Mountains, Trans-Pecos Texas," Bulletin 987, US Geological Survey, 98 pp.

Goddard, E.N., 1966, "Geologic Map and Sections of the Zuni Mountains Fluorspar District, Valencia County, New Mexico," Map I-454, Miscellaneous Geologic Investigations, US Geological Survey.

Gossling, H.H., 1972, "A Review of the World's Fluorspar Industry, with Particular Reference to South Africa," Report No. 1424, National Institute of Metallurgy, Johannesburg, 30 pp.

Gossling, H.H., 1978, "Fluorspar, 1973–1980: A Commodity Profile," Report No. 3, Mineral Bureau, Dept. of Mines, Republic of South Africa, p. 36.

Grogan, R.M., and Bradbury, J.C., 1967, "Origin of the Stratiform Fluorite Deposits in Southern Illinois," *Genesis of Stratiform Lead-Zinc-Barite-Fluorite Deposits,* Monograph 3, The Economic Geology Publishing Co., pp. 40–51.

Grogan, R.M., and Bradbury, J.C., 1968, "Fluorite-Zinc-Lead Deposits of the Illinois-Kentucky Mining District," *Ore Deposits of the United States, 1933–1967,* Vol. 1, J.D. Ridge, ed., AIME, New York, pp. 370–399.

Guccione, E., 1972, "What's Going on in the Fluorspar Industry," *Engineering & Mining Journal,* Vol. 173, Dec., pp. 64–72.

Hodge, B.L., 1970, "The U K Fluorspar Industry and Its Basis," *Industrial Minerals,* No. 31, Apr., pp. 23–37.

Hodge, B.L., 1973, "World Fluorspar Development, Part 1," *Industrial Minerals,* No. 68, May, pp. 9–25.

Hodge, B.L., 1973a, "World Fluorspar Developments: Part 2," *Industrial Minerals,* No. 69, June, pp. 9–21.

Hodge, B.L., 1981, "Fluorspar," *Mining Annual Review, 1981,* pp. 116–117.

Horton, R.C., 1961, "An Inventory of Fluorspar Occurrences in Nevada," Report 1, Nevada Bureau of Mines, 31 pp.

Horton, R.C., 1962, "Fluorspar Occurrences in Nevada," Nevada Bureau of Mines, Map 3.

Horton, R.C., 1964, "Fluorspar," *Mineral and Water Resources of Nevada,* 88th Congress, 2nd Session, US Government Printing Office, Washington, DC, pp. 198–202.

Jee, J.M., and Kye, J., 1971, "Report on the Fluorite Deposits of Korea," Bulletin No. 13, Geological Survey of Korea, pp. 7–368 (English Abstract).

Jewell, W.B., 1947, "Barite, Fluorite, Galena, Sphalerite Veins of Middle Tennessee," Bulletin 51, Tennessee Div. of Geology, 114 pp.

Johnson, R.C., et al., 1973, "Economic Availability of Byproduct Fluorine in the United States," Information Circular 8566, US Bureau of Mines, 97 pp.

Jones, W.R., and Wolter, F.J., 1973, "Chemical Markets for Hydrofluoric Acid," Preprint No. 73–H–32, Society of Mining Engineers of AIME, SME Fall Meeting, Pittsburgh, Sep.

Kim, H., et al., 1972, "Mineral Requirements for Korea's Industrialization," Korean Institute of Mining, pp. 80–84.

Kostick, D.S., and De Filippo, R.J., 1980, "Fluorspar," Preprint, 1978–1979, US Bureau of Mines, 17 pp.

Liddy, J.C., 1971, "Fluorspar in Australia," *Australian Mining,* Jan., pp. 34–38.

MacMillan, R.T., 1970, "Fluorine," *Mineral Facts and Problems,* Bulletin 650, US Bureau of Mines, pp. 989–1000.

Maersk, B., 1973, "Cryolite Concentrator in Copenhagen," *World Mining,* Vol. 26, No. 3, Mar., pp. 60–63.

McAnulty, W.N., 1970, "Evaluation of Fluorspar Deposits," Preprint 70–S–63, Society of Mining Engineers, AIME Annual Meeting, Denver, Feb.

Northholt, A.J.G., and Highley, D.E., 1971, "Fluorspar," Mineral Dossier No. 1, Mineral Resources Consultative Committee, Institute of Geological Sciences, London, 31 pp.

Ortel, M.K., 1966, "Fluorspar," South African National Resources Development Council, Dept. of Planning, Vol. 5, 31 pp.

Papke, K.G., 1979, "Fluorspar in Nevada," Bulletin 93, Nevada Bureau of Mines and Geology, 77 pp.

Peters, W.C., 1958, "Geologic Characteristics of Fluorspar Deposits in the Western United States," *Economic Geology,* Vol. 53, pp. 663–688.

Rothrock, H.E., et al., 1946, "Fluorspar Resources of New Mexico," Bulletin 21, New Mexico Bureau of Mines and Mineral Resources, 245 pp.

Sahinen, U.M., 1962, "Fluorspar Deposits in Montana," Bulletin 28, Montana Bureau of Mines and Geology, 38 pp.

Sheppard, R.A., and Gude, A.J., 1970, "Authigenic Fluorite in Pliocene Lacustrine Rocks Near Rome, Malheur County, Oregon," In *Geological Survey Research 1969,* Professional Paper 650-D, US Geological Survey, pp. D69–D74.

Snyder, K.D., 1978, "Geology of the Bayhorse Fluorite Deposit, Custer County, Idaho," *Economic Geology,* Vol. 73, No. 2, March-April, pp. 207–214.

Spada, A., 1969, "Il Giacimento de Fluorite e Baritina Esalativo-sedimentario in 'Facies' Lacustre, Intercalato nei Sedimenti Piroclastici della Zona di Castel Giuliano, in Province di Roma," *Industria Min.,* Vol. 20, pp. 501–518.

Staatz, M.H., and Osterwalld, F.W., 1959, "Geology of the Thomas Range Fluorspar District, Juab County, Utah," Bulletin 1069, US Geological Survey, 97 pp.

Steininger, E., 1972, "Making Fluorine Compounds from Waste Fluosilicic Acid," *Engineering & Mining Journal,* Vol. 173, No. 12, pp. 73–75.

Tabor, J.W., 1953, "Montana's Crystal Mountain Fluorspar Deposit is Big and High Grade," *Mining World,* Vol. 15, No. 7, pp. 43–46.

Thurston, W.R., et al., 1954, "Fluorspar Deposits of Utah," Bulletin 1005, US Geological Survey, 53 pp.

Van Alstine, R.E., 1964, "Fluorspar," *Mineral and Water Resources of Colorado,* 88th Congress, 2nd Session, US Government Printing Office, Washington, DC, pp. 159–165.

Van Alstine, R.E., 1965, "Fluorspar," in *Mineral and Water Resources of New Mexico,* Bulletin 87, New Mexico Bureau of Mines and Mineral Resources, pp. 260–267.

Van Alstine, R.E., 1969, "Fluorspar," in *Mineral and Water Resources of Arizona,* Bulletin 180, Arizona Bureau of Mines, pp. 348–357.

Van Alstine, R.E., and Schruben, P.G., 1980, "Fluorspar Resources of Africa," Bulletin 1487, US Geological Survey, 25 pp.

Weller, J.M., et al., 1952, "Geology of the Fluorspar Deposits of Illinois," Bulletin 76, Illinois State Geological Survey, 147 pp.

Wood, H.B., 1971, "Fluorspar and Cryolite," *Minerals Yearbook*, US Bureau of Mines, pp. 517–530.

Wood, H.B., 1972, "Fluorspar," *Engineering & Mining Journal,* Vol. 173, Mar., p. 152.

Worl, R.G., et al., 1973, "Fluorine," *United States Mineral Resources*, Professional Paper 820, US Geological Survey, pp. 223–235.

Zijl, P.J. van, 1962, "The Geology, Structure, and Petrology of the Alkaline Intrusions of Kalkfeld and Okorusu and the Invaded Damara Rocks," *Aunale Univ. Stellenbosch,* Ser. A., Vol. 37, No. 4, pp. 237–346.

Zurowski, M., 1972, "Barite-Fluorite Deposits of Lake Ainslie—An Appraisal From an Economic Viewpoint," *Transactions,* Canadian Institute of Mining & Metallurgy, Vol. 75, pp. 318–321.

Glauconite

Original by FRANK J. MARKEWICZ *
and
WILLIAM LODDING †
Revised by JOHN HOWER ‡

Greensand, greensand marl, and green earth are names given to sediments rich in the bluish green to greenish black mineral known as glauconite by the mineralogist. The word glauconite is from the Greek word *glaukos,* meaning bluish green. The term "greensand" as a rock name for a glauconite-bearing sediment is more appropriate than "greensand marl," a term that has been doggedly perpetuated in the literature.

Because of its potash and phosphate content, greensand was dug and marketed as a fertilizer and soil conditioner for more than 100 years. The advent of prepared fertilizers with adjustable nutrient ratios led to the decline of greensand for agricultural application. The material has since been recognized for use in water treatment. Unfortunately, in spite of large reserves and worldwide distribution, the mineral has not been utilized to any significant commercial extent because no major application has been found for a substance with its chemical composition and properties. This lack is probably due to a paucity of research on its potential commercial application.

Extraction of potash received considerable attention during and just after World War I. Because of relatively high extraction costs and generally low potash content (up to 8%), glauconite lost its appeal as a source of this commodity.

Historical Background

Greensand was used as a fertilizer in New Jersey in the latter part of the 1700s. During the early 1800s its use became more common; applications of up to 100 tons per acre were sometimes made, although recommendations for agricultural use suggested 20 to 50 tons per acre (Tedrow, 1957). Many crops, especially the forage type, were said to improve with greensand application; however, because of slow release of potash, large quantities were required. Certain greensands that contain sulfur and sulfide minerals are harmful to plant growth, and these were classified as "poison," "burning," or "black marls." The availability of higher-grade potash salts from other mineral sources, and the manufacture of prepared fertilizers, displaced the agricultural use of greensand during the later 1800s.

During the mid-1800s the domestic greensand industry, centered in a small section of the eastern United States, grossed more than $500,000 per year. Toward the end of the century, annual production had dwindled to less than $100,000. By 1910 there were only six or eight greensand producers, grossing less than $5000 (Tyler, 1934). There was a small revival in the industry during World War I because of the curtailment of foreign potash, especially from Germany.

During the later 1940s and early 1950s greensand was again recommended as a food nutrient for plants and farm crops. Agronomic studies discussed its potential as a soil additive that gradually releases potash and many trace-element nutrients essential for plant growth (Tedrow, 1957). Greensand was sold with the idea that it would condition soil and absorb and hold water while its base-exchange properties would release trace elements.

For a short time glauconite was utilized in certain parts of New Jersey as a binding additive in the brick industry, and in the 1800s for making green glass (Cook, 1868).

In the early 1900s the base-exchange property of glauconite was recognized for water treatment, and the mineral gained acceptance as a water softener. Mansfield (1922) does not mention base-exchange; however, the base ex-

* Acting State Geologist, New Jersey Geological Survey, Trenton, NJ.

† Formerly of Rutgers University, New Brunswick, NJ; now deceased.

‡ University of Illinois, Urbana, IL.

change phenomenon was known in 1916 or earlier. From 1916 through 1922, several patents for the use of glauconite as a water-softening agent were granted. A method was also patented for treating greensand to improve it for water softening and ready regeneration with common salt brine (Borrowman, 1920; Spencer, 1924; Kriegsheim and Vaughan, 1930).

Treated glauconite, on contact with water containing magnesia or lime, takes up the magnesium or calcium ions and releases sodium ions. This exchange is limited to the outer surface of glauconite grains, and when all the surfaces have absorbed their capacitive limit, the grains must be regenerated. Regeneration, simply stated, consists of treating or "backwashing" the glauconite charge with a common salt solution, which replaces the hard-water elements with sodium, thus reviving the glauconite. The process has become more sophisticated due to competition by the different companies in the water-softening business.

Greensand products for water softening generally consisted of several different grades, depending on the particular treatment the glauconite was given during processing. The standard greensand water softener was produced from natural glauconite after being washed and classified. Its characteristics for water softening are given in Table 1.

To increase its water-softening capacity, glauconite was treated by a process that made it more porous. Although most of the specifications remained the same as those of the washed greensand shown in Table 1, the porosity changed from 2–3% to 11–13%, and the exchange capacity as $CaCO_3$ from 2800–3000 grains to 3800–4500 grains per cu ft; maximum permissible raw water turbidity dropped from 10 to 5–7 ppm, and the maximum permissible iron in raw water decreased from 2.5 to 1.5 ppm. By increasing the glauconite porosity even more, the exchange capacity was also increased, with a corresponding change in some of the other characteristics shown previously. This special porosity process increased the salt requirement for regeneration from 1.4 to 2.5 lb per cu ft. Another method of processing greensand involved heating glauconite to a temperature of approximately 200° to 400°C to drive off water. After cooling it was treated with a salt brine, which made a more efficient greensand product because the softened water did not become cloudy.

During the later 1940s, greensand was displaced as a water-softening agent by phenol formaldehyde resin, which has twice the softening capacity, is not affected by water temperatures over 105°F, and requires the same amount of salt for regeneration. After 1950, phenol formaldehyde resin was displaced by styrene resins, which have even higher water-softening capacity, yet require no more salt for regeneration.

Current Use

Present-day use of glauconite is limited to two general applications: water treatment and soil conditioning. The industry is localized in the coastal plain of New Jersey, where greensand is extracted from Cretaceous and Tertiary sediments. A small amount of greensand for soil conditioning was dug in Maryland, but this operation was terminated in 1970, leaving New Jersey the sole producer in the United States.

TABLE 1—Greensand in Water Softening

Porosity, %	2 to 3
Effective size, mm	0.30 to 0.33
Uniformity coefficient	1.40 to 1.55
Screen analysis, mesh	18 to 60
Attrition loss per annum, %	2
Weight per cu ft, air-dried, lb	85
Recommended pH operating range	6.6 to 8.3
Maximum permissible temperature of raw water, °F	105
Maximum permissible turbidity in raw water, ppm	10
Maximum permissible iron in raw water as Fe, ppm	2.5
Bed expansion during backwash, %	30 to 35
Exchange capacity as $CaCO_3$, grains per cu ft	2800 to 3000
Salt requirement, lb per cu ft	1.4
Time required for regeneration in water softener, min	40
Recommended maximum softening rate, gpm per sq ft bed area	5
Recommended wash rate, gpm per sq ft bed area	7
Recommended minimum bed depth, ft-in.	2.5
Recommended maximum bed depth, ft	6

Data from New Jersey Geological Survey.

In water treatment, processed greensand is used to remove soluble iron and/or manganese salts. It is called manganese greensand "zeolite" as manufactured by the Inversand Co., Div. of Hungerford & Terry, Inc. The purple to almost black "zeolite" is used in two processes. One, rather old, involves utilizing it as a filter, regenerating it intermittently with a weak solution of potassium permanganate after an exchange cycle in which approximately 750 grains of iron per cu ft or 400 grains of manganese per cu ft has been removed.

A process developed and patented by Hungerford & Terry, Inc. about 1960 has replaced the older intermittent-regeneration process and is called "continuous regeneration" or "CR." Potassium permanganate is injected directly into the water ahead of the manganese oxide glauconite, which does not require an intermittent regeneration, only an intermittent wash. This process utilizes potassium permanganate more efficiently than the older process and has gained favor because of its ability to reduce the iron in water to extremely minute amounts, frequently to less than 10 parts per billion. When a water supply contains very little iron, but significant quantities of manganese, then the old IR, "intermittent regeneration," process is advantageously used.

For home and garden use, greensand is recommended as a mulch; top dressing mixture; and additive to soil for gardens, potted plants, and vegetable starting plots. The usual application is from 1 to 10 lb per 100 sq ft, although larger amounts may be applied depending on the nature of the soil. It is sold in bulk or bagged in 60-lb plastic bags at Sewell, NJ, by the Zeolite Chemical Co., Div. of Hungerford & Terry, Inc.

Geologic Occurrence and Character

Glauconite occurs in rocks ranging from Precambrian to Recent in age; however, when the terms glauconite greensand or greensand are used, a Tertiary or Cretaceous age is generally implied. The term greensand denotes an unconsolidated glauconite-rich sand. Glauconite frequently forms the major constituent in many sandy, silty, or even clayey formations of the Tertiary and Cretaceous. Greensand can occur as a solid massive bed, as disseminated grains, as glauconite-rich masses and clots, or as fillings in fossil worm tubes or in foraminifera. Brown ovoid phosphatic fecal pellets are common as accessory grains in greensand deposits.

Fresh glauconite is typically green to dark green in color and polylobate to rounded or tabular in shape. Irregularly shaped, vermiform to tabular sutured grains are usually authigenic in origin, whereas well-rounded to ovoid, well-polished grains found in some formations are detrital or second-generation (Figs. 1–4). Glau-

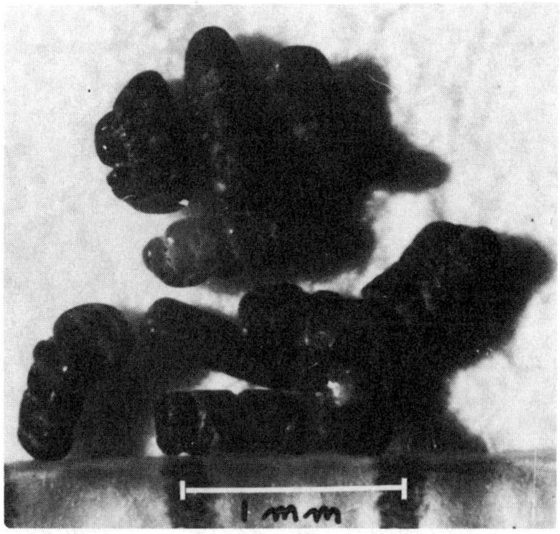

FIG. 1—*Vermiform pellets of glauconite. Note the elongated and segmented shape. X-ray diffraction and thin-section studies indicate that the segments are books of glauconite layers (see Fig. 3). The pellet at the far right has a longitudinal line dividing a more solid, perhaps more recently crystallized, portion from a rougher, perhaps older portion (epitaxial crystal growth?) (Tapper and Fanning, 1968).*

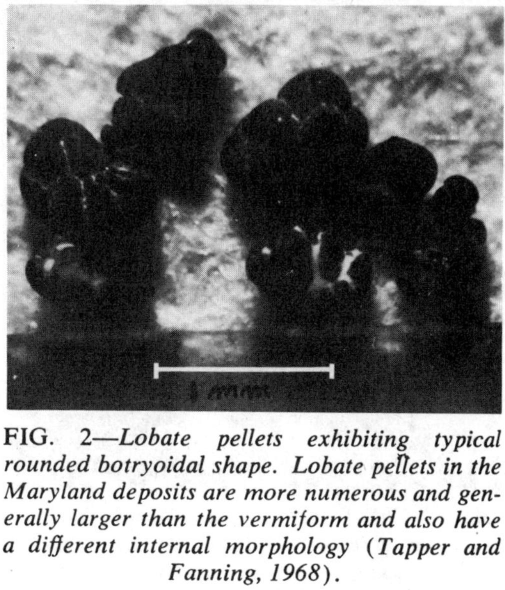

FIG. 2—*Lobate pellets exhibiting typical rounded botryoidal shape. Lobate pellets in the Maryland deposits are more numerous and generally larger than the vermiform and also have a different internal morphology (Tapper and Fanning, 1968).*

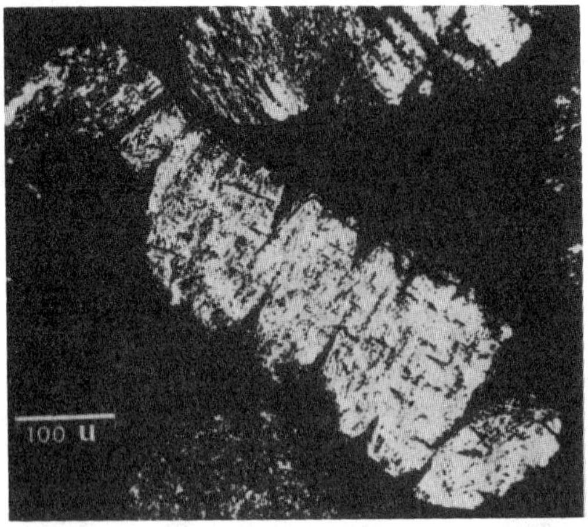

FIG. 3—*Longitudinal section of vermiform pellet showing glauconite layers crossing the pellet. (Thin section viewed with crossed nicols.) Note the appearance of a rectangular "crystal" outline within the pellet just above the center of the photograph. A portion of another pellet with a gross micaceous internal morphology is at the top (Tapper and Fanning, 1968).*

conite imparts a greenish color to the formation in which it occurs, depending on the amount present.

The greensand formations of New Jersey are well known and have long been utilized. So far as known, they are the only commercially productive strata in the United States today. Glauconite occurs as an accessory mineral in many of the New Jersey coastal plain formations, but greensand, as such, is best developed in two

FIG. 4—*Thin section, as viewed under crossed nicols, showing the "grainy" internal morphology that appears to be characteristic of the lobate pellets (Tapper and Fanning, 1968).*

units, the Navesink Formation of Late Cretaceous age and the Hornerstown Formation of Early Tertiary age. Both of these formations are inappropriately termed "marls" on the geologic map of New Jersey, and in Bulletin 50, "The Geology of New Jersey" (Lewis and Kummel, 1940). They crop out in a northeast-trending belt that extends across New Jersey from the vicinity of Sandy Hook to Delaware Bay near Salem. These formations or their equivalents extend into Delaware and beyond, but in general the glauconite content of the strata and the potash content of the glauconite decrease southward. In addition, both formations thin somewhat toward the southwest, and there are local bedding variations along the strike. At the north end in New Jersey, the Hornerstown rests on the Red Bank-Tinton Formation; at the south end, perhaps owing to nondeposition of the Red Bank, it lies directly on the Navesink Formation.

Hornerstown Formation

The Hornerstown Formation is a green medium-grained glauconite sand with a bright green clay cement and a little quartz sand. Locally it contains a few thin brown to reddish brown clayey or indurated limonitic bands. Thickness ranges from 15 to 30 ft. Dip is about 20 ft per mile to the southeast.

Navesink Formation

The Navesink Formation is a massive dark greenish-gray, medium-grained, poorly sorted glauconite sand, with varying amounts of quartz sand. It contains some calcareous clay toward the top, and limonitic bands or masses locally. The thickness ranges from 25 to 50 ft. The formation dips about 20 ft per mile to the southeast.

Table 2 lists some of the major greensand formations in the Coastal Plain of the United States.

Origin of Glauconite

The genesis of glauconite has been a puzzle for a long time, but facts are emerging that permit establishing a workable theory. The process now widely accepted (Hower, 1961; Ehlmann et al., 1963; Triplehorn, 1966; Grim, 1968; Bailey and Atherton, 1969; Zumpe, 1971; Odin, 1971; Odin and Giresse, 1972; Robert; 1972; Shutov et al., 1972) requires the presence of montmorillonite in a shallow marine environment (water depths 100 to 300 m), and very slow rates of sedimentation. Montmorillonite

TABLE 2—Major Greensand Formations of the United States

State	Formation	Age	Thickness, Ft
New Jersey	Hornerstown	Eocene	15 to 30
	Navesink	Cretaceous	25 to 50
Delaware	Aquia	Eocene	
Maryland	Aquia	Eocene	
	Nanjemoy	Eocene	20 to 30
Virginia	Aquia	Eocene	
	Nanjemoy	Eocene	
Alabama	Lisbon	Eocene	
	Eutaw	Cretaceous	
Tennessee	McNairy (?)	Cretaceous	
Mississippi	Winona	Eocene	
	Eutaw	Cretaceous	
Texas	Weches	Eocene	25+
	Reklaw	Eocene	
	Cook Mountain	Eocene	
	Kincaid	Paleocene	

Data from state geological surveys and US Geological Survey.

is converted to glauconite by substitution of Fe^{+3}, Fe^{+2}, and Mg^{+2} for Al^{+3} in the octahedral layer and the consequent fixation of K^{+}. The resulting mineral is a mixed-layer 10 Å hydrated clay. The 10 Å potassium-rich layers are those which have attained a sufficiently high negative charge (because of the substitution of Mg^{+2} and Fe^{+2} for Al^{+3}) to fix K^{+} and dehydrate. Aluminum ions, moving out of the octahedral layer, can form new montmorillonite on the surface of the potassium-iron-enriched globules, continuing the process of accretion. Most of the abundant ferric iron and silicon carried in solution by freshwater are chemically precipitated as hydroxide gels in a brackish environment (Jeans, 1971) and become available for glauconitization.

A suite showing alteration from (1) bleached clear biotite mica containing dark cores and swelled peripheral edges to (2) well expanded light olive-colored altered hydromica grains to (3) light green vermiform peripherally sutured early-stage glauconite has been observed in drill cuttings from several New Jersey coastal plain formations. The mica-glauconite diagenesis reported by Galliher (1939), Takahashi (1939), and Takahashi and Yagi (1929) finds support in these observations. Growth of the globules is limited by currents, which tend to move the newly formed glauconite pellets from areas of formation to those of deposition and burial. Diagenesis continues with the general substitution of hydrated layers by additional potassium, reducing the amount of mixed-layer material, and increasing the 10Å-mica component of mature glauconite.

These processes can adequately explain the formation of iron-potassium-silicate globules in a marine environment, and the accumulation of greensand strata of considerable thickness and extent. Glauconite has been found inside marine microfossils (foraminifera) and, occasionally, derived from fecal pellets (Haven and Morales-Alamo, 1968; Lamboy, 1967, 1968). The time of formation of various glauconites in the New Jersey and Maryland coastal plain was recently determined by K/Ar analysis to be very near 61.1 million years (Owens and Sohl, 1973).

Greensand Composition and Analyses

Most greensand deposits contain, in addition to glauconite, materials such as quartz, mica, pyrite, some heavy minerals, and calcium carbonate in the form of shell material or as a cement derived from marine organisms. Iron oxides, phosphatic nodules, and fecal pellets are common locally. Skeletal remains of marine vertebrates are occasionally found in the greensand deposits of southern New Jersey.

Glauconite grains range in size from 20 to 100 mesh. The largest percentage by weight is in the 20 to 40-mesh size. Grains of authigenic origin are commonly coarse. Many detrital grains are smooth and rounded, and easily pass through a 100-mesh sieve. Clay-size glauconite occasionally fills the interstices of coarse-grained sediments.

Glauconite is a hydrous iron potassium silicate, containing varying amounts of aluminum, magnesium, sodium, and trace elements such as beryllium, cobalt, chromium, nickel, molybdenum, vanadium, titanium, and uranium. Greensand analyses from ten localities in the United States are listed in Table 3.

The composition of glauconites was investi-

TABLE 3—Greensand Analyses (%)

	1	2	3	4	5	6	7	8	9	10
Silica (SiO_2)	51.83	46.28	63.55	77.93	77.80	73.00	35.18	53.61	30.00	36.44
Alumina (Al_2O_3)	6.23	5.42	4.41	7.48	7.86	9.60	5.30	9.56	14.11	29.02
Iron Oxide (Fe_2O_3)	20.08	23.64	15.74	3.54	5.42	7.20	17.35	23.04	25.09	13.36
Lime (CaO)	0.52	1.10	0.48	2.39	2.12	2.88	16.00	1.39	10.80	2.72
Magensia (MgO)	3.66	3.57	2.35	1.01	0.82	1.55	trace	2.87	3.46	1.88
Potash (K_2O)	6.60	7.50	6.24	2.30	3.60	3.30	1.69	3.49	0.80	—
Soda (Na_2O)	0.76	0.83	0.03	0.12	—	—	1.39	0.42	4.41	—
Phosphorus (P_2O_5)	0.31	0.52	0.19	0.22	—	—	3.30	—	0.44	—
Carbonic acid	—	—	—	—	—	—	8.00	—	—	—
Ignition loss	10.34	10.75	6.91	4.83	2.26	2.40	10.10	5.96	—	15.59

[1] Hornerstown greensand, Sewell, NJ (NJ Geological Survey).
[2] Navesink greensand, Sewell, NJ (NJ Geological Survey).
[3] Nanjemoy Formation, Maryland (Grim, 1968).
[4] Aquia Formation, Maryland (Glaser, 1971).
[5] Aquia Formation, Hop Yard, Rappahannock River, VA (Clark, 1912; Gildersleeve, 1942).
[6] Nanjemoy Formation, Woodstock, VA (Clark, 1912; Gildersleeve, 1942).
[7] Lower Eocene greensand, Gas Ridge anticline, Texas (Jones, 1936).
[8] Glauconite, McNairy County, TN.
[9] Weches Formation, Houston County, Texas (Fisher, 1965).
[10] Winona Formation, Montgomery County, MS.

gated by Foster (1969), who computed atomic ratios and layer charges for 32 samples from 12 countries. Table 4 shows the compositional range of these glauconites.

In the mica structure, according to Grim (1968), two silica tetrahedral sheets face one central octahedral sheet. The tips of the tetrahedra point toward the center unit, and the three sheets are combined into a single layer, with a suitable replacement of OH and O. Up to 25% of the silicons are replaced by aluminum, and the resultant charge deficiency is balanced by K-ions. In glauconite, aluminum can substitute 4.2 to 14.3% of silicon in the tetrahedral sheets. Some of the charge deficiency is balanced by Fe^{+2} and Mg^{+2} ions replacing Fe^{+3} and Al^{+3} ions in the octahedral sheet. Most glauconites are not pure dioctahedral micas, but a mixture of 10Å micas and 14Å expandable-lattice clays (Burst, 1958; Cimbalnikova, 1971).

Like other layer minerals, glauconite has the capacity to absorb ions from solutions. The exchangeable sites are on the outside of the silica-alumina-iron framework, and the exchange reaction generally does not affect the glauconite structure. Both cations and anions are exchangeable, but not much information is available on the latter. Ion exchange is important in chemical separations, water softening, and properties of soil (both agricultural and engineering). It is usually expressed in milliequivalents per 100 g; thus a typical cation exchange capacity for glauconite is approximately 20 milliequivalents per 100 g.

The compositional relationship of glauconite with other layer silicates is illustrated in the ternary plot by Yoder and Eugster (1955), Fig. 5. The exterior shapes of typical glauconite grains are shown in Figs. 1 and 2, and the internal morphology in Figs. 3 and 4.

TABLE 4—Compositional Range of Glauconites (%)

SiO_2	46.9-52.9
Al_2O_3	5.8-15.2
Fe_2O_3	9.3-24.1
FeO	1.0-6.3
MgO	2.3-4.6
CaO	Traces-1.9
Na_2O	Traces-1.6
K_2O	5.1-9.3
TiO_2	0 -1.8
H_2O	4.3-7.2

Source: Foster, 1969.

Purification of Greensand

Magnetic Susceptibility

Glauconite has marked magnetic susceptibility; magnetic separation is, therefore, a commonly used method to separate it from impurities. With a Frantz Isodynamic separator the following settings gave a clean separation of glauconite from quartz, coprolites, pyrite, and calcite particles:

side slope	20°
forward slope	15°
current	0.50 amp
mass magnetic susceptibility	$(K_m) = 25 \times 10^{-6} \pm 5$

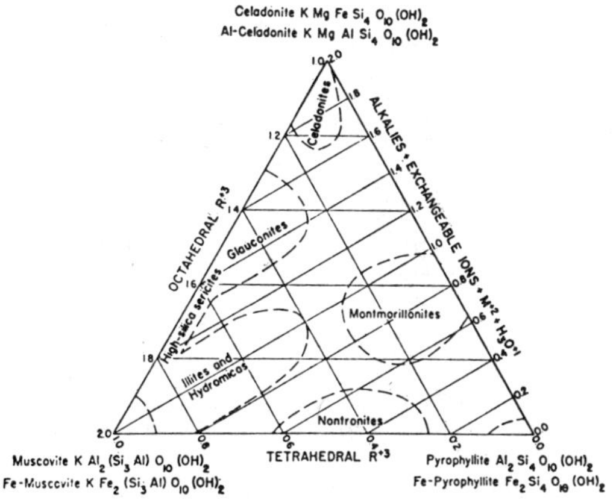

FIG. 5—*The relationships of dioctahedral mica-type phases, with fields indicated in which natural clay-mineral compositions fall (Yoder and Eugster, 1955).*

Three steps of medium-intensity magnetic separation, preceded by washing and removal of the +30, −200-mesh fractions, resulted in a purified glauconite.

Specific Gravity

The specific gravity of glauconites ranges widely, depending on the abundance of mixed-layer minerals and K_2O, the degree of substitutions in the tetrahedral and octahedral layers, and the state of weathering. Shutov et al. (1972) reported the results of careful fractionation of purified glauconites by means of a density column, which permitted separation in heavy liquids in steps of 0.020 ± 0.001 g per cc. The gravity spectra obtained from glauconites of various geologic ages and provenances are summarized in Table 5.

The wide range of specific gravities of glauconites limits heavy-liquid separation as a purification method. It can be used on some greensands with good results to separate glauconite mixtures of different provenances and different stages of diagenesis.

Mining and Treatment

Most greensand beds are unconsolidated sediments. Various adaptations of open-pit extraction have been utilized. These methods have ranged from pick and shovel in the earliest days to crane-operated jaw buckets, draglines, and hydraulicking. At Hungerford & Terry's Inversand operation in Sewell, NJ, which is the only commercially operating greensand company in the United States (perhaps in the free world), the hydraulicking method is used on the Hornerstown and Navesink Formations. The lower Hornerstown Formation, which forms the upper part of the deposit, is approximately 15 to 20 ft thick. Below this are 10 to 15 ft of nearly pure glauconite belonging to the Navesink Formation. After the beds are flushed with a high-pressure water jet, the loosened greensand is suction-pumped to a collecting tank. From this it flows through a complicated piping system containing special classifying and washing apparatus. After a thorough washing, the greensand is treated with chemicals such as aluminum sulfate, sodium silicate, sodium aluminate, and, for some applications, caustic soda and phosphoric acid. This treatment forms a very durable greensand having a high exchange capacity. At the Inversand operation the major application of treated greensand is for the removal of iron, manganese, and sulfide from water. Several grades of greensand are dried and bagged for shipment.

Domestic Production

New Jersey

In the late 1860s almost a million tons of greensand was dug every year for use as fer-

TABLE 5—Localities, Ages, and Specific Gravities of Glauconites

Greensand Locality	Mode, G per Cc	Range
California, Pacific, Recent	2.35	2.32-2.50
Akatosk trench, Pacific, Recent	2.68	2.46-2.78
Russian plateau, Tertiary	2.66	2.54-2.76
Baltic region, Cretaceous	2.68	2.54-2.76
Caucasus, Cretaceous	2.83	2.70-2.86
Basin of Laba, USSR, Jurassic	2.81	2.74-2.84
Baltic region, Ordovician	2.88	2.80-2.92
Pachelma, Precambrian	2.62	2.50-2.70
Ural, Precambrian	2.72	2.58-2.74
Ural, Sinian	2.71	2.66-2.76
Olenek, Sinian	2.70	2.66-2.74

Source: Shutov et al., 1972.

tilizer. In 1855 it brought 7¢ a bushel, and in 1877 it was selling for 60¢ a ton at the pit. Reports indicate that production was principally from New Jersey. By the 1870s it had dropped to less than 100,000 tons. Toward the turn of the century, demand for greensand improved, but in 1908 less than 10,000 tons was produced, at a total value of only $3500. The number of producers dropped from 13 in 1908 to 4 in 1912. The use of greensand for water softening in the early 1920s revived the industry. Table 6 shows tonnage and value, 1922 to 1951.

During 1926 the Inversand Co. started operations near Sewell, NJ. In the late 1940s the National Soil Conservation Co. started operations at Medford, NJ. The company produced greensand at this locality until the early 1960s. Since 1963 greensand production has been about 2000 tpy. Table 7 shows production for water treatment and agricultural uses.

Alabama

The only known commercial utilization of greensand in Alabama was in the production of an elixir made by passing water through the sands of the Lisbon Formation of Choctaw County. This operation was closed by the Pure Food and Drug Administration and no further attempts were made to use greensand commercially.

Delaware

There is no present-day commercial production. Research is being done on the possible economic use of glauconite, especially as a cation-exchange medium for the purification of polluted water.

Maryland

The Kaylorite Corp. mined greensand at Dunkirk, MD, for agricultural purposes and as a soil conditioner until 1970. It is not definitely known whether the operation was shut down because of the death of the owner or on account of difficulties in production and/or sales. Production was from the upper 20 ft of the Nanjemoy Formation. Two products were made, a finely ground greensand for use as a soil additive and nonpoisonous dusting agent for plants, and a −30-mesh granular greensand for agricultural application. Production in 1970 totaled 650 tons.

TABLE 6—Greensand Production in New Jersey

Years	Tons	Value, $
1922-23	18,500	188,000
1924-25	27,000	330,000
1926-27	24,000	337,000
1928-29	25,000	460,000
1930	12,700	225,000
1947-51	30,736	–
1967-71, avg	3,437	–

Data from New Jersey Geological Survey and US Bureau of Mines.

TABLE 7—Production of Greensand in New Jersey (Tons)

	Water Treatment	Agricultural Use
1959	2000	4600
1960	1900	3750
1961	1500	2900
1962	1900	2250
1963	2000	n.r.*

Data from New Jersey Geological Survey.

* n.r., not reported.

Tennessee

A small amount of greensand in McNairy County was used for direct application on agricultural land. No present-day commercial production is recorded.

Texas

Some limited past production is known from the Eocene horizon in Bexar County and from the Paleocene south of San Antonio. Greensand was mainly used for water purification and soil conditioning (Jones, 1936). No present-day commercial production is recorded.

Virginia

Used in the 1920s as an agricultural supplement because of its potash, phosphoric acid, and lime, greensand was dug from open pits and at some places along the edge of the Pamunkey River from horizontal drifts driven into the Eocene Aquia and Nanjemoy Formations. The potash content ranges from about 1 to 10 (?)% (Clark, 1912). There is no present-day commercial production in Virginia.

Greensand in Other Countries

Belgium

In Belgium the sands of Antwerp contain 40% or more of glauconite. Industrial exploitation has been suggested but there is no current utilization.

Great Britain

During the 19th century greensand was mined as a soil conditioner from pits located in Hampshire, Essex, and Kent. During World War II, the Eocene beds at Knaphill and Chobham Common in Surrey and at Hazeley Heath in Hampshire were explored as potential sources of potash; the beds are 8 to 10 ft thick, and reserves are estimated at 7,000,000 tons, with an average content of 50% glauconite (Oakley, 1943; Bessey, 1949). No greensand is being produced at present.

Netherlands

Glauconite sands occur in the Lower Tertiary and Upper Cretaceous sediments of the southwestern part of South Limburg, and in Lower Cretaceous sediments in the eastern part of Overijssel (Achterhock) (H. M. Harsveldt, 1972, private communication). Between 1940 and 1946 the usability of glauconite was investigated for (1) decalcifying of water, (2) manufacture of soluble potash salts for fertilizer, and (3) manufacture of green paint. Because of the low 2.6 to 6% potash content, its use as a source of potash salts proved uneconomic. Color and coarseness did not fulfill the paint requirements, and attempts to use the greensand to decalcify drinking water were of no avail.

Poland

Several Tertiary and Quaternary sediments have glauconite beds of variable purity. The "glauconite molding sand," from Zielnow near Grudziadz, at Zawichost near Sandomierz, and at Lechowka near Chelm Lubeski contain 13 to 16% glauconite; the remainder is mostly clay and quartz sand with minor muscovite, biotite, manganese, and iron aggregates (R. Osika, 1972, private communication).

USSR

Glauconitic sandstone is widespread in the northwestern region of the European USSR. It occurs in the Ukraine, in the Volga region, on the east slope of the Ural Mountains, in the Kazakhstan (Kazakh Soviet Socialist Republic), in the Uzbekistan (Uzbek Soviet Socialist Republic), and also in the Far East. These glauconitic deposits are mainly in Jurassic, Cretaceous, and Paleocene sands and sandstones. Glauconite content commonly reaches 50 to 80% (P. M. Tatarinov, 1972, private communication).

TABLE 8—Analyses of Glauconite from the USSR (%)

	I	II	III
SiO_2	48.80	44.46	47.92
Al_2O_3	9.25	10.52	5.20
Fe_2O_3	15.24	20.19	22.76
FeO	1.90	1.45	1.08
MgO	3.85	2.31	3.64
CaO	3.78	—	0.56
K_2O	8.20	4.85	5.88
Na_2O	—	11.38	0.18
H_2O	5.32	6.55	12.30
TiO_2	—	—	0.06
P_2O_5	—	—	0.16
Loss on ignition	4.20	—	—
	100.54	101.71	99.74

I. Glauconite, Saratovs deposit.
II. Glauconite sand, Lopatynsky deposit, Moscow region.
III. Glauconite, Changi deposit, Uzbekistan.

Data: Private Communication.

Glauconite is used in limited quantities for producing green mineral pigments and in quantity for softening hard water. It is also used for adsorption in the processing of petroleum products; for cleaning of sewage; and as a fertilizer for cotton, rice, and other agricultural crops. Table 8 gives analyses of glauconites from the USSR.

Australia

At Gingin, 50 miles north of Perth, there are thick greensand beds containing up to 40% glauconite. These beds are extensive and contain up to 3% potash, 1% phosphorus, and 15% calcium carbonate. Glauconite was obtained by electromagnetic separation for use as a water-softening agent, but production ceased in 1965 (McLeod, 1965). Table 9 shows greensand production in Australia.

Pricing

Increasing costs for materials, labor, and taxes have brought about a considerable increase in the cost of production during the past few years. Currently, greensand production for agricultural application is valued at $30 to $35 per ton. Quantities less than a ton are correspondingly higher. Glauconite produced for water purification is valued at $75 to $150 per ton depending on the treatment process it is given. Until more widespread uses are found, the present limited demand for glauconite will govern the price structure.

TABLE 9—Greensand Production in Australia

Year	Tons
1935 to 1950	3300
1950	324
1951	506
1952	230
1953	320
1954	258
1955	197
1956	85
1957	126
1958	112
1959	102
1960	111

Data from McLeod, 1965.

Potential Uses

In the light of declining use of greensand, recent efforts are under way to develop new uses. Its catalytic properties have been investigated in some detail in the United States, the USSR, Japan, Germany, and France (Hartough and Kosak, 1947; Kvirikashvili, 1962, 1964; Gornak, 1963). Acylating reactions, dehydration of cyclohexane to benzene, and condensation of aldehydes have been performed with glauconite catalysts. The relationship between catalytic activity and cation exchange capacity (CEC) has led to efforts to enhance the former by increasing the latter through heat treatment and chemical methods; since CEC increases with the ratio of mixed-layer component of glauconite to 10Å mica component, these efforts also are directed toward replacing more of the potassium in the glauconite structure with hydroxyl ions. Heating of the glauconite in a reducing atmosphere leads to the formation of metaglauconite; its absorbent properties are being studied.

An entirely different approach is the destructive dissolution of glauconite, and the recovery of at least three marketable products: high-purity silica, potash, and iron oxides. This process was first proposed by Turrentine (1925); at present, efforts are under way to prove economic feasibility with the help of up-to-date chemical engineering practice. Some extensive greensand beds are situated near population centers, and their capacity of absorbing waste fluids is being investigated.

A number of other uses have been suggested or locally tried. Finely ground glauconite has been used as a green paint pigment, but its low refractive index and tendency to weather to red iron oxide inhibits its use for this application. Pure glauconite has been used to a limited extent as a glass-polishing agent. Glauconite from the Weches Formation in Texas contains clayey beds that were formerly cut or sawed and dried to form a tough, reasonably durable building stone (Jones, 1936; Fisher, 1965).

Bibliography and References

Ashley, G.H., 1918, "Notes on the Greensand Deposits of the Eastern United States," Bulletin 660, US Geological Survey, pp. 27–50.

Bailey, R.J., and Atherton, M.P., 1969, "The Petrology of a Glauconitic Sandy Chalk," *Journal of Sedimentary Petrology,* Vol. 39, No. 4, Dec., pp. 1420–1431.

Bessey, G.E., 1949, "Potash from Greensand," Monograph, Mineral Supply, London, pp. 11–105.

Borrowman, G.L., 1920, "Water-Softening Compound and Method of Producing Same," US Patent 1,348,977, Aug. 10.

Bradley, W.F., 1945, "Diagnostic Criteria for Clay Minerals," *American Mineralogist,* Vol. 30, No. 11–12, Nov.-Dec., pp. 704–713.

Burst, J.F., 1958, "'Glauconite' Pellets: Their Mineral Nature and Applications to Stratigraphic Interpretations," *Bulletin of American Association of Petroleum Geologists,* Vol. 42, No. 2, Feb. pp. 310–327.

Cimbálníková, A., 1971, "Chemical Variability and Structural Heterogeneity of Glauconites," *American Mineralogist,* Vol. 56, No. 7–8, July-Aug. pp. 1385–1392.

Cimbálníková, A., 1971a, "Influence of 10 Å/14 Å Interlayering in Glauconites," *American Mineralogist,* Vol. 56, No. 7–8, July-Aug., pp. 1393–1398.

Clark, W.B., 1892, "A Preliminary Report on the Cretaceous and Tertiary Formations of New Jersey," Annual Report for 1892, Pt. 2, Geological Survey of New Jersey, 368 pp.

Clark, W.B., 1912, "The Physiography and Geology of the Coastal Plain Province of Virginia," Bulletin 4, Virginia Geological Survey, 274 pp.

Cook, G.H., 1868, "Geology of New Jersey," State Geological Survey, New Brunswick, 899 pp.

Ehlmann, A.J., et al., 1963, "Stages of Glauconite Formation in Modern Foraminiferal Sediments," *Journal of Sedimentary Petrology,* Vol. 33, No. 1, Mar., pp. 87–96.

Eilertsen, D.E., 1972, "Greensand," *Minerals Yearbook 1972,* US Bureau of Mines, p. 1359.

Fisher, W.L., 1965, "Rock and Mineral Resources of East Texas," Report of Investigations 54, Texas Bureau of Economic Geology, 439 pp.

Foster, M.D., 1969, "Studies of Celadonite and Glauconite," Professional Paper 614-F, US Geological Survey, 15 pp.

Galliher, E.W., 1935, "Geology of Glauconite," *Bulletin of American Association of Petroleum Geologists,* Vol. 19, No. 11, pp. 1569–1601.

Galliher, E.W., 1935a, "Glauconite Genesis," *Bulletin of the Geological Society of America,* Vol. 46, Sep., pp. 1351–1365.

Galliher, E.W., 1939, "Biotite-Glauconite Transformation and Associated Minerals," *Recent Marine Sediments,* P. Trask, ed., pp. 513–515.

Gildersleeve, B., 1942, "Eocene of Virginia," Bulletin 57, Virginia Geological Survey, 43 pp.

Glaser, J.D., 1971, "Geology and Mineral Resources of Southern Maryland," Report of Investigations 15, Maryland Geological Survey, 85 pp.

Goldman, M.I., 1919, "General Character, Mode of Occurrence and Origin of Glauconite," *Washington Academy of Science Journal,* Vol. 9, pp. 501–502.

Gornak, A.I., 1963, "Influence of Preliminary Treatment on the Exchange Capacity of Glauconite from the Loevsk Deposit of White Russia," SSR Ionobmen i Sorbtsiya iz Rastvorov, Akademii Nauk Belorussk, SSR, Inst. Obshch. i Neorgan. Khim.; Chemical Abstracts 61: 67h.

Grim, R.E., 1968, *Clay Mineralogy,* 2nd ed., McGraw-Hill, New York, 596 pp.

Gruner, J.W., 1935, "The Structural Relationship of Glauconite and Mica," *American Mineralogist,* Vol. 20, No. 10, Oct., pp. 699–714.

Hallimond, A.F., 1928, "The Formula of Glauconite," *American Mineralogist,* Vol. 13, No. 12, Dec., pp. 589–590.

Hartough, H.U., and Kosak, S., 1947, "Acylation Studies in the Tiophene and Furan Series, III, Natural and Synthetic Silica-Metal Oxide Catalysts," *Journal of American Chemical Society,* Vol. 69, pp. 1014–1016.

Haven, D.S., and Morales-Alamo, R., 1968, "Occurrence and Transport of Faecal Pellets in Suspension in a Tidal Estuary," *Sedimentary Geology,* Vol. 2, pp. 141–152.

Hendricks, S.B., and Ross, C.S., 1941, "Chemical Composition and Genesis of Glauconite and Celadonite," *American Mineralogist,* Vol. 26, No. 12, Dec., pp. 683–708.

Hower, J., 1961, "Some Factors Concerning the Nature and Origin of Glauconite," *American Mineralogist,* Vol. 46, No. 3–4, Mar.-Apr., pp. 313–334.

Jeans, C.V., 1971, "The Neoformation of Clay Minerals in Brackish and Marine Environments," *Clay Minerals* (London), Vol. 9, pp. 209–217.

Jones, C.A., 1936, "Report on a Mineral Resource Survey of Bexar County, Texas," Mineral Resource Survey Series, Circular 2, Texas Bureau of Economic Geology, 6 pp.

Kriegsheim, H., and Vaughan, W., 1930, "Water-Purifying Material and Process of Making the Same," US Patent 1,757,372, May 6.

Kvirikashvili, V.L., 1961, "Contact Catalytic Conversion of Cyclohexane and Methycylohexane in the Presence of Natural Glauconite Catalyst," Soobschch. Akademii Nauk Gruz. SSSR, Vol. 27, pp. 671–1; Chemical Abstracts 57: 700B.

Kvirikashvili, V.L., 1962, "The Conversion of *n*-Heptane in the Presence of Glauconite," *Transactions,* Inst. Khim. Akademii Nauk Gruz. Vol. 16, pp. 89–98; Chemical Abstracts, 60: 9131D.

Kvirikashvili, V.L., 1964, "Contact Conversions of Methylcylopentane in the Presence of a Natural Glauconite Catalyst," *Transactions,* Inst. Khim. Akademii Nauk Gruz. Vol. 17, pp. 201–205; Chemical Abstracts, 62:11697h.

Lamboy, M., 1967, "Répartition de la 'glauconie' sur le plateau continental de la Galice et des Asturies (Espagne)," *Comptes Rendus,* Academy of Sciences, Paris, 266-D, pp. 855–857.

Lamboy, M., 1968, "Sur un processus de formation de la glauconie en grains à partir des débris coquilliers. Rôle des organismes perforants," *Comptes Rendus,* Academy of Sciences, Paris, 266-D, pp. 1937–1940.

Lewis, J.L., and Kummel, H.B., 1940, "The Geology of New Jersey," Bulletin 50, New Jersey Dept. of Conservation & Development, 203 pp.

Light, M.A., 1950, "Glauconite of the New Jersey Coastal Plain," Ph.D. Thesis, Rutgers University, 244 pp.

Manghanani, M.H., and Hower, J., 1964, "Glauconites: Cation Exchange Capacities and Infrared Spectra—Part 1. The Cation Exchange Capacity of Glauconite," *American Mineralogist,* Vol. 49, No. 5–6, May-June, pp. 586–598.

Mansfield, G.R., 1922, "Potash in the Greensands of New Jersey," Bulletin 727, US Geological Survey, 146 pp.

McGrain, P., 1968, "Economic Geology of Calloway County, Kentucky," County Report No. 2, Series X, Kentucky Geological Survey, pp. 32–33.

McLeod, I.R., 1965, "Australian Mineral Industry: The Mineral Deposits," Bulletin 72, Bureau of Mineral Resources, 690 pp.

Nordell, E., 1935, "Zeolites: Mining, Processing, Manufacture, and Uses," Bulletin 61, Michigan State College Engineering Experiment Station, pp. 3–27.

Oakley, K.P., 1943, "Glauconite Sand of Brackesham Beds," Geological Survey of England & Wales, Vol. 33, 28 pp.

Odin, G.S., 1968, "Glauconie, glauconite et phyllosilicate verts," *Bulletin,* Gr. Franc. Argiles, Vol. 20, pp. 11–12.

Odin, G.S., 1971, "Sur la genèse des glauconies et leur signification sédimentologique d'après l'etude détaillée du sondage de Mont Cassel (Nord)," *Comptes Rendus,* Academy of Sciences, Paris, 272-D, pp. 697–699.

Odin, G.S., and Giresse, P., 1972, "Formation des minéraux phylliteux (berthiérine, smectites ferrifères, glauconite ouverte) dans les sédiments du Golfe de Guinée," *Comptes Rendus,* Academy of Sciences, Paris, 275-D, pp. 177–180.

Owens, J.P., and Sohl, N.F., 1973, "Glauconites from the New Jersey-Maryland Coastal Plain: their K/Ar ages and Application in Stratigraphic Studies," *Geological Society of America Bulletin,* Vol. 84, No. 9, Sep., pp. 2811–2838.

Robert, M., 1972, "Transformation experimentale de glauconites et d'illites en smectites," *Comptes Rendus,* Academy of Sciences, Paris, 275-D, pp. 1319–1322.

Ross, C.S., 1926, "The Optical Properties and Chemical Composition of Glauconite," *Proceedings,* US National Museum, Vol. 69, pp. 1–15.

Schneider, H., 1927, "A Study of Glauconite," *Journal of Geology,* Vol. 35, No. 4, May-June, pp. 289–310.

Shreve, R.M., 1921, "Greensand as a Source of Potash," *Chemical & Metallurgical Engineering,* Vol. 35, p. 1056.

Shutov, V.D., et al., 1972, "Crystallochemical Heterogeneity of Glauconite as Depending on the Conditions of its Formation and Post-Sedimentary History," *International Clay Mineral Conference,* Madrid, pp. 327–339.

Smulikowski, K., 1954, "The Problem of Glauconite," *Archiwum Mineralogiczne* (Polska Akademii Nauk, Komitet Geologiezny), Vol. 18, No. 1, 108 pp.

Spencer, A.C., 1924, "Process of Improving Glauconite," US Patent 1,491,561, Apr. 22.

Takahashi, J.I., 1939, "Synopsis of Glauconitization," *Recent Marine Sediments,* P. Trask, ed., pp. 502–512.

Takahashi, J.I., and Yagi, T., 1929, "The Peculiar Mud-Grains in the Recent Littoral and Estuarine Deposits with Special Reference of the Origin of Glauconite," *Economic Geology,* Vol. 24, No. 8, Dec., pp. 838–852.

Tapper, M., and Fanning, D.S., 1968, "Glauconite Pellets: Similar X-Ray Patterns from Individual Pellets of Lobate and Vermiform Morphology," *Clays and Clay Minerals,* Vol. 16, pp. 275–283.

Tedrow, C.F., 1957, "Greensand Soils Subject of Study by Station Scientists," *New Jersey Agriculture,* Nov.-Dec., 3 pp.

Triplehorn, D.M., 1966, "Morphology, Internal Structure, and Origin of Glauconite Pellets," *Sedimentology,* Vol. 6, pp. 247–266.

True, R.H., and Geise, F.W., 1918, "Experiments on the Value of Greensand as Source of Potassium for Plant Culture," *Journal of Agricultural Research,* Vol. 15, pp. 483–492.

Turrentine, J.W., 1925, "Potash from Greensand," *Industrial & Engineering Chemistry,* Vol. 17, Nov., pp. 1177–1181.

Tyler, P.M., 1934, "Greensand," Information Circular 6782, US Bureau of Mines, 8 pp.

Yoder, H.S., and Eugster, H.P., 1955, "Synthetic and Natural Muscovites," *Geochemica et Cosmochimica Acta,* Vol. 8, pp. 225–280.

Zumpe, H.H., 1971, "Microstructure in Cenomanian Glauconite," *Mineral Magazine,* Vol. 38, pp. 215–224.

Graphite*

GEORGE D. GRAFFIN†

The first use of graphite is lost in the mists of time. It was used by primitive man to make drawings on the walls of caves and by the Egyptians to decorate pottery. As early as 1400 A.D. graphite crucibles were being made in the Haffnerzell district of Bavaria.

Through the Middle Ages graphite was confused with other minerals, especially galena and molybdenite. Two common names that are still used for the mineral are "Plumbago," meaning leadlike, and "Black Lead." The latter name implied that graphite is either composed of lead or at least contains a large percentage of it. Gessner is credited with having recognized it as a separate mineral in 1565, but its composition was not determined until 1779, when Scheele demonstrated that graphite oxidized to carbon dioxide, thus proving its carbon constitution. In 1789, Werner named it "graphite" from the Greek word "graphein," meaning to write.

The United States has long relied on foreign countries for its graphite supplies. This reliance has been due more to the lack of suitable domestic graphites than to any particular preference for buying minerals overseas. Ceylon (now Sri Lanka) graphites were imported as early as 1820, and Madagascar graphites have been imported since the properties on that island were opened up about a hundred years later. Mexico has long supplied the United States with most of its requirements for amorphous graphite.

Physical Properties and Classification

Graphite is found in laminated, flaky aggregates disseminated in schistose rocks. It also occurs in veins and exhibits a foliated or fibrous structure.

Graphite is a black lustrous mineral that crystallizes in the hexagonal system, with rhombohedral symmetry. The crystals have tabular form and are six-sided; the faces are commonly striated. The flakes have perfect basal cleavage and are opaque. When well crystallized, the flakes have a black metallic luster while the amorphous material is black and earthy, with a microcrystalline compactness. The flakes feel greasy. Graphite is an excellent conductor of heat and electricity. It melts at approximately 3550°C at a triple point under 1250 psi. It sublimes between 3300–3500°C at 14.7 psi. At 3726°C a second triple point occurs at approximately 100,000 atmospheres. Thermal oxidation in the presence of oxygen begins at 300°C and the rate increases with temperature.

The three forms of carbon (charcoal, graphite, and diamond) are distinguished by chemical and physical tests. The specific gravity of charcoal is 1.3 to 1.9, of graphite 2.266 g/cm^3 (crystal density), and of diamond 3.5. Graphite has a hardness of 1 to 2 (Mohs' scale).

Natural graphite can be divided into three classes: disseminated flake, crystalline vein (fibrous or columnar), and amorphous. Flake graphite is a lamellar form found in metamorphic rocks, such as marble, gneiss, and schist. Each flake is separate, having crystallized as such in the rock. Crystalline vein graphite is found in the form of well defined veins or as pocket accumulations along the intrusive contacts of pegmatites with limestones. The graphite of such deposits is of two types, foliated and columnar. The important Sri Lanka (Ceylon) deposits are vein type.

Amorphous graphite is commonly found as minute microcrystalline particles more or less uniformly distributed in feebly metamorphic rocks, such as slates and shales, or in beds consisting practically entirely of graphite. The latter usually are metamorphosed coal seams and carry as much as 80–85% graphitic carbon, while the former, being altered carbonaceous sediments, commonly range from 25 to 60%

* The author has drawn heavily from the chapter on Graphite by Eugene N. Cameron in the third edition of *Industrial Minerals and Rocks*.

† Vice President, The Asbury Graphite Mills, Inc., Asbury, Warren County, NJ.

carbon. The graphite content of such amorphous deposits is dependent on the amount of carbon originally present in the sediments. There is no evidence of enrichment by the intrusive rocks. Certain amorphous graphite deposits have undoubtedly been formed by contact (thermal) metamorphism, while others are probably the result of dynamic (regional) metamorphism.

Manufactured (artificial) graphite is produced in electric furnaces from petroleum coke.

Definition of Terms, Grades, and Specifications

Through its long history of use graphite has been marketed as "graphite," "plumbago," or "black lead." Furthermore, industrial usage of terms may be confusing to one not intimately associated with the industry. For example, even though natural and manufactured (artificial) graphite are used extensively by industry, commercially the two main categories are *crystalline* and *amorphous*. Inasmuch as all graphite is truly crystalline and the distinction is only one of the size of the crystals themselves, the term "amorphous" is actually a misnomer.

"Amorphous," as applied to graphite, has been further complicated by long-standing industrial application of the term to very fine particles of crystalline flake graphite that can be sold only for low-value uses, such as foundry facings. Also, fine grained varieties of lump graphite that can be reduced easily to fine particle size by grinding are called *amorphous lump* to distinguish them from tough, platy, and acicular varieties, known as *crystalline lump,* that can be reduced in particle size with difficulty.

Crystalline flake graphite is well defined in paragraph 213 of the Tariff Act of 1930 as follows:

> The term "crystalline flake" means graphite or plumbago which occurs disseminated as a relatively thin flake throughout its containing rock, decomposed or not, and which may be or has been separated therefrom by ordinary crushing, pulverizing, screening, or mechanical concentration process, such as flake being made up of a number of parallel laminae, which may be separated by mechanical means.

Under the foregoing definition finely divided particles of crystalline flake graphite would be classified as crystalline graphite.

However, the Court of Customs Appeals has repeatedly held that commercial designations and not scientific terms must govern classification and that when a commercial meaning differs from the technical meaning the commercial designation must govern. Thus, large quantities of fine crystalline flake graphite are imported under the *amorphous* classification.

Sri Lanka lump graphite, one of the most important, is classified either as amorphous or crystalline. Each type is divided into a number of grades, depending upon the particle size (lump, ranging from the size of walnuts to that of peas; chip, from the size of peas to about that of wheat grains; dust, finer than 60 mesh), graphite carbon content, and degree of consolidation.

Amorphous graphite is graded primarily on graphitic carbon content. Commercial ores contain about 50 to 94% carbon, depending on their source.

Crystalline flake graphite from the Malagasy Republic is divided into two main grades, "flake" (coarse flake) and "fines" (fine flake). Malagasy crucible flake must have a minimum of 85% carbon and be essentially all −8 to +60 mesh. Other crystalline flake graphite is graded according to graphitic carbon content and particle size.

The terms "manufactured," "secondary artificial," "primary artificial," "electric-furnace," and "synthetic" are used to describe graphite produced from petroleum coke in electric furnaces.

The synthetic graphites generally are classified as electrodes, anodes, molded shapes, "secondary," and "primary" graphites. The term "secondary" artificial graphite mostly is used to describe the lathe and machine cuttings from formed shapes of artificial graphite. The term "primary" artificial graphite generally is used to describe granular particles of carbon which have been thermally converted to a graphitized state. Secondary artificial graphite generally is manufactured from calcined petroleum coke, coal tar pitch, iron oxide, and other minor ingredients. The ingredients are homogeneously blended and then extruded or molded into the desired shape. The shapes are commonly called "green" carbons. The "green" carbon shapes are then baked to a temperature of 810°C which usually requires two to three weeks, depending on the size and the mix composition. Small sizes may be baked in a matter of hours. "Green" carbons that have been heat treated are known as "baked" carbons. After the carbon shapes are heat treated, they are then processed in an electric furnace to a temperature of between 2,200°C and 2,800°C+. This may range from a few hours to several days, again depending upon the size and shape, base material, and mix formulation. These carbons are commonly known as "synthetic," "graphitized products," "electro-graphite" or graphite. The lathe and machine cuttings and turnings will generally range between 1.5 to 1.76 g/cm^3; special mix formulations may have a density of from 1.0 to 2.1 g/cm^3.

Primary artificial graphite is produced by heat treating carbon granulars in an electric furnace to a temperature of 2,200°C along with a catalyst in order to obtain higher purities and to lower the temperature of graphitization. The base material is generally calcined petroleum coke. The apparent density will range from 1.35 to 1.55 g/cm^3. The major difference between secondary and primary artificial graphite is the base raw material and purity.

The choice of graphites used is primarily dependent upon the application. In general, artificial graphite will have a lower density than natural graphite. It will have a higher electrical resistance and a higher porosity than natural graphite.

The one characteristic in natural graphite which has not been matched by artificial graphite is its excellent molding properties and high densities.

Mode of Occurrence

Graphite is widely distributed throughout the world. It occurs in many types of igneous, sedimentary, and metamorphic rocks. Many occurrences are, however, of little economic importance. The more important are those found in metasomatic-hydrothermal deposits and in sedimentary rocks that have been subjected to regional or thermal metamorphism.

Most, if not all, of the world's deposits of flake and crystalline graphite occur in metamorphic rocks of Precambrian age. Marble, gneiss, and schists are the most common types of rock in which the economic deposits of flake graphite occur. In many cases, the rocks have been intruded by pegmatitic veins. Vein graphite is normally found in rocks similar to these, but the enclosing wall rock is not necessarily graphitic. This type of deposit assumes the character of a true lode.

Economic deposits of graphite include five main geological types:

1) Flake graphite disseminated in metamorphosed, silica-rich sedimentary rocks.

2) Flake graphite disseminated in marble.

3) Deposits formed by metamorphism of coal or carbon-rich sediments (amorphous).

4) Veins filling fractures, fissures, and cavities in country rock.

5) Contact metasomatic or hydrothermal deposits in metamorphosed, calcareous sedimentary rocks.

Natural graphite in these deposits varies widely in physical appearance. Klar (1958) classified graphite occurrences on the basis of crystal characteristics into microcrystalline dense graphites and macrocrystalline silvery-bright graphites.

Microcrystalline graphites occur in deposits of high carbon content as exceptionally small crystals discernible only by high-powered microscopes, while the macrocrystalline types generally occur in lower concentrations and in a larger crystal form.

Following are descriptions of the five main geological types.

Flake Graphite Disseminated in Metamorphosed Silica-Rich Sedimentary Rocks

A large part of the world's total production of graphite is derived from rocks such as quartz-mica schists, feldspathic or micaceous quartzites, and gneisses. The graphite flakes in these rocks are oriented so that they lie parallel to the plane of foliation. The graphite varies widely in physical appearance and characteristics. The flakes may range in size from a fraction of a millimeter to an average of 16 in., and may vary from one deposit to another in thickness, toughness, density, and shape.

The principal deposits of flake graphite occur in lenses or layers that may reach as much as 100 ft in thickness and extend for many thousands of feet. The lenses have a variable graphite content both within themselves and from one lens to another. The content of an average deposit is about 25%. Some deposits containing as little as 2% graphite have been worked. In Madagascar one rich lens contains 60% graphite. Crystals in each deposit vary in size, usually reflecting the grain size in the parent rock. In a moderately good deposit crystals ranging up to ¼ in. are common.

Description of Deposits: *Alabama*—The deposits of northeastern Alabama, in Chilton, Coosa, and Clay counties, occur in two northeast-trending belts that are 1 to 2 miles in width and that have a combined length of about 55 miles. Worked intermittently since the late 1800s, the mines reached their peak production during World War I. The deposits, called "leads," are groups of parallel layers and lenses in the Ashland quartz-mica schist that contain

1 to 5% disseminated flake graphite. The leads mined range from 20 to more than 100 ft in thickness, dip gently to steeply, and have been traced up to 4000 ft along strike. Some are essentially constant in strike and dip over long distances; others are folded, faulted, or both. The deposits are weathered to depths as great as 100 ft. Weathered "ore" has furnished most of the graphite produced but unweathered ore has been worked at some mines. In general, the flake ranges from less than 0.05–0.20 in. in diameter, but most of it is less than 0.05 in. in diameter. Assays of samples from 32 properties in the northeastern belt by the US Bureau of Mines gave the following results:

Size	Lb of Flake per Ton of Ore
+50 mesh	6 to 14
−50 +100 mesh	7 to 38
−100 mesh	8 to 70
Total flake	35 to about 130

Reserves of flake graphite in unweathered rock (measured, indicated, and inferred) are estimated by the US Bureau of Mines at 14,410,000 tons. How much of each of the commercial grades that could be produced from the deposits, however, is uncertain. Most of the graphite produced in Alabama during and since World War II has been stockpiled, but Alabama graphite has been marketed at one time or another for most of the uses to which graphite is put.

Texas—The flake graphite deposits of Texas occur in the Precambrian Packsaddle schist in Llano and Burnet Counties. The deposits show a wide range of graphite content and flake size. The most important deposit is one in Burnet County operated in recent years by the Southwestern Graphite Co. The deposit consists of flake graphite disseminated in mica schist and micaceous quartzite. These rocks are cut by dikes of granite and granite pegmatite and contain abundant lenses and knots of similar material. Reserves of graphitic rock are substantial. The flake recovered is mostly small and has been marketed chiefly for foundry facings and for lubricants, packing materials, and pencils.

Norway—In Europe the principal productive graphite deposits of Norway, West Germany (Bavaria), and Czechoslovakia, and the graphite deposits in lower Austria, all consist of flake graphite disseminated in mica schist or mica gneiss, or are closely associated with these rocks. The productive deposits of Norway (A/S Skaland Grafitverk) are on Senja Island, North Norway (latitude 60°39′N; longitude, 17°E) and are lenses of graphitic rock enclosed in mica schists. The richest ore contains 25 to 30% graphite. The ore is hard and must be crushed to recover the graphite. The refined products are crystalline battery-grade concentrates and fine-ground graphite concentrates. Exploration has reportedly shown reserves amounting to one million tons of rock averaging 25% carbon. Reserves of 500,000 tons are also reported at Jennestad in the Lofoten Islands, a little southwest of Senja Island, but half the graphite is said to be amorphous and difficult to concentrate.

West Germany—Deposits east of Passau, Bavaria (West Germany, on the Austrian border), were known to prehistoric man and have been worked since the Middle Ages. They consist of crystalline graphite in seams, lenses, and disseminated flakes in gneiss and schist. Individual seams and lenses up to several tens of feet in thickness are mined for hundreds of feet. At Kropfmuhl, the only mine now operated, some 20 folded seams of graphite are interbedded with marble and micaceous gneiss in a zone that is 450 ft in stratigraphic thickness. Seams 1 to 5 ft thick are mined. Graphite forms 10 to 30% of the lenses and averages 20 to 25% after hand sorting. Associated minerals are mainly feldspar and calcite, with less quartz, pyrrhotite, pyrite, biotite, hornblende, sphalerite, and galena. Both weathered and unweathered rock have been mined. Two grades of crucible flake (92 to 95%), three of pencil flake (96 to 98%), and five of fines are produced. There is some question whether the deposits are epigenetic or due to metamorphism of carbonaceous sediments followed by hydrothermal alteration.

Madagascar—Probably the largest resources of high-grade, crystalline flake graphite in the world are on the Island of Madagascar. The deposits occur in belts of micaceous gneiss and schist over a distance of more than 500 miles in the eastern half of the island, from the latitude of Tamatave to the southern end. The principal producing area extends about 70 miles along the east coast, from Tamatave to Marovintsy. There are ten principal producers. Individual deposits are graphite-rich layers that range from 10 ft to more than 100 ft in thickness; some can be followed for thousands of feet. The graphite content reportedly ranges up to 60%, but ore averaging 4 to 11% disseminated graphite has been worked in recent years. Operations are confined to the weathered upper parts of the deposits, and reserves of weathered ore are considered ample for many years at current rates of production. Flake graphite from Madagascar is noted for its high proportion of coarse flake, for its uniform thickness, toughness, and cleanness, and for the care with which it is graded for export. It sets the world standard for high quality flake.

Brazil—In the area of Itapecerica, Minas Gerais, Brazil, graphite deposits with 20 to 25% carbon are mined in an open cut. Flotation and final cleaning yield flake and amorphous products with carbon contents up to 99%.

China—The deposits in China yield carbons 80 to 99% and have a flake up to ¼ in. in size, but the majority of the production is confined to −30 +80 mesh and −100 mesh.

India—The deposits in India yield carbons 70 to 88% and have flakes in the primary range of −30 +80 mesh and −100 mesh.

Flake Graphite Disseminated in Marble

Flake graphite disseminated in marble probably derives from carbonaceous impurities. It is commonly less than 1% of the rock, although in some localities it is as much as 5%. Deposits of this type exhibit much variation in grade and kind of accessory minerals over short distances and are structurally complex. Their contribution to world production is consequently much less than that from deposits in schists and gneisses.

Deposits Formed by Metamorphism of Coal or Carbon-Rich Sediments (Amorphous)

The graphite in deposits formed by the metamorphism of coal or carbon-rich sediments is almost invariably of the microcrystalline variety known to the graphite trade as amorphous graphite. A substantial part of the world production of amorphous graphite is from such deposits. The graphite occurs in seams, often distorted by folding and faulting, and is frequently intimately mixed with ungraphitized material. The ratio of graphite to ungraphitized material can vary widely. For example, some high-grade Mexican varieties contain as much as 95% graphite, while low grade Korean graphites are frequently burned for fuel because of the high nongraphitic carbon content.

Description of Deposits: *United States*—In the United States, coals in the Narragansett Basin of Rhode Island have locally been metamorphosed to graphitic anthracite. The material has been mined intermittently for many years as a source of low grade amorphous graphite for such uses as paint pigment. Amorphous graphite also occurs southwest of Raton, New Mexico, where Cretaceous coal beds have been converted to graphite adjacent to dikes.

Mexico—The most important deposits of amorphous graphite in the western hemisphere occur in the State of Sonora, Mexico, in an area 20 miles long by 10 miles wide, approximately 40 miles southeast of Hermosillo. The deposits were discovered in 1867. They were early described as beds of coal interlayered with sandstone, folded, and then intruded by granite, thereby being converted to graphite. Six beds of graphite are reported, the thickest averaging 9 to 10 ft but locally swelling to 24 ft because of folding. As the result of recent detailed examinations, however, the deposits at the Moradillas mine are now believed to be hydrothermal veins; no evidence has been found to indicate that they are metamorphosed coal beds. The veins are said to occupy fissures that cross the bedding of the country rocks at low angles. In the district as a whole, the graphite deposits are reported to be steeply dipping and are said to range from mere stringers to seams that are locally 12 to 15 ft wide. High grade Mexican graphite averages 80% carbon, but some contains as much as 95% carbon.

Italy—The principal Italian deposits, near Turin (Pinerolo) consist of material rich in finely crystalline graphite. The deposits are as much as 19 ft thick, consisting of layers or lenses enclosed in micaceous gneiss, phyllite, and schist. The graphite seams locally enclose masses of anthracite. The graphite content is 60 to 80%. The material is used chiefly for foundry facings and paint.

Austria—In southeastern Austria, large resources of graphite are present in a 30 mile-long belt of folded metasediments extending from Leoben to Rottenmann, in the Styrian Alps. There are two producing mines. At Kaiserberg, the principal mine, a series of graphitic beds with an aggregate thickness of 40 ft are enclosed in graphitic schists and quartzites. Crude ore contains 30 to 80% carbon and averages 50 to 60%. Refined graphites ranging from 66 to 95% carbon are produced by grinding and flotation. A new chemical process that will yield graphite with 99% carbon is reported to have been developed. The products are all amorphous graphite and are used for foundry facings, pencils, paint, and lacquer pigments and lubricants.

South Korea—The important amorphous graphite deposits of southern Korea, which have yielded about 1 million tons, occur as irregular lenses parallel to the structure of enclosing schists and phyllites of sedimentary origin. In the Kyeng-Sang district three seams having mineable widths of 78, 18, and 48 ft, respectively, have been reported. Neither the nature of the source rocks of these seams nor the origin of the seams is clear from published data. Deposits formed by dynamic and contact metamorphism of coal beds occur in southern Korea, but Overstreet (1947) reports that these are impure. Korean graphite is low in carbon content and is mostly exported to Japan for foundry facings and pigment; much of it is burned for fuel. Reserves of graphite in southern Korea are said to be large.

Other Areas—Elsewhere, deposits formed by the metamorphism of coal are worked in northern Italy, in the Styrian Alps of Austria, and at Undercliff in New South Wales, Australia. At Mt. Bopple, Collinsville, Cape Upstart, and Killarney, and in the Gympie goldfield, all in Queensland, igneous intrusions have locally altered Permian or Jurassic coals to amorphous graphite.

Vein Deposits Filling Fractures and Cavities

Vein deposits of graphite are typically fillings of fractures or cavities in country rocks. The veins are commonly sharply defined, and the graphite filling is typically banded and zoned. In the thinner veins, graphite occurs as crusts or layers of closely packed, coarse, elongate plates (needle lump) oriented perpendicular to the vein walls. Plates vary in length with vein thickness and may be as much as 4 in. long. Central portions of thicker veins may consist of platy layers parallel to the vein walls. Masses of coarse graphite and rosettes of radiating plates also occur. Gangue minerals are present in some deposits, totally absent in others.

The origin of vein deposits of graphite is uncertain. Winchell (1911), discussed various hypotheses and has concluded that the most probable mode of origin involved deoxidation of oxides of carbon.

Description of Deposits: *United States*—Deposits of this type at Sturbridge, MA, were worked in the 1640s and were the basis of the first graphite mining in the United States. Other deposits are known near Ticonderoga, NY. The only sizable domestic vein deposits, however, are those of the Crystal Graphite mine near Dillon, MT. There graphite fills networks of fractures in gneiss and pegmatite around the northwesterly plunging nose of an isoclinal fold, a structure that involves granite gneiss, schists, marbles, and quartzites of the Precambrian Cherry Creek series. The veins are fractions of an inch to 2 ft thick and range from a few feet to 50 ft in length. They pinch and swell rapidly both along strike and dip. No gangue minerals are present in the veins. The formation of the veins has been linked with the development of the pegmatites, but Ford (1954) has shown that fracturing and vein formation took place after formation of these rocks. He favors an origin from carbon originally present in the country rocks and introduced by gaseous emanations. Textural and structural features of the veins suggest deposition from fluids at no great depth.

Sri Lanka—The vein graphite deposits of Sri Lanka (Ceylon) have been worked systematically for more than 140 years, yielding more than 1.3 million tons of graphite since 1880. The main deposits are in Southern, Western, and Sabaragamuwa provinces. During some years more than 2000 pits have been active concurrently, and mining has reached depths as great as 1600 ft. The deposits occur in a complex series of Archean granulites, quartzites, garnet-sillimanite gneisses, and marbles in narrow, structurally controlled belts that persist for many miles along strike. The productive deposits range from simple fissure veins to branching veins or stockworks. Veins commonly range from fractions of an inch to several feet in thickness. Some of the larger deposits are tabular bodies parallel to foliation of enclosing rocks. Graphite filled "vugs" up to 80 ft in length and 6 x 5 ft in cross section are reported. The highest grades of Sri Lanka graphite carry 97 to 98% carbon. Lumps of pure, coarsely crystalline, platy graphite as much as a foot or more in diameter are found, and both needle and platy crystalline lump graphite are produced. A variety of minerals, chiefly quartz, pyrite, calcite, apatite, and pyroxene, with rutile, allanite, magnetite, and various lime-magnesia silicates, has been reported to be associated with graphite in the veins, but it is not clear whether these minerals were deposited with the graphite or are merely minerals of the wall rocks accidentally enclosed in the veins. Reserves of graphite in Sri Lanka are regarded as very large.

Mexico—The most important deposits of amorphous graphite in the western hemisphere occur in the state of Sonora, Mexico. Deposits of amorphous graphite occur in the area southeast of Navajoa approximately 25 miles from Empalme, but they are low grade and are not of economic importance at the present time.

The five companies that produce amorphous Mexican graphite are, in order of magnitude: Grafitos Mexicanos, S.A.; Compania Minera

Moraguirre; Grafitera de Sonora S.A. de C.V.; Grafito Superior S.A.; and Explotadora Sonorense.

Substantially more than half of the total production comes from the Lourdes Mine at Moradillas, owned and operated by Grafitos Mexicanos, S.A. These deposits, interlayered with sandstone, consist of steeply dipping N-S striking graphite veins ranging from small stringers to seams that at depth average 10 to 12 ft wide. Some veins more than 20 ft wide are being mined. The graphite content averages 95%. One mine, now defunct, averaged 90% carbon with pockets in excess of 95% carbon.

Other deposits were mined in the late 1940s in the Tonichi area, approximately 50 miles northeast of Guaymas. The graphite in this area was high grade, containing 85% carbon, and some veins as much as 5% carbon. However, the extraction process and the location of the mine, plus the long haulage to the railroad station at LaDura made it difficult to produce economically. After 1950, a dam was built in the Tonichi-Onavas-LaDura area, thereby causing all mining operations to cease.

Other—Vein deposits have also been reported from southern India (e.g., in the Madras province), from Madagascar, and from Brazil. Vein deposits also appear to be represented at the Sennotani mine, Toyama prefecture, Japan, and in certain deposits in Manchuria.

Contact Metasomatic or Hydrothermal Deposits in Marble

Concentrations of graphite occur in various parts of the world in silicified carbonate rocks. Some are clearly contact metasomatic; others show gradations to typical hydrothermal deposits. As a group, they are characterized by variable but generally small size, variability in flake size and content, and irregularity of form. The average grade of most deposits is low, and, consequently, they have accounted for only a minor part of total world production.

In the United States and Canada, deposits of this type occur at a number of places in the Grenville series of northern New York, southwestern Quebec, and southeastern Ontario. Small deposits west and northwest of Ticonderoga, NY were worked between 1850 and 1900. They consist of fine-to-coarse flake graphite irregularly concentrated in silicified marble adjacent to bodies of pegmatite. Some deposits in Quebec and Ontario are similar, but others appear to have been formed by metasomatic replacement of marble subsequent to silication. The most important Canadian graphite deposit, the Black Donald mine at Calabogie, Ont., which was worked until 1954, is an example. It is described as a thickened, northwesterly plunging lens of drag-folded ore more than 1000 ft long, 15 to 70 ft thick in one direction and 50 to 80 ft in another. The ore contained up to 55 to 65% graphite, but averaged 25%. The flake produced was mostly fine-grained, and 80% was sold to the foundry industry. Several grades of lubricant flake, two averaging 97 to 98% carbon were also produced.

Contact metamorphic deposits in silicified marbles also occur in Sosangun, Korea.

Regional Occurrence, Production, and Use

Asia

Republic of Korea: The Republic of Korea is the world's largest producer of natural graphite and produces both amorphous and flake. In 1961, the ratio was about seven of amorphous to one of flake.

Fifteen amorphous-graphite mines yielded 719,400 st in the seven-year period 1955 through 1961. Principal mining areas are in the provinces of Shung Chong Puk Do, Kiong Sang Puk Co, and Kang Won Do. Almost all the deposits are in the vicinity of anthracite coal mines, which lends support to the idea that amorphous graphite is derived from coal.

The refined amorphous graphite is 80% carbon or higher. One sample of amorphous lump was 87.86% carbon.

Flake graphite mines yielded 3400 st in the seven-year period 1955 through 1961. Of four mines, two are in the Seoul area. Material with 90% carbon remaining on a 60-mesh sieve has been mined.

The nongraphitic portion of Korean graphite is mainly silico-aluminous material. The iron content of associated minerals is the lowest of all known natural graphites; the ash is a distinctive white.

Exports were 131 kt of amorphous and 1.9 kt of flake for the period 1959 to 1961; 80% of exports were to Japan. The Korean government restricts the exportation of anthracite coal, and it is often shipped as graphite. Therefore, the actual tonnage of graphite exported is probably less than the reported figure.

Estimated reserves are 3.3 million st of amorphous graphite averaging 75% carbon and 1.8 million st of flake averaging 80% carbon.

Sri Lanka (Ceylon): Sri Lanka contains the largest known deposits of vein graphite. Very large deposits are located in the western and southwestern sections of the island. The Dutch government first reported graphite deposits in Sri Lanka in 1675.

The extent of the deposits has inspired much study, but, although the resulting literature is extensive, little is known about the origin of this vein graphite. According to Wadia (1943):

> Graphite occurs in a large area in the southwestern part of Ceylon in the extremely metamorphosed series of Archean sediments known as the Khondalite system. Disseminated graphite in granitic gneisses and in sillimanite-garnet rocks was probably formed by metamorphism of carbonaceous impurities, but vein graphite is secondary and fills cracks, fissures and foliation and joint planes. The vein graphite deposits are believed (by some) to have been formed by reduction, at high temperatures and pressures, of carbon dioxide liberated during the assimilation of limestones and dolomites by charnockitic and related igneous magmas. Pyrite, quartz, calcite and augite are the minerals most commonly associated with graphite; others include scapolite, wollastonite, forsterite, tremolite and phlogopite. Reserves of mineable graphite are believed to be large.

The structures contrast markedly with other forms of graphite and are typical of ores deposited as vein, dike, pegmatitic, or pneumatolytic occurrences. One feature is that the crystals grow inward from both margins of the vein toward the center. Changing conditions during the formation have also resulted in many parallel bands of columnar structure. These bands may vary considerably in width, texture, hardness, and purity. Many variations from the regular crystalline structure are seen. Often a jumbled maze of interlocking fibers is produced, or the fibers may radiate from centers, forming rosettes of graphite. Large flexible folia occur and bear the typical twinning striations so often seen on flake graphites. In places slippage has taken place, drawing out the fibers to form the so-called needle variety of graphite, or an amorphous dust may be formed.

The great bulk of high-quality vein graphite still comes from Sri Lanka. Despite rising labor costs, a considerable proportion of the production is still prepared by hand cobbing and hand sorting in order to satisfy those buyers who still retain a preference for lump grades. On the other hand, one of the larger operators has found it expedient to install complete mechanical beneficiation equipment for handling particle sizes below 6μ. Beneficiation in this case is based largely on the use of pneumatic tables, from which it is possible to obtain a wide range of carbon contents. Since most graphites are eventually utilized in industry in the form of powder, present operators anticipate a trend toward the direct supply of the powder from the mine to the importer. The requirements for hand sorting would be reduced progressively.

Sri Lanka graphite is highly graphitized; the best grades are completely graphitized. It is favored for lubrication purposes, pencils, electromotive brushes, and other uses requiring quality graphite.

The State Graphite Corp. of Sri Lanka is in charge of the mining operations and the sale of the Sri Lanka graphite.

Other Asian Sources: In 1847, Jean Pierre Alibert, a French merchant residing in Tawasthus, Asiatic Siberia, brought to Europe the knowledge of graphite in Siberia. He had discovered graphite in one of the mountain gorges near Irkutsk and traced the source to the summit of Mount Batougal (later renamed Mount Alibert) in the Saian Mountain range. In 1856, Alibert made an agreement with the German pencil company, A. W. Faber, that only Siberian graphite would be used in the production of Faber's pencils. The graphite is said to have been carried from the mountains to the ocean on the backs of reindeer.

One graphite deposit was worked on West Brother Island, off Hong Kong. The product is similar to Korean amorphous graphite.

The best graphite in India is found near Begumpet in Hyderabad. Deposits also occur near Bihar and Orissa. The Indian deposits are incompletely explored.

Europe

Austria: Two distinct mineralogical regions, Styria and Lower Austria, combine to make Austria the world's second largest known producer of natural graphite. As in all other European deposits, the graphite is thought to have originated by the metamorphism of carboniferous and bituminous substances under high pressure, temperature, and rock deformation. The deposits, therefore, occur in comparatively small lenses and, with one exception, must be mined underground.

Styria contains the more important deposits. A lode of graphitic phyllite extends from the Semmering over Aflenz and Bruck/Leoben into the Palten Valley. The major mines, some of them two centuries old, occur near Kaiserberg and Triben. Reserves are estimated at several million tons. Geologically, these Alpine deposits are embedded in a zone of fine grained, scaly schists and occur as lenses and individual beds up to 35 ft thick. The ore is black, soft, fine grained, and dense, with carbon content of 40–88%. It is low in sulfur (0.2–0.3%) and iron and practically free of carbonate and phosphate. The sorted ore is ground to 200 mesh and used in foundries and steel mills.

The first and only flotation plant for amorphous graphite is in Kaiserberg. The process was developed in the laboratories of Franz Mayr-Melnhof & Co.'s Grafitbergbau Kaisersberg, where fine (2μ average particle size) graphite of 92% carbon is produced, suitable for lubrication, pencils, pigment, and mud drilling. Work is under way to produce purities suitable for nuclear use.

Lower Austrian graphite is harder and more crystalline than that of Styria; moreover, it contains sulfides, carbonates, lime, and pyrrhotite. It is mostly used in foundries and in blast furnaces. It occurs in an area from Spitz, on the Danube, northward to the Bohemian massif in a belt of highly metamorphosed and coarse grained schists, partly of sedimentary origin and always accompanied by marble. Modern earthmoving equipment used in open pit mining has made these deposits economical to exploit. About 60,000 tons of graphite 40–45% carbon is produced annually for use in blast furnaces.

West Germany: The Passau district (Bavaria) has long produced flake graphite suitable for crucibles. Old documents frequently refer to this graphite, which probably furnished the graphite in the Middle Ages for alchemists' crucibles. Mines at Gruben (near Taxberg) are mentioned. By law the inhabitants of Pfaffenreuth in 1250 had to deliver graphite worth one-tenth of their income as a tax. In 1496, Deggendorf levied a tax of four Regensburger pfennigs on each 100 crucibles from the Passau district.

Today, Graphitewerk Kropfmuhl A.G. produces flake graphite exceeding 95% graphitic carbon suitable for crucibles, pencil leads, and lubricants. This company also chemically purifies graphite to 99.9+% carbon in several grades of fineness. The country rock is a part of the "kristallines Grundgebirge," the old gneissic and schistose rocks of the Bohemian massif; the gneissic rocks are considered to have been metamorphosed during the Carboniferous period.

Typically four or five graphitic beds averaging 5 ft thick are mined. Thicker layers occur where there is folding. Some mines have as many as eight layers. The ore runs 20–25% graphitic carbon. It is beneficiated and processed by flotation, grinding, and screening.

Other European Sources: Europe is more self-sufficient in natural graphite than the United States. In 1962, Austria, West Germany, and Norway were able to supply the full complement of kinds and purities of graphites required by all European industry.

Norway—A/S Skaland Grafitverk is the only operating mine in Norway. The deposit, on Senja Island, northern Norway, is of the vein type. One million tons of 25% carbon graphite is the estimated ore reserve. Flotation brings the carbon content up to 80–93%.

Italy—The principal Italian deposits are amorphous graphite associated with gneiss, slate, chalk, and marble. Several mines are located in the Germanasca and Chisone valleys in northern Italy near the French-Italian border. Another deposit is at Bormida, near Genoa. The main workings are southwest of San Germano, about 6600 ft above sea level; the graphite is brought down into the valley for processing.

Spain—Although Spain produces little graphite, there are large deposits in the province of Jaen and near Toledo. A graphite mine operates in Los Cotos de Guadamur.

North America

United States: Graphite production in the United States is inadequate to supply the demand. The one flake producer, Southwestern Graphite Co., of Burnet, TX, closed its mine operation in 1980. However, other mines could be reopened in an emergency. Some 20 to 25 states have deposits that were worked in the past. Most of the deposits are of either fine flake or low grade ores. A deposit of lump graphite is near Dillon, MT. The graphite plants in the United States process imported graphites that are beneficiated, ground, and blended as required to maintain the individual company grades.

Flake graphite deposits in both Alabama and Pennsylvania have been worked intermittently. The latest closing in Alabama was 1953 and in Pennsylvania, 1961. The only domestic production of amorphous graphite ceased in Rhode Island in 1959.

Canada: Canada is endowed with many deposits of commercially important flake graphite. It seems probable that many deposits are yet to be discovered because flake graphite is a common constituent of the Grenville-type Precambrian marble and gneiss of Ontario and Quebec.

Mining started in 1846 in Grenville, Que., and many small properties in the vicinity of

Buckingham were producers between then and 1936.

The only sizable Canadian operation was the Black Donald mine, about 65 miles west of Ottawa. It produced high quality flake graphite between 1897 and 1954, when it closed. The principal country rock is the Grenville marble, associated with which are bands of hornblende schist and pegmatite dikes.

In 1981, Asbury Carbons began production at its Canadian mine operation. It has the capability of producing −30 +80 mesh flake graphite ranging in carbon content from 75 to 95%. With the present know-how, it can be upgraded to a purity of 99.5% in carbon should demand increase.

Mexico: The state of Sonora contains extensive deposits of amorphous graphite and here are located Mexico's graphite mines and mills.

The Sonora deposits are homogeneous, contain about 80% graphitic carbon, and are an important source of amorphous graphite. They were discovered in 1867 and exploitation began in 1891. There are said to be as many as seven distinct beds of graphite within alternating layers of metamorphosed andalusite-bearing Triassic rocks.

The mines under active exploitation are located about 250 miles southeast of the US border in the region of Moradillas. This region is arid, and water for the operations is scarce. Trucking of graphite to the railroad during the brief rainy reason is sometimes interrupted for several days at a time.

Other amorphous graphite deposits are reported in the states of Baja California, and Coahuila in Northern Mexico, and in Hidalgo, Guerrero, and Oaxaca in central and southern Mexico.

The mineral suite for amorphous ore includes mica, clay minerals, tourmaline, and hematite with occasional pyrite and gypsum. The ore is prepared for shipment by hand sorting the admixed rock that comes from the hanging walls; then it is crushed, screened, and dried.

Africa

Malagasy Republic (Madagascar): The island of Madagascar is one of the most important producers of graphite because of the size and quality of its flake. Occurrences are widespread along the eastern coast near Tamatave, in the central section near Tananarive, and in a southern section between Betroka and Bekily. Reserves are undetermined, but considered to be exceptionally large. The first commercial interests were started by French settlers. Present production comes from the deposits along the eastern coast. The main producers are Société Minière de la Grande Ile (also manages the interests of Société Lyonnaise Agriole, Minière et Industrielle, SLAMI), Establissments Gallois, Establissments Izouard, Société Louys, and Establissments Rostaing (Pettifer, 1980). Each company markets its own graphite; graphites from different deposits are not blended.

The deposits are in a region where lateritic deposits of iron and bauxite also are found. The graphite, being resistant to weathering, is found among the weathered residue. Thus, quartz granules from the original crystalline rocks are common, the feldspathic constituents having been completely converted to clay. Some of the mica has survived in partly altered form and is usually found interlaminated between lamellae. The graphite occurs in highly metamorphosed schists and gneisses that have weathered deeply to soft, ferruginous clays. The graphitic content of the original rock has been raised through the natural leaching process.

The crude ore runs from 3 to 10% graphite, of which about two-thirds is large flake, and one-third fine flake. Commercial interest is primarily in the large flake.

The flake is large, strong, and flexible and is the best graphite for refractories. It is also favored for crucibles. The pyrometric cone equivalent (PCE) of the ash runs from cone 16 to cone 20 and is higher than the ash of other flake graphites.

The nongraphitic portion of the flake is uniform and consists (after oxidizing) chiefly of a clayey residue with some quartz. There exists a sprinkling of accessory minerals such as mica, zircon, rutile, and epidote. Sulfur compounds are absent and iron is low. Manganese may run higher than in most graphites. Potassium predominates in the alkali metal content. Carbonates are absent. Volatile matter, including water, is about 1.5%.

Other African Sources: South-West Africa (Namibia) produces an 85% carbon graphite in the district of Bethanien.

Deposits exist in the Transvaal at Carolina and Mopani.

Kenya graphite was mined in open pits near Kanziku. The ore runs about 10% graphitic carbon.

Two mines operate in the Republic of South Africa, at Cumbu and at Mutali. The ore is high in flake graphite, on the order of 20%, but the final product runs about 80% carbon.

Other Sources of Graphite

South America: Itapecerica in Minas Gerais, Brazil, is estimated to produce yearly some 1500 tons of a medium quality vein graphite and is the one commercially important source of Brazilian graphite. Some of this graphite is reportedly refined by flotation to 99% carbon.

Limited production in Argentina is reported. Some deposits are found in the mountains of Pie de Palo; the graphite mined there is shipped to Rosario for beneficiation. The final product is reported to be on the order of 65% carbon.

Australia and New Zealand: There has been small production from Mountain Bopple, Queensland; from Sleaford Bay, Eyre's Peninsula, South Australia; and from Undercliffe, New South Wales. There are also small deposits in West Australia. A deposit about 30 miles from Cloncurry, Queensland, is reported to have large reserves. Several occurrences are also reported in New Zealand.

USSR: Information on Russian deposits is still fragmentary. Resources of graphite, chiefly in Asiatic Russia, have repeatedly been stated to be large. Most of the principal types of deposits are probably represented. For example, thermally metamorphosed coal seams in the Tungusk coal basin west of Irkutsk yield amorphous graphite containing 87 to 94% carbon. Flake graphite deposits are present in the Ukraine, contact-metasomatic deposits have been reported from western Uzbekistan, and small vein deposits are reported from the Alibert mine and from the Krasnaya Polyana region, North Caucasus. The main Alibert deposits are unusual in that the graphite occurs as disseminated flakes and masses in nepheline syenite and in lenses along contacts of nepheline syenite and marble.

Exploration and Development

Prospecting has consisted primarily of outcrop examination, trenching, and sampling. This more-or-less standard procedure has from time to time been augmented by either churn drilling or diamond core drilling. However, the nature of disseminated deposits in particular makes accurate evaluation of grade and tonnage difficult to achieve by only drilling. The successful program of churn drilling near Burnet, Texas, by the Southwestern Graphite Co. in 1928 may have been one of the first to employ the drill as part of an exploration and development program. A diamond drilling program at Notre Dame du Laus in Canada in 1956 was abandoned when the test results showed widely varying percentages.

During World War II several more integrated programs involved geological mapping of specific areas both in Alabama and Pennsylvania in addition to trenching, pitting, and sampling. In 1942 there was some small production at one of these, the Benjamin Franklin Mine, Chester County, PA; the ore was of low grade (3%), however, and mining ceased in 1943.

Because of the high electrical conductivity of graphite, geophysical methods such as resistivity and self-potential have been employed in the search for deposits in Japan, Australia, India, and, to a limited extent, North America.

In the past it appears that the small economic return from graphite mining, at least in North America, has discouraged the launching of any large, integrated, or very sophisticated exploration programs.

Mining Methods

During the period 1890 to 1920, underground mining of graphite was practiced in New York and Pennsylvania. From 1942 until the end of World War II, open pit operation was the only method used, primarily because of the comparative ease of working in weathered rock. Graphite was mined underground at Dillon, MT during World War II and shortly thereafter until extraction of the ore became too costly to compete with Sri Lanka graphite and all operations ceased.

Madagascar operations are entirely open pit, but in Sri Lanka, Korea, Mexico, Bavaria, and Italy, underground mining is essential due to the depth and physical characteristics of the deposits.

Mexican underground mining operations range from a depth of 300 to 1100 ft below the surface, measured on the angle of the vein. Some of the older mines in Sri Lanka reached depths in excess of 1500 ft on a vertical plane.

For many years mining operations in Sri Lanka were primitive and ore extraction slow and cumbersome. The mines were fully mechanized after World War II.

Madagascar operations also were primitive because of the low labor cost. After 1938 the mines began to use mechanical equipment to remove the overburden, and the graphite-bearing schists were easily removed by bulldozers and tractors.

Preparation for Market

In Sri Lanka, the graphite ore is extracted in lumps and hand cobbed on a sorting patio to remove any quartz inclusions. The finished product is graded as follows large lump, 97 to 99% carbon; ordinary lump, 94 to 96% carbon; ordinary lump, 90% carbon; chips, 85 to 90% carbon; and chippy dust, 80% carbon. However, the chippy dust and flying dust are not produced in any quantity today.

In Madagascar the crude ore averages 10 to 12% carbon. It is crushed by a primary crusher and then conveyed to a series of roll crushers and classifiers to remove the oversizes and gangue. The −¼-in. product is then passed to rougher cells, which separate the graphite flakes from the clay and sand. The product of the rougher cells passes to the finishing cells where each particle of graphite is recovered by means of a pine oil frothing process. This yields an 80 to 84% carbon concentrate, which is dried; any residual pine oil is volatized. The run-of-mill product is upgraded by mechanical means, such as rod-mill grinding, to a 50 mesh US Standard sieve with 85 to 90% carbon. Higher carbon concentrates (up to 95%) are obtained by regrinding and flotation.

At the mill of Graphitwerk Kropfmuhl, S.A., in Bavaria, crude ores carrying 10 to 12% carbon are crushed, passed through a series of rolls, and finally a rod mill, which reduces the ore to 40 mesh. This product is fed to a series of flotation cells, and the graphite concentrate is filtered and kiln dried until the final product averages 90 to 93% carbon. For further refinement this product is milled and screened, producing grades with 95 to 96% carbon.

At Skaland, Norway, the crude ore averages 25 to 30% carbon. The crushing and flotation process used at this mine is similar to those already described. Grades of 80, 85, and 90% carbon are produced. The mill had a capacity of 12,000 tons in 1971.

Uses

Foundry Facings: The largest single use for natural graphite is foundry facings, which account for more than one-third of the tonnage and about one-fifth of the value of the total. This use, combined with graphite used in steelmaking (such as carbon raiser, hot topping, and ingot wash), accounts for about one-half of the total consumption of graphite and about one-third of the total value of all natural graphite used in the US.

Batteries: Natural graphite is suitable for batteries that are required to furnish small currents over long periods of time. However, it is not particularly efficient for use in the type of battery currently used in flashlights and has been replaced by a form of carbon black. With rural electrification and the declining use of large dry batteries, graphite consumption has declined. Since carbon black is satisfactory and cheap, there seems to be little opportunity for natural graphite to reverse this downward trend through product improvement.

Bearings: Graphite is used extensively in bearings but the reactions both of producers and consumers to the potential for increased use varies with respect to the type of bearings being produced. One bearing producer thinks the use of graphite will increase because of the increased sales of fully molded, self-lubricating, phenolic bearings. Graphite use might also be stimulated if graphite could be included in high-speed, high-load bearings.

Another producer of powder-metal bronze bearings uses graphite only when special formulations are specified by its customers. Despite this specialized use, graphite consumption has increased, particularly in ferrous-base formulations. This producer has found natural graphite better suited to his purpose than artificial.

The use of graphite in bearings subjected to extreme heat or low humidity causes problems. For example, if the humidity is low, graphite will act as an abrasive, scoring the surface with which it comes in contact.

Use of natural graphite in bearings should continue to increase, but the rate of increase might be accelerated if the problems of using it under conditions of high load, high speed, high temperature, and low humidity could be solved.

Brake Linings: As long as organic friction materials impregnated with graphite are used and accepted as satisfactory, the consumption of graphite should increase gradually in accordance with the increased number of vehicles on the road. Graphite is particularly useful in brake linings for heavy-duty vehicles as opposed to passenger vehicles.

Molybdenum disulfide is the only potential competitor, but it has been tried with only limited success in some applications and has a decided cost disadvantage.

The appearance of new, iron-based, sintered, friction materials containing up to 30% graphite may be significant. If these are utilized, the use of graphite in friction materials could increase more rapidly in the future. One consumer has suggested the desirability of a reduced content of abrasive alumina and silica in graphite but has emphasized that cost is more important than minor improvements along this line.

Crucibles: Consumers vary in their opinion as to the future trend of clay-graphite vs. carbon-bonded crucibles and the relative importance of fine vs. flake graphites. The disagreement centers primarily around how important the use of more oxidation-resistant, clay-graphite compositions with molten aluminum will become. The introduction of the clay-graphite bell in the immersion-bell technique for converting cast iron to ductile iron may create a new market.

The recent gradual decline in the use of graphite in crucibles may halt and a slight recovery is possible. On the negative side, the increase in ladle sizes will cause less use of stoppers and the consequent reduction in continuous casting might eliminate the need for stoppers. However, these negative trends might be balanced by the greater use of clay graphite for nozzles and perhaps for the well block.

One consumer has suggested the need for greater uniformity in graphite shipments, especially with respect to carbon content and particle size distribution, and a need for the reduction of impurities in graphite for stopper use.

Carbon Brushes and Other Electrical Uses: Natural graphite encounters considerable competition from artificial graphite for this use. Despite this competition, the tremendous growth of the electrical industry is expected to result in increased consumption of natural graphite. It seems that where artificial graphite has advantages it has already captured the market. For example, in the production of sliding contacts the variation in the ash content of natural graphite has proved unacceptable; now artificial graphite in which the abrasive materials can be controlled is being used.

Lubricants: This market is tremendously complex because of the variety of lubricating jobs that must be done and the imposing array of types of lubricants that are available. Graphite has been an important lubricant and will probably continue to be, but competition is stiff.

Graphite added to a soapy lubricant is considered to be a good die lubricant, particularly for galvanized wire, though the discoloration it produces is not a desirable feature. Natural graphite is a good antiseize agent in steel mill and railroad applications because it creates a solid lubricant film that retains its properties under pressure and high temperature. In this particular use the graphite content is specified by the customer; hence its use may not be based on strictly technical grounds. Natural graphite is used in greases where the grease application is such that the siliceous portion of the natural graphite is overshadowed by the indigenous contamination of dirt and sand, i.e., gear lubrication for mining machinery.

Graphite of colloidal size dispersed in kerosene, oil, or some other carrier can be used when a high-temperature film with a quick-drying characteristic is desired in high-temperature lubricating jobs, i.e., conveyors passing through ovens.

Packings: Fine flake graphite used with oils or greases is very satisfactory because of its resistance to heat up to 600°F. Closely controlled particle size is not too important for packing use. Use of graphite will tend to remain steady, but it may decline because equipment being manufactured today doesn't utilize old style packings. One consumer has mentioned that very small quantities of molybdenum offer excellent lubricating qualities.

Pencils: The use of pencils is increasing and, with it, the total market for graphite will increase. There is some shift from Sri Lanka graphite to Mexican and domestic amorphous graphite because of the grayish cast of the Sri Lanka variety. The pencil industry seeks to make improvements that would lead to greater jetness, higher carbon content, smaller particle sizes, and more unctuousness (opacity) in the graphite. However, such improvements must not add to the cost of graphite to the pencil manufacturer.

Refractories: The demand for graphite in linings that will impart non-wetting characteristics when in contact with molten metals should remain constant. Any improvement in oxidation resistance would be desirable. Though the quantity involved is small, the use of graphite should increase gradually in graphite magnesite refractories, plastic refractories, and patching materials.

Prices

The *Chemical Marketing Reporter,* formerly the *Oil, Paint and Drug Reporter,* quotes prices on an ex-warehouse basis. June 13, 1983 quotations:

	$ per Lb
No. 1 flake graphite, 90 to 95% carbon	0.45–0.75
No. 2 flake graphite, 90 to 95% carbon	0.45–0.75

	$ per Lb
Powdered crystalline graphite:	
88 to 90% carbon	0.25–0.60
90 to 92% carbon	0.35–0.65
95 to 96% carbon	0.45–0.75
Powdered amorphous graphite	0.14–0.38
Powdered amorphous or crystalline graphite, minimum of 97% carbon	0.60–1.00

April, 1983 prices, f.o.b. sources, were quoted in *Engineering & Mining Journal* for two major classifications of graphite imported by the United States:

	$ per Mt, 1983
Flake and crystalline graphite, bags:	
Sri Lanka	600–1700
Germany, West	350–3500
Malagasy Republic	250–700
Norway	300–900
China	250–1700
Amorphous, nonflake, cryptocrystalline graphite (80 to 85% carbon)	
Mexico (bulk)	70–100
South Korea (bags)	90–120

TABLE 1—Tonnage and Value of Graphite Imports, Exports, and Consumption in the US, and World Production, 1976-1980

		1976	1977	1978	1979	1980
United States:						
Apparent consumption*	short tons	†66,862	†73,773	90,396	77,562	52,438
Exports	do	12,236	13,783	9,595	8,623	8,880
Value	thousands	$2,388	$2,662	$2,304	$3,741	$3,695
Imports for consumption‡	short tons	79,098	87,556	99,991	86,185	61,318
Value	thousands	$6,753	$8,058	$11,700	$13,035	$15,765
World: Production	short tons	ʳ495,481	ʳ546,022	ʳ583,030	ʳ590,774	597,429

Source: Taylor, 1981.
ʳRevised.
*Excludes domestic production.
†Revised to include some manufactured graphite imported for consumption.
‡Includes some manufactured graphite.

TABLE 2—Consumption* of Natural Graphite in the United States, 1980, by Use

	Crystalline		Amorphous†		Total	
Use	Quantity (short tons)	Value	Quantity (short tons)	Value	Quantity (short tons)	Value
Batteries	W	W	W	W	1,737	2,178,963
Brake linings	933	959,438	2,013	1,534,062	2,946	2,493,500
Carbon products‡	182	243,258	408	360,378	590	603,636
Crucibles, retorts, stoppers, sleeves, nozzles	W	W	W	W	2,340	2,063,869
Foundries	1,366	1,092,086	6,466	2,411,549	7,832	3,503,635
Lubricants§	867	1,176,091	5,521	1,983,113	6,388	3,159,204
Pencils	977	1,502,433	340	337,236	1,317	1,839,669
Powdered metals	288	360,528	112	182,287	400	542,815
Refractories	1,062	224,887	11,577	2,161,400	12,639	2,386,287
Rubber	31	24,894	241	167,931	272	192,825
Steelmaking	386	164,875	9,373	4,732,611	9,759	4,897,486
Other††	6,928	5,471,546	805	693,470	3,656	1,922,184
Total	13,020	11,220,036	36,856	14,564,037	49,876	25,784,073

Source: Taylor, 1981.
ʳRevised. W Withheld to avoid disclosing company proprietary data; included with "Other."
*Consumption data incomplete. Small consumers excluded.
†Includes mixtures of natural and manufactured graphite.
‡Includes bearings and carbon brushes.
§Includes ammunition, packings, and seed coating.
†† Includes paints and polishes, antiknock and other compounds, drilling mud, electrical and electronic products, insulation, magnetic tape, small packages, and miscellaneous and proprietary uses.

TABLE 3—Graphite: World Production, by Country

(Short tons)

Country*	1976	1977	1978	1979[p]	1980[e]
Argentina	160	94	28	11	†13
Austria	36,439	38,898	44,645	44,664	44,000
Brazil (marketable)	6,634	10,127	11,417	11,979	13,200
Burma‡	177	106	309	295	300
China mainland[e]	55,000	66,000	88,000	110,000	110,000
Czechoslovakia[e]	49,600	49,600	49,600	49,600	49,600
Germany, Federal Republic of[e]§	[r]10,528	[r]9,178	7,034	4,047	4,000
India	42,189	53,412	70,310	56,141	†53,787
Italy	4,242	4,210	4,528	4,522	4,500
Korea, North[e]	[r]22,000	[r]22,000	[r]22,000	28,000	28,000
Korea, Republic of:					
Amorphous	42,193	68,904	59,288	59,789	†65,209
Crystalline flake	3,762	3,799	2,793	2,704	†1,575
Madagascar	19,193	17,336	18,326	15,699	16,000
Mexico:					
Amorphous	66,510	64,410	57,611	56,086	61,000
Crystalline flake	—	—	—	—	200
Norway	9,999	10,028	12,292	13,226	12,200
Romania[e]	6,600	6,600	6,600	6,600	6,600
Sri Lanka	[r]9,138	9,783	11,581	10,397	10,600
South Africa, Republic of	584	1,004	643	434	—
Thailand	33	33	25	25	25
Turkey	NA	NA	NA	231	220
USSR[e]	105,000	105,000	110,000	110,000	110,000
United States	W	W	W	W	—
Zimbabwe	[e]5,500	[e]5,500	[e]6,000	6,324	6,400
Total	[r]495,481	[r]546,022	583,030	590,774	597,429

Source: Taylor, 1981.

[e]Estimated. [p]Preliminary. [r]Revised. NA, Not available. W, Withheld to avoid disclosing company proprietary data.

In addition to the countries listed, Namibia may have produced graphite during the period covered by this table, but output is unreported and available general information is inadequate for formulation of reliable estimates of output levels.

†Reported figure.

‡Data are for fiscal year beginning Apr. 1 of that stated.

§Series revised; data now presented represents estimated marketable product derived from raw graphite mined indigenously, assuming that marketable output equals one-half of officially reported raw graphite production.

Graphite is granted a depletion allowance of 22%. Prices of imported graphite were generally higher, which indicated that importers were, for the time being, absorbing the cost increases and not passing them on to their customers.

Nevertheless, published prices for natural graphite merely represent the range of prices. Actual prices are negotiated between buyer and seller on the basis of a wide range of specifications.

Recent Production

Tables 1, 2, and 3 present basic data on the production, consumption, imports, and exports of graphite in the United States and on world production.

Manufactured graphite was produced at 28 plants in 1981 with additional production for in-house use likely. Table 4 lists US producers of manufactured graphite. Manufactured graphite was produced at other plants in the United States, but their output could not be included in this tabulation. Most of these plants produced specialty products, which were small in total tonnage, although their unit values were relatively high.

TABLE 4—US Producers of Manufactured Graphite

Company	Plant location	Product*
Airco Carbon, Div. of Airco, Inc.	Niagara Falls, NY	Anodes, electrodes, crucibles, motor brushes, refractories, unmachined shapes, powder.
Do	Punxsutawney, PA	
Do	St. Marys, PA	
Avco Corp., Avco Specialty Materials Div.	Lowell, MA	High-modulus fibers.
The Carborundum Co., Graphite Products Div.	Hickman, KY†	Electrodes, motor brushes, unmachined shapes, cloth.
Do	Niagara Falls, NY	
Do	Sanborn, NY	
Celanese Corp., Celanese Research Lab.	Summit, NJ	High-modulus fibers.
Fiber Materials, Inc.	Biddeford, Maine	Do.
BF Goodrich Co., Engineered Systems Div., Super Temp Operation.	Santa Fe Springs, CA	Other.
Great Lakes Carbon Corp.	Morganton, NC	Anodes, electrodes, powder.
Do	Niagara Falls, NY	
Do	Rosamond, CA	
Hercules Inc.	Salt Lake City, Utah	High-modulus fibers.
HITCO Materials Group ARMCO Co.	Gardena, CA	Cloth and high-modulus fibers.
Pfizer Minerals, Pigments & Metals Div.	Easton, PA	Other.
Poco Graphite, Inc.	Decatur, TX	Unspecified.
Polycarbon, Inc.	North Hollywood, CA	Cloth.
Sigri Carbon Corp.	Hickman, KY	Electrodes, other.
The Stackpole Corp., Carbon Div.	Lowell, MA	High-modulus fibers, anodes, motor brushes, unmachined shapes, powder.
Do	St. Marys, PA	
Superior Graphite Co.	Chicago, IL	Powder and other.
Do	Hopkinsville, KY	
Union Carbide Corp., Carbon Products Div.	Clarksburg, WV	Anodes, electrodes, unmachined shapes, motor brushes, powder, cloth, high-modulus fibers.
Do	Columbia, TN	
Do	Fostoria, Ohio	
Do	Greenville, SC	
Do	Niagara Falls, NY	
Do	Yabucoa, PR	

Source: Taylor, 1981.
*Cloth includes low-modulus fibers; electric motor brushes includes machined shapes; crucibles includes vessels.
†Plant sold to Sigri Carbon Corp., May 31, 1980.

Bibliography and References

Alling, H.L., 1917, "The Adirondack Graphite Deposits," Bulletin 199, New York State Museum, 150 pp.

Bascom, F., and Beach, L.M., 1922, "Graphite" and "History of Graphite Mining in Pennsylvania," *Mineral Resources of the United States, 1919*, Pt. 2, *Nonmetals*, US Geological Survey, pp. 309–324.

Bastin, E.S., 1912, "The Graphite Deposits of Ceylon, with a Description of a Similar Graphite Deposit near Dillon, Montana," *Economic Geology*, Vol. 7, No. 5, Aug., pp. 419–443.

Cameron, E.N., 1960, "Graphite," *Industrial Minerals and Rocks,"* 3rd ed., J.L. Gillson, ed., AIME, New York, pp. 455–468.

Cameron, E.N., and Weis, P.L., 1960, "Strategic Graphite, a Survey," Bulletin 1082–E, US Geological Survey, pp. 250–252.

Chelf, C., 1943, "Graphite in Llano County, Texas," Mineral Resources Circular 57, Bureau of Economic Geology, Texas University, 10 pp.

Clark, T.H., 1921, "The Origin of Graphite," *Economic Geology*, Vol. 16, No. 3, Apr.-May, pp. 167–183.

Cole, J.W., 1965, "Graphite," *Mineral Facts and Problems*, Bulletin 630, US Bureau of Mines, pp. 399–409.

Ford, R.B., 1954, "Occurrence and Origin of the Graphite Deposits near Dillon, Montana," *Economic Geology*, Vol. 49, No. 1, Jan.-Feb., pp. 31–43.

Gupta, B.C., 1949, "Graphite," *Indian Minerals*, Vol. 3, No. 1, p. 17–31.

Jones, W.B., 1929, "Summary Report on Graphite in Alabama," Circular 9, Alabama Geological Survey, 29 pp.

Kemberling, S.M., and Walker, P.L., Jr., 1961, "Progress Report on Natural Graphite Studies," Fuel Technology Dept., The Pennsylvania State University, University Park, pp. 1–6.

Klar, G., 1958, "The Important Graphite Deposits of the World," *Mining Magazine*, pp. 98, 137–142.

Lewis, R.W., 1970, "Graphite (Natural)," *Mineral Facts and Problems*, Bulletin 650, US Bureau of Mines, pp. 1025–1038.

Miller, B.L., 1912, "Graphite Deposits of Pennsylvania," Report 6, Pennsylvania Topographical and Geological Survey, 147 pp.

Murdock, T.M., 1963, "Mineral Resources of the Malagasy Republic," Information Circular 8196, US Bureau of Mines, pp. 112–121.

Needham, A.B., 1946, "Mining and Milling Operations of Southwestern Graphite Co., Burnet County, Texas," Information Circular 7339, US Bureau of Mines, 3 pp.

Overstreet, W.C., 1947, "Graphite Deposits of Southern Korea," *Economic Geology,* Vol. 42, No. 4, June-July, pp. 424–425.

Paige, S., 1911, "Mineral Resources of the Llano-Burnet Region, Texas," Geological Atlas, Folio 183, US Geological Survey, 16 pp.

Pallister, H.D., and Thoenin, J.R., 1948, "Flake Graphite and Vanadium Investigations in Clay, Coosa, and Chilton Counties, Alabama," Report of Investigation 4366, US Bureau of Mines, 84 pp.

Pettifer, L., 1980, "Natural Graphite—The Dawn of Tight Markets," *Industrial Minerals,* No. 156, Sept., pp. 19–39.

Pettifer, L., 1980a, "Synthetic Graphite—Electrodes Continue to Lead," *Industrial Minerals,* No. 158, Nov., pp. 19–31.

Piper, E.L., 1973, "Manufactured Graphites," SME Preprint 73H14, AIME Annual Meeting, Chicago, IL, Feb., 15 pp.

Rauch, H.W., Sr., 1971, "Graphite Fiber: Another Wonder Material," *Ceramic Age,* Vol. 87, No. 5, May, p. 24.

Schanz, J.J., Jr., 1960, "Natural Graphite Market Research," Progress Reports Nos. 1, 2, Jan. 1-Nov. 30, College of Mineral Industries, The Pennsylvania State University, University Park.

Shobert, E.I., II, 1964, "Natural and Synthetic Graphites," SME Preprint 64H56, AIME Annual Meeting, New York, Feb., 14 pp.

Seeley, S.B., 1963, "Natural Graphite," *Encyclopedia of Chemical Technology,* Vol. 4, 2nd ed., Kirk-Othmer, ed., Interscience Publishers, New York, pp. 304–335.

Spence, H.S., 1920, "Graphite," Canada Dept. of Mines, pp. 1–9.

Taylor, H.A., Jr., 1978–1979, "Graphite," Preprint, US Bureau of Mines, p. 9.

Taylor, H.A., Jr., 1981, "Graphite," *Minerals Yearbook, 1980,* US Bureau of Mines, pp. 375–384.

Tron, A.R., 1964, *The Production and Uses of Natural Graphite,* Her Majesty's Stationery Office, London, Feb., pp. 1–83.

Wadia, D.N., 1943, "Origin of the Graphite Deposits of Ceylon," Professional Paper 1, Records, Dept. of Mineralogy, Ceylon, Pt. 2, pp. 15–24.

Weis, P.L., 1973, "Graphite," *United States Mineral Resources,* Professional Paper 820, US Geological Survey, pp. 277–283.

Weis, P.L., and Salas, G.A., 1978, "Estimating Reserves of Amorphous Graphite in Sonora, Mexico," *Engineering and Mining Journal,* Vol. 174, No. 10, Oct., pp. 123–128.

Willard, D.G., 1971, "Graphite," *Minerals Yearbook 1971,* US Bureau of Mines, pp. 561–567.

Willard, D.G., 1971, "Graphite," *Minerals Yearbook 1972,* US Bureau of Mines, pp. 589–595.

Winchell, A.N., 1916, "A Theory for the Origin of Graphite as Exemplified in the Graphite Deposits near Dillon, Montana," *Economic Geology,* Vol. 6, pp. 218–230.

Gypsum and Anhydrite

FRANK C. APPLEYARD *

The two calcium sulfate minerals—gypsum and anhydrite—occur in many parts of the world, and gypsum has long been of economic importance in the family of industrial minerals. Gypsum, the dihydrate form of calcium sulfate ($CaSO_4 \cdot 2H_2O$) and anhydrite, the anhydrous form ($CaSO_4$) are frequently found in close association, and it is seldom that a calcium sulfate deposit will consist exclusively of one mineral or the other.

Although known gypsum deposits are extensive, anhydrite makes up the largest part of total calcium sulfate reserves. However, it has very minor economic use, and most of the following discussion will be devoted to gypsum.

Calcium sulfate is one of the principal constituents of evaporite deposits, and when pure, has the following composition:

	Lime (CaO)	Sulfur Trioxide (SO_3)	Combined Water (H_2O)
Gypsum	32.6%	46.5%	20.9%
Anhydrite	41.2%	58.8%	—

Deposits of pure gypsum or of pure anhydrite which are large enough to be considered commercial have never been found because of both the metastable relationship between the two minerals, and the presence of impurities such as calcium or magnesium carbonates, chlorides, other sulfate minerals, clay minerals, or silica. As a result most mine production of gypsum will range between 85 and 95% pure. Often it is used as mined, although in certain cases, one or more forms of mineral beneficiation are employed to upgrade the product.

End Uses

The largest use for gypsum is based upon the unique property which calcium sulfate has of readily giving up, or taking on, water of crystallization. With the application of a moderate amount of heat in a process known as calcining, gypsum is converted to plaster of paris (the hemihydrate of calcium sulfate, $CaSO_4 \cdot \frac{1}{2} H_2O$) which when mixed with water will set or harden as the calcium sulfate returns to the dihydrate form. This semifinished product, usually called stucco, is then manufactured into a large variety of plasters, wallboard, and block for construction use, or into plasters for industrial applications. About 75% of the gypsum used in the United States is calcined for these purposes.

Uncalcined uses of gypsum are principally as a retarder for portland cement, as a soil conditioner, as a mineral filler, and other minor industrial applications. About 25% of the gypsum mined in the United States goes into these markets; however, in less developed countries where construction and other industrial uses may be quite limited, the use of uncalcined gypsum for portland cement retarder may well be the dominant—if not the only—market for the mineral.

It has often been noted that calcium sulfate constitutes the world's largest reserve of sulfur, and minor use of both anhydrite and gypsum has been made to produce sulfuric acid or other sulfur compounds such as ammonium sulfate. However, because sulfur is readily available at lower cost from other sources, use of calcium sulfate has been limited to only a few locations where it could compete economically. No such use is reported in the United States at this time.

Mineralogy

Anhydrite, $CaSO_4$, because of its geologically rapid conversion to gypsum, and relatively high solubility (about 0.2 g per 100 g H_2O) is not often found outcropping in climates wet enough to support abundant vegetation except on steeply dipping slopes or other places where hydrated material is continuously removed. Anhydrite rock is most often light to bluish gray in color, and under the microscope, varies

* Formerly Vice President, Raw Material Resources, US Gypsum Co., Chicago, IL. Now retired, Tubac, AZ.

from granoblastic (ver Planck, 1952) to felty-lath crystal aggregates. Fig. 1 shows felty-lath anhydrite crystals from Shoals, IN.

Bassanite, $CaSO_4 \cdot \frac{1}{2}H_2O$, is a distinct phase intermediate between anhydrite and gypsum, but is identifiable only by X-ray diffraction or petrographic techniques using very carefully prepared samples. Bassanite is metastable under ordinary conditions (Wood and Wolfe, 1969); however, its occurrence in amounts under 1% is suspected to be widespread in calcium sulfate mineral deposits.

Gypsum, $CaSO_4 \cdot 2H_2O$, is the commonly observed calcium sulfate mineral in rock outcrops because of its diverse origins. It is easily distinguished from anhydrite by its inferior hardness (2.0 vs. 3 to 3.5) and lower specific gravity (2.2 to 2.4 vs. 2.7 to 3.0). Most gypsum is white to grayish white although the impurities in any given deposit frequently determine the color of the rock and resulting products. Petrographically, most rock gypsum is granoblastic (Goodman et al., 1957). Commercial deposits show a great variety of crystal sizes and textures, both from deposit to deposit and within a single deposit. Relatively undisturbed deposits often contain texturally consistent stratigraphic units, but in deposits that have been deformed the gypsum is often recrystallized and shows a variety of textures. Many deposits exhibit porphyroblastic textures wherein two distinct sizes and ages of crystals occur (Holiday, 1970), and in some deposits fibrous gypsum similar to felty-lath anhydrite makes up the mass (Van't Hoff et al., 1903).

Alabaster is a compact, very fine grained variety of rock gypsum prized by sculptors for its uniform workability under the chisel, and occasionally is found within commercial deposits. Fibrous gypsum composed of needle-shaped crystals in orientation parallel to the C-axis (Schmidt, 1914), is known as **satin spar** and is a type of stress mineral indicative of deformation. It is universally a secondary mineral and occurs widely as a fracture filling wherein the needle lengths are perpendicular to the fracture walls, and less commonly in shear zones wherein the needles are parallel to the direction of movement. Large euhedral gypsum crystals and large cleavages commonly known as **selenite** form in fluid-filled spaces or in an easily deformable host material, and are sometimes mistakenly identified as mica by the uninitiated. Both satin spar and selenite are of little or no economic importance other than being accessories in rock gypsum, although one deposit consisting primarily of coarse selenite crystals is being worked near Apex, NV (Bergstrom, 1961). Fig. 2 shows selenite from Sperry, Iowa.

Origin and Occurrence

The calcium sulfate minerals are deposited by precipitation from aqueous solution when the concentration of the components and the physical conditions are suitable, with the majority of the deposits originating from evaporation and concentration of marine brines in a dry climate, i.e., evaporite conditions. That this is the basic mechanism of formation of economic calcium sulfate deposits is unquestioned; however, there is considerable discussion in the literature as to the physical environment in which precipitation takes place, and as to whether the originally precipitated mineral is gypsum or anhydrite.

The older, well established concept of de-

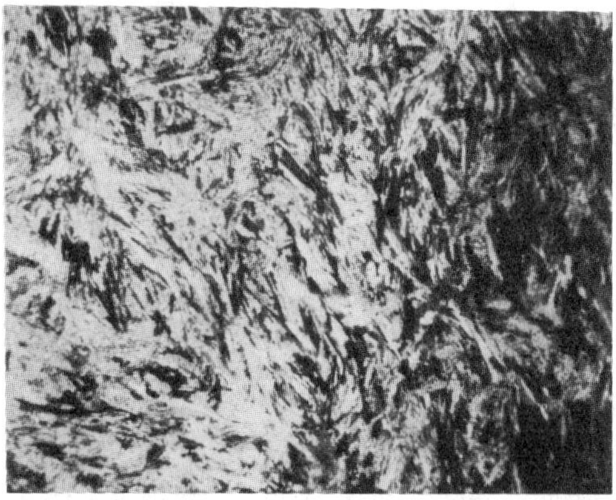

FIG. 1—*Felty-lath anhydrite from Shoals, IN. X39, polarized light.*

FIG. 2—*Selenite crystal from Sperry, Iowa.*

posit formation idealizes a brine-filled basin having restricted circulation (Branson, 1915; King, 1971) which permits limited replenishment of the brine as it evaporates. The resulting concentration causes precipitation of the contained salts in the inverse order of their solubility, and substantial thickness of essentially monomineralic deposits (e.g., calcium sulfate) are visualized as being derived from this type regime, an example of which is the A subzone of the Mississippian Windsor series of southeastern Canada.

A more recent concept of evaporite deposition has resulted from considerable work done in the Trucial Coast of Arabia (Butler, 1970; Kinsman, 1966, 1969) which describes the sabkha supratidal environment wherein calcium sulfate minerals are currently being deposited (*sabkha*, an Arabic word denoting salt-flat—Kinsman, 1966). These deposits are characterized by a distinctive sequence of sediments including lagoonal limestones, intertidal algal mat limestones, and precipitation of gypsum crystals and nodular anhydrite in the preexisting host sediments, which if originally calcium carbonate, become dolomitized. Clastic sediments, if present, are unaffected chemically, but are contorted and displaced as the gypsum crystals and nodular anhydrite grow in them. In the sabkha regime a basin is not required for deposition, although some concentration of the original seawater brine does take place in alongshore lagoons. Thick chloride accumulations would not be expected (Schroeder, 1970). The Mississippian evaporites of southwestern Indiana illustrate this type of deposit (Jorgensen and Carr, 1972).

The classical in-basin accumulations of sulfates and chlorides might be expected to have sabkha deposits along the margin of the basin, and the Silurian gypsum of Ohio, Ontario, and New York is possibly of this origin. The individual sulfate beds in sabkhas do not commonly exceed 20 to 40 ft in thickness, although 200-ft accumulations have been described in Jamaica, W.I. (Holliday, 1971).

Gypsum may also precipitate along fractures, bedding planes or in other available spaces where ground water carrying sulfate ions from the oxidation of sulfides comes into contact with carbonate rocks. Such deposits, however, are of limited extent, and rarely—if ever—could be expected to have economic importance.

Because gypsum will precipitate wherever brine concentration is appropriate, and recrystallizes or dissolves readily as the metastable or temporary environments change, interpretation of geologic evidence to determine exact depositional environments is difficult at best. Several recent works (Dean and Schreiber, 1978; Kendall, 1978; and others) are helpful in determining depositional environments.

Gypsite deposits usually occur in semiarid and arid climates, and result from the solution of existing gypsum deposits by ground water, which in turn is drawn toward the surface by capillary action where it evaporates to deposit the dissolved sulfate as a porous aggregate of gypsum containing considerable impurities. Also, in a few cases, reworking of existing deposits by wind erosion has concentrated gypsum in sand deposits, as at White Sands, NM or Cuatrocienegas, Coahuila, Mexico.

Of totally different origin are those calcium sulfate deposits which constitute part of the cap rock of salt domes in the Gulf Coast basin. The exact origin of these domes is not fully understood but most prevailing views involve flowage and upward movement of plastically deformed salt in response to overburden pressure from the overlying sediments (Martinez, 1974). Once the upward movement is initiated, it is possible for the top of the lighter salt mass to be injected into and through the overlying sediments.

The theory most widely accepted as to the origin of the cap rock is that it represents a residue of anhydrite and other relatively insoluble minerals which accumulate at the upper surface of the salt dome as it rises through water-bearing sediments and the salt is dissolved. The anhydrite crystals, which make up about 99% of the water insoluble residue (Walker, 1974), are then recrystallized and compacted to form a massive rock which may be locally hydrated to form gypsum.

The total height of a salt dome can exceed 10,000 ft, with the cap rock being well over 1000 ft. Most domes are roughly circular in plan view, and range in diameter from about 3280 ft to more than 4 miles, although there is great diversity in dome geometry and dome-to-cap-rock relationships. Most domes do not reach the present surface, although occasionally cap rock is found at or near the surface and might be of economic significance.

Distribution of Deposits

Calcium sulfate minerals have been found within every geologic system from Silurian through Quaternary. The largest production in North America is from Mississippian rocks in southeastern Canada, Michigan, Indiana, and Virginia, followed by that from Permian rocks in Texas, Oklahoma, Kansas, Colorado, Wyoming, Nevada, and Arizona. Production from Tertiary rocks is obtained in California, Jamaica, and Arizona, from Jurassic rocks in Iowa, New Mexico, Colorado, Utah, Wyoming, and Montana, and from Silurian rocks in New York, Ontario, and Ohio. Elsewhere, Tertiary rocks in Australia, France, and Italy, Permian and Jurassic rocks in England, and the Triassic in France, Germany, and England contain large commercial deposits. Commercial deposits range in thickness from 40 in. to over 100 ft, and reserves of individual deposits are usually measured in millions of tons. Production rates commonly average 200,000 to 300,000 tpy from each operation, but range from a few thousand tons up to about two million tons.

Although gypsum is produced from rocks of many ages, many deposits are Quaternary in age due to the recent and continuing conversion of anhydrite to gypsum that occurs under certain conditions. This conversion of anhydrite to gypsum, as well as the reverse reaction, has been extensively studied because of its importance in the original sequence of deposition from brines, as well as in metamorphism that occurs in calcium sulfate deposits after deposition. Several reviews have been made (Braitsch, 1971; Hardie, 1967; Mac Donald, 1953; Wood and Wolfe, 1969) of such variables as brine composition, temperature, water pressure, and overburden pressure. From these it is evident that no one set of conditions applies to all deposits. On the contrary, all these variables, plus structural, lithological, and topographical considerations which might affect water migration, must be examined in order to understand the degree of hydration of any given calcium sulfate deposit. To date, however, only limited attempts have been carried out in an effort to integrate theoretical models with field evidence of rock textures and structures.

Most commercial gypsum deposits are believed to have resulted from the action of surface and ground waters upon anhydrite, and hydration has been found at depths ranging from zero to more than 2000 ft below the surface. In many deposits, the degree of hydration is readily predictable once the variables controlling it are understood; however, in others the degree of hydration can vary widely, with corresponding reduction in the amount of available reserves, and increase in the cost of mining. In all cases, the operator is well advised to thoroughly explore and understand the anhydrite-gypsum relationship prior to development of a mine.

Impurities in Calcium Sulfate Deposits

The number of minerals which can occur in evaporite deposits is quite large (Braitsch, 1971), and many can be found as impurities in calcium sulfate deposits. Their occurrence is often dependent upon the mode of formation of the deposit. In varying degrees, most gypsum and anhydrite deposits contain clastic sediments, commonly clay minerals and fine sands, as well as chemical sediments such as limestones and dolomites. In sabkha-type deposits, these preexisting sediments will fill intermodular spaces of the gypsum and anhydrite, or form crude layers, and often appear to be replaced when they actually are displaced. Relatively insoluble evaporite minerals such as celestite, certain borates, some carbonate minerals, and silica may occur in calcium sulfate deposits as discrete crystals, crystal aggregates or nodules, and probably are in many cases original depositional features. However, strontium or boron, which may be present in only trace amounts in anhydrite, may migrate during gypsification to result in strontium and boron mineral accumulations.

Soluble evaporite minerals, e.g., halite, sylvite, mirabilite, epsomite, and others are also frequently found in calcium sulfate deposits. These too are expelled from anhydrite during gypsification, and most commonly, become associated with the clay minerals that may be present, probably being adsorbed on the clay mineral surfaces. They also are associated with the carbonate impurities where they may be absorbed into mineral surfaces, held as fluid inclusions, or included in a disordered lattice. In addition, these soluble minerals can occur as microfracture filling. The hydration halo of mixed gypsum-anhydrite commonly contains many times the soluble salt content of either unhydrated anhydrite or hydrated gypsym. Individual deposits show great variation in the quantity and type of soluble salts present, i.e., sulfates or chlorides, and it has been noted that, in general, chloride salts predominate in gypsum deposits of eastern North America, whereas sulfate salts exceed chlorides in western North American deposits.

In addition to the foregoing evaporite mineral impurities whose origin is generally a feature of original calcium sulfate deposition, impurities in any given mining operation may also be encountered as a result of structural features such as tight folding or faulting, or from recent erosional activity. These usually take the form of clays, sands, or gravels and occur most often in near surface deposits, although structural deformation can result in severe impurity problems at considerable depth. More often than not, the greatest quantity of impurity in a gypsum deposit is anhydrite (and vice versa), a result of the metastable relationship of these two minerals as previously discussed.

As utilized, most gypsum contains from 10 to 15% of impurities, although some deposits may be exceptionally pure (i.e., plus 95%) or somewhat impure (i.e., 80%). In general, the amount of impurity which can be tolerated depends upon: (1) the type of impurity, (2) the product being manufactured, and (3) the competitive situation.

Based upon the effect impurities may have upon the manufacturing process and finished products, impurities may be separated into three categories:

1) Insoluble or relatively insoluble minerals such as limestone, dolomite, anhydrite, anhydrous clays, silica minerals, etc.

2) Soluble chloride minerals such as halite, sylvite, etc.

3) Hydrous minerals such as the sulfate salts mirabilite and epsomite, and the montmorillonite group of clays.

The first category, to the extent that the minerals replace gypsum, reduce the strength of rehydrated stucco and increase weights of the finished plaster or wallboard, i.e., more pounds of an impure stucco are required to obtain a given strength. Nevertheless many commercial gypsum deposits contain as much as 10 to 15% of these insoluble impurities. The presence of the second category, chloride salts, affects calcining temperature, and stucco slurry consistency and set time. These impurities are usually limited to no more than 0.02 to 0.03%. The principal impact of the third category is in moisture pickup of the finished product and on bonding characteristics of the gypsum stucco core of wallboard to its paper covering. Hydrous sulfate salts must be limited to 0.02 to 0.03%; however, hydrous clays up to 1.0 to 2.0% may be tolerated.

Major North American Producing Areas

The principal gypsum producing areas in North America are shown on Fig. 3. Table 1 lists the North American gypsum production. In addition, there are many more known, but undeveloped gypsum deposits, especially in the midwestern and western states, southeastern Canada, and Mexico. Domestic reserves are geographically distributed in 23 states and have been estimated to contain 20 billion tons, or approximately 2000 years' supply at current production rates (Schroeder, 1970). Quantitative information is lacking on gypsum reserves in the rest of the world, but deposits are widespread, and inferred reserves are considered to be unlimited in relationship to requirements.

A brief description of the major North American producing districts follows.

Southeastern Canada

Marine evaporites of Mississippian (Visean) age underlie much of southern New Brunswick, northern Nova Scotia, and southwestern Newfoundland. The first production of gypsum in North America was recorded in Nova Scotia about 1735, and has grown to make this the largest gypsum mining region in the world, with an annual production exceeding 6 million tons.

Calcium sulfate is found in, or is correlative with, the Windsor Group which has been divided into five subzones: A, B, C, D, and E in ascending order. Although calcium sulfate is present in each of these subzones, the economic deposits of gypsum and anhydrite are confined to the Lower Windsor, i.e., subzones A and B. The most extensive is the A subzone, consisting of up to 1000 ft of anhydrite with occasional thin dolomite beds. Where topographic conditions are favorable, surface hydration, which rarely exceeds 30 to 40 ft, has produced several deposits of dense, white, high purity gypsum. Fig. 4 shows the operations in the Windsor B subzone of the Little Narrows Gypsum Co., Nova Scotia.

The B subzone, which varies in thickness from 250 to 500 ft, consists of interbedded calcium sulfate, limestones, and siltstones. The calcium sulfate, representing about 80% of the section, has been hydrated to depths varying from 150 to 700 ft, although occasional remnant anhydrite lenses with associated soluble salts are encountered. Unlike most gypsum deposits which have only minor structural features, if any, the rocks of the B subzone are frequently deformed by complex plastic folding and by occasional faults of substantial displacement, which make for difficult mining conditions. Nevertheless, the B subzone accounts for approximately 75% of the region's total production.

The southeastern Canada gypsum deposits are usually covered by glacial till which may be as thick as 200 ft, but averages 40 to 50 ft. All mining is by open pit methods, and in 1979 eight different companies operated in the region.

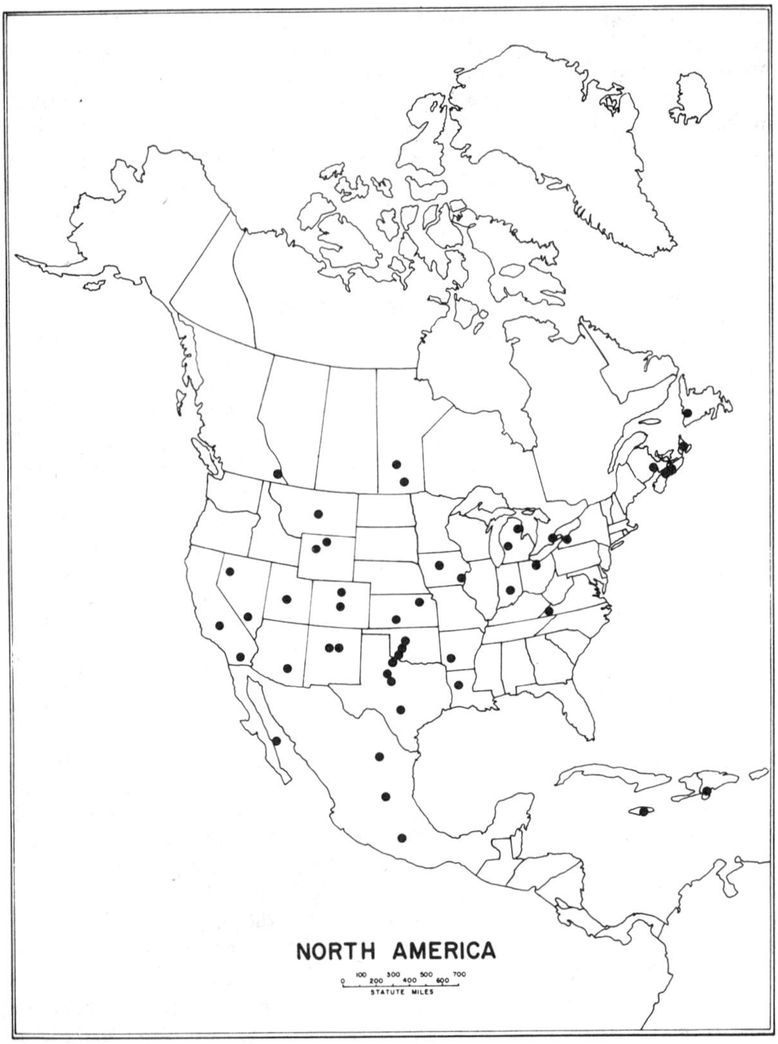

FIG. 3—*Location of North American gypsum-producing districts.*

Iowa

The largest single gypsum producing district in the United States is at Fort Dodge, Iowa, where mining began in 1872. Long considered to be an outlier of the Permian Basin, this deposit is now thought to be Jurassic in age (Cross, 1966). It consists of a single bed of gypsum up to a maximum of 30 ft in thickness, but which probably averages nearer 12 to 14 ft. Its areal extent of some 15 square miles appears to be limited by post-depositional erosion, particularly during the period between Late Cretaceous and the Pleistocene. Its continuity is interrupted by wide channels apparently formed prior to and/or during the glacial period, channels which cut through the gypsum into the underlying shales.

Overburden consisting of glacial till, varies from 20 to 60 ft, and the rock surface is commonly cut with mud-filled erosion crevices which must be cleaned out, thereby posing a mining problem; however, gypsum purity as used is usually plus 90%. Mining is by open pit methods, with four companies operating in the district. Total production in 1979 was approximately 1,250,000 tons.

Another gypsum operation at Sperry in southeastern Iowa went into production in 1961, and at 600 ft below the surface, is the deepest currently operating gypsum mine in North America. It is reached through a 15-ft-

TABLE 1—Gypsum Production, North America

Country	Thousand Short Tons 1978	1979ᵖ	1980
Canada:			
Nova Scotia	6,117	6,216	5,435
Newfoundland	892	884	733
New Brunswick	55	57	47
Ontario	817	845	946
Manitoba	201	141	150
British Columbia	819	790	825
Subtotal	8,901	8,933	8,136
United States:			
Arizona	184	231	209
California	1,578	1,624	1,640
Colorado	235	275	227
Iowa	1,602	1,695	1,468
Michigan	2,765	2,526	1,382
Nevada	1,335	1,075	852
New Mexico	263	251	182
Ohio	171	151	136
Oklahoma	1,398	999	1,326
Texas	1,864	1,903	1,681
Utah	316	772	287
Wyoming	370	366	312
Other States	2,812	2,762	2,674
Subtotal	14,893	14,630	12,376
Mexico	1,928	2,100	1,884
Dominican Republic	190	100	190
Jamaica	144	64	66
Total	26,056	25,827	22,652

Source: Pressler, 1979; Stonehouse, 1978, 1979, 1980.
ᵖPreliminary.
*Included with Other States.

diam concrete lined shaft, uses a room-and-pillar system in an average 9 to 10 ft of mining height, and is located in the Wapsipinicon formation of Devonian age. This calcium sulfate was apparently originally deposited as anhydrite, with hydration apparently confined to a zone paralleling the edge of the subcrop. Although the areal extent of the calcium sulfate is probably quite large, commercial gypsum reserves are restricted to this hydrated zone.

Michigan

Another important gypsum source in the midwest is the Michigan Basin, with two producing districts: Grand Rapids in the southwestern portion of the basin, and Alabaster-National City near the northeasterly edge on Lake Huron. These deposits are in the Michigan formation of Mississippian age, and consist of multiple units of gypsum 5 to 40 ft thick, separated by beds of shale varying from a few feet to 40 or 50 ft thick.

Three operations are active in the Alabaster-National City area, using open pit mining methods to recover three gypsum beds (separated by 5 ft of shale) at Alabaster, and a single bed at National City and Turner. This production serves seven manufacturing plants: one at National City, and after water shipment via the Great Lakes, plants at Waukegan, IL; East Chicago, IN; Detroit, MI; Lorain and Gypsum, Ohio; and Clarence Center, New York; as well as numerous cement plants. Overburden is glacial till ranging in thickness from 40 to 70 ft, and at Alabaster, where the gypsum grades into anhydrite as the overburden thickness increases, a minor amount of anhydrite is also produced.

Two operations are also active in the Grand Rapids district, both being shallow (less than 100 ft deep) underground mines employing the room-and-pillar system. Two different beds, 8 to 12 ft thick, are being mined with the production utilized by manufacturing plants at the mine sites.

New York—Ontario

Gypsum is found in a belt of Upper Silurian rocks which extends eastward from the Niagara Peninsula in Ontario across New York to the vicinity of Utica. The first recorded mining of gypsum in the United States took place in this belt in 1808 near Syracuse. In recent years, however, all production has come from the Oakfield area near Buffalo in the western part of the state, where one operation currently is active. The economic gypsum here consists of a 36- to 48-in. thick bed in the Salina Group of Upper Silurian age, overlain by 10 to 40 ft of thinly bedded shales and dolomites, covered by glacial till ranging up to 50 ft in thickness. Mining is by underground methods using the room-and-pillar system, and the gypsum averages about 88% purity.

In the westerly end of this belt, two underground mines in the Caledonia-Hagersville area of Ontario are important producers. The Hagersville mine also operates in a 36 to 48-in. bed which is correlative with the gypsum being mined in the Oakfield-Clarence Center district

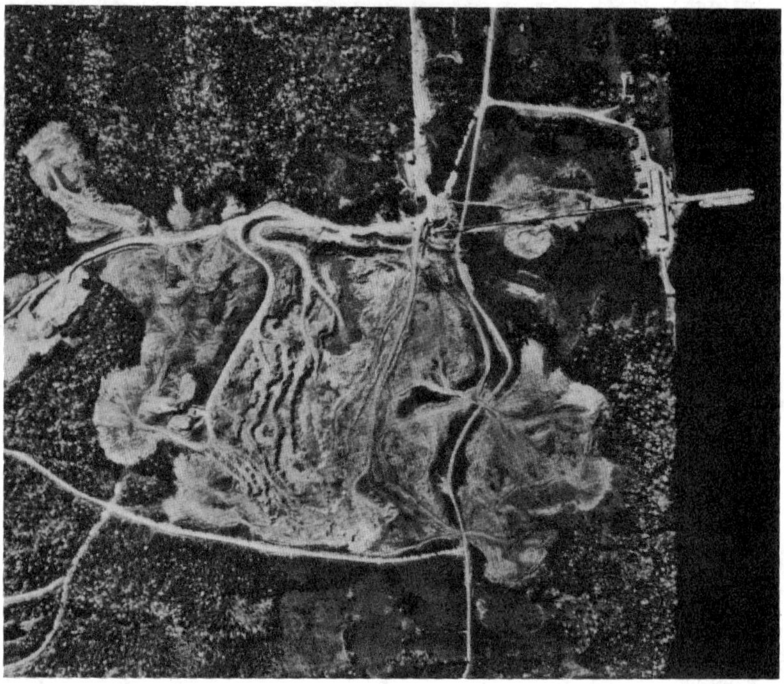

FIG. 4—*Operation of Little Narrows Gypsum Co., Cape Breton Island, Nova Scotia.*

of New York, giving these operations the dubious distinction of having the lowest mining heights in the gypsum industry of North America. However, the Caledonia mine is in a bed 8 to 10 ft thick that stratigraphically is approximately 200 ft below the Hagersville seam.

In addition to the Caledonia-Hagersville mines, a new operation came into production approximately 30 miles northwest of this area in 1978 near the town of Drumbo, mining a bed averaging 5.5 ft thick, about 380 ft below the surface.

Ohio

On the shores of Sandusky Bay of northern Ohio, gypsum has been continuously mined since 1849, with one open-pit mine being currently active. Several beds of calcium sulfate are found in the Tymochtee formation of Upper Silurian age, all of which are relatively impure due principally to inclusion of dolomite stringers and nodules, and to a lesser degree, anhydrite. As mined, this gypsum is probably the lowest grade of any commercial deposit in the United States (with the exception of gypsite in California) being in the 65 to 70% range; however, mine production is beneficiated by heavy media sink-float to produce an approximately 80 to 81% gypsum product.

Indiana

In 1952, gypsum was discovered in southern Indiana near the town of Shoals, and this has become one of the most important commercial gypsum producing districts in the United States. In part of the Illinois Basin, gypsum occurs in the St. Louis formation of Mississippian age, and lies 350 to 550 ft below the surface of this rather hilly area. The bed averages 14 to 16 ft thick, with room-and-pillar mining methods being employed by the two mines operating in the district.

The St. Louis limestones overlying the gypsum have been subjected to solution along bedding and joint planes to create a network of water-filled cavities of undetermined extent and location. This results in the creation of a flooding hazard to the mine workings, and one of the mines has been flooded on two different occasions. However, careful study of all the factors affecting the hydrology of the area, plus the adoption of precautionary operating measures has enabled mining to proceed with an acceptable level of safety.

Virginia

In the western tip of Virginia near Saltville, gypsum has been produced continuously since 1835, first from shallow pits or quarries and currently from one underground mine. Another mine located at the small community of Plasterco has the distinction of being the world's deepest gypsum mine, having reached a depth of 1300 ft below the surface. However, operations ceased in 1979 due to depletion of reserves and high operating costs. Fig. 5 shows a load-haul unit on the 1300-ft level, Plasterco, VA.

The calcium sulfate originated in the McCready formation of Mississippian age, but the deposits are structurally very complex as a result of Appalachian mountain building, and in particular, the Saltville fault, a major northeast-southwest trending thrust fault which in the Plasterco area has a vertical displacement of over 14,000 ft. As mined, the calcium sulfate is found in part as a more or less continuous bed, approximately vertical and varying in width from 8 to 10 ft up to 20 ft; and in part as remnants or fragments of an originally horizontal bed (or beds) known locally as "boulders." The host rocks are shales and limestones, and mining is done by developing horizontal levels at 100-ft intervals from which continuous prospecting is carried out to locate and provide access to either "boulders" or the vertical bed. A combination of sublevel and cut-and-fill mining methods are used to extract the mineral, which is then brought to the surface by conveyor belt (after underground crushing) up a 1330-ft 18° incline (Appleyard, 1965).

Hydration of the calcium sulfate is highly irregular, and as with most deposits, is a function of the availability of ground waters. As a general trend, there is less hydration at depth than near the surface; however, the ultimate limit of sufficient hydration to support a viable operation is as yet unknown.

The Permian Basin

One of the major geological features of the United States is the Permian Basin extending over parts of Texas, New Mexico, Oklahoma, and Kansas, and calcium sulfate is found extensively over this area. Where hydration has occurred, the resulting gypsum is usually of high (plus 92%) purity, and in fact most of the specialty gypsum products which require high quality raw material are made from deposits of Permian age. The best purity and whitest color gypsum being commercially exploited in the United States is found in the Blaine formation of the Permian at Medicine Lodge, KS and Southard, OK, and in the Easly Creek formation of the Permian at Blue Rapids, KS.

FIG. 5—*Load-haul-dump unit on 1300-ft level, Plasterco, VA.*

At the present time, there are seven operating mines in Texas, six in Oklahoma, and two in Kansas. All of the Texas and Oklahoma deposits are worked by open pit methods, but the two Kansas mines are underground and employ room-and-pillar mining methods.

Although considerable Permian calcium sulfate is found in southeastern New Mexico, none currently is being mined. There are, however, two operating mines (open pit) in the north central part of the state working gypsum of the Todilto formation of Jurassic age.

Rocky Mountain and Inter-Mountain States

This general area is particularly rich in calcium sulfate deposits, most of which are at, or just below, the surface and lend themselves to low cost open pit mining. As might be expected in this region, structural features play an important part in limiting the size of the deposits, and in mine design. In general, purity is good, but the extent of hydration is frequently difficult to predict. A major deterrent to development is the cost of transportation due in part to the distance to roads and railroads, and also, the long hauls to major markets.

Gypsum is currently produced in the Powder River Basin area of Wyoming from rocks of Upper Permian to Middle Jurassic age, and from the Front Range of Colorado in Larimer and Fremont Counties. Production has also been developed in Fergus County, MT (Piper formation, Jurassic age) and Sevier County, Utah (Arapien shale, Jurassic age).

In Nevada, two deposits are mined in Clark County and one in Pershing County, with most of the product from these operations being shipped to California markets. In Arizona, a deposit at Winkleman provides gypsum for a calcining and manufacturing operation in Phoenix, and three smaller operations in the state supply portland cement retarder and agricultural markets. All of these western operations are open pit mines with the exception of Montana where an underground mine using the room-and-pillar system operates with approximately 7 ft of mining height.

California

Gypsum rock for five of the seven calcining and manufacturing plants in California is shipped into the state from Mexico (four plants) and Nevada (one plant). The other two utilize gypsum from a deposit in Imperial County, which in 1979 produced approximately 900,000 tons, making it the largest operating gypsum mine in the United States. This deposit, of Miocene age, occurs as a remnant of a calcium sulfate bed well over 100 ft thick whose present extent is limited by structure and erosion. Nevertheless, it contains large reserves, in part because hydration has taken place to a greater degree than in many other cases. Little or no overburden is present, and mining is carried out by multi-bench open pit methods (ver Plank, 1952).

For many years, several low grade gypsite deposits have been worked in San Joaquin Valley to produce a 50 to 75% gypsum content material for use as a soil conditioner. These deposits are Holocene in occurrence and are of limited extent. Although of too low a purity for portland cement retarder or plaster products, they are important in treating the alkaline soils which predominate in the great agricultural valleys of the state.

Western Canada

Gypsum is produced from two open-pit mines, one in Manitoba, the other in British Columbia. These operations provide rock for eight calcining and manufacturing plants in the western provinces. Also, one plant in Vancouver is supplied with rock imported from Mexico.

Production from these deposits is used almost entirely in the manufacture of building products and as a portland cement retarder. In both cases, only mining and primary crushing operations are carried on at the mine site, with the crushed gypsum rock being shipped to plants located in or near the major cities of Winnipeg, Saskatoon, Calgary, Edmonton, and Vancouver.

Mexico

The primary use for gypsum in Mexico is as a retarder for portland cement and for building plaster, although in 1971 two small wallboard plants began operation in the city of Puebla in an attempt to introduce wallboard to the construction markets. Considerable gypsum is also exported to the West and Gulf Coasts of the United States, and on occasion, to the Far East.

Many large deposits are known to exist; however, at the present time the major producers are located on San Marcos Island in the Gulf of Baja California, and in the states of Puebla, San Luis Potosí, and Nuevo Leon. In general, the gypsum is found in rocks of Mesozoic and Tertiary age.

The largest volume of production in Mexico is from San Marcos Island where an open pit mine and deep water ship loading facilities are operated to produce and ship a crushed gypsum rock. Another large highly mechanized open pit mine is operated at Estación La Borreguita in the state of San Luis Potosí to produce a crushed gypsum which is shipped by rail to local cement companies and to the port of Tampico for export.

In the state of Puebla, from the town of Izucar de Matamoros on the east to the village of Axochiapan on the west, a zone of gypsum bearing rocks of Miocene-Pliocene age occurs in which several small quarries are operated. These provide gypsum for some eight or nine cement plants in the Mexico City-Puebla area, for two new gypsum board plants at Puebla, and for several small calcining operations making gypsum plaster. The highest purity gypsum presently mined in Mexico comes from near Monterrey, Nuevo Leon, where at least two calcining plants operate to produce plaster for both construction and industrial uses.

Caribbean Area

Two deposits are being worked in the Caribbean Area: one at Bull Bay, Jamaica, about 10 miles east of the capital of Kingston; and the other near Barahona, Dominican Republic, some 130 miles west of the capital city of Santo Domingo. In both countries only small domestic gypsum markets have been developed primarily for portland cement retarder, and the major part of the production serves the export trade.

In Jamaica calcium sulfate is of Lower Eocene age, and occurs as several separate bodies of relatively small size in a northwest-southeast trending basin known as the Wagwater Trough (Holliday, 1971). Perhaps 60 to 70% of the calcium sulfate mass has been hydrated, and gypsum is currently being mined from three different deposits, utilizing multi-bench open pit methods.

The Dominican Republic gypsum is of probable Upper Miocene age, and the largest outcrop is found in the Cerro de Sal, a 12-mile-long hogback trending northwest-southeast. Calcium sulfate is interbedded with limestones, claystones, sandstones, and salt of the Cerro de Sal formation with a total thickness of some 2000 to 2500 ft. Bentonite clays and soluble salts are a potential problem associated with some stratigraphic members. The geological structure has been complicated due to several repetitions of the stratigraphic column caused by faulting; however, large deposits of economic gypsum may exist. Mining is by open pit methods. At least three other gypsum outcrops are known in this general area, but to date have not been developed for production.

Gypsum Resources Outside North America

Reflecting its wide geologic occurrence and its basic use in the construction industry, gypsum is produced in some 72 countries around the world, as shown in Table 2. In 1980, with a total world production of 78,290,000 tons being reported, the US consumed 21,106,000 tons, or 27%. No statistics are available as to the potential world reserves of calcium sulfate, but it can be safely predicted that they are enormous both in terms of total tons and of years at the present rates of consumption.

In many of the countries reporting production, gypsum was used only for portland cement retarder, this being by far the most common use of the mineral on a worldwide basis. As a general rule, the calcining of gypsum for use as plaster and wallboard building materials, or for industrial plasters, is limited to the United States, Canada, Europe (including Russia), and Japan, although a few wallboard plants have recently been built or are being considered in certain South American and Asian countries. In these countries, the general practice of open structural frames, usually made of wood, to form a building has historically required the use of some type of covering material for walls and ceilings, and gypsum products have gained wide acceptance for this use. That is, gypsum in the form of either plaster or precast wallboad is a versatile, fireproof covering material, and these products account for over two-thirds of the gypsum used in the United States and Canada.

Conversely, in those countries where the usual methods of construction rely heavily on masonry products, such as concrete, concrete block, stone, and brick (including adobe), the need for fireproof covering materials is much less, and the use of gypsum for the manufacture of building materials will be very small, or even nonexistent. The production and use of gypsum in a given country, therefore, is usually a result of its mode of use rather than the availability of gypsum deposits.

Exploration Methods

As for most minerals, a thorough understanding of the origin and occurrence of gypsum is essential before undertaking an exploration program. This may be obtained by a review of published literature, and a basic reference would be "Selected Annotated Bibliography of Gypsum and Anhydrite in the United States and Puerto Rico" by C. F. Withington and M. C. Jaster (1960).

By studying the stratigraphic column in regions where sedimentary rocks occur, an indication can usually be obtained from the lithology whether or not an evaporite environment may have existed, this being, of course, a requisite for the deposition of calcium sulfate. If evaporite rocks are indicated, further details can be obtained from examination of outcrops of the particular formations, or if outcrops are not available, from logs of oil or gas wells, or from water wells which may have been drilled in the vicinity. Perhaps drill cuttings, or even core, might be available for inspection. As stratigraphic and lithologic details of an area are thus developed, structure should also be mapped so as to provide as detailed a picture of the area as the basic data will permit. From this, the possibility of calcium sulfate occurrence usually can be predicted, and also, it may be possible to judge the potential for hydration of the sulfate zones.

Sources of this type of information are published literature, unpublished university theses, geological surveys, geological departments of colleges and universities, oil companies, and water well drillers.

Geophysical work is only partially applicable to the search for calcium sulfate. The right combination of logs may indicate gypsum or anhydrite, but no log has yet been devised that will replace accurately a laboratory chemical or purity analysis. Geophysical logs are useful

TABLE 2—Gypsum: World Production, by Country

(Thousand Short Tons)

Country	1976	1977	1978	1979[p]	1980[e]
North America:					
Canada*	6,616	7,974	8,901	8,927	†7,947
Cuba[e]	94	100	105	100	100
Dominican Republic	243	249	190	193	190
El Salvador	[e]7	8	8	8	10
Guatemala	15	35	42	28	†37
Honduras	[e]11	20	[e]25	[e]25	25
Jamaica	279	237	148	64	66
Mexico	1,559	1,649	1,938	2,228	†1,884
Nicaragua[e]	33	40	40	40	44
United States‡	11,980	13,390	14,891	14,630	†12,296
South America:					
Argentina	559	603	674	648	†656
Bolivia	[e]1	—	1	1	†1
Brazil §	[r]568	[r]608	512	515	550
Chile	134	[r]224	246	240	242
Colombia	226	231	281	283	285
Ecuador	48	[r]46	38	39	39
Paraguay	18	15	10	12	11
Peru	189	237	263	239	242
Venezuela	122	172	404	464	485
Europe:					
Austria*	849	[r]892	844	880	†919
Belgium*	242	185	202	212	210
Bulgaria	[r]256	[r]325	375	341	†343
Czechoslovakia	728	752	768	809	834
France*	7,308	6,649	6,654	6,503	6,600
German Democratic Republic	[r]375	[r]375	385	400	410
Germany, Federal Republic of (marketable)*	2,315	2,445	2,467	2,481	2,480
Greece	490	452	474	496	510
Ireland	391	377	432	460	463
Italy	1,652	[r]1,674	1,624	1,630	1,810
Luxembourg	2	3	1	1	1
Poland[e]§	1,380	[r]1,480	1,490	1,500	1,430
Portugal	176	194	222	[re]220	220
Spain	[e]4,600	[r]6,042	5,918	5,815	6,060
Switzerland[e]	80	80	80	80	80
USSR[e]††	5,500	5,700	5,800	6,000	6,500
United Kingdom*	3,693	3,648	3,662	[e]3,600	3,600
Yugoslavia	[r]465	[r]532	554	626	635
Africa:					
Algeria[e]	190	190	190	210	220
Angola[e]	22	[r]22	[r]28	[r]28	28
Egypt	[r]819	[r]923	875	877	†761
Ethiopia	—	7	1	1	1
Kenya*	86	29	[e]33	[e]33	33
Libya	66	320	198	200	200
Mauritania	12	11	15	18	19
Niger	3	[e]3	3	[e]3	2
South Africa, Republic of	532	485	429	416	†478
Sudan*	20	17	22	33	33
Tanzania	[r]9	9	24	12	12
Tunisia	43	[e]44	[e]44	[e]44	44
Zambia	5	5	2	—	—

TABLE 2—Gypsum: World Production, by Country

(Thousand Short Tons)

Country	1976	1977	1978	1979ᵖ	1980ᵉ
Asia:					
Afghanistan	NA	NA	7	NA	NA
Burma	50	37	39	42	†40
China:					
Mainland[e]	1,100	1,100	[r]1,700	[r]2,200	2,200
Taiwan #	3	8	4	3	9
Cyprus	71	92	76	68	61
India	801	858	974	949	†943
Iran	7,165	7,600	8,800	[e]7,000	3,900
Iraq[e]	180	180	180	180	180
Israel	220	[e]220	[e]220	72	65
Japan #	5,581	[r]6,118	6,387	6,915	7,165
Jordan	23	24	40	40	†77
Korea, Republic of[e] #	550	660	680	680	700
Lebanon	[e]14	17	12	11	11
Mongolia[e]	28	[r]31	[r]31	[r]31	33
Pakistan	493	312	279	378	375
Philippines ‖	130	123	123	121	121
Saudi Arabia	19	22	231	331	331
Syrian Arab Republic	69	94	[e]95	[e]70	72
Thailand	295	419	310	388	†454
Turkey	36	72	67	[e]70	70
Vietnam[e]	11	13	15	15	17
Oceania: Australia	1,038	1,010	1,036	1,278	†1,420
Total	[r]72,888	[r]78,718	83,839	83,455	78,290

Source: Pressler, 1981.

[e]Estimated. [p]Preliminary. [r]Revised. NA, Not available.

*Includes anhydrite.

†Reported figure.

‡Excludes byproduct gypsum.

§Series revised to represent sum of (1) mine product sold without beneficiation and (2) output of concentrates.

Includes byproduct gypsum. (In the case of Japan, byproduct gypsum was virtually all gypsum consumed during 1976-1980.)

‖Series revised to include byproduct gypsum, which constitutes total output in 1979-1980 and virtually all output in 1976-1978.

for determining structural details and can be used, in conjunction with other tests, to help determine mining environment conditions. Geochemical methods are rarely applicable to calcium sulfate, although heavy concentrations of sulfate in ground waters might serve as a useful clue, and in a few instances differences in plant species may indicate changes in rock types that can aid in surface mapping.

Even though the extent of gypsum can be mapped from surface outcrops, drilling is necessary to predict the amount and regularity of hydration, i.e., how much of the calcium sulfate is gypsum, and how much is either only partially hydrated or is anhydrite. For deeper deposits, additional drilling data is almost always required to map projections of lithologies and/or structures. Core drilling is usually preferred, taking BX (or larger) size core to provide adequate material for sampling and to permit good core recovery. In some cases where it is only necessary to determine the depth of overburden, or to locate the contact between gypsum and anhydrite, rotary drilling is sufficient and is much less expensive. Also geophysical methods can be successfully used under some conditions to determine the depth of overburden.

Although calcium sulfate beds may sometimes be continuous over tens—or even hundreds—of miles, the exploration geologist should keep in mind that they may also occur as discontinuous lenses. Also, the extent of hydration is even less predictable, as is the occurrence of minor, but detrimental, amounts of impurities. Hence, a successful exploration program requires knowledgeable consideration and interpretation of all the factors involved, and sums of money which can run into six figures.

Evaluation of Deposits

The single most important factor in the evaluation of a gypsum deposit is its location with respect to markets, i.e., the logistical realities of the industry. The term "place value" is often used to describe this characteristic, which results because a local market demand will make nearby deposits more valuable even though they may not be as pure nor as easily mined as more remote deposits. Thus, the factor which more often than any other determines the value of one deposit as against another is the cost of transportation from the mine to the major market areas.

Also basic to the exploitation of gypsum is the fact that it is a relatively low priced commodity. This is due in part to the low cost inherent to extraction of the mineral from the ground and its preparation for market, and in part to the fact that in its major product areas, it must compete with relatively low priced materials. With this highly competitive pattern, which as a practical matter results in differing price structures from one country to another, or in the United States and Canada, from one section of the country to another, there is no set "market price" to measure against, and each deposit must be evaluated within the market area it might logically serve. Hence, early in the evaluation of any gypsum deposit, an investigation of marketing factors is essential.

The geologist and mining engineer should carefully study the geological environment of a deposit, looking at such features as overburden ratio (if a shallow deposit) or roof conditions if an underground mine is indicated. Also, thickness of the bed, structural features, and the possibility of unusual water conditions are important, as these translate into capital investment and operating costs. There is no simple rule as to how much overburden one can afford to strip for an open pit, or how much depth one might consider for underground mining. In general, the economics of the industry are such that mining costs must be relatively low, and many operating deposits today mine and deliver rock to the mill (on site) for a cost between $1.50 and $4.50 per ton (not including depreciation) although some others are in the $5.00 to $8.00 per ton range.

The grade of the deposit must also be assessed, particularly the type and amounts of impurities. As a rule of thumb, the higher the gypsum content the better, but in no case should the gypsum content be more than a percent or two lower than that of other deposits serving the same market. Impurities should be carefully analyzed, and if they are outside the limits, steps to eliminate or reduce them by beneficiation should be investigated. In such a case, the penalty may well eliminate the deposit from further consideration.

A separate chapter on utilization, discussing processing methods, markets, and price trends, provides further background for the evaluation of gypsum deposits.

Mining Methods

Gypsum is mined both from open pit and underground. Open pit mining of gypsum employs conventional methods and equipment, such as draglines and scrapers to remove overburden, and shovels or front end loaders to load rock. Rotary and/or auger drilling is almost universally used because of the relative softness of the rock, and penetration rates of 5 to 10 fpm are common. Drill holes are usually from 2 to 3.5 in. in diameter, and spaced relatively close together in order to distribute the explosive throughout the rock mass in a somewhat dense pattern. This is because gypsum tends to absorb explosive force without readily fracturing, and a hole pattern in excess of perhaps 12 × 12 ft may result in unsatisfactory fragmentation. Ammonium nitrate-fuel oil explosive is used where water conditions permit, and a ratio of three tons of rock broken per pound of explosive is about average.

Because little or no beneficiation is usually performed in the processing steps following mining, extra effort is required to remove all overburden, or to selectively mine a face in order to eliminate impurities or to blend varying grades of rock. Quality control is most important in the mining step, and may often result in somewhat higher per ton costs.

Underground mining commonly employs room-and-pillar methods with pillar design based upon the conditions of each mine; however, recovery of total rock is in the 65 to 80%

range. Auger drilling is standard practice, and loading is done with gathering arm type loaders, front end loaders or load-haul-dump units. Underground haulage may be by rail, or by diesel-powered trucks. In a few cases, conveyor belts are used. The use of diesel power has become quite common in gypsum mines, in part because explosive gas conditions do not exist, and in part because mine design usually can readily incorporate good ventilating conditions.

In general, gypsum mining operations are highly mechanized, although the limited volume of production, usually 500 to 1500 tpd, does restrict the size of equipment which can be utilized economically. The physical environment of most gypsum deposits favors good working and safety conditions, a major factor in contributing to low cost.

Future Considerations

As indicated throughout this discussion, there is no foreseeable shortage of either gypsum or anhydrite in the United States or in the world. Paradoxically, however, there are instances where it may be difficult to find gypsum which can be considered economic at a given time and location, a problem which has its roots in "place value." The best evidence of this situation is that the United States historically has imported 33 to 36% of its needs from Canada, Mexico, Jamaica, Dominican Republic, and Spain, as shown in Table 3.

All of this gypsum (with the exception of a few thousand tons of special grade rock) is used by plants on the Atlantic, Pacific, and Gulf Coasts of the United States. The basic

TABLE 3—US Imports of Crude Gypsum (Including Anhydrite)

(Thousand St)

Country	1977	1978	1979	1980
Canada	5,307	6,610	5,700	5,463
Mexico	1,448	1,610	1,815	1,565
Jamaica	238	167	5	11
Dominican Republic	70	144	80	69
Spain	—	198	125	250
Other countries	11	29	12	8
Total imports	7,074	8,308	7,773	7,366
Domestic production *	14,187	15,560	15,458	13,039
Total	21,261	23,868	23,231	20,405
% Imports	33	35	33	36

Source: Pressler, 1979; 1981.
*Includes byproduct gypsum.

reasons for this situation are: (1) large markets for gypsum products are concentrated in and around the coastal cities, (2) there are no known gypsum deposits along any of the three United States coastlines, and (3) large gypsum deposits exist on or near deep water of the exporting countries from which rock can be shipped to these major markets at lower cost than from inland United States deposits.

It is likely that these conditions will persist into the future, and the industry is built upon this premise, having made investments in offshore deposits, shipping facilities, and domestic port locations with appropriate marine rock handling equipment. There is no import duty on crude gypsum rock, but duties are placed upon processed gypsum.

Because of its widespread occurrence and huge potential reserves and also because its uses are such that it is not basic to survival in the event of a national emergency, gypsum is not considered to be a "strategic" mineral. This has permitted natural economic factors to prevail in the development of the mineral, literally on a worldwide basis, which overall is a healthy situation that hopefully will continue to prevail.

Future Trends and Problems

Competition between manufacturers of gypsum building products, as well as from manufacturers of substitute materials, has resulted in a continuing pressure to improve the quality of gypsum products. In turn, mine operators are being asked to reduce the average quantity of impurities, i.e., upgrade the gypsum purity, and maintain a uniform ton-to-ton level of impurities. To date, these objectives are being accomplished by selective mining techniques and screening during primary and secondary crushing stages to reject a size fraction in which impurities tend to concentrate. Other beneficiation techniques can be and are considered, but are slow to be adopted because of adverse economic impact.

Environmental concern in the mining of gypsum is limited almost exclusively to open pit operations, although in a few cases, shallow mines (less than 100 ft deep) may have an occasional subsidence problem that requires attention. Land reclamation of open pits has become a requirement as new laws and regulations take effect, and considerable ingenuity is needed to do a satisfactory job at a reasonable cost. It is believed that gypsum miners have as good, and in some cases, a better opportunity of doing satisfactory reclamation than miners of many other types of minerals, because there seldom are associated minerals which lead to such problems as acid waters. Also, the configuration of many gypsum open pits is such that surface restoration is readily feasible. In general, the water pumped from either open pit or underground gypsum mines is of a quality not considered as a pollutant; however, it is necessary to sample and analyze the pump discharge from time to time to develop adequate background data, and to assure that requirements are being met.

Processing of gypsum is energy-intensive; energy requirements represent the largest single cost in the production of gypsum wallboard. Future increases in energy cost will require innovation in conservation techniques and process technology to offset at least partially this adverse trend. For further discussion, see the section on Energy Considerations in the chapter "Gypsum and Anhydrite" under Construction Materials, Volume 1.

Synthetic or Byproduct Gypsum

Synthetic gypsum as a byproduct of various chemical operations has been available for some time in various parts of the world. Whether or not it might be substituted for natural gypsum requires analysis of the same factors that one would investigate in evaluating a calcium sulfate deposit, the most important being logistics, quality, and cost. In many cases, byproduct gypsum is of unacceptable quality for the majority of gypsum uses because of the type and amounts of impurities, or crystal size and shape. In all cases, it is a relatively fine, wet material which comes out of the chemical process as a filter cake or after settling in ponds or tanks.

Both the chemical and physical characteristics of byproduct gypsum are such that further treatment is necessary to make it into a useful product. The cost of such treatment varies widely with the type of byproduct and the use for which it is intended. By far the major source of byproduct gypsum at present is the manufacture of phosphoric acid. As pollution regulations are tightened, much additional byproduct material will also be developed by neutralization of the sulfur dioxide in stack gas and/or from waste sulfuric acid.

In the United States, use of synthetic gypsum is very small. It is not economical to go through the treatment required to make it suitable for major uses such as plasters or wallboard. On the other hand, minor quantities (828,000 tons in 1979) of synthetic gypsum, primarily "phosphogypsum," are being sold as a soil conditioner in California and the southeastern states, a use which tolerates the type and amount of impurities present and requires relatively low-cost processing to put the material in a useful form. Synthetic gypsum is being used in Japan for both portland cement retarder and wallboard. However, this is a unique situation; Japan's gypsum industry is well developed but there is a shortage of natural gypsum. The high cost of importing gypsum rock results in a condition where it is economical to upgrade the synthetic material. Additional use of synthetic gypsum to manufacture wallboard has recently been started in Australia and in Europe in situations where economics favor this use.

Bibliography and References

Adams, J.E., 1971, "Upper Permian Ochoa Series of Delaware Basin, West Texas and Southeastern New Mexico," *Origin of Evaporites,* American Association of Petroleum Geologists, Tulsa, OK, Reprint Series No. 2, pp. 60–89.

Appleyard, F.C., 1965, "The Locust Cove Mine," *Mining Engineering,* March, pp. 59–62.

Bergstrom, J.H., 1961 "Pabco's Gypsum Crystals Sparkle," *Rock Products,* April, pp. 90–95.

Borchert, H., and Muir, R.O., 1964, *Salt Deposits, The Origin, Metamorphism, and Deformation of Evaporites,* D. Van Nostrand, London, 338 pp.

Braitsch, O., 1971, *Salt Deposits, Their Origin and Composition,* Springer-Verlag, New York, 297 pp.

Branson, E.B., 1915, "Origin of Thick Gypsum and Salt Deposits," *Geological Society of America Bulletin,* Vol. 26, pp. 231–242.

Briggs, L.I., 1970, "Geology of Gypsum in the Lower Peninsula, Michigan," *Proceedings, Sixth Forum on Geology of Industrial Minerals,* Michigan Geological Survey, Misc. 1, pp. 66–76.

Butler, G.P., 1970, "Holocene Gypsum and Anhydrite of the Abu Dhabi Sabkha, Trucial Coast: An Alternative Explanation of Origin," *Proceedings, Third Symposium on Salt,* Northern Ohio Geological Society, Cleveland, Vol. 1, pp. 120–152.

Buzzalini, A.D., et al., 1969, ed. "Evaporites and Petroleum," American Association of Petroleum Geologists Bulletin, Vol. 53, No. 4, April, pp. 775–1011.

Cole, L.H., 1930, *The Gypsum Industry of Canada,* Canadian Dept. of Mines, Publication 714, pp. 1–164.

Conley, R.F., and Bundy, W.M., 1958, "Mechanism of Gypsification," *Geochimica et Cosmochimica Acta,* Vol. 15, pp. 57–72.

Cross, A.I., 1966, "Palynologic Evidence of Mid-Mesozoic Age of Fort Dodge (Iowa) Gypsum," Abstract, San Francisco Meeting, Geological Society of America, Nov. 14–16.

Dean, W.E., and Schreiber, B.C., eds., 1978, "Marine Evaporites," Mineral Short Course, *Notes,* No. 4, Society of Economic Paleontologists, 188 pp.

Friedman, G.M., and Sanders, J.E., 1978, *Principles of Sedimentation,* John E. Wiley & Sons, Inc., New York.

Gay, P., 1965, "Some Crystallographic Studies in the System $CaSO_4$—$CaSO_4 \cdot 2H_2O$, The Hydrous Forms," *Mineralogical Magazine,* Vol. 25, pp. 354–362.

Goodman, N.R., et al., 1957, "Gypsum," *The Geology of Canadian Industrial Mineral Deposits,* 6th Commonwealth Mining and Metallurgical Congress, pp. 111–137.

Ham, W.E., 1962, "Economic Geology and Petrology of Gypsum and Anhydrite in Blaine County," *Oklahoma Geological Survey Bulletin,* Vol. 89, pp. 100–151.

Ham, W.E., et al., 1961, "Borate Minerals in Permian Gypsum of West-Central Oklahoma," *Oklahoma Geological Survey Bulletin,* Vol. 92, 77 pp.

Hardie, L.A., 1967, "The Gypsum-Anhydrite Equilibrium at One Atmosphere Pressure," *American Mineralogist,* Vol. 52, Jan.-Feb., pp. 171–200.

Holliday, D.W., 1970, "The Petrology of Secondary Gypsum Rocks: A Review," *Journal of Sedimentary Petrology,* Vol. 40, No. 2, June, pp. 734–744.

Holliday, D.W., 1971, "Origin of Lower Eocene Gypsum-Anhydrite Rocks, Southeast St. Andrew, Jamaica," *Transaction Sec. B, Institution of Mining and Metallurgy,* Vol. 80, pp. B305–B315.

Jorgensen, D.B., and Carr, D.D., 1972, "Influence of Cyclic Deposition, Structural Features, and Hydrologic Controls on Evaporite Deposits in the St. Louis Limestone in Southwestern Indiana," *Proceedings, Eighth Forum on Geology of Industrial Minerals,* Iowa Geological Survey, 195 pp.

Keith, S.B., 1969, "Gypsum and Anhydrite," *Arizona Bureau of Mines Bulletin 180,* pp. 371–382.

Kelly, K.K., et al., 1941, "Thermodynamic Properties of Gypsum and Its Dehydration Products," *Technical Paper 625,* US Bureau of Mines, 73 pp.

Kendall, A.C., 1978, "Calcium Sulfate Diagenesis, Mississippean of Eastern Williston Basin," Abstract, *American Association of Petroleum Geologists Bulletin,* Vol. 62, No. 3, p. 530.

Kerr, S.D., and Thompson, A., 1963, "Origin of Nodular and Bedded Anhydrite in Permian Shelf Sediments, Texas and New Mexico," *American Association of Petroleum Geologists Bulletin,* Vol. 47, pp. 1726–1732.

King, R.H., 1971, "Sedimentation in Permian Castile Sea," *Origin of Evaporites, American Association of Petroleum Geologists,* Tulsa, OK, Reprint Series No. 2, pp. 90–97.

Kinsman, D.J.J., 1966, "Gypsum and Anhydrite of Recent Age, Trucial Coast, Persian Gulf," *Second Symposium on Salt,* Northern Ohio Geological Society, Cleveland, Vol. 1, pp. 302–326.

Kinsman, D.J.J., 1969, "Modes of Formation, Sedimentary Associations, and Diagenetic Features of Shallow Water and Supratidal Evaporites," *American Association of Petroleum Geologists Bulletin,* Vol. 53, No. 4, April, pp. 830–840.

MacDonald, G.F.J., 1953, "Anhydrite-Gypsum Equilibrium Relations," *American Journal of Science,* Vol. 251, pp. 883–898.

Martinez, J.D., 1974, "Tectonic Behavior of Evaporites," *Proceedings,* 4th Symposium on Salt, Northern Ohio Geological Society, Cleveland, Vol. 1, pp. 155–168.

McAide, H.G., 1964, "The Effect of Water Vapor Upon the Dehydration of $CaSO_4 \cdot 2\,H_2O$," *Canadian Journal of Chemistry,* Vol. 42, pp. 792–801.

Murray, R.C., 1964, "Origin and Diagenesis of Gypsum and Anhydrite," *Journal of Sedimentary Petrology,* Vol. 34, No. 3, Sep., pp. 512–523.

Newland, D.H., and Leighton, H., 1910, "Gypsum Deposits of New York," *New York State Museum Bulletin 143,* 94 pp.

Posnjak, E., 1938, "The System, $CaSO_4 \cdot H_2O$," *American Journal of Science,* Vol. 235A, pp. 247–272.

Posnjak, E., 1940, "Deposition of Calcium Sulfate From Sea Water," *American Journal of Science,* Vol. 238, pp. 559–568.

Pressler, J.W., 1979, Annual Advance Summary, "Gypsum in 1979," US Bureau of Mines, 17 pp.

Pressler, J.W., 1981, "Gypsum," *Minerals Yearbook 1980,* US Bureau of Mines, pp. 385–396.

Riley, C.M., and Byrne, J.V., 1961, "Genesis of Primary Structures in Anhydrite," *Journal of Sedimentary Petrology,* Vol. 31, pp. 553–559.

Schenk, P.E., 1969, "Carbonate-Sulfate-Redbed Facies and Cyclic Sedimentation of the Windsorian Stage (Middle Carboniferous), Maritime Provinces," *Canadian Journal of Earth Sciences,* Vol. 6, pp. 1037–1066.

Schmidt, R., 1914, "Uber die Beschaffenheit und Entstehung parallelfaseriger Aggregate von Steinsalz und Gips," *Kali.* 8, pp. 161, 218, 239.

Schroeder, Harold J., 1970, "Gypsum," *Mineral Facts and Problems,* Bulletin 650, US Bureau of Mines, pp. 1039–1048.

Shearman, D.J., 1966, "Origin of Marine Evaporites by Diagenesis," *Transactions Sec. B,* Institution of Mining and Metallurgy, Vol. 75, pp. B208 to B215.

Stone, R.W. et al., 1920, "Gypsum Deposits of the United States," *Bulletin 697,* US Geological Survey, 326 pp.

Stonehouse, D.H., 1970, "Gypsum and Anhydrite," *Canadian Minerals Yearbook,* Mineral Resources Branch, Dept. of Energy, Mines, and Resources, Ottawa, pp. 775–1011.

Stonehouse, D.H., 1972, "Gypsum and Anhydrite," *Canadian Minerals Yearbook,* Mineral Resources Branch, Dept. of Energy, Mines, and Resources, Ottawa, preprint, 7 pp.

Stonehouse, D.H., 1978, "Gypsum in Canadian Mineral Industry," Dept. of Energy, Mines and Resources, Ottawa, pp. 211–216.

Stonehouse, D.H., 1979, "Gypsum in Canadian Mineral Industry," Dept. of Energy, Mines and Resources, Ottawa, pp. 221–228.

Stonehouse, D.H., 1980, "Gypsum in Canadian Mineral Industry," Dept. of Energy, Mines and Resources, Ottawa, preprint, 8 pp.

Van't Hoff, J.H., et al., 1903, "Gips und Anhydrit," *Zeitschrift fuer Physik Chem.*, Vol. 45, pp. 257–306.

ver Planck, W.E., 1952, "Gypsum in California," *Bulletin 163,* California Div. of Mines, 151 pp.

Walker, C.W., 1974, "Nature and Origin of Cap Rock Overlying Gulf Coast Salt Domes," *Proceedings*, 4th Symposium on Salt, Northern Ohio Geological Society, Cleveland, Vol. 1, pp. 169–175.

Withington, C.F., 1962, "Gypsum and Anhydrite in the U.S.," Min. Inv. Res., Map MR-33 and text, US Geological Survey, 18 pp.

Withington, C.F., and Jaster, M.C., 1960, "Selected Annotated Bibliography of Gypsum and Anhydrite in the United States and Puerto Rico," Bulletin 1105, US Geological Survey, 126 pp.

Wood, G.V., and Wolfe, M.J., 1969, "Sabkha Cycles in the Arab-Darb Formation of Trucial Coast of Arabia," *Sedimentology,* Vol. 12, pp. 165–191.

Zen, E-An, 1965, "Solubility Measurements in the System $CaSO_4$-NaCl-H_2O at 35°, 50°, and 70° C and One Atmosphere Pressure," *Journal of Petrology,* Vol. 6, pp. 124–164.

Iodine

JOHN JAN *
Revised by L. A. ROE †

Iodine is a soft, lustrous, grayish-black nonmetallic element with a density of 4.9. It is the least active of the four members of the halogen family. The other members are, in order of increasing activity, bromine, chlorine, and fluorine. Iodine is a solid at ordinary temperatures, while bromine is a liquid and chlorine and fluorine are gases. Iodine melts at 113°C and volatilizes at 184.4°C to a blue-violet gas with an irritating odor. It occurs in nature only as iodates and iodides or other combined forms.

It was discovered by Bernard Courtois in France in 1811, who noticed its presence as an unknown substance in crude soda ash obtained from burning seaweed. Gay-Lussac recognized it as a new element and named it iode, from the Greek word "ioeides or iodes" for violet color (MacMillan, 1970).

The world production is not known with any great certainty, but it is probably between 25 and 35 million lb per year, of which about 30% is consumed in the United States.

Geology and Mineralogy

Iodine compounds occur as a minor constituent as a mineral in the Chilean nitrate deposits and in solution in brines and seawater. According to the US Geological Survey, it is the 47th most abundant element in the earth's crust. The minerals lautarite, $Ca(IO_3)_2$ (calcium iodate), and dietzeite, $Ca_2(IO_3)_2(CrO_4)$, (calcium iodate-chromate) are found in Chilean nitrate deposits which are located in Antofagasta and Tarapacá provinces on the eastern slope of the coast range in a desert area.

Various brines contain iodine compounds. Seawater contains about 0.05 ppm of iodine and some seaweeds will extract and accumulate it up to 0.45% on a dry basis. Some coals in West Germany also contain iodine compounds.

The known and potential sources of iodine reported in 1968 are given in Table 1. Table 2 lists estimated world iodine reserves.

Iodine has been recovered from brines in Java, Indonesia, France, England, and the USSR. Iodine has also been recovered from seaweed in Ireland, Scotland, France, Japan, Norway, and the USSR. The indicated reserves and future resources of iodine are large but have never been adequately measured.

Analysis

Iodine as the free element can be detected by the characteristic blue color it gives to a starch solution. Quantitatively it is determined as the free element by titration with standard thiosulfate solutions using starch as an indicator. Colorimetric methods are also applicable.

Markets and Prices

Iodine and its compounds are generally marketed in the form of: iodine, crude; iodine, resublimed; calcium iodates; calcium iodide; potassium iodide, sodium iodide, and numerous organic compounds.

Product Specifications

Crude iodine, World market: 99–99.5% I_2.

Crude iodine, US produced, average: 99.8% I_2.

Crude iodine, USP XVII Specs: not less than 99.8% percent I_2.

Resublimed iodine is usually 99.9% and ACS specifications call for not over 0.005% total bromine and 0.020% nonvolatile materials (Anon., 1971).

Prices, c.i.f. United States, for iodine have been steadily increasing from $1.18 per lb of crude in 1967 to $6.37–7.27 in January 1981 (Anon., 1981). In 1980 the price of iodine increased by 50%.

* Western Knapp Engineering Div., Arthur G. McKee Co., San Francisco, CA.
† Consultant, Downers Grove, IL.

TABLE 1—Known and Potential Sources of Iodine Reported in 1968

Location	Source	Estimated Iodine Content, % or Ppm	Estimated Reserves, Tons
Chile	Caliche nitrate deposits	Avg. 0.04%	400,000
Producer			
Sociedad Quimica y Minera de Chile			
Plants			
Maria Elena, 800 t/a			
Pedro de Valdivia, 2700 t/a			
Officina Victoria, 200 t/a			
Japan			
Producers			
AiOi Industrial Co.	Natural gas	0.01%	500,000-
Nippon Development Co.	well brines		1,500,000
Nippon Natural Gas Co.			
Daiichi Pharmaceutical Co.			
Ise Chemical Co.			
Chiba Iodine Co.			
US			
Anadarko Basin, OK	Oil field brine	Some contain over 500 ppm	
Los Angeles Basin, CA	Seal Beach		
Midland, MI,	Oil field brine	60-70 ppm	
Dow Chemical Co.	Well brine	38 ppm	
Shreveport, LA	Oil field brine	35 ppm	

Source: MacMillan, 1970.

End Uses

US consumption by principal end uses in the year 1980 is given in Table 3.

Other uses include iodine or its compounds used in lubricants for titanium and stainless steel, catalysts for producing such compounds as polybutadiene rubber, stabilized rosin, and tall oil products.

As of early 1981, extensive use was predicted for iodine or iodine derivatives as catalysts in plants using coal feedstock in the production of synthetic fuels and downstream products.

Iodine is also used in the modification of selenium to make semiconductors; manufacture of high-purity metals such as titanium, zirconium, boron, and hafnium; additives in rechargeable dry cells; smog inhibitors; cloud seeders to induce rainfall; and in certain dyes and food colorings.

An artificially prepared radioactive isotope I-131, with a half-life of eight days, is finding extensive use as a radioactive tracer (Anon., 1971; MacMillan, 1970).

Croft (1972) describes research efforts in Australia which point out advantages of cyclic iodination in upgrading ilmenite. Enriched titanium dioxide products were made by iodina-

TABLE 2—Estimated World Iodine Reserves
(Million pounds)

	Reserves*
North America: United States:	
Michigan	30
Oklahoma, California, and Louisiana	500
Total	530
South America: Chile	800
Europe:	
Federal Republic of Germany	10
Other†	400
Total	410
Asia:	
China, Mainland	NA
Japan	4,000
Indonesia	10
Total	4,010
World total	5,750

Source: Absalom, 1980.
NA, Not available.
*Except for Chile and Japan, figures denote only an order of magnitude.
†Includes the USSR, France, the United Kingdom, Norway, Italy, and Ireland.

tion of prereduced ilmenite to produce vapor phase iron iodide which can be treated for recovery of iodine which is recycled to the process. The upgraded titanium product consists of titanium oxide pseudomorphs of ilmenite grains.

The advantages of iodine in Croft's process over chlorine or bromine are numerous. Iodine is easily stored as a solid phase under normal conditions and it offers other advantages relating to vapor leakages, spillage, and corrosion.

New methods for the production of potassium iodide, potassium iodate, and ethylenediamine dihydriodide were reviewed by Garrison (1980) and their energy-saving features emphasized.

The US Bureau of Mines (1980) reported laboratory scale testing of a technique for recovery of silver from pregnant solutions generated in hydrometallurgical treatment of complex base-metal sulfide concentrates with oxidizing chloride media. The silver was precipitated as an iodide and then regeneration of the iodide salt recovered the iodide for reuse. Sodium or potassium iodide was used to precipitate silver iodide, which was then contacted with sodium sulfide to produce silver sulfide, with resulting regeneration of the iodide solution.

Principal Producing Countries

The production of iodine usually depends on production of other materials. In Chile, it is a byproduct of nitrate production. In Japan, it comes from natural gas well brines. Dow Chemical Co., recovers iodine along with salt, potassium, bromine, magnesium, and calcium

TABLE 3—US Consumption of Iodine in 1980

	Estimated Million lb	Quantities, %
Photographic chemicals, photolithographic supplies	1.71	19
Food supplements:		
animal & fowl	1.12	13
human (iodized salt)	0.13	1
Pharmaceutical	0.95	11
Disinfectants, industrial and household	1.75	20
Other (includes catalysts)	3.24	36
Total	8.90	100

Source: Absalom, 1980.

chlorides from natural brines at Midland, MI. A plant at Woodward, Oklahoma is under the control of PPG Industries and has a capacity of 2 million lb per year. In 1980, this plant was being expanded to an unstated capacity (Anon., 1980).

Chile was long the principal world producer of iodine from its nitrate fertilizer. In recent years, however, Japan has produced more iodine than Chile, as brines production has increased. In 1981, production in Chile was 5,800,000 lb and in Japan, it was 15,000,000 lb.

Aside from Chile and Japan, production in all other countries is considered minor. Available data for 1970 shows that West Germany produced 125 st of crude iodine. Other countries that are or may be iodine producers are: England, France, Indonesia, Java, Ireland, Norway, Scotland, and the USSR.

The two US producers are Dow Chemical Co. and PPG Industries. The US imports most of its requirements in the form of crude iodine. About 10,720 lb of resublimed iodine were also imported in 1969, of which Japan supplied 10,500 lb, Sweden 220 lb. Chile supplied about 40% of total imports. Kelp beds or brines continue as the best potential sources for iodine production (Anon., 1971; MacMillan, 1970). US demand for iodine is expected to double in this decade (Anon., 1980, 1980a).

Preparation for Market

Chilean Nitrate Deposits—The iodine-bearing nitrate ore is leached with an alkaline solution, with the iodine going into solution as sodium iodate (Anon., 1965). Sodium bisulfite is introduced to convert the iodate to iodide. Fresh leach solution with iodate is mixed with the iodide solution to precipitate iodine according to the equation:

$$5\ NaI + NaIO_3 + 3H_2SO_4 \rightarrow 3I_2 + 3Na_2SO_4 + 3H_2O.$$

The iodine is filtered, washed, and dried as crude iodine and sublimated to produce the purified grade. Only about 50 to 75% of the iodine in the ore is recovered. The recovery method for seaweed ash is similar to that used with Chilean nitrates.

Michigan Natural Brines—The natural brine is acidified, chlorinated, and passed downward through a packed tower, countercurrent to a stream of air. Iodine, displaced from the brine by chlorine, is blown to an absorption tower in which the iodine is absorbed in a solution of sulfuric and hydriodic acids. The solution is then treated with sulfur dioxide, after which iodine is precipitated with chlorine, filtered, and melted under a layer of sulfuric acid.

Japanese Gas Well Brines—In Japan iodine compounds are recovered from natural brine found in gas wells on the Chiba peninsula. The brines are first clarified to remove oil and suspended matter and then treated by one of the following processes:

1) Sulfuric acid, chlorine, and sodium nitrate are mixed with brine and the liberated iodine absorbed by activated carbon. The crude iodine is then dissolved in hot caustic solution and reprecipitated with chlorine to yield a refined iodine.

2) Iron sulfate and copper sulfate are added to the clarified brine, precipitating copper iodide which is oxidized to yield crude iodine.

3) Electrolysis of brines containing sulfuric acid, with absorption of the iodine on activated charcoal.

4) Brine containing iodine which has been liberated with chlorine is passed over copper wire to form insoluble cuprous iodide. The cuprous iodide is periodically washed off the copper wire, dried, and shipped as finished product.

Other Techniques—Although little used today, the recovery method from seaweed ash is similar to that used for Chilean nitrates. Silver nitrate was also used at one time in brine treatment. The brine was treated with silver nitrate solution to produce the highly insoluble silver iodide which was recovered by filtration. A suspension of silver iodide in water will react with iron to displace the silver. The resulting solution can be treated with chlorine to displace the iodine product.

Transportation

Crude iodine is shipped in double polyethylene plastic-lined fiber drums holding 100 to 200 lb each. Resublimed iodine is also shipped in the same type of container as well as in 25-lb glass jars and ¼-, 1-, and 5-lb bottles. Freight classification is "chemicals, not otherwise indexed by name" (NOIBN) and requires no special label.

Subsidies, Depletion, and Tariffs

On Nov. 30, 1981, the US government strategic stockpile contained 8,010,000 lb of iodine. The stockpile also contained 3,238 pounds of nonstockpile grade material. The stockpile objective established by the Office of Emergency Preparedness is 5 million lb.

Depletion allowances for domestic iodine producers were changed under terms of the Tax Reform Act of 1969. Effective after October 1969, the depletion allowance for iodine from both domestic and foreign production is 14%.

Tariffs on iodine and potassium iodide are as follows:

	Most Favored Nation (MFN)		Non-MFN
	1/1/82	1/1/87	1/1/82
Iodine, crude	Free	Free	Free
Iodine, resublimed	7.2¢/lb	6¢/lb	10¢/lb
Potassium iodide	3.2% *ad val.*	2.8% *ad val.*	7.5% *ad val.*

Bibliography and References

Anon., 1951, "Iodine—Its Properties and Technical Applications," Chilean Iodine Educational Bureau, Inc., New York, 74 pp.

Anon., 1965, "Iodine Into the Pool," *Chemical Week*, Vol. 97, No. 9, Aug. 28, pp. 127–128.

Anon., 1965a, "Iodine Lubricants Smooth the Way for Broader Use of Titanium," *Iron Age*, Vol. 196, No. 22, Nov. 25, pp. 68–69.

Anon., 1966, "Iodine Additives Reduce Friction," *Production Engineering*, Vol. 37, No. 20, Sep. 26, p. 40.

Anon., 1971, "Iodine in 1971," *Mineral Industry Surveys*, US Bureau of Mines, annual preliminary survey.

Anon., 1971a, "Minor Nonmetals in 1971," *Mineral Industry Surveys*, US Bureau of Mines, advance summary.

Anon., 1973, "Iodine 1972," *Commodity Data Summaries*, Division of Nonmetallic Minerals Bureau of Mines, January.

Anon., 1980, "Iodine Production is Planned for U.S. by Calabrian, With Start-Up in 1982," *Chemical Marketing Reporter*, Vol. 218, No. 18, Nov. 3, pp. 7, 23.

Anon., 1980a, "Growing Iodine Markets Spark Plant Expansions," *Chemical Week*, Vol. 127, No. 20, Nov. 12, p. 44.

Anon., 1981, "Iodine Price is Boosted Again, While Tight Market Eases Briefly," *Chemical Marketing Reporter*, Vol. 219, No. 4, Jan. 26, p. 24.

Absalom, S.T., 1980, "Iodine," Bulletin 671, Preprint, US Bureau of Mines, 9 pp.

Carnahan, T.G., et al., 1980, "Recovery of Silver from Chloride Leach Solutions by Iodide Precipitation," Report of Investigations 8428, US Bureau of Mines, 9 pp.

Collins, A.G., 1966, "Here's How Producers Can

Turn Brine Disposal into Profit," *Oil and Gas Journal,* Vol. 64, No. 27, July 4, pp. 112–113.

Collins, A.G., 1967, "Geochemistry of Some Tertiary and Cretaceous Age Oil-Bearing Formation Waters," *Environmental Science and Technology*, Vol. 1, No. 9, pp. 725–730.

Collins, A.G., and Eggleston, G.C., 1967, "Iodine Abundance in Oilfield Brines in Oklahoma," *Science,* Vol. 156, No. 377, May 19, pp. 934–935.

Collins, A.G., et al., 1967, "Bromide and Iodide in Oilfield Brines in Some Tertiary and Cretaceous Formations in Mississippi and Alabama," Report of Investigations 6959, US Bureau of Mines, 27 pp.

Croft, R.C., 1972, "Removal of Iron From Ilmenite by Cyclic Iodination," *International Symposium on Hydrometallurgy*, D.J.I. Evans and R.S. Shoemaker, eds., AIME, New York, pp. 920–942.

Dickson, T., 1979, "Iodine Ups and Downs," *Industrial Minerals*, No. 147, Dec., pp. 54–55.

Garrison, J., and Robe, K., 1980, "Hydrazine Reduction of KIO_3 Saves $5600/year, Ups Yield 5%," *Chemical Processing*, Vol. 43, No. 14, Dec., p. 154.

Hart, A.W., 1963, "Iodine and Iodine Compounds," *Encyclopedia of Chemical Technology,* Vol. 2, 2nd ed., Kirk-Othmer, ed., Interscience Publishers, New York, pp. 847–870.

Lyday, P.A., 1982, "Iodine," *Mineral Commodity Summaries 1982*, US Bureau of Mines, pp. 74–75.

MacMillan, R.T., 1970, "Iodine," *Mineral Facts and Problems*, Bulletin 650, US Bureau of Mines, pp. 1049–1058.

Palache, C., et al., 1951, "Iodates," *Dana's System of Mineralogy,* 7th ed., Vol. 2, pp. 312–319.

Park, F., 1967, "The Printed Word," *International Science and Technology,* Jan., pp. 24–26.

Park, W.G., 1969, "Iodine," *Minerals Yearbook,* US Bureau of Mines, pp. 1187–1191.

Shelton, R.A.J., 1968, "Thermodynamic Analysis of the Van Arkel Iodide Process," *Transactions,* Institution of Mining & Metallurgy, Vol. 77, No. 736, Sec. C., Mar., pp. C32–35.

Wang, K.P., 1972, "Iodine," *Minerals Yearbook 1972*, US Bureau of Mines, pp. 1359–1362.

Kyanite and Related Minerals

PAUL J. BENNETT *

JAMES E. CASTLE †

The sillimanite family of minerals, including kyanite, sillimanite, and andalusite, are anhydrous aluminum silicates with the formula $Al_2O_3 \cdot SiO_2$. Dumortierite and topaz are also included in this group because they are closely allied in composition and thermal behavior. However, neither are mined commercially today. They are typical metamorphic minerals which are found in metamorphic rocks on every continent. Sillimanite minerals are prized chiefly for their refractoriness and are important components in a broad range of acid refractory products, especially in mortars and castables. World production of sillimanite minerals is currently about 400,000 tpy and has been increasing at the rate of 3 to 8% a year for many years (Anon., 1972). While these minerals have widespread occurrence, the consumption of sillimanite minerals is concentrated in the relatively few highly industrialized areas where refractories are manufactured, and which in turn are typically close to the major iron and steel producing regions of the world. Thus northern Europe, England, the United States, and Japan are the principal consumers of refractories and sillimanite minerals. Of these countries only the US is a significant producer.

Uses of Kyanite

The following specifications are typical for the domestic kyanite industry:

Raw Kyanite Concentrates
Chemical Analysis

56% Al_2O_3 min	0.1% CaO max
42% SiO_2	0.1% MgO max
1% Acid soluble Fe_2O_3	0.3% combined alkali max
1.2% TiO_2 max	

Available Sizes

35, 48, 100, 200, and 325 mesh

* Bennett Mineral Co., Washington, GA.
† Consulting Mining Engineer, York, SC.

The largest use for domestic kyanite, both raw and calcined, is in the manufacture of refractory mortars, cements, castables, and plastic ramming mixes. In these applications kyanite constitutes from 10 to 40% of the mixture, the balance being refractory clays and coarser grog materials. A certain proportion of raw kyanite is used to offset the shrinking of the clay binder, whereas the calcined kyanite is used in the coarser sizes for body.

Certain high grade refractory shapes and kiln furniture are composed chiefly of graded sizes of calcined kyanite bonded with a little ball clay. In most applications, however, kyanite is used only to fortify the mix or as a minor ingredient. Some manufacturers of mullite brick use domestic calcined kyanite for the fine-grained portion of the brick, but seldom in quantities exceeding 10% by weight. Kyanite is used by ceramic manufacturers of wall tile and sanitary ware to offset shrinkage and cracking after firing.

Indian kyanite has traditionally been used to make the coarse grog sizes needed for the manufacture of mullite brick, and for the grog portions of mortar, cement, and castable mixes. It has been largely replaced, however, by synthetic mullite, high fired flint, and diaspore clay or calcined bauxite for other applications.

Indian kyanite has been available in lump as well as in granules and concentrates, as are sillimanite materials from India and South Africa. Andalusite is available as 8 mesh granules and as fines with chemical analyses not unlike those of domestic kyanites. From time to time and from place to place minor amounts of other aluminum silicate minerals such as dumortierite, topaz, and pyrophyllite have been made available at various prices and specifications, but have not been significant factors in the refractory industry. A mixture of kyanite and sillimanite of relatively fine grain-size and of low iron content was at one time offered to the molding sand industry by the Du Pont Co. as a byproduct from their deposits of heavy mineral sands in Florida. This product, "Kyasill," has been withdrawn from the market.

Sillimanite—Granular and massive forms of sillimanite are used in much the same way as granular and massive kyanite. Massive sillimanite from India and South Africa has been used for sawed block refractories principally in the glass industry in Europe. The massive varieties of kyanite and sillimanite have not enjoyed much use in the United States in recent years, but have been widely used by European refractory producers. Conversely, until recently, little granular domestic kyanite had found much application outside of North America. The use pattern is changing as a result of the depletion of the high grade massive deposits of the world, and of the increasing production of kyanite, andalusite, and sillimanite concentrates in India and Africa from deposits formerly thought to be unworkable. Andalusite concentrations are now being produced in France and are displacing or augmenting aluminum silicate sources from abroad.

The potential use of kyanite, sillimanite, and andalusite for the production of aluminum-silicon alloys has been investigated. To be an economic possibility, however, it appears that it would be necessary to produce aluminum silicate concentrates considerably purer than those currently produced and at a cost about half what the present market will bear. Short of some unforeseen and dire circumstances cutting off the free world trading and marketing of bauxite and silica, the likelihood of such production seems remote.

Other potential nonrefractory uses such as in wall and floor tile seem unlikely due to the high price of aluminum silicates compared to wollastonite, talc, and clays which are the raw materials currently used.

Prices

The average price for kyanite in the United States in 1981 ranged from $85 to $137 (depending upon grain size) per ton f.o.b. Georgia and Virginia. Aluminum silicate minerals of like specification are generally priced competitively throughout the world after freight considerations have been accounted for.

Geology

Mineralogy

The properties of the kyanite group of minerals are given in Table 1.

Classification of Deposits

Kyanite: *Kyanite Quartzite*—The kyanite produced in the United States comes from kyanite quartzites. Kyanite quartzites are rocks containing 15% to 40% kyanite and usually about 5% of other minerals such as pyrite, lazulite, rutile, and mica. The rock is characterized by an anomalously low content of potash and soda and the virtual absence of calcium and magnesium. The alumina content of kyanite quartzite ranges between 10% and 25% and averages about 18% Al_2O_3, which is generally similar to that of the schistose rocks which enclose the kyanite-bearing strata.

Kyanite quartzite deposits occur in the Piedmont region of the Appalachian Mountain system in a relatively narrow zone extending from northeastern Georgia through central South and North Carolina to southeastern Virginia. They are enclosed within a wider band of mildly metamorphosed acid metavolcanics and sediments known as the Volcanic Slate Belt in Virginia and the Little River Series in Georgia. The age of the rocks is uncertain, but several workers have suggested that they are lower Cambrian in age (Bennett, 1961).

More recent studies by Carpenter and Allard (1980) expand the mineralization and alteration theories of kyanite in the Georgia-South Carolina-McCormick district.

There are at least 13 distinct deposits of kyanite quartzite within this zone. All of them are known as "mountains" owing to their soli-

TABLE 1—Properties of Kyanite Group Minerals

	Andalusite	Kyanite	Sillimanite	Mullite
Formula	$Al_2O_3 \cdot SiO_2$	$Al_2O_3 \cdot SiO_2$	$Al_2O_3 \cdot SiO_2$	$3Al_2O_3 \cdot 2SiO_2$
Crystal system	Orthorhombic	Triclinic	Orthorhombic	Orthorhombic
Cleavage	{110} good {100} poor	{100} perf. {010} good {001} parting	{010} good	{010} perf.
Hardness	6 – 5 – 7	5.5 – 7*	6.5 – 7.5	6 – 7

* Varies with direction.

tary prominence rather than their relatively modest height. Graves Mountain in Georgia, and Willis Mountain, Baker Mountain, and East Ridge all in Virginia are being mined at the present time. Until 1970 the Henry's Knob deposit in South Carolina was also mined (Espenshade and Potter, 1960).

In southeastern California and extending into southwestern Arizona there is a zone of metavolcanics whose composition and lithology resembles that of the Carolina Slate Belt in which kyanite quartzite deposits are also found. However, the kyanite in these deposits, while averaging 25 to 35% of the rock in apparently large and recoverable crystals, has proven to be contaminated with extremely fine inclusions of quartz. To date, it has not been possible to produce a competitive kyanite product from these western kyanite quartzite deposits.

Kyanite quartzite has been described in several places in the world, and usually seems to occur within a geologic framework similar to that of the deposits of the southeastern United States. Such deposits have been described and explored in Surinam, Norway, Kenya, and Austria (Varley, 1968).

Kyanite Schist and Gneisses—Kyanite is very common in the highly metamorphosed schists and gneisses of the metamorphic regions of the world. Typically, the kyanite occurs in quantities ranging from a percent or two to as much as 25% in a gangue of biotite, feldspar, muscovite, garnet, and occasionally hornblende and other common rock forming minerals. Rocks containing a few percent of kyanite are extremely abundant and widespread. They are exposed over hundreds of square miles in the eastern and western metamorphic areas of North America and in the metamorphic rocks of other continents.

Repeated attempts to recover the kyanite from such rocks have been made and have been described in earlier editions of this volume. The most recent attempt was in the Timiskaming district in western Quebec by North American Refractories Co. of Cleveland, Ohio. An earlier effort was made in the late 1930s near Burnsville, NC. At the present time no kyanite schists are being mined for kyanite.

In order for such kyanite to be economic it is necessary that the region be deeply weathered and of gentle relief so that a mantle or segregation of resistant kyanite nodules, cobbles, and boulders can be accumulated at the surface. It is further necessary for labor to be abundant and cheap.

A great deal of literature has been published by the US Bureau of Mines, the US Geological Survey, and agencies of other governments describing investigations of kyanite-bearing schists. A recent exhaustive study was conducted by the USBM in Idaho on the huge deposits of kyanite, sillimanite, and andalusite at Goat Mountain and the kyanite deposit on Woodrat Mountain near Kemiah (Van Noy, 1970). Hundreds of millions of tons of 25% aluminum silicate ore available for open cast mining has been postulated. However, remoteness from the major markets and the difficulty of beneficiation have stalled development at this locality.

Massive Kyanite—Kyanite is found locally as nodules, knots, and huge boulder-sized segregations in very highly metamorphosed areas of aluminous sediments. This had been the principal source of kyanite from India for the past 40 years. Similar segregations were the basis for the kyanite production in Kenya which has been discontinued.

In several counties in Georgia, similar lumps of massive kyanite are found (Furcron and Teague, 1945). While the abundance and purity of the Georgia massive kyanite meet the requirements for commercial exploitation, high labor cost makes economic production by hand-sorting and gathering unfeasible and production has been limited to a few carloads during World War II.

Kyanite mineral segregations are probably the result of local pneumatolytic migration of silica and alumina during the late stages of regional metamorphism. Introduction of alumina does not seem to have been a factor since the overall composition of the segregations are similar to that of the country rock if the sample area considered is large enough to include the barren quartz segregations which invariably accompany the kyanite-corundum segregations. Aside from the size and abundance of the segregations in the Lapsa Bura deposits of India, they do not seem mineralogically dissimilar from kyanite segregations found occasionally in all kyanite schists.

Massive kyanite typically contains corundum and minor amounts of rutile. The kyanite is often felty and acicular and occurs in tightly interlocking aggregates. Kyanites from India are usually produced in lumps large enough to be hand-sorted according to kyanite and corundum content, and three grades are offered. Massive Indian kyanite has properties quite unlike that of large coarse kyanite crystals. Massive Indian kyanite is essentially volume-stable and calcines to a dense white aggregate that is much prized by European refractory manufacturers. On the other hand, the coarse kyanite produced from Georgia placer deposits in the 1940s, and more recently in Kenya, crumbles and loses much of its density and physical strength upon calcining. Apparently the interlocking acicular crystal mode of "Indian" kyanite prevents such expansion and consequent breakdown.

Sillimanite: *Sillimanite Schists*—Sillimanite is a very common rock-forming mineral in metamorphic rocks of relatively high rank. It is common in a series of metamorphosed rocks to find sillimanite, kyanite, and andalusite interchanging occurrences in given strata as the local conditions of temperature vary, as in the proximity to intrusives. For this reason the aluminum silicate minerals are often used to identify parameters of metamorphic intensity. Sillimanite, while common, seldom occurs as potentially exploitable crystals. The typical mode is what is often called "fibrolite" which is a felty aggregate of extremely fine "whiskers" of acicular sillimanite interlaced and interlocked with quartz, mica, and other minerals. Beneficiation is usually impossible. In some areas the sillimanite occurs as nodules and buttons which are marginally potential as in the Peltzer area of South Carolina (Espenshade and Potter, 1960).

In Hart County, Georgia, there is a northeasterly trending zone about ten miles long in which "matchstick" sillimanite occurs. Beneficiation tests have been conducted, from which it appears that a limited production of sillimanite in the 35 mesh range could be accomplished. However, the deposits are narrow and limited, and there is presently no existing market or incentive for production of sillimanite concentrates in the United States (Furcron and Teague, 1945).

Massive Sillimanite—Massive sillimanite has been produced for many years from the state of Assam, India. Production has been falling off in recent years as the cost of production has risen and as other materials are substituted for the sawed blocks which were formerly an important use of sillimanite. The deposits of Assam consist of huge segregations of sillimanite and corundum, often in intimate association, and weighing several tons. Considerable hand effort is employed to recover the boulders in a form suitable for sawing to refractory shapes, particularly for the English glass industry.

In the vicinity of Adelaide, South Australia, late-stage metasomatism resulted in a mixture of kaolin and included boulders and nodules of sillimanite. The sillimanite is recovered as a byproduct in the process of manufacturing refractory clays. Other deposits of residual boulders of massive sillimanite have been exploited in this region. Beneficiation has been tried with some success on the sillimanite itself and on the byproducts too low in grade to use directly. The Australian domestic market for such concentrates is limited, however, and thus far no large scale production has been reported.

The most important deposits of massive sillimanite-corundum occurs in the Republic of South Africa, in the Pella District near Pofadder, Namaqualand. Production reached a high of 56,000 tons in 1963, but has dropped steadily in recent years to an estimated 18,000 tons in 1971 (Anon., 1972).

Andalusite: *Andalusite*—Andalusite is a frequent constituent of metamorphic rocks, although it is not as abundant or common as sillimanite or kyanite. It is found in argillaceous and micaceous slates, and in schists and gneisses resulting from the contact metamorphism of intrusive rocks. Andalusite readily incorporates foreign matter in its crystals, and frequently grows around preexisting materials, including carbon. One variety, chiastolite, is so named for the crosslike inclusions of carbon oriented normal to the axis of the crystal.

In France, near Glomel in Brittany, an extensive, deeply weathered body of andalusite schist is being mined at the present time. Here the andalusite occurs as matchstick-sized crystals embedded in a fine-grained black groundmass composed of biotite, hornblende, muscovite, and feldspar. The andalusite is evenly disseminated and constitutes about 20% of the rock. The rocks in the area are very poorly exposed, and the geology of this occurrence is not well understood.

Near Canso, Nova Scotia, there is an extensive deposit of andalusite schist. Here the andalusite makes up about 15% of the rock and is evenly disseminated as large porphryoblasts averaging about 1 × ½ in. in cross section. The groundmass is principally muscovite, garnet, and feldspar. The crystals of black andalusite incorporate about 10% of finely disseminated magnetite and muscovite, however, and effective beneficiation is not practical.

Near Hillsboro, NC, a monadnock of andalusite-pyrophyllite-sericite rock is being mined by the Piedmont Minerals Co. The ore consists principally of pyrophyllite and quartz and contains 15 to 20% of disseminated pink andalusite. The andalusite is mined along with the pyrophyllite, and the resulting mixture is almost entirely consumed by the parent company in the manufacture of refractory products.

In the Goat Mountain deposit in Idaho, andalusite coexists with kyanite and sillimanite. No attempt has been made to separate it from the other aluminum silicate minerals in the tests made so far. The andalusite in this deposit typically incorporates a great many deleterious impurities. The Goat Mountain deposit is extremely large, but to date no commercial exploitation of the deposit seems feasible owing to beneficiation difficulties (Abbott and Prater, 1954).

Residual Andalusite—Alluvial deposits of andalusite sand occur in South Africa. These deposits have been worked on a very large scale, and current production is in the vicinity of 45,000 tpy. The source of the andalusite are shales of the Pretoria Series that have been intruded and metamorphosed by the Bushveld Complex. The andalusite has been weathered free from the parent rock and subsequently concentrated by the action of wind and water. Apparently concentration is still going on. The andalusite sand commonly contains 50% of recoverable andalusite. The reserves are estimated to be in the neighborhood of 800,000 tons of +50% andalusite. In 1964 a heavy media separation plant was installed, and since then the production of high grade concentrates has increased markedly.

Reserves

Domestic: The potential supply of kyanite minerals vastly exceeds the potential market. Therefore, what constitutes a reserve can not be defined without considering the probability of production. The two producing domestic kyanite deposits have enormous proven reserves. At least 65 million tons of kyanite quartzite containing at least 25% kyanite and amenable to open pit mining were indicated by an extensive US Bureau of Mines drilling program undertaken in 1949 at Willis Mountain in Virginia (Jones and Eiletsen, 1954). The potential reserves in this locality are probably twice the indicated reserves. At Graves Mountain, geologic mapping and diamond drilling has indicated a reserve of 25% ore in excess of 30 million tons. From these figures, which translate into many decades of production at current rates, it is readily apparent that the existing producers have little interest in acquiring new reserves unless a considerable economic advantage would result. If the +10% kyanite-bearing schists and gneisses are included, the reserves in the United States and Canada are truly vast. At current prices these reserves are obviously submarginal.

World: Most of the comments made about the reserves of kyanite minerals in the United States apply to other areas of the world as well. However, the massive kyanite reserves of India, being relatively rare and unique in occurrence, are more susceptible to reserve considerations. The Lapsa Buru area is credited with 700,000 tons of massive kyanite to a depth of 10 ft, and the Madhya Predesh area with 250,000 tons of massive sillimanite.

In the Republic of South Africa 800,000 tons of +50% andalusite sand are reported for the Transvaal and millions of tons of sillimanite reserves have been credited to Namaqualand in the Cape Province.

No figures have been reported from the Province of Brittany, France, where andalusite production recently began. However, the body of andalusite schist is reported to be at least 1 km wide and about 10 km long, and to contain an average of about 15 to 25% andalusite. This translates into many tens of millions of tons of potential reserves.

It is apparent that the existing producers of aluminum silicate minerals have adequate reserves, with the possible exception of the unique and relatively rare deposits of massive kyanite-corundum and massive sillimanite-corundum. It should be noted, however, that the primary use for the massive varieties is being increasingly challenged by the more consistent synthetic mullite materials which are produced from more available high alumina clays and bauxite.

The heavy mineral beach sand deposits of the world generally contain a significant proportion of andalusite, sillimanite, or kyanite. The US Bureau of Mines has described a method of producing a kyanite coproduct from the ilmenite sand treatment plants in Florida. The potential production from heavy mineral deposits throughout the world is large.

Exploration and Evaluation

Kyanite and sillimanite are fairly common minerals, and are found in the metamorphosed rocks on every continent. Andalusite is less abundant but still common. However, kyanite is consumed principally by the highly developed industrialized nations and its consumption is therefore restricted to certain areas. For example, the bulk of the American production of kyanite is consumed in a relatively small region extending from the northeastern part of the United States to central Missouri. A similar pattern exists in Europe, where consumption is concentrated in those parts of England, France, and Germany where the producers of refractories are concentrated. Kyanite and related minerals must be delivered to the consuming areas at competitive prices. Therefore, a very important preliminary consideration in any exploration project is the prospective cost of delivering the kyanite to the market. In the United States, typical freight costs for delivering kyanite concentrates from traditional sources to consumers ranges between $30 and $40 per ton. A relatively small market exists on the west coast of the United States where the freight is as high as $90 per ton.

The second problem to be considered is the probability of producing a salable product. Kyanite consumers are notoriously finicky about the specifications of materials used by them in the manufacture of refractory products. Thus any deviation from the commonly accepted specifications imposes a considerable burden on products attempting to enter the market. Probably the most important consideration in this regard when evaluating a raw prospect is the liberation size of the kyanite crystals. In the western United States there are several otherwise attractive kyanite-quartzite deposits which cannot be beneficiated to the required specifications without grinding the ore to about —200 mesh, which is quite fine considering that most existing kyanite deposits can produce concentrates in a size range of at least —35 to 28 mesh.

If the initial investigation indicates that both freight and quality of product can be competitive, then the normal exploration techniques apply. These consist of surface sampling, trenching, geologic mapping, and some limited diamond drilling to test the material in depth. Kyanite deposits are in reality kyanite-bearing rocks produced by regional metamorphism. They have inherently consistent compositions, unlike the hydrothermal deposits which are typical of the precious and base metals. Geochemical and geophysical exploration techniques are generally not useful for evaluating kyanite and other aluminum silicate deposits.

After a deposit has been explored, the potential markets determined, and studies indicate that a competitive product could be produced and delivered to the market, the next step should be a pilot plant production of the material intended to be marketed. Bench scale and laboratory testing are not sufficient in this regard. As an example, in recent years a major attempt was made to produce kyanite from schists and kyanite-bearing gneisses in Canada. After the construction of a plant and the expenditure of a great deal of money it was found that a salable concentrate could not be produced economically. Similar failures have occurred in the United States and in other countries. Therefore, it is most important to produce, by reproducible pilot plant techniques, a quantity of concentrates from a representative selection of ores sufficient to permit prospective consumers to test the new material extensively in their plants. There is no laboratory substitute for this kind of field testing in the plants of the intended consumers. Quoted physical and chemical specifications are to be used only as approximations of the suitability of a given kyanite or aluminum silicate material. New producers of aluminum silicates have found to their sorrow that kyanite conforming to specifications supplied by the prospective consumers may not in fact be what those consumers are really willing to use when confronted with the decision to utilize a new material from a new area. Competitive testing and specific approval is the only sure way that the marketability of a given kyanite concentrate can be assured.

Preparation for Markets

In the United States for the past 30 years kyanite concentrates have been produced by flotation and magnetic separation. The four kyanite-quartzite deposits currently being exploited, Willis Mountain, Baker Mountain, and East Ridge all in Virginia, and Graves Mountain in Georgia, are monadanock-type features which were created by differential weathering of the softer schists and phyllites surrounding the more resistant hard kyanite-quartzite ore bodies. As a result, the ore is exposed with little or no overburden.

The ore is drilled and blasted with ammonium nitrate. Secondary breaking is typically accomplished with a drop ball. The ore is picked up with diesel-powered shovels, loaded into trucks, and carried to the crushing plant, where it is reduced to approximately —1½ in. in size and stockpiled. The ore is withdrawn from the stockpile by conveyor belt controlled by an automatic feeder and fed into a rod mill in closed circuit with a rake classifier set to grind the ore to —28 mesh. The resulting slurry is carefully deslimed. This is an important prerequisite for successful flotation, but one which results in a considerable loss of fine-grained kyanite. The deslimed slurry is conditioned with pine oil and xanthate and an amine collector and passed through a series of flotation cells which remove the pyrite and micaceous contaminants from the slurry. The tailings from the pyrite circuit are again deslimed and conditioned with sulfuric acid and petroleum sulfonate and passed through a section of rougher flotation cells. The rougher concentrate goes to a two-stage recleaning circuit. The tails of the rougher circuit go to waste.

Flotation concentrate consists of about 91% kyanite, 2–5% iron oxides, and the balance quartz. The concentrates are dewatered by draining in open stockpiles or in drainage silos and are subsequently conveyed to a dryer where the moisture is reduced from 6–8% to essentially 0%. One American producer utilizes a reducing roast technique, elevating the temperature during the drying process to 900° F, followed by cooling in an oxygen-deficient atmosphere to render the iron oxide highly magnetic and to oxidize any residual pyrite. The other producer does not find this to be necessary. The dried concentrates are passed over high-intensity three-to-five-roll magnetic separators which reduce the acid-soluble iron in the concentrate to less than 1%. The resultant material is known as a raw 35 mesh kyanite and is ready for use. However, about 75% of the kyanite sold in the United States is further reduced in granite-lined pebble mills to −48, 100, and 200 mesh sizes. Until recently, about 30% of the kyanite was calcined in rotary kilns at a temperature of approximately 2900° F and sold as mullite or calcined kyanite. However, in the past year or two the demand for calcined kyanite leveled off and began to decline as cheaper calcined kaolin and bauxite became available. Approximately one-half of the kyanite sales made in the United States are shipped in bulk hopper cars, and the balance is packaged in 100-lb multiwall paper bags. Of the kyanite produced, 96% is shipped by rail to consumer's sidings.

Kyanite exports have increased over the past several years with an estimated tonnage of 25,000 exported in 1980 to Europe, Australia, Japan, South America, Mexico, and Canada.

Andalusite concentrates are produced near Gommel, in Brittany, France, by LaFerriere Aux E. Tangs. The ore is a black andalusite schist containing about 20% coarse euhedral andalusite porphyroblasts. The andalusite crystals are about the diameter of matchsticks and range from ¾ to 1 in. in length. The andalusite schist is deeply weathered and friable to at least a depth of 30 ft. The ore is drilled with a wagon drill, blasted, and transported with a rubber-tired Hough front-end loader with a 2½ to 3-yd bucket. It is fed into a surge hopper through an 8 x 8-in. grizzly and discharged onto a 24-in. conveyor belt by a reciprocating feeder. Oversized material is reduced with a sledge hammer to pass the 8-in. grizzly.

The conveyor belt carries the broken ore about 150 ft to an autogenous mill, which consists of a drum 15 ft in diameter revolving on eight truck tires, and which reduces the ore to 100% −⅛ in. The furnace is air-swept with heated air in closed circuit with two air classifiers. The sized ore is fed into a hopper, from which it is passed over a four-roll, high-intensity magnetic separator. The nonmagnetic fraction is conveyed to the heavy media separator and the magnetics are discarded.

The heavy media separator circuit consists of a hopper in which the andalusite ore is blended with a ferrosilicon slurry and then pumped through a hydrocyclone. The specific gravity of the slurry is controlled automatically by a water-metering device ahead of the blending hopper. The first stage of the two-stage heavy media separator circuit is controlled at about 2.3 specific gravity. The final effective specific gravity is adjusted by regulating the velocity of the slurry through the cyclone. The second stage is controlled in similar fashion. The sink product from the heavy media separator is screened at about 35 mesh to recover the media. About one-half of the andalusite is lost in the fine sizes at present. In the future, it is intended to install flotation cells to recover these fines.

The andalusite now being produced consists of dark brown to black, roughly equidimensional grains ranging in size from −8 mesh to +35 mesh, and containing about 59.2% Al_2O_3 and 1.0% Fe_2O_3. The plant is capable of producing 15 to 20,000 tpy.

The kyanite deposits of India have traditionally been exploited by labor-intensive techniques. In the beginning it was sufficient to simply gather the cobbles and boulders of kyanite which had been released and exposed by weathering. Later, primitive mining methods were initiated and the large segregations were drilled and blasted and hand-cobbed in preparation for shipment. Dozens of these small operations employed numerous men and women. As time passed and the availability of readily hand-picked material decreased, the beneficiation methods became more sophisticated. The picking belt is still an important beneficiation technique, however. High-grade kyanite nodules, and segregations too low in grade to be shipped directly, are now crushed and subjected to more sophisticated beneficiation techniques including flotation and magnetic separation. At the present time there are no large-scale integrated beneficiation plants in India of the kind employed in America.

In South Africa andalusite sands of the Marico district of the Transvaal are beneficiated with heavy media separators and high-intensity magnetic separators. A typical analysis of the product is as follows:

Al_2O_3	52-57%
SiO_2	35-44%
Fe_2O_3	1-4.5%
TiO_2	0.04-4%

Segregations of sillimanite-corundum rock in the Pellar District in the northwestern part of the Cape Province have been produced for many years by labor-intensive techniques similar to those employed in India.

In general, the large disseminated deposits containing 20 to 30% recoverable aluminum silicates are treated by modern beneficiation techniques, while the segregation-type deposits including the massive kyanites of India and the sillimanite-corundum segregations of South Africa are mined by more primitive labor-intensive techniques. As time passes the African and Indian deposits are increasingly subjected to the more sophisticated ore dressing techniques, and more aluminum silicate concentrates are being offered to the market from those areas which previously yielded only hand-selected products.

Future Considerations and Trends

The sillimanite family of minerals will be increasingly challenged in the future by synthetic mullite, calcined bauxite, and especially by the calcined kaolin which has only recently been produced and promoted on a large scale. The unique thermal expansion characteristics of kyanite at elevated temperatures should ensure its continued use in acid refractories designed for service in temperatures up to 3200° F. The increased use of castable and monolithic refractories also should dictate a correspondingly increased consumption of kyanite. The growth rate of the kyanite industry in the United States has been at a rate of 8 to 10% a year. Although there is a likelihood that this rate of growth will not be maintained indefinitely, a decline in absolute consumption is not anticipated.

Kyanite mining in the United States is subject to a depletion allowance of 22%. There is no tariff on imports of kyanite. There are no specific programs sponsored by the US government directed to the development or production of kyanite deposits, although the US Bureau of Mines has from time to time conducted research and development on particular aspects of production, and from time to time the US Geological Survey publishes exploration reports on promising deposits.

Recent reports by Coope (1980) indicate that kyanite production is on the rise in Brazil with a capacity of 40,000 tons per year from two operating mines. With this increase in world supply from Brazil, there is a decline in available material from India due to a governmental ban on export of high grade material and strict ceilings placed on allowable export tonnages of low grade product. In 1980 it was reported that 1,000 t/a were exported as compared to 70,000 t/a in 1970.

The major kyanite and kyanite-group mineral deposits of the world are located in relatively remote areas. The industry is relatively small in scope, and at the present and for the foreseeable future no special environmental problems are foreseen.

Bibliography and References

Anon., 1972, "Refractory Raw Materials," *Industrial Minerals*, No. 59, Aug., pp. 13–17.

Abbott, A.T., and Prater, L.S., 1954, "The Geology of Kyanite-Andallusite Deposits, Goat Mountain, Idaho, and Preliminary Beneficiation Tests on the Ore," Pamphlet 100, Idaho Bureau of Mines and Geology, 27 pp.

Bennett, P. J., 1961, "The Economic Geology of Some Virginia Kyanite Deposits," unpublished Ph.D. Thesis, University of Arizona, Tucson.

Browning, J.S., and McVay, T.L., 1963, "Flotation of Kyanite Quartzite Rock, Graves Mountain, Lincoln County, Georgia," Report of Investigations No. 6268, US Bureau of Mines, 9 pp.

Carpenter, R.H., and Allard, G.O., 1980, "Mineralization, Alteration and Volcanism in the Lincolnton-McCormick District, Georgia and South Carolina," Abstract, Bulletin of Geological Society of America, Nov., pp. 398–399.

Coope, B., 1980, "Patterns of Raw Material Supply to the Refractories Industry," Paper presented at ILAFA Conference, Lima, Peru.

Espenshade, G.H., and Potter, D.B., 1960, "Kyanite, Sillimanite and Andalusite Deposits of the Southeastern States," Professional Paper 336, US Geological Survey, 121 pp.

Furcron, A.S., and Teague, K.H., 1945, "Sillimanite and Massive Kyanite Deposits in Georgia," Bulletin 51, Georgia Dept. of Natural Resources, 76 pp.

Grametbaur, A.B., 1959, "Selected Bibliography of Andalusite, Kyanite, Sillimanite, Dumortierite, Topaz, and Pyrophyllite in the United States," Bulletin 1019–N, US Geological Survey, pp. 973–1046.

Jones, J.O., and Eiletsen, N.A., 1954, "Investigation of the Willis Mountain Kyanite Deposit, Buckingham County, Virginia," Report of Investigations No. 5075. U.S. Bureau of Mines, 41 pp.

Klinefelter, T.A., and Cooper, J.D., 1961, "Kyanite—A Materials Survey," Information Circular 8040, US Bureau of Mines, 54 pp.

Potter, M.J., 1982, "Kyanite and Related Minerals," *Mineral Commodity Summaries 1982,* US Bureau of Mines, pp. 82–83.

Van Noy, R.M., 1970, "Kyanite Resources in the Northwestern United States," Report of Investigations 7426, US Bureau of Mines, 81 pp.

Varley, E.R., 1965, *Sillimanite, Andalusite, Kyanite,* Overseas Geological Surveys, Mineral Resources Div., London, 165 pp.

Varley, E.R., 1968, *Sillimanite,* Chemical Publishing Co., New York, 165 pp.

Wells, J.R., 1972, "Kyanite and Related Minerals," *Minerals Yearbook*, US Bureau of Mines, pp. 689–693.

Williamson, D.R., 1960, "The Sillimanite Group," *Mineral Industries Bulletin,* Colorado School of Mines, Vol. 3, No. 4, July, 12 pp.

Lime

ROBERT S. BOYNTON *
KENNETH A. GUTSCHICK *
Revised by ROBERT C. FREAS †
and JEFFREY L. THOMPSON ‡

Lime, the "versatile chemical," is, generally speaking, a calcined or burned form of limestone commonly known as quicklime, calcium oxide or calcia, or, when water is added, calcium hydroxide or slaked lime. Almost 40 different lime products are available, a fact which has contributed to the rather loose use of the term *lime* as well as to much confusion and misunderstanding. The term is frequently, albeit erroneously, used to denote almost any kind of calcareous material or finely ground form of limestone or dolomite.

Lime is usually made from high-calcium or high-magnesium limestone, generally having a minimum of 97% combined carbonate content. Normally, high-calcium limes contain less than 5% MgO. When the lime is produced from a high-magnesium limestone, the product is referred to as dolomitic lime.

Calcination, or the production of lime, has its origins in the earliest days of alchemy, with the general reaction class having been identified in an Arabic text printed in 1000 A.D. It was much later, however, in the mid-1700s to the mid-1800s before this basic reaction became understood from a scientific perspective. The production of lime has been so basic and simple that its underlying scientific principles have over the years received only intermittent investigation. Rather, much of the thought and inquiry has been directed toward the development of kilns. Thus, it has only been in the last 15 to 20 years that lime has received any concentrated scientific investigation relative to the thermodynamics and kinetics of the calcination and hydration reactions.

Calcination refers to a broad class of reactions, of which the lime/limestone reaction is just one, wherein a substance is heated to less than its melting point, resulting in a weight gain or weight loss. In the calcination of limestone to produce lime, the basic chemical reaction is as follows:

$$\underset{100}{CaCO_3}\ (\text{limestone}) + \text{heat}\ (1000^\circ \text{ to } 1300^\circ C) \rightleftharpoons \underset{56}{CaO}\ (\text{quicklime}) + \underset{44}{CO_2} \uparrow$$

or

$$\underset{100}{CaCO_3} \cdot \underset{84}{MgCO_3}\ (\text{limestone}) + \text{heat}\ (900^\circ C \text{ to } 1200^\circ C) \rightleftharpoons \underset{56}{CaO} \cdot \underset{40}{MgO}\ (\text{quicklime}) + \underset{88}{2CO_2} \uparrow$$

While there is nearly universal agreement about the equilibrium conditions related to the limestone/lime reaction above, there have been numerous and varied calcination models developed for the reaction. Recent investigations have resulted in the development of the model shown in Fig. 1. From this model it can be seen that calcination is a function of both temperature and CO_2 pressure. It does not, however, provide any indication of the rate at which the reaction takes place.

Calcination is strongly time variant with different limestones. In a very broad sense this relates to the fact that the calcination reaction starts on the exterior surface of the limestone and then proceeds toward the center. As the calcination reaction takes place, the CO_2 released at the reaction interface must make its way through the lime to the exterior surface. Therefore, because calcination is limited by gas diffusion to the surface of the partially calcined limestone, natural impurities in the stone, differences in crystallinity, grain boundary chemistry, density variations, and imperfections in the atomic lattice all play a significant role in

* Article for 4th edition prepared by Robert S. Boynton, retired, formerly Executive Director, National Lime Assn., and Kenneth A. Gutschick, Technical Director, National Lime Association, Washington, DC.

† Limestone Products Corp., Sparta, NJ.

‡ Dravo Lime Co., Pittsburgh, PA.

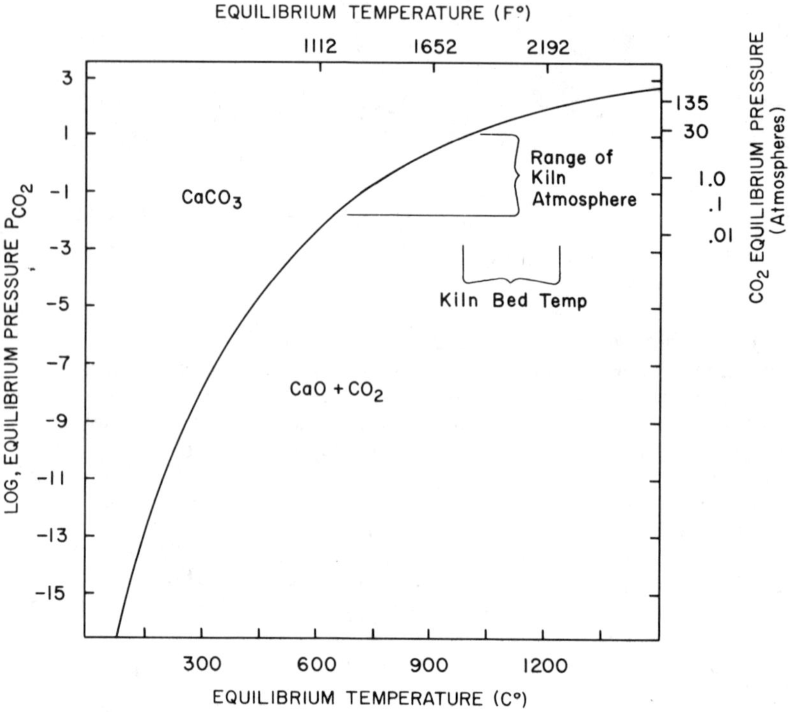

FIG. 1—*Equilibrium graph for* $CaCO_3 \rightleftharpoons CaO + CO_2$ *as a function of temperature and* CO_2 *pressure in atmospheres. Conditions below the curve cause the reaction to go toward lime (CaO), while above the curve the reaction proceeds toward calcite. Only on the curve is the equilibrium maintained and the reaction out does not proceed in either direction. Most like kilns have production bed temperatures of 1000 to 1150°C.*

calcination rate. The suitability of a given limestone as a source material for lime production can be determined only after completion of adequate burn tests designed to evaluate the various limiting factors. When a coal-fired kiln system is considered, the entire process of calcination is made even more complex. The reader is referred to the references for a more detailed description of the complete calcination reaction.

In the foregoing discussion it was noted that CO_2 is released during the reaction. This release results in a 44% weight loss during the complete calcination of a high calcium limestone, or a 48% weight loss for a highly dolomitic limestone. The trade term for this weight percent loss is *loss on ignition*, or LOI.

Because the calcination reaction is chemically reversible, *quicklime* or *burned lime* is frequently referred to as being highly reactive, or unstable. The more stable form of lime, *hydrated lime*, is commonly preferred and specified by the user. Hydrated lime is obtained by adding water to quicklime to produce a dry, fine powder. Quicklime's affinity for moisture is then satisfied, although it still retains a strong affinity for CO_2.

Hydration is the combining of calcium oxide with water in a reversible reaction to form calcium hydroxide:

$$CaO + H_2O \rightarrow Ca(OH)_2.$$

As a standard condition the reaction is exothermic for the formation of hydrate, with the accepted value for the heat of hydration being 15.456 kcal/g mol of CaO. This is a three-step reaction, with the first being the dissociation of CaO into a Ca++ ion and an $O^{=}$ ion:

$$CaO \rightarrow Ca^{++} + O^{=}.$$

Next, the water ionizes into two hydroxyl ions due to the strong attraction that an oxygen anion exerts on a water molecule:

$$H_2O + O^{=} \rightarrow 2(OH)^{-}.$$

Finally, the calcium and hydroxyl ions combine to form calcium hydroxide:

$$Ca^{++} + 2(OH)^{-} \rightleftharpoons Ca(OH)_2 + \text{heat}.$$

As noted this is a reversible reaction, and hydrated lime therefore can, when heated, revert to the original oxide form.

Analysis and Properties

The value of lime for most purposes is dependent upon its available lime content (CaO), or in the case of dolomitic lime its total oxide content. The purity of lime is influenced primarily by the purity of the limestone, and secondarily by its manufacture. The major impurities are silica, iron, alumina, and sulfur. During calcination these impurities react with some of the calcium oxide, yielding mineral forms (silicates, aluminates, ferrites, etc.) that reduce the available CaO by 3 to 5% or more. Even the purest calcium limes assay less than 95% in available CaO, several percentage points less than the total oxide content. In addition, uncalcined limestone particles may remain as core in the individual lime pieces. Typical chemical analyses of commercial quicklimes are presented in Table 1.

Impurities and core also affect the hydration reaction. As an example, sulfur in the form of calcium sulfate inhibits the hydration reaction out of all proportion to its actual weight percent concentration (perhaps more than 100 times). If the parent stone contains dolomite, this will also inhibit the reaction as its hydration may be nearly three orders of magnitude slower than that of calcium oxide. Early in the hydration reaction dolomite appears practically inert, but given sufficient time it will hydrate. The heat of reaction is nearly the same as for calcium oxide, but the time necessary for complete hydration may be hours, whereas for calcium oxide it is only seconds.

Important physical properties of lime are listed in Table 2. Lime is white in color of varying intensities, but some limes have a light cream, buff, or gray cast, depending upon the impurities present. The severity of calcination (time and temperature) determines the degree of porosity and chemical reactivity of the lime. When a lime is soft-burned, very little or no shrinkage occurs and a porous, softer, very reactive lime is produced.

When a lime is hard-burned, a denser, physically stronger, less reactive lime results. In either case the lime will readily hydrate in water except the rapidity of hydration is much greater with the former type than with the latter. In any event the heat of hydration is appreciable: 1134 and 886 kJ/kg of quicklime for high calcium and dolomitic types, respectively. This strong exothermic reaction will boil water easily and, under certain hydration conditions, temperatures of 290 to 315°C have been reached, causing dehydration of the freshly slaked lime. Quicklime can be made so reactive that it will literally explode on contact with water.

TABLE 1—Typical Analyses of Commercial Quicklimes

Component	High Calcium Quicklimes, Range,* %	Dolomitic Quicklimes, Range,* %
CaO	93.25-98.00	55.50-57.50
MgO	0.30-2.50	37.60-40.80
SiO_2	0.20-1.50	0.10-1.50
Fe_2O_3	0.10-0.40	0.05-0.40
Al_2O_3	0.10-0.50	0.05-0.50
H_2O	0.10-0.90	0.10-0.90
CO_2	0.40-1.50	0.40-1.50

Source: Anon, 1981.

*The values given in this range do not necessarily represent minima and maxima percentages.

Although limestone is almost totally insoluble, lime is slightly soluble in water in the range of 1.4 to 0.54 mg/L, depending on temperature. As water temperature rises, lime's solubility diminishes. The addition of sugar will increase these solubilities manyfold.

On slaking into a slurry or milk-of-lime, the saturated solution ionizes immediately into Ca^{++}, Mg^{++}, and OH^- ions, one of the strongest bases. Even a trace of lime will yield a pH of 11.2 and on up to nearly pH 13 at saturated solution and low temperature.

As an alkali, unlike other strong bases, (NaOH, KOH), $Ca(OH)_2$, and $Mg(OH)_2$ are diacid bases. Therefore, one molecule of lime will neutralize two molecules of acid (sulfuric, hydrochloric, etc.). For example, at equal weights CaO has 30% more neutralizing power than caustic soda (sodium hydroxide). However, dolomitic quicklime, because of its MgO content, has 16% greater neutralizing power than pure CaO.

Lime Manufacture

Production of lime involves three main processes—stone preparation, calcination, and hydration—although the latter is not necessary for several lime uses. The simplified flowsheet (Fig. 2) outlines the various production steps. Details of stone preparation, which comprises quarrying or mining, crushing, and screening to produce kiln-size stone, are covered in other chapters. However, because lime must meet existing chemical specifications, the various steps in stone production are carefully con-

TABLE 2—Properties of Typical Commercial Lime Products

Quicklimes

	High Calcium	Dolomitic
Primary constituents	CaO	CaO and MgO
Specific gravity	3.2-3.4	3.2-3.4
Bulk density (pebble lime), g/cm^3	0.88-0.96 (55-60 lb per cu ft)	0.88-0.96 (55-60 lb per cu ft)
Specific heat at 38°C (100°F), kJ/kg	0.4	9.94
Angle of repose	55° *	55° *

Hydrates

	High Calcium	Normal Dolomitic	Pressure Dolomitic
Primary constituents	$Ca(OH)_2$	$Ca(OH)_2 + MgO$	$Ca(OH)_2 + Mg(OH)_2$
Specific gravity	2.3-2.4	2.7-2.9	2.4-2.6
Bulk density, g/cm^3	0.40-0.56 (25-35 lb per cu ft) †	0.40-0.56 (25-35 lb per cu ft) †	0.48-0.64 (30-40 lb per cu ft) †
Specific heat at 38°C (100°F), kJ/kg	0.62	0.62	0.62
Angle of repose	70° *	70° *	70° *

Source: Anon., 1981.

* The angle of repose for both types of lime (hydrate in particular) varies considerably with mesh, moisture content, degree of aeration, and physical characteristics of the lime; e.g., for quicklime it generally varies from 50° to 55° and for hydrated lime it may range as much as 15° to 80°.

† In some instances these values may be extended. The Scott method is used for determining the bulk density values. In calculating bin volumes the lower figure should be used.

trolled to prevent contamination with undesirable impurities, particularly silica, alumina, and iron oxide. Accordingly, at several lime plants practices such as selective quarrying and washing are carried out.

Calcining

Although lime calcining (or burning) is a relatively simple operation chemically, a wide variety of kiln systems are used today. These include vertical, rotary, rotary with preheaters, multiple-shaft regenerative, and a variety of other types of kilns, in striking contrast to the portland cement industry which utilizes rotary kilns and rotary kilns with preheaters almost exclusively.

Lime kiln selection depends upon several factors, the most important ones being burning characteristics of the stone, fuel consumption, and capital equipment costs. Other factors include market requirements, fuel availability and air pollution regulations. Traditionally, vertical kilns have been considered to have the following advantages over rotaries: lower capital investment, more efficient fuel utilization, less attrition loss of stone and less refractory wear, and lower pollution control costs. Rotary kilns, on the other hand, have the advantages of being able to burn small stone readily, greater versatility (capable of producing a wide range of burn, from soft to hard, even dead-burned, and using all types of fuels, singly or in combination), more uniform product quality, greater capacity, and less manpower requirements. In addition, fuel efficiencies now being achieved with rotary kilns having preheaters and the adaptability of rotaries to the large size units (1000 tons or more per day) have had the effect of offsetting some of the advantages of vertical kilns. Today over 85% of the commercial lime produced in the US and Canada is from rotary kilns. However, current developments in calcining technology and the tremendous fuel cost increases experienced in the late 1970s and early 1980s have generated renewed interest in the US in the multiple-shaft regenerative kilns, as they are particularly fuel efficient.

Vertical Kilns: A revolution has taken place in recent years in vertical kiln design. These design changes have been largely the result of increasing fuel and capital costs, and are supplanting more traditional designs as new and replacement capacity is being brought on line. Among the newer vertical kilns are the European multiple-shaft parallel flow regenerative kilns (Fig. 3), the center burner gas-fired

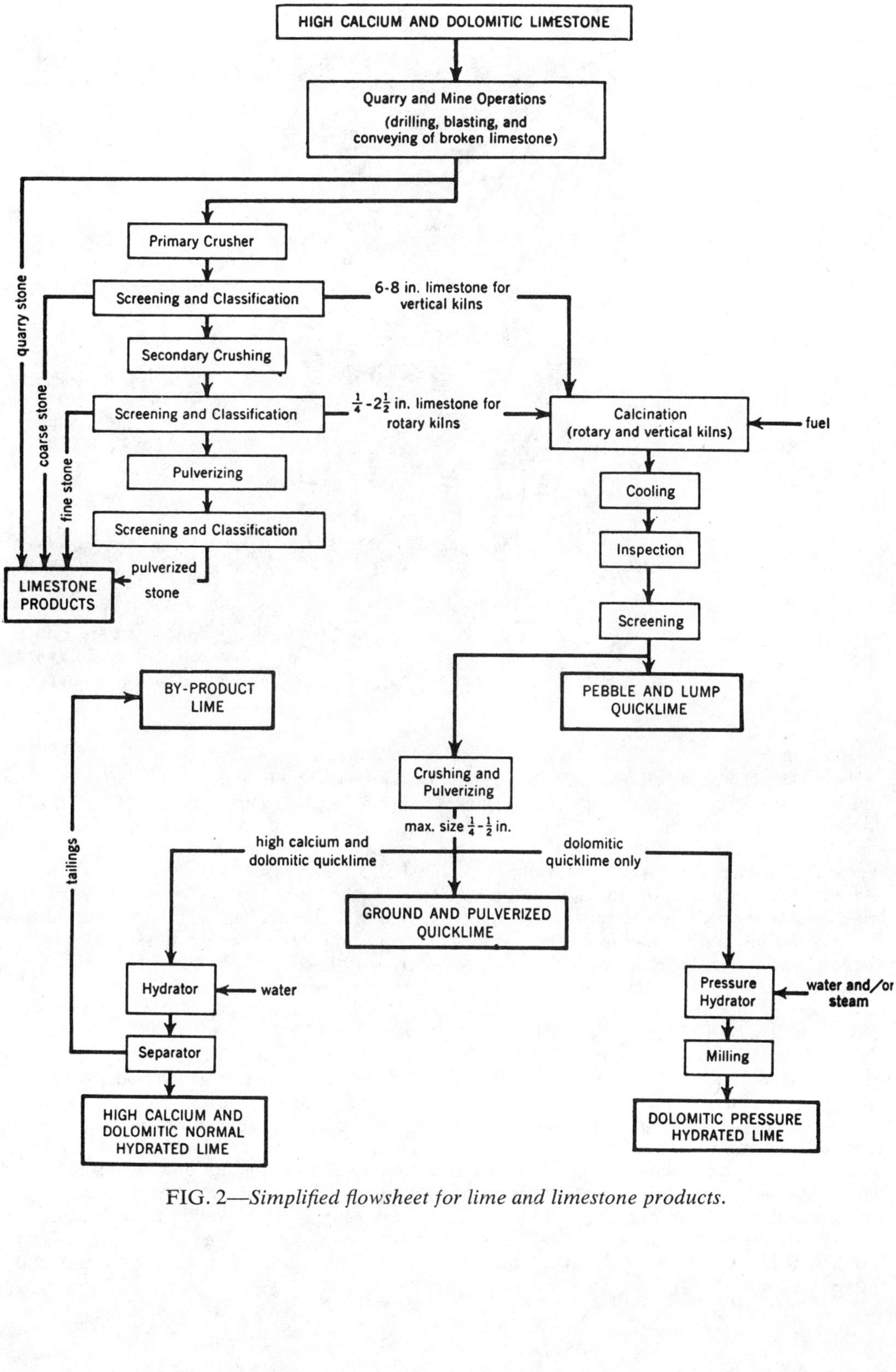

FIG. 2—*Simplified flowsheet for lime and limestone products.*

FIG. 3—*Multiple shaft parallel flow regenerative kiln.*

(Azbe, Union Carbide), oil-fired (West Catagas and Beckenback) annular, and the Beckenback double-inclined (cascade) kilns. The multiple shaft kilns are now available in the US through Warwick Furnace Co., Kennedy Van Saun, and the Fuller Co., while the annular kilns are available through Allis-Chalmers.

Vertical kilns traditionally have burned only larger stone (75 to 300 mm) with a size range of approximately 1:2. However, the newer designs can handle stone sizes as small as 20 mm and a widened size range to as much as 1:4. Fuel consumption has been a major advantage of shaft kilns, nominally requiring less than 5.2 GJ/t of lime produced. This has been further improved upon with the newer designs reportedly operating as low as 3.3 to 3.9 GJ/t of lime. Power requirements for these kilns will vary with stone size, but at nominal capacity will be in the range of 54 to 90 MJ/t. A disadvantage of traditional vertical kilns in today's energy environment is their fuel requirement of oil, natural gas, or coke. However, several recent developments in burner technology have come to the point where pulverized coal can now be utilized in a modern shaft kiln.

Vertical kilns may be of stone masonry, reinforced concrete, or boiler plate construction. The most widely used kiln has a refractory-lined steel shell and is usually circular in cross section. These kilns may be 2.7 to 7.3 m diam and 15 to 48.5 m high. Capacities vary from as low as 9.5 t/d to in excess of 700 t/d, with the larger tonnages being restricted to the newer generation of kilns.

Multiple Shaft Kiln: Developed to burn small stone, this kiln utilizes the parallel flow calcining principle in double- and triple-shaft units. The shafts are interconnected in the burning zone, and while one shaft is being fired (Fig. 4), the other is preheated. Fuel and combustion air are supplied to the burning shaft from above, ignite at the upper end of

FIG. 4—*Installation of fuel lances for natural gas and fuel oil in a multple shaft kiln.*

the burning zone, and calcine the lime in uniflow. The exhaust gases then pass into the second shaft, preheating the stone in counterflow. After a 10- to 15-min interval the shaft firing is reversed. Cooling air is blown into both shafts simultaneously. Because of the novel heat regeneration system, fuel consumption is 5.4 to 5.7 GJ/t of lime produced. Kilns vary in capacity from 90 to 550 t/d, and while nominally gas- or oil-fired, recent developments in burner technology have led to several installations now being equipped to fire pulverized coal. Kilns of this type have found worldwide acceptance. The recent installation of new multiple-shaft kilns in Japan and the United States brings their total number to over 185.

Annular Shaft Kiln: The burning process in this kiln is based upon counterflow for preheating and calcining and uniflow for residual calcination. Partitions provided by an inner cylinder and by staggered bridges in the burning zone permit even distribution of heat and uniform flow of material down the kiln. Stone as small as 25 x 75 mm can be calcined to produce a soft-burned lime, using either oil or gas firing. Capacities vary from 90 to 270 t/d, with fuel consumption reported under 6.2 GJ/t.

Rotary Kilns: Unlike vertical kilns, which operate fully charged, the rotary kiln has about 90% of its volume filled with flame and hot gases. As the kiln slowly rotates, new surfaces of stone are exposed to the hot gases, but there is little passage of the gases through the solids; hence radiation interchange between gas, solids, and refractory wall plays an important part in the overall heat transfer. Because the area of solids exposed is relatively small, a rotary kiln is less efficient than a shaft kiln. However, the rotary has the advantage of burning stone from as small as 6.5 mm to as large as 57 mm. Generally, though, the size ratio is 1:3 in order to minimize segregation, wherein the finer particles would sift to the bottom of the kiln and remain largely uncalcined.

Rotary kilns vary greatly in size, ranging from 2 x 25 m to 5.5 x 190 m, with capacities of 45 to 945 t/d. Because of the relatively high length to diameter ratios these kilns have come to be termed *long kilns.* In North America these long kilns are typically 2.5 to 4.0 m diam by 45 to 105 m long, producing 181 to 410 t/d of lime. While rotary kilns can burn a wide range of fuels, a major drawback of these long kilns has been their lack of fuel efficiency. Many of the earlier rotaries require in excess of 18.6 GJ/t of lime. This high fuel requirement has led to the development of more fuel efficient equipment such as external preheaters, internal heat exchangers and dams, recuperative-type coolers, more sophisticated and complex instrumentation, and improved refractories.

Preheaters: Of the several advances which have been made in improving the heat efficiency of rotary kilns, probably the most pronounced has been the trend away from long kilns to medium length kilns with external preheaters. Shaft-type preheaters (Fig. 5), available through Kennedy Van Saun, Fuller Co., and others, are the most widely employed. This preheater . a refractory-lined chamber located below the raw feed bin and ahead of the kiln. Exhaust gases from the kiln are drawn countercurrent through the stone, preheating it to 650 to 900°C. Retention time within the preheater is 1.0 to 1.5 hr, depending upon the system design, burning characteristics of the stone, etc. During preheating approximately 30% calcination is achieved before the stone is discharged to the kiln. Following preheating, the partially calcined stone is fed to the kiln at a predetermined rate by means of hydraulically actuated plungers. This type preheater has been adapted to the new large volume kilns which typically may be 5.2 m diam by 62.5 m long and have a capacity of 900 to 1000 t/d.

A second type preheater is the grate-type, developed by Allis-Chalmers as part of its grate-kiln system. This is a single pass, downdraft, enclosed traveling grate on which the finer stone sizes, placed on top of the coarser sizes are virtually calcined while the coarser sizes are preheated. Then, in the rotary kiln which follows, the coarse sizes are calcined, while the finer sizes are protected from overburning because they sink into the kiln load through a sifting action. Grate-kiln systems

FIG. 5—*Modern preheater rotary kiln system.*

of 272 to 545 t/d capacity are in operation in North America.

Heat exchangers and coolers which recuperate waste heat are also utilized to effect improved fuel efficiency. The heat exchangers, principally trefoil or quadrant sections, are mounted at the feed end. Their purpose is to turn over the stone more effectively and increase the turbulence of the hot gases. Dams and lifters may also be used to turn over the stone, create gas turbulence, or retard the passage of the stone. Coolers, on the other hand, are located at the discharge end of the kiln and serve to return hot gases to the kiln as secondary air. Three types of coolers are generally used with rotary kilns—the shaft or contact, the planetary, and the rotary. The most prevalent type is the Niem's contact (counterflow), which is refractory lined and mounted below the kiln hood. Cooling air is provided by a fan of sufficient size to reduce the lime temperature to 65°C or less. The planetary cooler consists of eight or more tubes approximately 4.2 x 6.2 m in size, mounted at the kiln discharge end. The lime is moved in each tube by conveyor flights counterflow to the cooling air. The rotary cooler is generally not as effective as the other two, and has the additional disadvantage of attrition loss during rotation.

The net effect of the use of both shaft and grate preheaters, and improved heat exchange and recovery within the kiln, has been the reduction of fuel ratios in rotary kilns to 6.9 to 8.5 GJ/t of lime produced. In addition, advances in instrumentation and monitoring (Fig. 6) have reduced manpower requirements and increased product quality and uniformity. Taken together, these changes have helped the large-capacity rotary kiln with preheater become the most commonly selected system today for new or expanded lime production facilities.

Other Kilns: The rotary hearth calciner, or Calcimatic kiln, consists of a preheater, circular hearth, and cooler, all refractory lined. Like the rotary, this kiln burns small stone which is typically sized in a 1:3 ratio. The stone is carried on the hearth in a thin layer, and one revolution of the hearth constitutes the calcining cycle. Numerous burners located inside and outside the hearth are used for firing, utilizing gas, fuel oil, coke-oven gas, and, most recently, pulverized coal. Fuel requirements approximate 7.7 GJ/t, with a wide range of burns possible because of the ease of operator control in calcining. One of the primary advantages of this kiln is the fact that attrition loss is negligible. Stone rests virtually motionless on the hearth, therefore permitting the utilization of soft limestones which would otherwise be subject to high mechanical breakage. A number of 90- to 273-t/d-capacity kilns of this design are in use in North America and Europe.

FluoSolids Kiln: For limestone that is friable or decrepitates during calcining, the Dorr-Oliver FluoSolids has been used to produce highly reactive lime. In this fluid bed process the material is maintained in a suspension in a rising current of hot gases until calcined. Kiln feed is normally 6 x 65 mesh. The system,

FIG. 6—*Modern control panel and instrumentation for a multiple kiln facility.*

which can be fired with gas or oil, incorporates a FluoDry unit (preheater) and FluoSolids reactor (calciner), the latter resembling a vertical kiln. Fuel consumption is approximately 7.8 GJ/t. Fluid bed calciners are available in sizes ranging from 45 to 226 t/d.

Flash Calciners: The flash calciner is one of the most recent developments in calcination, and in reality is not a kiln (Fig. 7). This system, developed by Fuller Co., is designed to calcine the undersized material, or fines, remaining after the primary kiln feed has been sized. The system includes preheater, flash calciner, and cooler. Preheating is achieved in a series of cyclones in which the heat of rising hot gases is absorbed in the counterflowing material which is fed from the top (Fig. 8) and heated to 980°C. As the preheated feed drops into the furnace or calciner, a forced air vortex flow patern pulls the particles into the center of the furnace and then up the sides before discharge. Within the furnace the temperature is 1200°C, but retention time is only a few seconds. A suspension-type cooler recovers heat from the calcined product before it is conveyed to the product bin. The system is designed in two modes depending upon what size feed is to be used. Where stone fines are to be calcined, 60 to 80% of the calcination occurs in the furnace. The final calcination takes place in a fluid bed located between the furnace and the cooler. In those systems where —28 mesh material is fed to the calciner no fluid bed reactor is necessary. This system can utilize gas, oil, or coal, and will normally require 6.2 to 8.5 GJ/t of product.

Milling and Hydrating

Following cooling, quicklime is carefully inspected to avoid *core*, or pieces of uncalcined rock. This is particularly true for shaft kilns burning large stone. The quicklime is then crushed and sized for shipment. Generally, the fines are converted to hydrated lime.

Hydrated lime, shipped in 23-kg bags or in bulk, is available in two principal forms: standard, where 85% passes a No. 200 sieve; and superfine or spray hydrate, where 98% or more passes a No. 325 sieve. These in turn are manufactured in two types, N (normal) and S (special) (see ASTM Designation C207–80), each including high calcium or dolomitic hydrates.

Type S limes are highly hydrated, generally by processes involving pressure hydration. They contain less than 8% unhydrated oxides and develop high and quick (15 min) plasticity and high water retentivity. Since World War II, the demand for type S lime has grown rapidly until currently it is the chief lime for structural uses. Some lime manufacturers now add entraining catalysts to lime to increase plasticity and durability (called S.A. and N.A. lime).

Mechanical hydrators used for type N lime are made in various designs, including batch hydrators like the Clyde and continuous hydrators like the Kritzer, Schaffer, Knibbs, Hardinge, and Kennedy Van Saun. The hydration process consists of adding water slowly to quicklime (usually in crushed or ground form), followed by agitation to produce intimate contact of lime with water. Accurate proportioning is essential, and this is accomplished by using flowmeters for water and weighing feeders for lime. More than the theoretical amount of water required should be used for hydration in order to allow for the water that is lost as steam caused by heat of hydration. Following hydration, the product is generally fed to a pulverizing or tube mill equipped with an air separator. The core, together with most of the silica and alumina, is rejected by the air separator. By adjusting the separator, hydrates of varying degrees of fineness can be made.

Type S hydrates are manufactured by various processes, the principal one being the Corson explosion method of continuous pressure hydration. In this process ground dolomitic quicklime and water are automatically fed in constant proportions to a high-speed slurry mixer. Hydration then takes place in an insulated autoclave operating at about 5.3 kg/cm^2 and 150°C. After about 30 min retention time, the product is discharged continuously (or exploded) through a small pipe at the rate of 3,050 m/s into a special cyclone collector at atmospheric pressure. There the hydrate and water are separated by the spray-drying principle. The product is then ground in a tube mill to increase the plasticity. Several US plants use the Corson process, marketing their product under the registered trade name Miracle Lime®.

Consumption

Chemical-Industrial Uses

Although lime to the average layman still connotes a mortar, plaster, or agricultural material and although it still enjoys these formerly predominant traditional uses, most lime today—over 90%—is utilized by the chemical and

FIG. 7—*Flash calciner.*

metallurgical process industries as a flux, acid-neutralizer, causticizing agent, flocculant aid, hydrolyzer, bonding agent, absorbent, and raw material. An allied product, dead-burned dolomite, is used as a refractory material. Lime has possibly the greatest number of diverse uses of any chemical or mineral commodity. However, by far its greatest tonnage lies in the metallurgical industries. Metallurgical uses, including MgO and Al_2O_3, account for 49% of total chemical-industrial usage, and of this amount, 83% is used for steel manufacture (nearly 50% of commercial noncaptive lime is for steel flux).

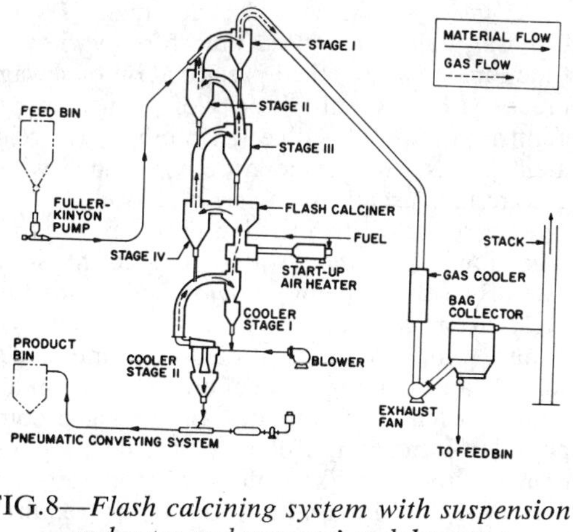

FIG.8—*Flash calcining system with suspension preheater and suspension deheater.*

Steel: Commencing during World War II, steel became lime's foremost market; however, it was not until the early 1960s that this use of lime surged ahead as a direct result of the basic oxygen furnace (BOF). Previously, lime was used as a flux in the open hearth steel furnace in varying amounts, usually with limestones as a coflux. But the factor of lime in a ton of open hearth steel averaged only 12.5 kg. With the BOF, only lime is used as the flux because the keynote of this process is speed. As a result, the lime factor in BOF steel markedly increased to 40 to 85 kg/t (64 kg/t average). Electric steel furnaces require an average of 30 kg of lime per ton of steel. As the BOF replaced the open-hearth and as overall steel production increased during the 1960s lime flux soared—from about 1.4 Mt in 1961 to 8.1 Mt in 1979.

Lime acts as a scavenger in purifying steel by fluxing out, into the molten slag, acid oxide impurities such as silica, alumina, phosphorus, and sulfur.

At first only a reactive high calcium quicklime of 90 to 93% available lime, with low core content and meeting strict tolerances on sulfur (0.03 to 0.06%), was used. Later it was discovered that if a pure dolomitic quicklime was substituted for a portion of the high calcium flux, greater refractory life was obtained, reducing the cost of BOF steel. As a result, most US steel plants use dolomitic lime as a 10 to 30% replacement of high-calcium lime—but this is not true in Europe or Japan.

Most steel companies use pebble or pelletized lime as flux, although the former type is preferred. Some European steel mills use pulverized quicklime, introducing it through their oxygen lances into the furnace. U.S. Steel Corp. developed a modification of the BOF, called the Q-BOP process, which also requires pulverized lime. There is also experimental usage of premixed fluxes in Europe, Japan, and jointly by Republic Steel Co. and Alabaster Lime Co. This involves blending and mixing by grinding high-calcium quicklime, ferrous flue dust from steel plants, alumina-bearing wastes such as red mud from alumina plants, with and without fluorspar, and then pelletizing the mixture into 25.4-mm pellets for charging into the BOF.

Minor lime uses in steel and steel fabricating plants are for wire drawing, neutralization of sulfuric pickle liquor wastes, water softening, and coating molds for "pig" casting. There has been experimental use of hydrated lime in place of pulverized limestone in making self-fluxing ferrous sinters.

Refractory lime, known as dead-burned dolomite, is universally used to line the bottom of open hearth steel furnaces, thereby extending the life of the costly refractory brick linings. With the growth of the BOF at the expense of the open hearth, this use has steadily declined from 2.2 Mt at its peak in 1956 to 0.7 Mt in 1979. This loss has been muted by the development of tar-bonded refractory brick made from dead-burned dolomite.

Nonferrous Metallurgy: *Copper*—Most domestic copper mining companies use lime as the principal reagent in beneficiating copper ores in flotation processes. Lime is used to neutralize the acidic effects of pyrites and maintain the proper pH. There is limited use of lime as a flux in smelting concentrated copper ores and to adsorb SO_2 from the stack gases of copper smelters.

Alumina—In the manufacture of alumina by the Bayer process, large quantities of lime are used in a few plants to causticize sodium carbonate solutions to regenerate sodium hydroxide for recyclical use. Nearly all plants will use some lime for secondary desilification in refining alumina. Some of the experimental processes developed by the US Bureau of Mines (USBM) for making alumina from clay or kaolin, instead of bauxite ore, require lime.

Magnesia—Most processes for the manufacture of magnesia and magnesium metal, such as the Dow seawater and natural brine processes, ferrosilicon process, and seawater magnesium process, require lime. Dolomitic lime, because of its magnesium content, is generally preferred, but high calcium lime has been successfully used. Two processes are diagrammed as follows:

$$CaCl_2 \cdot MgCl_2 + MgO \cdot CaO + 2CO_2 \rightarrow 2CaCO_3 + 2MgCl_2$$

$$MgCl_2 + \text{electrolysis} \rightarrow Mg + Cl_2 \quad (1)$$

$$Ca(OH)_2 + MgCl_2 \rightarrow Mg(OH)_2 + CaCl_2$$

$$Mg(OH)_2 + \text{heat} \rightarrow MgO + H_2O \quad (2)$$

Miscellaneous—In South Africa huge amounts of lime are used to recover uranium from the gangue (called gold slimes) of gold mining in flotation processes. In the US and Canada lime is employed to neutralize sulfuric acid waste liquors in acid uranium ore extraction plants. In the recovery of gold and silver by flotation processes, lime is used to curtail the loss of cyanides and for pH control. It is employed as a flux in the sintering of low-carbon chrome and in the recovery of nickel by precipitation.

Sanitation: *Water Treatment*—Over 1.5 Mt of lime are used annually in the treatment of municipal potable water and industrial process water. The former use exceeds the latter in tonnage.

Lime is used along or with soda ash to soften water in competition with ion exchange processes (zeolite, etc.). Lime is required to remove the temporary bicarbonate hardness from the water. The high pH of 11.5 induced by the lime is of value as a secondary sterilization agent to chlorine because retention for 3 to 10 hr at this pH will kill 99+% of the bacteria and most viruses. By introducing CO_2 into the lime-treated water, the pH is lowered to acceptable levels and most of the lime in solution is precipitated as a carbonate sludge.

An important application is in coagulation, where lime is employed with such coagulants as alum or iron salts to remove turbidity from river waters used for potable purposes or suspended solids from industrial water. Lime is used for pH control to negate the acid effect of the coagulants and to achieve optimum coagulating efficiency.

Subsidiary to its role in water softening, sterilization, and coagulation is its absorption of iron, manganese, and organic tannins from untreated water.

Sewage—For several decades many sewage disposal plants have used lime for a variety of functions: for chemical precipitation of sewage practiced by a small minority of plants; for pH control in the sludge digester in biological treatment plants; and for conditioning sludge when dewatering, usually with ferric chloride.

As a result of the growing problem of eutrophication of surface waters caused to some extent by sewage plant effluents, more lime will be used in the future. Sewage effluents will require much more treatment to remove dissolved and suspended solids which contain plant nutrients—phosphates and nitrogen compounds. Lime is useful in precipitating phosphates directly. When alum or iron salts are used as the coagulant, lime is generally required for pH control. It appears that a high pH of

10 to 11, inherent in lime treatment, is necessary for efficient removal of nitrogen. Therefore, the future of sewage treatment processes is swinging more from biological to chemical treatment processes that include lime, although there will be no standard treatment method because of the many divergent sewage conditions.

Pollution Abatement: Lime has been used for many years in treating liquid trade wastes and to a lesser extent industrial gaseous effluents to abate pollution of surface waters and air.

Lime is used to neutralize acid mine drainage and to precipitate iron; in coal washing plants, the acidic waste wash water is neutralized by lime. Wastes from sulfuric acid steel pickling plants and plating wastes are neutralized with lime in which heavy metal contaminants are also precipitated. Neutralization of chemical and pharmaceutical plant wastes is also achieved with lime. Clarification and color removal of paper mill wastes is obtained with lime alone or with supporting chemicals as well as other specialized applications.

A major new market for lime is air pollution control. Lime originally was envisioned as the primary reagent material for removing acidic gases from the stack gases of metallurgical and chemical plants. However, during the late 1970s it became apparent that the most significant tonnages of lime would be used by coal-fired electric utility generating stations for the removal of sulfur dioxide from flue gases (Fig. 9). Flue gas desulfurization, or scrubbing, currently requires nearly 1.8 Mt/a of lime, with this amount projected to double by 1990. The development of this market has been greatly enhanced by advances in scrubber hardware, including both wet and dry processes, and by the success achieved by Dravo Corp.'s Thiosorbic® lime. Lime scrubbing has been proven to be both an effective and viable means of meeting the US Environmental Protection Agency's air pollution control requirements.

Chemicals Manufacture: *Alkalies*—In the manufacture of soda ash and bicarbonate of soda by the Solvay process, lime is employed in large quantities to recover ammonia for recyclical use in the process; 635 kg of quicklime are required per ton of soda ash produced. In the process the lime reacts with the chloride ion to form calcium chloride, much of which is recovered and sold commercially. Another recovered byproduct is precipitated calcium carbonate. Additional lime is used to causticize sodium carbonate solutions to produce sodium

FIG. 9—*Sulfur dioxide scrubbers at a modern electric generating station.*

hydroxide (caustic soda). All alkali manufacturers make their own lime because they require an abundance of CO_2 in these as a coproduct. However, Solvay process ash is losing ground to natural soda ash (trona) and electrolytic caustic soda. Stringent pollution control regulations are imposing still greater economic pressure on this process. The future of the Solvay process is bleak, though undoubtedly its use will continue to some extent.

Carbide—Calcium carbide, formerly the main source of acetylene, is made in electric furnaces through the fusion of quicklime and coke in the presence of nitrogen at very high temperatures. About 1 t of quicklime is required to make 1 t of carbide. When water is added to carbide, acetylene gas is evolved, leaving a waste calcium hydroxide residue. The future of this product is poor because less costly processes for making acetylene from natural gas have been perfected. Furthermore, acetylene is losing ground to ethylene as a basic organic chemical building block. However, with increasing costs for natural gas this process may make a mild comeback.

Bleaches—High-test calcium hypochlorite (70% available chlorine) and chloride of lime, both dry sources of bleach, are made through the interaction of chlorine and hydrated lime.

Liquid calcium hypochlorite is made by paper plants for bleaching pulp.

Inorganics—Other inorganic chemicals made from lime include caustic magnesia by the brine and seawater processes (previously referred to); mono-, di-, and tricalcium phosphates; chrome chemicals; purifying salt brines to produce USP food grade salt; pesticides such as lime-sulfur sprays and powders; Bordeaux mixtures (copper sulfate and lime) and calcium arsenate; and paint pigments.

Organics—Lime is required in the manufacture of ethylene and propylene glycols in the chlorohydrin process; in calcium-based organic salts, such as calcium stearate, acetate, lactate, lignosulfonate, etc.; in the refinement and concentration of citric acid and glucose; and for dyes and dyestuff intermediates.

Pulp and Paper: In the manufacture of sulfate (kraft) paper pulp, lime is universally used to causticize the waste sodium carbonate (black liquor) to regenerate sodium hydroxide for reuse in this continuous chemical process. In spite of the fact that pulp mills in turn recover 90 to 96% of the lime employed by recalcining the dewatered calcium carbonate sludge precipitate, a substantial quantity of lime (over 680 kt) is required as makeup. Counting the regenerated lime, technically over 5.4 Mt/a of lime is actually consumed in the sulfate paper industry.

Secondary uses of lime by the paper industry include making calcium hypochlorite bleach; clarification of process water, including recyclical water; clarification and color removal of waste water effluents.

Building Products and Ceramics: Lime is used to make calcium silicate building products, such as dense sand-lime brick and block, cellular concrete block and insulation panels, and microporite; however, most of this is made in European and other foreign countries and very little in the US. West Germany alone consumes about 2.2 Mt/a of lime for these products. Lime is reacted with sand and fine forms of available silica under steam pressure in an autoclave to form stable, durable, and very strong cementing compounds (calcium silicates). To produce lightweight masonry units and insulation materials, aluminum powder is added to the mix which is attacked by the lime, producing a cellular effect of varying degrees of porosity.

Glass—In the manufacture of glass, dolomitic quicklime granules of Nos. 10 to 100 mesh are used interchangeably with limestone or dolomite of similar sizes as a flux raw material in the glass batch. Next to sand and soda ash, glass plants use far more lime and/or limestone than any other material in glassmaking. Most of the lime is used in bottle glass plants rather than plate glass.

Refractory Products—Silica refractory brick is made with hydrated lime, another calcium silicate reaction. Periclase (dead-burned MgO) is made from the aforementioned seawater magnesia processes as a raw material for basic refractory brick, and refractory brick is made from tar-bonded dead-burned dolomite.

Miscellaneous: *Sugar*—Lime is essential for sugar making in both beet and cane sugar refining processes. Large amounts of lime are required for beet sugar, averaging 0.2 t of quicklime per ton of sugar; with cane sugar only 1.9 to 4.9 kg/t of lime is required. Lime helps purify the sucrose juices by removing phosphatic and organic acid compounds as insoluble calcium compounds which are removed by filtration. Beet sugar plants invariably make their own lime in on-site kilns because they need a massive quantity of CO_2, which is obtained from the lime kiln stack gases. Cane sugar refiners purchase their much smaller requirements.

Produce Storage—Lime is used as a CO_2 adsorbant in controlled atmospheric storages for fresh fruit and certain vegetables to extend the freshness of the produce. This prevents CO_2 from building to abnormally high levels, which would accelerate rotting of produce. It is used to recover glue and gelatin from packinghouse byproducts. In citrus fruit processing, lime is used to treat the waste pulp, which after grinding and drying is sold as cattle feed. There are several other small isolated uses in the food industry.

Petroleum—Lime has limited use for neutralizing organic sulfur compounds (mercaptans) and SO_2 emissions in petroleum refineries. A common type of lubricating grease is made with lime acting as a saponifying agent of the petroleum oil. An ingredient of "red lime" drilling muds for oil drilling is hydrated lime. In a proprietary process, lime and pozzolans such as fly ash are used instead of cement in sealing producing offshore oil wells in the Gulf of Mexico.

Paints and Pigments—In addition to lime whitewash, proprietary water paints for masonry are made with hydrated lime, white portland cement, and pigments. Lime has minor uses in varnish manufacture, casein paints, and certain types of pigments.

Leather—Since biblical times, lime suspensions have been used for dehairing and plumping hides preparatory to leather tanning.

Construction

Stabilization: The total use of lime for soils stabilization for highways and off-highway uses now exceeds total structural lime consumption. In the construction of pavements for highways, city streets, airport runways, and parking lots, lime may be used to stabilize the clay subgrade, subbase, and/or base course. Clay must be present for lime to react with the available silica and alumina in the soil to form complex cementing compounds which bind the soil into a hard, stable mass that is not sensitive to water saturation. The plasticity of the soil, as well as shrinkage and swell, are markedly reduced. The amount of lime used generally ranges between 3 to 5% of the weight of the dry soil. Over 80% is hydrated lime, with the balance being pebble or ground quicklime. Lime is recognized by all federal agencies and most state highway departments, and has been used in at least 3200 km of interstate construction in 26 states. In the construction of the largest airport in the world in 1971–1974—The Dallas/Ft. Worth Airport—about 272 kt of hydrated lime was used in pavement and building areas, including access roads.

Lime also is used with pozzolan fly ash and varied aggregates in the preparation of base material at central-mix plants, which competes with crushed stone and soil-cement base courses. This process was introduced by the Corson Co. in a proprietary process called Poz-O-Pac. The Newark, NJ airport used lime-fly ash mixtures throughout its pavement system, with hydrated lime consumption of about 68 kt.

Use of lime continues to grow as its applications multiply into such diverse areas as building foundation construction, reinforcement of earth levees and dams, irrigation channels, railroad beds, race tracks, and even clay tennis courts. Contractors use it to expedite construction by drying up muddy subgrades and to construct temporary all-weather haul roads.

Asphalt: Increasingly, 1 to 2% of hydrated lime is added to asphalt hot mix in central-mix plants as paving material. The lime acts as an antistripping agent preventing the aggregate from raveling from the bituminous binder. In remote areas it permits use of marginal, unwashed aggregates and where needed serves as a filler. Based on the immersion compression test with some asphalts and aggregates, it markedly increases the wet/dry strength ratio and reduces volume change.

Mortar: For thousands of years lime has been employed as a mortar material—and it is still used today. Before the advent of portland cement in the latter part of the 19th century, mortars were composed only of lime and sand. Now, lime-cement mortars of varying proportions are used, ranging from 3 to 1/4 parts of lime by volume to 1 part cement. Lime is still the most dependable plasticizer for mortar, making it workable and more trowelable. Cement provides the rapid set.

In the US most hydrated lime is mixed at the job site. Rarely is quicklime slaked and used at the job site anymore, and what little is used is in pulverized form. Lime putty and ready-mixed lime-based mortars, with sand and even with cement, both in dry and wet form are also mixed at central plants and delivered in bulk. Additional lime is also mixed with cement or cement-blast furnace slag mixtures or with pozzolans such as fly ash, which is packaged for mortar mixing with sand at the job site. These proprietary compounds are known as mortar mixes or masonry cements. The biggest proportion of building lime used in the US is type S or S.A. dolomitic.

Lime encounters severe competition from masonry cements made by cement companies. Most masonry cements do not contain lime; instead they usually consist of a 50:50 mixture of pulverized limestone and portland cement, plus organic air entraining agents, the latter providing plasticity.

Plaster: In interior plastering today, lime is used only in the finish or white coat, usually gaged with gypsum or Keene's cement. Formerly it was used along with sand in the base coat, but it has been replaced by gypsum-aggregate plasters. Lime suffered a second loss as gypsum drywall construction replaced much conventional interior plaster. Some lime is also used in exterior plaster or stucco, particularly in California and Florida, again with mixtures of portland cement and sand in varying proportions. It is also purchased by stucco compounders for their proprietary stucco finishes of widely varying textures, white or pigmented.

Agriculture

Although most agricultural liming today is achieved with pulverized limestone, many farmers still use ground quicklime and hydrated lime, particularly in the middle Atlantic states. Many Amish farmers, in fact, are proponents of burned lime. Although lime costs more than limestone on an equivalent calcium oxide basis, it reacts faster, neutralizing soil acidity rapidly so crop yields do not suffer. In truck farming, use of burned lime for maintaining the optimum pH enables three crops to be grown per year. The added cost for lime is easy to justify in higher-value crops. In some sandy soil areas dolomitic lime only is used, due to persistent magnesium deficiencies. County agents who make pH tests recommend the amount of liming material to apply and the desired pH for a given crop. Most of these agents and state agronomists, however, promote the use of limestone, making it difficult for the lime industry to promote its product. Furthermore, the federal subsidies for liming markedly favor limestone over lime.

Some secondary farm uses of lime are for whitewashing dairy barns as a sanitation aid, for composting, in chicken litters, etc.

Hydraulic Lime

At one time there was a small production of hydraulic lime in the US. Currently, however, only one plant, located in Virginia, manufactures hydraulic lime, all of which is consumed internally. This product still flourishes in some foreign countries, notably Italy, Germany, and France. Chemically this product, which is high in silicates, aluminates, and ferrites, could be defined as an impure lime; its chemical composition is intermediate betweeen portland cement and lime. It is solely a structural lime, developing greater strength than pure lime and sand but much less than portland cement. It possesses limited hydraulic qualities, though much less pronounced than portland cement, in being able to harden under water.

A detailed statistical summary of lime uses for 1979 and 1980 is contained in Table 3. Note that refractory dolomite is included with lime, and that commercial and captive limes are differentiated.

Economics

Fig. 10 shows the trends in lime consumption since 1904, when the USBM first began compiling statistics; as in Table 3, four major end use categories are depicted. Three periods of growth are evident as well as the possible early phases of a fourth period. The three prominent growth periods include the late 1930s following the Great Depression, the decade following World War II, and the late 1960s and early 1970s, including the 1974 record consumption of slightly over 20 Mt of lime. The statistics for the late 1970s and early 1980s suggest that a fourth period may be starting, one which is underpinned with a strong new environmental applications demand including scrubbing, or flue gas desulfurization.

Further inspection of the trends depicted in Fig. 11 clearly shows that nearly all of the growth in lime since 1960 has been accounted for by chemical and industrial uses, particularly fluxing lime for steel. Assuming that environmental applications such as flue gas desulfurization are counted as chemical applications, this increasing tonnage trend for chemical and industrial uses would be expected to continue. Other growth areas for lime include water treatment, sewage treatment, and soil stabilization, while refractory dolomite and building lime continue to show little or no growth.

Competition

Traditionally the lime industry has been intensely competitive, on both an intra- and interindustry basis. As evidence of this, the number of lime plants has declined from about 1200 in 1910 to 154 commercial lime plants in 1980. This competitive pressure was further reflected in the average f.o.b. price of lime which only doubled from the 1920s to the early 1970s. In fact, only since the oil embargo of 1974 have truly significant increases in lime prices been effected. At one time lime was one of the commodities least affected by inflation. This is no longer the case; prices nearly tripled from 1972 to 1980.

Although cement, gypsum, caustic soda, and other chemical coagulants and neutralizers have provided serious competition in vying for common markets, limestone is still lime's most formidable competitor. This competition obviously has been fostered by the price differential between these two commodities as well as their many interchangeable applications such as open-hearth furnaces, alumina, glass, and a variety of environmental uses. As the price difference between lime and limestone continues to increase, the competitive pressure from limestone will be even more keenly felt by the lime industry.

Also of great importance in terms of competition and lime prices is the fact that there has always been a sizable percentage of captive lime produced, ranging between 26 and 40% of the total lime consumed. Traditionally much of this captive lime was produced in conjunction with alkali and beet sugar plants whose processes demand cheap CO_2 in order to be economically viable. However, large lime con-

TABLE 3—Lime (Quicklime, Hydrated Lime, and Refractory Dolomite), By Uses*, 1979 and 1980 (1000 st)

Use	1979 Sold	1979 Captive	1979 Total	1980 Sold	1980 Captive	1980 Total
Agriculture	71	—	71	79	—	79
Construction:						
Soil stabilization	695	—	695	170	—	170
Mason's lime	350	41	391	288	40	328
Finishing lime	195	—	195	99	—	99
Other uses	34	35	69	16	27	44
Total Construction	1,274	76	1,350	652	67	720
Chemical and Industrial:						
Steel, BOF	5,611	1,706	7,317	4,409	1,441	5,850
Steel, electric	964	28	992	755	34	789
Steel, open-hearth	603	49	652	564	38	602
Copper flotation	427	344	771	340	318	658
Aluminum and bauxite	162	111	273	160	114	275
Magnesium	W	W	177	W	W	187
Other metallurgy	55	3	58	31	4	35
Magnesia from seawater	W	W	682	W	W	648
Alkalies	6	1,252	1,258	6	1,167	1,173
Pulp and paper	1,149	109	1,258	1,039	116	1,156
Sugar refining	47	727	774	58	909	967
Calcium carbide	146	72	218	121	63	185
Glass	191	—	191	180	—	180
Other chemical	425	83	508	—	—	—
Other uses†	411	296	707	645	714	523
Total Chemical & Industrial	10,268	5,228	15,496	8,308	4,918	13,226
Environmental:						
Water treatment	1,846	79	1,925	1,487	9	1,496
Sewage	799	16	815	848	12	860
SO_2 removal	604	—	604	743	—	743
Total Environmental	3,249	95	3,344	3,078	21	3,099
Refractory dolomite	670	123	793	420	75	494
Total, All Uses‡	15,461	5,522	20,983	12,537	5,081	17,618

Source: Pressler, 1981.
*Excludes regenerated lime. Includes Puerto Rico.
†Includes petrochemicals, whiting, adhesives, plastics, explosives, and uses indicated by symbol W.
‡Data may not add to totals shown because of independent rounding.

sumers in steel, copper, alumina, and other chemical industries—and potentially even major utility companies—have turned to captive plants for a portion of their lime requirements. With the larger capacity kilns now coming on line this frequently leads to surplus capacity from captive plants finding its way into the open market. Consequently, the large percentage of captive lime produced has tended to have a depressant effect on lime prices.

Calcining Plants

Several decades ago nearly every lime plant operated its own adjacent or nearby limestone quarry. In the last few decades, particularly in the 1960s, there has been a marked increase in the number of plants that purchase their limestone kiln feed, usually transported by water, to their calcining plants. Fifteen such plants are located on the Great Lakes from Minnesota to Ohio, including seven captive operations that obtain limestone by boat from northern Michigan. Two calcining plants in Louisiana burn water-transported oyster and clamshell, dredged from the Gulf of Mexico and Lake Pontchartrain, respectively. Economically these plants are usually justified, in spite of much higher kiln feed costs, because they are located

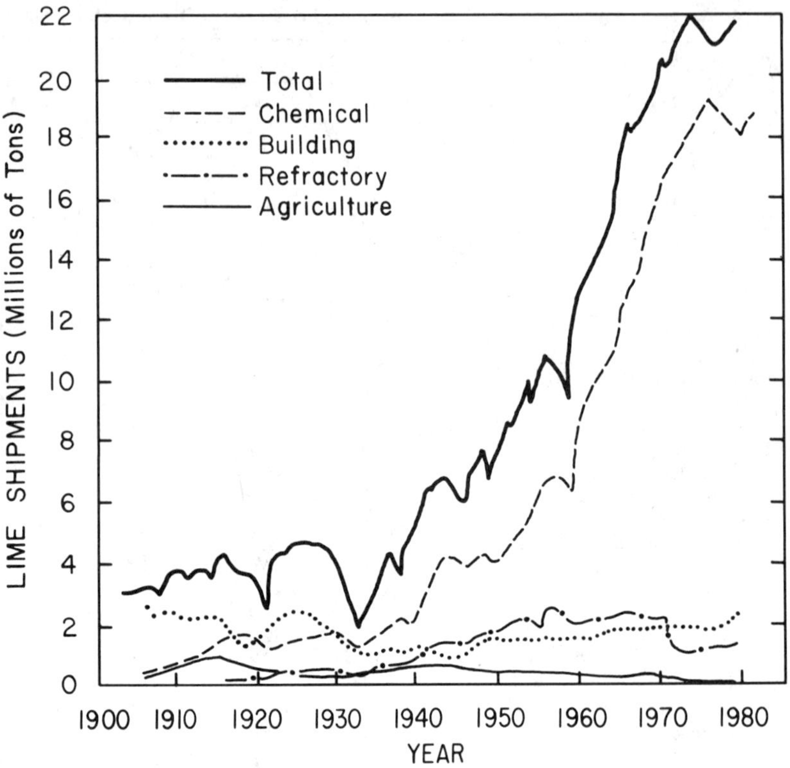

FIG. 10—*Trends in major uses of lime, 1904–1980. Sizable increase in 1959 was due largely to increased coverage of captive lime, with an additional 1.9 Mt added (based on US Bureau of Mines data).*

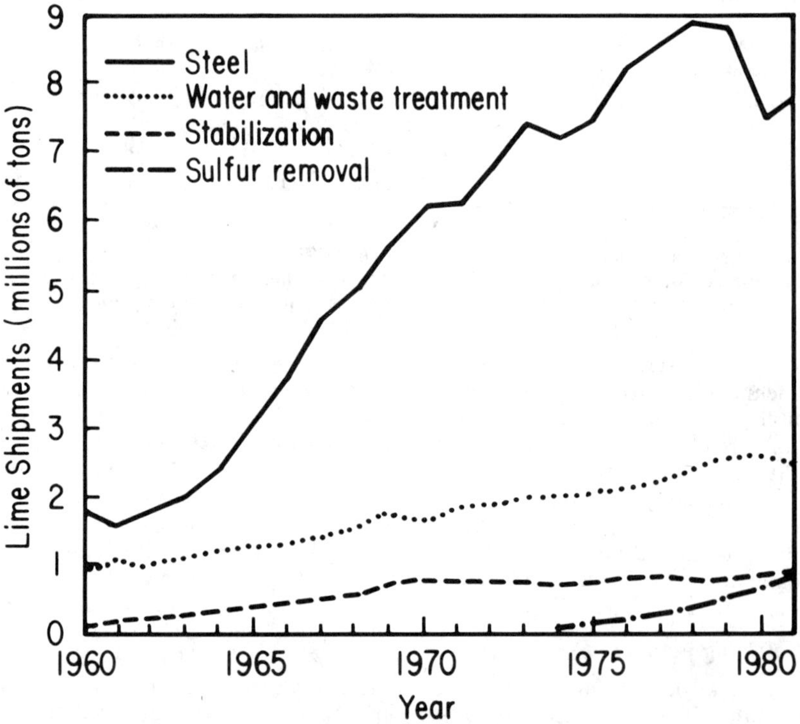

FIG. 11—*Trends in selected major uses of lime, 1960–1981.*

near large consuming markets. Similarly, two lime hydration plants have been built 200 or more miles from their company's calcining plant—in the Los Angeles and San Francisco bay areas. These plants also serve as distribution terminals for quicklime products, and are only the forerunners of other remote hydration and distribution facilities being considered by other lime producers.

Transportation

For many years railroads were the primary carrier of bulk lime. However, as rail service declined and the interstate highway system came into use, truck shipments became increasingly popular, to the point that in the late 1960s and early 1970s over 50% of lime shipped was moved by truck. While truck and rail shipments continue to predominate, an increasing tonnage of lime is moved by barge on the nation's inland waterways. Today over 15% of the lime produced in the US is moved via barge. The importance of this mode of transportation is seen in the fact that in addition to Mississippi Lime located at Ste. Genevieve, MO, two large new plants have been built on the Ohio River in northern Kentucky.

Distribution

Lime is produced in 40 states and consumed in all states. Table 4 lists the major consuming and producing states, including both commercial and captive lime as well as refractory dolomite. As would be expected, the two lists are similar but not identical.

Considering commercial lime only, the largest producing states in order of volume are Ohio, Missouri, Pennsylvania, Texas, Illinois, Alabama, Kentucky, Virginia, Michigan, and Arizona. Interestingly, as recently as 1977 neither Kentucky or Arizona were counted among the top producing states. Other locations which have experienced significant additions to their commercial production since 1978 include Tennessee, Utah, Texas, Michigan, and Illinois. Added to this are several plant modifications which have served to increase the capacity of existing facilities without the addition of new kilns.

On the consumption side, the leading quicklime-consuming states in 1979 were Pennsylvania, Ohio, Indiana, Michigan, and New York, each of which consumed more than 0.9 Mt. These five states accounted for 51% of the total quicklime consumed. Leading hydrate-consuming states were Texas, Pennsylvania, Illinois, Ohio, and Louisiana, each of which consumed more than 900 kt. These five states accounted for 52% of the total hydrate consumed.

Taxes and Prices

Since 1951 those lime companies which extract their own stone have had the benefit of percentage depletion, as applied to the federal income tax, on their kiln feed used to make lime. Until 1971 the percentage depletion rate was 15%, but then it was reduced to 14% along with similar reductions on most other minerals. The value of the kiln feed upon which percentage depletion is based has been subject to controversy and is often negotiated in accordance with various IRS interpretations on what to allow. Field market prices covering the quality and gradation of the stone and costs of producing the kiln feed (proportionate profits) are methods that have been applied.

The only statistics on prices are compiled by the USBM. The average mill price (bulk) for all types of lime is given in Table 5.

During the 1960s lime prices advanced very little, in spite of a moderate inflation rate. However, beginning in 1969 lime prices increased steadily due to fuel costs, air pollution and other government-mandated controls, escalating labor costs, inflation, etc. Then in 1974 the oil embargo initiated substantial increases in energy costs which have been clearly reflected in the cost of production and the cost of lime. These cost increases have been further exacerbated for the lime consumer by substantial increases in the cost of transportation. In addition to the cost of energy, the 1970s were a period of increased interest rates, which in 1980 reached and exceeded 20% for the first time in history. While these interest rates did not directly affect several producers, a large segment of the producers experienced major capital outlays in the period 1975 through 1980. Therefore, with the continuing impact of fuel costs, inflation, and high interest rates, annual price increases can be expected for the next several years.

Foreign Trade

Historically there has been relatively little foreign trade in lime in most countries, including the United States, because of the occurrence in most industrialized countries of limestone suitable for the production of lime, its

TABLE 4—Leading Lime Consuming and Producing States, 1979 and 1980* (1000 st)

Consuming	1979	1980	Producing	1979	1980
Pennsylvania	2662	2306	Ohio	3392	2786
Ohio	2380	1959	Pennsylvania	2153	1768
Indiana	2097	1699	Missouri	1790	1667
Michigan	1603	1355	Texas	1507	1515
Texas	1537	1535	Alabama	1273	1128
New York	1153	1077	Kentucky	1162	—
Illinois	1068	893	Michigan	1057	836
California	898	819	Illinois	W	—
Alabama	612	530	Virginia	872	824
Arizona	520	389	Arizona	673	514
Maryland	473	396	California	564	554
Florida	441	439	Indiana	W	—

Source: Pressler, 1981.
*Includes commercial lime, refractory dolomite, and captive lime.
W, Information withheld.

low cost, and the relative perishability of quicklime (and to a lesser extent, hydrate). US foreign trade in lime has included both exports and imports with Canada accounting for over 90% of this trade.

Exports of lime in recent years have generally been in the range of 36.2 to 45.3 kt and are normally measured against the record year of 1968 when nearly 63.5 kt of lime were exported. Of the total exports in recent years, Canada has received 60 to 65%, followed by Surinam, Guyana, and Mexico, each of which normally receives less than 10%. The remaining export tonnage is received by 30 to 40 other nations around the world, the exact number varying from year to year.

Imports of lime to the US have grown at an average rate of over 14% during the last 10 years. From 1976 to 1979 total imports, quicklime plus hydrate, have increased from 331 kt/a to 580 kt/a. The vast majority of this tonnage is imported from Canada, with the remainder coming from Mexico. In light of expanded production capacity in the northern half of the US and new lime production in Mexico, it is anticipated that the Mexican share of the import market will show modest gains.

The only worldwide foreign trade of consequence involves exports from Belgium and Germany into the Netherlands, where limestone production is virtually nil. Each of these countries exports 500 to 600 kt into the Netherlands annually. Belgium is also the main lime supplier for the Luxembourg steel industry and exports to France.

TABLE 5—Average Mill Price (Bulk), All Types of Lime

Year	Price Per Net st, $
1965	$13.87
1966	13.27
1967	13.42
1968	13.39
1969	13.94
1970	14.53
1971	15.78
1972	16.78
1973	17.42
1974	22.02
1975	27.45
1976	30.19
1977	33.42
1978	36.76
1979	41.18
1980	44.34
1981	46.20

Tests and Specifications

Because of the number of limes available, their many uses, and their wide variation in physical and chemical characteristics, there is no overall lime specification. Numerous specifications and buyer requirements for lime exists, not only for most of the major uses, but even among individual companies or plants.

The National Lime Assn., with its member lime companies, and a majority of the lime consumers, subscribe to the ASTM specifications on lime, which are promulgated by Committee C–7 on Lime. These specifications, which cover many uses and testing procedures, are:

ASTM Specifications

C25–79	Chemical Analysis of Limestone, Quicklime, and Hydrated Lime
C50–78	Sampling, Inspection, Packing, and Marking of Quicklime and Lime Products
C51–76	Terms Relating to Lime
C110–76	Physical Testing of Quicklime, Hydrated Lime, and Limestone
C400–76	Testing Quicklime and Hydrated Lime for Neutralization of Waste Acid
C593–76	Fly Ash and Other Pozzolans for Use with Lime
C602–75	Agricultural Liming Materials
C706–76	Limestone for Animal Feed Use
C737–78	Limestone for Dusting of Coal Mines
C821–78	Lime for Use with Pozzolans
C911–79	Quicklime, Hydrated Lime, and Limestone for Chemical Uses

Among the tests included in the C110–76 standard are residue (particle size), plasticity as determined by the Emley plasticimeter, soundness, water retention, settling rate, slaking rate, and air entrainment. The plasticity, water retention, soundness, and air entrainment tests apply especially to the mortar and finishing lime specifications. Slaking rate determination is of prime importance to many chemical lime uses, especially water and chemical treatment.

The C25–79 standard covers wet chemical tests for determining calcium and magnesium oxides, available lime content, CO_2 (loss on ignition or LOI), silica and other insolubles, sulfur, phosphorus, and various trace elements. In view of the many chemical uses for lime, the C25–79 standard is widely used in industry.

Another important lime specification is the AWWA Standard for Quicklime and Hydrated Lime (B202–77), which was promulgated by the American Water Works Assn. This specification includes a bonus and penalty arrangement based on available lime content, slaking rate, and slaking residue. It is also applicable to sewage and industrial waste treatment plants.

Bibliography and References

Because lime is one of mankind's oldest materials and its end-use pattern is diverse, the literature on lime is so extensive it would be impossible to give a complete bibliography. The following list of sources presents the more general references on lime, with emphasis on manufacture and major uses.

Among the principal sources of information are R. S. Boynton's textbook, *Chemistry and Technology of Lime and Limestones*, (2nd edition); publications of the National Lime Assn.; articles from *Pit and Quarry, Rock Products,* and *Zement-Kalk Gips* magazines; and publications of the US Bureau of Mines (particularly statistics) and the various state geological surveys (particularly Illinois, Indiana, Virginia, Pennsylvania, Ohio, and Missouri).

A selected bibliography follows.

Anon., 1920, "Lime—Its Properties and Uses," Circular 30, National Bureau of Standards, 25 pp.

Anon., 1927, "Manufacture of Lime," Circular 337, National Bureau of Standards, 104 pp.

Anon., 1940, *Symposium on Lime*, American Society for Testing and Materials, Philadelphia, 118 pp.

Anon., 1965, "Standard for Quicklime and Hydrated Lime," Standard B 202–65, American Water Works Assoc., 12 pp.

Anon., 1967, "Lime for Steelmaking: Tailored to Fit New Demands," *Metals Producing Magazine,* Vol. 33, Jan., 18 pp.

Anon., 1968, "Phosphate Extraction Process," Dorr-Oliver, Inc., 22 pp.

Anon., 1969, "Lime for Water and Wastewater Treatment," Reference No. I.22–24, B.I.F., 24 pp.

Anon., 1969a, *Reaction Parameters of Lime,* American Society for Testing and Materials, Symposium, Atlantic City, NJ.

Anon., 1969b, *Water Treatment Plant Design,* American Water Works Association, AWWA Inc., 359 pp.

Anon., 1970, "Lime," Bulletin T60.350–1, Wallace and Tiernan.

Anon., 1971, *Process Design Manual for Phosphorus Removal,* Black & Veatch, EPA Tech. Transfer, Oct.

Anon., 1981, *Chemical Lime Facts,* 4th ed., Bulletin 214, National Lime Assoc., Washington, DC, 44 pp.

Ahlberg, H.L., and Barenberg, E.J., 1965, "Pozzolanic Pavements," Engineering Equipment Station Bulletin 473, University of Illinois, 130 pp.

Azbe, V.J., 1946, *Theory and Practice of Lime Manufacture,* St. Louis, MO, 423 pp.

Azbe, V.J., 1954, "Rotary Kiln Performance, Evaluation and Development," National Lime Assn., pp. 89; also *Rock Products,* series, Feb. 1953-Oct. 1955.

Bishop, D.L., 1939, "Particle Size and Plasticity of Lime," *Journal of Research,* Vol. 23, No. 2, pp. 285–292.

Boynton, R.S., 1980, *Chemistry and Technology of Lime and Limestone,* John Wiley, New York, 578 pp.

Boynton, R.S., and Gutschick, K.A., 1968, *Building*

Lime, Its Properties, Uses, and Manufacture, United Nations Ind. Dev. Organization, New York, July, 104 pp.

Boynton, R.S., and Jander, F.K., 1952, "Lime and Limestone," *Encyclopedia of Chemical Technology*, R.E. Kirk, et al., eds., Interscience, New York.

Bowles, O., 1952, "The Lime Industry," Information Circular 7651, US Bureau of Mines, 43 pp.

Bowles, O., and Jensen, N.C., 1947, "Industrial Uses of Limestone and Dolomite," Information Circular 7402, US Bureau of Mines, 19 pp.

Burchard, E.F., 1912–1913, "Stone," *Mineral Resources of the U.S.*, US Geological Survey, Pt. 2, 1911, pp. 782–831; 1912, pp. 754–815; 1913, pp. 1335–1408. These contain a series of maps showing the location of every limestone quarry and lime-kiln in the United States.

Burchard, E.F., and Emley, W.E., 1913, "Source, Manufacture, and Use of Lime," *Mineral Resources of the United States,* US Geological Survey, Pt. 2, pp. 1509–1593.

Davison, J.L., 1969, "Curing of Lime Cement Mortars," STP 472, American Society for Testing & Materials, pp. 193–208.

Derge, G., and Shegog, J.R., 1969, "Observations on the Solution of Calcium Oxide in Slag Systems," STP 472, American Society for Testing & Materials, pp. 132–142.

Dickson, T., 1982, "North American Lime," *Industrial Minerals,* No. 177, June, pp. 51-63.

Eades, J.L., and Sandbert, P.A., 1969, "Characterization of the Properties of Commercial Limes by Surface Area Measurements and Scanning Electron Microscopy," STP 472, American Society for Testing & Materials, pp. 3–24.

Eckel, E.C., 1928, *Cements, Limes, and Plasters*, John Wiley, New York, 699 pp.

Emley, W.E., 1913, "Manufacture of Lime," Technical Paper 16, National Bureau of Standards, 130 pp.

Emley, W.E., 1920, "Measurement of Plasticity of Mortars and Plasters," Technical Paper 169, National Bureau of Standards, 27 pp.

Emley, W.E., 1927, "Manufacture of Lime," Circular 337, National Bureau of Standards, 104 pp.

Goudge, M.F., *Limestones of Canada, Their Occurrence and Characteristics*, Canada Dept. of Mines, published in five volumes, 1927 to 1938; Vol. 1, Introductory, 19; Vol. 2, Maritimes, 19; Vol. 3, Quebec, 19; Vol. 4 No. 781, Ontario, 362 pp.; Vol. 5, Western Canada, 19 pp.

Grancher, R.A., 1970, "Lime Marches On . . . Maybe," *Rock Products*, Mar., pp. 64–67.

Grancher, R.A., 1970a, "Past, Present and Future of Lime in the United States," International Lime Conference, Salzberg, Austria, May, 34 pp.

Hatmaker, P., 1931, "Utilization of Dolomite and High-Magnesium Limestone," Information Circular 6524, US Bureau of Mines, 18 pp.

Hatmaker, P., 1937, "Lime," *Industrial Minerals and Rocks,* 1st ed., AIME, New York, pp. 395–426.

Hedin, R., 1961, "Structural Processes in the Dissociation of Calcium Carbonate," Azbe Award No. 2, National Lime Assn., 12 pp.

Hedin, R., 1962, "Plasticity of Lime Mortars," Azbe Award No. 3, National Lime Assn., 11 pp.

Hirsch, A.A., 1962, "Dry Feed of Quicklime Without a Slaker," *AWWA Journal,* Dec., pp. 1531–1542.

Hoak, R.D., 1953, "How to Buy and Use Lime as a Neutralizing Agent," *Water and Sewage Works,* Dec., 6 pp.

Knibbs, N.V.S., 1924, *Lime and Magnesia*, Ernest Benn, Ltd., London, 306 pp.

Knibbs, N.V.S., and Gee, B.J., *Lime and Magnesia,* Pt. 1, H.I. Hall Corp., Ltd., 113 pp.

Lamar, J.E., 1961, "Uses of Limestone and Dolomite," Circular 321, Illinois Geological Survey.

Lamar, J.E., 1967, "Handbook on Limestone and Dolomite for Illinois Quarry Operators," Bulletin 91, Illinois Geological Survey, 119 pp.

Landes, K.K., 1953, *Metallurgical Limestone Reserves in the U.S.,* 2nd ed., National Lime Assn., July, 31 pp.

Lazell, E.W., 1915, *Hydrated Lime*, Jackson-Remlinger Ptg. Co., 95 pp.

Lewis, C.J., and Crocker, B.B., 1969, "The Lime Industry's Problem of Airborne Dust," *Journal of Air Pollution Control Assn.,* TI-2 Report No. 10, Jan.

Limes, R., and Russell, R.O., 1969, "Crucible Test for Lime Reactivity in Slags," STP 462, American Society for Testing and Materials, pp. 161–172.

Mayer, R.P., and Stowe, R.A., 1964, "Physical Characterization of Limestone and Lime," Azbe Award No. 4, National Lime Assn., Aug., p. 26.

McClellan, G.H., and Eades, J.L., 1969, "Textural Evolution of Limestone Calcines," STP 462, American Society for Testing and Materials, pp. 209–227.

McDonald, E.B., 1971, *Lime Stabilization Training Manual*, South Dakota Dept. of Highways, 79 pp.

McDowell, C., 1972, "Flexible Pavement Design Guide," Bulletin 327, National Lime Assn., 43 pp.

Miller, T.C., 1960, "A Study of the Reaction Between Calcium Oxide and Water," Azbe Award No. 1, National Lime Assn., p. 50.

Minnick, L.J., 1971, "Control of Particulate Emissions from Lime Plants—A Survey," *Journal of Air Pollution Control Assn.,* Apr., pp. 195–200.

Mullins, R.C., and Hatfield, J.D., 1969, "Effects of Calcination Conditions on Properties of Lime," STP 462, American Society for Testing and Materials, pp. 117–131.

Moyer, F.T., 1941, *Lime-fuel Ratios of Commercial Lime Plants in 1939,* Information Circular 7174, US Bureau of Mines, 9 pp.

Murray, J.A., 1956, *Summary of Fundamental Research at M.I.T. on Lime and Application of Results to Commercial Problems,* National Lime Assn., 19 pp.

Obst, K., et al., 1969, "Present Status and Technical Advances in Steelworks Lime for Basic Oxygen Furnaces in Germany," STP 462, American Society for Testing and Materials, pp. 173–192.

Parsons, W.A., 1965, "Chemical Treatment of Sewage and Industrial Wastes," Bulletin 215, National Lime Assn., 139 pp.

Philbrook, W.O., and Natesan, K., 1966, "Mixed-Control Kinetics in the Thermal Decomposition of Calcium Carbonate," Azbe Award No. 6, National Lime Assn., pp. 55.

Pressler, J.W., 1981, "Lime," *Minerals Yearbook, 1978–1979;* US Bureau of Mines, pp. 539–549; and 1980, pp. 507–518.

Pressler, J.W., 1982, "Lime," *Mineral Commodity Summaries 1982*, US Bureau of Mines, pp. 86–87.

Reed, A.H., 1971, "Lime," *Minerals Yearbook 1971,* US Bureau of Mines, pp. 695–702.

Riehl, M.L., 1970, *Water Supply and Treatment,* 10th ed., National Lime Assn., 220 pp.

Ritchie, T., 1969, "Influence of Lime in Mortar on the Expansion of Brick Masonry," STP 462, American Society for Testing and Materials, pp. 67–81.

Rockwood, N.C., 1922, "Lime and Tuberculosis," *Rock Products,* Jan., pp. 56–61.

Rockwood, N.C., 1949, "Lime," *Industrial Minerals and Rocks,* 2nd ed., AIME, New York.

Rooney, L.F., and Carr, D.D., 1971, "Applied Geology of Industrial Limestone and Dolomite," Geological Survey Bulletin 46, Indiana Dept. of Natural Resources, 59 pp.

Schlitt, W.J., and Healy, G.W., 1969, "A Comparison and Scaling Down of Coarse Grain Titration and ASTM Slaking Rate Test," STP 462, American Society for Testing and Materials, pp. 143–160.

Schwarzkopf, F., 1970, "A Comparison of Modern Lime Calcining Systems," *Rock Products,* July, pp. 68–71.

Searle, A.B., 1935, *Limestone and Its Products,* London, Ernest Benn, Ltd., 709 pp.

Staley, H.R., and Greenfeld, S.H., 1947, "Surface Areas of High Calcium Quicklimes and Hydrates," *Proceedings,* American Society for Testing and Materials.

Thompson, J.L., 1979, "Pebble Lime From Preheater Kilns," *Trans. SME-AIME,* Vol. 266, pp. 2041–2044.

Thompson, M.R., 1968, "Lime Stabilization of Soils for Highway Purposes," Series No. 25, Illinois Highway Engineering, Dec.

Timucin, M., and Morris, A.E., 1969, "Phase Equilibria and Thermodynamic Properties of Lime-Iron Oxide Melts," STP 462, American Society for Testing and Materials, pp. 25–31.

Voss, W.C., 1960, "Exterior Masonry Construction," Bulletin 324, National Lime Assn., 70 pp.

Wells, L.S., et al., 1936, "Differences in Limes as Reflected in Certain Properties of Masonry Mortars," Research Paper RP 952, National Bureau of Standards, 14 pp.

Wuhrer, J., 1965, "On the Reactivity of Lime from Different Kiln Systems," Azbe Award No. 5, National Lime Assn., p. 44.

The following are also good sources of information on lime:

Highway Research Board, National Academy of Sciences, Washington, DC, various bulletins on lime stabilization.

"Lime," *Minerals Yearbook.* These chapters contain statistical and general information, published each year by the US Bureau of Mines.

National Lime Association, Washington, DC:

- Bull. 192, *100 Questions and Answers on Liming Lawns,* 1955, 32 pp.
- Bull. 211, *Water Supply & Treatment,* 11th ed., 1976, 220 pp.
- Bull. 213, *Lime Handling, Application, and Storage in Treatment Processes,* 1976, 73 pp.
- Bull. 214, *Chemical Lime Facts,* 4th ed., revised, 1981, 42 pp.
- Bull. 215, *Chemical Treatment of Sewage & Industrial Wastes,* 1965, 139 pp.
- Bull. 216, *Acid Neutralization With Lime,* 1976.
- Bull. 217, *Lime in Municipal Sludge Processing,* 1980.
- Bull. 304-G, *Whitewash and Cold Water Paints,* 1955, 15 pp.
- Bull. 322, *Specifications for Lime and Its Uses in Building,* 1966, 16 pp.
- Bull. 324, *Exterior Masonry Construction,* 2nd ed., 1960, 70 pp.
- Bull. 325, *Hydrated Lime in Asphalt Paving,* 1962, 15 pp.
- Bull. 326, *Lime Stabilization Construction Manual,* 5th ed., 1972, 47 pp.
- Bull. 327, *Flexible Pavement Design Guide,* 1972, 43 pp.
- Bull. 328, *A Long Range Durability Study of Lime Stabilized Bases,* 1977.
- Bull. 400, *Lime Industry Safety Manual,* 2nd ed., 1970, 43 pp.
- Pamphlet, *Lime Dries Up Mud,* 1970, 4 pp.
- Pamphlet, *Masons Lime,* 1969, 12 pp.
- Map, *Commercial Plants in U.S. and Canada,* 1981.
- Lime Brochure (1980).

Pit and Quarry magazine, Chicago, IL, numerous articles on lime manufacturing, with May issue each year devoted to lime.

Pit and Quarry Handbook, Chicago, IL, annual handbook on production, including sections on lime plant equipment.

Rock Products magazine, Chicago, IL, numerous articles on lime manufacturing, with July issue each year devoted to lime.

Zement-Kalk-Gips, Wiesbaden, West Germany, numerous articles on lime manufacturing, in German.

Limestone and Dolomite*

DONALD D. CARR †

LAWRENCE F. ROONEY ‡

Perhaps no other mineral commodity in this volume has as many uses as limestone and dolomite. These carbonate rocks are the basic building blocks of the construction industry, the material from which aggregate, cement, lime, and building stone are made. Carbonate rocks, and their derived products, are used as fluxes, glass raw material, refractories, fillers, abrasives, soil conditioners, ingredients in a host of chemical processes, and much more. Although the use of limestone and dolomite goes back before recorded history, new uses and products are continually being found.

Carbonate rocks form about 15% of the earth's sedimentary crust and are widely available for exploitation. Found extensively on all continents, they are quarried and mined from formations that range in age from Precambrian to Recent. Reserves of carbonate rock are large and will last indefinitely, although high-purity deposits may be absent or have limited availability in certain areas.

Limestone and dolomite are so useful and so abundant that in 1981 about 644 million tons were produced in the United States. In fact, about 75% of all stone quarried or mined in the United States was carbonate rock. Sand and gravel was the only mineral commodity produced in greater quantity.

In this chapter we have sought to provide the general reader an overview of carbonate rocks, their composition, distribution, production, and uses, with emphasis on developments since 1960. In doing so we borrowed extensively from our report on the applied geology of limestone and dolomite (Rooney and Carr, 1971). The literature on the limestone and dolomite resources of the United States and Canada is so vast and diverse that one can easily become mired down in print. We tried to list those references that are easily accessible and that provide the economic geologist with a starting point for more intensive study. The table listing selected references on limestone and dolomite resources in Canada and the United States reflects our attempt to be selective rather than exhaustive and to direct the reader to the best sources of detailed information.

Geology

Mineralogy

The principal carbonate rocks used by industry are limestone and dolomite. Limestones are sedimentary rocks composed mostly of the mineral calcite ($CaCO_3$) and dolomites are sedimentary rocks composed mostly of the mineral dolomite ($CaCO_3 \cdot MgCO_3$). Aragonite ($CaCO_3$), which has the same chemical composition as calcite but a different crystal structure, is economically important only in modern deposits, such as oyster shells and oolites. Aragonite is metastable and alters to calcite in time. Some other carbonate minerals, notably siderite ($FeCO_3$), ankerite ($Ca_2MgFe(CO_3)_4$), and magnesite ($MgCO_3$), are commonly found associated with limestones and dolomites but generally in minor amounts.

Because of their similar physical properties, the carbonate minerals are not easily distinguished one from the other. Specific gravity,

* In addition to acknowledging the valued support of the Indiana Geological Survey and Flathead Valley Community College, we thank the many members of the Survey who provided assistance on particular problems. We extend special thanks to H. W. Allen, Chief Staff Geologist, Martin Marietta Cement Co., J. B. Patton, State Geologist, Indiana Geological Survey, and J. W. Baxter, Illinois Geological Survey, for their critical review of the manuscript.

† Indiana Geological Survey, Bloomington, IN.

‡ Formerly Flathead Valley Community College, Kalispell, MT; now US Geological Survey, Jeddah, Saudi Arabia.

TABLE 1—Physical Properties of Some Common Carbonate Minerals

Calcite: $CaCO_3$, hexagonal crystal system, commonly good rhombohedral cleavage. Mohs' hardness, 3. Specific gravity, 2.72. Commonly colorless or white but may be other colors due to impurities.
Dolomite: $CaCO_3 \cdot MgCO_3$, hexagonal crystal system, commonly good rhombohedral crystals with curved faces. Mohs' hardness, 3.5 to 4. Specific gravity, 2.87. Commonly white or pink.
Aragonite: $CaCO_3$, orthorhombic crystal system. Mohs' hardness 3.5 to 4. Specific gravity, 2.93-2.95. Commonly colorless, white, or yellow, but may be other colors due to impurities.
Siderite: $FeCO_3$, hexagonal crystal system, commonly distorted rhombohedral crystals. Mohs' hardness, 3.5 to 4. Specific gravity, 3.7 to 3.9. Commonly brown or black.
Ankerite: $Ca_2MgFe(CO_3)_4$, hexagonal system, commonly rhombohedral crystals. Mohs' hardness, 3.5 to 4. Specific gravity, 2.9. Commonly white, pink, or gray.
Magnesite: $MgCO_3$, hexagonal, usually in granular or earthy masses. Mohs' hardness, 3.5 to 4.5. Specific gravity, 2.96 to 3.1. Commonly white or yellowish but may be other colors due to impurities.

color, crystal form, and other physical properties (Table 1) are aids to mineral identification if the rock is relatively monomineralic and compact. The interested geologist can find further information on the chemical and physical properties of the carbonate minerals in Graf and Lamar's (1955) comprehensive review.

The rate of solubility of the different minerals in dilute hydrochloric acid is perhaps the most useful technique to identify them in the field. Calcite is much more soluble in dilute acid than dolomite; hence, if a fresh rock surface is etched, the amount of dolomite left standing in relief can be estimated by the use of a hand lens. Some staining techniques are based on differences in solubility of the carbonate minerals (in decreasing order of solubility: aragonite, calcite, and dolomite). These techniques are useful in the laboratory and have limited application in the field (Friedman, 1959; Högberg, 1971; Warne, 1962). Staining of thin sections is particularly effective in the laboratory (Dickson, 1965, 1966).

The X-ray diffractometer is commonly used in the laboratory for determining carbonate mineralogy of bulk samples. Techniques of determining calcite and dolomite ratios or the percentage of these minerals based on comparison of their diffraction intensities with those of known standards can be found in Azároff and Buerger (1958), Cullity (1956), Graf and Goldsmith (1963), Gulbrandsen (1960), Müller (1967), and Runnells (1970).

Thin section analysis may be a helpful adjunct to binocular examination of carbonate rocks. Although it is difficult to distinguish between calcite, dolomite, and ankerite in thin section unless staining techniques are used, identification of some other minerals, types of carbonate grains, fabrics, textures, and structures are sometimes facilitated by this method (see also "Physical Properties," p. 842). Appropriate parts of "Microscopic Sedimentary Petrography" (Carozzi, 1960) and "Petrology of Sedimentary Rocks" (Hatch and Rastall, 1971) provide introductions to microscopic investigation of carbonate rocks, as does the review on that subject by Gubler, et al. (1967). Good places for assistance in the identification of fossils and fossil fragments in thin section are "Introductory Petrography of Fossils" Horowitz and Potter (1971) and Scholle (1978).

Color, an important property of carbonate rocks, can be a rough guide to purity, but it can also be misleading. Only a small amount of noncarbonate material is necessary to produce a marked change in color. The famous building stone called Indiana Limestone, for example, with its distinct gray and buff colors, commonly contains less than 0.2% Fe_2O_3; "Carthage Marble," a fossiliferous dimension limestone from Missouri in shades of gray, generally has a total iron and aluminum oxide content less than 0.2%. Most high-purity limestones are shades of light brown and gray to white. Limestones in shades of gray or green generally indicate the presence of minerals containing ferrous iron oxides and (or) carbonaceous matter. As the state of oxidation increases, the colors change to yellows, browns, or reds. A color reference chart is useful in maintaining uniformity of rock descriptions. One of the most useful has been made available by the Geological Society of America (Goddard, 1963).

Impurities in carbonate rock vary considerably in type and amount but are important from an economic standpoint only if they affect the usefulness of the rock. Generally the two most important considerations of each impurity are how much is present and how it is distributed. A considerable amount of some impurities is tolerable in carbonate rock for some uses if the impurity is disseminated throughout the rock. On the other hand, if the impurity is concentrated in laminae, it may form planes of weakness that seriously affect the performance of the rock.

Perhaps the most common impurity in carbonate rocks is clay. The clay minerals—mainly kaolinite, illite, chlorite, smectite, and mixed-lattice types—may be either disseminated throughout the rock or concentrated in laminae or thin partings. The basic molecular building blocks of clay minerals are silica tetrahedra (a silicon atom and four oxygen atoms) and alumina and (or) magnesium octahedra (an aluminum or magnesium atom and six hydroxyl ions). Other chemical elements are incorporated into the structure, however, so that it is difficult to determine the type of clay mineral by chemical analysis alone. If clay mineral identification is important, other analysis techniques, such as X-ray diffraction, differential thermal analysis, and electron microscopy, may be employed. A comprehensive review of clay minerals can be found in the chapter of that title in this volume.

Chert is another common impurity in carbonate rocks and may be disseminated as grains throughout the rock or concentrated in nodules, lenses, or beds. It is composed mainly of very fine grained quartz (SiO_2) that may appear under the microscope as minute subequant crystals, usually 1 to 10 μ in diameter (microcrystalline quartz), or as radiating fibers (chalcedonic quartz). Chert easily incorporates impurities, including water, into its structure so that it may be found in almost all colors; its surface texture may range from dense or porcelaneous to porous or earthy. Dense cherts have a Mohs hardness of 7 and high impact toughness (greater than 6 in. per sq in.), which make them particularly abrasive to crushers and other processing equipment. Porous cherts, mainly because of their large surface area available for chemical reaction and moderate solubility in alkalies, are considered as deleterious components in aggregates used in concrete. Comprehensive coverage of the deleterious properties of cherts can be found in Dunn and Ozol (1962) and Kneller, et al. (1968), the latter of which contains a useful bibliography with 600 entries. A good place to start for further information on the texture and composition of chert is a paper by Folk and Weaver (1952).

Silica is also found in carbonate rocks as discrete silt- or sand-size grains of the mineral quartz. These grains may be disseminated throughout the rock or concentrated in laminae and beds. Detrital limestone especially may contain a considerable percentage of quartz silt and sand. These grains may act as the nuclei for coated carbonate grains, such as some ooliths and pisoliths.

Finely disseminated organic matter is a common constituent of limestones and dolomites and may give the rock a pronounced brown or black color. Bituminous material, an organic derivative of petroleum and residue of former pore fluids, may be present in sufficient quantity to make the rock undesirable for some uses.

Thin-section and insoluble-residue studies reveal trace amounts of a wide variety of other minerals in most carbonate rocks. Although these trace minerals may affect the economic usefulness of rocks used for chemical purposes, such as glass manufacture, they have little effect on rocks used principally for physical properties, such as construction materials.

Origin

Most limestones of economic importance were partly or wholly biologically derived from seawater and accumulated in a relatively shallow marine environment. The obvious skeletal material in limestones speaks of a biologic origin, but even nondescript, fine-grained material may derive from the life of organisms. Pellets in many instances are fecal material, and silt- and clay-size particles may be aragonitic-sheath crystals released upon the death of algae. Ooliths, which in the past have largely been thought to be the inorganic accumulation of calcium carbonate around a nucleus, may also depend in part on algal activity.

In some places, lime-secreting organisms, such as corals, calcareous algae, and mollusks, are able to erect large, wave-resistant structures called reefs. The biologically active parts of these structures are generally near the edge of shallow marine shelves where upwelling currents supply nutrients for the growth of the organisms. In other places along such shelves, small skeletal particles or other material may become coated by concentric layers of calcium carbonate to form ooliths. Oolites develop best in the high-energy zone of shelves where water currents agitate the grains, and as the oolites build up they form elongate lenticular bars that nearly reach the water surface. Very fine-grained carbonate muds (micritic limestone) derived from the comminution of coarser skeletal material or precipitated directly from seawater accumulate in low-energy environments, such as lagoons or deep water.

Environment of deposition is significant to the economic geologist because it determines the size, shape, purity, and other economically significant characteristics of the carbonate rock deposit. Limestones that form in high-energy zones generally contain little noncarbonate material and hence may be the source of high-purity carbonate material. Micrite, which accumulates in zones of low energy, is more likely to be diluted by clay and silt-size noncarbonate material.

Carbonate sediments are highly susceptible to postdepositional alteration and modification. The origin of dolomite is especially significant to the economic geologist. Although some dolomite may be precipitated directly from seawater, most dolomite is a result of the alteration of calcium carbonate sediments or rocks by hypersaline brines. Good examples are the almost-pure dolomite Silurian reefs in northern Illinois, Indiana, and Ohio, and in southern Michigan.

To understand the depositional environment and postdepositional history of carbonate rocks, economic geologists are well advised to study modern carbonate deposition. Good places to start for this information are the general reviews by Baars (1963), Ginsburg, et al. (1963), and Milliman (1974). For further information on locations and distributions of modern carbonate sediments, the summaries by Graf (1960a) and Taft (1967) are useful; the comprehensive discussions of the origin and occurrence of carbonate rock in the papers compiled by Jordan (1978), in the book by Wilson (1975), and the reviews of limestone by Sanders and Friedman (1967) and of dolomites by Friedman and Sanders (1967) fill the gaps of the other coverages. A good guide to the literature of both modern and ancient carbonate rock deposits is Potter's (1968) annotated bibliography of 202 references.

Classification

The explosive growth in the study of modern carbonate sedimentation from the late 1950s to the present has had a notable effect on carbonate-rock classification. Numerous classification schemes have emerged based on this new-found information with the result that carbonate rock descriptions are now more explicit and more conducive to genetic interpretation than before. It is unlikely that the ultimate objective of the industrial geologist is the origin of the rock; still a classification is useful in providing a written record that can be used to interpret the rock's physical and chemical properties.

Many aspects of carbonate rocks can be used as the basis of a classification scheme, but perhaps the most useful are composition and texture. Composition can be thought of in terms of mineralogy, types of fossils or grains, or chemical constituents. Texture refers to both depositional and post-depositional features, such as relative proportions of framework grains and lime mud, grain size, cement, and pores.

Carbonate rocks are rarely monomineralic in nature; thus, a mineralogical classification of these rocks needs to take into consideration the variation in amounts of calcite, dolomite, and noncarbonate materials (Fig. 1). Such a classification is useful in rock descriptions, especially when combined with textural parameters, but it commonly is not sufficient for industrial purposes. Although limestone and dolomite can be used equally well for a large number of uses, certain uses have special chemical requirements. These special requirements are stated in terms of chemical composition rather than mineralogical composition and specify the quantity of $CaCO_3$ (or CaO) and $MgCO_3$ (or MgO) or both in the rock along with the maximum percentage of impurities that can be tolerated. A practical chemical classification considers that ultrahigh calcium limestone is more than 97% $CaCO_3$, high-calcium limestone is more than 95% $CaCO_3$, high-purity carbonate rock is more than 95% combined $CaCO_3$ and $MgCO_3$, and high-magnesium dolomite is more than 43% $MgCO_3$ (theoretically pure dolomite is 45.7% $MgCO_3$).

A textural classification, along with a mineralogical classification, is fundamental to geologic studies aimed at determining the origin

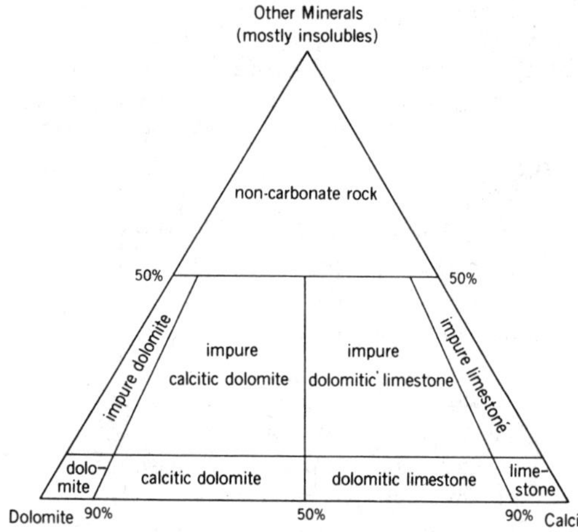

FIG. 1—*Mineralogical classification of carbonate rocks.*

of carbonate rocks. One such classification by Leighton and Pendexter (1962) considered that most limestones can be characterized by the types and relative amounts of four textural components: grains, lime mud (micrite), cement, and pores. The ratio of the relative proportions of grains to micritic material, which is the basis of their nomenclatural system (Fig. 2), gives some indication of water turbulence because muds cannot be deposited in areas of high bottom currents. Other classifications, such as the ones by Folk (1959, 1962), Dunham (1962), and Bissell and Chilingar (1967), make use of framework grains to mud ratios and have practical applications.

Dolomite presents a special problem in classification and may require separate handling from limestone. The textural classification can be used for secondary dolomite if the original depositional texture is preserved. Some dolomites, however, show only faint traces of original texture, called "ghosts" or "relics," and others may have had their original texture completely obliterated. For these cases, and for dolomite of primary origin, a classification based on crystal size may be required.

Distribution of Deposits

Carbonate rocks have been deposited from Precambrian to Recent time, and although they compose only about 0.25% of the volume of the crust of the earth (Parker, 1967), they comprise about 15% of the sedimentary rocks. In 1980, limestone and (or) dolomite were quarried or mined in all of the 50 states except Delaware, Louisiana, New Hampshire, and North Dakota; they were mined in all Canadian provinces except Saskatchewan.

Because carbonate rocks are widely distributed and differ in their geologic characteristics, each deposit must be considered on its own attributes. The best source of information for carbonate rock deposits is the state geological surveys, or their equivalents. Most state publications are oriented toward areal geology, commodities, or the two combined, rather than uses or methods. The US Bureau of Mines (USBM) and the US Geological Survey (USGS) also publish information on limestone and dolomite. The USGS's 7½-min geologic quadrangle maps are especially useful, as are the Bureau's circulars that contain cost analyses of limestone production (Eilertsen, 1964; Evans and Eilertsen, 1957; Kline, 1962; Kusler and Corre, 1968; Marshall, 1963; Riley and Schroeder, 1963). Table 2 lists the most useful publications on the limestone and dolomite resources of the different states and provinces. The annotated bibliographies of Gazdik and Tagg (1957) on high-calcium limestone deposits and of Davis (1957) on some high-magnesium dolomite deposits are helpful guides to the older literature, as is the report on high-grade dolomite deposits by Weitz (1942).

Although high-purity carbonate rock deposits are not overly abundant, they are by no means rare. Virtually all of the states except Delaware, Louisiana, New Hampshire, and Rhode Island, and all of the Canadian provinces have

Grain/ Micrite Ratio	% Grains	Grain Type					Organic Framebuilders	No Organic Framebuilders
		Detrital Grains	Skeletal Grains	Pellets	Lumps	Coated Grains		
9:1	~90%	Detrital Ls.	Skeletal Ls.	Pellet Ls.	Lump Ls.	Oolitic Ls. Pisolitic Ls. Algal Encrusted Ls.	Coralline Ls. Algal Ls. Etc.	Caliche Travertine Tufa
1:1	~50%	Detrital-Micritic Ls.	Skeletal-Micritic Ls.	Pellet-Micritic Ls.	Lump-Micritic Ls.	Oolitic-(Pisolitic-Etc.) Micritic Ls.	Coralline-Micritic Ls. Algal-Micritic Ls. Etc.	
1.9	~10%	Micritic-Detrital Ls.	Micritic-Skeletal Ls.	Micritic-Pellet Ls.	Micritic-Lump Ls.	Micritic-Oolitic (Pisolitic Etc.) Ls.	Micritic-Coralline Ls. Micritic-Algal Ls. Etc.	
		← Micritic Limestone →						

FIG. 2—*Textural classification of limestone. From Leighton and Pendexter, 1962.*

TABLE 2—Limestone and Dolomite Resources in Canada and United States—Information Supplied by State and Provincial Surveys in 1980

	Limestone and Dolomite Currently Being Produced	High-Calcium Limestone and/or High-Magnesium Dolomite Currently Being Produced	Potentially Commercial Deposits of High-Calcium Limestone and/or High-Magnesium Dolomite are Present	Selected References (Numbers Keyed to Bibliography) on Limestone and Dolomite Resources
CANADA				
Alberta	x	x	x	133,134,140,151,180,181,344,345
British Columbia	x	x	x	87,140,245
Manitoba	x	x	x	20,135,140
N. Bruns. and Nova Scotia	x	x	x	136,268,327
Nfld. and Labrador	x	x	x	56,91
Ontario	x	x	x	135,139,169,170
Quebec	x	x	x	135,137
Saskatchewan			x	123,140,214
UNITED STATES				
Alabama	x	x	x	3,225,358
Alaska	x		x	53,191,288,304,315
Arizona	x	x	x	202
Arkansas	x	x	x	46,178
California	x	x	x	40,41,42,154
Colorado	x	x	x	13,14,260,359,381
Connecticut	x	x	x	74,179,263
Delaware				
Florida	x	x	x	57,182,294,295,300,321,325,361,362,363,389,390
Georgia	x	x	x	47,250,287,290
Hawaii	x	x	x	374
Idaho	x	x	x	318,319
Illinois	x	x	x	186,216,218,219,220,221,382
Indiana	x	x	x	60,63,64,118,187,248,249,283,285,305,306,307,370
Iowa	x	x	x	100
Kansas	x		x	124,146,190,201,302,311,312,313,314,368
Kentucky	x	x	x	92,93,94,95,203,204,247,341,342,343
Louisiana				
Maine	x	x	x	4
Maryland	x	x	x	7,240,244
Massachusetts	x	x	x	12,67,112,284,297
Michigan	x	x	x	196,225,226,333
Minnesota	x		x	108,231,291,322,334,352,353
Mississippi	x	x	x	33,34,232,251,266
Missouri	x	x	x	9,49,50,175,176,185,238,239,377
Montana	x	x	x	66,81,227,228
Nebraska	x	x	x	54,73
Nevada	x	x	x	275,280,281
New Hampshire				
New Jersey	x	x	x	212,213,378,387
New Mexico	x	x	x	193,209,210,211
New York	x	x	x	19,155,194,195,274,292,356,366,391,392
North Carolina	x	x	x	2,29,32,75
North Dakota			x	10,55,58,152
Ohio	x	x	x	223,224,265,338,339,340,346
Oklahoma	x	x	x	8,28,30,109,149,150,197,309,348,369
Oregon	x	x	x	5,230,241,286,298,365
Pennsylvania	x	x	x	90,257,276,277,278,349
Rhode Island	x			284,296,297,379
South Carolina	x	x	x	71,167
South Dakota	x		x	145
Tennessee	x	x	x	168,351
Texas	x	x	x	21,69,106,126,234,270,303,323,337
Utah	x	x	x	51,264
Vermont	x	x	x	82,83,110,117,360,373
Virginia	x	x	x	22,23,76,77,104,105
Washington	x		x	31,86,259,261
West Virginia	x	x	x	215,246,376
Wisconsin	x		x	48,184,336
Wyoming	x	x	x	153,161,162,163,279

potentially commercial deposits of high-calcium limestone and (or) high-magnesium dolomite (Table 2). Twenty-one Mexican states had production of high-calcium limestone or high-magnesium dolomite in 1982 (Fig. 3). Because these deposits are present, however, does not necessarily mean that they can be exploited. In many areas competition is intense for potential mineral lands for construction sites, recreation areas, and highways; even nature has her own requirements for flowing streams and soil development. Society has also imposed environmental controls, now firmly ensconced in state and federal statutes, which prohibit or restrict mineral production in areas where it might significantly affect the quality of man's environment.

Exploration

Exploration for limestone and dolomite in North America is largely the detailed examination of known deposits. Because most limestone is a sedimentary rock, it occurs in strata generally of considerable extent. Some data as to chemical composition and physical character are available for most such strata in the published reports or files of state, provincial, and national geological surveys. Exploration for a new limestone deposit, therefore, in most instances begins with a search of these records to find the locations of deposits that satisfy the various economic factors and proceeds to a sampling program of favorable deposits. Geophysical techniques, if used at all, are used to determine the thickness of overburden. Geochemical techniques are not used.

All aspects of exploration are important, but the one most likely to be slighted is sampling. Yet it determines the validity of further study and may become the basis for hundreds of thousands and sometimes millions of dollars-worth of development work. Its goal must be the accurate representation of the limestone deposit. Coring, rock bitting, and surface (ledge) sampling are the most common sampling methods, and the choice among these depends on such matters as the geology of the deposit, the proposed use of the material, and the availability of equipment.

Special care should be taken when sampling weathered outcrops. In humid regions, the surface layer of a carbonate rock may be leached of calcite and dolomite and hence be less pure than the rest of the unit. On the other hand, in arid and semiarid regions, where evaporation exceeds precipitation for long periods of time, the surface layer may be enriched in calcite and dolomite. Thus if surface samples must be taken under these conditions, the geologist should be aware of a potential bias.

Coring

Coring is generally the best method of exploring a new deposit. A single core may not always be more representative than a section sampled in a quarry or in a natural exposure, where some judgment may be used as to what is representative, but cores taken on a grid pattern constitute a more representative and unbiased sample of a deposit than an equal number of surface sections. Coring avoids contamination by soil and weathered material and disproportionate sampling of different parts of a single unit, yet retains the surface material that may have worked down in solution cavities.

Initial drilling is generally widely spaced both to locate a potential deposit and to determine its potential size. Once an apparently large and suitable deposit is discovered it should be drilled in a more or less regular pattern. The core grid and type of drilling depends largely on the proposed use of the limestone, although other factors, such as the homogeneity of the deposit, geologic complexity, topography, and cost of drilling, are significant. In areas of steeply dipping strata, the grid spacing must take into account whether vertical or inclined drilling will be used. If stone is to be used as cement raw material and the magnesium content of the rocks is believed to be marginal and unpredictable, no greater than 100-ft centers would be required. If the deposit is relatively homogeneous in one direction and not in another, a rectangular rather than a square grid might be used. If the stone is to be used as aggregate and has proved relatively uniform in other deposits, generally only a few cores need to be taken in several hundred acres. We recommend that for most deposits of flat-lying strata in which the chemical composition of the rocks is important, cores be spaced on 100-ft centers until a pattern of uniformity indicates that the spacing can be increased safely. We know of one lime company, quarrying in a structurally complex area of Pennsylvania, that found coring on 50-ft centers was necessary for adequate quality control.

For most limestone exploration a BX core (1⅝ in. in diam.) suffices, but if physical tests are required, a core of larger diameter must be used. NX core (2⅛ in. in diam.) is adequate

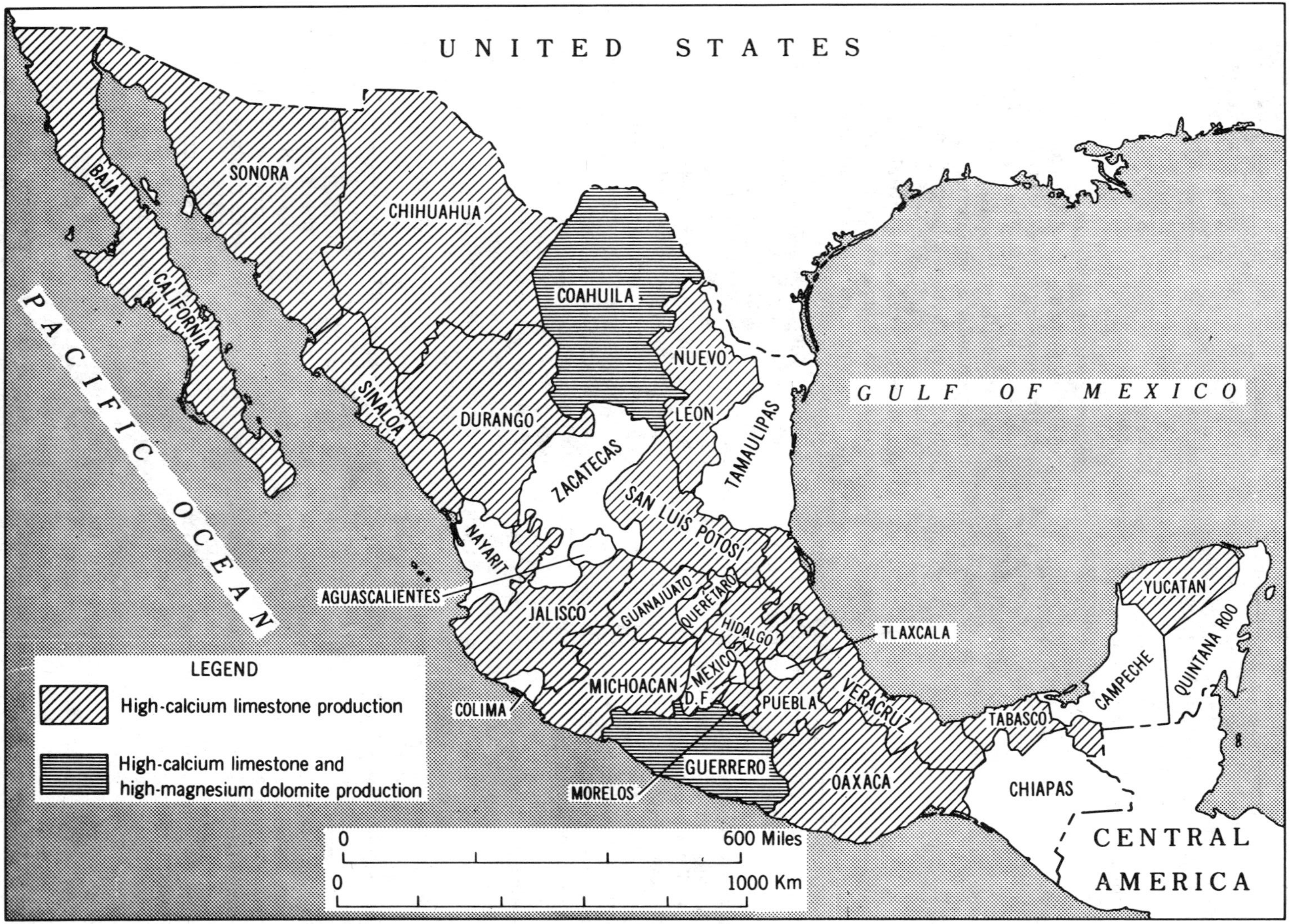

FIG. 3—*Map showing states in Mexico producing high-calcium limestone and high-magnesium dolomite in 1982. Data from the Consejo de Recursos Naturales No Renovables.*

for chemical analyses and limited physical testing, but too small for some physical tests.

Each common physical test of rock used for aggregate requires about 5 kg (11 lb) of rock. If tests for absorption, abrasion, and soundness are performed, more than 30 lb of rock is required. (Some useful data concerning weight and volume of limestone are given in Table 3.) If stone is to be tested for use as highway materials (ASTM D75–59 and AASHO T2–60), a minimum of 50 lb of rock is required. Highway commissions in some states do not perform physical testing of cores and accept only ledge samples. Testing agencies in other states will run physical tests on cores. For example, in prospecting for a quarry site near Skaneateles, NY, the General Crushed Stone Co. took a core 12 in. in diam and had the stone tested for quality by the New York State Dept. of Public Works before opening the quarry (Moore, et al., 1968).

Rock Bitting

If used alone, drill cuttings are probably the least reliable samples in exploration, but if used to supplement cores or information obtained from nearby quarries or outcrops, they are a cheap, rapid method of acquiring much information. If the drilling is done in a carefully cased hole so as to prevent contamination by overburden, if the cuttings are collected carefully, and if the geologist is experienced in interpreting well cuttings, these samples may be as reliable as core and surface sampling for some purposes. In fact, the percussion air drill is probably the fastest and least expensive method for preliminary sampling.

Because of the large amount of drilling for oil and gas during the past few decades, cuttings of thousands of wells are on file in government and other sample libraries. Although not intentionally so, these cuttings constitute the largest exploration program for industrial minerals that has ever been undertaken. Cuttings cannot be interpreted properly unless the interpreter understands drilling techniques, and the best interpretation of the well cuttings requires considerable well-site experience.

Surface Sampling

Chip samples taken carefully on a quarry face can provide a good representation of the limestone deposit. The geologist should first inspect the quarry face and divide the face into units of uniform lithology. He should mark tops of units by paint or flagging. He should subdivide thick homogeneous beds arbitrarily, so that no sample is more than about 5 ft thick. Starting at the base of the unit, he should take chips of uniform size as nearly as possible along a selected vertical line to the top of each unit. (He might start at the top and work down, but it is easier to work up than down.) This method of sampling is sometimes called "channel sampling." Chip samples should be taken from unweathered surfaces even if the weathered rind must be chipped away. Samples should be washed to remove some contaminants, such as lichens or soil, from the samples, but care should be taken not to wash out thin, interbedded shales.

Many channel or core samples weigh 10 to 20 lb, and this amount must be crushed, thoroughly mixed, and quartered several times to reduce the sample to the few ounces needed for chemical determination.

Evaluation and Testing

If given sufficient cores or exposures of rock, an experienced industrial minerals geologist

TABLE 3—Approximate Density of Limestone, Dolomite, Shale, and Cement

Material	Weight Lb, per Cu Ft	Weight Lb, per Cu Yd	Weight Tons, per Acre-Ft
Limestone, broken or crushed	95	2,570	
Limestone solid	160*	4,320	3,500
Dolomite, broken or crushed	105	2,840	
Dolomite solid	170	4,590	3,700
Shale, broken or crushed	100	2,700	
Shale solid	167	4,500	3,600
Portland cement	100	2,700	

*** 1 ft of 2 1/8-in. limestone core weighs about 3.7 lb.**

can evaluate a limestone deposit largely just by looking at it, using a hand lens, hammer, and weak hydrochloric acid as his only tools. He can generally appraise whether a stone will make class A aggregate and (or) cement raw material. He may not be able to determine whether it is of sufficient purity for some other uses, but he can eliminate much stone that is not of sufficient purity for those uses. Final evaluation of rocks suitable for aggregate, dimension stone, and similar uses requires physical testing. Rocks used for making lime, cement, or other products that depend on chemical purity should be chemically analyzed.

A list of laboratories capable of making chemical and physical determinations of rocks can be found in "Directory of Testing Laboratories, Commercial-Institutional," which is available from the American Society for Testing and Materials, 1916 Race St., Philadelphia, PA 19103.

Physical Properties

Procedures for physical testing can be obtained from two main sources: the American Society for Testing & Materials (ASTM) and the American Association of State Highway and Transportation Officials (AASHTO). Both organizations describe explicit procedures, in cookbook fashion, for testing limestone and dolomite. State highway commissions may specify slightly different procedures and should be consulted if stone is to be used as aggregate. In earlier years, several federal organizations developed physical tests for building stone, such as reported by the US Census Office (1884) and the US Bureau of Standards (Kessler and Sligh, 1927), but now building-stone testing generally follows procedure outlines by ASTM. Testing methods and procedures developed by USBM for testing diamond drill cores of mine rock are described in a report by Obert, et al., (1946).

Physical tests are designed to test how well a rock will perform for a particular use. As might be expected, a great number of tests have been developed in response to the many uses to which rock is put. The selected physical properties of carbonate rock shown in Table 4 are from reports of a lengthy study of the physical properties of mine rock by USBM. In addition to the physical properties listed in Table 4, many other properties have been reported in this same study, which include tensile strength, abrasive hardness, scleroscope hardness, Young's modulus, modulus of rigidity, Poisson's ratio, specific damping capacity, and longitudinal bar velocity. A large number of physical properties of limestones and other rocks have been collected and published in *Handbook of Physical Constants* (Birch, et al., 1950). This handbook contains data on bulk density, compressive strength, compressibility at high and low temperature, dielectric constant, electrical resistivity, porosity, thermal conductivity, thermal expansion, and other properties that are especially useful to the geophysicist. Considerable research on the properties of rock that have application to blasting and mining have been undertaken by USBM (Atchison, et al., 1964; Atchison and Pugliese, 1964; Bur, et al., 1969; Dick, et al., 1973; Merrill, 1956; Pugliese, 1972; Thill, et al., 1969; Willard and McWilliams, 1969). A tabulation of porosity and bulk density determinations of carbonate rock, as reported in the more accessible American, British, German, and Swiss literature, was published by Manger (1963).

A great many physical tests of carbonate rock used for aggregate have been performed in connection with state and federal road building programs, but much of this information is unpublished. In 1953, Woolf (1953) compiled results of physical testing of road-building aggregate as performed by the US Bureau of Public Roads up to 1951. Table 5 shows average values of physical properties for selected rock types from this compilation. Unfortunately data from later testing of aggregate in the federal highway program have not been tabulated. The Highway Research Board of the National Research Council has sponsored much research dealing with the relationships of the physical properties of carbonate rock to its serviceability as an aggregate. A report by Renninger and Nichols (1970) reviewed some of this work and gave the current status of aggregate research.

Through the years, the US Army Corps of Engineers has tested carbonate rock from many quarries in the continental United States in order to evaluate its potential use in constructing locks, dams, and other structures. A summary of this testing, which includes the standard tests for sulfate soundness, abrasion resistance, specific gravity, and absorption, can be found in Technical Memorandum No. 6–370, "Test Data—Concrete Aggregates in Continental United States" of the US Army Corps of Engineers, Waterways Experiment Station. This compilation, which occupies more than 2 ft of shelf space, cannot be purchased,

TABLE 4—Physical Properties of Some Carbonate Rocks

Age	Formation	Location	Rock Type	Apparent Specific Gravity	Apparent Porosity, %	Compressive Strength, Psi X 10^3	Modulus of Rupture, Psi X 10^3	Impact Toughness, In. per Sq. In.	Reference
Miocene	Unspecified	Eniwetok	Ls. (fossil.)	2.39	4.0	14.1	2.4	1.0	Blair (1956)
Eocene	Unspecified	Eniwetok	Ls. (dense)	2.53		17.7	3.2	2.9	Blair (1956)
Cretaceous	Niobrara	S. Dakota	Ls. (chalky)	1.81	8.3	3.7	0.6	1.2	Blair (1955)
Mississippian	St. Louis	Missouri	Ls. (oolitic, fossil.)	2.56		16.8	2.3	2.9	Blair (1956)
Mississippian	Maxville	Ohio	Ls. (fn.-grained)	2.41		15.8	2.5	3.1	Blair (1956)
Mississippian	Salem(Spergen)	Indiana	Ls. (fossil.)	2.37	11	10.9	1.6	1.9	Windes (1949)
Devonian	Columbus	Ohio	Ls. (v. fn.-grained)	2.60	5.4	17.9		3.6	Blair (1956)
Devonian	Columbus	Ohio	Ls. (fn.-grained)	2.69	0.7	28.5	2.9	8.6	Windes (1949)
Silurian	Brassfield	Ohio	Ls. (dolomitic)	2.8	1.3	26.	28	4.0	Windes (1950)
Silurian	Niagara	Ohio	Dol. (fn.-grained)	2.4	8.6	13.	11	1.8	Windes (1950)
Silurian	Clinton	Alabama	Ls. (cs.-grained)	2.83	0.9	24.0		6.6	Windes (1949)
Ordovician	Chickamauga	Tennessee	Ls.	2.73	3.4	>25.1	0.8	5.7	Blair (1956)
Ordovician	Lenoir	W. Virginia	Ls. (siliceous)	2.68	6.	23.0	1.9	2.5	Windes (1950)
Ordovician	Knox	Tennessee	Dol. (fn.-grained)	2.84	0.7	46.7	3.8	5.9	Windes (1949)
Cambrian	Bonne Terre	Missouri	Ls. (dolomitic)	2.66	3.3	25.4	1.8	4.8	Blair (1955)
Cambrian	Oro Grande	California	Marble	2.72	0.2	24.0	2.4	1.9	Blair (1955)
Precambrian	Cockeysville	Maryland	Marble (dolomitic)	2.87	0.6	30.8	2.8	2.7	Windes (1949)

TABLE 5—Average Values for the Physical Properties of Rocks Used for Aggregate*

Rock Type	Bulk Specific Gravity	Absorption, %	Loss of Abrasion: Deval Test, %	Loss of Abrasion: Los Angeles Test, %	Toughness†
Basalt	(229) 2.86	(228) 0.5	(203) 3.1	(24) 14	(203) 19
Chert	(74) 2.50	(74) 1.6	(78) 8.5	(6) 26	(29) 12
Diabase	(332) 2.96	(309) 0.3	(340) 2.6	(63) 18	(285) 20
Dolomite	(668) 2.70	(667) 1.1	(708) 5.5	(134) 25	(612) 9
Gneiss	(419) 2.74	(424) 0.3	(602) 5.9	(293) 45	(386) 9
Granite	(662) 2.65	(666) 0.3	(718) 4.3	(174) 38	(703) 9
Limestone	(1695) 2.66	(1673) 0.9	(1677) 5.7	(350) 26	(1315) 8
Marble	(184) 2.63	(162) 0.2	(175) 6.3	(41) 47	(188) 6
Quartzite	(208) 2.69	(204) 0.3	(233) 3.3	(119) 28	(161) 16
Sandstone	(716) 2.54	(707) 1.8	(699) 7.0	(95) 38	(681) 11
Schist	(297) 2.85	(296) 0.4	(314) 5.5	(136) 38	(212) 12

* Number of samples tested in parenthesis.
† Height in cm for a 2 kg steel plunger to cause a rock core about 1 in. in diameter and 1 in. high to fail.
Source: Woolf, 1953.

but it is available for inspection at the Corps of Engineers' libraries. Other more accessible publications that give physical test values and results of the Corps' extensive research relating to the use of carbonate rock as aggregates in portland cement concrete are those by Curry and Buck (1966) and Stowe (1969).

Some manufacturers of pulverized limestone, defined as limestone or dolomite having a minimum fineness of 97% passing a 325 mesh sieve, have established standards and test methods through the Pulverized Limestone Association. Test methods that are recommended include particle size, pH, dry brightness, water demand, and calcium and magnesium carbonate content.

A petrographic examination of carbonate rock can often foretell its physical properties and subsequent suitability for certain uses, but unfortunately too few people make use of this method. The definitive reference for this type of applied petrography is ASTM Standard C 295–65, "Petrographic Examination of Aggregates for Concrete," which is based on the work of Mather and Mather (1950). Techniques of petrographic examination and some applications of this kind of work can be found in Mather (1953), Mielenz (1954, 1955), and Newlon, et al., (1972).

Rippability is a physical property of rock that concerns mainly the construction industry. Limestones that are thinly bedded, low in compressive strength, or sufficiently inhomogeneous so as to contain horizontal planes of weakness can generally be ripped easily. Empirical testing by the Caterpillar Tractor Co. (1972) found a relationship between seismic wave velocities and rock rippability: the higher the wave velocity, the more difficult it is to rip the rock.

Various research workers have examined the relationships between different physical as well as chemical properties of carbonate rocks. These attempts have not met with much success because of the inhomogeneity of the rocks. Further information can be found in Judd and Huber (1961), D'Andrea, et al., (1964, 1965), West and Johnson (1965), Mutmansky and Singh (1968), Baxter and Harvey (1969), and Carr, et al., (1970).

Further information on physical requirements of carbonate rock can be found in the chapters of this book on "Crushed Stone," "Dimension and Cut Stone," and "Cement and Cement Raw Materials."

Chemical Properties

The techniques for making chemical analyses are beyond the scope of this chapter. Instrumental methods of analysis have become popular in recent years, and as a result a wide selection of instructional material is available on theory and techniques. Some of the most useful are those on atomic absorption spectroscopy by Angino and Billings (1967), Galle and Angino (1967), and Slavin (1968); and one on rapid chemical methods by Shapiro and Brannock (1962). Details of standard gravimetric and complexometric (EDTA titration) methods can be found in Hillebrand and Lundell (1953) and Diehl (1964), respectively.

The chemical and physical properties of carbonate rocks are interdependent. Pure calcite in the form of poorly cemented chalk is not only unique in its low strength and high absorption among the carbonate rocks, but it is also highly reactive chemically due to the large surface area of the component grains. Pure calcitic marble of the same chemical composition as the chalk is relatively strong, unabsorptive, and unreactive. Dolomite that contains quartz sand grains may have the same overall composition as dolomite that contains chert, but their suitability as aggregate differs widely due to the difference in reactivity of the two forms of silica. Processing also affects the degree of fracturing of stone and thus the surface area and chemical reactivity. Therefore, physical and mineralogic descriptions of carbonate rocks are of considerable use in predicting the chemical properties of the product from a deposit.

For some uses of carbonate rocks, chemical analysis may be of little or no help in estimating the suitability of a rock unit; for other uses various chemical determinations must be available to compare with specific optimum compositions. For example, in stone used for chemical purposes, such as glass raw material, flux, or cement, the percent of certain elements must fall within specified limits or ranges. On the other hand, the chemical content may or may not be important for stone that is used because of its physical properties, such as aggregate, building stone, or riprap. A thick section of rock that is almost pure dolomite is likely to be well-cemented reefal dolomite, and it can be predicted generally to make good aggregate. Rock that is pure calcite, however, may be a skeletal or oolitic limestone that is either well cemented or poorly cemented and thus might make excellent or only fair aggregate.

The proportion of alumina (Al_2O_3) and silica (SiO_2) in the rock may be helpful in determining the value of a carbonate rock for a use in which physical properties are important. Most of the silica in a carbonate rock is likely to be present as clay, silt- and sand-size quartz, or chert. For the common clay minerals found in limestones and dolomites, as much as 2% silica may be present for each 1% of alumina. The higher the alumina content, the more argillaceous the sample is likely to be. The alumina content can be multiplied by two to obtain an estimate of the amount of silica tied up in the clay. The way the remaining silica occurs can be determined only by reference to the sample description or by examining the sample itself.

The chemical analysis of a carbonate rock is essential for estimating the neutralizing value of agricultural limestone, which is usually expressed in terms of calcium carbonate equivalent (CCE). Pure calcite ($CaCO_3$) is assigned a CCE value of 100. Pure dolomite ($CaCO_3 \cdot MgCO_3$) has a theoretical CCE value of 108.6; that is, it is 8.6% more effective than pure limestone as a neutralizer. Such interpretation assumes that a molecule of $MgCO_3$, with a molecular weight of 84.32, is as effective a neutralizer as 1 molecule of $CaCO_3$, with a molecular weight of 100.09. Thus, a given weight of $MgCO_3$ is 1.19 times as effective as the same weight of $CaCO_3$. Because of differences in solubility, however, a dolomitic liming material will take longer to neutralize a given amount of acid than a limestone one, even though the CCE of the two is the same. A useful guide to selecting agricultural limestone materials has been prepared by Goodwin (1979).

Numerous chemical analyses of carbonate rocks have been published to show the varied compositions of limestones and dolomites (Clarke, 1924; Gillson, et al., 1960; Graf. 1960b; Siegel, 1967; Wedepohl, 1970), but these may not always be typical of the formation from which they were taken. Most sedimentary carbonate rocks vary in their impurities—including clay minerals, resistant minerals such as quartz, and organic material—because they were deposited in different environments. In addition, the rocks have evolved chemically as well as physically during compaction, dehydration, and lithification. Subsequent processes, including burial and exposure to percolation of water, provide for synthesis of authigenic minerals and alteration, such as oxidation of organic matter. Because of the limited extent of identical environments of deposition and subsequent postdepositional conditions, composition of any given rock unit is likely to be variable. Analyses of many samples taken at different sites are required to reveal the approximate composition of a particular unit of rock.

Good sources of chemical data on carbonate rocks are the state geological surveys or their equivalents. Many state surveys have files of chemical data obtained from quarry sampling and coring programs. Many of the references listed in Table 2 contain extensive lists of chemical analyses. The USGS published 1131 analyses of carbonate rocks from Montana, Wyoming, North Dakota, South Dakota, Nebraska, Colorado, and Kansas (Hill, et al., 1972) and 3585 analyses of carbonate rocks from Alaska, Idaho, Oregon, and Washington (Hill and Werner, 1972).

Specifications and Uses

The uses of limestone and dolomite depend largely on physical properties, chemical properties, or both. Physical properties are more important if stone is used "as is," such as for aggregate or building stone. Chemical properties are more important if stone undergoes changes from one form of matter to another, such as the manufacture of cement or lime. Chemical and physical properties are often interrelated; in whiteness, for example, the physical property of color is largely determined by the purity and chemical composition of the rock.

Specifications for carbonate rock should be considered generally as guidelines. For some limestones, minimum and maximum percentages of certain constituents are based on rea physical or chemical characteristics, but the precise figure may have been determined by custom, composition of the limestone being used, or "what the market will bear."

Published specifications may be based on a survey of manufacturers, but some are based on a survey of literature. For example, Bingham's (1916) recommended specifications for limestone used in manufacturing calcium carbide have been repeated often (Table 6). These repeated specifications are based on physical and chemical reactions during the manufacture and use of calcium carbide, but they have more latitude than the figures quoted from Bingham indicate or than Bingham himself suggested in his text. One should not interpret agreement in published specifications as corroboration so much as repetition of other published specifications.

As an example of "what the market will bear," a steel manufacturer in the midwest specified a maximum of 3% $MgCO_3$ and 0.04% sulfur in flux limestone obtained from a contract supplier in Indiana. When the magnesium and sulfur content of the initial material proved to be much lower, the steel company revised its specifications to 1.5% $MgCO_3$ and 0.025% sulfur.

Specifications for a particular limestone may be based on the average composition of the limestone being used regardless of the fact that limestone of a different composition might be cheaper and better. The cost of manufacturing is generally greater than the cost of the raw material, and a change in raw material might prove a costly mistake. Inadequate or unreliable supply could also be costly, and therefore a manufacturer is likely to be loyal to a dependable supplier.

Some specifications are primarily economic rather than physical or chemical. For example, a low silica content in limestone and dolomite used for flux stone is desirable, but the cost is an equally important factor. Generally a silica content lower than 2% is desired, but if only stone of much higher silica content (say, 8%) is available within an economical distance, it can be used. The drawback is that more high-silica stone than low-silica stone is required to flux the same amount of iron ore.

Some specifications may be in terms of tolerances from one shipment to the next rather than specific amounts. This is particularly true in glass manufacture where it is desirable to have uniform composition of raw materials from batch to batch. For example, one large US glass producer specifies that for glass-grade limestone and dolomite, the composition, except for iron content, must remain within specified tolerances (Kephart, 1972). Thus, a limestone containing 1% silica could be used by making adjustments for silica in the other batch components. The Fe_2O_3 content, however, must be less than 0.1% because of its colorant effect on glass.

Virtually hundreds of uses for limestone and dolomite exist, but for simplicity these can be grouped into about 30 uses that include more than 99% of the total crushed and dimension stone production (Table 7). Extensive lists of

TABLE 6—Some Published Specifications of Limestone Used for Manufacturing Calcium Carbide

$CaCO_3$, %	MgO, %	SiO_2, %	$Al_2O_3 + Fe_2O_3$, %	S	P, %	Reference
>97	<0.5	<1.2	<0.5	Tr*	<0.004	Bingham (1916)
>97	<0.5	<1.2	<0.5	Tr	<0.004	Mantell (1931)
>97	<0.5	<1.2	<0.5	Tr	<0.004†	Bowen (1957)
Very high	<0.5	<1.2	<0.5	Tr	<0.004	Gillson, et al. (1960)
>97	<0.5	<1.2	<0.5	Tr	<0.004	McGregor (1963)
Very high	<0.5-2.0	<1.5-3.0	Low	Tr	<0.05	Searle (1935)
>97	<0.5-2.0	<1.0-3.0	<0.05-0.75	Tr	<0.004-0.01	Lamar (1965)

* Tr—trace.

† The figure given in Bowen's publication is 1.2, a typographical error.

TABLE 7—Limestone and Dolomite Shipped or Used by Producers in the United States by Use in 1979

Crushed and Broken

Use	Quantity, 1000 St	Value, $1000
Agricultural limestone	32,902	116,457
Agricultural marl and other soil conditioners	436	2,136
Poultry grit and mineral food	2,442	19,567
Concrete aggregate	113,407	342,599
Bituminous aggregate	66,788	212,707
Macadam aggregate	22,646	59,884
Dense-graded roadbase stone	169,281	422,723
Surface treatment aggregate	38,517	122,515
Other construction aggregate and roadstone	134,839	371,024
Riprap and jetty stone	16,229	43,028
Railroad ballast	14,035	37,635
Filter stone	2,689	7,911
Manufactured fine aggregate (stone sand)	15,609	51,371
Terrazzo and exposed aggregate	624	7,531
Cement manufacture	104,908	223,603
Lime manufacture	34,054	98,042
Dead-burned dolomite	1,779	4,903
Flux stone	21,271	64,945
Refractory stone	20	64
Chemical stone for alkali works	1,966	6,409
Abrasives	141	656
Mine dusting	1,267	10,379
Asphalt filler	1,007	5,425
Whiting or whiting substitute	1,085	29,970
Other fillers or extenders	2,792	31,691
Acid neutralization	W	W
Building products	129	350
Other chemicals	41	152
Bedding materials	—	—
Dam construction	22	44
Drain fields	W	W
Fill	2,127	4,090
Glass manufacture	2,146	14,507
Paper manufacture	128	446
Roofing granules	376	2,307
Sugar refining	1,367	5,968
Waste material	39	76
Sulfur removal from stack gases	967	2,942
Other uses*	3,977	11,033
Total	812,054	2,335,089

Dimension Stone

Use	Quantity St	Quantity 1000 Cu ft	Value, $1000
Rough Stone:			
Rough blocks	207,843	2,672	$6,276
Irregular-shaped stone	5,785	74	172
Rubble	43,939	573	647
Flagging	26,645	357	414
Other rough stone	45	1	1
Dressed stone:			
Cut stone	55,884	754	10,866
Sawed stone	55,860	766	3,932
House stone veneer	64,569	853	3,063
Construction	5,823	69	223
Curbing	311	4	15
Flagging	2,679	35	117
Other uses	280	4	118
Total	469,663	6,161	25,845

Source: US Bureau of Mines *Minerals Yearbook.*

* Includes ferrosilicon, disinfectant and animal sanitation, magnesium metal manufacture, porcelain, and other uses.

uses can be found in Hatmaker (1931), Goudge (1937), Lamar and Willman (1938), Colby (1941), Bowles and Jensen (1941, 1947), Bowles (1956), Johnstone and Johnstone (1961), Lamar (1965), and Siegel (1967). Comstock (1963) listed many uses for magnesium compounds, some of which are derived from dolomite.

Preparation for Markets

Mining Methods and Costs

Open-Pit Mining: Most limestone is mined by open-pit methods, and most limestone producers will choose that method even where underground mining might be equally or slightly more economical.

Mining Methods—Limestone mining as compared to other mining does not require any special equipment. Overburden must be removed, whether by bulldozer, dragline, or shovel. The rock must be drilled and blasted. Oversized rock must be redrilled and blasted, broken by a drop ball, or otherwise disposed of. The broken stone must be loaded into a dump truck and hauled to the primary crusher (or it may be loaded onto a conveyor and carried directly to a portable or permanent crusher). And finally, for most uses of limestone, it must travel through a secondary crusher and be sized. For some uses limestone is ground or heated. For description of mining equipment and processing techniques the reader is referred to such trade publications as *Rock Products* and *Pit and Quarry,* the latter of which is published by a company that also publishes a useful handbook (Herod, 1980–81). For further information on mining and quarrying of stone, the reader may want to consult the proceedings of a symposium organized by the Institution of Mining and Metallurgy (1965) on "Opencast Mining, Quarrying and Alluvial Mining" or a compendium by John Sinclair (1969).

Overburden—Open-pit mining is less costly than underground mining if the overburden can be removed cheaply. We do not know the practical limit in thickness of overburden that can be removed. One large aggregate producer in Indiana has adopted a rule of thumb that overburden can be removed economically if the rock to overburden ratio is greater than three to one. In Indiana, that rule assumes that the overburden is unconsolidated material, such as till or residual clay. Where the overburden is extremely thick, however, this rule apparently does not apply. In some parts of Indiana the market could support a large quarry, but rock is not quarried even though the quarryable section is 300 ft thick and the drift thickness is about 100 ft. Where the drift is so thick, the length of a boom on a dragline, disposal of overburden and water, sloping of drift walls, and necessity to quarry high faces for volume production become limiting factors. On the other hand, in areas where thick limestone is not part of the stratigraphic succession, such as southwestern Indiana, overburden equalling or even exceeding the thickness of the limestone is removed.

At Marquette Cement Manufacturing Co.'s quarry near Superior, Ohio, as much as 90 ft of overburden, part of it bedrock, was removed to quarry an average 6 ft of limestone. The quarry was closed in 1965 (Bates, 1966).

In the 1960s, the Allegheny Minerals Corp. was mining 16 to 18 ft of limestone under 60 ft of overburden in two quarries in Pennsylvania (Herod, 1969). Moreover, the overburden consisted of 10 to 15 ft of clay and soil underlain by 12 ft of highly abrasive sandstone, 10 to 12 ft of soft shale, and 25 ft of hard shale. The hard shale and sandstone were blasted at the same time and the entire overburden stripped. In some areas the limestone is also immediately overlain by 2 ft of "ironstone" that had to be blasted separately and stripped. Even then, some clay had to be removed from solution channels in the limestone. Although the limestone was sold as portland cement raw material, bituminous concrete aggregates, road base, and aglime, clearly its market location and isolation from other sources of low-magnesium limestone must be the major factors in making the operation economical.

In Indiana about 40 or 50 ft is the maximum thickness of glacial drift that is removed consistently in any limestone quarry. In some quarries where the thickness of the drift averages less than 30 ft, more than 50 ft of drift has been removed in small areas. A few feet of rock overburden can be removed economically, especially if it is platy and weathered. Under exceptional circumstances—such as a small area of rock overburden, an exceptionally good market location, or a high-value product like dimension stone—a greater thickness of rock can be removed.

In most instances overburden is used as fill or is wasted, but sometimes it may have economic importance. Quarries that have sand and gravel overburden may use it for aggregate or other purposes (Bates, 1972; Rooney and Carr, 1971). In Indiana, one limestone aggregate quarry sells its shale overburden for cement raw material. In Ohio one cement company permits a brick company to remove and use the clay overburden from its limestone operation.

Underground Mining: The most important fact about underground mining of limestone is that it is done. In fact, 87 underground limestone mines were operating in the United States in 1979. In 1900, total production of limestone from all underground mines was 105,000 tons, or 0.37% of the total limestone production (Thoenen, 1926). In 1924, production was 5.8 million tons, or 4.5% of the total production. In 1965, it was 34.7 million tons (US Bureau of Mines, written communication, 1966), or about 6% of the total limestone production. In 1979, it was 36.1 million tons or about 4% of the total production. Many of the mines are in the midwest (Fig. 4), but their concentration reflects the abundance of high-quality limestone in horizontal beds in that part of the country as much as a concentration of population. We believe that this distribution pattern will change gradually as more mines are built to serve urban centers.

Mining Methods—Most limestone mines consist of horizontal drifts into the sides of hills (sometimes called tunnel mines) or into the walls of open-pit mines, but some consist of vertical or inclined shafts. Few limestone mines are more than 1000 ft deep, and most are only a few hundred feet deep. The deepest limestone mine known to us was the Pittsburgh Plate Glass Co.'s mine near Barberton, Ohio, in which the floor of the mined level was 2250 ft below the surface.

Most limestone mines are of the room-and-pillar type and, like salt mines, are characterized by cathedral-like dimensions. The rooms in the Mississippi Lime Co.'s tunnel mines near Ste. Genevieve, MO, are 90 ft high. The rooms in the Pittsburgh Plate Glass Co.'s mine near Barberton, Ohio, were 46 ft high. The inclined shaft that gives access to the Thomasville Stone and Lime Co.'s quarry near Thomasville, PA, is so large that a transcontinental bus can be driven down it.

Underground mining is generally more expensive than open-pit mining, but no single cost ratio applies. It is misleading to compare the relative cost of underground mining of one deposit in one part of the country with open-pit mining of another deposit in another part of the country. Some open-pit mining might prove more expensive than some underground mining. Perhaps the best comparison is of costs of simultaneous underground and surface mining in the same formation at the same general location. We know of no published costs of this type, but one operator reported that stone mined by the room-and-pillar method, at about 400-ft depth, in a 40-ft face, and hauled out by truck via an inclined shaft cost $1.05 per ton to produce, 57% more than the $0.67 per ton for the same company to produce stone in its adjoining quarry in 1969. Conversations with other mine superintendents have led us to believe that these costs were fairly representative.

Pros and Cons of Underground Mining—Assuming then that limestone costs about 50% more to produce underground, why is it economical to do so, even in conjunction with open-pit operations? Some of the reasons are:

1) Lack of surface deposits near the market. This is probably the most important reason. If no deposits of limestone suitable for a specific purpose are found at the surface, the additional cost of underground mining must be weighed against the cost of transporting stone into the area.

In some places the lack of surface deposits may be caused by urbanization of the area near a quarry, and thus the operator is prevented from acquiring new land or even from expanding the open pit on land already owned.

2) Unfavorable geologic structure. In some areas limestone may be found at the surface near markets, yet not be suitable for quarrying. Beds that dip steeply but are not vertical can be quarried, but the development of the quarry in two directions is limited to the thickness of the beds. In such places open-pit mines must become shaft mines so that the deposit can be worked downdip.

3) Efficient utilization of reserves. Many tunnel mines are simply the extension of surface quarries into adjoining hills where the overburden is too thick to remove. Thus, when the rock that can be quarried has been exhausted, the plant need not be moved to a new location. In some places tunneling and quarrying are carried on side by side to extend the life of the quarry or for the reasons given under 5 or 6 below.

4) Lowered cost by mining more than one commodity. In some places commodities other than limestone might be mined at different levels but serviced by a single shaft. Such mining would appreciably lower costs. In northern Indiana, both limestone suitable for cement raw material or aggregate and gypsum

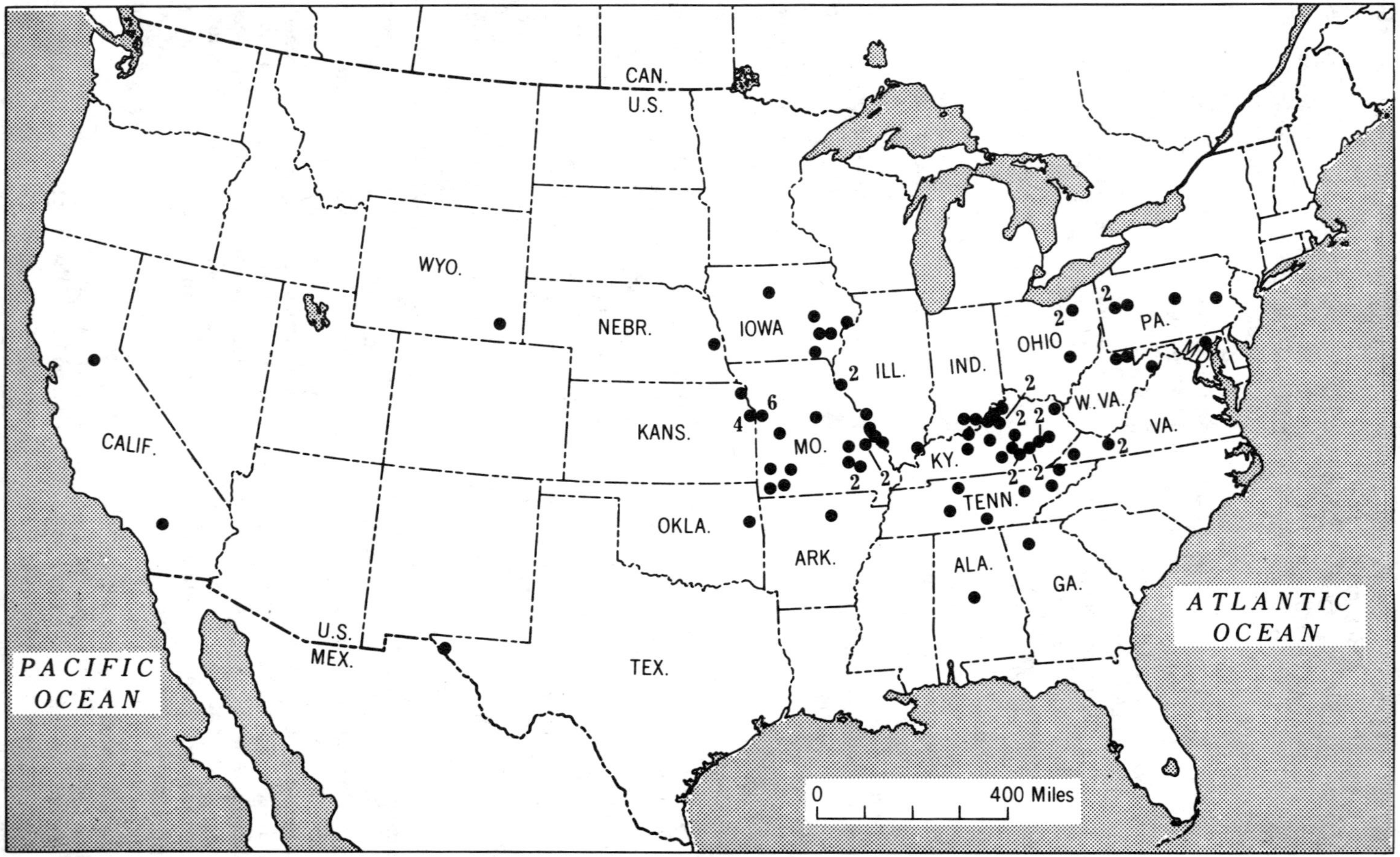

FIG. 4—*Map showing locations of active underground limestone mines in the United States. The figure beside a location symbol indicates the number of mines represented by the symbol. Information supplied by the US Bureau of Mines in 1972.*

could be produced by mining at different levels (Rooney and Ault, 1970).

5) Selective mining. Underground mining permits the selective removal of particular beds without removal of, or contamination by, the overlying materials or clay seams. Room-and-pillar mining of low-dipping beds also permits some horizontal selection of materials. Many underground operations in the United States selectively mine premium-grade material for chemical flux stone, portland cement raw material, or other purposes. The mines near Barberton, Ohio (now abandoned), Carntown, KY, and Riverside, CA are examples.

6) All-weather mining and uniform production. Underground mining is generally not affected by inclement weather. The operator of a shallow underground mine in Indiana reported that the temperature in the mine remains about 55° F summer and winter. The superintendent of a midwestern mine 350 ft deep reported a constant temperature of 59° F away from the air shafts. The superintendent of a mine 600 ft deep reported a temperature of 65° F. Deeper mines might be somewhat warmer, but few limestone mines are deep enough for the temperature to be above 70° F. The rock temperature in the Barberton mine, about 2250 ft deep, was 85° F, but the air was cooled to 80° by forced-air ventilation. A rule of thumb for increase in temperature with depth is 1° F for each 60 ft.

Some economies are thus possible through year-around operation. The workers have stable employment, stockpiles do not need to be large, rock does not freeze, and responses to unexpected demands of material are possible at any time. Some of the advantages of year-around operation, however, are lost in production of construction and agricultural materials because the maximum use of these materials is seasonal.

(7) Environmental problems. Underground mining is less conspicuous than open-pit mining and consumes less land. Most dust and noise are confined below ground, neighbors are not endangered by thrown rocks, and the choice of mining site depends less on surface topography and cultural features. If the roof is competent and sufficient limestone is left as pillars, say 30%, collapse of the roof causing subsidence at ground level should pose little threat. Underground mining, therefore, is favored where the surface is expensive or pre-empted, as near cities.

(8) Underground storage and waste disposal. Some operators have reported that the space created by mining can prove more profitable than the rock removed. Under some conditions the open space can be used for waste disposal or storage of gas or liquids. Mines above the water table with large rooms, solid roof, and level floor can be used for warehouses, frozen-food storage facilities, factories, office space, and fallout shelters (Dean, et al., 1969; Herod, 1972; Jewett, et al., 1967; Stearn, 1965). An abandoned mine in Indiana was even used commercially for raising poultry and one in Missouri is now used for a marina (Vineyard, 1969). Such facilities are less expensive than surface structures to build, and a constant temperature is easily maintained. Limestone mines are generally large enough to permit truck traffic and even railway spurs.

Inasmuch as most limestone is mined by the open-pit method, we feel that the disadvantages of underground mining compared with open-pit mining are well known and need to be cited only briefly:

1) A shaft mine requires a large initial investment before the first ton of rock is produced.

2) The size of equipment is controlled by the dimensions of the entrance and rooms, and less rock can be shot down at one time. Thus, an underground mine is not so adaptable to large-volume production.

3) About 30% of the reserves are left in the ground.

4) Equipment cannot be moved so easily in and out of the mine.

5) Water is more likely to be a threat, and its removal more costly. In areas of karst topography and subsurface drainage, rain storms may cause sudden flooding of the mine.

6) Safety precautions must be greater. Roof fall is a major threat.

7) Fresh air must be circulated through shaft mines.

8) State and federal regulations are more stringent for underground mines than for open-pit mines.

Only a few of the more than a hundred articles written on the underground mining of limestone are included in our bibliography. The US Bureau of Mines report on the underground mining of limestone (Thoenen, 1926) is out of date and has never been revised, but it probably remains the best single reference on this specific subject. More recently Loofbourow (1966) has written on underground mining of industrial minerals in general. The handbook by Lamar (1967) compiled for quarry operators gives a good account of the physical characteristics of limestone and dolomite deposits and their importance to open-pit mining.

Transportation

If a number of independent characteristics are essential to the value of a product, it is difficult to say which characteristic is most important. For most purposes to which limestone is put, however, location with regard to market is probably the prime variable. Cost of mining, except for thickness of overburden (discussed previously), and cost of exploration are much the same within any given region. Costs imposed by zoning restrictions are so arbitrary that it is not possible to generalize on them. Quality of stone is important; in fact, it is essential. (Dolomitic siltstone will not be used as concrete aggregate or the major cement raw material no matter how close to the market it may be.) But limestone and dolomite are so plentiful and their transportation relatively so costly that an inferior stone locally available is likely to be used rather than a superior stone shipped in by rail or barge.

Of course use is an important factor in determining how far limestone can be shipped economically. Ground white limestone used as a filler, say in putty, could probably be transported thousands of miles without adding appreciably to the cost of the finished product. High-calcium limestone used as a flux in open-hearth steel furnaces can bear transport more than 100 miles if none is available near the steel mill. Cement and lime plants are constructed generally, but not necessarily, at the raw material site.

Aggregate cannot be transported long distances economically, except under most unusual circumstances. Aggregate is carried as ballast in some ships and thus might travel thousands of miles. Limestone quarried on the fringe of a market area without limestone might bear long transportation, especially by water or rail. Limestone quarried on a waterway can be transported past producing areas that are not on a waterway but that are much closer to the market. Most limestone used as aggregate, however, is transported short distances and by truck.

In 1979, more than 80% of the crushed stone produced in the United States was transported by truck. The amount shipped by the next most important methods of transportation, rail and waterway, was about equally divided (Table 8). Trucks are generally used for transporting distances less than 50 miles, whereas rail and water are used for longer distances.

TABLE 8—Crushed Stone Shipped or Used in the United States in 1979, by Methods of Transportation

Method of Transportation	Thousand St	% of Total
Truck	900,707	82
Rail	86,201	8
Waterway	62,818	6
Other	47,381	4
Total	1,097,107	100

Source: US Bureau of Mines *Minerals Yearbook.*

Because so many factors influence the cost of transporting stone, it is difficult to determine average costs. A study of transportation of aggregate in Indiana was made by French (1969), and his generalizations probably hold true for the rest of the United States. He found that once trucks had replaced horse-drawn wagons and most rail gondola cars, most aggregate was hauled in trucks owned by the producing companies, later in trucks owned by the drivers, and now in trucks owned mostly by large contract haulers. Because these contract haulers are required in Indiana to file their rates with the Public Service Commission, he was able to estimate trucking costs, although actual rates may depart from the published rates. French found that one hauler charged ". . . approximately 5¢ per ton-mile for the first 6 miles, 3¢ per ton-mile for the next 18 miles, and 2½¢ per ton-mile thereafter." He also found that a haul of about 30 miles in rural areas would about double the cost of crushed rock, and that a haul of only about 9 or 10 miles might be required to double the cost in urban areas.

O'Donnell and Sliger (1972) conducted a more regional study than did French and obtained information from various limestone producers. They found that truck rates ranged from 5 to 10¢ per ton-mile for 10-mile shipments and 2 to 5¢ per ton-mile for hauls of 100 miles. In recent years, transportation costs have risen rapidly because of fuel prices. These costs vary so much, month by month and area by area, that average figures have little value. Nevertheless as a point of reference, in 1980 a trucking firm in a rural area of southern Indiana reported a price to ship aggregate of 15¢ per ton-mile for the first 10 miles and 5¢ per ton-mile for each additional mile. Costs in urban areas are generally higher.

Cost of transportation by rail is more difficult to determine. Because of the cost of loading and unloading, most stone shipped by rail is shipped long distances, and the rail cars are likely to be switched from one line to another, a practice which results in additional costs. But if a quarry is on a railway line and stone is shipped directly to the market on that one line, the only additional cost is the rehandling at the market end. In addition, if a large amount of stone is shipped, say on a unit train, the cost can be low. If the stone is shipped in cars owned by the aggregate company, the cost may be even lower. French (1969) found that the average charge for rail transport of aggregate in southern Indiana was 0.61¢ per ton-mile with a 99¢-per-ton minimum, but in northern Indiana the charge was an average 0.92¢ per ton-mile with a $1.84-per-ton minimum. These rates are similar to those rates published by the Interstate Commerce Commission in its 1% sampling of terminations in 1966 (Fig. 5).

Water transportation of crushed stone is the cheapest method, although for distances less than 300 miles railroads may be competitive (Fig. 5). Barging, which in 1965 accounted for 7.2% of the nation's freight movement exclusive of the Great Lakes (Fulkerson, 1969) is expected to increase in the future. Limestone shipments on the Great Lakes, which in 1966 ranked in tonnage only behind iron ore and bituminous coal, are expected to about double by 1995 (Aase, 1970). Fulkerson (1969) reported that the average revenue per ton-mile for carriers on the Mississippi River and its tributaries was 0.328¢. French (1969) reported one rate for transporting aggregate on the Ohio River to be 0.4 to 0.65¢ per ton-mile for minimum barge loads of 1200 tons (plus 50¢ to $1 per ton for loading and $1 to $2 per ton for unloading).

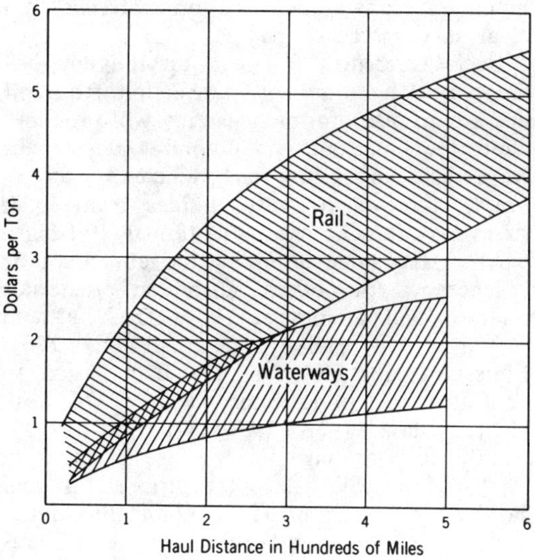

FIG. 5—*Graph showing range in rail and waterway transportation rates for shipments of crushed stone in the United States. Rail data from the Interstate Commerce Commission,* Carload Waybill Statistics, *1966 (1968). Waterway data from O'Donnell and Sliger (1972).*

US Production

The best source for statistics on limestone and dolomite production is the US Bureau of Mines *Minerals Yearbook,* which is published annually. The *Yearbook's* chapter on "Stone" gives national and state production figures for limestone, dolomite, marble, shell, and calcareous marl, as well as the amount and value of limestone and dolomite produced for different uses. The chapters on the mineral industry of the various states give data on

TABLE 9—Crushed and Broken Limestone and Dolomite Shipped or Used by Producers in the States That Account for About 90% of the Total United States Production in 1979

State	Quantity, 1000 St
1. Texas	70,661
2. Florida	63,609
3. Illinois	63,551
4. Pennsylvania	56,122
5. Missouri	54,246
6. Ohio	49,703
7. Tennessee	45,714
8. Michigan	39,721
9. Kentucky	39,298
10. Indiana	34,121
11. New York	32,578
12. Iowa	32,471
13. Oklahoma	27,649
14. Alabama	24,597
15. Virginia	22,689
16. Wisconsin	20,625
17. California	19,156
18. Kansas	18,853
19. Maryland	13,889
20. West Virginia	10,684
Total	739,937
Others	72,117
Total US	812,054

Source: US Bureau of Mines *Minerals Yearbook.*

production and uses of the rock in those individual states.

In 1979, twenty states accounted for about 90% of the total United States production of crushed limestone and dolomite (Table 9); seven states accounted for about 90% of the total United States production of dimension stone (Table 10); and one state, Louisiana, accounted for 71% of the total United States production of shell (Singleton, 1980). Nine states produced most of the calcareous marl (Table 11).

Between 1950 and 1979, the total production of carbonate rock in the United States increased a total of 357% (Table 12). In 1979, the tonnage of carbonate rock produced exceeded that of any other single mineral commodity except sand and gravel.

TABLE 11—Calcareous Marl Shipped or Used by Producers in the United States in 1979, by States

State	Quantity, 1000 St	Value $ 1000
1. North Carolina	260	957
2. Michigan	23	50
3. Indiana	13	19
Other States Florida, Maine, Mississippi, South Carolina, Texas, Virginia	2,354	3,507
Total	2,650	4,533

Source: US Bureau of Mines *Minerals Yearbook.*

Future Trends

According to the US Bureau of Mines (Singleton, 1980) the demand for dimension stone in the year 2000 will probably be between 1,000,000 and 2,500,000 st corresponding to an average annual growth rate of 1.4%; the demand for crushed stone in the year 2000 will probably be between 1,200,000,000 and 2,400,000,000 st corresponding to an average annual growth rate of 2.2%. The actual demand for limestone and dolomite—the nation's chief raw materials for dimension stone and crushed stone—could differ greatly from the predicted demand because of changes in population growth and in the gross national product.

These predictions take into consideration long-range economic forecasts and various contingencies such as technologic changes, competition from alternate raw materials, increasing labor costs, and new construction methods. But who can say how great these changes will be? As the frame house disappears from the construction scene, will concrete dwellings as well as mobile homes largely replace it? Will concrete poles replace the millions of wooden utility poles across the nation? As buildings and streets are demolished in the cities, will the aggregate be recycled? How will mass transit systems affect the demand for highway construction materials? No clear answers can be given.

Social pressures to clean up the environment and the imminent conversion from oil and gas to other forms of energy will probably affect the limestone and dolomite industry in ways that cannot now be predicted. For example, removal of sulfur oxides from stack gases may demand enormous tonnages of high-calcium limestone as either the absorbant or regenerant for other absorbant materials (Harvey and Steinmetz, 1971; Drehmel and Harvey, 1974; Harvey, et al., 1974, 1976). Limestone and dolomite are being used to neutralize acid mine waters (Ford, 1974; Hill, 1974; Mihok, 1968, 1970; Wilmoth, 1973), but the total tonnage for this use is relatively small. If the CO_2 Acceptor process for coal gasification should prove economically sound, it would create a new use for dolomite (Goodridge, 1973). Perhaps far in the future limestone may be used as raw material to produce hydrocarbon fuels and protein food supplements (Salotti, et al., 1970).

Whatever the new uses of limestone and dolomite, high purity deposits must be conserved through improved mineral extraction

TABLE 10—Dimension Limestone and Dolomite Shipped or Used by Producers in the States That Account for About 90% of the Total United States Production in 1979

State	Quantity, St
1. Indiana	337,502
2. Wisconsin	47,757
3. Iowa	10,197
4. Minnesota	9,832
5. Alabama	7,880
6. California	5,443
7. Texas	5,201
Total	423,812
Others	45,851
Total US	469,663

Source: US Bureau of Mines *Minerals Yearbook.*

TABLE 12—Carbonate Rock Shipped or Used by Producers in the United States, 1950 to 1979

Year	Crushed and Broken Limestone and Dolomite, St	Crushed and Broken Marble, St	Crushed and Broken Shell, St	Crushed and Broken Marl, St	Dimension Limestone and Dolomite, St	Dimension Marble, St	Total, St
1950	180,111,000	178,000	*	*	807,590	89,290	181,185,880
1951	204,673,000	169,000	*	*	806,842	87,191	205,736,033
1952	216,469,000	149,000	*	*	786,757	89,051	217,493,808
1953	223,915,000	378,000	*	*	798,911	76,255	225,168,166
1954	323,800,000	454,000	11,428,000	*	1,174,389	84,626	336,941,015
1955	362,196,000	976,000	15,130,000	*	1,182,459	116,029	379,600,488
1956	379,342,000	842,000	19,852,000	*	1,028,759	105,431	401,170,190
1957	384,295,000	1,274,000	18,510,000	1,917,000	1,131,376	149,000	407,276,376
1958	390,468,000	1,269,000	18,916,000	1,803,000	979,000	136,000	413,571,000
1959	433,003,000	1,758,000	20,180,000	2,043,000	952,000	137,000	458,073,000
1960	450,393,000	1,515,000	18,934,000	1,283,000	860,000	129,000	473,114,000
1961	437,398,000	1,435,000	20,973,000	1,099,000	855,000	157,000	461,917,000
1962	460,953,000	1,623,000	20,054,000	1,182,000	896,000	146,000	484,854,000
1963	488,348,000	1,752,000	19,019,000	1,164,000	895,000	150,000	511,328,000
1964	510,247,000	1,963,000	19,493,000	1,043,000	779,000	130,000	533,655,000
1965	554,034,000	2,046,000	21,560,000	1,291,000	732,000	126,000	579,789,000
1966	568,849,000	2,161,000	21,662,000	1,358,000	728,000	83,000	594,841,000
1967	568,902,000	2,158,000	22,026,000	1,227,000	561,000	74,000	594,948,000
1968	603,136,000	2,470,000	20,268,000	1,211,000	602,000	89,000	627,776,000
1969	628,362,000	2,271,000	19,731,000	2,490,000	574,000	77,000	653,505,000
1970	625,313,000	1,722,000	21,713,000	1,739,000	482,000	63,000	651,032,000
1971	628,035,000	1,641,000	18,537,000	3,459,000	468,000	75,000	652,215,000
1972	671,496,000	2,247,000	16,610,000	2,650,000	411.000	71,000	693,485,000
1973	774,397,000	2,023,000	19,896,000	2,327,000	370,000	48,000	799,061,000
1974	751,515,000	1,679,000	18,235,000	3,889,000	517,000	73,000	775,908,000
1975	664,820,000	1,461,000	15,453,000	3,504,000	420,000	50,000	685,708,000
1976	662,880,000	1,456,000	13,753,000	3,371,000	414,000	56,000	681,930,000
1977	706,521,000	1,540,000	13,492,000	2,517,000	443,000	97,000	724,610,000
1978	785,734,000	1,417,000	12,436,000	2,585,000	370,000	115,000	802,657,000
1979	812,054,000	1,461,000	12,177,000	2,650,000	470,000	80,000	828,892,000

Source: US Bureau of Mines *Minerals Yearbook.*

*Included with crushed and broken limestone and dolomite.

and processing techniques. Selective quarrying of specific lithologies is not new but neither is it widely practiced. The custom of "drill the blasthole to the bottom and shoot" should be replaced by benching based on composition of different strata. The use of selective recovery, such as described by Ames (1966) for portland cement raw materials, applies equally well to limestones for other uses and may allow a deposit of marginal quality to be utilized. High-purity deposits are uncommon and should not be wasted for uses not demanding such material. For example, impure limestone with a low-magnesium content can be used to manufacture portland cement. Thus some current practice of using high-calcium limestone and then adding clay, sand, and other materials to it in processing is comparable to flaring gas in an oil field. It may be desirable from the standpoint of present-day technology, but it is a waste of a valuable resource.

If future demands for limestone and dolomite are to be met at somewhat near the low costs that have prevailed, people must compromise their life styles. The desirability of sprawling urban areas and mini-farms, which cover potentially valuable deposits, will have to be placed in perspective against the need for building and construction materials. Underground mining must gain favor as a means of providing minerals while still allowing the land surface to be used for man's activities (Baxter, 1980; Carr and Ault, 1980). City and county planning must take into account the nature of mineral resources: economic deposits are rare, they are essential, and they should be produced before the land is urbanized.

Bibliography and References

1. Aase, J.H., 1970, "Transportation of Iron Ore, Limestone, and Bituminous Coal on the Great Lakes Waterway System, With Projections to 1995," Information Circular 8461, US Bureau of Mines, 61 pp.
2. Adair, R.B., et al., 1947, "Evaluation of North Carolina Raw Materials for the Production of Portland Cement, Part 1, Preliminary Laboratory Investigation," Bulletin 35, North Carolina State College, Dept. Eng. Research, 31 pp.
3. Alabama Geological Survey, see Publications List of County Reports and Maps.
4. Allen, H.W., 1953, "Progress Report of Limestone Survey, Knox County," Report of the State Geologist for 1951–1952, Maine Development Commission, pp. 11–30.
5. Allen, J.E., 1946, "Reconnaissance Geology of Limestone Deposits in the Williamette Valley, Oregon," Short Paper 15, Oregon Dept. of Geology and Mineral Industries, 15 pp.
6. Ames, J.A., 1966, "Limestone Deposits Vs. Beneficiation," *Ohio Journal of Science,* Vol. 66, No. 2, pp. 131–136.
7. Amsden, T.W., 1954, "Geology of Garrett County," *Geology and Water Resources of Garrett County,* Bulletin 13, Maryland Dept. of Geology, Mines and Water Resources, pp. 1–95.
8. Amsden, T.W., and Rowland, T.L., 1965, "Silurian Stratigraphy of Northeastern Oklahoma," Bulletin 105, Oklahoma Geological Survey, 174 pp.
9. Anderson, K.H., et al., 1979, "Geologic Map of Missouri," Scale: 1 in. = 8 miles, Missouri Geological Survey and Water Resources.
10. Anderson, S.B., and Haraldson, H.C., 1969, "Cement Rock Possibilities in Paleozoic Rocks of Eastern North Dakota," Report of Investigation 48, North Dakota Geological Survey, 62 pp.
11. Angino, E.E., and Billings, G.K., 1967, *Atomic Absorption Spectrometry in Geology,* Elsevier, New York, 144 pp.
12. Apfel, E.T., 1944, "Dolomite Marble in the Vicinity of Lee, Massachusetts, as an Available Source of Metallic Magnesium," Open File Report, US Geological Survey, 34 pp.
13. Argall, G.O., Jr., 1949, "Dolomite," in "Industrial Minerals of Colorado," *Colorado School Mines Quarterly,* Vol. 44, No. 2, pp. 144–149.
14. Argall, G.O., Jr., 1949, "Limestone," in "Industrial Minerals of Colorado," *Colorado School of Mines Quarterly,* Vol. 44, No. 2, pp. 254–274.
15. Atchison, T.C., et al., 1964, "Effect of Decoupling on Explosion-Generated Strain Pulses in Rock," Report of Investigations 6333, US Bureau of Mines, 49 pp.
16. Atchison, T.C., and Pugliese, J.M., 1964, "Comparative Studies of Explosives in Limestone," Report of Investigations 6395, US Bureau of Mines, 25 pp.
17. Azároff, L.V., and Buerger, M.J., 1958, *The Powder Method in X-Ray Crystallography,* McGraw-Hill, New York, 342 pp.
18. Baars, D.L., 1963, "Petrology of Carbonate Rocks," *Shelf Carbonates of the Paradox Basin,* Four Corners Geological Society 4th Field Conference, pp. 101–129.
19. Baird, G.C., 1979, "Sedimentary Relationships of Portland Point and Associated Middle Devonian Rocks in Central and Western New York," New York State Museum & Science Service Bulletin 433, 24 pp.
20. Bannatyne, B.B., 1975, "High-Calcium Limestone Deposits of Manitoba," Man. Mines Br., Pub. 75–1, Manitoba Mines Branch, 103 pp.
21. Barnes, V.E., and Bell, W.C., 1977, "The Moore Hollow Group of Central Texas," Report of Investigations 88, Bureau of Economic Geology, University of Texas, Austin, 169 pp.
22. Bartlett, C.S., Jr., and Biggs, T.H., 1980,

"Geology of the Abingdon, Wyndale, Holston Valley and Shady Valley Quadrangles, Virginia," Publication 16, Virginia Division of Mineral Resources, 39 pp.

23. Bassler, R.S., 1905, "Cement Materials of the Valley of Virginia," Bulletin 260, US Geological Survey, pp. 531–544.
24. Bates, R.L., 1966, "Geology of Cement Raw Materials in Ohio: A Summary," *Proceedings of 2nd Forum on Geology Industrial Minerals,* Indiana Univ., Indiana Geological Survey, pp. 23–29.
25. Bates, R.L., 1972, "Limestone Quarry in a Gravel Pit," *Proceedings Eighth Forum on Geology of Industrial Minerals,* Pub. Info. Series 5, Iowa Geological Survey, pp. 5–10.
26. Baxter, J.W., 1980, "Factors Favoring Expanded Underground Mining of Limestone in Illinois," *Mining Engineering*, Vol. 32, No. 4, pp. 1497–1504.
27. Baxter, J.W., and Harvey, R.D., 1969, "Alumina Content of Carbonate Rocks as an Index to Sodium Sulfate Soundness," Industrial Minerals Notes 39, Illinois Geological Survey, pp. 5–10.
28. Beach, J.O., and English, S.G., 1940, "Dolomite and Magnesium Limestone," Mineral Report 6, Oklahoma Geological Survey, 19 pp.
29. Beatly, K.O., and Adair, R.B., 1948, "Evaluation of North Carolina Raw Materials for the Production of Portland Cement, Part 2, Small-Scale Production of Portland Cement," Bulletin 42, North Carolina State College Engineering School, 51 pp.
30. Bellis, W.H., and Rowland, T.L., 1976, "Shale and Carbonate-Rock Resources of Osage County, Oklahoma," Circular 76, Oklahoma Geological Survey, 50 pp.
31. Bennett, W.A.G., 1944, "Dolomite Resources of Washington, Part 1," Report of Investigations 13, Washington Div. of Mines and Geology, 35 pp.
32. Berry, E.W., 1947, "Marls and Limestones of Eastern North Carolina," Bulletin 54, North Carolina Div. of Mineral Resources, 16 pp.
33. Bicker, A.R., Jr., 1970, "Economic Minerals of Mississippi," Bulletin 112, Mississippi Geological Survey, 80 pp.
34. Bicker, A.R., Jr., and May, J.H., 1976, "Agricultural Lime in Central Mississippi," Information Series MGS–77–1, Mississippi Geological Survey, 55 pp.
35. Bingham, C., 1916, *The Manufacture of Carbide of Calcium,* Raggett & Co., London, 219 pp.
36. Birch, F., et al., 1942 (reprinted 1950), "Handbook of Physical Constants," Special Paper 36, Geological Society of America, 325 pp.
37. Bissell, H.J., and Chilingar, G.V., 1967, "Classification of Sedimentary Carbonate Rocks," *Developments in Sedimentology, Carbonate Rocks,* Vol. 9A, G.V. Chilingar, H.J. Bissell, R.W. Fairbridge, eds., Elsevier, New York, pp. 87–168.
38. Blair, B.E., 1955, "Physical Properties of Mine Rock, Part III," Report of Investigations 5130, US Bureau of Mines, 69 pp.
39. Blair, B.E., 1956, "Physical Properties of Mine Rock, Part IV," Report of Investigations 5244, US Bureau of Mines, 69 pp.
40. Bowen, O.E., Jr., 1957, "Limestone, Dolomite, and Lime Products," *Mineral Commodities of California,* Bulletin 176, California Div. of Mines and Geology, pp. 293–306.
41. Bowen, O.E., Jr., 1966, "Limestone, Dolomite, and Lime Products," *Mineral Resources of California,* Bulletin 191, California Div. of Mines and Geology, pp. 221–233.
42. Bowen, O.E., et al., 1973, "The Mineral Economics of the Carbonate Rocks," *Limestone and Dolomite Resources of California,* O.E. Bowen, ed., Bulletin 194, California Div. of Mines and Geology, 60 pp.
43. Bowles, O., 1956, "Limestone and Dolomite," Information Circular 7738, US Bureau of Mines, 29 pp.
44. Bowles, O., and Jensen, M.S., 1941, "Limestone and Dolomite in the Chemical Processing Industries," Information Circular 7169, US Bureau of Mines, 15 pp.
45. Bowles, O., and Jensen, M.S., 1947, "Industrial Uses of Limestone and Dolomite," Information Circular 7402, US Bureau of Mines, 19 pp.
46. Branner, G.C., 1941, "Limestones of Northern Arkansas," Miscellaneous Publication, Arkansas Geological Commission, 24 pp.
47. Brantley, J.E., 1916, "Limestones and Marls of the Coastal Plain of Georgia," Bulletin 21, Georgia Geological Survey, 300 pp.
48. Buckley, E.R., 1898, "On the Building and Ornamental Stones of Wisconsin," Bulletin 4, Wisconsin Geological Survey, 544 pp.
49. Buckley, E.R., and Buehler, H.A., 1904, "The Quarrying Industry of Missouri," Missouri Geological Survey, Vol. 2, Series 2, 371 pp.
50. Buehler, H.A., 1907, "The Lime and Cement Resources of Missouri," Missouri Geological Survey, Vol. 6, Series 2, 255 pp.
51. Bullock, K.C., 1966, "Minerals of Utah," Bulletin 76, Utah Geological and Mineral Survey, 237 pp.
52. Bur, T.R., et al., 1969, "An Ultrasonic Method for Determining The Elastic Symmetry of Materials," Report of Investigations 7333, US Bureau of Mines, 23 pp.
53. Burchard, E.F., and Chapin, T., 1920, "Marble Resources of Southeastern Alaska," Bulletin 682, US Geological Survey, 118 pp.
54. Burchett, R.R., 1980, "Mineral Resource Map of Nebraska," Scale, 1:1,000,000, Nebraska Geological Survey.
55. Burr, A.C., et al., 1950, "Raw Materials for the Manufacture of Portland Cement in North Dakota," *1949 Proceedings,* North Dakota Academy of Science, Vol. 3, pp. 15–18.
56. Butler, J., and Bartlette, G., 1969, "Bibliography of the Geology of Newfoundland

and Labrador, 1814 through 1968," Bulletin 38, Newfoundland and Labrador Mineral Resources Div., 273 pp.

57. Calver, J.L., 1957, "Mining and Mineral Resources," Bulletin 39, Florida Geological Survey, 132 pp.

58. Carlson, C.G., 1964, "The Niobrara Formation of Eastern North Dakota; Its Possibilities for Use as a Cement Rock," Report of Investigation 41, North Dakota Geological Survey, 56 pp.

59. Carozzi, A.V., 1960, *Microscopic Sedimentary Petrography,* John Wiley, New York, 485 pp.

60. Carr, D.D., 1973, "Geometry and Origin of Oolite Bodies in the Ste. Genevieve Limestone (Mississippian) in the Illinois Basin," Bulletin 48, Indiana Geological Survey, 81 pp.

61. Carr, D.D., and Ault, C.H., 1980, "Potential for Deep Underground Limestone Mining in Indiana," Preprint 80–334, SME-AIME Fall Meeting, Minneapolis, Oct., 11 pp.

62. Carr, D.D., et al., 1970, "Relationships Between Physical and Chemical Properties of the Brassfield Limestone (Silurian) in Indiana, Ohio, and Kentucky," *Proceedings, 6th Forum on Geology of Industrial Minerals,* Miscellany 1, Michigan Geological Survey, pp. 127–137.

63. Carr, D.D., et al., 1971, "Crushed Stone Aggregate Resources of Indiana," Bulletin 42-H, Indiana Geological Survey, 38 pp.

64. Carr, D.D., et al., 1978, "Crushed Stone Resources of the Blue River Group (Mississippian) in Indiana," Bulletin 52, Indiana Geological Survey, 225 pp.

65. Caterpillar Tractor Co., 1972, "Handbook of Ripping, A Guide to Greater Profits," 4th ed., Caterpillar Tractor Co., Peoria, IL, 44 pp.

66. Chelini, J.M., 1965, "Limestone, Dolomite, and Travertine in Montana," Bulletin 44, Montana Bureau of Mines and Geology, 53 pp.

67. Chute, N.E., 1967, "Structural and Mineralogical Features of the Stockbridge Marble and Berkshire Schist Near Adams, Massachusetts," *Economic Geology in Massachusetts,* O.C. Farquhar, ed., Univ. of Massachusetts Graduate School, Amherst, pp. 169–180.

68. Clarke, F.W., 1924, "The Data of Geochemistry," Bulletin 770, US Geological Survey, 841 pp.

69. Cloud, P.E., Jr., and Barnes, V.E., 1946, "The Ellenburger Group of Central Texas," Publication 4621, University of Texas, Austin, 473 pp.

70. Colby, S.F., 1941, "Occurrences and Uses of Dolomite in the United States," Information Circular 7192, US Bureau of Mines, 21 pp.

71. Colquourn, D.J., and Duncan, D., 1966, "Geology of the Eutawville Quadrangle, South Carolina," MS-12, Map with Text, South Carolina Div. of Geology.

72. Comstock, H.B., 1963, "Magnesium and Magnesium Compounds," Information Circular 8201, US Bureau of Mines, 128 pp.

73. Condra, G.E., and Reed, E.C., 1959, "The Geological Section of Nebraska," Bulletin 14A, Nebraska Geological Survey, 82 pp.

74. Connecticut Geological and Natural History Survey, see Publications List of Geologic Quadrangle Maps and Quadrangle Reports.

75. Conrad, S.G., 1960, "Crystalline Limestones of the Piedmont and Mountain Regions of North Carolina," Bulletin 74, North Carolina Div. of Mineral Resources, 56 pp.

76. Cooper, B.N., 1944, "Industrial Limestones and Dolomites in Virginia: New River-Roanoke River District," Bulletin 62, Virginia Div. of Mineral Resources, 98 pp.

77. Cooper, B.N., 1945, "Industrial Limestones and Dolomites in Virginia: Clinch Valley District," Bulletin 66, Virginia Div. of Mineral Resources, 259 pp.

78. Cooper, J.D., 1970, "Stone," *Mineral Facts and Problems,* Bulletin 650, US Bureau of Mines, pp. 1219–1235.

79. Cullity, B.D., 1956, *Elements of X-Ray Diffraction,* Addison-Wesley, Reading, MA, 514 pp.

80. Curry, R.L., and Buck, A.D., 1966, "Development of Methods of Testing Aggregates Larger than 1½ Inch," Technical Report 6–747, US Army Engineer Waterways Experiment Station, 44 pp.

81. Czehura, S.J., and Berg, R.B., 1977, "Index of Graduate Theses on Montana Geology," Special Publication 77, Montana Bureau of Mines and Geology, 104 pp.

82. Dale, T.N., 1914, "The Commercial Marbles of Western Vermont," Report of the State Geologist on the Mineral Industries and Geology of Vermont, 1913–1914, Vermont Geological Survey, pp. 1–160.

83. Dale, T.N., 1914, "The Calcite Marble and Dolomite of Eastern Vermont," Report of the State Geologist on the Mineral Industries and Geology of Vermont, 1913–1914, Vermont Geological Survey, pp. 224–276.

84. D'Andrea, D.V., et al., 1964, "Prediction of Compressive Strength from Other Rock Properties," *Colorado School Mines Quarterly,* Vol. 59, No. 4B, pp. 623–640.

85. D'Andrea, D.V., et al., 1965, "Prediction of Compressive Strength from Other Rock Properties," Report of Investigations 6702, US Bureau of Mines, 23 pp.

86. Danner, W.R., 1966, "Limestone Resources of Western Washington," Bulletin 52, Washington Div. of Mines and Geology, 474 pp.

87. Danner, W.R., 1976, "Limestone Resources of Southwestern British Columbia," *Proceedings, 11th Forum on Geology of Industrial Minerals,* Special Publication 74, Montana Bureau of Mines and Geology, pp. 171–186.

88. Davis, R.E., 1957, "Magnesium Resources of the United States—A Geologic Summary and Annotated Bibliography to 1953," Bulletin 1019-E, US Geological Survey, pp. 373–515.

89. Dean, T.J., et al., 1969, "Underground Mining in the Kansas City Area," *Missouri*

Mineral Industry News, Vol. 9, Apr., Missouri Geological Survey, pp. 37–56.

90. Deasy, G., et al., 1967, "Supplement, Limestones and Dolomites of Pennsylvania," *Atlas of Pennsylvania's Mineral Resource, Pt. 1,* Mineral Resource Report M 50, Pennsylvania Geological Survey, 83 pp.
91. Degrace, J.R., 1974, "Limestone Resources of Newfoundland and Labrador," Report 74–2, Newfoundland Department Mines Energy, 117 pp.
92. Dever, G.R., Jr., 1973, "Stratigraphic Relationships in the Lower and Middle Newman Limestone (Mississippian), East-Central and Northeastern Kentucky," M.S. Thesis, University of Kentucky, Lexington, 121 pp; 1980, Thesis Series 1, Ser. 11, Kentucky Geological Survey, 49 pp.
93. Dever, G.R., Jr., 1974, "High-Carbonate Rock in the High Bridge Group (Middle Ordovician), Boone County, Kentucky," Information Circular 22, Ser. 10, Kentucky Geological Survey, 35 pp.
94. Dever, G.R., Jr., and McGrain, P., 1969, "High-Calcium and Low-Magnesium Limestone Resources in the Region of the Lower Cumberland, Tennessee, and Ohio Valleys, Western Kentucky," Bulletin 5, Ser. 10, Kentucky Geological Survey, 192 pp.
95. Dever, G.R., Jr., et al., 1978, "Industrial Limestone Resources along the Ohio River Valley of Kentucky," *Mining Engineering,* Vol. 30, No. 4, pp. 396–401.
96. Dick, R.A., et al., 1973, "A Study of Fragmentation from Bench Blasting in Limestone at a Reduced Scale," Report of Investigations 7704, US Bureau of Mines, 24 pp.
97. Dickson, J.A.D., 1965, "A Modified Staining Technique for Carbonates in Thin Section," *Nature,* No. 4971, Feb. 6, pp. 587.
98. Dickson, J.A.D., 1966, "Carbonate Identification and Genesis as Revealed by Staining," *Journal of Sedimentary Petrology,* Vol. 36, pp. 491–505.
99. Diehl, H., 1964, "Calcein, Calmagite, and o,o'-Dihydroxyazobenzene Titrimetric, Colorimetric and Fluorometric Reagents for Calcium and Magnesium," G. Frederick Smith Chemical Co., Columbus, Ohio, 124 pp.
100. Dorheim, F.H., 1970, "Mineral Resources of Iowa," Map with Text, Iowa Geological Survey.
101. Drehmel, D.C., and Harvey, R.D., 1974, "Carbonate Rock Properties Required by Desulfurization Processes," *Proceedings, 10th Forum on Geology of Industrial Minerals,* Miscellaneous Report 1, Ohio Geological Survey, pp. 59–66.
102. Dunham, R.J., 1962, "Classification of Carbonate Rocks According to Depositional Texture," *Classification of Carbonate Rocks,* W.E. Hamm, ed., American Association of Petroleum Geologists, Mem. 1, pp. 108–121.
103. Dunn, J.R., and Ozol, M.A., 1962, "Deleterious Properties of Chert," Physical Research Report No. 12, New York Dept. of Public Works, 121 pp.
104. Edmundson, R.S., 1945, "Industrial Limestones and Dolomites in Virginia: Northern and Central Parts of Shenandoah Valley," Bulletin 65, Virginia Div. of Mineral Resources, 195 pp.
105. Edmundson, R.S., 1958, "Industrial Limestones and Dolomites in Virginia: James River District West of the Blue Ridge," Bulletin 73, Virginia Div. of Mineral Resources, 137 pp.
106. Eifler, G.K., Jr., 1968, "Industrial Carbonates of the Texas Gulf Coastal Plain," *Proceedings, 4th Forum on Geology of Industrial Minerals,* Texas Bureau of Economic Geology, pp. 45–56.
107. Eilertsen, N.A., 1964, "Mining Methods and Costs, Kimballton Limestone Mine, Standard Lime, and Cement Co., Gile County, Va.," Information Circular 8214, US Bureau of Mines, 50 pp.
108. Emmons, W.H., and Grout, F.F., 1943, "Mineral Resources of Minnesota," Bulletin 30, Minnesota Geological Survey, 149 pp.
109. English, S.G., et al., 1940, "Limestone Analyses," Mineral Report 5, Oklahoma Geological Survey, 28 pp.
110. Erwin, R.B., 1957, "The Geology of the Limestone of Isle LaMotte and South Hero Island, Vermont," Bulletin 9, Vermont Geological Survey, 94 pp.
111. Evans, T.B., and Eilertsen, N.A., 1957, "Mining Methods and Costs at the Sunbright Limestone Mine, Foote Mineral Co., Sunbright, Virginia," Information Circular 7793, US Bureau of Mines, 44 pp.
112. Farquhar, O.C., 1967, "Drill Records of stone and Basalt in Western Massachusetts," *Economic Geology in Massachusetts,* O.C. Farquhar, ed., Univ. of Massachusetts Graduate School, Amherst, pp. 75–82.
113. Folk, R.L., 1959, "Practical Petrographic Classification of Limestones," *Bulletin,* American Association of Petroleum Geologists, Vol. 43, pp. 1–38.
114. Folk, R.L., 1962, "Spectral Subdivision of Limestone Types," *Classification of Carbonate Rocks,* W.E. Hamm, ed., American Association of Petroleum Geologists, Mem. 1, pp. 62–84.
115. Folk, R.L., and Weaver, C.E., 1952, "A Study of the Texture and Composition of Chert," *American Journal of Science,* Vol. 250, pp. 498–510.
116. Ford, C.T., 1974, "Selection of Limestones as Neutralizing Agents for Coal-Mine Water," *Proceedings, 10th Forum on Geology of Industrial Minerals,* Miscellaneous Report 1, Ohio Geological Survey, pp. 30–42.
117. Fowler, P., 1950, "Stratigraphy and Structure of the Castleton Area, Vermont," Bulletin 2, Vermont Geological Survey, 83 pp.
118. French, R.R., 1967, "Crushed Stone Resources of the Devonian and Silurian Carbonate Rocks of Indiana," Bulletin 37, Indiana Geological Survey, 127 pp.
119. French, R.R., 1969, "Transportation of Mineral Aggregates in Indiana," *Proceedings for 1968,* Indiana Academy of Science, Vol. 78, pp. 348–354.

120. Friedman, G.M., 1959, "Identification of Carbonate Minerals by Staining Methods," *Journal of Sedimentary Petrology,* Vol. 29, pp. 87–97.

121. Friedman, G.M., and Sanders, J.E., 1967, "Origin and Occurrence of Dolostones," *Developments in Sedimentology, Carbonate Rocks,* Vol. 9A, G.V. Chilingar, H.J. Bissell, R.W. Fairbridge, eds., Elsevier, New York, pp. 267–348.

122. Fulkerson, F.B., 1969, "Transportation of Mineral Commodities on the Inland Waterways of the South-Central States," Information Circular 8431, US Bureau of Mines, 88 pp.

123. Fuzesy, L.M., 1980, "Geology of the Deadwood (Cambrian), Meadow Lake and Winnepegosis (Devonian) Formations in West-Central Saskatchewan," Department Mineral Resources Report 210, Saskatchewan Geological Survey, 64 pp.

124. Galle, O.K., 1967, "The Geochemistry of the Oread Limestone," *Proceedings 3rd Forum on Geology of Industrial Minerals,* Special Distribution Publication 34, Kansas Geological Survey, pp. 97–109.

125. Galle, O.K., and Angino, E.E., 1967, "Determination of Calcium and Magnesium in Carbonate and Silicate Rocks by Atomic Absorption," Bulletin 187, Pt. 1, Kansas Geological Survey, pp. 9–11.

126. Garner, L.E., et al. (compilers), 1979, "The Mineral Resources of Texas," Map, Scale, 1:1,000,000, Bureau of Economic Geology, University of Texas, Austin.

127. Gazdik, G.C., and Tagg, K.M., 1957, "Annotated Bibliography of High-Calcium Limestone Deposits in the United States Including Alaska, to April 1956," Bulletin 1019-I, US Geological Survey, pp. 675–713.

128. Gillson, J.L., et al., 1960, "The Carbonate Rocks," *Industrial Minerals and Rocks,* 3rd ed., J.L. Gillson, ed., AIME, New York, 1960, pp. 123–201.

129. Ginsburg, R.N., et al., 1963, "Shallow-Water Carbonate Sediments," *The Sea,* Vol. 3, John Wiley, New York, pp. 554–582.

130. Goddard, E.N., et al., 1963, *Rock-Color Chart,* Geological Society of America, 16 pp.

131. Goodridge, E., 1973, "Status Report: the AGA/OCR Coal Gasification Program," *Coal Age,* Vol. 78, No. 1, pp. 54–59.

132. Goodwin, J.H., 1979, "A Guide to Selecting Agricultural Limestone Products," Illinois Mineral Note 73, Illinois Geological Survey, 7 pp.

133. Gotts, R.J., 1966, "The Cadomin Limestone Quarry," *Eighth Field-Trip Guidebook,* Edmonton Geological Society, pp. 17–18.

134. Gotts, R.J., 1980, "Minerals," *Industry and Resources 1980,* Alberta Department Economic Development, Edmonton, pp. 49–57.

135. Goudge, M.F., 1933, "Canadian Limestones for Building Purposes," Report 733, Canadian Bureau of Mines, 196 pp.

136. Goudge, M.F., 1934, "Limestones of Canada, Their Occurrence and Characteristics; Part 2, Maritime Provinces," Report 742, Canadian Bureau of Mines, 186 pp.

137. Goudge, M.F., 1935, "Limestones of Canada, Their Occurrence and Characteristics; Part 3, Quebec," Report 755, Canadian Bureau of Mines, 274 pp.

138. Goudge, M.F., 1937, "Limestone and Lime—Their Industrial Uses," *Mining and Metallurgy,* Vol. 18, No. 368, pp. 371–374.

139. Goudge, M.F., 1938, "Limestones of Canada, Their Occurrence and Characteristics; Part 4, Ontario," Report 781, Canadian Bureau of Mines, 362 pp.

140. Goudge, M.F., 1944, "Limestones of Canada, Their Occurrence and Characteristics; Part 5, Western Canada," Report 811, Canada Bureau of Mines, 233 pp.

141. Graf, D.L., 1960a, "Geochemistry of Carbonate Sediments and Sedimentary Carbonate Rocks, Part I, Carbonate Mineralogy, Carbonate Sediments," Circular 297, Illinois Geological Survey, 39 pp.

142. Graf, D.L., 1960b, "Geochemistry of Carbonate Sediments and Sedimentary Carbonate Rocks, Part IV-A, Isotopic Composition, Chemical Analyses," Circular 308, Illinois Geological Survey, 42 pp.

143. Graf, D.L., and Goldsmith, J.R., 1963, "Carbonate Mineralogy," *Subsurface Geology of Eniwetok Atoll,* by S.O. Schlanger, Professional Paper 260 BB, US Geological Survey, pp. 1048–1053.

144. Graf, D.L., and Lamar, J.E., 1955, "Properties of Calcium and Magnesium Carbonates and Their Bearing on Some Uses of Carbonate Rocks," *Economic Geology,* 50th Anniversary Vol., pp. 639–713.

145. Gries, J.P., 1964, "Limestone," *Mineral and Water Resources of South Dakota,* Bulletin 16, South Dakota Geological Survey, pp. 96–102.

146. Grisafe, D.A., 1977, "Kansas Building Limestones," Mineral Resource Series 4, Kansas Geological Survey, 42 pp.

147. Gubler, Y., et al., 1967, "Petrology and Petrography of Carbonate Rocks," *Developments in Sedimentology, Carbonate Rocks,* Vol. 9A, G.V. Chilingar, H.J. Bissell, R.W. Fairbridge, eds., Elsevier, New York, pp. 51–86.

148. Gulbrandsen, R.A., 1960, "A Method of X-Ray Analyses for Determining the Ratio of Calcite to Dolomite in Mineral Mixtures," Bulletin 1111-D, US Geological Survey, pp. 147–152.

149. Ham, W.E., 1949, "Geology and Dolomite Resources, Mill Creek-Ravina Area, Johnson County, Oklahoma," Circular 24, Oklahoma Geological Survey, 104 pp.

150. Ham, W.E., et al., 1943, "Geology and Chemical Composition of St. Clair Limestone near Marble City, Oklahoma," Mineral Report 16, Oklahoma Geological Survey, 24 pp.

151. Hamilton, W.N., 1973, "Economic Minerals," *Industry and Resources 1973,* Alberta Bureau of Statistics, Alberta Dept. of Industry & Commerce, pp. 39–60.

152. Hansen, M., 1953, "Geologic Report on Limestone Deposits in Stark County and Hettinger County, North Dakota," Report of Investigation 8, North Dakota Geological Survey, 71 pp.
153. Harrier, C.M., 1966, "Wyoming Iron-Ore Deposits," Information Circular 8315, US Bureau of Mines, 114 pp.
154. Hart, E.W., 1978, "Limestone, Dolomite, and Shell Resources of the Coast Ranges Province, California," Bulletin 197, California Div. of Mines and Geology, 103 pp.
155. Hartnagel, C.A., and Broughton, J.G., 1951, "The Mining and Quarry Industries of New York State, 1937 to 1948," New York State Museum & Science Service Bulletin 343, 130 pp.
156. Harvey, R.D., and Steinmetz, J.C., 1971, "Petrographic Properties of Carbonate Rocks Related to Their Sorption of Sulfur Dioxide," Environmental Geology Notes 50, Illinois Geological Survey, 37 pp.
157. Harvey, R.D., et al., 1974, "Lake Marls, Chalks, and Other Carbonate Rocks with High Dissolution Rates in SO_2-Scrubbing Liquors," *Proceedings, 10th Forum on Geology of Industrial Minerals,* Miscellaneous Report 1, Ohio Geological Survey, pp. 67–80.
158. Harvey, R.D., et al., 1976, "Behavior of Dolomite in Absorption of H_2S from Fuel Gas," *World Mining and Metals Technology,* AIME, New York, pp. 163–188.
159. Hatch, F.H., and Rastall, R.H. (rev. by J.T. Greensmith), 1971, *Petrology of the Sedimentary Rocks,* Vol. 2, Hafner, New York, 502 pp.
160. Hatmaker, P., 1931, "Utilization of Dolomite and High Magnesium Limestone," Information Circular 6524, US Bureau of Mines, 18 pp.
161. Hausel, W.D., and Glass, G.B., 1980, "Minerals and Coal Plate, Natrona County," Ser. 6, Wyoming Geological Survey.
162. Hausel, W.D., and Lageson, D.R., 1978, "Minerals Plate, Sheridan County," Ser. 5, Wyoming Geological Survey.
163. Hausel, W.D., et al., 1979, "Wyoming Mines and Minerals Map," Map MS-5, Scale, 1:500,000, Wyoming Geological Survey.
164. Herod, B.C., 1969, "Pennsylvania Producer Handles Tough Recovery Problem," *Pit & Quarry,* Vol. 62, No. 4, Oct., pp. 125–127, 129.
165. Herod, B.C., editor-in-chief, 1980–81, *Pit and Quarry Handbook,* 73rd ed., Pit and Quarry Publications, Chicago, 686 pp.
166. Herod, B.C., 1972, "Rock Removed = Valuable Space," *Pit & Quarry,* Vol. 64, No. 11, May, pp. 72–75, 82.
167. Heron, S.D., Jr., 1962, "Limestone Resources of the Coastal Plain of South Carolina," Bulletin 28, South Carolina Div. of Geology, 128 pp.
168. Hershey, R.E., and Maher, S.W., 1963, "Limestone and Dolomite Resources of Tennessee," Bulletin 65, Tennessee Div. of Geology, 231 pp.
169. Hewitt, D.F., 1960, "The Limestone Industries of Ontario," Industrial Mineral Circular 5, Ontario Div. of Mines, 177 pp.
170. Hewitt, D.F., and Vos, M.A., 1972, "Limestone Industries of Ontario," Industrial Mineral Report 39, Ontario Div. of Mines, 79 pp.
171. Hill, R.D., 1974, "Overview of Use of Carbonate Rocks for Controlling Acid Mine Drainage," *Proceedings, 10th Forum on Geology of Industrial Minerals,* Miscellaneous Report 1, Ohio Geological Survey, pp. 25–29.
172. Hill, T.P., and Werner, M.A., 1972, "Chemical Composition of Sedimentary Rocks in Alaska, Idaho, Oregon, and Washington," Professional Paper 771, US Geological Survey, 319 pp.
173. Hill, T.P., et al., 1967, "Chemical Composition of Sedimentary Rocks in Colorado, Kansas, Montana, Nebraska, North Dakota, South Dakota, and Wyoming," Professional Paper 561, US Geological Survey, 241 pp.
174. Hillebrand, W.F., and Lundell, G.E.F., 1953, (rev. by G.E.F. Lundell, H.A. Bright, J.I. Hoffman), *Applied Inorganic Analysis, with Special Reference to the Analysis of Metals, Minerals, and Rocks,* 2nd ed., John Wiley, New York, 1034 pp.
175. Hinchey, N.S., 1946, "Missouri Marble," Report of Investigation 3, Missouri Geological Survey, 47 pp.
176. Hinchey, N.S., et al., 1947, "Limestones and Dolomites in the St. Louis Area," Report of Investigation 5, Missouri Geological Survey, 80 pp.
177. Högberg, E., 1971, "Staining Method for Examination of Siliceous, Cretaceous Limestone II," *Geol. Fören. Stockholm Förh.,* Vol. 93, Pt. 4, pp. 707–713.
178. Holbrook, D.F., 1965, "High-Calcium Limestones in Independence and Izard Counties, Arkansas," Miscellaneous Publication, Arkansas Geological Commission, 21 pp.
179. Holmsley, F., 1935, "Marbles and Limestones of Connecticut," Bulletin 56, Connecticut Geological and Natural History Survey, 56 pp.
180. Holter, M.E., 1976, "Limestone Resources of Alberta," Alberta Research Council Economic Geology Report 4, Edmonton, 91 pp.
181. Holter, M.E., 1976, "Limestone Resources of Alberta," *Proceedings, 11th Forum on Geology of Industrial Minerals,* Special Publication 74, Montana Bureau of Mines and Geology, pp. 37–50.
182. Hopkins, R.H., 1942, "The Dolomitic Limestones of Florida," Report of Investigation 3, Florida Geological Survey, 105 pp.
183. Horowitz, A.S., and Potter, P.E., 1971, *Introductory Petrography of Fossils,* Springer-Verlag, New York, 302 pp.
184. Hotchkiss, W.O., and Steidtmann, E., 1914, "Limestone Road Materials of Wisconsin," Bulletin 34, Wisconsin Geological Survey, 137 pp.
185. Howe, W.B., et al., 1961, "The Stratigraphic Succession in Missouri," Missouri Geological Survey and Water Resources, Ser. 2, Vol. 40, 185 pp.
186. Illinois Geological Survey, see Publications List of Road Materials Resources Maps,

Regional and County Resource Reports, and Research Reports on Various Uses of Limestone.

187. Indiana Geological Survey, see Publications List of Regional Geologic Maps.

188. Institution of Mining & Metallurgy, 1965, *Opencast Mining, Quarrying and Alluvial Mining,* American Elsevier, New York, 772 pp.

189. Interstate Commerce Commission, 1968, "Carload Waybill Statistics, 1966, State-to-State Distribution, No. 32, Stone, Clay, and Glass Products," Interstate Commerce Commission, Washington, D.C., 58 pp.

190. Ives, W., and Runnels, R.T., 1960, "Lime Raw Materials in Kansas," Bulletin 142, *Pt. 3,* Kansas Geological Survey, pp. 123–148.

191. Jasper, M.W., and Mihelich, M., 1961, "Kings River Limestone Deposits, Anchorage Quadrangle," Report for the Year 1961, Alaska Div. of Mines and Minerals, pp. 40–48.

192. Jewett, J.M., et al., 1967, "Inland Underground Facilities," *Guidebook for Third National Forum on the Geology of Industrial Minerals,* Kansas Geological Survey, University of Kansas, April, 13 pp.

193. Jicha, H.L., Jr., 1956, "A Deposit of High-Calcium Lime Rock in Valencia County, New Mexico," Circular 36, New Mexico Bureau of Mines & Mineral Resources, 5 pp.

194. Johnsen, J.H., 1958, "Preliminary Report on the Limestones of Albany County, New York," Special Publication, New York State Museum & Science Service, 43 pp.

195. Johnsen, J.H., 1971, "Limestones (Middle Ordovician) of Jefferson County," Map and Chart Series 13, New York State Museum & Science Service, 88 pp.

196. Johnson, A.M., and Sorensen, H.O., 1978, "Drill Core Investigation of the Fiborn Limestone Member in Schoolcraft, Mackinac, and Chippewa Counties, Michigan," Report of Investigation 18, Michigan Geological Survey, 51 pp.

197. Johnson, K.S., 1969, "Mineral Map of Oklahoma (Exclusive of Oil and Gas Fields)," Map GM-15, Scale, 1:750,000, Oklahoma Geological Survey.

198. Johnstone, S.J., and Johnstone, M.G., 1961, "Limestone, Chalk and Whiting" in *Minerals for the Chemical and Allied Industries,* 2nd ed., John Wiley and Son, New York, pp. 293–326.

199. Jordan, C., (compiler), 1978, "Sedimentology Processes: Carbonate Sedimentology," Reprint Series 5, Society of Economic Paleontologists and Mineralogists, 235 pp.

200. Judd, W.R., and Huber, C., 1961, "Correlation of Rock Properties by Statistical Methods," *Proceedings, Symposium on Mining Research,* Vol. 2, Pergamon Press, New York, pp. 621–648.

201. Kansas Geological Survey, see Publications List of Regional and County Reports.

202. Keith, S.B., 1969, "Limestone, Dolomite, and Marble," *Mineral and Water Resources of Arizona,* Bulletin 180, Arizona Bureau of Mines, pp. 385–398.

203. Kentucky Dept. of Commerce and Kentucky Geological Survey, 1962, "Mineral Resources and Mineral Industry Map of Kentucky."

204. Kentucky Geological Survey, see Publications List of Geologic Quadrangle Maps.

205. Kephart, W.W., 1972, "Limestone and Dolomite Requirements for Glass Containers," *Proceedings of Eighth Forum on Geology of Industrial Minerals,* Public Information Series 5, Iowa Geological Survey, pp. 11–15.

206. Kessler, D.W., and Sligh, W.H., 1927, "Physical Properties of the Principal Commercial Limestones Used for Building Construction in the United States," Technical Paper 349, Part of Vol. 21, National Bureau of Standards, pp. 497–590.

207. Kline, H.D., 1962, "Methods and Costs of Mining and Crushing Limestone at Three Quarries, Anderson-Oxandale Rock Co., Kansas," Information Circular 8084, US Bureau of Mines, 15 pp.

208. Kneller, W.A., et al., 1968, "The Properties and Recognition of Deleterious Cherts which Occur in Aggregate Used by Ohio Concrete Producers," Final Report No. 1014, University of Toledo Research Foundation, 201 pp.

209. Kottlowski, F.E., 1957, "High-Purity Dolomite Deposits of South-Central New Mexico," Circular 47, New Mexico Bureau of Mines & Mineral Resources, 43 pp.

210. Kottlowski, F.E., 1962, "Reconnaissance of Commercial High-Calcium Limestones in New Mexico," Circular 60, New Mexico Bureau of Mines & Mineral Resources, 77 pp.

211. Kottlowski, F.E., 1965, "Limestone and Dolomite," *Mineral and Water Resources of New Mexico,* Bulletin 87, New Mexico Bureau of Mines & Mineral Resources, pp. 345–353.

212. Kümmel, H.B., 1901, "Report on Portland Cement Industry," Annual Report of the State Geologist for the Year 1900, New Jersey Geological Survey, pp. 9–101.

213. Kümmel, H.B., 1940, "The Geology of New Jersey," Bulletin 50, New Jersey Geological Survey, 203 pp.

214. Kupsch, W.O., 1953, "Ordovician and Silurian Stratigraphy of East Central Saskatchewan," Report 10, Saskatchewan Geological Survey, Dept. of Natural Resources, 62 pp.

215. Kusler, D.J., and Corre, H.A., 1968, "Limestone Resources in Western West Virginia," Information Circular 8369, US Bureau of Mines, 15 pp.

216. Lamar, J.E., 1957, "Chemical Analyses of Illinois Limestones and Dolomites," Report of Investigations 200, Illinois Geological Survey, 33 pp.

217. Lamar, J.E., 1961, reprinted 1965, "Uses of Limestone and Dolomite," Circular 321, Illinois Geological Survey, 44 pp.

218. Lamar, J.E., 1966, "High-Purity Limestones

in Illinois," Industrial Minerals Notes 27, Illinois Geological Survey, 20 pp.

219. Lamar, J.E., 1967, "Handbook on Limestone and Dolomite for Illinois Quarry Operators," Bulletin 91, Illinois Geological Survey, 119 pp.

220. Lamar, J.E., 1979, "Selected and Annotated List of Industrial Minerals Publications of the Illinois State Geological Survey," Industrial Minerals Notes 69, Illinois Geological Survey, 13 pp.

221. Lamar, J.E., et al., 1956, "Preliminary Report on Portland Cement Materials in Illinois," Report of Investigations 195, Illinois Geological Survey, 34 pp.

222. Lamar, J.E., and Willman, H.B., 1938, "A Summary of the Uses of Limestone and Dolomite," Report of Investigations 49, Illinois Geological Survey, 48 pp.

223. Lamborn, R.E., 1945, "Recent Information on the Maxville Limestone," Information Circular 3, Ohio Geological Survey, 18 pp.

224. Lamborn, R.E., 1951, "Limestones of Eastern Ohio," Bulletin 49, Series 4, Ohio Geological Survey, 377 pp.

225. Landes, K.K., 1963, "Metallurgical Limestone Reserves in the United States," 2nd ed., National Lime Assn., Washington, DC, 31 pp.

226. Landes, K.K., et al., 1945, "Geology of the Mackinac Straits Region," Publication 44, Michigan Geological Survey, 204 pp.

227. Landreth, J.O., 1968, "High Calcium Limestone Deposit in the Rattler Gulch Area, Granite County, Montana," Special Publication 44, Montana Bureau of Mines & Geology, 10 pp.

228. Lawson, D.C., 1980, "Directory of Mining Enterprises for 1979," Bulletin 111, Montana Bureau of Mines and Geology, 52 pp.

229. Leighton, M.W., and Pendexter, C., 1962, "Carbonate Rock Types," *Classification of Carbonate Rocks,* W.E. Hamm, ed., American Association of Petroleum Geologists Memoir 1, pp. 62–84.

230. Libby, F.W., 1957, "Limestone Resources of the Pacific Northwest," Resource Report 9, Raw Materials Survey, Inc., Portland, OR, 92 pp.

231. Lipp, R.J., 1979, "Minnesota Industrial Minerals Directory," Mineral Resources Research Center, University of Minnesota, Minneapolis, 41 pp.

232. Logan, W.N., 1916, "Preliminary Report on the Marls and Limestones of Mississippi," Bulletin 13, Mississippi Geological Survey, 82 pp.

233. Loofbourow, R.L., 1966, "Will Aggregate Shortages Drive Stone Producers Underground?" *Rock Products,* Vol. 69, No. 10, Oct., pp. 67–75.

234. Lozo, F.E., 1959, "Stratigraphic Relations of the Edwards Limestone and Associated Formations in North-Central Texas," *Symposium on Edwards Limestone in Central Texas,* Publication 5905, University of Texas, Austin, pp. 1–20.

235. Manger, G.E., 1963, "Porosity and Bulk Density of Sedimentary Rocks," Bulletin 1144-E, US Geological Survey, 55 pp.

236. Mantell, C.L., 1931, *Industrial Electrochemistry,* 1st ed., McGraw-Hill, New York, 528 pp.

237. Marshall, L.G., 1963, "Mining and Beneficiating Costs at Two Crushed Limestone Operations, Madison County, Iowa," Information Circular 8199, US Bureau of Mines, 18 pp.

238. Martin, J.A., 1967, "Limestone and Dolomite," *Mineral and Water Resources of Missouri,* Missouri Geological Survey & Water Resources, Vol. 43, Series 2, pp. 126–133.

239. Martin, J.A., 1967, "Cement and Cement Raw Materials," *Mineral and Water Resources of Missouri,* Missouri Geological Survey & Water Resources, Vol. 43, Series 2, pp. 172–183.

240. Maryland Geological Survey, see Publications List of County Reports.

241. Mason, R.S., 1969, "Stone," *Mineral and Water Resources of Oregon,* A.E. Weissenborn, ed., Bulletin 64, Oregon Dept. of Geology & Mineral Industries, pp. 246–268.

242. Mather, K., 1953, "Crushed Limestone Aggregates for Concrete," *Mining Engineering,* Vol. 5, No. 10, Oct., pp. 1022–1028; *Trans. AIME,* 1953, Vol. 196, 1954.

243. Mather, K., and Mather, B., 1950, "Method of Petrographic Examination of Aggregates for Concrete," *Proceedings,* American Society for Testing & Materials, Vol. 50, pp. 1288–1312.

244. Mathews, E.B., and Grasty, J.S., 1909, "The Limestones of Maryland," Vol. 8, Pt. 3, Maryland Geological Survey, pp. 227–484.

245. Mathews, W.H., and McCammon, J.W., 1957, "Calcareous Deposits of Southwestern British Columbia," Bulletin 40, British Columbia Dept. of Mines, 105 pp.

246. McCue, J.B., et al., 1939, "Limestones of West Virginia," West Virginia Geological Survey, Vol. 12, 560 pp.

247. McGrain, P., and Dever, G.R., Jr., 1967, "Limestone Resources in the Appalachian Region of Kentucky," Bulletin 4, Series 10, Kentucky Geological Survey, 12 pp.

248. McGregor, D.J., 1958, "Cement Raw Materials in Indiana," Bulletin 15, Indiana Geological Survey, 88 pp.

249. McGregor, D.J., 1963, "High-Calcium Limestone and Dolomite in Indiana," Bulletin 27, Indiana Geological Survey, 76 pp.

250. McLemore, W.H., and Hurst, V.J., 1970, "The Carbonate Rocks in the Coosa Valley Area, Georgia," Geology Dept., University of Georgia, Athens, 170 pp.

251. Mellen, F.F., 1942, "Mississippi Agricultural Limestone," Bulletin 46, Mississippi Geological Survey, 20 pp.

252. Merrill, R.H., 1956, "Roof-Span Studies in Limestone," Report of Investigations 5348, US Bureau of Mines, 38 pp.

253. Mielenz, R.C., 1954, "Petrographic Examination of Concrete Aggregate," *Proceedings,* American Society for Testing & Materials, Vol. 54, pp. 1188–1218.

254. Mielenz, R.C., 1955, "Petrographic Examination," *Significance of Tests and Properties of Concrete and Concrete Aggregates,* American Society for Testing & Materials, Special Technical Publication 169, pp. 253–273.

255. Mihok, E.A., 1968, "Mine Water Research-The Limestone Neutralization Process," Report of Investigations 7191, US Bureau of Mines, 20 pp.

256. Mihok, E.A., 1970, "Mine Water Research-Plant Design and Cost Estimates for Limestone Treatment," Report of Investigations 7368, US Bureau of Mines, 13 pp.

257. Miller, B.L., 1934, "Limestones of Pennsylvania," Mineral Resource Report M20, Pennsylvania Geological Survey, 729 pp.

258. Milliman, J.D., 1974, "Marine Carbonates," Part 1, Springer-Verlag, New York, 375 pp.

259. Mills, J.W., 1962, "High-Calcium Limestones of Eastern Washington," Bulletin 48, Washington Div. of Mines & Geology, 268 pp.

260. Mineral and Water Resources of Colorado, 1964, *Report of the US Geological Survey,* US Government Printing Office, Washington, DC, 302 pp.

261. Moen, W.S., 1966, "Stone, Including Limestone and Dolomite," *Mineral and Water Resources of Washington,* Reprint 9, Washington Div. of Mines & Geology, pp. 265–273.

262. Moore, F.C., et al., 1968, "Problems Encountered in the Design and Construction of a New Aggregate Plant," SME Preprint 68–H–324, 24 pp.

263. Moore, F.H., 1935, "Marbles and Limestones of Connecticut," Bulletin 56, Connecticut State Geological & Natural History Survey, 56 pp.

264. Morris, H.T., 1969, "Limestone and Dolomite," *Mineral and Water Resources of Utah,* Bulletin 73, Utah Geological and Mineral Survey, pp. 188–194.

265. Morse, W.C., 1910, "The Maxville Limestone," Bulletin 13, Ohio Geological Survey, 128 pp.

266. Morse, W.C., 1930, "Paleozoic Rocks," Bulletin 23, Mississippi Geological Survey, 212 pp.

267. Müller, G., 1967, *Methods in Sedimentary Petrology,* Pt. 1 (English trans. by Hans-Ulrich Schmincke), Hafner, New York, 283 pp.

268. Murray, D.A., 1975, "Limestones and Dolomites of Nova Scotia, Part II, Antigonish, Guysborough, Picton and Cumberland Counties," Bulletin 2, Nova Scotia Department of Mines, 155 pp.

269. Mutmansky, J.M., and Singh, M.M., 1968, "A Statistical Study of Relationships Between Rock Properties," *Status of Practical Rock Mechanics,* AIME, New York, pp. 161–177.

270. Nelson, H.F., 1959, "Deposition and Alteration of the Edwards Limestone, Central Texas," *Symposium on Edwards Limestone in Central Texas,* Publication 5905, University of Texas, Austin, pp. 21–95.

271. Newlon, H.H., Jr., et al., 1972, "An Evaluation of Several Methods for Detecting Alkali-Carbonate Reaction," Progress Report 5, Virginia Highway Research Council, 73 pp.

272. Obert, L., et al., 1946, "Standardized Tests for Determining the Physical Properties of Mine Rock," Report of Investigations 3891, US Bureau of Mines, 67 pp.

273. O'Donnell, J.J., and Sliger, A.G., 1972, "Availability of Limestones and Dolomites," Task 1 Final Report, M. W. Kellogg Co., National Technical Information Service PB–206–963, Feb. 1, 191 pp.

274. Oliver, W.A., Jr., 1956, "Biostromes and Bioherms of the Onondaga Limestone in Eastern New York," New York State Museum & Science Service Circular 45, 23 pp.

275. Olson, R.H., 1964, "Limestone," *Mineral and Water Resources of Nevada,* Bulletin 65, Nevada Bureau of Mines, pp. 217–220.

276. O'Neill, B.J., Jr., 1964, "Limestones and Dolomites of Pennsylvania," *Atlas of Pennsylvania's Mineral Resources,* Pt. 1, Mineral Resource Report M50, Pennsylvania Geological Survey, 40 pp.

277. O'Neill, B.J., Jr., 1975, "Potential High-Calcium Limestone Resources in the Mount Joy Area, Lancaster County," Information Circular 76, Pennsylvania Geological Survey, 18 pp.

278. O'Neill, B.J., Jr., 1976, "The Distribution of Limestones Containing At Least 90 Percent $CaCO_3$ in Pennsylvania," *Atlas of Mineral Resources,* Part 4, Mineral Resources Report M50, Scale, 1:500,000, Pennsylvania Geological Survey.

279. Osterwald, F.W., et al., 1966, "Mineral Resources of Wyoming," Bulletin 50, Wyoming Geological Survey, pp. 180–184.

280. Papke, K.G., 1973, "Industrial Mineral Deposits of Nevada," Map 46, Nevada Bureau Mines and Geology.

281. Papke, K.G., 1979, "List of Nevada Industrial Mineral Producers," Special Publication L-1, Nevada Bureau Mines and Geology, 3 pp.

282. Parker, R.L., 1967, "Composition of the Earth's Crust," *Data of Geochemistry,* 6th ed., M. Fleischer, ed., Professional Paper 440–D, US Geological Survey, 19 pp.

283. Patton, J.B., 1951, "Industrial Limestone in Indiana," *Scientific Monthly,* Vol. 72, No. 4, pp. 259–265.

284. Pearre, N.C., 1956, "Mineral Deposits and Occurrences in Massachusetts and Rhode Island, Exclusive of Clay, Sand and Gravel, and Peat," Mineral Investigations Resources Map MR 4, US Geological Survey.

285. Perry, T.G., and Smith, N.M., 1958, "The Meramec-Chester and Intra-Chester Boundaries and Associated Strata in Indiana," Bulletin 12, Indiana Geological Survey, 110 pp.

286. Peterson, N.V., and Mason, R.S., 1958, "Limestone Occurrences in Western Oregon," *The Ore Bin,* Vol. 20, No. 4, pp. 33–39.

287. Pickering, S.M., Jr., 1970, "Stratigraphy, Paleontology, and Economic Geology of Portions of Perry and Cochran Quad-

rangles, Georgia," Bulletin 81, Georgia Geological Survey, 67 pp.

288. Plafker, G., and Berg, H.C., 1964, "Limestone and Marble," *Mineral and Water Resources of Alaska*, US Geological Survey Report, pp. 135–138.

289. Potter, P.E., 1968, "A Selective, Annotated Bibliography on Carbonate Rocks," *Bulletin of Canadian Petroleum Geology*, Vol. 16, No. 1, pp. 87–103.

290. Power, W.R., 1978, "Economic Geology of the Georgia Marble District," *Proceedings, 12th Forum on Geology of Industrial Minerals*, Information Circular 49, Georgia Geological Survey, pp. 59–68.

291. Prokopovich, N., and Schwartz. G.M., 1956, "Minnesota Limestone Suitable for Portland Cement," Summary Report 8, Minnesota Geological Survey, 39 pp.

292. Prucha, J.J., 1953, "The White Crystal Dolomite Deposit near Gouverneur, New York," Report of Investigations 9, New York State Museum & Science Service, 13 pp.

293. Pugliese, J.M., 1972, "Designing Blast Patterns Using Empirical Formulas, A Comparison of Calculated Patterns with Plans Used in Quarrying Limestone and Dolomite, with Geologic Considerations," Information Circular 8550, US Bureau of Mines, 33 pp.

294. Puri, H.S., and Vernon, R.O., 1959, "Summary of the Geology of Florida and a Guidebook to the Classic Exposures," Special Publication 5, Florida Geological Survey, 255 pp.

295. Puri, H.S., et al., 1967, "Geology of Dixie and Gilchrist Counties, Florida," Bulletin 49, Florida Geological Survey, 155 pp.

296. Quinn, A.W., 1971, "Bedrock Geology of Rhode Island," Bulletin 1295, US Geological Survey, 68 pp.

297. Quinn, A.W., et al., 1949, "Bedrock Geology of the Pawtucket Quadrangle, Rhode Island-Massachusetts," Geological Quadrangle Map 1, Scale 1:31,680, US Geological Survey.

298. Ramp, L., and Peterson, N.V., 1979, "Limestone," *Geology and Mineral Resources of Josephine County, Oregon,* Bulletin 100, Oregon Dept. Geology and Mineral Industries, pp. 37–38.

299. Renninger, F.A., and Nichols, F.P., Jr., 1970, "Aggregate Technology—Wider Horizons Through Research," *Trans. SME/AIME,* Vol. 247, pp. 38–42.

300. Reves, W.D., 1961, "The Limestone Resources of Washington, Holmes, and Jackson Counties, Florida," Bulletin 42, Florida Geological Survey, 121 pp.

301. Riley, H.C., and Schroeder, H.J., 1963, "Crushed Limestone Operations, Watauga Quarry, Watauga Stone Co., Carter County, Tenn." Information Circular 8198, US Bureau of Mines, 21 pp.

302. Risser, H.T., 1960, "Kansas Building Stone," Bulletin 142, *Pt. 2,* Kansas Geological Survey, pp. 53–122.

303. Rodda, P.V., et al., 1966, "Limestone and Dolomite Resources, Lower Cretaceous Rocks, Texas," Report of Investigations 56, Texas Bureau of Economic Geology, 286 pp.

304. Roehm, J.C., 1946, "Some High Calcium Limestone Deposits in Southeastern Alaska," Pamphlet 6, Alaska Div. of Mines & Geology, 85 pp.

305. Rooney, L.F., 1970, "High-Calcium Limestone and High-Magnesium Dolomite Resources of Indiana," Bulletin 42–B, Indiana Geological Survey, 20 pp.

306. Rooney, L.F., 1970, "Dimension Limestone Resources of Indiana," Bulletin 42–C, Indiana Geological Survey, 29 pp.

307. Rooney, L.F., and Ault, C.H., 1970, "Potential Limestone and Dolomite Resources of Northern Indiana," *Proceedings Fifth Forum on Geology of Industrial Minerals,* Mineral Resources Report M64, Pennsylvania Geological Survey, pp. 179–224.

308. Rooney, L.F., and Carr, D.D., 1971, "Applied Geology of Industrial Limestone and Dolomite," Bulletin 46, Indiana Geological Survey, 59 pp.

309. Rowland, T.L., 1972, "General Survey of Carbonate Mineral Deposits in Oklahoma," *Oklahoma Geology Notes,* Vol. 32, No. 3, pp. 73–89.

310. Runnells, D.D., 1970, "Errors in X-Ray Analysis of Carbonates Due to Solid-Solution Variation in Composition of Component Minerals," *Journal of Sedimentary Petrology,* Vol. 40, pp. 1158–1166.

311. Runnels, R.T., 1951, "Some High-Calcium Limestones in Kansas," Bulletin 90, *Pt. 5,* Kansas Geological Survey, pp. 77–104.

312. Runnels, R.T., 1959, "Cement Raw Materials in Kansas," *1959 Reports of Studies, Pt. 2,* Bulletin 134, Kansas Geological Survey, pp. 105–124.

313. Runnels, R.T., and Dubins, I.M., 1949, "Chemical and Petrographic Studies of the Fort Hays Chalk in Kansas," Bulletin 82, *Pt. 1,* Kansas Geological Survey, 36 pp.

314. Runnels, R.T., and Schleicher, J.A., 1956, "Chemical Composition of Eastern Kansas Limestone," Bulletin 119, *Pt. 3,* Kansas Geological Survey, pp. 81–103.

315. Rutledge, F.A., et al., 1953, "Preliminary Report—Nonmetallic Deposits Accessible to the Alaska Railroad as Possible Sources of Raw Materials for the Construction Industry," Report of Investigations 4932, US Bureau of Mines, 129 pp.

316. Salotti, C.A., et al., 1970, "Limestone as a Raw Material for Hydrocarbon Fuels," *Proceedings, 6th Forum on Geology of Industrial Minerals,* Miscellany 1, Michigan Geological Survey, pp. 48–55.

317. Sanders, J.E., and Friedman, G.M., 1967, "Origin and Occurrence of Limestone," *Developments in Sedimentology, Carbonate Rocks,* Vol. 9A, G.V. Chilingar, H.J. Bissell, and R.W. Fairbridge, eds., Elsevier, New York, pp. 169–265.

318. Savage, C.N., 1965, "Economic Geology of Carbonate Rocks Adjacent to the Snake River South of Lewiston, Idaho," Mineral Resources Report 10, Idaho Bureau of Mines & Geology, 26 pp.

319. Savage, C.N., 1969, "Distribution and Economic Potential of Idaho Carbonate

Rocks," Bulletin 23, Idaho Bureau of Mines & Geology, 93 pp.

320. Scholle, P.A., 1978, "A Color Illustrated Guide to Carbonate Rock Constituents, Textures, Cements, and Porosities," American Association of Petroleum Geologists Memoir 27, 241 pp.

321. Schmidt, W., et al., 1979, "The Limestone Dolomite and Coquina Resources of Florida," Report of Investigation 88, Florida Bureau of Geology, 54 pp.

322. Schwartz, G.M., et al., 1959, "Investigation of the Commercial Possibilities of Marl in Minnesota," Office of Iron Range Resources and Rehabilitation, 190 pp.

323. Scott, A.J., 1968, "Environmental Factors Controlling Oyster Shell Deposits, Texas Coasts," *Proceedings, 4th Forum on Geology of Industrial Minerals,* Texas Bureau of Economic Geology, pp. 129–150.

324. Searle, A.B., 1935, *Limestone and Its Products; Their Nature, Production, and Uses,* Ernest Benn, London, 709 pp.

325. Sellards, E.H., et al., 1911, "Roads and Road Materials of Florida," Bulletin 2, Florida Geological Survey, 31 pp.

326. Shapiro, L., and Brannock, W.W., 1962, "Rapid Analysis of Silicate, Carbonate and Phosphate Rocks," Bulletin 1144–A, US Geological Survey, 56 pp.

327. Shea, F.S., and Murray, D.A., 1967, "Limestones and Dolomites of Nova Scotia, Part 1, Cape Breton Island," Bulletin 2, Nova Scotia Department of Mines, 300 pp.

328. Siegel, F.R., 1967, "Properties and Uses of the Carbonates," *Developments in Sedimentology, Carbonate Rocks,* Vol. 9B, G.V. Chilingar, H.J. Bissell, R.W. Fairbridge, eds., Elsevier, New York, pp. 343–393.

329. Sinclair, J., 1969, *Quarrying, Opencast and Alluvial Mining,* Elsevier, New York, 375 pp.

330. Singleton, R.H., 1980, "Stone," Preprint, Bulletin 671, *Mineral Facts and Problems,* US Bureau of Mines, 16 pp.

331. Slack, A.V., 1973, "Removing SO_2 from Stack Gases," *Environmental Science & Technology,* Vol. 7, No. 2, pp. 110–119.

332. Slavin, W., 1968, *Atomic Absorption Spectroscopy,* Interscience Publishers, New York, 307 pp.

333. Smith, R.A., 1915, "Limestones in Michigan," Publication 21, Pt. 2, Michigan Geological & Biological Survey, pp. 103–402.

334. Stauffer, C.R., 1950, "The High Magnesium Dolomites and Dolomitic Limestones of Minnesota," Summary Report 4, Minnesota Geological Survey, 29 pp.

335. Stearn, E.W., 1965, "Underground Storage: How Does It Look Today?" *Rock Products,* Vol. 68, No. 12, Dec., pp. 86–87, 96.

336. Steidtmann, E., 1924, "Limestones and Marls of Wisconsin," Bulletin 66, Wisconsin Geological Survey, 208 pp.

337. Stenzel, H.B., et al., 1948, "Geological Resources of the Trinity River Tributary Area in Oklahoma and Texas," Publication 4824, University of Texas, Austin, 252 pp.

338. Stith, D.A., 1969, "Potential Use of Ohio Limestones and Dolomites for Architectural Aggregate," Report of Investigations 73, Ohio Geological Survey, 14 pp.

339. Stith, D.A., 1972, "High-Calcium Limestone Facies of the Devonian Dundee Limestone," Report of Investigations 86, Ohio Geological Survey, 14 pp.

340. Stith, D.A., 1979, "Chemical Composition, Stratigraphy, and Depositional Environments of the Black River Group (Middle Ordovician), Southwestern Ohio," Report of Investigations 113, Ohio Geological Survey, 36 pp.

341. Stokley, J.A., 1949, "Industrial Limestones of Kentucky," Report of Investigations 2, Series 9, Kentucky Geological Survey, 51 pp.

342. Stokley, J.A., and McFarlan, A.C., 1952, "Industrial Limestones of Kentucky No. 2," Report of Investigations 4, Series 9, Kentucky Geological Survey, 95 pp.

343. Stokley, J.A., and Walker, F.H., 1953, "Industrial Limestones of Kentucky No. 3," Report of Investigations 8, Series 9, Kentucky Geological Survey, 62 pp.

344. Stonehouse, D.H., 1979, "Construction Materials and Other Industrial Minerals," *Canadian Minerals Survey 1979,* Energy, Mines and Resources Canada, Ottawa, pp. 63–67.

345. Stonehouse, D.H., 1979, "Lime—Report 26," Annual Review of the Canadian Mineral Industry 1979, Mineral Policy Sector of Energy, Mines and Resources Canada, Ottawa, 4 pp.

346. Stout, W., 1941, "Dolomites and Limestones of Western Ohio," Bulletin 42, Ohio Geological Survey, 468 pp.

347. Stowe, R.L., 1969, "Strength and Deformation Properties of Granite, Basalt, Limestone and Tuff at Various Loading Rates," Miscellaneous Paper C-69-1, US Army Engineer Waterways Experiment Station, 168 pp.

348. Suffel, G.G., 1930, "Dolomites of Western Oklahoma," Bulletin 49, Oklahoma Geological Survey, 155 pp.

349. Swain, F.M., et al., 1966, "High-Calcium Limestone Deposits of Cumberland Valley, Pennsylvania," *Ohio Journal of Science,* Vol. 66, No. 2, pp. 116–123.

350. Taft, W.H., 1967, "Modern Carbonate Sediments," *Developments in Sedimentology, Carbonate Rocks,* Vol. 9A, G.V. Chilingar, H.J. Bissell, R.W. Fairbridge, eds., Elsevier, New York, pp. 29–50.

351. Tennessee Division of Geology, see Publications List of Geologic Quadrangle Maps.

352. Thiel, G.A., and Dutton, C.E., 1935, "The Architectural, Structural and Monumental Stones of Minnesota," Bulletin 25, Minnesota Geological Survey, 160 pp.

353. Thiel, G.A., and Stauffer, C.R., 1947, "The High Calcium Limestones of Minnesota," Summary Report 1, Minnesota Geological Survey, 13 pp.

354. Thill, R.E., et al., 1969, "Correlation of Longitudinal Velocity Variation with Rock Fabric," *Journal of Geophysical Research,* Vol. 74, No. 20, pp. 4897–4909.

355. Thoenen, J.R., 1926, "Underground Limestone Mining," Bulletin 262, US Bureau of Mines, 100 pp.

356. Trainer, D.W., Jr., 1932, "The Tully Limestone of Central New York," New York State Museum & Science Service Bulletin 291, 23 pp.

357. US Census Office, 1884, "Report on the Building Stones of the United States and Statistics of the Quarry Industry for 1880," *10th Census,* Government Printing Office, Washington, DC, 410 pp.

358. US Geological Survey and US Bureau of Mines, 1968, "Mineral Resources of the Appalachian Region," Professional Paper 580, US Geological Survey, 492 pp.

359. Vanderwilt, J.W., 1947, "Mineral Resources of Colorado," Colorado State Mineral Resources Board, Denver, 547 pp.

360. Vermont Geological Survey, see Publications List of Quadrangle Reports.

361. Vernon, R.O., 1942, "Geology of Holmes and Washington Counties, Florida," Bulletin 21, Florida Geological Survey, 161 pp.

362. Vernon, R.O., 1943, "Florida Mineral Industry," Bulletin 24, Florida Geological Survey, 207 pp.

363. Vernon, R.O., 1951, "Geology of Citrus and Levy Counties, Florida," Bulletin 33, Florida Geological Survey, 256 pp.

364. Vineyard, J.T., ed., 1969, *Missouri Mineral Industry News,* Missouri Geological Survey, Vol. 9, July, pp. 102.

365. Wagner, N.S., 1958, "Limestone Occurrences in Eastern Oregon," *The Ore Bin,* Vol. 20, No. 5, pp. 43-47.

366. Walker, K.R., 1973, "Stratigraphy and Environmental Sedimentology of Middle Ordovician Black River Group in the Type Area—New York State," New York State Museum & Science Service Bulletin 419, 43 pp.

367. Warne, S.S.J., 1962, "A Quick Field or Laboratory Staining Scheme for the Differentiation of the Major Carbonate Minerals," *Journal of Sedimentary Petrology,* Vol. 32, pp. 29–38.

368. Waugh, T.C., 1966, "Analyses of High Calcium Chert-Free Beds in the Keokuk Limestone, Cherokee County, Kansas," Bulletin 180, *Pt. 2,* Kansas Geological Survey, 7 pp.

369. Wayland, J.R., and Ham, W.E., 1955, "General and Economic Geology of the Baum Limestone, Ravia-Mannsville Area, Oklahoma," Circular 33, Oklahoma Geological Survey, 44 pp.

370. Wayne, W.J., 1971, "Marl Resources of Indiana," Bulletin 42-G, Indiana Geological Survey, 16 pp.

371. Wedepohl, K.H., 1970, "Geochemische Daten von Sedimentären Karbonaten und Karbonatgesteinen in Ihrem Faziellen und Petrogenetischen Aussagewert," *Verhandl. Geol. Bundesanst,* (Austria), Vol. 4, pp. 692–705.

372. Weitz, J.H., 1942, "High-Grade Dolomite Deposits in the United States," Information Circular 7226, US Bureau of Mines, 86 pp.

373. Welby, C.W., 1961, "Bedrock Geology of the Central Champlain Valley of Vermont," Bulletin 14, Vermont Geological Survey, 296 pp.

374. Wentworth, C.K., 1939, "Physical Geography and Geology [of the Hawaiian Islands]," Hawaii Territorial Planning Board, 1st Progress Report, Honolulu, pp. 13–20.

375. West, T.R., and Johnson, R.B., 1965, "Analysis of Textural and Physical Factors Contributing to the Abrasion Resistance of Some Indiana Carbonate Aggregates," *Indiana Academy of Science Proceedings for 1964,* Vol. 75, pp. 153–162.

376. West Virginia Geological and Economic Survey, see Publications List of County Reports.

377. Wharton, H.M., et al., 1969, "Missouri Minerals—Resources Production and Forecasts," Special Publication 1, Missouri Geological Survey, 303 pp.

378. Widmer, K., 1964, *The Geology and Geography of New Jersey,* Van Nostrand, Princeton, NJ, 193 pp.

379. Willard, B., 1931, "Commercial Limestones of Rhode Island," *Pan-American Geologist,* Vol. 56, No. 2, pp. 116–122.

380. Willard, R.J., and McWilliams, J.R., 1969, "Microstructural Techniques in the Study of Physical Properties of Rock," *International Journal of Rock Mechanics and Mining Sciences,* Vol. 6, No. 1, pp. 1–12.

381. Williamson, D.R., and Burgin, L., 1960, "Limestone Occurrences in Colorado," Colorado School of Mines *Mineral Industries Bulletin,* Vol. 3, No. 1, 12 pp.

382. Willman, H.B., 1943, "High-Purity Dolomite in Illinois," Report of Investigations 90, Illinois Geological Survey, 89 pp.

383. Wilmoth, R.C., and Hill, R.D., 1973, "Mine Drainage Pollution Control Via Reverse Osmosis," *Mining Engineering,* Vol. 25, No. 3, pp. 45–47.

384. Wilson, J.L., 1975, *Carbonate Facies in Geologic History,* Springer-Verlag, New York, 471 pp.

385. Windes, S.L., 1949, "Physical Properties of Mine Rock, Part I," Report of Investigations 4459, US Bureau of Mines, 79 pp.

386. Windes, S.L., 1950, "Physical Properties of Mine Rock, Part II," Report of Investigations 4727, US Bureau of Mines, 37 pp.

387. Wolfe, P.E., 1948, "Agricultural Mineral Resources of New Jersey," Bulletin 2, Rutgers University, Bureau of Mineral Research, pp. 18–50.

388. Woolf, D.O., 1953, "Results of Physical Tests of Road-Building Aggregate," US Dept. of Commerce, Bureau of Public Roads, Washington, DC, 225 pp.

389. Yon, J.W., Jr., and Hendry, C.W., Jr., 1969, "Mineral Resource Study of Holmes, Walton and Washington Counties," Bulletin 50, Florida Geological Survey, 161 pp.

390. Yon, J.W., Jr., and Hendry, C.W., Jr., 1972, "Suwannee Limestone in Hernando and Pasco Counties, Florida," Bulletin 54, Pt. 1, Florida Geological Survey, pp. 1–42.

391. Zenger, D.H., 1965, "Stratigraphy of the Lockport Formation (Middle Silurian) in

New York State," New York State Museum & Science Service Bulletin 419, 210 pp.

392. Zenger, D.H., "Stratigraphy and Petrology of the Little Falls Dolostone (Upper Cambrian), East-Central New York," New York State Museum & Science Service Map and Chart Series 34, in press.

Lithium Raw Materials

IHOR A. KUNASZ *

Introduction

Lithium minerals occur predominantly in pegmatites which contain mineral assemblages derived from the crystallization of postmagmatic fluids or from the metasomatic action by residual pegmatitic fluids. They have been the traditional sources of raw materials for ceramic and chemical industries. With the discovery of Searles Lake, brines became new sources of lithium.

Indeed, the 1960s may well be referred to as the "brine decade" because several brine bodies of significant economic importance were discovered. In the United States, the brines of Clayton Valley, Nevada, are presently exploited by Foote Mineral Co. The brines of the Great Salt Lake of Utah, the Smackover formation, and the Imperial Valley geothermal field, have been defined as important resources of lithium. In South America, new occurrences have been identified in Bolivia and Argentina. At the Salar de Atacama, Chile, new production capacity is scheduled for 1984.

Following the termination of the AEC purchase program in 1959, several companies such as Maywood Chemical Co., American Lithium Chemicals, and Quebec Lithium Corp. were forced to close their operations. Nevertheless, with the development of new applications in glass ceramics, air conditioning systems, synthetic rubber, and metallurgy (aluminum potlines), the lithium industry grew steadily, and with the potential application of lithium metal in batteries and nuclear reactors, a healthy growth can be foreseen. New products and new applications are likely to stimulate exploration for expanded reserves as well as the search for new sources from which lithium could be extracted economically.

* Chief Geologist, Foote Mineral Co., Exton, PA.

Geochemistry of Lithium

Lithium is the third element in the periodic table. It is the lightest of all the metals, having an atomic weight of 6.938, an ionic radius of 0.68Å and a charge of +1.

The geochemistry of lithium has been extensively studied and has been summarized by Rankama and Sahama (1950), Goldschmidt (1937), and Hortsman (1957).

The distribution of lithium in igneous rocks is controlled by its size and its charge, and by the $(MgO+FeO)/Li_2O$ ratio. In the early stages of crystallization of a magma, that ratio is very large. Consequently, both magnesium and iron are removed by ferromagnesian minerals in preference to lithium which is then concentrated in the residual magma. This results in an enrichment of lithium in silicic rocks and pegmatites (Strock, 1936).

Pegmatites are coarse-grained igneous rocks formed by the crystallization of postmagmatic fluids. Minerals within pegmatites may also form by metasomatism (Jahns, 1955). Genetically, the pegmatites are associated with neighboring intrusives. Mineralogically, granitic pegmatites contain feldspar, quartz, and mica as the main constituents and a variety of exotic elements such as lithium, beryllium, tantalum, tin, and cesium, which may or may not occur in economically significant concentrations.

Detailed studies by numerous investigators (Cameron et al., 1949, 1954; Hanley et al., 1950; Jahns, 1953, 1955; Page et al., 1953) indicate that pegmatites often exhibit an internal zonal arrangement, with each zone containing a specific suite of minerals. The lithium minerals are usually found in the intermediate zones, and although as many as 13 zones have been recognized by Cameron, et al. (1949), a complete zonal arrangement is rarely found. Zoning of pegmatite bodies has also been observed on a regional basis. The regionally zoned pegmatite sequences exhibit mineral assemblages and complexity according to their respective distance from the granitic bodies to which they are genetically related.

Lithium is also found in small proportions in a variety of rocks. The average lithium content of igneous rocks is estimated at about 28 ppm Li. Sedimentary rocks contain an average of 53 ppm Li, the highest being recorded in shales (Hortsman, 1957).

Unusual amounts of lithium are found in the clay mineral, hectorite, which is expandable and belongs to the magnesian end member of the smectite group. Recent work by the US Geological Survey suggests that lithium-rich clays are ubiquitous in a number of clay environments.

Lithium is also present in significant amounts in waters associated with geothermal areas (White, 1957), in oil well brines (Mayhew and Heylmun, 1966), and certain brines of California (Searles Lake), Nevada (Clayton Valley), Utah (Great Salt Lake), and Chile (Salar de Atacama), Bolivia (Salar de Uyuni), and Argentina.

Lithium Minerals

Although lithium occurs in some 145 minerals, only spodumene, lepidolite, petalite, amblygonite, and eucryptite have been commercial sources of lithium.

Spodumene, a lithium aluminum silicate ($LiAlSi_2O_6$), is a monoclinic member of the pyroxene group. It has a very pronounced cleavage plane (110) which results in typically lath-shaped particles upon breaking. The color of spodumene is variable, being nearly white in low iron variety and dark green in iron-rich crystals.

When clear, spodumene is considered a gemstone. Three varieties are known: *hiddenite,* the green variety from Alexander County, North Carolina, first discovered in Brazil; *triphane,* the yellow variety also from Alexander County; and the lilac-colored *kunzite* from the Pala district, California, Brazil, and Afghanistan.

Spodumene undergoes pseudomorphic alteration to a variety of minerals. Norton and Schlegel (1955) have described spodumene replacement by quartz, albite, perthite, muscovite, beryl, amblygonite, apatite, and tourmaline. Weathering commonly alters spodumene to kaolinite and to montmorillonite.

Spodumene constitutes the most abundant commercial source of lithium. Theoretically, it may contain up to 3.7% Li, but the actual lithium concentrations vary from 1.35% to 3.56%, probably as a result of sodium and potassium substitution for lithium. Spodumene concentrates typically contain 1.9% to 3.3% Li.

Lepidolite is a phyllosilicate with the general formula $K_2(Li,Al)_{5-6}\{Si_{6-7}Al_{2-1}O_{20}\}(OH,F)_4$. The chemical variability expressed in the formula stems from a structural complexity attributed to a mixture of different polymorphs which include muscovite, lithium muscovite, and polylithionite (Winchell, 1942). On the other hand, Foster (1960) and Deer, et al. (1962) suggest that there is a continuous series between muscovite with a $2M_1$ structure to lepidolite, with 1M, $2M_2$, and 3T structures. The structural transition takes place when the lithia content in the mica reaches 1.53%.

The lithium concentration in lepidolite varies between 1.53% to a possible theoretical maximum of 3.6%. In commercial deposits, the concentrations are more normally 1.4% to 1.9% Li. In addition to lithium, lepidolites also carry substantial concentrations of rubidium and cesium (Deer, et al., 1962).

The major commercial occurrences of lepidolite are located in Zimbabwe (Bikita), Namibia (Karibib), Canada (Bernic Lake, Manitoba), and Brazil.

Petalite, $LiAlSi_4O_{10}$, is a monoclinic mineral with a framework silicate structure. Its color is grayish white and more rarely pinkish. It has two cleavage directions which form an angle of 38.5°. The basal cleavage is perfect.

The theoretical lithium content of petalite is 2.27%. In actual commercial deposits, the concentration varies from 1.6% to 2.1% Li. Sizable deposits of petalite occur with lepidolite in Zimbabwe (Bikita), Namibia (Karibib), Brazil (Araçuai), Australia (Londonderry), in the USSR (eastern Transbaikalia), and Sweden (Utö).

In certain pegmatites, there is evidence that petalite alters to a mixture of spodumene and quartz. In the Bernic Lake pegmatites of Manitoba, Cerny, et al. (1971) have described pseudomorphs of spodumene and quartz after petalite.

Eucryptite is also a lithium aluminum silicate which is deficient in silica. It has a formula $LiAlSiO_4$ and may contain 5.53% Li. The only large deposit of eucryptite is found in Zimbabwe (Bikita) where its occurrence with quartz suggests a spodumene origin (Westenberger, 1963). The grade of the eucryptite is 2.34% Li.

Amblygonite, with the generalized formula $LiAl(PO_4)(F,OH)$, is the fluorine-rich end member of a phosphate series, while montebrasite represents the hydroxyl-rich end member. It occurs in white to gray masses. Basal cleavage planes are pearly, others are vitreous. Amblygonite weathers to earthy apatite, wavellite, and other lithium deficient phosphates. Although amblygonite may contain as much as 4.74% Li, commercial ores usually carry 3.5% to 4.2% Li. Amblygonite has been mined in Canada, Brazil, Surinam, Zimbabwe, Ruanda, Mozambique, Namibia, and the Republic of South Africa.

Brines. Lithium is found in commercial quantities in certain brine deposits. The brines are present in desert areas and occur in playas and saline lakes where solutions have been concentrated by solar evaporation. In Searles Lake where production of dilithium phosphate began in 1938, the lithium concentration is 70 ppm Li. In Clayton Valley, Nevada, lithium-bearing brines contain 200 ppm Li. Smaller concentrations of lithium (28 to 60 ppm Li) are found in the Great Salt Lake. Rich concentrations of lithium have been identified at the Salar de Uyuni in Bolivia (100–700 ppm) and the Salar de Atacama in Chile (1000–5000 ppm).

Reserves

Projections of lithium metal requirements to satisfy the needs of the automotive and storage battery system as well as the potential use in thermal nuclear fusion energy generation resulted in concern regarding the availability of sufficient lithium. The US Geological Survey created the Lithium Exploration Group, which subsequently sponsored two symposia, one held in Denver, CO and another in Corning, NY. Concurrently, the Energy Research and Development Administration requested the National Academies of Sciences and Engineering to form a National Research Council Committee on Nuclear and Alternative Energy Systems (CONAES). One of the groups was the Lithium Subpanel consisting of experts from the industry and the government, formed to evaluate the lithium availability in the western world. The results of the Subpanel study were presented at the second lithium symposium held at Corning, NY in 1977. This document is considered the best reference regarding reserves and resources and the serious student of lithium should have this particular publication on hand (Penner, 1978). Many of the figures quoted in this review have been taken from the Subpanel study.

Major Producing Districts

At the present time, the United States and Zimbabwe supply the bulk of the free world's demand for lithium raw materials (Fig. 1), and the United States is the leading producer. Within the communist bloc, the Soviet Union probably represents an important producer; production figures, however, are not available. The People's Republic of China is reported to produce lithium compounds from ores mined in the northwestern part of the country.

United States

Lithium ores and lithium chemicals in the United States are supplied by three producing areas. Two of the areas are in the tin-spodumene belt of North Carolina, while brines are exploited in Clayton Valley (Silver Peak), Nevada.

The tin-spodumene belt of North Carolina constitutes the largest developed reserve of lithium in the free world. The area is characterized by numerous pegmatites that intrude amphibolites and schists along the eastern margin of the Devonian Cherryville quartz monzonite (Kesler, 1961).

The pegmatites are typically unzoned, and spodumene is distributed throughout the bodies. The contact zones and aplitic layers contain substantially lower percentages of spodumene. On the average, the pegmatites consist of 20% spodumene, 32% quartz, 6% muscovite, 41% feldspar (including albite and microcline), and about 1% trace minerals, such as beryl, lithiophilite, columbite, tantalite, pyrite, sphalerite, apatite, and rhodochrosite (Kesler, 1961).

Two companies exploit the spodumene-bearing pegmatites: Foote Mineral Co. and Lithium Corp. of America.

Foote Mineral Co., 83% owned by Newmont Mining Corp., mines spodumene from a pegmatite cluster located about one mile southwest of the town of Kings Mountain. Extensive drilling has outlined ore reserves of 29 million tons averaging 0.7% Li. At the present time, only the northern pegmatites are included in this estimate. Approximately one-half of the property where surface outcrops of spodumene-bearing pegmatites exist has not been evaluated by drilling. Geological inference permits an estimate of some additional 14 million tons of spodumene-bearing pegmatites for the unexplored portion of the property.

Lithium Corp. of America, a wholly-owned subsidiary of Gulf Resources and Chemicals Corp., has developed an open pit in the Long

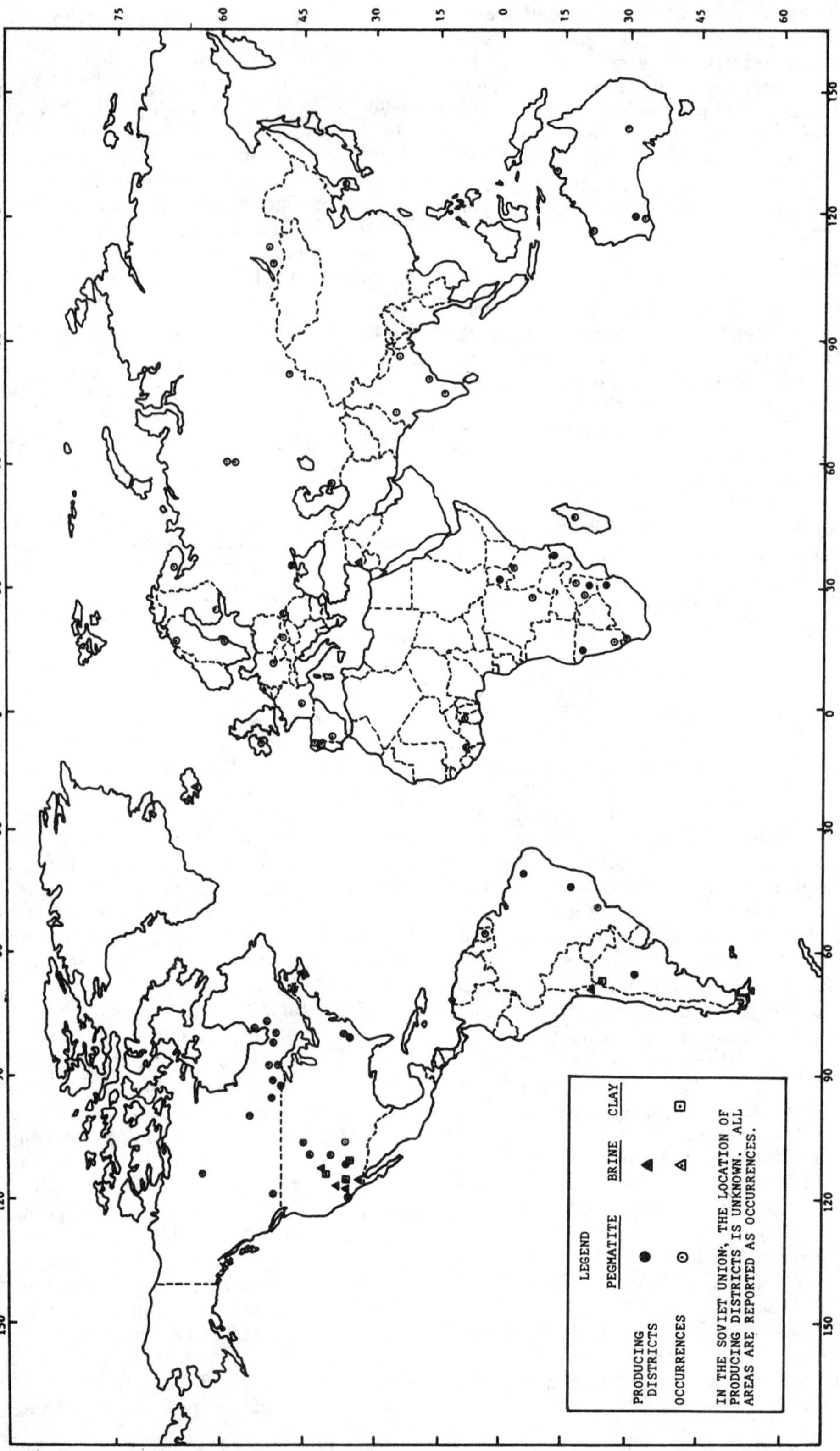

FIG. 1—*Lithium occurrences of the world.*

Creek area in western Gaston County. Proven and probable reserves on the Hallman-Beam property are reported at 25,700,000 tons of spodumene-bearing pegmatite grading 0.64% Li.

The tin-spodumene belt has a length of about 30 miles and the exploration activity has been restricted to the upper 984 ft. The geological setting and distribution of pegmatites suggest that pegmatites may well occur at greater depth. Kesler, in a study of the potential lithium contained in the total belt down to a 4921-ft level, has suggested the potential existence of a resource of 754 million tons at a grade typical for the area.

Other known pegmatite sources in the United States are not of great significance. These include the Black Hills of South Dakota, the Pala district of southwestern California, the White Picacho District of Arizona, the Harding and Pidlite mines in New Mexico, and several smaller areas in Colorado, Wyoming, Utah, and New England.

In addition to pegmatitic sources, significant amounts of lithium are produced from a brine deposit located in Clayton Valley (Silver Peak), Nevada. Lithium was also produced for a number of years from the brines of Searles Lake, where brines are presently beneficiated only for potash, sodium carbonate, and borate salts.

In Clayton Valley, located in Esmeralda County, NV, lithium-bearing brines occur in an undrained structural depression filled with Quaternary sediments composed mainly of clay minerals, including hectorite, volcanic sands, alluvial gravels, and saline minerals consisting of gypsum and halite (Kunasz, 1970).

The brine which saturates the sediments is chemically simple. It occurs as a concentrated sodium chloride solution containing subordinate amounts of potassium and minor amounts of magnesium and calcium. The lithium concentration is variable; the composite average pumped into the ponds is 200 ppm Li.

Foote Mineral Co. is presently operating a series of wells ranging from 500 to 1000 ft in depth. The brine is pumped through transite pipelines into evaporating ponds where it is concentrated by solar evaporation. The lithium resources, to a depth of 1000 ft, are estimated at 775,000 tons Li. The total recoverable reserves are estimated at 49,450 tons Li.

Zimbabwe

The largest lithium-bearing area in Zimbabwe is that of the Bikita tinfields, located about 45 miles east of Fort Victoria. Exploration has indicated the presence of an important mineralized zone within the Al Hayat, Bikita, and Southern sectors. The pegmatite is about 5100 ft long and its width varies from 95 to 210 ft. It strikes north-northeast and dips from 14° to 45° east. The property is operated by Bikita Minerals (Pvt.) Ltd., controlled by B.P. Minerals.

The pegmatite is asymmetrically zoned and contains a variety of commercially important lithium minerals as well as beryl and pollucite.

Drilling and development work have proved a reserve of 12 million tons of quarry ore grading 1.4% Li. Mining is presently done in the Al Hayat sector. The plant produces about 2,200 tons per month of petalite, 800 tons per month of lepidolite and 800 tons per month of spodumene. In the reduction plant, feldspar is separated from petalite and ground for sale (Wegener, 1980).

Other minor lithium-bearing occurrences are known in the Wankie, Salisbury, Umtali, Mtoko, Insizia, Matobo, and Mazoe districts (Toombs, 1962).

Minor Producing Districts

South America

Only two countries produce small amounts of lithium minerals in South America: Brazil and Argentina. The amblygonite deposit in Surinam has been essentially depleted.

In Brazil, lithium-bearing pegmatites are known in the districts of Minas Gerais and in the northeastern portion of the country which includes the states of Paraiba, Rio Grande do Norte, and Ceará (Heinrich, 1964).

The pegmatites which carry spodumene and amblygonite usually have been mined for cassiterite, tantalite, and beryl, while the lithium minerals have been sporadically recovered as byproducts.

An important occurrence of petalite has been reported in the tin-producing district near Araçuai, Brazil, located about 250 miles northeast of Belo Horizonte in the state of Minas Gerais. The pegmatite is reported to contain about 100,000 tons of petalite grading 2.0% Li.

Spodumene reserves have been estimated at 300,000 tons, while lepidolite reserves are considered to be nearly exhausted.

In 1976, production of spodumene was estimated at 900 tons and petalite at 2,500 tons (Afghouni, 1978).

In Argentina, lithium-bearing pegmatites occur in the western portion of the state in the Sierras Pampaneas, which include the productive districts of San Luis, Cordoba, and Catamarca. The pegmatites are zoned and contain spodumene. The reserves, considered to be small, total about 18,000 tons as spodumene (Angelleli and Rinaldi, 1963, 1965). Recently, the occurrence of brine has been reported in the salares of northwestern Argentina.

Namibia

Following a low production period, mining in the Karibib district was stimulated by the trade sanctions imposed on Southern Rhodesia (now Zimbabwe) in 1964. The district, located approximately 120 miles from Walvis Bay on the Atlantic Ocean, contains several strongly zoned pegmatites which contain lepidolite, petalite, and small amounts of amblygonite. In 1968, Klockner & Co. A.G. acquired the Rubicon and Helicon mines previously operated by S.W.A. Lithium Mines (Pty.) Ltd. Production has been expanded by the installation of a lepidolite flotation plant. Lepidolite reserves suitable as flotation feed material have been estimated at 1,000,000 tons (Anon., 1968a). Petalite reserves consist of some 200,000 tons of crude ore (Gevers, 1953).

Other African Deposits

In Africa, smaller productive areas occur in Mozambique, Uganda, and Ruanda.

In the Alto Ligonha area of Mozambique, lepidolite, amblygonite, and spodumene occur in zoned pegmatite bodies. Only lepidolite is exploited commercially. Amblygonite is rare, and spodumene is completely altered to kaolinite (Hutchinson and Claus, 1956).

Large amblygonite masses are found in pegmatite districts located west of the capital city of Kigal, in Ruanda (Valarmoff, 1954), and north of Kampala in Uganda (Roberts, 1948).

Other occurrences of lithium minerals include the Sudan, the Malagasy Republic, the Ivory Coast, and the Republic of South Africa (Nel, 1968).

Australia

The most important lithium-bearing districts occur in western Australia. Mining is active only in the Coolgardie district where small quantities of petalite and minor amounts of spodumene and amblygonite are produced at Londonderry. In the same area near Mt. Marion, the Western Mining Corp. has discovered a spodumene deposit. Drilling has indicated reserves in excess of 1,000,000 tons of ore (Anon., 1968).

Other lithium mineral occurrences include spodumene near Ravensthrope, lepidolite near Wodgina and Londonderry, and minor amblygonite near Euriowie, New South Wales (McLeod, 1965).

In addition to the producing areas described above, lithium minerals are known to occur in many European countries and in Ireland, India, Korea, and Japan. Sizable deposits exist in the Soviet Union but actual reserves data are not available.

Potential Producing Districts

Probably the largest lithium resources in the world are contained in the Manono and Kittolo pegmatites located in the Belgian Congo (Zaire). At the present time, only cassiterite and columbite are mined by Congo-Etain from the Manono pegmatite which has a length of 3 miles and a width ranging from 325 to 1300 ft. The adjoining Kittolo pegmatite has similar dimensions. Although the pegmatites are apparently zoned, their dimensions imply spodumene reserves which dwarf the presently known world reserves. However, the deposit may not have an economic value for years because of the very poor transportation facilities. The deposit is located about 1,360 miles from the Angolan port of Lobito.

Although many pegmatite districts have been described in Canada, only two are known to contain commercial quantities of lithium minerals: the Preissac-Lacorne district of western Quebec and the Bernic Lake area of Manitoba.

In the Preissac-Lacorne district, spodumene-bearing pegmatites intrude Precambrian metamorphic rocks. In addition to spodumene, they contain a variety of other minerals such as lepidolite, beryl, pollucite, molybdenite, and minor amounts of columbite and tantalite. The largest reserves occur on the property of the Quebec Lithium Corp. where drilling has indicated 15,000,000 tons of ore grading 0.56% (Mulligan, 1965). Although the company produced spodumene concentrates for a few years, the operation is presently inactive. Several other companies control spodumene-bearing properties in the district.

In the Bernic Lake area of southeastern Manitoba, the lithium-bearing pegmatites are part of a zoned complex, presently exploited for tantalum by the Tantalum Mining Corp. The property is located in the Cat Lake-Winnipeg River district which lies some 80 miles northeast of Winnipeg. The pegmatite is complexly zoned and contains several lithium-bearing units. It is a lenticular body about 3600 ft long, 1500 ft wide with a maximum thickness of 280 ft (Cerny and Turnock, 1971).

The spodumene reserves have been measured at 7,400,000 tons grading 1.34% Li. In addition to spodumene, three lepidolite zones averaging from 0.87% to 1.31% Li provide an additional 107,000 tons of ore (private communication).

The pegmatite also contains 300,000 tons of pollucite averaging 23% Cs_2O. 1,128,766 tons of tantalite grading 0.16% Ta_2O_5, and 1,000,000 tons of beryl ore averaging 0.22% BeO.

Lithium-bearing pegmatites are also known in the Herb Lake and East Braintree-West Hawk districts of Manitoba; the Yellowknife-Beaulieu district of the Northwest Territory; and the Nipigon, Dryden, Lac La Croix, Root Lake, Falcon Lake, and O'Sullivan Lake districts of Ontario (Mulligan, 1965).

Among the recently discovered brine deposits, the Salar de Atacama constitutes a very important future source of lithium. The salar is located in northern Chile, where it straddles the Tropic of Capricorn. The basin proper has a surface area of approximately 1200 square miles and the salt nucleus proper covers approximately 540 square miles. The salar, which lies at an elevation of 7500 ft, is bounded on the eastern side by the Andean Cordillera and on the western side by the Cordillera Domeyko. The salt nucleus consists almost exclusively of halite facies with a development of very narrow marginal facies of sulfate and carbonate.

The interstitial brine contains an average lithium concentration of 1500 ppm. Lithium resources are reported at 4.3 million tons.

In 1975, Foote Mineral Co. entered into an agreement to evaluate the feasibility of producing lithium carbonate.

Future Raw Materials

Several areas of the world carry potential lithium raw materials in the form of brines, geothermal brines, oil well brines, and clays. Among the existing brine deposits, the Great Salt Lake of Utah constitutes an important potential source of lithium chemicals. The lake holds a brine formed as a result of concentration by solar evaporation of waters contained in a much larger and more dilute Pleistocene lake—Lake Bonneville. In the Bear River Basin, the Great Salt Lake Minerals and Chemicals Corp. has constructed 14,000 acres of ponds and is producing potassium and sodium sulfates. The recoverable lithium resources are estimated at 286,000 tons (Evans, 1978).

A potential geothermal area lies within the Imperial Valley of southern California. The chemically complex brine contains large concentrations of sodium, calcium, and potassium. It also carries about 0.020% Li. Koenig (1970) indicates that the brine volume which exists under an area of 12 to 24 sq miles is in excess of 1 cu mile. This suggests a potential reserve exceeding 840,000 tons Li. Exploitation of lithium will depend upon whether or not some complex chemical engineering problems are solved.

Other geothermal areas include the Reykanes Field of Iceland where the visible active area covers an area of about 2 sq miles. Feasibility studies (Ludviksson and Hermannsson, 1970) indicate that a yearly production of 500 tons of lithium compounds is possible. The brine contains 8 ppm Li. Additional thermal areas are known at Reykir, Hveragardi, and Krysuvik (Karlsson, 1961; Prast, 1972).

A similar potential source of lithium exists in the geothermal waters at Wairakei, New Zealand. Potential production from bore waters containing 13 ppm Li are estimated at 2400 tpy as Li_2CO_3, assuming a discharge rate of 1,000,000 gal per hr (Kennedy, 1964).

Lithium-bearing clays are known to occur in several localities in the United States. The highest lithium value of 0.53% has been recorded at Hector, CA, where the Baroid Division of National Lead Co. mines hectorite for its swelling characteristics (Ames, et al., 1958). Hectorite has been identified in Clayton Valley where it occurs as an alteration product of volcanic ash along a fault zone and is one of the clay minerals in Tertiary and Quaternary lake sediments (Kunasz, 1970). The hectorite contains 0.24% to 0.35% Li. At Spor Mountain, Utah, 0.11% Li has been reported in beryllium-bearing tuffs (Shawe, 1968, Shawe, et al., 1964), where lithium probably also occurs in hectorite. The occurrence of lithium-bearing clays is also reported (Norton, 1965) in Yavapai County, AZ (0.10% Li), in Tertiary clays at Kramer, CA (0.19% Li) (Muessig, 1966), and in the Turilari playa (0.14% Li) in Argentina (Muessig, 1966). An extensive survey by the Lithium Exploration Group of the US Geological Survey has indicated that vast resources of lithium lie in a great number of sedimentary clays (Vine, 1976).

Although the clays contain lithium concentrations of the same order of magnitude as commercial pegmatites, they are not likely to become lithium sources in the near future because of technological problems associated with the extraction of lithium from the crystal lattice.

Production

The bulk of the world's lithium is manufactured in the United States by two companies: Foote Mineral Co., at Silver Peak, NV, Kings Mountain, NC, Sunbright, VA, and at Frazer, PA; and Lithium Corp. of America at Bessemer City, NC, and in the United Kingdom. In Europe, an important chemical producer is Metallgesellschaft A.G. in Germany. In addition, lithium chemicals are produced in Japan and Brazil. Significant quantities of lithium ores are produced in Zimbabwe by Bikita Minerals (Pvt.) Ltd. The USSR produces sizable quantities of chemicals. Lithium chemicals also have been supplied recently by the People's Republic of China.

Mining of lithium ores can be divided into three categories: underground, open-pit, and "mining" of brines.

Where practical, such as at Kings Mountain, NC, the open pit is the most economical method. Higher cost underground methods are used in Brazil and Namibia.

Although hand cobbing and screening are still used in Brazil and Africa, flotation is the main beneficiation method of lithium ores. In the production of spodumene concentrates, the mined ore undergoes several stages of crushing, grinding, and classification, followed by flotation with organic reagents. At Foote Mineral Co., the spodumene concentrate collected from the flotation cells is further processed into a chemical grade used in the manufacture of lithium chemicals. A low iron spodumene variety produced by the chlorination of β-spodumene (Fishwick, 1967; Richardson and Fishwick, 1968) is used as a substitute for petalite in glass-ceramic applications.

In addition to spodumene concentrates, the beneficiation of pegmatite ore includes several byproducts such as quartz, feldspar, and mica. Other potential byproducts are beryl, columbite, tantalite, cassiterite, and wolframite.

The extraction of lithium from spodumene can be achieved by several methods. Presently, the process used by Foote Mineral Co. and Lithium Corp. of America involves calcination to β-spodumene, which is then reacted with sulfuric acid to produce lithium sulfate. Conversion to lithium carbonate is accomplished by soda ash addition (Ellestad and Leute, 1950). Other processes have included the calcination of a spodumene-calcium carbonate mixture, followed by hot water leaching, evaporation-crystallization to produce lithium hydroxide as a final product. Another process, used in the past by the Quebec Lithium Corp., involved the calcination of spodumene and its reaction with an aqueous solution of sodium carbonate. Purification involved reaction with carbon dioxide, and then heating to produce a final lithium carbonate product (Archambault, 1963).

Lithium is produced from brines at only one location in the world—Clayton Valley, Nevada. Foote Mineral Co. "mines" by pumping saline solutions contained in sediments into large solar evaporation ponds (Barrett and O'Neil, 1970). Following a 20- to 30-fold concentration, the final lithium chloride solution is converted to lithium carbonate by addition of soda ash. Small impurities are removed within the pond system by the addition of lime (Shay, 1967).

At the Salar de Atacama, lithium rich brines will also be concentrated in solar evaporation ponds. Following the removal of calcium, magnesium, and sulfate impurities, the concentrated lithium chloride solution will be reacted with soda ash to produce a final lithium carbonate product. Production of lithium carbonate by the Sociedad Chilena de Lithio—a joint venture between Foote Mineral Co. (55%) and CORFO (45%)—is scheduled for 1984.

Uses

Lithium is marketed and used in three basic forms: as an ore and concentrate, as a metal, and as chemical compounds. As shown in Table 1, there are numerous applications.

Ores and concentrates are consumed by the glass, ceramic, and porcelain enamel industries. Petalite, lepidolite, and amblygonite can be used without prior beneficiation, except handcobbing, while spodumene must be concentrated and treated for removal of iron. Lithium is useful because it creates favorable internal nucleation conditions. By comparison, sodium favors external crystallization. Because of its small ionic radius and high field strength,

TABLE 1—Applications

Ores				
Spodumene	Petalite	Lepidolite	Amblygonite	Metal
Ceramics	Ceramics	Glasses	Li-Chemicals	Deoxidation
Glass ceramics	Glasses			Degasification
Enamels	Glazes			Alloys
Glasses	Enamels			Organic compounds
Li-Salts				Synthetic rubber
				Vitamins
				Nuclear energy
				Battery

Chemicals						
Lithium Carbonate	Lithium Hydroxide	Lithium Chloride	Lithium Fluoride	Lithium Bromide	Butyllithium	Other Lithium Compounds
Ceramics	Greases	Li-Metal	Porcelains	Absorption refrigeration	Polymerization	Electrolytes
Li-Chemicals	Li-Salts	Fluxing	Enamels		Pharmaceuticals	Catalysis
Fluxing	CO_2 absorption	Dehumidification	Glazes		Organic synthesis	
Pharmaceuticals		Synthetic fibers	Welding			
Aluminum electrolysis cells		Tracer	Brazing			
			Optics			

lithium imparts high mechanical strength, thermal shock resistance, as well as good chemical resistance to the product (Fishwick, 1966). In these applications, lithium is also introduced in the form of lithium carbonate.

In metal form, lithium is the lightest solid element, having an atomic weight of 6.94 and a specific gravity of 0.534 (20°C). It is manufactured by the electrolysis of a molten salt mixture of lithium and potassium chlorides. Lithium metal is used in the synthesis of butyllithium. In nonferrous metallurgy, the high reactivity of lithium with gases is used for scavenging oxygen and sulfur, converting them into stable compounds. Lithium is also used in lithium-aluminum and lithium-magnesium alloys where it imparts high temperature strength, improves elasticity, and increases the tensile strength (Bach, et al., 1967). An important development has been the production of lithium batteries used in pacemakers, electric calculators, and watches (Grady, 1980).

Among the various chemicals produced, the most important are: lithium carbonate, lithium hydroxide, lithium chloride, lithium bromide, and butyllithium. The remaining compounds constitute only a small proportion of the lithium market.

Lithium carbonate, in addition to its consumption by the ceramic industry, is being used at an increasing rate in the aluminum industry. When added to aluminum reduction cells, it increases the conductivity of the molten bath, reduces operating temperature, and results in higher production. It is also very effective in reducing fluorine emissions by retaining the fluorine as lithium fluoride in the bath. In purified form, lithium carbonate is being used in the chemotherapeutic treatment of manic depression.

Lithium hydroxide was first employed as an ingredient in alkaline storage batteries where its presence increased the life of the cells (Bach, et al., 1967). In the 1940s, a new application was discovered by Clarence Earl who developed a new grease by reacting lithium hydroxide with fatty acids. The resulting grease was found to retain its viscosity over a wide temperature range and remain stable in the presence of water. Today, lithium-based greases have become a standard product in military, industrial, and automotive lubrication fields. Anhydrous lithium hydroxide is capable of absorbing large quantities of carbon dioxide, a property that has been utilized in the air regeneration system of the Apollo command and lunar modules, as well as new applications in the mining industry for rescue-breathing apparatus.

Lithium chloride and lithium bromide brines have low vapor pressures and are used in absorption refrigeration systems. Lithium chloride also constitutes the feed material for the production of lithium metal. Lithium fluoride is used mainly as a flux in enamels, glasses, glazes, and in welding and brazing.

The most important organic compound is butyllithium. It serves as a catalyst in the polymerization of butadiene, isoprene, and styrene to produce polymers with special properties.

Numerous miscellaneous applications of lithium chemicals include sanitation and bleaching, hydrogen generation, oxygen generation, catalysis, and vitamin synthesis.

Prices

In the United States, the price of chemical spodumene is approximately $180 per net ton f.o.b. Freight rates for lithium ores vary from $25 to $35, according to the distance from the producing plants.

In other parts of the world, the 1983 prices of various ores (in dollars per long ton, c.i.f.) are as follows:

Spodumene—4 to 7% Li_2O: 315–485
Petalite —3.5 to 4.5% Li_2O: $315

Prices for lithium chemicals in 1983 are shown in Table 2.

TABLE 2—Typical Prices of Lithium Compounds in 1983

	$ per Lb
Lithium carbonate, powder	1.48
Lithium hydroxide, monohydrate, bulk	1.93
Lithium chloride, anhydrous, bulk	3.15
Lithium fluoride, bulk	4.50
Lithium metal, 1,000-lb lots	21.70
n-Butyllithium, 15% solution, bulk	13.39
Lithium bromide, anhydrous, bulk	6.51
Lithium bromide, brine, bulk	3.00
Lithium chloride, brine, bulk	2.67

Future Trends

Lithium is still a developing commodity. Its applications in various phases of the metallurgical and chemical industries have been rapidly diversifying and expanding in the past few years. In 1980, the estimated world consumption of lithium metal and chemicals was reported at 54.5 million pounds of lithium carbonate equivalents (Alexander, 1981). In addition, it was also estimated that the USSR consumes probably about 10 million pounds of lithium carbonate equivalents. At the present time, the industry has sufficient production capacity to supply the existing market, and the announced increases for new capacity appear to be sufficient to meet future growth.

A potential application for lithium is in nuclear power as an absorption blanket in fusion generation. In thermonuclear reactions, the most likely fuel will be a deuterium-tritium mixture because of its very large reaction rate (Bogart, 1980). Although tritium is naturally scarce, it can be artificially generated within the reactor itself by bombarding the lithium-6 isotope in the blanket with the neutrons created by the fusion reaction.

A promising application for lithium metal is in batteries. In the past decade, large efforts have been made to find a battery with a high energy-weight ratio. Because of its low weight and high electrochemical potential, lithium is a prime candidate (Grady, 1980). Primary lithium batteries have moved from the development laboratory into pilot plant production. Commercial utilization could require many thousands of pounds of lithium metal.

Singleton (1979) estimates that the world demand for lithium will quadruple by the end of the century. If the potential applications in nuclear power and batteries are considered, the growth rate may be considerably greater.

Presently known reserves exceed the cumulative conventional demands by about two orders of magnitude. Thus the amount of exploration has been justifiably limited. Should increasing demands be imposed on the lithium industry, vast resources can be tapped within a short period of time (Kunasz, 1980).

Bibliography and References

Anon., 1967, "Batteries," *Metals Week*, Sep. 25, pp. 11–27.

Anon., 1967a, "Lithium and the Electric Car," *Foote Prints*, Vol. 36, pp. 96–97.

Anon., 1968, "Lithium Find for Western Mining," *Mining Journal*, Mar. 8.

Anon., 1968a, "Klockner Buys SWA Lithium," *Metal Bulletin*, No. 5288, Apr. 5, p. 26.

Anon., 1976, "Lithium By the Year 2000," *Industrial Minerals*, No. 108, Sep., pp. 45–46.

Anon., 1977, "Lithium Chemical Expansion Insures Supply," *Industrial Minerals*, No. 122, Nov., pp. 67–73.

Afghouni, K., 1978, "Lithium Ores in Brazil," *Energy*, S.S. Penner, ed., Vol. 3, No. 3, pp. 247–253.

Alexander, J.H., 1981, Lithium Annual Review, *Engineering and Mining Journal*, pp. 111–113.

Ames, L.L., et al., 1958, "A Contribution on the Hector, California Bentonite Deposit," *Economic Geology*, Vol. 53, No. 1, Jan.–Feb. pp. 22–37.

Angelleli, V., and Rinaldi, C.A., 1963, "Yacimientos de Minerales de Litio de las Provincias de San Luis y Cordoba," Info. No. 91, Argentina, Com. Nac. Energia Atomica, 79 pp.

Angelleli, V., and Rinaldi, C.A., 1965, "Resena Acerca de la Estructura, Mineralizacion, y Aprovechamento de Nuestras Pegmatita Portadoras de Minerales de Litio," *Acta Geologica Lilloana*, Vol. 5, pp. 1–18.

Archambault, M., 1963, US Patent 3,112,171, Nov. 26.

Bach, R. O., et al., 1967, "Lithium and Lithium Compounds," *Encyclopedia of Chemical Technology*, A. Standen, ed., Vol. 12, 2nd ed., Interscience, New York.

Barrett, W.T., and O'Neill, B.J., Jr., 1970, "Recovery of Lithium from Saline Brines Using Solar Evaporation," *Third Salt Symposium*, J.L. Rau, and L.F. Dellwig, eds., Vol. 2, The Northern Ohio Geological Society, Inc., Cleveland, pp. 47–50.

Bogart, L.S., 1980, "Potential Lithium Requirement for Fusion Electric Power," *Foote Prints*, Vol. 43, No. 1, pp. 2–11.

Buckley, S., 1983, "Lithium—Recession Delays Market Lift Off," *Industrial Minerals*, No. 185, Feb., pp. 25–35.

Caldwell, H.C., et al., 1971, "A Pharmaco-Kinetic Analysis of Lithium Carbonate Absorption from Several Formulations in Man," *Journal of Clinical Pharmacology*, Vol. 11, No. 5, pp. 349–356.

Cameron, E.N., et al., 1949, *Internal Structure of Granitic Pegmatites*, Monograph 2, Economic Geology Publishing Co., Urbana, IL.

Cameron, E.N., et al., 1954, "Pegmatite Investigations, 1942–1945, New England," Professional Paper 255, US Geological Survey, 352 pp.

Cerny, P., and Turnock, A.C., 1971, "Pegmatites of Southeastern Manitoba," Special Paper No. 9, Geological Association of Canada, pp. 119–127.

Deer, W.A., et al., 1962, "Lepidolite," *Rock-Forming Minerals*, Vol. 3, J. Wiley and Sons, New York, pp. 85–91.

Eardley, A.J., 1970, "Salt Economy of the Great Salt Lake," *Third Symposium on Salt*, J.L. Rau and L.F. Dellwig, eds., Vol. 1, The Northern Ohio Geological Society, Inc., Cleveland, pp. 78–105.

Ellestad, R.B., and Leute, K.M., 1950, US Patent 2,516,109, July 25.

Fishwick, J.H., 1966, "Glass Ceramics," *Foote Prints*, Vol. 35, No. 1, pp. 20–23.

Fishwick, J.H., 1967, US Patent 3,394,988, Oct. 31.

Foster, M.D., 1960, "Interpretation of the Composition of Lithium Micas," Professional Paper 354–B, US Geological Survey, 146 pp.

Goldschmidt, V.M., 1937, "The Principles of Distribution of Chemical Elements in Minerals and Rocks," *Journal of the Chemical Society*, pp. 655–673.

Hanley, J.B., et al., 1950, "Pegmatite Investigations in Colorado, Wyoming, and Utah, 1942–1944," Professional Paper 227, US Geological Survey, 125 pp.

Hegelson, H.C., 1968, "Geological and Thermodynamic Characteristics of the Salton Sea Geothermal System," *American Journal of Science*, Vol. 266, March, pp. 129–166.

Heinrich, E.W., 1964, "The Tin-Tantalum-Lithium Pegmatites of Sao Joao Del Rei District, Minas Gerais, Brazil," *Economic Geology*, Vol. 59, No. 6, Sep.–Oct., pp. 982–1002.

Holdren, J.P., 1971, "Adequacy of Lithium Supplies as Fusion Energy Source," Hearing Before the Joint Committee on Atomic Energy, Congress of the United States, Nov. 10 and 11.

Hortsman, E.L., 1957, "The Distribution of Lithium, Rubidium, and Cesium in Igneous and Sedimentary Rocks," *Geochimica et Cosmochimica Acta*, Vol. 12, pp. 1–28.

Hutchinson, R.W., and Claus, R.J., 1956, "Pegmatite Deposits, Alto Ligonha, Portuguese East Africa," *Economic Geology*, Vol. 51, pp. 757–779.

Jahns, R.H., 1952, "Pegmatite Deposits of the White Picacho District, Maricopa and Yavapai Counties, Arizona," University of Arizona *Bulletin*, Vol. 23, No. 5, 105 pp.

Jahns, R.H., 1955, "The Study of Pegmatites, *Economic Geology*, 50th Anniversary Volume, Part II, pp. 1025–1130.

Karlsson, T., 1961, "Drilling for Natural Steam and Hot Water in Iceland," *Proceedings of the United Nations Conference on New Sources of Energy*, Rome, Vol. 3, pp. 215–221.

Kennedy, A.M., 1964, "Recovery of Lithium and Other Minerals from Geothermal Water at Wairakei," *Proceedings of the United Nations Conference on New Sources of Energy*, Rome, Vol. 3, pp. 502–511.

Kesler, T.L., 1961, "Exploration of the Kings Mountain Pegmatites," *Mining Engineering*, Vol. 13, No. 9, Sep., pp. 1062–1068.

Koenig, J.B., 1970, "Geological Setting of the Imperial Valley and its Geothermal Resources," Compendium of Papers Presented at the Imperial Valley—Salton Sea Area Geothermal Hearing, Oct. 22–23, Sacramento, CA, pp. E1–E5.

Kunasz, I.A., 1970, "Geology and Geochemistry of the Lithium Deposit in Clayton Valley, Esmeralda County, Nevada," Ph.D. Thesis, The Pennsylvania State University, 114 pp.

Kunasz, I.A., 1975, "Lithium Raw Materials," *Industrial Minerals and Rocks*, 4th ed., S.J. Lefond, ed., AIME, New York, pp. 791–803.

Kunasz, I.A., 1980, "Lithium, How Much?" *Foote Prints*, Vol. 48, No. 1, pp. 23–27.

Ludviksson, V., and Hermannsson, S., 1970, "Some Possibilities for Major New Industries in Iceland," National Research Council, Reykjavik, 26 pp.

Mayhew, E.J., and Heylmun, E.B., 1966, "Complex Salt and Brines of the Paradox Basin," *Second Symposium on Salt,* J.L. Rau, ed., The Northern Ohio Geological Society, Inc., Cleveland, pp. 221–235.

McLeod, I.R., 1965, *Atlas of Australian Resources Mineral Deposits,* 2nd ed., Department of National Development, Canberra.

Muessig, S., 1966. "Recent South American Borate Deposits," *Second Symposium on Salt,* J.L. Rau, ed., The Northern Ohio Geological Society, Inc., Cleveland, pp. 151–159.

Mulligan, R., 1965, "The Geology of Canadian Lithium Pegmatites," Economic Report No. 21, Geological Survey of Canada, 131 pp.

Nel, L.T., 1968, "Ore Deposits of Lithium in the Republic of South Africa," Atomic Energy Board, Republic of South Africa, PEL 23, 23 pp.

Norton, J.J., and Schlegel, D.M., 1955, "Lithium Resources of North America," Bulletin 1027–G, US Geological Survey, pp. 325–350.

Norton, J.J., 1965, "Lithium-Bearing Bentonite Deposit, Yavapai County, Arizona," Professional Paper 525–D, US Geological Survey, pp. D163–D166.

Page, L.R., et al., 1953, "Pegmatite Investigations, 1942–1945, Black Hills, South Dakota," Professional Paper 247, US Geological Survey, 228 pp.

Penner, S.S., 1978, "The Lithium Symposium 1977," *Energy*, Vol. 3, No. 3, 413 pp.

Prast, W.G., 1972, "Scope for Expanded Use of Iceland's Resources," *Mining Magazine,* Nov., pp. 465–468.

Rankama, K., and Sahama, T.G., 1950, "The Alkali Metals: Lithium, Sodium, Potassium, Rubidium, Cesium," *Geochemistry,* University of Chicago Press, Chicago, pp. 422–442.

Richardson, L.S., and Fishwick, J.H., 1968, "Removal of Iron from Spodumene Concentrates by Chlorination," *Glass Industry,* July, pp. 377–81.

Roberts, R.O., 1948, "Amblygonite and Associated Minerals from the Mbale Mine, Uganda," *Imperial Institute Bulletin,* Vol. 46, pp. 342–347.

Shawe, R.D., 1968, "Geology of the Spor Mountain Beryllium District, Utah," *Ore Deposits in the U.S., 1933/1967,* J. Ridge, ed., Vol. 2, AIME, New York, pp. 1143–1161.

Shawe, R.D., et al., 1964, "Lithium Associated with Beryllium in Rhyolitic Tuff at Spor Mountain, Western Juab County, Utah," Professional Paper 501–C, US Geological Survey, pp. 86–87.

Shay, F.B., 1967, "Low Cost Lithium from Brine at Silver Peak, Nevada," paper presented at AIME Annual Meeting, Los Angeles, Feb. 21.

Singleton, R.H., 1979, "Lithium," *Mineral Commodity Profiles,* US Bureau of Mines, 25 pp.

Strock, L.W., 1936, "Zur Geochemie des Lithiums," Nachrichten von der Gesellschaft der Wissenschaften zu Goeltingen, Mathematisch-Physikalische Klasse, IV, N.F. 1, No. 15, 171 pp.

Toombs, R.B., 1962, "A Survey of the Mineral Industry of Southern Africa," Mineral Information Bulletin MR 58, Department of Mines and Technical Survey, Ottawa, 275 pp.

Valarmoff, N., 1954, "Matériaux pour l'Etude des Pegmatites du Congo Belge et du Ruanda-Urundi," *Annales de la Société Géologique Belgique,* Vol. 78, pp. 1–25.

Vine, J.D., 1976, "Lithium Resources and Requirements by the Year 2000," Professional Paper 1005, US Geological Survey, 162 pp.

Wegener, J.E., 1980, "Lithium Minerals in Zimbabwe with Special Reference to the Operations at Bikita Minerals (Pvt) Ltd.," International Conference on Zimbabwe, Sep. 1–5, Salisbury.

Wegener, J.E., 1981, "Profile on Bikita—Processed Petalite, The New Priority," *Industrial Minerals,* No. 165, June, pp. 51-53.

Westenberger, H., 1963, "The Lithium Minerals, Their Formation and Occurrence," *Review of Activities,* No. 6, Metallgesellschaft A.G., Frankfurt/Main.

White, D.E., 1957, "Thermal Waters of Volcanic Origin," *Geological Society of America Bulletin,* Vol. 68, pp. 1637–1658.

Winchell, A.N., 1942, "Further Studies of the Lepidolite System," *American Mineralogist,* Vol. 27, pp. 114–130.

Magnesite and Related Minerals

O. M. WICKEN *

L. R. DUNCAN †

Magnesium, the eighth most abundant element in the earth's crust, is found widely distributed in a variety of minerals. Among the more commercially important ones are magnesite ($MgCO_3$), brucite ($Mg[OH]_2$), dolomite ($CaMg[CO_3]_2$), and the salts magnesium sulfate and chloride often found in natural brines. Among other commercially important magnesium-containing minerals are olivine ($[MgFe]_2SiO_4$), talc ($H_2Mg_3[SiO_3]_4$), and serpentine ($H_4Mg_3Si_2O_9$). These are valuable because of desirable characteristics given them by the magnesium content and its placement in the particular crystal structure.

These minerals are the raw materials for a host of products including magnesium metal and several grades of magnesia or magnesia-containing materials for refractories, fluxes, fillers, insulation, cements, decolorants, fertilizers, and chemicals.

Considering only those minerals valued for their magnesia or magnesium content, over 85% of the material tonnages are processed to be used as refractories. Dolomite, which is used in enormous tonnages for its physical properties in construction, has an important use as a refractory raw material in its own right and, after calcination, as a precipitant for magnesium hydroxide, which is an intermediate for the production of magnesia or magnesium metal. Magnesia from the reaction of calcined dolomite (or limestone) with seawater or magnesium-containing brines has supplanted much of the production of magnesia from mined magnesite, particularly in the United States.

* CEO, SRM Associates, Pittsburgh, PA.

† Vice President, Technical and Manufacturing Services, Harbison-Walker Refractories International, Div. of Dresser Industries Inc., Pittsburgh, PA.

The best known of the minerals directly and widely exploited for magnesia content is magnesite, one of the calcite group of rhombohedral carbonates which includes calcite ($CaCO_3$), siderite ($FeCO_3$), and rhodochrosite ($MnCO_3$) among others. The members of this group enter into a wide range of substitutional solid solutions when the positive ions have similar radii. The radii of magnesium and iron ions are within 6% of each other; hence, magnesite and siderite form a complete series of which breunnerite (ferroan magnesite) is a well-known end member. However, the radius of calcium ion is 36% larger than that of magnesium ion, and only limited substitution exists at each end of the $MgCO_3$-$CaCO_3$ series. Dolomite is not a member of the calcite group, but results when calcium and magnesium ions alternate in equal number in an ordered structure among carbonate ions. The result of these relationships is that calcite and dolomite, often found intermixed with magnesite, occur commonly as identifiable crystal entities which can often be separated to a varying degree from the magnesite by various beneficiation techniques.

Magnesite, when pure, contains 47.8% MgO and 52.2% CO_2. The pure mineral is sometimes, but rarely, found as transparent crystals resembling calcite, but, preponderantly, magnesite contains variable amounts of the carbonates, oxides, and silicates of iron, calcium, manganese, and aluminum. Magnesite may be either crystalline or amorphous (cryptocrystalline). The crystalline form has a hardness of 3.5 to 4.0. The color may range from white to black with shades of yellow, blue, red, or gray. The color is not a fundamentally significant indicator of purity, but in a given deposit, an experienced person can often roughly grade magnesite by assessing color and crystallinity.

The crystalline form of magnesite occurs in

relatively few but large deposits whereas the cryptocrystalline variety tends to occur in many but small deposits. Cryptocrystalline magnesite is massive with no cleavage and is sometimes descriptively called *bone magnesite*. The fracture is usually conchoidal, and hardness is 3.5 to 5.0. The color is normally white, but it can be tints of yellow, orange, or buff. Accessory siliceous minerals such as serpentine, quartz, or chalcedony are usually present. Calcium minerals are usually absent or in low concentration in cryptocrystalline magnesite as contrasted to their almost invariable presence and higher concentration in the crystalline variety.

The specific gravity of cryptocrystalline magnesite runs between 2.90 and 3.00, while that of pure, crystalline form is 3.02. In actuality, the crystalline form is often higher than 3.02 because of the presence of iron carbonate.

Magnesite dissociates upon heating to form magnesia (MgO) and carbon dioxide. When heated sufficiently, magnesia develops a crystal structure identical with that of the natural mineral periclase. The mineral periclase occurs only rarely in nature and not in any known, commercially workable deposits.

Brucite ($Mg[OH]_2$) is another mineral which can be directly reduced to periclase through the application of heat (Comstock, 1963). Brucite has been mined for the production of magnesia, but is no longer an important source. Theoretically it contains 69.1% MgO and 20.9% H_2O. The mineral is often associated with limestone, but most commonly with magnesite. The mineral has a translucent appearance and is relatively soft and lightweight with hardness of 2.5 and specific gravity of about 2.4. The color may be white, but blue or green with a gray cast is a more common coloration.

Nomenclature of Products

The word *magnesite* literally refers only to the natural mineral, but common usage has applied the name plus a prefatory word to two other types of materials. These are *dead-burned magnesite* and *caustic-calcined magnesite*. For the most part, these materials are essentially magnesia differing mainly in density and crystal development as the result of different levels of heat application. When magnesia produced from seawater or brines made its appearance, the products also received the appellations "dead-burned" or "caustic-calcined" magnesite, but of late years, the technical literature, particularly in the United States, increasingly refers to the materials as "refractory magnesia" or "refractory grade magnesia," and "caustic-calcined magnesia," as appropriate. These terms are now also increasingly applied to magnesias derived from the mineral magnesite, particularly for those materials with high MgO content. Nevertheless, the older terms are still widely used in the trade, particularly abroad.

The term *dead-burned magnesite* (refractory magnesia) refers to the granular product produced by firing magnesite, magnesium hydroxide, or other material reducible to magnesia at temperatures in excess of 1450°C and for enough time to produce a dense, weather-stable granule suitable for use in the manufacture of refractory materials.

The use of terms in describing magnesia products is somewhat confusing to those not closely connected with the industry. The vagueness of terminology may lead to misreading of statistics to a degree that simultaneous shortage and surplus of magnesia products can be indicated. Confusion has been further amplified by use of such undefined terms as "high-purity" or "super-high-purity" magnesias. In most cases, these terms have been used for magnesia being supplied to the refractory industry and refer to content and relationship of the accessory oxides rather than a specific amount of MgO.

The loose term, "high-grade," refers to refractory magnesia containing roughly over 96% MgO and having a density greater than 3.20 as a result of processing at high temperatures. The mineral name *periclase* is sometimes used in the trade as a catchall term for any refractory magnesia having high MgO content and density. The term *standard grade*, as distinguished from *high-grade*, applies to material of relatively lower purity.

Caustic-calcined magnesia is the product produced when magnesite, or material reducible to magnesia by heat, is heated to such degree that less than 10% ignition loss remains and the product displays adsorptive capacity or activity. The degree of sintering is much less than in the case of refractory magnesia, and commonly the firing temperature is lower than 900°C.

Origin and Occurrence of Magnesite

Crystalline Magnesite

Deposits of crystalline magnesite are usually found associated with dolomite, but some major deposits such as found in Brazil are in limestone measures. But even in these cases, the magnesite is not in direct contact with limestone, but is separated by a dolomitized zone. The major deposits of crystalline magnesite of the world are those in Austria, USSR, Korea, Manchuria, Brazil, Canada, Australia, Nepal, and the United States. All are in dolomite host measures or zones. Further, all are in regions which have had orogenic activity. Commonly, such areas also show igneous activity, and an early and strongly supported theory postulated that magnesite resulted from action of igneous intrusions and associated solutions on the dolomite.

The dolomite measures are widely viewed as being sedimentary (Dunbar and Rogers, 1957), and a few investigations have suggested a sedimentary origin for both the magnesite and the enclosing dolomite (Paone, 1970; Schroeder, 1948), but consensus favors secondary placement of magnesite in preexisting dolomite by hydrothermal action on a volume for volume basis (Bain, 1924; Bennett, 1943; Campbell and Loofbourow, 1962).

In the rather unique deposit of magnesite near Kilmar, Que., Canada, the intermixed rock of magnesite-dolomite is thought to be the product of hydrothermal dolomization of the limestone (Bray and Hilchey, 1957).

Magnesite is often closely associated both spatially and in time with intrusive activity (Bain, 1924; Bodenlos, 1954; Siegfus, 1927; Spencer, 1972). Igneous rocks associated with magnesite vary widely in composition. Bodenlos (1954) compiled a list which includes amphibolite, pyroxenite, diabase, peridotite, rhyolite, quartz, basalt, granite, and others.

The minerals associated with several crystalline magnesite deposits suggest relatively high temperature of formation (Niiomy, 1925). These associated minerals include dolomite, talc, serpentine, diopside, hematite, and enstatite. Olivine, pyrrhotite, and magnetite have been recognized in the Chewelah (Washington) deposits, while scapolite has been found in the Manchurian deposits. On the basis of associated minerals, Bodenlos (1954) has suggested a temperature of formation of 300°C or higher, and Schroeder (1948), a temperature within the hypothermal range of 300–500°C. It is reasonable to conclude on the basis of associated minerals that these magnesite deposits formed at a temperature in excess of 300°C.

Although the field evidence is clear and consistent, the nature of the hydrothermal solutions and the mechanism of replacement remains a problem. Mechanisms suggested for the hydrothermal emplacement of magnesite in dolomite involve reaction with Mg-rich, CO_2-bearing solutions. Although CO_2-bearing solutions may be conveniently derived from nearby intrusive bodies, it is not easy to account for the required Mg-enrichment of such solutions. An experimental investigation (Rosenberg and Holland, 1964; Rosenberg and Mills, 1966), throws new light on this problem. The investigation was carried out to conform with observed field relations and under conditions that obviated the necessity for Mg-rich solutions; the results were consistent with the presence of high-temperature mineral assemblages associated with magnesite. The proposed mechanism may be summarized as follows:

1) Attack on preexisting dolomite by CO_2-bearing solutions at temperatures below 200°C resulting in solutions with Ca:Mg ratio of unity.

2) Heating of these solutions to temperatures in excess of 200°C, resulting in the replacement of host rock dolomite by magnesite with the production of Ca-rich solutions.

In the discussion on the origin of the magnesite deposits occurring in the Kumaun Himalaya, India, the role of algae was presented (Valdiya, 1968). A chain of lentiform deposits of coarsely crystalline magnesite extends for 130 km northwest from the Kali Valley to the Alaknanda Valley. The Almora mining district immediately west of Nepal is within the mineralized zone. The deposits are of uniform thickness and chemical and mineralogical composition. The reported average chemical composition is $MgCO_3$, 90.2%; $CaCO_3$, 2.9%; and $SiO_2 + R_2O_3$, 6.9%. The magnesite is restricted to a narrow stratigraphic interval in the upper part of the Gangolihat Dolomites and is almost invariably associated with the stromatolite-bearing dolomite. Barred embayments, attributed to barrier building by algae, were formed in the back reef shelf of the ancient Gangolihat Sea. The Mg:Ca ratio of the basin waters progressively increased with the passage of time as a cumulative result of biogenic and organic precipitation of $CaCO_3$ during the aqueous stages of the sedimentary process. Accumulation of algoid debris further increased the concentration of magnesium. The Mg:Ca ratio became so high that during periods of high pH attributable to prolific growth of the algae, the carbonates formed earlier were converted into carbonate assemblages of higher magnesium content, including magnesite. Coarse granularity of magnesite is attributed to later recrystallization during regional metamorphism. The Sinian magnesite of Malyghingen of Manchuria and of northern China form irregular, though concordant, bedded deposits of a similar nature (Nishihara, 1956).

Cryptocrystalline Magnesite

Cryptocrystalline or "amorphous" magnesite is an alteration product of serpentine or allied magnesium rocks which have been subjected to the action of carbonate waters. The serpentine which lies near or surrounds the magnesite is itself an alteration product of the ultrabasic rocks. The mode of formation of the magnesite usually limits the amount of impurities to small amounts of iron, lime, and silica.

Occurrences of this type of magnesite are fairly common throughout the world, but because of their usually limited size, few—with the notable exceptions of those in Greece, Turkey, and Australia—are worked commercially. The action of surface water containing carbon dioxide percolating down through serpentinized fissures can convert serpentine to magnesite and other minerals. Also, waters rising through fissures could produce magnesite. In both cases, the relationship between temperature and carbon dioxide pressure is very critical to the formation of magnesite. The zones closer to the earth's surface are more favorably situated for the reaction. Therefore, most of the deposits of this type of magnesite are found near the surface and have limited extent in depth.

Distribution of Deposits

From the natural mineral magnesite, total world production (1980) of all grades of dead-burned and calcined magnesite was about 4.3 million st, representing about 10 million st of crude magnesite. The production came from many countries, with those of central Europe and Asia contributing more than 80%.

Prior to 1938 when the first large-scale seawater magnesia plant was started, these areas were the principal suppliers of refractory magnesia to the industrial nations of the world, including the United States. Even today, there is no major competition from seawater magnesia plants for markets in areas such as USSR, China, North Korea, and Czechoslovakia. The largest deposits of crystalline magnesite in the United States are those near Chewelah, WA (Bennett, 1941; Siegfus, 1927) and near Gabbs, NV (Vitaliano and Callaghan, 1956; 1957). Only the deposit at Gabbs is being worked commercially at the present time following the closure of the Chewelah operation of Northwest Magnesite Co. in 1969.

In Canada (Goudge, 1942) a deposit of crystalline magnesite near Kilmar, Que., is being mined commercially. Other Canadian deposits are in northern Ontario and in British Columbia. Three types of magnesite are found in British Columbia (McCammon, 1968). In the East Kootenays, beds of crystalline magnesite occur interlayered with sedimentary rocks. Pods of cryptocrystalline magnesite are common in the serpentine of the Clinton and Bridge River areas of central British Columbia. Veins of white magnesite are present in fault zones along the Yalakom River and Pinchi Mountain. A processing plant for commercial exploitation of the crystalline magnesite was being erected for possible operation by 1981.

The major producing countries of natural magnesite are USSR, North Korea, China, Czechoslovakia, Yugoslavia, Austria, and Greece. Brazil, Spain, Turkey, India, and South Africa produce important but lesser amounts.

The identified reserves of crystalline magnesite in the USSR are on the order of 400 million tons in the southern Ural Mountains and 2000 million tons in the Savan Mountains area of eastern Siberia. Most of the USSR production comes from the area of the Ural Mountains, particularly from the deposits near Satka in Ufa Province. The Satka deposits alone have more than 250 million tons of reserves. The ore has an average MgO content of 46%, and it is similar to Austrian magnesite in appearance and quality except for lower iron content. Production of crude magnesite—at one time as high as 3 million st—was estimated in 1980 to be approximately 1.8 million stpy. The deposits in the Savan Mountains have not been exploited to the same degree as those in the Ural Mountains; the deposits are remote from industrial centers and of relatively lower grade than those of the Urals.

North Korea has about 3000 million st of magnesite reserves. The main deposits are located in Kankyo Province and average about 45% MgO on a crude basis. Annual production of crude magnesite is in excess of 1.5 million st. The producers are beneficiating some of the crude ore to make a more acceptable product. Most of the dead-burned and caustic-calcined magnesite is exported.

China has 5000 million tons of reserves largely centered at the famous Manchurian deposits located near Ta-shin-chiao in the Province of Shengking. There are also 21 known deposits in Szechuan in an area of 500 sq km. Total production figures for China are difficult to verify, but from trading information, they have been placed at 1 million stpy of crude magnesite.

Czechoslovakia reserves are on the order of 100 million st. Dead-burned magnesite was produced at an annual rate of 0.6 million tons in 1979 from the extensive deposits of breunnerite in Slovakia. Large deposits are in the region of Kosice, Jelsava, and Lobinobana.

Yugoslavia has deposits of magnesite located in Serbia and Bosnia; these are operated as underground mines and open-pit quarries producing about 500,000 stpy. The magnesite deposits can be subdivided into hydrothermal-sedimentary and hydrothermal-vein types. Mines are located in the Kraljevo area at Bela Stena and Ilinjaca. Similar sedimentary deposits are located at Beli Kamen and Evati. Vein-type deposits are located at Sumadija, Goles, Dubovac, and Ibar. Recently deposits have been mined at Zlatibor, which run northwest from Kraljevo in Serbia. All the ore requires some beneficiation, such as heavy-media separation, to make it suitable for commercial purposes.

Austria, the oldest producer in western Europe, has a narrow belt of breunnerite magnesite extending westward from Semmering, Austria. The most important deposits are located at Semmering, Veitsch, Breitenau Trieben, Radentheim, and Dientin. These are important commercial deposits from which magnesia containing about 6% Fe_2O_3 is obtained. Austria produced 1.6 million st of crude magnesite in 1970 which was converted to 540,000 st of dead-burned magnesite and 180,000 st of caustic-calcined magnesite (Anon., 1971). By 1979, the output of crude magnesite had decreased to approximately 1 million stpy.

Greece has large deposits of cryptocrystalline magnesite which are being mined. A deposit located at Vavdos in the Khalkidiki region in southwest Thessaloniki was originally worked in 1957. Several other deposits have now been worked especially on the island of Euboea and in Ormylia in Khalkidiki. Greek production of crude magnesite increased from approximately 0.7 million st in 1970 to 1.3 million st in 1976. Since then it decreased, to under 0.8 million st in 1979. The magnesite in the deposits is of high purity, but the close intermixture with gangue makes beneficiation of all ore necessary before sintering.

Brazil has crystalline magnesite deposits which are among the largest in the world. Total known reserves are on the order of 500 million st. The two major deposits are located in the Alencar area of Ceará Province and in the Eguas range near the town of Brumado in southwestern Bahiá. Production of crude magnesite increased from approximately 0.3 million st in 1972 to over 0.7 million st in 1980.

Spain has magnesite deposits in Navarra Province, near the French border. Other deposits are located in northwest Spain. Total production for Spain is about 300,000 stpy of crude magnesite.

Turkey has several deposits of cryptocrystalline magnesite mainly located in the Eskisehir and Kutaya regions. Production of crude magnesite from 1977 through 1980 averaged 0.5 million stpy.

India, until recently, supplied crude, caustic-calcined and dead-burned magnesite from the white cryptocrystalline magnesites of the Salem district in Madras. Those deposits are essentially exhausted, and most of the mining is now centered in the Almora district in Uttar Pradesh. The Almora magnesite is coarsely crystalline and of white to yellowish color. Most is converted to dead-burned magnesite. Total crude magnesite production in India was about 0.4 million stpy from 1977 to 1979. The dead-burned product is standard grade material.

South Africa has been a producer of magnesite for more than 30 years, but the output rate has been irregular and generally declining. From the level of 100,000 st in 1966, output decreased to 40,000 st in 1979. The deposits exploited are in the eastern Transvaal near Olifantsfontein and Burgersfort. The cryptocrystalline magnesite often occurs in pockets at irregular intervals, making mine planning difficult.

Zimbabwe has become a moderately important producer during the past five years, and in 1979 produced as much as South Africa.

Australia has a deposit of cryptocrystalline magnesite that is being mined near Fifield in New South Wales. The magnesite occurs as small boulders rarely over 18 in. diam and randomly distributed in a layer of loose soil a few meters thick. Reserves are unknown, but are thought to be small. Another deposit is located near Ravensthorpe in Western Australia. The cryptocrystalline magnesite occurs in veins and stockwork in mineralized zones up to several meters thick. The only commercial production has been from the Fifield deposits; in 1980, output was approximately 30,000 st of crude magnesite.

Magnesite is found in many other countries of the world and some, such as Nepal, have large reserves awaiting development. Nepal's reserves are on the order of 300 million st of coarsely crystalline magnesite. Cryptocrystalline magnesite occurs in Baja California and in many countries of Central America. Guatemala has a deposit which may become commercially important.

A magnesite deposit in Saudi Arabia at Jabal Ar Rokhan, north of Jiddah, is a hydrothermal replacement of dolomite by magnesite, and inferred reserves are 40 million st. Other deposits are located at Zarghat and Jabal Wask in the Hijaz Mountains.

Prospecting and Exploration

The search for deposits of any type of magnesite should be guided by considerations of the probable origin of magnesite. In the case of crystalline magnesite, possible locations of deposits are areas of limestone or dolomite terrane that have been subjected to folding or igneous activity. In the case of cryptocrystalline magnesite, possible locations are areas of ultrabasic rocks which show extensive alteration to serpentine. Thus, in either case, the search can be confined within the boundaries of areas that show evidence of dynamic geologic activity. Areas such as midwestern United States would not be fruitful in spite of large deposits of limestone and dolomite.

Magnesite formations are generally more resistant to weathering than associated formations; consequently, bold outcrops characterize magnesite deposits. Outcrops are sampled by surface chipping and shallow trenching. Field differentiation between commercial and noncommercial magnesite can be aided by the use of cold, dilute hydrochloric acid; limestone will effervesce when treated with hydrochloric acid whereas magnesite will not. Sometimes the Jolly balance is used for the determination on the basis of specific gravity of the minerals (magnesite, 3.0+; dolomite, 2.85).

Following the preliminary investigation, a diamond drilling program may be carried out to assess the commercial potential of the deposit. The evaluation is based on the size and location of the deposit and on the quantity and distribution of undesired minerals containing silica, lime, iron-oxide, and alumina. Tests to determine whether the impurities can be removed by ore dressing aids the evaluation. Magnesite, to be commercially acceptable, should contain at least 95% magnesium carbonate, an amount which, depending on the accessory oxides, would result in a dead-burned product containing 90 to 94% MgO. Refractory magnesia of that quality would have some demand, but the present and growing demand is for a product with even higher MgO content. This demand places increasing emphasis on beneficiation of the ore.

Mining and Processing

The type of the deposit dictates the method of mining. Large, massive deposits are usually worked by open-pit methods; narrow and deep deposits by underground drifts and stopes or surface openings.

The mined ore is rarely shipped or used in crude form, but is processed to yield refractory magnesia or caustic-calcined magnesia. Invariably, some degree of sorting or beneficiation is given the ore prior to dead-burning or calcining.

Examples of mining and processing practices of varying degrees and complexity are found in many places around the world.

North America

United States: *Nevada*—The discovery of magnesite in the Gabbs area in 1927 followed the finding of deposits of brucite in terrane originally prospected for tungsten. Exploration delineated total magnesite reserves in excess of 25 million st. The ore bodies occur in Upper Triassic dolomites. The beds are folded, faulted, and intruded by a variety of dikes and granodiorite stocks.

The deposits are being worked by Basic Refractories Division of CE Minerals. The plant is located at 5000 ft and the open pit at 6000 ft in the foothills of the Paradise Range in Nye County.

All mining is by the open pit method, with 10- to 15-ft benches. From 6000 to 8000 stpd of ore and waste are selectively mined. The indicated bench height was found necessary to maintain proper ore control. Blast holes 2¾ in. in diameter are drilled 16 to 20 ft deep on an 8-ft diamond pattern to give good pit floor for the benches and good fragmentation. Blasting is done with fertilizer grade ammonium nitrate premixed with fuel oil, and charges are primed with ¾ lb booster and primadet attached to the trunk line. From 800 to 1500 holes are shot at a time, each hole giving approximately 61 st of rock. Secondary blasting is seldom necessary. The broken ore is trucked approximately two miles to the crushing plant (Jepsen, 1980).

The different grades of ore cannot be readily distinguished visually and sorting into grades is controlled by chemical analysis. All cuttings from drill holes are analyzed for lime and silica, and values are plotted on maps of the deposit. Contour lines are drawn through points of equal lime content on one map and of equal silica on another. Contour intervals are chosen to correspond to various grade limits. Superimposition of one map on the other allows selection of any particular grade of ore by following the colored area representing the lime limits around to where it is bounded by contours representing the desired silica limits. A grid is staked on the broken ore, which has not shifted appreciably during blasting, to show the shovel operator which portion of the rock to load. Loading is by two 4.5-yd power shovels and by two 5-yd front-end rubber-tired loaders. Haulage is by 35-ton trucks to the crushing plant, which has a capacity of 380 stph. Depending on the grade of the ore, the crushed material is placed on one of three stockpiles which respectively feed the kiln, flotation, and heavy media plants. Concentrates from the beneficiation plants also feed the kiln plant. The flotation plant has a capacity of approximately 42 stph; the heavy media plant, 125 stph.

The kiln plant feed goes into two separate circuits. One, for the production of dead-burned magnesite, uses two rotary kilns; the other, for caustic-calcined magnesite, uses two Herreshoff furnaces and a small rotary kiln. Some of the raw feed is mixed with flue dust and briquetted prior to being dead-burned in the rotary kiln.

Washington—Washington magnesite occurs as massive lenses in dolomite. Distributed throughout the magnesite are stringers of siliceous and calcareous impurities. The magnesite deposits are found within the Stensgar dolomite formation. While the magnesite outcroppings are numerous, only a few of the deposits have been worked, and with the discontinuance of the Red Marble Quarry operation in 1968, production in Washington ceased. However, the mining methods were typical of those used for similar magnesite deposits.

At this mine, the magnesite was removed in a series of 40-ft benches. Primary rock breakage was obtained by drilling 6 in.-diam holes with a rotary drill and blasting (Fisk, 1953). This was followed by some secondary breakage. The broken ore was put through a 42 in. gyratory crusher, loaded into aerial tramway buckets and transported to the mill, a distance of eight miles.

The rock from the quarry was crushed, scrubbed, and screened into two major fractions. The coarse fraction, 1½ × ⅛ in., was sent to the heavy media separation plant (Utley, 1954). The −⅛-in. material was stockpiled for further treatment by flotation. Separation of magnesite from the lighter impurities, consisting largely of dolomite and siliceous minerals, was carried out in ferrosilicon medium in a 20-ft-diam cone. The magnesite "sinks" were crushed, screened, and stored in bins, from which they were either transported 5 miles by aerial tramway to the calcining plant, or sent to the flotation mill for further treatment. Most of the magnesite was sent directly to the calcining plant where it was blended with a small amount of iron ore and fired in a rotary kiln at 2900°F to produce dead-burned magnesite of standard quality.

The magnesite which was conveyed to the flotation mill was reduced to minus 65-mesh (Tyler) by a ball mill in closed circuit with a classifier. In the two-stage flotation process, siliceous minerals were floated first, followed by flotation of magnesite from the calcareous minerals. The magnesite concentrate was thickened, filtered, dried in a rotary dryer, and transported to the calcining plant. Here it was fired in a rotary kiln to produce a high grade dead-burned magnesite.

Canada: *Quebec*—A deposit of high-lime magnesite is being mined at Kilmar, Que. (Bray and Hilchey, 1957). The magnesite ore body consists of a series of veins which dip about 85°. The footwall is quartz monzonite, and the hanging wall consists of serpentine, diopside, and quartzite.

The mine, from 1915 to 1936, was operated in narrow, deep opencuts, but since 1936 the mine has been operated by underground methods. Shafts are sunk to the desired level, and drifts and crosscuts are driven into the veins. The ore is removed from shrinkage stopes, hoisted to the surface in cars in a cage.

The ore is crushed, washed, and screened to separate it into two fractions. The coarse fraction, 1½ to 3⁄16-in., is fed to the heavy-media separation plant, and the −3⁄16-in. material is sent to a holding bin prior to transportation to the calcination plant for use as a blending material. The waste material, "floats," is discarded; the heavier "sinks" containing the magnesite goes to storage. The sinks are transported by truck to the calcination plant and placed in different piles according to chemical analysis. The ore is reclaimed from piles by a front-end loader which feeds a series of bins. Materials from the bins are reclaimed and proportioned through feeders onto a belt to give the required compositional blend. Blended ore goes to a ball mill where it is crushed and transported to blending silos for final compositional adjustments. The blended, fine feed is transported to a rotary kiln and calcined to dead-burned products. After dead-burning, the magnesia is crushed, loaded into railway hopper cars, and shipped either to the company's refractory brick plant or to customers.

Europe

Greece: The northern part of the island of Euboea consists of schists-hornstones and serpentines of the "oriental Greece" zone and is well known for important magnesite deposits. The magnesite is of the cryptocrystalline variety located in ruptures and stratified zones. Three types of magnesite deposits are present:

1) Vein type which may be from ¼ in. to 14 ft in width and of varying length.

2) The stockwork type which is a dense network of veins varying from a few inches to several feet in thickness.

3) Mass type, forming lenticular concentrations of magnesite with volumes of a few to several thousand cubic feet. This type is a local anomaly of the vein or stockwork magnesite types.

Among the operations of companies in Euboea, those of Financial Mining-Industrial and Shipping Corp. are the largest. The magnesite is exposed by extensive stripping of the overburden and then mined by open-pit methods. Benches are established at heights of 35 to 65 ft. Broken ore is loaded by shovel or front-end loaders into 35-ton capacity trucks and transported to the ore dressing plants. For one ton of dressed ore, more than 20 tons of crude ore and overburden must be moved.

At the dressing plants, the ore is washed, crushed, and beneficiated by a combination of optical, magnetic, and heavy-media separations. The beneficiated ore is stored in bins until transported to the calcining plant. An extensive program of sampling and chemical analysis facilitates separate storage of different grades of material at the calcining plant. The ore is reclaimed from the storage piles by a front-end loader and fed to six rotary kilns, which in total can produce 320 t/a of dead-burned magnesite (Manos, 1979). The dead-burned magnesite is cooled and transferred to the nearby port.

Output of raw magnesite by all producers was approximately 760,000 st in 1979, marking a steady decrease from 1,270,000 st in 1976.

Asia

Turkey: Magnesite is found in several places in Turkey. The main mining operations are near Eskisehir and Kutaya in the west-central part of the country. The magnesite is cryptocrystalline and is usually mined in narrow opencuts. The veins are more or less vertical, and in at least one operation near Kutaya, underground mining is practiced to some extent. The mined ore in all cases is sorted by hand and shipped to nearby calcining plants. The lump ore is calcined in shaft or rotary kilns.

Other Magnesium Minerals

Brucite is sometimes found associated with commercial deposits of magnesite as in Nevada and Washington. In those cases, it would be processed along with the magnesite to produce refractory magnesia.

By itself, as an exploitable mineral, there are no known operations for brucite. There was one at Wakefield, Que., but it closed down several years ago. At Wakefield, the brucite occurs as ricelike grains 1/16- to ⅛-in.-diam disseminated through a limestone matrix (Goudge, 1944). The process of winning the magnesia was based on controlled slaking of carefully sized and calcined ore. The system was controlled so that the discharge from the slaker consisted of an impalpable dust of hydrated lime mixed with magnesia granules. The granules were air separated, washed, and wet classified. The lime was sold directly, and the magnesia granules were further processed to refractory magnesia.

Olivine is composed of forsterite ($2MgO \cdot SiO_2$) and fayalite ($2FeO \cdot SiO_2$) in solid solution; in common olivine the latter is the minor constituent present to about 15%. Olivine is a common constituent of igneous rocks, but deposits of sufficient purity to be commercially attractive are not numerous.

North Carolina and Georgia reserves of olivine are estimated at 230 million tons averaging 48% magnesia (Hunter, 1941). In Washington, olivine occurs as dunite rock, a peridotite in which olivine is the dominant mineral. Reserves are estimated at 50 million st on Cypress Island and several million tons in the Twin Sister Region.

Small quantities of olivine were mined in the United States and Norway from 1930 to 1940, but it was not until the 1950s that world production reached 20,000 stpy. By 1974 world production was approximately 350,000 stpy, and in 1980 it was almost 1.5 million st. In addition, 400,000 st of dunite were produced. By far the largest producer of olivine is Norway; over 1 million st in 1980 (Watson). Other producing countries in order of rank are the US, Austria, and Italy. Dunite is produced mainly in Spain.

The largest Norwegian operation belongs to A/S Oliven, which operates a quarry at Aaheim. The ore is crushed, some is washed, and sized into various grain and flour categories. The other producer is K/S Norddal Olivin A/S & Co., one of the H. Björum group.

The Austrian production comes from a quarry at St. Lorenzen in Styria. Here Magnolithe GmbH, a subsidiary of Continental Ore Corp., produces forsterite brick, ramming mixtures, mortars, and foundry sand.

The only currently producing areas in the US are Washington State and North Carolina. Together they produce about 170,000 stpy, of which one company, International Minerals and Chemicals, accounts for more than 60%. In Washington, the company mines a talus accumulation at the base of the Twin Sister deposit. This material is washed, crushed, and transported to the processing plant at Hamilton. IMC has operations in North Carolina at Eddie and Burnsville. The ore varies in impurities content and must be beneficiated.

The second largest producer is National Olivine Co., which operates a wet processing facility at Dillsboro, near Asheville, North Carolina. Plant capacity is 100,000 stpy.

Synthetic Magnesite

Prior to 1938, natural magnesite was the source of essentially all grades of refractory magnesia. Since that time, the manufacture of refractory magnesia and caustic-calcined magnesia from other than natural magnesite has increased in importance until today over 85% of refractory magnesia produced in the US and suitable for brickmaking comes from nonmagnesite sources. Several basically different methods for production of magnesia from nonmagnesite sources have been studied, and at least four have been operated commercially. All center on the production and calcination of either magnesium hydroxide or magnesium chloride. The magnesium hydroxide may be obtained as a precipitate from a magnesium salt solution or as a residue remaining after the lime fraction of calcined dolomite is removed as a soluble compound by reaction with solutions of ammonium chloride or hydrogen sulfide. Magnesium chloride may be obtained as an end liquor after solar concentration of solutions of natural brines for production of salt or potash, or from well brines.

There are now 25 plants around the world producing refractory and caustic-calcined magnesia from magnesium hydroxide. In these plants, magnesium hydroxide is obtained by precipitation from liquids such as seawater, seawater bitterns, inland well brines, or end liquors from other chemical processes. One plant in Israel produces refractory magnesia by pyrolytic decomposition of magnesium chloride.

The commercial production of magnesia from seawater dates back to around 1885 when a small operation existed for a short time on the Mediterranean coast of France (Egleston, 1886). In that operation, milk of lime and seawater were combined in large masonry tanks which overflowed continuously into two similar tanks in series. From the last tank, the mixture of spent seawater and precipitated magnesium hydroxide was discharged into shallow troughs about 16 ft wide and 1000 ft long excavated in the beach sand. The sand acted as a filter for removal of the liquid. When the troughs became full of magnesium hydroxide mud, the inflow was stopped and after a few weeks in the sun, the material became sufficiently dry to allow excavation and handling. The broken lumps were then calcined. The plant shut down, presumably because it could not compete with mineral magnesite operations, but the principles of operation were similar to those now practiced by seawater-magnesia plants.

All the presently operating plants which make chemical magnesia by calcining magnesium hydroxide follow a basically similar process (Wicken, 1949). In all, the soluble magnesium salts are converted to insoluble magnesium hydroxide by treating with calcium hydroxide, which may come from oyster shells, limestone, or dolomite. Any of the common magnesium-bearing solutions such as seawater or brines may be used provided they can be treated to reduce carbonate and sulfate levels sufficiently so insoluble calcium compounds do not precipitate with magnesium hydroxide. The main considerations in selecting reactants are their purity and economic availability. After the reactants are established for any given plant, the limitations of pumping and thickening equipment will prevent that plant from changing to more dilute reactants without loss of capacity. A plant using seawater bitterns cannot change to using raw seawater, nor can one using dolomite change to using limestone without sacrifice of production. However, the reverse changes are feasible.

The stoichiometric requirements for production of one ton of magnesia are 2.3 st of dolomite or 2.5 st of limestone. Half of the magnesia comes from dolomite when it is used, but when limestone is used, all the magnesia comes from the seawater. Thus, when limestone is used, as compared to dolomite, twice the volume of seawater must be processed to obtain the same amount of magnesia.

In 1938 a commercial plant for refractory magnesia was built at Hartlepool, Durham, England (Gilbert and Gilpin, 1951). Calcined dolomite is used to treat the seawater to obtain magnesia. Dolomite is quarried, crushed, and sized and, then, calcined in a rotary kiln. The calcined dolomite is slaked to a fine, dry powder in hydrators. The hydrated product is made into a slurry to facilitate handling and mixing with seawater. The slurry is classified to remove impurities before being mixed with the seawater.

Incoming seawater is treated with sulfuric acid to remove carbon dioxide. The decarbonated water is pumped to reaction vessels and mixed with the slurry of calcined dolomite. Magnesium hydroxide is formed and the dilute suspension is delivered to large-diameter primary thickeners. The underflow of thickened magnesium hydroxide is repulped with treated seawater or, alternatively, freshwater and is then rethickened. The underflow from the secondary thickener is fed to a filter press operating at 2000 psi. The cake, containing about 70% solids, is fed directly into a rotary kiln for dead-burning.

The first commercial producer of magnesium compounds from seawater in the US was Marine Magnesium Products Corp., now part of the Merck & Co. group. In addition to the Merck & Co. operation, in south San Francisco, other plants in the US using seawater as a source of magnesium hydroxide are those of Harbison-Walker Refractories at Cape May, NJ; Kaiser Aluminum and Chemical Co. at Moss Landing, CA; Dow Chemical Co. at Freeport, Texas; Basic Refractories Div. of CE Minerals Co. at Port St. Joe, FL; and Corhart Refractories at Pascagoula, MS.

In all these operations, the processes are fundamentally alike. Seawater is treated with a small amount of slaked lime to precipitate the soluble bicarbonates as calcium carbonate or with acid to liberate carbon dioxide. The softened seawater is combined with either dry or slaked lime (or dolime). The resulting magnesium hydroxide is thickened and washed in a countercurrent system with freshwater in order to remove the calcium chloride reaction product and most of the seawater, and to concentrate the slurry. The washed and thickened magnesium hydroxide is then filtered, and the cake is either directly burned to produce refractory or caustic-calcined magnesia, or it may be calcined and pelletized prior to dead-burning to give the product desirable size and density characteristics.

In the Dow Chemical plant, the magnesium hydroxide is not dead-burned, but is converted to magnesium chloride for use in electrolytic cells to produce magnesium metal.

In Japan, seawater plants for magnesia production are at five locations. They are those of Ube Chemical Industries Ltd. at Ube, Yamaguchi; Shin Nihon Chemical Ind. Co. at Minamata, Kyushu; Nihon Kaisui Kako Co. Ltd. at Naoetsu; Asahi Glass Co. Ltd. at Iho; and the salt brine plant of Hokuriku Sein Kogyo KK at Hotsu. All these plants use calcined limestone as precipitant.

Two chemical magnesia plants owned by Industrias Peñoles are located in Mexico. Combined annual production in 1979 was 100,000 st. One, based on seawater, is that of Quimica del Mar, Ciudad Madero, Tampico; the other, based on end-liquor from a sodium sulfate operation, is that of Quimica del Rey, Laguna del Rey, Coahuila.

Sardamag SpA has a plant located at Sant'Antico in Sardinia; the plant produces refractory magnesia from seawater using calcined limestone as the precipitant. The production in 1973 was over 100,000 st of refractory magnesia. By 1978, production capacity had been increased to 180,000 st. Another Italian plant, Co.GE.MA SpA, with capacity of 60,000 stpy, is located in Sicily.

The Pfizer plant at Dungarvan, County Waterford, Eire manufactures refractory magnesia from magnesium hydroxide precipitated from seawater by calcined dolomite.

A plant reported at Sevash, Crimea (USSR), is said to be able to produce 100,000 stpy of magnesia from seawater.

A plant to produce magnesium hydroxide from seawater is operated by Norsk Hydro AS near Porsgrun, Norway. The magnesium hydroxide is used for the production of magnesium oxide for the pulp and paper industry.

The most recent entrants to the ranks of those producing magnesia from seawater or well brines are Premier Periclase, Ltd. and Magnesia International, BV. Premier Periclase at Drogheda in Ireland is owned by Cement Roadstone Ltd. Production of dead-burned magnesite from seawater and limestone commenced mid-1980 in a 100,000-stpy plant. Magnesia International, a subsidiary of Billiton International, which in turn is part of the Royal Dutch Shell group, has completed a 100,000-stpy plant at Veendam in the Netherlands. Imported calcined dolomite and the plant's own well brines are the raw materials for production, which began in early 1982.

Three US companies produce magnesium hydroxide from deep well brines. They are Dow Chemical Co. at Ludington and Midland, MI; Martin Marietta Chemicals at Manistee, MI; and Morton Chemical Co., a small operation also located in Manistee. The thickened product from Dow at Ludington is pumped to Harbison-Walker Refractories Div. of Dresser Industries, Inc., where the material is filtered and processed to refractory magnesia. The two other companies process their own material to yield caustic-calcined or refractory magnesia. A typical analysis of the Michigan brine is NaCl, 5.8%; $CaCl_2$, 17%; $MgCl_2$, 9.9%; KCl, 2.5%; Br_2, 0.1%; sp gr 1.29; and pH, 5 ± 0.5. The brine does not contain carbonate or sulfate; thus pretreatment ahead of magnesium hydroxide precipitation is not needed.

The brine is pumped to the surface from depths of 2500 to 3500 ft. The processes for recovery of magnesium hydroxide are based on the reaction between soluble magnesium salt and calcium hydroxide to produce magnesium hydroxide.

In 1937, Westvaco Chemical Co., a division of Food Machinery Co., started production of refractory magnesia at Newark, CA. Their process was based upon seawater bittern which was an end-liquor from a nearby solar salt-making operation. Seawater was concentrated by solar evaporation until most of the sodium chloride was crystallized leaving a bittern with nominal composition: NaCl, 12 to 16%; $MgCl_2$, 6 to 9%; $MgSO_4$, 4 to 5%; KCl, 1 to 2%; and $MgBr_2$, 0.16%. During the early years of the operation, calcined oyster shell was used as precipitant, but after 1943, calcined dolomite was used because it contained less sulfate than the oyster shell, thus resulting in less CaO in the final magnesia.

The bitterns were treated with calcium chloride solution to remove the sulfate ions as calcium sulfate before the magnesium hydroxide precipitation. The sulfate-free bittern was then reacted with finely ground, calcined dolomite to form magnesium hydroxide. The precipitate was washed, thickened, filtered, and fed into a rotary kiln for dead-burning.

The foregoing processes based on seawater or brines lead simply and directly to production of magnesium hydroxide and magnesia with no flowsheet interlocks for production of other products except that magnesium metal may be made from the magnesia. Other processes exist in which the production of magnesium hydroxide is less direct. These processes are of two general types: one in which relatively low-grade magnesium compounds or minerals are upgraded by insertion into a flowsheet primarily designed to produce something else; the other in which production of magnesia is locked in with the simultaneous production of a useful byproduct. Examples of these follow:

A process at one time but no longer used for the production of magnesia, is based on the substitution of calcined dolomite for calcined limestone in the Solvay process for making soda ash. The spent material from the ammonia stills consists of a slurry of dissolved calcium chloride and insoluble magnesium hydroxide rather than only a solution of calcium chloride as is characteristic of standard Solvay system. The slurry was thickened and filtered to recover the magnesium hydroxide which was converted to refractory magnesia.

The Dead Sea Works, a subsidiary of Israel Chemicals Ltd., has a plant at Arad to recover magnesium chloride from the Dead Sea brines. The magnesium chloride is converted to magnesia and hydrochloric acid by pyrolytic decomposition. The plant has an annual production of 60,000 st of refractory magnesia.

The world total annual capacity for chemical production of magnesia is about 2.5 million st. While there are many seawater magnesia plants, the choice of suitable sites is more limited than one would suspect. Large land masses can cause the seawater to be diluted because of freshwater runoff and tide action so it becomes merely brackish. The more dilute the seawater becomes, the less magnesia a given plant can produce. Normal seawater contains about 34 g of dissolved salts per kg; of this, 23 g reports as sodium chloride and the rest mainly as magnesium and potassium compounds. The amount of magnesium ion is equivalent to about 2.1 g of magnesia (MgO) per kg of seawater or over 17 lb per 1000 gal. The importance of seawater concentration as related to size of process equipment is evident. Other factors such as coastal topography, availability of high purity limestone or dolomite, fuel sources, and market locations must be closely studied before a site is selected.

Political and Commercial Control

The use of magnesia as a refractory material has a relatively short history when compared to that of other heat-resisting materials used in connection with metallurgy since antiquity.

The use of magnesite for metallurgical operations was first mentioned in 1866, and the first brick were manufactured in 1868 (Caron, 1868). Prior to World War I, magnesia was not produced regularly in the United States; industrial needs were largely met by imports from Central Europe. A small quantity of magnesite was mined in California as early as 1886, but output did not reach 10,000 st until 1910. Under the impetus of World War I, production of ore increased until in 1917, production reached 21,000 st. The state of Washington recorded its first production in 1916 and soon became the leading producer in the United States reaching a level of over 200,000 stpy. This rate was continued for many years until the increasing needs of the refractory industry for product of higher purity coupled with exhaustion of high grade ore forced the closure in 1968.

Depletion, Taxes, and Tariffs

Depletion allowance for magnesite, dolomite, and brucite mined from deposits in the United States is 14% and for olivine, 22%.

Soon after World War I, the infant magnesite industry in the United States needed protection to survive attack from low-cost imports. As a result, an import tariff of $11.50 per st of dead-burned magnesite was established (US Tariff Act, 1922, parg. 204). This rate held until October 1951 when it was decreased to $7.67 per st (Torquay). Subsequent differentiations as to types of refractory magnesia and inclusion of other refractory materials has resulted in the US tariff structure shown in Table 1 as it applies to countries in the "Most Favored Nation" category. The rates applied against imports from other countries are fixed at the higher rates of $15 per st and 30% ad valorem for the under- and over-4% lime types, respectively (Petkof, 1980). Other major countries around the world have followed a similar pattern of decreasing tariff for refractory magnesia.

Production and Consumption

About 75% of all refractory magnesia is used in one form or another by the steel industry, and any changes in technology or rate of production in that industry result in a change in the market for magnesite. Hence, trends in the total industrial use of magnesite can conveniently be expressed in terms relating to steel production. On that basis, the total apparent consumption of refractory magnesia by all industries in 1930 was equivalent to 4.1 lb per ton of steel produced. By 1958, the rate reached 11.5 lb as magnesia refractories supplanted other materials in steel and other industries. The advent of the basic oxygen process in 1954 began a new era in steelmaking technology (Price, 1969). The process was not fully exploited in the United States until the early 1960s, but it is now fully established as the major steel production method. The process changed the magnesia consumption pattern in that the new technology, as compared with open-hearth practice, requires much less magnesia per ton of steel produced. Some shops are now reporting under 5 lb of magnesia refractories consumed per ton of steel.

Development of the new steelmaking technology has resulted in heavy demand for sev-

TABLE 1—US Import Duties

Commodity	Most Favored Nation (MFN)		Non-MFN
	Jan. 1, 1982	Jan. 1, 1987	Jan. 1, 1982
Crude magnesite	$1.64/lt	Free	$10.50/lt
Refractory magnesia (including dead-burned and fused magnesite)			
4% or less lime	.18¢/lb	.16¢/lb	.75¢/lb
Over 4% lime	6% ad valorem	6% ad valorem	30% ad valorem
Caustic-calcined magnesite	$2.10/lt	$2.10/lt	$21/lt

eral grades and types of magnesia of high purity, controlled composition, and specific physical properties (Laming, 1971). The result of this proliferation of product requirement has been an emphasis on the production of chemical magnesia which is more amenable to compositional control than is that obtained from the mineral magnesite.

The net consumption of refractory magnesia in the US in 1979 was approximately 907,000 st, against a raw steel production of 139 million st. Domestic sources accounted for 814,000 st and imports the balance.

The US has long been an importer of refractory magnesia. The trade divides refractory magnesia into that containing over and under 4% lime. In 1972 the importation of the under-4% grade material was 128,000 st; of the over-4% grade material, 6,000 st (Chin, 1972). In 1979 the respective amounts were 65,000 and 28,000 st. The figures reveal that US producers have concentrated on producing more high-grade material but have left the less profitable standard-grade for foreign supply.

The market for caustic-calcined and technical grades of magnesia in the United States has grown from 100,000 to 130,000 stpy (1968–1972) to approximately 170,000 stpy in 1979. The country essentially is self-sufficient in these magnesias.

Marketing, Uses, and Prices

Crude Magnesite: Crude magnesite has been used in negligible quantities for the preparation of such chemicals as Epsom salts. Consumption for these purposes is extremely variable and probably has never exceeded a few tons per year in the United States. Crude magnesite has limited sale in Europe and the Middle East by producers who do not have their own processing facilities for conversion of the crude material to calcined products or chemicals. At present, in the United States, there is no known market for crude magnesite because manufacturers of chemicals prefer to use natural brines or caustic-calcined magnesia as starting material. The selling price for crude magnesite outside the United States ranges between $30 and $60 per short ton.

Dead-Burned Magnesite: Dead-burned magnesite is used almost entirely as a refractory material. It can be used directly as a grain product or as a constituent of brick, ramming mixes, gunning mixes, or castables. The refractories made from magnesia have a wide variety of uses but find their main use in the cement, glass, steel, and copper industries (McDowell and Howe, 1920). The great merit of magnesia-based refractories is their ability to resist basic slags at high temperatures (Spencer, 1972) and to lower the refractory cost per ton of product (Rigby, 1971).

Refractory magnesia is freely sold on the market, but many refractory producers have their own source of supply. The total production of refractory magnesia is generally adequate to meet market demands although shortages in certain special grades of high purity occur periodically.

The price of refractory magnesia is determined, in part, by the physical and chemical properties of the product. The magnesia content alone is not the sole control in establishing the price; the amounts and proportions of accessory oxides and the physical characteristics of the product are equally influential. Thus, different kinds of refractory magnesia, even though they contain the same amount of MgO, may be priced at different levels. The prices of refractory magnesias have increased greatly since 1973. In that year the range was $90 to $150 per st for the high-purity grades. In 1980 prices on the world market ranged up to over $500 per st. Similarly, prices for standard or maintenance grades have escalated from a range of $35 to $80 per st to more than $150 per st.

Caustic-Calcined Magnesia: Caustic-calcined magnesite has many uses, among which an important one is for the production of magnesium oxychloride and oxysulfate cements. The cement is prepared by mixing a solution of magnesium chloride or sulfate with ground caustic-calcined magnesite. The main use of the cement is as a resilient, fireproof, sparkproof, and vermin-proof flooring in industrial and institutional buildings. Additional large amounts of the cement are used for wall board, moldings, and acoustical tile (Anon., 1956).

Other major uses are found in the pulp and paper (Hull, et al., 1951), rayon, fertilizer, insulation, and chemical industries (Anon., 1977). It is used in stack gas scrubbing applications to remove sulfur dioxide and as an additive to decrease corrosion in furnaces burning high-vanadium oil.

The price of caustic-calcined magnesia varies widely with different grades. The grade used most widely in bulk for the paper, fertilizer, and chemical industries costs on the order of $300 per st.

Olivine: The patterns of use in Europe and the US are widely different in both markets and tonnages. In Europe about 70%, or about 1.2 million tons, of the total olivine and dunite produced is used as slag conditioner in blast furnaces. In the US, the blast furnace market in 1980 consumed about 50%, or 85,000 st of olivine produced. The second and third largest uses in Europe are as aggregates and as heat-storage blocks. The amounts used in the US for those purposes is essentially nil. The second largest use in the US is as a foundry sand (Watson, 1980).

The first commercial use of olivine was as a raw material for refractory brick in the early 1930s. Forsterite (2 $MgO \cdot SiO_2$) is the constituent in olivine that gives the material value as a refractory. However, after a period of slowly growing demand, largely for use in open-hearth steel operations through the 1950s, the use declined. In 1980 use as a refractory was around 15,000 stpy in the US. Olivine is also used in limited amounts for sandblasting, chemical manufacturing, and as an ingredient in fertilizer. Olivine is a relatively inexpensive material; prices in 1980 ranged from $25 to $90 per st depending on quality and size specifications.

Tests and Specifications

Crude Magnesite: Crude magnesite is used largely to produce dead-burned or caustic-calcined magnesite. The only characteristics of the raw magnesite that survive the calcining operation are those related to chemical composition; thus, the only tests regularly made on raw magnesite are for determining composition.

Refractory Magnesia: Refractory magnesia (dead-burned magnesite) is subjected to both physical and chemical tests. The nature of the tests and the product specifications are somewhat dependent upon the use to which the material is put. If the material is to be used for the manufacture of monoliths, certain characteristics are needed to assist sintering in service; these characteristics would not be necessarily needed nor desired in brick making mixes. Both services, however, would require dense, well-burned grain with a predominance of large grains.

Resistance to hydration is an important determination often made. The presence of free lime in dead-burned magnesite is deleterious and can result in shattered refractories as hydration takes place. In most dead-burned magnesites, lime is not free but is combined with other oxides in stable mineral combinations which are resistant to hydration. However, some combinations and magnesia itself can hydrate unless stabilization has been achieved during manufacture.

In synthetic and high purity magnesites, the size of the periclase crystal is considered important and is checked under the microscope. The presence and relationship of trace elements are important, as is the density of the grain of the final product.

Caustic-Calcined Magnesia: Most caustic-calcined magnesia is a commercial product sold on a tonnage basis in relatively large lots to chemical, paper, and fertilizer manufacturers.

The specifications for magnesia of this type are as varied as the uses. In general, specific requirements as to particle size, color, ignition loss, chemical analyses, bulk density, and activity may have to be met.

Magnesia for oxychloride cements is tested by making test slabs. Standards of setting time, dimensional change, and strength have been published by the American Society for Testing & Materials. The other requisites are that the magnesia have little or no free lime, have insufficient iron to color the magnesia, and be ground substantially through 200 mesh.

A class of magnesia products differing from commercial grades of caustic-calcined magnesia is that sold to specific technical and USP standards. These grades of magnesia are distinguished from others by being specially processed to meet rigorous physical and chemical standards. The products are used for medicinal purposes, specialized fillers, and various chemical needs. They may be referred to as "light" or "heavy" calcined magnesia.

When calcined at extremely high temperatures, the product finds use in refractory and electrical applications. These products are usually sold by the pound.

Environmental Considerations

Magnesite mining operations, being largely in mountainous regions, have little interface with urban society. Contestation as to land use is not a problem.

Mine drainage from underground or open-pit operations does not contribute significantly to pollution of streams because the drained area contains no significant amounts of soluble or oxidizable minerals.

Calcining plants, at one time, emitted substantial amounts of stack dust, but modern dust collectors and precipitators have removed most of these emissions and allowed the recovery of valuable products.

The greatest and most evident impact of magnesia operations on the environment has centered around seawater magnesia plants. Here, seawater, after having magnesia removed, is returned to the ocean. Recent innovations leading to greater control of the characteristics of the effluent have decreased the amount of turbid effluent.

Seawater magnesia plants calcine dolomite or limestone for use as a precipitant. The stack gas from the kilns is, in modern plants, controlled by suitable dust collecting devices.

During the calcination and dead-burning of magnesium hydroxide, substantial quantities of steam and some soluble salts are discharged. Wet scrubbers and electrical precipitants are used to control these discharges.

None of the discharges from either natural or synthetic magnesia plants has a noxious quality, and with modern remedial measures, even the cosmetic quality of the discharges is acceptable.

Bibliography and References

Anon., 1956, "Magnesium Chemicals," *Chemical Engineering,* Vol. 63, Aug., pp. 346–349.

Anon., 1960, "Basic Inc. Increases Magnesite," *Mining World,* Vol. 12, Nov., pp. 30–32.

Anon., 1971, "Magnesite," *Mining Annual Review,* 1971 ed., Mining Journal, London, pp. 115–117.

Anon., 1972, "Refractory Raw Materials—The Producers Reviewed," *Industrial Minerals,* No. 59, Aug., pp. 9–30.

Anon., 1977, "Magnesia—From Shortage to Surplus," *Industrial Minerals,* Sept., pp. 31–35.

Bain, G.W., 1924, "Types of Magnesite Deposits and Their Origin," *Economic Geology,* Vol. 19, No. 5, Aug., pp. 412–433.

Bennett, W.A.G., 1941, "Preliminary Report on Magnesite Deposits of Stevens County, Washington," Report of Investigation No. 5, Washington State Div. of Geology, 25 pp.

Bennett, W.A.G., 1943, "Character and Tonnage of the Turk Magnesite Deposit," Report of Investigation No. 7, Washington State Div. of Geology, 22 pp.

Bodenlos, A.J., 1954, "Magnesite Deposits in Serra Das Equas, Brumado, Bahiá, Brazil," Bulletin 975-C, US Geological Survey, pp. 87–170.

Bray, W.T., and Hilchey, G.R., 1957, "Magnesite Deposits of Kilmar, Quebec," *Geology of Canadian Industrial Mineral Deposits,* Canadian Institute of Mining & Metallurgy, Montreal, pp. 164–166.

Campbell, I., and Loofbourow, J.S., 1962, "Geology of Magnesite Belt of Stevens County," Bulletin 1142-F, US Geological Survey, pp. F1–F53.

Caron, M.H., 1868, "Preparation de La Magnesie Employee Comme Mature Refractaire," *Comptes Rendus,* Academie Science, Paris, Vol. 62, p. 839.

Chin, E.L., 1972, "Magnesium Compounds," *1972 Minerals Yearbook,* Vol. 1, US Bureau of Mines, pp. 747–755.

Comstock, H.B., 1963, "Magnesium and Magnesium Compounds," Information Circular 8201, US Bureau of Mines, 128 pp.

Coope, B., 1981, "Caustic Magnesia Markets—Agricultural Oversupply and Industrial Underdemand," *Industrial Minerals,* No. 161, Feb., pp. 43–51.

Dunbar, C.O., and Rogers, J., 1957, *Principles of Stratigraphy,* John Wiley and Sons, New York, 356 pp.

Egleston, T., 1886, "Basic Refractory Materials," *Trans. AIME,* Vol. 14, pp. 460–492.

Fisk, R.L., 1953, "Changes in Primary Drilling at Northwest Magnesite," *Mining Congress Journal,* Vol. 39, Feb., pp. 48–54.

Gilbert, F.C., and Gilpin, W.C., 1951, "Production of Magnesite from Seawater and Dolomite," *Research,* No. 8, pp. 348–353.

Goudge, M.F., 1942, "Sources of Magnesite and Magnesia in Canada," *Transactions,* Canadian Institute Mining & Metallurgy, Vol. 45, pp. 191–207..

Goudge, M.F., 1944, "Brucite Magnesia," *Journal of the American Ceramic Society,* Vol. 27, No. 1, pp. 8–10.

Hull, W.Q., et al., 1951, "Magnesia-Base Sulphite Pulping," *Industrial Engineering Chemistry,* Vol. 43, Nov., pp. 2424–2435.

Hunter, C.E., 1941, "Forsterite Olivine Deposits of North Carolina," Bulletin 41, Dept. of Conservation and Development, Div. of Minerals Research, 114 pp.

Krimball, H.M., 1972, "Magnesia Compounds," *Mining Engineering,* Vol. 24, Jan., pp. 46–47.

Ladoo, B., and Myers, W.M., 1951, *Nonmetallic Minerals,* 2nd ed., McGraw-Hill, New York, pp. 296–311; 431–436.

Laming, J., 1971, "Raw Materials and Refractories Performance," The Frank Scott Memorial Lecture, The Refractories Assn. of Great Britain, Mar. 24.

Lindgren, W., 1933, *Mineral Deposits,* 4th ed., McGraw-Hill, New York, pp. 388–390.

McCammon, J.W., 1968, "Magnesite, Hydromagnesite, and Brucite Occurrences in British Columbia," Information Circular, British Columbia Dept. of Mines and Petroleum Resources, Jan., 7 pp.

McDowell, J.S., and Howe, R.M., 1920, "Magnesite Refractories," *Journal of the American Ceramic Society,* Vol. 3, Mar., pp. 185–246.

Manos, A., 1979, "Greek Minerals—Development Through the 1970s," *Industrial Minerals,* Dec., pp. 37–53.

Niinomy, K., 1925, "The Magnesite Deposits of Manchuria," *Economic Geology,* Vol. 20, No. 1, Jan.-Feb., pp. 25–53.

Nishihara, H., 1956, "Origin of the Bedded Magnesite Deposits of Manchuria," *Economic Geology,* Vol. 51, No. 7, pp. 698–711.

Paone, J.O., 1970, "Magnesium," *Mineral Facts and Problems,* Bulletin No. 650, US Bureau of Mines, pp. 621–638.

Petkof, B., 1980, "Magnesium," *Mineral Facts and Problems,* Bulletin No. 671, US Bureau of Mines.

Petkof, B., 1982, "Magnesium Compounds," *Mineral Commodity Summaries 1982,* US Bureau of Mines, pp. 92–93.

Price, F.C., 1969, "An Industry in Transition: Iron and Steel Today," *Chemical Engineering,* Vol. 76, Aug., pp. 76–88.

Rigby, G.R., 1971, "Future Trends in Refractory Materials for Steel Production," *The Refractories Journal,* Vol. 46A, May, pp. 27–30.

Rosenberg, P.E., and Holland, H.D., 1964, "Calcite-Dolomite-Magnesite Stability Relations in Solutions at Elevated Temperatures," *Science,* Vol. 145, pp. 700–701.

Rosenberg, P.E., and Mills, J.W., 1966, "A Mechanism for the Replacement of Magnesite in Dolomite," *Economic Geology,* Vol. 61, No. 3, May, pp. 582–586.

Schroeder, N.C., 1948, "The Genesis of the Turk Magnesite Deposits, Stevens County Washington," *Compass,* Vol. 26, No. 1, Nov., pp. 37–46.

Siegfus, S.S., 1927, "Some Geological Features of the Washington Magnesite Deposits," *Engineering & Mining Journal,* Vol. 124, No. 22, Nov. 26, pp. 853–856.

Spencer, D.R.F., 1972, "Basic Refractory Raw Materials," *Transactions and Journal of British Ceramic Society,* Vol. 71, No. 5, pp. 123–134.

Utley, H.F., 1954, "Heavy-Media Separation Supplements Flotation in Magnesite Plant," *Pit and Quarry,* Vol. 46, June, pp. 90–92.

Valdiya, K.S., 1968, "Origin of the Magnesite Deposits of South Pithoragarh, Kumaun Himalaya, India," *Economic Geology,* Vol. 63, No. 8, Dec., pp. 924–934.

Vitaliano, C.J., and Callaghan, E., 1956, "Geologic Map of the Gabbs Magnesite and Brucite Deposits," Mineral Investigations Field Studies Map, MF 35, US Geological Survey.

Vitaliano, C.J., et al., 1957, "Geology of Gabbs and Vicinity, Nye County, Nevada," Mineral Investigations Field Studies Map, MF 52, US Geological Survey.

Watson, I., 1980, "Olivine and Dunite Slag Uses Foster Right Market," *Industrial Minerals,* Dec., pp. 57–63.

Wicken, O.M., 1949, "Production of Seawater Magnesite," *Electrical Furnace Steel Conference Proceedings,* Vol. 7, AIME, New York, pp. 212–217.

Willard, H.P., and Gates, R.W., 1963, "Selective Open Pit Mining Featured at Gabbs," *Mining Engineering,* Vol. 15, No. 10, Oct., pp. 44–46.

Manganese

CHARLES H. JACOBY *

In 1774 a Swedish chemist, C. W. Schule, first recognized manganese as an element. That same year Schule's associate, J. G. Gahn, isolated the element manganese for the first time. In 1856 the Bessemer process of steelmaking gave birth to the economic importance of manganese. Later, in 1882, Robert Hadfield discovered high manganese steels.

Although the primary use of manganese is in the ferroalloy industry, two additional important uses for manganese are making chemicals and dry cell batteries. Manganese is also vital to plant and animal life. Various chemical compounds of manganese are used in fertilizers, feeds, glass manufacture, paints, varnishes, and for numerous medicinal and chemical purposes.

For the study of industrial minerals and rocks, only the chemical and battery grade material is of concern, but for the sake of comparison, metallurgical aspects are included in this presentation. Although currently the world's largest consumer of manganese, the United States is producing only minor amounts of ore at the present time. This is due primarily to the fact that although huge deposits of manganese occur within our borders, all are of such low grade as to be economically unfeasible to mine.

For many years various government agencies and segments of private industry have devoted sizable sums of money and time in an attempt to upgrade ores from the various deposits, and also in an attempt to recover manganese from blast furnace slags. In view of current world prices, and the quantity of high grade material available, there is little likelihood that any of these approaches will be found to be economically feasible in the near future.

Geology and Mineralogy

Mineralogy

There are over one hundred minerals that contain manganese. These minerals vary from those with compositions that are predominantly manganese to those having only minor percentages.

Mineral	Chemical Composition	% Mn
Hausmannite	Mn_3O_4	72
Polianite	MnO_2	63.1
Pyrolusite	MnO_2 •	60–63
Cryptomelane	$KR_8O_{16}R$ (R=Mn)	Variable
Psilomelane	Ba Mn $Mn_8O_{16}(OH)_4$	45–60
Coronadite	Pb $R_8O_{16}R$ (R=Mn)	Variable
Hollandite	Ba $R_8O_{16}R$ (R=Mn)	Variable
Manganite	$Mn_2O_3 \cdot H_2O$	62
Braunite	3 $Mn_2O_3 \cdot MnSiO_3$	62
Tephroite	2 $MnO \cdot SiO_2$	54.3
Rhodochrosite	$MnCO_3$	47
Rhodonite	$MnSiO_3$	42
Spessartite	3 $MnO \cdot Al_2O_3 \cdot 3SiO_2$	33.3
Wad	Hydrous Mn Oxides	Variable
Franklinite	(Fe Mn Zn)O(Fe $Mn)_2O_3$	Variable
Asbolan	Cobaltiferous wad	Variable
Alabandite	MnS	63.14

Geology

From a geological standpoint most manganese deposits are complex. In general, it can be said that all primary deposits of manganese are carbonates or silicates. The most productive and profitable have been those of sedimentary origin and residual concentrations. Examples of these residual types of deposits are associated with the metamorphosed lodes of Madhya, Pradesh, India, and the lateritic deposits of Orissa and Bihar, India; nodules in the residual clays of the Philippine Islands, US southern Appalachians and Arkansas, Ghana, and Brazil.

Sedimentary manganese deposits are best exemplified by the Nikopol district of Russia and the Tchiatouri area of Georgia, USSR.

Hydrothermal replacements are characterized by the rhodochrosite deposits of Butte and Phillipsburg, MT. In general, the hydrothermal type deposit has not resulted in any large tonnages being produced, but may have given

* Jacoby & Co., Earth Science Consultants, Tempe, AZ.

rise to the formation of sedimentary deposits throughout geological history.

Some sedimentary and residual-type deposits have been metamorphosed, giving rise to small high-grade ore bodies. These deposits are regionally metamorphosed, occurring in marbles, slates, quartzites, schists, and gneisses. Some of these metamorphosed deposits, such as the Franklin, NJ, deposit, are rich enough to be commercial without secondary enrichment. However, most of the exploitable deposits have been secondarily enriched.

Manganese is widely distributed and, to a greater or lesser degree, replaces two sets of elements; first, alkaline earths, calcium, barium, and magnesium and, secondly, aluminum and iron. Due to the diversity and complexity of manganese deposits, both with respect to deposition and chemistry, a wide range of impurities is almost invariably present in the ores.

Distribution of Deposits

Distribution of manganese deposits can best be described by the term "scattered." They vary from occurrences such as the nodules on the bottom of Lake Michigan at Green Bay to the bog ores of Wickes, MT; the rhodochrosite of Phillipsburg, MT; the franklinite of Franklin, NJ, to the residual pyrolusite deposits of Cartersville, GA. More than 2100 manganese deposits are listed in 35 states. Currently the free world's largest deposits of manganese ore are in Gabon, Brazil, Republic of South Africa, Ghana, and India.

Principal reserves in foreign countries are as follows:

	Million Tons of Mn
USSR	Unknown-in excess of 200
Republic of South Africa	300
Gabon	96
Brazil	46
Australia	44
India	22
World reserves, total	700

Russia possesses what are probably the world's two largest manganese deposits. The largest of these is the flat-lying Eocene deposit at Tchiatouri in Georgia, south of the Caucasus Mountains. These are beds of oolitic pyrolusite interbedded with layers of clays, sandstones, and marls. Upon beneficiation, primarily washing, these ores produce concentrates containing 50% MnO_2, and from 1 to 2% iron. The second great deposit of Russia lies in the Nikopol District on the Dnieper River in the Ukraine. Again, these are flat-lying beds close to the surface and easily accessible to strip mining. The beds of pyrolusite are interbedded with sand and glauconitic clays. These beds of Oligocene age rest on crystalline rocks covering approximately 160 square miles.

The largest deposit, or at least one of the largest, in the United States is located at Chamberlain, SD. This deposit is sedimentary in origin and consists of manganiferous concretions in the Pierre shale which is Upper Carboniferous in age. This formation varies in thickness from a few feet to 38 ft at Chamberlain.

The Batesville, AR, deposit contains a sporadic scattering of pyrolusite and wad-type residual materials in clay. Also buttons and thin layers of oxides are found in the Cason Shale. This deposit rests on a fine grained weathered limestone which gives rise to pinnacles precluding the use of normal strip mining operations.

The Cuyuna iron range in Minnesota contains iron ore carrying varying quantities of manganese. The manganese content ranges from less than 1% to more than 17%. The manganese was originally sedimentary in origin but has undergone some secondary deposition in the filling of fractures and crevices.

Southern Appalachian deposits begin in the area of the Shenandoah Valley and extend south to the area of Cartersville, GA, and on into Alabama. While a wide variety of manganese minerals occur in lenses, beds, and residual pockets, only the oxides in residual pockets have been of any economic importance. In Virginia these occur as a series of small deposits extending for 80 miles along the foot of the Blue Ridge Mountains. In general, most of the mines which have produced in excess of one million tons of manganese concentrates have been small operations conducted by local operators. Exceptions to this are the Crimora mine, working residual layers and pockets in the Erwin sandstone and the Kendall and Flick mines. These deposits, which have proven of economic importance, are the result of deep weathering. While the geological features of the primary deposits are undoubtedly dissimilar, the secondary residual deposits have many similarities.

The deposit at Artillery Mountains in Arizona has been operated from time to time during national emergencies. It has been estimated that the deposit contains approximately 175 million tons of ore averaging from 3 to 4%

manganese, with zones of material carrying values as high as 20% manganese. Beneficiation studies have made concentrates from which ferromanganese has been produced. The average iron content of these stratified oxides is 3% with 0.08% phosphorous.

The Three Kids property near Henderson, NV, was operated by Manganese, Inc. during World War II. At that time it was the largest producer of manganese in the United States. The primary ore, containing about 18% manganese, was beneficiated to produce a concentrate averaging 45% manganese.

Numerous other districts supply varying quantities of manganese during times of national emergency and stockpiling. The Anaconda Co. has produced approximately four million pounds of manganese from the Butte District in Montana. Manganese, zinc, and silver have been deposited in the peripheral zone as rhodochrosite and manganosiderite. Almost all of the production has come from the mining of rhodochrosite. The Emma mine, located in the southernmost Anaconda vein, mined the "great bulge" at the east end. The Phillipsburg District gave rise to manganese, silver, lead, and zinc as replacement veins in limestones and shales. In part, the rhodochrosite of these deposits has been altered with the formation of high grade oxides. The primary source of the manganese was the Granite-Bimetallic vein, which produced silver and gold. Aroostook County, Maine, contains large reserves of manganese mixed with iron in a 1 to 2 ratio. The ore is a mixture of fine grained minerals including silicates and carbonates. These bedded deposits are steeply dipping slates which have been folded and faulted. Over 40 deposits are known to exist in an area 90 miles long by 30 miles wide.

Manganese Nodules on the Ocean Floors: The occurrence of large tonnages of manganese nodules in the three major oceans has been well established. Cardwell (1973) gives detailed chemical analyses for nodules taken from 54 different Pacific Ocean locations. The two major elements are manganese and iron. The minimal manganese analyses averaged 8.2% Mn, the maximum 50.1% Mn, and a median value of 24.2% Mn was determined. The percent nickel values on the same samples were 0.16%, 2.0%, and 0.99%, respectively.

The three manganese minerals found were todorokite, birnesite, and delta manganese dioxide. Todorokite varies in its chemical composition and can contain significant amounts of other elements substituted for manganese in the crystal lattice. It is basically a calcium, sodium, manganese, potassium, and magnesium oxide with three molecules of water. Birnesite contains less lattice substituted elements and has the typical formula:

$$(Na_7Ca_3)\ Mn_{70}\ O_{140}\ 28H_2O.$$

These three forms of manganese show the degrees of oxidation increasing from todorokite, through birnesite, to delta manganese dioxide. The only iron mineral which has been recognized in the nodules is goethite.

Archer (1973) in reviewing the progress and prospects of marine mining estimates that a commercial manganese nodule mining operation would operate at a rate between 1 and 3 million stpy (dry weights). Allowing for processing losses, a 1 million-stpy operation is likely to be capable of providing about 13,000 tons of nickel, 11,000 tons of copper, 2500 tons of cobalt, and 270,000 tons of manganese. It is highly probable that manganese and cobalt will be considered byproducts with the scale of mining depending mostly on the world demand for nickel.

Sources of Manganese: Table 1 (DeHuff and Jones, 1981) gives production statistics and sources of manganese for the years 1976–1980.

Exploration and Evaluation

Manganese deposits are one of the more difficult ores to evaluate. In general, the deposits are small and scattered and expensive geophysical means are not usually economically feasible. Also the geophysical exploration costs are often beyond the financial capacity of the small operator. However, since manganese has relatively high solubility, geochemical techniques can be an effective tool in an exploration program.

One of the main difficulties is securing representative samples of an ore body for proper evaluation and analysis. In the past, a large percentage of the holes drilled for samples were by cable or churn drill methods. These methods require extreme care in order to secure valid data. In most cases the data have been inaccurate.

Rotary drilling techniques are somewhat better, but again, extreme care must be exercised in order to secure representative samples. Where the deposit contains wad or soft oxides in clay, soil sampling methods can be used. Generally, where residual or detritus has occurred in a clay, the particle size of the manganese mineralization ranges from large boul-

TABLE 1—Manganese Ore: World Production, by Country

(Short tons)

Country*	% Mn[e]	1976	1977	1978	1979[p]	1980[e]
North America: Mexico[†]	35+	499,579	[r]536,409	576,692	543,108	[‡]492,874
South America:						
Bolivia[†§]	28-54	[r]13,520	9,464	1,364	11,574	4,960
Brazil[#]	38-50	1,869,738	1,670,741	2,113,239	2,490,483	2,400,000
Chile	36-40	26,058	19,843	25,621	27,524	24,900
Peru	26	676	—	—	—	—
Europe:						
Bulgaria	30-	44,100	44,100	44,100	46,300	44,100
Greece	48-50	9,075	8,631	7,727	6,338	6,600
Hungary	30-33	138,000	132,000	126,000	91,000	97,000
Italy	22+	4,917	10,267	10,738	9,921	[‡]10,103
USSR[‖]	35	9,520,000	9,470,000	9,984,000	11,292,000	[‡]11,300,000
Yugoslavia	30+	20,944	27,282	30,203	33,235	33,000
Africa:						
Egypt	28+	4,691	4,225	191	—	—
Gabon	50-53	2,443,556	2,039,857	1,830,959	2,535,417	[‡]2,366,386
Ghana	30-50	343,780	321,417	347,864	298,481	[‡]278,279
Morocco	53-50	129,305	125,164	139,112	149,017	165,000
South Africa, Republic of	30-48+	[r]6,010,079	5,564,411	4,758,721	5,712,615	[‡]6,278,125
Sudan	48	505	504	496	[e]500	400
Zaire	30-57	200,824	42,216	—	—	—
Asia:						
China, mainland[e]	20+	1,100,000	[r]1,200,000	1,400,000	1,650,000	1,750,000
India[¶]	10-54	[r]2,022,405	2,055,865	1,784,503	1,934,641	[‡]1,813,692
Indonesia	47-56	10,839	6,593	6,492	6,514	7,700
Iran**	33+	44,100	44,100	33,100	[e]25,000	22,000
Japan	26-28	156,244	[r]139,063	114,802	96,925	85,900
Korea, Republic of	23-40	1,524	732	823	39	[‡]89
Pakistan	35-	71	58	317	121	110
Philippines	35-45	11,658	22,706	4,311	5,508	5,500
Thailand	46-50	55,364	[r]84,836	79,599	38,984	[‡]54,299
Turkey	35-46	18,696	21,275	[e]22,000	[re]26,000	26,000
Oceania:						
Australia	37-53	2,374,560	1,531,113	1,376,699	1,836,752	[‡]2,161,511
Vanuatu (formerly New Hebrides)	40-44	[††]38,664	[‡‡]25,397	[‡‡]22,853	[‡‡]11,663	—
Total	NA	[r]27,113,472	[r]25,158,269	24,842,526	28,879,660	29,428,528

Source: DeHuff and Jones, 1981.

[e]Estimated. [p]Preliminary. [r]Revised. NA Not available.

*In addition to the countries listed, Colombia, Cuba, and the Territory of South-West Africa (Namibia) may have produced manganese ore and/or manganiferous ore, but available information is inadequate to make reliable estimates of output levels. Low-grade ore not included in this table has been reported as follows in short tons: Argentina (16% to 22% Mn) 1976—58,517, 1977—90,814, 1978—20,389, 1979—11,233, 1980—10,692; Czechoslovakia (about 17% Mn) 1976—1,211, 1977—1,003, 1978—[e]1,000, 1979—[e]1,000; 1980—[e]1,000; Malaysia (grade unspecified but apparently a manganiferous ferruginous ore) 1976—103,741, 1977—50,040, 1978—47,092, 1979—34,839, 1980—4,413; Romania (about 22% Mn) an estimated 90,000 in each year; the Republic of South Africa (15% to 30% Mn, in addition to material listed in table) 1976—56,178, 1977—266,930, 1978—105,490, 1979—1980—nil.

[†]Estimated on the basis of reported contained manganese.

[‡]Reported figure.

[§]Exports.

[#]Figures are the sum of (1) sales of direct shipping manganese ore and (2) production of beneficiated ore, both as reported in the 1977 through 1980 editions of Annuario Mineral Brasileiro.

[‖]Source: 1976-79, The National Economy of the USSR, Central Statistical Administration, Moscow; 1980, Pravda, Moscow. Grade represents the annual averages obtained from reported metal contents of the gross weights shown in the table for 1976-1979.

[¶]Much of India's production grades below 35% Mn; recent details on output by grade are not available, but in 1976, 65% of total exports of 787,533 short tons were below 35% Mn.

**Reported as if data are for calendar years, but may actually represent output for Iranian calendar years beginning March 21 of the year stated.

[††]Japanese imports.

[‡‡]Figures revised from Japanese imports to reported production.

ders to −200 mesh material. This makes soil sampling procedures difficult to employ.

In general the hard and deeper deposits are easier to sample with some degree of accuracy.

Although a great many manganese occurrences are known, the limited size of many of these deposits make them marginal to submarginal. Other large deposits are not economically exploitable due to the low concentration of manganese. The third consideration, and of prime importance, are the impurities associated with the manganese mineralization.

Due to the diversity and complexity of manganese formations, the impurities are many in number and complex in nature. Following is a broad generalization of the types of impurities:

1) Metallic impurities: iron, lead, zinc, copper, arsenic, and silver minerals.

2) Nonmetallic impurities: sulfur and phosphorous minerals.

3) Gangue: silica, alumina, lime, magnesia, and barium.

4) Volatiles: water, carbon dioxide, and organic matters.

Primary factors in the evaluation of deposits within the United States are the amenability of the material to beneficiation, and the price projection over the life of the necessary capital investment. Considerable effort has gone into the upgrading of ores from the various larger deposits within the continental limits of the United States.

Prior to World War II, few manganese operations had milling plants which were more complicated than washing, screening, jigs, tables, and, in a few instances, flotation. Now it is not only necessary to increase the manganese content, but also to decrease the percentage of impurities. Because each deposit is distinctly different from most other deposits, no single process is applicable to all.

End Use Specifications

The manganese industry in the US is confined to the processing of foreign ores having manganese content of 35% or greater. Low-grade ores containing from 5 to 35% manganese are produced and shipped in South Carolina, New Mexico, and Minnesota. South Carolina ore is used primarily as a pigment in the manufacture of brick in both North and South Carolina. Most of the ore produced in New Mexico and Minnesota is used for the production of pig iron. As of May 2, 1980, the Federal Emergency Management Agency (FEMA) established new stockpile goals: natural battery ore, 62,000 st; synthetic manganese dioxide, 25,000 st; chemical ore, 170,000 st; metallurgical ore, 2,700,000 st; and 439,000 st of high carbon ferromanganese.

The present processing plants in the US consist of 16 facilities owned by ten companies. Basically the plants may be expected to continue to diminish in number and capacity. Union Carbide has signed an agreement in principle to divest itself of all its ferromanganese alloy and metal plants to a group of investors. This group of Norwegians is relating to cheap hydroelectric power for processing. The most recent plant to begin production of ferromanganese is the plant at Theodore, Alabama. Tinsley (1974) reported on the close ties between manganese consumption and steel production. In 1977 the steel industry accounted for 80.2% of manganese consumption (DeHuff, 1977). Approximately 13 to 14 lb of manganese are used per ton of steel for deoxidation and control of sulfur.

Manganese dioxide (pyrolusite) is used in the manufacture of dry cell batteries. This market has been challenged in recent years by mercury, alkali, and lithium batteries. Additional inroads on the market can be expected with further technical advances.

Due to the complex nature of the geology of many manganese deposits, chemistry of the ores varies widely. Some analyses of ores from major areas of production are given in Table 2. In general, manganese ores can be classified as follows:

	% Mn
Manganese ore	+35
Ferruginous manganese ore	10 to 35
Manganiferous iron ore	5 to 10

The most promising segment of the industrial manganese picture is the heavy demand for synthetic manganese dioxide, which has prompted expansion of facilities in both Japan and South Africa. In most cases this expansion constitutes a doubling of capacity; in the case of Mitsui Mining & Smelting, however, capacity will be quadrupled.

The National Stockpiling Specifications of the General Services Administration are given in Table 3.

Only rich manganese ores can produce the standard 70 to 90% manganiferous alloys used in the manufacture of steel in the most industrially oriented countries. Standard grades of ferromanganese used in the United States are given in Table 4.

TABLE 2—Analyses of Manganese Ores

Country	Mining Area	Composition, %			
		Mn	Fe	SiO	P
Brazil	Miguel Burnier	50	–	1.0	0.03-0.05
Congo	Kisanga	50	–	–	1.5
Ghana	Nsuta	46-52	1.6-7.7	2.4-5.5	0.11-0.14
India	Balaghat	48-53	4.2-10.3	3.5-10.7	0.03-0.19
	Shivrajpur	46-50	4.7-5.5	6.0-9.3	0.24-0.25
	Bhandra*	52-58	1.5-5.1	0.6-2.6	0.23-0.38
Morocco	Imini	50-54	–	3-4	–
Republic of South Africa	Postmasburg	43-52	8.7-15.3	2.8-3.9	0.03-0.05
USSR	Chiatury	48-49	0-1.8	6-10	0.15-0.18
	Nikopol	48-51	0.7-1.2	7.5-10.5	0.16-0.33
United States	Montana	48-50	1-2.5	10-12	–
	Nevada	45-57	1.1	13-14	0.06

Source: Cole, 1960.

* Battery grade.

Chemicals and Glass Products

Manganese is used in the manufacture of numerous chemicals, disinfectants, glass, fertilizers, colored brick, tile, paints, and as a flux in the melting of gold and silver ore. As a chemical raw material manganese is used in both organic and inorganic chemicals as oxidizers, catalysts, and chemical intermediates. Small percentages of manganese dioxide are used to take the color out of glass which contains iron.

Manganese, as a base forming element, forms two series of salts. One series is bivalent and the other is tetravalent. Manganese also forms a series of compounds in which it acts as an acid radical. These compounds include the permanganates which are strong oxidants.

In the ceramic industry manganese gives a variation of colors from a bright reddish purple to a purplish brown or black.

Solution Mining.

Because of its position in the electromotive series, it will displace arsenic, antimony, bismuth, tin, lead, copper, iron, nickel, cobalt, cadmium, and zinc. It plays a role in the forming of cement metals in the solution mining of certain metalliferous deposits.

Fertilizers

With respect to fertilizers, manganese, because of its high solubility has often been depleted from the soil. Generally, this is replenished to the soil in the form of manganese sulfate either in its pure form or as an admix.

Paint

The paint industry uses manganese to promote the absorption of oxygen in the drying of paints and varnishes.

Batteries

The manufacture of batteries is the most important nonmetallurgical use of manganese. Its use as a solid cathodic material was a point of progress in battery development. Either the naturally occurring pyrolusite or synthetic MnO_2 may be used. Other systems, including the newly commercialized lithium batteries, may be expected to somewhat reduce the consumption of manganese for this purpose.

Preparation for Markets

Beneficiation

Both the Anaconda Co. at Butte, MT, and Manganese, Inc. at Three Kids, NV, have employed flotation techniques with either oil emulsions or conventional soap flotation.

In the case of Anaconda's treatment of manganese carbonates, rotary kilns were used to produce oxide nodules from the flotation concentrates. Some ores have been treated by heavy media separation, while others have been subjected to electromagnetic separation. Slime losses in the treating of soft ores such as pyrolusite and wads are always high, and recovery of the manganese from these slimes has been attempted with Humphreys spirals and various classifiers.

The Dean-Leute ammonium carbonate process is one of the methods which has been de-

TABLE 3—General Services Administration National Stockpiling Specifications

		% Wt
Metallurgical Grade*		
Manganese (Mn)		40.0
Iron (Fe)		16.0
Silica plus alumina ($Si_2O_3 + Al_2O_3$)†		
Phosphorus (P)		0.30 max
Copper plus lead plus zinc‡		1.00 max
Battery Grade		
Grade A (Military)		
Manganese dioxide in synthetic, ore, or concentrate form, suitable for use in the manufacture of dry cell batteries. This must conform to Signal Corp. Specifications SCL-3175.		
Grade A (Commercial)		
Manganese MnO_2		75 min.
Arsenic		— max
Copper		— max
Lead		0.5 max
Iron		3.0 max
Total heavy metals other than Fe & Pb		0.5
Total Insolubles		10.0
Total Mn		48.0 min.
pH		4 to 7.0
Chemical Grade		
Type A		
Manganese dioxide		80 min.
Iron		3 max
Type B	*P81*	*P-81-R*
Manganese dioxide	85.0 min.	82.0 min.
Manganese	— min.	53.0 min.
Iron	3.0 max	3.0 max
Silica	3.0 max	5.0 max
Alumina	3.0 max	3.0 max
Phosphorus	0.10 max	0.20 max
Arsenic	0.05 max	0.10 max
Ferromanganese		
Manganese		75.00 min.
Carbon		7.50 max
Silicon		1.25 max
Phosphorus		0.35 max
Sulfur		0.05 max
Iron		Balance

* In a separate category from the metallurgical grade ore is a National Stockpiling Specification P-30a-R of 3/4/58.
† No limit specified, but if over 15% purchased in exceptional cases only.
‡ Of which not more than 25% may be copper.

veloped for the processing of low grade manganese concentrates. This process has experienced some success in producing manganese dioxide from Cuyuna iron range ore. The process involves the roasting of the Cuyuna ore, leaching with ammonium carbonate solution to obtain a high grade carbonate, and the reduction of the carbonate to the oxide.

High manganese slags have been developed by the selective oxidization-reduction process using open hearth slags as feed stock. Manganiferous materials from Aroostook County, Maine, have been treated by chloride volatilization processes without economic success. The Sylvester process, in which limestone has been used to form a manganese clinker, has been tested on the siliceous ores of Aroostook as well as on open hearth slags. The material was treated in a ball mill and then fed to a magnetic separator to separate the high Mn-Fe material. Another process for treating siliceous ores was the Udy electric furnace process, in which two or three stages of reduction were employed to produce the desired quality of ferromanganese.

Fine-sized manganese ore particles have been successfully agglomerated into balls or pellets with an organic binder, and then introduced into a short stack blast furnace. In the presence of a high quality coke, which produces a reducing environment, 77% of the manganese is recovered in metal form having a composition: manganese, 92.0%; carbon, 6.8%; iron, 1.0%; and silicon, 0.2%.

In the Daugherty process of the Republic Steel Corp., low grade ores are ground, subjected to water and sulfur dioxide gas to produce $MnSO_3$ and $MnSO_4$, boiled by live steam injection into the solution, then filtered, and roasted in a kiln to produce manganese oxides. Similarly, the Welch process of Manganese Chemicals Corp. dissolves the manganese from low grade ores with sulfuric acid to form $MnSO_4$. This is introduced into an electrolytic cell to form MnO_2 of excellent battery grade.

The Bruce Williams process involves the roasting of a low grade manganese feed with ammonium salts to produce a manganese carbonate. The Sheer Korman process of Vitro Corp. treats siliceous ores with a high intensity arc.

The Nossen process recommends a regenerative nitric acid system to produce manganese dioxide from low grade manganous oxides.

In the evaluation of the economic feasibility of any low grade deposit, its amenability to processing, and the efficiency and cost of upgrading the ore will often be the deciding factor in commercial exploitation.

Transportation

The major centers of manganese consumption in the United States are east of the Mississippi River. During times of national emergency we are faced with long rail hauls from the low grade western sources.

Most of the foreign ore is transported by ocean freighters with the exception of imports

TABLE 4—Standard Grades of Ferromanganese Used in Steel Manufacture

	(1)	(2)	(3)	(4)	(5)	(6)
High-Carbon						
Manganese, %	78 to 82					
Iron, %	12 to 16					
Carbon, %	6 to 8					
Silicon, %	1.00 max					
Phosphorous, %	0.30 max					
Sulfur, %	0.05 max					
Low-Carbon						
Manganese, %	90.00	80 to 85	80 to 85	80 to 85	80 to 85	80 to 85
Carbon, %	0.07 max	0.10 max	0.15 max	0.30 max	0.50 max	0.75 max
Silicon, %	1.00 max	1.00 max	1.00 max	1.00 max	1.00 max	7.00 max
Phosphorus, %	0.06 max	0.20 max	0.20 max	0.20 max	0.20 max	0.25 max
Medium-Carbon						
Manganese, %	80 to 85					
Carbon, %	1.50 max					
Silicon, %	1.50 max					
Spiegeleisen or Spiegel						
Manganese, %	16 to 19	19 to 21	26 to 28			
Carbon, %	6.50 max	6.50 max	6.50 max			
Silicon, %	3.00 max	3.00 max	1.00 max			

from Canada and Mexico. In wartime our sea lanes are subject to severance, but with the value and strategic importance of manganese, it might be air-lifted from sources such as Brazil.

Markets and Consumption

By far the largest consumption of manganese is in the manufacture of manganiferous alloys used in the production of steel.

The sources of manganese in the production of ferromanganese, silicomanganese, spiegeleisen, and manganese metal are listed in Table 5. The quantities of ferromanganese produced in the US and the quantities of manganese ores consumed in manufacturing from 1973 to 1980 are listed in Table 6.

Table 7 shows consumption, and stocks of manganese ore in the US in 1980. Manganese consumption from 1960 to 1979 is given in Table 8.

Future Consideration and Trends

Substitutes and Synthetics

The substitution of other materials for manganese is very limited. In special cases, titanium, zirconium, and rare earth metals can replace manganese, but such substitutions are confronted with technological problems, adverse cost factors, and the availability of the material being used as a substitute. It might be said that because of its low cost, manganese is a substitute for several other alloying elements, including nickel.

Several substitutes exist for manganese in dry cells. Perhaps the substitute with the greatest future potential and the least strategic significance is lithium.

Synthetic manganese dioxide can be produced by electrolytic and chemical methods. Some of this material is of battery grade.

TABLE 5—Manganese Ore Used in Producing Ferromanganese, Silicomanganese, and Manganese Metal in the United States in 1980, by Source of Ore

Source	Gross weight (short tons)	Mn content, natural (%)
Domestic*	34,877	45
Foreign:		
Africa	288,991	46
Australia	49,021	52
Brazil	204,580	46
Chile*	18,754	45
India	39,676	48
Mexico	32,957	39
USSR*	7,452	48
Unidentified	49,819	—
Total or average	726,127	46

Source: DeHuff and Jones, 1981.

*Most, if not all, from US Government excess stockpile disposals.

TABLE 6—Ferromanganese Produced in the United States and Manganese Ore Consumed in Its Manufacture (+35%Mn)

	Ferromanganese Produced			Manganese Ore* Consumed (Short Tons)		
	Gross Weight,	Manganese Content		Gross Weight⁺		Gross Weight⁺ per Ton of Ferromanganese⁺ Made
Year	St	%	St	Foreign	Domestic	
1973	683,075	78.8	538,119	1,648,806	25,912	2.4
1974	544,361	78.0	424,405	1,348,425	55,822	2.5
1975	575,809	78.9	454,309	1,389,300	48,011	2.4
1976	482,662	79.0	381,328	1,208,336	53,632	2.5
1977	334,134	78.8	362,136	889,296	35,769	2.6
1978	272,530	80.6	219,707	740,906	90,660	2.7
1979	317,102	80.2	254,398	785,664	125,130	2.5
1980	189,472	79.7	150,982	691,250	34,877	1.9

Source: DeHuff, 1978-1979; DeHuff and Jones, 1981.
*Containing 35% or more manganese (natural).
⁺Includes ore used in producing silicomanganese and metal.

Subsidies

Where a national emergency exists or it is determined that it is in the nation's best interest to build stockpiles of strategic minerals, the United States Government has entered into extensive and effective subsidy programs for manganese ores. These programs have in the past varied widely in prices paid the operators as well as being realistic in the quality of ore that the districts could produce, keeping in mind the end use to which the material was needed. No such purchase program is currently active.

The US Geological Survey's Office of Mineral Exploration is encouraging the exploration for manganese with financial assistance. Since all manganese consumed in the United States at the present moment is of foreign import, the strategic importance is obvious.

Taxes and Depletion Allowance

Even though manganese bears a 22% depletion allowance, it has been insufficient tax incentive to encourage operators to mine even the best of the low grade properties in America. The 14% depletion allowance on foreign de-

TABLE 7—Consumption and Stocks of Manganese Ore* in the United States (Short Tons, +35% Mn)

	Consumption					
	1976	1977	1978	1979	1980	Stocks 12/31/80
By Use:						
Manganese alloys & metal	1,263,531	926,635	832,179	913,491	727,530	546,840
Pig iron & steel	143,761	200,803	219,663	230,742	131,516	158,422
Dry cells, chemicals & misc.	193,581	231,373	229,637	227,957	211,729	325,225
Total	1,600,873	1,358,811	1,281,479	1,372,190	1,070,775	1,030,487
By Origin						
Domestic	81,607	61,152	106,900	144,404	60,701	52,332
Foreign	1,519,266	1,297,659	1,174,579	1,227,786	1,010,074	978,155
Total	1,600,873	1,358,811	1,281,479	1,372,190	1,070,775	1,030,487

Source: DeHuff, 1978-1979; DeHuff and Jones, 1981.
*Containing 35% or more manganese (natural).

TABLE 8—US Manganese Consumption (Thousand short tons)*

Year	Consumption	Mine† Production	Imports	Exports	Inventories Private	Inventories Government	Price‡
1960	1,077	177	1,318	6	1,447	4,241	0.94
1961	842	78	1,172	7	1,278	4,811	0.94
1962	978	73	1,078	9	1,114	5,139	0.91
1963	1,096	94	1,125	6	1,043	5,327	0.81
1964	1,216	79	1,154	13	948	5,426	0.69
1965	1,373	83	1,432	12	977	5,527	0.73
1966	1,353	88	1,444	10	1,093	5,580	0.76
1967	1,207	73	1,171	11	1,171	5,528	0.67
1968	1,150	48	1,053	15	1,180	5,455	0.60
1969	1,317	93	1,249	14	1,241	5,405	0.50
1970	1,327	66	1,085	30	1,175	5,265	0.54
1971	1,170	38	1,150	30	1,281	5,147	0.60
1972	1,366	29	1,098	19	1,241	4,929	0.60
1973	1,554	31	1,058	40	978	4,687	0.65
1974	1,492	35	969	114	1,183	3,880	0.90
1975	1,133	19	1,112	131	1,359	3,571	1.38
1976	1,364	31	1,127	74	1,373	3,277	1.45
1977	1,523	27	935	78	1,092	2,919	1.48
1978	1,415	38	882	120	756	2,640	1.40
1979e	1,463	30	896	73	866	2,490	1.40

e Estimated.
* After DeHuff.
† Includes manganese in manganiferous ores.
‡ Average price for metallurgical ore containing 46% to 48% Mn in $ per 22.4 lb.

posits owned by US investors has been a factor in stimulating the development of deposits in developing nations. This incentive, plus the reduction in tariff from the 1967 level of 25¢ per lb of contained manganese, to 22¢ under the "General Agreement on Tariffs on Trade" of Jan. 1, 1968, undoubtedly helped to build the US supply from foreign sources.

Under this agreement the tariff has been steadily reduced so that on Jan. 1, 1972 the amount was 12¢ per lb. In June of 1964 the tax was completely suspended for three years. This suspension of duty was again initiated in 1967 for three years with no levy since that time.

Ecology

Since deposits are small and production is low even during times of national emergency, little or no damage is suffered even though the major portion of our ore is mined by the open pit method. Tailings ponds have effectively handled washing and milling rejects. A good example of such a tailings pond exists near Cushman, AR, in the Batesville District.

Bibliography and References

Anon, 1943, "Concentrating and Nodulizing 'Pink' Manganese at Anaconda," *Mining World*, Vol. 5, No. 1, Jan., pp. 3–8.

Archer, A.A., 1973, "Progress and Prospects of Marine Mining," *Mining Engineering*, Vol. 25, No. 12, Dec., pp. 31–32.

Barnard, P.G., et al., 1962, "Problems in Substituting Titanium for Manganese in Steel," Report of Investigations 6030, US Bureau of Mines, 27 pp.

Boardman, L.G., 1961, "Manganese in the Union of South Africa," Paper presented at the 7th Commonwealth Mining and Metallurgical Congress Johannesburg, Republic of South Africa, Preprint, 12 pp.

Brooke, D.B., 1966, *Low-Grade and Nonconventional Sources of Manganese*, Resources for the Future, Johns Hopkins Press, Baltimore, MD, 123 pp.

Caldwell, A.B., 1971, "Deepsea Ventures Readying Its Attack on Pacific Nodules," *Mining Engineering*, Vol. 23, No. 10, Oct., pp. 54–55.

Cardwell, P.H., 1973, "Extractive Metallurgy of Ocean Nodules," Preprint, American Mining Congress Meeting, Denver, Sep., 27 pp.

Chilton, C.H., 1958, "High Purity Manganese Via Electrolysis," *Chemical Engineering*, May 19, pp. 136–139.

Coetzee, J.J., 1961, "The Production of Ferro-Alloys," Paper presented at the 7th Commonwealth Mining and Metallurgical Congress, Johannesburg, Republic of South Africa, Preprint, 11 pp.

Cochran, A.A., and Falke, W.L., 1967, "A One-Step Operation for Recovery of Manganese as Chloride from Ores and Slags," Report of Investigations 6859, US Bureau of Mines, 22 pp.

Cole, S.S., 1960, "Manganese Ore," *Industrial Minerals and Rocks,* 3rd ed., J.L. Gillson, ed. AIME, New York, pp. 545–549.

DeHuff, G.L., 1960, "Manganese," *Mineral Facts and Problems,* Bulletin 585, US Bureau of Mines, pp. 493–510.

DeHuff, G.L., 1965, "Manganese," *Mineral Facts and Problems,* Bulletin 630, US Bureau of Mines, pp. 553–571.

DeHuff, G.L., 1971, "Manganese," *Minerals Yearbook—1971,* Vol. 1, US Bureau of Mines, pp. 717–729.

DeHuff, G.L., 1972, "Manganese," *1972 Minerals Yearbook,* Vol. 1, US Bureau of Mines, pp. 757–769.

DeHuff, G.L., 1977, "Manganese," *Minerals Yearbook 1977,* US Bureau of Mines.

DeHuff, G.L., 1978, "Manganese," Preprint, 1978–1979, *Minerals Yearbook,* US Bureau of Mines, 15 pp.

DeHuff, G.L., 1980, "Manganese in 1980," *Mineral Industry Surveys,* US Bureau of Mines.

DeHuff, G.L., and Jones, T.S., 1981, "Manganese," *Minerals Yearbook 1980,* US Bureau of Mines, pp. 543–553.

Dickson, T., 1980, "Manganese, Non-Metallic Market Development," *Industrial Minerals,* No. 155, Aug., pp. 21–29.

Douglas, W.S., et al., 1963, "The Production of Ferromanganese from Low-Grade Manganiferous Ores Using a Rotary Kiln and an Electric Smelting Furnace," *Canadian Mining & Metallurgical Bulletin,* Vol. 56, No. 5, May, pp. 387–389.

Dykstra, F.R., 1979, "Manganese—Its Strategic Implications," *Industrial Minerals,* No. 145, Oct., pp. 55–61.

Elkins, D.A., 1964, "Estimated Cost of Exploiting Enriched, Hard Manganese Ore from the Maggie Canyon Deposit, Artillery Mountains Region, Mohave County, Arizona," Report of Investigations 6438, US Bureau of Mines, 78 pp.

Falke, W.L., and Cochran, A.A., 1966, "Reduction-Volatilization Processes for Recovery of Manganese from Ores," Report of Investigations 6738, US Bureau of Mines, 19 pp.

Farnham, L.L., and Stewart, L.A., 1958, "Manganese Deposits of Western Arizona," Information Circular 7843, US Bureau of Mines, 87 pp.

Fillo, P.V., 1963, "Manganese Mining and Milling Methods and Costs, Mohave Mining Company, Maricopa County, Arizona," Information Circular 8144, US Bureau of Mines, 29 pp.

Huttl, J.B., 1955, "How Manganese, Inc. Upgrades Complex Three Kids Ore," *Engineering & Mining Journal,* Vol. 156, No. 11, Nov., pp. 88–93, 116.

Jacoby, C.H., 1975, "Manganese," *Industrial Minerals and Rocks,* 4th ed., S.J. Lefond, ed., AIME, New York, pp. 821–836.

Joyce, F.E., Jr. and Prasky, C., 1973, "Sulfatization-Reduction of Manganiferous Iron Ore," Report of Investigations No. 7749, US Bureau of Mines, 17 pp.

Johnson, A.C., and Trengove, R.R., 1956, "The Three Kids Manganese Deposit, Clark County Nevada: Exploration, Mining and Processing," Report of Investigations 5209, US Bureau of Mines, Apr., 31 pp.

Kline, H.D., 1958, "Methods and Costs of Mining and Washing Manganese Ore, Batesville District," Report of Investigations 5411, US Bureau of Mines, 46 pp.

Kline, H.D., 1962, "Methods and Costs of Mining and Washing Manganese Ore, Batesville District, Arkansas," Information Circular 8095, US Bureau of Mines, 22 pp.

Kraus, E.H., et al., 1959, *Mineralogy,* 5th ed., McGraw-Hill, New York, 686 pp.

LeVan, H.P., et al., 1965, "Extraction of Manganese from Georgia Umber Ore by a Sulfuric Acid-Ferrous Sulfate Process, Part One, Countercurrent Decantation Extraction and Agglomeration of Leached Residue," Report of Investigations 6692, US Bureau of Mines, 21 pp.

Lewis, W.E., et al., 1958, "Investigation of Cuyuna Iron Range Manganese Deposits, Crow Wing County, Minn., Progress Report One," Report of Investigations 5400, US Bureau of Mines, 49 pp.

Mero, J.L., 1961, "Economics of Deep-Sea Mining," *Mining Congress Journal,* Vol. 47, No. 9, Sep., pp. 52–56, 68.

Mero, J.L., 1968, "Seafloor Minerals: A Chemical Engineering Challenge," *Chemical Engineering,* July 1, pp. 73–80.

Nossen, E.S., 1951, "Manganese Concentration from Low-Grade Domestic Ore," *Industrial & Engineering Chemistry,* Vol. 43, No. 7, July, pp. 1695–1700.

Perkins, E.C., 1957, "Caustic Leaching of Manganese Flotation Concentrate from Artillery Peak, Arizona," Report of Investigations 5341, US Bureau of Mines, 16 pp.

Perkins, E.C., and Novielli, F., 1962, "Bacterial Leaching of Manganese Ores," Report of Investigations 6102, US Bureau of Mines, 11 pp.

Prasky, C., et al., 1961, "Evaluating Cuyuna Manganese Resources by Sulfatizing," Report of Investigations 5887, US Bureau of Mines, 27 pp.

Sears, C.E., 1957, "Manganese Deposits of the Appalachian Area of Virginia," *Mineral Industries Journal,* Virginia Polytechnic Institute, Vol. 4, No. 1, Mar., pp. 1–4.

Sheridan, E.T., 1970, "Manganese," *Mineral Facts and Problems,* Bulletin 650, US Bureau of Mines, pp. 315–331.

Sidwell, K.O.J., 1957, "The Woodstock, N.B. Iron-Manganese Deposits," *Canadian Mining & Metallurgical Bulletin,* Vol. 50, No. 7, July, pp. 411–416.

Sittig, M., 1968, *Inorganic Chemical and Metallurgical Process Encyclopedia,* Noyes Development Co., Park Ridge, NJ, pp. 437–444.

Sullivan, G.V., et al., 1962, "Extraction of Manganese from Low-Grade Dolomitic Materials by a Roast-Leach Process," Report of Investigations 6121, US Bureau of Mines, 24 pp.

Tinsley, C.R., 1973, "In Search for Commercial Nodules, Odds Look Best in Miocene-age Pacific Tertiary System," *Engineering & Mining Journal,* Vol. 174, No. 6, June, pp. 114–116.

Tinsley, C.R., 1973, "Mining of Manganese Nodules: An Intriguing Legal Problem," *Engineering & Mining Journal,* Vol. 174, No. 10, Oct., pp. 84–87.

Tinsley, C.R., 1974, "Manganese-Prices Boosted by Record Steel Output in 1973," *Engineering & Mining Journal*, Vol. 175, No. 3, Mar., pp. 81–84.

Weiss, S.A., 1980, "Manganese, the Other Uses," *Metals Bulletin Book*, London, 360 pp.

Williamson, D.R., et al., 1959, "Future United States Manganese Sources," *Colorado School of Mines & Mineral Industries Bulletin*, Vol. 2, No. 4, July, 12 pp.

Young, W.E., 1966, "Manganese Occurrences in the Eureka-Animas Forks Area of the San Juan Mountains, San Juan County, Colorado," Information Circular 8303, US Bureau of Mines, 52 pp.

Meerschaum

B. F. BUIE *

For over 200 years meerschaum has been a significant item of trade between the Near East and countries to the west. Best-known for its use in making smoking pipes and cigar and cigarette holders, it is also used in making a variety of decorative and ornamental items. It is easily carved, and is given an attractive, lustrous finish by rubbing. No published specifications are known, but compact form, purity, uniform whiteness or light color, and uniform, porous texture appear to be requisites. Fig. 1a shows the appearance of a piece of crude meerschaum from Turkey; Fig. 1b a carved item; and Fig. 1c, d, and e show the fibrous and porous texture as revealed by the scanning electron microscope.

Neither a major industrial mineral nor a precious stone, meerschaum nevertheless holds a record of long—and continuing—association with art and personal affairs of man. It is also a mineral to which modern methods of exploration and production probably could be profitably applied. An interesting account of the preparation and use of meerschaum in the early part of this century is given by Sterrett (1907).

Geology and Mining

Mineralogy

Meerschaum is recognized now to be the compact variety of sepiolite, the claylike variety of which is one of the minerals included in the chapter on clays. The word *meerschaum*, from the German meaning "sea foam," was in earlier times used synonymously with *sepiolite*, a word of Greek origin suggestive of the resemblance of the compact variety—then the only variety known—to cuttlefish bone. At one time *meerschaum* was used as the commercial name, and *sepiolite* for the scientific name. However, to think of meerschaum as the name by which sepiolite is known commercially is no longer valid, in view of the recent emergence of the earthy variety as a competitor of bentonite and other clay materials. Present tendency is to restrict the term *meerschaum* to the compact variety, and to use the term *sepiolite* as a more general name to include both the compact and earthy varieties. This is the terminology followed in this publication.

* Professor of Geology, Dept. of Geology, Florida State University, Tallahassee, FL.

Meerschaum typically is white to light gray, has a hardness of 2 to 2½, conchoidal to irregular fracture, and specific gravity varying with porosity from about 2 to less than 1. The composition is somewhat variable, but is near that determined by Schaller (1936) for sepiolite as $2MgO \cdot 3SiO_2 \cdot 4H_2O$. A more modern formula based on X-ray determination of the crystal structure is given by Caillère and Hénin (1961) as $(Si_{12})(Mg_9)O_{30}(OH)_6(OH_2)_4 6H_2O$. The reason for the firm, compact nature of meerschaum, in contrast with the earthy variety of sepiolite, is not fully understood. It may be due to the presence of a very minor amount of some bonding material such as silica. Scanning electron microscope (SEM) photographs of a specimen from Turkey (Figs. 1c-1e) indicate that the meerschaum is composed predominantly, if not entirely, of crystalline fibers which merge into sheetlike forms in a complex, intertwined mass.

Detailed X-ray and crystal structure data, as well as some data on thermal stability range, are given by Caillère and Hénin (1961) for the sepiolite family of minerals.

Mode of Occurrence and Origin

Most authors have attributed the origin of meerschaum to alteration of serpentine or other magnesian minerals. The occurrence in alluvium and other surficial deposits does not necessarily indicate that the alteration was supergene. In fact, hydrothermal origin appears more likely for some deposits, though not for all.

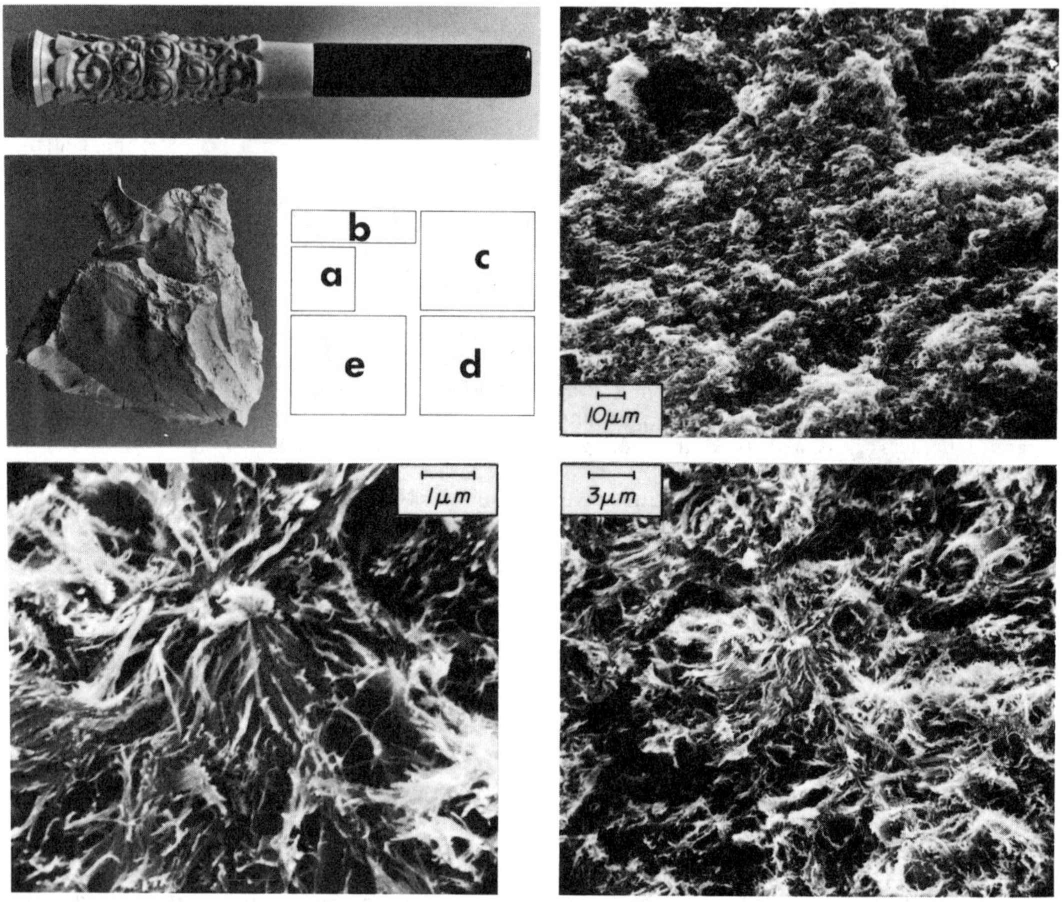

FIG. 1—*Meerschaum from Turkey. a—Broken piece of crude meerschaum showing characteristic fracture. b—Carved cigarette holder. c-e—Scanning electron microscope photographs at different magnifications showing porous texture and fibrous form.*

According to Bateman (1950) the Turkish meerschaum occurs as scattered nodules, from the size of an egg to that of a football, in a cemented valley fill that presumably came from the adjacent mountain sides. Oelsner and Krüger (1957) mention its occurrence in Eskishehir, Turkey, as bedded deposits very closely associated with serpentine, being an alteration product of olivine-rich rocks.

Sterrett (1907) describes the New Mexico meerschaum as occurring in veins, lenses, seams, and balls in limestone, in both nodular and massive form. Although he states that the deposits were not sufficiently studied to permit the expression of an opinion as to their origin, his description indicates hydrothermal action.

Sampson (1966) provides an excellent description of the meerschaum deposits in Tanzania and contributes much toward an understanding of their origin. The meerschaum occurs as irregular masses and as veins and films along bedding planes, faults, and fracture zones in dolomitic limestone, and to a lesser extent in an overlying unit of sepiolitic mudstone. The author believes that the sepiolitic mudstones and clays constitute the local source of the meerschaum. He states that the mode of emplacement of much of the meerschaum appears to have been simple deposition under wet conditions, by water percolating downward from the surface, and by pressure forcing the plastic meerschaum into cavities of the limestone. The lacustrine beds of limestone and mudstone have been subjected to folding, and the distribution of the meerschaum is partly attributed to structural control.

Concerning the origin of the highly magnesian sepiolitic mudstones and clays, Sampson notes that their color and fine grain, the absence of bedding, and their geographical position in relation to Kilimanjaro would allow for them to be volcanic ash. Lacking direct petrologic evidence, however, he declines to express an opinion as to the method of deposition.

Distribution and Description of Deposits

Meerschaum has been reported from many countries, including Turkey, Iran, India, Somali Republic, Kenya, Tanzania, France, Greece, Spain, Austria, Morocco, Republic of South Africa, and the United States. Only a few of these have been consistent producers.

Turkey: Turkey has been the major producing and exporting country for many years. Most production has been from the province of Eskishehir, in the vicinity of the town having the same name located about 200 km west of Ankara. All of the workings appear to be small; 31 separate localities are named in Eskishehir province. One deposit is in the province of Bilecik, near the town Inonu about 30 km northwest of Eskishehir. One is at the town of Kutahya in the province of the same name, adjacent to and west of Eskishehir province. Another deposit is reported near the town of Sharki-Karaagac, about 200 km south-southeast of Eskishehir. The deposit is in the Sultan Mountains, in the eastern part of Isparta province. Report No. 113 (Anon., 1963) of the Institute of Mineral Research and Exploration includes detailed information, in Turkish, on the location of each deposit. A less detailed summary in English is given by Ryan (1957).

New high-grade deposits of meerschaum were discovered 100 m below ground in Eskishehir in 1973.

Somali Republic: The deposits of meerschaum in Somali Republic first became recognized in international trade in 1965 (Woodmansee, 1967). The deposits are in the Mudugh region near El Bur, which is about 350 km north-northeast of Mogadiscio, and about 130 km from the coast. The Somali meerschaum is reported to be of good quality, with low density and high porosity, but little specific information is available. An early statement indicated that reserves were believed to total several million tons. It seems likely, however, that this included the earthy variety of sepiolite as well as meerschaum.

Tanzania: Little information has been found on the Tanzanian deposits other than the article by Sampson (1966). The deposits are located at Sinya, northwest of Kilimanjaro, near the northeastern boundary of Tanzania; they extend northward into Kenya. The Sinya meerschaum is light-colored, either white, light gray, light brown, or light green. It darkens quickly when wetted. Quality is variable. Sampson's data show that the bulk density ranges from 0.38 to 0.92 g per cc. Material below the water table is generally better quality than that above. Downward mining and exploration have been hindered by large amounts of ground water in the limestone, and reserves are not well-known. The average content of meerschaum in the rock mined is about ½ to ¾ lb per ton of rock.

United States: In the literature meerschaum is mentioned as occurring in several states, including Arizona, California, Pennsylvania, New Mexico, and Utah, but some of the reported occurrences are the earthy variety of sepiolite rather than meerschaum. Deposits in Arizona and New Mexico are the only ones which appear to be worthy of consideration for possible commercial development.

Arizona—Fibrous sepiolite, which from its description appears to be meerschaum, is the subject of a study by Kaufmann (1943). The material occurs in a contact zone in southwestern Yavapai County. According to Kaufmann the deposit was prospected thoroughly and found to be too limited in extent for commercial development.

New Mexico—The two known deposits in New Mexico are located in Grant County. In addition to the discussion by Sterrett (1907), the following is a summary given by Talmage and Wootten (1937):

> Two deposits of meerschaum from the same general locality in New Mexico have been described [by Sterrett], one about 12 miles northwest and the other about 24 miles north of Silver City. The meerschaum is described as occurring in veins in Tertiary igneous rocks. The veins carried the meerschaum in nodules or in blocks, some of them several feet across, but few if any of these nodules or blocks were clear and free from impurity. Nearly everywhere the meerschaum was shot through with crystals of quartz or calcite, so that grinding and washing were necessary to eliminate these crystals. Several shipments were reported to have been made some years ago, part of the material being used for pressing into tobacco pipes and part used as an absorbent for nitroglycerine.
>
> No detailed information is available as to the extent or success of the use of New Mexico ground meerschaum for these purposes, nor as to costs and prices at the time the deposits were worked. Either the quality of the material was too low, or more probably the cost of extraction and shipment

TABLE 1—Turkish Production of Meerschaum, 1969-1979

	1969	1970	1971	1972	1973	1974	1975	1976	1977	1978	1979
Kg	41,250	20,250	18,300	19,350	22,150	20,850	30,900	12,600	4,150	3,050	3,000
Lb	90,750	44,550	40,260	42,570	48,730	45,970	60,130	27,800	9,150	6,730	6,610

Source: *Minerals Yearbooks,* US Bureau of Mines.

was too high, to justify the continued production of this material, as the properties have been idle for many years.

Ford (1932), mentions meerschaum or sepiolite as occurring near Sapella, San Miguel County, New Mexico, but no other reference to this occurrence has been seen.

Production and Trade

Combined world production figures for meerschaum are not available, but some indication of their magnitude is given by the figures on production from Turkey, long the major producer. Data for production in Turkey for the years 1969 through 1979 are given in Table 1. No consistent trend is evident and production has dropped off considerably. For approximately the past decade, the Turkish government has discouraged, or prohibited, the export of crude meerschaum. The objective is to stimulate the manufacture of the finished products in Turkey.

Mining of meerschaum in what is now Tanzania began in 1954, with four tons reported for that year. Production increased to 110 tons in 1965, but subsequently declined. The indicated total through 1971 is over 320 tons; all apparently had gone to the pipe factory of the operating company, Tanganyika Meerschaum Corp., originally located in Nairobi but moved to Arusha in 1960. Pieces of meerschaum too small for whole pipes are used for making bowl inserts, or are ground and reconstituted into blocks. The reconstituted material is known as *Arcon.*

Other producing countries include Somali Republic, Kenya, France, India, and Iran. In 1965 about 2000 lb of meerschaum, 5% of the total imports, came into the United States from Somali Republic. This percentage increased, somewhat erratically, in subsequent years; in both 1970 and 1971 over 90% of US imports were from Somali Republic.

Total imports of meerschaum into the United States in 1971 amounted to 17,482 lb valued at $25,825; the quantity was 28% less (and the value 13% less) than in 1970. In 1972 US imports declined sharply for the second straight year, to a total of 11,137 lb, valued at $22,791. Sources of imports in 1972 were Belgium, Luxembourg, Japan, and Turkey.

No crude meerschaum was imported into the US during 1973 or 1974. In 1975 meerschaum imported for domestic consumption totaled 11,263 lb, valued at $20,337. The principal source of imports in 1975 was the Somali Republic (99.8%). In 1976 imports of crude meerschaum totaled 1200 lb, valued at $310 or $0.26 per lb; all from the Federal Republic of Germany (West Germany). In 1977 imports decreased to 485 lb, 68% from Spain and 32% from the Federal Republic of Germany. Customs value of the imported meerschaum was declared at $2.62 per lb. US imports of crude meerschaum in 1978 amounted to 14,055 lb, 78% from Somalia and 22% from the Federal Republic of Germany. Value of all imported meerschaum in that year was $35,405, or $2.52 per lb. The material was used by companies in New York and Ohio for the manufacture of smokers' pipes. No imports of meerschaum into the US were recorded for 1979. In 1980, imports of crude meerschaum totaled 3,793 lb, all from the United Kingdom. Customs declared value was $17,720, or $4.67 per lb (*Minerals Yearbook, 1971* through *1980*, US Bureau of Mines).

No domestic production has been reported by the Bureau of Mines since 1914. Bateman (1950) states that about 1000 tons has come from New Mexico.

Acknowledgments

The kind assistance of Yoldoray and Carolyn Buie Erdener, in supplying the specimens of meerschaum shown in Fig. 1 and in translating Turkish articles, is greatly appreciated. Credit and appreciation are due William Miller and Thomas Fellers for taking the SEM photographs, and to Dennis Cassidy for preparing Fig. 1.

Bibliography and References

Anon., 1963, Bilinen Maden Zuhurlari (No. 113), Maden Tetkik ve Arama Enstitüsünce, [Mineral Research and Exploration Institute of Turkey, Publication No. 113; in Turkish], Ankara.

Anon., 1973, "A New Meerschaum Deposit," *Industrial Minerals,* No. 66, Mar., p. 32.

Anon., 1981, "Meerschaum," *Minerals Yearbook, 1980,* US Bureau of Mines, p. 941.

Bateman, A.M., 1950, *Economic Mineral Deposits,* 2nd ed., John Wiley & Sons, New York, pp. 774–775.

Caillère, S., and Hénin, S., 1961, "Sepiolite," *The X-ray Identification and Crystal Structures of Clay Minerals,* Mineralogical Society, Clay Minerals Group, London, pp. 325–342.

Ford, W.E., 1932, *Dana's Textbook of Mineralogy,* 4th ed., John Wiley & Sons, New York, p. 679.

Kaufmann, A.J., Jr., 1943, "Fibrous Sepiolite from Yavapai County, Arizona," *American Mineralogist,* Vol. 28, Nos. 9–10, Sep.-Oct., pp. 512–520.

Nahai, L., 1958, "The Mineral Industry of Turkey," Information Circular 7855, US Bureau of Mines, p. 124.

Oelsner, O., and Krüger, E., 1957, *Lagerstätten der Stein und Erden,* Bergakademie Freiberg, p. 100.

Ryan, C.W., 1957, reprint 1960, "A Guide to the Known Minerals of Turkey," Mineral Research and Exploration Institute of Turkey, Ankara, 196 pp.

Sampson, D.N., 1966, "Sinya Meerschaum Mine, Northern Tanzania," *Transactions,* Institution of Mining & Metallurgy, Sec. B, pp. B23–B34.

Schaller, W.T., 1936, "Chemical Composition of Sepiolite (Meerschaum)," *American Mineralogist,* Vol. 21, No. 3, p. 202.

Sterrett, D.B., 1907, "Meerschaum in New Mexico," Bulletin 340, US Geological Survey, pp. 466–473.

Talmage, S.B., and Wootten, T.P., 1937, "Nonmetallic Mineral Resources of New Mexico," Bulletin 12, New Mexico School of Mines, State Bureau of Mines and Mineral Resources, p. 8.

Woodmansee, W.C., 1967, "The Mineral Industry of Other Areas of Africa," *Minerals Yearbook, 1965,* Vol. 4, US Bureau of Mines, p. 973.

Mica

GEORGE P. CHAPMAN *

Mica is a platy mineral occurring in a variety of complex hydrous aluminosilicate forms with differing chemical composition and physical properties. Principal minerals in the mica group include:

- Muscovite—potassium mica (colorless or pale green/ruby)
- Phlogopite—magnesium mica (dark brown or amber)
- Biotite—magnesium-iron mica (black or dark green)
- Lepidolite—lithium mica (lilac)

Selected properties are given in Table 1. Muscovite is most important commercially with phlogopite of lesser use. The laminated structure of mica enables it to be split into very thin films which are transparent and tough as well as having outstanding dielectric and insulating properties.

Commercially, mica is used in a number of forms (Chowdbury, 1941; Rajgarhia, 1951; Skow, 1962). *Sheet mica* consists of flat sheets mined from naturally occurring books of mica, free from defects and capable of being punched or stamped into required shapes. Sheet mica is separately classified into *blocks, films,* or *splittings* according to thickness.

Built-up mica, or *micanite,* is made by arranging overlapping splittings in layers cemented with a binder and pressed together at high temperature.

Reconstituted mica, or *mica paper*, is made by depositing fine flakes of high-quality scrap mica in a continuous film impregnated with binder.

Scrap and *flake mica* are normally unsuitable in quality or size for making sheet mica. Originally the term *scrap mica* referred to the waste byproduct of mining and fabricating sheet mica, and distinct uses were developed for this material. Subsequently it became economically feasible to mine lower quality mica for these uses, and the term *flake mica* was introduced. Although the two terms are still used somewhat synonymously, it is considered better, at least in the United States, to use the term *flake mica* in referring to mica mined for uses other than as sheet mica; and to restrict the term *scrap mica* to mica mined for or with sheet mica, but which will not meet the specifications.

Synthetic mica, such as fluorophlogopite, is produced by slow crystallization of a melted blend of pure raw materials.

Mineralogy

In the true sense the term *mica* does not relate to a particular mineral, but to a group or family of minerals of similar chemical composition and to some extent similar physical properties. These minerals are predominantly potassium aluminum silicates with varying amounts of magnesium, iron, and lithium. The precise formulas and isomorphic relationships of the various group members have been studied, but general agreement has not been reached (Grimshaw, 1971; Hurlbut, 1952; Skow, 1962). The general formula describing the chemical composition of micas is

$$X_2Y_{4-6}Z_8O_{20}(OH, F)_4,$$

where

X is mainly K, Na, or Ca;
Y is mainly Al, Mg, or Fe;
Z is mainly Si or Al

(Deer, et al., 1962).

These minerals have an internal structure of the layered lattice type where the silicon atoms are in the center of a tetrahedral grouping of oxygen atoms. The groups are linked together in a single plane by three oxygen atoms that lie within the common plane. Each of these oxygen atoms is shared by two tetrahedra. These linked tetrahedral groups, when continually extended, produce a hexagonal network within the plane. This internal structure of the mica has been used to explain the ex-

* Managing Director, Microfine Minerals and Chemicals Ltd., Derby, England.

TABLE 1—Selected Properties of Various Micas

Chemical Constituent	Muscovite	Phlogopite	Biotite
SiO_2	46	40	37
Al_2O_3	35	17	18
K_2O	10.5	10	9
MgO	0.5	26	8
Fe_2O_3	1	0.2	2
FeO	1	2.8	21
Na_2O	1	0.5	1
Minor	0.5	0.5	1
H_2O	4.5	3	3
Total	100	100	100
Specific gravity	2.77-2.88	2.76-2.90	2.70-3.30
Mohs' hardness	2½-3	2-2½	2½-4
Optic axial angle	30°-47°	0-15°	0-25°
Temp. of Decomposition	400-500°C	850-1000°C	
Dielectric constant	6.5-9.0	5.0-6.0	
Specific heat (25°C)	0.206-0.209	0.206-0.209	

ternal pseudohexagonal structure of a mica crystal. When a pair of silicon-oxygen sheets that have had about one-fourth of the silicon atoms replaced by aluminum to form a mica are oriented with the tetrahedra vertices pointing at each other, a firm double-layered structure is formed with hydroxyl groups and metallic atoms such as aluminum, magnesium, and lithium between them. These double-layered structures are joined together by potassium atoms. The cleavage plane of the mica is found between these double-layered structures.

In addition to the better known members of the mica group—muscovite, phlogopite, biotite, lepidolite—there are zinnwaldite, a lithium-iron mica, fuchsite, a chromium mica, and roscoelite, a vanadium mica. The mineral vermiculite, related to the mica group, is treated elsewhere in this volume. Of the known micas, only muscovite and phlogopite exhibit any large commercial demand.

These micas crystallize in the monoclinic system with crystals that usually form in hexagonal or rhomb-shaped scales, prisms, or plates, with plane angles on the base of about 60° or 120°. The crystal faces are rarely smooth or well defined except for the basal plane. When a mica cleavage plate is struck sharply by a blunt needle, a six-rayed percussion figure is formed. The figure's most prominent line is parallel to the crystal's plane of symmetry; the remaining two lines are almost parallel to the crystal's prismatic edges. The percussion figure, in conjunction with the position of the optic axes as seen in an interference figure under the polarizing microscope, is useful in making a quick distinction between muscovite and phlogopite, which may closely resemble each other visually. Phlogopite, like biotite, has the plane of the optic axes parallel with the plane of symmetry of the crystal; muscovite has the plane of the optic axes perpendicular to the plane of symmetry. (When a blunt punch is pressed on a mica cleavage plate, a pressure figure is produced. The lines of the pressure figure are perpendicular to those of the percussion figure, but the pressure figure generally develops only partially.)

Minerals of the mica group have the distinctive physical property of being both flexible and elastic. On this basis they are readily distinguished from chlorite, talc, and vermiculite, flakes of which will readily bend but will not snap back to their original position when released.

Muscovite has excellent basal cleavage that allows it to be split into very thin sheets exhibiting a high degree of flexibility, elasticity, and toughness. Very thin sheets of clear muscovite are transparent and colorless or almost colorless. In thicker sheets, it can be translucent and exhibit light shades of yellow, brown, green, or red. The mineral is strongly birefringent and exhibits a feeble pleochroism unless it is dark colored. The material is optically negative. The axial angle varies from 30° to 47° and is dependent on variation in the material's composition. The optic axes lie in a plane that is normal to the crystal's plane of symmetry. Upon heating, muscovite commences to lose water at about 500°C. Muscovite is decomposed only by hydrofluoric acid.

The utility and value of sheet mica depends on the freedom from certain physical defects that occur as a result of the environment and events that occurred during and after crystallization of the mica. These defects are structural imperfections of the cleavage surfaces due to mineral inclusions.

Some of the structural imperfections associated with mica are reeves or cross grains which are lines, striations, or sharp folds in the plane of cleavage; "A" structure, which usually consists of two sets of reeves intersecting on a cleavage surface at about 60°; herringbone structure formed by two sets of reeves intersecting at about 120° along a central line of reeves; wedge mica which consists of mica runs that are thicker at one end than the other; warping which causes a mica characterized by shallow waves or ridges; ruled mica which has regular, sharply defined parting planes that intersect the basal cleavage plane at about a 67° angle; and tangle-sheet mica which is formed by an intergrowth of parts of one mica crystal with another. Most of these defects preclude use of the mica as sheet mica but do not interfere with its use as flake mica.

Mineral inclusions in mica can also limit the utility of sheet mica. These inclusions and intergrowths of other minerals are broadly classed as stains. Some mineral stains appear parallel with the cleavage, but others can penetrate the cleavage plane and cause pinholes. Examples of staining are those formed during mica crystallization and include mottling, inorganic stains, mineral inclusions, and intergrowths. Additional staining can be caused by the circulation of ground water. Mica deposits that have been stained will yield little usable sheet mica.

Phlogopite has a pearly to submetallic luster and varies from translucent to transparent in thin sheets. Phlogopite has been identified in colors ranging from brownish red through yellowish brown, greenish brown, and dark pearl gray to an almost colorless pale green. The color intensifies as the iron content of the mineral increases. Darker colored samples of this mineral exhibit greater hardness than those of lighter color. Phlogopite can contain fluorine which is usually most prominent in reddish brown samples and least in greenish specimens. Phlogopite generally exhibits a distinct pleochroism.

The mineral is optically negative and its optic axial angle varies from 0° to 15°, apparently increasing directly as a function of the iron content. The plane of the optic axes is parallel to the crystal's plane of symmetry. When heated, phlogopite will not dehydrate greatly until about 1000°C. The reaction to acids is variable. Hydrochloric acid will attack the mineral mildly. Hot concentrated sulfuric acid decomposes phlogopite completely.

Historical Background

Sheet Mica

Evidence exists of the use of mica as far back as the earliest civilizations, especially in the Nile Valley and India prior to 2000 B.C.

One of the earliest uses of mica was as a medicine, and even today Hindu physicians use fused biotite. The lustre, iridescence and transparency of the larger sheets led to applications as windows, mirrors, adornments, and plates for painting mythological scenes and the like. The American Indians of the southern Appalachians are known to have used mica in ornamentation at grave sites in the 14th century.

The Romans were probably responsible for the name, from the Latin word *micare* meaning to shine or glitter. Pliny mentioned material scattered over the Circus Maximus to impart a shining whiteness, and window coverings of transparent stones; these may have been in reference to mica.

Muscovite probably derives from the Russian district of Moscovia where de Boodt found and identified potash mica in 1609. The Russians also used mica for medicine. Their men-of-war had mica portholes. Thin sheets of mica were used to protect the surface of icons from being kissed by worshippers.

Commercial mica mining was started in the United States in 1803 at the Ruggles Mines in Grafton County, NH. Primary uses of mica in the 1800s were for stove windows, shades for open-flame lights, and for furnace viewing glass. Although the United States originally was self-sufficient in its mica production, by 1885 India had become a major supplier of muscovite sheet.

The appearance of Thomas Edison's bipolar generator in 1878, his incandescent lamp in 1879, and his system of central-station power production in 1882 stimulated commercial development of electric power-generation.

Growth of the US electrical industry opened up many uses for sheet mica. Mica has proved to be the best insulation for use in commutator segments of large electric generators and motors. The US electrical industry grew so fast that the domestic reserves of large sheets were rapidly depleted.

A patent was issued in 1892 for the first built-up mica that maintained the dielectric and mechanical properties of natural mica. The use of splittings for built-up mica further depleted US reserves and led to increased imports of Indian mica.

Mica was used as spacers and insulators in the diode vacuum tube developed in 1904, and in the triode which followed in 1906. The vacuum tube industry grew in importance with the outbreak of World War I, and large quantities of mica were required to support the war effort. Mica retained its importance in the interwar period through use in the radio industry as well as in the growing electrical industry.

During World War II mica demand was heightened by the great increase in sophisticated electronic gear. The United States worked with the United Kingdom to stimulate production in India and other mica-producing countries. After the war, sheet mica production dropped off dramatically and the decline continues in the 1980s, caused by the rapid advancement of solid-state electronics.

Ground Mica

In 1890 J.S. Williams of the Richmond Mica Co. in Virginia produced ground mica as an ingredient for axle greases. Soon afterward the Vance brothers installed the first chaser mill in North Carolina producing wet ground mica for pigments, to be followed by wet-grinding by Williams.

From there the American industry grew. In 1908 Thomas English formed the English Mica Company in Spruce Pine, North Carolina. In the 1920s J.B. Preston Co. bought English Mica and merged it with the Biotite Mica Co. and Roofing Mica Co. Franklin Mica Products was started by John Davenport in Franklin, NC. In New Hampshire, what is now Concord Mica Co. was formed by Earl Moore. Diamond Mica Co. and Hayden Mica Co. were formed after World War II.

In the 1930s wet ground mica production started in Derby, England, and some dry grinding production began in Europe and India. Micronized mica emerged in Norway and England in the 1960s with further wet grinding plants in Norway, France, and South Africa.

The large growth in mica grinding came in dry ground mica from the 1940s onward, first with the uses in roofing felts and shingles in the US and, as that declined, with the emergence of drywall joint cements and oil-well lost circulation additives.

The 1970s saw the decline of water-grinding as rising energy costs made the products uneconomic and several plants in the US and Europe closed.

Countries of Origin

The most important countries producing mica are listed in Table 2, together with an estimate of their mica-scrap output. An attempt has been made to show how the total supply is distributed around the world for use in the manufacture of ground mica.

It will be seen that approximately 50% of the mica arises in the United States with four other countries—USSR, India, China, and Canada—making up the next 35%.

United States

Approximately 60% of US scrap is from North Carolina and consists of byproduct mica from the feldspar, kaolin, and lithium producers. Virtually all is ground in mills located near the mines. The remaining mica comes mainly from Alabama, Connecticut, Georgia, New Mexico, and South Carolina.

USSR

The pegmatite deposits of Karelia and Maura River regions appear to dominate production, but little information is published regarding mica-grinding. Apparently no exports are made outside Eastern Bloc countries.

India

High quality mica is mined in the States of Bihar, Rajasthan, and Andhra Pradesh for sheet mica production, and the consequent scrap is exported for mica-paper manufacture or ground in India for home and overseas consumption.

China

Mica is produced in a number of regions from pegmatite or as a feldspar byproduct. The most important area is Tientsin province.

Canada

Output consists of flake phlogopite mined in Quebec province and ground at a plant near Montreal. This is the only significant production of ground phlogopite on a world basis.

TABLE 2—Countries of Origin and Consumption of Mica, t/a

Country of Origin as Raw Material	Eventual Consumption as Mica Powder or Flakes: US and Canada	USSR and Eastern Bloc	Europe	Far East and Australia	Middle East	Africa	South America and West Indies	Total
USA	105,000	500	2,500	1,000	200	300	500	110,000
USSR		35,000						35,000
India	1,000	7,000	3,000	5,000	4,000	1,000		21,000
China			8,000	4,000				12,000
Canada	12,000		1,500	1,500				15,000
Brazil	1,000		5,000				1,000	7,000
France			6,000					6,000
Spain			4,000					4,000
South Africa			1,500	500		1,000		3,000
Korea				3,000				3,000
Norway			2,000					2,000
Argentina			1,000				1,000	2,000
Italy			1,000					1,000
Others		1,000	1,000	1,000				3,000
European Re-exports	500	1,000	(4,000)	500	800	700	500	
Total	119,500	44,500	32,500	16,500	5,000	3,000	3,000	224,000

Madagascar

High quality phlogopite is mined for use as sheet mica to satisfy the electrical and mica paper markets of Europe and America.

Geologic Features of Mica Deposits

Sheet Mica

Large crystals or books of muscovite and phlogopite mica are generally found in regionally metamorphosed rocks. Granitic pegmatites are the source of muscovite sheet. Phlogopite sheet is found in areas of metamorphosed sedimentary rocks into which pegmatite-rich granite rocks have intruded.

Pegmatites are generally the only source of usable quality sheet mica. Although there are many mica-bearing deposits throughout the world, only a few localities have deposits that are of economic importance. These pegmatites are light-colored, coarsely crystalline igneous rocks. They can be found as dikes or sills in metamorphic rocks and large granite intrusions. The mica crystals found in these deposits range from less than one inch to many feet in length. Variation of size within an individual deposit is not uncommon. Mica-bearing pegmatites have been known to exceed 200 ft in thickness and 1000 ft in length and have been worked to depths ranging from 200 to 500 ft. However, mica also has been produced from shallow surface deposits.

Pegmatites are composed primarily of feldspar, quartz, and mica. In many geologic situations accessory minerals such as garnet, tourmaline, and beryl occur with the primary pegmatite constituents. The distribution of minerals in pegmatites may be even, zoned, or segregated into layers.

Deposits of large books of phlogopite mica that can economically provide large quantities of sheet mica are available in only a few places. Skow (1962) has classified phlogopite deposits into vein, pocket, and contact deposits. Vein deposits, which are generally narrow and are enclosed in fine to medium grained pyroxenite, are the major source of phlogopite sheet mica. Pocket deposits have been found to be irregular in shape, size, course, and persistence. The surrounding pyroxenite of this type of deposit is usually more coarsely crystalline and open textured than that found with vein deposits. The phlogopite crystals may be irregularly distributed throughout the pocket along with crystals of other minerals such as pyroxene and calcite or may occur as very large solid masses of phlogopite.

Scrap and Flake Mica

Originally scrap mica was derived from the mining and processing of sheet mica and included poor quality sheet mica that did not meet the specifications for size, color, and quality. However, the industrial demand for scrap mica began to increase greatly in the 20th century, and mica deposits were mined specifically for the smaller size mica crystals or flake mica found in coarse-grained, weathered, granitic rocks such as alaskite and pegmatite, and in some schists. Flake mica is also recovered as a coproduct from the production of clay, feldspar, and spodumene.

Technology has been developed that permits the recovery of small particle size flake mica from various mica-bearing sources such as mica schist, graphitic mica schist, granite, silt, and recovery plant tailings. Good recovery rates for mica have been reported by the use of grinding and flotation techniques.

Exploration

Sheet Mica

Exploration for sheet mica is not amenable to most of the methodology used to search for other minerals. The sporadic occurrence of large runs of books of mica crystals in the host rock makes it uneconomic to drill a deposit to any great extent to determine the mica content. Any drilling that is done can only delineate the existence of pegmatite and may reveal very little information about the mica content.

Prospecting for sheet mica has failed to advance beyond the trial and error method. Geological evidence is generally insufficient to supply enough information for the evaluation of a deposit. The most practical prospecting method for sheet mica is to sink a test pit in an effort to determine the percentage, size, and quality of trimmed sheet that might be obtained from a deposit. A potential deposit may be stripped or trenched by mechanical or hand methods to outline the size, shape, and attitude of the pegmatite vein for the purpose of selecting the sites for test pits or to supplement the information obtained from test pits. A common method of further exploration for sheet mica consists of sinking a shaft either downdip in the pegmatite body or vertically to cut the pegmatite at the desired depth. Drifts can then be driven along the strike of the deposit in both directions. Any large mica crystals that are found are visually examined by experienced mica workers who determine their freedom from structural imperfection and their suitability for sale. Steeply dipping pegmatites can be explored by the construction of adits. As yet, geophysical and geochemical techniques have not been applied to prospect and evaluate deposits containing sheet mica. The search for mica sheet remains basically a pick and shovel operation with some limited usage of mechanized equipment.

Flake Mica

Small particle size flake mica that is distributed in deposits of pegmatite, schist, and clay can be evaluated by conventional methods. The location of a deposit can be determined from surface geology. If the surface outcrops are of sufficient size, surface sampling may be justified, but it is more likely that the deposit would be augered or drilled to a shallow depth to furnish samples for testing and delineation of the deposit.

Ore samples obtained in this manner are crushed to a range of sizes to determine the optimum crushing size at which the ore will release the maximum quantity of mica. Chemical analysis is not effective for mica determination because of the wide variation of the chemical composition of the various micas. Petrographic analysis is not completely satisfactory to evaluate the mica content of a sample. The mica industry has not established a standard method of analysis to determine the mica content of an ore sample.

Sink-float techniques can be applied to the crushed samples of pegmatites and schists to assess the capability of separating mica from its accessory minerals by the use of a heavy liquid such as tetrabromoethane or mixtures of tetrabromoethane and kerosene or trichloroethylene. Experiments can be carried out to determine if mica can be extracted from the ore sample by some method of flotation.

As a result of preliminary studies a flowsheet for the recovery of mica can be developed and a small pilot plant operation erected to determine the feasibility of the method of recovery.

The final test of any flake mica is reached when it is ground and/or classified to a particular particle size and accepted for an industrial end use.

Mining and Milling

Mining

Hydraulic mining methods are used where practical because screening and washing operations for the recovery of the mica require large quantities of water. One or more high-pressure streams of water are directed against the quarry face. The ore is broken up by the force of the water, washed from the working face into a sump, and then flumed or pumped to a processing plant. If necessary in partially weathered strata the quarry face is fractured by blasting prior to hydraulic mining.

Most flake mica ore is mined by the use of power-driven mechanical equipment. The exposed mica-bearing ore can be mined by power shovels, bulldozers, or tractor-drawn drag pans and moved to the processing plant by suitable means. Mining with power-driven equipment requires more equipment and labor than hydraulic methods but allows greater flexibility for mining deposits that are not suited for hydraulic methods. Power equipment allows the concentrating plant to be located favorably to a water supply, ore deposit, and transportation facilities. In addition, more small deposits can be worked economically.

Sheet Mica Mining: The recovery of sheet mica in all countries is dependent on the operation of a relatively large number of small mica mining operations that generally use little or no basic mechanical equipment such as compressors, air drills, and hoists. Mine workings are irregular in an effort to follow mica runs or books along rich shoots or from pocket to pocket in a pegmatite deposit. The methodology utilized must allow for the sporadic distribution of the mineral, recovery of undamaged crystals, and any other special condition of the deposit.

Where feasible, open pit, open cut, and stripping are the least expensive methods for the recovery of mica. These methods are used where the mica-bearing pegmatite lies in relatively flat or gently rolling country and appears as a surface outcrop or is covered by easily removed overburden.

In underground mining a main shaft is driven through the pegmatite deposit at suitable angles to the dip and strike of the deposit. Crosscuts and raises are developed to follow promising exposures of mica. Underground mining of pegmatites requires some use of drilling and explosives because the deeper pegmatites are harder than weathered surface pegmatites. Great care is taken to avoid drilling through mica crystals. Explosives are used only in small charges, consisting of low-velocity 40 to 60% strength dynamite, around pockets of mica. After mining, mica crystals are handpicked from the broken waste rock, boxed or bagged, and raised to the surface. In some operations where difficulties are encountered in hoisting, waste material is left in the mine for fill.

No current data relating to the cost of mining sheet mica are available because only small quantities of low-grade punch and circle mica have been domestically produced in recent years. However, if any significant quantity of good quality sheet mica were produced in the United States, its production cost would be much greater than that of imported foreign mica because of the higher cost of domestic labor.

Flake Mica Mining: Small particle size mica suitable for industrial use is recovered in the United States primarily from pegmatite and residual deposits formed by the weathering of mica-bearing rocks. Because of the regular distribution of the mica particles in the host rock, the mining of flake mica is amenable to conventional mining and beneficiation techniques. Most operations are open pit using conventional equipment to remove any overburden overlying a working face. The exposed mica ore is mined by hydraulic methods, power-driven equipment, or a combination of both methods.

Milling

Mica can be recovered from the mined ore by several general methods. The simplest method separates the mica from its host rock by differential crushing and screening in washer plants. Another method utilizes screens, classifiers, and Humphrey's spirals to concentrate differentially ground ore. The mica is also recovered by the use of screens, classifiers, and flotation techniques. Small particle size material such as sericite mica in clays can be concentrated by grinding, air classification, and drying.

Washer Plant Methods: Standard washer plants used to recover mica are low-cost units that are simple in design and can effectively recover +¼-in. or +3⁄16-in. mica. This type of plant cannot economically recover small-size mica, which is generally discarded. A washer plant generally has a series of roll crushers, trommel screens, and storage facilities through which material is moved by gravity flow or bucket elevators. The flake mica recovery process requires large quantities of water.

When entering the plant, mine-run ore is initially disintegrated by high-pressure water streams. The ore is then crushed in a jaw crusher and washed through a series of trommel screens and roll crushers. The process has little effect on the mica because of its platy and flexible characteristics. However, minerals such as the associated quartz and feldspar are reduced to fine grain sizes. The fine-sized gangue is removed by use of a trommel screen. The undersized gangue material consists of quartz, feldspar, clay, and fine mica. The coarse material is retained on the screen for the next operation. The crushing and screening operation is repeated as many times as necessary to remove the maximum quantity of gangue minerals. The wet, relatively clean coarse mica remaining from this process is then placed in a storage bin. Loss of mica from the original ore can be as high as 50% by this beneficiation process. The undersize discharge from this process can be stored for future recovery of the fine mica by other methods.

Humphrey's Spiral: The addition of Humphrey's spiral technique to a processing plant permits the recovery of a finer size mica than can be recovered from a roll and trommel washer plant. A spiral plant has a low initial and operating cost and minimal skilled labor requirements. Basic plant equipment consists of grinding, classification, and screening equipment, spirals, and storage facilities. In this type of plant, mine-run ore is washed by high-pressure hose into a bowl-rake classifier for initial desliming. Sand from the classifier is fed to a rod mill that discharges to a trommel screen. Any oversize material is returned to the rod mill for additional grinding. Undersize material from the trommel screen is directed to another bowl-rake classifier for additional desliming. The classifier sand is sent to a bank of Humphrey's spirals for initial concentration. The spiral rougher concentration is fed to the cleaner spirals, which produce a middling and a cleaner concentrate. The middling output is returned through the rougher spirals. The cleaner concentrate is passed over a series of screens to remove clay and fine-size gangue minerals. The screen oversize is fed to a hammer mill to delaminate the mica and remove fine-size quartz adhering to the mica. The concentrate is screened, centrifuged, and stored.

Flotation: Flotation methods for the recovery of mica from schists, pegmatites, and sediments have been developed by the Bureau of Mines during the past several years. This work led to the development of the acid cationic and alkaline anionic-cationic flotation methods for the recovery of mica from various types of ore.

The acid cationic flotation method allows the recovery of mica as coarse as 14 mesh. However, the ore must be completely deslimed at 150 to 200 mesh, allowing the loss of a significant quantity of fine mica. The ground ore pulp must be conditioned with sulfuric acid at 40 to 45% solids. Sulfuric acid is used for pH control and quartz depression. Optimum mica flotation was obtained in a pulp with a pH of 4.0. Cationic reagents, such as long-carbon-chain amine acetates, were the most effective collecting agents for floating mica.

The cationic method has been applied to the recovery of mica from pegmatite ores, and a flowsheet for the recovery of mica by this method is shown in Fig. 1.

The alkaline anionic-cationic method of mica flotation is effective for the recovery of mica when slimes are present. The ore is usually deslimed sufficiently to remove clay slimes, fine particle size mica, and any other granular material. This process allows the flotation of material as coarse as 20 mesh. The process requires conditioning the finely ground ore pulps at 40 to 45% solids with sodium carbonate and calcium lignin sulfonate and floating the mica with a combination of anionic and cationic collectors. Excellent recoveries were obtained in a pH range of 8.0 to 10.5. Separation was not greatly sensitive to pulp pH.

Continuous treatment of ore using the alkaline anionic-cationic method requires grinding, trommel screening, preconcentration, classification, conditioning, and flotation. A process flowsheet is shown in Fig. 1.

Recovery of mica is amenable to continuous flotation methods requiring relatively low capital investment in processing equipment and limited requirements for labor to operate the processing plant.

Pneumatic Processes: More recently, the US Bureau of Mines has concluded research on the recovery of flake mica using a dry pneumatic process. A system of crushers, screens, and zig-zag classifiers has been developed to concentrate mica particles. A generalized flowsheet is shown in Fig. 2. An alternative to the zig-zag classifier is the air-table or air jig. Such processes are in operation on pegmatite micas and mica schists in Europe and on phlogopite in Canada.

Preparation of Finished Products

Sheet Mica

The preparation of sheet mica for market relies entirely on hand labor that requires great expertise and experience to recover the maximum usable crystal area while developing minimal waste or scrap mica. Large quantities of scrap are produced during the preparation of

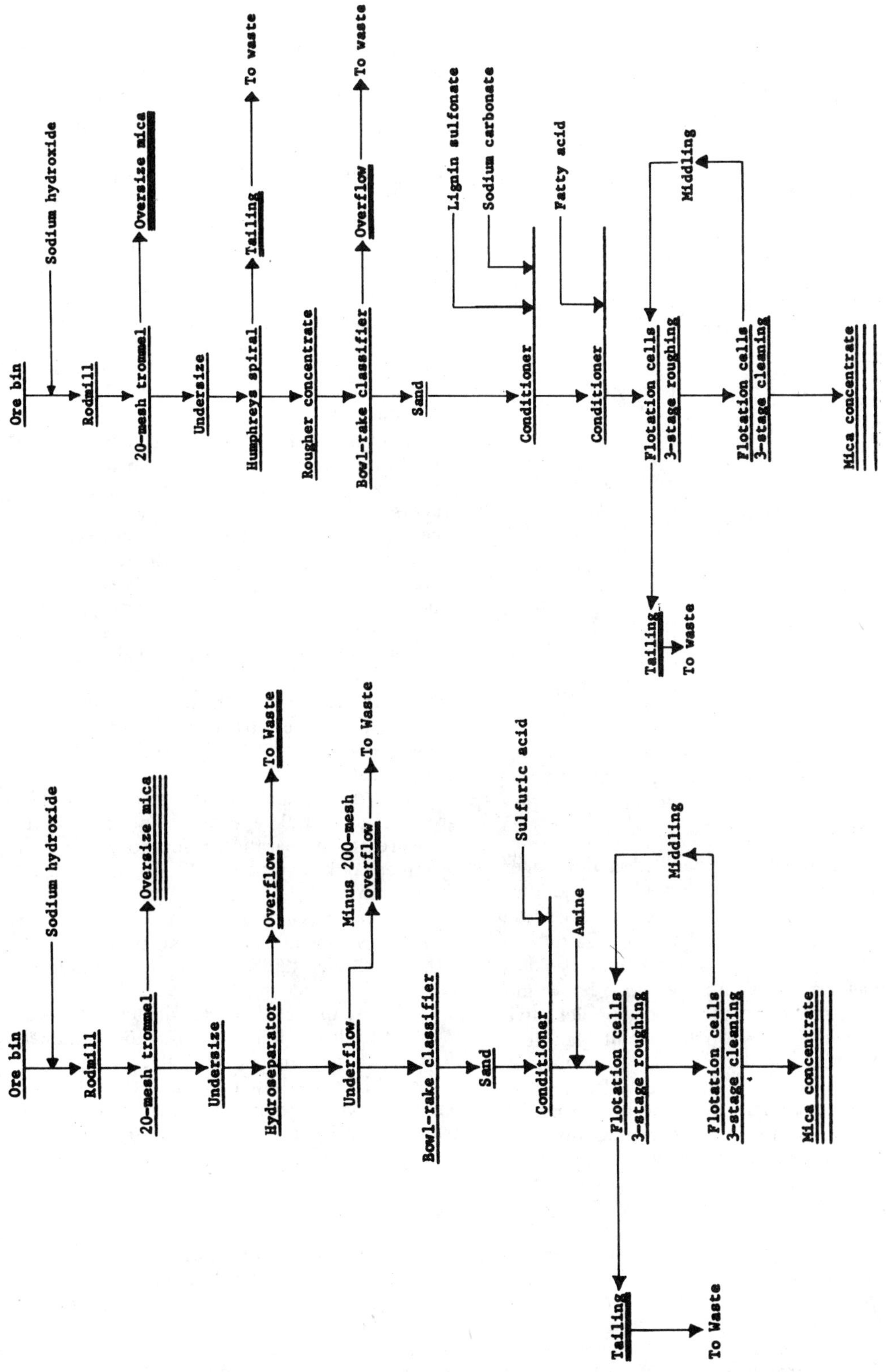

FIG. 1—*Flowsheet for recovery of mica; acid cationic method at left and alkaline anionic-cationic method at right.*

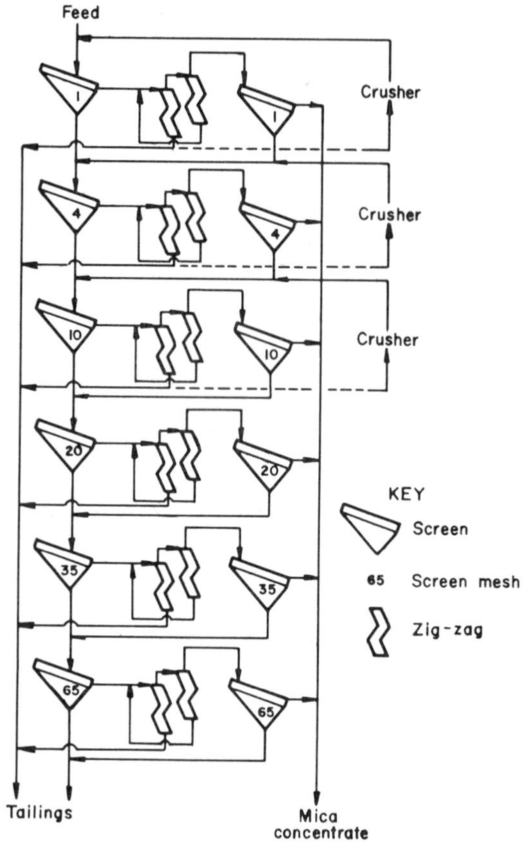

FIG. 2—*Generalized flow diagram of US Bureau of Mines pneumatic concentration method.*

the mica despite the use of skilled labor, proper preparation methods, and suitable precautions. The raw crystal is hand-cobbed and rifted using hammers, knives, and fingers. Trimming of the crystal is done with knives, sickles, shears, or fingers. Initially the rough books are cobbed to remove any dirt, rock, or defective mica and then rifted or split into suitable thicknesses. During these operations any obvious flaws in the crystals are removed. After rifting, the mica can be broadly classified as untrimmed sheet and scrap mica. Further processing removes broken and ragged edges, loose scales, and any other imperfections. When sheet mica was actively produced in the United States, the mica crystal was trimmed to a beveled-edge cut with a knife. The rifting and trimming operations required great care and judgment to determine what part of the mica should be removed to minimize waste and maximize the usable area of the crystal. The trimmed mica is classified according to grade (size), quality, and clarity. A visual determination, that can vary from one observer to another, is made of the quality of the natural sheet mica.

Additional splitting operations are required for the production of mica films and splittings. Films and splittings are not ordinarily produced in the United States. In other producing countries such as India and Brazil, splitting to the finer thickness is done by hand with a special, pointed double-edged knife with both edges sharpened near the point. A high degree of manual dexterity is necessary to split mica.

The fabrication of film and splittings demands highly skilled and experienced labor to split the mica to the necessary narrow thickness ranges. In India the preparation of splittings has remained a cottage industry.

All forms of sheet mica, especially in Brazil and India, are likely to move from the production area via a succession of mica buyers and traders until it eventually reaches a coastal city where sales are negotiated between exporters and importers from other countries. Most producing countries export their sheet mica to other consuming industrial countries. India does consume some of its own sheet production, but detailed data are not available.

Sheet mica is classified according to color, degree of preparation, thickness, visual examination, size and electrical quality according to a comprehensive and detailed series of descriptions and standards.

Color definitions include green, yellowish green, yellowish olive, brown, pinkish buff, etc.

Degree of preparation refers to crude, run-of-mine, book, hand-cobbed, untrimmed sheet, thumb-trimmed, half-trimmed, etc.

Block mica is not less than 0.007 in. thick; film mica in specified thickness from 0.0012 to 0.004 in.; "thick-thins" from 0.004 to 0.007 in., etc.

Visual quality is described in ten categories including clear, fair stained, good stained, densely stained, and black dotted.

Size is identified by the area of the minimum enclosed rectangle in square inches ranging from 8 x 12 in. down to 1½ x 2 in. plus circle and punch.

Ground Mica

Three different methods of grinding are employed each leading to products with distinct characteristics: water-ground mica, micronized mica, and dry-ground mica.

Water-Ground Mica: About half the total tonnage of mica is still made by traditional methods using equipment 20 to 30 years old. The process consists of grinding good quality scrap with up to 20% of moisture, in edge-runner mills followed by drying and classification to 100, 200, 325 mesh sizes. The resulting products have very thin and smooth particles providing sheen and good slip for use in wallpaper, tire manufacture, and paint.

Two manufacturers have introduced more up-to-date and economic proprietary processes. Their output, while not having the sheen required for the declining wallpaper industry, has the necessary properties for the bulk of the market, i.e., tires and paint.

Micronized Mica: First introduced in Europe to retain the paint markets previously held by water-ground mica, micronized mica involves grinding fine-grained byproduct scrap to sizes smaller than 40 microns. The method consists of accelerating the mica particles to very high speeds in a confined space so that they collide and disintegrate. A variety of such fluid energy micronizers exists, operating with superheated steam or compressed air. Particle size is controlled by adjusting the residence time in the mill and removing oversize by air classification.

Dry-Ground Mica: In general, US dry grinding practice differs from elsewhere as a consequence of the uniquely dominant requirement for mica in plasterboard joint-cements. Using minus 6-mesh scrap, the major production is by means of air micronizers; oversize (plus 60 or 100 mesh) is removed on dry screens.

Oil-well flakes, the second largest demand, are extracted from feldspar/kaolin byproducts by flotation or water classification. After drying, the various grades are sieved on multi-deck screens, any oversize (plus ¼ in.) being reduced by impact milling. Some oil-well flakes are extracted during the production of mica scrap of the mines by winnowing the output from the primary crusher through air boxes and cyclones. Other markets are satisfied by screening products from drilling flakes or joint-cement mica.

Dry-ground mica elsewhere in the world is made predominantly from prime scrap (plus 1 in.) fed through high-speed hammer-type mills, then topped and tailed through vibrating screens. The products generally are more highly delaminated and are therefore referred to as *high-aspect* ratio mica flakes and powders.

A wide variety of product specifications is necessary to meet user requirements. Examples of some of these are given in Table 3.

Mica Markets and Production Data

In production and marketing, as well as in usage, sheet mica and scrap and flake mica constitute almost entirely separate commodities. An exception is that scrap mica, recovered from block and film sheets from which fabricated shapes have been punched or stamped, is sold to the manufacturers of mica paper (reconstituted mica).

Sheet Mica

World production of sheet mica is principally from India, with lesser amounts from Brazil and Madagascar. US annual production for the past several years has been estimated at not more than 1000 lb, hand-picked from blast

TABLE 3—Particle Size Distribution of Ground Mica Products

Sieve Analysis Mesh	% Finer than Micron	Water-Ground Mica Rubber Grade	Micronized Mica Paint Grades	Dry Ground Mica: Oil Wells		Dry Ground Mica: Joint-Cement	Dry Ground Mica: Welding Electrodes	Plastics
¼-inch	6400			100				
7	2800			80	100		100	
14	1200			30	98		98	
30	500			10	50		60	
44	355				10		40	100
60	250					100	10	99
100	150	100				99		70
150	105	99.9				95		50
200	75	85				85		35
300	53	80	100			75		20
325	44	75	99.9			70		5
Bulk Density (kg/L)		0.18	0.22	0.30		0.29	0.26	0.24

TABLE 4—US Sheet Mica Production and Consumption

	Sheet Mica Production		Block and Film Consumption		Splitting	
	1000 lb	Value, $1000	1000 lb	Value, $1000	1000 lb	Value, $1000
1975	5	3	623	1608	4746	2634
1976	5	3	534	1413	5025	3226
1977	1	—	448	990	4144	2718
1978	—	—	247	1362	5537	3031
1979	1	—	282	1866	4877	3248
1980	NA	NA	147	1829	4383	3101

NA = Not available.
Source: Zlobik, 1980; Tepordei, 1981.

rubble at feldspar and mica quarries. Although world production of sheet mica has shown a moderate increase in the last five years, production declined from over 25 million lb in 1969 to 17 million lb in 1979. US demand during this period has remained fairly stable at approximately 5 to 7 million lb annually (Zlobik, 1980). The contrast between production and consumption of sheet mica in the US for 1975–1980 is shown in Table 4.

It is expected that growth in the use of substitutes will further reduce demand for sheet mica in the US and possibly elsewhere.

Most block and film mica is fabricated into mica parts by highly specialized mica fabricators for electrical, electronic, and other industrial users. Mica splittings are processed into built-up mica products such as molding plates, segment plates, heater plates and tape through the use of binders, adhesives, and backing materials.

Consumption of sheet mica products is divided into four categories: electrical, vacuum tubes, capacitors, and other uses. Demand pattern for these products is shown in Table 5.

Flake and Ground Mica

In comparison with the rest of the world, the United States has a disproportionate demand for plasterboard joint-cements resulting from building methods not normally found in other countries. In this context, it is interesting to note that the largest mica producer is also the largest gypsum and plasterboard producer. Apart from that, a higher-than-average proportion of water-ground mica goes to the American paint industry; elsewhere more micronized mica is used. The geographic demand for oil-well drilling flakes reflects the activity of exploration and development of the world's oilfields.

Quality and value of scrap and flake mica produced in the United States are shown in Table 6. An estimate of the world supplies of each type of ground mica is given in Table 7. The consumption of various types of ground mica produced outside the USSR is shown in Table 8.

TABLE 5—Sheet Mica Supply-Demand Relationships, 1975-1979 (1000 lb)

	1975	1976	1977	1978	1979
World Production	13,400	14,300	15,200	17,600	17,000
US Demand					
Electrical	4746	5025	4144	5537	4877
Vacuum tubes	375	318	241	152	170
Capacitors	7	8	7	8	5
Others	241	207	200	86	107
Total	5369	5558	4592	5783	5159

Source: Zlobik, 1980.

TABLE 6—US Production and Value of Scrap and Flake Mica and Ground Mica, 1975-1980 *

Year	Scrap and Flake		Ground Mica	
	Quantity (1000 st)	Value ($1000)	Quantity (1000 st)	Value ($1000)
1975	132	5,205	112	9,366
1976	123	5,667	115	10,027
1977	129	7,039	122	11,906
1978	139	7,916	124	12,979
1979	134	7,708	122	14,555
1980	117	5,296	109	12,992

* Excluding low-quality sericite.
Source: Zlobik, 1980; Tepordei, 1981.

The uses of the various types of mica are summarized as follows:

1) Water-ground mica:
   - a) wallpaper — to provide sheen; significant but diminishing market;
   - b) rubber tires — for mould lubrication and dusting; heavily dependent on automotive markets;
   - c) paint — several uses including anti-corrosion and anti-settling; largely being replaced by micronized mica because of cost-saving.

2) Micronized mica:
   used almost entirely in paints with a small quantity in phenolic moulding powders and adhesives; prices are highly competitive since production is from cheap mica byproducts from kaolin and feldspar.

3) Dry-ground mica:
   - a) Oil-well drilling — flakes used to seal well walls when drilling mud seeps into fractured strata; market follows pattern of oil development; prices normally very cheap and qualities variable;
   - b) Joint-cements — imparts crack resistance and assists smooth finishing; very large US market but less significant elsewhere; normally above-average bulk density;
   - c) Surface-coatings — medium and fine flakes used in house facade and interior textured plasters, textured stone paints, and roof paints;

TABLE 7—Estimated World Production of Ground Mica, Excluding USSR t/a

Country	Water-Ground Mica	Micronized Mica	Dry-Ground Mica	Total
US	13,000	10,000	87,000	110,000
India	—	1,000	14,000	15,000
UK	1,000	3,000	11,000	15,000
Canada	—	—	15,000	15,000
France	1,500	3,000	3,000	7,500
W. Germany	—	—	5,000	5,000
Japan	500?	1,000?	4,000?	5,500
Norway	—	3,000	2,000	5,000
Italy	—	1,000?	2,000?	3,000
China	—	—	3,000	3,000
South Africa	1,500	—	—	1,500
Brazil	—	—	1,000	1,000
Spain	—	500	500	1,000
Zimbabwe	500	—	500	1,000
Australia	—	—	1,000	1,000
Total	18,000	22,500	149,000	189,500

TABLE 8—Analysis of World Ground Mica Markets, Excluding USSR Products 1000 t/a

Consuming Nations	Water-Ground Mica	Micronized Mica	Dry-Ground Mica: Oil Wells	Dry-Ground Mica: Joint Cements	Dry-Ground Mica: Other	Total
US/Canada	13.3	9.0	25.2	66.0	11.0	123.5
USSR/Eastern Bloc	0.5	2.0	1.0	1.0	5.0	9.5
Europe	3.5	8.0	3.0	4.0	15.0	33.5
Far East/Australia	3.0	2.0	3.0	1.5	7.0	16.5
Middle East	0.5	0.5	3.5	—	0.5	5.0
Africa	0.5	0.5	1.0	—	1.0	3.0
South America/West Indies	0.2	0.5	1.3	—	1.0	3.0
Total	21.5	22.5	38.0	66.5	40.5	189.0
				155.0		

d) Insulation boards — a number of isolated uses in asbestos-free applications in low-density fire-resistant boards;

e) Welding electrodes — an essential ingredient in certain electrodes for control of flux and slag characteristics;

f) Plastics — growing use as reinforcement in rigidized plastics for automotive, electrical, and construction industries;

g) Miscellaneous — including roofing, rubber, foundry coatings, fire extinguishers, acoustic products, adhesives and lubricants.

Satisfying the market requirements creates a complexity of problems for the producer, arising from conflicts of specification between industries and even within the same industry. The examples in Table 3 do not reveal that a standard product list might well comprise three water-ground grades, three micronized grades, and over twenty dry-ground grades. Even within the dry-ground range, three or four products would be needed for oil-wells and six or seven for welding electrodes. In the latter case two or three different raw materials could be needed to satisfy the specified chemical analyses.

Health and Safety Requirements

Insufficient knowledge exists to state categorically whether dust from mica production is hazardous. It is accepted that if any danger exists it is confined to the inhalation of respirable dust. Various authorities have therefore fixed limits on the levels of such dust permitted in the workplace.

The American Conference of Government Industrial Hygienists has adopted the following Threshold Limited Values (TLV) for mica dust:

Total dust	—	6 mg/m^3
Respirable dust	—	3 mg/m^3

(TLV refers to airborne concentration of dust)

The Health and Safety Executive of Great Britain applies a standard of 10 mg/m^3 of total airborne dust for all grades of mica with an additional limit of 1 mg/m^3 for the respirable portion. Measurements are based on sampling carried out using a C.F. Casella size-selective cone sampler, and the "respirable" portion is that passing the cyclone and collected on the filter membrane.

Where TLV's are exceeded, the workers should wear approved masks to cover the nose and mouth.

Acknowledgments

This chapter has relied heavily on extracts from available published information included in the bibliography and references listed below.

The author particularly wishes to thank Mr. Benjamin Petkof for permission to reproduce significant portions of his chapter on Mica in the 4th edition of *Industrial Minerals and Rocks*, and the US Bureau of Mines for assistance in making data and publications available.

Bibliography and References

Adair, R.B., and Browning, J.S., 1969, "Flotation of Mica from Pegmatites of Randolph County, Ala.," Report of Investigation 7159, US Bureau of Mines, 9 pp.

Adair, R.B., and Browning, J.S., 1969a, "Flotation of Muscovite from Alabama Graphitic-Mica Schist Ore," Report of Investigation 7263, US Bureau of Mines, 7 pp.

Adair, R.B., and Crabtree, J.O., 1971, "Recovery of Mica from Silt Deposits in the Nolichucky Reservoir, Tenn.," Report of Investigation 7488, US Bureau of Mines, 9 pp.

Browning, J.S., 1973, "Mica Beneficiation," Bulletin 662, US Bureau of Mines, 21 pp.

Browning, J.S., and Adair, R.B., 1966, "Selective Flotation of Mica from Georgia Pegmatites," Report of Investigation 6830, US Bureau of Mines, 9 pp.

Browning, J.S., and Bennett, P.E., 1965, "Flotation of California Mica Ore," Report of Investigation 6668, US Bureau of Mines, 7 pp.

Browning, J.S., and McVay, T.L., 1963, "Concentration of Fine Mica," Report of Investigation 6223, US Bureau of Mines, 7 pp.

Browning, J.S., et al., 1965, "Anionic-Cationic Flotation of Mica from Alabama and North Carolina," Report of Investigation 6589, US Bureau of Mines, 9 pp.

Chapman, G.P., 1980, "The World Mica Grinding Industry and Its Markets," *Proceedings*, 4th Industrial Minerals International Congress, Atlanta, 7 pp.

Chowdhury, R.R., 1941, *Handbook of Mica*, Chemical Publishing Co., Brooklyn, NY, 340 pp.

Deer, W.A., et al., 1962, "Sheet Silicates," *Rock-forming Minerals, Vol. 3*, Longmans, London, 270 pp.

Eddy, W.H., et al., 1969, "Selective Flotation of Minerals from North Carolina Mica Tailing," Report of Investigation 7319, US Bureau of Mines, 10 pp.

Grimshaw, R.W., 1971, *The Chemistry and Physics of Clays and Allied Related Ceramic Materials*, Wiley-Interscience, New York, 1024 pp.

Hawley, G.C., 1981, "Expanding Markets and Technologies for Mica from Suzor Township, Quebec," Preprint 81–121, SME-AIME Annual Meeting, Chicago, 10 pp.

Hill, T.E., Jr., et al., "Separation of Feldspar, Quartz and Mica from Granite," Report of Investigation 7245, US Bureau of Mines, 25 pp.

Hurlbut, C.S., Jr., 1952, *Dana's Manual of Mineralogy*, 16th ed., John Wiley, New York, 530 pp.

Ivey, K.N., and Haskiel, R.S., 1969, "Fluorine Micas," Bulletin 647, US Bureau of Mines, 291 pp.

Jordan, C.E., Sullivan, G.V., and Davis, B.E., 1980, "Pneumatic Concentration of Mica," Report of Investigation 8457, US Bureau of Mines, 23 pp.

Lesure, F.G., 1973, "Mica," "United States Mineral Resources," Professional Paper 820, US Geological Survey, pp. 415–425.

Lusis, I., 1980, "Review of the Health Effects of Micas," *Industrial Minerals*, London, Oct., pp. 45–55.

Petkof, B., 1975, "Mica," *Industrial Minerals and Rocks*, S.J. Lefond, ed., 4th ed., AIME, New York, pp. 837–850.

Preston, J.B., 1973, "Aluminum Potassium Silicate—Mica," *Pigment Handbook, Vol. 1, Properties and Economics*, John Wiley & Sons, pp. 249–263.

Rajgarhia, C.M., 1951, *Mining, Processing and Uses of Indian Mica*, McGraw-Hill, New York, 388 pp.

Skow, M.L., 1962, "Mica: a Material Survey," Information Circular 8125, US Bureau of Mines, 240 pp.

Stanczyk, M.H., and Feld, I.L., 1972, "Ultrafine Grinding of Several Industrial Minerals by Attrition Grinding Process," Report of Investigation 7641, US Bureau of Mines, pp. 6–10.

Tepordei, V.V., 1981, "Mica," *Minerals Yearbook*, US Bureau of Mines, pp. 563–572.

Zlobik, A.B., 1980, "Mica," *Mineral Facts and Problems*, Bulletin 671, US Bureau of Mines, 16 pp.

Nepheline Syenite

D. GEOFFRY MINNES * †
Revised by STANLEY J. LEFOND ‡
and RAY BLAIR §

Nepheline syenite is a silica deficient crystalline rock consisting of albite and microcline feldspars and nepheline, together with varying but small amounts of mafic silicates and other accessory minerals. Nepheline-bearing rocks are widely distributed around the world but only in Canada, Norway, the Union of Soviet Socialist Republics, and the United States are deposits worked commercially.

Because of its low melting point and fluxing ability, nepheline syenite was investigated in the early 1900s by many students of glass and ceramics. Although the nepheline syenite deposits in Methuen Township, Ont., Canada, were discovered as early as 1897 it wasn't until 1935 that the first mill was erected to produce nepheline syenite products.

In Canada, nepheline syenite deposits are worked by Indusmin Ltd. and International Minerals and Chemical Corp. (Canada) Ltd. who have combined production capacity in excess of 1800 stpd. Between 1935 and 1972 seven million tons valued at $78 million were produced from the Methuen Township deposits. In Norway, the Norsk Nefelin Div. of Elkem Spigerverket has produced nepheline syenite from a deposit on Stjernöy Island in western Finland since early 1960. Production has increased from 8 kt in 1961 to over 200 kt in 1980.

Nepheline syenite has been under investigation in the Union of Soviet Socialist Republics (USSR) since at least 1928. In 1951 the first commercial alumina works was brought on stream at Volkhov, near Leningrad, utilizing nepheline concentrates from the Kola peninsula as mill feed. Since then two others have been completed and more are planned. Large quantities of portland cement, sodium carbonate, and potassium carbonate are produced in conjunction with the alumina.

A material called nepheline syenite, but actually a pulaskite, is produced in Pulaski County, Arkansas, by the McGeorge Construction Co. for construction aggregates and by the 3M Co. for roofing granules. The Dome Fill Co. purchases the fines and uses them in the preparation of enclosed construction fill. Production is over 2,000,000 stpy.

Chap. 16 on "Feldspar, Nepheline Syenite, and Aplite" in the 3rd ed. of *Industrial Minerals and Rocks* (Castle and Gillson, 1960) and *Nepheline Syenite Deposits of Southern Ontario* (Hewitt, 1960) contain excellent accounts of earlier commercial development of nepheline syenite as a glass and ceramic material. As in the 4th ed., this chapter deals mostly with the period since 1960 and with matters not previously reported in detail. An English monograph on *Nepheline-Syenite and Phonolite* (Allen et al., 1968) is a complete survey, at the time, of nepheline syenite which, with its very complete list of references, is an excellent source book.

End Uses

Nepheline syenite from the Canadian and Norwegian producers finds use primarily in the manufacture of glass and ceramics. One producer also makes several grades of nepheline syenite that are useful as extender pigments and fillers in paint, plastics, and rubber. In the Soviet Union nepheline is used in increasing quantities in the manufacture of alumina, alkali carbonates, and portland cement and to a lesser degree it is used in the manufacture of colored container glass.

* Geologist, Ontario Ministry of Natural Resources, Industrial Minerals Sec., Toronto, Ont., Canada.

† Acknowledgments: The writer is indebted to Hans-Peter Geis, D. F. Hewitt, and K. H. Teague for invaluable advice and assistance in the preparation of this chapter.

‡ Consultant, Evergreen, CO.
§ Consultant, Arvada, CO.

Glass

By far the largest use of nepheline syenite is in the manufacture of glass products including container glass, fiber glass, opal glass, plate glass, sheet glass, and tableware glass. Nepheline syenite provides necessary additions of alumina and alkalis in the glass batch; it is low in silica and contains no free quartz; it has a favorable ratio of sodium oxide to potassium oxide of 2/1. Nepheline syenite bearing glass batches have lower viscosity and easier workability compared with those containing potash feldspar.

The lower fusion point of nepheline syenite lowers the melting temperature of the glass batch with attendant faster melting, high productivity, and fuel savings.

In Canada about 70% of nepheline syenite is used for glass manufacture; over 75% of Canadian nepheline syenite is exported to the United States and much of this is used in glass manufacture. 85% of Norwegian production is believed to be used in the glass industry and 15% for ceramics.

Ceramic Ware

Nepheline syenite is being used for the manufacture of a wide variety of whiteware products including dinnerware, sanitaryware, floor and wall tile, electrical porcelain, art pottery, chemical porcelain, dental porcelain, porcelain balls, and mill liners. In ceramics, nepheline syenite serves as a vitrifying agent, contributing to the glassy phase which binds other constituents together and gives strength to the ware. The ready fusibility and abundant fluxing capacity of nepheline syenite benefits manufacturers by permitting reduced body flux content, lower firing temperatures, or faster firing schedules than could be attained with other raw material combinations.

Canadian consumption of nepheline syenite for ceramic applications in 1971 represented 12% of total consumption. Twelve percent of Norwegian production of nepheline syenite in 1971 was "ceramic grade." There is no direct statistical data to indicate the level of Canadian exports of ceramic grades of nepheline syenite but analysis of export data suggest the amount is not large, relative to total exports.

Extender Pigment and Fillers

Applications for nepheline syenite as extender pigments and inert fillers have been found and developed. These include exterior and interior latex and alkyd paint systems, traffic paints, metal primers, exterior wood stains, sealers, undercoats, and hardboard ground coats. Nepheline syenite for these products has desirable high dry brightness, high bulking value, low vehicle demand, extreme ease of wetting and dispersion, and a stabilizing pH value. In paint formulations nepheline syenite has exceptionally good tint retention characteristics.

In plastics nepheline syenite has found increasing use in rigid, flexible, and plastisol-type polyvinyl chloride, and in epoxy and polyester resin systems. It exhibits extremely low resin demand thus making high loadings possible. In PVC resins it exhibits extremely low tinting strength, has a refractive index close to that of vinyl resin, and has a very low optical dispersion. It has a specific gravity lower than calcium carbonate and talc and the properties described for paint applications are beneficial in plastics applications as well.

Nepheline syenite is also used as an inert filler in the manufacture of foam carpet backing.

Alumina

In the USSR nepheline syenite and nepheline are used to manufacture alumina for aluminum production, sodium and potassium carbonates, and portland cement. USSR production comes from Kola peninsula and Kiya Shaltyrsk deposits. The USBM (Anon., 1961–71) estimates that the present level of consumption of nepheline concentrates in the Soviet Union is 1 million ltpy. More recent data on nepheline syenite production in the USSR are not available. In addition, the Volkhov, Pikalevo, and Achinsk alumina plants require at least 3 million lt of nepheline syenite annually. Nowhere at this time outside the Soviet Union is nepheline or nepheline syenite used as the basis for such industries.

Product Specifications

Nepheline syenite used by the glass industry is characteristically a −30 mesh +200 mesh product having an alumina content in excess of 23%, alkali content in excess of 14%, and iron oxide content less than 0.1%. Other metallic contaminants must be very low. For the ceramic industry nepheline syenite is ground to typically −200 mesh and is characterized by its high fluxing and low level of iron and other impurities. For extender pigment and filler industries, nepheline syenite offers high dry

brightness, uniformity, chemical inertness, and specific index of refraction and is supplied in various finenesses.

Market requirements vary and while the European markets accept a 0.1% Fe_2O_3 content, North American markets require 0.08% **Fe_2O_3 or less. Norwegian nepheline syenite products are higher in alumina and alkalis than Canadian ones. Other contaminating metallic ions must be at acceptably low levels and refractory minerals must be absent.**

Complete product specifications for Canadian and Norwegian glass, ceramic, and filler grades of nepheline syenite are shown in Tables 1, 2, and 3, respectively.

Geology

Mineralogy

Nepheline syenite is a coarse to medium-grained crystalline *rock* generally considered of igneous origin in which the essential mineral constituents are microcline, orthoclase or albite feldspars, the feldspathoid nepheline, and ferromagnesium minerals of which the principal are hornblende, pyroxene, and biotite. The most frequently occurring accessory minerals include magnetite, ilmenite, calcite, garnet, zircon, and corundum. Nepheline syenites are quartz-free.

Commercial deposits of nepheline syenite contain at least 20% nepheline, at least 60% feldspar, and seldom more than 5% accessory minerals. For a deposit to be of interest, refractory minerals such as corundum and zircon must be absent or easily removed along with other accessory minerals and characterizing mafic silicates during processing.

Nepheline—Nepheline is the most common of the feldspathoid *minerals.* Its formula is $Na_3K\ Al_4\ Si_4\ O_{16}$. Potassium is always present in natural nepheline, most frequently in ratio amounts of Na/K: 3/1, although ratios of 4/1 and 6/1 are known. It crystallizes in the hexagonal system and frequently forms 6 or 12-sided prisms. It has a distinct and an imperfect cleavage and a subconchoidal fracture. It is brittle; the hardness is 5½ to 6 on Mohs' scale, the specific gravity is 2.5 to 2.7; the luster is vitreous to greasy, occasionally opalescent. It is colorless, white, or yellowish but when massive may be dark green, greenish, bluish-gray, brownish-red, and brick-red. It is also known as elaeolite. It gelatinizes readily in acid and because it is more soluble than associated minerals can be readily spotted in outcrops by the pitted appearance of the rock surfaces. Nepheline alters to various zeolites, analcite, sodalite or cancrinite, and the gieseckite micaceous materials.

TABLE 1—Nepheline Syenite—Typical Product Specifications, Glass Grades

	Indusmin		IMC		Norsk Nefelin
Product Designation Typical Chemical Analysis	330, %	333, %	Summit, %	Ridge, %	North Cape, %
Silica	59.9	60.0	60.2	60.1	55.9
Alumina	23.5	23.4	23.5	23.4	24.2
Ferric Oxide	0.08	0.35	0.07	0.5	0.1
Calcia	0.6	0.7	0.3	0.3	1.3
Magnesia	0.1	0.1	trace	trace	trace
Soda	10.2	9.9	10.6	10.5	7.9
Potash	5.0	4.8	5.1	4.9	9.0
BaO	—	—	—	—	0.3
SrO	—	—	—	—	0.3
P_2O_5	—	—	—	—	0.1
L.O.I.	0.6	0.7	0.4	0.3	1.0

Typical Sieve Analysis

US Sieve No.	330	333	Summit (−30)	Summit (−40)	Ridge (−30)	Ridge (−40)	Tyler Sieve No.	North Cape
On 25 mesh %	0.0	0.0					On 28 mesh %	0.0
30	0.1	0.1	0.2		1.3		32	0.1
40	14.5	14.0	15.7	1.0	19.5	0.3	35	4.9
50	48.0	46.0	—	—	—	—	48	30.0
60	—	—	50.8	32.1	43.3	40.1	65	52.0
100	86.0	84.0	20.1	31.9	13.0	38.7		
140	—	—	6.3	16.5	10.2	13.4		
200	98.0	97.2	3.9	7.9	5.4	3.5	200	89.0
pan	2.0	3.0	2.0	10.6	7.3	4.0	pan	11.0

TABLE 2—Nepheline Syenite—Typical Product Specifications, Ceramic Grades

	Indusmin			IMC				Norsk Nefelin	
Product Designation	A-200	A-270	A-400	Crest	Peak	Apex		North Cape	
Typical Chemical Analysis	%	%	%	%	%	%		%	
SiO_2	60.7	60.7	60.7	60.2	60.2	60.2		56.0	
Al_2O_3	23.3	23.3	23.3	23.5	23.5	23.5		24.2	
Fe_2O_3	0.07	0.07	0.07	0.07	0.07	0.07		0.1	
CaO	0.7	0.7	0.7	0.3	0.3	0.3		1.2	
MgO	0.1	0.1	0.1	trace	trace	trace		trace	
Na_2O	9.8	9.8	9.8	15.3-0.4	15.3-0.4	15.3-0.4		7.8	
K_2O	4.6	4.6	4.6	15.3-0.4	15.3-0.4	15.3-0.4		9.1	
L.O.I.	0.7	0.7	0.7	0.4	0.4	0.4		1.0	
Sieve Analysis US Sieve No.								Tyler Sieve No.	
On 70 mesh %	0.01	0.0		—	—			On 32 mesh %	—
100	0.05	0.01		—	—			35	—
140	0.20	0.05		—	—			48	—
170	—	—		—	—			65	—
200	0.70	0.15		0.3	0.3			200	0.1
270	2.00	0.40		—	—			270	—
325	5.25	1.75		4.7	2.9			325	0.5
pan	94.75	98.25		95.0	96.8			pan	99.4
						Apex			
Microns						400	700		
% Finer than 30	70.0	78.0	98.0			98.0		% Finer than 30	—
20	55.0	65.0	90.0			90.0	98.0	20	—
10	33.5	42.0	65.0			60.0	90.0	10	50
5	19.0	23.0	33.5			35.0	60.0	5	—
2.5	10.0	12.5	16.0			18.0		2.5	—

TABLE 3—Typical Properties of Nepheline Syenite for Extender Pigment and Filler Applications

Product Designation	Apex 400	Apex 700	Minex 2	Minex 3	Minex 4	Minex 7	Minex 10
Appearance				bright white powder			
Density, g per cc				2.61			
lb per US gal				21.7			
Bulking value, US gal per lb				0.046			
Hardness, Mohs' scale				5.5–6.0			
Refractive index, composite				1.53			
pH value, ASTM D-1208				9.9			
Particle shape				nodular			
Dry brightness, %, CIE Tristimulus	96	98	94	95	96	98	98
Oil Absorption,* ASTM D-281	22+	28+	21+	21+	22+	28+	28+
Particle size, μ							
mean	7.5	4.5	16	14	7.5	4.5	2.3
maximum	49	17	105	74	44	17	10
Specific surface area, sq m per g (Fisher Sub-Sieve Sizer)	0.95	1.45					
Specific resistance, ohm-cm, ASTM D-2448	3600	3300		4000	3600	3300	
Hegman grind	3.5-4.0	5.5-6.0		1.0	3.5-4.0	5.5-6.0	6.5-7.0
Screen Fineness							
thru 200 mesh, %	100	100	99.3	99.9	100	100	
thru 325 mesh, %	98.9	100	95.8	98.3	99.9	100	
thru 400 mesh, %							100

* Nepheline syenite tends to have a lower vehicle demand than indicated by its oil absorption value.

Feldspar—The major feldspars present are microcline and orthoclase with a theoretical composition SiO_2 64.7%, Al_2O_3 18.4%, and K_2O 16.9%; and albite with a theoretical composition SiO_2 68.7%, Al_2O_3 19.5%, and Na_2O 11.8%.

Pyroxene—The aegerine member of this series of minerals is customarily present in feldspathoid rocks but the theoretical composition ranges from aegerine SiO_2 52.0%, Fe_2O_3 34.6%, and Na_2O 13.4%; to that of diopside SiO_2 55.6%, CaO 25.9%, and MgO 18.5%.

Amphibole—A wide variety of amphiboles are present in nepheline syenites including hornblende, hastingsite, barkevikite, arfvedsonite, and riebeckite. These are calcium, magnesium, sodium, and iron aluminosilicates of various compositions.

Biotite—Two types of biotite are found in feldspathoidal rocks; the golden-brown variety, but more frequently the high-iron variety lepidomelane. It is a hydrated aluminosilicate of potassium, iron, and magnesium.

Accessory Minerals—Among the more frequently found accessory minerals are sodalite, noselite, hauynite, scapolite, cancrinite, calcite, apatite, magnetite, ilmenite, hematite, pyrite, zircon, sphene, pyrochlore, garnet, and corundum.

Nepheline syenites have a wide variety of textures which can be generally classified according to the origin of the rock; plutonic and hypabyssal nepheline syenites; nepheline syenite pegmatites, and the migmatic nepheline rocks or "nephelinized gneisses."

Classification of Deposits

Several methods can be used to classify nepheline syenite deposits—by the percentage of principal minerals present in the deposit, by their origin, and also by their color. Nepheline syenites have been studied intensively and works by Deer et al. (1963) and Eitel (1965) provide an introduction to complete studies of nepheline and nepheline syenite. Little information about worldwide nepheline syenite occurrences and their commercial potential was available in concise form until publication of *Nepheline-Syenite and Phonolite* by Allen and Charsley in 1968. The book is an excellent summary of the present state of commercial interest in this and allied rocks.

Although a wide variety of names have been given to nepheline-bearing rocks and the term nepheline syenite is often given to any such rock containing 5% or more nepheline, it is unlikely that a rock would be of commercial interest unless the nepheline content is at least 20%.

Nepheline Syenite Rocks: Some of the more common varieties of nepheline syenite are:

Congressite: mostly nepheline with minor orthoclase and albite.

Craigmontite: nepheline predominates, oligoclase, and minor corundum.

Ditroite: nepheline, microcline or microcline perthite, biotite, aegerine, and soda amphibole.

Fenite: nepheline, 70 to 90% perthite and 5 to 25% aegerine.

Foyaite: nepheline, orthoclase, biotite, hornblende, and augite-aegerine.

Ijolite: aegerine and from 50 to 70% nepheline.

Laurdalite: nepheline, cryptoperthite, pyroxene, biotite, and hornblende.

Litchfieldite: nepheline, albite, orthoclase, lepidomelane, and sodalite.

Melteigite: aegerine and less than 50% nepheline.

Miascite: nepheline, microperthite, biotite, and minor oligoclase.

Monmouthite: nepheline, hastingsite, and minor oligoclase.

Raglanite: nepheline with predominant oligoclase and minor corundum.

Rouvillite: nepheline, gabbro-nepheline, hornblende, augite, and labradorite.

Urtite: aegerine and more than 70% nepheline.

Feldspar-free feldspathoidal rocks are also classified on the basis of color index number with rocks of lower index number being of possible economic interest.

Nepheline Syenite Deposits: Deposits are rarely of great areal extent, the largest being those in the Kola peninsula of the Soviet Union (450 and 250 square miles), Brazil (300 and 100 square miles), and Greenland (80 square miles). Most others are typically much smaller, less than 20 square miles, and those in Blue Mountain, Ont. (3 square miles), and Stjernöy, Norway (0.2 square miles) are even smaller. Classification may also be made based on the mode of occurrence and origin of the nepheline syenite. Moorhouse (1959) for one has divided them into five groups:

1) Feldspathoidal rocks associated with undersaturated volcanics.

2) Differentiated ring complexes, often associated with carbonatite and usually characterized by metasomatism around their borders.

3) Layered intrusives, possibly related to the ring complexes.

4) Borders or satellitic stocks associated with syenites or granites.

5) Nephelinized gneisses, usually associated with nepheline pegmatites.

The first type has no economic significance. Examples of the others would be: group 2, Nemegos, Ont.; Mont St. Hilaire, Que.; and Kola peninsula, Karelian ASSR; group 3, Ice River, B. C.; Julienhaab, Greenland; group 4, Blue Mountain, Ont.; group 5, Stjernöy, Norway.

A variety of origins are possible (Turner and Verhoogen, 1951): from low temperature residual magmas which have been conditioned by volatile components; by fractional crystallization of undersaturated olivine-basalt magma; by desilication of granitic magmas, that might have invaded the limestone with evolution of a CO_2-rich gas phase which may be important; nepheline syenite magmas, and metasomatic alteration of various preexisting rocks, i.e., "nephelinization."

Distribution of Deposits

Nepheline-bearing rocks are widespread throughout the world and deposits of nepheline syenite are common. Those in Canada, the United States, Norway, and the Union of Soviet Socialist Republics have been well described and much is known about those in other countries of Asia, Africa, and the South Pacific. Two of the largest known are the Khibiny (450 square miles) and Lovozero (250 square miles) intrusives in the Karelian ASSR (Vlasov et al., 1966) of the USSR. Other large ones are located in Siberia. In Brazil, nepheline syenites occur in several extensive intrusive areas, the largest of which cover some 300 square miles in Minas Gerais and São Paulo, and 100 square miles in São Paulo. Other deposits are rarely large, seldom covering more than a few square miles or a few tens of square miles in area. Fig. 1 shows the location of many nepheline syenite occurrences in the world.

However size is not an overly important criterion in commercial considerations; more important are the purity and location of the deposits.

There are now seven known mining ventures in different parts of the world producing nepheline or nepheline syenite products; two in Canada, one in Norway, two in the USSR, and two in the United States

In Canada, nepheline syenite deposits are common in several provinces. Occurrences in the vicinity of Lake Albanel, Gouin, Labelle, Mont St. Hilaire, and Oka, Que. are well known. In Ontario, Hewitt (1960) has de-

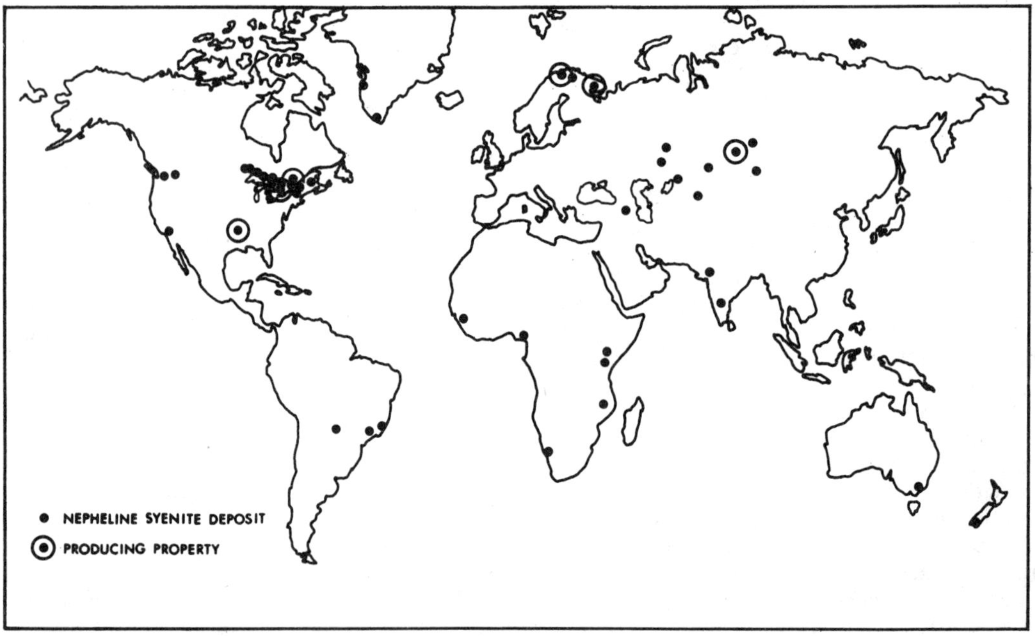

FIG. 1—*Location of principal nepheline syenite occurrences of the world.*

scribed deposits in the vicinity of Blue Mountain, Haliburton-Bancroft, French River, Callander Bay, Nemegos, and Port Caldwell as well as other minor occurrences. In British Columbia, the Ice River and Kruger Mountain complexes are well known. Much of the interest in Canadian deposits has stemmed from the successful development of the Blue Mountain intrusive in Methuen Township but none other has proven pure enough to warrant exploitation.

In Norway, nepheline syenite is known in the vicinity of Oslo and on several islands in western Finland, although only that on the southwest side of Stjernöy Island has the necessary characteristics for commercial development.

Reserves: The only known deposits of nepheline syenite suitable for low-iron products are being mined at existing operations in Norway and Canada. Reserves are considered limitless. In Norway at Norsk Nefelin, the volume of proven nepheline syenite exceeds 135 million cu m representing nearly 400 million tons of rock of a nature the company considers suitable for mining. In Canada, Indusmin has published a reserve figure of 14 million tons of proven ore, sufficient for 35 years at present mining rates. The company owns 85% of the Blue Mountain intrusive and believes potential reserves of minable rock are satisfactory for an indefinite period.

International Minerals and Chemical Corp. (Canada) Ltd. owns 500 acres of nepheline syenite at Blue Mountain, Ont. Since 1956 less than 50 acres of ground have been mined and reserves are considered to exceed 30 million tons, ample for any production requirements.

In the Soviet Union, there are no firm indications of reserves of nepheline-bearing rocks suitable for chemical conversion. The estimated production of apatite from the Kola peninsula mines exceeds 27 Mt/a and consequently the operation generates more than 35 Mt of nepheline-bearing tailings annually from which an estimated 15 Mt of nepheline concentrate could be recovered. In addition many square kilometers of nepheline syenite are present so reserves can be considered adequate for projected chemical needs. Nepheline syenite reserves at Goryachegorsk in central Siberia have been estimated at over 100 Gt. The planned production rate of 900 kt of alumina per year at Achinsk would require at least 3½ million tons of nepheline syenite rock from the Goryachegorsk deposit.

Principal Producing Countries: The principal producing countries of glass and ceramic grades of nepheline syenite are Canada and Norway. In the years between 1972 and 1981 combined production of the two countries has grown from 597 to 826 kt. Most of the Norwegian nepheline syenite was of glass grade destined for countries of western Europe, mainly the United Kingdom, West Germany, and the Netherlands. There is widespread acceptance of nepheline syenite in Europe as a feldspathic source. Canadian production of nepheline syenite has grown steadily from 438.4 kt in 1972 to 606 kt in 1981. Seventy percent of the total production was exported to the United States.

Canada—Indusmin Ltd. and International Minerals and Chemical Corp. (Canada) Ltd. operate the only two nepheline syenite mines in Canada. Although numerous nepheline syenite bodies have been examined in Ontario, Quebec, and British Columbia, only the Blue Mountain intrusive in Methuen Township, Ont., about 120 miles northeast of Toronto has the proven quality to produce uniform low-iron products.

The Blue Mountain intrusive is of Precambrian age having been emplaced, according to determinations, 1.3 billion years ago and was subsequently folded to its present attitudes some 300 million years later during the Grenville orogeny. Indicated temperature of the magma was 700° C and that of the country rock 400° C to 500° C.

The nepheline syenite intrusive consists of an elongate, irregular stock, having a length of 2½ miles in a north-south direction, and a width of 1¼ miles in an east-west direction. A long sill-like arm up to ¼ mile wide extends southwest from the main mass for 4 miles. The nepheline syenite intrudes and partially replaces para-amphibolite, paragneiss, and marble of the Blue Mountain metasedimentary band and is itself intruded and replaced by pink syenite and syenite pegmatite. The deposit appears to occupy the trough of a syncline plunging 20° southwest. Near the contacts the nepheline syenite is well foliated and within the main mass metamorphic lineation is readily observed. On vertical surfaces the usual plunge of the lineation is 10° to 30° SW. The accompanying map, Fig. 2, together with the table of formations, Table 4, shows the distribution of rocks and the relationship of the Blue Mountain intrusive with adjacent rocks.

The Blue Mountain nepheline syenite is a uniform, foliated, fine to medium-grained rock, composed essentially of: nepheline, 20 to 25%; albite, 48 to 54%; and microcline 18 to 23%.

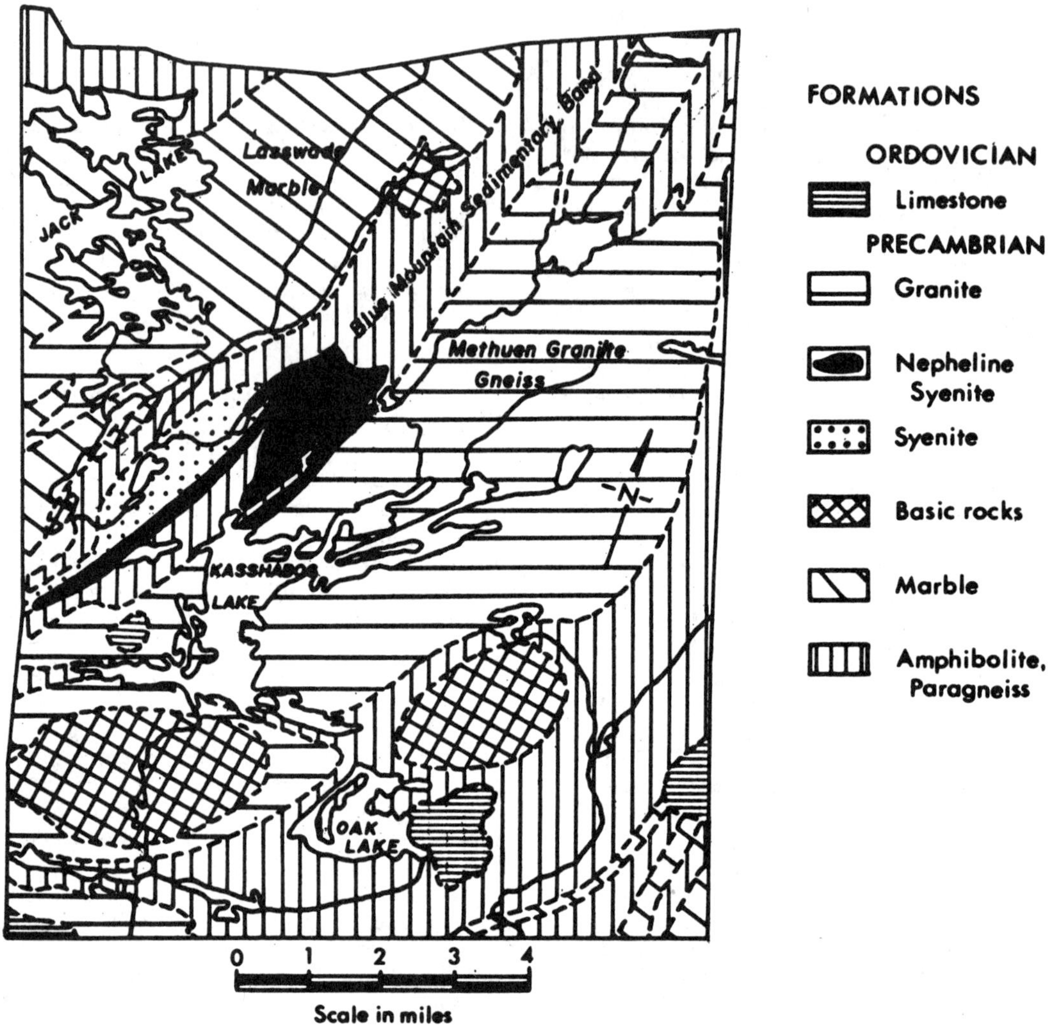

FIG. 2—*Geology of Methuen Township, Ont., Canada (after Hewitt, 1961; Payne, 1966).*

Characteristic accessory minerals are magnetite, 0.2 to 0.6%; biotite, 0 to 4%; hastingsite, 0 to 3%; muscovite, 0 to 2%; and aegerine; these in total seldom amount to more than 6%. Minor accessories encountered are: garnet, corundum, zircon, calcite, apatite, and titanite. Secondary hydrothermal and pegmatite minerals include epidote, riebeckite, natrolite, prehnite, analcite, hydronepheline, cancrinite, sodalite, and tourmaline. Recent studies (Guillet, 1962; Payne, 1966) indicate that a nepheline syenite magma intruded metasediments and hybrid syenite gneiss as a sill. The sill was fractioned during emplacement into five distinguishable zones: muscovite, muscovite-magnetite, biotite-muscovite, biotite, hornblende and pink syenite. Late-stage fluids of the magma were the source of pegmatites that intrude the sill as well as surrounding country rock. Lens, sheets, and pods of country rock that sloughed into magma are found in the nepheline syenite aligned parallel with the direction of flow. Some metasomatic and metamorphic effects of the intrusion developed above and to a minor extent below the sill. Analyses of typical nepheline syenites are shown in Table 5 and analyses of typical products derived therefrom are shown in Tables 1, 2, and 3.

Norway—By HANS-PETER GEIS, Elkem Spigerverket—The nepheline syenite deposit which is mined by Norsk Nefelin is situated within a Caledonian province of basic and

TABLE 4—Table of Formations, Methuen Township, Peterborough County*

Cenozoic Pleistocene Recent	Sand, gravel, clay, silt, till Great Unconformity	
Paleozoic Ordovician	Limestone, dolomitic limestone Red argillaceous limestone and shale Basal grit and arkose Great Unconformity	
Precambrian Plutonic Rocks	Granitic rocks: granite, granite gneiss, granite pegmatite Intrusive Contact Syenite rocks: syenite gneiss, hybrid syenite gneiss, syenite pegmatite Intrusive Contact Nepheline syenite Intrusive Contact Basic intrusives: diorite, gabbro, norite, pyroxenite, amphibolite Intrusive Contact	
Sedimentary and Volcanic Rocks	Southeast Methuen (stratigraphic succession) Vansickle formation (approximate thickness, 4500 ft + (1370 m +)] Upper schist Upper marble Middle schist Vansickle conglomerate Middle marble Lower schist Lower marble Lower conglomerate	Northwest Methuen Blue Mountain sedimentary band Casswade marble Apsley formation
	Oak Lake Formation [approximate thickness, 5000 ft (1525 m)] Pink arkose, quartzite, amphibolite schist, feldspar schist, some volcanic amphibolite	
	Big Island Beds: Interbedded marble and schist [approximate thickness, 2000 ft + (610 m +)].	
	Kosh Lake Beds: Epidotized amphibolite, paragneiss.	
	Troutling Bay Volcanics.	

Source: Hewitt, 1960.

ultrabasic rocks near the northern tip of Norway. This province has an areal extension of 30 sq km.

The deposit itself lies near the southern shore of the island of Stjernöy, cropping out at a height of 250 to 700 m above sea level. The outcrop is a lens-shaped form with a length of 1800 m and a maximum width of 300 m giving a total area of about 270,000 sq m.

The deposit has been mapped and sampled in great detail, both at the surface and underground. The wall rock on the northeastern side is an alternation of carbonatite and hornblendite; on the southwestern side it is gabbro gneiss. The accompanying map, Fig. 3, shows the different petrological types within the main nepheline syenite. It also shows a few minor bodies parallel to the main body which have not been studied in detail or mined.

There are two major types of nepheline syenite, viz., biotite type and hornblende type. Both types consist essentially of perthite feldspar (average 56%) and nepheline (average 34%), together with some minor minerals like plagioclase, calcite, clinopyroxene, hornblende, biotite, sphene, and magnetite. In the biotite type, the dominating dark mineral is biotite (2.5 to 6.0%), and in the hornblende type, common hornblende (0.3 to 2.5%) and aegerine (1.3 to 3.8%) together with small amounts of biotite. There is more sphene in the hornblende type than in the biotite type (0.2 to

TABLE 5—Typical Chemical and Mineralogical Composition of Some Nepheline Syenites, %

	Canada,* Methuen Township		Norway,† Stjernoy		Krasnoyarsk‡	United States,§ Litchfield, Maine	USSR,# Khibiny	Greenland,¶ Ilimaussaq	Congo**	Portugal††	Malawi‡‡	Brazil,§§ Cannon Mine
Chemical Composition	Hornblende	Biotite	Pyroxene	Biotite								
SiO_2	58.8	59.4	52.73	52.37	54.57	60.39	54.01	49.46	55.44	55.22	55.77	57.8
Al_2O_3	23.0	23.0	23.71	23.22	22.56	22.51	21.50	23.53	23.59	22.59	22.26	23.5
Fe_2O_3	0.8	0.7	1.9	1.1	3.44	0.42	2.60	3.04	0.44	1.14	1.33	0.1
FeO	1.4	1.5	1.89	1.14	—	2.26	1.80	1.02	1.42	1.17	2.50	—
TiO_2	0.004	0.004	0.51	0.61	0.15	—	1.20	0.16	0.20	0.59	0.50	—
Na_2O	9.4	9.5	7.78	6.87	9.69	8.44	9.50	14.71	10.20	8.76	8.05	10.9
K_2O	5.2	4.9	8.08	8.30	5.51	4.77	5.30	4.34	6.26	5.59	6.66	6.2
CaO	0.82	0.64	2.54	3.11	1.61	0.32	1.80	0.80	1.56	2.12	1.25	0.5
MgO	0.04	0.03	0.24	0.25	0.19	0.13	0.77	trace	0.14	0.28	0.58	0.05
MnO	0.049	0.045	0.06	0.09	—	0.08	0.17	0.17	0.15	0.13	0.19	—
P_2O_5	0.01	0.01	0.05	0.09	0.02	—	0.09	n.a.	0.18	—	0.21	—
CO_2	0.05	0.15	0.77	1.88	—	trace	—	—	—	nil	0.07	—
H_2O	0.3	0.4	0.26	0.26	—	0.57	1.14	1.38	1.07	2.16	0.50	—
BaO	—	—	0.40	0.47	—	—	0.12	—	—	—	—	—
Ignition Loss	—	—	—	—	—	—	—	—	1.62	—	—	0.9
Mineralogical Composition												
Albite	48.4	52.0	trace	6.0	20.3	47.8						
Microcline	22.7	18.9	—	—	} 37.8	16.5						
Microperthite	—	—	—	—	}	—						
Perthite	minor	minor	57.0	55.5	—	—						
Nepheline	24.9	24.1	37.0	29.0	26.9	23.7						
Biotite	0.1	2.2	trace	3.5	0.1	5.5						
Hornblende	3.0	—	1.0	—	—	—						
Pyroxene	0.2	—	1.0	—	6.1	—						
Magnetite	0.4	0.5	1.0	1.5	—	—						
Calcite	—	0.1	2.0	5.0	—	—						
Muscovite	—	0.7	—	—	—	—						
Zeolite	—	—	—	—	8.7	6.5						

Sources: * Payne, 1966, p. 140.
† Heier, 1966.
‡ Leucocratic nepheline syenite; Central Tartar Massif, 1968, Krasnoyarsk Territory, Siberia. Allen and Charsley, 1968, p. 109.
§ Allen and Charsley, 1968, p. 134.
# Gerasimovskii, 1956; Allen and Charsley, 1968, p. 101.
¶ Greenland—Ilimaussaq. Allen and Charsley, 1968, p. 127.
** Nepheline syenite Kirumba, Democratic Republic of The Congo. Allen and Charsley, 1968, p. 114.
†† Algarve Province. Allen and Charsley, 1968, p. 95.
‡‡ Biotite and hastingsite nepheline syenite, Lulwe Hill, Malawi. Allen and Charsley, 1968, p. 54.
§§ Harben, 1981.

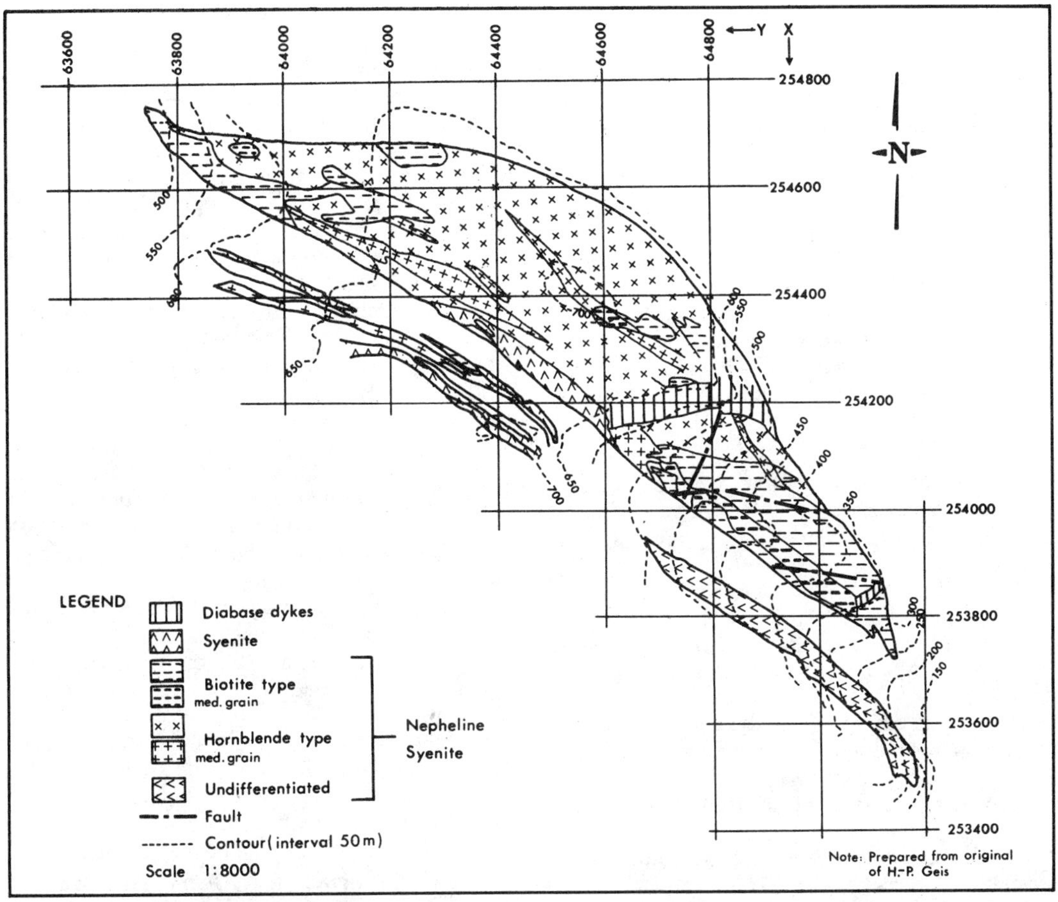

FIG. 3—*Surface geology of Stjernöy nepheline syenite deposit.*

1.2% as compared to trace). Traces of corundum and zircon have been observed but their content in the finished product are subject to routine control within the mine.

Under the microscope one can see that the rock consists of a network of feldspars in which there are veins containing the other minerals. The nepheline crystals contain more or less resorbed remainders of feldspar. The microscopic picture might, by these reasons, be interpreted as indicating a metasomatic origin of the nepheline syenite in its present form.

Geochemical studies have shown that the contents of Fe_2O_3 and CaO are lowest in the central part of the deposit near the southwestern contact and that they increase toward the northeast and toward both ends. Mining has been started in these central parts because the recovery here is highest. In the same direction as above, SiO_2 and Na_2O contents decrease.

A marked foliation has been observed in the nepheline syenite, caused by an elongated appearance of all minerals along strike and dip of the whole body. From the center of the body to the southeast end, foliation dips 70° SW. From the center toward the northwest, it dips steeper and at the northwest end it dips 70° NE. This foliation is so regular that by measurements at the surface it was possible to predict the contacts 500 m below with only very small deviation.

A number of diabase dykes from a few centimeters to tens of centimeters thick cross through the body. Their amount is estimated to represent from 6 to 7% of the whole nepheline syenite mass. A few fault zones intersect the nepheline syenite. Along them the rock is partly altered to zeolite, prehnite, and kaolin. For further reading one is referred to Barth (Barth, 1963) and Heier (Heier, 1961, 1964, 1965, 1966).

Union of Soviet Socialist Republics—The nepheline rock *urtite* is waste from the production of apatite at mines on the Kola peninsula. Production of apatite in 1970 was 27 Mt which would require mining an average of 62 Mt of ore and would make available over 15 Mt of nepheline concentrate. Production of nepheline concentrate was estimated by the US Bureau of Mines (Anon., 1961–71) to have grown from 350 kt in 1961 to 1 Mt by 1970.

Production is derived presently from the Khibiny pluton of probable Devonian Age at the west side of Lake Umpjaur in the central Kola peninsula. Seven phases of intrusion have been identified, and are listed in order of decreasing age (Polkanov, 1937):

1) Syenites and nepheline syenite. Adjoining Archean gneisses show evidence of both injection and metasomatism by the syenite magmas.

2) Coarse aegerine-nepheline syenites injected as a peripheral ring dyke.

3) Trachytoid aegerine-nepheline syenites forming an imperfect ring within and adjoining the coarse rocks of phase two.

4) An imperfect ring dyke of poikilitic micaceous nepheline syenites of variable composition. The assemblage of alkali feldspar, nepheline, mica, aegerine, astrophyllite is typical.

5) A massive cone sheet of silica-poor nepheline-rich rocks of the ijolite family.

6) Foyaites (alkali feldspars, nepheline, biotite, and hornblende or aegerine-augite), making the central core of the pluton.

7) Dikes of tinguaite, monchiquite, shonkinite, theralite, nepheline basalt, and leucite basalt.

The pluton is characterized by abundance of phosphorus, fluorine, titanium, zirconium, columbium, strontium, and rare earths, especially the cerium group. High grade deposits of apatite, sphene, and aegerine pegmatites rich in rare earths, are known. One apatite rock deposit is a flat lens 1½ miles long and 500 ft thick. Chemical analyses and mineral composition of some Soviet nepheline rocks are shown in Table 5.

Numerous other nepheline deposits in the Soviet Union have been described and at least one in central Siberia near Achinsk is mined to provide feed for alumina plants. Production of nepheline rocks is expected to continue to grow and eventually provide half the country's alumina requirements.

Minor Producing Countries:

United States—The nepheline syenite deposits of Pulaski, Saline, and Hot Spring counties in central Arkansas are well known. Attempts to produce ceramic and glass grade products from them have not been successful because the impurities are too finely divided within the feldspars for appropriate removal. Analyses of typical crude and treated rock, from Fourche Mountain in Pulaski County, are shown in Table 6.

The McGeorge Construction Co. operates a large construction aggregate quarry in the Fourche Mountain nepheline syenite intrusive near Little Rock. Nearby the 3M Co. has operated a roofing granule manufacturing plant since 1947. Its annual capacity is understood to exceed 1,000,000 st. Because the activities of these two companies are more properly dealt with in chapters dealing with construction materials and granules, they will not be considered further here.

Mexico—In Mexico, Industrias Peñoles, S.A. de C.V. is said to be developing reserves of alumina contained in 3 billion st of nepheline syenite near Cuidad Victoria, in the state of Tamaulipas.

Brazil—In 1981, Austral Ltd. (Austral Mineração Serviços Ltda.) of Brazil started a pilot plant with an annual capacity of 5,000 stpy to produce a glass-grade material from a nepheline syenite deposit near Rio de Janeiro.

Other Occurrences: Allen and Charsley (1968) in their monograph have assembled an impressive list of occurrences of nepheline syenite in many countries of the world together with initial information on the character of the rocks and the results of test work. In addition to the producing properties described herein they number deposits in 20 African countries, 5 Asian countries, Australia and New Zealand, 7 European countries, Mexico, 3 South American countries, and Greenland. The distribution of some occurrences is shown in Fig. 1. While few, if any described, could be used to prepare low-iron products, some might well be suited for the manufacture of colored glass and certain types of ceramic ware.

Exploration

As with most industrial minerals the "place value" of a nepheline syenite deposit is critical. In addition to the following comments on field examination, the exploration geologist should consider the overriding importance of costs of

TABLE 6—Analyses of Pulaski County, Arkansas, Nepheline Syenite*

	Crude Ore		Treated Ore	
	Hornblende Type Blue Syenite, %	Biotite Type Gray Syenite, %	Hornblende Type Blue Syenite, %	Biotite Type Gray Syenite, %
SiO_2	60.3	60.8	61.2	61.9
Al_2O_3	19.9	19.6	21.5	21.1
Fe_2O_3	4.7	4.3	0.9	1.1
TiO_2	1.1	1.2	0.3	0.2
CaO	1.3	0.8		
MgO	1.2	0.9		
Na_2O	6.3	6.7		
K_2O	5.3	5.9		
Ignition loss	0.1	1.3		

* Smothers et al., 1952.

production and the costs of transporting products to market.

Only by examining the interrelationship of all factors bearing on exploitation of a given nepheline deposit can one determine if development is warranted.

Field Techniques

The field geologist needs to search for a deposit that consists of a coarse to medium-grained rock. Crushing to minus US Standard 30 mesh should liberate all the impurities that need to be removed before a product is salable. Careful mapping of the deposit in detail will reveal textural, structural, and mineralogical changes that might affect mining. Samples need to be selected carefully to provide a representative suite on which initial microscopic, physical, chemical, and ceramic tests can be made.

If examination reveals that impurities are scattered through the feldspar crystals in a finely divided state, such a deposit would be unsuited for glass and ceramic uses. Only if impurities can be readily liberated by crushing should a deposit receive more detailed evaluation. One would normally expect to mine feldspathic materials by open-pit methods and due consideration needs to be given to bench heights, berms, and mining regulations when determining if a given deposit would be satisfactory. Ore reserves should be large enough to sustain an operation for a long time because customers demand assurance of ample supply and, in industrial mineral ventures, ten or more years may be needed to repay the invested capital. Normally one would not consider a deposit having ore reserves for less than 20 years operation.

Consideration must be given to possible methods that might be used to beneficiate nepheline syenite from a deposit. Present feldspar and nepheline syenite benefication plants have relatively simple schemes for flotation and electromagnetic removal of impurities, and flotation separation of quartz and feldspar. Flotation, electrostatic, electromagnetic, and other techniques can all be successfully used on different mineral assemblages but if the mineral suite in a given deposit is too complex, the combination of techniques needed to remove impurities could well be too expensive and involve too many operating variables to permit economical operation.

In areas where chemical weathering is a factor one must have fresh specimens with which to work. Nepheline dissolves from rock surfaces relatively easily, feldspars are altered to kaolin, iron oxides and sulfides may be leached away, all of which tends to distort results obtained when surface samples are examined.

Complex nepheline syenite deposits such as the ring-complex type and carbonatite type intrusives show little promise for utilization in glass and ceramics. They are too variable in normal feldspar and mafic constituents and furthermore may contain rare earths, halides, and numerous other trace elements that detract from the intended applications for the rock.

Nepheline syenite deposits of the sheet, gneissic, and border phase of granite and syenite intrusive types seem to offer the greatest chance of uniformity and quality needed.

Drilling

Deposits of merit need to be drilled to confirm the nature of the rocks and to provide samples for laboratory work. Depending on circumstances AX, BX, and NX diameter core may be recovered. In laying out a drill program, care must be taken that holes intersect all rock types at appropriate angles of inclination. Bands of biotite gneiss, for instance, might be erratic in distribution and unless drilling is adequate, erroneous conclusions might be drawn about the amount of such impurities present. A deposit may be zoned and although mapping indicates rock of one type, the core of the deposit may be another. There are always variations to be encountered in nepheline syenite deposits, as may be guessed from their mode of origin, and drilling a deposit provides assurance that surprises will be minimal.

Spacing of drill holes is determined by the uniformity of the rock and the stage of the investigation. Initial drilling is widespread, followed by grid drilling at perhaps 400-ft spacing, followed by spacing of 100 ft or less if conditions demand.

Geophysical and geochemical techniques are not a normal part of exploration for nepheline syenite although seismic surveys may be useful for helping to determine the depth of overburden.

Other Considerations

Consideration should be given to matters such as the size and character of the market, the mode and cost of transportation, availability of auxiliary services, labor supply, and laws that might be restrictive. Due concern for them at an early stage in exploration could provide a basis for judging which deposits have the greatest worth and save much time and wasted expense on further laboratory evaluation permit one to study the rock for signs of product warrants.

Evaluation of Deposits

Specifications—General

Deposits should lend themselves to preparation of products comparable to those in Tables 1, 2, and 3. Total waste should be as low as possible, preferably not more than 25%, with product in the −30 mesh +200 mesh range using US standard testing sieves. Impurities such as refractory minerals, sulfides, other oxides or ferromagnesian minerals, silicates other than feldspars and feldspathoids, should be absent or readily removed from material during processing. For filler products, whiteness is as important as the chemical composition. For production of chemicals from nepheline rocks, alumina levels should be 30% or higher but other elements are not as deleterious as in glass grade products. For all applications, a deposit must be consistently uniform in mineral content and character.

Testing and Evaluation

The first requirement is to characterize the nepheline syenite by describing as many of its properties as possible, then subject selected samples to laboratory-scale processing that attempts to reproduce techniques that would be used in an operating plant. Throughout the test work it is important to follow accepted mineral dressing practice.

Microscopic Examination: Much wasted time and expense can be avoided in studying nepheline syenites if thin sections and polished sections of the rock are made and the microscopic and macroscopic characteristics of the rock are determined. It is important to find out if the feldspars and feldspathoids can be liberated from mafic and other mineral impurities because, if not, the deposit cannot be considered further for most commercial uses.

Thin sections, polished sections, or appropriately sized fractions of nepheline syenite immersed in oils of specific indices of refraction, permit one to study the rock for signs of impurities within the feldspar and nepheline grains. They also permit one to estimate the degree of liberation of impurities from the desired minerals in the grain sizes to which the rock was crushed.

Sample Preparation and Analysis: Samples collected from a deposit are crushed and screened to provide −30 mesh +200 mesh material. Care should be taken to not overcrush material. While test work is carried out on this screen fraction some nepheline syenite is screened at intermediate size fractions. The −200 mesh fraction is kept for study if warranted. It is important to record the weight of all fractions so that material balances can be calculated. Specimens should be cut from the sample prior to any particular test so that their chemical and physical properties can be determined.

Physical Tests: The objectives of any sample testing program are, first, to determine if a commercially acceptable product can be made from the rock, and second, to establish if practical methods of treatment can be used to prepare those products.

Liberation of usable feldspathoid and feldspar minerals from coexisting contaminants is required. Because so much nepheline syenite is sold to glass manufacturers who purchase a coarsely granular product, commercial practice dictates that effective liberation should be achieved in size fractions coarser than 200 mesh.

It is customary to commence work on a sample that has been reduced in size to —30 mesh and from which the —200 mesh fraction has been removed. From this material any tramp iron introduced during crushing is removed completely either with a hand magnet, drum magnet, or low intensity electromagnet of adequate field strength. Carpco, Dings, Exolon, Jones, Rapid, and Stearnes type induced-roll, high-intensity, electromagnetic separators are available in models convenient for laboratory use. Samples are passed through such a unit one or more times to reduce the level of those impurities having medium to low magnetic susceptibility. Models of the Carpco and Jones separators are available that treat samples in a water medium which may be more successful than dry methods as dust will not coat the mineral grains in it. Minerals such as magnetite, hornblende, tourmaline, garnet, and biotite are readily removed by this magnetic method.

If the sample responds inadequately to magnetic treatment it may be subjected to flotation techniques. Sulfides, heavy minerals and nonmagnetic iron minerals can often be removed by flotation. Detailed procedures for treating samples can be developed from study of Taggart's *Handbook of Mineral Dressing* (1956) and AIME's *Froth Flotation* (Baarson et al., 1962).

Other methods of treatment could involve roasting, calcination, and various leaching media as well as combinations of methods. If impurities are present within the feldspar crystals then it is unlikely that suitable products can be prepared by normal mill practices.

Of course it is important during testing of nepheline syenite samples to perform some chemical tests on the crude sample, screen fractions of the sample as it is subjected to physical and chemical processing procedures, and on the derived products. Testing different screen fractions may provide a hint respecting any tendency of impurities to be present in disproportionate amounts in certain grain sizes and thus provide a basis for separation in a production flowsheet.

It is customary to analyse samples for SiO_2, Al_2O_3, Fe_2O_3, FeO, MgO, CaO, Na_2O, K_2O, and loss on ignition. One should check for chromium, manganese, titanium, copper, or other suspected metallic ions which, if present, would make the nepheline syenite unsuitable for glass or ceramic purposes.

It is useful to perform a heavy-media test on the nepheline syenite samples to detect the presence of minerals such as zircon, chromite, rutile, corundum, ilmenite, and sphene. This procedure is especially helpful in isolating nonmagnetic refractory minerals that may be present in small quantities but which adversely affect the value of a nepheline syenite deposit. Sufficient samples must be processed to detect their presence or absence. Acetylene tetrabromide (tetrabromethane) with a specific gravity of 2.96 is the commonly used separating medium. It can be diluted with carbon tetrachloride with a specific gravity of 1.59 to provide solutions of lower specific gravity if desired.

One may wish to prepare a small quantity of a typical glass composition in which the sample nepheline syenite is substituted for commercially available feldspathic materials, melt it, and compare the color and appearance with a regular glass melt. This test has the advantage that the effect of any previously undetected impurities will be readily apparent.

The foregoing tests are sufficient where a nepheline syenite is being considered for glass manufacturing applications. When applications for the ceramic, paint, and plastics industries are contemplated, further test work is needed.

Tests for Ceramic Applications: A sample of beneficiated nepheline syenite is ground to correspond in grain size to commercially available nepheline syenite or feldspar. The characteristics of nepheline syenite materials used in ceramic applications are shown in Table 2.

The fluxing character of nepheline syenites is important for ceramic applications. Typical Canadian ceramic grades have pyrometric cone equivalents ranging from three to six, depending on the fineness of the product. They start sintering at cone 08 and have a sintering range of 13 cones. This is attributed to the fact that the product consists of a mixture of the three minerals, albite, nepheline, and microcline, listed in order of abundance. Norwegian nepheline syenite consists essentially of perthite and nepheline. Standard cone fusion tests under controlled conditions of time and temperature are used to establish the fusion characteristics of samples. These are a function of the particle size, particle size distribution, and chemical composition of finely ground nepheline syenite products. A specially shaped cone of nepheline syenite is fired at a set rate with pyrometric cones 6, 7, and 8, and changes in the height and diameter of the fired cone are compared with established standards. Another test of fluxing ability is the "inclined plane" viscosity test wherein a given weight of nepheline syenite is packed in a specially designed wedge and fused under controlled conditions. The relative flows of the sample and standards are noted.

In addition to the cone fusion test, the +200 mesh residue from a wet sieve analysis of the specimen may be fused to provide a useful indication of the fired color and number of speck-producing impurities in the sample.

Of great importance in any test program are the grain size characteristics of the sample. Detailed description of test methods for determining these and discussion of the importance of grain size can be found in Grimshaw's (1971) text.

For screening of coarse fractions one can refer to the American Society for Testing & Materials (ASTM) specifications for sieves for testing purposes, ASTM designation E.11; sieve analysis of raw materials for glass manufacture, ASTM designation C-429; and wire cloth sieve analysis of nonplastic pulverized ceramic materials, ASTM designation 371.

For determining the size of subsieve particles the Andreasen method is used frequently. This is a refinement of the ordinary pipette method for the determination of subsieve grain sizes. For substances of common specific gravities, when water is used as a sedimentation liquid, the apparatus allows determination of fineness of substances, the grains of which range from 30 to 0.5 μ. The method is based on Stoke's law relating to the rate of fall of particles in a liquid medium. The procedure is described in detail in Grimshaw (1971). For rapid size determination the Bahco method is often employed. It involves the use of a Bahco Micro-Particle Classifier which is a combination air centrifuge elutriator in which various size fractions may be obtained by controlled air velocities. The classifier has been calibrated against the Andreasen Pipette Sedimentation method to determine the grain size of the various fractions. The method is described in a paper by Taylor and Wilson (1962).

Determination of the pH of a given nepheline syenite and water mixture provides an indication of the presence and solubility of impurities.

In addition, the following are typical tests that may be carried out on ceramic bodies made from nepheline syenite and in combination with other raw materials to determine its relative usefulness.

1) Deflocculation and flocculation curves.
2) Casting or extrusion properties.
3) Body preparation.
4) Thermal gradient analysis.
5) Dry properties—shrinkage, ASTM designation C326; modulus of rupture, ASTM designation C369; bulk density, ASTM designation C373; linear thermal expansion, ASTM designation C372.
6) Fired properties—shrinkage, ASTM designation C326; modulus of rupture, ASTM designation C369; adsorption, ASTM designation C373; bulk density, ASTM designation C373; linear thermal expansion, ASTM designation C372.

Body compositions and test methods are discussed in both Grimshaw (1971) and *Industrial Ceramics* (Singer and Singer, 1963).

When consideration is being given to nepheline syenite for extender pigment or filler applications, useful and important tests of samples include determination of dry brightness, refractive index, and oil absorption as well as grain size distribution, pH, and chemical composition. Samples should be compared with commercially available products to provide a basis for judgment of their value.

Evaluation of Results

The test results must indicate that nepheline syenite products can be prepared that are as good as or better than those being offered to consuming glass, ceramic, and other industries. From the test data, engineering analyses will show if a plant could be designed to produce products at competitive prices.

Other Important Considerations

An undeveloped nepheline syenite deposit must be assessed for its "place value." Transportation costs to carry products to market, the nature of markets, the specific product demands of individual companies, the capital and operating costs, and the effect of jurisdictional regulations imposed by government must all be considered in a deposit evaluation. It goes without saying that the field and laboratory evaluation of samples are only two of the many aspects of the overall feasibility study that must be made to assess the merit of a given nepheline syenite deposit. The overwhelming interdependency of the information developed in a study is evident only when favorable facts spur, or unfavorable ones hinder, a project.

Preparation for Markets

Canada

Indusmin Ltd.: Indusmin operates a mine at Nephton, Ont., some 120 miles northeast of Toronto (Fig. 4). Canadian Nepheline Ltd., the predecessor of Indusmin, was the first company in the world to produce commercial grades of nepheline syenite for the glass and ceramic markets. Production began in 1935 and has been continuous to the present. Credit must be given to Ernest Craig, Harold Deeth, C. J. Koenig, and C. L. Cruickshank for their enormous pioneering effort during the first 30 years of nepheline syenite utilization for they were the first to realize its potential and guide nepheline syenite to full industrial acceptance.

Mining—Indusmin mines rocks from two quarries, situated three miles apart on the Blue Mountain ridge, at the rate of 1800 stpd, a single shift. The ore to waste ratio is approximately 4/1. There is practically no overburden on the deposit because of the scouring effect of numerous continental ice masses that passed over the area. What little dirt there is can be readily removed by bulldozer and Grade-All equipment. Typical rock waste consists of biotite gneiss bands that parallel the lineation of ore bands and are scattered through it.

A Joy Air-Trac crawler-mounted drill, powered by a Gardner Denver 600 rotary compressor is used in rough terrain and to develop new benches. Drilling pattern in new areas is 7 x 7 ft and hole diameter is 3 in. Developed benches are drilled on an 11 x 13-ft pattern by a Joy Challenger rubber-mounted, self-propelled drill, powered by a 1200-cfm stationary compressor located at the mill. Holes are 4½ in. in diameter drilled to a depth of up to 41 ft. Bench heights may be 30 to 40 ft. Amex II ammonium nitrite primed with 60% Forcite is detonated with short-period electric caps; 2½ x 8 in. sticks of 60% Forcite are spaced through the Amex II to ensure complete detonation. A 5500-lb drop ball is used for secondary breaking.

Two front-end loaders place the pit-run ore into four 35-ton capacity off-the-highway rear-dumping trucks.

Milling and Processing—Pit-run ore is discharged into a surface bin feeding a 42 x 48 in. Allis-Chalmers jaw crusher set to reduce rock to −7 in. This is further reduced in a 4½-ft Symons standard cone crusher to −2 in. and transferred to a 60 x 96-ft 4500-st-capacity crude ore storage building. The primary and

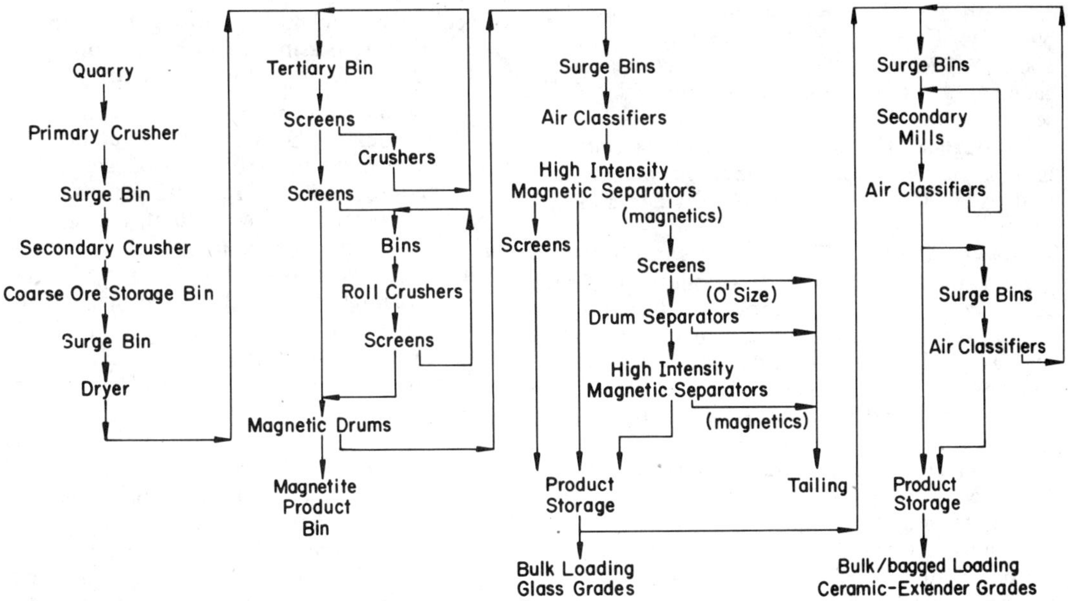

FIG. 4—*Production flow chart of Indusmin Ltd. plant, Nephton, Ont. Canada.*

secondary crushing is carried out on one shift-per-day.

Minus 2-in. ore is dried in 70-in. by 50 ft parallel flow oil-fired rotary dryers and conveyed to a 300-st-capacity bin. Ore then passes over double-deck mechanical screens, the first of many different screens in the plant. Plus 3/16 in. material goes to two 3-ft Symons short-head cone crushers or one impact crusher and the discharge is returned to the mechanical screens. Minus 3/16 in. material is fed to six 4 x 10 ft Tyler Hummer double-deck screens. Minus 3/16 in. + 30 mesh sand is fed to roll crushers for further reduction. Minus 30 mesh sand is fed to low intensity magnetic drums and other screens prior to being directed to banks of Exolon and Lurgi type high-intensity electromagnetic separators. Discharge from the separators constitutes the glass-grade nepheline syenite having an Fe_2O_3 content of 0.07%. There are four mill units consisting of feed bins, roll crushers, screens, magnetic drums, high intensity electromagnetic separators, and scalping screens together with necessary materials handling equipment. Rejects from the magnetic circuits are mixed with water and pumped to a tailings pond. Magnetite is collected from the magnetic drums and sold as iron ore. Total mill and mine recovery ranges from 75 to 80%.

Ceramic and extender grades of nepheline syenite are ground in pebble mills operating in closed circuit with air classifiers.

Finished product storage consists of concrete bulk silos—four of 800-st capacity, four of 600-st capacity, and three of 150-st capacity each. In addition, warehousing is provided for several hundred tons of bagged products.

Throughout the entire mill particular attention is paid to dust collection. In the dryer building 30,000 cfm of air are provided and in the main mill building 400,000 cfm.

Quality Control—The quality control program starts with the mine development drilling and carries through to sampling drill cuttings, mining faces, numerous checkpoints in the mill, and as products are loaded into conveyances for delivery to customers. Routine chemical analyses include determination of Al_2O_3, Na_2O, K_2O, CaO, MgO, Fe_2O_3; determinations of heavy mineral content, screen analyses, and subsieve particle size distribution; cone fusion tests and dry brightness. The company maintains a technical and research center in Toronto, Ont., responsible for new product development, customer service, and quality control.

The mine is served by rail and all-weather highway. Products are shipped from the mine to more than 21 countries around the world.

International Minerals and Chemical Corp. (Canada) Ltd.: Although the Blue Mountain property was explored as early as 1912 by Norman B. Davis and staked by him in 1936 it was developed in 1956 by International Minerals and Chemical Corp. (Canada) Ltd. (IMC) who erected a mill to produce glass and ceramic grades of nepheline syenite. The mine and plant are four miles northeast of Nephton, Ont., at the northeast end of the Blue Mountain Intrusive.

Mining—Mining conditions are similar to those elsewhere on the Blue Mountain ridge. Maximum relief of the nepheline syenite deposit controlled by IMC is approximately 300 ft above local lake level. Three benches are under development at the 25, 70, and 100-ft levels. To date less than 10% of the IMC property has been opened for mining. Quarry faces are generally advanced either parallel or normal to the strike of the gneissic rock, taking advantage where possible of the joint patterns. Normal quarry methods are employed with the drill pattern being usually 6 x 6 ft. Individual shots range up to 20,000 tons of rock. Broken rock is reduced in size, drop-balling where necessary, loaded with a front-end loader, and moved to the primary jaw-crusher at the plant by truck. Quality control in the quarry is practiced by close sampling of drill cuttings. Grade of rock to be broken is known prior to shooting; thus it is possible to blend for optimum mill feed and ultimate product control.

Milling and Processing—Dry processing involves various methods of size reduction, including rod mills, to —30 and —40 mesh, followed by iron mineral removal with various types of magnetic-type separators, further sizing, fine grinding of ceramic grades, and preparation for either bagged or bulk shipment. Frequent sampling throughout the plant of in-process material and of finished products is practiced to permit size and chemical control. Fig. 5 provides a detailed listing of equipment and the sequence of operations.

IMC produces eight main products; low iron "Summit" glass-grade nepheline syenite and "Crest" and "Peak" ceramic grades. "Ridge" grade in two screen sizes is produced for colored glass. *Apex* 400 and 700 are produced for the filler, extender, and ceramic markets. The present plant capacity is in the 350,000 stpy range of product.

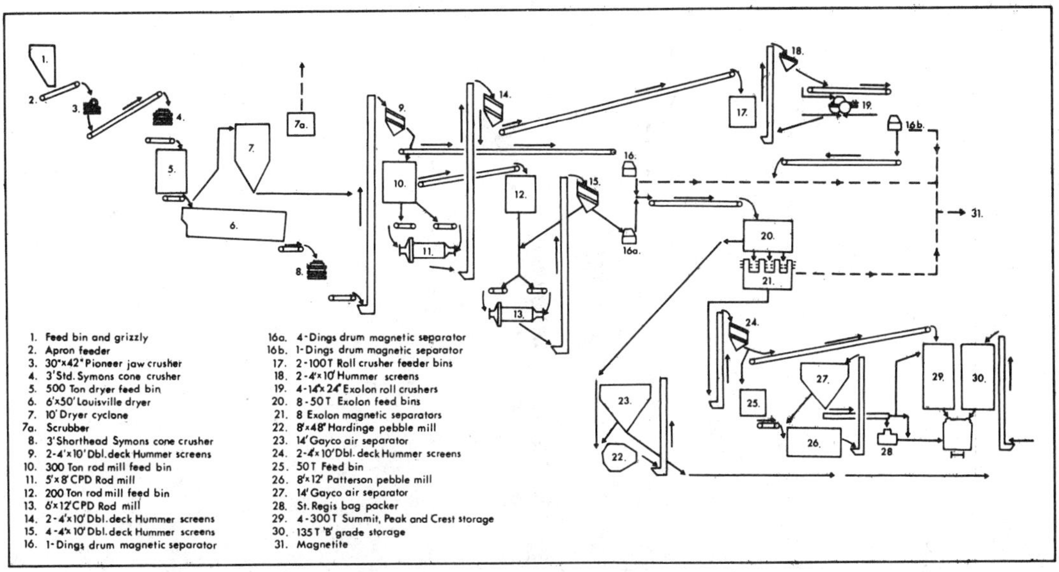

FIG. 5—*Production flow chart of International Minerals & Chemical Corp.* (*Canada*) *Ltd. plant, Blue Mountain, Ont., Canada* (*Source: Anon., 1973*).

Rail and an all-weather highway serve the IMC mine. More than 90% of shipments are in bulk.

Norway

Norsk Nefelin Division of Elkem S/A: Evaluation of the nepheline syenite deposits in northern Norway was started in 1951 by the Norwegian Geological Survey. Christiania Spigerverk undertook, in 1959, to build a 45-kt-capacity plant at Stjernöy at a reported cost of $1.1 million to exploit the deposit its geologists had defined. Commercial samples were available in early 1960 and production has increased annually since then.

In 1972 the Norsk Nefelin Division of Christiania Spigerverk became part of the enlarged corporation, Elkem Spigerverket, now Elkem S/A.

Mining—The mine and plant are situated on the southwest coast of the island of Stjernöy. The island is 250 miles north of the Arctic Circle some 25 miles northwest of Alta. The deposit lends itself to underground mining methods. Additionally, the arctic climate, heavy snowfall, prevalence of avalanches, and rugged topography all helped to dictate the mining method chosen by Norsk Nefelin. Fig. 6 shows schematically the mine layout in relation to the Nabbaren Mountain and the plant on Lillebukt Inlet. Mining is by room-and-pillar method. Ore is loaded on Kiruna 25- and 33-ton-capacity trucks by Caterpillar 966 front-end loader. Cutler-Sater filters are fitted to the trucks to reduce the amount of gases emitted within the mine. The excellent natural ventilation helps keep the mine air fresh also. Ore is trucked to the 28 x 40 in. Blake-type jaw-crusher where it is reduced to —4 in., thence transported 1600 yd by truck to the 4-ft Symons standard cone crusher for reduction to —1½ in. Ore may be stored in a 2500-t underground bin or a 250-t steel bin on the hill slope above the mill.

Milling and Processing—Ore is drawn into the plant by conveyor and passes over a 6 x 20-ft single deck screen. Oversize passes through a 5½-ft Symons shorthead cone crusher and the —⅜ in. stone passes to a 7½ x 45 ft oil-fired rotary dryer where moisture is reduced from 1.5% to 0.04%. The ore then is distributed to one of five 200-ton-capacity steel bins. Further size reduction is handled in circuits consisting of four 4 x 10-ft double-deck Hummer screens, 16 double-deck Rotex screens, 18 rolls crushers with 14 x 24 in. rolls, two 24 x 80 in. drum magnets, and one 2 x 20 ft screen at which point 90% of the nepheline syenite has been reduced to 30 mesh (Fig. 7).

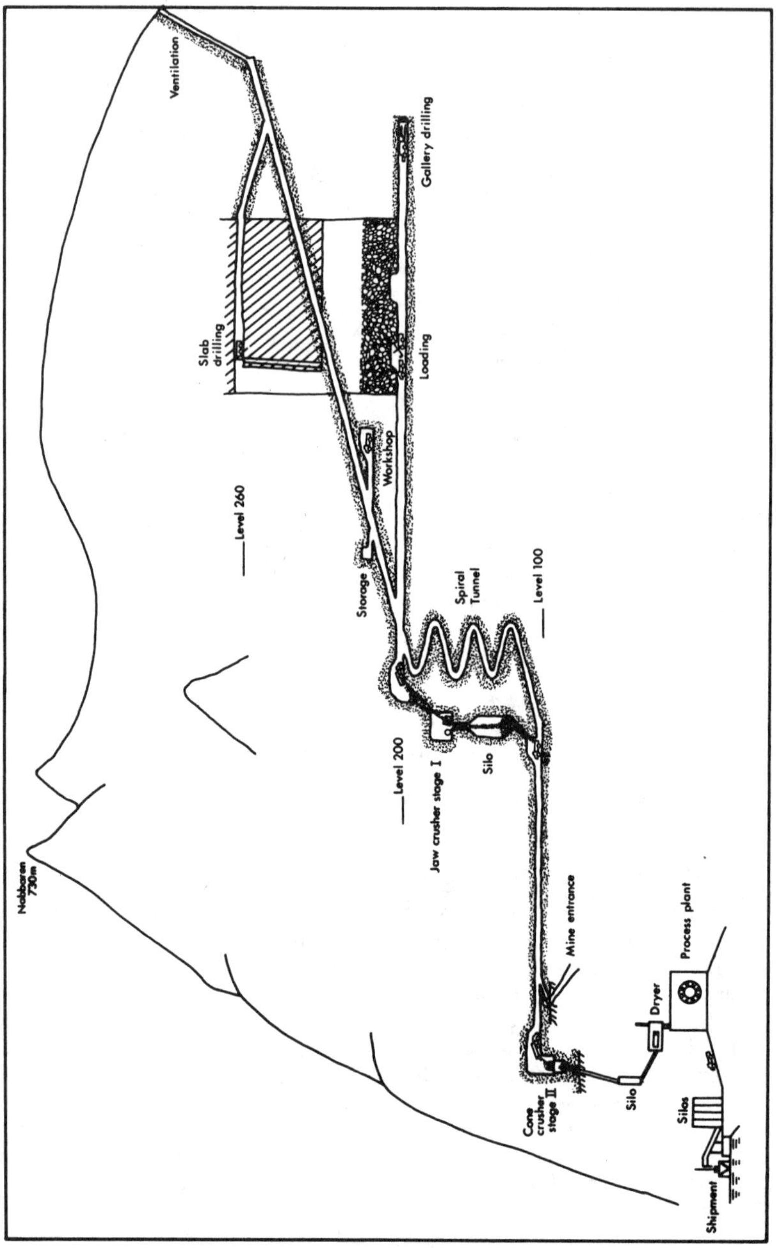

FIG. 6—Schematic diagram of mine *workings, Stjernöy, Norway.*

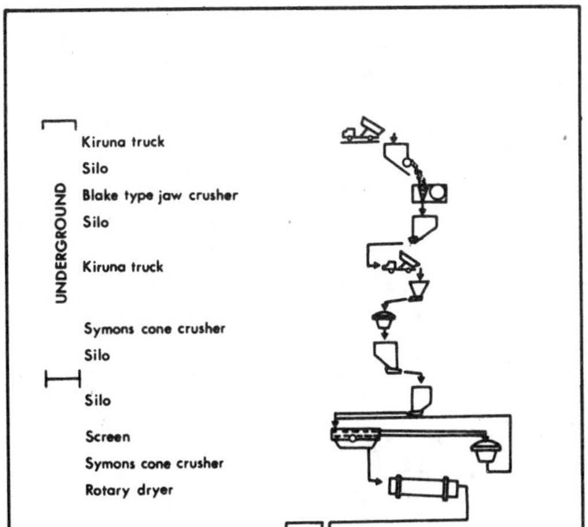

FIG. 7—*Flowsheet for production of nepheline syenite, Stjernöy, Norway.*

The sand is passed through high intensity magnetic separator circuits consisting of 3 single roll Lurgi separators followed by double roll Lurgi separators followed by 14 three-roll Exolon separators during which the iron oxide level is reduced from 1.5% to 0.1%.

The sand then passes over 12 Sweco 48-in. screening units where the remaining +30 mesh material is removed. Oversize provides mill feed to the Unidan 7 x 39 ft pebble mill. Undersize passes through an 8-ft-diam Wedag air classifier from which the oversize passes to the glass grade finished product bins and the undersize is fed to the pebble mill for grinding into ceramic grade nepheline syenite.

The original plant capacity in 1961 was 45 kt of glass-grade nepheline syenite annually but this has been increased several times to the present 200 kt of glass grade and about 25 kt of ceramic grade. The company has the capacity to increase production further, if market conditions warrant it.

Quality control consists of complete analyses of ore faces in the mine, drill cuttings, and preproduction diamond drill samples and mill samples. Size analysis and Fe_2O_3 determinations are carried out on plant samples. Regular particle size determinations of the ceramic grades of nepheline syenite are made to ensure the consistency that is so important in this material. Other chemical analyses are carried out as requested on a Perkin-Elmer atomic absorption spectrophotometer.

Between the mine and Alta, Norsk Nefelin operates a fast ferry service for its employees that makes at least four round trips each day throughout the year.

The ocean-side location and remoteness dictates the need for water transport of products to markets in 17 countries, principally those in western Europe. Glass-grade North Cape nepheline syenite is shipped in bulk loads of about 850, 4000, and 6000 tons. Norsk Nefelin has built or leased modern storage silos at several ports (Anon., 1968)—5000-ton-capacity silos at Birkenhead, near Liverpool in England; a 6000-ton silo at Terneuzen in the Netherlands; and a 5000-ton silo at Duisburg in West Germany. Stockpiles are also maintained in France and Spain.

Ceramic-grade North Cape nepheline syenite is packed in 50-kg bags and shipped in loads of 700 to 850 tons to stockpiles in western Europe. Distribution of both grades to customers from the stockpiles is by truck or covered hopper car. Large consignments may be discharged from the freighters onto barges for direct delivery to customers with plants on European waterways.

Union of Soviet Socialist Republics

As early as 1931 a process was advanced by Vladavits for the preparation of alumina from nepheline concentrates. Prior to that various workers had tested its value in glass, ceramics, and agriculture. A cement plant was built in

1938 at Volkhov near Leningrad to use nepheline slurry and in 1941 a plant to produce alumina utilizing the same material. Actual commercial production began at Volkhov in 1951 utilizing nepheline concentrate from the Kola peninsula as mill feed. A second and larger plant was completed at Pikalevo, also near Leningrad, in 1959. The third and largest alumina works built to date was operational in 1970 and is situated at Achinsk in central Siberia. Plant feed is nepheline syenite from Goryachegorsk about 90 miles southwest of Achinsk.

Mining: The processes of conversion of nepheline concentrates and nepheline syenite to alumina, portland cement, and alkali carbonates (Fig. 8) have been well described by a number of authors, among whom are Allen and Charsley (1968); Bozhenov and Kavalerova (1966); Dudkin, Kozyreva, and Pomerantseva (1964); Smirnov (1971); and Talmud (1957, 1961). Brief outlines also appeared in *Engineering and Mining Journal* (Baer, 1959) and in *Phosphorus and Potassium* (Anon., 1971). The following account was prepared largely from these sources.

Preliminary Treatment: The urtite rock tailings, from the production plants at Kirovsk and Apatity in the Kola peninsula, consist largely of nepheline (65 to 75%), apatite (6 to 10%), and titanite (1 to 3%), having a chemical analysis of 23.8% alumina, 6.8% Fe_2O_3, and 4.5% P_2O_5. This is concentrated by flotation and 44% of the mill feed is recovered as nepheline concentrate, 12.9% as apatite concentrate, and the remainder is waste. The nepheline concentrate is 90 to 93% nepheline and has an approximate composition of 29.3% alumina and 7% other compounds including iron oxide.

The dry concentrate is transported by rail to Volkhov and Pikalevo near Leningrad where it constitutes the feed for complex plants that convert it.

Processing: Raw Materials—The nepheline concentrate is about 50% −170 mesh and is mixed with appropriate portions of preground limestone which are then pulped and reground in a wet state. Suitably pure limestone is available in the area of the alumina works, and it must be less than 1% in MgO and low in SiO_2, otherwise treatment costs rise.

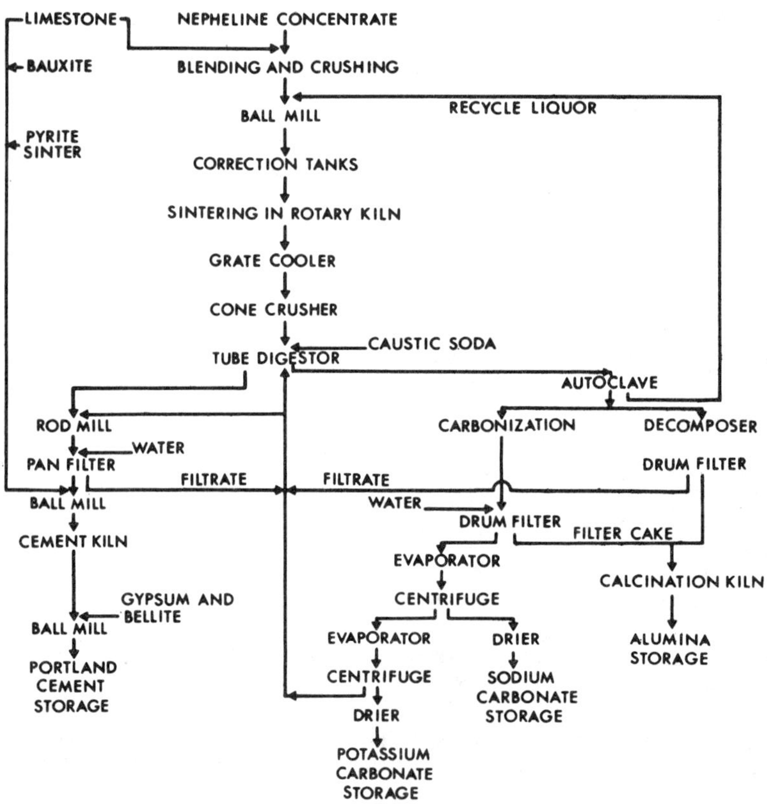

FIG. 8—*Simplified flowsheet for production of alumina, alkali carbonates, and portland cement from nepheline syenite.*

The nepheline-limestone mixture is ground to less than 5% on 170 mesh.

Production of Alumina—The pulp passes through a coal-fired sinter kiln where the intake temperature ranges from 1500°C to 1600°C. Reaction between the limestone and nepheline takes place at 1300°C and the product is a sinter cake of essentially beta dicalcium silicate and sodium and potassium aluminate together with any impurities present. The chemical reaction is:

$$4CaCO_3 + (Na,K)_2O \cdot Al_2O_3 \cdot 2SiO_2 \rightarrow (Na,K)_2O \cdot Al_2O_3 + 2Ca_2SiO_4 + 4CO_2.$$

The cake is cooled, then crushed in a cone crusher, and wet ground in a ball mill to a specific grain size and distribution being about 3% +10 mesh, 30% −200 mesh, and the balance as much as possible within the size range −10 +40 or +50 mesh. The pulp is leached in a tube digestor with a solution of caustic soda and mother liquor derived from other circuits in the plant, and passes continuously to a special vacuum filter-condenser to remove the alkali aluminates. The dicalcium silicate slurry is washed and thickened and is available for manufacture into portland cement.

Contaminating silica is removed from the aluminate in pressurized autoclaves and the solution filtered. The indicated reactions are:

$$6NaAlO_2 + 5Na_2SiO_3 + Ag \rightarrow (3\text{–}4)Na_2O \cdot 3Al_2O_3 \cdot 5SiO_2 \cdot nH_2O + (8\text{–}10)NaOH + Ag.$$

Filtrate from the desilication circuit is returned to the primary nepheline-limestone wet grinding circuit. The potassium reaction is similar. The solutions then pass to the carbonization towers where the alkalis remain as soluble carbonates in solution and the alumina is precipitated as aluminum hydroxide. The hydroxide is calcined to form alumina and the carbonates are separated by a process of selective crystallization.

The reactions are $2(Na_2K)AlO_2 + CO_2 + 3H_2O \rightarrow 2Al(OH)_3 + (Na,K)_2CO_3$.

The indicated product characteristics in 1972 were:

Alumina

Al_2O_3	99.39%
Fe_2O_3	0.001%
Na_2O	0.37%
SiO_2	0.062%
Ignition loss	0.22%
Alpha alumina content	0.35%

Soda (Approx.)

Na_2CO_3	97.1%
K_2CO_3	2.1%
K_2SO_4	0.6%
KCl	0.07%
Fe_2O_3	0.01%

Potash

K_2CO_3	98.0% (min.)
Na_2CO_3	0.9% (max)
KCl	0.15% (max)
K_2SO_4	0.73% (max)
Fe_2O_3	0.005% (max)
Residue	0.1% (max)

Production of Portland Cement—The dicalcium silicate from the alumina plant is known as "bellite." The typical chemical composition of bellite used in the Volkhov works in 1961 was SiO_2, 29.85%; Al_2O_3, 3.57%; Fe_2O_3, 2.7%; CaO, 56.58%; Na_2O, 2.21%; MgO, 1.27%; SO_3, 0.34%; ignition loss, 3.61%. Since then improved recoveries in the alumina circuits have resulted in bellite in 1972 of typical composition: SiO_2, 31.5%; Al_2O_3, 1.8%; Na_2O, 0.9%; K_2O, 0.3%; ignition loss 3.0%. This material constitutes about 40% of the cement mix. It is combined in slurry form with additional limestone, low-grade bauxite, and pyrite clinker in appropriate amounts for the manufacture of normal type-one portland cement and ground in a ball mill.

The ground mixture is fired in a rotary kiln at about 1600°C, then cooled. Coal is the customary heat source. The clinker is mixed with about 15% of dry bellite and 5% gypsum and ground to the required fineness to make finished portland cement.

Costs and Efficiency (Anon., 1969)—In the USSR, the raw materials required for, and products obtained from, the lime-sinter process are:

Raw Materials:

1) Nepheline concentrate or rock	3.9 to 4.3	t
2) Limestone for alumina	6.0 to 7.8	t
for cement	5.0 to 6.0	t
3) Fuel (min. 7000 kcal/kg)		
for alumina	1.67 to 1.70	t
for cement	1.3 to 1.6	t
4) Steam heat	4.68 to 4.12	g-cal
5) Electric power		
for alumina	1050 to 1190	kw-hr
for cement	700 to 860	kw-hr

Products:

1) Alumina	1.0	t
2) Sodium carbonate	0.62 to 0.76	t
3) Potassium carbonate	0.18 to 0.28	t
4) Portland cement	9.0 to 11.0	t

Studies indicate that the operating costs for the manufacture of alumina, portland cement, soda, and potash in an integrated plant are from 10 to 15% lower than costs of producing these in separate plants where alumina is made by the Bayer process, soda by the ammoniated method, potash from other raw materials, and cement from limestone and clay. The capital cost of the necessary plant however is somewhat higher than indicated for other methods of manufacturing these products.

Production Costs: No cost information has been published by the producers of nepheline syenite and one is left to estimate costs on the basis of the published flowsheets of each company.

Markets

Consumption—There are no complete data on world consumption of nepheline syenite. Between 1961 and 1982 world production of high quality nepheline syenite (excluding the USSR) rose from 250,000 to 911,000 st. Major expansion of markets in Europe and the United States were responsible for much of the growth in demand. In 1981, nearly 70% of the Canadian nepheline syenite production was exported to the United States.

Of the 240,000 tons of nepheline syenite produced by Norsk Nefelin, about 85% is used in the glass industry and 15% in ceramics. Only about 1000 tons per year are used internally, with the balance being exported to France, the U.K., West Germany, Belgium, the Netherlands, Italy, and Greece. In addition, some material is exported as far away as Thailand and Singapore.

The principal use of nepheline syenite is in the manufacture of glass of all types, but especially for glass containers. Glass containers typically contain 1 to 3% alumina, and feldspathic materials are the chief source of this important oxide. Statistical information on the quantities of nepheline syenite sold to individual industries in the United States is not available. Table 7 shows the consumption of feldspar in the United States and the consumption of nepheline syenite by industry in Canada.

It is apparent that nepheline syenite markets are heavily weighted in favor of glass, whereas feldspar markets are more evenly distributed between glass, whitewares, and other markets. Table 8 shows the amount of feldspar or nepheline syenite that may be used by segments of the ceramic manufacturing industry in typical product compositions.

Only small quantities of nepheline syenite are used in extender pigment and filler applications by paint, plastics, and rubber manufacturers. In 1970 in Canada, the latest year for which figures are available, 5000 st, representing 6% of nepheline syenite consumption, were sold for all nonglass and nonceramic applications including the aforementioned. No comparable figures are available for consumption of nepheline syenite in the United States for these purposes although the figure for feldspar was 7%.

Prices—The prices for nepheline syenite have been very low since 1955 when glass grade was sold for US $15 per st f.o.b. plant. The comparable price for feldspar was nearly US $14 per ton. Because of the intense competition between producers of feldspathic materials the price of glass-grade feldspar reached a low of US $8 per ton in 1961 while that of nepheline syenite dropped to US $9 per ton. By 1970 prices had recovered to US $11 for feldspar and Canadian $12 for nepheline syenite. The United States average import value of nepheline syenite was US $10 in 1966, and rose to US $11.60 in 1970, and US $11.88 in 1971. Today's

TABLE 7—Percentage Consumption of Feldspar and Nepheline Syenite By Industry

	Feldspar Consumption			Nepheline Syenite Consumption
	1968 US, %	1970 US, %	1980 US, %	1970 Canada, %
Flat glass	11	–		–
Container glass	44	–	57	–
All glass	–	55		70
Whiteware	33	36	39	12
Mineral wool	–	–	–	12
Porcelain enamel	3	2		0.3
Paint	–	–	4	2
Other	9	7		3.7
Total	100	100	100	100

feldspar costs vary between \$27.50 to \$64 per st, f.o.b. plant.

Norwegian nepheline syenite is exported mainly to the United Kingdom, West Germany, and the Netherlands. The average import value of Norwegian nepheline syenite in the United Kingdom remained constant between 1966 and 1970 at \$22.70 per st. Most is used in the manufacture of glass although some glass grade material is ground by customers for ceramic applications. Prices of nepheline syenite in 1982 varied from \$22 to \$47 (Canadian) for Canadian nepheline syenite to \$76 to \$116 c.i.f. for Norwegian products.

Competition—Nepheline syenite competes with feldspar, aplite, certain slag derivatives, and silica-feldspar sand mixtures in glass applications.

Several methods are used to determine the relative value of these feldspathic materials to container and other glass manufacturers. In Tooley (1953) and Feitler (1967) are to be found complete descriptions of various procedures. Using that described by Feitler (1967) to calculate raw material batch compositions for desired glass composition, 90 lb of feldspar, 79 lb of aplite, and 73 lb of Canadian nepheline syenite would be required in a given container glass batch. In Table 9 is shown a comparison of the chemical values of the three principal feldspathic materials used in eastern North America and the basis whereby glass container companies can calculate the relative values of each.

The total weight of the oxides is not identical with the indicated weight of raw material needed to contain unit weight of alumina due to minor constituents not reported in the chemical composition, i.e.; loss on ignition. With the data in Table 9, and knowing the delivered cost of silica, feldspathic materials, soda ash, limestone, and alumina, and by crediting each raw material with its SiO_2, $Na_2O + K_2O$, and CaO content, a glass company can calculate which of the available feldspathic materials provides the lowest net cost of alumina and is consequently a best buy. The same methods can be helpful in estimating if a nepheline syenite deposit has potential merit. The value of higher alumina and alkali in nepheline syenite is apparent in calculations such as the foregoing and indicates why nepheline syenite has successfully competed with other comparable raw materials.

Since 1970 several large United States feldspar producers have completed major plant expansions allowing them to meet growth in domestic demand close to their plants. Canadian nepheline syenite producers will find competition keen in traditional market areas.

TABLE 8—Weight of Feldspathic Material Used in Ceramic Ware

	Feldspar or Nepheline Syenite, %
Sanitary ware	25 to 35
Hotel ware	15 to 35
Chemical porcelain	15 to 30
Semi-vitreous chinaware	15 to 30
Floor and wall tiles	15 to 55

Future Considerations and Trends

It is expected that the glass and ceramic industries will continue to be the principal users of nepheline syenite. New types of better, stronger, and cheaper glass containers and ceramic ware will undoubtedly be available from manufacturers and a healthy demand for feldspathic materials should continue.

Nepheline syenite competes directly with feldspar, aplite, feldspar-silica sand mixtures, and slag as a source of alumina and alkalis for the glass and ceramic industries. Because they are the cheapest available forms of the needed compounds, nepheline syenite, feldspar, and similar materials are not threatened by mineral substitutes or synthetics. Suppliers are well situated to satisfy market needs and competition between them will remain strong.

TABLE 9—Calculation of Chemical Values

Raw Material Chemical Composition			
Oxides	Aplite	Feldspar	Nepheline Syenite
SiO_2	63.5%	67.9%	60.0%
Al_2O_3	22.4	19.2	23.7
$Na_2O + K_2O$	8.58	11.2	15.4
CaO	5.2	0.01	0.03
Fe_2O_3	0.1	0.07	0.07
Raw Material Needed for Unit Alumina Weight			
	Aplite	Feldspar	Nepheline Syenite
Weight of Raw Material	4.46	5.21	4.22
Weight of Contained Oxides			
Al_2O_3	1.00	1.00	1.00
SiO_2	2.84	3.54	2.53
$Na_2O + K_2O$	0.38	0.58	0.65
CaO	0.23	–	–
Total weight	4.45	5.12	4.18

Nepheline syenite sold as an inert filler or extender pigment competes to a limited degree with other materials having similar properties, and even dissimilar materials. Demand for all types of fillers is strong but increased sales of nepheline syenite will depend on more widespread acceptance of the advantages it offers over less inert materials.

Trends

North America: United States glass container production grew at an average rate of 8% during the period 1966 to 1971. The US Bureau of Mines (Cooper and Wells, 1970) report suggests a median growth rate for glass containers of 4% per year through to the year 2000. The same report suggests median annual growth rates for feldspar demand of from 3.6 to 5.1%. Using this same method of calculation for nepheline syenite, the demand in the year 2000 would be between 1.3 million and 2.1 million st for Canadian products.

The consumption of nepheline syenite depends to a significant degree on transportation rates to the principal consuming industries in Canada and the United States. In the 1950s and 1960s new glass manufacturing plants were built in the northeastern mid-Atlantic states and midwestern states which could be supplied equally well by most feldspathic material producers. In the 1970s there was an apparent trend toward erection of glass plants away from the northeast which favors feldspathic materials other than nepheline syenite.

The glass container industry needs a lightweight, unbreakable bottle that will compete with other containers of similar use. One move in this direction is the bulb-shaped glass container enclosed in plastic which serves to reinforce and keep the bulb upright. These can be manufactured at the rate of 600 to 800 units per min.

New machinery for filling glass containers at the rate of 2000 bottles per min, more than three times the conventional rate, is available. Nepheline syenite helps impart greater strength to such containers and will certainly be needed for these and other improved containers.

For the ceramic industry the US Bureau of Mines report (Cooper and Wells, 1970) projected a growth in demand for feldspar of 4% and 1.75% per year for use in whiteware and porcelain enamel, respectively, to the year 2000. For other uses, such as extender pigments and fillers, a rate of 4% was used also. Similar rates should apply to demand for nepheline syenite in these applications, although actual performance of nepheline syenite will depend on developments in its traditional marketing areas.

Europe: Using growth rates of 3.6 to 5.1% demand for Norwegian nepheline syenite would amount to between 245 kt and 390 kt by the year 2000. The rate of growth of Norsk Nefelin since 1961 has been much higher than this.

Norwegian nepheline syenite is readily available at moderate cost to consuming industries in Europe and other parts of the world. It competes with feldspars from other European countries, with Cornish Stone in the United Kingdom (Johnstone and Johnstone, 1961), and with North American nepheline syenite in certain of the glass and ceramic producing areas. Norwegian nepheline syenite can be shipped in almost any quantity, bagged or bulk from the ocean-side plant at Stjernöy, and stocks of bulk syenite are maintained at major European ports for truck delivery to customers.

Although the Norwegian plant has been expanded several times, expansion of other European feldspar producers and the entry of new Finnish and Norwegian feldspar producers make the markets highly competitive. If all the planned expansions are completed European producers face a serious state of overcapacity.

Substitutes: Nepheline syenite and other feldspathic materials are the cheapest sources of necessary alkalis and alumina for many glass and ceramic manufacturers and consequently are not vulnerable to competition from higher priced and purer alternatives such as alkali carbonates and alumina, in most applications.

Trends to thin-walled, stronger, and lighter bottles as well as changes in the percentage of returnable and nonreturnable bottles produced will certainly affect nepheline syenite requirements in years to come. How, is not clear. Plastic and metal containers emerged during the 1960s to capture many traditional glass markets and can be expected to continue the challenge in the 1980s. The changing character of the container industry certainly adversely affects demand for nepheline syenite.

One reason why nepheline syenite has not been used more for extender pigment and filler applications is the ready availability of excellent alternative materials such as calcium carbonate, feldspar, kaolin, silica, and talc. Increased utilization of nepheline syenite will depend wholly on increased demand for its special properties.

Transportation: Any inequality in the application of freight rate increases tends to work to the detriment of users of Canadian nepheline syenite. Rail and truck rates for international movements are generally higher than rates for comparable domestic movements and for comparable movements of feldspar within the United States. Consequently the numerous freight rate increases by United States and Canadian carriers in the 1970s and 1980s tend to hurt producers of nepheline syenite. Similar trends apply in Europe. However, because Norwegian nepheline syenite is waterborne for most of the journey to customers, the effect of increased rail and truck rates will be less on users of nepheline syenite than users of feldspar in Europe. Increases in freight rates the world over are bound to affect traditional supply patterns and cause reorientation in favor of local producers.

Government Control, Zonal Restrictions

There are no specially inhibiting government controls affecting producers of nepheline syenite. In the case of Norsk Nefelin, its efforts are accorded full government encouragement because of the expressed desire to develop industry in the Norwegian far north. Canadian producers of nepheline syenite are subject to normal business regulations. Various acts are of particular significance to IMC and Indusmin, among which are The Mining Act, The Act to Protect the Natural Environment, The Municipal Act, and the Planning Act. These acts bear on land use in Ontario and directly affect the way nepheline syenite producers operate.

Taxes, Depletion and Tariffs

Taxes and Depletion—Since 1970 Canadian federal taxes on mining companies have tended to increase and much of the encouragement granted mine developers is in the process of being withdrawn. These changes affect nepheline syenite producers and inhibit development of many mining properties. Effective Jan. 1, 1972, depletion allowances granted shareholders of mining companies were withdrawn. Depletion allowances granted operators of mines in the past were progressively reduced until 1976 when they ceased and were replaced by new allowances based on profits and eligible capital expenditures. The three-year exemption from federal income taxes of new mines terminated on Dec. 31, 1973, and purchases and sale of mining properties after Dec. 31, 1971 are treated as income transactions and taxable as such. After 1976 a 15% abatement of federal tax was allowed in respect to mining profits.

The maximum federal tax rate to which Canadian nepheline syenite producers would be subject is 48%. Mineral rights in Canadian provinces are normally owned by the province and licenses to explore for and develop minerals are granted by them for fixed periods. Nepheline syenite is subject to provincial regulations and provincial taxes. Currently there is a trend in Canada, generated by provincial authorities, encouraging companies to develop mining properties held under license or else to abandon the license. There is an equally strong trend to increased taxation of mining companies.

The Norwegian government has indirectly encouraged development in its northern regions by granting tax incentives and quick write-offs for capital investments.

Tariffs—Nepheline syenite, whether processed or unprocessed, enters the United States duty-free from all countries. The United Kingdom levies no duty on nepheline syenite from Commonwealth and European Free Trade Area countries. Nepheline syenite enters European Economic Community countries free of duty but EEC members may levy value added taxes. Exports of Canadian nepheline syenite are permitted to all countries but export permits are required for shipments to Communist countries.

Recycling

Public interest and concern in glass containers as sources of waste quickened in the latter part of the 1960s and will be of significant importance to raw material producers in the 1980s. Although glass represents only about 6.5% of municipal waste, it is highly visible and consequently is the focus of legislative reaction against the growing quantities of waste generated by society.

Throughout the United States and Canada, glass container manufacturers have established depots in major traffic areas where people may return clean glass containers for recycling. In 1970 the Glass Container Manufacturers Institute (Anon, 1972a) sponsored a pilot program for bottle reclamation at the plants of eight member companies in the Los Angeles area. Subsequently the project was expanded to nearly 100 collection centers at bottle factories in 25 states. Glass container manufacturers pay an average of 1¢ per lb for used glass containers brought to collection centers, the only stipulation being that the glass be free of metal contamination and sorted as to color. Properly prepared waste glass (cullet) can constitute up to 50% of the glass batch. In many glass plants 10 to 15% waste glass is being used, up from 2 to 3% only a few years ago.

Throughout the United States and Canada, governments are considering ways to encourage the reuse and recycling of all types of materials including plastic, metal, and glass beverage containers.

The glass recycling program is too recent for significant effects to be felt by producers of raw materials but manifestly the recycling of glass into new containers, fiber glass, and other products will affect the markets for nepheline syenite and other feldspathic materials.

Ecology

The direct impact of current international interest in ecology on nepheline syenite producers is minimal. All three operating nepheline syenite plants prepare products by "dry" methods with attendant need for large and efficient dust collection systems to keep plants clean and to prevent atmospheric emissions. Between 1966 and 1972 Norsk Nefelin spent over $1 million at Stjernöy on improved dust control equipment and between 1966 and 1972 Indusmin spent $750,000 at Nephton to improve dust control and tailings disposal systems.

All industry, including mining, can expect to spend significant sums in the 1980s to make their activities correspond to the realities of environmental protection. Reports by the American Chemical Society (Reese, 1969, 1971) attempt to catalog the environmental conditions in North America and indicate the magnitude of the task ahead. *Chemical Engineering* (Anon., 1972) in 1972 summarized international activity in the field of pollution control.

More and more public and governmental interest and intervention in the affairs of nepheline syenite producers can be expected. While the results will mean cleaner, better, and more efficient operations, the cost will be high and will to some degree tend to affect the competitive position of producers in world markets.

Nepheline Syenite in Other Countries

There is little likelihood that nepheline syenite from the Soviet Union will be a competitive factor in world glass and ceramic markets in the near future. The emphasis has been on the internal utilization of nepheline syenite for the manufacture of alumina, alkali carbonates, and portland cement, and only small quantities have been used for the production of drainage tile, floor tile, and green glass. Indications are that Soviet deposits are not suited for economic production of products comparable to those from Canadian and Norwegian sources.

Soviet technology for the chemical conversion of nepheline-bearing rocks may well be of interest to other nations in the near future. While similar technology has been developed in the United States (Lundquist and Singleton, 1962; Johnson and Peters, 1968; Archibald and Nicholson, 1948) and Canada (Archibald, 1942), the only commercial operations based on nepheline and feldspar minerals are located in the Soviet Union. The trend in Soviet plants is to larger and larger units to take advantage of any economies that result from such scale-up. The annual production capacity of nepheline-based alumina plants has grown from the original Volkhov plant (estimated 50 kt/a) to the Pikalevo plant (estimated 200 kt) to the Achinsk plant (800 kt) to the currently planned 1.5 Mt plant near the Sayano-Shushenskaya dam. Present methods of production of alumina and cement in non-Soviet countries are low cost and nepheline syenite will not be used for such production until comparable costs are attained. Many such deposits would be satisfactory for applications requiring good fluxing character where iron impurities are not detrimental such as colored glass, drainage and sewer pipe, brick, floor tile, and backup wall tile.

Elsewhere in the world no nepheline syenites have been found that lend themselves to production of low-iron products. Some nations in Africa, Europe, and South America may satisfy local industrial needs for feldspathic raw materials by mining domestic nepheline-bearing deposits.

Bibliography and References

Anon., 1967, "Mining and Milling Nepheline Syenite," *Canadian Clay and Ceramics*, Vol. 40, No. 5, p. 12.

Anon., 1968, "Nepheline Syenite 30 Years Rapid Growth," *Industrial Minerals*, Metal Bulletin Ltd. London, England, No. 7, Apr., pp. 9–19.

Anon., 1969, *Alumina from Nepheline*, V. O. Licensintorg, Moscow, 8 pp.

Anon., 1971, "Potash: A By-Product of Soviet Alumina from Nepheline," *Phosphorus and Potassium*, May/June, pp. 40–41.

Anon., 1972, "Environmental Engineering," *Chemical Engineering,* Vol. 79, No. 10, May 8, p. 259.

Anon., 1972a, *Glass Containers,* 1972 ed., Glass Container Manufacturers Institute, Inc., New York, 20 pp.

Anon., 1973, "Ceramic Manufacturing Sales Estimates," *Ceramic Age,* Jan. p. 12.

Anon., 1961–1971, "Nepheline Syenite," *Canadian Minerals Yearbook,* Mineral Resources Div., Dept. of Energy, Mines and Resources, annual publication.

Anon., 1961–1971a, "Feldspar, Nepheline Syenite and Aplite," *Minerals Yearbook,* US Bureau of Mines, annual publication.

Allen, J.B., and Charsley, T.J., 1968, *Nepheline-Syenite and Phonolite,* Institute of Geological Sciences, Minerals Resources Div., London, England, 169 pp.

Archibald, F.R., and Nicholson, C.M., 1948, "Alumina from Clay by the Lime-Sinter Method II," AIME, Tech. Pub. 2390, 25 pp.

Archibald, F.R., 1942, "Report of Investigation on Production of Alumina and Alkalis from Nepheline Syenite," unpublished report to Bureau of Mines, Ottawa, 40 pp.

Baarson, R.E., et al., 1962, "Plant Practice in Non-Metallic Mineral Flotation," Chap. 17, *Froth Flotation,* D.W. Fuerstenau, ed., AIME, New York, pp. 427–453.

Baer, F.H., 1959, "Soviets Push Ambitious Aluminum Plans," *Engineering and Mining Journal,* Vol. 160, No. 5, May, 1959, pp. 102–105.

Barth, T.F.W., 1963, "The Composition of Nepheline," *Schweiz, Min. Petr. Mitt.* 43/1, pp. 153–164.

Bozhenov, P.I., and Kavalerova, V.I., 1966, *Nepheline Slurries,* Editor of Literature for the Building Industry, Moscow-Leningrad, (in Russian), 243 pp.

Castle, J.E., and Gillson, J.L., 1960, "Feldspar, Nepheline Syenite and Aplite," *Industrial Minerals and Rocks,* J.L. Gillson, ed., 3rd ed., AIME, New York, pp. 339–362.

Cooper, J.D., and Wells, J.R., 1970, "Feldspar," *Mineral Facts and Problems,* Bulletin 650, US Bureau of Mines, pp. 977–988.

Deer, W.A., et al., 1963, *Rock-Forming Minerals,* Vol. 4, Longmans Green and Co. Ltd., London, England, pp. 231–269.

Dudkin, O.B., Kozyreva, L.V., and Pomerantseva, N.G., 1964, *Mineralogy of Apatite Occurrences of the Khibinsk Tundra,* Academy of Sciences, Moscow-Leningrad.

Eitel, W., 1965, *Silicate Science,* Vol. 3, Academic Press, New York, pp. 236–264.

Feitler, S.A., 1967, "Feldspar Resources and Marketing in Eastern United States," Information Circular 8310, US Bureau of Mines, 41 pp.

Gerasimovskii, V.I., 1956, "Geochemistry and Mineralogy of Nepheline Syenite Intrusions," *Geochemistry,* No. 5, pp. 494–510.

Grimshaw, R.W., 1971, *The Chemistry and Physics of Clays and Allied Ceramic Materials,* 4th ed., John Wiley and Sons Inc., New York, pp. 372–442.

Guillet, G.R., 1962, "A Chemical and Inclusion Study of Nepheline Syenite for Petrogenetic Criteria," unpublished M.A. Thesis, University of Toronto, Canada, 75 pp.

Harben, P.W., 1977, "Raw Materials for the Glass Industry," *Industrial Minerals,* Metal Bulletin, Ltd., London, pp. 6–14.

Heier, K.S., 1961, "Layered Gabbro, Hornblende, Carbonatite and Nepheline Syenite on Stjernöy, North Norway," *Norsk Geol. Tidsskr 41,* pp. 109–155.

Heier, K.S., 1964, "Geochemistry of the Nepheline Syenite on Stjernöy, North Norway," *Norsk Geol. Tidsskr 45,* pp. 205–215.

Heier, K.S., 1965, "A Geochemical Comparison of the Blue Mountain (Ontario, Canada) and Stjernöy (Finnmark, North Norway) Nepheline Syenites," *Norsk Geol. Tidsskr 45,* pp. 41–52.

Heier, K.S., 1966, "Some Crystallo-Chemical Relations of Nepheline and Feldspars on Stjernöy, North Norway," *Journal of Petrology,* Vol. 7, Pt. 1, pp. 95–113.

Hewitt, D.F., 1960, "Nepheline Syenite Deposits of Southern Ontario," Ontario Dept. of Mines, Vol. 69, Pt. 8, 194 pp.

Huvos, J.B., 1980, "The Mineral Industry of Norway," Preprint, US Bureau of Mines, 14 pp.

Johnson, P.W., and Peters, F.A., 1968, "Methods for Producing Alumina from Anorthosite, and Evaluation of a Lime-Soda Sinter Process," Report of Investigation 7068, US Bureau of Mines, 42 pp.

Johnstone, S.J., and Johnstone, M.G., 1961, *Minerals for the Chemical and Allied Industries,* Chapman and Hall, London, England, pp. 188–194.

Lundquist, R.V., and Singleton, E.L., 1962, "Some Characteristics of Iron in the Lime-Soda Sinter Process for Recovering Alumina from Anorthosite," Report of Investigation 6090, US Bureau of Mines, 13 pp.

Merrett, A.J., and Sykes, A., 1963, *The Finance and Analysis of Capital Projects,* Longmans Green and Co. Ltd. London, England, 539 pp.

Minnes, D.G., 1975, "Nepheline Syenite," *Industrial Minerals and Rocks,* S.J. Lefond, ed., 4th ed., AIME, New York, pp. 861–894.

Moorhouse, W.W., 1959, *The Study of Rocks in Thin Section,* Harper and Brothers, New York, pp. 302–312.

Payne, J.G., 1966, "Geology and Geochemistry of the Blue Mountain Nepheline Syenite Body," Ph.D. Thesis, McMaster University, Canada, 183 pp.

Polkanov, A.A., ed., 1937, *The Northern Excursion, Kola Peninsula,* 17th International Geological Congress, Moscow, 119 pp.

Reese, K.M., ed., 1969, "Cleaning Our Environment, the Chemical Basis for Action," report by the Sub-Committee on Environmental Improvements, Committee on Chemistry and Public Affairs, American Chemical Society, Washington, DC, 249 pp.; 1971 Supplement, 20 pp.

Robie, E. H., ed., 1959, *Economics of the Mineral Industries,* AIME, New York, 755 pp.

Singer, F., and Singer, S.S., 1963, *Industrial Ceramics,* Chemical Publishing Co. Inc., New York, 1455 pp.

Smirnov, M.N., 1971, *Physical-Chemical Fundamentals of Alumina Production from Nepheline,* Vol. 3, Proceedings, 2nd International Symposium of ICSOBA, pp. 337–345.

Smothers, W.J., et al., 1952, "Ceramic Evaluation of Arkansas Nepheline Syenite," Arkansas Resources and Development Commission, Research Series No. 24, pp. 21.

Taggart, A.F., 1956, "Electrical Concentration," Sec. 13, *Handbook of Mineral Dressing,* John Wiley and Sons Inc., New York, pp. 13-01—13-47.

Talmud, I.L., 1957, "Aluminum from Nepheline," *Science and Life,* No. 4, Moscow, April (in Russian).

Talmud, I.L., 1961, "Complex Treatment of Nepheline Ore," *Chemical Industry,* No. 4, April, (in Russian), pp. 226–232.

Taylor, G.H., and Wilson, R.C., 1962, "Particle Size Control of Nepheline Syenite for Whitewares," *Bulletin,* American Ceramic Society, Vol. 41, No. 1, pp. 12–13.

Tooley, F.V., 1953, *Handbook of Glass Manufacture,* Vol. 1, Ogden Publishing Co., New York, pp. 75–79.

Turner, F.J., and Verhoogen, J., 1951, *Igneous and Metamorphic Petrology,* McGraw Hill, New York, pp. 338–342.

Vlasov, K. A., et al., 1966, *The Lovozero Alkali Massif,* Oliver and Boyd, Edinburgh and London, 627 pp.

Watson, I., 1981, "Feldspathic Fluxes—The Rivalry Reviewed," *Industrial Minerals,* No. 163, Apr., pp. 21–45.

Watson, I., 1981a, "The Industrial Minerals of Scandinavia—Norway," *Industrial Minerals,* No. 171, Dec., pp. 34–45.

Nitrogen Compounds

E.A. HARRE and R.D. YOUNG *

Nitrogen exists in two broad categories commonly designated as elemental nitrogen and fixed nitrogen. Elemental nitrogen is found in nature as a diatomic molecule and constitutes about 78%, by volume, of the earth's atmosphere. Fixed nitrogen is a common term used for nitrogen that is chemically bound to other elements. Although elemental nitrogen is very abundant, nonleguminous plants cannot utilize nitrogen unless it is in the chemically combined form. The content of this chapter is confined to fixed nitrogen since the uses and properties of elemental nitrogen and the fixed nitrogen compounds are different.

Certain bacteria associated with legumes are able to fix atmospheric nitrogen, but not to the extent required by succeeding nonleguminous crops. The need to use fertilizers containing fixed nitrogen compounds to grow food crops has been recognized for hundreds of years, and from antiquity farmers have used organic wastes such as animal manures, vegetable wastes, sewage sludge, fish scraps, etc., containing these compounds for fertilizers. These organic materials have a low nitrogen content, however, and cannot support large-scale food production.

Known commercial-size deposits of natural nitrates containing much higher nitrogen content exist only in Chile, and the possibility of the discovery of other significant deposits is remote. The Chilean deposits were the major source of nitrogenous fertilizers until the Haber-Bosch process for the production of synthetic ammonia was developed. In the Haber-Bosch process, nitrogen reacts with hydrogen at high temperatures and pressures and in the presence of a catalyst to form ammonia. Modern plants use air, water, and natural gas or other hydrogen sources to produce the ammonia that is now the key building block of the fixed nitrogen industry.

Several major problems that developed in the 1970s continue with the nitrogen industry into the 1980s. These are (1) the continuing sharp escalation in the cost of feedstocks for ammonia production as well as all energy sources, and (2) the very steep increase in the investment cost of production facilities.

More than 75% of the fixed nitrogen produced domestically is consumed as fertilizers. The types of materials used as fertilizers have changed significantly during the past years. There has been a trend from the use of low-analysis fertilizers to types which contain high percentages of plant nutrients. The average plant nutrient content ($N + P_2O_5 + K_2O$) of fertilizers used in the United States has increased from 18% in 1930 to 32% in 1960 and today approaches an average of 44%.

In the case of nitrogen fertilizers, the low analysis natural nitrates (16% N) gave way to ammonium sulfate (20% N) after about 1920. After World War II, ammonium nitrate (33.5% N) previously used in munitions, became established as the leading nitrogen fertilizer. Urea use as a fertilizer (46% N) began in the 1960s, and in the past 10 years has become the world's leading solid nitrogen fertilizer.

Nitrogen Materials

Anhydrous Ammonia—Anhydrous ammonia is a colorless gas at ambient temperatures and pressures and has a sharp pungent odor. Anhydrous ammonia is stored and transported as a liquid in high-pressure cylinders and tanks at ambient temperature, and in low-pressure tanks, barges, and ships at −28°F. Anhydrous ammonia is applied directly at 4 to 7-in. depths in the soil as a high-pressure liquid by special applicators, after which it expands to a gas that dissolves in the soil moisture or is absorbed on soil colloids. In other systems ammonia is metered into irrigation water and is thereby introduced into the soil.

* Office of Agricultural and Chemical Development, Tennessee Valley Authority, Muscle Shoals, AL.

Aqua Ammonia—Aqua ammonia is obtained by reacting anhydrous ammonia with water. Commercial aqua ammonia solutions usually contain between 22 and 25% nitrogen and are strongly alkaline. Aqua ammonia has a significant vapor pressure and must be contained in pressure vessels to prevent loss of ammonia. Equipment for the direct application of aqua ammonia is similar to that used for the application of anhydrous ammonia.

Ammonium Nitrate—Fertilizer ammonium nitrate is sold as a solid, normally in small round pellets called prills or granules, or dissolved in water as a component of nitrogen solutions. Solid ammonium nitrate is hygroscopic and it is usually treated with a conditioning agent, such as kaolin clay, to prevent caking. In addition to its use as fertilizer, large amounts of ammonium nitrate are used as blasting agents when sensitized with fuel oil.

Ammonium Nitrate-Limestone—This product is a mixture of ammonium nitrate and limestone, and is marketed as a solid in prilled or pellet form. Its soil reaction is alkaline or neutral. Used quite widely in Europe, Asia, and Latin America, where it is usually called calcium-ammonium nitrate (CAN), it is of little importance in the United States.

Urea—Prilled urea is marketed as products containing 45 and 46% nitrogen. Urea is popular as a direct application fertilizer because of its high nitrogen content, and it is also used extensively in nitrogen solutions. Urea is used as a reactant in many industrial chemical applications, including the production of plastics and synthetic foams.

The rapid growth of urea to its place as the world's leading nitrogen fertilizer resulted to a large extent from its being a "natural" for production from ammonia and carbon dioxide, the coproduct from ammonia synthesis.

Nitrogen Solutions—The nutrient content of nitrogen solutions can vary from 19 to 59%. Nitrogen fertilizer solutions are prepared, for the most part, from various mixtures of ammonium nitrate, urea, ammonia, and water. There are, of course, an infinite number of possible combinations. The vapor pressure ranges from nearly zero for the low-analysis solutions to high vapor pressures for the high-analysis materials. Vapor pressure increases with increasing ammonia content. The nonpressure mixtures are easy to handle, store, ship, and apply. The grade of the very popular urea-ammonium nitrate solutions (UAN) ranges from 28 to 32% N, depending on ambient temperature.

Ammonium Sulfate—The nitrogen content of ammonium sulfate averages about 20.5%. Ammonium sulfate also contains a large percentage of sulfur which is of value for sulfur-deficient soils. Ammonium sulfate is marketed as a white crystalline salt and is often an ingredient of mixed fertilizers. It is a byproduct of coke oven and steel industries, the fibers industry, and is also manufactured synthetically. Its importance as a fertilizer in the United States and the world continues to decline in comparison with higher analysis nitrogen materials.

Calcium Nitrate—Calcium nitrate (15.5% N) is extremely hygroscopic and, if exposed to sufficiently humid air, may liquefy. Its main source is as a byproduct from some nitric phosphate fertilizer plants located in Europe. A paraffin coating is used to provide moisture protection.

Sodium Nitrate—Only about 550,000 tons of this once important fertilizer was produced worldwide in 1980. Sodium nitrate contains 16% nitrogen, is produced synthetically in small amounts, but is obtained primarily from natural nitrate deposits in Chile. Sodium nitrate has a residual alkaline effect on the soil. It is also used as a food preservative.

Calcium Cyanamide—Calcium cyanamide for fertilizer use contains about 21% nitrogen, which means that it has a low purity, since pure calcium cyanamide contains about 35% nitrogen. Calcium cyanamide has a high liming effect and is equivalent to about 2 lb of calcium carbonate per pound of nitrogen applied.

Nonfertilizer Uses of Nitrogen

Other significant end uses for fixed nitrogen compounds are for plastics, textiles, and elastomers, some of which are described in the following (Billmeyer, 1971).

Nitrile Rubbers—Nitrile rubbers are polymers of butadiene and acrylonitrile and are used in applications which require resistance to oil, such as in gasoline hoses and fuel tanks. Also, nitrile rubbers are used in adhesives.

Acrylic Fibers—Acrylic fibers are polymers containing 85% or more of acrylonitrile along with copolymers such as vinyl acetate, acrylic esters, and vinyl pyrrolidone. Acrylic fibers exhibit high strength, abrasion resistance, resilience, and have good moisture and chemical resistance.

Nylon—Nylon is a generic term used for synthetic polyamides. The nylons are perhaps the most familiar of the plastics and synthetics.

Nylon 6/6 is the polymer of hexamethylenediamine and adipic acid. About half of the nylon fiber produced is consumed as tire cord, but other end uses include rope, thread, cloth, and garments. Nylon plastics are used as metal substitutes in bearings, gears, rollers, slides, etc.

Other nylon polymers are made from dimerized vegetable oil acids and diamines such as ethylene diamine. These materials are used in paints, for paper coatings, and as curing agents for epoxy resins.

Polyurethanes—Polyurethanes are typically formed through the reaction of a diisocyanate and a glycol. The urethane polymers are used to produce foams, fibers, elastomers, and coatings. Polyurethane elastomers are noted for their abrasion resistance and hardness combined with a resistance to oils and solvents. The materials are used in tire treads, heel lifts, and small industrial wheels.

Urethane foams can be made in either flexible or rigid foams. Flexible urethane foams are used for furniture and automobile cushions, while rigid foams are used for thermal insulations and for imparting buoyancy to boats.

Amino Resins—The two important classes of amino thermosetting resins are the condensation products of urea and of melamine with formaldehyde. Melamine is a trimer of cyanamide. Almost all urea molding compounds are cellulose-filled while the melamine molding resins are filled with cellulose, asbestos, glass, or cotton fabric. The resins are used, among other things, for appliance housings, hardware, and dinnerware. Also, amino resins are used widely for adhesives such as the adhesives used in the lamination of plywood and the fabrication of furniture. Melamine resins are often used for the production of laminates for counter, cabinet, and table tops. Urea-based enamels are used for refrigerator and kitchen appliances, and melamine enamels are used in automobile finishes.

Ammonium Nitrate-Fuel Oil—In recent years ammonium nitrate sensitized with fuel oil has dominated the industrial explosives market. The primary reason for this is the relatively low cost of the ammonium nitrate explosives.

Black Blasting Powder—Black blasting powder is perhaps the oldest of the blasting agents and consists of a mixture of sodium nitrate, sulfur, and charcoal.

Dynamite—Dynamite is a combination of nitroglycerine absorbed in kieselguhr and ammonium nitrate in various proportions.

Other Nitrogen-Containing Explosives—Other industrial or military explosives include nitrocellulose, nitroglycerine, nitrostarch, trinitrotoluene (TNT), pentraerythritol tetranitrane, cyclonite, trinitrobenzene, ammonium picrate, and picric acid. Nitrogen-containing detonators include lead azide and mercury fulminate.

Animal Feed—Urea is used as a supplement in the feed of ruminant animals, and in 1980 about 250,000 tons of urea was produced for use in feed compounds.

There are also other uses of fixed nitrogen compounds. Weak solutions of ammonia in water are used in industrial and home cleaning agents, usually in combination with detergents. Ammonia is used as a curing agent in leather-making, and ammonia and certain amines are used as depilatories in the leather industry. Many pharmaceuticals are chemicals containing fixed nitrogen. Ammonia is used in commercial and industrial refrigeration units. Nitrogen tetroxide and hydrazine are sometimes used as rocket propellants. Nitric acid has a wide range of industrial applications, but is usually an intermediate chemical for some other end use.

Synthetic Nitrogen Fixation Processes

Production of Anhydrous Ammonia

The fixed nitrogen industry of today is largely based upon the production of synthetic anhydrous ammonia. Although the process for producing ammonia has been known for 70 years, the technology has continued to improve through the years, with major changes occurring in the technology since the early sixties. One of the most significant changes was the development of high-pressure centrifugal compressors that paved the way for single-train plants of much higher production capacity.

Engineering firms completely redesigned the ammonia plant to utilize the new compressors and to achieve efficient utilization of plant energy. These changes resulted in a capacity of 1000 tpd becoming the standard and a few plants of 1500 tpd were built by 1980.

Two chemical elements, nitrogen and hydrogen, are combined to synthesize ammonia (NH_3). Nitrogen is readily available from the air but providing hydrogen is more difficult and expensive. About 94% of the ammonia plants in the United States use natural gas to obtain the process hydrogen, while in other parts of the world, naphtha has been used in a number of plants instead of natural gas. Some ammonia plants have used coke-oven gas, a wide range of hydrocarbons, or pulverized coal to produce hydrogen. By 1980 the interest in use of coal had increased because of increased cost and uncertainty of resources of natural gas in many areas. A demonstration plant (135 tpd NH_3) was in test operation in the United States in 1980 (Anon., 1980c).

In a typical new ammonia plant, a feed gas such as methane is compressed, if necessary, to 550–600 psi and desulfurized before it enters the primary reformer. Steam is added and the mixture of steam and hydrocarbon is passed through a series of tubes containing a nickel catalyst. In the primary reformer tubes, which are heated externally, the hydrocarbon and steam react to form hydrogen, carbon monoxide, and carbon dioxide. The gas exit temperature from the primary reformer ranges from 1400° to 1600°F.

Next, the gas mixture enters a secondary reformer, which also contains a nickel catalyst. Compressed air enters the secondary reformer and the oxygen in the air is completely consumed in exothermic reactions with hydrogen, carbon monoxide, and residual methane from the primary reformer. Exit gas temperature in the secondary reformer is from 1700° to 1900°F. The temperature of the outlet stream from the secondary reformer is reduced in a tubular waste heat boiler in which high pressure steam is generated for use in the plant.

From the secondary reformer, the process stream enters a two-stage shift converter where most of the carbon monoxide is converted to carbon dioxide. Steam and/or water are added to a catalyst bed in the shift converter to effect the conversion of carbon monoxide to carbon dioxide. In the first stage, shift conversion is carried out over a chromium-promoted iron catalyst at temperatures of 600° to 800°F. The second-stage shift conversion is carried out over a copper-zinc-alumina catalyst at temperatures of 400° to 500°F.

The next step in the process is the removal of carbon dioxide from the gas stream. A number of processes have been used, all based on the scrubbing of the gas with various solvents. One of the popular processes uses ethanolamines which have a high solubility for carbon dioxide. The process is relatively simple, consisting of countercurrent extraction in the absorber and subsequent regeneration of the ethanolamines in a reactivator by steam stripping and heating.

Before the synthesis gas is sent to the ammonia converter, the carbon dioxide and carbon monoxide content must be reduced to very low levels. In the widely used methanation process, carbon dioxide and carbon monoxide are reacted with hydrogen over a nickel catalyst to form methane and water.

After the previous steps, the gas mixture consists of the proper ratio of hydrogen and nitrogen and is ready for the ammonia-synthesis process. The synthesis gas must be compressed for the ammonia reaction, and the newer plants use centrifugal compressors. The more economical centrifugal compressors can be driven by electric motors, gas turbines, or steam turbines. Pressure of 600 psig is used in the modern 1000- to 1500-tpd plants.

In the new plants of 600 tpd and larger, all of the compressors are centrifugal including both the process air and ammonia refrigerant compressors. Pressures in the ammonia converter have historically covered a wide range from 100 to 1000 atm. The ammonia synthesis is generally carried out at a temperature of about 950°F over an iron oxide catalyst promoted by aluminum oxide, potassium oxide, calcium oxide, and/or magnesium oxide.

The size of the ammonia converters has increased dramatically in recent years and 1000 tpd single-train units are now common with a few 1500 tpd plants now onstream. At the other extreme, small skid-mounted converters with a capacity of 30 tpd are also available.

The reaction of hydrogen and nitrogen in the presence of a catalyst to form ammonia is highly exothermic and means must be provided in the converter for dissipating the excess heat generated in the system. Starting in the mid-1970s, the pressure toward, and incentives for, increased energy efficiency have resulted in design innovations that will be beneficial in conservation of energy. These generally consist of improved heat transfer for steam generation and a series of small improvements in the efficiency of various components in the system.

Liquid ammonia is withdrawn from a separator at about 20°F. The ammonia stream is passed to a flash drum where the pressure is reduced to about 14 atm. In the flash drum, dissolved inerts are flashed and vented. Product ammonia from the flash drum is passed into a weighing tank and then into storage or transportation lines or vessels for shipment (McVickar, 1966; Noyes, 1967; Sittig, 1965; Slack and James, 1979; Anon., 1979). Fig 1 is a diagram of the Kellogg ammonia process.

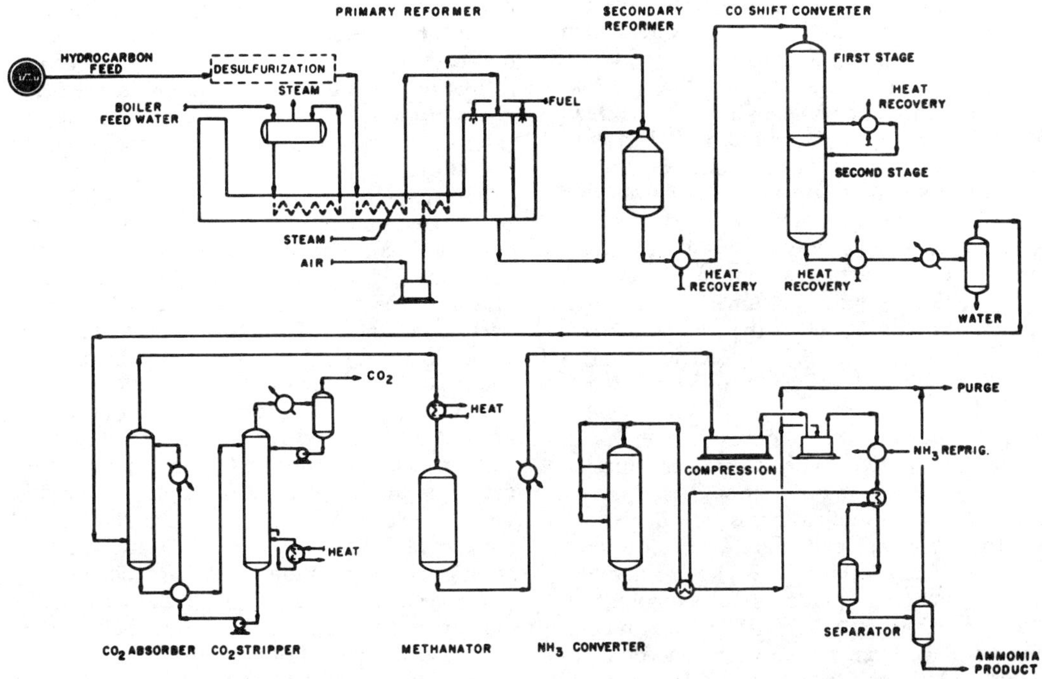

FIG. 1—*The Kellogg ammonia process* (*courtesy of the M.W. Kellogg Co.*).

Production of Nitric Acid

Ammonia and air are the raw materials in all modern plants for the manufacture of nitric acid. Gaseous ammonia is mixed with air and converted to nitric oxide in a high-temperature reaction which is usually catalyzed by a noble metal catalyst. The conversion reaction is rapid and essentially complete with about 96% conversion of ammonia to nitric oxide.

Next, the nitric oxide is further oxidized with air to form nitrogen dioxide which is then dissolved in water in a specially designed absorption tower to produce nitric acid of 52 to 64% concentration. There is an increasing demand for concentrated, about 98%, nitric acid. Early methods to produce concentrated acid used dehydrating agents such as sulfuric acid or magnesium. Some newer methods produce high-concentration nitric acid directly by oxidation of ammonia with oxygen or air. In the new methods used to produce high-concentration acid, the option of producing dilute acid also can be included.

Although the basic chemistry of the nitric acid processes is the same, there are several different processes which are used and there is a constant effort within the industry to lower capital and production cost and to improve energy efficiency. Nearly all nitric acid plants in the US are of the high-pressure (7 atm) type for conversion and absorption. In Europe and some other areas most nitric acid plants are of the low-pressure type or operate at low pressure for oxidation and high or medium pressure for absorption. The high-pressure (7 atm) design saves considerably on investment because of the smaller equipment. Conversion is more efficient in a larger catalyst bed at atmospheric pressure.

One widely used process is the Chemical Construction (Chemico) process. In the past Chemico has designed and installed low and medium-pressure nitric acid plants, but the more recent plants operate at what is called high pressure, 7 atm. Air is compressed in the oxidation section and about 80% of the hot compressed air is filtered and mixed with superheated and filtered ammonia vapor. The ammonia and air mixture flows to the ammonia burner where it passes through a series of platinum-rhodium catalyst gauzes which convert the ammonia to nitric oxide. Hot gases leave the burner at 1650°F and pass through a number of heat exchangers before entering the absorption tower. Nitrogen dioxide, which is formed on cooling, is absorbed in the water and weak acid of the absorption tower. Pure steam condensate ordinarily is used as feed water. The nitric acid produced is then contacted with compressed air to remove dissolved oxides before going to storage. Tail gas leaving the tower is under pressure and can be used to provide most of the energy for the air compressor and to generate additional steam.

The major power consumer in a high-pressure nitric acid plant is the air compressor, which can be operated by an electric motor, a steam turbine, or by a gas turbine. If a gas turbine is used and the tail gas utilized, there is a saving in power but an increase in investment. There is also a reduction in steam consumption that can be used in other processes. Several options in energy utilization and recovery systems for nitric acid plants are available.

The nitric acid industry was hit hard in meeting environmental standards for emissions of oxides of nitrogen (NO_x). After experimentation and limited experience with catalytic combustion and ion exchange, the predominant effective control by the late 1970s was "extended absorption," essentially a smaller second stage of absorption.

The relative importance of nitric acid worldwide as an intermediate for production of ammonium nitrate fertilizer continues to decrease as urea dominates in new facilities. It is used in substantial amounts in the production of ammonium nitrate for use in blasting agents, and for explosives and a variety of industrial chemicals (Anon., 1979).

Production of Urea

Urea is synthesized when ammonia and carbon dioxide combine in an exothermic reaction to form ammonium carbamate and in a subsequent reaction, which is endothermic, the ammonium carbamate is dehydrated to form urea and water. The urea process is credited as being the first industrial application of organic synthesis from inorganic chemicals.

In a typical plant the reaction takes place at about 4000 to 5000 psi and 400°F. Reactor effluent containing excess ammonia and a portion of the unconverted carbamate is reduced to a pressure of about 250 psi and flashed in a primary decomposer. The residual liquid is then heated to decompose most of the remaining carbamate to ammonia and carbon dioxide for recycling to the process. Gases from the flash stage and the carbamate decomposition are combined and sent to a primary absorber for recovery of the reactants. To minimize the formation of biuret, an undesirable urea decomposition product, the residence time at elevated temperature in the primary decomposer is held to a minimum.

The aqueous urea solution is further depressurized and heated in a secondary decomposer at atmospheric pressure to achieve the optimum recovery of ammonia and carbon dioxide. The vacuum flash reduces the temperature of the urea solution to further inhibit biuret formation. The off-gas from the primary decomposer is contacted in countercurrent flow with an aqueous ammonia solution in the primary absorber to recover all of the carbon dioxide and a portion of the ammonia for recycle to the reactor. After all of the carbon dioxide is removed, the bulk of the ammonia is condensed with cooling tower water without additional compression or refrigeration.

Inert gases introduced with the carbon dioxide feed stream leave the top of the absorber and pass through the wash tower where they are scrubbed with the process water used in the primary absorber for final recovery of a small amount of ammonia. The overhead vapors from the secondary decomposer and flash tower are sent to the secondary absorber where they are absorbed in a water solution.

The aqueous urea solution from the synthesis unit has a concentration of about 75%. It can be used at that concentration for preparation of urea-ammonium nitrate solution, but for most uses is concentrated, by evaporation, to about 99.5% urea melt. The steam-heated evaporators operate under vacuum to lower the operating temperature and are designed for a short retention of urea to minimize the formation of undesirable biuret. Stainless steel or similar alloys are used for construction of the falling-film or forced-circulation types of evaporators. A flow diagram of a typical urea synthesis process is shown in Fig. 2.

The 99+% concentrated urea solution or melt is formed into small round pellets called prills in most facilities. The concentrated melt is sprayed downward from a spinning perforated basket or from shower-head sprays at the top of towers 60 to 100 ft tall. The droplets of urea-melt cool and solidify as they fall through upward moving air from large fans. The product is cooled and the small proportion of offsize prills is removed by screening and redissolved or remelted. Since urea is hygroscopic, the product is treated with a surface

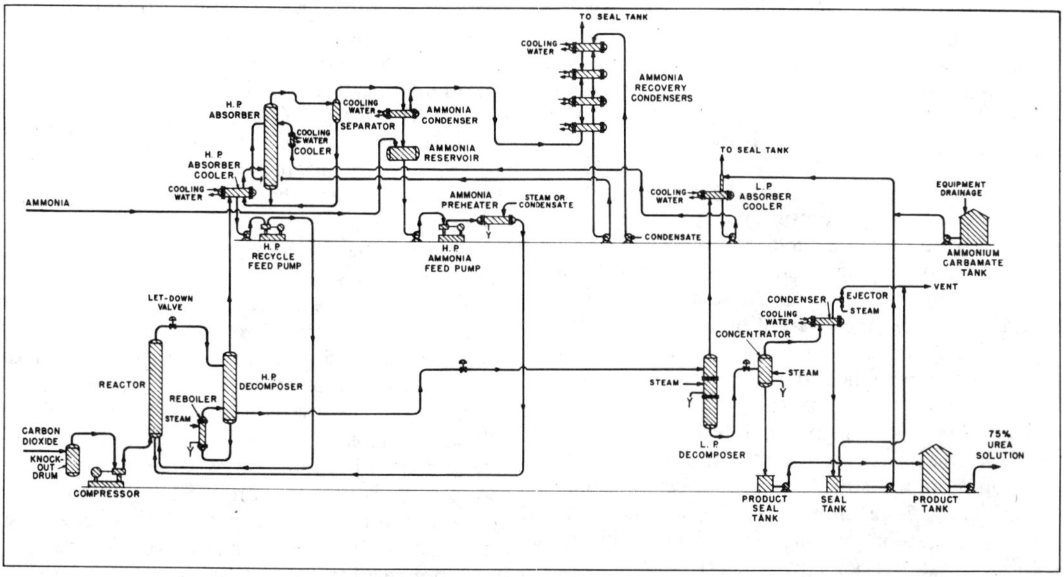

FIG. 2—*Typical total-recycle urea process* (*TVA plant*) (*Anon., 1979*).

coating or a chemical additive to minimize caking.

During the 1970s processes were commercialized for the production of urea in granular form by spray-drum granulation (*spherodizing*) and by inclined-pan granulation. Granules have the advantages of greater crushing strength (hardness) as compared with prills and the granulation processes can be readily controlled to produce granules that are much larger than prills. The larger granules are desired for aerial application on forests and for other uses.

There were no major improvements in the basic urea synthesis process in most of the 1970s, but engineering firms and process designers developed innovations late in the decade to decrease energy consumption and in some cases to increase production capability. These innovations are mainly related to designs for improved heat transfer and for more efficient pumps, motors, etc. (Anon., 1979; Young, 1974; McCamy and Norton, 1977; Anon., 1980c).

Production of Ammonium Nitrate

Nitric acid (HNO_3) and ammonia (NH_3) are the intermediates needed for the production of ammonium nitrate. It has been produced industrially since before World War I; its main use then was in munitions. The importance of ammonium nitrate and its production increased sharply after its use as a good high-analysis fertilizer was demonstrated in the United States and Europe in the early 1950s.

Production of ammonium nitrate is not particularly difficult. It consists of (1) neutralization of nitric acid with anhydrous ammonia, (2) concentration of the solution to either 96% or 99+% by evaporation, and (3) forming the product into prills or granules.

There are a number of designs of systems for neutralization and concentration. They all take into account the potential explosive hazard of ammonium nitrate, so temperatures and retention of material at high temperature are minimized. Care also is exercised to prevent contamination with oils and other organic materials that can sensitize ammonium nitrate to detonation.

Concentrators are operated under vacuum like those described earlier for urea. Prilling also is by far the main method used for forming ammonium nitrate, and design and operation are about the same as described earlier for urea. Spherodizing and pan granulation of ammonium nitrate came into the commercial picture during the 1970s (Young and McCamy, 1967; Young, 1980).

Ammonium nitrate is produced in two prilled forms: (1) *high-density prills* made from 99+% melt in a short prill tower (60 ft) for fertilizer and other industrial uses, and (2) *low-density prills* made from 95–96% solution in a tall (100–110 ft) prill tower. Porosity is provided by voids in the low-density prills to facilitate incorporation of oil added to produce blasting agents.

Nitrogen Solutions

Nitrogen solutions containing various combinations of urea, ammonium nitrate, and in some cases ammonia, have continued to grow in popularity in the United States and to a lesser extent in a few other countries. The greatest use of these solutions (nonpressure type) is in formulating fluid mixed fertilizers and for direct application.

Urea-ammonium nitrate (UAN) solutions are usually made from the intermediate solutions produced as the first step in making urea and ammonium nitrate. The most popular combination is a mixture of urea and ammonium nitrate in proportion to give maximum solubility. It contains by weight about 45% ammonium nitrate, 35% urea, and 20% water. Its crystallization (salting-out) temperature is 32°F and it contains 32% nitrogen. Solutions of 30 and 28% nitrogen content can be handled at temperatures below freezing (Hignett, 1971; Anon., 1979).

Ammonium Sulfate

Ammonium sulfate continues as an important industrial chemical product and in some countries remains important as a fertilizer for direct application. The main source of ammonium sulfate is byproduct material from the coke oven gas and steel industries. It also is produced from ammonia and sulfuric acid, although the cost usually is higher than for byproduct ammonium sulfate. When ammonium sulfate is produced directly, ammonia and sulfuric acid react to form the product which is crystallized from solution. The ammonium sulfate crystals can be recovered by recycling a side stream of the reaction solution through centrifuges, or the ammonium sulfate crystals can be collected batchwise for centrifuging. The crystals are usually washed with ammonia or water and dried in a rotary drum dryer and then screened before being put in bulk storage or bagged.

More refined types of crystallizers are used to grow large "grains of rice" crystals that are preferred for bulk blending with other fertilizer materials and for direct application.

Another major source of byproduct ammonium sulfate since the 1960s is from the process for production of caprolactam, an intermediate used in production of nylon. About 4 tons of ammonium sulfate is obtained per ton of caprolactam produced. In some cases the ammonium sulfate is fed directly to a fertilizer process as a 40% solution rather than preparing solid product. When produced as a solid, the ammonium sulfate is crystallized from solution and separated by centrifuging as in the other processes.

A method has been used for the manufacture of ammonium sulfate from gypsum ($CaSO_4 \cdot 2H_2O$) which, in the finely divided state, is agitated in a concentrated solution of ammonium carbonate formed by introducing ammonia and carbon dioxide into the aqueous medium. Calcium carbonate precipitates and ammonium sulfate remains in solution. Ammonium sulfate can then be obtained by the standard procedures. This process is not economically suitable for operation in the United States. It has found limited use in other countries where sulfur is not readily available or is very expensive.

Ammonium sulfate is a particularly good fertilizer for paddy rice. It is superior to ammonium nitrate and sometimes measurably better than urea, depending on application methods and water management practices. Another substantial use for ammonium sulfate is as an additive to supply nitrogen in fermentation processes. Much smaller amounts are used in medicinal and industrial chemical applications (Anon., 1979).

Ammonium Phosphates

Ammonium phosphates are generally classified as phosphates rather than nitrogen compounds. However, since these compounds contain 11 to 18% nitrogen (46–54% P_2O_5) and are so important in the world as fertilizers, they are covered here.

The ammonium phosphates are in two main forms, monoammonium phosphate (MAP) and diammonium phosphate (DAP). They were of comparatively minor importance until the simple and dependable process for their production for use as fertilizers from economical materials was developed and introduced to industry by the Tennessee Valley Authority (TVA) in 1961. By 1970, ammonium phosphate production had grown rapidly to become the leading phosphate fertilizer in the United States and reached this status in the world by the mid-1970s.

The first industrial production of ammonium phosphates was as compounds made in crystalline form from high purity phosphoric acid and ammonia. Later, there was some production of crystalline ammonium phosphates as by-products from the coke-oven gas and steel industries that were used mainly as fertilizers. The TVA process, which is easily adaptable to production of either DAP or MAP, utilizes much lower cost and more readily available "wet-process" phosphoric acid. The wet-process acid is made by acidulating pulverized phosphate rock with sulfuric acid followed by filtration and concentration. It is the major phosphate intermediate for fertilizer production. Partial neutralization of the acid with ammonia in a single reactor is a simple operation. The remaining ammonia is fed during granulation in a conventional rotary drum of the type widely used in industry. Product drying, cooling, and sizing are conventional steps. The granules do not require conditioning and have excellent bulk and bagged storage and transporting properties. Diammonium phosphate (18% N and 46% P_2O_5) is most popular, but production and use of monoammonium phosphate (12% N and 53% P_2O_5) is growing (Young, 1962; Young and Hicks, 1967; Anon., 1979).

As is the case with most fertilizer and other chemical processes, the main improvements in granular ammonium phosphate production starting in the mid-1970s were directed toward energy conservation and lower capital investment. A main innovation in ammonium phosphate and NPK fertilizer granulation units is the comparatively simple pipe reactor for acid(s) and ammonia. Effective use of the heat of reaction evaporates most of the water in the acid and drying of the product is eliminated or decreased substantially. Total investment is lower for new plants with pipe reactors and environmental controls are simpler. At the end of 1980, there were about 35 pipe-reactor installations in operation or nearing completion. Seven are in countries other than the United States (Achorn and Salladay, 1976; Shields, 1975).

Nonsynthetic Sources of Nitrogen

Biological Nitrogen Fixation

It has been known for many years that certain types of bacteria are capable of producing fixed nitrogen directly from elemental nitrogen. Total world biological nitrogen fixation was estimated at 100 million tpy (Postgate, 1971). The mechanisms of bacterial fixation have intrigued scientists for years, and for good reason. The biological nitrogen fixation takes place at atmospheric pressure and ambient temperature while the present synthetic processes take place at high temperatures and pressures and require large amounts of energy. If scientists are able to understand and duplicate the process of bacterial nitrogen fixation, major changes possibly could result in the fixed nitrogen industry.

There is a symbiotic relationship between members of the bacterial genus Rhizobium and leguminous plants. Some of the most familiar of the legumes include alfalfas, clovers, peas, beans, lupines, soybeans, and cowpeas; and nitrogen-fixing bacteria are found, for the most part, in nodules on the roots of the leguminous plants. Botanists estimate that there are from 10,000 to 12,000 species of leguminous plants, most of which are found in the tropics. About 200 species are cultivated by man and about 50 species are grown commercially in the United States. The planting and plowing under of legumes was recommended in the United States in many areas until about 1950 as a major practice in nitrogen fertilization.

In addition to the rhizobia, some types of free-living bacteria are capable of nitrogen fixation. Most of these bacteria occur in shallow muddy ponds or estuarine muds, or in environments with a relatively high content of organic matter. Estimates indicate that no more than 6 lb of nitrogen per acre is added to the soil each year through the combined activities of free-living nitrogen-fixing microorganisms.

There is evidence that more than a dozen species of blue-green algae are capable of direct nitrogen fixation. The algae occur in almost every environmental situation where sufficient sunlight is available for photosynthesis, including barren rock surfaces and frozen wastelands. They form symbiotic relationships with a variety of organisms, the most important being the lichen fungi. The blue-green algae may play an important role in the nitrogen economy of rice soils in India and in the Peoples Republic of China; for example, the ability to produce at least a modest yield of rice on the same land for many years without applying nitrogen fertilizer is believed to be due to nitrogen fixation by algae.

There are other bacteria which cause denitrification, a process just the opposite of nitrogen fixation, and the gaseous loss of nitrogen from soil. Biological denitrification is of considerable importance in well-drained and actively nitrifying soils which become partially anaerobic during wet periods, or through additions of large quantities of decomposable organic matter. The ability to reduce nitrates and nitrites to gaseous products fortunately appears to be limited to a relatively few bacteria (Postgate, 1971; Sauchelli, 1964).

Production of Nitrogen Compounds from Coke-Oven Gases

Coal carbonized at high temperature releases fixed nitrogen in the form of ammonia, which coke-oven operators recover as anhydrous ammonia, aqua ammonia, ammonium sulfate, or diammonium phosphate. The ammonia must be removed from the gas prior to further processing; otherwise it would form corrosive salts which would damage equipment or, if released as a waste material, would create pollution problems.

The ammonia formed during coking exists in both the water and gas that form part of the volatile products. The recovery of this ammonia can be accomplished by several methods. The most widely used method is called the semidirect process in which the ammonia in the liquor produced during coal carbonization is removed by distillation and alkali treatment and added to the gas stream. Then the gas containing all of the ammonia is passed through an absorber containing dilute sulfuric acid that reacts with the ammonia.

In an Otto-type ammonia absorber, coke-oven gas enters near the bottom and is sprayed with a dilute solution of sulfuric acid as it rises to the top. Ammonia in the gas reacts with the acid to form ammonium sulfate. The resulting solution drains into a crystallizer from which it is recirculated to the absorber, and a constant flow of sulfuric acid is added to the ammonia absorber to replace the acid neutralized by the ammonia in the coke-oven gas. After the solution becomes supersaturated, crystals of ammonium sulfate are precipitated in the crystallizer and accumulate as a slurry in the bottom. A portion of this slurry is removed and pumped to the slurry tank where the salt settles and the liquid overflows and returns to the ammonia absorber. The concentrated slurry is withdrawn from the bottom of the tank and is fed continuously or in batches to centrifuges that spin at high speed to separate and dewater the crystals. The partially dried ammonium sulfate is conveyed to heated rotary-drum dryers for final drying to a moisture content of about 0.1%.

Another process used for the recovery of ammonia is the Wilputte process for producing ammonium sulfate. Coke-oven gas is passed through a spray-type absorber over which is circulated a 6% solution of sulfuric acid nearly saturated with ammonium sulfate. Leaving the absorber, the solution is delivered to the solution-circulating system of a crystallizer in which crystallization takes place by the combined cooling and concentration effects of vacuum evaporation. By variation of the circulating rate and the degree of concentration, the size range of the ammonium sulfate crystals can be controlled within narrow limits. As the crystals grow in size, they settle to the bottom of the suspension tank and are delivered to a continuous centrifuge or filter, followed in either case by a dryer, as in the other processes. Essentially all of the byproduct ammonium sulfate produced is sold for use as fertilizer (McGannon, 1971; Sauchelli, 1964).

In the US Steel "Phosam" process, coke-oven gas is treated to recover anhydrous ammonia as the product. After the gas has passed through coolers and conventional gas cleaning equipment which removes entrained solids, water, and tar, ammonia is scrubbed from the gas by countercurrent contact with low-pH ammonium phosphate solution. Gas leaves the top of the absorber with 98 to 99% of the ammonia removed.

The ammonia-rich solution from the absorber is pumped through heat exchangers and into the stripper, recovering heat enroute. In the stripper the solution is contacted with steam at elevated pressure, stripping out the absorbed ammonia and regenerating the lean solution. The lean solution (low pH) is cooled and returned to the absorber. The overhead vapor from the stripper is condensed to form an aqueous ammonia feed to the fractionator where anhydrous ammonia is produced by steam fractionization. All the ammonia recovered is normally produced as liquid anhydrous ammonia, although gaseous ammonia can be produced if desired.

Deposits of Natural Nitrates

Deposits of natural nitrates, primarily the sodium and potassium salts, occur in the United States, Mexico, South Africa, Egypt, and Chile. The Chilean deposits of natural nitrates are the only deposits of commercial significance with estimated reserves of 200 million tons. The Chilean deposits and mining operations have been described in several publications (Anon., 1964; Graham, 1949; Sauchelli, 1964).

These naturally occurring inorganic nitrates were an important source of nitrate for fertilizers and explosives in the western hemisphere until about 1930. Their history of use goes back about 150 years. At one time the Chilean nitrates accounted for 40% of world nitrogen production. As production of synthetic nitrogen grew, their importance has waned. Total world production of Chilean nitrates in 1980 was only about 90,000 tons of nitrogen, a very small fraction of total world nitrogen production of 60 million tons. Although quite unimportant in the total fertilizer picture, these naturally occurring nitrates still are preferred for specific crops in some locations.

The nitrate fields in Chile are situated in the northern desert regions of the country in an area covering approximately 20,000 km^2. Several theories have been put forward to explain the origin of these deposits, and it seems likely that the nitrate content was derived from several sources including electrical discharge, atmospheric oxidation, and the action of nitrifying bacteria. After the formation of the nitrate, a secondary process peculiar to the area was probably responsible for concentrating the salts into economically workable deposits.

The nitrate-bearing ore is known as caliche. It is a hard rock-like layer varying from 0.5 to 5 m in thickness and occurring 1 to 3 m below the surface of the soil. In addition to a sodium nitrate content of from 5 to over 20%, caliche also contains chlorides, sulfates, borates, and iodates, as well as other salts. Certain of the deposits also contain potassium in workable amounts.

The mining and processing of the deposits is by the state-owned Sociedad Quimica y Minera de Chile SA (SQM), which operates two major plants having a combined capacity of up to 800,000 tons of product per year. The two production sites, Pedro de Valdivia and Maria Elena, are near the port of Tocopilla.

Open cast mining methods are used to remove the crude ore from the ground. The overburden is stripped away and the caliche is then blasted free and transported by electric railway to the processing plant. Here crushing and sieving are the primary operations and the screened caliche is then conveyed to extracting tanks, using water. The extraction process depends on the differing water solubilities of the salts present in the ore. By dissolving them as far as possible, the insoluble or slightly soluble substances such as clay or gypsum can be separated, and sodium chloride also can be largely eliminated. Fractional crystallization leads to the various salts being separated by centrifuges. A fine crystalline powder is the end result, and since this had a tendency to cake during storage, the product is melted and prilled in large towers similar to those used for ammonium nitrate and urea.

Various methods of processing the Chilean nitrate have evolved over the years, with the object of increasing the yield and decreasing the cost of operation. Solar energy makes a very real contribution to fuel requirements. Huge evaporation ponds are used to concentrate the weaker nitrate solutions. By this means economies can be made, since thousands of tons of fuel oil would otherwise be needed. Further work along these lines is continuing, with the objective of further use of solar energy and more efficient overall heat utilization.

Chilean nitrate is used primarily as a fertilizer for particular crops and in particular agronomic situations. Sugar beets, for example, have a definite sodium demand if maximum yields are to be obtained, and in European countries the sugarbeet crop is a major user of Chilean nitrate. Use in the United States is primarily limited to tobacco and cotton fertilization in the south Atlantic states.

Sodium nitrate is widely used in the explosives industry, especially in the manufacture of slurry and gel explosives and also in the production of SANFO, a mixture of sodium nitrate, ammonium nitrate, and fuel oil (Anon., 1980).

Although world production of Chilean nitrate was only about 90,000 tons of nitrogen in 1980 with 13,000 tons consumed in the United States, this naturally occurring material continues to be of some world significance and its export is important to the economy of Chile.

Ecology

Some controversy has developed concerning the possible accumulation of nitrate in soil, water, and plants. Some environmentalists and ecologists claim that we are faced with a crisis from the use of commercial nitrogen fertilizers and the release of nitrogen oxides by high temperature combustion in power plants and internal combustion engines.

The US Dept. of Agriculture published a report in 1971, giving an appraisal on the accumulation of nitrate (Viets and Hageman, 1971). The authors concluded that, on the basis of available information, the potential for nitrate accumulation does not pose a threat of an environmental crisis, and there is no indication of widespread upward trends of nitrate concentrations in foods, feeds, surface or ground water. The report concluded that the levels of nitrate currently found in foods pose no major health hazard to human adults, but, in isolated instances, the nitrate content of water and foods has been toxic to infants. In some of the cases, the toxicity resulted from microbial contamination that converted nitrate to nitrite. Future cases can be eliminated by more extensive testing for nitrate in water and foods consumed by humans. These findings generally have been substantiated in later studies.

A comprehensive research program is underway at the Tennessee Valley Authority (TVA) National Fertilizer Development Center at Muscle Shoals, Alabama, to study the fate of nitrogen fertilizers applied to the soil. Fertilizer nitrogen-soil interactions are being studied in cooperation with land-grant universities to determine effects on the environment and efficiency of nitrogen use for crop production.

To isolate the fertilizer nitrogen from that already present in nature, the scientists *tag* the fertilizer nitrogen by changing its isotope composition. In nature, nitrogen is composed of one Nitrogen-15 atom for every 273 Nitrogen-14 atoms.

TVA makes fertilizers with "depleted" material, in which the proportion of Nitrogen-14 and Nitrogen-15 is very different from that in the soils, plants, waters, and other natural substances. When this nonradioactive tagged fertilizer is added to the soil and taken up by the plant, the isotope composition of the plant changes. Then the amount of nitrogen in the plant that came from the fertilizer can be measured by instrumental analysis (mass spectrometry). Or the amount of fertilizer nitrogen in the drainage water can be determined. How much of the fertilizer remains in the soil, how far it moves in the soil, or whether some of it is available for plant use the next growing season can be determined.

TVA research scientists point out that "tracer" technology has not identified all the problems, nor has it helped develop many solutions as yet. But nitrogen tracers have given more accurate information about nitrogen in the biosphere and have proven to be valuable tools in tracing the movement of nitrogen. This in turn will help in the long-range search for ways to improve efficiency of plants in their use of nitrogen and, as a consequence, will help protect the quality of the environment (Anon., 1976).

Besides making tagged nitrogen fertilizers available to university scientists, the TVA program helps graduate students and senior researchers gain experience in using the nitrogen isotopes. University students and professors work in the nitrogen isotope laboratories at Muscle Shoals for periods of a few days to several months for training and experience.

Much work remains to be done to better determine the fate of soluble nitrogen in fertilizers and to assess any effect on our environment. But work underway at several locations is beginning to provide some answers.

Recent Changes in Design and Production Capacity of Manufacturing Processes

Since about 1975 there have been no major breakthroughs in basic process designs for ammonia, urea, nitric acid, and ammonium nitrate. There have been a number of improvements related mainly to conservation of energy and emphasis will continue in this area well into the 1980s. Some innovations have resulted in decreases of 15 to 30% in energy consumption for various processes without major changes in the basic processes. A large nitrogen fertilizer complex due to come onstream in Canada in 1985 will have several energy saving innovations. The total natural gas consumption per ton of NH_3 will be 25 to 26 million Btu's as compared with an average of 33 million Btu's in existing plants. Together with other features, this will cut total energy requirements by 16% (Anon., 1981e). Use of a physical solvent in CO_2 removal and recovery greatly decreases the heat required for regeneration.

Changes in various facilities include a series of minor modifications that allow greater recovery and use of process heat. Also, changes in pumps and compressors to combine some duties have decreased utility requirements. Engineering rationalization of the use of process-generated steam drivers in a complete production complex concept rather than piecemeal has also lowered energy requirements. Efforts in this direction will continue as energy costs increase.

There also have been important developments in the control of atmospheric emissions and aqueous effluents within required limits. In a number of cases this has been accomplished without massive expenditures by the application of sound engineering and practical judgment. One of the main problem areas is keeping the emission of brown oxides of nitrogen fumes from nitric acid plants within required limits. After trial of costlier and more difficult schemes, the technique of "extended absorption" has generally proved best. This involves use of a second, shorter absorption stage operating with cooled, dilute-process liquid.

Impoundment and reuse of process and cooling water has necessitated soundly engineered systems and tight operating controls for meeting essentially "zero discharge" of pollutants and limited heat in effluent water. In production units for prilled or granular urea and ammonium nitrate, dust collection is now required. A major step has been the successful use of bag filter units for dust collection. Good design and rigid operating discipline are keys to the use of these very effective collectors.

Product quality has been improved substantially by development of new or modified process schemes for preparation of finished solid urea and ammonium nitrate products. Pan granulation allows a wider range of particle sizes and better quality of products. This system is used in a few commercial plants. "Spray-drum" granulation (spherodizing) is used in several plants. A recent development is TVA's new "curtain-granulation" process that was demonstrated in a 2-tph pilot plant in late 1980. A decrease of about 40% in energy requirements for cooling is realized and very hard spherical product with an unlimited selection of particle size range is produced (Anon., 1980c).

By the mid-1970s the move to larger production units had leveled off, and the more-or-less standard size for large units now is 1000 tpd for ammonia, with a companion 1520-tpd plant for urea. Ammonium nitrate units had already leveled off at 900 to 1100 tpd. A few ammonia plants have been built for capacity of 1500 tpd, but most engineering firms are not venturing into this area. A major factor is the continuing sharp escalation in the investments required for production units. Units of 1500-tpd capacity put even greater pressure on the operation at a high percentage of the time (90% or more) to control the investment-related costs per ton of product. Depending on feedstock costs, investment costs can represent more than half of the total production cost.

World Review

There have been dramatic changes in nitrogen fertilizer production and trade patterns. A much greater interdependence among regions has developed with countries no longer isolated from world market trends as they were 20 years ago. These changes are indicated by the production, consumption, and trade statistics for the major regions presented in Table 1. In 1980, for the fertilizer year ending June 30, Asia and East Europe were the leading consumers of nitrogen fertilizers. Both of these regions have increased nitrogen consumption at a faster rate than North America and West Europe—the leading nitrogen use areas in 1967. Table 1 also indicates production and use statistics for developed, developing, and the Central Planned Economies of East Europe, USSR, and the Peoples Republic of China (PRC). The developed countries remain net exporters of nitrogen fertilizers. The developing countries, while increasing fertilizer use at a rapid rate, have done so by increasing their dependence on imported materials. The Central Planned Economies remain largely self-sufficient, with the exception of large nitrogen imports by the PRC.

Total world nitrogen use was estimated to be about 57 Mt of N in 1980. From 1967 to 1980, world nitrogen use increased at an average rate of more than 7% per year. By 1985, world use is expected to exceed 70 Mt of N—an average growth of about 4% per year. This lower average annual increase reflects the maturing of the nitrogen market as application rates on major agricultural and cash crops come closer to optimum rates given today's crop varieties, management levels, and commodity prices.

In 1967, world ammonia plant capacity was just under 37 Mt of nitrogen. By 1980 total world ammonia capacity had increased to more than 92 million tons and indications are that capacity will exceed 110 million tons by 1985 (Table 2).

World nitrogen fertilizer production is dominated by urea and ammonium nitrate. The remainder consists of ammonium sulfate, and other lower analysis materials such as calcium and sodium nitrate, along with ammonia for direct application and in combination with phosphate and potash materials. Urea has emerged as the leading nitrogen fertilizer material in many parts of the world. Capacity is about 34 million tons of nitrogen—an increase of more than 27 million tons since 1967. Ammonium nitrate capacity doubled during the same period but should increase very little as compared to the continued expansion in urea capacity over the next five years.

Most nitrogen production has been located in the developed regions: North America, West Europe, and Japan. However, the rapidly changing world energy supply situation and rising energy costs are shifting the location of nitrogen supplies. As indicated in Table 3, natural gas is the most common feedstock for ammonia production. Developed regions, however, account for less than 20% of total world reserves while the developing regions are estimated to control almost half of all natural gas reserves. The share of world ammonia capacity in the developing regions is less than

TABLE 1—World Nitrogen Fertilizer Production, Trade, and Consumption by Region
Mt, N

Region		1966-1967	1972-1973	1977-1978	1979-1980	Forecast 1984-1985 #
North America	Production	6.02	9.20	11.07	12.86	—
	Export*	0.87	1.16	2.55	3.32	—
	Import	0.63	0.64	1.79	2.48	—
	Consumption	5.74	7.92	9.74	11.15	13.07
Latin America	Production	0.50	0.84	1.33	1.50	—
	Export	0.21	0.27	0.18	0.20	—
	Import	0.53	1.07	1.41	0.14	—
	Consumption	0.81	1.63	2.52	2.67	4.07
Western Europe	Production	7.03	8.95	9.73	11.25	—
	Export	2.65	2.83	3.36	3.96	—
	Import	0.83	1.44	2.15	2.65	—
	Consumption	5.02	6.99	8.36	9.54	10.05
East Europe-USSR	Production	5.07	10.68	15.09	15.51	—
	Export	0.24	1.27	2.35	2.37	—
	Import	0.33	0.45	0.30	0.33	—
	Consumption	4.76	9.29	12.11	12.45	15.00
Africa	Production	0.24	0.55	0.77	0.84	—
	Export	—	0.03	0.06	0.04	—
	Import	0.32	0.65	0.64	0.71	—
	Consumption	0.55	1.06	1.33	1.52	2.27
Asia	Production	4.03	8.16	13.20	17.46	—
	Export	0.97	2.16	1.76	2.12	—
	Import	2.34	3.26	3.87	4.78	—
	Consumption	5.30	9.31	15.54	19.56	25.60
Oceania	Production	0.04	0.18	0.22	0.21	—
	Export	—	0.02	—	—	—
	Import	0.06	0.05	0.06	0.07	—
	Consumption	0.12	0.21	0.26	0.27	0.34
Developed Countries†	Production	15.09	21.17	3.33	26.73	—
	Export	4.45	5.72	6.85	8.16	—
	Import	1.62	2.20	4.07	5.34	—
	Consumption	12.06	16.47	19.85	22.60	25.60
Developing Countries‡	Production	1.29	3.86	6.23	8.12	—
	Export	0.23	0.75	1.09	1.51	—
	Import	2.01	3.50	4.27	5.24	—
	Consumption	2.98	6.39	9.59	11.33	15.90
Central Planned Economies§	Production	6.56	13.56	21.83	24.78	—
	Export	0.26	1.27	2.33	2.35	—
	Import	1.41	1.87	1.89	1.83	—
	Consumption	7.25	13.54	20.43	23.22	28.90
World	Production	22.95	38.58	51.39	59.64	—
	Trade	4.94	7.74	10.27	12.03	—
	Consumption	22.29	36.40	49.86	57.16	70.40

Source: Anon., 1968, 1974, 1980b, 1981c.

* Export and import data for finished nitrogen products only and does not include anhydrous ammonia used for production of other fertilizer materials.

† North America, West Europe, Oceania, Japan, Israel, and Republic of South Africa.

‡ Latin America, Africa excluding Republic of South Africa, and Asia excluding Japan, Israel, and Central Planned Economies.

§ East Europe, USSR, Peoples Republic of China, Vietnam, North Korea, and Mongolia.

\# Forecasts from Harris and Harre (1979) revised to reflect historical data revisions for USSR and China.

TABLE 2—World Nitrogen Fertilizer Product Capacity, By Region
Mt, N

Region	Product	1966-1967	1972-1973	1977-1978	1979-1980	Forecast 1984-1985
North America	Ammonia	11.08	14.03	18.47	17.66	18.88
	Ammonium Nitrate	2.88	3.00	3.47	3.70	3.70
	Urea	1.34	1.96	3.49	3.78	4.38
Latin America	Ammonia	1.02	1.76	3.54	4.03	6.84
	Ammonium Nitrate	0.23	0.26	0.44	0.44	0.54
	Urea	0.18	0.38	1.21	1.29	2.74
Western Europe	Ammonia	9.22	14.15	14.85	15.62	15.30
	Ammonium Nitrate	2.69	4.15	4.70	5.08	5.41
	Urea	1.36	3.45	4.52	5.00	5.41
East Europe-USSR	Ammonia	8.01	15.61	22.80	29.75	37.48
	Ammonium Nitrate	3.32	6.11	8.27	8.53	8.78
	Urea	1.96	4.78	7.83	9.83	12.91
Africa	Ammonia	0.35	0.93	1.21	1.83	3.14
	Ammonium Nitrate	0.04	0.21	0.34	0.66	0.79
	Urea	0.09	0.28	0.57	0.88	1.37
Asia	Ammonia	7.01	13.62	18.81	23.26	28.31
	Ammonium Nitrate	0.62	0.96	0.97	1.13	1.09
	Urea	2.33	5.43	10.02	13.78	17.34
Oceania	Ammonia	0.13	0.45	0.45	0.45	0.52
	Ammonium Nitrate	0.03	0.11	0.11	0.11	0.11
	Urea	0.12	0.10	0.10	0.10	0.18
Developed Countries	Ammonia	23.07	33.53	38.11	38.00	39.57
	Ammonium Nitrate	5.81	7.76	8.86	9.56	10.04
	Urea	3.80	7.58	9.99	10.26	11.57
Developing Countries	Ammonia	2.71	6.99	12.21	15.77	23.66
	Ammonium Nitrate	0.26	0.57	0.81	1.18	1.39
	Urea	1.01	3.27	7.36	10.29	14.61
Central Planned Economies	Ammonia	11.03	20.03	29.80	38.82	47.24
	Ammonium Nitrate	3.74	6.48	8.64	8.89	9.01
	Urea	2.58	5.54	10.38	14.10	18.15
World	Ammonia	36.82	60.55	80.12	92.59	110.46
	Ammonium Nitrate	9.81	14.80	18.31	19.63	20.43
	Urea	7.39	16.39	27.73	34.65	44.33

Source: Capacity data from the records of Marketing and Distribution Economics Section, Tennessee Valley Authority, National Fertilizer Development Center, Muscle Shoals, AL, May 1981.

20% of total world capacity, but will gain in the share of total capacity, as the developed nations continue to decline in relative importance.

This shift in nitrogen capacity to areas having large reserves of natural gas will mean an increasing level of nitrogen trade as greater amounts of ammonia are shipped to developed regions.

Transportation

Nearly all forms of transportation are used to move nitrogen fertilizer materials from production points to the market. Anhydrous ammonia is transported by barges, railroad tank cars, pipelines, transport trucks, and nurse tanks (Table 4). Large tonnages of anhydrous ammonia move by barge up the Mississippi River and along other inland waterways. Barge transportation is relatively low-cost for those areas having access to waterways. Barges for anhydrous ammonia are of two types: high-pressure (250 psig), capable of transporting anhydrous ammonia at ambient temperatures and a fairly high vapor pressure; and low-pressure (10 to 55 psig), with insulated storage tanks and mechanical refrigeration units to

TABLE 3—Estimated Proved Reserves of Natural Gas

Region	Natural Gas	Share of World Natural Gas Reserves	Share of Ammonia Capacity 1980	Share of Ammonia Capacity 1985	Ammonia Capacity Based on Natural Gas (1980)
	(10^9 cu ft)	(%)	(%)	(%)	(%)
North America	278,300	11	19	18	96
Latin America	159,811	6	4	6	87
West Europe	157,815	6	17	14	67
East Europe	930,900	35	32	35	85
Africa	208,470	8	2	3	45
Asia	867,105	33	25	27	35
Oceania	36,100	1	—	—	42
Developed Countries	472,815	18	41	36	75
Developing Countries	1,210,286	46	17	21	58
Central Planned Economies	955,400	36	42	43	72
World	2,638,501	—	—	—	71

Source: Auldridge, 1980; and capacity data from the records of Marketing and Distribution Economics Section, Tennessee Valley Authority, National Fertilizer Development Center, Muscle Shoals, AL, April 1981.

keep the vapor pressure of the anhydrous ammonia below the maximum allowable design pressure.

Two types of rail tank cars are used in ammonia transport. Standard tank cars are of 11,000-gal capacty and transport 25 to 26 tons of liquid ammonia. These cars have an inner high-pressure (300 psig) tank covered by a layer of insulating material with an outer light steel shell. Jumbo tank cars of 30,000-gal, 70-ton capacity now move most of the rail-shipped ammonia. The jumbo tank cars have noninsulated tanks designed for 400 psig test pressure.

Anhydrous ammonia is also delivered through long-distance pipelines from major producing areas to major consuming areas. In the early 1970s Gulf Central placed in operation a 1700-mile-long pipeline system that carries liquid anhydrous ammonia from Louisiana to Iowa, Nebraska, Illinois, Indiana, Missouri, and other midwest consuming areas. The Gulf Central pipeline cost about $59,000 per mile and has a capacity of 3 million st per year. Gulf Central offers shippers the option of contract storage and has constructed refrigerated storage tanks at strategic points along the line.

A second long-distance anhydrous ammonia pipeline has been operating since 1967. Built by MidAmerica Pipeline Co. (MAPCO), it stretches 720 miles from Borger, Texas, to distribution points in Kansas, Nebraska, and Iowa. The MAPCO pipeline has a design capacity of 3,000 to 3,500 tpd.

A third ammonia pipeline in Florida with a capacity of 600,000 tpy began operation in December 1979. The capacity of the pipeline is leased by two phosphate fertilizer producers and extends from the partners' respective anhydrous ammonia terminals on Tampa Bay to their phosphate manufacturing complexes, as well as other manufacturing facilities in central Florida.

Nitrogen solutions containing 29% N are being transported by pipeline by the Williams Pipe Line Company of Tulsa, Oklahoma. The line extends from Tulsa into the midwest and north through Iowa, Minnesota, Nebraska, and North and South Dakota. Materal can be accepted at three locations and unloading facilities are available at nine other locations. The minimum shipment is 6,000 tons and the carrier does not provide storage at either point of origin or destination. As petroleum products are shipped in the same line, shipments of nitro-

TABLE 4—Fertilizer Transportation by Mode, 1978, Percentage

Product	Rail	Truck	Water	Pipeline
Anhydrous ammonia	27.2	42.8	11.7	18.3
Nitrogen solutions	19.8	45.4	23.8	11.0
Urea	46.2	25.6	28.2	—
Ammonium nitrate	66.6	33.4	—	—

Source: Anon., 1980a.

gen solutions in the winter months are scheduled only when temperatures and operating conditions are satisfactory and no damage will occur to the pipeline or the other products being transported. The route of the long-distance ammonia and nitrogen solutions pipelines in relation to US ammonia plant locations is shown in Fig. 3.

Highway transport trucks are still another means used for transporting anhydrous ammonia. These trucks are usually semitractor trailer rigs with high-pressure, 265 psig, tanks varying in size from 6000 to 9000 gal and which haul from 12 to 19 tons of anhydrous ammonia per trip. Some transport trucks are equipped with vapor compressors or liquid pumps for unloading. Other trucks have no pumping equipment and must be unloaded by pumps or compressors located at the receiving plant.

Barges, rail cars, and transport trucks are often used for both liquefied petroleum gas and ammonia service. For dual service, however, all fittings must be of steel and approved for ammonia service. For direct farm fertilizer applications, anhydrous ammonia is moved from the bulk plant or storage to the farmer's fields in high-pressure (up to 265 psig) tanks, usually 1000-gal capacity, mounted on heavy duty, two- or four-wheel wagons. Tanks are generally not filled to more than 85% of capacity as a safety precaution. A 1000-gal tank contains about 2.2 tons of ammonia.

In international trade, modern ships are designed to handle both liquefied petroleum gas and anhydrous ammonia. A typical vessel is designed for refrigeration down to −45° C to keep the vapor pressure of the cargo at a low level. World ammonia trade has increased from 1.7 Mt N in 1972 to an estimated 5.3 Mt N in 1980. Leading exporters include the United States, Mexico, Trinidad, Venezuela, the Middle East, and the USSR, which represents one-third of all ammonia exports.

United States: Production, Consumption, and Trade

Nitrogen Fertilizer Consumption

Total nitrogen fertilizer consumption in the United States has increased from 2.7 million st in 1960 to 11.4 million st in 1980—or four times the level of 20 years ago. The rate of increase in nitrogen use has been declining from an average of about 10% per year between 1960 and 1970 to slightly more than 4% from 1970 to 1980.

Marketing patterns for nitrogen fertilizers have changed dramatically since 1960 when mixed fertilizers accounted for 37% of total nitrogen use. In 1980, mixtures were down to 22% of total consumption. Direct application of ammonia increased from less than 600,000 tons N in 1960 to 4.5 million tons in 1980. Similar gains also have been made by nitrogen solutions and urea while lower analysis materials have either declined in tonnage or shown little gain. Ammonium nitrate use increased from 415,000 tons N in 1960 to a peak of 1.1 million tons in 1973. Since then use has slowly declined to just under 900,000 tons N in 1980. Consumption of nitrogen fertilizers as mixtures and materials over the past 20 years is shown in Table 5.

While consumption of most nitrogen fertilizers has increased or remained relatively stable over the past 20 years, there have been some dramatic changes in the share of the market for the major nitrogen materials. Anhydrous ammonia has moved from 20% of total nitrogen use in 1960 to almost 40% of today's market. Urea has increased from only 3% in 1970 to about 9% in 1980 and nitrogen solutions have expanded from 7% to more than 16% since 1960. The nitrogen in diammonium phosphate for direct application today represents more than 5% of total use compared to less than 1% in 1960. Losers in market share include ammonium nitrate, ammonium sulfate, and the dry mixed fertilizers containing nitrogen.

Nitrogen fertilizer consumption is centered primarily in two regions of the United States—the East and West North Central states. In 1980 almost 57% of all nitrogen use was in this area, which has 59% of the harvested acreage and 43% of the nation's total crop and livestock receipts. Nitrogen use in these two regions has increased at more than 10% per year for the past 20 years and still leads the country with a 5% average gain per year since 1970. The West South Central states and the Pacific Coast states are also large consumers of nitrogen fertilizers, followed by the South Atlantic and East South Central area. Regional use patterns for nitrogen fertilizers in the United States during the past 20 years are shown in Table 6.

Industrial Explosives

Consumption of explosives and blasting agents in the United States totaled 4.55 billion lb in 1980. In the past 20 years the use of explosives has increased at about the same annual

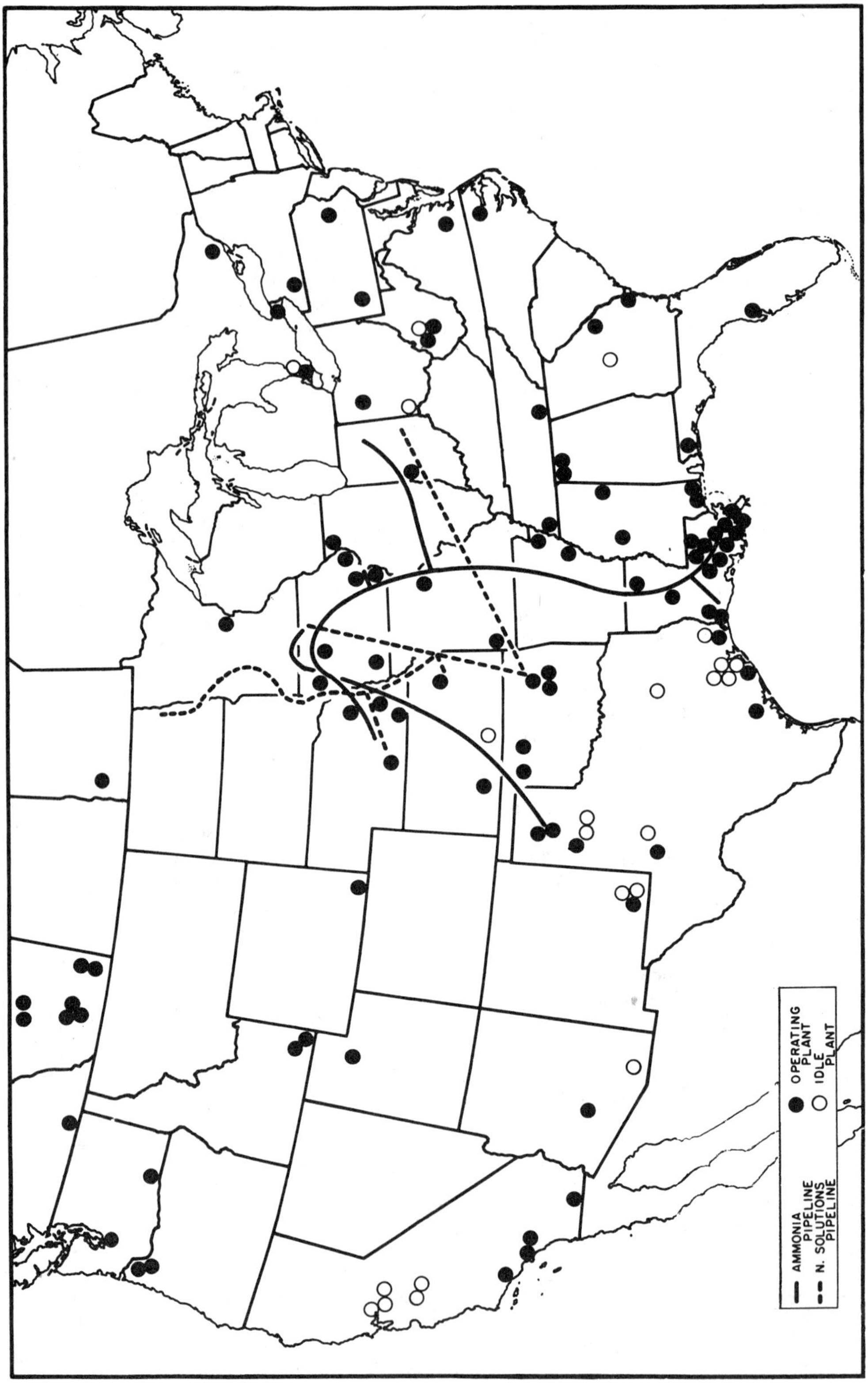

FIG. 3—*North American ammonia plants and pipeline systems, 1981.*

TABLE 5—Nitrogen Fertilizer Consumption in the US

		Direct Application Materials* St, N						
Year†	Nitrogen in Mixtures	Anhydrous Ammonia	Nitrogen Solutions	Ammonium Nitrate	Urea	Ammonium Sulfate	Total	Total Nitrogen Consumption
1960	1,017,415	581,924	194,740	415,855	64,596	106,959	1,720,632	2,738,047
1961	1,071,224	666,234	292,566	446,585	92,448	115,715	1,959,564	3,030,788
1962	1,147,266	767,425	371,379	468,523	132,804	116,687	2,222,714	3,369,980
1963	1,263,641	1,006,762	471,392	499,378	165,198	147,121	2,665,448	3,929,089
1964	1,377,033	1,148,071	526,755	555,320	184,327	148,865	2,975,776	4,352,809
1965	1,452,084	1,281,968	595,859	547,488	193,713	161,848	3,186,454	4,638,538
1966	1,591,927	1,606,872	712,445	610,705	211,615	165,797	3,734,376	5,326,303
1967	1,764,372	1,973,596	784,392	710,503	227,952	174,834	4,262,625	6,026,997
1968	1,867,091	2,457,261	803,092	786,346	243,359	167,306	4,920,474	6,787,565
1969	1,901,393	2,576,431	839,925	863,792	264,755	157,042	5,056,207	6,957,600
1970	1,939,077	2,844,058	969,882	952,861	242,758	160,911	5,519,927	7,459,004
1971	2,062,782	3,254,160	1,036,498	969,091	273,607	185,386	6,070,824	8,133,606
1972	2,131,639	2,982,274	1,009,119	1,015,548	357,386	192,205	5,890,627	8,022,266
1973	2,391,300	2,700,704	1,144,407	1,095,651	432,420	197,088	5,903,805	8,295,105
1974	2,431,387	3,427,672	1,184,507	1,061,743	467,934	193,025	6,725,821	9,157,208
1975	2,098,763	3,294,794	1,191,691	941,881	524,183	170,614	6,502,056	8,600,819
1976	2,417,282	4,025,258	1,620,308	985,555	729,250	219,760	7,994,302	10,411,584
1977	2,551,338	4,040,328	1,685,028	936,863	855,344	220,226	8,096,065	10,647,403
1978	2,374,999	3,721,289	1,591,935	820,356	880,753	188,208	7,589,620	9,964,619
1979	2,573,382	4,003,845	1,723,860	841,272	967,040	157,846	8,141,337	10,714,719
1980	2,501,355	4,499,947	1,908,917	884,718	946,359	175,566	8,898,153	11,399,508

Source: Anon., 1981a.

*Includes some materials reported prior to blending with phosphate and potash materials for delivery to farm.

†Fiscal year ending June 30.

growth rate as nitrogen fertilizers—7.0% per year vs. 7.4% for fertilizer. Unlike fertilizer use, the rate of increase has been maintained throughout the period and has averaged 6.6% from 1970 to 1980 and 7.4% per year from 1960 to 1970.

Coal mining remained the single largest consumer of industrial explosives accounting for 56% of total use in 1980. Use in other mining and construction declined in 1980 reflecting economic and market conditions. Consumption by major use category is shown in Table 7.

In 1979 seven states accounted for 59% of all explosives used in the United States. Kentucky was the leading market followed by Pennsylvania, Alabama, Arizona, Ohio, West Virginia, and Indiana.

Of the 4.55 billion lb of explosives used in 1980, half was unprocessed ammonium nitrate, while 36% was mixed ammonium nitrate-fuel blasting agents. The principal distinction between explosives and blasting agents is sensitivity to initiation. Explosives are cap-sensitive while blasting agents are not and require a primer.

Fertilizer Trade

Since 1960 US foreign trade in nitrogen fertilizer materials has grown in importance as world events have had an ever-increasing impact on the domestic nitrogen market. In the early 1960s the United States was a net importer of nitrogen; the trade balance however, seldom exceeded 200,000 tons of nitrogen in any given year. In 1966, the nation became a net exporter with annual trade balances reaching as high as 860,000 tons of nitrogen in 1969. This rapid rise in nitrogen exports was brought about by large surpluses in the United States and a worldwide program to increase food supplies in developing countries.

Between 1975 and 1979 the nation once again was a net importer of nitrogen materials. Since then import and export tonnages have been expanding rapidly but the nation's nitrogen fertilizer trade has remained in balance.

As energy costs have increased sharply, ammonia plants have been forced to close, little capacity has been added, and ammonia imports have increased from 400,000 tpy in 1970 to

TABLE 6—Nitrogen Fertilizer Consumption in the US, by Region
1000 St, N

Year*	Northeast	South Atlantic	East North Central	West North Central	East South Central	West South Central	Mountain	Pacific
1960	166.2	462.9	423.2	494.8	315.0	314.7	126.7	382.1
1961	167.7	476.8	484.0	645.0	317.1	340.3	149.7	392.7
1962	179.3	496.3	558.7	700.0	344.2	442.4	160.6	421.9
1963	201.9	539.8	674.0	882.9	364.5	539.4	172.3	466.3
1964	217.0	582.2	848.3	973.3	385.4	588.0	193.0	504.1
1965	228.3	581.8	951.4	1,075.5	385.2	612.7	207.4	537.1
1966	249.8	628.6	1,174.4	1,345.2	417.3	667.6	240.5	549.3
1967	261.4	687.6	1,308.0	1,655.4	452.5	725.1	275.3	609.3
1968	270.4	722.6	1,319.7	2,139.8	442.4	880.4	323.9	637.2
1969	265.9	772.7	1,354.8	2,074.8	486.7	986.3	366.7	601.8
1970	277.5	806.1	1,419.0	2,337.5	487.0	1,059.9	374.7	645.9
1971	310.1	833.0	1,692.1	2.567.1	532.7	1,080.9	386.8	678.2
1972	280.7	834.8	1,415.1	2,545.4	577.3	1,167.0	412.3	738.2
1973	330.7	882.3	1,396.8	2,646.5	593.8	1,154.9	457.4	784.0
1974	321.2	901.6	1,676.5	2,906.5	644.8	1,317.2	488.8	848.8
1975	310.0	853.2	1,696.0	2,784.7	562.4	1,053.7	458.2	836.4
1976	326.2	1,008.8	2,063.6	3,620.7	641.1	1,258.7	539.2	911.0
1977	354.7	1,090.6	2,258.1	3,490.3	692.6	1,302.5	514.9	898.4
1978	333.6	937.0	2,074.2	3,364.2	616.8	1,179.9	534.9	878.5
1979	351.4	963.1	2,247.3	3,743.2	569.5	1,285.8	559.7	951.6
1980	356.8	944.3	2,437.6	4,032.0	595.9	1,347.8	605.1	1,036.4

Source: Hargett and Berry, 1981.
*Fiscal year ending June 30.

almost 2 million tons in 1980. Imports of other nitrogen products have generally been on the decline from their peaks of 1977 and 1978.

While the nation has been importing more ammonia in the past few years, it has been able to maintain a favorable trade balance by rapidly increasing trade in ammonium phosphates and urea. In 1980 the nitrogen content of ammonium phosphates accounted for 35% of total nitrogen exports. Urea and ammonia each accounted for an additional 25% while ammonium sulfate and nitrogen solutions made up most of the remainder. Total nitrogen exports represented about 17% of total nitrogen production in 1980. Exports of nitrogen products for 1960 to 1980 are shown in Table 8 and imports for the same period in Table 9.

Nitrogen Production

In 1980, the US ammonia industry operated at more than 90% of total capacity, producing just over 19 million tons of material. Ammonia production has increased steadily at an annual growth rate of about 7% per year. Only twice in the past 20 years has production declined from the previous year, once in 1968 in response to oversupply, and again in 1978 primarily because of feedstock shortages (Table 10). Nitrogen production is expected to continue to expand but the rate of increase will be relatively slow.

Anhydrous ammonia is produced primarily in areas having large reserves of natural gas, and where water transport is available for both domestic shipments and export of nitrogen products to world markets. This represents a shift from the concept of establishing production units in market areas that prevailed through the early years of the US nitrogen industry. In 1980, Louisiana accounted for 33% of US ammonia capacity followed by Oklahoma with 11% and Texas with 9%. Other major producing states are Mississippi, Iowa, Alaska, and Arkansas. These seven states make up about 70% of the total US ammonia capacity. North American ammonia capacity for the years 1967, 1973, 1978, and 1980 is shown in Table 11. Anticipated additions and expansions of ammonia capacity by 1985 are also indicated.

In the past 10 years urea production has increased faster than any of the other nitrogen fertilizer materials. Total urea production in

1980 exceeded 7.2 million tons of primary solution, of which 3.15 million tons was solid (prilled or granular) urea. Production of solid urea has been expanding at more than 10% per year over the past decade. In 1980 total US urea capacity was estimated to be 7.8 million tons of material yielding a 90% operating rate for the year.

Ammonium nitrate, the other leading solid nitrogen fertilizer, has experienced declining production levels in the past few years. From a high of almost 4 million tons of material in 1973, production reached a low of 3 million tons in 1978 and in 1980 was about 3.2 million tons. Production of ammonium nitrate solution has continued to increase for use as an intermediate for nitrogen solutions and in nonfertilizer materials. During the 1970s production levels averaged a 3% annual growth rate. In 1980, total ammonium nitrate capacity was estimated to be 10.9 million tons of material, with a total production of almost 8.6 million tons.

Nitric acid production, primarily as an intermediate for producing ammonium nitrate, has been steadily increasing from 3.3 million tons of material in 1960 to almost 9 million tons in 1980. About 75% of the nitric acid produced is used in producing ammonium nitrate.

Since 1964, ammonium sulfate production has remained relatively constant, ranging between a high of 2.8 million tons in 1966 and 1978 to a low of 2.3 to 2.4 million tons in the early 1970s and in 1980. Of the total production in 1980, byproduct ammonium sulfate production from coke ovens amounted to 400,000 tons, while the remainder consisted of 640,000 tons of synthetic material and 1.45 million tons of byproduct—58% of total production—from the fiber industry. While much of the production is involuntary, the total estimated capacity for ammonium sulfate production in the United States was more than 3.1 million tons in 1980.

Nitrogen solutions have gained in importance in the nitrogen fertilizer market in recent years. Production has increased from 800,000 tons of nitrogen equivalent in 1960 to more than 3 million tons in 1980. During the 1970s, production increased at an average rate of almost

TABLE 7—US Consumption of Industrial Explosives by Use
1000 lb, Gross Wt.

Year	Coal Mining	Metal Mining	Quarrying and Nonmetal Mining	Railway and Other Construction Work	All Other Purposes	Total
1960	414,667	221,186	262,749	209,739	64,657	1,172,998
1961	414,697	234,155	268,479	209,796	76,497	1,203,624
1962	437,895	245,187	287,609	273,412	68,304	1,312,407
1963	506,045	252,800	320,632	305,837	70,608	1,455,922
1964	549,286	318,299	361,246	332,120	104,602	1,665,553
1965	594,964	356,715	371,034	411,384	150,803	1,884,900
1966	618,158	353,331	396,283	437,020	165,364	1,970,156
1967	661,581	327,596	409,525	412,447	93,564	1,904,713
1968	684,166	403,444	397,998	411,664	50,466	1,947,738
1969	820,114	470,791	438,789	411,784	55,000	2,196,478
1970	962,331	479,508	455,424	446,568	49,660	2,393,491
1971	1,071,309	457,286	489,572	478,317	57,534	2,554,018
1972	1,212,585	430,686	493,677	467,273	65,568	2,669,789
1973	1,177,062	495,879	643,292	393,821	44,892	2,754,946
1974	1,186,614	465,490	551,380	362,617	196,189	2,762,290
1975	1,652,251	449,271	493,125	328,337	196,043	3,119,027
1976	1,798,873	488,653	493,656	339,237	208,110	3,328,529
1977	2,093,312	446,406	522,678	350,222	297,132	3,709,750
1978	2,168,630	574,213	604,955	416,609	164,782	3,929,189
1979	2,237,393	612,820	653,033	474,122	113,090	4,090,458
1980*	2,559,142	552,944	576,086	412,038	451,961	4,552,171

Source: Anon., 1976, 1981.
* Preliminary.

TABLE 8—US Exports of Nitrogen Fertilizer Materials
1000 St, N

Year*	Ammonia	Ammonium Nitrate	Ammonium Sulfate	Urea	Nitrogen Solutions	Ammonium Phosphates	Other	Total
1960	66.0	21.0	51.0	16.0	—	8.0	26.0	188.0
1961	72.0	11.1	43.9	43.5	—	21.4	23.4	215.3
1962	66.0	12.4	90.4	41.7	—	13.6	10.5	234.6
1963	41.2	8.8	102.0	11.1	—	22.0	11.9	197.0
1964	66.9	12.9	86.8	20.0	—	49.4	30.5	266.5
1965	101.7	38.6	135.3	18.8	—	66.4	35.8	396.6
1966	107.8	28.1	296.5	17.2	—	83.1	18.2	550.9
1967	211.7	24.7	271.1	30.6	—	172.0	25.6	735.7
1968	382.0	16.2	257.6	67.1	—	273.0	31.4	1,027.3
1969	818.3	36.3	248.9	254.3	—	174.7	34.0	1,566.5
1970	627.0	26.8	111.0	301.9	—	177.5	50.2	1,294.4
1971	490.7	19.3	126.1	168.4	—	204.3	46.3	1,055.1
1972	345.1	11.1	117.1	209.0	—	277.5	53.9	1,013.7
1973	569.0	7.1	102.0	235.3	—	370.9	46.6	1,330.9
1974	436.3	12.2	117.1	145.1	—	387.7	52.6	1,151.0
1975	211.4	7.4	117.7	202.5	—	403.8	60.4	1,003.2
1976	276.3	20.3	157.9	261.2	—	489.8	28.6	1,234.1
1977	365.6	3.7	103.1	165.6	—	570.2	29.8	1,238.0
1978	468.5	10.5	136.9	412.7	2.5	730.6	29.2	1,790.9
1979	455.5	27.7	231.6	697.3	101.9	900.2	48.4	2,462.6
1980	638.8	24.8	176.6	660.3	164.3	924.2	41.0	2,630.0

Source: Anon., 1981d.
* Fiscal year ending June 30.

6% a year, exceeded only by the urea products. Total nitrogen solution capacity in the United States in 1980 was more than 3.1 million tons of nitrogen as 32% solution.

Challenges for the Industry

During the 1980s and beyond, the nitrogen industry will continue to face the severe challenges that arose or became much more severe during the 1970s.

There are four general areas of concern:

1. Meeting and maintaining environmental standards and continuing to improve the industrial work environment.
2. Continuing the substantial work already accomplished in energy conservation in existing and new production facilities.
3. Adapting processes and developing new schemes for conservation of key natural resources, including natural gas and petroleum derivatives.
4. Coping with the continuing sharp escalation in the investment for new facilities and plant modifications.

The US fertilizer industry, and in particular the important nitrogen segment, made great strides in meeting rather rigid environmental standards during the 1970s. Considerable work remains to be done, and much effort and money will be needed to maintain established environmental standards.

Energy conservation quickly became a major endeavor of industry in the early 1970s as the cost of natural gas, fuel oil, and electric power increased sharply. Major improvements were made by repairing steam leaks, operating equipment only when needed, and enforcing other good housekeeping practices. Tight surveillance on the nonprocess use of fuels such as natural gas became essential as its cost rose from 30¢ per million Btu's in the 1950s to $3 and higher in the late 1970s. Higher costs led to innovations in process design that allow greater heat recovery and consume less total energy. A large part of the improvements so far involve redesign of systems and components to better use and conserve heat energy from chemical reactions. These efforts must continue and will become even more important as the cost of natural gas, fuel oil, and electricity continues to increase. Unless proven reserves of natural gas are extended well past the turn of the century, the situation will become critical.

Considerable effort is underway in adapting processes to the use of coal instead of natural gas or petroleum fuels as the process feedstock for preparation of hydrogen. The United States and other countries have abundant coal reserves that will last for hundreds of years. Use of coal for producing hydrogen, however, is untested in most countries. The technology is difficult and investment in process systems is much higher. The experience gained in using coal as process feedstock in Germany and South Africa, and in the past few years in India, can be drawn on. The Tennessee Valley Authority began testing a demonstration plant in late 1980 using coal gasification to provide hydrogen equivalent for 135 tpd of ammonia to an existing plant. This is the first large facility of this type in the western hemisphere.

The sharp escalation in the investment required for new and modified production facilities throughout the 1970s had an immense effect on the production cost of nitrogen products. The cost of materials, process equipment, and total construction doubled between 1970 and 1980. The increase in construction cost indices showed no tendency to slow down in 1981.

Investment related costs comprise a large percentage of the total production costs for primary nitrogen compounds such as ammonia and urea. Estimates (1981 basis) indicated that \$87 of the total production cost of \$154 for urea was related directly to investments. This is about 57% of the total manufactured cost of urea in a facility with a 1000-tpd ammonia plant and a 1520-tpd urea plant. A high rate of capacity utilization for those large facilities is important because investment related costs continue whether the facilities are operating or standing idle.

Since a large part of the sharp escalation in investment costs throughout the 1970s was tied to general inflation in the economy, there is not much that can be done unless inflation is brought under control. However, the trend to larger production units has leveled off, raising the question of a new trend toward smaller facilities with lower capital outlay.

Not all opportunities for effective use of resources and controlling costs are on the supply

TABLE 9—US Imports of Nitrogen Fertilizer Materials
1000 St, N

Year*	Ammonia	Ammonium Nitrate	Ammonium Sulfate	Urea	Nitrogen Solutions	Ammonium Phosphates	Other	Total
1960	—‡	78.0	54.0	12.0	6.0	22.0	126.0	298.0
1961	—‡	80.0	40.7	40.6	19.3	18.4	128.3	327.3
1962	—‡	90.6	56.7	61.5	23.9	26.1	144.3	403.1
1963	—‡	101.1	47.4	96.2	23.5	27.7	114.4	410.3
1964	74.5	84.7	47.8	126.2	26.3	19.2	102.9	481.6
1965	145.8	59.4	40.5	110.9	23.6	20.1	93.2	493.5
1966	211.7	59.3	32.2	90.7	25.7	32.8	96.1	548.5
1967	321.9	57.9	35.8	123.8	26.4	34.9	86.8	687.5
1968	344.5	74.2	30.1	108.5	22.3	40.4	67.1	687.1
1969	348.6	77.7	28.3	113.0	25.9	49.9	62.7	706.1
1970	391.3	101.0	37.7	190.6	31.2	71.2	63.0	886.0
1971	411.2	120.8	45.9	148.3	62.2	84.9	74.5	947.8
1972	322.2	128.8	55.3	164.3	38.3	88.0	67.9	864.8
1973	281.3	108.7	58.0	302.3	46.3	78.1	60.1	934.8
1974	358.9	153.7	57.3	300.7	53.2	71.4	140.4	1,135.6
1975	490.6	153.7	52.1	365.3	29.3	44.5	130.3	1,265.8
1976	628.7	103.2	88.3	237.4	60.1	61.1	92.5	1,271.3
1977	793.4	130.3	95.4	659.7	142.2	69.7	88.4	1,979.1
1978	864.3	103.0	69.2	643.6	172.1	61.7	92.9	2,006.8
1979	1,429.4	90.5	54.0	490.5	59.9	54.0	83.2	2,261.5
1980	1,819.4	87.2	58.1	511.2	44.0	57.2	89.9	2,667.0

Source: Anon., 1981d.
* Fiscal year ending June 30.
† Includes materials for nonfertilizer use.
‡ Included in other nitrogen materials.

TABLE 10—Production of Nitrogen Fertilizer Materials in the US
1000 St

Year	Ammonia	Nitric Acid	Ammonium Nitrate Total	Ammonium Nitrate Solid*	Urea Total	Urea Solid*	Nitrogen Solutions†	Ammonium Sulfate
1960	4,817.7	3,315.0	3,121.7	1,572.0	746.8	301.9	804.0	1,491.7
1961	5,206.7	3,379.7	3,235.3	1,653.4	914.7	421.3	812.6	1,545.4
1962	5,809.7	3,669.6	3,405.5	1,668.2	1,029.4	466.0	891.2	1,697.5
1963	6,693.0	4,242.4	3,989.9	1,926.8	1,105.8	478.5	1,107.3	1,823.1
1964	7,634.2	4,732.5	4,543.0	2,161.5	1,243.6	546.1	1,156.1	2,306.1
1965	8.869.4	4,898.0	4,663.1	2,287.6	1,399.5	546.4	1,111.5	2,656.1
1966	10,604.7	5,513.8	5,116.8	2,516.8	1,768.4	693.5	1,198.5	2,859.5
1967	12,194.1	6,462.7	6,005.0	2,851.8	2,179.8	875.6	1,517.0	2,824.3
1968	12,119.9	6,992.0	5,737.1	2,951.7	2,436.4	985.1	1,149.8	2,723.3
1969	12,768.8	7,223.4	5,891.2	3,213.5	2,976.1	1,351.8	1,359.2	2,563.7
1970	13,824.0	7,602.8	6,456.0	3,314.6	3,250.0	1,201.6	1,720.6	2,489.4
1971	14,537.6	7,638.2	6,635.3	3,566.0	3,181.1	1,050.2	1,499.5	2,359.8
1972	15,169.3	7,981.2	6,863.2	3,813.5	3,424.2	1,317.0	1,464.9	2,419.1
1973	15,208.1	8,397.7	7,235.2	3,995.0	3,486.9	1,456.2	1,882.9	2,587.8
1974	15,732.5	8,119.7	7,541.9	3,858.5	3,664.5	1,435.1	2,101.8	2,658.0
1975	16,419.0	7,527.4	7,087.6	3,320.6	3,564.5	1,497.5	2,067.8	2,595.3
1976	16,716.1	7,791.1	7,185.5	3,574.2	3,928.2	2,433.1	2,117.1	2,507.2
1977	17,764.6	7,987.1	7,177.3	3,148.3	4,445.5	2,684.1	2,468.1	2,733.9
1978	17,119.3	7,933.7	7,210.1	3,000.7	6,272.6	3,161.6	2,286.2	2,877.1
1979	18,523.3	8,464.6	7,542.8	3,280.5	7,027.1	3,673.0	2,244.9	2,818.3
1980‡	19,035.9	8,932.5	8,589.5	3,249.6	7,218.2	3,150.7	3,031.4	2,493.8

Source: Anon., 1981b; Bridges, 1979.
* Fertilizer material only.
† Tons of nitrogen.
‡ Preliminary.

TABLE 11—North American Ammonia Plant Capacity
1000 St

Company	Location	1967	1973	1978	1980	1985
Agrico Chemical Co.—Williams	Blytheville, AR	340	390	407	407	407
	Donaldsonville, LA	—	400	468	468	468
	Verdigris, OK	—	—	840	840	840
Air Products & Chemicals, Inc.	New Orleans, LA	210	210	210	210	210
	Pace Jct, FL	100	100	100	100	100
Allied Corp.	Laplatte, NE	200	200	172	172	172
	Hopewell, VA	400	340	340	340	425
	Geismar, LA	—	340	340	340	340
	South Point, OH	240	240	160	—	—
	Helena, AR	—	—	210	210	210
American Cyanamid	Fortier, LA	394	340	340	580	580
Amoco Oil Co.	Texas City, TX	198	720	522	522	—
Apache Powder Co.	Benson, AZ	15	15	15	—	—
Apple River Chemical	East Dubuque, IL	230	230	—	—	—
Arkla Chemical Corp.	Helena, AR	210	210	—	—	—
Atlas Chemical (Tyler)	Joplin, MO	136	136	136	136	136
Beker Industries	Conda, ID	—	—	100	100	100
Borden Chemical Co.	Geismar, LA	—	340	340	340	400
	San Jacinto, TX	40	—	—	—	—
Calumet Nitrogen	Hammond, IN	140	—	—	—	—
Car-Ren, Inc.	Columbus, MS	—	—	68	68	68
CF Industries, Inc.	Donaldsonville, LA	—	750	1,590	1,590	1,590
	Fremont, NE	48	48	48	48	48
	Terra Haute, IN	150	150	150	150	150
	Tunis-Ahoskie, NC	—	210	210	210	210
	Tyner, TN	170	170	170	170	170

TABLE 11—North American Ammonia Plant Capacity (Continued)

Company	Location	1967	1973	1978	1980	1985
Chemical Distributors	Chandler, AZ	—	—	—	33	33
Chemicals, Inc.—IMCC	Bartow, FL	105	—	—	—	—
Chevron Chemical Co.	Pascagoula, MS	—	530	530	530	530
	Richmond, CA	130	130	130	—	—
	Fort Madison, IA	95	95	95	95	95
	El Segundo, CA	—	4	20	20	20
Cities Service	Lake Charles, LA	140	—	—	—	—
	Tampa, FL	120	120	—	—	—
Columbia Nitrogen	Augusta, GA	122	122	122	510	510
Cominco-American	Borger, TX	—	400	400	400	400
Diamond Shamrock	Deer Park, TX	35	—	—	—	—
	Dumas, TX	160	160	160	160	160
Dow Chemical Co.	Freeport, TX	115	115	115	115	115
	Midland, MI	34	—	—	—	—
	Plaquemine, LA	60	—	—	—	—
	Pittsburg, CA	12	—	—	—	—
E. I. du Pont de Nemours & Co.	Beaumont, TX	—	340	340	340	340
	Belle, WV	275	340	340	—	—
	Gibbstown, NJ	75	—	—	—	—
	Victoria, TX	100	100	100	100	100
Duval Corp. (Oxy)	Hanford, CA	—	21	42	—	—
El Paso Products	Odessa, TX	135	115	115	115	115
Estech General Chemical	Beaumont, TX	—	—	300	—	—
Farmland Industries	Fort Dodge, IA	210	210	210	210	210
	Dodge City, KS	—	210	210	210	210
	Hastings, NE	140	140	140	140	140
	Enid, OK	—	—	840	840	840
	Lawrence, KS	190	340	340	340	340
	Plainview, TX	26	—	—	—	—
	Pollock, LA	—	—	420	420	420
Felmont Oil Corp.	Olean, NY	85	85	85	85	85
First Mississippi Corp.	Fort Madison, IA	—	340	365	365	365
First Mississippi Corp. (Ampro)	Donaldsonville, LA	—	—	—	400	400
FMC Corp.	S. Charleston, WV	24	24	24	24	24
Gardinier	Tampa, FL	—	—	120	120	120
Georgia Pacific	Plaquemine, LA	—	—	—	196	196
Goodpasture, Inc.	Dimmitt, TX	31	31	71	40	40
Grace-Oklahoma Nitrogen	Woodward, OK	—	—	400	400	400
W. R. Grace & Co.	Woodstock, TN	275	340	340	340	340
	Big Springs, TX	100	100	—	—	—
Green Valley Chemical	Creston, IA	35	35	35	35	35
Gulf Oil Corp.	Pittsburg, KS	189	—	—	—	—
	Henderson, KY	107	—	—	—	—
	Vicksburg, MS	81	—	—	—	—
Hawkeye Chemical Co. (Getty)	Clinton, IA	138	138	138	138	220
Hercules, Inc.	Hercules, CA	70	70	—	—	—
	Louisiana, MO	70	70	70	70	70
Hooker Chemical Co.	Tacoma, WA	28	28	28	—	28
International Minerals	Sterlington, LA	140	340	770	400	770
Jupiter Chemical	Lake Charles, LA	—	—	78	78	78
Kaiser Agricultural Chemicals Co.	Savannah, GA	150	150	150	100	100
	Pryor, OK	—	—	—	105	105
Ketona Chemical Corp.	Ketona, AL	51	—	—	—	—
Mississippi Chemical Corp.	Yazoo City, MS	453	340	393	393	393
	Pascagoula, MS	175	175	175	175	175
Mobil Chemical Co.	Beaumont, TX	300	300	—	—	—
Monsanto Co.	El Dorado, AR	280	—	—	—	—
	Luling, LA	450	450	850	850	850
	Muscatine, IA	100	—	—	—	—
New Jersey Zinc-Gulf & Western	Palmerton, PA	35	35	35	35	35
Nipak, Inc. (Enserch)	Pryor, OK	105	105	105	—	—
	Kerens, TX	96	96	115	—	—

TABLE 11—North American Ammonia Plant Capacity (Continued)

Company	Location	1967	1973	1978	1980	1985
Nitrin, Inc.	Cordova, IL	140	—	—	—	—
Northern Chemical Industries	Searsport, ME	40	—	—	—	—
N-Ren Corp.	East Dubuque, IL	—	—	238	238	238
	Pine Bend, MN	90	—	—	—	30
	Carlsbad (Hobbs) NM	—	—	68	68	68
	Pryor, OK	55	55	94	94	94
Occidental Agricultural Chemical Co.	Taft, LA	—	—	90	90	90
	Lathrop, CA	120	120	160	—	—
	Plainview, TX	52	52	52	—	—
Olin Corp.	Lake Charles, LA	588	490	490	490	490
Pennwalt Chemicals	Wyandotte, MI	34	—	—	—	—
	Portland, OR	8	8	8	8	8
Phillips Pacific Chemical Co.	Kennewick, WA	155	155	155	155	155
Phillips Petroleum	Beatrice, NE	210	210	210	210	210
	Etter, TX	210	210	—	—	—
	Pasadena, TX	230	230	—	—	—
PPG Industries	Natrium, WV	50	50	50	50	50
Reichhold Chemicals	St. Helens, OR	—	90	90	90	90
Reserve Oil & Gas	Hanford, CA	21	—	—	—	—
Rohm & Haas	Deer Park, TX	45	45	45	—	—
Shell Chemical Co.	St. Helens, OR	90	—	—	—	—
	Pittsburg, CA	110	—	—	—	—
	Ventura, CA	165	—	—	—	—
J. R. Simplot	Pocatello, ID	54	54	108	108	108
	El Centro, CA	—	—	—	—	210
Southwest Nitrochem	Chandler, AZ	40	—	—	—	—
Sun Oil Co.	Marcus Hook, PA	133	133	—	—	—
Tenneco Chemical	Houston, TX	210	210	210	—	—
Tennessee Valley Authority	Muscle Shoals, AL	45	74	74	74	74
Terra Chemicals	Port Neal, IA	—	210	210	210	210
Texaco, Inc.	Lockport, IN	77	—	—	—	—
Triad Chemical	Donaldsonville, LA	—	340	340	340	340
Union Carbide Co.	Texas City, TX	88	—	—	—	—
Union Chemical Co.	Kenai, AK	—	510	1,020	1,020	1,020
	Brea, CA	380	260	280	280	280
U.S. Industrial Chemicals	Tuscola, IL	80	—	—	—	—
USA Petrochem Corp.	Ventura, CA	—	—	60	60	60
U.S.S. Agri-Chemicals	Clairton, PA	—	325	325	325	325
	Crystal City, MO	98	—	—	—	—
	Cherokee, AL	175	175	175	175	175
	Geneva, UT	70	70	70	70	70
Valley Nitrogen Producers	El Centro, CA	—	210	210	210	—
	Helm, CA	164	164	164	—	—
	Chandler, AZ	—	—	33	—	—
	Hercules, CA	—	—	70	—	—
Vistron Corp.	Lima, OH	136	450	475	475	475
Vulcan Materials	Wichita, KS	23	23	35	—	—
Wycon Chemical Co.	Cheyenne, WY	33	167	167	167	167
Total US		13,287	17,373	22,028	20,935	21,068
Canada						
Canadian Ind., Ltd.	Courtright, Ont.	340	340	400	400	810
	Millhaven, Ont.	66	—	—	—	—
Canadian Fert., Ltd.	Medicine Hat, Alta.	—	—	800	800	800
Cominco, Ltd.	Calgary, Alta.	125	125	125	125	125
	Trail, B.C.	155	70	70	70	70
	Carseland, Alta.	—	—	400	400	400
Cyanamid of Canada	Welland, Ont.	250	250	250	250	250
Dow Chemical of Canada	Sarnia, Ont.	140	—	—	—	—
Esso Chemicals	Redwater, Alta.	—	210	260	260	860

Compnay	Location	1967	1973	1978	1980	1985
Genstar Chemical	Maitland, Ont.	125	88	88	88	88
N. W. Nitro Chemicals	Medicine Hat, Alta.	66	—	—	—	—
Sherritt-Gordon Mine	Ft Saskatch, Alta.	160	160	160	160	572
J. R. Simplot Co.	Brandon, Manitoba	110	110	110	110	195
Western Coop Fertilizer	Calgary, Alta.	70	70	70	70	70
	Medicine Hat, Alta.	—	66	66	66	66
Total Canada		1,607	1,489	2,799	2,799	4,306
Total North America		14,894	18,862	24,827	23,734	25,374

Source: Capacity data from the records of Marketing and Distribution Economics Section, TVA-NFDC, Muscle Shoals, AL, May 1981.

side of the equation. It is estimated that as much as 50% of the nitrogen applied to the soil is unused by the crop and is eventually lost to the atmosphere or in surface and subsurface water runoff. Potential savings with increases in efficiency of as little as 10% has brought about a renewed research effort to find ways of reducing losses and developing new products that effectively control nitrogen release.

The demand for nitrogen fertilizers, essential to the future world food supply, will continue to expand as population and the demand for improved living standards continue to pressure the world's food supply. The challenge facing the nitrogen industry, both in the public and private sector, is to continue to meet these growing needs using the limited resources available as effectively as possible.

With an expanding market for its products and a sound technology base to work from, the world nitrogen market is in a good position to meet the challenge that it will face in the 1980s. Since the alternative is world famine and economic chaos, the nitrogen industry must continue the research and development programs that have been so important to agriculture.

Bibliography and References

Anon., 1964, "Chilean Nitrate Modernizes to Stay," *World Mining,* Vol. 17, No. 10, Sep., pp. 32–36, 89.

Anon., 1968, *Fertilizers: An Annual Review of World Production, Consumption, and Trade 1967,* Food and Agriculture Organization, United Nations, Rome, Italy, 211 pp.

Anon., 1974, *Annual Fertilizer Review 1973,* Food and Agriculture Organization, United Nations, Rome, Italy, 190 pp.

Anon., 1976, "Apparent Consumption of Industrial Explosives and Blasting Agents in the United States, 1912–1975," *Mineral Industry Surveys,* Aug. 21, US Bureau of Mines, 15 pp.

Anon., 1976a, "Nitrogen Fertilizer—A Potential Polluter," *Tennessee Valley Perspective,* Vol. 6, No. 4, Summer, pp. 15–19.

Anon., 1979, *Fertilizer Manual,* Ref. Manual IFDC–R–1, IFDC/UNIDO, International Fertilizer Development Center, Muscle Shoals, AL, 353 pp.

Anon., 1980, "Chilean Nitrate in the 1980s," *Nitrogen,* No. 28, Nov./Dec., pp. 25–28.

Anon., 1980a, *Fertilizer Reference Manual,* June, The Fertilizer Institute, Washington, DC, p. 86.

Anon., 1980b, *FAO Fertilizer Yearbook 1979,* Food and Agriculture Organization, United Nations, Rome, Italy, 143 pp.

Anon., 1980c, "New Developments in Fertilizer Technology," 13th Demonstration, Bulletin Y–158, National Fertilizer Development Center, Tennessee Valley Authority, Muscle Shoals, AL, 93 pp.

Anon., 1981, "Apparent Consumption of Industrial Explosives and Blasting Agents in the United States," *Mineral Industry Surveys,* annual reports, 1976–1981, US Bureau of Mines.

Anon., 1981a, *Commercial Fertilizers,* annual reports, 1969–1980, US Department of Agriculture Statistical Reporting Service, Washington, DC.

Anon., 1981b, "Inorganic Fertilizer Materials and Related Products," *Current Industrial Reports,* monthly, 1979–1980, Bureau of the Census, US Department of Commerce, Washington, DC.

Anon., 1981c, "Fertilizers: Production, Trade, and Consumption of Nitrogen, Phosphate, and Potash," *Monthly Bulletin of Statistics,* Vol. 4, No. 3, Mar., Food and Agriculture Organization, United Nations, Rome, Italy, pp. 13–20.

Anon., 1981d, *The Fertilizer Supply,* annual reports, 1960–1980, US Department of Agriculture, Agriculture Stabilization and Conservation Service, Washington, DC.

Anon., 1981e, "New Ammonia Technology Saves Energy," *Farm Chem.,* Vol. 144, No. 2, Feb., pp. 48, 50.

Achorn, F.P., and Salladay, D.G., 1976, "Pipe-Cross Reactor Eliminates the Dryer," *Farm Chem.,* Vol. 139, No. 7, July, pp. 34, 36, 38.

Auldridge, L., 1980, "World Oil Flow Slumps, Reserves Up," *Oil and Gas Journal,* Vol. 78, No. 52, Dec. 29, pp. 75–82.

Billmeyer, F.W., Jr., 1971, *Textbook of Polymer Science,* 2nd ed., Wiley-Interscience, New York, 598 pp.

Bridges, J.D., 1980, "Fertilizer Trends 1979," Bulletin Y–150, Jan., Tennessee Valley Authority, Muscle Shoals, AL, 48 pp.

Briggs, T.C., 1975, "Nitrogen Compounds," *Industrial Minerals and Rocks,* 4th ed., S.J. Lefond, ed., AIME, New York, pp. 895–920.

Graham, H.R., 1949, "Nitrates and Nitrogenous Compounds," *Industrial Minerals and Rocks,* 2nd ed., AIME, New York, pp. 643–660.

Hargett, N.L., and Berry, J.T., 1981, "1980 Fertilizer Summary Data," 12th ed., Bulletin Y–165, Apr., Tennessee Valley Authority, Muscle Shoals, AL, 136 pp.

Harris, G.T., and Harre, E.A., 1979, "World Fertilizer Situation and Outlook—1978–85," Technical Bulletin IFDC–T–13, Mar., International Fertilizer Development Center, Muscle Shoals, AL, 27 pp.

Hignett, 1971, "Liquid Fertilizer Production and Distribution," TVA Circular Z–27, National Fertilizer Development Center, Tennessee Valley Authority, Muscle Shoals, AL, 38 pp.

McCamy, I.W., and Norton, M.M., 1977, "Have You Considered Pan Granulation of Urea?" *Farm Chem.,* Vol. 140, No. 2, Feb., pp. 61, 64–65, 68.

McGannon, H.E., 1971, *The Making, Shaping and Treating of Steel.* 9th ed., US Steel Corp., Pittsburgh, PA, pp. 168–170.

McVickar, M.H., et al., 1966, "Agricultural Anhydrous Ammonia Technology and Use," Agricultural Ammonia Institute, Memphis, TN, 314 pp.

Noyes, R., 1967, "Ammonia and Synthesis Gas," Chemical Process Monograph No. 26, Noyes Development Corp., Park Ridge, NJ, 177 pp.

Postgate, J.R., 1971, *The Chemistry and Biochemistry of Nitrogen Fixation,* Plenum Press, New York, 326 pp.

Sauchelli, V., 1964, *Fertilizer Nitrogen—Its Chemistry and Technology,* Reinhold, New York, 424 pp.

Shields, J.T., et al., 1975, "An Appraisal of the Fertilizer Market and Trends in Asia," Bulletin Y–95, National Fertilizer Development Center, Muscle Shoals, AL, 141 pp.

Sittig, M., 1965, *Nitrogen In Industry,* Van Nostrand, Princeton, NJ, 278 pp.

Slack, A.V., and James, G.R., 1979, *Ammonia* (four parts) Fertilizer Science and Technology Series, Part IV, Marcel Dekker, Inc., New York, pp. 16–17.

Viets, F.G., Jr., and Hageman, R.H., 1971, "Factors Affecting the Accumulation of Nitrate in Soil, Water, and Plants," *Agriculture Handbook* No. 413, US Dept. of Agriculture, Washington, DC, Nov. 63 pp.

Young, R.D., et al., 1962, "TVA Process for Production of Granular Diammonium Phosphate," *Journal of Agriculture Food Chemistry,* Vol. 10, Nov./Dec., pp. 442–447.

Young, R.D., and Hicks, G.C., 1967, "Production of Monoammonium Phosphate in a TVA-Type Ammonium Phosphate Granulation System," *Commercial Fertilizer,* Vol. 114, No. 2, Feb., pp. 26–27.

Young, R.D., and McCamy, I.W., 1967, "TVA Development Work and Experience with Pan Granulation of Fertilizers," *The Canadian Journal of Chemical Engineering,* Vol. 45, Feb., pp. 50–56.

Young, R.D., 1980, "TVA Development of Pan Granulation Processes for Nitrogen Fertilizers," Bulletin Y-160, Dec., National Fertilizer Development Center, Tennessee Valley Authority, Muscle Shoals, AL, 16 pp.

Olivine

KEFTON H. TEAGUE *

Olivine is a mineral containing a mixture of forsterite (Mg_2SiO_4) and fayalite (Fe_2SiO_4) in solid solution. The name *olivine* was first applied by Werner in 1790 (Hunter, 1941) because of the olive-green color of the mineral.

Olivine is the principal component of the rock dunite, which itself is a member of the peridotite group of ultrabasic igneous rocks. European uses of the terms *olivine* and *dunite* are somewhat different than US usage. In commerce European usage defines *dunite* as containing 36 to 42% MgO, 36 to 39% SiO_2, and loss on ignition of approximately 10%. The term *olivine* is used to designate a material containing approximately 85% forsterite, with a chemical composition of approximately 45 to 50% MgO, 40 to 43% SiO_2, 5 to 8% Fe_2O_3, and ignition loss of 1 to 2%.

Uses

The principal use of olivine is in various applications involving hot metal. Table 1 shows the best estimate available as to the amount of olivine consumed by various industries worldwide. In excess of 90% of world uses involve hot metal, with approximately 75% as a slag conditioner in the blast furnace production of pig iron. Approximately 15% of total olivine usage is as a special foundry sand in mold making for the brass, aluminum, magnesium, and manganese steel foundries (Schaller, 1957, 1958; Snyder, 1957; Anon., 1977).

Olivine was first used as an industrial mineral in the early 1930s for a refractory material (Anon., 1970; Hunter, 1941). As a refractory raw material, olivine was first introduced in the United States as hand-cobbed, selected, shaped blocks of crude olivine. This use met with limited success. More recently, finely ground olivine blended with MgO and pressed into bricks, which are then fired, has found use in glass tank furnaces and open-hearth furnaces. Ramming or gunning mixes for basic furnace linings also utilize olivine. Olivine has been used in ladle linings with varied success. In Europe, substantial tonnages of olivine are utilized in refractory brick for night storage heaters (Anon., 1970).

A limited amount of olivine has been used in the past as a fertilizer (magnesium source) and has been fused with rock phosphate to produce a magnesium phosphate as a plant food. The relatively high magnesia (MgO) content of olivine also attracted attention to it as a potential source of both magnesium compounds and as a source of metallic magnesium (Bengston, 1956; Hunter, 1941). The use of olivine as an additive in the blast furnace is relatively new, having been developed during the last decade. J.W. Currier (Private communication, 1978) describes the behavior of olivine in the blast furnace as follows:

> Olivine (Mg, Fe_2, SiO_4) represents a unique lime-free source of magnesia and silica as a blast furnace charge component. Only recently has the magnesia-rich mineral been recognized as beneficial cleaner for blast furnace experiencing the problems associated with excessive alkali inputs and/or low coke stabilities.
>
> For some time it has been known that alkalis tend to recycle and accumulate in blast furnaces. In the high temperature region of volatilization the elemental alkalis formed attack the integrity of the graphite structure. This alkali attachment is enhanced when poor coke is used. Higher in the furnace stack volatile alkali cyanides condense and oxidize. This accumulation of alkalis tends to agglomerate the burden materials, producing scabs and scaffolds. These structures not only inhibit the downward movement of burden materials and the upward passage of wind, but also lead to erratic furnace performance

* Senior Geologist, Industry Group, International Minerals & Chemical Corp., Mundelein, IL.

TABLE 1—World Olivine Usage, 1978

Use	Short Tons
Blast furnace	3,000,000
Foundry sand	600,000
Heat storage blocks	75,000
Refractories	75,000
Abrasives	50,000
Miscellaneous	100,000
Total	3,900,000

such as irregular cast temperatures, off-quality hot metal, burden slips, broken tuyeres, coke messes, and excessive coke consumption and restricted production. In short, under an alkali problem a furnace fails to function and respond to control predictably.

Since alkali problems can occur at alkali inputs as low as 4 lb total alkali oxides per net ton of hot metal, the restriction of alkali inputs is not always practical. Indeed, the reduction in the quality of ores and coals available mandates higher alkali inputs. In order to handle these larger inputs special cleaning practices have been developed to cope with the problem. Calcium chloride has been found to be slightly effective in alleviating some of the symptoms associated with high alkali loadings. The charging of calcium chloride does not alleviate the problem; however, more effective in coping with the alkali problems is a reduction of the slag base/acid ratio in order to allow the slag to carry and purge more of the basic alkali components. This has several limitations, however. The acidic slags usually cannot produce the appropriate quality hot metal. Therefore, most operators resort to the intermittent charging of quartzite to purge the alkalis on an intermittent basis. While this can be shown to have some effect on the quantities of alkali purged, it serves to accentuate the nonsteady rate of the furnace operation and performance.

Comparatively recently it has been found that continuous charging of olivine can have a marked effect on a blast furnace alkali problem. An appropriate charging of olivine in the region of 20 to 25 lb/NTHM has been shown to be effective even at alkali oxide loadings as high as 30 lb/NTHM. This charge rate corresponds to 25 tons of olivine per day for an average sized furnace. The olivine is charged as continuously as practicable for the particular charge sequence and system. For the most part the olivine should be toward the outside of the furnace and segregated from all lime components as much as possible. The tendency toward higher MgO contents in the slag can be compensated for by reduced additions of dolomite without diminishing the beneficial effect.

Under the above continuous charging conditions smoother alkali purging results and the tendency for alkali accumulation is markedly reduced. As a result, erratic furnace performance, slips, peels, coke messes, and the like are dramatically reduced. It becomes possible to achieve a steady state furnace performance in which the normal controls predictably influence the independent variables of temperature, hot metal quality, permissible wind rates, and burden movement.

The blast furnace tests with olivine to date have been reasonably consistent. Production increases of as much as 10% have been documented. Coke savings of the order of 4% are common and provide an adequate economic incentive to use this practice. In addition, uniform hot metal quality within the limits that can be expected of this slag operating practice are attained without exception.

As a foundry sand olivine is used in mold making for the brass, aluminum, magnesium, and manganese steel foundries. It has a fusion point of approximately 1816°C (3300°F) and lower thermal expansion than silica; thus, it exhibits less defects in the castings such as "rat tails," "scabs," and "buckles." Little or no free silica occurs with olivine, so the silicosis hazard is reduced as compared to the use of normal silica sand in foundry applications.

Normally, coarse sand is used in manganese steel foundries and the finer sands in aluminum, magnesium, and brass foundries. Fine sand is used in the manufacture of "hot tops" where quartz sand is not permitted because of health hazards.

Industries using olivine have not standardized specifications. Normally, producers supply a product which meets individual customer requirements. Such requirements involve type and amount of impurities, grain size of product, refractory characteristics, chemical analysis, etc.

Geology

Olivine occurs commonly as accessories in basic igneous and basic metamorphic rocks. Economic deposits are of magmatic origin and are restricted to essentially the dunite variety of peridotite. Dunites are medium to coarse-grained crystalline rocks, generally reddish brown on weathered outcrop, and are composed primarily of olivine.

Mineralogy

The name *olivine* is a generic term used to indicate a group of orthosilicate minerals in isomorphous series, with magnesium-rich forsterite (Mg_2SiO_4) and iron-rich fayalite (Fe_2SiO_4) as end points (Ramberg and DeVore, 1951; Reed, 1959). In commercial practice, use of the term olivine applies to deposits containing a mixture of forsterite and fayalite in solid solution with fayalite content usually restricted to less than 15%. Accessory constituents may include the primary minerals: ilmenite, magnetite, chromite and garnet; and various secondary minerals as alteration products.

The chemical composition of typical olivine may be expressed either as $(Mg,Fe)_2SiO_4$ or as $2(Mg,Fe)O.SiO_2$. The ratio of Mg:Fe varies from 16:1 to 2:1, passing from forsterite on the one hand to fayalite on the other. Chemical analyses of some commercial olivines are noted in Table 2.

Mode of Occurrence and Origin

Commercial olivine occurs as alpine-type and zoned dunite bodies. Controversy exists among various researchers concerning the origin and mode of emplacement (Astwood et al., 1972; Bennett, 1940; Bowen and Schairer, 1935, 1936; Gaudette, 1963; Mossman, 1972; Ragan, 1959, 1961; Ross et al., 1954; Yudin, 1959). There is general agreement, however, that most olivines have been emplaced in a partially crystallized condition (mush) and that in part, at least, foliation present in many unmetamorphosed dunites results from flowage during emplacement prior to complete crystallization.

All commercial bodies are of igneous origin and may be in part magmatic segregates from basic magmas. Domestic olivine bodies, of present or potential economic importance, range in age from pre-Ordovician for the North Carolina deposits, to post early-Tertiary age for the Washington dunites.

TABLE 2—Chemical Analyses of Some Commercial Olivines, %

	1	2	3	4	5	6
MgO	47.5	43-44	49.0	47.7	46.4	46.9
SiO_2	40.4	24-35	42.6	40.8	42.5	40.8
Fe_2O_3	9.0	7.6-7.7	6.0	7.5	8.0	9.4
Other oxides	2.5	0.7-0.8	1.8	1.9	2.4	3.2
LOI	0.8	?	0.6	2.0	0.5	0.6

1 Ste. Anne des Monts, P. Q., Canada
2 Leoben, Austria
3 Aaheim, Norway
4 Norddal, Norway
5 Burnsville, NC
6 Hamilton, WA

Source: Anon., 1970.

Distribution of Deposits

Large deposits of olivine-bearing dunites crop out in Norway, Sweden, USSR, Austria, Japan, New Zealand, Zimbabwe, South Africa, United States, New Caledonia, Italy, Greece, Spain, India, Brazil, and Canada (Anon., 1970; Bennett, 1940; Brothers, 1960; Du Rietz, 1935; Francis, 1956; Gwinn, 1943; Hunter, 1941; Ragan, 1961; Roberts, 1930; Yudin, 1959).

In the United States numerous lenslike bodies extend in a belt from northeastern Georgia, northeastward across western North Carolina. Some 25 of these deposits are relatively fresh, with each containing up to five million tons of recoverable olivine.

Many small to large dunite bodies occur in northwestern Washington, with the Twin Sisters deposit being the largest.

Taylor (1967) describes eight zoned ultramafic complexes in southeastern Alaska containing cores of dunite. Some of these olivine cores are as great as one mile in diameter.

Reserves

The largest body of olivine in the United States is the Twin Sisters deposit in Whatcom and Skagit Counties, Washington. This dunite has an outcrop area of approximately 36 square miles and has a relief of about 5000 ft.

Other countries containing large reserves of olivine include Norway, where the Aaheim deposit is reported to contain two billion tons. Several other Norwegian deposits contain up to five million tons. In Sweden the largest deposit is at Arutats and is reported to be comparable in size to the Twin Sisters deposit. In Austria large deposits of serpentinized-olivine occur in the Province of Styria. Scotland, Spain, and Italy contain large deposits of serpentinized olivine. In New Zealand the Dun Mountain dunite is reported to be approximately 1½ miles in diameter. This deposit is located near Nelson, New Zealand (Gwinn, 1943). A dunite deposit in the coastal area of southern New Zealand has been described by Mossman (1972) "as exceeding 600 meters in thickness." This occurrence apparently is a magmatic differentiate in the Greenhills Ultramafic Complex.

Unaltered dunite crops out as a lenslike body approximately 25 miles south of Ste. Anne des Monts on the Gaspé Peninsula of Quebec, Canada. Dunite is also reported in Newfoundland.

Exploration

Most of the dunites containing economic olivine outcrop as prominent ridges or domes. Soils developed from the weathering of dunites support limited vegetation, thus many deposits crop out as "balds."

As a first step in exploration, an olivine deposit should be thoroughly investigated by reconnaissance techniques followed by detailed topographic and geologic mapping. Grab samples, representing various varieties of dunite which occur in the deposit, should be collected for petrographic and chemical evaluation. Grain size, alteration products, and distribution of impurities should be noted, both in samples and throughout the deposit. Portions of the deposit covered with soil should be trenched to determine whether or not the bedrock is olivine. For example, some of the North Carolina olivines, originally regarded as being relatively continuous, have shown by trenching and drilling that areas covered by soil are underlain by non-olivine material that has resulted from alteration of the olivine.

After the surface extent of the olivine deposit is established, details of vertical variation should be determined. As most olivine bodies in the Southeastern deposits are rather closely fractured and jointed and may contain alteration products considerably softer than olivine, core drilling is not recommended for determining vertical continuity. Some variety of percussion drilling, such as Air-Trac drilling or down-the-hole drilling is normally superior to core drilling in olivine deposits. The driller can note variations in drilling rates as well as character of cuttings. If the drill is equipped with a vacuum pump, continuous samples of cuttings can be collected and evaluated.

Evaluation of Deposit

A commercial olivine should contain in excess of 40% MgO, before any beneficiation. Many olivine deposits, such as Twin Sisters in Washington, and those in Norway and Sweden, are relatively pure and are amenable to evaluation by dry crushing, followed by screen analysis of the crushed product and chemical and petrographic analysis of the various size fractions. Refractory characteristics also should be determined on the various screen sizes.

Olivine samples which contain objectionable impurities can be evaluated by crushing to a size that will permit removing these impurities and evaluating the clean olivine. Perhaps the most efficient laboratory method to accomplish removal of impurities from crushed or broken samples is to use heavy liquids. Most objectionable impurities which occur with olivine "float" on a liquid of specific gravity of about 3.0. Accessory chromite and olivine report as "sinks." If complete liberation is obtained prior to separation with heavy liquid, the resultant olivine product can be evaluated as outlined previously. Olivine deposits, which require beneficiation in order to produce a usable olivine concentrate, should yield at least 65% product to be considered commercial.

Most accessory minerals associated with olivine lower its refractory characteristics. For example, serpentine, chlorite, and vermiculite contain 12 to 17% water of crystallization, and unless these materials are removed from the olivine, they will contribute to a high loss-on-ignition of the product. Also, the platy minerals, chlorite and vermiculite, react explosively when heat shocked. Olivine sands which are to be coated should contain a minimum amount of platy minerals because of the surface area-weight ratio.

Preparation for Market

Mining Methods

Methods employed in the United States and in Europe in mining olivine are similar. Where possible, open cut methods are used with bench heights ranging from 12 to 20 ft.

North Carolina: There is considerable variation in accessory and alteration minerals in the North Carolina deposits; therefore, prior to mining, rather detailed quality data is needed for mine planning. Percussion drill holes, up to 70 ft deep, spaced on centers up to 50 ft, are drilled in advance of mining. Samples of the drill cuttings are collected and are evaluated for olivine and impurity content. Data developed from this exploration are used in mine planning.

Face heights range up to 20 ft and blasting charges are relatively small, generally yielding less than 10,000 tons of crude olivine. Holes are drilled on a 4 to 6-ft grid pattern which normally results in reasonably good breakage. Some secondary drop-ball work is required at times.

Washington: One of the Twin Sisters olivine deposits presently being worked is a talus accumulation along the south side of the Twin Sisters body. Olivine fragments have accumulated at the base of the slope. It is loaded by shovel and trucked to a portable crushing-washing plant. This plant consists of a jaw crusher, which reduces the olivine to approximately —4 in., screening facilities where the —⅛-in. material is removed, and a picking belt from which the non-olivine material is removed. The olivine is trucked approximately 25 miles to the processing plant at Hamilton, WA.

At one time underground mining methods were attempted on the Twin Sisters deposit. A high quality product was produced; however, the nature of the rock, especially its closely spaced fractures and joints, resulted in heavy, unsafe ground. Ore costs from underground mining were substantially greater than presently experienced utilizing the talus material.

In the near contact zone of the Twin Sisters deposit, relief is severe. It would be possible to mine this rock by open cut methods, but mining cost again would exceed present cost experienced by mining talus accumulation of olivine. One producer active on the Twin Sisters deposit is currently producing material from bedrock in an area on the northwestern side, where relief is relatively moderate.

In the production of olivine for use as a slag conditioner, processing involves crushing and sizing to normal sizes of —1¼ +¼ in. Depending on the amount of impurities in the —¼ in., this may or may not be used as a sand source. In production of fine sizes (sand), again crushing and sizing are involved. Dry processing is utilized where possible, such as for the Washington, Norway, and Sweden operations.

Mining cost of the ore ranges between $3 and $7 per ton. The uniformity and purity of the ore and topography of the deposit are major factors in mining cost.

Milling and Processing Techniques

Olivine is reduced by jaw crushers, cone crusher, and impact or roll crushers to the desired size. In addition to a dry operation, the A/S Olivine operation at Aaheim, Norway, is reported to process the sand through a washing procedure. This process is described as follows (Anon., 1970):

> Part of the material fed through the primary crusher is damp. This is passed through a washing plant, which comprises the second line at Aaheim. Before being washed, the olivine is crushed in the Prall-Mühle impact mill to give a suitable grain size. Washing takes place in two Callow cones placed in line. From the washing plant the olivine—which is now mainly free from particles less than 0.02 mm in size, is conveyed to a screw classifier for primary dewatering. Simultaneously, with the removal of water, some of the remaining fine particles (fines) are eliminated. In the next stage the water content of the sand is reduced to 8% by passing it through a centrifuge. The moisture content is finally reduced to maximum 0.5% in a vertical dryer.

Because of the relatively high content of undesirable impurities in the North Carolina olivine, it is necessary to beneficiate it in order to produce a high quality foundry sand. Flotation was used when beneficiation was first commenced in the 1940s. Later gravity methods, using Humphrey's spiral or tables, replaced flotation. Processing in North Carolina is essentially as follows:

Quarry rock in the —2 ft range is loaded, using either shovels or front-end loaders, trucked from quarry to the primary crusher, reduced to —4 in., and stage crushed by jaw or shorthead crushers to —1 in. Fines (—20 mesh) are removed, slurried, deslimed at 200 mesh, sized, and sent to either concentrating tables or flotation. The plus 20 mesh material is wet rod milled, deslimed, sized, and moved either to concentrating tables or flotation.

Concentrate is recombined and dewatered, dried, and sized. Foundry sand may be shipped either in bulk or bagged.

Milling cost in the production of olivine relates closely to the purity of the olivine ore and beneficiation requirements. In general, cost of milling ranges between $12 and $20 per ton.

Normally, the finer sands have a f.o.b. price approximately $10 greater per ton than does the coarser sands. Price ranges for sand are generally between $40 and $60 per ton (bagged) and somewhat less per ton for bulk. Special sands with special size distribution command higher prices.

Transportation

In the United States olivine is transported from plant to market by either truck or rail. Transportation cost to the main consuming markets are substantial, frequently equaling the f.o.b. plant cost of the olivine.

In Europe truck, rail, and water transportation is used for moving the olivine to market. These transportation costs are thought to be substantially less than those experienced by US producers.

Consumption

Industries utilizing olivine generally are located in those countries which have major production or have access to this production. Industries in Japan utilizing olivine represent the largest market.

Table 3 represents the best available data of current production figures for the principal producing countries.

In Europe major consuming industries are located in Norway, Sweden, West Germany, France, Spain, Italy, and England. In these countries substantially larger tonnages of olivine enter the foundry sand application than in the United States. This situation is brought about, in part, by the higher cost of good silica foundry sand and the desire to avoid all risks of silicosis.

TABLE 3—World Olivine Production, 1978

Country	Short Tons
Japan	2,100,000
Spain	700,000
Norway	400,000
United States	240,000
Italy	100,000
Sweden	50,000
Austria	30,000
Other	300,000
Total	3,900,000

Future Considerations and Trends

As mentioned under uses, perhaps 90% of olivine is used in conjunction with the production and utilization of hot metal. Therefore, demand for olivine is closely related to the activity in this industry. When blast furnaces are operating at a high level, a correspondingly high demand for slag conditioner is to be expected. The requirements for foundry sand fluctuate with the activity in the nonferrous and manganese steel foundries.

Palmour, et al. (1979) indicates the demand for high heat capacity refractory ceramics utilizing olivine for use in heat storage furnaces in the United States may offer a substantial market for olivine.

TABLE 4—Comparison of Applications, Olivine vs. Other Refractory Materials

Advantages	Disadvantages
Resistance to most basic slags.	Resin-bonded foundry cores cannot be made with olivine sand unless calcined.
No silicosis hazard.	Attacked by acid slags.
Price advantage over similar materials (chromite & zircon) for use in applications requiring low expansion, high cooling, and penetrating resistance.	Melting point is lowered in a reducing atmosphere (ferric to ferrous iron).
Requires smaller amounts of bentonite as binder (disperse grain-size distribution).	Hydrated contaminant causes high loss on ignition.
Can be recycled, owing to its long life and grain-size durability.	Requires larger amounts of liquid binders such as self-curing oil; also, cold-setting resins must be neutral, hardened by an isocyanate and a catalyst to equal the characteristics attained with silica sands.
Washed sand product is more concentrated and has lower content of fines (suitable for fluid binders used in core production and in resin-coated sand for shell moldings).	High chilling capacity of olivine linings and limited resistance to thermal shocks.
Grain surfaces are more angular than silica sands.	Relatively low durability of forsterite bricks compared to chromite bricks, because of spalling.
Improved finish of foundry castings (no scabs or metal penetration).	Not as plentiful or readily available as silica sand.
High mechanical strength from the texture and fabric of dunite (irregular intergrowth of olivine crystals).	
Very accurate dimensions of castings (free from microporosity) produced in nonferrous foundries (aluminum, bronze, etc.).	

Source: Beckius, 1970.

For foundry sand uses olivine competes with quartz, chromite, and zircon. The price of quartz sand is less than that of olivine; however, chromite and zircon sands command substantially higher prices than olivine. There is general concern in foundries where silica sand is used because of the silicosis hazard; in those applications where olivine is substituted for silica sand, the silicosis hazard is removed. Future increased uses of olivine as a foundry sand may be at the expense of silica sand because of the silicosis problem.

Table 4 summarizes the advantages and disadvantages of olivine as a foundry sand with competitive materials.

Olivine is permitted a 22% depletion allowance for tax purposes. It may enter either the United States or Canada duty free.

As has been mentioned previously, almost all of the United States production of olivine is by open cut methods. State or federal regulations now require that all disturbed areas in the course of open cut mining of olivine be reclaimed. Also all dissolved and suspended solids must be removed from process water before it is returned to streams. These and possible other regulations pertaining to environmental protection and land use must be considered concerning their effect on the economics and hence market position of olivine. On the positive side is the health and safety factor of elimination of silicosis when olivine is substituted for silica sand.

Bibliography and References

Anon., 1970, "Opportunities for Increasing Olivine Output," *Industrial Minerals,* No. 29, Feb., pp. 11–21.

Anon., 1976, "The Industrial Minerals of Spain," *Industrial Minerals,* Apr., pp. 15–51.

Anon., 1977, "Olivine and Dunite—Blast Furnace Usage Adds New Dimension," *Industrial Minerals,* May, pp. 39–50.

Ashby, G., 1976, "Minerals in the Foundry Industry," Second Industrial Minerals International Congress, Munich.

Astwood, P.M., et al., 1972, "A Petrofabric Study of the Dark Ridge and Balsam Gap Dunites, Jackson County, North Carolina," *Southeastern Geology,* Vol. 14, No. 3, Sep., pp. 183–194.

Beckius, K., 1970, "Olivine: Its Properties and Uses," *Industrial Minerals,* No. 29, Feb., pp. 22–26.

Bengston, K.B., 1956, "Magnesium from Olivine *via* Chlorination: A Possibility, Trend in England," University of Washington, Vol. 8, Jan., pp. 21–26, 35–36.

Bennett, W.A.G., 1940, "Ultrabasic Rocks of the Twin Sisters Mountains, Washington," *Bulletin,* Geological Society of America, Vol. 51, p. 2019.

Bowen, N.L., and Schairer, J.F., 1935, "The System MgO, FeO, SiO_2," *American Journal of Science,* 5th Ser., Vol. 29, pp. 151–217.

Bowen, N.L., and Schairer, J.F., 1936, "The Problem of the Intrusion of Dunite in the Light of the Olivine Diagram," *Proceedings,* 16th International Geological Congress, Vol. 1, pp. 391–396.

Brothers, R.N., 1960, "Olivine Nodules from New Zealand," *Proceedings,* 21st International Geological Congress, Pt. 13, pp. 68–81.

Dana, E.S., and Ford, W.E., 1940, *Textbook of Mineralogy,* John Wiley, New York, 851 pp.

Du Rietz, T., 1935, "Peridotites, Serpentines, and Soapstones of Northern Sweden," *Geologiska Foreningen i Stockholm,* Forh. 401, Vol. 57, pp. 133–260.

Francis, G.H., 1956, "The Serpentine Mass of Glen Urquahart, Iverness-Shire, Scotland," *American Journal of Science,* Vol. 254, pp. 201–226.

Gaudette, H.E., 1963, "Geochemistry of the Twin Sisters Ultramafic Body, Washington," unpub. Ph.D. Thesis, University of Illinois, Urbana, 104 pp.

Goldschmidt, V.M., 1938, "Olivine and Forsterite Refractories in Europe," *Industrial & Engineering Chemistry,* Vol. 30, pp. 32–33.

Gwinn, G.R., 1943, "Olivine," Information Circular 7239, US Bureau of Mines, 11 pp.

Hunter, C.E., 1941, "Forsterite Olivine Deposits of North Carolina and Georgia," Bulletin 41, North Carolina Dept. of Conservation & Development, Div. of Mineral Resources, 117 pp.

Kaufman, A.J., 1952, "Industrial Minerals of the Pacific Northwest." Information Circular 7641, US Bureau of Mines, p. 37.

Lamont, W.E., et al., 1977, "Olivine Foundry Sand from North Carolina Dunite by Differential Grinding," Preprint 77–H–369, SME-AIME Fall Meeting, St. Louis, 22 pp.

Misra, K.C., and Keller, F.B., 1978, "Ultramafic Bodies in the Southern Appalachians: A Review," *American Journal of Science,* Vol. 278, pp. 389–418.

Mossman, D.J., 1972, "The Geology of the Greenhills Ultramafic Complex, Bluff Peninsula, Southern New Zealand," *Abstracts with Programs,* Vol. 4, No. 7, Oct., p. 605.

Palmour, H., III, et al., 1979, "Ceramics for Energy Storage Units: Bricks from North Carolina Olivine for Heat Storage Furnaces," North Carolina Minerals Research Laboratory Report No. 79–16–P.

Ragan, D.M., 1959, "The Mode of Emplacement of the Twin Sisters Dunite, Washington," *Bulletin,* Geological Society of America, Vol. 70, pp. 1742–1743.

Ragan, D.M., 1961, "Geology of the Twin Sisters Dunite in the Northern Cascades, Washington," unpub. Ph.D. Thesis, University of Washington, 88 pp.

Ralston, O.C., 1937, "Annual Report of the Nonmetals Division, Fiscal Year 1937," Information Circular 6974, US Bureau of Mines, p. 9.

Ralston, O.C., et al., 1938, "Annual Report of the Nonmetals Division, Fiscal Year 1938," RI 3427, US Bureau of Mines, pp. 5–9.

Ralston, O.C., et al., 1939, "Annual Report of the Nonmetals Division, Fiscal Year 1939," Report of Investigations, 3473, US Bureau of Mines, pp. 27–28.

Ramberg, H., and DeVore, G.W., 1951, "The Distribution of Fe and Mg in Coexisting Olivines and Pyroxenes," *Journal of Geology,* Vol. 59, pp. 193–210.

Reed, J.J., 1959, "Chemical and Modal Composition of Dunite from Dun Mountain, Nelson," *New Zealand Journal of Geology and Geography,* Vol. 2, No. 5, pp. 916–919.

Roberts, M., 1947, "Washington's Vast Olivine Deposit," *Mining Congress Journal,* Vol. 33, No. 6, pp. 29–32.

Ross, C.S., et al., 1954, "Origin of Dunites and of Olivine-rich Inclusions in Basaltic Rocks," *American Mining,* Vol. 39, pp. 693–737.

Schaller, G.S., 1957, "New Foundry Sand," *Modern Metals,* Vol. 13, No. 9, Oct., pp. 82–86.

Schaller, G.S., and Snyder, W.A., "Industrial Applications of Olivine Aggregate," *Modern Castings,* Vol. 33, No. 6, June, pp. 99–104.

Snyder, W.A., 1957, "How to Use Olivine Sand," *Foundry,* Vol. 85, No. 9, pp. 100–105.

Taylor, H.P., Jr., 1967, "The Zones Ultramafic Complexes of Southeastern Alaska," *Ultramafic and Related Rocks.*

Yudin, M.I., 1959, "Dunite of the Boris Mountain Range and Their Origin," *Academy of Science Bulletin,* USSR, Geol. Ser. 2., pp. 47–62.

Perlite

FREDERIC L. KADEY, JR.*

Perlite, as a volcanic glass, has been recognized since the Third Century, B.C. (Langford, 1978). The precise details of discovery often become lost in antiquity, and the variations among the stories pertaining to the more recent discovery of perlite as a material of commerce are no exception. Credit in the United States is given to a dentist who, while experimenting with tooth enamels about 1941, discovered that perlite—the rock—intumesced when subjected to heat. At about the same time it is reported that the chief geologist of Silver and Barytes Ores Mining Co. attempted to put out a picnic bonfire on the shores of Milos Island, Greece by throwing beach sand on it. The ensuing pyrotechnic display immediately conjured up in that man's mind the possibility of a new use for the volcanic rock that constituted most of the island. Very little was done with this discovery either here or abroad until after World War II.

Today the name perlite is applied to both the hydrated volcanic glass, generally of rhyolitic composition, and to the lightweight aggregate that is produced from the expansion of the glass after it has been crushed and sized. Petrologically, it is defined as a glassy rhyolite that has a pearly luster and concentric, onionskin parting. Occurrences of perlite are restricted to several Tertiary to Quaternary age rhyolitic belts that trend in a generally north-south direction around the world. Commercially suitable deposits generally occur as domes of several hundred feet in height, although glassy zones in welded ash-flow tuffs and others associated with dikes and sills also have been reported. Mining is by ripping and blasting from open pits. Because of weight considerations, perlite usually is shipped to the local market area for subsequent expanding. In the United States, New Mexico leads in production with Arizona, California, Nevada, Idaho, and Colorado following in approximately that order. The principal use for expanded perlite is as a lightweight insulating aggregate in cryogenics, in plaster, concrete, and in loose fill insulation. Expanded perlite is also used in horticultural applications, and after subsequent milling and classification, as a filter aid.

The United States is the world's largest producer and consumer of perlite. Table 1 shows the world production of perlite and Table 2 shows the perlite mined, processed, expanded, and sold or used by producers in the United States.

Geology

Composition and Morphology

Any discussion of perlite must take into consideration its dual nomenclature, for it is known by the same name as both the naturally occurring rock and, after processing and expansion, as the lightweight aggregate of commercial significance.

In its naturally occurring form, perlite is a rhyolitic glass that contains from 2 to 5% combined water. While perlite also can occur as andesitic or dacitic glass, these latter types are of negligible commercial significance. Table 3 lists the chemical composition of a few typical perlites (Anderson, et al., 1956; Langford, 1979).

What sets perlite of commercial significance apart from other volcanic glasses is the fact that under the proper conditions of preparation—crushing and sizing—it will, when rapidly introduced into a flame of sufficient temperature, expand or "pop." All of the elements of composition contribute to the expansibility of the rock. The role of the combined water, however, is the most significant because it is believed not only to produce a fluxing effect in the softening of the highly siliceous glass prior to expansion, but it is also responsible for the explosive force of expansion through volatilization during heating. The current theory of the origin of the water in perlite is now less con-

* Retired. Formerly Exploration Manager, Manville Products Corp., Denver, CO.

TABLE 1—World Production of Perlite*, 1969-1979
(Thousand short tons)

Country	1969	1970	1971	1972	1973	1974	1975	1976	1977	1978	1979[e]
North America:											
United States[†]	471	456	432	545	544	555	512	553	597	641	660
Mexico	12	14	12	14	15	13	21	16	5	27	28
Total	483	470	444	559	559	568	533	569	622	668	668
Europe:											
Greece[‡]	108	118	105	120	139	126	125	[r]140	163	148	150
Hungary[§]	67	66	67	94	106	103	79	106	114	102	103
Italy[e]	80	90	90	95	105	110	100	105	100	100	100
USSR[e]	125	150	200	250	300	320	340	360	380	400	400
Others[e]	26	31	38	45	52	55	60	62	62	60	60
Total	406	455	500	604	702	714	704	[r]773	819	810	813
Asia:											
Japan[‡e]	40	45	50	55	60	63	68	72	77	80	83
Philippines	—	13	1	1	1	1	1	2	2	2	2
Turkey	3	4	16	32	16	19	13	27	33	30	30
Others[e]	—	—	33	33	35	30	28	28	28	30	30
Total	43	62	100	121	112	113	110	129	140	142	145
Africa: Total[e]	#	#	1	1	#	#	#	#	1	1	1
Oceania:											
Australia	1	1	2	2	3	2	[e]2	4	2	2	3
New Zealand	[e]1	2	2	3	3	3	[e]2	2	1	[e]1	1
Total	[e]2	3	4	5	6	5	[e]4	6	3	[e]3	4
World total‖	934	990	1,049	1,290	1,379	1,400	1,351	[r]1,477	1,585	1,624	1,651

Source: Meisinger, 1980.
[e] Estimated. [r] Revised.
* Perlite mined and/or processed.
[†] Processed ore sold or used by producers.
[‡] Processed ore.
[§] Mined ore.
Less than ½ unit.
‖ Does not include minor quantities of perlite mined from several deposit areas in South America.

TABLE 2—Perlite Mined, Processed, Expanded, and Sold and Used by Producers in the United States
(Thousand short tons and thousand dollars)

		Processed perlite					Expanded perlite		
	Perlite mined*	Sold to expanders		Used at own plant to make expanded material		Total quantity sold and used		Sold and used	
Year	Quantity	Quantity	Value	Quantity	Value		Quantity produced	Quantity	Value
1977	871	298	5,514	299	5,239	597	504	498	53,600
1978	939	320	6,813	321	6,927	641	553	546	64,300
1979	847	322	7,996	338	8,439	660	[r]551	[r]543	[r]61,200
1980	824	334	9,053	304	7,447	638	544	537	69,200
1981	710	324	9,888	267	7,530	591	484	475	65,900

Source: Meisinger, 1982.
* Crude ore mined and stockpiled for processing.
[r] Revised.

TABLE 3—Some Typical Perlite Chemical Compositions (Percent)

	No Agua, NM	Superior, AZ	Pioche, NV	Big Pine, CA	Milos, Greece	Akita, Japan	Bulgaria	Argentina	Hungary
SiO_2	72.1	73.6	73.1	73.6	74.2	74.2	73.8	72.3	73.5
Al_2O_3	13.5	12.7	12.8	13.2	12.3	12.9	12.8	13.4	13.0
Fe_2O_3	0.8	0.7	0.7	0.8	0.95	0.68	0.56	1.0	1.8
TiO_2	0.06	0.1	0.08	0.07	0.08	0.06	0.07	0.08	—
CaO	0.89	0.6	0.9	0.6	0.85	0.45	0.50	0.59	1.5
MgO	0.50	0.2	0.2	0.1	0.13	0.05	0.03	0.3	0.4
Na_2O	4.6	3.2	3.0	4.1	4.0	4.1	3.0	3.4	3.5
K_2O	4.4	5.0	4.7	4.1	4.4	4.0	4.9	4.7	3.8
H_2O+	3.0	3.8	3.9	3.3	2.8	3.3	4.0	3.7	3.0

troversial than it was a decade earlier. The 2 to 5% range for total combined water in perlite was originally thought to have been chilled or frozen into the glass as it was injected under an ice sheet or into a lake (Huntting, 1949). Now, there is universal acceptance of the theory that perlite was formed by the secondary hydration of obsidian after its emplacement (Ross and Smith, 1955; Friedman, et al., 1966; Jesek and Noble, 1978). Studies have shown that the combined water in perlite exists in at least two forms (Lehmann and Knauf, 1973; Lehmann and Rossler, 1974): one as molecular water and the rest as hydroxyl water. The ratio of one kind of water to the other is different for perlites of different origins. The composition of the perlite, particularly the amount of MgO and CaO, is believed to influence the rate of hydration of the obsidian.

Perlite of commercial significance occurs in a variety of textures that appear to be related to the depth of emplacement. An emplacement model for perlitic domes suggests that the outermost material will be pumiceous, grading inward into increasingly more compact textures and finally into a felsitic core. The following textures that fit the above model have been observed at No Agua Mountain, New Mexico and elsewhere (Whitson, 1981). While these textures also may be known by other names, the following seems to be a convenient and logical nomenclature.

Pumiceous: Near surface, this is a lightweight and frothy perlite where vesiculation is less confined by lithostatic and hydrostatic pressures. The vesicularity and the degree of distortion of the vesicles is a function of local confining pressures and of the amount of viscous flow during vesiculation of the super-cooled liquid. Flattening of vesicle walls increases with depth while elongation occurs during flow. Hence, at depth, the vesicles of pumiceous perlite when viewed with a 10 X hand lens appear elongated and taffy-like. At No Agua Mountain, this taffy-like texture is referred to as "shardy." Pumiceous perlite is usually light grey but may be buff.

Although the expansibility of pumiceous perlite may be quite satisfactory, its usefulness as a commercial grade is limited by its pronounced friability during milling and subsequent handling. Not only are excessive quantities of undersize material produced but suitably sized material is less likely than other textures to survive handling during transit to the expanding location and may arrive out of specification by virtue of excessive fines content. During exploration, near surface pumiceous perlite is difficult to core drill. In the field, the more compacted pumiceous varieties of perlite—shardy—are characteristically difficult to sample because they tend to powder when struck by a geologist's pick and much of the energy is absorbed rather than expended in breaking a chip off the outcrop.

Granular: This texture is found adjacent to and deeper than pumiceous perlite in the emplacement model. It is more dense than the overlying textures, has a "sugary" or saccharoidal appearance and commercially is highly satisfactory from the milling, classification, and expansibility standpoints. It cores well during drilling and blasts easily during mining. Color usually is buff but varies from grey to brown. It often displays flow banding (see Fig. 1).

Classical: This is the typical pearl grey material with "onionskin" concentric parting. Classical texture also may be dark grey to black. The name is derived from the classical description of perlite in the literature (Johannsen, 1939). Classical perlite is found stratigraphically below the granular perlite in the

FIG. 1—*Hand specimen of granular perlite showing flow banding.*

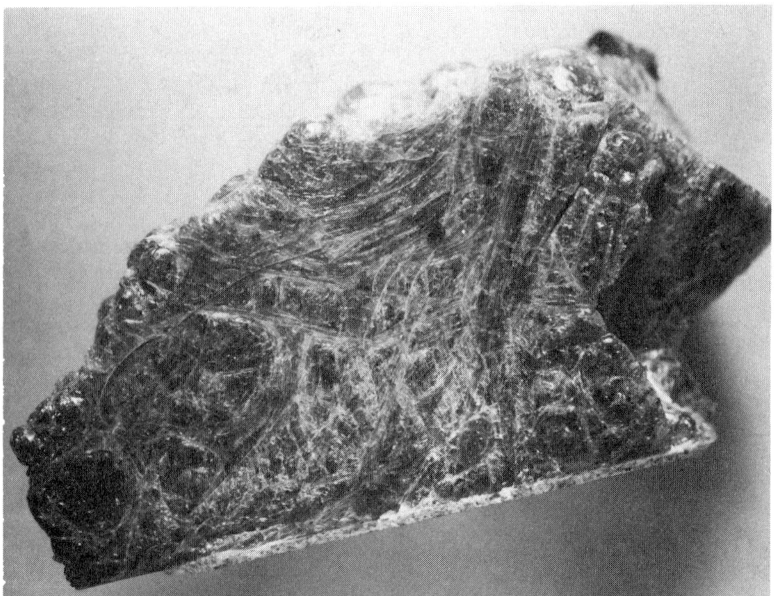

FIG. 2—*Hand specimen of classical perlite. Concentric onionskin-like cracks are evident. The name is derived from the classical description of perlite rock.*

FIG. 3—*Hand specimen of perlite breccia consisting of granular perlite in felsitic matrix. Granular fragments may vary from microscopic to several centimeters in cross-section.*

dome. Obsidian, whenever present, is always encased in classical perlite lending substance to the theory that perlite is formed from the hydration of obsidian. The concentric rings around each obsidian nodule were probably formed as the result of volume increase due to the hydration which advanced only as far as the perlite-obsidian interface (see Fig. 2). Toward the interior of the flow, the amount of obsidian in classical perlite tends to increase. Also found in perlite deposits are a rhyolitic or felsitic core, flow breccia (Fig. 3), and other materials and structures associated with volcanic flows and domes.

In addition to obsidian which may contain what have been identified as tridymite inclusions, often such nonexpansibles as quartz, feldspar, biotite, magnetite, and other accessory minerals, as well as products of devitrification, may be present.

Thin section studies of expanded perlite particles reveal the internal structure to be cellular and quite variable depending on the source (Bailey and Kadey, 1962). A comparison of Figs. 4, 5, and 6 explains why some perlites are considered *hard* and have high compaction resistance while others have low compaction resistance and are known as *soft* perlites. Hard perlites are particularly suited for plaster and concrete aggregate. Soft perlites which are not suited for these applications will mill readily to produce good filter aids. The size of the internal cell compartments and the thickness of the individual cell walls all depend on the expansibility of the particle. These in turn are thought to be a function of the composition and in particular the amount and distribution of the combined water.

Obsidian which contains less than one percent water can also be made to expand, but at a higher temperature (see Fig. 7).

Distribution of Deposits

Perlite deposits are restricted to Tertiary or younger volcanics of rhyolitic composition. Because rhyolitic glasses are unstable, they devitrify with age. Very young acidic volcanics have not had sufficient time to hydrate, hence they are still essentially obsidian and not expansible within the temperature range of perlite. There is a belt of Tertiary and Quaternary rhyolites that begins in Iceland, extends south into Ireland, Scotland, through the Massif Central of France into some of the Aegean Islands, through Sardinia and mainland Italy and farther south into Morocco, Algeria, and into South Africa. A Pacific belt splits, with a western limb extending through Japan, the Philippines, China, New Zealand, and Aus-

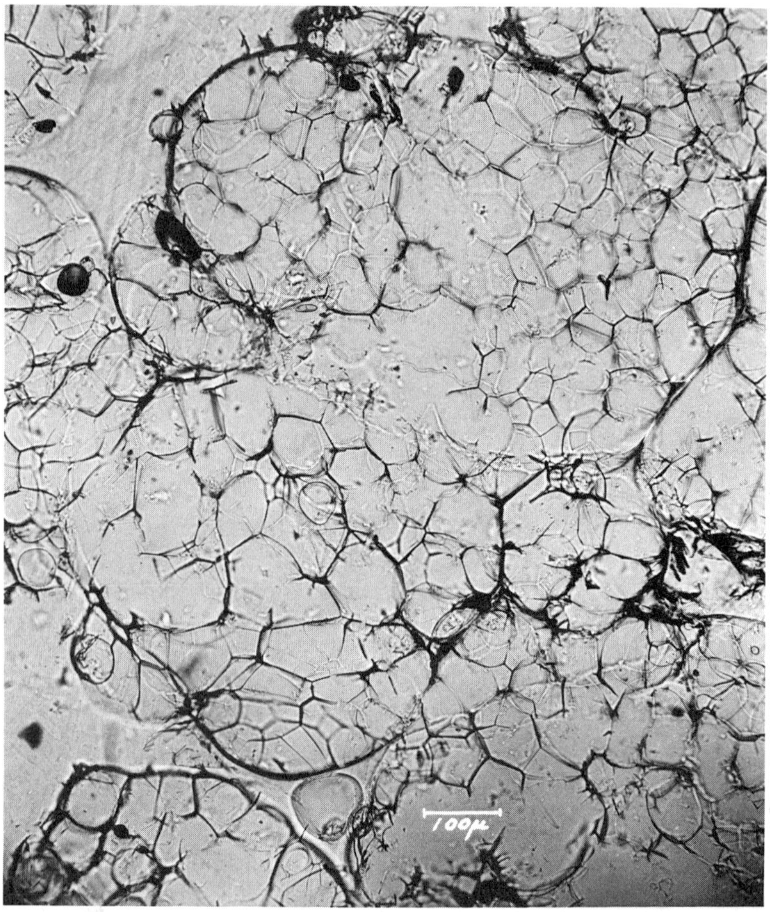

FIG. 4—*Thin section micrograph through perlite aggregate expanded to low density (less than 2 lb per cu ft). The thin fragile walls will break under pressure. This material, while not having sufficient compaction resistance for plaster or concrete aggregate, is quite suitable for milling into a filter aid.*

tralia. The eastern part of the circum-Pacific belt extends from Alaska (Plafker, 1963) through western Canada, down through the western United States into Mexico, and emerges in parts of the Andes in Chile and Argentina. Fig. 8 shows the perlite resources of the United States.

North America: While perlitic occurrences have been noted in Alaska and western Canada, the continental United States has the largest known reserves and commercial production in the world (Jaster, 1956; Meisinger, 1980).

There are 12 operations in six of the western states. In 1982, New Mexico accounted for approximately 83% of US production. No Agua Mountain in Taos County mined by both Manville Products Corp. and Grefco, Inc. is considered to be the largest commercial deposit in the world and accounts for most of this production. Fig. 9 shows the perlite operations at No Agua, New Mexico. Other producing locations in the state are Brushy Mountain about 20 miles northeast of No Agua, mined by Silbrico Corp., a deposit in Socorro County operated by Grefco (Weber, 1963), and one near Grants that is mined by U.S. Gypsum Co. for use in plaster (Bassett, et al., 1963). Elsewhere in the southwestern part of the state are smaller and lower quality occurrences.

Main production in Arizona, which centers

around the area near Superior and the southern half of the state, is accounted for by Filters International Corp. and by Harborlite Corp. Much of the perlite from the Superior area displays classical texture. Commercial grade perlite produced by Delamar-Mackie and U.S. Gypsum comes from Nevada. Perlite is also produced from deposits in California by American Perlite Corp., in Idaho by Oneida Perlite Corp. (Staley, 1962), and in Colorado by Persolite Products, Inc. Utah, Oregon, and Washington report lesser occurrences with negligible current production (Huntting, 1949; Jaster, 1956).

Mexican production is centered in the state of Sonora by Aislantes del Pacifico, S.A., in Pueblo by Cia. Minera Oriental, S. A., in Durango by Termolita, S.A., and in Pueblo by Dicalite de Mexico, S.A. (a Grefco, Inc. affiliate).

South America: Argentina is the major producer of South American perlite. Perfiltra S.A. and Minaclar, S.A. mine perlite from deposits in Salta Province near Salar de Pocitos (Quartino, 1975). The only other significant source of perlite in South America is in Chile where Cia Perlite Ltda. mine from a deposit in the Lagana de Maula area.

Europe: Greece is the largest western European producer of perlite. Major production comes from the Island of Milos where Silver and Barytes Ores Mining Co., General Enterprises Sarides, S.A., Mycobar Mining Co. S.A. (a subsidiary of Dresser Industries), and N. Bouras and Co. are the principal producers. The Greek Islands of Kos and Lesbos in the

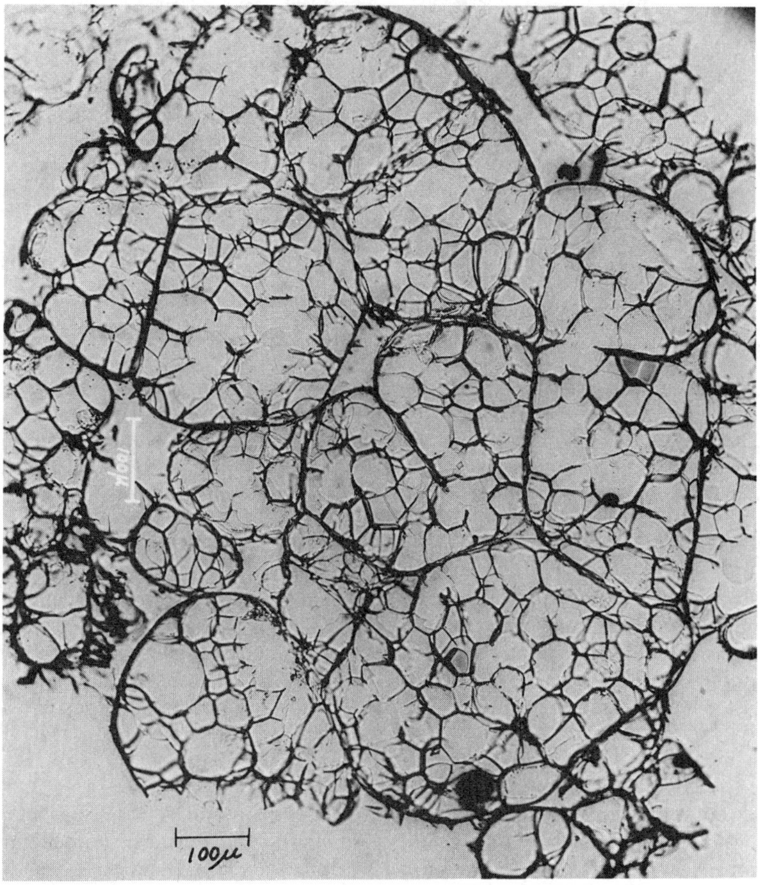

FIG. 5—*Thin section micrograph through perlite aggregate expanded to intermediate density (approximately 4½ lb per cu ft). Note the relatively coarse cell wall structure compared with Fig. 4.*

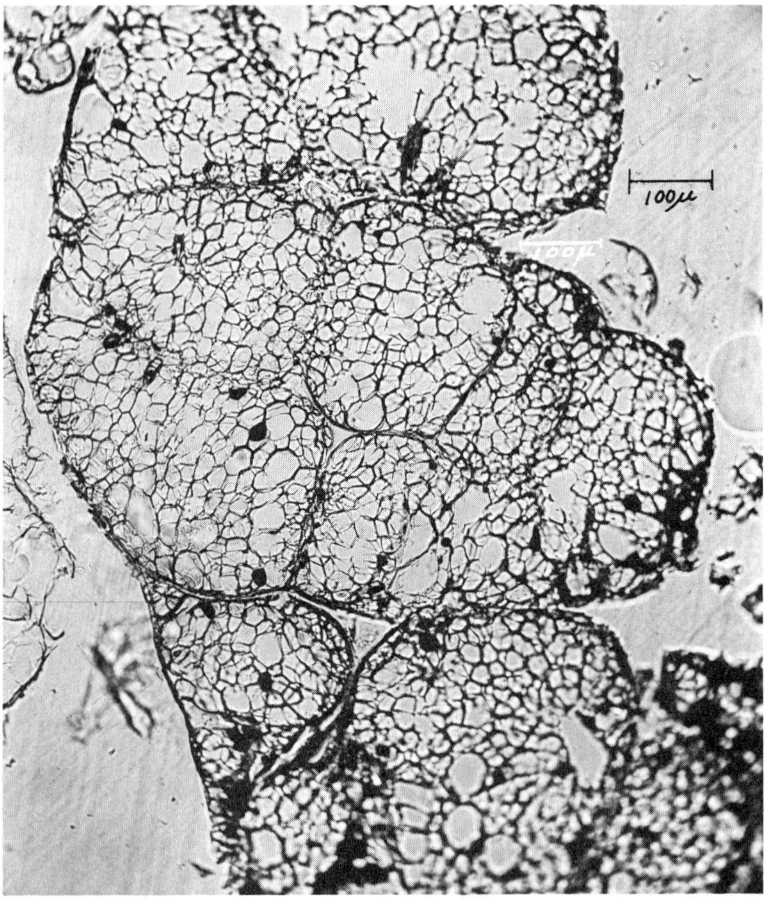

FIG. 6—*Thin section micrograph through perlite aggregate expanded to relatively high density (approximately 8 lb per cu ft). Note the small cells and coarse wall structure. This perlite would produce excessive floaters content if milled into a filter aid. The apparent strength of the aggregate would be ideal for plaster and concrete aggregate.*

Aegean Sea also have commercial grades of perlite from which there has been some production. Numerous other Greek islands contain perlite (Anastopoulos, 1963). Turkey has substantial reserves of high quality perlite in the western part of the country. There are also occurrences in central and eastern Turkey as well. Etibank, the government controlled mineral producing company, has substantial reserves near Cumoavasi in western Turkey and is the only source to date with processing facilities (Anon., 1977). Turkey could become a major producer of high quality perlite for both filter aid and aggregate use if the owners of the various deposits could attract the necessary capital to construct crushing and screening facilities. Some of these producers are shipping mined crude elsewhere for processing.

Italy is an important source of perlite with the most important deposits located on the island of Sardinia near the town of Uras (Carta, 1976). Perlite SpA is jointly owned by British Gypsum Ltd. CECA (Carbonisation et Charbons Actifs.), also mines from a deposit on Sardinia. Perlite is mined on the island of Ponza by Vic Italiana SpA. In Eastern Europe, Hungary (Perlaki and Szoor, 1973) and the Soviet Union are significant producers of perlite (Rudnyanszky, 1978). Bulgaria, Czechoslovakia, and Yugoslavia have smaller but growing industries (Vladimirov, 1975). Iceland has a commercially challenging but as yet unproductive perlite deposit at Priest Mountain (Prestahnukur) located about 100 km northeast of Reykjavik. The remoteness of the deposit, the relatively high concentrations of obsidian, and the short four-month mining season offset the otherwise good quality of perlite in many parts of the mountain. A smaller occurrence at Lodmundarfjordur has been investigated, but the location and quality do not justify commercial consideration. Northern Ireland did at one time have limited production from deposits in the Sandy Braes area of County Antrim.

Small deposits in France and Germany are devitrified or have not been economical to mine.

Africa: The only known commercial production of perlite is in southern Africa by Pratley Perlite Mining Co. (Pty.) at the Nxwala Estate, Zululand and by Perlite Industries (Pty.) Ltd. from a deposit in the Lebombo Mts. area both in northern Natal, South Africa (Coetzee, 1976; Anon., 1977a).

Asia: Japan is the most significant producer of perlite in Asia, which because of the volcanic nature of the islands has numerous perlite deposits. Mitsui Mining and Smelting Co. Ltd., the dominant Japanese perlite producer, operates mines at Kitakata, Ippongisita, and at Arita, Tomioka. Other deposits are located at the Kushiro prefecture of Hokkaido; at Akita; in Yamagata; in Nagano; in Chiba prefecture; in Kawasaki and Osaka; in Yamaguchi; in Kumamoto and in Fukuoka.

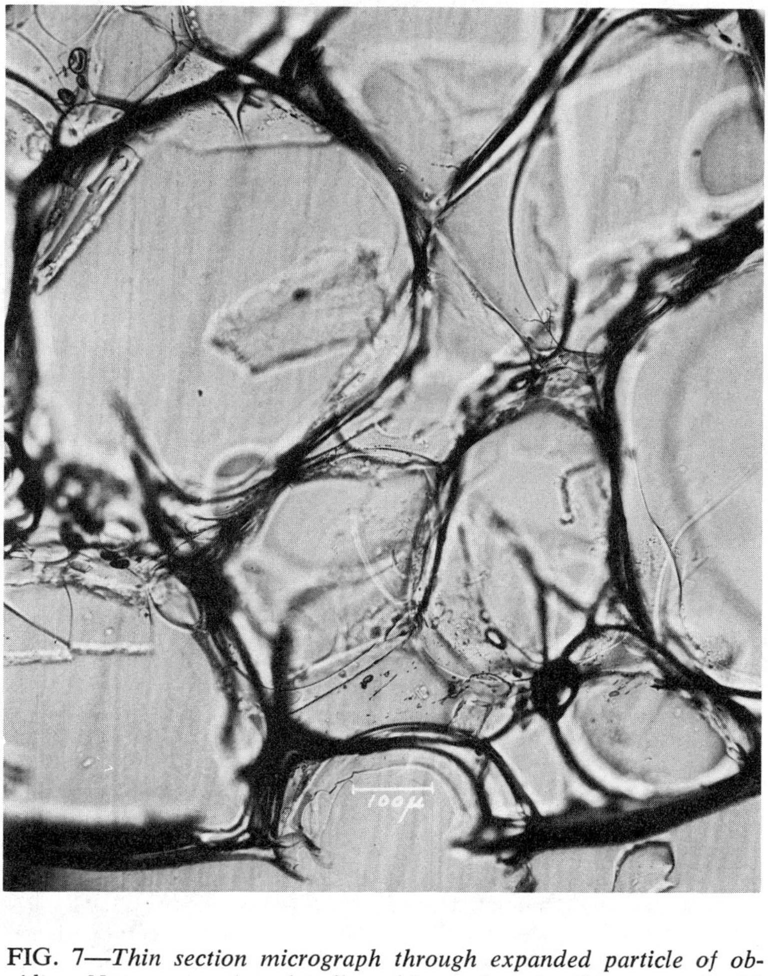

FIG. 7—*Thin section micrograph through expanded particle of obsidian. Note coarse size of walls and large size of cells compared with those aggregates in Figs. 4, 5, and 6.*

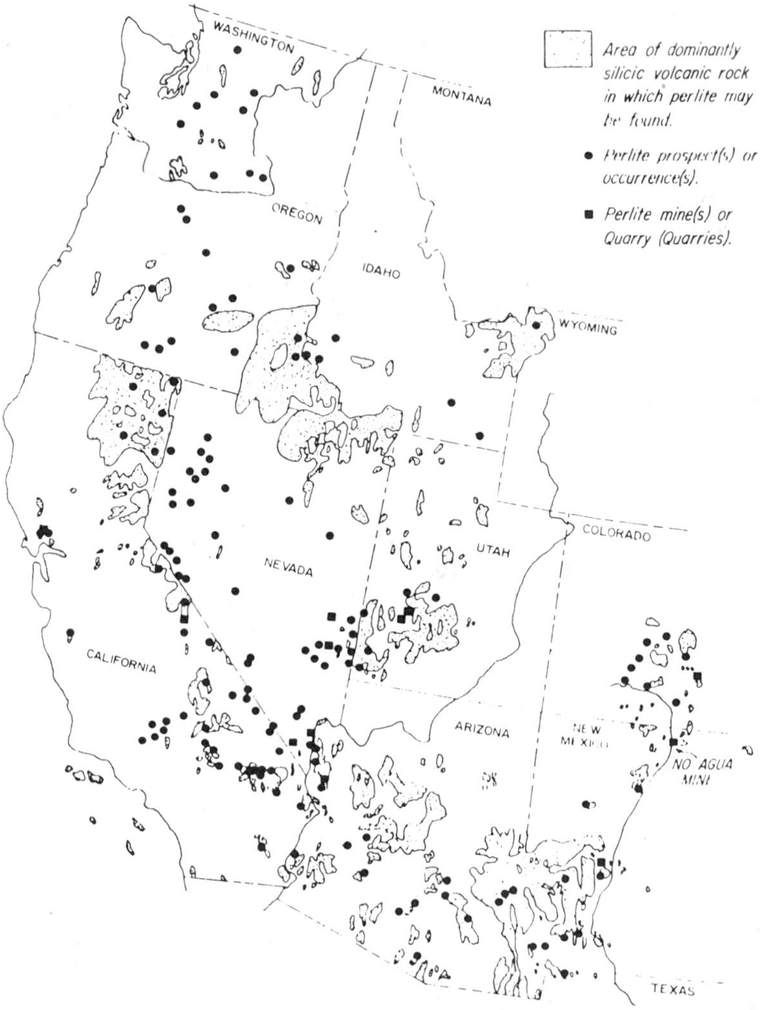

FIG. 8—*Perlite resources of the United States (after Jaster, 1956; updated, 1977, by A. L. Bush, US Geological Survey).*

New Zealand has a deposit at Kenlieth which is mined by New Zealand Perlite, Ltd. A substantial deposit occurs on Great Barrier Island, located a few miles off the east coast northeast of Auckland. Several other deposits in New Zealand have been described (Thompson and Reed, 1954). Perlite in Queensland, Australia, is mined and processed by Australian Gypsum Industries, Ltd.

The Philippines have perlite on Luzon Island. A deposit near Pagaspi is operated by Perlite Industries and Minerals Corp. Grefco, Inc. mines from a deposit in the Albay area of Luzon and Trinity Lodge Mining Corp. also mines ore from the Albay area.

Reserves

There does not appear to be any accurate or complete estimate of total world ore reserves. The US Bureau of Mines has estimated US reserves to be approximately 700,000,000 tons (Meisinger, 1979). Depending on the quality for specific applications and on the growth of the industry, this is enough to supply domestic needs for approximately 500 to 1,000 years. Although foreign reserves are even less known, it is reasonable to assume that on a worldwide basis, there is enough perlite for the foreseeable future.

Exploration

Exploration for perlite is restricted to those areas of rhyolitic volcanism between Tertiary and Quaternary age. Any glassy rhyolites of earlier than the Tertiary age have long since devitrified and are not of commercial interest. Younger acidic volcanics will not as yet have hydrated from obsidian. Perlite is recognized in the field by its characteristic appearance, whether it be a rather light grey pumiceous to classical onionskin or the grey to buff granular texture. The presence of marekanite or obsidian nodules (*Apache teardrops*), while detrimental in commercial perlite, are a good field indicator that the prospector may not be far from "pay dirt." Once grab samples have been taken from outcrops or shallow trenches, and upon receipt of encouraging test results, the next step includes the careful mapping of the dome or flow. Careful field work should ascertain the nature and distribution of rock textures and identify other such features as flow banding, joint systems, faults, etc.

The next step, diamond core drilling, is one in which considerable difficulty may be encountered. The friable nature of some perlitic textures, together with fracturing and jointing, may result in poor core recovery. Often only sludge samples can be obtained. In many cases such mechanical difficulties as hung bits, lost drill steel, and loss of circulation of drilling fluid may be encountered. Such problems can be minimized with the use of a triple tube wire line core barrel and NQ size core. On some occasions and in less accessible terrain, portable drilling equipment will produce adequate results and often can furnish information not otherwise attainable. The spacing between drill holes and the intervals into which the core is to be split for testing depends on the variability of quality. In many occurrences, variability is so great that the economics of the matter determines the closeness of the drilling and sampling

FIG. 9—*Aerial view of No Agua Mountain, New Mexico. The Manville Products Corp. North Hill Quarries may be seen in the center foreground with the crushing and screening plant in the lower right foreground. Manville is also mining the southern flank of the south hill (rear left peak). The east peak is out of view, to the left. Grefco, Inc. is mining a flow that is hidden in this photo by the west hill (rear right peak).*

pattern. In general multistage drilling programs, in which drilling is begun on wide spacings, then tightened up in subsequent stages, are most economical in time and costs.

Evaluation of Deposits

As is the case with so many industrial minerals, laboratory evaluation is based on tests designed to simulate commercial conditions of processing and application. Foremost among such tests is one that measures the expansibility of the sample. This can be accomplished in its simplest form by dropping dry crushed perlite into a bunsen flame and observing the resulting popping. A more refined apparatus is required to obtain some degree of control. One of the best laboratory expanders is the test furnace manufactured by the Perlite Corporation of Chester, PA. The basic design consists of a 48 in. long (40 in. effective length from feed port to flame) vertical tube, about 4 in. in diameter. It has a cyclone type collecting device fitted to the upper end, with a burner at the lower end. The crushed and sized test sample can be fed into the tube at any one of three side ports, whereupon it drops into the flame and upon expansion rises in the hot draft and passes over into the cyclone collecting system. The furnace can be modified and the design embellished so that either liquid propane or natural gas can be used, and the feed rate, draft, and other parameters monitored. One of the restrictions of laboratory expansion, however, is that the short tube length requires the furnace feed to have a very narrow size gradation. The relative expansibility of an unknown sample can be judged against that of a reference standard that is run at the same time. A 50/100 mesh gradation works moderately well, although narrow distributions of coarser gradations can be expanded with careful control of temperature and draft. There is no known suitable scale-down to laboratory size to simulate the expansibility of perlite in a horizontal furnace. Whether a crude perlite sample is worth laboratory testing can be ascertained by a petrographic examination for percent contamination, refractive index of the glass, and other criteria (Kadey, 1963).

The properties of the laboratory expanded sample can be measured in the same way as are properties of perlite expanded in a commercial size furnace. Loose density and compacted density in pounds per cubic foot, percent yield, percent nonexpansibles, compaction resistance, etc., are all standard tests that have been described in the Perlite Institute handbook of tests (Perlite Institute, 1977). For filter aid evaluation, further treatment beyond the expansion of the granule must be performed. Laboratory milling and filter aid testing that simulates commercial filter aid performance is required. Evaluation of the data depends on comparison of test properties against those of the reference standard and/or against the specifications required by the marketplace.

Preparation for Markets

Mining Methods: Open pit mining is the only method of ore removal that is economically feasible and practical for the extraction of perlite. With pumiceous textures, the ore is friable enough that rippers on the back of bulldozers are used to loosen the rock which is then pushed into hoppers or bins (traps), under which trucks are loaded for transit to the stockpile or mill. With harder ore, blasting is necessary and the ore is mined from benches. It is then loaded into trucks by front-end loaders. The variable nature of perlite deposits often necessitates selective mining techniques to avoid clay seams, obsidian, or other non-perlitic areas. The nature of some ore is such that it doesn't even have to be ripped. The Priest Mountain, Iceland deposit and parts of the Sardinian deposits are fractured enough in place that front-end loaders can break and load what otherwise appears to be solid rock. Rock of this type, however, is difficult to core drill and has some of the same commercial drawbacks as pumiceous textured perlite.

Assuming satisfactory quality, the part of the flow that is overlain by the least amount of overburden and in which maximum advantage can be taken of topography in the development of the mine plan is usually selected as the first mine site.

Milling Techniques: Perlite is usually used in the expanded form and the applications for which it is intended dictate the particle range into which it must be crushed and sized prior to expansion. Theoretically, equidimensional particles, which when expanded, fall between a maximum of ⅜ in. for horticultural grades and a minimum of 200 mesh for most other applications, are ideal. Axiomatically, the coarser the required product, the coarser the corresponding furnace feed must be. The size range of the expanded perlite dictates the size range of the furnace feed. This relationship, however, is not necessarily the same for all perlites and must be determined for each source since no two deposits are alike.

The crushing and sizing of crude perlite is designed with this objective in mind. The first step is to reduce the stockpile to lumps of less than 3 in. because much of the mine run ore is of boulder size. This is done by a jaw crusher. Further reduction to less than ⅝ in. is done by cone crusher, whereupon reduction to minus 8 mesh is accomplished by impact milling. Vibratory screening and air classification separates the milled perlite into basic grades for storage in silos, whereupon it is blended to meet specifications or to meet the customer's requirements. Undersize material is wasted, but research in recent years has been directed toward finding uses for undersize material.

Processing Techniques: The processing of crushed and sized perlite consists essentially of its expansion or "popping" into a lightweight, cellular aggregate. This is accomplished at temperatures ranging between 1400°F and 2100°F depending on the composition of the perlite (Sharps, 1961). As has been previously noted, the name perlite is also used for the expanded material. The largest volume of perlite is used in this form; however, the application of perlite as a filter aid requires a subsequent milling and classifying stage.

Although there have been many styles of furnaces patented for the expansion of perlite, two basic types have survived—the vertical (Fig. 10) and the horizontal furnace (Fig. 11). Both types depend on the relatively rapid introduction of the crushed and sized perlite into a hot zone during which the particle reaches its softening range, coincident with the volatilization of combined water. In the case of the vertical furnace, perlite is gravity fed through a side port into the hot zone (see Fig. 10).

Upon expansion, the density of the expanding particle decreases to a point where the downward force of gravity on its mass is exceeded by the updraft which carries it over into the cyclone. Large particles, therefore, fall further down into the hot zone than do finer particles before expanding and rising in the draft. Nonexpansible particles like obsidian, felsite, and other contaminants continue to drop out the bottom whereas the shattered fines, etc., are carried upward and trapped in the bag house.

In the case of a horizontal furnace, the furnace feed is introduced at the hot end through an arrangement of baffles which cause the perlite to cascade into the hot zone. It travels down the length of the rotating furnace during which it is expanded and is then drawn up the stack by the draft and over into the collecting system. The expansibility of some perlites can be enhanced by preheating below 800°F (Sharps, 1961). In vertical furnacing, this is accomplished by passing the furnace feed through a separate heater at about 700°F. Horizontal furnaces preheat the feed by passage between an inner and outer shell of the furnace before it cascades into the hot zone (see Fig. 11).

There must be a good balance between the softening of the glass and the volatilization of the combined water. Excessive combined water will cause some perlites to explode or shatter with the subsequent production of excessive fines. Insufficient water or an overly viscous glass, on the other hand, will result in partially expanded, heavy density perlite. The former is often termed *lively* and the latter *dead*. Lively perlites can be modified by preheating to reduce some of the combined water. Some dead perlites may be helped by preheating. Otherwise, they are less suitable commercially and must be relegated to applications where a low density thin-walled product is not needed.

Expansion for the various aggregate uses requires that the perlite be expanded to a specific size range and density. Processing into a filter aid, in addition to expansion to a specified size and density, requires a subsequent milling

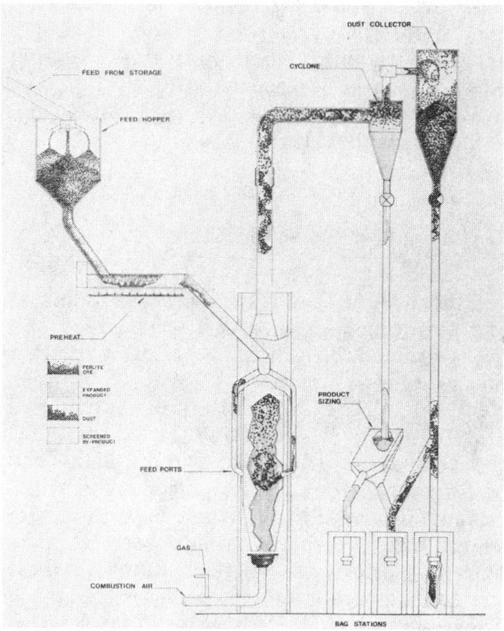

FIG. 10—*Vertical expander and handling system.*

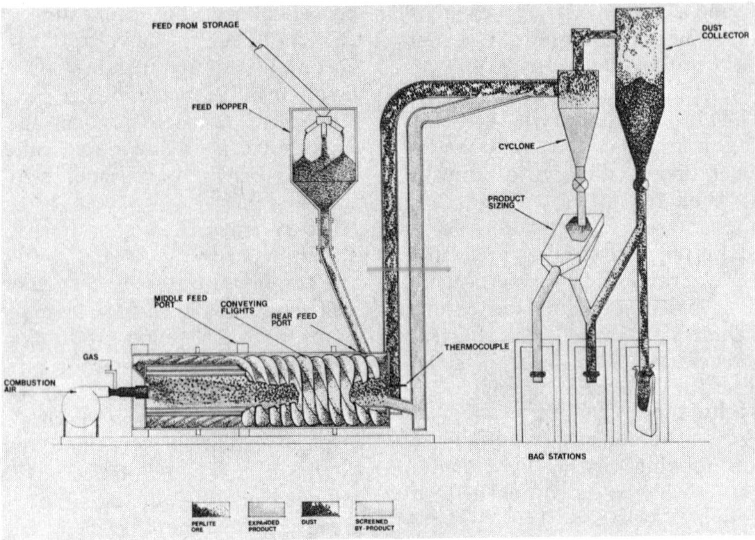

FIG. 11—*Horizontal expander and handling system.*

and classifying operation. The perlite bubble aggregates must be broken into curved glass plates and bubble junctions. A filter aid requires the right proportion of these with a minimum of unmilled cells. Residual unmilled cells, or "floaters," are undesirable in a filter aid. There are, of course, additional requisites and specifications for both aggregates and filter aids. These will be discussed later.

Table 4 shows typical size distributions for a range of perlite ores.

Testing and Specifications

Guidance for the maintenance of production and performance standards for expanded perlite is handled through the auspices of the Perlite Institute with offices at 45 West 45th Street, New York. The Perlite Institute publishes a *Manual of Test Methods* and related standards that control every phase of perlite quality and performance from the shipment of the unexpanded material in box and hopper cars to its ultimate incorporation into such applications as plaster and concrete. Many of these test methods are the creation of the Technical Committee of the Perlite Institute whereas others are selected specifications of the American Society for Testing & Materials (ASTM) applicable to the perlite industry. Following is a list of current Perlite Institute tests:

PI 108–77 Method for Bulk Box Car Sampling of Sized Perlite Ore
PI 109–77 Method for Sampling Sized Perlite Ore from Hopper Cars
PI 110–77 Sampling of Perlite
PI 111–77 Preparation of Perlite Sample for Testing by Coning and Quartering Method
PI 112–77 Preparation of Perlite Sample for Testing by Splitter or Riffle Method
PI 113–77 Sieve Analysis of Fine Materials
PI 114–77 Sieve Analysis of Expanded Perlite, Volume Basis
PI 115–77 Sieve Analysis of Expanded Perlite, Weight Basis
PI 116–77 Fractional Density of Expanded Perlite
PI 117–77 Wet Screen Analysis
PI 118–77 Determination of Free and Combined Moisture in Perlite
PI 200–77 Loose Weight Determination of Expanded Perlite
PI 201–77 Compacted Density
PI 202–77 Test for pH Value of Expanded Perlite
PI 300–77 Determination of *Floats* and *Sinks* in Expanded Perlite
PI 303–77 Water Repellency Test
PI 305–77 Yield Test for Perlite Aggregate
PI 306–77 Compaction Resistance Test
PI 307–77 Perlite Ore Expansibility Test
PI 400–77 Setting Time of Gypsum Plaster
PI 401–77 Plaster Coverage Test

The following specifications of the ASTM are applicable to the perlite industry:

C 28–76 Specification for Gypsum Plasters
C 29–76 Test Method for Unit Weight of Aggregate
C 35–76 Specification for Inorganic Aggregates for Use in Gypsum Plaster
C 136–76 Test Method for Sieve or Screen Analysis of Fine and Coarse Aggregates
C 331–77 Specification for Lightweight Aggregates for Concrete Masonry Units
C 332–77 Specification for Lightweight Aggregates for Insulating Concrete
C 472–73 Physical Testing of Gypsum Plasters and Gypsum Concrete
C 495–77 Test Method for Compressive Strength of Lightweight Insulating Concrete
C 513–69 Method for Securing, Preparing, and Testing Specimens from Hardened Lightweight Insulating Concrete for Compressive Strength
C 520–65 Test Method for Density of Granular Loose Fill Insulations
C 549–73 Specification for Perlite Loose Fill Insulation
E 11–70 Specification for Wire-Cloth Sieves for Testing Purposes
E 380–76 Standard for Metric Practice

Marketing

Because of its low bulk density, perlite is generally shipped in the unexpanded form to the local marketing area, and expanded at that point. Perlite mined from twelve operations in six western states is shipped to 73 expanding locations in 32 states (Table 5). Foreign producers likewise ship crushed and sized perlite all over their marketing areas to the local expanding points for redistribution. The foreign distances are considerably greater. For example, Greek perlite is shipped throughout Europe. There was considerable interest in the early 1960s in importing Greek and Icelandic perlite to the East Coast of the United States. The attempt failed for a number of reasons, one of which was the lack of sufficient warehousing space for boatloads of crude perlite. The United States, therefore, continues to supply all its own needs.

Because of its unique internal cellular structure, perlite offers many advantages as an insulating material for the construction industry. Perlite aggregate combined with portland cement and water produces a lightweight concrete that is used for lightweight floor and roof fill, lightweight structural roof decks, precast components, and numerous other insulating applications. In roof deck construction, in

TABLE 4—Typical Size Distribution for Some Perlite Ores

	Manville Grade				
	PA-1000	PA-115	PA-220	PA-420	PA-610
	Application				
Sieve Size	Plastic and Resin Filler	Cryogenics, Acoustical Tile, Filter Aid	Plaster Aggregate	Concrete Aggregate, Masonry, Loose Fill	Horticulture, Foundry Slag
+8				0-2	10-25
8-16				22-30	
8-12					40-60
12-16					25-42
+20			7-16		
16-20					0-8
+30		0-2			
16-30				53-65	
20-30			26-35		
-20					0-4
30-50		4-12	40-52	10-21	
50-100		30-55	6-13		
-50				0-4	
-100			0-4		
+100	0-8				
100-200	20-47	25-45			
-200	50-75	0-20			

Source: Manville Bulletin FF 391, January, 1980.

TABLE 5—Producers of Expanded Perlite in the United States, 1979

Company	Address
Airlite Processing Corp. of Florida	Vero Beach, FL
American Perlite Products, Inc.	Gilliam, LA
Armstrong Cork Co.	Lancaster, PA
Aztec Perlite Co.	Escondido, CA
Brouk Company	St. Louis, MO
Buffalo Perlite Div. of Pine Hill Concrete Mix Corp.	Buffalo, NY
Carolina Perlite Co.	Gold Hill, NC
C-E Refractories	Port Kennedy, PA
Chemrock Corp.	Nashville, TN
Cleveland Gypsum Co.	Cleveland, OH
Conwed Corp.	St. Paul, MN
Filter Media, Inc.	Houston, TX
Filter Products Corp.	Lake Zurich, IL
Georgia-Pacific Corp.	Portland, OR
Grefco, Inc.	Los Angeles, CA
Harborlite Corp.	Escondido, CA
Manville Products Corp.	Denver, CO
Lite Weight Products, Inc.	Kansas City, KS
Mica Pellets, Inc.	DeKalb, IL
Midwest Perlite Co.	Appleton, WI
National Gypsum Co.	Buffalo, NY
Oneida Perlite Corp.	Malad City, ID
Pacific Coast Products	Sebastopol, CA
Pamrod Products	McQueeney, TX
Paramount Perlite Co., Inc.	Paramount, CA
Pennsylvania Perlite Corp.	Lehigh Valley, PA
Perlite Manufacturing Co.	Carnegie, PA
Perlite of Houston, Inc.	Houston, TX
Perlite Popped Products	Santa Fe Springs, CA
Perlite Processing Co.	Santa Fe Springs, CA
Persolite Products, Inc.	Florence, CO
Redco, Inc.	North Hollywood, CA
Schundler Co.	Metuchen, NJ
Scolite International Corp.	Troy, NY
Silbrico Corp.	Hodgkins, IL
Sil-Flo, Inc.	Port Jefferson, NY
South Texas Perlite	San Antonio, TX
Strong-Lite Products	Pine Bluff, AR
Supreme Perlite Co.	Portland, OR
The Celotex Corp.	Tampa, FL
The Pax Co.	Salt Lake City, UT
Thermo-O-Rock Div. of Allied Block Chemical Co.	New Eagle, PA
United States Gypsum Co.	Chicago, IL
Whittemore Perlite Co., Inc.	Andover, MA
W. R. Grace & Co.	Cambridge, MA

Source: Meisinger, 1980.

addition to its insulating capacity, it is stronger, more rigid, and more fire safe than other roof insulations. On flat roofs, it can be applied to provide the proper drainage slopes and it makes an excellent base for conventional built-up roofing. It is also nailable. Perlite plaster consists of a blend of expanded perlite aggregate and neat gypsum (or portland cement) properly mixed with water for application to wall and ceiling surfaces or for fireproofing structural steel. Perlite plaster walls have low sound transmission. Silicone treated loose fill insulation is used for cavity wall and concrete or cinder block wall construction. Its use will minimize winter heat loss and summer heat gain. The silicone treatment renders it water repellent to improve its insulating capacity. Perlite is rot-proof, nonsettling, and noncombustible. Its insulating capacity is used in cryogenic insulations. It fills the space between double shelled tanks for liquid oxygen and similar chemicals. In this and other loose fill applications, it is easy to apply. It has horticultural use as a soil conditioner—an application that is more apparent to the general public, particularly to the home gardener, than is its more commercial uses.

As a filter aid for the separation of suspended solids from fluids, in addition to use with pressure leaf filters, it has particular application in rotary precoat filtration. The criteria for a good filter aid are as follows:

1) Inert in the liquid to be filtered
2) Unaffected by temperature
3) High permeability
4) Low wet density
5) Ability to remove nonsettleable solids
6) Easily kept in suspension
7) Precoatability
8) Neutral

A properly processed perlite filter aid will meet all these criteria. It has established itself particularly in the filtration of sugar, of alginates, and of pharmaceuticals. Unexpanded perlite has demonstrated suitability in certain specialty uses such as scouring compounds, blasting sands, and some filler applications. Table 6 shows the amount of expanded perlite sold or used in the United States by producers.

Pricing

The US Bureau of Mines reported (Meisinger, 1982) that the total value of processed perlite sold or used by producers in 1981 was \$17.4 million compared with \$13.7 million in 1978. The average price, f.o.b. plant for raw, crushed grades in 1980 was \$25.86 per ton, and for 1981, was \$29.47 per ton. Transportation costs can more than double the delivered price to the East Coast. The trend for previous years is shown in Table 7. The price of expanded perlite is variable depending on use, grade, and competitive market location.

The range for expanded perlite in 1979 was \$75 to \$275 (loose fill) with the average price \$127.79.

Health and Ecology

Perlite per se is not known to present a health hazard either in the mining, milling, or processing of the material. As a glass, it is amorphous and regarded as innocuous. Most environmentally conscious companies, however,

TABLE 6—Expanded Perlite Sold and Used by Producers in the United States, by Use (Short tons)

Use	1979	1980	1981
Concrete aggregate	37,000	29,800	21,800
Fillers	9,000	10,000	6,200
Filter aid	96,500	102,300	94,400
Formed products*	291,000	289,900	256,000
Horticultural aggregate†	41,400	40,900	40,200
Low-temperature insulation	6,200	7,700	5,900
Masonry and cavity fill insulation	21,800	20,900	20,000
Plaster aggregate	23,200	24,000	16,700
Other‡	16,900	11,200	14,100
Total §	543,000	537,000	475,000

Source: Meisinger, 1982.
* Includes acoustic ceiling tile, pipe insulation, roof insulation board, and unspecified formed products.
† Includes fertilizer carriers.
‡ Includes fines, high-temperature insulation, paint texturizer, refractories, and various nonspecified industrial uses.
§ Data may not add to total shown because of independent rounding.

consider all dusts to be a nuisance, and in this regard, protect their workers with dust masks and practice dust control at considerable cost. Furthermore, as an abrasive and glassy material, crude and expanded perlite dust can be a serious irritant to the eyes and safety glasses are a necessity in the mine and in the plant.

With the recent emphasis on pollution of the environment, any operation that emits dust is suspect. Blasting, crushing, screening, and expansion of perlite are all dusty operations. Most perlite deposits are located in remote places and are not likely to cause complaints. The expansion of perlite is usually carried out in populated and industrial locations so dust collection and waste disposal problems must be recognized and dealt with effectively.

Further Considerations and Trends

Substitutes

Just as the advent of perlite produced a competitive threat to other industrial minerals, some of these have in turn replaced perlite in the marketplace or have joined perlite to share the market. Vermiculite probably ranks highest on the list of perlite competitors. Both are used in plaster and in loose fill insulation. In loose fill application, vermiculite is less amenable to beneficiation because it cannot be rendered water repellent with a monomolecular silicone coating. This treatment is preferable to asphalt treatment which can be applied to vermiculite. Vermiculite is also used for soil conditioning. From an aesthetic viewpoint, the white color of perlite which is so much of an advantage otherwise is less preferable than the color of the vermiculite, which does not stand out as white aggregates in the darker soil. Dark coatings can be applied to the perlite but at added cost. While perlite invaded the realm of filtra-

TABLE 7—Time-Price Relationships for Crude Perlite

	Average annual price, dollars per ton*	
Year	Actual prices	Based on constant 1978 dollars
1958	8.44	19.42
1959	8.43	18.98
1960	8.54	18.91
1961	8.59	18.85
1962	8.31	17.91
1963	8.39	17.82
1964	8.78	18.36
1965	8.55	17.49
1966	9.67	19.16
1967	9.62	18.51
1968	9.87	18.18
1969	10.82	18.97
1970	10.75	17.89
1971	11.44	18.12
1972	11.44	17.39
1973	10.28	14.77
1974	12.66	16.59
1975	14.22	17.01
1976	16.99	19.32
1977	18.01	19.33
1978	21.44	21.44
1979	24.90	22.88

Source: Meisinger, 1980.
* F.o.b. processing plant.

tion long held by diatomite and considered to be superior in this application, each of these filter aids appears to have its own area of influence. The particulate structure of some diatomite filter aids is such that perlite performs better in rotary precoat filtration. On the other hand, where high clarity is required, the ratio of clarity to flow rate is higher for diatomite. Since most filter aids are sold on a weight basis and used on a volume basis, the lower density of perlite can be considered a competitive advantage.

Tariffs and Taxes

There is no tariff imposed on the importation of perlite into the country. The depletion allowance for perlite is 10%.

Acknowledgments

The author gratefully acknowledges the assistance of J. M. Sharratt, D. N. Whitson, R. O. Y. Breese, and J. A. Sarles of the Manville Exploration Department for their review and comments on the manuscript; and of Ann B. Michael, Kathy Hodges, and Judith Neale of the Manville Corporate Information Center in the compilation of the bibliographical references. The patience and expertise of S. J. Lefond, editor-in-chief, in the final draft is also greatly appreciated.

Bibliography and References

With some exceptions, the references listed herein are dated 1970 and later. The reader is referred to the Perlite chapter in the 4th edition of *Industrial Minerals and Rocks* for additional bibliographic information prior to 1970.

Anon., 1972, "Perlite: New Sources to Meet Growing Demand but Consumption Undergoing Change," *Industrial Minerals,* June, pp. 9–26.

Anon., 1977, "Perlite—Market Patterns and Future Potential," *Industrial Minerals,* May, pp. 17–37.

Anon., 1977a, "Perlite in North America," *Industrial Minerals,* July, pp. 31–35.

Anderson, F.G., et al., 1956, "Composition of Perlite," Report of Investigations 5199, US Bureau of Mines, 13 pp.

Austin, G.S., and Weber, R.H., 1976, "Perlite Operations in New Mexico," *Geological Society of America,* Vol. 8, No. 5, pp. 564–565.

Bailey, D.A., and Kadey, F.L., 1962, "Petrographic Thin-Section Study of the Structure of Expanded Perlite," *Trans. SME-AIME,* Vol. 223, pp. 37–40.

Bassett, W.A., et al., 1963, "Potassium-Argon Ages of Volcanic Rocks Near Grants, New Mexico," *Geological Society of America Bulletin,* Vol. 74, pp. 221–226.

Bush, A.L., 1974, "National Self Sufficiency in Lightweight Aggregate Resources," address given at the Perlite Institute Annual Meeting in Colorado Springs, CO, Apr. 22, pp. 94–99.

Carmichael, I.S.E., 1979, "Glass and the Glassy Rocks," Chap. 8, *The Evolution of the Igneous Rocks,* Princeton University Press, pp. 233–244.

Carta, M., et al., 1976, "The Industrial Minerals of Sardinia: Present Situation and Future Prospects," Second Industrial Minerals International Congress, Munich, pp. 41–55.

Chesterman, C.W., 1954, "Genesis of Perlite" (abstract), *Geological Society of America Bulletin,* Vol. 65, No. 12, p. 1336.

Chesterman, C.W., 1975, "Perlite," *Industrial Minerals and Rocks,* 4th ed., S.J. Lefond, ed., AIME, New York, pp. 927–934.

Christiansen, R.L., and Lipman, P.W., 1966, "Emplacement and Thermal History of a Rhyolitic Lava Flow Near Fortymile Canyon, Southern Nevada," *Geological Society of America Bulletin,* Vol. 77, pp. 671–684.

Coetzee, C.B., 1976, "Perlite in Lenses and Intermittent Layers," *Perlite—South Africa Geological Survey Handbook,* pp. 397–398.

Dunn, L.R.L., and Billinghurst, W.M., 1954, "Perlite Deposits in New Zealand," Part II, Evaluation, *New Zealand Journal of Science and Technology,* Nov., pp. 218–226.

Elevatorski, E.A., 1975, "Perlite in Pinal County near Superior," *Arizona Industrial Minerals,* Minobras, 73 pp.

Friedman, I., et al., 1966, "Hydration of Natural Glass and Formation of Perlite," *Geological Society of America Bulletin,* Vol. 77, Mar., pp. 323–328.

Huntting, M.T., 1949, "Perlite and Other Volcanic Glass Occurrences in Washington," Report of Investigation No. 17, Division of Mines and Geology, State of Washington, 77 pp.

Jaster, M.C., 1956, "Perlite Resources of the United States," Bulletin 1027–1, US Geological Survey, pp. 375–403.

Jesek, P.A., and Noble, D.C., 1978, "Natural Hydration and Ion Exchange of Obsidian: An Electron Microprobe Study," *American Mineralogist,* Vol. 63, pp. 266–273.

Johannsen, A., 1939, "A Descriptive Petrography of Igneous Rocks," University of Chicago Press, 318 pp.

Kadey, F.L., 1963, "Petrographic Techniques in Perlite Evaluation," *Trans. SME-AIME,* Vol. 229, pp. 332–336.

King, C.R., 1948, "Pumice and Perlite as Industrial Materials in California," *California Journal of Mines and Geology,* Vol. 44, No. 3, July, pp. 293–319.

Lacroix, A., 1916, "Volcanic Glasses of the Central Massif," *Comptes Rendus Hebdomadaires des Seances; Memoires de l'Academie de Sciences de l'Institut de France,* Vol. 163, No. 12, pp. 406–411.

Langford, R.L., 1979, "Mineral Dossier No. 21," Institute of Geological Sciences, London.

Lehmann, H., and Knauf, A., 1973, "Entwicklung Einer Pruemethode 2UR Beuteilung Der Blahfahigkeit Von Perlitgestein," *Tomind-Ztg.,* Vol. 97, No. 3, pp. 65–66.

Lehmann, H., Rossler, M., 1974, "A Contribution to the Nature of Water-Binding in Perlites,"

Thermal Analysis, Vol. 2, Proceedings, Fourth ICTA, Budapest, pp. 619–628.

Lehmann, H., and Rossler, M., 1976, "Thermoanalytische Untersuchungen An Perliten Und Perlitartigen Gestein," *Tomind-Ztg.,* Vol. 100, No. 7, pp. 271–274.

Lipman, P.W., and Mehnert, H.H., 1979, "The Taos Plateau Volcanic Field, Northern Rio Grande Rift, New Mexico," *Rio Grande Rift: Tectonics and Magmatism,* R. E. Riecker, ed., American Geophysical Union, Washington, DC, pp. 289–309.

Loney, R.A., 1969, "Flow Structure and Composition of the Southern Coulee, Mono Craters, California, A Pumiceous Rhyolite Flow," Memoir 116, Geological Society of America, pp. 415–440.

Meisinger, A.C., 1979, "Perlite," *Mineral Commodity Profiles,* US Bureau of Mines, 14 pp.

Meisinger, A.C., 1980, "Perlite," *Mineral Facts and Problems,* Bulletin 671, US Bureau of Mines, pp. 1–12.

Meisinger, A.C., 1982, "Perlite in 1981," *Mineral Industry Surveys,* US Bureau of Mines.

Meisinger, A.C., 1982a, "Perlite," *Mineral Commodity Summaries, 1982,* US Bureau of Mines, pp. 110–111.

Perlaki, E.I., and Szoor, Gy., 1973, "The Perlites of the Tokaj Mountains," *Acta Geologica Academiae Scientiorum Hungaricae,* Vol. 17 (1–3), pp. 85–106.

The Perlite Institute, *Perlite Design Manual.*

The Perlite Institute, 1977, *Manual of Test Methods.*

Plafker, G., et al., 1963, "Investigations for Perlite in the Alaska Range," Bulletin 1155, US Geological Survey, pp. 49–66.

Quartino, B.J., Zardini, R.A., and Llorente, R.A., 1975, "Estudio Geologico Economico de los Yacimientos de Perlita Taurus Y Anfitrite Salar De Pocitos, Prov. de Salta," *Ed. Cientificas Arg Librart* (ECAL), pp. 337–351.

Ricker, K., 1960, "Perlite with Special Reference to Icelandic Occurrences," *Zeitschrift der Deutschen Geologischen Gesellschaft,* Vol. 112, pp. 197–207.

Ross, C.S., and Smith, R.L., 1955, "Water and Other Volatiles in Volcanic Glasses," *American Mineralogist,* Vol. 40, Nos. 11, 12, Nov.-Dec., pp. 1076–1089.

Rudnyanszky, P., 1978, "Development and Uses of Perlite in Hungary," Perlite Institute Annual Meeting, Dubrovnik, Yugoslavia, May, pp. 135–138.

Sharps, T.I., 1961, "Perlite in Colorado and Other Western States," *Mineral Industries Bulletin,* Vol. 4, No. 6, Nov., Colorado School of Mines, pp. 1–15.

Staley, W.W., 1962, "The Oneida Perlite Deposit," Idaho Bureau of Mines and Geology Resources Report 9, 6 pp.

Stamboliev, H.T., and Sapunov, P., 1973, "Perlite-Vorkommen in Mazedonien/Jugoslawien," *Tomind-Ztg.,* Vol. 97, No. 3, pp. 63–64.

Stein, H.A., and Murdock, J.B., 1955, "The Processing of Perlite," *California Journal of Mines and Geology,* Vol. 51, No. 2, Apr., California Div. of Mines, pp. 105–116.

Stewart, D.B., 1979, "The Formation of Siliceous Potassic Glassy Rocks," Chap. 11, *The Evolution of the Igneous Rocks,* Princeton University Press, pp. 339–350.

Szep, E., 1976, "Die Forderung und Aufbereitung Des Perlits in Ungarn," *Tomind-Ztg.,* Vol. 100, No. 8, pp. 297–299.

Thompson, B.N., and Reed, J.J., 1954, "Perlite Deposits in New Zealand," *New Zealand Journal of Science and Technology,* Series B., Vol. 36, No. 3, pp. 208–218.

Toth, K., 1972, "Untersuchungen Zur Bestimmung Des Blahuermogens Von Perlitgestein," *Tomind-Ztg.,* Vol. 96, No. 6, pp. 137–142.

Toth, K., 1973, "Ein Mikrover Fahren Zur Bestimmung Der Eigenfestigkeit Von Perlithauewerken," 67 pp.

Varju, G., 1976, "Perlitrohstoffe und Einege Problem Ihrer Genese," *Tonindustrie Zeitung,* 100 Jahrg, Nr. 8, pp. 285–312.

Vassiliou, B.E., 1975, "Perlite in Greece," address given at Perlite Institute Annual Meeting, Athens, Greece, pp. 33–37.

Vassiliou, B.E., 1975, "The Market Outlook for Perlite Today," Industrial Minerals International Congress, *Proceedings,* No. 1, pp. 190–195.

Vgenopoulos, A., 1977, "Der Chemismus Einiger Sauren Vulkaniten Aus Dem Evros Gebiet Westthrakiens/Griechenland," Inst. Geol. Min. Res., Athens, Greece, *Proceedings,* 6th Colloquium on the Geology of the Agean Region, Vol. 3, pp. 945–954.

Vlaointirov, I., 1975, "The Perlite Industry in Bulgaria," Address given at the Perlite Institute Annual Meeting, Athens, Greece, pp. 39–46.

Weber, R.H., 1955, "Processing Perlite, the Technologic Problems," *Mining Engineering,* Vol. 7, No. 2, pp. 174–176.

Weber, R.H., 1957, "Geology and Petrography of the Stendel Perlite Deposit, Socorro County, New Mexico," Circular 44, New Mexico Institute of Mining and Technology, 22 pp.

Weber, R.H., 1963, "Geologic Features of the Socorro Perlite Deposit," New Mexico Bureau of Mines and Mineral Resources, New Mexico Institute of Mining and Technology, New Mexico Geological Society, 14th Field Conference, pp. 144–145.

Whitson, D.N., 1981, "Geology of the No Agua Peaks, New Mexico Perlite Deposit," *Proceedings,* 17th Forum on Geology of Industrial Minerals (Paper Presented May 15).

Williams, H., 1932, "The History and Character of Volcanic Domes," Bulletin of the Dept. of Geological Sciences, Vol. 21, No. 5, University of California, pp. 51–146.

Williams, H., 1941, "Calderas and Their Origin," Bulletin of the Dept. of Geological Sciences, Vol. 25, No. 6, University of California, pp. 239–346.

Phosphate Rock

G. DONALD EMIGH *

Nothing is more important to life—plant and animal—than phosphate. Its compounds are essential to the energy functions of all living systems and for the formation of bones and teeth. Animals get their phosphate from eating plants and other animals, or domestically from feed supplements. Plants get their phosphates from the soil.

Man's most important use of phosphate is for fertilizer; approximately 95% of the world's phosphate rock production is consumed by fertilizers. Most of the balance is processed in electric furnaces into elemental phosphorus, the important raw material for making industrial phosphates. Recovery of uranium from phosphates has been possible but not widely used. However, recent significant developments have generated interest in such recovery in the US and elsewhere.

Phosphate occurs in all igneous and sedimentary rocks, and in all fresh and salt water. However, economical recovery is limited to deposits where natural concentration of the phosphate mineral has occurred. Occasionally natural concentration is great enough that the material can be used as mined; generally, however, the ore is low grade and must be concentrated.

Phosphate rock is produced in 31 countries. Fig. 1 shows the location of major phosphate rock producers as well as some of the known phosphate deposits not yet in production.

In the last decade the world phosphate industry has experienced steady growth. World production rose from 98 Mt of phosphate rock in 1973 to 140 Mt in 1982. Fig. 2 shows the growth curve of rock production and graphically illustrates the almost straight-line rise since 1945 in both US and world production. Table 1 gives the breakdown, by countries, of world rock production, 1974 through 1979. Table 2 shows US rock production by producing areas.

* Consultant, Burley, ID.

Definitions of Terms Used in the Phosphate Industry

Phosphate Rock: Commonly called *rock* in sedimentary deposits and *apatite* in igneous deposits. Those expressions generally include any mined, or mined and beneficiated, fluorine-containing calcium phosphate used as the raw material for the next stage of manufacturing. The average phosphate content in rocks is 0.1 to 0.2% P_2O_5, as documented by McKelvey (1973). About 200 minerals contain more than 1% P_2O_5. However, the important mineral in igneous rocks is fluorapatite, $[Ca_5(PO_4)_3F]$, containing about 42% P_2O_5 and 3.8% F_2; in sedimentary rocks the important mineral is francolite, a carbonate fluorapatite containing up to 2% molecular CO_2. Both are in the apatite family of minerals.

Most phosphate rock, whether beneficiated or not, is a fine-grained material. The expression *phosphate rock* has no relation to its phosphate content. For example, phosphate rock from Idaho used in the production of phosphorus may contain about 24% P_2O_5, whereas rock from Morocco may contain 36.6% P_2O_5.

Phosphorite: A deposit of phosphate directly or indirectly of sedimentary origin, which is of economic interest.

Grade of Phosphate Rock: The calcium phosphate content of phosphate rock is expressed in different world areas by one of the following terms:

BPL (bone phosphate of lime)
TPL (triphosphate of lime)
P_2O_5 (phosphorus pentoxide)
P (phosphorus—not commonly used)

An illustration of relationship is:

80% BPL = 80% TPL = 36.6% P_5O_2 = 16% P

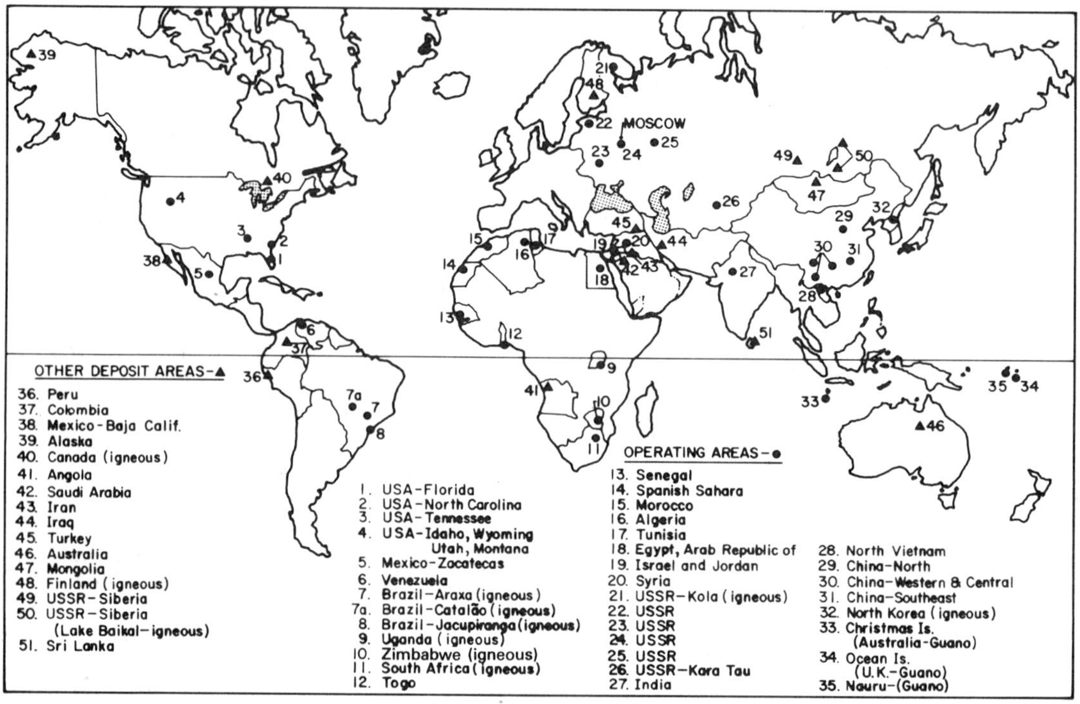

FIG. 1—*Location of the major phosphate rock-producing areas of the world and some of the known deposits not now in operation. All deposits are pellet phosphates except where indicated in legends to be igneous.*

Units of Weight:

metric ton (t) = 2205 lb
long ton (lt) = 2240 lb
short ton (st) = 2000 lb

Ore or Matrix: Used to denote naturally occurring phosphatic material to be upgraded into phosphate rock.

Wet-Process Acid: Phosphoric acid produced by treating phosphate rock with acid. It is sometimes called *green acid*.

Food Grade Acid: A very pure phosphoric acid produced from elemental phosphorus.

Processing of Phosphate Rock

Phosphate rock is commercially processed in three ways:

Acidification Treatment

1) With sulphuric acid to make:

Normal superphosphate—a fertilizer.

Concentrated superphosphate—a fertilizer.

Phosphoric acid—a fertilizer or the base for making ammonium phosphate fertilizers or calcium phosphate for animal feed supplements. Phosphoric acid to be purified for production of industrial chemicals.

2) With nitric acid to make nitric phosphate fertilizers.

3) With hydrochloric acid to make phosphoric acid. Hydrochloric acid treatment has had only limited use.

Electric Furnace Treatment

The electric furnace treatment of rock produces elemental phosphorus which is converted largely to very pure phosphoric acid used in the production of such industrial chemicals as phosphates of sodium, calcium, potassium, and ammonia. The phosphoric acid is also called food-grade phosphoric acid when further purified.

Simple Physical Treatment

1) Fine grinding for use as a fertilizer by direct application to acid soils.

2) Removal of fluorine by heat to produce calcium phosphate for animal feed supplement.

3) Heat treatment of wavellite (aluminum

TABLE 1—World Phosphate Rock Production, 1000 t

Country	1974	1975	1976	1977	1978	1979	% Change, 1974-1979 (Selected)
Western Europe	104	125	138	150	100	60	−42
France	19	20	28	19	20	0	
West Germany	85	75	85	81	0	0	
Sweden	—	30	25	50	80	58	
Finland	—	—	—	—	—	2	
Eastern Europe							
USSR	22 500	24 155	24 230	24 375	24 000	25 000	+11
North America							
United States	41 445	44 450	44 670	47 256	50 037	50 996	+21
Central America	179	153	147	150	150	171	
Curacao	107	83	77	80	80	47	
Mexico	72	70	70	70	70	124	
South America	432	471	555	661	1 164	1 699	
Brazil	281	345	465	608	1 094	1 694	+503
Colombia	10	10	10	10	10	5	
Venezuela	141	116	80	43	60	0	
Africa	33 089	25 555	25 545	29 661	32 637	33 829	+2
Algeria	802	669	820	1 001	1 136	1 073	
Egypt	550	536	550	580	642	645	
Morocco	19 326	13 548	15 285	17 285	19 278	20 175	
Morocco (Sahara)	2 385	2 682	172	232	400	—	
Zimbabwe (Rhodesia)	130	140	140	140	100	140	
Senegal	1 878	1 677	1 580	1 802	1 843	1 619†	
South Africa	1 550	1 651	1 631	2 403	2 699	3 221	+108
Toga	2 553	1 161	2 068	2 857	2 827	2 916	
Tunisia	3 903	3 481	3 294	3 614	3 712	4 040	
Uganda	11	10	5	5	0	0	
Asia	9 906	9 515	10 317	11 401	13 217	15 643	+58
China	3 000	3 400	3 750	4 000	4 300	5 500	+83
Christmas Island	1 809	1 343	1 037	1 260	1 386	1 367	
India	431	429	613	635	690	514	
Indonesia	6	5	5	4	5	—	
Israel	1 000	660	741	1 232	1 759	2 216	+122
Jordan	1 600	1 353	1 702	1 782	2 223	2 826	+77
North Korea	400	420	450	450	450	550	
Philippines	11	6	9	13	4	—	
Syria	500	500	511	425	800	1 170	+134
North Vietnam	1 150	1 400	1 500	1 600	1 600	1 500	
Oceania	2 838	2 178	1 446	2 054	2 677	2 248	
Australia	1	124	254	496	210	0	
Nauru	2 288	1 534	757	1 146	1 999	1 828	
Banaba (Ocean Island)	549	520	438	419	468	420	
Total World	110 498	106 602	107 049	115 959	123 982	128 157	+16

Source: The British Sulphur Corp., London.
* Compiled largely from *Phosphate Rock Statistics*, ISMA, 1979.
† Excluding aluminum phosphates.

TABLE 2—US Production of Phosphate Rock, Mt

State	1974	1975	1976	1977	1978	1979	1979-1980*
Florida and North Carolina	40.0	40.7	41.5	40.6	43.3	44.3	47.1
Tennessee	2.4	2.3	1.8	1.7	1.7	1.9	2.1
Western states	6.3	5.8	5.9	4.9	5.1	5.5	5.3

Source: US Bureau of Mines.
* Fertilizer year July 1979 through June 1980.

phosphate in Senegal) to produce fertilizer and animal food supplement.

4) Heat treatment with alkaline rocks, or other alkaline materials, to produce fused phosphate fertilizer. The process has only limited use.

Geology of Phosphate Deposits

Phosphate of economic importance occurs in two types of rocks—igneous and sedimentary. A comprehensive and detailed report on the world's phosphate deposits by the British Sulphur Corp. (Anon., 1971) covers nearly all the known phosphate occurrences and describes their geology, mining, and beneficiation. Among other information are maps and chemical analyses of the phosphate rock. Barr (1960) gives a good short description of the world's major phosphate areas. Emigh (1973) describes phosphate deposits in the United States and some in other parts of the world.

Igneous Phosphate Deposits

About 20% of the world's phosphate rock now comes from the mining and beneficiation of igneous deposits, but this percentage will drop as production from sedimentary deposits increases.

Igneous apatite in relatively small quantities has been recovered from veins and from magnetite ore bodies where it occurs in minor amounts.

The important igneous apatite deposits of commercial interest are found in certain intrusive complexes of alkalic rocks. They generally occur as ringlike structures with limited areal extent (generally 1 to 20 square miles). Commonly near the center are veins and larger bodies rich in carbonates (calcite, dolomite, siderite, and ankerite) emplaced either in a molten condition or by metasomatic processes. These carbonate deposits are called carbonatites. Apatite of economic interest can be in the carbonatites or in the alkalic rocks located outward from the central carbonatite. In carbonatites, it may be in a weather-enriched surface mantle, or in the unweathered rock. Carbonatites are well described by Heinrich (1966) and by Tuttle and Gittins (1966).

The following description of igneous apatite deposits associated with alkalic rocks are excerpts from Deans (1968).

> The most outstanding apatite deposit in the world occurs in the Khibina nepheline syenite massif near Kirovsk, in the Kola Peninsula, USSR. . . . The Khibina massif is a circular pluton about 25 miles in diameter, consisting of foyaite surrounded by concentric inward-dipping sheets of other varieties of nepheline syenites characterized by eudialite, aegirine, arfvedsonite or mica. The apatite deposits are restricted to a phase of layered ijolitic (nepheline-aegirine-apatite-sphene) rocks which occupy only four percent of the area of the complex, mainly in an arcuate zone parallel to, and about two to four miles within, the southwestern boundary. . . .
>
> Carbonatitic alkaline igneous complexes, once regarded as rare and isolated phenomena, are now known to be widely scattered over all the continents, except (as yet) Australasia and Antarctica. They constitute a distinctive subvolcanic and volcanic rock association, characteristic of deeply faulted regions in continental shield and platform areas, but not strictly confined to these areas. Eruptions of this type are believed to originate at deep subcrustal levels, and have occurred at many periods extending back from the present to the Precambrian of about 2000 million years ago. The carbonatites, low-temperature magmatic and in part metasomatic in origin, are typically associated with feldspathoidal or alkaline ultrabasic rocks, and the surrounding country rocks display a characteristic form of metasomatic alteration termed fenitization, often accompanied by widespread breccia-

tion. Commonly they form rather small intrusive ring complexes, about one to five miles in diameter, although the pattern and suites of rocks exposed varies considerably, depending partly on the depth to which they have been uncovered by erosion, and probably partly on conditions at the deeper point and time of origin. The surface volcanic manifestations of this activity are gas-drilled explosion craters of the Laacher See (Germany) type, or nephelinitic volcanoes, such as Oldoinyo Lengai in Tanzania, whose eruptive cycle includes nephelinite lavas and tuffs, sodic carbonatite lavas and mixed carbonatitic and nephelinitic ashes. The older eroded complexes may reveal a core of calcitic or dolomitic carbonatites surrounded by syenitic rocks, usually nepheline syenites (ijolites and/or foyaites), often forming ring dykes or cone sheets, or in other cases by ultrabasic suites comprising dunites, pyroxenites or biotite or vermiculite-rich rocks. In other cases the carbonatites are emplaced along major rift faults, forming long dykes or zones of metasomatism. The petrology of these complexes is highly varied, often including extremely calcic, sodic or potassic suites, but a general characteristic is enrichment in minor elements such as fluorine, phosphorus, strontium, barium, niobium and the cerium-group of the rare earths. In association with the carbonatites there are several examples of workable deposits of these elements. Among the latter some nine apatite deposits have been worked successfully in South Africa, Rhodesia [Zimbabwe], Uganda, and Brazil, and several others appear to be of potential economic importance.

Carbonatites have been found in Australia in 1968, New Zealand in 1970, and recently in Paraguay.

World production of apatite (igneous) concentrates in 1980 was about 25 Mt. Table 3 lists the operating apatite mines in the world in 1980. Brazil is actively developing new apatite mines; by the mid-1980s its production of phosphate rock, largely igneous apatite, should be 5.8 Mt, compared to 1.7 Mt currently. The USSR continues to develop new operations in the Kola area and soon will have a new mine operating on apatite in the Lake Baikal region in Siberia. Foskor in South Africa is planning another large mine-beneficiation operation on the Phalaborwa carbonatite complex.

Some of the USSR's apatite mines in the Kola Peninsula are underground, but the majority of its mines, as well as all other apatite mines in the world, are surface operations (Anon., 1979, 1979a).

Sedimentary Phosphate Deposits

Sedimentary phosphorites are of two types: guano and pelletal. A suggested classification is shown in Table 4.

Guano deposits result from the accumulation of animal excrement, principally that of large sea birds. They are of historical importance and, although they still contribute significant tonnage, now account for only a small percentage of total production. Guano deposits are mined on two islands—Nauru and Christmas—where combined production is about 3 Mt/a. Christmas Island contains the largest deposit of guano-type rock known, with reserves of over 200 Mt. More detailed information on origin and geology, which can be complex, may be found in the works of Hutchinson (1950), Warin (1968), and Barrie (1967).

TABLE 3—World Apatite Mines (1980)

Country	No. Mines	No. Beneficiation Plants	Conc., 1000 t/a	Grade % P_2O_5	Type Deposit
USSR (Kola)	6	2	20	17	Nepheline syenite (unweathered)
South Africa	2	1	3.5	7	Pyroxenite and foskorite (unweathered)
Brazil	5	3	1.7	13.5-15	1 carbonatite (unweathered) 4 carbonatite weathered residual

TABLE 4—Classification of Sedimentary Phosphatic Formations as Proposed by Emigh

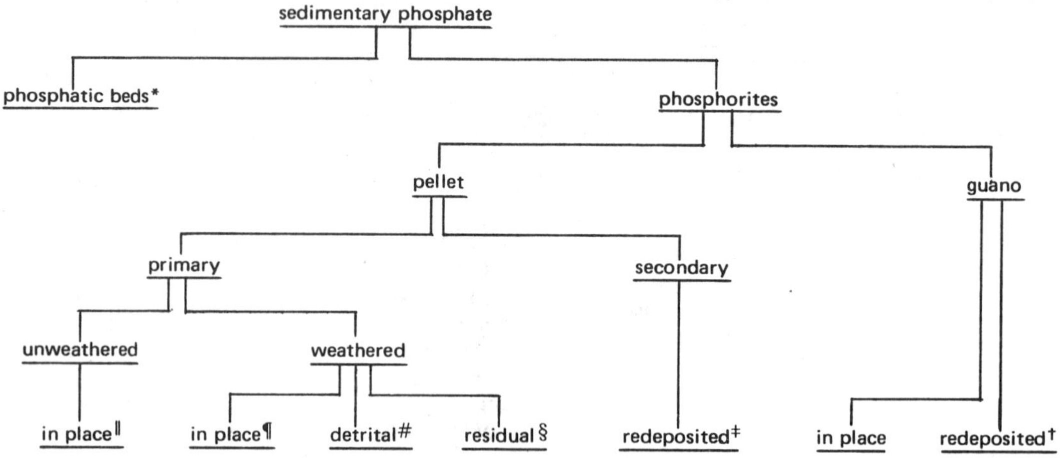

* Marine sediments in which the phosphate is believed present always as pellets—examples, Hawthorne formation, Florida, and Bigby limestone, Tennessee.

† Phosphate from guano—examples, Curacao, Nauru, and Christmas Island.

‡ Phosphate from pellet phosphorites—examples, white rock of Tennessee, hard rock of Florida, nonpelletal rock of Queensland, Australia.

§ Derived from phosphatic beds—example, brown rock of Tennessee.

\# Examples, pebble field of Florida and other similar fields in the southeastern United States.

¶ Surface and near surface weathered parts of all marine phosphorites—examples, Phosphoria formation of western USA, presently worked beds of North Africa (one exception is Kalaa-Djerda, Tunisia), Beetle Creek formation in Queensland, Bu-Craa phosphorite of Spanish Sahara.

‖ Unweathered parts of all marine phosphorites—examples, Phosphoria formation, La Caja formation of Mexico, lower unworked beds in Morocco, the "black" rock found at lower levels at Youssoufia, Morocco, and the bed at Kalaa-Djerda, Tunisia.

The greatest phosphate deposits are pelletal phosphorites. These are in marine formations—commonly associated with dolomite or limestone. They furnish over 80% of the world phosphate rock requirements and this proportion will increase in the future. As previously noted, the phosphate is present as the mineral francolite—a carbonate fluorapatite. In 1870, this mineral was named collophane, and it was believed to be isotropic. Subsequently, it was found to be anisotropic, conforming to the mineral francolite. Unfortunately, the word collophane still appears occasionally in modern literature.

The francolite of pelletal phosphorites was originally emplaced as small pellets, and most of it is still in this form. These are commonly ovoid in shape although some skeletal pellets, such as broken shells, can be elongate. Emigh (1967) describes six pellet types. In any one bed pellets are about the same size, but in differing beds they can be of different size. The sizes of pellets in some phosphorites are shown in Emigh (1967); an average is in the range of 0.25 to 0.35 mm. Practically all pellets represent original aragonite pellets of diverse origin, later phosphatized by phosphate ions in seawater.

Under certain weathering conditions, the phosphate of pellets can move and be emplaced elsewhere. Examples include the white rock of Tennessee, the hard rock of northern Florida, and the nonpelletal parts of the phosphorites of Queensland, Australia, as pointed out for the latter by Cook (1972).

Highly phosphatized marine formations can be very large in areal extent. For example, the Phosphoria formation of Montana, Idaho, Wyoming, Utah, and Nevada originally covered 135,000 square miles and is being mined in several places along a 450-mile belt across this area.

The thickness of a bed of phosphorite may vary from a few inches up to tens of feet. Beds may extend several miles laterally, maintaining fairly equal thickness and phosphate content.

The phosphate content of a phosphorite bed ranges from a few percent P_2O_5 up to practically solid francolite pellets—about 35% P_2O_5. Dark brown to black hydrocarbons commonly are in the phosphorite beds together with varying amounts of detrital materials—largely quartz. Such beds may even be referred to as black shales. Chert is commonly associated with phosphorites, sometimes in the matrix containing the phosphate pellets, but more commonly as either thin or thick marine beds located stratigraphically close to the phosphorite—either above or below. In the famous Rex chert of the Phosphoria formation of the western United States, the chert appears to be derived from siliceous sponge spicules (skeletons). This is also true of thin chert beds associated with the phosphorites of middle Tennessee.

Sheldon et al. (1967) describes, for the Phosphoria formation, the theory of deep upwelling ocean currents depositing phosphate on and near outer continental shelves. This theory has been applied elsewhere in the world. Not everyone believes this condition is necessary for the deposition of phosphatic marine formations of economic interest, but there is generally common agreement that moving seawater is necessary for providing the phosphate.

Phosphorites occur widely scattered throughout geological history. Table 5 shows the ages of various phosphorites.

Among the many descriptions of pellet phosphorites, the following excerpts are taken from an article by Notholt (1968):

> Most marine phosphorites appear to have been formed in somewhat restricted basins of deposition, occupying arms or

TABLE 5—Age of Phosphate in Some Pellet Phosphorites*

Location	Formation or Group	Age
Canada, Alberta	Rocky Mountain	Permian
Western USA, MT, ID, UT, WY, NV	Phosphoria	Permian
Western USA, Utah	Brazer	Mississippian
California	Monterey	Miocene
Arkansas, Searcy Co.	Morrow	Upper Pennsylvanian
Iowa	Maquoketa Shale	Upper Ordovician
Tennessee, Brown Rock	Bigby	Ordovician
Tennessee, Blue Rock	Chattanooga Shale	Lower Mississippian
North Carolina	Pungo River	Lower Miocene
Florida, Pelletal Type	Hawthorne	Middle Miocene
Mexico, Zacatecas and Coahuila	La Caja	Upper Jurassic
Venezuela, Western		Upper Cretaceous
Venezuela, Falcon State	Riecito	Lower Miocene
Colombia	Guadalupe	Upper Cretaceous
Peru, Sechura Desert	Zapayal	Upper Miocene
Brazil, Pernambuco	Gramamme	Upper Cretaceous
Angola, Cabinda Enclave		Upper Cretaceous
Togo		Middle Eocene
Senegal, Taiba		Eocene
Spanish Sahara, Bu-Craa		Upper Cretaceous/Lower Tertiary
Morocco		Upper Cretaceous/Lower Tertiary
Algeria		Upper Cretaceous/Lower Tertiary
Tunisia	Metlaoui	Upper Cretaceous/Lower Tertiary
Egypt		Upper Cretaceous
Israel		Upper Cretaceous
Jordan	Belqa	Upper Cretaceous/Lower Tertiary
Saudi Arabia	Hibr	Lower Tertiary
	Aruma	Upper Cretaceous
Iran	Pabdeh	Lower Tertiary
	Geirud	Upper Devonian
Iraq		Upper Cretaceous/Lower Tertiary
Lebanon		Upper Cretaceous
Syria		Upper Cretaceous/Lower Tertiary
Turkey	Germav	Upper Cretaceous
	Karabogaz	
	Karababa	
USSR		Upper Precambrian, Cambrian, Ordovician Jurassic, Lower Cretaceous
India		Precambrian, Jurassic, Upper Cretaceous
North Vietnam	Coc Xan	Lower Cambrian
Australia	Beetle Creek	Middle Cambrian

* Source: Anon., 1971b.

gulfs of preexisting seas and developed either on the neritic shelf areas of geosynclines or as subsiding areas on relatively stable continental platforms. Marine phosphate deposits have thus been classified into geosynclinal and platform varieties, each being recognizable in a very general way by the differences in petrography of the phosphorite beds themselves and of the associated sediments and in various structural relationships. The classification is unsatisfactory in many ways, however, and at best the two groups, where typically developed, can only be regarded as representing end members of a continuous series. Most phosphate deposits show features characteristic of both geosynclinal and platform-type deposits. Nevertheless, in the absence of a much-needed classification based on precise stratigraphical and structural criteria, it seems that a separation into the two groups of deposits can usefully be made for preliminary exploration purposes, the distinction being modified subsequently where their gross geological characteristics are known.

The phosphorite lithofacies was first recognized as such by Russian geologists during the early 1930s, the frequent association of phosphorite with black, carbonaceous or bituminous shales or mudstones, cherts (or other siliceous sediments), as well as various carbonate rocks probably providing to date one of the most effective aids to phosphate exploration. Although some important variations exist, the black shale-phosphorite-chert-carbonate rock assemblage characterizes geosynclinal phosphate deposits in many parts of the world. Its recognition in some cases after examining samples obtained in the course of oil-drilling operations, has led in recent years to the successful discovery of promising deposits.

The Permian Phosphoria Formation of the western United States, embracing one of the largest phosphate fields in the world, contains a classic example of what may be termed the *geosynclinal phosphorite lithofacies.* This facies occurs in the western part of the phosphate field where the Phosphoria and its partial equivalents form part of a thick sedimentary pile deposited in the Cordilleran miogeosyncline. It is characterized by black, high-grade phosphorite, phosphatic and carbonaceous mudstone, cherty mudstone and chert. Calcareous rocks, as well as sandy beds, are generally absent. Accordingly, McKelvey has concluded that almost any marine black bituminous shale or bedded chert (except for those occurring with volcanic rocks or other thick eugeosynclinal sediments) may be suspected of being associated with phosphorite or phosphatic shale, although the latter rocks may not necessarily contain sufficient amounts of phosphate to be of commercial interest. The close association of bedded chert and phosphorite was not recorded by Kazakov although radiolarian and diatomaceous cherts had been known for many years previously from the North African phosphate fields and elsewhere. Of interest is the association of Miocene phosphorites with diatomite in the Monterey formation of California and the Zapallal formation underlying large areas of the Sechura Desert, northwestern Peru.

The British Sulphur Corp., in "World Survey of Phosphate Deposits" (Anon., 1971b), includes phosphate reserves; however, it issues a word of caution in the following statement:

> Quantitative information on reserves of phosphorus-containing minerals is most unsatisfactory. Published data are seldom given in sufficient detail that their significance can be evaluated, and even where this detail is available it may only serve to reflect the widely different methods of assessment used to calculate and classify reserves throughout the world.

Some phosphate reserves are given in Table 6, although this list does not represent total reserves. Reserves of phosphate rock are for all practical purposes unlimited (Emigh, 1972).

Weathering

The wide range of geologic time over which phosphorite deposits have been deposited has been noted. Over this long period of time tectonic movements have moved the original flat-lying phosphorite beds into all attitudes—from gently dipping to vertical and to overturned, all further complicated by faulting. Such rearrangement of beds, with its attendant faulting, complicates exploration and mining, but it does aid substantially that all important factor—surface and near-surface weathering.

With a few notable exceptions, such as the unconsolidated phosphorite deposits along the coast of North Carolina and the Sechura Desert of Peru, all phosphate deposits were originally hard rock. Certainly this is true of igneous deposits. The originally unconsolidated pellet phosphorite beds became dense hard rocks as they were subjected to lithification. In these phosphorites, the hard cementing matrix is commonly dolomite, but sometimes calcite. Less commonly the cementing matrix can be cryptocrystalline quartz, or chert, derived from

TABLE 6—Summary of Uraniferous Phosphate Resources of the Free World

Area	Phosphate Resource (millions of st)	Typical P_2O_5 Content of Phosphate Resource	Average Uranium Content of Phosphorite (ppm U)	Potentially Mineable Phosphate Product (~30% P_2O_5)* (millions of st)	Estimated Recoverable Phosphate Product (~30% P_2O_5)† (millions of st)	Average Uranium Concentration of Phosphate Product (ppm U) e = estimated	Uranium Contained in Potentially Mineable Phosphate (thousands of st)	Estimated Recoverable Uranium‡ (thousands of st)
United States								
Southeastern								
Central Florida	9,760	15-18%	65	1,737	1,129	110	191	124
Southern Extension	25,854	10-15%	36	5,084	3,050	110	560	336
North Florida-South Georgia	19,110	8-10%	64	3,230	1,764	70	225	123
East Florida	25,590	8-16%	51	4,437	2,662	102	453	272
East Georgia-South Carolina	18,179	4-15%	30-60	3,469	1,787	60-90	276	142
North Carolina	71,761	8-18%	30	15,086	9,429	60	906	566
Others	272	16-23%	20-33	144	69	25	3	2
Subtotal	170,526			33,187	19,890		2,614	1,565
Western Phosphate Field								
Idaho	324,904§	24-26%	96	203,065∥	84,072	124	25,180	10,425
Utah	154,479§	24-26%	76	96,204∥	35,171	98	9,428	3,447
Wyoming	287,573§	24-26%	64	178,778∥	67,397	83	14,839	5,594
Montana	58,306§	22-24%	65	35,861∥	15,684	84	3,012	1,317
Subtotal	825,262			513,908	202,324		52,459	20,783
Other¶	9,412	12-28%	30-76	2,257∥	1,128	98	221	111
Total United States	1,005,200			549,352	223,342		55,294	22,459
Rest of Free World								
Algeria	1,100	25%		700	525	100	70	52
Australia	4,000	17%		1,600	1,200	68	110	82
Brazil	2,200	5-15%		422	316	e 65	30	22
Egypt	1,645	20-26%		1,000	750	100	100	75
Iraq	2,360	18%		770	579	130	100	75
Jordan	16,000	15-33%		4,000	3,000	120	480	360
Mexico	51,190	3-27%		5,534	4,150	e100	553	415
Morocco	88,000	25-34%		60,000	45,000	110	6,600	4,950
Peru	10,000	8-12%		3,000	2,250	e100	300	225
South Africa	105,300	7-16%		50,770	3,600**	12	3,147	43
Syria	900	19-28%		540	405	130	70	52
Tunisia	1,300	26-30%		910	682	60	55	41
Western Sahara	11,000	31-33%		7,400	5,550	100	740	555
Others	4,941	13-30%		1,834	1,378	30-160	167	126
Total, Rest of Free World	299,936			138,480	69,385		12,522	7,073
Total, All of Free World	1,305,136			687,832	292,727		67,816	29,532

Source: Anon., 1979b.

* Includes the phosphate product (of approximate 30% P_2O_5 content) that could be producible from the total phosphate resource, without accounting for a mining recovery factor, but accounting for an unavailability factor. In the United States, resources unavailable for mining development because of federal land classifications, cultural features, or other environmental factors are included in the total phosphate resource column but are excluded from the potentially mineable phosphate product column. Only the portion of RARE II lands recommended for wilderness designation are considered unavailable for purposes of this table.

† Average mining recovery factors generally of 50 to 75% for surface mineable resources, depending on the district, 50% for underground resources to a depth of 5,000 feet and subocean resources, and 35% for deep underground (greater than 5,000 foot depth) resources have been applied to the "Potentially Mineable Phosphate Product" to determine this "Estimated Recoverable Phosphate Product."

‡ The total uranium contained in the estimated recoverable phosphate product.

§ Includes ore-grade ($\geq$18.0% P_2O_5) resources in all resource depth categories.

∥ A beneficiation recovery factor of 1.6:1, the operating experience in the Western Phosphate Field, has been applied to the phosphate resources in order to obtain the potentially mineable phosphate product. In other words, 1.6 tons of phosphate resource must be mined and beneficiated to produce 1.0 ton of approximately 32% P_2O_5 product.

¶ Includes all other marine phosphorites of the United States, but not any of the monazite placer or igneous apatite deposits, because of their clear lack of economic viability.

** The offshore phosphorites off South Africa, which are lithified to a great extent, are considered to be not mineable with current technology.

Note: tons are short tons (2,000 pounds)

siliceous skeletons of tiny marine animals and plant life. Commonly also dark brown-black hydrocarbon is present.

Surface weathering, with its solubilizing and leaching away of cementing carbonates, is a most important factor in the development of economic phosphate deposits. The removal of carbonates in phosphorites and carbonatites softens the formation, upgrades the phosphate content, removes carbonate material which is deleterious to the treatment of the final concentrate, and makes mining and beneficiation easier. In pellet phosphorites, weathering also changes the originally finely dispersed pyrite into iron oxide and reduces the quantity of hydrocarbons, changing the original black color to gray or brown.

Hale (1967) describes the effects of weathering on the Phosphoria formation of the western United States, and Emigh (1973) describes the effects on pellet phosphorites in general.

Exploration and Development

Phosphate deposits of economic interest, both igneous and sedimentary, are difficult to recognize in the field. Commonly, even the discrete apatite crystal, or the small phosphorite pellet, cannot be positively identified. But a simple field test is possible by placing a drop of acid (sulfuric, nitric, or hydrochloric, not weaker than approximately 20% concentrated acid in water by volume) on the questionable area with a medicine dropper and then adding a small crystal, or a small amount of powder, of ammonium molybdate off the end of a knife blade. The rapid development of a bright yellow color indicates phosphate. This field test is very sensitive but positive results, however, do not mean a commercial deposit has been found.

Exploration is generally directed toward: (1) finding new deposit areas and (2) searching for new deposits associated with known fields. In the case of (2) exploration will, of course, be guided by the practical knowledge acquired in that field before the new exploration begins.

The report of the United Nations (Anon., 1968d) contains over 50 articles by authoritative authors on the geology and exploration of world phosphate deposits as well as the use of rock.

Igneous Deposits

Although the world's great phosphate deposits are of the pellet phosphorite type, igneous deposits are also important as they yield about 20% of the present world production of phosphate rock.

Deans (1968) states the number of known alkalic complexes favorable for apatite has grown from 20 in 1945 to about 200 in 1967. At least 20 of these probably contain apatite deposits. The following are excerpts from Deans' chapter:

> It may be useful to outline practical steps which can be taken to initiate a wider search for carbonatites in the Economic Commission for Asia and the Far East (ECAFE) region. The new books on carbonatites should be studied to provide a factual background. Then, before field work begins, there must be a critical search for possible clues in the regional literature, maps, files and rocks and mineral collections of all the geological departments in each country so that specific areas may be selected for reconnaissance inspection. The clues may be sought under three headings: petrological, mineral records, and geological and geophysical information.
>
> Petrological information: Occurrences of alkaline rocks obviously deserve consideration since carbonatites are genetically linked with these rocks, but it should be noted that many occurrences have no carbonatites associated with them. . . .
>
> Having selected localities or small districts for examination on the basis of the more promising clues found in this search, a round of geological field inspections can begin. The objective is now to locate carbonatites, 'promising' alkaline or ultrabasic rocks, or fenites etc. for further study, and to eliminate fruitless clues and if possible explain them so that similar mistakes are not repeated unnecessarily. The methods and intensity of search are matters for the geologist to decide in the particular circumstances, but collecting material for identification or confirmation in the laboratory is the major task, and in addition to rock specimens, samples of residual soils are most important. This is because they are often more representative of the chemistry and mineralogy of the bedrock as a whole than are the resistant and often heterogeneous rocks which form outcrops. Normally they will be enriched relative to bedrock in some of the diagnostic elements of carbonatites, notably in phosphorus, barium, rare-earths and niobium (strontium is leached out). . . .
>
> In the field it is useful to carry a strong hand-magnet, conveniently shielded in a thin aluminum cup, for detecting magnetite-rich soils, as magnetite and apatite are commonly associated in carbonatites and phosphorites, and both are enriched in the residual soils overlying them. . . .
>
> Having located carbonatitic alkaline or ultrabasic complexes, the chances of apatite discoveries may still be uncertain, and finding them may not be easy. Apatite-rich material may be quite nondescript and variable in appearance, especially when fine-grained or weathered, and chemical or microscopic checks should be applied to all doubtful or suspect rocks, coupled with soil sampling. In general the most deeply eroded complexes appear to be most favourable for apatite deposits of the phosphorite or pyroxenite type. In these, carbonatites may be relatively minor constit-

uents of the complex (at Palabora they form only about 2 percent of the outcrop area), so a complex should not be underrated just because the area of carbonatite is small. The maximum development of carbonatites appears at a somewhat higher level of intrusion in the form of large carbonatite plugs, and these may yield important residual deposits of apatite, provided climatic and erosional factors are favourable. In the less eroded volcanic complexes one may find apatite associated with the highly potassic feldspathic breccias and trachytic rock, as exemplified by the Tundulu deposits and several minor occurrences. The grade of deposit which may be regarded as of possible practical interest depends on many factors, but present practice suggests that a recoverable apatite content of about 15 percent (6 percent P_2O_5) may prove economic, provided there are adequate reserves, and location and working costs are favourable.

A brief description of prospecting for igneous phosphate deposits in the USSR is given by the USSR Delegation (Anon., 1968b), both for those associated with carbonatites and those associated with alkaline rocks such as the nepheline syenites of the Kola Peninsula. The following is an excerpt from the article:

> In the first stage of prospecting, on the basis of data of the Complex Geological Survey and other sources, general small-scale forecast maps are compiled for the areas of possible occurrence of these deposits. For carbonatites, they are the regions of ancient platforms and areas of completed folding; while for magmatic deposits associated with basic rock, they are the areas of a wide development of ancient basic gabbro-anorthosite and gabbro-diorite complexes.
>
> At the initial stage of search for carbonatites, the usual aeromagnetic and aeroradiometric surveys are particularly useful, as they reveal anomalies that are distinctly outlined in the regions of development of normal sediments. At the ground control of these anomalies, the areas of possible apatite occurrence are selected where reconnaissance prospecting is carried out in order to reveal their possible commercial importance.

The USSR article refers to the use of boreholes (presumably diamond drill holes). The article states in the reconnaissance survey that the exploration network of holes is roughly 500 × 200 m for simple deposits and 250 × 150 m for complex deposits. The maximum depth of drilling during the preliminary survey is about 700 m. In the following detailed survey, category A reserves are those drilled out on grids of about 125 × 100 m and category B on grids of about 250 × 200 m. Cores are sectionally sampled in lengths of 2.5 to 3 m and analyzed for P_2O_5 and Al_2O_3. Other chemical analyses are performed on groupings of eight or ten of the individual samples.

Pellet Phosphorites

Only in the last 25 years has there been concerted worldwide search for pellet phosphorites with the greatest emphasis occurring in only the last 15 years.

Fig. 2 shows the rapid increase in demand for phosphate rock since the mid-1940s. Prior to that time, the relatively modest world rock requirements came from deposits that had been long known and worked. These deposits had been found largely by chance.

The surge in world demand for rock sparked a rapid buildup in capacity in the producing fields and attendant exploration worldwide.

Especially for areas beyond known fields, planned exploration applies new ideas about the origin and emplacement of pellet phosphorite. Of special value was the recognition of the importance of the movement of phosphate-containing seawater in large basins.

The exploration and development of pellet phosphorites in new areas might be separated into four phases as follows:

Phase 1: The search for areas of marine sedimentation.

Phase 2: Within areas of marine sedimentation, search for beds of chert, or marine diatomite and/or black shales.

Phase 3: Search for beds of phosphorite in or near the black shales or cherts. Also search for residual phosphatic soil areas associated with beds of phosphatic dolomite or limestone.

Phase 4: Determine thickness, grade and extent of weathering by trenching and drilling.

The thickness of chert beds, so commonly associated with pellet phosphorites, may be from a few inches to over 200 ft. Pellet phosphorite commonly contains from 0.001% to 0.02% U_3O_8. Where the higher percentage is present in exposed beds, airborne scintillator equipment can also be employed.

Commonly, an outcrop of phosphorite will have patches of a thin skin of "phosphate bloom" where exposed at the surface. Its characteristic light gray-blue color is readily visible up to distances of several feet. A 10 to 20-power

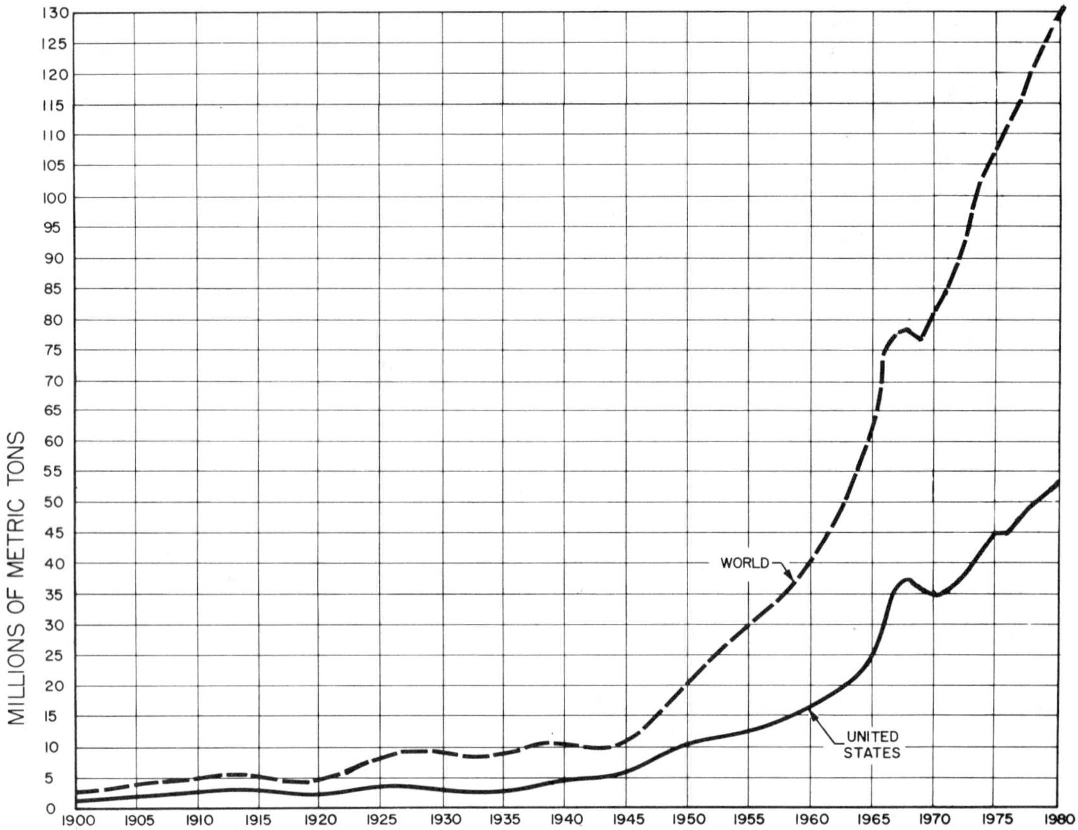

FIG. 2—*Production of phosphate rock in metric tons for the world and the United States, 1900–1980. Through 1965 curves are drawn on five-year averages; after 1965 they are drawn on a yearly basis.*

hand lens is indispensable to aid in identifying the small ovoid-shaped phosphate pellets.

In areas of marine sedimentation which appear to be a favorable locale for phosphorite, but where no outcrops are present, previously drilled holes, if present, can prove a valuable aid. For example, in the mid-1950s water wells in North Carolina when flushed out revealed phosphate pellets. Later, preliminary and successful exploration of the field was conducted by gamma-ray testing of about 100 existing holes drilled for water.

The important pellet phosphorites of the Georgina Basin in the Queensland-Northern Territory of northeastern Australia were found in 1966. This discovery resulted from a study of cores of holes drilled in a search for oil. It was then established that phosphate beds could be identified on the gamma-ray logs of the holes. The phosphate beds so identified were then found in outcrops some distance away (Howard, 1971, 1972).

Early in 1962, chemical testing of sludge obtained from many holes drilled for oil in Turkey showed a few holes with enough phosphate to merit interest. Later gamma-ray logs of such holes were examined for phosphate. The results obtained from the drill holes led to the discovery of pellet phosphorite outcrops.

Other than gamma-ray measuring techniques, there has been to date very little application of geophysical methods in the exploration of pellet phosphorites. Cooksley (1967) describes some work on the Phosphoria formation in southeastern Idaho made in connection with obtaining additional information on previously discovered phosphorite deposits. Limited seismic work in northern Florida and Tennessee has been unproductive. Except for the possibilities outlined by Cooksley, and the use of gamma-ray measurements, it appears probable that the geophysical approach to phosphorite exploration has only a minor role to play.

To the writer's knowledge, geochemical prospecting has not been used in the exploration for phosphate deposits. Presumably it might have some value, especially in exploration for residual phosphorites of the Tennessee type.

Under Phase 3 of exploration mention was made of residual phosphorites represented by deposits in Tennessee. The brown-rock pellet phosphorites of Tennessee are residual deposits formed by weathering of the underlying phosphatic limestone. The brown-colored phosphorite is covered with up to several feet of brown-colored soil. The soil mantle, coupled with the low uranium content of the phosphorite (less than 0.001% U_3O_8), makes airborne scintillator prospecting useless. Road cuts and an occasional gully will reveal the underlying phosphorite. Even at such places the inexperienced eye probably would not distinguish between the soil and the phosphorite. Prospecting for isolated phosphate deposits is simply by blind drilling at the proper horizon, i.e., on top of the phosphatic limestone. Individual economic deposits may contain from 20 kt to over 1 Mt of matrix (ore). It seems probable that throughout the world there are many areas of residual phosphorites of the Tennessee type. They are difficult to recognize and, in all likelihood, generally have not been looked for.

Phosphorite beds vary laterally in thickness and phosphate content. In some places, as in western United States' Phosphoria formation, this lateral variation is small for long distances—several tens of miles—but in other places, and other formations, more extreme variations do occur. In exploration therefore, extensive lateral exploration should be made even though preliminary exploration may detect only low grade ore. The economically important detrital (reworked residual) phosphorites of the Bone Valley formation in middle and northern Florida occur as unevenly distributed beds over large areas. Their lateral variation in thickness and grade requires extensive drilling to ascertain the sometimes tortuous outlines of minable material.

After a bedded primary pellet phosphorite has been found, it is commonly difficult to follow the bed in outcrop because weathering has softened the formation releasing loose soil-like material that masks the beds. Even when the exact location is known, pellets or small pieces of phosphorite are difficult to find or recognize. Therefore, harder "marker" beds stratigraphically above or below the phosphorite often can be better used to map for location of a phosphorite bed than the phosphate bed itself.

Utilization of existing holes drilled for oil in exploration for pellet phosphorite has been mentioned. However, to date, present economic conditions would not justify extensive drilling for the primary purpose of finding phosphate. But in certain areas, shallow drilling has proven itself where the objectives are near-surface deposits. Residual phosphate deposits of Florida-Georgia-South Carolina-North Carolina and Tennessee are representative. Drilling in those areas is generally limited to depths that could be open pit mined. For the past several years, exploration drilling for phosphate deposits in selected areas of Florida and North Carolina has reached depths of 300 ft or greater. Present exploration drilling for both shallow and deep deposits is done with a truck mounted rotary rig collecting a 4 to 5 in. core with a 10 ft single tube core barrel.

In wet places, tests can be made by washing cuttings from the hole. Cuttings are caught in tubs on the surface. This type drilling commonly results in loss of slimes and, consequently, under-reporting of the quantity of slimes in the final laboratory tests.

In Tennessee exploratory drilling is most economical by using powered 4-in. augers. One type is a two-wheel trailer with weight of the drill only being applied to the drill stem. But a preferred type is mounted on the back of a truck and the drill stems are forced downward hydraulically. Again, field inspection suffices to determine whether phosphate of interest has been found. Even until recently, hand drilling was done in Tennessee and Florida using a cup auger drilling head mounted on a 1-in. pipe. Diameter of the hole was 4 in., and the head was pulled out and cleaned every 8 in. Some hand-drilled holes require casing through the soil overburden. Here the casing is forced ahead of the drill.

The discovery of the economically important Bu-Craa phosphate in Spanish Sahara in the early 1960s is an example of drilling being used in the exploration phase. As described by ENMINSA (Anon., 1967), geological reconnaissance work started at the already known low-grade phosphate horizon in the Izic district. These studies, in a flat desert terrain with poor rock exposures, indicated a richer phosphate concentration 60 miles away. Drilling in this area on a 1-km grid found the important Bu-Craa deposit. This drilling was done with a rotary rig using a tricone bit. Diameter was 11.34 cm (4.5 in.) with cuttings air-blown to the surface where they were collected as samples.

Phase 4 in the exploration and development of pellet phosphorites involves the determination of the thickness, grade, and extent of weathering by surface work which may include one or more of the following: trenching, shafts, or drilling. Trenching can be by hand or with mechanical equipment—dozers or backhoes. Its purpose is to expose the phosphatic beds for measurement and sampling. Depths of 15 ft or more may be necessary to get through loose overburden and minimize the effects of surface creep. Down-slope surface creep can significantly string out and thin the phosphatic beds and change the apparent dips of the beds. Surface slopes as low as 5° can result in significant creep. Where the material moved is dry and relatively incoherent, deep trenches pose a real danger to men and equipment from sudden slumping of the vertical walls.

A brief description has already been given of the kinds of drills used for exploration of residual pellet phosphorites in Florida and Tennessee. The same Phase 3 equipment is used for the additional holes required for Phase 4. In the western US Phosphoria field, diamond drilling (commonly NX) is used as well as rotary drilling with air return of cuttings. When the rotary drilled hole has reached the phosphate zones (and this is determined visually), the cuttings are collected as samples. Coring can, of course, also be done with a rotary drill rig, but this is not common nor is drilling with water common with these rigs. Air return of cuttings works very well, and without casing, up to depths of several hundred feet; however, encountering wet material usually prevents recovery of cuttings.

Rotary rigs may average 400 ft per 8-hr shift vs. perhaps 30 to 40 ft for the diamond drill. Nevertheless, diamond drilling has certain advantages, i.e., ability to make deeper holes needed to determine the limits of alteration, wet ground does not stop drilling, holes can be angled to better intersect dipping beds, and a core is a more accurate sample for geological study. A common criticism of coring in soft shales is that core recovery is not good. Proper drilling, however, results in good core recovery—largely due to not forcing the bit.

In Phase 4 exploration for a pellet phosphorite bed such as present in the Phosphoria formation of western United States there is no established rule for the lateral spacing of holes, but trenches and drilled holes about 1000 ft apart are generally sufficient to establish thicknesses and grades. As the time approaches for mining, however, additional information is required, but this is more for the purpose of better defining local structures such as faulting and change of dip of the beds. These local structural features dictate the amount of additional drilling; however, different operators have different approaches. One may be satisfied with holes before mining being spaced several hundred feet apart, where another, in a similar situation, may want 50-ft spacing.

In the Florida fields, the matrix (phosphate ore) is commonly separated every 5 ft into an individual sample. Each sample is processed in laboratories by bench tests to simulate plant operations, i.e., a weighed sample is pulped with water and deslimed; slime is any material less than 150 mesh. The deslimed material is then screened at 14 mesh to obtain coarse phosphate aggregate called pebble, and the undersize −14 +150 mesh portion is floated to recover a phosphate concentrate (rock). The pebble and flotation feed portions furnish the sample for assaying and not the original matrix.

In Tennessee, the matrix obtained by drilling is separated into 4-ft sections, sacked, and sent to a laboratory. The sample is split and analysis made for P_2O_5. The other split is saved until a minable deposit has been drilled. Then one composite is made, proportioning weights by feet of core in each sample. This composite sample is split down to a working sample of about 6000 g. One half is analyzed for P_2O_5, SiO_2, Al_2O_3, Fe_2O_3, and CaO. The other half is wet screened. There may be one screen size used (325 mesh) or three, i.e., 200, 325, 400. All products, including undersize, are weighed and analyzed for P_2O_5, SiO_2, Al_2O_3, Fe_2O_3 and CaO. As in Florida, Tennessee samples are handled to simulate plant beneficiation practices.

In the western Phosphoria field, about half of the phosphorite is utilized without beneficiation into the elemental phosphorus industry. Samples from trenches and drilling are taken at about 5-ft intervals, dried, and analyzed for P_2O_5, SiO_2, Al_2O_3, Fe_2O_3, and CaO.

In the United States, analyses on samples are performed by the old established wet chemical procedures or by X-ray fluorescence. The X-ray process, however, is not always accurate. It depends on the establishment of standard curves but between deposits or even within one deposit variations frequently do not fit these curves. For exploration and development, therefore, the X-ray procedure should not be used. A titration process is used for P_2O_5, Fe_2O_3 and CaO, but SiO_2 and Al_2O_3 analyses are determined gravimetrically.

Evaluation of Deposits

Even before exploration begins, a search for phosphate deposits has an end-use objective. Objectives include export of rock, export of fertilizer, export of phosphorus, or satisfying domestic markets for fertilizer or industrial chemicals.

With the exception of an occasional unusual deposit such as the wavellite (aluminum phosphate) of Senegal, and guano, the world's phosphate deposits contain the mineral fluorapatite. Deposits containing this mineral, whether of igneous or sedimentary origin, can be processed to make both phosphorus and fertilizer.

Where igneous apatite is contained in unweathered alkalic rocks, such as in the Kola Peninsula area of the USSR and in South Africa, the absence of carbonates permits flotation of apatite after fine grinding. Where the igneous apatite is in a carbonatite, the presence of the large quantities of carbonates in the unweathered rock heretofore has prevented the recovery of rock by flotation because flotation reagents work on the calcium end of the mineral molecule and calcium carbonate will float along with calcium phosphate. This condition has been overcome in Brazil and will be commented on later.

With pelletal phosphorites, impurities contained in the phosphate ore may influence end use of the rock. Such impurities include detrital silica grains, cryptocrystalline silica derived from siliceous skeletons of small organisms, clay, iron oxide, calcite, dolomite, iron sulfide (pyrite), and hydrocarbons. Most of these impurities occur in the ore as discrete particles separate from the phosphate; however, a certain amount is present in the individual phosphate pellet or particle.

Impurities present as discrete particles in weathered phosphorites can be partially, or nearly wholly, removed by simple washing or by washing followed by flotation. However the small particles of impurities present in a phosphate pellet generally cannot be economically removed, except that calcining, generally in the range of 900° to 950°C, removes the CO_2 of carbonates and the hydrocarbons. The remaining CaO can be removed by slaking and screening. This is done at Oron in Israel.

Because under normal conditions phosphate rock for any use has a relatively low sales value, the evaluation of phosphate deposits from the start is involved with limiting cost factors. Among these are the need for low-cost open pit mining, large volume operations with their cost benefits, low-cost large supplies of water for beneficiation, and low-cost transportation to markets. If a large, high-grade phosphate deposit were found in the middle of the Sahara Desert it would have no value for the foreseeable future because of transportation costs to say nothing of the possible lack of water.

The mining and milling techniques for the preparation of rock are essentially the same whether the rock is used to make phosphorus or fertilizer. Rock for phosphorus production can be as low-grade as 24% P_2O_5; therefore, it may require no beneficiation, or, if beneficiated, a low-grade concentrate is acceptable. On the other hand, for fertilizer use, the rock (concentrate) should be as high-grade as economically possible.

Mining

About 15% of the world's phosphate rock production in 1981 was mined underground, coming largely from the USSR, Morocco, and Tunisia, and minor quantities from the Arab Republic of Egypt and the state of Montana, USA. China also produces a significant amount of phosphate ore from underground operations. In Morocco and Tunisia underground mining employs room-and-pillar methods. The relatively soft ore does not require drilling and blasting but large quantities of timber posts are used to prevent roof slabs from falling. Underground mining of the USSR Kola ore is done by sublevel open stoping. (Anon., 1968a, 1979, 1979a; Woodrooffe, 1972).

The bulk of the world's phosphate rock is mined open-pit for the simple reason that it is the least expensive way to move large volumes of material—both overburden and ore. Unit costs are largely affected by the thickness, and hardness, of the overburden. Thus larger equipment with its lower unit cost of operation permits removal of greater thicknesses of overburden. Both overburden and phosphate ore are moved by (1) shovels and trucks, (2) draglines and trucks, (3) rubber-tired scrapers, (4) draglines and slurry pipelines, (5) bucket wheel excavators, and (6) grabbing cranes (guano).

Where the overburden is hard, such as the thin-bedded carbonate formations in Morocco and Utah and the chert bed in Idaho, drilling with rotary rigs and shooting with ANFO (ammonium nitrate fuel oil) may be required. Rippers on heavy-duty dozers are also useful.

Generally, more efficient use of equipment, and therefore lower overall costs, results from using the same equipment to both strip and mine. This is the practice in Florida where the mining operations use 30- to 65-cu-yd walking draglines. The same machine strips the sandy overburden, then picks up the matrix (ore) and dumps it into a pump-sump from where it is pumped to beneficiation plants. Stripping and mining of the residual brown-rock deposits of Tennessee is done with the same dragline equipment with the matrix loaded into trucks. These trucks either deliver directly to the phosphorus furnace plants or they dump into open gondola railroad cars for final delivery to the plants. Small crawler draglines with 1.5 to 3.5-cu-yd buckets are used because the small buckets are needed to dig matrix out of frequent narrow and deep troughs called cutters. One operator in Idaho uses rubber-tired scrapers both for removing overburden and picking up and delivering the ore to a truck-loading tipple. The scrapers are assisted in loading by pusher dozers. These dozers are equipped with ripper teeth used to loosen both overburden and ore. Another operator in Idaho in moving overburden uses both scrapers and 4½-cu-yd shovels loading 35-ton end-dump trucks.

Because of the nature of the deposits, other operators find it best to use different size equipment in their operations. The North Carolina operation uses two hydraulic dredges to remove the overburden. The area is drained, and a 50- and a 72-cu-yd dragline remove the remaining 50 ft of overburden to expose 40 ft of sandy phosphorite. The draglines remove the ore and place it on the surface where a 20-cu-yd dragline feeds the ore into sumps from which slurried ore is pumped to the beneficiation plant.

At some open-pit operations in Morocco larger draglines are used on the overburden and smaller ones on the matrix. Belts deliver matrix to beneficiation plants. The overburden consisting of thin-bedded carbonate rocks is drilled and blasted, then removed with draglines up to 65 cu yd in size. The matrix draglines are about 10-cu-yd size. In Togo, overburden is stripped with crawler-mounted bucket wheel excavators, each with a capacity of 2600 cu yd per hr. Matrix is mined with a 600-tph bucket wheel excavator.

At the Fosfago mine at Catalão, Brazil, mining is conducted by open pit techniques using rippers, loaders, and trucks on an altered carbonatite with ore grading from 13 to 15% P_2O_5. Beneficiation is similar to Jacupiranga except the major contaminate, magnetite, is removed by magnetic separation prior to fine grinding and flotation. Current production is 500 kt/a of 36 to 38% P_2O_5 apatite concentrate.

The examples given here of equipment used in surface mining of phosphate rock do not, of course, cover all world phosphate mining. They do indicate types and sizes commonly used. In soft material the large draglines move material at the lowest cost—several cents a cubic yard. With harder material and smaller machines, of whatever type used, costs are much greater.

For details on earth-moving equipment, refer to Pfleider (1968). This excellent volume covers not only equipment, but other matters relevant to surface mining, including mine design, drilling, blasting, excavation, transportation, and maintenance. The importance of maintenance cannot be overemphasized. For those contemplating a new operation, and without previous experience in maintenance of heavy equipment, use of an experienced contractor can be well worthwhile. The manufacturers of earth-moving equipment have a wealth of accumulated information on the performance of their equipment. They are glad to make this information available to a potential client.

Brief summaries of mining methods and equipment for most of the world's phosphate mining operations are given in Anon., 1971b.

Beneficiation

Igneous Deposits

At present, there are four areas producing substantial quantities of apatite—USSR, South Africa, Brazil, and Zimbabwe.

USSR: In 1972 production of Kola concentrates was about 12 Mt, containing 39% P_2O_5 (Woodrooffe, 1972). By 1980 this had increased to about 20 Mt. In the Kola area the unweathered nepheline syenite ore containing about 17% P_2O_5 is crushed, ground, and floated.

South Africa: In South Africa the unweathered apatite-bearing ore of the Phalaborwa alkalic igneous complex (pyroxenite and foskorite) is crushed, ground, and floated. The flotation concentrate contains about 36% P_2O_5. The ore contains about 7% P_2O_5. Production in 1980 was about 3.5 Mt.

Brazil: At the Jacupiranga mine in Brazil, mining operations are on an unweathered carbonatite containing 5% P_2O_5. Beneficiation consists of crushing, grinding, and floating an apatite concentrate containing 36 to 38% P_2O_5.

The operations are described by Silvia and Andery (1972).

In addition to Brazil's long established apatite mines and plants at Jacupiranga and Aráxa, several other apatite-bearing carbonatites have been developed in recent years. These are all in southwestern Brazil in the adjoining states of Sao Paulo, Minas Gerais, and Goias. By 1979 Brazil's yearly production of apatite rock (concentrates) had increased to 1.7 Mt. In 1980 production was from four mines on three carbonatites (Jacupiranga, Aráxa, and Catalão). By 1985 five more mining and beneficiating operations will raise Brazil's yearly production of rock to 5.8 Mt. Four of these will be on carbonatites (Tapira, Ipanema, and Santa Catarina). The fifth will be on a pellet phosphorite at Patos de Minas. With the exception of Jacupiranga all apatite is on soft residual ore resulting from weathering. Grade of ore is 7 to 15% P_2O_5 and beneficiation is crushing, grinding, magnetic separation, desliming, and flotation. Brazil's rapidly developing phosphate industry is described in Anon., 1978 and Breathitt and Finch, 1977.

Zimbabwe: At the Dorowa alkalic complex in Zimbabwe, weathered surface material is mined at about 7% P_2O_5. The beneficiation plant grinds, washes, and floats the ore to produce about 90 kt/a of apatite concentrate.

Pellet Phosphorite Deposits

Table 7 shows the chemical composition of beneficiated pellet rock for some of the grades of many of the world's phosphorites.

United States: Florida, with some 20-odd beneficiation plants, is the largest phosphate rock producing area of the world. One operation is in northern Florida, but the others are in the vicinity of the city of Bartow in central Florida. The rock-producing capacity, all

TABLE 7—Chemical Analyses of Some Pellet Phosphorites*†

Location Code	P_2O_5	CaO	Fe_2O_3	Al_2O_3	MgO	Na_2O	K_2O	MnO	Cr_2O_3	TiO_2	V_2O_5	SiO_2	CO_2	SO_3	F_2	Cl_2	Org. C	Insol.
1	30.31	45.52	1.54	1.31	0.52	0.24	0.16	0.05	0.02	0.07	0.01	10.25	4.02	0.91	3.62	0.01	0.30	
2	34.42	49.34	1.20	0.95	0.17	0.22	0.12	0.02	0.01	0.15	0.01	4.21	3.12	0.68	3.84	0.01	0.18	
3	32.95	46.45	0.55	1.50	0.25	0.50	0.15					4.25	3.40	0.90	3.60		0.15	4.50
4	30.08	49.00	0.45	0.40	0.50										3.53		1.20	2.28
5	33.10	52.10	0.75	0.45	0.60										3.90		0.04	2.00
6	34.26	52.78	0.26	0.37	0.48	0.84	0.09					1.36	3.59	1.59	3.05	0.03		
7	37.21	54.24	0.13	0.39	0.10	0.27	0.06					0.97	2.64	0.62	4.17	tr		
8	33.05	51.95	0.21	0.43	0.40	0.80	0.07					2.46	4.48	1.22	2.49	0.03		
9‡	36.60	51.90	0.14	0.35	0.06	0.24	0.14					4.52	1.83	0.44	3.93	0.036	tr	
10	37.68	51.08	1.02	1.15	0.04	0.17	0.03					2.95	1.45	0.25	3.32	0.03		
11	36.85	51.69	1.30	1.00	0.03	0.27	0.05					2.99			3.75	0.12		
12	30.23	48.92	0.37	0.77	0.76	1.50	0.08					3.12	5.18	3.42	3.26	0.09		
13	34.58	54.25	0.45	0.46	0.77	0.60	0.02					2.35	1.25	1.60	4.00	tr		
14	29.86	48.96		1.50	0.98							4.98	6.09	2.51	2.95	0.40		
15	35.10	55.60	0.10	0.30	0.20	0.40	0.03					1.60	1.40	1.80	4.00	0.07		
16	31.20	47.70	0.35	0.27	0.35	0.58	0.06					3.00		1.16	3.00	0.23		
17	34.60	53.60										0.83			4.05	0.10		
18	34.50	42.50	1.75	3.50	0.40			0.40				9.50	0.95		3.05			
19‡	29.00	46.14	2.15	1.96	0.77	0.73	0.32	0.03				6.78	5.33	1.51	2.95		0.33	
20‡	30.10	54.04	0.37	0.65	0.16				0.006		0.07	8.68	2.95		3.25		0.04	
21‡	31.30	43.86	2.10	3.05	0.10	0.34	0.46	0.11				12.66	1.76	0.52	4.06		0.07	
22‡	29.90	47.04	3.56	1.31	0.91	1.46	0.35	0.01	0.06			3.70	5.41	2.22	3.49		1.04	
23‡	32.30	45.85	0.95	2.13	0.17	0.46	0.61	0.01	0.09	0.05	0.25	9.87	2.10	0.84	2.99		2.50	
24‡	33.12	47.70	0.51	1.26	0.30													
25‡	25.20	37.33	1.63	4.36	0.25	0.52	1.48	0.02	0.14	0.09	0.20	20.56	0.45	2.62				
26‡	33.00	47.86	0.75	1.28	0.36	0.81	0.77		0.25		0.39	5.06	2.15	1.68	3.28	2.53		

KEY TO LOCATION CODES

1. Central Florida (66% BPL grade from one producer)
2. Central Florida (75% BPL grade from one producer)
3. North Florida
4. North Carolina (Texasgulf Inc.)
5. North Carolina (Texasgulf Inc., calcined)
6. Morocco (Khouribga area, 75% TPL grade)
7. Morocco (Khouribga area, 80-82% TPL grade)
8. Morocco (Youssoufia)
9. Spanish Sahara (80% TPL grade)
10. Senegal (Taiba)
11. Togo
12. Tunisia (65-68% TPL grade)
13. Algeria (Djebel-Onk)
14. Arab Republic of Egypt (Kosseir area)
15. Israel (Oron)
16. Syria (Kneifess area)
17. Jordan (El Hassa)
18. North Vietnam (Lao-Kay)
19. Baja California, Mexico (beach sands, beneficiated, not calcined)
20. Mexico (Saltillo area, La Caja formation, calcined, washed)
21. Tennessee, Maury Co. (washed brown rock)
22. 40-mile bank off coast of Southern Calif. (sea floor nodule)
23. Idaho, Caribou Co. (Phosphoria Frm., main bed—high grade, 6-ft thick)
24. Idaho, Caribou Co. (washed, calcined)
25. Idaho, Caribou Co. (electric furnace shale rock—not beneficiated)
26. Tennessee, Maury Co. (washed electric furnace brown rock)

* Selected from Anon., 1971b.
† All are beneficiated concentrates except where otherwise indicated.
‡ Obtained from various sources other than *.

grades, of these plants is about 47 Mt/a.

Basically the beneficiation flowsheets are about the same. The matrix is pumped several miles as water slurry containing about 35% solids from the mines to the plants in steel pipes. At the plants the slurry passes over coarse screens to remove trash and waste rock, then over about 14 mesh screens to remove pebble rock. The −14 mesh material is deslimed at 150 mesh. The −14 +150 mesh material is called flotation feed, and its grade is 6.4 to 11.4 P_2O_5. It is separated into two sizes and each is floated to recover the concentrate (rock). References for more details on beneficiation practices are: Beall and Merritt (1966), Aparo (1969), Anon. (1971b), and Lawver et al. (1978). Beall and Merritt not only describe beneficiation operations in Florida, but mining as well. Additionally, they describe mining and beneficiation in North Carolina. Beneficiation of the North Carolina pellet phosphorite is described also by Wright (1971) and Breathitt and Finch (1977).

Rock production in the western United States was 5.5 Mt in 1979. About 50% goes to electric furnaces without beneficiation except screening to remove unweathered carbonate rock and unweathered phosphorite. Of the 2.75 Mt going to fertilizer plants, about 2.5 Mt is beneficiated in four plants—a calcining plant and a washing-flotation plant in Utah, and a washing-calcining plant and a washing plant in Idaho. For description of past and present beneficiation plants, refer to Clitheroe and Mele (1967).

Morocco: In 1979 Morocco's phosphate rock capacity (phosphate pellets) was 24 Mt. Capacity will be over 37 Mt by 1985, and planned rock capacity by the year 2000 is 100 Mt/a.

The major part of the known phosphate reserves are located in marine pelletal phosphorite districts. The three northern deposits are 50 miles inland from the Atlantic Ocean along a northeast-southwest trend 150 miles long. Within this district are the two established (1920s) rock-producing districts of Khouribga and Youssoufia. At the southern end is the newly established Meskala area. Still further south in former Spanish Sahara (now Western Sahara) is the Bu-Craa mine complex built by the Spanish government, which was taken over by Morocco in 1976.

The three northern phosphate areas contain four to seven phosphate beds of economic importance with thicknesses of 3 to 6 ft and grades of 24 to 34% P_2O_5. In the fourth area, the Bu-Craa mine in western Sahara, the phosphate beds total 18 ft thick and contain 31 to 33% P_2O_5. The Bu-Craa complex operated at near capacity for one year, 1975, then was transferred to Morocco. Thereafter, Polisario guerrilla activity disrupted operations by sabotaging the 60-mile belt transportation system connecting the mine to the plant and living quarters complex on the coast at El Aaium.

Morocco's past rock production has been from the two northern areas of Oulad Abdoun (Khouribga) and Ganntour (Youssoufia) from both underground and surface mines. The main producing area is Khouribga, where for many years the soft, flat upper bed was high grade enough (about 33% P_2O_5) that beneficiation was not required. Where beneficiation is now required the ore is crushed, ground, and separated by elutriation practices. Over the years the ore has become less weathered and therefore is harder and lower grade, due to cementing calcium carbonate. This ore will require new beneficiation techniques.

The ore at Bu-Craa is weathered and soft and is crushed, washed with seawater, then finally washed with fresh water to remove chlorides.

The Meskala area has not been developed but tentative plans provide for the USSR to develop a 10 Mt/a rock operation for a period of 30 years.

Morocco's energetic program to develop its export rock and finished fertilizer industry has been well described in Anon. (1975), Anon. (1980), and Anon. (1980a).

Tunisia: Tunisia produced 4 Mt in 1979 from about seven underground mines and one surface mine. The ore was beneficiated (in 1977) in four mills using both wet and dry elutriation methods to concentrate the pellet phosphate. Expansions are planned, including a second open-pit mine. Part of the rock is processed in fertilizer plants in the cities of Sfax and Gabes and part is exported by ship.

Israel: In 1980 there were four operating plants in Israel—Oron, Arad, Makhtesh Qatan, and Nahal Zin. 1980 production at Oron was about 450 kt. At the Oron phosphate deposit in Israel the phosphate pellets are associated with soft carbonate rock; grade of the ore is about 24% P_2O_5. Shortage of water originally led to beneficiation by air separation. Later flotation was tried, but both processes still yielded relatively low-grade concentrates, i.e., about 30% P_2O_5 concentrate. The process consists of calcining at 950°C, slaking with controlled amounts of water to convert CaO to $Ca(OH)_2$, and air separation at 325 mesh to remove part of the $Ca(OH)_2$. Finally the coarse product is washed in hydrocyclones. The beneficiation process is described in Anon. (1968c).

The production of the newest mine in Israel, Nahal Zin, was about 1.7 Mt of concentrates in 1980. The surface-mined ore is crushed and wet-screened to remove +1 mm waste. Hydrocycloning removes —0.1 mm slimes. The pulp is washed with fresh water to reduce chlorides, then dried. A description of Israel's deposits and mining/beneficiation operations is given in Anon. (1980b).

Jordan: Phosphates, one of Jordan's few valuable mineral resources, were discovered in 1934. The most important deposits are found in an area about 25 miles wide which extends from north of Amman south along the Hegas railroad to El Hasa, about 80 miles. Deposits include Ruseija, El Hasa, Naki Musa, Es Salt, and Qatrana. The phosphate occurs in the Upper Cretaceous to Eocene sequence that is about 1300 ft thick. Today Jordan produces about 4.5 Mt of phosphate rock. Most of this is exported as internal requirements are small.

As an additional beneficiation process, calcining of pellet phosphate rock, is being increasingly used. When the objective is elimination of hydrocarbons, temperature on the order of 800°C is adequate, but removal of CO_2 from carbonates requires higher temperatures, from 900 to 950°C.

There are a number of advantages in using calcined rock in the production of wet process phosphoric acid. Among them is the ability to produce clear acid (light green to amber in color), which is a more acceptable component of liquid fertilizers.

Worldwide, phosphate calcining capacity in 1981 is about 14 Mt/a. About 75% of world calcining is treated in Dorr-Oliver FluoSolids systems (Table 8). Allowing 85% equipment availability, the yearly design capacity is 8.9 Mt/a. Actual production is probably 120% of design capacity, or 10.70 Mt/a for Dorr-Oliver equipment and 14.3 Mt/a for all calcining installations. These installations are designed to reduce the organic content of phosphate rock. However, in Algeria, the calciners, which are operated at higher temperatures, are used to drive off CO_2 from carbonates.

Phosphate Rock Use

Fertilizer Industry

Fertilizers consume 95% of the world's rock production. The future of the phosphate rock industry is, therefore, directly related to the future of phosphate fertilizer; accordingly, it is pertinent that any description of the rock industry be tied in with comments on the fertilizer industry. Table 9 shows world phosphate consumption by region.

The United States is the world's largest producer and user of fertilizer and is the leader in developing new types of fertilizers and their uses. Particularly in the last 15 years there has been a rapid switch to mixed and higher grade fertilizers, both granular and liquid types. The world trend is also in this direction.

Most world fertilizer manufacturing processes use sulfuric acid, although some, mostly in Europe, use nitric acid. Where acid soil conditions favor its use, some finely ground rock is applied directly to the soil, but the quantity used is minor and in the United States continues to decline.

A continuing trend is to produce phosphoric acid (52 to 54% P_2O_5) which is then used in the formulation of final fertilizers. A more recent trend is to use superphosphoric acid (67 to 76% P_2O_5) for the same purpose.

Export of phosphoric acid continues to grow. Most involves ocean shipment from countries with indigenous rock resources. These include the United States, Morocco, Tunisia, and South Africa. Other exporters, based on imported rock, include Spain, Mexico, France, The Netherlands, and Belgium (Anon., 1978a).

The grade of a fertilizer is denoted by the use of three numbers such as (18–46–0). The first number defines the percent elemental nitrogen (N), the second, P_2O_5, and the last is K_2O. In recent years there has been a trend to using numbers referring to the element only as (N-P-K). Chief among the types of phosphate fertilizers are the following:

Normal superphosphate (0–18–0) to (0–20–0) (solid).

Concentrated superphosphate (0–42–0) to (0–50–0) (solid).

Diammonium phosphate (18–46–0) (solid).

Monoammonium phosphate (16–48–0) and (11–48–0) (solid).

Granular mixed fertilizers (combination of foregoing types and K_2O where needed).

Liquid mixed fertilizers (phosphoric acid

TABLE 8—World Phosphate Calcining Capacity, 1981

Installation	Location	Units	Design Capacity, t/d
Djebel Onk	Algeria	3	3 000
OCP	Morocco	2	3 700
Beker	Idaho	4	5 830
Simplot	Idaho	4	5 000
Stauffer	Wyoming	3	2 000
Texas Gulf	North Carolina	6	9 000
Total		22	28 530

neutralized with ammonia and nitrogen solution and potash added where needed).

As pointed out earlier, any phosphate rock, igneous or sedimentary, can be used to meet any of man's required end products; however, economics help govern the selection. Grindability and reactivity in acid influence costs. For example, the fluorapatite crystals in igneous apatite are denser than the francolite crystals of pellet phosphorites. Igneous apatite is, therefore, not as easy to grind nor is it as reactive to acid. Conversely, igneous apatite does not have the hydrocarbon content which generally occurs in all pellet phosphorites and which may impose a penalty during acidification. Robinson (1972), Hill (1960), Everhart (1971), and Pacl et al. (1978) describe the physical and chemical characteristics of phosphate rock which influence its reaction to acidulation.

The following is quoted from Everhart:

> In the acidulation of phosphate rock, the presence of too much combined iron and aluminum oxide (known as "I and A" in the industry) is a serious trouble-maker in chemical plants and is the factor in ruling out many deposits as ore. A combined iron and aluminum oxide content of greater than 2.5% to 4.0% (depending on the end use) is generally considered prohibitive. Another critical factor is the weight ratio of CaO to P_2O_5 content in the rock. If this ratio is above 1.6 to 1, the cost of sulphuric acid consumption is generally so prohibitive as to eliminate the candidate rock. Inasmuch as the origin of phosphorite is marine, many such formations in the world have a substantial chloride content, if the geologic history of the marine sediments under consideration has involved the subsequent formation of evaporites in the area. Acidulation of phosphate rock in wet acid plants commonly can tolerate no more than 0.13% of chlorine in chlorides, because of the resulting severe corrosion problems in processing equipment.
>
> Still another deleterious element in the chemical processing of phosphate is magnesium oxide. No more than 0.25% is desirable if fluid fertilizers produced from superphosphoric acid are the principal end products. Even small quantities of MgO in the raw phosphate rock will result in an intolerable viscosity index in superphosphoric acid produced by the wet acid process.
>
> Fluorapatite is the principal phosphate mineral in marine phosphorites, and so deposits of this origin characteristically have a constant P_2O_5 to F_2 ratio. The range is 8 to 11:1 and tends to be about the same within an individual deposit but differs from district to district. A P_2O_5 to F_2 ratio of less than 8:1 would indicate possible troublesome fluorine problems in processing. Even organic matter in phosphate is undesirable since it leads to discolored acid that is very difficult to sell. Pyrite in the phosphate rock can lead to danger in a wet acid plant because of the generation of lethal H_2S during acidulation.

Phosphorus (Industrial Chemicals) Industry

Phosphorus is produced in electric furnaces charged with calcined phosphate rock, silica, and coke. It is used to produce high purity compounds—mostly sodium, calcium, potassium, and ammonium phosphates. The manufacturing steps begin with burning the phosphorus in air and dissolving the resultant oxide in water to make a relatively pure phosphoric acid suitable for the manufacture of industrial phosphates. They include detergents, industrial cleaners, and forest fire retardants. Further purification with hydrogen sulfide to remove traces of arsenic yields a very pure "food grade" acid which is used extensively in the manufacture of toothpaste and food additives. Phosphorus is also used in the manufacture of organophosphorus compounds for such uses as flame-retardant additives for fabrics and plastics, and insecticides.

Until about 30 years ago the United States produced the majority of the world's phosphorus with lesser amounts largely from England, France, Canada, and Germany. Since ɛn, however, production capacity has increased in many countries so that by 1972 the United States accounted for about half the world production.

A rough estimate of the yearly world production of elemental phosphorus in 1979 was 3,400 million pounds. Of this, 1,500 million pounds was attributed to Russia (for fertilizer), 1,300 million pounds to North America, and 600 million pounds to Western Europe. Except for Russia, the world's phosphorus is used largely in the production of detergents.

Based on the foregoing statistics, in 1979 about 11% of the world's rock production was used to make elemental phosphorus and 6% went into detergents. In the US about 12% of the rock consumed went into elemental phosphorus.

Phosphorus electric furnaces (enclosed) operate in a reducing atmosphere so that the vaporized phosphorus can be pulled off in the gases without contacting oxygen in the air. The gases are mainly CO and vaporized phosphorus. They are first cleaned of entrained particulate matter, then sprayed with water which cools them, causing condensation of the phosphorus into droplets which settle and are collected underwater.

In the phosphorus furnace coke is the reducing agent. Silica in the melt reacts with the calcium end of the phosphate mineral to produce a calcium silicate slag. Any other calcium, magnesium, or aluminum mineral, such as calcite, dolomite, or clay also reacts with silica and adds to the slag. Part of any iron oxide present in the phosphate rock is reduced to molten iron which settles to the bottom of the furnace under the silicate melt. As the iron is formed it reacts with phosphorus at a ratio of about three parts iron to one part phosphorus to make an alloy called ferrophosphorus. It is tapped separately from tapholes located below the silicate-melt tapholes.

Calcium, magnesium, and clay minerals, therefore, are undesirable in the phosphate rock, as they consume electricity. Iron is undesirable for the same reason, but, even worse, it ties up phosphorus in a low-value end product—ferrophosphorus.

The calcined phosphate rock charged to the furnace cannot be fine-grained because a porous bed must be maintained in the furnace to permit circulation of gases. The rock is charged as prepared material about an inch in size. This prepared material is made by one of two routes—nodulizing and calcining at about 1350°C in nearly horizontal kilns or calcining rolled flakes, or round pellets, in grate kilns.

Phosphate rock used to produce phosphorus can be as low grade as 24% P_2O_5. The CaO/P ratio should be in the range of 3.30 to 3.60—preferably the lower figure. An acceptable

TABLE 9—Summary of P_2O_5 Consumption, Mt, World and Five Regions

Year	North America	Western Europe	East Europe & USSR	Asia	Remainder	Total
1962*	2.71	3.68	1.75	1.07	1.38	10.58
1963	2.97	3.84	1.82	1.24	1.44	11.31
1964	3.29	4.12	2.12	1.52	1.63	12.67
1965	3.44	4.33	2.61	1.60	1.98	13.95
1966	3.86	4.38	2.95	1.66	2.11	14.95
1967	4.28	4.51	3.21	1.91	2.20	16.11
1968	4.44	4.78	3.48	2.01	2.24	16.96
1969	4.54	4.92	3.85	2.43	2.43	18.17
1970	4.43	5.19	4.05	2.68	2.47	18.82
1971	4.68	5.49	4.41	2.68	2.53	19.79
1972	4.76	5.72	4.80	3.09	2.73	21.10
1973	5.02	5.84	5.04	3.41	3.13	22.44
1974	5.11	5.99	5.51	3.90	3.65	24.16
1975	4.59	5.04	6.09	3.81	3.15	22.69
1976	5.25	4.91	6.89	3.84	3.24	24.13
1977	5.64	5.29	7.19	4.45	3.93	26.49
1978	5.15	5.60	7.50	4.75	4.25	27.25
1980	5.80	6.00	8.50	5.15	4.80	30.25
1985	6.50	6.50	10.50	6.50	6.00	36.00

Source: Douglas, 1979 and F.A.O. Fertilizer Statistics.
* 1962-1977 actual, 1978 estimated, 1980 and 1985 projected.

CaO/P ratio is, however, directly related to the cost of power. About 6 kw-hr is required in a phosphorus plant for each pound of phosphorus produced. The iron content of phosphate rock should not exceed a few percent Fe_2O_3.

In the early 1930s while planning for the new phosphorus industry in the US, pilot plant furnace plants of up to 1500-kw capacity were built and operated. Since that time, however, with the experience gained in operating large commercial plants, pilot plants have not been considered necessary.

Cost calculations for the production of phosphorus are based on chemical analyses. Pilot plant work is required, however, in determining the manner of preparing the calcined rock product to be added to the furnace.

In the United States phosphorus produced at an electric furnace plant is shipped in closed and insulated railroad tank cars to manufacturing plants located near markets for the end product. These tank cars, with capacities of 8,000 to 22,000 gal (holding 117,000 to 317,000 lb of phosphorus), are jacketed so that if the temperature of the phosphorus drops to 44°C (when the liquid phosphorus becomes a solid), the contents can be remelted by steam at destination for pumping to the burning towers. In the burning tower oxygen in air reacts with the phosphorus to make P_2O_5 powder, which is mixed with water to make a very pure phosphoric acid containing 54% P_2O_5.

Outside the US, most phosphorus plants are located near markets for end products and the phosphorus therefore is converted to industrial chemicals at the plant site. However, a phosphorus plant built in Newfoundland to take advantage of low-priced electricity, ships phosphorus overseas in two specially designed 5000-st vessels. The USSR is reported to have six phosphorus plant installations. Rail transportation of some of the phosphorus may be involved, but there is no readily available information on such movements.

Phosphorus is not inexpensive. The principal production cost factors are electricity, coke, and rock coupled with heavy capital investment. Its production is justified by its high purity which is required to meet strict specifications of end products.

Very little has been published, at least in English, on the production of phosphorus. A detailed description was compiled by Burt and Barber (1952) and reprinted in 1967. This TVA report mentions furnaces of 20,000 to 25,000-kw size; however, since 1952, many of the more modern furnaces are over double these sizes. In Waggaman (1952), there is a chapter on the manufacturing of phosphorus and other chapters on the use and manufacturing of industrial phosphates produced from phosphorus. Other publications by TVA personnel are Barber and Marks (1962) and Hignett and Stripling (1967). Several articles on phosphorus production including a description of world furnace plant installations in 1968 can be found in Anon., 1966, 1966a, and 1968.

Phosphate Fertilizer Industry *

The world supply-demand situation has rapidly come back toward a balanced position since 1975. This was accomplished as a result of many changes. Demand has increased; many older plants closed as the chaotic financial situation led to economic obsolescence. Planned plants were postponed indefinitely, and some were scrapped altogether. The result has been a rapid recovery of the remaining wet phosphoric (W-P) acid and nitric phosphates (N-P) segments of the industry. In 1978 they were able to operate at almost 90% of the rate at which they operated during the shortage of 1972–1974—a real recovery from the 81% level in 1975.

Further improvement is expected through 1980. If we accept the statistics projected for new plants, by 1980 the relative operating rate will be about 94%—possibly even higher, because some of the new plants announced are in regions that historically have time slippages on construction schedules. Regardless, the phosphate supply-demand situation should become progressively better through 1980.

Many new plants are proposed for 1981 and 1982. The new plants planned and announced are sufficient to lead to another serious downswing if all plants are completed as planned. Based on history, this seems unlikely.

On the other hand, only one new W-P acid plant has been announced for North America, though there have been recurring rumors of four or five other major expansions which could lead to another 2 million st capacity by 1982–1983. If these plans are completed, in conjunction with other plants already announced, the industry by 1983 could well re-

* Summarized from J.R. Douglas, 1979, "World Phosphate Supply Plagued by Questions," Tennessee Valley Authority National Fertilizer Development Center, presented at the 7th Phosphate-Sulphur Symposium, Innsbruch, FL, Jan. 18–19.

vert to a position comparable to 1968 and 1975.

Western Europe: The recovery of the phosphate industry in Europe has been remarkably rapid but at great sacrifices on the part of much of the industry. Many plants have been closed altogether. Less basic slag is being sold—and less NSP. Imports of finished phosphates or intermediate wet-process acids have increased. The net result has been a rapid recovery of the remnants of the industry, but no new plants are operating or scheduled. Further progress in attaining a balanced supply-demand situation can be expected unless imports are increased.

Eastern Europe and the USSR: A casual glance at the supply-demand situation for the phosphate industry at first appears to indicate that this region is planning on building sufficient N-P and W-P acid capacity to remain, as in the past, relatively self-sufficient. At the same time, the region is embarking on long-range trade which will result in annual importation of about 700,000 st of P_2O_5. We don't know if this will be a net import or if the area will in turn begin exporting finished phosphate fertilizers to the world market. If the importation of the wet-process acid works out satisfactorily, the USSR could delay some of its proposed high-cost construction and divert funds to other uses. Or it might decide to substitute finished phosphate fertilizer for some of the million tons or so of P_2O_5 it now uses in the form of ground rock phosphate or the large amounts of NSP now used.

Asia: We know too little of the fertilizer industry—past, present, and future—in communist Asia to attempt a prediction. In the past, the Asian industry has been relatively self-sufficient. There is some conjecture that, at least temporarily, Asia will begin importing relatively large amounts of finished phosphate fertilizers while building its own industry. Such an approach would not be abnormal in a planned development program, and would only tend to improve the world demand situation.

A quick glance at the figures might lead one to believe that noncommunist Asia is rapidly building toward self-sufficiency by the early 1980s. However, most of the new capacity planned is to be in the Near East. Much of the subcontinent will remain major importers. Some few nations who now import P_2O_5 may become self-sufficient. Others will need to import more. The major question here is: Who will furnish this material.

Central and South America: Although there are many new P_2O_5 plants proposed for Central and South America between now and 1985, it appears certain that this region will remain a major net importer—even if all plants now planned are built, which is doubtful. The region could become an even larger net importer over the next five to seven years.

Africa: A plethora of plants were announced and built in Africa in the past ten years. Many more are planned. Many of these will be delayed; some, however, will be built, and Africa can well be expected to become an increasingly strong competitor for sales of world traded finished phosphate materials.

Oceania: This region has always been relatively self-sufficient with a major reliance on NSP. No changes are foreseen.

North America: The North American phosphate industry has pulled out of the doldrums of 1975 almost entirely by rapid increases in exports. Net exports have increased from less than 1 million st P_2O_5 in the 1960s to more than 1.5 st in 1975 and an estimated 2.75 million st in 1978. In 1979, exports were considerably above those of previous years. A new major export has been developed to the USSR which is to reach a reported 700 kt annually by 1980. Domestic use has increased about as expected —with some annual decreases offset by a continuing upward trend in use—but exports have skyrocketed. As a result, the operational rates of North American W-P acid plants have increased and last year approached the relative rates of the shortage period of 1972–1974. If the total demand for domestic use and exports is to be met, plant operations will have to exceed the highest rates of any past date. Inventories could be pulled down to very low levels. Even then we could very easily run into a shortage of material during the middle of the spring rush reason.

If we assume exports will continue in the future and will be expanded by the new trade with the USSR, this North American industry as presently configured will not be able to meet domestic and export demand for long, if at all. More plants will be needed, or considerable debottlenecking of existing plants. We will need more production capacity—again—if the export demand holds up.

There have been rumors of four or five new expansions for the 1980–1981 period. These, if accomplished, could add 2 million st capacity or more, almost overnight. Even this 2 million st capacity would not throw the North American industry into a large oversupply position if domestic demand increases as expected and exports continue at the current projected levels.

If, as historically often happens, this capacity becomes available all at once, and if at the same time there should be a reduction of exports of 1.5 million tons from projected levels, then our industry would face the same degree of problems it faced in 1968 and 1975. With rising fertilizer prices offsetting relatively stable world prices in grain and farm products, at some point exports will drop. It is hoped that the industry can more adequately judge what this price level is and thus avoid a recurrence of the chaos of past cycles.

The future of the North American industry is as bright as we make it. We have a natural competitive advantage in producing finished phosphate fertilizer. We are the only region of the world with ample supplies of phosphate rock in close conjunction economically with ample supplies of sulfur—and both close to the world's sea lanes. In addition, we have leading technology. With such a combination, we have a real competitive advantage over other areas of the world and should, if proper decisions are made, remain the world's major trader in finished phosphates.

World Phosphate Fertilizer Production *

The phosphate market differs from the nitrogen market in that there are two basic sources of finished phosphate materials: products produced with phosphoric acid including TSP, most nitric phosphates, ammonium phosphates, and other complex fertilizer materials, and products produced without phosphoric acid. Normal superphosphate, made by treating phosphate rock with sulfuric acid, is the major product. Others include basic slag (a byproduct of steel production) available for fertilizer use in Europe and the portions of concentrated superphosphate and nitric phosphate that come directly from phosphate rock. This second group of products is declining in importance and, as the scheduled expansion in phosphoric acid comes into production, the decline should accelerate.

Total phosphate fertilizer capacity is not very meaningful because capacity of NSP and TSP usually is well above actual production levels. Thus fertilizer capacity is considerably larger than most estimates if market conditions become very favorable.

One of the most important changes taking place in the phosphate industry is the trend toward phosphoric acid-based fertilizers with the phosphate rock-producing countries producing and exporting greater amounts of phosphate fertilizers and phosphoric acid. Normal superphosphate will continue to face increased competition from higher analysis products. Its usage, primarily in the production of low-analysis mixed fertilizers, is expected to remain at about today's levels through 1985; however, its market share will decline. Normal superphosphate still comprises a large portion of the phosphate market in all regions except North America. Also, NSP contains sulfur, which is needed in some soils. Future production will be concentrated in only a few areas. An increase is expected in the total world production of TSP by 1985 with the largest increase in Asia. The ammonium phosphates and other high-analysis complex fertilizers are fast becoming the most popular way of providing phosphates to the farmer.

Basic slag production is centered in Western Europe. Usage of this product as a source of phosphate has been declining with changes in the steelmaking process. Because this product is directly related to the region's steel industry, it appears that a shift to new steelmaking processes or use of different ores would be necessary to bring about any significant changes from the current production pattern.

The future of the world phosphate industry will depend on developments in phosphoric acid and to a lesser extent in nitric phosphates. World nitric phosphate capacity is expected to climb from the 1978 level of about 3.2 Mt of P_2O_5 to almost 4.5 Mt of P_2O_5 by 1985. Production will center in Eastern and Western Europe, which together will have 3.7 Mt of capacity by 1985. Nitric phosphates will be of minor importance in other regions of the world.

Phosphoric acid is sometimes used in the manufacture of complex fertilizers by the nitric phosphate process. Only that portion of the P_2O_5 from the nitric acid acidulation of phosphate rock was considered to avoid double counting.

Phosphoric Acid Capacity: World phosphoric acid capacity in 1977 was 26.4 Mt, a 92% increase over eight years ago (Table 10). Capacity is expected to increase 30% during the next eight years and reach 34.2 Mt by 1985. The largest percentage increase will occur in

* Excerpted from G.E. Harris and E.A. Harre, "World Fertilizer Situation and Outlook," International Fertilizer Development Center, and National Fertilizer Development Center, TVA, Muscle Shoals, AL, March, 1979.

TABLE 10—World Phosphoric Acid Capacity, Mt, P_2O_5, by Region

Region	1976	1977	1980*	1985*	Rate of Change, % 1969-1977	Rate of Change, % 1977-1985
North America	9.1	9.3	9.4	9.4	48	1
Latin America	0.9	1.0	1.4	2.2	54	122
Western Europe	4.7	4.7	4.7	4.7	54	1
Eastern Europe	1.5	1.6	2.1	2.5	166	56
USSR	4.0	4.5	5.4	6.8	276	51
Asia	2.4	3.1	3.5	5.3	154	73
Africa	1.4	2.0	2.5	3.2	290	60
Oceania	0.3	0.3	0.3	0.3	14	0
World	24.2	26.4	29.3	34.2	92	30

Source: Harris and Harre, 1979.
* Numbers after 1977 are estimates.

Latin America, but the largest absolute increase will be in Asia and the USSR, where capacity in both areas will increase by over 2 Mt. These two regions will have 35% of the world's phosphoric acid capacity by 1985, compared to 29% currently. All regions except North America, Western Europe, and Oceania are projected to increase their phosphoric acid capacity by 50% or more between 1977 and 1985. North America, which now has 35% of the world's capacity, will have only 27% by 1985.

In order to analyze the future supply capability of the phosphate industry, it is necessary to look at the raw materials important in the manufacture of phosphate fertilizers. World production of phosphate fertilizers is given in Table 11.

Byproducts from the Use of Phosphate Rock

Gypsum: Gypsum, $CaSO_4 \cdot 2H_2O$, is produced as a byproduct in making wet-process phosphoric acid. Above five tons of gypsum results from the production of a ton of P_2O_5—equivalent to roughly 1.5 tons of gypsum for each ton of rock used. Commonly, the gypsum is stockpiled near the fertilizer plant. Because of the increasing use of rock, with the trend to wet-process phosphoric acid production, together with the trend to larger plants, the quantity of gypsum accumulating in stockpiles is becoming large.

Much attention has been given to putting this byproduct gypsum into the main market—plasterboard. For a number of reasons, efforts to use any significant quantity of byproduct gypsum have not been successful.

Appleyard, in the chapter "Gypsum and Anhydrite," in Part 4, comments on byproduct gypsum resulting from production of phosphoric acid.

Uranium: Uranium occurs in minor quantities in the mineral apatite $[3Ca_3(PO_4)_2CaF_2]$, which is the source of phosphorous in phos-

TABLE 11—World Phosphate Fertilizer Production, Mt, P_2O_5, by Region

Region	1969	1977	1980*	1985*	Rate of Change, % 1969-1977	Rate of Change, % 1977-1985
North America	5.1	7.8	7.8	7.9	53	1
Latin America	0.3	1.2	1.3	2.1	319	84
Western Europe	5.6	5.5	6.1	6.0	−2	9
Eastern Europe	1.8	2.9	4.0	4.4	65	52
USSR	1.9	4.4	5.9	7.5	127	70
Asia	2.0	3.6	4.7	6.8	83	92
Africa	0.7	0.9	2.2	2.8	26	205
Oceania	1.2	1.1	1.4	1.4	−8	30
World	18.5	27.3	33.3	38.9	47	42

Source: Harris and Harre, 1979.
* Based on 90% operating rates in developed countries and 70% rate in developing countries. Numbers after 1979 are estimates.

phate ore. After mining, phosphate ore is crushed and ground, leaving an apatite concentration with a grade of 30 to 34% P_2O_5. If the apatite is derived from a marine deposit, as most are, this concentrate generally carries 50 to 200 ppm U_3O_8; and lower if the apatite is derived from an igneous deposit.

During the last ten years, there has been much interest in the United States in the recovery of uranium from wet-acid by solvent extraction methods. Several plants are now recovering uranium from Florida wet acid; more are planned both in the US and elsewhere. A good summary of the worldwide status of uranium recovery from pellet phosphate rock is given in Anon., 1980c.

Because uranium is recovered as a byproduct from existing operations, the environmental difficulties are significantly less troublesome than those associated with new mining and milling facilities. Given the environmental attractiveness and competitive position with conventional ore deposits, it is likely that more extraction facilities will be constructed in the US and other countries, and that henceforth these facilities may supply about 10% of the world's uranium requirements.

Vanadium: Pellet phosphorites commonly contain significant quantities of other metals besides uranium. Those of the Phosphoria formation of the western US in places contain considerable vanadium. In the 1930s and 1940s, rock mined in Idaho was processed into wet-process acid at a plant in Montana and red cake, sodium vanadate, recovered. The ferrophosphorus from an electric furnace operation in Idaho is processed in a separate plant for recovery of vanadium.

Fluorides: The world's phosphate rock deposits represent the largest known resources of the all-important element fluorine. Much of the fluorine in rock is evolved in gaseous form in the production of fertilizer. Plants in Florida have long recovered H_2SiF_6, used by municipalities to fluorinate water. In recent years, fluorides recovered from Florida fertilizer plants have been converted to synthetic cryolite and aluminum fluoride in two plants, one in Florida and one in Louisiana. The products are used by the aluminum industry. Other plants around the world either in production or planned to recover fluorides from rock are located in Canada, Mexico, USSR, India, Japan, West Germany, China, and Austria. In all likelihood the future will see a growing recovery of fluorine from this source.

Other Byproducts: Nepheline syenite tailings from the Kola apatite operation in the USSR are utilized in the production of aluminum and ceramics.

In Brazil, a plant makes cement from calcite tailings resulting from the flotation of apatite (igneous) from a carbonatite phosphate deposit.

The calcium silicate slag resulting from production of electric furnace phosphorus is used for road metal, railroad ballast, and for concrete aggregate. It has a ready market in Florida, where there is a shortage of natural stone. Where natural stone is readily available, as in Tennessee and the western United States, the slag has limited markets.

Over the past few years governmental agencies have expressed concern that the small amount of uranium in the slags might be harmful to humans.

Governmental and Environmental Influences

In the United States, federal and state laws and regulations affect the mining of rock. The phosphorite beds of the Phosphoria formation occur largely in Montana, Idaho, Wyoming, and Utah. In the early 1900s, the federal government withdrew nearly all the phosphate land from the location of mining claims for phosphate. In 1920, Congress passed a law, commonly referred to as the Mineral Leasing Act of 1920, retaining all federal ownership of phosphate and certain other minerals. The great majority of the Phosphoria formation is, therefore, owned by the federal government with lesser amounts on Indian, state, and private lands. Federal and Indian phosphate deposits can be leased. Federal lands are leased by the Department of the Interior for an indefinite time period; however, the government retains the right to make reasonable changes in lease terms and conditions every 20 years. Indian lands are leased by the tribes involved and are for 10 years. Leases are administered by the US Department of the Interior. The Mineral Leasing Act of 1920 is administered under regulations of the Department of the Interior. Numerous changes in regulations have been made, particularly since 1966, designed for protection of the environment, so that as of 1972, prior to surface mining operations, mining and reclamation plans would have to be approved by the Dept. of the Interior. Environmental impact statements may be required. Beginning in 1969, the western phosphate states of Montana, Idaho, and Wyoming have enacted surface mining laws. These require that operations have state permits, approval of reclamation plans, and bonding for reclamation.

In the late 1960s, a company obtained federal phosphate prospecting permits in the Los Padres National Forest of California. Test drilling proved the presence of phosphate, and a preference right phosphate lease application was made to the Department of the Interior. All of this was prior to Congress passing the National Environmental Policy Act of 1969. Environmental groups protested issuance of the federal leases, and public hearings were held in July 1971. As of July 1983, The Department of Interior had not issued a preference right lease to the holders of phosphate prospecting permits.

A somewhat similar situation exists for phosphate lease applications on the Osceola National Forest of northern Florida. There the issuance of phosphate leases after phosphate was found under federal permits also has been held up by a law suit filed in a federal court by the State of Florida through the Attorney General in 1971. The suit requests that the Secretaries of the Interior and of Agriculture be enjoined from approving federal leases for various alleged causes including threatening the existence of endangered species of wildlife, possibly destroying the freshwater aquifer, etc. The courts have held that the holders of federal permits in the Osceola National Forest have the right to convert to federal leases. Early in 1983, the Dept. of the Interior denied the lease applications on the Osceola National Forest, stating that the lands, if mined, could not be properly reclaimed. One of the parties applying for leases has filed suit in federal court, and others may do the same.

Nearly all of the phosphate-bearing lands of Florida, North Carolina, and Tennessee are in private ownership. Much of that in Tennessee is in areas of rapid housing developments and has for years been subjected to city and county zoning ordinances. Starting July 1, 1971, the phosphate rock industry in Florida was subjected to a 3% severance tax on the value of the rock which in 1975 was raised to 5% with certain credits for other taxes and cost of land reclamation. The tax was increased to 10% of value of wet rock in 1981. By 1982, the severance tax on Florida rock had increased from $0.16 per ton in 1973 to $2.03.

Tennessee passed surface mining legislation in 1967 requiring, among other things, yearly permits, publishing of plans to mine, and bonding for reclamation.

The generation of large quantities of byproduct gypsum from phosphoric acid production has been mentioned. Sweeney and Timmons (1972) list existing stockpiles of this gypsum in Florida in 1972 at 153 million tons covering 1800 acres with a yearly addition rate of 21 million tons. They point out that it is not an environmental threat, but does present a potential aesthetic problem.

Beginning in the late 1960s concern developed, especially in the United States, over the possibility that phosphate in freshwater was the cause of excessive growth of algae—particularly in some lakes. Phosphate enters freshwater through many sources, predominately natural leaching of phosphate from rocks and soils, animal excrement (largely human waste through sewage systems), phosphate detergents used for cleaning, and runoff from fields fertilized with phosphate. It quickly became popular to believe that elimination of phosphate detergents was desirable and, by the early 1970s, hastily devised laws were enacted by several states, counties, and cities. Some of these laws set lower phosphate content in detergents, some eliminated it entirely. By 1980 public emotionalism over phosphates in detergents had subsided from that of the early 1970s; however, damage has been done to the detergent industry markets. By the late 1970s detergent markets were holding steady. Similar public emotionalism about phosphate fertilizers had also subsided by that time.

Tests by researchers from Purdue University revealed that the Indiana ban on phosphate detergents made no difference in the water quality of 15 Indiana lakes (Anon., 1978b). Another group from Cornell University has preliminary results from ten lakes in New York which confirm the findings of the Purdue team.

Phosphate fertilizer plants, rock defluorination plants, and electric furnace plants emit gaseous fluorides in the rock treatment process. By early 1970s a number of states had passed laws setting maximum allowable limitations on fluorine emission. The principal purpose of such laws is to limit excess buildup of fluorine in forage plants, thereby preventing fluorosis primarily in cattle, although there has been concern, also, over certain flower and orchard crops.

Various state laws establish limits in any, or all, of the following: accumulation of fluorine in animal forage, the quantity of HF in ambient air around industrial plants, and the quantity of fluorine emitted by the plant operation. Allowable maximums differ from state to state but that for fluorine in forage is generally about 40 ppm, ambient air fluorine is generally 1 to 4 ppb. Plant emissions are stated in parts per billion over certain periods of time or by fraction of a pound of fluorine per ton of P_2O_5 equivalent produced.

In complying with laws on fluorine emission, the phosphate industry is faced with substantial capital costs for control equipment and the continuing costs to operate such equipment.

Phosphate rock is a large commodity in world export markets. World rock production in 1982 was 140 Mt. Of this, about 60 Mt were exported, largely by ship. The balance was processed domestically into individual uses or exported.

With a few minor exceptions, the movement of rock is not hampered by tariffs or other restrictions. There are some tariffs on the international movement of phosphate fertilizer, although the United States has none. An example of such tariffs is the European Economic Community (EEC), which imposes a duty of 13.2% c.i.f. (cost, insurance, and freight) on phosphoric acid and 8% c.i.f. on diammonium phosphate (DAP). These EEC duties do not apply to Tunisia and Morocco, as they are Common Market associates.

Costs, Prices and Taxes

In this chapter there has been no reference to the cost of producing phosphate rock. Producers consider such costs confidential, and the same is true for the production of phosphate fertilizer.

Until 1974, finished phosphate rock was a low-value material. In 1971, for example, the rate price of 72% BPL Florida rock f.o.b. Tampa was only $6 per long ton. This quickly changed in October, 1973, when Morocco announced export prices effective January 1, 1974 which nearly tripled existing prices. Shortly thereafter, the other ex-US rock exporters went along with the Morocco increase. In November 1973, the Phosphate Rock Export Association (Phosrock), made up of most of the Florida rock producers, announced increased prices of export rock from the United States effective Jan. 1, 1974.

Domestic and export prices fluctuate. The US Bureau of Mines periodically publishes rock prices. One of these reports on "Marketable Phosphate Rock," February 1983, gives the following information on 1982 prices per metric ton for rock from Florida and North Carolina, f.o.b. mines: domestic rock prices for grades of 60 to 74% BPL were $23.22 and $30.93; export rock prices for grades of 60 to 74% BPL were $26.63 and $29.08. The shipping cost from mines to ports averaged $5.24/t.

For federal income tax purposes a phosphate rock producer in the United States can take either cost depletion or percentage depletion. If percentage depletion is used, the producer is allowed a depletion deduction of 14% of gross income from the mining property, but limited to 50% of net income (gross income less costs). The dollar amount of the allowable deduction is subtracted from income, thereby reducing the taxable profit. Put another way, the allowable deduction can be considered as an expense and, added to other expenses, establishes the profit subject to federal income tax.

Future of the Phosphate Rock Industry

The future of the industry is largely tied to the future of the fertilizer industry. Phosphate rock, used in fertilizer, is necessary to the life of modern man. There is no substitute. In today's economics only a minor quantity is recovered, from municipal sewage, but it appears likely the future will see more phosphate recovered from this source.

The rapid growth of the world rock industry since 1945 is shown in Fig. 2, but whether future growth can be projected from this past growth is questionable.

Harre et al. (1972) states: ". . . In many areas of the world, developed and developing, the fertilizer market is maturing to a point where increased productivity will depend more heavily on an optimum combination of farm inputs—fertilizer, water, pesticides and herbicides, varieties, etc.—rather than just the result of increased use of one input factor only. Fertilizer has been the key to acceptance of these new management techniques; thus, its major impact in some areas may have already been achieved, and a slowing in demand growth can be expected in the next decade."

As Fig. 2 graphically shows, US and world rock production since 1945 has grown rapidly, almost along a straight line. The 1980s will see expansion of existing rock-producing capacity as well as production from new areas. A straight-line projection of the growth curves in Fig. 2 indicates a world production of 180 Mt in the year 2000, with a US production of 72 Mt.

In mid-1980, the producers of rock and fertilizer in Florida began reducing operating rates to correct imbalances between supply and demand in the United States. Such reductions continued through 1981 and by the end of 1982 had resulted in a number of mines and plants stopping operations as well as others curtailing production. Phosphate rock production in the US was 37.414 Mt in 1982, 30% less than 1981. By early 1983, it appeared rock and fertilizer production for 1983 may exceed 1982 by 5%. Complete recovery to normal will be a slow process, but should take place between 1985 and 1987.

The rock production for Florida and North Carolina accounts for over 85% of US production. This large production raises questions about life of the phosphate rock reserves, especially for the main producer—Florida.

Government and other surveys go back over many years; since 1978, there have been a number of new estimates published. Four of these are: Zellar-Williams, Inc., 1978, done for the US Bureau of Mines; Anon., 1979b, done for the US Dept. of Energy; Sandvick, 1979; and Mayberry, 1981.

The US Bureau of Mines in using its report by Zellar-Williams tends to tie reserves into past historically low costs and prices for rock. This limits life of reserves, especially for Florida.

The other three estimates take the approach that continually improving methods of producing and using marketable rock will allow for developing more reserves as well as recognizing that costs and prices will adjust for production as long as there is a market for the rock. These three estimates paint a brighter picture for a long reserve life.

Table 6 (Anon., 1979b) shows known estimated phosphate rock reserves in the noncommunist world.

Bibliography and References

Anon., 1965, "Phosphate Rock Mining and Treatment Plant at Phalaborwa," *Phosphorus & Potassium,* No. 19, The British Sulphur Corp., London, October, pp. 12–18.

Anon., 1966, "Elemental Phosphorus in the United States," *Phosphorus & Potassium,* No. 24, The British Sulphur Corp., London, August–September, pp. 12–24.

Anon., 1966a, "World Elemental Phosphorus—Part II," *Phosphorus & Potassium,* No. 25, The British Sulphur Corp., London, October–November, pp. 10–16.

Anon., 1967, *A Study of the Bu-Craa Phosphate Deposit,* Empresa Nacional Minera del Sahara, S.A. (ENMINSA), Madrid, 28 pp. (English translation).

Anon., 1968, "Elemental Phosphorus Developments," *Phosphorus & Potassium,* No. 36, The British Sulphur Corp., London, July–August, pp. 16–25, 33.

Anon., 1968a, "Kola," *Phosphorus & Potassium,* No. 37, The British Sulphur Corp., London, September–October, pp. 13–20.

Anon., 1968b, "Methods for Prediction, Search and Prospecting of Agrochemical Raw Materials in the USSR," *Proceedings on Sources of Mineral Raw Materials for the Fertilizer Industry in Asia and the Far East,* United Nations Mineral Resources Development Series, No. 32, pp. 119–124.

Anon., 1968c, "Phosphate Rock Calcination in Israel," *Phosphorus & Potassium,* No. 35, The British Sulphur Corp., London, May–June, pp. 12–15, 21.

Anon., 1968d, *Proceedings of the Seminar on Sources of Mineral Raw Materials for the Fertilizer Industry in Asia and the Far East,* United Nations Mineral Resources Development Series, United Nations, No. 32, 392 pp.

Anon., 1969, "Calcined Phosphate Rock," *Phosphorus & Potassium,* No. 41, British Sulphur Corp., London, May–June, pp. 13–17, 19.

Anon., 1971, *Fluorides,* Committee on Biologic Effects of Atmospheric Pollutants, National Academy of Sciences, Washington, 295 pp.

Anon., 1971a, "The Phosphate Fertilizer Industry of the U.S.," *Phosphorus & Potassium,* No. 55, The British Sulphur Corp., London, September–October, pp. 21–28.

Anon., 1972, "Developments in Morocco's Calcined Rock Capacity," *Phosphorus & Potassium,* No. 62, The British Sulphur Corp., London, November–December, pp. 24–26.

Anon., 1975, "Morocco Plans for the Future," *Phosphorus & Potassium,* The British Sulphur Corp., Ltd., London, March-April, pp. 25–36.

Anon., 1978, "Brazil's Phosphate Rock Industry," *Phosphorus & Potassium,* The British Sulphur Corp., Ltd., London, January-February, pp. 21–24.

Anon., 1978a, "Phosphorus Acid Trend in 1977," *Phosphorus & Potassium,* The British Sulphur Corp., Ltd., London, July-August, pp. 17-21.

Anon., 1978b, "Does Banning Detergent Phosphate Make Any Significant Difference?" *Phosphorus & Potassium,* The British Sulphur Corp., Ltd., London, January-February, p. 42.

Anon., 1979, "Soviet Phosphates in the Tenth FYP-I (Five-Year Plan)" *Phosphorus & Potassium,* The British Sulphur Corp., Ltd., London, July-August, pp. 16–20.

Anon., 1979a, "Soviet Phosphates in the Tenth FYP-II (Five-Year Plan)" *Phosphorus & Potassium,* The British Sulphur Corp., Ltd., London, September-October, pp. 21–23, 43.

Anon., 1979b, "Uraniferous Phosphate Resources and Technology and Economics of Uranium Recovery from Phosphate Resources, United States and Free World," Open File Report GJBX-110 (79), Earth Sciences, Inc., US Dept. of Energy, Grand Junction, CO.

Anon., 1980, "OCP Unveils Its Plans for the Future, Part I," *Phosphorus & Potassium,* The British Sulphur Corp., Ltd., London, January-February, pp. 20–26.

Anon., 1980a, "OCP Unveils Its Plans for the 1980s, Part II," *Phosphorus & Potassium,* The British Sulphur Corp., Ltd., London, March-April, pp. 34–39.

Anon., 1980b, "Negev Phosphates Comes of Age," *Phosphorus & Potassium,* The British Sulphur Corp., Ltd., May-June, pp. 27–32.

Anon., 1980c, "Uranium Extraction from H_3PO_4—A Review," *Phosphorus & Potassium,* The British Sulphur Corp., Ltd., London, July-August, pp. 20–23.

Altschuler, Z.S., 1973, "The Weathering of Phosphate Deposits, Geochemical and Environmental Aspects," *Environmental Phosphorus Handbook,* John Wiley—Interscience, New York, pp. 33–96.

Aparo, S.J., 1969, "Current Flotation Practices in the Beneficiation of Florida Phosphate," Preprint 69-H-319, Society of Mining Engineers of AIME, New York, 13 pp.

Barber, J.C., and Marks, E.C., 1962, "Phosphorus Furnace Operations," *Journal of Metals,* December, pp. 1–5.

Barr, J.A., 1960, "Phosphate Rock," *Industrial Minerals and Rocks,* 3rd ed., J.L. Gillson, ed., AIME, New York, pp. 649–668.

Barrie, J., 1967, "The Geology of Christmas Island," Australia Department of National Development, Bureau of Mineral Resources, Geology and Geophysics, Canberra, 46 pp.

Beall, J.V., and Merritt, P.C., 1966, "Phosphate and Potash," *Mining Engineering,* Vol. 18, No. 10, October, pp. 83–89.

Breathitt, H.W., and Finch, E.P., 1977, "Phosphate Rock Beneficiation," Preprint 77–H–134, SME-AIME Annual Meeting, Atlanta, 31 pp.

Burt, R.B., and Barber, J.C., 1952, *Production of Elemental Phosphorus by the Electric-Furnace Method,* Chemical Engineering Report No. 3, Tennessee Valley Authority, Muscle Shoals, AL, (republished 1967), 312 pp.

Clitheroe, J.B., and Mele, S., 1967, "Beneficiation Techniques for Western Phosphate Rock," *Anatomy of the Western Phosphate Field,* L.A. Hale, ed., Intermountain Association of Geologists, Salt Lake City, Utah, pp. 225–235.

Cook, P.J., 1972, "Petrology and Geochemistry of the Phosphate Deposits of Northwest Queensland, Australia," *Economic Geology,* Vol. 67, No. 8, pp. 1193–1213.

Cooksley, J.W., Jr., 1967, "Application of Geophysical Methods in Phosphate Exploration Southeastern Idaho," *Anatomy of the Western Phosphate Field,* L.A. Hale, ed., Intermountain Association of Geologists, Salt Lake City, Utah, pp. 161–166.

Deans, T., 1968, "Exploration for Apatite Deposits Associated with Carbonatites and Pyroxenites," *Proceedings of the Seminar on Sources of Mineral Raw Materials for the Fertilizer Industry in Asia and the Far East,* United Nations Mineral Resources Development Series, No. 32, pp. 109–119.

Douglas, J.R., 1979, "World Phosphate Supply Plagued with Questions," prepared for presentation at the 7th Phosphate-Sulphur Symposium, Innisbrook, FL, January 18–19, TVA National Fertilizer Development Center, Muscle Shoals, AL, 10 pp.

Emigh, G.D., 1967, "Petrology and Origin of Phosphorites," *Anatomy of the Western Phosphate Field,* L.A. Hale, ed., Intermountain Association of Geologists, Salt Lake City, Utah, pp. 103–114.

Emigh, G.D., 1972, "World Phosphate Reserves—are there really enough," *Engineering and Mining Journal,* Vol. 173, No. 4, April, pp. 90–95.

Emigh, G.D., 1973, "Economic Phosphate Deposits," *Environmental Phosphorus Handbook,* John Wiley—Interscience, New York, pp. 97–116.

Emigh, G.D., 1975, "Phosphate Rock," *Industrial Minerals and Rocks,* 4th ed., S.J. Lefond, ed., AIME, New York, pp. 935–962.

Everhart, D.L., 1971, "Evaluation of Phosphate Rock Deposits," Preprint No. 71–H–52, Society of Mining Engineers of AIME, New York, 7 pp.

Hale, L.A., 1967 "Phosphate Exploration Using Gamma-Radiation Logs, Dry Valley, Idaho," *Anatomy of the Western Phosphate Field,* L.A. Hale, ed., Intermountain Association of Geologists, Salt Lake City, Utah, pp. 147–159.

Harben, P., 1980, "Where Is Florida's Phosphate Industry Going?" *Industrial Minerals,* No. 148, London, January, pp. 48–55.

Harre, E.A., et al., 1972, *Estimated World Fertilizer as Related to Future Demand,* Tennessee Valley Authority, Muscle Shoals, AL, August, 25 pp.

Harris, G.E., and Harre, E.A., 1979, "World Fertilizer Situation and Outlook," International Fertilizer Development Center and National Fertilizer Development Center, TVA, Muscle Shoals, AL, March, 27 pp.

Heinrich, E.W., 1966, *The Geology of Carbonatites,* Rand McNally & Co., Chicago, 607 pp.

Hignett, T.P., and Striplin, M.M., Jr., 1967, "Elemental Phosphorus in Fertilizer Manufacture," *Chemical Engineering Progress,* Vol. 63, No. 5, May, pp. 85–92.

Hill, W.L., 1960, "Phosphate Rock: Reactivity Scales for Ground Rock," *Chemistry and Technology of Fertilizers,* V. Sauchelli, ed., Reinhold Publishing Corp., New York, pp. 116–138.

Howard, P.F., 1971, "The Discovery of Phosphorites in Australia—A Case History," Preprint 71–H–97, SME-AIME Annual Meeting, 11 pp.

Howard, P.F., 1972, "Exploration for Phosphorite in Australia—A Case History," *Economic Geology,* Vol. 67, No. 8, pp. 1180–1192.

Hutchinson, G.E.. 1950, "Survey of Contemporary Knowledge or Biogeochemistry: 3. The Biogeochemistry of Vertebrate Excretion," *Bulletin,* American Museum of Natural History, Vol. 96, 554 pp.

Jayawardena, D., 1979, "The Occurrence of Rock Phosphate Deposits in Sri Lanka in Relation to the Precambrian Basement," Paper presented at I.G.C.P. Project 156, East-West Center, Hawaii, August 9.

Lawver, J.E., et al., 1978, "New Techniques in Beneficiation of Phosphate Rock," Preprint 78–B–331, SME-AIME Fall Meeting, Orlando, FL, 36 pp.

Mayberry, R.C., 1981, "Phosphate Reserves, Supply and Demand—Southeastern Atlantic Coastal States, 1980–2000 A.D.," Preprint 81–409, SME-AIME Fall Meeting, Denver, Nov., 39 pp.

McKelvey, V.E., 1973, "Abundance and Distribution of Phosphorus in the Lithosphere," *Environmental Phosphorus Handbook,* John Wiley —Interscience, New York, pp. 13–31.

Merriman, G.M., and Hobbs, C.S., 1962, "Bovine Fluorosis from Soil and Water Sources," *Tennessee Agricultural Experiment Station Bull. 347,* University of Tennessee, Knoxville, 46 pp.

Notholt, A.J.G., 1968, "Phosphate Exploration Techniques," *Proceedings of the Seminar on Sources of Mineral Raw Materials for the Fertilizer Industry in Asia and the Far East,* United Nations Mineral Resources Development Series, No. 32, pp. 214–228.

Pacl et al., 1978, "The Reactivity of Natural Phosphates," *Phosphorus & Potassium,* The British Sulphur Corp., Ltd., London, November-December, pp. 20–25.

Pfleider, E.P., ed., 1968, *Surface Mining,* AIME, New York, 1061 pp.

Robinson, N., 1972, "The Assessment of Phosphate Rocks for Phosphoric Acid Ammonium Phosphate Manufacture," *Phosphorus & Potassium,* No. 57, The British Sulphur Corp., London, January–February, pp. 31–36.

Sandvik, P.O., 1979, "US Phosphates—Abundant Resources Will Last for Hundreds of Years," *Engineering & Mining Journal,* October, pp. 99–101.

Sheldon, R.P., et al., 1967, "Sedimentation of Rocks of Leonard (Permian) Age in Wyoming and Adjacent States," *Anatomy of the Western Phosphate Field,* L.A. Hale, ed., Intermountain Association of Geologists, Salt Lake City, Utah, pp. 1–13.

Silvia, A.F., and Andery, P.A., 1972, "Mining and Beneficiation of Apatite Rock at the Jacupiranga Mine, Brazil," *Phosphorus & Potassium,* No. 57, The British Sulphur Corp., London, January–February, pp. 37–40.

Sweeney, J.W., and Timmons, B.J., 1972, "Availability and Potential Utilization of Byproduct Gypsum in Florida Phosphate Operations," Eighth Forum on Geology of Industrial Minerals, Iowa City, Iowa, April 13, pp. 89–97.

Stowasser, W.F., 1980, "Phosphate Rock," Preprint, *Mineral Facts and Problems, 1980,* Bulletin 671, US Bureau of Mines, 20 pp.

Stowasser, W.F., 1980a, "Phosphate Rock," Preprint, *Minerals Yearbook, 1978–1979,* US Bureau of Mines, 21 pp.

Stowasser, W.F., 1982, "Phosphate Rock," *Mineral Commodity Summaries 1982,* US Bureau of Mines, pp. 112–113.

Tuttle, O.F., and Gittins, J., eds., 1966, *Carbonatites,* John Wiley—Interscience, New York, 610 pp.

Waggaman, W.H., 1952, *Phosphoric Acid, Phosphates and Phosphoric Acid,* 2nd Ed., Reinhold Publishing Corp., New York, 683 pp.

Warin, O.N., 1968, "Deposits of Phosphate Rocks in Oceania," *Proceedings of the Seminar on Sources of Mineral Raw Materials for the Fertilizer Industry in Asia and the Far East,* United Nations Mineral Resources Development Series, No. 32, pp. 124–132.

Woodrooffe, H.M., 1972, "Phosphate in the Kola Peninsula, USSR," *Mining Engineering,* Vol. 24, No. 12, December, pp. 54–56.

Wright, T.J., 1971, "Texas Gulf Sulphur Beneficiation of North Carolina Phosphate," Preprint No. 71-H-116, Society of Mining Engineers of AIME, New York.

Zellar-Williams, Inc., 1978, "Evaluation of the Phosphate Deposits of Florida Using the Minerals Availability System," Open File Report 112–78, US Bureau of Mines, 199 pp. Available from NTIS PB 286 648/AS.

Potash

SAMUEL S. ADAMS *†
Revised by ROBERT J. HITE ‡

Potash, the generic term for a variety of potassium-bearing minerals, ores, and refined products (Table 1), owes its importance as an industrial mineral to the potassium requirement of growing plants. The term *potash* originally referred to potassium carbonate produced by the leaching of wood ashes. This process, for which the first US patent was issued in 1790, was the basis for the largest United States chemical industry in the early nineteenth century. The agricultural use of potash soon eclipsed the chemical applications following the German discovery in 1840 that potassium is essential to plant growth. The initial discoveries of large deposits of soluble potassium minerals, essential for agriculture applications, were made in 1857 in Germany and in 1914 in the United States. More than 90% of the 27.5 Mt K_2O produced in the world in 1980 (Searls, 1981) was derived from marine evaporite deposits and approximately 90% was used in fertilizers. The potassium content of an ore, mineral, or product is customarily expressed as % K_2O, a unit of measure which is neither a natural nor synthetic compound of potassium (1.0% K is equivalent to 1.2046% K_2O).

Geology

Potassium, the seventh most abundant element in the earth's crust, is present in silicate minerals in igneous, sedimentary and metamorphic rocks. Potassium is also a major constituent of many surface and subsurface brines (Table 2). Economic potash deposits are essentially restricted, however, to widespread, thick accumulations of marine chloride evaporite deposits. These deposits yield high grade, large tonnage ore bodies, many of which are amenable to low cost mining and beneficiation. The products from these deposits are, moreover, ideal for use in fertilizer because of the high relative solubility of the potassium chloride and sulfate evaporite minerals. World potash production and reserves for such deposits so vastly overshadow other sources of potassium that the following discussion will be limited largely to potash deposits in marine evaporites with brief reference to certain potash-rich brines. Approximately 85% of current domestic production is from bedded salt deposits and much of the balance is from brines.

Marine evaporite deposits are accumulations of salt minerals deposited in structural sedimentary basins through the evaporation of seawater or mixtures of seawater and other brines. The salt minerals are precipitated in assemblages and sequences of assemblages as determined by the compositions of the brines and the solubility relations in the salt systems (D'Ans, 1933). The evaporation of seawater normally yields carbonate minerals as the first precipitates (Table 3). Further evaporation and concentration of the brine leads to the precipitation, in succession, of sulfate, chloride, and supersaline mineral assemblages including the salts of potassium and magnesium. The common minerals, ores, and products of potassium deposits are listed in Table 1. Many potash deposits contain accessory evaporite minerals together with a variety of detrital and authigenic silicates (Braitsch, 1971).

The structural basins within which marine evaporites occur are of two main types: (a) cratonic basins formed by broad epirogenic movements (for example, the Michigan Basin of the midwestern United States and the Permian Zechstein Basin of eastern Europe), and (b) grabens, commonly related to continental

* Retired, formerly Chief Geologist-Uranium and Aluminum, The Anaconda Co., Salt Lake City, Utah.

† The author gratefully acknowledges the assistance of T. L. Britt, D. L. Everhart, H. A. Gorrell, R. J. Hite, K. O. Linn, E. C. Pendery, G. P. Salas, and A. T. Wells in preparing this manuscript. The staff of International Minerals and Chemical Corp. provided considerable data. The Anaconda Co. kindly granted the author permission to prepare the manuscript.

‡ US Geological Survey, Denver, CO.

TABLE 1—Some Minerals, Ores, and Products of Potash Deposits*

Name	Formula	K_2O, Equivalent Wt %	Remarks
	Marine Evaporite Deposits		
MINERALS			
Chlorides			
Sylvite	KCl	63.17	Principal ore mineral
Carnallite	$KCl \cdot MgCl_2 6H_2O$	16.95	Ore mineral and contaminant
Kainite	$4KCl \cdot 4MgSO_4 \cdot 11H_2O$	19.26	Important ore mineral
Bischofite	$2MgCl_2 \cdot 12H_2O$	0	Accessory contaminant
Halite	$NaCl$	0	Principal ore contaminant
Sulfates			
Polyhalite	$K_2SO_4 \cdot MgSO_4 \cdot 2CaSO_4 \cdot 2H_2O$	15.62	Ore contaminant
Langbeinite	$K_2SO_4 \cdot 2MgSO_4$	22.69	Important ore mineral
Leonite	$K_2SO_4 \cdot MgSO_4 \cdot 4H_2O$	25.68	Ore contaminant
Schoenite (Picromerite)	$K_2SO_4 \cdot MgSO_4 \cdot 6H_2O$	23.39	Accessory
Glaserite (Aphthitalite)	$3K_2SO_4 \cdot Na_2SO_4$	42.51	Accessory
Syngenite	$K_2SO_4 \cdot CaSO_4 \cdot H_2O$	28.68	Accessory
Bloedite	$Na_2SO_4 \cdot MgSO_4 \cdot 4H_2O$	0	Accessory
Loeweite	$6Na_2SO_4 \cdot 7MgSO_4 \cdot 15H_2O$	0	Accessory
Vanthoffite	$3Na_2SO_4 \cdot MgSO_4$	0	Accessory
Kieserite	$MgSO_4 \cdot H_2O$	0	Common ore contaminant
Hexahydrite	$MgSO_4 \cdot 6H_2O$	0	Accessory
Epsomite	$MgSO_4 \cdot 7H_2O$	0	Accessory
Anhydrite	$CaSO_4$	0	Common ore contaminant
ORES			
Sylvinite†	$KCl + NaCl$	10-35	Canada, USA, USSR, Brazil, Congo, Thailand
Hartsalz	$KCl + NaCl + CaSO_4 + (MgSO_4 \cdot H_2O)$	10-20	Germany
Carnallitite‡	$KCl \cdot MgCl_2 \cdot 6H_2O + NaCl$	10-16	Germany, Spain, Thailand
Langbeinite§	$K_2SO_4 \cdot 2MgSO_4 + NaCl$	7-12	USA, USSR
Mischsalz	Hartsalz + Carnallite	8-20	Germany
Kainite	$4KCl \cdot 4MgSO_4 \cdot 11H_2O + NaCl$	13-18	Italy, Ethiopia
PRODUCTS			
Potassium chloride	KCl	63.17	Principal potassium product
Potassium sulfate	K_2SO_4	54.05	Artificial product
Potassium-magnesium-sulfate (langbeinite)	$K_2SO_4 \cdot 2MgSO_4$	22.69	Natural product (New Mexico)
Manure salts	$KCl + NaCl$	40-60	USSR and East Germany
	Other Sources		
Alunite	$K_2Al_6(SO_4)_4(OH)_{12}$	11.4	
Niter (caliche)	$KNO_3 + NaNO_3 + Na_2SO_4 + NaCl$	0.6-1.9	Chile
Potassium nitrate	$KNO_3 + NaNO_3$	10-14	

* Source: Modified after Adams, 1968.
† May contain one or more sulfate minerals or carnallite.
‡ May contain sylvite.
§ May contain sylvite and leonite, kainite or other sulfate minerals.

TABLE 2—Potassium Content of Selected Rocks and Waters

	K_2O, Equivalent Wt %
Rocks*	
Average for earth's crust	3.01-3.13
Continental crust	2.10-3.00
Average for igneous rocks	3.12-3.13
Basalts	1.00
Granites	3.02-5.07
Average for sediments	1.90
Clays and shales	2.74-3.20
Sandstones	1.29
Carbonate rocks	0.35
Marine evaporites†	0.0029-63.17
Waters	
Oceans†	0.0458
Rivers (world average)‡	0.0003
Searles Lake, CA§	1.86-3.17
Great Salt Lake, UT#	0.506-0.846
Salton Sea, CA‡	0.0135
Mono Lake, CA‡	0.135
Dead Sea, Israel‡	0.45-0.758
Sechura Desert, Peru¶	0.35-0.65
Lake Eyre, Australia**	0.18
Lake McLeod, Australia¶	0.41
Lake bed, Wendover, UT††	0.63
Subsurface thermal waters‡‡	0.0012-0.2310
Salton Sea geothermal water, CA§§	3.0
Subsurface brines associated with salt deposits‡‡	0.25-3.12
Subsurface oil field brines‡‡	0.0005-0.475
Subsurface brine, Dariala, West Pakistan¶¶	4.100

* Source: Parker, 1967, except as noted
† Source: Stewart, 1963
‡ Source: Livingstone, 1963
§ Source: Ryan, 1951
\# Source: Hahl and Handy, 1969
¶ Source: W. A. Seedorff, personal communication, 1972
** Source: Johns and Ludbrook, 1963
†† Source: M. W. Lallman, personal communication, 1973
‡‡ Source: White et al., 1963
§§ Source: White, 1965
¶¶ Source: Anon., 1966

margin rift systems (for example, the deposits of the Danakil Depression, Ethiopia, and Sergipe, Brazil). Significant concentrations of potassium minerals are commonly restricted to certain parts of those halite deposits of greatest thickness and areal extent. The potash deposits of the cratonic basins are generally of Paleozoic age and are characterized by assemblages containing sylvite and potassium sulfate-bearing phases, commonly of a diagenetic or metasomatic origin. The deposits of the continental margin rift systems are related to the early stages of crustal rifting and are of Triassic or younger age. These deposits commonly have experienced less post-depositional metasomatism, hence primary carnallite-bearing assemblages are more commonly preserved.

Marine evaporite deposits of one age or another are present on all continents with the possible exception of Antarctica (Table 4). Evaporite deposits of various ages are, however, unevenly distributed between continents and within the individual continents. Proterozoic and Paleozoic deposits of the cratonic basins (Fig. 1), for example, are widely distributed within the continental masses with the notable exception of Africa. It may be inferred that Africa did not experience this type of Paleozoic tectonism and sedimentation. Mesozoic, Cenozoic and Recent marine evaporites (Fig. 2), on the other hand, occur at the edges of continents or in linear zones such as that extending from England and Spain through Turkey to Thailand. These distributions apparently reflect a different tectonic pattern related, in part at least, to rifts associated with crustal separations.

Marine evaporite deposits are composed of sequences of tabular beds which are characterized by the proportions of the minerals they contain as well as by color, texture, and grain size. Mappable beds may be a few inches to several feet in thickness and may be present over hundreds of square miles. These strati-

TABLE 3—Natural Evaporite Salt Sequences Compared with Sequences for the Theoretical, Static Evaporation of Seawater*

Rock Type (in order of precipitation)	Approximate Thickness Normalized to 100 M of Halite			
	Zechstein Evaporites, Germany	Average of Other Evaporites	Theoretical From Seawater	Equivalent Layer in Table 5
Carbonate	4	10	0.5	–
Sulfate	20	40	5	–
Halite	100	100	100	100 (A)
Na-Mg sulfates	1	0.3	7	10.8 (B)
Potash salts	3	2	15	17.3 (C, D)

* Source: Adapted from Borchert and Muir (1964) and Braitsch, (1971).

TABLE 4—World Occurrences of Halite-Bearing, Marine Evaporite Deposits.*

Country	Geological Age	Geographical Region	Geological Designation	Potash Occurrences	Remarks
NORTH AMERICA					
Canada	Cambrian (?)	Great Bear Lake, N.T.	MacDougal Fm.	None reported	3000-5000 ft deep
	Ordovician (?)	Great Slave Lake, N.T.		None reported	1000-3000 ft deep
	Cambrian ?-Ordovician?	Arctic Islands	Cornwallis	None reported	Salt structures
	Devonian	Saskatchewan, Manitoba, Alberta	Elk Point group	Sylvite, carnallite	10 potash mines
	Mississippian	Maritime Provinces	Windsor group	Sylvite, carnallite	Salt deformed, new mine development
Cuba	Jurassic-Triassic (?)	Cunagua	Punta Alegre Fm.	None reported	Salt domes
Dominican Republic	Miocene (?)	Las Salinas		None reported	
Mexico	Permian (?)	Chihuahua	Cuchillo Parado	None reported	Possible Salado equivalent
	Jurassic	Monterrey	Minas Viejas group	None reported	
	Jurassic	Isthmus of Tehuantepec	Salina Fm., Isthmian salt	Carnallite, sylvite	
	Jurassic-Cretaceous	Tehuantepec-Guatemala	Chiapas salt	None reported	Chiapas Basin
United States	Silurian	Michigan, New York, Pennsylvania, Ohio, W. Virginia, (Ontario, Canada)	Salina Fm.	Sylvite, carnallite polyhalite	Potash beds below 8300 ft in Michigan Basin
	Devonian	Michigan	Detroit River	None reported	Michigan Basin
	Devonian	North Dakota, Montana	Prairie evaporite	Sylvite, carnallite	Williston Basin
	Mississippian	Michigan	Michigan	None reported	
	Mississippian	Virginia	McCrady Fm.	None reported	Salt deformed
	Mississippian	North Dakota, Montana	Charles Fm., Mission Canyon Fm.	None reported	Williston Basin
	Pennsylvanian	SE Utah, SW Colorado	Hermosa Fm.	Sylvite, carnallite	Salt deformed, 1 mine Paradox Basin
	Pennsylvanian	Colorado	Eagle Evaporite	Sylvite, langbeinite	
	Permian	SE New Mexico, West Texas	Castile, Salado, and Rustler Fms.	Sylvite, carnallite langbeinite, polyhalite	7 potash mines
	Permian	Western Oklahoma	Blaine Fm., Clear Fork group, Flower Pot shale	None reported	
	Permian	Western Kansas	Wellington Fm., Salt Plain Fm. (?)	Polyhalite	
	Permian	Nebraska, Wyoming	Goose Egg Fm.	None reported	
	Permian	North Dakota, Montana	Opechi Fm.	None reported	Williston Basin
	Permian	Arizona	Supai Fm.	Sylvite	
	Permian-Jurassic (?)	North Dakota, Montana	Pine salt	None reported	
	Jurassic	North Dakota, Montana	Dunham salt	None reported	

TABLE 4—(Continued)

Country	Geological Age	Geographical Region	Geological Designation	Potash Occurrences	Remarks
	Jurassic	Idaho, Utah, Wyoming	Preuss sandstone	None reported	
	Jurassic	Gulf Coast	Louann salt, Werner Fm.	Sylvite	Salt domes
	Jurassic	Texas, Alabama	Buckner	None reported	Salt domes
	Jurassic	Utah	Arapien shale	None reported	
	Cretaceous	Florida	Comanchean, Trinity	None reported	
	Cretaceous	Texas	Fredericksburg	None reported	
SOUTH AMERICA					
Argentina	Cretaceous	Neuquen	Salina Fm.	None reported	
Bolivia	Cretaceous	Chiquisaca	Margas-Multicolors	None reported	
Brazil	Pennsylvanian	Amazon Basin	Novo Olinda	Sylvite	
	Cretaceous	Sergipe	Muribeca	Carnallite, sylvite, new mine development	
Chile	Cretaceous	Salinas de Purilactis	Purilactis	None reported	
Colombia	Cretaceous	Cordillera Oriental	Caqueza, Villeta groups	None reported	Salt deformed
Peru	Permian	Eastern Cordillera	Mintu group	None reported	
	Triassic	Eastern Cordillera	Pucara	None reported	
	Jurassic	Utcumbamba	Saraxaquillo	None reported	
	Miocene-Pliocene	Cerros de la Sol	San Pedro	None reported	
EUROPE					
Austria	Permian(?)-Triassic(?)	Nordliche Kalkalpen		None reported	
Bulgaria	Permian(?)-Triassic(?)	Provadiar		None reported	
Cyprus	Miocene(?)	Xeri		None reported	
England, Ireland	Triassic	Cheshire, Carrickfergus	Keuper	None reported	
	Permian	Yorkshire, Cleveland	Zechstein	Sylvite	1 mine
East Germany, West Germany, Netherlands, Denmark, Poland, England	Permian	Western Europe	Zechstein Fm.	Sylvite, carnallite, kainite, polyhalite	Mines in East and West Germany
East Germany, West Germany	Triassic		Bundsandstein, Rot Salz	Polyhalite	
France	Triassic	Lorraine, Franche-Compte	Keuper-Muschelkalk	None reported	
	Triassic	Dax	Keuper	Sylvite	Salt domes
Italy	Miocene	Sicily	Tortorian	Kainite, carnallite sylvite	Mines

TABLE 4—(Continued)

Country	Geological Age	Geographical Region	Geological Designation	Potash Occurrences	Remarks
	Miocene	Lungro		None reported	
	Miocene	Volteranno		None reported	
Netherlands	Triassic	Hengalo	Rot Salz	None reported	
Portugal	Triassic	Coastal Provinces		None reported	
Rumania,	Miocene		Tortorian	None reported	Several salt basins
West Ukraine,				Sylvite, langbeinite	
Poland		Bachnia-Wieliczka		None reported	
Czechoslovakia		Presov-Trebisov		None reported	
Russia	Cambrian	Angara-Lena	Ussolskaja Series	Sylvite	
	Devonian	Byelorussia, Latvia	Fammenian	Sylvite, carnallite	
	Permian	Embya, Manych, Dneiper, Donetz	Kungurian	None reported	Several basins; salt domes
	Permian	Urals, Upper Kama	Kungurian	Sylvite, carnallite	Several mines
	Jurassic	Kugitang, Gaurdak	(?)	None reported	
	Oligocene-Miocene	Nakhichevan		None reported	
	Miocene	West Ukraine		Sylvite, kainite, langbeinite	
		Yakutsk Krasnoyarsk		Sylvite, kainite langbeinite	
Spain	Triassic-Jurassic	Eastern Province	Keuper	None reported	
	Eocene-Oligocene	Catalonia and Navarra	Ludian	Sylvite, carnallite	Several potash mines
Switzerland	Triassic	Zurgach-Bex	Muschelkalk	None reported	
West Germany	Permian	Hamburg	Rotliegendes	None reported	
	Triassic		Keuper		
	Triassic	Southern Area	Werfener	Polyhalite, sylvite, langbeinite, kainite	
	Triassic		Muschelkalk	None reported	
	Jurassic		Munder	None reported	
West Germany, France	Oligocene	Buggingen, Alsace		Sylvite, carnallite	Several potash mines
Yugoslavia	Eocene-Miocene(?)	Tuzla		None reported	
AFRICA					
Algeria,	Triassic	Constantine-Laghourt		None reported	Salt domes
		Jeffara Plains	Bu Sceba		

TABLE 4—(Continued)

Country	Geological Age	Geographical Region	Geological Designation	Potash Occurrences	Remarks
Algeria	Miocene	Ouled Kebbeb		None reported	
Cabinda				None reported	
Egypt	Miocene	Gulf of Suez	South Gharib Fm.	Polyhalite	
Ethiopia	Quaternary	Danakil Depression	Houston Fm.	Sylvite, kainite, carnallite	
Gabon, Congo, Angola	Cretaceous(?)	Atlantic Coast		Carnallite and sylvite	Abandoned mine
				None reported	
Libya	Tertiary	Edri, Pisida, Marada		Carnallite and brine	Shallow marsh deposits
Morocco	Triassic	Mogador	(?)	None reported	Salt domes
	Triassic	(?)	(?)	Carnallite, sylvite	Salt domes
Tanzania	Pre-Jurassic	Mandawa	(?)	None reported	
Tunisia	Triassic	Djebel Hadifa	(?)	None reported	
ASIA					
Afghanistan		Northern Provinces		None reported	Salt domes
China					No information
India		Mandi District	Krol Series or Subathu group	None reported	
Iran, Saudi Arabia	Cambrian(?)	Persian Gulf	Hormuz	None reported	Salt domes
Iran, Saudi Arabia, Kuwait	Jurassic	Persian Gulf	Riyadh group	None reported	Salt domes
Iran	Oligocene	Tehran	Lower Red Fm.	None reported	Salt domes
	Miocene	Tehran	Asmari	None reported	Salt domes
Iran, Iraq	Miocene	Persian Gulf	Fars Fm.	None reported	
Israel, Jordan	Tertiary-Pleistocene	Dead Sea Graben	Dead Sea group	None reported	
Saudi Arabia	Miocene	Red Sea Coast		None reported	Salt structures
Syria	Miocene	Eastern Syria	Fars Fm.	None reported	
Thailand, Laos	Cretaceous	Khorat Plateau	Maha Sarakham Fm.	Sylvite and carnallite	

TABLE 4—(Continued)

Country	Geological Age	Geographical Region	Geological Designation	Potash Occurrences	Remarks
Turkey	Oligocene-Miocene	Anatolia		None reported	
West Pakistan	Eocene (Cambrian ?)	Punjab Ranges		Sylvite, kainite carnallite, langbeinite	
	Tertiary(?)	Kohat		None reported	
Yemen	Miocene	Red Sea Coast		None reported	Salt structures
OCEANIA					
Australia	Precambrian	Amadeus Basin	Bitter Springs Fm.	None reported	
	Precambrian-Lower Paleozoic(?)	Officer Basin	Browne Beds(?)	None reported	Salt domes
	Cambrian	Amadeus Basin	Chandler Limestone	None reported	
	Ordovician(?)-Devonian(?)	Bonaparte Gulf	Not known	None reported	Salt dome offshore
	Ordovician(?)-Devonian(?)	Canning Basin	Carribuddy Fm.	None reported	
	Silurian	Carnarvon Basin	Dirk Hartog Fm.	None reported	
	Devonian	Adavale Basin	Etonvale Fm.	Minor occurrences	

* (Compiled from references in bibliography, in particular; Benavides, 1968; Lefond, 1969; Liechti, 1968; Pierce and Rich, 1962; Rios, 1968; Stocklin, 1968; Tortochaux, 1968); A. T. Wells, personal communication, 1972; H. A. Gorrell, personal communication, 1972; R. J. Hite, personal communication, 1973.

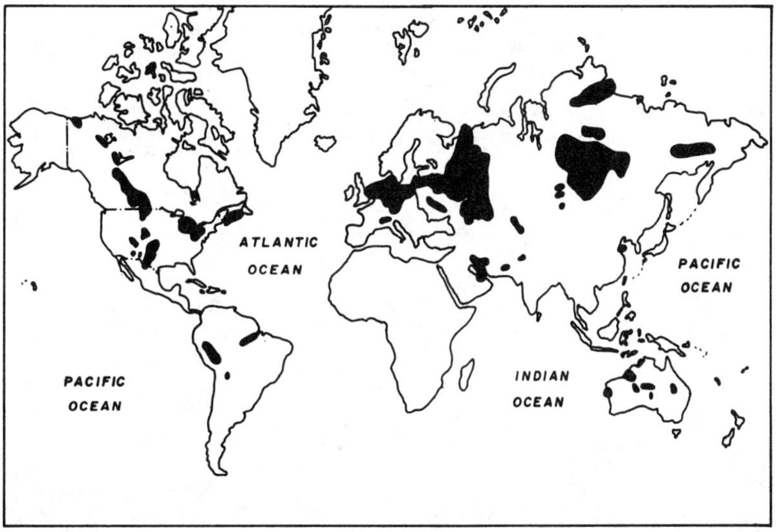

FIG. 1—*Approximate distribution of Proterozoic and Paleozoic halite-bearing marine evaporite deposits (adapted from Lefond, 1969; Pendery, 1970; Kozary et al., 1968; and Lotze, 1957).*

graphic features reflect relatively rapid precipitation in comparatively stable structural settings.

Theoretical and experimental salt assemblages bear sufficient resemblance to natural assemblages that they are useful in the study of marine potash deposits. Study of the origin of these deposits is aided by a knowledge of the original composition of the ore forming fluid (seawater) and the duplication of most salt assemblages in the laboratory. As an example, Table 5 lists the sequence of chloride-bearing salt assemblages which precipitate from seawater under conditions of static evaporation at 25°C. Fig. 3 traces the changes in brine composition which accompany the precipitation of the assemblages in Table 5. Similar data for a variety of temperatures and initial brine compositions may be prepared from the data of D'Ans (1933). The proper interpretations of field relations in the context of these chemical data improve the evaluation of known potash deposits and the exploration for new ones. However, most salt deposits were not formed from the isothermal, static evaporation of seawater and are, therefore, more complex than the sequence in Table 5. Whereas precipitation generally results from evaporation at the brine-air interface, it may also occur with temperature changes, the mixing of brines of different compositions, and the reaction of brines with previously precipitated salt. Repeated brine in-

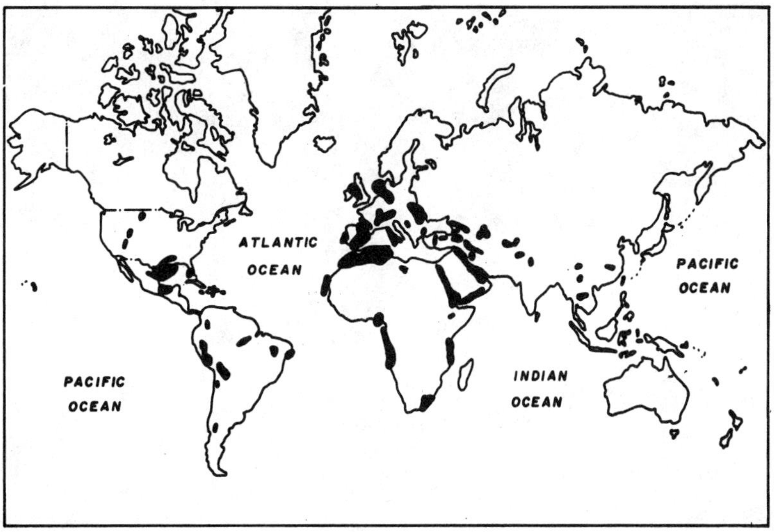

FIG. 2—*Approximate distribution of Mesozoic, Cenozoic, and Recent halite-bearing marine evaporite deposits (adapted from Lefond, 1969; Pendery, 1970; Kozary et al., 1968; and Lotze, 1957).*

TABLE 5—Theoretical Primary Precipitation of Stable Chloride Assemblages from Normal Seawater + Static Evaporation at 25°C (without Reaction at Transition Points)*

Layer	Mineral Assemblages: Mineral	%	Thickness of Layer, M
A	Halite	100	100.0
B_1	Halite	72	6.3
	Bloedite	28	
B_2	Halite	20	4.5
	Epsomite	80	
C_1	Halite	29	6.3
	Epsomite	30	
	Kainite	41	
C_2	Halite	21	5.7
	Hexahydrite	3	
	Kainite	76	
C_3	Halite	11	1.7
	Kieserite	4	
	Kainite	85	
D_1	Halite	12	3.6
	Kieserite	40	
	Carnallite	48	
E	Halite	0.5	38.0
	Kieserite	1	
	Carnallite	0.5	
	Bischofite	98	

* Source: Adapted from Braitsch, 1971.

fluxes of differing composition and incomplete evaporation lead to random and/or cyclic repetition of assemblages. Post-depositional metamorphism and metasomatism may substantially alter the mineral assemblages. The study and exploration of an evaporite deposit should be based, therefore, on (1) the thorough study of its stratigraphy and mineralogy, and then (2) on the interpretation of these features in the context of the chemistry of the salt systems.

Exploration

The search for new evaporite deposits requires the use of subsurface and indirect surficial evidence because the soluble chloride minerals rarely crop out at the surface. Evaporites of cratonic basins are commonly associated with extensive chemical, biochemical, and clastic sediments such as shelf carbonates, fringing reefs, and red beds. Evaporite deposits at the continental margins occur where crustal extensions developed during the Mesozoic and Cenozoic periods. The proximity of buried evaporites may be inferred from saline waters in springs and wells, thick outcrops of gypsum or anhydrite, and domal or collapse structures or geophysical anomalies such as those associated with the piercement salt structures under and around the Gulf of Mexico (Halbouty, 1967). Surface geophysical techniques, particularly gravity and seismic methods (Sawatzky et al., 1960), may be useful in determining the distribution and shape of evaporite bodies.

Initial exploration of an evaporite deposit may require the evaluation of geophysical logs from oil and gas exploration drill holes. Gamma-ray, neutron, sonic, caliper, density,

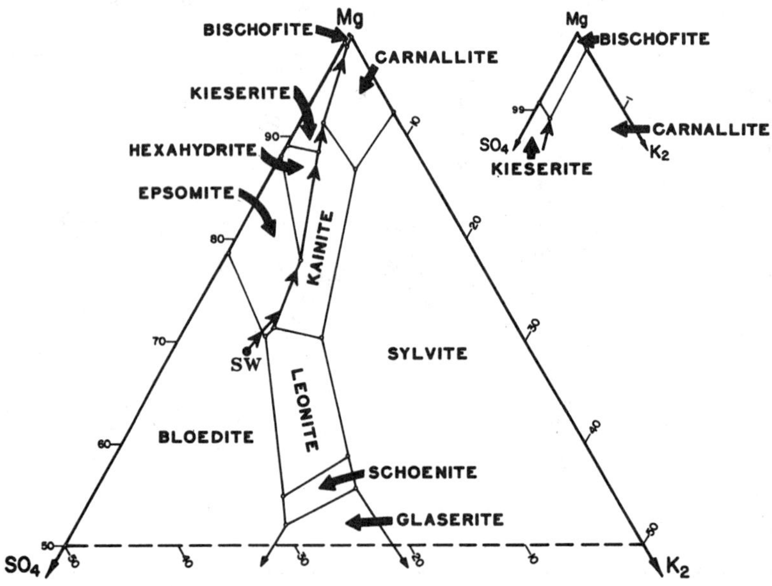

FIG. 3—*Magnesium-rich portion of six component (Na-K-Mg-Cl-SO_4-H_2O) system at 25°C and NaCl saturation (from Braitsch, 1971, Table 6, Fig. 16). Diagram calculated for solution composition where $Mg+K_2+SO_4=100$. Transition boundaries for minerals shown in fine lines; crystallization path for static evaporation of seawater (SW) without reaction at transition points shown in heavy lines with arrows.*

and resistance logs are useful in the interpretation of evaporite lithology, stratigraphy, and mineralogy (Alger and Crain, 1968). The gamma-ray log is particularly useful as it detects the gamma radiation from the natural potassium isotope K^{40}, hence it provides a continuous measure of the potassium content of the salt section. Initial exploration core holes are drilled on several mile centers using a sodium chloride saturated brine. If a natural brine is not available, rock salt can be added to the drilling fluid. Where potash salts are present, the brine must be saturated with respect to all the readily soluble minerals with due allowance for possible higher temperatures (hence greater solubilities of all evaporite minerals) in deep holes. If the composition of the brine cannot be rigorously controlled, a petroleum-based drilling fluid is used.

The ratio of sodium to potassium in seawater is 27.7 to 1. Minable potash beds, therefore, are generally accompanied by considerably thicker halite deposits. Furthermore, the potassium mineral assemblages commonly occur in the younger, upper portions of the thick halite deposits because the potash salts are precipitated late in the evaporation sequence.

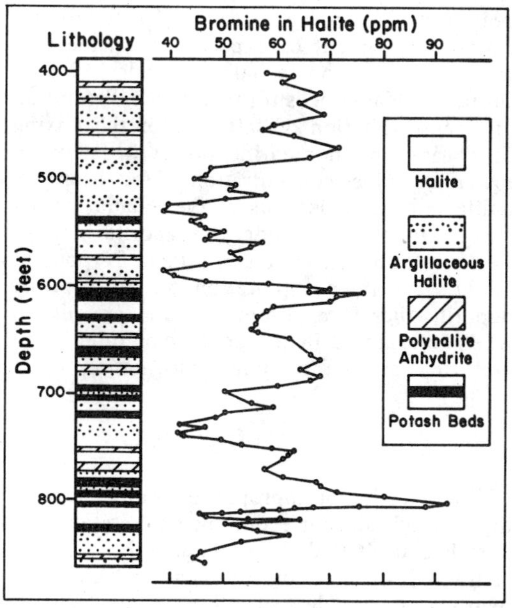

FIG. 4—*Stratigraphy and bromine content of halite in potash-bearing upper Salado Formation, core test IMC 115, Carlsbad potash district, New Mexico. Bromine curve smoothed by averaging successive groups of three analyses (modified after Adams, 1968).*

Multiple potash horizons may occur in the upper one-quarter to one-third of major halite deposits reflecting the cyclic nature of precipitation (Fig. 4).

The principal potash deposits of a district may be associated with areas of greatest halite thickness, peripheral to them or in nearby subbasins, depending on the tectonic history of the area. Figs. 5 and 6 are simplified maps for the Carlsbad, New Mexico, and Saskatchewan, Canada, potash districts showing the gross salt thickness, the areas of significant potash deposits, and the location of mining operations. Thick halite deposits in some basins do not have associated potash deposits, the potash-rich brines having been lost by reflux before potash precipitation or the precipitated salts having been subsequently dissolved by refreshened brines.

The exploration for potash deposits in known evaporite basins is based on studies of the chemical and physical characteristics of the salt beds. Stratigraphic criteria favorable to the occurrence of economic potash deposits include: (a) halite sequences in excess of 200 to 300 ft in thickness, (b) the presence of potassium in sulfate beds (possibly as polyhalite) or as disseminated grains of potassium chloride or sulfate minerals in halite beds, and (c) the presence of clastic zones of mixed

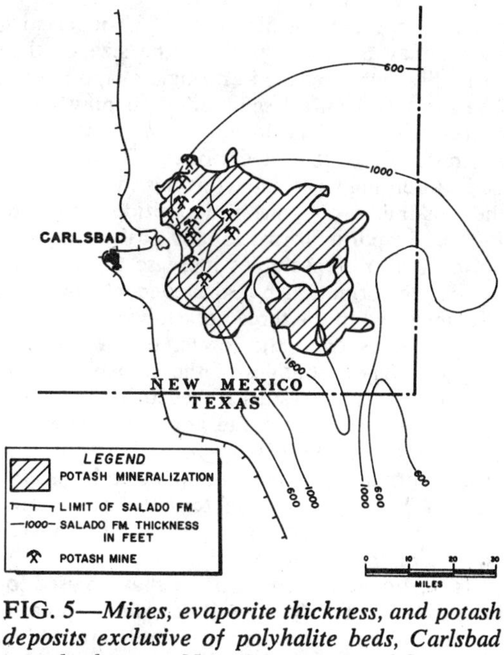

FIG. 5—*Mines, evaporite thickness, and potash deposits exclusive of polyhalite beds, Carlsbad potash district, New Mexico (modified after Pierce and Rich, 1962; and Adams, 1970).*

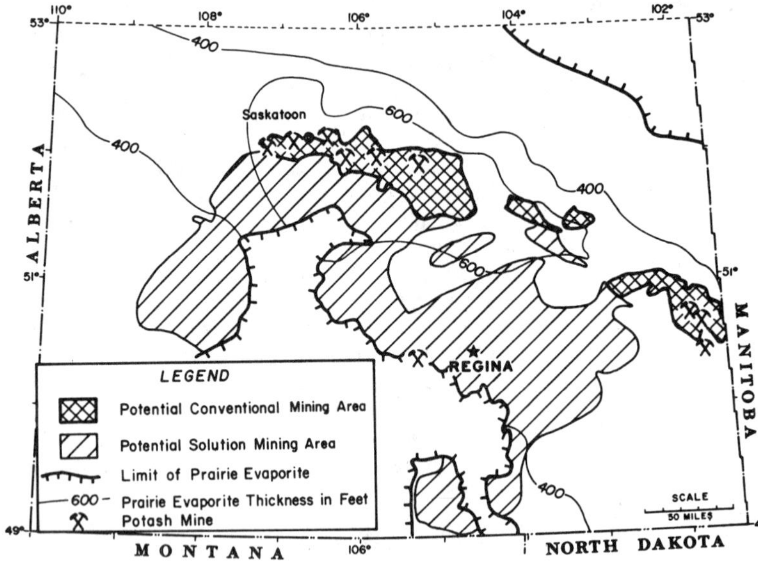

FIG. 6—*Potash deposits, mines, and evaporite thickness, Saskatchewan potash district, Canada (modified after Holter, 1969).*

layer clays, particularly in the upper part of the evaporite section. Bedded or disseminated potassium chloride and sulfate minerals are clearly the product of more concentrated brines. The presence of clays interbedded with, and disseminated within, the salt beds, on the other hand, indicates a closer source for detrital material and a decreasing rate of chemical relative to clastic sedimentation. The clastic source may be closer because the size of the basin has been reduced through evaporation. The rate of chemical sedimentation may have decreased due to the lower vapor pressure of the residual concentrated brines. Both conditions infer higher brine concentration, hence the possibility of potash deposition. In the Prairie Evaporite formation of Saskatchewan, not only do the insolubles increase toward the top of the formation but the concentration of insolubles within the three potash-bearing members increases from the lowest to the highest (Fig. 7). The insolubles of the potash beds in the Yorkton area of Saskatchewan include anhydrite, gypsum, dolomite, quartz, hematite, and micas or illite (Streeton, 1967). In the Salado formation of New Mexico, both insoluble-rich beds and polyhalite beds are more common with the potash deposits in the upper portion of the formation (Fig. 4).

The bromine content of halite may be used to infer the favorability of a salt deposit for potash horizons. The substitution of bromine for chlorine in the halite crystal lattice increases as the concentration of the brine increases, hence halite in and near potash beds contains more bromine. In deposits formed from the simple evaporation from seawater the absolute bromine content of halite is an indication of the stage of evaporation, hence the likelihood of a potash deposit being present (Holser, 1968). Potash beds might be expected when the bromine concentration in halite exceeds 150 ppm bromine. Halite beds formed by the dissolution and reprecipitation of salt beds or the mixing of marine and nonmarine brines will contain lower bromine concentration. For example, halite in potash deposits in the Salado formation, largely a secondary evaporite deposit (Adams, 1968), contains as little as 70 ppm Br (Fig. 4). In the exploration of an evaporite deposit, therefore, bromine data are useful when interpreted in the context of other mineralogical and stratigraphic information.

Evaluation of Deposits

The selection of appropriate mining and milling methods and the economic exploitation of a potash deposit will depend upon the accurate characterization of the physical and chemical features of the deposit. Similarities between deposits render much of this evaluation routine. The recognition and evaluation of features peculiar to a deposit may, however, determine the success of the mining operation. Some general guidelines are listed in Table 6.

NX or NC cored drill holes are the basis for

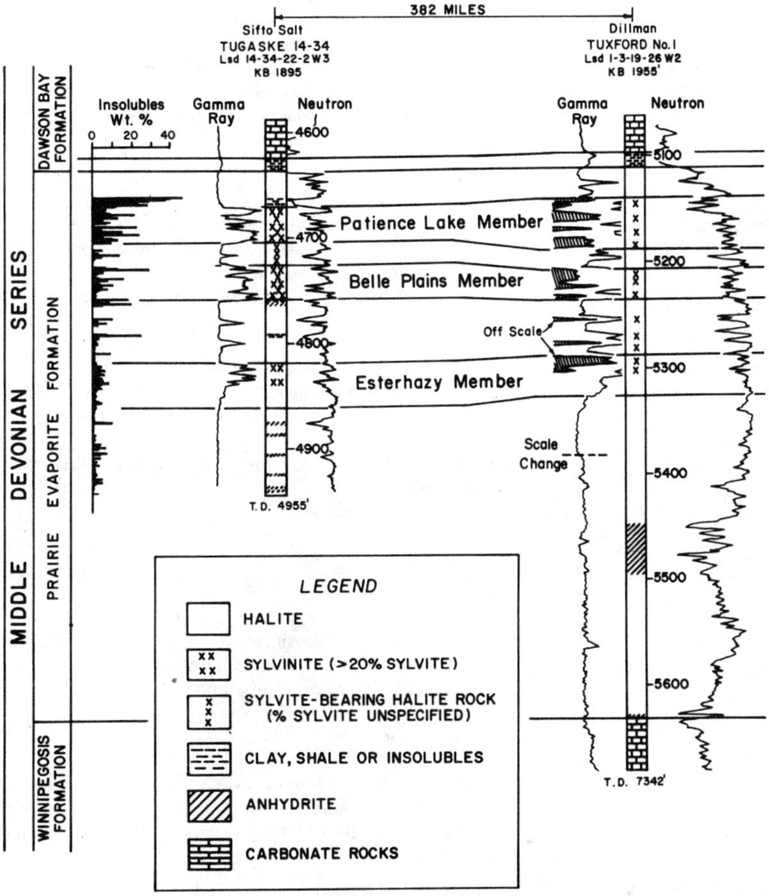

FIG. 7—*A section from the Prairie Evaporite Formation, Saskatchewan. Geophysical logs not necessarily at same scale (modified after Holter, 1969).*

the initial appraisal of a deposit. Due to the coarse grain size of salt assemblages, smaller core is generally unrepresentative. Development drill holes are spaced on ¼ to 1 mile centers, depending on the depth and complexity of the deposit. The salt cores are logged with attention to all mappable stratigraphic, lithologic, and mineralogic features (a logging scale of 1 in. = 10 ft is appropriate for the halite section; 1 in. = 1 ft for potash horizons). The potash horizons should be sampled by individual geological units for chemical analysis and determinative mineralogy without compositing dissimilar mineral assemblages. It is generally sufficient to analyze for K, Na, Ca, Mg, Cl_2, SO_4, and H_2O after drying to constant weight at 90°C. The accuracy of the sampling and analytical procedures may be confirmed if the megascopic mineralogy of the sample, estimated from hand specimen and thin section, can be calculated from the chemical analyses within the accuracy of the estimates and the analytical methods. The geological interpretation of these data in terms of the factors listed in Table 6 should indicate the potential for a minable deposit or the need for additional drilling or metallurgical tests.

World Potash Resources

The reserves of potash recoverable with existing technology from known deposits are estimated to be approximately 135 billion short tons K_2O. Russia and Canada have, respectively, 40 and 37% of the total. The United States' share is 4%. The resource estimates compiled in Table 7 are based largely on an evaluation of published information. For the most part, the reserves may best be referred to as "probable ore," that is to say, "ore whose occurrence is to all essential purposes reasonably assured but not absolutely certain." There is some question, for example, as to the reliability of the reserves reported for Russia as well as the portion of the Canadian solution mining reserve that will actually be recoverable.

TABLE 6—Guides to the Evaluation of Potash Deposits

Factor to be Considered	Potential Problem	Observation to be Made
Structure of ore horizon	Faulting, folding, or flowage of ore horizon complicating or precluding room-and-pillar mining method	a) Correlation of stratigraphy between drill holes b) Examination of core for evidence of deformation and recrystallization
Stratigraphy	Incorrect correlation of ore horizons; salt thickness above ore insufficient to protect deposit from water; slabbing and ore dilution from clay beds above ore horizon	Detailed core logging and correlation of beds between drill holes
Ore depth	Low mine extraction or excessive development and mining costs.	Mining method and size of mine openings designed on basis of rock mechanics and geology
Ore thickness	Local systematic or random thinning of ore zone	a) Isopach maps of stratigraphic units and mineralization in ore zone b) Inspection of core for evidence of metasomatism and potash leaching, i.e., recrystalized halite, clay in intergranular clots, and horizontal mineral zoning
Ore mineralogy	Assemblages incompatible with proposed mill circuit; certain assemblages reduce stability of mine openings; high insoluble content requires extra desliming capacity.	Determinative mineralogy of core
Ore grade	Grade may be too low or erratic to maintain acceptable mill feed; recoverable K_2O may be significantly less than total K_2O	Correlation and interpretation of drill hole data; comparison of chemical analyses with determinative mineralogy
Ore continuity	Erratic barren or thin areas in ore zone	Correlation of stratigraphic units and mineralization in ore zone; inspect core for evidence of recrystallization and horizontal mineral zoning
Ore crystallinity	Inclusion of contaminants within potash minerals; fine-grained assemblages requiring finer grind	Correlation of microscopic examination of ore samples with chemical analyses of mineral separates
Ore reserves	Insufficient reserves or excessive mining and milling costs	Accurate geological description of deposit for mining and metallurgical studies

A more conservative estimate of world reserves, perhaps more closely approaching "proven ore," yields about 50 billion tons K_2O (Adams, 1968).

The distribution of potash reserves and production by continent and the geological age of the host rock is presented in Table 8. Paleozoic evaporites account for 98% of the potash production plus reserves. This is particularly striking when one notes that the principal Paleozoic deposits occur on the continents of North America and Europe which represent only 23% of the earth's crustal area. Despite their widespread occurrence, Mesozoic and younger evaporites (see Fig. 2) account for only 2% of potash production plus reserves.

North America

Canada has the largest reserves of high grade sylvinite ore in the world. Reserves recoverable by underground mining are now placed at

TABLE 7—Estimated World Potash Resources*

Country	Potash Mineralogy	Grade, % K_2O	Depth, Ft	K_2O Equivalents, Millions St Subtotal	Breakdown	Remarks
NORTH AMERICA						
Canada				55,000		
					5,000	Conventional mining ore
Saskatchewan and Manitoba	Sylvite and carnallite	14-32	3000 – 8000		50,000	Solution mining ore
Maritime Provinces	Sylvite and carnallite	5-25	800 – 4000		Potential	Salt deformed
United States				500		
Arizona	Sylvite	14-16	1800		Potential	Holbrook Basin
California	Brine	3.0	Lake bed		20	Searles Lake
Michigan	Sylvite	4.0	8300		Potential	Michigan Basin
New Mexico	Sylvite and langbeinite	10-22 7-12	600 – 2000		90	Additional potential
North Dakota	Sylvite and carnallite	NA	5500–12,500		Potential	Williston Basin
Utah	Sylvite and carnallite	20-25	1000–10,000		280	Paradox Basin, Deformed and undeformed
	Brine	0.5	Surface		100	Great Salt Lake
	Brine	0.7	Lake bed		10	Wendover
Mexico	Carnallite and sylvite	NA	200 – 2000		Potential	Tehuantepec and Yucatan
SOUTH AMERICA				58		
Brazil	Carnallite and sylvite	0-25	1500 – 2000		30	Sergipe
Chile	Nitrates	1.0	Surface		15	
Peru	Brine	0.35-0.65	Shallow		13	Sechura Desert
EUROPE (WEST)				12,420		
Denmark	Sylvite and carnallite	10-20	700 – 3000		Potential	
England	Sylvite	15-30	3200 – 4500		340	
France	Sylvite and carnallite	15-25	1400 – 3400		400	Alsace
West Germany			1300 – 3500		11,000	
	Sylvite	15-20				
	Hartsalz	9-15				
	Carnallite	9-15				
Italy			1000 – 2600		170	Sicily
	Kainite	12-18				
	Carnallite	11-13				
	Sylvite	18-29				

TABLE 7—(Continued)

Country	Potash Mineralogy	Grade, % K_2O	Depth, Ft	K_2O Equivalents, Millions St: Subtotal	K_2O Equivalents, Millions St: Breakdown	Remarks
Netherlands		15	3200		10	Zechstein salt
Spain			1000 – 4000		500	Catalonia and Navarra
	Sylvite	15-25				
	Carnallite	14				
EUROPE (EAST)				65,180		
East Germany	Sylvite	13-17	1000 – 3200		10,000	
	Carnallite	12				
Poland	Carnallite	8	600 – 900		180	Zechstein Salt
Russia					55,000	
West Ukraine			250 – 2200			Stebnik-Kalush
	Hartsalz	16				
	Sylvite	8-30				
	Kainite	10				
	Langbeinite	NA				
Byelorussia	Sylvite	9-30	1000 – 3100			Starobin-Soligorsk
Urals	Sylvite	15-30	200 – 1300			Solikamsk-Brezhniki
Siberia						
Irkutsk	Sylvite	NA	2300 – 3400			
Krasnoyarsk	Sylvite, kainite, langbeinite	NA	4000 – 4600			
AFRICA				134		
Congo			850 – 1300		44	Holle
	Sylvite	15-20				
	Carnallite	15				
Ethiopia			80 – 400		50	Danakil Depression
	Sylvite	20-30				
	Kainite	NA				
Libya	Brine and carnallite	1-4	Shallow		10	Marada
Morocco			1600 – 2600			Khemisset Basin
	Sylvite	10-11			4	
	Carnallite	11-13			26	

TABLE 7–(Continued)

Country	Potash Mineralogy	Grade, % K_2O	Depth, Ft	K_2O Equivalents, Millions St: Subtotal	Breakdown	Remarks
ASIA				2,000		
Israel and Jordan	Brine	0.7	Surface		2,000	Dead Sea
Pakistan	Brine	4.2	4000 – 8000		Potential	
	Sylvite, carnallite, and langbeinite	NA	NA		Potential	Punjab Range
Thailand and Laos	Carnallite and sylvite					Khorat Plateau
OCEANIA						
Australia	Brine	0.41	Surface		Potential	Lake McLeod, W.A.
			Total	135,292		

* Source: Anon., 1968; Adams, 1968; Koepke, 1971.

TABLE 8–Potassium Production and Reserves for Marine Evaporites by Continent and Geological Age of Host Rock*

	Cumulative Area of Continents		Paleozoic Potash Deposits			Mesozoic and Younger Potash Deposits		
	Square Miles, Millions	% of Total	Cumulative Production† Millions St K_2O	Reserves, Millions St K_2O	% of Production Plus Reserves	Cumulative Production† Millions St K_2O_3	Reserves, Millions St K_2O	% of Production Plus Reserves
North America	9.4	16	77	55,370	42			
Europe‡	3.8	7	197	75,530	58	67	2,070	90
Subtotal	13.2	23	274	130,900	100%	67	2,070	93%
South America	6.9	12					30	2
Asia	17.1	30						
Africa	11.7	20					124	5
Oceania	3.3	6						
Antarctica	5.1	9						
Subtotal	44.1	77					154	7
Grand Total	57.3	100%	274	130,900	100%	67	2,224	100%

*Source: Adapted from US Bureau of Mines Mineral Yearbooks, Anon., 1966, and Table 7 of this paper.
† Figures are approximations only and may include some production from sources other than marine evaporites.
‡ Includes portion of Russia west of Ural Mts.

5 billion tons K_2O (Holter, 1969; reserve calculations based on minimum 20% K_2O ore grade, minimum ore thickness 8 ft, 3500 ft maximum depth, minimum 20 ft of salt above ore, 40% mine extraction, and 90% mill recovery). In-place solution mining reserves are estimated to be 65 billion tons K_2O, 50 billion tons of which may be recoverable (Holter, 1969; calculation of in-place reserves based on minimum depth of 3500 ft, minimum of 20 ft of salt above ore, and minimum ore thickness of 30 ft).

Nine mines and one solution operation currently produce from the Prairie Evaporite of the Mid-Devonian Elk Point cratonic basin (Fig. 6). Three potash horizons, the Esterhazy, Belle Plains, and Patience Lake members, occur in the upper part of the evaporite sequence (Fig. 7). The thickness and continuity of the ore permit the use of continuous mining machinery, which minimizes rock stability and ventilation problems, in a room-and-pillar mining method. Canada will play a major role in world potash production and marketing, a fact apparent since 1970 when the Saskatchewan Provincial Government imposed production limitations in an effort to stabilize prices and bring potash supply and demand into closer balance.

Exploration continues (1981) in the Mississippian evaporite basin of the Maritime Provinces. Scattered, low grade sylvite and carnallite mineralization has been encountered in the deformed evaporites of Nova Scotia. Recent drilling by the government of New Brunswick, however, has encountered high grade (21 to 25% K_2O) mineralization in zones 9 to 32 ft thick at depths of 850 to 1000 ft. If minable reserves are developed, they will enjoy a significant transportation advantage in world trade. Potash Corp. of America has completed sinking a 2,000-ft shaft and currently is exploring this deposit with expectations of production sometime in early 1983. Shaft sinking at a second mine is underway by Denison Mines Ltd. (Denison/Potacan), with first production scheduled for 1984.

United States potash production since 1931 has come principally from the Permian evaporites of the Carlsbad District, southeast New Mexico. By 1961 gross revenue for the district had exceeded one billion dollars. The current reserves of 90 million tons K_2O include sylvinite, argillaceous sylvinite (more than 3% insoluble material) and mixed ores containing sylvite, langbeinite and lesser amounts of other sulfate minerals. Substantial additional reserves are available in thin and lower grade beds. The average grade of sylvinite mined in 1981 was about 16% K_2O.

The thick evaporite deposits of the Carlsbad district occur in the basin and shelf provinces of the cratonic Permian Delaware Basin. The potash salts occur at 12 horizons irregularly over an area of approximately 2000 square miles. Post-depositional metasomatism has locally reconstituted the ore mineralogy and produced barren areas within the ore zones. These barren areas, referred to as salt horses, complicate mining operations (Linn and Adams, 1966). Seven companies currently operate seven mines which produce from 5 of the 12 potash horizons.

Domestic potash production from brines is increasing. In 1972, the Texasgulf potash mine at Moab, Utah, then the only underground potash mine outside the Carlsbad district, was converted to a solution mining operation. Deformation of the ore zone precluded successful underground mining. Potash brines are being pumped from the flooded mine workings, concentrated through solar evaporation, and processed in the existing plant. Recently Buttes Oil and Gas Co. has been exploring a deep solution-mining prospect 30 miles north of Moab. Several other potential solution-mining prospects are known in this region.

In 1971, potassium sulfate production began from the Great Salt Lake, Utah. Initial production was less than anticipated due reportedly to a poor evaporation year and harvesting problems. Potash production from other brines continues at Searles Lake, California; Bonneville, Utah; and in Michigan.

Argillaceous sylvinite has been delineated in the Permian Supai formation of northeastern Arizona and the Dow Chemical Co. has discovered potash minerals at depths of 8000 ft in the Michigan Basin (Matthews, 1970). These deposits are not likely to be mined in the foreseeable future.

South America

Significant proven potash reserves in South America occur only in the marine evaporite deposits of Sergipe, Brazil, the nitrates of Chile, and the brines of the Sechura Desert, Peru. The continent has but few marine evaporites and thus far no known large, epicontinental basin with extensive, high grade potash deposits, such as occur in North America and Europe, have been discovered. The very large Amazonas

Basin of *Brazil* contains sylvite-bearing evaporite deposits of Permian age (Szatmari et al., 1979). Reportedly, these deposits are now being explored by Petrobras Mineracão.

The sylvite and carnallite deposits of Sergipe, Brazil, occur in subbasins of a coastal Cretaceous evaporite basin. Sylvinite reserves, estimated at 30 million tons K_2O, could be substantially increased by mining multiple horizons. The mining of these reserves, however, may be complicated by the occurrence of tachyhydrite ($CaMg_2Cl_6 \cdot 12H_2O$), an extremely hygroscopic mineral, and by faulting. A Brazilian company, Kalium Mineracão S.A., is undertaking the development of the potash and tachyhydrite deposits, the latter reportedly containing reserves of about 1.5 billion tons $MgCl_2$. Shaft sinking by Petrobras Mineracão with technical assistance by Potasse d'Alsace began in 1979, near Santa Rosa de Lima. Production, expected to start in 1983, will come from a high grade sylvite deposit. There are no plans for exploiting the large tachyhydrite deposits in this area.

Polyhalite ($K_2SO_4MgSO_42CaSO_4 \cdot 2H_2O$) occurs in the nitrate ores of *Chile*. Minor amounts of potassium borate and perchlorates are also found.

The brines of the Sechura Desert, *Peru*, are a potential source of potash. The brines occur in shallow porous sediments and are essentially concentrated seawater. The reserves, however, are limited and, although the brines have been under investigation for several years, there are no current plans for their production.

Europe

Europe has dominated potash statistics since the initial exploitation of the Zechstein deposits in 1861. European countries, including Russia, have accounted for 77% of world production and now contain 58% of known world reserves.

The Boulby Mine, in *England*, began producing ore in 1973 (Woods, 1979). Reportedly the great depths (in excess of 1000 m) of the ore, bad roof conditions, erratic nature of the ore, and methane gas have caused difficult mining conditions. Potash reserves are present in two main sylvinite beds which occur at depths between 3200 and 9000 ft in the Permian Zechstein. The beds average 20 to 25 ft in thickness at grades of 10 to 30% K_2O. Two other potash ventures in Yorkshire have been temporarily deferred.

The *French* potash mines of the Mulhouse area in Alsace (Fig. 8) are operated by state-owned monopoly, Mines de Potasse d' Alsace. During recent years the potash industry has been modernized, resulting in the expansion of some mines and plants and the closing of others. The longwall rather than the room-and-pillar mining method is generally used because bedding dips in excess of 10° are common.

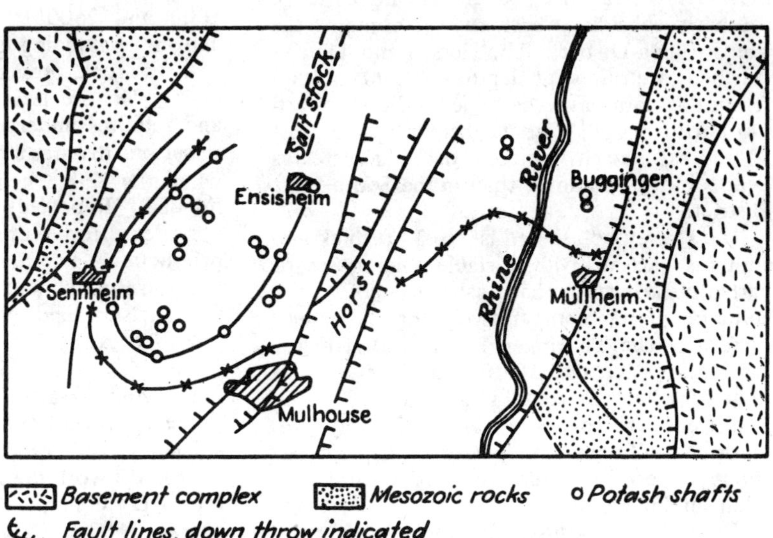

FIG. 8—*Location map for the potash mines of the Rhine Graben, in France and West Germany. (Smith, 1950).*

The evaporites, which contain the principal potash reserves of France, were deposited in the Oligocene Rhine Graben. Similar deposits are mined nearby in *West Germany*. Two ore zones have been mined; the lower contains 15 to 20% K_2O over 6 to 15 ft and the upper zone 20 to 25% K_2O over 3 to 6 ft. The district is presently faced with the difficult problem of sodium chloride disposal. Previous use of the Rhine River for this purpose is no longer environmentally acceptable. Some curtailment of potash production may result. The erratic and irregular sylvinite beds of the Triassic evaporites near Dax in southwestern France contain minor reserves in zones up to 30 ft thick with grade of 12 to 19% K_2O.

The salt cycles and potash beds of the Permian Zechstein of Middle Europe have been defined by Richter-Bernberg (1955) as follows:

Evaporite Cycle	*Series*	*Potash Horizons*
Zechstein 4	Aller	
Zechstein 3	Leine	Riedel
		Ronnenberg
Zechstein 2	Stassfurt	Stassfurt
Zechstein 1	Werra	Hessen
		Thuringen

Zechstein 2 through Zechstein 4 are present in the main evaporite basin which includes the Hannover, Stassfurt, and South Harz potash mining districts (Fig. 9). The potash beds of the older Werra series occur only in the more southerly subsidiary basin which contains the Werra-Fulda District. The Hessen and Thuringen beds are mined at depths of 1200 to 3000 ft. In the main basin the regional dip is toward the north so that, except where raised by salt structures, potash horizons occur at greater depths in the Hannover than in the South Harz District.

The Hessen potash bed is 7 to 35 ft thick and contains hartsalz with variable thicknesses of carnallite. Hartsalz (hard salt) is a potash ore containing the comparatively hard mineral kieserite. The Thuringen bed is 180 ft below the Hessen, averages 10 ft in thickness and is mined for hartsalz containing 10 to 15% K_2O. The beds are flat-lying and relatively undeformed with uniform potash grade and distribution permitting the use of the room and pillar mining method.

The potash deposits of the main basin, principally the Hannover district, are on the steeply dipping flanks of conformable and piercement salt structures. Mining is by stoping and backfilling and is concentrated between depths of 1000 and 2500 ft. Above 1000 ft there is danger of flooding from ground water; below 2500 ft rock pressures and temperatures are high. The ore is irregular and discontinuous and requires underground exploration drilling. A serious disaster struck the Hannover District when the Kali Chemie Ronnenberg mine was flooded. Reportedly this important mine is considered to be a total loss in terms of future production. The Stassfurt, and to a lesser extent the Ronnenberg and Riedel bed, are mined for hartsalz and smaller amounts of other potash ores.

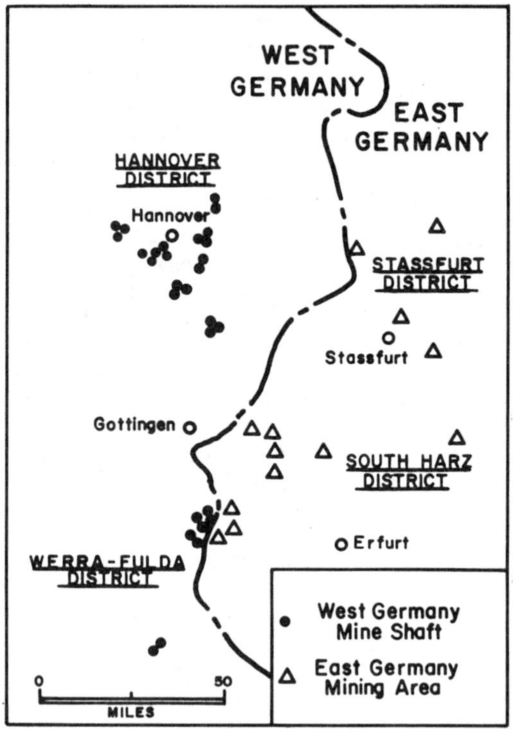

FIG. 9—*Location map for the potash mining areas of the Permian Zechstein, West and East Germany (modified after Anon., 1971).*

The *East German* potash industry exploits the sylvinite reserves of the Zechstein salt. Recent modernization has increased the industry's efficiency and production. A new mine at Zielitz will work the Ronnenberg horizon which reportedly contains 13 million tons K_2O at an average 16 to 17% K_2O. In addition, the Zechstein of East Germany contains large carnallite reserves.

The *West German* potash industry has exploited the famous potash beds of the Zechstein for over one hundred years. Recently the industry has undergone modernization similar to that in France and East Germany. Ownership of the industry has been consolidated in such manner that Kali and Salz AG, an amalgamation of Wintershall and Salzdetfurth, contributes 85% of the production from 11 mines. The remaining 15% of production is from two mines owned by Kali-Chemi AG. Currently 30 shafts service a district which once contained 220 operating mines. Ores containing sylvite, carnallite, and sulfate minerals are mined at grades between 9 and 17% K_2O. At Buggingen, near the Alsace deposits of France, sylvinite ores containing 18 to 20% K_2O are mined from the Oligocene evaporites.

In recent years *Italy* has essentially doubled its potassium sulfate mining and production capacity. The mines exploit the kainite reserves of Sicily which, with the exception of Ethiopia, are unique in the world. The kainite reserves occur, with lesser amounts of sylvite and carnallite, at four horizons in beds 6 to 100 ft in thickness in Miocene rocks which dip up to 60°.

In *Poland,* the eastern extension of the Zechstein evaporites contain low grade potash beds. Attempts have been made to exploit carnallitic reserves near Klodawa which reportedly contain 8% K_2O at depths of 1500 to 2500 ft. There is now some interest in the deep hartsalz mineralization which extends under the Baltic Sea near Gdansk.

Russia contains 40% of the world potash reserves and has considerable potential for new discoveries. From 1969 to 1970, production increased about 25%, on the strength of major expansions in the Urals and in Byelorussia, two of the three principal potash districts (Fig. 10). This expansion has been accompanied by a greater use of mining machines, with a significant increase in productivity per man shift (Anon., 1972).

The deposits of the Urals reportedly contain the vast majority of Russia's reserves. The deposits are generally thick, high grade, occur at depths of about 800 ft, and are undeformed. Three zones of mineralization containing multiple potash horizons occur over large areas.

The sylvinite deposits of Byelorussia contain 1.5 to 3.0 billion recoverable metric tons of K_2O in three sylvinite horizons and one sylvinite-carnallite bed. The deposits are high grade (20 to 25% K_2O) but also contain 4 to 10% water insoluble material.

The deformed Miocene deposits of Stebnik

FIG. 10—*Location map of the potash mining districts of Russia (Anon., 1972a, modified).*

and Kalush in the West Ukraine contain hartsalz, sylvinite, kainite, langbeinite and up to 12 to 18% insoluble material in at least five potash zones. The deposits are lower grade than those in the Urals and mining in steeply dipping beds is common.

The potash industry in *Spain,* as elsewhere in Europe, has been undergoing expansion. Oligocene evaporites are mined in the Navarra and Catalonia Districts of northern Spain (Fig. 11). The deposits in the Navarra District occur in two beds between depths of 400 and 2000 ft. Sylvinite ore contains 18 to 20% K_2O. The upper bed is principally carnallite with an average grade of 14% K_2O.

The evaporites of Catalonia have been deformed through flowage, folding, and faulting such that mining dips of 30° are common.

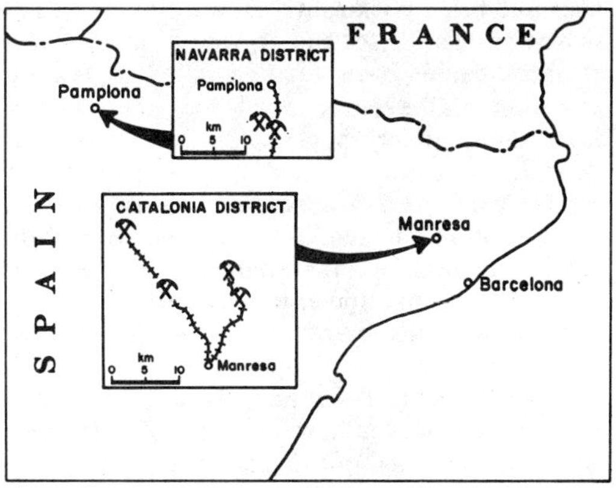

FIG. 11—*Location map of the potash mining districts of Spain (Anon., 1971, modified).*

Carnallite and sylvite ores occur in up to four beds with grades between 15 and 29% K_2O.

Africa

The potash reserves of Africa are estimated at 134 million short tons K_2O, approximately one-tenth of 1% of the total world reserves. Production has been equally small by world standards, the first mine opening as recently as 1969.

The *Congo* (Brazzaville) began potash production in 1969 with completion of the facilities at Holle 25 miles northeast of Point Noíre. Irregularities in the potash beds have reportedly caused initial difficulties for the boring machines. In 1977, the Holle potash mine was abruptly flooded and lost. Since then, there have been no attempts to renew production from this district (De Ruiter, 1979). The mining grades, on the other hand, reportedly have been high, in some cases up to 39% K_2O. Sylvinite ore bodies are inferred to be small and erratic in grade, distribution and thickness. Carnallite, by contrast, averages 16% K_2O and occurs in beds 10 to 40 ft thick.

***Ethiopia* contains significant potash deposits unique for their young (Quaternary) age and extensive kainite beds. The deposits occur in a rift valley characterized by recent vulcanism and high heat flow (Holwerda and Hutchinson, 1968). The Houston formation contains, from the top, three potash-bearing members: the sylvinite member (up to 30 ft thick), the intermediate members containing carnallite throughout, sylvite at the top and kainite at the base (10 to 80 ft thick), and the lower kainite member (13 to 43 ft thick). Ground water within clastic sediments interbedded with the evaporites and brine pockets within the evaporites are potential hazards for underground mines. Several companies have been involved in the exploration and evaluation of the deposits. A shaft and underground workings are now flooded.**

At Marada, *Libya,* potassium-bearing brines **occur within a shallow, porous carnallite-bearing formation in a marshlike depression. These brines are not exploited although their development has been considered for several years.**

Deposits near Khemisset, *Morocco,* **reportedly contain 40 million tons of sylvinite and 200 million tons of carnallite ore. At the** grades given in Table 7 the total inplace K_2O reserves would be about 30 million tons. The deposits occur over an area of at least 30 sq km at depths between 1600 and 2600 ft. The sylvinite occurs around the edges of the two carnallite beds which average 3 to 30 ft in thickness. The sylvinite bodies are reportedly low grade, erratic, and probably irregular. Exploration continues for sylvinite reserves adequate to support a mine and mill complex.

Asia

The potash potential of *China* is essentially unknown. It seems probable that the country contains one or more continental evaporite basins with their characteristic high grade, large tonnage potash deposits. For many centuries, the Chinese have produced sodium chloride from brine wells in the provinces of Szechwan and Yunnan. There are believed to be extensive deposits of Jurassic evaporites in this area. Recently it has been reported that potash deposits have been located by the Chinese government at Jiang Cheng, Yunnan. The balance of Asia does contain scattered Paleozoic evaporites and several Mesozoic deposits. Potash salts, however, are limited thus far to the Eocene (Cambrian?) salts of the Punjab Ranges of *West Pakistan* where no minable deposits are yet known to exist. Evaluation of a deep brine occurrence at Dariala, West Pakistan, is reportedly continuing. The Geological Survey of *India* has identified an area of over 30,000 square miles in western Rajasthan as a potential area for potash exploration. Exploration in the Naguar Basin, which began in 1974, has established the presence of large potash resources.

A very large potash deposit was discovered in 1973, at the Khorat Plateau of *Thailand* (Hite and Japakasetr, 1979). This Cretaceous deposit consists primarily of carnallite, although sylvinite and tachyhydrite are also present. Locally the carnallite deposit is as much as 95 m thick and at shallow depths (90 to 600 m). Lenticular high-grade bodies of sylvite are found associated with the carnallite deposit on the flanks of salt anticlines. The deposits also extend north across the Mekong River into Laos. A large-scale exploration program by the Thailand government is still continuing.

Israel's production from the Dead Sea remains the only significant source of potash in **Asia. The production facilities at Sodom, expanded in 1969, have a production capability of 1.1 million short tons of K_2O annually.**

Construction of facilities for potash recovery from Dead Sea brine on the *Jordan* side is nearing completion. Production is scheduled for sometime in 1983.

Oceania

Minable marine potash deposits have not as yet been discovered in *Australia* although occurrences of potash minerals are known in the Adavale Basin in Queensland. Potash production was scheduled by Texada to begin in 1973 from the brines of Lake McLeod in Western Australia.

The program was curtailed in 1974. In 1978, Texada was acquired by Dampier Salt Co., Ltd. In 1981, construction of test crystallizers and pilot plant tests commenced with production scheduled by 1985. Production rates are scheduled to be in the region of 60 to 120 kt/a.

The brines occur in porous halite and gypsum sediments approximately 2 ft below the dry lake surface. The brines will be concentrated by solar evaporation and then processed to yield a synthetic langbeinite product.

Production

Mining

Potash mines are serviced by vertical shafts similar to many other underground mines. The shafts are constructed either by drilling and blasting or with large diameter drilling machines. Shafts in the Carlsbad District, New Mexico, are 600 to 1650 ft deep and encounter no particular problems in construction. In Saskatchewan, on the other hand, shafts exceed 3000 ft in depth and, due to high water pressure in the Blairmore Formation, are constructed with freezing and tubbing techniques. In other districts shaft depth and design are determined by the particular geology features of the deposits.

Mining in the Carlsbad District is by the room-and-pillar method using conventional drilling and blasting and, to a limited extent, continuous mining machines. Conventional mining with jumbo drills, undercutters, ammonium nitrate explosives, mucking machines, and shuttle cars is applicable to all the deposits. The continuous mining machines do not perform well in irregular ore containing barren areas or in ores with appreciable amounts of the hard sulfate minerals. Initial mine extraction in the district is 60 to 70%. Final extraction after pillar robbing is about 85%.

The uniform deposits of Saskatchewan are mined entirely by continuous mining machines (with the exception of one solution mine) which produce about 4000 tons per day. Ore from these machines is moved to the shafts on extensible and fixed belts. Initial mine extraction is held to 35 to 40% and conventional blasting is avoided so as to diminish the risk of fracturing through the overlying salt and flooding the mines. Final mine extraction may ultimately reach 50%.

The annual production capacity of the Carlsbad District is about two million K_2O t/a. Current mine productivity averages 75 to 100 tons per man shift.

Past production in the Carlsbad District has been largely from ores which contain 15 to 25% K_2O and 1 to 1.5% insoluble material, principally as clays. In 1981, most production was from ores with lower potash grades (14 to 17% K_2O) and higher insoluble contents (3 to 5%). The ores of Saskatchewan are uniformly higher in grade (20 to 30% K_2O) than those of the Carlsbad District. Furthermore, the reactive insoluble material in the Carlsbad ores causes desliming problems not experienced in the Saskatchewan ores (3–4% insoluble material).

Potash mining methods outside North America are similar to those in Saskatchewan and New Mexico except in those deposits which have been deformed. In some of the deposits of Germany and Sicily, for example, mining is by stopes rather than room-and-pillar, generally with somewhat higher mining costs.

Processing

Most marketable potash products are concentrated by (a) flotation, (b) selective dissolution ("washing") of the gangue minerals, or (c) precipitation of the potash mineral from a hot brine. The majority of sylvinite ores mined in New Mexico (White and Arend, 1950) and Saskatchewan are processed in flotation mills equipped with grinding, desliming, flotation, and drying circuits. The ore is first pulped in a saturated sodium chloride-potassium chloride brine and ground in ball and rod mills to separate the sylvite from the gangue minerals. The fine-grained contaminants, largely clays and quartz, are removed in the desliming circuit. The slurry of sylvite and halite is then passed through froth flotation cells where amine reagents attach the sylvite grains to bubbles in the froth causing the sylvite to float away from the halite. The sylvite is drawn off the top of the cells, dried, and sent to storage.

In the past langbeinite of the langbeinite and mixed langbeinite-sylvite ores of the Carlsbad

District has been concentrated by dissolving or "washing" and removing the readily soluble halite and sylvite gangue minerals. One company now processes ores containing sufficient sylvite (so-called mixed ores) in a heavy media circuit whereby the langbeinite and sylvite can both be physically separated and concentrated.

Carnallite ores and sylvinite ores with abundant insoluble material or impurities within the sylvite grains are processed by selectively dissolving the potassium minerals in a sodium chlorite brine, desliming the brine, and precipitating a pure sylvite product by controlled variation of the composition and temperature of the brine.

Potassium sulfate does not occur as a natural mineral. It may be produced, however, from certain natural brines (for example, Great Salt Lake and Searles Lake) or by mixing brines produced by dissolving potassium and potassium-sulfate minerals. The potassium-sulfate minerals commonly used in this process are langbeinite (Carlsbad District) and kainite (Sicily). The extra potassium required in the reaction is introduced as a brine generally prepared by dissolving sylvite.

Solution mining is applicable only to the readily soluble potassium chloride minerals and is probably profitable only for thick, high-grade sylvinite ores. Only two large scale solution mining operations are currently in production. The first, located in Saskatchewan, has operated since 1964 and in 1969 increased its annual capacity from 600,000 to 900,000 tons K_2O. The nature of the operation has not been made public, but it is likely that the thick, undeformed, high-grade ore horizons have contributed substantially to the success of the operation. The second solution operation began in 1972 when an unsuccessful conventional mine in Utah was flooded and converted to a solution mining-solar evaporation process. Conventional mining was precluded by the deformation of the ore horizon. The solution mining operation is reported to be proceeding successfully.

Product Specifications

The principal potash products for fertilizers [muriate of potash (KCl), potassium sulfate (K_2SO_4) and potassium-magnesium sulfate ($K_2SO_4 \cdot 2MgSO_4$)] are sold on the basis of chemical compositions and particle size. Products with higher potash grades and coarser particles are generally more desirable. Most products are sold on the basis of a guaranteed minimum potash content in standard, coarse, and granular grades, in order of increasing particle size (Table 9). Muriate of potash, the highest grade potash product, accounted for 80% of domestic potash consumption in 1981. The low chlorine content and lower solubility of the potassium sulfate products make them desirable for certain crops and soil conditions. Products produced by solution and precipitation (some muriate and all potassium sulfate) are often purer, hence are preferred for certain applications including the chemical industry.

Transportation

Potash mines in the United States are located distant from oceans and areas of high fertilizer consumption, hence transportation represents a major portion of the delivered price of potash products. East of the Mississippi, for example, average rail freight charges for Carlsbad muriate in 1981 were approximately 50% of the delivered price. Carlsbad production destined for export, although subject to lower freight rates than domestic shipments, incurs freight charges of about $10 per ton to the west coast or $8 per ton to Houston (1975). These rates include the cost of transfer to ships at about $1 per ton.

Imports from Europe and Canada are competitive with domestic production or hold a freight advantage over considerable portions of the United States. In the Great Lakes region, which accounts for more than 40% of domestic consumption, freight costs from Carlsbad and Saskatchewan are comparable (Lewis, 1971). Along the eastern seaboard, potash from Europe holds a freight advantage for up to 150 miles inland.

Transportation costs are equally significant to the Saskatchewan potash industry. In 1980, for example, only a small amount of Saskatchewan's production was consumed in Canada while 67% was exported to the United States. In spite of lower production costs, freight rates to Vancouver and the Great Lakes region of the United States become a burden in world competition during periods of excess production.

The importance of transportation costs is exemplified by current attempts to develop potash production from the deposits in New Brunswick and Nova Scotia. These deposits may have relatively higher mining costs because of geologic complexity, but since they are close to deep water ports on the Atlantic coast, they have a decided transportation cost advantage over deposits in the Saskatchewan District.

Alternate methods of transportation such as slurry pipelines, have been considered, particu-

TABLE 9—Typical Specifications for Standard Grade Potash Products*

	Muriate of Potash, KCl	Sulfate of Potash, K_2SO_4	Potassium, Magnesium, Sulfate, $K_2SO_4 \cdot 2MgSO_4$
Chemical Specifications, %			
K_2O	60.0 – 61.5	50.0 – 52.4	22.0 – 22.9
MgO			18.0 – 18.8
S			22.3 – 22.6
Cl		1.0 – 2.5	1.0 – 2.3
H_2O	0.06 – 0.12	0.10 – 0.12	0.12 – 0.15
Physical Specifications, % Cumulative			
+ 8 mesh			0 – 3
+10			3 – 16
+14	0 – 6	0 – 15	16 – 38
+20	17 – 32	20 – 45	39 – 63
+28	42 – 63	65 – 85	60 – 83
+35	66 – 90	80 – 95	75 – 92
+48	84 – 98	90 – 98	85 – 96
+65	93 – 100	95 – 100	94 – 100
Typical Bulk Density (loose, lb per cu ft)	66 – 68	88	92
Typical Chemical Analysis, %			
K	50.5	41.9	18.5
Mg	0.17	0.9	11.1
Ca	0.02	0.3	0.05
Na	1.05	0.2	0.76
Cl	47.86	1.5	
SO_4	0.04	51.5	67.4
Insol.	0.30	0.7	0.33

* Source: Adapted from product specifications published by International Minerals and Chemical Corp.

larly in Canada. It is probable, however, that railroads will continue to move the production of Carlsbad and Saskatchewan and it is unlikely that the costs will be significantly reduced.

Supply and Demand

The estimated world production of potash for 1979 was 26.35 Mt of K_2O equivalent. Of this, the USSR, Canada, and the German Democratic Republic were the world's leading potash producers, as shown in Table 10.

Table 11 shows salient statistics of US potash production. Approximately 85% of US production comes from New Mexico. Utah and California supply the balance.

Potash prices are quoted on the basis of the short ton unit, f.o.b. minesite. Canadian prices (Lyon, 1981) are shown in Table 12. Published prices, f.o.b. Carlsbad, NM, are about 5¢ per unit higher than the Canadian prices for the grades shown in the table.

Potash sales are highest in late spring in anticipation of the spring fertilizer season (Fig. 12). To encourage consumers to purchase and store products in the off season, quoted prices in late winter and spring are a few cents per unit higher.

Potash supply and demand ended the 1980 fertilizer year in fairly close balance. The fertilizer year starts July 1 of each year and ends June 30 the following year. Table 13 (Lyon, 1981) shows the North American potash supply and demand picture for 1978–1979 with an estimate of the supply and demand for 1980–1981.

In 1981, the Government of Saskatchewan, through the Potash Corp. of Saskatchewan (PCS), a Crown corporation, controlled five separate operations with a total capacity of about 2.9 Mt K_2O or about 39% of the Province's total capacity (Singleton and Searls, 1981). The PCS is now the largest potash producer in North America. PCS also is planning a new 2.0 Mt potash mine in the Bredenbury area and indicated it will increase potash

TABLE 10—Marketable Potash: World Production by Country*

(1000 t of K_2O equivalent)

Country†	1976	1977	1978	1979^p	1980^e
Canada (sales)‡	ʳ5,215	ʳ5,764	6,340	7,074	§7,532
Chile #	ʳ17	ʳ11	15	21	21
China, mainland‖	ʳ12	ʳ18	ʳ21	ʳ16	12
Congo	254	ʳ136	—	—	—
France	1,603	ʳ1,580	1,795	1,920	§1,939
German Democratic Republic	3,161	3,229	3,323	3,395	§3,422
Germany, Federal Republic of	2,036	2,341	2,470	2,690	§2,674
Israel	680	ʳ707	695	730	§797
Italy	ʳ140	ʳ151	196	182	185
Spain	630	ʳ562	722	781	770
USSR	8,310	8,347	8,193	6,635	8,000
United Kingdom	ʳ46	81	150	264	280
United States	2,177	2,229	2,253	2,225	§2,239
Total	ʳ24,281	ʳ25,156	26,173	25,933	27,871

Source: Searls, 1981a.

^e Estimated. ^p Preliminary. ʳ Revised.

* Table includes data available through Apr. 27, 1981.

† In addition to the countries listed, Australia apparently produced small quantities of marketable potash during 1976-1980, but output was not reported quantitatively, and general information was inadequate for the formulation of reliable estimates of output levels.

‡ Official Government figures.

§ Reported figure.

Series revised; new data represent officially reported output of potassium nitrate product (gross weight basis) converted assuming 14% K_2O equivalent.

‖ Series revised to reflect officially reported Chinese data on production of potassic fertilizers in terms of nutrient content; small additional quantities may be produced and used by the nonfertilizer chemical industry.

production to an estimated 6.9 Mt by 1990. The Government of Saskatchewan has announced plans to eventually acquire 50% of the Saskatchewan potash industry, apparently by internal expansion.

Other Considerations

World food requirements are projected to increase, hence so also will the demand for fertilizers and potassium. The principal market

TABLE 11—Salient Statistics on Potash, United States

(1000 t)

	1977	1978	1979	1980	1981^e
Production: Marketable	2,229	2,253	2,225	2,239	2,150
Imports for consumption	4,605	4,707	5,165	4,972	5,200
Exports	846	809	635	762	500
Apparent consumption	5,992	6,205	6,918	6,427	6,700
Price: Muriate, ¢/t unit of K_2O, f.o.b. mine*	71	76	95	130	140
Stocks, producer, yearend	467	414	251	250	420
Employment: Mine	2,100	1,900	1,800	1,800	1,750
Employment: Mill	1,300	1,400	1,450	1,450	1,400
Net import reliance† as a percent of apparent consumption	63	64	66	65	68

Source: Searls, 1981.

^e Estimate.

* Average prices based on actual sales, excluding soluble and chemical muriates.

† Net import reliance = imports − exports + adjustments for government and industry stock changes.

TABLE 12—Canadian Muriate of Potash Prices
(Dollars per unit K_2O)*

Grade	March 1, 1980	March 1, 1981	% Change
Standard	0.96	1.16	+20.8
Coarse	1.10	1.25	+13.6
Granular	1.12	1.27	+13.4

Source: Lyon, 1981.
* A unit is equivalent to 20 lb or 1% K_2O short ton.

of potash is assured because potash is essential to plant metabolism even if substitutions should occur in some of the industrial and chemical applications.

By 1985, potash consumption is expected to be about 33 million short tons of K_2O per year. The US demand is expected to increase to about 8.6 Mt of K_2O equivalent by 1990. The estimated US demand by the year 2000 is 10.1 Mt of K_2O equivalent.

Adequate reserves of low grade potash are available to meet the world's needs for the next few hundred years. Deposits with lower production costs or advantages in location may even see expansion toward the end of the decade at the expense of less profitable operations. Beyond the protection of some portion of domestic production capacity, countries are likely

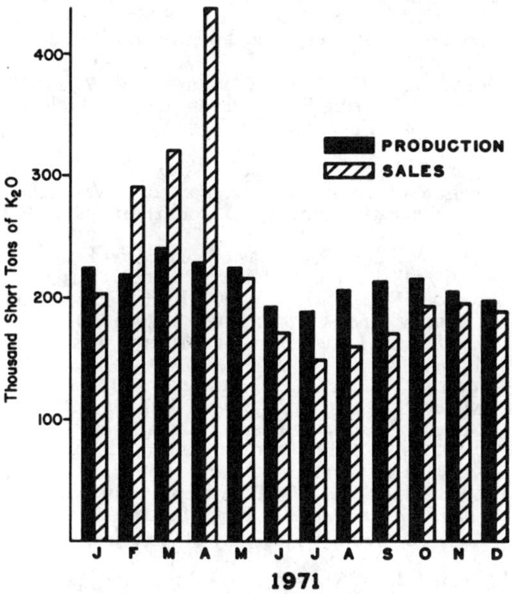

FIG. 12—*Monthly total of US potash production and sales, 1971 (data supplied by International Minerals and Chemical Corp.)*.

TABLE 13—North American Potash Supply and Demand
(1000 St, K_2O)

	Actual* 1978-1979	Actual* 1978-1980	Estimated† 1980-1981
Supply			
Opening inventory	1,622	1,155	1,445
Production (US and Canada)	9,296	10,265	10,800
Imports	268	301	350
Less closing inventory	1,155	1,445	1,445
Total Supply	**10,031**	**10,276**	**11,150**
Demand			
Agricultural			
US	6,405	6,548	6,900
Canada	381	401	420
Total	6,786	6,949	7,320
Industrial	382	423	430
Export	2,829	2,847	3,200
Total demand	**9,997**	**10,219**	**10,950**

Source: Lyon, 1981.
* Potash-Phosphate Institute.
† CF Industries Inc.

to permit world trade to continue with minimal restrictions.

The potash industry is subject to taxes and royalties similar to other extractive industries. Domestic production from federal lands is administered by the US Geological Survey and the Bureau of Land Management insofar as potash is a leasable mineral. Domestic production is also subject to a depletion allowance of 14% of gross income and in New Mexico a state severance tax is based on an ore valuation formula. Production from the Saskatchewan District is subject to a Provincial mining tax and a royalty payable to the Crown. Tax consideration, similar to the depletion allowance, is permitted but is under revision.

The extraction of potash ores and the application of fertilizers constitute alteration of the natural environment as, indeed, does the planting and harvesting of crops. Whereas it is unlikely that either the extraction or the application of potash has any serious deleterious effect on the environment, both are receiving study. Potash mines in North America are remote from population centers. The deposits of New Mexico are in low density, semiarid grazing land where brackish lakes and marshes are common. Mining operations disturb only sufficient surface area for mining and milling facilities and tailings ponds. The tailings ponds are the only potential long range problem as they are too saline to support life and must be stabilized against wind erosion. The proper application of potash in fertilizer is not known to constitute a hazard of any kind. Hearings by the Illinois Pollution Control Board identified potential hazards from the misapplications and buildup of nitrogen and phosphorus but noted no such concern for potash (Aldrich, 1972).

Bibliography and References

Anon., 1965, *Salt Basins Around Africa,* Institute of Petroleum and the Geological Society, London, Elsevier, Amsterdam, 122 pp.

Anon., 1966, "World Survey of Potash, 1966," British Sulphur Corporation Ltd., London, 93 pp.

Anon., 1968, "World Supply and Demand for Potash and its Impact upon the Saskatchewan Industry," Saskatchewan Potash Committee, 86 pp.

Anon., 1970, "The United States Position and Outlook in Potash," Information Circular 8487, US Bureau of Mines, 47 pp.

Anon., 1971, "New Name in the World Potash Industry," *Phosphorus and Potassium,* No. 51, January-February, pp. 47–49.

Anon., 1971a, "Spanish Potash—1970 and After," *Phosphorus and Potassium,* No. 53, May-June, pp. 36–39.

Anon., 1972, "The Technical Advances by the USSR Potash Industry," *Phosphorus and Potassium,* No. 59, May-June, pp. 41–45.

Anon., 1972a, "The Canadian Potash Industry: A Study in Government Cooperation," International Minerals and Chemical Corp., Libertyville, IL, 24 pp.

Adams, S.S., 1968, "Potassium Reserves in the World," *The Role of Potassium in Agriculture,* V.J. Kilmer, S.E. Younts, and N.C. Brady, eds., American Society of Agronomy, Crop Science Society of America, Soil Science Society of America, Madison, pp. 1–21.

Adams, S.S., 1969, "Bromine in the Salado Formation, Carlsbad Potash District New Mexico," Bulletin 93, New Mexico Bureau of Mines and Mineral Resources, 122 pp.

Adams, S.S., 1970, "Ore Controls, Carlsbad Potash District, Southeast New Mexico," *Third Symposium on Salt,* Rau, J.L., and Dellwig, L.F., eds., Vol. 1, Northern Ohio Geological Society, Cleveland, pp. 246–257.

Aldrich, 1972, "Opinion of the Board in the Matter of Plant Nutrients," R 71–15, Illinois Pollution Control Board, 7 pp.

Alger, R.P., and Crain, E.R., "Defining Evaporite Deposits with Electrical Well Logs," *Second Symposium on Salt,* J.L. Rau, ed., Vol. 2, Northern Ohio Geological Society, Cleveland, 1966, pp. 116–130.

Benavides, V., 1968, "Saline Deposits of South America," *Saline Deposits,* R.B. Mattox, ed., Special Paper 88, Geological Society of America, pp. 249–290.

Bentor, Y.K., 1968, "Salt Deposits of the Dead Sea Region," *Saline Deposits,* R.B. Mattox, ed., Special Paper 88, Geological Society of America, pp. 139–156.

Borchert, H., and Muir, R.O., 1964, *Salt Deposits,* Van Nostrand, London, 330 pp.

Braitsch, O., 1971, *Salt Deposits, Their Origin and Composition,* Springer-Verlag, New York, 297 pp.

D'Ans, J., 1933, *Die Losungsgleichgewichte de System der Salze Oceanischer Salzablagerungen,* Hrsq. von Kali-Forschungs-Anstalt, Verlagsgellschaft fur Ackerbau, Berlin, 254 pp.

De Ruiter, P.A.C., 1979, "The Gabon and Congo Basins Salt Deposits," *Economic Geology,* Vol. 74, No. 2, Mar.-Apr., pp. 419–431.

Eilertsen, D.E., 1972, "Potash," *Minerals Yearbook 1972,* US Bureau of Mines, pp. 1055–1067.

Eilertsen, D.E., 1972, "Potash in 1971," *Mineral Industry Surveys,* US Bureau of Mines, July 5, 4 pp.

Goudarzi, G.H., 1981, "Potash—Libya," abstract, 17th Annual Forum on the Geology of Industrial Minerals, Albuquerque, NM.

Hahl, D.C., and Handy, A.H., 1969, "Great Salt Lake: Chemical and Physical Variations of the Brine, 1963–1966," Water Resources Bulletin 12, Utah Geological and Mineralogical Survey, 33 pp.

Halbouty, M.T., 1967, *Salt Domes—Gulf Region, United States and Mexico,* Gulf Publishing Co., Houston, 425 pp.

Hite, R.J., 1982, "Progress Report on the Potash Deposits of the Khorat Plateau, Thailand," US Geological Survey Open File Report 82–1096, 70 pp.

Hite, R.J., and Japakasetr, T., 1979, "Potash Deposits of the Khorat Plateau, Thailand and Laos," *Economic Geology,* Vol. 74, No. 2, Mar.-Apr., pp. 448–458.

Holser, W.T., 1968, "Bromide Geochemistry of Salt Rock," *Second Symposium on Salt,* J.L. Rau, ed., Vol. 1, Northern Ohio Geological Society, Cleveland, pp. 248–275.

Holter, M.E., 1969, "The Middle Devonian Prairie Evaporite of Saskatchewan," Report No. 123, Saskatchewan Department of Mineral Resources, 134 pp.

Holwerda, J.G., and Hutchinson, R.W., 1968, "Potash-Bearing Evaporites in the Danakil Area, Ethiopia," *Economic Geology,* Vol. 63, pp. 124–150.

Johns, R.K., and Ludbrook, N.H., 1963, "Investigation of Lake Eyre," Report of Investigation 24, Geological Society of South Australia, 102 pp.

Keyes, W.F., 1973, "Potash," *Minerals Yearbook 1973,* US Bureau of Mines, Preprint, 12 pp.

Knudson, R.H., 1972, "World Potash Production Capacity and Consumption 1970–1980," Saskatchewan Department of Mineral Resources, 14 pp.

Koepke, W.E., 1971, "Fertilizers and Fertilizer Minerals in Canada," Mineral Bulletin MR 115, Mineral Resources Branch, Department of Energy, Mines and Resources, Ottawa, 157 pp.

Kozary, M.T., Dunlap, J.C., and Humphrey, W.W., 1968, "Incidence of Saline Deposits in Geologic Time," *Saline Deposits,* R.B. Mattox, ed., Special Paper 88, Geological Society of America, pp. 43–57.

Lefond, S.J., 1969, *Handbook of World Salt Resources,* Plenum Press, New York, 384 pp.

Lewis, R.W., 1970, "Potassium," *Mineral Facts and Problems, 1970 Edition,* Bulletin 650, US Bureau of Mines, pp. 1157–1170.

Liechti, P., 1968, "Salt Features of France," *Saline Deposits,* Mattox, R.B., ed., Special Paper 88, Geological Society of America, pp. 83–106.

Linn, K.O., and Adams, S.S., 1968, "Barren Halite Zones in Potash Deposits, Carlsbad, New Mexico," *Second Symposium on Salt,* J.L. Rau, ed., Vol. 1, Northern Ohio Geological Society, Cleveland, pp. 59–69.

Livingstone, D.A., 1963, "Chemical Composition of Rivers and Lakes," *Data of Geochemistry,* Professional Paper 440-G, US Geological Survey, 60 pp.

Lotze, F., 1957, *Steinsalz und Kalisalze,* 2nd ed., Vol. 1, Gebruder Borntraeger, 465 pp.

Lotze, F., 1963, "The Distribution of Evaporites in Space and Time," *Problems in Paleoclimatology,* A.E.M. Nairn, ed., Proceedings of NATO Paleoclimates Conference, University of Newcastle-Upon-Tyne, pp. 491–507.

Lyon, F.D., 1981, "Major Expansions Programmed at Canadian Mines and Plants," *Engineering and Mining Journal,* Vol. 182, No. 3, Mar., pp. 95–99.

Matthews, R.D., 1970, "The Distribution of Silurian Potash in the Michigan Basin," *Forum on Geology of Industrial Minerals,* 6th, Proceedings Michigan Geological Survey Div., Misc., No. 1, p. 20–23.

Parker, R.L., 1967, "Composition of the Earth's Crust," *Data of Geochemistry,* Professional Paper 440-D, US Geological Survey, 19 pp.

Pendery, E.C., 1970, "Distribution of Salt and Potash Deposits: Present and Potential Effect on Potash Economics and Exploration," *Third Symposium on Salt,* J.L. Rau, and L.F. Dellwig, eds., Vol. 2, Northern Ohio Geological Society, Cleveland, pp. 85–95.

Pierce, W.G., and Rich, E.I., 1962, "Summary of Rock Salt Deposits in the United States as Possible Storage Sites for Radioactive Waste Materials," Bulletin 1148, US Geological Survey, 91 pp.

Richter-Bernberg, G., 1955, "Stratigraphische Gliederung des deutschen Zechsteins," *Zeitschrift der Deutschen Geologischen Gessellschaft,* Vol. 105, p. 843.

Rios, J.M., 1968, "Saline Deposits of Spain," *Saline Deposits,* R.B. Mattox, ed., Special Paper 88, Geological Society of America, pp. 59–74.

Rittenhouse, P.A., 1979, "Potash and Politics," *Economic Geology,* Vol. 74, No. 2, Mar.-Apr., pp. 353–357.

Ryan, J.E., 1951, "Industrial Salts: Production at Searles Lake," *Trans. SME/AIME,* Vol. 190, pp. 447–452.

Sawatzky, H.B., Agarwal, R.G., and Wilson, W., 1960, "Helium Prospects of Southwest Saskatchewan," Report No. 49, Saskatchewan Department of Mineral Resources, 26 pp.

Searls, J.P., 1981, "Potash," *Mineral Commodity Summaries, 1981,* US Bureau of Mines, pp. 116–117.

Searls, J.P., 1981a, "Potash," *Minerals Yearbook, 1980,* US Bureau of Mines, pp. 651–661.

Singleton, R.H., and Searls, J.P., 1980, "Potash," Preprint, 1978–1979, US Bureau of Mines, 15 pp.

Smith, J.P., 1950, "Notes on the Potash Deposits of Germany, France, and Spain," *Trans. AIME,* Vol. 187, No. 1, pp. 117–121.

Stewart, F.H., 1963, "Marine Evaporites," *Data of Geochemistry,* Professional Paper 440-Y, US Geological Survey, 52 pp.

Stocklin, J., 1968, "Saline Deposits of the Middle East," *Saline Deposits,* R.B. Mattox, ed., Special Paper 88, Geological Society of America, pp. 157–181.

Streeton, D.H., 1967, "The Geology of the Prairie Evaporite Formation of the Yorkton Area of Saskatchewan," M.A. Thesis, University of Saskatchewan, 81 pp.

Szatmari, P., et al., 1979, "A Comparison of Evaporite Facies in the Late Paleozoic Amazon and Middle Cretaceous South Atlantic Salt Basins," *Economic Geology,* Vol. 74, No. 2, Mar.-Apr., pp. 432–447.

Tortochaux, F., 1968, "Occurrence and Structure of Evaporites in North Africa," *Saline Deposits,* R.B. Mattox, ed., Special Paper 88, Geological Society of America, pp. 106–138.

White, D.E., Hem, J.D., and Waring, G.A., 1963, "Chemical Composition of Subsurface Waters," *Data of Geochemistry,* Professional Paper 440-F, US Geological Survey, 67 pp.

White, D.E., 1965, "Saline Waters of Sedimentary Rocks," *Fluids in Subsurface Environments,* A. Young, and J.E. Galley, eds., Memoir 4, American Association of Petroleum Geologists, pp. 342–366.

White, N.C., and Arend, C.A., Jr., 1950, "Potash Production at Carlsbad," *Chemical Engineering Progress,* pp. 523–530.

Woods, P.J.E., 1979, "The Geology of the Boulby Mine," *Economic Geology,* Vol. 74, No. 2, Mar.-Apr., pp. 409–418.

Pumice, Pumicite, and Volcanic Cinders

N. V. PETERSON *

R. S. MASON †

The violent explosion of Mount St. Helens in the state of Washington on May 18, 1980 showered volcanic ash over vast areas of the Northwest. It was another eruptive episode in a long history of similar occurrences in the region, which extends from Canada to California and reaches several hundred miles inland from the Pacific Ocean.

Pumice, pumicite, and cinders are products of explosive volcanic eruptions. Pumice and pumicite are produced by the violent expansion of dissolved gases in a viscous silicic lava such as rhyolite or dacite. Scoria is the mafic counterpart of pumice.

Pumice is a light-colored, cellular, almost frothy rock made up of glass-walled bubble casts. It may occur as coherent, massive blocks composed of highly vesicular glassy lava in either a flow or vent filling, or it may be more or less fragmented by violent eruption. Pumicite, the diminutive of pumice, has the same origin, chemical composition, and glassy structure, differing only in particle size. Particles less than 0.16 in. in diam are designated pumicite.

Pumice usually is found relatively close to the vent from which it was expelled, while pumicite may be carried by winds for great distances before settling as an accumulation of fine-grained ash or tuffaceous sediment.

Mount St. Helens lofted an estimated 5 billion cu yd of ash and pumicite. The volcano, unlike most of the High Cascade peaks, is composed largely of dacitic material with only minor amounts of andesite and basalt.

Volcanic cinders are the reddish to black vesicular fragments which pile up during explosive eruptions of volcanoes of basaltic composition. Most cinder deposits occur as cones or mounds of stratified fragments ranging from a fraction of an inch to several inches in diameter. Individual cones or mounds may be several hundred feet in diameter and up to 500 ft high.

* Retired, formerly District Geologist, Grants Pass Office, State of Oregon Department of Geology and Mineral Industries, Portland, OR.

† Retired, formerly State Geologist, State of Oregon Department of Geology and Mineral Industries, Portland, OR.

Properties and Uses

Low bulk density, good heat and sound insulating properties, and excellent abrasive capabilities make pumice and pumicite useful in industrial applications. Cinders, though less cellular and somewhat heavier, also have a variety of industrial uses.

The main use of pumice and cinders continues to be in the construction industry as road-surfacing material, railroad ballast, and building block aggregate. Pumicite is used in sizable quantities as a pozzolanic additive to monolithic concrete where it increases workability, strength, and durability of the concrete and reduces the heat of hydration.

Volcanic cinders generally are denser and more coarsely cellular than most pumice. Markets for cinders include aggregate for road construction and surfacing and use in lightweight concretes, particularly blocks. Some producers having access to both pumice and cinders offer a variety of blends of the two materials to block manufacturers. Table 1 lists the principal applications of industrial pumice. Tables 2 and 3 show US sales and uses of volcanic products.

Most pumice deposits are unconsolidated and usually have a minimum of overburden. Mining is by open pit and can be easily carried out with conventional loading equipment. Sometimes air drying precedes hauling to a crushing and screening plant which yields the size product desired. Multiple bins at railroad hauling facilities permit blending if desired. Pumice for abrasive use may require more sophisticated processing by fine grinding and air classification before being bagged for market.

TABLE 1—Principal Applications of Industrial Pumice

Market	Application	Grade
Paint manufacture	Nonskid coatings	Coarse
	Acoustic insulating ceiling paints	Coarse
	Fillers for textured paints	Intermediate
Chemical industry	Flattening agents	Extra fine
	Filtration media	Coarse
	Chemical carriers	Coarse
	Sulfur matches and strikers	Intermediate
Metal and plastic finishing	Cleaning and polishing	Extra fine
	Vibratory and barrel finishing	Extra fine-Intermediate
	Pressure blasting	Intermediate
	Electroplating	Extra fine-Intermediate
Compounders	Cleaning lithographic plates	Extra fine
	Powdered hand soaps	Intermediate
	Glass cleaners	Extra fine
	Polishing natural teeth and dentures	Fine
	Smoothing rough skin	Lump
Rubber	Erasers	Intermediate
	Mould release agents	Extra fine
Glass and mirror	TV tube processing	
	Glass buffing and polishing	Fine
	Bevel finishing	Extra fine
	Cut glass finishing	Extra fine
Furniture	Hand-rubbed satin finishing	Extra fine
	Piano keys	Extra fine
	Picture frame gold-leafing	Extra fine
Leather	Buffing	Intermediate
Electronics	Cleaning circuit boards	Extra fine
Pottery	Filler	Extra fine-fine
Agricultural	Soil substitute and additive	Coarse

Source: Meisinger, 1977.

Block pumice in economical quantity and quality is limited to a few geologically young deposits located in California and Oregon. The California deposits are discussed by Chesterman (1957). Production in California has come mainly from Glass Mountain and Little Glass Mountain in eastern Siskiyou County just south of the Oregon-California border, and at Mono Craters, Mono County. In Oregon, block pumice occurs principally at Newberry Crater and at Rock Mesa, both in Deschutes County just south of the South Sister, a 10,000-ft peak in the Cascade range. There has been minor production of lump pumice from the Newberry Crater site for many years. The Rock Mesa deposit covers roughly 1,200 acres, but has not been developed.

Deposits containing block pumice typically are composed of a mixture of rhyolitic, dacitic, or rhyodacitic volcanic glasses ranging in texture from glassy, solid obsidian to highly vesicular pumice.

Block pumice sizes range from 2 in. or more in one dimension to chunks 5 ft or larger in diam. The smaller pieces have a variety of uses including abrasive blocks for the removal of callouses, for cleaning barbecue grills, and for dressing metal polishing belts. Pumice blocks are also used for filtering vinegar.

Larger blocks may be sold either rough or with one or, at most, two edges sawed. The latter are used architecturally for wall or other surface coverings, including suspended ceilings. The rough chunks are often used in landscape

TABLE 2—Pumice and Volcanic Cinders Sold and Used by Producers, US

	1976	1977	1978	1979	1980	1981
Pumice and pumicite (1000 st)	906	1,178	1,208	1,173	543	499
Value* ($1000)	3,830	4,625	4,836	5,008	4,267	4,311
Average value per st ($)	4.23	3.93	4.00	4.27	7.86	8.64
Volcanic cinder and scoria (1000 st)	3,228	2,831	3,549	3,241	3,236	3,667
Value* ($1000)	6,636	7,340	9,619	10,958	11,258	13,400
Average value per st ($)	2.06	2.59	2.71	3.38	3.48	3.65
Exports (1000 st)	1	2	2+	2+	1†	1†
Imports for Consumption (1000 st)	81	253	216	62	194	90
World production, pumice and related volcanic materials (1000 st)	16,800‡	18,000	19,600¶	19,500¶	17,712	17,790

Source: Meisinger, 1978-1979, 1980, 1981.
*Value f.o.b. mine and/or mill
† estimated
‡ revised
¶ preliminary

TABLE 3—Volcanic Material Sold and Used by Producers in the United States, By Use

Use	1978		1979		1980		1981	
	Quantity, 1000 st	Value, $1000	Quantity, 1000 st	Value, $1000	Quantity, 1000 st	Value, $1000	Quantity, 1000 st	Value $1000
Volcanic Cinder and Scoria								
Concrete admixture and aggregate	726	2,289	744	3,066	514	3,316	534	4,020
Landscaping	352	2,253	184	2,538	209	2,513	184	2,568
Railroad ballast	199	405	193	400	140	377	31	50
Road construction (including ice control and maintenance)	2,139	3,176	1,848	3,831	2,292	4,268	2,856	6,230
Other uses*	133	496	272	1,118	82	426	63	582
Total	3,549	9,619	3,241	10,953	3,236	11,258	3,667	13,400
Pumice and Pumicite								
Abrasives (including cleaning and scouring compounds)	15	448	28	649	27	568	19	486
Concrete admixture and aggregate	1,153	2,968	1,094	3,254	459	2,515	404	2,469
Landscaping	10	484	25	196	19	249	34	370
Other uses†	30	936	26	909	38	935	42	986
Total, Pumice and Pumicite	1,208	4,836	1,173	5,008	543	4,267	499	4,311

Source: Meisinger, 1978-1979, 1980, 1981.
* Includes absorbents, asphalt mix, roofing granules, and miscellaneous uses.
† Includes decorative building block, heat- or cold-insulating media, pesticide carriers, soil conditioners, roofing granules, and miscellaneous industrial uses.

TABLE 4—Pumice, Pumicite, Volcanic Cinder, and Scoria Sold and Used by Producers in the US, by State

State	1975		1976		1977		1978		1979		1980[†]		1981[†]	
	Quantity, 1000 st	Value, $1000	Quantity, 1000 st	Value, $1000	Quantity, 1000 st	Value, $1000	Quantity, 1000 st	Value, $1000	Quantity, 1000 st	Value, $1000	Quantity, 1000 st	Value, $1000	Quantity, 1000 st	Value, $1000
Arizona	856	1,294	802	1,240	621	1,226	1,135	3,130	940	2,367	990	3,228	1,088	3,189
California	348	2,762	705	3,245	636	3,838	831	3,458	800	3,973	568	3,159	770	4,462
Colorado	—	—	—	—	—	—	—	—	—	—	W	W	107	615
Hawaii	318	912	330	636	260	574	272	658	359	1,240	W	W	373	1,364
Idaho	111	187	W	W	—	—	—	—	—	—	—	—	—	—
Kansas	—	—	—	—	—	—	—	—	—	—	—	—	W	W
Montana	—	—	5	8	5	7	—	—	—	—	—	—	—	—
Nevada	W	W	388	763	656	1,154	706	1,282	W	W	—	—	—	—
New Mexico	397	1,280	486	1,560	457	1,835	631	2,706	604	3,550	448	3,028	538	3,810
Oklahoma	1	W	1	W	1	W	1	W	1	W	—	—	2	W
Oregon	1,470	3,937	1,125	2,311	1,083	2,429	915	2,016	781	1,644	1,090	2,734	878	1,547
Utah	17	23	164	264	W	W	28	270	28	280	35	347	—	—
Washington	—	—	—	—	W	W	50	63	—	—	—	—	—	—
Wyoming	—	—	—	—	—	—	7	W	—	—	—	—	—	—
Other states*	374	808	128	439	290	902	181	872	901	2,907	647	3,029	410	2,724
American Samoa	—	—	—	—	1	10	4	24	2	15	3	32	—	—
Total	3,892	11,203	4,134	10,466	4,010	11,975	4,761	14,479	4,416	15,976	3,781	15,557	4,166	17,711

Source: Adapted from Meisinger, 1978-1979, 1980, 1981.

*Colorado, Idaho, Kansas (1978), Nevada (1979), Oklahoma (value only), Utah (1977), Washington (1977), and Wyoming (1978 value only).

[†] Volcanic cinders and scoria reported separately after 1979 by US Bureau of Mines, but combined here.

TABLE 5—Pumice and Related Volcanic Materials: World Production, by Country
(1000 st)

Country*	1976	1977	1978	1979[p]	1980[e]
Argentina[†]	63	72	24	27	28
Austria: Pozzolan	13	10	10	9	8
Cape Verde Islands: Pozzolan[e]	17	17	NA	NA	NA
Chile: Pozzolan	109	175	201	242	275
Costa Rica[e]	1	1	2	2	2
Dominica: Pumice and volcanic ash[e]	120	120	120	120	120
Egypt[e]	(‡)	(‡)	(‡)	(‡)	(‡)
France: Pozzolan and lapilli	703	774	648	[e]650	660
Germany, Federal Republic of:					
Pumice (marketable)	2,422	1,928	2,294	1,579	1,440
Pozzolan	110	131	192	215	220
Greece:					
Pumice	441	626	827	692	695
Pozzolan	1,081	1,385	1,565	1,235	1,650
Guadeloupe: Pozzolan	220	209	220	220	220
Guatemala:					
Pumice	NA	NA	21	[r e]20	20
Volcanic ash	26	29	39	41	40
Iceland	2	8	9	27	26
Italy:					
Pumice and pumiceous lapilli[e]	[r]949	[r e]825	[e]860	[r e]940	880
Pozzolan[e]	[r]6,600	6,300	6,400	6,500	6,600
Martinique: Pumice	[r e]330	316	183	172	165
New Zealand	55	31	44	[e]44	45
Spain[§]	133	1,027	759	854	860
United States (sold or used by producers):					
Pumice and pumicite	906	1,178	1,208	[r]1,172	543 #
Volcanic cinder (including scoria)[‖]	3,275	2,832	3,553	3,243	3,215 #
Total	[r]17,576	[r]17,994	19,179	[r]18,004	17,712

Source: Meisinger, 1980.
[e] Estimated. [p] Preliminary. [r] Revised. NA Not available.
* Pumice and related volcanic materials are also produced in a number of other countries, including (but not limited to) Iran, Japan, Mexico, Turkey, and the USSR, but output is not reported quantitatively and available information is inadequate for the formulation of reliable estimates of output levels.
[†] Unspecified volcanic materials produced mainly for use in construction products.
‡ Less than 1/2 unit.
[§] Includes Canary Islands.
Reported figure.
[‖] Includes American Samoa.

architecture and for sculptured pieces. The low specific gravity of block pumice makes it feasible to use large boulders; pieces 3 to 4 ft diam can be hand-trucked to the site and no special support is necessary.

Substitutes are available for most or all uses of pumice, pumicite, and volcanic cinder, so a market exists only when a combination of cost, ease of handling and processing, and quality are favorable. Prices for crude or processed pumice usually are determined by negotiation.

Adverse economic conditions in the late 1970s reduced residential and industrial construction uses and slowed demand for natural lightweight aggregate. In the years 1974 through 1979 volume increased only 17%, although value rose 43%. Average prices for all volcanic aggregates rose from $2.32 per ton f.o.b. mine in 1974 to $3.46 per ton in 1979.

Lightweight naturally expanded volcanic aggregates appeared to be gaining a stronger foothold in competition with commercially expanded shales at the end of the 1970s. Greatly increased fuel costs forced some kilns to shut down; producers of naturally expanded materials promptly penetrated the new market.

A volcanic environment is necessary for sizable high quality deposits, and this pretty well confines the domestic supply and markets to the western United States. Table 4 shows production and value of pumice, pumicite, volcanic cinder, and scoria by state.

About 57,000 tons of pumice are imported, primarily from Italy and Greece. Table 5 shows world production of pumice by country.

References and Bibliography

Anon., 1977, "Pumice—A Dual Role in Industry," *Industrial Minerals,* No. 120, Sept., pp. 15–29.

Chesterman, C.W., 1957, "Pumice, Pumicite, Perlite, and Volcanic Cinders," *Mineral Commodities of California,* Bulletin 176, California Division of Mines and Geology, pp. 443–448.

Chesterman, C.W., 1966, "Pumice, Pumicite, Perlite, and Volcanic Cinders," *Mineral Resources of California,* Bulletin, 191, California Division of Mines and Geology, pp. 336–341.

Eckel, E.B., 1960, "Pumice and Pozzolan Deposits in the Lesser Antilles," Misc. Paper No. 6–400, US Geological Survey, prepared on behalf of the International Cooperation Administration by US Army Engineer Waterways Experiment Station, Vicksburg, MS, Sept., 80 pp.

Meisinger, A.C., 1974, "Pumice and Volcanic Cinder," *Minerals Yearbook,* Vol. I, US Bureau of Mines, pp. 1111–1115.

Meisinger, A.C., 1975a, "Pumice and Volcanic Cinder," *Minerals Yearbook,* Vol. I, US Bureau of Mines, pp. 1209–1213.

Meisinger, A.C., 1975b, "Pumice and Volcanic Cinder," *Mineral Facts and Problems,* US Bureau of Mines, pp. 871–880.

Meisinger, A.C., 1976, "Pumice and Volcanic Cinder," *Minerals Yearbook,* Vol. I, US Bureau of Mines, pp. 1159–1163.

Meisinger, A.C., 1977, "Pumice and Volcanic Cinder," *Minerals Yearbook,* Vol. I, US Bureau of Mines, pp. 763–767.

Meisinger, A.C., 1978, "Pumice and Volcanic Cinder," Preprint, 1978–1979, *Minerals Yearbook,* US Bureau of Mines, 5 pp.

Meisinger, A.C., 1980, "Pumice and Volcanic Cinder," *Minerals Yearbook,* Vol. I, US Bureau of Mines, pp. 663–667.

Meisinger, A.C., 1981, "Pumice and Pumicite," *Minerals Yearbook,* Vol. I, US Bureau of Mines, pp. 693–696.

Wagner, N.S., 1969, "Perlite, Pumice, Pumicite and Cinders," *Mineral and Water Resources of Oregon,* Bulletin 64, Oregon Dept. of Geology and Mineral Industries, pp. 222–228.

Wentworth, C.K., and Williams, H., 1932, "The Classification and Terminology of Pyroclastic Rocks," *Report of Commission of Sedimentation,* Bulletin 89, National Research Council, pp. 19–53.

Williamson, D.R., and Burgin, L., 1960, "Pumice and Pumicite," *Mineral Industries Bulletin,* Colorado School of Mines, Vol. 3, No. 3, May, 12 pp.

Pyrophyllite

B. E. CORNISH *

Most technical and statistical data published on pyrophyllite relating to production figures, uses, markets and sales, have in the past traditionally linked the mineral with talc and soapstone. This is due to several common superficial physical properties resulting in substitute uses and applications. However, these minerals are not similar in chemical composition and do not generally occur in a similar geological environment.

Pyrophyllite now stands on its own in the marketplace—particularly in ceramic applications—due to the combination of excellent heat shock and creep resistance properties of the mineral. The increase in pyrophyllite use in the field of ceramics indicates the continuing realization of these marked unique properties. Several new fields of ceramic applications, previously the preserve of high-alumina materials, have been developed on the intrinsic properties of pyrophyllite.

Pyrophyllite is an aluminum silicate with the molecular formula of $Al_2O_34SiO_2H_2O$. It would be more appropriate to group pyrophyllite with such materials as kyanite, diaspore, andalusite, and certain high-alumina clays in relation to its general applications (Stuckey, 1928). These materials all have good heat shock resistance properties at high temperatures and are widely used for the manufacture of refractories. For those reasons they are interchangeable in some applications. Pyrophyllite also has many uses in nonceramic applications, the most important to date being as a substitute for talc in fillers.

The major pyrophyllite mining operations are located in the US, Japan, Korea, Canada, and Australia, where the mineral is widely used as a raw material for general ceramic, mineral filler, and dilutant applications.

* Group Development and Exploration Manager, Steetley Industries Limited, Sydney, Australia.

Commercial Applications

Pyrophyllite was first used in the last century for lining stoves and fireplaces, and also in the carving and polishing of ornamental objects. The rise in industrial demand for new products and applications resulted in the introduction of pyrophyllite as an alternative, mainly for talc and soapstone. Industrial filler applications followed for paint, paper, cosmetics, rubber, pesticides, crayons and pencils, and bleaching soap.

The first use of pyrophyllite in ceramic applications was prior to World War II, in the manufacture of ceramic cladding tiles in the US. This was followed by the production of kiln furniture refractories and whiteware when pyrophyllite was offered by suppliers in North Carolina as an alternative to some local clays.

The largest tonnages of pyrophyllite used today are in refractory linings in steel ladles in Japan, as seen in Table 1. The consumption of pyrophyllite in Japan is second only to clay and chamotte among refractory raw materials used. Recently this technology has been adopted in Australia. Japanese and Australian practice has resulted in appreciable technical and economic benefits, particularly when pyrophyllite is used in combination with zircon. The potential growth for this application is substantial if US and European steelmakers follow.

Investigation and research into the properties of pyrophyllite determined the following main advantages of pyrophyllite-based ceramics: high corrosion resistance for molten iron, steel, and slags; good heat shock resistance; good indices of deformation under load and hot creep resistance (refractories); and increase in mechanical strength of whiteware products promoted by a better distribution of mullite in finished products (also obtained at lower firing temperatures than normally required in production of triaxial ceramics). White cement

TABLE 1—Consumption of Refractory Raw Materials In Japan, t

Fiscal year	Chrome ore	Magnesia clinker	Synthetic magnesia dolomite clinker	Dolomite clinker	Fused Alumina and sintered alumina	Bauxite	Synthetic mullite
1965	72,878	166,494	—	57,588	6,818	22,236	—
1966	82,230	204,474	—	62,023	8,986	31,839	—
1967	94,850	259,079	—	83,216	11,747	41,413	—
1968	73,112	242,738	—	72,412	17,210	59,641	—
1969	78,122	285,046	—	89,303	25,746	63,728	—
1970	84,602	327,196	49,468	64,987	36,207	65,451	24,428
1971	71,331	261,829	66,853	53,470	31,811	58,720	21,777
1972	86,235	311,995	74,928	56,778	45,244	77,767	33,977
1973	94,498	367,713	89,911	49,282	59,641	101,520	45,351
1974	89,860	375,488	79,475	39,930	76,262	96,710	46,684
1975	59,660	315,100	59,030	28,620	74,120	75,120	51,490

Fiscal year	Sillimanite group	Clay and chamotte	Pyrophyllite	Quartzite	Carborundum	Zircon and zirconia	Alumina
1965	18,639	803,896	345,091	154,260	10,750	7,143	9,763
1966	22,363	907,596	378,777	171,156	13,749	9,292	10,687
1967	28,907	1,145,738	478,930	206,542	19,736	17,287	14,866
1968	32,094	1,321,310	515,257	274,164	27,360	24,243	19,093
1969	28,207	1,362,362	589,024	287,168	33,757	37,471	22,546
1970	31,017	1,518,989	673,402	356,423	44,922	49,462	28,554
1971	17,470	1,227,790	628,105	303,015	42,058	52,005	26,089
1972	17,738	1,168,504	649,430	226,520	46,787	79,086	38,833
1973	15,189	1,204,117	700,160	250,180	57,796	106,980	46,890
1974	15,529	1,121,108	636,340	251,710	56,564	105,670	47,393
1975	15,528	839,960	515,600	211,700	48,420	69,600	42,860

Source: Nameishi, 1976.

clinker formulations based on pyrophyllite generally have an advantage of a low iron content when compared with batches made up from silica and clay mixtures. A lower temperature of clinker sintering of batches is possible due to the higher reactivity of the pyrophyllite. White road aggregate made from calcined pyrophyllite rock exhibits high polished stone values (PSV), high aggregate impact values (AIV), and high luminance factors (LF).

These intrinsic properties of pyrophyllite disclosed potential ceramics uses which subsequently have been commercially exploited in industrial countries around the world in contact refractories for lining steel ladles; kiln furniture for tunnel kilns; refractory mortars; ingredients for vitreous china, stoneware, chinaware, and wall tile bodies; manufacture of white cement; calcined pyrophyllite for use in white road-making aggregate; and refractories for use in furnaces, melting ceramic and enamel fritts.

Contact Refractories

In the late 1960s the Japanese steel industry successfully developed ladle lining refractories incorporating Japanese and Korean pyrophyllite. In the 1970s, after receiving confirmation of availability of sufficient reserves of similar grade Australian pyrophyllite, the Australian steel industry decided to fully evaluate refractories based on this material. The successful performance of trial linings in 250-t ladles was reported at the 2nd (1976) Industrial Minerals International Congress in Munich and laid the foundation for the introduction of pyrophyllite ladle bricks into the Australian steel industry. The industrial application of pyrophyllite in ladle linings over the past five years has confirmed two things: first, that linings made from pyrophyllite and pyrophyllite/zircon increase ladle life significantly; and second, that the use of the new refractories gave adequate ladle availability. This in turn saved substantial capital investment for ladle maintenance facilities and reduced overall manpower requirements. In terms of actual performance, these new refractories proved in service that the ladle life increases twofold when lined with pyrophyllite bricks and 4½ times when lined with pyrophyllite/zircon bricks, compared to previously used bloating clay bricks. This was achieved at 75% of the cost per ton of steel produced using the traditional bloating brick.

Refractory plants which had supplied bloating ladle bricks installed new equipment for shaping the pyrophyllite and pyrophyllite/zircon bricks. These are now produced by quick-action friction presses and replace the old and cumbersome process of plastic extrusion followed by a repress. The full details of this development and the introduction of the refractories into steel plant practice were presented at the AIME conference in Detroit (Pauline and Jones, 1979). In fact, this successful introduction of pyrophyllite-based refractories for the steel industry is a confirmation of a similar development which took place ten years ago in Japanese refractory practice (Nameishi, 1976).

Recently Australian Industrial Refractories Ltd., a subsidiary of the Broken Hill Proprietary Co. Ltd. (Australia's largest steel manufacturer), officially opened the Pyrophyllite Zircon Ladle Brick Plant in Newcastle, NSW, with an initial capacity of 27 kt/a of finished brick.

Kiln Furniture

Pyrophyllite initially was used in the preparation of kiln car furniture in the US between 1930 and 1940. Subsequently, with the major progress in development of tunnel kiln technology, pyrophyllite-based tunnel-kiln top car furniture has proven particularly economical in side-fired automated brick kilns. The US and more recently Australia (Kaiser Refractories, Ltd.) are currently major producers of kiln car furniture shapes.

Pyrophyllite is used as the coarse fraction instead of a grog for kiln furniture shapes such as slabs, saggers, blocks, and pins, and has been in continuous and successful use in western industrial countries and ceramic plants producing bricks and whiteware. Many advantages have been confirmed in this field of application, including improved life and greatly lowered maintenance costs; excellent heat shock resistance; low coefficient of thermal expansion (see Fig. 1); minimal shrinkage; high resistance to deformation under load; moderate to high PCE (30–32); direct firing of the material without preliminary grog preparation in the processing and manufacture of the kiln car-top furniture (which results in appreciable savings of energy and ancillary processing expenses); and excellent creep resistance when compared with zircon refractory materials (Fig. 2).

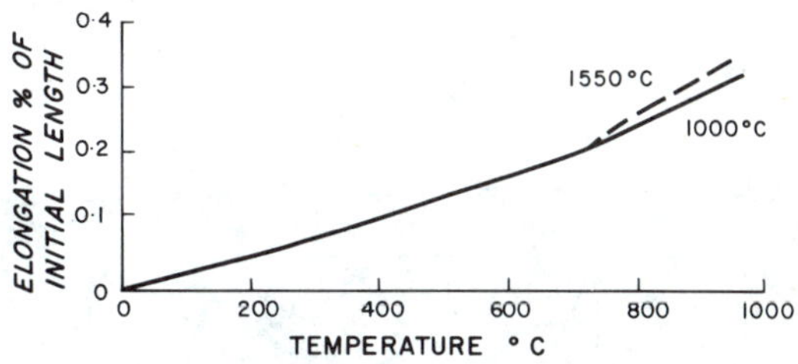

FIG. 1—*Coefficient of thermal expansion.*

Refractory Mortars

Formulations made from fine fractions of pyrophyllite are used for the manufacture of mortar. Such mortars have satisfactory plasticity but low drying and firing shrinkages when compared with clay-based compositions.

Whiteware

Pyrophyllite in stoneware and earthenware bodies including floor tiles, wall tiles, sanitary ware, crockery, and electrical porcelain is in commercial production in Japan and the US and is approaching general introduction to the industry in Australia and the Pacific region. For instance, a series of stoneware compositions was investigated and described in Australia (Taubar et al., 1973; Table 2) confirming early suggestions by Vanderbilt Co. in the US (Table 3). Pyrophyllite lowers the firing temperature, suppresses dunting and cracking, increases whiteness of the bodies, lowers firing shrinkage, and improves thermal shock resistance. Lower maintenance expenses produce cost savings in the slip house and extruding equipment.

White Cement

Pyrophyllite with a low iron content, when used as a component in the manufacture of white cement, assists in maintaining whiteness levels in excess of 90%. This was proven in preliminary laboratory tests where clinkerization processes were successfully achieved. Py-

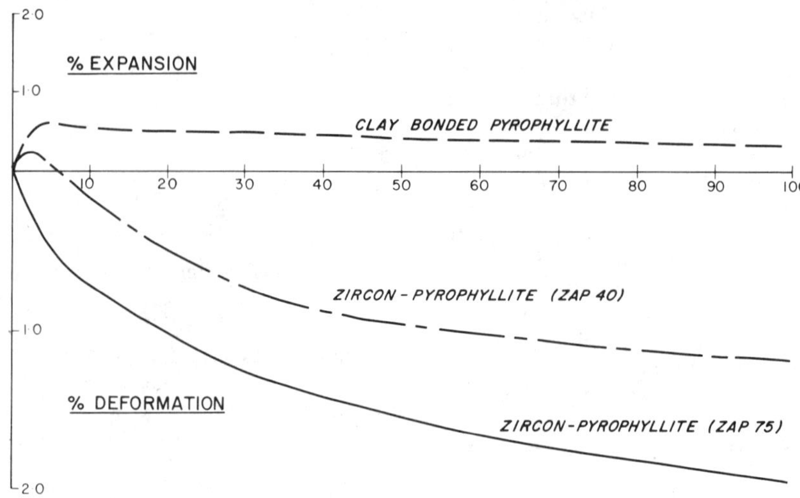

FIG. 2—*100 hr creep test, 1400°C, 2 kg/cm² load. Source: Australian Industrial Refractories, Ltd.*

rophyllite is used in the commercial production of substantial tonnages of white and off-white cement in Japan and is under consideration in Australia.

Safety Road Surface Aggregates

Research has been carried out on the application of calcined pyrophyllite in freeway center stripes, pedestrian crossings, intersection curves, malls, and highly reflective surfaces. The material shows many advantages after calcining the natural crushed rock, among them high antiskid properties, high polished stone value (PSV), high aggregate impact value (AIV), high luminance factor (LF), lower cost of natural crushed rock over synthetic materials, reduced road lighting costs, and excellent safety silhouette delineation in poor lighting conditions.

Melting Furnaces for Ceramic Glazes

An interesting new and successful use for pyrophyllite is the manufacture of ceramic glazes in melting furnaces. Trials in rotary and

TABLE 2—Chemical and Batch Composition of Pyrophyllite Bodies in Stoneware, Percentages

Body No.	SiO_2	Al_2O_3	Na_2O	K_2O	CaO	Pyrophyllite	Ball Clay	Nepheline Syenite	Potash Feldspar
1N	67.93	28.67	1.85	0.86	0.27	75	10	15	
2N	66.96	28.02	3.00	1.38	0.26	65	10	25	
3N	67.87	28.58	1.88	0.91	0.27	65	20	15	
4N	66.89	27.92	3.03	1.43	0.25	55	20	25	
5N	67.82	28.48	1.90	0.95	0.26	55	30	15	
6N	66.85	27.82	3.06	1.48	0.25	45	30	25	
7N	66.34	27.44	3.66	1.75	0.24	35	35	30	
8N	66.40	27.55	3.62	1.71	0.25	45	25	30	
1F	68.62	27.41	0.79	2.35	0.36	70	10		20
2F	68.33	26.30	1.12	3.45	0.39	60	10		30
3F	68.64	27.32	0.80	2.39	0.35	60	20		20
4F	68.29	26.20	1.14	3.52	0.38	50	20		30
5F	68.60	27.21	0.82	2.45	0.34	50	30		20
6F	68.24	26.08	1.16	3.59	0.37	40	30		30
7F	68.15	25.75	1.29	4.00	0.40	55	10		35
8F	68.06	25.53	1.33	4.14	0.39	35	30		35

TABLE 3—Typical Whiteware Bodies Containing Pyrophyllite

Product	1	2	3	4	5	6	7	8
Flint	10.0		20.0			5.0	8.0	10.0
Feldspar		52.0	3.0	35.0	33.0	20.0	33.0	9.0
HS *Pyrax* pyrophyllite	40.0	16.0	13.0	19.0	22.0	20.0	10.0	27.0
KY-TN Clay Co. No. 4 ball	18.0	7.0	16.0	8.0	10.0	20.0	9.0	10.0
Wade No. 5							10.0	15.0
No. 12 ball				8.0				
KY Clay Mining Co. No. 5 ball			16.0					
KCM					8.0	15.0	6.0	10.0
Cherokee GA kaolin	18.0		25.0	15.0				
Peerless SC kaolin		10.0			10.0	10.0	10.0	18.0
Edgars E.P.K. FL kaolin		15.0		15.0	10.0	10.0		
Harris *Lunday* NC kaolin					7.0		8.0	
Loomis, NY, talc	12.0		6.0					7.0
Whiting	2.0		1.0					
Moore and Munger No. 27, GA							6.0	

Source: R.T. Vanderbilt Co., North Carolina.
1. Wall-tile body, Cone 1 to 5.
2. Floor-tile body, Cone 9 to 10.
3. Earthenware body, Cone 6 to 8.
4. Dry-pressed insular body, Cone 11 to 12.
5. Plastic molded insulator body, Cone 10 to 12.
6. Plastic molded insulator body, Cone 11 to 13.
7. Sanitary-ware body, Cone 11 to 12.
8. Artware body, Cone 1 to 3.

continuous flake melting fritts and furnaces confirm that pyrophyllite/zircon refractories have better corrosion resistance in contact with molten glazes than the previously used aluminosilicate materials and shaped zircon refractories.

Other General Industrial Applications

Paints and Wallboard: Pyrophyllite requires the finest grinding or smallest particle size for use in paints and wallboard. In paint, it acts as low-priced extender of expensive pigments. In wallboard it has the effect of making the plaster flow more smoothly. It also is used as a filler in crack fillers for use with wallboard.

Insecticides: As a dilutant, extender, vehicle, or carrier in agricultural insecticides, finely ground pyrophyllite is excellent. Its practically neutral pH, inertness, good flowability, and nonhygroscopic and fluffy characteristics make it ideal for this purpose. It is compatible with both acid and alkaline poisons. It allows longer storage of prepared insecticides without loss of effectiveness. It readily settles and covers the foliage of the plants. In passing through the blowers of dusting machinery it picks up an electrostatic charge and, with the plants acting as lightning rods, it is attracted to the underside of the leaves, as well as settling onto the surfaces.

Miscellaneous: Finely ground pyrophyllite has also been used in soap, textiles, cosmetics, rubber, composition battery boxes, in welding rod coating, and in numerous other minor usages. An interesting use of South African pyrophyllite (wonderstone) is as a material in the manufacture of high-temperature furnaces for growing diamonds, where high temperature and creep resistance are of primary importance.

Geology

Mineralogy

Pyrophyllite is an hydrated aluminosilicate which has a phyllosilicate sheet or layer structure. This structure consists of two (Si_2O_5) nets between which are interposed a layer of groups of octahedral aluminum cations surrounded by oxygen and hydroxyl anions. The crystalline system of pyrophyllite is monoclinic.

The structural formula of pyrophyllite, $Si_4O_{10}Al_2(OH)_2$, also could be presented in the oxide form $4SiO_2 \cdot Al_2O_3H_2O$. The centidecimal theoretical pure composition of pyrophyllite is SiO_2, 66.7%; Al_2O_3, 28.3%; and H_2O, 5%.

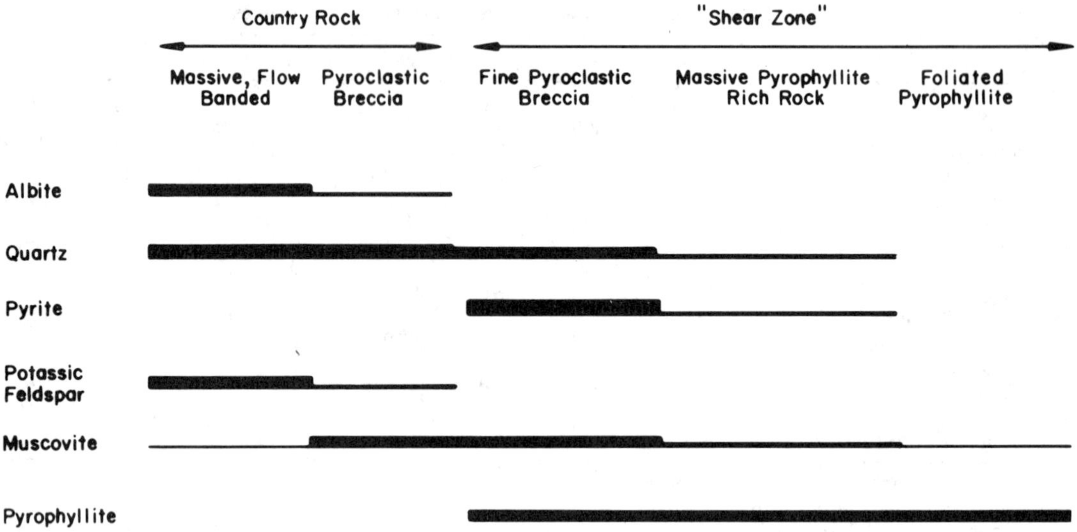

FIG. 3—*Relationship between structural features and mineral assemblages in regionally altered acid volcanics.*

Pyrophyllite ore bodies often contain the following characteristic mineral assemblage: diaspore, corundum, pyrophyllite, kaolinite, alunite, silica (quartz), sericite, and montmorillonite (Iwao, 1953). A diagrammatic representation of Iwao's concept of mineral segregation within hydrothermal deposits derived from volcanic rocks in Japan is given under "Classification of Deposits." The mineral assemblage at Pambula, NSW, Australia is somewhat similar: diaspore, chlorite-cookeite, pyrophyllite, kaolinite, chalcedony (quartz), and sericite.

Diaspore occurs commonly as a minor constituent in most pyrophyllite deposits. Less than 20% of the samples contain diaspore, with minor amounts of kaolinite in the massive pyrophyllite zones.

Chlorite-cookeite occurs near the central shear zone of the ore body where quartz and sericite are virtually absent. Appreciable quantities of chlorite and, to a lesser extent, diaspore are associated with the pyrophyllite. The chlorite has a dioctahedral x-ray diffraction pattern and is characterized by abundant aluminum and a small significant lithium content, and has been identified as the mineral cookeite ($Al_{3.93}Fe_{0.02}Mg_{0.02}Li_{0.81}Al_{0.74}Si_{3.26}O_{10}(OH)_8$).

Pyrophyllite occurs as a solid mass varying in color from green to yellow to white. It has physical properties similar to talc. The mineral has a low hardness and a good mechanical strength.

Kaolinite often occurs in associated shear zones and is probably developed during a retrogressive stage, possibly by the silicification of diaspore but more likely from cookeite through the loss of lithium.

Chalcedony appears throughout pyrophyllite ore bodies as in Australian and Korean deposits, with the exception of the pure pyrophyllite lenses ranging in size from large inclusions of quartz to small grains.

Sericite occurs in solid lensoid masses similar to pyrophyllite and can generally be separated by visual inspection. The percentage of alkalis in sericite lenses is in the range of 0.05 to 5%. Sericite is primarily a potash alkali.

Figs. 3 and 4 show the distribution of the minerals albite, quartz, pyrite, galena, potash, feldspar, muscovite, and pyrophyllite in relation to the degree of alteration taking place in the acid volcanics as described by the mapping of the main zones at Pambula.

Classification of Deposits

A twofold classification of pyrophyllite deposits is evident based on geological mode of occurrence.

Hydrothermal Deposits: These are the most common and are associated with acid-volcanic complexes. Hydrothermal solutions originating within the complex passed along channelways

ZONES RECOGNIZED	ZONES MAPPED	MINERALOGY	TRENDS	
CENTRAL PYROPHYLLITE ZONE	MASSIVE PYROPHYLLITE	Pyro. + Dias. + Cook	W.F. - S.F.	
	SILICEOUS PYROPHYLLITE	Pyro. + Minor Qtz.	W.F.	
sharp contact				
INNER SILICEOUS ZONE		Qtz. + Minor Pyro.	N.F.	
poorly defined contact				FeS Content Increases
MUSCOVITE ZONE	MUSCOVITE MILL ROCK	Pyro. + K'Mica. + Qtz.	M.F.	
		(K'Mica. + Qtz. + Pyro)	S.F.	
gradational contact				
OUTER SILICEOUS ZONE	SILICEOUS MILL ROCK	Qtz. + K'Mica. + Alb.	N.F. - W.F.	
	SILICEOUS FRAGMENTAL		N.F.	
gradational contact				
FELDSPAR ZONE		Qtz. + K'Spar. + Alb. + K'Mica. + Chl.		Decreasing Intensity
gradational contact				
RELATIVELY UNALTERED COUNTRY ROCK		Qtz. + K'Spar. Alb. + K'Mica. + Chl. Ep. + Pyrite + Galena		

FIG. 4—*Alteration zones and related mineralogy associations, Pambula, NSW. Foliation for "Trends": S.F. = Strong, M.F. = Moderate, W.F. = Weak, and N.F. = No.*

formed by faults or shears and transformed the country rock from rhyolite or feldspar-quartz to pyrophyllite-quartz. Hydrothermal alteration is indicated by dilution of alkali and iron oxide values and abnormally high lithium oxide values. The Pambula pyrophyllite deposit near Eden, NSW, is such an example.

Figs. 5 and 6 show diagrammatic representation of Iwao's (1953) concept of mineral segregation within hydrothermal deposits derived from hydrothermal rocks in Japan showing an hypothetical erosion surface and subsequent surface-exposed mineralogical relationship.

Metamorphic Deposits: These are less common and are found to be associated with metamorphosed volcanic ashes. Pyrophyllite-schist is developed and typically grades irregularly into other schistose rocks. In this case dilution of K_2O values is difficult to explain in terms of isochemical metamorphism. The recording of pyrophyllite-schist near Peak Hill, NSW, is an example of this metamorphic origin.

Principal Producing Countries

Japan: Pyrophyllite was first discovered in Japan in 1797 on Mount Omotoyama, at the site of the Mitsuishi mine of the Shinagawa Mining Co. In the early 1800s sawn blocks of pyrophyllite were used for carving images and in the manufacture of slate pencils, signature seals, and firebricks. In 1885 deposits in the Mitsuishi district were surveyed by the Imperial Geological Survey, and this work established the existence of abundant reserves of usable quality ore.

The most important pyrophyllite production is in Okayama. This prefecture is near the south coast of southwest Honshu Island, approximately 90 miles west of Osaka. Fifty-seven of the 70 mines in Japan are known to have produced various quantities of pyrophyllite. Of these 57 principal mines, 27 are located in Okayama prefecture.

The largest production from Okayama has been from the Shinagawa-Mitsuishi Ohira, Kawabe, and Kato mines. Significant production has also been recorded from the Shinagawa-Makata, Fukuyama, and Kiyotaki mines in Hyogo prefecture. The Shokozan mine in Hiroshima prefecture has also produced significant quantities of pyrophyllite. In Hiroshima and Hyogo prefectures, massive pyrophyllite deposits in rhyolite and porphyrite include the accessory minerals diaspore, kaolinite, corundum, pyrite, and alunite.

The second important pyrophyllite-producing area in Japan is in Kanzaki gun, Hyogo prefecture. Six mines are known in this area, which is located 70 miles northeast of Okayama.

Three important mines also exist in the Kinosaki gun, Hyogo prefecture. These mines occur between the Okayama and Kanzaki districts, in which the largest reserves of pyrophyllite in the world occur. In 1974, 3.5 Mt of reserves had been estimated in Okayama prefecture, with 33 Mt in Kanzaki gun, Hyogo

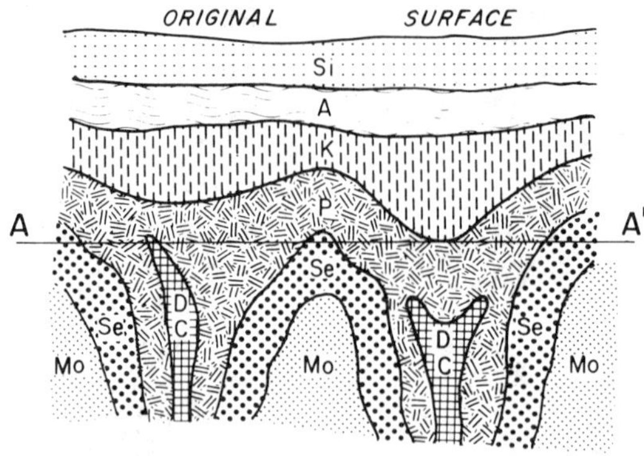

FIG. 5—*Diagrammatic representation of Iwao's concept of mineral segregation within hydrothermal deposits derived from volcanic rocks in Japan. Source: Modified from Iwao (1953).*

prefecture, and 17.5 Mt in Kinosaki gun, Hyogo prefecture.

Reserves of pyrophyllite in Japan were estimated in 1977 to be 179.4 Mt, of which 151.8 Mt was recoverable. In the periods 1965–1970 and 1975–1980 respectively, 92 and 108 pyrophyllite mines were in production in Japan.

The Japanese pyrophyllite deposits were formed by the hydrothermal alteration of porphyry and liparite. In Okayama prefecture pyrophyllite occurs as veins and irregular masses within sedimentary and volcanic rocks. Kaolin accompanies pyrophyllite within the upper parts of the ore bodies, with nodules of diaspore occurring in the central part of the ore mass, and pyrite and marcasite in the lower parts. The deposits were probably formed by the replacement of andesites, shales, agglomerates, rhyolites, and porphyries which were permeated by hot hydrothermal solutions associated with volcanism.

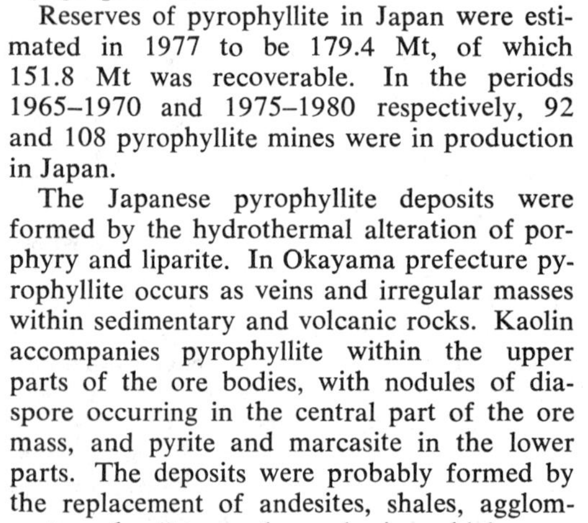

FIG. 6—*Diagrammatic representation of Iwao's concept of mineral segregation. D = diaspore, C = corundum, P = pyrophyllite, K = kaolin, A = alunite, Si = silica, Se = sericite, Mo = montmorillonite. A-A' is a hypothetical erosion surface.*

South Korea: Pyrophyllite is mined in two major areas of South Korea, both near the coast in the southern part of the peninsula. The largest production is from the Wan-Do mine at Haenam. Major mines in the Haenam area are the Sungsan, Hwansan, Okmesan, Dae-Do, and Chin-Do mines. The deposits mined are tabular, and are associated in places with diaspore.

The second major pyrophyllite-producing area in South Korea is at Dong Nae where the Nilyang, Yangsan, Kimhae, Pusan, and Kyongnam mines are worked. The deposits in the Dong Nae area were formed by the alteration of intermediate volcanic rocks including andesite, trachyte, trachytic andesite, and quartz porphyry. The host rocks of the Pusan and Kyongnam mines consist of quartz-porphyry and feldspar porphyry, while trachytic andesite is the host rock in the Dong Nae, Imgi, and Nilyang mines. Pyrophyllite in the Dong Nae area is thought to have been formed by hydrothermal alteration of volcanic rocks by granite intrusives of Cretaceous age, from which hydrothermal solutions filled fissures in tuffs and volcanics. The Yangsan and Un-Yang granite are thought to be the principal mineralizers in the Dong Nae area. In the Haenam area pyrophyllite deposits resulted from fissure-filling tuffs and rhyolites.

The major reserves of pyrophyllite in South Korea are primarily located at Wan-Do, where inferred ore reserves total approximately 30 Mt.

United States: The production of pyrophyllite in the US is limited to the states of North Carolina and California. Annual US production is approximately 100 kt, and reserves of various grades of pyrophyllite are estimated at 1.5 Mt.

Principal production in the US is in Moore and Randolph Counties in North Carolina. Other occurrences in North Carolina are in Granville, Orange, Alamance, and Montgomery Counties. The North Carolina belt actually starts near the Virginia line and extends across the state through South Carolina to Graves Mountain, Georgia. Other occurrences of pyrophyllite in the US have been reported in the South Mountain area of Pennsylvania and in California.

Pyrophyllite occurs in irregular, lenticular, or bedded deposits. The pyrophyllite deposits of North Carolina occur within rhyolite volcanic rocks and form oval or lenticular bodies up to 500 ft wide and several hundred feet long, located within fracture zones. According to Stuckey (1928) the North Carolina deposits were formed through metasomatic replacement

TABLE 4—Chemical Composition of Pyrophyllite-Bearing Rocks from Pioneer Mine Area, Percent

	A*	B*	C*	D*	E†	F
SiO_2	80.13	65.58	77.16	66.74	66.80	66.70
Al_2O_3	11.79	24.07	18.29	26.68	28.20	28.30
Fe_2O_3	0.28	0.38	0.16	0.44	0.14‡	
MgO	0.06	0.45	0.05	0.06	Tr	
CaO	0.74	0.30	0.10	0.05	Tr	
Na_2O	n.d.	n.d.	n.d.	n.d.	0.05	
K_2O	n.d.	n.d.	n.d.	n.d.	Tr	
H_2O	3.10	3.43	3.13	4.65	5.00	5.00
H_2O-	1.00	0.46	0.21	0.16	n.d.	
TiO_2	1.85	0.68	0.65	0.70	0.02	
P_2O_5	0.24	0.07	0.05	0.02	0.06§	
Total	99.19	95.42	99.80	99.50	100.27	100.00

* Eilun H. Kane, analyst.

†Reported by L. M. Richard, "Pyrophyllite in San Diego County, California," *Am. Ceramic Soc. Bull* (Oct., 1935) Vol. 14, p. 353.

‡Includes 0.04 % MnO.

§Reported as S; P_2O_5 not determined.

A. Silicified and slightly pyrophyllitized dacite.
B. Moderately pyrophyllitized dacite.
C. Pyrophyllite-quartz schist.
D. Pyrophyllite schist.
E. Pyrophyllite schist.
F. Theoretically pure pyrophyllite.

of Precambrian acid tuffs and breccias of both dacitic and rhyolitic composition. Pyrophyllite occurs with quartz and sericite, forming the central part of these lenses and grading outward into impure zones. The bulk of production comes from a few large deposits mined by open-cut and underground methods and used in ceramic filler dilutant and carrier applications.

The other district in the US where pyrophyllite has been produced commercially is in California. Sericitic pyrophyllite is mined in Inyo and San Diego Counties for ceramic applications. The San Diego County pyrophyllite deposit was described by Jahns and Lance (1950). An important deposit called the Pioneer mine occurs in the San Deiguito area northeast of La Jolla and 7½ miles from Escondido, the shipping point by rail. The deposits occur in the Santiago Peak volcanic series, which comprises miscellaneous rocks of Jurassic age. The pyrophyllite deposits, which occur within an area of about one square mile, represent progressive stages in the alteration of the volcanic rocks. Nearly all the pyrophyllite mined has been obtained from open pits in a single mass of high-grade schist 150 ft long and 15 ft wide. Chemical composition of typical material shipped from the Pioneer mine is given in Table 4.

Australia: The main pyrophyllite deposit in Australia is located at Pambula near the deep water port of Eden on the southeast coast of New South Wales. The stratigraphic sequence is described in Fig. 7. An intrusive and extrusive volcanic assemblage formed in an active tectonic environment during the Upper Devonian period provided the basis for extensive hydrothermal activity and burial metamorphism. This hydrothermal activity associated with the volcanism formed irregular and lenticular mineralization, rich in aluminous phases, by the remobilization and the dissipation of the alkali components of the host rock (Taylor, 1978). Suitability of this host rock and structural competence of the volcanic sheet control the geometry and distribution of the deposits, with optimal conditions favoring large, strongly cleaved, broadly altered lensoid bodies (Figs. 8, 9, 10, 11).

It is probable that the mineralization at Pambula originated by a similar process to the zonal distribution of mineralization in the Japanese deposits as explained by Iwao (1953). However, the segregation process of diaspore-pyrophyllite-kaolinite-alunite-silica was achieved

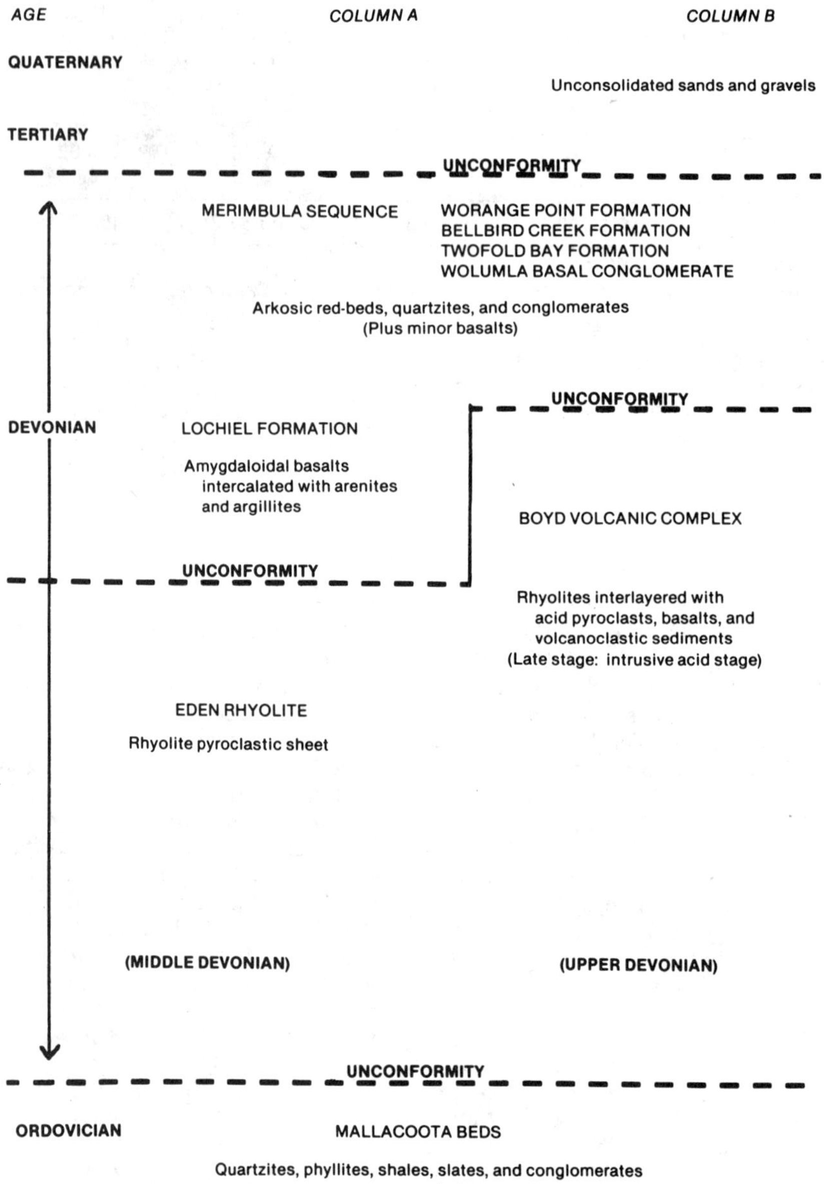

FIG. 7—*Stratigraphy of the Pambula region, NSW, Australia .*

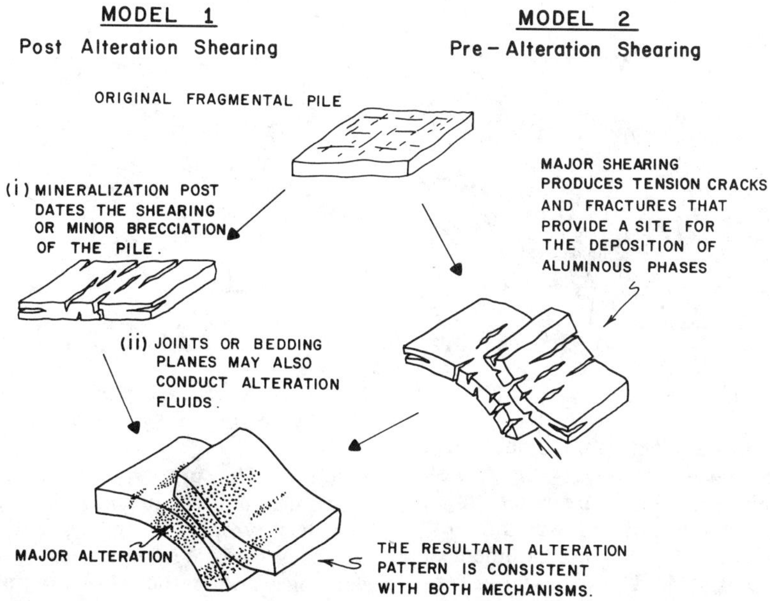

FIG. 8—*Mechanisms of alteration of the Back Creek Deposit, Pambula, NSW.*

through the alteration of lavas and pyroclastic rocks at shallow depth by ascending acid solutions of volcanic and subvolcanic derivation. During ascent the temperature decreased from 600 to 150°C and the solution became progressively more acidic, selectively mobilizing the alkalis which hydrated and laterally giving rise to zones enriched in sericite and montmorillonite. Simultaneously the residue (mainly alumina and silica) reacted with the components of the solution.

Recoverable reserves of all grades of pyrophyllite total approximately 10 Mt, with an estimated additional 20 Mt in place.

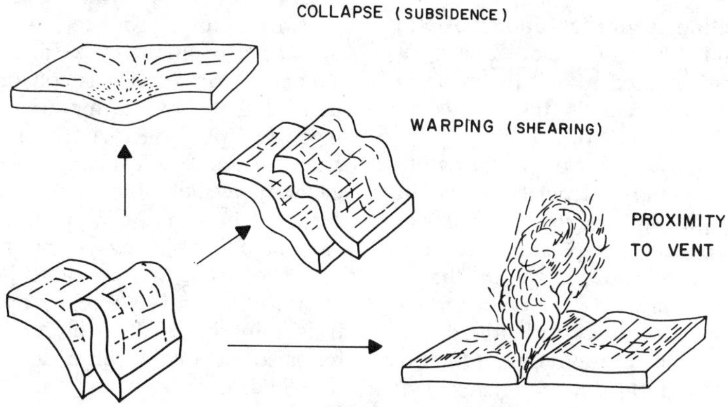

FIG. 9—*Pyroclast bedding irregularities.*

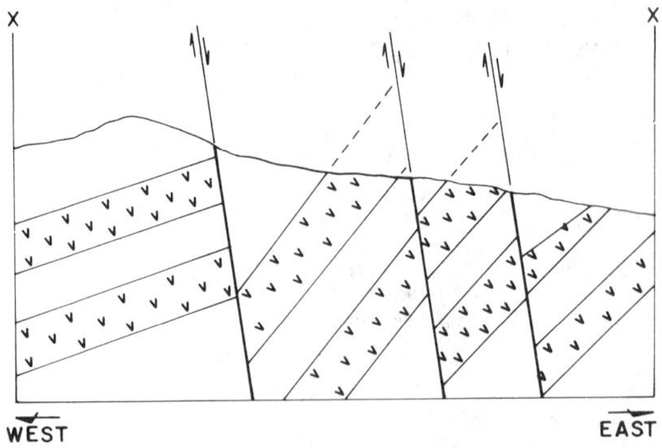

FIG. 10—*Step faulting.*

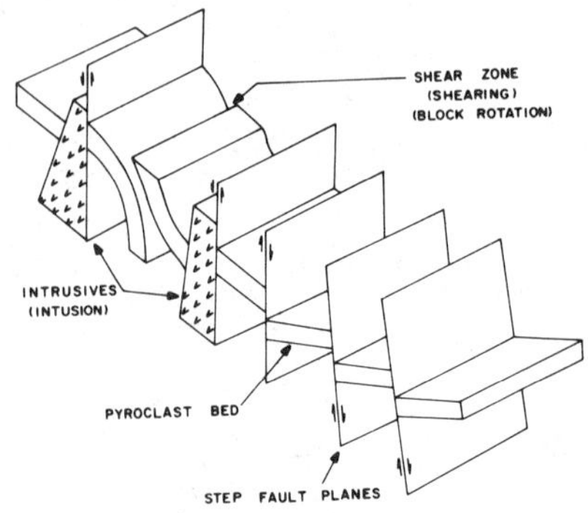

FIG. 11—*Structural features of a portion of the Back Creek Mine Sequence.*

Canada: The only current commercial source of pyrophyllite in Canada is on the Avalon Peninsula of Newfoundland, near the small town of Toxtrap. Pyrophyllite was first extracted in the area in 1904 when a shipment of 1,750 tons was sent to the US. Thereafter production continued intermittently, the most consistent period being 1938–1948 when Industrial Minerals Co. of Newfoundland mined and milled 6,500 tons. However, Newfoundland Minerals Ltd., a wholly owned subsidiary of American Olean Tile Co. Inc. of the US, acquired the operation in 1956. Since then production has been continuous.

The pyrophyllite occurs in a narrow belt running south of Conception Bay in eastern Newfoundland and along the east side of the Precambrian Holyrood granite batholith. It is thought that hydrothermal solutions, moving up via faults and shear zones from the granite, may have recrystallized rhyolitic rocks into the fine-grained assemblage of quartz, sericite, and pyrophyllite. Rhyolitic rocks remote from the granite showed no signs of pyrophyllite mineralization.

Current production from the 450- by 300-m open pit is around 30 to 40 kt/a. The mineralized material is crushed to —7 in. prior to shipping to American Olean's tile plants in Lansdale, PA and Jackson, MS.

A future Canadian source of pyrophyllite may well be the Senneterre deposit located in Carpentier Township, Abitibi region, Quebec, which has been actively prospected since the 1930s. It was not until 1962, however, that a mapping project conducted by the Quebec Department of Natural Resources identified the rock in the shear zone as pyrophyllite. Domtar Ltd. optioned the property between 1964 and 1974. Since then it has been in the hands of Descarreaux & Mousseau Tremblay Inc. of Quebec. The property, which is within the mineral-rich Precambrian Shield area of Canada, has pyrophyllite mineralization in a series of zones associated with sheared metamorphic volcanics and pyroclastics ranging in composition from andesites to rhyolites. Again, a metasomatic origin has been assigned to the deposit. The present owners hope further tests will prove pyrophyllite won from the ore is suitable for present markets.

Some occurrences of pyrophyllite have been reported in western Canada. For example, the northeastern portion of Kyuquot Sound, a small inlet on the west coast of northern Vancouver Island, has been described as a pyrophyllite area. In addition, Mountain Minerals Ltd. of Lethbridge, Alberta, has been actively investigating several pyrophyllite claims in western Canada.

Brazil: The Onca de Pitangui and Pitangui deposits in the state of Minas Gerais and the Santano de Parnaiba deposit in the state of Sao Paulo contain more than 1 Mt measured reserves of pyrophyllite (Table 5). Brazil's vast refractory producer, Magnesita SA, has an output of about 60 kt/a at Belo Horizonte. The entire production is consumed by the company, largely for the manufacture of ladle linings. Finapa Assessoria Commercial Industrial Ltda. of Sao Paulo claims to have a 100 kt/a capacity pyrophyllite mine which supplies the ceramics industry. Cia. Minera Agregados Calcareous SA of Lima, Peru, has been mining and processing a mineral similar in chemical composition and uses to pyrophyllite, although it is mineralogically different. The company has also isolated a large deposit of pyrophyllite—the only one on the Pacific coast of South America—which it hopes to exploit in the near future.

TABLE 5—Brazilian Pyrophyllite Reserves

State	Region	Quantity (tons)	
		Proven	Indicated
Minas Gerais		687,613	310,000
	Onca di Pitangui	434,555	310,000
	Pitangui	253,058	—
Sao Paulo		412,140	—
	Santana do Parnaiba	412,140	—
Total		1,099,753	310,000

Minor Producing Countries

India: Three deposits at Madhya, Aradesh, Rajasthan, and Uttar Pradesh produce a total of approximately 15 kt of pyrophyllite annually.

Argentina: Argentinan pyrophyllite is mined principally from the Juanita del Puerto mine in La Rioja province and is used in the manufacture of electrical porcelain and vitreous enamel. It produces less than 10 kt/a.

South Africa: Pyrophyllite, locally referred to as *wonderstone,* is mined near Ottosdal in the Lichtenburg district. The deposit is of the hydrothermal type and crops out in a series of elongate lenses associated with two lineaments which transgress rhyolite and acid types of the Dominion Reef series of the Witwatersrand System. Table 6 shows the chemical composition of *wonderstone* and indicates that it is not a pure pyrophyllite. The material mined in fact is a metasedimentary rock made up of pyrophyllite (89%), chloritoid, epidote, and rutile. The product, with an average chemical composition of 55.5% SiO_2, 35.24% Al_2O_3, 0.31% MgO, 2.8% TiO_2, and 6.15% loss on ignition, is different from a standard pyrophyllite in having a Mohs hardness of 8. Naturally, the end uses are therefore different, the main ones being bushes for submerged pumps, inserts for sluiceway nozzles, spigots for coal washing plants, and abrasion- and acid-resistant linings for pipes and coal chutes. Sales of wonderstone are handled by Ore and Metal Co. Ltd. of Johannesburg.

Swaziland has been producing small tonnages of pyrophyllite for some years now, particularly from the Usetu area between the Mkhondvo and Great Usutu Rivers. However, production is only on the order of a few hundred tons a year.

Thailand: Minor production of pyrophyllite occurs in central Thailand for general ceramic uses in Southeast Asia. The mining capacity of the Thailand operation is on the order of 12 kt/a. It is apparently limited.

Following is an analysis of the Thailand material.

SiO_2	67.8%
Al_2O_3	22.8%
Fe_2O_3	0.12%
K_2O	0.15%
Na_2O	0.15%
Loss on ignition	6.7%

Mineralogically it is a mixture of pyrophyllite, kaolinite, and diaspore according to X-ray diffraction analysis.

Turkey: Turkey has been a commercial producer of pyrophyllite since 1976 when Endustrie Mineralleri ve Taslari Sanayi (Industrial Minerals of Turkey Ltd.) started up its produc-

TABLE 6—Chemical Composition of Wonderstone

Sample Nos.	1.	2.	3.	4.	5.	6.
Silica (SiO_2)	56.76	55.50	56.71	56.84	54.96	54.20
Alumina (Al_2O_3)	32.39	34.30	33.27	33.57	34.03	34.94
Ferric oxide (Fe_2O_3)	1.54	0.82	1.48	0.75	0.86	1.06
Titanium oxide (TiO_2)	2.80	2.60	2.30	2.25	2.75	2.50
Lime (CaO)	0.25	0.16	0.14	0.07	0.18	0.12
Magnesia (MgO)	0.31	0.28	0.26	0.29	0.13	0.21
Loss on ignition	6.37	6.71	6.15	6.42	7.63	7.40
Total	100.42	100.37	100.31	100.19	100.54	100.43

TABLE 7—Pyrophyllitic Analyses, Turkey

Chemical Analysis	Massive Type, %	Natural Powder, %	Sericete Contaminated (High aluminous type)
Loss on ignition	4.26	5.60	4.23
SiO_2	65.88	61.54	42.70
Al_2O_3	26.39	28.59	43.48
TiO_2	1.00	0.60	0.60
Fe_2O_3	0.60	0.50	0.43
CaO	0.67	0.79	0.45
MgO	0.17	0.78	0.78
K_2O	0.33	0.33	4.66
Na_2O	0.70	1.27	2.67

tion in the Malatya Province of eastern Turkey. Approximately 6 Mt of pyrophyllite are contained in hydrothermally altered dacite tuffs found in lenses scattered over a wide area. Three main grades are mined: a green colored massive block variety; a white soft clay material; and a sericite-contaminated dark green pyrophyllite. The typical chemical compositions are given in Table 7.

Current production is around 3 kt/a, although there are plans for a substantial increase. The entire production is currently being consumed by Canakkale Ceramic Refractories, part owners of Industrial Minerals of Turkey, in its production of wall tiles and cordierite products, and as a natural grog for kiln furniture such as setters, slabs, pins, and posts. There are plans to produce pyrophyllite for the production of pyrophyllite/zircon ladle lining bricks. Later the market for pyrophyllite as a paper filler will be examined.

Exploration

Exploration programs for pyrophyllite are usually conducted in areas where geological evidence indicates that underlying rocks belong to acid-volcanic complexes. Pyrophyllite is found in hydrothermal alteration zones associated with faults or shears which transect these felsic-igneous rocks.

Field Techniques

Aerial photographs are most useful in reducing the area to be explored to a manageable size. This is achieved by plotting on the photographs linear features which may represent faults or shear zones. The geologist then examines each of these lineaments on the ground and undertakes reconnaissance work along favorable zones.

Pyrophyllite is generally identified in the field by its translucent green color and waxy feel. In addition, outcrops of chalcedonic pyrophyllite tend to be bolder than the enclosing rocks and display a characteristic weathering pattern, whereas outcrops of chloritic pyrophyllite are typically subdued and feature jagged boulders.

When a discovery has been made, drilling or trenching is performed to provide fresh samples for preliminary laboratory evaluation.

Geochemical

The pattern of alteration in pyrophyllite deposits is usually zoned, with a central core of chloritic pyrophyllite surrounded by imperfectly segregated zones of chalcedonic pyrophyllite and sericitic pyrophyllite which grade into unaltered host rocks. Geochemical techniques have been found to be successful in delineating the zonal pattern and defining the margins of the ore body.

Surface samples could be collected on a 10-m^2 grid basis. In areas of good outcrop conditions, clean representative samples can be obtained without difficulty. However, in areas of poor outcrop conditions shallow drilling or trenching may be necessary.

Contoured plots of alumina and total-alkali values are prepared and superimposed on a geological plan. The results not only define the zonal pattern but also provide a good basis for designing detailed drilling programs. As an example, in the evaluation of the Pambula deposit 3,000 samples, both surface and subsurface, were analyzed for potash alkalis and alumina values. These were used to prepare histogram plots (Figs. 12 and 13) and cumulative frequency distribution curves (Figs. 14 and 15). These plots confirmed the bimodal distribution of both Al_2O_3 and K_2O populations, probably representing both a primary and a secondary alteration phase and two periods of mineralization. The primary mode has a definite positive skew with the greater proportion of the mineralization showing alkali values at less than 0.5% K_2O. The mean values of K_2O and Al_2O_3 suggest a composition for the pyrophyllite mineralization occurring during the primary alteration phase of 12.7% Al_2O_3 and 0.57% K_2O.

Fig. 16 is a bivariate scattergram depicting grade lines and sample frequency distribution based on Al_2O_3 v. K_2O plots.

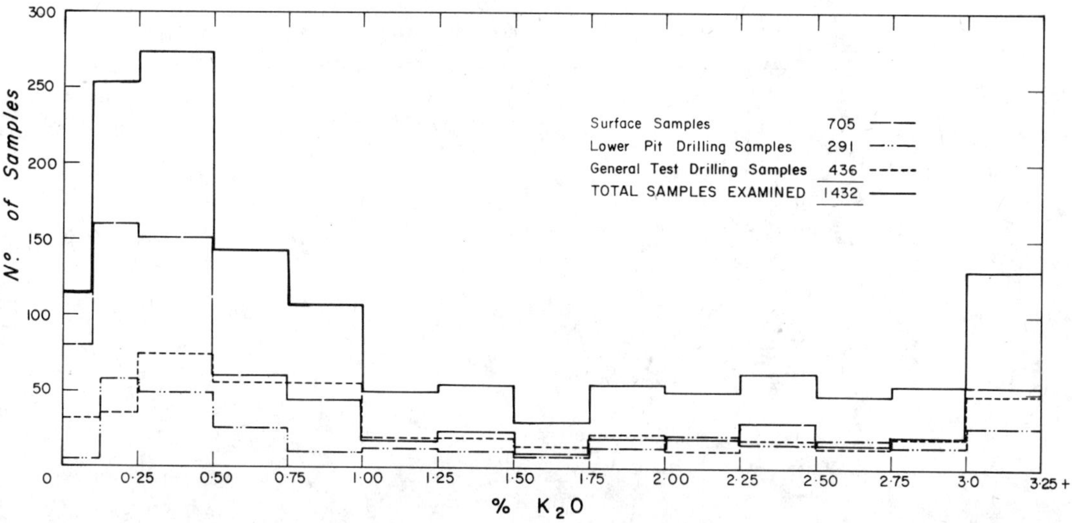

FIG. 12—*K_2O distribution curves—histograms.*

Drilling

Two kinds of drilling operations generally apply to pyrophyllite exploitation, diamond drilling and percussion drilling.

Diamond drilling provides core samples to enable the geologist to predict the subsurface behavior of the ore body and to find and correlate the zonal pattern determined from surface observations. Diamond drill holes are usually inclined at low angles and designed to penetrate the entire width of the steep dipping ore bodies.

Percussion drilling is carried out as a forerunner to production operations for detailed grade evaluation. Vertical holes are usually drilled on a square grid spacing of between 1 and 3 m and chip samples are collected at 1- or 2-m intervals to depths of up to 10 m. Compressed air raises drill chips to the surface

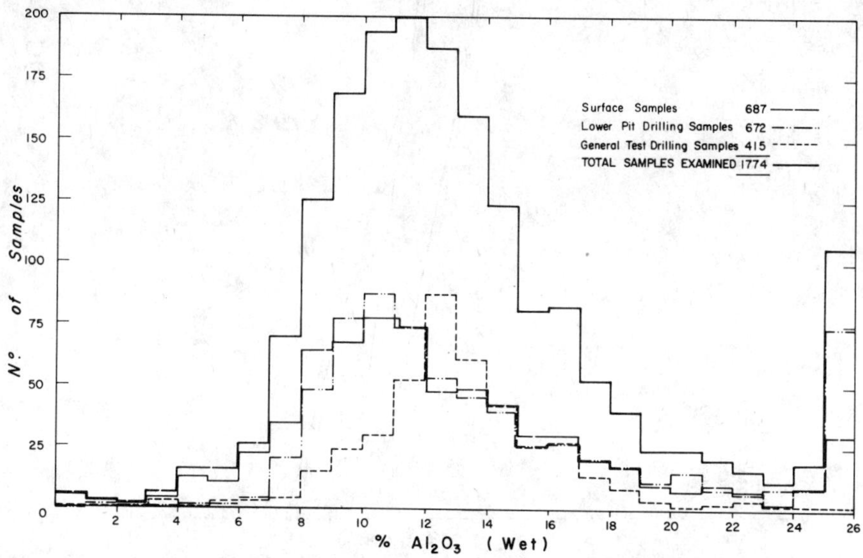

FIG. 13—*Al_2O_3 distribution curves—histograms.*

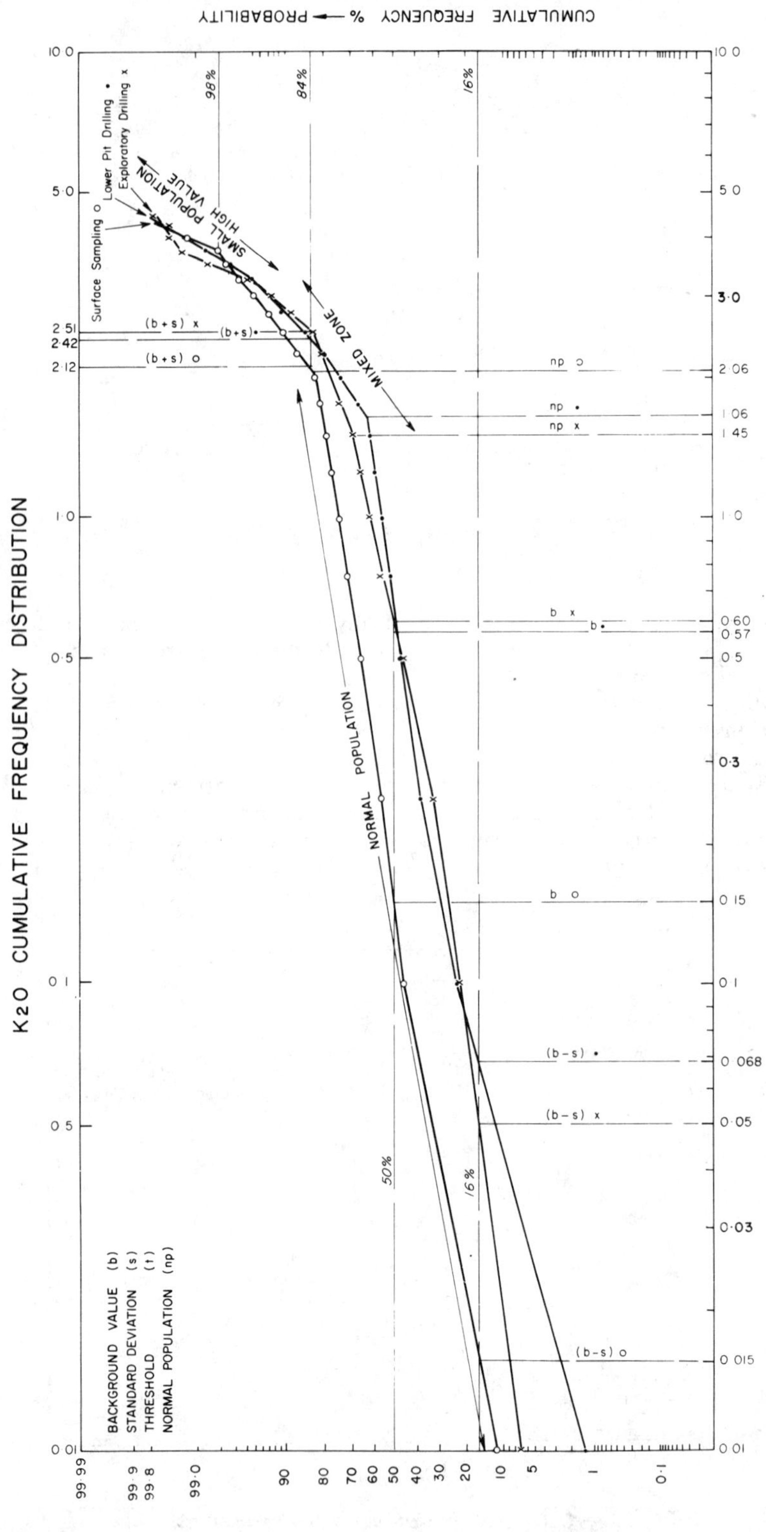

FIG. 14—K_2O cumulative frequency distribution.

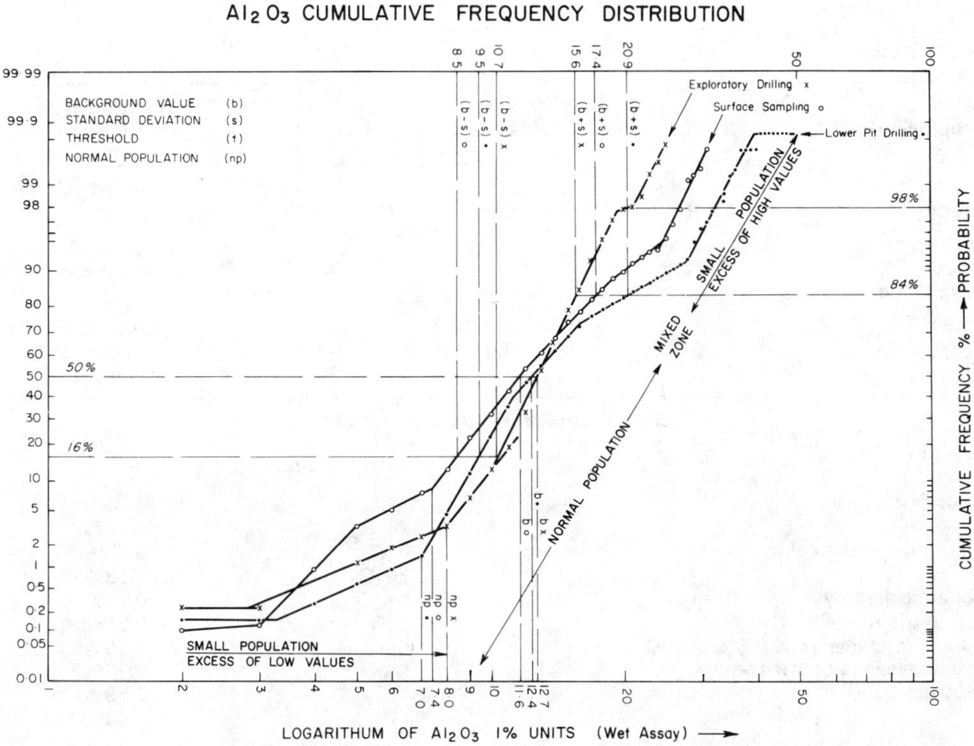

FIG. 15—*Al_2O_3 cumulative frequency distribution.*

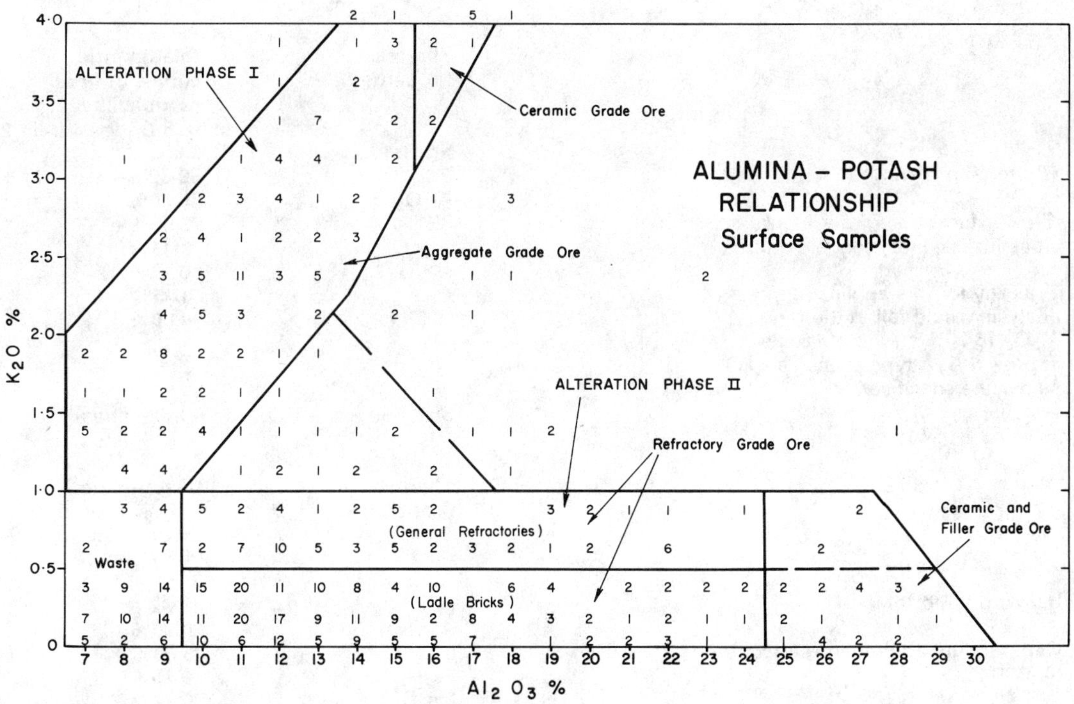

FIG. 16—*Bivariate scattergram depicting grade lines and sample frequency distribution.*

TABLE 8—Main Grades and Broad Specifications of Australian Pyrophyllite

Raw Material	Specification (%) Al_2O_3	K_2O	Fe_2O_3	Application
Chloritic pyrophyllite	24-28	0.2-0.5	0.2	Paint filler Plastics filler Rubber filler
Chalcedonic pyrophyllite*	13-15	0.2-0.5	0.4	Contact refractories
Chalcedonic-sericitic pyrophyllite*	15-20	0.4-0.8	0.5	Tunnel kiln furniture
Sericitic pyrophyllite	18-22	0.5-3.0	0.5	Whiteware ceramics, floor tiles, electrical porcelain

Typical Australian Product Specifications

	Specification	
Contact Refractories	PCE	28-30
	Al_2O_3	15.80%
Metallurgical and steel industries	SiO_2	78.50%
	Fe_2O_3	0.38%
The chemical analysis specification occurs in the range Al_2O_3 14-20% Na_2O and K_2O less than 0.5% and a typical analysis is as per the specification.	Na_2O	0.11%
	K_2O	0.24%
	CaO	0.30%
	MgO	0.12%
	TiO_2	0.39%
	LOI	3.90%
	Size Range	1.5 cm) ± 30# & fines 1.0 cm) 10#)
	Thermal Properties	Dilatometric curves of fired pyrophyllite PCE Cone 31 (Min 2)
	Sp Gr	2.65
Tunnel Kiln Furniture	PCE	30-32
	Al_2O_3	20.60%
Kiln car furniture, girder blocks, slabs, decks, pins, saggers, etc.	SiO_2	73.90%
	Fe_2O_3	0.45%
	Na_2O	0.12%
Typically for this application, the analysis would fall in the range Al_2O_3 18-20% and total alkalis approx. 1%. A typical analysis is as per the specification.	K_2O	0.44%
	LOI	4.30%
	Sp Gr	2.65
	Size Range	10# + fines 30# + fines
	Thermal Properties	Dilatometric curves of fired pyrophyllite PCE Cone 31
Refractory Mortars	PCE	28-32
	Al_2O_3	20.60%
General duty air setting refractory mortars.	SiO_2	73.90%
	Fe_2O_3	0.45%
	Na_2O	0.12%
	K_2O	0.44%

TABLE 8—Continued

Typical Australian Product Specifications—Continued

		Specification
	Size	10# + fines
Whiteware	PCE	16-33
	Al_2O_3 22.0%	28.20%
Applications stoneware, earthenware floor tiles, wall tiles, sanitary ware, electrical porcelain.	SiO_2 70.0%	65.80%
	Fe_2O_3 0.5% max.	0.11%
	Na_2O 0.5% max.	0.10%
	K_2O 2.0%	0.04%
	CaO 0.3% max.	0.01%
	MgO 0.2% max.	0.01%
	TiO_2 0.5% max.	0.18%
	LOI 4.0%	5.10%
	PCE Cone 16	33
	Sp Gr 2.70	2.80
	Screen Analysis:	
	−100# + 200#	Trace
	−200# + 300#	0.30%
	−300#	99.70%
	Thermal Properties: Dilatometric curves of fired pyrophyllite (PCE Cone 33).	
White Cement	Al_2O_3	15-20%
	Fe_2O_3	10-20%

For use in manufacture of *White Cement* a low iron pyrophyllite component can achieve the necessary specifications

Safety Road Surface Aggregate

Freeway stripes, pedestrian crossings, intersections, curves, malls, and general high reflective surfaces. The material obtains the following properties after calcining the natural crushed rock: high antiskid properties; high PSV; high AIV; high luminance factor; lower cost of using natural crushed rock over synthetic materials; reduced road lighting costs; excellent safety silhouette lighting.

Source of Silica and Alumina

A range of applications exist in the manufacture of fiberglass and asbestos cement.

Pyrophyllite may be a successful alternative to any application which requires either silica or alumina, or both, as raw material components.

Mineral Fillers	Al_2O_3	28.20%
	SiO_2	65.80%
Filler in rubber, plastics, paints, cosmetics, and general replacement for talc.	Fe_2O_3	0.22%
	Na_2O	0.10%
	K_2O	0.04%
	CaO	0.28%
	MgO	0.01%
	TiO_2	0.18%
	LOI	5.10%

TABLE 8—Continued

Typical Australian Product Specifications—Continued

	Specification	
	Brightness	80-90
	Water Absorption	50%
	Sp Gr	2.74
	Size range	Various
	Screen Analysis:	
	−300#	99.70%
	−200# + 300#	0.30%
	−100# + 200#	Trace

* Chalcedonic and chalcedonic-sericite pyrophyllite are often referred to as refractory-grade material and must be low in K_2O and Fe_2O_3. Requirements for this grade are most rigid.

and into a cyclone where they are removed from the airstream and discharged into a sample collecting tube.

Evaluation of Deposits

Specifications

A variety of grades is available from each deposit. Grade classification takes account of market requirements on the one hand and mineralogy of the ore on the other. Ore is graded according to its pyrophyllite mineral content and the nature of abundance of other minerals in the ore. Mineralogy is determined indirectly by chemical analysis, principally for Al_2O_3, K_2O, and Fe_2O_3 contents. The main grades and specifications of Australian pyrophyllite from Pambula are listed in Table 8 and described in detail. Analyses of pyrophyllite from North Carolina showing the relationship between chemical analysis, mineralogy, and grades are shown in Table 9.

Testing

Physical Tests: When pyrophyllite is being considered for refractory use, samples are tested for refractoriness (PCE) in firing trials at a range of temperatures using subsequent dilatometric methods.

Chemical Tests: Routine analysis of the major components is carried out on samples generated during both exploration and production phases.

Evaluation

Grades and reserves are calculated by conventional geological techniques and the results used to optimize quarry design.

Preparation for Markets

Mining Methods

Pyrophyllite is usually mined by conventional open-pit methods, employing the most modern types of mining, earthmoving, and bulk handling equipment. Some underground mining operations remain in production in North Carolina. Open-pit operations commence with clearing of trees and natural vegetation. Topsoil is removed and carefully stockpiled for use during site rehabilitation. Standard drilling and blasting techniques are used to establish and develop pit faces according to a predetermined quarry design. Pit faces usually range from 6 to 10 m high and benches vary from 4 to 8 m wide.

Ore blasted onto the quarry floor can be sorted not only according to size (because oversize rock must be redrilled and blasted), but also according to grade from preliminary visual sorting. Lump pyrophyllite ore is generally hauled to the processing plant by road trucks.

The following is a history of developed mining techniques as described by Chappell (1960).

Early mining methods consisted of conventional open cuts. As quality requirements and product control became more exacting, and as the pits were deepened, it was found necessary to go to underground operations, or to extremely deep open pits.

The oldest Glendon operations were started as open pits, later turned into underground mining, and later reverted to open pit operations, which are presently being operated by the Glendon Pyrophyllite Co. Standard Mineral Co. also has an open-pit operation at Glendon adjacent to the other operations.

The Standard Mineral Co.'s operations at

Robbins have been principally underground for a number of years. Standard also operates open pits in this same area. The underground operations consist of shafts, drifts, and crosscuts to reach the ore. The main drifts are driven in the footwall side of the deposit, with crosscuts through the ore body. The crosscuts are expanded to open square-set stopes. Broken ore is loaded from square-set stopes through chutes into mine cars and conveyed by battery locomotives to the shaft, where it is hoisted in a skip directly to the crushing plant. When extraction in a stope progresses near the level above, the stope is caved, in order to provide support to the pillars on either side.

The mine at Staley, known as the Gerhardt mine when first opened and currently being operated by the Carolina Pyrophyllite Co., was started with a tunnel and raise. The top of the raise was widened into a glory hole. As extraction progressed, this operation grew into a very large open quarry.

The Snow Camp mine operated by the North State Pyrophyllite Co. is operated exclusively as an open pit mine, the product going to North State's own refractory.

Processing

Many processing methods apply to pyrophyllite and each operation has a processing plant layout based on its individual ore characteristics and local market requirements. Nonetheless, the principal processing methods comprise crushing and screening to provide a series of size fractions.

Flowsheets from pyrophyllite processing plants of Steetley at Pambula, NSW, Australia, and North State Pyrophyllite at Hillsboro and Standard Mineral Co. at Robbins, North Carolina are shown in Figs. 17, 18, and 19.

Strict quality control is essential during processing because pyrophyllite ore commonly consists of a mixture of soft pyrophyllite mineral and relatively harder chalcedony and, during crushing and screening, the softer pyrophyllite tends to become concentrated in the finer size fractions.

Transportation

Pyrophyllite is a commodity of relatively low unit value and, therefore, the cost of transportation to market is an important consideration. Products are generally transported by road in bulk tipper-trucks and by rail and ship in bulk carriers. Often bulk milled pyrophyllite is freighted in road and rail tankers.

Consumption of Pyrophyllite

Most major present production and consumption areas lie in the Pacific Basin region where the principal consuming industries are located in the refractories and ceramics fields. The main exception is the industry in the northeastern US, which is dependent on supplies of pyrophyllite from North Carolina and Newfoundland.

TABLE 9—Analysis of Pyrophyllite from North Carolina, US

	Composition, %.						
Constituent	(1) *	(2) *	(3) *	(4) *	(5) *	(6) †	(7) ‡
Silica	76.32	73.50	70.26	69.90	69.38	83.34	77.22
Alumina	19.80	22.53	24.95	25.13	26.02	13.93	15.95
Iron	0.18	0.09	0.08	0.07	0.08	0.13	0.35
Calcium	0.14	0.08	0.16	0.16	0.14	0.42	Tr
Magnesium							
Potash	0.27		0.13	0.00	0.00	0.00	3.54
Soda	0.07	0.06	0.31	0.08	0.24	0.18	0.93
Loss	3.44	3.95	4.32	4.67	4.50	2.20	2.08
Total	100.22	100.21	100.21	100.01	100.36	100.24	100.07
Pyrophyllite	66	79	85	88	89	48	5
Quartz	31	21	12	11	9	51	57
Sericite	3		3	1	2	1	38

* Pyrophyllite
† With granular quartz
‡ Sericite

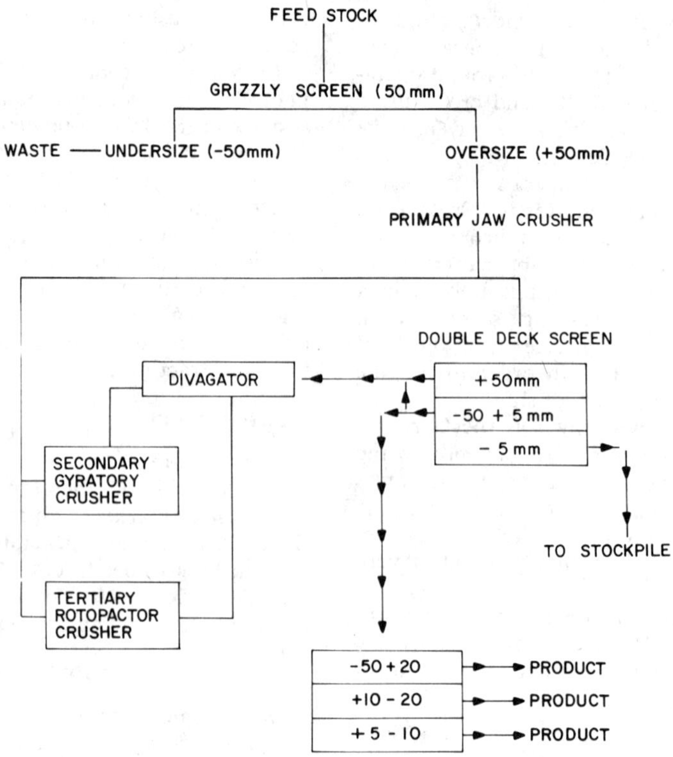

FIG. 17—*Flowsheet for the crushing and screening of pyrophyllite from Pambula pyrophyllite deposit, Australia.*

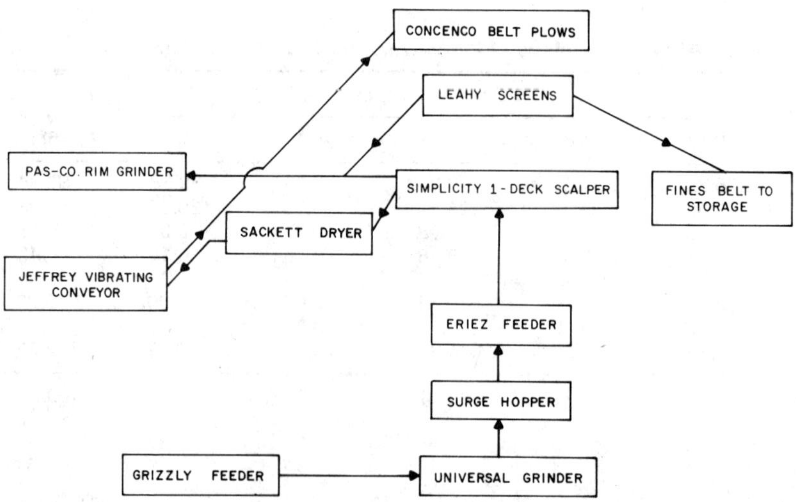

FIG. 18—*Flowsheet of automated grinding plant at North State Pyrophyllite Co., Greensboro, NC.*

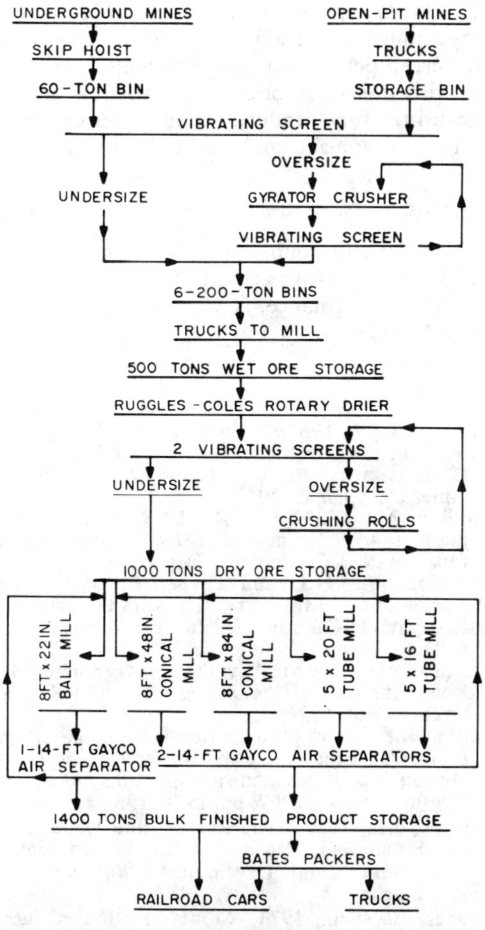

FIG. 19—*Flowsheet, Standard Mineral Co., Inc., Robbins, NC.*

Separate production statistics for pyrophyllite have seldom been available, because production figures are commonly incorporated in the statistics for talc. Nonetheless it is evident that Japan is the leading producer with about 70% of world production, followed by South Korea and the US with about 14 and 6%, respectively, of world production. Four countries—Australia, Brazil, Canada, and India—produce between 1 and 5% of world production, and five countries—Argentina, Peru, South Africa, Thailand, and Turkey—produce less than 1% of world production. World production is shown in Table 10.

Pacific Markets for Pyrophyllite

The production of pyrophyllite has shown steady increases over the last decade and world consumption now totals almost 2 Mt/a. Japan is the world's leading producer and consumer of pyrophyllite and approximately 1.5 Mt/a are used for various refractory and ceramic applications.

In Japan, recoverable pyrophyllite ore reserves in 1978 stood at 60 Mt. However only a very limited percentage of these reserves are suitable in the application of refractories to the steel industry. Thus Japan is becoming increasingly dependent on the external supply of higher quality pyrophyllite with the consequence that imports of Korean pyrophyllite are running in excess of 150 kt/a.

Korean pyrophyllite production approached 400 kt/a in 1980, placing Korea among the top three world producers.

TABLE 10—World Production of Pyrophyllite, 1971-1976 (t)

	1971	1972	1973	1974	1975	1976
Argentina	7,263	8,325	6,313	5,412	8,357	NA
Australia	1,577	8,288	7,804	11,478	13,589	10,320
Brazil	6,356	33,136	40,058	45,813	65,778	NA
India	11,780	15,086	14,912	15,562	14,994	14,807 *
Japan	1,434,286	1,365,131	1,372,834	1,396,184	1,073,111	NA
South Korea†	142,335	163,396	304,842	238,418	223,000	349,000
Peru	4,528	8,944	8,343	10,968	NA	NA
Thailand	50	1,550	9,550	1,640	10,300	NA
Turkey	—	—	—	—	—	3,000
South Africa‡	3,333	2,059	4,743	8,510	6,782	5,784
Swaziland	204	108	126	36	—	NA
USA	91,653	NA	NA	95,892	83,364	NA

Source: Institute of Geological Sciences, Mineral Statistics and Economics Unit, London, U.K.
Canada averages 30 to 40 kt/a.
NA—not available.
* Jan.-July only.
† Includes the following output of *agalmatolite* (a massive pyrophyllite): 1973-224 kt; 1974-205 701 t; 1975-196 239 t; 1976-108 025 t.
‡ Wonderstone.

Production and sales of pyrophyllite in the US have remained fairly static over the period 1970 and 1980 at approximately 100 kt/a. The main production is primarily limited to five mining operations in North Carolina with intermittent production in California.

Substitutes

High alumina products are not necessarily the best choice for many refractory applications as previously had been perceived. Pyrophyllite brick linings containing approximately 15% Al_2O_3 are often superior to high alumina clay linings in certain applications. The superior creep performance at high temperatures of pyrophyllite refractories may make it a suitable substitute for products manufactured from kyanite, andalusite, and calcined bauxite.

Future Considerations and Trends

The use of pyrophyllite in its industrial application is increasing and the mineral stands on its own performance. New uses will continue to be found.

Bibliography

Bowman, R., and Tauber, E., 1979, "Potential Applications of Two High Grade Kaolinitic Clays and a Pyrophyllite Deposit," *Journal Australian Ceramic Society,* Vol. 15, No. 1, pp. 5–8.

Chappell, F., 1960, "Pyrophyllite," *Industrial Minerals and Rocks,* 3rd ed., J. L. Gillson, ed., AIME, NY, pp. 681–686.

Cornish, B.E., 1980, "Australian Pyrophyllite and Its Growing Influence in the World Markets," 4th Industrial Minerals International Congress, Atlanta, Georgia, pp. 179–184.

Cornish, B.E., Tauber, E., and Nichol, D., 1980, "Pambula Pyrophyllite: Production and Applications," *Proceedings,* 9th Australian Ceramic Conference, pp. 132–134.

Fergusson, C.L., et al., 1979, "The Late Devonian Boyd Volcanic Complex, Eden, N. S. W.: A Reinterpretation," *Journal Geological Society Australia,* Vol. 26, pp. 87–105.

Harben, P., 1978, "Pyrophyllite Stands on Its Own," *Industrial Minerals,* No. 129, June, pp. 19–25.

Iwao, S., 1953, "Ceramic Minerals of Japan," Society Mineral Geology, Japan, Vol. 1, No. 2, pp. 105–124.

Jahns, R.H., and Lance, J.F., 1950, "Geology of the San Dieguito Pyrophyllite Area, San Diego County, Calif.," Special Report 4, California Div. Mines, 32 pp.

Jolly, W.T., and Smith, R.E., 1971, "Degredation and Metamorphic Differentiation of the Keweenawan Tholeiitic Lavas of Northern Michigan, U. S. A., *Journal of Petrology,* Vol. 13, Part 2, pp. 273–309.

Lipman, P.W., et al., 1972, "Cenozoic Volcanism and Plate Tectonic Evolution of the Western United States," Phil. Trans. Royal Society, London, A., pp. 217–248, 271.

Loughnan, F.C., and Steggels, K.R., 1976, "Cookeite and Diaspore in the Back Creek Pyrophyllite Deposit near Pambula, N. S. W.," *Mineralogical Magazine,* Vol. 40, pp. 765–772.

Nameishi, N., 1976, "Recent Status of Steel Plant Refractories in Japan," XIX International Refractory Colloquium, Aachen, Germany, pp. 209–271.

O'Brien, R.J., and Tauber, E., 1976, "The Uses of Zircon in Steel Ladle Linings," 2nd Industrial Minerals International Congress, Munich, Germany.

Parsons, W.H., 1969, "Criteria for the Recognition of Volcanic Breccias: Review," *Memoir 115,* L. Larson et al., ed., Geological Society of America, pp. 263–304.

Pauline, M., and Jones, R., 1979, "Zircon-Pyrophyllite—An Economical Steel Ladle Refractory," AIME Conference, Detroit, March.

Pepplinkhouse, H.J., and Tauber, E., 1976, "The Formation of Mullite from Kaolin-Gibbsite and Pyrophyllite-Gibbsite Mixtures," *Journal Australian Ceramic Society,* Vol. 13, No. 1.

Phillips, B.L., 1976, "Synthetic Aggregates for Road Surfaces," *Proceedings,* 8th ARRB Conference, Vol. 8, No. 1, p. 1.

Smith, R.E., 1968, "Redistribution of the Major Elements in the Alteration of Some Basic Lavas During Burial Metamorphism," *Journal of Petrology,* Vol. 9, Part 2, pp. 195–219.

Stuckey, J.L., 1928, "The Pyrophyllite Deposits of North Carolina," North Carolina Department of Conservation and Development *Bulletin,* Vol. 37, 62 pp.

Tauber, E., et al., 1973, "Stoneware Bodies Based on Pyrophyllite," *Journal Australian Ceramic Society,* Vol. 9, No. 2, pp. 47–51.

Tauber, E., et al., 1974, "Earthenware Tile Body Based on Pyrophyllite and Orthophosphoric Acid," *Journal Australian Ceramic Society,* Vol. 19, pp. 46–49.

Tauber, E., et al., 1976a, "Development of a China (Porcelain) Body from Australian Minerals," *Interceram,* Vol. 26, p. 211.

Tauber, E., et al., 1976b, "Replacement of Silica by Pyrophyllite in Vitreous China Products," *Interceram,* Vol. 25, p. 195.

Tauber, E., and Pepplinkhouse, H.J., 1972, "Ceramic Properties of Pyrophyllite from Pambula, N.S.W.," *Journal Australian Ceramic Society,* Vol. 8, No. 3, pp. 62–64.

Taylor, J.R.P., 1978, "Pambula Eden Regional Geology," Unpub. Thesis, Monash University, Melbourne, Australia.

Wall, V.J., 1976, "Gold and Pyrophyllite Mineralization in the Devonian Acid Volcanics of the Yalwal-Eden Belt," *Bulletin,* Australian Society Exploration Geophysics, Vol. 7, No. 1.

Wall, V.J., and Kesson, S., 1969, "Pyrophyllite-Bearing Rocks in a Regionally Altered Volcanic Sequence," *Abstracts,* Part 7, Geological Society Australia, 232 pp.

Rare Earths and Thorium

SPENCER S. SHANNON, JR.*

The lanthanide elements from lanthanum (atomic number 57) to lutetium (71) plus yttrium (39) are called the rare-earth elements; scandium (21) is chemically similar to yttrium and the lanthanides, but differs in occurrence. The rare-earth metals were divided into three categories by the chemist, Berzelius, in the early 1800s; however, modern usage defines only two groups. The cerium group is named after the most abundant element and comprises lanthanum (57), cerium (58), praseodymium (59), neodymium (60), promethium (61), samarium (62), europium (63), and gadolinium (64). Traces of natural promethium were discovered in apatite in Finland (Erametsa, 1965). The yttrium group, also named after its most abundant representative, comprises terbium (65), dysprosium (66), holmium (67), erbium (68), thulium (69), ytterbium (70), and lutetium (71). Yttrium most closely resembles holmium in chemical properties, and occurs most commonly with the heavy lanthanides. The rare-earth minerals may be divided into cerium-group, yttrium-group, or mixed-group classes.

A number of rare-earth minerals contain thorium and uranium, but these are variable in content and are not essential to the composition of the minerals. In monazite, a mineral that is a principal source of rare earths, the worldwide average content of thorium oxide is 7.2%; however, varieties of monazite contain as little as 0.001% and as much as 31.5% thoria. Monazite is the chief ore of thorium and, until recently, thorium has been the main product; the rare-earth metals were in fact byproducts. However, because of the great demand for rare-earth metals by the petroleum refining and the iron and steel industries since the early 1970s, the roles have been reversed so that thorium is now the byproduct.

* Staff Geologist, Los Alamos National Laboratory, Los Alamos, NM.

Uses

A 9 to 1 mixture of yttrium and thorium oxides is used for high-intensity incandescent lamps and for high-temperature-resistant windows and lenses. Yttrium oxysulfide and orthovanadate, doped with europium oxide, are used in color television tube phosphors. Yttrium, as well as cerium, praseodymium, and samarium, is alloyed with cobalt to form strong permanent magnets. It hardens magnesium alloys and retards the oxidation of iron-chromium alloys. Yttrium-iron garnets are used for transmitting low microwave frequencies. Yttrium-aluminum garnets are used as host crystals for lasers and as artificial gems. Yttria also forms strong, stable, high-temperature refractories. Gadolinium-gallium garnets are used in substrate wafers on which bubble-memory films are deposited.

Rare-earth chlorides, particularly of lanthanum, neodymium, and praseodymium, are used as catalysts for petroleum cracking. Mischmetal is used for lighter flints and to improve the ductility and impact strength of iron and steel alloys. Mischmetal also reduces the fatigue resistance of magnesium and aluminum alloys. Rare-earth oxides are used for polishing glasses and lenses. Individual or specific combinations of high-purity oxides are used as additives in glass to maintain or change its color, dispersion, or refractive index. High-purity oxide combinations are also used to color tile and to opacify porcelain enamel. Rare-earth fluorides and oxides are used in carbon-arc electrodes to provide incandescent white light. Lanthanum or gadolinium oxysulfide phosphors, activated with terbium, are used as x-ray screen intensifiers. Hydrogen can be stored by sorption on lanthanum-nickel alloys (Anon., 1979).

Thorium^{232} is bombarded with slow neutrons in breeder reactors to create fissionable thorium^{233} for nuclear fuel cells. The 330-MW, high-temperature gas-cooled reactor at Fort St.

Vrain, CO, is an example of this process in operation. An experimental light-water breeder reactor at Shippingport, PA, uses a thorium-uranium^{233} fuel system to generate electricity.

The chief nonenergy use for thorium is in the manufacture of Welsbach incandescent gas mantles. From 20 to 40% thorium is used in thorium-magnesium hardener alloys for making magnesium alloys containing 2 to 3% thorium. Addition of small amounts of thorium oxide to metals such as nickel and tungsten increases their strength and resistance to corrosion.

Product Specifications

Commonly, monazite concentrates must contain at least 55% rare-earth oxides. The specifications for particular rare-earth metals and oxides depend upon their intended uses, complicated by the cost and difficulty of totally separating specific elements. Requirements for production chemicals are comparatively low, but standards are more stringent for specific purposes. Rare-earth oxides, from which most of the cerium oxide has been removed, are referred to commercially as didymium oxide. Mischmetal, a mixture of rare-earth metals, can be cerium-free or contain as much as half cerium.

Thoria or thorium for use in atomic reactors must have a purity of 99.9%. Furthermore, either must contain less than 1 ppm of boron, cadmium, dysprosium, europium, gadolinium, and samarium, which absorb high-energy neutrons. Standard metallurgical-grade thoria must contain at least 99.8% ThO_2. The standard specification for high-purity yttrium is 99.9%.

Prices

In 1980, the nominal price for monazite concentrates was $0.34 per lb of contained rare-earth oxides. Calcined bastnaesite concentrates sold for $0.90 per lb of contained rare-earth oxides. Prices of high-purity rare-earth oxides and metals in 1979 are shown in Table 1. Thorium oxide was quoted at $17.27 per lb as of Jan. 1, 1981 (Kirk, 1980b).

TABLE 1—Prices of High Purity Oxides and Metals in 1979*

Element	Oxide, $/kg	Metal, $/kg
Cerium	18	108
Dysprosium	100	270
Erbium	120	450
Europium	1,650	6,500
Gadolinium	120	430
Holmium	375	1,100
Lanthanum	17	108
Lutetium	4,200	13,200
Neodymium	65	250
Praseodymium	110	290
Samarium	110	280
Terbium	825	2,000
Thorium†	26	33
Thulium	2,650	7,000
Ytterbium	180	720
Yttrium	74	320

* Moore, 1980a.
† Kirk, 1980a.

Geology

Mineralogy

Rare-earth and thorium minerals occur chiefly in association with alkalic plutons and in placers derived from them. Commonly, cerium- or yttrium-group lanthanide elements partially or wholly replace calcium in minerals such as fluorite and apatite.

Discussion of specific minerals will be restricted to those that have been mined for their rare-earth or thorium content in the past. Other exotic minerals are cited by Parker and Baroch (1971).

Monazite (Ce-La-Nd-Pr phosphate) and bastnaesite (Ce-La-Nd-Pr fluorcarbonate) are the chief minerals in rare-earth and thorium ores. Other ore minerals of local significance include yttrofluorite and two phosphates, xenotime and apatite; two silicates, cerite and gadolinite; and various multiple oxides such as betafite, brannerite, euxenite, fergusonite, and samarskite. Although thorite and thorianite, the silicate and oxide of thorium, contain more thorium than monazite, they are less important commercial sources of that metal.

Monazite contains rare earths of the cerium group, plus an average of 7.2% thorium and minor yttrium. It is a yellowish to reddish brown monoclinic mineral having both a hardness and a specific gravity of 5. It occurs in commercial concentrations in beach and stream placers and in lesser amounts in veins. It also occurs as an accessory mineral in igneous and metamorphic rocks.

Bastnaesite contains as much as 75% rare-earth oxides of the cerium group. It is a light yellow to brown hexagonal mineral having a hardness of 4.5 and a specific gravity of 5. It occurs chiefly in carbonatite plutons and subordinately in veins, pegmatites, and skarns.

Yttrofluorite and yttrocerite, the yttrium- and cerium-group analogs of fluorite, occur in pegmatites. Xenotime is an yttrium-group phosphate that occurs in igneous and metamorphic rocks, pegmatites, and placers. It is a pale yellow to brownish green tetragonal mineral having both a hardness and specific gravity of 4.5. Rare-earth elements replace part of the calcium in some apatite, particularly in some carbonatite plugs, alkalic magnetite deposits, and marine phosphate deposits.

Cerite is a brown to gray, calcium, cerium-group hydroxyl silicate, having a hardness of 5.5 and a specific gravity of 4.9. It occurs in carbonatites, pegmatites, and skarns. Gadolinite is a ferrous, beryllium, yttrium-group orthosilicate. It is a brown to black monoclinic mineral having a hardness of 6.7 and a specific gravity of 4.4 which occurs chiefly in pegmatites.

Multiple oxide minerals occur chiefly as brown to black, heavy (4.0 to 5.7), hard (4.5 to 6.5), radioactive, metamict minerals in pegmatites, alkalic igneous rocks and related veins, and placers. The five commercial minerals (betafite, brannerite, euxenite, fergusonite, and samarskite) are niobate-tantalate-titanates. Samarskite, however, is low in titanium, while brannerite contains titanium only. All contain uranium, and all but betafite contain appreciable thorium and yttrium-group rare earths. Betafite, brannerite, and euxenite contain calcium, while brannerite and samarskite contain iron. Euxenite and samarskite also contain cerium-group rare earths; fergusonite and samarskite contain erbium.

Both thorite and thorianite are radioactive and have a hardness of 5 to 5.5. Thorite is a reddish brown to black analog of zircon, whereas thorianite is a heavy isometric dark gray to black mineral.

Classification of Deposits

Rare-earth and thorium deposits have an ultimate genetic relation to alkalic igneous rocks, especially carbonatites. For example, in the Iron Hill district of Gunnison County, CO, such ores are found in association with a late carbonatite pluton and in the veins that fill the fractures that accompanied its emplacement (Armbrustmacher, 1980). The bastnaesite-rich carbonatite pluton at Mountain Pass, CA, was overlooked until 1949 (Olson et al., 1954). Carbonatite stocks, therefore, are excellent exploration targets for large deposits of rare-earth and thorium ores in economic concentrations.

Rare-earth elements also occur both as discrete minerals and in apatite in association with high-temperature, metamorphic, nontitaniferous magnetite deposits. The large contact metasomatic rare-earth-bearing iron ore deposits at Bayan Obo in Inner Mongolia adjoin an alkalic granitic pluton (Argall, 1980). Bastnaesite and monazite are important accessory minerals associated with hematite, magnetite, and martite. Rare-earth concentrates are coproducts derived from the beneficiation of iron ores of this type. Less commonly, rare-earth minerals such as allanite can occur in economic concentrations in skarns, such as at the Mary Kathleen mine in Queensland (Parker and Baroch, 1971).

Economic amounts of thorite, monazite, bastnaesite, and xenotime have been found in veins in a few places. Thorite occurs in veins in the Lemhi Pass district of Idaho (Staatz, 1979). Thorite also occurs in veins associated with alkalic plutons in the Iron Hill district and the Wet Mountains of Colorado. Bastnaesite is associated with barite, galena, and pyrite in epithermal veins in New Mexico and Burundi. Monazite, bastnaesite, and xenotime occur locally in tin-tungsten veins and greisens in Bolivia and South Africa (Parker and Baroch, 1971).

Small quantities of exotic rare-earth minerals occur in igneous pegmatites formed from residual fluids that were derived from nearly complete crystallization of presumably alkalic or subalkalic magmas. As with beryl, such pegmatitic occurrences are of only academic concern, unless labor costs are low enough to permit them to be economic.

Ancient and modern littoral and fluviatile placers are major loci for large deposits of rare earths and thorium minerals. The radioactive pyritic conglomerates of Aphebian age (early Proterozoic) in Quebec and Ontario, and similar deposits in the Northwest Territories, Canada, are perhaps the largest sources of potentially economic concentrations of thorium (as monazite and brannerite) in the world. According to Roscoe (1969), the deposits at Elliot Lake, Ontario, were formed under a reducing atmosphere. Such deposits extend across the Grenville Front into higher temperature and higher pressure, isochemical, metamorphic facies, yet still retain their radioactive elements. The Precambrian reefs of the Witwatersrand in the Transvaal are metamorphosed placers that contain thorium minerals.

Tertiary and Recent beach placers in Brazil, India, Australia, Egypt, the Malagasy Republic, and the United States are major sources of monazite recovered as a coproduct from the mining of magnetite, ilmenite, and rutile sands (Overstreet et al., 1968). Euxenite and brannerite have been mined from Recent alluvial placers in south-central Idaho (Savage, 1961).

Yttrium occurs in certain marine phosphatic shales such as the Phosphoria formation of Permian age in Idaho. Although the apatite in these rocks has a maximum content of 1000 ppm of yttrium, it might be feasible to extract an yttrium concentrate as a byproduct during the beneficiation of the phosphate rock.

Carbonatite plutons, aureoles surrounding alkalic plutons, and ancient and modern placers derived from them should continue to be the most attractive places to prospect for deposits of rare-earth and thorium ores during the remainder of the century.

Reserves

World yttrium reserves in 1979 were estimated at 38,200 st (Moore, 1980b). They are estimated by country in Table 2. It is anticipated that approximately 90% of the metal may be recovered from monazite ores; the remainder may be derived from xenotime, bastnaesite, euxenite, and uranium minerals.

Rare-earth oxide reserves are estimated at 17,730,000 st (Moore, 1980b; Argall, 1980) and are shown by country in Table 2. More than half of the published resources are attributed to bastnaesite-bearing alkalic plutons. The rest comprise monazite sands, plus a minor amount of euxenite sands.

World reserves of thorium oxides that could be produced by existing processes recently

TABLE 2—Published Reserves of Lanthanide Oxides, Yttrium, and Thorium Oxide, St

Country / Continent	Lanthanide Oxides*	Yttrium*	Thorium Oxide†
Egypt	10,000		
Malagasy Republic	20,000		
Malawi	15,000		
Nigeria			15,000
South Africa	5,000		12,000
Other	20,000		10,000
Africa	70,000		37,000
China‡	10,000,000		
India	1,000,000	20,000	350,000
Korea	50,000	1,000	
Malaysia	30,000	600	
Sri Lanka	10,000	200	
Other	10,000	200	9,000
Asia	11,100,000	22,000	359,000
Australia	400,000	6,000	35,000
Russia	500,000	1,600	40,000
Scandinavia	60,000	200	146,000
Europe	560,000	1,800	186,000
Canada	250,000	2,400	202,000
United States	5,000,000	3,500	192,000
North America	5,250,000	5,900	394,000
Brazil	350,000	2,500	75,000
Other			10,000
South America	350,000	2,500	85,000
World	17,730,000	38,200	1,096,000

* Moore, 1980b.
† Kirk, 1980b.
‡ Argall, 1980.

were estimated at 1,096,000 st (Kirk, 1980b). Distribution by countries is summarized in Table 2.

Most low-cost thorium reserves are in modern or ancient placers. Reserves in the thorite veins in the Lemhi Pass district of Idaho may be several hundred thousand tons (Staatz, 1979).

Principal Producing Countries

Australia, India, Brazil, and Malaysia are the chief producers of monazite concentrates. Domestic output is company confidential. Typical monazite concentrates contain approximately 90% monazite. Production of monazite concentrates by country during 1979 is summarized in Table 3.

The United States and China are the chief producers of bastnaesite concentrates. In 1979, the concentrates produced from the Molybdenum Corp. of America mine at Mountain Pass, CA, contained 18,205 st of rare-earth oxides. Malaysia is the chief source of yttrium from xenotime recovered as a byproduct of alluvial tin placers.

Other Countries and Potential Sources

Given favorable market conditions, thorite can be recovered as a coproduct from mining the Huronian uranium conglomerates at Elliot Lake, Ontario. Carbonatites in Kenya, Tanzania, South Africa, Malawi, and Norway are large potential sources of rare-earth-bearing pyrochlore and monazite. Monazite is also a possible byproduct of Bolivian tin ores. Placers in Argentina, Indonesia, the Malagasy Republic, Malawi, and Uruguay may be future sources of monazite concentrates.

TABLE 3—Production of Monazite Concentrates by Country During 1979

Country	Monazite Concentrates, Short Tons
Australia	17,000
Brazil	2,700
India	3,100
Korea	10
Malaysia	2,200
Sri Lanka	220
Thailand	800
United States	na
Zaire	85
	26,115

Source: Moore, 1980a; na—not available.

Exploration

Standard techniques of geophysics, geochemistry, trenching, and drilling are used to search for and outline bastnaesite deposits. These methods, plus panning and jigging, can be used to investigate the economic potential of monazite in beach and stream placers.

Field Techniques

Gravity separations are useful field techniques for preliminary evaluation of placer sands. Random samples obtained from auger holes or pits can be screened and panned or jigged. Minerals in the concentrate can then be examined under a binocular microscope or checked radiometrically.

Geophysics

Airborne, surficial, and borehole radiometric surveys with scintillation equipment can detect rare-earth and thorium minerals associated with small amounts of uranium in veins or plutons. Because the same resistant minerals exist in many sands, such techniques work equally well to detect rare-earth or thorium minerals in ancient and modern beach or stream placers. Furthermore, because these minerals are commonly associated with magnetite and ilmenite, airborne or surficial magnetometric surveys can be used to find favorable target areas for placer concentrations of rare-earth and thorium minerals.

Geochemistry

Field tests have been devised for the detection of the rare-earth elements (Rose, 1976). Multielemental hydrogeochemical and stream-sediment reconnaissance data from the National Uranium Resource Evaluation can be used effectually for resource studies on rare-earth minerals (Bolivar, 1980; Price et al., 1980). Carbonatite plutons in eastern Uganda have been investigated by means of geochemical soil and stream-sediment surveys. Niobium, beryllium, and rare earths are pathfinder elements for carbonatites.

Drilling

Conventional core drilling (Parker and Baroch, 1971) is used to search for plutonic bastnaesite deposits. Drivepipe and hand augers are used to drill shallow holes in sampling beach placers down to depths of 30 ft. Waterjet drilling, using rotary drill rigs, has also been used to evaluate beach placers. Churn drills are used to test stream placers, particularly the deeper parts where the heavy minerals are concentrated just above bedrock. Typically, a 6-in. casing is driven and samples are recovered at 2.5- to 5-ft intervals.

Evaluation of Deposits

Tonnage and grade are the key factors in determining the economic value of rare-earth and thorium deposits. The geomorphology and paleogeomorphology of stream and beach deposits determine the ultimate limits of possible placers. Detailed geologic mapping by aerial photography may indicate possible targets. Drilling or pitting on 500- to 2000-ft centers should outline the general boundaries of the deposits. Detailed borders can be delimited by closer spacing of sample sites.

The shape of bedrock deposits depends upon the extent and configuration of favorable host units. Careful geologic mapping, followed by geochemistry, geophysics, and diamond drilling, should allow an estimate of the probable tonnage of ore-bearing rock.

Rare-earth and thorium minerals occur in the heavy fraction of littoral and fluviatile placers and in conjunction with the heavy accessory minerals in igneous rocks. Therefore, gravity separation by panning samples will concentrate the desired minerals in the heavy black-sand residue. In bedrock sampling, the rock must first be crushed and ground to a sufficient fineness in order to free the accessory minerals sought.

Magnetite and ilmenite can be removed from dried heavy concentrates by means of a strong magnet. Other minerals may be collected by use of a variable-intensity magnetic separator. The mineral grains can be identified under a microscope and their relative proportions can be counted. Rare-earth minerals containing thorium or uranium can be detected with a Geiger counter. Those containing more than 5% of combined praseodymium and neodymium can be identified with a portable spectroscope (Parker and Baroch, 1971).

A rapid field test is based on the relative insolubility of oxalates of thorium, yttrium, and rare-earth elements. The sample is boiled in sulfuric acid, then cooled and diluted with cold water, filtered, and neutralized with ammonia. It is boiled again and a saturated solution of oxalic acid is added until a light-colored curdy precipitate forms. Later, if the precipitate is ignited, a buff to tan color indicates yttrium-group rare earths, and a dark brown color indicates cerium-group rare earths. To test for thorium, the oxide should be dissolved in nitric and oxalic acids and added to a potassium iodate solution. White thorium iodate forms if thorium is present (Parker and Baroch, 1971). Individual rare-earth elements can be identified best by spectrophotometry, x-ray fluorescence, flame photometry, neutron-activation analysis, or emission spectroscopy.

Preparation for Markets

Mining Methods

The technology for mining monazite beach placers is similar to that employed for diamonds, gold, or cassiterite (Parker and Baroch, 1971). Offshore operations use floating dredges having either suction or bucket elevators. Slurry is carried hydraulically from the dredges to the beneficiation plant.

Subaerial deposits are worked by draglines or power shovels. However, in places where scattered pockets of ore must be mined selectively, bulldozers and scrapers are used, and the ore is trucked to the concentrator.

In underdeveloped nations where capital is scarce and labor cheap and abundant, primitive methods are used, in which hand-dug pits and simple sluice boxes recover black-sand concentrates.

Conventional open-pit mining methods are employed in the bastnaesite deposit at Mountain Pass, CA. The ore is drilled, blasted, and loaded into trucks by power shovels, then hauled to the mill.

Milling Techniques

Pure monazite contains approximately 70% rare-earth oxides. Standard acceptable grades for monazite-sand concentrates are 55, 60, and 66% rare-earth oxides.

Primary beneficiation of beach sands is effected either on floating dredges or at terrestrial plants. After oversized particles have been screened, a black-sand concentrate is recovered mechanically by means of jigs, sluice boxes, shaking tables, and/or Humphreys spiral concentrators. Through the use of induced-roll electromagnetic separators and high-tension electrostatic roll separators, a monazite concentrate, which may represent only 1% of the total black sands, can be recovered in the magnetic nonconductive fraction. Flotation cells containing oleic acid are used to enrich Indian monazite sand concentrates.

The crude ore at Mountain Pass, CA, contains 9% rare-earth oxides. (Pure bastnaesite contains approximately 75% rare-earth oxides.) Following primary and secondary crushing, the ore is passed through a rod mill and a classifier. The slurry is then heated and passed through flotation cells that depress the barite gangue and yield a 55 to 60% rare-earth oxide concentrate. This concentrate, in turn, is leached by hydrochloric acid and countercurrent decantation to remove calcite, thereby upgrading the concentrate to 68 to 72%. Finally, this concentrate is calcined to remove carbon dioxide from the carbonates, yielding a 92% concentrate of rare-earth oxides and fluorides (Parker and Baroch, 1971).

In the uranium mills at Elliot Lake, Ontario, thorium concentrates were recovered from waste solutions by solvent extraction to obtain a 35% thorium oxide precipitate (Shortt, 1970).

Processing Techniques

The caustic process and the acid process are two common methods for treating monazite concentrates (Parker and Baroch, 1971). In the caustic process, monazite is digested in hot sodium hydroxide and filtered. The insoluble thorium and rare-earth hydroxides are separated by treatment with weak hydrochloric acid, which dissolves the rare-earth hydroxides and leaves solid thorium hydride. The thorium hydride is then dissolved in nitric acid, and thorium is recovered by solvent extraction.

In the acid process, monazite is digested in hot sulfuric acid. Rare-earth and thorium sulfates are dissolved and removed by filtration. Thorium is then precipitated as a pyrosulfate, leaving the rare-earth ions in solution. Next, rare-earth elements are precipitated as oxalates or as sodium-rare-earth sulfates. These, in turn, can be roasted to form oxides, which are then dissolved in nitric acid. The rare-earth elements are then separated from each other by solvent extraction.

Four products may be recovered from the treatment of bastnaesite concentrates at Mountain Pass, CA: europium oxide; a lanthanum-rich mixture of rare-earth metals; a heavy subgroup mixture chiefly composed of samarium and gadolinium; and technical grade cerium. The bastnaesite concentrates are roasted and leached with hydrochloric acid. Cerium oxide is filtered from the solution. The europium and heavy rare-earth elements are separated from the solution by solvent extraction. Then the lanthanum-rich product is precipitated. After further solvent extraction, europium sulfate is precipitated, leaving samarium and gadolinium in solution.

Thorite concentrates are digested in hot nitric acid and filtered. The thorium is removed from solution by solvent extraction and purified by countercurrent solvent extraction. Such extraction yields a high-quality thorium nitrate from which pure thorium oxide, tetrachloride, or tetrafluoride may be produced.

Thorium metal may then be made by metallothermic reduction of thorium oxide, thermal decomposition of thorium tetrachloride through reduction of thorium halide or oxide with calcium, or fused-salt electrolysis. Metallic yttrium is produced by direct reduction of yttrium trifluoride (Anon., 1979).

Separation and purification of the lanthanide elements is a major problem. Ion exchange is an effective way to separate individual rare-earth elements. EDTA (ethylene diamine tetraacetate) solution is the best elutant for removing the lanthanide elements and yttrium from the resin in that order. Solvent extraction can be used effectively to separate rare-earth groups, and to isolate yttrium. Cerium and europium can be separated from other lanthanide elements by valency-change reactions. Lanthanide sulfates can be subdivided by adding cold aqueous solutions containing excess alkali sulfates to alkali-rare-earth sulfates. The heavy lanthanides and yttrium remain in solution. Fractional recrystallization permits the selective separation of lanthanum, praseodymium, and neodymium.

Rare-earth metals are produced most successfully by electrolysis or by metallothermic reduction of rare-earth halides. Mischmetal is produced by electrolytic fusion of mixed anhydrous rare-earth chlorides.

Transportation

Monazite concentrates are transported by ship, rail, or truck from primary beneficiation plants to chemical processing plants. Bastnaesite concentrates are shipped to processing plants in steel drums by truck or rail from the mill at Mountain Pass, CA.

Markets, Consumption

The United States imported 6,931 st of monazite concentrates valued at $1,677,000

TABLE 4—Domestic Imports of Rare-Earth and Thorium Concentrates, Metals, Alloys, and Chemicals in 1979

Commodity	Amount, lb	Value, $	Chief Source
Monazite concentrates	13,862,000	1,677,000	Australia
Cerium oxide	13,668	115,852	France
Other rare-earth oxides	848,081	15,153,469	France
Rare-earth metals (alloys)	144,864	577,162	United Kingdom
Rare-earth metals (+ scandium)	8,975	186,259	Russia
Ferrocerium	123,218	679,612	France
Thorium nitrate	47,415	162,837	France
Thorium oxide	31,509	160,490	France
Thorium oxide in gas mantles	2,867	476,842	Malta
Other thorium chemicals	181	33,688	Switzerland

Sources: Moore, 1980a; Kirk, 1980a.

during 1979. Of this amount, 90% was obtained from Australia and 9% from Malaysia. Average grade of the concentrates was 55% rare-earth oxides and 6% thoria. Domestic imports of concentrates, metals, alloys, and chemicals in 1979 are shown in Table 4.

In 1979, domestic stocks of monazite increased markedly, whereas stocks of bastnaesite concentrates decreased. Stocks of rare-earth compounds increased during that same year.

Consumption of bastnaesite and monazite concentrates increased 7 and 10%, respectively, during 1978. Domestic shipments from processors to consumers contained 11,000 st of rare-earth oxides. The foreign and domestic supply of thorium has exceeded demand because of increased requirements for coproduct rare earths derived from monazite concentrates. Consumption of nonenergy thoria equivalent in 1979 was estimated at 32 st.

In 1979, the United States exported 84,100 lb of ferrocerium, of which Korea was the chief recipient. Domestic exports of uranium and thorium metals, as well as of alloys, were not subdivided for 1979. Total shipments were 10,651 lb, valued at $216,130, of which 64% went to Sweden. Similarly, shipments of uranium and thorium compounds were undifferentiated for 1979. A total of 37,367 lb (chiefly uranium), valued at $430,472, was exported, of which 54% went to Chile (Kirk, 1980a).

Future Considerations and Trends

The future demand for thorium hinges primarily on the successful use of thorium as a nuclear fuel in power reactors. Probable domestic consumption in the year 2000 has been estimated at 155 st. Probable nonenergy consumption of thorium in 2000 should be approximately 85 st, of which one-third will be used in the aerospace industries in magnesium-base alloys and one-fourth in a thorium-copper catalyst to be used in the production of methanol (Kirk, 1980b). It is anticipated that all requirements can be satisfied from domestic operations.

Because domestic thorium output is a byproduct of the output of rare-earth concentrates from monazite, the accumulation of excessive stockpiles may halve the price of thorium by the year 2000. National producers may have to reduce mining costs to compete successfully with higher grade foreign operations. Development of plutonium or other types of breeders could make obsolete the high-temperature gas reactors fueled by thorium-uranium. Furthermore, the failure to make aggressive improvements in thorium breeders could render them uneconomic (Shortt, 1970).

The annual demand for yttrium by the year 2000 is expected to be 73 st, chiefly used as phosphors for cathode-ray picture tubes and as electronic components. Sufficient domestic ores exist to meet anticipated demands, but the unit price may rise somewhat.

Moderate growth of the rare-earth industry is anticipated by 2000. All demands should be met from domestic sources without difficulty. Probable annual demand for all rare earths by 2000 is expected to be 57,000 st, of which 25,000 st will be used in iron and steel foundries and 15,000 st will be used in petroleum refining.

Substitutes, Synthetic Competition

Palladium can replace rare-earth oxides in petroleum cracking. Selenium and arsenic compounds can substitute in part for rare-earth

TABLE 5—Tariffs on United States Imports of Rare-Earth and Thorium Commodities

Commodity	Most Favored Nation (MFN)		Non-MFN
	1/1/81	1/1/87	1/1/81
Rare-earth ore and concentrates	Free	Free	Free
Cerium oxide and compounds	13.1% ad val.	7.2% ad val.	35% ad val.
Other rare-earth oxides	4.7% ad val.	3.7% ad val.	25% ad val.
Rare-earth metals	4.7% ad val.	3.7% ad val.	25% ad val.
Rare-earth alloys (including mischmetal)	45¢/lb	32¢/lb	$2/lb
Ferrocerium and other pyrophoric alloys	43¢/lb + 5.1% ad val.	22¢/lb + 2.6% ad val.	$2/lb + 25% ad val.
Thorium ore	Free	Free	Free
Thorium nitrate and oxide	15.1% ad val.	7.8% ad val.	35% ad val.
Unwrought thorium metal, waste, and scrap	5.6% ad val.	4.2% ad val.	25% ad val.
Thorium alloys	6.4% ad val.	3.0% ad val.	25% ad val.

Sources: Kurtz and Stephenson, 1981; Kirk, 1981.

oxides as glass decolorizers. In the production of ductile nodular iron, magnesium-nickel alloys and magnesium are more effectual and less expensive than cerium alloys. Alumina and rouge can replace rare-earth oxides for optical polishing. Boron is preferable to rare earths as a thermal neutron absorber for control rods because it generates no secondary alpha radiation and is less expensive.

Thorium breeder reactors may in time be competitive with uranium breeder reactors. However, thorium for nonnuclear uses is likely to be replaced by less expensive metals. Yttria and beryllia can replace thoria in refractories. Titanium and zirconium can substitute for thorium in electronic tubes.

Governmental Control

On a worldwide basis, most governments control exports of thorium concentrates and metal. Some treat thorium production as a governmental monopoly. India initiated the first embargo on monazite concentrates in 1946; Brazil followed in 1950.

As of Dec. 31, 1980, the entire United States stockpile of rare earths and all but 408 st of the thorium nitrate stockpile were authorized as available for disposal by the General Services Administration.

Taxes, Depletion, and Tariffs

The allowable depletion on domestically mined monazite and other thorium-bearing minerals in the United States is 22%. The rate for domestic rare-earth minerals that do not contain thorium and for foreign production of monazite is 14%. Depletion for other rare-earth minerals produced outside the United States is 14%. New tariff agreements with developed countries were set for mineral commodities in Tokyo in 1979. Rates for imports from most-favored nations were to be lowered at the beginning of each year from 1980 through 1987. Table 5 shows the present 1981 and ultimate 1987 end member of declining tariff structures for imports of various rare-earth and thorium products into the United States.

Ecology

The accepted radiation protection standard for thorium is 50 microcuries. Thorium and thorium hydride powders must be treated carefully because spontaneous ignition of dust layers may cause fire or explosions.

Conservationists are concerned about the destruction of beaches through placer mining. For this reason, many mining applications have been rejected in Australia. Reclamation has been successful at mined-out monazite placers in Georgia and Florida.

Acknowledgment

The author thanks Sam Rosenblum, US Geological Survey, for critically reading the manuscript, making suggestions, and especially for preparing a new introduction. Christine M. Moore and William S. Kirk of the US Bureau of Mines kindly provided preprints and other current information on rare earths and thorium. Aaron Waters of the Los Alamos National Laboratory made helpful editorial suggestions that improved the paper. This article was prepared under the auspices of the US Department of Energy.

Bibliography and References

Anon., 1979, "Rare Earths: Industry Profile and Market Review," *Industrial Minerals,* No. 138, pp. 21–59.

Argall, G.O., Jr., 1980, "Three Iron Ore Bodies of Bayan Obo," *World Mining,* Vol. 33, No. 1, pp. 38–41.

Armbrustmacher, T.J., 1980, "Abundance and Distribution of Thorium in the Carbonatite Stock at Iron Hill, Powderhorn District, Gunnison County, Colorado," Professional Paper 1049B, US Geological Survey, pp. B1–B11.

Bolivar, S.L., 1980, "An Overview of the National Uranium Resource Evaluation Hydrogeochemical and Stream Sediment Reconnaissance Program, Los Alamos Scientific Laboratory," LA–8457–MS, Los Alamos National Laboratory, Los Alamos, NM, 24 pp.

Erametsa, O., 1965, "Separation of Promethium from a Natural Lanthanide Mixture," *Acta Polytech. Scand., Chem. and Met. Series,* No. 37, 21 pp.

Kirk, W.S., 1980a, "Thorium," *Minerals Yearbook, 1978–79,* US Bureau of Mines, pp. 907–913.

Kirk, W.S., 1980b, "Thorium," *Mineral Facts and Problems, 1980,* Bulletin 671, US Bureau of Mines, pp. 937–945.

Kirk, W.S., 1981, "Thorium," *Mineral Commodity Summaries 1981,* US Bureau of Mines, pp. 164–165.

Kurtz, H.F., and Stephenson, P.A., 1981, "Rare-Earth Metals," *Mineral Commodity Summaries 1981,* US Bureau of Mines, pp. 124–125.

Moore, C.M., 1980a, "Rare-Earth Minerals and Metals," *Minerals Yearbook, 1978–1979,* US Bureau of Mines, pp. 735–742.

Moore, C.M., 1980b, "Rare-Earth Elements and Yttrium," *Mineral Facts and Problems, 1980,* Bulletin 671, US Bureau of Mines, pp. 737–752.

Olson, J.C., et al., 1954, "Rare-Earth Mineral Deposits of the Mountain Pass District, San Bernardino County, California," Professional Paper 261, US Geological Survey, 75 pp.

Overstreet, W.C., et al., 1968, "Fluvial Monazite Deposits in the Southeastern United States," Professional Paper 568, US Geological Survey, 85 pp.

Parker, J.G., and Baroch, C.T., 1971, "The Rare-Earth Elements, Yttrium and Thorium; A Materials Survey," Information Circular 8476, US Bureau of Mines, 92 pp.

Price, V., Jr., et al., 1980, "Use of NURE HSSR Data for Resource Studies on Rare-Earth Minerals," (abstract) *Abstracts with Programs,* Vol. 12, No. 5, Geological Society of America, p. 205.

Roscoe, S.M., 1969, "Huronian Rocks and Uraniferous Conglomerates in the Canadian Shield," Paper 68–40, Geological Survey of Canada, 205 pp.

Rose, E.R., 1976, "A Field Test for Rare-Earth Elements," Paper 75–16, Geological Survey of Canada, 10 pp.

Savage, C.N., 1961, "Economic Geology of Central Idaho Black-Sand Placers," Bulletin 17, Idaho Bureau of Mines and Geology, 160 pp.

Shortt, C.E., 1970, "Thorium," *Mineral Facts and Problems,* Bulletin 650, US Bureau of Mines, pp. 203–218.

Staatz, M.H., 1979, "Geology and Mineral Resources of the Lemhi Pass Thorium District, Idaho and Montana," Professional Paper 1049–A, US Geological Survey, pp. A1–A90.

Salt

STANLEY J. LEFOND *

CHARLES H. JACOBY †

Salt, or halite, has a long and most varied history. While we know the Chinese were producing salt as early as 3000 B.C., the first written reference to salt appears in the book of Job recorded about 2250 B.C., which reads:

"Can nothing which is unsavory be eaten without salt?"

We do not know when man first used or realized that he needed salt. Perhaps one of our early Eolithic ancestors noted a piece of salt shining along the shore of a salt lake or a bay. He may have been attracted by its shining luster and tasted it. He found that it had a tart, zesty, distinctive flavor and perhaps chewing a large piece, became ill. However, by experimentation he soon learned that taken in the proper proportions salt was a very pleasant additive to his food. He may have noticed, too, that it increased his feeling of well-being. Thus, when the curtain of recorded history rises, we find man using salt for his health's sake, seasoning his food with it, feeding it to his animals, and using it in many superstitious and religious ceremonies. These uses, however, are a far cry from the uses for salt today.

The first attempts of white man to make salt in continental United States were reported in the year 1614, while the first commercial production was in 1753 from the Kanawha licks. Practically all this salt was for human consumption, preserving food, and curing hides. Prior to the coming of the white man, Indians are known to have made salt from salt springs such as those that existed at Avery Island, LA, Charleston, WV, Saltville, VA, and Onondaga, NY. In 1862 the first rock salt was mined in the US from the Avery Island salt dome. Almost immediately thereafter, rock salt was mined at Saltville, VA and then at Retsof, NY in 1885.

Of the more than 39 million tons produced annually in the United States, about 58% or 23 million tons are consumed by the chemical industry in the production of chlorine, caustic soda, hydrochloric acid, sodium metal, and some 30 other basic chemicals. These basic chemicals, in turn, are used in the preparation of approximately 14,000 chemicals which range from soap and detergents, to such chlorinated hydrocarbons as DDT, BHC, carbon tetrachloride, etc.

Utilization of cavities created in salt beds and domes has grown rapidly since 1947 until now the waste salt is creating an ecological problem. The uses for both dry- and solution-mined cavities range from the storage of liquid hydrocarbons to containerized storage of toxic wastes, from compressed air for the peak shaving of power to storage of nuclear wastes.

Geology

Mineralogy

Properties: Because of its simple structure, halite, the mineralogical name for salt, was the first structure to be analyzed by X-rays. It has the following properties:

Formula: $NaCl$ (Na-39.34%, Cl_2-60.66%)
Crystallography: isometric, hexoctahedral, $4/m\ \bar{3}2/m$
Space group: Fm3m
A_0: 5.627 kX
Cell contents: Na_4Cl_4
Habit: usually cubic, rarely octahedral, massive, granular to compact
Physical properties:
Refractive index: 1.554
Twinning: observed on {111}
Cleavage: {001} perfect
Fracture: conchoidal, brittle

* Consultant, Industrial Minerals, Evergreen, CO.

† Jacoby and Co., Earth Science Consultants, Tempe, AZ.

Hardness: 2, specific gravity, 2.168
Melting point: 804°C; boiling point: 1413°C
Luster: vitreous, normally colorless to white but occasionally red, yellow, blue, and purple
Solubility: 0°C: 35.7 parts per 100 parts of water; 100°C: 39.8 parts per 100 parts of water

Classification of Deposits

Geologically salt is found in solution or in a solid state.

Solution: Salt is found in solution in oceans, lakes, and springs.

Oceans—It is generally conceded that the bulk of the world's salt supply is in solution in the oceans. The most often quoted figure is that the oceans contain 4.5 million cubic miles of rock salt—a volume half again as much as the entire North American continent above sea level.

The actual salt content of the oceans varies from 1 to 5% but averages about 3.5%. This variation, which exists from place to place and by depth, is due to the influence of inflowing waters, variations in the evaporation and freezing rates, and man. Today, in areas along the coasts where conditions are ideal for solar evaporation, salt is produced from ocean waters. The facilities vary from extremely crude (Fig. 1) to highly efficient and well-operated solar salt plants capable of producing upwards of 4,000,000 tpy of high quality salt (Fig. 2).

Marginal salt pans and marine salinas (natural solar salt pans) are found along the coasts, usually in the more arid countries. A marginal salt pan is a shallow depression, usually covered only during periods of abnormal high tides or during storms. The water is evaporated during dry spells leaving a thin salt crust. Many of these are exploited along the northwestern coast of Mexico, as well as other areas.

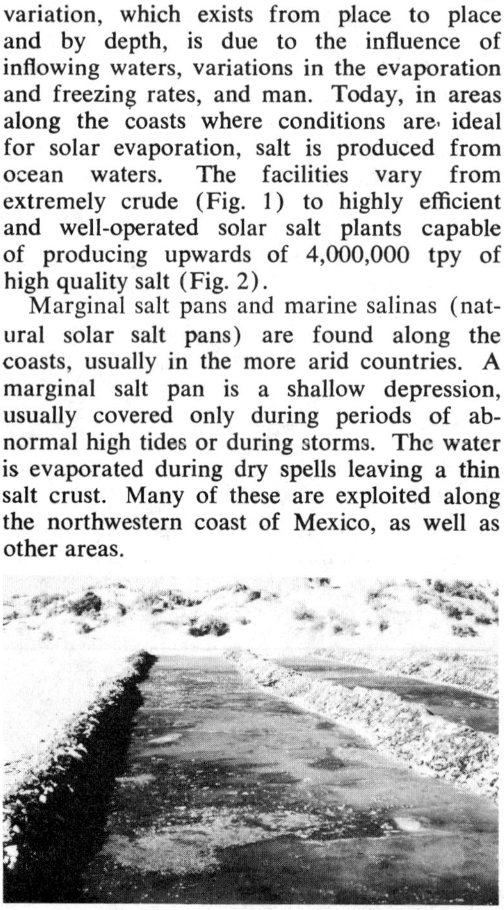

FIG. 1—*Crude solar salt pond.*

FIG. 2—*Modern solar salt operation.*

A marine salina is a salt lake which is fed by underground seepage from the ocean. The salt lake at Larnaca, Cyprus, is a typical example.

Lakes—Lakes are formed when water accumulates in a topographic depression. Waters flowing into the depression carry material in solution. The composition of this material will depend upon the type and solubility of rocks in the watershed traversed by the meteoric waters. Thus it is possible to have lakes that vary greatly in composition. Clarke (1924) classifies mineralized lakes as follows:

1) Sodium chloride lakes
2) Natural bittern lakes—magnesium salts dominate
3) Sulfate water lakes
4) Sulfate-chloride lakes
5) Alkaline lakes: carbonate lakes, carbonate-chloride lakes, chloride-sulfate-carbonate lakes

Normally, mineralized lake deposits contain little or no potassium. Sodium and calcium are the predominant cations and the most common minerals are the sodium sulfates, mirabilite and thenardite, burkeite, gaylussite, trona, and halite. Landes (1960) feels these sodium chloride lakes are "either separated bodies of ocean water or they owe their composition to the presence of salt within the rocks that floor the watershed."

Ground Water—Ground waters or their surface expression "springs" vary in salinity from practically pure water to those that are almost completely saturated. Ground waters are classified as either connnate or meteoric water. Connate waters are those trapped in the rock during the time of formation. Meteoric (rain) water is fresh initially but picks up soluble material as it travels through air, soil, rocks, etc. Therefore the composition of ground water varies according to the environment encountered. Connate waters usually reflect the composition of the water in which the rock was formed. For example, most sedimentary rocks were deposited under the sea, so much connate water is entrapped seawater. How-

TABLE 1—Analysis of Natural Brines, Ppm

	1	2	3	4	5
Ca	37,832	40,567	22,400	850	60
Mg	3,824	3,092	—	5,980	15,800
Na	67,966	79,253	49,000	96,400	34,940
SO_4	180	115	508	14,270	5,920
HCO_3	148	186	116	120	—
Cl_2	182,660	202,810	—	156,930	41,960
I_2	7	7	—	—	—
Br_2	4,570	4,600	—	—	208,020
K	—	—	1,380	trace	7,560
Fe	—	—	138	20	—
CO_3	—	—	—	—	540
Total solids	292,670	326,023	204,400	—	—

1. Smackover Brine, Arkansas
2. Smackover Brine, Arkansas
3. Mt. Simon fm, Indiana
4. Tashkurghan, Afghanistan
5. Dead Sea, Israel

ever, sedimentary rocks are also formed in freshwater lakes and interior salt lakes.

Clark (1924) has classified ground waters into the following groups: chloride, sulfate, carbonate, mixed (chloride-sulfate, chloride-carbonate, sulfate-carbonate, chloride-sulfate-carbonate), siliceous, borate, nitrate, phosphate, and acid.

Ground waters range in chemical content from practically nothing to as much as 254,000 ppm. Some brines are exploited for their bromine and magnesium contents. Table 1 lists analyses of typical brines.

Solid Salt Deposits: Solid salt is found in playa deposits, as bedded salt deposits, and diapirs and domes.

Playa Salts—According to Grabau (1913) a playa is a sandy, salty, mud-covered floor of a desert basin. During the rainy season a playa may be covered with rainwater which dissolves part of the soluble material. When the water evaporates, it leaves behind the soluble material, which may be concentrated in one area of the lake. This body of soluble material will shift around the lake and many will say that a new crop of crystals "grows" after each wet cycle. Oftentimes, too, a brine reflecting the composition of the crystallized crust will be found just a short distance beneath the surface of the playa. In a few instances a crystal body is formed, which reflects the composition of the surrounding brine. Searles Lake in southern California is probably the most famous example of a crystal body in a playa lake.

In Mexico, tequesquite, a mixture of salt, sodium sulfate, and sodium carbonate, forms a crust on many playa lakes and is used by natives in cooking and cleaning. Its composition varies greatly as shown in Table 2 (Flores, 1918). Figs. 3 and 4 show tequesquite being harvested.

Playa deposits are fairly common in arid regions and saline playas result from the evaporation of a mineralized lake. The dissolved minerals in the lake result from the leaching of the rocks in the area surrounding the enclosed basin. Salt in the playas results from the solution of sodium chloride-bearing sediments. The presence of borates suggests fairly recent volcanism in the area.

The composition of material in a playa lake varies greatly. For example, in California, Cadiz Lake is predominantly calcium chloride. Bristal Lake is predominantly salt, and Owens

TABLE 2—Tequesquite Analyses, Mexico

Playa Lake	NaCl, %	Na_2CO_3, %	$NaHCO_3$, %	Na_2SO_4, %
Sta. Clara, Lake Texcoco	37.5	19.4	9.1	0.60
Los Leyes, Lake Texcoco	31.5	18.3	1.15	0.56
El Carmen, Llanos de San Juan	11.7	30.1	5.7	1.6
Cuitsea (Michoacán)	4.6	2.45	0.9	2.45
Sayula (Jalisco)	25.0	12.3	6.5	5.6
San Luis Potasí	9.8	31.6	17.5	21.6

Source: Flores, 1918.

FIG. 3—*Harvesting tequesquite, Mexico.*

FIG. 4—*Harvesting tequesquite, Mexico.*

Lake is sodium carbonate. Others contain sodium sulfate, while still others contain mixtures of many different compounds.

Bedded Salt Deposits—Bedded salt deposits are true sedimentary rocks and as such are found in sedimentary sequences with shales, limestones, gypsum, anhydrite, etc. Thus, salt beds have been found in every geologic period from the Cambrian to the Tertiary, and there is evidence that some deposits are Precambrian in age. A list of salt deposits in the various geological periods is delineated in Table 3 (Lefond, 1969).

Salt beds can be as much as 1300 ft thick. However, due to the manner in which they are formed, thick salt beds characteristically contain thin bands or lenses of anhydrite and/or clay, with a salt interval 2 to 4 in. thick. In some areas as much as 2000 ft of salt are known. Fig. 5 shows bedded salt and a "salt horse," in the Detroit mine of the International Salt Co.

The formation of such thick salt beds has intrigued geologists for many years. Hence many theories and ideas have been advanced to explain the formation of thick salt beds which extend over hundreds of square miles.

TABLE 3—Ages of Salt Deposits

Cenozoic Era

Recent. Solar salt operations, Playa Lakes.

Pleistocene. California, Nevada, Russia, Mexico, Israel

Pliocene. Nevada, Utah, Italy, Jordan

Miocene. Dominican Republic, Cyprus, Czechoslovakia, Poland, Spain, Turkey, Russia, Algeria, Egypt, Sudan, Iran, Iraq, Syria, Trucial States, Morocco

Oligocene. France, Germany, Spain, Turkey, Iran, Iraq

Eocene. Green River Basin, WY, British Honduras, Pakistan(?), Morocco, Iran

Mesozoic Era

Cretaceous. Florida, Bolivia, Brazil, Colombia, Peru, Russia, Angola, Zaire, Congo Republic (Brazzaville), Gabon, Morocco, Libya, Senegal, Nigeria, Mexico

Jurassic. Gulf Coast area, Idaho, Cuba, Chile, Germany, Aden, Kuwait, Tanzania

Triassic. Isthmus of Tehuantepec-Mexico, Bulgaria(?), France, Greece(?), Germany, Netherlands, Portugal, Spain, Switzerland, United Kingdom, Algeria, Libya, Morocco, Tunisia, Ethiopia, Peru, Bolivia

Permian. Permian Basin-US, Supai Basin, Williston Basin, Mexico, Peru, Germany, Greece(?), Netherlands, Poland, United Kingdom, Russia, Denmark, Australia (?), Brazil

Pennsylvanian. Colorado, Paradox Basin, Brazil

Mississippian. Virginia, Williston Basin, New Brunswick, Nova Scotia

Devonian. Williston Basin, Russia, Australia, Canada

Silurian. Salina Basin-US, and Canada

Ordovician. Williston Basin, WY, Bolivia(?)

Cambrian. Northwest Territories, Australia, Russia, Iran, Pakistan(?)

Precambrian

Iran, Pakistan(?), Australia

Source: Modified from Lefond, 1969.

FIG. 5—***Bedded salt and a "salt horse," Detroit mine, International Salt Co.***

Landes' (1960) classification of the various theories follows:

(I) Terrestrial type. Evaporating interior sea becomes saline due to:
  (1) Primary sodium chloride produced as a weathering product.
  (2) Flushing of connate water from sedimentary rocks cropping out above water level in drainage basin.
  (3) Leaching of salt from sedimentary rocks cropping out in drainage basin.
(II) Marine type. Concentration of ocean water in cutoff sea:
  (1) Without further enrichment.
  (2) With enrichment by
    (A) Wind-blown salt
    (B) Ocean water
      (a) Marginal salt pans
      (b) Marine salinas
      (c) Barred lagoons
      (d) Silled downwarping basins

Today all acceptable theories involve the precipitation of salt from an evaporating body of salt water. Landes (1960) suggests the following stages for the formation of a major bedded salt deposit:

> *Generalized Chronological History Proposed for Thick Salt Deposits.* The following stages are suggested in the natural history of a major bedded salt deposit:
> (1) Peneplanation of the continental platform so that it is a nearly flat land of very little elevation above the sea.
> (2) Submergence of wide areas beneath the sea. Shallow ocean waters spread over a considerable part of the continent.
> (3) Beginning or renewed downwarping of a basin marginal to this epicontinental sea.
> (4) Sedimentation of both basin and epicontinental sea. Mainly clastic and organic (reef) deposits.
> (5) Lowering of sea to approximately sill level. Period of evaporite deposition. Enrichment by seawater flowing across sill. Because of the shallowness of the water, and the distance from the open ocean in such areas as the Michigan Basin, the brine crossing the sill may have been abnormally strong. Some enrichment could also take place by wind-blown salt, for the flat topography and scantiness of land vegetation no doubt resulted in incessant winds of considerable magnitude.

Thus, as more and more sediments and/or salt is loaded onto the mass of salt crystals on the sea floor the water is squeezed out and the salt crystals are compacted or recrystalized into massive salt beds.

Salt Domes—Salt is very vulnerable to pressure. Thus, as pressure is exerted on a salt bed, due to the static weight of overlying sediments, or tectonic forces, the salt "flows" plastically, first bulging to form salt anticlines such as those that occur in the Utah-Colorado salt basin and in Europe. If the pressure increases, the overlying rock is ruptured and salt is squeezed into, and through, the overlying sediments to form *salt domes* or *diapiric folds.* Such features are known in the Gulf Coast of the US, in the Paradox Basin of Colorado-Utah, in Mexico, Romania, Iran, Germany, Tunisia, Algeria, and elsewhere. Fig. 6 shows an outcrop of a salt-gypsum diapir in Rheous, Tunisia. Fig. 7 shows mining in the Avery Island salt dome of the International Salt Co.

In some instances the salt has reached the surface (Avery Island dome, LA), or actually flows like a glacier onto the surface. Kent (1958) says of the Kuh-i Namak (Dashti) salt plugs:

> The parent anticline is a high whaleback fold about 30 miles long and 8 miles wide, with the salt mass piercing the culmination and forming a vast dome 4,000 feet above plains level. From this dome broad salt glaciers flow between flanking limestone crags down the mountain sides northeast and southwest. Where the salt overhangs rock cliffs the precipitous slopes are at times swept by salt avalanches—a complete parallel with conditions on glaciated mountains.

The distribution of salt and salt domes in the United States is shown in Fig. 8 (Lefond, 1969). The greatest concentration in the US is in an area that extends from western Alabama to the Texas-Mexican border—and perhaps even on into northeastern Mexico. The Gulf Coast salt bodies are vertical, or nearly so, and are circular to elliptical in plan. The diameters range from about 3000 ft to as much as 4 miles. The vertical dimension is not ac-

FIG. 6—*Salt-gypsum diapir, Rheous, Tunisia.*

FIG. 7—*Mining in salt dome, Avery Island mine, International Salt Co.*

curately known, but may be as much as 40,000 ft.

The salt anticlines of the Paradox Basin range in length from 30 to 70 miles, and contain salt cores that range from about 4,100 to 13,700 ft in thickness.

Distribution of Deposits

Salt deposits and/or solar salt operations are found in nearly every country of the world. However, in a few areas such as Japan, Norway, Sweden, Finland, and parts of the US, salt has to be imported to furnish raw material for the manufacture of chemicals. In other areas, for example, in northwestern Mexico, Inagua, Long Island, Bonaire, etc., large solar plants exist primarily to furnish salt for export to large chemical complexes of the more highly industrialized nations.

The United States has tremendous salt reserves in the Salina Basin of the northeast, the Permian Basin, Gulf of Mexico Salt Basin as well as smaller areas in Colorado, Idaho, Arizona, Wyoming, Virginia, California, Utah, and elsewhere (Fig. 8). The deposits are large, of good grade, and in most instances are exploited either by solution mining and/or conventional underground methods. Analyses of salt from these areas are shown in Table 4.

While large areas of the US are underlain by salt deposits, several areas, notably the northwest and most of the eastern states, are without salt. Large amounts of salt are imported into these areas to supply raw material for caustic-chlorine manufacturing and for the fishing industry.

While Canada has large reserves, the principal areas of salt production are in Ontario and Nova Scotia, with 85% of the production coming from Ontario. Recently, an extension of the Nova Scotia deposit was discovered in Newfoundland. There are plans to exploit this deposit to produce salt for a soda ash and/or chlorine-caustic soda company. Salt deposits are also found in New Brunswick.

In Mexico rock salt deposits exist near Monterrey, Coatzacoalcos, and Punta Gorda. The deposits of Coatzacoalcos are a series of salt domes and salt anticlines, several of which have been drilled for sulfur. At the present time salt is being produced from the Parajitos

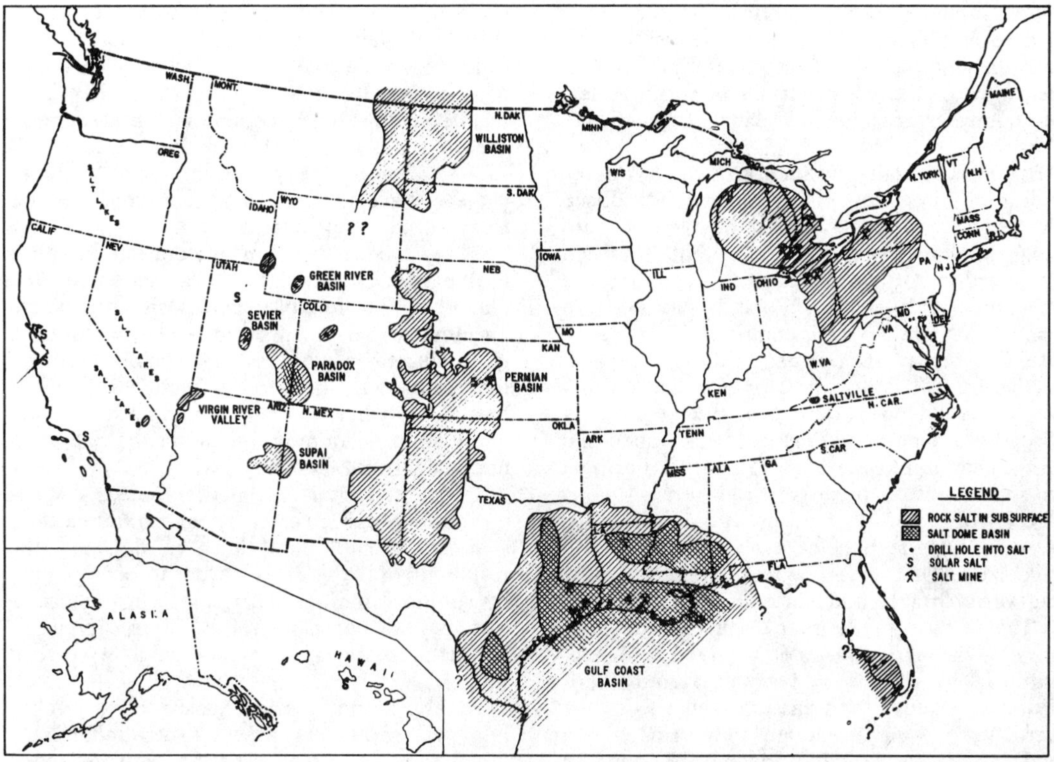

FIG. 8—*Distribution of salt and salt domes, United States (Lefond, 1969, courtesy Plenum Press).*

dome for use in a chlor-alkali plant at Coatzacoalcos. The salt deposit near Monterrey is exploited for use in a soda-ash plant. Although these deposits contain huge reserves, the major tonnage of Mexican salt is solar evaporated, coming from the Black Warrior Lagoon area.

Large solar salt plants exist in many areas of the Caribbean, principally Long Island, Bonaire, and the Bahamas. Production from these operations is exported primarily to the United States. Although salt production in the other Latin American countries is relatively small, it should be noted that Brazil has huge reserves in the Amazon Basin, Sergipe, and the newly discovered offshore domes.

The primary salt-producing countries of

TABLE 4—Analyses of Rock Salt, %

	1	2	3	4	5	6
NaCl	92.08	76.10–99.80	99.25	96.150	98.220	98.150
$Fe_2O_3 + Al_2O_3$	3.52	— —	—	0.015	—	—
$CaCl_2$	2.04	0.50–0.20	0.04	0.018	0.053	0.011
$CaSO_4$	0.71	0.01–0.03	0.69	3.252	0.634	0.639
$CaCO_3$	0.15	1.10–0.39	—	—	—	—
$MgCl_2$	0.08	— —	0.01	0.342	0.052	0.027
SiO_2	1.42	— —	—	—	—	—
$MgSO_4$	—	— —	—	—	—	—
H_2O insoluble	—	22.80–0.30	—	0.223	1.041	1.173

1. Louann Salt, Union County, AL
2. Haynesville Salt, Se4, T5N, R2E, AL
3. International Salt Co., Avery Island, LA
4. Hutchinson Salt, Carey Salt Co.
5. International Salt Co., Detroit, MI
6. International Salt Co., Cleveland, OH

South America are Argentina, Brazil, Chile, and Colombia. With the exception of the underground salt mine at Zipaquira, Colombia, most of the salt produced in South America comes from solar salt pans or playa lake deposits.

In Europe salt production is one of the oldest of industries. The Romans introduced the open-pan evaporation of brines into England as early as 300 B.C. Salt has been produced in Austria since the 8th century, and the mine at Wieliczka, Poland, has been in operation since the 13th century. Today, Russia, France, West Germany, Italy, England, Netherlands, Poland, and Romania are the major salt-producing countries of Europe. Newly discovered deposits in Portugal near Lisbon and those near Loule are expected to lift Portugal into the ranks of the larger producers.

One-fourth of the salt produced in Europe is rock salt; the balance is salt brine used in chlor-alkali and soda ash plants and for making various table and agricultural salt.

The Zechstein (Permian) salt extends from eastern England to Western Poland and constitutes one of the major salt resources of Europe. Keuper (Triassic) salt beds are particularly well developed in Germany. In addition, extensive salt horizons are found in the Miocene, Oligocene, and Jurassic regions of several European countries. Solar salt is produced along the coasts of France, Portugal, Spain, Italy, and Greece (Lefond, 1969).

While there are large rock salt resources in many countries of Africa, most of the salt production comes from the solar evaporation of seawater and salt lakes. Egypt, the major producer, produced only 770,000 tons in 1980. In fact, the total salt production of Africa in 1980 was slightly over 3 million tons. Over two-thirds of this was produced in only five countries and the balance in the remainder of Africa.

Salt deposits in Asia have been exploited for hundreds of years. Solar salt operations and drilling for salt brines have been an industry in China since 2000 B.C. Today, China, the second largest salt-producing country in the world, operates solar salt plants in the Coastal provinces from Southern Manchuria to Kwangsi Province. In the inland areas, salt is produced from brine wells, salt lakes, and a mixture of salt and soils, called "salt earth" (Lefond, 1969).

Exploration

As a result of oil and potash exploration, many new salt deposits have been found in various countries such as France, Portugal, Brazil, Thailand, Morocco, and others.

For example, the recent discovery of salt in Morocco's Mohammedia area was the result of the search for a commercial-grade potash ore in the Khemmiset-Berrechid Basins by a United Nations exploration team. Similar projects have defined, in rather broad general terms, other occurrences of salt.

Thus, the work of an exploration geologist looking for salt is usually a matter of the "final" delineation of the deposit and its environment. The manner in which exploration is conducted is governed to a large extent by the planned method for extracting the salt—that is, by either normal dry mining techniques or by solution mining, including the projected use of the resultant cavity.

If the deposit is to be dry mined, the exploration program must be much more extensive and greater care taken in detailing the results, than if the deposit were to be exploited by solution mining. This care is made necessary by the comparatively greater capital investment in dry mining than in a brine field (Fig. 9).

In evaluating a salt deposit for dry mining, the geological aspects surrounding the deposit are almost as important as the deposit itself. The primary aspects are:

1. Ground Water: The amount of ground water and its hydrostatic head will have a direct influence on the method and cost of shaft sinking. If this ground water is a potable water supply for a community, the protection of the aquifer will require careful hydrological studies. Shaft sinking difficulties will materially increase if an aquifer contains a brine or significant amounts of hydrogen sulfide, ammonia, or hydrocarbon gases or if the hydrostatic pressure is very high.

2. Overburden Characteristics: Considering the overburden as the alluvium plus the rock sequences above the salt deposit, it is essential that the physical characteristics of these horizons be known as fully as possible. It is of particular importance that this information be gathered from the area in which the shaft or shafts will be sunk. These data should include the thickness and hardness of the beds, drillability, fracture pattern, friability, and tendency to oxidize. Shaft sinking contractors generally require this information and occasionally make a physical examination of the cores themselves. Usually NX size "slim" hole develops a core sufficiently large enough for chemical, hardness, drillability, and other required tests (Read and Jacoby, 1957; Severy, 1946).

FIG. 9—*Modern salt plant.*

3. Physical Characteristics of the Rocks Associated With the Salt: In sedimentary layered evaporite deposits, the structural competency of both the rocks underlying and overlying the bed to be mined are of particular importance. Roof spans, pillar size, percentage of extraction, mine layout, and the roof bolting program are some of the mine features which will be dictated by rocks surrounding the deposit.

4. Faults: Fault systems which are common to many salt deposits can create numerous serious problems and should be defined as carefully as possible during the original exploration program (Jacoby, 1969). Domes in southern United States have numerous faults associated with the spines that characterize their upper reaches. These vertical or near vertical fractures often contain brine, oil and/or gases. Failure to develop mining patterns to avoid creating differential stresses across these features often results in the expenditure of large sums of money and energy to grout off fluid intrusions.

Downdropped graben blocks are often the result of post-depositional water intrusions along faults. The dissolving action of the water on one or more of the salt beds may cause a collapse that can take the form of a single block fault or echelon step faults. Either of these may cause expensive realignment of the mining pattern.

5. Facies Changes: Both minor and major facies changes may occur in salt beds over a relative short distance. This has been demonstrated in the salina salts of Michigan, the potash salts of the Khemmiset Basin in Morocco, and the salts of Realmonte, Sicily. In Michigan the A-1 salt grades from a pure anhydrite on the basin's edge near Detroit to an almost pure salt in East China Township, MI, and then to a heavily contaminated potash salt in the area of Midland. In the Khemmiset Basin, salts grade from a lean potash ore at Berrechid to a 98.3% NaCl in Mohammedia. The salt at Realmonte, Sicily, although metamorphosed, shows a similar gradational change from lean potash mineralization to a pure halite. Since most customers desire a product of constant uniform chemical composition, the exploration program must define the limits of various salt grades both vertically within the bed as well as horizontally and the reserves of a standardized or uniform product.

6. Attitude and Position: The attitude and position of the salt beds or, in the case of domes, the flowage lines, influence the mining system, the percentage of extraction, the depth of the mining level, the selection of a shaft vs. an inclined entry, and the rock mechanics of superimposing pillars where multilevel systems are contemplated. In certain cases where internal features of the deposit or the continuity of the quality of salt is in question, then a mill sample should be secured by a bored shaft or inclined entry.

There are a number of advantages to this approach. In addition to allowing an in-situ study of the geology, this test entry can serve to (a) secure samples for mill tests, (b) evaluate the customer acceptance of the salt product, and (c) be a facility for ventilation and/or escapeway.

Where solution mining is contemplated, exploration is largely concomitant with the expansion or replacement of the present brining facilities. Usually regional geology leads to the selection of the brine field site. Based on previously published information, a decision is made on whether or not geophysical information would be useful in guiding subsequent drilling. In cases where domes are involved, or the beds are known to have been altered by diapirism or faulting, geophysical studies may be of distinct value (Mattox, 1968).

7. Chemical Composition: In general, it may be said that the deposit should have a minimum sodium chloride content of 95% or greater to be commercial. In the northeastern portion of the United States the average sodium chloride content required is 97%. Where salts must be competitive with either rock salt from southern salt domes or solar evaporated salt, the sodium chloride content must approach 99%. Depending upon the end use of the salt, various trace elements such as copper, vanadium, boron, chromium, and iron as well as the calcium sulfate, ammonia, and water insolubles must be considered.

Drilling and Coring

Most operators prefer to core at least the first hole. If seismic or other geophysical studies show the level of the top of the salt to be uniform and previous data on the deposit is accurate, a changeover from water to a saturated brine drilling fluid can be made at an elevation in the overlying sediments above and close to the top of the salt deposit. Where zones of lost circulation have been encountered "up the hole," a casing point is usually selected above the top of salt and a string of protective casing installed. Depending upon the design of the well and the economics, the casing is either set temporarily on a packer or permanently cemented in the hole. Where drill holes expose shales which swell and spall or slough in the presence of freshwater, early conversion to a saturated brine drilling fluid often is remedial.

The most important aspect of coring salt is to have a clean, fully or supersaturated brine drilling fluid which will not leach or mechanically etch the salt. In general, even a brine which is fully saturated must be passed through a salt lixator (mechanical saturator) to produce a supersaturated drilling fluid. It can be expected that if the temperature of the drilling fluid on the surface is less than that which is experienced down the hole, even a supersaturated solution will leach the salt from the core and the walls of the hole. A heated saturated solution will help to solve these problems. Table 5 shows the solubility of salt at various temperatures.

Hydrocarbon and trade-named drilling

TABLE 5—Solubility of Salt at Various Temperatures

Temperature		%	Temperature		%
°F	°C	Salt	°F	°C	Salt
− 6	−21.11	23.31*	70	21.11	26.45
0	−17.78	23.83	80	26.67	26.52
+10	−12.22	24.70	100	37.78	26.68
20	− 6.67	25.53	125	51.67	26.92
30	− 1.11	26.16	150	65.56	27.21
32	0.0	26.29	175	79.44	27.52
32.3	+ 0.1	26.31†	200	93.33	27.91
40	4.44	26.33	212	100.00	28.12
50	10.00	26.36	220	104.44	28.29
60	15.56	26.395	227.7	108.7	28.46‡

* Eutectic point
† Transition point
‡ Boiling point at one atmosphere pressure

fluids which will not leach the salt are available. However, these usually are either a fire hazard or expensive and in most cases contaminate the core to such a degree as to hinder chemical analyses.

Failure to use a clear brine will result in the mechanical erosion of the soft halite. One should start with clean equipment and use clean, high purity rock salt of a "CC" grade size. Recycling the drilling fluid through a lixator helps to filter out finely divided cuttings and mud.

After the protective casing is set where needed, the coring is started with a 50 to 90-ft oil field type double-tube core barrel. Even though a double-tubed barrel is used, care should be taken to keep the drill fluid free of mechanically suspended particles.

Cores are usually cut into 1 ft lengths and halved. One portion is halved again to form a quarter which is crushed and ground for chemical analyses. Physical testing is usually accomplished on the second quarter of the core.

When the core hole is completed, it should then be logged with a suite of logs such as gamma-neutron, sonic, or possibly the 3-D, all of which are available from various well logging service companies. These logs are then correlated with the core log, and the geolograph which records the rate of drill penetration. Since faults, particularly low angle thrust faults, are difficult to recognize in cores of salt beds, this correlation with logs will often show repeat sequences of rock strata. Subsequently, a reexamination of the core may show a fine grained granular halite section of core which was previously unrecorded. Sonic logging differentiates anhydritic-dolomite from dolomitic-anhydrite, while gamma-neutron will clearly depict salt from anhydrite (Leroy, 1951).

After one hole has been cored and a suite of logs obtained, most operators consider it unnecessary to core subsequent holes. If the geology is uniform and relatively undisturbed, such a procedure is often satisfactory. However, when the salt deposit has been subjected to tectonic forces, metamorphosed, or affected by ground water solution, cores should be taken from every, or every other, hole depending on the complexity of the geology. It may be expected that coring will increase the cost of the well by approximately 10%, depending on various conditions. When there are plans to use the resultant cavity, coring and the permanent retention of the cores is recommended.

Evaluation of Deposits

Procedure—The procedure for the evaluation of a salt deposit with respect to its exploitation by solution mining is outlined as follows (Jacoby, 1972):

1. Preliminary Considerations
   - A. Plant Requirements
     1. Quantity of brine
     2. Quality of brine
     3. Degree of saturation required
   - B. Subsurface Factors
     1. Geology
     2. Rock mechanics
     3. Hydrology
     4. Previous extractive operations and environment
   - C. Surface Factors
     1. Environment
     2. Topography
     3. Property boundaries
     4. Plant—brinefield relationships
   - D. Feasibility Study
     1. Preliminary brinefield layout
     2. Capital investment
     3. Cost of capitol
     4. Operating cost
     5. Brine treatment costs
     6. Taxes and tax incentives
     7. Return on investment
2. Preliminary Engineering
   - A. Property
     1. Acquisition
        - a. Option of surface and mineral rights
        - b. Rights of well sites
        - c. Ingress and egress rights
        - d. Right-of-way for roads, pipelines, and power lines
        - e. Storage rights
        - f. Conditions of abandonment
   - B. Water Supply
     1. Environmental Considerations
        - a. Ground water table
        - b. Salt water encroachment
        - c. Effect on local agriculture, industry, and domestic use
     2. Supply characteristics
        - a. Volume (continuity and cyclic variations)
        - b. Quality
   - C. Salt Supply
     1. Exploration
        - a. Geophysical surveys
          - (1) Seismic
          - (2) Gravity
        - b. Drill holes
          - (1) Cores
          - (2) Geophysical logs
          - (3) Hydrological tests
     2. Design data development
        - a. Analyses of well logs and cores
        - b. Physical testing of cores
        - c. Water supply evaluation
        - d. Equilibrium studies
3. Design and Cost Analysis
   - A. Brinefield Layout and Cost
     1. Salt wells design and cost
     2. Subsidence grid system design
     3. Roadways and well sites
     4. Pipeline and power line
     5. Cathodic protection
     6. Pump and pump facilities
     7. Feed water treatment facilities
     8. Reservoirs and aeration
     9. Security
   - B. Supervision Costs
   - C. Operating Costs
   - D. Maintenance Costs
   - E. Taxes and Insurance
   - F. Depletion and Depreciation
   - G. Cost of Capital
   - H. Return on Investment

Specifications

"With regard to any given salt occurrence, the geologist or mining engineer must know the location of the market to be served, the ultimate use of the salt, and the customer's requirements. This will determine whether the prospective deposit is to be dry-mined or solution-mined and whether, if mined by solution, the resultant brine is to be evaporated or electrolyzed" (Jacoby, 1972).

Three types of rock salt are normally produced in the US. They are based on screen sizes and are fine crystal (FC), coarse crystal (CC), and grade No. 1. Typical analyses of these grades are shown in Table 6.

TABLE 6—Typical Analyses of Rock Salt Products, New York, %

	FC*	CC†	No. 1
Moisture	0.038	0.023	0.022
Water insolubles	1.306	1.227	1.613
Calcium sulfate ($CaSO_4$)	0.533	0.473	0.639
Calcium chloride ($CaCl_2$)	0.022	0.022	0.030
Magnesium chloride ($MgCl_2$)	0.006	0.006	0.008
Sodium chloride (NaCl)	98.095	98.249	97.688
Sodium chloride (dry basis)	98.133	98.272	97.710

* FC, "fine crystal."
† CC, "coarse crystal."

These analyses demonstrate the resistance of hard, insoluble particles of anhydrite, dolomite, and shale contained in coarse No. 1 to degradation, resulting in a corresponding lowering of the sodium chloride content in these coarser fractions.

After completing chemical and physical testing on the cores, the crushing and screening test may be conducted on the residual portion of the core to determine the chemical composition of the products which might be formed and to roughly ascertain the percentage of FC (fine crystals) and/or −10 mesh material which may be obtained during operations.

The purity of rock salt products derived from bedded deposits is not only related to the deposit itself, but also to the particle sizes that are manufactured. Typical screen analyses of the three principal rock salt products marketed in the US are shown in Table 7.

TABLE 7—Typical Screen Analyses of the Three Rock Salt Products Marketed in the United States

Sieve No.	Cumulative % Retained on	Component % of Aggregate
Screen Analysis: Fine Crystal (FC)		
4	0.0	0.0
8	0.0	0.0
12	12.2	12.2
20	52.9	40.7
40	72.1	19.2
70	87.1	15.0
80	89.3	2.2
Pass		
Screen Analysis: Coarse Crystal (CC)		
3/8 in.	0.0	0.0
4	24.0	24.0
8	82.0	58.0
10	89.7	7.7
12	94.7	5.0
16	98.2	1.8
Pass		
Screen Analysis: No. 1		
3/4 in.	0.0	0.0
3/8 in.	5.0	5.0
4	90.3	85.3
8	98.7	8.4
Pass		1.3

Mining

Dry Mining

All presently operating rock salt mines in the United States and Canada use the room-and-pillar mining method. The height of the pillars is limited by the vertical section of commercial quality salt that is being mined and in some cases the thickness of salt required for a back or roof. The room width is a function of the thickness of the overlying incompetent rocks and the physical characteristics and thickness of the salt which forms the roof.

Pillar widths are governed by the percentage of extraction permissible at various depths and room widths. Rooms widths are basically related to the equipment used in the mining process and more particularly to the size and requirements of the haulage units (Fig. 10). Because of the adverse economics involved in the construction of salt fillets or curved wall-roof intersections, mining procedures dictate the formation of 90° angles between the roof and pillars. This procedure occasionally leads to severe pillar spalling. In bedded deposits the mining height is usually not sufficient that these slabs create safety hazards, but in dome deposits serious problems do sometimes develop.

Most mines in the bedded deposits of the northeastern portion of the United States and Canada roof bolt their haulageways and permanent work areas. The diameter, length, and spacing of the bolts vary with circumstances. Most operations use bolts that vary from 4 to 10 ft in length on patterns of 4-ft centers. Rusting often occurs where connate waters are encountered.

Due to the difficulties experienced in setting bolts in salt, only minor roof bolting is done in salt domes. Bolt holes in salt have a smooth slick surface which forms poor seats for bolt anchors and which deform plastically in time under pressure. Wedge-type slotted bolts are virtually impossible to set.

Many operators drill vertical roof holes at periodic intervals to test the thickness and character of the roof salt. This is made necessary by the fact that in bedded deposits solution erosion usually occurs at the top of the deposit, causing rather rapid changes in the thickness and nature of the roof salt. These vertical holes also allow any fluids or gases to escape, thus preventing a pressure buildup as minor roof bed separation occurs. Occasionally shales, exposed by the bolt borehole, oxidize, causing adverse effects.

In mines in the salt domes of the southern United States, pillar and room sizes approach a near constant width of 100 ft. Recently some deviation from this historical pattern has

FIG. 10—*Haulageway and pillars, underground salt mine.*

occurred. At Avery Island rooms monitored for rock movement show that mining heights of 135 ft are feasible. Continued advances in rock mechanics led to the establishment of a room-and-pillar system where the room width and height were designed for 150 ft. Here fillets were left at the intersection of the roof and rib line (Halbouty, 1967).

Equipment used in the mining of salt varies more than perhaps with any other segment of the mineral industry (Fig. 11). Because of the softness of salt, abrasion is at a minimum. There is practically no corrosion to equipment in a salt mine. Mine temperatures vary from a year around average of about 60°F in the Detroit salt mine to 75°F at the Avery Island salt mine. With such favorable factors, power shovels which were first installed in 1935 in the Detroit mine are still operating efficiently today. With the continued expansion of the tonnage demand experienced by the salt industry, new and more modern equipment is added, with a reluctance to discard the older, yet still operable, machines.

Depending upon the height and width of the rooms, the depth of the mining level, the tonnages produced, cost of labor, and specific operating techniques used, direct operating costs may vary from about $0.90 per ton to $1.70 per ton. The single most important factor in these costs is labor.

Solution Mining

Solution mining in the last 20 years has undergone radical changes. Prior to 1955 all solution mining was conducted by either "annulus injection" or "tubing injection" of solvent with relatively small tonnages being produced by the Trump method. At this time the system of hydraulic fracturing was developed which had the following effects (Bays, 1963; Bays, et al., 1960):

1) Altered the cavity configuration from the "morning glory" shape to hemispheres and ellipsoids, thus increasing the stability of openings and reducing the likelihood of subsidence.

2) Increased the percentage of extraction from less than 5% to something in excess of 40%.

3) Reduced brining costs in most operations by 50%.

FIG. 11—*Mining equipment in an underground salt mine, Retsof mine, International Salt Co.*

4) Increased the productive capacity of wells from approximately 50 gpm to about 500 gpm or more.

5) Reduced well maintenance costs to less than 5% of their former amount.

Most companies now favor using a contract driller instead of operating their own rigs with a company drill crew. Practically all new brine wells are constructed with rotary drilling equipment. Steel casing is centralized in the drilled hole and cemented back to surface.

Underground storage of hydrocarbons has lead to utilization of the cavity created by brining. This procedure started in Texas and has gradually spread throughout the US at a much slower pace than the acceptance of hydraulic fracturing techniques. In conjunction with brining operations the storage of hydrocarbons has a tendency to lower brining costs or even to show a profit over and above the initial cost of producing salt.

Recently, plant wastes were injected into abandoned brine cavities. At Watkins Glen, NY, pan bleeds from the evaporators and residue from brine purification are injected into an old cavity and the resultant salt effluent from the cavity reprocessed. As the feasibility of this process proves itself to environmental authorities this practice should spread rapidly. Once again, this will have the effect of reducing the actual cost of solution mining.

Brining costs vary tremendously from one area to another, depending largely on the capital costs involved in well construction. Brine wells in some areas are as shallow as a few hundred feet and in others are over 10,000 ft in depth. Thus, individual well construction costs vary from about \$20,000 to over \$250,000. The operating technique, together with the thickness of salt in the deposit which is to be exploited, will determine the tonnage of salt produced by a given well and thus, the capital cost per ton of salt. Operating cost of a brine well using the hydraulic fracturing technique is primarily a power cost. Thus, the major cost of brine well operation is a function of depth of the well and power costs. In the United States the cost of producing a ton of salt in brine varies from a low of about \$0.45 per ton to approximately \$2.00 per ton.

Solution mining in the United States in 1980 extracted some 22,231,000 tons of salt in the form of brine for chlor-alkali production (Kostick, 1981). Added to this was salt produced in the brine for the manufacture of 6,314,000 tons of evaporated salt. Thus, the chemical specifications and geological circumstances surrounding the solution mined production of about 62% of salt is of major importance.

It is gradually becoming apparent that the insoluble rocks which are exposed to the cavity during brining operations are in some cases major contributors of impurities to the brine being produced. Brine-soluble and water-soluble components of an otherwise insoluble sedimentary rock are occasionally present in quantities sufficiently high as to cause significant contamination of the brine being produced. In bedded deposits these rocks often contain connate waters which have high concentrations of minerals other than sodium chloride.

Generally, it is necessary to purify the brine before it can be used in the chlor-alkali cells or soda ash plants. Some types of electrolytic cells require solid salt for the resaturation of the spent cell liquor. As the spent cell liquor is recirculated, the metal ions or other impurities, unless removed, build up in the brine. Eventually the whole system has to be purged. The effect of trace amounts of soluble metal ions in the electrolysis process is very complex. Vanadium, molybdenum, chromium, and titanium are detrimental to the process since they cause hydrogen to be liberated in the chlorine. These metals also plate out on the amalgam in mercury cells and cause difficulties with thick mercury and formation of mercury "butter." The ions of copper, zinc, and lead are undesirable as they contaminate the finished caustic. Calcium and magnesium have to be removed from diaphragm cell feed to prevent blockage of the diaphragm by precipitation of lime and magnesia. Sulfate ions cause the oxidation of the graphite electrodes.

The exact specifications of the feed brine either for electrolytic cells or vacuum pans vary with individual operators. In general, however, specifications for maximum allowable concentration of impurities in cell feed brine are:

	Mg per L
SO_4	4000
Ca^{++}	100
Mg^{++}	1
ClO_3	2000
Fe^{+++}	1
Cu^{++}	0.5

The more contaminated the raw brine, the more complex and costly is its treatment. This is true of brine used not only for cell feed, but also for the production of granular salt. These additional costs are experienced not only in additional chemicals for treatment and additional costs for operating and maintenance labor, but also for the investment needed for larger treatment facilities (Kauffman, 1960).

Solar Salt

Solar evaporation operations may have a variety of brine sources—seawater, salt springs, salt lakes, or the shallow wells of the original Syracuse, NY, operations. Even playa deposits themselves or the recrystallized salts from the solution of weathering salt outcrops may be catagorized as naturally occurring solar salt evaporation operations. Salt lakes furnish the feed brine for operations around their shores, while in other areas artificial brines produced from bedded rock salt deposits constitute the brine feed.

The construction of solar salt ponds is a matter of good engineering. Levees should be built of an impervious clay. Pond floors should be free of vegetation, sand lenses, and coral knobs which might give rise to seawater "boils" during operations. Such boils can be a major source of dilution as well as an escape route for brines which have been concentrated depending on the balance of the hydrostatic heads. Not only do these boils lead to a dilution and/or loss of concentrate, but, in the case of harvesting ponds, to a contaminated salt.

In most solar evaporation operations salt water is introduced by pumping or by high tides into large concentrating ponds. As the gravity of the brine in the concentrating pond increases it flows by gravity toward the pond's outlet and into the "lime" pond where calcium sulfate and other more insoluble compounds begin to precipitate out of solution. These lime ponds usually constitute about 10% of the total area of the concentrating ponds.

Prior to this transfer point, that is, before it enters the lime pond, the liquor is about 20°Be (20% dissolved solids). In the concentrator pond a part of the calcium sulfate, and all of the calcium, magnesium, and iron carbonates are precipitated. When the density in the lime pond has reached 26 to 26.5°Be, the fluid is either drained or pumped into the crystallizer or harvesting ponds where large quantities of calcium sulfate are deposited on the floor of the pond. Some salt starts to crystallize out of solution at 25°Be.

After its transfer into the crystallizer ponds the salinity continues to increase to 29.5 to 30.0°Be. In the range of salinity from 26.5 to 30.0°Be very pure halite is deposited on the floor of the crystallizer. The volume of salt deposited in this range of salinites leaves approximately 12% NaCl in the bitterns or "tails." Continued evaporation causes contamination of salt with magnesium sulfate, magnesium and potassium chlorides, sodium and magnesium bromides, together with other trace elements.

The crystallizing or harvesting pond is the most important area in the operation for many reasons. It should be rectangular in shape to accommodate modern harvesters. The material underlying the pond floor must be dense enough to support heavy equipment such as the harvester (Fig. 12). These crystallizer ponds should be close to the storage and dock areas to keep haulage costs to a minimum. The tonnage per acre of area employed in the total operation should be in excess of 40 tons and harvesting ponds should produce approximately 600 tons per acre.

In comparison to the lime ponds, the harvesting ponds usually occupy one-half to two-thirds the area. Because of the obvious economic advantage of gravity flow, ponds are arranged in such a manner as to take maximum advantage of natural contours and gradients. Fig. 13 illustrates harvesting ponds near the dock at International Salt Co.'s operation at Bonaire, Netherlands Antilles.

Although in some underdeveloped areas of the world salt is still manually harvested from the crystallizer ponds, all modern operations use mechanical harvesters. They are of various designs, but generally handle between 100 to 250 tph. Narrow gage rail haulage systems are giving way to truck haulage.

Raw salt is hauled to a wash plant where it is washed with a brine from the concentrating ponds and then finally sprayed with freshwater. The brine wash removes any dirt, clay, particles of calcium sulfate, and any organic material. The freshwater removes the occluded bitterns.

The chemical composition of a typical bittern is shown in Table 8 (ver Planck, 1957).

These salts or bittern are usually pumped back into the sea at a point where it will not become a contaminant of the raw brine feed.

Where sufficient quantities of bitterns are available the liquor is subjected to further

TABLE 8—Composition of a Typical Bittern

	%
NaCl	12.5
$MgCl_2$	8.7
$MgSO_4$	6.1
KCl	1.9
$MgBr_2$	0.18

Source: ver Planck, 1957.

FIG. 12—*Harvesting pond and ship-loading facilities, International Salt Co., Bonaire.*

FIG. 13—*Solar salt harvesting.*

processing for the removal of other salts such as those of magnesium and potassium. The cycle time for the concentration of raw seawater to a precipitated salt and a waste or byproduct bittern may be from a few months to several years, depending upon winds and various climatic factors. For example on Taiwan the cyclic time from raw seawater to bittern is about four to five years. Heavy rains or hurricanes can rupture the levees or cause the brines to become diluted, thus increasing the time span required to complete the cycle from seawater to salt.

Processing

Rock salt as dry mined is generally subjected only to crushing and screening. Exception to this procedure is the use of the Sortex process and/or the Thermoadhesive process to upgrade bedded rock salt products from an average NaCl content of 97% to a product of 99.0%+. The Sortex process depends on the translucence of salt and a jet of compressed air. The thermoadhesive process depends on the absorption of light by dark particles of anhydrite, shale, and dolomite. These dark heated particles adhere to a resin-coated belt, separating them from the translucent halite. In most instances the thermoadhesive process will remove 40% of the water insolubles from the crushed salt in a single pass through the process.

The purest grades of commercial salt are produced by the treatment of FC (fine crystal, 10 mesh) rock salt in a recrystallizer. In this process the fine granular rock salt is dissolved in high temperature brine. Due to the inverse solubility of calcium sulfate, the calcium sulfate is not taken into solution resulting in the production of a very pure hot brine. The salt produced by the recrystallizer may be as pure as 99.99% NaCl. The water balance in the recrystallizer system results in no net evaporation capacity. Direct injection and condensation of steam through a direct contact heater into the system dissolves the fine rock salt and heats the brine at the same time. Salt is produced in the evaporator by flash evaporation and by cooling. The vapors released in the lower effects are then used to augment the high pressure steam and thus recover as much of the heat as possible. Steam from the last effect is condensed in a barometric leg condenser (private communication).

Evaporated salt had its beginning before recorded history. Modern evaporation technique had its birth in England in 1812, when a Mr. Howard built a single-effect vacuum evaporator for the sugar industry. The first multiple-effect system was built in 1834, again for the sugar industry. It was not until 1887 that the first single-effect evaporator was used in the United States for salt production at Silver Springs, NY. In 1899 a multiple-effect vacuum pan was built at Manistee, MI.

Basically this is the principle for our current technology. The most common system used today is triple and quadruple multiple-effect system constructed of either Monel or Monel-clad steel. From the design standpoint, there are dozens of different original designs many of which have undergone numerous modifications during their productive lifetime. Some of these have progressed from the vertical tube type first built in 1850 to basket type used first in 1877 to the forced circulation of the 1890s.

Standard means for producing granulated salt for human consumption is by either the enclosed vacuum pans or in open pans. The most common process is the multiple effect evaporators. In this process the vapors from the boiling brine in the first vessel or effect are used to boil the brine in the second effect. This second effect is at a lower pressure than the first with the first effect vapors circulating through heat exchange tubes. The vapors from each succeeding effect are circulated through the following effect, which in turn is at a lower pressure and enables it to boil at a correspondingly lower temperature. In general, the multiple effect system has three or four of these vacuum pans.

Other means by which closed evaporators operate are by passing hot brine into a series of vessels or effects without the addition of heat to the effect. These depend on a low vacuum to "flash" the water from the hot brine.

Open pans or grainers consist of long shallow pans where the brine is heated by steam that is injected into the brine from below its surface. In this process the brine is never brought to the boiling point, with the resultant formation of a flake-type salt rather than the typical cube of halite produced by the vacuum pan. Gradually this process of producing salt is being discarded due to the high energy consumption per unit weight of salt produced.

All of these methods can operate on raw brine of varying degrees of purity. Depending on the purity of product required, the raw brine can be treated chemically before evaporation to remove the calcium and magnesium salts. Again, the brine can be treated with various additives to modify the crystal habit of the salt produced.

Transportation, Handling, and Storage

Rock salt, evaporated salt, and solar salt are transported from producing locations to customers and redistribution points by water, rail, and highway. Under normal conditions of distribution, the cost of transportation often exceeds the value of any large shipment of salt that travels over a few hundred miles.

Waterborne Salt

Ocean Movements: Although all types of salt are transported by oceangoing vessels, the major tonnage of salt moved is solar salt or rock salt. Salt moves in bulk freighters which range in size from 6,000 to over 100,000 tons, and salt may comprise a full or partial cargo.

Present major ocean movements of salt are as follows:

Solar Salt:

From the Bahamas, Netherlands Antilles, and Venezuela to eastern United States.

From Mexico to both eastern and western United States and to Japan.

From Australia to Japan.

From the Mediterranean, particularly North African coasts, to Europe and the United States.

Rock Salt:

From Chile to the United States.

From central Europe to the rest of Europe and the United States.

From Canada to the United States.

Great Lakes and St. Lawrence Seaway: Salt movements on the Great Lakes and the western regions of the St. Lawrence are mainly between the United States and Canada and vice versa, and between major United States east coast ports. In recent years some imported salt from Europe and the Caribbean has entered the region via the St. Lawrence Seaway. The main commodity shipped is bulk industrial salt and highway ice control salt.

Salt moves from the major producing facilities on the shores of the Great Lakes to major redistribution points by bulk and/or self unloading freighters, or barges, thus taking advantage of the economics of large volume waterborne transportation and the strategic value of lake ports for serving their respective areas.

As the economics of bulk transportation favor large cargoes and handling systems, the trend on the Great Lakes is to larger self-unloading vessels, requiring deeper water and more extensive dock loading and handling facilities, with a gradual demise of the smaller freighters. The normal navigational season on the Great Lakes is from April to December. Traffic generally stops in mid-December due to the ice buildup in the harbors. These harbors reopen in mid-April. Plans are currently being formulated to create ice-free ports with the use of large ice breakers.

US River and Canal Systems: Movement by barge is the lowest cost transportation for those commodities for which it is adaptable. Thus, salt moves by barge on all navigable sections of the central and eastern river and canal systems in the United States, on certain navigable rivers on the west coast, and between ports on the Gulf of Mexico. Barges range in size from 900 tons or less to massive 10,000-ton oceangoing units that are normally constructed with a double shell hull. The outer shell forms a floating hull and the inner shell forms an obstruction-free hopper or cargo hold. Salt moves almost exclusively in covered dry cargo barges, with either roll top or lift top covers.

Barge movement of salt in the southern United States begins at the producing mines, most of which are located south and west of New Orleans, in salt domes along the low lying southern Louisiana and Texas coasts, close to the Intracoastal Waterway system. The salt-laden barges move via the Intracoastal Waterway to join larger tows proceeding north up the Mississippi River. Barged salt reaches Kansas City, MO, Omaha, NE, Sioux City, IA on the Missouri River system; as far north as St. Paul and Minneapolis on the Mississippi River and Chicago, via the Illinois River. On occasion, river barges enter Lake Michigan via the Calumet River and proceed to destinations on the southern shores of the lake. Barges of southern salt also reach Pittsburgh, PA, and both Charleston and Fairmont, WV, via the Ohio, Allegheny, Kanawha, and Monongahela Rivers. Nashville, TN, is supplied with bulk salt via the Cumberland River; Chattanooga and Knoxville receive barged salt from movements on the Tennessee River.

In addition to these major movements, barged salt reaches Mobile and southern Alabama via the Intracoastal Waterway. The west coast of Florida is also served by gulf barges. In the Pacific northwest, limited quantities of salt move in barges on the navigable sections of the Columbia River system.

Normal winter conditions in midwestern United States result in the closing of certain sections of the Missouri and Mississippi River systems due to their freezing over and, hence, barge movements cease. The Missouri River above Kansas City, MO, and the Mississippi River above Davenport, IA, are usually closed to traffic from November through April. Except in extremely severe cold weather, traffic on the lower stretches of the Illinois and the Ohio Rivers and their tributaries is not interrupted in the winter months.

Rail Shipment of Salt

Bulk salt is carried in both boxcars and hopper cars, although by far the largest tonnage moves in the latter. Invariably covered hopper cars are used for bulk salt, as vagaries of weather preclude the successful use of the open hopper car.

Covered hoppers, originating in the south, particularly those loaded on hot humid days, experience problems with condensate collecting on the internal surfaces of the car. Where finer sizes are involved i.e., CC size and smaller, any moisture such as condensate will produce "caking." In order to minimize this effect, free flowing agents such as YPS (yellow prussiate of soda) are added. Salt shipments by rail or truck which contain any free moisture and where shipping during cold weather is anticipated are treated with an antifreeze formulation. Shipments are normally made in multicar lots (3–12 cars) to take advantage of substantial savings through special rate for unit trains.

High quality rock salt generally requires specialized equipment such as rail cars fitted for pneumatic loading and unloading and handling systems at destination. Large volumes of packaged salt products move in box cars, largely on pallets, and in many instances utilize specialized DF (damage free) and DFB (damage-free-bulkheaded) equipment.

Truck Shipments of Salt

Both bulk and packaged salt moves on trucks, primarily on short haul routes, generally less than 200 miles, from producing locations or from warehouses and stockpiles direct to customers. Bulk salt is carried in dump trucks or specialized pneumatic and tank trucks, depending on its grade, purity, and end use. It is important in shipping bulk salt that the truck be properly tarped. Any moisure inherent in the salt from production, or gathered during loading or hauling, will freeze in cold weather due to the supercooling effect of wind on the metal body of the truck. Certain industries take delivery of high purity salt in small specialized transportable bins carried on trucks. Packaged salt is moved on flatbed or van-type units.

Storage and Handling

Economics of supply and distribution dictate that salt be transported in bulk quantities to points as close to the area of consumption as possible and distributed from this point to the consumer in as large a quantity as the transportation system will allow. These deliveries should be in as large a quantity as the receiving point can handle without overtaxing the handling systems or causing undue delay in unloading the hauling equipment.

Apart from direct shipments of salt by one of the transportation systems discussed, both bulk and package salt are distributed through an extensive network of stockpiles and warehouses. In almost all cases the redistribution point is supplied from the producing plant in bulk quantities, or in quantities sufficient to enable the shipper to take advantage of freight concessions based on volume.

The maintenance of grade or quality from the producing plant to redistribution point or to the customer is of major importance in the salt industry. This concern arises from the normally high degradation rate of the halite crystal. The fewer the number of times salt is rehandled in the distribution process the better the retention of particle sizes within the grade specifications of the product being shipped. Hence the distribution system utilized must be designed to minimize the numbers of rehandling and transfer points.

Off-Loading Waterborne Salt

The most common method of unloading bulk salt from oceangoing vessels and barges is by clamshell bucket and crane. The unloading equipment is often an integral part of oceangoing vessels. However, when bulk freighters do not have self-unloading gear, cranes on shore or on barges are brought alongside for the transfer of the salt from freighter to dockside storage or to a transportation system. On the river systems where barges are moved to the discharging facility, either barge-mounted or shore-mounted cranes unload the cargo. In both cases, salt is normally dropped into a hopper which in turn feeds a conveyor system or trucks for continued movement to the customer and/or storage.

The majority of the salt moving on the Great Lakes is now carried in self-unloading vessels. Conveyor systems built into the boats in tunnels beneath the load carry the cargo to an onboard discharging conveyor system supported on a boom. The booms vary in length, with many now over 200 ft long with up to 120° travel on either side of the ship's centerline. This versatility enables bulk salt to be off-loaded at convenient dockside locations and placed in windrow piles.

Off-Loading Rail Salt

Unloading from boxcars requires bulk salt to be brought to doorway by hand shovels, miniature front-end loaders, or other mechanical means, then dropped into a hopper, chute, or dissolving tank alongside the boxcar.

Hopper cars are provided with release doors or gates beneath the car. Movement away from the car is accomplished by either an undercar pit and conveyor system, or a portable overtrack conveyor which is placed between bottom of car and top of rails. Salt can be conveyed directly to trucks or into a warehouse or stockpiles. Certain covered hopper cars are unloaded pneumatically, and salt is conveyed from rail car to storage by flexible hoses and pipes. This requires specialized handling and receiving equipment and is installed only where a constant salt supply is required.

Off-Loading Trucked Salt

Depending on grade size, purity, and use, bulk salt is carried from the producing point in either dump or tank trucks. Dump trucks drop loads either directly into hoppers or on a surface pad, so that salt can be loaded into hopper by front-end loaders. On occasion the salt is dumped directly into hoppers equipped with slingers. Salt is then conveyed into warehouse or stockpile. Dump trucks also off-load into pits, particularly where there are brine manufacturing operations.

Tank trucks, depending on type, can discharge either directly through bottom doors into pits, or by pneumatic methods into dissolving tanks or dry storage bins.

The selection of locations for warehouses and stockpiles for the secondary redistribution of salt products, as in other commodities, depends primarily on the marketing and servicing requirements, mix of products, the economics of transportation both to and from the storage and handling facility, and the costs of handling salt through the facility.

Storage Facilities

Bulk Outside Storage: Bulk storage is common for highway salt and certain grades of salt used by industry. Salt is brought to strategically located storage areas by water, rail, and truck, depending on accessibility from producing point and market requirements. The salt pile is usually built with a conveyor, although bulldozers, clamshell buckets, and cranes are also employed at some locations. Of prime importance in this type of storage is the control of salt dust during the period when the pile is being constructed. Piles are normally developed as cones or flat-topped windrows, both straight or curved; and every effort is made to produce even geometric shapes to minimize size of stockpile base and to simplify covering.

Environmental considerations dictate that salt piles must be placed on impervious, well-drained asphalt or concrete pads, and be fully covered with a suitable waterproof material. If moisture is allowed to enter a stockpile, brine leaches out, and drains away from the pile, to create unsightly white residues, and also to pose a threat to surface waters and potable ground water aquifers. Many types of covers are available ranging from treated burlap and canvas to high strength plastics and/or combinations of these materials. They are used mainly to prevent rainwater from entering the pile and to prevent fine salt from being blown away from the pile. Good housekeeping is a necessity at the edges of the piles so that rainwater runoff does not come in contact with salt while flowing away from pile base. Mesopiestic water will be drawn into the pile by capillary action if any portion of the toe of the pile is in contact with free water. This moisture will, through evaporation, cause a hard crust to form on the pile. In addition, moisture in bulk stored materials will cause caking and freezing problems in transport during winter months. Finer particle sizes draw more moisture than coarser sizes. Covered material lasts one to two years if undisturbed. Covers are not normally reusable if a pile has been depleted.

The selection of locations for stockpiles and warehouses for the redistribution of salt products depends primarily on the marketing requirements, product mix, and economics of transportation and handling. Environmental aspects which govern the design and construction may, depending on local conditions, make a site untenable from the standpoint of economics.

Bulk Inside Storage: Normally, industrial salt is stored inside, in either surface bins or silos. Salt which is to be used for bagging is usually stored inside. The main reasons for inside storage are to keep salt dry and uncontaminated. In some areas, highway ice control salt is stored inside and, as environmental factors become more rigid, the tonnages stored in this manner will increase. Most warehouses separate salt grades with walls or dividers in order to avoid mixing.

Two of the problems associated with storage of salt are listed below with the most common means of prevention and methods to overcome the problems:

Caking—Caking is the greatest problem in all types of salt storage. The degree of caking depends primarily on transit time, changes in moisture content, particle size, storage method and duration, chemical composition, and temperatures. Long transit times with the vibration of loads and inherent segregation of fines toward the bottom encourages caking, and long storage times allow salt to pass through several caking cycles, thus increasing overall caking.

Coarse salt, greater than +⅜ in., seldom gives trouble in storage. Bins for storage of fine salts should be built with poke holes so that chunk or caked salt can be displaced. Salt should be kept moving through bins as rapidly as possible with frequent salt additions and discharges.

Transit time from the production facility to the consumer should be at a minimum. Salt should be unloaded as soon as possible after being received and in-plant storage time should be kept at a minimum.

Degradation—Salt tends to degrade during loading, conveying, discharging from heights, and through handling by any and all types of equipment. Salt therefore must be handled as gently and as few times as possible between producing facility and customer.

Package Salt Storage: Standard warehouse procedures are used for inside storage of all forms of packaged salt. For the most part, packaged salt is palletized when received at the warehouse, or immediately after receipt, and is placed in storage areas with forklift trucks. The container shape and strength—either bag or carton—determines the stacking pattern of units on each pallet, and also determines the number of pallet loads which can be safely stacked one on another. The use of slip sheets and the shrink wrap techniques are gaining favor.

Storage areas must be clean and dry, as packaged salt is susceptible to moisture damage. Development of plastic bags is helping to alleviate this problem.

Marketing

A large percentage of the nations of the world have their own source of salt. Of the more than 181,608,000 tons of salt produced worldwide in 1980, the United States produced 41,480,000 tons (Kostick, 1981). Because of the wide distribution of salt either in the form of rock salt or solar-evaporated salt, prices for many years have been low and stable. Recently rock salt prices have as much as quintupled partially because of the inflated energy cost. As one of the five building blocks of the chemical industry, industries requiring salt have generally located near a source of salt and cheap power.

In most cases, irrespective of the low freight rates of water transportation, the cost of transportation is often more than the f.o.b. price of the salt and in some cases is as much as 75% of the delivered value.

Because the United States is the world's largest producer and consumer of salt, salt economics are governed to a large extent by US industrial and governmental agency purchasing. In the purchase of salt by governmental agencies for snow and ice control, contracts are awarded on a bid basis without regard to point of origin and, in some cases, require two or more sources of supply.

Use of salt for ice and snow control by federal, state, and local agencies was 6.4 million tons in 1980, down from 9.3 million tons in 1972. Practically all this tonnage was rock salt. During 1980 US rock salt producers mined 12.3 million tons, down from 14.4 million tons in 1972. Another 1,087,000 tons of snow and ice control salt was imported. Salt imports by end use are shown in Table 9 (Kostick, 1981).

Prices as quoted by the May 9, 1983 *Chemical Marketing Reporter* are as follows:

Salt, evaporated, common 80-lb bags, carload lot, North works	$3.00
Salt, 80-lb bag chemical grade, same basis	3.20
Salt, rock, medium coarse, same basis, 80 lb	2.05

The average value of evaporated salt reported by producers to the US Bureau of Mines during 1980 was $76.44 per ton, up from $21.26 per ton in 1972. On the same basis, the average value of rock salt in 1980 was $14.65, up from $6.58 in 1972. Salt in brine form was $6.50 per ton in 1980, up from

TABLE 9—Salt Imports Into US by End Use (Thousand Short Tons)

Use	1979	1980
Government (highway use)	2,396	1,087
Chemical industry	762	803
Water conditioning service	148	179
Other	388	260
	3,695	2,330

Source: Kostick, 1981.

$3.61 per ton in 1972. The energy-intensive evaporated salt product demonstrates the most dramatic price increase. Tariff rates for salt are as follows:

Item	Most Favored Nation		Non-Most Favored Nation
	Jan. 1, 1983	Jan. 1, 1987	Jan. 1, 1983
In brine	4.8% ad val.	3.7% ad val.	20% ad val.
In bulk	2.6% ad val.	Free	26% ad val.
Other	Free	Free	11¢/100 lb

For tax purposes a depletion allowance of 10% is allowed for both foreign and domestic deposits.

Five states contributed a total of 33,703,000 tons, or 81.3% of total US production (Table 10).

The average value per ton of salt established for each state (Table 11) is a composite of types of salt. Various other factors influence price. Where a large percentage of the salt produced is in the form of brine, as in Texas, costs are low. Those states producing large tonnages of salt by evaporation reflect increased energy costs in a higher cost per ton. New York, Michigan, and Ohio are situated close to population centers and therefore have less freight absorption with respect to making their products competitive in the marketplace.

In 1980, 11 of the 47 salt-producing companies accounted for 82% of US production. Each of these companies produced in excess

TABLE 10—Major Salt-Producing States of the US 1979-1980 (Thousand Short Tons, Thousand Dollars)

	1979		1980	
	Quantity	Value	Quantity	Value
Louisiana	14,207	113,167	12,662	132,182
Texas	11,283	67,602	9,978	93,414
New York	6,387	77,751	5,509	99,395
Ohio	4,135	79,598	3,228	87,371
Michigan	3,080	82,540	2,406	104,842

Source: Kostick, 1981.

TABLE 11—Average Value of Salt Produced for the Chemical Industry, 1980

State	Average Value $ per ton of all salt	Evaporated, Tons
Michigan	$43.58	1,133,000
Ohio	27.07	NA
New York	18.04	638,000
Louisiana	10.44	280,000
Texas	9.36	NA

NA—Not available.

of 1 million tons. Brine produced in 1980 constituted 55% of the total salt produced while mined rock salt accounted for 30%. Vacuum pan and grainer-evaporated types of salt amounted to 9% of the total, with solar-evaporated furnishing only 6% of domestic production.

Leaders in world salt production are shown in Table 12.

United States exports of salt amounted to 831,000 tons with a value of $12,829,000 in 1980. Over 95% of this salt was shipped to Canada.

Imports totaled 5,263,000 tons with a valuation of $44,071,000 or $8.37 per ton (Table 13). The majority of the tonnage was purchased by government agencies for snow and ice control. While the value per ton of exports was $15.43 as opposed to import value of $8.37 per ton, the deficit trade balance for the world's largest salt producer was $31,242,000.

The increase in US imports of salt from 2.494 million tons in 1972 to 5.263 million tons in 1980 was due in part to US companies acquiring foreign sources of cheap solar-evaporated salt. This salt, loaded on oceangoing freighters, can be landed close to the points of consumption along the eastern coast. This, together with inherently low labor costs, adds to the competitive edge of the off-shore producers. As some of these third world governments and labor unions become more sophisticated, this present advantageous position will diminish. Some US companies have already found their Caribbean investments less than attractive.

TABLE 12—Salt Production (tons) by Country, 1980

United States	41,480,000
China	20,000,000
USSR	15,980,000
Germany (West)	14,300,000
United Kingdom	7,300,000
France	7,830,000
Canada	7,748,000
Australia	5,859,000
Mexico	6,600,000
Romania	6,200,000
Italy	5,806,000
Poland	3,700,000
Netherlands	3,818,000
India	8,000,000

Source: Kostick, 1981.

TABLE 13—US Imports for Consumption of Salt, by Country (Thousand Short Tons, Thousand Dollars)

	1979		1980	
Country	Quantity	Value	Quantity	Value
Bahamas	528	3,985	531	5,573
Brazil	197	1,625	62	608
Canada	2,057	15,580	2,089	16,515
Chile	244	1,699	341	2,689
Colombia	41	480	273	2,280
Italy	42	1,205	—	—
Mexico	1,649	11,282	1,457	10,216
Nepal	—	—	22	161
Netherlands	57	960	104	2,034
Netherlands Antilles	175	1,597	193	2,031
Spain	252	1,745	99	831
Tunisia	33	250	60	530
Yemen Arab Republic	—	—	31	163
Other	—	452	—	439
Total	5,275	40,860	5,263	44,071

Source: Kostick, 1981.

A determining factor in world demand for salt is its use in the chlor-alkali and soda ash industries. In the five years from 1966 to 1970, the consumption of salt for this use fell from 20% to 14% of total salt consumed in the United States, although use of salt for other purposes increased. In 1972 salt used for chlorine-caustic soda and soda ash was 26,517,000 tons. By 1980 this total tonnage had fallen to 23,941,000, largely because of the impact of Green River trona, which is progressing as expected.

The demand for chlorinated solvents, chlorofluorocarbons, and chlorinated hydrocarbons is the dominant factor in controlling the demand and hence the price of chlorine. The growth of chlorine consumption in countries where large quantities of pulp and paper are produced is higher than that experienced by other chlorine users. The aluminum industry and manufacturers of certain synthetic fibers also consume a vast amount of salt-based chemicals.

TABLE 14—End Uses of Salt

SALT NaCl

Nutrient or Flavor	Preservative	Food Processing Material	Chemical Manufacturing	Freezing Point Depressant	Metallurgical Processing	Miscellaneous Processing
baking	cheese making	blanching seafood & vegetables	Calcium Hypochlorite	coal antifreeze	chloride roasting	artificial seawater
breakfast cereals	cucumber salting	chicken deboning	Chlorine Dioxide	highway deicing	drawing lubricant	coal briquets
butter and cheese	fish bait curing	crabmeat pickling	Sodium Chlorate	ice cream making	foam killer	dehydrating agent
canning	fish curing	egg preservative	Sodium Fluosilicate	ice manufacture	heat treating baths	dye processing
cattle blocks	hay preserving	fish striking agent	Sodium Hypochlorite	iron ore antifreeze	iron ore cementation	dyestuff carrier
flour mixes	hide curing	gravity separation	Sodium Perchlorate	refrigerating brines	metallurgical flux	electrolytic milling
heat tablets	meat curing	oyster shucking		refrigerating cars	mill scale remover	emulsion breaker
isotonic solutions	sausage casings	wine stabilization			molten metal cover	etching aluminum foil
livestock feeds	well drilling fluids	yeast processing			rare metal refining	herbicides
oleomargarine					sink and float baths	ion exchange regeneration
pickles						leather tanning
potash substitute						rubber coagulant
salted nuts						soap salting-out agent
table salt						soil stabilizer
						starch manufacture
						textile dyeing
						tile glazing
						water softening
						weed killing

Soda Ash, Na_2CO_3

- abrasives
- adhesives
- batteries
- ceramics
- cleansers
- cosmetics
- degreasers
- dyes
- explosives
- fats and oils
- fertilizers
- fire extinguishers
- inhibitors
- insecticides
- leather
- metal fluxes
- ore refining
- paint removers
- paper
- petroleum
- pigments
- soap
- textiles
- water softeners

Sodium, Na

- bactericides
- case hardening
- cosmetics
- detergents
- dye fixation
- dyes
- flour conditioning
- fumigation
- heat transfer
- ore refining
- organic syntheses
- paints
- pharmaceuticals
- photography
- pigments
- plating salts
- pulp bleaching
- starch conversion
- tetraethyl lead
- textile bleaching
- titanium metal
- zirconium metal

Sodium Sulfate, Na_2SO_4

- ceramics
- detergents
- dyes
- explosives
- fertilizers
- metal fluxes
- paper
- pharmaceuticals
- photography
- pigments
- plating salts
- rubber
- soap
- textiles

Hydrogen, H_2

- alcohol
- ammonia
- cooking fats
- high energy fuels
- hydrochloric acid
- metallurgy
- meteorology
- organic syntheses
- petroleum products
- pharmaceuticals
- synthetic fibers
- synthetic gems
- welding

Hydrochloric Acid, HCl

- adhesives
- ceramics
- dyes
- engraving
- inks
- leather
- metal cleaners
- ore refining
- perfumes
- pigments
- printing
- rubber
- soldering flux
- textiles

Chlorine, Cl_2

- anaesthetics
- bleaches
- ceramic colors
- cleansers
- disinfectants
- dyes
- explosives
- fertilizers
- fire extinguishers
- fungicides
- insecticides
- leather
- paint removers
- paper
- plastics
- refrigerants
- rubber
- sewage treatment
- solvents
- synthetic fibers
- textiles
- water treatment
- weed killers

Caustic Soda, NaOH

- adhesives
- batteries
- building materials
- ceramics
- cosmetics
- dyes
- explosives
- fruit peeling
- inks
- ion exchange
- laundering
- leather
- lubricants
- ore refining
- pharmaceuticals
- pigments
- plastics
- rayon
- refractories
- rubber
- soap
- synthetic fibers
- water treatment
- wood processing

Of tremendous importance to the rock salt mining industry is the use of salt for snow and ice control. Approximately 40% of the rock salt mined and about 60% of the profit in its mining are related to the de-icing of roads and highways. Vagaries in the weather plus environmental considerations make this use cyclical. Eventually advanced technology will eliminate this use.

Uses

Salt is the most readily available and abundant source of sodium and chlorine ions. These ions are used in making a number of chemicals such as soda ash, chlorine, caustic soda, and metallic sodium.

Although it is said that there are over 14,000 uses for salt, only a few consume a major portion of the salt production. Table 14 gives seven general categories of uses with a breakdown of the major chemicals derived from salt, and a listing of more important products of these chemicals.

Table 15 shows classes of consumption in the United States. Currently the chemical industry consumes 53% of all domestic salt produced, with chlorine accounting for 86% of this amount. Soda ash now consumes only a small amount of chemical salt.

A new use of salt is in-situ creation of cavities for storage and other uses. Some of the uses will allow dry or solution mining and primary use of the salt as a consumer item, with use of the cavity as a byproduct of the operation. In other cases, the cavity itself is the principal product, with the salt extracted forming a waste with environmental consequences.

New uses for salt cavities during the last three decades, including those currently in the design stage, are storage of liquid hydrocarbons (propane, butane, kerosene, fuel oil and jet fuel); storage of hazardous wastes (arsenides, cyanide, pesticides, herbicides, toxic wastes, and outdated drugs); storage of nuclear or radioactive wastes; storage of compressed air (peak shaving of power in electrical generation); geothermal energy; and wasting of insoluble materials (fly ash and other plant waste in abandoned brine cavities and mines).

Future Considerations and Trends

Salt is relatively inexpensive, readily accessible, and abundant, with no substitute either natural or synthetic. Because of its abundance there are no government programs to stimulate its production or use.

Environmental problems have always plagued the soda ash and chlor-alkali industry as well as rock salt mines and evaporator plants. For many reasons, one by one the Solvay soda ash plants have closed; as of July 1983 only one Solvay soda ash plant was operating. It is scheduled to be closed soon. Eventually all soda ash production in the US will come from the Green River area of Wyoming and Searles Lake in California.

Salt is widely used for the removal of snow and ice from our highways in the north and northeastern portions of the United States. Although this use is a source of some water pollution during the snowstorms of the winter months, the deleterious effects are far overshadowed by the safety benefits of ice-free roads. Stabilization of the subgrade of highways, airports, and parking lots by the use of brine in the compaction and hardening of the subgrade will see expanded use.

Some economists have suggested that due to the eventual shortage of paper pulp and ecological objections to salt's use for snow and ice removal the consumption will level off and/or decrease. The substitution of trona for soda ash will be another prime factor in the reduction of tonnages of salt produced by US industry.

It is the opinion of the authors that the trend of a gradual increase in salt consumption will prevail over the long term. Nothing has been found to replace salt for de-icing highways. Other scientific advances will lead to increases in salt consumption. One of these will be the use of salt in the control and reduction of coal dust in mines. The process of "salt crust" has long been known in European coal centers but because of the bureaucratic regulations and union politics, salt use as a dust collector and substitute for limestone dusting has been blocked. Although salt could be used in a similar manner in other dusty mines, such use in coal mines alone would consume millions of tons.

Another scientific advance has shown that sodium chloride added to certain solvents will enhance their ability to dissolve various metallic sulfides. As solution mining increases in the extraction of metalliferous values, there will be an increase in the use of salt. Adding salt to sulfuric acid dramatically improves the solubility of copper sulfides. This process for

TABLE 15—Distribution of Salt Sold or Used by Producers in the United States, By Use '(Thousand Short Tons)

Consumer or use	1979 Evap-orated	1979 Rock	1979 Brine	1979 Total*	1980 Evap-orated	1980 Rock	1980 Brine	1980 Total*
Chlorine, caustic soda, soda ash	557	1,819	23,824	26,200	682	2,103	21,156	23,941
All other chemicals	446	625	150	1,222	396	505	119	1,020
Textile and dyeing	134	53	—	188	124	53	—	177
Meatpackers, tanners, casing manufacturers	259	287	—	546	278	268	—	546
Dairy	78	7	—	85	79	5	—	84
Canning	181	99	†	280	182	104	†	287
Baking	109	10	—	119	105	8	—	113
Flour processors (including cereal)	70	25	†	95	68	24	†	92
Other food processing	204	56	†	261	197	40	14	251
Feed dealers	688	506	—	1,194	732	440	—	1,172
Feed mixers	364	359	—	723	337	326	—	662
Metals	70	286	†	356	56	215	†	272
Rubber	W	9	W	99	W	5	W	95
Oil	228	103	218	550	345	100	264	709
Paper and pulp	W	134	W	194	W	135	W	230
Water softener manufacturers and service companies	464	345	†	810	443	234	8	686
Grocery stores	887	253	†	1,140	825	168	†	995
Highway use	308	8,433	†	8,742	252	6,137	†	6,389
US government	20	58	†	78	39	82	†	121
Distributors (brokers, wholesalers, etc.)	588	W	W	1,249	529	W	W	989
Miscellaneous and undistributed‡	603	1,430	491	§1,714	637	1,320	501	§1,810
Total*	6,260	14,901	24,684	45,844	6,306	12,272	22,063	40,641

Source: Kostick, 1981.
W Withheld to avoid disclosing company proprietary data; included with "Miscellaneous and undistributed."
* Data may not add to totals shown because of independent rounding.
† Less than 5 units; included with "Miscellaneous and undistributed."
‡ Includes withheld figures and some exports and consumption in overseas areas administered by the United States.
§ Incomplete totals; withheld totals are included with total for each specific use.

mining deep-seated and/or marginal deposits has gained considerable acceptance by the mining industry.

Thus, with our advancing technology and increase in the consumer demands of our society, salt and its sodium and chlorine components will continue to play an increasingly important role in man's life.

Acknowledgment: This is to acknowledge, with deep appreciation, the efforts of John A. C. Atkins in respect to the section on "Transportation, Handling, and Storage."

Bibliography and References

Anon., 1978, "New Problems Arise for Nuclear Waste Storage," *Chemical and Engineering News,* Vol. 56, No. 24, June 12, p. 28.

Anon., 1978a, "Pakistan Opens New Salt Mine," *Mining Journal,* Vol. 290, No. 7446, May, p. 337.

Adams, J.E., 1968, "Permian Salt Deposits, West Texas and Eastern New Mexico," Special Paper No. 88, Geological Society of America, p. 407.

Al Naqib, N.M., 1970, "Geology of Jabal Sanam, South Iraq," *Journal,* Geological Society of America, Vol. 3, No. 1, pp. 9–36.

Aleksin, A.G., Kozhevnikov, I.I., and Sokolin, K.G., 1968, "Geologicheskoye Stroyeniye i Neftegazonosnost' Solyanokupol'nykh Basseynov Zapadnoy Afriki (Gas and Oil-Bearing Salt Domes of West Africa)," *Geologiya Nefti i Gaza,* Vol. 12, No. 12, pp. 47–51.

Andreyeva, R.I., and Kabyshev, V.P., 1970, "Kharakternyye Osobennosti Solyanoy Tektoniki Dneprovsko-Donetskoy Vpadiny (Characteristics of Salt Tectonics of the Dnieper-Donets Basin)," *Sovetskaya Geologiya,* No. 2, pp. 135–142.

Antoine, J.W., and Bryant, W.R., 1969, "Distribution of Salt and Salt Structures in Gulf of Mexico," *Bulletin,* American Assn. of Petroleum Geologists, Vol. 53, No. 12, pp. 2543–2550.

Atwater, G.I., 1968, "Gulf Coast Salt Dome Field

Area," Special Paper No. 88, Geological Society of America, pp. 29–40.

Auzende, J.M., et al., 1971, "Upper Miocene Salt Layer in the Western Mediterranean Basin," *Nature,* Vol. 230, No. 12, pp. 82–84.

Ballard, J.A., and Feden, R.H., 1970, "Diapiric Structures on the Campeche Shelf and Slope, Western Gulf of Mexico," *Bulletin,* Geological Society of America, Vol. 81, No. 2, pp. 505–512.

Bays, C.A., 1963, "Significant Uncertainties in Current Salt Solution Extraction Operations," *1st Symposium on Salt,* Northern Ohio Geological Society, Cleveland, pp. 467–481.

Bays, C.A., et al., 1960, "Solution Extraction of Salt Using Wells Connected by Hydraulic Fracturing," *AIME Transactions,* Vol. 217, pp. 266–277.

Benavides, V., 1968, "Saline Deposits of South America," Special Paper No. 88, Geological Society of America, pp. 249–290.

Benavides, G.L., and Sansores, E., 1968, "Salt Deposits of Southern Mexico," Special Paper No. 88, Geological Society of America, pp. 407–408.

Bentor, Y.K., 1968, "Salt Deposits of the Dead Sea Region," Special Paper No. 88, Geological Society of America, pp. 139–156.

Bleimeister, W.C., 1970, "Mining the Salar Grande," *3rd Symposium on Salt,* Vol. 2, Northern Ohio Geological Society, Cleveland, pp. 479–486.

Bornhauser, M., 1971, "Salt Piercement versus Downbuilding on Shallow Gulf Coast Salt Domes," Discussion, *Bulletin,* American Assn. of Petroleum Geologists, Vol. 55, No. 8, pp. 1360–1361.

Braitsch, O., 1971, *Salt Deposits, Their Origin and Composition,* Vol. 4, "Minerals, Rocks and Inorganic Materials," Springer-Verlag, New York-London-Heidelberg, 297 pp.

Briggs, L.T., 1968, "Evaporite Deposits of the United States," Abstract, Special Paper No. 88, Geological Society of America, p. 408.

Brinkman, R., and Logsters, H., 1968, "Diapirs in Western Pyrenees and Foreland, Spain," *Diapirism and Diapirs,* Geological Memoir No. 8, American Assn. of Petroleum Geologists, pp. 275–292.

Brognon, G., 1971, "The Geology of the Angola Coast and Continental Margin," *The Geology of the East Atlantic Continental Margin,* Vol. 4, "Africa, Great Britain," Report No. 70/16, Institute of Geological Science, pp. 143–152.

Brunstrom, R.G.W., 1967, "Origin of the Keuper Salt in Britain," *Nature,* Vol. 215, No. 5109, p. 1474.

Brunstrom, R.G.W., and Walmsley, P.J., 1968, "Permian Evaporites in North Sea Basin," *Bulletin,* American Assn. of Petroleum Geologists, Vol. 52, No. 3, p. 522.

Bush, V.A., et al., 1970, "Solyanokupol'nyye Struktury Vostochnoy Chasti Severo-Germanskoy Vpadiny i Osobennosti ikh Razmeshcheniya (Salt Dome Structures in the Eastern Part of the North German Basin and Their Distribution)," *Geologia i Razvedka,* Izvestiya Vysshikh Uchebnykh Zavedeni, No. 3, pp. 27–34.

Cater, F.W., and Craig, L.C., 1970, "Geology of the Salt Anticline Region in Southwestern Colorado," Professional Paper No. 637, US Geological Survey, 80 pp.

Clarke, F.W., 1924, "The Data of Geochemistry," Bulletin 770, US Geological Survey, 841 pp.

Clarke, O.M., Jr., et al., 1970, "Mineral Resources of Wilcox County, Alabama," Circular No. 66, Alabama Geological Survey, 26 pp.

Contreras, V.H., and Castillon, B.M., 1968, "Morphology and Origin of Salt Domes of Isthmus of Tehauntepec," *Diapirism and Diapirs,* Geological Memoir No. 8, American Assn. of Petroleum Geologists, pp. 244–260.

Cordini, I.R.A., 1967, "Reservas Salinas de Argentina (Salt Resources of Argentina)," *Annales,* Inst. Nac. Geol. Mineria, No. 13, 108 pp.

DeMille, G., and Schouldice, J.R., 1968, "Saline Deposits of Alberta and Saskatchewan," Special Paper No. 88, Geological Society of America, pp. 408–409.

Dingman, R.J., 1967, "Geology and Ground-Water Resources of the Northern Part of the Salar de Atacama, Antofagasta Province, Chile," Bulletin No. 1219, US Geological Survey, 49 pp.

Dunnington, H.V., 1968, "Salt-Tectonic Features of Northern Iraq," Special Paper No. 88, Geological Society of America, pp. 183–227.

Eargle, D.H., et al., 1971, "Uranium Geology and Mines, South Texas," Guidebook No. 12, Bureau of Economic Geology, University of Texas, 59 pp.

Eaton, G.P., et al., 1970, "Luke Dome, a Halokinetic Salt Mass Near Phoenix, Arizona," *Abstracts,* Geological Society of America, Vol. 2, No. 5, p. 332.

Fernandes, G., 1968, "O Sal-Gema de Formacao Nova Olinda, Bacia do Medio Amazonas (Rock Salt in the Nova Olinda Formation, Middle Amazonas Basin)," *Bol. Tec. Petrobras,* Vol. 11, No. 2, pp. 245–248.

Fischer, G., 1967, "Salzbergwerk Berchtesgaden (Berchtesgaden Salt Mine)," *Fortschritte der Mineralogie,* Vol. 45, No. 1, 6 pp.

Flores, T., 1918, "El Tequesquite del Lago de Texcoco," *Annales,* Mexican Institute of Geology, No. 5, pp. 1–14.

Foster, R.J., 1980, "Salt," Preprint, US Bureau of Mines, 12 pp.

Frey, M.G., and Grimes, W.H., 1969, "Bay Marchand-Timbalier, Bay-Caillou Island Salt Complex, Louisiana," in "Geology of the American Mediterranean," *Transactions,* Gulf Coast Assn. of Geological Societies, Vol. 19, p. 266.

Galitskiy, I.V., and Pistrak, R.M., 1970, "Role of Rhomboid Block Fields in the Tectonics of the Dnieper-Donets Depression," *Doklady,* Earth Sciences Sections, Akademiya Nauk SSSR, Vol. 187, pp. 23–25.

Gardner, L.S., et al., 1967, "Salt Resources of Thailand," Report of Investigations No. 11, Thailand Dept. of Mineral Resources, 100 pp.

Garlicki, A., 1970, "Zloze Soli Kamiennej Moszczenica-Lapczyca na Zachod od Bochni (Rock Salt Deposit in the Moszczenica-Lapczyca Area West of Bochnia)," *Kwart. Geol.,* Polish Institute of Geology, Vol. 14, No. 2, pp. 350–360.

Garlicki, A., 1979, "Sedymentacja Soli Mioceńskich W Polsche, (Sedimentation of Miocene Salts in Poland)," *Prace Geologiczne 119,* Polska Akademia Nauk-Oddzial W Krakowie Komisja Nauk Geologicznych, 67 pp.

Gawel, A., 1967, "Salt Deposit at Wieliczka," *Bulletin,* Polish Institute of Geology, No. 211, Pt. 2, pp. 325–328, 337–339.

Geyer, H., 1973, "How Hydraulic Excavators Cut Costs, Raise Efficiency at German Salt Mine," *World Mining,* Vol. 9, No. 3, May, pp. 41–43, 66.

Gimm, W., ed., 1968, *Kali- und Steinsalzbergbau (Potash and Rock Salt Mining),* Band 1, "Aufschluss und Abbau von Kali- und Steinsalzlagerstatten (Vol. 1, Exploration and Exploitation of Potash and Rock Salt Deposits)," Deutsche Verlag Grundstoffind., 600 pp.

Gorfunkel', M.V., and Slepakova, G.I., 1970, "Pervichnaya Moshchnost' Solyanoy Tolshchi v Prikaspiyskoy Vpadine po Geofizicheskim Dannym (The Primary Thickness of Salt Series in the Caspian Basin Based on Geophysical Data)," *Sovetskaya Geologiya,* No. 4, pp. 166–169.

Gorfunkel', M.V., and Slepakova, G.I., 1971, "Original Thickness of Salt in the Caspian Depression, from Geophysical Data," *International Geology Review,* Vol. 13, No. 3, pp. 419–421.

Gorrell, H.A., and Alderman, G.R., 1968, "Elk Point Group Saline Basins of Alberta, Saskatchewan and Manitoba, Canada," Special Paper No. 88, Geological Society of America, pp. 291–317.

Gussow, W.C., 1965, "Energy Source of Intrusive Masses," Abstract, *Bulletin,* American Assn. of Petroleum Geologists, Vol. 49, No. 3, Mar., p. 343.

Halbouty, M.J., 1967, *Salt Domes,* Gulf Publishing Co., 425 pp.

Hamilton, W.N., 1971, "Salt in East-Central Alberta," Bulletin No. 29, Research Council of Alberta, 53 pp.

Hauber, L., et al., 1971, "Bericht uber die Exkursion der Schweizerischen Geologischen Gesellschaft in das Gebiet der Rheintaflexur und des Tafeljuras bei Basel vom 19. und 20. Oktober 1970 (Report on the Field Trip of the Swiss Geological Society in the Area of the Rhine Valley Flexure and the Jura Plateau near Basel, October 19–20, 1970)," *Eclogae Gol. Helvetica,* Vol. 64, No. 1, pp. 204–214.

Hermann, A., et al., 1967, "Die Halotektonische Deutung der Elfas-Uberschiebung im Sudniedersachsischen Bergland (Salt Tectonics of the Elfas Overthrust Zone in the Mountainous Region of Southern Lower Saxony)," *Geologisches Jahrbuch,* Vol. 84, pp. 407–460.

Hills, J.M., 1968, "Permian Basin Field Area, West Texas and Southeastern New Mexico," Special Paper No. 88, Geological Society of America, pp. 17–27.

Hite, R.J., 1968a, "Salt Deposits of the Paradox Basin, Southeast Utah and Southwest Colorado," Special Paper No. 88, Geological Society of America, pp. 319–330.

Hofrichter, E., 1967, "Subrosion und Bodensenkungen am Salzstock von Stade (Subsurface Leaching and Subsidence on the Salt Dome of Stade)," *Geologisches Jahrbuch,* Vol. 84, pp. 327–340.

Hofrichter, E., 1978, "Geologie, Tektonik, Mineralogie, Geochemie," *V. International Salz-Symposium in Deutschland,* Kali Steinsalz, Vol. 7, No. 8, pp. 338–344.

Holser, W.T., 1971, "Geologic Significance of the Minor-Element Composition of Marine Salt Deposits," *Economic Geology,* Vol. 66, No. 5, p. 813.

Jacoby, C.H., 1969, "Faults and Their Impact on (Salt) Mining," *3rd Symposium on Salt,* Vol. 2, Northern Ohio Geological Society, Cleveland, pp. 447–452.

Jacoby, C.H., 1969a, "Correlation, Faulting, and Metamorphism of Michigan and Appalachian Basin Salt," *Bulletin,* American Assn. of Petroleum Geologists, Vol. 53, No. 1, pp. 136–154.

Jacoby, C.H., 1972, "Evaluation of Salt Deposits," *Trans. SME/AIME,* Vol. 252, pp. 118–122.

Jacoby, C.H., 1973, "Salt," *Mining Engineering,* Vol. 25, No. 1, Jan., pp. 46–47.

Jacoby, C.H., 1977, "Geology—Hydrology of Avery Island Salt Dome," July, Y/OWI/SUB.

Jacoby, C.H., 1977a, "Scoping Report of Various Salt Mines in the United States," July, Y/OWI/SUB and Y/OWI/SUB, 77/16523/1.

Jaritz, W., 1968, "Einige Bemerkungen uber die Entstehung der Salzstrukturen Nordwestdeutschlands (Remarks on the Genesis of Salt Structures in Northwestern Germany)," *Erdol Kohle,* Vol. 21, No. 9, pp. 519–520.

Johnson, H.A., and Bredeson, D.H., 1971, "Structural Development of Some Shallow Salt Domes in Louisiana Miocene Productive Belt," *Bulletin,* American Assn. of Petroleum Geologists, Vol. 55, No. 2, pp. 204–226.

Johnson, K.S., 1970, "Salt Produced by Solar Evaporation on Big Salt Plain, Woods County, Oklahoma," *Geological Notes,* Geological Survey of Oklahoma, Vol. 30, No. 3, pp. 47–54.

Johnson, K.S. and Gonzalez, S., 1978, "Salt Deposits in the United States and Regional Geologic Characteristics Important for Storage of Radioactive Wastes," March, Y/OWI/SUB–74141.

Kapustin, N.N., 1970, "Rol'tektonicheskogo Faktora v Formirovanii Solyanokupol'nykh Struktur, na Primere Severo-Germanskoy Vpadiny (The Role of Tectonics in the Formation of Salt Dome Structures, North German Basin)," *Geologia i Razvedka,* Izvestiya Vysshikh Uchebnykh Zavedenii, No. 12, pp. 42–48.

Kaufman, D.W., 1960, *Sodium Chloride,* Reinhold, New York, 743 pp.

Kent, P.E., 1970, "The Salt Plugs of the Persian Gulf Region," *Transactions,* Philosophical Society, Vol. 64, pp. 55–58.

Khrushchov, D.P., and Gavrish, V.K., 1968, "Pro Deyaki Strukturni Osoblivosti Solyanikh Porid Dniprovs'ko-Donets'koi Zapadini (Some Structural Features of Halite in the Dnieper-Donets Basin, USSR)," *Dopovidi,* Akademii Nauk Ukrainskoi SSR, Series B., No. 1, pp. 53–58.

Kirkland, D.W. and Evans, R., 1973, *Marine Evaporites, Origin, and Diagenesis and Geochemistry,* Dowden, Hutchinson and Ross, Inc., Stroudsburg, PA, 426 pp.

Klingsberg, C. and Duguid, J., 1980, "Status of Technology for Isolating High Level Radioactive Waste in Geologic Repositories," D.E.E./T.I.C. 11207, pp. 77–85.

Koch, K., et al., 1968, "Geochemische Untersuchungen an Salzen und Laugen von Salzlagerstatten in der Deutschin Demokratischen Republik (Geochemical Investigations of Salts and Brines of the Salt Deposits of the German Democratic Republic)," *Abstract Volume,* 23rd International Geological Congress, Prague, Czechoslovakia, 161 pp.

Kostick, D.S., 1981, "Salt," *Minerals Yearbook, 1980,* US Bureau of Mines, pp. 683–694.

Kostick, D.S., 1982, "Salt," *Mineral Industry Surveys,* US Bureau of Mines, pp. 132–133.

Krishnan, M.S., 1966, "Salt Tectonics in the Punjab Salt Range, Pakistan," *Bulletin,* Geological Society of America, Vol. 77, No. 1, pp. 115–122.

Krishnan, M.S., 1968, "Geology of the Salt Deposits in the Punjab Salt Range, Pakistan," Special Paper No. 88, Geological Society of America, pp. 410–411.

Krishnan, M.S., 1968a, "Saline Lake Deposits of Rajasthan, India," Special Paper No. 88 (abstract), Geological Society of America, p. 411.

Kupfer, D.H., ed., 1970, *Geology and Technology of Gulf Coast Salt,* School of Geoscience, Louisiana State University, Baton Rouge, 183 pp.

Kupfer, D.H., 1970a, "Mechanism of Intrusion of Gulf Coast Salt," *Geology and Technology of Gulf Coast Salt,* Louisiana State University, School of Geosciences, Baton Rouge, pp. 25–65.

Landes, K.K., 1960, "Geology of Salt Deposits," *Sodium Chloride,* D.W. Kaufman, ed., Reinhold, New York, pp. 28–69.

Lawrence, C.R., 1971, "Occurrence and Production of Halite and Gypsum in Victoria," *Mining Geology Journal,* Mines Dept., Victoria, Vol. 7, No. 1, pp. 17–19.

Lefond, S.J., 1965, "Salt—The Universal Mineral," *The Explorer,* Cleveland Natural Science Museum, Vol. 7, No. 4, pp. 11–15, 34–35.

Lefond, S.J., 1969, *Handbook of World Salt Resources,* Monographs in Geoscience, Plenum Press, New York, 384 pp.

Leite, D.C., 1968, "Investigacoes Sobre as Possibilidades de Salgema na Parte Sudoeste da Bacia Sedimentar do Reconcavo (Investigations on the Possibilities of Rock Salt in the Southwestern Part of the Reconcavo Sedimentary Basin, Bahia)," *Bol. Tec. Petrobras,* Vol. 11, No. 2, pp. 231–242.

Leroy, L.E., 1951, *Subsurface Geologic Methods,* 3rd ed., Colorado School of Mines, Golden, 1166 pp.

Liechti, P., 1968, "Salt Features of France," Special Paper No. 88, Geological Society of America, pp. 83–106.

Lotze, F., 1968, "Salt Deposits of Europe (Including the USSR)," Special Paper No. 88, Geological Society of America, pp. 411–412.

Lukac, M., 1969, "Estratigrafia y Genesis de la Sal Gema en Punta Alegre y en Loma Cunagua, Provincia de Camaguey (Stratigraphy and Genesis of Rock Salt in Punta Alegre and Loma Cunagua, Camaguey Province)," *Rev. Tecnol.* (Havana), Vol. 7, No. 5–6, pp. 20–42.

MacMillan, R.T., 1972, "Salt," *Minerals Yearbook, 1972,* US Bureau of Mines, pp. 1093–1102.

Martinez, J.D., 1969, "The Impact of Salt on Man's Environment," in "Geology of the American Mediterranean," *Transactions,* Gulf Coast Assn. of Geological Societies, Vol. 19, pp. 49–62.

Martinez, J.D., 1971, "Environmental Significance of Salt," *Bulletin,* American Assn. of Petroleum Geologists, Vol. 55, No. 6, pp. 810–825.

Maslankiewicz, K., 1970, "Historical Survey of Concepts on the Geological Structure of the Wieliczka Rock-Salt Deposit (South Poland)," *History of Concepts on Mineral Deposits,* Symposium, International Commission on History of Geological Sciences, Freiberg, summary paper, p. 88.

Mattox, R.B., ed., 1968, "Saline Deposits," Special Paper No. 88, Geological Society of America, 701 pp.

McNaughton, D.A., et al., 1968, "Evolution of Salt Anticlines and Salt Domes in the Amadeus Basin, Central Australia," Special Paper No. 88, Geological Society of America, pp. 229–247.

Multhauf, R.P., 1978, *Neptune's Gift,* The Johns Hopkins Press, Baltimore and London, 325 pp.

Neal, J.T., 1975, "Playas and Dried Lakes, Occurrence and Development," *Benchmark Papers in Geology/20,* Dowden, Hutchinson and Ross, Inc., Stroudsburg, PA, 411 pp.

Norris, S.E., 1978, "Hydrologic Environment of the Silurian Salt Deposits in Parts of Michigan, Ohio and New York," US Geological Survey, Open File 78–684, 57 pp.

Notholt, A.J.G., and Highley, D.E., 1973, "Salt," Mineral Dossier No. 7, Mineral Resources Div., Institute of Geological Sciences, London, 36 pp.

Osika, R., and Poborski, J., 1970, "Halogenic Map," Polish Geological Institute, sheet 4.

Oxley, P., 1971, "Salt Ridge-Salt Dome Origins in Gulf Coast Area," *Bulletin,* American Assn. of Petroleum Geologists, Vol. 55, No. 2, p. 357.

Pauca, M., 1968, "Beitrage zur Kenntnis der Miozanen Salzlagerstatten Rumaniens (Contributions to Knowledge of the Miocene Salt Deposits of Romania)," *Geologische Rundschau,* Vol. 57, No. 2, pp. 514–531.

Read, L.E., and Jacoby, C.H., 1957, "Exploring and Mining for Salt," *Mining Engineering,* Vol. 9, No. 5, pp. 538–541.

Richter-Bernburg, G., 1964, "Vergleichende Betrachtung Deutscher und Iranischer Salztocke (Comparative Study of German and Iranian Salt Stocks)," *Zeitschrift fuer Deutschen Geologischen Gessellschaft,* Vol. 114 (1962), No. 3, pp. 692–693.

Rios, J.M., 1968, "Saline Deposits of Spain," Special Paper No. 88, Geological Society of America, pp. 59–74.

Rona, P.A., 1969, "Possible Salt Domes in the Deep Atlantic off North-West Africa," *Nature,* Vol. 224, No. 5215, pp. 141–143.

Sannemann, D., 1968, "Salt-Stock Families in Northwestern Germany," *Diapirism and Diapirs,* Memoir No. 8, American Assn. of Petroleum Geologists, pp. 261–270.

Schauberger, O., 1967, "Die Geologische Kartierung der Alpinen Salzlagerstatten und ihre Wirtschaftliche Bedeutung fur den Salzsoleberg-bau (Geologic Mapping of Alpine Salt Deposits and Its Economic Significance)," *Berg und Huettenmaennische Monatshefte,* Vol. 112, No. 6, pp. 184–190.

Severy, C.L., 1946, "Salt Cored Successfully with Oil Field Equipment," *Engineering & Mining Journal,* Vol. 147, No. 8, pp. 85–87.

Shaw, W.S., and Blanchard, J.D., 1968, "Salt Deposits of the Maritime Provinces of Canada," Special Paper No. 88, Geological Society of America, pp. 414–415.

Smith, D.A., and Reeve, F.A.E., 1970, "Salt Piercement in Shallow Gulf Coast Salt Structures," *Bulletin,* American Assn. of Petroleum Geologists, Vol. 54, No. 7, pp. 1271–1289.

Smith, D.A., and Reeve, F.A.E., 1971, "Salt Piercement versus Downbuilding on Shallow Gulf Coast Salt Domes," Reply, *Bulletin,* American Assn. of Petroleum Geologists, Vol. 55, No. 8, pp. 1361–1362.

Stocklin, J., 1968, "Salt Deposits of the Middle East," Special Paper No. 88, Geological Society of America, pp. 157–181.

Svitoch, A.A., 1971, "Solevaya Struktura Prikaspiyskoy Vpadiny; Nekotoryye Voprosy Stroyeniya, Razvitiya i Prichiny Obrazovaniya (Salt Structure of the Basin North of the Caspian; Questions of Composition, Evolution, and Reasons for Formation)," *Byulleten Otdel Geologicheskii,* Moskovskii Obshchest. Ispyt., Vol. 46, No. 1, pp. 25–40.

Tashchi, V. M., 1968, "Nekotoryye Voprosy Gidrogeologii Solotvinskogo Mestorozhdeniya Kamennoy Soli (Hydrology of the Solotvino Rock-Salt Deposit)," *Geologiya Geokhimiya Goryuch. Iskop.,* Akademii Nauk Ukrainskoi SSR, No. 17, pp. 66–73.

Teixeira, A.A., and Saldanha, L.A., 1968, "Bacia Salifera Aptiana de Sergipe/Alagoas; Ocorrencias de Sais Soluveis (The Sergipe-Alagoas Salt Basin, Occurrence of Soluble Salts)," *Bol. Tec. Petrobras,* Vol. 11, No. 2, pp. 221–228.

Tkhorzhevskiy, S.A., 1970, "O Solyanoy Tektonike v Prikaspiyskoy Vpadine (Salt Tectonics in the Caspian Basin Area)," *Sovetskaya Geologiya,* No. 10, pp. 131–139.

Turk, L.J., 1973, "Hydrogeology of the Bonneville Salt Flats, Utah," Utah Geological and Mineral Survey Water Resources Bulletin 19, 80 pp.

Ujueta, L.G., 1968, "Sal en la Cordillera Oriental de Colombia (Salt in the Cordillera Oriental of Colombia)," *Bol. Geol.,* Univ. Ind., Santander, No. 21 (1965), pp. 5–18.

Ujueta, L.G., 1969, "Salt in the Eastern Cordillera of Colombia," *Bulletin,* Geological Society of America, Vol. 80, No. 11, pp. 2317–2320.

Vancea, A., 1968, "Le Role du Sel dans la Formation des Domes de la Curvette de Transylvanie, (The Role of Salt in the Formation of Domes in the Transylvania Basin)," *Revue Roumaine de Geologie, Geophysique et Geographie,* Serie de Geophysique, Vol. 12, No. 2, pp. 203–217.

ver Planck, W.E., 1957, "Salt in California," Bulletin 175, California Div. of Mines and Geology, 168 pp.

Vos, M.A., 1970, "Salt in Ontario; Southern Ontario," in "Summary of Field Work," Miscellaneous Paper No. 43, Ontario Dept. of Mines, pp. 94–96.

Wagenbreth, O., 1966, "Entwurf einer Abgedeckten Geologischen Karte des Stassfurt-Oscherslebener Salzattels und Seiner Tertiaren Randsenken (Sketch of a Geologic Map of the Stassfurt-Oschersleben Salt Structure (With the Sedimentary Cover Removed) and Its Tertiary Marginal Troughs)," *Geologie* (Berlin), Vol. 15, No. 9, pp. 1009–1022.

Wagner, G., et al., 1971, "Der Salzstock von Cardona in Nordostspanien (The Cardona Salt Dome in Northeastern Spain)," *Geologische Rundschau,* Vol. 60, No. 3, pp. 970–996.

Webb, G.W., 1970, "Salt Structures East of Nova Scotia," *Marine Sediments,* Vol. 6, No. 3, p. 138.

Weidie, A.E., and Martinez, J.D., 1970, "Evidence for Evaporite Diapirism in Northeastern Mexico," *Bulletin,* American Assn. of Petroleum Geologists, Vol. 54, No. 4, pp. 655–657.

Zaykov, V.V., et al., 1967, "Novyye Dannyye o Solenosnosti Devonskikh Otlozheniy Tuvy (New Information on the Salt Content of the Devonian Deposits of Tuva)," *Geologiya i Geofizika,* Akademiya Nauk SSSR, Sibirskoe Otdelenie, No. 8, pp. 21–27.

Zbyszewski, G., and De Faria, J.B., 1971, "O Sal-Gema em Portugal Metropolitano; Suas Jazidas, Caracteristicas e Aproveitamento (Rock Salt of Metropolitan Portugal; Deposits, Properties, and Utilization)," *Notas Trab.,* Serv. Fom. Mineiro, Estud., Portugal, Vol. 20, No. 1–2, pp. 5–105.

Zholtayev, G.Z., 1968, "O Terminakh "Solyanoy Kupol," "Solyanoy Shtok," "Solyanoy Massiv," i "Solyanaya Antiklinal" (The Terms "Salt Dome," "Salt Stock," "Salt Massif," and "Salt Anticline")," *Izvetiya, Seriya Geologicheskaya,* Akademiya Nauk SSSR, No. 8, pp. 103–109.

Sand and Gravel

HAROLD B. GOLDMAN *

DON REINING †

The sand and gravel industry is the largest nonfuel mineral industry in the nation. In 1981, the production of sand and gravel totaled 755 million tons valued at $2.3 billion. California, which leads the nation with more than 110 million tons, together with Alaska, Texas, Michigan, and Ohio, account for 36% of the total production in the nation (Table 1).

In commercial usage, *sand* applies to rock or mineral fragments ranging in size from particles retained on a No. 200 sieve (0.074 mm openings) to those passing a No. 4 sieve (4.76 mm openings). *Gravel* consists of rock or mineral fragments larger than 4.76 mm, ranging up to 88.9 mm (3½-in.) maximum size.

The construction industry consumes 97% of the sand and gravel produced; the remainder is sand used for specialized products such as glass (see the chapter on "Silica and Silicon").

Utilization

The building industry uses sand and gravel chiefly as aggregate in portland cement concrete, mortar, and plaster; the paving industry uses sand and gravel in both asphaltic mixtures and portland cement concrete. Aggregate is commonly designated as the inert fragmental material which is bound into a conglomerate mass by a cementing material such as portland cement, asphalt, or gypsum plaster.

Portland Cement Concrete Aggregates

Portland cement concrete consists of sand and gravel surrounded and held together by hardened portland cement paste. Concrete mixes commonly contain 15–20% water, 7–14% cement, and 66–78% aggregate. Sand and gravel used as concrete aggregate have to meet many requirements (Goldman, 1956). Premature deterioration of concrete has been traced in many instances to the use of unsuitable aggregates.

Asphaltic Aggregate

Asphaltic mixtures used predominantly for paving consist of combinations of sand, gravel, and mineral filler (material finer than 0.003 in.), uniformly coated and mixed with asphalt produced in the refining of petroleum. Except for the addition of mineral filler, sand and gravel used as asphaltic aggregate must meet the same general physical requirements as materials used for portland cement concrete aggregate.

General Requirements of Aggregates

Construction aggregate has many requirements that are difficult to meet if only unprocessed material from natural deposits is used. Suitable material is composed of clean, uncoated, properly shaped particles which are sound and durable. Soundness and durability are terms used to denote the ability of aggre-

TABLE 1—Total Produced (Million Tons) Five Largest Producing States

State	1978	1979	1981	% of US Total
California	115	129	112	14.8
Alaska	69	51	46	6.1
Texas	57	52	45	6.0
Ohio	47	45	36	4.8
Michigan	48	50	33	4.4

Source: US Bureau of Mines.

* Consulting Geologist, San Francisco, CA.

† President, Southern California Rock Products Assn., South Pasadena, CA.

gates to retain a uniform physical and chemical state over a long period of time so as not to cause disruption of the concrete when exposed to weathering and other destructive processes. To have these attributes, individual particles must be tough and firm, possessing the strength to resist stresses and chemical and physical changes, which may cause swelling, cracking, softening, and leaching. The aggregate should not be contaminated by much clayey material, silt, mica, organic matter, chemical salts and surface coatings.

The quality of aggregate depends upon its physical and chemical properties. These, in turn, may be inherent mineralogical and textural features of the rock or may be the effects of later changes such as tectonic fracturing, mechanical or chemical weathering, or incrustations.

The physical properties most significant with regard to concrete are: (1) abundance and nature of fractures and pores, (2) particle shape and surface texture and (3) presence of material which may cause volume change. An aggregate is considered to be physically sound if it is adequately strong and capable of resisting the agencies of weathering without disruption or decomposition. Minerals or rock particles that are physically weak, extremely absorptive, and easily cleavable are susceptible to breakdown by weathering. The use of such materials in concrete reduces strength or leads to early deterioration by promoting weak bond between cement and aggregate, or by inducing cracking, spalling, or popouts. Severely weathered, soft, micaceous, or porous materials may cause localized stresses to develop in concrete by swelling and shrinking during wetting and drying or freezing and thawing cycles.

Suitability of the Various Rock Types: Sedimentary rocks show a wide range in physical qualities and suitability. Sandstones and limestones, if hard and dense, are ordinarily satisfactory, but many sandstones are friable and excessively porous and commonly are clay-bearing. Shales generally make poor aggregate material, being soft, light, weak, and absorptive. Most igneous rocks are satisfactory, being normally hard, tough, and dense. Tuffs and certain flow rocks may be extremely porous and have high absorption and low strength. Metamorphic rocks differ in character. Most marbles and quartzites are usually massive, tough, and dense. Gneisses are ordinarily very tough and durable. Some schists contain micaceous minerals which are undesirable because they are soft, laminated, and absorptive. Micaceous minerals are susceptible to weathering along cleavage planes and thereby impair strength and durability. Some schists and slates in particular are thinly laminated and tend to assume flat slabby shapes which lack strength.

Any or all of these rock types may be rendered undesirable because of harmful exterior coatings. Weathering processes, particularly the action of ground waters, deposit these coatings. The most common coatings are calcium carbonate, clay, silt, opal, chalcedony, iron oxide, manganese oxide, and gypsum. Particles with these coatings are generally undesirable as aggregates. The bond between particle and coating may be weak and decrease the strength of the aggregate-cement bond.

The chemical properties which may affect service life of concrete are (1) reaction of certain rocks and minerals with high-alkali cement (alkali-aggregate reactivity), (2) leaching of water soluble substances, (3) solution of certain secondary minerals, such as the zeolites, to release sodium and potassium which aids in attacking susceptible aggregate particles, and (4) oxidation by weathering to produce compounds that may retard cement hydration.

Alkali-aggregate reactivity has been discussed at length in many publications. A reactive aggregate is any rock, gravel, or sand that contains one or more constituents that react chemically with the alkalies (sodium and potassium) in some types of portland cement. This reaction causes expansion, cracking, and deterioration of concrete and arises from osmotic pressures produced by the formation and hydration of alkali-silica gels. The gels are formed through interaction between reactive silica in the mineral aggregate and the alkalies which are liberated by the cement during hydration. Opal (amorphous hydrous silica) is the most conspicuous aggregate material reacting in this manner. Other rocks and minerals known to be reactive are: glassy volcanic rocks of medium to high silica content (andesite and rhyolite), chalcedonic rocks, certain phyllites which contain a hydromica, and the minerals tridymite and heulandite. Any rock containing a significant proportion of reactive substances may be deleteriously reactive; thus normally nonreactive sandstone, shale, basalt, granite, and other rock types may be harmful if impregnated or coated with opal, chalcedony, or other reactive substance.

Certain sulfide minerals, such as the iron sulfides, pyrite, and marcasite, oxidize and cause unsightly rust stains or generate acidic compounds injurious to the concrete matrix and cause popouts. Chemical salts such as sulfates, chlorides, carbonates, and phosphates also occur in some aggregates. Some of these substances dissolve easily or react to impede the setting of the cement.

A chemical alteration contributing to physical unsoundness is hydration. Shales, clays, and some rock nodules are examples of materials which expand when they absorb water and shrink as they dry. This absorptive character increases the rock's susceptibility to disruption by weathering.

Most of the foregoing features can be observed in the field, and on the basis of these observations, laboratory test procedures can be set up to further evaluate the properties of the aggregate.

Geologic Occurrence

In the United States, sand and gravel is obtained commercially from rock units of many types and ages. The principal sources are along existing or ancient river channels and in glaciated areas; marine and lake deposits and older geologic formations are significant but of lesser importance.

Classification of Deposits

Stream Deposits: *Stream Channel Deposits*—Stream channel deposits consist of sand and gravel deposited in stream beds along present or former stream courses. Most channel deposits are generally easily accessible and easily mined.

Materials in these deposits are desirable as aggregate for many reasons. The natural abrasive action of stream transport has removed most of the soft weak rocks, leaving only the harder and firmer particles. These latter particles have undergone some degree of rounding and are subrounded to well-rounded, a desirable aspect for use in concrete.

Most channel sand and gravel deposits are replenished by material carried by seasonal floods, except in portions of the streams downstream from dams. Overburden is rarely present, but high flood waters may leave silt, clay, and wood debris covering parts of some channel deposits. The size of gravel gradually decreases downstream in the streams with long reaches. Commercial production is concentrated in the deposits where a proper blend of sand and gravel can be obtained. Fortunately, the favorable portions of many streams occur in flat-lying areas near population centers.

Channel deposits normally are free of excessive amounts of silt and clay and most deposits contain sand and gravel in the size gradations necessary for concrete design. Mining operations are often relatively simple, consisting of no more than washing and screening to obtain suitable aggregate. In spite of these qualities some deposits are unsuitable because they contain harmful ingredients such as physically unsound or chemically reactive rocks. The nature of the material in the stream channel is determined in large part by the nature of the source rocks within its drainage area. Different geological formations in a drainage basin contribute a variety of rock types which show a wide range in chemical composition and degree of weathering.

An example of the influence of source rocks on the quality of materials in stream deposits is provided by the Colorado River between Hoover and Parker Dams. The Colorado River in the vicinity of Hoover Dam contains a complex assemblage of rock types of which only a small proportion are chemically reactive. Hoover Dam concrete made with this aggregate shows no sign of harmful chemical reactivity after 20 years of service. Below Hoover Dam the Colorado River flows over predominantly volcanic rocks and the content of reactive volcanic rock increases progressively downstream.

At Davis Dam, 67 miles below Hoover Dam, the content of reactive ingredients increases to the point where it was unwise to use the aggregate unless special precautionary measures were taken to forestall alkali-aggregate reaction. Farther downstream, the tributary Bill Williams River contributes large amounts of highly reactive andesites and rhyolites to the main stream. Parker Dam built with this aggregate exhibited signs of alkali-aggregate reaction within 2 years after completion.

Flood Plain Deposits—True flood plain deposits consist of material deposited on plains bordering streams by periodic overflow of the streams from their channels. The sediments deposited are normally composed of silt and sand grains. However, fine materials may mantle usable deposits of sand and gravel, particularly in areas where, in the geologic past, the streams were more vigorous and transported greater volumes of coarser material. The sand and gravel in these older, deeper flood plain deposits is similar to that in channel deposits and is suitable for use after overlying flood plain silt layers are removed.

Terrace Deposits—Stream terrace deposits are benchlike deposits of sand and gravel which border a stream but lie above the level of the present flood plain. These deposits are remnants of older flood plains through which the stream has cut. Terrace deposits, because they are above the stream level, may be more desirable than stream channel deposits if water tables are shallow and abundant ground water makes stream channel and flood plain operations difficult. However, there is little possibility of replenishment of terrace deposits. The materials in these deposits have the general properties of stream channel materials, but weathering processes may have diminished the quality of some of their constituents by converting certain minerals to clay. Not all terrace deposits can be used for aggregate. Some deposits are only thin veneers of sand and gravel deposited on a stream-cut bedrock surface. The materials in the deposit may not be suitable if they have undergone post-depositional weathering or if the grains have become coated with harmful substances due to the action of ground waters.

Alluvial Fans: An alluvial fan is formed when streams carrying large volumes of sand and gravel down a steep mountain slope enter an adjacent valley or plain. The abrupt change in slope causes a decrease in the speed of the stream. This change causes the stream to deposit the sand and gravel it has been carrying. The deposited matter spreads in a gently sloping fan-shaped mass from the mouth of the canyon onto the valley floor. The heavier, coarser material is deposited near the mouth of the valley, while the finer material is carried out toward the edges of the fan. Alluvial fan deposits have been found to reach depths up to and occasionally more than about 200 ft.

Fan deposits, because of the frequent shifting of the stream channel, ordinarily contain lenticular beds or tongues of poorly sorted sand and gravel interbedded with varying proportions of silt and clay. The particles are subangular to angular in the fans built by streams with small drainage areas. The particles in large fans, covering several square miles, are generally subangular to subrounded. Suitable aggregate is obtained from alluvial fans which are free from thick clay lenses.

In the past, the deposits were replenished during periods of flooding until the time that flood control dams were built to control the rivers.

Glacial and Fluvial-Glacial Deposits: Sand and gravel is produced commercially from glacial deposits in the midwestern and northeastern parts of the United States. The bulk of the sand and gravel was deposited directly or indirectly from continental ice sheets which originated in Canada and spread south into the United States. Glacial deposits that have been subjected to stream action are called fluvial-glacial deposits.

The glaciation period consisted of advances and recessions of the continental ice sheets which were several thousand feet thick, and moved roughly north to south.

Fragments of the rocks in the areas over which the glaciers passed, such as the igneous rocks in Canada, were picked up and incorporated in the glacier. When the glacial ice melted, the debris of clay, silt, sand, and pebbles and boulders were deposited as glacial till. When the advance and melting of the ice were in balance the debris deposited along the margin of the glacier formed irregular ridges termed moraines. As the water flowed away from the melting glaciers, the glacial debris was transported and deposited in outwash plains or elongate valley trains which are now major sources of sand and gravel (Ekblaw and Lamar, 1961). In addition, kames (hills) and eskers (ridges) of sand and gravel were deposited by melt waters within the ice sheet.

The fluvial-glacial deposits exhibit a better degree of sorting and contain less clay than the true glacial deposits and are of high enough quality to yield satisfactory aggregate comparable to true stream sand and gravel. Generally, the fluvial-glacial deposits are thinner than stream deposits in regions of considerable topographic relief.

Outwash Deposits—Outwash debris varies in size with the distance from the former glacial ice. The coarsest materials, pebbles, cobbles, and boulders, were deposited near the glacier, and the finer materials, sand, silt, and clay, were carried successively farther away from it. Since the water velocities fluctuated, the maximum size of transported particle varied and the deposits are characterized by great variation in particle sizes. Instances are cited where layers of gravel abruptly terminate, to be replaced by thick accumulations of silt or clay.

Kames and Eskers—Kames are hills composed of material deposited by glacial melt water in vertical crevasses or holes in the glacier, generally at or near its front. Eskers consist of material deposited in the beds of streams that flowed under or in the glacier. When the ice melted, the kames remained as rounded hills, and the eskers were left as ridges. The materials in these deposits, which show a wide variation in size and physical characteristics, are a common source of sand and gravel.

Beach and Dune Deposits—During the last stage of glaciation, the Lake Michigan basin was occupied by a larger lake whose level varied with the advances and recessions of the glacier. Each stage is marked by beach ridges of gravelly sand. In places the beach sand has been blown into dunes. These deposits are suitable as sources of sand, although urbanization and preservation as natural and recreational areas have removed many of them from consideration as minable reserves.

Moraines—True glacial deposits such as moraines have been transported solely by glacial ice with little or no reworking by streams. Morainal debris contains material of diverse size, shape, and quality ranging from pulverized rock flour to erratic boulders and generally is usable only after expensive processing.

Dredge Tailings: As a result of bucket dredge operations for gold, huge furrowlike ridges of gravels have been accumulated as tailings along many of the western state streams. A normal gold-dredging operation involves washing and screening the stream deposit. The oversize gravel, which ordinarily consists of everything over ½ or ¾ in., is washed and sent to tailing piles behind or alongside the dredge. The undersize fine gravel and sand are then processed to remove the gold and returned to the dredge pond and subsequently covered by other oversize tailings. This results in segregated deposits of fine gravel and sand which underlie coarser gravels, requiring that considerable material be processed by aggregate producers to obtain the desired gradings of sand and gravel. Most commercial operations employ equipment to crush the oversize tailings. Tailing piles which have not been subjected to excessive weathering contain material as suitable as the related stream deposit.

Older Geologic Formations: Pre-Quaternary formations, particularly partly consolidated sedimentary beds of sandstone and conglomerate, afford usable sources of aggregate when post-depositional weathering has not been too complete. Ordinarily, these formations have been subjected to long periods of weathering and are either too well cemented or contain too much clayey material to be processed economically. Accessibility of the deposit and depth of overburden also present problems.

The pre-Quaternary formations which are composed of well indurated sedimentary rocks or hard crystalline rocks require expensive quarrying operations to obtain suitable material for aggregate. Quarrying rock outcrop has some advantage over developing stream deposits in that once the physical and chemical characteristics of the rock are established, production can proceed without the danger of encountering harmful ingredients. Production can be closely controlled in all grade sizes.

Beach Deposits: Sand and gravel formed by the winnowing action of currents and waves on a beach make excellent aggregate. The gravel and coarse sand size particles are generally well-rounded, hard, and firm. The sands are commonly composed predominantly of resistant quartz and feldspar. Most beach deposits, however, are thin and lack proper size gradation.

Distribution of Deposits

The occurrence of sand and gravel in the United States is related to geologic processes. On a geographic-geologic basis, in the northern states, the principal sand and gravel resources are various types of glacial and outwash glacial deposits (Cooper, 1970). Marine terraces, ancient and recent geologically, are the major sand and gravel sources in the Atlantic and Gulf Coastal Plains. River deposits are the prime sources in several of the southeastern and south-central states. In the Great Plains, sand and gravel are mainly stream deposits. On the West Coast, deposits are principally alluvial fans, stream deposits, beach and dune sand, and fluvial-glacial deposits.

Offshore deposits, although of minor importance tonnagewise, are mined along the east coast from Washington to Boston, and in the Great Lakes.

Reserves

Any estimate of potential resources of sand and gravel in the United States is of little value if the trend toward urban expansion with its land conflicts and stronger environmental controls continues. In 1965, the National Sand and Gravel Assn. estimated the reserves held by producers at 16 billion tons (Cooper, 1970). There are enormous undeveloped deposits in the United States that are uneconomical to mine today because of the high cost of transportation to market. These would constitute reserves if some means can be found to set aside these lands for future use and avoid the conflicts of urbanization and the mineral industry (Goldman, 1959).

The heavily populated and urbanized Middle Atlantic, New England, and Pacific Coast states now are facing rapid depletion of deposits, loss of mineral lands to urbanization, and stiff opposition from environmental groups to development of new sources.

The United States is by far the largest producer and consumer of sand and gravel, accounting for an estimated one-sixth of the world's consumption.

Supplies in the rest of the world may be considerable, but as is the case with the United States, the deposits are not located in the areas of greatest demand. The more highly developed countries also are faced with the problem of loss of resources to urban encroachment (Pajalich, 1976).

The United States' foreign trade in sand and gravel is mostly with Canada. Exports are about 2.5 million tons and imports less than 1 million tons annually.

Exploration

Exploration for sand and gravel is generally guided by the needs of the potential producer. There are certain criteria that a deposit should meet in order to be of economic interest for a commercial producer, such as proximity to market, ease of obtaining the necessary mining permit, and a deposit of sufficient quantity to amortize the cost of a modern processing plant.

As sand and gravel is a low-cost, high-volume commodity, it cannot be transported far before the cost of transportation equals or exceeds the cost of the processed material at the plant. Thus, in most urbanizing areas the proximity to market would determine the target area for exploration. In aggregate-deficient parts of the country where materials are transported hundreds of miles, the target area naturally would be much larger.

Another important criteria to be met before field exploration begins is the socioeconomic restrictions on a potential mining operation. If the governing bodies have adopted a policy of refusing to grant operating permits, there would be no purpose to exploring in that jurisdiction. Also, there should be an awareness of the opposition that would be generated by conservation and ecology minded groups to granting of mining permits.

The size of the deposit is critical. In some parts of the country, where large volumes of sand and gravel are consumed, new sand and gravel plants cost upward of a million dollars. In order to amortize the life of such a plant, the deposit should contain sufficient reserves for at least 20 years. In California, a deposit should contain more than 10 million tons to be of commercial interest.

Naturally, in prospecting for sand and gravel for noncommercial production, such as for use in dams, the foregoing criteria rarely apply. In this instance, the quantity and quality are the prime factors to consider.

Field Technique

Preliminary field evaluation by a geologist or by an engineer with some geologic training may often suffice to evaluate a deposit as a possible source of aggregate material. The characteristics that bear upon the usefulness of a deposit as a source of aggregate material have been determined by the geologic processes by which the deposit was formed and subsequently modified. A basic understanding of the processes, therefore, is helpful to the person evaluating it. The important characteristics are types and physical condition of the rock, grading, rounding and degree of uniformity, particle size and shape, location, thickness and type of overburden, and ground water level.

By the judicious use of geologic maps a geologist can determine the type of materials to be encountered in a certain drainage area and predict the type of alluvial material to be found in a given stream channel. He thereby can recognize likely sources of aggregate and often can rule out areas unlikely to contain suitable materials. The geologist uses commercial sand and gravel deposits as a yardstick to guide him in evaluating new deposits. By comparing service records and laboratory test results with the petrographic character of commercially proven aggregates, the geologist builds up a background of data from which he can extrapolate to predict the behavior of untested deposits of similar petrographic character. This technique has been used successfully by the US Bureau of Reclamation in preliminary planning studies for large structures that require large amounts of suitable aggregates. In many instances field evaluation has obviated the need for expensive and time-consuming laboratory tests.

Target Areas: As outlined in the section on geology, those areas to be prospected for sand and gravel are predominantly those associated with either the present stream drainage or ancient drainages. Clues to buried or older deposits may be gleaned from well-drilling logs, geophysical data, and from geologic reports.

Aerial photographs, coupled with topographic maps, are essential tools. Any geologic information in the form of maps or reports should be researched. In most instances, the state geological surveys can provide the necessary maps.

The initial exploration can be greatly expedited with the use of airplanes, preferably a helicopter, to observe promising geologic features. Preliminary investigations, which include determining such features as crude estimates of volume of available material, thickness and type of overburden, water and power availability and access, are based primarily upon visual examination, either by a geologist or someone with experience in aggregate production.

Detailed Exploration: The method of exploration will depend upon the particular set of conditions, the objective being to obtain representative samples of the entire deposit and to delineate the dimensions of the deposit.

No matter what method of exploration is considered, ultimately representative samples must be obtained so that standard laboratory acceptance tests can be performed on the samples.

Where the deposit is shallower than 30 ft, test pits may be excavated with bulldozers, backhoes, clamshells, or dragline. The location of the water table often will determine the type of test pit.

For deeper deposits, truck-mounted drill rigs should be used. Ordinarily, truck-mounted bucket augers, which use buckets from 18 to 36 in. in diam, perform satisfactorily when large boulders are not present in the deposit. When drilling beneath the water table, steel casing is used to prevent the hole from caving.

Other drilling rigs, such as clamshells and reverse circulation drills, perform satisfactorily. Choice of drilling method will depend upon the availability of competent drillers in the area.

Geophysical methods can be used in conjunction with a drilling program. Geophysics may be used to check continuity of geologic units between drill sites or beyond a known deposit. For example, electrical resistivity techniques may be used to locate thick zones of clay interbedded in the sand and gravel.

The depth of a sand and gravel deposit resting on a hard bedrock surface can be determined by means of refraction seismic methods.

The cost of geophysical methods equals the cost of drilling, on an hourly basis. However, much more information is obtained within the same time interval through geophysics.

Obviously, the geophysical methods are useful exploration tools but cannot substitute for the other methods when samples must be obtained for laboratory testing.

The standard laboratory acceptance tests are described more fully in the chapter on uses. Equally important as laboratory testing is the sampling in the field. Where possible, large bulk samples should be obtained and quartered in the field. If it is possible, enough material should be stockpiled and trucked or railed to an existing plant to determine size gradation and quality of the finished product.

As an integral part of the exploration program, cross sections should be drawn of the deposit and estimates made of the quantity of material available. This figure should be reduced by the percentage of waste material, such as beds or lenses of clay and silt, and the loss of material to setbacks and finished slopes required by mining permits.

Evaluation of Deposits

General Specifications

The study of the results of laboratory tests of aggregates that have good service records in concrete has led to the establishment of certain minimum requirements or specifications to which aggregates are expected to conform. These specifications are designed so that completely serviceable concrete may be made, using any aggregate that meets the requirements. Most specifications are written by government agencies, engineering societies, and concrete technologists and attempt to conform to one standard set of specifications, those set up by the American Society for Testing and Materials. Modifications of these standards for certain types of concrete work make it difficult to compare individual requirements of the various organizations. Therefore, to evaluate the suitability of a deposit by judging the test results of selected samples is a difficult task. Some deposits which may not meet certain required specifications may have to be utilized because of other outside factors, such as the greater expense of hauling a more suitable aggregate.

In general, aggregate from an untried deposit will be satisfactory for most uses if it meets the following minimum standards (these specifications are a general average of the basic requirements recommended by the ASTM, California Division of Highways, US Army Corps of Engineers, and the US Bureau of Reclamation):

Abrasion—The abrasion loss should be less than 30%.

Soundness—The loss in the sodium sulfate test should be less than 10%.

TABLE 2—Laboratory Methods of Determining Suitability of Sand and Gravel for Use as Aggregate in Portland Cement Concrete

Property of Aggregate	Importance in Concrete	Test Methods	Reference to Tests*
Hardness and durability (resistance to abrasion)	Affects strength, resistance to wear	"Los Angeles Rattler;" measure proportion of fine material produced by abrasion in revolving metal drum after 100 and 500 turns	ASTM Test C131-51 (ASTM 1954, p. 40)
Soundness (lack of fissures in particles)	Affects strength, susceptibility to frost damage from expansion of absorbed water	Alternately soak in sodium or magnesium sulfate solution and dry; crystallization of absorbed solution forces open invisible cracks. Subject test beams made with aggregate to alternate cycles of freezing and thawing	ASTM Test C88-46T (ASTM 1954, p. 76) ASTM Test C290-52T (ASTM 1954, p. 191)
Specific gravity—dry and with absorbed liquid	Determine mass (specific gravity commonly specified 2.5 or more); absorption affects bond of cement paste to particles	Compare oven-dry weight with immersed weight, and weight after surface re-dried	ASTM Tests C127-42, and C128-42 (ASTM 1954, pp. 82,84)
Size grading characteristics	Affects flowability, residual void spaces, strength	Standard sieve analysis; screen in standard-size screens; weigh various fractions; plot on appropriate graphs	ASTM Test C136-46 (ASTM 1954, p. 69)
General characteristics: 1. Particle shape 2. Character of surface 3. Grain size 4. Texture (e.g., pore space, grain packing, cementation) 5. Color 6. Mineral composition 7. General physical condition (e.g., weathering) 8. Presence of potentially deleterious chemical substances (e.g., gypsum, zeolite, pyrite, opal, chalcedony, volcanic glass)	Different effects on strength, hardness, color, and permanence of concrete	Examine by naked eye, hand lens, and under petrographic microscope	ASTM Test C295-54 (ASTM 1954, p. 97)

Potential chemical reactivity	Affects permanence of concrete; reactive substances cause "popouts" and failures due to expansion	Weigh silica dissolved in sodium hydroxide solution; measure reduction in alkalinity caused by immersion of sample in standard sodium hydroxide solution	ASTM Test C289-54T (ASTM 1954, p. 57)
		Measure expansion of mortar bars made with aggregate over lengthy periods; 1-2 years	ASTM Test C227-52T (ASTM 1954, p. 296)
Lack of organic matter (coal, lignite, organic impurities)	Affects strength, resistance to wear	Separate material lighter than 2.0 specific gravity in heavy liquid, and weigh; compare color of sample with standard color solution—dark color assumed due to organic material	ASTM Tests C123-53T and C40-48 (ASTM 1954, pp. 51,56)
Cleanness (lack of dirt, clay, or silt finer than 200 mesh)	Determines quality of bond with cement	Measure material passing 200-mesh sieve; measure suspended material after shaking in water	ASTM Test C117-49 (ASTM 1954, p. 47)
Unit weight	Determines mass	Weigh aggregate contained in standard cubic foot measure	ASTM Test C29-42 (ASTM 1954, p. 90)
Lack of soft or friable fragments	Affects strength, resistance to wear	Scratch test using brass rod of Rockwell hardness B65 to B75; rock softer than rod is unsatisfactory	ASTM Test C235-54T (ASTM 1954, p. 74)
Toughness	Affects strength, resistance to wear	Impact; measure distance a standard-size hammer drops on specimen to fracture it	ASTM Test D3-18 (ASTM 1954, p. 88)

* Standardized tests as outlined by the American Society for Testing and Materials: *ASTM Standards on Mineral Aggregates, Concrete and Nonbituminous Highway Materials*, Dec. 1954.

Specific Gravity—The specific gravity should be greater than 2.55.

Size and Grading—

a) The deposit has proper grading so that the fine aggregate should contain no more than 45% of the material between two consecutive sieve sizes.

b) The fineness modulus should be between 2.3-3.1.

c) No more than 5% of the material should pass the No. 200 sieve.

Reactivity—A mortar bar containing the aggregate should have an expansion less than 0.10% in one year with a 0.8% alkali content cement.

Absorption—The absorption should not exceed 3%.

Durability—The concrete containing the aggregate should not have a loss in the modulus of elasticity exceeding 50% in the freeze-thaw test.

Sand Equivalent—The fine aggregate should have a sand equivalent of not less than 75.

Testing and Evaluation

Laboratory testing is a means of scientifically evaluating the suitability of aggregate material. In an attempt to forecast the behavior of the aggregate in concrete, numerous tests have been devised, many of which are complicated and require expensive equipment and trained technicians. Several of these tests have been used for many years and are familiar to those in concrete construction work. A strong effort is being made to standardize testing procedures throughout the nation, and many laboratories use, with little or no modifications, test methods as set forth in detail by the American Society for Testing and Materials. The principal tests performed on aggregates are for toughness and abrasion resistance, soundness, organic content, grading, specific gravity, absorption, alkali-aggregate reactivity, and thermal incompatibility. Petrographic examination in greater detail than possible during field examination supplements the laboratory tests, which are outlined in Table 2.

Petrographic Examination: Petrographic examination is a laboratory method for determining physical and chemical properties of aggregates primarily by visual appraisal. Standard techniques for mineral identification, which commonly involve the use of the petrographic microscope, and microchemical tests are employed. Representative samples containing hundreds of particles are carefully examined and classified as to (1) rock type, (2) physical condition, (3) chemical stability—presence of reactive or otherwise deleterious ingredients, (4) shape of particles present, and (5) nature of particle coatings. Reports are tabulated to show percentage composition, condition of the rock types present, and percentage of reactive ingredients, as well as diagnostic features of the particular constituents and the relative qualities of the different size fractions. Examination by petrographers experienced in concrete work permits interpretation of routine laboratory tests and aids in comparing new aggregates with other aggregates for which service data are available. On the basis of petrographic observations supplementary tests other than the usual laboratory tests may sometimes be recommended.

Depletion Allowance

The depletion allowance for common sand or gravel is 5%; for silica sand or pebbles, the allowance is 14%.

Preparation for Markets

Mining Methods

Sand and gravel deposits are mined with power shovels, draglines, slackline cable draglines, bulldozers, or dredges. The choice of excavating equipment depends upon whether the deposit is mined dry or wet.

In a dry-pit operation, shovels, loaders, or draglines load the sand and gravel into trucks or onto conveyor belts for transfer to the processing plant. The description of the CONROCK plant in a subsequent section (processing) provides an example of a large, dry-pit operation.

A wet-pit operation recovers sand and gravel from deposits that are below the water table. The sand and gravel may be excavated by a land-based dragline or by a floating dredge on a man-made or natural lake. Dredges are also used to recover sand and gravel from river channels.

Dredging: Some dredges are equipped for complete processing of material; others only scalp off unwanted sizes, and processing is completed on shore.

Dredging, especially in rivers, has had a long and stable background in the mining of sand and gravel. Now torn between two forces (a forward impetus due to local use restriction of land-based deposits, and increasingly restrictive federal and state regulations), dredge operations still account for up to 50% of the nation's sand and gravel production. This figure includes such forms of dredging as barge-mounted draglines, straight suction pumping, hydraulic cutterhead techniques, and bucket ladder methods.

Stearn (1970) stated that in 1968 there were 603 sand and gravel dredge operations in 37 of the 50 states. Nebraska headed the list with 129 dredges working. Kansas ranked second with 55. Florida, Michigan, Indiana, and Louisiana followed, with 34, 33, 31, and 29 dredges, respectively.

A brief survey of rock products operators utilizing dredges for all or part of their production revealed that some were dredging aggregates as far back as 100 years ago. Roughly 45% of the dredges covered by the survey were operating in a pond or wet pit. Rivers and lakes accounted for 25% each, and the remainder operated in bays.

Maximum working depths reported ranged from a shallow 15 ft in a river to 125 ft in a lake. The overall average was 50 ft. Bay dredges worked at depths ranging from 20 to 100 ft, with an average of 60 ft. In lakes, dredges reached depths from 20 to 125 ft, averaging 80 ft.

Sixty-five percent of the operators were working mixed deposits of sand and gravel. Straight sand, straight gravel, shell, and silica sand were the other raw materials mentioned. Sizes ranged from 8 in. boulders to 200 mesh sand, with roughly half those reporting handling topsized material in the 6 to 8 in range. Slightly more than 60% of the producers processed material ashore, but there appeared to be an increasing trend toward furnishing dredges with processing equipment.

Production capacities varied widely, from 20 to more than 2000 tph, for an average of approximately 350 tph. Principal product categories were specification sand and gravel, concrete aggregates, industrial and mason sand, roofing gravel, and ballast.

Diesel and electric power accounted for 45% each among the engines chosen by these dredge operators. Steam, natural gas, and fuel oil were in limited use. Pump sizes ranged from 4 to 18 in., with 12 in. average.

Ocean Mining: Ocean mining may supply future aggregate needs in metropolitan coastal areas as local deficiencies develop in construction aggregate supplies from nearby land deposits. Sand and gravel probably constitutes the most widespread resources of the sea floor of the US continental shelf. Although no large-scale commercial exploitation of offshore sand and gravel deposits is known to be underway at present on the continental shelf proper, tidal rivers and estuaries have been dredged commercially for many years.

Marine technology today is essentially an extension of conventional dredging systems which have been used successfully in rivers and estuaries. These include clamshell, dragline, and dipper dredges for relatively shallow water (generally under 100 ft), bucket ladder types (150-200 ft depth), and hydraulic systems (generally equipped with a rotary cutter) capable of pumping dredged material in slurry form.

Processing Techniques

Commercial operations are located on portions of deposits where the proper blend of sizes of sand and gravel can be obtained in order to produce a variety of products. Noncommercial production by highway contractors generally is obtained from portions of deposits where there is an abundance of +1½-in. gravel to provide crushed materials for use as road base or bituminous aggregate.

Ideally a commercial deposit would contain about 60% gravel and 40% sand (−¼-in.), enough +1½-in. gravel to crush for use as road base or bituminous aggregate, and sand in the correct sizes and proportions for use in concrete. Few deposits meet these requirements and plants are designed to cope with natural deficiencies. Plant capacities can range from 30 to 1000 tph of processed material.

Most plants are departmentalized to produce different products. There is usually a "dry" side, where pit run material is screened and crushed for use as road base, or in some instances as bituminous aggregate, and a "wet" side where sand and gravel are washed and screened for use as concrete aggregate. Various stages of crushing are used to reduce the gravel fraction to the necessary sizes. Ordinarily, a jaw crusher is used for primary reduction, particularly where boulders greater than 3 in. are present. Gyratory and roll crushers further reduce the gravel to −1 in., and −¾ in. Rod mills are used to manufacture sand from ¼ to ⅜-in. "pea" gravel to supplement natural sand that is deficient in fine sizes.

The sand fraction ordinarily is washed and classified in spiral classifiers (sand screws). Plants processing sand deficient in fine sizes usually employ wet cyclone-type separators to recover fine sand which may be present in the overflow from a classifier. In a number of plants, hydraulic settling tanks are used to separate the sand into various sizes. The desired blend of sand sizes is drawn off through gates in the bottom of the tank and dewatered in spiral classifiers.

Plants processing clayey material use revolving scrubbers or log washers to clean the pit run material before further processing.

Standard screens which are mechanically vibrated separate the gravel into the necessary sizes.

Some plants beneficiate their raw material by removing soft particles, such as shale, in a heavy media separation which utilizes the "sink-float" principle. The desired material sinks and the undesirable lighter fraction is floated off. The heavy liquid is made up of a mixture of finely ground magnetite and ferrosilicon, with a specific gravity that can be varied from 1.0 to 3.4. Jigs are also used to remove light particles in the granule size range. (Goldman, 1962).

Perhaps the best example of new and modern methods for the harvesting and processing of sand and gravel can be found in the newest facility of a company in southern California.

In the fall of 1972, CONROCK Co. began operation of its newest and most modern sand and gravel plant, which is located in the City of Irwindale, CA, on the alluvial fan of the San Gabriel River.

The plant was designed to operate initially in the 1200–1400 tph range, but provisions were made in the design to permit expansion to 2000–2400 tph without major modifications.

Material is harvested dry in the pit using a 9-cu-yd electric shovel. This material, of about 48-in. maximum diameter, is then reduced in a 48 x 60-in. primary jaw crusher mounted on a portable crusher car frame.

Material from the pit is then transported to the plant site via 42-in. wide conveyor belts and deposited in a 100-ft high surge pile.

Material is drawn from the surge pile using vibrating feeders, and the feed to the plant is automatically controlled by a conveyor belt scale. The initial plant feed is split on two 6 x 16-ft primary screens. The 1½-in. material goes to the wet side of the plant where it is scrubbed in a 7 x 35-ft trommel scrubber. The gravel portion is then washed and sized on 5 x 16-ft horizontal doubled deck screens. Concrete sand is washed in a scalping-classifying tank and then partially dewatered in 54-in. diam by 34-ft long sand screws. Further dewatering is accomplished by depositing the sand in large stockpiles for drainage.

The +1½-in. material from the primary screens is reduced, using three 7-ft secondary cone crushers in a closed circuit crushing system. Crushed materials are screened into various size groups and then combined into products by blending, using an elaborate chute system. With an almost unlimited number of combinations available for combining sizes, and with provisions for recrushing any of the unwanted larger sizes, it is possible to make only those products that are in demand at any respective time. Also, gradations of crushed products can be made coarser or finer in the blending process so as to meet exacting specifications. Provisions have also been incorporated to permit certain percentages of the crushed sizes to be used for supplementing natural gravel products.

Natural gravel and crushed products, together with washed concrete sand, which is conveyed back to the plant by belt conveyor, are stored in large overhead reinforced concrete bins. Each bin holds either 500 or 1000 tons of live storage. The total overhead live storage for the plant is 25,000 tons.

Marketing

Commercial interests produce approximately 85% of the sand and gravel supply, with the remaining 15% produced by government and/or contractors. In the latter category, it is apparent that such production operations are set up to handle extremely large projects (such as dams, rural highways, etc.), or in remote or rural areas where no commercial operations exist. The work is done with semi-mobile plants which remain only until completion of a specific project and then relocate for the next job.

In or around metropolitan areas where continuing work and markets for sand and gravel exist, and where transportation networks (rail, water, highways) are adequate, most production is done by commercial companies with permanent production facilities. Since transportation of sand and gravel is a disproportionate share of selling cost, it is not unusual for asphaltic concrete and ready-mix concrete plants to be located on the sand and gravel production site. Indeed, some producers are in the ready-mix business, some also in the asphaltic concrete business, and there are some producers who are also construction contractors.

Because of the broad distribution and general availability of sand and gravel, there are many commercial operations throughout the US. The number of commercial operations declined from 7291 in 1978 to 6836 in 1979.

The US Bureau of Mines reveals that in 1979 there were 5447 commercial operations with production of less than 200,000 tons, accounting for approximately 30% of total US production. Some 860 operations with annual production between 200,000 and 500,000 tons produced 28% of the total US production, and 369 operations with production over a half million tons accounted for the remaining 42%.

Price Trends: The price of sand and gravel as used at its ultimate location is a function of two elements—first, the cost of the material itself at the production location and, second, the cost of transport to the using site. These two elements must be considered when evaluating industry trends. Publications of the US Bureau of Mines reveal that although the unit price of sand and gravel in actual dollars has increased since 1954, in constant dollars the price has decreased from $1.55 per ton in 1954 to $1.34 per ton in 1974. Such a low price has been maintained (in spite of rising labor costs, higher land values, and increasing restrictions) through the development and use of larger and more efficient mining and processing equipment. Such improvements do have limits, and it is expected that prices will rise slowly during the 1980s to the year 2000.

Average f.o.b. plant prices for various types of gravel are shown in Table 3 for the years 1971 and 1979.

TABLE 3—Average F.O.B. Plant Prices for Various Types of Gravel

Type of Gravel	Average Price Per Ton, $ 1971*	1979†
Building	1.55	2.59
Paving	1.19	2.65
Fill	0.69	1.55

*Pajalich, 1972.
†Tepordei, 1980.

Major technological developments which have been instrumental in maintaining adequate production of sand and gravel at stable or slightly declining real cost include: use of larger operating units, more efficient portable and semi-portable plants, unitized plants for versatility of plant capacity, new prospecting methods utilizing aerial and geophysical surveying methods, and greatly increased rehabilitation and resale of mineral areas where advantageous.

Transportation: The proximity of market to the resource to date has maintained a relatively stable cost index for most regions. The cost of transporting material is based on production cost, equipment used, and manpower. A good rule-of-thumb estimate on rates for transporting sand and gravel is approximately $0.10–$0.15 per ton for each mile hauled, though in some areas costs of $0.20–$0.25 or more per ton-mile are realized.

The single factor which most escalates the delivery price of sand and gravel is the distance of haul under present practices and regulations. At 10¢ a ton-mile an increase of 20 miles in the hauling distance would increase the price of aggregate $2 per ton. This could easily represent an increase of 50 to 100% on the delivered cost of sand and gravel to the consumer. Some idea of the effect distance has on the price of aggregate delivered to the consumer is indicated in Fig. 1.

When aggregates are moved great distances, the final cost for the finished product is not the production cost but transportation. In some metropolitan areas, public utility commission truck zone rates apply; in other more outlying areas truck rental rates apply. The rate is highly variable depending on method of transportation, governmental regulation, labor contracts, and area. Other factors include haul distance, legal load limits, on-road or off-road operation, and, certainly, competition.

Truck haulage is the predominant form of transportation used by the industry, both as the only method or often in conjunction with an alternate method. Rail hauls are used in many areas particularly when spurs are available, quantities are large, and distances from production to destination are long. Rehandling of material is undesirable because of added costs and breakdown of the materials during handling. In coastal areas and areas on inland waterways, barge hauling is a readily accepted method. Together, rail and barge transportation account for approximately 10% of the total sand and gravel movement.

Consumption: The American Road Builder's Assn. estimates that highway consumption (construction and maintenance) will require 1970 million tons in 1985. In 1902, output was 1.848 million tons, and 35 years ago, 254.1 million tons.

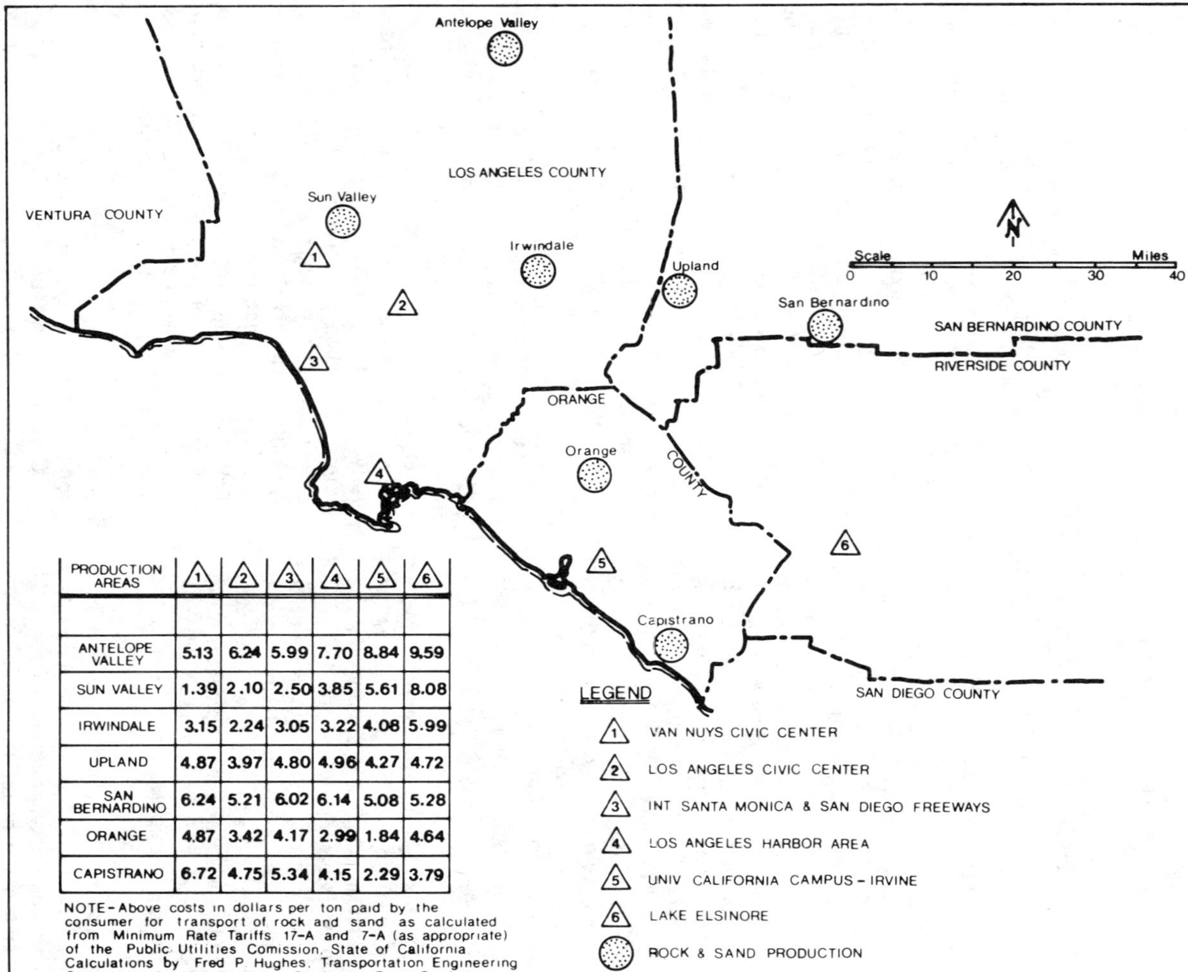

PRODUCTION AREAS	1	2	3	4	5	6
ANTELOPE VALLEY	5.13	6.24	5.99	7.70	8.84	9.59
SUN VALLEY	1.39	2.10	2.50	3.85	5.61	8.08
IRWINDALE	3.15	2.24	3.05	3.22	4.08	5.99
UPLAND	4.87	3.97	4.80	4.96	4.27	4.72
SAN BERNARDINO	6.24	5.21	6.02	6.14	5.08	5.28
ORANGE	4.87	3.42	4.17	2.99	1.84	4.64
CAPISTRANO	6.72	4.75	5.34	4.15	2.29	3.79

NOTE-Above costs in dollars per ton paid by the consumer for transport of rock and sand as calculated from Minimum Rate Tariffs 17-A and 7-A (as appropriate) of the Public Utilities Comission, State of California. Calculations by Fred P. Hughes, Transportation Engineering Consultant for the Southern California Rock Products Association, May 1981

FIG. 1—*Typical variations in transport cost per ton of rock and sand between production areas and selected locations.*

The average annual growth rate of domestic demand for sand and gravel since 1954 has been about 2.8%. The estimated annual growth rate for the balance of the century, obtained by contingency forecasting based on anticipated requirements for individual end use, is expected to fall somewhere between 2.5 and 4.7%. The forecast demand for sand and gravel in the United States in the year 2000, corresponding to this growth range, covers a range from 2.0 to 2.6 billion tons (Pajalich, 1976).

Pattern of Use: Within the construction industry, highway construction or road building consumes the major volume of sand and gravel, used as aggregate for both concrete and asphalt pavements; in concrete for bridges, tunnels, and appurtenant structures; in treated and untreated road base material; and in both structural and nonstructural fill. The next largest segment of use is in building construction, where practically all of the volume is used in the manufacture of concrete for both cast-in-place and precast structural and cladding elements. The remaining quantities go into fill, septic fields, and other construction-related uses.

Specialty sand uses demand more rigid specifications for special types and gradings. Such specialty sands are classified by the US Bureau of Mines as "industrial sand" and are discussed in the chapter on "Silica and Silicon."

Future Consideration and Trends

Substitutes

As in the case with most materials, sand and gravel are not immune to substitution for various reasons—economic, application, or even environmental. Perhaps the most widely used substitute or alternate material is crushed stone (see "Sand and Gravel" in Part 2, *Uses*). It competes in almost every category with sand and gravel and is produced in a wide variety of types, sizes, and gradation.

A growing competitor to sand and gravel in the concrete aggregate market area is lightweight aggregate. This material is manufactured by a sintering process in which appropriate clays, shales, slate, or fly ash are passed through rotary kilns or traveling grate furnaces to produce a bloated aggregate product. These materials are used extensively to produce a lightweight concrete used in cast-in-place structures, in precast elements, and in concrete block. Other lightweight materials for concrete aggregate include natural pumice, expanded perlite, expanded (or exfoliated) vermiculite, and iron blast-furnace slag. These materials find use in insulating fills, roof decks, and similar basically nonstructural uses.

Urbanization, Government Control, and Zone Restrictions

The sand and gravel industry, more often than any other major segment of the mineral industry, is beset with land use conflicts and environmental problems associated with rapid urban expansion. The problems are complicated because urban growth depends upon centrally located sources of low-cost construction materials. It is an economic fact not understood by the consuming public.

As a city or county expands, the growth process itself may cut off the vital supply of industrial minerals which are needed for this growth. This usually happens in the following sequence. As a community or county begins to grow, a demand or market is created for industrial minerals, such as sand, gravel, crushed rock, and clay for brick products. Since sand and gravel form a high-bulk, low-cost product, the extraction operations begin as close as possible to the consuming areas—the population centers. As the community or county grows, more and more homes are built in the outlying or rural areas. This urban expansion gradually approaches and then surrounds the established gravel pits. The new homes may cover sand and gravel deposits; this makes it impossible to expand the sand and gravel extraction. In addition to covering potential reserves, the new homeowners may resent the existing extraction operations. Complaints are often made to the local government authorities which allege excessive noise, dust, and hazards to children. Continued pressure may result in ordinances which restrict operating conditions, truck routes, etc. Finally, as a result of depleted reserves and/or harassment, the pit operator may be forced to move to more remote locations. Ultimately, this reduces the available supply and increases construction costs (Goldman, 1959).

Rapid population growth and urbanization have had a decided impact upon the sand and gravel industry. Their ability to continue operations and supply the needed aggregates to maintain the present and projected construction needs of our society is imperiled. Local government has imposed planning and zoning restrictions that are inconsistent and usually only the zoning needs for that particular city are considered. There is little uniformity or consistency in the laws. Little thought is given to those areas in the total region served by the mineral resources. Prime mineral resources are lost to development and urban growth.

No uniform mining and mineral policy exists. In most cases, legislative bodies usually respond to demands of local voters who prefer that the sand and gravel operators go elsewhere. To date, 15 states have passed environmental laws which will affect sand and gravel operations.

Closely associated with land use conflicts are problems of environmental contamination, such as water pollution, dust, solid waste, noise, and unsightly conditions that sometimes accompany sand and gravel operations.

Bibliography and References

Anon., 1957, "Standard Specifications for Construction of Roads and Bridges on Federal Highway Projects," FP-57, US Dept. of Commerce, Washington, DC, 363 pp.

Anon., 1958, "Bibliography on Mineral Aggregates," No. 23, Highway Research Board, Washington, DC, 111 pp.

Anon., 1961, "Mineral Aggregates," Pt. 4, *ASTM Standards,* American Society for Testing & Materials, Philadelphia, PA, pp. 491–657.

Anon., 1962, "Engineering Problems of Sand and Gravel Production," NSGA Circular No. 88, National Sand and Gravel Assn., Silver Spring, MD, 34 pp.

Anon., 1963, *Concrete Manual,* 7th ed., US Bureau of Reclamation, Denver, CO, 642 pp.

Anon., 1964, *Pit and Quarry Handbook,* Pit and Quarry Publications, Chicago, IL.

Ahearn, V.P., Jr., 1964, "Land Use Planning and the Sand and Gravel Producers," National Sand and Gravel Assn., Silver Spring, MD, 30 pp.

Cooper, J.D., 1970, "Sand and Gravel," *Mineral Facts and Problems,* Bulletin 650, US Bureau of Mines, pp. 1185–1199.

Drake, H.J., 1972, "Stone," *Minerals Yearbook, 1972,* US Bureau of Mines, pp. 1153–1173.

Dunn, J.R., and Cutcliffe, W.E., 1971, "Selecting Aggregate Deposits: A Geologic View," *Rock Products,* Mar., pp. 75–79.

Dunn, J.R., and Hudec, P.P., 1972, "Frost and Sorption Effects in Argillaceous Rocks," No. 393, Highway Research Board, Washington, DC, pp. 65–78.

Ekblaw, G.E., and Lamar, J.E., 1964, "Sand and Gravel Resources of Northeastern Illinois," Circular 359, Illinois State Geological Survey, 8 pp.

Goldman, H.B., 1956, "Sand and Gravel for Concrete Aggregate," *California Journal of Mines and Geology,* Vol. 52, No. 1, pp. 79–104.

Goldman, H.B., 1959, "Franciscan Chert in California Concrete Aggregate," Special Report 55, California Div. of Mines and Geology, 28 pp.

Goldman, H.B., 1959a, "Urbanization and the Mineral Industry," *Mineral Information Service,* California Div. of Mines and Geology, Vol. 12, No. 12, Dec., pp. 1–5.

Goldman, H.B., 1961, "Sand and Gravel in Northern California," Bulletin 180A, California Div. of Mines and Geology, 38 pp.

Goldman, H.B., 1961a, "Urbanization—Impetus and Detriment to Mineral Industry," *Mining Engineering,* Vol. 13, No. 7, July, p. 717.

Goldman, H.B., 1962, "Aggregate from 'Fossils,'" *Rock Products,* Vol. 65, No. 11, pp. 65–68.

Goldman, H.B., 1964, "Sand and Gravel in Central California," Bulletin 180B, California Div. of Mines and Geology, 58 pp.

Goldman, H.B., 1968, "Sand and Gravel," in *Mineral and Water Resources of California,* Bulletin 191, California Div. of Mines and Geology, pp. 361–369.

Goldman, H.B., 1969, "Sand and Gravel in Southern California," Bulletin 180C, California Div. of Mines and Geology, 56 pp.

Goldman, H.B., and Klein, I.E., 1958, "Sand and Gravel Resources of Cache Creek in Lake, Colusa, and Yolo Counties, California," *California Journal of Mines and Geology,* Vol. 54, No. 2, Apr., pp. 237–296.

Goldman, H.B., and Klein, I.E., 1961, "Sand and Gravel Resources of the Kern River, near Bakersfield, California," Special Report No. 70, California Div. of Mines and Geology, 33 pp.

Pajalich, W., 1972, "Sand and Gravel," *Minerals Yearbook 1972,* US Bureau of Mines, pp. 1103–1121.

Pajalich, W., 1976, "Sand and Gravel," *Mineral Facts and Problems, 1975,* US Bureau of Mines, pp. 933–946.

Schellie, K.L., and Rogier, D.A., 1963, "Site Utilization and Rehabilitation Practices for Sand and Gravel Operations," National Sand and Gravel Assn., Silver Spring, MD, 80 pp.

Stearn, E.W., 1970, "Dredging in the Aggregates Industry," *Rock Products,* Vol. 73, No. 9, Sep., pp. 74–76.

Tepordei, V.V., 1980, "Sand and Gravel," *Minerals Yearbook, 1978–79,* US Bureau of Mines, 26 pp.

Yeend, W., 1973, "Sand and Gravel," *U.S. Mineral Resources,* Professional Paper 820, US Geological Survey, pp. 561–565.

Silica and Silicon

T. D. MURPHY *

G. V. HENDERSON †

The element silicon, with its usual partner, oxygen, plays the same role relative to inorganic materials as carbon and hydrogen play with respect to living organisms. The crystallographic structure of silicon dioxide (SiO_2) consists of one atom of silicon well-nigh inseparably bonded to four contiguous atoms of oxygen, a gas of but slightly more than half its atomic weight, forming a three-dimensional network of SiO_4 tetrahedra (Dietz, 1968). This wispy arrangement is the fabric of the mineral quartz which is one of the harder, more abrasive, and chemically stable raw materials to be found in Mother Nature's cupboard. Most of us are vaguely familiar with the observation that nearly 60% of the lithosphere to an approximate depth of 10 miles is composed of these two elements, which proportion actually rises to about 75% if the atmosphere, hydrosphere, and biosphere are included (Parker, 1967).

Elemental silicon is a brittle steel-gray metalloid with a density of 2.42 and does not occur in nature. In its limited commercial form, which is still brittle and metallic in appearance, its density drops to 2.33. It combines with oxygen to form the gaseous oxide SiO, or the solid dioxide SiO_2 (analogous to its congener carbon in CO and CO_2); a tetrachloride $SiCl_4$ (analogous to CCl_4); or, combining with oxygen and one or more metal ions, forms the largest rock-making family of all, the silicates. Further, next to the feldspars, quartz is the most abundant known mineral and, in some form or another, accounts for roughly 12% of surficial terrestrial rock. It is this so-called "free silica" to which your attention is chiefly directed in this chapter.

In any such consideration of silica, its several crystalline phases with their attendant differing physical properties appear worthy of delineation. Alpha quartz is the only one of this interesting polymorphic family that is thermodynamically stable at normal pressures and temperatures up to about 573°C. The low phases of tridymite and cristobalite can exist in the so-called metastable state below 117° to 200°C but seldom occur in nature in these two different crystal forms (Sosman, 1965). Opaline silica, now considered by most researchers to be a variety of cristobalite, occurs in abundance and is the only exception. The three remaining polymorphs, keatite, coesite, and stishovite, also are metastable under ambient conditions. Keatite, however, is not known to occur naturally and has only been synthesized in the laboratory. The other two have been identified in submicron sizes and in microscopic amounts, occurring naturally as a result of high order shock-wave pressures generating quite elevated temperatures—such as might accompany the impact of a large meteorite on a hard sandstone terrain (Chao et al., 1962).

An intriguing characteristic of these silica minerals, and one which has very practical applications, is their extremely variable densities. The classic 2.65 specific gravity of alpha quartz drops to 2.20 in tridymite, cristobalite, and lechatelierite (natural silica glass), as well as in many man-made glasses of commerce which are merely noncrystalline or amorphous solids in the form of super-cooled liquids which can no longer change their shapes (Pauling, 1950). It rises again to 2.50 and 2.93 for keatite and coesite, respectively, and finally becomes a reported 4.35 for stishovite—nearly a 50% range in density for the known physical combinations that silica naturally assumes (Frondel, 1962; Konnert et al., 1973; Pecora, 1960).

Pursuing its esthetic aspects, briefly, silica is well-represented among gem stones which, to qualify as such, must have the attributes of beauty, durability, and rarity (Jahns, 1983). As quartzose gems usually have but one or, at

* Deceased, 1980.

† Chapter modified for 5th edition by G.V. Henderson, Associate Professor, Earth Sciences Dept., California State Polytechnic College, deceased 1981.

most, two of these necessary attributes in any single physical form, they qualify only in the semiprecious category. Among the crystalline forms may be mentioned the well-known amethyst, citrine, "tiger's-eye," aventurine, rock crystal, and rose or smoky quartz. Among the amorphous (so-called opaline) or microcrystalline forms are agate, carnelian, chrysoprase, bloodstone, and, of course, the true opal in its several manifestations. Quartz is one of the gem stones which, together with emerald, ruby, spinel, and sapphire, have been successfully synthesized by man.

From a strictly utilitarian point of view, the bulk of the free silica of commerce consists of quartzites together with the detrital quartzose sands and gravels and the weakly consolidated sandstones and conglomerates resulting from erosive and sedimentary processes. These are the "bread-and-butter" types that furnish most of the high-purity siliceous raw materials going into the ceramic, refractory, and heavy chemical trades as well as to foundry, steel, and other metallurgical outlets. In addition, but to a lesser extent, respectable tonnages of massive quartz in the form of pegmatitic replacements and vein types, together with tripoli, ganister, chert, novaculite, and quartz-mica schist, find their way to market for the production of specialty items or ultra high-purity products such as vitreous silica, chemical ware, and TV tubes.

Uses and Specifications

Siliceous raw materials are often used in industry directly as finished products, shipped without recourse to elimination of natural contaminants. Such an item as naturally bonded molding sand, in which case the mined material is merely dried before shipment to preserve the natural interstitial clay bond, is an example of this, although that used for very heavy steel castings is shipped crude or green with contained moisture. Runner sands for the blast furnace casting floor and furnace bottom sand for acid open hearth practice both fit in this category. Carefully graded quartz sand filter media is an example where the exact size frequency distribution is of paramount importance. This is also especially true of hydraulic fracturing sand wherein the grain shape, in terms of a high Krumbein roundness factor, is of critical significance. Other examples could be cited, such as blasting sands and pulverized silica flour, as these materials are largely purchased on very precise physical and chemical specifications (Wilborg and Henderson, 1983). The use of selected natural quartz crystals as well as cultured or man-made quartz crystals is a strategic as well as a growing business (Hale and Blair, 1983). Specialty silica products, in both powdered and liquid forms, are being increasingly used in industry. These synthetic silicas referred to as xerogels and hydrogels, together with pyrogenic and precipitated silica as well as colloidal silica sols which are used as binder, bonding, and antislip agents, are described by Teicher (1974). Another specialty item is the commercially prepared cristobalite that C-E Minerals and Harbison-Walker Refractories now manufacture. It is 60% − 100 mesh and is used in investment casting to compensate for metal shrinkage and to impart increased dimensional accuracy of cast parts.

However, the greater tonnage is purchased for its use as an additive to some manufacturing process where, in this capacity, it qualifies only as a semifinished product. Thus, while physical characteristics are significant, a particular material might chiefly be used because of its relatively unique chemical properties or composition. These products demand even greater care in preparation and must often meet very rigorous chemical specifications. Such prepared silica materials as electrometallurgical gravel for production of elemental phosphorus or silicon alloys, granular mix for acid refractory shapes, and specially prepared sand for manufactured abrasives, soluble silicates, and glassmaking are the principal tonnage items, the last, by far, the largest.

To generalize, a naturally light-colored quartzite, sandstone, or quartz sand or one which readily breaks down and washes to a light color, having at least 95% of its grains between US Standard Sieves of 20 mesh (850 μm) and 140 mesh (106 μm), and with a fairly uniform size frequency distribution, is, from a textural standpoint, a potential multipurpose industrial silica raw material, per se. If further inspection reveals that the individual quartz grains are relatively free of identifiable blemishes, e.g., black iron-bearing mineral inclusions, hard, tight irony-clay crusts and coatings, or a noticeable percentage of deeply colored grains, the chances are that with specialized preparation the material might qualify for higher grade use from a compositional standpoint also. The degree to which it would qualify is tightly controlled by the type and percentage content of discrete particles native to the raw material. These normally consist of high gravity detrital minerals, which are black or opaque and usually high in iron, and/or semitransparent minerals which are extremely refractory, and/or miscellaneous rock particles. However, bulk color is not an absolute criterion for the judgment of purity in a silica sand. Needlelike graphitic inclusions in the quartz grain, quantitatively sufficient to impart a drab gray or even a blackish color to a commercially prepared sand that is otherwise acceptable, do occur sporadically (Heinrich, 1974).

The aforementioned defects that generally plague most silica source materials are not only known to most producers but also to most potential users, particularly in the glass industry. Even the smaller glass producers have their own laboratory and glass technologist. Because they know the trouble they can get into almost overnight, they protect themselves against this by carefully sampling and testing the incoming raw materials on a very stringent specification basis. Although explicit ranges and tolerances of unwanted contaminants are furnished to the supplier, shipments of unfailing uniform quality, year in and year out, are most sought after by the enlightened purchasing agent. He knows that he can keep his batch and furnace supervisor the most content with a raw material that does not have erratic and continuing fluctuations in the sieve and chemical analyses even though the top sizes or the fines are sometimes a little excessive or the contaminants are on the high side of the specifications limits.

This is axiomatic for any raw material consumer, of course, but it is particularly true where trace quantity limitations are critical, as with silica. For instance, it is well-known in the silica industry that nickel, chromium, copper, and cobalt are not normally indigenous to the usual silica deposit, per se. For glass-melting sand use, however, even trace amounts of these four metallic ions are deleterious because of their high colorant or defect-producing properties when dissolved in molten glass. While it is true these elements are often added to the glass batch this is done under highly controlled conditions by the batchmaster (Mills, 1983). As a consequence, most glass producers put very stringent particulate limits for these on their purchase specifications. The producer, therefore, must be conscious of any widespread secondary manganese mineralization in his deposit which could harbor trace quantities of cobalt; must eliminate all danger of copper contamination in his quarry blasting by substituting aluminum caps and iron leg wires for the more conventional copper blasting accessories, or, better, go entirely to "Primacord" for the initiation of the blast; and must be sure that no nickel or chromium alloys are used where any wear surfaces are involved in his mill flowsheet. If these precautions are not strictly enforced, the penalty can be the eventual loss of valuable orders.

While it is difficult to generalize on tolerances and specifications of a sand suitable for flint glass melting or as a base for chemical manufacture, the total acceptable iron content usually ranges from 0.015 to 0.030%; alumina should not exceed 0.20% or fluctuate; calcium and magnesium oxides combined should be uniform and not exceed 0.05% for glass and 0.15% for chemical use. Alkalies and titania are tolerated in trace amounts (0.01% or less) but must not vary; cobalt and chromium—none. Sand with 0.05 to 0.15% iron is used for the production of amber container ware and flat glass. Specifications for other than ceramic raw materials—which are constantly being tightened—have not changed materially in the past 20 years, so will not be recited here (Murphy, 1960, 1960a).

Geology, Distribution, and Occurrences

Silica, from the standpoints of its genesis and geographic distribution, is indeed ubiquitous. In North America, as elsewhere, silica has been exposed in massive weathered outcrops and reaccumulating in fresh detrital deposits during practically every era and period of recorded geologic time. Current witnesses to this range from impressive Canadian operations in the Lorrain quartzite of ancient Huronian time (Hewitt, 1951) to our own Holocene beach sand deposits even now being winnowed by nature for future exploitation on the Gulf Coast of Florida's Panhandle. As an illustration of the versatility of this mineral, it may be pointed out that a major silica producer in this country currently utilizes raw materials ranging in geologic age from late Precambrian to mid-Tertiary. These run the lithologic gamut from pegmatitic quartz through tightly cemented vitreous quartzite to loose, unconsolidated quartz sands.

Eastern United States

The bulk of this country's silica production lies in the populous east where most of the consumers are and hence the demand for silica-based products is highest. Twenty years ago a study of source distribution by the writer showed that 145 viable silica mining companies, regardless of the quality or type of product sold, were operating in the silica industry east of the Mississippi River (Murphy, 1960).

The census of silica-bearing formations in the 26 eastern states referred to indicated the surficial occurrence of approximately 67 mapped, named, and formally accepted stratigraphic units that might have an economic potential in this industry. Lithologically, these break down into quartz-mica schists, quartzites, sandstones (variable in degree of consolidation), quartz conglomerates, quartz gravels, and several widely separated tripoli districts. Also included were many geologically undesignated dune sand occurrences, terrace sands and gravels, commercial chert deposits, massive quartz pegmatites, hydrothermal replacement bodies, and silicified zones, some of which have been and probably are viable from an economic point of view.

In this connection there are many scattered, very pure, silica sand deposits of Holocene age along the coastal plains of South Carolina, Georgia, and Florida, some of which have been and are being reworked by wind and the present river systems. Another interesting and valuable source type may be referred to as the St. Lucie soil zone which is well-developed along a stretch of east Florida coast particularly in St. Lucie county (Hudson, 1946), the type locality, as well as inland in this state and in south Georgia (Teas, 1921). The key to the purity of these deposits relates to a fluctuating ground water table and an excess of azohumic and other organic acids resulting from decaying swamp vegetation which leaches the iron normally present in the surficial sand. This organic chemistry develops the so-called Van Dyke or sap-brown ore at the base of the leached zone which was much sought after as an organic dye for paint, wood stains, and inks before the advent of aniline dyes. Most of these types of deposits are from 4 to possibly 12 ft thick on flat terrain with consequent disappointingly low tonnages. However, they are commercially viable at a few points in each of the states mentioned and can supply local flint glass sand markets for indeterminant periods of time if extensive enough, which several have proved to be. The inevitable silica sand rings which partially surround the puzzling "Carolina Bays," are believed to be a manifestation of this leaching and in one instance were commercially exploited for several years in Williamsburg County, South Carolina (Buie, 1958).

The Carolina Silica Co. has a major silica sand production in North Carolina at Marston, which has been in production for about nine years. Pennsylvania Glass Sand has a major silica sand production in South Carolina near Columbia. Martin Marietta has a silica operation at Camden. Jessie Morrie Co. has a major silica sand production near Junction City, Georgia. In Virginia, the major silica sand production is from Gore, where Unimin Corp. has an operation.

Two formational units for high grade ceramic products exposed in the eastern and a few middle western states warrant detailed citation as the most successfully exploited silica raw material sources on the basis of tonnage and profit in this country. They are the St. Peter sandstone together with its stratigraphic equivalents of Lower Ordovician age, and the Oriskany sandstone or orthoquartzite of Lower Devonian age. The St. Peter is exposed and exploited principally in northern Illinois and, to a lesser degree, in southern Wisconsin and Missouri. Sandstones of the same time equivalency also are exploited in Arkansas and Oklahoma (Lamar, 1927; Ham, 1945). These formations are relatively flat-lying and massively bedded materials of weak to moderate consolidation that lend themselves equally well to open pit or underground mining recovery, both methods being currently employed. Physical recovery is high and a minimum of plant preparation is necessary because of its extreme natural purity. Reserves are large and nearly completely controlled by current producers. Even deeply covered areas, now unused because of excessive stripping costs or otherwise marginal because of local factors, have been explored and acquired by operating companies for possible future use. What little St. Peter sandstone that remains under private ownership will soon be, if it has not already been, completely zoned out of the market by urbanization pressures.

The Oriskany quartzite, more precisely the Ridgeley sandstone (Harris, 1972), is a ridge maker, to a large degree, because of its greater induration and consequent structural attitudes. For the most part it is found in strongly dipping beds ranging from 45° to the vertical and even overturned. In a few areas it is as flat-bedded as the St. Peter, whereas, locally, it may be preserved intact in homoclinal, anticlinal, and synclinal postures (Bates, 1960). Being an integral part of the Appalachian Highlands, its economic occurrences are pretty well limited to the tri-state area of Virginia, West Virginia, and Pennsylvania. The texture, purity, and degree of induration are all subject to extreme variability which is much less seldom encountered with the St. Peter. However, by and large, it shares the same perfection of grain that the St. Peter enjoys and most of the defects are secondarily induced and can be mitigated and often nearly eliminated by suitable plant beneficiation. Here again, all known reserves in the technical sense and much that could be classed as potential resource acreage is believed to be owned by, or under long-term lease to, local operating companies. A substantial portion of what is already controlled will doubtless never be completely exploited because of the ever-increasing zonal restrictions due to urbanization.

Although these two formations are the premier sources for all-purpose silica products in the United States, several others bear mention. The Jordan sandstone of Upper Cambrian age in Minnesota and Wisconsin has a preferred grain shape for specialty use end products but is limited to very restricted areas of exposure. Over most of its near-outcrop area it is covered with either thick glacial drift or heavy dolomite beds which make it completely uneconomic for commercial use. As it is a massive and moderately cemented sandstone, mining by room-and-pillar method has been carried on sporadically along the Mississippi River bluffs near St. Paul, producing chiefly foundry sand products.

The Cohansey sand, of probable Miocene age, is a strong local factor in the foundry and ceramic supply business of southern New Jersey where it is well located with respect to a heavy concentration of consumers. This is a very weakly consolidated sand of extremely erratic quality, both physical and compositional, but selected deposits can be and are upgraded by suitable processing into salable products, mostly for ceramic and refractory uses. Exploitable reserves are strictly limited and those that are known are jealously held by mining companies who operate in the outcropping areas.

A long-time supplier of industrial-type silica products, especially for the foundry trades, is the Pottsville formation of lower Pennsylvanian age. This formation includes the Homewood and Connoquenessing sandstone members in Pennsylvania, and the Sharon and Olean quartz-conglomerate and sandstone members in Ohio, Pennsylvania, and New York State (Bowen, 1953). These several stratigraphic units supply a large portion of the metallurgical pebble used in silicon alloys in the northeastern states, as well as amber quality melting sand and several types of sand for refractory use. Most of the coarse pebble sizes in the conglomerate phase have been exhausted through the years, forcing the metallurgical trade to switch, in part, to silica pebble from the Sewanee conglomerate of Pennsylvanian age in east-central Tennessee or the Pee Dee type of Pliocene to Holocene terrace gravels produced in both the Carolinas and to a lesser extent in Alabama, Georgia, and Florida. In Maryland the Bryn Mawr gravel of Pliocene age is a source material for this type of end use.

Western United States

In 1960 there were 48 producers west of the Mississippi, which represented roughly 25% on a national basis. This situation remains virtually unchanged at present writing, and breaks down into 22 producers in California and Nevada; five in the Pacific Northwest (which includes Oregon, Washington, Idaho, and Montana). This leaves an additional 21 producers for the remaining 16 states in the conterminous United States.

West Coast deposits, by and large, are inferior as to grade, nationally, necessitating expensive beneficiation which raises the ex-mine price substantially. However, except for certain items such as pulverized silica and a few quality specialty products obtainable only from eastern or mid–continent producers, the delivered price insures the use of western silica products in this region.

About the same situation exists with respect to the Pacific Northwest. Here, in addition to relatively second-grade raw materials, the additional burden of poor location with respect to transportation exists, with all known deposits being anywhere from 1 to 40-odd miles from rails. Since most of these deposits, geologically, are Tertiary to Holocene in age, they consequently have not undergone the natural winnowing of nonquartz particles which results from the recycling that the older Paleozoic and Mesozoic rocks east of the Continental Divide have undergone. Most of these far western deposits have been derived from the breakdown of simple binary granites, relatively free of ferromagnesian and ore minerals. Hence they are either only feldspathic, clayey, or both.

Production of feldspathic sand from Pacific Grove, CA, ceased several years ago. However, production of a feldspathic sand near Byron, CA (Martin Marietta) was begun about 1974. This operation serves the glass container industry in the Bay area and the Central Valley of California. This is an amber sand used principally for wine bottles and fiberglass production.

Martin Marietta has significant production of feldspathic sand from deposits near Emmett, Idaho. These sands are used in glass container manufacturing, sand blasting, filtering, building products, and other applications.

Probably the most successfully utilized material for production of glass container ware in northern California has been exploited since 1955 in Amador County from the Ione formation of Eocene age. The only significant production in this area today is the production of a high silica sand by Owens Illinois. The workable member consists of relatively clean quartz sand mixed with residual kaolin and anauxite in roughly a 60% to 40% ratio. The operation is close to captive San Francisco Bay area markets and the product is sold also to independent consumers. A similar venture was initiated in 1956 by competitive interests in an extension of the Ione into adjoining Calaveras County. A marketable quality product was made and less waste was involved here because the raw material, while quite feldspathic, was essentially free of clay, as such. However, the operation was forced into abandonment shortly after startup because the property was deeply flooded by the construction of a new dam to form the present Camanche Reservoir on the North Fork of Mokelumne River—an unfortunate but increasingly common victim of environmental pressures.

A detrimental impediment to the use of Ione sand (or, for that matter any other far western sand exploited to date) is that even though a completely adequate flint-quality glass sand can be produced at a high cost, it cannot be ground into an acceptable color quality for silica flour in the burgeoning commercial cleanser market which, perforce, must import silica flour from mid-continent or eastern producers. This is because the Ione formation contains residual particles of Mariposa slate and other dark-colored rock fragments which, because of the formation's geological youth, have not been eliminated by recycling. An exception to this in California is the availability of a limited supply of cleanser-grade silica flour with acceptable brightness (>85% dry reflectance and > 56% wet reflectance based on US Bureau of Standards MgO comparisons). This is produced from selected vein quartz pebble accumulations resulting from early placer hydraulic tailings that are dredged from the Bear River near Colfax between Nevada and Placer counties.

In Southern California, the Paleocene Silverado and Eocene Tejon formations are, respectively, mined and processed for both flint and amber glass at Rancho Mission Viejo in Orange County and near Oceanside in San Diego County. These outlets supply both glass and foundry silica raw materials as well as a great many specialty products such as blast sands, filter media, traction sand, soluble silicates, and ganister mix, all of which are principally used in the Los Angeles Basin area (Gay, 1957).

In the contiguous state of Nevada, Clark County contains substantial deposits of mostly second-grade raw silica materials in the form of quartzites, sandstones, and dune sands. The most heavily exploited of these is the Baseline sandstone of Longwell, so-called, who correlated these beds with early Upper Cretaceous formations in Wyoming and Idaho. It is uniform in composition and highly quartzose, contaminated only by a few clay pockets and laminations; the cementing material is weak, producing a friable sandstone easily quarried and beneficiated. It is produced and shipped into Southern California where it is utilized for flint and amber container ware and several grades of foundry sands. Other potential silica sources here include the Eureka quartzite of middle Ordovician age, very pure in most exposures but also tightly cemented and indurated. This has discouraged its use as a ceramic sand product for cost or production reasons but it could well be utilized for refractory and metallurgical uses. The Supai formation is classified as Pennsylvanian or Permian in age with restricted outcrops and, although sandy and friable, is plagued with highly colored beds of hematitic and limonitic-stained beds which lower its potential purity, productwise. The Aztec sandstone of Jurassic age is exposed extensively in imposing outcrops but suffers from red and yellow coloration that, again, would pose expensive cost problems in processing, especially for white glass use. Finally, a plethora of eolian sands of Pleistocene and Holocene age occur in sheets and dunes that are the end products of wind scour on exposed outcrops of the Supai and Aztec sandstones which crop out sporadically throughout the vicinity of Overton. The finer grades of this crude sand are suitable for medium to light corework in iron, brass, bronze, magnesium, and aluminum, but are more commonly utilized as a base in molding light steel, brass, and aluminum castings. In summary, some of the Clark County sands—chiefly those washed and graded products of the Baseline sandstone—are favored by virtue of market specifications and delivered price in the Southern California foundry trade, to a limited extent by the glass container trade, and much more for special purpose work such as, for example, manufacture of Transite pipe for which the mix requires up to 30% SiO_2 ground to a −140-mesh size (Murphy, 1954).

Some rather competent silica natural resources occur in the Pacific Northwest but the sparse population of this region, together with the relative isolation of most of the deposits with regard to rail transportation and end-use points has resulted in retardment of exploitation up to the present time. However, the Lane Mountain silica deposit in Stevens County, WA, is the major producer of silica sand in the Northwest and the only producer of flint quality glass sand. Lane Mountain Silica Co. has been a most successful operation for a number of years.

Results of a study (Carter et al., 1962) based on field and laboratory examinations, furnished data for commercial evaluations of some 82 silica deposits in this region. Of the 37 deposits found to be of satisfactory quality, for some sort of exploitation, only 16 were determined to have sufficient tonnage to be commercially viable. The investigators concluded there were two outstanding supply points: one was the Quartz Mountain deposit in Spokane County, Washington, which is an excellent source for silicon metal and ferrosilican alloy raw materials even though the operation is 20 miles from Mead, the nearest rail point. The deposit is huge and consists of massive quartz intruded into the local granitic country rock. The other was the Bovill clay-sand deposit in Latah County, Idaho, with accessibility to rails only a mile from the deposit. It was felt that this material should be a source for low-iron sand in the northwest for many years. The Lane Mountain deposit in Stevens County, Washington, shows promise but more field and laboratory work must be done. The Addy quartzite (age unstated) shows great textural variations throughout its outcrop area and contains argillaceous impurities. Rail transportation is available from 1 to 12 miles, depending on the outcrop locations.

Most high-purity silicon metal (96–99% Si) produced in the United States comes from raw materials located in the Pacific Northwest (including British Columbia) induced by the availability of cheap hydroelectric power according to Richardson et al. (1974). Therefore, this region appears to have a good potential for further eventual exploitation of its silica raw materials.

Middle United States

The 16-state region of the United States formerly alluded to represents a relatively sparse market for silica products except for certain areas. Again, demand based on population plays a large part in this. It is largely an agrarian region bordering on the Missisippi River with relatively few scattered market centers for silica raw materials. Those that do exist are amply serviced by the St. Peter sandstone and its stratigraphic correlatives which are chiefly exploited in the tier of states bordering on the Mississippi. A plexus of high-grade silica reserves of St. Peter age exists in Oklahoma which, together with a few stratigraphically younger and generally second-grade deposits in Texas, Kansas, and South Dakota furnishes silica demand west to the Rockies. Major production of hydra-fracturing sand comes from deposits near Brady, Texas (Pennsylvania Glass Sand and Texas Mining Co.). These deposits provide a major portion of the frac sand used in the US.

Unimin Corp. has recently completed a new plant facility for production of frac sand near Le Sueur, Minnesota. Ottawa Silica Co. operates a silica sand-kaolin operation near Kosse, Texas. This plant has shipped over 300,000 stpy of product for the past 15 years. The deposit is on the Simmsboro sand/kaolin member of the Wilcox formation. Ottawa Silica Co. also has a silica operation near Dubberly, LA, which supplies silica to the glass and chemical industries.

The five mountain states, together with Utah and Arizona, either lack sufficient population demand to warrant local silica production or lack silica raw materials of sufficient quality and accessibility to make exploitation profitable. Arizona is the only exception with its long-established production of excellent hydra-frac sand from the Bidahochi formation of Pliocene age in Apache County. Production of post-Carboniferous pegmatitic quartz for high-grade ceramic use and cleanser-grade silica flour for California markets comes from Mohave County.

Canada

Eastern Canada: The overwhelming proportion of Canadian silica and silicon alloy business is limited to two provinces, Ontario and Quebec, mainly because there are nearly twice as many people living here as in the other eight contiguous and more agriculturally oriented provinces. Of equal importance, the cheap hydroelectric power available at Welland, Thorold, and Niagara Falls, Ont., and at Beauharnois, Que., make these points the centers of important ferroalloy production.

This comprises the principal foreign outlet for one of Canada's prime silica raw materials, the Precambrain Lorrain quartzite cropping out in the Georgian Bay area of Ontario. Early quarries were opened on Manitoulin Island and later on nearby Badgely Island. As this is highly metamorphosed paraquartzite which was derived from an argillaceous sandstone, much trouble has been encountered in keeping alumina to acceptable limits which occurs as sericitic partings in some of the beds. Quarrying is also carried on along the northern shore of Georgian Bay in the Killarney area which can be reached only by boat from Little Current. Most of the product goes into ferroalloys but some is used as a metallurgical flux stone for slagging iron and basic oxides into silicates in the production of nickel and copper matte. These shipments go to the Sudbury district in Ontario and the Noranda district in Quebec and some are captive quarries such as the Lawson operation at Whitefish Falls.

Production was initiated at St. Donat in a friable orthoquartzite of probable Grenville age which is now beneficiated into an acceptable grade of glass sand at a plant 50 miles southeast at St. Canut. Portions of the product are sold for silicon carbide manufacture and foundry sand, as well as pulverized silica for use as filler in pipe and concrete blocks (Minnes, 1967).

The only other potential silica raw materials that occur in eastern Canada are: pegmatitic and vein quartz which are mostly handcobbed as waste in small feldspar operations and used locally for such items as pottery flint, architectural veneers, or flux stone; unconsolidated glacial and river sands which are generally loaded with rock fragments and used principally for construction aggregate; and several miscellaneous sandstones which, for one reason or another, are incompetent for highgrade industrial use without a great deal of expensive upgrading. Hewitt (1951) lists the latter as being of Pleistocene age.

The Potsdam sandstone, also referred to as the "Nepean" sandstone, is either late Cambrian or early Ordovician in age, and is exposed over wide areas as discontinuous patches of relatively flat-bedded deposits with up to 40° dips, having a maximum known thickness of 280 ft (Keith, 1949). The formation has been quarried in several areas in eastern Ontario and southern Quebec with questionable success. This stems from its pyritiferous nature which also condemns the formation in all personally investigated exposures in upper New York State. The pyrite occurs not only as loose, intergranular mineralization but also as tight coatings on the quartz grains themselves. Beneficiation to date has proven partially successful but costly and the Potsdam should doubtless be relegated to the role of a second-grade ceramic raw material acceptable on a delivered price basis only if it meets competition from other imported US sands.

The Medina sandstone of lower Silurian age crops out in several counties west of the Niagara escarpment in Ontario and is principally quarried for building stone. Hewitt (1951) concludes that its potential as a marketable silica product is low although it has been used in a ganister mix for cupola and ladle lining in the past and is an attractive stone for architectural veneer.

Oriskany sandstone is exposed as an extensive outlier in Haldimand County, Ontario, with a reported maximum thickness of about 20 ft. It is described as a coarse-grained, well-rounded, friable, white sandstone poorly bonded by lime cement. Although glass and foundry sands reportedly were produced from this area many years ago, the high iron content (0.15%–0.49%) rules it out as a first-class raw material (Cole, 1923).

The Sylvania sandstone of lower Devonian age crops out modestly near Sylvania in Lucas County, Ohio, this being the type locality. It is mined extensively in the vicinity of Rockwood, Munroe County, Michigan, and furnishes a fair proportion of the glass sand exported to Canada from the United States. Hewitt (1951) correctly points out that this is a potential glass sand in the Windsor-Amherstburg area of Essex County, Ontario, even though subsurface methods to win it would be required because of a strong northeasterly dip. He reports that the top of this sandstone was encountered in a well in Anderdon Township at a depth of 150 ft. However, the writer has examined undoubted Sylvania sand exposed in the bottom of an operating dolomite quarry in the overlying Detroit River formation near Amherstburg with no more than 30 to 60 ft of overburden. This is not an insurmountable factor using modern methods of stripping. The Sylvania, however, requires careful processing as it is high in base oxides which must be carefully controlled by washing and leaching to insure the necessary uniformity, day in and day out, that any commercial ceramic product requires.

Collings (1965) pinpoints the occurrences of several deposits of unconsolidated sands in the Maritime Provinces of eastern Canada; specifically, the beach sands near Souris on Prince Edward Island and those at Barrington Bay and Port Mouton on the southeastern coast of Nova Scotia. These are Holocene sands, feldspathic and impure, and would have to be thoroughly beneficiated to become saleable. The economics are further complicated by their distance to markets.

Goudge (in Cole, 1928) describes additional silica potential in Nova Scotia as consisting of two deposits in Cape Breton County. One is a deposit of clayey sand derived from a weathered orthoquartzite of Precambrian George River age. It has been both quarried and mined and at the time examined was still in the initial stages of development. A fair market was assumed for use as a foundry sand for iron and brass casting in Sydney and New Glasgow. The other deposit had been opened in a fine-grained, brittle paraquartzite facies of the same formation fractured or faulted with serpentinite and chlorite fillings. Production at the time of examination was intermittant, the product from this captive operation going into silica brick lines for the open hearth furnaces in nearby Sydney, N.S.

Unconsolidated bar and bank sands are exposed along the Mattagami and Missinaibi rivers in northeastern Ontario which are relatively pure or could be economically upgraded. However, their great distance from present population and consumption centers prevents their exploitation at the present time. In Newfoundland, paraquartzite produced at Villa Marie on the Avalon Peninsula is shipped to Long Harbor where it is used as a flux in the reduction process for the production of elemental phosphorous in the electric furnace.

Western Canada: The best known and most exploited material in Manitoba—as well as one of the most poorly situated—reportedly occurs in the Winnipeg formation of probable Cambro-Ordovician age (Cole, 1928; Collings, 1965; Spiece, 1980). It is exposed on three islands and two promontories in Lake Winnipeg. As it is mined principally on the southeast shore of Black Island, it takes its name in the trade from this source. The sandstone is described as being very loosely consolidated with argillaceous-ferruginous cementing material that washes off readily leaving a fine-grained, subrounded to rounded grain, white, and relatively free of heavy minerals. Economically, the principal drawback is its isolated location as it has to be barged 80 miles to Fort Selkirk where it is washed, dried and sized, and shipped 20 miles farther on to Winnipeg for use as foundry sand. Some of the material is even shipped as far as Redcliff, Alberta, for container-ware use.

Two other deposits of silica raw material with potential are: (1) A deposit occurring near the town of Beausejour on the Canadian National Railway, 36 miles northwest of Winnipeg. Believed to be of Pleistocene age and fluvioglacial origin, it contains many rock fragments and ferromagnesian minerals. It was opened for foundry materials but is nonoperative at present. Although location-wise it is quite well situated, the necessary ore dressing required to produce a ceramic raw material might well price it out of the market. (2) In contrast, unconsolidated silica sand deposits occurring surficially in the Pine River district are reportedly relatively pure. The chief impurity is a light clay coating on the grain which is readily removed by attrition scrubbing according to Collings (1965). However, these deposits are 200 miles, straight-line, from the nearest use points in the Winnipeg area and nearly 500 miles from Redcliff.

Both Cole (1928) and Collings (1965) speak highly of the great purity of an extensive white sandstone that crops out on the south shore of Wapawekka Lake in east-central Saskatchewan. One sample submitted for analysis showed 98.50% SiO_2, 0.21% Al_2O_3, and 0.008% Fe_2O_3, presumably after washing only, with 91% of the grains on the 35 and 48 mesh sieves. The grains are described as being subangular. If representative, this is a remarkable accumulation of silica, but unfortunately it will probably never be exploited for it is over 100 miles from the nearest point on the Canadian National Railroad through lake-studded, true wilderness, and bush country. Two other potentially commercial silica deposits in Saskatchewan are the Flin Flon sand 125 miles east of Wapawekka Lake and near that town, and the Red Deer River sandstone 150 miles due south of Flin Flon. These raw materials apparently qualify for ceramic use after suitable beneficiation but although both are close to rails, they are too distant from present markets to be other than potential resources.

Excessive rail distance from present markets of known silica deposits again limits exploitation in Alberta, even though glass container ware is produced at Redcliff nearly 600 miles southeast of the only known potentially competent deposit in this province. Collings (1965) refers to this as the Peace River deposit of friable sandstone that occurs practically on rails south of the town of that name. The material apparently qualifies as marketable ceramic-grade material, after attrition scrubbing, but the nearest potential market area is at Edmonton about 275 miles to the southeast. This area has been periodically investigated for manufacture of fiberglass, using this deposit as a raw material.

Cole (1928) has examined the bituminous sands of the Fort McMurray district at rail's end about 250 miles north of Edmonton. Laboratory tests on a composite sample taken from six representative outcrops indicated a bitumen content of about 15% and a sand recovery of about 83% analyzing 95.50% SiO_2, 2.25% Al_2O_3, and 0.35% Fe_2O_3 with less than 1% total base oxides and alkalies. The silica grains are clear quartz and sharply angular to round, with sizing predominantly in the 40 to 80 mesh range. It was concluded that with proper treatment, a green bottle glass might be made of the sand tailings. McConville (1975) has given a comprehensive description of these so-called "Athabasca tar sands" and the current development being initiated for energy purposes. No mention is made, however, of the sand tailings' potential as a source for commercial silica sand. The sand is lower Cretaceous in age.

One other silica deposit, worthy of mention, is a large deposit of well-rounded quartzose pebbles exposed near Cypress Hills 22 miles off the Canadian Pacific Railway. Grinding tests made from representative samples indicated that the work done by these pebbles compared favorably with that of expensive imported Danish pebbles.

Although Cole (1928) cites and describes three bona fide silica deposits in British Columbia, the descriptions are not impressive and lead one to the conclusion that they qualify as good prospects only. They are all quartz bodies of the vein or igneous type which generally, but not always, are restricted as to tonnage and pinch out on strike. Collings (1965) calls attention to the shutdown of an operation near Oliver, in which igneous quartz or quartzite was being shipped for ferrosilicon and silicon carbide manufacture. He points out that much of the British Columbia silica market is currently supplied by the Lane Mountain quartzite deposits near Valley, WA. However, he speaks optimistically of the silica potential of deposits at both Golden and Canal Flats in southeastern British Columbia, suggesting that future glass or fiberglass raw materials could be developed; or that the Alberta markets, which already exist at Redcliff and Edmonton, might be supplied from here.

A potential is also believed to exist in some beach and dune sands at the southern tip of Vancouver Island which bear similarity to the Oregon State coastal deposits between Fort Stevens and Coos Bay. Evaluation of the stateside sands indicated that even with high-intensity magnetic separation followed by acid-leaching the remaining iron oxides precluded use for high-grade ceramic products because of low recovery of product and excessive reagent costs (Carter et al., 1964).

Summing up, the Canadian Mineral Resources Branch gives production statistics for the Dominion for the record year of 1979 (Boyd, 1979). It shows a total domestic production of 2 368 497 t of silica, of which 60 823 t of quartzite were exported. Imports of silica sand and quartz amounted to 1 651 890 t, of which nearly all came from the United States. This gave Canada a total silica consumption of 2 987 735 t. Domestic production figures included 400 316 t produced in western Canada from the provinces of Manitoba, Saskatchewan, and British Columbia.

Table 1 lists a few dependable analyses of representative silica formations exploited throughout the United States and Canada together with their geologic age.

Mexico *

The following companies account for about 90% of the Mexican production of about 400 kt of silica sand:

Company	
Materias Primas Monterrey, S.A.	164 990
Materias Primas Minerales San José, S.A.	94 630
Silice Pizzuto, S.A.	64 440
Arena Silica Industrial de México, S.A.	17 280
Silice Potosi, S.A.	15 630
Cia Minera y Mercantil "El Palizar," S.A.	13 600
	370 570

* Courtesy Guillermo P. Salas, Chief Economic Studies, Consejo de Recursos Naturales no Renovables, Mexico City, D.F., Mexico.

TABLE 1—Representative Silica Samples from the United States and Canada
Percent

Oxides	1	2	3	4	5	6	7	8	9	10	11	12	13	14	15
SiO_2	99.07	99.76	99.59	99.70	99.77	98.93	99.42	96.62	99.16	99.52	99.83	99.45	98.61	96.71	99.27
Al_2O_3	0.56	0.22	0.20	0.08	0.15	0.65	0.16	1.72	0.43	0.27	0.05	0.05	0.47	1.71	0.26
Fe_2O_3	0.03	0.03	0.03	0.02	0.01	0.03	0.03	0.38	0.03	0.01	0.04	0.02	0.25	0.17	0.05
TiO_2	Nd	0.02	0.06	0.02	0.02	0.03	0.02	0.12	0.03	0.01	Nd	0.01	Nd	Nd	0.04
CaO	0.03	0.01	Nd	0.01	Tr	0.01	0.07	Nil	0.08	0.01	0.02	0.04	0.13	Tr	0.02
MgO	0.09	Tr	Nd	0.01	Tr	0.04	Tr	0.08	Tr	0.01	0.01	0.02	0.05	0.05	0.02
Na_2O+K_2O	Nd	Nd	Nd	0.01	Nd	0.40	Nd	0.24	Nd	Nd	0.12	Nd	Nd	0.34	0.54
Fusion loss	Nd	0.12	0.12	0.10	0.08	0.24	0.12	Nd	0.18	0.10	0.09	0.07	0.35	Nd	Nd
% total	99.78	100.16	100.00	99.95	100.03	100.33	99.82	99.16	99.91	99.93	100.16	99.66	99.86	98.98	100.20

No.	Formation	Geologic Age	County and State	Type Sample	Reference
1	Bovill clay-sand	Holocene	Latah, ID	Laboratory processed	Carter, 1962
2	Cohansey sand	Tertiary-Miocene	Cumberland, NJ	Laboratory processed	Lefond, private comm.
3	Ione clay-sand	Tertiary-Eocene	Amador, CA	Commercial product	Mills, 1983
4	Silverado sandstone	Tertiary-Paleocene	Orange, CA	Commercial product	Mills, 1983
5	Tuscaloosa sand	Upper Cretaceous	Marion, GA	Laboratory processed	Lefond, private comm.
6	Baseline sandstone	Upper Cretaceous	Clark, NV	Laboratory processed	Murphy, 1954
7	Trinity sandstone	Lower Cretaceous	Coleman, TX	Laboratory processed	Lefond, private comm.
8	Sharon conglomerate	Basal Pennsylvanian	Portage, OH	Laboratory processed	Bowen, 1953
9	Oriskany quartzite	Lower Devonian	Frederick, VA	Commercial product	Lowery, 1954
10	Eureka quartzite	Middle Ordovician	Clark, NV	Laboratory processed	Murphy, 1954
11	McLish sandstone	Lower Ordovician	Pontotoc, OK	Commercial product	Ham, 1945
12	St. Peter sandstone	Lower Ordovician	LaSalle, IL	Commercial product	Lamar, 1927
13	Potsdam sandstone	Upper Cambrian	Leeds, Ont.	Laboratory processed	Keith, 1946
14	Chickies quartzite	Lower Cambrian	Lancaster, PA	Commercial product	Harris, 1965
15	Lorrain quartzite	Precambrian-Mid. Huronian	Manitoulin, Ont.	Laboratory processed	Hewitt, 1951

Nd, not determined. Tr, Trace.

However, as they produce only about 58% of the total requirements, the balance is imported as shown in the following:

Years	Imports, t
1977	303 500
1978	319 200

Materias Primas Monterrey, S.A., the largest silica sand producer in Mexico, operates in the Jaltipan area of Vera Cruz. Silica sand is also produced by Materias Primas Minerales de San José, S.A., at San Jose de Iturbide in Guanajuato, and Silice Pizzuto, S.A., and Silice Potosi, S.A., near the town of San Luis Potosi in the state of San Luis Potosi.

The principal uses of silica sand in Mexico are as follows:

	Domestic Production	Imports
Glassmaking	290 000	247 150
Foundry	80 700	8 620
Asbestos cement	13 600	——
Chemicals (silicates)	12 000	8 620
Ceramics	——	20 120
Others	15 100	2 881
Total, t	411 400	287 391

Some silica sand is exported, primarily to Brazil, Venezuela, Argentina, and Peru.

Exploration

Potential source grounds for commercial silica, other than those used for construction aggregates, have been mapped and reported on in the literature for so many years, that a certain degree of serendipity enters into the finding of a completely unknown and competent deposit to fit the need in a new target area today. All successful silica-producing companies either have a trained geological specialist to oversee the prospecting and exploration or can hire an experienced consultant to do this.

Because of the extensive amount of published information relating to commercially usable silica deposits, one is indeed lucky to find a deposit that has been overlooked, particularly within marketable distance of a potential customer. Even so, the ecological impact of encroaching urbanization may well have lowered such a deposit's potential for exploitation and although it may qualify technically, it cannot now be exploited and will lie fallow indefinitely.

However, in the unlikely event of a new surficial deposit being discovered which is well-located with respect to markets, the usual standard sampling and exploration can be carried out after arrangements to option or purchase the property have been completed. The methodology of doing this is so well-known by people in this business that it would be superfluous to go into any detail regarding it. There are many sources in the literature that outline such drilling procedures in detail, a few of which may be found in the bibliography.

Field and Laboratory Evaluation

Field Evaluation

All samples, either outcrop or drill hole, should be initially judged by the geologist or mining engineer in charge of the project before remission to the laboratory for final evaluation. Sampling procedures may be modified as more knowledge is gained about the raw material being sought. Pitting and trenching may be utilized in areas with a depressed ground-water table and judgments made which should and can be just as competent as using the drill. Dozers and backhoes, together with hand labor where it is economic, have all been used with complete success.

Buried and ancient dune sands as well as developing fore dunes and beach sands are used for glass raw materials throughout the world (Bird, 1968). In the eastern United States much drilling has been done in reworked sand in the so-called Sandhills Belt developed on the Upper Cretaceous Tuscaloosa formation in the Carolinas, Georgia, Alabama, Tennessee, and Kentucky. Good results have been obtained by hand auger drilling, utilizing a standard portable 3-in. auger set, consisting of 2½-in. steel casing, a 2-in. pump, and 5-ft aluminum extension rods. This same equipment can be used with a 7½-hp motor as a so-called "portable" drill which requires at least a three-man crew in rough country.

Another primitive drilling tool used in shallow, unconsolidated sand, preferably where surface conditions are not conducive to mobility such as wet, heavily wooded swampy ground, is a modification of the ancient Banka drill. This drill is standard in Indonesia where prospecting for alluvial tin is done by cheap labor. An adaptation of this tripod drilling method has sometimes been used in the United States where ground conditions will not accommodate a heavy mechanical drill. Three men and a foreman are required to accomplish this task which is not only slow but tedious, and consequently expensive in these days of high labor costs.

However, if a relatively high water table exists and the terrain is solid and self-supporting, which is usually the case, a heavy machine drill usually can be employed. Nearly any standard type of foreign or American drill, properly adapted to a sandy material, can be used with successful results except for a variable, though low (up to 5%), loss of fines. A standard water jet or percussive type of rig using casing will give good results in water-saturated sand formations; so will a machine auger drill if proper care is taken to avoid cavitation. These methods have all been amply described (Wells, 1973).

Sandstone, if not too highly cemented, can be successfully probed with a heavy auger drill and good results obtained. A quarry-type air-track drill can be used successfully for prospecting as well as a down-the-hole percussion-type rig if larger samples are desired. As far as silica exploration is concerned, using both of these types of drills to depths exceeding 80 to 100 ft becomes time-consuming because of hole clean-out, especially if the material is full of pore water. The more the sandstone is saturated, the slower the progress made so that the time element eventually becomes an unbearable cost factor unless one is on a footage contract. With any of these drilling methods—air-track, machine auger, or percussion—it is mandatory to collect the samples at once, dry them thoroughly, and demagnetize the cuttings so all drill bit or surface casing iron is removed before it gets a chance to oxidize and distort the true quantity of iron native to the raw material—possibly the most significant impurity to be guarded against in silica used for glassmaking.

If the formation to be tested is quartzitic, particularly a vitreous quartzite, the only option of sampling it (short of an adit which is just as expensive and delineates a much more restricted tonnage) is by the diamond core drill. This is the ultimate tool in hard rock as one gets a complete cross section of the formation either at right angles to the bedding or at any angle thereto. The only drawback is that, in a friable orthoquartzite, ofttimes in certain areas, the original calcareous cement has been leached out and it has not yet had time to become vitrified through infiltration by silicious-bearing waters. In this instance core recovery becomes very poor, sometimes yielding almost nothing, particularly on angle holes. It is a truism in this business that every drilling method known has its good and bad features and only experience will show which is the best type of rig to use on such and such a deposit.

Careful meticulous sampling is about as important a consideration as any phase of establishing the competency of a deposit of silica or any other natural material. However, such an abundance of information exists in the literature on procedural tactics that it would be redundant to dwell in much more detail in a chapter of this type (Lake, 1973; Payne, 1973; Waterman and Hazen, 1968).

Laboratory Evaluation

Glass producers, whether they specialize in container, pressed and blown ware, or flat-ground and floated products, all demand precise chemical and particle size distribution specifications that must be met day in and day out by the supplier. One good reason for this, in the case of quartz sand, is that it represents nearly three-quarters of the total batch mix which may contain as many as 6 to 13, or even more, other raw material components, most of which can and do contain contaminants. This is detailed in the chapter on Glass Raw Materials by Mills, in Volume 1.

There are few, if any, silica deposits that do not contain detrital particles in the form of so-called heavy and refractory minerals, together with chert or other rock fragments, because of their resistance to decay or abrasion in the natural reduction of sandstones or quartzites to sand-size particles. The primordial igneous rocks, from whence originated the present quartz-bearing formations, contained variable amounts of ferromagnesian and ore minerals. The extent to which these are present in currently exposed siliceous formations depends largely on their resistance to mechanical and chemical weathering. Some are relatively unstable such as andalusite, sillimanite, titanite, and kyanite; others are quite stable such as rutile, zircon, ilmenite, spinels, magnetite, and tourmaline; others, equally undesirable, are of secondary origin and relatively newly formed such as pyrite, marcasite, glauconite, siderite, leucoxene, and certain manganese and iron minerals (Pettijohn et al., 1973).

Even the smallest silica producer today has some sort of laboratory to determine whether or not his products meet specifications, which is usually done before samples are submitted to a prospective customer. The standard determination of the nature and content and the degree of removal of these heavy detrital minerals, as well as those which are refractory, together with resistant rock particles, can be made at bench level, employing attrition scrubbing, flotation, or, if required, magnetic or electrostatic methods (McQuiston and Bechaud, 1968).

Development, Mining, and Processing

If the raw material is proven potentially competent, and adequate reserves have been demonstrated, development of an operational site is usually promptly initiated. This is especially true if the deposit is discovered in a new marketing area. Since the term "development" has been used, historically, in several different senses by mining engineers in the past, it is being employed here in the evolutionary sense of being the intermediate step following exploration and investigative evaluation, but preceding actual production mining (Banfield, 1972).

Problems relating to terrain, proximity of rail haulage, condition of nearby roads and highways—and, in rare instances, even waterways—are always carefully considered. In addition, the type and accessibility of a good freshwater supply, what sort of power and fuel is available, together with sufficient storage space for overburden, waste material, tailings, and settling ponds must be evaluated. The topographic characteristics of the deposit may influence or even strongly control some of these factors, as well as the lithologic type, attitude, and minable thickness of the deposit. Unfortunately for the modern producer, the proximity, density, and customs of the local population must be assessed and goodwill established, all of which is becoming ever more difficult to do.

If all the foregoing factors appear to be satisfactory, or at least surmountable, a pit, a face, or an adit is opened, dependent on the type of raw material and the amount of burden involved. Usually, large bulk samples are removed for trial processing. If this step is successful, actual mining is initiated, which consists of dredging or hydraulic jetting in soft rock deposits, and open cutting or underground mining in rock that is harder and more competent.

Good planning dictates that the mill or processing plant has been built, shaken down, and is ready to produce by the time the quarry or mine is up to production capacity.

Conventional practice in processing embraces coarse and fine crushing often followed by tertiary grinding. Following this, stage washing and even high-density attrition scrubbing is done to remove surficial impurities containing clay or iron. If necessary, the remaining impurities consisting of detrital minerals can be reduced through the use of froth flotation, low (or high) intensity magnetic separation, or high-tension electrostatic separation. Such beneficiation is then followed by drying in steam-static, rotary kiln, or fluid-bed types of equipment. The dried and cooled material is then screened into such commercial sizes as may be required. Some of the dried material can be bypassed to compeb mills (pebble mills) where it is further pulverized into whatever grade of silica flour may be desirable.

Such processing requires sophisticated and expensive equipment backed by technical know-how gained only by years of experience. For these reasons, only a relatively few companies are able to successfully compete in the silica producing business today. A few of the literature sources embodying the processing referred to are as follows: Gaudin and Fuerstenau, 1955; Humphreys et al., 1973; Taggart, 1945; and Tyler, 1945. A series of papers in the May 1973 issue of *Minerals Processing* is also of interest.

Economic Considerations

Place Value

Even though the amount of free silica or quartz is statistically placed at something over 10% of our so-called earth's crust, we must not deduce this to mean that such a stupendous tonnage is available for potential exploitation. In all probability something like 99.9% of this is tied up in the quartz phenocrysts of acid igneous rocks and locked away, possibly forever, from commerce and industry during mankind's tenure on this planet. But the small remaining fraction is still a huge amount and much of it, because of its relative insolubility and inertness, is constantly being reworked, refined, and segregated by natural processes into a multitude of diverse repositories. This is the principal reason silica products are relatively low-priced, ex-mine, and consumers are willing to pay relatively high rates to transport it from its source to its point of use. The delivered price of the product controls the purchasing transaction even though cost of transportation may nearly equal cost of product—or even exceed it in many situations.

The f.o.b. shipping price of flint grade glass sand varies widely, from approximately $6.25 to $15/t. It averages $9 to $10/t, considerably less than the per-ton value of soda ash, feldspar, etc. used in a glass batch.

The very fact of this mineral's relative abundance makes its production highly competitive which the consumer is quick to take advantage of. Given sources A and B, both furnishing a product equal to or better than specification and both capable of meeting shipments on schedule, the source with the cheaper delivered price will become the prime supplier, and rightly so. However, assume that sources A and B have practically the same total cost of production but that source A is 100 miles closer to the customer's raw material receiving bins. This, at least theoretically and without considering equalization of freight rates or favored price adjustments, can give source A a definite delivered price advantage. In effect, it places a higher intrinsic value on A's silica reserves regardless of their otherwise computed book worth. Ladoo (1964) coined a phrase to describe this situation and referred to it as "place value." Since then, many writers have recognized the importance of this term and referred to it, particularly with respect to industrial minerals.

The term is particularly relevant when used for high-bulk, low unit-value commodities of which silica is a good example. Relatively small amounts of specialized silica products are bagged, whereas in the case of glass sand, a large tonnage item for example, most of it is shipped in covered hopper railroad cars (sometimes as unit trains), pneumatic tank trucks, or dual purpose vans in bulk form. With respect to unit value, the average ex-mine price of top grade flint sand is nearly a dollar a ton cheaper than that of any other high tonnage glassmaking raw material commonly utilized (Mills, 1983).

Unfortunately for newly aspiring entrepreneurs, the older, well-entrenched silica producers have long since explored for, located, and purchased or leased the best situated silica properties with respect to quality and market position, leaving little for the newcomer. As a consequence, barring a fortuitous find, new venture capital must, perforce, buy out an already established producer in order to obtain a competent supply of raw material which is reasonably well-situated with regard to markets, or stay out of the business.

Marketing

Prices, uniformity of product, customer services, research facilities, and supplier reliability are all ingredients that make up marketing in the silica business today. Prices are the controlling factor, basically, but other aspects play as important a role at times. For example, even if the deposit is practically next door to a large consumer and enjoys such a high *place value* advantage that the cost to the consumer is much below that of his other potential suppliers, the ensuing price advantage would be of little value if the customer could not get long-term adequate service. Interrupted shipping schedules, off-grade materials, and similar problems would soon drive this customer into the arms of a more distant but reliable competitor whose delivered price might be somewhat higher. This the enlightened purchasing agent would cheerfully pay, as his unit cost for bulk raw materials is a relatively small item in his final product cost analysis.

It is often stated that market demand combined with solid sales expertise is of possibly greater worth to a business than a sound deposit of raw material, especially in the field of industrial minerals which does not enjoy the almost automatic market that metals do. (This is obviously debatable as there would be no business at all without an adequate raw material source.) However, some considerations of this point are in order: a market generally exists for most elemental metals at the going price; which price, of course, is dependent on a chain of complex factors. This certainly does not follow for any industrial mineral or nonmetal unless it has been refined to its purest state, either by man, or by nature, as in the case of native sulfur. All or most industrial mineral products are refined mineral compounds, many of them oxides or carbonates. The physical properties and quite often even the chemical compositions of a given industrial mineral vary dramatically from deposit to deposit, and silica is no different in this vital respect. It is therefore very difficult and often practically impossible to introduce a new industrial mineral into a consumer field already satisfied with its present supply. Only unusual selling ability backed by comprehensive research effort can get the potential customer to accept a trial shipment on its own merits; and it had better be right.

For these reasons, good management, enlightened foresight, and unwavering integrity in production and sales are watchwords today for success in the silica business. Confidence built up in the customer, accruing from these tenets, is an invaluable asset to a supplier and for lack of a better term it might be spoken of as "sales credibility." With it, a strong bond of business relationship is built up that can last a generation or more. Without it, a vital and very necessary ingredient for sucess is irretrievably lost—and with it, usually the customer.

Government Controls and Restrictions

Restrictions on silica producers from the environmental and ecological viewpoints are proliferating yearly. A few of the smaller producers have had to shut down or at least curtail portions of their business for lack of capital to make the improvements or changes demanded. As with most industrial mineral producers who must include stripping, washing, and drying in their flowsheet, silica producers cannot now, or soon will not be able to, spoil their overburden where they please simply because they own or lease the property involved; cannot dump their mill tailings or acidulated waste water from their flotation circuits into the nearest flowing stream; cannot allow the excessive stack losses or fumes from their dryer operations to pollute the air. If pulverization of the sand into silica flour is done, very low levels of toxic dust must be maintained to prevent silicosis in those exposed in the working areas and this involves expensive dust collection systems. Even noise pollution from heavy equipment, both inside and outside the plant, which also includes the noise and vibration of heavy blasting and that of abnormal truck movements, is beginning to be noticed, particularly in or near populated areas. Literally millions of dollars annually are being, or soon will be, spent in the silica industry to abate these nuisances or risk being shut down if repeated warnings from state or federal authorities are not met (Banfield, 1973).

As industry management is forced more and more to comply with these new governmental edicts just to stay in business, they are, or should be, taking steps to prepare for these new regulations. A good start would be instilling an awareness of what good environmental practice means from the top plant management level to the workmen themselves. Further, it is believed by some that industrial mineral companies affected by this new governmental pressure should not be satisfied by simply lobbying through their trade associations against the new regulations. Instead, they should take an active part in drawing up the legislation to make these new environmental laws more practicable and functional, especially in the field of land reclamation (Appleyard, 1974). At the very least, in this matter, silica companies must start planning ahead on waste disposal and mining practice or they will have it done for them by governmental agencies with no concern for the financial burden that would be suddenly thrust upon the industry. A good paper reflecting on these problems as they relate to a sister industry, sand and gravel, is that by Davison (1974).

State and Growth of the Industries

The silica and silicon industries, so far, appear capable of keeping up with the natural growth and expansion of the ceramic, foundry, metallurgical, and allied industries they supply. Competition is keen and the silica producers are constantly going through a natural process of attrition with the smaller, less efficiently equipped companies falling by the wayside as their deposits become depleted or the specifications for their products become too stringent to meet from year to year.

A new silica operation requires a competent deposit with at least a 50-million-ton reserve to fully amortize the multimillion-dollar investment needed for a plant required to meet present competitive pressures. Also, as with everything else, cost of production in the silica supply business has risen drastically in recent years even though the operations are now highly mechanized compared to their predecessors of yesteryear.

It is difficult to determine the amount of high-purity manufactured silica sand and flour consumed in the United States for all purposes. These statistics are hidden in the US Bureau of Mines annual state reports to protect the production and sales of an individual producer by including them in the figures given for construction sand and gravel aggregates. However, in 1979, reportedly 13,870,000 st of glass sand valued at $111,870,000 was sold or used in the United States and a total of only 423,000 st valued at $1,179,000 was imported—possibly Canadian quartzite for ferroalloy production (Tepordei, 1978–79). US Bureau of Mines reports are believed to reflect about 75% of the sand produced and used by US industry.

Silicon is a relatively low-priced commodity whose cost is influenced by availability of cheap energy and convenient transportation. Production capacity is geared mainly to the requirements of the steel and aluminum industries. Demand is expected to increase at an annual rate of about 2.1%. Silicon's low cost, high availability, and inertness make it a prime material for research to develop substitutes to replace more expensive materials (Shekarchi, 1974).

Richardson et al. (1974) supports this thesis by pointing out that the growth of silicon metal and its ferroalloys should parallel the growths of steel castings as well as carbon and special steels with demand for the silicon metal becoming even greater for aluminum castings and silicones. He, therefore, predicts the need for a total of 525,000 st of silicon units for ferroalloy use and requirements of 160,000 st of silicon as a metal. This he translates into a demand ot the annual production of nearly two million tons of metallurgical-grade quartz or quartzite with a typical specification of SiO_2 99.5%, Al_2O_3 < 0.15%, Fe_2O_3 < 0.1%, CaO < 0.1%, LOI <0.2%. Size requirements must not deviate from a range of 4 x ½ in. because fines in the charge are undesirable. The stone also must not decrepitate upon heating. A *kop* or ridge of igneous quartz, intruded into Precambrian country rock and located in the northern Transvaal region of South Africa near Pietersburg, has recently been recognized as meeting these rigid requirements. The deposit reportedly will supply a substantial demand for this type of raw material for at least 20 years and exploitation has already begun.

Bibliography and References

Anon., 1970, "Few Sources of Glass-Grade Silica Sand, But Reliance on Continental Suppliers Reduced," *Industrial Minerals,* No. 35, Aug., pp. 9–25.

Anon., 1974, "Glass-Grade Silica Sand in Western Europe," *Industrial Minerals,* No. 77, Feb., pp. 9–24.

Appleyard, F.C., 1974, "Surface Mining of Industrial Minerals: A Positive Approach to Land Reclamation," SME Preprint 74H52, AIME Annual Meeting, Dallas, TX, Feb., 14 pp.

Banfield, A.F., 1972, "Ore Reserves, Feasibility Studies, and Valuations of Mineral Properties," SME Preprint 72AK87, AIME Annual Meeting, San Francisco, Feb., 30 pp.

Banfield, A.F., 1973, "Surface and Facility Requirements, Pollution and Environment," *SME Mining Engineering Handbook,* Vol. 1, Sec. 8, A.B. Cummins and I.A. Given, eds., AIME, New York, pp. 8–11–8–19.

Bates, R.L., 1960, "Sedimentary Rocks," *Geology of the Industrial Rocks and Minerals,* Chap. 5, Harper & Brothers, New York, pp. 98–117.

Bird, E.C.F. 1968, "Beaches, Spits and Barriers," and "Coastal Dunes," *Coasts,* Vol. 4, Chaps. 5 and 6, Australian National University Press, Canberra, pp. 81–146.

Bowen, C.H., 1953, "Petrology and Economic Geology of the Sharon Conglomerate in Geauga and Portage Counties, Ohio," Engineering Experiment Station, Ohio State University, Bulletin 153, Vol. 22, No. 3, 58 pp.

Boyd, B.W., 1981, "Silica," Preprint, *Canadian Minerals Yearbook 1979,* Minerals Resources Bureau, Department of Energy, Mines and Resources, Ottawa, 5 pp.

Broadhurst, S.D., 1954, "High-Silica Sand Resources of North Carolina," Information Circula 11, North Carolina Division of Mineral Resources.

Buie, B.F., and Robinson, G.C., 1958, "Silica for Glass Manufacture in South Carolina," Bulletin 23, State Development Board, Columbia, SC, pp. 28–29.

Carter, G.J., et al., 1962, "Industrial Silica Deposits of the Pacific Northwest," Information Circular 811E, US Bureau of Mines, 57 pp.

Carter, G.J., et al., 1964, "Benefication Studies of the Oregon Coastal Dune Sands for Use as Glass Sand," Report of Investigations 6484, US Bureau of Mines, 21 pp.

Chao, E.C.T., et al., 1962, "Stishovite, SiO_2, A Very High Pressure New Mineral from Meteor Crater, Arizona," *Journal of Geophysical Research,* Vol. 67, No. 1, pp. 419–421.

Clarke, G., 1980, "Cheshire Silica Sand—Vital to the U.K. Industry," *Industrial Minerals,* No. 151, Apr., pp. 19–33.

Clarke, G., 1980a, "Mexico's Industrial Minerals—Gathering Momentum," *Industrial Minerals,* No. 153, June, pp. 21–53.

Cole, L.H., 1923, "Silica in Canada, Its Occurrence, Exploitation and Uses," Memoir 555, Pt. 1, Eastern Canada, Mines Br., Dept. of Mines & Technical Surveys, Ottawa, 126 pp.

Cole, L.H., 1928, "Silica in Canada, Its Occurrence, Exploitation and Uses," Memoir 686, Pt. II, Western Canada, Mines Br., Dept. of Mines & Technical Surveys, Ottawa, 59 pp.

Collings, R.K., 1965, "Silica Sand—Canadian Sources of Interest to the Domestic Glass Industry," Technical Bulletin 69, Mines Br., Dept. of Mines and Technical Surveys, Ottawa; reprinted from *Journal of Canadian Ceramic Society,* Vol. 32, 1963, pp. 39–45.

Davison, E.K., 1974, "Land Use and Environmental Problems of the Sand and Gravel Industry," SME Preprint 74H27, AIME Annual Meeting, Dallas, Texas, Feb., 9 pp.

Dietz, E.D., 1968, "The Glassy State," *Science & Technology Journal,* Nov., pp. 10–21.

Dunn, J.R., 1974, "Appraisal of High Bulk, Low Unit Value Mineral Deposits (tentative)," *Suggested Practices and Guides,* American Institute of Professional Geologists, Denver, CO, 25 pp.

Dyer, W.S., 1930, "Sylvania Sandstone Deposit at Amherstburg," *Investigations of Non-Metallic Mineral Resoutces of Ontario 1928,* Vol. 38, Pt. 4, 38th Annual Report, Ontario Dept. of Mines, pp. 41–46.

Edwards, P.K., and Rose, D.C., 1974, "Geology, Mining, Milling and Marketing of Silica Sands for the Foundry Industry," SME Preprint 74H45, AIME Annual Meeting, Dallas, TX, Feb., 8 pp.

Frondel, C., 1962, "Silica Minerals," *Dana's The System of Mineralogy,* 7th ed., Vol. 3, John Wiley, New York, pp. 251–318.

Gaudin, A.M., and Fuerstenau, D.W., 1955, "Quartz Flotation with Anionic Collectors," *Trans. AIME,* Vol. 202, pp. 66–72.

Gay, T.E., Jr., 1957, "Specialty Sands," *Mineral Commodities of California,* Bulletin 176, California Div. of Mines & Geology, Sacramento, pp. 547–564.

Hale, D.R., and Blair, R.E., 1983, "Electronic and Optical Uses," *Industrial Minerals and Rocks,* 5th ed., S.J. Lefond, ed., AIME, New York, pp. 213–231.

Ham, W.E., 1945, "Geology and Glass Sand Resources, Central Arbuckle Mountains, Oklahoma," Bulletin 65, Oklahoma Geological Survey, Norman, OK, 103 pp.

Harris, A.T., Jr., and Miller, W.T., 1965, "Quartzite Mining and Processing Methods and Costs at the Honey Brook, Pa., Plant of Geo. F. Pettinos, Inc.," Information Circular 8248, US Bureau of Mines, 21 pp.

Harris, W.B., 1972, "High-Silica Resources of Clarke, Frederick, Page, Rockingham, Shenandoah, and Warren Counties, Virginia," Report 11, Virginia Div. of Mineral Resources, Charlottesville, VA, 42 pp.

Heinrich, E.W., 1974, "The Sands of Glass," SME Preprint 74H73, AIME Annual Meeting, Dallas, TX, Feb., 13 pp.

Hewitt, D.F., 1951, "Silica in Ontario," Industrial Mineral Circular No. 2, Ontario Dept. of Mines, 16 pp.

Hewitt, D.F., 1963, "Silica in Ontario," Industrial Mineral Report No. 9, Ontario Dept. of Mines, 36 pp.

Hudson, W.C., 1946, "Investigations of the Miami-West Palm Beach Belt of Silica Sand in Florida," Report of Investigation 3865, US Bureau of Mines, 5 pp.

Humphreys, K.K., et al., 1973, "Mineral Processing," *SME Mining Engineering Handbook,* Vol. 2, Sec. 27, A.B. Cummins and I.A. Given, eds., AIME, New York, pp. 27–2 – 27–94.

Jahns, R.H., 1983, "Gem Materials," *Industrial Minerals and Rocks,* 5th ed., S.J. Lefond, ed., AIME, New York, pp. 279–338.

Jaster, M.C., 1957, "Selected Annotated Bibliography of High-Grade Silica of the United States and Canada Through December 1954," Bulletin 1019–H, US Geological Survey, pp. 609–673.

Keith, M.L., 1949, "Sandstone as a Source of Silica Sands in Southeastern Ontario," Annual Report, Vol. 55, Pt. 5, Ontario Dept. of Mines, 36 pp.

Konnert, J.H., et al., 1973, "Crystalline Ordering in Silica and Germania Glasses," *Science,* Vol. 179, No. 4144, Jan., pp. 177–179.

Ladoo, R.B., 1964, "Marketing of Industrial Minerals," *Economics of the Mineral Industries,* 2nd ed., Chap. 7, Pt. 2, E.H. Robie, ed., AIME, New York, pp. 304–317.

Lake, J.L., and Perry, J.K., 1973, "Exploration for Mineral Deposits," *SME Mining Engineering Handbook,* Vol. 1, Sec. 5, A.B. Cummins and I.A. Given, eds., AIME, New York, pp. 5–70 – 5–74.

Lamar, J.E., 1927, "Geology and Economic Resources of the St. Peter Sandstone of Illinois," Bulletin 53, Illinois State Geological Survey, Urbana, 175 pp.

Lowry, W.D., 1954, "Silica Sand Resources of Western Virginia," Vol. 47, No. 12, Virginia Polytechnic Institute, Blacksburg, 63 pp.

McConville, L.B., 1975, "The Athabasca Tar Sands," *Mining Engineering,* Vol. 27, No. 1, Jan., pp. 19–38.

McQuiston, F.W., Jr., and Bechaud, L.J., Jr., 1968, "Metallurgical Sampling and Testing," *Surface Mining,* Chap. 3.2, E.P. Pfleider, ed., AIME, New York, pp. 103–121.

Merrill, C.W., 1973, "Exploration for Mineral Deposits," *SME Mining Engineering Handbook,* Vol. 1, Sec. 5, A.B. Cummins and I.A. Given, eds., AIME, New York, pp. 5–74 – 5–77.

Mills, H.N., 1983, "Glass Raw Materials," *Industrial Minerals and Rocks,* 5th ed., S.J. Lefond, ed., AIME, New York, pp. 339–347.

Minnes, D.G., 1967, "Silica, Past, Present and Future for Canada," Industrial Minerals of Canada, Ltd., Toronto, Oct., 40 pp.

Minnes, D.G., Lefond, S.J., and Blair, R.E., 1983, "Nepheline Syenite," *Industrial Minerals and Rocks,* 5th ed., S.J. Lefond, ed., AIME, New York, pp. 931–960.

Murphy, T.D., 1954, "Silica Resources of Clark County, Nevada," Bulletin 55, Nevada Bureau of Mines, Reno, 28 pp.

Murphy, T.D., 1960, "Distribution of Silica Resources in Eastern United States," Bulletin 1072–L, US Geological Survey, pp. 657–665.

Murphy, T.D., 1960a, "Silica Sand and Pebble," *Industrial Minerals and Rocks,* 3rd ed., J.L. Gillson, ed., AIME, New York, pp. 763–772.

Murphy, T.D., 1975, "Silica and Silicon," *Industrial Minerals and Rocks,* 4th ed., S.J. Lefond, ed., AIME, New York, pp. 1043–1060.

Pajalick, W., 1972, "Sand and Gravel," *Minerals Yearbook 1972,* US Bureau of Mines, pp. 1003–1121.

Parker, R.L., 1967, "Composition of the Earth's Crust," *Data of Geochemistry,* 6th ed., Professional Paper No. 440–D, US Geological Survey, pp. D1–D195.

Pauling, L., 1950, "The Chemistry of Silicon," *College Chemistry,* 2nd ed., Chap. 31, W.H. Freeman, San Francisco, CA, pp. 620–635.

Payne, A.L., 1973, "Exploration Drilling," *SME Mining Engineering Handbook,* Vol. 1, Sec. 5, A.B. Cummins and I.A. Given, eds., AIME, New York, pp. 5–58 – 5–63.

Pearse, G.H.K., 1971, "Silica," *Canadian Minerals Yearbook 1971,* Minerals Resources Br., Dept. of Energy, Mines and Resources, Ottawa, 5 pp.

Pecora, W.T., 1960, "Coesite Crators and Space Geology," *Geotimes,* Vol. 5, No. 2, Feb., pp. 16–20.

Pettijohn, F.J., et al., 1973, "Production and Provenance of Sand," *Sand and Sandstone,* 1st ed., Chap. 8, Springer-Verlag, New York, pp. 294–326.

Richardson, L.S., et al., 1974, "Silicon Metal and Ferrosilicon Alloys," SME Preprint 74H11, AIME Annual Meeting, Dallas, TX, Feb., 15 pp.

Shekarchi, E., 1974, "Silicon," *Commodity Data Summaries,* "Appendix to Mining and Minerals Policy," US Bureau of Mines, pp. 150–151.

Shufflebarger, T.E., Jr., 1974, "The Evaluation of Silica Deposits," SME Preprint 74H76, AIME Annual Meeting, Dallas, TX, Feb., 17 pp.

Shufflebarger, T.E., Jr., 1977, "Economics of Glass Sand Prospects," *Industrial Minerals,* No. 120, Sep., pp. 57–63.

Sosman, R.B., 1965, "The Crystalline Phases of Silica," *The Phases of Silica,* 2nd ed., Chap. 4, Rutgers University Press, New Brunswick, NJ, pp. 34–72.

Spiece, E.L., 1980, "Manitoba Silica Sands," *Industrial Minerals,* No. 154, July, pp. 41–47.

Taggart, A.F., ed., 1945, "Flotation," *Handbook of Mineral Dressing,* 2nd ed., Vol. 1, Sec. 12, John Wiley, New York, pp. 12–01 – 12–130.

Teas, L.P., 1921, "Preliminary Report on the Sand

and Gravel Deposits of Georgia," Dept. of Mines, Mining & Geology, Atlanta, GA, pp. 373–375.

Teicher, H., 1974, "Specialty Silica Products," SME Preprint 74H17, AIME Annual Meeting, Dallas, TX, Feb., 29 pp.

Tepordei, V.V., 1978, 1979, "Sand and Gravel," Preprint, 1978–1979, US Bureau of Mines, 29 pp.

Tyler, P.M., 1945, "Industrial Minerals," *Handbook of Mineral Dressing,* 2nd ed., Vol. 1, Sec. 3, A.F. Taggart, ed., John Wiley, New York, pp. 3–82 – 3–87.

Vos, M.A., 1978, "Silica in Ontario—Supplement," Open File Report 5236, Ontario Geological Survey, 50 pp.

Waterman, G.C., and Hazen, S., 1968, "Development Drilling and Bulk Sampling," *Surface Mining,* Chap. 3.1, E.P. Pfleider, ed., AIME, New York, pp. 69–102.

Waters, A.B., 1980, "Stimulation of Hydrocarbon," *Industrial Minerals,* No. 157, pp. 57–65.

Wells, J.H., 1973, "Special Exploration Techniques—Placer Deposits," *SME Mining Engineering Handbook,* Vol. 1, Sec. 5, A.B. Cummins and I.A. Given, eds., AIME, New York, pp. 5–44 – 5–50.

Wilborg, H.E., and Henderson, G.V., 1983, "Foundry Sand," *Industrial Minerals and Rocks,* 5th ed., S.J. Lefond, ed., AIME, New York, pp. 271–278.

Wimpfen, S.P., and Severinghaus, N., Sr., 1968, "Applications and Economics—Industrial Minerals," *Surface Mining,* Chap. 13.2, E.P. Pfleider, ed., AIME, New York, pp. 849–873.

Also useful:

Minerals Processing, 1973: Walvoord, O.W., "Mill Design of the Future," pp. 4–6; Salat, S.J., "Planning Aspects of Minerals Processing Plants," pp. 7–10; Seiler, F.J., "Design of a Sand-Gravel Aggregates Plant," pp. 11–14; Cruz, L.D., et al., "Computer Techniques in the Design and Operation of Multi-Stage Crushing Plants," pp. 15–19; Anderson, R.P., and Tanner, J.A., "Design Considerations for Minerals Plants," pp. 20–25.

Sodium Carbonate Deposits

L. E. MANNION *

Sodium carbonate (soda ash) is one of two principal commercial alkalis. Its principal competitor is sodium hydroxide. The use of sodium carbonate is recorded in ancient Egypt, where naturally occurring brines and solid salts provided impure soda for early glassmaking. Until the 19th century, soda ash (as well as potash) was recovered mainly by leaching wood or other plant ashes. In 18th-century Europe, for example, leaching seaweed ash with hot water produced a brown laundry lye. One of the better grades of this material was a Spanish product, *barilla,* which analyzed 24 to 30% Na_2CO_3 (Hou, 1942).

The modern chemical industry was born in 1691 with the LeBlanc process for making sodium carbonate from salt, sulfuric acid, and lime. English industry used this process extensively in the 17th and 18th centuries. The next major advance was in 1863, when Ernst Solvay devised the ammonia-soda, or Solvay process. The first of many Solvay plants was built in 1874. These supplanted the LeBlanc method and dominated worldwide soda ash production until 1974, when mined sodium carbonate became predominant.

Although considerably less important until recently, natural sodium carbonate deposits historically have supported small soda ash operations throughout the world. Shortages during World War II stimulated production of "natural" soda ash from such deposits, particularly at Searles Lake, CA and Lake Magadi, Kenya. More spectacularly, increased production from the vast trona deposits in Wyoming has virtually supplanted the Solvay process in the US. Principal producers of natural soda ash as well as the character of the source deposits are listed in Table 1.

* Chief Geologist, Stauffer Chemical Co., Richmond, CA.

This chapter was reviewed by Dr. H.P. Eugster, Dept. of Earth and Planetary Sciences, Johns Hopkins University, Baltimore, MD.

TABLE 1—Principal Producers of Soda Ash from Natural Sodium Carbonate Sources

Company	Location	Production Started and 1982 Capacity t/a	Type of Deposit	Remarks
Allied Chemical Corp.	Green River, WY	1969 2,000,000	Horizontal 0.6-3 m trona bed at 460 m depth	Underground room-and-pillar mine; refinery; same bed as FMC
FMC Corp.	Green River, WY	1953 2,360,000	Horizontal 2.5-3 m trona bed at 460 m depth	Underground room-and-pillar mine; refinery
Kerr-McGee Chemical Corp.	Searles Lake, CA	1916 1,270,000	Subsurface complex brine in permeable, varied salt layers which include trona; depth 0 to 75 m	Brine pumped from wells; sodium carbonate recovered plus other salts
The Magadi Soda Co. Ltd. (ICI)	Lake Magadi, Kenya	1917 250,000	Surface layers of trona in shallow alkali lake	Trona dredged, crushed, dried, and calcined
Sosa Texcoco, S.A.	Lake Texcoco, near Mexico City, Mexico	1947 230,000	Subsurface residual brine from former alkaline lake	Brine pumped to solar ponds for concentration. Half production converted to caustic soda
Stauffer Chemical Co.	Green River, WY	1962 1,450,000	Two horizontal beds about 3 m thick at 260 m depth	Underground room-and-pillar mine; refinery
Tenneco Inc.	Green River, WY	1982 900,000	Horizontal 3 m trona bed at 460 m depth	Underground room-and-pillar mine; same bed as FMC
Texasgulf Inc.	Green River, WY	1976 900,000	Two horizontal beds about 3 m thick at 400 m depth	Underground room-and-pillar mine; refinery

Use and Specifications

Sodium carbonate (Na_2CO_3) is commonly produced as a white, crystalline hygroscopic powder which forms a strongly alkaline aqueous solution. From the Solvay plants two forms are commonly sold: a fine, open crystalline material, called light ash, and a more solidly crystalline dense ash. Their respective bulk densities are 560 and 960 to 1060 kg/m^3. The Wyoming trona plants produce a dense ash, 960 kg/m^3, which dominates the US market. Some Wyoming producers make a somewhat lighter ash, weighing 720 to 800 kg/m^3.

Most commercial soda ash must meet high standards of purity and uniformity. The close quality correspondence between Solvay process and Wyoming natural soda ash is shown in Table 2. The major difference between them is the higher sodium chloride content of Solvay ash. Minor differences important to some consumers are the slightly higher iron and silica content of Wyoming soda ash.

World use of sodium carbonate now exceeds 29 Mt/a, of which the US produces about 7.5 Mt. Fig. 1 shows percent consumption by product of soda ash in Europe, Japan, and the United States. The market for soda ash is dominated by the glass industry, which uses about half the supply. Inorganic chemicals, including phosphates and silicates, consume about 30%. Other principal users are manufacturers of soaps and detergents, petroleum, caustic soda, organic chemicals, and paper, and in ferrous and nonferrous metallurgy.

The price of soda ash stayed relatively stable during the 1960s at about $30/t. By 1975 it reached $38/t f.o.b. Green River, and since then the price has escalated to a current (1983) value of $84/t.

Geology

Mineralogy

Table 3 lists the more common sodium carbonate-bearing minerals. Anhydrous sodium carbonate is almost never found in nature. Only natron, thermonatrite, and trona—minerals without admixture of other saline elements—have been used directly as sources of sodium carbonate.

Mode of Occurrence

Sodium carbonate commonly occurs as a precipitate in nonmarine alkaline lakes and marshes, but such deposits rarely are economic.

TABLE 2—Typical Analyses of United States Soda Ash

Constituent	Typical Analysis, % Wyoming	Solvay
Na_2CO_3	99.8	99.5
NaCl	0.04	0.25
Na_2SO_4	0.04	0.015
Fe_2O_3	0.004	0.002
SiO_2	0.02	0.003
CaO	0.015	0.016
Insol	0.010	0.010

Nonmarine alkaline bodies of water characteristically are shallow, and many undergo vast seasonal changes in area and concentration. Most such deposits, because of their susceptibility to destruction by erosion, are young. Pre-Tertiary deposits are virtually unrecorded.

Some of the more common forms of occurrence include: (1) natron crystals in cool, wet environments on the bottom of ponds or at shallow depths in salt marshes, as at Lake Goodenough, B.C. (Cummings, 1949) and Owens Lake, CA (Smith and Friedman, 1975); (2) powdery surface efflorescences of thermonatrite around alkaline lakes or marshes, as in the alkaline soils of the lower Danube River valley and at Lake Nyasa, Tanzania (Orr and Grantham, 1931); (3) relatively hard but porous deposits of trona, either around the shores or at the bottom of shallow alkaline lakes or playas, as at Lake Magadi, Kenya (Baker, 1958; Eugster, 1970), Owens Lake, CA (Dub, 1947), and Lake Chad basin, Africa (Eugster and Maglione, 1979); (4) older beds of buried solid trona, nahcolite, or other sodium carbonate-bearing minerals, as the Green River formation in Wyoming (Bradley and Eugster, 1969; Culberson, 1971; Deardorff and Mannion, 1971) and in Colorado (Smith and Milton, 1966; Anon., 1972) and at Rail-

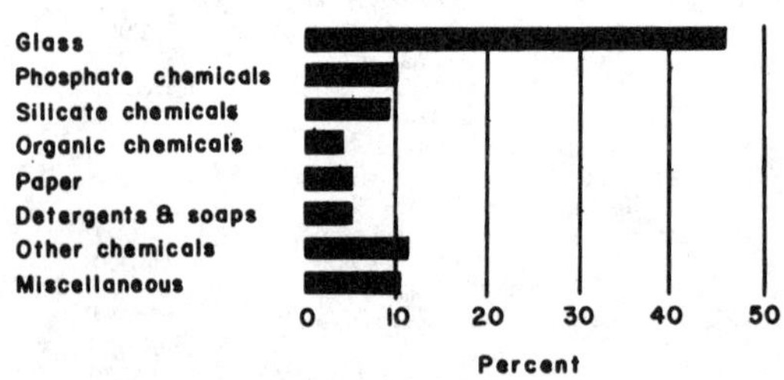

FIG. 1—*Consumption of soda ash in Europe, Japan, and the United States by percentage.*

TABLE 3—Sodium Carbonate Bearing Minerals

Mineral	Composition	% Na_2CO_3 *
Natron (sal soda)	$Na_2CO_3 \cdot 10H_2O$	37.1
Thermonatrite (monohydrate)	$Na_2CO_3 \cdot H_2O$	85.5
Nahcolite	$NaHCO_3$	63.1
Trona (sesquicarbonate)	$Na_2CO_3 \cdot NaHCO_3 \cdot 2H_2O$	70.4
Gaylussite	$Na_2CO_3 \cdot CaCO_3 \cdot 5H_2O$	35.8
Pirssonite	$Na_2CO_3 \cdot CaCO_3 \cdot 2H_2O$	43.8
Shortite	$Na_2CO_3 \cdot 2CaCO_3$	34.6
Burkeite	$Na_2CO_3 \cdot 2Na_2SO_4$	27.2
Hanksite	$2Na_2CO_3 \cdot 9Na_2SO_4 \cdot KCl$	13.5
Northupite	$Na_2CO_3 \cdot NaCl \cdot MgCO_3$	42.6
Dawsonite	$NaAl(CO_3)(OH)_2$	35.8

*Including bicarbonate converted to carbonate.

road Valley, NV (Free, 1913); and (5) in solution as a constituent of brine, either surface or subsurface, as at Searles Lake, CA (Smith, 1979) and Makgadikgadi, Botswana (Massey, 1973; Lefond, 1981). In all these occurrences the sodium carbonate may be mixed with other salts, notably sulfates and chlorides.

Origin

The origin of sodium carbonate in natural deposits has been attributed to various processes, including volcanic spring activity, reactions between sodium carbonate waters and sodium sulfate, and ion exchange in sodium-bearing soils. Most ground waters in volcanic, igneous, or metamorphic terrains yield alkaline solutions on evaporation (Hardie and Eugster, 1970). The relative absence of chloride and sulfate in these rocks permits solutions to become predominantly sodium-bearing and carbon dioxide-bearing. Chemical fractionation of inflowing waters and brines within closed basins can result in different minerals accumulating in separate areas. The fractionation mechanisms may be mineral precipitation, selective dissolution of efflorescent crusts, degassing of brines, and/or redox reactions (Eugster and Jones, 1979).

The sodium may be derived from leaching of sodic feldspars or volcanic ash deposits, and the carbon dioxide from biogenic processes and the atmosphere. Ground water percolating through siliceous soils, e.g., those developed on granite or windblown sands, can become highly alkaline and enriched in sodium carbonate (Garrels and Mackenzie, 1967). Very high CO_2 contents—up to 10% of soil gas—are known to occur in soils as a result of biological activity. This can produce CO_2 in soil water in excess of 200 mg/L (Jakucs, 1977). The Pretoria pan, South Africa, and the sandhill regions of Nebraska and the Sind are examples of sodium carbonate in soil waters (Dutoit, 1948; Barbour, 1916; Cotter, 1923). In lakes, deposition of monomineralic beds of trona may be aided by the continual generation of CO_2 by algae and bacteria in the lake water (Eugster, 1971). Perelman (1967) states that semiarid rather than arid climates are most favorable for producing alkaline waters, as in severely arid regions water evaporation is too rapid to stabilize soda solutions.

Drainage from regions of sedimentary rocks commonly contains too much chloride, sulfate, and calcium to produce either brine or solid salts of sodium carbonate. Even though sodium carbonate waters arise from igneous and volcanic rock weathering, these same waters may interact with other waters or with other rock and soil types in the area of accumulation so that mixed chloride and sulfate salts are the end product (Risacher, 1978).

Occurrences of Sodium Carbonate

Numerous occurrences of sodium carbonate are known in the world. Some are described briefly in Table 4, where they are numbered to correspond to numbered locations in Fig. 2. Data pertaining to Africa, Australia, Europe, and North and South America are reasonably adequate. Information on Asian deposits is less so, and doubtless many Chinese and Siberian occurrences are not recorded here. Western North America has an abundance of sodium carbonate deposits and many are well known. Deposits may contain solid salts, solutions, or both. Many have produced for local and small markets, but only a few have been long-term suppliers of commercial grade soda ash.

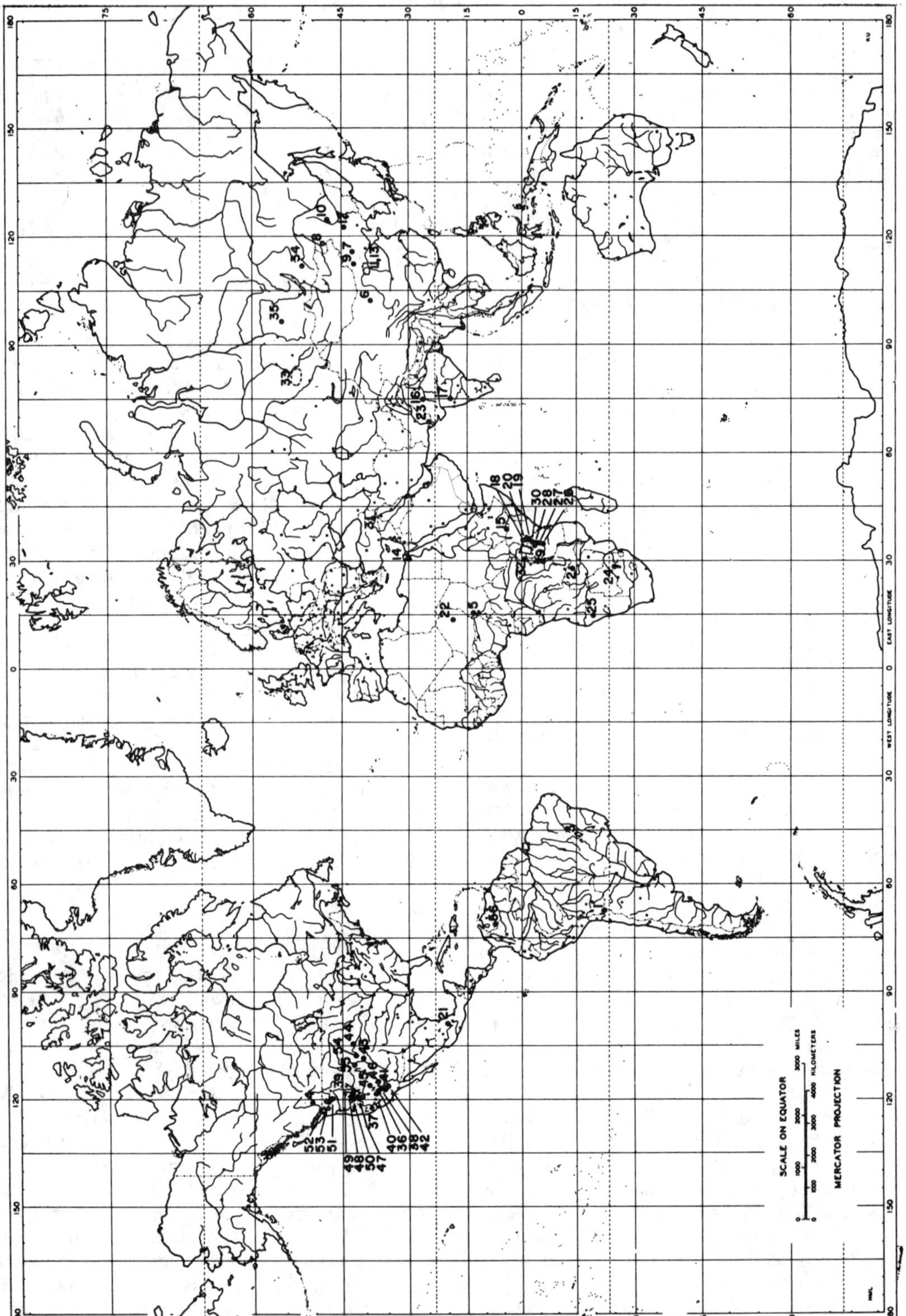

FIG. 2—*Location of major natural sodium carbonate deposits. Numbers are keyed to Table 4.*

TABLE 4—Occurrences of Sodium Carbonate in Various Parts of the World

Name and Location; No. Refers to Location on Fig. 2	Physical Character	Geologic Character	Chemical Composition*	Remarks and References
BOLIVIA				
(1) Collpa Laguna, southwest corner of Bolivia (Lipez) just east of Salar de Chalvira	Shallow lake 800 m in diameter	Salt crust and subsurface brine in mud	Brine Na_2CO_3 6.0% NaCl 2.6 Organic 0.5	Ahlfeld and Schneider-Scherbena, 1964; Risacher, Echenique, and Ballivian, 1978
(1) Cochi Laguna, southwest Bolivia in northern Lipez, 50 km south of Chiguana	Shallow lake 1000 m across, located south of the volcano Tapoquilaha	Brine in mud, also 6-10 cm crust with trona and thermonatrite	Solid salts dried at 80° Na_2CO_3 47.3% $HaHCO_3$ 11.0 NaCl 7.1 Na_2SO_4 11.0 Insol 1.0	Soda ash produced from brine in past, not economic; Ahlfeld and Schneider-Scherbena, 1964; Risacher, Echenique, and Ballivian, 1978
BOTSWANA				
(2) Sua Pan, Makgadikgadi basin in northeast Botswana	Low area in a vast drainage area of flat terrain; Sua Pan is eastern of two	Brine in sand aquifer overlain by sand and clay layers totaling 30 m thick; aquifer underlies 900 km²	Avg. composition of brine Na_2CO_3 26.0 g/L NaCl 140.0 Na_2SO_4 14.5	Investigation and test work on possible commercial use since 1963; Pelletier, 1964; Lefond, 1981
BRAZIL				
(3) Upper Sao Francisco drainage, northeast of Brasilia	Numerous salt pans up to 175,000 m² area in plateau region 900-1100 m in elevation	Salt crusts and shallow brine in area of alkaline igneous rocks	Range of 5 crust analyses Na_2CO_3 23-41% NaCl 1.5-10 Na_2SO_4 1-7 $MgSO_4$ 0.4-2.2 H_2O 40-50	Freise, 1932
CANADA				
(4) Eighty-Three Mile Lake, Green Timber Plateau, south-central B.C., north of Clinton	Area 0.1 km²; max depth brine 0.6 m; in level terrain.	Alkaline lake with a mud bottom on glacial drift and basalt. Crystals and brine present.	Brine Na_2CO_3 94.0 g/L NaCl 1.3 Na_2SO_4 Trace	Many other small lakes and ponds in region also contain high carbonate content; Cummings, 1949; Livingston, 1963
(4) Goodenough Lake, Green Timber Plateau, south-central B.C., north of Clinton	Area 0.05 km² dries up	Crust of salines 2 cm thick on natron containing mud 3 m thick.	Brine Na_2CO_3 208 g/L NaCl 11.0 Na_2SO_4 71.0 KCl 8.7	Solid Crust Na_2CO_3 97.0% $NaHCO_3$ 0.7 NaCl 0.8 Na_2SO_4 00.5 $MgCO_3$ 0.35 Cummings, 1949
(4) Last Chance Lake, Green Timber Plateau south-central B.C., north of Clinton	Area 0.1 km²; mostly dry	Lake contains permanent crystal layers in numerous bowl-shaped depressions separated by mud	Brine Na_2CO_3 180.0 g/L NaCl 12.0 Na_2SO_4 37.0 KCl 3.3	Cummings, 1949
CHAD				
(5) Lake Chad, near Bo, Chad	Lake highly variable in area 10,000-26,000 km²; depth is 3-8 m; vast drainage area	Lake held in by sand dunes northeast shores; layers of natron with halite found 1.5 m below surface	Solids as mined are impure sodium carbonate mixed with sodium chloride	Long continued export of 5-7 kt/a to Nigeria; also Lake Orari 60 km northeast of Bedao, and Lake Tellie near Ounianga Kebir, northern Chad; Eugster and Maglione, 1979; Maglione, 1972

TABLE 4—Continued

Name and Location; No. Refers to Location on Fig. 2	Physical Character	Geologic Character	Chemical Composition*	Remarks and References
CHINA				
Inner Mongolia (6) Alashan and Ehtingol areas	Alkali lakes abundant			Produced "soda ash"; Torgasheff, 1929
(7) Chalar Region, Dolon Nor, 250 km north of Peking	Alkali lakes abundant			Produced "soda ash"; Torgasheff, 1929
(8) Hailar Area (Hailaerh), 150 km east of Manchouli	Two groups of lakes north and south of Hailar; aggregate 500 km²	Salt crust and brine in steppe	Brine Na_2CO_3 115 g/L NaCl 213 Na_2SO_4 71 Salt crust Na_2CO_3 29% Na_2SO_4 12 H_2O 58	Include Chassun-Nor and Bain-Tsugan-Nor; production from south group; Hou, 1942; Torgasheff, 1929
Inner Mongolia, LioO-Ning (9) Polishan (Liao-Ning)	One of large group alkali lakes		Salt crust Na_2CO_3 12.0% NaCl 1.5 Na_2SO_4 1.0	Also Tafus Lake; Torgasheff, 1929
Heilungkiang Province (10) Tsitsihar (Ch'ich'haerh)	Soda lakes in steppe in valley of Nonni River			Torgasheff, 1929; Hou, 1942
Kiangsi Province (11) Ning-Kian-Hsien	Soda lakes in steppe		Natural solid salts Na_2CO_3 81% $NaHCO_3$ 7	Other soda lakes in region; Hou, 1942
Kirin Province (12) Taboos Nor, 320 km north of Shenyang, near Fuyu	Elliptical lake, 8 km across in flat plain surrounded by sand dunes	Generally dry lake with 1 cm crust of thermonatrite on 30 cm mud layer; mud contains natron 0-30%, which lies on gaylussite-bearing muds up to 2 m depth.	Solids Na_2CO_3 35% H_2O 30	Possibly a million tons of sodium salts; numerous other lakes including Fu-U-Hsien; Hou, 1942; Niinomy, 1930
Shansi Province (13) U-Tsu-Hsien	Dry lake		Natural solid salts Na_2CO_3 54.0% $NaHCO_3$ 16.0 NaCl 3.5 Na_2SO_4 17.0 H_2O 12.0	Hou, 1942
Shensi Province (13) Shen-Mu-Hsien	Soda lakes in steppe		Natural solid salts Na_2CO_3 78.3% $NaHCO_3$ 0.5 NaCl 5.0 Na_2SO_4 0.9 H_2O 15.0	Other soda lakes in region; Ordos area in adjacent Inner Mongolia incl. Cha-Han-Nor, Nalin-Nor, and Pa-Yen-Nor; Hou, 1942
EGYPT				
(14) Wadi El-Natron, 97 km northwest of Cairo	Chain of 10 or so small, partly ephemeral lakes; central ones carry natron, others sodium sulfate	Brine of varying composition; bottom and bordering solid salts	Brine Na_2CO_3 52.3 g/L Na_2SO_4 67.1 NaCl 221.2 Product impure carbonate-bicarbonate with mud, NaCl, and Na_2SO_4	Probably oldest known soda sources; probably fed by seepage from Nile River; other occurrences at El Barnougi in Baheira Province, 50 km north of Wadi El-Natron; and in small lakes at El-Mahamid, near Edfo; Lucas, 1912; Said, 1962

TABLE 4–Continued

Name and Location; No. Refers to Location on Fig. 2	Physical Character	Geologic Character	Chemical Composition*	Remarks and References
ETHIOPIA				
(15) Lake Magado and El Soda, near Mega, 500 km south of Addis Ababa	Salt pans in volcanic craters.	Brine plus solid salts.	Brine a mixture of sodium chloride and sodium carbonate.	Exploited on local scale. Murdock, 1949
INDIA				
(16) Sambhar Lake in northwest Rajasthan (Rajputana), 310 km southwest of New Delhi	Variable area, max 230 km² and 4 m deep	Brine with salt layers and saline muds to 85 m depth	Brines predominantly NaCl with lesser Na_2SO_4 and Na_2CO_3.	A significant salt producer, remaining bittern contains Na_2CO_3 3.8%, $NaHCO_3$ 1.0, Na_2SO_4 7.7. Roy, 1959; Sahni, 1951
(17) Lonar Lake, Buldana District of Berar, Lat. 20°N, Long. 76½°E	Circular shallow lake 150 m deep, 1 km², variable amounts of water	Impact crater in basalt plateau; crusts of trona around edge, mixed trona and halite on bottom; spring-fed.	Brine Na_2CO_3 25.0 g/L NaCl 55.0 Na_3SO_4 2.0 K 0.2	Christie, 1912; Fredriksson, et al., 1973
KENYA				
(18) Lake Bogoria, southeastern Kenya	Alkaline lake in Rift Valley		Brine $NaHCO_3$ 48.0 g/L NaCl 5.8 Na_2SO_4 0.3	Livingston, 1963
(19) Lake Magadi, southeastern Kenya	Trona salt flat 65 km², in a low part of Rift Valley	Trona surface with subsurface trona layers and brine, in alkalic volcanic terrane	Brine, variable Na_2CO_3 100 g/L $NaHCO_3$ 20 NaCl 60 Na_2SO_4 1-2 NaF 0.01-0.2	Trona dredged as raw material for soda ash production since 1914; Baker, 1958; Eugster, 1970, 1980
(20) Lake Nakuru, southeastern Kenya, northwest of Nairobi	Shallow brine lake in Rift Valley, modest relief	Brine with shore crusts of trona until lake level rose	Brine $NaHCO_3$ 17.0 g/L NaCl 2.0 Na_2SO_4 0.4	Abundant organisms in water; Livingston, 1963
MEXICO				
(21) Lake Texcoco, west of Mexico City	Low point in Valley of Mexico, subsurface brines	Residual waters from drying up of former Lake Texcoco; now subsurface brines dilute	Water Na_2CO_3 8.0 g/L NaCl 8.6 Na_2SO_4 0.1 $Na_2B_4O_7$ 0.1 KCl 0.8	Solar evaporation in spiral-shaped ponds; much of soda ash production converted to caustic soda; Durand-Chastel, 1968; Gonzales-Reyna, 1956
NIGER				
(22) Bilma oasis in central Sahara, 350 m north of Lake Chad	Shallow saline lakes	Trona precipitated from salt lakes	Dominant sodium chloride	Significant trade in salt
PAKISTAN				
(23) In southern Sind, 100-110 km northeast of Hyderabad	Hundreds of shallow lakes mostly a kilometer or less in max dimension in sand dunes, level terrain	Brine lakes with clay bottom, fed by ground water; trona crusts deposited with scarce natron.	Solids Na_2CO_3 24.0% $NaHCO_3$ 15.0 NaCl 3.0 Na_2SO_4 7.5	Both saline and carbonate lakes present; much organic matter in carbonate lakes; past production at several thousand t/a; Cotter, 1923
SOUTH AFRICA				
(24) Pretoria salt pan, 40 km north of Pretoria	Crater 1 km diam, 130 m deep in granitic terrane, salt pan in bottom	Salts including trona and halite, underlain by laminated clays containing brine; volcanic rocks occur in crater	Brine Na_2CO_3 37.4 g/L NaCl 56.9 SO_4 0.30 SiO_2 0.07 CaO 0.02 MgO Trace	Worked in past; proportion of Na_2CO_3 to NaCl is greater in outer parts of pan; brine likely represents highly concentrated normal ground water; Bond, 1946; Dutoit, 1948

TABLE 4—Continued

Name and Location; No. Refers to Location on Fig. 2	Physical Character	Geologic Character	Chemical Composition*	Remarks and References
SOUTH-WEST AFRICA				
(25) Otjiwalunda Pan just west of Etosha Pan in northern South-West Africa	Alkali pan in nearly flat terrain	Trona and halite precipitated from brines		Possible production of soda investigated in 1964; Pelletier, 1964
TANZANIA				
(26) Bahi Swamp, central Tanzania	Salt pan underlain by 110 m of clays and sandy clays	Alkaline waters contained in sediments. Drill holes had some with high concentrations sodium carbonate	Variable brine composition Na_2CO_3 20.0 g/L $NaHCO_3$ 17.5 Na_2SO_4 9.0 NaCl 28.0 F_2 0.15	Wells can be pumped at rates in excess of 2500 L/min Fawley, 1958
(27) Lake Balangida (Balangda), north-central Tanzania, 80 km south of Lake Eyasi	Salt pan, 30 km², at foot of escarpment in Rift Zone	Spring-fed lake; crust forms only 10 cm thick on mud; volcanic terrane	Specimen of crust Na_2CO_3 22.3% $NaHCO_3$ 13.8 NaCl 28.2 Na_2CO_4 21.1 Insol. 10.	Dries up in summer; only 140 kt sodium carbonate and bicarbonate in crust; Orr and Grantham, 1931
(28) Lake Eyasi (Nyarasa), north-central Tanzania	Saline, sandy pan 1000 km², dry, at foot of escarpment in Rift Zone	Principal salts occur as thin crusts and in brine pools along Sibiti River bed in southwest part of pan	Variable composition, both brine and crust; a supersaturated brine from pool shows: TDS 432.0 g/L Na_2CO_3 3.6% $NaHCO_3$ 1.5 NaCl 51.2 Na_2SO_4 9.1	Sample of alkali salt below a top crust of NaCl Na_2CO_3 32.5% $NaHCO_3$ 25.7 NaCl 15.8 Na_2SO_4 10.5 Moisture 11.1 Insol. 1.3 Organic matter 3.1 Orr and Grantham, 1931
(29) Lake Lagarja, Olduvai Valley, north-central Tanzania	Rift Valley Lake	Mostly dry salt pan in volcanic terrane	Crust Na_2CO_3 44.0% $NaHCO_3$ 17.0 NaCl 13.9 Na_2SO_4 9.8	Guest, 1956
(30) Lake Natron, northeast Tanzania on Kenya border	570 km², large shallow lake in Rift Valley; northwest part contains salts	At times as much as 200 km² are underlain by a meter of salts	Salt crusts contain Na_2CO_3 53-60% NaCl 20-30 Na_2SO_4 1.5-2.0	Estimated 65 Mt of total salts in mud layers; Orr and Grantham, 1931
TURKEY				
(31) Lake Van, southeast Turkey	3730 km², avg. depth 55 m, appears on 1700 m high plateau	Dilute alkaline water; may be density stratified	Water Na_2CO_3 8.6 g/L NaCl 9.4 Na_2SO_4 3.3 K_2SO_4 0.5	Also some lakes in the Araxes plain are carbonate; Lake Urmi is a sodium chloride lake; Karajian, 1920; Tulus, 1944
UGANDA				
(32) Lake Katwe, southwest Uganda	In a volcanic crater, near Lake Edward, salt lake 0.6 km in diam		Salt from evaporation of brine Na_2CO_3 2.46% Na_2SO_4 5.3 NaCl 81.7 K_2SO_4 8.4 Fe_2O_3 0.15	Investigation made toward commercial extraction from brine similar to Searles Lake, CA; Barnes, 1961

TABLE 4—Continued

Name and Location; No. Refers to Location on Fig. 2	Physical Character	Geologic Character	Chemical Composition*	Remarks and References
USSR Altay Region				
(33) M. Khailor Lakes	Two lakes, in semiarid steppe.	Trona and brine	Brine Na_2CO_3 5-11% $NaHCO_3$ 1-2 NaCl 3-7 Na_2SO_4 1-3	Current production from trona at Mikhaylovskiy; Bukhshtein and Zakina, 1949
(33) Petukhov Lakes (Siberia), 210 km from Slavgorod	One large, several small, interconnected lakes.	Brine lake with alkali crusts in steppe region; when temperature drops below 0°C, natron precipitates in 8-15 cm layers	Solid salts Na_2CO_3 36.4% NaCl 0.39 Na_2SO_4 Trace H_2O 62.16	Estimate Na_2CO_3 at 2.4 Mt in lake water, 4.5 Mt precipitated in muds; natrum mined and dehydrated to 99% Na_2CO_3; Niinomy, 1930
Chita Region (34) Doronin Lake, 400 km east of Lake Baikal	Alkaline lakes in semiarid steppe		High in sulfate as well as carbonate	Produced "soda ash"; crystalline salts amount to 350 to 450 kt; Torgasheff, 1929
(35) Tura salt lakes in Krasnoyarsk Region, Lat. 56°N, Long. 90°E	Various lakes in steppe	Brines and solid sulfate salts	Brines containing sulfate and some with high carbonate and bicarbonate content	These are mostly sulfate lakes; some have large quantities of thenardite and mirabilite
UNITED STATES California				
(36) Black Lake, north of Benton, Mono County	Small lake 1.6 km long, up to 150 m wide and up to 20 m deep	Carbonate-sulfate lake colored with organic matter	Dilute lake water Na_2CO_3 12.0 g/L NaCl 2.0 Na_2SO_4 3.4 KCl 0.7	Clarke, 1924
(37) Borax Lake, southeast of Clear Lake, Lake County	0.8 to 1.2 km² in small basin adjacent to Clear Lake	Carbonate water with notable borate content; recent volcanic rocks adjacent	Variable composition Na_2CO_3 18.5 g/L NaCl 1.7 KCl 1.7 $Na_2B_4O_7$ 0.5 Na_2SO_4 0.01 Misc. and Organic 0.5	1.4 Mt of Na_2CO_3; lake supports a notable amount of life; Averill, 1947
(38) Deep Springs Lake, northern Inyo County	Mountain valley, 21 ×7 km, contains saline playa 2.6 km² in area	Shallow brine pond underlain and bordered by saline crust with thenardite, burkeite, trona and nahcolite	Solid crust contains mainly sodium carbonate, sulfate, and chloride	Jones, 1965
(39) Middle Alkali Lake, western Modoc County	One of three shallow lakes in downfaulted valley	Dilute brine in region dominated by Cenozoic volcanic rocks	TDS 9.3 g/L Na_2CO_3 3.5 NaCl 5.4 Na_2SO_4 0.9	Livingston, 1963
(40) Mono Lake, Mono County	Lake area variable 190-220 km², mean depth 18 m	Rather complex saline water in volcanic terrane	Dilute lake water Na_2CO_3 46.0 g/L NaCl 23.0 Na_2SO_4 10.0 $Na_2B_4O_7$ 1.0 KCl 2.1	84 Mt salts; Livingston, 1963
(41) Owens Lake, west-central Inyo County	Up to 260 km², very shallow	Brine on surface and near surface, also trona, burkeite, and halite as surface salts and in subsurface saline muds up to 3 m thick	Brine composition variable Na_2CO_3 8.5% NaCl 17.0 Na_2SO_4 4.5 KCl 0.7 $Na_2B_4O_7$ 0.5	Present drying up of lake resulted from 1917 diversion of Owens River water; intermittent soda ash producer; contains several tens of millions of tons of Na_2CO_3; last plant shut 1968; Dub, 1947

TABLE 4—Continued

Name and Location; No. Refers to Location on Fig. 2	Physical Character	Geologic Character	Chemical Composition*	Remarks and References
(42) Searles Lake, northwest San Bernardino County	83 km² saline playa in mountain valley.	Complex salt body with interstitial brine, underlies 83 km² of surface	Concentrated high-quality brine Na_2CO_3 4-5% NaCl 16 Na_2SO_4 7 KCl 3-5 $Na_2B_4O_7$ 1.5 Spec. Grav. 1.3	Soda ash together with sodium sulfate, borate potash, lithium, and bromine extracted from brines; contains several hundred million tons of sodium carbonate; Smith, 1966, 1979
Colorado				
(43) Piceance Basin, northwest Colorado	Saline part of Green River formation in center part of basin.	Beds of nahcolite and disseminated nahcolite and dawsonite in oil shale	Several thousand feet of thickness contain about 10% dawsonite and nahcolite	Possible future extraction as byproduct from shale oil production; Smith and Milton, 1966; Anon., 1972
Nebraska				
(44) Jess Lake and other alkaline lakes in northwest Nebraska, chiefly in Cherry and Sheridan Counties	Many lakes among sand hills, more than 1 km², rolling terrain	Natural drainage through sandy soils accumulates in hollow with impervious mud bottom	Jess Lake (near Antioch) TDS 190.0 g/L K_2SO_4 26.5 K_2CO_3 21.5 Na_2CO_3 36.2 $NaHCO_3$ 9.2 NaCl 5.8 SiO_2 0.2	Potash produced briefly; Barbour, 1916
Nevada				
(45) Soda Lakes (Ragtown Lakes)	One lake about 1.2 km²; smaller lake about 0.8 km²	Lakes in cinder cone of sandy pyroclastic debris surrounded by sandy lake beds; trona precipitated.	Dissolved solids include: Na_2CO_3 20% NaCl 80 Na_2SO_4 20	Had some production between 1893 and 1969; est. 0.5 Mt Na_2CO_3 present; Chartard, 1890; Morrison, 1964
(46) Railroad Valley, northeast Nye County	Buried salts in lake beds at 200-300 m below playa lake in broad valley	Beds of both solid gaylussite and of gaylussite crystals in clay of Plio-Pleistocene lake; brine and solids, trona gaylussite, and natron	Gaylussite $Na_2CO_3 \cdot CaCO_3 \cdot 5H_2O$	Large reserves of a material with no known commercial utilization; Free, 1913
Oregon				
(47) Lake Abert, Lake County, southeast Oregon	Area 130-150 km²; 1.5-1.8 m deep; dries up	Alkaline semipermanent lake in downfaulted block in volcanic rocks; inflow mostly the Chewaucan River	Brine Na_2CO_3 10-20 g/L $NaHCO_3$ 4-13 NaCl 17-38 Na_2SO_4 1-1.5 KCl 1-1.5 pH 9.0+	Up to 141 TDS in playa interstitial water; Allison, 1947; Phillips and Van Denburgh, 1971
(48) Alkali Lake 31 km north-northeast of Abert Lake	Area 13-15 km², water in very shallow pond	Playa containing "pot holes" containing mostly natron and brine; recharged by ground water; local trona crust	Brine HCO_3+CO_3 171 g/L NaCl 76 Na_2SO_4 65 K_2SO_4 29	Allison, 1947; Phillips and Van Denburgh, 1971
(49) Harney Lake, central Harney County	120 km² lake, max depth 1.4 m; dries up	Surface water dilute subsurface interstitial brine up to 50 g/L dissolved solids; no solid salts	Brine Na_2CO_3 3.6 g/L Na_2SO_4 1.2 NaCl 4.8 KCl 0.4 $Na_2B_4O_7$ 1.9	TDS recorded 1912: 22.0 g/L; Phillips and Van Denburgh, 1971

TABLE 4—Continued

Name and Location; No. Refers to Location on Fig. 2	Physical Character	Geologic Character	Chemical Composition*	Remarks and References
(50) Summer Lake, Lake County, southeast Oregon	Area 260-470 km^2, 0.5-1 m deep; dries up	Alkaline semiperennial lake in downfaulted block; inflow maintained by ground-water recharge and rain	Brine Na_2CO_3 8-14 g/L $NaHCO_3$ 3-7 NaCl 5-9 Na_2SO_4 1-2 KCl 0.5-1	Up to 130 g/L total solids in playa mud water, pH 9.0; Allison, 1947; Phillips and Van Denburgh, 1971
Washington				
(51) Carbonate Lake southeast Grant County	0.6 km^2, very shallow in small valley rimmed by basalt flow	Brine in pools and under ground; crystalline salts occurred in pans or "pot holes"	Concentrated brine Na_2CO_3 128.0 g/L $NaHCO_3$ — NaCl 55 Na_2SO_4 38 KCl 21 Na 91 Sp. Gr. 1.3	Destroyed by construction of dam; former efforts to produce soda ash; Bennett, 1962
(52) Omak Lake, south-central Okanagan County	15 km^2 lake, 30-90 m deep in steep-sided valley in crystalline rocks	Stream fed in abandoned valley, vertically stratified	Very dilute water Na_2CO_3 4.0 g/L NaCl 0.3 Na_2SO_4 1.9 K 0.4	3.6 Mt of Na_2CO_3 present in water; Bennett, 1962
(53) Soap Lake, north-central Grant County	3.5 km^2 lake, avg. depth 7.5 m at Grand Coulee in abandoned valley of Columbia River, floored by basalt	The last of a chain of four lakes fed by sub-surface flow; density stratified	Rather dilute and variable Na_2CO_3 21 g/L NaCl 8 Na_2SO_4 10	570 kt of Na_2CO_3 present in brine; contains water-soluble heavy oil; Bennett, 1962
Wyoming				
(54) Berthaton and Yale Lakes (claims), 50 km southwest of Casper	Alkaline lakes 0.3 to 216 km^2 in area	Lakes in area of Tertiary rocks; brine and crystal-line salts present	Old analyses Brine Na_2CO_3 66% NaCl 14 Na_2SO_4 20 Crystal Deposit Na_2CO_3 50 $NaHCO_3$ 6 NaCl 6 Na_2SO_4 37	Schultz, 1910
(55) Green River trona, Sweetwater County	Buried trona beds of 120-900 m underlie 2800 km^2 in Green River Basin	40 trona beds up to 10 m thick in organic marl-stone and oil shale deposited in Eocene saline lake	Mined trona $Na_2CO_3 \cdot NaHCO_3$ 90.26% NaCl 0.040 Na_2SO_4 0.024 SiO_2 0.003 Organic Trace Water insol 10.0	Five producers, 6.7 Mt/a soda ash capacity; Bradley and Eugster, 1969; Deardorff and Mannion, 1971
VENEZUELA				
(56) Lago de Laguinillus State of Merida	Small pond which may dry up	In an area of alkalic rocks	Trona and gaylussite form from evaporation	Not an important source; Liddle, 1946

*Analytical data in some instances have been recalculated to chemical compounds; most values have been rounded.

Africa

Africa has a number of recorded sodium carbonate occurrences, some with a long history of production. Wadi Natron, about 80 km northwest of Cairo, contains small lakes which probably were the earliest worked deposits in the world. Impure sodium carbonate is obtained from crystallizing pans among sand dunes in the northeast part of Lake Chad (Eugster and Maglione, 1979). This material is exported to Nigeria at the rate of a few thousand tons per year.

Trona has been reported from near Fezzan in the Sahara (Chatard, 1890). A near-surface brine deposit of large size occurs in the Makgadikgadi basin in Botswana. Investigation of its commercial potential was underway for several years, but no significant developments occurred. The Otjevalunda pan (Dutoit, 1948) of South-West Africa contains sodium carbonate of potential value.

The largest deposits are present in east Africa. Among these the most remarkable is at Lake Magadi in southern Kenya, which contains a very large deposit of trona of Recent

TABLE 5—Analyses of Magadi Area Waters in Ppm

	Spring 16 South End of Lake Magadi	Spring 20 at Edge of Lake Magadi	Little Lake Magadi 1.6 km North of Lake Magadi	Lake Magadi, Bird Rock Lagoon	Brine in Borehold B, Lake Magadi, 6.1-m Depth
Na_2CO_3	15,480	5,900	93,000	46,800	200,000
$NaHCO_3$	6,300	20,550	32,000	11,100	1,200
NaCl	10,250	9,550	62,000	19,400	109,000
Na_2SO_4	290	250	ND	ND	ND
NaF	240	330	2,100	620	3,200
Total dissolved solids	30,350	29,130	ND	ND	ND
Temp.	40°C	81°C	–	–	–
Sp. Gr.	1.026	1.026			1.30

Source: Baker, 1958.

age. It is the most soda-rich of a number of alkaline lakes in the Rift Valley of Kenya and Tanzania, where surrounding terrain is largely underlain by faulted alkalic volcanic rocks of Tertiary and Pleistocene age. The climate near Magadi is semiarid, with rainfall of about 360 mm per year, concentrated in two rainy seasons.

Lake Magadi is essentially a trona salt pan, some 65 km^3 in area, whose surface is largely dry. Most of the lake bed is underlain by as much as 45 m of relatively pure trona, plus interbedded trona and trona-bearing muds. Brine is present at or within a meter of the surface. The trona flats are bordered locally by alkaline lagoons fed by numerous warm springs, the combined flow of which approaches 280,000 m^2/d. Analyses of the water of two springs and of a lagoon water produced by a group of springs are given in Table 5. As the aggregate inflow of Na_2CO_3 from the springs amounts to 3600 t/d, it appears that the present rate of soda accumulation greatly exceeds extraction (Baker, 1958).

The alkaline spring water which supplies sodium carbonate to Lake Magadi is considered by Baker (1958) to be the result of weathering and leaching of igneous silicate minerals. Eugster (1970) points out that the dilute ground water from the igneous highland terrain recharges a deep reservoir, which in turn supplies the warm springs with concentrated carbonate waters. The same process appears to be operating elsewhere in the Rift Valley region to produce alkaline waters at Lake Natron, Lake Nakaru, Lake Elmenteita, and Lake Bogorio (Orr and Grantham, 1931).

Lake Magadi has been a producer of sodium carbonate since 1916. Raw trona is dredged from the salt flats, crushed, washed, and dewatered. It has a composition of about 45% Na_2CO_3, 35% $NaHCO_3$, 1.7% NaCl, 0.06% Na_2SO_4, and 0.9% NaF. The trona is calcined to drive off CO_2 and convert it to soda ash, which is then pulverized and screened to produce a dense ash product analyzing about 97% Na_2CO_3, 0.5% $NaHCO_3$, 0.5% NaCl, 0.4% Na_2SO_4, and 0.9% NaF. Production in 1966 was nearly 100 kt. Present plant capacity is 227 kt/a and could be expanded readily. The soda ash is marketed principally in the Far East. Salt for local markets is also produced by brine evaporation.

Australia

Australia appears devoid of significant surface sodium carbonate concentration. Although salt pans are common, they contain chiefly sodium chloride and sodium sulfate. Buried deposits may exist, particularly in the extensive areas of dominantly crystalline rock.

Asia

Sodium carbonate occurrences apparently are abundant in southern Siberia, Mongolia, and northern and western China (Hou, 1942; Torgasheff, 1929). Unfortunately, geologic information and other data are not readily available. Significant alkali lakes and salt pans are known and used in Manchuria and Inner Mongolia, on both sides of the Greater Khingan Range, which is reported to be largely granitic in character. The western part of Inner Mongolia and the Tarim basin (Takla Makan) of Sinkiang are reported to contain surface deposits of soda. Soda-ash plants, producing by means of a trona process, are reported at three localities in Inner Mongolia. These are O-T O-K O, Erhlien, and Pa-Yen Nao.

In Siberia, the regions of Kazakhstan and the Kulundinsk depression of Altay near Slovgorod and Barnaul have produced sodium carbonate from alkali lakes. Such lakes are also reported in eastern Siberia around Chita. Soda-ash plants based on a trona process exist at Makhaylovskiy in the Altay and at Temir-Tau, Karaganda. Presumably these utilize natural materials from soda lakes.

Table 4 presents some data on alkali lakes in China and Siberia, but even their locations are not readily identified on maps. It appears, however, that numerous and locally economically significant deposits occur as a broad, diffuse belt running roughly east-west from eastern Kazakhstan through Mongolia and adjacent parts of the USSR and China to northern Manchuria.

Europe

The present climate of Europe generally is not conducive to the development of significant lake salines. Sodium carbonate efflorescences have long been reported in the Danube Valley of Hungary, and saline lakes which may include alkaline varieties exist in Spain.

North America

North America appears unique in the number, size, and variety of known sodium-carbonate deposits. This may be in part a result of more extensive exploration. The deposits at Searles Lake, CA and Green River, WY account for the bulk of the natural soda ash produced in the world and will likely continue to do so.

Searles Lake: Searles Lake is a large salt flat surrounded by mud. It occupies the lowest part of Searles Valley about 226 km northeast of Los Angeles. The salt surface is the topmost part of a series of permeable crystalline saline lenses which contain interstitial brine and are interbedded with saline-bearing muds. The principal solid salt is halite. Also present are hanksite, trona, and borax, and lesser amounts of burkeite. The most highly saline-bearing part of the deposit extends from the surface to a depth of about 180 m and has an extent of about 100 km^2 (Smith, 1979).

The chief production intervals are termed "upper and lower salt structures." Almost immediately below the surface is the upper structure, a complex lens of salts with a maximum thickness of about 27 m. The lower structure consists of a series of six salt lenses and five mud layers and is separated from the upper by about 3 m of mud. Somewhat deeper are trona beds containing brine with high concentrations of sodium carbonate. Searles Lake had its origin as the principal terminal sump of a chain of lakes in which water overflowed from one to the other until the Searles basin accumulated a brine and salt deposit almost free of detrital material.

Although the crystal bodies contain much trona and other potentially valuable solid salts, the contained brine is the present source of commercial products. Its composition varies both laterally and vertically, and brine density ranges between 1.25 and 1.3. A characteristic composition is given in Table 4. The principal recovered salts are sodium sulfate, sodium carbonate, potassium chloride, and borax. Lithium carbonate, bromine, and phosphoric acid also are produced.

Green River Basin, WY: Production of soda ash in the United States is dominated by the trona from southwest Wyoming. Numerous and extensive layers of trona were formed in a vast Eocene lake which is now represented by deposits of the Green River formation. In the Green River basin, some 3100 to 3600 km^2 are underlain by nearly flat trona beds at depths of 100 to 1070 m (Fig. 3). In just 11 of these beds, which average 2 m or more in thickness, trona reserves in place total more than 47 Gt (Culbertson, 1971).

Within the Green River basin, the Wilkins Peak member of the Eocene Green River formation consists chiefly of nondetrital lacustrine sediments deposited in ancient Lake Gosiute (Bradley, 1964). The saline-bearing facies is composed largely of organic marlstone, oil shale, claystone, and tuff, together with trona and halite. Most of the Wilkins Peak member is characterized by abundant crystals of shortite and is notable for an extraordinary suite of rare minerals (Fahey, 1962; Milton, 1971). A playa lake model for the deposition of Wilkins Peak trona as well as the surrounding and enclosing rocks has been proposed and fits many of the observed features (Eugster and Surdam, 1973; Surdam and Wolfbauer, 1975).

During the existence of Lake Gosiute, conditions for the deposition of trona evidently were repeated many times. More than 40 beds of trona, and trona plus halite, are distributed through about 300 m of evaporite-bearing strata. The position of many trona

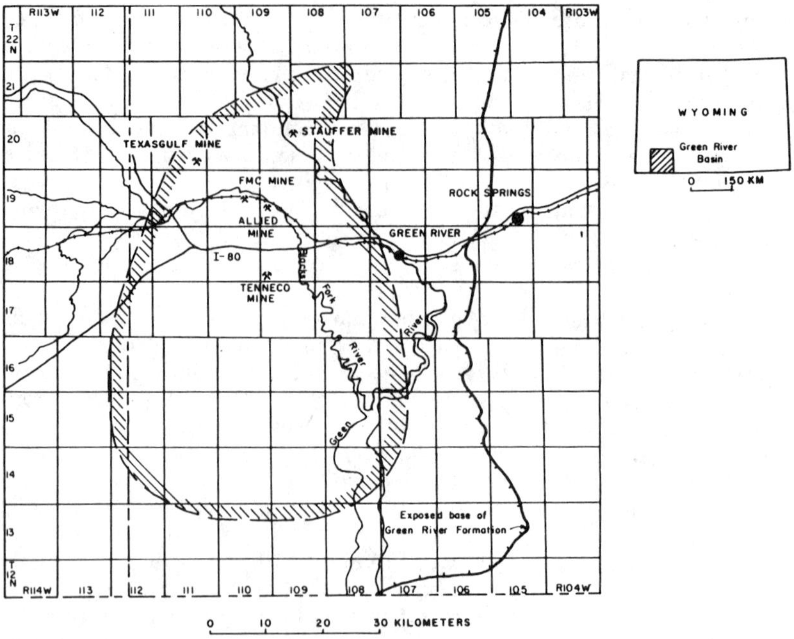

FIG. 3—*Trona Basin, Green River, Wyoming.*

beds on and within oil shales is noteworthy (Culbertson, 1971; Deardorff and Mannion, 1971). Trona beds range up to 12 m in thickness, and the most extensive layer has an area of almost 2600 km^2. The 12 deepest and stratigraphically lowest trona beds occur in the southern half of the area. Some of these beds contain much halite in the central portions, whereas the highest beds are essentially devoid of sodium chloride. These higher strata extend farther north and are less deep, the shallowest known being in the northeastern part of the basin in the vicinity of the Stauffer Chemical Co. mine.

Minable Wyoming trona consists of light to dark brown, almost monomineralic, massive beds. The brown color is due to small amounts of organic matter. The trona has a bladed habit; crystals range in size from needles less than 0.5 mm long to laths 100 mm long that commonly occur in radiating or sheaflike aggregates. A commercial bed normally contains about 90% trona and has a minable height of 2 to 3 m. Localities are known where minable thicknesses consist of as much as 97% trona. Mined trona contains less than a tenth of a percent each of sodium chloride and sodium sulfate. Water-insoluble waste consists of shortite and varied amounts of dispersed oil shale and marlstone as well as thin seams of the same material.

Currently (1981) four shaft mines are operating in the Wyoming trona district and a fifth is being developed. FMC Corp., Tenneco, Inc., and Allied Chemical Co. are mining different parts of the same extensive trona bed at a depth of about 460 m. Texasgulf operates a mine on two beds between 400 and 425 m. Stauffer Chemical Co. is mining the two youngest beds at about 260 to 275 m. All mines employ a multiple-entry, room-and-pillar method of mining. Both continuous mining machines and conventional drilling, blasting, and loading techniques are used. Ore movement underground is by shuttle car and belt haulage. Each of the mines supplies an on-site refinery which produces finished soda ash.

South America

Despite the presence of some of the world's largest salt pans or salars, South America appears lacking in significant resources of natural sodium carbonate. Salt pans at high and low elevations are known, but the great bulk of known surface salts are sulfate or sulfate-chloride varieties. Occurrences have been reported in southern Peru, in western Minas Gerais, Brazil, and in the province of Los Andes in northwestern Argentina. The sodium carbonate deposits which occur in the region where Chile, Bolivia, and Argentina meet are

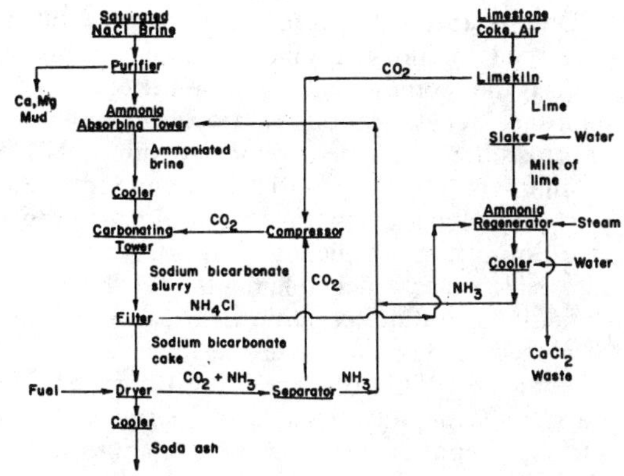

FIG. 4—*Flow diagram, Solvay soda ash process.*

small and economically unimportant (Catalano, 1926; Cordini, 1967).

Processing

World production of soda ash in 1979 was about 29 Mt (Watson, 1980). Of this, all but about 7.7 Mt was produced in Solvay process plans in 33 countries. Solvay plants continue to be built in countries where it is the most economical method of manufacture.

Solvay Process

Virtually all artificially maufactured soda ash is derived from the Solvay or modified

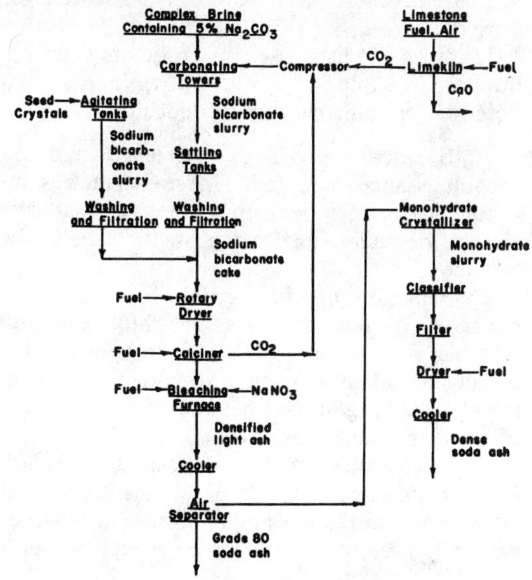

FIG. 5—*Flow diagram, Searles Lake simple carbonation process.*

Solvay process (Rau, 1969) (Fig. 4). In this method saturated sodium chloride brine is first purified of metal ions, notably calcium and magnesium, then saturated with recycled NH_3 gas. The ammoniated brine is carbonated at 300 to 400 kPa pressure, using gas from a lime kiln. Sodium bicarbonate is thus precipitated and filtered. The filtered sodium bicarbonate is calcined to drive off H_2O, NH_3, and CO_2, thus producing a light ash.

The ammonia-bearing brine filtrate is recycled to an ammonia recovery system. In it the ammonia chloride solution is heated with milk of lime to drive off the NH_3 while at the same time producing large quantities of calcium chloride, which is mostly unsalable. A modified Solvay process produces marketable ammonium chloride rather than recycling the ammonia and wasting the calcium chloride.

Searles Lake Brine Process

At Searles Lake, brine feed of differing compositions is obtained from a complex of wells that tap various depths and locations in the deposit. Soda ash is produced through carbonation-precipitation of brines with high sodium carbonate content. At the West End plant the CO_2 gas for carbonation is produced from lime kilns (Fig. 5). At the much larger Argus plant, CO_2 is obtained from power plant gases.

The Argus plant uses brine produced by solution mining trona with 107 m deep wells. Spent warm liquors from the plant are recycled by injecting them into the ore zone to dissolve the trona. The brine is stored, cooled, and concentrated in open-air solar-evaporation ponds. Increased brine concentration helps improve the yield of the plant process. The brine from the ponds is then carbonated to produce a bicarbonate crystal slurry. This slurry is filtered, dried, and calcined to produce a light soda ash. The light ash can be converted to dense ash, if desired, by dissolution, reprecipitation as monohydrate, and drying to remove the free water.

Wyoming Trona Process

Production of soda ash from trona is somewhat simpler than the ammonia soda process and can be done at lower cost. The energy requirements are also lower. Fig. 6 shows a flow diagram of the Wyoming trona process. Trona ore as mined contains the equivalent of about 60% sodium carbonate; approximately

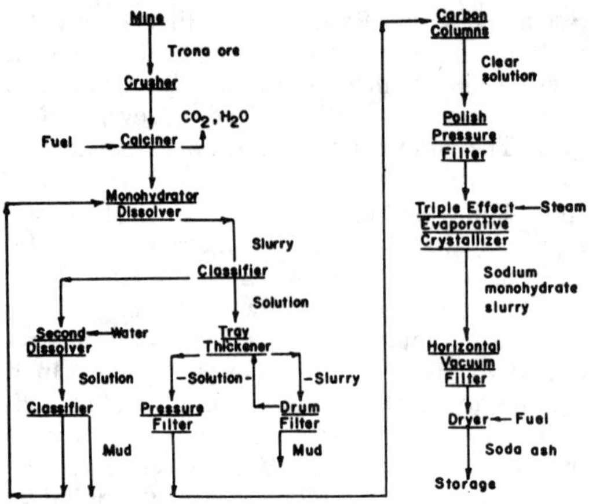

FIG. 6—*Flow diagram, Wyoming trona process.*

1.8 t of trona are necessary to make 1 t of soda ash.

Two variations of the process are used (Frint, 1971). The trona process begins by dissolving trona ore; the monohydrate process first calcines the trona and dissolves the crude soda ash thus produced. In the trona process, the hot solution from dissolving trona is clarified by removal of solids and organic matter. The solution is cooled to precipitate trona, which is then calcined to produce soda ash. In the monohydrate process the conversion of trona to soda ash requires a temperature of about 200° C to achieve the following reaction within a reasonable length of time:

$$2(Na_2CO_3 \cdot NaHCO_3 \cdot 2H_2O) \xrightarrow{\text{Heat}} 3Na_2CO_3 + CO_2\uparrow + 5H_2O\uparrow$$

The solution derived from the crude soda ash is evaporated to produce monohydrate crystals, which in turn are heated to convert them into soda ash.

Exploration

Commercial concentrations of sodium carbonate are confined to nonmarine geologic environments, modern or ancient. The drier areas of the world with interior drainage contain numerous salt pans and salt lakes where alkaline salts may occur. However, most are dominated by sodium chloride and sodium sulfate rather than carbonate. Because saline lakes are conspicuous and accessible, some knowledge of their composition is commonly available.

Despite the widespread occurrence of playa lake type sodium carbonate, notable problems attend the commercial development of such deposits. These problems are the small size of most deposits; their poor location relative to markets for a low-priced commodity; the presence of large amounts of difficult-to-separate impurities (typically chlorides and sulfates but also such elements as fluorine); difficult operating conditions because of the dispersed nature of the deposit and wet ground; variability of composition of brine and salts; and the adverse effects, particularly in surface deposits, of weather changes, e.g., changes in temperature which cause dissolution or precipitation of salts and change in concentration of brine because of rain and increased surface drainage.

Conditions favoring development of economical concentrations of sodium carbonate are: semiarid rather than a severe desert climate; drainage dominated by crystalline or volcanic rocks rather than sedimentary rocks; and a reasonably constant and sufficient inflow of alkaline water to provide large amounts of alkali salts.

In looking for ancient or buried deposits, interpretation of geologic history and paleoenvironments is valuable in determining whether the appropriate conditions existed. The presence of peripheral salines and dolomitic strata in exposed lake beds indicates ancient alkaline lakes. Smith (1966) suggests additional criteria to be used in the search for buried salines.

Many geologists use the following procedure when exploring for, sampling, and testing potential sodium carbonate deposits:

1) Surface deposits over a large area are reconnaissance sampled. Mere surface sampling may not be representative; samples should be taken at least a meter below the surface.

2) The size of the deposit is noted, particularly its compactness or thickness and whether of brine or solids. In sampling brine at substantial depths, care should be taken to avoid obtaining mixed samples from aquifers of varying concentration and composition.

3) Uniformity of the deposit is verified. Both crystalline and liquid constituents of ancient and modern alkaline lakes and marshes can vary greatly from place to place and with depth.

4) If solar evaporation and concentration of brines is to be used, local deposits of clay

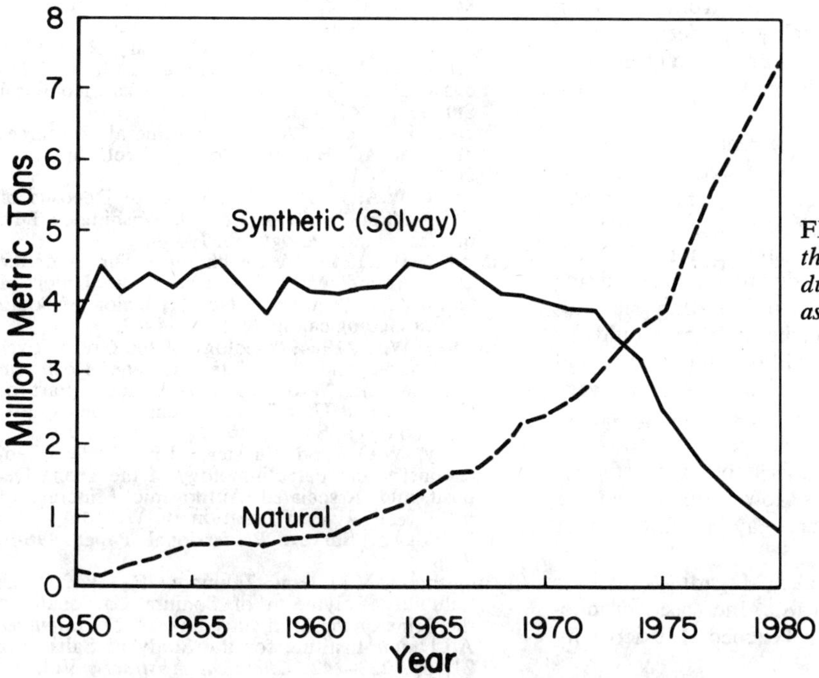

FIG. 7—*Solvay (synthetic) soda ash production vs. natural soda ash production in the United States.*

must be proven adequate for construction of the evaporation lagoons.

5) Climatic variables which might affect the proposed method of operation, particularly the composition of the raw material, are studied. Many ores contain hygroscopic and differentially soluble minerals. In some areas seasonal rains dilute ground and surface brines.

6) If a brine is to be processed, an inexpensive and effective method for the collection and recovery of the brine must be found. Wells, for example, may not always function satisfactorily or produce adequate amounts of brine. Tests are essential to ensure an effective operation.

7) If near-surface solids are to be mined, stability and effectiveness of excavations must be determined. Will contamination be a problem?

8) The capacity of enclosing rocks to support normal mine openings must be determined if underground solids are to be mined. It may be necessary to keep such openings confined to the saline deposit itself when the enclosing rock is weak.

9) Particularly if underground mining will be used, the possibility of invasion of the deposit by extraneous water must be determined.

Future of the Industry

Sodium carbonate is an essential bulk chemical and increases in use generally parallel the growth of the economy. Growth has been about 2% per year for the past 20 years and will probably continue at that rate.

In the United States, Wyoming soda ash has had rapid development, while Solvay process ash has almost disappeared (Fig. 7). The foremost contributing factors are the high quality and size of the Wyoming deposits and the high quality of the product. These deposits can be developed economically and, despite the distance from eastern markets, can compete effectively. Environmental difficulties have beset Solvay plants, which have, among other things, the problem of disposing of large amounts of calcium chloride. The one remaining Solvay plant, located in New York, will probably continue to be competitive.

Some problems exist in the mining of trona. One of these is maintaining integrity of mine openings where the enclosing strata are weak. The floor and roof strata of some mines tend to fail locally when the supporting trona is removed. Handling, recycling, and disposal of waste brine and tailings solids may present additional problems.

Finally, vast amounts of trona exist in Wyoming, but are probably too deep for conventional mining. Some form of solution mining by means of wells may be feasible for recovery of this resource.

The vast quantities of nahcolite in the Piceance Basin of Colorado, occurring mostly as disseminated material in oil shale but also as bedded deposits, represent a great potential source of sodium carbonate (Anon., 1972a). If soda ash is recovered during oil shale mining in the saline-bearing zone of the Piceance Basin, the impact on the market would be great. However, the additional environmental problems attendant upon shale oil recovery from the saline-zone oil shale make saline-free oil shale a more attractive prospect. Mining of bedded nahcolite is another possibility, although ore grade is lower than that of Wyoming trona and there may be some mining problems.

Production from Lake Magadi, Kenya, is planned to double in 1983. Increased production is also planned at Texcoco, Mexico City (Watson, 1980).

Attempts have been made to market Wyoming soda ash in Europe but the special conditions of the European market, particularly governmental concern, have prevented significant inroads.

Bibliography and References

Anon., 1959, *Mineral Resources of the Union of South Africa,* 4th ed., Union of South Africa Geol. Surv., pp. 554–555.

Anon., 1970, *Minerals Yearbook,* Vol. III, Area Reports, International, US Bureau of Mines, p. 959.

Anon., 1972, "An Economic Analysis of a White Nahcolite Installation in Colorado," Option I, Open File Report 31–72, July, National Technical Information Service, PB-212 046, US Bureau of Mines, 69 pp.

Anon., 1972a, "An Economic Analysis of an Oil Shale, Nahcolite, Dawsonite Complex in Colorado," National Technical Information Service Publications, PB-212 046, US Bureau of Mines, 3 vol.

Anon., 1980, "Trona Soda Ash and the Argus Facility," Bulletin 1400, Kerr-McGee Chemical Corp., 23 pp.

Ahlfeld, F., and Schneider-Scherbena, A., 1964, "Los Yacimientos Minerales y de Hidrocarbures de Bolivia," Bolletin No. 5 Minister de Minas v Petroleo, La Paz.

Allison, I.S., 1947, "Sodium Salts of Lake County, Oregon," GMI Short Paper No. 17, Oregon Dept. of Geol. and Mineral Industries, 12 pp.

Averill, C.V., 1947, "Mines and Mineral Resources of Lake County, Cal.," *California Journal of Mines and Geology,* Vol. 43, No. 1, pp. 16–40.

Baker, B.H., 1958, "Geology of the Magadi Area," Report 42, Kenya Geol. Surv., 81 pp.

Barbour, E.H., 1916, "A Preliminary Report on the Alkali Resources of Nebraska," Miscellaneous Papers, Vol. 4, Pt. 28, Nebraska Geological Survey, pp. 405–438.

Barnes, J.W., ed., 1961, "The Mineral Resources of Uganda," Bulletin No. 4, Geol. Surv. of Uganda, 46 pp.

Bennett, W.A.G., 1962, "Saline Lake Deposits of Washington," Bulletin No. 49, Washington Div. of Mines and Geology, pp. 19–49.

Bond, G.W., 1946, "A Geochemical Survey of the Underground Water Supplies of the Union of South Africa," Memoir No. 41, Union of South Africa Geological Survey, pp. 44–47.

Bradley, W.H., 1964, "Geology of the Green River Formation and Associated Eocene Rocks in Southwestern Wyoming and Adjacent Part of Colorado and Utah," Professional Paper 496-A, US Geological Survey, 86 pp.

Bradley, W.G., and Eugster, H.P., 1969, "Geochemistry and Paleolimnology of the Trona Deposit and Associated Authogenic Minerals of the Green River Formation of Wyoming," US Geological Survey, Professional Paper 496-B, 71 pp.

Bukhshtein, V.M., and Zakina, A.E., 1949, "The Solubility Polytherm of Sodium Carbonate in the Brine of the Mikhailov Lakes," *Transactions,* All-Union Institute for the Study of Salts, No. 21, pp. 223–242; *Chemical Abstracts,* Vol. 46, 1952, p. 8911.

Catalano, L.R., 1926, "Geologia Economica de Los Yacimientes de Borate y Materiales de las Cuencas—Solar Cauchari," Publication No. 23, Direccion General de Minas Geologia y Hidrologica, Buenos Aires, 110 pp.

Chatard, T.M., 1890, "Natural Soda; Its Occurrence and Utilization," Bulletin 60, US Geological Survey, pp. 162–163, 237–243.

Christie, W.A.K., 1912, "Note on the Lonar Soda Deposit," *Records,* Geol. Surv. of India, Pt. 4, Vol. 41, pp. 276–285.

Clarke, F.W., 1924, "The Data of Geochemistry," Bulletin 770, US Geological Survey, 841 pp.

Cordini, I.R., 1967, "Reservas Salinas de Argentina," Anales No. 13, Instituto Nacional de Geologia y Mineria, Buenos Aires, 108 pp.

Cotter, G.deP., 1923, "The Alkaline Lakes and the Soda Industry of Sind," Memoir 47, Pt. 2, Geological Survey of India, pp. 202–311.

Culbertson, W.C., 1966, "Trona in the Wilkins Peak Member of the Green River Formation, Southwestern Wyoming," Professional Paper 550-B, US Geological Survey, pp. B159–164.

Culbertson, W.C., 1971, "Stratigraphy of the Trona Deposits in the Green River Formation, Southwest Wyoming," *Contributions to Geology,* Vol. 10, No. 1, R.B. Parker, ed., University of Wyoming, Laramie, pp. 15–28.

Cummings, J.M., 1949, "Saline and Hydromagnesite Deposits of British Columbia," Bulletin No. 4, 1940, British Columbia Dept. of Mines, 160 pp.

Deardorff, D.L., and Mannion, L.E., 1971, "Wyoming Trona Deposits," *Contributions to Geology,* Vol. 10, No. 1, R.B. Parker, ed., University of Wyoming, Laramie, pp. 25–27.

Dub, G.D., 1947, "Owens Lake—Source of Sodium Minerals," Technical Publication No. 2235, AIME, 13 pp.

Durand-Chastel, M.H., 1968, "L'exploitation de Saumures de Carbonate de Soude au Mexique," *Science et Techniques,* No. 12, pp. 23–30.

Dutoit, A.L., 1948, *The Geology of South Africa,* Hafner, New York, p. 428.

Eugster, H.P., 1970, "Chemistry and Origin of the Brines of Lake Magadi, Kenya," Special Paper 3, Mineralogical Society of America, pp. 213–235.

Eugster, H.P., 1971, "Origin and Deposition of Trona," *Contributions to Geology,* Vol. 10, No. 1, R.B. Parker, ed., University of Wyoming, Laramie, pp. 15–23.

Eugster, H.P., 1980, "Geochemistry of Evaporitic Lacustrine Deposits," *Annual Review,* Earth and Planetary Sciences, Vol. 8, pp. 35–63.

Eugster, H.P., 1980a, "Lake Magadi, Kenya, and Its Precursors," *Hypersaline Brines and Evaporitic Environments,* Elsevier, Amsterdam, pp. 195–232.

Eugster, H.P., and Jones, B.F., 1979, "Behavior of Major Solutes During Closed-Basin Brine Evolution," *American Journal of Science,* Vol. 279, pp. 609–631.

Eugster, H.P., and Maglione, G., 1979, "Brines and Evaporites of the Lake Chad Basin, Africa," *Geochimica et Cosmochimica Acta,* Vol. 43, pp. 973–981.

Eugster, H.P., and Smith, G.I., 1965, "Mineral Equilibria in the Searles Lake Evaporites," *Journal of Geology,* Vol. 6, pp. 473–522.

Eugster, H.P., and Surdam, R.C., 1973, "Depositional Environment of the Green River Formation of Wyoming: A Preliminary Report," Geological Society of America Bulletin, Vol. 84, pp. 1115–1120.

Fahey, J.J., 1962, "Saline Minerals of the Green River Formation," Professional Paper 405, US Geological Survey, 50 pp.

Fawley, A.P., 1958, "Diamond Drilling in the Bahi Swamp," *Records,* Tanganyika Geological Survey, Vol. 6 (1956), pp. 47–51.

Fredriksson, K. et al., 1973, "Lonar Lake, India: An Impact Crater in Basalt," *Science,* Vol. 180, No. 4088, pp. 862–864.

Free, E.E., 1913, "Progress in Potash Prospecting in Railroad Valley, Nevada," *Mining and Scientific Press,* Vol. 107, No. 5, August, pp. 176–178.

Freise, F.W., 1932, "Beobachtungen au Brasilianischen Salzsumpfen," *Chemie der Erde,* Band 7, Heft 1, pp. 24–34.

Frint, W.R., 1971, "Processing of Wyoming Trona," *Contributions to Geology,* Vol. 10, No. 1, R.B. Parker, ed., University of Wyoming, Laramie, pp. 25–27.

Garrels, R.M., and Mackenzie, F.T., 1967, "Origin of the Chemical Composition of Some Springs and Lakes," *Equilibrium Concepts in Natural Water Systems,* Advances in Chemistry Series 57, American Chemical Society, Washington, pp. 222–242.

Garrett, D.E., and Phillips, J.F., 1960, "Sodium Carbonate from Natural Sources in the United States," *Industrial Minerals and Rocks,* 3rd ed., J.L. Gillson, ed., AIME, New York, pp. 799–809.

Gonzalez-Reyna, J., 1956, "Riqueza Minera y Yacimiento," *Minerales de Mexico.* 3rd ed., 20th Congreso Geologico Internacional, Banco de Mexico S.A. Dept. de Investigaciones Industriales, Mexico, pp. 431–439.

Guest, N.J., 1956, *Records,* Geological Survey of Tanganyika, Vol. 4, p. 7.

Hardie, L.A., and Eugster, H.P., 1970, "The Evolution of Closed-Basin Brines," Special Paper 3, Mineralogical Society of America, pp. 273–290.

Horino, F.G., and Hooker, V.E. 1978, "Mechanical Properties of Cores Obtained from the Unleached Saline Zone, Piceance Creek Basin, Rio Blanco County, Colo., Report of Investigations 8297, US Bureau of Mines, 21 pp.

Hou, T.P., 1942, *Manufacture of Soda,* Monograph Series No. 65, American Chemical Society, Reinhold, New York, pp. 14–30.

Jakucs, L., 1977, *Morphogenetics of Karst Region* (*Variants of Karst Evolution*) translated from the Hungarian by B. Balhay, Halsted Press, New York, pp. 29–33.

Jones, B.F., 1965, "The Hydrology and Mineralogy of Deep Springs Lake, Inyo County, California," Professional Paper 502-A, US Geological Survey, 55 pp.

Jones, B.F., 1966, "Chemical Evolution of Closed-Basin Water in the Western Great Basin," *Second Symposium on Salt,* Vol. 1, J.L. Rau, ed., Northern Ohio Geological Society, Cleveland, pp. 181–200.

Karajian, H.A., 1920, *Mineral Resources of Armenia and Anatolia,* Armen Technical Book Co., New York, pp. 121–125.

Klingman, C.L., 1972, "Sodium and Sodium Compounds," *Minerals Yearbook 1972,* US Bureau of Mines, pp. 1147–1151.

Kostick, D.S., and Foster, R.J., 1979, "Soda Ash (Sodium Carbonate), Sodium Sulfate, and Sodium," *Mineral Commodity Profiles,* US Bureau of Mines, 14 pp.

Kuchin, M.I., 1928, "Salt Lakes of the Slavgorod Region in Siberia," *Journal of Chemical Industry* (Moscow), Vol. 5, pp. 292–297; *Chemical Abstracts,* Vol. 22, 1928, p. 3963.

Lanbein, W.B., 1961, "Salinity and Hydrology of Closed Lakes," Professional Paper 412, US Geological Survey, 20 pp.

Lefond, S.J., 1981, "Botswana—An Opportunity," Preprint 81–73, SME-AIME Annual Meeting, Chicago, Feb., 3 pp.

Liddle, R.A., 1946, *The Geology of Venezuela and Trinidad,* 2nd ed., Paleontological Research Institution, Cayuga Press, Ithaca, p. 589.

Livingston, D.A., 1963, "Chemical Composition of Lakes and Rivers," Chapter G, *Data of Geochemistry,* 6th ed., M. Fleischer, ed., Professional Paper 440-G, US Geological Survey, pp. 617, 633.

Lucas, A., 1912, "Natural Soda Deposits in Egypt," Paper No. 22, Ministry of Finance, Survey Dept., Government Press, Cairo, 29 pp.

Maglione, G., 1972, "New Data on the Processes at Present Giving Rise to Saline Concentrations on the North-East Periphery of Lake Chad," *Geology of Saline Deposits,* G. Richter-Bernberg, ed., UNESCO, Paris, pp. 211–213.

Mannion, L.E., 1975, "Sodium Carbonate Deposits," *Industrial Minerals and Rocks,* 4th ed., S.J. Lefond, ed., AIME, New York, pp. 1061–1079.

Massey, N.W.D., 1973, "Resources Inventory of Botswana: Industrial Minerals and Rocks," Mineral Report No. 3, Geological Survey Dept., Lobatse, Botswana, 39 pp.

Milton, C., 1971, "Authigenic Minerals in the Green River Formation," *Contributions to Geology,* Vol. 10, No. 1, R.B. Parker, ed., University of Wyoming, Laramie, pp. 57–63.

Morrison, R.B., 1964, "Lake Lahontan: Geology of Southern Carson Desert, Nevada," Professional Paper 401, US Geological Survey, p. 116.

Murdock, T.G., 1949, "Industrial Minerals of Ethiopia," *Trans. AIME,* Vol. 181, pp. 376–384.

Niinomy, K., 1930, "A New Locality of Gaylussite in Eastern Mongolia with Associated Natural Soda," *Economic Geology,* Vol. 25, No. 7, pp. 758–763.

Orr, D., and Grantham, D.R., 1931, "Some Salt Lakes of the North Rift Zone," Short Paper No. 8, Geological Survey Dept., Tanganyika Territory, 21 pp.

Pelletier, R.A., 1964, *Mineral Resources of South-Central Africa,* Oxford University Press, Oxford, p. 131.

Perel'man, A.I., 1967, *Geochemistry of Epigenesis,* trans. from Russian by N. N. Kohanowski, Plenum Press, New York, pp. 176–181.

Phillips, K.N., and Van Denburgh, A.S., 1971, "Hydrology and Geochemistry of Abert, Summer, and Goose Lakes and Other Closed-Basin Lakes in South-Central Oregon," Professional Paper 502-B, US Geological Survey, 86 pp.

Rau, E., 1969, "Sodium Compounds," *Encyclopedia of Chemical Technology,* Vol. 18, 2nd ed., Kirth-Othmer, ed., Interscience, New York, pp. 458–468.

Rettig, S.L., and Jones, B.F., 1965, "Determination of Carbonate, Bicarbonate, and Total CO_2 in Carbonate Brines," Professional Paper 501-D (1964), US Geological Survey, pp. D134–137.

Risacher, F., 1978, "Le Cadre Geochemique des Bessins a Evaporites des Andes Boliviennes," Vol. X, No. 1, Cah O.R.S.T.O.M., Ser. Geol., pp. 37–48.

Risacher, F., Echenique, A, and Ballivian, O, 1978, "Informe Geologico—Economico de la Laguna Collpa y de la Laguna Hedronda," unpublished paper, 19 pp.

Roy, B.C., 1959, "The Economic Geology and Mineral Resources of Rajasthan and Ajmer," *Memoirs,* Geological Survey of India, Vol. 86, pp. 271–273.

Sahni, M.R., 1951, "Alkaline Soils and Lakes of India," *Records,* India Geological Survey, Vol. 81, Pt. 2, pp. 241–296.

Said, R., 1962, *The Geology of Egypt,* Elsevier Publishing Co., Amsterdam, p. 272.

Schultz, A.R., 1910, "Deposits of Sodium Salts in Wyoming," Bulletin 430, US Geological Survey, pp. 570–589.

Singewald, J.T., 1949, "Mineral Resources of Colombia," Bulletin 964-B, US Geological Survey, pp. 186–192.

Smith, G.I., 1964, "Character and Distribution of Non-clastic Minerals in the Searles Lake Evaporite Deposit, California," Bulletin 1181-P. US Geological Survey, 58 pp.

Smith, G.I., 1966, "Geology of Searles Lake—A Guide to Prospecting for Buried Continental Salines," *2nd Symposium on Salt,* Vol. 1, J.L. Rau, ed., Northern Ohio Geological Society, Cleveland, pp. 167–180.

Smith, G.I., 1979, "Subsurface Stratigraphy and Geochemistry of Late Quaternary Evaporites, Searles Lake, California," Professional Paper 1043, US Geological Survey, 130 pp.

Smith, G.I., and Friedman, I., 1975, "Chemical Sedimentation and Diagenesis of Pleistocene Evaporites in Searles Lake, Calif.," 9th International Sedimentation Congress, Nice, pp. 137–140.

Smith, J.W., and Milton, C., 1966, "Dawsonite in the Green River Formation of Colorado," *Economic Geology,* Vol. 61, No. 6, Sep.-Oct., pp. 1029–1042.

Surdam, R.C., and Wolfbauer, C.A., 1975, "Green River Formation, Wyoming, A Playa Lake Complex," Geological Society of America, Bulletin, Vol. 86, pp. 335–345.

Teale, E.O., and Oates, F., 1946, "The Mineral Resources of Tanganyika Territory," Bulletin No. 16, 1946, Geol. Div., Dept. of Lands and Mines, Tanganyika Territory, p. 150.

Torgasheff, B.P., 1929, "Soda in China, Manchuria, and Neighboring Countries," *Chinese Economic Journal,* Bureau of Industrial & Commercial Information, National Government of the Republic of China, Vol. 5, No. 2, Aug., pp. 662–681.

Tulus, R., 1944, "Die Analyse des Wassers vom Van-see (Ostanatolium)," *Review Faculte Science,* University of Istanbul, Vol. 9, pp. 61–63.

Watson, I., 1980, "Soda Ash," *Industrial Minerals,* No. 154, July, pp. 17–31.

Watson, I., 1981, "U.S. Soda Ash—Green River's Solution—Europe's Problem," *Industrial Minerals,* No. 162, Mar., pp. 37–38.

Young, N.B., and Smith, O.W., 1970, "Dawsonite and Nahcolite Analyses of Green River Formation Oil-Shale Sections, Piceance Creek Basin, Colorado," Report of Investigations 7445, US Bureau of Mines, 22 pp.

Sodium Sulfate Deposits

WM. I. WEISMAN *

SID McILVEEN, JR.†

Sodium sulfate is an important industrial chemical, being one of perhaps a dozen or so chemical commodities produced and consumed in the United States in quantities exceeding 1 Mt/a. In recent years approximately half that production has come from natural sources, whereas 27 years ago only 25% was derived from these sources. Table 1 illustrates the history of production of natural sodium sulfate in the United States. Production of natural sodium sulfate from various types of deposits is the main source of this chemical in Canada and Mexico, and probably in Argentina, Chile, Spain, and the USSR.

Mineralogy and Physical Properties

Sodium sulfate is widespread in occurrence and is a common constituent of many mineral waters, as well as seawater. Atmospheric precipitation contains sulfate; it is one of the major dissolved constituents of rain and snow (Davis and DeWiest, 1966). Many of the saline lakes throughout the world contain varying amounts of sodium sulfate. Because sodium is usually the dominant cation, some workers make an anionic distinction, referring to lakes containing predominately sulfate as bitter lakes and those containing predominately carbonate as alkali, or soda lakes (Bateman, 1950). Sodium sulfate in its natural form is found as the hydrous salt *mirabilite* (commonly called Glauber's salt), and as *thenardite,* the anhydrous variety. The largest quantities occur in the form of mirabilite. Sodium sulfate is found in varying degrees of purity, from theoretically pure efflorescent crystals of mirabilite to combinations and admixtures of other salts and impurities. It is a common constituent of some brines; from this source much is extracted commercially.

TABLE 1—Production of Sodium Sulfate in the United States, Manufactured and Natural (1000 t)

Year	Production, 100% Basis[1]	Sold or Used by Producers, Natural Only[2]
1953	903	225
1954	877	227
1955	982	259
1956	998	302
1957	949	300
1958	855	315
1959	986	366
1960	974	409
1961	1030	423
1962	1108	416
1963	1119	395
1964	1194	522
1965	1275	563
1966	1312	581
1967	1238	578
1968	1346	635
1969	1338	610
1970	1246	543
1971	1231	625
1972	1205	636
1973	1305	610
1974	1224	621
1975	1114	606
1976	1118	602
1977	1088	577
1978	1121	549
1979	1071	484
1980	1142	529
1981	1043	540
1982	896	W

[1] Current Industrial Reports, Inorganic Chemicals, US Dept. of Commerce, Bureau of Census.

[2] US Dept. of Interior, *Minerals Yearbook*, and USBM data.

W—Withheld to avoid revealing individual company data.

* President, Ozark-Mahoning Co., Tulsa, Oklahoma.

† Geologist, Ozark-Mahoning Co., Brownfield, Texas.

Sodium sulfate also is found in compounds, such as the minerals *glauberite,* the double salt of anhydrous sodium and calcium sulfate, *bloedite,* the hydrous double-salt of sodium sulfate and magnesium sulfate, and *burkeite,* the anhydrous double-salt of sodium carbonate and sodium sulfate.

Over 40 minerals contain sodium sulfate in varying proportions; many are of special interest because of their frequent occurrence. Table 2 lists some sodium-sulfate bearing minerals. The reader is referred to other publications (Cole, 1926; Dana, 1932; Grabau, 1920; and Dietrich, 1969) for descriptions of these minerals. Only mirabilite and thenardite will be described herein.

Mirabilite, $Na_2SO_4 \cdot 10H_2O$, contains 55.9% water of crystallization. It is noted for its efflorescence or spontaneous loss of water. On dehydration it changes to the anhydrous form, Na_2SO_4. Mirabilite is an opaque to colorless, water-soluble mineral that tastes first cool, then slightly bitter. It has a specific gravity of 1.48. It frequently forms as efflorescent, needlelike monoclinic crystals, but generally is found in the massive form.

Thenardite, the anhydrous mineral, Na_2SO_4, contains 43.68% Na_2O and 56.32% SO_3. It ranges from colorless to white, and may be tinted shades of gray or brown. It is a water-soluble mineral with a slightly salty taste. Its specific gravity (2.67) and hardness (2.5 to 3) exceed those of mirabilite. It commonly occurs in the massive form without visible crystals. Its crystals are frequently tabular pyramids of the orthorhombic system.

Sodium sulfate also occurs as a heptahydrate, containing seven molecules of water, but this is unstable and has not been found in the natural environment.

The solubility of sodium sulfate has an important effect on the crystallization of the salt in nature, as well as in its production. Its solubility in water generally increases as a

TABLE 2—A List of Minerals That Contain Sodium Sulfate

Mineral	Composition	%Na_2SO_4
Thenardite (anhydrous)*	Na_2SO_4	100
Hanksite	$9Na_2SO_4 \cdot 2Na_2CO_3 \cdot KCl$	81.7
Lecontite	$(Na,NH_4,K)_2SO_4 \cdot 2H_2O$	<79.8
Vanthoffite	$3Na_2SO_4 \cdot MgSO_4$	78
Sulphohalite†	$2Na_2SO_4 \cdot NaCl \cdot NaF$	73.9
Burkeite‡	$Na_6(SO_4)_2(CO_3)$	72.8
Darapskite	$NaNO_3 \cdot Na_2SO_4 \cdot H_2O$	58.0
Glauberite	$Na_2SO_4 \cdot CaSO_4$	51.1
Loeweite	$MgSO_4 \cdot Na_2SO_4 \cdot 2\frac{1}{2}H_2O$	46.2
Ferrinatrite	$3Na_2SO_4Fe_2(SO_4)_3 \cdot 6H_2O$	45.6
Mirabilite (Glauber's salt)	$Na_2SO_4 \cdot 10H_2O$	44.1
Bloedite (astrakanite)	$MgSO_4 \cdot Na_2SO_4 \cdot 4H_2O$	42.5
Kroehnkite	$CuSO_4 \cdot Na_2SO_4 \cdot 2H_2O$	42.1
Nickelbloedite§	$Na_2Ni(SO_4)_2 \cdot 4H_2O$	40.3
Sideronatrite‡	$Na_2Fe(SO_4)_2(OH) \cdot 3H_2O$	38.9
Caracolite	$Pb(OH)Cl \cdot Na_2SO_4$	35.4
Palmierite	$(K,Na)_2Pb(SO_4)_2$	<31.9
Tychite	$2MgCO_3 \cdot 2Na_2CO_3 \cdot Na_2SO_4$	27.2
Aphthitalite (glaserite)‡	$(K,Na)_3Na(SO_4)_2$	21-38
Tamarugite¶	$Na_2SO_4 \cdot Al_2(SO_4)_3 \cdot 12H_2O$	20.3
Natrochalcite	$Cu_4(OH)_2(SO_4)_3 \cdot Na_2SO_4 \cdot 2H_2O$	18.8
Almeriite	$Na_2SO_4 \cdot Al_2(SO_4)_3 \cdot 5Al(OH)_3 \cdot H_2O$	15.9
Mendozite (soda alum)¶	$Na_2SO_4 \cdot Al_2(SO_4)_3 \cdot 24H_2O$	15.5
Natrojarosite	$Na_2Fe_6(OH)_{12}(SO_4)_4$	14.7
Noselite	$3Na_2Al_2Si_2O_8 \cdot Na_2SO_4$	14.3
Slavikite	$(Na,K)_2SO_4 \cdot Fe_{10}(OH)_6(SO_4)_{12} \cdot 63H_2O$	<4.6

Note: Formulae are taken from Dana and Ford, 1932, unless otherwise noted.
* Salt cake is anhydrous sodium sulfate with a purity lower than 98%.
† Kogarkoite—Na_3SO_4F (monoclinic), schairerite—$Na_{21}(SO_4)_7F_6Cl$ (trigonal), and galeite—$Na_{15}(SO_4)_5F_4Cl$ (trigonal) are minerals related to sulphohalite (cubic) (Fanfani, et al., 1975).
‡ Dietrich, 1969.
§ Nickel and Bridge, 1977.
¶ Dennis S. Kostick, Sodium Compounds Commodity Specialist, Bureau of Mines, private communication.

nonlinear function of temperature. Below 1.2° C, ice and mirabilite form. As the temperature is increased above 0° C, increasing amounts of sodium sulfate become soluble. At 32.4° C, a transition point on the solubility curve is reached, as the decahydrate melts in its own water of crystallization and the anhydrous form crystallizes. With increasing temperatures, solubility decreases somewhat. The presence of other dissolved salts lowers the transition temperature and changes the solubility characteristics of sodium sulfate.

Classification of Deposits

Essentially all commercial deposits of sodium sulfate have resulted from accumulation and evaporation of surface and ground water in basins with interior drainage, called playas. They are found in arid to semiarid regions. Thus these deposits fall within the broad classification of evaporites. They can be further classified in two categories: those occurring as crystalline beds of mirabilite or glauberite and/or brines within or underlying playa lakes; and those found as buried beds of thenardite, glauberite, and associated minerals. The first category includes the principal deposits of the world. Notable among these are the lake deposits of western Canada, Great Salt Lake, Utah, the vast mirabilite-glauberite deposits of the Kara-Bogaz-Gol Gulf on the eastern shore of the Caspian Sea, and the brine deposits of the western United States and Mexico. These require pumping, dredging, or reservoiring for exploitation.

Commercial deposits of the second category are relatively rare. Deposits of thenardite and glauberite are presently being mined in Spain, and thenardite has been mined in the past at Camp Verde, AZ and at Rhodes Marsh near Mina, NV. Conventional mining techniques are used for extraction.

Origin of Sodium Sulfate Deposits

Evaporation of surface waters, with consequent concentration of salts, is the principal agent in the formation of alkali and bitter lakes from which sodium sulfate and other salts are extracted. Because of the great difference in composition between seawater salts and those of playa lake deposits, it is generally accepted that these deposits are Recent accumulations from nonmarine sources.

The brines of playa lake deposits are more diverse in composition than the oceans because each deposit contains soluble salts from discrete sources within its drainage area. The salinity of the brines is much greater, because of the high rates of evaporation characteristic in arid interior basins where the deposits occur.

The evolution of brine and subsequent precipitation of minerals involve complex geochemical factors. These include the weathering of solutes from the watershed, their transport to the playa basin, and their evaporative concentration. Jones (1966) points out the importance of lithology and weathering processes to the dominant anion in playa lakes of the western United States. Krauskopf (1967) discusses the fate of various elements and radicals during transport and the eventual arrival of sodium as the dominant cation and either sulfate or chloride as the chief anion. Hardie and Eugster (1970) discuss the evolution of dilute waters in closed basins into four groups: Group A (Na-CO_3-SO_4-Cl brines), Group B (Na_2SO_4-$NaCl$ brines), Group C (Na-Mg-Ca-Cl brines) and Group D (Na-Mg-SO_4-Cl brines). The precipitation of calcite, sepiolite, and gypsum are important points of bifurcation during the evolutionary process.

In nature two processes, evaporation and cooling, cause the deposition of sodium sulfate. Evaporation decreases the volume of water, and cooling decreases the solubility of the salt. Below the freezing point of water, ice decreases the available water for solution and the solubility of sodium sulfate minerals, causing mirabilite to form. This natural phenomenon accounts for the accumulation of intermittent beds of crystalline sodium sulfate in the bitter lakes of Canada, Russia, and the United States from which sodium sulfate may be harvested during winter. Crystallization from cooling also results from day and night temperature changes, particularly at high elevations where these changes can be extreme.

In the warmer climates, evaporation is the principal agent of precipitation of sodium sulfate. In those climates natural cooling is not employed in the production of sodium sulfate, but concentrated brines are pumped to the surface and cooled by refrigeration to yield mirabilite crystals.

Most sodium sulfate deposits of commercial significance contain permanent crystalline beds and brines at or near the surface. The crystalline beds can range in thickness from a few millimeters to several meters. They usually are interstratified with other salts, clay, silt, and organic matter. The impervious clays and silts help preserve a portion of the intermittent beds from redissolving. Brines generally occur within crystal beds, but in some deposits crystal beds are lacking.

In most deposits sodium sulfate predominates, with sodium and magnesium chlorides, magnesium sulfate, and carbonates also present. In more complex deposits, concentrations of potassium chloride and borates can be appreciable, depending on the location of the deposit in relation to the source of soluble salts.

The origin of sodium sulfate deposits has prompted the development of many theories. Variations in these theories mainly concern the source of alkalic salts that make up the deposits. Other factors considered are the origin of the basin and the agent of transportation and deposition of the sodium sulfate (Ruffel, 1970).

Early theories generally attributed the source of the salts to decomposition of granitic rocks. It was assumed that the feldspars in granites yielded soluble salts that were transported by surface waters, and that pyrite and other sulfides in these rocks were oxidized to sulfuric acid, which immediately dissolved some of the basic oxides to produce soluble sulfates (Wells, 1923).

Following an extensive study of the deposits of western Canada, Cole (1926) considered four possible sources for the sodium sulfate. These are: (1) underlying sands and shales, which contain small amounts of alkalies, and interbedded volcanic-ash beds, which contain an appreciable percentage of these salts; (2) connate waters in underlying strata; (3) springs in or near the deposits; and (4) surface deposits of bentonitic glacial drift. Cole concluded that the probable source was the salts contained in the bentonites, which are widely distributed in the glacial drift. Surface waters carrying calcium salts in solution react by base exchange with the alkali silicates, releasing the salts contained in the bentonite in the form of soluble sulfates, etc. These in turn are concentrated and deposited in undrained lake basins.

More recent investigators have attributed the source of some sodium sulfate to the decomposition of rocks and sediments, such as bentonite, that also contain sulfates, and from the leaching of buried beds of sulfates. Organic rich clay and shale also would produce sulfate through the oxidation of pyrite and marcasite (Davis, 1966). A subaerial means of transport is provided because many playa lakes are hydrologically coupled with the surrounding water table. Sulfides are found in the organic rich muds (containing stromatolites, etc.) and vegetative debris of some playa-lake sediments. The fermentation of organics under anaerobic conditions would produce hydrogen sulfide which could oxidize to sulfate (Reeves, 1968).

Grossman (1968) studied the sodium sulfate deposits of the Northern Great Plains of Canada and the United States and attributed the origin of those deposits to the solution of deeply buried marine evaporites. In his model, ground water from the Rocky Mountains moved downward and eastward into the Saskatchewan Sub-Basin during the Quaternary, leaching the buried evaporites of the Prairie Formation (Devonian). Ascending mineralized ground water during the late Pleistocene discharged into stratified drift where freezing segregated pure crystals of sodium sulfate in meromictic (chemically stratified and stagnant) lakes.

Occurrence, Reserves, and Production

Sodium sulfate deposits of sufficient size and purity for commercial mining are found in several countries (Fig. 1). The principal producing countries are Canada, Mexico, the United States, the USSR, and Spain. Less important are Argentina, Chile, Iran, and Turkey.

Canada

Numerous saline lakes containing sodium sulfate are known on the western plains of Canada but, at present, only nine lakes and five companies account for essentially all sodium sulfate production (personal communication). One deposit is in Alberta; the remainder are in Saskatchewan. Four companies use dredges to mine the permanent crystal beds (Watson, 1980). One additional lake in Saskatchewan is worked only intermittently.

The deposits occur in basins and depressions in glacial drift and lack apparent outlets. In general, the deposits are crystalline beds that underlie saturated brine lakes containing a few millimeters to approximately a meter of brine in the spring and summer. The deposits contain two recognizable phases of crystalline beds: intermittent and permanent.

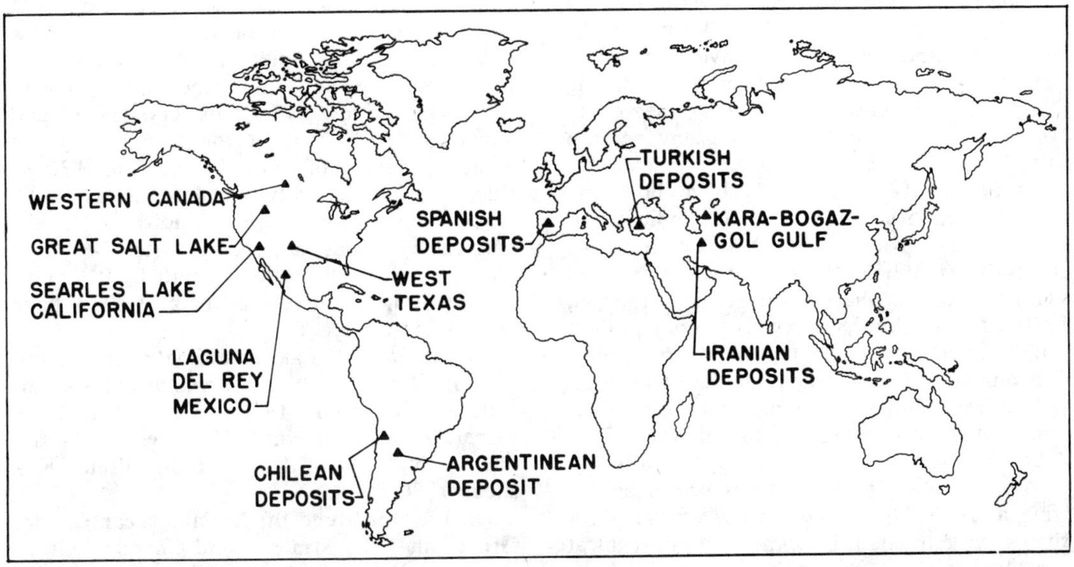

FIG. 1—*Principal producing sodium sulfate deposits of the world.*

The intermittent beds form on the bottom and along the shores of the lakes. They crystallize and redissolve with fluctuations in temperature. This can occur in a matter of hours, but as winter sets in less solution of the crystals occurs, causing the beds to become thicker and more compact.

The permanent beds underlying the lake usually are interstratified with thin layers of mud, clay, and organic matter. These beds range in thickness from less than one to several meters. At Ingebrigt Lake, permanent beds are reported (Cole, 1926) to cover an area of 2.7 km^2 with an average thickness of 6.7 m. In one part of the deposit a thickness of 42.4 m was measured. Cole estimated the tonnage of hydrous salt in Ingebrigt Lake to be 22.7 Mt. A more recent evaluation is 8.2 Mt of anhydrous sodium sulfate (Broughton, 1977).

Most of the deposits presently in production contain several million tons of sodium sulfate. The bulk of the Canadian reserve occurs in Saskatchewan. One deposit in Alberta contains 2.7 Mt (Watson, 1980). Estimates of total reserves in Saskatchewan range between 54.4 and 181 Mt. A more recent evaluation places the total anhydrous sodium sulfate reserve at 54.4 to 63.5 Mt and evaluated commercial reserves at 27.2 to 36.2 Mt (Broughton, 1977).

Production of natural sodium sulfate in 1979 was over 440 kt and is expected to be about 490 kt in 1980. Of this quantity nearly 200 kt was exported.

Current capacity is reported at approximately 742 kt/a. Production has never been close to reported capacity. About 270 kt/a of capacity belongs to Saskatchewan Minerals, a government corporation (Watson, 1980).

United States

Deposits of sodium sulfate are found in Arizona, California, Colorado, Idaho, Montana, Nevada, New Mexico, North Dakota, Oregon, Texas, Utah, Washington, and Wyoming. At the present time, production is chiefly from brine deposits at Searles Lake, CA, at Great Salt Lake, UT, and in West Texas. Small quantities have been produced from saline lake deposits in Wyoming. Most of the southwest brines are contained in dry lakes or playas underlain by beds of mirabilite, thenardite, and other evaporite salts. In the northwestern United States, the deposits associated with saline or intermittent dry lakes are similar in origin and composition to the Canadian deposits. Except for the brine deposits in West Texas, the deposits of the southwest are chemically more complex than the northern deposits.

California: California has several sodium sulfate deposits. Dale Lake, Durmid Hills, Soda Lake, and Danby Lake have a record of past production but are not now producing (Smith, 1966). Owens Lake has reserves of 10.9 Mt but has not produced any sodium sulfate (Kostick, 1980).

At Searles Lake, minerals have been extracted from brines since 1873. Sodium sulfate was first produced in 1914. Searles Lake currently is the leading source of natural sodium sulfate in the United States (Moulton, 1980). It probably has produced more sodium sulfate than any other deposit in the world. The annual value of the minerals production to date exceeds $30 million and total production to date exceeds $1 billion (Smith, 1979).

Sodium sulfate and other usable salts are produced at Searles Lake from brines underlying a large playa. The surface of the playa covers an estimated 155 km² but the central part, which forms the main deposit, is about 30 km² in area (Graubau, 1920). A hole drilled by Kerr-McGee Corp. in 1968 encountered 930 m of basin fill and bottomed in quartz-monzonite (Moulton, 1980). The salt deposits vary in areal extent and are listed in Table 3.

Brine occurs in the interstices of the salt deposits. Ions present in the brine are sodium, potassium, carbonate, bicarbonate, sulfate, chloride, and borate. The Upper Salt, Lower Salt, and Mixed Layer are the productive zones, each contributing a distinctive brine to meet the individual requirements of the plants. In some instances, solar concentration of the produced brine is used to improve plant efficiency (Moulton, 1980).

A plant expansion underway in 1977 was to bring capacity to 590 kt/a (Parkinson, 1977). The plant never operated as planned and in fact, in 1982 Kerr-McGee shut down about half of their sodium sulfate capacity, so output dropped to about 200 kt/a.

TABLE 3—Searles Lake Salt Deposits

Unit	Typical Thickness, m
Overburden mud (surface)	7
Upper salt	15
Parting mud	4
Lower salt	12
Bottom mud	30
Mixed layer	200+

Source: Smith, 1979.

Utah: Sodium sulfate is produced from natural brine at Great Salt Lake, Utah. Thick beds of mirabilite, intercalated with thin clay layers, have been found beneath the lake at several localities. Mirabilite crystallizes and sinks to the bottom of the lake during the winter. A layer of mirabilite 0.15 to 0.30 m thick occurs seasonally over large areas of the lake. Under favorable conditions, it is carried by winds and currents to the shore where it accumulates to a depth of 0.3 m or more. Strand deposits as thick as 3 m have been observed.

The Great Salt Lake is estimated to contain 3.8 Gt of salt, 342 Mt of which are sodium sulfate (Cohenour, 1966). Great Salt Lake Minerals and Chemicals Corp. estimates the playa contains 408 Mt of sodium sulfate (Kostick, 1980).

On the north end of the lake, near Ogden, Great Salt Lake Minerals and Chemical Corp. operates 69 km² of evaporation ponds and a plant complex for producing potassium sulfate and magnesium chloride. Sodium sulfate and common salt are byproducts. First production was reported in 1970. The operation has a sodium sulfate capacity between 22.7 and 32.9 kt/a.

Texas: Four sodium sulfate deposits are presently being exploited in Gaines and Terry Counties, by the Ozark-Mahoning Co. The company has produced sodium sulfate from brine in West Texas continuously since 1933. The Texas deposits underlie playas and occur as brine and crystalline mirabilite. The brine contains mainly sodium chloride and sodium sulfate, with lesser amounts of magnesium and potassium salts, and is produced from 30.5 m deep wells. A fifth deposit, Soda Lake, near Monahans in Ward County, was abandoned in 1970 after nearly 40 years of continuous operations, during which time over 1 Mt of anhydrous sodium sulfate were produced.

Cedar Lake deposit, near Seagraves, covers a surface area of about 21 km². Beds of crystalline salts are minimal, but strata containing saturated brines are found from near the surface to a depth of 30.5 m. The shallower strata contain brines of low salt content and are less productive. Richer brines come from beds 12 to 30.5 m below the surface. Brines from the Cedar Lake deposit contain about 10.5% sodium sulfate, 14% sodium chloride, and 1% each of magnesium and potassium.

Brownfield Lake deposit, 16 km east of

Brownfield, covers an area of about 4 km^2. The main deposit lies outside the boundary of the present dry lake surface. Saturated brine is produced from crystal beds at depths from 9 to 30.5 m. These beds, which contain thin layers of clay and silt, range from a meter to as much as 4 m in thickness. The brine contains 15% sodium sulfate and 10% sodium chloride. A solution mining project to extract the crystalline reserve has been developed.

Rich Lake deposit, 19 km north of Brownfield Lake, produces sodium sulfate bearing brine from a crystal bed averaging 2 m in thickness. The brine is transported by pipeline to the Brownfield Lake plant for processing. Ozark-Mahoning operates plants at Brownfield Lake and Cedar Lake. The combined capacity of the two plants is 159 kt/a.

A fourth deposit, Mound Lake, is located 13 km north of the Brownfield Lake plant. The lake has an area of 5 km^2 and lacustrine sediments occur to a depth of 24 m. The major portion of the reserve is brine similar in composition to that of Cedar Lake. A significant tonnage of mirabilite also exists. Mound Lake began producing in April 1981 and the brine also is transported to the Brownfield plant for processing.

Arizona: At the Camp Verde deposit in Yavapai County, solid beds of thenardite of varying thickness were mined for many years by Western Chemical Co. The deposit is unusual, as pure anhydrous sodium sulfate occurring in solid crystal form is rare. The deposit consists of layers of thenardite and other sodium salts, interbedded with clay, which form a part of the Verde Formation. The total thickness of the beds is at least 46 m. Mining was done in a large quarry and by underground room-and-pillar methods. The last mining was done by Arizona Chemical Co., a subsidiary of American Cyanamid Co., in 1934 (Ralston, 1949).

Nevada: Nevada's principal producing deposit in years past was Rhodes Marsh, near Mina in Mineral County. The deposit covers an area of 28 km^2. Production was confined mainly to the central portion of the deposit, where a 1- to 2-m thick bed of solid thenardite was mined (Ralston, 1949).

New Mexico: A sodium sulfate deposit is known near Alamogordo in Dona Ana County. In Eddy County, near the potash mines, brine wells containing appreciable amounts of sodium sulfate and magnesium sulfate have been reported (Lang, 1941). These wells were produced by Ozark-Mahoning from 1951 through 1957; the brines were trucked to Ozark's plant near Monahans, Texas, for processing. The wells depleted rapidly and were abandoned.

North Dakota: The sodium sulfate deposits of northwestern North Dakota are located in Divide, Williams, Mountrail, and Ward Counties. They are a southeasterly extension of the deposits in Saskatchewan. Sodium sulfate in North Dakota was first noted by the Lewis and Clark Expedition. Several lakes are known. Most of them were investigated and prospected by FERA (Federal Emergency Relief Administration) in 1934. An important reevaluation, based on drilling, was made by the US Bureau of Mines in 1948 (Binyon, 1952).

The sodium sulfate content is generally high; sodium chloride, magnesium sulfate, and calcium sulfate are the main impurities. The US Bureau of Mines reported tonnages occurring in 18 lakes. Nine of the largest were estimated to each contain over 1 Mt of Glauber's salt. The major portion of the reserve occurs in the permanent and intermittent crystal beds and only a minor portion occurs in the form of brine. Total reserves are reported to be 25 Mt of Glauber's salt or 11 Mt of anhydrous sodium sulfate (Binyon, 1952; Crosby, 1973).

Grenora Lake No. 2, the largest deposit, covers 2 km^2. Drilling done by FERA showed the maximum thickness of the permanent crystal bed to be 24 m and the average thickness to be 3.66 m. The deposit was estimated by the US Bureau of Mines to contain 5 Mt of Glauber's salt or 2.2 Mt of anhydrous sodium sulfate (Binyon, 1952). Drilling conducted by Ozark-Mahoning in 1977 generally corroborated the Bureau's estimate. Grenora Lake No. 2 is owned by Ozark-Mahoning, but is not favorably located for economical exploitation.

Other States: A number of deposits in Washington, Wyoming, and Montana are not considered economically important because of their location and size, even though their composition is similar to the Canadian deposits.

Mexico

A large evaporite deposit containing a high concentration of sodium sulfate is located in northern Coahuila, approximately 160 km north of Torreon. The evaporite lens consists of glauberite and bloedite, this last one as thin beds and irregular aggregates.

The crystal bed is 20 km long and 5 km

wide, with 30 m thickness at the center. The main concentration of bloedite (a double sulfate salt of sodium and magnesium) is found as a layer 1 to 2 m thick, located about 8 m below the top of the evaporite mass. Mirabilite and epsomite occur in minor proportions. The formation underlying the crystal bed is impervious clay and silt containing some sodium chloride. Glauberite occurs as thin tabular crystals, while bloedite is found in massive concentrations. The intercrystalline spaces, amounting to as much as 25% of the evaporite mass, are partially filled with clay, and the rest with a brine in equilibrium with the solid phase.

A composite average analysis of brines from a depth of 20 m from several wells in the lake bed showed 18.46% sodium sulfate, 5.91% sodium chloride, and 5.28% magnesium sulfate.

Analysis of several cores of the crystal bed showed 35 to 36% sodium sulfate and only 1.5 to 1.8% sodium chloride. Magnesium sulfate ranged from 4.7 to 12.7%.

Estimated reserves are approximately 160 Mt of anhydrous sodium sulfate.

Sodium sulfate is produced at Laguna del Rey by Quimica del Rey, S. A. a subsidiary of Industrias Peñoles, S. A. de C. V. The plant was built in 1963 with a capacity of 30 kt/a. Present capacity is 440 kt/a.

Other deposits are known to occur in northern Coahuila and in the state of Chihuahua, but in size and sodium sulfate content they are smaller than the deposit at Laguna del Rey.

Sulfato de Viesca recovers 20 kt/a of sodium sulfate as a byproduct from production of sodium chloride from natural brine. The plant is near Viesca in the southwestern part of Coahuila.

Deposits of thenardite occur in the state of Sonora near the municipality of Bacadehuachi some 190 km northeast of Hermosillo. Outcroppings of thenardite and efflorescences in the soil occur discontinuously along the east side of the Bacadehuachi Valley in an area approximately 6 km in length and 4 km in width. The deposits occur within the Baucarit Group (Oligocene). The outcroppings are covered by unconsolidated surface material, which varies in thickness between 10 and 30 m.

Four workings have been developed within this area. In the largest pit, the Oro Blanco, the thenardite is reported to attain a thickness of 20 m. Analyses of selected samples from the Oro Blanco ran between 95 to 99% Na_2SO_4 (Anon., 1981).

During 1982, Para Mex, the Mexicanized affiliate of the Spanish company Tolsa, had conducted some exploration of the Oro Blanco and was evaluating the commercial possibilities of producing sodium sulfate from this area (personal communication).

According to US Bureau of Mines reports, Mexico produced 128 kt of sodium sulfate during 1978. Quimica del Rey, the country's main producer, states it produced 298 kt in 1978. Approximately 30% of 1980 production is destined for South American markets (Watson, 1980).

Union of Soviet Socialist Republics (USSR)

Many deposits containing sodium sulfate occur in the USSR, in a large area from the Black Sea to western China and farther north into west-central and eastern Siberia. The economically important deposits are in the Aral and Caspian depressions and in the Kalunda Steppe in Kazakhstan. While literature on sodium sulfate deposits in the USSR is reportedly fairly extensive, it is not readily available and requires translation. The reader is referred to a paper (Dort and Dort, 1970) that contains a summary of several publications on Russian deposits.

The largest known occurrence of sodium sulfate in the world is in the Kara-Bogaz-Gol Gulf, an embayment on the eastern side of the Caspian Sea. It is separated from the Caspian by a bar with a narrow strait through which water flows from the sea into the embayment. Sixty-five years ago the gulf covered 18 000 km² and had a depth of 12 m. Owing to evaporation, the Gulf has shrunk in area to 10 000 km² and in depth to about 3 m and has a much higher saline content (Garbell, 1963).

Kolosov (1974) reported that Kara-Bogaz-Gol experienced a final phase of intense desiccation between 1939 and 1948. Seawater now moves across the bar from the Caspian Sea at a reduced rate of 9 to 11 km³ per year. The movement of seawater into the bay has apparently reached equilibrium with evaporation. Since 1957 or 1958 no decline in water level or constriction of the bay's area or appreciable change in brine composition was observed. Gypsum and glauberite were being precipitated out of the saturated brine throughout the bay. Bloedite (astrakanite) and halite were being deposited locally. Some carbonates were being deposited in the zone of mixing adjacent to the strait.

It has been reported very recently that the

completion of a dam across the mouth of the Kara-Bogaz-Gol may halt the future supply of sodium sulfate bearing water from the Caspian Sea (Anon., 1983).

In 1924, Kara-Bogaz-Gol was estimated to contain 1 Gt of sodium sulfate (Grabau, 1920). More recent estimates of 2 Gt have been reported (Goudge, 1960; Hemy, 1969).

The Aral Sea is considered to be another rich source of sodium sulfate. Some sulfate is produced in the central Asian republics that adjoin it. Lakes in western Siberia and in Kazakhstan are good sources, but are not presently produced except on a local scale. The steppe region around Kalunda, in Alta Kray, has a number of lakes rich in sodium sulfate. A sulfate combine has been established on Lake Kuchuk, the largest of these (Hemy, 1969).

Sodium sulfate and magnesium salts are produced from the Kara-Bogaz-Gol Gulf by Karabogazgolsulphate Combine. It is impossible to determine the amount of natural sodium sulfate produced in the USSR. An estimate is that 318 kt are produced annually, mostly from the Kara-Bogaz-Gol deposit (USBM data). Watson (1980) reports capacity at 600 kt/a.

Spain

Sodium sulfate from deposits of solid thenardite is produced in Spain. The El Castillar deposit is located in Villarrubia de Santiago, Toledo Province, where beds of thenardite 5 to 6 m thick are mined by room-and-pillar methods. The deposit extends over an area of 30 km^2 and contains probable reserves of 200 Mt. The composition of the ore is reported as 70 to 75% sodium sulfate, 11 to 17% calcium sulfate, 0.4 to 0.6% sodium chloride, 3 to 9% clay, and 1 to 5% moisture (Rovira, 1960). The mine, and a plant producing anhydrous sodium sulfate, is operated by Union Salinera de Espana S.A.

Criaderos Minerales y Derivados S.A. (Crimidesa) mines glauberite at Cerezo del Rio Tiron in Burgos Province.

Another company, S.A. Salquisa, recently began production of sodium sulfate by solution mining of glauberite. It is a joint venture of Industrias Penoles S.A. de C.V. of Mexico and Spanish mining interests. Rated capacity is 75 kt/a.

Total production of natural sodium sulfate from Spain is approximately 150 kt/a and has shown a slight increase over the last few years (Watson, 1980).

Turkey

Natural sodium sulfate is produced in Turkey, although output is not large. First production recorded was in 1963, when 1264 t were produced. Output has increased steadily, and production in 1978 was about 80 kt.

Several saline lakes in southwestern Turkey contain appreciable amounts of sodium sulfate. Tersakan Lake is located about 160 km south of Ankara, near the town of Cihanbeyli. The lake, 13 km long and 4 km wide, covers an area of 52 km^2. Its sodium sulfate content is in excess of 8%. Sodium chloride and magnesium sulfate are present in amounts less than 5 and 2%, respectively, and the calcium sulfate content is less than 1%. Where the lake water has been impounded to induce evaporation, the sodium sulfate content of the brine increases. In the bottom of the ponds crystalline layers over 95% pure have formed.

At Aci Gol, a large lake 56 km east of Denizli, the saline content is similar to that of Tersakan Lake, although the sodium content is slightly lower and that of sodium chloride slightly higher. Aci Gol is over 26 km long, nearly 10 km wide, and has a surface area of about 200 km^2.

Three companies currently produce sodium sulfate—Alkim Alkali Kimya A.S., Sodas Sanayii A.S., and Otuzbir Kimya ve Sanayii Ltd.

Argentina

Many salt lakes and playas, some of which contain rich deposits of sodium sulfate, are located in Argentina, on the arid plateaus and in the intermontane valleys of the Andes (Wells, 1923).

A large playa deposit of mirabilite is located in northwestern Argentina in Jujuy Province. The deposit is in the Andes at an elevation of 3353 m. A permanent bed of mirabilite covering 53 km^2 occurs in the middle of a dry lake bed. The permanent bed, covered by a thin layer of sodium chloride, calcium sulfate, and sand, is reported to be 4.6 m thick at the center of the deposit and gradually thinner toward the edges. Subsurface waters from melting snow in the surrounding mountains feed into the lake and thus furnish a supply of brine year round.

Production of sodium sulfate in Argentina was 43 kt in 1978. Sulfarentina SAMIC, a subsidiary of CADIPSA, produces sodium sulfate at LaLoberia, Santa Cruz Province, and at Salar del Rio Grande, Salta Province.

Total output from Sulfargentina is about 18 kt/a (Watson, 1980).

Another producer of natural sodium sulfate in Argentina is Pagrun S.A.M.I.C.

Chile

Small amounts of natural sodium sulfate are produced from playa deposits in northern Chile. The deposits are in the Atacama Desert, which lies along the west side of the Andes between the mountains and the Pacific Ocean, at an elevation of about 1067 m.

Chile produced 20 kt of natural sodium sulfate in 1978. Some additional natural output is byproduct from the nitrate industry. Total production from Chile in 1979 was nearly 64 kt of natural sodium sulfate from Sociedad Quimica y Miniera de Chile S.A. (SOQUIMICH). A new anhydrous sodium sulfate plant is reported to be opened in Maria Elena in the Antofagusta region (Watson, 1980).

Africa

A series of salt pans, some of which contain sodium sulfate, occurs in the deserts and velds of the South African uplands. To date, none have been commercially exploited. In 1968, however, Associated Sulfate (Pty.) Ltd. was established to build a plant to produce sodium sulfate, using raw materials from solar salt pans and effluent from the Rustenburg platinum plant.

About 110 Mt of sodium sulfate reserves exist in the Makgadikgadi Basin of Botswana. Some Japanese developers were contemplating the development of the deposit for sodium sulfate and sodium carbonate (Kostick, 1980; Massey, 1973). However, by early 1981 no plans to develop the deposit had been formalized (Lefond, 1981).

No production of natural sodium sulfate in South Africa was reported between 1975 and 1979 (USBM data).

Antarctica

Since mirabilite is the first mineral, after ice, to appear in freezing seawater, it is not surprising that several occurrences have been noted in Antarctica. Mirabilite efflorescences commonly occur on the surfaces of rocks, glaciers, and sea ice; some occurrences are also reported farther inland.

Sodium sulfate has been described in coastal lakes which have been elevated and separated from the sea due to isostatic rebound. Other occurrences are reported in impoundments behind glaciers. The inland deposits are probably a result of seawater-derived salts transported with snow and accumulated in local depressions.

Though these occurrences are not economic, they are of geological interest. The reader will find them described by Dort and Dort (1970) along with a discussion of the low temperature origin of sodium sulfate.

Iran

Iran produced 25 kt of sodium sulfate in 1976 and 1977 and 22.7 kt in 1978 (Watson, 1980). Production is presumably from Urmiah Lake (also called Rezaiieh) located in northwestern Iran (Jolly, 1976).

Byproduct Sodium Sulfate

For many years, manufactured sodium sulfate has been a byproduct or coproduct from the production of other chemicals. Important sources are:

1) Manufacture of viscose rayon. Sodium sulfate is generated in an acid spin bath by the neutralization of sodium hydroxide used to prepare the viscose solution. In the past this source has been of almost equal importance with natural sodium sulfate on a worldwide basis. The growing popularity of other synthetic fibers, however, has resulted in reduced production. Of particular significance has been the replacement of rayon tire cord by other synthetic fibers, fiberglass, and steel. Nevertheless, rayon manufacture continues to be an important source of supply.

2) Manufacture of hydrochloric acid from sodium chloride and sulfuric acid in the Mannheim furnace, and similar processes. In the Hargreaves process, sulfur, steam, and air are substituted for sulfuric acid. Many years ago the Mannheim furnace was the main source of sodium sulfate, and this chemical was truly a coproduct rather than a byproduct. Because the salt discharged from the furnace tended to form cakes, it was named "salt cake." This name remains as a trade designation for low-grade sodium sulfate. Large amounts of hydrochloric acid are produced today as a byproduct in the manufacture of chlorinated hydrocarbons. As a result, the Mannheim furnace process is generally considered to be obsolete; however, in certain geographical areas and in particular raw-material situations it

continues to be an economically significant source of sodium sulfate.

3) As a byproduct in numerous other processes wherein certain sodium salts are converted to acids by reaction with sulfuric acid, or sodium alkalies are neutralized with sulfuric acid as part of the process. Examples are the processing of chromium ores to produce sodium dichromate, and the production of lithium salts, boric acid, phenol, formic acid, and various catalysts.

As noted, natural sodium sulfate has captured about half the market in the United States and had become increasingly important worldwide during the past two decades. Although vast resources are available, producers of the natural material always are threatened by byproduct sources. With the greatly increased emphasis on pollution control many dilute waste waters containing sodium sulfate can no longer be discharged into streams and rivers. Hence treatment processes usually attempt to recover any materials of value, such as sodium sulfate. The material is then sold at whatever price it will bring, whether or not the price covers the cost of recovery. Conceivably the removal of sulfur from fuels or combustion products can generate sulfur compounds, which may compete with sodium sulfate for certain uses. The outlook for long-term growth in the natural sodium sulfate business is not particularly promising, although natural sources will continue to be vital.

Worldwide total production of sodium sulfate is reported at 4153 kt for 1978, whereas productive capacity was listed at 5 Mt (Kostick, 1980). Natural production represents about 41% of world total output (Watson, 1980).

End Uses

The principal use of sodium sulfate is in the manufacture of kraft paper pulp, also known as sulfate pulp. In the process, the sodium sulfate is reduced chemically to sulfide forms, which are the active constituent of pulping liquor. In the second edition of this book (1949), Ralston reported that 179 kg of sodium sulfate were required to produce 1 t of kraft paper pulp. Since that time, process improvements have reduced this figure to approximately 40 kg/t. Technological developments in the paper industry indicate that the amount of sodium sulfate required will continue to decline.

Certain new pulp-bleaching processes utilize raw materials that produce waste streams containing sodium and sulfur, which may be utilized instead of sodium sulfate for kraft pulping. Likewise the neutral soda semi-chemical pulping process generates a liquor which may be utilized for kraft pulping in place of sodium sulfate. Some newer paper mills combine both processes economically (Anon, 1971a). Waste streams from caustic washing of catalytically cracked gasolines can be converted readily to sodium sulfide compounds, used as replacement for sodium sulfate in kraft mills. The authors estimate that perhaps 20% of the requirements of the kraft paper industry in the southeastern United States are now supplied from various waste streams.

Even though new pulping methods and recovery techniques will reduce the amount of sodium sulfate required per ton of kraft paper, total usage of sodium sulfate may not decrease because the sulfate-pulp process is a major paper-making technique. Kraft pulp produces the strongest paper, its production continues to grow worldwide, and it is by far the dominant type of pulp being produced. We expect it to occupy this position in the foreseeable future.

Another important consumer of sodium sulfate is the glass industry. Soda ash is the major chemical raw material for glass manufacture, and sodium sulfate may supply some of its sodium needs. For certain types of glass the sulfur content of sodium sulfate is useful; it prevents scum formation on the surface of molten glass in the refining stage. For technical reasons most sodium sulfate consumed in the glass industry is used in the production of flat glass. Lesser quantities are used in the production of containers and other glass products.

For the detergent industry, sodium sulfate is an ideal filler and diluent, as it is noncorrosive, neutral, and cheap. In addition, it is claimed to have very mild detergent properties. In 1971 and 1972 the detergent industry went through a revolution in the United States as governmental pressure to reduce phosphate discharges in municipal sewage required the reformulation of detergents. Nonphosphate and low-phosphate detergents are now on the market. Further changes may be expected in the future. The result was an increase of sodium sulfate in detergents. Currently, about one-third of dry home laundry detergent is sodium sulfate.

A number of other uses of sodium sulfate are significant. In the United States it is an ingredient of various stock foods and remedies, probably because of its mild laxative properties. It is useful in the production of synthetic sponges. In Europe and Japan it is used in dyeing and for the production of sodium sulfide and related compounds. In the United States these uses are minor.

New uses for sodium sulfate may have the potential to become significant in the future. A joint industry-government project sponsored by EPA (Environmental Protection Agency) discovered that small amounts of sodium sulfate added to low-sulfur coal before burning improve the performance of high-temperature electrostatic precipitators (Anon., 1980). The project was conducted at a power plant of the Gulf Power Co. near Panama City, FL. Partners in the investigation were the Industrial Environmental Research Laboratory of EPA as well as Southern Company Services, Birmingham, AL, and The Electric Power Research Institute, Palo Alto, CA.

Another new use involves sodium sulfate as a constituent in molten salt used to decompose pesticide wastes. A third new use may have a greater potential for the future. This use takes advantage of the phase change properties of anhydrous sodium sulfate to Glauber's salt as a storage material for solar heat. Of nine commercial units being offered in 1980, four utilize sodium sulfate (Eissenberg, 1980; Holland, 1980).

It is difficult to estimate worldwide consumption patterns because of great variations from one country to the next. Canada, Sweden, and Finland are large producers of kraft paper pulp, and perhaps 95% of the sodium sulfate consumed in those countries is used for this purpose. In the United Kingdom, France, West Germany, Japan, and many other countries, paper production is small, so the major consumption of sodium sulfate is in detergents, glassmaking, production of other chemicals, dyeing processes, and miscellaneous uses.

The authors' estimate for consumption in the United States is as follows:

Kraft paper pulp	55%
Detergents	35%
Glass	5%
Other uses	5%

The foregoing does not include the use of materials which contain both sodium and sulfur values that are used interchangeably with sodium sulfate in the paper industry. Worldwide consumption patterns probably are not far from the estimate shown for the United States.

Product Specifications

The old salt cake nomenclature continues to be used for material of lower quality. The product of the Mannheim furnace, for example, may contain only 97% Na_2SO_4. Some natural material from Canadian deposits may be as low in grade, although most of the plants are now producing higher quality sodium sulfate. The salt cake grade is quite acceptable to the paper industry.

The other designation is anhydrous sodium sulfate, technical grade, which usually has a purity in excess of 99% Na_2SO_4, and frequently may be specified as 99.5%. Different consumers for the same end use may have quite different specifications. In some cases particle size and bulk density are important.

The decahydrate, known as Glauber's salt (mineralogically, mirabilite), at one time was an item of commerce. In virtually all uses anhydrous sodium sulfate is preferred today because of savings in freight and ease of handling. Small quantities of Glauber's salt are still sold for some special uses.

Processing

Natural sodium sulfate is produced by five companies in Canada (Ruffel, 1970, 1972; Broughton, 1977; Watson, 1980). Saskatchewan Minerals, Sodium Sulphate Division, a corporation owned by the Province of Saskatchewan, operates plants at Chaplin Lake, Bishopric, Sybouts, and Ingebrigt Lake. The Bishopric plant is at Frederick Lake and has operated on an intermittent basis in recent years. Ormiston Mining and Smelting Co., Ltd., has a plant at Ormiston on Shoe Lake; Midwest Chemicals, Ltd., at Palo, produces from Whiteshore Lake; and Francana Minerals, at Cabri, operates on Snakehole Lake and at the Alsask Lake deposit, which formerly was operated by Sodium Sulphate (Saskatchewan, Ltd.). All of the foregoing are located in Saskatchewan. In Alberta the only operation is the Metiskow Lake plant of Alberta Sodium Sulphate, Ltd.

Many Canadian producers utilize seasonal temperature differences to assist in the processing of the sodium sulfate deposits. Brine is allowed to form on the lake bed and becomes nearly saturated in the late summer months. At that time it is pumped to large reservoirs which may be over 3 m deep. When cold weather sets in, Glauber's salts crystallize from the saturated brine. The residual mother liquor is drained off; it contains most of the magnesium sulfate and sodium chloride impurities, which may be returned to the lake bed. After draining, the Glauber's salt is harvested and accumulated in stockpiles at the plant sites.

At some of the plants of Ormiston Mining and Smelting, Francana Minerals, Saskatchewan Minerals, and Alberta Sodium Sulphate, the supply of Glauber's salt crystals from the reservoirs is supplemented by salt dredge mined from the permanent bed and pumped to the plant. At the Metiskow Lake operations of Alberta Sodium Sulphate, solution mining also is practiced. A heated solution is injected into the crystal bed to generate a brine, which is pumped to the reservoirs for subsequent chilling and crystallization. In some of the Saskatchewan lakes the reserves are present principally as brine, which can be pumped directly to the reservoir.

Obviously the Canadian plants are somewhat dependent on nature to produce the crystals of Glauber's salt. Drought or other unusual conditions can affect the production cycle. The plants try to keep more than one year's supply of Glauber's salt in their stockpiles to keep production from being interrupted by such conditions. This method has proved very successful in Canada, producing a salt cake of high quality.

To prepare a salable anhydrous product the Glauber's salt must be dehydrated. Because of the inverse solubility of sodium sulfate (i.e., less soluble at higher temperatures), this is not a simple straightforward operation. In the early days dehydration was done in large rotary kilns, some of which were fired by lignite. The Sybouts plant of Saskatchewan Minerals continues to use this process.

When Glauber's salt is melted, the solubility of anhydrous sodium sulfate in the water of crystallization is exceeded, causing precipitation. The balance of the water must be evaporated to produce an anhydrous product. Ozark-Mahoning submerged combustion evaporators are used for this purpose by Ormiston Mining and Smelting, Midwest Chemicals, Alberta Sodium Sulphate, and Saskatchewan Minerals at the Ingebrigt Lake plant. Francana Minerals utilizes triple effect vacuum evaporators. Another type, called the Holland evaporator, provides for intimate contact between the solution and hot products of combustion by means of mechanical agitation. These are in use at the Bishopric plant of Saskatchewan Minerals and the Alsask plant of Francana Minerals.

At all plants, the anhydrous salts precipitated in the evaporators may be separated from the saturated solution by mechanical means, including centrifuges, and the wet salt finally dried in direct-fired rotary kilns. Parallel-flow dryers are now preferred for ease of control and fuel economy, whereas in the older processes countercurrent flow was used in the dryers.

United States

Each US producer of sodium sulfate has a different process dictated by the nature of the deposit being worked. Two plants operated by Ozark-Mahoning from brine deposits in West Texas, one near Brownfield and the other near Seagraves, use a process developed in the early 1930s for a plant near Monahans, Texas (Weisman, 1953). The Monahans plant closed in 1970 because the deposit was essentially exhausted. The Ozark submerged combustion burner used at the old Monahans plant probably represented the first commercial application of submerged combustion in the United States.

The Texas brines contain both sodium chloride and sodium sulfate, as well as smaller amounts of magnesium and potassium salts. At the Brownfield plant a portion of the brine is pumped through the Permian salt bed to introduce sodium chloride. The high sodium chloride content depresses the solubility of sodium sulfate, improving its extraction. At the Seagraves plant sufficient sodium chloride is present in the brine; additional salt is not needed. The Texas climate does not lend itself to the production of Glauber's salt by atmospheric chilling, so mechanical refrigeration and heat exchange are required. The Glauber's salt formed in this manner is filtered and washed, yielding a product of high purity.

Melting and dehydration of Glauber's salt is accomplished by the Ozark submerged combustion system, which utilizes gas burners submerged under the surface of the solution, with products of combustion in direct contact with the liquid phase. Centrifuges separate the anhydrous crystals from the saturated solution, which is returned to the evaporators. The crystals are dried in parallel-flow rotary kilns.

Use of the submerged combustion system largely eliminates scaling problems, which can occur in tubular evaporators from the inverted-solubility character of sodium sulfate solutions. Some plants operate successfully with evaporators that transfer steam heat through tubes to the saturated solution; however, as noted, many Canadian plants have adopted the Ozark submerged combustion method. Most byproduct plants use multiple-effect vacuum evaporators.

Newer plants use vapor recompression evaporators to achieve high fuel economy. High velocities in the tubes minimize scaling problems. The Brownfield plant was converted to vapor recompression evaporation in 1983 because of the high cost of natural gas.

The largest production of natural sodium sulfate in the US still comes from Searles Lake, CA, even though output is much less than in the past. Prior to 1982, the Kerr-McGee Chemical Corp. plants made about 500,000 tons of sodium sulfate annually. As mentioned earlier, a planned plant expansion did not result in increased production and the Trona plant, representing about half of the total sodium sulfate capacity, was shut down in 1982. The West End plant now is Kerr-McGee's only source of sodium sulfate.

The Searles Lake brines are complex and a number of salable products are extracted. At Kerr-McGee's West End plant, sodium sulfate is recovered along with soda ash and borax. Mixed brines are first carbonated with carbon dioxide to precipitate sodium bicarbonate, which is removed by filtration. The decarbonated brine is cooled to crystallize borax. This borax is separated from the brine, which is then cooled further to produce a crop of Glauber's salt. After it is removed, a final cooling recovers a second crop of borax. By heating, the sodium bicarbonate is converted to soda ash and the borax is either crystallized as a hydrate or dehydrated to the anhydrous form. The Glauber's salt is washed, melted, and recrystallized as anhydrous sodium sulfate. The following description of the Trona plant is included because it is of interest, even though the sodium sulfate part is shut down.

In the Trona main plant cycle there were three major process plants. The brines from the lake were cooled to produce a crop of Glauber's salts which were removed and dried to sodium sulfate. The resultant mother liquor from this step was processed further to recover potash and borax. A different brine was treated in a liquid-liquid solvent extraction plant to recover boron values as boric acid. After it was removed, mixed sodium and potassium sulfates were precipitated by evaporation and these mixed sulfates were fed to the potassium sulfate plant. The Trona carbonation plant was somewhat similar to the West End plant in that carbon dioxide precipitated sodium bicarbonate which was filtered and removed. Borax was then precipitated by cooling. Burkeite, a double salt of sodium sulfate and sodium carbonate, finally was recovered and fed to the Argus process.

In the Argus plant, which is the newest addition, brines from the lake are mixed with burkeite solutions and streams from other plants to recover additional soda ash. Nothing has been published since Moulton (1980), so the details of present plant practice at Trona have not been revealed.

The other producer of natural sodium sulfate in the United States is Great Salt Lake Minerals and Chemicals Corp. The plant began production late in 1970. The initial rated capacity announced was 15 kt/a of sodium sulfate, plus 220 kt of potassium sulfate and 90 kt of anhydrous magnesium chloride. Solar evaporation is an important step in the processing, with several complex salts harvested from evaporation ponds for further refining in the plant by proprietary processes. Early operations were beset with difficulties but these problems were overcome. Very little has been published on production processes used by the company. According to company advertising literature the process used in 1980 was dependent on harvesting Glauber's salts from solar ponds where salts were deposited during the winter months, similar to the Canadian deposits. The Glauber's salt is melted and anhydrous sodium sulfate precipitated by adding NaCl to decrease the solubility of the sodium sulfate. A high-purity sodium sulfate is the final product.

Great Salt Lake contains tremendous reserves of sodium sulfate and no doubt will be a factor of some importance in the future. A railway causeway across the lake, which affects the concentration of salts on both sides, has created controversy. When this controversy is resolved, the lake's concentration of sodium sulfate and the production of this and other salts could be drastically altered.

Mexico

The brine processed at Quimica del Rey is very strong and probably saturated. It contains more than 26% sodium sulfate and appreciable amounts of chlorides and magnesium. About 100 wells produce the brine from a depth of approximately 12 m. The brine temperature is 25°C before precooling with the spent brine. The cooled brine is fed to a vacuum crystallizer where it is cooled to 13°C by evaporative cooling, causing Glauber's salt to crystallize. The slurry is thickened, filtered, and melted. Evaporation is accomplished in triple-effect evaporators. Final drying is in countercurrent-fired rotary kilns. Part of the mother liquor is processed further for recovery of magnesia products while the balance goes to waste.

The most recent expansion of this plant occurred in 1980. Six separate production lines are capable of producing 440 kt/a. It is expected that one line will be converted to recovery of magnesium sulfate from the spent brine.

Other Countries

Little is published about processing in other countries. It can be reasonably assumed that methods used are similar to those used in the United States, Mexico, and Canada, wherein chilling by natural or mechanical means precipitates Glauber's salt, allowing separation from other salts. The Glauber's salt would then be dried to anhydrous sodium sulfate.

In Argentina saturated brine from a lake bed is pumped to crystallization tanks, where great night and day temperature variation results in precipitation of Glauber's salt. This is possible because the plant site is at an elevation of 3353 m. The Glauber's salt is melted by heating; the precipitated anhydrous sodium sulfate is then separated and dried. There is no evaporation of water, so the saturated solutions probably are recycled through the system.

Deposits in Spain, one of the few countries where thenardite is the starting material, are mined underground and present special problems because the mineral contains an appreciable amount of calcium sulfate in the form of glauberite (Na_2SO_4, $CaSO_4$). Countercurrent leaching at temperatures between 35 and 40°C dissolves most of the sodium sulfate, leaving the calcium sulfate undissolved. The solids, which probably contain other insoluble impurities, are settled in large Dorr thickeners. A second leaching step with water at 29°C results in maximum recovery of sodium sulfate. Evaporation is carried out in a single-effect vacuum evaporator with forced circulation. Final drying is done in a fluid bed dryer with hot air heated by steam (Rovira, 1960).

Prices

Published price information for sodium sulfate usually only approximates the true market price. In the United States prices usually are based on quotations from byproduct sources located near consuming points. Because producers of the natural salt are located at great distances from most customers, the f.o.b. plant value for natural material may be much less than the published price. On the other hand, in periods of shortages and rising prices, published information often lags behind actual selling levels.

The May 16, 1983 *Chemical Marketing Reporter* quoted sodium sulfate in bulk carload lots, works, freight equalized, in the East, at $90 to $96 per short ton. It also quoted sodium sulfate, West, bulk, carload lots, works, freight equalized, at $90 to $96 per short ton.

United States tariff rates for sodium sulfate are as follows:

	Most Favored Nation		Non-Most Favored Nation
	Jan. 1, 1982	Jan. 1, 1987	Jan. 1, 1982
Crude (salt cake)	Free	Free	Free
Anhydrous	40¢/t	34¢/t	$3.05/t
Crystallized	3% ad val.	2.5% ad val.	4% ad val.

For tax purposes a depletion allowance of 14% for both domestic and foreign deposits is permitted by the Internal Revenue Service.

Acknowledgment

The authors acknowledge the valuable assistance and suggestions received from Dennis S. Kostick, Sodium Compounds Commodity Specialist, US Bureau of Mines, and Charles W. Tandy, Vice President, Ozark-Mahoning Co.

Bibliography and References

Anon., 1967, "The Sodium Sulfate Industry in Saskatchewan," Unpublished Report, Industrial Minerals Div., Saskatchewan Dept. Mineral Resources, 8 pp.

Anon., 1970, "Great Salt Lake Starts $30 Million Pond-Plant Complex," *Chemical & Engineering News,* Vol. 48, No. 51, Dec. 7, p. 17.

Anon., 1970a, "Utah Company Gets Set to Tap Mineral Wealth of Great Salt Lake," *Engineering & Mining Journal,* Vol. 171, No. 4, Apr., pp. 67–70.

Anon., 1971, "Chemical Profile, Sodium Sulfate," 2nd Rev., *Chemical Marketing Reporter,* Oct. 1.

Anon., 1971a, "1971 World Review," *Pulp and Paper,* World Review ed., pp. 150–154.

Anon., 1972, "Business Newsletter," *Chemical Week,* Vol. 110, No. 20, May 17, p. 17.

Anon., 1972a, "Harvest of Sulfate of Potashes to be 3–4 Times Greater in 1972," *The Mining Record,* Oct. 11, p. 3.

Anon., 1972b, "Petrochemicals Get the Emphasis in Mexican Buildup," *Chemical Week,* Vol. 110, No. 20, May 17, p. 33.

Anon., 1972c, "Sodium Sulphate: World Survey of Production, Consumption and Prices with Special Reference to Future Trends," Roskill Information Services, Ltd., London, England.

Anon., 1980, *Industrial Research and Development,* Vol. 22, No. 10, Oct., pp. 67–69.

Anon., 1981, "Evaluacion Preliminar del Yacimiento de Sulfato de Sodio en Bacadehuachi, Son.," Informe de Avance, Direccion de Mineria, Geologica y Energeticos, State of Sonora, Mexico, Oct.

Anon., 1983, "Mineral Notes," *Industrial Minerals,* No. 187, Apr., p. 68.

Ashley, B.E., 1972, "Mineral Industry of Mexico," Preprint, *Minerals Yearbook,* US Bureau of Mines, 13 pp.

Bateman, A.M., 1950, *Economic Mineral Deposits,* 2nd ed., John Wiley and Sons, pp. 194–196.

Binyon, E.O., 1952, "North Dakota Sodium Sulfate Deposits," Report of Investigations 4880, US Bureau of Mines, 41 pp.

Broughton, P.L., 1977, "Natural Salt Cake in Canada," *Industrial Minerals,* No. 121, Oct., pp. 51–59.

Cohenour, R.E., 1966, "Salt Inventory of Great Salt Lake," *Second Symposium on Salt,* Vol. 1, Northern Ohio Geological Society, p. 212.

Cole, J.W., 1970, "Mineral Industry of Chile," *Minerals Yearbook,* Vol. 3, US Bureau of Mines, pp. 197–206.

Cole, L.H., 1926, "Sodium Sulphate of Western Canada," No. 646, Canada Dept. of Mines, pp. 26–27, 74–78, 113.

Corrick, J.D., 1972, "Mineral Industry of Spain," Preprint, *Minerals Yearbook,* US Bureau of Mines, 21 pp.

Crosby, E.J., 1973, "Sodium Sulfate," Mineral and Water Resources of North Dakota, North Dakota Geological Survey Bulletin 63, pp. 151–154.

Dana, E.S., 1932, *A Textbook of Mineralogy,* 4th ed., John Wiley and Sons, pp. 809–810.

Davis, S.N., and DeWiest, R.J.M., 1966, *Hydrogeology,* John Wiley and Sons, pp. 108–109.

Dewey, L.S., 1972, "Mineral Industry of Argentina," Preprint, *Minerals Yearbook 1972,* Vol. 3, US Bureau of Mines, 11 pp.

Dietrich, R.V., 1969, *Mineral Tables,* McGraw-Hill Book Co.

Dort, W., Jr. and Dort, D.S., 1970, "Low Temperature Origin of Sodium Sulfate Deposits, Particularly in Antarctica," *Third Symposium on Salt,* Vol. 1, Northern Ohio Geological Society, pp. 181–203.

Eissenberg, D., and Wyman, C., 1980, "What's in Store for Phase Change?" *Solar Age,* Vol. 5, No. 5, May, pp. 12–16.

Fanfani, et al., 1975, "The Crystal Structure of Schairerite and Its Relationship to Sulphohalite," *Mineralogical Magazine,* Vol. 40, June, pp. 131–139.

Fletcher, S.W., Jr., 1979, Talk Presented to the Fall 1979 Seminar, Raw Material Section, American Paper Institute, Oct.

Garbell, M., 1963, "Sea that Spills into Desert," *Scientific American,* Vol. 209, Aug., pp. 94–100.

Goudge, M.F., and Tomkins, R.V., 1960, "Sodium Sulfate from Natural Sources," *Industrial Minerals and Rocks,* 3rd ed., J.L. Gillson, ed., AIME, New York, pp. 809–818.

Grabau, A.W., 1920, "Geology of the Non-Metallic Mineral Deposits," *Principles of Salt Deposition,* 1st ed., Vol. 1, McGraw-Hill, New York, pp. 29–40, 277–279.

Grossman, I.G., 1949, "Sodium Sulfate Deposits of Western North Dakota," Rept. of Investigation No. 1, North Dakota Geological Survey, 66 pp.

Grossman, I.G., 1968, "Origin of the Sodium Sulfate Deposits of the Northern Great Plains of Canada and the United States," Professional Paper 600-B, US Geological Survey, pp. B104–B105.

Hardie, L.A., and Eugster, H.P., 1970, "The Evolution of Closed-Basin Brines," Special Paper 3, Mineralogical Society of America, pp. 273–290.

Havighorst, C.R., 1963, "AP&CC's New Process Separates Borates From Ore by Extraction," *Chemical Engineering,* Vol. 70, No. 23, Nov. 11, pp. 228–232.

Hemy, G.W., 1969, "Chemical Minerals in the U.S.S.R.," Part I, *Chemical and Process Engineering,* Vol. 50, No. 8, Aug., pp. 63–68.

Holland, E., 1980, "The Mattapoisett House," *Solar Age,* Vol. 5, No. 5, pp. 24–28.

Jolly, J.L.W., 1976, "The Mineral Industry of Iran," *Minerals Yearbook,* US Bureau of Mines, pp. 12–13.

Jones, B.F., 1966, "Geochemical Evolution of Closed Basin Water in the Western Great Basin," *Second Symposium on Salt,* Vol. 1, Northern Ohio Geological Society, pp. 181–200.

Klingman, C.L., 1972, "Sodium and Sodium Compounds," *Minerals Yearbook 1972,* US Bureau of Mines, pp. 1147–1151.

Koelling, G.W., 1971, "Mineral Industry of Argentina," *Minerals Yearbook,* Vol. 3, US Bureau of Mines, pp. 87–98.

Kolosov, A.S., Pustyl'nikov, A.M., and Fedin, V.P., 1974, "Precipitation of Glauberite and Conditions of Deposition of Glauberite-Bearing Sediments in the Kara Bogaz Gol," *Doklady Akademii Nauk USSR,* Vol. 219, No. 6, pp. 1457–1460.

Kostick, D.S., 1978–1979, "Sodium and Sodium Compounds," Preprint, *Minerals Yearbook, 1978–1979,* US Bureau of Mines, 8 pp.

Kostick, D.S., 1982, "Sodium Sulfate," *Mineral Commodity Summaries, 1982,* US Bureau of Mines, pp. 144–145.

Krauskoph, K.B., 1967, *Introduction to Geochemistry,* McGraw-Hill, Inc., p. 320.

Lang, W.B., 1941, "New Sources of Sodium Sulphate in New Mexico," American Association of Petroleum Geologists, *Bulletin,* Vol. 25, Jan., pp. 152–160.

Lefond, S.J., 1981, "Botswana—An Opportunity," Preprint 81–73, SME–AIME Annual Meeting, Chicago, Feb., 3 pp.

Massey, N.W.D., 1973, "Resources Inventory of Botswana: Industrial Rocks and Minerals," Min. Res. Report No. 3, Geological Survey Dept., 39 pp.

Moore, L., 1972, "Mineral Industry of Chile," Preprint, *Minerals Yearbook,* Vol. 3, US Bureau of Mines, 12 pp.

Moulton, G.F., 1980, "Compendium of Searles Lake Operations," Preprint 80–133, SME-AIME Annual Meeting, Las Vegas, Feb., 6 pp.

Nickel, E.H., and Bridge, P.J., 1977, "Nickelbloedite, $Na_2Ni(SO_4)_2 \cdot 4H_2O$, A New Mineral from Western Australia," *Mineralogical Magazine,* Vol. 41, Mar., pp. 37–41.

Parkinson, G., 1977, "Kerr-McGee Expands Soda

Ash Output Nine-fold from Searles Lake Brines," *Engineering and Mining Journal,* Vol. 178, No. 10, Oct., pp. 71–75.

Rathjen, J.A., 1972, "Minerals Industry of Canada," Preprint, *Minerals Yearbook 1972,* Vol. 3, US Bureau of Mines, 21 pp.

Ralston, O.C., 1949, "Natural Sodium Carbonate and Sodium Sulphate," *Industrial Minerals and Rocks,* 2nd ed., S.H. Dolbear, ed., AIME, New York, pp. 945–962.

Reeves, C.C., Jr., 1968, *Introduction to Paleolimnology,* Elsevier Publishing Co., pp. 68–69.

Rovira, J.M.R., 1960, "Sodium Sulfate: Its Recovery from the Minerals Containing It," *ION,* No. 233, Dec.

Ruffel, P.G., 1970, "Natural Sodium Sulfate in North America," *Third Symposium on Salt,* Vol. 1, Northern Ohio Geological Society, pp. 429–451.

Shekarchi, E., 1972, "Mineral Industry of Turkey," Preprint, *Minerals Yearbook,* Vol. 3, US Bureau of Mines, 11 pp.

Smith, G.I., 1966, "Sodium Sulfate," Mineral Resources of California, California Div. of Mines and Geology Bulletin 191, pp. 389–392.

Smith, G.I., 1979, "Subsurface Stratigraphy and Geochemistry of the Late Quaternary Evaporites, Searles Lake, Calif.," Professional Paper 1043, US Geological Survey.

Watson, I., 1980, "Sodium Sulphate: Present Shortage on Short-term Basis," *Industrial Minerals,* No. 152, May, pp. 61–77.

Weisman, W.I., and Anderson, R.C., 1953, "The Production of Sodium Sulfate from Natural Brines at Monohans, Texas," *Mining Engineering,* July.

Weisman, W.I., and Tandy, C.W., 1975, "Sodium Sulfate Deposits," *Industrial Minerals and Rocks,* 4th ed., S.J. Lefond, ed., AIME, New York, pp. 1081–1093.

Wells, R.G., 1923, "Sodium Sulphate: Its Sources and Uses," Bulletin 717, US Geological Survey, pp. 17–18.

Staurolite

ROBERT B. FULTON, III *

Staurolite, an iron aluminum silicate mineral, is used industrially as a high value-in-use sand-blasting agent, as a premium grade foundry sand, and as the source of aluminum in portland cement manufacture in areas where the aluminum constituent is not economically available from shale or argillaceous limestone.

It is produced by E. I. du Pont de Nemours & Co. as a coproduct in titanium mineral production in Florida where it was first recovered on a commercial scale in June 1952 from the Pleistocene sand deposit known as Trail Ridge. In the Trail Ridge deposit it comprises about one-fifth of the heavy minerals, the principal other heavy mineral constituents being ilmenite, leucoxene, rutile, zircon, kyanite, sillimanite, and tourmaline. Staurolite and the other silicate nonconductive heavy minerals are separated from the conductive titanium minerals on high-tension separators. The silicates are then separated from one another magnetically, the staurolite being relatively magnetic. In the total processing, 200 tons of raw sand yield 7 tons of heavy mineral concentrate from which about one ton of staurolite is recovered.

This process yields staurolite concentrates of the characteristics given in Table 1.

These concentrates are sold as three products with the size distribution of subrounded particles shown in Table 2.

All three grades are used in sand-blasting for cleaning metal, for paint removal, and for cleaning buildings. Staurolite compares favorably with lower unit cost abrasives, such as boiler slags and mineral aggregates, in applications where the abrasive is used only one time and is not recovered. Because it is hard, fine, clean, and dense it can reduce blasting time, with less material reportedly being required

TABLE 1—Characteristics of Staurolite Concentrates

Typical Mineral Composition, %	
Staurolite minerals	89
Titanium minerals	4
Kyanite	1
Zircon	2
Quartz	4
Typical Chemical Composition, %	
Al_2O_3	45 (min.)
Fe_2O_3	18 (max)
ZrO_2	3 (max)
TiO_2	5 (max)
Free silica	<5 (max)
Physical Characteristics	
Bulk density	143-150 lb per cu ft
Specific gravity	3.1-4.6
Hardness (Mohs')	7
Melting point	2800° F (1537° C)
Coefficient of expansion	7.8×10^{-6}

TABLE 2—Size Distribution of Subrounded Particles in Three Staurolite Products

	% Retained on Sieve		
US Standard Sieve No.	Coarse	Biasill®	Starblast®
20	1	–	–
30	3	–	–
40	26	<1	3
50	58	5	11
70	9	20	19
100	–	53	45
140	–	19	19
200	–	3	3
270	–	<1	<1
Pan	3	Trace	none

®Registered trademarks for staurolite and abrasive sand products of E.I. du Pont de Nemours & Co., Inc.

* Basic Materials Planning, Materials and Logistics Dept., E.I. du Pont de Nemours & Co., Wilmington, DE.

TABLE 3—Cleaning Rate and Profile of Abrasives for Blast Cleaning

Abrasive	Cleaning Rate, Sq Ft per Hr	Abrasive Flow Rate, Lb per Hr	Profile, Mils
Starblast®	240	600	1.1
Biasill®	240	580	0.9
Coarse Staurolite	200	535	1.5
Rotoblast® G80A*	150	700	1.2
Rotoblast® G40A	120	750	2.1
Garnet RT-80	224	640	1.0
Garnet RT-60	194	525	1.0
Aluminum Oxide 30 FDT	170	665	1.5
54/70	224	575	1.3
80/100	250	540	1.1
Black Beauty® 50†	160	750	1.5
Flintbrasive® #3‡	120	520	1.7
Flintshot®¶	200	700	1.7
Flint Silica®§	220	730	1.4

Tests conducted at Du Pont Engineering Test Center, 7/1/70.
100% mill scale on New Steel Plate. All surfaces to Class I white metal.
Nozzle, $^{3}/_{8}$-in. 92-95 psi.

* Registered trademark for a metal blasting grit of The Pangborn Corp.
† Registered trademark for an abrasive slag of H. B. Reed Corp.
‡ Registered trademark for a blasting abrasive of the Independent Gravel Co.
¶ Registered trademark for a blasting abrasive of Ottawa Silica Co. and United States Silica Co.
§ Trademark for a silica sand of Ottawa Silica Co.

TABLE 4—Results of Field Trials

Abrasive	Degree of Cleaning	Cleaning Rate, Sq Ft per Hr	Abrasive Flow Rate, Lb per Hr	Profile, Mils
		New Pipe—Intact Millscale		
Polygrit®*	Commercial	176	1,360	1.75
Flintshot®	Near white	308	845	1.29
Biasill®	White	295	568	1.31
Coarse Staurolite	Commercial/ near white	313	645	1.14
Starblast®	White	265	632	1.18
		New Pipe—Millscale and Rust		
Polygrit®	Commercial	222	1,650	1.92
Flintshot®	Commercial	235	880	1.44
Biasill®	Near white/ white	204	545	1.36
Coarse Staurolite	Commercial	228	868	1.48
Starblast®	Near white	233	640	1.46

Tests conducted at William W. Pearson, Inc., Chester, PA, 1/19/71-1/21/71.
Conditions: ½-in. nozzle; 75-80 psi nozzle pressure.

* Registered trademark for an abrasive slag of MDC Industries.

TABLE 5—Breakdown Factors of Abrasives

Abrasive	Number of Uses	Breakdown Factor*
Aluminum Oxide 30 FDT	1	0.80
	3	0.61
Garnet RT 60	1	0.61
	3	0.38
Starblast®	1	0.77
	3	0.45
Flintshot®	1	0.68
	3	0.36
Flintbrasive® #3	1	0.58
	3	0.32

* Breakdown factors can range from 1 to 0. A material showing no change after blasting has a factor of 1, while one which breaks completely into dust has a factor of 0.

Tests conducted at Du Pont Engineering Test Center, 7/1/70.
¼-in. nozzle.
92-95 psi nozzle pressure.
¼-in. steel plate target.

to clean a specific area. Compared to higher unit value abrasives such as aluminum oxide and garnet, which are recycled, it costs significantly less per pound and cleans at equal or faster rates. The breakdown factor for staurolite, which is a measure of recycling performance, is intermediate between those for aluminum oxide and garnet.

Tables 3 and 4 show performance of the three staurolite products compared to other abrasives. Table 5 compares longevity of premium priced abrasives in recycling.

Fringe benefits claimed for staurolite's use as an abrasive are:

1) Good flow properties and freedom from lumps or slivers found in silica sands or boiler slag abrasives.

2) Low dusting which affords good visibility, permits faster working, and reduces cleanup problems.

3) Heated storage is not required in freezing conditions because it is nonhygroscopic.

Bibliography and References

Anon., 1972, Technical Bulletin "Starblast" Abrasive, E. I. du Pont de Nemours & Co.

Huskonen, W.D., 1970, "Spotlight on Specialty Sands," *Foundry Magazine,* Oct., pp. 40–45.

Laux, J.P., and Lipton, J.M., 1972, "An Introduction to Staurolite Sand," *Foundry Magazine,* May, pp. 158, 160.

Wagner, A.J., 1972, "Biasill Staurolite Sand," *Modern Casting Magazine,* Feb., pp. 64, 65.

Strontium

ROBERT B. FULTON, III *

Commercially, celestite ($SrSO_4$) is the predominant strontium mineral. Among other strontium-bearing minerals, only strontianite ($SrCO_3$) occurs commonly; however, it is rarely an item of commerce.

Production of celestite is principally in Mexico, Spain, and Turkey. An attempt to produce celestite in Canada was unsuccessful. The United States, which is wholly dependent on imports, chiefly from Mexico (99%), is the largest consumer of strontium. Lesser quantities are consumed in Japan and Europe. The Federal Republic of Germany is the principal source of strontium compounds imported to the US.

Uses

The principal uses of strontium are for the manufacture of strontium carbonate and nitrate. Carbonate is used in glass for TV tube face plates and in ceramic ferrites for magnets. Nitrate is used in pyrotechnics, mainly in safety flares, to which it imparts a characteristic brilliant red color. Minor amounts of strontium compounds are used in electrolytic zinc manufacturing where the carbonate is used to remove unwanted lead. Small quantities are consumed in greases, ceramics, soaps, alloys, and pharmaceuticals.

Specifications

Typical celestite specifications for the carbonate and nitrate markets are shown in Table 1. Specifications of manufactured carbonate, sold in two grades, one to the glass industry and the other for ceramic ferrites, are shown in Table 2. Both grades of carbonate are shipped in palletized 50-lb bags. Glass grade material is shipped in bulk by rail.

TABLE 1—Celestite Specifications

	For Carbonate	For Nitrate
$SrSO_4$	90% min.	95% min.
$CaSO_4$	—	1.5% max
$BaSO_4$	2% max	2% max
F	0.1% max	—
Size	6 in. max	minus 6 in., plus ¼ in.

Manufacturing Processes

The following description is adapted from Anon. (1980).

The manufacturing process for strontium carbonate and strontium nitrate chemicals is conceptually simple. Strontium nitrate is made from strontium carbonate in two additional steps. The production process begins with celestite, the natural form of strontium sulfate. In the first step the ore is crushed, ground, and mixed with coke. This mixture is then fed into a kiln where the ore is reduced at high temperatures to strontium sulfide; the coke is added to the ore as a source of energy and carbon. The carbon combines with the oxygen

TABLE 2—Specifications for Manufactured Carbonate Made from Celestite

	Glass Grade	Ceramic Grade
$SrCO_3$	96% min.	96% min.
$BaCO_3$	3% max	1.5% max
$CaCO_3$	0.5% max	—
Total S	0.4% max	0.4% max
Fe_2O_3	0.01% max	—
Na_2CO_3	1.0% max	—
Size (Tyler mesh)		
on 10 mesh	0.0%	no size specification
on 14 mesh	2.0% max	
on 100 mesh	70-85%	
on 150 mesh	95% min.	
thru 150 mesh	5% max	

* Basic Materials Planning, Materials and Logistics Department, E.I. du Pont de Nemours & Co., Wilmington, DE.

of the sulfate group and escapes as carbon dioxide. The sulfide is removed from the kiln, purified, and dissolved in water. The sulfide solution is then reacted with either sodium carbonate (soda ash) or carbon dioxide to produce the carbonate. The reactions are as follows:

The soda ash method:

$$SrS + Na_2CO_3 = SrCO_3 + Na_2S$$

Carbon dioxide method:

$$SrS + CO_2 + H_2O = SrCO_3 + H_2S$$

In both reactions, the carbonate forms as a solid which is dried and screened. The reaction conditions and drying techniques determine the particle size and, thus, whether the product will be ceramic or glass grade.

The coproducts obtained in both methods are marketable. The sodium sulfide solution obtained in the soda ash method is concentrated, dried, and packaged in 50-lb palletized bags for shipment. The hydrogen sulfide obtained in the carbon dioxide method can be converted to sodium sulfide, sodium sulfahydrate, or ammonium sulfide for market.

Strontium nitrate is produced by mixing strontium carbonate with concentrated nitric acid. The product is then crystallized and dried.

The production process for strontium carbonate is basically continuous. Production is controlled either by how long the equipment is run or by the flow rate used. The equipment in use generally dates back to the 1940s and 1950s. Although improvements in operating efficiency are being made, there have been no substantial technological improvements since then.

Markets

In the United States, principal consumers of celestite for the manufacture of carbonate are Chemical Products (CPC) of Cartersville, GA, and FMC of Modesto, CA, each with an estimated capacity of about 15,000 stpy. Companies which have discontinued carbonate manufacture include Foote Mineral and Sherwin-Williams, the latter having shut down in 1972 after Kaiser Chemical opened a new facility in Canada. US production of carbonate dropped from 20,000 st in 1971 to a low of about 15,000 st in 1975 at the time of maximum impact of Kaiser's Canadian production, then rose to about 30,000 st in 1978 after Kaiser ceased operation. Consumers of carbonate for glassmaking have included Corning, Owens-Illinois, and RCA. Consumers of electronic grade to make ferrites and titanates include Arnold Engineering, Alan Bradley, Colt Industries, Indiana General, and Stackpole Carbon.

Since Du Pont withdrew in mid-1975, the only US consumer of celestite for the manufacture of nitrate is FMC. FMC sales increased in 1978, dropped slightly in 1979, then declined sharply in January through June 1980 compared to the same period in 1979, while end-of-year inventories quadrupled from 1977 to 1979. US consumption of nitrate is about 15,000,000 lb per year, of which 12,000,000 lb are produced in the US.

Ground celestite is consumed at a rate of about 2,000 stpy by electrolytic zinc manufacturers in the US and Canada.

Kali-Chemie AG is a major producer of strontium carbonate in West Germany, accounting for essentially all US imports since 1978, following the closure of Kaiser's Canadian plant in late 1976. A subsidiary of Kali-Chemie, Societa Bario e Derivati SpA (SABED), produces strontium nitrate in Italy, as does Ammi Bario. SABED has accounted for virtually all US nitrate imports since it started production in 1978, making Kali-Chemie in effect the main foreign supplier of both strontium chemicals to the United States. SABED exports nearly all of its production to the US as a supplier to Olin since mid-1978 when Olin's plant at Peru, IN, switched from domestic to imported nitrate. Olin's other Signal Products Division plant at Morgan Hill, CA, continues to be supplied by FMC.

Imports of carbonate from West Germany rose sharply from 1977 to 1979, then declined steeply in January through June 1980, increasing from 2,300,000 lb in 1977 to 7,700,000 lb in 1979, but dropping from 2,600,000 lb January through June 1979 to 400,000, January through June 1980. The value rose from \$364,000 in 1977 (\$0.159 per lb) to \$1,498,000 in 1979 (\$0.195 per lb) and dropped from \$472,000 January through June 1979 (\$0.181 per lb) to \$86,000 January through June 1980 (\$0.217 per lb).

US imports of nitrate have switched from West Germany to Italy, dropping from 201,000 lb in 1977 to 2,000 lb in 1979 from West Germany, and rising from zero in 1977 to 3,100,000 lb in 1979 from Italy. Value of West German imports was US \$61,000 in 1977 (\$0.303 per lb). Value of Italian imports was

US $128,000 in 1978 ($0.249 per lb) and US $792,000 ($0.257 per lb) in 1979.

US exports of strontium chemicals is a very small percentage of production, essentially all by FMC.

Consumption

US consumption of strontium (as Sr) was about 40,000,000 lb, roughly divided as follows:

Glass	65%
Magnets	5%
Pyrotechnics	15%
Metal refining	4%
Pigments and Fillers	4%
Other	7%

North America probably accounts for three-fourths of world consumption, the rest being divided between Europe and Japan.

Substitutes and alternative materials have not been the subject of scrutiny because the total strontium mineral market is small and supplies are adequate, leaving little incentive for attempts at penetration by other materials.

Production of Celestite

World production of strontium minerals is shown in Table 3.

England: In England, mining is in the vicinity of Yate, near Bristol, where lumps of celestite are found near the ground surface in a clay matrix derived from the Keuper marl formation. Commercial development of these deposits began in the 1880s and reached a peak during the purchases by the US government for stockpiles in the early 1960s. English China Clays, a large producer of china clay, succeeded Bristol Minerals Co. as the producer from these deposits in the late 1960s. Mining methods introduced by Bristol Minerals and continued by English China Clays include stripping of top soil, followed by excavation using front-end loaders, hand-picking of celestite lumps, and transportation to a wash plant for removal of clay and sizing. Final product resulting from this simple operation is 95 to 96% $SrSO_4$. Export is through nearby Bristol Channel ports.

R. F. Bennett of Albright & Wilson, Ltd. and T. G. Pleasants of English China Clays, Ltd., in an informal paper, reviewed the history of strontium mineral markets in England, citing a number of potential uses which failed to materialize or which enjoyed a period of prosperity then declined. An example of the latter case of a prosperous market was the early use of strontium compounds in recovery of sugar from beet molasses, particularly in Germany in the period following 1890. During World War I production was stimulated briefly for pyrotechnics. Use in glass to replace red lead was the subject of an early patent but did not become a significantly large use. World War II gave a brief stimulus for pyrotechnics, similar to that of World War I. Use of strontium chloride in the 1950s for case hardening

TABLE 3—Strontium Minerals: World Production, by Country (Short Tons)

Country*	1976	1977	1978	1979 p	1980 e
Algeria	7,147	r5,622	6,418	e6,000	6,000
Argentina	2,264	r925	1,317	134	†209
Canada e	13,200	—	—	—	—
Iran e‡	6,000	11,000	16,535	r11,000	5,500
Italy	e770	e770	402	1,866	1,300
Mexico	24,424	50,302	36,563	re38,500	33,000
Pakistan	665	402	239	680	670
Spain	re9,100	12,120	15,430	19,840	20,000
Turkey	e7,000	e18,300	16,038	9,058	10,000
United Kingdom	5,952	5,622	4,740	6,600	5,500
Total	r76,522	r105,063	97,682	93,678	82,179

Source: Taylor, 1981a.

e Estimated. p Preliminary. r Revised.

* In addition to the countries listed, mainland China, the Federal Republic of Germany, Poland, and the USSR produce strontium minerals, but output is not reported quantitatively and available information is inadequate for formulation of reliable estimates of output levels.

† Reported figure.

‡ Year beginning March 21 of that stated.

of steel was short-lived, as was use of the sulfide for luminous paints. The sulfide was also used as a depilatory but was superseded by thioglycollic acid. An abortive attempt was made to employ strontium compounds to impart a bright flame color to coal briquettes but high cost discouraged this use. Use in ferrites is growing. Europeans do not use it in TV tube face-plate glass because of a finding by European TV manufacturers that x-ray absorbent glass is not required if the circuitry is properly designed. A minor but growing use for strontium chromate has been found in anticorrosive paint, especially for aircraft. Expectations for its use in water-soluble primers for automobiles have not been realized. In Europe strontium compounds await a major development, meanwhile experiencing a slow growth. Future of the mines is clouded by the environment of suburban developments and consequent zoning restrictions.

Mexico: Mexico supplied 99% of US imports of celestite from 1976 through 1980. The main producer is Minera Valenciana, headquartered in Torreon, with mines in Paila district where the celestite occurs as lens-shaped bodies concordant with bedding of the gently dipping Cretaceous limestone host rock. Lenses vary in thickness, averaging about 2 m. The ore is mined underground where horizontal tunnels connect to room-and-pillar stopes in the ore lenses. Ore is trucked to the railhead at Marte which is about 50 km south of the mines and 200 km east of Torreon. In contrast to the situation in England, described above, there has been little attempt to develop markets for celestite within the country. Essentially all of the sales are to US consumers, with FMC and Chemical Products being the principal buyers. About 2,000 stpy are ground at Brownsville for sale to electrolytic zinc manufacturers.

Spain: In Spain, celestite is produced from the Aurora mine at Montevive, near Granada, by Bruno SA, a company headquartered in Madrid. The mine is open pit and the celestite is selectively mined from the baritic gypsiferous host rocks. The principal products are -4 in., -2 in. lumps, and smaller quantities of ground-bagged material. The products are trucked a distance of 85 km to the port of Motril for export, currently two-thirds to Japan and the rest mainly to West Germany, France, and the US. The Aurora mine first produced during World War II, after which it remained idle until 1967 when it was reopened with initial shipments to the United States. Recent expansion has increased its capacity to a reported 40,000 stpy, typically assaying 93 to 94% $SrSO_4$.

Bruno SA has announced construction of a plant for manufacturing strontium chemicals with a capacity of between 3,000,000 and 4,000,000 lb per year of strontium carbonate located near the port of Motril using ore from the Aurora mine and operating under a separate company, Proimsur SA, of the same ownership as Bruno SA.

A report by the Spanish government refers to another deposit near Fortuna in Murcia, called Nuestra Senora, where prospect pits indicate possibilities of extracting ore grading 98% $SrSO_4$.

United States: In the United States, all known deposits are uneconomic in today's markets. Historically production has been sporadic and short-lived. Surplus from the US stockpiles of material purchased mainly from England and Mexico has been sold from time to time by the General Services Administration to domestic consumers. The US imported about 21,750 st of strontium mineral (as Sr) in 1981, which equates to about 54,120 st of celestite ore, up from 15,670 st in 1976 (about 39,000 st celestite).

It is noteworthy that US manufacturers of strontium compounds are wholly dependent on foreign sourecs of celestite and have significant competition from Italy and West Germany in the domestic market for finished strontium compounds.

Canada: In Canada at Loch Lommond, Cape Breton County, Nova Scotia, Kaiser Chemical Co. attempted to develop an open pit mine in ore averaging 50 to 60% $SrSO_4$, upgrading it by froth flotation to a concentrate assaying more than 88% $SrSO_4$ for use in making strontium compounds in a 60,000,000 lb-per-year facility at Point Edward.

The mineralization occurs in rocks of the lower Windsor group of Mississippian age as gently dipping mantos concordant to bedding of the limestone-gypsum host rocks into which the ore mantos grade laterally. Open-pit mining started in September 1969. The concentrator first produced celestite concentrates in March 1971 and continued until its closure in August 1976.

Other: In Turkey, the largest celestite producer, Barit Moden Turk, is expanding production at its Silvas operation from the current 20,000 stpy to 44,000 stpy capacity.

In Iran, Nakhjir, a new celestite deposit located about 190 km southeast of Tehran in the

Dasht-e-Kavir desert, is being developed by the Strontium Co., a wholly owned subsidiary of the Simiran group of companies. Production is estimated to be about 5,500 stpy of hand-sorted celestite.

Elsewhere in the world, celestite has been offered from time to time from production in Algeria, Argentina, Italy, Morocco, and Pakistan,

Summary

In summary, several points of interest are:

1. Celestite is important in glass manufacturing for TV face plates in the US because of the x-ray shielding qualities of strontium carbonate, while European TV manufacturers have claimed such shielding is unnecessary if circuitry is properly designed.

2. A growing market for strontium carbonate is provided by its use in making ceramic ferrites which are demonstrably superior in performance and are achieving acceptance by industry.

3. Use of celestite in nitrate for pyrotechnics (mainly for signal flares) grew from 1977 to 1978, but dropped sharply by 1980, declining with the business recession.

4. Numerous other uses of strontium compounds are identified, none of which are presently commercially important.

5. On balance, the trend of celestite consumption is expected to be modestly upward as strontium ferrites gain acceptance and TV manufacturers for the US market, both in the US and Japan, continue to use strontium carbonates in face plates.

6. While deposits in the US are not economic, mining of celestite from well-established sources abroad is adequate for present world needs and other deposits are known which could be used to meet new or expanded requirements.

Bibliography and References

Anon., 1969, "New Look Strontium Industry," *Industrial Minerals,* No. 27, Dec., pp. 11–20.

Anon., 1973, "The Economics of Strontium," Roskill Information Services, Ltd. report, Mar., 91 pp.

Anon., 1980, "Barium Carbonate and Strontium Carbonate from the Federal Republic of Germany and Strontium Nitrate from Italy," Investigation Nos. 731-TA-31-33 (Preliminary), USITC Publication 1105, US International Trade Commission, Oct.

Anon., 1980a, "Bruno Opens Strontium Compounds Plant," *Industrial Minerals,* No. 154, July, p. 11.

Crowell, G.D., 1973, "Mineralogy of Strontium," *Proceedings,* International Conference on Strontium Containing Compounds, Nova Scotia Technical College, pp. 245–255.

Fulton, R.B., 1971, "World Economics of Selected Industrial Minerals," *Centennial Volume,* AIME, New York, pp. 74–87.

Ireland, J.R., 1973, "Hard Ferrites in P.M. Motors," *Proceedings,* International Conference on Strontium Containing Compounds, Nova Scotia Technical College, pp. 33–46.

Massone, J., 1982, "Technology and Uses of Barium and Strontium Compounds," *Industrial Minerals,* No. 177, June, pp. 65–69.

Schiebel, W., 1978, "New Strontium Deposit in Iran," *Industrial Minerals,* No. 132, Sept., pp. 54–57, 59.

Stein, D.L., 1973, "Extraction of Strontium Values from Celestite Concentrate at the Kaiser Plant in Nova Scotia," *Proceedings,* International Conference on Strontium Containing Compounds, Nova Scotia Technical College, pp. 1–9.

Taylor, H.A., 1981, "Strontium," *Mineral Commodity Summaries, 1981,* US Bureau of Mines, pp. 152–153.

Taylor, H.A., 1981a, "Strontium," *Minerals Yearbook, 1980,* US Bureau of Mines, pp. 944–947.

Thomas, I.A., 1973, "Celestite," Mineral Dossier No. 6, Her Majesty's Stationery Office, London, 26 pp.

Wininger, D.C., 1970, "Strontium," *Minerals Yearbook, 1970,* US Bureau of Mines, Vol. 1, pp. 1233–4.

Sulfur

JAMES M. BARKER *

Sulfur is a nonmetallic element of great physical and economic importance to the world. It is widely but sparingly distributed throughout the hydrosphere, lithosphere, and biosphere. Sulfur is the tenth most abundant element in the universe and eighth most abundant in the sun. In the earth's crust, which is mainly igneous, it ranks 14th, but is eighth or ninth in sediments (Kaplan, 1972; Stanton, 1972). It apparently is very abundant in the core of the earth, because it is a strong chalcophile, and it is fifth in abundance in stony and iron meteorites (Field, 1972) where it occurs as trolite (FeS).

In ancient times sulfur was called *brimstone*, literally "burning stone." Today the term *brimstone* is used interchangeably with the term *elemental sulfur*. Shelton (1979) summarized most of the common definitions of terms, grades, and specifications for the sulfur industry as follows (with minor additions):

Sources of Sulfur:

Combined sulfur—Sulfur that occurs in nature combined with other elements.

Cupriferous pyrites—Pyrite containing relatively minor amounts of copper sulfide minerals.

Hydrogen sulfide—A toxic gas (H_2S) occurring in natural gas and petroleum.

Involuntary sulfur—Sulfur produced primarily as a result of legislative or process mandates.

Native sulfur—Sulfur that occurs in nature in the elemental (uncombined) form.

Nonferrous metal sulfides—Copper, lead, zinc, nickel, and molybdenum sulfides that are processed for their metal content.

Organic sulfurs—Complex organic sulfur compounds occurring in petroleum, coal, oil shale, and tar sands.

Pyrites—Iron sulfide minerals that include pyrite, marcasite, and pyrrhotite.

Sulfate sulfurs—Sulfur in anhydrite and gypsum.

Voluntary sulfur—Sulfur produced in response to market forces.

Basic Sulfur Products Produced and Marketed:

Acid sludge—Contaminated sulfuric acid usually returned to acid plants for reconstitution.

Brimstone—Synonymous with crude sulfur.

Bright sulfur—Crude sulfur free of discoloring impurities and bright yellow in color.

Broken sulfur—Solid crude sulfur crushed to -8 in.

Byproduct sulfuric acid—Sulfuric acid produced as a byproduct of a metallurgical or industrial process, generally relating to combined sulfur sources.

Crude sulfur—Commercial nomenclature for elemental sulfur.

Dark sulfur—Crude sulfur discolored by minor quantities of hydrocarbons ranging up to 0.3% carbon content.

Elemental sulfur—Processed sulfur in the elemental form produced from native sulfur or combined sulfur sources, generally with a minimum sulfur content of 99.5%.

Formed sulfur—Elemental sulfur cast or pressed into particular shapes to enhance handling and to suppress dust generation and moisture retention.

Frasch sulfur—Elemental sulfur produced from native sulfur sources by the Frasch mining process.

Hydrogen sulfide—Associated with natural gas or petroleum and produced at refineries and coke oven plants; used to make sulfuric acid or recovered sulfur.

Liquid sulfur—Synonymous with molten sulfur.

Liquid sulfur dioxide—Purified sulfur dioxide compressed to the liquid phase.

Molten sulfur—Crude sulfur in the molten phase.

* Industrial Minerals Geologist, New Mexico Bureau of Mines and Mineral Resources, Socorro, NM.

This chapter is based in part on Gittinger, 1975.

Prilled sulfur—Solid crude sulfur in the form of pellets produced by cooling molten sulfur with air or water.

Recovered sulfur—Elemental sulfur produced from combined sulfur sources by any method.

Slated sulfur—Solid crude sulfur in the form of slatelike lumps, produced by allowing molten sulfur to solidify on a moving belt.

Specialty sulfurs—Prepared or refined grades of elemental sulfur that include amorphous, colloidal, flowers, precipitated, wettable, flour, or paste sulfur.

Sulfur ore—Unprocessed ore containing native sulfur.

Sulfuric acid—Sulfuric acid of commerce as produced from all sources of sulfur, generally reported in terms of 100% H_2SO_4 with a 32.69% sulfur content.

Sulfur is produced commercially from one or more sources in about 70 countries (Shelton, 1980b). World production of sulfur in all forms totaled 58 Mt * in 1980 (Barker and Cochran, 1981). The per capita consumption of sulfur often is used as a reliable index of the level of industrial development and the activity of a nation's economy (Field, 1972).

Sulfur resources are abundant and exist throughout the world, but the extent to which they can be classified as reserves is greatly circumscribed by prevailing prices and technology. The 1980 world estimate of sulfur reserves totals 1.765 Gt (Shelton, 1980a), about ten times that calculated for the United States (Table 1).

Resources of elemental sulfur in evaporite and volcanic deposits and sulfur associated with natural gas, petroleum, tar sands, and metal sulfides amount to about 6.385 Gt. The sulfur contained in gypsum and anhydrite is almost limitless and some 600 Gt are contained in coal, oil shale, and shale rich in organic matter, although low-cost methods have not been developed to recover sulfur from these sources. The domestic resource is about 5% of the world total.

Native deposits, sulfides, oil, and gas contributed, respectively, 32, 36, 30.1, and 0.9% to world sulfur production in 1973 (Anon., 1974a). Of the sulfur produced from native deposits, salt domes accounted for 53% and evaporite basin (stratiform deposits), 45%. Volcanic deposits (2%) are of little significance and are mined mainly for local consumption in Italy, Turkey, and the Andes Mountains of South America.

The past two decades have seen continued shifts in the world's sulfur sources. In 1950, native deposits and sulfides each supplied about one-half the world's production. During the 1950s, sulfur recovered from oil and natural gas grew rapidly, representing about 10% of the total by 1960. Since then, the petroleum industry has continued to increase its relative importance as a source of sulfur, representing 18% in 1965 and 30% in 1973 for the world. It comprised more than 34% for 1979 in the United States. The proportion of sulfur produced by the Frasch or similar processes has declined in importance to 52% of all sulfur in the United States for 1980. The economics of the impact of recovered sulfur on the US Frasch sulfur industry is covered in detail by Rieber et al. (1981).

Elemental sulfur is produced from salt dome deposits in the Gulf Coast region of the United States and the Isthmus of Tehuantepec in Mexico, and from stratiform evaporite deposits in the United States (west Texas), Poland, the USSR, and Iraq. It is recovered from native sulfur ore of volcanic origin in about eight countries and is recovered in large amounts from sour natural gas in ten countries, of which Canada and the United States are the principal producers with Saudi Arabia due to become a major producer in the 1980s. Sulfur is derived from petroleum refinery gases in about 50 countries, of which the United States and Japan are the largest producers. The leading 15 countries produce 95% of total output.

About 30 countries produce pyrites, of which the USSR, Japan, Spain, Italy, and South Africa are the largest. About the same number of countries recover sulfur largely in the form of byproduct sulfuric acid from nonferrous smelter gases or petroleum refinery gases. The USSR, Japan, and the United States are the leading producers. Sulfate minerals are utilized as a source of sulfur in a few countries.

Elemental sulfur represented 64.5% of the 56.1 Mt of sulfur produced in 1980. The western world accounted for 63.9% of all sulfur forms and 71.5% of elemental sulfur. World production of all forms grew at a rate of 6% per year from 1950 to 1971 and elemental sulfur grew at a rate of 6.5% per year. Growth in the western world was about 5% per year for all forms and 5.5% per year for elemental sulfur. The Communist world recorded growth of 11% per year for all forms and 16.5% per

* Sulfur statistics in this chapter are given in metric tons (t). t × 1.102 311 4 = st (short tons); t × 0.984 206 = lt (long tons).

TABLE 1—Identified World Sulfur Resources,* Mt

Type of deposit	Reserves	Other	Total
United States			
Salt domes and evaporites	90	25	115
Nonferrous metal sulfides	55	30	85
Natural gas	20	10	30
Petroleum	10	10	20
Pyrites	—	50	50
Volcanic	—	20	20
Tar sands	—	10	10
Total US	175	155	330
World			
North America			
United States	175	155	330
Canada	250	2,000	2,250
Mexico	90	60	150
Other	5	—	5
Total	520	2,215	2,735
South America, Total	30	30	60
Europe			
USSR	250	450	700
Poland	150	450	600
France	30	10	40
Germany, Federal Republic of	30	5	35
Spain	30	450	480
Italy	15	25	40
Other	185	285	470
Total	690	1,675	2,365
Africa, Total	20	—	20
Asia			
Japan	10	40	50
Iraq	150	50	200
Near East	250	400	600
China, Mainland	25	50	75
Other	50	200	250
Total	485	740	1,175
Oceania, Total	20	10	30
World Total	1,765	4,670	6,385

Source: Shelton, 1980a.
* Derived in consultation with US Geological Survey.

year for elemental sulfur. Fig. 1 shows the trend in production for the western and Communist worlds from 1950 to 1981.

Elemental sulfur and, to a lesser extent, pyrites are internationally traded so that the geographic pattern of consumption differs somewhat from that of production. Table 2 shows total sulfur production, from 1976 to 1980, by major countries that produce over 0.5 Mt of sulfur from all sources.

Commercial grades of elemental sulfur are classified as *crude sulfur* (run-of-mine, bright or dark); and *recovered sulfur*. Both grades are shipped in liquid or solid form. All grades are guaranteed 99.5% pure, and the only allowable impurity of any significance is usually carbon in the form of hydrocarbons in solution in the sulfur. Dark sulfurs may contain up to 0.3% carbon. All grades are free of arsenic, selenium, and tellurium.

Some sulfur is solidified (formed) and shipped as slates or prills rather than in the more common broken lumps in an attempt to minimize dust problems during handling.

Specialty grades of sulfur are prepared by treating commercial grades to impart desirable

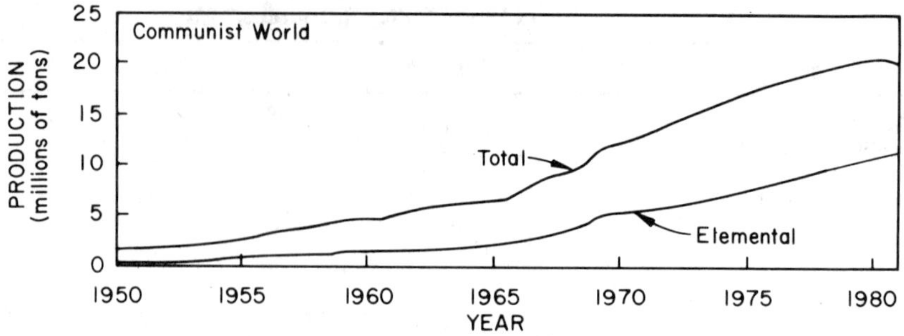

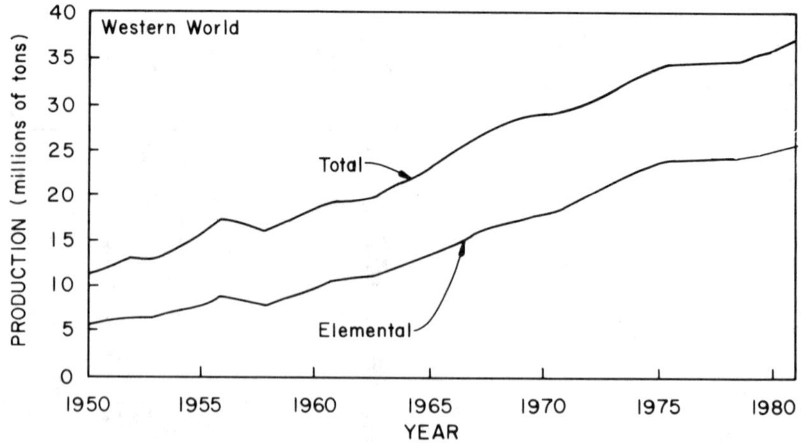

FIG. 1—*Sulfur production for the western and Communist worlds, 1950 to 1981. Source: updated from Gittinger, 1975.*

characteristics. Some special grades such as refined candles, cones, sticks, and sublimed sulfur are prepared from distilled crude sulfur.

Finely-ground crude sulfur is sold as "flour sulfur," "rubber-makers sulfur," "agricultural dusting sulfur," etc. Conditioning agents may be added to produce wettable grades. Special processing is involved in the production of insoluble sulfur, a material which contains up to 90% of the amorphous allotrope, which is insoluble in carbon disulfide.

Commercial grades of pyrite usually contain a minimum of 45% sulfur unless the copper content is at least 1%. If the pyrite price is predicated on realizing values for iron content, impurities such as arsenic and selenium should be at very low levels or absent. Residual copper, lead, and zinc contents should not exceed 0.05% if the iron values are to find a ready market in the United States. Some European and Japanese steel mills accept material with combined residuals of 0.3 to 0.4%, at a discount.

The sulfur content of pyrite sells at a considerable discount from the brimstone price level because capital and operating costs of acid plants based on pyrite are much higher than those for brimstone plants.

Mineralogy

Elemental sulfur occurs in subsurface sedimentary deposits associated with gypsum and anhydrite in salt domes and stratiform evaporites. It also is common in volcanic, especially solfataric, associations and at mineral or thermal springs.

Sulfur occurs in compounds as sulfates (Staples, 1972a) such as gypsum or anhydrite. It is also found in sulfides (Staples, 1972b), both ferrous (pyrites, phyrrhotite, etc.) and nonferrous (sulfides of copper, lead, zinc, niçkel, etc.). Organic materials often contain sulfur, e.g., hydrogen sulfide in petroleum and natural gas, organic compounds in oil and oil

TABLE 2—Major World Production of Sulfur In All Forms, By Country and Source (1000 t)

Country and source	1976	1977	1978	1979	1980
Canada					
Pyrite	15	12	5	12	12
Byproduct					
Metallurgy	705	736	676	667	903
Natural gas	6,241	6,475	6,248	5,935	6,000
Petroleum	200	160	200	200	190
Tar sands	100	100	118	213	300
Total	7,261	7,483	7,247	7,027	7,405
China					
Mainland					
Native	150	200	200	200	200
Pyrite	900	1,252	1,605	1,682	1,700
Byproduct, all sources	300	300	350	400	400
Total	1,350	1,752	2,155	2,282	2,300
France					
Byproduct					
Natural gas	1,737	1,911	1,900	1,940	1,839
Petroleum	88	89	86	90	88
Unspecified	143	160	160	160	150
Total	1,968	2,160	2,146	2,190	2,077
Germany, Federal Republic of					
Pyrite	233	235	221	203	200
Pyproduct					
Metallurgy	390	385	348	361	380
Natural gas	460	624	666	690	700
Petroleum	119	186	190	214	220
Unspecified	161	197	195	343	300
Total	1,363	1,627	1,620	1,811	1,800
Iraq					
Frasch	582	620	600	550	600
Byproduct, petroleum and natural gas	40	40	40	70	60
Total	622	660	640	620	660
Italy					
Native	35	36	16	16	23
Pyrite	366	371	330	330	331
Byproduct, all sources	211	259	299	225	250
Total	612	666	645	571	604
Japan					
Pyrite	471	389	327	303	300
Byproduct					
Metallurgy	1,252	1,336	1,296	1,350	1,300
Petroleum	926	1,100	1,105	1,241	1,300
Total	2,649	2,825	2,728	2,894	2,900
Mexico					
Frasch	2,054	1,723	1,818	2,025	2,102
Byproduct					
Metallurgy	75	80	100	100	150
Petroleum and natural gas	96	146	168	249	300
Total	2,225	1,949	2,086	2,374	2,552
Poland					
Frasch	4,341	4,321	4,546	4,310	4,000
Native	550	450	505	520	500

TABLE 2—Major World Production of Sulfur In All Forms, By Country and Source (1000 t) (Cont'd)

Country and source	1976	1977	1978	1979	1980
Byproduct					
Metallurgy	239	314	315	310	300
Petroleum	25	35	35	35	30
Gypsum	55	30	20	20	20
Total	5,210	5,150	5,421	5,195	4,850
Romania					
Pyrite	375	395	400	425	450
Byproduct, all sources	98	110	120	130	140
Total	473	505	520	555	590
Saudi Arabia					
Native	1	1	1	1	1
Byproduct, petroleum and natural gas	12	12	14	125	700
Total	13	13	15	126	701
Spain					
Pyrite	1,052	1,102	1,071	1,019	1,100
Byproduct					
Metallurgy	123	129	117	120	125
Petroleum	4	5	10	10	12
Coal (lignite) gasification	1	2	3	3	3
Total	1,180	1,238	1,201	1,152	1,240
United States					
Frasch	6,365	5,915	5,648	6,357	6,390
Pyrite	291	169	301	400	322
Byproduct					
Metallurgy	957	960	1,103	1,167	1,003
Natural gas	1,298	1,426	1,753	1,760	1,730
Petroleum	1,890	2,198	2,309	2,310	2,316
Unspecified	78	59	61	107	78
Total	10,879	10,727	11,175	12,101	11,839
USSR					
Frasch	500	500	800	800	900
Native	2,200	2,400	2,700	2,700	2,800
Pyrite	3,300	3,500	3,500	3,500	3,550
Byproduct					
Coal	40	40	40	40	40
Metallurgy	2,040	2,180	2,210	2,210	2,310
Natural gas	870	920	1,100	1,100	1,100
Petroleum	190	200	200	200	200
Total	9,140	9,740	10,550	10,550	10,900

Source: Shelton, 1980b.

sands, and pyrite and organic compounds in coal and oil shale.

The atomic weight of sulfur is 32.064, with an atomic number of 16. It has four stable isotopes—S^{32}, S^{33}, S^{34}, and S^{36}—with percent abundances of 95.1, 0.74, 4.2, and 0.016, respectively. The terrestrial S^{34}/S^{32} ratio varies naturally by at least 4.5% from a mean (0.0) close to that of meteors. Sulfur also has four oxidation states (+6, +4, 0, and −2), of which the S^{-2} sulfide ion (ionic radius 1.84 Å) and the S^{+6} sulfate cation (0.30 Å) are most important in minerals.

Sulfur occurs as both orthorhombic crystals and amorphous masses. Crystals commonly are dipyramidal and sometimes thick tabular on {001} or sphenoidal. Other habits are irregular masses of imperfect crystals, incrustations, stalactitic/stalagmitic, spherical, reniform, or earthy masses. Cleavage is imperfect to none, and twinning is rare. Sulfur is brittle to slightly sectile, with conchoidal to uneven fracture and resinous to greasy luster. It has a hardness of 1.5 to 2.5 and a density of 2.07 (range: 2.05 to 2.09). Color varies widely, from straw yellow through red to gray with a white streak. Pure, cool sulfur is yellow, but it is often discolored by high temperatures or impurities, especially bitumen or clay. Orthorhombic sulfur [alpha (α) sulfur] is stable to 95.5°C whereupon it slowly converts to monoclinic [beta (β) or gamma (γ) sulfur]. The melting point is 118.9°C and the boiling point is 444.6°C, with viscosity increasing markedly with increasing temperature over this range. Sulfur is easily ignited in air, burning at 270°C with a blue flame and evolution of sulfur dioxide. It is nearly insoluble in water, is nonhydroscopic, and is unaffected by most acids. It is soluble in carbon disulfide and some oils (Berry and Mason, 1959).

Solid sulfur has a number of physical properties of interest to industry. The explosive limits of sulfur dust in air are about 35 g/m^3 (minimum) to about 1400 g/m^3 (maximum). The auto-ignition temperature of dust in air is 190°C (dust clouds) and 220°C (undispersed dusts). Corrosivity is nil for dry sulfur with slow generation of sulfuric acid when wet. Theoretical density is 1922 kg/m^3 with practical bulk densities ranging from 120 to 1393 kg/m^3. Specific heat is 0.67 kJ/kg at −17.8°C, increasing to 4.98 kJ/kg at 93°C. Vapor pressure is less than 0.1 μm Hg at 20°C (Anon., 1968).

Molten sulfur properties at atmospheric pressure include some of interest to industry (Tuller, 1954). Specific gravity at 132°C is 1.79 (basis: H_2O at 20°C = 1) and relative viscosity, same basis, is 9.0 at 132°C. The flash point (modified Cleveland open cup) is 168.3 to 187.3°C (Anon., 1966).

Historical Aspects

Production of sulfur on a commercial basis commenced in Sicily early in the 15th century. For 150 years Sicilian brimstone had a virtual monopoly on the world supply. In 1838, French interests gained control of the Sicilian sulfur industry and attempted to triple sulfur's price, whereupon countries dependent on Sicilian sulfur developed other sources. Pyrites gradually displaced Sicilian brimstone for the manufacture of sulfuric acid, except in the United States where import duties on the iron and copper in pyrites held back their use. These duties were removed in the 1890s, and by the turn of the century pyrites had become the principal acid-making material in the United States. At the same time, the United States was becoming an important producer of sulfur with the successful development of the Frasch hot-water process and its application to mining sulfur from salt dome deposits of the Gulf Coast. However, pyrites held the dominant position as an acid-making material for the US chemical industry until shipping problems during World War I forced a major conversion to Frasch sulfur. The development of the contact sulfuric acid process also favored the use of brimstone.

World War II disrupted sulfur production outside North America and recovery after the war was slow. Many foreign countries turned to US Frasch sulfur for rehabilitation and expansion of their acid industries. As foreign consumers became more dependent on US sulfur, the requirements of the US acid industry were also expanding. The US Frasch industry expanded production but in addition was forced to depend upon inventories to meet demand.

By 1950, when the Korean War broke out, inventories of Frasch sulfur had been reduced to about a six-month supply. The industry was producing enough sulfur to supply the needs of US consumers, but not enough to supply consumers in allied countries. In 1951, the US government ordered producers to ship specified allocations to its allies. Later the government limited domestic consumption, adopted a policy of maintaining existing inventories of mined sulfur, and controlled prices. The sulfur shortage of the Korean War period did not last long. After the cessation of hostilities, controls on the industry were removed and, stimulated by shortages and high prices, new sources of sulfur were developed both in the United States and abroad.

The US Frasch producers brought in new mines in 1953 at Garden Island Bay and Damon Mound with Chacahoula and Nash Dome underway. Italy was second in world production, but competition from low-cost mines was causing a rapid decline there.

By 1954, Mexico had entered the Frasch market with the opening of San Cristobal and continued construction at Salinas and Jaltipan. Of even greater significance were the first expansions of Canada's recovered sulfur capacity, which began about 1950 and accelerated growth in recovered sulfur.

All this activity caused supplies to outstrip demand by the mid-1950s with downward pressure on prices. This was enhanced by Mexican increases, the appearance of recovered sulfur from France in 1957, and the work begun on gas pipelines in Canada.

Freeport commenced construction at Grand Isle in 1958 and the Mexican Frasch industry reached maturity. Recovered sulfur output continued to expand in North America. The most significant event in 1958 was the opening of Texasgulf's liquid sulfur terminal, the world's first. This began a strong trend in the direction of liquid sulfur handling. The long-term contract became common owing to high inventory costs of liquid sulfur as compared to storage of solid sulfur.

The massive increases in North American Frasch output coupled with increases in recovered sulfur output triggered a slack world market in 1959. Pyrite sulfur/acid producers suffered further declines. Domestic pyrites were important locally, such as in Japan and east Europe, but the trend was clear with the beginning of Polish development of Frasch deposits in the Tarnobrzeg region. French recovered sulfur reached world markets via the new terminal at Bayonne.

The first Polish sulfur from the Machow open pit appeared on the world market in 1960 with French and Canadian recovered output increasing. US producers restricted output in the face of oversupply, but the market remained soft. By 1963, Canadian and French recovered sulfur output each exceeded 1 Mt and the US share of world exports had fallen 19% in four years to 31%. International trade in liquid sulfur commenced with construction of terminals, for US output, at Rotterdam and Dublin.

A fertilizer and industrial boom led to tight sulfur supplies and higher prices, particularly in western Europe and the United States. Mexico surpassed the US as the dominant exporter in 1964. Non-Frasch sulfur reached 4 Mt in 1965 compared to 1.25 Mt in 1953, largely due to pollution control requirements.

Poland took a large step toward export dominance by opening Frasch mines at Grzybow in 1966 and Jesiorko in 1967. A large deposit in Culberson County in west Texas was discovered by Duval in 1967.

Increased recovered sulfur in the United States, Japan, West Germany, Iran, and Iraq helped boost world supplies past consumption. Producer stocks jumped 41% in 1969 alone. Recovered sulfur output exceeded Frasch output for the first time in 1970. In the 1970s recovered sulfur, whose output was related to nonmarket forces, dominated the sulfur market.

The early 1970s were characterized by oversupply somewhat mitigated by growth in demand and the forced vatting of Canadian sulfur via nondust legislation by the Vancouver Port Authority. Thus, stocks actually fell in 1972 even though Iraq brought the Frasch mine at Mishraq on stream.

By 1973, increased demand coupled with the high cost of centralizing scattered recovered sulfur output led to major price increases, even though supply statistically matched demand. Tight supply continued in 1974 with demand up 11%, mainly from additions to consumer stocks. For the first time in many years, the pyrite decline was reversed by increased consumption in a few pyrite-producing countires, primarily Spain, Portugal, and South Africa.

The first major cartel price offensives led to significant changes in the 1970s. The oil and phosphate cartels increased prices sharply in 1974 and 1975, causing industrial and fertilizer recessions. Consumers began drawing down the sulfur stocks they had built up in the previous two years. The combined result was plummeting demand with major cutbacks in the United States and Mexico. This was the beginning of the nearly continual decline in North American Frasch caused by aging mines and lack of exploration success.

A gradual market recovery began in 1977 while sulfur production stagnated. Slower growth cut demand as well as supply of recovered sulfur. After a period of adjustment in world economies, growth in fertilizer use built up the demand in sulfur to previous levels and beyond by 1979.

The major exporters of the late 1970s and early 1980s were Poland and Canada. Very high prices have developed since demand outstripped supply in 1979–1981. These high prices have reversed, at least for the short term, the decline in US Frasch production, but the United States remains a net importer. The net reliance on imports began in 1977. Canada began drawing down its vast reserve in vats in 1980, a trend likely to continue into the 1990s. Major sour gas discoveries in the Overthrust Belt in the United States could make it a net exporter by the mid-1980s. New output in recovered sulfur in Saudi Arabia coupled with that in the US should ease tight supplies by 1983–1985. The conversion to liquid sulfur has peaked and new forms, primarily prills, are now encroaching on the liquid market.

Exploration

Exploration for sulfur deposits is a difficult task, especially for Frasch deposits. They are scattered and give little surface indication of their presence. While the presence of salt domes seems to narrow the area, most are barren. Even when sulfur is present, it tends to be very steeply dipping and hence presents a small, easily-missed drilling target.

Stratiform deposits, while often larger and flatter than domes, do not have the easily found dome structure to guide exploration. In fact, searching for a stratiform sulfur deposit is as close to oil and gas exploration as mineral exploration allows. This derives from the genetic relationship between petroleum and biogenetic sulfur. Exploration centers on areas containing anhydrite with petroleum in or below it. Within the basins, traps identical to many petroleum traps are located and drilled. Surface venting of hydrogen sulfide has led to discovery. Hollister (1977), with expansion by Barker et al. (1979), summarized the generally known conditions for sulfur deposits as: (1) sulfate-bearing rocks, preferably thick and composed of anhydrite; (2) proximity to petroleum deposits; (3) hydrodynamic communication between sulfates and petroleum, which is often accomplished via faulting, jointing, or solution-derived porosity; (4) a stratigraphic or structural trap to contain the precursors of the sulfur-forming reaction; (5) a reducing environment where the petroleum and sulfate are biochemically metabolized; and (6) an oxidizing environment where the metabolic products are altered, yielding sulfur.

Tools useful in locating sulfur deposits are few. Gravity studies continue to draw interest both for stratiform and dome (including offshore) exploration. Anhydrite has a density of about 2.96 while sulfur bodies vary from about 1.80 to 2.20, yielding a potentially measurable negative gravity anomaly. The variation depends on interaction between sulfur content, void development, water, and the type of limestone matrix. Actual use has yielded few if any exploration successes. Refer to Haye (1969), Heintz and Alexander (1978), Schmoker and Robbins (1979), and Schmoker (1979) for discussions and additional references.

Other methods popular in varying degrees include traditional geologic modeling and mapping (Ruckmick et al., 1979) photogeology (structure and geochemical coloration), thermal IR (heat of reaction), seismic (especially shallow high resolution), geochemistry (including airborne spectrometry of gases), resistivity, induced polarization and, most successfully, the drilling rig and downhole logging. Surface geology and stratigraphy are useful at times, but many deposits have very subtle to no surface indications.

Exploration for volcanic sulfur follows the more traditional style. Exploration expressly for pyrites is very uncommon, because they are generally a byproduct. When done, however, methods useful to sulfide exploration in general are used.

Elemental Sulfur

Elemental sulfur occurs in and is produced from deposits of three general types: cap rock over salt domes, evaporite-basin (stratiform) deposits, and volcanic deposits.

Salt Domes

Salt domes and similar piercement structures are known to occur in various regions of the world, but sulfur deposits of economic interest have been discovered only in the salt domes of the Gulf Coast region.

The salt dome sulfur deposit typical of the Gulf Coastal Plain is depicted in Fig. 2. It shows an intrusive salt plug capped successively by barren anhydrite, sulfur-bearing limestone, and barren limestone cap rock, all enclosed in unconsolidated sediments.

Numerous references that discuss the origins of the salt dome, its cap rock, and associated sulfur may be found in the bibliographies of previous editions of this work (Lundy, 1949; Fogarty and Mollison, 1960; Gittinger, 1975). Interlocking scientific efforts of geologists, geophysicists, engineers, chemists, biologists, and nuclear chemists have contributed concepts or confirming data to the origin of salt dome sulfur deposits.

The salt that forms the domes was derived by evaporation of seawater to form a thick sedimentary deposit of salt, gypsum, and anhydrite many thousands of feet below the present surface of the Gulf Coast area. Paleontological studies of pollen and spores found in the salt indicate that the salt was deposited during the Late Triassic to Early Jurassic periods, about 180 million years ago. It is a layered deposit composed primarily of soluble NaCl (halite), with 5 to 10% insoluble material. The insoluble material is 99% $CaSO_4$ (anhydrite) and most of the remaining 1% is dolomite. Calcite, pyrite, quartz, limonite, hematite, celestite, and barite are present in trace amounts. Scarcity of potassium compounds, which concentrate in the last stage

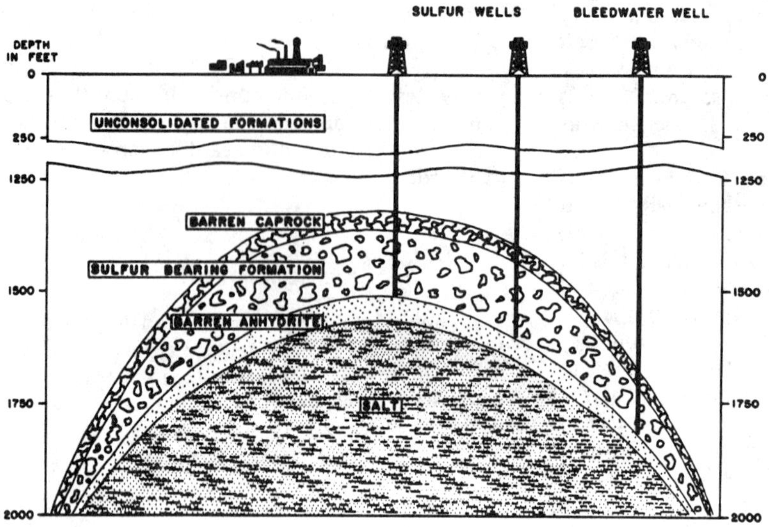

FIG. 2—*Typical cross section of a sulfur-bearing salt dome (Gittinger, 1975).*

brines before desiccation, indicates that evaporation probably was never complete.

Explanation for the formation of the domes from the original bed of salt is derived from a fluid-mechanics theory proposed by Nettleton (1934). The salt bed was deposited with a specific gravity of about 2.2, and because salt is only slightly compressible, this gravity would not have changed appreciably with burial. During subsequent geological time, the salt was buried beneath many layers of sediment. Burial and compaction of the sediments increased their specific gravity until at lower depths they had a specific gravity greater than the salt. Salt will flow when subjected to differential stress and Nettleton proposed that the salt acted like a liquid and flowed up through the overlying sediments driven by its lower specific gravity. An overburden of between 300 and 600 m was necessary before the gravity differential was great enough to initiate movement. Actual start of movement may have resulted from variations or weaknesses in the overburden, structures in the salt, or tectonism. There is a critical depth of overburden which, if exceeded, will stop the growth of a dome. A compressed sediment increases in strength as well as density. If the strength of the compressed sediment exceeds the buoyant force available for upward movement of salt, no movement is possible unless faulting or fracturing provides an avenue of escape.

The layered cap rock on top of the salt dome was originally thought to be of sedimentary origin, a remnant of overlying beds pushed up on the rising finger of salt. Work by Taylor (1938) and Goldman (1952) showed that the cap rock resulted from the residual accumulation of anhydrite present in the salt. As the salt dome rose through the basin sediments, it came into contact with zones of circulating ground water which dissolved away the salt, leaving the insoluble crystals of anhydrite to be compacted into a cap over the salt plug. Locally, the anhydrite may be hydrated to form gypsum.

Geologists noticed that sulfur and limestone occur as replacements of the anhydrite of the cap rock. The common association of oil and sulfur was also noted. This led to the hypothesis that hydrocarbons reduced the anhydrite to form sulfur and limestone, as represented by the following equations:

$$CaSO_4 + CH_4 \rightarrow CaS + CO_2 + 2H_2O \quad (1)$$

$$CaS + CO_2 + H_2O \rightarrow CaCO_3 + H_2S \quad (2)$$

$$3H_2S + SO_4^{=} \rightarrow 4S + 2H_2O + 2OH^- \quad (3)$$

Although these reactions have permissive thermodynamics, the rate of reaction is very slow for temperatures that can reasonably be expected in Gulf Coast sediments. For example, at 100°C, 150 million years would not suffice for development of a cap rock deposit. This is impossibly slow considering that the salt was originally deposited only 180 million years ago.

It is now generally recognized that a biogenic, epigenetic origin of sulfur in salt domes and stratiform deposits is most likely. Many microbiologists have since studied the subject and greatly refined the knowledge of anaerobes (Postgate, 1972, 1979; Silverman, 1972). These bacteria are highly versatile, such that a strain collected from saltwater can be acclimated to live in brackish water and then fresh water, and vice versa. At present, two genera and at least three species are commonly accepted: *Desulfovibrio desulfuricans,* the most widely distributed and most active form at atmospheric temperatures; *Desulfovibrio orientis,* a mesophilic (medium temperature) form; and *Clostridium nigrificans,* a thermophilic (heat-loving) form. All of these anaerobic bacteria consume hydrocarbons as a source of energy, but use sulfur instead of oxygen as a hydrogen acceptor to produce hydrogen sulfide, calcite, and water. The hydrogen sulfide may be oxidized to a colloidal form of sulfur or, in an anaerobic environment, it can react with excess calcium ions to form calcium sulfide and by hydrolysis to calcium hydrosulfide. Calcium hydrosulfide solutions will react with colloidal sulfur in previously oxidized zones to produce calcium polysulfides. In the presence of carbon dioxide generated by bacteria the polysulfide precipitates crystalline sulfur and coprecipitates secondary calcite. If hydrogen sulfide is able to escape from the system, a limestone cap rock may be free of sulfur. Sulfur may also be removed by sulfur-oxidizing bacteria (Ehrlich, 1972).

Further investigations support the role of sulfate-reducing bacteria in the formation of sulfur. Although all four sulfur isotopes chemically react alike, the lighter isotopes enter some phases more readily than do the heavier ones (Thode, 1972). This results in fractionation or a concentration of lighter isotopes in some compounds and a residual enrichment of heavier isotopes in others. Feely and Kulp (1957) and others produced evidence to support the theory of sulfate-reducing bacteria. Their laboratory evidence showed that sulfate-reducing bacteria fractionate the sulfur isotopes S^{32} and S^{34} in the reduction of sulfate ions to sulfide ions comparable to that found in sulfur from salt dome cap rock. The sulfur is enriched in the S^{32} isotope while the surrounding gypsum and anhydrite are enriched in the S^{34} isotope. Feely and Kulp (1957) also showed that the carbon in the cap rock is isotopically similar to the carbon in Gulf Coast crude oil and that the C^{12} : C^{13} ratio was 4.5% higher for cap rock carbon than the ratio for sedimentary limestone.

Nearly all salt dome sulfur deposits were discovered during petroleum and natural gas exploration. Some earlier discoveries were made by drilling, while others were recognized by sulfur water seeps or low mounds. Various geophysical and geological methods are used in prospecting. Because sulfur is not uniformly distributed throughout the cap rock, numerous boreholes must be drilled to determine the extent and richness of the sulfur area for evaluating its commercial potential.

Prospect drilling has proved the existence of over 400 salt dome structures in the coastal and offshore area of the Gulf of Mexico, most of which contain no commercial sulfur. In November 1980, there were seven salt dome sulfur deposits in commercial operation: one onshore in Louisiana, one offshore Louisiana in the Gulf, three in Texas, and two in Mexico. These seven mines were producing at an annual rate of around 8 Mt. They have produced about 159 Mt in their life. Production history of all Frasch process sulfur mines (including stratiform deposits) in the United States and Mexico is shown in Table 3.

The Frasch Process:

Herman Frasch's efforts to develop the hotwater process for mining sulfur from the first discovered salt dome deposit in Calcasieu Parish, LA, is described by Haynes (1942; 1959). The hotwater process now used in extracting sulfur from salt dome deposits in the United States and Mexico and from the evaporite basin deposits in west Texas, Poland, the USSR, and Iraq is only slightly modified. Requisites for the economical production of sulfur by the Frasch process are: a porous, sulfur-bearing formation; an impervious stratigraphic seal; a large, dependable supply of water; and a cheap fuel supply. The heart of the operation is the power plant, where water is heated, air is compressed for pumping the liquid sulfur to the surface, and electric power is generated for drilling, mining, loading, and auxiliary operations.

A typical setting of equipment for a Frasch installation is shown in Fig. 3. A well consists of four concentric pipes varying in diameter. The outer pipe acts as a protective casing, extending to the barren cap rock. The next pipe carries the flow of hot water, which is forced out through holes at the bottom into the sulfur-bearing limestone formation. As sulfur melts it percolates down through the porous limestone

TABLE 3—Production of Frasch Sulfur in the United States and Mexico by Mine and Company, 1894 through 1979

Dome/Mine	Company	Years Operated	Total Output, 1000 lt*	Total Output, kt *
United States				
Sulphur Mine, LA	Union	1894-1924	9,412	9,563
Bryan Mound, TX	Freeport	1912-1935	5,001	5,081
Gulf (Big Hill), TX	Texasgulf	1919-1936	12,350	12,548
Hoskins Mound, TX	Freeport	1923-1955	10,895	11,070
Big Creek, TX	Union	1925-1926	2	2
Palangana, TX	Duval	1928-1935	237	241
Boling, TX	Union	1928-1929	8	8
Boling, TX	Texasgulf	1929-	75,556	76,768
Long Point, TX	Texasgulf	1930-1938	402	408
Jefferson Island, TX	Jefferson Lake	1932-1936	431	438
Grande Ecaille, LA	Freeport	1933-1978	40,398	41,046
Boling, TX	Duval	1935-1940	571	580
Boling, TX	Baker-Williams	1935-1935	1	1
Clemens, TX	Jefferson Lake	1937-1960	2,976	3,024
Orchard, TX	Duval	1938-1970	5,495	5,583
Long Point, TX	Jefferson Lake	1946-	8,129	8,259
Moss Bluff, TX	Texasgulf	1948-	8,847	8,989
Starks, LA	Jefferson Lake	1951-1960	840	853
Spindletop, TX	Texasgulf	1952-1976	9,837	9,995
Bay Ste. Elaine, LA	Freeport	1952-1959	1,131	1,149
Damon Mound, TX	Standard Sulphur	1953-1957	140	142
Garden Island Bay, LA	Freeport	1953-	18,522	18,819
Nash, TX	Freeport	1954-1956	151	153
Chacahoula, LA	Freeport	1955-1962	1,199	1,218
Fannett, TX	Texasgulf	1958-1977	3,477	3,533
High Island, TX	US Sulfur	1960-1962	37	38
Grand Isle, offshore LA	Freeport	1960-	19,414	19,726
Lake Pelto, LA	Freeport	1960-1975	5,622	5,712
Gulf (Big Hill), TX	Texasgulf	1965-1970	502	510
Sulphur Mine, LA	Allied Chemical	1966-1970	68	69
Nash, TX	Phelan Sulphur	1966-1969	171	174
Heiner, west Texas	Duval	1967-1970	579	588
Bryan Mound, TX	Hooker Chemical	1967-1968	5	5
Chacahoula, LA	John W. Mecom	1967-1970	154	156
Ft. Stockton, west Texas	Atlantic Richfield	1968-	1,467	1,491
Lake Hermitage, LA	Jefferson Lake	1968-1972	441	448
Caminada, offshore LA	Freeport	1968-1969	325	330
Bully Camp, LA	Texasgulf	1968-1978	1,725	1,753
High Island, TX	Pan Am Petroleum	1968-1971	110	112
Culberson, west Texas	Duval	1969-	17,027	17,300
Comanche Creek, west Texas	Texasgulf	1975-	679	690
	Total United States		264,334	268, 573
Mexico				
San Cristobal	Azufre Mexicana	1954-1957	162	165
Jaltipan	Azufrera Panamericana	1954-	25,961	26,378
Salinas	Azufre Veracruz	1956-1969	3,869	3,931
Nopalapa	Exploradora del Istmo	1957-1960	322	327
		1967-1971	348	354
Texistepec	Central Minera	1959-1961	9	9
Texistepec	Exploradora del Istmo	1971-	5,059	5,140
	Total Mexico		35,730	36,304
	Total US and Mexico		300,063	304,877

* 1 long tons (lt) × 1.016047 = metric tons (t).
1000 t = 1 kt.

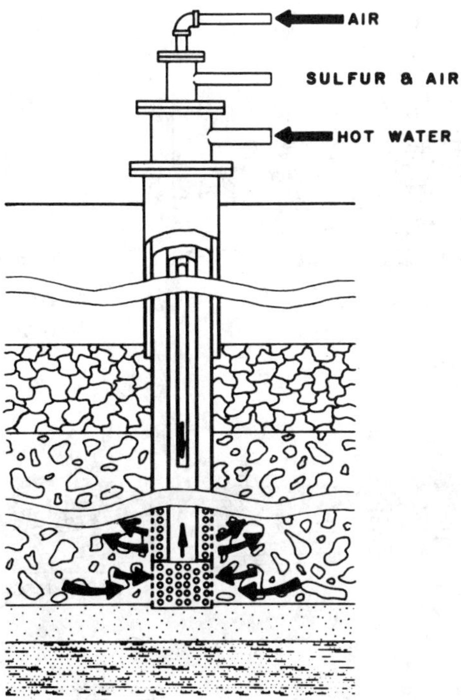

FIG. 3—*Typical sulfur well equipment for the Frasch process (Gittinger, 1975).*

to form a liquid sulfur pool at the bottom of the well. Of the two inner pipes, the smaller one carries compressed air which acts as an air lift to raise the liquid sulfur to the surface through the larger pipe. The volume of compressed air is regulated so that the production rate equals the sulfur melting rate in order not to deplete the sulfur pool and cause the well to produce water. If the pool does become depleted, air injection is stopped and additional hot water is injected down through the sulfur pipe, as well as the normal water annulus, until a sulfur pool is reestablished.

On the surface, the liquid sulfur moves through steam-heated lines to a separator where the air is removed prior to shipment by pipeline, thermally insulated barges, or rail cars to a central shipping terminal, customers' plants, or into storage. The sulfur may be transported as either a liquid or a solid.

Hot water for sulfur mining by the Frasch process is produced in power plants with water-heating capacities that range from 3.7 to upward of 53 ML/d. Until 1952, all mines used fresh water obtained from wells, rivers, or other surface sources. In that year Freeport pioneered the use of seawater for mining at its Bay Ste. Elaine, LA mine, located some 48 km from the nearest source of fresh water. Since then, Freeport has brought four additional mines into operation using seawater. Two of these mines, Grand Isle and Caminada, located approximately 11 km off the Louisiana coast in the Gulf of Mexico, are built on mining platforms. Caminada is presently on a standby basis because of the depressed condition of the sulfur market. A third mine, Lake Pelto, now closed, utilized a barge-mounted power plant originally used at Bay Ste. Elaine and now used at Caillou Island, the fourth and latest seawater Frasch mine (Anon., 1979a, 1980i, 1980j).

The hot water pumped into the formation to melt the sulfur must be withdrawn at approximately the same rate at which it is injected in order to prevent pressure in the formation from increasing to the point where further water injection would be impossible. Bleed wells which extract water from the formation usually are located on the flanks of the dome, away from the mining area.

Stratiform Deposits

Elemental sulfur occurs also in subsurface deposits which are similar to salt dome deposits in that the sulfur with secondary limestone is associated with bedded gypsum and anhydrite (Fig. 4). Deposits of this type are found in Italy, Sicily, France, Spain, Poland, the USSR, Iraq, Iran, west Texas, and southeast New Mexico.

The origin of sulfur in these deposits is generally believed to be derived by hydrocarbon reduction of sulfates, assisted by bacterial action. Dessau et al. (1962), in a study of Sicilian sulfur deposits, suggests a biogenic origin for H_2S and a subaqueous sedimentary origin for the sulfur. According to this source, H_2S was formed in the bottom oozes of lagoons and was oxidized by atmospheric oxygen in the water; sulfur was precipitated as a sediment on the lagoon floor.

Polish scientists Czerminski (1960, 1968) and Pawlowski (1968) propose for their deposits, which are concentrated in the Miocene marine sediments in the Carpathian foredeep, that beds of gypsum were reduced to H_2S and $CaCO_3$, probably by biogenic activity. Bacteria as the active agents then precipitated sulfur from the H_2S. A number of bacteria including the genera *Beggiatoa* and *Chromatium* consume H_2S to produce sulfur as a waste which accumulates in the body of the bacteria. On death of the bacteria, the sulfur remains. The

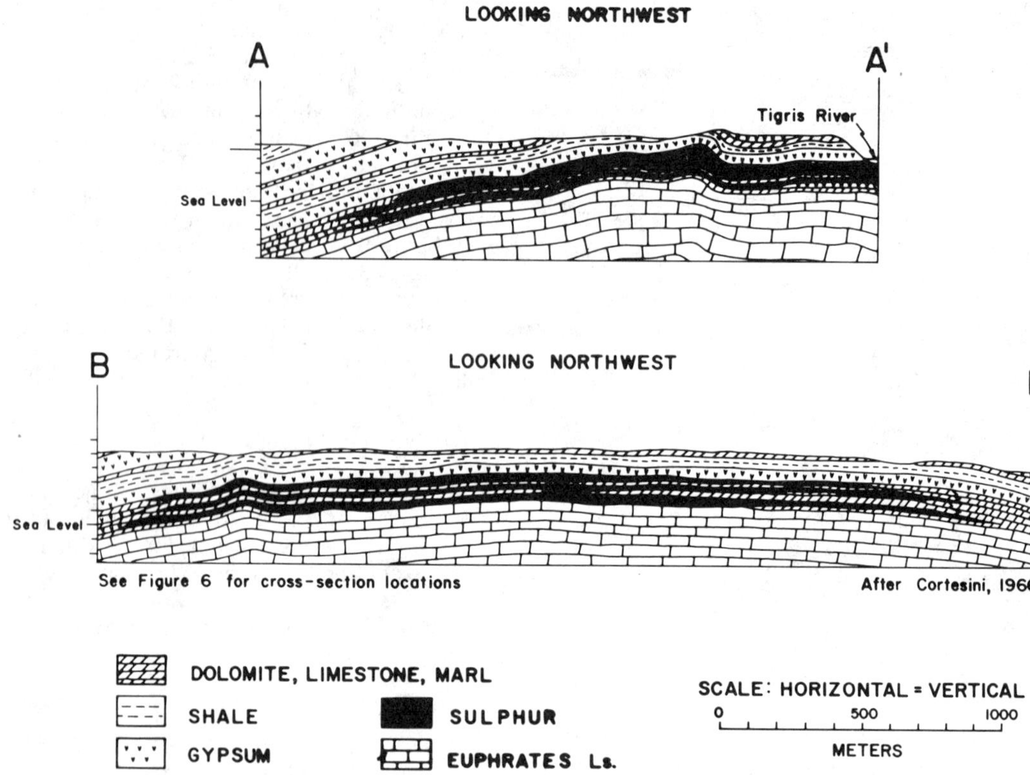

FIG. 4—*Subsurface deposits of elemental sulfur at Mishraq, Iraq.*

sulfur is available for solution, transport, and redeposition, or it may stay as microcrystalline grains disseminated in the calcite of the sulfur-bearing bed.

The Polish sulfur deposits, which are located in the upper valley of the Vistula River near Tarnobrzeg, were discovered in 1953. The initial mine, the Piaseczno open pit, was brought into production in 1960 and operated until 1971. The Machow open pit mine was brought into production in 1969. The ore from the open pit operations is beneficiated by flotation. The flotation concentrate is then melted in autoclaves and filtered to yield pure sulfur or burned to make sulfuric acid. A modified Frasch process [the hydrodynamic process of Zakiewicz (1975)] is being used at the Gryzbow and Jeziorko mines (Bixby, 1968; Anon., 1967, 1968c; Singhal, 1969). Production began at Gryzbow in 1966, and about a year later at Jeziorko. Poland is now second only to the United States in production of Frasch-process sulfur, and fourth after the United States, USSR, and Canada in production of elemental sulfur.

The Polish deposit extends into the west Ukraine region where the USSR has operations utilizing both open pit and hot water mining methods at Rozdol. Additional Frasch-process open-pit mines are in operation, one in the Ukraine and the other in southeastern Turkestan at the Gaurdak deposit in Upper-Jurassic limestone. The USSR also mines sulfur at the Kuybyshev complex from Permian limestone deposits in the middle Volga region (Shabad, 1969).

The sulfur mines of west Texas are in secondary limestone pods within cyclic evaporite deposits generally of Permian Age. Sulfur of commercial occurrence is located mostly in formations along the western edge of the Delaware Basin and on the south end of the Central Basin Platform in Culberson, Reeves, and Pecos counties (Zimmerman and Thomas, 1969). Cap rock as described for salt domes of the Gulf Coast area does not exist in west Texas. The sulfur occurs commercially in sections of evaporite beds where the gypsum/anhydrite has been biogenetically converted to limestone and sulfur. However, cores from the sulfur zones are very similar to those of Gulf Coast salt dome deposits. Based on stratigraphic, mineralogical, chemical, and isotopic analyses of core samples, the source of sulfur and its host limestone is considered to be epigenic and formed biogenetically within a calcium sulfate environment (Davis and Kirkland, 1970; McNeal and Hemenway, 1972). Native sulfur was derived from H_2S generated by bacterial reduction of anhydrite in the presence of dilute hydrocarbons. This bacterial action was also associated with oxidation of hydrocarbons to form CO_2, followed by the deposition of calcium carbonate. Subsequently, oxidation of H_2S by oxygenated ground waters which penetrated the porous formation yields sulfur and water. The H_2S may migrate considerable distances prior to oxidation. It must also be trapped by mechanisms identical to hydrocarbon traps; sulfur from H_2S frequently is associated with hydrocarbons.

The USSR has large reserves of sulfur, but grades are low and operating costs high. Estimated production of contained sulfur totaled 8.9 Mt, of which 3.8 Mt were recovered from pyrite, 2.6 Mt from native sulfur, and 2.5 Mt from other elemental sulfur. Estimated levels of total sulfur output for 1977 and 1980 are 9.5 and 11 Mt, respectively. Sulfur exports, mainly to Comecon countries, decreased from 463.8 kt in 1970 to 441 kt in 1975. Imports of sulfur increased from 216.7 kt in 1970 to 690 kt in 1975.

The dressing mills of all three main Russian complexes are treating ore with a sulfur content from 8 to 40% and produce a flotation concentrate with 70 to 85% sulfur content. Sulfur production by the Frasch process at the Gaurdak complex has increased rapidly since the second installation (100 kt/a) went into service in 1975. The No. 3 (100 kt/a) installation using the Frasch process at this complex was under construction in 1976. The Rozdol complex produced concentrates with a 72 to 75% sulfur content from ore with a 22 to 23% sulfur content. It is planned to recover 895 kt of sulfur from the Orenburg sour gas. Agreements have been signed for the supply by Poland of three sulfuric acid production units to the USSR with a capacity of 1515 t/d each. The Soviet Union has already purchased 25 sulfuric acid plants from Poland (Strishkov, 1977).

In late 1971, the state-owned National Iraqi Minerals Company started sulfur production at the Mishraq mine. The Mishraq native sulfur deposit is in northern Iraq about 40 km southeast of Mosul and 315 km north of Baghdad. This deposit is the largest known occurrence of stratiform bioepigenetic sulfur, containing at least 100 Mt of elemental sulfur. It underlies a 10 km^2 portion of the southeast end of a northwest-trending, doubly plunging anticline on the west bank of the Tigris River at its confluence with the Great Zab River. The deposit is in one anticline among many in the folded portion of the Mesopotamian depression and represents part of what may be the largest reserve of elemental sulfur in the world (Anon., 1974).

Small, tight, en echelon folds are superimposed on the southwest limb of the Mishraq anticline with axial trends similar to the main fold (320°). Faulting is normal and reverse with dominant northwest and subordinate northeast trends. Mining-induced subsidence of 4 to 5 m causes fractures which tend to follow these preexisting directions (Zackiewicz, 1975).

The sulfur mineralization is in three main zones (Chebanenko, 1969) of vuggy and bituminous bioepigenetic limestone in the basal Lower Fars Formation of middle Miocene age and marine origin (Fig. 5). The Lower Fars consists of intercalated bituminous marine carbonates (dolomite, limestone, marl) and anhydrite/gypsum with minor shale and sandstone interbeds. The depositional environment was of periodic marine influxes followed by evaporation in a partially barred basin with southern access to the sea (Al-Sawaf, 1977).

The sulfur was formed by oxidation of hydrogen sulfide produced during metabolysis of petroleum and anhydrite/gypsum by anaerobic bacteria such as *D. desulfuricans* (Al-Sawaf, 1977), a conclusion supported by the S^{34} and C^{13} fractionation data (Lein et al., 1975). Sulfur mineralization is thickest in areas of intense folding and faulting which enhance solution, movement, and mixing of reactants. Combined thickness (Fig. 6) of the three sulfur-bearing zones is from 2 to 123.9 m, with a maximum ore section of 107.9 m and a grade of 23.14 wt % sulfur (Chebanenko, 1969). Seven ore types have been recognized, of which coarse-crystalline, secondary sulfur alternating with bands of secondary limestone predominates (Niec and Al Nouri, 1976).

Hydrologic conditions at the mine are complex with subsurface drainage via karst features, fractures, faults, and vuggy porosity within the ore zones of the productive member (Niec and Al Nouri, 1976; Featherstone and Al-Samarrie, 1975; Al-Sawaf, 1977; Zakiewicz, 1975). The productive member contains the

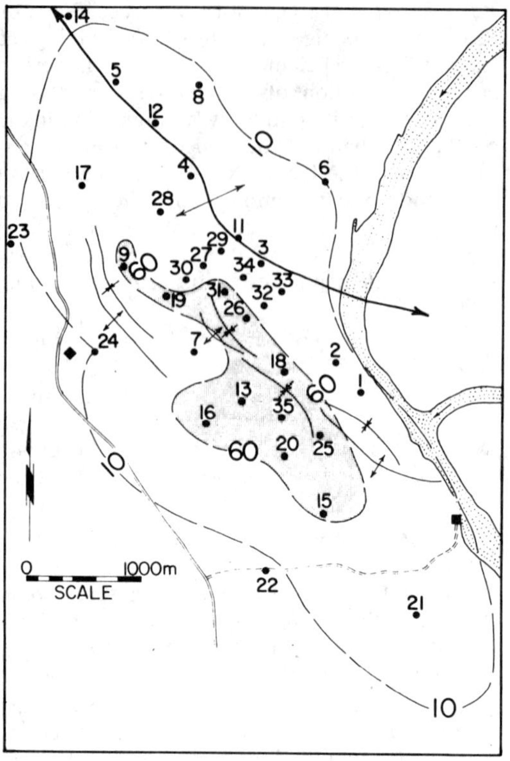

FIG. 5—*Isopach of total solid feet of sulfur, zones 1 to 3, as a function of folding at Mishraq, Iraq (based solely on early Russian mapping and drill holes 1–9, 11–35; Cortesini et al., 1966).*

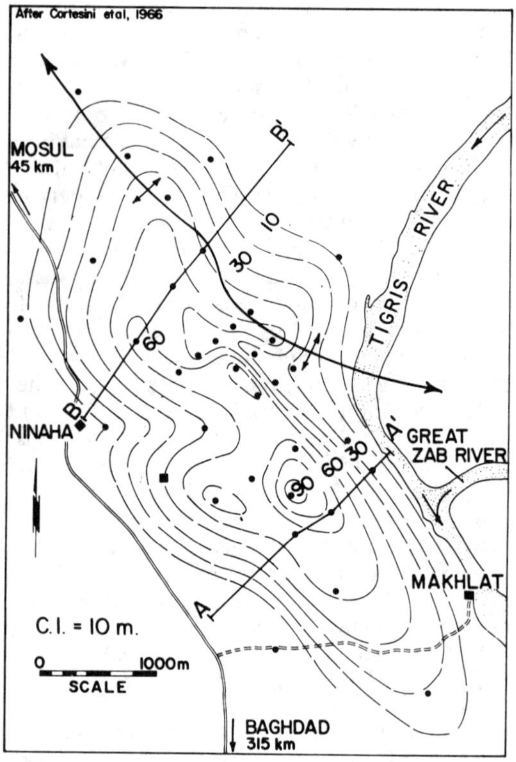

FIG. 6—*Isopach of total solid feet of sulfur, zones 1 to 3, Mishraq, Iraq (based solely on early Russian data; Cortesini et al., 1966).*

bulk of the ground water, although up to 13 aquifers have been recognized throughout the section, and it is intersected by the Tigris River. A very steep hydraulic gradient exists with flow southeastward toward the Tigris where it is discharged from springs in and beside the river. Hydrogen sulfide-carbonate waters predominate. Recharge is mainly at the northwest portion of the anticline where the Lower Fars Formation crops out. The ground water connection and flow to the river is important in sulfur generation as it is the pathway by which oxygen is carried into contact with hydrogen sulfide at depth yielding sulfur and water.

The deposit is currently being mined by the Polish hydrodynamic (modified-Frasch) process (Zackiewicz, 1975) at a rate of about 600 kt/a out of a total capacity of 1 Mt/a. The maximum capacity has not been reached because of infrastructural constraints related to transportation and because the mine is difficult to operate. Large injection water losses and excessive bitumen contamination are paramount. Water losses are mitigated by the installation of subsurface impermeable screens. Bentonite/sulfuric acid filtration removes the excess bitumen but at a cost of 8 to 25% loss of sulfur in filter cake and skim foam (Ali and Al-Shahwani, 1975).

Initially, sulfur was vatted at the mine, crushed, and shipped via rail to the port of Um Qasr about 1,000 km south of Mishraq (Anon., 1974). Liquid sulfur is now shipped to the port where vatting is done as necessary.

The feasibility of using the Frasch process for mining the bedded sulfur deposits of west Texas was first proved by Duval Corp., the mining subsidiary of Pennzoil Co., when it began operations in 1967 in Pecos County, near Fort Stockton (Zimmerman and Thomas, 1969). Atlantic Richfield Co. also operated a small Frasch mine near Fort Stockton (McNeal and Hemenway, 1972). In 1968, Duval discovered and delineated a large sulfur deposit in the Rustler Hills area of Culberson County and brought the mine into operation in 1969 (Aalund, 1970). Duval closed its Fort Stockton mine in 1970. In 1979 and 1980 production at Duval's Culberson property exceeded 2 Mt/a and a small Frasch mine was opened at Phillips Ranch, about seven miles from Culberson (Anon., 1979b; Barry and Davis, 1969; Ellison, 1971).

Volcanic Deposits

Sulfur of volcanic origin is probably the most widely distributed of native sulfur deposits. Deposits of this type are known in most volcanic regions of the world but mainly throughout the mountain ranges which border the Pacific Ocean, particularly in South and Central America, Japan, the Philippine Islands, the Kamchatka Peninsula, the Kurile Islands, Taiwan, and New Zealand. Most currently known reserves occur within this region (White, 1968). Mining of volcanic deposits is limited primarily to Japan, Turkey, Mexico, Italy, and the Andean countries of South America. Detailed discussion of deposits in Japan and Chile are available in Anon. (1966b).

Volcanic sulfur deposits include those originating from all forms of volcanic activity and are generally classified by mode of emplacements: (1) impregnation and replacement deposits, usually formed by the action of hot acid springs, which are rich in sulfurous gases; (2) sedimentary deposits, formed by precipitation of elemental sulfur in hot crater lakes, usually in the latter stages of volcanic activity; (3) sublimation deposits, formed by gaseous or fumarolic activity near volcanic craters; and (4) sulfur flows, the sulfur probably originating from previously formed deposits which have been remelted. Volcanic deposits are usually found in tuffs, lava flows, and other volcanic rocks, as well as sedimentary and intrusive formations.

Mining methods used are tunneling, room-and-pillar, open pit, cut-and-fill systems, and various forms of stoping. Processing of ore to separate the sulfur from the volcanic ash is by various combinations of melting, distillation, agglomeration, flotation, solvent extraction, or any one of these (Anon., 1960; Anon., 1968d; Anon., 1969; and White, 1968). Commercial interest in volcanic deposits is high only when premium or politically protected markets exist.

Japanese volcanic deposits have supplied that country's elemental sulfur requirements for many years, but the growing supplies of recovered sulfur from petroleum refining and byproduct sulfuric acid from nonferrous smelting are forcing a major contraction of the Japanese native sulfur mining industry, as are dwindling reserves.

Turkey obtains most of its elemental sulfur requirements from the Keciborlu mines in the southwestern part of the country. Mexico has a small mine, Los Cerritos, in San Luis Potosi. In the United States where many small deposits of this type are found in several western states, ore has been mined only intermittently and generally in small tonnages for local use. One exception is the Leviathan deposit in Alpine County, CA, which was mined by The Anaconda Co. from 1953 until 1962 to supply 20% sulfur ore for a sulfuric acid plant at its Yerington, NV copper operation (Gould, 1955).

Recovered Elemental Sulfur

The principal sources of recovered sulfur are the hydrogen sulfide contaminants in sour natural gas and the organic sulfur compounds contained in crude oil. Recovery is mainly in the elemental form, although some is converted directly to sulfuric acid. Smaller quantities of sulfur are recovered from tar sand, oil shale, and coal.

Sour Natural Gas and Crude Oil

The growth in the recovery of sulfur from natural gas and crude oil since the mid-1950s has been one of the most significant trends in world sulfur. Rising from less than 0.5 Mt in 1950 to about 2.5 Mt in 1960, production at 1972 year-end reached an annual rate of nearly 15 Mt of which over 11 Mt were recovered from sour natural gas. The Province of Alberta, Canada, accounted for about two-thirds of this 11 Mt; France, the second largest producer, for a sixth. Canadian production of sulfur from sour gas wells totaled 6 Mt in 1980. The hydrogen sulfide content of the sour gas in the Alberta foothills of the Rocky Mountains ranges from trace amounts to more than 50%.

Recovery of elemental sulfur from sour natural gas in France is principally from the Lacq field 24 km northwest of Pau in the Department of Aquitaine. This deposit was discovered in 1951 and brought into production in 1957. In the early 70s, the Lacq plant had eight sulfur recovery units and processed gas from Pont d'As, Meillon/St. Faust, and Ucha gas fields in addition to that from the Lacq field. Gas from the Lacq field has a hydrogen sulfide content of 15.2%; the other fields are less sour. Sulfur production was about 1.75 Mt/a, but has begun to decline.

Sulfur was first recovered from sour natural gas in the United States in 1944, and until 1952, the United States was the sole producer. Today, more than 60 plants located in 10 states produce about 1 Mt/a, with Texas accounting for more than half.

Several other countries recover sulfur from natural gas, of which Iran, West Germany, Iraq, and Mexico are the most important. The USSR installed a gas processing and sulfur recovery plant to exploit gas from the Orenburg field in the southern Urals which was discovered in 1959 (Gardner, 1973). The gas is reported to have a hydrogen sulfide content of about 3%.

About 3.25 Mt/a of sulfur are recovered from crude oil in 46 countries, of which the United States is the largest producer. Sixty-eight plants located in 19 states recovered 1.2 Mt of sulfur, primarily brimstone, from crude oil in 1972. Texas and California accounted for 59% of the total. Japan, the second largest producer, recovers just over 1 Mt.

Sulfur values associated with crude oil are recovered during the refining process and the amount of sulfur recovered depends not only upon the crude throughput, but on the source of the crude, the refining process, and the products produced. The average sulfur content of the world's crude oil reserves is approximately 1.7%. In general, crudes in the Middle East, Venezuela, and the USSR have a higher sulfur content than those of the United States and North Africa.

Refineries in general are equipped to remove the sulfur only from the lighter refinery products and much of the sulfur in crude oil remains in the refinery emission gases or the residual fuel oils and is ultimately emitted to the atmosphere when these oils are burned. For the western world as a whole some 10 to 15% of the sulfur contained in crude oil is recovered. In the United States where residual oil is a lesser part of refinery output, recovery is higher—probably more than one-third. Low-sulfur fuel oils are growing in demand as an anti-air-pollution measure and facilities have been and are being built in the United States, Japan, the Caribbean, and the Middle East for reducing the sulfur content of fuel oil products. Although direct desulfurization of high-sulfur residual oils is costly, several processes have been developed. However, to date most desulfurization is done on distillate fractions which are blended back with a part of the residual. As an alternative to low-sulfur fuels, the SO_2 released from the burning of high-sulfur fuel oils may be removed from the combustion gases prior to their emission into the atmosphere. Processes for controlling SO_2 emission are discussed in a subsequent section.

In general, the process for recovery of sulfur from sour natural gas and crude oil consists of first separating the H_2S from the other gases, then converting the H_2S to elemental sulfur. The absorbent normally used in processing the natural gas is monoethanolamine (MEA) although other absorbents are used and have been proposed (Anon., 1968a; Sweeney and Valentine, 1970). In processing the H_2S-laden refinery gases diethanolamine (DEA) is the usual absorbent because it is less susceptible to contamination by heavy hydrocarbon fractions.

Unlike sour natural gas where the sulfur occurs naturally as hydrogen sulfide, the sulfur compounds in crude oil must be removed from the refinery feed and converted to hydrogen sulfide by a hydrogenation process before the sulfur can be recovered. It is beyond the scope of this chapter to discuss all of the various types of desulfurization processes used in refineries.

C.F. Claus developed the basic process for converting H_2S to elemental sulfur in Germany about 1880 (Sands and Schmidt, 1950). Over the years, the process has been improved and is now generally referred to as the modified-Claus process. The concentrated H_2S is fired in a combustion chamber connected to a waste heat boiler as shown in the generalized flowsheet of Fig. 7. Air is regulated to the combustion chamber so that one-third of the H_2S is burned to produce SO_2, water vapor, and sulfur vapor. Retention time in the combustion zone controls the amount of sulfur formed there, conversion increasing with increased retention. The high temperature gases generate steam in the waste heat boiler and are further cooled in a condenser for sulfur removal. Usually an in-line acid-gas burner is used to reheat the total gas stream before introduction to the catalytic converter where H_2S and SO_2 react to produce sulfur vapor and water vapor. The sulfur vapor produced is condensed and removed from the gas stream in a second condenser. A second converter and scrubber follow. Finally, the total gas stream is incinerated to convert all remaining H_2S, CS_2, and COS to SO_2 before emission to the atmosphere. Two converter plants may attain recoveries approaching 95 to

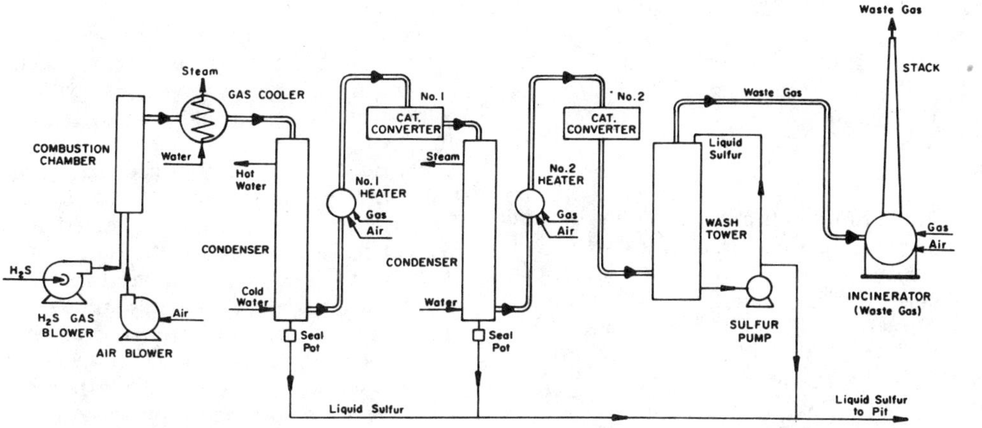

FIG. 7—*Generalized flowsheet, recovery of sulfur from hydrogen sulfide (Gittinger, 1975).*

96%. Air pollution regulations force most new plants to have three converters to attain up to 98% conversion; the Province of Alberta is pressing for 99.5% conversion (Rowland, 1972). A number of new processes have been announced for cleaning up the residual sulfur values from tail gases in recovery processes (Anon., 1972, 1972b; Beavon, 1971; Davis, 1972; Groendaal and Van Meurs, 1972; Hirai et al., 1972).

Oil Sands

Suncor, Inc., a subsidiary of Sun Oil Co., commenced production of synthetic crude oil from the Athabasca oil sands deposit in the northeastern corner of Alberta in September 1967. Hydrogen sulfide derived from upgrading the bitumen product is converted into elemental sulfur. Overcoming early technical difficulties, the plant now has a capacity of about 58,000 bbl of synthetic crude oil and 325 t of elemental sulfur per day.

Other groups have proposed similar projects and a consortium called Syncrude Canada Ltd. (a joint company owned by Atlantic Richfield Co., Cities Service Co., Gulf Oil Corp., and Imperial Oil Ltd.) was granted a permit by the Alberta government and a plant was completed in 1978. The plant has a daily capacity of 129,000 bbl of synthetic crude and 800 t of sulfur.

A number of other companies are conducting research programs in the Athabasca area, including research on methods for extracting the bitumen from the sands other than by surface mining.

The Athabasca deposits are estimated to contain more than 700 billion bbl of raw oil in place, with 400 billion bbl recoverable. The sulfur content, primarily in the form of organic compounds, ranges from 4 to 6%. Other large deposits of tar sands are located in Venezuela and Colombia. The United States has deposits in Utah, Kentucky, and California but they are relatively small.

Oil Shale

More important than tar sands as a potential source of synthetic crude oil for the United States are its oil shale deposits, particularly those of Colorado, Utah, and Wyoming. In general, the oil shale lands most attractive for commercial development are owned by the federal government and, under a proposed leasing program, industry would be permitted to develop a small part of the resources on public lands in those three states.

US oil shales have a sulfur content of less than three-fourths of 1%, of which 80% is in the pyritic form. Under a conventional mining and retorting process for recovery of the oil, the sulfur in the shale feed will be distributed among the oil, the retort off-gases, and the spent shale. An alternative process under study for recovery of the oil is based on an in-situ combustion principle in which thermal decomposition takes place underground.

Outside the United States, the largest reserves of oil shales are in Brazil, the USSR, and Africa. Large-scale production of shale oil is presently limited to the USSR and the People's Republic of China although no sulfur is apparently recovered from these operations.

In Brazil, a shale-oil prototype complex was

completed in late 1972. The plant is designed to treat 2 kt/d of shale and produce 1,000 bbl a day of oil with a sulfur content of 1.06%, 36,245 m^3 of fuel gas, and 14 t of elemental sulfur (Franco, 1972).

Coal

Very little of the sulfur contained in coal and lignite is recovered in a usable form. In 1981, the world mined over 2.3 Gt of bituminous coal and lignite. Sulfur content is estimated to average over 1%, equivalent to more than 30 Mt, of which less than a 0.5 Mt were recovered, largely from coke oven gases and mainly as sulfuric acid. Coal presently mined in the United States contains around 2% sulfur which is higher than the average of US reserves.

Some 40 to 80% of the sulfur in coal occurs as pyrite (see "Combined Sulfur") which is physically mixed with other constituents. The remaining sulfur is present mainly as complex organic compounds in the coal itself. Existing coal-cleaning technology can remove on an average about half the pyrite, but it appears organic sulfur can be removed only by processes of hydrogenation, liquefaction, or gasification which change the form of the coal. A number of techniques are in various stages of development. Processes for desulfurization of coal, including processes for gasification, are reviewed in Anon., (1972a), Chopey (1972, 1972a), Mills (1971), and Sisselman (1972).

Combined Sulfur

Pyrites and Other Metal Sulfides

Metal sulfides, found widely distributed throughout the world, have long been an important source of sulfur. Sulfide ores of commercial interest for their sulfur value belong to the ferrous group known as *pyrites*. Most important of these are iron pyrites (pyrite, FeS_2); white iron pyrites (marcasite, FeS_2); and magnetic pyrites (pyrrhotite, $Fe_n S_n + 1$).

The most important nonferrous metal sulfides include those of copper as chalcopyrite ($CuFeS_2$) and chalcocite (Cu_2S); lead as galena (PbS); zinc as sphalerite or blende (ZnS); nickel as pentlandite $(Fe, Ni)_9 S_8$; and arsenic as arsenopyrite ($FeS_2 \cdot FeAs_2$).

Pure pyrite contains 53.4% S and 46.6% Fe. Commercial grade ores are usually between 40 and 50% S and frequently contain small amounts of arsenic, lead, copper, cobalt, gold, and zinc. Pyrite occurs in isometric crystal forms having a brass-yellow color, a metallic luster, and a specific gravity of 4.95 to 5.10. Marcasite, which has the same chemical composition and hardness as pyrite, occurs as orthorhombic crystals having a pale yellow color, metallic luster, and a specific gravity of 4.85 to 4.90.

Pyrrhotite has varying chemical compositions ranging from Fe_5S_6 to $Fe_{16}S_{17}$, with $Fe_{11}S_{12}$ being the generally accepted formula. When pure, it contains 38.4% S and 61.6% Fe. Pyrrhotite is hexagonal in crystal form, with a bronze color and metallic luster. It is softer than pyrite, is weakly to strongly magnetic, and has a specific gravity of 4.58 to 4.65. Cobalt, nickel, and copper are frequently found in association with pyrrhotite. Silver, gold, and platinum are found less frequently.

Sulfide ore deposits often occur in massive lenses, but may occur in tabular masses, in veins, and in disseminated zones. Metal sulfide deposits are formed under a wide variety of geological conditions ranging from those which are purely of igneous origin to those formed solely by sedimentary processes. In both cases, metamorphism has, to a varying degree, also played a part in the further concentration of the metal sulfide minerals to the extent that they constitute ore. Some interesting hypotheses on the role of certain bacteria in the formation of some sedimentary deposits of metal sulfides are discussed in Jensen (1962).

In 1979, the world produced an equivalent of 8.9 Mt of sulfur from pyrites. The USSR is the world's largest producer of pyrites with an annual production exceeding the equivalent of 3 Mt of contained sulfur. Japan and Spain, the world's second and third largest producers, each produce in excess of 1 Mt. Other important producers include Italy, Cyprus, Finland, Norway, and Sweden.

Sulfur is recovered from pyrites primarily by roasting to secure sulfur dioxide gas as a feed for the manufacture of sulfuric acid or for liquid sulfur dioxide. The iron sinter from the roasting operation may be leached for the recovery of byproduct nonferrous metals and the final residue sold as iron ore. Sulfur may also be recovered from pyrites in the elemental form. Of three commercial processes, only the Outokumpu process is still being used.

Outokumpu Oy, Finland's largest mining and metallurgical company, developed its process to recover sulfur and iron cinder from pyrite concentrates obtained from beneficiation of a deposit of complex ore at Pyhasalmi. Dried pyrite concentrate is blown into the top of a flash smelter. Here, oxygen-free combustion gases obtained from burning oil heat and decompose the pyrite to elemental sulfur and an FeS matte which is continuously tapped from the bottom of the furnace. The sulfur-laden gases from the furnace pass through a waste heat boiler, an electrostatic precipitator for dust removal, and a series of catalytic converters, and finally the sulfur vapor is condensed to liquid sulfur. Sulfur is shipped in liquid form or as prills. The FeS matte is roasted in fluidized-bed roasters to obtain SO_2 for sulfuric acid manufacture and iron cinder. The heat generated is recovered as steam and utilized for electric power production (Anon., 1964; Anon., 1966a; Argall, 1967; Guccione, 1966). Outokumpu's plant at Kokkola has been in operation since August 1962 and produces 100 kt of elemental sulfur and an additional 75 kt in the SO_2 gas.

The Orkla process, developed in the early 1930s by Orkla Mining Co. of Norway, used a shaft furnace for direct smelting of pyritic copper ore in the presence of a carbonaceous reducing agent. Heat in the upper zone was sufficient to drive off the labile atom of sulfur. The FeS formed was oxidized in the bottom zone to convert the remaining sulfur to sulfur dioxide, which was reduced to sulfur by the coke in the middle zone (Kiaer, 1954). Plants using this process were also operated in Portugal and Spain (Potts and Lawford, 1949).

The Noranda process developed by Noranda Mines Ltd. also controls temperature to drive off about half of the sulfur in the elemental form and the remainder as sulfur dioxide for sulfuric acid manufacture (Anon., 1954).

In the smelting of nonferrous sulfide ores, the sulfur is converted into sulfur dioxide which can be recovered from the stack gas in the form of sulfuric acid. Minor quantities are recovered as liquid sulfur dioxide or utilized directly in pulping mills. The amount of sulfur which can be recovered by conventional copper smelting is limited by the fact that a portion of the off-gases within a smelter contain too low a concentration of SO_2 for the economic manufacture of sulfuric acid; other copper smelting processes such as flash-smelting and electric furnace smelting yield gases with higher concentrations of SO_2 from which sulfur can be recovered more economically (Semrau, 1971). Hydrometallurgical processes for copper production are also under study (Gardiner and Warwick, 1971).

Sulfur Dioxide Emission Control: The problems associated with the emission of gaseous pollutants into the atmosphere have received considerable attention during the past decade. The smelting of nonferrous sulfide ores has been estimated to account for 12% of annual worldwide emissions of SO_2 into the atmosphere. Although sulfur compounds are but one of numerous air pollutants, they have been one of the most topical. Sulfur enters the atmosphere as sulfur dioxide, hydrogen sulfide, sulfuric acid, and various sulfates. On a world basis the equivalent of about 220 Mt of sulfur are discharged into the atmosphere annually, of which two-thirds comes from natural processes. The remaining third comes from air pollution sources, mostly in the form of SO_2 (Robinson and Robbins, 1968). The burning of coal and petroleum products, mainly residual oils, has been estimated to account for 86% of worldwide emissions of SO_2 to the atmosphere from pollution sources (Robinson and Robbins, 1968).

Emphasis on control of air pollution and more stringent standards for flue gas emissions, as a means for meeting ambient air quality standards, have spurred research and development of techniques for removal and possible recovery of sulfur from SO_2 emissions. Many processes have been actively investigated and a number are available to industry through various engineering firms (Table 4).

Control processes may be classified broadly as regenerative processes or throwaway processes depending upon whether the sulfur values are recovered in a useful form or lost in output intended for disposal as solid waste (Gittinger, 1975; Moser, 1979). Both recovery methods can be carried out in wet or dry systems. Process chemistry may be absorption, adsorption, or catalysis. Operational processes utilize lime/limestone, sodium compounds, or magnesium compounds. Processes under development are much more broadly based.

Laws for the control of atmospheric pollution are likely to require nonferrous smelters to install some type of control technique, but the ultimate choice will vary with conditions at individual smelters. Where they are favorable, smelters are likely to install sulfuric acid plants. Some of this acid can be used for leaching copper oxide ores and additional quantities may find local markets. In some cases, however, excess acid may have to be neutralized with limestone to be disposed of as gypsum. The Smelter Control Research Assn. was formed by a number of copper-smelting companies to study processes for reducing smelter SO_2 (Campbell and Ireland, 1972). American Smelting and Refining Co. and Phelps Dodge Corp. conducted a joint pilot plant test at

TABLE 4—Summary of Flue Gas Desulfurization Methods Through 1978

Method	Class	Type	Main Output*	Remarks
Operational (% of Total)				
Calcium-based (90%)				
Slaked lime or milled limestone	W	T	$CaSO_4 \cdot 2H_2O$ (10%) $CaSO_3 \cdot \frac{1}{2}H_2O$ (90%)	waste or gypsum product via forced oxygenation
Limestone + $MgSO_4$ (sol.)	W	T		
Sodium-based (7%)				
Sodium carbonate	W	T		
Wellman-Lord (sodium sulfite)	W	R	H_2SO_4, S, SO_2 (liq.)	little or no waste
Double alkali (sodium sulfite + lime)	W	T	$CaSO_3$	waste
Magnesium-based (3%)				
Magnesium oxide	W	R	H_2SO_4	little or no waste
Developmental				
Saarberg-Holter (clear $CaCl_2$)	W	T	$CaSO_4 \cdot 2H_2O$	gypsum product
Sodium citrate	W	R	S	little or no waste
Ammonia	W	R	S	blue fume problems
Potassium thiosulfate	W	R	S	little or no waste
Aqueous sodium carbonate	D	T or R	Na_2SO_3 or S	waste or product
Activated carbon	D	R	S	little or no waste
Cupric oxide	D	R	S, H_2SO_4, SO_2 (liq)	little or no waste
Dry nahcolite	D	T	dust	waste

Source: R. Moser, 1979.
W = wet process, D = dry process; T = throwaway, R = regenerative
* Not including flyash.

Asarco's El Paso, Texas smelter of that company's process for reducing SO_2 to elemental sulfur (White, 1971).

Sulfates

Sulfate minerals represent the world's largest, yet virtually untapped, sulfur resource. Deposits of naturally occurring anhydrite and gypsum are widely distributed geographically and large amounts of gypsum are produced as byproduct waste material from several industrial processes, notably the phosphogypsum generated by manufacture of phosphoric acid from phosphate rock and sulfuric acid (Sweeney and Timmons, 1973). Furthermore, dissolved sulfates in oceans, saline lakes, and brines are almost inexhaustible.

Owing to the large amount of capital required and the high cost of operating the plants, sulfate resources have been little utilized as a source of sulfur. Production or significant development has occurred primarily in the United Kingdom, United States, Austria, East Germany, Italy, India, and the USSR. India was particularly active in development during the 1970s.

Sulfur values can be recovered in three ways: as sulfuric acid, as ammonium sulfate, or as sulfur. Only the first two are currently operational although work on processes with sulfur output continue. The established methods for elemental sulfur recovery are either microbial (Postgate, 1979) or thermochemical in nature. The latter process is the only one attempted commercially in recent times (by the Elcor Co. in west Texas in the late 1960s). This involved a two-step process (Anon., 1980e) in which calcium sulfide produced in a kiln is hydrolyzed to hydrogen sulfide which can yield sulfur by the Claus process. This very energy-intensive process was the primary reason for the Elcor failure, along with technical problems.

One of the oldest industrial processes for use of gypsum or anhydrite was developed in Germany during World War I. Generally referred to as the Muller-Kuhne process, it produces sulfuric acid and cement clinker from anhydrite (Hull, 1957; Ramirez, 1968), or from byproduct phosphogypsum derived from wet process phosphoric acid manufacture (Anon., 1968e).

The Merseberg process for the manufacture of ammonium sulfate from either natural gypsum, anhydrite, or byproduct gypsum from phosphoric acid is based on reacting calcium sulfate with ammonium carbonate liquor to produce ammonium sulfate and byproduct calcium carbonate (Anon., 1952; Hardy, 1957).

Processes to recover elemental sulfur from gypsum or anhydrite have been studied by the US Bureau of Mines and others (Anon., 1968). A bacteriological system has also been studied as an alternative to conventional physiochemical recovery methods (Rosenweig, 1968).

Formed Sulfur

Sulfur forming (Anon., 1979g) first developed in the early 1950s on a very small scale at refineries. Flakes of sulfur were usually produced but some slates production occurred. Increased sour gas production in the 1960s resulted in higher sulfur output. The generation of sulfur dust increased accordingly, as did the handling problems related to flakes in large volume. This led to development of many forming processes during that decade (Table 5), especially prilling by air, water, or combination techniques. Large-scale movement into prilling occurred in the early 1970s, especially after the bans on handling of bulk solid sulfur in the United States and Canada.

The Canadians initially converted to slates from crushed bulk sulfur derived from vats. While a great improvement, slates were still dusty, retained moisture, and were relatively difficult to handle. Other forms were developed and by the end of the 1970s, four wet-pelleting and four granulating plants were operating or under construction and one air prilling plant was under construction.

Production

Sulfur forming techniques were adopted to overcome the handling difficulties and attrition losses inherent in solid sulfur operations. Basically, liquid sulfur is formed, upon leaving the source, into moderately-sized particles of selected shapes. These particles are designed to resist attrition, be unwettable, and handle easily during stockpiling, loading, transport, and unloading.

There are four primary forms into which sulfur can be made to improve or relieve the problems associated with its transport and handling: flakes, slates, prills (or pellets), and granules (or nuggets). Flakes are perhaps the easiest, but least popular, of the four forms to produce. Sulfur is cooled and solidified on the outside of large rotating drums. This gives a thin skin or coating on the drum, which eventually peels off into small flakes.

Slating, the least expensive process, is one of the most widely used forming techniques. Molten sulfur is cast onto a continuous conveyor belt and is cooled, directly or indirectly, with water so that it solidifies into a thin sheet, usually between 3 to 19 mm in thickness. As it reaches the end of the belt, this sheet is broken, or breaks naturally, into smaller pieces that resemble slates.

Prilling sulfur is a more complicated—and correspondingly more expensive—forming method. There are two fundamental types of sulfur prilling: air and water. In air prilling, molten sulfur is sprayed from the top of a tower against an upward flow of air. As it falls the sulfur breaks into small droplets, which cool and solidify into prills between 1and 2 mm in diam. In water prilling, the sulfur is sprayed into tanks containing water, which quenches and solidifies the sulfur droplets. The prills formed are removed from the bottom of the quench tank and usually must be dried. It is not uncommon to find the prilling technique, in particular the water prilling method, referred to as pelleting—a legacy from the production of lead shot in towers, the process on which all sulfur prilling is based.

The last of the four common forming techniques is granulation or melt agglomeration. This involves applying successive coats of sulfur to solid particles of sulfur in a granulator until the particle size reaches the required value. A similar technique is widely used in fertilizer production.

Characteristics

A number of characteristics of a sulfur form have a direct bearing on the suitability of the form for transport. These generally can be termed friability, moisture performance, and handling characteristics.

The purpose of a sulfur form is to reduce the amount of dust produced in handling and transport, so the friability of the form (the degree to which it is degraded) is perhaps the most important of its characteristics. This is influenced by a number of factors including particle size, surface texture, internal voids, and crystallinity.

Moisture performance is defined by the moisture content of the newly formed product and the moisture uptake and retention when it is stockpiled. Those forms produced by water quenching have a higher initial moisture con-

TABLE 5—Forming Processes

Process Type/Name	Developer	Country of Origin	Date of First Use	Prime Mover	Heat Sink	Shape	Bulk Density, (Kg/m³)	Angle of Repose
Slating								
V & E Process	Vennard & Ellithorpe	Canada	1970	rubber belt	air or water	slate or tiles		
Sandvik	Sandvik Steel	Sweden	1954	steel belt	water	slate or tiles		
Candybar						bevelled slate		
Water Pelleting								
Sulpel	Elliot Assoc. Devel. Ltd.	Great Britain	1964	gravity or pressure	water	sphere	1280-1400	30-35°
Capsul	Cambrian Engineering	Canada	1975	gravity or pressure	water	sphere	1152-1169 (dry)	31° (dry) 38° (wet)
Kaltenbach	Kaltenbach and Cie.	France	1973	gravity	water	sphere		
Chemsource	Chemsource	USA	—	gravity/ pressure	water	sphere	1121-1281	35°
Fletcher	R. B. Fletcher & Assoc.	Canada	1978	gravity	water	semi-flat, irregular	1121	25-30°
P.V. Commodity System	P.V. Container Systems	Canada	—	gravity	water	nuggets	1121-1217 (compact.)	32° +
Sulfur Nuggeting	Liquid Terminals, Inc.	USA	—	pressure		nuggets	1025	41°
Air Prilling								
Outokumpu	Outokumpu Oy	Finland	1962	pressure/ gravity	water	sphere	1200-1300	33-35°
Ciech	Ciech	Poland	1966	pressure/ gravity	air	sphere		
Stearns-Roger	Stearns-Roger Canada, Ltd.	Canada	1977	gravity	air	sphere	1120-1280	20°
Misc. Prilling								
Hydroprill	Ross Thermal Systems	Canada	—	pressure/ gravity	air & water	sphere		
Popcorn Sulfur	Union Oil Co. of California	USA	—	high pressure	air & water	"popcorn"	1000	37°
Granulation								
Perlomatic	APC & PEC Engineering	France	early 1960s	pressure	air	granules	1000	20-32°
Procor GX	Sulphur Services Div. of Procor	Canada	1970s	pressure	air	granules	1150	30°
Briquettes	Komarek	USA	—		air	"egg"		
Bulk Sulfur			—		air	mixed fragments		

tent than forms produced by dry processes. The moisture content of a form can be reduced, at extra cost, by mechanical dewatering and drying. Moisture uptake and retention are particularly important because sulfur is usually exposed to the atmosphere during storage. Formed sulfur, especially sulfur formed with silicon additives, usually has reasonable water rejection properties because sulfur itself is water-repellent. The spherical forms have the better water rejection properties; water will often penetrate only the first few inches into a stockpile and will then drain away through this surface layer. The handling characteristics

TABLE 5—Forming Processes

Friability	920 Tumbles	Hardness	Stability	Level 1 Stress Test	Level 2 Stress Test	Remarks
high	30+	10±				Water immersion of precooled sulfur on moving belt; produces 3-19 mm thick slates.
high	30+	10				Water and sulfur kept separate by cooling bottom of belt with sulfur on top; produces 3-19 mm thick slates.
low	10 (5-10)			6.7	5.3	Fed via special nozzles to rotating quench water with silicones added; 0.5-6 mm diam pellets with 0.25% internal moisture.
low	5.30	5.8	8.0	3.2	4.7	Vortex chamber nozzle ejects liquid sulfur into quench water causing turbulence. Additives are surfactants, pH controllers, and others; no commercial plants as of June, 1979.
						No commercial plants as of June, 1979.
low	10-30			5.8	6.2	Sprayed from special nozzle plate into quench water; 3-4% surface moisture on prills.
moderate	10 (5-8)					Fed into hot quench water through steam heated perforated plate: pH control additives used; 3-6 mm pellets produced.
	7.1	2.6	1.7	5.6	5.9	Sulfur discharged into agitated quench water via perforated plate; pH control additives used; 0.5-1% inherent moisture in pellets; not available for license (June, 1979).
moderate	10-30			8.5	6.0	Spray nozzle immersed in quench water; 0.5-1% internal moisture in nuggets.
						Spray of air/water/sulfur mixture downward into rising air; sprayed from side; 0.5-3 mm diam prills.
	tr.-1.76	2.07	1.51	0.4	1.1	Spray from top down into tower full of seed particles in updraft, 2-3.5 mm diam prills
						Sulfur cooled to near freezing in hydrostatic head tanks, then fed into tower with air updraft; not commercial as of June, 1979.
low	10-30			6.3	5.3	Spray of sulfur into jets of water in air updraft; no large scale plants as of June, 1979.
moderate	10-30			14.3	12.7	Spray sulfur/water mixture (1:1) directly onto storage pile; high residual moisture in prills (1-5%).
low	1-4	0.6	0.5	0.8	1.7	Spouting head feeds sulfur/air mixture into bottom of cyclically rotating conical-based vessel with seed sulfur.
low	10-30			9.6	5.4	Sulfur is sprayed into rotating granulation drum with seed material.
high	10-30					Powdered sulfur is compressed into shapes mechanically.
very high						Liquid sulfur is fed into bounded storage areas where it solidifies; these masses are broken for shipment or remelting.

of a sulfur form depend on particle size, uniformity, surface texture, density, and hardness. In fact, it is difficult to relate some of these to field experience. More directly identified with the way a form handles are the properties of bulk density (and the extent to which the form can be compressed), the angle of repose, and the flowability. The last of these can be calculated, as can the compressibility of the form. The highest compressibility and poorest flowability are shown by slates, with better flowability shown by prills.

Probably the most important of the handling characteristics are the bulk density and the angle of repose. Certainly these are normally stated in a description of the properties of the form. Normal dry bulk sulfur has a bulk density between 1.345 and 1.440 kg/m^3 and an angle of repose of about 35°. Thus any form should have a bulk density as near to this as possible and as high an angle of repose as possible. These vary considerably for the different types of forms and even for forms produced by the same technique. Additionally, two samples of a form produced by the same process can vary depending on, for example, the moisture content of the form, which increases both the bulk density and angle of repose. General figures are shown in Table 5 for most of the different types of forms.

Refined Sulfur

Of all the brimstone consumed in the western world in 1978, only about 400 to 450 kt of the total 25.6 Mt were used as refined sulfur or special grades—less than 2%. Thus, the ground sulfur business lies on the periphery of the huge world sulfur industry. Nevertheless, it is a business important in its own right by supplying essential and often highly specialized products for agriculture and industry (Anon., 1979c).

The category includes a wide range of manufactured sulfur products of which ground sulfur is the most prominent. The category also includes wettable, insoluble, precipitated (also referred to as sublimed or flowers of sulfur), colloidal, micronized, and double-refined sulfur, as well as special grades used in manufacturing pharmaceutical and photochemical products.

Although it is the oldest end use of brimstone, refined sulfur having been used since antiquity particularly in pharmacology and alchemy, demand for refined sulfur continues to grow. The manufacture of refined sulfur is confined to a relatively small number of enterprises most of which are located in Western Europe, and are engaged in sulfur grinding. Outstanding amongst them are Stauffer in the United States and Raffineries des Soufres Reumines, with its associated subsidiary companies, in France.

In agriculture, refined sulfur in powdered (dry or wettable), micronized, sublimed, and other forms often combined with metal compounds, is used in direct application on crops, in viticulture, and in horticulture as, first and foremost, a protection against mildew, powdery mildew, and other damaging fungi. However, in both these end uses crude sulfur increasingly is used simply formed into the traditional shape of sulfur rolls without being refined, as recovered sulfur with a purity of 99.9% meets this specification. Refined sulfur, mainly ground, is also used in wine-making and sugar refining, as a pesticide (an ancient end use of sulfur), in fertilizer as a soil conditioner, and as an additive to compound animal feeds. However, owing to cost, fertilizer sulfur now tends to be crude sulfur formed into granules or pellets prior to grinding, or is the fines separated from bulk sulfur to be used in single superphosphate mixtures.

In industry, the rubber and tire sector is the prime user of special sulfur, making use of ventilated, micronized, and insoluble forms, all of which are necessary in the vulcanization process. Much smaller amounts of special grades are used in the manufacture of such diverse items as pharmaceuticals, matches, fireworks, and textiles.

Although ground and refined sulfur are used in some proportion almost everywhere in the western world, production of the various types and grades generally is concentrated in those areas where the level of economic development is already high. Companies involved in manufacture tend to be located in North America, Europe, and Japan, with other producers to be found where there is significant domestic demand as in Australia, South Africa, India, Mexico, Brazil, and Chile.

Characteristically, individual firms making ground and refined sulfur are set up initially to satisfy the demands of the domestic economy (this is particularly true for agricultural sulfur). This impetus then leads to a search for export markets. An exception to this is the relatively new sulfur-grinding industry in Kuwait, which makes use of the easy availability of recovered sulfur from petroleum refining, and has been from the beginning, largely export-oriented.

A healthy international trade in refined sulfur exists because almost all economies need some, but not enough to justify establishing a domestic industry. Certain developing countries have a significant potential need for ground sulfur, but lack financial and entrepreneurial resources to set up their own plants. In general, the pattern of trade seems to be from North to South America in the Western Hemisphere, and from Europe and Japan to nonproducing countries in the Eastern Hemisphere.

Transportation

Sulfur can be transported either in liquid or in solid states. Conversion to sulfuric acid, the main end use, is easiest using liquid sulfur so this has become the preferred form, especially when environmental effects are considered. Stockpiling of sulfur is cheapest by far in large solid blocks (vats) which are rebroken or remelted prior to shipment. Because sulfur block remelting is expensive, as is liquid sulfur-handling equipment, transport and handling of formed solid sulfur is now more economical except where liquid-handling facilities are in place.

End Uses

Sulfur is consumed mainly in the form of acid or other intermediate chemicals used to manufacture a final product that often does not contain much, if any, sulfur. Domestically, 85% of sulfur supplies are utilized in the form of sulfuric acid with similar proportions worldwide. The balance is used for a variety of nonacid industrial purposes. These are detailed for the United States in Figs. 8 and 9 and Tables 6 and 7.

The largest single use is in phosphatic fertilizer production followed by petroleum and coal refining, copper production, and industrial organic chemicals. Combined agricultural uses account for over 60% of domestic sulfur use. Shelton (1980a) describes details of major end-use sulfur requirements and modes of sulfur utilization.

Sulfuric Acid

World sulfuric acid production totaled 135 Mt in 1979 (Table 8), and domestic sulfuric acid production totaled 38 Mt from 120 plants operating in 1979 (Table 9). Producers in the southern states rely on Frasch sulfur as feedstock; western states producers rely on by-product sulfuric acid produced by smelters and on recovered sulfur from natural gas in Canada (Shelton, 1980a).

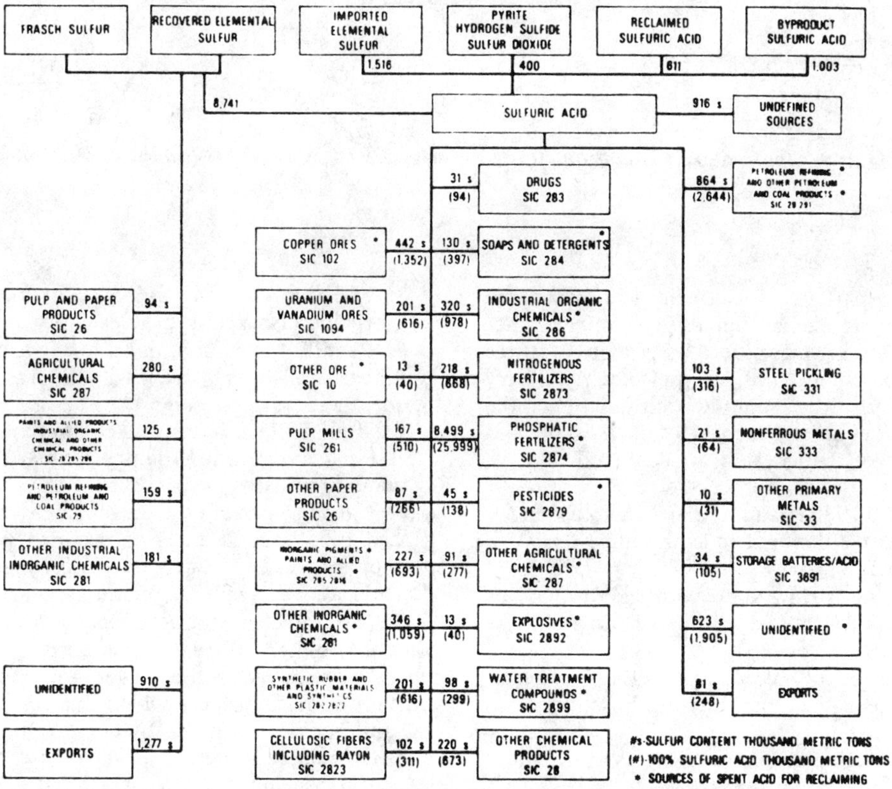

FIG. 8—*Domestic sulfur-sulfuric acid supply/end-use relationship, 1980 (Shelton, 1980b).*

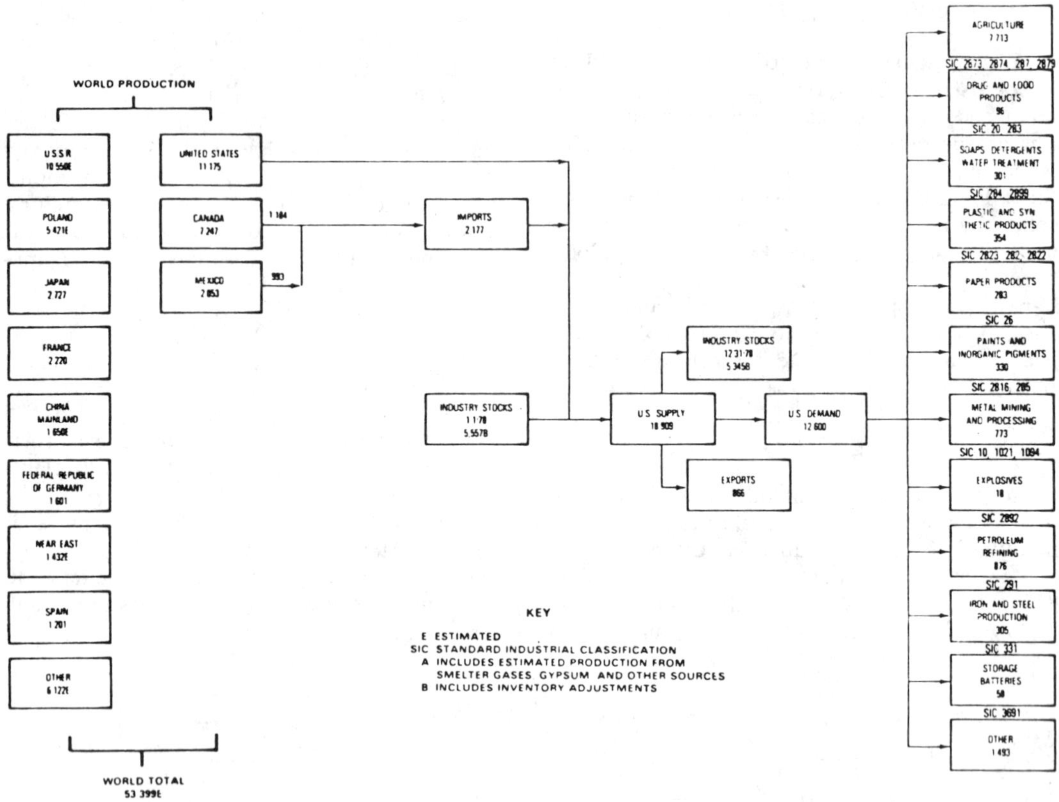

FIG. 9—*Supply-demand relationships for sulfur, 1978 in 1000 t S (Shelton, 1980a).*

A large part of the sulfuric acid consumed by industry is used as a process chemical rather than for its sulfur value. In certain of these applications, particularly in petroleum refinery alkylation and the manufacture of alcohols, the sulfur values in the acid are not used up and the spent or sludge acid is recovered and reused as such or is reprocessed.

Shelton (1980a) summarized the reasons for sulfuric acid's dominance as the most desirable mineral acid as follows:

1. Relatively low cost both as to the raw material (sulfur) and sulfuric acid plant operating costs.
2. Less severe shipping, handling, and storage problems compared with those of other acids.
3. Wide adaptation to numerous acidulations and neutralization processes.
4. Excellent dehydrating properties.
5. Effective catalysis in many hydrocarbon and organic chemical syntheses.
6. Readily forms organic sulfates with many hydrocarbons, which are easily hydrolized to form other basic products.
7. High boiling point, which limits volatilization losses due to high temperatures required in many processes.
8. Can be partially used in one process, reused in a second or even a third process, and/or be readily regenerated, fortified, and concentrated for still further uses.

Sulfuric acid is produced near consumption sites by either of two processes; the older chamber process or the newer and more important contact process which has lower capital and operating costs. In the production of sulfuric acid, elemental sulfur or any number of sulfur-bearing materials are burned to produce SO_2. The contact process requires pure

TABLE 6—Sulfur Supply-Demand Relationships, 1969-1979 (1000 t)

	1969	1970	1971	1972	1973	1974	1975[†]	1976	1977	1978	1979
World Production											
Production:											
United States	9,698	9,710	9,734	10,382	11,096	11,602	11,440	10,879	10,727	11,175	12,101
Rest of world	28,099	32,227	33,001	35,113	37,102	39,551	39,238	39,660	41,773	42,224	[p]42,733
Total	37,797	41,937	42,735	45,495	48,198	51,153	50,678	50,539	52,500	53,399	[p]54,834
Components and Distribution of US Supply											
Domestic mines	9,698	9,710	9,734	10,382	11,096	11,602	11,440	10,879	10,727	11,175	12,101
Imports	1,824	1,694	1,452	1,207	1,242	2,184	1,927	1,755	2,009	2,177	2,494
Industry stocks, Jan. 1	2,697	3,392	3,890	4,186	3,857	3,990	4,020	5,208	5,652	5,557	5,345
Total	14,219	14,796	15,076	15,775	16,195	17,776	17,387	17,842	18,388	18,909	19,940
Distribution of US supply:											
Industry stocks, Dec. 31	3,392	3,890	4,186	3,857	3,990	4,020	5,208	5,652	5,557	5,345	4,239
Exports	1,576	1,456	1,561	1,882	1,804	2,706	1,374	1,290	1,197	866	2,044
Demand	9,316	9,375	9,320	10,012	10,399	10,991	10,773	10,941	11,657	12,600	13,739
Apparent surplus (+), deficit (−) of supply*	−65	+75	+9	+24	+2	+59	+32	−41	−23	+98	−82
US Demand Pattern											
Agriculture (fertilizers)	4,537	4,755	4,877	5,294	5,609	6,076	5,977	6,164	7,225	7,713	8,142
Drugs and food products	—	—	—	—	—	—	44	42	47	96	101
Soaps, detergents, and water treatment	—	—	—	—	—	—	196	197	237	301	323
Plastic and synthetic products	579	503	523	549	589	620	357	314	455	354	328
Paper products	371	356	325	345	376	396	243	276	345	283	372
Paints	452	427	386	396	427	447	341	229	324	330	343
Metal mining and processing	376	396	417	437	508	569	510	656	701	773	869
Explosives	264	259	259	264	284	294	42	33	20	18	19
Petroleum refining	193	198	203	223	234	244	732	686	879	876	890
Iron and steel production	127	122	107	112	112	112	106	97	130	305	288
Storage batteries	—	—	—	—	—	—	33	33	44	58	51
Other	2,417	2,359	2,223	2,392	2,260	2,233	2,192	2,214	1,250	1,493	2,013
Total US primary demand	9,316	9,375	9,320	10,012	10,399	10,991	10,773	10,941	11,567	12,600	13,739

Source: Shelton, 1980a.
[p]Preliminary.
* The difference between total US distribution of supply and total US supply.
[†]Data since 1975 not comparable to data for prior years.

gas stream which can be obtained from a brimstone burner without the purification equipment required with a pyrites roaster. The chamber process produces rather weak acid, about 70% H_2SO_4; the contact process produces 98% acid and stronger grades containing excess SO_3, called oleums. The chamber process is characterized by large, lead reaction chambers and the use of gaseous nitrogen oxides as the catalyst in a complex reaction to form sulfuric acid. The contact process uses a solid contact substance as a catalyst to convert SO_2 and excess O_2 into SO_3. Platinum, once used widely as the catalyst, has been replaced by V_2O_5, which is less susceptible to poisoning by arsenic and other similar impurities found in metal sulfides and sulfur of volcanic origin.

An innovation in the contact process was introduced commercially by Farbenfabriken Bayer of Germany in 1964. Plants of this new type are variously referred to as "double-contact," "double-absorption," or "double-catalysis" plants. The Bayer process increases yield by using thermodynamic equilibrium by removing a product of reaction. In the Bayer process, after a substantial proportion of the SO_2 is converted to SO_3 in two or three conversion stages, the SO_3 is removed from the gas stream by absorption. The remaining uncoverted gas is returned to additional catalyst stages for conversion of SO_2 to SO_3, which is absorbed in acid in a secondary absorption tower. Besides the additional absorption tower, gas-to-gas heat exchangers are required for cooling and reheating the gas stream as shown in the simplified flowsheet of Fig. 10. Overall conversion of 99.5% or better is possible by this process. New sulfuric acid plants are likely to use this process in order to meet air pollu-

TABLE 7—Sulfur and Sulfuric Acid Sold or Used in the United States, By End Use (1000 t sulfur content)

SIC	Use	Elemental sulfur* 1978	Elemental sulfur* 1979	Sulfuric acid (sulfur equivalent) 1978	Sulfuric acid (sulfur equivalent) 1979	Total 1978	Total 1979
102	Copper ores	—	—	626	693	626	693
1094	Uranium and vanadium ores	—	—	103	128	103	128
10	Other ores	—	—	10	8	10	8
20	Food and kindred products	W	W	—	—	W	W
261, 26	Pulpmills and paper products	96	124	187	248	283	372
2816, 285 28, 286	Inorganic pigments, paints and allied products, industrial organic chemicals, and other chemical products	214	166	203	205	417	371
281	Other inorganic chemicals	167	192	375	379	542	571
2822, 2823, 282	Synthetic rubber, cellulosic fibers, other plastic materials and synthetics	W	W	321	294	321	294
283	Drugs	—	—	94	96	94	96
284	Soaps and detergents	—	—	137	121	137	121
286	Industrial organic chemicals	—	—	195	385	195	385
2873	Nitrogenous fertilizers	—	—	412	179	412	179
2874	Phosphatic fertilizers	—	—	6,854	7,581	6,854	7,581
2879	Pesticides	—	—	34	48	34	48
287	Other agricultural chemicals	287	272	126	62	413	334
2892	Explosives	—	—	18	19	18	19
2899	Water treating compounds	—	—	164	202	164	202
28	Other chemical products	—	—	383	166	383	166
281, 29	Petroleum refining and other petroleum and coal products	108	103	768	787	876	890
30	Rubber and miscellaneous plastic products	17	18	16	16	33	34
3111	Leather tanning and finishing	—	—	5	W	5	W
331	Steel pickling	—	—	305	288	305	288
333	Nonferrous metals	—	—	29	34	29	34
33	Other primary metals	—	—	5	6	5	6
3691	Storage batteries	—	—	58	51	58	51
	Exported sulfuric acid	—	—	4	29	4	29
	Subtotal	889	875	11,432	12,025	12,321	12,900
	Unidentified	985	952	401	453	1,386	1,405
	Total	1,874	1,827	11,833	12,478	13,707	14,305

Source: Shelton, 1980a.
W Withheld to avoid disclosing individual company proprietary data; included with "Unidentified."
*Does not include elemental sulfur used for production of sulfuric acid.

tion control regulations. Modern plant design is discussed by Phillips (1977), May (1980), Srinivason (1980), and Appl and Neth (1979). New plants often recover the large heat credit inherent in sulfuric acid manufacturing (Anon., 1978d; Anon., 1980d).

New Uses for Sulfur

Various industrial, academic, and governmental groups are researching new uses for sulfur in an effort to absorb the growing supplies of this basic chemical raw material, although some uses are so valuable as to compete for current supplies. Among new uses with potential large-volume markets are (1) sulfur-asphalt compositions for paving applications to give increased durability and solvent resistance, (2) as an extender or recycling agent for used asphalt; (3) a form of plasticized sulfur for paving (Sulphlex); (4) foamed sulfur with high compressive strengths and insulating properties; (5) impregnation of a variety of materials to improve strength and water and wear resistance; (6) sulfur concretes, mortars, and aggregates with enhanced corrosion and salt resistance; (7) sewage and waste-water treatment; (8) a sulfur electrode in alkali metal batteries; and (9) plant and soil treatment (Davis, 1972; Shelton, 1980a). See also Anon., 1979d; Bourne, 1978; McBee and Sullivan, 1979; Raymont, 1978b; Rennie, 1977; West, 1975; Raddatz et al., 1981; McBee et al., 1981; and Anon., 1980c.

In some places, sulfuric acid often is disposed of at a loss. Some new or expanded uses are possible (Habashi, 1979).

TABLE 8—World Sulfuric Acid Production (1000 t 100% H_2SO_4)

	1977	1978	1979
World Total*	123,030	131,295	135,246
Western World	85,794	91,351	94,703
Communist World	37,236	39,944	40,543
West Europe	26,640	27,376	27,741
Belgium	2,030	2,109	2,309
France	4,382	4,487	4,957
West Germany	4,236	4,591	4,458
Italy	3,200	3,150	2,950
Netherlands	1,638	1,858	1,786
Spain	2,997	2,965	2,950
United Kingdom	3,405	3,453	3,498
Others	4,752	4,763	4,833
Africa	6,210	6,367	6,764
Morocco	1,326	1,271	1,435
South Africa	2,341	2,525	2,644
Others	2,543	2,571	2,685
Asia	11,411	12,342	12,382
India	1,850	2,311	2,574
Japan	6,392	6,437	6,582
South Korea	988	1,400	1,450
Others	2,181	2,194	1,777
Oceania	2,493	2,570	2,773
Australia	1,803	1,890	2,058
New Zealand	690	680	715
North America	34,417	37,712	39,211
Canada	3,431	3,527	3,030
United States	30,986	34,186	36,181
Latin America[†]	4,623	5,156	5,832
Mexico	2,295	2,372	2,430
Argentina	244	249	282
Brazil	1,348	1,650	2,234
Chile	350	355	351
Others	386	530	535
Communist World	37,236	39,944	40,543
Poland	3,293	3,300	3,500
USSR	21,100	22,400	22,400
Others	12,843	14,244	14,643

Source: British Sulphur Corp., Statistical Supplement No. 21, July/August 1980.
*Totals and subtotals may not add due to rounding.
[†] Latin America excluding Cuba.

Outlook for Demand

Demand for sulfur has continued to grow steadily, although dominance of agricultural end uses has caused flattening when fertilizer slumps for reasons exogenous to the sulfur producers. Shelton (1980a) predicts 3% growth through the year 2000 for traditional uses but an overall 4.6% growth rate driven by sharp increases in new sulfur-based technologies (Tables 10 and 11). These rates are lower than 1950 to 1970 levels and imply a domestic demand of 34 Mt and world demand of 142 Mt by 2000 under probable conditions. This is compared to 1981 levels of about 12 Mt and 58 Mt, respectively.

Shorter-term demand and supply predictions by Price (1980) are shown in Tables 12 and 13.

Outlook for Supply

Shelton (1980a) predicts chronic shortfalls in domestic production of from 1.6 to 4.0 Mt annually through the year 2000, with a world cumulative shortfall of 255 Mt. Clearly, higher prices in the early 1980s should shift resources into reserves, a process that can be rapid.

An additional problem in prediction is the rapidly changing balance between sulfur sources. The interaction between Frasch, recovered, and pyrite sulfur is complex and dominated by price and geology. Both Frasch and pyrite sulfur are largely voluntary sulfur, and thus respond to price. Recovered sulfur is primarily involuntary and, in fact, is dominated by forces exogenous to the world sulfur market which are political, legislative, and environmental in nature.

Examples of expected large changes are the decline in Canadian export dominance, especially for offshore. A rapid rise in Middle East recovered sulfur should offset this and more. Other factors, such as interaction between byproduct and brimstone sulfuric acid, may cause problems for virgin acid producers (Anon., 1979).

High prices will continue until shortfalls are rectified. Regionalization of supply may occur, with associated regional independence and market penetration of smaller disseminated sulfur sources.

TABLE 9—Sulfuric Acid Sold or Used in the United States, By End Use (1000 t 100% H_2SO_4)

SIC	Use	Quantity 1978	Quantity 1979
102	Copper ores	1,915	2,119
1094	Uranium and vanadium ore	314	391
10	Other ore	32	25
261	Pulp mills	480	683
26	Other paper products	92	77
285, 2816	Inorganic pigments and paints and allied products	622	628
281	Other inorganic chemicals	1,147	1,159
282, 2822	Synthetic rubber and other plastic materials and synthetics	582	647
2823	Cellulosic fibers including rayon	400	252
283	Drugs	287	293
284	Soaps and detergents	418	370
286	Industrial organic chemicals	595	1,178
2873	Nitrogenous fertilizers	1,262	546
2874	Phosphatic fertilizers	20,968	23,192
2879	Pesticides	103	148
287	Other agricultural chemicals	385	189
2892	Explosives	56	57
2899	Water treating compounds	503	617
28	Other chemical products	1,171	509
29, 291	Petroleum refining and other petroleum and coal products	2,351	2,407
30	Rubber and miscellaneous plastic products	48	49
3111	Leather tanning and finishing	15	*
331	Steel pickling	932	880
333	Nonferrous metals	90	105
33	Other primary metals	14	20
3691	Storage batteries/acid	177	157
	Unidentified	1,228	1,385
	Total domestic	36,187	38,083
	Exports	12	88
	Total	36,199	38,171

Source: Shelton, 1980a.
* Included in Unidentified.

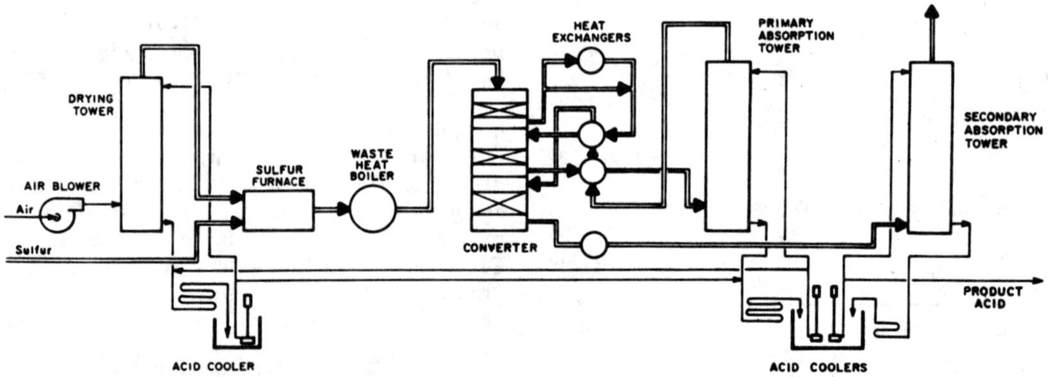

FIG. 10—*Simplified flowsheet, double contact sulfuric acid plant (Gittinger, 1975).*

TABLE 10—Projections and Forecasts for US Sulfur Demand By End Use (1000 t)

End use	1978	2000 Statistical projections *	Contingency forecast for United States: Forecast range Low	Forecast range High	Probable
Agriculture (fertilizers)	7,713	18,000	13,000	15,000	14,000
Drug and food products	96	†	115	125	125
Soaps, detergents, and water treatment	301	†	360	400	400
Plastic and synthetic products	354	†	1,200	1,350	1,300
Paper products	283	0	300	600	400
Paints	330	100*	—	—	—
Metal mining and processing	773	1,700	1,400	1,800	1,800
Explosives	18	†	20	40	30
Petroleum refining	876	†	1,500	1,900	1,900
Iron and steel production	305	0*	—	—	—
Storage batteries	58	†	115	125	125
Other	1,493	2,600*	3,800	4,200	3,900
Potential new uses	—	—	5,000	30,000	10,000
Total	12,600	—	26,800	55,500	34,000

Source: Shelton, 1980a.
* Statistical projections, provided by the Branch of Economic Analysis, are derived from regression analysis based on historical time series data and forecasts of economic indicators such as GNP, FRB index. Projection equations with a coefficient of determination (R squared) less than 0.70 are indicated by an asterisk (*).
† On basis of new series, data are not available for sufficient number of years to make meaningful statistical projections.

TABLE 11—Summary of Forecasts of US and Rest-of-World Sulfur Demand (1000 t)

	1978	2000 Forecast range Low	2000 Forecast range High	Probable 1990	Probable 2000	Probable average annual growth rate 1978-2000 (%)
United States:						
Total	12,600	27,000	56,000	21,600	34,000	4.6
Cumulative	—	420,000	660,000	200,000	480,000	—
Rest of world:						
Total	40,000	85,000	175,000	68,600	108,000	4.6
Cumulative	—	1,340,000	2,100,000	650,000	1,540,000	—
World:						
Total	52,600	112,000	231,000	90,200	142,000	4.6
Cumulative	—	1,760,000	2,760,000	850,000	2,020,000	—

Source: Shelton, 1980a.

TABLE 12—US Sulfur Outlook, Mt

	1979	1980	1985
Production			
Frasch	6.4	6.8	6.8
Recovered elemental	4.1	4.4	6.5
Total Elemental	10.5	11.2	13.3
Sulfur in other forms	1.7	1.6	2.0
Total production	12.2	12.8	15.3
Consumption			
Sulfuric acid			
Phosphates	8.5	9.0	9.7
Other	3.7	3.6	3.8
Total	12.2	12.6	13.5
Nonacid	1.1	1.0	1.5
Total	13.3	13.6	15.0

Source: Price, 1980, via Texasgulf, Inc.

TABLE 13—World Sulfur Outlook, Mt

	1979	1980	1985
Sulfur Production			
Elemental			
Frasch	8.9	9.5	9.8
Recovered	16.5	17.5	21.5
Total	25.4	27.0	31.3
Sulfur in other forms	10.2	10.3	12.0
Total	35.6	37.3	43.3
Consumption			
US	13.3	13.6	15.0
Other Western Hemisphere	4.4	4.7	6.0
Western Europe	11.4	11.8	13.0
Africa	2.6	2.9	4.0
Asia/Oceania	6.2	6.5	8.0
Total	37.9	39.5	46.0
Supply/Demand			
Western world consumption	37.9	39.5	46.0
Imports from East Bloc	2.0	2.0	2.5
Exports to East Bloc	0.4	0.5	0.7
Demand to be met by western world suppliers	36.3	38.0	44.2
Western world production	35.6	37.3	43.3
Inventory change	(0.7)	(0.7)	(0.9)

Source: Price, 1980, via Texasgulf, Inc.

Bibliography and References

Anon., 1952, "Ammonium Sulfate from Gypsum," *Chemical Engineering*, Vol. 59, No. 6, June, pp. 242–245.

Anon., 1954, "Noranda's Elemental Sulfur Plant Slated to Begin Operation October 1," *Engineering & Mining Journal*, Vol. 155, No. 9, Sep., p. 142.

Anon., 1960, "Hot Water Separates Volcanic Sulphur," *Engineering & Mining Journal*, Vol. 161, No. 1, Jan., p. 96.

Anon., 1964, "Outokumpu Process for the Production of Elemental Sulphur From Pyrites," *Sulphur*, No. 50, Feb./Mar., pp. 33–38.

Anon., 1966, "Finland—New Technical Development by Outokumpu Oy," *Sulphur*, No. 67, Oct./Nov., pp. 17–22.

Anon., 1966a, "Handling Liquid Sulfur," Data Sheet, Engineering and Research Commission, Fertilizer Section, National Safety Council, Chicago, 8 pp.

Anon., 1966b, *World Survey of Sulphur Resources*, 1st ed., British Sulphur Corp., pp. 42–44, 61–72.

Anon., 1967, "Polish Sulphur," *Sulphur*, No. 72, Sep./Oct., pp. 12–14.

Anon., 1968, "Handling and Storage of Solid Sulfur," Data Sheet, Engineering and Research Commission, Fertilizer Section, National Safety Council, Chicago, 4 pp.

Anon., 1968a, "Gypsum Ready to Fill the Sulphur Gap?", *Chemical Engineering*, Vol. 75, No. 10, May 6, pp. 94–96.

Anon., 1968b, "Natural Gas Desulfurization Looks to a New Absorbent," *Chemical Engineering*, Vol. 75, No. 4, Feb. 12, p. 66.

Anon., 1968c, "Polish Sulphur Deposits and Their Exploitation," *Sulphur*, No. 78, Sep./Oct., pp. 10–13.

Anon., 1968d, "Processes for the Recovery of Sulfur Values from Calcium Sulphate," *Sulphur*, No. 77, pp. 36–39.

Anon., 1968e, "Route to Sulphur Via Volcanics and Gypsum," *Engineering & Mining Journal*, Vol. 169, No. 7, July, pp. 69–76.

Anon., 1968f, "Sulphuric Acid and Cement from Phosphoric Acid By-Product Phospho-Gypsum," *Sulphur*, No. 74, Jan./Feb., pp. 27–29.

Anon., 1969, "New Thermal Process for Sulfur Recovery," *Engineering & Mining Journal*, Vol. 170, No. 1, Jan., p. 36.

Anon., 1972, "Aquitaine Cleans Claus Tail Gas With Sulfreen Unit," *Oil & Gas Journal*, Vol. 70, No. 26, June 26, pp. 85–88.

Anon., 1972a, "Coal Gasification—Vital Part of the US Energy Programme," *Sulphur*, No. 99, Mar./Apr., pp. 30–37.

Anon., 1972b, "New Process Removes Sulfur From Gas," *Oil & Gas Journal*, Vol. 70, No. 27, July 3, p. 47.

Anon., 1973a, "Duval's Rustler Springs Mine Reaches Design Capacity," *Sulphur*, British Sulphur Corp., Ltd., Reprint, 4 pp.

Anon., 1973b, "World Sulphur Supply and Demand, 1960–1980," Industrial Development Organization, United Nations, New York, 165 pp.

Anon., 1974, "Iraq Emerges as Major Sulphur Exporter," *Sulphur*, No. 111, pp. 36–40.

Anon., 1974a, *World Survey of Sulphur Resources*, 2nd ed., British Sulphur Corp., London, 183 pp.

Anon., 1976, "Compliance with RRC Rule 36; Safe Operations in Handling Hydrogen Sulfide," Railroad Commission of Texas, Oil and Gas Division, Austin, 80 pp.

Anon., 1977, "World Sulphur Supply and Demand, 1970–1982," British Sulphur Corp., London, 82 pp.

Anon., 1977a, "The North American and Caribbean Sulphur Outlook 1977–1985," Manderson Assn., Inc., Winter Park, 213 pp.

Anon., 1978a, "Sulphur—The Past 25 Years," *Sulphur,* No. 136, pp. 22–26.

Anon., 1978b, "Sulphur—The Next 25 Years," *Sulphur,* No. 136, pp. 28–33.

Anon., 1978c, "Frasch Sulphur Production Down the Years: Past, Present—and Future?" *Sulphur,* No. 136, pp. 45–47, 67.

Anon., 1978d, "Outline of Recent Experiences in Sulphur Melting and Burning," *Sulphur,* No. 136, pp. 54–65.

Anon., 1979a, "Freeport to Produce Sulphur at Caillou Island," *Freeporter,* Vol. 26, No. 6, p. 7.

Anon., 1979b, "Duval to Open Sulfur Mine in Culberson County, Texas," *Skilling's Mining Review,* Vol. 68, No. 47, Nov. 24, p. 10.

Anon., 1979c, "Refined Sulphur—A Small Industry with Growing Prospects," *Sulphur,* No. 141, pp. 19–22.

Anon., 1979d, "SUDIC's 1978 Highlights," *Sulphur,* No. 141, pp. 23–27.

Anon., 1979e, "Improving the Quality of Solid Sulphur; Silicon Additives May Be the Answer," *Sulphur,* No. 141, pp. 29–31.

Anon., 1979f, "Flue Gas Desulphurization in the United States," *Sulphur,* No. 141, pp. 34–39.

Anon., 1979g, "Sulphur Forming Processes," *Sulphur,* No. 142, pp. 30–39.

Anon., 1979h, "Aznalcollar; Start-up of Phase I," *Sulphur,* No. 143, pp. 26–28.

Anon., 1979i, "New Frasch Mine for Freeport," *Sulphur,* No. 144, p. 10.

Anon., 1979j, "Polish Sulphur Industry Revisited; Immediate Outlook Not Promising," *Sulphur,* No. 144, pp. 17–20.

Anon., 1979k, "Sulphur Grinding—The Art of Avoiding Explosions," *Sulphur,* No. 144, pp. 33–38.

Anon., 1979l, "By-product Acid versus Brimstone Acid: The US Dilemma," *Sulphur,* No. 145, pp. 40–43.

Anon., 1980a, "Florida, the World's Largest Sulphur Market—and Still Growing," *Sulphur,* No. 146, pp. 18–21.

Anon., 1980b, "Tharsis Pyrite Mines Slated for Expansion," *Sulphur,* No. 147, p. 20.

Anon., 1980c, "New Uses of Sulphur—Current Progress and Problems," *Sulphur,* No. 147, pp. 29–31.

Anon., 1980d, "Waste Heat Recovery from Sulphuric Acid Plants; Uses in Water Heating, Power Generation, and Desalination," *Sulphur,* No. 147, pp. 32–33, 39–43.

Anon., 1980e, "Elemental Sulphur Production, Recovery of Low-grade Pyrites and Phosphogypsum," *Sulphur,* No. 147, pp. 36–38.

Anon., 1980f, "Expansion Plans at Aljustrel," *Sulphur,* No. 148, pp. 18–19.

Anon., 1980g, "The Enormous Potential of the Overthrust Belt," *Sulphur,* No. 148, pp. 29–30.

Anon., 1980h, "West German Sulphur Recovery Running Strong," *Sulphur,* No. 149, pp. 24–26.

Anon., 1980i, "Freeport Commences Production at Caillou Island Sulfur Mine," *Skilling's Mining Review,* Vol. 69, No. 47, Nov. 29, p. 4.

Anon., 1980j, "Caillou Island, A New Mine Is Born," *Freeporter,* Vol. 27, No. 6, Sept./Oct., pp. 10–11.

Anon., 1981a, "Brazil Approves Commercial-Scale Production of Oil Shale in Parana," *Engineering & Mining Journal,* Vol. 182, No. 3, pp. 43, 47.

Anon., 1981b, "Spiraling Costs Slow Rundle Oil Shale Development," *Engineering & Mining Journal,* Vol. 182, No. 5, p. 54.

Anon., 1981c, "Crested Butte Begins Exploration of Montana Oil Shale/Minerals Area," *Engineering & Mining Journal,* Vol. 182, No. 6, p. 57.

Anon., 1981d, "The Future of Heavy Crude and Tar Sands," United Nations Institute for Training and Research, US DOE and Alberta OSTRA, *Proceedings,* available from *Engineering & Mining Journal* Mining Information Service, New York, 968 pp.

Aalund, L., 1970, "New Sulfur Region Has Modern Plant," *Oil & Gas Journal,* Vol. 68, No. 18, May 4, pp. 125–129.

Aalund, L., 1972, "Hydrodesulfurization Technology Takes on the Sulfur Challenge," *Oil & Gas Journal,* Vol. 70, No. 37, Sep. 11, pp. 79–104.

Alexis, J., 1980, "Cleaning Coal and Refuse Fines with the Humphreys Spiral Concentrator," *Mining Engineering,* Vol. 32, No. 8, pp. 1224–1228.

Ali, L.H., and Al-Shahwani, K.I., 1975, "Method for the Purification of Raw Frasch Sulfur from Bitumen and Ash Impurities," *Fuel,* Vol. 54, No. 3, 221 pp.

Al-Samarrie, A.M., 1980, "Hydro-Dynamic Barrier for Sulphur Mines," *Sulphur,* No. 149, pp. 30–31.

Al-Sawaf, F.D.S., 1977, "Sulfate Reduction and Sulfur Deposition in the Lower Fars Formation, Northern Iraq," *Economic Geology,* Vol. 72, pp. 608–618.

Anderson, C.J., 1979, "A Comparative Look at Overseas Sulphur Distribution Systems," *Sulphur,* No. 140, pp. 29–31.

Appl, M., and Neth, N., 1979, "The Catalytic Oxidation of Sulphur Dioxide: Problems and Advances," *Sulphur,* No. 145, pp. 26–39.

Argall, G.O., 1967, "Outokumpu Adds Second Catalyzer to Raise Pyrite-to-Sulphur Conversion to 91 Percent," *Mining World,* Mar., pp. 42–46.

Baillie, R.A., and Mertes, T.S., 1968, "Development and Production of Oil from the Athabasca Tar Sands," *Exploration and Economics of the Petroleum Industry,* V.S. Cameron, ed., Vol. 6, Mathew Bender, New York, pp. 241–271.

Ban, T.E., Johnson, E., and Marlow, W.H., 1981, "Devleopment of the Clean Pellet Fuel Process—A Progress Report," *Mining Engineering,* Vol. 33, No. 1, pp. 71–74.

Barker, J.M., and Cochran, D.E., 1981, "Sulfur," Annual Commodity Review, *Mining Engineering,* Vol. 33, No. 5, pp. 589–591.

Barker, J.M., et al., 1979, "Economic Geology of the Mishraq Native Sulfur Deposit, Northern Iraq," *Economic Geology,* Vol. 74, No. 2, pp. 484–495.

Barry, C.B., and Davis, G.R., 1969, "Sulphur Mining in West Texas," *Journal of World Sulphur,* No. 83, July/Aug., pp. 22–29.

Baughman, G.L., and Sladek, T.A., 1981, "Shale Oil Recovery Methods," *Mining Engineering,* Vol. 33, No. 1, pp. 43–47.

Beard, H.R., et al., 1979, "Gas Analysis Procedures Applicable to Flue Gas Desulfurization by the Citrate Process," Information Circular 8793, US Bureau of Mines, 15 pp.

Beavon, D.K., 1971, "Add-On Process Slashes Claus Tailgas Pollution," *Chemical Engineering,* Vol. 78, No. 28, Dec. 13, pp. 71–73.

Berry, L.G., and Mason, B., 1959, *Mineralogy,* W.H. Freeman and Co., San Francisco, 630 pp.

Bixby, D.W., 1968, "The Sulphur Mining Industry of Poland," Tour of Polish Sulphur Mining Facilities by US Sulphur Delegation, Apr. 21-30, 1967, Sulphur Institute, Washington, DC, May, 29 pp.

Bourne, D.J., ed., 1978, "New Uses of Sulfur—II," Advances in Chemistry Series 165, American Chemical Society, Washington, DC, 282 pp.

Boyle and Assoc., 1965, "Surface and Shallow Oil Impregnated Rocks and Shallow Oil Fields in the US," Monograph 12, US Bureau of Mines, 375 pp.

Campbell, I.E., and Ireland, J.D., 1972, "Status Report on Lime and Wet Limestone Scrubbing to Control SO_2 in Stack Gas," *Engineering & Mining Journal,* Vol. 173, No. 12, Dec. pp. 78–85.

Carrales, M., Jr., and Martin, R.W., 1975, "Sulfur Content of Crude Oils," Information Circular 8676, US Bureau of Mines, 61 pp.

Chebanenko, V.V., 1969, "Geology of the Mishraq Native Sulfur Deposit (Iraq)," Geolgiia mestorzhdenii samorodnoi sery, Izdatel 'stro "Nedra," Moscow, pp. 374–379.

Chopey, N.P., 1972, "Coal Gasification: Can It Stage a Comeback?," *Chemical Engineering,* Vol. 79, No. 7, Apr. 3, pp. 44–46.

Chopey, N.P., 1972a, "Taking Coal's Sulfur Out," *Chemical Engineering,* Vol. 79, No. 16, July 24, pp. 86–88.

Cooper, B.R., and Petrakis, L., eds., 1980, "Chemistry and Physics of Coal Utilization—1980," American Institute of Physics, New York, 476 pp.

Cortesini, A., et al., 1966, "Mishraq Sulfur Project Geologic Report," unpublished, Mideast Industries, Ltd., Baghdad, Iraq, 20 pp.

Crane, S.R., et al., 1981, "Hydrogen Sulfide Generation by Reaction of Natural Gas, Sulfur, and Steam," Report of Investigations 8539, US Bureau of Mines, 54 pp.

Crocker, L., et al., 1979, "Citrate Process Pilot Plant Operation at the Bunker Hill Company," Report of Investigations 8374, US Bureau of Mines, 77 pp.

Czerminski, J., 1960, "Microorganogenic Structures in Sulphur in the Tortonian," *Kwartalnik Geologiczny,* Warsaw, Vol. 4, No. 2.

Czerminski, J., 1968, "Epigenic Processes Within Tortonian Sulphur-Bearing Series," *Proceedings* of Section 8, International Geological Congress, Czechoslovakia, pp. 121–127.

Dale, J.M., 1981, "Sulfur from Surface Ores," *Mining Engineering,* Vol. 33, No. 9, pp. 1340–1341.

Davis, J.B., and Kirkland, D.W., 1970, "Native Sulfur Deposition in The Castile Formation, Culberson County, Texas," *Economic Geology,* Vol. 65, No. 2, Mar. Apr., pp. 107–121.

Davis, J.B., et al., 1970, "Evidence Against Oxidation of Hydrogen Sulfide by Sulfate Ions to Produce Elemental Sulfur in Salt Domes," *Bulletin,* American Assn. of Petroleum Geologists, Vol. 54, p. 2444.

Davis, J.C., 1972, "Sulfur: New Uses Needed," *Chemical Engineering,* Vol. 79, No. 17, pp. 30–32.

Dayton, S., ed., 1981, "Mining Oil Shale," *Engineering & Mining Journal,* Vol. 182, No. 6, pp. 61–135.

Dessau, G., et al., 1962, "Geology and Isotopic Studies of Sicilian Sulfur Deposits," *Economic Geology,* Vol. 57, No. 3, May/June, pp. 410–438.

Dresher, W.H., and Rodolff, D.W., 1981, "Smelter Pollution Abatement; How the Japanese Do It," *Engineering & Mining Journal,* Vol. 182, No. 5, pp. 67–74.

Dunnington, H.V., 1958, "Generation, Migration, Accumulation and Dissipation of Oil in Northern Iraq," *Habitat of Oil,* L.G. Weeks, ed., American Assn. of Petroleum Geologists, Tulsa, pp. 1194–1251.

Ehrlich, H.L., 1972, "Sulfur Oxidation—Bacterial," *Encyclopedia of Geochemistry and Environmental Sciences,* Encyclopedia of Earth Sciences Series, Vol. IVA, R.W. Fairbridge, Jr., ed., Van Nostrand Reinhold, New York, p. 1158.

Eliot, R.C., ed., 1978, "Coal Desulfurization Prior to Combustion," Hayes Data Corp., Park Ridge, 307 pp.

Ellison, S.P., 1971, "Sulfur in Texas," Handbook No. 2, Bureau of Economic Geology, University of Texas, Austin, 48 pp.

Enright, R.J., 1979, "Saudi Desert Sprouts Gas Plants," *Oil and Gas Journal,* Vol. 77, No. 28, July 9, pp. 63–122.

Featherstone, R.E., and Al-Samarrie, A.M., 1975, "Geohydrology of Mishraq—Groundwater Movement Study with Electrical Analogue Simulation," *Sulphur,* No. 120, pp. 44–49.

Feely, H.W., and Kulp, J.L., 1957, "Origin of Gulf Coast Salt-Dome Sulphur Deposits," *Bulletin,* American Assn. of Petroleum Geologists, Vol. 41, No. 8, Aug., pp. 1802–1853.

Field, C.W., 1972, "Sulfur: Element and Geochemistry," *Encyclopedia of Geochemistry and Environmental Sciences,* Vol. IVA, R.W. Fairbridge, Jr., ed., Van Nostrand Reinhold, New York, pp. 1142–1148.

Fjerdingstad, E., 1979, "Sulfur Bacteria," STP650, American Society for Testing & Materials, 121 pp.

Fogarty, C.F., and Mollison, R.D., 1960, "Sulfur and Pyrites," *Industrial Minerals and Rocks,* 3rd ed., J.L. Gillson, ed., AIME, New York, pp. 819–833.

Franco, A., 1972, "Brazil Tries New Shale-Oil Process," *Oil & Gas Journal,* Vol. 70, No. 37, Sep. 11, pp. 105–111.

Freeman, M.R., 1979, "The Fertilizer and Sulfur Outlook in the USSR," Preprint, Annual Conference, Agricultural Chemistry Division, American Marketing Assn., Alexandria, VA, Oct. 20, 21 pp.

Gardiner, S.A., and Warwick, G.C.I., 1971, "Pollution-Free Metallurgy: Copper via Solvent-Extraction," *Engineering & Mining Journal,* Vol. 172, No. 4, Apr., pp. 108–110.

Gardner, F.J., 1973, "Soviet Master Plan for Gas Growth Falling Apart," *Oil & Gas Journal,* Vol. 71, No. 1, Jan. 1, pp. 6–8.

Gittinger, L.B., Jr., 1975, "Sulfur," *Industrial Minerals and Rocks,* 4th ed., S.J. Lefond, ed., AIME, New York, pp. 1103–1125.

Goldman, M.I., 1952, "Deformation Metamorphism and Mineralization in Gypsum-Anhydrite Cap Rock," *Memoir 50,* Geological Society of America, 169 pp.

Gould, A.J., 1955, "Fluosolids Roasting of Sulphur Ore," *Mining Congress Journal,* Vol. 41, No. 6, June, pp. 70–73.

Grimm, C., et al., 1978, "The Colstrip Flue Gas Cleaning System," *Chemical Engineering Progress,* Vol. 74, No. 2, pp. 51–57.

Groendaal, W., and Van Meurs, H.C.A., 1972, "Shell Launches Its Claus Off-Gas Desulfurization Process," *Petroleum & Petrochemical International,* Vol. 12, No. 9, Sep., pp. 54–58.

Guccione, E., 1966, "From Pyrite: Iron Ore and Sulfur via Flash Smelting," *Chemical Engineering,* Vol. 73, No. 4, Feb. 14, pp. 122–124.

Habashi, F., 1979, "Hydrometallurgical Applications of Sulphuric Acid," *Sulphur,* Vol. 140, pp. 37–42.

Hardy, W.L., 1957, "Ammonium Sulfate by the Gypsum Process," *Industrial & Engineering Chemistry,* Vol. 49, No. 2, Feb., pp. 57–58.

Haye, E.F., 1969, "Application of Gravity to Sulfur Exploration—Delaware Basin," Photo-Gravity Co., Inc., Houston, 4 pp.

Haynes, W., 1942, *The Stone That Burns,* D. Van Nostrand, New York, 345 pp.

Haynes, W., 1959, *Brimstone, The Stone That Burns,* D. Van Nostrand, New York, 308 pp.

Hazelton, J.E., 1970, *The Economics of the Sulphur Industry, Resources for the Future,* Johns Hopkins Press, Washington, DC, 172 pp.

Heintz, K.O., and Alexander, M., 1978, "Sulfur Exploration with Core Hole and Surface Gravity," Preprint, SEG Meeting, San Francisco, 17 pp.

Heintz, K.O., and Alexander, M., 1979, "Sulfur Exploration with Core Hole and Surface Gravity," abstract, *Geophysics,* Vol. 44, p. 370.

Hirai, M., et al., 1972, "Solvent Catalyst Mixture Desulfurizes Claus Tailgas," *Chemical Engineering,* Vol. 79, No. 8, Apr. 17, pp. 78–79.

Hollister, V.F., 1977, "Potential for the Occurrence of Sedimentary Sulphur Deposits in Northeastern British Columbia," *CIM Bulletin,* Vol. 70, No. 777, pp. 106–108.

Hull, W.Q., et al., 1957, "Sulfuric Acid From Anhydrite," *Industrial & Engineering Chemistry,* Vol. 48, No. 8, Aug., pp. 1204-1214.

Hunt, J.M., 1976, "Origin of Athabascan Oil," *Bulletin,* American Assn. of Petroleum Geologists, Vol. 60, No. 7, p. 1112.

Idemura, H., et al., 1978, "Jet Bubbling Flue Gas Desulfurization," *Chemical Engineering Progress,* Vol. 74, No. 2, pp. 46–50.

Israel, D.H., 1981, "The Government Plays a Prominent Role in Leasing and Developing Oil Shale," *Mining Engineering,* Vol. 33, No. 1, pp. 52–53.

Ivanof, M.V., 1964, "Microbiological Processes in the Formation of Sulphur Deposits," Academy of Sciences, USSR; Israel Program for Scientific Translation, 1968, 298 pp.

Jensen, M.L., ed., 1962, "Biogeochemistry of Sulfur Isotopes," *Proceedings,* National Science Foundation Symposium, Apr. 12-14, Yale University, New Haven, CT.

Johnson, J.L., 1979, *Kinetics of Coal Gasification,* John Wiley and Sons, New York, 324 pp.

Kaplan, I.R., 1972, "Sulfur Cycle," *Encyclopedia of Geochemistry and Environmental Sciences,* Encyclopedia of Earth Sciences Series, Vol. IVA, R.W. Fairbridge, Jr., ed., Van Nostrand Reinhold, New York, p. 1158.

Kiaer, T., 1954, "Smelter Gases Yield Elemental Sulfur at Orkla-Grube Plant in Norway," *Engineering & Mining Journal,* Vol. 155, No. 7, July, pp. 88–90.

Komarnicki, J.J., 1980, "Strachan Prilling Unit Starts Up," *Sulphur,* No. 147, pp. 22–27.

Krienberg, M., 1978, "Handling Molten Sulfur," *Chemical Engineering,* Vol. 85, No. 27, pp. 125–126.

Laseke, B.A., and Devilt, I.W., 1979, "Status of Flue Gas Desulfurization," *Chemical Engineering Progress,* Vol. 75, No. 2, pp. 37–50.

Laulajainen, R., 1978, "Spatial Dynamics in the U.S. Sulphur Industry, 1955–1975," Swedish School of Economics and Business Administration, Helsingfors, Sweden, 62 pp.

Lein, A.Y., et al., 1975, "Origin of Native Sulfur in Mishraq Deposit (Iraq)," *International Geology Review,* Vol. 17, No. 8, pp. 881–885,

Leonard, J.W., ed., 1979, *Coal Preparation,* 4th ed., AIME, New York, 1204 pp.

Lundy, W.T., 1949, "Sulfur and Pyrites," *Industrial Minerals and Rocks,* 2nd ed., S.H. Dolbear, ed., AIME, New York, pp. 989–1017.

Marchant, L.C., and Jackson, D., 1981, "U.S. Tar Sands: Drawing Interest But No Mega-Projects Are on the Horizon," *Engineering and Mining Journal,* Vol. 182, No. 6, pp. 129–133.

Matveyev, A.K., ed., 1974, *Deposits of Fossil Fuels,* Vol. 4, *Oil Shale Deposits Outside the Soviet Union,* G.K. Hall and Co., Boston, 120 pp.

May, S.K., 1980, "Sulphuric Acid Plant Operation with Reference to Catalyst Performance," *Sulphur,* No. 146, pp. 30–34.

McBee, W.C., and Sullivan, T.A., 1979, "Development of Specialized Sulfur Concretes," Report of Investigations 8346, US Bureau of Mines, 21 pp.

McBee, W.C., et al., 1981, "Modified-Sulfur Cements for Use in Concretes, Flexible Pavings, Coatings, and Grouts," Report of Investigations 8545, US Bureau of Mines, 24 pp.

McConville, L.B., 1975, "The Athabasca Tar Sands," *Mining Engineering,* Vol. 27, No. 1, pp. 5, 19–38.

McIntyre, H., 1976, "Heavy Oil Recovery in the Alberta Oil Sands," *Engineering Journal,* Vol. 59, No. 1, pp. 61–65.

McIntyre, H., 1977, "Heavy Crude Will Backstop Regular Oil," *Canadian Chemistry Process,* Vol. 61, No. 6, pp. 34–35.

McIntyre, H., 1981, "Oil Sands Rounte Looks for Jobs in Canada, U.S.," *Chemical Engineering,* Vol. 17, No. 88, p. 59.

McNeal, R.P., and Hemenway, G.A., 1972, "Geology of Fort Stockton Sulfur Mine, Pecos County,

Texas," *Bulletin,* American Assn. of Petroleum Geologists, Vol. 56, No. 1, Jan., pp. 25–37.

Mehrotra, V.P., et al., 1980, "Oil Agglomeration Offers Technical and Economic Advantages," *Mining Engineering,* Vol. 32, No. 8, pp. 1230–1235.

Meyer, B., 1977, *Sulfur, Energy and Environment,* Elsevier Scientific Publ. Co., New York, 448 pp.

Mills, G.A., 1971, "Gas From Coal, Fuel of the Future," *Environmental Science & Technology,* Vol. 5, No. 12, Dec., pp. 1178–1182.

Moser, R., 1979, "Flue Gas Desulphurization in the United States," *Sulphur,* No. 141, pp. 34–39.

Nettleton, L.L., 1934, "Fluid Mechanics of Salt Domes," *Bulletin,* American Assn. of Petroleum Geologists, Vol. 18, No. 9, Sep., pp. 1175–1204.

Niec, M., and Al Nouri, M., 1976, "Zloze Siorki Mishraq w Polnocnym Iraku," *Przegl. Geol.,* Vol. 24, No. 5, pp. 272–276.

Nissen, W.I., and Madenburg, R.S., 1979, "Citrate Process Demonstration Plant Design," Information Circular 8806, US Bureau of Mines, 16 pp.

O'Brien, E., 1980, "Fine Coal Cleaning in the 28 × 0 Fraction: A Technical Overview," *Mining Engineering,* Vol. 32, No. 8, pp. 1213–1214.

Opychral, A.M., and Wang, K., 1981, "Economic Significance of the Florida Phosphate Industry: An Input-Out Analysis," Information Circular 8850, US Bureau of Mines, 62 pp.

Palm, J.W., 1979, "Sulphur Recovery Plant Operation; Amoco Reviews 30 Years' Experience," *Sulphur,* No. 143, pp. 34–39.

Pawlowski, S., 1968, "Geology of Sulphur Deposits in Poland," *Proceedings* of Section 8, International Geological Congress, Czechoslovakia, pp. 249–267.

Pfeiffer, J.B., ed., 1975, "Sulfur Removal and Recovery from Industrial Processes," Advances in Chemistry Series, No. 139, American Chemical Society, Washington, DC, 221 pp.

Phillips, A., 1977, "The Modern Sulphuric Acid Process," *The Modern Inorganic Chemicals Industry,* R. Thompson, ed., Special Publication No. 31, The Chemical Society, London, pp. 183–200.

Phizackerley, P.H., and Scott, L.O., 1967, "Major Tar Sand Deposits of the World," *Proceedings,* Vol. 3, 7th World Petroleum Congress, pp. 551–571.

Plouf, T.M., 1980, "Froth Flotation Techniques Reduce Sulfur and Ash," *Mining Engineering,* Vol. 32, No. 8, pp. 1218–1223.

Postgate, J.R., 1972, "Sulphate Reduction—Microbial," *Encyclopedia of Geochemistry and Environmental Sciences,* Encyclopedia of Earth Sciences Series, Vol. IVA, R.W. Fairbridge, Jr., ed., Van Nostrand Reinhold, New York, pp. 1127–1129.

Postgate, J.R., 1979, *The Sulphate-Reducing Bacteria,* Cambridge University Press, Cambridge, MA, 151 pp.

Potts, H.R., and Lawford, E.G., 1949, "Recovery of Sulfur From Smelter Gases by the Orkla Process at Rio Tinto," *Transactions,* Institute of Mining & Metallurgy, Vol. 58, pp. 1–36.

Pratt, L., 1976, *The Tar Sands,* Hurtig Publ., Edmonton, 197 pp.

Price, W., 1980, "Sulfur: Tight Supplies May Turn Into Glut," *Chemical Engineering,* Mar. 24, pp. 64–68.

Princiotta, F.T., 1978, "Advances in SO_2 Stack Gas Scrubbing," *Chemical Engineering Progress,* Vol. 74, No. 2, pp. 58–64.

Pryor, W.A., 1962, *Mechanisms of Sulfur Reactions,* McGraw-Hill, New York, 241 pp.

Raddatz, A.E., et al., 1981, "Laboratory Investigation of Sulfurous Acid Leaching of Kaolin for Preparing Alumina," Report of Investigations 8533, US Bureau of Mines, 15 pp.

Ramirez, R., 1968, "Gypsum Finds New Role in Easing Sulfur Shortage," *Chemical Engineering,* Vol. 75, No. 24, Nov. 4, pp. 112–114.

Raymond, W.J., 1978, "The Kellog-Weir Scrubbing System," *Chemical Engineering Progress,* Vol. 74, No. 2, pp. 75–80.

Raymont, M.E.D., 1978a, "Engineering and Economic Analysis of Sulphur Forming Processes," *Sulphur,* No. 135, pp. 32–36.

Raymont, M.E.D., 1978b, "Sulphur Concretes and Coatings," Sulphur Development of Canada, New Uses for Sulphur, No. 4, 43 pp.

Redford, D.A., and Winestock, A.G., eds., 1978, "The Oil Sands of Canada—Venezuela 1977," Canadian Institute of Mining and Metallurgy, Montreal, 784 pp.

Rennie, W.J., 1977, "Sulphur Asphalts," Sulphur Development Institute of Canada, New Uses for Sulphur, No. 2, 52 pp.

Rieber, M., et al., 1981, "Studies on Sulfur Pollution Control—The Impact of Stack Gas Cleanup on the Sulfur Mining Industry of Texas and Louisiana," Reports OFR94(1)–81, OFR94(2)–81, OFR94(3)–81, and OFR94(4)–81, US Bureau of Mines.

Robinson, E., and Robbins, R.C., 1968, "Source, Abundance, and Fate of Gaseous Atmospheric Pollutants," SRI Project PR-6755, American Petroleum Institute, New York, Feb.

Rochille, G.T., and King, C.J., 1978, "Alternatives for Stack Gas Desulfurization by Throwaway Scrubbing," *Chemical Engineering Progress,* Vol. 74, No. 2, pp. 65–69.

Rogers, K.A., and Hill, R.F., 1980, "Synfuels Processing Technology," *Mining Engineering,* Vol. 32, No. 2, pp. 167–170.

Rosenweig, M.D., 1968, "Sulfur Shortage Spawns New Technology," *Chemical Engineering,* Vol. 75, No. 1, Jan. 1, pp. 16–18.

Rowland, L., 1972, "Sulphur Recovery Plants in Alberta Could Face $500 Million Clean Up Bill," *Oilweek,* Vol. 23, No. 36, Oct. 23, pp. 9–12.

Ruckmick, J.C., et al., 1979, "Some Aspects of the Genesis of Bioepigenetic Sulphur Deposits," *Economic Geology,* Vol. 74, No. 2.

Russell, P.L., 1980, "History of Western Oil Shale," Center for Professional Advancement, New Brunswick, 152 pp.

Russell, P.L., 1981, "An Oil Shale Perspective," *Mining Engineering,* Vol. 33, No. 1, pp. 29–32.

Sakseev, G.T., 1973, "Geologic Structure and Genesis of the Teisarov Sulfur Deposit," *Lithology and Mineral Resources,* Vol. 8, No. 5.

Sands, A.E., and Schmidt, L.D., 1950, "Recovery of Sulfur From Synthesis Gas," *Industrial & Engineering Chemistry,* Vol. 42, No. 11, Nov., p. 2281.

Schamm, L.W., 1979, "U.S. Tar Sand Oil Forecasts (1985–1995)" DOE/EIA Report 0183/15, 44 pp.

Schmoker, J.W., 1979, "Interpretation of Borehole Gravity Surveys In a Native-Sulfur Deposit, Culberson County, Texas," *Economic Geology,* Vol. 74, No. 6, Sept./Oct., pp. 1462–1470.

Schmoker, J.W., and Robbins, S.L., 1979, "Borehole Gravity Surveys in Native Sulfur Deposits, Culberson and Pecos Counties, Texas," Open File Report 79–361, US Geological Survey, 23 pp.

Schwaderer, R., ed., 1980, *Synfuel Handbook* (including the Yellow Pages of Synfuels) *Coal Week,* Washington, DC.

Scott, T.F., 1974, "Athabasca Oil Sands to A.D. 2000," *CIM Bulletin,* Oct., p. 98–102.

Selmeczi, J.G., and Steward, D.A., 1978, "The Thiosorbic Flue Gas Desulfurization Process," *Chemical Engineering Progress,* Vol. 74, No. 2, pp. 41–45.

Semrau, K.T., 1971, "Sulfur Oxides Control and Metallurgical Technology," *Journal of Metals,* Vol. 23, No. 3, Mar., pp. 41–47.

Shabad, T., 1969, *Basic Industrial Resources of the U.S.S.R.,* Columbia University Press, New York, 393 pp.

Shafer, J.R., et al., 1978, "Sulfuric Acid Plants for Handling H_2S," *Chemical Engineering Progress,* Vol. 74, No. 12, pp. 62–65.

Shelton, J.E., 1979, "Sulfur," *Mineral Commodity Profile,* US Bureau of Mines, 25 pp.

Shelton, J.E., 1980a, "Sulfur," Preprint, Bulletin 671, US Bureau of Mines, 22 pp.

Shelton, J.E., 1980b, "Sulfur and Pyrites," Preprint, *Minerals Yearbook 1980,* US Bureau of Mines, 19 pp.

Silverman, M.P., 1972, "Sulfide Mineral Oxidation—Microbial," *Encyclopedia of Geochemistry and Environmental Sciences,* Encyclopedia of Earth Sciences Series, Vol. IVA, R.W. Fairbridge, Jr., ed., Van Nostrand Reinhold Co., New York, pp. 1132–1134.

Singhal, R.K., 1969, "Sulphur Mining In Poland," *Mining & Minerals Engineering,* March, pp. 35–39.

Sisselman, R.T., 1972, "Coal Gasification a Partial Solution to the Energy Crisis," *Mining Engineering,* Vol. 24, No. 10, Oct., pp. 71–78.

Skolnik, E., 1980, "Heavy Medium Cleaning of –28 Mesh Coal," *Mining Engineering,* Vol. 32, No. 8, pp. 1235–1237.

Slock, A.V., 1978, "Lime-Limestone Scrubbing: Design Considerations," *Chemical Engineering Progress,* Vol. 74, No. 2, pp. 71–74.

Smith, J.W., 1981, "Alumina From Oil Shale," *Mining Engineering,* Vol. 33, No. 6, pp. 693–697.

Smith, M., 1982, "Sulfur—Pyrites Set to Return," *Industrial Minerals,* No. 182, Nov., pp. 49–65.

Spencer, G.B., et al., 1969, "Domestic Tar Sands and Potential Recovery Methods," *Bulletin,* Vol. 11, No. 2, Interstate Oil Compact Committee, pp. 5–12.

Srinivason, G., 1980, "The Modern Wet Gas Sulphuric Acid Plant," *Sulphur,* No. 148, pp. 32–36.

Stanwood, R.M., 1981, "Environmental Issues: Still Clouded in Uncertainty," *Mining Engineering,* Vol. 33, No. 1, pp. 49–51.

Stanton, R.L., 1972, "Sulfides in Sediments," *Encyclopedia of Geochemistry and Environmental Sciences,* Encyclopedia of Earth Sciences Series, Vol. IVA, R.W. Fairbridge, Jr., ed., Van Nostrand Reinhold Co., New York, pp. 1134–1140.

Staples, L.W., 1972a, "Sulfosalts," *Encyclopedia of Geochemistry and Environmental Sciences,* Encyclopedia of Earth Sciences Series, Vol. IVA, R.W. Fairbridge, Jr., ed., Van Nostrand Reinhold Co., New York, pp. 1141–1142.

Staples, L.W., 1972b, "Sulfides (with Selenides, Tellurides, Arsenides, Antimonides)" *Encyclopedia of Geochemistry and Environmental Sciences,* Encyclopedia of Earth Sciences Series, Vol. IVA, R.W. Fairbridge, Jr., ed., Van Nostrand Reinhold Co., New York, pp. 1129–1132.

Strishkov, V.V., 1977, "Mineral Industries of the U.S.S.R.," *Mineral Perspectives,* Vol. 2, US Bureau of Mines, 19 pp.

Sweeney, J.W., and Valentine, J.P., 1970, "Physical Solvent Stars in Gas Treatment Purification," *Chemical Engineering,* Vol. 77, No. 19, Sep. 7, pp. 54–56.

Sweeney, J.W., and Timmons, B.J., 1973, "Availability and Potential Utilization of Byproduct Gypsum in Florida Phosphate Operations," *Proceedings,* Vol. 811, Public Information Circular No. 5, Forum on Geology and Industrial Minerals, 1972, Iowa Geological Survey, pp. 89–97.

Taylor, R.E., 1938, "Origin of the Caprock of Louisiana Salt Domes," Bulletin No. 11, Louisiana Geological Survey, 191 pp.

Thode, H.G., 1972, "Sulfur Isotope Fractionation —In Biological Processes," *Encyclopedia of Geochemistry and Environmental Sciences,* Encyclopedia of Earth Sciences Series, Vol. IVA, R.W. Fairbridge, Jr., ed., Van Nostrand Reinhold Co., New York, pp. 1152–1157.

Tiernon, C.H., 1980, "Concentrating Tables for Fine Coal Cleaning," *Mining Engineering,* Vol. 32, No. 8, pp. 1228–1230.

Tost, D., 1976, "Syncrude Vows Environmental Responsibility," *Hydrocarbon Process,* Vol. 55, No. 10, p. 102.

Trevoy, L.W., et al., 1978, "Development of the Heavy Minerals Potential of the Athabasca Tar Sands," *CIM Bulletin,* Vol. 71, No. 3, pp. 175–180.

Tuller, W.N., 1954, *The Sulphur Data Book,* McGraw-Hill, New York, 143 pp.

Utter, S., 1981, "A Range of Mining Techniques to Meet Site-Specific Conditions," *Mining Engineering,* Vol. 33, No. 1, pp. 39–43.

Van den Berkhof, B., 1980, "Sulphuric Acid as a By-product of the European Nonferrous Metals Industry," *Sulphur,* No. 146, pp. 24–28.

Weeks, L.G., 1976, "Origin of Athabasca Oil," *Bulletin,* American Assn. of Petroleum Geologists, Vol. 60, No. 7, pp. 1110–1111.

West, J.R., ed., 1975, "New Uses of Sulfur," *Advances in Chemistry Series,* Vol. 140, American Chemical Society, Washington, DC, 236 pp.

White, J.A.L., 1968, "Native Sulfur Deposits Associated With Volcanic Activity," *Mining Engineering,* Vol. 20, No. 6, June, pp. 47–51.

White, L., 1971, "SO_2 Laws Force US Copper Smelters Into Russian Roulette," *Engineering & Mining Journal,* Vol. 172, No. 7, July, pp. 61–71.

White, L., 1975, "Texasgulf Readies for '75 Startup at Comanche Creek Frasch Mine in Texas,' *Engineering & Mining Journal,* Vol. 176, No. 8, pp. 83–88.

White, L., 1980, "Mining in Mexico," *Engineering & Mining Journal,* Vol. 181, No. 11, Nov., p. 70.

Wideman, F.L., 1957, "A Reconnaissance of Sulfur Resources in Wyoming, Colorado, Utah, New Mexico and Arizona," Information Circular 7770, US Bureau of Mines, 61 pp.

Yen, T.F., and Chilingarian, G.V., eds., 1976, *Oil Shale,* Elsevier Scientific Publ. Co., New York, 292 pp.

Zakiewicz, B., 1975, "Exploitation of Bedded Sulphur Deposits by the Hydrodynamic Method," *Sulphur,* No. 120, pp. 35–43.

Zimmerman, J.B., and Thomas, E., 1969, "Sulfur in West Texas: Its Geology and Economics," Geological Circular 69–2, Bureau of Economic Geology, University of Texas, Austin, 35 pp.

Talc

LAWRENCE A. ROE *

RICHARD H. OLSON †

Talc, when it can be isolated as a pure mineral, has a composition of 63.36% SiO_2, 31.89% MgO, and 4.75% H_2O. However, as an industrial commodity, talc rarely approaches theoretical purity. Nevertheless, the impure talc products of commerce find a multitude of uses and few substitutes in many industrial applications.

Minerals commonly associated with and sold together in talcose mixtures are tremolite, chlorite, dolomite, mica, and magnesite. *Steatite* was originally a mineralogical name applied to pure talc. In today's commerce *steatite* generally refers to the massive variety of talc suitable for electrical insulator manufacture. Block steatite ore can be machined into various shapes. Impure varieties of massive or block talc are still commonly referred to as soapstone. Soft massive talc, suitable for crayon manufacture, has been referred to as French chalk.

It has been common practice to discuss talc, soapstone, and pyrophyllite under the same general heading. In the case of soapstone and talc, this is a natural thing since many different types of platy, soft minerals exhibiting a high degree of lubricity (commonly referred to as "slip") have been grouped together and called soapstone or talc.

Further, there is a mineralogical relationship of pyrophyllite to montmorillonite and talc to hectorite. Thus, talc and pyrophyllite are sometimes referred to as clay minerals. When finely divided talc or pyrophyllite are combined with water in proportions to make a slurry, the end product does have the appearance of a clay-water mixture.

Soapstone for utensils and ornaments was mined by prehistoric Indians on Santa Catalina Island, CA. In the mid-1800s, soapstone from deposits along the western foothills of the Sierra Nevada mountains was used by white settlers for building and ornamental stone and in the linings and foundations of furnaces (Anon. 1956). Previous to 1916, the annual recorded production in California did not exceed 2000 tons. In the period 1912 to 1918, output rose sharply when the Talc City, Western, and Silver Lake mines were put into operation. From 1916 to 1935, the state's annual output of talc was in the range of 9,000 to 20,000 tons. In the mid-1930s, the use of talc in wall tile grew rapidly. In California, the production of talc grew to 63,000 stpy in 1943. The post-war building boom helped California production to grow to 120,000 tons in 1951. In 1968, production was 165,000 tons, 1969 was 145,000 tons, 1970 was 185,000 tons, and 1972 was 155,000 tons.

Talc mining in New York state dates to about 1878 when a Colonel Palmer and associates opened the first commercial talc mine on the Nelson Freeman farm near Talcville, New York. In 1893, this operation was sold to the International Pulp Co. which changed its name to the International Talc Co. in 1944. The company was acquired by R.T. Vanderbilt Co., Inc. in 1974.

A new talc operation, the Gouverneur Talc Co. owned by the R.T. Vanderbilt Co., began operations near Balmat, NY, in 1948. The initial capacity of the processing plant was 200 stpd (Gillingham, 1950). Subsequent expansions have increased capacity to over 600 stpd.

The Vermont talc industry began just after the start of the 20th century. Talc was discovered in the area of the present Johnson, VT, talc mine in 1902. The American Minerals Co. initiated plant operations at Johnson in 1904. The Magnesia Talc Co. opened a plant at Waterbury, VT, in 1913 (Burmeister, 1963; Trauffer, 1964) and acquired American Minerals Co. in 1923. The Waterbury mine has a long history as a producer of talc crayons.

The Eastern Magnesia Talc Co. was formed in 1924 by a merger of the Eastern Talc Co.

* Consultant, Downers Grove, IL.

† Consultant, Golden, CO.

The Geology section of this chapter was written by Richard H. Olson.

and Magnesia Talc Co. In 1956, the Vermont Mineral Co., with a talc mine at Hammondsville was also acquired. A new plant was then built at Gassetts to produce roofing products.

In order to produce high-grade products for several markets, the Eastern Magnesia Talc Co. began operating a modern froth flotation plant at West Windsor, VT, in 1964. The ore is provided by the Hammondsville mine. As of 1972, this operation is known as Windsor Minerals, Inc., a totally owned subsidiary of the Johnson and Johnson Co.

The Texas talc deposits were first mined commercially in 1952 and by 1970 ranked fourth in US production behind New York, California, and Vermont. Principal use of this talc is in the ceramics industry. Lundquist (1970) reports on a process for acid treatment, calcination, and grinding Texas talc and talc-kaolin mixtures. By his process, Lundquist increased the brightness value from 65 to over 94.

End Uses

Talc is an extremely versatile mineral. In spite of the relatively low purity of most ores mined, talc has found a steadily increasing number of uses. With the exception of pure steatite grades, hand-picked platy cosmetic talcs, and a few products from wet processing plants, the industrial products are really mixtures of many minerals. For example, much of the talc used by the ceramic industry is a mixture of platy talc and tremolite; most of the filler-grade talc sold to the paper, plastics, and rubber industry is, at best, about 90% talc mineral with the balance being dolomite, calcite, chert, clays, serpentine, chlorite, actinolite, iron and manganese-containing minerals, and carbonaceous material.

Table 1 lists chemical analyses of typical talc ores and products.

Talc, along with many of the filler-type industrial minerals, is subject to steadily increasing demands for higher purity and better quality control. A good example is the talc sold to the cosmetic and toilet goods industry. Prior to the end of 1960, a wide variety of talc and talcose products were accepted for cosmetic use if they met rather loose and uncomplicated specifications. A good cosmetic talc had to be reasonably light-colored, have good "slip," be free of gritty or hard mineral particles, and measure up to a certain fragrance-retention standard. Now, the cosmetic talc market is a demanding one. Not only has the industry itself set higher and new standards, but new rulings from federal

TABLE 1—Typical Chemical Analyses of Talc Ores and Products, %

	Pure Talc (Theoretical)	1	2	3	4	5	6
SiO_2	63.36	35.98	59.15	47.92	62.65	59.80	54.92
MgO	31.89	32.95	31.34	26.00	30.23	27.45	27.20
Fe_2O_3	–	0.65	3.36	6.82	1.51	0.05	0.46
TiO_2	–	0.02	–	0.15	–	–	–
Al_2O_3	–	0.43	0.26	7.35	0.31	0.57	–
CaO	–	0.00	0.15	4.14	trace	6.80	5.76
K_2O	–	0.00	–	0.00	0.05	–	–
Na_2O	–	0.00	–	0.00	0.15	–	–
CO_2	–	20.45	1.76	–	0.27	1.18	–
H_2O	4.75	2.73	4.30	0.05	4.87	–	–
MnO	–	0.41	–	0.00	–	0.39	–
S	–	0.06	–	0.09	–	(SO_3) 0.07	–
NiO	–	0.21	–	–	–	–	–
Cr_2O_3	–	0.18	–	–	–	–	–
CoO	–	0.01	–	–	–	–	–
FeO	–	5.96	–	–	–	0.15	–
Less O for S	–	0.05	–	–	–	–	–
P_2O_5	–	0.01	–	0.00	–	–	–
LOI	–	–	–	7.51	–	4.75	10.76
	100.00	100.10	100.32	100.03	100.04	101.21	99.10

Source: 1-5, Chidester, 1964; 6, Pence, 1955.

1. Average Vermont carbonate ore.
2. Flotation talc, Johnson Mine, Vermont.
3. Roofing granules, Cohutta Talc Co., Georgia.
4. Steatite, Yellowstone mine, Montana.
5. Average talc ore, Talcville, Gouverneur District, New York.
6. Texas Talc.

agencies have made heavy demands on talc product quality.

Table 2 lists the various properties of talc which are important in specific markets or industries.

Talc in Paint

In 1979, an estimated 237,000 tons of talc were used as extender and filler pigments in the paint industry. This is 24.7% of the total US talc production of about 960,000 tons for that year. This paint pigment extender tonnage was exceeded only by calcium carbonate.

Some of the important properties of talc used in paint are color, fineness, oil absorption, chemical inertness, and optimum viscosity. The measurement of oil absorption is important. This property is measured by ASTM Method D 281-31 which measures the number of parts of acid-refined linseed oil required to produce a coherent paste with 100 parts of talc pigment.

Talc helps reinforce paint films and also prevents sagging of paint films. It prevents settling of solids to the bottom of the paint container. Talc is very soft and thus contributes minimal abrasion to process and application equipment. It disperses easily in both aqueous and solvent-based paints. Talc is used for gloss control in paints where an exact degree of sheen or luster is required. Gloss is usually determined with a reflectance meter at 60° inclination. Sheen values are determined with the same meter and with the film at 85° inclination.

A very useful, uncomplicated method used for testing the fineness and degree of dispersion of paint and other wet ground products involves use of the Hegman gage. This gage can also be called a "fineness of grind" gage (Anon., 1970). It is a precision-machined, smooth, flat, hardened steel block with one or two grooves or "paths" about 5¼ in. long and ½ in. wide. The groove depth ranges from 100μ (about 4 mils) at one end to zero depth at the other end. A scraper blade is the second component of the gage. In use (on talc products), an aliquot of a mixture of talc and linseed oil (about 1 part talc to 4 parts of oil) is placed in the tapered part of the groove. Then the mixture is wiped down toward the "zero" depth end of the gage with the stainless steel wiper blade. The fineness is determined by holding the gage to a familiar light source and observing the area where the coarse particles begin to appear. The angle between the face of the gage and the line of vision should be not more than 30° nor less than 20°. The fineness readings can be given

TABLE 2—Talc Properties Important to Specific Industries

Paint Industry
- Color (whiteness)
- Particle shape
- Packing quality
- Oil absorption
- Fine particle size (Hegman gage rating)
- Opacity

Paper Industry
- Free of grit (low Valley Abrasion value)
- Color (with MgO as 100, prefer 90 or higher)
- Opacity
- Particle size (less than 5μ)
- Low alkali content
- Effective in controlling pitch, oil, or other oleoresinous substances
- Talc pigment gives lower wax pick values than clay pigments
- Talc gives lower ink receptivity than clay

Ceramic Industry
- Uniform chemical composition
- Constant amount of shrinkage on firing
- Fired color
- Particle size distribution

Cosmetic Industry
- Contain only traces of dolomite, tremolite, quartz, or any harsh minerals
- Some chlorite may be acceptable
- No color change after heating
- Odorless; have good fragrance retention when compounded
- 100% through 100 mesh and 98% minimum through 200 mesh; finer grades as specified by buyer
- Neutral to litmus paper
- Water-soluble substances, 0.1% maximum
- Acid-soluble substances, 2.0% maximum
- Loss on ignition, 5.0% maximum
- Arsenic (as As); 3 ppm maximum
- Lead (as Pb), 20 ppm maximum
- Total aerobic plate count including yeast and mold, 100 per g maximum
- Gram negative bacterial plate count, less than 10 per g
- Good "slip" and "unctuosity" when applied to human body, also the white color should disappear as the talc is rubbed into the skin
- Good deodorizing ability

Plastics Industry
- Low iron content
- Particle shape
- Reinforcing ability
- Compatibility with resins and other components (talc is inert)
- Superfine particles
- Resistivity

Roofing Industry—Asphalt Backing and Surfacing
- Minimum oil absorption
- Color
- Particle size consist
- Brightness
- Particle shape

Petroleum and Automotive Industries—Lubricants, Body Putty, Asphalt Undercoating
- Free from grit (pure platy talc for lubricants)
- Chemically inert
- Nonwicking (undercoating)

Rubber Industry
- Good lubricity
- Free of grit
- Color (only in white rubber and latex)
- Resistivity
- Must be chemically compatible when used as latex filler

in microns or mils. A reading of 6 on the gage is about equal to 25μ.

Talc in Paper

One of the fastest growing uses for talc has been in the coating and filling of paper. As a filler it is effective as an inexpensive titanium dioxide extender and is outstanding as a pitch control additive. Talc also assists in improvement of gloss, opacity, and brightness. Its retention in the paper fibrils usually approaches 100%. In 1970 (Cooper and Hartwell, 1970), a forecast of a 3.75% annual increase in demand for talc paper filler was made. Present indications are that this is a conservative estimate and the annual increase may reach as much as 5%.

The main reason for the rapid growth of talc consumption in paper is the availability of high-quality micronized talc manufactured to exacting specifications. Talc filler should not contain more than 2 to 5% calcium carbonate, or similar minerals which react with the alum used in paper formulations. Particle size ranges from 5μ to less than ½μ with surface areas from 4 to 25 m^2/g. Generally, the Valley Abrasion * number will be very low, from 2 to 6; thus, abrasion of paper making equipment and printing equipment is minimized.

When talc is used as a titanium dioxide extender in paper, it has a distinct weight advantage (talc specific gravity 2.8 vs. 4.2 for TiO_2). This feature grows in importance as postal rates increase.

High-purity talc products are finding a good market in competition with titanium dioxide fillers. Considerable history has already been made in the paper industry where talc fillers have found an excellent market for use as titanium dioxide extender agents. In 1979, titanium dioxide (anatase) sold for about 53¢ per pound in slurry shipments and 57¢ per pound in bags.

The best grades of talc sold, in 1979, for only 5 to 10¢ per lb. This indicates that talc, when it can be substituted, even partially, will continue to be a formidable competitor for substantial titanium dioxide markets.

* The Valley Abrasion tester measures wear (in milligrams), of a wire cloth specimen, caused by a circulating suspension of the talc and the rubbing action of an oscillating Micarta block. After 6000 cycles, the wire cloth is washed, dried, and weighed. The loss in milligrams is the Valley Abrasion number.

The unique property of preferentially wetting oily materials in the presence of water makes talc extremely effective in "pitch control." Pitch and other oleoresinous components of paper pulps cause serious manufacturing problems if they are not controlled. Talc absorbs the pitch and helps prevent accumulations which would otherwise deposit on the rolls, wire, and other parts of the paper machine. Competitive materials such as diatomaceous earth and clay usually are either too abrasive or much less effective in the presence of water.

The ultrafine particle size talc products (maximum particle size of 5μ) now finding use in paper coating colors have similar ranges of particle size distribution as coating clays and precipitated calcium carbonate. The talc products can be used to pigment latex, starch, or alpha protein coating colors. The ultrafine talc products can be used along with clay, calcium carbonate, and titanium dioxide pigments to control rheological properties, calendering, gloss, ink holdout, opacity, and brightness.

Talc in Ceramics

Talc finds use in many ceramic products. It is a versatile component having beneficial effects relating to the control of thermal expansion of ceramic bodies. Talcs low in accessory carbonate minerals exhibit low ignition losses. A typical New York talc will have an ignition loss, at 2100°F, of 5 to 6%. In comparison, many Texas and California talcs lose 8 to 13% of their weight when fired.

Talc is used in cordierite (talc-clay mixture fired to Cone 12) bodies such as developed for use in electrical insulators. Cordierite has a very low thermal expansion. New York and Texas talcs are used in the semivitreous whiteware field, especially wall tile and in dinnerware bodies. These have been found to be ideal raw materials for the following reasons:

1) The formation of enstatite produces high thermal expansion bodies resulting in glazes being put in compression which in turn tends to prevent crazing (hairlike cracking in a glaze).

2) Low moisture expansion bodies are produced, resulting in good resistance to delayed crazing.

3) The hard, massive nature of the ores aids in either dry-pressing or in the making of good casting slips.

4) Low firing temperatures are possible.

5) Fast firing schedules are possible—as fast as 1 hr cold to cold.

6) Shrinkage and absorption of bodies are fairly constant over a long temperature range.

7) Good white-fired bodies are possible.

8) Glazes of unusual brilliance and attractiveness can be fitted readily to high talc bodies.

As far back as 1930 reports began to appear praising the improved craze resistance of semi-vitreous bodies when New York talc was used.

A minor use of talc is as an addition to the clay-flint-feldspar type of dinnerware bodies. Talc is added up to 6% of the body and results in greatly improved craze resistance.

Talc is used as a glaze ingredient as it provides a low-cost source of MgO.

Talc is used as a mold dusting compound to help the release of large castware pieces.

Synthetic lava talc is made from ground steatite-grade talcs. In these applications, as much as 80% talc is used in the mixtures. When steatite-grade talcs are used for the manufacture of high-frequency insulators, limits of 1.5% CaO, 4% Al_2O_3, and 1.5% iron oxides are usually imposed. This extrapolates to a maximum of about 5 to 10% nontalc minerals. Preferably, this should be only 1 to 2% nontalc components.

Talc can be used to replace clay in many ceramic bodies. It promotes translucency of the finished ware and makes it more durable.

Specifications for talc used in the ceramic industry are increasingly more demanding. The uniformity of most ceramic talcs supplied by producers using hand-sorting or other primitive beneficiation methods is only fair at best. As more sophisticated ore processing methods are applied to talc ores, the ceramic industry is beginning to receive higher quality talcs with very uniform chemical and physical properties.

The use of talc in ceramic tile has not even approached the increases in residential and commercial construction. For example, a 50% increase in housing starts in 1970 to 1971 gave only a 10% increase in tile consumption (Perry, 1972). New grades, sizes, and colors of tile are being developed to help increase the market. The Tile Council of America forecasts replacement of portland cement as a setting material in the future (Bennett, 1972). Also, increased use of edge-bonded panels and colored grouts is predicted.

Pence (1955) gave an early report on white-firing talc for ceramic use from west Texas. The talc described was from claims located in Hudspeth County about 23 miles west of Van Horn, Texas. The ore analysis was 54.92% SiO_2, 27.20% MgO, 5.76% CaO, 1.36% Al_2O_3, 0.46% Fe_2O_3, and 10.76% ignition loss. By firing standard 4¼ x 4¼ x ⅜-in. hand-pressed tile, Pence obtained satisfactory results in color, density, glaze fit, and other properties required in ceramic wall tile.

At 750° to 900°C talc loses its one molecule of combined water; at 900° to 1000°C it dissociates into the metasilicate 3 $MgO \cdot SiO_2$ and silica. At 1200° to 1300°C it converts to clinoenstatite, $MgO \cdot SiO_2$. The final products of the thermal decomposition of talc are clinoenstatite and cristobalite.

In pottery, talc may constitute 40% of the body composition. The thickness of the object being made, the texture, porosity required, and the firing conditions all influence the amount of talc used. Tableware often contains 15% talc.

Talc in Plastics

Talc is an important filler material for plastics. It improves chemical and heat resistance, impact strength, dimensional stability, stiffness, hardness, thermal conductivity, tensile strength, creep resistance, and electrical insulation properties. In some cases, the processability, especially faster molding cycles, of the plastic is also improved. Talc can be used in both thermoplastics and thermosets.

Talc is used as a reinforcing extender filler in many thermoplastics. It controls melt flow, reduces creep in molded parts, increases molding cycles, increases heat deflection temperature, and dimensional stability. When platy varieties of talc are used, a beneficial lubricating effect is noted on molding machine parts. And perhaps most important of all, the use of talc decreases the overall cost of the plastic product. Talc can be used as a filler in quantities from 1 up to 50% by weight when polyolefin compositions are involved (Poppe et al., 1972). In a rare case, up to 70% talc has been used in a cellulose ether that is both thermoplastic and water soluble.

In all filler applications, where talc is added to an organic component, a new dimension is rapidly developing. This is the chemical bonding of the talc filler to the resin matrix (in the case of plastic composites). The technology involved is old since initial successes in bonding glass fibers to various plastics were reported in the early 1950s.

The development of processes for the chemical coating of talc fillers is now reaching a considerable degree of sophistication. Silane coupling agents are used to improve performance of talc fillers in plastic materials. Ranney, et al. (1972) report that the tensile strength of a 50% (by weight) loaded polyethylene plastic can be increased 25 to 30% over the non-silane-containing standard.

In both rubber and plastics, the use of filler minerals was originally for the sole purpose of reducing cost. Thus such minerals were called extenders or extender pigments. With the advent of products such as silane-coated clays in the rubber industry, it became evident that filler minerals could be functional as well as reduce overall cost.

Silane coupling agents are used with clays, talc, and other mineral fillers. These agents are a chemical family of organosilicon monomers which characteristically possess two different kinds of chemical functionality. The general formula for the molecule is: $R—Si(OX)_3$ where R is the organofunctional group attached to the silicon atom in a thermally and hydrolytically stable manner. In this formula, OX designates hydrolyzable groups upon silicon. These groups are generally believed to be the means whereby the silane coupling agents interact with the siliceous inorganic surfaces (e.g., talc) to form a more durable bond with the filler mineral. With the increased bonding, it is often possible to increase the filler loading without sacrificing the properties required in the polymer system.

Talc in Roofing Products

Talc is used both as a filler and as a surfacing material in the manufacture of roofing. When used as a filler, the talc acts as a stabilizer for the melted asphalt constituents. It helps increase stability and weather resistance. When talc is sprinkled on the surface of either asphalt shingles or roll roofing, it prevents the roofing from sticking during manufacture and storage before use.

Typical size specifications for talc used for stabilizing roofing products are:

Tyler Mesh	Range, % Wt.
+ 35 mesh	none
+ 65 mesh	10-20
− 65, + 100 mesh	15-30
− 100, + 200 mesh	25-35
− 200 mesh	20-25

The market for low-grade talc used for surfacing, filler, and backing for smooth surface roofings and for filler and backing of asphalt roofing shingles is growing rapidly. When used for surface coatings, the color and staining properties of the product are important. When talc is used as backing, the primary qualifications are particle size range, particle shape, absorbency, and above all, low cost.

Miscellaneous Uses of Talc

Talc is used in a wide variety of textile materials. Textile sizing compositions include binders and mineral fillers. Some textiles use sizings for temporary stiffeners and appearance for sales appeal. The degree of whiteness is often very important and thus favors talc or other extremely white minerals. Where extreme whiteness of color is important, talc often competes with titanium dioxide at a considerable cost advantage.

A relatively new market for talc and pyrophyllite fillers is in the integral foamed latex rubber backing for carpets, rugs, and parquet hardwood floor panels. When used with fabric, the main purpose of the latex is to provide a flexible backing for the fabric and, at the same time, lock the pile material to its own backing. Impurities in the filler such as calcium, magnesium, iron, and manganese which can chelate with the latex are undesirable. Specially coated fillers with additional latex reinforcement properties are under development.

Upholstery fabric backing also uses substantial tonnages of mineral fillers. The filler loadings run about 75 to 100 parts per 100 parts of latex. Major fillers are clay, talc, pyrophyllite, and calcium carbonate. Similar back coatings are used in drapery materials, but the filler loadings are lower than for upholstery. Talc is a favored material because of its ability to meet the demand for whiteness of color.

The lubrication properties of talc are unique and found in very few minerals. The history of talc as a component in extreme temperature range greases is extensive and significant. The dispersion characteristics so important to lubricants which operate over a wide temperature range without freezing or running off the bearing surface are assisted by use of talc. A representative grease capable of use over an extreme temperature range might include a silicone oil, colloidal silica thickener, molybdenum disulfide, talc, and several chemical additives to prevent corrosion of the bearing surface or decomposition of the lubricant (Elstron, 1961).

Both platy and nodular talc powders are covered by US Patents for use in corrosion proofing compositions such as those used in automotive undercoating (Miller and Holtzapfel, 1970). The talc is about 50% by weight finer than 10μ and no more than 3% coarser than 325 mesh. Talc plays an important part in the prevention of "wicking" action in the composition. Undercoating products containing fillers such as asbestos are often subject to failure resulting from a wicking action which causes moisture to travel through the corrosion proofing coatings.

The use of talc in dry fire-extinguishing powders has been patented (Cawood, 1962). Talc is used in quantities ranging from 10 to 15 parts per 100 parts of product. The talc is used primarily to maintain the desired free-flowing condition so essential in dry powders used in fire extinguishers.

Talc is used in the textile industry both for loading and "bleaching" materials such as cotton sacks, cordage, rope, and string. For such purposes, a good grade of low-abrasive white talc is required. Talc used for backfilling textiles can be of lower quality.

Other uses for talc include:

- Cereal polishing (rice, corn, barley)
- Bleaching agents
- Odor absorption from foods
- Floor wax
- Water filtration
- Leather treatment (oil absorption)
- Joint fillers and grouts
- Insecticides
- Textiles
- Shoe polishes
- Welding rod coatings
- Printing inks
- Encapsulant for acceleration testing artillery shells up to 50,000 g
- Coating for iron ore pellets in direct reduction processes
- Source of magnesium in plant foods

Mineralogy

Talc is a hydrated magnesium silicate with the theoretical formula $Mg_6[Si_8O_{20}](OH)_4$. The chemical composition of talc found in nature, however, is quite variable. It is an extremely soft mineral and is rated No. 1 in Mohs' scale of hardness. It varies in color from snow-white to greenish gray and various shades of green. Its specific gravity ranges from 2.58 to 2.83. Talc usually derives as a secondary mineral by alteration of other magnesium silicates such as serpentine, pyroxene, or siliceous dolomite.

In industry the word talc covers a broad variety of rocks and rock products.

Talc deposits can be classified into four major categories:

1) Steatite—compact, massive, cryptocrystalline, can be sawed, drilled, or machined to required shapes. Steatite converts on firing at 1800°F for 6 hr to interlocking crystals of clinoenstatite—then called "lava." This has good electrical insulating properties.

2) Soft platy talc—is an alteration product of sedimentary magnesium carbonate rocks. Chlorite is a common accessory mineral. This is the most important type of talc. It possibly has more uses and more potential than any other talcose material.

3) Tremolite talc—sometimes called "hard" talc. It is massive or laminated rock composed of varying percentages of tremolite, anthophyllite, calcite, dolomite, serpentine, and true soft talc. It is characterized by calcium oxide contents of 6 to 10%.

4) Mixed talc ores—includes the so-called "soft talc," a friable, white schistose rock composed of platy talc, dolomite, calcite, serpentine, and many other trace minerals. In some low-grade deposits, mixtures of talc, chlorite, and dolomite are common.

Talc is a mineral that meets the criteria of a "sheet silicate" mineral. Its structure is like that of pyrophyllite except that the octahedral sites in the layers or "sheets" are occupied by magnesium instead of aluminum. Also, none of the octahedral sites are vacant (which is not the case with pyrophyllite). Chemically pure talc is rarely found in nature in commercial quantities. Practically all of the talc produced is an impure product. Even the hand-sorted, selectively mined cosmetic talcs contain many extraneous mineral components.

When the major talc products in world trade are examined, it becomes obvious that only a miniscule percentage of world reserves are "pure" talc. When impure talc products are reprocessed by froth flotation, or similar selective separatory methods, very high purity talc mineral can be produced. It would seem objective to call a talc "pure" if it contains upward of 95% $Mg_6[Si_8O_{20}](OH)_4$.

Table 3 lists the approximate composition of the minerals present in talc ores.

The main isomorphous components of the tremolite group are ferric iron, manganese, and aluminum. Tremolite can contain 2 to 3% sodium. Tremolite includes three varieties

TABLE 3—Approximate Composition of Minerals Present in Talc Ores, %

	SiO_2	MgO	CaO	CO_2	Fe	Al_2O_3	K_2O	Na_2O	H_2O	TiO_2
Talc	63	32							3-7	
Serpentine (antigorite)	44	43							8-13	
Chlorite	33	36				18			5-14	
Anthophyllite	58	30	2						15-2.2	
Tremolite	57	28	13						15-2.3	
Actinolite	52	5	9		34				3	
Diopside	56	18	26							
Feldspar	65					18	17			
Magnesite		48		52						
Dolomite		22	30	48						
Calcite			56	44						
Quartz	100									
Muscovite	46	0.1				39	10	0.6	4.3	
Magnetite (Fe_3O_4)					72					
Ilmenite ($FeTiO_4$)					33					48.0
Pyrite (FeS_2)					47					
Tourmaline*	36	11	0.5			33	0.6	2.3	3.8	0.2
Graphite (100% C)										

* 10% B_2O_3

—iron tremolite, manganese tremolite (hexagonite), and fluorine tremolite. Tremolite, which has the theoretical composition $2CaO \cdot 5MgO \cdot 8SiO_2H_2O$, is commonly the major component in talc ores, at the expense of the mineral talc itself. This relationship may hold throughout large talc-mining districts, e.g., the New York and California-Nevada districts. Other talclike minerals which may constitute various proportions of talc ores are anthophyllite, antigorite, chlorite, pyrophyllite, and serpentine.

Povarennykh (1972) includes pyrophyllite, talc, minnesotaite, and willemseite in the pyrophyllite group of minerals. In his description of the group varieties, he includes Mg-pyrophyllite, Fe-pyrophyllite, Fe^{2+}-talc, $Fe^{2+}Fe^{3+}$-talc, Ni-talc, and Mg-willemseite. Talc can contain up to 1% nickel and pyrophyllite can contain up to 9% magnesium. These chemical variations of the pyrophyllite group show the wide variety and possibilities of the occurrence of "talclike" pyrophyllite.

Barnes (1958) reported weak to medium infrared luminescence of talc minerals from various parts of the world. Steatitic talc from Manhattan, NV, showed medium luminescence. Tremolite from Kragero, Norway, and Outokumpu, Finland, also exhibited medium reaction.

At the Jatoba talc mine near Ponta Grossa, Brazil, a claylike antigorite has found commercial acceptance as a diluent for insecticides and as "talc" for pharmaceutical uses (Brindley and Santos, 1971). This is another example of a talclike mineral finding commercial acceptance as "talc."

Faust and Fahey (1962) provide a good review of extensive studies by others relating to the distribution of minor elements in magnesium-rich minerals and rocks. There is good agreement by the scientists that "the rocks of the ultrabasic suite and their hydrothermally altered forms rich in serpentine contain an unusually high concentration of nickel, chromium, cobalt, and scandium. Moreover, some of the investigators have observed that the carbonate-rich rocks, in contrast, contain very small to practically negligible concentrations of these elements." This is in agreement with the high nickel content of some Vermont talcs and its complete absence in some western carbonate-talc ore bodies.

Differential thermal analysis (DTA) and X-ray diffraction both provide good analytical methods for identification of talc and also many of the impurities associated with talc ores. A large endothermic peak at 965°C is typical of relatively pure talc mineral analyzed by DTA methods.

It is of considerable geological significance that tremolite from some metamorphosed carbonate rocks contains fluorine. On the basis of fluorine analyses of 22 talcs from worldwide deposits, it was noted talc from metamorphosed sedimentary rocks contained from 0.11 to 0.48% fluorine. Talcs derived from ultramafic rocks contained less fluorine (Ross and Smith, 1968).

Later work by Troll and Gilbert (1972) on fluorine-bearing tremolites points out that the theoretical value for pure fluoro-tremolite is 4.65% fluorine. One sample of tremolite used in this work from Richville, NY, contained

1.51% fluorine. Analysis of a sample of lavender-colored tremolite showed 0.30% fluorine. The US Bureau of Mines (USBM) and the National Institute of Environmental Health Sciences (NIEHS) have reported on the characterization data of fine mineral test samples, including nonfibrous tremolite (Campbell, et al., 1980).

Occurrence and Reserves

Talc deposits are found in many countries. Good quality talc ores are found in Africa, the Americas, Australia, Republic of China, Finland, France, Greece, India, Italy, North and South Korea, Manchuria, Norway, Romania, Sardinia, and the USSR.

Table 4 is a summary of geologic and commercial data for the principal talc districts of the United States (Chidester, et al., 1964).

As governmental regulations concerning exposure to mineral dusts increase in frequency and severity, those talc deposits containing asbestiform minerals may decrease in importance. The filler industry shows preference for very high-purity platy talc which has minimal health hazards upon exposure to its dust.

Geology

Talc deposits occur on every continent and the geology of these deposits is roughly similar to those of the United States. The most common host rocks for the formation of talc are dolomite and ultramafic rocks. Chidester et al. (1964) consider that there are three major types of talc deposits in the United States: deposits associated with sedimentary rocks; deposits associated with ultramafic igneous rocks; and deposits associated with mafic igneous rocks.

Good examples of talc deposits associated with sedimentary rocks are those of the California-Nevada, Montana, New York, and Texas districts.

Good examples of talc deposits associated with ultramafic igneous rocks are those of the Vermont district.

Talc deposits associated with mafic igneous rocks occur in the central Appalachian Mountains and are the source of most of the soapstone quarried in the United States. These rocks, which contain relatively minor amounts of the mineral talc, were probably formed by the hydrothermal alteration of gabbroic rocks. As they are not a source of talc products, they will not be discussed here. The first two types of talc deposits, however, are important as a source of talc products.

The easiest method of forming pure or relatively pure talc deposits is shown in the following reaction as presented by Winkler (1974):

$$\text{3 dolomite} + \text{4 quartz} + H_2O = \text{talc} + \text{3 calcite} + 3\ CO_2.$$

The geological settings of US talc deposits are as varied and diverse as would be expected for any naturally occurring mineral commodity of secondary origin, but there are nevertheless some broad similarities. First, talc is invariably a secondary mineral formed in preexisting rocks either directly from those rocks, or through the introduction of new material. Second, all economically important deposits have been formed under conditions of low-grade regional metamorphism. Third, commercial bodies are generally tabular and concordant, being molded after the shape of the parent material, which is most commonly a sedimentary or metasedimentary rock unit. Fourth, most of these talc ore bodies occur within Precambrian rock units; the few which do not are found within rock units whose age is no younger than Early Paleozoic. Finally, in almost every instance where the conditions of genesis can be either firmly established or confidently inferred, the time of the talc formation is Precambrian as well.

The reasons for this almost total restriction of talc to the Precambrian in both space and time are presently unclear. It may be easier to comprehend in the case of ultramafic rocks, for these may indeed be much more abundant in the Precambrian than in younger geological systems. It is, however, somewhat more difficult to understand with regard to dolomitic rocks, for these are certainly much more common in the Cambrian and younger geological systems than they are in the Precambrian. It may be that deep burial and the consequent conditions of dynamic metamorphism could have attained sufficient intensities only during Precambrian time; such a theory would perhaps explain the formation of talc in Montana and New York, for instance, but would probably not be applicable to some of the other districts, which admittedly have undergone deep burial but do not show conditions of extreme dynamic metamorphism and isoclinal structure, such as California-Nevada. It seems safe to say, how-

TABLE 4—Summary of Geologic and Commercial

State and District	Form and Size of Talc Body	Country Rock and Parent Rock	Paragenetic-Sequence in Alteration of Parent Rock to Talc
California: Silver Lake—Yucca Grove	Lenses, mainly of massive tremolite rocks, as much as 800 ft long, 20 ft thick, and 200-300 ft deep; schistose/talc rock forms layers and lenses 1 ft or more thick, mostly along the footwall.	Feldspar-quartz-diopside-carbonate hornfels, commonly separated from the talc body by phlogopite-feldspar-tremolite schist a few inches to several feet thick. Tremolite and talc apparently derived from beds of sedimentary dolomite 5-20 ft thick that form parts of pendants in bodies of granitic rock.	Dolomite → tremolite → talc.
Southern Death Valley, Kingston Range.	Tabular to lenticular bodies as much as 5000 ft long, 80 ft thick, and 400 ft downdip; most are of schistose to massive talc rock, but some contain interlayered tremolite rock in various proportions, and some are entirely of tremolite rock.	Cherty and noncherty dolomite, diabase, and argillite, commonly silicated at the contact. In most deposits, one kind of rock forms the hanging wall, another the footwall. Talc and tremolite derived from generally massive and commonly cherty or siliceous beds of dolomite.	Carbonate+chert → tremolite → serpentine → talc. Intermediate steps in the process did not take place in many deposits.
Inyo Range—Northern Panamint Range.	Small pods to lenses and tabular bodies as much as 500 ft long, 50 ft wide, and 400 ft downdip. Some bodies show large-scale variations related to differences in parent rock, permitting selective mining.	Principally dolomite and quartzite, in a few places granitic rock and limestone. Talc derived principally from massive beds of dolomite and quartzite, locally from granite or diorite.	Dolomite and quartzite → talc; also granite and diorite → talc.
Georgia: Chatsworth	Lenticular bodies as much as 5000 ft long, 150 ft thick, and 300 ft downdip. Lenticular and spindle-shaped bodies of massive talc rock (crayon talc) are dispersed through talc-carbonate rock that shows large-scale variations in color, texture, and mineral composition.	Schist, slate, granite, and granite gneiss. Mode of origin doubtful; parent rock either sedimentary carbonate rock, mafic igneous or volcanic rock, or ultramafic igneous rock.	Probably: dolomite → serpentine → (chlorite) → talc. Possibly: ultramafic or mafic igneous rock → serpentine → (Chlorite) → talc+carbonate.
Maryland	Irregular tabular masses adjacent to pegmatite dikes in serpentinite; irregular envelopes surrounding a serpentinite core; and lenticular masses.	Schist, gneiss, pegmatite, and serpentinite. Talc derived by replacement of parts or the whole of serpentinite bodies that vary widely in size.	Dunite and peridotite → serpentine → talc+carbonate.
Montana: Dillon-Ennis	Small veins and pods to lenses as much as 700 ft long, 100 ft thick, and more than 200 ft in vertical extent.	Dolomite and quartz-mica schist. Talc derived from beds of dolomitic marble that range widely in size.	Dolomite →talc; also dolomite → tremolite → talc.
Nevada: Palmetto-Oasis	Elongate zones of discontinuous pods and lenses to tabular masses more than 5-500 ft long; generally 1-15 ft, but as much as 50 ft thick; and 140 ft downdip.	Chiefly dolomitic marble, hornfels, and diabase; locally granitic rock, phyllite, and schist, commonly one kind of rock forms the hanging wall, another kind the footwall. Talc derived chiefly from extensive beds of dolomitic marble and hornfels, but locally from phyllite, schist, and granitic rock; commonly occurs at fault contacts.	Chiefly: dolomite → talc. Locally: hornfels, granite, schist, and phyllite →talc.
New Mexico: Hembrillo Canyon—San Andres Mountains.	Lenses as much as 300 ft long, 25 ft thick, and of undetermined depth.	Chiefly argillite; locally carbonate rock and silica-carbonate rock. Talc probably derived from extensive sedimentary units of argillite.	Not determined
New York	Sheetlike to lenticular zones as much as several miles long and 300 ft wide.	Dolomitic and calcitic marble, diopsidic marble, and thin quartz-diopside layers. Talc derived chiefly from beds of dolomite and siliceous dolomite that form units as much as 5000 ft thick and several miles long; some talc derived from granite, syenite, quartzite, and amphibolite.	Dolomite + quartz → diopside → tremolite → anthophyllite → serpentine → talc. One or more minerals may be omitted in the sequence.
North Carolina: Murphy	Small pods to lenses as much as 700 ft long, 50 ft thick, and 200 ft downdip. Many deposits exhibit large-scale variations in color, texture, and distribution of impurities, permitting selective mining.	Calcite marble, locally highly siliceous and commonly tremolitic at the borders of the talc deposits.	Dolomite+quartz →silicated rock →talc.

Data for the Principal Talc Districts of the United States

Mineral Composition and Impurities	Physical and Chemical Characteristics of the Commercial Talc	Commercial Use	Production through 1955, References, and Remarks
Chiefly tremolite and talc with minor forsterite, serpentine, and carbonate. Impurities include phlogopite, chlorite, quartz, carbonate, and mafic dike rocks.	Tremolite rock mostly massive; talc rock commonly schistose.	Wall tile, pottery, rubber, insecticides.	About 320,000 tons (Wright, 1954; 1957).
Principally talc; tremolite mostly-subordinate, but in places predominant. Impurities chiefly carbonate, iron oxide stains, and wall-rock inclusions.	Talc rock blocky and massive to schistose or laminated. Tremolite rock massive to laminated. Particle shape acicular to platy.	Paint, ceramics, paper, textiles.	About 820,000 tons (Wright, 1957).
Chiefly talc; tremolite rare. Impurities generally very minor; chiefly chlorite, carbonate, quartz, pyrite, and iron oxide.	Talc rock mostly massive and blocky; ranging in color from dark gray to white or pale green; shows large-scale variations in color and texture.	Steatite, ceramics, cosmetics, pharmaceuticals, paints.	About 340,000 tons (Page, 1951; Wright, 1957).
Talc rock principally talc, minor magnetite; talc-carbonate rock chiefly talc and carbonate with minor magnetite, pyrite, chlorite, quartz, feldspar, actinolite, and sericite.	Talc rock varies from massive to schistose, translucent medium green. Talc-carbonate rock varies from medium to dark gray, and varies widely in texture and composition.	Crayons, cosmetics, paint, rubber, roofing, insecticides, lubricants, foundry facings, dusting agents.	Probably about 850,000 tons (Furcron and others, 1947).
Talc rock chiefly talc; impurities, chiefly chlorite, variable. Talc-carbonate rock essentially talc and carbonate, with chlorite, magnetite, serpentine, and other impurities.	Talc rock associated with pegmatites massive, fairly high in iron. Other talc rock and talc-carbonate rock commonly schistose.	Nonsteatite block talc, low-grade ground talcs.	About 320,000 tons (Pearre and Heyl, 1960, p. 795).
Essentially talc with minor amounts of impurities such as chlorite, graphite, dolomite, manganese and iron oxides, and inclusions of country rock.	Massive and blocky to schistose nearly pure talc; ranges in color from white to pale green and buff; particles range from subequant to flaky.	Steatite, possibly block steatite, cosmetics, paints, paper, ceramics.	About 200,000 tons (Perry, 1948).
Chiefly talc with minor chlorite and carbonate, locally some tremolite. Much of the material is nearly pure talc.	Massive to schistose talc rock, ranging in color from dark gray to pale green and white. Talc particles commonly finely platy.	Cosmetics, pharmaceuticals, paints, paper, insulators. Much probably suitable for steatite.	About 110,000 tons.
Chiefly talc with minor chlorite; locally stained with iron oxide.	White to gray talc rock with platy particles.	Cosmetics. Some hand-sorted material probably suitable for steatite.	Probably less than 10,000 tons. Incorporated in White Sands Military Reservation and withdrawn from mining in 1942.
The proportion of tremolite, anthophyllite, talc, and serpentine varies widely; any may predominate, but tremolite rock is most common. Quartz, carbonate, diopside, manganese and iron oxides, and gypsum are minor impurities.	Varieties of commercial talc range from schistose to massive. Particles range from nearly equant to bladed and fibrous. Very white.	Largely paint and ceramics; also rubber, insecticides, foundry facings, plastics, and other uses.	About 6,500,000 tons up to July 1961.
Essentially talc. Impurities are chiefly minor amounts of quartz, carbonate, iron oxide stain, and rarely tremolite.	Massive to schistose talc rock, ranging in color from light gray to white. Particles subequant to flaky.	Crayons, cosmetics, paint, paper, textiles, ceramics; some suitable for steatite.	Probably more than 100,000 tons (Van Horn, 1948).

TABLE 4—Continued

State and District	Form and Size of Talc Body	Country Rock and Parent Rock	Paragenetic-Sequence in Alteration of Parent Rock to Talc
Texas: Allamoore	Largely unknown, but some tabular masses are several thousand feet long and several hundred feet wide, with tabular and irregular inclusions of carbonate rock, phyllite, and masses of chert that range widely in size.	Cherty dolomite; some conglomerate, phyllite, and diabase. Talc probably derived from extensive beds of sedimentary or pyroclastic rock such as dolomitic marl or magnesium-rich tuff.	Not determined
Llano	Mostly small lenses and pods in schist or along margins of serpentinite masses as much as 4 miles long and 1 mile wide.	Schist, gneiss, and serpentinite. Talc derived from lenticular bodies of serpentinite and from tabular and lenticular units of schist, gneiss, mafic volcanic rock, and carbonate rock.	Details unknown. Serpentinite and metamorphosed sedimentary and volcanic rock altered variously to one or more of talc, carbonate, tremolite, and anthophyllite.
Vermont	Shell of talc rock 1-3 ft thick surrounds a zone of talc-carbonate rock that ranges in size from a shell 1-20 ft thick around a serpentinite core to the entire mass of the original serpentinite body.	Schist, greenstone, and serpentinite. Talc rock and talc-carbonate rock derived almost wholly from serpentinite bodies that range in size from small pods to lenses several miles long and a mile wide; a small proportion of the talc rock is derived from schist.	Dunite and peridotite →serpentinite. Serpentinite+CO_2 →talc+carbonate. Serpentinite+SiO_2 →talc. Schist or greenstone →talc+chlorite.
Virginia: Schuyler	Lenticular bodies of the same extent as the original igneous rock body.	Gneiss, schist, and amphibolite. The talc (soapstone) was derived from lenticular bodies, probably originally hypersthene gabbro, as much as 180 ft thick and 1500 ft long.	Hypersthene gabbro →amphibole+chlorite →talc+carbonate.

Source: Chidester et al., 1964.

ever, that even though conditions of extreme dynamic metamorphism have not necessarily been attained in all six talc-mining districts of the United States, conditions of burial beneath miles of rock cover have been attained in every one of them.

The six major talc-mining districts in the United States are New York, Vermont, the southern Appalachians, Texas, California-Nevada, and Montana.

New York

All of the known commercial talc here occurs within the Balmat-Edwards or Gouverneur mining district. A comprehensive study of talc in the United States by the US Geological Survey (Chidester, et al., 1964) indicates that New York had at that time almost 75% of the United States' "measured and indicated" talc ore and about 40% of its "inferred" talc ore. With only relatively minor exceptions, all known talc in this district lies within a narrow belt, which probably averages only 500 ft or so wide and is only about eight miles long.

The New York talc district was long the most important in the United States in terms of total annual production, being rivaled or surpassed only by Vermont. Over the past couple of decades, however, both Montana and Texas have also pulled ahead of New York. It is most peculiar, therefore, that the New York district, although dominant in production for so much of the past century, and even more dominant in terms of proven and potential ore reserves, is geographically the smallest in area of all six talc-mining districts in the United States.

The bedrock complex of the Gouverneur mining district consists exclusively of metamorphic rocks of Precambrian age (the Grenville Series), but two specific lithologic types, gneiss and marble, dominate. The gneiss is thought to be the older unit and is about 3,000 ft thick; the marble, thought to be the younger unit, has a maximum known thickness of about 2,000 ft. Thin sheets and lenses of amphibolite occur within both the marble and the gneiss, but are relatively minor in total volume. Brown and Engel (1956) subdivided the marble sequence into 15 units, of which only Unit 13, the "talc unit," is known to contain commercial deposits of talc, tremolite, or zinc minerals. Unpublished geological investigations indicate that the composition of most of the marble within this mining district is relatively simple and probably does not and did not originally differ greatly along strike. The talc unit, a dolomitic marble, shows evidence of being only 250 ft or so thick in its least contorted areas; greater thicknesses are probably due to structural flowage.

TABLE 4—Continued

Mineral Composition and Impurities	Physical and Chemical Characteristics of the Commercial Talc	Commercial Use	Production through 1955, Reference, and Remarks
Essentially talc. Impurities consist of varying amounts of chert, carbonate, wall rock inclusions, surface weathering products, and organic material.	No information	Chiefly wall tile; also insecticides.	120,000 tons through 1958 (King and Flawn, 1953; Flawn, 1958).
Talc; talc and tremolite; talc and anthophyllite; anthophyllite. Commonly contains impurities of quartz, magnetite, chlorite, and other materials.	Variable in composition and physical properties. Small amounts of massive talc rock.	Chiefly low-grade uses. Some suitable for crayons and for carvings.	Production unknown, but very small (Dietrich and Lonsdale, 1958).
Talc rock is essentially talc with minor chlorite. Talc-carbonate rock essentially talc and magnesite with minor serpentine, magnetite, chlorite, and sulfides.	Talc rock generally schistose, talc-carbonate rock schistose to massive; both are medium to light gray. Talc ranges from fine to coarsely flaky.	Paper, rubber, textiles, paint, roofing, cosmetics, filler, foundry facings, crayons.	About 1,500,000 tons (Chidester and others, 1951, 1952a, b).
Talc, chlorite, carbonate, amphibole, and magnetite.	Massive, felted texture	Sawed and shaped slabs, insecticides.	No reliable production figures available, but total production probably about 2 million tons, of which probably more than 80% was in sawed and shaped slabs (Hess, 1933a; Spence, 1940).

The zinc and talc ores occur within the same stratigraphic unit, but not together (except at a few localities). Locally, the siliceous dolomite of Unit 13 has been converted to tremolite under conditions of dynamic metamorphism and/or metasomatism; subsequently, talc, anthophyllite, and serpentine have been formed locally from the tremolite. In the northeasternmost part of the district, anthophyllite locally may be the major ore constituent.

New York talc ores are commonly more than half tremolite. Tremolitic ores have been proven by underground workings to downdip depths of at least 1,100 ft and by core drilling (generally in search of commercial zinc deposits) to downdip depths of more than 3,000 ft. Bodies of almost tremolite-free talc (platy or scaly variety) have been discovered in surface and underground mines near Talcville; these relatively small deposits, however, are at or near the hanging-wall contact and are difficult to mine by themselves. Field mapping by US Geological Survey workers has indicated that tremolite, talc, and associated secondary minerals were formed prior to 10^9 years ago.

The talc ore zones locally pinch and swell, but they are on the whole conformable with the dolomitic marble layers. The regional dip averages about 45° to the northwest; departures from this are only locally significant. Tight and contorted cross-folds superimposed upon the main talc belt have resulted locally in the thickening of the talc sequence by at least 50%. All of the rocks here are complexly folded; metamorphism to the amphibolite facies is common, if indeed not ubiquitous.

Both open-pit and underground mining methods are employed. A large crushing facility is installed within one underground mine.

The deposits of the New York talc-mining district may well constitute the largest reserves of tremolitic-type talc ore in the world, for nothing described in the literature at the present time is of similar magnitude. The US Bureau of Mines groups New York's production in with six other states; this production in 1979 was 224,000 tons, with most of this obviously coming from New York.

New York talc ores are milled as is, i.e., without any prior beneficiation. Milling is accomplished in ball mills and in fluid energy mills and the product is then air-classified.

Vermont

The district which is commonly known as the Vermont talc-mining district runs through the central part of that state, but also extends southward into Massachusetts and northward into Quebec.

Bodies of ultramafic rock have been intruded into a sequence of phyllite, schist, gneiss, greenstone, and amphibolite rocks. These ultramafic bodies subsequently have been locally serpentinized and partly converted into talc. The talc deposits are commonly associated with dunite, peridotite, and serpentinite, which were probably their parent rock.

These ultramafic rocks form part of a belt more than 2,000 miles long, which extends from Alabama to Newfoundland (Chidester, et al., 1951). The talc ore itself is more of a talc-carbonate complex, which locally grades into a carbonate complex commonly rich in magnesite. The talc ore is relatively dark-colored, has schistose structure, and is relatively high in iron content, particularly at the surface where the iron has been enriched by weathering processes. Although the mineral talc itself is a relatively major constituent, the grade of Vermont talc ores generally is considerably lower than those of the other districts in the United States.

The ultramafic bodies are as much as 1 mile wide and as much as 3.5 miles long (Chidester, et al., 1951). The commercial talc bodies, which occur in those ultramafic bodies known as the *verde antique* type, are zoned around cores of serpentinite. The talc-containing zone (steatite zone) forms the outer shell or zone of the body; between it and the serpentinite core is the *grit zone,* which is commonly several feet to several tens of feet thick. *Grit* is the miners' term for rock which is composed of talc and carbonate minerals only.

There are few masses of high-purity talc in the Vermont district large enough to mine as such. Tremolite is not known and talc itself is the only talclike mineral contained within the ultramafic bodies.

In northern Vermont, entire ultramafic bodies have been converted into grit and have widths on the order of 100 ft or so. In southern Vermont one similar body has a width of about 175 ft. At least 145 talc occurrences are known in Vermont within an area having a strike length of about 150 miles and a width ranging from five to 25 miles. Homoclinal folding is the regional structure; isoclinal folding is thought to be uncommon, for little repetition due to folding has been noted.

Few workers have attempted to date either the host rocks of the Vermont talc-mining district or the time of steatization, but studies by Christman (1959) and Christman and Secor (1961) suggest that the age of the parent rocks is Early Paleozoic, probably Ordovician.

Open-pit mining is relatively minor, with most of the production coming from underground workings, some of which extend to vertical depths of more than 1,000 ft.

The ore reserve situation in the Vermont district has never been seriously investigated by anyone who would be at liberty to disclose the results of such a study. Judging from the geographical distribution of the known talc bodies, however, the total ore reserves must be extremely large.

Producers historically have had the choice of grinding the talc ore as mined and simply size-classifying it for relatively low-priced dusting compounds and fillers or of beneficiating it by flotation or other means in order to derive high-value products containing major amounts of the mineral talc. To manufacture the most profitable product line in this district, an operator must employ wet-processing beneficiation techniques. The Eastern Magnesia Talc Co., Inc. (now Engelhard Minerals and Chemical Corp.) has been using wet-beneficiation methods on ores near Johnson, VT since 1937. The ore is subjected to froth flotation, then dewatered in a rotary vacuum drum filter. The filter cake is then dried, pulverized, and air-classified (Harrah, 1956). Eastern Magnesia Talc Co., Inc. established another, more modern froth flotation plant at West Windsor, VT in 1964; since 1972 this operation has been known as Windsor Minerals, Inc., a wholly owned subsidiary of Johnson & Johnson Co. (Trauffer, 1964).

Vermont's talc production in 1979, according to the US Bureau of Mines, was 346,000 tons with a value of $2,755,000.

Southern Appalachians

Although numerous deposits of talc and soapstone are known in the Appalachian Mountain province between Vermont and western North Carolina, none of them have such a history of past or present production, nor in the writer's opinion the future potential, to even suggest the possibility of their becoming major talc-mining districts. These occurrences in Massachusetts, Connecticut, Rhode Island, New Jersey, Pennsylvania, Maryland, and Virginia are described by Chidester, et al. (1964).

Several more attractive talc occurrences and ore deposits, however, are known between Murphy in western North Carolina and Dadeville in eastern Alabama. Compared with those of the other five major talc-mining districts in the US, these are relatively small deposits, but some of them are quite important in that they are of exceptionally high purity and serve specialized consumer needs.

Production figures issued by the US Bureau of Mines do not differentiate this area from the New York district, but it is likely that the total annual talc production of the southern Appalachian area does not exceed 50,000 tons.

North Carolina: The Murphy talc district is about 85 miles long, extending from southwesternmost North Carolina into northern Georgia. Although talc is found in sedimentary rocks and is also associated with ultramafic bodies in North Carolina, only that talc in sedimentary rocks is presently of commercial interest. The host rock for these ore bodies is the Murphy Marble, which is generally more dolomitic than calcitic. Although the marble has been reported to be as thick as 400 ft or so, such thicknesses may be caused by structural deformation and therefore may be extremely anomalous. The Murphy Marble in the vicinity of Murphy, NC has been divided into as many as eleven zones, only one of which (about 45 ft thick) is known to contain commercial talc deposits. Little information is available about the geology of the marble belt and its talc zones in areas away from mining activity, probably because climatic conditions here cause outcrops to be few and far between.

Despite the relatively small size of the North Carolina talc deposits, operations there can be profitable because the purity and color of the talc are extremely high-grade. Many of the resulting ground products can compete with those manufactured from Montana talc ores and from the ores of a few deposits in the Death Valley area of California-Nevada as the only pure US talc products of high color produced without the use of wet beneficiation methods. Tremolite is minor; indeed, even talc formed after tremolite, which is so common in most other districts, is rare here.

The talc appears to have originated by relatively simple metasomatic processes (Chidester, et al., 1964). Although the white talc-bearing zone in the Murphy area is dolomitic marble which is locally slightly sandy, Van Horn (1948) does not believe that the parent material contained sufficient magnesia and silica to permit the talc to form by dynamic metamorphism alone. Owens (1968), however, on the basis of a petrologic study of diamond drill cores, feels that the talc is of metamorphic origin. The age of the Murphy Marble has not yet definitely been established. The general consensus, however, is that it is either of latest Precambrian or Cambro-Ordovician age.

Both open-pit and underground mining methods have been employed in the past, but recently the trend has been toward deeper underground mining. The developmental costs of shaft sinking to relatively small ore bodies are among the highest such costs known to the writer, further testifying to the superior quality of these carbonate-derived ores.

North Carolina is presently the only domestic producer of talc crayons or welders pencils, a relatively minor item tonnage-wise, but an extremely lucrative product line.

Georgia: A few occurrences of talc are known in the Canton area in northern Georgia, which is the southwesternmost portion of the Murphy Marble Belt, but Georgia's principal talc deposits are in Murray County. The individual talc deposits there may be tens of feet wide and several thousands of feet long and may extend for several hundred feet downdip (Chidester, et al., 1964). Furcron, et al. (1947), in the most detailed investigation to date, conclude that the talc has been formed by the alteration of dolomitic portions of the Cohutta Schist of Precambrian age; earlier workers, however, believe that it has been derived from peridotite. Needham (1972) considers that the soapstone and other talcose rocks of Murray County have been formed by serpentinization, followed by metasomatism and metamorphism. Chidester, et al. (1964) state that ultramafic bodies are not rare in Georgia, but that no talc is known to be associated with them.

Talc in Murray County was discovered about 1872 and mining was begun shortly thereafter. Mining operations are now conducted underground. The principal product lines have been crayons and ground products of both light and dark colors, but at the present time only the latter product line is being produced.

Alabama: Pure white talc has been mined near Winterboro in Talladega County, Alabama. As far as is known, all of this production went to cosmetic and pharmaceutical uses. The parent rock is a carbonate sequence of probable Cambro-Ordovician age which underlies the area, but the geological relationships are far from clear, according to McMurray and Bowles (1941). Blount and Vassiliou (1980) feel that the two ore bodies near Winterboro are associated with Cambrian dolomite, which has undergone little if any metamorphism. Their work suggests that hydrothermal solutions and metasomatic processes produced talc in and near fractures in the dolomite.

Talc and anthophyllite asbestos deposits, first

noted in 1873, are present in Tallapoosa and Chambers Counties, Alabama, where they are associated with ultramafic bodies (Neathery, 1968). The talc content ranges from 3 to 26%, probably averaging 20% for the better deposits; annual production has been about 5,000 tons (Neathery, 1970). Neathery, et al. (1967) describe flotation-concentration feasibility studies, which show that good-grade talc concentrates can be made from these ores, but the economics of such an operation have yet to be proven.

Texas

Texas has two talc districts—the Llano district in the central part of the state and the Allamoore district to the northwest of Van Horn in the western part of the state. The latter district is the one most recently put into production and is now the only one of any importance.

The Allamoore district extends about 20 miles east-west and is as much as 5 miles wide. Its ore deposits are in the Allamoore Formation, of Precambrian age, which consists of thousands of feet of carbonate rocks, volcanic rocks, and phyllite (King and Flawn, 1953). The talc appears to the writer to be associated with dolomite, but it also appears to interfinger with phyllite in certain parts of the district. Edwards (1980) feels that phyllite is invariably the host rock, but Bourbon (1981) believes that the talc formed in Precambrian time by the replacement of primary sedimentary magnesite. Field relationships indicate that carbonate rock other than calcite is the host rock for these talc deposits.

The ore-bearing strata, which are in the lower plate below a large overthrust fault (the upper plate having been almost entirely eroded), dip either vertically or steeply to the south. No ore body of which the writer is aware has ever been bottomed-out by either drilling or mining. Large iron-oxide-stained "horses" of either carbonate or volcanic rock are common within the talc zones, but generally are easily eliminated by the use of selective mining methods. Talc is commonly found from the grass roots down and the soil cover is generally less than 1 ft thick. Caliche is abundant in the uppermost 5 ft or so of these dark-colored ore bodies, but seldom causes any problems at greater depths.

Initial but far from complete petrographic and X-ray studies of the ores indicate that tremolite is not present to any significant degree. Dolomite, however, is so finely admixed with the talc as to render it extremely unlikely that pure talc can ever be produced in significant quantities from known ore bodies by presently known beneficiation techniques. There is little indication of a hydrothermal origin for these ores.

The talc ore (ceramic-type) is generally strongly foliated, with the individual foliae ranging from paper-thin to as thick as ¼ in. This dark Texas crude looks little like talc ore, even to the trained observer, when one first visits the district. The first impression is to identify it as black shale or perlite or some other similar nontalc-bearing rock. However, despite its dark—or indeed even black—color and its striking untalclike appearance, this talc ore is a superior ceramic crude with superb pressing qualities and excellent firing characteristics.

Flawn (1958) describes the district in a stage of infancy and King and Flawn (1953) describe in detail the overall geology of the district and surrounding areas.

All of the production has been from large open pits, with the exception of minor amounts of light-colored (generally pink to white) talc which have been mined in small underground workings and then finely ground into filler and extender products. Almost all of the talc ore produced in the Allamoore district is shipped out of the district to the east, and even into Mexico, for use in the ceramic industry, particularly for the manufacture of wall tile.

Mining costs probably are the lowest anywhere for talc ores. Flawn (1958) gave the costs of mining, transportation, and loading into cars as ranging from \$1.75 to \$3.00 per ton.

This is by far the youngest of all the major US talc-mining districts for production commenced only in 1952. The US Bureau of Mines credits Texas with the production of 207,000 tons of talc ore in 1979, but this is sharply down from immediately prior years.

California-Nevada

Of the major talc-mining districts in the United States, the one which combines the largest geographical area with the greatest number of productive commercial deposits is that of southeastern California, which barely extends northward into Nevada. It is about 200 miles long and has an average width of 30 miles, but is locally as much as 75 miles wide. Engel and Wright (1960) divide this long belt into three separate areas, each of which contains a slightly different characteristic type of deposit. See Wright (1968) for a detailed and comprehensive description of the geology and talc deposits in the central portion of this belt.

The deposits in the southern two areas occur in Precambrian sedimentary rocks. They were probably formed in Precambrian time and are spatially associated with a thick diabase sill which has been intruded near the base of a widespread and thick carbonate unit. In the northernmost of the three areas—which historically has not been as important a producer as the other two areas—the deposits occur in Lower Paleozoic formations and are thought to have been formed by replacement processes, probably in Cretaceous or Tertiary time.

The deposits come in all sizes. The largest is thought to be the Western-Acme mine complex near Tecopa in San Bernardino County, CA, which extends more than 5,000 ft along strike (although locally severely complicated by cross-faulting) and is as much as 80 ft wide. Portions of this zone have been mined downdip for at least 350 ft.

This district is too large and its geology too complicated to be satisfactorily summarized here. In general, however, all of the deposits have been formed by the metamorphism and hydrothermal replacement of siliceous and silicated magnesian marbles and limestones (Engel and Wright, 1960).

There are two main types of ore in this district: the "hard" ore of the miners is tremolitic and the "soft" ore is talc schist. Many deposits contain both types, but their relative abundance and distribution is erratic and unpredictable. Most of the ore is markedly tremolitic, but there are some relatively small bodies of high-grade talc. The problem with most of these high-purity soft talc bodies, however, is that they commonly cannot be mined by themselves, but for practical economic reasons must be extracted along with the accompanying tremolitic ores. In the past the production of high-grade talc ore in the California-Nevada district generally had to be a function of the production of tremolitic ores; this is not so true today, for several mines now produce only the talc schist, or soft ore.

Papke (1975) describes 34 deposits in southern Esmeralda County, Nevada, along the California border. Sixteen of these have talc as the principal ore mineral and nine have mixed talc and chlorite; the others consist of chlorite and sericite. Dolomite is obviously the preferred host rock for the talc-rich ore bodies. The host rocks range in age from Precambrian to Early Cambrian, but the age of mineralization is believed to be Late Jurassic (Papke, 1975).

In the past, most of the mining in the California-Nevada district was done underground; in such operations relatively little mechanization is employed and timbering costs are extremely high. In recent years, however, an increasingly larger proportion of the district's production has been obtained by open-pit mining methods.

In no other talc-mining district anywhere in the United States are the mines so far from rail transportation as are those in the California-Nevada district. Mills and other processing facilities are not located at or near the mines, but are established either in the Los Angeles area or at various points along two railroad lines in this area which serve Los Angeles. Truck haulage routes either to a processing facility or to the railroad are commonly 100 miles or more.

Soapstone was mined from small deposits along the foothills of the western Sierra Nevada in the mid-1880s, but the current California-Nevada talc-mining district was established during World War I. At that time, the area around Darwin in Inyo County, California supplied block-grade steatite talc to compensate for the temporary elimination of foreign supplies. This production slackened off in the postwar period but was rejuvenated during World War II. Since then, however, California has not been a significant factor in the supply of steatite-grade talc.

In the mid-1930s the use of California tremolitic ores for the manufacture of wall tile began to come into its own and the ceramic industry became the major market. The talc content of such ores is not critical; a considerable amount of carbonate material can be tolerated and, indeed, may even be desirable.

The current production from this district, however, appears to be fairly well balanced between the ceramic industry and the paint and paper trade.

Montana

Next to the Allamoore talc-mining district in western Texas, the talc-mining district in the southwestern corner of Montana is the youngest in the United States. It is the most important US source of tremolite-free talc products which may be produced directly from crude ore without the need for processing.

The pioneering work on Montana talc was

done during the 1940s by Henry Mulryan, then with Sierra Talc Co., and Walter K. Skeoch, then with Southern California Minerals Co.

The earliest significant work in the geological literature is that of Perry (1948), an excellent field study which was compiled in the very early days of the Montana talc industry. Subsequent works worthy of consultation for further and more timely details on this district are Olson (1976), whose work mainly concerns the Ruby Range, and Berg (1979), whose work covers the remainder of the district.

High-purity talc deposits in southwestern Montana are totally restricted in their occurrence to dolomitic marble in the Prebeltian Cherry Creek Series. The talc, which is generally white to pale green, commonly occurs as conformable lenses and stringers within the marble; if the talc is anywhere discordant, such a relationship is usually at low angles to the foliation of the bedrock. Structural control in the strictest sense has not been widely recognized, but there is an apparent restriction of talc mineralization to that part of the dolomitic marble which is spatially close to large bodies of the Dillon Granite Gneiss.

The following reaction is thought to account for the formation of these talc deposits:

$$3 \text{ dolomite} + 4 \text{ quartz} + H_2O = \text{talc} + 3 \text{ calcite} + 3\ CO_2.$$

Inasmuch as the host dolomite normally is not notably siliceous adjacent to the talc, it is likely most of the silica required for formation of the talc was metasomatically introduced. Although this is merely the writer's opinion, it is in part substantiated by Garihan (1973).

All of the pre-Cherry Creek rocks, the rocks of the Cherry Creek Series, and most of the rock types of the Dillon Granite Gneiss have been subjected to upper amphibolite grades of metamorphism, part of a long-lived metamorphic event in Prebeltian time, which included the development of intense isoclinal deformation. It is the writer's opinion that the talc most likely was formed during the period of regional metamorphism of greenschist grade, which postdated the amphibolite period and predated the intrusion of the diabase dikes. These vertical or near-vertical dikes, which are now best referred to as metadiabase, have a consistent northwesterly strike which coincides with the main direction of high-angle faulting.

Precambrian structural deformation was extreme and is characterized by isoclinal folding and refolding and the flowage of less competent rocks, rather than by faulting. Post-Precambrian folding has been negligible. The characteristic Laramide structure is high-angle block-faulting along northwest trends.

The talc was almost certainly formed in Precambrian time. Evidence is the relationship of the talc to the metadiabase and the fact that no talc whatsoever is known to have been formed in the thick dolomite units of the Cambrian and Devonian systems in the northern third of the Ruby Range, even though some of these dolomite units are within only a mile or so of talc deposits in the Precambrian dolomitic marble. Garihan (1973) feels that this lack of talc formation in the Paleozoic dolomites, plus the fact that the Paleozoic rocks are completely unmetamorphosed, speaks strongly for the talc having been formed in Precambrian time. Okuma (1971) also feels that the most likely age of the talc mineralization is Precambrian.

Several major talc deposits are known in the Ruby Range and one major deposit is known in the Gravelly Range. One relatively small deposit and several minor occurrences are known in the Greenhorn Range; several minor occurrences are also known in the Tobacco Root Mountains. Virtually all of Montana's talc production comes from three large mines—two in the Ruby Range and one in the Gravelly Range.

Ross, et al. (1955) have mapped the unit in which all of the known commercial talc deposits occur as covering an area about 180 miles long in an east-west direction (from Red Lodge to Bannack) and 85 miles wide (from Three Forks south to the Idaho border). The dolomitic marble which is the only host rock for the formation of the talc, however, does not occur over the entire extent of this unit. As known today, admittedly from still incomplete and extremely fragmental information, the dolomitic marble is probably significant only in the Ruby Range, the Gravelly Range, the Greenhorn Range, the Tobacco Root Mountains, and possibly in small areas of the Madison Range.

The mechanism of the formation of Montana talc bodies was probably similar to that postulated by Van Horn (1948) for the Murphy, NC talc ores. The geology of the southwestern Montana talc district is similar in many ways to that of the Murphy area and many of Van Horn's conclusions concerning the genesis of the Murphy talc ores also appear to apply to the genesis of Montana talc ores.

The dynamic metamorphism of the earlier upper amphibolite period may have formed zones of weakness and allowed the introduction of siliceous material (probably from the Dillon Granite Gneiss) and possibly some magnesia also (probably remobilized from amphibolite and other ultramafic rocks). After this metasomatic event had afforded the opportunity for the introduction of all of the constituents necessary for the formation of talc which had been lacking in the original dolomite, it is thought that the talc itself could have been formed during the retrograde phase of this major metamorphic event.

Montana's talc production in 1979, according to the US Bureau of Mines, was 343,000 tons.

Mining

It is estimated that over three-fourths of domestic talc production comes from open-pit mines. The major underground mines are located in the eastern United States.

Talc mining operations are generally carried out on a small scale in which an "average" mine produces less than 300 stpd.

The equipment used in underground mines is generally unsophisticated and capable of producing only a few tons per hour. Some underground operations have vertical shafts over 1000 ft deep. Mining methods range from conventional shrinkage stoping operations to room-and-pillar mine workings. Timber is often required in mines producing high quality soft, platy talc.

Open-pit mines utilize conventional stripping, loading, and haulage equipment. The degree of drilling and blasting effort is directly related to the type of talc ore being mined. Only the massive, blocky, and hard talc ores require extensive drilling and blasting.

One unusual aspect of talc mining is the extreme necessity for good housekeeping practice. This is important because many of the talc products are sold on the basis of a good white color and absence of abrasive materials.

It is difficult to give examples of the cost of mining talc ore. Since talc mine output may range from 2,000 or 3,000 stpy up to 150,000 stpy, there is no such thing as an average mining cost.

In New York State both open pit and underground mining methods are used. One of the largest underground talc mines in North America is operated by the Gouverneur Talc Co. (Erdman, 1972; Gillingham, 1950). This mine has an 1100-ft shaft carrying a 6-ton skip. Open stope mining is practiced and hard rock mining methods are required due to the hardness of the ore. Jackleg drills are used and blasting is done with gelatin dynamite and ANFO. Ore transport is either by gravity drawpoint loading or slusher hoist to a rail car system which moves the broken rock to the central underground crushing system on the 700-ft level. A 24x36-in. jaw crusher reduces the ore to suitable size for loading into the 6-ton skip. In the mine headframe, a 36-in. gyratory crusher reduces the ore to —¾-in. for transfer to the grinding plant.

Mining operations at the Engelhard Minerals and Chemicals Corp. talc mine near Johnson, VT, in Lamoille County are underground. The inclined shaft is driven at an angle of 45° to a depth of over 1000 ft below the surface (Anon., 1971d; Burmeister, 1963). Ore transport is by slusher and gravity to rail haulage drifts below the slusher drifts. The crushed ore is hauled by truck to the processing plant. Fig. 1 is a sketch of stoping operations at the Johnson mine. The ore is about 55% talc, 40% magnesite, and 5% other minerals.

The Vermont Talc Co. which was acquired by Vermont Marble Co. in 1969 operates an underground mine only 85 ft deep at Windham, VT. The mine consists of 11 headings, 300 ft wide and about 1000 ft long (Anon., 1971f). A modified shrinkage stope mining method is used. The ore is hauled to a plant 11 miles away at Chester, VT for processing.

The talc mining operations of Cyprus Industrial Minerals are carried out in California, Montana, and Texas. The Yellowstone and Beaverhead mines in Montana are both open-pit operations. These are conventional drill, blast, and shovel operations. Some overburden can be removed by ripping methods. At the Yellowstone mine some hand-sorting is carried out to remove carbonate and other gangue minerals. About ten tons of rock and ore are handled to obtain one ton of acceptable crude talc ore (Mulryan, 1971). Pfizer, Inc. operates the Treasure mine (open pit) less than a mile north of Cyprus's Beaverhead mine in the Ruby Range, Montana.

California talc mines include both open pit and underground operations. The underground mine workings are usually developed and carried out in ore with square set timbered stopes. This underground mining method is required because of the small size and irregular nature of the talc occurrences.

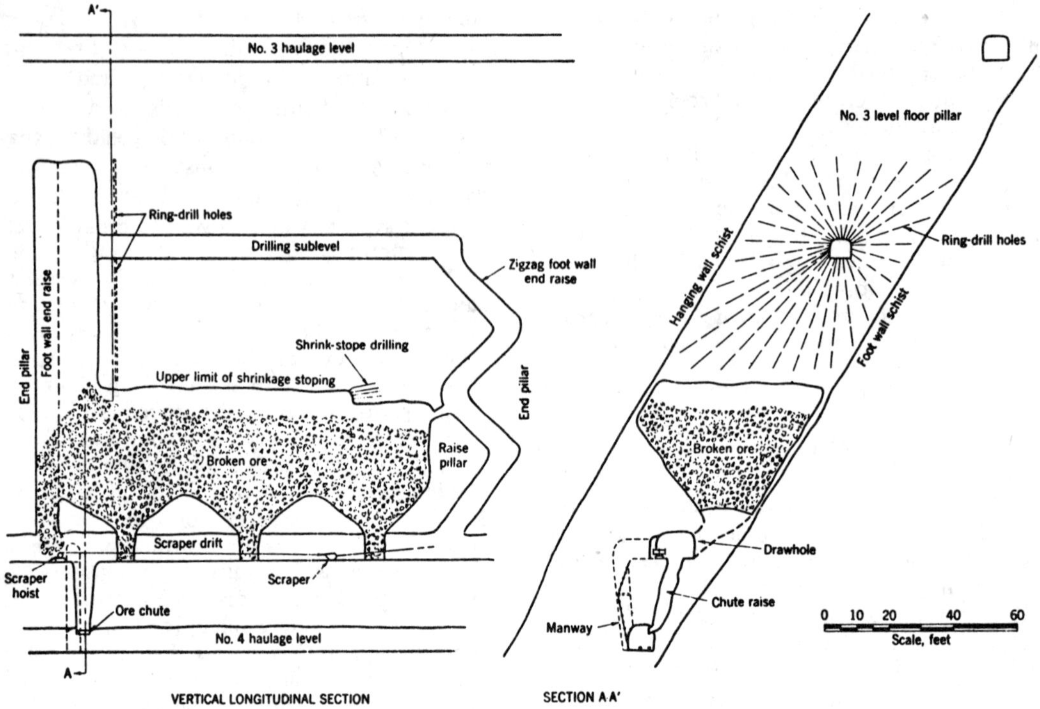

FIG. 1—*Drifts, raises, and combined shrinkage and sublevel stopes (Burmeister, 1963).*

The larger California talc mines are being converted to open pit mining.

Processing

Talc milling has traditionally involved dry-processing operations. Since much of the talc market is supplied with a relatively impure talc product, beneficiation is not usually carried out. The talc milling industry is slowly changing from uncomplicated grinding plants to sophisticated processing operations.

The new generation of talc mills includes complex froth flotation, sedimentation, hydrocycloning, dry and wet magnetic separation, centrifugal sizing, spray drying, and new grinding techniques. It should be noted that grinding is required both to prepare the ore for processing as well as to finish off the final product which must usually meet stringent particle size demands.

The processing of talc ores is complicated by the fact that extreme white color is a very desirable feature. Therefore, the grinding equipment used should not discolor the talc in any way. This generally rules out conventional rod and ball milling with steel grinding media. Fortunately, most talc ores are very soft and reasonably good grinding rates can be obtained with ceramic grinding media.

Some very sophisticated grinding facilities are often required for final preparation of processed talc sold to specific markets. For example, when talc is sold for paper filler and coating applications, it is generally designed to meet an average particle size of less than 5μ. In some special products this may be lowered to less than 1μ. This ultrafine grinding requirement is met with vertical-shaft pulverizing mills (3500 rpm) and jet-milling equipment. The latter utilizes the fluid energy principle with either compressed air or steam providing the energy. Fig. 2 is a sketch of a fluid energy type grinding mill.

In pioneer plants talc was ground in flour-grinding mills equipped with buhrstones and then sized on silk bolting reels (Parsons, 1948). The operation of such a mill grinding both wheat and talc was recorded as late as 1922.

Porcelain-lined flint-pebble mills were used at Gouverneur, NY, as early as 1885. In Vermont, where softer talc ore was ground, vertical emery mills, disintegrators, pulverizers, and roller mills were used. Cosmetic-grade talcs were sized by bolting through cloth.

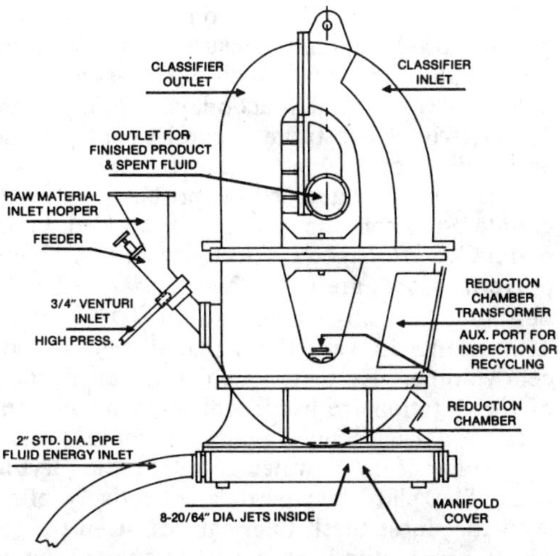

FIG. 2—*Sketch of fluid energy grinding mill.*

At present, little information has been published on wet grinding of talc ores. A pertinent reference to the subject appeared in Russian literature in 1971 (Karmazin, et al., 1972). The authors found that grinding in rod and ball mills did not result in "opening" of the mineral intergrowths even when grinding to a 5 to 10μ particle size. Jet mill grinding did accomplish intramineral grinding without the overgrinding resulting from impact mill grinding.

Windsor Minerals, Inc., a totally owned subsidiary of the Johnson and Johnson Co., operates two talc mines in Vermont. The Chester mine and the Hammondsville mine are both in Windsor County. The two talc processing plants serving these mines are located at Gassetts and West Windsor, VT. The plant at West Windsor utilizes sophisticated wet-process technology and has contributed toward better understanding of the basic problems involved in the beneficiation of talc ores.

Ore is delivered by trucks to the flotation plant where several high-grade talc products are manufactured. Fig. 3 is a flowsheet of the plant as reported in 1964 (Trauffer, 1964).

A Canadian talc processing facility using modern wet-processing equipment is operated by Baker Talc Ltd. Ore is produced from an underground operation at the Van Reet mine, in Potton Township about 80 miles southeast of Montreal. Talc was first mined here in 1871 (Anon., 1969).

The Baker talc plant utilizes froth flotation and high-intensity magnetic separation to produce high quality talc products. This ore could not be sufficiently beneficiated by froth flotation alone due to the presence of iron-bearing minerals which discolored the products. The magnetic separator in use is a modified high-intensity Jones wet magnetic separator which is now also finding major application in the beneficiation of many industrial minerals.

Eastern Magnesia Talc Co., Inc. (now Engelhard Minerals and Chemicals Corp.), was a pioneer in the Vermont talc industry. After a 1902 discovery of an ore body, a dry mill was constructed in 1904 to process 25 stpd (Harrah, 1956). This plant operated until 1920 when a 100-stpd dry plant was placed in operation. In 1932, wet methods of beneficiation were investigated. This work, along with US Bureau of Mines and the Missouri School of Mines development work, led to installation of a flotation pilot plant in 1937. Modification and enlargement of the pilot plant in 1938 resulted in commercial application of the froth flotation process. The ore is ground to 72 to 77% −200 mesh in dry pebble mills prior to flotation and shaking table concentration. The tables remove a high-gravity product containing a small quantity of nickel, iron, and cobalt minerals. A talc-magnesite mixture is floated at 18 to 21% solids in 21 Fahrenwald-type flotation cells.

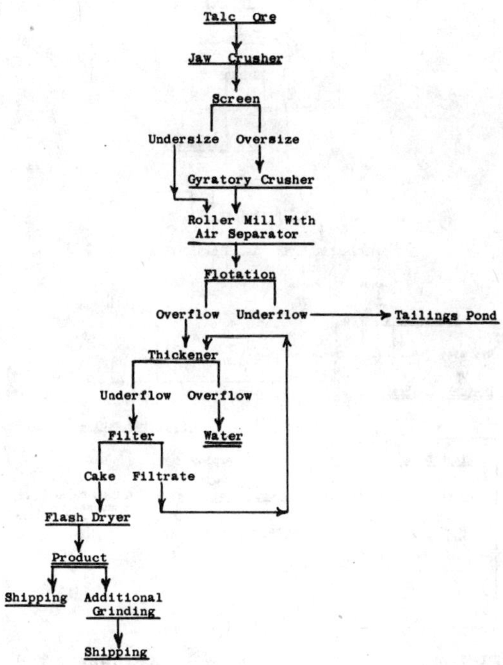

FIG. 3—*Flowsheet for Windsor Minerals, Inc., Vermont.*

The talc froth product contains talc particles of sufficiently coarse size that a rotary vacuum drum filter can be used for dewatering to 18% moisture. The filter cake is flash dried, pulverized, and air classified prior to shipping. Fig. 4 is a mill flowsheet as of 1962.

Naturally hydrophobic minerals such as talc, sulfur, graphite, and molybdenite will often respond positively to froth flotation without the addition of collector reagents. Lepetic (1972) has reported on the flotation of copper and molybdenum sulfides without collector reagents. Klassen and Mokrousov (1963) point out the striking similarity of talc and molybdenite as regards water wettability of the cleavage planes and the faces of these layered minerals.

A detailed report on a laboratory test program involving a Dadeville, AL, talc-asbestos deposit was issued in 1967 by Neathery, et al. (1967). The report concludes, "it is technically feasible to obtain good grade talc concentrates by flotation concentration of each of the five samples examined." Pine oil was the favored flotation reagent, and at least four flotation cleaning stages were necessary for optimum results. It was not considered feasible to produce a commercial anthophyllite concentrate for use as an asbestos product.

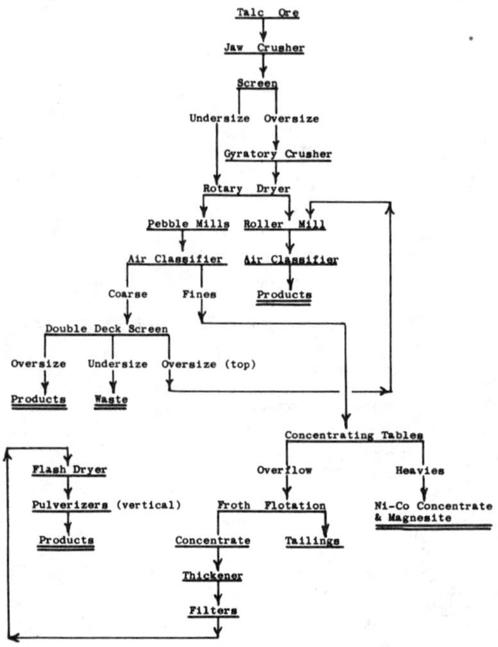

FIG. 4—*Flowsheet for Eastern Magnesia Talc Co., Vermont.*

The Johnson and Johnson Co. is an important producer and consumer of cosmetic-grade talc. In 1963, they were issued two patents (Brown and Macdonald, 1963; Chase, 1963), which constitute a good review of the art of flotation of platy talc ores. While platy talc is easily floated, many problems are encountered when the process is applied commercially. For good selectivity, when finely pulverized ores are used, the flotation cell pulp density is usually around 6 to 10% solids. This results in very low capacity per unit of cell volume. Further, when a major portion of the flotation froth concentrate product consists of particles finer than 325 mesh (44μ), it is difficult to dewater prior to the drying step. The platy talc particles effectively blind even the finest mesh filler fabrics. Centrifugal dewatering methods have proved effective when all conventional filtration methods fail.

There is no publication of commercial applications of the froth flotation process to tremolitic-talc ores. Norman, et al., (1939) published results of a detailed laboratory investigation on the flotation of talc ores from Gouverneur, NY, in 1939. These data indicate that a clean separation of soft, platy talc from tremolite is difficult. With a feed containing 36% tremolite, a cleaner concentrate containing 13.2% tremolite was floated. The underflow, or rougher tailing from this test, contained 47.3% tremolite.

Fig. 5 is a flowsheet showing typical dry-processing of New York tremolitic-talc ores.

Adaptation of a noncomplicated sedimentation process to the beneficiation of certain soft, platy talc ores was announced by Roe in 1972. A pilot plant operation on a talc-dolomite ore from the Death Valley-Kingston Range district, California, was successful in decreasing the carbonate mineral content from over 40% to less than 4%. The beneficiated product had a Valley Abrasion number of 4, a G.E. brightness of 93 to 95, 3.3% acid solubles, and 1.6% carbon dioxide. This, and similar processes, hold promise of the future development of similar low-grade talc ores and other filler minerals.

As wet processing is accepted by talc producers, the interest in specialized delamination equipment increases. Some talc deposits include a small percentage of naturally occurring, discrete, thin plates. For the most part, talc is in the form of booklets of plates which require physical effort to split into sheets for greater utility in paint, paper, plastics, and cosmetics.

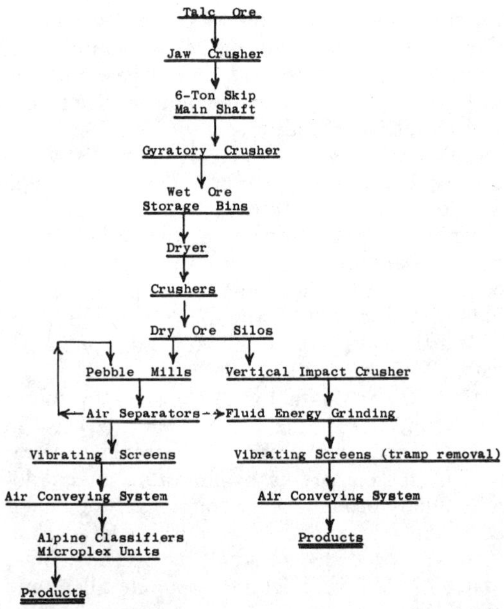

FIG. 5—*Flowsheet for Gouverneur Talc Co., Inc., New York.*

The mica industry has a long history of evaluating delamination processes. Mica used in joint cement (for wallboard, etc.) should have a low bulk density; the degree of delamination is generally directly proportional to the bulk density. A good mica will have a bulk density of less than 10 lb per cu ft. Several types of high-shear wet process equipment for delamination have been tested on mica and talc. In general, the processing usually involves the use of high-pressure pumps, such as those used on mineral-slurry pipelines, to force the mica or talc through a restricted orifice thus causing delamination. The same process is also used for size reduction of sulfur particles. It is reported that a pressure homogenizer can produce sulfur dispersions containing 99% of the particles in the 1μ range, with none larger than 3μ (Anon., 1966).

Ashton and Russell (1972) describe a useful method for study of delaminated talc platelets in two dimensions, area and thickness. In this method a thin film of talc particles is dusted onto the surface of distilled water and a section of the monolayer is picked up on a microscope glass slide. A precisely trimmed 10 cm^2 area of the talc film is examined photographically. By subtracting the void areas a total area of talc is determined on a projected image. The weight of the talc film is determined to 0.10 mg on a microbalance. The average thickness of the talc particles was calculated from the foregoing data as follows:

$$\text{Volume } (V) = \frac{\text{weight } (w)}{\text{density } (d)}$$

$$\text{Volume } (V) = \text{thickness } (t) \times \text{area } (a)$$

$$t = \frac{w}{ad}$$

$$t \times 10^{-4} = \text{thickness } (t) \text{ in microns}$$

where d is density of talc in grams per cubic centimeters = 2.7, w is weight of talc on slide in grams, a is actual area covered by talc in square centimeters, and t is average thickness of talc in centimeters. By using the described technique, talc platelet thicknesses of 1.4 to 10.1μ were measured.

Quality control in many talc plants consists of uncomplicated test procedures. For example, the former Eastern Magnesia Talc Co. reported (Burmeister, 1963), "Quality control is largely based on percentage of contained talc (insolubility test in hydrochloric acid) and particle size and brightness or color."

In larger talc plants (+100,000 stpy production) very sophisticated laboratory procedures are used to control product quality. Those talc plants supplying the larger segments of the talc market, such as paint, paper, and cosmetics, will have facilities for chemical and instrumental analysis, surface area determination, color, oil absorption, abrasivity, particle size and shape determination, packing quality, resistivity, lubricity, bacterial count, melting point, vitrification range, and fired shrinkage.

Production and Markets

The total world production figures for talc, soapstone, and pyrophyllite have generally been compounded. This addition of quantities results from lack of both mineralogical and definitive production data.

As the talc industry grows, uses more sophisticated beneficiation methods, and meets stringent product specifications, there is evolving a more logical statistical reporting system. Production statistics should list, for example, figures for paper filler grade talc, cosmetic talc for face and body powders, cosmetic talc for use in aerosol applications, talc used in many types of lubricants, and talc used in agricultural applications.

US mine production of talc exceeded 1 million tons for the first time in 1969. World production averaged 1.7 million tons for the period 1945–1960 but by 1969 world production rose to 5 million tons.

Many of the important large tonnage uses for talc are closely related to the degree of sophistication of industrialization of a given country. Over the 15-year period 1945 to 1960, the United States produced about 38% of the total world production of talc, soapstone, and pyrophyllite; by 1962, this had dropped to 26%. For the ten years preceding 1973, US production stabilized to average about 22% of the total world production.

In 1980, talc was produced by 21 companies from mines in 12 states. Vermont, Montana, New York, and Texas were the top producing states, accounting for 90% of the year's output. Georgia, North Carolina, Arkansas, Nevada, Oregon, Virginia, and California produced the remaining tonnage. Iowa is listed as a producer but actually, it has only a grinding facility processing ores from other areas. The 1979 US production of talc was 1,268,000 tons.

Table 5 shows tonnages of ground talc sold or used by US producers in 1978, 1979, and 1980.

Future Trends

The uses for talc in putties, caulks, and many types of sealants and adhesives are rapidly increasing. This is especially true in those products requiring close control of color (white and tinted) and dependence on high quality through adherence to strict control of talc specifications.

TABLE 5—End Uses for Ground Talc

Use	Thousand short tons 1978	1979	1980
Ceramics	257	260	282
Cosmetics*	69	74	59
Insecticides	13	13	11
Paint	192	237	197
Paper	87	105	102
Plastics	147	112	110
Refractories	6	6	2
Roofing	18	19	20
Rubber	36	39	37
Other uses+	92	95	83
Total‡	917	960	903

Source: Clifton, 1980, 1981.

* Incomplete data. Some cosmetic talc known to be included in "Other uses."

+ Includes art sculpture, asphalt filler, crayons, floor tile, foundry facings, rice polishing, stucco, and other uses not specified.

‡ Data may not add to totals shown because of independent rounding.

Talc has several competitive minerals which are sold for similar uses. Feldspar, wollastonite, and nepheline syenite compete in ceramics. Mica competes in the roofing markets. Titanium dioxide is losing ground as a competitive filler in some markets due to its high cost. Specially purified, calcined, and compounded clays compete with talc in paint, paper, and plastics.

One important factor to consider in future competition for the talc filler markets in paint, paper, and plastics is the rapid penetration of these markets by limestone and dolomite-filler products. Calcium carbonate is often considered to be the most versatile commercially produced mineral filler. The United States market for calcium carbonate fillers and extenders exceeds the total domestic market for talc and pyrophyllite. The world market for carbonate fillers is growing at a faster rate than any other mineral filler. The widespread availability of large carbonate deposits strategically located for low-cost transportation will encourage continuing competition in the white-mineral filler market. A detailed comparison of calcium carbonate and talc fillers is presented by Roe (1980).

As more industrial minerals operations change over to wet beneficiation methods, especially flotation, there may be considerable amounts of byproduct talc produced. The very successful introduction of wet high-intensity magnetic separation in the beneficiation of kaolin clays and other industrial minerals has added another dimension to the options in talc ore processing.

Bibliography and References

Anon., 1956, "Talc," California Mineral Information Service, Vol. 9, No. 11, Nov. 1, 8 pp.

Anon., 1966, "An Alternate Route to Size Reduction," *Chemical Processing*, Aug., p. 82

Anon., 1969, "Process Break-Through Revitalizes 100 Year Old Mine," *Mining in Canada*, Oct., pp. 27–30.

Anon., 1970, "Standard Method of Test for Fineness of Dispersion of Pigment-Vehicle Systems," ASTM Designation D1210–64 (reapproved 1970).

Anon., 1971a, "The Operations of A/S Norwegian Talc," *Industrial Minerals* (London), Jan., pp. 27–28.

Anon., 1971b, "Pyrenean Producer: S.A. des Talcs

de Luzenac," *Industrial Minerals* (London), Jan., p. 25.

Anon., 1971c, "Talc: Micronized Grades Lead the Way," *Industrial Minerals* (London), Jan., pp. 9–18.

Anon., 1971d, "Talc Operations of Engelhard Minerals and Chemicals," *Industrial Minerals* (London), No. 41, Feb., p. 46.

Anon., 1971e, "Talc: UK Consumption Shows Gradual Rise," *Industrial Minerals* (London), Jan., pp. 19–24.

Anon., 1971f, "Windham Plant Grinds Talc for Specialized Applications," *Rock Products,* Vol. 74, No. 6, June, p. 59.

Anon., 1972, "Chemical Profile—Titanium Dioxide," *Chemical Marketing Reporter,* Vol. 202, No. 24, Dec. 11, p. 9.

Anon., 1982, "Talc, Stability in a Soft Market," *Industrial Minerals,* No. 183, Dec., pp. 59–73.

Ashton, W.H., and Russell, R.S., 1972, "Talc Beneficiation," US Patent 3,684,197, Aug. 15.

Barnes, D.F., 1958, "Infrared Luminescence of Minerals," Bulletin 1052-C, US Geological Survey, p. 113.

Bennett, F.E., 1972, "Tile Council of America Research in the Seventies," American Ceramic Society, Fall Meeting, Sept. 27–29, Bedford, PA.

Berg, R.B., 1979, "Talc and Chlorite Deposits in Montana," *Memoir 45,* Montana Bureau of Mines and Geology, 65 pp.

Blount, A.M., and Vassiliou, A.H., 1980, "The Mineralogy and Origin of the Talc Deposits Near Winterboro, Alabama," *Economic Geology,* Vol. 75, No. 1, pp. 107–116.

Bourbon, W.B., 1981, "The Origin and Occurrences of Talc in the Allamoore District, Texas," in Program with Abstracts, 17th Annual Forum on the Geology of Industrial Minerals, Albuquerque, NM, May.

Brindley, G.W., and Santos, P. de S., 1971, "Antigorite—Its Occurrence as a Clay Mineral," *Clays and Clay Minerals,* Vol. 19, pp. 187–191.

Brown, J.S., and Engel, A.E.J., 1956, "Revision of Grenville Stratigraphy and Structure in the Balmat-Edwards District, Northwest Adirondacks, New York," *Bulletin,* Geological Society of America, Vol. 67, No. 12, pp. 1599–1622.

Brown, W.E., and Macdonald, R.D., 1963, "Talc Beneficiation," US Patent 3,102,855, Sept. 3.

Burmeister, H.L., 1963, "Mining Methods and Costs, Eastern Magnesia Talc Co., Johnson Mine, Johnson, Vermont," Information Circular 8142, US Bureau of Mines, 44 pp.

Campbell, M.E., 1972, "Solid Lubricants, A Survey," NASA SP-5059(01) publication prepared under contract by Midwest Research Institute, Kansas City, MO, 126 pp.

Campbell, W.J., et al., 1980, "Chemical and Physical Characterization of Ampite, Chrysotile, Crocidolite, and Nonfibrous Tremolite for Oral Ingestion Studies by the National Institute of Environmental Health Sciences," Report of Investigations 8452, US Bureau of Mines, 63 pp.

Cawood, E.E.C., 1962, "Fire Extinguishing Composition," US Patent 3,063,940, Nov. 13.

Chase, W.E., 1963, "Platy Talc Beneficiation," US Patent 3,102,856, Sept. 3.

Chidester, A.H., Billings, M.P. and Cady, W.M., 1951, "Talc Investigations in Vermont," Preliminary Report, Circular 95, US Geological Survey, 33 pp.

Chidester, A.H., et al., 1964, "Talc Resources of the United States," Bulletin 1167, US Geological Survey, pp. 1–61.

Christman, R.A., 1959, "Geology of the Mount Mansfield Quadrangle, Vermont," Bulletin 12, Vermont Geological Survey, 70 pp.

Christman, R.A., and Secor, D.T., Jr., 1961, "Geology of the Camels Hump Quadrangle, Vermont," Bulletin 15, Vermont Geological Survey, 70 pp.

Clifton, R.A., 1980, "Talc and Pyrophyllite," Preprint 1978–1979, US Bureau of Mines, 7 pp.

Clifton, R.A., 1981, "Talc and Pyrophyllite," *Minerals Yearbook 1980,* US Bureau of Mines, pp. 815–820.

Cooper, J.D., and Hartwell, J.W., 1970, "Talc, Soapstone and Pyrophyllite," *Mineral Facts and Problems,* Bulletin 650, US Bureau of Mines, pp. 1267–1281.

Dickson, T., 1982, "North American Talc," *Industrial Minerals,* No. 183, Dec., pp. 75–78.

Dillender, R.D., Jr., and Gower, I.W., 1953, "Preliminary Report on the Froth Flotation of Wake County Talc," Bulletin No. 7, Dept. of Engineering Research, North Carolina State College, Raleigh, NC, Nov., 16 pp.

Drake, M.J., and Brett, B.A., 1962, "Talc, Soapstone and Pyrophyllite," *Minerals Yearbook 1962,* US Bureau of Mines, pp. 1189–1198.

Edwards, G., 1980, "Tumbledown Mountain Talc Deposit, Allamoore District, Culberson County, Texas," *Guidebook,* 31st Field Conference, New Mexico Geological Society, pp. 245–250.

Elstron, L.W., 1961, "Slipping and Sliding," *Research Engineer,* Georgia Institute of Technology, Feb., pp. 15–18.

Engel, A.E.J., 1949, "New York Talcs, Their Geological Features, Mining, Milling and Uses," *Transactions AIME,* Vol. 184, pp. 345–348.

Engel, A.E.J., and Wright, L.A., 1960, "Talc and Soapstone," *Industrial Minerals and Rocks,* 3rd ed., J.L. Gillson, ed., AIME, New York, pp. 835–850.

Erdman, G.R., 1973, "Dust Control at Gouverneur Talc Company," *Trans. SME-AIME,* Vol. 254, No. 2, June, pp. 161–165.

Faust, G.T., and Fahey, J.J., 1962 "The Serpentine-Group Minerals," Professional Paper 384–A, US Geological Survey, pp. 80–82.

Flawn, P.T., 1958, "Texas Miners Boost Talc Output," *Engineering and Mining Journal,* Vol. 159, No. 1, pp. 104–105.

Frommer, P.W., and Fine, M.M., 1956, "Laboratory Flotation of Talc from Arkansas and Texas Sources," Report of Investigations 5241, US Bureau of Mines, June, 5 pp.

Furcron, A.S., Teague, K.F., and Calver, J.L., 1947, "Talc Deposits of Murray County, Georgia," *Bulletin 53,* Georgia Geological Survey, 75 pp.

Garihan, J.M., 1973, "Geology and Talc Deposits of the Central Ruby Range, Madison County, Montana," PhD dissertation, Pennsylvania State University, 209 pp.

Gillingham, W.P., 1950, "Grinding Talc to Superfine Size," *Compressed Air Magazine,* Feb., pp. 32–37.

Harrah, H.W., 1956, "Eastern Magnesia Talc

Company, Inc." *Deco Trefoil,* Denver Equipment Co., May–June, pp. 7–14.

Harrell, G.O., and Harrell, D.E., 1971, "Survey of Mineral Fillers in Selected Industries, Phase I—Plastics and Textiles," The Industrial Extension Service, School of Engineering, North Carolina State University, Raleigh, NC, Apr., 105 pp.

Herod, B.C., 1954, "Pioneer Talc Leads Way in West Texas Field," *Pit and Quarry,* Vol. 54, Apr., pp. 134–135, 156.

Heystek, H., and Planz, E., 1964, "Mineralogy and Ceramic Properties of Some California Talcs," *Bulletin,* American Ceramic Society, Vol. 43, No. 8, pp. 555–561.

Hunt, R.E., 1969, "How to Select Fillers and Reinforcements for Thermoplastics," *Plastics Technology,* Nov., pp. 38–43.

Karmazin, V.I., et al., 1972, "Grinding and Concentrating Capacity of Talc-Magnesite Rocks of the Pravda Deposit of the Ukranian SSR," Obogasheh, Polez, Iskop, 1971, No. 9, p. 28–29 (Russ.); *Chemical Abstracts,* Vol. 77, 92190c, p. 262.

Katz, H.S., and Mileski, J.V., 1978, *Handbook of Fillers and Reinforcements for Plastics,* Van Nostrand Reinhold Co., New York, 652 pp.

King, P.B., and Flawn, P.T., 1953, "Geology and Mineral Deposits of Precambrian Rocks of the Van Horn Area, Texas," No. 5301, Texas University Publications, 218 pp.

Klassen, V.I., and Mokrousov, V.A., 1963, *An Introduction to the Theory of Flotation,* London, 493 pp.

Lamb, F.D., and Ruppert, J., 1950, "Flotation of a North Carolina Pyrophyllite Ore," Report of Investigation 4674, US Bureau of Mines, Apr., 7 pp.

Lepetic V.M., 1972, "Flotation of Copper and Molybdenum Sulfides Without Collector," Preprint 72B308, SME-AIME Fall Meeting, Birmingham, AL, Oct. 18–20, pp. 1–23.

Lundquist, J.D., 1970, "Talc Treatment and Talc-Containing Pigments," US Patent 3,533,821, Oct. 13.

McMurray, L., and Bowles, E., 1941, "The Talc Deposits of Talladega County Alabama," Circular 16, Geological Survey of Alabama, 31 pp.

Miller, C.R., and Holtzapfel, P.J., 1970, "Corrosion Proofing Composition and Method," US Patent 3,549,391, Dec. 22.

Mulryan, H.T., 1971, "World-Wide Operations of United Sierra," *Industrial Minerals* (London), No. 41, Feb., pp. 43–45.

Neathery, T.L., 1968, "Talc and Anthophyllite Asbestos in Tallapoosa and Chambers Counties, Alabama," Bulletin 90, Geological Survey of Alabama, 98 pp.

Neathery, T.L., 1970, "Geology and Mining of Low Grade Talc Deposits, Tallapoosa County, Alabama," Preprint 70H311, SME-AIME. Fall Meeting, St. Louis, MO, 15 pp.

Neathery, T.L., et al., 1967, "Talc and Asbestos at Dadeville, Alabama," Report of Investigations 7045, US Bureau of Mines, 57 pp.

Needham, R.E., 1972, "The Geology of the Murray County, Georgia Talc District," Master's thesis, Pennsylvania State University, 107 pp.

Norman, J.E., et al., 1939, "Froth Flotation of Talc Ores from Gouverneur, New York," *Bulletin,* American Ceramic Society, Vol. 18, No. 8, pp. 292–297.

Okūda, S., et al., 1969, "Negative Surface Charges of Pyrophyllite and Talc," *Proceedings of the International Clay Conference,* Tokyo, Vol. 1, pp. 31–44, Vol. 2, pp. 14–16.

Okuma, A.F., 1971, "Structure of the Southwestern Ruby Range near Dillon, Montana," PhD dissertation, Pennsylvania State University, 122 pp.

Olson, R.H., 1976, "The Geology of Montana Talc Deposits," Spec. Publ. 74, Montana Bureau of Mines and Geology, pp. 99–143.

Owens, M.H., 1968, "Petrologic Study of Talc Mineralization in the Murphy Marble in Southwestern North Carolina," Master's thesis, Tennessee University, 64 pp.

Papke, K.G., 1975, "Talcose Minerals in Nevada—Talc, Chlorite, and Pyrophyllite," Bulletin 84, Nevada Bureau of Mines and Geology, 62 pp.

Parsons, A.B., 1948, *Seventy-Five Years of Progress in the Mineral Industry, 1871–1946,* AIME, New York, p. 355.

Pence, F.K., 1955, "Commercially Proven White-Firing Talc Occurring in West Texas," *Bulletin,* American Ceramic Society, Vol. 34, No. 4, pp. 122–123.

Perry, A., 1972, "Supply and Demand for Ceramic Tile in the Seventies," American Ceramic Society, Fall Meeting, Sept. 27–29, Bedford, PA.

Perry E.S., 1948, "Talc, Graphite, Vermiculite and Asbestos in Montana," Memoir 27, Montana Bureau of Mines and Geology, 44 pp.

Poppe, W., et al., 1972, "Talc Filled Metallizable Polyolefins," US Patent 3,663,260, May 16.

Povarennykh, A.S., 1972, *Crystal Chemical Classification of Minerals,* Vol. 1, trans. from Russian by J.E.S. Bradley, Plenum Press, New York, pp. 415–416.

Ranney, M.W., et al., 1972, "Silane Coupling Agents in Particulate Mineral-Filled Composites," 27th Reinforced Plastics Technical and Management Conference, SPI, Washington, DC, Feb. 8–11, pp. 1–29.

Rayner, J. H., and Brown, G., 1966, "Structure of Pyrophyllite," *Clays and Clay Minerals,* Proceedings, 13th National Conference on Clays and Clay Minerals, Pergamon Press, New York, pp. 73–84.

Roe, L.A., 1972, "High-Purity Talc from Western Ores," Preprint 72H312, SME-AIME Fall Meeting, Birmingham, AL, Oct. 18–20, 16 pp.; US Patent 3,806,043, Apr. 23, 1974.

Roe, L.A., 1980, "A Comparison of Calcium Carbonate and Talc Fillers," Preprint 80–301, SME-AIME Fall Meeting, Minneapolis.

Ross, C.P., Andrews, D.A., and Witkind, I.J., 1955, "Geologic Map of Montana," US Geological Survey.

Ross, M., and Smith, W.L., 1968, "Triclinic Talc and Associated Amphiboles from Gouverneur Mining District, New York," *American Mineralogist,* Vol. 53, May-June, pp. 751–769.

Seymour, R.B., et al., 1972, "Fibrous Reinforcements for Polymers," *1972–73 Modern Plastics Encyclopedia,* pp. 365–394.

Trauffer, W.E., 1964, "New Vermont Talc Plant Makes High-Grade Flotation Product for Special Uses," *Pit and Quarry,* Vol. 57, No. 6, Dec., pp. 72–76, 101.

Troll, G., and Gilbert, M.C., 1972, "Fluorine-Hydroxyl Substitution in Tremolite," *American Mineralogist,* Vol. 57, pp. 1386–1403.

Van Horn, E.C., 1948, "Talc Deposits of the Murphy Marble Belt," Bulletin 56, North Carolina Dept. of Conservation and Development, Division of Mineral Resources, 54 pp.

Wells, J.R., 1973, "Talc, Soapstone, and Pyrophyllite in 1973," Annual Advance Summary, US Bureau of Mines, 2 pp.

Winkler, H.G.F., 1974, *Petrogenesis of Metamorphic Rocks,* Springer-Verlag, New York, 320 pp.

Wright, L.A., 1950, "California Talcs," *Trans. AIME,* Vol. 187, pp. 122–128.

Wright, L.A., 1968, "Talc Deposits of the Southern Death Valley-Kingston Range Region, California," Special Report 95, California Division of Mines and Geology, 79 pp.

Zen, E-an, 1961, "Mineralogy and Petrology of the System AL_2O_3-SiO_2-H_2O in Some Pyrophyllite Deposits of North Carolina," *American Mineralogist,* Vol. 46, Jan.-Feb., pp. 52–66.

Titanium Minerals*

LANGTRY E. LYND †

STANLEY J. LEFOND ‡

Elemental titanium has become famous as a space age metal, because of its high strength/weight ratio and resistance to corrosion. However, the major use is in the form of titanium dioxide pigment, which because of its whiteness, high refractive index, and resulting light-scattering ability, is unequaled for whitening paints, paper, rubber, plastics, and other materials. A relatively minor use is in welding rod coatings, in the form of the mineral rutile. United States consumption of titanium dioxide in concentrates for these applications in 1982 was as follows: pigments, 630,000 st; titanium metal, 31,000 st; and welding rod coatings, 5,000 st. The only commercially important titanium ore minerals at the present time are ilmenite and its alteration products, and rutile.

Titanium was discovered by Gregor in 1790, as a white oxide which he recovered from menaccanite, a variety of ilmenite occurring as a black sand near Falmouth, Cornwall. Barksdale (1966) stated that the fundamental chemical reactions on which the present day titanium industry is based were known before 1800, although it was not until 1918 that these pigments were available commercially on the American market.

Pings (1972) outlined the early history of the titanium industry in the United States, referring to the work of Guise (1964) and others:

> Mining of titanium minerals in the United States began sometime between 1880 and 1900, in Chester County, Pennsylvania. Small quantities of rutile were also produced during that time in North Carolina and Georgia. In 1901 rutile was mined from a deposit near Roseland, VA, and was used in making titanium chemicals and for coloring ceramics. Ilmenite in the deposit was produced as a separate item in 1913, and rutile and ilmenite were obtained from this deposit through 1921.
>
> The mining of titanium bearing beach sands began in 1916 near Mineral City and Pablo Beach, FL, for the purpose of making titanium tetrachloride to be used in tracer bullets, flares, and smokescreens. The titanium pigment industry was founded in 1918. By 1928 a large part of the domestic production of rutile, ilmenite, and zircon came from Florida. However production ceased in 1929 in Florida with the mining of newly discovered deposits in Virginia which were mined until 1968. The discovery of new deposits in Florida, and the development of better mining and concentrating methods for low-grade sands, led to a return of activities in this area. The large deposits near Tahawus, NY, were brought into production in 1942 by National Lead Co. (now NL Industries, Inc.), and by 1949 this company was the leading producer of ilmenite in the world. In 1948 the extensive deposits of ilmenite and ilmenite-bearing iron ores were discovered in eastern Quebec.

The large Trail Ridge sand deposit in Florida was discovered by E. I. du Pont de Nemours & Co., Inc. geologists in cooperation with the US

* The writers are indebted to many sources of information for material used in this chapter. We especially wish to thank Stanford O. Gross for geological, mining, milling, and production information on the MacIntyre Development, and H. Leroy Scharon for his contribution on methods of exploration for hard-rock ilmenite deposits. Special thanks are also due the US Bureau of Mines; NL Industries, Inc.; W. P. Dyrenforth of Colorado School of Mines Research Institute, Golden, CO; A. G. Naguib, Vice President, Mineral Development International, Orange Park, FL; and Thomas E. Garnar, Jr., E. I. du Pont de Nemours & Co., Inc., for their cooperation and valuable assistance in preparing the manuscript.

† Physical Scientist, US Bureau of Mines, Division of Nonferrous Metals, Washington, DC.

‡ Industrial Minerals Consultant, Evergreen, CO.

Bureau of Mines and the Florida Geological Survey, and was developed by Du Pont in the late 1940s (Garnar, 1980).

The beginning of the modern titanium metal industry was in 1948, when Du Pont produced the first metal. US Bureau of Mines reports, which gave details of the Kroll process, together with the attractive properties of the metal for military aircraft, led to a concerted effort by industry and government to develop a large-scale titanium metal industry, which reached a peak capacity of over 36,000 stpy from six producers by 1958 (Pings, 1972a).

The major developments affecting the titanium metal industry since the late 1950s have been fluctuations in the production rate of military aircraft, and the increasing use of the metal in commercial aircraft and for industrial applications. A major change in the titanium dioxide pigment industry starting in 1957 was the development and expansion of the chloride process, which generally uses higher grade TiO_2 raw materials such as rutile in contrast to the relatively low TiO_2 ilmenite feed used in the older sulfate process. The resulting increased demand for rutile led to higher prices, which in turn stimulated the development of processes for making rutile substitutes by beneficiation of ilmenite.

Since 1960, new deposits of sand ilmenite were developed in New Jersey, Georgia, and Florida. Annual domestic titanium dioxide pigment production was 635,000 st in 1982, 19% below the peak of 787,000 st reached in 1974.

Production and Consumption of Titanium Concentrates and Products

United States production and consumption of ilmenite plus slag in 1982 were down appreciably from 1978 levels. Consumption of rutile also decreased, resulting in a 9% drop in United States pigment production from 1978 through 1982. World production of ilmenite plus slag has declined since 1978, and rutile production has increased (Table 1).

TABLE 1—Salient Titanium Statistics

	1978	1979	1980	1981	1982
United States:					
Ilmenite concentrate:					
Mine shipments, st	580,878	646,399	593,704	523,681	233,063
Value, 1000 $	$25,628	$32,965	$32,041	$37,013	$19,093
Imports for consumption, st	308,671	184,478	357,488	236,217	348,366
Consumption, st	792,289	791,063	848,607	856,116	583,250
Titanium slag:					
Imports for consumption, st	149,172	111,210	194,994	268,825	247,845
Consumption, st	128,826	144,708	181,582	252,826	225,541
Rutile concentrate, natural and synthetic:					
Imports for consumption, st	289,617	283,479	281,605	202,373	163,325
Consumption, st	263,184	313,761	297,582	285,371	238,937
Sponge metal:					
Imports for consumption, st	1,476	2,488	4,777	6,490	1,354
Consumption, st	19,854	23,937	26,943	31,599	17,328
Price, Dec. 31, per lb	$3.28	$3.98	$7.02	$7.65	$5.55
Titanium dioxide pigments:					
Production, st	700,755	742,081	727,245	r761,190	635,061
Imports for consumption, st	117,708	104,968	97,590	124,906	138,922
Apparent consumption, st	801,728	837,042	753,480	r806,040	722,362
Price, Dec. 31, cents per lb:					
Anatase	46.0	53.0	57.0	69.0	69.0
Rutile	51.0	59.0	63.0	75.0	75.0
World production:					
Ilmenite concentrate, st	r3,874,661	r3,874,586	4,015,772	p4,009,737	e3,371,090
Titaniferous slag, st	1,037,193	r842,038	1,343,200	p1,245,000	e1,170,000
Rutile concentrate, natural, st	*332,690	*r393,807	*459,634	*p409,220	*e381,253

Source: US Bureau of Mines.

e Estimated. p Preliminary. r Revised.

* Excludes US production data to avoid disclosing company proprietary data.

TABLE 2—Consumption of Titanium Concentrates in the United States, by Products
Short Tons

Year and Product	Ilmenite*		Titanium slag		Rutile (natural and synthetic)	
	Gross weight	TiO_2 content[e]	Gross weight	TiO_2 content[e]	Gross weight	TiO_2 content[e]
1978	792,289	475,448	128,826	91,490	263,184	245,184
1979	791,063	487,228	144,708	106,346	313,761	292,912
1980	848,607	513,315	181,582	133,933	297,582	277,882
1981:						
Alloys and carbide	§	§	#	#	—	—
Pigments	843,055	501,301	252,826	186,020	[1]206,257	[1]192,779
Welding-rod coatings and fluxes	§	§	—	—	7,389	6,944
Miscellaneous ‖	13,061	9,721	—	—	71,725	66,873
Total	856,116	511,022	252,826	186,020	[1]285,371	[1]266,596
1982:						
Alloys and carbide	§	§	#	#	—	—
Pigments	574,634	345,618	225,541	168,433	[1]194,994	[1]184,403
Welding-rod coatings and fluxes	§	§	—	—	5,607	5,275
Miscellaneous ‖	8,616	6,775	—	—	38,336	35,435
Total	583,250	352,393	225,541	168,433	[1]238,937	[1]225,113

Source: US Bureau of Mines.
[e] Estimated. [r] Revised.
* Includes a mixed product containing rutile, leucoxene, and altered ilmenite.
† Includes estimate of imported ilmenite used to make synthetic rutile in the United States.
‡ Includes imported synthetic rutile, but excludes synthetic rutile made in the United States from imported ilmenite.
§ Included with "Miscellaneous" to avoid disclosing company proprietary data.
\# Included with "Pigments" to avoid disclosing company proprietary data.
‖ Includes ceramics, chemicals, glass fibers, and titanium metal.
[1] Includes synthetic rutile made in the United States.

Table 2 shows that about 94% of total United States TiO_2 consumption in 1982 was used in titanium dioxide pigment manufacture. Table 3 shows that paper and plastics are the fastest growing uses for titanium pigments. As shown in Table 4, the major imports of titanium concentrates are ilmenite and rutile from Australia, and titanium slag from Canada.

World production of titanium concentrates, shown in Table 5, comes mainly from four major producing countries, Australia, Norway, the United States, and the USSR, with about 230,000 to 1,300,000 stpy each, and from Finland, India, and Malaysia, each producing 120,000 to 200,000 stpy. Australia is by far the largest producer of rutile, while Canada and the Republic of South Africa produce nearly all of the world's titaniferous slag.

Prices of titanium concentrates and products (Table 6) generally increased significantly in the years 1978–1981, due to increased production costs and to development of a tight metal supply in 1979–1980 caused mainly by sharply increased demand for commercial aircraft. The prices of rutile and titanium sponge metal dropped in 1982 because of the economic recession, and collapse of the commercial aircraft market.

Titanium Dioxide Pigments

Titanium dioxide pigments are made by two basic processes. In the older sulfate process, ilmenite (43 to 65% TiO_2) or slag (70 to 74% TiO_2 or 85% TiO_2) is converted to water-soluble sulfates, and the TiO_2 is recovered by hydrolysis of high-TiO_2 solutions. In the chloride process, rutile (95% TiO_2), high-TiO_2 ilmenite (60 to 70% TiO_2), or rutile substitutes (85 to 95% TiO_2, including 85% TiO_2 slag), are converted to volatile chlorides, and the TiO_2 is recovered by vapor phase oxidation of titanium tetrachloride.

Until the early 1950s, titanium dioxide pigments were produced entirely by the sulfate

TABLE 3—Distribution of US Titanium-Pigment Shipments, by Industry, TiO_2 Content Percent

Industry	1978	1979	1980	1981	1982
Paints, varnishes, lacquers	47.9	47.4	44.1	43.4	43.3
Paper	20.8	21.8	24.3	23.8	24.6
Plastics (except floor covering and vinyl-coated fabrics and textiles)	11.6	11.8	10.6	11.4	11.4
Rubber	2.8	2.9	2.1	2.2	2.3
Printing ink	2.0	1.9	2.8	1.3	0.9
Ceramics	2.1	1.9	1.7	1.4	1.1
Other	6.7	7.1	8.2	8.6	6.4
Exports	6.1	5.2	6.2	7.9	10.0
Total	100.0	100.0	100.0	100.0	100.0

Source: US Bureau of Mines.

TABLE 4—US Imports for Consumption of Titanium Concentrates by Countries Short Tons

Country	1978	1979	1980	1981	1982
Ilmenite:					
Australia	308,649	184,478	338,676	234,562	342,279
Finland	—	—	27	—	—
Germany, Federal Republic of	—	—	—	—	24
India	—	—	18,739	—	—
Netherlands*	—	—	46	—	—
Norway	22	—	—	1,656	—
Sri Lanka	—	—	—	—	6,063
Total†	308,671	184,478	357,488	236,217	348,366
Titanium slag:					
Canada	149,172	81,289	145,475	246,137	201,168
South Africa, Republic of	—	29,921	49,519	22,685	45,685
Other	—	—	—	3	992
Total	149,172	111,210	194,994	268,825	247,845
Rutile, natural:					
Australia	242,505	140,291	143,038	88,345	74,501
Malaysia	—	—	267	11	—
Sierra Leone	—	7,980	40,900	25,236	53,308
South Africa, Republic of	5,453	10,819	18,907	47,406	11,320
Sri Lanka	6,063	6,305	—	—	—
Thailand	—	—	197	—	—
Other	8	18	33	25	2
Total	254,029	165,413	203,342	161,022	139,131
Rutile, synthetic:					
Australia	23,546	72,218	60,962	39,708	22,744
Germany, Federal Republic of*	—	—	2	—	—
India	11,011	22,134	10,471	440	—
Japan	675	1,243	6,590	1,200	1,450
Taiwan	356	22,471	238	—	—
Other	—	—	—	3	—
Total†	35,588	118,066	78,263	41,351	24,194
Titaniferous iron ore‡					
Canada	51,640	153,714	10,185	12,271	6,996

* Country of transshipment rather than country of production.
† Data may not add to totals shown because of independent rounding.
‡ Includes materials consumed for purposes other than production of titanium commodities, principally heavy aggregate and steel furnace flux.

TABLE 5—World Production of Titanium Concentrates—Ilmenite, Leucoxene, Rutile, and Slag, by Country,* St

Concentrate type and country	1978	1979	1980	1981 [p]	1982 [e]
Ilmenite and leucoxene[†]					
Australia:					
Ilmenite	1,383,400	[r]1,267,656	1,442,924	1,452,033	#1,276,463
Leucoxene	[r]17,752	[r]24,001	26,393	21,657	#22,198
Brazil	22,131	[r]14,541	18,562	16,631	17,000
China	NA	NA	NA	150,000	150,000
Finland	145,395	131,947	175,267	178,023	176,000
India[§]	178,063	161,867	185,078	208,147	209,000
Malaysia #	205,929	220,262	208,470	190,432	121,000
Norway	845,461	903,690	912,508	724,907	#608,215
Portugal	358	295	258	368	370
Sri Lanka	36,421	61,035	37,430	88,197	88,000
USSR[e]	450,000	450,000	460,000	470,000	475,000
United States[‖]	589,751	639,292	548,882	509,342	#227,844
Total	[r]3,874,661	[r]3,874,586	4,015,772	4,009,737	3,371,090
Rutile:					
Australia	283,376	[r]307,435	323,801	263,729	#243,343
Brazil	402	484	472	190	220
India[§]	6,239	5,445	5,908	9,647	8,800
Sierra Leone	—	[r]8,267	52,356	55,992	#52,590
South Africa, Republic of[e]	20,000	46,000	53,000	55,000	52,000
Sri Lanka	12,673	16,176	14,097	14,662	14,300
USSR[e]	10,000	10,000	10,000	10,000	10,000
United States	W	W	W	W	W
Total	332,690	[r]393,807	459,634	409,220	381,253
Titaniferous slag:					
Canada[¶]	937,000	[r]525,840	964,200	837,000	750,000
Japan[¶]	193	198	—	—	—
South Africa, Republic of[e**]	100,000	316,000	379,000	408,000	420,000
Total	1,037,193	[r]842,038	1,343,200	1,245,000	1,170,000

Source: US Bureau of Mines.

[e] Estimated. [p] Preliminary. [r] Revised. NA Not available. W Withheld to avoid disclosing company proprietary data.

* Table excludes production of anatase ore in Brazil (4,298,731 tons produced prior to 1979 and apparently largely mined in 1978; 7,373,074 tons mined during 1979; and unreported quantities mined in 1980 and 1981), all of which was stockpiled without beneficiation. This material reportedly contains 20% TiO_2. The table includes data available through June 15, 1983.

[†] Ilmenite is also produced in Canada and in the Republic of South Africa, but this output is not included here because an estimated 90% of it is duplicative of output reported under titaniferous slag and the rest is used for purposes other than production of titanium commodities, principally as steel furnace flux and heavy aggregate.

[‡] Reported figure.

[§] Data are for fiscal year beginning Apr. 1 of year stated.

\# Exports.

[‖] Includes a mixed product containing ilmenite, leucoxene, and rutile.

[¶] Contains 70% to 72% TiO_2.

** Contains 85% TiO_2.

process. However, by 1981, as shown by the listings of US and world production capacities in Tables 7 and 8, the proportion of pigment produced by the chloride process was about 78% in the United States, and was substantial in the rest of the world as well. Total world production capacity in 1982 was about 1,900,000 stpy chloride process pigment and 900,000 stpy sulfate process pigment. Future annual growth in consumption of TiO_2 pigments has been estimated at 2 to 3% in the United States, and at 3 to 4% in the rest of the world.

The chloride and sulfate process TiO_2 pigments have each demonstrated superior prop-

TABLE 6—US Prices of Titanium Concentrates and Products

	Yearend prices				
	1978	1979	1980	1981	1982
Concentrates: $ per st:					
Domestic ilmenite, 64% TiO_2	—	—	35*	35*	40*
Imported ilmenite, 54% TiO_2, f.o.b. Atlantic ports	45	50	50	65	65
Imported rutile, 95% TiO_2, f.o.b. Atlantic ports:					
Small lots	325-350	425-450	425-450	450-475	450-475
Large lots, long term, estimate	250-275	350-375	350-375	325-350	275-300
Titanium slag:					
70-72% TiO_2, f.o.b. Quebec	100	100	104	121	134
85% TiO_2, f.o.b. Republic of South Africa	—	120	124	156	156
Synthetic rutile, f.o.b. Mobile, AL	—	—	310	340	350
Products: $ per lb					
TiO_2 pigments, regular grade:					
Rutile, carlots, f.o.b. plant	0.51	0.59	0.63	0.75	0.75
Anatase, carlots, f.o.b. plant	0.46	0.53	0.57	0.69	0.69
Titanium sponge metal:					
Domestic, f.o.b. plant	3.28	3.98	7.02	7.65	5.55
Imported, Japanese, c.i.f. US ports, not including import duty	3.20	3.60	7.50-8.70	7.50-8.50	No quotation

* Bulk, large lots, f.o.b. Green Cove Springs, FL.
Open market price not available in previous years.

erties over the other in various applications. The chloride process, being a continuous process is inherently simpler, consisting of fewer steps, and has less waste material to dispose of, as it uses feed materials with higher TiO_2 content and lower iron content than the sulfate process. Chloride process manufacturers are increasing their ability to use lower grade feed materials, at the expense of having to dispose of more iron chloride or iron oxide. The sulfate process has the advantage of lower cost raw materials, but has the greater quantity of waste material for disposal.

TABLE 7—United States Titanium Dioxide Pigment Producers in 1982.

	Estimated Dec. 31, 1982 Annual Capacity, st	
Company and plant location	Sulfate process	Chloride process
American Cyanamid Co., Savannah, GA	64,000	46,000
E. I. du Pont de Nemours & Co., Inc.:		
Antioch, CA	—	35,000
De Lisle, MS	—	150,000
Edge Moor, DE	—	110,000
New Johnsonville, TN	—	228,000
Gulf + Western Natural Resources Group, Chemicals Div.:		
Ashtabula, OH	—	35,000
Gloucester City, NJ	44,000	—
Kerr-McGee Chemical Corp., Hamilton, MS	—	56,000
NL Industries, Inc., Sayreville, NJ *	—	—
SCM Corp., Glidden Pigments Group, Chemical/Metallurgical Div:		
Ashtabula, OH	—	42,000
Baltimore, MD	66,000	42,000
Total	174,000	744,000

* Closed 100,000 stpy sulfate process plant in September 1982.

TABLE 8—World Titanium Dioxide Pigment Production Capacity, Dec. 31, 1982

Country	Producing Companies	Estimated Annual Capacity, st	
		Sulfate	Chloride
North America:			
Canada	Canadian Titanium Pigments, Tioxide Canada	76,000	—
Mexico	Pigmentos y Productos Quimicos	33,000	—
United States	American Cyanamid, Du Pont, Gulf & Western, Kerr-McGee, NL Industries, SCM	174,000	744,000
South America:			
Brazil	Titanio do Brasil (Tibras)	33,000	—
Europe, Western:			
Belgium	Kronos SA-NV, Bayer SA	72,000	—
Finland	Kemira Oy	88,000	—
France	Thann et Mulhouse, Tioxide SA	182,000	—
Germany, Federal Republic of	Bayer, AG, Kronos Titan GmbH, Sachtleben Chemie	298,000	66,000
Italy	Montedison	44,000	—
Netherlands	Tiofine	39,000	—
Norway	Kronos Titan A/S	28,000	—
Spain	Dow-Uniquinesa, Titanio SA	77,000	—
United Kingdom	BTP Tioxide, Laporte Industries	143,000	100,000
Europe, Eastern:			
Czechoslovakia	Prerovske Chemiske	22,000	—
Poland	ZPN	40,000	—
USSR	State Authority	137,000	—
Yugoslavia	Cinkarna Celje	22,000	—
Asia:			
China, Mainland	State Authority	20,000	—
Taiwan	ISK Taiwan	10,000	—
India	Travancore Titanium Products	15,000	—
Japan	Fuji Titanium, Furukawa Mining, Sakai Chemical, Teikoku Kaku, Titan Kogyo, Tohuku Chemical, Titanium Industries	224,000	26,000
Korea, Republic of	Hankuk Titanium, Donhwa Titanium	10,000	—
Africa:			
South Africa, Republic of	South African Titan Products	35,000	—
Australia	Tioxide Australia, Laporte Titanium (Australia)	75,000	—
Totals		1,897,000	936,000
Total, chloride plus sulfate		2,833,000	

Source: Modified from Coope, 1982.

Sulfate Process

Specifications for Sulfate Plant Feed: There are no rigid specifications for feed material to be used in the sulfate process, but certain impurities such as chromium, vanadium, columbium, manganese, and phosphorus are known to seriously impair pigment properties, so the degree of freedom from these elements is an important factor in selection of concentrates for this process. The TiO_2 content must be high enough to be recovered economically, and capable of dissolution in sulfuric acid at practical temperatures. Some typical commercial concentrates with their chemical and mineralog-

TABLE 9—Composition of Typical Commercial Ilmenite Concentrates and Titaniferous Slag Weight Percent

	United States		Australia			India		Malaysia	Canada	Republic of South Africa
	New York	Florida	Cable Bunbury	Western Titanium	Norway	Quilon	MK	Amang	QIT Slag	RB Slag
TiO_2 (total)	46.1	64.00	54.4	55.4	45.0	60.6	54.2	53.1	70-74	85.0 (min)
Ti_2O_3									10-15	25.0*
Fe_2O_3	6.7	28.48	19.0	11.1	12.5	24.2	14.2	8.7		
FeO	39.3	1.33	19.8	22.5	34.0	9.3	26.6	33.6	12-15	
Al_2O_3	1.4	1.23	1.5		0.6	1.0	1.3		4-6	
SiO_2	1.5	0.28	0.7	1.4	2.8	0.7			3.5-5	
CaO	0.5	0.007	0.04		0.25				1.2*	0.15*
MgO	1.9	0.20	0.45		5.0	0.9	1.0		4.5-5.5	1.3*
Cr_2O_3	0.009		0.2	0.03	<0.076	0.12	0.07	0.005	0.25	0.3*
V_2O_5	0.05		0.12	0.13	0.16	0.15	0.16	0.02	0.5-0.6	0.6*
ZrO_2	0.01					0.9	0.8			
S	0.6		<0.01		<0.05	0.21	0.12		0.03-0.10	
P_2O_5	0.008	0.12	0.02		<0.04			0.085	0.025*	
MnO	0.5		1.4		0.25	0.4	0.4	4.0	0.2-0.3	2.5*
H_2O (loss on ignition)	1.3		0.4			2.0	0.3			
Rare earths						trace	0.12			
C	0.22				<0.055				0.03-0.10	
Reference	(1)	(2)	(3)	(3)	(4)	(3)	(3)	(3)	(5)	(6)

References: (1) NL Industries; (2) DuPont; (3) Anon., 1971c; (4) Anon., 1973c; (5) Guimond, 1964; (6) Anon., 1977, chloride grade.
* Maximum

ical compositions are listed in Table 9. It will be noted that the TiO_2 contents range from about 45% for unaltered ilmenite concentrates to 85% for slag.

Processing to Pigment: In the sulfate process finely ground ilmenite or high-TiO_2 slag is digested with strong sulfuric acid, forming a solid, porous cake which is dissolved in dilute acid and water to yield a solution of titanyl sulfate ($TiOSO_4$) and iron sulfate. Any ferric iron present is reduced to the ferrous state by adding scrap iron. This is done to avoid precipitation of ferric iron later in the process and to facilitate washing the precipitated titania, since ferrous iron is less strongly adsorbed. After reduction, the solution is clarified by settling and filtration. With ilmenite solutions, some of the iron is usually crystallized out as ferrous sulfate heptahydrate (copperas). The solution is then concentrated, and the titania is precipitated by hydrolysis, filtered, washed, calcined at 900° to 1000°C to the oxide, and further treated prior to the finished pigment stage. The nature of the various special treatments determines the grade and type of the finished pigment, i.e., whether it is anatase or rutile, chalking or nonchalking, oil or water dispersible, etc.

The estimated tons of raw material required to produce one ton of TiO_2 are as follows:

Ilmenite or titanium slag	1.5–2.8
Sulfuric acid	3.0–4.0
Iron scrap	0.1–0.2

Chloride Process

In the chloride process, titanium tetrachloride ($TiCl_4$) is produced by chlorinating titaniferous, higher-grade materials at about 850° to 900°C in the presence of petroleum coke. The estimated tons of raw materials required to produce one ton of TiO_2 are as follows:

Rutile	1.1–1.2
Chlorine	0.1–0.2
Petroleum coke	0.1–0.2
Oxygen	0.4–0.5
Aluminum chloride	0.03

All the commercial $TiCl_4$ plants in the United States use fluid-bed chlorinators but static-bed chlorinators also can be used.

Titanium tetrachloride is a highly reactive liquid at room temperatures and reacts violently with moisture. Hence, all handling of the $TiCl_4$ is carried out under inert conditions to avoid contamination by oxygen or moisture in the air.

Unfortunately, the inherent impurities in the ore also form chlorides which have to be removed by fractional condensation before the next step, otherwise they are precipitated with the titanium dioxide and cause off-color pigments. Iron, vanadium, and silicon are the more deleterious impurities. Trace impurities are removed by distillation and/or chemical treatment.

Processing to Pigment: The purified titanium tetrachloride is converted directly to titanium dioxide by oxidizing the $TiCl_4$ with air or oxygen at a high temperature. The resulting fine grained TiO_2 is calcined at about 500° to 600°C to remove any residual chlorine and hydrochloric acid that may have been formed in the reaction. The calcination also helps to improve some of the pigment characteristics. Aluminum chloride is added to the titanium tetrachloride to assure that virtually all of the titanium is oxidized in the rutile crystalline form. The chlorine formed during oxidation is recovered and recycled.

The process can produce pigments in either the anatase or rutile form. The cost per ton of pigment is believed to be less by the chloride process than by the sulfate process.

Titanium Metal

Titanium has two major advantages over other structural metals—its high strength/weight ratio, and its outstanding corrosion resistance. Disadvantages are its relatively high cost and difficulty in fabrication.

Applications and Consumption

The major applications of titanium metal have been in the aerospace industry, and consumption has been strongly dominated by military aircraft demands. The proportion of titanium consumption used for military planes has decreased somewhat but in 1981 was still estimated to be 35% compared to 40% for commercial planes, and 25% for nonaerospace applications. The heavy dependence on aerospace markets has led to severe fluctuations in demand. The most serious setback was in 1958–1959, when military aircraft production was cut back in favor of missile development, with a drop in titanium sponge production from about 17,000 tons in 1957 to 4600 tons in 1958 and 4000 tons in 1959. Gradual recovery and further fluctuations in demand led to successive new peaks in mill product shipments of 16,000 tons in 1969, 17,000 tons in 1974, and about 27,000 tons in 1980, with intervening lows coinciding approximately with cancellation of the supersonic transport program in 1971 and of the B-1 bomber program in 1977. The strong market in 1978 through 1981 resulted mainly from sharply increased rates of ordering for commercial aircraft such as the Boeing 747, Lockheed L1011, and McDonnell-Douglas DC10 in 1978–1979, and for the Boeing 757 and 767 in 1979–1981, as well as moderately increased demand for tubing to be used in nuclear power plant condensers in 1979.

Of the titanium used for airplanes in 1980–1981, about 55% was for commercial, and 45% for military aircraft. The main use in engines is for compressor disks and blades, and other compressor components. In airframe structures titanium is used to reduce the weight of various parts such as frames, bulkheads, longerons, skin, fasteners, firewalls, ducts, tubes, fairings, landing gear assemblies, and for parts subjected to high temperature. Titanium is also used extensively in missile and space vehicle components.

Nonaerospace consumption was only 5 to 6% of mill product shipments for many years, but reached 20% of shipments or about 3,000 tons in 1973, and about 25% of shipments, or 6,700 tons in 1980. Besides the use in power plant condensers, the largest industrial application, nonaerospace applications for titanium metal include heat exchangers in the chemical and petroleum industries; anodes in electrochemical processing; evaporators for water desalination; and pumps, valves, tanks, piping, and other chemical processing equipment. Anticipated new growth areas include power plant gas turbine blades; salt water environment uses, including offshore drilling platforms, titanium-hulled submarines, and ocean temperature energy conversion; and possible automotive applications.

Processing to Metal

Kroll Process: Commercial production of titanium metal involves the reduction of $TiCl_4$ with magnesium or sodium metal, using various modifications of the Kroll process (Kroll, 1940).

The titanium tetrachloride used for manufacture of titanium metal is produced by chlorination of rutile, blends of rutile with altered ilmenite, or other high-TiO_2 concentrates, using fluid-bed chlorinators. Coke is used as a reductant, and the reaction is generally carried out at temperatures of 850° to 950°C. The liquid $TiCl_4$ from the chlorinator is purified by chemical treatment and distillation.

Titanium sponge producers in the United States and their 1982 production capacities are listed in Table 10.

At the Titanium Metals Corp. of America plant at Henderson, NV (Lloyd, 1956), reduction of $TiCl_4$ is carried out with magnesium in steel pots in gas-fired furnaces, under a helium

TABLE 10—1982 United States Titanium Sponge Capacity

Sponge Producers	Annual Capacity, St
Titanium Metals Corp. of America (NL Industries, Inc. and Allegheny International, Inc.), Henderson, NV	15,000
RMI Co. (US Steel Corp. and National Distillers Corp.), Ashtabula, OH	9,500
Oregon Metallurgical Corp. (Publicly owned, with Armco Steel Co. as major stockholder), Albany, OR	4,500
International Titanium, Inc. (Wyman-Gordon Co.; Ishizuka Research Institute and Mitsui & Co., Ltd., Japan; and other US and Japanese interests)	2,500
Teledyne Wah Chang Albany, Albany, OR (Teledyne, Inc.)	1,500
Western Zirconium Co. (Westinghouse Electric Corp.)	500
Total	33,500

atmosphere, at a temperature above the melting point of magnesium chloride, but below the melting point of the titanium metal, which forms as a spongelike mass. The sponge is removed from the reactor in a dry room to avoid absorption of moisture and is crushed and leached with acid to remove residual magnesium and magnesium chloride. The magnesium chloride drained from the reactors is transported molten to the magnesium plant, where magnesium and chlorine are recovered by electrolysis.

Oregon Metallurgical Corp. (Oremet) and Teledyne Wah Chang Albany also use magnesium to reduce the $TiCl_4$ to metal, while RMI Co. uses sodium for reduction. Oremet uses an inert gas sweep to remove nearly all of the magnesium chloride from the sponge, ITI and Teledyne use vacuum distillation, and RMI uses dilute acid leaching.

Titanium sponge is converted to ingot by compacting and double or triple vacuum arc-melting, along with scrap and alloying constituents, as a consumable electrode, using a water-cooled copper crucible. Vacuum melting removes hydrogen, residual magnesium chloride, and other volatile impurities.

All of the sponge manufacturers except ITI are integrated producers. Producers of ingots from titanium sponge and scrap in 1982, besides the sponge producers, included Martin-Marietta Aluminum, Inc., Teledyne Allvac, Howmet Corp., and Lawrence Aviation Industries, Inc.

A National Materials Advisory Board report (Anon., 1983) on availability of titanium provides a broad, integrated overview of the titanium field.

Electrolytic Titanium: Electrolytic procedures for producing titanium metal have been extensively investigated with the object of developing a continuous process which would be more economical than the batch Kroll process (Cobel et al., 1980; Leone et al., 1967; Priscu, 1966; Ramsdell and Mathews, 1960; Snyder, 1966). Early work was done by the US Bureau of Mines in Boulder City, NV. National Lead Co. was also involved in electrolytic titanium research as early as 1950. Titanium Metals Corp. of America built a semi-works plant and produced about 68,000 kg (150,000 lb) of electrolytic sponge titanium, but discontinued the operation in 1968 because of overcapacity for making sponge by the Kroll process. In 1980 D-H Titanium Co., a joint venture of Dow Chemical Co. and Howmet Turbine Components Corp., announced startup of an electrolytic titanium demonstration plant in Freeport, TX. The D-H process, which involves reduction of titanium tetrachloride in a fused salt bath, is based on early work done by the Bureau of Mines (Leone et al., 1967), research work by Dow, and pilot plant operation by Howmet at Whitehall, MI (Cobel et al., 1980). It is claimed that electrolytic processes will produce a higher purity product, require less capital investment, and consume less energy than plants using magnesium or sodium reduction (Minkler, 1980). The D-H venture was terminated in late 1982, mainly because of adverse market conditions.

Production and Prices

United States annual sponge and ingot production up to 1963, and ingot production from 1964–1980 are given in Table 11. Recovery to the 1957 peak of 17,249 st was not achieved until 1966, and two subsequent peaks were reached in 1969, 1974, and 1981.

From 1948 until 1954 the price of domestic titanium sponge metal was $5.00 per lb. Since then the price dropped many times, reaching a low of $1.32 in 1964 through 1972, before rising to reach $2.70 in 1975, $3.98 in 1979, $7.02 in 1980, and $7.65 per lb in mid-1981.

TABLE 11—United States Production* of Titanium Sponge and Ingot (Short tons)

Year	Sponge	Ingot
1950–54, average	1,851	NA
1955	7,398	4,573
1956	14,595	11,688
1957	17,249	10,009
1958	4,585	5,408
1959	3,898	6,017
1960	5,311	8,297
1961	6,727	9,371
1962	6,730	10,400
1963	7,879	11,138
1964	—	13,964
1965	—	15,294
1966	—	24,253
1967	—	25,960
1968	—	19,234
1969	—	28,490
1970	—	24,331
1971	—	18,450
1972	—	20,267
1973	—	28,932
1974	—	36,132
1975	—	25,560
1976	—	21,614
1977	—	26,302
1978	—	31,385
1979	—	37,414
1980	—	41,864
1981	—	45,923
1982	—	25,236

Source: *Minerals Yearbooks* and *Mineral Industry Surveys,* US Bureau of Mines.

*After 1963, the US Bureau of Mines no longer reported data on sponge production, to avoid disclosing company proprietary information. Data for ingot production, 1964-1980, include alloying constituents and the use of imported as well as domestically produced sponge.

NA–Not available.

Miscellaneous Uses

Fairly large amounts of rutile concentrate (about 7,000 st US consumption in 1980) are used for welding-rod coatings. Ilmenite, slag, or manufactured TiO_2 may also be used as a source of TiO_2 for welding-rod coatings, but rutile is considered to be the most desirable for this purpose (Stamper, 1970). The titanium dioxide stabilizes the electric arc, reduces the viscosity of the slag formed, and decreases the surface tension of the metal droplets on the electrodes, resulting in a smooth weld (Peterson, 1966).

Titanium dioxide also is used in producing ceramic capacitors and electromechanical transducers, in producing glass fibers, and for making titanium carbides. Commercial carbide cutting tools contain 8 to 85% titanium carbide, with or without tungsten carbide, in a matrix of molybdenum, nickel, or cobalt. Organo-titanium compounds such as alkyl and butyl titanates, derived from alcohols by reaction with titanium tetrachloride, and titanium esters are used as catalysts for various polymerization processes, as water repellents, and in dyeing processes (Stamper, 1970).

Titanium, formerly used mainly as a deoxidizer, has assumed considerable importance in the manufacture of stainless steels as a carbon- and nitrogen-stabilizing element to inhibit intergranular corrosion. The titanium is usually added as ferrotitanium, which is generally made by melting titanium scrap and iron together in an electric furnace. Ferrotitanium contains 15 to 70% titanium and 7 to 0.03% carbon. US consumption of titanium in steel and other alloys in 1982 was about 2,000 tons.

Large crystals of rutile, and certain titanates, especially strontium titanate, have been produced commercially, mainly by NL Industries, Inc., using a flame fusion process (Lynd and Merker, 1955; Moore, 1955). Because the refractive index of strontium titanate (1.41) is almost identical to that of diamond, strontium titanate gemstones look very much like diamonds, except that strontium titanate has more fire (flashes of spectral color) due to its higher dispersion. Strontium titanate also has been used in optical applications, such as for infrared detection devices (Beals and Merker, 1960).

Geology

Mineralogy

Although titanium is the ninth most abundant element of the lithosphere, comprising an estimated 0.62% of the earth's crust, there are only a few minerals in which it occurs in major amounts: rutile, anatase, and brookite (which are polymorphs of TiO_2), ilmenite and its alteration products, including leucoxene, perovskite ($CaTiO_3$), and sphene ($CaTiSiO_5$). Anatase may be emerging as a significant ore mineral of the future, but ilmenite, altered ilmenite, leucoxene, and rutile have been the only large volume ore minerals through 1980.

Ilmenite: The chemical formula of theoretically pure ilmenite is $FeO.TiO_2$. It was shown by Ramdohr (1950) that up to 6% Fe_2O_3 may be dissolved in solid solution, and at 1050°C a continuous solid solution series exists between ilmenite and hematite (Nicholls, 1955). Hematite may, and often does, occur with ilmenite as minute exsolution lamellae. Magnesium and manganese may substitute for the ferrous iron in ilmenite, which can produce the rare end-members $MgTiO_3$ (geikelite) and $MnTiO_3$ (pyrophanite), but usually these two elements are present as minor impurities. Magnetite is a common associate of ilmenite in igneous and metamorphic rocks, and in such coexisting pairs chromium, nickel, and vanadium tend to concentrate in magnetite while manganese concentrates in ilmenite.

In basic igneous rocks, notably anorthosites, gabbros, and basic lavas, ilmenite frequently occurs in intimate intergrowths with magnetite. The ilmenite forms lenses following octahedral parting planes in the magnetite host, and magnetite may, in turn, form crystallographically oriented inclusions within the ilmenite lenses.

Altered Ilmenite and Leucoxene: In sand deposits ilmenite frequently exhibits a degree of alteration caused by oxidation and removal of iron. The end product is essentially TiO_2. The process was described by Temple (1966) as follows:

> Alteration is initiated along grain boundaries and structural discontinuities within the grain. After going through an amorphous stage, oxidation and partial removal of iron from the ilmenite lattice results in an intermediate iron titanate of definite structure for which the name pseudorutile has been proposed. The alteration product at the stage of complete oxidation of the original iron in the ilmenite analyses 65–70% TiO_2 as compared with the 52% TiO_2 of ilmenite. Complete removal of iron from the pseudorutile lattice results in a grain composed of crystallites of the mineral rutile. Only the three mineral phases: ilmenite, pseudorutile and rutile have been identified in the commercial titanium mineral concentrates studied. Natural occurrence, in sand deposits, indicates that in weathering final removal of iron from pseudorutile to give the rutile end product takes place only either above or in the zone of the fluctuating water table.
>
> The alteration products of the mineral ilmenite are herein termed "altered ilmenite." Many commercial ilmenite concentrates actually consist of altered ilmenite. However, the term "leucoxene" is also applied to high-TiO_2 products of alteration. Commercial justification for distinguishing between the high- and low-iron products of alteration lies in the fact that the former are used in the sulfate process by pigment manufacturers whereas the latter are not economically soluble in sulfuric acid but are used, as is rutile, for the production of titanium tetrachloride used for pigment and titanium metal manufacture.

Alteration is an extremely slow process which is aided by elevated temperature, so that older sand deposits in temperate and tropical regions of the world generally contain high-TiO_2 ilmenite. Younger deposits, for example, those found on modern beaches, and those in the higher latitudes usually contain unaltered ilmenite with a TiO_2 content around 50%, near the theoretical level for pure ilmenite.

Rutile: Rutile, the high pressure, high temperature polymorph of TiO_2, is the commonest form in nature and is a widespread accessory mineral in high grade metamorphic gneisses and schists and in igneous rocks. It is also a common detrital mineral.

Commercial rutile concentrates run 95% TiO_2 or more, with SiO_2, Cr_2O_3, V_2O_3, Al_2O_3, and iron oxides comprising the remainder. Analyses of rutile from other occurrences may show major amounts of tantalum and columbium, which can enter titanium minerals because of the close similarity in ionic radius between Ti^{+4} and both Cb^{+5} and Ta^{+5}. There is also a high iron variety termed ferroan rutile.

Rutile may form by alteration from ilmenite or anatase, and while it is very stable over a broad range of geologic conditions, occasionally processes may be reversed so that rutile alters to sphene, possibly ilmenite, and more rarely anatase.

The characteristic color of rutile is reddish brown, but it may bé black, violet, yellow, or green.

Classification of Deposits

Titanium minerals have been mined from both rock and sand deposits. Until about 1942, nearly all of the ilmenite and rutile produced commercially came from sand deposits, but about one third of the world's ilmenite in 1982 came from rock deposits. Rutile, however, is now produced exclusively from sand deposits.

Rock Deposits: *Anorthositic Deposits*—Nearly all of the known commercially important rock deposits of titanium minerals are associated with anorthositic or gabbroic rocks, and are of three main types: ilmenite-magnetite (titaniferous magnetite), ilmenite-hematite, and ilmenite-rutile. Ilmenite-magnetite deposits usually contain ilmenite and magnetite as granular intergrowths that can be separated rather readily to yield concentrates of ilmenite and magnetite which may be essentially homogeneous minerals, or may consist of intimate intergrowths of one mineral in the other. Ilmenite-hematite deposits usually contain these minerals as intimate intergrowths and yield an ilmenite-hematite, or hemo-ilmenite concentrate rather than a separate concentrate of each mineral. Ilmenite-rutile deposits contain rutile and ilmenite either as separate concentrations or occurring together.

Miscellaneous Deposits—Other types of rock deposits in the United States that have been mined, or seriously considered as sources of titanium, include a deposit of ilmenite disseminated in schist in Yadkin Valley, North Carolina (Broadhurst, 1955), and a complex deposit of rutile, anatase, and brookite in a pegmatitic phase of alkalic rocks surrounding sediments at Magnet Cove, Arkansas (Fryklund and Holbrook, 1950; Fryklund et al., 1954).

A perovskite deposit in southwestern Colorado owned by Buttes Gas & Oil Co. was estimated to contain about 50 million tons of TiO_2 (Thompson, 1977). Processes to convert perovskite to titanium dioxide were being investigated through 1982.

The US Bureau of Mines (Llewellyn et al., 1980) and the US Geological Survey (Force, 1980) investigated the porphyry copper ores and mill tailings as a possible source of rutile. It was concluded that this material, which contains about 0.3% of potentially recoverable fine-grained rutile, could constitute a sizeable domestic resource, but more work is needed to develop an economic recovery process.

Major occurrences of anatase and ilmenite in weathered carbonatite bodies at Tapirá, Salitre, and Catalão in Minas Gerais, Brazil, are under investigation as new raw material sources. Feasibility and pilot plant studies to determine grades and recoveries of possible future commercial titanium mineral concentrates have been carried out. Modifications in pigment manufacturing processes may be necessary to accommodate potential production as a feed, or beneficiation or slagging may be used to make concentrates acceptable to existing pigment plants, but successful exploitation by any avenue will cause these and possibly other carbonatite occurrences to be classed as an important rock ore type.

Occurrences of titanium minerals in other types of rocks are known, but are not considered to be of commercial significance (Force et al., 1976).

Sand Deposits: Beaches, bars, dunes, and stream sands in many parts of the world are enriched by gravity segregation of the heavy minerals that are chemically resistant to weathering and physically hard enough to withstand considerable abrasive action. The energy of currents, waves, and to a lesser extent, winds, mobilizes the sand grains and permits them to behave as individuals within a fluid medium. Where the energy decreases, the heavier particles fall out while the lighter ones are carried farther. The composition of the heavy mineral concentrate depends on the nature of the geologic terrane being subjected to weathering and erosion and supplying materials to transporting streams. If available at the source, titanium-bearing minerals, zircon, magnetite, chromite, rare earth minerals such as monazite, staurolite, kyanite, sillimanite, garnet, xenotime, precious metals, and diamonds may be concentrated in this way. The resultant gravity segregations are usually dark colored and are called "black sands." Not all black sands are ore mineral occurrences; those found on the beaches of volcanic islands, for example, may be mostly amphibole and pyroxene.

There is a wide variation in titaniferous black sands in terms of heavy mineral concentration, percentage of titanium minerals within the heavy mineral suite, and TiO_2 content of the titanium mineral concentrate. All of these factors interact in rating deposits according to quality and as to whether they are today's reserves or possible resources of the future. Furthermore, the presence or absence of valuable coproduct or byproduct minerals such as zircon is a complicating factor. A classification on the basis of mineralogy, as suggested in the following section, is only broadly applicable.

Sand deposits in which rutile is the only economically important titanium mineral occur along the eastern shore of Australia. Ilmenite, altered ilmenite, and rutile form inland elevated strand-line deposits in Western Australia and in older sands of the Atlantic Coastal Plain of the United States. Ilmenite and altered ilmenite are the principal titanium ore minerals in other Western Australian districts; in Kerala, India; in deposits north of the Black Sea in the USSR; and in Florida and Georgia. Relatively unaltered ilmenite is found in large beach and dune occurrences along the northeastern coast of South Africa, in the Nile Delta of Egypt, and in still other Western Australian deposits, those closest to the present coast. Sand deposits of titaniferous iron ores occur as dune and beach deposits in many volcanic areas, of which those in New Zealand are the outstanding examples.

Mode of Occurrence and Origin

Rock Deposits: The anorthositic deposits contain titanium minerals either as massive ore, or disseminated in rock ranging from anorthosite to gabbro in composition, and varying in grade from solid ore to almost barren rock. While there are metamorphic and hydrothermal features in some parts of these deposits, for the most part, they seem to have formed at magmatic temperatures by magmatic processes.

The only other type of ilmenite rock deposit to be exploited commercially was the Yadkin Valley deposit in Caldwell County, North Carolina (Broadhurst, 1955), which was operated by a Glidden Co. subsidiary from 1942 to 1952. The ore consists of small masses of ilmenite disseminated throughout a talcose body which lies conformably with the enclosing quartzite and mica schist.

The Hot Spring County, Arkansas, deposits have been studied extensively for possible commercial development, and some rutile was recovered from the Magnet Cove deposit from 1932 to 1944. The deposits occur in a complex mixture of alkalic igneous rocks intruding folded sedimentary and metamorphic rocks. Much of the titanium is in the form of rutile or brookite, but many other titanium minerals have been identified. The ore minerals are intimately associated with gangue (Fryklund and Holbrook, 1950; Fryklund et al., 1954; Toewe et al., 1971).

Sand Deposits: Titanium-bearing black sands are found mainly in ancient or modern ocean and sea beaches around and occasionally within continental land masses. They frequently form highly visible surficial layers between the high and low water marks which may extend intermittently along coasts for miles, but such concentrations, containing perhaps 80% heavy minerals, are not mined on a large scale because they are usually too shallow and narrow to represent major reserves. Minable bodies are multilayered occurrences of a similar nature left behind by retreating seas, or coastal dunes formed when heavy minerals from black sand beaches were being transported inland by wind action. Heavy minerals tend to be disseminated within such dunes rather than layered as in beach-type deposits.

The history of a black sand ore body may be simple or complex. The essential elements are: (1) a "hinterland" of crystalline rocks in which the heavy minerals were accessory constituents, (2) a period of deep weathering, (3) uplift with rapid erosion and quick dumping into the sea of the products of stream erosion, and (4) emergence of the coastline with longshore drift and high-energy waves acting during the process of shoreline straightening. There may be intermediate stages such as partial concentration of the heavy minerals in a coastal plain sediment and subsequent elevation, erosion, and reconcentration. The sand brought to the sea by rivers is picked up and carried away from their mouths by longshore currents, forming offshore bars and filling in bays between headlands, particularly during storms. Where bars are formed, the sand-carrying waves drag bottom and lose their energy so that the heavy minerals fall on the seaward side while the light minerals are cast over the bar and into the quieter water beyond. Layer upon layer of varying concentrations of heavy minerals accumulates on the growing bar in this way. Where bays are being filled with sand, both heavy and light minerals are churned from the bottom by landward-rushing waves and are hurled up the beach slope. The smoother, slower retreat of each wave mobilizes the uppermost layer of sand deposited there, and draws away the light minerals, to be picked up again and again by waves as currents move them along the coast, while leaving the heavy minerals behind. Alternating periods of stormy and calm weather leave alternating layers of high and low concentrations of heavy minerals in the beach sand as it advances toward the sea.

Distribution of Deposits

Reserves

Ilmenite resources in rock and sand deposits in the United States, recoverable under the economic and technological conditions of 1981, were estimated to contain 18 million tons of titanium dioxide. Rock deposits in New York account for 29% of the total. The remainder is in beach and river sands in Florida (Garnar, 1980), Georgia, New Jersey (Markewicz, 1969), and Tennessee. An additional 94 million tons of titanium dioxide in ilmenite and 48 million tons in perovskite (Thompson, 1977), not recoverable under present conditions, occur in identified resources, widely dispersed throughout the United States. Major occurrences lie in Colorado, Minnesota, New Jersey, New York, and Wyoming.

Reserves of rutile in the United States are in sand deposits in Florida, Georgia, and Tennessee. The rutile in these deposits contains about 1.7 million tons of titanium dioxide.

TABLE 12—World Reserves of Ilmenite and Rutile
Thousands of Short Tons of TiO_2 Content

Country	Ilmenite	Rutile
Arab Republic of Egypt	—	—
Australia	27,000	10,000
Brazil	2,000	67,000*
Canada	26,000	—
China	34,000	—
Finland	3,400	—
India	22,000	3,000
Norway	52,000	—
Sierra Leone	—	3,000
South Africa, Republic of	57,000	5,500
Sri Lanka	21,000	5,000
United States	18,000	1,700
USSR	7,000	3,000
Total	270,000	140,000

* Mainly anatase.

Identified but presently subeconomic deposits in Arizona, Arkansas, California, Florida, North Carolina, South Carolina, Tennessee, Utah, and Virginia contain an estimated 13 million tons of titanium dioxide in the form of rutile. The rutile in the Arizona and Utah resources occurs mainly as an accessory mineral in porphyry copper ores and mill tailings (Force, 1980; Llewellyn and Sullivan, 1980).

Estimates for world titanium reserves and resources are shown in Table 12. World reserves of ilmenite are estimated to contain 266 million tons of titanium dioxide, of which US reserves account for 7%. The titanium dioxide content of world rutile reserves is estimated at 97 million tons; US reserves represent 1.6% (Lynd, 1983).

Rock Deposits—Principal Producing Countries

United States: *Sanford Lake District, New York*—The Sanford Lake deposits (Gross, 1968), are located in the heart of the Adirondack Mountains. Discovery of these titaniferous magnetite deposits dates back to 1826. Several attempts to exploit the ores for iron prior to 1942 proved uneconomical because of difficulties with the associated titanium, and the isolated location (Bachman, 1914; Masten, 1923). Since 1942, National Lead Co. (now NL Industries, Inc.) has produced ilmenite concentrates for the titanium pigment industry, and a magnetite byproduct used in the steel and refractory industries and as heavy media in coal cleaning operations. The district is within the large anorthosite massif making up the central high peak area of the Adirondacks. All of the various low silica rock types associated with the massif are found within the boundaries of the district. These consist of both the Marcy and Whiteface types of anorthosite and of gabbroic anorthosite, gabbro, and different grades of titaniferous magnetite ores. All of the rocks contain the same minerals and differ only in the percentages of these constituents.

There are four mineralized areas where an economic grade of ore has been found. Three of these have both gabbroic-type ore and anorthositic-type ore. The fourth has only gabbroic ore. The TiO_2 contents of the ores range from 9.5% to over 30.0% TiO_2. Ore bodies of both types are related to gabbro and conform to the configuration of the gabbro bodies within the anorthosite.

Except for a few thousand tons mined before 1900, all ore production has come from the Sanford Hill-South Extension ore body. Mining began on Sanford Hill in 1942, and as a result of further exploration and development work, was transferred in the early 1960s to the South Extension part of the ore body, which was overlain by glacial till and Sanford Lake.

General Geology—Rocks of the Sanford Lake district are regarded as members of a genetically related anorthositic series, the whole being part of the large Adirondack anorthosite massif. Locally, anorthosite grades into gabbro by an increase in the content of mafic minerals. Buddington (1939) divided this sequence into four rock types depending upon the amount of ferromagnesian minerals present: anorthosite with 0 to 10%, gabbroic anorthosite with 10 to 22.5%, anorthositic gabbro with 22.5 to 35%, and gabbro with over 35% mafic minerals.

For mine mapping purposes, rock types are designated as anorthosite, gabbroic-anorthosite, gabbro, and the various grades of ore. The relationship of rock and ore type to TiO_2 assay is shown in Table 13.

Economic Geology—The titaniferous magnetite ore bodies of the district are of two types. One type, referred to locally as "anorthositic ore," is associated with anorthosite waste rock as coarse-grained massive lenses of irregular shape, with little structure and having sharp contacts with the anorthosite. The second ore type, "gabbroic ore," occurs as fine to medium grained oxide-enriched bands within the gabbro, with well defined structures similar to those in the gabbro. Contacts with the gabbro range from distinct to gradational.

In the Sanford Hill-South Extension ore body, both types of ore occur. The anortho-

TABLE 13—TiO_2 Assay of Various MacIntyre Mine Rock Types

Classification	%TiO_2
Anorthosite	0–5.4
Gabbro	5.5–9.4
Low grade protore	9.5–13.4
Medium grade ore	13.5–17.4
High grade ore	17.5+

sitic type forms a footwall ore body and the gabbroic ore forms a hanging wall ore body. These are separated by various widths of anorthosite and/or gabbro rock. In some instances the two types of ore are in direct contact.

The ore body is designated by two names because it has been developed in two pits. Mining started where ore was exposed in outcrops on the west side of Sanford Hill. Diamond drilling indicated the ore lenses as becoming very narrow where they extended underneath the basin of Sanford Lake. Later magnetometer results, along with aerial work in 1955 and ground followup in 1959, revealed a change in strike and width of the magnetic anomalies. Diamond drilling proved the South Extension of the Sanford Hill ore body to be of commercial grade and tonnage. At Cheney Pond, about 2 km (1½ miles) west of the South Extension pit, a development drill program proved commercial tonnages of gabbroic ore, and revealed none of the massive anorthositic type. Oxide rock bands up to 30 m (100 ft) thick are found within a synclinal body of fine grained gabbro, which is in sharp contact with the underlying anorthosite. An extensive study of the petrology of the Cheney Pond deposit was carried out by Sun (1971).

Ore of both types occurs in two smaller ore bodies: the Mount Adams ore body, about two miles northeast of Sanford Hill, and the Upper Works or Calamity-Mill Pond ore body about three miles north of Sanford Hill.

MINERALOGY OF THE ORE BODIES—The ores of the Sanford Lake district contain both ilmenite and magnetite as granular aggregates and disseminated grains. The ratio of total Fe: TiO_2 is generally 2:1 or greater in the anorthositic ore, and is less than 2:1 in the gabbroic ore. Gabbroic ore is finer-grained than anorthositic ore.

The magnetite grains are seldom homogeneous, and normally contain some ilmenite in solid solution and up to 35% of ilmenite as exsolution intergrowths. The intergrown ilmenite occurs as tabular plates oriented parallel to the octahedral planes of the magnetite or along the boundary between magnetite grains.

A second highly magnetic phase, ulvospinel (Fe_2TiO_4), has been identified as a fine network within the magnetite of the Sanford Hill-South Extension ores. This ulvospinel was first identified by Ramdohr (1956), using high magnification metallography. Kays (1965) used X-ray techniques along with chemical assay data to estimate that the magnetite grains may have an ulvospinel content of 34%.

About 0.5% vanadium occurs in solid solution within the magnetite. No separate vanadium mineral has yet been identified.

Ilmenite occurs as a matrix around magnetite and is observed under the microscope to be corroded by magnetite. It is finer-grained than magnetite and can be distinguished megascopically in coarse-grained anorthositic ore by its high luster and conchoidal fracture, compared to the dull luster and parting planes in magnetite. In finer-grained ore, the ilmenite and magnetite are more equigranular and cannot be recognized individually in hand specimens.

The typical ore and gangue minerals and their weight percentages in both ore types are shown in Table 14.

Other minor minerals identified in the Sanford Hill-South Extension ore body include

TABLE 14—Typical Mineral Analysis of Ore Types in Sanford Hill Ore Body

	Gabbroic or Hanging Wall Ore, wt %	Anorthositic or Footwall Ore, wt %
Black opaques, total	61.5	77.3
Magnetics (magnetite)	25.7	40.8
Nonmagnetics (ilmenite)	35.8	36.5
Feldspar	19.2	10.3
Garnet	8.1	3.4
Pyroxene (mostly clinopyroxenes)	6.6	4.7
Amphiboles (green and brown hornblende)	2.3	1.5
Sulfides (pyrrhotite and pyrite)	1.5	1.7
Apatite, spinel, chlorite, calcite	<1.0	<1.0

chalcopyrite, sphalerite, molybdenite, prehnite, barite, leucoxene, scapolite, epidote, orthoclase, and quartz.

In his extensive study of mineral paragenesis, Stephenson (1945) concluded that the large plagioclase phenocrysts were first to develop, followed by the finer-grained plagioclase, and later by the sequence: apatite, hypersthene, augite, hornblende, garnet, and ore minerals. The chlorites, carbonates, and scapolite formed as later alteration minerals as a result of deformation and minor hydrothermal activity. Kays (1965), in a study of Sanford Hill pit, agreed in general with the sequence given by Stephenson (1945), but pointed out that no one single paragenetic sequence holds rigorously for all rock types in the Sanford Lake district.

Ore Genesis—Investigators have generally considered the Sanford Lake ores to have formed by magmatic segregation, as a result of such processes as simple segregation, filter-pressing of residual liquid with injection into the wall rock, and gravitational purification of the oxide melt by floating out of crystallized silicates with later injection, with replacement playing a minor role (Balsley, 1943; Bateman, et al., 1951; Buddington et al., 1955; Evrard, 1949; Osborne, 1928; Stephenson, 1945). Gillson (1956), however, related the variations in the anorthosites and the origin of the gabbro and ore to a series of pneumatolylic replacements involving andesination of Marcy anorthosite. Later solutions were assumed to be the source of the ferromagnesian and ore minerals.

Heyburn (1960) made reference to two stages of gabbro emplacement, one of them associated with the anorthosites, and the second a gabbroic intrusive rich in iron and titanium with sufficient volatiles to allow partial replacement of anorthosite by ore minerals.

Kays (1965) proposed that during granulation and shearing of the anorthosite, calcium and aluminum were released from the original laboradorite plagioclase, resulting in andesinization, but by a different means than proposed by Gillson (1956). Creation of pressure gradients by rock fracturing was then postulated to cause iron, magnesium, and titanium to migrate to the low pressure fractured areas, where they reacted with the calcium and aluminum released by andesinization to form ferromagnesian silicates, garnet, ilmenite, and magnetite. The resulting replacement could be partial, as in gabbro, or complete over large volumes, as in anorthosite.

Sun (1971) concluded that the Cheney Pond deposit is more likely to be of magmatic rather than replacement or metamorphic origin, stressing evidence of magmatic crystallization temperatures and the layered nature of the ore bands, which resembles the structure of ultramafic stratiform sheets.

Gross (1968), after working for several years at the MacIntyre Development, concluded:

> No one theory can explain satisfactorily all of the rock and mineral relationships now in evidence.
>
> It is quite certain that all the rocks of the district are genetically related as shown by their mineral constituents. The primary mode of emplacement is partially or wholly obscured by subsequent metamorphic processes. It is likely that certain portions of the rocks have been remobilized one or more times by deformation pressures. Different degrees of plasticity are evident . . .
>
> There is evidence that volatiles were present at some point in the history of the deposits, probably in both early and late stages. The faults and joints now exposed show movement along many surfaces. Along the major breaks there are breccia fragments of ore or gabbro in a matrix of carbonates. These are also the areas of recrystallized magnetite, large ilmenite crystal masses, and minor sulfide minerals . . .
>
> The present faults and associated hydrothermal minerals represent the last minor sequence in a very long and complicated series of geologic events.

Virginia Titanium Deposits—The titanium ores of Virginia are also associated with anorthosite, and are of two types: (1) rutile and ilmenite disseminated in the border facies of the anorthosite and bordering gneiss, and (2) rutile and/or ilmenite with apatite in the form of dikelike masses which Watson and Taber (1913) named nelsonite after the county in which they are most common. Deposits of both types have been mined commercially, utilizing mainly the soft, saprolite portions of the ore bodies which form an upper layer about 9 to 37 m (30 to 120 ft) thick.

Rutile was first mined commercially at Roseland, VA, in 1900 by the American Rutile Co. The ore was obtained by open cut mining of both saprolite and hard rock. The mine was operated intermittently until 1949 when increased labor costs, low grade ore, and low market prices for rutile led to a shutdown of operations. The ore consists almost exclusively of disseminated rutile in the anorthosite (Fish, 1962; Hillhouse, 1960).

Commercial surface mining for ilmenite began in 1930 from a large saprolite ore body along the Piney River on the old Warwick

Tract. This mine was first operated by the Vanadium Corp. of America, and later by the Southern Mineral Products Corp. (Hillhouse, 1960) until it was acquired by the American Cyanamid Co. in 1944. In 1958 the company transferred operations to the deposit on the S.V. Wood property near Lowesville about 5 km (3 miles) to the West (Fish, 1962), supplying ilmenite to its nearby sulfate process pigment plant until 1971, when both mine and pigment plant were closed down, mainly because the supply of readily mined soft ore at this location was running out.

Fish (1962) estimated the saprolite deposits in the area to contain reserves of titanium-bearing material in excess of 20 million tons, averaging 7.0% TiO_2. Unweathered rock beneath the saprolite contains comparable titanium values, but mining and processing this ore would be more expensive. The ore is derived from two types of rocks; one, containing ilmenite as the dominant titanium mineral, is a diorite, and the other, containing ilmenite and rutile, is an anorthosite.

The nelsonite deposits are higher grade, but of limited size. Many of them are lenticular and pinch out at shallow depths (Fish and Swanson, 1964).

Davidson et al. (1946) studied the Piney River deposit and concluded that it was emplaced as a large dike fingering into its walls. Replacement was agreed to have occurred, but this was not regarded as proof that the main development of ore was by replacement.

Theories proposed regarding the origin of the Virginia titanium ores in general include magmatic segregation from the anorthosite, favored by Watson and Taber (1913), differentiation from a granodiorite magma, suggested by Moore (1940), and replacement by invading solutions, proposed by Ross (1941, 1947); and Hillhouse (1960).

Yadkin Valley Deposit, North Carolina—The Yadkin Valley deposit is about 21 km (13 miles) north of Lenoir, in Caldwell County. The ore occurs as a series of narrow, close-spaced lenses which form a nearly continuous vein about 305 m (1000 ft) long. The ore consists of small masses of ilmenite disseminated throughout a talcose body which lies conformably with the enclosing quartzite and mica schist (Broadhurst, 1955). Some rutile occurs as fine inclusions in the mica, but was not recovered.

The ore zone is parallel to the gneissic structure, is about 9 m (30 ft) thick, and was mined by quarrying, running about 30 to 35% TiO_2. The mine was operated from 1942 to 1952 by the Yadkin Mica and Ilmenite Co., a subsidiary of Glidden Co. A flotation ilmenite concentrate was produced which contained 49 to 52% TiO_2 (McMurray, 1944), unusually high grade for a rock ilmenite.

Hot Spring County, Arkansas Deposits, (Magnet Cove, Christy, Hardy-Walsh)—These deposits occur in a complex mixture of alkalic igneous rocks intruding folded sedimentary and metamorphic rocks, many of which contain various quantities of titanium-bearing minerals (Fryklund and Holbrook, 1950). Much of the titanium is in the form of rutile or brookite, but many other titanium minerals have been identified. Rutile was recovered commercially from the Magnet Cove (Reed, 1949a) deposit from 1932 to 1944. A considerable amount of work has been done on ore dressing techniques for these ores (Fine and Frommer, 1952; Fine et al., 1949). The features contributing most to the difficulty of utilizing these deposits are their comparatively low grade, the larger deposits averaging 3 to 6% recoverable TiO_2 (Fryklund et al., 1954), the formation of large amounts of slimes during grinding, and the intimate association of ore minerals and gangue, resulting in low recoveries and comparatively low grade concentrates (Toewe et al., 1971). No commercial output of titanium minerals has been reported from the Christy (Reed, 1949) or Hardy-Walsh deposits.

Other US Deposits—Other deposits of potential commercial importance include the titaniferous magnetite deposits in the Laramie Range, Wyoming (Diemer, 1941; Frey, 1946; Pinnell and Marsh, 1954). Metamorphic rocks were intruded by anorthosite, which in turn was cut by gabbro and titaniferous magnetite dikes. The main dike at Iron Mountain consists of a granular aggregate of homogeneous ilmenite and magnetite containing very small intergrown ilmenite inclusions. In the Taylor deposit, as much as 60% apatite is locally present. Resources of high grade ore containing about 45% Fe and 20% TiO_2 may amount to about 30 million tons. Further metallurgical research would be needed, and the geographical location is at present unattractive for commercial development.

Other large deposits occur in the San Gabriel Mountains in Los Angeles County, California, as disseminations in a gabbroic facies of an anorthosite (Moorhouse, 1938). Very large tonnages may be available, but the average grade is only about 4.5% TiO_2.

Smaller deposits occur in nearly every one of the western states, for example in Montana (Wimmler, 1946); Minnesota (Broderick, 1917); Wichita Mountains of Oklahoma (Merritt, 1939); Boulder County, Colorado (Jennings, 1913); and the San Juan district, Colorado (Singewald, 1913).

Canada: *Quebec*—The geology and general characteristics of the Canadian titanium deposits were well described in an extensive and thorough report by Rose (1969). He concluded that the titaniferous magnetite and ilmenite deposits in eastern Canada were formed by magmatic differentiation and injection, mainly in Precambrian time, presumably from 850 million to 1,500 million years ago. There is an unmistakable genetic relationship between these deposits and anorthositic rocks. The anorthosite bodies, as exemplified by the Morin, St. Urbain, Lac St. Jean, Sept-Iles, and Lac Allard anorthosites, appear in general to be composite, multiple intrusions composed mainly of anorthositic and gabbroic (noritic) rock. According to Rose, (1969):

> True anorthosite characteristically is host to the massive ilmenite-hematite lodes, whereas gabbroic anorthosite characteristically is host to the extensive deposits of low-grade, disseminated, titaniferous magnetite containing mixtures of titanomagnetite, ilmenite, ilmenite-hematite, and other intermediate members that possibly form a solid solution series of iron-titanium oxide minerals. Although generally separate, in places the two types of deposits, as well as the host rocks, appear to be transitional.
>
> Although a few ilmenite and titaniferous magnetite occurrences have been found in rocks that were invaded by anorthositic magma, the major deposits are almost always within the intrusion itself. Low-titanium magnetite deposits are commonly found in Grenville-type rocks outside the anorthosite massifs.

The main deposits include: the sill-like Lac Tio ilmenite-hematite deposits being mined by QIT-Fer et Titane, Inc. (formerly Quebec Iron and Titanium Corp.), estimated to hold about 100 million tons of high grade open pit material; the dikelike Magpie Mountain medium grade titaniferous magnetite deposits, averaging 43% iron and 6% titanium, probably containing over 250 million tons of open pit material; the St. Urbain deposits with over 20 million tons of high grade ilmenite-hematite; and several million tons of slightly lower grade material near Ivry, and in the Lac du Pin-Rouge area, near St. Hippolyte-de-Kilkenny. Substantial amounts of low-grade titaniferous magnetite and ilmenite occur at many other localities in the Morin, Lac St. Jean, Sept-Iles, Lac Allard, and St. Urbain anorthosite areas (Rose, 1969).

Kish (1972) studied the chemical composition of the Quebec deposits, particularly with regard to the vanadium content, and reported that vanadium concentrations of economic importance have so far been found only in the titaniferous magnetite deposits. The composition and origin of selected titaniferous deposits were discussed by Lister (1966).

ALLARD LAKE (LAC TIO) DEPOSITS—The geology of the area in which the Allard Lake deposits occur was investigated by Retty (1944), who noted numerous concentrations of ilmenite within a mass of anorthosite, along the shores of several lakes. These ilmenite showings, while not large enough to warrant commercial development, led to detailed exploration of the anorthosite area in 1946 by Kennco Explorations, Ltd., which resulted in the discovery of the large Lac Tio deposit. The deposit lies between Allard Lake and Puyjalon Lake, about 40 km (25 miles) north of Havre St. Pierre on the north shore of the St. Lawrence River.

The geology of the area and the nature of the ilmenite deposits have been well described by Hammond (1949, 1952), and by Hargraves (1959). The most important lithological unit in the area is the Allard Lake anorthosite. It is one of several anorthosite masses occurring at intervals in the southeastern part of the Precambrian shield, in a line trending northeast from the Ontario-Quebec boundary to the Labrador coast. The Allard Lake anorthosite mass is about 145 km (90 miles) long, and 32 to 48 km (20 to 30 miles) wide, its length paralleling the Gulf of St. Lawrence. The several facies of the anorthosite range from almost pure feldspar rock through anorthositic gabbro, ilmenite-rich anorthosite, and norite. The norite occurs as steeply dipping sheets, as much as 6 km (4 miles) long and 914 m (3000 ft) thick, intruded into the anorthosite, and is rich in hemo-ilmenite and magnetite.

The Lac Tio deposit is a flat-lying, tabular body about 1097 m (3600 ft) long, and 1036 m (3400 ft) wide. With an estimated tonnage of 125,000,000 st of ilmenite averaging 32% TiO_2

and 36% Fe, it is the largest known ore body of its type in the world.

The ore consists of exsolution intergrowths of ilmenite and hematite, with coarse-grained ilmenite containing numerous blades and lenses of hematite up to 0.3 mm wide, which in turn may contain similarly shaped but smaller inclusions of ilmenite, in parallel orientation. The typical high grade ore contains about 75% ilmenite, 20% hematite, and 5% gangue minerals consisting of pyroxene, feldspar, and minor amounts of pyrite, pyrrhotite, and chalcopyrite.

The ore occurs in anorthosite and anorthositic gabbro and is identical in character with that in other deposits in the area. Inclusions of anorthosite are found in the ore, and do not show evidence of replacement by ilmenite. The ilmenite is very coarse-grained along its contacts with anorthosite, and the contacts are very sharp. Tiny dikelets of coarse granular ilmenite commonly cut the anorthosite at contacts with the ore bodies. On a basis of such direct field evidence, Hammond (1949, 1952) concluded that the Allard Lake ores are late magmatic, probably emplaced by a process of late gravitational liquid accumulation, with injection of the oxides into fractures within the anorthosite.

Hargraves (1959) recognized a genetically significant relationship between the ore deposits and the oxide-rich norite phases of the anorthosite rocks with which they are associated. The usual occurrence of the isolated ilmenite masses in anorthosite below norite sheets, with hemo-ilmenite concentrated at the base of the sheets, suggests that the norite sheets may have been the original source of the iron-titanium oxides. Hargraves (1959) presented evidence suggesting immiscibility between the oxide fraction and the silicate fraction from which pyroxenes crystallized, and emphasized that the crystallization of the oxide fraction was subsequent to all other primary crystallization. He suggested that:

> Gravitational concentration by downward settling within the sheets could have been accentuated by contemporaneous deformation ("filter pressing") which in some places, caused it to be squeezed out of the sheets, and injected into surrounding anorthosite. In this way the isolated deposits in anorthosite are postulated to have been formed.

Based on his own extensive study of the St. Urbain deposits, and the similarity between these and the Allard Lake deposits, Gillson (1932, 1949) felt that the St. Urbain and Allard Lake ore bodies are pneumatolytic replacement deposits.

The ilmenite-hematite ore produced from the Lac Tio operation contains about 34% TiO_2 and 40% Fe. Because of its high iron content it is not used directly for titanium dioxide pigment manufacture, but is subjected to further concentration and electric furnace smelting at Sorel, Que., to produce iron metal, and a high-TiO_2 slag which is widely used as a feed for sulfate process pigment plants.

China: The largest titanium deposit in China is reportedly the 1.1 billion ton Panzihua titaniferous magnetite deposit containing about 7% titanium in the form of ilmenite, near Dukou, Sechuan province (Brady, 1981).

Norway: The iron-titanium provinces of Norway have been described by Geis (1971), Vokes (1968), Carstens (1957), Michot (1956), Hubaux (1956), and others. There are three types of ilmenite deposits: (1) ilmenite, (2) vanadium-bearing magnetite-ilmenite, and (3) apatite-bearing magnetite-ilmenite. Production of ilmenite has come only from Type 1 deposits (Storgangen and Tellnes), except for a small amount produced from a Type 2 deposit (Rodsand), as a byproduct from magnetite production. These three commercial deposits all occur in the Egersund anorthosite in southwestern Norway.

Storgangen Deposit—The Storgangen ore body located near Hauge, was deposited in a fracture in anorthosite, forming a dike about 1585 m (5200 ft) long and 49 m (160 ft) wide, of unknown depth, dipping 45° to 50° N. The ore contains about 40% ilmenite and 9% magnetite with a gangue that consists mainly of hypersthene and plagioclase, with accessory biotite, pyrite, chalcopyrite, and spinel. The ilmenite contains about 13% hematite as exsolution lenses and lamellae in parallel orientation, ranging up to a few microns in width. The magnetite is essentially homogeneous, containing very little exsolved ilmenite. The ilmenite concentrate was produced by a combination of gravity, magnetic, and flotation methods. From this deposit, Titania A/S (NL Industries, Inc., subsidiary) mined 9 million st of ore from 1916 to 1966.

Tellnes Deposit—Around 1954, it became apparent that because of the increased demand for ilmenite, the amount of ore accessible at currently mined levels in the Storgangen deposit would be soon depleted. By 1956, two alternatives had been investigated: to extend operations to deeper levels at Storgangen, or to discover and develop a new ore source in nearby areas considered favorable to the occurrence of titanium ore. Diamond drilling at Storgangen showed that the ore available at depth was less than expected, and at current production rates would have lasted only another 10 to 12 years. At the Tellnes deposit which was discovered by an aeromagnetic survey in 1954, drilling showed ore reserves of 220 million st, with another 110 million st probably present, based on projection to depth. After extensive laboratory and pilot plant work to demonstrate that a satisfactory concentrate could be made from the Tellnes ore, which is much finer grained than the Storgangen ore, it was decided to develop the Tellnes deposit as an open pit mining operation (Brun, 1957).

From 1960 to 1966 ilmenite was produced from both the Storgangen and Tellnes deposits. In 1966 production at Tellnes reached about 330,000 stpy, and production ceased at Storgangen. Since that time, production capacity was increased in three steps, and in 1974 production of ilmenite flotation concentrate reached a peak of 935,000 st.

The Tellnes ore body is about 40 km (25 miles) east of the Storgangen deposit, and consists essentially of an ilmenite-rich norite intruded into anorthosite. The ore body has a known outcrop length of 2682 m (8800 ft), and a surface area of about 56 hm^2 (140 acres).

The ore is magmatic in character, as proved by its texture, by apophyses of ilmenonorite into the surrounding rocks, by xenoliths of anorthosite in the ore itself, and by the presence of eruptive breccia (Dybdahl, 1960).

The Tellnes ore is much finer-grained than the Storgangen ore, the intergrowth of ilmenite and magnetite is more intimate, and the amount of magnetite is lower in the Tellnes ore—only 2% compared to about 7% in the Storgangen ore. Dybdahl (1960) gave the following average mineral composition for Tellnes ore: ilmenite, 39%; plagioclase, 36%; hypersthene, 15%; biotite, 3.5%; and accessories, 3.5%, including a little apatite, and some pyrrhotite containing nickel and copper. As at Storgangen, the ilmenite contains minute exsolution lenses and lamellae of hematite.

Because of the fineness of grind needed to liberate the ore minerals, and the relatively high phosphorus content present as apatite, extensive changes in the ore dressing procedure had to be made, resulting in an all-flotation ilmenite concentrate, and an acid washing step to remove apatite. The ilmenite concentrate contains 44.5 to 45.5% TiO_2, about 34.0% FeO, and 12.5% Fe_2O_3.

Tellnes ilmenite concentrate is the major feed material for a large number of European sulfate process pigment plants, and is in a particularly strong market position because of relatively stable North Sea freight rates, and its proximity to the European market (Anon., 1973c; Anon. ,1978a).

Finland: *Otanmäki Deposit*—The ilmenite-magnetite deposit at Otanmäki is located almost at the geographical center of Finland. The Otanmäki deposit and a neighboring deposit at Vuorokas were discovered in 1938 as a result of magnetic surveys by Veikko Okko of the Geological Survey of Finland (Harki et al., 1956).

As described by Harki et al. (1956), the deposit at Otanmäki extends about 2 km (1.2 miles) within an east-west zone which curves northward in a semicircle at its east end. This circle is the eastern boundary of a mass of amphibolite occupying most of a roughly elliptical area within which the ore bodies are found. The ore occurs as lenses which are usually located along contacts between amphibolite and anorthosite, or between anorthosite and rocks that occur in a heterogeneous band along the southeast border of the ore zone. The anorthosite appears as continuous conforming lenses in the innermost border of the curve. A mass of gabbro borders the southwest part of the ore zone. According to Harki et al. (1956), it has been established that the ore is intrusive in nature, except possibly in the case of lower grade stratified ores.

The ore minerals, magnetite and ilmenite, occur as independent grains which after fine grinding are separated to produce magnetite and ilmenite concentrates. There are no microscopic intergrowths of magnetite and ilmenite, although the ilmenite contains minor amounts of intergrown hematite (Harki et al., 1956; Vaasjoki, 1947). The main gangue minerals are chlorite, hornblende, and basic plagioclase, with accessory spinel and sphene.

Reserves are about 28 million st containing 40% magnetite, 30% ilmenite, 1 to 2% pyrite, and 0.25 to 0.3% vanadium associated with the magnetite (Anon., 1978b).

Rock Deposits—Potential Sources

Brazil: *Tapira and Salitre Titanium Deposits*—These deposits are unique in that their major titanium mineral is the anatase (or octahedrite) polymorph of TiO_2, rather than rutile or ilmenite. The deposits occur in an alkaline pipe, 6 km (4 miles) in diameter, about 48 km (30 miles) southeast of Araxá in

Minas Gerais. According to the National Department of Mineral Production, proved and indicated reserves contain 108 million st of TiO_2 (Anon., 1980a). In places, anatase may make up as much as 70% of the ore. In preliminary tests concentrates containing up to 86% TiO_2 were produced with very low chromium and vanadium content, said to be suitable for producing TiO_2 pigments by the chloride process. These concentrates have some solubility in sulfuric acid (Anon., 1972).

Through 1979, about 12 million tons of anatase-bearing ore had been mined and stockpiled in conjunction with phosphate production.

Mexico: *Pluma Hidalgo Deposit, Oaxaca*—The Pluma Hidalgo Deposit occurs in an area known for some time to have scattered occurrences of rutile, and is located about 125 km (78 miles) south of the City of Oaxaca, the state capital. Starting in 1953, Republic Steel Corp. carried out an extensive program of exploration, but did not find high grade ore bodies large enough to justify a mining operation. In 1957, when the price of rutile dropped substantially, the project was discontinued.

Paulson (1964) described the dominant country rock as a quartz feldspar gneiss, or granulite, which may be classed as an anorthosite by analogy with the anorthosite associated with the similar Virginia titanium deposits. The common mineral association in ore zones is ilmenite, rutile, and apatite in a green rock that is mostly chlorite.

Regarding possible economic exploitation of the Pluma Hidalgo deposit, Paulson was of the opinion that instead of trying to find large zones of high grade ore, it might be better to block out a large tonnage of the irregularly distributed ore which, including the barren rock that would have to be removed with the ore, might have an overall grade that would permit mining by open pit methods.

Sand Deposits—Principal Producing Countries

United States: All commercially important titanium mineral sand deposits in the United States are within the Atlantic and Gulf Coastal Plain geologic provinces. States which have established or potential exploitable heavy-mineral-bearing sands are Florida, Georgia, New Jersey, and Tennessee.

Concentrations of heavy minerals in the coastal areas of the southeast are related to both recent and ancient marine shorelines, the latter of Pleistocene age. Up to seven old shorelines have been identified by different investigators (Cooke, 1941, 1945; Flint, 1940, 1942, 1947; MacNeil, 1949; Parker and Cooke, 1944), but the most widely recognized are: (1) Okefenokee, 46 m (150 ft), (2) Wicomico, 31 m (100 ft), (3) Pamlico, 8 to 11 m (25–35 ft), and (4) Silver Bluff, 2 to 3 m (8–10 ft) above sea level. All are regarded as lines of farthest marine transgression during interglacial and postglacial periods, and all have produced commercial heavy mineral deposits. Mining began where black sands were exposed on modern beaches, and when these deposits were exhausted operations moved inland to the larger and harder-to-find deposits of the older elevated strands. Florida has three heavy mineral mines operating with possibilities for more in the future; Georgia's only deposit exploited to date recently has been worked out but, again, other occurrences of possibly economic size and grade are known.

In central New Jersey two operations have recovered titanium minerals and byproducts from Pliocene sands of the Cohansey formation, about 24 km (15 miles) inland from the present shore.

The host for heavy mineral deposits in western Tennessee which have been considered for mining is the Cretaceous McNairy sand, outcropping within the Mississippi embayment of the Gulf Coastal Plain. Past, present, and possible future ore bodies in each of the three areas of occurrence are considered in turn below.

Florida and Georgia—Storm-line concentrations of heavy minerals in modern beach sands have been mined for ilmenite at two locations on Florida's eastern beaches. Mining began at Pablo Beach near Ponta Vedra in 1916 (Martens, 1928), and Riz Mineral exploited a deposit on the ocean front near Vero Beach in later years. Similar modern beach, bar, and barrier island black sands are found on the northeast and west coasts of the peninsula and along the shore of the western panhandle. The ratios among species in heavy mineral suites in the western Florida occurrences are different from those in eastern deposits, probably because of dissimilar ultimate sources on opposite sides of the southern Appalachian Mountains. The potential ore reserves in the known near-shore concentrations are small, with the possible exception of Amelia Island in extreme northeastern Florida, Cumberland Island just to the north in Georgia, and several others of the Sea Island chain formed as barrier islands during the development of the Silver Bluff shoreline. Land values in the islands have increased because of their growing popularity as resort areas over approximately three decades since the heavy mineral deposits were first drilled, so that mining may not be economically competitive with tourism and recreation.

Elevated sand bars were mined by Riz Mineral south of Vero Beach and by Humphreys Gold Corp., on behalf of Titanium Alloy Manufacturing Co., in a Pamlico shoreline feature just west of Jacksonville Beach. These deposits are now mined out.

Just south of the St. Marys River, which forms Florida's northern boundary, and northeast of the community of Yulee, ITT-Rayonier owns land containing an unmined heavy mineral deposit which has been drilled by several organizations, most recently by Pennsylvania Glass Sand Corp., the ITT-Rayonier subsidiary. Ore grade lenses form low north-south ridges, interspersed with shallow swamps, and rest on clay and shell beds. The Yulee deposit is also a Pamlico shoreline development.

In Georgia other black sand concentrations, all related to the Pamlico shoreline, are located north and south of Brunswick and south of Savannah. Potential ore reserves are modest in all known cases, as they are at Yulee, but exploitation might be possible if the same mining equipment could be used at several deposits sequentially, and if a heavy mineral concentrate could be economically transported to a central separating plant.

TRAIL RIDGE, FL—The broad sand ridge that extends from the southern parts of Clay and Bradford counties in north-central peninsular Florida for 201 km (125 miles) northward into southeastern Georgia, called Trail Ridge, is thought to have formed by one or more of several processes in late Miocene or Pliocene time. It has been suggested that it developed as:

1) An accumulation of deltaic sediments from southward flowing rivers which was reworked by encroaching seas and partly washed away (Bishop, 1956).

2) A prograding bar built up on the seaward side of finer marine to estuarine deposits accumulating on its protected Gulf side (Brooks, 1966).

3) A residual sand ridge resulting from the weathering in situ of underlying sediments.

4) A barrier island originating as a ridge built up landward of a shoreline and cut by tidal inlets during coastal subsidence or sea level rise, forming lagoons behind the ridge (Hoyt, 1967).

5) A reworking, through the action of winds, waves, and currents, of the northern part of the remnant Lake Wales ridge, a possible deltaic deposit extending more than 241 km (150 miles) farther to the south (Pirkle and Yoho, 1970).

6) Sediments transported from the high terrace sands to the west of Trail Ridge (Pirkle, 1975).

Du Pont's Trail Ridge ore body occupies the southern 29 km (18 miles) of the western part of the feature, and is from 2 to 3 km (1 to 2 miles) wide. The base of the ore body is at an elevation of 44 m (145 ft) to more than 61 m (200 ft) above sea level, and the average thickness is 11 m (35 ft). It overlaps barren coarse sand on the east. An indurated layer of concentrated organic material comprised of decomposed remnants of roots, branches, and trunks of trees underlies the central portion, and the western margin lies on pre-Pleistocene clayey sand in many places (Grogan et al., 1964).

The heavy minerals are thinly layered and disseminated in brown, oxidized, cross-bedded sand. In random lenticular areas at various depths organic materials and clayey minerals cement the sand grains to form a poorly consolidated sandstone locally called "hardpan." The average grade in heavy minerals is low, only about 4%, but there are about 45% high-TiO_2 minerals in the heavy mineral suite (Pirkle and Yoho, 1970).

The titanium ore minerals are altered ilmenite, leucoxene, and a very little rutile. Staurolite, zircon, tourmaline, spinel, kyanite, sillimanite, monazite, corundum, and topaz are also present. Authigenic pyrite occurs in small amounts associated with present-day swamps. The absence of monazite, garnet, and epidote, which are common in the underlying formations and in other southeastern heavy mineral deposits, implies different source rocks and a different age for the Trail Ridge minerals. The absence of these three minerals makes the separation and marketing of staurolite possible (see "Staurolite" chapter).

Alteration of the titanium minerals by oxidation and leaching of iron has been extensive. Grogan et al. (1964) wrote:

> The TiO_2 content of the mineral grains is highest near the land surface, decreases rapidly from the surface to a depth of about 10 feet, and then remains essentially constant to the bottom. The iron content varies inversely with the TiO_2 content. The particle size of the titanium mineral grains ranges between 48 and 200 mesh, with the average size near 80 mesh. The grains are spherical, and some are porous as a result of leaching.

Their color ranges from black through brown, gray, and tan to light yellow or yellowish-white depending on the degree of alteration.

The term "ilmenite" is applied to the more magnetic of the titanium minerals. These particles are black with a metallic luster and range from 62 to 70% TiO_2.

The term "leucoxene" is applied to the less magnetic opaque titanium minerals which contain from 70 to 95% TiO_2. They have the same size and shape as the ilmenite grains but are lighter in color, and represent more advanced stages of alteration. Their magnetic susceptibility ranges from moderate for the darker grains to feeble for the very light colored grains.

Unaltered crystalline rutile is common in the nonmagnetic fraction. The grains are dark red in color and are mostly minus 150 mesh in size.

GREEN COVE SPRINGS, FL—About 32 km (20 miles) east-southeast of the south end of Trail Ridge an ilmenite-leucoxene-zircon-monazite heavy mineral deposit was mined by Titanium Enterprises, a joint venture of American Cyanamid and Union Camp Corp., until 1978 when mining ceased for economic reasons. In 1980 the property was purchased and mining resumed by Associated Minerals (U.S.A.) Ltd., Inc., a subsidiary of the Australian firm, Associated Minerals Consolidated Ltd. The ore body is near the eastern margin of the Duval Upland, which is thought to be a regressional beach ridge plain, and is probably related to an ancient shoreline at an elevation of 27 to 31 m (90 to 100 ft) above present sea level or to the Wicomico shoreline of MacNeil (1949).

The ore zone at Green Cove Springs is 16 to 19 km (10 to 12 miles) long, 1 km (¾ mile) wide, and it averages about 6 m (20 ft) thick. The heavy minerals comprise 3 or 4% of loose to slightly consolidated quartz sands which are underlain by subgrade quartz sands, by brown and gray sands containing some clay, by shell beds, and finally, at a depth of 31 m (100 ft) or more below surface, by limestone and dolomite.

The grain size of both quartz and heavy minerals is smaller than at Trail Ridge. Also, the heavy mineral suite, containing a significant percentage of monazite together with minor amounts of epidote and garnet, is different. It has been suggested (Pirkle et al., 1974) that while the Trail Ridge concentration of heavy minerals was derived from the Northern Highlands of the northwestern part of peninsular Florida, the Green Cove Springs deposit was formed from the sands of the Duval Upland to the east.

BOULOUGNE, FL—Humphreys Mining Co. in 1974 to 1979 mined a heavy mineral deposit about 3 km (2 miles) south of the town of Boulougne on the St. Marys River, which forms the boundary between extreme northern Florida and Georgia. It was located near the eastern edge of the Duval Upland at an elevation above sea level the same as that at Green Cove Springs, and had similarly fine sands containing monazite, garnet, and epidote. The two deposits probably had a common genesis.

The Boulougne ore body was 4 to 5 km (2½ to 3 miles) long north-south, 0.8 to 1.2 km (½ to ¾ mile) wide, and 2 to 8 m (5 to 25 ft) thick. The ore-bearing surface sands were underlain by sand with a diminished heavy mineral content, then by various colors of clayey sand, then by a thin shell bed, and finally by the sediments of the Hawthorne formation of middle to late Miocene age.

FOLKSTON, GA—Du Pont, in a program of systematic examination of elevated bars and old high level shorelines along the Atlantic Coast, discovered the Folkston heavy mineral deposit in 1952. It was situated in a flat, broad elevated area a few miles north of the Boulougne, FL, ore occurrence and the two may at one time have been one, later cut by the St. Marys River. The flat area is better described as a marine terrace rather than a sand bar or old shoreline, as it is up to 6 km (4 miles) wide and over 32 km (20 miles) long. The zone in which commercial deposits of heavy minerals were found, 8 km (5 miles) east of Trail Ridge, is not geomorphically different from the remainder of the terrace. It is a relatively thin layer of sand resting on clay. Three lenses were outlined in development drilling: the Main Area, the West Extension, and the North Extension. Each of the lenses was bounded by swamps, underlain by barren sand and clay, which were probably original depressions rather than erosion channels. Humphreys Mining Co. completed mining of the Main Area for Du Pont in 1974 and moved mining and primary concentrating equipment to Boulougne, FL.

Titanium minerals comprised 56% of the heavy mineral suite in the Main Area, and the average TiO_2 content of a combined titanium mineral concentrate was about 72%. The grain size averaged 90 to 94% −100 mesh. Byproducts were zircon and monazite.

New Jersey—The New Jersey Geological Survey, in an exploration program which grew out of observations of minor titanium mineral concentrations during investigation of monazite placers as sources of radioactive materials, identified the upper Tertiary sediments in the northern part of New Jersey's Coastal Plain as an ilmenite province in 1956 (Markewicz, 1969). The main heavy mineral formations were found to be the Miocene Kirkwood marine micaceous sand, silt, and clay, the Pliocene (?) Cohansey fluvial poorly sorted quartz sand, and the Pleistocene Cape May sands and gravels derived from the Kirkwood and Cohansey and from sediments farther inland. The Cohansey formation, which contains the greatest concentrations among these three, is ilmenite-rich in the northern third of its extent in New Jersey, in the vicinity of Lakehurst, about 48 km (30 miles) southeast of Trenton. Exploratory drilling by private companies resulted in the discovery of four ore bodies, and two heavy mineral mines were established.

LAKEHURST, NJ—The deposit of Glidden Pigments Group of SCM Corp. was located at Legler, about 3 km (2 miles) due north of Lakehurst, and was mined from 1962 to 1978. A mantle of Pensauken sand and gravel from 0.3 to 3.0 m (1 to 10 ft) thick overlay the Cohansey ore body, which was from 6 m (20 ft) to more than 12 m (40 ft) thick, with the average being 8 m (25 ft). A barren red sandstone 6 to 12 m (20 to 40 ft) thick underlay the ore, and beneath it was a zone in the Kirkwood formation, more than 12 m (40 ft) thick in places, averaging 4.5% heavy minerals. Only the Cohansey ore was mined.

The Cohansey sand is highly variable in color and in ilmenite, clay, and ironstone content. Occasional pebbles and limited gravel layers are present, and a coarse, iron stained, barren sand unit forms a bottom marker horizon. Cross-bedding, slump structures, roots and other pieces of organic material, and clay bodies up to 152 mm (6 in.) in diam are fairly common. Partially and wholly indurated iron oxide lenses and masses occur randomly above and below the water table (Anon., 1974a). A black, silty, lignite-charcoal band up to 5 m (15 ft) thick, found within the ore in the northwestern portion of the deposit, was interpreted by Markewicz as possibly the remains of a partially burned wooded swamp.

Heavy mineral content of the ore varied from 3% to more than 15% over distances of a few feet, with 5% being the average. The composition of the heavy mineral suite was reported by Markewicz as follows:

Mineral	Grain Count, %
Ilmenite-leucoxene	85–90
Zircon	2–4
Kyanite-sillimanite	1–2
Staurolite	tr-1
Rutile, anatase	tr-1
Tourmaline	tr-2+
Garnet, monazite, epidote, andalusite, hypersthene	tr

Glidden produced about 90,000 stpy of 61.5% TiO_2 ilmenite-leucoxene concentrate for its own consumption from 1962 to 1978.

MANCHESTER MINE—The Manchester mine of Asarco, Inc., also in the Cohansey formation, was about 97 km (60 miles) south of New York City, near Lakehurst, NJ. Geological features are similar to those at the Glidden operation at Lakehurst. Reserves were reported to be 180 million tons of sand averaging about 4% heavy minerals (Anon., 1974a) and 1.95% TiO_2 (Li, 1973). The operation was expected to have a 20- to 22-year life from 1973, when production began, based on a planned initial rate of 155,000 tons and a possible ultimate rate of 185,000 tpy of ilmenite, averaging 63% TiO_2. All ilmenite produced was under a sales contract to Du Pont. The mine was closed in March 1982 because of escalating costs and the prospect of a long-term oversupply situation. Figs. 1 to 4 are scenes of the operation.

Tennessee—A reconnaissance exploration program conducted by Du Pont to investigate reports of surface heavy minerals in 1957–1958 led to discoveries of ore grade occurrences of titanium minerals in the Cretaceous McNairy formation. McNairy is a poorly consolidated, clay-bearing fine-grained sandstone, about 91 m (300 ft) thick, which outcrops in a north-south belt just west of Kentucky Lake. The dip is very gentle to the west. Heavy minerals, including ilmenite, leucoxene, rutile, zircon, monazite, and minor amounts of kyanite, staurolite, tourmaline, and xenotime, are concentrated in the lower portion of the McNairy above the underlying Coon Creek formation. The ilmenite is somewhat altered, and concentrates have shown an average of about 62% TiO_2.

Ethyl Corp. and Kerr-McGee carried out exploratory drilling programs during the period 1970–1972, and both companies outlined several ore bodies. Production plans were deferred because of environmental problems and changing markets.

Australia: Australia is the world's most important heavy mineral sand mining country, producing 64% of the rutile mined in 1982, and 38% of the ilmenite.

FIG. 1—*General view of Asarco's dredge, pond, slurry pipeline, and pumping station at Manchester, NJ (courtesy of Asarco).*

FIG. 2—*Suction dredge recovered 20,000 gpm (20% solid) of titanium-bearing sands. The lake moved forward as the dredge advanced and closed in behind as clean sand from the concentration plant was returned (courtesy of Asarco).*

FIG. 3—*Asarco's custom-built Dixie CS–20 suction dredge was designed to excavate sands from a depth of 20 m (65 ft). The slurry was pumped by pipeline to a terminal for processing (courtesy of Asarco).*

FIG. 4—*View of a portion of the 1524 Humphrey spirals. They rejected over 95% of the slurry intake as waste in rougher, cleaner, and finisher cycles (courtesy of Asarco).*

Heavy mineral concentrations occur widely around the Australian coast, and small black sand deposits have been reported in Tasmania, Victoria, South Australia, and the Northern Territory. Those at Nepean Bay on Kangaroo Island have been mined in the past. In addition to onshore beach-type deposits, offshore occurrences have been discovered and explored in shallow water along portions of the New South Wales-Queensland coasts. In 1982, however, mining was restricted to three areas: (1) the coast of New South Wales and Queensland from Newcastle to Gladstone, (2) near Bunbury, Capel, and Busselton on the west coast of Western Australia south of Perth, and (3) near Eneabba 29 km (18 miles) inland and about 225 km (140 miles) north of Perth (Fig. 5).

As elsewhere, the heavy mineral grains were derived from the erosion of granites, intrusives, quartz reefs, and sandstones, and transported to the coast and concentrated by water and wind action. They occur in present-day beaches, fossil beaches, buried strand lines, and coastal dunes, and may be up to nearly 32 km (20 miles) inland as in Western Australia.

Ratios among individuals in Australian heavy mineral suites vary geographically. For example, between Sydney and Newcastle, rutile and zircon comprise about 90% of the concentrates, while farther north in New South Wales the rutile plus zircon content drops to 60 to 70%. Continuing northward, in southern Queensland the ilmenite level increases to about 60% of the heavy minerals. In the area between Busselton and Bunbury in Western Australia ilmenite makes up about 90% of the heavy mineral fraction, and near Eneabba concentrations contain 40 to 60% ilmenite, and about 10% rutile.

There is also a geographical variation in the chemical composition of ilmenite. Along the east coast ilmenite is generally too high in chromium to be suitable as a raw material for pigments manufacture by the sulfate process, and has found only a limited market in the Japanese steelmaking industry. The chromium content does, however, decrease from south to north, and the ilmenite is apparently suitable for upgrading to rutile grade material by some beneficiation processes. In Western Australia ilmenites from the Bunbury-Busselton area average 54 to 57% TiO_2, while Eneabba ilmenite grades around 61% TiO_2.

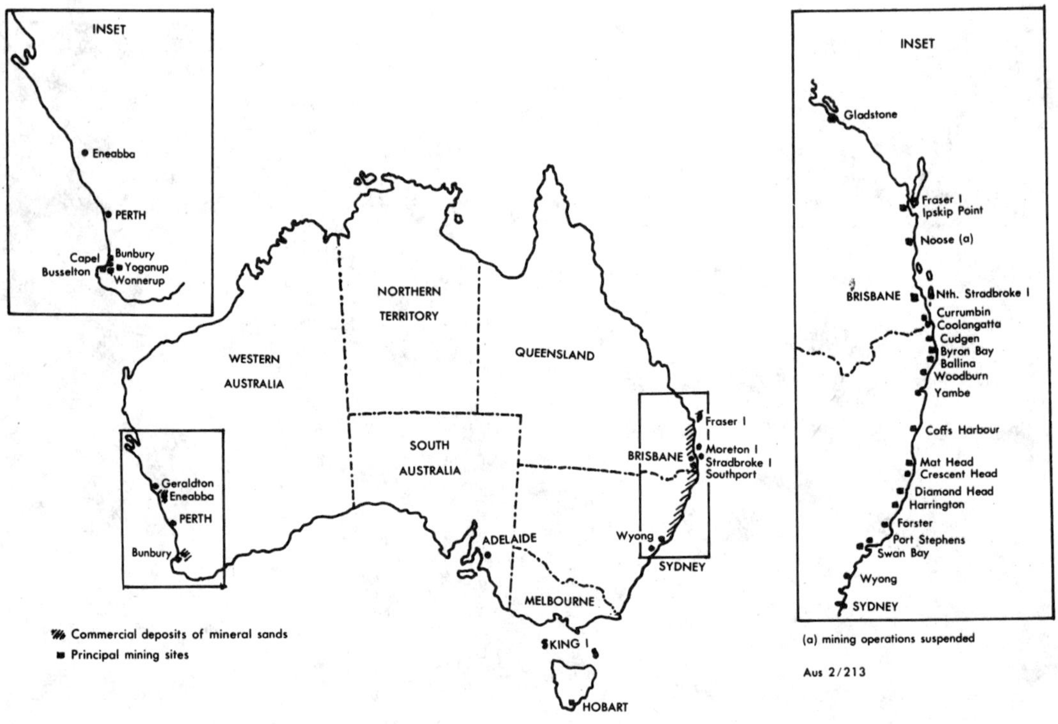

FIG. 5—*Location of principal ilmenite and rutile mineral sand reserves and mining operations in Australia, 1980 (Anon., 1980c, and Ward, 1972).*

Titanium mineral mining began in eastern Australia in 1934. Ore bodies were modern beach placers containing up to 50% heavy minerals, mostly rutile and zircon. All these high grade deposits have been exhausted over the past 40 years, and the average grade in the east by 1980 was about 3% heavy minerals. The increased cost of working lower grade deposits farther inland has been offset by improved techniques for mining and processing, and some ore bodies containing less than 0.25% TiO_2 as rutile (plus at least as much zircon) are now mined successfully on the east coast. In Western Australian ore bodies titanium minerals must be more highly concentrated because of the relative scarcity of rutile, and although there are wide variations within individual deposits, the average is between 5 and 10% heavy minerals. Mining of heavy mineral sands began in Western Australia in 1956.

The development of the large and important Eneabba ore field began in 1970. Rutile occurs in the heavy mineral suite there with ilmenite and leucoxene, so that with full-scale production commencing in 1974 and 1975, Western Australia became an important source of rutile for the first time.

Australia's titanium reserves are shown in Table 15. Table 16 lists company names, locations, estimated reserves, and 1980 production of rutile and ilmenite.

Eastern Australia—Ore bodies are in ancient beaches and bars in New South Wales and parts of Queensland. On North Stradbroke and Fraser Islands they are in high coastal dunes. A variety of mining equipment including dredges, bulldozers, front-end loaders, and bucket wheel excavators is employed by the seven producers.

Associated Minerals Consolidated Ltd. (AMC) is the world's leading producer of rutile and zircon, operating the world's largest mobile sand mineral concentrator on North Stradbroke Island off Brisbane. The suction-cutter dredge is 25 m (81 ft) long and 12 m (40 ft) wide. It is followed by a 34 x 34-m (110 x 110-ft) pontoon-mounted concentrator with a capacity of 1500 stph.

The Forster, N.S.W., mining plant of Mineral Deposits, Ltd., the second largest mineral sand producer in Australia, has a capacity ot 900 to 1500 stph. It consists of a suction-cutter dredge, a 400-ton pontoon-mounted surge bin, and a concentrator mounted on a 24 x 31-m (80 x 103-ft) pontoon.

Consolidated Rutile Ltd. mines 91-m (300-ft) dunes on North Stradbroke Island by bulldozer, frequently moving a sand-receiving trommel and slurrying unit to keep it close to the mining site. The sand is pumped to a nearby wet concentrator built in several basic units for rapid dismantling and moving a few times a year. Heavy mineral concentrates are barged to the dry mill on the mainland at Meeandah near Brisbane.

TABLE 15—Titanium Reserves of Australia, Thousands of Short Tons

	Rutile	Ilmenite
East coast	6,500	14,400
West coast	3,500	36,000
Total	10,000	50,400

Source: Anon., 1980b.

The opposition of conservationists to sand mining on Fraser Island led to the withdrawal of export licenses for Fraser Island concentrates at the end of 1976, and to restrictions on sand mining in other areas in Queensland and New South Wales.

Western Australia—Heavy mineral concentrations occur in three separate strand lines in the Bunbury-Busselton area: the present beach, the Capel line, and the Yoganup line, with the Yoganup line being oldest and 16 km (10 miles) inland. The TiO_2 content of the titanium mineral fraction of the heavy minerals increases with distance from the present coast.

Longshore drift in the area is southward and westward toward Cape Naturaliste, and it appears that longshore currents acting in the past sorted sands brought to the coast by northwestward-flowing rivers, creating heavy mineral deposits along beaches near their mouths. The existence of the old Yoganup and Capel lines indicates two past episodes of shoreline standstill.

There are no mining operations now on the beach. Cable Sands Pty. Ltd. and AMC are mining the Capel line, and Westralian Sands is working the Yoganup line. Both dredging and dry mining with scrapers, bulldozers, draglines, front-end loaders, and backhoes are practiced. In some deposits iron oxide-cemented, hard lenses near the bottom of the ore, which contain up to about 70% heavy minerals, are drilled, blasted, removed from the pits in blocks, and stored for possible future milling.

The Yoganup line sand has a high clay content, requiring dewatering of tailings in a series of large settling ponds. Water is scarce, and is conserved by recycling.

Commercial concentrates of ilmenite, leu-

TABLE 16—Australian Producers of Titanium Minerals, 1980

Company	Location: Mines (m) and Separation plants (s)		Reserves, Thousand st: Rutile	Reserves, Thousand st: Ilmenite	Production, Thousand st: Rutile	Production, Thousand st: Ilmenite	Associated Companies and Controlling Interests
East coast							
Associated Minerals Consolidated Ltd.	N. Stradbroke Island, Qld.	m,s	800	2,000	71	NA	Consolidated Goldfields Australia Ltd., 61.7%
	Southport, Qld.	s					
	Jerusalem Creek, N.S.W.	m					
	Kempsey, N.S.W.	m					
	Hexham, N.S.W.	s					
Consolidated Rutile Ltd.	N. Stradbroke Island	m	1,100	2,400	60	NA	Cudgeon RZ, 50%, and Union Corp., Republic of South Africa
	Meeandah, Qld.	s					
Currumbin Minerals Pty. Ltd.	Currumbin, Qld.	m,s	200	200	4	NA	Neumann Associate Companies Pty. Ltd.
	Kingscliff, N.S.W.	m					
Mineral Deposits Ltd.	Crescent Head, N.S.W.	m,s	900	2,000	50	NA	Utah Mining Australia Ltd., 84%
	Myall Lakes/Hawks Nest, N.S.W.	m,s					
Rutile and Zircon Mines (Newcastle) Ltd.	Harrington, N.S.W.	m,s	300	700	39	NA	Peko Wallsend Ltd., 50%, Kathleen Investments Australia Ltd. 50%
	Tomago, N.S.W.	m,s					
Reserves unavailable for mining because of environmental considerations			3,100	7,100	—	—	
Total east coast			6,400	14,400	224	60	
West coast							
Allied Eneabba Ltd.	Eneabba, W.A.	m,s	2,000	10,000	60	325	E. I. du Pont de Nemours, 59%
Associated Minerals Consolidated Ltd.	Capel, W.A.	m,s	1,600	13,000	40	433	Consolidated Goldfields Australia Ltd., 61.7%
	Eneabba, W.A.	ms					
Cable Sands Pty. Ltd.	Capel, W.A.	m	—	2,000	—	191	Kathleen Investments, principal shareholder.
	Bunbury, W.A.	s					
Westralian Sands Ltd.	Yoganup, W.A.	m	—	11,000	—	434	Tioxide Australia Ltd., 40%
	Capel, W.A.	s					
	North Capel, W.A.	m,s					
Total west coast			3,600	36,000	100	1,383	
Total Australia			10,000	50,400	324	1,443	

Sources: Anon., 1980b; and industry contacts.
NA = Not available

coxene, zircon, and monazite are produced from this region of Western Australia.

Exploration northward toward Perth and beyond resulted in the discovery of heavy mineral occurrences in old strand lines at Boyanup, Waroona, Bull's Brook, and Gin Gin. Rutile appears significantly in the heavy mineral suites at Bull's Brook and Gin Gin. The ilmenite is reported to be high in chromium content and it could be more suitable as a raw material for beneficiation than for direct sale.

ENEABBA—It was suggested by Lissiman and Oxenford (1973) that the heavy mineral deposits at Eneabba were formed by agencies acting similarly to those which formed the ore bodies in the Bunbury-Busselton area. Rapid erosion of a Mesozoic sedimentary rock highland contributed sand to a westward-flowing stream which emptied into a bay partially protected on the southwest from south-to-north longshore currents by a headland. Longshore drift produced countercurrents within the bay which concentrated black sands on the beaches south of the mouth of the stream. There are several distinct strand lines marked by north-trending bands of high heavy mineral concentration at the bottom of the ore zone, and they are thought to have been formed during relatively brief periods of sea level standstill. The strand lines are at successively lower levels from east to west. The geologic history of the ore field is made complex by evidence of reworking of the higher strand lines by waves which must have been associated with a sea which rose again, perhaps repeatedly, and by wind which created disseminated heavy mineral sands between and over the strand line deposits. Ore bodies in the northern portion of the district, which is about 16 km (10 miles) long and over 2 km (1½ miles) wide, are thought to have been formed by wind concentration of heavy minerals from black sand dunes then exposed in areas to the south.

Total reserves of the Eneabba district are in excess of 20 million tons of heavy minerals. Rutile, ilmenite, leucoxene, zircon, and monazite are present in variable proportions from one area to another, but the average ratios may be 1:5:0.2:2:0.2, respectively. Shipments of concentrates are from the port of Geraldton.

The development of Eneabba for mining presented considerable challenge because of its remoteness and relative lack of labor, power, water, transportation, and nearby port facilities. Furthermore, a portion of the ore, which occurs both in free-flowing sand and in harder, clay-rich sands, is bound up in nodules and in cellular and solid masses of iron oxide-cemented material locally called *laterite*. Extensive pilot testing of concentrating and separating methods and pilot mining were carried out to prove the physical and economic feasibility of heavy mineral operations. Production from Eneabba will make an important contribution to world titanium mineral and zircon markets for the next 15 years and more, and could compensate for diminishing production of rutile and zircon from the east coast.

KING ISLAND—Beach placers containing heavy minerals are located along the central part of the east coast from Naracoopa some 11 km (7 miles) along Sea Elephant Bay north to Cowper Point. The heavy mineral content is high, up to about 60%. Rutile and zircon occur in about equal proportions. Accessory minerals are leucoxene, ilmenite, magnetite, garnet, and cassiterite.

Naracoopa Rutile Ltd., a subsidiary of New Mount Costigan Mines Ltd., mined the sands by dragline and used trucks to haul to a wet plant stockpile. A dry plant had a production capacity of 10,000 stpy of rutile and 10,000 stpy of zircon. In early 1972 Buka Minerals N.L. acquired Naracoopa's mining properties and plant through its subsidiary, Kibuka Mines Pty. Ltd., and recommenced production after a reconstruction program.

China: Coastal sand deposits in Guandong and Guangxi provinces contain ilmenite, zircon, and other heavy minerals (Brady, 1981).

India: At one time India was a leading producer of ilmenite from the state of Kerala (formerly Travancore-Cochin). The beach sands were mined in the Manavalakurichi (M.K.) area and later the Quilon deposit of ilmenite near Chavra was put into production. These deposits supplied the bulk of the titanium ore used by the US prior to World War II.

The two deposits have more differences than similarities. The ilmenite in the M.K. deposit analysed only 54% TiO_2 and the sand was rich in garnet and monazite. The ilmenite in the Quilon deposit analyzes about 60% TiO_2. The sand carried almost no garnet and is high in monazite in only two places. Analyses of ilmenite from these areas are shown in Table 17.

The M.K. deposit consisted of buried seams of rich black sands at or just above the present ocean level. The beach sands were rich only when the monsoon storms sorted out the light minerals from the heavy ores.

TABLE 17—Ilmenite Analyses, India

	1, %	2, %
TiO_2	54.1	60.4
FeO	25.6	9.55
Fe_2O_3	15.3	24.6
Cr_2O_3	0.09	0.17
V_2O_5	0.23	0.36

Source: Gillson, 1959.

1. Travancore, "M.K."
2. Travancore; average "Quilon."

The Quilon deposit is not only much larger but the ilmenite is of a better grade. The deposit is a barrier beach extending parallel to the old shore in front of the mouths of two large rivers, the Panalur at Neendakara and Pallikal Todu. The ocean beach is black with layers carrying 80% heavy minerals and dunes on the beach are gray with 40 to 50% heavy minerals. Below them are old beaches now buried under dunes formed before the shore retreated, which are as rich as those now being washed by the waves. The extent of the old buried beaches has never been completely determined.

Malaysia: Substantial tonnages of ilmenite are produced in Malaysia as a byproduct of tin mining either as rough concentrates in mills using magnetic separation techniques or in the form of *Amang,* a crude mixture of heavy minerals which has to be further treated to recover the ilmenite content. The main centers of production are at Perak and Ipoh. The deposits are alluvial in nature.

Sierra Leone: A deposit along the Sherbro River was originally developed as a joint venture of Pittsburgh Plate Glass Co. and British Titan Products. Although relatively high grade (±1.5% TiO_2) the sands were extremely fine and disseminated in lumpy lateritic clay. Reserves are estimated to be between 3 and 30 million tons of contained rutile. The ore problems plus mechanical and political difficulties caused Sherbro Minerals to shut down in 1971 after only two years of operation.

Bethlehem Steel acquired 85% ownership of the property in 1976 (15% Nord Resources) and conducted an extensive prospecting and geological investigation that extended into 1977, finding higher grade deposits (±2.5% TiO_2) than that mined by Sherbro. Improvements were made in both the wet and dry flowsheets and as of 1981 the operation now called Sierra Rutile Limited was approaching specifications and design capacity of 110,000 stpy.

Rutile grades are highest in the topsoil, averaging 2.5% TiO_2, and in the basal sands and gravels, with up to 3.0% TiO_2. The average grade of the Mogbwemo deposit where Sierra Rutile began mining in 1979 is over 2.0% TiO_2 (Anon., 1981).

South Africa, Republic of: Shipments of rutile and zircon were started in 1977 from the Richards Bay mineral sands operation in Natal. Richards Bay Minerals manages the combined affairs of Tisand Pty. Ltd. and Richards Bay Iron and Titanium Pty. Ltd., the companies responsible for the operation of the mine and the smelter, respectively. A high-TiO_2, low-magnesium-plus-calcium slag (RB slag) containing 85% TiO_2 has been produced from ilmenite since 1978, and is reportedly suitable for use in either the sulfate or chloride titanium dioxide pigment processes. Since 1980, production rates have approached planned annual capacities of 440,000 tons of RB slag, 220,000 tons of low-manganese iron, 53,000 tons of rutile, and 110,000 tons of zircon. Proven reserves were estimated at 770 million tons grading 5% ilmenite, 0.3% rutile, and 0.65% zircon. Principal shareholders at the end of 1980 were QIT-Fer et Titane, Inc. (32%), General Mining Union Corp. (30%), and the Industrial Development Corp. of South Africa Ltd. (16%).

Operations in 1980 were in an area 17 km (11 miles) long and 1 km (0.6 miles) wide adjacent and parallel to the shoreline. The ilmenite, rutile, and zircon, along with magnetite, are evenly distributed throughout the sands which occur as dunes up to about 38 m (125 ft) in height on a clay base (Anon., 1980d).

Sri Lanka: Sri Lanka contains extensive beach deposits of titanium-bearing sands at Pulmoddai, Tirukkovil, Kelani River, Kalu River, Modoragam River, Kudremalai Point, Negombo, and Induruwa.

The Pulmoddai area contains 5.6 million st of titaniferous material with 2.451 million st of contained TiO_2. The deposit extends for a distance of 7 km (4½ miles), has a maximum width of about 91 m (300 ft), and a thickness of about 2.4 m (8 ft). There is no overburden. The deposit contains about 80% ilmenite and rutile. The average mineralogical composition of the black sand is as follows (Anon., 1974):

Ilmenite	70–75%
Rutile	10–12%
Zircon	8–10%
Sillimanite	±1%
Monazite	±0.4%

Other minerals present include spinels, garnet, tourmaline, etc., with quartz and shells as the main waste materials.

The separation of rutile has been adversely affected by the presence of excessive amounts of residual ilmenite and quartz in the tailings. The separation of zircon has been hampered by inadequate water and insufficient wet tabling equipment to handle the extremely fine-grained Pulmoddai ore.

Sand Deposits—Potential Sources

Brazil: Brazilian deposits of heavy minerals on ocean beaches occur in a zone about 161 km (100 miles) long, extending north from the northeast corner of the state of Rio de Janeiro up into the state of Espirito Santo as far north as the Rio Doce. There is also a zone in southern Bahia. Deposits are known to exist near Natal in the state of Rio Grande do Norte. Fig. 6 shows the location of the ilmenite deposits of Brazil (Gillson, 1950). Rutile deposits also are found in the state of Goias.

The minerals of the Brazilian beaches are:

Mineral	% Heavy Mineral
Magnetite	tr to 3
Ilmenite	35–75
Monazite	1–20
Zircon	5–35
Rutile	1/2–5
Sillimanite	1–1/2–5
Kyanite	1/2–2
Corundum	tr-1/2
Spinel	1/4–5
Staurolite	tr-3
Others	tr-3

The ilmenite assays from 39.7 to 61.6% TiO_2. Table 18 lists Brazilian ilmenite and rutile analyses.

Egypt: Rich deposits of black sands occur along the northern beaches of the Nile Delta for about 241 km (150 miles). There are two types: one is dark in color and contains about 70 to 90% heavy minerals and a second which is grayish-yellow to dark gray and contains about 40% heavy minerals. The deposits cover areas 10 km (6 miles) long and a few meters (yards) wide and up to 0.5 m (1½ ft) in thickness.

The black sand concentrate is composed of ilmenite, magnetite, zircon, rutile, monazite, and garnet. Analysis of typical concentrates follows:

Ilmenite—55%
Magnetite—15–20%
Zircon—7–8%
Garnet—4–6%
Rutile—1.5%
Monazite—0.5–1%

FIG. 6—*The heavy mineral deposits of Brazil (Gillson, 1950).*

New Zealand: Ilmenite and associated heavy minerals are found in New Zealand on the west coast of South Island from Jackson Bay to Karamea and on the west coast of North Island at the mouth of Waikato River, Muriwai, and at Manukan Heads. Total reserves near Westport are estimated to be between 17 to 31 million tons. Tests of the sand from Westport indicate the titanium content of the ilmenite averages about 47%.

Approximately one billion tons of sand are available between Karamea and Jackson Bay from which ilmenite could be recovered. These deposits on the west coast of South Island are believed to be one of the world's major reserves of low chromium ilmenite.

South Africa: Ilmenite-bearing sands occur in scattered deposits along the east coast of the Republic of South Africa from the area of East London to the Mozambique border and northward. Ilmenite sands also occur along the west coast particularly in the Vannhyrodorp district.

Titanium minerals are found on the east coast in three types of deposits:

1) Older red and brown coastal sands.
2) Recent dunes and beach sands.
3) Alluvium in coastal lagoons.

Older red and brown coastal sands occur in a belt of undulating fixed dunes which is roughly parallel to the coast. The belt ranges from a few meters (yards) to 6 km (4 miles) or more in width. Some dunes are as much as 152 m (500 ft) high but the average height is much less.

The heavy mineral content varies from 2 to 25%. The heavy mineral content is about:

Ilmenite—70–80%
Leucoxene—2–5%
Rutile—2–5%
Zircon—8–10%
Magnetite—2–8%
Monazite—tr-0.3%
Garnet, pyroxene, etc. 1–5%

Recent dunes and beach sands are partially derived from the older sands and contain the same mineral suites.

TABLE 18—Rutile and Ilmenite Analysis, %, Brazil

	1	2	3	4	5	6	7	8	9	10	11	12	13	14	15
TiO_2	58.5	61.6	55.6	38.9	66.5	70.5	79.3	69.0	75.2	96.7	92.9	93.5	97.8	97.1	95.6
FeO	35.4	32.0	36.6	—	26.2	24.6	19.8	30.9	21.7						
Fe_2O_3				44.2	1.8					2.0	6.8	6.6	2.3	3.0	3.4
SiO_2	0.2	2.5		4.0	5.2					0.5	0.4	0.1	0.2		0.7
CaO	0.9														
MnO	0.1														
MgO	2.5														
H_2O–110°	0.1														
H_2O+110°	2.7														

Source: Abreu, 1962.

1. Ilmenite from coastal sand, S.G.
2. Ilmenite and rutile, Bom Jardin, M.G.
3. Ilmenite from Pissaguera, P.R.
4. Ilmenite from São Sebastiao, S.P.
5. Ilmenite and rutile, Bom Jardin, M.G.
6. Ilmenite and rutile, Anápólis, G.O.
7. Material from riverbed, Rio Grande, M.G.
8. Ilmenite and rutile, Santa Quitéria, C.E.
9. Ilmenite and rutile, Neopolis, S.E.
10. Rutile, Mossâmedes, G.O.
11. Rutile, Mossâmedes, G.O.
12. Rutile, Corumba, G.O.
13. Rutile, Pirenopolis, G.O.
14. Rutile, Santa Luzia, G.O.
15. Rutile, Average of shipment from G.O.

Ilmenite concentrates occur in the alluvium in coastal lagoons, i.e., lagoon sands and in the bars at river mouths.

On the west coast, ilmenite-bearing sands exist in the Vannhyrodorp District between Strandfontein and the mouth of the Zout River. One deposit, a recent dune, skirts the coast for 6 km (4 miles) and another on the Geelwae Karoo is found in a narrow modern beach about 8 km (5 miles) long. The sands are unconsolidated and fine to medium grained.

Analyses of ilmenites from South Africa are shown in Table 19.

Titaniferous sandstone containing up to about 50% ilmenite with lesser amounts of rutile and zircon occur in the Ecca series of the Karoo system. They outcrop in several areas, principally near Bothaville about 483 km (300 miles) from the east coast in the Orange Free State. Total reserves are estimated at 85 million tons of material, some of which might be recovered with future technological improvements.

The Cape Morgan Titanium Mining Co. has produced a small amount of ilmenite in the Komga District from coastal sands.

Uruguay: Black sands are found northeast of Montevideo at Agua Dulces, in the state of Rocha. The sands extend along the coast for a distance of 11 km (7 miles) with an average depth of 5.5 m (18 ft). Reserves are estimated at 3.3 million tons of heavy minerals, with an average 2.5% heavy mineral concentrate. The composition of the heavy mineral fraction is approximately:

Ilmenite	60%
Zircon	5%
Rutile	1%
Monazite	0.6%

In addition, titanium sand deposits are known to exist in the following areas:

China: At Luanping and Chente, and the coast of Jiangsu province.

Korea: Resources of 2,535,000 tons with 507,000 tons of contained TiO_2.

Mozambique: Rutile and ilmenite are known to exist on the coastal beaches of Mozambique north of 16° 30′ latitude and near Vila Luisa, about 40 km north of Lourenco Marques.

TABLE 19—Analyses of South African Ilmenites

	1, %	2, %	3, %
TiO_2	50.5	46.6	41.6
FeO	38.2	39.51	34.1
Fe_2O_3	9.2	9.11	19.7
Cr_2O_3	0.2	0.10	0.0
V_2O_5	0.12	2.76	0.92

Source: Anon., 1959.

1. Ilmenite concentrate, Umgababa, Natal; Analysis Government Laboratory, Malaya.
2. Ilmenite concentrate, Isipingo, Natal; analysis, Div. of Chemical Services, Pretoria.
3. Ilmenite fraction, Vanrhynsdorp District, Cape Province; analysis, Div. of Chemical Services, Pretoria.

USSR: Black Sea deposits; Aldan River; Boludka River; Pit River in Siberia; Sylvista River; Chusovaya River; and Vizhai River in the Urals.

Exploration and Evaluation

Hardrock Ilmenite Deposits

Hardrock ilmenite deposits, because of their inherent magnetic properties, are readily amenable to the application of aero and ground magnetic geophysical surveys. With few exceptions, these deposits respond to such an application by reflecting abnormally positive magnetic intensities (gammas).

Examples can be cited, however, where negative magnetic anomalies are associated with such deposits. In such cases, the ore occurs either as a titaniferous-hematite concentration or a titaniferous-magnetite occurrence when the titanium content is greater than the iron content. Theoretically, it would seem possible that a commercial deposit may exist where no magnetic response would be obtained. Such a deposit has not yet been found.

Once such anomalies are mapped, further exploration may take place in the form of detailed surface geological observations and ultimate drilling to first test the anomalies and hopefully to delineate a viable ilmenite deposit.

MacIntyre Development: The original discovery of magnetite-ilmenite deposits at MacIntyre Development, Tahawus, NY, was made from surface outcrops. Extension of these outcrop locations has been made through use of geophysical methods and diamond drilling.

Geophysical—Since the ore is magnetic, the geophysical methods used have been for the purpose of outlining magnetic anomalies. Early work was by dip needle readings over a grid pattern extending outward from surface outcrops. With the advent of the aerial magnetometer the entire property was flown. This produced anomalies well beyond any previous work extending to areas of deep overburden and under-lake area.

Aerial work was followed with detailed ground magnetometer readings taken on 15-m (50-ft) spacings along section lines 30 m (100 ft) apart. This served to guide the location of diamond drill holes for exploratory drilling on one or two sections with drill holes spaced 122 m (400 ft) apart on section.

The initial ground magnetometer work on any anomaly is a section across the center of the anomaly at right angles to the long axis indicated from aerial work. The magnitude and gradient of the ground anomaly compared to the magnitude and gradient of the aerial anomaly, along the same section, gives an indication of depth and configuration of the source of the anomaly.

Drilling—If initial diamond drilling indicates an economic deposit, development drilling proceeds on a set grid pattern of 61 or 91 m (200 or 300 ft). Intermediate holes are drilled where necessary to fill in major gaps in geologic interpretation of structure, or ore continuations.

Core Sampling—Drill cores are visually logged and split for chemical assay. In ore areas, samples are taken of the entire core and composited in 1.5-m (5-ft) intervals for assay. In waste areas samples may be taken every 1.5 m (5 ft) and composited for a maximum of 8 m (25 ft) of core. Routine assaying is done to determine TiO_2 and total iron content. Based upon past experience, the nominal cutoff for ore is 13.5% TiO_2.

Mill Testing—Metallurgical laboratory work is done using core assay rejects to determine milling characteristics. Concentrates from the laboratory work are used for testing end use performance. Complete analyses are run on concentrates to determine presence and quantity of detrimental elements.

Tellnes Deposit, Norway: The Tellnes ilmenite deposit near Hauge-I-Dalane in Norway was discovered in 1954 as a result of an airborne magnetometer survey carried out for Titania A/S, with the object of locating new sources of ilmenite to provide an alternative to developing additional underground ore at Storgangen. Anomalies were also found at four other locations, so that ground geological study and diamond drilling had to be done at all of these sites to be certain that the best deposit was selected for devlopment. A total of 10,058 m (33,000 ft) of drill holes was completed in 1956, and showed the total amount of ore in the Tellnes deposit to be about 300 million tons, averaging 18% TiO_2 (Brun, 1957; Anon., 1978a).

Allard Lake (Lac Tio) Deposit, Quebec: *Ground Surveys*—Preliminary investigation of the original ilmenite discoveries at Allard Lake (Hammond, 1949), indicated that all of these were too small to be of commercial importance. A detailed exploration program was carried out in 1946 to more thoroughly explore the area of 647 km^2 (250 square miles) where ilmenite had been found along certain of the lakeshores.

Geologic prospecting parties carried out traverses originating from the shores of the larger lakes, at 0.4-km (¼-mile) intervals in an east-west direction, normal to the strike of the previously known ilmenite occurrences. Mapping along and on both sides of the line of traverse was supplemented by dip needle readings at 61-m (200-ft) intervals or less. This work resulted in the discovery of eight ilmenite deposits, among them the very large Lac Tio deposit.

Drilling—A diamond drilling program followed to explore the Lac Tio deposit in depth. A series of vertical holes was drilled at 61-m (200-ft) intervals along five east-west sections to an average depth of 91 m (300 ft). Cores were sampled in 8 m (25-ft) lengths where ore intersections were sufficiently long. The specific gravity was determined on all samples before shipment for analysis. Later drilling on the portions of the ore body to be mined was carried out on a 15-m (50-ft) grid pattern.

Core Testing and Grade Control—Because of the marked difference in specific gravity between ore and gangue, it was found that density measurements gave a reliable measure of ore content. The accuracy of the method is about 2%, sufficient for the method to be used for controlling grade during mining.

Aeromagnetic Survey—An airborne magnetometer survey was carried out to explore the remainder of the anorthosite mass for other ilmenite deposits closer to the coast which might be developed more economically. Flight traverses were flown in a northwest-southeast direction at 152-m (500-ft) elevation, completing some 7200 km (4500 miles) of air traverse in a two-month period. The survey was first carried out over the known ore occurrences to obtain an indication of what results to expect from a new ore body.

No new deposits were discovered as a result of the aeromagnetic survey, but strong negative anomalies were obtained over all the known major ore bodies. Areas of disseminated ilmenite in anorthosite generally showed positive magnetic anomalies.

Otanmäki Deposit: The Otanmäki ilmenite-magnetite deposit and a neighboring deposit at Vuorokas were discovered in 1938 as a result of magnetic surveys by the Geological Survey of Finland (Harki et al., 1956).

Sand Deposits

Exploration: There are only a few large areas of the world where the granite-clan rocks and high-grade metamorphic gneisses which are likely to contain ilmenite (not titaniferous-magnetite) and rutile are close enough to continental margins to have contributed their erosion products to the sediments of coastal plains. Well-sorted sands are much more likely hosts than unsorted sands. These are the areas on which exploration efforts should be focused. Since the alteration of ilmenite to remove iron is aided by humic acid developed by the decomposition of organic material near the water table in hot and humid climates, it follows that the highest TiO_2 ilmenites are more likely to be found in the tropical and temperate regions of the world.

Titanium minerals are dark-colored and their concentration, as in black beach sands, tends to be fairly readily noticeable against the light brown or white quartz. Many sand ore bodies, therefore, have been discovered through surface observation of high-grade placer zones formed on beaches and along the courses of streams, and by following their traces into the larger, lower grade concentrations which constitute economic ore bodies.

There are areas in which potential heavy mineral concentrations in ancient beach sands may be masked by younger sand, gravel, or soil. Exploration under these circumstances then involves interpretation of geomorphic and subsurface geologic data to define areas which could have been beaches or dunes in the past, and then drilling to obtain samples.

Field Techniques—In areas where heavy mineral concentrations are suspected but where the concentration level is difficult to estimate because of dissemination, as in dunes, hand panning of samples is an excellent method of rapidly producing a clean concentrate. A hand magnet gives a quick identification between ilmenite and magnetite or titaniferous-magnetite. The presence of potentially valuable by-product zircon, monazite, kyanite-sillimanite, etc., can be determined with a hand lens. A Geiger counter or Scintillometer also helps in identifying radioactive minerals such as monazite and zircon. Wind and current action can develop very high-grade layers of heavy minerals on the surface which have no economic significance; thus samples for panning and examination should be taken from the maximum convenient depth. Often an idea of the vertical distribution of heavy minerals can be gained from observation of wave-cut cliffs in sands behind the beaches, cutbanks in stream margins, road cuts, construction excavations, and material encountered in sinking or drilling wells. Dark, heavy minerals may even be observed in the sand brought to surface by burrowing insects and small animals.

Hand augering with a two-man crew is possible to a depth of 6 m (20 ft) or more with a jointed auger stem and is reasonably rapid and inexpensive. Some heavy-mineral bearing sands contain so much clay that they are difficult to hand auger and a mechanical auger mounted on a light off-the-road vehicle (like a post-hole digger) is preferable. The samples so obtained are not precise because of the contamination of the hole by sand falling into it. The hole cannot be advanced nor reliable samples obtained below the water table when the material is free-flowing and easily disturbed by the auger.

Drilling—For the first phase of development, drilling equipment should have a depth capability of at least 15 m (50 ft) and preferably 30.5 m (100 ft). The selection of the type of drill will depend upon the "stiffness" of the ground as influenced by clay, hardpan, caliche, indurated and iron-oxide cemented layers; the presence of roots, stumps, and other organic material; as well as upon the elevation of the water table.

The truck-mounted jet drill is a popular exploration and development tool of the southeast US. One-inch flush-jointed steel pipe, with a chisel lower end or bit and perforations near the bit to permit water to jet forward and downward, is attached to a hammer of about 136 kg (300 lb) which is activated in 0.3-m (1-ft) strokes by a hand-held rope about a capstan winch. Water is supplied to the bit by a swivel connection above the hammer and by a duct through it. As the hammer rises, falls, and turns, it causes the bit to chop and churn into the ground and drives a 59-mm (2-in.) casing downward. The water pipe is within the 59-mm (2-in.) casing. The bit of the water pipe is about 25.4-mm (1-in.) or less ahead of the casing when casing and hammer are in contact. The sand sample is flushed out of the hole continuously by water rising in the annulus between water pipe and casing, and is collected at the surface in vessels or tubs. A jet drill under favorable conditions with short moves between holes can drill about 61 m/d (200 fpd) of hole. Its disadvantages lie in its inability to penetrate very hard layers, and the casing can stick in tenacious clay.

The auger-within-an-auger type of drill has good depth and penetration capabilities, particularly in the less free-flowing sands. The internal auger is advanced about 1.5 m (5 ft) to collect a sample on its flights, then the external auger is put down to the same depth to form a casing, and the internal auger is withdrawn for sample collection. There is some possibility of particularly fluid sands below the water table invading the casing when the internal auger is raised.

In Australia extremely light, gasoline-powered augers, portable and operable by two-man crews, have been developed and used for rapid, low-cost depth investigations. They are particularly successful in high, clean dunes.

A truck-mounted rotary drill driven by rapid strokes of a hydraulic hammer is also very useful. The drill string is a double pipe, and compressed air forced down the inner pipe returns the sample in the annulus between the drill-pipe and casing. A drag bit with hardened cutting edges may be used.

It is important in all types of machine drilling to avoid contamination of samples by material originating higher in the walls of the hole. The hole and casing should be flushed free of loose sand after collecting each sample, and before the bit is advanced farther.

The nature of the sand and other materials penetrated by the drill is usually logged by a geologist, who makes a rough estimate of heavy mineral content and variation by panning a small portion of each sample.

Depending on hole diameter, a 0.9 or 1.5-m (3 or 5-ft) sample may be too large for laboratory requirements. If so, a simple heavy duty riffle may be used to reduce the weight to about 2 kg (5 lb). The bulk of the samples may then be bagged or drummed and left at the hole site for future reference.

Geophysical Techniques—Some titanium mineral ore bodies contain sufficient iron to be detectable by magnetometers in ground or airborne surveys, but many are too dilute and shallow to respond.

If monazite or radioactive zircon is present, low-level radiometric surveys may help in exploration. Scintillometers and Geiger counters are sometimes useful on the ground to define horizontal limits of ore and to detect internal variations in heavy mineral concentrations close to the surface.

The problem of downward contamination during drilling can lead to a misrepresentation of the total depth of ore because the heavy minerals, being higher in specific gravity, are hardest to remove by flushing with air or water and may be reported even when the hole has been advanced into barren underlying sand. A scintillometer probe with a surface recording instrument can show a sudden decrease in radioactivity from the walls of the hole at the true bottom of the ore. Radiometric surveying of all or at least some of the holes drilled during development of an ore body is a highly desirable check on the fidelity of drilling and sampling.

Evaluation of Deposits: An economic titanium mineral deposit must have reserves large enough to support depreciation over a period of at least 10 to 20 or more years. The capital investment in 1980 was in the range of \$75 to \$80 million in the US for a mine and mill plant with an output of 100 to 200 thousand stpy of ilmenite (or equivalent rutile) with given "normal" geologic parameters. Significant contributions can be made by zircon and other byproducts. Another general rule is that a new and separate ore body, if its production is to be all ilmenite which cannot be treated in an existing mill, should have a minimum reserve of about 1 million tons of recoverable TiO_2 in the titanium minerals. Small, high-grade concentrations are uneconomic under the present conditions.

The definition of economic reserves depends, of course, upon many factors, among them:

1) Cost of mining and milling, as influenced by depth of overburden (if any); cost of surface and mineral rights; and availability of water, power, labor, and transportation facilities for bulk shipments.

2) Recoverability in mining and milling.

3) Cost of treatment and disposal of waste slimes.

4) Cost of waste water treatment and land reclamation.

5) Distance to markets and cost of transport.

6) Ability of markets to absorb the type of titanium minerals to be produced, and prevailing prices for titanium minerals and byproducts.

Specifications—In the titanium mineral industry, as in some others, grade and reserve do not completely define the quality of a deposit. Titanium minerals for pigments manufacture must meet specifications regarding those chemical elements which would reduce the whiteness of the titanium pigment. These include chromium, manganese, vanadium, and columbium. To be considered economic for the chloride process, the ilmenite product should contain at least 60% TiO_2.

There are no industry-wide specifications for ilmenite in the pigments industry. Usually consumers and producers establish typical, maximum, and minimum levels, for TiO_2 and other oxides in the concentrates, in separate purchase contracts. In general terms, ilmenites containing not more than a few tenths of 1% of the elements—chromium, manganese, vanadium, and columbium—may be marketable. A few tenths of 1% of the oxides of aluminum and phosphorus may also be acceptable.

In ilmenite the ratio of ferric to ferrous oxide is important. Ferric oxide is objectionable in the sulfate process because it must be reduced to the ferrous condition in the processing, and this requires more handling and increases cost.

TiO_2 concentrates used in the chloride process (including ilmenite) must be low in Ca, P, and Mg. During chlorination, calcium and magnesium form gummy substances which cause plugging difficulties and which cannot be removed easily.

Thus, it is extremely difficult to generalize, since conditions and costs vary widely in different parts of the world, but some *minimum* guidelines for the industry in typical locations are about as follows:

Southeastern United States

Recoverable reserves, contained TiO_2: 1 million tons.

Average heavy mineral content: 3–4%, cutoff of 2%.

Average TiO_2 content of raw ore: 1% or a little less.

Average depth of ore: about 5 m (15 ft) for dredging, with a cutoff at 1.5 m (5 ft).

Eastern Australia (Rutile and Zircon)

Recoverable reserves, rutile: about 300 thousand tons.

Average heavy mineral content: 1%, cutoff around 0.5%.

Average TiO_2 content of raw ore: about 0.5%, cutoff around 0.2%.

Western Australia (Ilmenite, Altered Ilmenite, Rutile, and Zircon)

Recoverable reserves, heavy minerals: about 3 million tons.

Average heavy mineral content: about 5%, cutoff around 2.5%.

Clay minerals or slimes in heavy mineral sands are important because they must be removed. In addition, the disposition often calls for impoundment, sometimes with flocculation, and dewatering. Clay and other slimes require extra equipment and cost for removal and disposal, obscure the vision of operators who keep the equipment adjusted, and reduce the specific gravity contrast. Particle coatings make high tension separation more difficult, and in the case of zircon, may affect the sales value of the mineral. It is not unusual to find clay contents of 15% by weight, and some ore sands, notably at Eneabba in Western Australia, contain up to about 30% slimes in some places.

Particle size, shapes, and grain size distribution of the ore minerals can affect both mill recovery and marketability. Relatively coarse, well-rounded, uniformly sized grains are generally most desirable. A titanium mineral concentrate with the majority of the grains passing a 200-mesh screen may have limited market appeal, particularly to pigments manufacturers using the chloride process. The finer the particle size the more difficult it is to achieve good recovery in the three kinds of concentrating equipment normally used: gravity, high-tension, and magnetic.

Coating of particles by iron oxide or by secondary silica can change their natural magnetic and electrostatic properties, and such coatings may require removal prior to separation. Iron oxide cementation may produce nodules, cellular or even uniform masses of material so hard that the deposit requires drilling and blasting in mining. Liberation of grains can be poor in milling.

Angular minerals are more troublesome to separate in both wet and dry equipment and increase maintenance costs due to abrasion wear in pumps, screens, and piping.

Deposits in which a high proportion of the titanium content is in the form of secondary anatase present recovery problems. The anatase is the result of natural weathering of ilmenite and is both fine-grained and friable.

Testing and Evaluation—The number of drill holes required to give confidence in ore reserve calculations will vary directly with the irregularity of the deposit in depth, heavy mineral content, clay content, degree of concentration or development of hardpan or caliche, and abundance ratios among the members of the heavy mineral suite. Ore bodies tend to be several times greater in length than in width, reflecting their origins in narrow beach zones. The frequency of holes across the strike in the direction of most rapid change in physical characteristics should be much greater than along the strike. Where physical conditions such as the absence of swamps or particularly steep dune slopes permit, a regular grid system of drilling leads to simpler and probably more reliable ore reserves estimates. Holes at 30.5 m (100-ft) intervals along lines spaced 91 m (300 ft) apart have been found to make a good pattern in some deposits. Reserves in other deposits have been defined with holes at 15-m (50-ft) intervals on lines 152 m (500 ft) apart.

There are two common tests for ore or non-ore indications: weight percent of heavy minerals as determined by separation in bromoform or acetylene tetrabromide (which has a slightly higher specific gravity and separates shell fragments from minerals more effectively) and weight percent TiO_2 in the raw sand by wet chemical analysis. Both imply a knowledge of the mineralogical composition of the heavy mineral suite. Thus the total heavy minerals or total TiO_2 can be more or less directly related to weight percent of titanium minerals in the sand. In new areas either or both of these tests should be preceded by petrographic examination of heavy mineral concentrates from panned samples.

Once the presence of ore has been indicated and drilling changes from exploration to development, material handling and rapid analysis become critical, as up to about 50 samples a day, each representing 1.5 m (5 ft) of hole, can be generated by a single drill. Each sample should be dried, weighed, and screened to remove coarse particles. It is then slurried, agitated, decanted to remove slimes, dried, and weighed again. The original weight minus the weight of the coarse fraction and the weight of the dry sample after desliming gives the weight of slimes. The sample may then be split to a convenient size, say 50 g, for heavy mineral determination. The "sink" fraction is washed, dried, and weighed, and its weight related to that of the original dried sample.

Heavy mineral concentrates from samples from one out of every few holes should be petrographically examined and their mineralogical compositions determined by grain counting. Magnetite, which can be mistaken for ilmenite under the microscope, and ilmenite, which can be difficult to differentiate from rutile, may be removed as discrete fractions with different settings of a magnetic separator prior to petrographic study. Grain counting should be by skilled technicians, and all laboratory personnel should be alert to the possibility of the introduction of error due to coatings of iron oxide, silica, and clay. An acid or caustic attrition scrub of the heavy mineral concentrate can improve the accuracy of identification. Data gathered in such detailed studies of both titanium minerals and byproducts can be valuable in guiding mining operations. The information is highly susceptible to computer analysis for projections of production rates and economics.

The precise methods of laboratory tests outlined can produce determinations of heavy mineral contents higher than those obtained in later production. The difference is due in part to recoverability in the concentrator phase of milling. The addition of a mechanical heavy mineral concentrator such as a jigging table or a spiral to the sample analysis flow pattern after desliming and before separation in a heavy liquid can lead to a better early knowledge of recoverable reserves.

The weight of sand per unit volume varies among ore bodies, and the volume to weight conversion factor for reliable ore reserve calculations should be established by careful measurement of excavated volumes of representative sand types, and by averaging their dried weights. In the southeastern United States the factor varies from 41 to 45 kg (90 to 100 lb) per 0.03 m^3 (cu ft). Exceptionally high grade sands have higher densities.

There is enough difference in physical characteristics of heavy mineral deposits that the best mining and milling methods for a new discovery may not be obvious. Pilot tests, either by commercial consulting organizations specializing in heavy mineral treatment and mill flowsheet design, or in a pilot plant erected at the site, are extremely valuable.

Garnar (1978) outlined a laboratory procedure for determination of separation characteristics, using a sample large enough to furnish at least 2 kg (5 lb) of heavy mineral concentrate. The heavy minerals are concentrated on a laboratory size shaking table, then split magnetically at 7,000 gauss to separate the ilmenite. The magnetic portion is passed over a laboratory high tension unit and the conductor fraction cleaned. The final conductor fraction is examined under the microscope, and if it contains 98% or more opaques, it is probably free from surface coatings that could interfere with dry milling. If coatings are present they may be removed by high energy attritioning. However, some ores may require treatment with acid or sodium hydroxide to remove coatings.

Material for pilot tests may be gathered by excavation with standard earth moving equipment or, preferably, by sinking numerous regular, vertical pits through the ore body by backhoe and/or by hand. Cribbing or concrete culvert sections may be necessary to keep the pits from caving. Pit walls can be channel sampled and the collected sand assayed to check results from nearby drill holes. This provides confidence in exploration and development drilling and sampling.

The degree of alteration of ilmenite in the deposit, together with the presence or absence of rutile, will determine the TiO_2 level of the concentrates that can be produced. Pilot plant work should be directed to production of various concentrates and testing the marketability by close communications and providing bulk samples, up to a few thousand tons, to consumers. Technical liaison between would-be producer and consumers is vital at this stage.

Complete chemical analysis should be performed on all potential products. The most sensitive elements in titanium mineral concentrates which can affect the whiteness of the pigment or interfere with the pigment manufacturing process include chromium, calcium, and magnesium. Some pigments manufacturers permit only a few tenths of 1% of these elements in their raw materials. The sum of the oxides of iron and titanium is employed as a measure of the purity of titanium mineral concentrates in some cases.

Because titanium minerals have various physical and chemical properties, each mineral product is not always suitable for utilization by each consumer. There is considerable latitude and flexibility in some instances because of highly developed technology, but most pigments manufacturing plants operate within narrow limits as far as the quality of the raw material is concerned. Usually both the raw material and the plant have to be tailored to fit one another, and in all cases the chemical and physical properties of a titanium mineral product have to be well-known to enable a manufacturer to judge its suitability. In a relationship such as this, naturally the best way to effect accurate evaluation of a titanium mineral deposit is through very close cooperation between would-be producer and would-be consumer, through all stages of exploration, bulk sample testing, development, and plant design and operation.

Evaluation of Results: The titanium mineral industry is restricted, with relatively few producers worldwide supplying relatively few consumers. Evaluation of a newly discovered and developed deposit is therefore undertaken within a narrow framework of economic and materials supply considerations. Present and future prices for products and byproducts under long-term purchase contracts have to be established with pigment manufacturers, and production rates forecast from pilot plant information. All aspects of production costs including those involved with pollution control and reclamation are taken into account in the economic model. Again, close liaison with consumers is essential as these companies, dependent on steady supply, are reluctant to enter long-term arrangements unless they can have reasonable assurance that the project is economically viable.

Preparation for Markets

Hardrock Ilmenite Deposits

MacIntyre Development, NY: *Mining*—Geologic sections and plans, produced from drill results, are used to determine open pit mining plans including waste and ore tonnages, mining ratios, and mining sequence. These, in turn, determine equipment and manpower requirements, and overall mining costs.

Mining is carried out on 14-m (45-ft) benches in an open pit using large blasthole drills, electric shovels, and diesel-electric trucks.

Milling—Ore from the mine is reduced in three stages of crushing to −14-mm (−9⁄16-in.) size. At the −64-mm (−2½-in.) size there is a magnetic separation step which discards about 20% by weight as a waste. Ore is reduced in size to −208 μm (−65 mesh) in rod mills and ball mills. Magnetic separators remove the magnetite fraction of the ore, and the nonmagnetics are treated by flotation to produce an ilmenite concentrate and tailing. A flowsheet of this milling process is shown in Fig. 7.

Prior to 1972, the ore was ground to −590 μm (−28 mesh) and was then fed to wet magnetic separators to remove magnetite. The nonmagnetic fraction was sized and fed to tables. Finer products went to a flotation plant to produce a flotation ilmenite concentrate. The wet table concentrate was dried and subjected to high intensity magnetic separation, producing a coarser ilmenite concentrate. A flowsheet of this earlier milling process is shown in Fig. 8.

Magnetite concentrate is dewatered on a filter and stockpiled. Ilmenite is filtered and then dried prior to shipment.

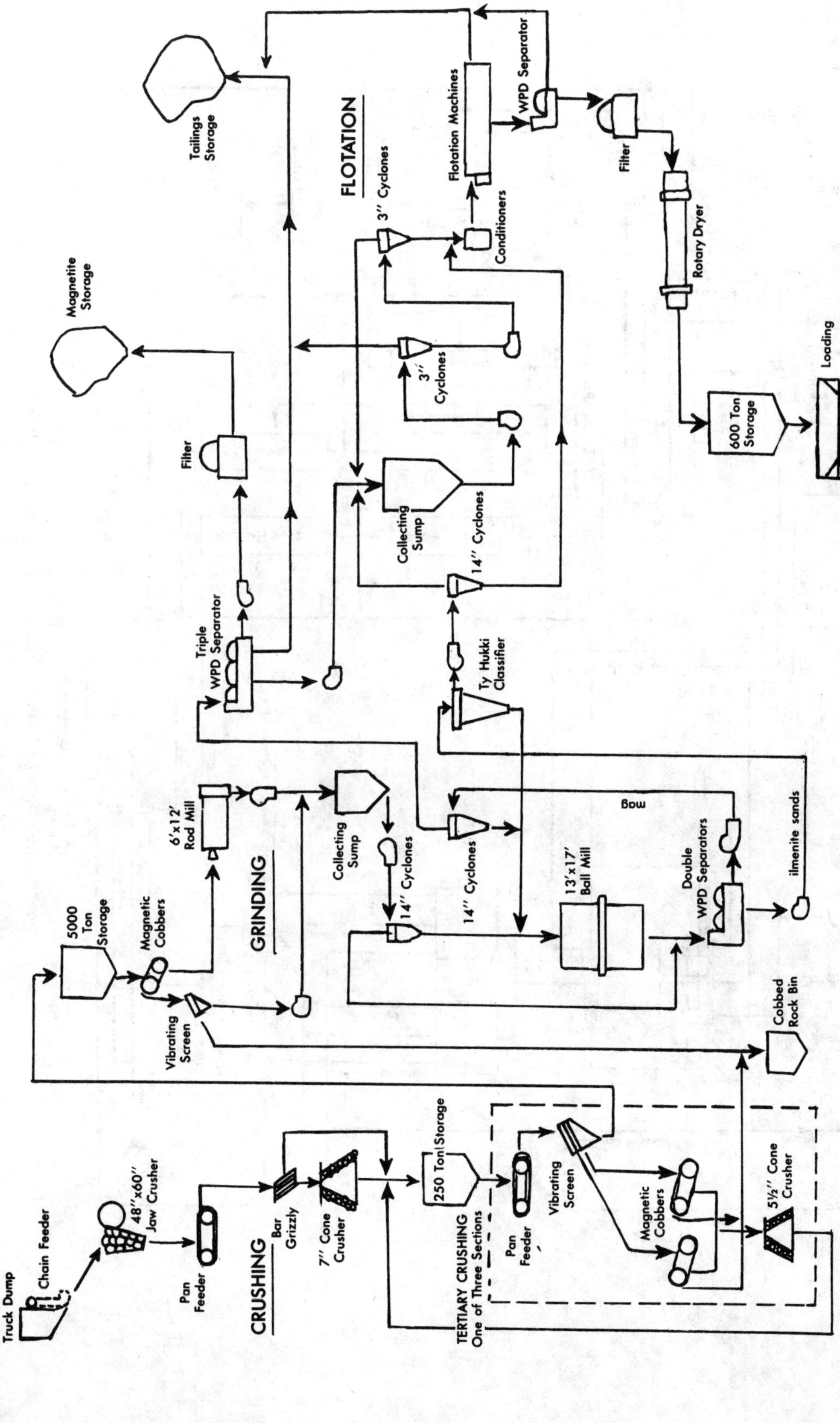

FIG. 7—*Flowsheet for all-flotation ilmenite, MacIntyre Development, NL Industries, Tahawus, NY (courtesy of NL Industries, Inc.).*

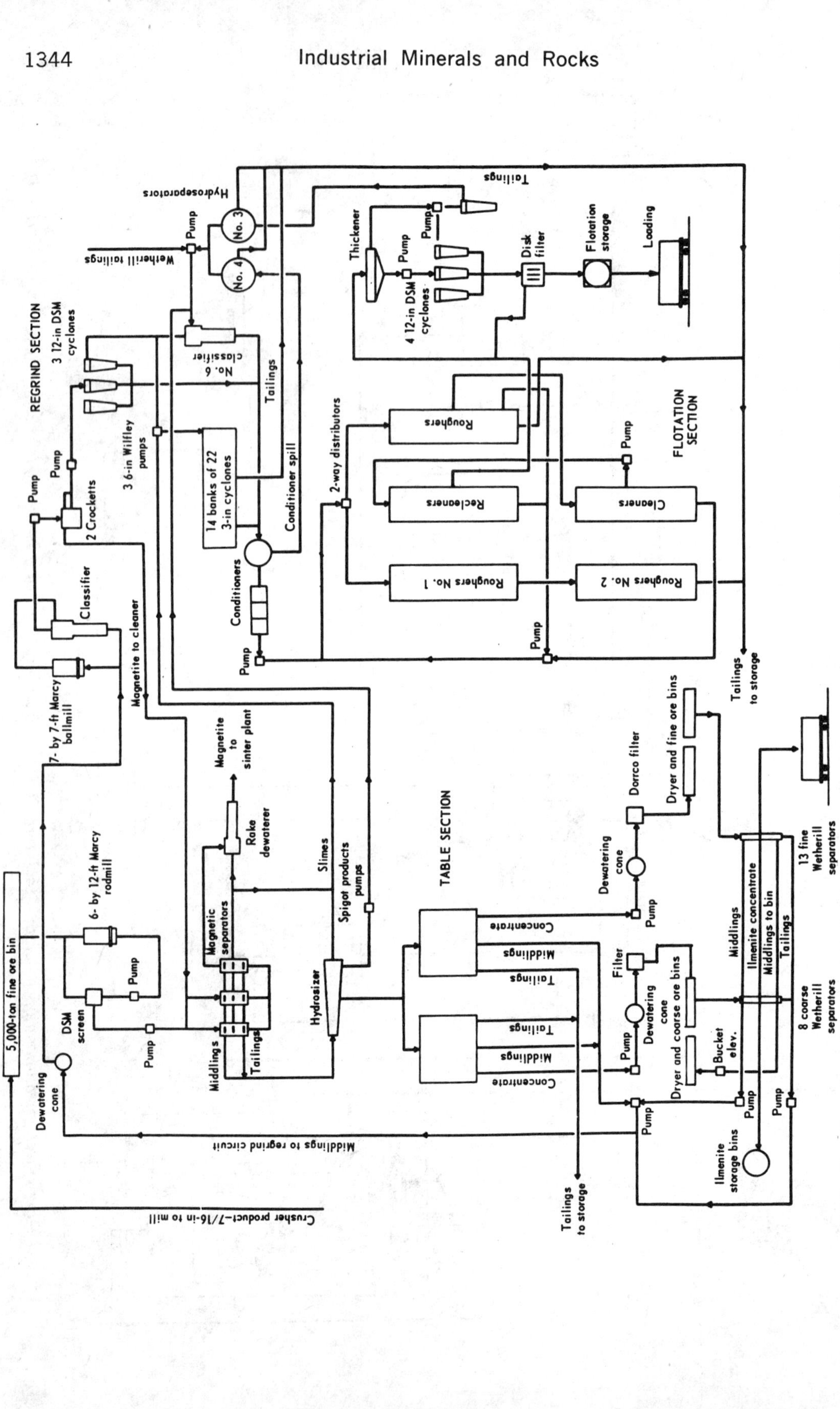

FIG. 8—*Flowsheet for magnetic and flotation ilmenite concentrates, MacIntyre Development, Tahawus, NY (Peterson, 1966).*

Transportation, Markets, and Consumption—Ilmenite and magnetite concentrates are shipped by rail in open cars. Ilmenite went mainly to the NL Industries TiO_2 pigment plant at Sayreville, NJ. A relatively new use for ilmenite is as a substitute for barite in well-drilling muds. The magnetite has been used as blast furnace feed, in refractories, and for heavy media separation in the coal industry.

Tellnes Deposit, Norway: *Mining and Milling*—The ore at the Tellnes deposit (Anon., 1978a; 1973c; Brun, 1957) is mined in an open pit using 12 to 14-m (40 to 46-ft) benches. The primary processing plant, including crushing, grinding, magnetic separation, and flotation is located close to the mine. The ilmenite flotation concentrate passes by gravity in slurry form through a pipeline in an underground tunnel to the final processing plant near Jossingfjord, where leaching and drying are carried out. The leaching is done with dilute sulfuric acid to remove apatite. The magnetic concentrate is transported by means of a smaller pipeline through the same tunnel. The ilmenite is dewatered on disk filters, to about 7% moisture, and dried in a rotary kiln.

The magnetite slurry is partially dewatered and subjected to flotation treatment to remove sulfides. The sulfide-free magnetite is dewatered, filtered on drum filters, and stored. The sulfide fraction is combined with a sulfide product from the ilmenite flotation circuit, thickened, and filtered. Production of these byproducts has been about 40,000 stpy of magnetite concentrate (65 to 65.5% Fe) and 10,000 stpy of sulfide concentrate (4 to 4.5% Ni and 2 to 2.5% Cu).

Transportation, Markets, and Consumption—Fully processed concentrates are stored in underground silos beneath the leaching and drying plant. A belt conveyor in a tunnel takes material from the silos directly onto ships in Jossingfjord.

Titania A/S is by far the largest producer of titanium concentrates in Europe and is the only important exporter in Europe. Even Spain and Finland, which produce ilmenite themselves, import some from Norway. Besides its proximity to the European market, Titania A/S has the advantage of relatively stable North Sea freight rates.

Tellnes ilmenite is the major feed material for the Kronos (NL Industries) group of sulfate plants in Europe, including plants in West Germany and Belgium as well as the Frederikstad plant in Norway. In addition, Titania has supplied companies in the United Kingdom, Italy, Spain, and other European countries, including Czechoslovakia (Anon., 1973c).

Allard Lake Deposite, Quebec: *Mining*—The ore at the Allard Lake deposit is mined by open pit methods. The mined ore has a grade of 32 to 36% TiO_2 and 39 to 43% iron. It is shipped by rail and boat via Havre St. Pierre to the company's beneficiation and smelting plant at Sorel, Que., for further processing (Elliot, 1959; Hatch and Cuke, 1956; Guimond, 1964; Hammond, 1949; Peirce, 1949).

Beneficiation—The ilmenite-hematite ore is crushed, and beneficiated at Sorel by heavy media (magnetite) separation of the −6.4-mm + 1400 μm (−¼-in. + 14-mesh) fraction, and Humphreys spiral concentration of the −1400 μm (−14-mesh) portion. Sulfur is eliminated from the combined gravity concentrates by heating in rotary kilns, using as fuel excess carbon monoxide from the electric smelting furnaces.

Because of its intimate intergrowths of ilmenite and hematite, the physically beneficiated concentrate, averaging 36.8% TiO_2 and 41.8% Fe, is still not high enough in TiO_2 content for direct use as a pigment plant feed. It is, therefore, smelted by a process developed by QIT-Fer et Titane, Inc. for the purpose of utilizing this type of high iron titanium ore.

Smelting—The smelting process used by QIT uses little or no flux, the furnace charge consisting of ilmenite-hematite concentrate with 0 to 10% lime and 8 to 14% low ash coal. Due to the extremely corrosive nature of titanium slag, it is necessary to keep the melt within solid banks of ore to prevent the slag from attacking the refractory lining. The average furnace measures 15 × 6 × 6 m (50 × 20 × 20 ft), and has 610-mm (24-in.) diam graphite electrodes on 2-m (6-ft) centers, heating the charge to 1500° to 1700°C, to produce a titanium-rich slag containing 70 to 74% TiO_2, and a form of pig iron known as Sorelmetal. QIT planned to increase the TiO_2 content of the slag to 80% in 1984. A flowsheet of the Sorel beneficiation and smelting process is shown in Fig. 9.

Transportation, Markets, and Consumption—QIT slag is widely used as sulfate pigment plant feed and is shipped by rail and by boat to various plants in Canada, the United States, and Europe. Production capacity has been periodically increased from 240,000 stpy of slag in the mid-1950s to a capacity of about 960,000 stpy slag (2.0 million tons ore) in 1980.

Otanmäki Deposit, Finland: The ilmenite is produced as a flotation concentrate, amounting

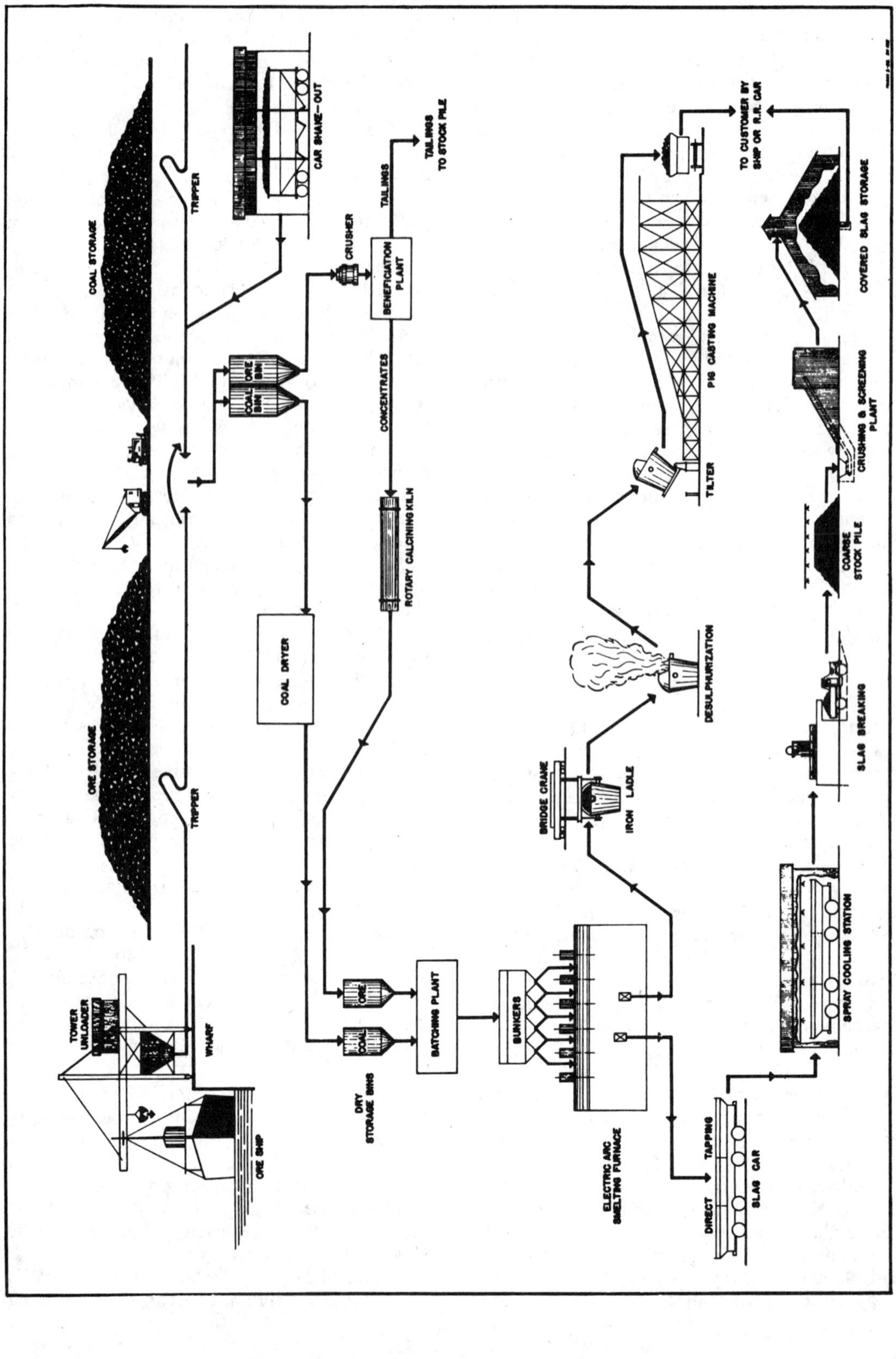

FIG. 9—*Flowsheet for Quebec Iron and Titanium Corp. smelter, Sorel, Que., Canada (Lynd, 1960).*

to about 160,000 stpy. Most of this is consumed domestically, but some is exported to other European countries. A vanadium-bearing magnetite is produced as a byproduct (Anon., 1971c; Anon., 1978b; Harki et al., 1956; Runolinna, 1957).

Beach and Alluvial Sand Deposits

Mining Methods: Sand titanium mineral deposits are most commonly mined by suction dredge where the ground is not excessively "stiff" because of clay, hardpan, or induration of heavy mineral layers due to cementation by iron oxide. Dredges of widely different power and production capacity are used, depending on local conditions and desired production rates (see Figs. 1–4). The minimum dredging depth is about 1.5 m (5 ft), the maximum about 21 m (70 ft). The elevation of the dredge pond is adjusted to compensate as much as possible for variations in topography. The lowest practicable cutbank above the water level in the pond is maintained to avoid wave-producing shocks caused by sudden collapse of large volumes of sand into the pond, but high coastal dunes on North Stradbroke Island in Queensland, Australia, are successfully dredged with sloping banks over 15 m (50 ft) high.

Some deposits are mined by dragline, some by bulldozer, some by front-end loader, and some by bucket wheel excavator. Occasionally a combination of equipment types is employed. Particularly rich but highly cemented layers near the bottoms of some Western Australian deposits are mined in their softer portions by backhoe.

Clearing and stumping of the land surfaces precedes mining, and it is now becoming standard in the industry to remove any topsoil which may be present and to pile it aside for spreading over the refilled excavation after mining has passed through.

In all cases the object of mining is to excavate the heavy mineral sand and transport it to a "wet plant," which makes the first step of milling through gravity concentration. Large trommels, mounted either on dredges or on mobile supports in dry mining operations, remove roots, lumps of clay, and oversize gravel, boulders, and cemented nodules or fragments before the slurried sand is pumped to the wet mill. In dredging operations the wet mill floats on pontoons moored to the bank and moved periodically to keep up with dredge advances. In dry mining the gravity concentrator is located close to the mine. In both cases the tailings of quartz and other light materials are returned to the mined excavation, so that at any given time part of the ore body is being prepared for mining, part is being mined, part is being filled, and part is being reclaimed.

There is some expansion of the ore sand in mining due to disturbance of grain packing, so that the volume returned to the excavation may be 4% or so greater than that removed. The creation of unfilled excavations is not unavoidable in normal heavy mineral sand mining, although it is possible to resculpture the surface by planned distribution of tailings to leave low hills and small lakes or ponds if desired. Replanting can be accomplished by seeding fast-growing, hardy grasses and fertilizing, or simply by respreading the stockpiled topsoil with its retained fertile seeds of native plants. In the pine forests of the southeastern United States, where trees are valuable for papermaking, it has been demonstrated that pine seedlings planted in areas where sand dredging has been done grow at a natural rate.

Processing Beach and Alluvial Sand Deposits §**:** Throughout the world, there are mining companies producing titanium, zirconium, and rare earth minerals from marine beach deposits. To concentrate and separate these sandlike minerals, a combination of gravimetric, high tension (electrostatic), and magnetic separations is the standard method. The metallurgy is straightforward and well-understood, and the valuable minerals are produced from the sands with relative ease.

When the minerals occur in stream placer deposits the production of high-grade products along with a high recovery presents a more difficult task of ore dressing.

Physical Characteristics of the Deposits and Minerals—BEACH DEPOSITS—In the marine beach deposits, the sands are very well-sized and the entire mineral suite is usually well-classified. The deposits are, to a large degree, unconsolidated free grains of sand. In some cases, there is cementation of grains due to organic matter, iron, or carbonates. The mining engineer must be fully aware of these conditions as the presence of these coatings not only creates mining but recovery and grade problems as well when the marine deposit is mined and concentrated. This type of deposit being formed on a beach tends to be long and shallow and may not be uniform in mineral composition due to the concentration characteristics of the waves and of the wind. The evalua-

§ This portion of the chapter was written by W. P. Dyrenforth, formerly Vice President, Mountain States Engineers, Tucson, AZ, now Marketing Executive, Colorado School of Mines Research Institute, Golden, CO.

tion of this type of deposit cannot be done on the basis of heavy minerals alone (those heavier than quartz), since the distribution of rutile and ilmenite and other commercial minerals in relation to noncommercial heavy minerals may vary to a large degree throughout the deposits.

Besides being well-sized and classified, usually unconsolidated, and easy to mine, these deposits have presented the ideal situation of a maximum degree of weathering. This means that there will be very few grains that aren't clearly a single mineral and that usually the high iron materials such as magnetite have almost completely disappeared. Also the ilmenite will usually be weathered to the point that the iron has been sufficiently leached away to leave an ilmenite high in TiO_2. It is interesting to note that many of the beach deposits from which monazite and zircon are produced are worked primarily because of the quality of the ilmenite which they can produce.

ALLUVIAL DEPOSITS—Alluvial deposits formed in stream beds present an entirely different problem to the mining engineer. Instead of uniform grain size and the classified characteristics of a beach deposit, the commercial mineral values may vary from -75 μm to 2 mm (−200 mesh to 10 mesh) in size. Any one mineral such as monazite in an alluvial deposit may vary from -75 μm to $+600$ μm (-200 mesh to $+30$ mesh) in size. The deposits are usually a heterogeneous mixture from clay on up to boulders depending on the age and source material of the stream.

These deposits, of course, present an entirely different mining and concentration problem. Extensive weathering and smoothing of the grains has not taken place so that one may have to contend with cemented grains of two or more minerals, and/or the grains are more likely to be coated with iron or clay.

Mining and Concentration Methods—The beach-type deposits, providing they are deep and extensive enough, are usually exploited by suction dredging. This has been considered the least expensive way of mining, provided the material is unconsolidated and that the cross section of the deposit lends itself to maintaining a dredge pond. Due to irregular cross section of some deposits in Africa, dredging was found to be impractical, so they are being mined by draglines. Dragline-type mining is useful where comparatively rich deposits may be worked, in that this method allows selective winning of rich streaks and material in pockets. Draglines also are able to remove consolidated as well as unconsolidated overburden simultaneously.

A bucket-line dredge was planned for use on a marine beach deposit in the southeastern United States. It was believed that the higher capital cost of the equipment would be offset by more efficient mining of the ore, and that the ability of the large bucket-line dredge to carry its own concentration equipment should keep costly pumping of sand and water to a minimum. The plant was never built due to a titanium recession in the late 1950s.

Bucket-line dredges are almost a necessity for most stream-type deposits. They are required to break up the clay and other forms of consolidation as well as to handle the large stones, boulders, and even stumps and logs. Where only small gravel and stones occur, suction dredges have been used but they present a problem of lower mining recovery because the largest stones are left in the bottom of the pond. This is a mining recovery problem created by the loss of heavy minerals which are concentrated by water action into the interstices of the porous rocks.

Concentration of the minerals from sands is to date best done in small plants by use of wet shaking tables. For larger production, the spiral concentrator has been standard over a period of years. Flotation has been used to some degree. At the present time, there is a great deal of interest and work being done on various types of gravity concentrators to provide large tonnage rates of feed in comparatively light weight and low cost concentrating devices.

Jigs have been studied and adapted for this kind of concentration in many places. They were actively used in the British Titan Products' operation in Africa and are still being studied in a number of places. There are a number of separators being developed and used that incorporate the sluicing principle. In the United States, examples of this type of separation action are found in the Cannon ring-conformation pinched-sluice concentrators, the Lamflo sluice concentrator developed by Carpco, and the pan concentrator developed by Du Pont. In Australia the Reichert cone concentrator, which is also operated on the sluice principle, has been very successful for large tonnage operations (Paterson, 1977). All of the separators based on the sluice principle require a feed density of 55 to 60% solids which results in quite a saving in water when compared to other gravity methods.

For the separation of the minerals from

stream-type deposits, the jig is usually chosen. The jig has long been used on alluvial deposits containing gravel and tests have indicated that they are the concentrating device least sensitive to extreme grain size variation. The jigs also allow large throughput capacity in a small space and can be mounted on the dredge. Mounting the concentration equipment on the dredge allows maximum maneuverability for the stream deposits that may be comparatively narrow and crooked.

Separation of the Valuable Heavy Minerals from Each Other—Two quite different flowsheets are required to separate the minerals and to produce commercial products from the two types of deposits. High-purity minerals are particularly difficult to produce at high recovery from stream deposits due primarily to the grain size distribution and also due to the unweathered characteristic of the deposits which means there are many middling products to be dealt with. These middlings create heavy circulating loads, are contaminants to finished products, and their disposal can cause losses in recovery.

Minerals concentrated from stream deposits will often require attrition scrubbing to remove surface coatings of organics, clay, and iron. The sands from beach deposits will often be uncoated. A beach mineral concentrate, however, that has been produced with the use of saltwater, may require a freshwater wash to remove salt prior to the dry separation steps.

Once the minerals are concentrated and the surfaces of the grains are clean, the next problem that confronts the ore dresser is the size range of the minerals. Table 20 shows the size range distribution of the heavy mineral concentrate from a typical Florida beach-type deposit as compared to the screen analysis of a concentrate from an alluvial deposit.

TABLE 20—Comparison Chart of Screen Analyses

Alluvial Concentrate			Beach Concentrate		
Screen Size, Mesh		%	Screen Size, Mesh		%
+20		9.6	+20		–
−20	+28	9.8	−20	+28	–
−28	+35	14.2	−28	+35	–
−35	+40	9.5	−35	+40	–
−40	+48	14.5	−40	+48	–
−48	+65	18.6	−48	+65	0.5
−65	+100	13.9	−65	+100	7.8
−100	+150	6.1	−100	+150	37.0
−150	+200	1.9	−150	+200	45.0
−200		1.0	−200		9.7
		100.0			100.0

Since the minerals from beach deposits are very well-sized and classified usually no further sizing is considered necessary. Normally grade of product and recoveries have been acceptable as produced without sizing. Further study seems to point out that even in beach-type deposits, sizing or classification may eventually be incorporated for optimum recoveries.

Sizing of the minerals concentrated from stream deposits is a must and can be accomplished in two ways. The entire feed to the dry separation plant can be sized into fractions which are handled separately in the mill or they may be introduced into the mill as unsized feed wherein the natural sizing effects of the various separation equipment are employed to do simultaneous separation and sizing.

A good example of taking advantage of the sizing effect of separating devices appears in nearly all of the plants producing rutile and its associated minerals. Grains of coarse quartz and/or staurolite or garnet are thrown off the rotors of both the high tension (electrostatic) and magnetic separators, since their mass and kinetic energy cannot be overcome by high voltage or magnetic fields present which are adjusted to the highest recovery of the bulk of the finer minerals present. These coarse grains report finally with the nonmagnetic conductor fraction which is typically rutile or cassiterite. It is considered a very simple system to purposely allow these coarse grains to follow through to the end of the flowsheet where they can be scalped off by a small screen rather than to attempt to screen the entire feed to the mill. At some plants, however, the occurrence of coarse-grained rutile requires a modification of this method. Screening is employed at two points in the circuit in conjunction with high tension sizing: (1) between the rougher and thrown cleaner rolls and (2) between the final pinned product of the pinned cleaner rolls and the primary magnetic separators used for zircon-quartz, monazite-staurolite separation. In the first case, coarse staurolite is rejected as oversize with some coarse rutile. This oversize product is then treated separately by magnetic and high tension separators at special settings to recover the rutile. In case two, the valuable minerals, zircon and monazite, are -500 μm (-35 mesh). Thus, screening at this point eliminates coarse quartz and staurolite, reduces the load on the magnets, and improves the overall efficiency. Figs. 10a and b are typical flowsheets for processing beach deposits and alluvial deposits.

If the entire feed to a plant is sized, it

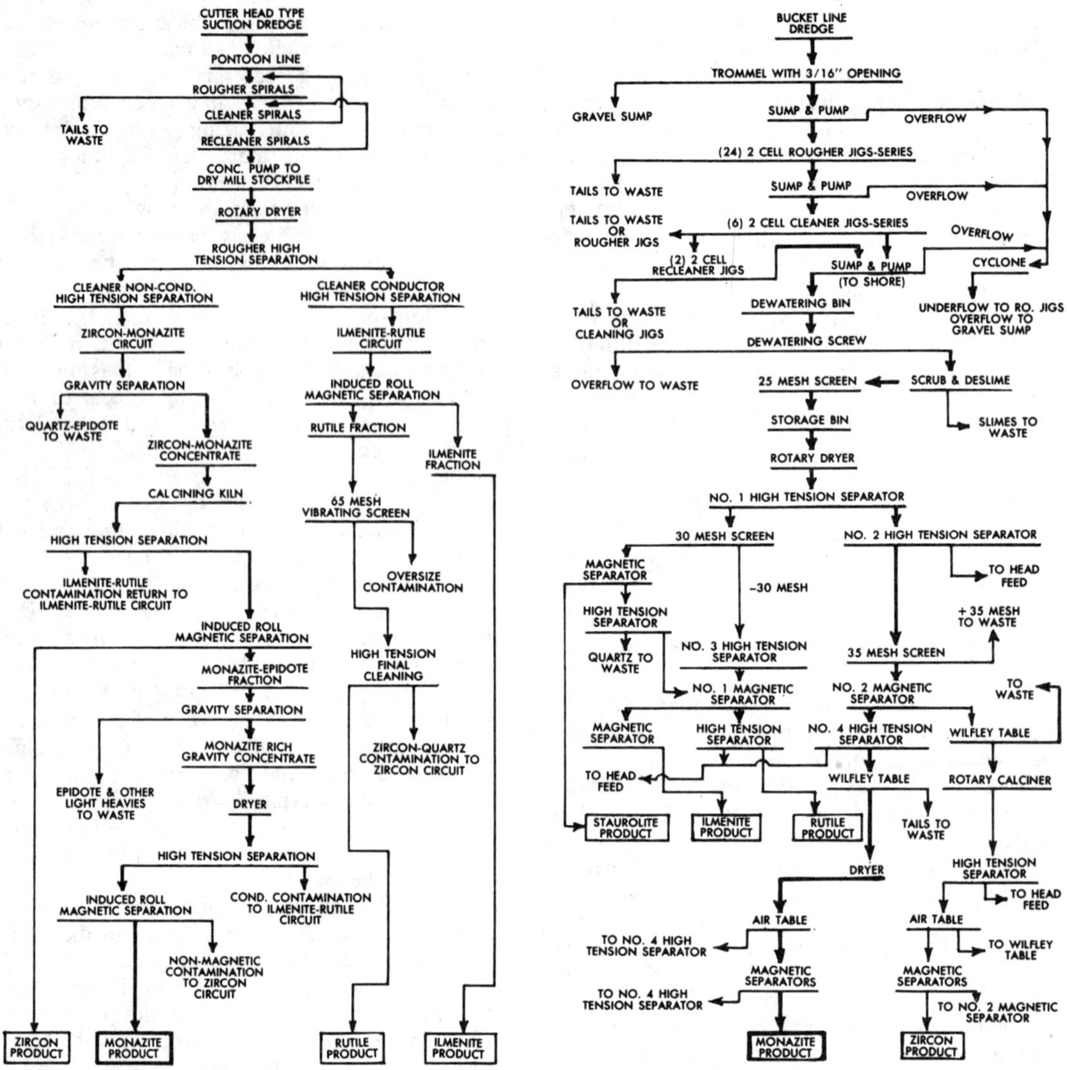

FIGS. 10a, b—*Typical flowsheets for processing beach sands (left) and alluvial deposits (right).*

creates the problem of multiple lines of separation equipment or an alternate of bulk storage and batch processing which requires a large amount of operational attention and resetting of the circuit. Accordingly, an alluvial flowsheet is usually designed to accept an unsized feed on a continuous basis and to make use of the sizing effect occurring during separation. At first glance this gives the appearance of a complicated flowsheet but after the operation and technical personnel have had time to balance the flow of the various minerals in accordance with their size, the production of finished mineral products is accomplished with a minimum of operational changes. Highest recoveries can be gained by continual study and development of this type of return products system.

The rather simple flowsheet as commonly used for the production of titanium minerals and byproducts from beach sands is shown along with a flowsheet used for the production of same from an alluvial deposit. Notes on the various types of separation equipment are given, particularly on the alluvial flowsheet. The relative difference in the flowsheets required to produce valuable minerals from the two types of deposits is very apparent.

The production of high-grade minerals from a stream-type deposit is a more complicated problem than production from marine beach sands. The major differences in the metallurgical flowsheet required are due to the uneven grain size distribution. It is necessary to remove clay and iron coatings and the larger amount of middling minerals occurring due to the lack of weathering in stream-type deposits.

Future Considerations and Trends

With present-day technology, there are no totally competitive substitutes for titanium minerals as raw materials for pigments manufacture. There are, however, potential substitutes for rich sand deposits of ilmenite, altered ilmenite, and rutile, and these are in the form of extensive deposits of lower TiO_2 grade ilmenite in several parts of the world including South Africa, New Zealand, and possibly Egypt. Beyond these, there are widespread deposits of titaniferous magnetite sands which could produce concentrates averaging 10 to 20% TiO_2. Exploitation of the latter would involve development of technology to produce separate iron and titanium dioxide concentrates by smelting or hydrometallurgy. Weathered carbonatites in Brazil, India, and Africa have already demonstrated possible large reserves of anatase which could produce high TiO_2 concentrates if difficulties in recovery and concentration can be overcome.

Substitutes, Synthetic Rutile

Pigments: Titanium dioxide pigments have largely displaced white lead and other white pigments in the United States, and the question comes to mind whether titanium dioxide itself may some day be replaced by some new material with still better pigment properties. Some idea of the likelihood of this happening may be gained by a consideration of what other materials exist or might be synthesized that can match or surpass the physical and chemical properties that make titanium dioxide such an outstanding pigment: high refractive index (2.61 to 2.90 for rutile, 2.49 to 2.56 for anatase), high brightness and whiteness, chemical inertness, and the ability to be produced in the optimum particle size to develop maximum hiding power.

Materials having high refractive indices comparable to those of titanium dioxide include calcium, strontium and barium titanates, diamond, and silicon carbide. It seems unlikely that any of these would offer any substantial advantage in properties or cost, even if they could be made in the required amounts in pigmentary form.

A number of other materials, such as talc, clay, silica, and alumina, are white and can be used as pigment, but in most applications because of their low refractive indices they either do not provide the desired performance in terms of hiding power, brightness, etc., or are uneconomical because of the relatively large amounts needed. These materials are, therefore, used mainly as fillers rather than as pigments.

The use of microvoids to obtain opacity by means of a negative refractive index difference between the suspending vehicle and the pigment (microscopic gas bubbles in this case) has been found promising for dry hiding applications, such as in paper and in matte latex paint, rather than in glossy paint films (Kershaw, 1971).

At present, there seems to be no serious rival for titanium dioxide pigments. On the contrary, consumption of titanium dioxide pigments has increased at an average rate of about 2% per year since 1965.

Rutile Substitutes: Rutile concentrates containing about 95% TiO_2 have been the preferred feed material for chloride process pigment plants. As demand for rutile has grown, rutile prices have increased in line with the increasing cost of producing rutile from lower grade or more remote deposits as the older higher grade deposits become depleted. There has, therefore, been considerable cost incentive to develop processes for making rutile substitutes from the much more abundant, less costly ilmenite. Another approach to lower chloride feed material cost is to learn to handle lower grade materials, such as the rutile-leucoxene-ilmenite blends used by Du Pont, containing about 75 to 85% TiO_2. Extensive reviews of processes that have been investigated for making rutile substitutes from ilmenite have been published by the US Bureau of Mines (Henn and Barclay, 1970) and by the National Materials Advisory Board (Schlechten, 1972).

The main efforts to make a rutile substitute have been directed toward beneficiating ilmenite by using some combination of oxidation, reduction, leaching, and/or chlorination to remove most of the iron and produce a high-TiO_2 concentrate. Because this concentrate is intended for use as feed to fluid bed chlorinators, it is important that the particle size be closely controlled, with a minimum of fines. The sand ilmenite concentrates are generally of suitable size for this purpose, so that most beneficiation processes are designed to utilize sand ilmenites, preserving their original particle size as much as possible. The processes which are most favored for commercial applications are discussed briefly below, in approximately the order of their degree of development. The compositions of typical products from several of the processes are given in Table 21.

Ishihara Process—In 1971 Ishihara Sangyo Kaisha began operating a 30 kt/a synthetic rutile plant at Yokkaichi, Japan, and expanded production to 53 kt/a in 1974 (Anon., 1971b; Kataoka and Yamada, 1973; Yamada, 1976).

The process involves sulfuric acid leaching of reduced ilmenite in the presence of hydrated titanium dioxide, which is said to act as a seed to increase the rate of precipitation of titanium salts, allowing leaching of iron at relatively low temperature. The original particle size of the ilmenite, which may be either a sand ilmenite or a rock ilmenite ground and sized to about 840 μm to 75μm (20 to 200 mesh), is maintained in the leached product.

Ilmenite is mixed with coke and reduced in an oil-fired rotary kiln for about 1 hr at about 900°C, converting over 95% of the iron in the ilmenite to the ferrous state. After cooling, the residual coke is separated magnetically and the reduced ore is leached with waste acid from a sulfate process TiO_2 pigment plant. By adding the seed TiO_2 and keeping the leaching temperature at 130°C for 8 hr, iron and other impurities are selectively dissolved. The upgraded ilmenite is filtered, washed, and calcined at about 900°C, yielding an SO_3-free product containing over 95% TiO_2. Magnetic or gravity separation may be used to upgrade the washed concentrate before calcining, recycling the lower grade fraction.

Waste acid from the leaching step is either used as raw material for an ammonium sulfate plant, or is recycled in the leaching process with the addition of makeup acid.

Benilite Process—Benilite Corp. of America developed an ilmenite beneficiation process which involves partial reduction of iron oxides and hydrochloric acid leaching (Iammartino, 1976). Under Benilite licensing agreements, Ballarpur Paper and Straw Mills Ltd., of India, and Taiwan Alkali Co. Ltd., of Taiwan, each built beneficiation plants having 25,000 to 30,000 stpy production capacity (Anon., 1971; 1973a).

In a Benilite patent example, ilmenite is prereduced 40 min at 900°C with coke, and is then leached 5 hr in each of two steps at 100°C with 2.6 parts 20%HCl to give, after separation and drying, an ilmenite concentrate containing 92.1% TiO_2, 3.1% Fe_2O_3, and 0.25% FeO_3 (Chen, 1971). The particle size of the concentrate is essentially the same as that of the original ilmenite.

The proposed Taiwan plant was claimed to be the first closed-cycle effluent-free ilmenite upgrading plant ever to be built on a commercial scale. The effluent problem is avoided by regenerating and recycling the hydrochloric

TABLE 21—Composition of Rutile Substitutes and Typical Australian Natural Rutile Weight Percent

Constituent	Kerr-McGee	Ishihara	Benilite	Dhrangadhra	Western Titanium	Chlorine Technology	Mitsubishi Metal	Summit	Murso	Natural Rutile
TiO_2 (total)	90.5-93.0	96.1	93.0	90-92	92.0	97.5	96.7	94.0	96.2	95.2
Ti_2O_3					10.0					
Fe (total)	1.5-2.6		2.0	3.0	3.6	0.46				
Fe (metallic)					0.2					
FeO	0.1-0.6							2.0		
Fe_2O_3	2.0-3.0	1.30				0.66	0.4		1.5	0.9
SiO_2	1.5-2.5	0.50	1.6	1.5	0.7	0.46	0.1	0.5		1.0
Al_2O_3	<1.0-1.5	0.46	0.42	1.0	0.7	0.49	0.7	1.7	0.2	0.2
CaO	<0.35	0.01		0.05	0.03	0.065		0.7		0.02
MgO	0.05-0.40	0.07	0.05	0.05	0.15	0.027			0.04	0.07
Cr_2O_3	0.05-0.12	0.15	0.04	0.2		0.11		0.10	0.15	0.18
V_2O_5	0.10-0.25	0.20	0.06	0.25	0.12	0.04	tr	0.15	0.04	0.6
MnO	0.05-0.15	0.03	0.22	0.1	2.0	<0.01	0.06	1.7	0.05	0.01
S	0.04-0.07		0.01		0.15	0.006		0.02		0.008
Na_2O	0.0034-0.0047					0.05				<0.1
C	0.04-0.07				0.15	0.09				0.04
P_2O_5	0.05-0.09	0.17		0.2		0.072		0.1		0.03
ZrO_2	0.20-0.40	0.15	0.12	1.0		0.19		0.2		0.8
Ignition loss				3.8						0.2
Moisture	0.05-0.40			1.5						0.1
Nb_2O_5	0.20-0.40	0.25	0.6			0.07				0.3
SnO_2			0.14			0.003				0.03
Reference:	(1)	(2)	(3)	(4)	(5) (6)	(7)	(8)	(5)	(9)	(7)

References: (1) Kerr-McGee analysis; (2) Yamada, 1976; (3) Benilite analysis; (4) Anon., 1973; (5) Roberts, 1971; (6) Bracanin, et al., 1972; (7) NL Industries analysis; (8) Anon., 1972a; (9) Sinha, 1972.

acid, using Woodall-Duckham's spray-roasting process.

In 1977, Kerr-McGee Chemical Corp. began operation of a $53 million synthetic rutile plant at Mobile, AL, using the Benilite process. Following a shutdown in 1978 to devise process improvements, the plant reopened in early 1980, and in late 1981 was approaching its nominal production capacity rate of 110,000 stpy.

Dhrangadhra Process—The Dhrangadhra process involves hydrochloric acid leaching of ilmenite supplied by Indian Rare Earths Ltd., of Quilon in Kerala State. The hydrochloric acid requirements are met by an adjacent caustic soda plant. Capacity of the beneficiated ilmenite plant at Sahupuram in Tamil Nadu State was to be increased beyond the initial 25 kt/a (Anon., 1973).

Western Titanium Process—Western Titanium NL (now part of Associated Minerals Consolidated Ltd.) began operation of a semicommercial ilmenite beneficiation plant at Capel in 1968. A full-scale commercial plant with 30,000 stpy initial capacity began production at Bunbury in 1975, and was later expanded to about 64,000 stpy. The Western Titanium process involves high temperature oxidation to form a pseudobrookite structure, and reduction with carbon to metallize the iron, which is then oxidized in an aerated water slurry and separated as a hydrated iron oxide suspension, yielding a beneficiated ilmenite containing about 92% TiO_2. By adding as little as 2% of sulfur to the ilmenite it was found that iron and manganese contents of the upgraded ilmenite are lowered still more, raising the TiO_2 content to as high as 94% (Anon., 1973e, 1978c; Bracanin, 1972; Roberts, 1971).

Chlorine Technology Process—Chlorine Technology Ltd., a subsidiary of Rutile and Zircon Mines Ltd., developed a novel chlorine beneficiation process for upgrading ilmenite. This process was developed through the pilot plant stage at Mount Morgan, Australia.

In the Chlorine Technology process, ilmenite is selectively chlorinated to remove iron and other impurities as volatile chlorides, using a fluid bed reactor under reducing conditions at about 1000°C, with the novel feature of continuously removing a portion of the partially chlorinated bed, and separating it magnetically into an essentially iron-free fraction, and a second iron-oxide-containing fraction, which is recycled to the chlorinator. Recycling of the iron-containing part of the bed lowers TiO_2 losses, and produces less porous, harder, more abrasion-resistant grains in the final concentrate than when there is no recycling or when beneficiation is done by leaching. Chlorine can be recovered by oxidizing the ferric chloride formed during chlorination. The final product contains 95 to 98% TiO_2 and 0.1 to 0.5% Fe_2O_3 (Dunn, 1972, 1973).

Mitsubishi Process (Mitsubishi Metal Corp.)—In this chlorine beneficiation process, which has been developed through the pilot plant stage, ilmenite is first oxidized at 950° to 1000°C to convert all the iron to the ferric form. The oxidized material is then fed to a fluid bed chlorinator along with petroleum coke, where the iron is volatilized as ferric chloride. The ferric chloride is then oxidized to yield ferric oxide, and chlorine gas which is recycled to the chlorinator. The solids from the chlorinator are subjected to electrostatic and magnetic separation to remove unreacted ore, coke, and other impurities. About 10% of the original charge is returned to the fluidized reactor in the form of unreacted material. The final product contains 95 to 97% TiO_2 (Anon., 1972a).

Halomet Process—The Halomet process (Anon., 1971a; Othmer and Nowak, 1972) is based on a chemical displacement series that ranks metals according to the halogen affinities of their oxides, somewhat analogous to the electrochemical displacement series. The Halomet process uses dual chlorination—a reduction-chlorination followed by an oxidation-chlorination to take advantage of the ability of a metal chloride to react with the oxide of a metal above it in the chloride affinity series. The oxidation-chlorination reaction is therefore an effective separation technique.

The Halomet process was originally conceived as a method of using low grade ores for the production of iron. For application to ilmenite upgrading, partially reacted ilmenite is subjected to reduction-chlorination, with CO and Cl_2, forming $TiCl_4$ and $FeCl_3$, which pass to the oxidation part of the chlorinator where fresh ilmenite reacts with the $TiCl_4$ to form more $FeCl_3$ and TiO_2. Proper metering of chlorine permits isolation of high purity ferric chloride at one end of the reactor and a titanium dioxide concentrate at the other.

Halomet and Tellus, A.G., of Frankfurt, West Germany, collaborated in the operation of a large pilot plant to produce ferric chloride from low grade ores. The ferric chloride was oxidized in air to form ferric oxide and chlorine.

Summit Process—In this process ilmenite ore is oxidized in air or other oxygen-containing gases at 750° to 1200°C to convert all the iron to the ferric state. The oxidized ore is then reduced at 750° to 1200°C, using solid carbon fuels, carbon monoxide, hydrogen, or other reducing agents, reducing over 90% of the iron content to metal. The metallic iron is removed from the reduced ore by leaching about ½ hr with an aqueous solution of ferric chloride. The residue from the leaching process is washed and dried to yield a final concentrate containing up to about 94% TiO_2 (Roberts, 1971; Shiah, 1966). The spent leach solution is regenerated by sparging with air or oxygen to convert part of the ferrous chloride to ferric oxide, and the remainder to ferric chloride which is recycled to the leaching step.

Murso Process—This process was developed in Australia jointly by Murphyores Inc. Pty. Ltd. and the CSIRO Division of Mineral Chemistry. Ilmenite is first oxidized at 900° to 950°C with air, then reduced with hydrogen or hydrogen-containing gas in a fluid bed at 850° to 900°C, reducing the ferric iron to the ferrous state with a minimum formation of metallic iron. This results in formation of a very reactive synthetic ilmenite product which is leached in 20% HCl solution at 108° to 110°C at atmospheric pressure for 3 to 4 hr, yielding a concentrate containing over 95% TiO_2. Magnetic separation may be used to remove chromium and other impurities from the final product. HCl can be regenerated from the $FeCl_2$ spent leach liquor by known commercial processes, and recycled (Sinha, 1972). Planning was begun to build a commercial plant in Australia using the Murso technique, following a pilot plant development program carried out jointly with Mitsubishi Chemical Industries Ltd. in Japan (Anon., 1973d).

Other Ilmenite Beneficiation Processes—High TiO_2 concentrates have also been produced by direct leaching of ilmenite with sulfuric or hydrochloric acid, and by a variety of other processes, some of which are similar to those described above. A few of these processes are discussed in the following paragraphs.

Direct acid leaching of ilmenite requires more severe leaching conditions and more finely ground ore than when the ilmenite is oxidized and reduced before leaching. Leaching time may be shortened by use of high temperature and pressure, but the resulting concentrate is fine grained and would require agglomeration treatment to make it suitable for feed to the fluid bed chlorinators currently used in the chloride pigment process. Examples of processes of this type are those of Anderson and Rowe (1956), using sulfuric acid; and Leddy and Schechter (1962) and Marshall and MacMahon (1972), using hydrochloric acid.

In leaching tests on ilmenite and partially reduced ilmenite, Mackey found that leaching at atmospheric pressure takes place very slowly and is only partially effective, even with concentrated acids. After reduction for 6 hr with hydrogen at 900° C to reduce all of the ferric oxide to ferrous oxide, leaching times of 10 hr in two stages at 105° to 125°C and 10 to 20 psig resulted in leached concentrates containing about 90% TiO_2 (Mackey, 1974).

Electric furnace smelting processes have been investigated by the US Bureau of Mines, and by NL Industries, with the object of producing, besides pig iron, high-TiO_2 slags which would be suitable for chloride process and/or sulfate process pigment plant feed (Armant and Cole, 1949; Armant and Sigurdson, 1954, 1956; Banning et al., 1955; Cole, 1953; Moore and Sigurdson, 1949; Sigurdson and Cole, 1949; Stoddard et al., 1950). Concentrates used were classed as low, medium, or high grade, depending on the ratio of TiO_2 to iron oxides. The lower grade materials, such as the blend of magnetite and ilmenite concentrates produced at the MacIntyre Development, Tahawus, NY, which contains 20 to 25% TiO_2 and 60 to 70% iron oxides, yield a very large amount of pig iron, but the gangue oxide content is increased severalfold in the slag, compared with the original smelter feed. Gangue oxides such as magnesia and lime form low-melting chlorides which cause sticking of the charge when such slags are chlorinated.

Slags made from this low-grade blend contain 55 to 60% TiO_2 compared with 65 to 70% and 75 to 80% in slags made from medium-grade concentrate (MacIntyre ilmenite, 45% TiO_2) and high-grade concentrate (Quilon ilmenite, 59% TiO_2). The high-grade concentrate gives a slag with lower gangue content, but yields less iron product.

In general, slags made from low-grade concentrates are not suitable for manufacture of $TiCl_4$, since they contain too much magnesia and lime. The best type of slag for chlorination would be obtained from high-grade ilmenite. Slags from all three types of concentrate are digestible in sulfuric acid, provided impurities are low, and over-reduction of titanium has not occurred.

In processes involving reduction of the iron in ilmenite to the metallic state, followed by acid leaching, the reduction may be carried out with solid carbon or with hydrogen, and the leaching may be done with either sulfuric or hydrochloric acid to yield concentrates containing 90 to 95% TiO_2 (Michal and Nilsen, 1969; Nilsen, 1972; Volk, 1968). The acid used for leaching can be more dilute and leaching is faster than when reduction of iron is only to the ferrous state. When hydrogen is used for reduction, it is regenerated during acid leaching of the metallic iron, and can be recycled.

Chlorine beneficiation processes other than those discussed previously have been developed by other investigators (Daubenspeck and Toomey, 1956; Hughes and Arkless, 1965; Love et al., 1960). For example, Daubenspeck and Toomey (1956) patented a process in which the iron in ilmenite is selectively chlorinated with $TiCl_4$ under oxidizing conditions, comparable to the oxidation-chlorination part of the Halomet process.

The US Bureau of Mines has developed some novel processes for beneficiating ilmenite. One of these (Elger, Kirby, and Rhoads, 1976) is based on oxidation and fluxing of titania slag produced by electric furnacing of ilmenite. Phosphorus pentoxide flux additions enhance conversion of the titaniferous phase to rutile, forming rutile crystals of 70 to 150 μm in size in slags heated to temperatures of 800° to 1550°C. The synthetic rutile crystals are liberated from ground slag by treating with phosphoric acid to dissolve the glass matrix, followed by tabling to recover concentrates analyzing as high as 96.8% TiO_2. Another USBM process involves metallization of ilmenite and removal of the iron by reaction with CO at 110° to 130°C and 1000 to 1400 psig for 2 hr to form iron pentacarbonyl, which is removed and decomposed by heating to yield iron metal powder and CO which is recycled. The TiO_2 product is porous, has rutile structure, and is suitable for chlorination (Visnapuu et al., 1973).

The US Bureau of Mines also developed a process in which titanium carbide was recovered from domestic perovskite and ilmenite concentrates, and was chlorinated in fluid bed tests with good results (Elger, Hunter, and Mauser, 1980). The Bureau has recently been investigating treatment of titanium slags with mixtures of sulfur dioxide and oxygen, followed by leaching to remove calcium, magnesium, and manganese impurities, to render such slags usable as rutile substitutes in fluid bed chlorinators (Elger et al., 1981).

Government Control, Tariffs, Depletion Allowance

Sand titanium mineral deposits occur on or near ocean beaches for the most part. Government control of mining is in a sense exercised through establishment of seashore parks and wildlife agencies in some areas. In 1980 about 50% of the economic ilmenite and rutile resources on the east coast of Australia were unavailable for mining because of environmental considerations, which had led to withdrawal of export licenses for Fraser Island concentrates at the end of 1976.

In the United States the regulations imposed by the Environmental Protection Agency could have direct bearing on the economic viability of sand mining. Government regulations are beoming more strict especially regarding the waste disposal of effluents from pigment-producing processes and other pollution-producing operations. There are also land-use conflicts where black sand deposits exist, principally along the Atlantic littoral. In addition the high cost of land in ocean shore areas has been a major drawback to mining in coastal Georgia and Florida.

The US government strategic stockpile goal for titanium sponge metal was increased 48% in May 1980 to 195,000 st, while the goal for rutile was lowered 39% to 106,000 st. Government stockpile inventories in December 1982 were 32,331 st of titanium sponge metal, including 10,866 st of nonstockpile material, and 39,186 st of rutile.

The Australian government has involved itself with prices of titanium minerals for export under long-term purchase contracts in an effort to insure that as much value as possible will accrue to that country. There is a growing trend among governments of exporting nations to seek revenues to oblige consumers to undertake at least the first stages of pigment or metal manufacture in the country where the raw material originates, and this trend can be expected to continue.

Tariffs

US import duties on titanium concentrates, metal, pigment, and other compounds are given

TABLE 22—US Import Duties

Tariff Item	Number	Most Favored Nation (MFN) Jan. 1, 1983	Most Favored Nation (MFN) Jan. 1, 1987	Non-MFN Jan. 1, 1980
Ilmenite	601.5120	Free	Free	Free
Titanium slag	603.6200	Free	Free	Free
Rutile	601.5140	Free	Free	Free
Synthetic rutile	603.7010	6.3% ad val	5.0% ad val	30% ad val
Waste and scrap metal	629.1200	12.6% ad val	7.2% ad val	25% ad val
Unwrought metal, sponge	629.1420	17% ad val	15% ad val	25% ad val
Other unwrought metal	629.1460	17% ad val	15% ad val	25% ad val
Wrought metal	629.2000	17% ad val	15% ad val	45% ad val
Titanium dioxide pigments	473.7000	6.8% ad val	6.0% ad val	30% ad val
Titanium compounds except pigment grade	422.3000	6.3% ad val	4.9% ad val	30% ad val

in Table 22. The duties shown reflect results of the Tokyo round of negotiations completed in 1979, under which tariffs on many items were scheduled to be reduced in several stages over the period January 1, 1980 to January 1, 1987.

Depletion Allowance

Titanium ores carry a 22% domestic depletion allowance and a 14% foreign allowance.

Transportation

Shipment of ilmenite and altered ilmenite is in bulk, in lots of 15,000 or 20,000 tons if ocean transport is involved. Rutile may move in bulk shipments of 5,000 tons or more, but many shipments, especially those for spot market sales, involve bagged rutile. Shipping port facilities should include covered storage for finished concentrates and channels and berths for vessels of at least 15,000 and preferably about 30,000 tons.

Recycling and Ecology

Mining

Sand mineral mining presents few serious ecological hazards. The land is restored and can even be improved in low, flat areas. Dune areas can be left in nearly their original form and the sand can be stabilized by seeding and replanting. Nothing of an unnatural chemical character is added to the quartz sand tailings.

Disturbance and aeration of material during mining can cause bacterial activity in clay, and the bacteria can contribute to ground-water acidity, but careful control of pH in settling ponds can eliminate any hazard. In New South Wales, Australia, rutile and zircon are mined from sand in a government-controlled freshwater reserve.

Clay slimes may present some problem, but careful monitoring of impoundments to avoid breaching by erosion and of overflow water to ensure its clarity and nonacid nature has proved successful. The effluent water stream from a Florida operation enters a creek, and fish reportedly have found the area particularly to their liking, to the delight of local anglers.

Sand mineral mining has some advantage because no open excavation remains after mining has passed through, and the land can be returned to its original form and use at reasonable cost. Companies and private owners with large holdings have begun to appreciate that their properties may have both mineral and land value, and that two profits might be made: the minerals first and the land afterward.

Sulfate Process

The sulfate process results in relatively large quantities of spent acid and iron sulfate [up to about 2 kg (4 lb) of waste per 0.5 kg (lb) of titanium dioxide]. Some of the iron sulfate is crystallized out as copperas and is utilized for manufacturing iron oxide pigments and in water treatment and sewerage. The remainder generally has been disposed of, along with the spent sulfuric acid solution obtained on hydrolyzing the titanium, in waterways or the ocean. Even though an independent study (Anon., 1973b) found no ill effects from ocean disposal of such wastes, federal and state environmental regulations require the phasing out of all ocean dumping by 1990. Although the practice of disposing of the spent acid and iron sulfate results in a loss of sulfur and iron, to date all methods to recover these materials have proven uneconomic.

The sulfate process also has a considerable potential for causing air pollution during the vigorous digestion reaction of sulfuric acid with ground ilmenite or slag. To minimize such pollution, some producers use gas scrubbers on digestion tank stacks. NL Industries developed a modified digestion process which it called liquid phase digestion, and applied it to its Sayreville, NJ pigment plant in early 1982.

There are two general methods which are favored for recycling or neutralizing the acid in pigment waste liquor:

1) Concentration of H_2SO_4 by evaporating water, and roasting of the ferrous sulfate precipitated during evaporation, along with excess copperas, forming SO_2 for manufacture of new acid, and iron oxide for use as pigment, or as raw material for iron powder or steel manufacture.

2) Neutralization of H_2SO_4, and reaction of $FeSO_4$, with lime or calcium carbonate, converting the acid and sulfates to gypsum, and the iron to iron oxide.

The first method minimizes the amount of material to be finally sold or disposed of (only iron oxide) but requires large amounts of energy for evaporation and roasting and poses some problems of technology. The second method, neutralization, requires less energy, but uses large amounts of limestone and yields a gypsum byproduct (and CO_2 if limestone is used) for which markets must be established, or disposal areas provided.

Since 1975 American Cyanamid Co. has been operating a waste treatment facility at its Savannah, GA pigment plant, using limestone and lime to neutralize spent sulfuric acid-ferrous sulfate solution, forming gypsum and iron orides. Glidden Pigments, SCM Corp., has been operating a similar plant at its Baltimore sulfate process plant since 1977. In 1979 these pigment producers reported arrangements for the sale of byproduct gypsum. Gypsum from American Cyanamid's plant was to be converted by an adjacent Lemco, Inc. plant into briquettes for use by the cement industry. SCM was selling its byproduct gypsum to a local wallboard manufacturer. In 1982, G&W Industries was preparing to build waste neutralization facilities at its Gloucester City, NJ sulfate process plant. It was estimated in 1977 that the use of such neutralization processes would add about $90 per st to pigment cost, and that sale of all of the byproduct gypsum would recover only 25% of this added cost.

Chloride Process

The chloride process produces only about 0.5 kg (1.2 lb) of waste for each pound of titanium dioxide produced. The waste material is predominently ferric chloride which is usually dumped into the ocean. Du Pont has developed a process for recovering ferric chloride from the waste for use in the treatment of waste and drinking water. Tests are also being conducted on a process to recover the chlorine for reuse and to produce an iron oxide for the steel mills. However, large markets have to be developed for the waste products. In addition, even if markets were developed there is still a disposal problem for the miscellaneous metal chlorides that make up about 25% of the waste.

Bibliography and References

Anon., 1959, "The Mineral Resources of the Union of South Africa," *Handbook, Geological Survey*, 4th ed., US Geological Survey, pp. 312–318.

Anon., 1971, "Benilite Comes to Agreement on Upgrading Process," *Industrial Minerals,* No. 49, Oct., p. 28.

Anon., 1971a, "Chlorination Process Upgrades Low-Grade Ores," *Chemical and Engineering News,* Nov. 29, pp. 31–32.

Anon., 1971b, "Ishihara Takes the Plunge—40,000 Tons a Year of Synthetic Rutile," *Industrial Minerals,* No. 50, Nov., pp. 29–32.

Anon., 1971c, "Titanium Minerals 1. The Producers Reviewed," *Industrial Minerals,* No. 43, Apr., pp. 9–23.

Anon., 1972, "Brazilian Titanium," *Mining Journal,* Vol. 278, No. 7121, Feb. 11, pp. 118–119.

Anon., 1972a, "Mitsubishi Process for Upgrading Ilmenite," in *Processes for Rutile Substitutes, National Materials Advisory Board NMAB No. 293,* Mitsubishi Metal Corp., PB 212 898, June, 193 pp.

Anon., 1973, "Dhrangadhra Set to Expand Beneficiated Ilmenite Capacity," *Industrial Minerals,* No. 71, Aug., p. 35.

Anon., 1973a, "Effluent-Free Ilmenite Upgrading Plant," *Mining Journal,* Oct. 26, pp. 343–345.

Anon., 1973b, "Heavy Going Ahead for Waste Discharging at Sea," *Chemical Week,* June 27, pp. 45–47.

Anon., 1973c, "Norwegian Ilmenite: A/S TITANIA's Major Expansion at Tellnes," *Industrial Minerals,* No. 72, Sep., pp. 24–28.

Anon., 1973d, "Plan New Murphyores Plant," *American Metal Market,* Aug. 2.

Anon., 1973e, "30,000 Ton/Year Ilmenite Upgrading Plant," *Mining Journal,* June 8, p. 470.

Anon., 1974, "Pulmoddai's Mineral Sands," *Industrial Minerals,* No. 77, Feb., p. 27.

Anon., 1974a, "U.S. TiO_2 Mine on Stream," *Mining Magazine,* Vol. 130, No. 1, Jan., p. 7.

Anon., 1977, "RBM Progress Report," Sep., Richards Bay Minerals, 4 pp.

Anon., 1978a, "Titania: The Largest Producer of Titanium Minerals in Europe," *Mining Magazine,* Vol. 139, No. 4, Oct., pp. 365–371.

Anon., 1978b, "Rautauruukki—A Major Force in World Vanadium Supplies Is Still Expanding," *World Mining,* Mar., pp. 44–46.

Anon., 1978c, "The Hockin Process," *Industrial Minerals,* No. 93, Sep., pp. 55–57.

Anon., 1980a, "Titanio, Anuário Mineral Brasileiro," Brasilia, Vol. IX, p. 358.

Anon., 1980b, "Australia's Mineral Resources: Mineral Sands," Australian Department of Trade and Resources, 10 pp.

Anon., 1980c, "Australian Mineral Sands Processing Industry—Potential for Expansion," Commonwealth/State Joint Study Group on Raw Materials Processing, Australian Government Publishing Service, Canberra, pp. 17–18.

Anon., 1980d, "South Africa—Mining at Richards Bay," *Mining Journal,* Vol. 295, No. 7579, Nov. 21, pp. 411–413.

Anon., 1981, "Sierra Rutile," *Mining Magazine,* Vol. 144, No. 6, June, pp. 458–465.

Anon., 1983, "Titanium: Past, Present, and Future," National Materials Advisory Board, Commission on Eng. and Tech. Systems, Nat. Res. Council, Publication NMAB-392, National Academy Press, Washington, DC, 226 pp.

Abreu, S.F., 1936, "Rutilo no Brasil," Pub. 19, Instituto Nacional de Technologia, 32 pp.

Abreu, S.F., 1962, *Recursos Minerais do Brasil,* Vol. 1, Instituto Nacional de Technologia, pp. 566–579.

Anderson, W.W., and Rowe, L.W., 1956, "Method for Preparing Chlorination Feed Material," US Patent 2,770,529, Nov. 13.

Armant, D.L., and Cole, S.S., 1949, "Laboratory Smelting of Titaniferous Ores," *Trans. AIME,* Vol. 185, pp. 909–913; *Journal of Metals,* Dec.

Armant, D.L., and Sigurdson, H., 1954, "Preparation of Titanium Slag Composition," US Patent 2,680,681, June 8.

Armant, D.L., and Sigurdson, H., 1956, "Method for Producing Titanium Concentrates," US Patent 2,751,307, June 19.

Bachman, F.E., 1914, "The Use of Titaniferous Ores in the Blast Furnace," *Iron and Steel Industry Yearbook,* pp. 370–419.

Balsley, J.R., Jr., 1943, "Vanadium-Bearing Magnetite-Ilmenite Deposits Near Lake Sanford, Essex County, New York," Bulletin 940-D, US Geological Survey, pp. 99–123.

Banning, L.H., et al., 1955, "Electric Furnace Smelting of Ilmenite Concentrates from Valley County, Idaho," Report of Investigations 5170, US Bureau of Mines, 18 pp.

Barksdale, J., 1966, *Titanium, Its Occurrence, Chemistry, and Technology,* 2nd ed., Ronald Press, New York, 691 pp.

Bateman, A.M., et al., 1951, "Formation of Late Magmatic Oxide Ores," *Economic Geology,* Vol. 46, No. 4, June-July, pp. 404–426.

Beals, M.D., and Merker, L., 1960, "Three New Single Crystal Materials," *Materials in Design Engineering,* Jan., pp. 12–13.

Bishop, E.W., 1956, "Geology and Ground-Water Resources of Highlands County, Florida," Report of Investigation 15, Florida Geological Survey, 115 pp.

Bracanin, B.F., et al., 1972, "The Development of a Direct Reduction and Leach Process for Ilmenite Upgrading," *Light Metals,* W. C. Rotsell, ed., AIME, New York, pp. 209–259.

Brady, E.S., 1981, "China's Strategic Minerals and Metals—Titanium," *The China Business Review,* Vol. 8, No. 5, Sep.-Oct., pp. 62–65.

Broadhurst, S.D., 1955, "The Mining Industry in North Carolina from 1946 through 1953," Economic Paper No. 66, North Carolina Dept. of Conservation and Development, Div. of Min. Resources, pp. 26–27.

Broderick, T.M., 1917, "The Relation of the Titaniferous Magnetites of Northeastern Minnesota to the Duluth Gabbro," *Economic Geology,* Vol. 12, No. 8, Dec., pp. 663–696.

Brooks, H.K., 1966, "Geological History of the Suwanee River," *Geology of the Miocene and Pliocene Series in the North Florida-South Georgia Area,* N.K. Olson, ed., Atlantic Coastal Plain Geological Assn., 7th Field Trip, Southeastern Geological Society, 12th Field Trip, Guidebook, pp. 37–45.

Brun, R.M., 1957, "The Tellnes Story," *Ilmenitten* TITANIA, A/S, Norway, Summer issue.

Buddington, A.F., 1939, "Adirondack Igneous Rocks and Their Metamorphism," *Geological Society of America Memoir* 7, pp. 19–48.

Buddington, A.F., et al., 1955, "Thermometric and Petrogenetic Significance of Titaniferous Magnetite," *American Journal of Science,* Vol. 253, pp. 497–532.

Chen, H., 1971, "Hydrogen Chloride Leaching of Ilmenite," German Offen. 2,004,878, May 6.

Carstens, H., 1957, "Investigations of Titaniferous Iron Ore Deposits, Part I Gabbros and Associated Titaniferous Iron Ore in West-Norwegian Gneisses," *K. Norske Vidensk. Selsk. Skr.,* No. 3, 67 pp.

Cobel, G.B., et al., 1980, "Electrowinning of Titanium from Titanium Tetrachloride: Pilot Plant Experience and Plant Projections," *Titanium '80, Science and Technology, Proceedings,* 4th International Conference on Titanium, Kyoto, Japan, May 19–22, H. Kimura and O. Izumi, eds., Vol. 3, AIME, New York, pp. 1969–1976.

Cole, S.S., 1953, "Titanium Concentrates," US Patent 2, 631,941, Mar. 17.

Cooke, C.W., 1941, "Two Shore Lines or Seven?," *American Journal of Science,* Vol. 239, No. 6, pp. 457–458.

Cooke, C.W., 1945, "Geology of Florida," Bulletin 29, Florida Geological Survey, 339 pp.

Coope, B., 1982, "Titanium Dioxide Pigments," *Industrial Minerals,* No. 181, Oct., pp. 47–51.

Daubenspeck, J.M., and Toomey, R.D., 1956, "Separation of Iron from Titaniferous Iron Ores," US Patent 2,758,019, Aug. 7.

Davidson, D.M., et al., 1946, "Notes on the Ilmenite Deposit at Piney River, Virginia," *Economic Geology,* Vol. 41, No. 7, Nov., pp. 738–748.

Diemer, R.A., 1941, "Titaniferous Magnetite Deposits of the Laramie Range, Wyoming," Bulletin No. 31, Geological Survey of Wyoming, 23 pp.

Dunn, W.E., Jr., 1972, "High Temperature Chlorination of TiO_2 Bearing Minerals II," AIME Annual Meeting, San Francisco, Feb.

Dunn, W.E., Jr., 1973, "Process for Recycle Beneficiation of Titaniferous Ores," Canadian Patent No. 932,964, Sep. 4.

Dybdahl, I., 1960, "Ilmenite Deposits of the Egersund Anorthosite Complex," *XXI International Geological Congress,* Copenhagen, Guidebook to Excursion No. C-10, pp. 49–53.

Elger, G.W., et al., 1976, "Producing Synthetic Rutile from Ilmenite by Pyrometallurgy," Report of Investigations 8140, US Bureau of Mines, 31 pp.

Elger, G.W., et al., 1980, "Preparation and Chlorination of Titanium Carbide from Domestic Titanium Ores," Report of Investigations 8497, US Bureau of Mines, 20 pp.

Elger, G.W., et al., 1982, "Utilization of Domestic Low-Grade Titaniferous Materials for Producing Titanium Tetrachloride," *Light Metals,* The Metallurgical Society of AIME, pp. 1135–1147.

Elliot, R.A., 1959, "Beneficiation of Titanium Ores with Particular Reference to Canadian Ores," *CIM Transactions,* Vol. 62, pp. 90–95; *CIM Bulletin,* Mar., pp. 186–191.

Evrard, P., 1949, "Differentiation of Titaniferous Magmas," *Economic Geology,* Vol. 44, No. 3, May, pp. 210–232.

Fine, M.M., and Frommer, D.W., 1952, "Mineral Dressing Investigation of Titanium Ore from the Christy Property, Hot Spring County, Arkansas," Report of Investigations 4851, US Bureau of Mines, 7 pp.

Fine, M.M., et al., 1949, "Titanium Investigations . . . The Laboratory Development of Mineral Dressing Methods for Arkansas Rutile," *Mining Engineering,* Vol. 1, No. 12, pp. 447–452.

Fish, G.E., Jr., 1962, "Titanium Resources of Nelson and Amherst Counties, Virginia (In Two Parts) 1. Saprolite Ores," Report of Investigations 6094, US Bureau of Mines, 44 pp.

Fish, G.E., Jr., and Swanson, V.F., 1964, "Titanium Resources of Nelson and Amherst Counties, Virginia (In Two Parts) 2. Nelsonite," Report of Investigations 6429, US Bureau of Mines, 25 pp.

Flint, R.F., 1940, "Pleistocene Features of the Atlantic Coastal Plain," *American Journal of Science,* Vol. 238, No. 11, pp. 757–787.

Flint, R.F., 1942, "Atlantic Coastal 'Terraces'," *Washington Academy of Sciences Journal,* Vol. 32, No. 8, pp. 235–237.

Flint, R.F., 1947, *Glacial Geology and the Pleistocene Epoch,* John Wiley, New York, 589 pp.

Force, E.R., 1980, "Is the United States Geologically Dependent on Imported Rutile?" Presented at 4th Industrial Minerals International Congress, Atlanta, GA, 4 pp.

Force, E.R., et al., 1976, "Geology and Resources of Titanium," Professional Paper 959–A through F, US Geological Survey.

Frey, E., 1946, "Exploration of Iron Mountain Titaniferous Magnetite Deposits, Albany County, Wyoming," Report of Investigations 3968, US Bureau of Mines, 37 pp.

Fryklund, V.C., Jr., and Holbrook, D.F., 1950, "Titanium Ore Deposits of Hot Spring County, Arkansas," Bulletin No. 16, Arkansas Research and Development Comm., Arkansas Div. Geology, 173 pp.

Fryklund, V.C., Jr., et al., 1954, "Niobium and Titanium at Magnet Cove and Potash Sulphur Springs, Arkansas," Bulletin 1015-B, US Geological Survey, pp. 23–57.

Garnar, T.E., Jr., 1978, "Geological Classification and Evaluation of Heavy Mineral Deposits," 12th Forum on the Geology of Industrial Minerals, Information Circular No. 49, Georgia Geological Survey, Atlanta.

Garnar, T.E., Jr., 1980, "Heavy Minerals Industry of North America," Presented at 4th Industrial Minerals International Congress, Atlanta, GA, 13 pp.

Geis, H.P., 1971, "A Short Description of the Iron-Titanium Provinces of Norway, with Special Reference to Those in Production," *Minerals Science Engineering,* Vol. 3, No. 3, pp. 13–24.

Gillson, J.L., 1932, "Genesis of the Ilmenite Deposits of St. Urbain, County Charlevoix, Quebec," *Economic Geology,* Vol. 27, No. 6, Sep. pp. 554–577.

Gillson, J.L., 1949, "Titanium," *Industrial Minerals and Rocks,* 2nd ed., S.H. Dolbear, ed., AIME, New York, pp. 1042–1069.

Gillson, J.L., 1950, "Deposits of Heavy Minerals on the Brazilian Coast," *Trans. AIME,* Vol. 187, pp. 685–693.

Gillson, J.L., 1956, "Genesis of Titaniferous C and Their Origin," *Trans. AIME,* Vol. 205, 296–301; *Mining Engineering,* Vol. 8, Mar. I

Gillson, J.L., 1959, "Sand Deposits of Titanium Minerals," *Trans. SME-AIME,* Vol. 214, pp. 421–429; *Mining Engineering,* Vol. 11, No. 4.

Grogan, R.M., et al., 1964, "Milling at Du Pont's Heavy Mineral Mines in Florida," *Milling Methods in the Americas,* N. Arbiter, ed., Gordon and Breach, New York, pp. 205–229.

Gross, S.O., 1968, "Titaniferous Ores of the Sanford Lake District, New York," *Ore Deposits in the United States, 1963/1967,* John D. Ridge, ed., AIME, New York, Vol. 1, pp. 140–153.

Guimond, R., 1964, "Quebec Iron and Titanium Corporation, A Study in Growth," *Canadian Mining Journal,* Vol. 85, No. 11, pp. 47–53.

Guise. F.P., et al., 1964, "Titanium in the Southeastern United States," Information Circular 8223, US Bureau of Mines, 30 pp.

Hammond, P., 1949, "Allard Lake Ilmenite Deposits," *Canadian Mining & Metallurgical Bulletin,* Vol. 42, pp. 117–121.

Hammond, P., 1952, "Allard Lake Ilmenite Deposits," *Economic Geology,* Vol. 47, No. 6, Sep.–Oct., pp. 634–649.

Hargraves, R.B., 1959, "Petrology of the Allard Lake Anorthosite Suite and Paleomagnetism of the Ilmenite Deposits (Quebec)," Ph.D. Thesis, Princeton University, Princeton, NJ, May, 193 pp.

Harki, I., et al., 1956, "Discovery and Mining Methods at Finland's Largest Fe-Ti-V Mine," *Mining World,* Vol. 18, Aug., p. 62.

Hatch, G.G., and Cuke, N.H., 1956, "Iron Operations of the Quebec Iron and Titanium Corporation," *CIM Transactions* Vol. 59; *CIM Bulletin 533,* pp. 619–622.

Henn, J.J., and Barclay, J.A., 1970, "A Review of Proposed Processes for Making Rutile Substitutes," Information Circular 8450, US Bureau of Mines, 27 pp.

Heyburn, M.M., 1960, "Geological and Geophysical Investigation of the Sanford Hill Ore Body Extension, Tahawus, New York," Unpublished M.S. Thesis, Syracuse University, Syracuse, NY, 48 pp.

Hillhouse, D.M., 1960, "Geology of the Piney River-Roseland Titanium Area, Nelson and Amherst Counties, Virginia," Unpublished Ph.D. Thesis, Virginia Polytechnic Institute, Blacksburg, VA, 169 pp.

Hoyt, J.H., 1967, "Pleistocene Shore Lines: Guide to Tectonic Movements, Northern Florida and Southern Georgia," *Abstracts, 1967 Annual Meeting,* Geological Society of America, New Orleans, LA, p. 104.

Hubaux, A., 1956, "Various Types of Black Ores of the Egersund Norway Region," Bulletin 79, Ann. Soc. Geol. Belg., pp. 203–215.

Hughes, W., and Arkless, K., 1965, "Titanium Ore Beneficiation Process," British Patent 992,317, May 19.

Iammartino, N.R., 1976, "Beneficiated-Ilmenite Process Recycles HCl Leach Liquor," *Chemical Engineering,* Vol. 83, No. 11, May 24, pp. 100–102.

Jennings, E.P., 1913, "A Titaniferous Iron Ore Deposit in Boulder County, Colorado," *AIME Trans.,* Vol. 44, pp. 14–25.

Kataoka, S., and Yamada, S., 1973, "Acid Leaching Upgrades Ilmenite to Synthetic Rutile," *Chemical Engineering,* Mar. 19, pp. 92–93.

Kays, M.A., 1965, "Petrographic and Modal Relations, Sanford Hill Titaniferous Magnetite Deposit," *Economic Geology,* Vol. 60, No. 6, Sep.-Oct., pp. 1261–1297.

Kershaw, K.W., 1971, "A New Class of Pigments," *Australian OCAA Proceedings and News,* Aug., pp. 4–9.

Kish, L., 1972, "Vanadium in the Titaniferous Deposits of Quebec," *CIM Bulletin,* Mar., pp. 117–123.

Kroll, W.J., 1940, "Method for Manufacturing Titanium and Alloys Thereof," US Patent 2,205,854, June 25.

ddy, J.J., and Schechter, D.L., 1962, "Pressure Leaching of Titaniferous Material," US Patent 3,060,002, Oct. 23.

Leone, O.Q., et al., 1967, "High-Purity Titanium Electro-won from Titanium Tetrachloride," *Journal of Metals,* Vol. 19, No. 3, pp. 19–23.

Li, T.M., 1973, "Startup of Manchester Mine and Mill Boosts U.S. Production of Primary Ilmenite," *Engineering & Mining Journal,* Dec., pp. 71–75.

Lissiman, J.C., and Oxenford, R.J., 1973, "The Allied Mineral N.L. Heavy Mineral Deposit in Eneabba, W.A.," *Conference Volume,* Australasian Institute of Mining & Metallurgy, pp. 153–161.

Lister, F.G., 1966, "The Composition and Origin of Selected Iron-Titanium Deposits," *Economic Geology,* Vol. 61, No. 2, Mar.-Apr., pp. 275–310.

Llewellyn, T.O., and Sullivan, G.V., 1980, "Recovery of Rutile from a Porphyry Copper Tailings Sample," Report of Investigations 8462, US Bureau of Mines, 18 pp.

Lloyd, R.R., 1956, "Production of Titanium Metal at Henderson, Nevada," Rocky Mountain Minerals Conference, Salt Lake City, Utah, Sep. 28.

Love, F.E., et al., 1960, "Beneficiation of Titaniferous Iron Ores," US Patent 2,933,373, Apr. 19.

Lynd, L.E., 1960, "Titanium," *Industrial Minerals and Rocks,* 3rd ed., J. L. Gillson, ed., AIME, New York, pp. 851–880.

Lynd, L.E., 1980, "Titanium," *Mineral Facts and Problems,* Bulletin 671, US Bureau of Mines, 18 pp.

Lynd, L.E., 1983, "Titanium," *Mineral Commodity Profile,* US Bureau of Mines, 17 pp.

Lynd, L.E., and Merker, L., 1955, "Metal Titanate Composition," US Patent 2,723,916. Nov. 15.

Mackey, T.S., 1974, "Acid Leaching of Ilmenite into Synthetic Rutile," Ind. Eng. *Chem. Prod. Res. Develop.* Vol. 13, No. 1, pp. 9–18.

MacNeil, F.S., 1949, "Pleistocene Shore Lines in Florida and Georgia," *Shorter Contributions to General Geology,* Professional Paper 221-F, US Geological Survey, pp. 93–106.

McMurray L.L., 1944, "Froth Flotation of North Carolina Ilmenite," *Trans. AIME,* Vol. 173, 1947; *Mining Technology,* Jan. 1944.

Markewicz, F.J., 1969, "Ilmenite Deposits of the New Jersey Coastal Plain," *Geology of Selected Areas of New Jersey and Eastern Pennsylvania and Guidebook of Excursions,* S. Subitzky, ed., Rutgers University Press, New Brunswick, NJ, pp. 363–382.

Marshall, J. and MacMahon, D.M., 1972, "Ilmenite Dressing," German Offen. 2,200,954, Dec. 7.

Martens, J.C.H., 1928, "Beach Deposits of Ilmenite, Zircon, and Rutile in Florida," 19th Annual Report, Florida Geological Survey, pp. 124–154.

Masten, A.H., 1923, *The Story of Adirondac,* Princeton Press, Princeton, NJ, 199 pp.

Merritt, C.A., 1939, "Iron Ores of the Wichita Mountains, Oklahoma," *Economic Geology,* Vol. 34, No. 3, May, pp. 268–286.

Michal, E.J., and Nilsen, A.E., 1969, "Titanium Dioxide Concentrate and Method for Producing the Same," US Patent 3,446,590, May 27.

Michot, P., 1956, "The Deposits of Black Ores of the Egersund Region," Bulletin 79, Ann. Soc. Geol. Belg., pp. 183–201.

Minkler, W.W., 1980, "Titanium in 1980," *Journal of Metals,* Vol. 33, No. 4, Apr., pp. 41–44.

Moore, C.H., Jr., 1940, "Origin of the Nelsonite Dikes of Amherst County, Virginia," *Economic Geology,* Vol. 35, No. 5, Aug., pp. 629–645.

Moore, C.H., Jr., 1955, "Rutile Boule and Method of Making Same," US Patent 2,715,070, Aug. 9.

Moore, C.H., Jr., and Sigurdson, H., 1949, "Petrology of High Titanium Slags," *Trans. AIME,* Vol. 185, pp. 914–919; *Journal of Metals.* Dec.

Moorhouse, W.W., 1938, "Some Titaniferous Magnetites in the San Gabriel Mountains, Los Angeles County, California," *Economic Geology,* Vol. 33, No. 7, Nov., pp. 737–748.

Nicholls, G.D., 1955, "The Mineralogy of Rock Magnetism," *Advances in Physics* (Supplement to *Philosophical Magazine*), Vol. 4, p. 113.

Nilsen, A.E., 1972, "Extraction of Iron from Titaniferous Ores," US Patent 3,647,414, Mar. 7.

Osborne, F.F., 1928, "Certain Magmatic Titaniferous Ores and Their Origin," *Economic Geol-*

ogy, Pt. 1, Vol. 23, No. 7, Nov., pp. 724–761; Pt. 2, Vol. 23, No. 8, Dec., pp. 895–922.

Othmer, D.F., and Nowak, R., 1972, "Halogen Affinities—A New Ordering of Metals to Accomplish Difficult Separations," *AICHE Journal,* Vol. 18, No. 1, Jan., pp. 217–220.

Parker, G.G., and Cooke, C.W., 1944, "Late Cenozoic Geology of Southern Florida," Bulletin 27, Florida Geological Survey, 119 pp.

Paterson, O.D., 1977, "How Many Low Grade Ores Are Now Being Recovered by Gravity Concentration," *World Mining,* Vol. 30, No. 8, July, pp. 44–49.

Paulson, E.G., 1964, "Mineralogy and Origin of the Titaniferous Deposit at Pluma Hidalgo, Oaxaca, Mexico," *Economic Geology,* Vol. 59, No. 5, Aug., pp. 753–767.

Peirce, W.M., et al., 1949, "Titaniferous Material for Producing Titanium Dioxide," US Patent 2,476,453, July 19.

Peterson, E.C., 1966, "Titanium Resources of the United States," Information Circular 8290, US Bureau of Mines, 65 pp.

Pings, W.B., 1972, "Titanium, Pt. 1," *Colorado School of Mines Industries Bulletin,* Vol. 15, No. 4, July, 13 p.

Pings, W.B., 1972a, "Titanium, Pt. 2," *Colorado School of Mines Industries Bulletin,* Vol. 15, No. 5, Sep., 17 pp.

Pinnell, D.B., and Marsh, J.A., 1954, "Summary Geological Report on the Titaniferous Iron Ore Deposits of the Laramie Range, Albany County, Wyoming," *Mines Magazine,* Vol. 44, No. 5, p. 30.

Pirkle, E.C., and Yoho, W.H., 1970, "The Heavy Mineral Ore Body of Trail Ridge, Florida," *Economic Geology,* Vol. 65, No. 1, Jan.-Feb., pp. 17–30.

Pirkle, E.C., et al., 1974, "The Green Cove Springs and Boulougne Heavy Mineral Sand Deposits of Florida," *Economic Geology,* Vol. 69, No. 7, Nov., pp. 1129–1137.

Pirkle, F.L., 1975, "Evaluation of Possible Source Regions of Trail Ridge Sands," *Southeastern Geology,* Vol. 17, No. 2, Dec., pp. 93–114.

Priscu, J.C., 1966, "Electrolytic Cell for the Production of Titanium," US Patent 3,282,822, Nov. 1.

Ramdohr, P., 1950, *Die Erzmineralien und ihre Verwachsungen,* Akademie Verlag, Berlin, 875 pp.

Ramdohr, P., 1956, "Die Beziehungen von Fe-Ti Erzen und Magmatischen Gesteinen," Bulletin No. 173, Comm. Geol. Finlande, pp. 1–18.

Ramsdell, J.D., and Mathews, D.R., 1960, "Effect of Impurities on Mechanical Properties of Electrolytic Titanium," Report of Investigations 5701, US Bureau of Mines, 12 pp.

Reed, D.F., 1949, "Investigation of Christy Titanium Deposits, Hot Spring County, Arkansas," Report of Investigations 4592, US Bureau of Mines, 10 pp.

Reed, D.F., 1949a, "Investigation of Magnet Cove Rutile Deposits, Hot Spring County, Arkansas," Report of Investigations 4593, US Bureau of Mines, 9 pp.

Retty, J.A., 1944, "Lower Romaine River Area, Saguenay County, Quebec," Report 19, Quebec Dept. of Mines & Geology, pp. 3–29.

Roberts, J.M.C., 1971, "Ilmenite Upgrading," *Mining Magazine,* Vol. 125, No. 6, Dec., pp. 543–551.

Rose, E.R., 1969, "Geology of Titanium and Titaniferous Deposits of Canada," Economic Geology Report No. 25, Geological Survey of Canada, 177 pp.

Ross, C.S., 1941, "Occurrence and Origin of the Titanium Deposits of Nelson and Amherst Counties, Virginia," Professional Paper No. 198, US Geological Survey, 59 pp.

Ross, C.S., 1947, "Virginia Titanium Deposits," *Economic Geology,* Vol. 42, No. 2, Mar.-Apr., pp. 194–198.

Runolinna, U., 1957, "How Otanmäki Floats Ilmenite in Finland," *Mining World,* Vol. 17, Apr., pp. 49–55.

Schlecten, A.W., 1972, "Processes for Rutile Substitutes," National Materials Advisory Board, Publication No. NMAB-293, National Academy of Sciences, June.

Shiah, C.D., 1966, "Process for Producing Titanium Dioxide Concentrate and Other Useful Products from Ilmenite and Similar Ores," US Patent 3,252,787, May 24.

Sigurdson, H., and Cole, S.S., 1949, "Melting Points in the System TiO_2-CaO-MgO-Al_2O_3," *Trans. AIME,* Vol. 185, pp. 905–908; *Journal of Metals,* Vol. 1, Dec.

Singewald, J.T., Jr., 1913, "Titaniferous Iron Ores of the United States, Their Composition and Economic Value," Bulletin 64, US Bureau of Mines, 145 pp.

Sinha, N.H., 1972, "Ilmenite Upgrading by the Murso Process," *Light Metals, 1972,* W. C. Rotsell, ed., AIME, New York.

Snyder, L.E., 1966, "Electrolytic Production of Titanium," US Patent 3,274,083, Sep. 20.

Stamper, J.W., 1970, "Titanium," *Mineral Facts and Problems,* Bulletin 650, US Bureau of Mines, pp. 773–794.

Stephenson, R.C., 1945, "Titaniferous Magnetite Deposits of the Lake Sanford Area, New York," Bulletin No. 340, NY State Museum, 95 pp.

Stoddard, C.K., et al., 1950, "Pilot Plant Smelting of Ilmenite in the Electric Furnace," Report of Investigations 4750, US Bureau of Mines, 15 pp.

Sun, S.S., 1971, "Fission Track Study of the Cheney Pond Titaniferous Iron Ore Deposit, Tahawus, N.Y.," Ph.D. Thesis, Washington University, St. Louis, MO, June, 134 pp.

Temple, A.K., 1966, "Alteration of Ilmenite," *Economic Geology,* Vol. 61, No. 4, June-July pp. 695–714.

Thompson, J.V., 1977, "Appraising Large Diameter Core and Percussion Drilling for Bulk Samples," *Engineering and Mining Journal,* Vol. 178, No. 8, Aug., pp. 80–82.

Toewe, E.C., et al., 1971, "Evaluation of Columbium-Bearing Rutile Deposits, Magnet Cove, Arkansas," Prepared for US Department of the Interior, Office of Minerals and Solid Fuels, Contract No. 14–01–0001–1738, Battelle Memorial Institute, Columbus, OH, Sep., 134 pp.

Vaasjoki, O., 1947, "Microstructure of Titaniferous Iron Ore at Otanmäki," Bulletin No. 40, Comm. Geologique de Finlande, pp. 107–114.

Visnapuu, A., et al., 1973, "Conversion of Ilmenite to Rutile by a Carbonyl Process," Report of Investigations 7719, US Bureau of Mines, 20 pp.

Vokes, F.M., 1968, "Forelesninger i Malmgeologi I," *Noregs Malmgeologi,* Hösten, Vol. 99, p. 488.

Volk, W., 1968, "Reduction of Metallic Oxides," US Patent 3,383,200, May 14.

Ward, J., 1972, *Australian Mineral Industry Quarterly Review,* Bureau of Mineral Resources, Geology and Geophysics, Vol. 25, No. 1, Sep., pp. 12–22.

Watson, T.L., and Taber, S., 1913, "Geology of the Titanium and Apatite Deposits of Virginia," Bulletin 3A, Virginia Geological Survey, 308 pp.; *Metallurgical and Chemical Engineering,* Vol. 13, 1915, p. 573.

Wimmler, N.M., 1946, "Titaniferous Magnetite Deposits in Montana," Report of Investigations 3981, US Bureau of Mines, 12 pp.

Yamada, S., 1976, "Ilmenite Beneficiation and Its Implications for Titanium Dioxide Manufacture," *Industrial Minerals,* No. 100, Jan., pp. 33–40.

Tripoli

JAMES C. BRADBURY *

HENRY P. EHRLINGER, III †

Tripoli is a naturally occurring, very finely divided form of silica found chiefly in some midwestern and southeastern states and used commercially as fillers and abrasives.

Definitions

Tripoli is a microcrystalline, finely particulate, more or less friable form of silica that appears to be the product of leaching of siliceous limestone or calcareous chert. The term was originally applied to a deposit near Seneca, MO, that resembled material of that name found near Tripoli in North Africa (Hovey, 1894). The two deposits were subsequently found to have different physical characteristics —the North African tripoli, now called *tripolite,* is a diatomite, composed of the siliceous skeletons of microscopic marine plants called diatoms; the Missouri tripoli contains no diatom remains. The term *tripoli,* however, continued to be used for the Missouri material and is now applied in a general way to most silica deposits similar to it in characteristics and geologic origin. In commercial trade *tripoli* is commonly understood to designate material from the Missouri-Oklahoma field; silica from southern Illinois, the other major producing area, is referred to commercially as *amorphous silica.* Each term, however, may be used commercially for material, regardless of place of origin, that has the characteristics or uses commonly ascribed to the material from one or the other major producing area. For example, in the commodity price listings, *Engineering & Mining Journal* routinely lists under *Tripoli,* prices "f.o.b. Elco, IL" and "f.o.b. Seneca, MO, and Rogers, AR," and under *Silica, amorphous,* prices "f.o.b. IL," and "f.o.b. Dierks, AR."

Amorphous silica, the commercial designation given to southern Illinois tripoli, is, like the term *tripoli,* a misnomer. Mineralogically, the southern Illinois deposits are composed of microcrystalline quartz, and no amorphous material has been detected during X-ray and scanning electron microscope investigations (Leamnson, Thomas, and Ehrlinger, 1969; Thomas et al., 1970; Keller, 1978).

Soft silica is another term that has been applied to Illinois and Tennessee tripoli, presumably to distinguish it from the more compact Missouri material. It has also been used in the ceramic trade to distinguish tripoli from "hard" silica, or silica flour, produced by pulverizing silica sand (Heinz, 1937).

Rottenstone, produced commercially in Pennsylvania, is similar to tripoli in that a major constituent is finely particulate silica, and its uses are much the same as those of tripoli— as abrasives and fillers. The origin of rottenstone also is similar to that of tripoli—decomposition of a siliceous sedimentary rock. The parent rock is defined as a siliceous limestone (Gary, McAfee, and Wolf, 1972; Thrush, 1968) or as shale. Heinz (1937) quotes a brief statement describing the Pennsylvania deposits —"a black impure shaly limestone grading into black shale"—but does not identify the source of the statement. For convenience, rottenstone is included with tripoli in the reporting of production statistics in US Bureau of Mines (USBM) *Minerals Yearbooks.*

Other materials that have been called tripoli include siliceous deposits in San Bernardino County, California (USBM *Minerals Yearbooks,* 1934–1941) and in Nevada (Chandler, 1960). These, however, differ in composition and/or origin from tripoli and, therefore, should not be identified by the name *tripoli.*

* Principal Geologist and Head, Geological Group, Illinois State Geological Survey, Champaign, IL.

† Senior Supporting Development Engineer, DRAVO Research Center, Pittsburgh, PA.

Uses

Tripoli is used as an abrasive, as a component of buffing compounds, and as an inert filler and extender in paints, plastics, caulking compounds, and rubber. The properties of tripoli from each of the several areas of production are different and unique; consequently, producers within an area are competitors, but producers in different areas service different markets.

Physical Characteristics of Tripoli

Tripoli, as broadly defined, is composed of extremely fine-grained (microcrystalline) quartz in various stages of aggregation from minute particles to porous masses. In the friable silica from southern Illinois, particle sizes are mostly between 1 and 5 μm (Lamar, 1953), but electron micrographs show that particles from 0.2 to 0.1 μm are common (Thomas et al., 1970). Single euhedral crystals of quartz up to 10 μm are also present (Thomas et al., 1970; Keller, 1978). Particle sizes in other soft silicas (Arkansas, southeastern states) may be somewhat finer or coarser but are comparable to those in southern Illinois silica (Heinz, 1937). Electron micrographs of tripoli from the Arkansas novaculite belt show particle sizes of 1 to 10 μm (Keller, 1978). In the firmer, more coherent type from the Missouri-Oklahoma field, aggregates of particles tend to be more firmly bonded, but electron micrographs show that the sizes of individual particles generally are comparable to those in the other tripolis (Keller, 1978).

Tripoli may be white or some shade of yellow, brown, or red, depending on the degree of iron oxide staining in the deposit. Missouri-Oklahoma tripoli is generally colored and is marketed as "rose" and "cream." The Illinois variety typically is white, and deposits or parts of deposits that are iron-stained are generally regarded as noncommercial.

Different varieties of tripoli have different use characteristics. The whiteness of the Illinois type makes it preferred for most filler uses. The Missouri-Oklahoma tripoli is sold chiefly for use in abrasives; its abrasiveness is probably related to its degree of aggregation and, possibly, to the manner in which the aggregates break down during processing and use. Another characteristic ascribed to the Missouri-Oklahoma product is that in buffing compounds it "sticks to the buff" better than other silicas (Heinz, 1937). However, no comprehensive studies relating physical characteristics to use characteristics of the different types of silica are known to have been made, and preference for one or the other appears to be based largely on known or presumed behavioral characteristics.

Chemical Composition

Commercial tripoli generally contains 98% to 99% silica and minor amounts of alumina (as clay) and iron oxide. Analyses of Missouri-Oklahoma tripoli, which is normally marketed as cream or rose, commonly show very close to 98% SiO_2 (Heinz, 1937; Perry, 1917). Illinois amorphous silica is marketed as white and typically contains 99% SiO_2 (Lamar, 1953). Iron oxide, which imparts the color to the Missouri-Oklahoma product, is generally less than 1%, and in the Illinois material is less than 0.1%. Alumina content varies somewhat but is generally less than 1%.

The SiO_2 content of Pennsylvania rottenstone is approximately 60% and Al_2O_3 about 18% (personal communication). However, much of the silica is combined with the alumina as clay minerals, and the proportions of clay and free (particulate) silica are probably about the same. The balance of rottenstone is chiefly iron oxide (about 10%) and calcium and magnesium carbonates (about 5%).

Geologic Occurrence and Distribution

Tripoli occurs in geologically conformable deposits in siliceous, calcareous sedimentary strata and is believed to be the result of leaching over an extended period of time of siliceous limestone or calcareous chert or both. The deposits are limited essentially to Paleozoic strata but range in age from Precambrian in the southeastern states to Pennsylvanian in Texas (Fig. 1). The most favored parts of the geologic column from the standpoint of commercial production are: (1) lower and middle Mississippian strata (Fort Payne Chert of the western Tennessee Valley, beds of approximately equivalent age in the Boone formation of the Missouri-Oklahoma district, and the upper division of the Arkansas Novaculite) and (2) strata of Lower Devonian age in southern Illinois. Locations of known domestic deposits are shown in Fig. 2.

Missouri-Oklahoma District

Although widely scattered deposits of tripoli are known to be present over an area of 1300 km² (500 sq miles) centered in Wyandotte, OK, tripoli production in the Missouri-Oklahoma district has come chiefly from mines located in

ERA	PERIOD	ROCK UNIT	LOCATION
PALEOZOIC	PENNSYLVANIAN	Marble Falls Limestone	Texas
	MISSISSIPPIAN	"Boone Formation" Fort Payne Chert Bangor Limestone Arkansas Novaculite	Missouri-Oklahoma NW Arkansas W Tennessee NE Alabama NW Georgia W Cen Arkansas
	DEVONIAN	Clear Creek Chert Grassy Knob Chert	S Illinois
	SILURIAN		
	ORDOVICIAN	Knox Dolomite (Upper)	E Tennessee NE Alabama NW Georgia
	CAMBRIAN	Rome Formation Shady Limestone	E Tennessee NW Georgia
PRECAMBRIAN		Murphy Marble	NW Georgia

ISGS 1981

FIG. 1—*Stratigraphic distribution of tripoli in the United States. Heavy lines in rock unit column indicate large or numerous deposits; light lines indicate horizons of lesser importance.*

Ottawa County, OK and adjacent Newton County, MO, all within a 16-km (10-mile) radius of Seneca, MO (Fellows, 1967). Present producers are American Tripoli Division of Carborundum Co., with a processing plant at Seneca and a mine in Ottawa County, and Midwestern Materials Co., with a mine and plant in Ottawa County.

The tripoli occurs in flat-lying deposits up to 6 m (20 ft) thick under broad, flat-topped hills. Overburden, ranging in thickness from about 1 to 3 m (3 to 10 ft), generally consists of "rotten" tripoli grading upward to a red, cherty clay at the surface (Perry, 1917). Since mining rarely, if ever, extends to the base of a deposit, the character of the strata underlying the deposits is not well known. Hovey (1894) mentions a drill hole in one of the deposits that penetrated a layer of red clay beneath the deposit, and Perry (1917) cites "reliable descriptions" in making the statement that the tripoli deposits are underlain by "several feet of red clay filled with many fragments of greatly decomposed chert" on top of solid rock. The deposits occur in the Mississippian age "Boone formation," a composite of lithologically similar units of limestone, cherty limestone, and chert. In places the Boone can be subdivided into five recognizable formations; its lower and middle parts (St. Joe Limestone, Reeds Spring Limestone, Burlington Limestone, and Keokuk Limestone formations) have been assigned to the Osage Series, and its upper part (Warsaw Limestone) is assigned to the Meramec Series (Schoff, 1955). The tripoli deposits appear to occur at stratigraphic positions both below and above the Short Creek Oolite Member of the Keokuk formation and are tentatively referred, respectively, to the Keokuk and Warsaw Limestones (Fellows, 1967).

Missouri-Oklahoma tripoli may occur as

FIG. 2—*Tripoli deposits in the United States.*

loose, unconsolidated material (Heinz, 1937), but that mined and processed commercially is a relatively firm, coherent variety. During quarrying, the tripoli is drilled and shot into chunks that are hand-sorted by color and grade, chert nodules being discarded in the process. After being stored in drying sheds for about six months, the crude tripoli is crushed, ground, sized by screening or air classification, and bagged for shipment.

Southern Illinois District

The silica deposits of the southern Illinois district occur in Alexander and Union counties in extreme southwestern Illinois. The area of principal deposits lies in northern Alexander County and adjacent southern Union County near the towns of Tamms and Elco in Alexander County and Mill Creek in Union County (Lamar, 1953). Past production has also come from areas near Wolf Creek in northern Union County and Olive Branch in southern Alexander County. Present producers are Illinois Minerals Co. and Tammsco, Inc., with processing plants at Elco and Tamms, respectively, and mines in northern Alexander County.

The major source of the silica deposits is the 90-m (300-ft) thick Lower Devonian Clear Creek Chert, which normally consists of white to mottled chert associated with smaller amounts of siliceous gray limestone (Weller, 1944). The Grassy Knob Chert, which directly underlies the Clear Creek in much of the silica-producing area (Lamar, 1953; Hood and Levine, 1973), contains scattered deposits of silica, some of which have supported commercial operations. The Olive Branch Minerals Co. produced silica for many years from deposits in the Grassy Knob. In northern Union County the Backbone Limestone intervenes between the Grassy Knob and Clear Creek Cherts.

The silica consists of friable, easily powdered silica; firm, porous silica; and nodules of chert (Lamar, 1953). The soft, pulverulent material commonly dominates. Scattered nodules of chert are normally mined and processed along with the other silica, but the parts of the deposits that contain large amounts of chert are bypassed during mining.

The silica is normally white, but iron oxide staining, resulting in yellows, browns, and reds, is common in portions of the deposits. The stained material is particularly prevalent at the outcrop, at the tops and bottoms of the deposits, and locally along fractures and more permeable beds and is generally avoided by selective mining.

The commercial deposits are generally 2.4 to 12 m (8 to 40 ft) thick. Bedding is horizontal to gently inclined but shows local waviness and distortion. Loess and Tertiary gravels generally cap the hills under which the silica is found. Since iron staining is common in the bottoms of the deposits, downward progress of mining normally ceases well above the actual base of the tripoli, and therefore little is known concerning the strata directly underlying the deposits.

Arkansas

Silica occurs chiefly in two areas of Arkansas: the northwestern corner of the state in Benton County, and west-central Arkansas in Garland, Hot Springs, Howard, Montgomery, Pike, and Polk counties (Mather, 1951). Occurrences have also been reported in Washington County in northwestern Arkansas and in Baxter County in north-central Arkansas (Branner, 1959). Production in recent years has been recorded for west-central Arkansas only and tripoli mining in Benton County has been, at best, intermittent.

The most productive deposits, those in west-central Arkansas, occur in the upper division of the Arkansas Novaculite, which is Mississippian in age. The upper division is described as "a massive, highly calcareous novaculite," which gives rise on weathering to a "white or cream-colored" soft residue of fine-grained silica (Miser, 1920). Tripoli deposits in the novaculite have been reported up to 33 m (100 ft) in thickness (Mather, 1951).

The Benton County deposits are associated with the Boone formation of Mississippian age and are believed to have been derived from siliceous limestone. Thicknesses of tripoli in Benton County are said to reach 12 m (40 ft) (Mather, 1951).

The tripoli from the novaculite belt differs from that in Benton County in character and shape of grain. Electron micrographs (Keller, 1978) illustrate the unique particle shape—polygonal triple-point—of the novaculite tripoli. The Benton County tripoli, derived from the Boone formation, may be assumed to have particle shapes more like the Missouri-Oklahoma tripoli. It is probably these differences that have led to the separate listings of past years in *Engineering & Mining Journal.*

The only consistent Arkansas producer for many years has been Malvern Minerals Co., with a plant at Hot Springs, Garland County.

West Tennessee River Valley

The most important silica-producing area of the southeastern states, from the standpoint of both past and potential production, is the region of the western Tennessee River valley in Tennessee, northwestern Alabama, and northeastern Mississippi, comprising the counties of Decatur, Dickson, Hardin, Humphreys, Lewis, Perry, and Wayne in Tennessee; Lauderdale and Colbert in Alabama; and Tishomingo in Mississippi (Clark, 1966; Spain, 1938; Whitlatch, 1937). The most productive part of the region has been a belt extending from northeastern Wayne County through Hardin County in Tennessee into Lauderdale and Colbert counties in northwestern Alabama and Tishomingo County in northeastern Mississippi. The Waynesboro-Collinwood district in Wayne County is thought to have the largest reserves of high-quality tripoli in the western Tennessee Valley Region (Spain, 1938; Whitlatch, 1937).

Most of the deposits are in the upper part of the Mississippian Fort Payne Chert, although some deposits are in the basal part of the overlying Warsaw formation. The tripoli is similar to the Illinois material in particle size and in that it is white (except where stained yellow or brown by iron oxides), largely incoherent, and easily powdered between the fingers. It is commonly referred to as the *Illinois-Tennessee type* (Spain, 1938; Whitlatch, 1937). The tripoli derived from the Warsaw formation is said to be finer and more uniform in particle size and less cherty than that in the Fort Payne (Spain, 1938).

The deposits are for the most part horizontally bedded, but in places the bedding becomes wavy or otherwise distorted by slumping related to solutional removal of carbonate minerals. The thickness of the deposits averages 7.5 m (25 ft), but one exposure in Wayne County was 18 m (60 ft) thick (Spain, 1938).

Except for a small amount produced near Florence, AL, in 1965 for the attempted production of silica brick, no production has been reported for the western Tennessee Valley Region since 1939. In previous years production had been centered chiefly around Waynesboro and Collinwood in Tennessee, Waterloo and Riverton in Alabama, and Iuka in Mississippi. A plant near Parsons in Decatur County, Tennessee, was reportedly in operation in the late 1930s (Whitlatch, 1937).

Other Areas in the Southeastern United States

Deposits in the southeastern United States are also known along the belt of folded Appalachians in northeastern Alabama, northwestern Georgia, and eastern Tennessee. These deposits occur chiefly in the Cambro-Ordovician Knox Dolomite, but some may be found in other formations both older and younger than the Knox.

Northeastern Alabama: Small scattered deposits of tripoli are known in Talladega and Calhoun counties in the northeastern part of Alabama (Pallister, 1955). These deposits occur in the Knox Dolomite and in the Mississippian Age Fort Payne Chert. No production has been recorded from northeastern Alabama. Industrial Fillers, Inc. developed a deposit near Piedmont in northeastern Calhoun County in the mid-1960s but failed to achieve significant commercial production.

Northwestern Georgia: Tripoli occurs in the northwest corner of Georgia in the counties of Bartow, Catoosa, Chatooga, Floyd, Gilmer, Murray, Polk, Walker, and Whitfield. Most of the deposits are in the upper part of the Knox Dolomite, but others have been found in the Precambrian Murphy Marble (Gilmer County), in the Cambrian Shady Limestone (Bartow County), and in the Mississippian Bangor Limestone (northern Chatooga County—deposits in southern Chatooga County are in the Knox) (Crickmay, 1937). Most of the deposits appear to have been formed by the leaching of siliceous limestone; those in the Shady Limestone evidently have been derived from material called novaculite by Crickmay (1937).

In grain size and shape, most of the Georgia material is reported to resemble the Missouri-Oklahoma tripoli more closely than it does the material from Illinois and Tennessee. However, the Georgia tripoli is generally lighter in color than the Missouri product and many deposits contain large amounts that are ivory to white (Crickmay, 1937).

Tripoli has been mined chiefly in Chatooga, Walker, and Whitfield counties (Crickmay, 1937). No production has been reported for Georgia since 1935.

Eastern Tennessee: Tripoli has been produced in the past near Cleveland, Bradley County, in southeastern Tennessee, and has been mined to a limited extent in Johnson County in the northeast corner of the state (Whitlatch, 1937). The Bradley County tripoli is described as "white to off-color" and is found in the Knox Dolomite. That in Johnson County is reported to be "bright yellow, porous, and incoherent" and appears to have been derived from siliceous limestones in the Cambrian Watauga Shale (now called the Rome formation). Whitlatch (1937) refers to the Johnson County material as *rottenstone* but gives no reasons for preferring this particular term to the name *tripoli* for the Johnson County deposits.

Texas

Tripoli has been produced from deposits northwest of Austin in Lampasas, Burnet, and San Saba counties (Dietrich and Lonsdale, 1958; Barnes et al., 1947). The tripoli is a "white, finely divided material consisting almost entirely of siliceous fragments of sponge spicules" (Evans, 1946). The deposits, given the rock term *spiculite* (Damon, 1946), occur in limestone, from which they evidently were derived by leaching and residual concentration. Their distribution is erratic, and deposits of nearly pure silica may grade abruptly into limestone (Damon, 1946). Spiculite is currently produced by Texas Architectural Aggregates, Inc., San Saba County, who market the spiculite as corkstone, a lightweight building stone used chiefly as rubble and field stone. Until recently, Southwest Portland Cement Co., Lampasas County, used spiculite as a source of silica in its cement plant at Odessa, Texas, and material from Burnet County was at one time used in special drilling muds (Dietrich and Lonsdale, 1958). Although the origin of the silica in spiculite (organic) is different from that of the silica in ordinary tripoli (largely inorganic), the mode of origin of the deposits (residual concentration from the leaching of siliceous, calcareous sedimentary strata) and their general character are the same as those of tripoli; therefore, it is logical and desirable to include spiculite under the general rock and commodity term *tripoli.*

Other Deposits Reported As Tripoli

Tripoli production was reported in California for the years 1933–1941, and deposits of tripoli are said to occur in Nevada (Chandler, 1960).

A deposit near Barstow, San Bernardino County, mined by Western Talc Co., was listed in *Minerals Yearbooks,* US Bureau of Mines, as the source of the California tripoli for the years 1934–1936. No deposits of tripoli are known in San Bernardino County or elsewhere in California. Several deposits near Barstow, ranging from siliceous marl to altered volcanic rock and referred to as mineral fillers (Bowen, 1954), were suggested as possible sources of the material reported as tripoli. Heinz (1937) stated that silica of volcanic origin was produced in Inyo and San Bernardino counties but declined to call it tripoli. Inasmuch as none of these materials appears to meet the definition of tripoli, in either composition or origin, they should not be called tripoli.

Deposits in Nevada of white, fine-grained silica, "consisting of opal and occasionally some cristobalite" and "formed by leaching of rhyolitic volcanic rock in a shallow hydrothermal environment," have been suggested as the occurrences that have been reported as tripoli. Because of the volcanic-hydrothermal origin of the material, it is not considered tripoli by the Nevada Bureau of Mines and Geology, an opinion in which the present authors concur.

Rottenstone

Rottenstone is being produced from one location in Pennsylvania. Keystone Filler and Manufacturing Co. produces from the Devonian Mahantango formation in Northumberland County. The material quarried is reported to be weathered shale and is used as abrasives and fillers.

Origin

In general, tripoli deposits are thought to be the result of a long period of leaching of siliceous limestones and calcareous cherts under surface or near-surface conditions. In a discussion of the origin of the southern Illinois deposits, Lamar (1953) concluded that the microcrystalline silica composing those deposits is the residue remaining after the removal of carbonate minerals from siliceous Devonian rocks; the character of the tripoli, soft and powdery to firm and porous, was believed related to the character of the parent rock—the powdery material having been derived from a siliceous limestone in which the individual silica particles were physically separated, and the more coherent silica from a calcareous chert in which the grains were at least partially interlocked. A more recent study (Hood and Levine, 1973) reached essentially the same conclusions. In none of the Illinois deposits that have been exposed by mining has there been observed a transition from tripoli to parent rock. Lamar (1953), however, traced in a roadcut a 152-mm (6-in.) bed, which displayed progressively greater leaching from siliceous limestone (24% SiO_2) at its origin to typical tripoli (98% SiO_2) some 107 m (350 ft) away.

Silica deposits of the southeastern states are, likewise, ascribed to leaching of siliceous limestone or calcareous chert (Jones, 1926; Miser, 1921; Pallister, 1955; Spain, 1938; Whitlatch, 1937). Crickmay (1937) states that the origin of the Georgia deposits is debatable but admits that the "most plausible" mode of origin is leaching of siliceous limestones.

In the Arkansas novaculite belt, field observations have indicated that the silica deposits were developed by leaching of the novaculite. Miser (1920) noted tripoli grading into normal novaculite. Keller (1978) found that both novaculite and the tripoli derived from it were composed of quartz crystals with a polygonal triple-point shape, whereas quartz crystals in the Illinois and Missouri-Oklahoma tripolis exhibited the typical hexagonal forms. Keller interpreted the triple-point texture as having formed from recrystallization in a confined space during metamorphism of a precursor siliceous rock to novaculite.

In the Missouri-Oklahoma district, however, there has been somewhat less agreement over the genesis of the tripoli. While it appears that the generally accepted theory of origin of tripoli (residual concentration of silica from leaching of siliceous limestone or calcareous chert) is favored for the Missouri-Oklahoma deposits (Fellows, 1967), some observers have held that these deposits are sufficiently unlike other silica deposits that they must have had a different mode of origin. According to the latter viewpoint, the Missouri-Oklahoma deposits were laid down in essentially their present form by direct precipitation of silica, with later modifications by the leaching of coprecipitated alkaline salts or by the recrystallization of originally colloidal silica (Heinz, 1937).

Exploration

Exploration for deposits is deceptively simple but requires a thorough knowledge of the local terrain and long experience in judging the quality of the material in the deposits. Outcrops are located by field exploration, commonly aided by aerial photographs, in an area known to contain tripoli-bearing strata. The common exploration drilling techniques are unsatisfactory for tripoli exploration because of the abrasiveness of the siliceous host rocks and the difficulty of recovering samples in the pulverulent, incoherent tripoli. Consequently, a preliminary evaluation of the deposit is carried out by drifting into the hillside or by trenching and test pitting, depending on accepted practice in the particular district. On the basis of experience, a judgment is generally made on the spot regarding color and texture, and either the prospect is abandoned or development work is continued.

Development

In the case of underground mines, ore bodies must be developed months or years ahead of production. The exploration drift is driven on a level near the floor of the deposit and the thickness of the deposit is determined by raising. As the drift advances, crosscuts are driven to determine the width of the deposit and to evaluate the deposit with respect to discolored or iron-stained material, excessively cherty zones, or clay inclusions, which must be left in the mines as pillars or sold as a low-value product. The development of a block of ore is labor-intensive and expensive. Continual appraisal is required to determine if the development is a potential source of high-grade ore that has sufficient tonnage to justify mining and has strength of rock to reduce the need for artificial support.

Mining

Because of differences in the physical characteristics of the deposits, the purity of the material, and depth beneath the surface, different systems of mining are used in the various tripoli districts.

In Illinois, underground mining is practiced utilizing a room-and-pillar mine plan. Insofar as possible, the off-specification parts of the mine are used as pillars in order to maximize the higher grades of material for plant feed. The mines are small, and two or three are continually in production so areas which must be bypassed can be accommodated without changing the feed rate at the plants. Each of the mines is normally run on a one-shift-per-day basis. Each mine is provided with an electrical generator and air compressor at the portal, the generator to provide light in working areas and haulageways, and the compressor to provide air for drills and ventilation. Mining is done with a jackleg drill. After drilling, the ore is blasted, picked up with a front-end loader, and transferred to a truck which delivers the ore to the crusher at the centrally located plant. As in many room-and-pillar mines, less than 50% of the ore encountered in place is delivered to the plants. No pillar recovery program has been instituted because of accessibility to additional ore and the desire to avoid subsidence problems.

The mines in Missouri and Arkansas use an open-cut method of mining. Extraordinary care must be taken to prevent contamination by earth or clay. The ore is highly absorbent and leaves the mine with 25 to 35% moisture content. It is stored in drying sheds for extended periods to reduce this moisture prior to going to the plant.

Mineral Preparation

We refer to the plant as a mineral *preparation* plant rather than a mineral *beneficiation* plant because beneficiation implies upgrading by the rejection of a lower grade or tailings product. Such is not the case in the tripoli plant. The flowsheet includes crushing, drying, and grinding. Everything delivered from the mine, with the exception of contained moisture, ends up in the final plant product.

The ore from the mine goes through a jaw crusher, a surge bin, a gyratory crusher, another surge bin, a rotary dryer, and is delivered into a fine ore bin. Ore from the fine ore bin is ground in a jasper-lined pebble mill in closed circuit with a bank of mechanical classifiers. The jasper lining and pebble media maintain the purity of the product as iron contamination would greatly reduce the marketability. The mechanical separators have the capability of cutting off any desired size fraction by proper setting, so feed rate is a function of the product being prepared at any given time.

The materials from the mechanical classifiers may be sold as several sizes of air-floated grades or become feedstock for several grades of micronized materials. These ultrafine grades are prepared in steam- or air-powered micronizing mills, in which they are ground and classified simultaneously. The finest product is sold at 100% minus 10 μm and 96% minus 5 μm. The average particle size is 1.12 μm and the

Hegman gage, a measure of the coarsest particle, is greater than 7. The ultrafine grinding is unique in that it does not involve the reduction of discrete crystalline particles to ever finer fragments. The individual particles in the feed to the micronizing mills are clusters of quartz crystallites, many of which are less than 1 μm in size, held together at their points of contact. The required size reduction is accomplished by a breaking of these cohesive bonds.

Other plants use hammermills and screens, coupled with or bypassing the technological steps indicated above.

Uses

Tripoli is defined in the sales brochure of one of the companies that processes and sells it as a "very uncommon variety of a very common mineral." Tripoli was first used as shaped products for kitchen scrubbing and water filters. Presently, virtually all of the Missouri, Arkansas, and Illinois tripoli is ground for the market. This market consists of abrasives, buffing and polishing compounds, inert mineral fillers, and more recently, surface-treated silica used in both thermoplastic and thermosetting resins for molded engineering plastics, casting compounds, adhesives, and coatings. Each of these uses will be discussed separately.

Abrasives

Tripoli at the present time is used in abrasive soaps, cleaners, powders, etc. Well-known examples are teeth-cleaning preparations and the hand soaps used in shop areas.

Buffing and Polishing Compounds

The uses of tripoli in buffing and polishing compounds are a function of the maximum and average particle sizes of the particular grade of the tripoli. In the grades normally used in buffing and polishing compounds, 99.5% of the particles are less than 10 μm in size. The average particle size is 5 to 7 μm and the Hegman grind is 4 to 7. The products are used in metal finishing. The coarser grades act as a mild abrasive, being much milder than ground silica sand, which is sold for similar purposes. The finer grades are used in buffing out hairline scratches and are used in liquid cleaner-polisher formulations, in which the thixotropic properties of the tripoli are beneficial.

Mineral Fillers

Because of the many degrees of fineness that the several companies market, the prices and applications are numerous. Probably the largest use is as an extender and filler in paint. The material required for trade sales paints (exterior latex coatings) is a low-micron grade of micronized tripoli, which yields a Hegman grind of about 6. It aids in tint retention, durability, leveling, and flowability. In baked finishes, enamels, etc., it permits higher loadings with no appreciable increase in vehicle demand and improves sheen. Ease of dispersion and film uniformity are a result of its controlled particle size. In many products its abrasiveness yields significant wear resistance. The paints in which silica filler is used are much more resistant to chemicals than those paints in which water-ground whitings and other reactive fillers are used.

A use of tripoli that has experienced rapid growth is as a filler and extender in plastics. Ground silica is used extensively in plastics for electrical use because of the excellent dielectric properties and the flexural and compressive strength it imparts. Its low oil absorption, coupled with wetting and rapid dispersion, permits high loading in most compounds. The only reported drawback in the use of tripoli as a filler in plastics comes from the degree of wear which the extruding nozzles and molds endure as a result of this use. Research by one operation has indicated that products with finer particle sizes would reduce this wear; however, they would command a higher price.

The properties which make tripoli a valuable additive to plastics are also important in the rubber industry, where it and pyrogenic or precipitated silica are used as fillers, the pyrogenic silica serving additionally as a reinforcing pigment. Tripoli is also utilized in epoxy sealants for electrical applications.

Surface-Treated Silica

Recent surface treatment of very finely sized tripoli products have broadened their uses in the fields of thermoplastic and thermosetting resins. Examples are the surface treatment with an organofunctional group such as aminos, mercaptos, epoxies, and a hydrolyzable alkoxy group. One end of the chain hydrolyzes with the surface of the silica; the organofunctional group is left attached but exposed and unreacted so that it will become reactive when added to the resin.

Comparing surface-treated and nontreated fillers at a constant loading, the treated silica would be expected to increase the mechanical strength and electrical properties and make significant improvements in chemical resistance, weatherability, and salt spray resistance.

Other improvements in favor of the use of surface-treated tripoli are a reduction in viscosity and an increase in dispersion rate in a medium.

About 30 chemicals are used commercially as adjuncts to the thermosetting resins, thermoplastic resins, and elastomers. The details are long and highly technical. It is the belief of the writers that only the concepts have been touched. The uses of organically surface-treated tripoli will be extensive and expanding in many directions before the end of the decade. None will be large-tonnage applications, but the market for special use items will continue to grow.

Miscellaneous Uses

There are numerous other uses of tripoli that involve significantly lower tonnage consumption but are important and should not be slighted. These include use in adhesives; in insecticides, both as fillers and as carriers; in dry cleaning and laundering formulations; in refractories and ceramic glazes; in foundry facings and injection thermoset moldings; in cosmetics; as wallboard filler; and as plastic wood filler. Lower grades have found uses in construction.

Substitutes

In the filler uses cited, ground calcium carbonate, ground silica sand, pyrogenic and precipitated silica, calcined kaolin, and talc are reasonable substitutes for tripoli. Materials that may substitute for tripoli as an abrasive and polishing agent include pumice, diatomite, ground feldspar, ground calcium carbonate, ground silica sand, and precipitated silica.

Prices, Production, and Forecast

Prices in 1983 ranged from $71 to $196 per ton, in direct proportion to the degree of fineness. Missouri-Oklahoma tripoli, which is used mainly as abrasives, was quoted in *Engineering & Mining Journal* in mid-1983 as follows: Once-ground, 2.9¢ per lb; double-ground, 2.9¢; and air-float, 3.15¢. At a per-ton rate, these prices are $58 and $63. Tripoli f.o.b. Elco, IL was quoted at $3.55 per lb for air-floated through 200 mesh. Prices of amorphous silica, f.o.b. Elco, IL, ranged from $71 per ton for 90 to 95% through 200 mesh to $196 per ton for the finest grade, 99% below 10 μm. Intermediate grades had prices between these limits. After being listed for a number of years, amorphous silica f.o.b. Dierks, AR was last quoted by *Engineering & Mining Journal* during 1977 at $40 per ton for 200 mesh and $50 per ton for 325 mesh, prices comparable to those listed at that time for Illinois silica.

Production statistics for the years 1970 to 1980 are shown in Table 1. Noteworthy is the almost annual increase in the use of tripoli in fillers, a trend continued from the previous decade (Bradbury and Ehrlinger, 1975). This increase may be assumed to reflect, at least in part, its increasing use in plastics, as indicated in the section on uses.

The conventional uses of tripoli are expected

TABLE 1—Processed Tripoli* Sold or Used by Producers in the US, 1970-1980

Year	Abrasives		Fillers		Other		Total	
	Short Tons	Value, †$	Short Tons	Value, †$	Short Tons	Value, †$	Short Tons	Value, †$
1970	41,703	1583	18,093	545	1134	28	60,930	2156
1971	44,899	1692	20,457	681	1327	32	66,683	2406
1972	47,321	1918	25,973	847	1584	43	74,878	2807
1973	55,420	2233	32,407	1158	2105	62	89,932	3453
1974	50,651	2251	33,361	1346	2025	66	86,000	3665
1975	38,815	1518	27,630	1205	1739	60	68,124	2783
1976	68,874	2525	40,247	1811	5000	175	114,121	4511
1977	70,631	2805	42,599	2212	2689	119	115,919	5136
1978	75,574	3709	36,505	2220	2190 e	97e	114,269	6026
1979	53,600	2468	62,409	3811	—	—	116,009	6279
1980	39,352	2253	59,909	4025	—	—	99,261	6277‡

Source: *Minerals Yearbooks,* US Bureau of Mines.
* Includes tripoli, amorphous silica, and rottenstone.
† Thousands of dollars.
‡ Data may not add to total because of independent rounding.
e Estimated.

to grow at a rate comparable to the national economy. The uses of exotically treated silicas will grow at a substantially greater rate because of new, previously unavailable feedstocks.

Environmental Considerations and Government Controls

No serious dislocations have been reported by tripoli producers with respect to land use, probably because most tripoli mines are in lightly populated areas. Safety regulations in mining must be met and are evidently being taken in stride for the most part. Stringent air quality regulations in processing plants have been difficult to meet and have caused temporary plant closings on occasion. Producers indicate that 80% of the recent increases in market prices of tripoli, particularly the ultrafine grades, are directly attributable to the increase in production costs caused by the enforced addition of new dust control measures. The silica industry has been exercising careful dust control for years and feels present air quality limits are unrealistic.

Acknowledgments

We wish to acknowledge the cooperation of the late Jay Seymour; John Norton; Joseph K. Davis; John Dillingham; Robert B. McElwaine; R.H. Dobbs; and J.E. Moreland. Information on silica deposits and the industry in Alabama has been provided by Michael W. Szabo; in Arkansas by Norman F. Williams; in California by Fenelon F. Davis; in Nevada by Keith G. Papke; in Pennsylvania by Arthur A. Socolow; in Tennessee by James L. Moore; and in Texas by Roselle Girard. The late J.E. Lamar was of valuable help both as a source of information and as a reviewer of the manuscript. Without the generous aid of these people, we would not have been able to develop a feeling for the past, present, and future of the tripoli industry in the United States.

Bibliography and References

Anon., 1981, "Abrasive Materials," *Minerals Yearbook 1980,* US Bureau of Mines, p. 54.

Bain, H.F., 1907, "Analyses of Certain Silica Deposits in Southern Illinois," *Year-Book for 1906,* Bulletin 4, Illinois State Geological Survey, pp. 185–186.

Barnes, V.E., Dawson, R.F., and Parkinson, G.A., 1947, "Building Stones of Central Texas," Publication 4246, University of Texas, p. 157.

Barrett, N.O., 1922, "Mineral Resources of Illinois in 1917 and 1918," *Year-Book for 1917 and 1918,* Bulletin 38, Illinois State Geological Survey, pp. 25–112.

Baskin, G.D., 1980, "Abrasive Materials," *Minerals Yearbook 1978–1979,* US Bureau of Mines, pp. 27–41.

Bowen, O.E.,Jr., 1954, "Geology and Mineral Deposits of the Barstow Quadrangle, San Bernardino County, California," Bulletin 165, California Div. of Mines, pp 150–160.

Bradbury, J.C., and Ehrlinger, H.P., III, 1975, "Tripoli," *Industrial Minerals and Rocks,* 4th ed., S.J. Lefond, ed., AIME, New York, p. 1217.

Branner, G.C., 1959, "Mineral Resources of Arkansas," Bulletin 6, Arkansas Geological Commission, pp. 78–79.

Chandler, H.P., 1960, "Tripoli," *Industrial Minerals and Rocks,* 3rd ed., J.L. Gillson, ed., AIME, New York, pp. 881–888.

Clarke, O.M., Jr., 1966, "Clay and Shale of Northwestern Alabama," Circular 20B, Alabama Geological Survey, 88 pp.

Clarke, R.G., 1973, "Abrasive Materials," *Minerals Yearbook 1971,* US Bureau of Mines, pp. 135–136.

Clarke, R.G., 1974, "Abrasive Materials," *Minerals Yearbook 1972,* US Bureau of Mines, pp. 125–126.

Crickmay, G.W., 1937, "Tripoli Deposits of Georgia," Information Circular 9, Georgia Div. of Geology, 7 pp.

Damon, H.G., 1946, "The Origin and Distribution of Spiculite near Lampasas, Lampasas County, Texas," *Texas Mineral Resources,* University of Texas Publications 4301, pp. 271–282.

Dietrich, J.W., and Lonsdale, J.T., 1958, "Mineral Resources of the Colorado River Industrial Development Association Area," Report of Investigations No. 37, Texas Bureau of Economic Geology, pp. 71–74.

Evans, G.L., 1946, "Mineral Abrasive and Polishing Materials in Texas," *Texas Mineral Resources,* University of Texas Publications 4301, pp. 245–248.

Fellows, L.D., 1967, "Tripoli," *Mineral and Water Resources of Missouri,* Vol. 43, 2nd Series, Missouri Div. of Geological Survey and Water Resources, pp. 220–223.

Gary, M., McAfee, R.,Jr., and Wolf, C.L., eds., 1972, *Glossary of Geology,* American Geological Institute, Washington, DC, 805 pp.

Harben, P., 1983, "Tripoli and Novaculite—The Little Known Relations," *Industrial Minerals,* No. 184, Jan., pp. 28–32.

Heinz, C.E., 1937, "Tripoli," *Industrial Minerals and Rocks,* 1st ed., S.H. Dolbear, ed., AIME, New York, pp. 911–922.

Hood, W.C., and Levine, C.R., 1973, "Tripoli Deposits of Alexander County, Illinois," *Guidebook to 37th Annual Tri-State Field Conference,* F.G. Ethridge, et al., eds., Southern Illinois University, Carbondale, pp. 128–138.

Hovey, E.O., 1894, "American Tripoli," *Scientific American Supplement,* Vol. 38, No. 969, July 28, p. 15487.

Jones, W.B., 1926, "Index to the Mineral Resources of Alabama," Bulletin 28, Alabama Geological Survey, p. 245.

Keller, W.D., 1978, "Textures of Tripoli Illustrated by Scanning Electron Micrographs," *Economic Geology,* Vol. 73, No. 3, May, pp. 442–446.

Lamar, J.E., 1953, "Siliceous Materials of Extreme Southern Illinois," Report of Investigations 166, Illinois State Geological Survey, 39 pp.

Leamnson, R.N., Thomas, J.,Jr., and Ehrlinger, H.P.,III, 1969, "A Study of the Surface Areas of Particulate Microcrystalline Silica and Silica Sand," Circular 444, Illinois State Geological Survey, 12 pp.

Mather, W.B., 1951, "Nonmetalliferous Mineral Resources of Arkansas," *Trans. AIME,* Vol. 187, p. 581.

Miser, H.D., 1920, "Geology and General Topographic Features of Arkansas," *Outlines of Arkansas Geology,* J.G. Ferguson, ed., Arkansas Bureau of Mines, Manufactures, and Agriculture, Little Rock, pp. 21–139.

Miser, H.D., 1921, "Mineral Resources of the Waynesboro Quadrangle," Bulletin 26, Tennessee Geological Survey, pp. 129–131.

Pallister, H.D., 1955, "Index to the Minerals and Rocks of Alabama," Bulletin 65, Alabama Geological Survey, p. 39.

Perry, E.S., 1917, "Tripoli Deposits of Oklahoma," Bulletin 28, Oklahoma Geological Survey, 32 pp.

Schoff, S.L., 1955, "Geologic Formations and Their Water-Bearing Character," *Groundwater Resources of Ottawa County, Oklahoma,* Bulletin 72, Oklahoma Geological Survey, pp. 48–52.

Spain, E.L., 1938, "Tripoli Deposits of the West Tennessee Valley," *Trans. AIME,* Vol. 129, pp. 501–515.

Thomas, J.Jr., et al., 1970, "Colloidal-Size Silica Produced from Southern Illinois Tripoli," Industrial Minerals Note 40, Illinois State Geological Survey, 6 pp.

Thrush, P.W., ed., 1968, *A Dictionary of Mining, Mineral, and Related Terms,* US Bureau of Mines, Washington, DC, 1269 pp.

Weller, J.M., 1944, "Devonian System in Southern Illinois," Bulletin 68, Illinois State Geological Survey, pp. 89–102.

Whitlatch, G.I., 1937, "Tripoli," Markets Circular No. 1, Tennessee Div. of Geology, 12 pp.

Vermiculite

Original by PHILIP R. STRAND *
Revised by O.F. STEWART †

Vermiculite is the name generally applied to the group of hydrated ferromagnesian aluminum silicates that are characterized by the ability to expand when heated. This process, called exfoliation, results in a lightweight product of commercial value. Most uses of vermiculite are for the expanded material. The chief markets are in construction, agriculture, and horticulture, with lesser uses in general industry.

Composition and Properties

Vermiculite, in its natural state, has the characteristic micaceous habit, a perfect basal cleavage which causes splitting into thin laminae that are soft, pliable, and inelastic. The structure of vermiculite is basically that of a talc. The prominent monoclinic crystal faces are often marked by lines at 60° and 120°. Hardness varies from 1.5 to 2 or more; specific gravities are between 2.1 and 2.8; color varies from almost clear to amber, bronze, brown, green, or black. Vermiculite feels like talc, especially when wet.

Although much research has been performed on the chemical and structural composition, there is not yet complete agreement on the exact formula. This is to be expected when different workers have examined the many different varieties. Vermiculite is not considered to be a single mineral species but a family of related minerals. The structural formula for a trioctahedral vermiculite may be written:

$$(H_2O)-(Mg,Ca,K)-(Al_2,Fe,Mg)-(Si,Al,Fe)_4O_{10}(OH)_2$$

Hydrobiotite also occurs with vermiculite and is usually considered a vermiculite for commercial uses.

When heated quickly to elevated temperatures, vermiculite expands by exfoliating at right angles to the cleavage into wormlike particles. The name vermiculite is derived from the Latin *vermiculare,* to breed worms. This characteristic of expansion is the result of the mechanical separation of the layers by the rapid conversion of contained water to steam. The decrease in bulk density of commercial grades is usually approximately 10 times, from 50 to 5 lb/ft^3, but varies depending on the quality, size, and furnace efficiency. Individual flakes may expand up to 30 times. Vermiculite may also be expanded by soaking in chemicals such as hydrogen peroxide, weak acids, and other electrolytes. Color change during expansion is dependent upon the type of vermiculite and furnace conditions. Heating in an oxidizing atmosphere produces a dull gray or tan color, whereas a reducing atmosphere can produce a bronze or gold color.

The expansion of the vermiculite crystal results in large pores being formed between the platelets. Thus, exfoliation makes available a large increase in void volume which is important in the application of vermiculite as a chemical carrier.

Mathieson and Walker (1954) state vermiculite must be regarded as a true clay mineral. The characteristic properties of the mineral, such as high cation exchange capacity, organic complexing ability, and variable interlamellar distance are very similar to those of montmorillonite. The cation exchange capacity of vermiculite is one of the highest of all the clay minerals. The interplatelet space is accessible to penetration by some electrostatically neutral molecules. Water and glycerine are two common substances whose molecules may be so imbibed.

In the natural state and under normal atmospheric conditions, water occupies the spaces between the silicate layers. The crystal *d*-spacing is near 14.2Å. By differential thermal analysis, it has been determined that the water

* Deceased, formerly Director of Advertising and Public Relations, Construction Products Div., W.R. Grace and Co., Cambridge, MA.

† Geologist, Construction Products Div., W.R. Grace and Co., Enoree, SC.

is released at three temperature ranges. The "unbound water" is released near 300°F. This water is reversible and comes to equilibrium with the environment. The second water, designated "bound water," is removed at about 500°F. This is the water necessary for exfoliation. The third water is released at approximately 1600°F and is probably hydroxyl. Little of this water is released by commercial exfoliation and when it is, a very noticeable change occurs in the color and physical characteristics of the product.

Vermiculite is closely related to biotite and phlogopite. The essential difference is that the unit cell of vermiculite contains a layer of water and the biotite contains a layer of potassium. Biotite or phlogopite are almost always associated with vermiculite in commercial deposits and are sometimes intermixed within crystals or across a single crystal face. The term *hydrobiotite* has been used for those varieties where the analysis indicates that there is a layered mixture of the two minerals in some definite proportion.

Table 1 shows chemical analyses typical of the three major vermiculite deposits.

TABLE 1—Chemical Analyses of Vermiculite, %

	Libby, Montana	Enoree, South Carolina	Palabora, South Africa
SiO_2	40.16	39.77	39.37
MgO	20.63	18.32	23.37
Al_2O_3	12.01	13.88	12.08
Fe_2O_3	13.00	12.84	5.45
FeO	—	—	1.17
K_2O	5.93	5.11	2.46
CaO	1.54	1.02	1.46
TiO_2	1.44	2.07	1.25
H_2O	5.29	6.99	11.20
Total	100.00	100.00	97.81

Origin

Detailed investigations have been made of the three largest commercial deposits of vermiculite. According to Boettcher (1966), the Rainy Creek alkaline-ultramafic complex near Libby, MT, represents a composite of successive intrusions of igneous rocks emplaced into the Precambrian metasedimentary rocks. Most of the biotite in the inner body of pyroxenite has been altered to hydrobiotite and vermiculite. From his studies, Boettcher suggests the vermiculite is a product of leaching of the biotite by ground waters, whereas the hydrobiotite may represent a higher temperature alteration product.

Stewart (1949) in his report on the vermiculite deposits at Tigerville, SC, states that hydrothermal activity was necessary in the formation of the high grade deposits only to the extent of furnishing the biotite, which was later altered to vermiculite by meteoric waters. Buie and Stewart (1954) determined the paragenetic sequence as:

pyroxene→amphibole→biotite→vermiculite
(hypogene) (supergene)

Libby (1975) suggests that the potassic ultramafic plutons of the Enoree, SC, vermiculite district probably were intruded into Late Precambrian to Cambrian sedimentary rocks and subsequently were metamorphosed. Studies indicate that both the vermiculite and hydrobiotite form under weathering conditions.

The Palabora deposit in South Africa at Loolekop is located in a carbonatite complex. Both the hydrothermal and weathering theory have been proposed for the formation of the vermiculite. There is a gradation of vermiculite to phlogopite or biotite with increasing depth.

The Palabora carbonatite complex (Dekun, 1965) is about 4 by 1½ miles in size. The dolomite core is surrounded by a thin inner ring of altered phoscorite (serpentine-apatite-magnetite) and an outer ring of diopside pyroxenite. Outward from this is a discontinuous 1½ mile-wide ring of fenitized gneisses. Numerous alvikite and orthoclasite veins have been injected into the pyroxenite ring. Serpentine, apatite, and magnetite occur in both inner zones, and a pegmatitic pyroxenite zone occupies the center of the northern ultrabasic ring. Vermiculite occurs only in the ultrabasics, especially close to serpentinized patches. Vermiculite also occurs disseminated with apatite. The principal vermiculite areas are located (1) in the north-central pegmatic pyroxenite one half mile from the limit of the carbonatite plug; (2) three quarters of a mile south of the core; and (3) in a zone near the southeastern rim of the pyroxenite.

A vermiculite deposit near Louisa, VA, occurs in a body of basic pyroxenite, about 4 to 5 miles in diameter, surrounded by gneisses and granitoid type rocks typical of the Piedmont. The vermiculite occurs in varying flake size but nearly all are less than 10 mesh. The grade of vermiculite varies considerably within a few feet both laterally and with depth. The vermiculite appears to have been altered from biotite by surface weathering. However, little, if any, geological investigation has been conducted on this deposit.

Distribution of Deposits

Vermiculite occurrences have been reported from many countries of the world. As the mineral has become more widely recognized new occurrences have been reported. Market conditions, location, size, grade, quality, and economics are some of the factors affecting the commercial value and development of a deposit.

In the United States vermiculite deposits or occurrences are found in Alabama, Arizona, Arkansas, California, Colorado, Georgia, Idaho, Kansas, Montana, Nevada, New Mexico, North Carolina, Pennsylvania, South Carolina, Texas, Virginia, and Wyoming. Production has been reported from California, Colorado, Georgia, Montana, Nevada, North Carolina, South Carolina, Texas, Virginia, and Wyoming. However, production is reported today only from mines in Montana, South Carolina, and Virginia.

Other countries where deposits have been reported are Argentina, Australia, Brazil, Canada, China, Egypt, Finland, India, Japan, Kenya, Korea, Mexico, Morocco, Pakistan, Zimbabwe, Republic of South Africa, Spain, Tanzania, Uganda, and the USSR. Minor production has been reported from Argentina, Brazil, China, Canada, Egypt, India, Kenya, Korea, Mexico, Spain, and Tanzania. Some vermiculite is produced in the USSR, but quantitative data are not available.

Prospecting and Exploration

In the United States vermiculite deposits of consequence are located in two principal areas: the Piedmont region from Alabama to Pennsylvania and in the Rocky Mountain Range from Montana southward into New Mexico, including the central mineral region of Texas. A few deposits occur in eastern Canada in the Precambrian shield.

Vermiculite deposits usually are covered by vegetation because the minerals are soft and when weathered have a considerable water-holding ability. Outcrops showing vermiculite are rare, but in some areas the mineral can be recognized in road cuts. Outcrops of associated rocks such as biotite, alkalic pyroxenites, and dikes of carbonatites or syenite, generally are more noticeable. In most instances, the only field evidence will be the presence of vermiculite flakes in the soil or in stream beds.

The larger commercial size deposits usually are found associated with ultrabasic rock, commonly pyroxenite. The pyroxenite may be intruded by numerous dikes such as pegmatites, syenites, and carbonatites. Numerous veins containing vermiculite are found in ultramafic intrusives such as dunites, peridotites, and pyroxenites, but these do not often represent commercial size deposits.

In a vermiculite-bearing formation the rock will vary from barren to a high content of vermiculite. The bulk of the material containing vermiculite will fall in the 20 to 30% content range. Drilling or trenching is necessary to prove the presence of ore in economic quantities and grade. Since the preservation of flake size is most important, auger or large rotary hole drilling is necessary with special attention to bit design. Diamond core drilling is most impractical.

The primary requirement of a commercial deposit is that the vermiculite be of acceptable quality. The concentrate must expand to a high degree without decrepitating, and the expanded particles must be strong enough to withstand handling. The biotite or phlogopite content must be low, or occur in such a form that it can be kept from mixing with the mill feed. The deposit predominantly must contain flake size material larger than 65 mesh, as there is little demand for the minus 65 mesh size.

Another characteristic of a commercial deposit is that the ore body be large enough to be mined with modern earthmoving equipment. Vermiculite production is very energy-intensive, as a high percentage of the commercial cost is for mining, reagentizing, drying, expanding, and transportation.

Deposits are considered to be high grade if their content of plus 65 mesh vermiculite is over 30%, and uneconomical or at least borderline if the content is 20% or below.

Preparation for Markets

Mining and Milling Methods

The two larger operations in the United States are described as examples of the production now in use. Fig. 1 shows the typical flowsheet from mine to shipping.

Montana: In the pyroxenite near Libby the vermiculite occurs in high grade pods and disseminated form. Mining starts on top of the mountain and progresses downslope by slicing off the top of the hill in approximately 27 ft benches. In the exposed faces of the benches,

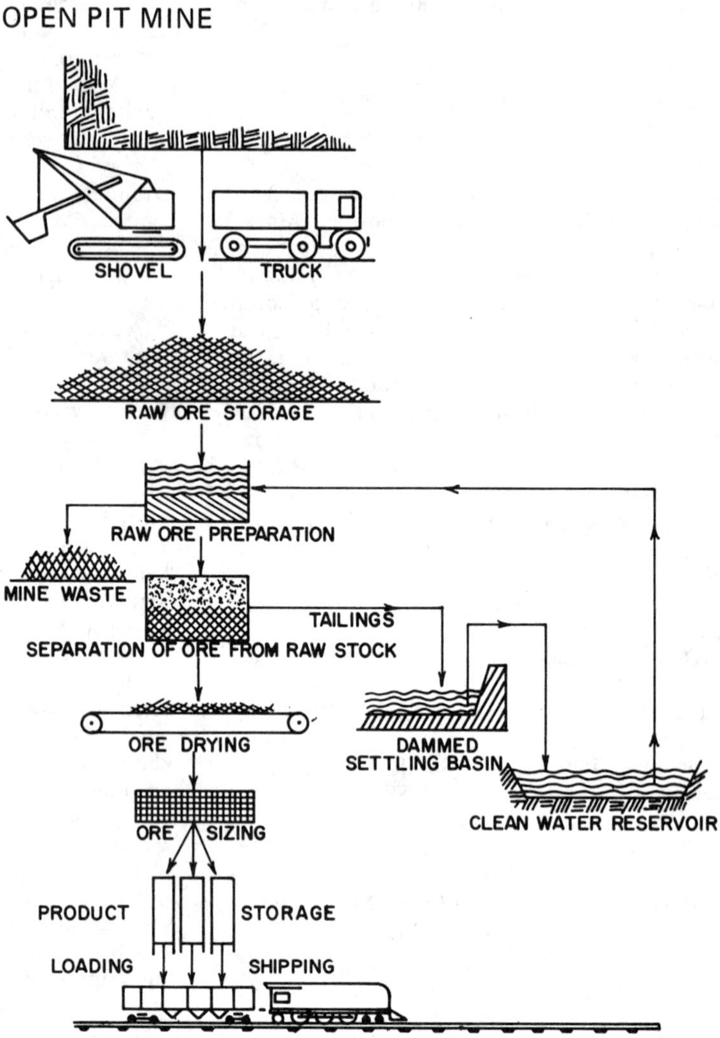

FIG. 1—*Flow diagram of vermiculite ore processing (Strand, 1975).*

vermiculite-bearing ore and waste are removed at the same time. Most of the material is loosened by rotary drilling 6-in. holes and shooting with ammonium nitrate. Large front-end loaders are used to load 85-ton off-highway trucks. Conventional mining and construction equipment, such as dozers, graders, drills, trucks, and loaders, is used. Considerable road building, using the syenite waste rock, is necessary because of the large truck size and the soft pyroxenite. The ore is hauled to a primary screening plant located between the mine and mill areas. At this point the +⅝-in. waste rock is removed and the ore is conveyed to a large storage dome.

Waste from the mine is hauled by truck to the edge of the pyroxenite and dumped downslope into the valley.

The ore is fed to the mill at a closely controlled rate where it is separated into size fractions and beneficiated. All concentrates pass through a dewatering step and then are dried. The minus 65 mesh fraction goes to a tailings pond where the material is settled and the clear water recirculated back to the process. All air or dryer exhausts are passed through bag dust collectors.

Concentrates from the mill are conveyed to a loading station and then hauled by truck to a plant near the railroad where screening separates the standard commercial sizes for storage and shipment.

South Carolina: The largest of the South Carolina operations is located near Enoree, approximately 35 miles southeast of Greenville. The ore reserves consist of a group of deposits which are mined and blended for feed to the centrally located mill. The mines are relatively small open pit type with sloped benched walls. The overburden and wall material are removed by self-loading scrapers and stacked near the pit for use in reclamation. Weathering is deep and no blasting is required. The ore is usually pushed by dozer to a point for loading by small draglines. Some blending of the ore is accomplished by the dozer as it moves across the area. Hauling is done with tandem rear-axle diesel tractor-trailer dump units. All equipment must be of the size that can travel or be hauled over public highways. Dozers and graders are used to build and maintain the mine roads. Upon completion of mining, the pit is reclaimed by pushing in the walls, returning the waste, terracing, and establishing a cover of vegetation.

The ore from the mines is dumped into piles on the rim of a pit at the head of the mill. The various grades of ore (usually four) are blended and pushed into the pit by a dozer. Two hydraulic monitors are used to break up the lumps and wash the ore into the mill as a wet slurry. The ore goes through a series of scrubbing, desliming, and beneficiating steps. The concentrate is dewatered, dried, screened into the standard sizes, and stored for later shipment. Ponds are maintained for the collection of all mill waste. The clarified water is returned to the mill process in a closed system.

Processing

Vermiculite is sold mostly in the expanded form which has been exfoliated by heating. The expanded material weighs from 4 to 8 lb/ft^3 depending on the particle size and type of furnace used. In the United States the unexfoliated vermiculite concentrate, 50 to 60 lb/ft^3, is usually shipped in 100-ton bulk rail cars from the mills to expanding plants located throughout the United States and Canada.

There are five standard sizes produced by the mills so that when the concentrate is exfoliated at the plants it will be a finished product. The standard commercial screen sizes (Tyler Standard) for the concentrates are −3 +10, −6 +14, −8 +28, −20 +65, −50 mesh. At the expanding plant the bulk hoppers usually are unloaded by conveyors and bucket elevators and stored in silos or bins. From the storage areas, the ore is fed at controlled rates into vertical furnaces fired by gas or oil. Processing flame temperature varies from approximately 2000° to 3600°F. The time-temperature relationship is critical in obtaining maximum expansion and good quality. Usually a shorter retention time and higher temperature will produce better expansion. After leaving the furnace the expanded material is passed across a separator where rock impurities are removed and the product is cooled. The product is normally packaged in 3- or 4-ft^3 paper or plastic bags and shipped to customers by truck. Both bulk and bagged material are shipped by rail. Most furnaces are equipped with bag collectors for dust control. Various products are manufactured at selected plants, using the expanded vermiculite as an ingredient.

Tests and Specifications

At the mills samples of drill cuttings, mined ore, mill feed, and concentrates are continuously tested to maintain quality control. Concentrates are checked for expansion loss (water content), impurities of nonexpandable minerals, screen size, and volume yield.

The expanded products must meet many specifications depending on use. Specifications have been established by Underwriter's Laboratories, Inc., the American Society for Testing and Materials, the Food and Drug Administration, the producers, and the customers. Numerous methods for quality control have been devised for testing the expanded vermiculite or manufactured products to ensure that these specifications are met.

Production

Although the mineral vermiculite was described in 1824 the first report of production was in 1913 in the Turrett Mining District near Salida, CO, where eight carloads were produced. In the 1920s commercial production was started at Libby, MT, and grew slowly until 1946. Production was started in South Carolina in 1946, which helped to develop the market in the eastern United States. Production in Virginia commenced in 1978. South Africa began production in the late 1940s.

Table 2 shows the trend of production since 1940, and Table 3 lists vermiculite exfoliating plants in the US in 1980.

TABLE 2—World Vermiculite Production Since 1940 (Thousand Short Tons)

Year	United States	South Africa	Others*	World
1940	22	0	0	22
1950	208	47	0	255
1960	199	69	1	269
1970	285	134	12	431
1971	301	145	13	459
1972	337	163	12	512
1973	365	172	12	549
1974	341	201	15	557
1975	330	210	28	588
1976	304	245	28	576
1977	359	182	29	570
1978	337	230	49	616
1979	346	211	50	607
1980	337	205	41	583
1981	320	210	46	576
1982[e]	310	200	40	550

Source: US Bureau of Mines.
[e] Estimate
* Exclusive of Central Economy countries, for which data are not available.

Markets and Uses

The principal markets for vermiculite are in construction, agriculture, horticulture, and general industry. It is estimated that approximately 70% of the production is used in construction and perhaps 30% in various agriculture and horticulture uses. Most of the products sold to construction are used in nonresidential building.

Construction Uses

These include loose fill used as a pour-in insulation for attics, safes, and block walls. A special water-repellent product is used as a fill in masonry walls.

Vermiculite is used as an aggregate to produce a lightweight insulating concrete roof deck which is poured in place.

Unexpanded concentrate is used extensively in the production of fire-retardant gypsum wallboard.

Special sprayed-on products, in which vermiculite is mixed with gypsum and other ingredients, protect structural steel and concrete members against fire damage.

Agricultural Uses

Manufacturers of lightweight high analysis fertilizers for the home lawn and garden market use vermiculite as the base carrier. It also is used as the carrier for other agriculture chemicals, including pesticides and weed killers. During the past five years the market has been increasing for the use of vermiculite as a growing medium. Mixtures of vermiculite, peat moss, and other ingredients make excellent artificial soil such as potting soils, soil substitutes, soil conditioners, and soils for plant propagation.

The average price per short ton of unexpanded vermiculite concentrate, f.o.b. mine, in the United States for the past five years is given by Meisinger (1981) as:

Year	*Price*
1977	$51.81
1978	58.46
1979	63.58
1980	69.73
1981	81.88
1982 (estimate)	90.00

TABLE 3—Vermiculite Exfoliating Plants in the United States in 1980

Company	City	State
Brouk Co.	St. Louis	Missouri
Cleveland Builders Supply Co.	Cleveland	Ohio
Diversified Insulation, Inc.	Minneapolis	Minnesota
J. P. Austin, Assoc., Inc.	Beaver Falls	Pennsylvania
W. R. Grace & Co.	Irondale	Alabama
	Phoenix	Arizona
	North Little Rock	Arkansas
	Newark	California
	Santa Ana	California
	Denver	Colorado
	Pompano Beach	Florida
	Jacksonville	Florida
	Tampa	Florida
	West Chicago	Illinois
	Newport	Kentucky
	New Orleans	Louisiana
	Muirkirk	Maryland
	Easthampton	Massachusetts
	Dearborn	Michigan
	Minneapolis	Minnesota
	St. Louis	Missouri
	Omaha	Nebraska
	Trenton	New Jersey
	Weedsport	New York
	High Point	North Carolina
	Oklahoma City	Oklahoma
	Portland	Oregon
	New Castle	Pennsylvania
	Enoree	South Carolina
	Travelers Rest	South Carolina
	Nashville	Tennessee
	San Antonio	Texas
	Dallas	Texas
	Milwaukee	Wisconsin
International Vermiculite	Girard	Illinois
Koos, Inc.	Kenosha	Wisconsin
Mica Pellets, Inc.	Dekalb	Illinois
Patterson Vermiculite Co.	Lanford	South Carolina
Robinson Insulation Co.	Great Falls	Montana
	Minot	North Dakota
The Schundler Co.	Metuchen	New Jersey
O. M. Scott	Marysville	Ohio
Strong-Lite Products	Pine Bluff	Arkansas
Verlite Co.	Tampa	Florida
Vermiculite of Hawaii, Inc.	Honolulu	Hawaii
Vermiculite-Intermountain, Inc.	Salt Lake City	Utah
Vermiculite Products, Inc.	Houston	Texas

Bibliography and References

Bassett, W.A., 1959, "The Origin of the Vermiculite Deposits of Libby, Montana," *American Mineralogist,* Vol. 44, No. 3–4, Mar.-Apr., pp. 282–299.

Boettcher, A.L., 1966, "The Rainy Creek Igneous Complex Near Libby, Montana," Ph.D. Thesis, Pennsylvania State University, 70 pp.

Buie, B.F., and Stewart, O.F., 1954, "Origin of Vermiculite at Tigerville, South Carolina" (abstract), *Geological Society of America Bulletin,* Vol. 65, No. 12, Pt. 2, pp. 1356–1357.

Dekun, N., 1965, *Mineral Resources of Africa,* Elsevier, Amsterdam, pp. 440–441.

Haines, S.K., 1978, "Vermiculite," *Mineral Commodity Profiles,* US Bureau of Mines, 10 pp.

Heinrich, E.W., 1966, *The Geology of Carbonatites,* Rand McNally & Co., Chicago, 555 pp.

Libby, S.C., 1975, "Origin of Potassic Ultramafic Rocks in the Enoree Vermiculite District, South Carolina," Ph.D. Thesis, Pennsylvania State University.

Mathieson, A.L., and Walker, G.F., 1954, "Crystal Structure of Magnesium Vermiculite," *American Mineralogist,* Vol. 39, pp. 231–255.

Meisinger, A.C., 1979, "Vermiculite," *Minerals Yearbook, 1979,* US Bureau of Mines.

Meisinger, A.C., 1981, "Vermiculite," Preliminary Annual Report, US Bureau of Mines.

Meisinger, A.C., 1982, "Vermiculite," *Mineral Commodities Summaries 1982,* US Bureau of Mines, p. 170.

Stewart, O.F., 1949, "Origin and Occurrence of Vermiculite at Tigerville, South Carolina," M.S. Thesis, University of South Carolina.

Strand, P.R., 1975, "Vermiculite," *Industrial Minerals and Rocks,* 4th ed., S.J. Lefond, ed., AIME, New York, pp. 1219–1226.

Wollastonite

E. A. ELEVATORSKI *
Revised by L. A. ROE †

Wollastonite, named after William H. Wollaston, an English chemist, is a calcium metasilicate, $CaSiO_3$; CaO: 48.30%, SiO_2: 51.70%. It has a short history as an industrial mineral.

The earliest production of wollastonite is reported to be from a deposit near Code Siding, located north of Randsburg, CA. At this locality small tonnages of wollastonite were quarried during 1933–34 and 1938–41, and processed into mineral wool. This operation was largely experimental and virtually no United States production was again reported until the 1950s when a large deposit near Willsboro, NY, was developed by the Cabot Corp. A processing plant was placed onstream in 1953, with nearly continuous production to date. It is currently operated by NYCO, a division of Processed Minerals, Inc. Since 1958, wollastonite deposits in the Little and Big Maria Mountains of Riverside County, and in the Panamint Range of Inyo County, both in California, have operated intermittently for production of both ornamental and commercial wollastonite.

During 1980, the United States was the major producing country, furnishing about 75% of the world's output. Current production comes from Finland, Mexico, India, and Kenya. Small amounts have been shipped intermittently from the USSR, New Zealand, Republic of the Sudan, Republic of South Africa, and Namibia (South-West Africa).

The principal use of wollastonite is in the manufacture of plastics. Other uses are for paints, ceramics, adhesives, fluxes, glazes, thermal insulation board, and refractory products.

Mineralogy

Pure wollastonite, $CaSiO_3$, has the composition of 48.3% CaO and 51.7% SiO_2. However, it is seldom found in the pure state due to the ease with which it takes into solution the metasilicates of manganese, magnesium, iron, and strontium. Predominantly, wollastonite occurs as a contact metamorphic deposit forming between limestones and igneous rocks. Commonly associated minerals are garnet, diopside, epidote, calcite, and quartz.

It has a specific gravity of 2.8 to 3.0, and hardness of 4.5 to 5 on Mohs' scale. When pure, it has a brilliant white color, but with impurities it may be grayish or brownish. Luster is vitreous to pearly. Melting point of wollastonite is about 1540° C.

Wollastonite occurs in coarse-bladed masses, rarely showing good crystal form. It is usually acicular or fibrous, even in the smallest of particles. The most unique property of crushed and ground wollastonite is its cleavage. Fragments of crushed wollastonite tend to be needle-shaped, imparting a high strength, and this property is the basis for many of its uses. The fiber lengths are commonly in the ratio of 7 or 8 to 1, length to diameter. The average diameter of wollastonite is 3.5 μm. Some crystals of wollastonite fluoresce under short-wave or long-wave ultraviolet light, or both, with colors ranging from yellow-orange to pink-orange. Specimens may also show phosphorescence.

The acicularity of wollastonite is a property of considerable importance to the marketplace. The plastics industry makes utilization of high aspect ratio grades of wollastonite (20:1) for reinforcing thermoplastics and thermoset polymer compounds.

The naturally high pH of 9.9 (10% water slurry) is a prime property to the coatings industry. The coatings industry uses milled grades of wollastonite as a pH stabilizer in interior and exterior PVA and acrylic latex systems. Processed wollastonite can have a G.E. brightness of 90 to 93.

Chemically, wollastonite is inert and this property makes it useful as a filler and reinforcing agent.

There are two polymorphs of calcium silicates: wollastonite, a low temperature form, and pseudowollastonite, a high temperature

* Minobras, Dana Point, CA.
† Consultant, Downers Grove, IL.

form. Inversion of wollastonite to pseudowollastonite occurs at about 1120° C, resulting in an increase in the coefficient of expansion and a color change. Pure white wollastonite, on inversion, may change to a cream tint, or various shades of red or brown. The color change is thought to be due to the presence of iron and strontium.

Mode of Occurrence and Origin

Wollastonite is a contact metamorphic mineral, occurring within impure limestones near intrusive bodies of granite or other acidic rocks. It can also be formed by the metasomatism of calcareous sediments and by crystallization of certain magmas.

In contact metamorphism, limestones are recrystallized by the heat of intrusive rocks. Silica emanations from igneous rocks yield calc-silicate hornfels, along with "skarn" rocks formed by the transfer of manganese, silicon, aluminum, and iron from the magma to the limestone. Skarn minerals of wollastonite, garnet, and diopside form in the adjacent limestone, usually in a well-defined zone and probably at the end stages of igneous activity.

The reaction between the silica emanations of a granitic intrusion and calcite from limestone is as follows:

$$\underset{\text{Calcite}}{CaCO_3} + \underset{\text{Silica}}{SiO_2} \rightleftharpoons \underset{\text{Wollastonite}}{CaSiO_3} + \underset{\text{Carbon dioxide}}{CO_2}$$

Studies by Winkler (1965) indicate that when the temperature rises to about 400° to 450° C, the reaction begins and continues until the supply of calcite or silica is exhausted.

Deposits

United States

As the largest producer in the world, the United States output comes from New York and California.

New York: An extensive deposit is located on the western side of Lake Champlain near Willsboro. Wollastonite-bearing rocks outcrop in a belt of contact metamorphosed limestones and metasomatized sediments that is about 6 miles long and ¼ mile wide. There are three main deposits, known as Willsboro, Lewis, and Deerhead, with proven reserves in excess of 10 million tons containing 55 to 65% wollastonite.

The largest wollastonite rock band is between 30 to 75 ft thick, interbedded with iron garnets, mostly almandite, and iron diopside. Host rock is a Precambrian limestone partly replaced by skarn bands, varying in composition from pure wollastonite to pure garnet. Both garnet and diopside are feebly magnetic, and are easily separated from the wollastonite by magnetic separators. A typical analysis of wollastonite-bearing rocks from the Willsboro deposit consists of: SiO_2, 47.7%; CaO, 37.8%; Al_2O_3, 3.1%; FeO, 6.6%; MgO, 3.6%; and MnO, 1.2%.

Also, in New York, an occurrence of wollastonite has been reported in a contact zone of skarn rocks, between marble and granite-gneiss near the Clifton mine, located about 80 miles west of Willsboro. Wollastonite-bearing rocks up to several feet in thickness are reported.

California: Many occurrences of wollastonite are reported by Troxel (1957), in the desert areas, but only a few have produced commercial-grade or ornamental wollastonite.

Wollastonite is produced from the Little and Big Maria Mountains, located about 20 miles northwest of Blythe. It is found as high-grade pods in a Paleozoic crystalline limestone and in a lower zone of metamorphic rocks beneath the limestone. In the lower zone, the wollastonite is fine-grained, hard, and intimately associated with diopside, making separation difficult. The fibers are interlaced, tough, and difficult to grind. In addition, some of the ore is high in magnesium and iron. Some wollastonite is selectively mined by Pfizer Inc., and milled at Victorville, where it is beneficiated and ground to —200 and —325 mesh products. It is marketed under the trade name Wolcon and sold to the ceramic and coatings industries.

Talus material consisting of thin slabs of weathered wollastonite is collected on the slopes of the Big and Little Maria Mountains and Arica Mountains in Riverside County. Uneven erosion has caused irregular surfaces to develop on the wollastonite rocks, with brownish-gray subparallel streaks, resembling driftwood. This material is sold as ornamental stone for decorative stonework, landscaping, and roofing material.

A large deposit of high-grade wollastonite is mined in the Panamint Range, about six miles southeast of Ubehebe Peak, just outside of Death Valley National Monument. It occurs in calc-silicate rocks produced by contact metamorphism of coarse-crystalline Permian or Devonian limestones by a quartz-monzonite intrusive. Diopside, idocrase, tremolite, quartz, and calcite are the contaminants present. Reserves are very large and the wollastonite is both fine-grained and acicular.

The earliest mining of wollastonite is reported to have occurred near Code Siding, north of Randsburg, about midway between Randsburg and Ridgecrest. The host rock is a Paleozoic metasedimentary rock, intensely folded and containing wollastonite-garnet-diopside layers. The wollastonite is medium to fine-grained, gray in color, and interstratified with nearly equal amounts of diopside and garnet. The outcrop is about 8000 ft long, 400 ft wide, and from 10 to 30 ft in thickness. Reserves are reported to be about 12 million tons.

White, crystalline wollastonite outcrops in Warm Springs Canyon, on the east slope of the Panamint Range in Inyo County. The deposit is an elongate lens, about 750 ft long and 35 ft in thickness. Some of the wollastonite layers are of high-grade; however, some layers are of siliceous and calcareous material with constituents of diopside, quartz, and calcite.

Other California localities where wollastonite occurs are: Hunter Mountain near Darwin, in the Argus Range east of Darwin, both in Inyo County; at Sheep Creek in the Avawatz Mountains, in the western foothills of the Shadow Mountains, about 22 miles northeast of Victorville, both in San Bernardino County; and in the Cargo Muchacho Mountains of Imperial County.

Arizona: In the Mineral Hill area of the Sierrita Mountains, and near Rosemont, both in Pima County, wollastonite is locally abundant, where large masses of limestone have been metamorphosed. A deposit is also reported in the vicinity of Tank Pass, about 10 miles northwest of Salome, in Yuma County. Other occurrences of wollastonite are reported in the Dripping Springs Mountains near Christmas; and in the Tombstone District, Cochise County.

Nevada: Large masses of wollastonite-garnet-diopside occur near Yerington in Lyon County, where limestones have been metamorphosed into calc-silicate rocks. At this location, Triassic limestones have been intruded by granodiorite and quartz-monzonite.

Other States: In Idaho, a wollastonite-marble deposit is located in the Alder Creek mining area near Mackay, Custer County. Wollastonite occurs in metamorphosed limestone xenoliths of the South Canyon and Organ Districts of Dona Aña County, New Mexico. In Utah, wollastonite, in association with garnet and diopside, is reported in the Bingham area.

Mexico

Extensive deposits exist in the states of Chiapas and Zacatecas, where wollastonite was first mined during the 1950s in the La Blanca District of Zacatecas.

The large "Santa Fe" deposit in the state of Chiapas, near the town of Pichucalco, is unusually pure and free of garnet and calcite. Extensive mine workings have shown the deposit to be at least 300 ft thick, elliptical, and dome-shaped, being about 1400 ft long and 400 ft wide. Associated minerals total less than 1%. The wollastonite is a brilliant white and has typical acicular cleavage. Near the edge of the deposit, contaminants of bornite, chalcopyrite, and garnet are found. Compañia Minera de Cerralvo, S.A., operates the project and most of the output is consumed by the domestic ceramics industry.

In the La Blanca District, about 30 miles east of Ciudad Zacatecas, state of Zacatecas, wollastonite is produced by Wollastonia de Mexico. Irregular tabular bodies, about 8 ft in thickness, occur in folded Cretaceous limestones, intruded by granite. A very pure and bright white wollastonite is found. Also, a gray wollastonite-marble band, hosting thin layers of garnet, is found adjacent to a granite contact. Reserves of the deposit are estimated to be in excess of 40 million tons. Output from the La Blanca District is shipped to tile manufacturers in Monterrey and Mexico City.

Near Xalostoc, Sierra Tlayacac, in the state of Morelos, ornamental wollastonite is produced from a metamorphosed limestone near its contact with granitic rocks. The material polishes well and the brilliant white wollastonite contrasts markedly with pink garnet and yellow idocrase.

Additional occurrences of wollastonite appear in skarns over large areas of the Naica area in Chihuahua, mostly as small pods that are coarsely crystalline and white to pale pink in color. At another location, limestones in the Sierra Magistral area of the Llanos District, north of Puebla, have been metamorphized, forming masses of wollastonitic-bearing rocks.

USSR

Many occurrences of wollastonite are known and deposits in Uzbekistan and Tadzhikistan were being worked in the mid-1960s.

In central Asia, the Uzbekistan deposit is located in the Nuratau Mountains about 45 miles north of Samarkand. Calc-silicate hornfels containing 30 to 40% wollastonite occur at the contact between a granodiorite intrusion and argillaceous limestones. The wollastonite has a radial-fibrous structure and contains appreciable calcite with small quantities of feldspar, garnet, diopside, and pyroxene.

There has been some production of wollastonite from a deposit located 10 miles north of Leninabad in northern Tadzhikistan. Calc-silicate hornfels are characterized by wollastonite-garnet-pyroxene bands. The wollastonite is white to pink, fine-grained, and radial to fibrous in structure.

A large deposit with a thickness of about 200 ft is located between the cities of Almalyk and Tashkent, in Central Asia. The wollastonite rock has an average grade of 50%, and reportedly was formed in limestone horizons by the reaction of post-magmatic solutions.

Wollastonite rocks are also common in the Aktau Range of the Nuratau Mountains, south of Samarkand; in northwest Russia at Kalkkitekhda, just outside of Leningrad; and in southwest Russia at Tyrnyauz.

Finland

In southwest Finland, wollastonite is quarried from Archean limestones about 2 miles south of Lappeenranta. This elliptically shaped deposit is about 3 km long and 1 km wide and is surrounded by a large granite mass. The main wollastonite zone is about 150 ft thick and 2000 ft long. It contains bands of wollastonite and diopside interbedded with leptite, a quartz-feldspar rock, and calcite. The host rocks reportedly contain an average of 20%, ranging upward to 60%, wollastonite. Wollastonite is mined selectively, crushed, and the calcite and iron-bearing minerals removed by flotation. The deposit is worked by Pargas Kalkbergs Ab, and the output used almost entirely by a subsidiary company in the manufacture of wall tiles.

Kenya

Wollastonite-marble rocks were discovered in 1965 at Lolkidongai, about 50 miles south of Nairobi. These rocks reportedly outcrop along several hills forming a three-mile long arc. Wollastonite comprises 40 to 50% of the limestone host rock. Chief impurity is quartz, although calcite, and garnet and diopside minerals are present. Flotation concentrates, of about 85% pure wollastonite are dried and the −60 mesh material bagged and exported to European countries.

Namibia (South-West Africa)

Extensive bands of wollastonite marble are found at Usakos and in a belt extending northwest of Damaraland for nearly 40 miles. Composition of the wollastonitic marbles varies greatly. The higher grades are utilized for industrial material with the lower grades used for building stone and ornamental rock. Marble was formed by regional metamorphism and the subsequent intrusion of a pegmatitic granite may have formed the wollastonite. Some ceramic grade material is also produced and exported to Holland.

Other Areas

Large deposits of wollastonite are known in the northeastern Red Sea Hills about 70 miles from Port Sudan, Republic of the Sudan. The wollastonite occurs as lenticular masses up to 100 ft long in a skarn assemblage that includes idocrase, garnet, and diopside. In the Republic of South Africa, wollastonite occurs near Garies, Namaqualand, in a belt of limy metamorphic rocks, enclosed by granite-gneisses. Grade of the ore varies considerably, averaging 50% wollastonite. Chief contaminants are a green diopside and a brownish garnet.

A high-grade wollastonite deposit is located at Merida, in Badajoz Province, near the Portuguese border of Spain. It is located between bands of Devonian limestone, and offers promise in the manufacture of dielectric porcelains. Calc-silicate rocks containing wollastonite-diopside-garnet are found in the northern foothills of Sudety Mountains, near Strzelin, Poland. In Morocco, wollastonite beds over 30 ft thick are reported at Azegour, southwest of Marrakech, in an area between the northern flanks of the Atlas Mountains and the east coast.

Wollastonite occurs in a number of areas of Canada; however only the Fintry deposit, located 10 miles north of Kelowna, in British Columbia, has been explored. Steeply dipping limestone beds contain wollastonite-calcite lens and stringers averging 30% wollastonite. Mineralization occurs as a contact aureole of a granitic intrusion into Paleozoic sediments.

In 1969 an extensive deposit near the village of Khila, Rajasthan State of India, was being developed. Estimated reserves, to a depth of 50 m, are about 5 million tons.

Occurrences of wollastonite have been noted in limestone blocks on the islands of Santorini, Greece, and on Monte Somma, Vesuvius, Italy, where it probably was formed at depth and ejected during volcanic activity.

Mining and Processing

At Willsboro, NY, wollastonite is mined by open stope methods. Stopes are about 30 ft wide, varying from 15 ft at the lower level to 80 ft at the surface. At the mill lump ore is dumped into a crusher pit, scraped to a pan feeder, and discharged through a jaw crusher. The dry beneficiation process is depicted by the flowsheet in Fig. 1. Unique in the operation is the high intensity magnetic separation of both garnet and diopside. The beneficiated wollastonite is ground to four product sizes in pebble mills and a high aspect ratio product produced in an attrition mill. Finished products are sold under the trade names of Nycor® and Nyad®.

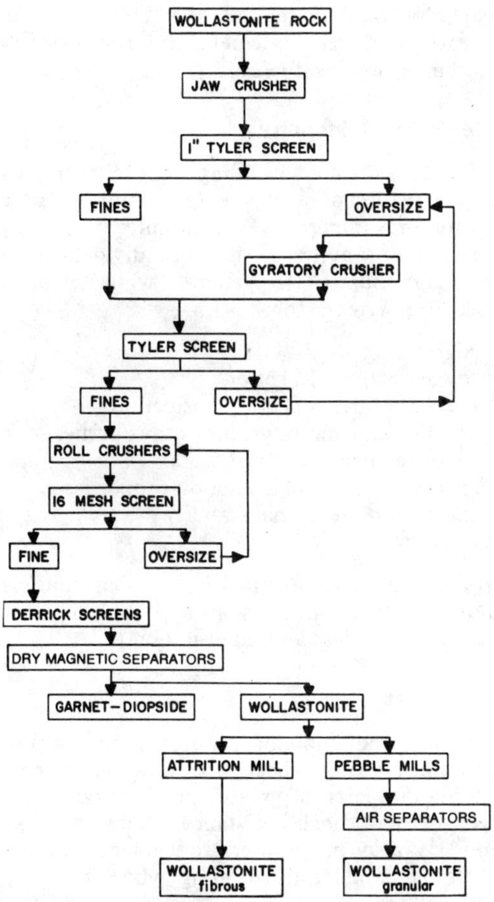

FIG. 1—*Flowsheet of wollastonite processing, Willsboro, NY.*

Quarrying is the method used in mining wollastonite at Lappeenranta, Finland. Here selective mining is necessary because the wollastonite occurs in thin bands up to 1 m in thickness. After primary and secondary crushing, a two-stage flotation process is employed. Calcite, the main impurity, is depressed into the underflow and the wollastonite and other silicates report to the first froth product. Then the unwanted silicate minerals are floated from the first froth product by using a combination anionic-cationic reagent as the collector. The final underflow is the wollastonite product. After drying, the flotation concentrates are ground in a pebble mill to produce two principal grades, one −50 mesh and the other −200 mesh. Concentrates are composed of SiO_2, 51 to 53%; CaO, 43 to 45%; Al_2O_3, 0.5%; Fe_2O_3, 0.2%; and MgO, 0.3%.

Reserves

World reserves of proven wollastonite-bearing rocks exceed 90 million tons, with probable reserves about three times this figure. Many countries have extensive deposits that are unsurveyed. In addition to the size of the deposit, the economic feasibility of producing wollastonite is greatly affected by the purity of the crude ore and the difficulties encountered in removing the contaminating minerals.

Consumption, Trade, and Production

Estimated world consumption of wollastonite approached 140,000 tons in 1980. In the United States, the plastics and thermal insulation board industries were the major consumers followed closely by the ceramics and refractory industries, with the coatings industry a medium-size consumer. Plastics and board account for 25 to 30% of the total consumption.

In Germany, Denmark, and Finland the coatings and ceramic industries are the major consumers. In Italy the ceramic industry is the major consumer, while the plastics and coatings industries in Japan are the primary consumers.

Ornamental wollastonite, used as decorative rock and in crushed form for terrazzo and roofing material, is produced in Mexico and in Namibia (South-West Africa).

The US production of 135,000 tons of industrial wollastonite comes from producers in

New York and California. Production of 20,000 to 40,000 tons is available in Mexico, 40,000 tons in Finland, and an unknown volume in India.

In trade, the United States and Finland consume 80 to 90% of their respective production, with the remainder exported to Denmark, Germany, Italy, Great Britain, and Canada.

Specifications and Prices

Wollastonite is a unique mineral due to its acicular particle shape. Its markets are directly related to its properties since wollastonite is not a widely-occurring mineral and freight costs are high. Commercial products fall into two broad classifications: milled grades and attrition milled grades. Milled grades are —200 mesh, —325 mesh, —400 mesh, and a new 10 μm top size product. Attrition grades are currently limited to one—a 20:1 L/D product.

Since the United States is the largest producing and consuming country, current US dollar prices will be listed. The milled grades cost between 2¢ and 5¢ per lb, the attrition grade is 7¢ per lb (1981). Price lists are published by all producers.

Uses

The major potential markets for wollastonite are the plastics and coatings industries, followed closely by thermal insulation board. Other significant consumers are ceramics and refractories. Wollastonite has minor application in metal casting plasters, fluxes, matchheads, abrasives, pesticide carrier, and friction papers.

A major general application area is as a partial substitute for fiberglass and for asbestos in the plastics and coatings industries.

The Ceramic Industry

The advantages of using wollastonite in ceramics are twofold: it improves the mechanical properties of the ceramic ware, and it greatly reduces warping and cracking for ceramic materials that must be rapidly fired. Wollastonite fluxes readily with silica and alumina at low temperatures, reducing thermal expansion and thereby minimizing cracking. Tiles have higher green strengths and better pressing qualities because of the needlelike shape of wollastonite particles. Also, drying rates are improved because the wollastonite particles provide openings for rapid passage of moisture through the body. Since wollastonite contains no chemically bound water nor carbonates, its use in ceramic materials minimizes the liberation of gas. Without its use, gases commonly cause cracks and laminations. Apart from its other merits, wollastonite is almost 50% silica and its adoption in place of free silica-bearing materials such as sand, flint and china stone, reduces the risk of silicosis of workers in the ceramic industry.

The Coatings Industry

For coatings, the high pH of wollastonite allows its wide use as a pH stabilizer (replacing ammonia and chemical buffers) in PVA and acrylic latex paint systems. Most trade sales paints have ¼ lb per gal of wollastonite to prevent can and lid corrosion, and to keep the pigments in suspension. Exterior acrylic house paints (containing zinc oxide) utilize the buffering action of wollastonite for stabilization and weathering.

Wollastonite's properties of acicular particle shape, low oil demand, and good color allow its use for film reinforcement, improved scrub resistance, and weatherability.

The Plastics Industry

For plastics compounding, the acicular particle structure of wollastonite identifies it as a reinforcing mineral. Wollastonite has many individual properties related to individual polymer compounds. In general, wollastonite is used in polymers for:

- high purity
- low moisture adsorption
- excellent heat deflection/distortion
- good tensile, modulus, impact properties
- reinforcement
- excellent electrical insulation properties
- excellent dimensional stability
- economics

Wollastokup®, surface modified wollastonite, is finding wide application as a partial replacement for fiberglass and other minerals.

Other Uses

In the production of ceramic-bonded abrasives and abrasive wheels, the addition of wollastonite imparts a greater fluxing rate and high thermal shock resistance. Some wollastonite is also being used to replace natural cryolite as a filler in grinding wheel bonding formulations.

At Lappeenranta, Finland, some wollastonite is used to produce mineral wool. Dolomite, wollastonite, and coke are finely ground, mixed, and melted in furnaces. At about 1500°C the melt is poured on to fast spinning wheels. The fibers produced are blown onto a conveyor where blankets of material are formed. A variety of acoustic and thermal insulation products are made from mineral wool. These include ceiling tiles, cork and mineral wool boards, tubes, and blankets interlayed with wire screening.

Synthetics

Synthetic calcium metasilicates have been commercially produced for some years in the United States, Denmark, Italy, Germany, and the Soviet Union. Most of the synthetics are hydrated, and none of the anhydrous compounds have the crystal form of natural wollastonite. Unlike wollastonite, the hydrated calcium silicates have a high absorptivity.

Danish synthetic calcium metasilicates are made from sand and chalk, which are abundant in Denmark, and marketed under the trade name of Synopal. Chalk and sand are mixed into a slurry with small amounts of dolomite which acts as a flux. This slurry is heated in a kiln to a temperature of about 1560°C, forming a viscous mixture that is water-cooled. When solidified, the material is graded and then reheated in a second kiln to a temperature of 1250°C, sufficient to permit crystallization to occur. The finished product is subangular, white, and opaque, consisting of about 50% wollastonite and the remainder gehlenite ($Ca_2Al_2SiO_7$) and akermanite ($MgCa_2Si_2O_7$). Uses of Synopal are for road surfacing, mosaic tiles, flooring, and roofing felts.

A product known as Wollanita is made in Italy by heating a mixture of silica, sand, chalk, and dolomite. Production methods are similar to those used in Denmark. Wollanita is used for road material, as an abrasive, and in ceramics.

Both hydrated and anhydrous calcium metasilicates are manufactured in the United States, as Silene by Pittsburgh Plate Glass Co. and as Micro-Cel by Manville Products Corp. Many grades of both products are produced by subjecting lime, sand, and dolomite slurries to high pressure and temperature.

Actual and potential uses of synthetic calcium silicates are about the same as those of natural wollastonite. In countries where wollastonite is abundant and can be produced cheaply, it is preferred over the synthetic calcium metasilicates. Common uses of both wollastonite and its synthetic equivalents are in tile making and as extenders in the paint industry. Applications of the synthetic material are as a flow-aid agent in powdered materials, in foam, and in mineral wool.

Tonnages of synthetic wollastonite produced are unavailable; however, they are known to be minor in comparison to natural wollastonite production. Prices of the synthetic equivalents are about double the price of natural wollastonite.

Bibliography and References

Anon., 1969, "Wollastonite Comes of Age," *Industrial Minerals,* Apr., pp. 8–13.

Anon., 1975, "Wollastonite: U.S.A. Dominates Both Production and Consumption," *Industrial Minerals,* No. 94, July, pp. 15–23, 29.

Anon., 1980, "Wollastonite Facility Set By Processed Minerals for Willsboro Plantsite," *Chemical Marketing Reporter,* Vol. 217, No. 21, May 26, pp. 3, 23.

Amberg, C.R., et al., 1969, "Wollastonite, An Industrial Mineral," *Ceramics Bulletin,* New York State College, No. 4, 60 pp.

Andrews, R.W., 1970, *Wollastonite,* Monograph, Her Majesty's Stationery Office, Great Britain, 114 pp.

Burnham, K., and Broughton, J.G., 1955, "Occurrences and Uses of Wollastonite from Willsboro, New York," *Mining Technology,* No. 8, July, pp. 1–8.

Clarke, G., 1980, "Mexico's Industrial Minerals—Gathering Momentum," *Industrial Minerals,* No. 153, June, pp. 21–53.

Copeland, J., and Rush, O.W., 1978, "Wollastonite, Short Fiber Filler Reinforcement," *Plastic Compounding,* Nov.-Dec., pp. 26–41.

Copeland, J., and Rush, O.W., 1979, "Reinforcing with Wollastonite Filler Makes for a Tougher Polypropylene," Preprint, *Modern Plastics,* Mar., 2 pp.

Doctorman, V.C., 1958, "Wollastonite Spun into Rock Wool," *Rock Products,* Aug., pp. 80–81.

Engelhardt, C.L., 1979, "Calcium Metasilicate—An Extender Pigment," Preprint, *American Paint and Coatings Journal,* Sept. 10, 7 pp.

Hall, A.L., et al., 1952, "Enter Wollastonite—A New Commercial Nonmetallic Mineral," *Mining Engineering,* Vol. 4, No. 10, pp. 952–953.

Ladoo, R.B., 1960, "Wollastonite," *Industrial Minerals and Rocks,* 3rd ed., AIME, New York, pp. 897–899.

Mason, J.E., 1966, "The Lolkidongai Wollastonite Occurrence, Kajiado District, Kenya," Information Circular No. 3, Kenya Dept. of Mines and Geology, 28 pp.

Neely, J.R., 1954, "Properties of Ceramics Containing Wollastonite," M.S. Thesis, University of California at Los Angeles.

Neely, J.R., and Knapp, W.J., 1964, "California Wollastonites," *Ceramic News,* Vol. 14, No. 5, pp. 12–13.

Shvetsov, V.Y., and Tsapkov, N.T., 1964, "Producing Wollastonite Concentrates for Ceramics," *Glass Ceramics,* Vol. 21, No. 6, pp. 32–33.

Simpson, W. and deJager, D.H., 1962, "Wollastonite near Garies, Namaqualand," *Annals of the Geological Survey of South Africa,* Vol. 1, pp. 127–135.

Thorndike, J.D., 1936, "Mineral Wool from Wollastonite," *Mining and Metallurgy,* March, pp. 133–136.

Tolliday, J.M., 1958, "The Crystal Structures of Parawollastonite and Wollastonite," Ph.D. Thesis, Birbeck College, London, 110 pp.

Troxel, B.W., 1957, "Wollastonite," Bulletin 176, Mineral Commodities of California, California Div. of Mines and Geology, pp. 693–697.

Vukovich, M., et al., 1957, "The Use of Wollastonite in Ceramic Bodies," *Engineering Experiment Station Bulletin No. 164,* Ohio State University, Vol. 26, No. 2, 76 pp.

Winkler, H.G.F., 1965, *Petrogenesis of Metamorphic Rocks,* Springer-Verlag, pp. 28–30.

Zeolites

Introduction

RICHARD H. OLSON *

"Rarely in our technological society does the discovery of a new class of inorganic materials result in such a wide scientific interest and kaleidoscopic development of applications as has happened with the zeolite molecular sieves." That opening sentence in a landmark volume on zeolite molecular sieves by D. W. Breck (1974), is, if anything, an understatement.

Zeolites were recognized as a new group of minerals in the 1750s, but it was not until 1930 that the first analysis of the crystal structure of a zeolite mineral was made. In the late 1940s research workers at Union Carbide Corp. began a program of zeolite synthesis and study which has resulted in one of the major research achievements of all time.

Naturally, with the vast research and development efforts devoted to the synthetic molecular sieves, Union Carbide became concerned in the mid-1950s about its position should minable deposits of natural analogues be discovered.

Zeolite minerals have diverse origins, and even a brief discussion of these is beyond the scope of this paper. The early work on zeolites was confined to igneous rock occurrences (e.g., amygdules and veins) which were formed by "space-filling;" these crystals tend to be euhedral, thereby rendering them appealing to both mineralogists and collectors. During the first year or so of Union Carbide's exploration (which began in mid-1957) attention was devoted solely to such igneous occurrences.

A landmark paper by Bramlette and Posnjak (1933) heralded the potential significance of the occurrence of zeolite minerals as alteration products of vitric ash and similar materials. In the summer of 1958, full attention was directed by Union Carbide to the search for deposits of this type in the western United States. Since then, such exploration has been continuous by both commercial firms and private parties. The future economic potential for natural zeolite deposits seems to be restricted to those of the "open-and-closed system" types as described by Sheppard (this chapter). Therefore, this and the following discussions will concentrate upon these types; the other types of zeolites in sedimentary rocks (e.g., deep-sea sediments, burial metamorphic, and hydrothermal) will be, for the most part, ignored.

Despite the relative cheapness of natural zeolite deposits as compared with the cost of synthetic molecular sieves, there has been minimal development of zeolite mining in the US over the past 25 years or so. Mumpton (this chapter) discusses the present commercial utilization of natural zeolites and their future potential in detail, but perhaps the following will serve as useful background. In the early days of synthetic molecular sieve research and production, the efforts were concentrated largely upon such high-value applications as physical separation of components in a stream (based upon molecular size and geometry) and in catalysis. Due to a number of reasons (some of which are difficult for an outsider to understand), natural zeolite ores (beneficiated or not) have not made substantial inroads upon the use of the higher-priced synthetic molecular sieves for the foregoing applications, but this is not to say that such a condition will exist forever. Furthermore, in view of the many other uses which are now known, or which loom in the future, the competitive position of the natural zeolites vs. the more expensive

* Consultant, Industrial Minerals Evaluations, Inc., Golden, CO.

synthetic molecular sieves is, in the writer's opinion, bound to improve markedly.

It has certainly been to the advantage of the higher-priced synthetic molecular sieves to see that they are used in preference to natural zeolite products; therefore, much which could have been stated about the relative merits of natural zeolites has understandably remained unsaid. Consequently, most of those who would mine or use the natural varieties have been handicapped by a lack of knowledge with regard to their comparative efficiencies. Time, in its inexorable manner, will increasingly serve to diminish the inhibiting nature of this barrier to knowledge.

In addition, who can foresee which zeolite minerals, with known and well-demonstrated economic potential but which are not yet known to exist in commercial quantities, might be found as ore bodies in the future. In 1957 there was probably less than a kilogram of erionite in all the museums and mineral collections of the world; now several deposits are known in the US alone, aggregating many millions of tons!

Sheppard (1971) has plotted the known occurrences of the six most common zeolites occurring in sedimentary deposits in the conterminous United States. Fig. 1 shows analcime occurrences, while Table 1 lists those occurrences. Fig. 2 shows the occurrences of chabazite, erionite, mordenite, and phillipsite, while Table 2 lists those occurrences. Fig. 3 shows clinoptilolite occurrences, while Table 3 lists those occurrences.

The use of natural zeolites in environmental cleanup and many fields of pollution control is only now beginning to receive concerted research efforts. This field alone and its obvious potential would appear to justify a close look at natural zeolites by many firms who are apparently not yet so interested. But there are many other uses which also should be explored, and this is well-documented by the strength of the zeolite-mining industry in Japan as contrasted with that of the United States and the rest of the world.

Commanding positions in the zeolite mining industry in the future will be held by those who are willing to prove new uses and develop markets now, rather than by those who wait for others to do so and then hurriedly embark upon "me-too" exploration programs.

In the 3rd edition of this volume (1960), not only was there no chapter on zeolites, but the word did not even appear in the Index! There was good reason for this, because the first search for zeolite ores had only begun in mid-1957. The past 25 years or so have seen the appearance of hundreds of papers and volumes regarding zeolite minerals and synthetic molecular sieves, but it is significant that the four most comprehensive contributions which deal with both natural zeolites and synthetic molecular sieves, and also touch upon their relative merits (however lightly), have all been presented or published within a 12-month period in 1973–74. These four contributions, i.e., Breck (1973, 1974), Mumpton (1973), and Sheppard (1973), have been revised or enlarged to furnish the material for the parts of the chapter that follow.

TABLE 1—Analcime Occurrences in Sedimentary Rocks

Locality No. on Fig. 1	Locality	Occurrence
1	Near Vaughn, Cascade County, MT	Siltstone and sandstone in the Taft Hill Member of the Blackleaf Formation of Cretaceous age and tuff in the Bootlegger Member of the Blackleaf Formation of Cretaceous age
2	Near Twin Creek, Bear Lake County, ID	Tuff in the Twin Creek Limestone of Jurassic age
3	Near Gros Ventre River, Teton County, WY	Ocher oolitic beds in the Popo Agie Member of the Chugwater Formation of Triassic age
4	Near Dubois, Fremont County, WY	Ocher oolitic beds in the Popo Agie Member of the Chugwater Formation of Triassic age
5	Near Lander, Fremont County, WY	Ocher oolitic beds in the Popo Agie Member of the Chugwater Formation of Triassic age
6	Near Thermopolis, Hot Springs County, WY	Purple and ocher units of the Popo Agie Member of the Chugwater Formation of Triassic age
7	Near Hyattville, Big Horn County, WY	Bentonite in the Mowry Shale of Cretaceous age
8	South Fork of the Powder River, Natrona County, WY	Bentonite in the Mowry Formation of Cretaceous age
9	Near Casper, Natrona County, WY	Bentonite in the Mowry Formation of Cretaceous age
10	Near Lysite Mountain, Hot Springs County, WY	Tuff in the Tepee Trail Formation of Eocene age
11	Beaver Rim, Fremont County, WY	Tuff in the Wagon Bed Formation of Eocene age
12	Near Green River, Sweetwater County, WY	Tuff in the Green River Formation of Eocene age
13	Near Ludlow, Harding County, SD	Lignite in the upper member of the Tongue River Formation of Paleocene age
14	Cathedral Bluffs, Rio Blanco County, CO	Tuff in the Green River Formation of Eocene age
15	Near Piceance Creek, Rio Blanco County, CO	Oil shale in the Green River Formation of Eocene age
16	Anvil Points, Garfield County, CO	Oil shale in the Green River Formation of Eocene age
17	Along Piceance Creek, about 20 miles west of Meeker, Rio Blanco County, CO	Tuff in the Parachute Creek Member of the Green River Formation of Eocene age
18	Lone Tree Mesa, Montrose County, CO	Tuffaceous mudstone in the Brushy Basin Member of the Morrison Formation of Jurassic age
19	Near Slick Rock, San Miguel County, CO	Tuffaceous mudstone in the Brushy Basin Member of the Morrison Formation of Jurassic age
20	Near Vernal, Uintah County, UT	Ocher oolitic beds in the Chinle Formation of Triassic age
21	White River Canyon, Uintah County, UT	Tuff in the Green River Formation of Eocene age
22	Near Two Water Creek, Uintah County, UT	Tuff in the Parachute Creek Member of the Green River Formation of Eocene age
23	Near Duchesne, Duchesne County, UT	Dolomitic oil shale of the Green River Formation of Eocene age
24	Near Currant, Nye County, NV	Tuff in Horse Camp Formation of Miocene and Pliocene age
25	Nevada Test Site, Nye County, NV	Tuff and lapilli tuff of Tertiary age
26	Teels Marsh, Mineral County, NV	Tuff in lacustrine deposit of Quaternary age
27	Near Silver Peak, Esmeralda County, NV	Tuff in the Esmeralda Formation of Miocene and Pliocene age
28	Deep Springs Lake, Inyo County, CA	Saline crusts of Holocene age
29	Saline Valley, Inyo County, CA	Mud of Holocene age
30	Owens Lake, Inyo County, CA	Tuff and tuffaceous sediments of Pleistocene age
31	Lake Tecopa, Inyo County, CA	Tuff in lacustrine rocks of Pleistocene age
32	Searles Lake, San Bernardino County, CA	Tuff and mudstone of Quaternary age
33	Mojave Desert, eastern Kern County and San Bernardino County, CA	Tuff and mudstone of late Tertiary and Quaternary age

Table 1 (cont'd)

Locality No. on Fig. 1	Locality	Occurrence
34	Near Delano, Kern County, CA	Pond series soil
35	Near Wikieup, Mohave County, AZ	Tuff in Big Sandy Formation of Pliocene age
36	Maggie Canyon, Mohave County, AZ	Sandstone of the Chapin Wash Formation of Pliocene(?) age
37	Near Horseshoe Reservoir, Maricopa County, AZ	Tuff in the Verde Formation of Pliocene(?) or Pleistocene age
38	Near Eloy, Pinal County, AZ	Silty claystone of late Tertiary age
39	Willcox Playa, Cochise County, AZ	Mudstone of Pleistocene age
40	Along San Simon Creek, Cochise and Graham Counties, AZ	Tuff in unnamed lacustrine formation of late Cenozoic age
41	Near Nutrioso, Apache County, AZ	Sandstone in unnamed formation of Tertiary age
42	Near Red Wash, San Juan County, NM	Tuffaceous mudstone of the Brushy Basin Member of the Morrison Formation of Jurassic age
43	About 2.5 miles southeast of Senorito, Sandoval County, NM	Siliceous tuff in the Brushy Basin Member of the Morrison Formation of Jurassic age
44	Wichita Mountains, Kiowa County, OK	Arkose in the Tepee Creek Formation of Permian age
45	Near Terlingua, Brewster County, TX	Black tarry shale of late Mesozoic or early Tertiary age
46	Near Yardley, Bucks County, PA	Argillite in the Lockatong Formation of Triassic age
47	Near Frenchtown, Hunterdon County, NJ	Argillite in the Lockatong Formation of Triassic age
48	Near Pursglove, Monongalia County, WV	Concretion in the Pittsburgh coal bed of the Monongahela Formation of Pennsylvanian age

Source: Sheppard, 1971.

FIG. 1—*Analcime occurrences in sedimentary rocks—United States; see Table 1 (courtesy USGS).*

TABLE 2—Chabazite, Erionite, Mordenite, and Phillipsite Occurrences in Sedimentary Rocks

Locality No. on Fig. 2	Locality	Zeolites	Occurrence
1	Near Bearbones Mountain, Lane County, OR	Mordenite	Tuff and lapilli tuff in the Little Butte Volcanic Series of Oligocene and Miocene age
2	Vicinity of Stein's Pillar, Crook County, OR	Mordenite	Tuff in the John Day Formation of Oligocene and Miocene age
3	Near Durkee, Baker County, OR	Erionite	Tuff of Tertiary age
4	Near Rome, Malheur County, OR	Erionite, mordenite, phillipsite	Tuff and tuffaceous sandstone in an unnamed lacustrine formation of Pliocene age
5	West face of Hart Mountain, Lake County, OR	Mordenite, phillipsite	Tuff and tuffaceous sedimentary rocks of late Oligocene or early Miocene age
6	Near Harney Lake, Harney County, OR	Erionite, phillipsite	Tuff and tuffaceous sedimentary rocks in the Danforth Formation of Pliocene age
7	Beaver Rim, Fremont County, WY	Chabazite, erionite, phillipsite	Tuff in the Wagon Bed Formation of Eocene age
8	Near Split Rock, Natrona County, WY	Phillipsite	Tuff in the Moonstone Formation of Pliocene age
9	Near Green River, Sweetwater County, WY	Mordenite	Tuff in the Tipton Shale Member of the Green River Formation of Eocene age
10	Near Mud Buttes, Butte County, SD	Phillipsite	Bentonite in the Gammon Ferruginous Member of the Pierre Shale of Cretaceous age
11	Sheep Mountain Table, Shannon County, SD	Erionite	Tuff in the Arikaree Formation of Miocene age
12	Near Creede, Mineral County, CO	Mordenite	Tuff in the Windy Gulch Member of the Bachelor Mountain Rhyolite of Oligocene age
13	Pine Valley, Eureka, County, NV	Erionite, phillipsite	Tuff in the Hay Ranch Formation of Pliocene and Pleistocene age
14	West flank of the Shoshone Range, Lander County, NV	Erionite	Tuff in unnamed lacustrine formation of Pliocene age
15	Reese River, Lander County, NV	Erionite	Tuff in unnamed lacustrine formation of Pliocene age
16	Jersey Valley, Pershing County, NV	Erionite, phillipsite	Tuff in unnamed lacustrine formation of Pliocene age
17	Near Lovelock, Pershing County, NV	Mordenite	Tuff in unnamed lacustrine formation of late Tertiary age
18	Near Copper Valley, Churchill County, NV	Mordenite	Tuff in unnamed lacustrine formation of late Tertiary age
19	Near Eastgate, Churchill County, NV	Erionite	Tuff in unnamed lacustrine formation of late Tertiary age
20	Teels Marsh, Mineral County, NV	Phillipsite	Tuff in lacustrine deposit of Quarternary age
21	Near Silver Peak, Esmeralda County, NV	Mordenite, phillipsite	Tuff in the Esmeralda Formation of Miocene and Pliocene age
22	Nevada Test Site, Nye County, NV	Chabazite, mordenite	Tuff and lapilli tuff of Tertiary age
23	Owens Lake, Inyo County, CA	Erionite, phillipsite	Tuff and tuffaceous sediments of Pleistocene age
24	Lake Tecopa, Inyo County, CA	Chabazite, erionite, phillipsite	Tuff and tuffaceous rocks of Pleistocene age
25	Searles Lake, San Bernardino County, CA	Phillipsite	Tuff of Quaternary age
26	Mojave Desert, eastern Kern County and San Bernardino County, CA	Chabazite, erionite, mordenite, phillipsite	Tuff and tuffaceous rocks of late Tertiary and Quaternary age
27	Near Nipomo, San Luis Obispo County, CA	Mordenite	Tuff in the Obispo Formation of Miocene age

Table 2 (cont'd)

Locality No. on Fig. 2	Locality	Zeolites	Occurrence
28	Union Pass, Mohave County, AZ	Mordenite	Tuff and lapilli tuff in the Golden Door Volcanics of Tertiary age
29	Near Wikieup, Mohave County, AZ	Chabazite, erionite, phillipsite	Tuff in Big Sandy Formation of Pliocene age
30	Near Horseshoe Reservoir, Maricopa County, AZ	Erionite, phillipsite	Tuff in the Verde Formation of Pliocene(?) or Pleistocene age
31	Near Morenci, Greenlee County, AZ	Mordenite	Tuff and lapilli tuff in unnamed formation of Tertiary age
32	Near Bear Springs, Graham County, AZ	Chabazite, erionite, phillipsite	Tuff in unnamed lacustrine formation of late Cenozoic age
33	Along San Simon Creek, Cochise and Graham Counties, AZ	Chabazite, erionite	Tuff in unnamed lacustrine formation of late Cenozoic age

Source: Sheppard, 1971.

FIG. 2—*Chabazite, erionite, mordenite, and phillipsite occurrences in sedimentary rocks—United States; see Table 2 (courtesy USGS).*

TABLE 3—Clinoptilolite Occurrences in Sedimentary Rocks

Locality No. on Fig. 3	Locality	Occurrence
1	Near Vaughn, Cascade County, MT	Tuff and tuffaceous siltstone and sandstone in the Taft Hill, Vaughn, and Bootlegger Members of the Blackleaf Formation of Cretaceous age
2*	Near Livingston, Park County, MT	Tuffaceous mudstone, siltstone, and sandstone in the Livingston Group of Cretaceous age
3	Near Preston, Franklin County, ID	Tuff in the Salt Lake Group of late Tertiary age
4	Near Renton, King County, WA	Sandstone and conglomerate in unnamed marine formation of Oligocene age
5	Near Bearbones Mountain, Lane County, OR	Tuff and lapilli tuff in the Little Butte Volcanic Series of Oligocene and Miocene age
6	Near Stein's Pillar, Crook County, OR	Tuff in the John Day Formation of Oligocene and Miocene age
7	Near Deep Creek, Wheeler County, OR	Tuff in the lower part of the John Day Formation of Oligocene and Miocene age
8	Near the Painted Hills, Wheeler County, OR	Tuff and claystone in the lower part of the John Day Formation of Oligocene and Miocene age
9	Sucker Creek, Malheur County, OR	Tuff and tuffaceous sandstone in the Sucker Creek Formation of Miocene age
10	Near Sheaville, Malheur County, OR	Tuff probably equivalent to part of the Sucker Creek Formation of Miocene age
11	Near Rome, Malheur County, OR	Tuff and tuffaceous sandstone in unnamed lacustrine formation of Pliocene age
12	East face of Steens Mountain, Harney County, OR	Tuff in the Pike Creek Formation of Oligocene(?) and Miocene age
13	Near Harney Lake, Harney County, OR	Tuff and tuffaceous sedimentary rocks in the Danforth Formation of Pliocene age
14	West face of Hart Mountain, Lake County, OR	Tuff and tuffaceous sedimentary rocks of late Oligocene or early Miocene age
15	Near Pedro, Weston County, WY	Bentonite in the Pierre Shale of Cretaceous age
16	Near Lysite Mountain, Hot Springs County, WY	Tuff in the Tepee Trail Formation of Eocene age
17*	Snake River Canyon, Lincoln County, WY	Shale in the Aspen Formation of Cretaceous age
18	Beaver Rim, Fremont County, WY	Tuff in the Wagon Bed Formation of Eocene age
19	Near Cameron Spring on Beaver Rim, Fremont County, WY	Tuffaceous sandstone in the White River Formation of Oligocene age
20	Near Split Rock, Natrona County, WY	Tuff in the Moonstone Formation of Pliocene age
21	Near Green River, Sweetwater County, WY	Tuff in the Tipton Shale Member of the Green River Formation of Eocene age
22	Near Twin Buttes, Sweetwater County, WY	Tuff and tuffaceous sandstone in the Bridger Formation of Eocene age
23	Near Chamberlain, Buffalo County, SD	Bentonite in the Sharon Springs Member of the Pierre Shale of Cretaceous age
24	Sheep Mountain Table, Shannon County, SD	Tuff in the Arikaree Formation of Miocene age
25	Near Vermillion Cliffs, Moffat County, CO	Tuff in the Bridger Formation of Eocene age
26	Near Creede, Mineral County, CO	Tuff in the Windy Gulch Member of the Bachelor Mountain Rhyolite of Oligocene age
27	Near Mountain Green, Morgan County, UT	Tuff in the Salt Lake Group of Tertiary age
28	Northern part of the Markagunt Plateau, Iron County, UT	Tuffaceous sandstone of Oligocene and Miocene(?) age
29	Near Elko, Elko County, NV	Oil shale in unnamed formation of Oligocene age
30	Near Carlin, Eureka County, NV	Tuff in the Safford Canyon Formation of Oligocene(?) or Miocene(?) age and the Carlin Formation of Pliocene age
31	West flank of the Shoshone Range, Lander County, NV	Tuff in unnamed lacustrine formation of Pliocene age
32	Reese River, Lander County, NV	Tuff in unnamed lacustrine formation of Pliocene age
33	Jersey Valley, Pershing County, NV	Tuff in unnamed lacustrine formation of Pliocene age
34	Near Lovelock, Pershing County, NV	Tuff in unnamed lacustrine formation of late Tertiary age

Table 3 (cont'd)

Locality No. on Fig. 3	Locality	Occurrence
35	Near Eastgate, Churchill County, NV	Tuff in unnamed lacustrine formation of late Tertiary age
36	Teels Marsh, Mineral County, NV	Tuff in lacustrine deposit of Quaternary age
37	Near Silver Peak, Esmeralda County, NV	Tuff in the Esmeralda Formation of Miocene and Pliocene age
38	Near Goldfield, Esmeralda County, NV	Tuffaceous sandstone in the Siebert Formation of Miocene(?) age
39	Nevada Test Site, Nye County, NV	Tuff and lapilli tuff of Tertiary age
40	Near Bullfrog Hills, Nye County, NV	Tuff of Tertiary age
41	Death Valley, Inyo County, CA	Tuff in the Furnace Creek Formation of Pliocene age
42	Lake Tecopa, Inyo County, CA	Tuff in lacustrine rocks of Pleistocene age
43	Owens Lake, Inyo County, CA	Tuff and tuffaceous sediments of Pleistocene age
44	Mojave Desert, eastern Kern County and San Bernardino County, CA	Tuff and tuffaceous rocks in numerous formations of late Tertiary and Quaternary age
45	Near Branciforte Creek, Santa Cruz County, CA	Tuffaceous sandstone in the Santa Margarita Formation of Miocene age
46	Near Nipomo, San Luis Obispo County, CA	Tuff in the Obispo Formation of Miocene age
47*	Near Oakview, Ventura County, CA	Bentonite in the Modelo Formation of Miocene age
48	Near San Pedro, Los Angeles County, CA	Dolomitic sandstone in the Monterey Formation of Miocene age
49	Near Wikieup, Mohave County, AZ	Tuff in Big Sandy Formation of Pliocene age
50	Near Dome, Yuma County, AZ	Bentonite of Tertiary(?) age
51	Near Horseshoe Reservoir, Maricopa County, AZ	Tuff in the Verde Formation of Pliocene(?) or Pleistocene age
52	Near Nutrioso, Apache County, AZ	Tuff and sandstone in unnamed formation of Tertiary age
53	Near Morenci, Greenlee County, AZ	Tuff and lapilli tuff in unnamed formation of Tertiary age
54	Along San Simon Creek, Cochise and Graham Counties, AZ	Tuff in unnamed lacustrine formation of late Cenozoic age
55	Near Bayard, Grant County, NM	Tuff in the Sugarlump Tuff of Oligocene age
56	Near Coy City, Karnes County, TX	Tuff and tuffaceous sandstone in the Jackson Group of Eocene age
57	Near Tilden, McMullen County, TX	Tuff in the Jackson Group of Eocene age
58	Near Meridian, Lauderdale County, MS	Tuffaceous sandstone in the Meridian Sand of Eocene age
59	Near Nettleboro, Clarke County, AL	Tuffaceous sandstone in the Meridian Sand of Eocene age
60	Near McKenzie, Butler County, AL	Tuff and tuffaceous claystone in the Tallahatta Formation of Eocene age
61	Near Paducah, McCracken County, KY	Claystone in the Clayton(?) Formation of Paleocene age and clay in the Porters Creek Clay of Paleocene age
62*	Near Jackson, Madison County, TN	Fossiliferous rock of Paleocene age
63*	Near Caryville, Washington County, FL	Suwanee Limestone of Oligocene age
64	Near Coosawhatchie, Jasper County, SC	Clay in the Hawthorn Formation of Miocene age and the Santee Limestone of Eocene age
65	Central SC	Mudstone in the Black Mingo Formation of Paleocene and Eocene age
66	Near Eward, Beaufort County, NC	Phosphorite in the Pungo River Formation of Miocene age

Source: Sheppard, 1971.

***Zeolite was identified as heulandite.**

FIG. 3—*Clinoptilolite occurrences in sedimentary rocks—United States; see Table 3 (courtesy USGS).*

Synthetic Zeolites: Properties and Applications

D. W. BRECK *

Zeolites were first recognized as a new group of minerals by Cronstedt with the discovery of stilbite in 1756. The word zeolite was coined from the two Greek words meaning "to boil" and "a stone" because of the loss of water when heated in the mineralogist's blowpipe (Breck, 1974).

In 1845 it was discovered that certain soils have the power of retaining ammonium salts (Way, 1850). These were the first ion exchange experiments. Later it was discovered that it was the hydrated silicates in the soil that produced this phenomenon. Several years later a paper was published dealing with the action of dilute salt solutions on silicates which showed that this base-exchange principle, which we now call cation exchange, is reversible. The quantitative cation exchange behavior of zeolite minerals, such as chabazite, was studied. As a result any aluminosilicate which exhibited the property of ion exchange was referred to as a "zeolite"; even today the term has been applied to clay minerals as well as synthetic organic ion exchange resins.

* Deceased, formerly of Union Carbide Corp., Tarrytown, NY.

The first synthetic aluminosilicates prepared specifically for their cation exchange properties were the permutites which are amorphous, gel-like, hydrous aluminosilicates. In the early part of this century extensive effort was devoted to the preparation of permutite materials for use in water softening (Shreve, 1930). These have been replaced by the organic exchange resins. In recent years some of the synthetic crystalline zeolites are finding ion exchange applications due to their unusual cation selectivities.

The term "molecular sieve" as it is applied to the crystalline zeolites was originated by J. W. McBain (1932) to define porous solid materials which exhibit the property of acting as sieves toward gas molecules. Dehydrated zeolite crystals act as sieves by the selective adsorption or rejection of molecules due to differences in their sizes and other structural factors. They belong to the class of microporous adsorbents; they have uniform pores permeating the solid with diameters of as much as 20Å. Although other microporous molecular-sieve-type adsorbents have been studied in laboratories, only the crystalline zeolites have achieved commercial importance as adsorbents.

The first experimental observations of the adsorption of gases on zeolites and their behavior as molecular sieves were conducted on zeolite minerals. The first definitive experiments on the separation of mixtures using a dehydrated zeolite, chabazite, as a molecular sieve were performed by Barrer in 1945 (Barrer, 1945). He classified the zeolites into three groups as based upon their ability to adsorb or exclude molecular species of different sizes. The classification defined the approximate intrachannel dimensions.

Beginning in 1948, the pioneering research efforts by R. M. Milton and his associates at Union Carbide Corp. resulted in the first synthesis and later the commercial manufacture of molecular sieve zeolites. This synthesis has been acknowledged as a major research achievement. The properties and uses of zeolites are now being explored by many diverse scientific disciplines. A wide variety of applications includes the separation and recovery of normal paraffin hydrocarbons, catalysis of hydrocarbon reactions, many uses in drying such as the drying of refrigerants, adsorbents in the separation of air components and the production of oxygen-enriched streams, the recovery of radioactive ions from radioactive waste solutions, the removal of carbon dioxide and sulfur compounds from natural gas, and the control of air pollution.

Zeolite Structural Chemistry

Along with the feldspars, the zeolites are framework silicates; their structures consist of a three-dimensional framework of SiO_4 and AlO_4 tetrahedra. As in the feldspars, the trivalent aluminum in these tetrahedral positions requires the presence of additional positive charge such as alkali metal or alkaline earth ions in order to maintain electrical neutrality. In the zeolites, for stability reasons, the maximum substitution of aluminum for silicon is the ratio of 1:1. In this event, a complete ordering of the aluminum and silicon is generally present. Recently zeolites with a Si/Al ratio of up to 100:1 have been reported. However, unlike the feldspars which have tight and dense structures, zeolite structures are open and contain large cavities filled with water molecules. The cavities may be interconnected in one, two, or three directions producing, upon dehydration, a crystal permeated with channel systems in one, two, or three directions. The metallic ions which are needed for the charge compensation occupy sites in the channel or adjacent to the cavities and are generally available for exchange by other cations.

Many zeolite framework structures consist of simple polyhedra, and each polyhedron is a three-dimensional array of the individual tetrahedra in a simple geometric form. These arrangements of polyhedra are referred to as secondary building units (Meier, 1968). Examples include single rings of four or six, cubic arrays, and hexagonal prism arrays or double six rings. For illustration, in the structure of the synthetic zeolite A, units of double four rings or cubes are joined together. This cubic arrangement produces a larger polyhedron known as the truncated octahedron (Breck, et al., 1956), Fig. 4. Internally, in the middle of the cube there is another type of polyhedron referred to as the truncated cube octahedron. This internal cavity is 11Å in diameter and is entered through six circular apertures formed by a regular ring of eight oxygens giving a free diameter of 4.2Å. It is these apertures which upon dehydration of the zeolite produce the molecular sieve character of this well known synthetic zeolite. In the crystal these cavities are stacked in a continuous three-dimensional pattern forming a system of unduloidlike channels with 4.2Å periodic re-

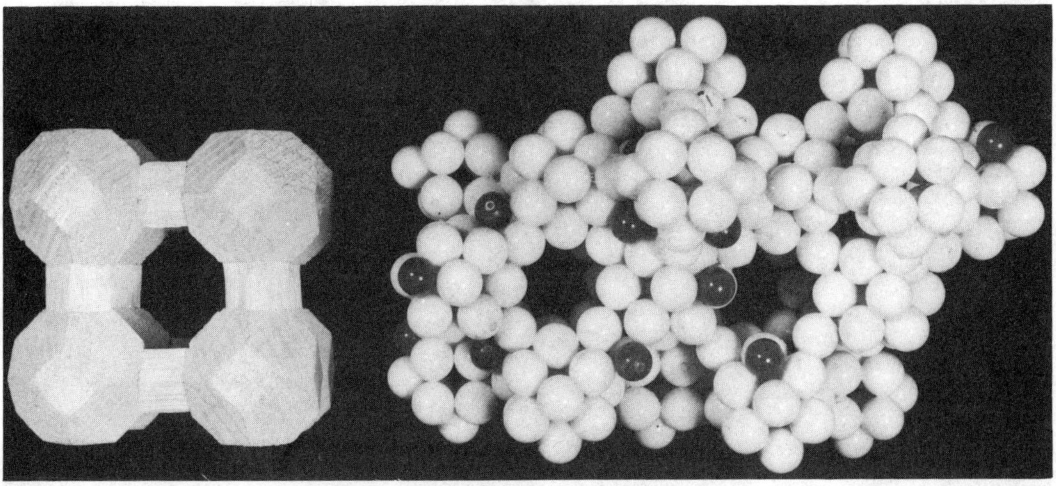

FIG. 4—*Model of the crystal structure (unit cell) of zeolite A showing the cubic array of truncated octhedra r.h., surrounding one adsorption cavity. A packing model at the left shows the apertures into the large cages and some of the exchange cations, Na+, near the main 4.2Å apertures. The truncated octahedra, each of which consists of 24 tetrahedra, are joined by cubic units of eight tetrahedra.*

strictions. Zeolite A has the following chemical composition:

$$Na_{12}Al_{12}Si_{12}O_{48} \cdot 27H_2O.$$

Eight of the twelve sodium ions lie in the center or near the center of the six rings in the hexagonal faces, and four occupy positions adjacent to the larger eight-membered ring. Dehydration of the zeolite leaves these four cations in positions near the eight-membered ring where they protrude into the apertures and may interfere with the movement of molecules. When replaced by other cations, such as calcium, the adsorption behavior changes. Inasmuch as one calcium ion replaces two sodium ions, the cations are removed from the eight-membered rings, and the zeolite has a larger effective pore size. Replacement of the sodium by larger univalent ions, such as potassium, restricts the eight-membered ring even more and produces a zeolite of a smaller pore size.

A second structure which illustrates some of the structural principles is the faujasite-type structure which is found in the synthetic zeolites X and Y (Breck and Flanigen, 1968). This framework is based upon a different configuration of the truncated octahedra. Eight of these are arranged in a unit cell in a tetrahedral configuration which produces eight super-cages of about 13Å in diameter. The cage is entered through four, 12-membered rings which produce a pore size of about 8Å. In this structure the cation positions are even more complex; there are five types of positions. Some of the typical cation positions are shown in Fig. 5. This complex cation siting accounts for the variation in the adsorptive and catalytic properties as a result of cation exchange. Zeolites A and X and Y have three-dimensional channel systems.

The mineral zeolite, mordenite, is different: its structure is more complex and provides for a one-dimensional channel system. The main channels do not intersect but are parallel to one particular crystallographic direction; they have a cross-sectional diameter of about 6 x 7Å (Fig. 6). The basic framework consists of chains of five-membered rings which are cross-linked by the sharing of neighboring oxygen ions. Perpendicular to the main channels there are smaller side channels providing pockets having a diameter of 3.9Å. These side channels, however, are blocked by cations (Meier, 1961). The crystal structure of mordenite provides for large channels, but the mineral does not exhibit this large-pore character in adsorption. A synthetic type, known as "Zeolon," does, however, have this large pore character. The reason for the difference is not yet clear but is probably due to some form of intercalated matter which blocks the channels.

The basic structures of most of the mineral zeolites and of the important synthetic zeolites are known. A structural classification is shown

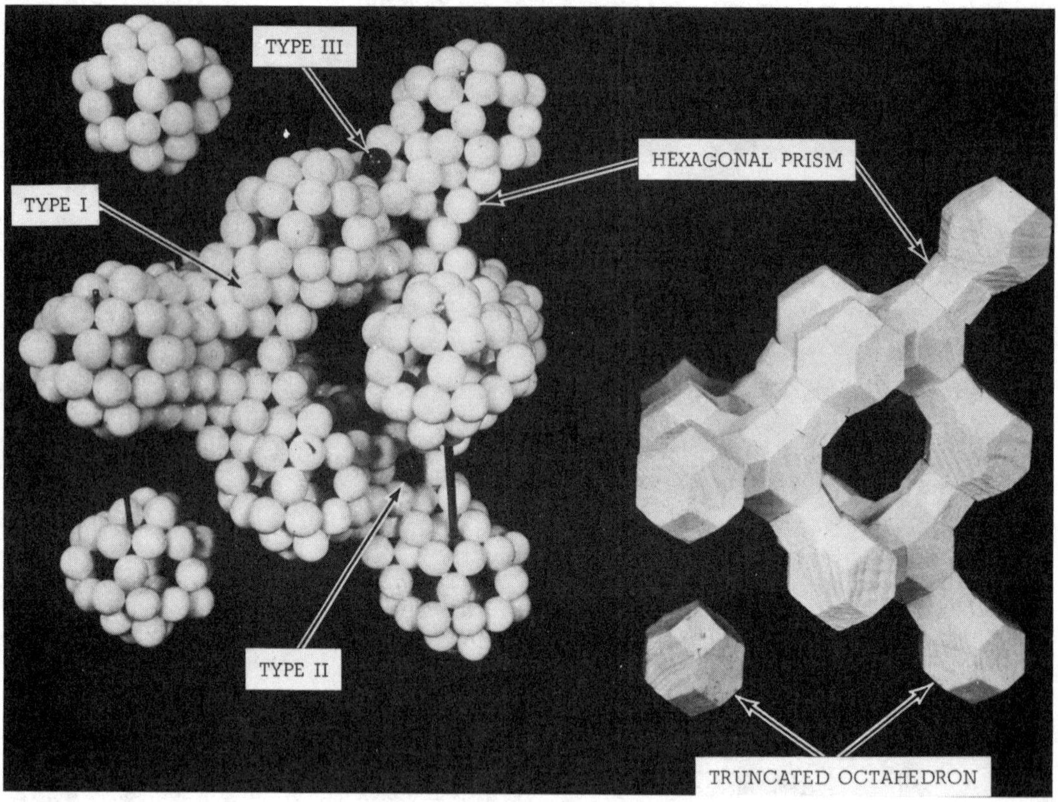

FIG. 5—*Model of the crystal structure (unit cell) of zeolites X and Y. On the right is shown the tetrahedral arrangement of truncated octahedra surrounding one large cavity. On the left is shown the packing model. The 12-membered ring is visible as well as the smaller 6-rings. Cation positions for three types of* Na^+ *ions in zeolite X are illustrated.*

in Table 4. The zeolites are arranged in groups based upon the secondary building unit dominant in the structure. There are now about 90 synthetic zeolites; some have structures related to known minerals. However, there are many more for which the structures are not known.

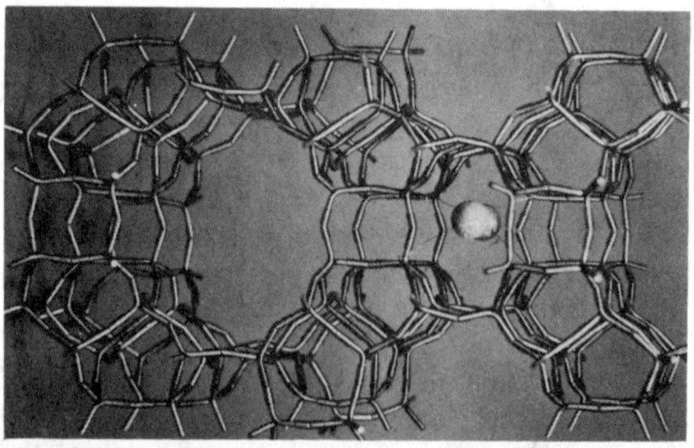

FIG. 6—*Model of the crystal structure of mordenite showing the 7Å channel and two sodium ions located in side pockets adjacent to the channels.*

Because no simple chemical system can be used to name synthetic complex aluminum silicates, there is a nomenclature problem. As defined by chemical composition, cation structure, framework structure, and structurally related physical properties, there are 34 species of zeolite minerals. There are established rules for the naming of new minerals. In the naming of synthetic zeolites the following practices are generally applicable (Breck, 1971):

1) The synthetic zeolite is designated by the letter(s) assigned to the zeolite by the original investigator (Law of Priority), for example, zeolite A, zeolite K-G, zeolite α, zeolite ZK-5, etc.

The designation of some synthetic zeolites by Greek letters and other combinations has been followed, e.g., synthetic zeolites α, β, ZK-4, and ZK-5.

2) The letters designate the zeolite as syn-

thesized. Thus, zeolite A designates the synthetic zeolite, $Na_{12}[(AlO_2)_{12}(SiO_2)_{12}]\cdot 27H_2O$ (pseudo-cubic unit cell) as prepared in the Na_2O, Al_2O_3, SiO_2, H_2O system, and zeolite L designates the zeolite $K_9[(AlO_2)_9(SiO_2)_{27}]\cdot 22H_2O$ as prepared in the K_2O, Al_2O_3, SiO_2, H_2O system. These refer to typical compositions.

3) In some cases, investigators have referred to a synthetic zeolite by the name of a related mineral, e.g., "synthetic analcime," "synthetic mordenite," etc. This approach is inaccurate and inadequate. The terms "mordenite-type," "analcime-type," etc., are preferable for indicating that the synthetic zeolite is related structurally to a mineral. In the light of current knowledge of the effect of cation type and location, silicon and aluminum distribution, and Si/Al ratio on the properties of the zeolite, a statement of identity is not warranted when only similarity is established.

4) Unfortunately, some confusion is unavoidable. The same letter has been used to name different species. Thus, Na-B refers to a synthetic analcime-type zeolite and Na-D indicates a mordenite-type. The letter B refers to the synthetic zeolite phases which Barrer and others have referred to as "P." Since the use of P for these phases has precedence in the published literature, it is preferred. Where a letter has been employed to refer to more than one synthetic zeolite, additional letter(s) are necessary; in these examples, the symbol for the predominant alkali metal is used.

5) A problem arises if the synthetic zeolite contains tetrahedral atoms other than aluminum and silicon, e.g., phosphorus, gallium, germanium. The element symbol is used as a prefix to indicate other tetrahedral atoms. Thus, P-L refers to zeolite L with phosphorus substitution in the framework.

6) It is necessary to indicate the particular composition in those cases where the letter(s) refer to a synthetic zeolite which can vary in framework composition. This is accomplished by giving the Si/Al ratio, unit cell contents, etc.

7) When a different cationic form of a synthetic zeolite is prepared by ion exchange, it may be so referred to, i.e., calcium exchanged (zeolite) A abbreviated as $Ca^{ex}A$ or CaA. A hyphen between Ca and A, Ca-A, could refer to a completely different zeolite. Thus:

$$\text{Ca exchanged A} = Ca^{ex}A = CaA \neq \text{Ca-A}.$$

It is assumed that the major cation component is, in this example, Ca^{2+} (greater than 50% exchange). In many cases, it is necessary to specify additionally the degree of exchange as percent of exchange equivalents, or in terms of the unit cell contents. Thus, $Ca_2Na_8[(AlO_2)_{12}(SiO_2)_{12}]\cdot XH_2O$ is equal to 33% exchange.

The designation of a zeolite as a separate species must be based on characteristic properties. Identification of a synthetic zeolite with a mineral species based upon a general similarity in X-ray powder diffraction patterns alone is inadequate. The properties to be included for complete characterization of a zeolite are the framework structure, cation content, chemical composition, and various structure-related properties such as stability, ion exchange behavior, and physical adsorption characteristics.

There are excellent reviews which cover the occurrence of mineral zeolites (Hay, 1966; Mumpton, 1973; Sheppard, 1971). The sedimentary zeolite minerals which occur in large deposits have not found extensive application as commercial molecular sieve materials. There are reasons for this. The use of zeolites as adsorbents and catalysts frequently requires materials of very uniform composition and purity. Variations in cation content and type dramatically affect the adsorption and catalytic characteristics; impurities likewise may affect catalysis. It is difficult to achieve a high degree of reproducibility in important industrial applications with a material that varies widely in mineral content, chemical composition, and impurities.

Some Aspects of Zeolite Chemistry

Dehydration—Zeolites useful as molecular sieve adsorbents and catalysts do not exhibit any change in structure when they are dehydrated under very severe conditions. Although in many zeolites water molecules affect the locations of the exchange cations, they do not have a primary structural function. In the stable zeolites the desorption of water is exhibited as a continuous plot when shown as a thermogravimetric curve of loss in weight vs. temperature. When hydrated the zeolite is a polyanionic framework which surrounds a water solution of positive ions. After dehydration the stable zeolite can withstand a temperature of about 700°C.

In the laboratory, zeolites are dehydrated before use by heating to a temperature of at least 350°C in a vacuum. In commercial practice, dehydration is brought about using a dry purge gas such as air, nitrogen, or methane in order to reduce the partial pressure of water vapor in contact with the zeolite.

The cation exchange behavior of several zeolites has been studied in detail (Rees, 1970;

TABLE 4—Classification of Zeolites

Name	Typical Unit Cell Contents	Type of Polyhedral Cage*	Framework Density, G per CC†	Void Fraction‡	Type of Channels§	Free Aperture of Main Channels, Å#
Group 1 (S4R)						
Analcime	$Na_{16}[(AlO_2)_{16}(SiO_2)_{32}]\cdot 16\,H_2O$		1.85	0.18	One	2.6
Phillipsite	$(K,Na)_{10}[(AlO_2)_{10}(SiO_2)_{22}]\cdot 20\,H_2O$		1.58	0.31	Two	4.2–4.4
Paulingite	$K_{68}Na_{13}Ca_{3.6}Ba_{1.5}[(AlO_2)_{152}(SiO_2)_{520}]\cdot 700\,H_2O$	α, γ, δ (10-hedron)	1.54	0.49	Three	3.9
Yugawaralite	$Ca_2[(AlO_2)_4(SiO_2)_{12}]\cdot 8\,H_2O$		1.81	0.27	Two	3.5
Group 2 (S6R)						
Erionite##	$(Ca,Mg,K_2Na_2)_{4.5}[(AlO_2)_9(SiO_2)_{27}]\cdot 27\,H_2O$	ε, 23-hedron	1.51	0.35	Three	3.6 × 5.2
Offretite##	$(K_2,Ca)_{2.7}[(AlO_2)_{5.4}(SiO_2)_{12.6}]\cdot 15\,H_2O$	ε, 14-hedron (II)	1.55	0.40	Three	3.6 × 5.2, ∥a 6.4, ∥c
Omega¶	$Na_{6.8}, TMA_{1.6}[(AlO_2)_8(SiO_2)_{28}]\cdot 27\,H_2O$	γ, 14-hedron (II)	1.65	0.38	One	7.5
T	$Na_{1.2}K_{2.8}[(AlO_2)_4(SiO_2)_{14}]\cdot 14\,H_2O$	ε with D6R	1.50	0.40	Three	3.6 × 4.8
Group 3 (D4R)						
A	$Na_{12}[(AlO_2)_{12}(SiO_2)_{12}]\cdot 27\,H_2O$	α, β	1.27	0.47	Three	4.2
P	$Na_6[(AlO_2)_6(SiO_2)_{10}]\cdot 15\,H_2O$		1.57	0.41	Three	3.5
Group 4 (D6R)						
Faujasite	$(Na_2,K_2,Ca,Mg)_{29.5}[(AlO_2)_{59}(SiO_2)_{133}]\cdot 235\,H_2O$	β, 26-hedron (II)	1.27	0.47	Three	7.4
X	$Na_{86}[(AlO_2)_{86}(SiO_2)_{106}]\cdot 264\,H_2O$	β, 26-hedron (II)	1.31	0.50	Three	7.4
Chabazite	$Ca_2[(AlO_2)_4(SiO_2)_8]\cdot 18\,H_2O$	20-hedron	1.45	0.47	Three	3.7 × 4.2
Gmelinite	$Na_8[(AlO_2)_8(SiO_2)_{16}]\cdot 24\,H_2O$	14-hedron (II)	1.46	0.44	Three	3.6 × 4.9, ∥a 7.0, ∥c
ZK-5	$(R_2,Na_2)_{30}[(AlO_2)_{30}(SiO_2)_{66}]\cdot 98\,H_2O$	α, γ	1.46	0.44	Three	3.9
L**	$K_9[(AlO_2)_9(SiO_2)_{27}]\cdot 22\,H_2O$	ε	1.61	0.32	One	7.1
Group 5 (T_6O_{10})††						
Natrolite	$Na_{16}[(AlO_2)_{16}(SiO_2)_{24}]\cdot 16\,H_2O$		1.76	0.23	Two	2.6 × 3.9
Thomsonite	$Na_4Ca_8[(AlO_2)_{20}(SiO_2)_{20}]\cdot 24\,H_2O$		1.76	0.32	Two	2.6 × 3.9
Edingtonite	$Ba_2[(AlO_2)_4(SiO_2)_6]\cdot 8\,H_2O$		1.68	0.36	Two	3.5 × 3.9
Group 6 (T_8O_{16})‡‡						
Mordenite	$Na_8[(AlO_2)_8(SiO_2)_{40}]\cdot 24\,H_2O$		1.70	0.28	Two	6.7 × 7.0, ∥c 2.9 × 5.7, ∥b
Dachiardite	$Na_5[(AlO_2)_5(SiO_2)_{19}]\cdot 12\,H_2O$		1.72	0.32	Two	3.7 × 6.7, ∥b 3.6 × 4.8, ∥c
Epistilbite	$Ca_3[(AlO_2)_6(SiO_2)_{18}]\cdot 18\,H_2O$		1.76	0.25	Two	3.2 × 5.3 ∥a 3.7 × 4.4 ∥c

Group 7 ($T_{10}O_{20}$)§§					
Heulandite	$Ca_4[(AlO_2)_8(SiO_2)_{28}] \cdot 24\ H_2O$	1.69	0.39	Two	4.0 X 5.5 ‖a 4.0 X 7.2 ‖c
Stilbite	$Ca_4[(AlO_2)_8(SiO_2)_{28}] \cdot 28\ H_2O$	1.64	0.39	Two	4.1 X 6.2 ‖a 2.7 X 5.7, ‖c

Source: Breck and Flanigen, 1968

* Of the five space filling solids, three (cube, hexagonal prism, and truncated octahedron) are found as polyhedral units in zeolite frameworks. The cube is the double four-ring (D4R) as shown here. The double six-ring (D6R) is the hexagonal prism or 8-hedron. The α cage is the Archimedean semiregular, solid, truncated cuboctahedron referred to also as a 26-hedron, Type 1. The β cage is the truncated octahedron or 14-hedron, Type 1. The γ cage is the 18-hedron and the ϵ cage the 11-hedron.

† The framework density is based on the dimensions of the unit cell of the hydrated zeolite and framework contents only. Multiplication by 10 gives the density in units of tetrahedra per 1000 $Å^3$.

‡ The void fraction is determined from the water content of the hydrated zeolite.

§ Refers to the network of channels which permeate the structure of the hydrated zeolite. Considerable distortion may occur in the Group 6 and 7 zeolites upon dehydration.

Based upon the structure of the hydrated zeolite.

Erionite and offretite may also be considered to consist of double 6-rings linked by single 6-rings.

¶ Zeolite Ω may be considered to consist of single 6-rings linked by double 12-rings.

** Zeolite L consists of double 6-rings linked by single 12-rings.

†† The T_5O_{10} refers to the unit of 5 tetrahedra as given by Meier for the 4–1 type of SBU. See Meier, 1968.

‡‡ The T_8O_{16} unit refers to the characteristic configuration of tetrahedra shown in Fig. 8b of Meier, 1968.

§§ The $T_{10}O_{20}$ unit is the characteristic configuration of tetrahedra shown in Fig. 8a of Meier, 1968.

‖ Dimension of channel parallel to either the a, b, or c axis.

Sherry, 1971). Typical types of ion exchange isotherms are shown in Fig. 7 (Breck, 1974). The isotherms for four univalent ions show a variation in selectivity ranging from high selectivity for Ag^+ to the low selectivity for Li^+. For some ions, such as K^+, the selectivity varies from positive to negative. In other cases, such as Ca^{++}, the isotherm does not terminate at the theoretical limit of complete substitution but at a lower point. This type of isotherm is commonly observed in zeolite Y for rare earth cations as well as for calcium and cesium. Certain sites are inaccessible to large cations and some polyvalent ions, so complete exchange cannot occur. In a few cases ion exchange hysteresis is observed, such as for Sr^{2+} sodium exchange in zeolite X.

Various methods of achieving hydrogen or proton exchange have been developed. In most instances the treatment of zeolites with a strong acid solution results in complete decomposition. In zeolites a certain degree of hydrogen exchange can be achieved by mineral acids and ultimately the aluminum is removed from the framework tetrahedral sites. This dealumination has been achieved in mordenite and clinoptilolite (Barrer and Makki, 1964; Kranich, et al., 1971).

The formation of hydroxyl groups in zeolites is important because hydroxyls play an important role in hydrocarbon catalysis (Bolton and Bujalski, 1971). Hydroxyl groups probably exist on the terminal surfaces of the zeolite crystals, but little is known about the terminal surfaces. Investigation of the formation and nature of hydroxyl groups in zeolites has been extensive and has utilized various techniques including thermal analysis and infrared spectroscopy (Ward, 1971). Hydroxyl groups are formed by two methods:

1) Deammoniation of NH_4^+ exchanged zeolites.

2) Hydrolysis of polyvalent cations (Rabo and Poutsma, 1971). These methods are illustrated in Fig. 8.

After dehydration, cations on the surface of the large cavities in the zeolites X and Y are not well shielded and have a low coordination

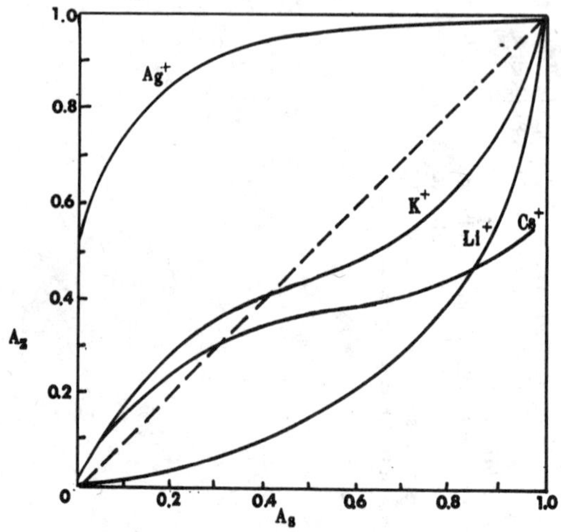

FIG. 7—*Exchange isotherms for univalent cation exchange in zeolite X at 25°C.* $Na^+ \rightarrow Ag^+$, *0.2 N;* $Na \rightarrow K^+$, *0.2 N;* $Na^+ \rightarrow Li^+$, *0.2 N; and* $Na^+ \rightarrow Cs^+$, *0.1 N. The equivalent fraction of exchange ion in the zeolite at equilibrium for the reaction* $Na^+_{zeolite} + A^+_{solution} = A^+_{zeolite} + Na^+_{solution}$ *is given by* A_Z *and in solution by* A_S. *The* CS^+ *isotherm terminates at* $Cs_Z \simeq 0.5$.

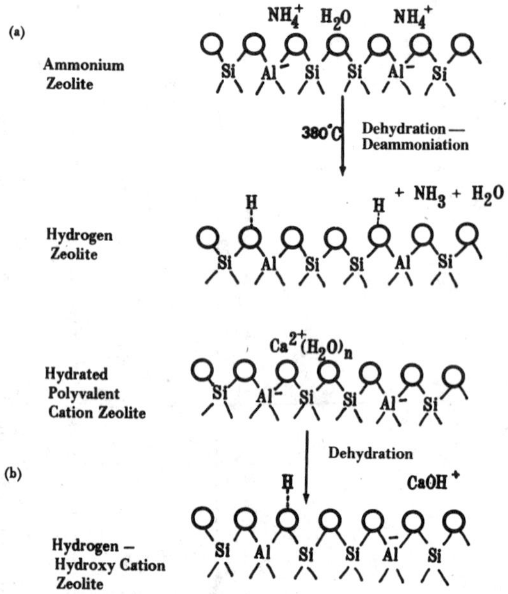

FIG. 8—*Schematic representation of the formation of hydroxyl groups in zeolites. (a) The example is zeolite Y, Si/Al = 2. After* NH_4^+ *exchange in water solution, the zeolite is dehydrated. The temperature is raised,* NH_3 *is evolved, leaving hydrogen atoms in place of cations but attached to framework oxygen atoms forming hydroxyl groups. The acidic nature of these hydrogens is responsible for activity in hydrocarbon reactions. (b) During dehydration of polyvalent cation forms of zeolite Y, some hydrolysis of the cation occurs, resulting in formation of a univalent hydroxy cation and hydroxyl groups on the framework. Trivalent rare earth cations produce more hydroxyl groups.*

with the framework oxygens. Thus, electrical fields extend into the main cavities and may interact with adsorbed species. The effect of these fields varies with the silicon/aluminum ratio of the zeolite. Increasing the silicon/aluminum ratio from 1 to 2 increases the distance between cation sites. This increase results in the location of the divalent ions, such as calcium, in a nonsymmetrical position nearer one negative site than the other. The effect of this is to create a strong electrostatic field. Calculated electrostatic fields show that in the vicinity of trivalent cations, the electrostatic fields are as great as one volt per Angstrom. This field can cause substantial shifts of bonding electrons and adsorbed molecules. It contributes to the formation of hydroxyl groups on the polyvalent ion (Fig. 8) and acidic hydrogens on the oxygen framework. This is particularly true of divalent ions such as calcium and of trivalent ions such as those of the rare earths. During dehydration, hydrolysis of the polyvalent ions, in this case illustrated by calcium, results in the formation of a hydroxyl group for every two exchanged cation sites.

Synthesis of Zeolites

Extensive studies have been carried out in an attempt to synthesize "minerals" (Morey and Ingerson, 1937). In the preparation of zeolites early investigators attempted to duplicate what was then supposed to be a natural process without much success. In the late 1940s the initial work carried out by Union Carbide scientists was based on a new departure —a low temperature process using reactive ingredients. The results were very successful and resulted in the preparation of many different types of synthetic zeolites.

This approach was primarily based on the use of a freshly prepared, highly reactive aluminosilicate gel, or hydrogel. The term "gel" in this sense means a hydrous metal aluminosilicate mixture which is prepared from either aqueous solutions or reactive solid phases. Typical gels are prepared from aqueous solutions of sodium aluminate, sodium silicate, and sodium hydroxide or perhaps another alkali metal hydroxide such as potassium hydroxide. The zeolites form when the gels are crystallized at temperatures ranging from room temperature to 200°C at atmospheric or autogeneous pressure. This method is well suited for the alkali metals since they form soluble hydroxide, aluminates, and silicates; and it is possible to prepare reactive and homogeneous gel mixtures (Breck, 1974; Breck and Flanigen, 1968).

The gel structure is produced by the polymerization of the aluminate and the silicate anions. The composition and structure of this hydrous polymer gel seems to be controlled by the size and structure of the polymerizing species. Differences in the chemical structure and molecular weight distribution of the starting species in the silicate solutions lead to differences in the gel structures and major differences in the zeolite phases produced. During the crystallization of the gel the sodium ions, aluminate, and silicate components undergo a rearrangement into the zeolite crystalline structure. This comes about by a depolymerization of the gel by the hydroxyl ions present in the reaction mixture. Following this, the nuclei formed as small crystallites grow into the zeolite crystals.

A typical crystallization curve is shown in Fig. 9. Large numbers of crystallites nucleate from the supersaturated gel and the final product consists of a finely divided white crystalline powder that is a few microns in particle size. By controlled variation of parameters such as composition, temperature, and type of reactant, many different zeolite species have been prepared at Union Carbide and by others (Barrer, 1966; Breck and Flanigen, 1968). Some of these synthetic species may be structurally related to mineral zeolites; others seem to have no known mineral analogue.

Hydrogel Process

The process to make a zeolite by the hydrogel process is shown in Fig. 11 (Breck, 1974).

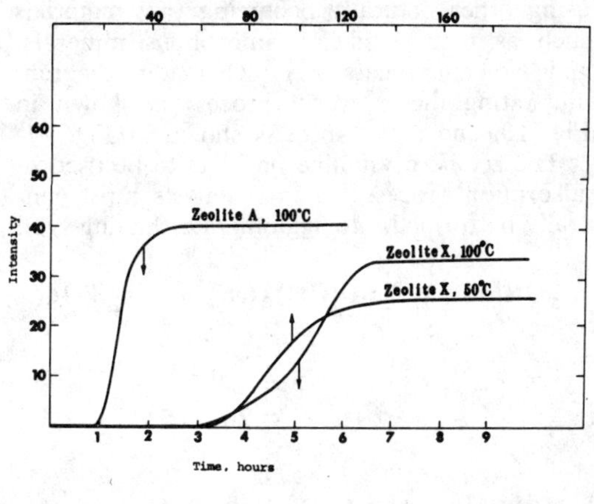

FIG. 9—*Crystallization of zeolite A and zeolite X at 50°C and 100°C as a function of time. Intensity, in arbitrary units, indicates the degree of crystallization as determined by X-ray powder methods.*

TABLE 5—Molecular Sieve Zeolite Preparation Processes

Process	Reactants	Products
Hydrogel	Reactive oxides Soluble silicates Soluble aluminates Caustic	High-purity powders Gel preform Zeolite in gel matrix
Clay conversion	Raw kaolin Meta-kaolin Calcined kaolin Acid treated clay Soluble silicate Caustic Sodium chloride	Low-to-high purity powder Binderless, high-purity preform Zeolite in clay-derived matrix
Other	Natural SiO_2 Amorphous minerals Volcanic glass Caustic $Al_2O_3 \cdot 3H_2O$	Low-to-high purity powder Zeolite on ceramic support Binderless preforms

Alumina trihydrate, $Al_2O_3 \cdot 3H_2O$, is dissolved in hot caustic and mixed with sodium silicate solution in the gel makeup tank until a homogeneous gel is formed. This gel is pumped to an ambient age tank, if required, and then subsequently crystallized after several hours at 200°F to form the zeolite. Progress of crystallization is followed by X-ray diffraction.

The first commercial processes were based on the original laboratory synthesis using amorphous hydrogels. Processes now used for the manufacture of commercial molecular sieves may be classified into three groups (Table 5). They are (1) the preparation of zeolites in high purity from reactive aluminosilicate gels; (2) the conversion of minerals of the clay group, kaolin in particular; and (3) processes using other naturally occurring raw materials such as natural silicas, amorphous minerals, and volcanic glass. A schematic diagram illustrating the hydrogel process is shown in Fig. 10, and a flowsheet is shown in Fig. 11.

The zeolite crystalline powders to be used in adsorption processes or as catalysts must generally be formed into agglomerates having high physical strength and high attrition resistance. Methods for forming the powders into agglomerates include the addition of an inorganic binder, generally a kaolin-type clay, to the high-purity zeolite powder in a wet mixture. The clay-zeolite mixture is extruded into cylindrical pellets or formed into beads which are subsequently calcined to convert the clay to an amorphous binder of considerable mechanical strength. The preparation of zeolite binder agglomerates as spheres or cylindrical pellets which have high mechanical attrition resistance alone is not difficult. However, in order to use these materials in adsorption or catalytic processes, the diffusion characteristics of the zeolite must not be altered. Consequently, the binder component should be such that a macroporosity is maintained that does not increase the diffusion resistance of the zeolite crystal itself.

The second method involves hot pressing the zeolite powder into binderless particles. When subject to high pressures at an elevated temperature, a mass of the zeolite crystals may self-bond into a 100% zeolite pellet. This method has not achieved extensive industrial use.

$$\mathrm{NaOH(aq) + NaAl(OH)_4(aq) + Na_2SiO_3(aq) \xrightarrow{t_1} [Na_a(AlO_2)_bSiO_2)_c \cdot NaOH \cdot H_2O]\ gel}$$

$$\downarrow t_2$$

$$\mathrm{Na_x[(AlO_2)_x(SiO_2)_y] \cdot MH_2O + solution}$$

Zeolite crystals

t_1 = 20°C
t_2 = 20–175°C

FIG. 10—*Scheme showing zeolite crystallization from typical reactants by the hydrogel process. Parameters which must be controlled within critical limits include the temperature* t_1 *and* t_2, *the composition of the gel, the nature of the reactants, and the time.*

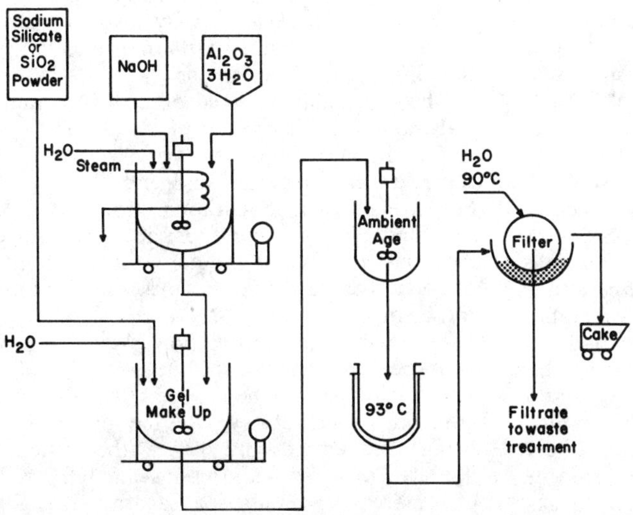

FIG. 11—*Hydrogel process* (source: Breck, 1974).

	Wt (To Produce 1000 Lb, Dry Basis)		
	4A	13X	Y
Raw Materials			
Sodium silicate*	1350	2000	——
SiO_2 powder†	——	——	1450
Alumina trihydrate‡	575	500	340
Caustic, 50% NaOH	870	1600	1400
Water	3135	7687	5300
Gel Composition (Moles)			
Na_2O	2.04	4.09	4.0
Al_2O_3	1.0	1.00	1.0
SiO_2	1.75	3.0	10.6
H_2O	70	176	161

* 9.4% Na_2O, 28.4% SiO_2.
† 95% SiO_2.
‡ 65% Al_2O_3, 35% H_2O.

A third method is the "preform" approach. The zeolite precursor reactant is formed into pellets, either as a gel or as a clay agglomerate. The reactants in these pellets are then converted *in situ* to zeolite crystals producing an agglomerate which has a high zeolite content, as high as 95%.

Application of Synthetic Zeolites as Molecular Sieves

The two major applications of synthetic zeolites are in adsorption and catalysis.

Adsorption

The zeolites are high capacity, selective adsorbents for two reasons:

1) They separate molecules based upon molecular size and configuration of the molecule relative to the size and geometry of the zeolite structure.

2) They adsorb molecules, in particular those with a permanent dipole moment and other interaction effects, with a selectivity that is not found with conventional adsorbents.

Two types of separation may occur: one based upon the molecular sieve effect and the second upon preferential adsorption. The sodium form of zeolite A will readily adsorb materials such as water, carbon dioxide, sulfur dioxide, and all hydrocarbons containing one or two carbon atoms. Propane and higher hydrocarbons are physically excluded with the exception of propylene which is adsorbed slowly. When the sodium ions are exchanged by calcium the effective pore size increases to a point corresponding to about 30% exchange by Ca^{++}. This zeolite will then adsorb compounds such as straight-chain paraffins, olefins, alcohols, and cyclopropane but not isoparaffins such as isobutane or aromatics. If the sodium ions are exchanged by potassium, the opposite effect is observed, i.e., the effective pore size is decreased and the potassium-exchanged-type A zeolite will not adsorb any hydrocarbon or even oxygen at low temperatures. Because the effective pore sizes based upon the molecular diameters are about 3Å for the potassium A, 4Å for the sodium A, and 5Å for the calcium A, these materials are commercially known as type 3A, type 4A, and type 5A, respectively. For the sodium X zeolite, all molecules with diameters up to about 8Å are adsorbed. The exchange of the sodium by calcium ions in this zeolite reduces the effective pore size by about 1Å. By convention these zeolites are termed type 13X for the sodium form and 10X for the calcium form.

If two molecules are capable of entering the micropore system, then the zeolite retains one preferentially to the other on the basis of polarity or other molecule-zeolite interaction effects. Polar and unsaturated molecules are selectively adsorbed. Water, a strongly polar molecule, is very selectively adsorbed. The zeolite cavities are essentially filled with water at very low concentrations and, in the case of zeolite A, each cavity retains 20 molecules at vapor pressures as low as 1/1000 of an atmosphere at room temperature. The cavity is essentially filled. For gases such as nitrogen and oxygen, nitrogen is preferentially adsorbed due to the interaction of the quadrupole of the nitrogen molecule with the cations of the zeolite.

Although the N_2 molecule is only 0.2Å larger than the O_2 molecule, this difference is sufficient for rejection by zeolite NaA at cryogenic temperatures. Argon is intermediate in its behavior. At cryogenic temperature, nitrogen and argon diffuse with considerable difficulty, and in any reasonable period of time, true adsorption equilibrium cannot be reached. In addition to this process of activated diffusion, there is also the effect of temperature on the thermal vibration of the oxygen atoms which surround the aperatures. A variation in vibrational amplitude of 0.1–0.2Å can be expected over the temperature interval of 80–300°C.

Normal paraffins have an effective cross-sectional diameter of 4.3Å whereas a branched chain hydrocarbon is 5Å or greater. Aromatic hydrocarbons are also too large for adsorption in the 5A (calcium A) zeolite. This provides an efficient way for recovering normal paraffin hydrocarbons from various hydrocarbon mixtures.

The use of other adsorbents such as carbon and silica gel in practical separations has not generally involved the adsorption of one component from a mixture to the exclusion of all others. The use of zeolites as selective adsorbents makes such separations possible.

For partial molecular sieve action, it is important to know the relative adsorptivities of the various components. At present there is no good correlation between the adsorption of a single component on a zeolite and its behavior in a mixture although attempts at predicting relative adsorptivities from the individual adsorption isotherms of the separate components have been made. It is necessary to study the equilibria of the mixture experimentally.

The use of molecular sieves in many separations is based upon partial molecular sieve action where the dimensions of the molecular sieve pores are of the same magnitude as the dimensions of the molecule and an activation energy for diffusion is required. In the case of the larger pore zeolites such as the sodium X type, the channels are large, 7.4Å, compared to most molecules, and diffusion effects such as those that are encountered in type 4A or 5A are not as important. In some cases total molecular sieve behavior may revert to partial behavior as the result of thermal effects upon the diffusional activation energy.

The molecular sieve adsorbents in commercial use today are primarily those based on the A, X, Y, and mordenite-type structures. To some extent, chabazite- and erionite-type zeolites are utilized. Separation processes utilize the fundamental factors previously discussed. Adsorption processes require regeneration of the adsorbent after it becomes saturated with the adsorbate.

In general, adsorption processes utilize two or more beds in unison (Barry, 1960; Collins, 1968). While one bed is being used for the adsorption part of a cycle the other bed is being regenerated and hence the system is operated in a semi-continuous manner. A list of some typical commercial separations is given in Table 6. Many other separations are possible (Milton, 1968).

The preferred material for the dehydration of unsaturated hydrocarbon streams is the zeolite KA. Its pore size of 3Å excludes all hydrocarbons including ethylene and other olefins which might tend to undergo secondary polymerization reactions.

A molecular sieve effect is utilized in the drying of the refrigerants. NaA is used because the size of refrigerant molecules is such that they are not adsorbed by the zeolite. NaA is also utilized in the prepurification of natural gas to remove impurities such as water and CO_2; it excludes hydrocarbon molecules of the size of propane and larger.

CaA is uniquely suited to the separation and recovery of normal paraffin hydrocarbons from various hydrocarbon feedstocks such as natural gasoline and kerosene. Larger molecules such as the isomeric paraffin and cyclic hydrocarbons are excluded. More than a half dozen plants are now in operation to recover normal paraffins which are subsequently utilized in the manufacture of biodegradeable detergents.

Many important commercial separations are based upon differences in the relative adsorption selectivity. An outstanding example of this is the use of zeolites such as CaA and CaX in the removal of nitrogen from air to produce an oxygen enriched gas (Batta, 1972). Oxygen produced by this selective adsorption process is used in the treating of secondary sewage effluent.

The high selectivity shown by zeolites for sulfur compounds such as hydrogen sulfide and mercaptans is utilized in the removal of these

TABLE 6—Molecular Sieve Adsorbents

Molecular Sieve Basic Type	Molecules Adsorbed*	Molecules Excluded	Typical Applications
3A	H_2O, NH_3, He (molecules with an effective diameter < 3Å)	CH_4, CO_2, C_2H_2, O_2, C_2H_5OH, H_2S (molecules with an effective diameter > 3Å)	Drying cracked gas, ethylene, butadiene, and ethanol
4A	H_2S, CO_2, C_2H_6, C_3H_6, C_2H_5OH, C_4H_6 (molecules with an effective diameter < 4Å)	C_3H_8, compressor oil (molecules with an effective diameter > 4Å)	Drying natural gas, liquid paraffins, and solvents. CO_2-removal from natural gas
5A	*n*-Paraffins, *n*-Olefins, *n*-C_4H_9OH (molecules with an ellective diameter < 4A)	Iso-compounds, all 4-carbon rings (molecules with an effective diameter > 5Å)	*n*-Paraffin recovery from naphtha and kerosene
10X	Iso-paraffins, iso-olefins (molecules with an effective diameter < 8Å)	Di-*n*-butylamine and larger (molecules with an effective diameter > 8Å)	Aromatic separation
13X	Di-*n*-butylamine (molecules with an effective diameter < 10Å)	$(C_4F_9)_3$-N (molecules with an effective diameter > 10Å)	Desulfurization, general drying, simultaneous H_2O and CO_2-removal

* Each type adsorbs listed molecules plus those of preceding type.

sulfur compounds from reformer recycle hydrogen. Water, carbon dioxide, and sulfur compounds are likewise removed from natural gas.

The zeolite CaX shows a selectivity for certain individual components of aromatic hydrocarbon mixtures. In practice, various types of cycles are used which differ in terms of the desorption part of the process cycle. Generally, the adsorption part of the process cycle is rapid (Barry, 1960).

Molecular Sieve Zeolites in Catalysts

As in adsorption, catalytic reactions take place within the cavities of crystalline zeolites—hence, the voids must be accessible to the reactants. The zeolites important in catalysis are those which have the largest pore sizes and a maximum of available void volume. Inasmuch as zeolites of this type are either rare or nonexistent in nature, molecular-sieve catalysts are based upon the synthetic zeolites. (All of the extensively occurring zeolite minerals are of the small-pore variety and at best adsorb normal-paraffin hydrocarbons with some difficulty.) Sieving effects occur in catalysis as in adsorption. Not only must the reactants permeate the zeolite structure, but the products likewise must exit.

There may be a minor role played by the external surface which is about 1% of the total intracrystalline zeolite surface available. In the cracking of gas-oil molecules, the very large molecules cannot enter the intracrystalline cavities and therefore must crack on the external surface. Subsequently, the smaller molecules may enter into secondary reactions in the intracrystalline cavities. The reduction of olefin content in the product using zeolite cracking catalysts is attributed to factors of this type.

Some of the features of zeolite catalysts and the factors which are concerned with catalytically active sites include: (1) the zeolite framework structure, since it determines the pore system; (2) the cation type (size, charge), since it influences the electrostatic fields and the active catalytic sites; (3) the cation locations in zeolites; (4) the Si/Al ratio which may affect activity and stability; (5) the presence of proton donors in hydroxyl groups; and (6) the presence of active elemental metals in a highly dispersed state.

Molecular sieves catalyze other organic reactions such as dehydration (Venuto and Landis, 1968). In this instance an alcohol may be dehydrated to form an olefin with a zeolite acting as a dehydration catalyst. Zeolite catalyst activity for dehydration depends on the cation size and charge. A higher activity is found in zeolite systems with a greater cationic charge density and therefore higher electrostatic field strength. A second type of reaction is polymerization. This is important in catalyst aging since an olefin may polymerize within the pore structure and plug the pores and consequently not desorb, as has been observed in the alkylation of benzene with ethylene.

The hydrogen form of a synthetic type of mordenite catalyst is used to react propylene with carbon monoxide. Using zeolite X, ammonia reacts with aromatics such as toluene to form benzonitrile.

Zeolite catalysts are best known in hydrocarbon reactions, in particular cracking (Rabo and Poutsma, 1971). Although several mechanisms have been proposed for the carbonium ion reactions involving hydrocarbons and zeolite catalysts, there is no agreement on the source of carbonium-ion activity although it may be related to a combination of effects. These include the interaction of Lewis sites (due to three coordinate aluminum) with available hydroxyl groups (Bronsted sites). The presence of polyvalent cations, in particular the rare earths, enhances the acidity of the hydrogen in the hydroxyl group by an inductive effect. In one proposal the hydroxyl groups are said to play a role which involves consumption of the hydrogens through reaction with olefins. The protons are consumed during reaction, and when depleted the reaction ceases. In this sense the hydrogen is a reactant (Bolton and Bujalski, 1971).

As exemplified in catalytic cracking, catalysts containing polyvalent ion-exchange X or Y exhibit higher activity and better selectivity as compared with conventional silica-alumina cracking catalyst. Typically, up to 20% higher gasoline yields are obtained. More than 95% of installed cracking catalyst capacity is now based on molecular sieve catalysts.

In isomerization reactions, noble metal containing polyvalent ion exchanged Y catalysts are used to upgrade low octane number normal paraffiins. These catalysts have a long life and resistance to sulfur poisoning and, unlike conventional catalysts, are noncorrosive. In hydrocracking, similar molecular sieve catalysts increase the refinery gasoline yield and octane number. The performance in a typical hydrocracking process is excellent from both product selectivity and rate of deactivation standpoints. In reforming petroleum naphthas, the molecular sieve catalysts exhibit excellent yield-octane performance and show good resistance to nitrogen poisoning. Anticipated competitive features of molecular sieves in reforming as in

TABLE 7—Potential Molecular-Sieve Catalyst Applications

Process	Feed/Products	Competitive Features
Catalytic reforming	Naphtha/gasoline	Activators unnecessary; gasoline with reduced sensitivity; feed pre-treatment minimized.
Polymerization	Low molecular weight olefins/gasoline	Noncorrosive.
Alkylation	Aromatic and low-value olefinic streams/valuable alkylated aromatics	Noncorrosive; feed pre-treatment minimized.
Hydrodealkylation	Toluene/benzene	High activity; improved selectivity.
Hydrogenation	Benzene/cyclohexane	Improved resistance to sulfur poisoning.
Hydrogenation of fats, oils	Unsaturated oils/saturated oils	High selectivity; low isomerization.
Selective hydrogenation	Straight and branched olefins/ n-alkanes and branched olefins	Separation problems minimized.
Methanation	Synthesis gas/methane	High yields; resistant to poisons.
Dehydrogenation	Ethylbenzene/styrene	Improved selectivity.
Dehydration	Alcohols + acids/esters	Improved rates and yields.
Dehydrohalogenation	Alkylhalides/olefins	Molecular size; selectivity.

TABLE 8—Comparison of Natural Zeolites vs. Synthetic Zeolites

Property	Zeolite Minerals	Synthetic Zeolites
Availability	Restricted to 8–9 species: analcime clinoptilolite, mordenite, phillipsite, erionite and chabazite. Also laumontite and ferrierite.	Many types are available by manufacture from inexpensive raw materials.
Purity	Generally impure. Bedded deposits are mixtures. Deposits of the useful zeolites in high purity are limited. Cation contents are complex. Undesirable impurities such as iron are common.	Manufactured as pure species. Other cationic forms prepared by cation exchange.
Pore size	Limited. Largest pore size in chabazite and erionite limits adsorption to *n*-paraffins.	Available from 3Å up to 8Å. Adsorb large molecules in catalysis and adsorption, or reject small molecules in adsorptive separation.
Pore volume (adsorption capacity)	Limited. Only chabazite and erionite have good pore volume.	Up to 50% by volume.
Catalytic activity	Limited due to small pore sizes and impurities.	Wide applicability.

numerous other catalytic process are summarized in the accompanying Table 7 (Collins, 1968). Potential catalytic applications range from methanation to aromatic alkylation.

Synthetic Zeolites—Advantages

Important aspects of zeolite minerals and synthetic zeolites (which must be considered in major applications) are compared in Table 8.

The use of molecular sieves in adsorption and catalytic processing usually requires that they meet closely controlled specifications in terms of purity, mechanical properties, and porosity. Although some zeolite minerals have been utilized to a limited extent, synthetic zeolites are better suited to meet the stringent requirements imposed on molecular sieve adsorbents in adsorption and catalytic processes.

Zeolites in Sedimentary Rocks *

RICHARD A. SHEPPARD †

Zeolites—crystalline hydrated aluminosilicates of the alkalis and alkaline earths—are important rock-forming constituents in sedimentary rocks and are potentially valuable industrial minerals. They have a framework structure that encloses interconnected cavities occupied by the relatively large cations and water molecules. The cations, chiefly sodium, potassium, and calcium, and the water have considerable freedom of movement within the structure and give the zeolite their cation-exchange and reversible-dehydration properties. Some zeolites have essential contents of barium, strontium, or magnesium. The basic structure

* Reprinted courtesy US Geological Survey, from Professional Paper 820.

† US Geological Survey, Denver, CO.

of all zeolites consists of $(Si,Al)O_4$ tetrahedra wherein each oxygen is shared between two tetrahedra; thus, the atomic ratio O:(Si+Al) is 2. The net negative charge of the structure is balanced by the cations. The porous framework of the zeolites enables them to act as molecular sieves for the separation of molecular mixtures according to the size and shape of the molecular compounds.

More than 30 distinct species of zeolites occur in nature. Numerous zeolites have also been synthesized, but many of these have no natural counterparts. Most zeolites have a specific gravity of 2.0–2.3 and indices of refraction of 1.44–1.52. Zeolites rich in barium or strontium commonly have higher specific gravities and indices of refraction. Because the zeolites have similar physical properties, X-ray diffraction techniques are generally used for identification.

Synthetic zeolites have been produced commercially in the United States since the early 1950s. These synthetic zeolites are used principally as catalysts, selective sorbents, and desiccants; they are now utilized in about 90% of the petroleum catalytic cracking installations and have greatly increased the recovery of gasoline. Natural zeolites, however, have found only limited use in this country. Clinoptilolite is used to remove cesium from radioactive wastes (Brown, 1962), and chabazite is used to desiccate mildly acidic natural gas. Zeolitic tuff has been used as a pozzolan in cement.

In Japan, on the other hand, zeolites have been extensively mined and utilized since their discovery there in sedimentary deposits in 1949. Clinoptilolite and mordenite are the only zeolites currently mined in Japan, and they are utilized chiefly in the fields of agriculture and animal husbandry. Minato and Utada (1969) described the following uses: desiccant for gases, separator of oxygen from air, adsorbent for obnoxious odors in farmyards, filler and whitening agent for paper, and a soil conditioner to increase the effectiveness of chemical fertilizers. More clinoptilolite is mined in Japan for the last use listed than for any of the others.

Some of the natural zeolites may find applications in pollution control. For example, laboratory and pilot-plant studies by Pacific Northwest Laboratory, Battelle Memorial Institute, Richland, WA (Mercer et al., 1970) indicate that clinoptilolite is effective in the removal of ammonia from waste water. Also, recent studies suggest that mordenite may be suitable for the removal of SO_2 from stack gases.

The United States is self-sufficient with respect to the present fledgling zeolite industry. Because of the large potential zeolite resources and the diverse zeolite mineralogy, the United States will not only remain self-sufficient when new markets develop but will probably become a major exporter.

Exploitation

Production figures have not been published for natural zeolites in the United States; however, small tonnages were mined from upper Cenozoic tuffs. Several hundred tons of chabazite, clinoptilolite, and erionite have been mined each year since the mid-1960s; the mined deposits are near Bowie, AZ, near Hector, CA, and in Jersey Valley, NV, respectively.

Exploration and sampling of bedded zeolite deposits by commercial organizations have been active since the late 1950s in Arizona, California, Nevada, Oregon, Texas, and Wyoming. The exploration peaked in the early 1960s, when the search concentrated on natural molecular-sieve zeolites that might compete with the synthetic zeolites. Since 1970, exploration has accelerated again. This time the emphasis is on the location of zeolite deposits suitable for pollution-control processes.

Japan is probably the world's leading producer of natural zeolites. Since the mid-1960s, Japan has had an annual production of about 100,000 tons. About half of this production is from one mine, at Itaya, Honshu.

Geologic Environment

Zeolites have been known since the mid-1750s, and they occur in rocks that are diverse in age, lithology, and geologic setting. Zeolites have been recognized in sedimentary deposits since 1891, when Murray and Renard (1891) described phillipsite in deep-sea deposits. Prior to the early 1950s, however, most zeolite occurrences were reported from fracture fillings and vesicle fillings in igneous rocks, particularly in basaltic rocks. Most of the large attractive zeolite specimens in museum collections were obtained from igneous rocks. In recent years, zeolites have been recognized as important rock-forming constituents in low-grade metamorphic rocks and in a variety of sedimentary rocks (Hay, 1966). The zeolites in sedimentary rocks are very finely crystalline and do not appeal to mineral collectors; however, deposits of this type are voluminous and have great economic potential. The remainder of this section summarizes those zeolite resources in sedimentary deposits.

Of the more than 400 published reports that describe zeolites in sedimentary rocks throughout the world, more than 75% were published in the last 20 years. The factors chiefly responsible for this recent surge of reports are (1) the widespread use of X-ray powder diffraction techniques in the study of fine-grained sedimentary rocks, (2) the exploration for zeolite deposits suitable for commercial use, and (3) the early review papers by Coombs et al., (1959) and Deffeyes (1959), both of which emphasized the widespread and relatively common occurrences of diagenetic or authigenic zeolites in sedimentary rocks.

Zeolites are among the most common authigenic silicate minerals that occur in sedimentary rocks, and they have formed in sedimentary rocks of diverse lithology, age, and depositional environment. Those zeolites given in Table 9 have been reported from sedimentary rocks; however, some, such as faujasite, gismondine, gonnardite, scolecite, and thomsonite, generally occur in trace amounts and are rarely reported. Only nine zeolites commonly make up the major part of zeolitic rocks. These are analcime, chabazite, clinoptilolite, erionite, ferrierite, heulandite, laumontite, mordenite, and phillipsite. Analcime and clinoptilolite are by far the most abundant zeolites in the sedimentary rocks. All nine of these common zeolites in sedimentary rocks show a considerable range in Si:Al ratios and cation contents. Except for heulandite and laumontite, they are generally alkalic and more siliceous than their counterparts that occur in mafic igneous rocks. Summaries of their chemistry are given by Deer, Howie, and Zussman (1963), Hay (1966), Sheppard (1971), and Utada (1970).

TABLE 9—Zeolites Reported from Sedimentary Rocks

Zeolite	Formula
Analcime	$NaAlSi_2O_6 \cdot H_2O$
Chabazite	$(Ca,Na_2)Al_2Si_4O_{12} \cdot 6H_2O$
Clinoptilolite	$(Na_2,K_2,Ca)_3Al_6Si_{30}O_{72} \cdot 24H_2O$
Epistilbite	$(Ca,Na_2)_3Al_6Si_{18}O_{48} \cdot 16H_2O$
Erionite	$(Na_2,K_2,Ca)_{4.5}Al_9Si_{27}O_{72} \cdot 27H_2O$
Faujasite	$(Na_2,Ca)_{1.75}Al_{3.5}Si_{8.5}O_{24} \cdot 16H_2O$
Ferrierite	$(K,Na)_2(Mg,Ca)_2Al_6Si_{30}O_{72} \cdot 18H_2O$
Gismondine	$(Ca,Na_2,K_2)_4Al_8Si_8O_{32} \cdot 16H_2O$
Gonnardite	$Na_2CaAl_4Si_6O_{20} \cdot 5H_2O$
Harmotome	$(Ba,Na_2)_2Al_4Si_{12}O_{32} \cdot 12H_2O$
Heulandite	$(Ca,Na_2)_4Al_8Si_{28}O_{72} \cdot 24H_2O$
Laumontite	$Ca_4Al_8Si_{16}O_{48} \cdot 16H_2O$
Mordenite	$(Na_2,K_2,Ca)Al_2Si_{10}O_{24} \cdot 7H_2O$
Natrolite	$Na_4Al_4Si_6O_{20} \cdot 4H_2O$
Phillipsite	$(K_2,Na_2,Ca)_2Al_4Si_{12}O_{32} \cdot 12H_2O$
Scolecite	$Ca_2Al_4Si_6O_{20} \cdot 6H_2O$
Stilbite	$(Ca,Na_2)_4Al_8Si_{28}O_{72} \cdot 28H_2O$
Thomsonite	$NaCa_2Al_5Si_5O_{20} \cdot 6H_2O$
Wairakite	$CaAl_2Si_4O_{12} \cdot 2H_2O$
Yugawaralite	$CaAl_2Si_6O_{16} \cdot 4H_2O$

Most zeolites in sedimentary deposits formed after burial of the enclosing sediments by the reaction of aluminosilicate materials with the pore water. Silicic volcanic glass is the aluminosilicate material that most commonly served as a precursor for the zeolites, although materials such as clay minerals, feldspars, feldspathoids, and gels also have reacted locally to form zeolites. Hay (1966) showed that authigenic zeolites and associated silicate minerals can be correlated with the following factors: (1) composition, grain size, permeability, and age of the host rocks; (2) composition of the pore water including pH, salinity, and proportion of dissolved ions; and (3) depth of burial of the host rock. Except for laumontite and possibly some heulandite, the common zeolites generally occur in tuffaceous sedimentary rocks that have not been deeply buried or exposed to hydrothermal solutions.

Classification of the diverse zeolitic sedimentary rocks is difficult, but the following tenuous classification is based on geologic setting and is offered as a basis for further discussion: (1) hydrothermal, (2) burial metamorphic, (3) weathering, (4) open system, and (5) closed system. The hydrothermal type includes those zeolites associated with metallic deposits such as those at the East Tintic district, Utah (Lovering and Shepard, 1960), and especially with hot-spring deposits. Well-known examples of the latter are at Yellowstone National Park, WY (Fenner, 1936; Honda and Muffler, 1970); Wairakei, New Zealand (Steiner, 1953, 1955); and Onikobe, Japan (Seki et al., 1969). Zeolites of geothermal areas commonly show a vertical zonation, and the downward succession of mineral assemblages seems to correlate with an increase in temperature. At Wairakei and Onikobe, for example, the downward succession is characterized by mordenite, laumontite, and then wairakite.

Zeolites of the burial metamorphic type were originally recognized by Coombs (1954) in Triassic sedimentary rocks of the Southland syncline, New Zealand. Coombs et al., (1959) demonstrated a vertical zonation of mineral assemblages that is characterized by a downward succession of clinoptilolite-heulandite-analcime, laumontite-albite, and then prehnite-pumpellyite-albite. Rocks rich in prehnite and pumpellyite commonly grade downward into rocks typical of the greenschist metamorphic facies. Locally, a zone rich in wairakite occurs between the laumontite-albite and prehnite-pumpellyite-albite zones or overlaps these zones. The zeolites and associated silicate minerals of the burial metamorphic type commonly occur in marine volcaniclastic strata that are more than 10,000 ft thick and locally are as much as 40,000 ft thick. The vertical succession of mineral assemblages is one of decreasing hydration with depth and is generally thought to be temperature dependent; however, chemical variables may prove to be equally important. Besides occurrences in New Zealand, zeolite rocks of the burial metamorphic type have been recognized in Australia (Packham and Crook, 1960), Puerto Rico (Otálora, 1964), USSR (Kossovskaya and Shutov, 1963), Japan (Seki, 1969; Utada, 1970), British Columbia, Canada (Surdam, 1968), and the United States. Laumontite-bearing rocks of the burial metamorphic type occur in central Oregon (Dickinson, 1962; Brown and Thayer, 1963), in Mount Rainier National Park, WA (Fiske et al., 1963), and near Cache Creek, CA (Dickinson et al., 1969).

Zeolites of the weathering type are volumetrically rather minor, but many deposits have probably been overlooked. Analcime was recently reported from alkaline saline soils of the eastern San Joaquin Valley, CA (Baldar and Whittig, 1968). The analcime was detected to a depth of about 4 ft, and its abundance decreased with depth. Hay (1963a) recognized analcime, chabazite, natrolite, and phillipsite in alkaline saline soil profiles at Olduvai Gorge, Tanzania. Surface concentrations of analcime were also reported in reddish, root-marked claystones on the Luboi Plain, Kenya (Hay, 1970).

The most voluminous and potentially valuable zeolite deposits belong to the open- and closed-system types. The terms "open system" and "closed system" are used in a hydrologic sense rather than in a thermodynamic sense. Deposits of the open-system type formed by the reaction of volcanic glass with subsurface water that originated as meteoric water. The original volcanic material commonly was deposited in marine or fluviatile environments or was air-laid on the land surface. Deposits of the closed-system type formed by the reaction of volcanic glass with the connate water trapped during sedimentation in a saline alkaline lake.

Deposits of the open-system type commonly formed in thick tuffaceous strata and show a vertical zonation of authigenic silicate minerals. Hay (1963a) proposed hydrolysis and solution of silicic glass by subsurface water to account for the formation of clinoptilolite in tuff and tuffaceous claystone in the lower part of the Tertiary John Day Formation in central Oregon. The upper part of the formation contains unaltered glass or montmorillonite. The authigenic mineral zonation is more complex in the Tertiary tuffs at the Nevada Test Site (Hoover, 1968). An upper zone consists of unaltered glass with local concentrations of chabazite and clay minerals. Zeolitic tuff continues downward for as much as 6000 ft and is characterized by a downward succession of zones rich in clinoptilolite, mordenite, and then analcime. The zeolite zones of the open-system type commonly cut across stratigraphic boundaries.

Zeolite deposits of the closed-system type formed during diagenesis in alkaline saline lakes commonly of the sodium carbonate-bicarbonate variety. Brine of this composition generally has a pH greater than 9, which probably accounts for the relatively rapid solution of vitric material and precipitation of zeolites. The authigenic silicate mineralogy can be correlated with the salinity in deposits of the closed-system type. The Pleistocene deposits of Lake Tecopa, CA (Sheppard and Gude, 1968) are characteristic of the closed-system type. Vitric material is unaltered or partly altered to clay minerals in tuff deposited in freshwater near the lake shore and inlets; however, the tuffs consist of zeolites where deposited in moderately saline water and of potassium feldspar where deposited in the highly saline and alkaline water of the central part of the basin. Thus, individual tuffs show a lateral zonation in a basinward direction of unaltered glass to zeolites and then to potassium feldspar. Zeolitic tuffs at Lake Tecopa consist chiefly of phillipsite, clinoptilolite, and erionite. Chabazite is a minor constituent of tuffs at Lake Tecopa, but it is locally the major constituent in zeolitic tuffs of other saline lacustrine deposits. In some deposits of the closed-system type, such as the Miocene Barstow Formation of California (Sheppard and Gude, 1969) and the Eocene Green River Formation of Wyoming (Surdam and Parker, 1972), a zone of analcime separates the other zeolites from the zone of potassium feldspar. In addition to clay minerals, zeolites, and potassium feldspar, deposits of the closed-system type locally contain opal or chalcedony, searlesite ($NaBSi_2O_6 \cdot H_2O$), fluorite (CaF_2), or dawsonite ($NaAl(CO_3)(OH)_2$) of authigenic origin.

Prospecting Techniques

Prospecting for bedded zeolite deposits is difficult because the zeolites are finely crystalline and resemble bedded diatomite, feldspar, or bentonite in the field. Zeolitic tuffs generally have an earthy luster and are resistant. Although some zeolitic tuffs are pastel shades of yellow, brown, red, or green, many are white or pale gray. If the zeolitic tuff is nearly monomineralic, certain gross physical properties of the rock may aid field recognition (Sheppard and Gude, 1969). Utilizing the ion-exchange and molecular sieve properties of zeolites, Helfferich (1964) designed a "field" test for the recognition of zeolites. Helfferich's test distinguishes zeolites from clay minerals, feldspars, and volcanic glass, but the test will not identify the zeolite species. My experience with Helfferich's test suggests that it is better suited to the laboratory than to the field.

X-ray powder diffraction analysis of bulk samples is the technique generally used for identification of the zeolites in sedimentary rocks. This method also permits a semiquantitative estimate of the abundance of zeolites and associated minerals in the samples. Tuffaceous strata are sampled, and then the samples are brought to the laboratory for examination by X-ray diffraction. Fresh tuff is generally distinguishable from altered tuff in the field, so only the altered parts of the tuffaceous rocks are sampled in both vertical and lateral directions. Once zeolites have been identified by X-ray diffraction, an additional detailed sampling is necessary to ascertain the distribution and abundance of the zeolites and associated authigenic minerals. Inasmuch as the samples must be returned to the laboratory for X-ray study, the availability of a truck-mounted X-ray diffractometer unit, suitable for field use, would facilitate the location of favorable targets and provide a guide for a meaningful sampling program.

Potential Resources

Evaluation of zeolite resources is hampered by the paucity of data on the extent and grade of the identified deposits. Although more than 100 occurrences of zeolites in sedimentary rocks of the United States have been recorded (Sheppard, 1971), information such as the grade, mineralogy, and vertical and lateral extent of the zeolitic rock and the thickness of overburden are lacking except for several deposits. The occurrence of zeolites in a certain sedimentary unit is commonly recognized from the examination of only one sample. Nevertheless, very large tonnages of zeolitic tuff of high purity are known from the Cenozoic deposits of the western United States and the Gulf Coastal Plain.

The only published estimate of zeolitic resources in the United States was made by Deffeyes (1968), who estimated that about 120 million tons of clinoptilolite, chabazite, erionite, mordenite, and phillipsite occur in near-surface deposits of the Basin and Range province. This estimate is probably conservative for the zeolite deposits of the Basin and Range province, and the potential zeolite resources of the United States are probably several orders of magnitude greater. If the restraints of grade and depth of overburden are ignored, the total of identified, hypothetical, and speculative zeolite resources in the United States is conservatively estimated at 10 trillion tons. The nine common zeolites listed in approximate order of decreasing abundance are clinoptilolite (including heulandite), analcime, mordenite, erionite, phillipsite, chabazite, laumontite, and ferrierite. The United States probably has the world's largest potential resources of high-grade chabazite, erionite, and phillipsite. The only high-grade deposits of ferrierite in the world have been reported from central Nevada (Regis, 1970).

If an estimate of zeolite resources in the United States is difficult to make, a meaningful estimate of zeolite resources for the other countries of the world is impossible to make. Zeolites have been reported in sedimentary rocks from many other countries, and their occurrences were summarized by Hay (1966) and Iijima and Utada (1966). However, no resource estimates have been published. In addition to the foreign zeolite deposits mentioned in these summaries and those foreign deposits mentioned previously in this chapter, apparently extensive and relatively high-grade zeolite deposits occur in France (Estéoule et al., 1971), Germany (F. A. Mumpton, written communication, 1972), Italy (Sersale, 1960; Alietti and Ferrarese, 1967; Alietti, 1970), Yugoslavia (Stojanović, 1968), Hungary (Nemecz and Varju, 1962), and Bulgaria (Alexiev, 1968). Most of the occurrences are in Cenozoic tuffaceous rocks, and the major zeolite is clinoptilolite.

Additional deposits of zeolites will undoubtedly be discovered in the western United States and in other countries of the world. Zeolites can form from a variety of aluminosilicate materials during diagenesis, providing the interstitial water has a relatively high pH and high concentration of alkalis. The purest zeolite deposits, however, form from volcanic ash that lacks crystal and rock fragments. In the Western Hemisphere, zeolites have not been reported from countries south of the United States except Argentina, Chile, and Mexico. An examination of upper Mesozoic and Cenozoic tuffaceous rocks will probably reveal large zeolite deposits in Mexico and many of the countries of Central America and western South America (Mumpton, 1973). Future exploration will also probably identify bedded zeolite deposits in the volcanic terranes of southern Europe, eastern Africa, and southern Asia.

Research Needed

The genesis of zeolites in sedimentary rocks is understood in general terms; however, several problems merit additional study. Phillipsite and clinoptilolite are the chief zeolites in young deep-sea deposits, but the relative importance of the precursor materials and the interstitial fluids in their formation is unknown. Tuffs of originally rhyolitic composition in deposits of alkaline, saline lakes characteristically contain a variety of zeolite minerals, but the factors that control which particular zeolite forms are poorly understood. Low-temperature experimental studies combined with studies of the zeolites and associated interstitial fluids in Holocene tuffs presently undergoing alteration may provide some answers. Another problem that warrants investigation is the relationship of the chemistry of a certain zeolite species to the geologic setting of the host rock.

Other research of a technological nature that deserves consideration is (1) development of techniques for the separation of zeolites from the gangue, (2) development of techniques for the separation of one zeolite from another, and (3) chemical or structural modifications of natural zeolites to increase their usefulness. Natural zeolites or zeolitic rocks will undoubtedly be used in many industrial and agricultural processes; however, additional applied research is necessary to achieve these goals.

Commercial Utilization of Natural Zeolites

FREDERICK A. MUMPTON *

For more than 200 years zeolites have been familiar minerals to geologists and mining engineers as minor, but ubiquitous constituents in vugs and fractures of most basalt and traprock formations. More than 35 different species have been recognized including several, such as chabazite, erionite, mordenite, and faujasite, whose adsorption properties rival those of many synthetic molecular sieves. Igneous occurrences of this type have never been mined on a commercial scale; they are generally small and low grade, and they commonly contain as many as four zeolite minerals, along with small quantities of calcite, quartz, and other silicate minerals. Their economic beneficiation to monomineralic zeolite products has also not been achieved and, for the most part, industrial applications of zeolites have relied primarily on synthetic molecular sieves that were crystallized from hydrous aluminosilicate gels.

During the middle 1950s, when Union Carbide Corp. and other groups were developing processes and markets for their fledgling synthetic zeolite businesses, natural zeolites

* State University of New York College, Brockport, NY.

FIG. 12—*Zeolitic tuff, 1 to 2 m thick, in lacustrine deposit near Buckhorn, Grant County, NM.*

were discovered as major constituents of numerous volcanic tuffs in saline-lake deposits of the western United States and of massive, marine deposits in Japan and Italy. Since that time, more than 1000 occurrences have been reported from sedimentary rocks of volcanic origin in more than 40 countries of the world. The flat-lying nature and high purity of the natural deposits have aroused considerable commercial interest both here and abroad, and today efforts are being made to utilize "sedimentary" zeolites in many of the same ways as their synthetic counterparts, as well as to develop other applications which take advantage of the low mining costs of the bedded, near-surface deposits. Although the use of natural zeolites is still in its infancy, more than 300 kt/a of zeolitic tuff are mined in the United States, Japan, Italy, Hungary, Yugoslavia, Bulgaria, Mexico, and Germany, and used for filler in the paper industry, in pozzolanic cements and concretes, as lightweight aggregate, in fertilizer and soil conditioners, as ion exchangers in pollution abatement processes, as dietary supplements in animal nutrition, in the separation of oxygen and nitrogen from air, and as acid-resistant adsorbents in gas drying and purification.

Types of Deposits and Properties

Based upon their geologic setting, mineralogical composition, and origin, zeolite deposits in sedimentary rocks have been classified into several types, including (1) deposits formed from volcanic material in "closed" saline-lake systems, (2) deposits formed in "open" freshwater-lake or ground water systems, (3) deposits formed in marine environments, (4) deposits formed by low-grade burial metamorphism, (5) deposits formed by hydrothermal or hot-spring activity, and (6) deposits formed without direct evidence of volcanic precursor material (Mumpton, 1973a; Sheppard, 1973; Munson and Sheppard, 1974). Commercial interest is directed primarily toward deposits of the first three types. Zeolitic tuffs in saline-lake deposits (Figs. 12–14) are usually only a few centimeters to a few meters thick, but they commonly contain nearly monominerallic zones of the large-pore zeolites, chabazite and erionite, which generally do not occur in other types of deposits. Open-system and marine deposits are characterized by clinoptilolite and mordenite and may be several hundred meters thick (see Fig. 15). Small to moderate amounts of montmorillonite and opal are also present,

FIG. 13—*Erionite-rich tuff, 2 to 4 m thick, capping series of fluvial-lacustrine sedimentary rocks, near Rome, Malheur County, OR.*

FIG. 14—*"Skyline" tuff of Barstow Formation, Rainbow Basin Natural Area, Barstow, San Bernardino County, CA. Locally altered to analcime and K-feldspar (Sheppard and Gude, 1969).*

along with varying quantities of crystal and lithic pyroclastic material.

All commercial applications of zeolite minerals make use of one or more of several attractive physical or chemical properties which characterize both the natural and synthetic materials. These include (1) ion exchange, (2) adsorption and related molecular-sieve phenomena, (3) dehydration and rehydration, and (4) a siliceous composition. In addition, certain extrinsic properties such as the tendency of "sedimentary" zeolites to occur as light-colored, lightweight, porous aggregates of micrometer-size crystals, have contributed to their past and present-day use in industrial and agricultural technology. Brief descriptions of many important uses of natural zeolites, including several that are of special significance in the current climate of pollution control and energy conservation, are discussed below.

Dimension Stone

Historically, devitrified volcanic ash and altered tuff have been used for more than 2000 years as lightweight dimension stone. Only within the last 25 years, however, has the zeolite content of many of these materials been recognized. Zeolitic tuffs are generally characterized by low bulk densities, high porosities, and homogeneous, close-knit textures. They are lightweight and easily cut or sawed into expensive building blocks. Many buildings associated with the Mayan centers at Mitla and Monte Alban in southern Mexico were constructed of blocks of massive green tuff recently found to consist of 85 to 90% mordenite and clinoptilolite (Mumpton, 1973). This tuff and others like it are still being quarried today at Etla, Tecoatlan, and Tejupan in Oaxaca for local use as dimension stone in walls, buildings, and other structures.

At Otsunomiya City, Tochigi Prefecture, Japan, a 100-m-thick zeolite tuff has been used in a similar manner for several hundred years (Ota and Sudo, 1949). This pumiceous marine formation shown in Fig. 15 is part of the extensive Green Tuff region which contains most of the major zeolite deposits of that country (Utada, 1970; Iijima and Utada, 1972) and consists of 80 to 85% clinoptilolite with small amounts of montmorillonite, celadonite, and glass. The easily cut and fabricated Neopolitan yellow tuff in central Italy has also been used as a common dimension stone for many years, and to the casual visitor it seems as though the entire city of Naples is built out of it. According to Norin (1955) and Sersale (1958) this material contains major quantities of chabazite and phillipsite and was formed by the alteration of volcanic glass in ground

FIG. 15—*Clinoptilolite-rich marine tuff, 100 m thick, near Otsunomiya City, Tochigi Prefecture, Japan (Ota and Sudo, 1949).*

water systems. Numerous cathedrals and public buildings in central Europe were constructed with blocks of zeolitic tuff from the Laacher See area of Germany. TUBAG Trasszement and Steinwerke AG of Kruft continues to quarry this tuff and to supply carved dimension stone for replacement blocks in these edifices.

Early ranch houses (Fig. 16) in the American West were built with blocks of erionite quarried in the vicinity. The structures were cool and apparently did not crumble in the arid climate, despite the friable nature of the tuff. In the 1950s geologists exploring for zeolites would commonly examine all known building-block quarries in these regions, often with considerable success. Similar structures are constructed of zeolitic tuff blocks at Kurdžali, Bulgaria; Tokaj, Hungary; and in the vicinity of almost every other large zeolite deposit in Europe and Japan. The total tonnage of zeolites mined each year for this purpose is not known; however, based upon the size of the operations examined by this author, this figure is probably 50 to 100 kt. The only known zeolite deposit being quarried for dimension stone in the United States crops out a few miles southwest of Rome, OR. Here, a few hundred tons of impure erionite-rich tuff is cut into facing stone each year for local consumption.

Pozzolanic Cements and Concrete

The first and most popular pozzolanic raw material used by the ancient Romans in their highways, aqueducts, and public buildings was also obtained from the Neopolitan yellow tuff near Pozzuoli, Italy (Norin, 1955; Sersale, 1958). Similar altered tuffs and devitrified ash in other parts of Italy and in the Eifel region of Germany were also exploited by the Romans during this period and have been used continuously for cement production throughout Europe. The specific reactions of zeolite minerals in pozzolans are not well understood, but it is likely that their high silica content allows them to combine with and neutralize excess lime produced by setting concrete in much the same way as does finely powdered pumice or fly ash from coal-burning power plants. Zeolitic pozzolans have important applications in hydraulic cements where the final concrete must withstand continuous underwater corrosion. The sodium and potassium content of many zeolites from sedimentary deposits, however, may contribute to increased alkali-aggregate reactions.

FIG. 16—*Early ranch house constructed of erionite blocks, Jersey Valley, Pershing County, NV.*

Deposits of zeolitic tuff are undoubtedly used locally as pozzolanic raw materials in many other parts of the world; however, the mineral content of these materials is probably unknown to the user. In recent years, many of these tuffs were found to be zeolite-rich. In northern Yugoslavia several thousand tons of zeolite marine tuff are quarried each month near Celje as a pozzolan and as a source of lightweight aggregate. Both the clinoptilolite-rich tuff at Zaloska Gorića and the analcime-rich tuff at Gorenje are about 100 m thick and contain a small amount of montmorillonite which limits their pozzolanic activity (Grimšičar, 1967). An exceptionally large deposit of clinoptilolite-rich lithic tuff is currently mined near Vranska Banja, Serbia, at a rate of more than 100 kt/a for exclusive use in the production of pozzolanic portland cement.

Five separate trass deposits in the Eifel area are being worked as pozzolans. Ludwig and Schwiete (1962) indicated that these materials contain about 6% chabazite and 15% analcime. One such deposit near Kretz is 8 to 14 m thick and is interbedded with other zeolitic and barren ash formations (Sersale and Aiello, 1965). A massive marine tuff, nearly 100 m thick, is being mined in an open pit in the northeastern Rhodopes, near Kurdžali, Bulgaria. The tuff crops out over a 50-km² area and is remarkably homogeneous. Alexiev (1968) found that this material contains about 80% clinoptilolite, with admixed cristobalite and a small amount of montmorillonite. It is currently mined solely as a pozzolan; however, new uses are actively being sought.

In the United States nearly $1 million was saved in 1912 during the construction of the 386-km-long Los Angeles aqueduct by replacing up to 25% of the required portland cement with a zeolitic tuff mined near Tehachapi, CA (see Mielenz et al., 1951; Drury, 1954). This clinoptilolite-rich ore was until recently mined from a massive, ash-flow tuff by Monolith Portland Cement Co. and was a principal constituent of their pozzolanic cement products. Similar bodies are abundant in the western United States and eventually may compete locally with volcanic ash as a source of pozzolans in the construction of dams, highways, and irrigation canals.

Lightweight Aggregate

A third construction use of zeolitic tuff, just now being investigated in Europe, is in the production of expanded, lightweight aggregate. Much as perlite and other volcanic glasses are frothed into low-density pellets for use in concrete, so can zeolite-rich tuff be "popped" by calcining at high temperature. Clinoptilolite from the Slovenian deposit described previously and from several occurrences in southern Serbia yield excellent aggregate of this type on firing between 1200° and 1400°C. Stojanović (1972) reported densities as low as 0.8 g/cm^3 and porosities of up to 65% for products made by expanding clinoptilolite from the Katalanac, Zlatokop, Toponica, and Meškovac deposits. None of these deposits is currently mined for this use; however, the abundance of zeolitic tuffs in Yugoslavia should give rise to greater effort in the development of such aggregate products.

Filler in Paper

High-brightness zeolite ores are finding more and more application as fillers in the paper industry. Approximately 3000 tons of clinoptilolite are mined each month from a massive marine tuff at Itaya, Japan, and used in the production of high-quality paper (Minato, 1968). Fine grinding followed by cyclone classification yields a soft —10-μm-size product having an abrasion index of less than 3% and a brightness of about 80. The exact function of the zeolite in the paper-making process is not fully understood, but it is not unlikely that the adsorptive nature of the zeolite provides a means of controlling pitch, or that the zeolite simply replaces higher priced kaolinite as a filler and bulking agent. According to Kobayashi (1970) Kraft papers filled with clinoptilolite are bulkier, more opaque, easier to cut, and less susceptible to ink blotting than comparable clay-filled products. The Itaya deposit is one of the largest zeolitic tuffs being mined in Japan. It is flat-lying, about 100 m thick, and outcrops over a 5-km^2 area. The deposit is currently being worked by both open pit and underground methods by Zeeklite Chemical & Mining Co. Ltd., and yields products commanding prices of up to $150 per ton.

Another clinoptilolite deposit, several tens of meters thick, is also being mined as a paper filler in the Tokaj district of eastern Hungary. This and other zeolitic tuffs are widespread in the region and have been described in detail by Nemecz and Varju (1962). Although only a few hundred tons are produced each month, many uses are actively being investigated for this material. One potential use is in the preparation of a commercial adsorbent called "clinosorb," apparently an acid-leached, hydrogen form of clinoptilolite. Tokaj clinoptilolite is also being tested as a purification agent for wine.

Ion-Exchange Processes

Large-scale ion exchange processes utilizing natural zeolites were first developed in the 1960s at the Hanford Laboratory of Battelle Northwest in Richland, WA, under the auspices of the US Atomic Energy Commission. Using clinoptilolite from the Hector, CA, deposit, Ames (1959, 1960) demonstrated a method of concentration and isolation for radioactive strontium and cesium from waste streams of nuclear facilities. Once saturated with these ions, the clinoptilolite can be stored indefinitely or the ions removed by chemical means and the zeolite reused. Scale-up and supply problems have slowed the introduction of this process into existing installations; however, the technique is being used to concentrate cesium from low-level effluents. With an exchange capacity of greater than 150 milliequivalents per 100 g, clinoptilolite loaded with Cs^{137} might well provide an easily handled source of heat to fill some of the nation's ever-expanding energy requirements.

Similar studies involving natural zeolites as collectors of radioactive wastes have been and are being carried out in France, Italy, Great Britain, Hungary, Bulgaria, Mexico, Canada, and Japan. In a related area, Nishita and Haug (1972) showed that the addition of clinoptilolite to soils contaminated with Sr^{90} markedly reduced the strontium uptake by plants; thus, natural zeolites may be an important means of limiting the effect of radioactive fallout in crop and pasture lands.

As a spin-off of their work on the removal of radioactive ions from low-level wastes, Ames and Mercer et al. showed that clinoptilolite is an effective agent for extracting ammonium ions from sewage and agricultural effluents (Ames,

1967; Mercer et al., 1970). Using a mobile exchange unit at a Lake Tahoe test site, they were successful in removing 97% of the ammonium from sewage streams. The ammonia was later discharged harmlessly into the atmosphere, and the clinoptilolite regenerated for further use. Based upon this process, several large sewage-treatment plants are currently being designed in several parts of the United States. A 2.3ML/d unit in the Minneapolis-St. Paul area went onstream in February 1974, and uses six 8.5 m³ columns of 850 μm x 300 μm clinoptilolite. The initial 90 tons of zeolite were obtained from the Hector deposit operated by Baroid Div. of NL Industries at a cost of about $0.40/kg. The zeolite is expected to last five to ten years. Two larger plants with capacities of 205 and 38 ML/d are in the design stages for the Alexandria and Reston areas in Virginia, respectively. The Reston operation will require 1800 tons of clinoptilolite at a cost of more than $700,000. A smaller unit has been designed for communities of about 10,000 population by US Energy Corp. in Riverton, WY, and will use clinoptilolite from the Beaver Rim deposit near Sand Draw, WY. Several other clinoptilolites, said to be superior to the Hector material for ammonium removal, are also being readied for production by Double Eagle Petroleum & Mining Co. in Wyoming and other western states.

In a related area of sewage treatment, Sims (1972) reported that nitrification can be enhanced in activated-sludge processes by the addition of clinoptilolite. The zeolite selectively exchanges NH_4^+ from the waste water and provides an ideal surface for the attachment of nitrifying bacteria which then oxidize the ammonium ions to nitrate. The total tonnage of natural zeolites used today in sewage treatment applications is still small; however, if the stringent requirement of less than 1 ppm total nitrogen is imposed on sewage effluent by the Environmental Protection Agency, several hundred thousand tons of clinoptilolite may be required by the year 2000.

Air Separation

An exciting new use for natural zeolites was developed in the late 1960s in Japan for the production of high-purity oxygen for secondary-smelting operations in the iron and steel industry (Tamura, 1971a, 1972). As shown by Domine and Haÿ (1968), nitrogen gas is easily separated from air by any of several synthetic zeolites, yielding products containing up to 95% O_2 and a few percent argon. A small plant at Toyohashi City, a few miles north of Osaka, has used a pressure-swing adsorption process with natural mordenite since 1968 to produce 500 m³ of 90% O_2 gas per hr. The plant consists of three towers, each filled with 13 tons of acid-washed mordenite from the Itado mine, Minase, Akita Prefecture, and operates at ambient temperature on a 9-min cycle of adsorption-desorption-standby. Nitrogen may also be prepared by this process with purities up to 99.95%. According to Tamura (1971), the method is competitive with air distillation for relatively small-scale oxygen production. Clinoptilolite is also useful in air separation; however, the nitrogen-adsorption capacities of the several samples Tamura tested are lower than those of mordenite.

Zeolite-adsorption processes may provide inexpensive sources of oxygen or nitrogen where large-scale liquefaction facilities are not warranted. Nitrogen could be produced for wineries or breweries in on-site adsorption units when needed, thereby reducing their dependence on costly liquid or cylinder gas. As claimed by the German chemical-engineering company, J.F. Mahler (Anon., 1970), such a process is capable of producing 60% O_2-enriched air at a cost of $0.11/2.8 m³, thereby filling a need of users who now stand the cost of cryogenic distillation to produce pure oxygen, only to have to dilute it down to their required levels. Several commercial processes using synthetic zeolites have been developed and seem to be economically favorable over liquefaction at throughputs of less than 25 t/d (see Batta, 1971, 1972; Lee, 1972). It is not unlikely that mordenite and other zeolites will also find application in this area, although final acceptance of the natural materials depends upon the total cost of the operating system.

Lee (1973) listed several potential markets for a cheap oxygen source, including river and pond aeration, pollution control in the paper and pulp industry, feed gas to ozone generators, and chemical oxidation processes. In Japan small, freestanding zeolite adsorption units have been designed to produce oxygen-enriched air for hospitals, for use in fish-breeding and transportation, as well as for operation in areas of high air pollution. Sheppard (private communication) has suggested that a natural-zeolite adsorption plant might be a means of furnishing oxygen for the in-place gasification of low-grade coal deposits. Current gasification methods have been stymied by the lack of an inexpensive oxygen source (Abelson, 1973). Mordenite or clinoptilolite appear to be best suited for this process, but other zeolites, such as the Bowie, AZ, chabazite/erionite might be employed to purify the gases produced by the controlled burning process (vide infra).

Animal Nutrition

Experiments have been in progress since 1965 in Japan on the use of natural zeolites as dietary supplements for swine and poultry. Based on the use of bentonite in the normal protein diet of chickens (Quisenberry, 1968), clinoptilolite and mordenite have been added in amounts up to 10% to the feed of pigs and chickens. Test animals generally experienced faster growth than control groups, with simultaneous decrease in the amount and cost of the feed. In one 11-week experiment, young and mature pigs were fed a diet containing 5% clinoptilolite. Both groups gained an average of 16% more weight than animals fed a normal diet (Kondo and Wagai, 1968). Severe cases of scours were also relieved when the afflicted animal was placed on a diet containing 15% of this zeolite. Onogi (1968) found that additions of clinoptilolite or mordenite to the diets of chickens produced similar results. Test birds gained up to 8% more weight than did control animals, with no adverse effects. In both cases the animals' excrement was less odoriferous due to the zeolitic adsorption of ammonia. Zeolites are commonly used in Japan (in areas close to sources of zeolitic tuff) to control the moisture content and odor of animal droppings.

Similar studies have been carried out in several agricultural experiment stations in the United States. Full experimental data are not yet available, but initial experiments have met with limited success, especially in the prevention of intestinal diseases in young swine. Currently heavy doses of prophylactic antibiotics are used to control these diseases; however, if federal regulations prohibit such use, other means must be found. Zeolites may well be the answer. Little work has been done on the biochemical function of zeolites in nutritional processes; however, it is generally thought that the adsorptive nature of the zeolite allows nutrient molecules to be retained in the animal's digestive system for a longer time, thus allowing more efficient use of the feed. The result is larger animals for the same feed cost. The ammonium selectivity of clinoptilolite suggests that it may also be useful as an ammonia reservoir in ruminant feeding of cattle.

The use of zeolites in animal nutrition is still in the experimental stage, but offers attractive possibilities of increasing the world's food supply. The potential market in the United States alone exceeds several hundred thousand tons per year, and it may be even larger in the rest of the world. Before zeolites become a dietary staple, however, considerable experimental work is necessary. Special attention must also be paid to the long-range pathological effects of zeolites in animals, especially of needle-shaped erionite and ultrathin fibers of mordenite (see Figs. 17–20).

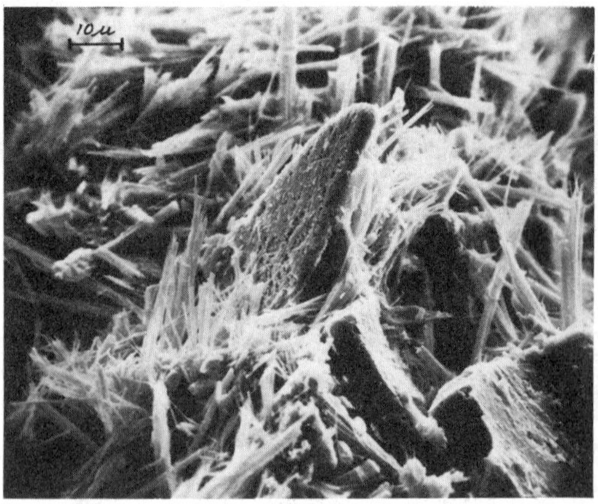

FIG. 17—*Scanning electron micrograph of erionite surrounding a pitted fragment of volcanic glass from zeolitic tuff, Jersey Valley, Pershing County, NV. X900*

Agricultural Products

Although little has been published on the subject, farmers in many parts of Japan have

FIG. 18—*Scanning electron micrograph of filiform mordenite spanning a crevice in a clinoptilolite-rich tuff, 3 m thick, Toponica, Yugoslavia. X5000.*

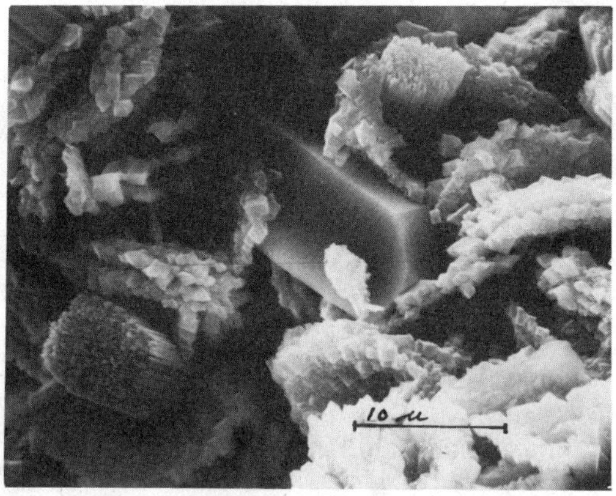

FIG. 19—*Scanning electron micrograph of pseudocubic chabazite, sheafs and rods of erionite, and laths of clinoptilolite from a 1-m-thick tuff, Bowie, Graham County, AZ. X2600.*

added pulverized zeolitic tuffs to their fields for more than 100 years in an attempt to neutralize acidic volcanic soils. This may be the most common use of natural zeolites in that country today, although new uses are actively being investigated (Minato, 1968). Zeolite soil conditioners have been shipped from Japan to Taiwan and will undoubtedly be exported to many other parts of southeastern Asia within a few years. Both mordenite and clinoptilolite are used as agglutinizing agents for mixed fertilizers. The zeolites control the release of ammonium and other cations from the fertilizer, as well as retain desired cations in the soil for longer periods of time.

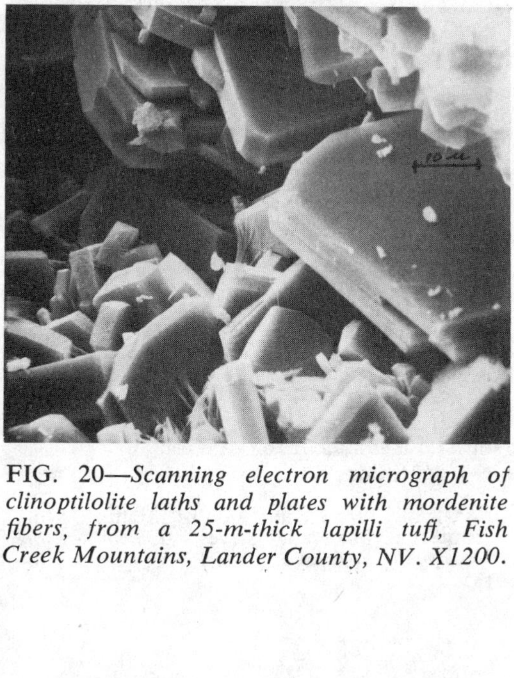

FIG. 20—*Scanning electron micrograph of clinoptilolite laths and plates with mordenite fibers, from a 25-m-thick lapilli tuff, Fish Creek Mountains, Lander County, NV. X1200.*

Similar to their synthetic counterparts, the high ion exchange and adsorption capacities of many sedimentary zeolites suggest that they may be useful carriers of fungicides and pesticides. Yoshinga et al. (1973) found that clinoptilolite is an effective fungicide carrier in the control of stem blasting in rice.

Gas Adsorption and Catalyst

Two principal uses of synthetic molecular sieves are the purification of gaseous hydrocarbons under controlled conditions of adsorption and desorption, and the preparation of catalysts for petroleum refining. In general, natural zeolites cannot compete with their synthetic counterparts in these areas due to their inherent lower adsorption capacities and, to some extent, to the presence of trace amounts of iron and other cationic "poisons." Some synthetics also have much larger pore openings. Despite the low cost of the natural materials (a few cents per 0.5 kg), the economics of hardware construction, activation, and regeneration favor the more expensive synthetics, even at \$2/kg. Certain natural zeolites, however, have made headway in these areas, especially in the drying and purification of acidic gases. Mordenite and a few other species are capable of withstanding the rigors of continuous cycling in acid environments and have been used extensively to remove water and carbon dioxide from sour natural gas. Between 1970 and 1972 approximately 120 204 kg of chabazite/erionite were sold by Union Carbide Corp. from their Bowie, AZ, deposit (see Figs. 19 and 21) under the trade name AW-500 (Langerhans, 1973). Union Carbide Corp. (Anon., 1962) reported this material to be stable at pH's as low as 2.5 and therefore capable of removing HCl from reformer hydrogen streams, water from chlorine, and CO_2 from stack-gas emissions.

FIG. 21—*Chabazite/erionite-rich tuff, 152-mm thick, being stockpiled north of Bowie, Graham County, AZ.*

N R G Corp. recently introduced a package gas-purification system for the removal of H_2O, CO_2, and H_2S from natural gas to meet pipeline specifications. Using the Bowie zeolite, they claim extractions of up to 25% CO_2 from wellhead products. This company has also developed a zeolite-adsorption process for purifying methane gas produced by decaying garbage in landfills. A pilot plant began operations in late 1974-early 1975 at the Palos Verdes landfill near Los Angeles, and produces over 28 m^3/min of methane. Such a process may also be useful in the removal of noncombustible CO_2 and corrosive H_2S from methane gas produced during the anaerobic digestion of sludge in municipal sewage-treatment plants. Much of this gas is currently burned to raise the sludge to its 32.2°C activation temperature and other in-house heating, but the impurities present take their toll in equipment breakdown. Similar digestion of animal wastes to yield methane is also feasible. The product gas purified, represents an enormous source of untapped energy. According to Jewell (1974) once the impurities are removed, the methane gas produced by the digestion of the organic wastes from a New York State dairy farm of 60 cows would be equivalent to the farm's entire fossil fuel requirements. Zeolitic purification may be the key to the economic feasibility of this process.

Although not yet implemented, Munson and Clifton (1971) suggested that adsorption on zeolites may be a means of increasing the on-vehicle storage capacity of trucks and autos using natural gas or other condensed gas fuels. The system would eliminate the need of expensive and heavy pressure tanks which heretofore have stymied the widespread application of this form of motor-vehicle fuel. This and other applications listed previously will undoubtedly expand in the next few years in the United States and in all other energy-conscious nations of the world.

In the area of catalyst preparation, the selective-forming process recently developed by Mobil Oil Corp. is thought to make use of catalysts derived from an erionite-clinoptilolite ore mined from their Jersey Valley, NV, deposit (Chen, 1971; Wilson, 1972). Here, tilted erionite beds crop out over a 2 023 500 m^2 area and have been mined by both open-pit and underground methods for the past ten years. In Japan a hydrocarbon-conversion catalyst for the disproportionation of toluene to benzene and xylene has been patented by Ohtani et al. (1972) based upon a hydrogen-exchanged natural mordenite. Clinoptilolite from the Tokaj district of Hungary was found by Papp et al. (1971) to be an effective catalyst for the hydrodemethylation of toluene after ion-exchange treatment of the zeolite.

Conclusions

If the efforts made on the occurrences, properties, and uses of natural zeolites in the past two decades are paradigms for the future, the United States will experience a minor "zeolite rush" during the next decade. The third edition of *Industrial Minerals and Rocks* made no mention of these intriguing minerals. The fourth edition was published at a time when zeolites were just beginning to find important applications in many areas of industrial and agricultural technology. Scores more have been on the drawing boards of R&D organizations throughout this country and in many other nations of the world. As this and other countries turn their research efforts toward the protection of their environments and the conservation of their mineral and energy resources, even greater efforts will undoubtedly be made to harness the unique and attractive properties of this abundant mineral commodity.

Bibliography and References

Introduction

Bramlette, M.N., and Posnjak, E., 1933, "Zeolitic Alteration of Pyroclastics," *American Mineralogist,* Vol. 18, No. 4, Apr., pp. 167–171.

Breck, D.W., 1973, "Synthetic Zeolites, Properties and Applications," SME Preprint 73H36, AIME Annual Meeting, Chicago, Feb., 32 pp.

Breck, D.W., 1974, *Zeolite Molecular Sieves: Structure, Chemistry and Use,* John Wiley, New York, 771 pp.

Mumpton, F.A., 1973, "Worldwide Deposits and Utilisation of Natural Zeolites," *Industrial Minerals,* No. 73, Oct., pp. 30–45.

Papke, K.G., 1972, "Erionite and Other Associated Zeolites in Nevada," Bulletin 79, Nevada Bureau of Mines & Geology, 32 pp.

Sheppard, R.A., 1971, "Zeolites in Sedimentary Deposits of the United States—A Review," *Molecular Sieve Zeolites—I,* R.F. Gould, ed., Advances in Chemistry Series 101, American Chemical Society, pp. 279–310.

Sheppard, R.A., 1973, "Zeolites in Sedimentary Rocks," Professional Paper 820, US Geological Survey, pp. 689–695.

Synthetic Zeolites: Properties and Applications

Barrer, R.M., 1945, "Separation of Mixtures Using Zeolites as Molecular Sieves," *Journal of the Society of Chemical Industry,* Vol. 64, pp. 130–135.

Barrer, R.M., 1957, "Some Researches on Silicates: Mineral Syntheses and Metamorphoses," *Transactions,* British Ceramic Society, Vol. 56, pp. 155–189.

Barrer, R.M., 1966, "Mineral Synthesis by the Hydrothermal Technique," *Chemistry in Britain,* Vol. 2, pp. 380–394.

Barrer, R.M., and Makki, M.B., 1964, "Molecular Sieve Adsorbents from Clinoptilolite," *Canadian Journal of Chemistry,* Vol. 42, pp. 1481–1487.

Barry, H.M., 1960, "Fixed-Bed Adsorption," *Chemical Engineering,* Vol. 67, No. 2, pp. 105–120.

Batta, L.B., 1972, "Adiabatic Selective-Adsorption Gas-Separation Process," US Patent 3,636,679, Jan. 25.

Bolton, A.P., and Bujalski, R.L., 1971, "The Role of the Proton in the Catalytic Cracking of Hexane using a Zeolite Catalyst," *Journal of Catalysis,* Vol. 23, pp. 331–339.

Breck, D.W., 1971, "Recent Advances in Zeolite Science," *Molecular Sieve Zeolites–I,* Advances in Chemistry Series 101, American Chemical Society, pp. 1–9.

Breck, D.W., 1974, *Zeolite Molecular Sieves, Structure, Chemistry, and Use,* John Wiley, New York, 771 pp.

Breck, D.W., and Flanigen, E.M., 1968, "Synthesis and Properties of Union Carbide Zeolites L, X and Y," *Molecular Sieves,* Society of Chemical Industry, London, pp. 47–61.

Breck, D.W., et al., 1956, "Crystalline Zeolites I. The Properties of a New Synthetic Zeolite, Type A;" "Crystalline Zeolites II. Crystal Structure of Synthetic Zeolite, Type A," *Journal of the American Chemical Society,* Vol. 78, pp. 5963–5977.

Collins, J.J., 1968, "Molecular Sieves in the Process Industries," *Chemical Engineering Progress,* Vol. 64, No. 8, pp. 66–82.

Hay, R.L., 1966, "Zeolites and Zeolitic Reactions in Sedimentary Rocks," Special Paper 85, Geological Society of America, 130 pp.

Kranich, W.L., et al., 1971, "Properties of Aluminum Deficient Large-Port Mordenites," *Molecular Sieve Zeolites-I,* Advances in Chemistry Series 101, American Chemical Society, pp. 502–513.

Morey, G.W., and Ingerson, E., 1937, "The Pneumatolytic and Hydrothermal Alteration and Synthesis of Silicates," *Economic Geology,* Vol. 32, Sup. to No. 5, pp. 607–761.

McBain, J.W., 1932, *The Sorption of Gases and Vapors by Solids,* George Rutledge and Sons Ltd., London, Chap. 5.

Meier, W.M., 1961, "The Crystal Structure of Mordenite," *Zeitschrift für Kristallographie,* Bd. 115, pp. 439–450.

Meier, W.M., 1968, "Zeolite Structures," *Molecular Sieves,* Society of Chemical Industry, London, pp. 10–27.

Milton, R.M., 1968, "Commercial Development of Molecular Sieve Technology," *Molecular Sieves,* Society of Chemical Industry, London, pp. 199–203.

Mumpton, F.A., 1973, "Worldwide Deposits and Utilisation of Natural Zeolites," *Industrial Minerals,* No. 73, Oct., pp. 30–45.

Rabo, J.A., and Poutsma, M.L., 1971, "Structural Aspects of Catalysis with Zeolites: Cracking of Cumene and Hexane," *Molecular Sieve Zeolites–II,* Advances in Chemistry Series 102, American Chemical Society, pp. 284–314.

Rees, L.V.C., 1970, "Ion Exchange in Zeolites," *Annual Reports on the Progress of Chemistry,* Vol. 67, pp. 191–212.

Sheppard, R.A., 1971, "Zeolites in Sedimentary Deposits of the United States—A Review," *Molecular Sieve Zeolites—I,* R.F. Gould, ed., Advances in Chemistry Series 101, American Chemical Society, pp. 279–310.

Sherry, H.S., 1971, "Cation Exchange in Zeolites," *Molecular Sieve Zeolites–I,* Advances in Chemistry Series 101, American Chemical Society, pp. 350–379.

Shreve, R.N., 1930, "Greensand Bibliography to 1930," Bulletin 328, US Bureau of Mines, 78 pp.

Venuto, P.B., and Landis, P.S., 1968, "Organic Catalysis over Crystalline Aluminosilicates," *Advances in Catalysis,* Academic Press, New York, Vol. 18, pp. 259–371.

Ward, J.W., 1971, "Infrared Spectroscopic Studies of Zeolites," *Molecular Sieve Zeolites–I,* Advances in Chemistry Series 101, American Chemical Society, pp. 380–404.

Way, J.T., 1850, "Power of Soils to Adsorb Manure," *Journal of the Royal Agricultural Society,* Vol. 11, pp. 313–379.

Zeolites in Sedimentary Rocks

Alexiev, B., 1968, "Clinoptilolite des Rhodopes du Nord-Est," *Comptes Rendus,* Bulgare Akademii Nauk, Vol. 21, No. 10, pp. 1093–1095.

Alietti, A., 1970, "I Minerali di Neeformazione dei Monti Berici," *Mineralog. et Petrog. Acta,* Vol. 16, pp. 27–32.

Alietti, A., and Ferrarese, G., 1967, "Clinoptilolite, Na-Montmorillonite e Ossididi Manganese in Una Formazione Sedimentaria a Zovencedo (Vicenza)," *Mineralog. et Petrog. Acta,* Vol. 13, pp. 139–145.

Baldar, N.A., and Whittig, L.D., 1968, "Occurrence and Synthesis of Soil Zeolites," *Proceedings,* Soil Science Society of America, Vol. 32, pp. 235–238.

Brown, C.E., and Thayer, T.P., 1963, "Low-Grade Mineral Facies in Upper Triassic and Lower Jurassic Rocks of the Aldrich Mountains," *Journal of Sedimentary Petrology,* Vol. 33, pp. 411–425.

Brown, R.E., 1962, "The Use of Clinoptilolite," *Ore Bin,* Vol. 24, No. 12, pp. 193–197.

Coombs, D.S., 1954, "The Nature and Alteration of Some Triassic Sediments from Southland, New Zealand," *Transactions,* Royal Society of New Zealand, Vol. 82, pp. 65–109.

Coombs, D.S., et al., 1959, "The Zeolite Facies, With Comments on the Interpretation of Hydrothermal Syntheses," *Geochimica et Cosmochimica Acta,* Vol. 17, pp. 53-107.

Deer, W.A., et al., 1963, "Framework Silicates," Vol. 4, *Rock-Forming Silicates,* John Wiley, New York, 435 pp.

Deffeyes, K.S., 1959, "Zeolites in Sedimentary Rocks," *Journal of Sedimentary Petrology,* Vol. 29, pp. 602–609.

Deffeyes, K.S., 1968, "Natural Zeolite Deposits of Potential Commercial Use," *Molecular Sieves,* Society of Chemical Industry, London, pp. 7–9.

Dickinson, W.R., 1962, "Petrology and Diagenesis of Jurassic Andesitic Strata in Central Oregon,"

American Journal of Science, Vol. 260, pp. 481–500.

Dickinson, W.R., et al., 1969, "Burial Metamorphism of the Late Mesozoic Great Valley Sequence, Cache Creek, California," *Geological Society of America Bulletin,* Vol. 80, pp. 519–526.

Estéoule, J., et al., 1971, "Sur la Présence de Clinoptilolite dan les Dépôts Marno-Calcaires du Crétacé Supérieur de l'Anjou," *Comptes Rendus,* Academie des Sciences, Vol. 272, Ser. D, pp. 1569–1572.

Fenner, C.N., 1936, "Bore-Hole Investigations in Yellowstone Park," *Journal of Geology,* Vol. 44, pp. 225–315.

Fiske, R.S., et al., 1963, "Geology of Mount Rainier National Park, Washington," Professional Paper 444, US Geological Survey, 93 pp.

Hay, R.L., 1963, "Zeolite Weathering in Olduvai Gorge, Tanganyika," *Geological Society of America Bulletin,* Vol. 74, pp. 1281–1286.

Hay, R.L., 1963a, "Stratigraphy and Zeolitic Diagenesis of the John Day Formation of Oregon," *Geological Science,* California University Publications, Vol. 42, No. 5, pp. 199–262.

Hay, R.L., 1966, "Zeolites and Zeolitic Reactions in Sedimentary Rocks," Special Paper 85, Geological Society of America, 130 pp.

Hay, R.L., 1970, "Silicate Reactions in Three Lithofacies of a Semiarid Basin, Olduvai Gorge, Tanzania," Special Paper 3, Mineralogical Society of America, pp. 237–255.

Helfferich, F., 1964, "A Simple Identification Reaction for Zeolites (Molecular Sieves)," *American Mineralogist,* Vol. 49, pp. 1752–1754.

Hoover, D.L., 1968, "Genesis of Zeolites, Nevada Test Site," *Nevada Test Site,* E.B. Eckel, ed., Memoir 110, Geological Society of America, pp. 275–284.

Honda, S., and Muffler, L.P.J., 1970, "Hydrothermal Alteration in Core from Research Drill Hole Y-1, Upper Geyser Basin, Yellowstone National Park, Wyoming," *American Mineralogist,* Vol. 55, pp. 1714–1737.

Iijima, A., and Utada, M., 1966, "Zeolites in Sedimentary Rocks, with Reference to the Depositional Environments and Zonal Distribution," *Sedimentology,* Vol. 7, pp. 327–357.

Kossovskaya, A.G., and Shutov, V.D., 1963, "The Correlation of Zones of Regional Epigenesis and Metagenesis in Terrigenous and Volcanic Rocks," *Doklady, Earth Sciences Secs.,* Akademii Nauk SSSR, Vol. 139, pp. 732–736.

Lovering, T.S., and Shepard, A.O., 1960, "Hydrothermal Alteration Zones Caused by Halogen Acid Solutions, East Tintic District, Utah," *American Journal of Science,* Vol. 258–A, pp. 215–229.

Mercer, B.W., et al., 1970, "Ammonia Removal from Secondary Effluents by Selective Ion Exchange," *Water Pollution Control Federation Journal,* Vol. 42, No. 2, Pt. 2, pp. R95–R107.

Minato, H., and Utada, M., 1969, "Zeolite," *The Clays of Japan—International Clay Conference,* Japan Geological Survey, Tokyo, pp. 121–134.

Mumpton, F.A., 1973, "First Reported Occurrences of Zeolite in Sedimentary Rocks in Mexico," *American Mineralogist,* Vol. 58, No. 34, Mar.-Apr., pp. 287–290.

Murray, J., and Renard, A.F., 1891, "Report on Deep-Sea Deposits," *Report on the Scientific Results of the Voyage of H.M.S. Challenger During the Years 1873–1876,* London, 520 pp.

Nemecz, E., and Varju, G., 1962, "Sodium Bentonitization, Clinoptylolitization and Adularization in the Rhyolitic Tuffs of the Szerencs Piedmont Area," *Acta Geologica,* Vol. 6, pp. 389–427.

Otálora, G., 1964, "Zeolites and Related Minerals in Cretaceous Rocks of East-Central Puerto Rico," *American Journal of Science,* Vol. 262, pp. 726–734.

Packham, G.H., and Crook, K.A.W., 1960, "The Principle of Diagenetic Facies and Some of Its Implications," *Journal of Geology,* Vol. 68, pp. 392–407.

Regis, A.J., 1970, "Occurrences of Ferrierite in Altered Pyroclastics in Central Nevada," *Abstracts with Programs,* Geological Society of America, Vol. 2, No. 7, pp. 661.

Seki, Y., 1969, "Facies Series in Low-Grade Metamorphism," *Geological Journal,* Geological Society of Japan, Vol. 75, pp. 255–266.

Seki, Y., et al., 1969, "Zeolite Distribution in the Katayama Geothermal Area, Onikobe, Japan," *Japanese Journal of Geology & Geography,* Vol. 40, pp. 63–79.

Sersale, R., 1960, "Genetic and Constitution Analogies Between Volcanic Tuffs with Pozzolanic Activity," *Silicates Industry,* Vol. 25, pp. 499–509.

Sheppard, R.A., 1971, "Zeolites in Sedimentary Deposits of the United States—A Review," *Molecular Sieve Zeolites–I,* R.F. Gould, ed., Advances in Chemistry Series 101, American Chemical Society, pp. 279–310.

Sheppard, R.A., and Gude, A.J., III, 1968, "Distribution and Genesis of Authigenic Silicate Minerals in Tuffs of Pleistocene Lake Tecopa, Inyo County, California," Professional Paper 597, US Geological Survey, 38 pp.

Sheppard, R.A., and Gude, A.J., III, 1969, "Diagenesis of Tuffs in the Barstow Formation, Mud Hills, San Bernardino County, California," Professional Paper 634, US Geological Survey, 35 pp.

Steiner, A., 1953, "Hydrothermal Rock Alteration at Wairakei, New Zealand," *Economic Geology,* Vol. 48, pp. 1–13.

Steiner, A., 1955, "Hydrothermal Rock Alteration," Industrial Research Bulletin 117, New Zealand Dept. of Science, pp. 21–26.

Stojanović, D., 1968, "Vulkanski Tufovi i Sedimentne Stene u Srbiji sa Sadržajem Zeolita," *Srpsko Geološko Drustvo,* 15 pp.

Surdam, R.C., 1968, "Low-Grade Metamorphism of the Karmutsen Group, Central Vancouver Island, British Columbia, Canada," *Abstracts for 1967,* Special Paper 115, Geological Society of America, pp. 217–218.

Surdam, R.C., and Parker, 1972, "Authigenic Alumino-Silicate Minerals in the Tuffaceous Rocks of the Green River Formation, Wyoming," *Geological Society of America Bulletin,* Vol. 83, pp. 689–700.

Utada, M., 1970, "Occurrence and Distribution of Authigenic Zeolites in the Neogene Pyroclastic Rocks in Japan," *Science Papers Collection General Education,* Tokyo University, Vol. 20, pp. 191–262.

Commercial Utilization of Natural Zeolites

Anon., 1970, "Molecular Sieves Offer Low-Cost Oxygen Source," *Chemical Engineering,* Vol. 77, No. 12, pp. 54–56.

Anon., 1972, "Linde Molecular Sieve Type AW-500," Linde Molecular Sieves Bulletin F-1617, Union Carbide Corp., Aug. 15, 1962.

Abelson, P.H., 1973, "Underground Gasification of Coal," *Science,* Vol. 182, p. 1297.

Alexiev, B., 1968, "Clinoptilolite des Rhodopes du Nord-Est," *Comptes Rendus,* Bulgare Akademii Nauk, Vol. 21, pp. 1093–1095.

Ames, L.L., 1959, "Zeolitic Extraction of Cesium from Aqueous Solutions," Unclassified Report HW-62607, US Atomic Energy Commission, 23 pp.

Ames, L.L., 1960, "The Cation Sieve Properties of Clinoptilolite," *American Mineralogist,* Vol. 45, pp. 689–700.

Ames, L.L., 1967, "Zeolitic Removal of Ammonium Ions from Agricultural Wastewaters," *Proceedings,* 13th Pacific Northwest Industrial Waste Conference, Washington State University, Pullman, pp. 135–152.

Batta, L.B., 1971, "Adiabatic Selective-Adsorption Gas-Separation Process," US Patent 3,636,679, Jan. 25.

Batta, L.B., 1972, "Separation of Gas Mixtures by Adiabatic Adsorption Involving Pressure Changes," German Patent 2,153,808.

Chen, N.Y., 1971, "Shaped Selective Transition Metal Zeolite Hydrocracking Catalysts," US Patent 3,630,966, Dec. 28.

Domine, D., and Haÿ, L., 1968, "Process for Separating Mixtures of Gases by Isothermal Adsorption: Possibilities and Applications," *Molecular Sieves,* Society of Chemical Industry, London, pp. 204–216.

Drury, F.W., 1954, "Pozzolans in California," *Mineral Information Service,* Vol. 7, No. 10, pp. 1–6.

Grimšičar, A., 1967, "Zeoliti v Oligocenskih Tufih med Mozirjem in Celje," *Razprava Procila,* Vol. 10, pp. 239–245.

Iijima, A., and Utada, M., 1972, "A Critical Review on the Occurrence of Zeolites in Sedimentary Rocks in Japan," *Japanese Journal of Geology & Geography,* Vol. 42, pp. 61–84.

Jewell, W.J., 1974, "Methane . . . The Energy-Sufficient Farm," *The News,* State University of New York, Vol. 3, No. 2, p. 4.

Kobayashi, Y., 1970, "Natural Zeolite-Fillers for Paper," Japanese Patent 70–41,044.

Kondo, N., and Wagai, B., 1968, "Experimental Use of Clinoptilolite-Tuff as Dietary Supplements for Pigs," *Yotonkai,* May, pp. 1–4.

Langerhans, R., 1973, "United States of America, Contestant, vs. Union Carbide Corporation, Contestee," Testimony in Hearing A-7345, Dept. of the Interior, Office of Hearings & Appeal, Denver, CO, July 17, pp. 125–134.

Lee, H., 1972, "Pressure Swing Oxygen—A New Source of Oxygen for Waste Water Treatment," speech at Symposium on Application of Commercial Oxygen to Water and Waste Water Systems, University of Texas, Austin, Nov. 13–15.

Lee, H., 1973, "Applied Aspects of Zeolite Adsorbents," *Molecular Sieve Zeolites–I,* R.F. Gould, ed., Advances in Chemistry Series, Vol. 121, American Chemical Society, pp. 311–318.

Ludwig, U., and Schwiete, H.W., 1962, "Beitrag zur Konstitution Einiger Rheinischer Trasse," *Zement-Kalk-Gips,* Vol. 15, pp. 160–165.

Mercer, B.W., et al., 1970, "Ammonia Removal from Secondary Effluents by Selective Ion Exchange," *Water Pollution Control Federation Journal,* Vol. 42, pp. R95-R107.

Mielenz, R.C., et al., 1951, "Natural Pozzolans for Concrete," *Economic Geology,* Vol. 46, No. 3, May, pp. 311–328.

Minato, H., 1968, "Characteristics and Uses of Natural Zeolites," *Koatsugasu,* Vol. 5, pp. 536–547.

Mumpton, F.A., 1973, "First Reported Occurrence of Zeolites in Sedimentary Rocks of Mexico," *American Mineralogist,* Vol. 58, pp. 287–290.

Mumpton, F.A., 1973a, "Worldwide Deposits and Utilisation of Natural Zeolites," *Industrial Minerals,* No. 73, pp. 30–45.

Munson, R.A., and Clifton, R.A., 1971, "Natural Gas Storage with Zeolites," Technical Progress Report No. 38, US Bureau of Mines, 9 pp.

Munson, R.A., and Sheppard, R.A., 1974, "Natural Zeolites: Their Properties, Occurrences and Uses," *Mineral Science and Engineering,* Vol. 6, No. 1, pp. 19–34.

Nemecz, E., and Varju, G., 1962, "Sodium Bentonization, Clinoptylolitization and Ardularization in the Rhyolitic Tuffs of the Szerencs Piedmont Area," *Acta Geologica,* Vol. 6, pp. 389–426.

Nishita, H., and Haug, R.M., 1972, "Influences of Clinoptilolite on Sr-90 and Cs-137 Uptakes by Plants," *Soil Science,* Vol. 114, pp. 149–157.

Norin, E., 1955, "The Mineral Composition of the Neopolitan Yellow Tuff," *Geologische Rundschau,* Vol. 43, pp. 526–534.

Ohtani, S., et al., 1972, "Conversion Catalyst for Hydrocarbon," Japanese Patent 72,46,667, Nov. 24.

Onogi, T., 1966, "Experimental Use of Zeolite-Tuffs as Dietary Supplements for Chickens," Report, Kamagata Stock Raising Institute, pp. 7–18.

Ota, S., and Sudo, T., 1949, "Studies on 'Oya-ihi,' Part II, Mineralogical Composition," *Journal of the Geological Society of Japan,* Vol. 55, pp. 242–246.

Papp, J., et al., 1971, "Hydrodemethylation of Toluene on Clinoptilolite," *Journal of Catalysis,* Vol. 23, pp. 168–182.

Quisenberry, J.H., 1968, "The Use of Clay in Poultry Feed," *Clays and Clay Minerals,* Vol. 16, pp. 267–270.

Sersale, R., 1958, "Genesi e Constituzione del Tufo Giallo Napolitano," *Rendiconti della Accad. Sci. Fische Mat. (Napoli),* Vol. 25, pp. 181–207.

Sersale, R., and Aiello, R., 1965, "Ricerche sulla Genesi, sulla Constituzione e sulla Reattivita del 'Trass' Ranano," *Silicates Industriels,* Vol. 30, pp. 1–11.

Sheppard, R.A., 1973, "Zeolites in Sedimentary Rocks," Professional Paper 820, US Geological Survey, pp. 689–695.

Sims, R.C., 1972, "Enhancement of Nitrification in Activated Sludge by Addition of Hector

Clinoptilolite," *Environmental Science Engineering Notes,* Vol. 9, pp. 2–4.

Stojanović, D., 1972, "Zeolite-Containing Volcanic Tuffs and Sedimentary Rocks in Serbia," *Proceedings for 1968–1970,* Serbian Geological Society, pp. 9–20.

Tamura, T., 1971, "Gas Adsorption Properties and Industrial Applications of Japanese Tuff," unpublished abstract, Seminar on Occurrence and Mineralogy of Sedimentary Zeolites in the Circum-Pacific Region, Nikko, Japan, US-Japan Cooperative Science Program.

Tamura, T., 1971a, "Oxygen Concentration Process," British Patent 1,258,417, Dec. 30.

Tamura, T., 1972, "Adsorption Material and Method for Gas Separation," German Patent 2,214,820, Oct. 12, 41 pp.

Torii, K., et al., 1969, "Gas Chromatography with Natural Mordenite as Column Packing," *Kogyo Kagaku Zasshi,* Vol. 72, pp. 661–664.

Torii, K., et al., 1971, "Adsorption on Zeolite Tuff," *Kogyo Kagaku Zasshi,* Vol. 74, pp. 2018–2024.

Utada, M., 1970, "Occurrence and Distribution of Authigenic Zeolites in the Neogene Pyroclastic Rocks of Japan," *Science Paper, College of General Education,* Tokyo University, Vol. 20, pp. 191–262.

Wilson, R.C., 1972, "Clinoptilolite Blends with Shape-Selective Catalysts," US Patent 3,640,905, Feb, 8.

Yoshinago, E., et al., 1973, "Organophosphate-Containing Agricultural and Horticultural Granule Formation," US Patent 3,708,573.

Additional General References

Anon., 1977, "Cation Exchange Capacity of Clinoptilolite Rocks—A Study on Selective Absorption of Potassium," *Scientia Geologica Sinica,* No. 3, Mineralogy Group, Institute of Geology, Academia Sinica, pp. 284–288.

Anon., 1980, "Natural Zeolites in Agriculture," *Proceedings,* Symposium on the Utilization of Natural Zeolites in Agriculture, Georgian SSR Academy of Sciences, Sukhumi, Oct. 16–21, 1978, 254 pp.

Anon., 1980a, "Zeolites—Gaining Ground as Replacement for Phosphates in Detergents," *Journal,* American Oil Chemists Society, Vol. 57, No. 2, pp. A228-A229.

Aiello, R., and Porcelli, C., 1975, "Chabazitic Tuff Deposits of Possible Economic Value in the Campanian Region," *Rendiconti della Accad. Sci. Fische Mat.,* Naples, Vol. 41, pp. 426–434.

Aleksiev, B., and Djourova, E.G., 1974, "Zeolite Rocks: Classification and Nomenclature," *Bulg. Akad. Nauk Dokl.,* Vol. 27, No. 3, pp. 373–374.

Ataman, G., 1977, "Zeolite Occurrences in West Anatolia," *Yerbilimleri,* Vol. 3 (Turkish with English abstract) published by Institute of Earth Sciences of Hacettepe University, pp. 85–94.

Ataman, G., 1980, "The Role of Erionite (Zeolite) on the Development of Pulmonary Mesothelioma," *Comptes Rendus Academie des Sciences,* Ser. D, Vol. 291, No. 2, pp. 167–169.

Barrer, R.M., 1978, *Zeolites and Clay Minerals,* Academic Press, New York, 497 pp.

Barrer, R.M., 1979, "Chemical Nomenclature and Formulation of Compositions of Synthetic and Natural Zeolites," Commission on Colloid and Surface Chemistry, pp. 1093–1100.

Boreskov, G.K., and Minacher, K.M., eds., 1979, "Application of Zeolites in Catalysis," First All-Union Conference on Molecular Sieves in Catalysis, Novosibirsk, USSR, Akademiai Kiado, Budapest, 179 pp.

Castro, M., and Elias, A., 1978, "Effect of the Inclusion of Zeolite in Final Molasses-Based Diets on the Performance of Growing-Fattening Pigs," *Cuban Journal of Agricultural Science,* Vol. 12, No. 1, pp. 69–76.

Dyer, A., 1978, "Separation of Closely Related Systems by Molecular Sieve Zeolites," *Separation Science and Technology,* Vol. 13, No. 6, pp. 501–516.

Eyde, T.H., 1978, "Bowie Zeolite, an Arizona Industrial Mineral," *Field Notes,* Vol. 8, No. 4, Arizona Bureau of Geology and Mineral Technology, pp. 1–5.

Franco, E., and Aiello, R., 1969, "Zeolitization of Natural Glasses," *Rendiconti della Accad. Sci. Fische Mat.,* First Mineral. Chim. Ind., Univ. Napoli, Naples, Italy, Vol. 36, pp. 174–197.

Frost, B.R., 1980, "Observations on the Boundary Between Zeolite Facies and Prehnite-Pumpellyite Facies," *Contr. Mineralogy and Petrology,* Vol. 73, No. 4, pp. 365–373.

Goto, I., and Ninaki, M., 1979, "Agricultural Utilization of Natural Zeolites as Soil Conditioners. I. Effects of the Zeolite Application to Soils on the Inhibition of Nitrification and the Prevention of Leaching Out of Bases," *Shuko,* Vol. 24, No. 2, Tokyo Nogyo Daigaku Nogaku, pp. 164–183.

Goto, I., and Ninaki, M., 1980, "Agricultural Utilization of Natural Zeolites as Soil Conditioners. II. Changes in Physical and Chemical Properties of Upland Soils With the Application of Natural Zeolites," *Shuko,* Vol. 24, Nos. 3–4, Tokyo Nogyo Daigaku Nogaku, pp. 305–312.

Gottardi, G., and Obradovic, J., 1978, "Sedimentary Zeolites in Europe," *Fortschr. Mineral.,* Vol. 56, No. 2, pp. 316–366.

Hayhurst, D.T., 1980, "Gas Adsorption by Some Natural Zeolites," *Chemical Engineering Commun.,* Vol. 4, No. 6, pp. 729–735.

Imafuku, M., et al., 1979, "Removal of Environmental Pollutants," Japan Kokai Tokyo Koho 79,152,661 01 Dec.

Kashkay, M.A., and Babayev, I.A., 1976, "Clinoptilolite, Its Physical Properties and Genesis," *Izd. Nauka,* P.M. Tatarinov, ed., Leningrad, USSR, pp. 76–90.

Katzer, J.R., ed., 1977, "Molecular Sieves—II," American Chemical Society Symposium Series, No. 40, 732 pp.

Kim, J.T., et al., 1979, "A Study on Natural Zeolites. I. The Chemical Treatment and Adsorption Capacities of Natural Zeolites," *Hwahak. Kognhak,* Vol. 17, No. 5, pp. 331–344.

Kirov, G.N., 1974, "Zeolites in Sedimentary Rocks of Bulgaria," *God. Sofii.,* University Geol.-Geogr. Fak., Vol. 66, No. 1, pp. 171–185.

Klieve, J.R., and Semmens, M.J., 1980, "An Evaluation of Pretreated Natural Zeolites for Ammonium Removal," *Water Resources,* Vol. 14, No. 2, pp. 161–168.

Koellike, J.K., et al., 1980, "A Zeolite Packed Air Scrubber to Improve Poultry House Environments," *Trans. ASAE,* Vol. 23, No. 1, pp. 157–161.

Lee, Y.C., et al., 1979, "Experiments Using Natural Zeolite Powder to Replace Antibiotic Additives for Fattening Pigs," *Chung-hua Min Kuo Shou I Hseuh Hui Tsa Chih,* Vol. 5, No. 2, pp. 129–131.

Leonard, D.W., 1979, "The Role of Natural Zeolites in Industry," Preprint 79–380, SME-AIME Fall Meeting, Tucson, AZ, Oct., 21 pp.

Lisowski, A., and Kurowski, Z., 1979, "The Use of Clinoptilolite for Ammonia and Nitrogen Removal in Water Purification," Metody Fizyochem. Oczyszczania Wody Sciekow, Mater. Konf. Nauk. Miedzynar Konf. 2nd, 3, Paper No. 61, 10 pp.

Ma, Chueng-Shyang, et al., 1979, "Effect of Zeolite Feeding of Pregnant Pigs on the Litter Size at Birth," *Ki'o Hsueh Nung Yei* (Taipeh), Vol. 27, Nos. 5–6, pp. 189–192.

Makhmudov, F.T., 1979, "Kinetics of Sorption of Silver and Nickel Ions from Solutions on Modified Natural Zeolites," Mater. Konf. Molodykh Uch. Aspir., 13.

McCaslin, B.D., and Boyle, F.W., 1980, "Report of Research on Soil Conditioners," Res. Rept.-N.M., Agric. Exp. Stn. 411, 44 pp.

Meier, W.M., and Olson, D.H., 1978, *Atlas of Zeolite Structure Types,* Structure Commission of the International Zeolite Assn., 99 pp.

Mikhaylov, A.S., 1976, "Zeolite Rocks of Transcaucasia," *International Geological Review,* Vol. 18, No. 10, pp. 1201–1207.

Mikhaylov, A.S., et al., 1975, "Distribution of Zeolites in Volcanic Sedimentary Formations of the USSR and Some Physical Methods for Their Study," *Kristallokhim. Miner. Geol. Probl.,* pp. 177–184.

Mimura, H., et al., 1979, "Dynamic Properties of Ion Exchange of Cesium into Zeolites," *Tohoku. Daigaku Senko Seirea Kenkyusko Iho,* Vol. 35, No. 1, pp. 19–26.

Minato, H., and Watanabe, M., 1978, "Adsorption of CO_2 and N_2 Gases on Natural Zeolites and Its Theoretical Interpretations," *Scientific Papers of the College of General Education,* Vol. 28, No. 1, University of Tokyo, pp. 135–141.

Minato, H., and Watanabe, M., 1978a, "Adsorption of CO_2 and N_2 Gases on Natural Zeolites and Their Ion-Exchanged Forms," *Scientific Papers of the College of General Education,* Vol. 28, No. 2, University of Tokyo, pp. 215–220.

Mumpton, F.A., ed., 1977, "Mineralogy and Geology of Natural Zeolites," *Short Course Notes,* Vol. 4, Mineralogical Society of America, 233 pp.

Mumpton, F.A., 1979, "Reconnaissance Study of the Association of Zeolites with Mesothelioma Cancer Occurrences in Central Turkey," Preprint 79–332, SME-AIME Fall Meeting, Tucson, AZ, Oct., 22 pp.

Mumpton, F.A., and Ormsby, W.C., 1976, "Morphology of Zeolites in Sedimentary Rocks by Scanning Electron Microscopy," *Clays and Clay Minerals.*

Nathan, Y., and Flexer, A., 1977, "Clinoptilolite, Paragenesis and Stratigraphy," *Sedimentology,* Vol. 24, No. 6, pp. 845–855.

Negishi, T., 1972, "On the Measurement of Cation Exchange Capacity of Zeolitic Tuffs," *Nendo Kagaku,* Vol. 12, pp. 23–30.

Nikishina, V.A., et al., 1976, "Study of the Selectivity of Different Types of Zeolites Towards Some Non-Ferrous Metals," *Chromatography,* Vol. 120, No. 1, pp. 155–158.

Rabo, J.A., ed., 1976, "Zeolite Chemistry and Catalysis," Monograph 171, American Chemical Society, 796 pp.

Rees, L.V., ed., 1980, *Proceedings,* Fifth International Conference on Zeolites, held in Naples, Italy, June, Heyden, London, 902 pp.

Semmens, M.J., 1979, "The Regeneration of Clinoptilolite by Biologically Restored Brine," Order No. PB-296507, 218 pp., available from NTIS.

Semmens, M.J., and Porter, P.S., 1979, "Ammonium Removal by Ion Exchange, Using Biologically Restored Regenerant," *Journal,* Water Pollution Control Fed., Vol. 51, No. 12, pp. 2928–2940.

Shabtai, J., 1979, "Zeolites and Cross-linked Silicates as Media for Selective Catalysis," *La Chimica e L'Industria,* Vol. 61, No. 10, pp. 734–741.

Sherman, J.D., 1979, "Application of Molecular Sieve Zeolites to Pollution Control," *Ion Exchange Pollution Control,* Vol. 2, pp. 227–236.

Sherry, H.S., 1979, "Ion-Exchange Properties of the Natural Zeolite Erionite," *Clays and Clay Minerals,* Vol. 27, No. 3, pp. 231–237.

Shigeishi, R.A., et al., 1979, "Solar Energy Storage Using Chemical Potential Changes Associated with the Drying of Zeolites," *Solar Energy,* Vol. 23, No. 6, pp. 489–495.

Sims, R.C., and Hindin, E., 1978, "Use of Clinoptilolite for Removal of Trace Levels of Ammonia in Reuse Water," *Chemical Wastewater Technology,* pp. 305–323.

Surdam, R.C., and Eugster, H.P., 1976, "Mineral Reactions in the Sedimentary Deposits of the Lake Magadi Region, Kenya," *Geological Society of America Bulletin,* Vol. 87, pp. 1739–1752.

Townsend, R.P., ed., 1979, *The Properties and Applications of Zeolites,* Proceedings of a conference held at The City University, London, April, Special Publication No. 33, The Chemical Society, 430 pp.

Tsitsishvili, G.V., ed., 1977, *Clinoptilolite,* Proceedings of the Symposium on the Problems of Clinoptilolite Studies and Applications, held at Tbilisi, Nov. 1974, 243 pp.

Tsitsishvili, G.V., ed., 1979, *Natural Zeolites,* Proceedings of the Soviet-Bulgarian Symposium on Studies of Physiochemical Properties of Natural Zeolites, held at Tbilisi, Oct. 1976, 333 pp.

Tsitsishvili, G.V., et al., 1979, "Ethanol Drying by Natural Clinoptilolite," *Izv. Akad. Nauk. Gruz. SSR,* Ser. Khim., Vol. 5, No. 3, pp. 267–269.

Urotadze, S.L., et al., 1975, "Adsorption of Carbon Dioxide on Clinoptilolite," *Journal of Applied Chemistry,* Vol. 48, No. 12, Pt. 1, pp. 2760–2762.

Valyon, J., et al., 1979, "Investigation of the Gas-Adsorption Characteristics of Hungarian Natural Clinoptilolite," *Magyar Kemiai Folyoirat,* Vol. 85, No. 2, pp. 55–60.

Veitser, Y.I., and Sterina, R.M., 1979, "Removal of Ammonium Nitrogen from Municipal Wastewaters Using Clinoptilolite," *Nauch. Tr. Akad. Kommun. Khoz-va,* Vol. 164, pp. 43–50.

Willis, W.L., Quarles, C.L., and Fagerberg, D.J., 1980, "Zeolites Fed to Male Boiler Chickens," (abstract) *Poultry Science,* Vol. 59, No. 7, p. 1673.

Zirconium and Hafnium Minerals

THOMAS E. GARNAR, JR.*

Zirconium and hafnium are curious elements because they are almost always found together in nature. Zirconium was discovered by Klaproth in 1789 and isolated 35 years later by Berzelius. Because hafnium's chemical properties are so close to those of zirconium, it was not discovered until 1923 when Coster and Von-Heresey detected it using x-ray spectrographic analysis. These elements occur most commonly in nature as the mineral zircon ($ZrSiO_4$) and less commonly as the oxide baddeleyite (ZrO_2). They also are found as a variety of other silicates. Zircon was identified as a component in alluvial and beach sands in 1895, but it was not produced in any quantity until 20 years later. Zircon is always a coproduct from TiO_2 mining and processing. During World War I it was produced as a coproduct of beach sand mining for titanium minerals just south of Jacksonville Beach, FL, and was patented as a refractory. It was not until the 1930s when Zircon Rutile, Ltd. began mining at Byron Bay on the east coast of Australia that zircon was first used as a foundry sand. Later in the 1940s, NL Industries, Humphreys Gold Corp., and E. I. du Pont de Nemours & Co., Inc. began production of zircon sands from fossil beaches in northeast and northcentral Florida. Baddeleyite first became available as a commercial product in 1916, but never in the quantity of zircon.

Zircon holds a unique position as an industrial mineral because it is used for both its physical and chemical properties as well as an ore of zirconium and hafnium metals. Australia is the major zircon producer followed by South Africa and the US. Zircon is also produced in India, Sierra Leone, Sri Lanka, Malaysia, China, Thailand, and Brazil. Undeveloped heavy mineral beach sand reserves containing zircon are known to occur in Egypt, Malawi, Senegal, and Tanzania. Locations are shown in Fig. 1.

* E. I. du Pont de Nemours & Co., Inc., Starke, FL.

Mineralogy

Zirconium and hafnium are always present together in naturally occurring compounds. Most commonly they form as the silicate sometimes containing iron, calcium, sodium, manganese, and other elements. Less commonly they are found as oxides in combination with titanium, thorium, calcium, and iron. Zircon is the most common form. It is extremely resistant to weathering, therefore, often it is present in ancient beach sands and placer deposits. Although resistant to alteration from external sources, it is vulnerable to internal alteration as the result of thorium and uranium substituting for zirconium either in the zircon lattice or in solid solution. Alteration to the metamict state takes place as radioactive emanations from these elements disorder the crystal lattice, accompanied by hydration, reduction in specific gravity, and changes in color. The mineralogy of zircon and baddeleyite and variation of physical and chemical properties are shown in Table 1.

The effects of zircon's internal alteration and darkened color can be corrected by heating to 1000°C for 30 min or so. A certain time-temperature relationship makes it possible for decolorization and reordering of the crystal lattice to take place when heating at higher temperatures for shorter periods or at lower temperatures for longer periods. After heating, x-ray diffraction patterns are sharper, the grains may be harder, and except for particles of malacon, the grains will be white or colorless in the absence of grain surface coatings. The color change is only temporary and the particles slowly darken with time.

Zircon fluoresces light yellow in shortwave ultraviolet light. If exposed to ultraviolet light for extended periods of time, white (calcined) zircon turns purple. With time the induced color begins to fade and eventually disappears (e.g., calcined Florida zircon exposed to ultraviolet light for 24 hr still retains faint, but discernible, purple color after six months).

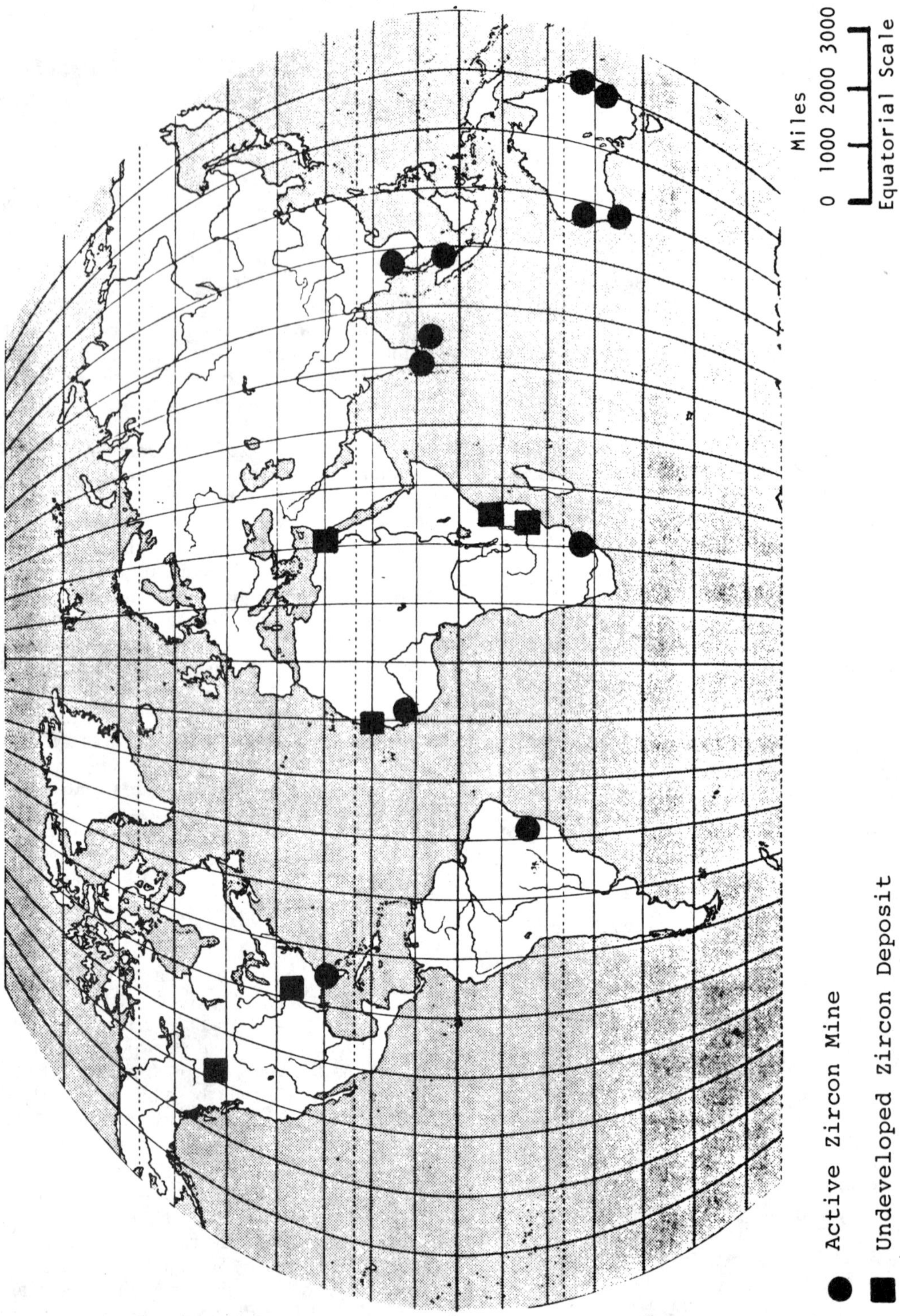

FIG. 1—*Location of world zircon deposits and mines.*

TABLE 1—Mineralogy of Zircon and Baddeleyite

	Normal Zircon	Altered Zircon (Hyacinth)	Much Altered Zircon (Malacon)	Baddeleyite
Crystal	Ditetragonal Dipyramidal	Ditetragonal Dipyramidal	Amorphous	Monoclinic Prismatic
Specific gravity	4.6-4.7	4.2-4.6	3.9-4.2	5.4-5.7
Mohs hardness	7.5	7	6-7	6.5
Color	White	Purplish	Dark	Variable
Index of refraction	1.92-1.96	1.90-1.92	1.76-1.90	2.19
Cleavage	110 poor	110 poor	None	001 Perfect
Loss on ignition	0.1%	0.5-1%	10%	

Occurrence of Zircon and Baddeleyite

Commercial zircon is available only in the form of sand mined, for the most part, from ancient beach sand deposits. In the beginning, zircon occurs as an accessory mineral in a variety of igneous and metamorphic rocks, especially those containing sodic feldspars such as granite, syenite, diorite, etc. It is one of the earliest minerals to crystallize from a cooling magma and frequently incorporates inclusions (e.g., apatite, magnetite). The zircon in these rocks often crystallizes out in tetragonal prisms with pyramidal terminations but it also commonly occurs as rounded grains in igneous rocks. It is sometimes found in metamorphic rocks such as gneiss and schists. One occurrence has been reported in a meteorite.

Before the zircon particles can become part of a beach sand deposit their host rocks must undergo a series of events that will ultimately liberate the zircon for transportation to a seacoast. This begins with the exposure of the host rocks to subaerial weathering. The rocks are broken down into smaller fragments and transported downhill by rainwater and gravity. The rocks further decompose to the point at which the smaller zircon grains are liberated from the enclosing feldspars and quartz. The mineral particles are transported by streams and ultimately end up along a marine shoreline. Here the action of waves, tidal currents, and wind may remove lighter quartz, forming a heavy mineral deposit rich in zircon. Many of the older beach sand deposits have been consolidated and the grains cemented together to form sandstones containing concentrations of titanium minerals, zircon, monazite, and other heavy minerals. Deposits such as these occur in the four-state area of Utah, Wyoming, New Mexico, and Colorado. Since this type of deposit requires more costly mining methods, crushing, and grinding, they are less attractive for mining and exploitation than are the more recent less consolidated deposits. All of the commercial zircon products today are mined and separated from relatively young beach sand deposits occurring on or near active coast lines.

Australian Zircon Deposits

Australia is the major world producer and exporter of zircon sand. Deposits are found on both the east and west coasts of Australia. In eastern Australia major heavy mineral sand deposits lie between Broken Bay, New South Wales, to the south and Cape Clinton north to Rock Hampton, Queensland. The richest occurrences are concentrated in the area between Yamba and North Stradbroke Island. Although there had been production of zircon from a mine at Ponte Vedra, FL between 1918 and 1929, the east Australian mines were the first major producers of commercial zircon sands beginning in the 1930s.

The east Australian heavy mineral beach sands deposits are of both Pleistocene and Holocene age. Zircon and its coproduct rutile have been derived from the breakdown and erosion of well sorted silica sandstone in the late Paleozoic and Mesozoic basins of eastern Australia. Zircon-bearing sands as mined contain less than 0.5% zircon. Zircon from the east coast of Australia ranges in color from dark tan to pale lavender. Because of iron-bearing grain surface coatings, these zircon sands become brownish orange to orange on heating to 1000°C. Grain size and shape vary from deposit to deposit. The eastern Australia zircon is characterized by subangular to rounded particles with many elongated prismatic grains. Particle size ranges from 90 to 100 μm depending upon which mine produced the zircon. Until 1983, five companies produced zircon from beach sands along the east coast of Australia. Two have since shut down.

Mineral sand deposits of Western Australia are found along ancient beaches which lie at different elevations ranging from 10 to 200 m above present sea level. The strand line deposits are part of a series of Pleistocene and Recent sediments which exist as a thin veneer throughout the Perth basin. These overlie much older marine and continental deposits. Shoreline deposits formed during stable Pleistocene sea level stands. The age of these deposits ranges from fairly young to perhaps over a hundred thousand years. Heavy mineral sand deposits along the southern coast of Western Australia near Cable and Bunbury are very rich in the mineral ilmenite ($Fe_2O_3 \cdot 3TiO_2$) but contain some zircon.

Zircon has become a major export of Western Australia since the discovery of the deposits near the town of Eneabba. The Eneabba heavy minerals occur at different strand lines ranging from 82 to 120 m above sea level. The presence of clay, induration, and grain surface coatings created difficult mining and mineral separation problems; however, these have largely been corrected by the producers. The mineral grains are extensively coated with ferrugenous aluminum silicates as a result of laterization. The grain surface coatings cause the zircon to take on a light orange color when heated to 1000°C. Australian zircon grains are shown in Fig. 2.

South African Deposits

Heavy minerals are produced from a high dune paralleling the Indian Ocean about 160 km north of Durban. The ore body being mined is 17 km long and 2 km wide and is reported to contain 700 million tons of ore. Life of the present ore body is estimated at 30 years.

As shown in Fig. 3, the zircon grains are rounded to irregular or prismatic in shape and are slightly finer than Du Pont Florida zircon but coarser than Associated Minerals, US. Like Australian zircon the grains have surface coatings which turn a dark orange color on heating.

There are reported to be similar concentrations of heavy minerals in dunes to the north of Richards Bay.

Baddeleyite is recovered from mill tailings at the Palabora copper mine in the Transvaal. The ZrO_2 and HfO_2 range from 97 to 99%.

Baddeleyite may occur either as a primary or secondary mineral. It is sometimes found in alluvial deposits associated with ilmenite, zirkelite (mixture of zircon and baddeleyite), apatite, and perovskite. It is also found in Brazil, Sri Lanka, Italy, and the United States.

US Deposits

A small amount of zircon was produced from a beach sand mine located near Ponte Vedra, FL between 1918 and 1929. There were no companies producing heavy minerals in Florida from 1929 to 1939. In 1940 a small quantity of heavy minerals including zircon was produced from surface concentrates found in the beach sand near Melbourne, FL; this operation was later moved to a dune area near Vero Beach and operated until about 1963. Between 1943 and 1963 zircon was produced from a deposit at Arlington, a suburb of Jacksonville, FL. In 1963 production began from a heavy mineral deposit near Folkston, GA. When this deposit was mined out in 1974, the mining and

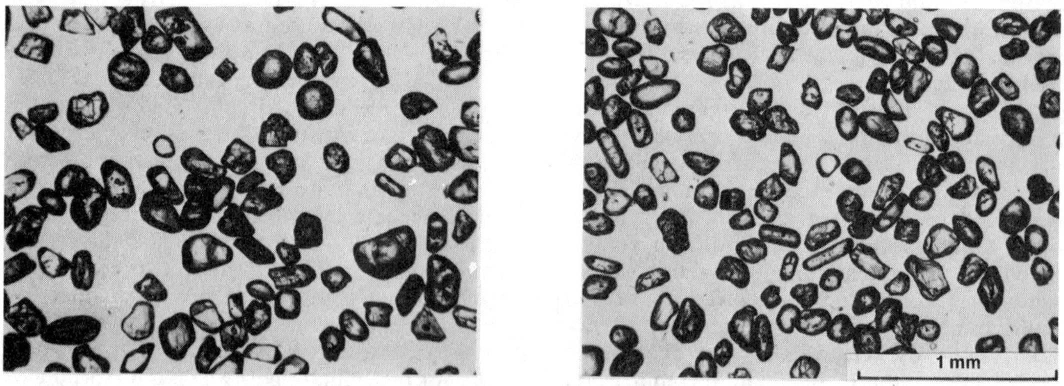

FIG. 2—*Australian zircon sands. Left, Western Australia; right, Queensland.*

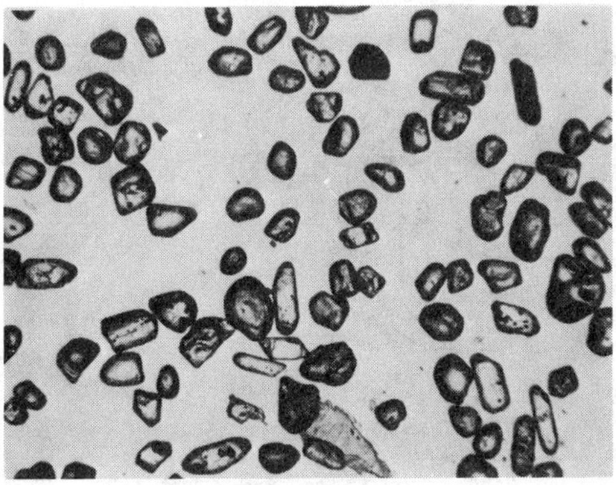

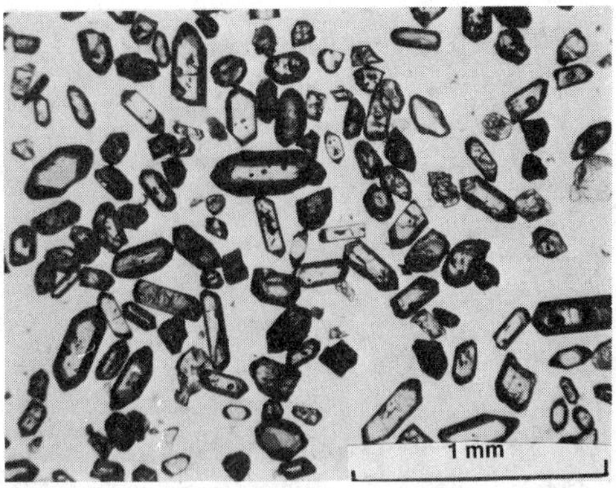

FIG. 3—*African and Malaysian zircon sands. Left, Richards Bay, South Africa; right, Malaysia.*

concentrating equipment was moved a few miles south to a deposit near Boulougne, FL where mining continued until the ore was mined out in 1979. Zircon production began from an old shoreline deposit near Green Cove Springs, FL in 1973 (Fig. 4). Zircon from this deposit averages about 98 μm in size. This is similar to the zircon produced from the Jacksonville deposit during World War II.

The largest domestic supply of zircon sand comes from the Trail Ridge deposit in north central Florida. This large deposit was discovered by Du Pont in 1946 and has been in continuous production of zircon since 1950. The deposit was formed at the height of Florida's submergence during Pleistocene times. Reworked sands transported from older sand features were redeposited forming the ridge. Waves, currents, and wind action removed part of the lighter silica sand, leaving an enrichment of heavy minerals in sand dunes on the western flank of this prominent geographic feature. After deposition of the heavy mineral bearing sands the ridge was covered with wind blown sand, forming an overburden.

Zircon from the Trail Ridge deposit is extremely uniform with respect to particle size and grain morphology as well as chemical purity (Fig. 4). The grains are free of surface coatings and have been slightly etched as a result of wind action during formation of the deposit. The grains have no surface coatings, therefore, the zircon turns white on heating.

Undeveloped zircon deposits are also known to exist in western Tennessee.

Mining and Processing

Most of the heavy mineral operations in the world are similar. The sand ores are mined with dredges, both cutterhead-suction or in some cases, bucketline. In areas where there is insufficient water for dredging, the ores are mined using elevating scrapers and bulldozers.

The heavy minerals are concentrated by removing the quartz and other light minerals using gravity concentrators (spirals, sluices, or cones). The gravity concentrates are scrubbed, if necessary, to remove any surface coatings that might interfere with dry processing. These are dried and separated into ilmenite, leucoxene, and rutile by magnetic and high tension separation. The remaining materials are further treated to produce zircon using some combination of spirals, gravity tables, high tension separators, magnets, and air tables. A flow diagram is shown in Fig. 5.

In the absence of surface coatings, the zircon concentrates may be calcined to produce a white color. The white zircon has long served as a trademark for sands produced in the southeastern United States. Zircon from the Australian and South African deposits turns a yellowish orange color when heated because of ferruginous coatings on the grain surfaces.

Zircon Quality Standards

When zircon was originally marketed in the 1950s, only one grade was available from each producer. As competition increased and the consuming markets became more sophisticated, a number of different grades evolved with names selected by each producer to describe the product. Sometimes, however, the same name did not mean the same thing to different producers. For example, *premium* zircon may

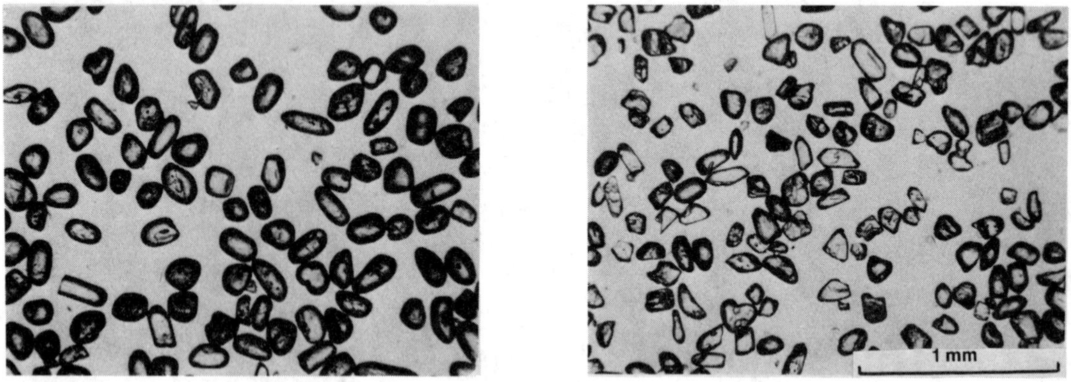

FIG. 4—*US zircon sands. Left, Trail Ridge Deposit, Starke, FL; right, Green Cove Springs, FL.*

mean one or more of the following: low TiO_2, low Fe_2O_3, low Al_2O_3, high ZrO_2, calcined, classified, etc. In addition, such terms as foundry grade, standard grade, ceramic grade, etc., are also applied to zircon products.

Typical zircon sand specifications are shown below:

$ZrO_2 + HfO_2$	65–66% (min.)
Fe_2O_3	0.02–0.10% (max.)
TiO_2	0.10–0.35% (max.)
Al_2O_3	0.20–2.00% (max.)

Du Pont also markets two special grades of zircon called *Zircon T,* a foundry grade containing up to 2% rutile and *Zircon M,* a mixture of zircon and magnetic heavy minerals.

Zircon particle size distribution and shape are important in most foundry applications. Zircon sands from each deposit are different in particle size (Table 2). The Eneabba ores produce the coarsest zircon particles presently available. The finest particle size comes from the Green Cove Springs Deposit in Florida. Photomicrographs illustrating the variation in zircon particle size and shape are shown in Figs. 2 through 4.

Markets and Uses

Zircon sands are used in a variety of markets ranging from foundry sands to zirconium metal manufacture. The major uses in the United States are:

Foundry sands	30%
Flour	30%
Refractories	30%
Abrasives	8%
Other	2%

Foundry Use

Widespread use of zircon as a foundry sand began about 1950 when large quantities became available after Du Pont opened its mines near Starke, FL. Zircon is ideal for foundry use because of the following properties:

1. Very low thermal expansion on heating.
2. Very high thermal conductivity and bulk density giving about four times the cooling rate of quartz. These properties provide good "chill" in the casting process.
3. Unwetted by molten metal.
4. Chemically nonreactive with metals.
5. Clean rounded grains which readily accept any binder.
6. Uses less binder than other sands.
7. Volatiles removed by heating during processing.
8. Superior dimensional and thermal stability at elevated temperatures.
9. pH is neutral or slightly acid.

TABLE 2—Typical Particle Size Distribution of Commercial Zircon Sands

Wt % Retained, US Standard Series	Opening in μm	Eneabba Deposit, Western Australia	Capel, Western Australia	Queensland, Australia	Starke, Florida	Green Cove Springs, Florida	Richards Bay, South Africa
50	300	Trace	1	—	1	—	—
70	212	14	3	1	1	Tr	Trace
100	150	46	30	6	13	1	8
140	106	30	52	47	48	25	42
200	75	9	13	44	36	63	46
270	53	1	1	2	2	10	4
Pan	53	—	—	1	Trace	1	Trace
Average Particle size in μm		162	141	111	120	98	111

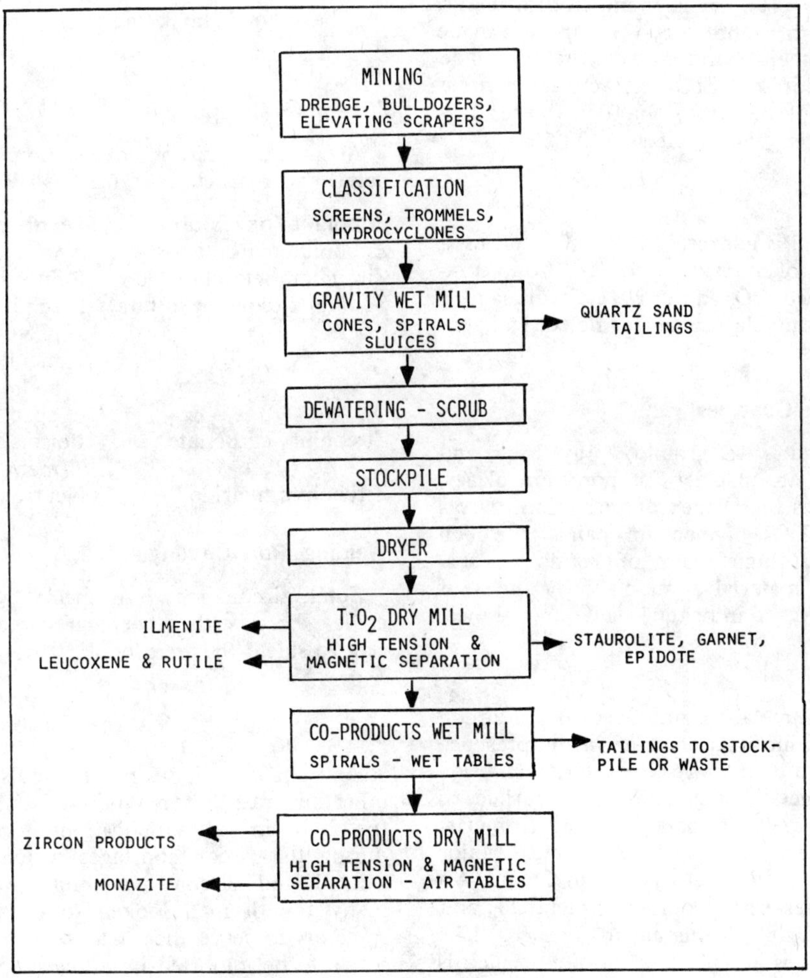

FIG. 5—*Simplified flow diagram of a heavy mineral processing plant.*

Zircon sand is used in the foundry for shell cores and mold linings. It is also dry ground to flour, then used to make mold washes which are painted on mold facings to give castings smoother surface finish. The investment casting industry uses zircon as a stucco grain and the flour for smoother finishes.

Basic Steel Production

Zircon sand is used in basic steel production in ladle brick because it reduces erosion and extends ladle lining life. It is used in coatings, mortars, and as ladle nozzle fill.

Refractory Use

Zircon sands are fused to make refractory brick. These are used to line kilns, glass melting furnaces, etc. High temperature boats, dishes, and other laboratory ware are also made
dishes, and other laboratory ware are also made
from fused zircon. ZrO_2 refractories can be used at temperatures to 3600°F where they begin to soften, then melt at 4300°F.

Abrasives

Zircon sand is converted into ZrO_2 and used in a number of abrasive products. Four companies produce ZrO_2 in the US. Zircon sand is also used to sandblast turbines in electrical generating plants.

Zircon Flour Opacifier

Zircon sand is wet ground, spray dried, and marketed as an opacifier for porcelain glazes. Because of its high index of refraction, it acts much like TiO_2 pigment in paint in which material with high index of refraction suspended in a material of much lower index of refraction gives it an opaque white appearance.

Zirconium Metal

Zirconium metal is produced from zircon sand by reacting with chlorine in the presence of carbon and then converting the tetrachloride to sponge metal which is further worked to produce a variety of shapes ranging from zirconium plate to zirconium bolts, etc. Major metal producers are Teledyne Wah Chang, Albany, OR, Western Zirconium, Ogden, UT, and Pechniney Ugine Kuhlman in France. The sponge contains about 2.5% hafnium which is closely related to the zirconium and very difficult to separate. The commercial metal usually contains hafnium but reactor-grade zirconium for use in atomic work must be hafnium-free.

Commercially pure zirconium is not a high-strength metal but has properties similar to those of iron. It is valued for atomic construction purposes because of its low neutron-capture cross section, thermal stability, and corrosion resistance. The metal is 99.99% pure and is used to make rods, plates, sheets, foil, and wire. The unalloyed metal is difficult to roll and is usually worked at temperatures to 900°F. The metal is not attacked by nitric, sulfuric, or hydrochloric acids, but is dissolved by hydrofluoric acid.

Small amounts of zirconium are used in steel. It is a powerful deoxidizer, removes the nitrogen, and combines with the sulfur reducing hot-shortness and giving ductility. Steels with small amounts of zirconium are fine grained and have good shock and fatigue resistance.

Zirconium Compounds

A number of zirconium compounds are prepared and marketed for uses as shown below:

Compound	Use
Zirconium oxychloride	Textile coatings
Zirconium carbide	Abrasive
Zirconium tetrachloride	Refine Al and Mg
Sodium zirconium sulfate	Precipitation of proteins, pigment stabilizer, and paper opacifier
Zirconium carbonate	Poison ivy ointment
Zirconium hydride	Neutron moderator

Welding Rod Coatings

Some zircon goes into special welding rod coats where, with other ingredients, it forms slag to shield the weld pool.

Substitutes for Zircon

As with most industrial minerals, there are other minerals that may be used as a substitute for zircon in some applications. For example, during the zircon shortages, when prices escalated and zircon availability was limited, many foundrymen looked to more available minerals to serve their needs. Chromite sand began to be imported in increasing amounts to replace zircon. The foundrymen found, however, that chromite contained serpentine and other hydrated alteration products that caused casting defects as volatiles were released during the casting process. In addition, chromite is a very basic mineral that requires the addition of acid in certain binder mixes to lower the pH. Other drawbacks of chromite include the irregular particle shape resulting from crushing to sand size and expansion on heating. Many of the foundries that used chromite returned to zircon after it became more available and less expensive. Some foundrymen tried olivine, but it also has many undesirable characteristics that limit its use in the foundry industry.

During periods of shortages in the late 1960s Du Pont extended its Florida zircon supplies by offering *Zircon T* which had all of the foundry properties of regular zircon, but contained more than normal TiO_2 (2% max.).

More recently Du Pont has offered a specialty foundry sand made up of a mixture of zircon and aluminum silicates (kyanite and sillimanite) which is marketed under the trade name *Zircore®*. This highly refractory sand has some advantages over pure zircon. For example, its coarser particle size gives cores and molds more porosity to vent off gases evolved from some binders, thus reducing surface defects on the castings.

Hafnium Metal

Hafnium, though relatively abundant in nature, is sparsely disseminated and very costly to extract. Zircon sand is the only commercial source of this metal. Hafnium's melting point is 4000°F, which is higher than the melting point of zirconium. Heat-resisting parts for special purposes have been made by compacting hafnium powder to a density of 98%. The metal has a close-packed hexagonal structure. It has excellent resistance to a wide range of corrosive environments. Because of its high thermal neutron-capture cross section and excellent strength up to 1000°F, hafnium is useful in unalloyed form in nuclear reactors.

Hafnium Compounds

Hafnium oxide is a better refractory ceramic than zirconia, but is more costly. The inversion of the crystal from the monoclinic to the tetragonal form occurs at 3100° F with a 3.4% expansion compared to 2000°F and 7.5% in zirconia.

Hafnium carbide is produced by reacting hafnium oxide and carbon at high temperatures. The resulting crystals have a melting point of 4160°C, making hafnium carbide one of the most refractory materials known.

Zircon Production, Pricing, and Supply

There has been a continuing demand for zircon since 1950. Until recently, price and demand varied inversely. In the mid-1960s and again in the early 1970s, zircon was in short supply and prices rose. In the intervening years oversupplies caused prices to decline. More recently, demand and prices both dropped. Foreign zircon production peaked at over 600,000 tons in 1980 but dropped to 564,000 tons in 1982, according to the US Bureau of Mines trade statistics shown in Table 3. Fig. 6 shows that US imports almost doubled between 1977 and 1980 but have declined sharply since 1980.

TABLE 3—Zirconium Concentrate Production by Country*
(Thousands of Short Tons)

	1975	1976	1977	1978	1979	1980	1981	1982
Australia	421	463	439	432	488	506	468	456
South Africa	13	12	19	40	90	88	110	88
India	11	11	12	11	11	14	16	13
Other (Brazil, Malaysia, Sri Landa, Thailand)	7	6	6	10	11	7	9	7
Total (Excluding US)	452	492	476	493	600	615	603	564

Source: US Bureau of Mines *Minerals Yearbooks.*
* US production figures are withheld to avoid disclosing company proprietary data. US production is estimated at 120,000 stpy.

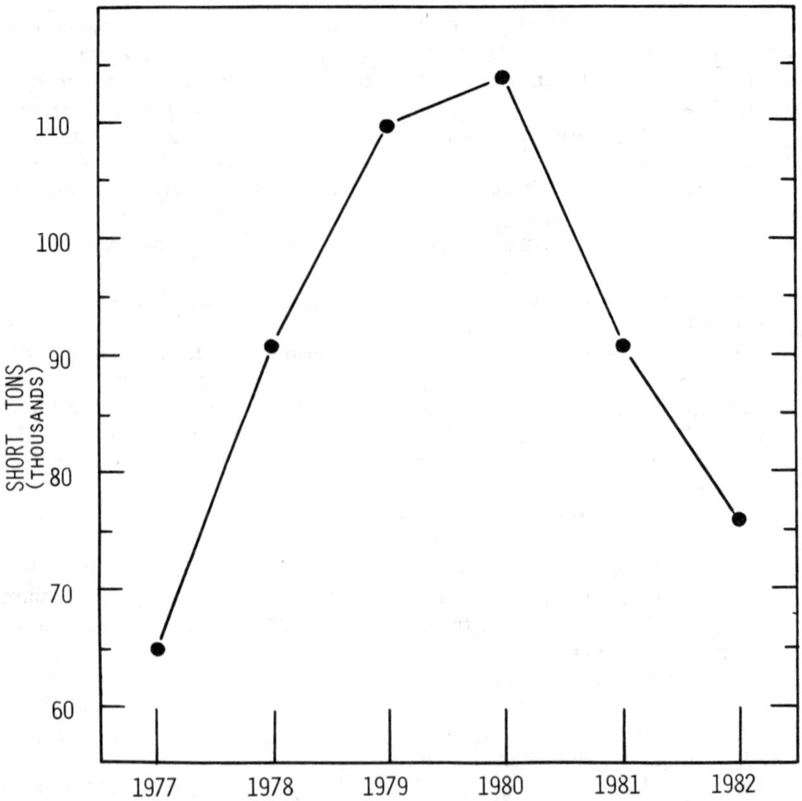

FIG. 6—*Zirconium ore imports for use in the United States. (Source: US Bureau of Mines Minerals Yearbooks.)*

Australian zircon prices dropped from an all-time high of about $275 per net ton in 1973 to less than $80 in 1980. US zircon prices vary somewhat with demand but not as much as Australian prices do.

The cost to produce each heavy mineral increases with the amount of processing required. Ilmenite costs are lowest, zircon or monazite (where present) costs are significantly higher.

In the past, world prices for heavy minerals have not reflected their true costs of production but rather their supply vs. demand. In particular, zircon has most often sold at prices below actual cost.

In 1980, Humphreys Mining Company shut down after mining out their deposit near Boulougne, FL and Associated Minerals (US) purchased the Titanium Enterprises Plant near Green Cove Springs, FL. In 1981, zircon production began from the heavy mineral sand operation near Richards Bay, South Africa.

Australia continues to lead the world in zircon production with 456,000 tons produced in 1982. Australian producers are under increasing pressures from conservation groups and the government to stop mining present deposits and to discontinue development of new reserves. Declining east coast Australian zircon production will be partially offset by increased production from South Africa, Western Australia, and the US. Major world zircon producers are shown in Table 4.

World Zircon Reserves

The US Bureau of Mines reports identified world resources of zircon exceed 60 million tons; however, only a small fraction is economic to mine. Of this 60 million tons, the US has about 14 million tons. Studies of zircon associated with phosphate, sand, and gravel have been carried out by several US governmental agencies and private interests. In all cases the tonnage of zircon and other heavy minerals produced in those industries is insufficient to be considered economic at this time. An important future source may be Canada tar sands. Although the percent heavy minerals in the tar sands is less than 1, about 41,000 st of zircon

TABLE 4—Major World Zircon Sand Producers

Producer	Location	Remarks
Australia		
Allied Eneabba Pty., Ltd.	Eneabba, W.A.	60% Du Pont 40% Allied Minerals
Associated Minerals Consolidated, Ltd.	Stradbroke Island, Qld.	Subsidiary of Consolidated Goldfields
	Capel, W.A. Geralton, W.A. Eneabba, W.A.	Formerly Western Titanium
Cable Sands Pty., Ltd.	Bunbury, W.A.	Subsidiary of Kathleen Investments, Ltd.
Consolidated Rutile, Ltd.	Stradbroke Island, Qld.	Subsidiary of Cudgen RZ, Ltd.
Cudgen RZ, Ltd.	Kingscliff, NSW	General Mining (49%)
Rutile and Zircon Mines	Newcastle, NSW	Peko-Wallsend, Ltd. Kathleen Investments
Westralian Sands, Ltd.	Capel, W.A.	
Brazil		
Nuclemon	Bahia and Espirito Santo States	Subsidiary of Nuclebras
India		
Indian Rare Earths Ltd.	Manavalakurichi, Tamil Nadu, Chavara, Quilon	
Kerala Minerals & Metals	Chavara, Quilon	Owned by the Government of Kerala
South Africa		
Richards Bay Minerals (Tisand Pty, Ltd.)	Richards Bay	Quebec Iron & Titanium, General Mining, Industrial Development Corp., S.A. Mutual Life Assurance Soc., Southern Life Association
Sri Lanka		
Ceylon Mineral Sands Corp.	Pulmoddai China Bay	
United States		
E. I. du Pont de Nemours & Co., Inc.	Starke, FL Lawtey, FL	
Associated Minerals Consolidated, Inc.	Green Cove Springs, FL	Subsidiary of Consolidated Goldfields

could be produced from the 95 million st of tar sands mined annually.

The Future of Zircon in North America

The demand for zircon will be set by the needs of the various consuming industries and by the availability of equal or lower cost alternative products.

The supply of zircon will be determined by demand for titanium minerals and availability of deposits. It is doubtful that new heavy mineral mines will open until the current oversupply of rutile and zircon passes and prices stabilize at higher levels. Several known deposits have already been put off limits. Others may be eliminated if governmental attitudes toward mining and the environment follow the trend of the 1970s.

Additional North American reserves are available for development to meet domestic needs. These will be developed if the delivered cost of the zircon and other heavy minerals is competitive with that from foreign sources and yet high enough to provide producers a reasonable return on their investment.

Bibliography and References

Anon., 1967, "Contribuição do Departamento Nacional da Produção Mineral no Desenvolvimento Geo-econômico do Nordeste Brasileiro," Publicação Especial 4, Brazil Departamento Nacional da Produção Mineral, 125 pp.

Anon., 1969, "Pattern of Zircon Use May Change," *Industrial Minerals,* No. 16, Jan., pp. 9–13, 15–17.

Anon., 1969a, "Titanium Minerals—Uruguay," *Mineral Trade Notes,* US Bureau of Mines, Vol. 66, No. 6, pp. 25–26.

Anon., 1969b, "World Zircon Producers—Present Production and Future Potential," *Industrial Minerals,* No. 16, Jan., pp. 19, 21, 23–27.

Anon., 1970, "On the Trail of Zirconium Oxides," *South African Mining and Engineering Journal,* Vol. 81, Pt. 1, No. 4034, pp. 1085, 1087, 1089, 1091.

Anon., 1970a, "South Africa—Foskor's Huge Phosphate Reserves," *Industrial Minerals,* No. 28, Jan., p. 36.

Anon., 1971, "Titanium Minerals—Part 1, The Producers Reviewed," *Industrial Minerals,* No. 43, pp. 9–13, 15–23.

Anon., 1973, "Zirconium and Hafnium," *Commodity Data Summaries,* US Bureau of Mines, pp. 166–167.

Angelelli, V., and Chaar, E., 1967, "Los Depositos de Titanomagnetita, Ilmenita y Zircon de la Bahia San Blas (Tramo Baliza la Ballena-Faro Segunda Baranca), Partido Carmen de Patagones, Provincia de Buenos Aires," (Informe) CNEA 210, Argentina Republica, Comisión Nacional de Energía Atómica, Buenos Aires, 53 pp.

Brooks, C.K., 1969, "On the Distribution of Zirconium and Hafnium in the Skaergaard Intrusion, East Greenland," *Geochimica et Cosmochimica Acta,* Vol. 33, No. 3, Mar., pp. 357–374.

Brooks, C.K., 1970, "The Concentrations of Zirconium and Hafnium in Some Igneous and Metamorphic Rocks and Minerals," *Geochimica et Cosmochimica Acta,* Vol. 34, No. 3, Mar., pp. 411–416.

Calver, J.L., 1957, "Mining and Mineral Resources," Geological Bulletin No. 39, Florida Geological Survey, pp. 15–31.

Carpenter, J.H., and Griffith, R.H., 1960, "Production of Monazite from Alluvial Concentrates," Carpco Technical Bulletin, Carpco, Jacksonville, FL, 10 pp.

Chao, E.C.T., and Fleischer, M., 1960, "Abundance of Zirconium in Igneous Rocks," *Report,* 21st International Geological Congress, Copenhagen, Part 1, pp. 106–131.

Chessex, R., and Delaloye, M., 1965, "Données sur les Teneurs en Hafnium et en Yttrium des Zircons," *Schweizerische Mineralogische und Petrographische Mitteilungen,* Vol. 45, No. 1, pp. 295–315.

Clarke, R.G., 1970, "Zirconium and Hafnium," *Minerals Yearbook 1970,* US Bureau of Mines, Vol. 1, pp. 1205–1211.

Coope, B., 1982, "Titanium Minerals—Focus on Production," *Industrial Minerals,* No. 178, July, pp. 27–35.

Dana, E.S., 1932, *A Textbook of Mineralogy,* W.E. Ford, ed., 4th ed., John Wiley & Sons, New York, 851 pp.

Degenhardt, H., 1957, "Untersuchungen zur geochemischen Verteilung des Zirkoniums in der Lithosphäre," *Geochimica et Cosmochimica Acta,* Vol. 11, No. 4, pp. 279–309.

Dow, V.T., and Batty, J.V., 1961, "Reconnaissance of Titaniferous Sandstone Deposits of Utah, Wyoming, New Mexico, and Colorado," Report of Investigations 5860, US Bureau of Mines, 52 pp.

Fleischer, M., 1955, "Hafnium Content and Hafnium-Zirconium Ratios in Minerals and Rocks," Bulletin 1021-A, US Geological Survey, pp. 1–13.

Franco, R.R., and Loewenstein, W., 1948, "Zirconium from the Region of Poços de Caldas," *American Mineralogist,* Vol. 33, No. 3–4, pp. 142–151.

Frondel, C., 1957, "Zirconium—Mineralogy and Geochemistry," *Advances in Nuclear Engineering,* Vol. 2, *Proceedings,* 2nd Nuclear Engineering and Science Congress, Philadelphia, Pt. 2, pp. 305–312.

Gadsden, J., 1972, "Zirconium and Hafnium," in "Mining Annual Review," *Mining Journal,* London, p. 94.

Garnar, T.E., Jr., 1978, "Geologic Classification and Evaluation of Heavy Mineral Deposits,"

12th Forum on the Geology of Industrial Minerals, Information Circular No. 49, Georgia Geological Survey, Atlanta, pp. 25–36.

Garnar, T.E., Jr., 1980, "Heavy Minerals Industry of North America," Presented at 4th Industrial Minerals International Congress, Atlanta, GA, 13 pp.

Gerasimovskii, V.L., 1956, "Geochemistry and Mineralogy of Nepheline Syenite Intrusions," *Geochemistry,* No. 5, pp. 494–510.

Gottfried, D., and Waring, C.L., 1964, "Hafnium Content and Hf/Zr Ratio in Zircon from the Southern California Batholith," Professional Paper 501-B, US Geological Survey, pp. B88–B91.

Grogan, R.M., et al., 1964, "Milling at Du Pont's Heavy Mineral Mines in Florida," *Milling Methods in the Americas,* N. Arbiter, ed., VII International Mineral Processing Congress, Gordon and Breach, New York, pp. 205–229.

Guimarães, D., 1948, "The Zirconium Ore Deposits of the Poços de Caldas Plateau, Brazil, and Zirconium Geochemistry," Boletim 6, Minas Gerais Instituto de Tecnologia Industrial, pp. 1–40 (Portuguese), 41–79 (English).

Hansen, J., 1968, "Niobium Mineralization in the Ilimaussaq Alkaline Complex, South-West Greenland," *Proceedings,* 23rd International Geological Congress, Prague, Sec. 7, "Endogenous Ore Deposits," pp. 263–273; reprinted as Miscellaneous Papers No. 60, Grönlands Geologiske Undersögelse.

Hess, H.D., 1962, "Hafnium Content of Domestic and Foreign Zirconium Minerals," Report of Investigations 5856, US Bureau of Mines, 62 pp.

Horn, M.K., and Adams, J.A.S., 1966, "Computer-Derived Geochemical Balances and Element Abundances," *Geochimica et Cosmochimica Acta,* Vol. 30, No. 3, Mar., pp. 279–297.

Kauffman, A.J., Jr., and Holt, D.C., 1965, "Zircon—A Review, with Emphasis on West Coast Resources and Markets," Information Circular 8268, US Bureau of Mines, 69 pp.

Klemic, H., et al., 1973, "Zirconium and Hafnium," US Mineral Resources, Professional Paper No. 820, US Geological Survey, pp. 718–722.

Klemic, H., 1975, "Zirconium and Hafnium Minerals," *Industrial Minerals and Rocks,* 4th ed., S.J. Lefond, ed., AIME, New York, pp. 1275–1283.

Kramer, J.W., et al., 1976, "Survey of Heavy Minerals in Surface Mineable Area of Athabasca Oil Sand Deposit" *Canadian Mining and Metallurgical Bulletin,* Vol. 69, No. 776.

Lewis, R.M., 1978, "Possible Recovery of Heavy Minerals from Phosphate Tailings," Preprint No. 78B300, SME-AIME Fall Meeting, Orlando, FL, 7 pp.

Lissiman, J.C., and Oxenford, R.J., 1973, "The Allied Mineral N.L. Heavy Mineral Deposit in Eneabba, W.A.," Conference Volume, Australian Institute of Mining and Metallurgy, pp. 153–161.

Lustman, B., and Kerze, F., Jr., 1955, *The Metallurgy of Zirconium,* McGraw-Hill Book Co., Inc., New York, 776 pp.

Lyakhovich, V.V., and Shevaleyevskii, I.D., 1962, "Zr:Hf Ratio in the Accessory Zircon of Granitoids," *Geochemistry,* No. 5, pp. 508–524.

Lynd, L.E., 1978, "US Dependence on Foreign Sources of Heavy Mineral Concentrates," Preprint No. 78H366, SME-AIME Fall Meeting, Orlando, FL, 17 pp.

Lynd, L.E., 1980, "Zirconium and Hafnium," *Minerals Industry Surveys,* US Bureau of Mines, 1 p.

Lynd, L.E., 1981, "Zirconium," *Mineral Commodity Summaries,* US Bureau of Mines, pp. 182–183.

Martens, J.H.C., 1928, "Beach Deposits of Ilmenite, Zircon, and Rutile," 19th Annual Report of the Florida State Geological Survey, pp. 124–125.

Middleton, J.M., 1969, "Zircon-Its Applications in the Foundry," *Industrial Minerals,* No. 16, Jan., pp. 29–31.

Palache, C., Berman, H., Frondel, C., 1951, "Baddeleyite," *Dana's System of Mineralogy,* John Wiley & Sons, New York, pp. 608–610.

Pirkle, E.C., Pirkle, W.A., and Yoho, W.H., 1974, "The Green Cove Springs and Boulogne Heavy Mineral Sand Deposits of Florida," *Economic Geology,* Vol. 69, pp. 1129–1137.

Pirkle, F.L., 1975, "Evaluation of Possible Source Regions of Trail Ridge Sands," *Southeastern Geology,* Vol. 17, No. 2, Duke University, pp. 93–114.

Sørensen, H., 1970, "Low-Grade Uranium Deposits in Agpaitic Nepheline Syenites, South Greenland," *Uranium Exploration Geology, Proceedings,* International Atomic Energy Agency Panel on Uranium Exploration Geology, Vienna, pp. 151–159.

Stamper, J.W., and Chin, E., 1970, "Hafnium," *Mineral Facts and Problems,* Bulletin 650, US Bureau of Mines, pp. 587–594.

Stamper, J.W., and Chin, E., 1970a, "Zirconium," *Mineral Facts and Problems,* Bulletin 650, US Bureau of Mines, pp. 825–835.

Stow, S.H., 1968, "The Heavy Minerals of the Bone Valley Formation and Their Potential Value," *Economic Geology,* Vol. 63, No. 8, pp. 973–975.

Sullivan, G.V., Browning, J.S., 1970, "Recovery of Heavy Minerals from Alabama Sand and Gravel Operations," Technical Progress Report No. 22, US Bureau of Mines.

Taylor, S.R., 1966, "The Application of Trace Element Data to Problems in Petrology," *Physics and Chemistry of the Earth,* Vol. 6, Pergamon Press, Oxford, England, pp. 133–213.

Tolbert, G.E., 1966, "The Uraniferous Zirconium Deposits of the Poços de Caldas Plateau, Brazil," Bulletin 1185–C, US Geological Survey, 28 pp.

Turekian, K.K., and Wedepohl, K.H., 1961, "Distribution of the Elements in Some Major Units of the Earth's Crust," *Bulletin,* Geological Society of America, Vol. 72, No. 2, pp. 175–191.

Vainshtein, E.E., et al., 1959, "The Hf/Zr Ratio in Zircons from Granite Pegmatites," *Geochemistry,* No. 2, pp. 151–157.

Vlasov, K.A., 1966, *Geochemistry and Mineralogy of Rare Elements and Genetic Types of Their Deposits—Vol. 1, Geochemistry of Rare Elements,* IPST No. 2123, Israel Program for Scientific Translations, Jerusalem, 945 pp.

Weast, R.C., et al., 1965, *Handbook of Chemistry and Physics,* 46th ed., The Chemical Rubber Co., Cleveland, Ohio, pp. B-3, B-27, B-28, B-63, and B-64.

Wedow, H., Jr., 1967, "The Morro do Ferro Thorium and Rare-Earth Ore Deposit, Poços de Caldas District, Brazil," Bulletin 1185-D, US Geological Survey, 34 pp.

Wessel, F.W., 1958, "Zirconium Raw Materials Supply," *Survey of Raw Material Resources, Proceedings*, 2nd International Conference on Peaceful Uses of Atomic Energy, Geneva, Vol. 2, pp. 17–20.

Williams, L., 1964, "Titanium Deposits in North Carolina," Information Circular 19, Div. of Mineral Resources, North Carolina Dept. of Conservation and Development, 51 pp.

Winchell, A.N., 1951, "Zircon," *Elements of Optical Mineralogy. Part II—Description of Minerals*, John Wiley & Sons, New York, pp. 494–495.

Index

A

Abandoned mines, use of, 851, 1132
Abietic acid, 260
Abrasives, 11–31, 521
  abrasive value, 11
  action of, 11
  alumina, 16, 29, 521
  analyses, 29
  artificial abrasives, 27–31
  bauxite, 35, 503, 515, 518, 521
  beach pebbles, 27
  bonded abrasives, 16, 17
  borazon, 30
  boron carbide, 12, 29, 30
  boron nitride, 29, 30, 43, 313
  brass wool, 30
  ceramic pebbles, 27
  cerium oxide, 16, 31
  chemical precipitates, 31
  chert, 15
  choice of, 11
  chromium oxide, 31
  classification of, 18–31
  clay, 31
  coated, 17, 18
  copper wool, 30
  corundum, 15, 16
    synthetic, 313–315
  crocus, 31
  crystal size, importance, 15
  definition, 11
  density, bulk testing, 13, 14
  diamonds, 16, 19, 20, 654–657
    see also Diamonds.
    manufactured, 19, 20, 30
    wheel marking system, 20
  diatomite, 16, 25, 26, 38, 699
  electric furnace, used in, 28
  emery, 16, 18, 19
  feldspar, 16, 18, 26, 721
  flint, 17, 18, 27
  fluorspar, 724
  friability test, 12, 13
  fuller's earth, 16
  fused alumina oxide, 29, 521
  garnets, 21–25
  glass, 31
  grade sizes, 17
  grinding pebbles, 27
    lime, 31
  iron oxide, 30, 31
Abrasives, hardness, 12
    Knoop, 12
    Mohs' scale, 292
  kaolin, 26
  lampblack, 31
  lime, 31
  loose-grains, 15–18
  loose-pack density test, 13, 14
  magnesia, 31
  manufactured abrasives, 28–31
  metal finishing, 16, 31
  metallic abrasives, 28, 30
  monocrystalline count, 15
  mullite, 27, 34, 376
  natural abrasives, 18–27
    corundum, 15, 16, 18, 19, 313
    diamonds, 19, 20, 654–656
    diatomite, 16, 25, 26, 38, 699
    emery, 16, 18, 19
    garnets, 21–25
    pumice, 26
    quartz, 25
    silica, 25, 26
    staurolite, 26
    tripoli, 16, 18, 26, 1364, 1371
  nonsiliceous, 26
  polishing, 30, 31
  production
    alumina oxide, 28
    metallic abrasive, 28
    putty powder, 31
    silicon carbide, 28
    special silica stone product, 28
  properties, physical, 11
  pumice, 16, 18, 26
  putty powder, 31
  rottenstone, 26
  rouge, 30
  sand, crushed, 25
  shape, grain, 13, 14
  sharpening stones, 26, 27
  silica, microcrystalline, 26
  silica sand, 16, 17, 25
  siliceous powder, 25, 26
  silicon carbide, 28, 29
Abrasives, soft, 25, 26
  steel shot, 30
  steel wool, 30
  strength of grains, 13
  tests, 12, 13, 14, 15
  tripoli, 16, 18, 26, 34, 39, 1364, 1371
  tube-mill linings, 27
  tungsten carbide, 30
  types of abrasives, 15–18, 19
  uses, 15, 16, 17, 18, 24, 1388
  value, abrasive, 11
  Wig-L-Bug, use of, 12, 13
  weight per cubic foot, test, 14, 15
  whiting, 16, 26
  zircon-alumina, 29
Absorbance, 226
Absorbents, minerals, 255
Accelerators, 189
Acrylic fibers, 962
Actinolite, 435, 440–443, 456, 473
Activated clay, 589
Addy quartzite, 1173
Adobe, 639
Afghanistan
  chromite, 578
  gems, 302
  lithium, 870
Africa
  See separate countries.
  gems, 302
  geological agencies, 407–410
Aggregates, 59–131, 1151–1166
  alternate choices, 69
  asphaltic, 1151
  barite, 485
  byproduct, 1042
  cinders, 81, 85, 86
  consumption, 64
  crushed stone, 60–80, 847, 852
    alternate design choices, 69
    capacity of plant, 69, 70
    chert, 63

Aggregates, crushed stone
competition, 74–76
consumption, 61–64
costs, 64, 73–76
crushing, 72
definition, 60
demand, 74, 75
depletion allowance, 78
factors affecting competition, 75, 76, 77
future trend, 78, 79
general plant considerations, 70–73
government controls, 70, 77
gravel, 1151–1166
gradation, acceptable, 68
highway construction, 63–65
industry, 75, 76, 83
labor requirements, 75
land use, 77, 78
limestone, 63, 68, 844
location requirements, 61–63, 69, 70
nonconstruction use, 62
physical properties, rocks, 63
plant considerations, 69, 70, 71–73, 75, 76
prices, 77, 78
production, 61, 69
profitability, 77
qualities, 60
reserves, 79
rock types used, 62, 63, 844
physical properties, 63, 844
shape, particle, 68
sand, 1151–1166
shipments, 852
slag, as an aggregate, 111
specifications, 60, 61, 65–69
structure, industry, 75, 76
substitutes, 69
supply, 79
taxation, 78
tests, 65–68, 100, 119, 1158
transportation, 73, 74, 852
use, 59, 61, 62
value, 77, 78
zoning regulations, 78
definition, 60, 61, 1151
demand, 107, 108
dolomite, 63
lightweight aggregates, 81–95
bituminous paving mixes, 120

Aggregates, lightweight
byproducts, 81, 85, 111–131
calcilux process, 128
cinders, 81, 85
classification, 81
clay, 81
coke breeze, 81
commodities, used for, 82
consumption, 85–87
costs, operating, 85
definition, 81
deleterious substances, 87, 1152
economic factors, 89, 90
expanded slag, 85
government considerations, 90
grading requirements, 88
marketing, 90
operating costs, 85
outlook, 90, 91
perlite, 59, 81, 86, 88
prices, 86, 89
production, 81–85, 86
properties of, 82
pumice, 81, 85
scoria, 81
shale, 82, 1152
silicone treated, 86
slag, 81, 85, 86, 111–131
slate, 85
specifications, 82, 87
substitutes, 89
transportation, 89, 90
tuff, 81
types, 81
uses, 82
value, average, 86
vermiculite, 81, 86, 87, 1375–1381
volcanic cinders, 81, 85, 1079–1184
weight of, dry, 81, 87
zeolite, use in, 1422
limestone, aggregate, 63
pumice, 81, 85
rock types, suitability of, 1152
sand and gravel, 96–110, 1151–1166
abrasion tests, 65, 66, 100
conservation, 105–107
consumption, 96, 97, 1163
control, governmental, 103, 104, 1065
definition, 96
demand, 106, 107
deleterious material, 1152
depletion allowance, 104, 1160
deposits, 1153–1155
dredging, 1160, 1161

Aggregates, sand and gravel
economic factors, 101–103
environmental considerations, 104, 107, 1165, 1166
evaluation, 1157–1160
exploration, 1156, 1157
exports, 1156
geologic occurrences, 1153–1155
government considerations, 103, 104, 1065, 1160
grading, 1158, 1160
growth rate, 96, 1165
imports, 1156
land use, multiple, 106
marketing, 103, 124, 126, 1162
microscope examination, 1160
mining, 1160, 1161
prices, 101, 1163
processing, 1161, 1162
production, 101, 1151
properties, 1151, 1152
reactivity, 1160
requirements, general, 1151
reserves, 1155, 1156
resource availability, 105, 106
safety regulations, 107
specifications, 96, 97, 109
substitutes, 100, 101, 1165
supply-demand considerations, 107, 108
taxes, 103, 104
technology, changes in, 105, 127–129
tests, 98–100, 1152, 1157–1160
transportation, 73, 103, 124, 1163
transportation costs, 103, 1163, 1164
trend, 101, 1163, 1165
utilization, 1151
uses, 96, 1151–1153
value, 101
zoning, 104, 106, 1165
shale, 81, 82
slag, 111–131
air-cooled blast-furnace slag, 114, 115, 118–122
aggregates, lightweight, 82–95
ballast, railway, 121
bituminous paving mixes, 120, 121

Aggregates, slag, blast furnace slag, production, 120
chemical composition, 110
coarse slag, 120
concrete making properties, 87
consumption, 112, 113
composition, chemical, 119
definition, 111
deleterious substances, 87
employment in, 112
environmental considerations, 127
expanded slag, 116, 117, 119, 122, 123
fire resistance ratings, 123
future prospects, 127
government considerations, 126, 127
gradation requirements, 121
granulated slag, 115–117, 119, 121, 122
granules, roofing, 203–212
history, early, 111
marketing, 124–126
mixes, 120–122
pavements, use in, 121
physical properties, blast furnace slag, 119
plants, 112
pozzolan, 143
price trends, 123, 124
prices, 113, 115, 117, 123
producers, 125
production, 112, 113
production centers, 82, 83, 109, 110
properties, 119
sales, 113, 124–126
slag, use in cement, 143
specifications, slag, 118–123
steel slag, 118, 119, 122
substitutes, 122, 123
transportation, methods of, 124
types of slag, 114–122
uses, 114, 118, 119, 121
value, slag, 114, 118, 119
wool, slag, 115, 121
Air-cooled blast furnace slag, 114, 115, 117, 119, 120, 121
analyses, 119
production, 113, 115, 125
prices, 113, 123
sales, 114
specifications, 118–120
strength, 120, 122
uses, 115, 120
weight, 120
Aglime, 237
Air separation, zeolite used in, 1423, 1424
Akermanite, 1389
Alabama
asphalt, 530, 531
barite, 492
bauxite, 515
bentonite, 599
clay, 374, 375
fuller's earth, 599
gannister, 377
glauconite, 752, 765
granite, 170
graphite, 759, 760
kaolin, 622, 638
lime, 827, 828
limestone, 853
manganese, 898
marble, 164, 170
marl, 854
mica, 918
monument stone, 164, 854
pegmatites, 711
plastic clay, 375
silica, 1171
slag, 112, 125
talc, 1288, 1289, 1296
tripoli, 1368
zircon, 1368
Alabaster, 776
Alaska
barite, 492, 498
chromite, 575
fluorspar, 728
jade, 303
perlite, 1002
Alaskite, 711
Albania
asphalt, 530
chromite, 381, 576, 578
Alberta
bentonite, 601
silica, 1176
sodium sulfate, 1210, 1218
sulfur, 1251, 1253
tar sand, 1253
Albertite, 530, 531
Alexandrite, 295, 312
synthetic, 312, 316
Algeria
barite, 493
bentonite, 605
celestite, 1233
diatomite, 687
phosphate, 1035
Alkali-dolomite reactions, 99
Alkali-flux, 262
Alkali-halide flux, 263
Allard Lake, 1337, 1345
Alluvial deposits, 1154
processing, 1347–1351
flowsheet, 1350
Almandite, 22, 226
Alpha-hemihydrate, 186, 189, 191
Alpha quartz, 1167
Alumina, 29
abrasives, 35
analysis, 29
bricks, 376, 377
cement, use in, 521
ceramics, 33, 35
chemical analysis, 29
fused, 29
hydrate, 633
lime, used in, 820
nepheline syenite, 932, 942–953, 954
processing, 29, 519–523
production, 28
screen analysis, 29
sintered, 376
Alumina hydrate, 633
Aluminum chemicals, 521
Aluminum
kaolin, source of, 629
nepheline syenite, source of, 932, 953, 954, 1042
Aluminum oxide, fused, 29, 417
Alunite, 417–434
alunitization, 422
Alumet process, 419–421, 428–430
alumina source, 417, 427
analyses, 420
bacteria, role of, 423
bochones, 427
boehmite, synthetic, 428
bullseye, 422
composition, 419, 420
deposits
Argentina, 423
Australia, 417, 419
Chile, 423
China, 417
France, 417
Italy, 417, 418
Japan, 417
Spain, 417
types of, 418–422
United States
Arizona, 418, 423
California, 425
Colorado, 425
Nevada, 418, 422
New Mexico, 423
Texas, 425
Utah, 417–419, 423, 430
Washington, 425
USSR, 418
fertilizer, 418
genesis of, 421–423
geology, 418–423
historical background, 417, 418
hypogene ore deposits, 430, 431
jarosite, 419–421

Alunite, Kalunite process, 428
Marysvale, 417–419, 425
mining, 430
natroalunite, 419
photomicrographs, 422
potash, source of, 417
processes
Alumet, 418, 419, 421, 428–430
Kalunite, 428
Soviet process, 430
UG process, 428
prospecting for, 431
resources, 423–426
Argentina, 427, 430
Asia, 426
Bolivia, 427
Europe, 426, 427
Iran, 427
Pakistan, 427
South America, 427
Turkey, 427
United States, 423–426
Arizona, 424, 425
Colorado, 423, 425
Nevada, 424
New Mexico, 425
Utah, 423, 424
USSR, 426
Roman alum, 417, 418
technology, 428–430
UG process, 428
uses, 417
zoning, of deposit, 420, 421
Amargosite, 598
Amber, 279, 281, 291, 301, 315, 316
Amblygonite, 870, 877
see also Lithium raw materials.
Amino resins, 963
Ammonia
see also Nitrogen compounds.
coke oven byproduct, 970
synthetic, 963–970
Ammonia, anhydrous, 961
fertilizer, 234
Ammonium nitrate, 234
Ammonium sulfate, 197, 234, 981
Ammonium thiosulfate, 239
Amorphous graphite, 758
Amorphous silica, 1363
see also Tripoli.
Amosite, 437–443, 455–457, 461, 475, 476
Amphibole asbestos, 435, 444, 456
Analcime, 1393, 1394
Anatase, 1311, 1323
Anauxite, 1172
Andalusite, 34, 35, 376, 799, 800, 802, 803
Andreasen size determination, 946
Angola
diamonds, 302
fluorspar, 734
gems, 302
kaolin, 626
Angular steel grit, 30
Anhydrite, 183–199, 775–777
portland cement material, 141
sulfur source, 1256, 1257
use, 775
Animal feed, urea, 963
Ankerite, 833, 834
Anorthite, 36
Antarctica, sodium sulfate, 1216
Anthrophyllite, 435, 439, 440, 441, 443, 444, 456, 483
see also Asbestos.
Antigorite, 1282
Apache tears, 1007
Apatite, 1017, 1021
see also Phosphate rock.
Aplite, 710, 716, 719–721, 955
see also Feldspar.
production, 644, 716
Aragonite, 35, 134, 301, 833, 834
Arcon, 912
Arc welding, 265, 266
Argentina
barite, 492
bentonite, 602
borates, 549, 550
celestite, 1233
diatomite, 688
fire clay, 638
fluorspar, 732
garnet, 23
glance pitch, 531
kaolin, 624, 638
lithium, 869, 870, 873, 876
perlite, 999, 1003
rare earths, 1113
refractory clay, 638
salt, 1126
soda ash, 1200
sodium sulfate, 1207, 1215, 1221
vermiculite, 1377
zeolite, 1418
Aridized anhydrite, 188, 189
Arizona
asbestos, 454
barite, 492
bentonite, 598
diatomite, 25, 685
feldspar, 711
fluorspar, 728
gem, 303, 319
gypsum, 778, 783
hectorite, 600
kyanite, 801
lime, 827, 828
Arizona, manganese, 898
meerschaum, 911
perlite, 997, 999, 1003
potash, 1066
salt, 1124
silica, 1173
sodium sulfate, 1209, 1213
titanium, 1316
vermiculite, 1377
wollastonite, 1385
zeolites, 1414, 1424, 1425
Arkansas
asphalt, 530
barite, 489, 491, 496, 497
bauxite, 508, 515, 519
bromine, 48
diamonds, 303
fuller's earth, 591
halloysite, 635
kaolin, 618–620, 623, 638
kimberlite, 662
manganese, 897, 898
nepheline syenite, 508, 931, 942, 943
novaculite, 27, 1369
pyrophyllite, 1097
rock asphalt, 530
silica, 1170
talc, 1298
titanium, 1316, 1320
tripoli, 26, 1367, 1369
vermiculite, 1377
whetstones, 27
Arkose, 166, 170
Artificial abrasives, 27–31
Asbestos, 435–484
actinolite, 435, 440–444, 456, 473
amosite, 437–443, 455–457, 461, 475, 476
amphibole asbestos, 435, 444, 456
anthophyllite, 435, 439–444, 456, 483
bulk sampling, 465
Cape Blue, 476
chemical composition, 436–440
chemical resistance, 443
chrysotile, 435–437, 439–444, 446, 454, 455, 474–481
classification, 474–476
composition, 436–440
consumption, 481
crocidolite, 437, 440–443, 455–457, 475
cross fiber, 444, 446, 462, 463
definition, 435
deposits, 435, 444, 446, 454, 455
distribution, 446–454

Asbestos, dolomite, deposits in, 454, 455
- drilling, 462, 463
- electron micrographs, 436–439
- exploitation, 466–473
- exploration, 461–466
- exports, 479
- evaluation, 463–466
- face reading, 465
- fiber evaluation, 437, 444, 446, 462–464
- filler, 250, 479
- geology, 444–446
- grading of, 474, 475, 476
- health hazards, 481
- heat, effect on, 440
- markets, 481
- mass fibers, 441, 442, 444
- mill, flowsheets, 469, 470, 471, 472
- milling, 467–473
- mining, 466, 467
- montasite, 455
- occurrence, 446–454
- ores, 437, 444, 446, 461–466
- prices, 478
- production, 456–461
- properties, 435–444
- prospecting, 461–466
- Quebec grading, 474
- Quebec standard, 473, 474
- reserves, calculation of, 463–466
- slip fibers, 444, 463, 464, 466
- solubility of, 443
- specifications, 473, 474
- spinning, 441, 442
- stabilizer, use as, 209
- specifications, 473, 474
- strength units, 466
- structures, 444
- supply-demand relationship, 480
- tensile strength, 466, 473, 474
- tests, 473, 474
- tremolite, 435, 440–444, 456, 461, 1282
- types of, 440
- uses, 250, 479, 480
- varieties, 435
- veining, 444–446
- zeta potential, 430

Ashlar, 162

Asia
- *see also* individual countries.
- geological agencies, 411–413

Asphalt, native, 529, 530
- aspect ratio, 925
- stabilizer, 190, 209

Asphalt roofing, development of, 204

Asphaltites, 530, 531
- gilsonite, 530
- glance pitch, 531
- grahamite, 530, 531
- petroleum asphalts, 529

Asterism, 295, 296, 312

AT-cut quartz, 213, 216, 217

Attapulgite
- *see also* Fuller's earth.
- absorbent, 255
- analysis, 596
- deposits, 591
- palygorskite, 591
- uses, 591–594

Australia
- asbestos, 454
- barite, 493
- bauxite, 508, 511–513, 523
- bentonite, 606
- chromite, 578
- coral, 318
- crocidolite, 454
- diamonds, 658
- diatomite, 689
- fire clay, 638
- fluorspar, 736
- fuller's earth, 606
- gems, 302, 317, 318
- geological agencies, 413
- glauconite, 753
- gypsum, 778
- graphite, 767
- kaolin, 627, 638
- lithium, 870, 874
- magnesite, 883, 884, 885
- monazite, 1113
- pearls, 318
- perlite, 1006
- phosphate, 1028
- potash, 1071
- pyrophyllite, 1093–1095
- rare earths, 1111
- refractory clay, 638
- refractory kaolin, 638
- sillimanite, 802
- titanium, 1305, 1315, 1322, 1333, 1339, 1340, 1352
- vermiculite, 1377
- zeolites, 1416
- zircon, 273, 1433, 1435, 1443

Austria
- barite, 492
- graphite, 761, 762, 764
- iron oxide pigment, 355
- kaolin, 624
- kyanite, 801
- magnesite, 883, 884, 885
- meerschaum, 911
- olivine, 889, 991
- sulfur, 1256

Autoclave, quartz synthesis, 216, 217

Aventurine, feldspar, 296

Aztec sandstone, 1172

B

Bahco classifier, 946

Bacteria, 1249

Baddeleyite, 1433, 1435, 1436

Baquette, 33

Bahamas
- aragonite, 101
- salt, 1125

Balangero area, 452

Balds, 992

Ball clay, 35, 38, 612, 633, 634
- ceramics, 35, 38
- composition, 633
- definition, 35, 633
- depletion allowance, 641
- deposits, 620, 633
- distribution, 633, 634
- evaluation, 634
- future consideration, 634
- geology, 633, 634
- history, 633
- impurities, 633
- milling, 634
- mining of, 634
- mineralogy, 633
- occurrence, 633, 634
- origin, 633, 634
- preparation for market, 634
- production, 586, 633
- reserves, 634
- tariff, 642
- trends, 634
- uses, 633

Ballas, 19, 292, 296, 316

Ballast, railroad, 61

Bank sand, 271

Banka drill, 1178

Barbados, glance pitch, 531

Barilla, 1187

Barite. *See* Barium minerals.

Barium chemicals, 486, 496, 497

Barium minerals, 485–501
- aggregate, 486
- barium chemicals, 486, 496, 497
- beneficiation, 495, 496
- ceramics, 486
- chemicals, 486, 496
- consumption, 497, 498
- depletion allowance, 499
- deposits, distribution, 490–493
- deposits, types of, 487–490, 495, 496, 498
- ecological impact, 499
- evaluation of deposits, 495
- exploration, 493–495
- filler, 244, 245, 247, 250
- flowsheet, 495, 496
- future considerations, 498, 499

Barium minerals, geology, 486–495
- imports, 498
- marketing, 497, 498
- milling, 495–497
- mineralogy, 487
- mining, 495–497
- plants, chemical, 497
- preparation of chemicals, 496, 497
- prices, 498
- processing, 496, 497
- production, 494
- recycling, 499
- reserves, 493, 494, 498
- shielding, 37, 485, 486
- specifications, 486
- substitutes, 499
- tariff, 499
- taxes, 499
- transportation, 497
- trends, 499
- uses, 35, 388, 389, 485, 486
- witherite, 485, 499

Barium titanate, 313

Barytes. *See* Barium minerals.

Basalt, 165, 168, 169, 205, 209

Base exchange, 745

Base granules, 206

Base, molding sands for, 272

Baseline sandstone, 1172

Basic brick, 381–383

Basic oxygen furnace, 122, 382, 819

Basic oxygen process, 122, 741

Bassanite, 776

Bastnaesite, 1110–1114, 1116, 1117

Bath brick, 26

Bath rouge, 359

Batteries
- graphite in, 768
- lithium, 878
- manganese, 902

Bauxitic raw materials, 503–527
- abrasive material, 35, 521, 522
- abundance, 503
- activated, 521
- age, of deposits, 508, 510
- alumina, 519, 520
- analyses, 508, 509, 518, 519
- bauxite, 504, 508, 509
- Bayer process, 418, 520
- boehmite, 504, 515
- cement, 521
- chemicals, 521
- classification of deposits, 504, 505
- composition, 508, 509, 518, 519
- consumption, 521
- deposits, 504–511
- diaspore, 504, 515, 637, 638
- environmental considerations, 524
- evaluation, 517–521
- exploration, 516, 517
- ferruginous bauxite, 515
- fused aluminum oxide, 29
- future considerations, 523, 524
- geology, 503–508
- gibbsite, 504, 515
- government control, 523, 524
- grades, 517–521
- high alumina bricks, 520, 521
- imports, 522
- impurities, 503
- milling, 519
- mineralogy, 503, 504
- mining, 519
- mixed ore, 503
- nepheline syenite, 506
- origin, 506–508
- preparation for market, 519–521
- prices, 521–523
- processing, 519–521
- producing countries, 511–516
- production, 512, 522
- properties, 503, 504
- refractories, 520, 521
- reserves, 509–511
- specifications, 35, 517, 518
- substitutes, 523
- testing, 517, 518
- transportation, 519
- uses, 35, 518, 519, 520, 521, 522

Bayer process
- aluminum process, 418, 519, 520
- lime used in, 820
- sulfuric acid, 1263

Beach pebbles, 27

Beach sand deposits, 1347, 1348

Beam and cylinder test, 68

Beam splitter, 226

Bearings, graphite for, 768

Belgium
- flint pebbles, 27
- glauconite, 752
- kaolin, 626
- lime, 828
- quartz, 27
- silex, 27

Beneficiation. *See* preparation under names of rocks and minerals.

Benitoite, 303

Bentone, 611

Bentonite, 38, 587–612
- absorbent granules, 255, 589, 594
- activated clay, 589
- analyses, 596
- beneficiation, 609
- bleaching clay, 589
- bond, foundry sand, 593
- claiming land, 641
- classification, 587–589
- definition, 587, 588
- demand, 611
- depletion allowance, 641
- deposits, distribution, 598–606
- drilling, 607
- drying, 609, 640
- electron micrograph, 595
- evaluation, 606
- exploration, 606, 607
- fillers, 244
- fuller's earth, 587–612
- future considerations, 611, 612
- geology, 594–599
- history, 589, 590
- impurities, 595
- industry growth, 589
- ion exchange, 594
- iron ore pelletizing, 593
- K-bentonite, 588
- metabentonite, 588
- milling, 609
- mineralogy, 594–597
- mineralogical investigation, 608
- mining methods, 608, 609
- mixed bentonites, 588
- montmorillonite, 594
- occurrence, 597, 599–606
- organic-clad bentonite, 610, 611
- origin, 588, 591–593, 597, 598
- plants, 598, 601, 602
- plasticizer, 38
- potash-bentonite, 588, 589
- preparation for market, 608, 609
- processing techniques, 609
- production, 585–587, 589, 590
- results, appraisal, 608
- size, ore body, 597
- smectite, 588, 594
- southern bentonite, 588
- specifications, 593, 594
- sub-bentonite, 588
- swelling bentonite, 588
- synthetic, 610, 611
- tariff, 643
- testing, 607, 608
- transportation, 609, 610
- trend, 611, 612
- use, 589–591, 593

Bentonite, western bentonite, 588
  Wyoming bentonite, 588
Beryl, 35, 37, 42, 292, 294, 296
BET, surface area, 350
Beta-hemihydrate, 186, 188
Bicarbonate of soda, 388, 390
Bidahochi formation, 1173
Biotite, 915, 916
Bitumens, fillers for, 248, 249
Bituminous material, 529–532
  albertite, 529, 531
  asphalt, 529
  asphaltic pyrobitumens, 529
  asphaltites, 530–532
  bituminous limestone, 530
  ceresine, 531
  definition, 529
  elaterite, 531
  first used, 529
  gilsonite, 530
  glance pitch, 531
  grahamite, 530, 531
  impsonite, 531
  limestone, rock asphalt, 530
  manjak, 531
  mineral waxes, 529
  mining, 530
  montan wax, 531
  native asphalt, 529, 530
  nonasphaltic pyrobitumens, 529
  ozokerite, 531
  pavement bituminous mixes, 120
  petroleum asphalts, 529
  processed rock asphalt, 530
  production, 531
  pyrobitumens, 529, 531
  rock asphalt, 530
  sands, bituminous, 529, 1176
  uses, 529
  uintahite, 529
  varieties, 530, 531
  wurtzilite, 529, 531
Black granite, 165, 169
Black marl, 745
Black powder, 963
Black rouge, abrasive, 31
Black sand, 1316
Blanc fixe, 486
Blast furnace slag, air-cooled, 114, 115, 135, 143, 344, 346
Blast sand, 1225
Bleaching clay, 523
  see also Fuller's earth.
Block mica, 915
Blockspar, 711
Blue ground, 301
Blue Mountain syenite, 937, 948
Bluestone, 166
Boehmite, 504
Bochones, 427
Boeton, 530
BOF. *See* Basic oxygen furnace.
BOP. *See* Basic oxygen process.
Bolivia
  bastnaesite, 1111, 1113
  borates, 548
  crocidolite, 455
  lithium, 869, 870, 871
  monazite, 1111, 1113
  soda ash, 1200
Bonaire
  salt, 1125
Bonded abrasives, 16, 17
Bonding clay, 589
Bone magnesite, 882
Bone phosphate of lime, 1017
Boracite, 550
Borax
  see also Boron and Borates.
  Boron, CA, 539, 541
  brazing flux, 36
  brines, 553, 554
  composition, 536
  deposit size, 552
  flux, 36
  growth potential, 48
  origin, 41, 536–553
  prices, 48
  processing, 553, 554
  production, 48
  Turkey, 543–546
  uses, 48
Borazon, 29, 30, 313, 671
Bordeaux mixture, 822
Boric acid, 547, 555
Boron and borates, 45–48, 533–560
  borax, 36, 41, 48, 539, 541, 553, 554
  boron chemicals, 41, 43, 44, 46, 47
  classification, 536
  colemanite, 5, 43, 536, 552
  composition, minerals, 536
  consumption, 45, 534, 555
  depletion allowance, 555
  deposits, 536–543, 552
  environmental considerations, 555
  evaluation, 552
  exploration, 550–552, 554, 555
  exports, 534
  fertilizer, 239
  flux, 45, 262, 263
  future considerations, trends, 48, 544, 555
  geology, 535–538
  government controls, 555
  imports, 555
  marine deposition, 537
  markets, 555, 556
  marsh deposits, 537
  micronutrient, 223, 239
Boron and Borates, milling, 553
  mineralogy, 45, 535, 536
  minerals, 536
  mining, 552, 553
  neutron absorption, 37
  nonmarine deposition, 537, 538
  ores, 43
  origin, 41, 536–538
  preparation for market, 552–554
  prices, 48
  processing, 553, 554
  producers, 539
  production, 48, 533, 556
  products, 533–535
  refining, 554
  regions
    Argentina, 549, 550
    Chile, 550
    China, 550
    Germany, 550
    Italy, 547
    Mexico, 550
    Peru, 550
    Tibet, 550
    Turkey, 543–546
    US, 539–543
      California, 539
      Nevada, 539
    USSR, 546, 547
  reserves, 539–546, 552
  resources, map, 539
  sodium sulfate from, 553
  test for borates, 551
  testing, 552
  transportation, 555
  trends, 554–558
  uses, 36, 38, 43, 44, 47, 533–535, 557
Boron, CA, 539–541
Boron carbide, 12, 29, 30
Boron nitride, 29, 30, 43, 313, 320, 671
Bort, 19, 292, 296
Botswana
  diamonds, 302
  soda ash, 1216
Boule, 311, 312
Brake linings, graphite for, 768, 769
Brass wool, abrasive, 30
Brazil
  asbestos, 451
  baddeleyite, 1436
  barite, 492
  bauxite, 508, 514
  bentonite, 602
  bort, 19
  carbonados, 19
  chromite, 575
  diamonds, 653, 658
  diatomite, 689

Brazil, fire clays, 638
fluorspar, 732
gems, 302
graphite, 761, 763, 767
kaolin, 624, 638
kyanite, 806
lithium, 870, 871, 873
magnesite, 882, 884, 885
manganese, 898
monazite, 1113
oil shale, 1253
nepheline syenite, 935, 936, 942, 1042
olivine, 991
phosphate, 1021, 1032, 1033, 1042
potash, 1051, 1066, 1067
pyrophyllite, 1096, 1097
quartz, 213
rare earths, 1112
salt, 1125, 1126
soda ash, 1200
sulfur, 1260
talc, 1282
titanium, 1315, 1323, 1334
tourmaline, 302
vermiculite, 1377
yttrium, 1113
zirconium, 1443
Brazil twinning, 218
Brazing, 259, 261–263
Breeder reactors, 1117
Breunnerite, 881, 885
Brick, bath, 26
Bricks, refractory, 373–378, 381–383
Brilliance, 292–294
Brimstone, 1235
see also Sulfur.
Brine, artificial, 42
Brines, natural, 41, 44, 871, 873, 875
borax source, 541, 542
bromine, 41, 45, 561
chart, uses, 44
chlorine, 41, 45
iodine, 41, 793, 796
lithium, 41, 869, 871, 873, 876
magnesium, 881, 889
potassium, 41
sodium, 41
types of, 44
Briquetting, 723, 725, 739, 740, 887
British Columbia
barite, 492
bentonite, 601
chrysotile, 450
diatomite, 685
fluorspar, 728
gypsum, 784
magnesite, 884
British Columbia, nepheline syenite, 936
perlite, 682, 685
silica, 1176
soda ash, 1188
wollastonite, 1386
zeolite, 1416
Bromine, 48, 561–565
brines, 41, 45
depletion allowance, 563
exploration guide for potash, 1060
prices, 48
production, 48
salt, content, 562
sources, 41
uses, 48
Brookfield viscosity, 616
Brownstone, 166
Bruce Williams process, 903
Brucite, 382, 881, 882, 888
Bryn Mawr gravel, 1711
Buckles, 274
Buhrstone, 27
Building stone, 161, 164
Built up mica, 915
Bulgaria
amphibole asbestos, 461
barite, 493
chromite, 578
fluorspar, 734
kaolin, 625
perlite, 999, 1004
zeolite, 1417, 1419, 1421, 1422
Bulk sampling, asbestos, 465, 466
Bullseye alunite, 422
Burkeite, 1189, 1199
Burma, gems, 19, 302
Burnishing sand, 25
Burundi, bastnaesite, 1111
Butyllithium, 877, 878

C

Calcareous marl, 854
Calcined clay, 614, 636
Calcined kaolin, 614, 628
Calcite
fertilizer, 233, 237, 238
optical, 221, 297, 834
Calcium, 237, 238
Calcium carbide, 724
lime, for use in, 821
limestone, for use in, 846
Calcium carbonate
abrasive, 19
precipitated, 31, 633
rubber filler, 251
substitute for, 1298, 1372
Calcium cyanamide, 962
Calcium nitrate, 962
Calcium zirconate, 35
Caliche, nitrates from, 971
California
asbestos, 451
asphalt, 530
ball clay, 633, 634
barite, 488, 489, 491, 492
bastnaesite, 1113, 1114, 1115
bentonite, 597
borax, 537, 539–543
bromine, 48
chromite, 575
clay, 633, 634, 638
diatomite, 25, 26, 682–685, 693
dimension dolomite, 854
dimension limestone, 854
feldspathic sand, 713
fire clay, 638
flint pebbles, 27
fluorspar, 729
gems, 303, 318, 319
glass sand, 639
granite, 171
gypsum, 778, 784
hectorite, 598, 600
iodine, 794
kaolin, 618, 619, 622, 638
kyanite, 801
lime, 828
limestone, 853
lithium, 870, 875
magnesite, 890, 891
pegmatites, 997, 999, 1003
perlite, 1003
phosphate, 1043
potash, 1073
pumice, 171, 1079, 1080
pyrophyllite, 1093
rare earths, 1111
salt, 1121, 1124
sandbornite, 487
silica, 1171, 1172, 1173
slag, 125
soda ash, 1187, 1188
sodium sulfate, 1212, 1220, 1290, 1291, 1293, 1294, 1298
sulfur, 1251, 1252
talc, 1275, 1283, 1284, 1285
tar sands, 1253
thorium, 1113–1115
titanium, 1316
tripoli, 1363, 1369
vermiculite, 1377
witherite, 487
wollastonite, 1383–1385
zeolites, 1414, 1416, 1421
Cameos, 333
CAN, 962
Cana Brava area, 451
Canada
asbestos, 445, 446–451, 456, 460
barite, 492
bentonite, 601, 602

Canada, blasting sand, 25
  see also individual provinces.
  brucite, 382
  celestite, 1032, 1229
  chromite, 575
  diatomite, 685
  fluorspar, 728
  graphite, 765
  gypsum, 184, 777, 779, 784
  kyanite, 803
  lime, 828
  lithium, 870, 871
  magnesite, 883, 884
  mica, 918
  monazite, 1113
  nepheline syenite, 931, 933, 937, 938, 947–949
  olivine, 991
  perlite, 1002
  potash, 1062
  pyrophyllite, 1096
  quartzite pebbles, 27
  silica sand, 25, 1174–1176
  sodium sulfate, 1207
  tar sand, 1442
  titanium slag, 1305
  vermiculite, 1377
  zeolites, 1422
Cap rock, 1243, 1244
Cape Blue. *See* Crocidolite.
Carat, unit of weight, 323
Carbolon, 28
Carbon
  see also Graphite.
  ceramic raw material, 35
  forms of, 757
  refractory, 379
Carbonada, 19, 292, 296, 316
Carbonate rocks. *See* Dolomite, Limestone, Marble.
Carbonatites
  apatite in, 1020
  fluorspar in, 727
  phosphates in, 1020
  rare earths in, 1111, 1112
Carborundum, 28
Carlsbad area, 1059–1066, 1071
  bromine, 1059
  potash, 1059, 1060
  stratigraphy, 1059
Caribbean Islands, geological agencies, 402
Carnelian, 1168
Carnallite, 1050
  see also Potash.
Carthage marble, 834
Castable refractories, 380
Casting plaster, 194
Catalyst, 636, 1425, 1426
Cat's eye, 296
Caustic, calcined magnesite, 882
  see also Magnesite.
Caustic soda
  see also Sodium hydroxide.
  production, 49
  use, 346
Cawk, 487
CCE, 845
Celestite. *See* Glass raw materials, raw materials composition; Strontium.
Cement, 37, 133–159
  aggregates for, 37, 59–131, 1151–1153
  alternate raw materials, 147
  capacity of industry, 146, 150
  chemical requirements, 144–146
  clinker, 139–141, 155
  composition, 144
  comment, general, 157, 158
  competition, foreign, 152
  compounds, 141
  consumption, 146–148, 150
  control, manufacturing, 141, 142, 156
  costs, 147, 148
  definition, 133
  deleterious material, 135
  depletion allowance, 151
  developments, notable, 155, 156
  dry blending, 138
  dry process plants, 138, 139
  drying, 155
  dust control, 141
  economic factors, 135, 148–151
  environmental considerations, 152–154, 156
  evaluation, 135, 136, 247
  gypsum, cement from, 1256
  high-alumina, 521
  history, 133, 134
  hydraulic cement, 157
  kilns, 135, 139, 140
  location, plants, 134, 135
  manufacture, steps in, 137–139, 141, 142, 155
  marketing, 149, 150, 155
  physical requirements, 144–146
  plants, 146, 150
  pollution control, 153
  processing, 136–141, 155
  price trend, 151, 152
  pozzolan cement, 143
  profits, 151, 152
  production, 146
    US, 134
    world, 157
  products, types of, 142, 143
  pyroprocessing, 139
  quality control, 138
  quarry operations, 136, 155
Cement, raw materials, 134–136, 141, 143, 147, 635, 784, 785, 931, 953, 1256, 1279, 1421
  requirements, 144–146
  rotary kiln, 140
  specifications, 144–146
  subsidies, 151
  substitutes, 147
  summary view, 154
  tests, 68
  transportation, 148, 149
  types, 142, 148
  underground mines, 136
  units, 133, 134
    costs, 148
  use of, 147
  weight, 133, 134
  wet process plants, 138, 139
  white, 1087, 1088, 1103
Central African Republic, diamonds, 661
Ceramic, 35
Ceramic pebbles, 27
Ceramic raw materials, 33–39
  anadalusite, 34, 35, 376, 799, 802, 803
  barium ceramics, 35, 486
  dielectric materials, 38, 213
  feldspar, 33, 36, 719, 721
  fluorspar, 35, 724, 725
  function of, 34–39
  kaolin, specifications, 617, 618
  lime, 34, 37, 822
  nepheline syenite, 36, 932, 945, 946
  olivine, 35, 989
  silica, 33, 34, 36, 39, 1170
  specifications, 617, 618
  talc, 35, 1277
  titanium, 1312
Cerium, 1110, 1116
Cerium oxide abrasive, 16, 31
Cermets, 37
Ceylon. *See* Sri Lanka.
Chabazite, 1395, 1396
Chalk, 167
  abrasive, 26
  cement raw materials, 135
  flint, 25
Chamottes, 636, 1085
Chaser stones, 18
Chatham process, 313
Chatoyance, 296
Chemical industry, 41–58
  brines, salt, 41, 42, 44
  chlor-alkali industry balance, 48

Chemical industry, consumption, 43
  industrial minerals, 41–43
  major elements, 43
    boron, 36, 38, 43–48, 533–560
    bromine, 46–49, 561–565
    caustic soda, 52
    chlorine, 48–50
    consumption, 51, 52
    fluorine, 50–52, 723, 724
    hydrofluoric acid, 50–52, 723
    producers, 51
    sodium carbonate, 52, 1077, 1187–1206
    sodium chemicals, 52
    sodium hydroxide, 52, 53
    sodium sulfate, 52, 1207–1223
    sulfur, 41, 43, 53–55, 1235–1273
    sulfuric acid, uses, 41–43, 54, 55
Chemical precipitates, abrasives, 30, 31
Chemical reactivity tests, sand and gravel, 99, 100
Chert, 15, 68, 835
Chile
  barite, 492
  borates, 550
  diatomite, 689
  fluorspar, 732
  gems, 302
  iodine, 795
  kaolin, 624, 638
  lithium, 869, 871
  lazurite, 302
  lithium, 869, 871
  nitrates, 970, 971
  perlite, 1003
  potash, 1066, 1067
  salt, 1126
  soda ash, 1200
  sodium sulfate, 1207, 1216
  sulfur, 1251, 1260
  zeolite, 1418
China
  asbestos, 454
  barite, 493
  bastnaesite, 1113
  bauxite, 458, 515
  borates, 550
  calcined bauxite, 518
  chromite, 578
  diaspore, 515
  diatomite, 689
  flint clay, 638
  fluorspar, 735
  gems, 302
  graphite, 761
  kaolin, 627
China, magnesite, 884
  mica, 918
  oil shale, 1253
  potash, 1070
  refractory clay, 638
  salt, 1126
  talc, 1283
  titanium, 1336
  vermiculite, 1377
  zirconium, 1433
China clay, 26, 612
Chlorastrolite, 303
Chlor-alkali industry, 48–51
Chloride process, TiO_2 manufacture, 49
Chlorine, 48–50
  fertilizer, 233
Christmas Island
  guano, 1021
  phosphate, 1021
Chromite, 567–584
  analyses, 569, 571
  beneficiation, 572, 573
  ceramic, 35
  chemical grade, 568, 569
  composition, 569–571
  consumption, 569, 578
  deposits, types, 572
  definition, 567
  economic factors, 567, 579–581
  evaluation, 573, 574
  exploration, 573, 574
  foundry sand, 274, 568, 1440
  future trends, 581, 582
  gangue, 571, 572
  geology, 569–573
  grades, 35, 274, 567–569
  imports, 578, 579–581
  metallurgical grade, 567, 568
  milling, 572, 573
  mineralogy, 569–571
  mining, 572, 573
  molding sand, 274, 568
  olivine, competition, 994
  podiform deposits, 572, 576
  prospecting, 573, 574
  prices, 274, 578–581
  production, 578, 579
  properties, 274
  ratios, Cr:Fe, 567, 568
  refractories, 379–381, 568
  refractory filler, 35
  reserves, estimating, 574
  resources, world, 574–578
    Africa, 576, 577
      Madagascar, 576
      South Africa, 576
      Zimbabwe, 576
    Asia, 577, 578
      China, 578
      India, 577
      Philippines, 577
Chromite, resources, Asia, Turkey, 577, 578
    Central America, 575
    Europe, 576
      Albania, 576
      Finland, 576
      USSR, 576
      Yugoslavia, 576
    North America, 575
      Cuba, 575
      Greenland, 575
      United States, 575
    South America, 575, 576
  specifications, 567–569
  stockpiling, 567
  stratiform deposits, 572, 576
  uses, 35, 274, 567–569
Chromium oxide, abrasives, 31
Chrysoberyl, 279, 282, 291
  synthetic, 313, 314
Chrysophase, 1168
Chrysotile
  see also Asbestos.
  composition, 440
  electron micrograph, 436, 439
  geology, 444–446
  heat, effect on, 440
  occurrences, 454, 455
  origin, 436
  pH, 437
  production, 444
  properties, table, 441, 442
  solubility, 443
  structure, 435
  thermal analysis, 437
  uses, 474–481
  zeta potential, 439
Cinders
  as aggregate, 81
  coal, 85
  volcanic, 81, 85, 86, 1079
Circum-Pacific belt, 1001, 1002
Claus process, sulfur from H_2S, 1252, 1253
Clays, 585–651
  abrasive, 31
  absorbent granules, 593, 594
  analyses, table of, 596
  attapulgite. *See* Fuller's earth.
  ball clay, 35, 38, 586, 612, 620, 633, 634, 641, 642
  bentonite, 38, 587–612
  calcined clay, 614, 628, 636
  ceramic, 33, 34
  definition, 585
  depletion allowance, 641
  drying, 609
  ecological considerations, 641, 642
  electron micrographs, 595, 619
  fillers, 244, 250–252

Clays, fire clay, 34, 373–376, 576, 586, 641
- firing, shrinkage, 640
- fuller's earth, 16, 38, 587–612, 641
- glass, use in, 346
- granule, 209, 593, 594
- green strength, 640
- halloysite, 596, 612, 613, 620, 634–636
- hectorite, 594, 596, 598, 600, 607–609, 611
- illite, 34, 633
- kaolin, 26, 34, 35, 38, 375, 376, 586, 596, 599, 612–633, 635, 641, 642
- lightweight aggregate, 82
- opal claystone, 596
- palygorskite, 596
- porcelain, clays for, 614
- pottery, clays for, 639
- production, 586, 590, 614
  - Argentina, 624
  - Australia, 627, 628
  - Austria, 624
  - China, 627
  - Czechoslovakia, 625
  - Denmark, 625
  - France, 625
  - Guyana, 624
  - Hungary, 625
  - India, 626
  - Italy, 625
  - Japan, 672
  - Korea, South, 627
  - Malaysia, 627
  - Mexico, 624
  - New Zealand, 628
  - Philippines, 627
  - Spain, 625
  - UK, 623, 624
  - US, 622, 623
  - USSR, 624
  - West Germany, 626
- refractory clay, 586, 620, 636–639
- reserves
  - Brazil, 624
  - Bulgaria, 625
  - Czechoslovakia, 625
  - Iran, 627
  - New Zealand, 628
  - Poland, 625
  - Sri Lanka, 626
  - Turkey, 627
  - US, 623
  - USSR, 624
- saponite, 594, 600
- sepiolite, 596
  - *see also* Meerschaum.

Clays, smectite, 594, 597, 599, 604, 605, 633, 640, 870
- societies, 586
- uses, 588–592, 599, 602, 606, 633–636
- vitrification range, 640
- whiteware, 1089

Clayton Valley, 871, 873
Clinker, 139–141
Clinoenstatile, 1279
Clinoptilolite, 1397, 1398
Coal, sulfur from, 1254
Coated abrasives, 17, 18
Ccated optics, 298
Coatings
- calcium carbonate, 633
- decorative, protective, 252
- kaolin, 616
- talc, 1278

Coesite, 1167
Coke, 35
Coke oven gases, 970
Colemanite, 5, 43, 536
- *see also* Boron and borates.

Coalinga area, 451
Colloidal silica, 1168
Colombia
- asbestos, 451, 452
- barite, 492
- chromite, 578
- diatomite, 689
- emeralds, 302
- glance pitch, 531
- kaolin, 624
- salt, 1126
- tar sands, 1253

Colorado
- barite, 492
- clay, 374, 375, 637
- cryolite, 741
- feldspar, 711
- fluorspar, 727, 729
- gannister, 377
- gems, 303, 319
- grahamite, 530
- granite, 171
- gypsum, 778, 782
- lithium, 873
- oil shale, 1253
- pegmatites, 711
- perlite, 997, 1003
- salt, 1123, 1124
- sand-gravel demand, 108
- slag, 125
- soda ash, 1188
- thorium, 1111
- titanium, 1321
- vermiculite, 1377
- zirconium, 1435

Composite gems, 334
Computers, use of, 142

Concrete
- aggregates, 60–131
- air-entrained, 142

Congo (Brazzaville), potash, 1070
Connecticut
- feldspar, 711
- fuller's earth, 591
- granite, 170, 711
- mica, 918
- pegmatites, 711
- talc, 1288
- trap rock, 62

Construction materials
- aggregates
  - crushed stone, 60–80
  - lightweight aggregates, 81–95
  - sand and gravel, 96–110
  - slag, 111–131
- cement, cement raw materials, 133–159
- dimension, cut stone, 161–181
- gypsum, anhydrite, 183–198
- insulation materials, 199–201
- roofing granules, 203–212

Contact process, sulfuric acid, 1266
Copper, 233, 240
Copper wool abrasive, 30
Copperas red, 361, 364
Copperopolis area, 451
Coquina, 167, 170
Coral, gem, 291, 301, 303, 318
Cordierite, ceramic, 34, 35, 1278
Corduroy spar, 710
Corkstone, 162
Cornwall stone, 36
Corundum, 15, 16, 18, 292
- synthetic, 312, 313, 319

Cosmetics, 52, 1277
Costs
- open pit, 74
- transportation, 73, 853

Cristobalite, 1167, 1168
Crocidolite, 437, 440–443, 457
- *see also* Asbestos.
- uses, 475

Crocus, abrasive, 31
Cross fiber, 444, 463
Crucibles, graphite for, 769
Crushed steel, 30
Crushed stone, 60–80, 847
- aggregates, 60–80
- capacity, plant, 69, 70
- chert, 63
- competition, factors affecting, 74–76
- consumption, 61–64
- costs, 64, 73, 75, 76
  - transportation, 73, 74, 852, 853
- crushing, 72
- definition, 60

Crushed stone, deleterious material, 68
- demand, 74
- depletion allowance, 78
- distribution, 69
- future trend, 78, 79
- government controls, 70, 77
- gradation, acceptable, 68
- grading requirements, 67, 68
- highway construction, 63–65
- industry, 75, 76, 83
- labor requirements, 75
- land use, 77, 78
- limestone, 62, 63, 844
- location requirements, 61–63, 69, 70
- nonconstruction use, 62
- physical properties, rocks, 63
- plant considerations, 69, 70–73, 75, 76
- prices, royalty, 78
- production, 61–69
- profitability, 77
- qualities, 60
- reserves, 79
- rock types used, 62, 63, 844
- shape, particle, 68
- shipments, 852
- slag aggregate, 111
- specifications, 60, 61, 65–69
- substitutes, 69
- supply, 79
- taxation, 78
- tests, 65–69, 98, 100, 1157, 1158
- transportation, 73, 74, 852, 853
- uses, 59, 61, 62, 847
- value, 77, 78
- zoning regulations, 78

Crushing
- asbestos, 468
- borates, 553
- stone, 72

Cryolite, 741, 742, 1042
- wollastonite, substitute for, 1388

Crystalline graphite, 757

Crystallite, 350, 351

Crystals
- *see also* Electronic and optical uses.
- producers, 214, 226
- synthetic, 214, 226

Crystolon, 28

Cuba
- barite, 492
- chromite, 575
- glance pitch, 531
- grahamite, 530

Cuicatlan area, 451

Cullet, 346

Cultured quartz, 38, 210, 213, 214, 216–221
- *see also* Quartz.

Curbing block, 164, 165

Cured sands, 272

Curvettes, 333

Cut stone, 162, 163

Cuttables, diamonds, 19, 296

Cutting sand, 25

Cyprus
- asbestos, 452
- bentonite, 602
- chromite, 578
- iron oxide pigments, 355
- sulfur, 1254

Czechoslovakia
- barite, 493
- bentonite, 602
- clay, 625, 636, 638, 683
- magnesite, 884
- ozokerite, 531
- perlite, 1004

D

Danakil Depression, 1051

Darling Range, 512, 513

Daugherty process, 903

Dauphine twinning, 218

Dawsonite, 523

Dead-burned gypsum, 189

Dead-burned magnesite, 381, 882, 887, 893
- *see also* Magnesite.

Dead gems, 295, 332, 654

Dead perlite, 1010

Death Valley, 542, 543

Decorative coatings, 252

Delamination, 630, 631, 632

Delaware, glauconite, 752

Delay line, 226, 227

Denmark
- diatomite, 687
- flint pebbles, 27
- kaolin, 625
- quartzite, 27

Density test, loose grains, 20

Desulfurization, 1256

Deval abrasion test, 100, 844

Diabase, 314

Diamite, 314

Diamonair, 314

Diamond drilling, asbestos, 462, 463

Diamonds, 19, 20, 653–676
- *see also* Gem materials.
- abrasive, 15, 16, 19, 20
- accessory minerals, 658
- alluvial occurrences, 667, 668
- ballas, 19, 657
- black diamond, 656
- boart, 656
- boort, 656
- bort, 19, 656
- bortz, 656
- bowr, 656
- brilliance, 653
- carbonado, 19, 656, 657
- classification, 656, 657
- composition, 659
- consumption, 656
- cuttables, 298
- cutting, 653
- dead stone, 295, 332, 654
- diamite, 314
- diamonair, 314
- discoveries, 653
- dispersion, 294, 653
- ecology, 673
- evaluation, 663
- exploration, 662, 663
- fancies, 295
- fire, 295, 653
- flowsheets, 666, 667
- future of, 673
- gems, 279, 283, 295, 332, 653, 654
- geology, 657, 658, 662
- government control, 672
- grade, 663
- grits, 657
- hardness, 291–293
- inclusions, 299
- index of refraction, 653–660
- industrial, 19, 296, 654–657
- kimberlite, 657, 658, 662
- manufactured, 16, 19, 20, 30, 662, 668–670
- marketing, 668
- metal coating, 30
- milling, 310, 311, 665–668
- mining, 663–665
- powders, grade of, 658
- price/time, 673
- processing, 665–668
- production, 19, 654, 662, 672
- producing controls, 668
- properties, 659–661
- reserves, 658, 661
- scintillation, 653
- secondary recovery, 672, 673
- shapes, 654
- short-bort, 657
- sources, 654
- stockpile, 672
- substitutes, 671, 672
- synthetic, 313, 318, 668–670
- tariffs, 672
- taxes, 672
- transportation, 668
- trends, 668
- trimond, 314
- value, 321
- uses, 653

Diaphragm cells, chlorine production, 50

Diaspore, 34, 376, 377, 504, 515, 637, 638, 1090
Diatomite, 16, 25, 26, 38, 677–708
  abrasive, 16, 25, 26, 38, 699
  absorbent, 699
  acid treatment, 692
  aggregate, lightweight, 89
  analyses, 678, 679, 694
  calcining, 692
  ceramic, 36, 39
  chromatographic support, 699
  classification of deposits, 682, 683
  composition, 677–681, 694
  consumption, 695
  definition, 677
  density, 678, 690
  depletion allowance, 701
  deposits, 682–691
  ecology, 701
  electron micrographs, 680, 681, 683, 684–687, 693
  evaluation, 690, 691
  exploration, 689, 690
  exports, 696
  filler, 254, 697–699
  filter, 254, 696, 697
  flow diagram, 692, 693
  flux calcining, 692
  geology, 677–683
  glass former, 36
  impurities, 677
  lacustrine deposits, 682
  marine deposits, 682
  marketing, 696
  microstructure, 678
  milling, 691–694
  mining, 691
  morphology, 677–681
  occurrences, 681, 682
  origin, 681, 682
  packaging, 694
  preparation, for market, 691–699
  processing, 692, 693
  producers, 658, 662
  production, 695
  properties, 659–661, 699, 700
  reserves, 689
  sampling, 663
  specifications, 694, 695
  structure, diatom, 677
  substitutes, 699–701
  taxes, 701
  tariffs, 701
  testing, 694, 695
  transportation, 696
  types of, 678
  uses, 244, 254, 388, 677
    cement, 143
    ceramic, 36, 39
    filler, 697–699
Diatomite, uses, filter, 696, 697
    properties of, 698
    well drilling fluids, 388
Didymium oxide, 1110
Dielectric materials, 38, 213, 227
Dimension stone, 161–181
  ashlar, 157, 158, 162
  basalt, 168, 169
  building stone, 161–181
  color, 161
  consumption, 162, 847
  curbing block, 164, 165
  cut stone, 162, 163
  cutting, 173–175
  definition, commercial, 165–167
  depletion allowance, 180, 181
  deposits, distribution, 169–171
  dolomite, 167
  end use, 161–165
  evaluation of deposits, 179, 180
  finishing, 164, 177
  flagstone, 166
  fleuri pattern, 167
  geologic provinces, 169–171
  granite, 161–166
  greenstone, 168
  limestone, 162, 163, 166, 167, 847
  lithologic classification, 165–169
  marble, 162, 167–173
  marketability, 180
  mill-stock slate, 164
  monument stone, 164
  obsidian, 167, 168
  onyx, 167
  pattern, quality of, 163
  paving blocks, 164
  polish, 163, 164
  preparation for markets, 171–177
  preservation of stone, 177, 178
  producers, 162
  production, 162, 847, 854
  pumice, 168
  quarrying, 171–177
  reserves, 169–171
  rock types, commercial definition, 165–167
  roofing slate, 164
  rough construction stone, 161, 162
  rubble, 161, 162
  sandstone, 166
    abrasive use, 25
    bluestone, 166
    brownstone, 70
    freestone, 166
    reeds, 166
    refractory brick, 165
Dimension stone, sandstone, rift, 166, 176
  sawing, 176, 177
  slate, 162–164, 167, 168
    filler, use as, 245
    granules, 209
  soundness, 180
  underground quarries, 172, 173
  use, 847
  zeolites, as, 1420
Discrete particle, definition, 350
Dispersion, 294
Dolomite, 167, 833–868
  aggregate, 61
  burnt dolomite, 269
  ceramic, 33, 35
  composition, 269
  fertilizer, 238
  flux, 33, 269
  refractories, 381, 382, 881
  smelting use, 269
  specifications, 770–772
  uses, 33, 238, 269, 342, 345, 381, 382, 881
Domes, salt, 1123–1125
Dominican Republic
  bauxite, 513, 523
  gypsum, 784
Dredges, 1155, 1160, 1328, 1348
Drilling
  asbestos, 462, 463
  bauxite, 456, 457
Drilling muds. *See* Well drilling fluids.
Dry press process, brick, tile, 375
Dumortierite, 35, 799
Dunite, 989
Durability test, sand, gravel, 98
Dysprosium, price, 1110

E

Earth diatomaceous, 653–676
  see also Diatomite.
Ecuador, kaolin, 624
Egypt
  chromite, 578
  clay, 626, 638
  gems, 303, 316
  nitrates, 970
  phosphate, 1031
  rare earths, 1112
  salt, 1126
  soda ash, 1197
  titanium, 1315, 1334
  vermiculite, 1377
  zirconium, 1433
Elaeolite, 933
Elastomic products, 49
Elaterite, 531
Electric furnace production, 28, 1018
Electrical twinning, 218

Electrochemical unit, 50
Electronic uses, 213–231.
    see also Quartz.
Emeralds, 313, 317, 318
Emery, 16, 18, 19
Endellite. *See* Halloysite.
Eneabba, 1332, 1333
Epsom salts, 893
Erbium, price, 1110
Erionite, 1395, 1396
Esker, 1154, 1155
Ethiopia
    kaolin, 626
    potash, 1051, 1070
Ethylene dichloride, 49
Eucryptite, 870
Eureka sandstone, 1172
Europe, geological agencies, 404–407
Europium, price, 1110
Expanded aggregates, 86
Expanded slag, 85, 116, 117, 119, 122
Extenders, 253
    see also Mineral fillers.

F

Fancies, 295
Fat clay, 639
Feathers, in quarrying, 175
Feldspar, 709–722
    abrasive, 16, 18, 26
    alaskite, 709–711
    aplite, 709, 712, 716, 721
    block spar, 711
    bonding agent, 16
    cement, use in, 134
    ceramics, use in, 33, 36, 716, 721
    depletion allowance, 720
    feldspathic sand, 710
    filler, 244
    glass, use in, 716, 719, 721
    government control, 721, 722
    granules, 204
    history, 718, 719
    markets, 719, 720
    milling, 717, 718
    mining, 717
    nepheline syenite, 719, 721
    pegmatites, 710, 711
    perthite, 710
    prices, 722
    processing, 716–718
    producers, 717, 719
    producing areas, 711, 713, 715, 716
    production, 713, 714, 716
    reclamation, 720
    reserves, 712, 713
    sand, 710
Feldspar, substitutes, 719, 720
    uses
        ceramic, 33, 36, 721
        fillers, 721
        glass, 344, 721
Feldspathic sand, 710
Ferrierite, 1417
Ferrites, 367, 1229
Ferroalloys, 904, 1116, 1173
Ferroelectric crystal, 227
Ferromanganese, grades, 904
Ferrotitanium, 1313
Fertilizer minerals, 233–241
    ammonia, 234
    boron, micronutrient, 239
    calcium, 233, 237, 238
        see also Limestone, Dolomite.
    consumption, 234
    copper, micronutrient, 233, 240
    importance, 233
    industry scope, 233
    iron, micronutrient, 239
    magnesium, 238, 239
        see also Limestone, Dolomite.
    micronutrients, 239, 240
    nitrogen, 233, 234, 977
        see also Nitrogen compounds.
    nutrient consumption, 234
    olivine, 989
    phosphorus, 233, 235, 236
        see also Phosphate rock.
    potassium, 233, 236, 237
        see also Potash.
    primary plant nutrients, 233–237
    secondary plant nutrients, 233, 236, 237
    sulfur, 233, 238, 239
        see also Sulfur.
    tagged nitrogen, 972
    trace elements, 233, 239, 240
    zinc, micronutrient, 240
Fiber glass, production, 339
Fibrolite, 802
Fiji, barite, 493
Fillers
    see also Mineral fillers.
    ceramic, 34
    clinoenstatite, 1279
    diatomite, 245, 247
    feldspar, 245, 721
    kaolin, 245, 246, 247
    nepheline syenite, 931, 934, 944
    perlite, 1013
    tripoli, 245
    wollastonite, 245
    zeolite, 1422
Filter aids
    see also Mineral filters.
    diatomite, 254
Filter stone, 61
Fining agent, 720
Finland
    anthophyllite, 461
    chromite, 576
    sulfur, 1254, 1255
    talc, 1282, 1283
    titanium, 1338
    vermiculite, 1377
    wollastonite, 1383, 1386, 1388
Fire clay, definition, 373, 374, 613
    see also Refractory clay.
Fire, of gem, 286, 292
Flagging, 164, 166
Flagstone, 162, 164
Flake graphite, 758
Flake mica, 915, 919–921
Flame cutting, 173–175
Flame treatment, 167, 339
Fleuri pattern, 167
Flint, 17, 18
Flint, abrasive, 17, 18
Flint clays, 374
Flint kaolin, 618
Flint paper, 17
Flint pebbles, 27
Florida
    attapulgite, 591
    clays, 591, 599, 622
    diatomite, 685
    kyanite, 800, 803
    limestone production, 62
    magnesite, synthetic, 890
    marl, 854
    phosphate, 1030, 1032–1034, 1039, 1043, 1045
    shells, 62
    silica, 1170, 1171
    staurolite, 26, 27, 1225
    titanium, 1303, 1304, 1316, 1324–1326
    zircon, 274, 1433, 1436–1438
Flour, silica, 1168, 1172
Flour, sulfur, 1238
Flue gas, 1256
Fluid-energy grinding mill, 1295
Fluorine, 50, 51, 723–725
    phosphates, from, 740, 1041
    tremolite, 1283
Fluorinated hydrocarbons, 723, 724
Fluorite. *See* Fluorspar.
Fluorocarbons, 51, 52
Fluorescence, 227
Fluorspar, cryolite, 269, 723–744
    beneficiation, 737, 738
    ceramics, use in, 34, 35, 734
    consumption, 725, 738, 739
        fluorocarbons, 52, 724

Fluorspar, consumption
fluxes, types of, 33, 267, 268, 739
glass, 724
hydrofluoric acid, 51, 724
steel, 724, 725
fluorides from phosphates, 740
flux, 33, 267, 268, 739
gem, 297
occurrences, mode of, 726–728
reserves
Alaska, 728
Italy, 733
Ontario, 728
world, 736
uses, 723, 724
abrasives, 724
ceramics, 34, 35, 724, 734
chemicals, 723
fluorinated hydrocarbons, 52, 724
flux, 33, 267, 268, 739
glass, 724
hydrofluoric acid, 56, 724
optics, 298
steel, 724, 725
welding flux, 269
Flux factor, 377
Fluxes, 33, 259–270
agents, 26, 36, 269
borates, 45, 46, 262, 263
ceramic fluxes, 36
economy of, 36
flux modifiers, 37
lithium, 36
nepheline syenite, 36
salt cake, 36
Fly ash, 85, 135, 209
Forsterite, 35
Forum on Geology of Industrial Minerals, 394, 395
Foundry facings, graphite, 768
Foundry sand, 271–278
olivine, 274–278, 989–996
staurolite, 274, 1225–1227
zircon, 273, 274, 1433–1446
France
andalusite, 802
asphalt, 530
barite, 492
bauxite, 514, 515
bromine, 48
buhrstone, 27
clay, 602, 625, 638
chalk, 25, 26
chrysotile, 457
diatomite, 686
feldspar, 715
flint pebbles, 27
fluorspar, 732, 733
gems, 302
gypsum, 184, 778
iodine, 795
iron oxide pigments, 355
France, kyanite, 803
meerschaum, 911, 912
mica, 918
perlite, 1005
potash, 1067
salt, 1126
slag, 113
sulfur, 1247, 1251
talc, 1283
zeolite, 1417, 1422
Fraccing sand, 1168
Francolite, 1022
Frasch process, 53, 1235, 1241, 1242, 1245–1247
Freestone, 166
Freeze-thaw test, 98
French chalk, 25, 26
Friability tests, 12, 13, 15
Frits, 36
Fuchite, 916
Fuller's earth, 12, 16, 587–612
abrasive, 16
absorbent, 255
ceramic, 35
distribution, 598
filler, 241
processing aid, 38
tariff, 642
uses, 38, 591, 592, 594
Fur, asbestos, 444
Fused alumina oxide, 29
Fusion cast products, 379

G

Gabon, manganese, 898
Gadolinium, price, 1110
Gadolinium-gallium garnet, 314
Galena, 388
Gannister, 363, 374, 379, 1168
Garnets, 17, 21–25
yttrium-aluminum, 314
Gas, natural, 976
Gehlenite, 1389
Geikelite, 1389
Gem materials, 279–338
see also names of stones.
ballas, 19, 296
bort, 10, 292, 296
carbonado, 19, 292, 296
control, political, 321–323
corundum, 15, 16, 18
synthetic, 312, 313, 319
diamonds, 19, 20, 314, 653–676
garnet, 24, 315
marketing, preparation for, 23, 323–330
quartz, 1167, 1168
uses
decorative, 291
gems, 653, 654
history, 279
industrial, 654, 655
Geological agencies
see also Sources of information.
Africa, 407–410
Asia, 411–413
Australia, 413
Caribbean Islands, 402
Central America, 402
Europe, 404–407
Middle East, 410, 411
North America, 393–402
Pacific Islands, 413
South America, 402–404
Georgia
attapulgite, 592
barite, 491, 492
bauxite, 515
clay, 592, 599, 613, 618, 621, 622, 638
feldspar, 711
granite, crushed, 62, 170
halloysite, 634, 635
iron oxide pigments, 355
kyanite, 376, 800, 801
manganese, 898
marble, 164, 170
mica, 918
monumental stone, 164
olivine, 888, 889, 991
pegmatite, 711
silica, 26, 1170, 1171
sillimanite, 802
titanium, 1316, 1324, 1326
talc, 1284, 1285, 1289, 1298
tripoli, 26, 1368
vermiculite, 1377
zircon, 1436
Germany, East
barite, 492
borates, 550
fluorspar, 733
kaolin, 625
potash, 1068
sulfur, 1256
Germany, West
asphalt, 530
barite, 492
bromine, 562
clay, 492, 604, 626, 638
diatomite, 686
feldspar, 715
fluorspar, 733
graphite, 760, 765, 768
gypsum, 778
iodine, 795
lime, 828
lithium, 876
perlite, 1005
potash, 417, 1068, 1069
quartz, 214
salt, 1126
slag, 113
sulfur, 1242, 1252

GGG, 314
Ghana
  bauxite, 508, 513, 523
  diamonds, 302
  gems, 302
  manganese, 897
Gibbsite, 376, 377, 504, 635
Gilsonite, 530
Glacial deposits, source of aggregates, 1154, 1155
Glance pitch, 531
Glass, 31, 33, 36
  raw materials for, 344, 902
Glass batch gypsum, 196
Glass paper, 31
Glass raw materials, 339–347
  abrasive, 17, 31
  batch requirements, 339, 340
  ceramic, 33, 36
  feldspar, 721
  fiber glass, 339
  formers
    diatomite, 36
    silica, 33, 36
  gypsum, used in, 196, 340
  imitation gems, 315
  industrial minerals, consumption, 33, 340, 341
  natural, 279
  processing, 33
  production, 340, 955
  raw materials composition, 340, 341, 344
    borax, 340, 341
    celestite, 1229, 1231
    clay, 33, 346
    dolomite, 33, 340
    feldspar, 340, 341, 344
    lime, 822
    limestone, 33, 341, 344, 846
    nepheline syenite, 341, 344, 955
    silica sand, 33, 36, 340, 341, 344
    soda ash, 340, 341, 344, 1217
Glass wool, 199–201
  see also Insulating materials.
Glauberite, 1209
Glauber's salt, 1207
Glauconite, 745–756
Glossary, electronic, optical, 226–228
Gneiss, 165
Goethite, pigment, 352
Gosiute, Lake, 1199
Gove Peninsula, 512
Grading requirements
  lightweight aggregates, 88
  slate, 164, 168
Grahamite, 530, 531
Grainers, 1136
Granite, 161, 164, 165, 167, 168
  aggregate, 61
  composition, 712
  crushed, 62
  cutting, 177
  definition, 167, 168, 712
  flame treatment, 173, 174
  graphic, 710, 711
Granulated slag, 115–117, 121, 122
  base rock, 204, 205, 208
Granules, roofing, 203–212
  absorbent, characteristics of, 594
Graphic granite, 710, 711
Graphite, 5, 757–773
  refractories, 379, 769
  reserves
    Alabama, 760
    Korea, 763, 764
    Norway, 765
  types of, 757, 758, 759
  uses, 388, 757, 768, 769
Gravel, definition, 1151
Gravel spar, 730
Graywacke, 166
Great Britain. *See* United Kingdom.
Great Dyke, 572
Great Salt Lake
  lithium in, 875
  potash, 1072
  sodium sulfate, 1212, 1220
Greece
  asbestos, 452
  barite, 492
  bauxite, 514
  chromite, 578
  clay, 602, 625
  fluorspar, 733
  magnesite, 382, 884, 885, 888
  meerschaum, 911
  olivine, 991
  perlite, 999, 1003
  pumice, 1084
  salt, 1126
  talc, 1283
  wollastonite, 1387
Green earth. *See* Glauconite.
Green River Basin, 1199, 1200
Green rouge, 31
Greenland
  chromite, 575
  cryolite, 741
  flint pebbles, 27
  nepheline syenite, 935, 936
Greensand. *See* Glauconite, Foundry sands.
Greenstone, 168, 205
Grinding pebbles, 27, 31
Grog, 638
Guano, 234, 1021, 1031
Guatemala
  chromite, 575
Guatemala, fluorspar, 732
Guillotine, 162
Guinea
  bauxite, 508, 513, 523
  diamonds, 661, 662
  gems, 302
Guyana
  bauxite, 5, 513, 514
  calcined bauxite, 458
  diamonds, 658
  kaolin, 624
Gypsite, 777
Gypsum, anhydrite, 183–199, 755–792, 1256, 1257
  ammonium sulfate, 968
  anhydrite, 775, 776, 1256
  anhydrous, 195
  cement raw material, 134, 141, 775, 784, 785, 1256
  fertilizer, 238
  glass batch gypsum, 196, 344
  major uses, 775
    agriculture, 196, 238, 968
    construction, 192–194
    industrial, 194–196
  phosphate, byproduct, 238
  portland cement material, 134, 141, 784, 785, 1256
  processing, 185–196
  secondary plant nutrients, 238
  sulfur, source of, 1256, 1257
  superphosphate, processing of, 238, 1041
  uses, 196, 197, 775
    agriculture, 196, 238, 1256, 1257
    cement, use in, 134, 141, 775
    construction group, 190, 194, 195
    fertilizer, 238
    filler, 244, 247
    industrial, 194–196
    stucco, 188, 189

H

Hafnium, 1441
  see also Zirconium.
Haiti, bauxite, 513, 523
Halloysite, 612, 613, 620, 634–636
    analyses, 596, 612
Hand molding, 375
Hanksite, 1199
Hard kaolin, 614
Hardness, scales
  Knoop, 12, 659
  Mohs', 292, 293
Hardness, selected materials, 19, 284, 285
Harmonic generation, 227

Hawaii
bauxite, 515
coral, 291, 301, 303
Headlap granules, 209
Hectorite, 594, 596, 598, 600, 607, 611, 875
Hegman test, 617
Helfferichs test, 1417
Hematite, pigment, 351, 352
Hercules viscosity, 629
Hiddenite, 870
see also Lithium raw materials.
High alumina brick, 374, 520, 521
High alumina cement, 521
High grading, 311
Highway construction, aggregates, 63–65
Hispaniola, bauxite, 453, 504
Histogram, 1099
Holland
glauconite, 753
salt, 1126
Holography, 227
Honduras, fluorspar, 732
Hong Kong, graphite, 764
Horse, salt, 1122
Howlite, 550
Humphrey's spiral, 922
Hungary
bauxite, 514
clays, 603, 625, 638
perlite, 999, 1004
zeolites, 1417, 1419, 1421, 1422, 1426
Hydraulic cement, 1957
Hydraulic lime, 824
Hydrocarbons, 529–532, 1132
see also Bituminous material.
Hydrochloric acid, flux, 260
Hydrofluoric acid, 50–52, 260
Hydrogel process, 1407–1409
Hydrogen sulfide, 489
Hysteresis loop, 351

I

Iceland
diatomite, 686
geothermal sources, 875
lithium, 875
perlite, 1001, 1005
Iceland spar, 297
Idaho
andalusite, 803
barite, 492
clay, 597, 618–620, 623, 635, 638
diatomite, 685
feldspar, 711
fluorspar, 729
garnet, 23, 24
gems, 303
Idaho, glass sand, 639
kyanite, 801
monazite, 1111
pegmatite, 711
perlite, 997, 1003
phosphate, 1022, 1031, 1034, 1042
salt, 1124
silica, 1171, 1172, 1173
thorium, 1111
travertine, 171
vermiculite, 1377
wollastonite, 1385
yttrium, 1112
Illinois
barite, 492
clay, 596, 599, 623
dimension stone, 62
fluorspar, 726, 729, 730
lime, 827, 828
limestone, 836
limestone production, 62, 853
silica, 1170
slag, 112, 125
tripoli, 26, 1363, 1364, 1366, 1367
Ilmenite, 1313, 1314
see also Titanium.
Illite, ceramic, 34
Imitation gems, 315
Impsonite, 531
Inagua, salt, 1124
Inder Lake, 507
India
amosite, 455
asbestos, 454, 455
barite, 493
bauxite, 508, 515
chromite, 577, 581
clay, 597, 605, 626, 634, 638
crocidolite, 455
diamonds, 302
fire clay, 638
garnet, 28
gems, 302
graphite, 761, 764
kyanite, 376, 799, 801, 803
magnesite, 883, 884, 885
manganese, 897
meerschaum, 911, 912
mica, 5, 917
monazite, 1113
olivine, 991
pyrophyllite, 1097
rare earths, 1113
sillimanite, 802
sulfur, 1256, 1260
talc, 1283
thorium, 1113
titanium, 1333
vermiculite, 1377
wollastonite, 1383, 1386
yttrium, 1113
zircon, 1443
Indiana
dimension stone, 62, 854
dolomite, 62, 854
gypsum, 125, 777, 782
halloysite, 635
lime, 828, 836
limestone, 62, 854
marl, 854
slag, 125
whetstones, 27
Indonesia
bauxite, 515, 523
chromite, 578
diatomite, 689
gypsum, 795
iodine, 793
kaolin, 626, 627
monazite, 1113
Industrial minerals, 1–6, 42, 43, 393–414
Information, sources of, 292–414
Insoluble anhydrite, 189
Insulating materials, 199–201
bricks, 377, 378
Intaglios, 333
Iodine, 41, 793–797
Ion exchange, clay, 594, 595
Ione formation, 1172
Iowa
dimension stone, 854
dolomite, 854
gypsum, 778, 780, 781
limestone, 854
Iran
barite, 493
chromite, 577
kaolin, 627, 638
meerschaum, 911, 912
sodium sulfate, 1216
sulfur, 1242, 1252
Iraq
asphalt, 530
sulfur, 1242, 1249, 1250, 1252
Ireland, Northern, perlite, 1001
Ireland, Republic of
barite, 493
bauxite, 505
diamonds, synthetic, 662
iodine, 795
magnesite, synthetic, 890
Iron, 233, 239
Iron oxide
abrasives, 30, 31
ceramics, 30, 36, 38
natural, 351–359
synthetic, 359–368
Isotopes
carbon, 1240
iodine, 794
sulfur, 1240
Israel
bromine, 562
diaspore, 638
clay, 638

Israel, magnesium chloride, 891
phosphate, 1031, 1035
potash, 1070
Italy
asbestos, 452, 457, 461
asphalt, 530
baddeleyite, 1436
barite, 492
borates, 547, 548
celestite, 1233
clay, 603, 638
diatomite, 686
feldspar, 713
fluorspar, 728, 733
gems, 302
graphite, 761, 765, 767
gypsum, 778
iron oxide pigments, 355
kaolin, 625
magnesite, 890
marble, 171
olivine, 991
perlite, 1004
potash, 1069
pumice, 1084
pyrite, 1236, 1254
salt, 1126
slag, 113
sulfur, 1236, 1241, 1247, 1251, 1254
tremolite, 461
wollastonite, 1387
zeolite, 1417, 1419, 1420, 1422
Ivory Coast
diamonds, 661, 662
lithium, 874

J

Jade, 299, 302, 303, 314, 319
Jamaica
bauxite, 513, 523
gypsum, 778
Japan
asbestos, 454
barite, 493
bromine, 48
chromite, 578
clay, 605, 627, 636, 638
cultured quartz, 214
diatomite, 689
fluorspar, 735
graphite, 763
garnet, 23
iodine, 793, 795, 796
magnesite, synthetic, 890
olivine, 991
perlite, 999
pearls, 318
pyrite, 1236, 1254
pyrophyllite, 1085, 1091
refractories, consumption, 1086
Japan, silica brick, 377
slag, 113
sulfur, 1236, 1242, 1251, 1254, 1260
vermiculite, 1377
zeolites, 1414, 1419, 1420, 1421, 1422, 1423, 1424
Java, iodine, 795
Jet, 292, 301, 315
Jet channelling, granite, 175
Jetty stone, 61
Jordan
bromine, 563
potash, 1070
phosphates, 1035
Jugoslavia, diatomite, 687

K

K-bentonite, 588
synthetic, 610, 611
Kainite, 1070
Kames, source of aggregate, 1154
Kansas
diatomite, 685
gypsum, 783
vermiculite, 1377
Kaolin, 26, 34, 35, 38, 375, 376, 586, 594, 596, 599, 612–633, 635, 641
definition, 612
processing, 629, 630, 631, 632
production, 586, 590, 614
Argentina, 624
Australia, 627
Austria, 624
Czechoslovakia, 625
Denmark, 625
France, 625
Guyana, 624
Hungary, 625
India, 626
Italy, 625
Japan, 627
Korea, South, 627
Malaysia, 627
Mexico, 624
New Zealand, 628
Philippines, 627
Spain, 625
UK, 623, 624
US, 622, 623
USSR, 624
West Germany, 626
reserves
Brazil, 624
Bulgaria, 625
Czechoslovakia, 625
Iran, 627
New Zealand, 628
Poland, 625
Sri Lanka, 626
Turkey, 627
Kaolin, reserves
US, 623
USSR, 624
saponite, 594, 600
sepiolite, 596
see also Meerschaum.
smectite, 594, 597, 599, 604, 605, 633, 640
see also Bentonite, Fuller's earth.
specifications, 616–618
uses, 26, 34, 35, 38, 614–617
Kaolinite. *See* Kaolin.
Kara-Bogaz-Gol, 1209
Kasota vein, creamy, 122, 167
Keatite, 1167
Keenes cement, 189, 823
Kellogg, M.W., ammonia process, 964
Kentucky
asphalt, 530
barite, 492
clay, 374, 375, 599, 623, 624, 633, 634, **637**, **638**
fluorspar, 729, 730
lime, 827, 828
limestone, 853
rock asphalt, 530
silica, 1178
slag, 125
tar sand, 1253
Kenya
bentonite, 605
diatomite, 687
fluorspar, 734
graphite, 766
kaolin, 626
kyanite, 801
meerschaum, 911, 912
monazite, 1113
soda ash, 1187, 1197
vermiculite, 1377
wollastonite, 1383, 1386
zeolite, 1416
Kernite, 45, 536
Kettle, calcining, 63–65
Keuper, 1125
Kieselguhr. *See* Diatomite.
Kilns, 81–83, 85, 139, 140, 186–188, 812, 813
Kimberlite, 308
Knoop hardness number, 12
Kokoman, 734
Kop, 1183
Korea, North
barite, 493
chromite, 578
fluorspar, 735
magnesite, 884
talc, 1283
Korea, South
barite, 493
chromite, 578

Korea, South, fluorspar, 735, 736
 graphite, 761–763, 767
 halloysite, 636
 kaolin, 627
 pyrophyllite, 1092, 1107
 magnesite, 883
 talc, 1283
Kramer. *See* Boron, CA.
Kroll process, titanium, 1311, 1312
Kunzite, 303, 318, 870
Kyanite, 376, 799–807
Kyasill, 800

L

Lake sand, 271
Lampblack, abrasive, 31
Land plaster, 183, 186, 187
Langbeinite, 236, 238, 1071
Lapis lazuli, synthetic, 314
Laponite, 611
Larvikite, 171
Las Brisas, area, 451
Lasca, 214, 215
Laser, 227
Laterite, 35
Lazurite, 292, 298, 300
Lead, flux, 36
Lean clay, 639
Lepidocrocite, 352
Lepidolite, 915
Lechatelierite, 1167
Leucoxene, 1314
Liberia, barite, 493
Libya, 1070
Light-sensitive materials, ceramic, 37
Lightweight aggregates, 81–95
 bituminous paving mixes, 120
 properties, desirable, 82
 types, 81
 zeolite, use in, 1442
Lime, 809–831
 abrasive, 31
 analyses, 269, 811
 burned lime, 269, 801
 burnt dolomite, 269
 composition, 269, 824
 flux, 269
 uses, 31, 34, 37, 134, 267, 388, 821–824
Limestone, 62, 833–868
 see also Crushed stone, Dimension stone.
 aggregate, 60, 61
 agricultural use, 237
 bituminous, 530
 cement, use in, 135
 ceramic, 33, 35
 chalk, 167
 commercial definition, 163
 composition, 61, 269, 839, 855
Limestone, consumption, agricultural, 237
 crushed stone, 60–80, 847, 852
 cut stone, 162, 163
 definition, commercial, 163
 density, 63, 842, 843
 dimension stone, 166, 167, 847
 evaluation, 841–848
 filler, 244, 847
 flux, 35, 269
 glass, use in, 344
 oolitic, 167, 835
 production, 847, 849, 852, 853
 production, dimension stone, 162, 166, 854
 shells, consumption, 62, 855
 uses, 62, 846–848
  aggregates, 62, 847
  agriculture, 62, 237, 847
  calcium carbide, 846
  cement, 135, 847
  ceramic, 33, 35
  flux, 269
  glass, 344, 847
  well drilling fluids, 388
Limestone rock asphalt, 530
Lithium, 869–880
 batteries, 902
 brines, 41, 871, 873, 875
 flux, 36
 minerals, 870, 875, 877
 potential districts
  Chile, 875
  Utah, 875
  Zaire, 874
 producing districts, 871–874
  Argentina, 873, 874
  Australia, 874
  Brazil, 873
  Canada, 874, 875
  Chile, 875
  Mozambique, 874
  Namibia, 874
  Ruanda, 874
  South Africa, 874
  Uganda, 874
  US, 871, 872
 reserves, 871–874
  Argentina, 873, 874
  Australia, 874
  Brazil, 873
  Canada, 874, 875
  Chile, 875
  Namibia, 874
  US, 871, 872
  Zaire, 874
  Zimbabwe, 873
Lithopone, 486
Lively perlite, 1010
Loading materials, 243–257
 see also Mineral fillers.
Long Island, salt, 1124
Loose abrasive grains, 15–17
Loose-pack density test, 13, 14
Lorrain quartzite, 1174
Lord Rayleight waves, 220
Los Angeles rattler test, 65, 100, 119, 844, 1158
Louisiana
 asphalt, 530
 bentonite, 599
 iodine, 794
 salt, 1119, 1123, 1140
 shell, 62, 854
 silica, 1173
 sulfur, 1245
Lubricants, graphite for, 769
Lump graphite, 758
Luster, 294, 295
Lutetium, price, 1110

M

MacIntyre mine, 1317–1319, 1343, 1344
Magadi, Lake, 1187, 1197, 1198
Magnesia, 31, 34
Magnesite, 881–896
 ceramic, 35
 dead-burned, 882, 893
 deposits, 884–886
 flux, 35
 production, 892, 893
 seawater as source, 381
 synthetic, 889–892
 talc with, 1145
 uses, 35, 883, 893, 894
Magnesium
 brines, 881, 889–891
 secondary fertilizer, 238
Magnesium metal, 890
Magnesium sulfate, soundness test, 66, 98
Magnet Cove, 1320
Magnetic surveying, asbestos, 461, 462
Magnetite pigment, 352, 353
Magnetoptic, 227
Maine
 diatomite, 685
 feldspar, 711
 gems, 303, 319
 granite, 170
 graphite, 762
 manganese, 899
 marl, 854
 pegmatites, 639
Make, of a gem, 280, 329
Makgadikgadi, 1197
Malagasy Republic (Madagascar)
 bauxite, 515
 chromite, 576
 clays, 605, 626
 garnet, 23

Malagasy Republic (Madagascar)
  gems, 302
  graphite, 758, 759, 760, 761, 766, 768
  mica, 918
  monazite, 1112, 1113
  rare earths, 113
  thorium, 1112
Malawi
  fluorspar, 734
  monazite, 1113
  zirconium, 1433
Malaysia
  kaolin, 627
  titanium, 1333
  zirconium, 1433
Manchuria
  graphite, 763
  magnesite, 883
  soda ash, 1198
  talc, 1283
Manganese, 897–908
  deposits, 897, 898
  fertilizer, 239, 240, 902
  micronutrient, 233, 239
  uses, 233, 239, 901, 902
Manganese dioxide
  abrasive, 31
  battery use, 905
  synthetic, 904
Manganese silicate, flux, 268
Manitoba
  bentonite, 601
  chromite, 575
  gypsum, 784
  lithium, 874, 875
  silica, 1175
Manjak, 531
Man-made quartz
  see also Cultured quartz.
Manto, 726
Manufactured abrasives, 27–31
  boron carbide, 12, 29
  diamonds, manufactured, 16, 19, 20, 30
  silicon carbide, 16, 17, 29
Manufactured diamonds, 30, 668–670
  see also Diamonds.
Marble, 167, 168, 173, 835
  types, 166, 167
Marcasite, 321, 323
Marekanite, 1007, 1003
Marinas, salt, 1120
Marl, calcareous, 854
  cement raw materials, 135
Marsh borate deposits, 537
Maryland
  chromite, 575
  clay, 374, 375
  diatomite, 684
  glauconite, 752
  limestone, 853
  silica, 1171
Maryland, slag, 112
  talc, 1284, 1285, 1288
Mass fiber, 441, 442, 444
Massachusetts
  clay, 596
  diatomite, 685
  feldspar, 711
  granite, 170
  pegmatites, 711
  talc, 1287, 1288
Matrix, 1018, 1030
Medina sandstone, 1174
Meerschaum, 89, 909–914
Mercury cells, chlorine production, 50
Merseber process, ammonium sulfate from, 1256
Met-spar, 725
Meta-bentonite, 588
Metal finishing abrasive, 16, 31
Metahalloysite, 612
Metallic abrasives, 28, 30, 31
Metallurgical pebbles, 1171
Methyl bromide, use, 48
Mexico
  asbestos, 451
  barite, 492
  borates, 550, 551
  celestite, 1229
  chromite, 578
  clay, 602, 624, 638
  diatomite, 685, 686
  feldspar, 715
  fluorspar, 727, 728
  glance pitch, 531
  graphite, 761, 762, 766, 767
  gypsum, 777
  limestone, 840
  magnesite, synthetic, 890
  nepheline syenite, 942
  nitrates, 970
  perlite, 1003
  salt, 1121, 1124, 1125
  silica, 1176, 1177
  sodium sulfate, 1213, 1220
  sulfur, 1236, 1241, 1251, 1252, 1260
  tequesquite, 1121
  titanium, 1324
  vermiculite, 1329, 1377
  wollastonite, 1383, 1385
  zeolite, 1418–1420, 1422
Mica, 203, 209, 915–929
  deposits, 919, 920
  filler, 244, 252, 927
  flake, 915, 919, 920
  processing, 921–923
  scrap, 915, 918, 919, 920, 921
  substitute for talc, 1298
  synthetic, 915
  uses, 209, 388, 917, 927, 928
Micanite, 915
Michigan
  bromine, 48, 562
  dolomite, 85
  gypsum, 777, 781
  iodine, 795, 796
  lime, 827, 828
  limestone, 836
  magnesite, synthetic, 891
  marl, 854
  potash, 1066
  salt, 1140
  slag, 112, 125
Micrite, 837
Micro-Cel, 1389
Microcrystalline silica, 26
Microporite, 822
Midland type brine, 44
Mill scale, 360
Mill-stock slate, 164
Mine, salt, temperature, 1311
Mineral absorbents, 255
Mineral fillers, 243–257
  ceramics, 24, 25
  nepheline syenite, used as, 245, 247, 932, 934
  paints, 252, 931
  plastics, use in, 249, 250, 931
  properties, 244, 245, 246, 247
  rubber, fillers used in, 251, 931
  uses, 243–245
  zeolite, 1422
Mineral filters, 253–255
  properties, table of, 698
Mineral oils, bleaching clays, processing, 594
Mineral pigments, 349–372
  definition, 349
  processing, 353–356, 360–363
  synthetic, 359, 360, 366–368
  titanium, 1303, 1304
Mineral wax, 529, 531
Mineral wool, 199–201
  see also Insulating materials.
Minnesota
  dimension stone, 854
  dolomite, 854
  kaolin, 623
  limestone, 854
  manganese, 898
  quartzite, 27
  silica, 1173
  titanium, 1321
Mirabilite, 1207
Miridas, 313
Misch metal, 1110
Mississippi
  asphalt, 531
  bauxite, 515
  clays, 599, 633
  magnesite, synthetic, 890
  marl, 854

Missouri
- barite, 490
- bauxite, 515
- clay, 374, 375, 599, 637
- diaspore, 515
- granite, 170
- lime, 827, 828
- limestone, 62
- silica, 1170
- tripoli, 26, 1361, 1364–1366, 1369

Modulation, 227
Modules of rupture, 617
Mohs' scale of hardness, 292, 293
Molded stone, 325
Molding plaster, 194
Molding sand
- *see also* Foundry sand.
- chromite, 568
- natural, 271
- olivine, 898, 990
- silica, 271, 1168

Molecular sieves, 1400, 1407, 1408, 1411–1413
- *see also* Zeolites.
- definition, 1400
- uses, 1411–1413

Moler, 687
Molybdenum, 233, 240
Monazite, 1109–1117
- *see also* Rare earths.
- deposits, 1112
- uses, 1109, 1110

Mongolia
- fluorspar, 734
- rare earths, 1111
- soda ash, 1198

Monocrystalline count, 15
Montana
- barite, 492
- chromite, 575
- clays, 587, 600
- fluorspar, 730, 731
- gems, 303, 318
- graphite, 762, 765, 767
- gypsum, 778, 783
- jade, 303
- manganese, 898, 902
- phosphate, 1022, 1031, 1042
- silica, 1171
- sodium sulfate, 1213
- talc, 1283, 1284, 1285, 1291–1293, 1298
- titanium, 1321
- vermiculite, 200, 1376, 1377, 1378

Montasite, 455
Montmorillonite, 255
- *see also* Clays.

Montan wax, 529, 531
Monumental stone, 164
MOR, modules of rupture, 617
Mordenite, 1395, 1396, 1419
Morocco
- barite, 493
- celestite, 1233
- clays, 605, 626, 636
- fluorspar, 734
- meerschaum, 911
- phosphate, 1031, 1032, 1034
- potash, 1070
- salt, 1126
- vermiculite, 1377
- wollastonite, 1386

Mortar, 823
Mother-of-pearl, 301
Mountain leather, 446
Mountain Pass, 1113, 1114, 1115
Mozambique
- anthophyllite, 456
- clays, 605, 626
- fluorspar, 734
- lithium, 871, 874
- monazite, 1112, 1113
- titanium, 1336

Muds, drilling, 387–390
- *see also* Well drilling fluids, 387–390.

Muller-Kuhne process, 1256
Mullite, 24–27, 34, 376, 377, 520, 799, 800, 806, 1085
Muriate of potash, 236
Muscovite, 915, 916

N

Nahcolite, 1203
Namibia
- amblygonite, 874
- diamonds, 302, 658
- fluorspar, 727, 735
- gems, 302
- graphite, 766
- lepidolite, 870, 874
- lithium, 874
- petalite, 874
- sillimanite, 799, 802
- soda ash, 1063
- sodium sulfate, 1197
- wollastonite, 1383, 1386

Native asphalts, 529–532
Natroalunite, 419
Natron, 1188
Natural abrasives, 15, 16, 18–27
Natural glass, 279
Natural granules, 209
Natural iron oxides, 351
- goethite, 352
- hematite, 351
- lepidocrocite, 352
- magnetite, 352, 353

Natural molding sand, 271
Natural pigments, 351–359
Nauru, phosphate, 1021
Nebraska, soda ash, 1189
Neodymium, price, 1110
Nepal, magnesite, 883, 885, 886
Nepheline syenite, 506, 931–960
- alumina, from, 932, 953, 954, 1042
- deposits, 935, 936
- flowsheets
  - IMC, Ont., 949
  - Indusmin, Ont., 947
  - Stjernöy, Norway, 951
- granule, 205
- reserves
  - Canada, 937
  - Norway, 937
  - Siberia, 937
  - USSR, 937
- talc, substitutes for, 1298
- uses, 931, 932

Nevada
- barite, 489, 490, 491, 497
- bentonite, 590, 597, 598
- borates, 537
- brines, 869, 871
- brucite, 382
- diatomite, 685
- fluorspar, 727, 731, 739
- gems, 319
- gypsum, 783
- halloysite, 635
- lithium, 869, 870, 871, 876
- magnesite, 382, 886, 887
- manganese, 899, 902
- perlite, 997, 999, 1003
- phosphate, 1022
- silica, 272, 1171, 1172
- soda ash, 1189
- sodium sulfate, 1209, 1213
- tripoli, 1369
- vermiculite, 1377
- wollastonite, 1385
- zeolite, 1414, 1417, 1426

New Brunswick
- diatomite, 685
- gypsum, 779
- potash, 1059, 1072

New Caledonia
- chromite, 578
- olivine, 991

New Guinea, chromite, 578
New Hampshire
- diatomite, 685
- feldspar, 711
- granite, 170, 711
- mica, 917
- pegmatites, 711
- scythestones, 27

New Jersey
- clays, 591, 636
- feldspar, 711
- foundry sand, 271
- glauconite, 751
- magnesite, synthetic, 890
- manganese, 898
- silica sand, 25

New Jersey, titanium, 1316, 1324, 1326, 1327
traprock, 62
New Mexico
barite, 492
bastnaesite, 1111
feldspar, 711
fluorspar, 727, 731
gems, 303
graphite, 761
gypsum, 778, 783
kaolin, 623
meerschaum, 910, 912
mica, 918
pegmatites, 711
perlite, 997, 999, 1002
potash, 1059, 1060, 1066, 1071, 1073, 1075
sodium sulfate, 1213
sulfur, 1247
talc, 1284, 1285
vermiculite, 1377
wollastonite, 1385
zirconium, 1435
New York
corundum, 18
diatomite, 685
emery, 18
feldspar, 711
foundry sand, 273
garnet, 18, 22, 23
gem, 303
granite, 763
gypsum, 781, 782
limestone, 62
salt, 1129, 1132, 1133, 1135, 1140
sand and gravel, 101, 102
sandstone, 170
silica, 1171
slag, 112, 125
slate, 170
talc, 1275, 1276, 1283–1285, 1286, 1287, 1293, 1296, 1298
titanium, 1303, 1316, 1317, 1336, 1342
wollastonite, 1384, 1387
New Zealand
asbestos, 454
barite, 493
clays, 606, 628, 636, 638
diatomite, 689
geothermal sources, 875
graphite, 767
lithium, 875
olivine, 991
perlite, 1006
sulfur, 1251
titanium, 1314, 1335
wollastonite, 1383
zeolites, 1415
Newfoundland
asbestos, 444, 450, 457
Newfoundland, fluorspar, 728
gypsum, 779
olivine, 992
pyrophyllite, 1094, 1105
salt, 1124
Nigeria
cryolite, 741
fluorspar, 734
kaolin, 626
thorium, 1112
Nikkagel, 611
Nitrates, Chilean, 961, 970, 971
Nitric acid, 963, 965, 966
Nitrile rubber, 962
Nitrogen compounds, 234, 961–988
ammonia, synthetic, 234, 963–965, 978, 982–987
application, 234
composition, 234
fertilizer, 234, 961, 968, 974, 977, 979
uses, 962, 963
Noble gems, 279
Norbide, 29
North Carolina
alaskite, 713
andalusite, 802
anthophyllite, 456
barite, 492
clays, 613, 618, 619, 623
feldspar, 711
gems, 303
granite, 27, 67
hiddenite, 870
kyanite, 800
lithium, 870, 871, 872, 876
marl, 854
mica, 918
olivine, 274, 888, 889, 991, 992
phosphate, 1028, 1032, 1034, 1043
pyrophyllite, 1092, 1105–1107
quartzite, 27
silica, 170
talc, 1284, 1285, 1288, 1289, 1298
titanium, 1315, 1320
vermiculite, 1377
North Dakota, sodium sulfate, 1213
Northwest Territories
bastnaesite, 1111
bentonite, 601
thorium, 1111
Norway
feldspar, 715
flint pebbles, 27
fluorspar, 733
graphite, 760, 765
iodine, 795
kyanite, 801
Norway, larvikite, 171
magnesite, synthetic, 890
mica, 918
monazite, 1113
nepheline syenite, 933, 935, 936, 938–942, 949–951
olivine, 888, 889, 991
pyrite, 1254, 1255
sulfur, 1254, 1255
talc, 1282, 1283
titanium, 1322, 1337, 1345
Nossen process, 903
Nova Scotia
andalusite, 802
barite, 492
celestite, 1229, 1232
diatomite, 685
fluorspar, 728
gypsum, 779
potash, 1066, 1072
salt, 1124
silica, 1175
Novaculite, 204, 1367
Nylon, 962, 963, 968

O

Obsidian, 168, 1001
Ocean floor deposits, manganese, 899
Ocean mining, 1161
Oceania, phosphate, 1039
Ohio
bromine, 562
clays, 374, 375, 637
gannister, 277
gypsum, 778, 782
lime, 828
limestone, 62, 836, 853
salt, 1140
sandstone, 170
silica, 1171
slag, 112, 125
Oil absorption, definition, 350
Oil shale, sulfur from, 1253, 1254
Oklahoma
asphalt, 530
barite, 492
clays, 599
feldspar, 709
grahamite, 530
granite, 170
gypsum, 778, 783
iodine, 795
limestone, 170, 853
silica, 1170, 1173
tripoli, 26, 1363, 1364–1366, 1369
Olivine, 33, 238, 274, 888, 889, 893, 894, 989–996
ceramics, 35

Olivine, deposits, 991
fertilizer, 238, 989
foundry sand, 274, 990
gem, 318
magnesium, source of, 238, 989
processing, 993
properties, 274
sources, 274, 991
uses, 989, 990
ceramic, 35
fertilizer, 238, 989
foundry sand, 274, 990
heat storage, 894, 994
refractory, 893, 894, 921
Ontario
asbestos, 450
bastnaesite, 1111, 1113, 1115
blasting sand, 25
fluorspar, 728
graphite, 763
gypsum, 781, 782
kimberlite, 662
nepheline syenite, 936, 937, 938, 947–949
salt, 1124
silica, 25, 1174, 1175
Onyx, 167, 171
Oolites, 167, 835
Opal, 279, 296, 301, 314, 320, 1168
Opal claystone, 596
Opalescence, 296
Opaline quartz, 1167
Open pans, 1136
Optical calcite, 221
Optical materials, 221–231
Oregon
chromite, 575
clays, 597, 638
diatomite, 25
ferruginous bauxite, 515
fluorspar, 732
gems, 319
perlite, 1003
pumice, 1080
silica, 1171
sodium sulfate, 1211
talc, 1298
traprock, 62
zeolites, 1421
Organoclad bentonite, 610, 611
Oriskany sandstone, 1170, 1174
Ornamental stones, 279
Otanmaki deposit, 1345
Otjevalunda pan, 1197, 1198
Owens Lake brine, 44
Oxychloride cement, 894
Ozokerite, 531

P

Packing, graphite for, 769
Paints, 614–616
lime, use in, 822
manganese, use in, 902
pyrophyllite, use in, 1089
talc, use in, 1277
wollastonite, use in, 1383, 1387, 1388
Pakistan
celestite, 1233
chromite, 578
clays, 606, 627
fluorspar, 736
potash, 1070
vermiculite, 1377
Palygorskite, 591
Pandermite, 535
Pans, salt, 1120
Paper
coatings, 614, 616
fillers used in, 614, 616, 1422
lime, used in, 822
sodium sulfate, 1217, 1218
talc, use in, 1277
titanium, used in, 1305, 1306
zeolite, 1422
Paradox Basin, 1123, 1124
Paraguay, kaolin, 624
Particle size, 350, 351
paste, 315
Paving blocks, 164, 165
Pay streak, 308
Pearl grain, unit of weight, 323
Pearls, 301, 315, 322, 323
Pebbles, ceramic, 27
Pee-Dee gravel, 1171
Pegmatites, 299, 300, 710, 728, 869, 870, 919
see also Feldspars, Aplite, Mica.
gem stone source, 299, 300
thorium in, 1111
Pencils, graphite for, 769
Pennsylvania
barite, 492
bauxite, 515
chromite, 575
clays, 374, 375, 637, 638
diaspore, 515
gannister, 374
granite, 170
graphite, 765
lime, 827, 828
limestone, 62
pegmatites, 71
refractory, 637
rottenstone, 26
sandstone, 170
silica, 1170
slag, 122, 125
slate, 170
talc, 1288
vermiculite, 1377
Periclase, 381, 382, 384
abrasive, 31
Peridot. *See* Olivine.
Perlite, 81, 86, 87, 199–201, 244, 378, 997–1015
see also Lightweight aggregates.
analyses, 999
definition, 997
deposits, 1001–1006
evaluation, 1008
processing, 1009, 1010
uses, 997, 1013
Permanent vs. portable plants, 71, 72
Perovskite, 1313
Perthite, 710
Peru
barite, 492
borates, 548, 550
clays, 602, 624, 638
diatomite, 689
phosphate, 1024
potash, 1066, 1067
soda ash, 1200
Petalite, 870, 877
Petite granite, 171
Petrochemicals, fluorspar, uses in, 724
Petroleum
asphalt, 529
source of sulfur, 1251–1253
Phianite, 314
Philippines
barite, 493
chromite, 381, 577
clays, 627, 636
manganese, 897
perlite, 1006
sulfur, 1251
Phillipsite, 1395, 1396
Pholgopite, 915, 917, 919, 922
Phosam process, 970
Phosgene, 49
Phosphate rock, 1017–1047
deposits, 1018, 1020, 1021
diammonium phosphate, 969
evaluation, 1031
fertilizers, 235, 236, 1035, 1036, 1038–1040
fluorine in, 1042
glauconite, source of, 745
processing, 1018–1020
rare earths in, 1112
terminology, 1017, 1018
yttrium, in, 1112
Phosphogypsum, 1256
Phosphoric acid, 1040, 1041
Phosphorus, 1018, 1031, 1036–1038
see also Phosphate rock.
fertilizer, 235, 236
uses, 1035–1040
Phosphoria formation, 1022–1024
Piezoelectricity of quartz, 228

Pigments. *See* Mineral pigments.
Pigment volume concentration (PVC), 252
Pipe clay abrasive, 26
Pitch bonded refractories, 381
Pitch impregnated refractories, 381
Place value, 4, 788, 942, 946, 947, 1180, 1181
Plant nutrients. *See* Fertilizer minerals.
Plaster
  gypsum, 193, 194
  lime in, 823
  vermiculite, use in, 1380
Plaster of paris, 34, 183
Plastic clays, 380
Plastics, 249–251
  asbestos, 250
  barite, 250
  clay, 250, 251
  fillers used in, 249, 251
  kaolin, 616, 617
  talc used in, 1277
  titanium, in, 1305
  use as imitation gems, 315
  wollastonite used in, 1388
Playa lake, 1121, 1122
Pleochroism, 294
Plumbago, 757
Plumping hides, 822, 823
Poland
  barite, 493
  clays, 603, 625, 638
  glauconite, 753
  gypsum, 184
  ozokerite, 531
  potash, 1069
  salt, 1126
  sulfur, 1242, 1247, 1248
  wollastonite, 1385
Polaroid, 297
Polishing, abrasives for, 30, 31
Porcelain blocks and pebbles, 27
Porcelain, clays, 615
Portable vs. permanent plants, 71, 72
Portland cement. *See* Cement, Cement raw materials.
Portugal
  barite, 493
  chromite, 578
  kaolin, 625
  marble, 171
  salt, 1126
Potash, 236, 237, 1049–1078
  brines, 41, 1957
  bromine, use in exploration for, 1060
  Carlsbad district, 1059, 1060, 1066, 1071
  deposits, 1052–1057
Potash, fertilizer, 236, 237
  glauconite, source of, 745
  nepheline syenite, source of, 951–953
  processing, 1071, 1072
  reserves, 1065, 1066
    Africa, 1070
    Brazil, 1067
    Canada, 1062, 1066
    Morocco, 1070
    New Mexico, 1066
    Russia, 1069
    South America, 1066, 1067
    world, table, 1063–1065
  resources, 1061, 1063–1065
  uses, 236, 237, 1049
  Zechstein, 1068, 1069
Potash bentonite, 588
Potassium
  see also Potash.
  brines, 41
  carbonate, from nepheline syenite, 951–953
  fertilizer, 236, 237
  oxide. *See* Potash.
  sulfate, 419, 1066, 1072
Potsdam sandstone, 1174
Pottery, clays for, 639
Pottsville formation, 1171
Powders, abrasives, 16
Pozzolan, 134, 135, 143
  zeolite used in, 1421
Prairie Evaporite formation, 1060
Praseodymium, price, 1110
Precious stones. *See* Gem materials.
Pressure blasting, 16, 27
Priceite, 435
Prills, 967
Protective coatings, 252
Puerto Rico, zeolites, 1416
Pulaskite, 931
Pulp, paper, lime used in, 822
Pumice and pumicite, 199, 999, 1079–1084
  abrasive, 16, 18, 25, 26, 1081
  aggregate, 84, 1083
  consumption, 85
  lightweight aggregate, 82, 1083, 1084
  mining, 86, 1079
  uses, 1079
    abrasive, 16, 18, 25, 26, 1081
    aggregate, 1081
    building stone, 1081
    filler, 244
Pumicite. *See* Pumice.
Putty powder, abrasive, 31
Putunig area, 458
PVC (pigment volume concentration), 252
Pyke area, 454
Pyrite, 353
  source of iron, 135
  source of sulfur, 1254
Pyrobitumens, 531
Pyrolusite, 897, 902
Pyrometric cone equivalents (PCE), table, 613
Pyrophanite, 1313
Pyrophyllite, 1085–1108
  analyses, 1093, 1097, 1098, 1102, 1105
  deposits
    Argentina, 1096
    Australia, 1093, 1094
    Brazil, 1096
    Canada, 1094, 1095
    India, 1097
    Japan, 1091
    South Africa, 1097
    South Korea, 1092
    Turkey, 1097
    United States, 1092
  evaluation, 1104
  processing, 1105, 1106
  substitutes, 1085, 1108
    for feldspar, 720
  uses, 1085, 1086, 1087, 1280
Pyropissite, 531
Pyroprocessing, 139, 140
Pyrrhotite, source of sulfur, 1254–1256

Q

Quarrying, 136
  dimension stone, 171–175
  granules, 207
  underground, 172, 173
Quartz, 213–221
  see also Silica and silicon.
  abrasive, 25
  cultured, 38, 214
  dielectric material, 38, 213, 227
  gems, 1167, 1168
  glossary, 226–228
  synthesis, 216–218, 320
  uses, 219–221
Q-spoiler, 228
Q-switch, 228
Q of quartz, 217, 218
Quartzite, 27
  pebble, metallurgical, 1171
Quicklime. *See* Lime.
Quebec
  asbestos, 445, 446–451
  bastnaesite, 1111
  blasting sand, 25
  brucite, 382, 888
  chrysotile, 445, 446–451
  graphite, 763, 765
  kyanite, 801
  lithium, 874
  magnesite, 883, 884, 887, 888

Quebec, mica, 918
nepheline syenite, 936
olivine, 992
pyrophyllite, 1096
silica, 1050
talc, 1287, 1295
thorium, 1111
titanium, 1321, 1337, 1345
Quebec standard test, 464

R

Radiation shielding, ceramics, 37
Ramming, 375, 379
mixes, 380, 381, 799, 989
Rare earths, 1109–1118
bastnaesite, 1110
deposits, 1111, 1112
exploration, 1113, 1114
monazite, 1110
processing, 1114, 1115
uses, 1109, 1110
Rattails, 274
Reaction flux, 263
Reclamation costs, 642
Reconstituted mica, 915
Reconstructed gems, 311
Recovered sulfur from
coal, 1254
crude oil, 1251, 1252
oil sand, 1253
oil shale, 1253, 1254
sour gas, 1051, 1052
sulfides, 1254, 1255
tar sands, 1253, 1254
Red mud, 740
Reeds, 166
Reefs, 835, 836
Refractories, 373–385
alumina, 376, 377
chromite, 381
clay, 373–376
lime, 822
magnesite, 381, 382
mullite, 24–27, 34, 376, 377, 520, 799, 800, 1085
olivine, 989
periclase, 381, 382, 882
raw materials
dolomite, 167, 382
graphite, 379
mullite, 27, 376, 377, 799
olivine, 989
periclase, 381, 382
zircon, 378
talc, 1277, 1279
types, 373–378, 380, 381–383
Refractory clay, 636–639
ultrahigh alumina, 379
uses, 588–592, 599, 602, 606, 633–636
Resonator, 228
Retarders, in cement, 189, 190
Rhinestone, 315
Rhode Island
granite, 170
graphite, 761
talc, 1288
Rhodesia. *See* Zimbabwe.
Rift, 166
and grain, 175
Riprap, 61
Rochelle salt, 226
Rock asphalt, 530
Rock hounds, 308
Rock salt. *See* Salt.
Rocks used for aggregates, 1444
see also Crushed stone.
physical properties, 846–848
Rock wool, 199–201
Roman alum, 417, 418
Romania
barite, 493
chromite, 578
clays, 603, 626
diatomite, 687
fluorspar, 734
ozokerite, 531
salt, 1126
talc, 1283
Roofing materials, 203–210
see also Granules.
Rosin solders, flux, 260
Rottenstone, 1363, 1364, 1369
abrasives, 16, 26
composition, 26, 1364
deposits, 26, 1364
sales, 26
uses, 1369
Rouge, 31
Ruanda, amblygonite, 871, 874
Rubber
fillers in, 251, 614
wollastonite use in, 1383
Rubble, 161, 162
Ruby, 279
asterism, 295
occurrence, 300, 304
production, 320
synthetic, 311–313
use, 296
Runner sand, 1168
Russia. *See* USSR.
Rutile, 1314
see also Titanium minerals.
gem, 313
mineralogy, 1314
synthetic, 313, 321

S

Sabhka, 777
Salado formation, 1059, 1060
Salinas, 1120
Salt, 98, 1119–1149
brines, derivatives, chart, 44
cap rock, 1243, 1244
ceramic use, 38
Salt, deposits, 1120, 1121, 1123, 1124
domes, 1123, 1124, 1243–1245
cap rock, 777
evaluation, 1129
evaporated, 1135
mining, 1130, 1131, 1133–1135
origin, 1122, 1123
solar, 1133–1135
solution mining, 1131–1133
uses, 1142, 1143, 1144, 1145
Salt cake, 419, 1217, 1218
see also Sodium sulfate.
ceramic use, 36, 38
detergents, 1217
flux, 36
glass, use in, 344
paper, use in, 1217
Salt earth, 1126
Salt lakes, 1120, 1121
Salt pan, 1120
Samarium, price, 1110
Sampling, silica, 1178–1180
Sanbornite, 487
Sand and gravel, 96–110, 1151–1166
deposits, types of, 1154, 1155
evaluation, 1157–1160
sand equivalent, 1160
tests, 98, 99, 1157, 1158, 1160
uses, 96
Sand blasting, 1125
Sand, foundry. *See* Foundry sand.
Sandhills Belt, 1178
Sandpaper, 17
Sandstone, 61, 166
Sanford Lake, 1317–1319
Saponite, 594, 600
Sapphire, 279
asterism, 295
production, 312, 320
synthetic, 311, 312
uses, 296, 297
Sardinia
fluorspar, 733
talc, 1283
Saskatchewan
bentonite, 601
potash, 1059, 1061, 1071–1074
quartzite pebbles, 27
silica, 1175, 1176
sodium sulfate, 1210, 1218
Sassolite, 535
Satin rouge, 31
Satin spar, 296
Saudi Arabia, magnesite, 886
Saws, stone cutting, 173, 174
Scabs, 274
Scapolite, 883
Scattergram, 1101

Scintillation, 654
Scoria, 1079
  aggregate, 81
Scotland
  diatomite, 687
  flint clay, 638
  iodine, 795
  olivine, 991, 992
  perlite, 1001
Scrap mica, 915, 919, 920
Scratch test, 11
Screw dislocation, 218
Scythestones, 27
Sea foam, 909
  see also Meerschaum.
Searles Lake, 44, 48, 533, 537, 541, 542, 553, 873, 1072, 1121, 1187, 1199, 1201, 1212, 1220
Seawater, source of magnesia, 382, 891
Sechura Desert, 1024, 1067
Secondary plant nutrient, 237–239
Seed quartz, 216, 217
Selenite, 221, 298, 776
Selenitza, 530
Self-fluxing ore, 268
Semi-precious stones, 279
Senegal
  bentonite, 597, 605
  fuller's earth, 597
  wavellite, 1031
  zirconium, 1433
Sepiolite. *See* Meerschaum.
Sergipe, 1066, 1067, 1125
Serpentine, 446
  dimension stone, 168
Shale, 81, 82, 639–641
  aggregate, 81, 82, 86
  cement material, 134, 639
  properties, 639
Shape, particle, definition, 351
Sharpening stones, 26
Shells, 61–63, 204, 854
  use in cement, 134, 135
Shielded arc welding, 265
Shot steel, 30
Siberia
  bauxite, 508
  graphite, 764
  nepheline syenite, 936, 937
  soda ash, 1198
Sicily
  potash, 1071
  sulfur, 1241, 1247
Siderite, 353, 834
Sierra Leone
  bauxite, 513, 523
  chromite, 578
  diamonds, 302, 658
  titanium, 1333
Silex, 27
Silica, silicon, 1167–1185
  see also Tripoli.
  abrasive, 16, 17, 25, 27
  bricks, 377
  ceramics, 33, 34, 35, 39
  filler, 244
  flux, 269
  foundry sand, 271–278
  gannister, 363, 374, 377, 1168
  glass former, 33, 36
  glass, use in, 344, 1179
  granules, 204
  microcrystalline, 26
  molding sand, 271
  skeleton former, 34
  specifications, 344, 1168, 1169
  tripoli, 1363–1374
  uses, 1168, 1169, 1173
  zonal restrictions, 106, 107
Silicate cotton, 199
Siliceous powder, 25, 26
Silicon carbide, 12, 16, 17, 27–29, 378, 379, 1174
Silicon metal, 1167, 1173
Silicone, 86
Sillimanite, 376, 799, 800, 802
  ceramics, 34
  porcelain, use in, 27
  refractories, 376
Silver Peak Marsh, 44, 871–873
  see also Lithium.
Skeleton formers, 34
Slag, 81, 82, 89, 111–131
  cement, raw materal for, 134, 143
  deleterious substances, 87
  expanded slag, 85, 116, 117, 122, 123
  glass, raw material for, 344
  granules, roofing, 203–212
  steel, 118, 122
  types, 115–122
  uses, 120–122
Slate, 163, 164, 168
  filler, use as, 245
  granule, 209
  lightweight aggregate, 82
  mining, 82
  soundness, 180
  specifications, 162–164
  travertine, 167
  underground quarries, 172, 173
  uses, 164
Slip casting, 375
Slip dislocation, 218
Slip fiber, 464
Smackover formation, 869
Smectite, 594, 640, 870
  see also Bentonite, Fuller's earth.
Soap, scouring, 18
Soapstone, 165, 175
  see also Talc.
Soda ash, 52, 1187–1206
  see also Sodium carbonate.
  ceramic, 36
  flux, 36, 38
Sodium, 49
Sodium carbonate, 52, 1187–1206
  ceramic, 36
  flux, 36
  geology, 1188, 1189
  Green River, 1199, 1200, 1201, 1202
  nepheline syenite, 931
  Searles Lake, 48, 533, 537, 541, 542, 1072, 1121, 1187, 1199, 1201, 1212, 1220
  Solvay process, 52, 1201
  uses, 52, 344, 1188
Sodium chloride, 38, 98
  see also Salt.
Sodium hydroxide, 42, 48, 52, 99
Sodium sulfate, 52, 1207–1223
  glass use in, 346
  producers
    Canada, 1210, 1211, 1218
    Mexico, 1220
    US, 1211–1213, 1218–1220
  production, 1217
    Argentina, 1215
    Canada, 1210, 1211
    Chile, 1216
    Mexico, 1213, 1214
    Russia, 1214, 1215
    Spain, 1215
    Turkey, 1215
    US, 1207, 1211–1213
  soundness test, 98
  uses, 36, 38, 1217, 1218
Soffioni. *See* Sassolite.
Soft kaolin, 614
Soft silica, 1363
Soft siliceous powder, 25
Soil stabilization, lime in, 823
Solar salt, 1133–1135
Soldering, 259, 261, 262
Soluble anhydrite, 188
Solution mining, 902
  potash, 1072
  production, 1132
  salt, 1127, 1131
  wells, 1132
Solvay process, 52, 1201
Somali Republic, meerschaum, 911, 912
Sorel slag, 1345, 1346
Sorption sensitivity, sand, gravel, 99
Sour gas, 1251–1253

Sources of information, 393–414
Africa, geological agencies, 407–410
Asia, geological agencies, 411–413
Australia, geological agencies, 413
Caribbean Islands, geological agencies, 402
Europe, geological agencies, 404–407
Forum on Industrial Minerals publications, 394, 395
North America, geological agencies, 397–402
Pacific Islands, geological agencies, 413
periodicals, 402
Public Inquiries Office, USGS, 396
publications
Canada, 397
US, 402
schools of mines
Canada, 397
US, 401, 402
societies, mining, geological, 401, 402
South America, geological agencies, 402–404
state agencies, US, 399–401
state mineral, water resource publications, 395
trade associations, 414
US Bureau of Mines offices, 398
US Geological Survey offices, 398, 399
South Africa
amblygonite, 871
amosite, 453, 455
andalusite, 803
apatite, 1021
bastnaesite, 1111, 1113
bort, 19
carbonado, 19
chromite, 274, 381, 576
chrysotile, 453
clays, 605, 612, 636, 638
corundum, 806
crocidolite, 455
diamonds, 302
diaspore, 572, 658
fluorspar, 726, 727, 734
gems, 302
graphite, 766
kimberlite, 658
kyanite, 799, 803, 806
lithium, 874
magnesite, 884, 885
manganese, 898
meerschaum, 911
South Africa, mica, 918
monazite, 1113
nitrates, 970
olivine, 991
pyrite, 1236
pyrophyllite, 1097
rare earths, 1111, 1113
silica, 1183
sillimanite, 802
slag, 113, 1305, 1315, 1334, 1335
soda ash, 1193
sulfur, 1260
titanium, 113, 1305, 1315, 1334, 1335
vermiculite, 200, 1376
wollastonite, 1383, 1386
zirconium, 1433, 1436, 1437
South America
see individual countries.
bauxite, 504
diatomite, 688, 689
feldspar, 715
gems, 302
South Carolina
barite, 492
clay, 596, 599, 613, 618, 622, 638
feldspar, 711, 713
granite, 62, 170
kyanite, 376, 800, 801
marl, 854
South Carolina, mica, 918
silica, 1170
sillimanite, 802
vermiculite, 200, 1376, 1379
South Dakota
amblygonite, 873
feldspar, 711
granite, 170
manganese, 898
silica, 1173
spodumene, 873
South West Africa. *See* Namibia.
Southern bentonite, 588
Spain
barite, 493
celestite, 1229
clays, 597, 598, 603, 625, 636
cryolite, 741
diatomite, 686
fluorspar, 726, 727, 733
graphite, 765
magnesite, 884, 885
meerschaum, 911
olivine, 991
potash, 1069, 1070
pyrite, 1236, 1254
salt, 1126
sepiolite, 592
sodium sulfate, 1207, 1215, 1221
sulfur, 1236, 1247, 1254
Spain, vermiculite, 1377
wollastonite, 1386
Spanish Sahara, phosphates, 1029
Spar
block, 711
feldspar, 709–722
heavy. *See* Barite.
Iceland, 297
satin, 296
Spath, 709
Specifications. *See* individual commodities.
Sphalerite, dispersion, 294
Spiegeleisen, 904
Spinel, synthetic, 311, 312
Splittings, 915
Spodumene, 870, 877
see also Lithium raw materials.
ceramics, 36
Spring apron, 537
Sri Lanka (Ceylon)
baddeleyite, 1436
gem, 302
graphite, 5, 758, 762, 763, 764, 767, 768
kaolin, 626
monazite, 1113
titanium, 1334
yttrium, 1112
zirconium, 1433, 1443
St. Lucie zone, 1170
St. Peter sandstone, 272, 1173
Stabilization, land, 823
Stabilizer, asphalt, 209
Staurolite, 27, 274, 1225–1227, 1325
abrasive, 26
foundry sand, 274
use, 274, 1225
Steatite, 1275, 1281
Steel shot, abrasive, 16, 18, 30
Steel slag, 122, 126
Steel wool, 30
Stishovite, 1167
Stjernöy, Norway, 938–942, 949–951
Stockpiling
fluorspar, 740
iodine, 796
Stokes diameter, 350
Stone
see also names of rocks.
crushed stone, 60–80
dimension stone, 161–181
sand, gravel, 96–110
solidifier, 178, 179
Stoneware, clays for, 612, 617
Storgangen, 1322
Strass, 315
Stratiform deposits, sulfur, 1247
Strontianite, 1229

Strontium, 1229–1233
- specifications, 1229
- titanate, gem, 313
- uses, 1229, 1231

Stucco, 188, 189
Submerged arc fluxes, 267, 268
Sudan
- chromite, 578
- fluorspar, 735
- lithium, 874
- wollastonite, 1383, 1386

Sugar industry, lime use in, 822
Sulfates, 1256, 1257, 1308–1310
Sulfide ores, sources of sulfur, 1238
Sulfur, 238, 239, 1235–1274, 1297
- alunite, with, 421
- ceramic, 35
- consumption, fertilizer, 239
- deposits, 1245–1251
- economic factors, 68, 1268
- fertilizers, 238, 239
- plant nutrient, 233, 238, 239
- secondary plant nutrient, 233, 238, 239
- sources of, 1235, 1236, 1238, 1243, 1251
- uses, 1261–1264

Sulfur dioxide emission control, 1255, 1256
Sulfuric acid, 1261–1266
- superphosphates, use of, 238

Sunstone, 296
Supai formation, 1172
Superphosphates, 238
Surface area, definition, 350
Surinam
- bauxite, 513, 523
- kyanite, 801
- lithium, 871

Surveying, airborne for asbestos, 462
Swallow nest workings, 310
Swaziland
- chrysotile, 453
- kaolin, 626

Sweden
- clay, 626, 638
- feldspar, 715
- fluorspar, 733
- lithium, 870
- olivine, 991
- petalite, 870
- pyrite, 1254
- sulfur, 1254

Switzerland
- asphalt, 530
- bentonite, 603
- fluorspar, 733

Sylvania sandstone, 1174
Sylvanite, 1072
- *see also* Potash.

Synopal, 1389
Synthetic materials
- corundum, 311, 312
- diamonds, 313, 318, 668–670
- gems, 311–315
- iron oxides, 359–368
- magnesite, 889–892
- mica, 915
- quartz, 213–221, 314
- rutile, 313
- zeolites, 1399–1414

Szaibelyite, 535, 536

T

Taiwan, kaolin, 627
Talc, 203, 209, 1275–1301
- ceramic use, 34, 35, 38, 1277
- deposits, 1283–1293
- filler, use as, 33, 35, 38, 250, 251, 1280
- paint, use in, 1277, 1278
- processing, 1294–1297
- properties, 1277, 1281
- specifications, 1276, 1277

Tanganyika
- bentonite, 605
- meerschaum, 912

Tanzania
- clays, 605, 626
- diamonds, 302, 658
- garnet, 23
- gems, 302
- meerschaum, 910, 911
- monazite, 1113
- soda ash, 1198
- vermiculite, 1377
- zeolites, 1416
- zirconium, 1433

Tap density, 20, 21
Tar sands, 1235
Taylorite, 587
Tecopa, Lake, 1416
Tejon formation, 1172
Tellnes, deposit, 1337, 1345
Tennessee
- barite, 491, 492
- bauxite, 515
- clays, 589, 633, 634, 638
- fluorspar, 732
- foundry sand, 271
- glauconite, 752
- lime, 827
- microcrystalline silica, 26
- phosphate, 1029, 1030, 1043
- quartzite, 27
- silica, 26
- titanium, 1316, 1324, 1327
- tripoli, 26, 1364, 1367, 1368

Tequesquite, 1121
Terbium, price, 1110
Terra alba, 195, 196
Tetraethyl lead, 48
Texas
- asphalt, 530, 531
- barite, 492
- clays, 375, 589, 590, 599, 623, 633, 634
- dimension stone, 170
- feldspar, 711
- flint, 27
- fluorspar, 732
- gems, 319
- glass sand, 639
- glauconite, 752
- granite, 170
- graphite, 750, 765
- gypsum, 778, 783
- lime, 827, 828
- limestone, 62, 854
- magnesite, synthetic, 890
- marl, 854
- pegmatite, 711
- rock asphalt, 530, 531
- salt, 1132, 1140
- sand, 108
- silica, 1173
- slag, 125
- sodium sulfate, 1212, 1219
- sulfur, 1236, 1248, 1250, 1252, 1256
- talc, 1275, 1283, 1286, 1287, 1290, 1298
- tripoli, 1368, 1369
- vermiculite, 1377
- zeolites, 1414

Thailand
- barite, 493
- fluorspar, 736
- gems, 302
- kaolin, 627
- potash, 1070
- pyrophyllite, 1097
- salt, 1126
- zirconium, 1433

Thenardite, 1207
Thermonatrite, 1188
Thetford area, 446, 447
Thin sections, use of, 944
Thorite, 1111
Thorium, 1109–1112, 1114, 1116, 1117
Thulium, price, 1110
Tibet, borates, 550
Tiff, 487
Tiger's-eye, 1168
Tile, 375
Tincal, 45, 535
- *see also* Borax.

Tincalconite, 45, 536
- *see also* Borax.

Tin oxide
- abrasive, 16, 31
- ceramic, 35, 37

Titangem, 313
Titania, 313, 520, 671

Titanium carbide, 1313
Titanium dioxide pigments, 1304
Titanium minerals, 1303–1362
alluvial deposits, 1348–1351
analyses, 1335
Brazil, 1335
Sanford Hill, 1318
South Africa, 1336
beach deposits, 1347–1351
capacity, industry
US, 1308
world, 1309
distribution of deposits, 1316–1336
Australia, map, 1330
Brazil, map, 1335
evaluation, 1339–1342
gem, 313
processes
chlorite, 149, 1310, 1311
sulfate, 1305–1308
reserves
Australia, 1331
world, 1317
sponge, producers, 1312
substitutes, 1278, 1351
uses, 1303, 1306
Tofa, 417, 418
Topaz, 291, 292, 799
Tourmaline, 297
synthetic, 314
use, 297
Trace elements fertilizers, 239
Trade associations, 414
Trail Ridge, 1437, 1438
Transparency, 292
Transportation
costs, 73, 103, 104
lightweight aggregates, 89
molding sands, 274, 275
Transvaal, South Africa
amosite, 455, 457
chromite, 576
chrysotile, 455
crocidolite, 455, 457
diatomite, 687
fluorspar, 727, 734
graphite, 766
thorium, 1111
Traprock, 61, 164
Travertine, 167, 171
Tremolite, 435, 440–444, 456, 461, 1281
Triamond, 314
Tridymite, 1167
Trinidad
asphalt, 529
grahamite, 530
Triphane, 870
Tripoli, 26, 1363–1374
abrasives, 16, 18, 26, 1364, 1371
ceramics, 34, 39
occurrence, 26, 1364–1369
Tripoli, uses
abrasive, 16, 18, 26, 1364, 1371
ceramics, 34, 39
fillers, 1371, 1372
Tripolite, 1363
Tritium, 878
Trona, 1188, 1197, 1198, 1199, 1200
see also Sodium carbonate.
Troodos area, 452
Tube-mill linings, 27
Tuff, aggregates, 81
Tumbling, 332
Tungsten carbide, 30
Tunisia
fluorspar, 735
phosphate, 1031, 1034
salt, 1123
Turilari playa, 876
Turkey
bauxite, 515
borates, 5, 535, 543–546
celestite, 1229
chromite, 381, 577, 578
clays, 606, 627
diatomite, 687
fluorspar, 735
magnesite, 884, 885, 888
meerschaum, 910, 911
perlite, 1004
phosphate, 1028
pyrophyllite, 1097
sodium sulfate, 1215
sulfur, 1236, 1251
Turquoise, stabilized, 314
Twinning defects, 218

U

Uganda
apatite, 1021
lithium, 874
vermiculite, 1377
Uintahite, 530
Ulexite, 534, 536, 537, 542, 544, 548
Ultrahigh-alumina bricks, 376
Umber, 355
Underground mining, 849
Underground storage, 1132
Ungava, asbestos, 444
United Kingdom
barite, 493
bath brick, 26
bromine, 48
celestite, 1231, 1232
chalk, 26
flint, 25
clays, 603, 608, 623, 624, 634, 638
cultured quartz, 216
diaspore, 638
diatomite, 687
flint clay, 638
flint pebbles, 27
fluorspar, 723, 726, 733, 734
glauconite, 753
gypsum, 778
iodine, 795
magnesite, synthetic, 890
mica, 918
potash, 1067
quartz, 216
salt, 1135
slag, 113
witherite, 485
zeolites, 1422
United States
see also individual states.
andalusite, 802, 803
bauxite, 515, 516
bentonite, 599–601
bromine, 561, 562, 563
chromite, 384
chrysotile, 451, 457
clays, 585, 599–601, 612, 622–624
diamonds, synthetic, 662, 670
diatomite, 685, 686
feldspar, 709, 713–715, 718
fluorspar, 728–732
gems, 303, 318–321
gypsum, 184
kyanite, 800–802
lime, 827
magnesite, 883
mica, 981
monazite, 1114, 1115
nepheline syenite, 931, 942
nitrates, 970
olivine, 991, 992, 993
perlite, 998
phosphate, 1030, 1033, 1034
potash, 1059, 1060, 1066, 1071, 1073, 1095
pyrophyllite, 1092, 1093
rare earths, 1114, 1115
salt, 1125
sillimanite, 799, 802
slag, 111, 112
soda ash, 1187, 1188
sulfur, 1236
vermiculite, 1377
wollastonite, 1384, 1385
zeolites, 1393–1399, 1414
Uranium in phosphate, 1041
Urea, 234, 962, 963, 966
Urite, 942, 952
Uruguay
fire clay, 638
monazite, 1113
rare earths, 1113
titanium, 1336
USSR (Russia)
apatite, 1021
asbestos, 452, 453, 479

USSR (Russia), barite, 492
bauxite, 515
borates, 546, 547
bromine, 48
chromite, 381, 576
chrysotile, 452, 453, 479
clays, 597, 604, 638
cryolite, 741
cultured quartz, 216
diamonds, 653, 662
diatomite, 687
fluorspar, 734
garnet, 23
gems, 303
glance pitch, 531
glauconite, 753
graphite, 767
iodine, 795
kaolin, 624
lithium, 870
magnesite, 883, 884, 890
manganese, 897, 898
mica, 918
nepheline syenite, 931, 935, 951–954
oil shale, 1253
olivine, 991
perlite, 1004
phosphate, 1020, 1027, 1031, 1032, 1037, 1039
potash, 1069
salt, 1126
slag, 113
sodium sulfate, 1207, 1209, 1214
sulfur, 1236, 1249, 1252, 1256
synthetic diamonds, production, 662
talc, 1283
thorium, 1112
titanium, 1336
vermiculite, 1377
wollastonite, 1383, 1385
yttrium, 1112
zeolites, 1416
zirconium, 1416
Utah
asphalt, 530
brines, 869
diatomite, 685
fluorspar, 732
gems, 303
gilsonite, 530
glance pitch, 531
gypsum, 778
halloysite, 635
hectorite, 875
lime, 827
lithium, 870, 875
oil shale, 1253
ozokerite, 531
perlite, 1003
phosphate, 1022, 1031, 1034, 1042
Utah, potash, 1066, 1073
salt, 1123
silica, 1173
slag, 125
sodium sulfate, 1209, 1212
tar sand, 1253
titanium, 1316
wollastonite, 1385
wurtzilite, 531
zeolites, 1415
zirconium, 1435

V

Valley abrasion test, 1278
Van Dyke zone, 1170
Venetian red iron oxide, 365
Venezuela
bauxite, 514
diamonds, 302, 658
gems, 302
iron oxide pigments, 355
kaolin, 624
sulfur, 1252
tar sand, 1253
Verde antique, 164, 167, 170
Vermiculite, 81, 85, 87, 90, 199, 200, 916, 1375–1381
consumption, 87
deposits, 1377
evaluation, 1377
expansion, 86, 1375
filler, 245
grading requirements, 88
lightweight aggregates, 81, 86, 87, 1380
marketing, 90
prices, 86, 1380
production, 84, 1379, 1380
specifications, 1379
uses, 81, 87, 1380
value, unit, 86
weight, dry, 87
Vermont
asbestos, 451
granite, 170
kaolin, 623
marble, 73, 164, 170
monument stone, 164
serpentine, 170
slate, 170
talc, 1275, 1276, 1282, 1286–1288, 1293, 1295, 1298
Verde antique, 170
Verneuil process, 311–313
Vienna lime, 31
Viet Nam, North, chromite, 578
Viet Nam, South, kaolin, 627
Virginia
aplite, 709, 716, 718, 719
barite, 492
bauxite, 515
Virginia, feldspar, 712, 716, 718, 719
gem, 303
glauconite, 752
granite, 170
crushed, 62
gypsum, 778, 782
kaolin, 623
kyanite, 376, 801
lime, 824, 827, 828
limestone, 853
marble, 164
marl, 854
pegmatites, 711
salt, 1119, 1124
serpentine, 170
silica, 1170
slate, 170
talc, 1286, 1287, 1288, 1298
titanium, 1319
vermiculite, 1377
Viscosity, 616, 629
Volcanic
cinder, 81–86, 1081
see also Pumice.
well drilling fluids, 388
dust, 26
glass, 997

W

Wallboard, 194, 195, 1089
Washington
barite, 492
brucite, 382
chromite, 575
diatomite, 685
feldspar, 711
ferruginous bauxite, 515
fire clay, 638
fluorspar, 732
granite, 171
magnesite, 382, 883, 887
olivine, 272, 274, 991, 993
perlite, 1003
silica, 1171, 1173
sodium sulfate, 1213
traprock, 62
zeolite, 1416
Waste treatment
glauconite, 746, 747
lime use, 820, 821
Wavellite, 1031
Waxes, mineral, 529
Weight-per-cubic-foot test, 14, 15
Weights, units, gems, 323
Welch process, 903
Welding rod coatings, 265, 268, 269
Well drilling fluids, 387–390
Western bentonite, 390

West Virginia
- clays, 374, 375, 637
- grahamite, 530
- limestone, 853
- salt, 1119
- silica, 1170
- white cement, 1087, 1088, 1103

Wheel abrasive, 16, 17
Whiting abrasive, 16, 26
Wicking, 1281
Wig-L-Bug mill, 12, 13
Winnepeg formation, 1175
Wire saws, 173, 174
Wisconsin
- barite, 492
- dolomite, 31, 854
- feldspar, 711
- gannister, 377
- granite, 171
- limestone, 854
- manganese, 898
- quartzite, 27
- silica, 1170

Witherite, 485, 487, 499
Wollanita, 1389
Wollastonite, 800, 1383–1390
- ceramics, 34, 35, 1388
- production, 22, 1387, 1388
- refractory filler, 35
- uses, 1383, 1388
  - ceramic, 34, 35

Wonderstone, 1089, 1097
Woodsreef area, 454
Wool
- blanket, 201
- brass, 30
- mineral, 200, 201, 1388
- slag, 115, 121
- steel, 30

Wurtzilite, 531
Wyoming
- bentonite, 589, 595, 600
- chromite, 575
- feldspar, 711
- fluorspar, 732
- gypsum, 778, 783
- jade, 303
- lithium, 873
- oil shale, 1253
- pegmatites, 711

Wyoming, phosphate, 1022, 1042
- salt, 1124
- silica, 1172
- soda ash, 1188, 1199, 1200, 1201
- sodium sulfate, 1213
- tar sand, 1253
- titanium, 1320
- trona, 1199, 1200
- vermiculite, 1377
- zeolites, 1423
- zirconium, 1435

X

Xenotime, 1111
X-ray diffraction
- clay identification, 594
- gems, color changes, 333
- zeolite identification, 1415

Xerogels, 1168

Y

Yadkin valley, 1320
YAG, 314, 671
Y-bar seed, 210, 217
Yellow ground, 301
Yellow iron oxide, flow diagram, 362
YGG, 314
Yttrium, 1112
- *see also* Rare earths, Thorium.
- growth rate, 1116
- price, 1110

Yttrium-aluminum-garnet, 314
Yttrocerite, 1111
Yttrofluorite, 1111
Yugoslavia
- asbestos, 452
- barite, 492
- bauxite, 513, 523
- chromite, 576
- clays, 604, 626, 638
- diatomite, 687
- fluorspar, 734
- magnesite, 884, 885
- perlite, 1004
- zeolites, 1417, 1419

Yukon, chrysotile, 450

Z

Zaire, diamonds, 302, 653, 658
Zechstein, 1049, 1067–1069, 1126
Zeolites, 1391–1431
- deposits, 1419–1421
- minerals, 1415
- processing, 1407–1413
- reserves, 1414, 1417, 1418
- structure of, 1400–1403, 1413
- synthetic, 1399–1413, 1414
- uses, 1411, 1412, 1420, 1425, 1426

Zeta potential, chrysotile, 439, 440
Zidani Kozani area, 452
Zimbabwe
- amblygonite, 871
- apatite, 1021, 1033
- asbestos, 453
- chromite, 576, 577, 579, 581
- eucryptite, 870
- ferroalloy, 581
- gems, 302
- lepidolite, 870
- lithium, 870, 871, 873, 874
- magnesia, 885
- olivine, 991
- rhodocrosite, 881
- vermiculite, 1377

Zinc, 38
- fertilizer, 233
- micronutrient, 240

Zinnawaldite, 916
Zircon, 27, 273, 314, 378
- alumina abrasive, 29
- ceramic use, 35
- distribution, 273, 274
- foundry sand, 273, 274, 1433
- prices, 1441, 1442
- refractories, 378, 1438
- uses, 273, 274

Zirconia, 34, 35, 314
Zirconium minerals, 273, 274, 1433–1446
- *see also* Titanium.
- abrasive, 29
- ceramics, use in, 35
- foundry sand, 273
- gems, 314
- refractories, 378
- uses, 1086, 1438, 1440

Zoning, 35, 271, 272
- crushed stone, 78
- sand and gravel, 106, 107
- silica, 1177